Fundamental Physical Constants

Quantity	Symbol	Value
Gas constant	R	$8.314\,510(70)$ J K^{-1} mol^{-1}
		$8.205\,783(69) \times 10^{-5}$ m^3 atm K^{-1} mol^{-1}
Avogadro constant	N_A	$6.022\,136\,7(36) \times 10^{23}$ mol^{-1}
Faraday constant	F	$9.648\,530\,9(29) \times 10^4$ C mol^{-1}
Permeability of vacuum	μ_0	$4\pi \times 10^{-7}$ H m^{-1} (exactly)
Speed of light in vacuum	c	$299\,792\,458$ m s^{-1} (exactly)
Permittivity of vacuum	$\varepsilon_0 = 1/\mu_0 c^2$	$8.854\,187\,817\ldots \times 10^{-12}$ F m^{-1}
	$4\pi\varepsilon_0$	$1.112\,650\,056\ldots \times 10^{-10}$ F m^{-1}
Elementary charge	e	$1.602\,177\,33(49) \times 10^{-19}$ C
Electron rest mass	m_e	$9.109\,389\,7(54) \times 10^{-31}$ kg
Proton rest mass	m_p	$1.672\,623\,1(10) \times 10^{-27}$ kg
Neutron rest mass	m_n	$1.674\,928\,6(10) \times 10^{-27}$ kg
Atomic mass unit	$m_u = 1$ u $= 1$ amu	$1.660\,540\,2(10) \times 10^{-27}$ kg
Planck constant	h	$6.626\,075\,5(40) \times 10^{-34}$ J Hz^{-1}
	$\hbar = h/2\pi$	$1.054\,572\,66(63) \times 10^{-34}$ J s
Boltzmann constant	k_B	$1.380\,658(12) \times 10^{-23}$ J K^{-1}
Gravitational constant	G	$6.672\,59(85) \times 10^{-11}$ N m^2 kg^{-2}
Standard acceleration of free fall	g	$9.806\,65$ m s^{-2} (exactly)
Fine structure constant	$\alpha = \mu_0 e^2 c/2h$	$7.297\,353\,08(33) \times 10^{-3}$
	α^{-1}	$137.035\,989\,5(61)$
Bohr radius	$a_0 = 4\pi\varepsilon_0\hbar^2/m_e e^2$	$5.291\,772\,49(24) \times 10^{-11}$ m
Hartree energy	$E_h = \hbar^2/m_e a_0^2$	$4.359\,748\,2(26) \times 10^{-18}$ J
Rydberg constant	$R_\infty = E_h/2hc$	$1.097\,373\,153\,4(13) \times 10^7$ m^{-1}
Bohr magneton	$\mu_B = e\hbar/2m_e$	$9.274\,015\,4(31) \times 10^{-24}$ J T^{-1}
Nuclear magneton	$\mu_N = (m_e/m_p)\mu_B$	$5.050\,786\,6(17) \times 10^{-27}$ J T^{-1}
Electron magnetic moment	μ_e	$9.284\,770\,1(31) \times 10^{-24}$ J T^{-1}
Proton magnetic moment	μ_p	$1.410\,607\,61(47) \times 10^{-26}$ J T^{-1}
Free electron g value	$g_e = 2\mu_e/\mu_B$	$2.002\,319\,304\,386(20)$

Source: E. R. Cohen and B. N. Taylor, *The 1986 Adjustment of the Fundamental Physical Constants*, CODATA Bulletin 63, Pergamon, Elmsford, N. Y. (1986); *Rev. Mod. Phys.* **57**, 1121 (1987).

Notation: $1.234\,5(67)$ specifies a standard deviation uncertainty of 67 in the final two digits.

PHYSICAL CHEMISTRY

PHYSICAL CHEMISTRY

John S. Winn
Dartmouth College

HarperCollins*CollegePublishers*

Executive Editor: Doug Humphrey
Project Coordination and Text Design: Elm Street Publishing Services, Inc.
Cover Design: Kay Fulton
Compositor: Monotype Composition Company, Inc.
Printer and Binder: R.R. Donnelley & Sons Company
Cover Printer: The Lehigh Press, Inc.

Source acknowledgments:

In addition to those sources specifically cited in the text, grateful acknowledgment is due the following sources of figures and data tables for permission to reproduce them or extracts from them in this text: Academic Press (Figures 14.18 and 14.21), Dr. B. J. Alder (Figure 17.4), Butterworth (Figure 16.26), Dr. M. W. Chase, Jr. (Tables 7.6–7.9), Prof. K. A. Jackson (Figure 6.27), MIT Press (Figure 8.5), National Institute of Standards and Technology (Tables 13.3–13.4), Prof. R. H. Soderberg (Figure 16.16b), Prof. G. A. Somorjai (Figures 16.23 and 16.25), Van Nostrand Reinhold (Tables 19.2 and 19.3), John Wiley and Sons (Figure 16.16a), and Wiley Interscience (Table 7.2).

Physical Chemistry

Copyright © 1995 by HarperCollins College Publishers

All rights reserved. Printed in the United States of America.
No part of this book may be used or reproduced in any manner whatsoever without written permission, except in the case of brief quotations embodied in critical articles and reviews. For information address HarperCollins College Publishers, 10 East 53rd Street, New York, NY 10022.

Library of Congress Cataloging-in-Publication Data
Winn, John S.
   Physical chemistry / John S. Winn.
      p. cm.
   Includes bibliographical references and index.
   ISBN 0-06-047148-4
    1. Chemistry, Physical and theoretical.   I. Title.
QD453.2.W56   1994
541--dc20   94-17318
                                                    CIP

For John Christopher

CONTENTS

Preface xix

CHAPTER 1

Equations of State, Thermodynamic Variables, and Gas Behavior 1

1.1 The Ideal Gas Equation as an Equation of State 2
1.2 Real Gases—The Origins of Nonideal Behavior 13
1.3 The van der Waals Equation of State 17
    A Closer Look at the van der Waals Equation of State 23
1.4 Corresponding States 25
    Summary 28
    Further Reading 28
    Practice with Equations 28
    Problems 29

CHAPTER 2

Energy, Work, and Heat 32

2.1 What Is Energy? 33
2.2 The Transfer of Energy Called Work 34
2.3 The Variation of Work with the Path 38
    A Closer Look at Work Calculations 41
2.4 The Transfer of Energy Called Heat 42
2.5 Simple Consequences of the First Law 44
    The Joule Expansion 46
    General Adiabatic Processes 48
    Adiabatic Reversible Processes in an Ideal Gas 49
    Paths Equivalent to an Adiabatic Path 50
    A Closer Look at the Meaning of Adiabaticity 52
2.6 A New State Function, Enthalpy 52
    The Joule–Thomson Effect 54
    A Closer Look at the Joule–Thomson Effect 56
    Summary 58
    Further Reading 59
    Practice with Equations 59
    Problem 59

CHAPTER 3

Spontaneity, Equilibrium, and Entropy 63

3.1   A Statistical View of Entropy 64
3.2   A Thermodynamic View of Entropy 67
3.3   A Mathematical View of Entropy 71
         A Closer Look at Temperature Scales 72
3.4   The Second Law of Thermodynamics 73
3.5   General Properties of Entropy Changes 75
3.6   Consequences of Entropy as a State Function 80
         A Closer Look at the Derivation of $(\partial U/\partial V)_T$ 82
         Summary 85
         Further Reading 85
         Practice with Equations 86
         Problems 86

CHAPTER 4

Heat Capacities, Absolute Zero, and the Third Law 88

4.1   The Variation of Entropy with Temperature 88
4.2   Life at Absolute Zero 89
         A Closer Look at the Two-Level System 93
4.3   The Meaning of Heat Capacity 94
         Solids 95
         Liquids 97
         Gases 98
         A Closer Look at C_P and C_V for Gases 100
4.4   A First Look at Phase Transitions 103
4.5   The Third Law and Absolute Entropy 105
4.6   The Calculation of Absolute Entropies 107
         Summary 111
         Further Reading 111
         Access to Data 111
         Practice with Equations 112
         Problems 112

CHAPTER 5

Stability, Free Energy, and Thermodynamic Potentials 116

5.1   Conditions for Stable Equilibrium 117
         A Closer Look at Stability 119
5.2   Free Energy 120
         A Closer Look at the Meaning of Work, w 122
5.3   Multicomponent Systems and Chemical Potential 124
         A Closer Look at the Uniformity of Chemical Potential 126
5.4   The Calculation of Chemical Potentials 127
         One Component Ideal Gas—In General 129
         One Component Ideal Gas—The Standard State 130
         One Component Real Gas—Fugacity 130
5.5   A Summary of Thermodynamic Calculus 134
         Internal Energy 135
         Enthalpy 135

Helmholtz Free Energy 135
Gibbs Free Energy 136
Summary 136
Further Reading 137
Access to Data 137
Practice with Equations 137
Problems 137

CHAPTER 6

Phase Equilibria, Mixtures, and Solutions 141

6.1 Why Do Phase Transitions Occur? 142
6.2 The Coexistence of Phases in Pure Substances 147
  A Closer Look at the Clapeyron Equation 151
  Vapor–Liquid or Vapor–Solid Coexistence Line (Vaporization and Sublimation) 155
  Solid–Liquid Coexistence Line (Fusion) 155
  Solid–Solid Coexistence Line 157
6.3 Composition Variables in Mixtures 157
6.4 Differences between Solutions and Mixtures 159
  Ideal Mixing and Partial Pressure 159
  A Closer Look at Partial Pressure 160
  Nonideal Mixing 163
  Excess Functions and Activity 165
  Ideal Solution 165
  Phase Separation and Spontaneous Un-mixing 166
6.5 Colligative Properties 170
  Vapor Pressures Over Solutions 171
  Henry's Law and the Ideal Dilute Solution 173
  Freezing Point Depression and Boiling Point Elevation 176
  Osmotic Pressure 179
6.6 Many Compounds, Many Phases—The Phase Rule 181
6.7 Phase Diagrams 186
  Cu/Ni 188
  Cd/Bi 189
  Sn/Pb 190
  Summary 191
  Further Reading 192
  Access to Data 192
  Practice with Equations 192
  Problems 193

CHAPTER 7

Chemical Reactions and Chemical Equilibria 200

7.1 Chemical Reactions as a Thermodynamic Path 201
7.2 Energy and Entropy Changes in Chemical Reactions 203
7.3 Named Enthalpies and Enthalpy Diagrams 206
  Standard Enthalpies of Formation 206
  Bond Enthalpies and Energies 209
  Other Named Enthalpies 212

CONTENTS

- 7.4 Chemical Reaction Equilibria 217
  - Ideal Mixture of Reacting Ideal Gases 220
  - Real Reactive Mixtures 220
- 7.5 The Stability of Chemical Equilibria 222
  - Temperature Stability 222
  - Adiabatic Reactions 223
  - Pressure Stability 225
  - Stability in Open Systems 227
- 7.6 Methods of Calculating Equilibrium Constants 228
  - Summary 242
  - Access to Data 243
  - Practice with Equations 243
  - Problems 244

CHAPTER 8

Thermodynamics of Surfaces and Interfaces 253

- 8.1 The Energy Contributions of Interfaces 254
- 8.2 Small Drops and Bubbles 257
- 8.3 The Nucleation of Phases 259
- 8.4 Equilibrium Forces among Interfaces 261
  - Wetting and Capillarity 263
  - Summary 264
  - Further Reading 265
  - Access to Data 265
  - Practice with Equations 265
  - Problems 265

CHAPTER 9

Gravitational, Magnetic, and Electric Fields 269

- 9.1 Equilibrium in Gravitational Fields 270
- 9.2 Centrifugal Fields 274
- 9.3 Magnetic and Electric Field Energies 277
  - Electric Field Phenomena 278
  - A Closer Look at Polarization 288
  - Magnetic Field Phenomena 290
  - A Closer Look at Magnetization 296
  - Summary 298
  - Further Reading 299
  - Practice with Equations 299
  - Problems 300

CHAPTER 10

Ionic Solutions and Electrochemistry 307

- 10.1 Aqueous Solutions of Ions—Extreme Nonideality 308
  - A Closer Look at the Debye–Hückel Theory 316
- 10.2 Chemical Reactions that Transfer Electrons 323
- 10.3 The Electrochemical Potential 326
- 10.4 Electrochemical Devices 327

10.5 Using Electrochemical Data 333
   Summary 341
   Further Reading 341
   Access to Data 341
   Practice with Equations 342
   Problems 342

CHAPTER 11

The Need for Quantum Mechanics 349

11.1 Basic Quantum Mechanical Phenomena 350
11.2 What to Expect from Quantum Mechanics 357
11.3 How to Think Quantum Mechanically 371
   The Uncertainty Principle 371
   Discrete Energy Levels 376
   Continuous Energy Levels 380
   The Correspondence Principle 380
   Degeneracy 381
   Summary 381
   Further Reading 382
   Practice with Equations 382
   Problems 383

CHAPTER 12

Quantum Mechanical Model Systems 388

12.1 The One-Dimensional Free Particle 389
12.2 The One-Dimensional Trapped Particle 394
12.3 Multidimensions 409
   A Closer Look at Separation of Variables 411
12.4 Angular Momentum and the Hydrogen Atom 415
12.5 "It Might As Well Be Spin" 435
   Summary 440
   Further Reading 442
   Practice with Equations 442
   Problems 443

CHAPTER 13

The Structure of Atoms 450

13.1 Basic Ideas—A Review of the Periodic Table 451
13.2 The Problems of Too Many Particles 452
13.3 An Interlude on Approximation Methods 456
   Variation Theory 456
   Perturbation Theory 462
   A Closer Look at Perturbation Theory 466
13.4 The Wavefunctions of Atoms 468
   A Closer Look at the "r_{12} Integral" 469
   A Closer Look at the Exchange Integral 475
13.5 The Description of Atomic Energy Values 479
   Summary 490

Further Reading 491
Practice with Equations 492
Problems 492

CHAPTER 14

The Origins of Chemical Bonding 498

14.1 The Simplest Molecules: H_2^+ and H_2 499
    The Closer Look at Molecular Electronic Angular Momentum 506
    A Closer Look at Spin Effects on Bonding 517
14.2 Approximate Descriptions of Covalent Bonds 519
14.3 Named Bonds and the Concept of Delocalization 525
14.4 The Forces that Shape Molecules 536
    A Closer Look at Hybrid Orbitals 545
14.5 Taking Molecules Apart 547
    Summary 553
    Further Reading 553
    Practice with Equations 554
    Problems 554

CHAPTER 15

Electric and Magnetic Properties of Molecules 559

15.1 Basic Properties of Molecular Charge Distributions 560
    A Closer Look at Parity 565
    A Closer Look at the Multipole Expansion 568
    A Closer Look at the Hellmann–Feynman Theorem 576
15.2 Special Consequences of Unpaired Electrons 576
15.3 The Motion of Molecules in External Fields 580
    Summary 582
    Further Reading 582
    Practice with Equations 582
    Problems 583

CHAPTER 16

The Solid State 588

16.1 Periodicity and Symmetry in Solids 589
    The Rare Gases 589
    Metallic Elements 594
16.2 Bonding Types in Solid Phases 596
    Ionic Solids 597
    A Closer Look at the Madelung Constant 604
    Metals 605
    Free Electron Model of Metals 605
    The Engel–Brewer Model 607
    Covalent Solids and Band Theory 611
    Molecular Crystals 613
16.3 Crystallography: Measuring Solid Structures 616
16.4 Amorphous Solids and Crystal Defects 625
16.5 Solid Surface Structures 627

Summary 634
Further Reading 634
Practice with Equations 634
Problems 635

CHAPTER 17

The Liquid State 642

17.1 Why Liquids Are Messy 643
17.2 The Connections to Solids and Gases 643
17.3 Simple Liquids 649
  A Closer Look at Computer Simulations 650
17.4 Ordinary Liquids—Water 655
17.5 Liquid Crystals 657
  Summary 662
  Further Reading 662
  Problems 662

CHAPTER 18

Molecules and Radiation 665

18.1 What Is Electromagnetic Radiation? 666
18.2 How Spectroscopies Work—Absorption and Emission 669
  A Closer Look at Blackbody Radiation 670
18.3 Atomic Spectroscopy and Selection Rules 675
  A Closer Look at Spontaneous Emissions 677
18.4 Too Much Energy—Photoionization and Photodissociation 682
  Photoionization 682
  Photodissociation 683
  Summary 686
  Further Reading 686
  Practice with Equations 686
  Problems 687

CHAPTER 19

Molecular Spectroscopy of Small Free Molecules 691

19.1 The Classification of Spectroscopies by Molecular Motions 692
19.2 Small Molecule Energy Levels 696
  Diatomic Molecules 697
  A Closer Look at + and − States 698
  Polyatomic Molecules 712
19.3 Molecular Structure as Deduced by Spectroscopy 720
  Diatomic Rotational Spectroscopy 720
  Diatomic Vibrational Spectroscopy 724
  Diatomic Electronic Spectroscopy 727
  Polyatomic Spectra 731
19.4 Potential Energy Surfaces 733
  A Closer Look at the Morse Potential 734
19.5 Line Shapes and Spectral Congestion—What's Observable? 738
  A Closer Look at the Doppler Effect 741

Summary 746
Further Reading 746
Practice with Equations 747
Problems 747

CHAPTER 20

Molecular Spectroscopy of Large Molecules 752

20.1 Why a Triatomic May Be a Large Molecule 753
20.2 Electronic Structure in Large Molecules 755
20.3 Radiationless Transitions—Where Did the Energy Go? 762
20.4 Atomic Chaos at Large Energies 765
20.5 Spectroscopy in Condensed Media 770
    A Closer Look at Tunable Dye Lasers 771
    Summary 773
    Further Reading 773
    Practice with Equations 774
    Problems 774

CHAPTER 21

Spectroscopy in Magnetic and Electric Fields 779

21.1 The Zeeman and Stark Effects 780
    A Closer Look at the Landé g Factor 783
    A Closer Look at Atomic Time Standards 789
21.2 Spin Resonance Spectroscopies 791
    Nuclear Magnetic Resonance 791
    A Closer Look at Fourier Transforms 807
    Electron Spin Resonance 809
    Summary 812
    Further Reading 812
    Practice with Equations 812
    Problems 813

CHAPTER 22

The Dynamic Nature of Equilibrium 818

22.1 The Meaning of Distribution Functions 819
    A Closer Look at FD and BE Statistics 827
22.2 The Statistical Necessity of Equilibrium 832
22.3 The Role of Intermolecular Forces in Maintaining Equilibrium 842
    Summary 848
    Further Reading 848
    Practice with Equations 849
    Problems 849

CHAPTER 23

The Molecular Basis of Equilibrium 852

23.1 Partition Functions—How Many States Contribute? 853
    Translational Motion 854
    A Closer Look at the Translational Partition Function 856
    Vibrational Motion 857

                Electronic Motion 859
                Nuclear Spin Motion 859
                Rotational Motion 859
                A Closer Look at the Symmetry Number 863
                Spectroscopic Quantities 864
23.2   From Microscopic Individuality to Macroscopic Average 864
                Molecular Distribution Functions 864
                The Equipartition Theorem 868
                Statistical Thermodynamics 870
23.3   Statistical Thermochemistry 881
                Summary 890
                Further Reading 891
                Practice with Equations 891
                Problems 892

CHAPTER 24

The Kinetic Theory of Gases 899

24.1   Speeds and Velocities of Gases 900
24.2   The Dynamic Nature of Pressure 910
                A Closer Look at an Effusive Molecular Beam 917
24.3   Intermolecular Forces and Nonideal Gases 919
                The Intermolecular Potential Energy 921
                The Molecular Description of Nonideal Gases 936
                Summary 938
                Further Reading 939
                Practice with Equations 940
                Problems 941

CHAPTER 25

Nonequilibrium Dynamics 947

25.1   The Nonequilibrium Distribution Function 948
25.2   Phenomenological Transport—Forces and Fluxes 957
                Diffusion 958
                A Closer Look at Concentration Gradients 963
                Thermal Conduction 965
                Electrical Conduction 968
                Viscosity 970
25.3   Microscopic Transport Coefficients 975
                Diffusion 976
                Thermal Conductivity 978
                Electrical Conductivity 979
                Viscosity 981
25.4   Transport in Condensed Media 984
                Summary 989
                Further Reading 989
                Practice with Equations 990
                Problems 991

CHAPTER 26

The Phenomenology of Chemical Reaction Rates 997

26.1 Basic Ideas—Reactants Go, Products Appear 998
26.2 Mechanisms and Integrated Rate Expressions 1002
　　　First-Order Reactions 1002
　　　Second-Order Reactions 1003
　　　Sequential Reactions 1007
　　　Branching Reactions 1010
　　　Complex Mechanisms—Chain Reactions 1012
　　　Branching Chain Reactions—Explosions 1016
26.3 The Effects of Thermodynamic Variables on Reaction Rates 1018
　　　A Closer Look at the Arrhenius Plot 1020
26.4 The Attainment of Chemical Equilibrium 1022
　　　Detailed Balance 1022
　　　Chemical Relaxation 1025
　　　Chemical Oscillations 1028
　　　Summary 1031
　　　Further Reading 1032
　　　Access to Data 1032
　　　Practice with Equations 1032
　　　Problems 1033

CHAPTER 27

Elementary Processes and Rate Theories 1039

27.1 What Is an Elementary Process? 1040
　　　Unimolecular Processes 1040
　　　Bimolecular Processes 1041
　　　Termolecular Processes 1043
　　　Photochemical Processes 1045
27.2 Simple Collision Theories of Reactions 1049
　　　Gas-Phase Reactions 1049
　　　Solution Reactions 1052
27.3 The Transition State 1058
　　　Activated-Complex Theory 1063
　　　A Closer Look at Activated-Complex Theory 1067
　　　A Closer Look at the Transition State 1072
　　　Thermochemical Kinetics 1073
27.4 Unimolecular Reactions—The Decay of Energy 1077
27.5 Catalysts and the Design of Potential Energy Surfaces 1081
　　　Enzyme Catalysis 1081
　　　Homogeneous Catalysis 1085
　　　Heterogeneous Catalysis 1085
　　　Summary 1089
　　　Further Reading 1090
　　　Practice with Equations 1091
　　　Problems 1092

CHAPTER 28

Chemical Reaction Dynamics 1099

28.1 Where Does Chemical Energy Come From and Where Does It Go? 1100
28.2 State-to-State Kinetics 1109
      Elastic Scattering 1109
      Inelastic Scattering 1119
      Reactive Scattering 1126
28.3 Models for Elementary Reaction Dynamics 1128
28.4 Is There Microscopic Control over Reactions? 1136
      Summary 1141
      Further Reading 1142
      Practice with Equations 1142
      Problems 1142

APPENDIX

Units, Dimensions, and Constants 1147

Index 1153

PREFACE

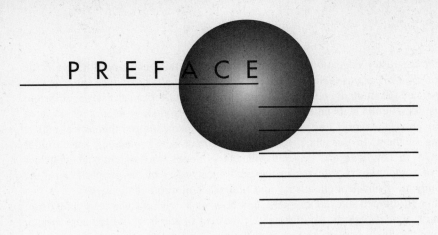

This is a vast field for employing the zeal and abilities of young chemists, whom I would advise to endeavor to do well . . . .

Antoine Lavoisier, *Elements of Chemistry* (1789)

Chemistry works with an enormous number of substances . . . ; it is an extensive science. Physics on the other hand works with rather few substances . . . ; it is an intensive science. Physical chemistry is the child of these two sciences; it has inherited the extensive character from chemistry. Upon this depends its all-embracing feature, which has attracted so great admiration. Physical chemistry may be regarded as an excellent school of exact reasoning for all students of natural sciences.

Svante Arrhenius, *Theories of Solutions* (1912)

Honk If You Passed PChem

Bumper sticker sold by the American Chemical Society (latter quarter of the twentieth century)

TO THE STUDENT

Do not call me Ishmael.

There is a suspicion among students that physical chemistry is the "Great White Whale" of the chemistry curriculum. This is not true. Physical chemistry will call upon your prior knowledge of chemistry, your knowledge of physics, and your knowledge of math, but physical chemistry is neither impossibly difficult nor frighteningly challenging.

There was a time when most physicists could not get more than about three elements deep into the Periodic Table (H, He, and maybe Na), and most chemists could not get more than about one differential equation deep into physics (maybe $F = m\, d^2x/dt^2$), but that time has long past. The vast interface of molecular science bridging chemistry and physics, which is the realm of physical chemistry, is not only vital to a chemist's knowledge, it is also vital to those branches of physics and chemistry that have evolved as new and growing disciplines themselves: materials science, biophysical chemistry, geochemistry, atmospheric chemistry, astrochemistry, and so forth.

Physical chemistry is usefully divided into three areas of focus, and this book (along with the courses in which you will use it) is organized following those areas. The order in which you first learn them may not follow their order here, but your physical chemistry education is not complete until you touch upon them all: the study of *equilibrium chemical systems,* the study of *molecular structure and chemical bonding,* and the study of *chemical dynamics* in both equilibrium and nonequilibrium settings. This study will take you to new places, introduce you to new ways of thinking about the world, and show you new phenomena and our theoretical understanding of them. In this book you will read about new molecules, see familiar molecules in new settings, travel to planetary moons and supernovae, consider topics of contemporary research in the field, meet many of the people behind the development of physical chemistry, and find your intuition challenged.

Although divided into 28 chapters, this book is a coherent whole. Topics from Chapter 1 appear periodically throughout the rest of the book where appropriate, including the final chapter. As you read the book in your assigned order (and it is unlikely that you will be assigned every chapter in your courses—read the remainder at your leisure!), pay attention to this interconnectedness. Ideas from thermodynamics apply to reaction kinetics; ideas from molecular spectroscopy apply to statistical mechanics; ideas from quantum chemistry apply to the reactivity of bulk metals.

Take notes as you read. The size of this text reflects the concern that has gone into careful explanations more than it does the quantity of material to learn. As a key result is derived, do not worry over the mathematical details behind, for instance, the evaluation of an integral or some other mathematical step. Pay more attention to the *physical* reason for an integration (or other mathematical operation) than to its *mathematical* evaluation, at least at first. The analogy between learning a foreign language and learning a new branch of science is stressed several times in this text, and it is an apt one. With a new language, a basic vocabulary and a feel for the spoken and written language come before mastering the fine points of the grammar. So it is here, but the fine points—the mathematical techniques—take you toward fluency in physical chemistry.

Many chapters have short discussions of advanced topics or supplementary material. These discussions carry the title "A Closer Look at ..." and they may be skipped on first reading. They should not be ignored completely, however, since they fill in details of derivations or otherwise explain a concept in greater depth.

Some Thoughts on Problem Solving

Mastering physical chemistry takes practice, and in addition to the experimental practice you will have in a laboratory course, you must practice the reasoning and mathematics of physical chemistry through problem solving. To aid you, each chapter has a number of example problems worked in some detail. Each example ends with a list of related problems that appear at the chapter's end. These problems are grouped according to the chapter section for which they provide practice, ending with a number of general problems that are often more challenging or otherwise greater in scope than those grouped into a single section's collection.

Equally important, however, are the short questions called *Practice with Equations* that precede each collection of problems. A reference to a numbered equation in the chapter follows each of these questions, along with the answer. They are designed to make you think about the physical quantities that go into key equations, the ways any one equation might be rearranged in a new context, and the dimensions, magnitudes, and common units used to express important physical quantities. These questions should not take you long to answer, and you should attempt many of them before moving into the problems themselves.

Before you begin working your first problem, however, you should read the Appendix at the end of the book on units, dimensions, and constants. This material will teach you the conventions used in this book to write physical quantities and their all-important physical units. *There is no more common mistake a beginning student of physical chemistry makes than to neglect units and proper orders of magnitude.* There is also no mistake that is easier to correct than this one. There are many important tables of data in nearly every chapter, and some of these are taken from important contemporary data sources. Many chapters have a list of references that give you access to data in a particular area (but every problem states all the information you will need to solve it).

As you work a problem, look in the text for analogies to the mathematical techniques you might need, and try to see the *physical* basis of the question rather than just the numerical answer. Why does an answer have the magnitude (or even the algebraic sign) it has? Do these make physical sense? If a problem asks for an energy and your answer ends with units of meters per second, for example, then you have made a mistake somewhere. If a related example in the text yields an answer of a few hundred kJ mol^{-1} and your answer is many orders of magnitude smaller, then perhaps you forgot a factor of Avogadro's constant.

Make sure you have mastered your computational tools before you begin a problem. Do you know what all those keys on your calculator do? Is a list of fundamental constants and a Periodic Table handy? (Both are inside the front cover of the book.) Is a sketch of a graph called for somewhere in the problem? If you have access to a personal computer spreadsheet program, learn it well and use it as you work. With practice, it will be as easy to use as a sheet of paper. The same is true of a symbolic algebra computer program, but remember that you may be alone with your calculator on exams!

If you cannot answer every part of any one problem, do not be discouraged. (On the other hand, if you can answer every problem with ease, then congratulations! You are ready to move into the references listed for further reading at the end of each chapter.) Think of the problems as a training ground that not only tests your skills but shows you the concepts, techniques, and ideas you should continue to review. Your instructor will probably not assign all of the problems in any one chapter (and she or he may well supplement these with additional ones), but there is never a penalty for doing more problems than those assigned.

I wish you well in your study of physical chemistry, and I welcome your comments about this book. I have enjoyed writing it, and I have learned some new things myself in doing so. This book is only an introduction to physical chemistry, but in its a comprehensive one. I hope you continue to find it useful beyond the courses for which you first bought it.

PREFACE

TO THE INSTRUCTOR

As you read the table of contents, you will see that the standard mix of topics common to physical chemistry courses at the sophomore through senior undergraduate level are included here. Perhaps a few are lacking that you will miss (polymers and group theory are two that do not have their own chapter), and for this I apologize. What the table of contents alone will not tell you is the level of the treatment or the innovations that have been brought to the presentation of some familiar key concepts.

The level assumes the chemical, physical, and mathematical preparation appropriate to the serious advanced undergraduate student for whom a professional career in science or medicine is the goal. Difficult topics are introduced with considerable care and discussion in the hope that the student will see not only the mathematical result but also the physical reasoning behind that result. Virtually all the figures were drawn explicitly for this text, and most were generated through personal computer graphics programs that faithfully reproduce the mathematics and physics rather than qualitatively approximating them. From the van der Waals isotherms in Chapter 1 to the collinear reaction trajectories in Chapter 28, care has been taken to show phenomena accurately.

There has also been a conscious effort to integrate material throughout the book. This integration takes place in several ways: a uniform system of notation (including notation prevalent in research journals in areas such as spectroscopy), cross-referencing equations and ideas, and the use of a small set of compounds as paradigms. Hydrogen, argon, water, and several other key substances are used to illustrate ideas such as equilibrium reaction thermodynamics, phase diagrams, atomic structure, molecular structure, reaction kinetics, and statistical mechanics. Often, the student is reminded of the behavior of one of these key substances from an earlier chapter's discussion, connecting that discussion to the current one. Similarly, several physical models are used again and again, such as the hard-sphere potential and the two-level system. Each new use of a model builds on the previous uses but not in a way that precludes teaching topics in a different order from that presented here.

An effort has been made to introduce important data from contemporary sources and to teach students how to use these sources. These include JANAF Thermochemical Tables, the NIST Atomic Energy Level tables, the diatomic spectroscopic constant tables of Huber and Herzberg, and other more general sources included in sections called *Access to Data*. The tabulated data themselves generally include the same species no matter what the table (wherever appropriate).

Worked examples point to related problems at the end of each chapter, and a series of short questions, called *Practice with Equations,* is collected before each chapter's problems. These questions include a reference to an equation, table, or idea presented in the chapter, along with the question's answer. They are intentionally brief, since their purpose is to guide students through the quantities that enter a particular equation without adding the extra layer of reasoning that the problems typically invoke. The problems themselves are grouped according to chapter section, and they end with a collection of general problems that are often more challenging. The majority of the problems are designed to allow the student to test her or his understanding in a particular context. Often, they make a point or tell a story that helps to carry the discussion beyond the text. A solutions manual for these problems is available from the publisher.

I have tried to present physical chemistry in its contemporary setting. Several new derivations, presentations, or topics themselves have been included in order to achieve this goal. An additional benefit of some of these presentations has been to ease the unification of several topics that I had felt were somewhat disconnected by discussions that over-simplified some derivations. The first ten chapters consider equilibrium phenomena, building chemical thermodynamics from one-component, one-phase systems, through many-component, many-phase systems, adding chemical reactions in Chapter 7, surfaces in Chapter 8, external fields in Chapter 9, and electrolytes in Chapter 10. These latter four chapters can be used or ignored as the instructor needs without loss of continuity.

Quantum phenomena, molecular structure, and spectroscopies follow in Chapters 11–21. Again, there is considerable flexibility possible in one's choice of chapters for a particular course. Spectroscopies are emphasized in Chapters 18–21, while molecular and condensed-phase structures are emphasized in Chapters 14–17.

Chapters 22–28 cover dynamic or nonequilibrium situations. Chemical kinetics and reaction dynamics fall into Chapters 26–28, and each chapter is somewhat more sophisticated than the previous one. Statistical mechanics, the kinetic theory of gases, and transport phenomena fall in Chapters 22–25. These use the idea of a distribution function as their unifying theme. Selected sections in these chapters can stand alone, depending on one's need for a more or less microscopic discussion.

ACKNOWLEDGMENTS

Any book of this size and scope benefits from the advice and criticism that only its intended audience can offer. I have been fortunate to have had an excellent set of reviewers of all or selected portions of this book over the years during which it evolved, and I include among them the many students at the University of California at Berkeley and at Dartmouth College who have taken courses from me and helped me to develop the presentations that appear in this book. I particularly thank the following expert reviewers for their time and comments: S. L. Bernasek (Princeton University), L. K. Brice (Virginia Polytechnic Institute and State University), A. Campion (University of Texas at Austin), E. D. Cater (University of Iowa), M. J. Coté (Bates College), W. R. Gilkerson (University of South Carolina), D. Gilson (McGill University), L. P. Gold (Penn State University), G. A. Kenney-Wallace (University of Toronto), J. Laane (Texas A&M University), J. R. Marquart (Eastern Illinois University), J. H. Moore (University of Maryland), N. Muller (Purdue University), G. A. Raiche (Hamilton College), R. L. Scott (University of California–Los Angeles), P. Smith (Duke University), and C. A. Trapp (University of Kentucky–Louisville). I also thank my colleagues at Dartmouth for their interest in and contributions to this book. In particular, Professor J. E. G. Lipson kindly helped field-test the first ten chapters in her course, and Professor W. H. Stockmayer read and provided many valuable comments to many chapters. Professor W. Hubble (University of California–Los Angeles) kindly shared with me his copy of the 1828 chemistry text quoted in the introduction to Chapter 6.

I also thank my publishers for their expert help and advice. Special thanks go to Jane Piro and Doug Humphrey at HarperCollins and to Nancy Shanahan at Elm Street Publishing Services.

The Third Law of Thermodynamics says that a state of perfect order is attained at the absolute zero of temperature and that this state can never be attained. The publishing corollary to that law states that a book of this size and complexity can probably never be completely free of errors. In spite of extensive proofreading, I expect a typographical error or two may still remain. I hope these are few and harmless, and I welcome comments that will in any way improve this book.

Finally, to everyone who has lived with me over the years during which I wrote this book, it is now safe to ask me, "When's the book gonna be finished?"

John S. Winn
Hanover, New Hampshire
jwinn@dartmouth.edu
December 1994

CHAPTER 1

Equations of State, Thermodynamic Variables, and Gas Behavior

THE ideal gas equation, Eq. (1.1), is probably familiar to you:

$$PV = nRT .\qquad(1.1)$$

It expresses what is intuitive to everyone who has played with balloons or inflated tires. In a tied-off balloon, the amount of gas is constant, and if the temperature is constant, one must squeeze the balloon (increase the pressure) to decrease its volume. Mathematically, if the right-hand side of Eq. (1.1) is constant, a change in pressure, P, can only be accomplished by a simultaneous change in volume, V, such that the product of pressure and volume before the change is numerically equal to the PV product after the change.

In this chapter we will examine this equation, its variables, the assumptions tacitly made when it is written, and the changes it must undergo to widen its predictive ability and accuracy. We begin this text with gases at equilibrium not only for the inherent interest in gaseous behavior but also for the wide-ranging introduction to the methods of physical chemistry that this topic affords. In many ways, this first chapter is among the most important. Many definitions, concepts, types of equations, and mathematical procedures will be introduced here and used again and again later in the text. To begin this foundation, we examine the variables of the ideal gas equation.

1.1 The Ideal Gas Equation as an Equation of State

1.2 Real Gases—The Origins of Nonideal Behavior

1.3 The van der Waals Equation of State

1.4 Corresponding States

1.1 THE IDEAL GAS EQUATION AS AN EQUATION OF STATE

One of the hallmarks of the gas phase is that a gas occupies all of any volume enclosing it. Usually, the walls of the container will be the boundaries of the volume, but on occasion, we will find it convenient to consider a region in space bounded by imaginary surfaces as our volume of interest. Dimensionally, volumes are cubed lengths,[1] such as the cubic centimeter, cm^3, or the cubic decimeter, dm^3, which is the same as the liter, L.

The amount of gas is measured by the number of moles, n. Note that the mass of the gas (and for that matter, its chemical identity) does not directly enter the discussion. All that matters is the number of molecules, and these molecules need not be all the same.

The ratio V/n is the *molar volume* $\overline{V}$ with units of $cm^3\,mol^{-1}$, for example. It is a different type of variable than is V or n alone, since if we double V and *simultaneously* double n, $\overline{V}$ does not change. This means we can rewrite Eq. (1.1) as

$$\frac{PV}{n} = P\overline{V} = RT \,. \tag{1.2}$$

Note that we are *not* restricting ourselves to just one mole of gas in Eq. (1.2). With $\overline{V}$, we need not state how much gas is present, but we consider instead the fate of each and every mole of gas, no matter how many.

EXAMPLE 1.1

Which of the following pairs of situations are characterized by equal molar volumes?

(a) $V = 0.1\,dm^3$, $P = 1.5$ atm, $T = 300$ K and $V = 0.2\,dm^3$, $P = 3.0$ atm, $T = 600$ K

(b) $V = 0.1\,dm^3$, $P = 1.5$ atm, $T = 300$ K and $V = 0.1\,dm^3$, $P = 1.0$ atm, $T = 300$ K

(c) $V = 0.1\,dm^3$, $P = 2.0$ atm, $T = 450$ K and $V = 0.1\,dm^3$, $P = 2.0$ atm, $T = 250$ K

SOLUTION We rewrite Eq. (1.2) in the form $\overline{V} = RT/P$, which shows that ideal gas situations that have the same value of RT/P (or simply T/P, since R is a constant) have the same molar volumes. Pair (a) satisfies this requirement since 300 K/1.5 atm = 600 K/3.0 atm. Pair (b) has common temperatures, but differing pressures, while pair (c) has common pressures, but differing temperatures. Neither of these would have a common molar volume.

➠ RELATED PROBLEMS 1.1, 1.2

Pressure has the dimensions of force per unit area. This must be kept in mind even when units for pressure such as the atmosphere or bar are used. Since energy has the dimension of force times length, pressure could also be expressed as energy per unit volume: pressure = force/area = (force · length)/(area · length) = energy/volume. Thus, the product PV has the dimensions of energy, and so must the product nRT that it equals.

Much of this text will discuss energy, and the product nRT becomes our first opportunity to do so. We begin first with a brief discussion of temperature. Temperature is a common quantity in everyday life, but in spite of our constant use of it, temperature is one of the more difficult variables to understand and interpret correctly. For now we take the following definition of the absolute temperature: T will be

---

[1] We will not limit ourselves to one particular unit of length (or anything else) in this text, since many systems of units are in common use. A detailed discussion of dimensions, units, and conversion factors is given in the Appendix at the end of the book. Familiarize yourself with the material there at your earliest convenience.

measured in degrees Kelvin (symbolized K, rather than °K) where 273.16 K is the difference between the coldest temperature, which we assign the value 0 K (the absolute zero of temperature), and the temperature at which water, ice, and water vapor coexist alone in equilibrium (the so-called triple point of water).

At first glance this may seem to be a strange way to define so fundamental a quantity and, in a sense, it is. The definition is partly historical accident, as was the ancient definition of the yard as the length from the King's nose to the tip of his outstretched arm. You should also note that we have not given any argument as to why an absolute zero of temperature should exist, why there is one and only one temperature at which water, ice, and water vapor can coexist, or how we measure temperature in general. These arguments will come in due time and will be seen to be independent of our choice of temperature scale.

The dimensions of R, the universal gas constant, can be deduced by rearranging the ideal gas equation:

$$R = \frac{PV(\text{energy})}{n(\text{mol})\,T(\text{K})}\ .$$

Dividing energy by moles gives us a *molar energy,* and R has the dimensions of molar energy per unit temperature.[2] If we measure energy in units of joules, R has the numerical value 8.314 51 J mol^{-1} K^{-1} or about 8.3 J mol^{-1} K^{-1}, an approximate value worth remembering. If pressure is measured in atm units and volume in cm^3 units, then R = 82.057 83 cm^3 atm mol^{-1} K^{-1} $\cong$ 82 cm^3 atm mol^{-1} K^{-1}. It is vital to select the correct value of R according to the mix of units required in any calculation.

EXAMPLE 1.2

What is the molar volume of a room temperature ideal gas at 1 atm pressure? What is its $\overline{PV}$ product molar energy value?

SOLUTION Since $\overline{V} = RT/P$, we only need T (which is roughly 300 K at room temperature) and P to find the molar volume. With P in atm units, the appropriate value for R is 82.0 cm^3 atm mol^{-1} K^{-1}. We find

$$\overline{V} = \frac{(82.0\ \text{cm}^3\ \text{atm mol}^{-1}\ \text{K}^{-1})\,(300\ \text{K})}{(1\ \text{atm})} \cong 25\,000\ \text{cm}^3\ \text{mol}^{-1}$$

which, in English units, is roughly 1 ft^3 mol^{-1} (1 ft^3 = 28 317 cm^3), a value that may be easier to visualize. The $\overline{PV}$ product equals RT; with R in SI units, RT = (8.3 J mol^{-1} K^{-1}) (300 K) $\cong$ 2.5 kJ mol^{-1}. It remains for us to see whether 2.5 kJ mol^{-1} is a large or small molar energy value.

➡ *RELATED PROBLEMS* 1.1, 1.2, 1.3

There is one other condition necessary to forge a link between energy and temperature. This is the condition of *equilibrium*. When an object is in equilibrium, the physical parameters that specify the state of the object (such as P and T) are uniform throughout the object.[3] Thermal equilibrium means that the object is uniformly at one temperature and that the temperature is the same as that of any matter

---

[2] We will see again and again that the ratio of molar energy to R is a common one in physical chemical calculations. This ratio has the dimensions of temperature.

[3] This statement is modified slightly in the presence of external force fields, such as gravity, as we will see in Chapter 9. In Chapter 5, we will prove the conditions for equilibrium in the absence of such fields.

immediately surrounding and in contact with the object. In describing phenomena that are not in a state of equilibrium, temperature may not be so easy to define.

This is one of the characteristic traits of classical thermodynamics, the topic of the first third of this text. *Classical thermodynamics treats only equilibrium situations.* We will see that thermodynamics can predict the composition of a reacting mixture, for instance, but *only* when the mixture has reached equilibrium. Thermodynamics has nothing to say about how long we must wait for equilibrium to be attained, in spite of the "dynamics" part of its name.

We can make one further classification of thermodynamic variables. At equilibrium, certain quantities are the same at every point. These include temperature, pressure, and molar volume, in the sense that we can specify these quantities without specifying the physical extent of the system in question. Such variables are called *intensive variables*. In contrast, *extensive variables,* such as volume, mass, or number of moles, require (or by themselves give) information as to the system's size.

EXAMPLE 1.3

Classify the following variables as either intensive or extensive:

(a) the molar energy of an ideal gas
(b) the surface area of a raindrop
(c) the density of the isotope ^{235}U
(d) the pH of pure water
(e) the volume of 1 M HCl(*aq*) needed to lower the pH of water to pH = 5.

SOLUTION (a) Molar energy, like all molar quantities, is intensive. It is the ratio of two extensive quantities: the system's energy and amount (number of moles).

(b) Surface area depends on the total volume of the system as well as the *shape* of the system. It is thus extensive (and requires a rather detailed description of the system for its evaluation).

(c) Density (which usually refers to *mass* density) is an intensive variable, but for the spontaneously fissionable isotope ^{235}U, which is not actually in equilibrium due to its radioactivity, the density can change rather rapidly and dramatically if the critical fissionable mass is exceeded.

(d) The pH expresses the $H^+(aq)$ *concentration,* which is an intensive quantity. Every point in any solution has the same pH.

(e) Here, the amount of acid needed to change the pH of water depends on the amount of water at hand. Thus, this amount (volume) is an extensive quantity.

➡ RELATED PROBLEM 1.4

If we know two of the variables of Eq. (1.2), all of which are intensive, we can calculate the third. Depending on which two we know, we write one of the following:

$$P = P(\overline{V}, T) = \frac{RT}{\overline{V}} \tag{1.3}$$

$$\overline{V} = \overline{V}(P, T) = \frac{RT}{P} \tag{1.4}$$

$$T = T(P, \overline{V}) = \frac{P\overline{V}}{R} . \tag{1.5}$$

In each of these equations, we have chosen two known variables to be *independent,* such as $\overline{V}$ and T in Eq. (1.3), and the third becomes the *dependent* variable, such as P in Eq. (1.3), a function of the chosen two.

These equations (really all three the same) are called *equations of state* for the ideal gas. *An equation of state is the mathematical relationship among the relevant thermodynamic variables of an equilibrium system.*

Note the term *system*. In thermodynamics, the system is "that part of the physical world under consideration." We specify the extent of the system by locating some kind of *boundary* between the system and the rest of the world, the *surroundings*, where we usually will be found. The location of the boundary is usually obvious, but it can sometimes be quite subtle when we need to construct an abstract mathematical surface as our boundary. For instance, given a bulb of gas, the bulb may be of less interest to us than the gas, and we place the boundary infinitesimally inside the bulb so that it encloses (in our mind) all the gas, leaving us and the bulb in the surroundings.

Next we consider how the chosen dependent variable will change when we alter, in one way or another, the independent variables. These changes are expressed by the appropriate partial derivatives of the equation of state.

A partial derivative is the rate of change of a function of many independent variables when all but one of these variables are held constant. For example, if we keep the molar volume constant by enclosing a fixed amount of gas in a rigid container, then the rate of change of pressure with temperature is

$$\left(\frac{\partial P}{\partial T}\right)_{\overline{V}} = \left(\frac{\partial (RT/\overline{V})}{\partial T}\right)_{\overline{V}} = \frac{R}{\overline{V}}. \tag{1.6}$$

Note carefully the location and meaning of the symbols in Eq. (1.6). The independent variables that are held constant (just $\overline{V}$ here) are written as subscripts to the partial derivative symbol. Each of these subscripted variables is considered a constant when taking the derivative. Similarly we can write

$$\left(\frac{\partial P}{\partial \overline{V}}\right)_{T} = \left(\frac{\partial (RT/\overline{V})}{\partial \overline{V}}\right)_{T} = -\frac{RT}{\overline{V}^2}. \tag{1.7}$$

The minus sign tells us (since $\overline{V}$, T, and R are inherently positive numbers) that pressure decreases as molar volume increases at constant temperature, just as we expected.

EXAMPLE 1.4

Match the symbolic partial derivatives on the left to the correct corresponding expression on the right, all of which are based on the ideal gas equation of state.

(a) $(\partial \overline{V}/\partial T)_P$ (i) $-RT/P^2$
(b) $(\partial T/\partial P)_{\overline{V}}$ (ii) $\overline{V}/R$
(c) $(\partial \overline{V}/\partial P)_T$ (iii) R/P

SOLUTION Derivatives (a) and (c) come from Eq. (1.4), which expresses $\overline{V}$ as a function of P and T:

(a) $\left(\dfrac{\partial \overline{V}}{\partial T}\right)_P = \left(\dfrac{\partial (RT/P)}{\partial T}\right)_P = \dfrac{R}{P}$ **(choice iii).**

(c) $\left(\dfrac{\partial \overline{V}}{\partial P}\right)_T = \left(\dfrac{\partial (RT/P)}{\partial P}\right)_T = -\dfrac{RT}{P^2}$ **(choice i).**

Derivative (b) comes from Eq. (1.5):

(b) $\left(\dfrac{\partial T}{\partial P}\right)_{\overline{V}} = \left(\dfrac{\partial (P\overline{V}/R)}{\partial P}\right)_{\overline{V}} = \dfrac{\overline{V}}{R}$ (choice ii).

⇒ **RELATED PROBLEMS** 1.5, 1.6

We can use Eq. (1.7) to calculate the infinitesimal change in pressure experienced by a gas when the molar volume is changed by the infinitesimal amount $d\overline{V}$ at constant temperature. We write

$$\text{at constant temperature:} \quad dP = \left(\dfrac{\partial P}{\partial \overline{V}}\right)_T d\overline{V} = -\dfrac{RT}{\overline{V}^2} d\overline{V} \quad (1.8)$$

but how do we use this expression in a calculation?

Consider a gas with molar volume 24.6 dm³ mol⁻¹ at 300 K and 1.0 atm. The value of $(\partial P/\partial \overline{V})_T$ at this particular molar volume and temperature is

$$\left(\dfrac{\partial P}{\partial \overline{V}}\right)_T = -\dfrac{RT}{\overline{V}^2} = -\dfrac{(0.082 \text{ dm}^3 \text{ atm mol}^{-1} \text{K}^{-1})(300 \text{ K})}{(24.6 \text{ dm}^3 \text{ mol}^{-1})^2} = -4.1 \times 10^{-2} \text{ atm mol dm}^{-3}.$$

In Eq. (1.8), this number gives us the amount, dP, by which the pressure changes when the molar volume is changed from 24.6 dm³ mol⁻¹ to $(24.6 + d\overline{V})$ dm³ mol⁻¹. The quantity $d\overline{V}$ represents an infinitesimal amount, but it can be approximated by a small, finite quantity. Suppose $d\overline{V} = +0.1$ dm³ mol⁻¹. Then Eq. (1.8) predicts

$$dP = (-4.1 \times 10^{-2} \text{ atm mol dm}^{-3})(0.1 \text{ dm}^3 \text{ mol}^{-1}) = -4.1 \times 10^{-3} \text{ atm}.$$

The pressure would change from 1.0 atm to $1.0 + dP = 0.9959$ atm. (The question of significant figures in our answer will be ignored for a moment in order to make a point.) We can, of course, turn around and calculate the pressure for $T = 300$ K and $\overline{V} = 24.7$ dm³ mol⁻¹ directly from Eq. (1.3). This calculation gives $P = 0.9960$ atm (to four significant figures). The point here is not that the two values differ slightly ($d\overline{V}$ should represent an infinitesimal amount, but we used a finite amount for it in the calculation) but rather that the calculation based on the partial derivative is sometimes the *only* one we can make, especially whenever we lack a complete equation of state.

To indicate that $d\overline{V}$ is finite, we replace it by the symbol $\Delta \overline{V}$ and rewrite Eq. (1.8) as the approximation

$$\Delta P \cong \left(\dfrac{\partial P}{\partial \overline{V}}\right)_T \Delta \overline{V}.$$

This notation is very useful. Whenever we write Δ(something) we will *always* mean

$$\Delta(\text{something}) = (\text{final value of something}) - (\text{initial value of something}).$$

The exact way to use Eq. (1.8) and equations like it with an explicit expression for the derivative is to integrate the equation directly. Since integration and differentiation are inverse operations, we will recover Eq. (1.3) when we integrate Eq. (1.8). Thus, the result will not be new, but the method will be.

We write

$$\text{at constant temperature:} \quad \int_{P_i}^{P_f} dP = \int_{\overline{V}_i}^{\overline{V}_f} \left(\dfrac{\partial P}{\partial \overline{V}}\right)_T d\overline{V} \quad (1.9)$$

where the subscripts i and f denote initial and final values, respectively. The left-hand side is

$$\int_{P_i}^{P_f} dP = P_f - P_i \equiv \Delta P,$$

and the right-hand side is

$$\int_{\overline{V}_i}^{\overline{V}_f} \left(\frac{\partial P}{\partial \overline{V}}\right)_T d\overline{V} = -\int_{\overline{V}_i}^{\overline{V}_f} \frac{RT}{\overline{V}^2} d\overline{V}$$

$$= -RT \int_{\overline{V}_i}^{\overline{V}_f} \frac{d\overline{V}}{\overline{V}^2}$$

$$= RT \left(\frac{1}{\overline{V}_f} - \frac{1}{\overline{V}_i}\right).$$

Equating the two sides gives

$$\Delta P \quad = \quad \frac{RT}{\overline{V}_f} \quad - \quad \frac{RT}{\overline{V}_i}.$$

$$\uparrow \qquad \uparrow \qquad \uparrow \qquad \qquad (1.10)$$

Change in pressure — Final pressure — Initial pressure

as expected.

Next we consider pressure changes when the temperature and molar volume *both* change. There are again three approaches: (1) approximation for small, finite ΔT and $\Delta \overline{V}$ from partial derivatives at the initial T and $\overline{V}$ values, (2) direct evaluation from the equation of state, and (3) integration of the partial derivative expression. The last two are equivalent, and the third is important more for the *method* than for the result.

The infinitesimal change in pressure, dP, due to infinitesimal changes in both temperature and molar volume is simply the sum of the pressure changes caused by each. We write a combination of Equations (1.6) and (1.7):

$$dP = \left(\frac{\partial P}{\partial T}\right)_{\overline{V}} dT + \left(\frac{\partial P}{\partial \overline{V}}\right)_T d\overline{V} = \frac{R}{\overline{V}} dT - \frac{RT}{\overline{V}^2} d\overline{V} \qquad (1.11)$$

called the *total derivative* of P. When we write $P = P(T, \overline{V})$, we are associating a pressure with each pair of T and $\overline{V}$ values. The total derivative is the general way of expressing how pressure will change as we move from one $(T, \overline{V})$ pair to a second infinitesimally nearby.

There are two convenient graphical representations of a function of two variables such as this. The first uses three Cartesian axes, P, T, and $\overline{V}$. Values for $P = P(T, \overline{V}) = RT/\overline{V}$ produce a two dimensional surface in this three dimensional space. This axis system and a section of the surface $P = RT/\overline{V}$ are shown in Figure 1.1.

A second kind of graph plots the surface as a contour map, just as is done for geographical maps. We plot lines of constant pressure in a T, $\overline{V}$ coordinate system with contours labeled by the pressures they represent. Similarly, we can change point of view and choice of independent variables and plot contours of constant T in a P, $\overline{V}$ coordinate system or contours of constant $\overline{V}$ in a P, T coordinate system. These three contour maps derivable from Figure 1.1 are shown in Figure 1.2. Contours of constant temperature, Figure 1.2(a), are called *isotherms;* those of

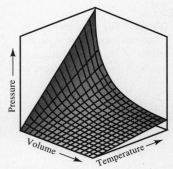

FIGURE 1.1 The equilibrium states of an ideal gas lie on this surface: $P = P(\overline{V}, T) = RT/\overline{V}$.

constant pressure, Figure 1.2(b), are *isobars;* and those of constant molar volume, Figure 1.2(c), are *isochores*.

If we expand a portion of the surface in Figure 1.1 and look in the vicinity of the point $P_i = P(T_i, \overline{V}_i)$, as in Figure 1.3, we can see a graphical interpretation of the total derivative. Suppose we keep the temperature fixed at T_i and change $\overline{V}$ from $\overline{V}_i$ to $\overline{V}_i + d\overline{V}$ *moving along the isotherm* $T = T_i$. Then the pressure would change from P_i to $P_i + dP_T$ where

$$dP_T = \left(\frac{\partial P}{\partial \overline{V}}\right)_T d\overline{V}$$

FIGURE 1.2(a) Isotherms of the ideal gas equation of state are graphs of $P = RT/\overline{V}$ as a function of $\overline{V}$ for various values of T. Equilibrium states of the gas at any one temperature lie on the isotherm for that temperature.

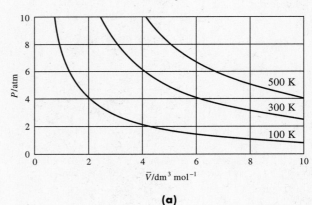

(a)

FIGURE 1.2(b) Isobars of the ideal gas equation of state are graphs of $\overline{V} = RT/P$ as a function of T for various values of P.

FIGURE 1.2(c) Isochores of the ideal gas equation of state are graphs of $P = RT/\overline{V}$ as a function of T for various values of $\overline{V}$.

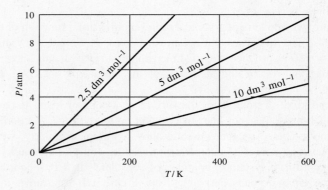

(and the subscript T on dP_T reminds us that the temperature is held constant at T_i). This isothermal change is shown in Figure 1.3. Similarly, we could move along the isochore $\overline{V} = \overline{V}_i$ and find the change in pressure due to a change in temperature of amount dT from

$$dP_{\overline{V}} = \left(\frac{\partial P}{\partial T}\right)_{\overline{V}} dT \; .$$

If we change both T and $\overline{V}$, the pressure would change from P_i to $P_i + dP$ where $dP = dP_T + dP_{\overline{V}}$. This is just the total derivative expression, Eq. (1.11), and it shows us how to calculate a change from one state of the gas to another infinitesimally nearby:

$$P(T_i + dT, \overline{V}_i + d\overline{V}) = P(T_i, \overline{V}_i) + dP \; . \tag{1.12}$$

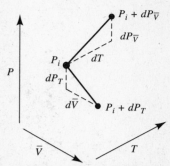

FIGURE 1.3 A close-up view of a point on the P, $\overline{V}$, T surface of Figure 1.1 shows how pressure changes as either T or $\overline{V}$ changes infinitesimally.

To move a finite distance along this surface, we need to specify the initial coordinates of the change, $(T_i, \overline{V}_i)$, the final coordinates, $(T_f, \overline{V}_f)$, and *the way in which temperature and molar volume are to change*. In other words, do we change T first, then $\overline{V}$, or do we change both at arbitrary rates, or does it matter how T and $\overline{V}$ change in detail? If we imagine a machine that is changing T and $\overline{V}$, its operation may suggest how the changes are made; we are about to see if our calculation is forced to follow such changes.

In the terminology of thermodynamics, the changes in the independent variables are called the *path*. The path includes, by definition, all the intermediate states of the system between $(T_i, \overline{V}_i)$ and $(T_f, \overline{V}_f)$, even if they are not equilibrium states, as well as any changes in the *surroundings*. (The reason for considering the surroundings will become clearer in subsequent chapters.) When we specify the system, the initial state, the final state, and the path, we have described a thermodynamic *process*.

We can express the process mathematically in an abstract way through

$$P_f = P_i + \int_{\text{path}} dP \; . \tag{1.13}$$

The integral over the path in Eq. (1.13) is a *line integral*, since the path can frequently be represented by a line in the $(P, T, \overline{V})$ space of Figure 1.1. In general, this line will *not* lie on the surface $P = P(T, \overline{V})$ shown in the figure, since *this surface represents only equilibrium states* and the *path may include non-equilibrium states* not on the surface. For example, consider popping a balloon in an otherwise evacuated and closed chamber. The molar volume would suddenly increase when the balloon popped, but for an instant or two, as the gas rushed out of the balloon, the system would be in disequilibrium and would *not* be represented by a point on the surface in Figure 1.1.

Now we make Eq. (1.13) concrete with two examples. In both the change is from the initial state $P_i = P(T_i, \overline{V}_i)$ to the final state $P_f = P(T_f, \overline{V}_f)$, but the processes will have different paths.

The first (path A) has two parts:

(A1) change the temperature from T_i to T_f keeping the volume fixed at $\overline{V}_i$
(A2) then change $\overline{V}$ from $\overline{V}_i$ to $\overline{V}_f$ while keeping T fixed at T_f.

Since the path has two parts, so will the path integral:

$$P_f = P_i + \int_{A1} dP + \int_{A2} dP \; .$$

Since path A1 is at constant $\overline{V}$, the infinitesimal $d\overline{V} = 0$, and dP is just $dP_{\overline{V}}$. Similarly, along path A2, $dT = 0$ and $dP = dP_T$. Therefore

$$P_f \;=\; P_i \;+\; \underbrace{\int_{T_i}^{T_f} \left(\frac{R}{\overline{V}_i}\right) dT}_{\uparrow} \;+\; \underbrace{\int_{\overline{V}_i}^{\overline{V}_f} -\left(\frac{RT_f}{\overline{V}^2}\right) d\overline{V}}_{\uparrow} \;. \qquad (1.14)$$

$\uparrow$ Final pressure $\qquad$ $\uparrow$ Initial pressure $\qquad$ Path A1 $\qquad$ Path A2

Note that $\overline{V}_i$ appears in the integrand for the path A1 integral, since $\overline{V} = \overline{V}_i$ along that path. Similarly, T_f appears in the second integral, since $T = T_f$ along path A2. Integration yields

$$P_f = P_i + \frac{R}{\overline{V}_i}(T_f - T_i) + RT_f\left(\frac{1}{\overline{V}_f} - \frac{1}{\overline{V}_i}\right)$$

$$= P_i - \frac{RT_i}{\overline{V}_i} + \frac{RT_f}{\overline{V}_f}$$

$$= \frac{RT_f}{\overline{V}_f} \;.$$

Now consider a second path (path B) for which the two variables change in a different order:

(B1) change $\overline{V}$ from $\overline{V}_i$ to $\overline{V}_f$ while keeping T fixed at T_i
(B2) then change T from T_i to T_f while keeping $\overline{V}$ fixed at $\overline{V}_f$.

The path integrals become

$$P_f \;=\; P_i \;+\; \underbrace{\int_{\overline{V}_i}^{\overline{V}_f} -\left(\frac{RT_i}{\overline{V}^2}\right) d\overline{V}}_{\uparrow} \;+\; \underbrace{\int_{T_i}^{T_f} \left(\frac{R}{\overline{V}_f}\right) dT}_{\uparrow} \;. \qquad (1.15)$$

$\uparrow$ Final pressure $\qquad$ $\uparrow$ Initial pressure $\qquad$ Path B1 $\qquad$ Path B2

Before we evaluate these integrals, you should compare them, term by term and symbol by symbol, with those of Eq. (1.14). Make sure you can relate the algebra to the verbal description of the path.

Integrating gives

$$P_f = P_i + RT_i\left(\frac{1}{\overline{V}_f} - \frac{1}{\overline{V}_i}\right) + \frac{R}{\overline{V}_f}(T_f - T_i)$$

$$= P_i - \frac{RT_i}{\overline{V}_i} + \frac{RT_f}{\overline{V}_f}$$

$$= \frac{RT_f}{\overline{V}_f} \;.$$

Note that the final answer, $P_f = RT_f/\overline{V}_f$, is the same for both paths. This is both a completely expected result and a profoundly important result. It is expected because we know the gas must *always* exhibit the same P_f for any given $(T_f, \overline{V}_f)$ no matter how these values were achieved. The result is profound because the next chapter has differential expressions for quantities other than pressure which, when integrated between the same initial and final limits but *along different paths,* give *different results.*

Whenever the path integral of a differential expression does not depend on the details of the path, we call the dependent variable a *state variable* or a *state function.* The pressure of the system is a state variable, as our discussion above has shown. We can write an equation of state for this variable ($P = RT/\overline{V}$), and we can differentiate the equation to find the total differential, Eq. (1.11).

Sometimes we will find the total differential first and will want to know if an equation of state can be found from it. As an example, suppose we have two arbitrary independent variables, x and y, a dependent variable z, and some total differential of the form

$$dz = f_x(x, y)\, dx + f_y(x, y)\, dy$$

where $f_x(x, y)$ and $f_y(x, y)$ are two functions of x and y. For instance, consider the two (arbitrarily chosen) functions

$$f_x(x, y) = xy \quad \text{and} \quad f_y(x, y) = x^2 y^2 \;.$$

The total derivative is

$$dz = xy\, dx + x^2 y^2\, dy \;.$$

Now we ask if there exists some function (a state function) $z(x, y)$ for which this dz is its total derivative. If there is, then it must be true that (compare to Eq. (1.11))

$$\left(\frac{\partial z(x, y)}{\partial x}\right)_y = f_x(x, y) = xy$$

and

$$\left(\frac{\partial z(x, y)}{\partial y}\right)_x = f_y(x, y) = x^2 y^2 \;.$$

If the state function $z(x, y)$ exists, then it must be true that

$$\frac{\partial}{\partial x}\left(\frac{\partial z}{\partial y}\right) \equiv \left(\frac{\partial^2 z}{\partial x \partial y}\right) = \frac{\partial}{\partial y}\left(\frac{\partial z}{\partial x}\right) \equiv \left(\frac{\partial^2 z}{\partial y \partial x}\right)$$

since the order of differentiation does not matter. We are asking

$$\text{does} \quad \frac{\partial f_y(x, y)}{\partial x} = \frac{\partial f_x(x, y)}{\partial y} ?$$

This is known as the *cross-derivative equality test.* If and only if this equality holds can we expect the function $z(x, y)$ to exist. In our arbitrary example,

$$\frac{\partial f_x}{\partial y} = \frac{\partial (xy)}{\partial y} = x$$

but

$$\frac{\partial f_y}{\partial x} = \frac{\partial (x^2 y^2)}{\partial x} = 2xy^2 \;.$$

Since $x \neq 2xy^2$ for all x and y, there is *no* function $z(x, y)$ for which $xy\, dx + x^2 y^2\, dy$ is the total differential. We can still integrate $dz = xy\, dx + x^2 y^2\, dy$ along any path, but *the value of the path integral will depend on the details of the path.*

EXAMPLE 1.5

Integrate the example total differential $dz = xy\, dx + x^2 y^2\, dy$ from $(x, y) = (0, 0)$ to $(x, y) = (1, 1)$ over the following two paths:

(a) change x from 0 to 1 with $y = 0$, then change y from 0 to 1 with $x = 1$
(b) change y from 0 to 1 with $x = 0$, then change x from 0 to 1 with $y = 1$

SOLUTION For path (a), the first step is to evaluate $\int xy\, dx$ from $x = 0$ to $x = 1$, but with $y = 0$. Since y multiplies the entire integrand, this integral is zero. The second step involves the integral $\int x^2 y^2\, dy$ from $y = 0$ to $y = 1$ with $x = 1$. Thus the integral is simply $\int_0^1 y^2\, dy = \frac{1}{3}$. Adding both steps together gives $\Delta z = \int dz = \frac{1}{3}$. For path (b), the first step is an integral over y, but again this integral ($\int x^2 y^2\, dy$) is zero since $x = 0$ in the first step. The second step integrates over x from $x = 0$ to $x = 1$ with $y = 1$. We find $\int_0^1 x\, dx = \frac{1}{2}$, the value of Δz for this path. Note that Δz is *different for the two different paths* as predicted by the analysis in the text.

▸ **RELATED PROBLEMS** 1.7, 1.8

Whenever a total differential expression passes this test, we say that the differential is *exact*. Only exact differentials can be integrated without regard to the path. (Those that fail the test are called *inexact differentials*.) You can now see why exact differentials and state functions are so important. We can pick *any* path for the calculation, as long as the initial and final states are correct; we are not tied to the path that the real physical process might follow. Such process paths are often very complicated, and the ability to pick any convenient path (convenient in a mathematical sense) is a very great advantage.

EXAMPLE 1.6

Which of the following are exact differentials?

(a) $dT = \dfrac{V}{R} dP + \dfrac{P}{R} dV$

(b) $dP = -\dfrac{RT}{(V - b)^2} dV + \dfrac{R}{(V - b)} dT$ (b is a constant)

(c) $dP = -\dfrac{RT}{V(V - b)} dV + \dfrac{R}{(V - b)} dT$ (b is a constant)

SOLUTION We apply the cross-derivative equality test to each in turn.

(a) Does $\left(\dfrac{\partial (V/R)}{\partial V}\right) = \left(\dfrac{\partial (P/R)}{\partial P}\right)$? Yes; thus, (a) is an exact differential. (It is the total differential for $T = PV/R$.)

(b) Does $\left(\dfrac{\partial \left(\dfrac{-RT}{(V-b)^2}\right)}{\partial T}\right) = \left(\dfrac{\partial \left(\dfrac{R}{(V-b)}\right)}{\partial V}\right)$? Yes, and again we have an exact differential. (The equation of state is $P = RT/(V - b)$, which we will encounter in Problems 1.18 and 1.19.)

(c) Does $\left(\dfrac{\partial\left(\dfrac{-RT}{V(V-b)}\right)}{\partial T}\right) = \left(\dfrac{\partial\left(\dfrac{R}{(V-b)}\right)}{\partial V}\right)$? No, and thus (c) is *not* an exact differential.

⟹ RELATED PROBLEM 1.8

This brings us to the end of a very important section. We have defined many thermodynamic terms, developed mathematical techniques to express them, and formulated some mathematical rules of the road. Make sure you understand *equation of state, process, path, path integral, total differential, state variable, exact differential,* and *system*.

1.2 REAL GASES—THE ORIGINS OF NONIDEAL BEHAVIOR

The ideal gas equation gives us a qualitative feeling for gas behavior and a simple equation to apply and explore in various contexts, but it is not always sufficiently accurate. Nature's real gases are only approximated by ideal gas behavior.

Later on, we will introduce ideal solutions, ideal mixtures, etc., always with the expectation that most of the essential physical phenomena will be reproduced, if not the numerically exact details observed in a laboratory (or in day-to-day life, for that matter). Once the essential characteristic features of idealized situations are understood, we will adjust our ideal equations so that they conform more closely to reality. Our first such adjustment concerns the observed effects not described by $PV = nRT$.

Real gases turn into liquids and ultimately solidify as the temperature is lowered, but the ideal gas is always gaseous. Real gases have finite size molecules; the ideal gas can be compressed into an arbitrarily small volume. Real gases approach the ideal gas equation of state *only* in the limit of high temperatures and large molar volumes.

Condensation and solidification are possible because real molecules stick to one another, if perhaps only weakly, at sufficiently low temperatures. There exists a *real intermolecular attractive force* between any two real molecules. When we say a molecule has a size, we mean that two molecules can approach each other only so closely. There exists a *real intermolecular repulsive force between molecules* at distances that define molecular sizes. When we say real gases approach ideal behavior "at large T and $\overline{V}$," we can crudely measure the range and strength of intermolecular forces by specifying how "large" are "large T and $\overline{V}$."

We gauge the ways real gases deviate from ideal behavior with a simple experiment. Suppose we fix the temperature of a real gas and measure its molar volume as a function of pressure. The ratio of the real molar volume to the ideal molar volume

$$Z = \frac{\overline{V}(\text{real})}{\overline{V}(\text{ideal})} = \frac{P\overline{V}(\text{real})}{RT} \quad (1.16)$$

is a measure of non-ideality known as the *compressibility factor*. A graph of Z at one temperature for H_2 and CO_2 is shown in Figure 1.4. If H_2 and CO_2 were truly ideal, then at all pressures we would find $Z = 1$. Notice that H_2 has a molar volume greater than the ideal value ($Z > 1$) while CO_2 has a molar volume less than the ideal value ($Z < 1$) at elevated pressures.

Without worrying about the molecular origin of non-idealities, we can take as an experimental fact that the data for H_2 and CO_2 in Figure 1.4 fall on lines that

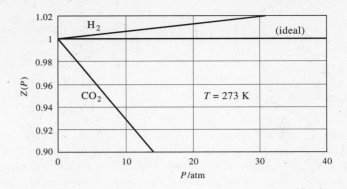

FIGURE 1.4 The ideal compressibility factor is independent of pressure, but Z for real gases can be either greater or lesser than 1.00 at elevated pressures. The near linearity of $Z(P)$ is the experimental basis for the pressure form of the virial expansion.

are fairly straight, at least over the pressure range of that figure. We can construct accurate empirical equations of state if we fit such data to an equation of the form

$$Z(T, P) = 1 + A_1(T)P + A_2(T)P^2 + \cdots \quad (1.17)$$

which is a power series expansion of Z in pressure with coefficients $A_1(T)$, $A_2(T)$, etc., each of which may be positive or negative, and each of which depends on temperature. Each gas will have a unique set of such coefficients.

Alternately, we could express Z as a power series in $1/\overline{V}$, writing

$$Z(T, \overline{V}) = 1 + \frac{B(T)}{\overline{V}} + \frac{C(T)}{\overline{V}^2} + \cdots \quad (1.18)$$

Equation (1.18) is known as the *virial equation of state*. Equation (1.18) is more interesting than (1.17) since there are theories for calculating the coefficients $B(T)$, $C(T)$, etc., directly from intermolecular force information. These coefficients are known as the *second virial coefficient*, the *third virial coefficient*, and so on, and each depends on temperature and the molecular identity of the gas. We will pay closest attention to $B(T)$, the second virial coefficient.

EXAMPLE 1.7

What error is made if the ideal gas equation is used to calculate the number of moles of Ar gas at 10.0 atm, 200 K, and 15.0 dm³ when compared to a more accurate virial equation calculation?

SOLUTION The ideal gas amount is

$$n = \frac{PV}{RT} = \frac{(10.0 \text{ atm})(15.0 \text{ dm}^3)}{(0.082 \text{ dm}^3 \text{ atm mol}^{-1} \text{ K}^{-1})(200 \text{ K})} = 9.2 \text{ mol.}$$

The virial equation can be written

$$PV = nRT + \frac{n^2 BRT}{V},$$

which is a quadratic expression for n. We find $B(T = 200 \text{ K}) = -47.4 \times 10^{-3} \text{ dm}^3 \text{ mol}^{-1}$ in Table 1.1 (converting from the tabulated cm³ units to the dm³ units of the problem). Solving the quadratic expression yields $n = 9.4$ mol, which is about 2% greater than the ideal gas value.

➠ RELATED PROBLEMS 1.9, 1.10

The quantity $B(T)$ has units of molar volume, since Z itself is dimensionless. The values for $B(T)$ for the rare gases at several temperatures are given in Table 1.1. Near room temperature $B(T)$ for He is about 12 cm³ mol⁻¹. Recall that 1 mol

TABLE 1.1 $B(T)/\text{cm}^3\,\text{mol}^{-1}$ for the Rare Gases at Selected Temperatures

T/K	He	Ne	Ar	Kr	Xe
20	−3.34	—	—	—	—
50	7.4	−35.4	—	—	—
100	11.7	−6.0	−183.5	—	—
150	12.2	3.2	−86.2	−200.7	—
200	12.3	7.6	−47.4	−116.9	−276.
300	12.0	11.3	−15.5	−50.5	−133.
400	11.5	12.8	−1.0	−22.0	−69.4
600	10.7	13.8	12.0	1.7	−19.4

of an ideal gas at 300 K and 1 atm exhibits a molar volume of ~25 000 cm³ mol⁻¹, a value considerably larger than $B(T)$. However, at 4.2 K and 1 atm, He liquifies. The density of liquid He is 0.15 g cm⁻³, and since the mass of 1 mol of He is 4.0 g, the molar volume of He(l) is

$$\overline{V}(\text{He}(l)) = \frac{(4.00\text{ g mol}^{-1})}{(0.15\text{ g cm}^{-3})} = 27.0\text{ cm}^3\text{ mol}^{-1}\;.$$

This value is similar in magnitude to $B(T)$, and with good reason. The molar volume of a condensed phase is a rough measure of size, since condensed phases consist of molecules packed closely together. If one assumes a gas to consist of very small marbles (more correctly called *hard spheres*), each of volume v, the second virial coefficient of such a gas, B_{hs}, (which, since the molecules have a finite volume, is a non-ideal gas) can be calculated by the theories alluded to above. (See Problem 1.13 for a particularly simple theory.) The result is

$$B_{hs} = 4N_A v \tag{1.19}$$

where N_A is Avogadro's constant.

The phrase "hard spheres" specifies the intermolecular force. At all center-to-center distances greater than the diameter of the hard sphere, there are no attractive or repulsive forces. But when two spheres' centers are brought to a separation of one diameter, the spheres just touch and they can be brought no closer together. In other words, an infinite repulsive force is felt at the contact distance.

Note, however, that Eq. (1.19) shows that B_{hs} is inherently positive and temperature independent. Thus, the hard-sphere model cannot be a useful representation of a real gas at *low* temperatures where Table 1.1 shows $B(T)$ to be negative and strongly temperature dependent, but at *high* temperatures, the model is at least in accord with the positive, nearly constant values of $B(T)$.

EXAMPLE 1.8

From X-ray diffraction measurements, it is found that solid He at 2 K has atoms that are 3.57 Å apart. How does this measure of helium's size compare to the hard-sphere model for gaseous helium?

SOLUTION Table 1.1 lists high-temperature B values of around 12 cm³ mol⁻¹. From the hard-sphere model, $B_{hs} = 4N_A v$, and the effective hard-sphere volume of He is thus (12 cm³ mol⁻¹)/(4N_A) = 5.0 × 10⁻²⁴ cm³. This corresponds to a radius ($v = 4\pi r^3/3$) of 1.1 Å, or a diameter of 2.2 Å. Thus, the hard-sphere model gives only a crude approximation to atomic size when compared to measures made by more direct methods.

➡ RELATED PROBLEMS 1.12, 1.13

FIGURE 1.5 The potential energy function for two Ar atoms decreases slowly as the two atoms are brought together (implying the atoms attract one another) until the atomic separation is 3.76 Å. For shorter distances, the potential energy rises rapidly, corresponding to a strong repulsive force between the atoms.

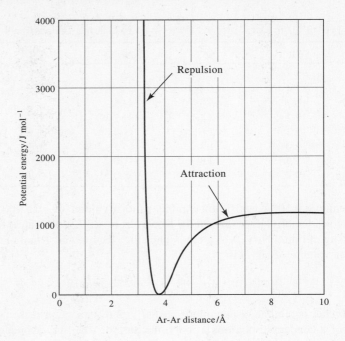

This comparison of real $B(T)$ behavior to predict hard-sphere behavior suggests repulsive forces dominate the high-temperature behavior of $B(T)$. The low-temperature behavior is dominated by attractive forces. Gases condense at low temperatures in ways that repulsive forces alone cannot duplicate,[4] and we cannot expect to model real gas behavior without considering all aspects of intermolecular forces.

It is the *intermolecular potential energy* rather than the force that is usually discussed. The two are related by

$$Force = -\frac{d(Potential\ energy)}{d(Distance)}.$$

For example, Figure 1.5 plots the intermolecular potential energy between two Ar atoms as a function of the Ar–Ar distance. The large *negative slope* of the potential energy at small distances implies a large *positive force*—a repulsion. Likewise, the gentle *positive slope* at large distances implies a small *negative force*—an attraction.

If we can estimate a typical Ar–Ar separation in, say, a 1 atm, room-temperature sample, we can use Figure 1.5 to estimate the type and magnitude of force the atoms feel. In such a gas, the volume of the container that is available on average to each molecule is the ratio of the molar volume to Avogadro's constant:

$$\frac{25\,000\ cm^3\ mol^{-1}}{6.02 \times 10^{23}\ mol^{-1}} = 4.2 \times 10^{-20}\ cm^3\ molecule^{-1}.$$

If we imagine each molecule to be in the center of a cube of this volume, then the average atom–atom separation is approximately the cube root of this number:

$$\sqrt[3]{4.2 \times 10^{-20}}\ cm = 3.5 \times 10^{-7}\ cm = 35\ Å.$$

---

[4]The "hard-sphere fluid" as a model of *condensed* phases is discussed in Chapter 17.

This distance is so large it is off the scale of Figure 1.5; two Ar atoms 35 Å apart exert essentially no force on each other. Thus, we can conclude that even at 1 atm pressure and ordinary temperatures, gases behave ideally to a good approximation because the molecules are too far apart to interact appreciably. As the molar volume is decreased, the average separation will decrease and attractive forces will become more important.

If the high temperature behavior of $B(T)$ is dominated by repulsive forces and the low temperature behavior by attractive forces, then there should reasonably be an intermediate temperature at which the effects of attraction and repulsion on nonideal behavior are in some sense balanced. One such temperature is called the *Boyle temperature*, T_B. At the Boyle temperature, the second virial coefficient is zero:

$$B(T_B) = 0 \ .$$

If we set $B = 0$ in the virial expansion of Eq. (1.18), we see that at large $\overline{V}$ (low pressures), $Z(T_B, \overline{V}) \cong 1$, and the gas appears to behave ideally, if only at this one temperature. In a more precise definition of the Boyle temperature (which includes the previous one), T_B is the temperature for which a plot of Z versus P has a zero slope as P approaches 0:

$$\lim_{P \to 0} \left(\frac{\partial Z}{\partial P} \right)_{T_B} = 0 \ .$$

Before we can use the Boyle temperature to tell us something about this balance of forces, we need a way of measuring intermolecular attractions. Repulsive forces can be gauged by size considerations, as in the hard-sphere model, but attractions are somewhat more subtle. In the next section we discuss a simple and remarkable equation of state that incorporates the most rudimentary effects of attraction and repulsion on the state variables of a real gas.

1.3 THE VAN DER WAALS EQUATION OF STATE

For his PhD thesis in 1873, Johannes van der Waals proposed an equation of state for real gases that has proven to be remarkably insightful. Van der Waals introduced into the ideal gas equation two empirical parameters, to be determined by experiment but amenable to physical interpretation. One parameter, symbolized a, represents the major effects of intermolecular attractions, while the second, symbolized b, represents the effects of repulsive forces.

Recall from the end of the last section that gas molecules at low densities are so far apart, on average, that to a good approximation they behave as free particles. This is true even near the inside surface of the gas container where the attractive force exerted by the myriad of molecules at the container's surface averages to zero as any one gas molecule first approaches the surface, then strikes it, and then rebounds from it. It is, of course, the "striking" interactions that produce the pressure as we ordinarily measure it.[5]

Since pressure is proportional to the effects of *each* surface collision *times the number of such collisions per second*, non-ideality represents a *change in the rate of gas–surface collisions from the ideal gas standard rate*. If the rate increases, the pressure increases. Finite molar volumes (from repulsive intermolecular forces) do

---

[5]Pressure is a force per unit area, and force is the time rate of change of momentum; by approaching, striking, and rebounding, the gas molecule's component of momentum towards the surface is reversed over the time of the collision. This is the molecular basis for pressure.

just that. As van der Waals correctly argued, a certain fraction of the geometric volume of the container is *excluded* by the volume occupied by the molecules themselves. Thus we replace the ideal molar volume with a corrected volume:

$$\text{the \textit{free molar volume}} = \overline{V} - b$$

where b is the second van der Waals parameter.

Attractive forces *reduce* the surface collision rate, but it is more difficult to see how this reduction occurs on a molecular level than it was for repulsions. Attractions lower the pressure, and the effect depends on gas density since it is the result of two molecules colliding with each other. Such collisions are more probable at greater gas densities, and the number of them increases as the *square* of the gas density. Since density is proportional to $1/\overline{V}$, the effect is proportional to $1/\overline{V}^2$, and we modify the ideal pressure by

$$\text{the \textit{internal pressure}} = -\frac{a}{\overline{V}^2}$$

where a is the first van der Waals parameter.

Incorporating these two terms into the ideal gas equation yields *the van der Waals equation of state*:

$$P = \underbrace{\frac{RT}{\overline{V} - b}}_{\substack{\uparrow \\ \frac{RT}{\text{Free molar volume}}}} - \underbrace{\frac{a}{\overline{V}^2}}_{\substack{\uparrow \\ \text{Internal pressure}}}. \qquad (1.20)$$

(Real gas pressure ↑)

Values for a and b for a number of gases are given in Table 1.2. Note that b increases with atomic or molecular size, as expected. (The large b values for the small species H_2 and He can be traced to the fact that these species exhibit hardly any attractive interactions so that repulsive forces dominate at unusually large distances.) Note also that a increases as one goes from gases that condense at the lowest temperatures (He, Ne, H_2) toward gases that condense at higher temperatures (NH_3, H_2O, Cl_2). Normal boiling points are also controlled in part by attractive intermolecular forces.

EXAMPLE 1.9

Xe is the most expensive rare gas; it is produced from fractional distillation of air which contains Xe at about 90 parts per billion. One company sells Xe in a 0.44 dm³ cylinder with a 15.6 atm pressure at 295 K for $385 or with a 24.8 atm pressure for $448. Find the cost per mole of each using the van der Waals equation.

SOLUTION If Xe were ideal, we would easily find that the first cylinder holds 0.28 mol and the second holds 0.45 mol. But writing the van der Waals equation explicitly in terms of moles

$$P = \frac{nRT}{(V - nb)} - \frac{n^2 a}{V^2}$$

leaves us with a cubic equation for n when expanded. This type of problem is well suited to an iterative solution with a pocket calculator or programmable computer. We iteratively substitute guesses for n (starting with the ideal gas value) in the right side of the expression above until the computed pressure agrees with the desired pressure. Such a solution gives 0.31 mol (instead of 0.28 mol) and 0.53 mol (instead of 0.45 mol). Thus Xe costs about \$1240 mol^{-1} (for the low pressure cylinder) or about \$845 mol^{-1} (for the high pressure

TABLE 1.2 van der Waals parameters and Boyle Temperatures

	a/dm^6 atm mol^{-2}†	b/cm^3 mol^{-1}	(a/Rb)/K‡	T_B(expt)/K
H_2	0.2444	26.61	112	110
He	0.03412	23.70	17.5	25
Ne	0.2107	17.09	150	127
Ar	1.345	32.19	509	410
Kr	2.318	39.78	710	577
Xe	4.194	51.05	1001	775
N_2	1.390	39.13	433	327
O_2	1.360	31.83	521	405
Cl_2	6.493	56.22	1407	—
CO	1.485	39.85	454	—
CO_2	3.592	42.67	1026	713
OCS	3.933	58.17	824	—
CS_2	11.62	76.85	1843	—
H_2O	5.464	30.49	2184	—
H_2S	4.431	42.87	1260	—
NH_3	4.170	37.07	1371	—
CH_4	2.253	42.78	642	509
C_2H_2	4.390	51.36	1042	—
C_2H_4	4.471	57.14	954	—
C_2H_6	5.489	63.80	1048	—
C_6H_6	18.00	115.4	1901	—
C_6H_{12}	22.81	142.4	1952	—

†To convert from dm^6 atm mol^{-2} to cm^6 atm mol^{-2} units, multiply by 10^6.
‡Boyle temperatures, T_B, predicted by van der Waals parameters.

cylinder). The moral is twofold: the ideal gas equation can be seriously in error at high pressures, and one should always buy rare chemicals in quantity!

➡ RELATED PROBLEM 1.15

If we rewrite the van der Waals equation as a virial expansion, we can identify the van der Waals expressions for the virial coefficients. We start with the van der Waals expression for Z:

$$Z = \frac{P\overline{V}}{RT} = \frac{\overline{V}}{(\overline{V} - b)} - \frac{a}{RT\overline{V}} = \frac{1}{1 - \frac{b}{\overline{V}}} - \frac{a}{RT\overline{V}}.$$

At large $\overline{V}$ (where $b/\overline{V} \ll 1$), we can use the expansion

$$\frac{1}{1 - \frac{b}{\overline{V}}} \cong 1 + \frac{b}{\overline{V}} + \frac{b^2}{\overline{V}^2} + \cdots$$

which yields the virial expansion

$$Z = 1 + \left(b - \frac{a}{RT}\right)\frac{1}{\overline{V}} + \frac{b^2}{\overline{V}^2} + \cdots \quad (1.21)$$

The van der Waals expressions for the second virial coefficient is the coefficient of $1/\overline{V}$,

$$B(T) = b - \frac{a}{RT} ,\qquad (1.22)$$

and the third virial coefficient is

$$C(T) = b^2 .\qquad (1.23)$$

Note that $C(T)$ is independent of temperature in this approximation. The expression for $B(T)$, however, does have the qualitative properties observed experimentally. At large T, Eq. (1.22) approaches the constant value b. At small enough T, such that $a/RT > b$, $B(T)$ becomes negative, as observed experimentally. The van der Waals expression for the Boyle temperature is

$$T_B = \frac{a}{Rb} \quad \text{since} \quad B(a/Rb) = 0 .\qquad (1.24)$$

This is the temperature in the next-to-last column of Table 1.2. Experimental values of the Boyle temperature (the last column) are usually lower, but the van der Waals expression at least gives the correct trends and approximate magnitudes of T_B. Except for He, Ne, and H_2, T_B is well above room temperature. This means that $B(T)$ will be negative at or near room temperature, and, therefore, *the non-ideal behavior of almost all gases at room temperature can be traced to the effects of intermolecular attractions.*

If we measure $B(T)$ in multiples of b, and T in multiples of T_B, we can rewrite Eq. (1.22) in the dimensionless form

$$\frac{B(T)}{b} = 1 - \frac{a}{RbT} = 1 - \frac{T_B}{T} .$$

This very simple equation is graphed in Figure 1.6 along with observed data for the rare gases taken from Tables 1.1 and 1.2 (using the experimental T_B). Note that all the rare gases except He have remarkably similar values when scaled in this way.

While the empirical van der Waals parameters allow us to write a reasonably accurate equation of state for a real gas, another valuable property of Eq. (1.20) is that it anticipates and approximates other phenomena observed in real gases but not predicted by the ideal gas equation. It is helpful to rewrite Eq. (1.20) in a way that frees us from the need to specify the identity of the gas. We can do this by a change

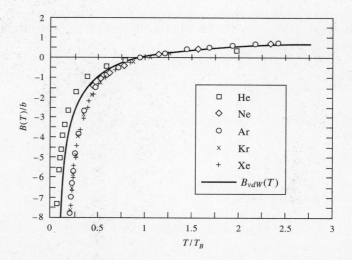

FIGURE 1.6 The van der Waals expression for $B(T)$ (the solid line) gives an approximate agreement with the measured values of $B(T)$ for the rare gases. Note that He is distinctly different from the others.

from the real variables, P, T, and $\overline{V}$, to appropriately scaled *dimensionless* variables $\tilde{P}$, $\tilde{V}$, and $\tilde{T}$ that are proportional to the real variables.

A way to scale $\overline{V}$ is suggested immediately by noting that $\overline{V}$ and b have the same dimensions; thus, we define

$$\tilde{V} = \frac{\overline{V}}{b}.$$

Next, note from Eq. (1.20) that $a/\overline{V}^2$ has the dimensions of pressure, and so does a/b^2. We therefore define

$$\tilde{P} = \frac{P}{(a/b^2)} = \frac{Pb^2}{a}.$$

We can define a dimensionless temperature as we did for Figure 1.6, scaling the real T by the Boyle temperature:

$$\tilde{T} = \frac{T}{T_B} = \frac{T}{(a/Rb)} = \frac{TRb}{a}.$$

These three variable changes can be written as

$$\overline{V} = \tilde{V}b, \quad P = \tilde{P}a/b^2, \quad \text{and} \quad T = \tilde{T}a/Rb$$

and substituting these into Eq. (1.20) yields, after simplification,

$$\tilde{P} = \frac{\tilde{T}}{(\tilde{V} - 1)} - \frac{1}{\tilde{V}^2}. \tag{1.25}$$

This is the equation we were seeking: it has the algebraic properties of the van der Waals equation without requiring us to specify the individual gas parameters a and b.

Isotherms of Eq. (1.25) are plotted in Figure 1.7 for values of $\tilde{T}$ less than unity, that is, for real temperatures below the Boyle temperature. Three features of these curves are striking. The first is the appearance of negative pressures at sufficiently low temperatures. More will be said of this in Chapter 6 where we will see that negative pressure does have physical meaning. The second is the appearance of the high-temperature isotherms. They have the qualitative appearance of the ideal gas isotherms in Figure 1.2a. But the third and most striking point to notice is the radically different shape of the isotherms for $\tilde{T} < 8/27$. These isotherms have their unusual shape over real temperatures and molar volumes that are low, characteristic of conditions under which a real gas liquifies.

We will concentrate on these isotherms and see what clues they have to indicate that the gas is changing to a liquid.

There are three clues. The first is that these isotherms have temperatures that are fairly small fractions of the Boyle temperature so that non-ideality is dominated strongly by intermolecular attractions. The second clue is the range of $\tilde{V}$ over which an isotherm suddenly changes shape. This range is roughly $1.5 < \tilde{V} < 5$, characteristic of condensed media. The third clue lies in their slope.

The slope is $(\partial \tilde{P}/\partial \tilde{V})_{\tilde{T}}$, but its reciprocal, which measures how rapidly volume changes with pressure, is more intuitive. The quantity

$$\kappa \equiv -\frac{1}{V}\left(\frac{\partial V}{\partial P}\right)_T \tag{1.26}$$

is called the *isothermal bulk compressibility* (not to be confused with the compressibility factor, Z). The derivative in Eq. (1.26) is always negative; the volume always decreases when the pressure is increased. Hence, a negative sign is incorporated in the definition to make κ positive for all substances under all conditions. One divides by V to make κ the decrease in volume *per unit volume* per unit pressure increase.

EXAMPLE 1.10

Find an expression for κ for the ideal gas.

SOLUTION From Eq. (1.26), we see we need the partial derivative $(\partial V/\partial P)_T$ for the ideal gas. Writing $V = nRT/P$, we find

$$\left(\frac{\partial V}{\partial P}\right)_T = nRT\left(\frac{\partial (1/P)}{\partial P}\right) = -\frac{nRT}{P^2}$$

so that

$$\kappa \equiv -\frac{1}{V}\left(\frac{\partial V}{\partial P}\right)_T = -\frac{P}{nRT}\left(-\frac{nRT}{P^2}\right) = \frac{1}{P}\;.$$

Thus, κ for the ideal gas is simply the reciprocal of the pressure.

➡ *RELATED PROBLEMS* 1.18, 1.19

Gases are certainly more readily compressed than are liquids or solids; hence, gaseous κ values are much larger than those for liquids or solids. A portion of any isotherm that is relatively flat, as are those at large $\tilde{V}$, have small negative slopes and large compressibilities. Conversely, at *small* $\tilde{V}$, the isotherms have large *negative* slopes, implying *small* compressibilities.

Now notice an embarrassing defect in Figure 1.7. There are portions of the low temperature isotherms with *positive* slopes and thus *negative* values of κ. Such behavior is not physically possible in an equilibrium system, and thus these portions of the isotherms cannot be correct.

One last feature of Figure 1.7 is worth our attention. What is so special about the isotherm $\tilde{T} = 8/27$ (the dashed line isotherm in Figure 1.7)? And why 8/27? This isotherm divides those with a non-physical positive slope region (at lower $\tilde{T}$) from those with *only* negative slopes (at higher $\tilde{T}$). It is the only isotherm with a horizontal inflection point—a single point where both the first and second derivatives of the isotherm are zero.

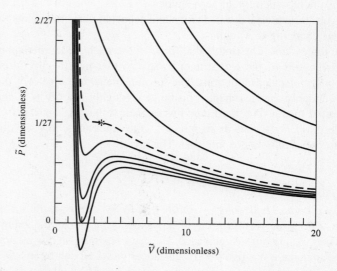

FIGURE 1.7 The isotherms of the van der Waals equation of state, plotted in terms of the dimensionless variables of Eq. (1.25), have very different shapes from those of the ideal gas shown in Figure 1.2(a). The isotherms are shown for $\tilde{T} = $ 20/27, 15/27, 10/27, 8/27 (the critical temperature, shown by the dashed line), 7.5/27, 7/27, 6.75/27, and 6.5/27. The critical point, denoted by the *, is at $\tilde{T} = 8/27, \tilde{P} = 1/27$, and $\tilde{V} = 3$.

The first derivative of any isotherm, based on Eq. (1.25), is

$$\left(\frac{\partial \tilde{P}}{\partial \tilde{V}}\right)_{\tilde{T}} = -\frac{\tilde{T}}{(\tilde{V}-1)^2} + \frac{2}{\tilde{V}^3}$$

and the second derivative is

$$\left(\frac{\partial^2 \tilde{P}}{\partial \tilde{V}^2}\right)_{\tilde{T}} = \frac{2\tilde{T}}{(\tilde{V}-1)^3} - \frac{6}{\tilde{V}^4} .$$

Setting each to zero and considering them along with Eq. (1.25) as a system of three equations in three unknowns allows us to find the coordinates of this unique point. The algebra leads to

$$\tilde{T} = \frac{8}{27}, \quad \tilde{P} = \frac{1}{27}, \quad \text{and} \quad \tilde{V} = 3 .$$

This point is called the *critical point*. The real coordinates of the critical inflection point, T_c, P_c, and $\overline{V}_c$, for the van der Waals gas are

$$T_c = \frac{8}{27} T_B = \frac{8}{27} \frac{a}{Rb} \qquad (1.27a)$$

$$P_c = \frac{1}{27} \frac{a}{b^2} \qquad (1.27b)$$

and

$$\overline{V}_c = 3b . \qquad (1.27c)$$

Since the critical isotherm is horizontal at the critical point, the bulk compressibility is infinite at this one point. Also, all isotherms above T_c lack a region of nonphysical slope, and we will see in Chapter 6 that this is an indication of a system that does not undergo a phase change from gas[6] to liquid as V is decreased. This remarkable behavior is brought forth most clearly if we consider a gas at some temperature below T_c, but at large $\overline{V}$ where we would all agree the system is a gas in the ordinary sense of the word. This gas can be heated at constant $\overline{V}$ to some temperature above T_c, compressed to some $\overline{V} < \overline{V}_c$ at this temperature, and then cooled at constant $\overline{V}$ to the original temperature. This path is shown in Figure 1.8. At the end of the path, we would have something we would all agree is a liquid, but *nowhere along the path would we find a two-phase (liquid and gas) condensation*! We have smoothly converted the gas to a liquid in an unusual way. Running the path backwards, we can take a liquid and produce a gas without ever seeing the liquid boil.

A Closer Look at the van der Waals Equation of State

The total differential of the van der Waals pressure is not difficult to calculate:

$$dP = \left(\frac{\partial P}{\partial T}\right)_{\overline{V}} dT + \left(\frac{\partial P}{\partial \overline{V}}\right)_T d\overline{V}$$

$$= \left(\frac{R}{\overline{V}-b}\right) dT + \left(\frac{2a}{\overline{V}^3} - \frac{RT}{(\overline{V}-b)^2}\right) d\overline{V} .$$

---

[6]This would be a very dense gas, perhaps, which is why it is better to call a system above T_c simply a "fluid," a term that encompasses gases and liquids.

FIGURE 1.8 If a gas is subjected to a path that goes around the critical point (*), such as that shown in the figure, the gas can be converted to a liquid without ever passing through a state in which gas and liquid coexist as separate phases in equilibrium.

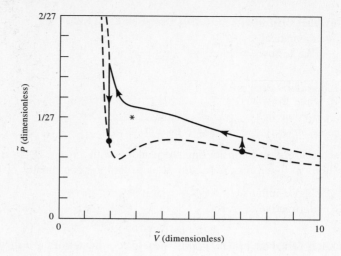

But suppose we need the total differential for $\overline{V}$, or were simply given P and T and asked to find the corresponding $\overline{V}$ from the equation of state. We would quickly confront the fact that the van der Waals equation is cubic in $\overline{V}$ when expanded:

$$P\overline{V}^3 - (Pb + RT)\overline{V}^2 + a\overline{V} - ab = 0 \ .$$

Thus we cannot write a simple analytic expression of the form $\overline{V} = \overline{V}(P, T)$. Cubic equations can be solved exactly (see most mathematical handbooks for details), but the solutions are very unwieldy. It would thus seem that an expression such as

$$d\overline{V} = \left(\frac{\partial \overline{V}}{\partial P}\right)_T dP + \left(\frac{\partial \overline{V}}{\partial T}\right)_P dT$$

would be out of reach. But we can use the identity

$$\left(\frac{\partial \overline{V}}{\partial P}\right)_T \equiv \left(\frac{\partial P}{\partial \overline{V}}\right)_T^{-1}$$

and we know $(\partial P/\partial \overline{V})_T$. Likewise, $(\partial \overline{V}/\partial T)_P = (\partial T/\partial \overline{V})_P^{-1}$, and $(\partial T/\partial \overline{V})_P$ can be computed.

Another valuable technique that can be applied to difficult differentials comes from recognizing that in an expression like $(\partial \overline{V}/\partial T)_P$, the subscript P is saying "$dP = 0$." Thus, we can use the total differential for pressure to note that

$$dP = 0 \quad \text{implies} \quad \left(\frac{\partial P}{\partial T}\right)_{\overline{V}} dT = -\left(\frac{\partial P}{\partial \overline{V}}\right)_T d\overline{V}$$

Rearranging this equality shows that

$$-\frac{(\partial P/\partial T)_{\overline{V}}}{(\partial P/\partial \overline{V})_T} = \left(\frac{\partial \overline{V}}{\partial T}\right)_P ,$$

which gives us a second route to the difficult quantity $(\partial \overline{V}/\partial T)_P$.

Note how the *constraint*, $dP = 0$, is expressed in the subscript to $(\partial \overline{V}/\partial T)$. These important mathematical techniques in the calculus of thermodynamics will be used frequently in the following chapters.

We will leave the van der Waals equation at this point, but only temporarily. There are many features of this remarkable equation left to explore. While the van der Waals equation is quantitatively poor in many ways—real systems have far more complexity than can be expressed by only two parameters—its qualitative predictions are well worth our study. They show in a simple way the influence of intermolecular forces in generating non-ideal behavior, and they allow a rough separation of these influences into those dominated by repulsive forces and those dominated by attractive forces.

1.4 CORRESPONDING STATES

Figure 1.9 compares the experimental isotherms of Xe in the vicinity of the critical point to the van der Waals isotherms. The van der Waals parameters for this figure have been adjusted to give the observed critical point (as discussed in Problem 1.16). The agreement is not what one might have wanted, but whenever a real gas is held at the critical coordinates, whatever these may be, the phenomena predicted by the van der Waals equation, such as an abnormally high isothermal compressibility, are observed. Moreover, real gases do exhibit single fluid phase behavior at temperatures above T_c for a wide range of P and $\overline{V}$.

The van der Waals expressions for the critical coordinates, Eq. (1.27), indicate that critical behavior results from a fairly subtle interplay between intermolecular attractions and repulsions. As with the Boyle temperature, it is tempting to say that these forces are in some sense "balanced," but critical behavior is more complicated than this.

While it is difficult to *predict* critical point coordinates, *measured* values can be used to predict other behavior. A number of measured values are given in Table 1.3. The last column of this table gives the critical point compressibility factor:[7]

$$Z_c = \frac{P_c \overline{V}_c}{RT_c},$$

which varies from about 0.24 to 0.31 for the species tabulated. For an ideal gas, Z_c would, of course, equal 1. For a van der Waals gas, the expressions in Eq. (1.27) yield $Z_c = 3/8 = 0.375$, also larger than is observed.

If we measure the pressure, molar volume, and temperature of a real gas in *multiples* of the critical coordinates, then we have scaled the real coordinates in much the same way we scaled the van der Waals equation in deriving Eq. (1.25). We define

$$p = \frac{P}{P_c}, \quad \overline{v} = \frac{\overline{V}}{\overline{V}_c}, \quad \text{and} \quad t = \frac{T}{T_c}. \tag{1.28}$$

The hope is that these scaled variables will be in some sense universal—independent of the chemical identity of the gas. Whenever two gases are in states characterized by the same set of these scaled coordinates, they are said to be in *corresponding states*.

---

[7]Remember: the *isothermal compressibility*, κ, is infinite at the critical point, but the *compressibility factor*, Z, is finite.

FIGURE 1.9 The experimental isotherms for Xe in the vicinity of the critical point (located by the dashed lines) show that the corresponding van der Waals isotherms (solid lines) are in only qualitative agreement with observation.

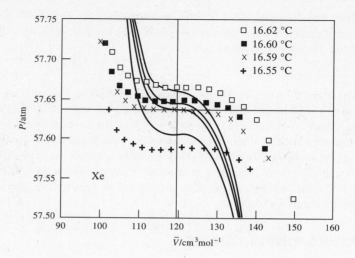

EXAMPLE 1.11

What are the scaled coordinates of CO_2 with $P = 5.00$ atm, $\overline{V} = 6.00$ cm^3 mol^{-1}, and $T = 400$ K? What state of Xe is a corresponding state to this one?

SOLUTION We look up the critical coordinates of CO_2 in Table 1.3 and compute $p = (5.00 \text{ atm})/(72.7 \text{ atm}) = 6.88 \times 10^{-2}$, $\overline{v} = (6.00 \text{ cm}^3 \text{ mol}^{-1})/(94 \text{ cm}^3 \text{ mol}^{-1}) = 6.4 \times 10^{-2}$, and $t = (400 \text{ K})/(304.2 \text{ K}) = 1.31$. Next, we seek the real coordinates of Xe that have these scaled coordinates. Using the Xe critical coordinates from the table, we find $P = pP_c = (6.88 \times 10^{-2})(57.6 \text{ atm}) = 3.96$ atm, $\overline{V} = \overline{v}\overline{V}_c = (6.4 \times 10^{-2})(119 \text{ cm}^3 \text{ mol}^{-1}) = 7.6$ cm^3 mol^{-1}, and $T = tT_c = (1.31)(289.7 \text{ K}) = 380$ K. Thus, CO_2 at (5.00 atm, 6.00 cm^3 mol^{-1}, 400 K) is in a state corresponding to Xe at (3.96 atm, 7.6 cm^3 mol^{-1}, 380 K).

➡ **RELATED PROBLEMS** 1.21, 1.22

TABLE 1.3 Critical Point Coordinates and Critical Compressibility Factors

	P_c/atm	T_c/K	$\overline{V}_c$/cm^3 mol^{-1}	$\dfrac{P_c \overline{V}_c}{RT_c}$
H_2	12.8	32.99	65.5	0.309
He	2.24	5.19	57.3	0.301
Ne	27.2	44.40	41.7	0.312
Ar	48.1	150.8	74.9	0.291
Kr	54.3	209.4	91.2	0.288
Xe	57.6	289.7	119.	0.288
N_2	33.5	126.2	89.5	0.289
O_2	49.7	154.6	73.4	0.288
Cl_2	76.1	417.	124.	0.276
CO	34.5	133.	93.1	0.295
CO_2	72.7	304.2	94.0	0.274
CH_4	45.5	190.5	99.	0.288
H_2O	218.3	647.3	59.1	0.243
NH_3	111.5	405.4	72.5	0.243

Systems in corresponding states have properties that are very often the same to a high degree of accuracy. The phrase "properties of the two gases" means not only the physical state of the samples, such as both gaseous or both liquid, but also quantities such as the compressibility factor, the isothermal bulk compressibility, and many others.[8]

Figure 1.10 shows the compressibility factor for several gases plotted as a function of the scaled pressure, p, for several scaled temperatures, t. Given the critical coordinates of any gas, we can calculate reduced variables and read from the graph its equation of state. For very accurate calculations, this procedure is not good enough, but if errors of a few percent can be tolerated, then the principle of corresponding states provides a fast way to estimate an enormous amount of information on real gas behavior. Moreover, strong deviations from corresponding-states behavior is often a clue that something of an inherently chemical nature is going on in the gas. For example, HF gas has appreciable deviations that are related to the formation of HF clusters (stable species such as $(HF)_2$, $(HF)_3$, etc.) at high density and low temperature. In fact, it can be shown that the corresponding-states principle is exact for molecules that interact *pairwise* (which means the interaction between molecules 1 and 2 is independent of the location of molecules 3, 4, etc.) through a simple intermolecular potential energy that depends only on the molecular separation.

EXAMPLE 1.12

Predict the density of water vapor (steam) at $T = 700$ K and $P = 300$ atm.

SOLUTION Using data from Table 1.3, we calculate the scaled coordinates to which these real coordinates correspond: $t = (700 \text{ K})/(647.3 \text{ K}) = 1.08$ and $p = (300 \text{ atm})/(218.3 \text{ atm}) = 1.37$. From Figure 1.10, we interpolate a value for Z at these (p, t) coordinates of about 0.49. Thus, $\overline{V} = ZRT/P = (0.49)(82 \text{ cm}^3 \text{ atm mol}^{-1} \text{ K}^{-1})(700 \text{ K})/(300 \text{ atm}) = 94 \text{ cm}^3 \text{ mol}^{-1}$. Since the molecular mass of H_2O is 18.0 g mol^{-1}, the mass density of the gas, ρ, is approximately $\rho = (18.0 \text{ g mol}^{-1})/(94 \text{ cm}^3 \text{ mol}^{-1}) = 0.19 \text{ g cm}^{-3}$. The experimental value is 0.188 g cm^{-3}.

⟹ RELATED PROBLEMS 1.23, 1.24

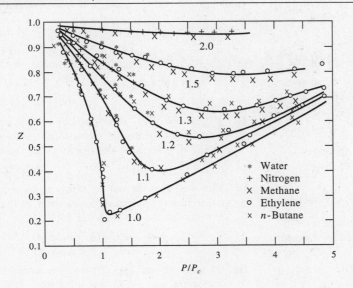

FIGURE 1.10 Compressibility-factor isotherms of a variety of gases lie on common lines when plotted as a function of reduced variables. Each line represents a particular reduced temperature: $T/T_c = 1.0, 1.1, 1.2, 1.3, 1.5,$ and 2.0.

---

[8]Van der Waals was the first to introduce the concept of corresponding states, by the way, but only some eight years after he first derived his equation of state.

CHAPTER 1 SUMMARY

This chapter has used the behavior of gases to introduce several important definitions—system, state, equation of state, path, process, etc.—several important mathematical methods—the total differential, path integrals, exact differentials, scaled coordinates, etc.—and several important physical concepts—intensive and extensive variables, intermolecular forces and potential energies, physical models (ideal gas, hard sphere gas, van der Waals gas), critical point behavior, and the principle of corresponding states. The important equations are collected below for quick reference, but you should make sure you understand the conditions under which each is applicable, the physical meaning of each symbol in them, and, with the experience of using them, the order of magnitude of their variables and parameters.

ideal gas equation of state: $\dfrac{PV}{n} = P\overline{V} = RT$

compressibility factor: $Z = \dfrac{\overline{V}(\text{real})}{\overline{V}(\text{ideal})} = \dfrac{P\overline{V}(\text{real})}{RT}$

virial expansion:
$$Z(T, \overline{V}) = 1 + \dfrac{B(T)}{\overline{V}} + \dfrac{C(T)}{\overline{V}^2} + \cdots$$

hard-sphere second virial coefficient: $B_{hs} = 4N_A v$

van der Waals equation of state:
$$P = \dfrac{RT}{(\overline{V} - b)} - \dfrac{a}{\overline{V}^2}$$

van der Waals second virial coefficient:
$$B(T) = b - \dfrac{a}{RT}$$

van der Waals Boyle temperature: $T_B = \dfrac{a}{Rb}$

van der Waals critical coordinates:
$$T_c = \dfrac{8a}{2Rb},\ P_c = \dfrac{a}{27b^2},\ \overline{V}_c = 3b$$

FURTHER READING

Van der Waals's thesis has been published in translation (J. D. van der Waals, "On the Continuity of the Gaseous and Liquid States," edited with an introductory essay by J. S. Rowlinson [Elsevier Science, New York, 1988]).

The virial expansion is discussed in detail in *The Virial Equation of State*, by E. A. Mason and T. H. Spurling (Pergamon Press, New York, 1969). Similarly, the reference work, *The Virial Coefficients of Gases and Mixtures: a Critical Compilation*, by J. H. Dymond and E. B. Smith (Clarendon Press, Oxford, 1980), contains a discussion of virial equations of state along with extensive tables of data.

PRACTICE WITH EQUATIONS

1A What is the rate of increase of pressure with temperature for 5.00 mol of an ideal gas confined to a volume of 150 dm³? (1.6)

ANSWER: 2.73×10^{-3} atm K^{-1}

1B What is the rate of change of pressure with molar volume for 1.00 mol of an ideal gas at 500 K and 2.00 atm? (1.2), (1.7)

ANSWER: -9.76×10^{-2} atm mol dm^{-3}

1C What is the total derivative of the equation of state $P = RT/\overline{V} + BRT/\overline{V}^2$ where B is a function of temperature only? (1.11)

ANSWER: $P = (R/\overline{V} + BR/\overline{V}^2 + BRT/\overline{V}^2 \, (dB/dT))\, dT$
$- (RT/\overline{V}^2 + 2BT/\overline{V}^3)\, d\overline{V}$

1D What is the molar volume of a gas with $Z = 0.856$ at $P = 37.9$ atm and $T = -25.0\ °\text{C}$? (1.16)

ANSWER: 0.460 dm³ mol^{-1}

1E If 2.50 mol of Cl_2, represented by the van der Waals equation, is confined in a container of volume 85.0 dm³ at 325 K, what is the pressure? (1.20), Table 1.2

ANSWER: 0.780 atm

1F How well does the van der Waals equation predict the experimentally observed second virial coefficient for Xe at room temperature? (1.22), Tables 1.1, 1.2

ANSWER: predicted: -119 cm³ mol^{-1},
observed: -133 cm³ mol^{-1}

1G At what pressure does CO have the same critical-point scaled pressure as CO_2 has at 2.4 atm? (1.28), Table 1.3

ANSWER: 1.14 atm

PROBLEMS

SECTION 1.1

1.1 (a) Modern surface science research (discussed in Chapter 16) is often carried out in ultra-high vacuum (UHV) apparatus that can attain pressures on the order of 10^{-11} torr $\cong 10^{-14}$ atm. How many gas molecules are there in each cubic centimeter in such an apparatus at room temperature? What is the molar volume? (b) The average density of interstellar space is roughly 1 H atom per cm^3, and the temperature is typically on the order of 10 K. What is the pressure of this interstellar medium? What is its molar volume?

1.2 Write the ideal gas equation in terms of P, T, molecular mass M, and mass density, ρ. What is the mass density of He(g) with a molar volume of 50 dm^3 mol^{-1}? What is the pressure of this gas if the temperature is 275 K?

1.3 It is found that the molar volume of a gaseous saturated hydrocarbon C_nH_{2n+2} is 30.61 L mol^{-1} if 0.432 g of the hydrocarbon fills a 300.0 mL container at a constant temperature and pressure. What is the hydrocarbon? What can you say about the temperature and pressure of the experiment?

1.4 An intensive variable such as mass density or molar volume is the ratio of two extensive variables: ρ = system mass/system volume and $\overline{V}$ = system volume/amount of system in moles. Is the ratio of two extensive variables *always* an intensive variable?

1.5 Calculate the value of $(\partial P/\partial \overline{V})_T$ for an ideal gas at 300 K and 1 atm pressure. Then, graph the ideal isotherm for 300 K in such a way that the point on the isotherm with $P = 1$ atm is roughly in the middle of your graph. Draw a straight line that passes through this point and has a slope given by your value for $(\partial P/\partial \overline{V})_T$, and verify visually that this line is tangent to the isotherm as it should be.

1.6 Find the numeric values for $(\partial \overline{V}/\partial T)_P$, $(\partial \overline{V}/\partial P)_T$, and $(\partial P/\partial T)_{\overline{V}}$ for an ideal gas at 1.00 atm and 298 K. Then state a (P, T) condition that would double each derivative.

1.7 (a) Starting from $T = T(P, \overline{V}) = P\overline{V}/R$, write the total differential for dT.
(b) Integrate this total differential from $(P_i, \overline{V}_i)$ to $(P_f, \overline{V}_f)$ over the two paths

(i) P_i to P_f at $\overline{V}_i$ followed by $\overline{V}_i$ to $\overline{V}_f$ at P_f
(ii) $\overline{V}_i$ to $\overline{V}_f$ at P_i followed by P_i to P_f at $\overline{V}_f$

and verify that both calculations give the same value for ΔT: $P_f\overline{V}_f/R - P_i\overline{V}_i/R$.

1.8 Is there any n for which $dz = (xy)^n \, dx + xy \, dy$ is an exact differential? How about for $dz = (nx^2y + y^n) \, dx + (x^n + nxy^2) \, dy$?

SECTION 1.2

1.9 The measured values for Z for H_2 at 35 K and a variety of molar volumes are tabulated below:

Z	$\overline{V}/cm^3 \, mol^{-1}$
0.99863	50 000.
0.98875	8 332.
0.97798	5 130.
0.95916	3 390.
0.92845	2 381.

First, use the virial expansion in $1/\overline{V}$ through just the second virial coefficient and find the best value for B at this T. A graphical determination, plotting Z versus $1/\overline{V}$, will suffice, but if you have access to a least-squares fitting program, by all means use it. Then assume that both the second and third virial coefficients are needed, and determine both B and C. (*Hint:* $Z = 1 + B/\overline{V} + C/\overline{V}^2$ can be rewritten in the form of a straight line of slope C and intercept B.)

1.10 A colleague of yours calibrates a pressure gauge for a 1.00 L cylinder that will hold Ar at 150 K. The calibration run, however, used He instead of Ar and assumed the ideal gas equation of state. How many moles of Ar can be placed in the cylinder before the pressure reading is in error by 10%?

1.11 Since Z can be written as either a pressure expansion or as an expansion in $1/\overline{V}$, there must be a relationship between the coefficients of each type of expansion. Find the relationship between $A_1(T)$ of Eq. (1.17) and $B(T)$ of Eq. (1.18). (*Hint:* Note that $Z = P\overline{V}/RT = 1 + B/\overline{V} + \cdots$ can be solved for P, which can in turn be substituted into Eq. (1.17).) Calculate A_1 for He at 300 K where $B = 12.0$ cm^3 mol^{-1}. At what pressure would He have $Z = 1.01$?

1.12 (a) Figure 1.5 plots a real intermolecular potential energy function. How would the intermolecular potential energy function look for the interaction of two hard spheres each of volume v?
(b) The quantity B_{hs} is temperature independent, but real $B(T)$ values tend to approach a maximum as T is increased and then slowly fall at higher temperatures. (See, for instance, the He data in Table 1.1.) Use features of Figure 1.5 and your graph from (a) to explain the real high-temperature behavior of $B(T)$. Take into account the fact that molecules collide with greater average energy as the temperature is increased.

1.13 The relationship $B_{hs} = 4N_A v$ can be derived from a general theory of the second virial coefficient that can treat any pairwise intermolecular potential energy function, but this result for hard spheres can be derived without the general theory. Let the spheres have volume v and corresponding radius r. Then the center of any two spheres can approach no closer than $2r$. When the two spheres touch, one sphere excludes a certain volume from accessibility by the center of the other. Calculate this volume, divide it equally between each sphere, and show that you have derived the value $4v$ per molecule, or $4N_A v$ per mole of molecules.

1.14 Estimate the Boyle temperature for He from the values for $B(T)$ in Table 1.1, recalling that $B(T_B) = 0$. How well does your answer compare to the experimental value $T_B = 25$ K?

SECTION 1.3

1.15 A commercial cylinder of $N_2(g)$ contains 6.5 kg in a 28 L volume. Calculate the cylinder pressure at 20 °C using both the ideal gas and the van der Waals equations of state. You should find very close agreement. Can you explain why? (*Hint:* What are the individual values of each of the two pressure contributions in the van der Waals equation?)

1.16 Heretical behavior is more often justifiable in science than in other disciplines. For instance, suppose we drop the notion that R is a universal constant and instead view its role in the van der Waals equation as that of another parameter, along with a and b. Use the critical coordinates of the rare gases to deduce values for a, b, and R, and compare your answers to the a and b values in Table 1.2 and the real value of R.

1.17 Many empirical equations of state have been proposed since van der Waals first wrote his. One of the more successful was suggested in 1949 by Redlich and Kwong. It is a simple modification of the van der Waals equation, and it also has two parameters, a and b, both independent of temperature and volume:

$$P = \frac{RT}{(\overline{V} - b)} - \frac{a}{T^{1/2} \overline{V} (\overline{V} + b)}.$$

Show that the second virial coefficient for this equation of state is

$$B = b - \frac{a}{RT^{3/2}}.$$

1.18 The hard-sphere gas equation of state comes from the van der Waals equation with the a parameter set equal to zero. How do isotherms of this equation of state differ from those of the ideal gas?

1.19 A "hard-sphere gas" sounds like something less compressible than an ideal gas. Does a hard-sphere gas (with the equation of state discussed in Problem 1.18) in fact have a smaller bulk isothermal compressibility than an ideal gas at the same $(P, T, \overline{V})$ point?

SECTION 1.4

1.20 Look up the normal boiling points of the gases listed in Table 1.3 and plot their critical temperatures versus their boiling points. Can you see a correlation? Can you derive a useful rule of thumb relating these two temperatures?

1.21 Consider N_2 at room temperature and pressure. What are the corresponding-states scaled coordinates for N_2 in this state? What are its van der Waals reduced coordinates, $\tilde{P}$, $\tilde{T}$, and $\tilde{V}$? Now turn this type of calculation around. Using Figure 1.10 and data from Table 1.3, find the *real P, T,* and $\overline{V}$ for N_2 with $t = 1.1$ and $Z = 0.5$.

1.22 Show that the virial expansion, Eq. (1.18), when carried through the term $C/\overline{V}^2$, has $Z = 1/3$ at the critical point. (The method of Problem 1.25 is especially useful here. With it, you will find you do not need explicit expressions for P_c, $\overline{V}_c$, and T_c to complete this problem.)

1.23 Calculate the density of Kr and Xe at their critical points. You should find values quite close to the density of water at ordinary conditions, about 1 g cm^3. If the pressure on water is raised a few atmospheres, its density does not change appreciably. Is the same true of a critical fluid? Estimate the densities of Kr and Xe at 100 atm and the critical temperature.

GENERAL PROBLEMS

1.24 A chemist is given a small cylinder of gas by her colleague. The label on the cylinder reads: "$P = 23.5$ atm, $V = 10.0$ cm^3, 0.81 g of CF...," but at that point, a drop of acetone has obscured the label. The chemist knows that the gas is either CF_4 or CF_3H. Thus, she uses the ideal gas relationship (with $T = 300$ K, the lab's temperature) and finds $\overline{V} = 1050$ cm^3 mol^{-1}, which lets her calculate the molecular mass as $(0.81 \text{ g})(1050 \text{ cm}^3 \text{ mol}^{-1})/(10.0 \text{ cm}^3) = 85$ g mol^{-1}. This is closer to the molecular mass of CF_4 (88 g mol^{-1}) than to that for CF_3H (70 g mol^{-1}), and CF_4 is what she assumes she has. Is she justified? Critical point data for CF_4 and CF_3H are shown below:

	CF_4	CF_3H
T_c	−45.7 °C	25.9 °C
P_c	41.1 atm	46.9 atm

1.25 In the text, we used derivatives of the equation of state to find the critical-point coordinates of the van der Waals equation. Here is an alternate way. At the critical point, $\overline{V} = \overline{V}_c$, but since the equation is cubic in $\overline{V}$, it must also be true that $(\overline{V} - \overline{V}_c)^3 = 0$, since the cubic equation has three equal roots at the critical point. Expand this expression, then write the van der Waals equation as a cubic equation in $\overline{V}$ of the form $\overline{V}^3 + (\cdots)\overline{V}^2 + (\cdots)\overline{V} + (\cdots) = 0$ so that the coefficient of $\overline{V}^3$ is unity, and equate coefficients of equal powers of $\overline{V}$ in both expressions. You will then have three equations in the three unknowns, $\overline{V}_c$, P_c, and T_c. Verify that these equations give the same expressions for $\overline{V}_c$, P_c, and T_c as are in the text.

1.26 Use the method of Problem 1.25 to find the critical point coordinates of the Berthelot equation of state:

$$P = \frac{RT}{(\overline{V} - b)} - \frac{a}{T\overline{V}^2}.$$

Introduced in 1907, this expression modifies the van der Waals equation by allowing the internal pressure term to depend on temperature. Otherwise, a and b are parameters independent of volume and temperature. You should find

$$\overline{V}_c = 3b, \quad P_c = \frac{1}{6}\left(\frac{aR}{6b^3}\right)^{1/2}, \text{ and } T_c = \frac{4}{3}\left(\frac{a}{6Rb}\right)^{1/2}.$$

Does this equation of state have an improved value for the compressibility factor at the critical point over that predicted by the van der Waals equation?

CHAPTER 2

Energy, Work, and Heat

2.1 What is Energy?

2.2 The Transfer of Energy Called Work

2.3 The Variation of Work with the Path

2.4 The Transfer of Energy Called Heat

2.5 Simple Consequences of the First Law

2.6 A New State Function, Enthalpy

A large part of physical chemistry is concerned with the distribution of energy among chemical systems. Energy is an attribute of everything. It can be transferred from one object to another, redistributed over and over, and made known to us in many ways. Our recognition of energy is a remarkable and fairly recent occurrence in human affairs, yet we use the term in our day-to-day life with hardly a second thought—an energetic person, a high energy food, an energy crisis.

The term is not only a part of our daily language, but, usually without noticing it, we produce or encounter a huge variety of energies in the course of daily life. From a bowl of corn flakes in the morning (about 350 kJ per cup of flakes), through a noon milkshake (1.4 MJ per 10 oz chocolate shake), and on to a lean 3 oz sirloin steak (750 kJ) at dinner, we consume foods with varying useful energy contents. We accelerate cars on the highway to a kinetic energy of about 0.4 MJ (much less energy than the milkshake!) and we benefit in many ways from the 10^{23} J or so of solar energy received per day on Earth.

In this chapter, we will define energy in a way most useful to thermodynamics, and we will begin to learn how to use thermodynamic concepts to calculate the changes in energy that accompany physical processes. Only these *changes* in energy are calculable.

We will have no need to know the absolute amount of energy[1] contained by a system. We will see how work and heat are defined, measured, and calculated, and how they are related to energy changes. This will give us our first "Law of Thermodynamics," a general statement that has universal applicability.

2.1 WHAT IS ENERGY?

Energy is an important concept because all experimental evidence (and all of human experience) leads us to believe that the total intrinsic energy of the entire universe is a fixed quantity that does not change with time. The universe of today is the way it is only because this fixed amount of energy has been redistributed over the eons without being added to or subtracted from. Why should this be so? Why does the total energy have whatever value it has? We have no idea. We do not even know if these questions are answerable. But by accepting the *postulate* that energy exists in a universally constant amount, we have been enormously successful in constructing predictive theories, such as thermodynamics.

Whenever the total value of a physical quantity cannot change with time (even though it may be distributed among several systems in various ways), we say that the quantity is *conserved*. It is the conservation of energy (in this technical sense of the word "conservation" rather than the ecological sense) that constitutes the First Law of Thermodynamics. It must always be remembered that this law, like the other laws of thermodynamics, is an *assumed* universal principle that summarizes observations and experience. It is *not* derived from more fundamental concepts.

The exchange or transfer of energy is manifest in a number of phenomena, many of which have common names. For instance, glowing wires in electric heaters are said to warm us by radiant energy. We are conscious of the calorie (energy) content of the foods we eat (or refuse to eat), and we work (expending energy) when we do manual labor or play at sports. But energy is energy, and to retain so many names for the same quantity is unnecessarily confusing and detailed. Thermodynamics concentrates on energy *transfers* and divides them into only two categories given the generic names *work* and *heat*.

These words have been in our languages far longer than the words "energy" and "thermodynamics." It took an extraordinary effort by many talented people to organize and refine the meanings of work and heat into the precise (and related) forms we use in contemporary science. For the most part, these people lived during the rise of industrialization—the golden age of the steam engine, the blast furnace, and our exploding use of fossil fuel. Many of the premier physicists and chemists of the last century—Joule, Kelvin, Clausius, Maxwell, Boltzmann, Gibbs, and many others—were involved in establishing the principles of thermodynamics. James Prescott Joule was one of those who spent the majority of his scientific career worrying about the meanings of work and heat and the way in which they become unified under the general term, energy.

The following passage, taken from Jacob Bronowski's book, *The Ascent of Man*, gives a good impression of Joule's fervor. Keep in mind that science in Joule's day was quite a bit different from the science of today. But also keep in mind that his experiments, caricatured somewhat in this passage, were every bit as significant to

---

[1]Of course, the Einstein relation, $E = mc^2$, is a valid measure of absolute energies. It is of little use to us here, however, since chemical energy changes produce unmeasurably small mass changes.

nineteenth-century physics as the multi-million-dollar accelerator experiments of today are to twentieth-century physics.

In the summer of 1847, young William Thomson (later to be the great Lord Kelvin, the panjandrum of British science) was walking—where does a British gentleman walk in the Alps?—from Chamonix to Mont Blanc. And there he met—whom does a British gentleman meet in the Alps?—a British eccentric: James Joule, carrying an enormous thermometer and accompanied at a little distance by his wife in a carriage. All his life, Joule had wanted to demonstrate that water, when it falls through 778 feet, raises one degree Fahrenheit in temperature. Now on his honeymoon he could decently visit Chamonix (rather as American couples go to Niagara Falls) and let nature run the experiment for him. The waterfall here is ideal. It is not all of 778 feet, but he would get about half a degree Fahrenheit. As a footnote, I should say that he did not—of course—actually succeed; alas, the waterfall is too broken by spray for the experiment to work.

Fortunately, Joule, Thomson, and their contemporaries persevered. The result of their efforts, classical thermodynamics, has remained unchanged since the early part of this century.

2.2 THE TRANSFER OF ENERGY CALLED WORK

Thermodynamic systems do not "have work" in the sense that they "have mass" or "have a temperature." In fact, work is neither an extensive nor an intensive quantity. *Work is only associated with the performance of a thermodynamic process.* Something has to happen in order for work to have any meaning. Therefore, work is *not* a quantity associated with an equilibrium state of a system. This means that work must be calculated by a procedure that explicitly describes the path for the process. As we saw in the last chapter, this means that a differential amount of work, dw, cannot be treated by the special and helpful mathematics reserved for exact differentials and state functions.

It is convenient to describe work in terms of the various phenomena that produce it. These include

1. mechanical work—such as the turn of a shaft, the motion of a piston, the compression of a spring;
2. electrical work—such as the flow of a current through a circuit, the charging or discharging of a capacitor or a battery;
3. magnetic work—such as the macroscopic effect of magnetizing the system;
4. gravitational work—such as the raising or lowering of a weight; and
5. surface work—such as the formation of bubbles or drops, or the change in shape of a soap film.

We will concentrate on mechanical work in this chapter, but we will have occasion to invoke gravitational effects to produce it.

Mechanical work requires the action of an external force (located in the surroundings) doing something to the system. The force may push in one side of the system (as when a gas is compressed) or, in general, deform the shape of the boundary. Or, the force may rotate a shaft. If the shaft passes into the system (for instance, a motor-driven stirrer), the action of the shaft and its attachments transfer the effects of the external force into the system. A simple case we will examine in detail is the compression (and expansion) of a gas by a piston. External forces will control the position and motion of the piston.

Force is a *vector quantity*. Force has both a particular magnitude and a particular direction, and both are important. Similarly when something is moved by a force, we must consider not only how far it is moved, but also the direction of motion. Thus, displacement is also a vector quantity. We need both concepts—magnitude and direction—when we calculate work. Mechanical work is defined as

$$\text{work} = \int (\text{force}) \times (\text{displacement in the direction of the force}).$$

One can easily imagine processes and paths for which the force varies in magnitude and/or direction from one incremental displacement to the next, such as rearranging furniture in a room. Consequently, we write

$$\text{work} = \int_{\text{path}} \mathbf{F}(\mathbf{r}) \cdot d\mathbf{r} \ . \tag{2.1}$$

How the (vector) force $\mathbf{F}$ varies with location (the vector $\mathbf{r}$) is a statement of the path. The dot between $\mathbf{F}$ and the displacement $d\mathbf{r}$ is mathematical shorthand for the expression "displacement in the direction of the force." If the force and displacement are directed along the same line throughout the path, even if they are in different directions, this equation can be written in terms of the magnitudes of the force and displacement.

EXAMPLE 2.1

Calculate the work for the following two processes:

(a) a mass M is raised a distance h from a table, and
(b) the same mass is moved, without friction, a horizontal distance h across the table.

SOLUTION The force due to gravity acting on a mass M is just Mg in magnitude, where g is the acceleration of gravity, about 9.8 m s^{-2}. This force acts in a vertical direction pointed towards the center of the Earth. Thus for part (a):

the displacement is "up," in a direction *opposite* that of the gravitational force, and it has magnitude h. The work done *by the gravitational force* is therefore $-Mgh$, since the displacement is in a direction opposite that of the force.

But for part (b):

the displacement is *at right angles to the direction of the gravitational force*. There is thus *no* component of displacement along the gravitational-force direction, and therefore the *work associated with a frictionless displacement along a straight line perpendicular to the gravitational force's direction is zero.*

We will use these results in the discussion below.

➠ RELATED PROBLEMS 2.2, 2.3

Imagine an ideal gas confined to a cylinder by a piston of area A as shown in Figure 2.1(a). The piston is massless, frictionless, and leak-proof. The entire cylinder is immersed in a large water bath to ensure that the temperature of the gas remains fixed. The initial gas pressure, P_i, is 1 atm. Since the piston is massless and initially stationary, the external pressure must also be 1 atm. The initial volume is V_i. We define the x axis to point up and down, along the direction of gravity. We take $x = 0$ to be at the bottom of the cylinder, and we take x to increase upward. The initial location of the piston is at the point $x = x_0$.

To one side of the cylinder, we imagine a platform at x_0 holding a brick of mass M. If we slide the brick over to the piston, the external force on the piston will suddenly increase due to the weight of the brick, and the piston will move down, compressing the gas to a final pressure P_f and moving the piston a distance h, as shown in Figure 2.1(b). This process has a certain amount of work associated with it which we will now calculate.

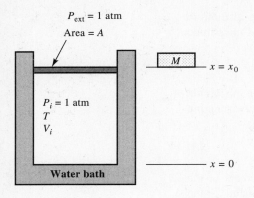

FIGURE 2.1(a) A gas is confined by a massless piston of area A at an initial pressure of 1 atm. A brick of mass M is poised at the height $x = x_0$, ready to be moved horizontally onto the piston, increasing the external force on the gas. The cylinder is immersed in a large water bath in order to keep the gas temperature constant. **(b)** When the brick is placed on the piston, the external pressure jumps from 1 atm to 1 atm + Mg/A. The piston moves down a distance h, and equilibrium is reestablished with the gas in the state P_f, T, V_f.

First, we compute P_f. At the end of the process, the gas pressure and the external pressure will again balance:

$$\underset{\substack{\uparrow \\ \text{External equilibrium} \\ \text{pressure}}}{P_{\text{ext}}} = \underset{\substack{\uparrow \\ \text{Final process} \\ \text{pressure}}}{P_f} = \underset{\substack{\uparrow \\ \text{Initial process} \\ \text{pressure}}}{P_i} + \underset{\substack{\uparrow \\ \text{Pressure added} \\ \text{by process}}}{\frac{Mg}{A}} = 1 \text{ atm} + \frac{Mg}{A} \, . \quad (2.2)$$

Here, g is the gravitational acceleration constant, and Mg is the total force exerted on the area A, so that Mg/A is the pressure component due to the brick.

Now we compute the work. The *magnitude* of the external force is

$$F(x) = P_i A + Mg = \text{a constant, independent of position.}$$

The *direction* of the force is downward, which is a *negative* direction in our coordinate system. Eq. (2.1) is

$$w = \int_{\text{path}} \mathbf{F(r)} \cdot d\mathbf{r} = -\int_{\text{path}} F(x) \, dx = -\int_{x_0}^{x_0 - h} (P_i A + Mg) \, dx \, . \quad (2.3)$$

Note how the path is specified; the initial coordinate of the piston becomes the lower limit of the integral and the final coordinate of the piston becomes the upper limit. Since the force is a constant, we take it out of the integral and find

$$w = (P_i A + Mg)h \, . \quad (2.4)$$

Next we look at the volume change, in order to simplify this otherwise perfectly valid expression for the work. We have

$$\Delta V = V_f - V_i = A(x_0 - h) - Ax_0 = -Ah \, .$$

Thus $P_i Ah = -P_i \Delta V$ and $Mgh = -Mg\Delta V/A$. Eq. (2.4), with the help of Eq. (2.2), becomes

$$w = -\left(P_i + \frac{Mg}{A}\right)\Delta V = -P_{\text{ext}} \Delta V \, . \quad (2.5)$$

This is the equation we can most readily interpret. It states that the mechanical work depends on the *external pressure* and the *volume change of the system*. Only the external pressure appears in this expression since *only external forces figure in the calculation of work*. Moreover, the system pressure is not even well-defined during the process, since the piston causes turbulence as it flies down to its final position; the gas is not at equilibrium during the entire process.

Now consider the algebraic sign of the work in Eq. (2.5). The quantity P_{ext} is the magnitude of the external pressure and is positive. The quantity ΔV, however, is *negative*. Thus the work is *positive* in this process. This leads to a generalization: *if the external force "does work" on the system, as it does here, the work is a positive quantity*. On the other hand, if the system does work on the surroundings (as it would if we removed the brick and the gas expanded against the 1 atm external pressure) then ΔV would be positive, and the work would be negative. Engineering applications of thermodynamics have historically used the opposite definition for the algebraic sign of work. Their definition arose from the design of motors, engines, and the like, systems "designed to do work." It was natural to assign a positive value to work done by such devices. (Imagine trying to sell a motor rated at -10 horsepower!) For scientific discussions, the sign of work as we have defined it here is the more relevant. We, with our external forces, are located in the surroundings, and we experiment with systems, doing work on them.

EXAMPLE 2.2

Calculate the work required to compress 1 mol of gas at 100 atm pressure and 300 K in a single step, at constant temperature, to a final volume that is half the initial volume. Assume first that the gas is ideal, then assume it is a hard-sphere gas, $P = RT/(\overline{V} - b)$, with $b = 40$ cm^3 mol^{-1}.

SOLUTION First, the ideal gas. Because T is constant, and $2V_f = V_i$, the final pressure is 200 atm. Since the process was a one-step compression, the (constant) external pressure is $P_{ext} = P_f = 200$ atm. We find ΔV from

$$\Delta V = V_f - V_i = V_f - 2V_f = -V_f = -\frac{nRT}{P_f}$$

and, from Eq. (2.5),

$$w = -P_{ext}\,\Delta V = P_f\left(\frac{nRT}{P_f}\right) = nRT = 2.46 \text{ kJ}.$$

Note that the work is independent of the initial pressure here, a somewhat counter-intuitive result. The same amount of work is associated with halving the volume starting from a pressure of 100 atm, 1 atm, or whatever (for a constant amount of gas).

Next, the hard-sphere gas. We use the equation of state with $b = 40$ cm^3 mol^{-1} to find $V_i = 286$ cm^3. Thus, $V_f = 143$ cm^3, which we can use to find $P_f = 239$ atm $= P_{ext}$. As before, $\Delta V = -V_f$, but here P_{ext} is greater. We find

$$w = -P_{ext}\,\Delta V = P_{ext}\,V_f = 34.2 \times 10^3 \text{ cm}^3 \text{ atm} = 3.46 \text{ kJ}.$$

Note that this is a greater amount of work than for the ideal gas. The increase reflects the repulsive force between the hard-sphere molecules—they pushed back more than the ideal gas.

➡ RELATED PROBLEMS 2.4, 2.5

The derivation of Eq. (2.5) involved several algebraic steps. A graphical method, shown in Figure 2.2, can arrive at the same result. First, we locate the equilibrium initial state of the gas (P_i, V_i) and the final state (P_f, V_f) on a PV diagram. Note that the water bath keeps these two points on the same isotherm (which is not shown in the figure). Next we graph the path. Initially, the external pressure is just P_i. When the brick is placed on the piston, P_{ext} rapidly jumps to its constant, final value. This part of the path is denoted by the vertical broken line. The final portion of the path is the horizontal broken line that ends at (P_f, V_f) and represents compression

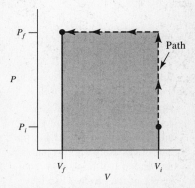

FIGURE 2.2 To graph the Figure 2.1 process, we locate the initial and final equilibrium states on a (P, V) diagram. These states are connected by the path (the dashed line) that describes the change in the external pressure with the system volume. The *magnitude* of the work associated with the process is given by the shaded area: $P_f(V_i - V_f)$.

of the gas at a constant external pressure. The shaded rectangle in the figure has a height P_{ext} and a width $V_i - V_f = -\Delta V$. Thus *the area of the shaded rectangle is* $-P_{ext} \Delta V$ *which is the work as calculated by Eq. (2.5)*.

This graphical calculation of work is very useful. Note that it involves three steps:

1. locate the initial and final equilibrium states (along with any intermediate equilibrium states encountered during the process);
2. connect these states by lines representing the way P_{ext} varies with the volume of the system (this is the path); and
3. find the total area under the path lines.

This area will always give the magnitude of the work. The algebraic sign can be deduced from our definition that *work is positive when it results in energy being transferred to the system*.

2.3 THE VARIATION OF WORK WITH THE PATH

Now we imagine taking the gas from (P_i, V_i) to (P_f, V_f) by a different process and ask if the work involved is different from that discussed in the last section. Suppose we break the brick into halves with one half on the platform at $x = x_0$ and the other on a new platform positioned to be at the height of the piston with the first half brick on it. Figure 2.3(a) describes this set-up. We compress the gas in two steps: first one half brick and then the other is placed on the piston. The new feature of this process is the intermediate equilibrium state reached when the first half brick is alone and at rest on the piston.

The work associated with this process is illustrated graphically in Figure 2.3(b). The algebraic calculation gives

$$w = -\int P_{ext}(V)\, dV = \int_{V_i}^{V'} \left(P_i + \frac{Mg}{2A}\right) dV - \int_{V'}^{V_f} \left(P_i + \frac{Mg}{A}\right) dV$$

$$= -P_i \Delta V - \frac{Mg}{A}\left(V_f - \frac{V_i}{2} - \frac{V'}{2}\right)$$

(2.6)

where V' is the intermediate volume. Note that the first step is to integrate $dw = -P_{ext}(V)\, dV$ where $P_{ext}(V)$ means "how the external pressure varies with the system volume along the path." If we compare Equations (2.5) and (2.6) or if the graphical areas of Figures 2.2 and 2.3(b) are compared, it is clear that the two values for the work are different, even though the initial and final states of the system are the same.

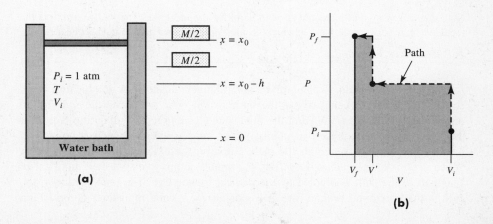

FIGURE 2.3(a) In a two-step compression, the brick of Figure 2.1 is broken in half and the half bricks are placed so that the second half-brick is at the height of the piston when the gas is compressed by the first half-brick. The combined result of both bricks is to lower the piston to the same height, $x = x_0 - h$, as in Figure 2.1(b). **(b)** The graphical representation of the two-step process shown in Figure 2.3(a) shows that the work (the shaded area) has a magnitude different from that associated with the one-step process of Figure 2.2.

This is a specific demonstration of the general result that *work is a path dependent quantity*. In particular, the work associated with a two-step compression is *less* than that associated with a one-step compression. This is most readily seen by comparing the graphical areas of Figures 2.2 and 2.3(b). We now make an inductive leap—if we had broken the brick into three equal parts (a three-step compression), the work would have been even less. One can continue this way and surmise that the *minimum work* would involve a process with infinitely many steps. How can we approach this process in reality? Suppose we pulverize the brick and carefully stack all the brick powder as shown in Figure 2.4(a). To compress the gas, we slide grain after grain over from the stack onto the piston, slowly increasing P_{ext}.

As each grain of mass dM is added to the piston, the external pressure increases by $(g/A)dM$. Since we started with $P_{ext} = P_{sys} = P_i$, this path is characterized by

$$P_{ext} = P_{sys} + dP = P_{sys} + \frac{g}{A} dM \tag{2.7}$$

where P_{sys} is the system pressure, a defined and meaningful quantity throughout the path.

The work is

$$w = -\int_{V_i}^{V_f} P_{ext}(V)\, dV = -\int_{V_i}^{V_f} (P_{sys} + dP)\, dV = -\int_{V_i}^{V_f} P_{sys}\, dV - \int_{V_i}^{V_f} dP\, dV$$

or simply

$$w = -\int_{V_i}^{V_f} P_{sys}\, dV \tag{2.8}$$

since $\int dP\, dV$ is infinitesimal. The system pressure appears in Eq. (2.8) *only* because of the special path we have chosen. Since the gas is ideal, and since the temperature and amount of gas are constant, $P_{sys}(V) = nRT/V$ and

$$w = -\int_{V_i}^{V_f} P_{sys}(V)\, dV = -nRT \int_{V_i}^{V_f} \frac{dV}{V} = -nRT \ln\left(\frac{V_f}{V_i}\right). \tag{2.9}$$

(Since $V_f < V_i$, $\ln(V_f/V_i) < 0$, and w is still positive.)

We can again use a graphical method for this process, as is shown in Figure 2.4(b). *For the first time*, the path follows the system isotherm. At each point in the path, the system is only infinitesimally displaced from equilibrium. This type of process is clearly very slow, even neglecting the time it takes to stack up

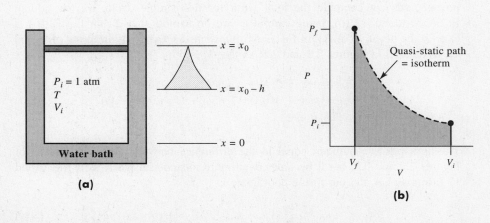

FIGURE 2.4(a) The minimum work that will compress the gas isothermally from V_i to V_f comes from powdering the brick and stacking the brick grains as shown schematically in this figure. As each grain is moved horizontally onto the piston, the gas is slowly compressed. **(b)** If each brick grain is infinitesimally small, the process is quasi-static and the path is the equilibrium system isotherm. The work (the shaded area) is the minimum possible for this isothermal compression from (P_i, V_i) to (P_f, V_f).

all those grains of powdered brick! Such processes are termed *quasi-static,* since equilibrium (a static situation, macroscopically) is closely approximated throughout the process. Quasi-static processes are important limiting cases of real processes.

EXAMPLE 2.3

Compare the work in a *quasi-static* process to that in a *one-step* process both of which compress isothermally 1 mol of an ideal gas from 100 atm, 300 K, to a final volume that is half the initial volume.

SOLUTION   The one-step work was found in Example 2.2 to be $w = nRT = 2.49$ kJ. For the quasi-static process, we use Eq. (2.9) with $V_f = V_i/2$:

$$w = -nRT \ln(V_f/V_i) = -nRT \ln(1/2) = 0.693\, nRT = 1.73 \text{ kJ} \ .$$

This is a smaller amount of work, and, in fact, it is the minimum amount that could possibly do the job.

➠ *RELATED PROBLEMS*   2.6, 2.7, 2.8

Next we look at the work associated with isothermal gas expansions. An expansion will be associated with a *negative* amount of work, but the work is otherwise conceptually the same to calculate as for a compression. However, a combination of an expansion and a compression can lead to a process which is *cyclic*—it ends where it begins. *Any process that takes a system from one state through one or more intermediate states and returns the system to the initial state is termed a cyclic process.*

We take the configuration of Figure 2.1(b) as the initial state of an expansion process. We slide the brick horizontally from the piston and allow the gas to expand against the 1 atm constant P_{ext}. Clearly, the final state of the *system* will be identical to that of Figure 2.1(a), but the brick will be at $x = x_0 - h$ at the end. We can readily compute the work from the graphical construction shown in Figure 2.5; note how the path here differs from that of the compression in Figure 2.2. The work is

$$w = -P_{\text{ext}} \Delta V = -(1 \text{ atm}) \Delta V \tag{2.10}$$

but, in contrast to compression, ΔV is *positive* and w is *negative.* Comparing Equations (2.5) and (2.10) shows that this expansion work differs not only in sign but also in magnitude from the one-step compression work between the same two states.

If we combine the one-step compression and expansion paths into one overall process, we can compute the total work for this simple cycle, w_{cy}, by adding Eq. (2.5) to Eq. (2.10), being careful as we do to note that ΔV in Eq. (2.10) is exactly the negative of ΔV in Eq. (2.5). We write ΔV explicitly in terms of V_i and V_f, as defined in Figures 2.2 and 2.5, so that $V_i > V_f$:

$$\begin{aligned} w_{\text{cy}} &= w_{\text{compression}} + w_{\text{expansion}} \\ &= -(1 \text{ atm} + Mg/A)(V_f - V_i) - (1 \text{ atm})(V_i - V_f) \\ &= \frac{Mg}{A}(V_i - V_f) \ . \end{aligned} \tag{2.11}$$

This is a positive number, equal to the difference between the shaded areas of Figures 2.2 and 2.5, but it becomes even more meaningful when we write V_i and V_f in terms of A and our piston coordinate x:

$$w_{\text{cy}} = \frac{Mg}{A}[Ax_0 - A(x_0 - h)] = Mgh \ . \tag{2.12}$$

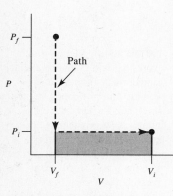

FIGURE 2.5   The path of a one-step isothermal expansion indicates that the external pressure rapidly drops to P_i and that the gas expands against this constant pressure. Once again, the magnitude of the work is given by the shaded area.

The net work for the cycle equals the change in potential energy of the brick when it is moved a vertical distance h in the gravitational field.

EXAMPLE 2.4

Construct an isothermal cyclic process for an ideal gas that starts at (V_i, P_i), begins with a one-step compression at constant P_{ext}, and ends with an expansion (or a series of expansions) so that w_{cy} is maximized.

SOLUTION A graphical solution is easiest here. The first step is simply the compression diagrammed in Figure 2.2. We want to expand from the end point of this path back to (V_i, P_i) in such a way as to minimize the energy loss from the system, since such a loss would be subtracted from w_{cy}. The way to do this is to expand with *no* work involved at all. A free expansion (against no external pressure) is such an expansion. We can imagine putting stops to catch the piston at the end of the free expansion so that the process stops at V_i again, and we return to (V_i, P_i), since the process is isothermal. A free expansion path is drawn (see diagram) as a vertical line from the starting point down to $P = 0$ and vertically (along the V axis) to the final volume. Since this path has no area under it, free expansions involve no work.

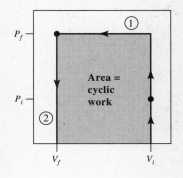

⇒ RELATED PROBLEMS 2.9, 2.25

This cyclic process is interesting for several reasons. We have returned the gas to its initial state by a process that involved a non-zero amount of work. But note that while the system winds up where it started, *the surroundings are different* at the end of the cycle. The brick has moved down. We can imagine many other types of cyclic processes based on the other compressions (two-step, etc.) we have discussed, but only one cyclic process will return the system *and* the surroundings to their initial state. Such a cycle would have *both the compression and expansion steps done quasi-statically* and is termed a *reversible cycle*. One of the important features of running an isothermal cycle reversibly is that w_{cy} is zero. This may be seen by referring to Figure 2.4(b) and noting that the path line for a quasi-static isothermal expansion is the same as for a quasi-static isothermal compression. The magnitude of the work in each will be the same, but the signs will be opposite.[2]

A Closer Look at Work Calculations

If you refer back to Figure 2.1 and imagine doing the experiment it represents, you will quickly recognize that a few simplifications were made in its discussion. First, as the brick is placed on the piston and the piston flies down, it will not gracefully stop in the position of Figure 2.1(b). Instead, it will fly past the point $x = x_0 - h$ until the gas is over-compressed to some $P \geq P_f$. The piston will then be accelerated upward by the over-compressed gas and will continue to oscillate. Since the piston was frictionless by assumption, we might expect the piston to oscillate forever. In fact, it would not. The source of "friction," the damping force that finally brings the piston to rest at $x = x_0 - h$, is provided by the gas itself through two complex nonequilibrium phenomena. The first is the viscous nature of the gas—the inherent resistance to motion of a body through the gas. The second is thermal conduction of energy through the gas, through the walls of the container, and into the water bath. Calculations of either of these effects are complicated problems of hydrodynam-

---

[2]This statement carries the following corollary. We can conclude that not only is a quasi-static process the one that *requires the least amount of work* for an isothermal *compression,* but also it is the one that *produces the greatest magnitude of work* on isothermal *expansion.*

ics. (And we should also remember that the air above the piston will be turbulent as the piston flies up and down. The external pressure due to the atmosphere is only approximately constant.)

In spite of these complications, our discussion of isothermal compression work is essentially complete. On the one hand, the oscillatory motion of the piston about the point $x = x_0 - h$ will contribute a net work of zero as the piston first overcompresses the gas and then the gas expands against the piston. On the other hand, we could imagine a mechanical stop at $x = x_0 - h$ that would catch the piston on its way down. This stop would have to absorb the kinetic energy of the piston and it would have to be in thermal contact with only the gas in order to include this energy component in the total energy transfer of the process. Such a stop would perform the role of the viscosity and thermal conduction properties of the gas.

2.4 THE TRANSFER OF ENERGY CALLED HEAT

Equation (2.12) states that the one-step cyclic process involves a net amount of work equal to the change in gravitational potential energy of the brick when it is lowered a distance h. Now we must confront and answer an important question: where did this energy go?

We can approach this question by considering just the one-step compression. If we took the cylinder out of the water bath and repeated the experiment, we would probably find that the product $P_f V_f$ was different from that observed when the cylinder was in the water bath. If we made the cylinder walls out of different materials—metals, glass, plastics, etc.—we would find a variety of different $P_f V_f$ products. Each would be *larger* than the $P_f V_f$ product observed when the system was placed in the water bath.

The implication is clear. As we go from a system constrained (by the water bath) to be at one temperature to systems with walls that more or less isolate them from the surroundings, we should not be surprised to find the temperature of the gas changing as a result of the process. In this example, isolation has produced an *increase* in temperature, since $P_f V_f$ is larger when the system is isolated than when it is not. We are using the ideal gas equation of state, $PV = nRT$, as a thermometer.

Recall our definition of temperature in Section 1.1. If we construct a reference system to be a cylinder in a bath of water and ice in equilibrium with pure water vapor, then, by definition, the temperature[3] is 273.16 K. Suppose we measure the PV product of a gas (real or ideal) in such a reference system. Call this product $(PV)_{273.16}$. If we measure the PV product of the same amount of gas in any other system, then we can say that the temperature of that system is

$$\frac{T}{273.16} = \lim_{P \to 0} \frac{(PV)_T}{(PV)_{273.16}} \tag{2.13}$$

where the limit $P \to 0$ is taken to ensure ideal behavior. This is the basis of the *ideal gas temperature scale*.

Now we return to the question posed at the beginning of this section. If our system is isolated from thermal contact with the surroundings, experiment will show

---

[3]The external pressure will no longer be 1 atm, however, since we require the external atmosphere to be pure water vapor. This equilibrium "triple point" for water, as it is called, only exists when the water vapor pressure is 4.58 torr, or (4.58 torr)/(760 torr atm^{-1}) = 6.03×10^{-2} atm. We can always add weights to the piston to make the initial pressure of the system 1 atm.

2.4 THE TRANSFER OF ENERGY CALLED HEAT

that the one-step compression raises the temperature of the gas. This rise must be a consequence of the energy transferred as work from the surroundings to the isolated system. We call a process done to a thermally isolated system an *adiabatic* process. We still allow mechanical contact with the surroundings, but we prohibit mass flow between the system and surroundings, and we construct the boundaries of the system out of materials that we recognize to be thermal insulators, such as plastic foam or the vacuum jacket of a "thermos bottle."

We can do more experiments with our adiabatic system. There are many paths that will take it from a given initial state to a given final state. Experimentation will show that the work associated with all of these paths is the same. This result appears to be in direct conflict with our previous conclusion that work is a path-dependent property. We must now modify our conclusion in light of such experiments and say that *work in an adiabatic process is not path-dependent. Adiabatic work must therefore be a state function.*

This is a pivotal result. We now postulate a new property of all thermodynamic systems, the *internal energy,* symbolized U. This property must be extensive, but we will find it useful to talk of an intensive quantity, the molar internal energy, $\overline{U} = U/n$, just as we used molar volume, $\overline{V}$, in Chapter 1. Moreover, we can state that changes in the internal energy of a system are measured by the amount of adiabatic work, w_{ad}, required to transform the system from one state to another:

$$\Delta U_{ad} = U_f - U_i = \int_{U_i}^{U_f} dU = \int_{\text{adiabatic path}} dw = w_{ad} \ . \tag{2.14}$$

This equation is a real step forward. It has a path that lets us relate a calculable quantity, adiabatic work, to a state function, the internal energy of the system. But Eq. (2.1) is incomplete. Adiabatic paths take us from one state to many other final states, but not to *all* final states. For example, the final state of our gas will *always* exhibit an increased temperature when adiabatically compressed. We need to extend Eq. (2.14) to include different phenomena (different paths) so that we can go from one state to *any* final state.

As we implied in the introduction to this chapter, the way thermodynamics makes this extension is to lump all forms of energy transfer not due to work (not produced by external forces) under the umbrella term *heat,* symbolized q. Imagine taking our gas cylinder and gluing the piston to the walls so that the system volume remains constant and holding the cylinder over a flame. The temperature of the gas will, of course, increase. We have seen that an increase in temperature brought about by adiabatic work implies an increase in internal energy. Similarly, the increase in internal energy due to a temperature rise in the *absence* of work is said to be *due to a quantity of heat being associated with the process*. We can write

$$\Delta U_V = U_f - U_i = \int_{U_i}^{U_f} dU = \int_{\text{constant volume path}} dq = q_V \tag{2.15}$$

in analogy with Eq. (2.14) with q_V reminding us the heat is associated with a constant volume process.

This is our first expression involving heat, and it is one that shows heat equals a change in a state function, ΔU. But, just as only an adiabatic path makes work a path-independent quantity, so, too, is a constant volume path a special one for heat. In general, *the heat associated with a process will depend on the details of the path*. It is *meaningless* to say a system "has a certain amount of heat." Heat, like work, is *only* associated with the execution of a thermodynamic process, and it is *not* an equilibrium property of a system.

We have not yet discussed how to calculate q directly; that comes in the next section. For now, note that q has an algebraic sign which has a similar interpretation to the algebraic sign of work. *Heat is taken to be positive when it represents a transfer of energy from the surroundings to the system.*

Internal energy, however, is a quantity associated with *any equilibrium state of a system.* Thus, we combine the ideas behind Equations (2.14) and (2.15) and state, first in words and then in equations, the *First Law of Thermodynamics:*

> The internal energy of a system is changed by the additive effects of the heat and the work associated with a thermodynamic process.

In differential form, these words become

$$dU = dq + dw \tag{2.16}$$

and in an integrated form, they become

$$\Delta U = \int_{U_i}^{U_f} dU = \int_{\text{path}} dq + \int_{\text{path}} dw = q + w \ . \tag{2.17}$$

These equations nearly end our introductory look at the First Law. Next, we will see how to calculate q and probe the consequences of this Law. Behind the mathematics, we will always seek physically meaningful interpretations. The way to master thermodynamics (and virtually every other branch of science) is to understand the connection between these physical interpretations and the mathematics that represent them.

2.5 SIMPLE CONSEQUENCES OF THE FIRST LAW

The transfer of energy by heat alone often (but not always) results in a change in temperature, as we have just seen. Experiments have shown that the relationship between an amount of heat, dq, and the temperature change it produces, dT, can be written as a simple proportionality:

$$dq = C \, dT \ .$$

Experiments also show that the proportionality factor, C, depends on the materials from which the system is constructed *and on the details of the path.* (Since temperature is a state variable and q is not, C must depend on the path.) This quantity, C, is called the *heat capacity*. We append to this symbol some description of the path (and the materials that constitute the system, if need be), such as C_V, the heat capacity at constant volume, and C_P, the heat capacity at constant pressure, the two most important paths.

EXAMPLE 2.5

One way of measuring heat capacities involves measuring the temperature rise in a system for a known heat transfer. A common and quite accurate way of transferring a known amount of energy as heat uses electrical heating. The wattage of a heater expresses the rate of energy transfer: 1 watt = 1 joule per second. For example, if a 50 W heater is in contact with a 75 g block of Cu and the heater is turned on for 19 s, the Cu block's temperature will increase by 33 K. Calculate the molar heat capacity of Cu.

SOLUTION Running a 50 W heater for 19 s corresponds to the energy transfer as heat $q = (50 \text{ J s}^{-1})(19 \text{ s}) = 950$ J. A 75 g block of Cu contains $(75 \text{ g})/(63.5 \text{ g mol}^{-1}) = 1.2$ mol Cu. The (extensive) heat capacity (at constant pressure) of the block is thus

$$C_P = \frac{q}{\Delta T} = \frac{950 \text{ J}}{33 \text{ K}} = 29 \text{ J K}^{-1}$$

and the (intensive) molar heat capacity is

$$\overline{C}_P = \frac{29 \text{ J K}^{-1}}{1.2 \text{ mol}} = 24 \text{ J K}^{-1} \text{ mol}^{-1} \; .$$

The magnitude of molar heat capacities will be discussed in detail in Chapter 4, but here we can note that 24 J K^{-1} mol^{-1} is the expected value for the molar heat capacity of metallic elements near room temperature. Metallic elements have roughly the same heat capacities due to the similarity of metallic bonding in all such solids. In later chapters, we will see how various molecular models—ideal gas, metallic bonding, real gases of various types—can be used to calculate heat capacities.

⟹ RELATED PROBLEMS 2.13, 2.14

Heat capacities are extensive quantities, but we can and often will speak of intensive molar heat capacities. Heat capacity has the units of energy divided by temperature; these are also the units of nR. In fact, one can show (by methods due to statistical mechanics but *not* by thermodynamics directly) that, *for an ideal gas*[4]

$$C_V = \frac{3}{2}nR \quad \text{or} \quad \frac{\overline{C}_V}{R} = \frac{C_V}{nR} = \frac{3}{2} \; . \qquad (2.18)$$

From this result, thermodynamics will let us deduce that

$$C_P = \frac{5}{2}nR \quad \text{or} \quad \frac{\overline{C}_P}{R} = \frac{C_P}{nR} = \frac{5}{2} \qquad (2.19)$$

for an ideal gas. (We will see one way thermodynamics relates C_P to C_V in the next section.) In general, heat capacities have to be measured by experiment or calculated by theories other than classical thermodynamics.

We can incorporate heat capacity and the relationship $dw = -P_{\text{ext}}\, dV$ into Eq. (2.17):

$$\Delta U = \int_{\text{path}} dq + \int_{\text{path}} dw = \int_{\text{path}} C_{\text{path}}\, dT - \int_{\text{path}} P_{\text{ext}}\, dV \; . \qquad (2.20)$$

The integral of $P_{\text{ext}}\, dV$ is under control, since we explored work in detail in the last section, but we have yet to consider the heat capacity integral. Before doing that, however, it is instructive to consider and exploit the rather natural way T and V, state variables of the system, have popped into the equation for ΔU.

Equation (2.20) seems to imply that U is naturally a function of T and V in some sense. The quantity U is also a function of n, but we will assume n is constant. Thus, following our notation in Equations (1.3)–(1.5), we write $U = U(T, V)$. If such a function exists, then its total differential is

$$dU = \left(\frac{\partial U}{\partial T}\right)_V dT + \left(\frac{\partial U}{\partial V}\right)_T dV \; . \qquad (2.21a)$$

Compare this expression with the differential form of Eq. (2.20):

$$dU = C_{\text{path}}\, dT - P_{\text{ext}}\, dV \; . \qquad (2.21b)$$

---

[4]We are using the term "ideal gas" in a more expansive way here than we did in Chapter 1. The ideal gas is a specific molecular model—point non-interacting particles characterized by only their mass and kinetic energy. This model has certain consequences; one is $PV = nRT$, of course, but another is Eq. (2.18). From now on in this text, whenever we use the term "ideal gas," we will mean not only $PV = nRT$ but also $C_V = 3nR/2$ and other consequences discussed in later chapters. It is correct to say that gases with $C_V \neq 3nR/2$ can behave "ideally" at low pressures, meaning that they follow $PV = nRT$, as long as it is understood that the genuine ideal gas model encompasses more than this equation of state. Real monatomic gases follow the ideal gas model closely.

We *cannot* say that the coefficients of dT and dV in each equation are equal. In general, they are not. But suppose $dV = 0$. Then

$$dU_V = \left(\frac{\partial U}{\partial T}\right)_V dT$$

from Eq. (2.21a), and also

$$dU_V = C_V \, dT$$

from Eq. (2.21b). We can make the connection

$$C_V = \left(\frac{\partial U}{\partial T}\right)_V. \qquad (2.22)$$

Can we find an expression for $(\partial U/\partial V)_T$, the other coefficient in Eq. (2.21a)? If we suppose $dT = 0$, then Eq. (2.21a) retains only the second term on the right, but Eq. (2.21b) *must still have both terms* since we know that *the path that makes the first term, $C\,dT$, equal to zero is an adiabatic path so that $dU = -P_{ext}\,dV$, rather than an isothermal one.*[5] If we suppose an adiabatic path, Eq. (2.21b) becomes $dU = -P_{ext}\,dV$, but now we must keep *both* terms in Eq. (2.21a) since both T and V might change in an adiabatic path.

The Joule Expansion

If we cannot find an expression for $(\partial U/\partial V)_T$, can we at least imagine an experiment that would measure it?[6] Our eccentric friend, James Joule, was one of the first to try. His experiment is shown schematically in Figure 2.6. He took a two-chamber vessel that had one side evacuated and the other enclosing a gas (the system) at a fairly high pressure (about 22 atm). When he allowed the gas to flow from this chamber to the other, he clearly increased the volume of the system. Moreover, the entire apparatus was immersed in a water bath (which itself was thermally insulated from the rest of the surroundings) equipped with one of Joule's thermometers. Joule reasoned as follows. When the gas expanded, there would be no work associated with the process. Why? Because $P_{ext} = 0$; the expansion was into a vacuum. (This is called a *free expansion*.) But if the internal energy of the gas changed because $(\partial U/\partial V)_T$ was non-zero, then the gas should transfer this energy change to the water bath (as heat) either warming the water or cooling it, depending on the sign of

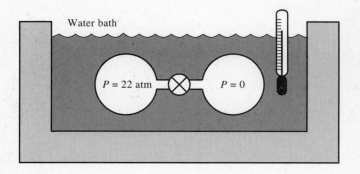

FIGURE 2.6 In the Joule experiment, a gas at high pressure (22 atm in the diagram) expands into an adjoining container that is initially evacuated. Both containers are immersed in a water bath. If the internal energy of the gas should change as a result of the expansion, the water-bath temperature would change when the expanded gas and the water return to thermal equilibrium.

---

[5]As an aside, things are even worse. The heat capacity along a constant temperature path is not well-defined!

[6]We will find an expression for $(\partial U/\partial V)_T$ eventually, but equations based on the First Law alone are not sufficient.

$(\partial U/\partial V)_T$. Any internal energy change for the gas would change the water's temperature.

Joule did the experiment and found no temperature change, implying $(\partial U/\partial V)_T = 0$. In fact, he should have measured a small temperature change, but his experiment was not sensitive enough—too much water and too little temperature rise. But his conclusion is still valid as a limiting result for real gases and a rigorous result for ideal gases:

$$\textbf{real gas:} \quad \lim_{P \to 0} \left(\frac{\partial U}{\partial V}\right)_T = 0 \; ; \quad \textbf{ideal gas:} \quad \left(\frac{\partial U}{\partial V}\right)_T = 0 \; .$$

We can see why real gases approach the ideal gas value as a limit if we invoke Chapter 1's discussion of intermolecular forces. Real molecules change their mutual intermolecular potential energy as their separation changes. This potential energy component is lacking in an ideal gas, but it is a component of the gases' internal energy. As the volume of the real gas is changed, the average distance between molecules changes, and thus the small intermolecular potential energy component of U changes as well. Once again we have a mechanism based on intermolecular forces that accounts for deviations from the limiting ideal behavior.

EXAMPLE 2.6

Derive a general expression for $\Delta \overline{U}$ for an ideal gas.

SOLUTION Since $(\partial U/\partial V)_T = 0$, $\overline{U}$ for an ideal gas is a function of T only. Thus, Eq. (2.21a) becomes

$$d\overline{U}(\text{ideal gas}) = \left(\frac{\partial \overline{U}}{\partial T}\right)_V dT = \overline{C}_V \, dT = \frac{3}{2} R \, dT \; ,$$

which can be integrated easily to yield

$$\Delta \overline{U}(\text{ideal gas}) = \frac{3}{2} R \, \Delta T \; .$$

▶ RELATED PROBLEM 2.15

In the next chapter, we will derive an expression for $(\partial U/\partial V)_T$ that, when applied to the van der Waals equation of state, yields

$$\left(\frac{\partial U}{\partial V}\right)_T = \frac{an^2}{V^2} = \frac{a}{\overline{V}^2} \; . \tag{2.23}$$

(Recall that $a/\overline{V}^2$ is the internal pressure term; $\partial U/\partial V$ has the dimensions of pressure.) Note that this expression approaches zero as P approaches zero, as expected from the ideal gas limiting value. Since $a/\overline{V}^2$ is always positive, U increases during an isothermal expansion and decreases during compression, since $dU_T = (\partial U/\partial V)_T \, dV$ = (a positive number) dV.

EXAMPLE 2.7

Lower the temperature of an ideal gas 1 K at constant volume and its molar internal energy falls by $\Delta \overline{U} = \overline{C}_V \, \Delta T = (1.5 \, R)(-1 \, \text{K}) = -12.5 \, \text{J mol}^{-1}$. If Xe is taken to be a van der Waals gas, to what final molar volume should it be compressed from an initially large molar volume to cause an equivalent drop in $\overline{U}$?

SOLUTION We integrate $(\partial U/\partial V)_T$ in Eq. (2.23) to get a general expression for ΔU_T for an arbitrary volume change between any $\overline{V}_i$ and $\overline{V}_f$:

$$\Delta \overline{U}_T = \int_{\overline{V}_i}^{\overline{V}_f} \frac{a}{\overline{V}^2} d\overline{V} = \left(\frac{a}{\overline{V}_i} - \frac{a}{\overline{V}_f}\right),$$

which, if $\overline{V}_i$ is very large, is approximately $\Delta \overline{U}_T = -a/\overline{V}_f$. For $\Delta \overline{U}_T = -12.5$ J mol^{-1} = -123 cm^3 atm mol^{-1} and with $a = 4.194 \times 10^6$ cm^6 atm mol^{-2} for Xe (Table 1.2), $\overline{V}_f = -(4.194 \times 10^6$ cm^6 atm mol$^{-2})/(-123$ cm^3 atm mol$^{-1}) = 34\,100$ cm^3 mol^{-1}. Recall from Chapter 1 that this is about 1.4 times the molar volume at room temperature and 1 atm. Thus, real gases at ordinary densities have internal energy components from intermolecular attractions that are small fractions of the internal energy controlled by temperature.

➥ **RELATED PROBLEM** 2.16

We now understand (or have the promise of understanding) all of the terms in Equations (2.21a) and (2.21b). We can integrate these equations and calculate ΔU values for various processes. For those involving an ideal gas, the key step will be to find ΔT, since ΔT alone governs ΔU.

General Adiabatic Processes

We begin with adiabatic processes, for which $dq = 0$ and $C\, dT = 0$. (Even though $dT \neq 0$ for an adiabatic process, C_{ad}, the adiabatic heat capacity, is zero, since $C_{ad} = (dq_{ad}/dT)$ and $dq_{ad} = 0$.) Equation (2.21b) becomes

$$dU_{ad} = -P_{ext}\, dV = dw_{ad}. \tag{2.24}$$

For an ideal gas, Eq. (2.21a) is always

$$dU = \left(\frac{\partial U}{\partial T}\right)_V dT = C_V\, dT = \frac{3}{2} nR\, dT. \tag{2.25}$$

Combining these equations and integrating yields

$$\int dU_{ad} = \Delta U_{ad} = w_{ad} = \int C_V\, dT = \frac{3}{2} nR\, \Delta T. \tag{2.26}$$

Suppose P_{ext} is constant. Then $w_{ad} = -P_{ext}\, \Delta V$, and Eq. (2.26) shows that

$$\Delta T = -P_{ext} \frac{\Delta V}{C_V} = -\frac{2}{3} \frac{P_{ext}\, \Delta V}{nR}.$$

Imagine next that $P_{ext} = 0$. Then $w_{ad} = 0$ and $\Delta T = 0$; this is the Joule experiment run adiabatically. Next, suppose P_{ext} is not zero, but the process is a compression so that ΔV is negative, w_{ad} is positive, and ΔT is positive. The positive work represents a flow of energy into the system. Since the process is adiabatic, this energy cannot escape as heat. But, *even though there is no heat,* the heat capacity of the gas controls the observed temperature rise.

EXAMPLE 2.8

One mole of an ideal gas at 500 K and 20 atm is expanded adiabatically against a constant external pressure of 1 atm. What is the maximum amount of work that can be done *by* this system in such a process?

2.5 SIMPLE CONSEQUENCES OF THE FIRST LAW

SOLUTION Since P_{ext} is constant, w_{ad}, the work done *on* the system, is $-P_{ext} \Delta V$ so that the work done *by* the system is $-w_{ad} = P_{ext}(V_f - V_i)$. This is the quantity we want to maximize. From Eq. (2.26), we can also write

$$-w_{ad} = -\frac{3}{2}nR \Delta T = -\frac{3}{2}nR(T_f - T_i) = P_{ext}(V_f - V_i) \;.$$

To maximize $-w_{ad}$, we should make T_f as small as possible. For an ideal gas, we can imagine T_f dropping all the way to absolute zero. This limit has a physical interpretation. Expanding the gas adiabatically until its temperature drops to zero extracts all the extractable internal energy there is—the gas "runs out of gas," so to speak. The energy extracted as work is $3nRT_i/2$.

➠ RELATED PROBLEMS 2.17, 2.18

Adiabatic Reversible Processes in an Ideal Gas

This very restricted process—adiabatic, reversible, ideal gas—is valuable to consider as a limiting process on real gases. Since the process is reversible, $P_{ext} = P_{sys} + dP$ (Eq. (2.7)), and a combination of Equations (2.24) and (2.25) gives (neglecting the term $-dP\, dV$)

$$C_V\, dT = -P_{sys}\, dV = -\frac{nRT}{V}\, dV \;. \tag{2.27}$$

To integrate this equation, we collect each independent variable by itself on either side of the equality, integrate:

$$\int_{T_i}^{T_f} C_V \frac{dT}{T} = -nR \int_{V_i}^{V_f} \frac{dV}{V}$$

and find

$$C_V \ln\left(\frac{T_f}{T_i}\right) = -nR \ln\left(\frac{V_f}{V_i}\right) \;.$$

A little rearrangement, using $C_V = 3nR/2$, gives

$$\ln\left(\frac{T_f}{T_i}\right) = \frac{2}{3} \ln\left(\frac{V_i}{V_f}\right) = \ln\left(\frac{V_i}{V_f}\right)^{2/3}$$

or, exponentiating each side,

$$T_f = T_i \left(\frac{V_i}{V_f}\right)^{2/3} \;. \tag{2.28}$$

We can rewrite this equation in an illuminating way using the equation of state to introduce pressure. First, we rearrange Eq. (2.28) to

$$T_i V_i^{2/3} = T_f V_f^{2/3}$$

and then substitute PV/nR for T, yielding

$$P_i V_i^{5/3} = P_f V_f^{5/3} \;. \tag{2.29a}$$

This is an equation for the *path of a reversible, adiabatic process involving an ideal gas,* and we can generalize this to

$$P_{sys} V^{5/3} = P_{ext} V^{5/3} = \text{constant} \tag{2.29b}$$

where the constant is either $P_i V_i^{5/3}$ or $P_f V_f^{5/3}$, since these are equal.

This equation is often written

$$PV^\gamma = \text{constant}$$

where γ is the *heat capacity ratio*:

$$\gamma = \frac{C_P}{C_V},$$

which is 5/3 for an ideal gas. For a real gas at low pressure, which may follow the ideal gas equation of state, but may not have ideal gas heat capacities, this form of the adiabatic reversible path equation can be used with whatever value of γ is appropriate to the gas.

Equation (2.29b) is graphed in Figure 2.7 (curve I, the *adiabat*) along with the isotherm $PV = nRT_i$ (curve II). The graph is drawn for an expansion ($T_f < T_i$ along curve I), and the isochore (curve III) at $V = V_f$ is explained below.

Paths Equivalent to an Adiabatic Path

It is not uncommon for someone learning thermodynamics to look back at these equations, especially Eq. (2.27), and ask how C_V, a constant volume quantity, got mixed into an equation describing a process in which the volume is most definitely not constant! This seemingly contradictory state of affairs is a consequence of dealing with path-independent state functions such as U. Equation (2.27) equates the internal energy change in a constant volume process ($C_V \, dT$) to the change in an adiabatic, reversible process ($-P_{\text{sys}} \, dV$). Since the internal energy is a state function, the equality holds as long as the initial and final states of both processes are the same.

Figure 2.7 shows how these two paths can be used to connect the same two states. First, imagine calculating ΔU for path I, the adiabatic, reversible expansion. Then imagine a process that first expands the gas *isothermally* to V_f (path II) followed by an isochoric cooling to T_f (path III). Along path II $\Delta U_{\text{II}} = 0$, since it is isothermal and the gas is ideal. Finally, along path III, $\Delta U_{\text{III}} = C_V (T_f - T_i)$. Since U is a state function and since path I = path II + path III,

$$\Delta U_{\text{I}} = \Delta U_{\text{II}} + \Delta U_{\text{III}} = 0 + C_V (T_f - T_i) \tag{2.30}$$

or, with the help of Eq. (2.28) to express T_f,

$$\Delta U_{\text{I}} = C_V T_i \left[\left(\frac{V_i}{V_f} \right)^{2/3} - 1 \right]. \tag{2.31}$$

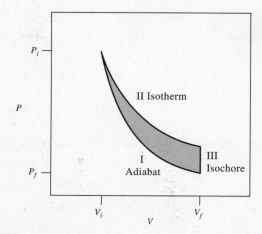

FIGURE 2.7 Curve I shows the path for the *adiabatic*, reversible expansion of an ideal gas from (P_i, V_i) to (P_f, V_f). Curve II is the ideal gas isotherm for temperature T_i. It is also the path for the *isothermal*, reversible expansion of an ideal gas from V_i to V_f. Curve III is a constant volume path connecting states (V_f, T_i) and (V_f, T_f).

We can also calculate ΔU_1 from w_{ad} and Eq. (2.29):

$$w_{ad} = -\int_{V_i}^{V_f} P_{ext}\, dV = -\int_{V_i}^{V_f} \frac{P_i V_i^{5/3}}{V^{5/3}}\, dV$$

$$= \frac{3}{2} P_i V_i^{5/3} (V_f^{-2/3} - V_i^{-2/3})$$

$$= \frac{3}{2} (P_f V_f - P_i V_i)$$

$$= \frac{3}{2} nR (T_f - T_i) = C_V (T_f - T_i)$$

in agreement with Eq. (2.30).

EXAMPLE 2.9

A quantity of He (assumed to be an ideal gas) is contained at 300 K in a 10 dm³ volume at 1 atm. The container is thermally insulated, but fitted with a piston. A weight is suddenly placed on the piston, compressing the gas in one step to a final pressure of 5 atm. What is the work associated with this process? What is the final temperature of the gas?

SOLUTION We know $T_i = 300$ K, $V_i = 10$ dm³, $P_i = 1$ atm, and $P_f = 5$ atm. The process is an adiabatic compression under a constant external pressure of 5 atm. Thus

$$\Delta U = C_V (T_f - T_i) = w_{ad} = -P_{ext} (V_f - V_i).$$

We don't know T_f or V_f, but we know they are related via the equation of state

$$V_f = \frac{nRT_f}{P_f}.$$

Substitution gives

$$C_V (T_f - T_i) = -P_{ext} \left(\frac{nRT_f}{P_f} - V_i \right)$$

but we don't know n. We could find n from the relationship $n = P_i V_i / RT_i$, but this step turns out to be unnecessary. If we solve the above for T_f and use $P_{ext} = P_f$ and $C_V = 3nR/2$, we obtain

$$T_f = \frac{P_f V_i + \frac{3}{2} nRT_i}{\frac{5}{2} nR}.$$

Next, we make use of the identity $P_f V_i = P_f (P_i V_i)/P_i = (nRT_i)(P_f/P_i)$, which gives an expression for T_f that we can evaluate:

$$T_f = T_i \left[\frac{(P_f/P_i) + \frac{3}{2}}{\frac{5}{2}} \right] = 2.6\, T_i = 780 \text{ K} . \text{ (Quite an increase!)}$$

We can find w_{ad} most simply from

$$w_{ad} = C_V (T_f - T_i)$$

$$= 1.5\, nR\, (2.6\, T_i - T_i)$$

$$= (1.5)(1.6)\, nRT_i = (1.5)(1.6)\, P_i V_i$$

$$= (1.5)(1.6)(10 \text{ dm}^3 \text{ atm}) = 24 \text{ dm}^3 \text{ atm} .$$

Note that we never calculated n, although we could have ($n = 0.41$ mol), nor did we need a numerical value for R.

➠ RELATED PROBLEMS 2.18, 2.26

A Closer Look at the Meaning of Adiabaticity

A few more words about adiabatic processes are in order before we move on. As we have described them so far, we have required a thermally insulating boundary for adiabatic processes to ensure $dq = 0$. One other way of defeating energy transfer as heat recognizes that this transfer is not instantaneous. Thus, if we carry out a process fast enough, it will appear to be adiabatic, at least momentarily. Better yet, a fast, cyclic process (compression, expansion, compression, expansion, etc.) will appear to be purely adiabatic. An example of this type of process is sound. Sound waves are rapid gas compression–expansion cycles (tens to several thousand per second, for sound humans can hear) that can be described as a cyclic adiabatic process.

A second process of great practical interest is based on a slow, atmospheric phenomenon that closely approaches adiabaticity simply because air is not a very efficient thermal conductor. As air rises from near sea level, it expands due to the general pressure drop with increasing altitude. Such an expansion is accompanied by a temperature drop as the vast amounts of atmospheric air act as their own thermal insulators. This phenomenon is discussed in detail in Chapter 9.

2.6 A NEW STATE FUNCTION, ENTHALPY

Constant external pressure processes are very common and easy to arrange. Typically, one need only keep the system open to the external 1 atm pressure of the surroundings, as in an uncovered beaker. The relevant heat capacity is C_P, and we calculate the heat involved in a constant pressure process by integrating $dq_P = C_P \, dT$.

For such paths, we will now show that it is helpful to consider the sum $U + PV$. This sum is a new state function, since it is the sum of a state function, U, and a product of state variables, PV. We can write the differential for this sum as

$$d(U + PV) = dU + P \, dV + V \, dP$$
$$= dq - P_{ext} \, dV + P \, dV + V \, dP \; .$$

At constant pressure, $P = P_{ext}$ and $dP = 0$, and we are left with

$$d(U + PV)_P = dq_P \; . \quad \textbf{(2.32)}$$

This result is so helpful and useful that the sum $U + PV$ is given a special name and symbol:

$$U + PV = H = enthalpy. \quad \textbf{(2.33)}$$

Enthalpy is another extensive state function defined for any equilibrium state of any system. Just as systems have and can transfer energy, they also have and can transfer enthalpy.

From Equations (2.32) and (2.33), $dH_P = dq_P = C_P \, dT$. Following arguments very similar to those leading to Eq. (2.22), we can show that

$$\left(\frac{\partial H}{\partial T} \right)_P = C_P \; .$$

From the ideal gas equation of state, we have

$$H(\text{ideal gas}) = U(\text{ideal gas}) + PV = U + nRT .\qquad(2.34)$$

Since U(ideal gas) is a function of T only (at constant n), so is H(ideal gas). For an ideal gas, $\Delta U = \Delta H = 0$ for *all* isothermal processes. The differential form of Eq. (2.34) is

$$dH = dU + nR\, dT$$

or

$$C_P\, dT = C_V\, dT + nR\, dT$$

so that

$$C_P - C_V = nR \quad \text{for an ideal gas} \qquad(2.35)$$

as indicated by Equations (2.18) and (2.19). We have implicitly used $(\partial U/\partial V)_T = 0$ and $(\partial H/\partial P)_T = 0$ for an ideal gas here. We will find a more general expression for $C_P - C_V$ in Section 3.7.

EXAMPLE 2.10

What is the difference between the molar enthalpy and the molar internal energy of a room temperature ideal gas at 1 atm pressure? At 10 atm?

SOLUTION Here we need only apply the definition of enthalpy. Since $H = U + PV$, we have $\overline{H} - \overline{U} = P\overline{V} = RT$. At room temperature, $RT \cong 2.5$ kJ mol^{-1}. We encountered this number in Example 1.2 in Section 1.1, and now we can see that $P\overline{V}$ measures the difference between the molar enthalpy and molar internal energy of a gas—*any* gas, but only for the ideal gas does $\overline{H} - \overline{U}$ also equal RT. Since neither H nor U depends on pressure for an ideal gas, the answer here is the same for 1 atm, 10 atm, or any pressure at all.

➞ RELATED PROBLEM 2.19

Although enthalpy was invented for constant pressure applications, the change in enthalpy may be calculated for any process. For example, we know how U varies with T at constant volume: $(\partial U/\partial T)_V = C_V$; we can also ask how H varies with T at constant volume. We start with H as a function of its natural variables, T and P: $H = H(T, P)$, and write the total differential in the usual way:

$$dH = \left(\frac{\partial H}{\partial T}\right)_P dT + \left(\frac{\partial H}{\partial P}\right)_T dP = C_P\, dT + \left(\frac{\partial H}{\partial P}\right)_T dP .\qquad(2.36)$$

Next we divide by dT holding V constant to find an expression for $(\partial H/\partial T)_T$:

$$\left(\frac{\partial H}{\partial T}\right)_V = C_P + \left(\frac{\partial H}{\partial P}\right)_T \left(\frac{\partial P}{\partial T}\right)_V .\qquad(2.37)$$

How can we interpret or even use this expression? We can measure C_P and calculate $(\partial P/\partial T)_V$ from an equation of state, but what can we do with $(\partial H/\partial P)_T$? If we set $dH = 0$ in Eq. (2.36) we can arrange the result as

$$\left(\frac{\partial H}{\partial P}\right)_T = -C_P \left(\frac{\partial T}{\partial P}\right)_H .\qquad(2.38)$$

(Note how the constraint $dH = 0$ is expressed as the subscript to $\partial T/\partial P$.) Substituting this into Eq. (2.37) yields

$$\left(\frac{\partial H}{\partial T}\right)_V = C_P - C_P \left(\frac{\partial T}{\partial P}\right)_H \left(\frac{\partial P}{\partial T}\right)_V = C_P \left[1 - \left(\frac{\partial T}{\partial P}\right)_H \left(\frac{\partial P}{\partial T}\right)_V\right]. \qquad (2.39)$$

Compare this expression for $(\partial H/\partial T)_V$ to that in Eq. (2.37) where $(\partial H/\partial P)_T$ is unknown. In Eq. (2.39), we have replaced it with $(\partial T/\partial P)_H$, a derivative involving T and P, but one which is to be *evaluated along a path of constant enthalpy*. To make Eq. (2.39) useful instead of just another abstract expression, we must find a process that occurs *at constant enthalpy* (an *isenthalpic* process).

The Joule–Thomson Effect

Joule and William Thomson found such a process. They confined a quantity of gas in an adiabatic cylinder, as shown schematically in Figure 2.8. The cylinder contained two moveable pistons with a porous wall between them. Initially the gas was all on the left of the wall as shown in Figure 2.8(a) in the state (P_i, V_i, T_i). The left piston then drove the gas through the porous wall (maintaining P constant at P_i on the left) while the right piston was withdrawn so as to maintain a constant final pressure in the right hand chamber throughout the process. The final position of both pistons is shown in Figure 2.8(b), with the gas in the final state (P_f, V_f, T_f).

It is not obvious that this process is isenthalpic, but we can show very easily that it is. First, $q = 0$, since the cylinder walls are adiabatic. Second, $\Delta U = w_{ad} = P_i V_i - P_f V_f = \Delta(PV)$ since the work consists of two parts: $P_i V_i$ is the work done compressing the gas to zero volume on the left under a constant $P_{ext} = P_i$ and $-P_f V_f$ is the work associated with the gas expansion on the right from zero volume to V_f against $P_{ext} = P_f$. We next write $\Delta U = \Delta(PV)$ as

$$U_i + P_i V_i = U_f + P_f V_f$$

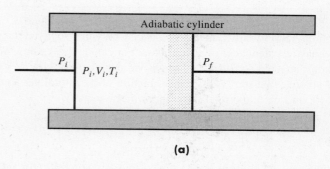

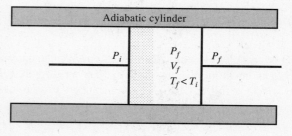

FIGURE 2.8(a) In the initial state of the Joule–Thomson experiment, the gas is confined to the volume V_i on the left side of a porous plug mounted in an adiabatic cylinder. The gas is compressed through the plug by the constant external pressure P_i on the left and against the constant external pressure P_f on the right. **(b)** The final state of the Joule–Thomson experiment has the gas transferred to the right-hand side of the porous plug. Since the temperature and pressure on either side of the plug are both different, in general, the final volume V_f will be different from the initial volume.

and recognize that the left-hand side is H_i, the initial enthalpy, while the right is H_f. Thus, $\Delta H = H_f - H_i = 0$, and the process is isenthalpic.

The important measurement in this experiment is the final temperature at various final pressures. (In fact, the most important result is that the temperature *does* change for a real gas.) Suppose we pick a convenient P_i, T_i starting point and plot measured T_f values for various experiments done at different P_f values. If we connect those data by a smooth line, as is shown in Figure 2.9, we draw a curve that connects states having the same enthalpy; we draw an *isenthalp*. The slope of this curve, $(\partial T/\partial P)_H$, is the quantity we need to allow us to use Eq. (2.39). It is called the *Joule–Thomson coefficient*:

$$\mu_{JT}(T, P) = \left(\frac{\partial T}{\partial P}\right)_H . \qquad (2.40)$$

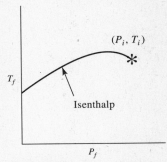

FIGURE 2.9 In a series of Joule–Thomson experiments, a gas at the initial state (P_i, T_i) is expanded to a series of final pressures $P_f < P_i$. For each experiment, the final temperature T_f is measured, and the results are plotted as a series of (T_f, P_f) points. When these points are connected, the resulting line shown in the figure locates states of the gas with the same enthalpy. This line is an *isenthalp*. The slope of an isenthalp on such a plot is μ_{JT}, the Joule–Thomson coefficient.

It is a function of both T and P, and it may be positive, negative, or zero. Since H(ideal gas) is a function of temperature only, ideal gas isenthalps are horizontal lines (zero slope) on a graph like Figure 2.9, and thus μ_{JT}(ideal gas) = 0.

In the next chapter, we will derive the following expression for μ_{JT}:

$$\mu_{JT}(T, P) = \frac{1}{C_P}\left[T\left(\frac{\partial V}{\partial T}\right)_P - V\right] . \qquad (2.41)$$

This is a general expression, but suppose the gas follows a virial equation of state, Eq. (1.18), through the second virial coefficient. Then we can use Eq. (2.41) to find an explicit expression for μ_{JT}. The low pressure ($\overline{V} \gg B$) approximation to the exact expression is revealing:

$$\mu_{JT} \cong \frac{1}{C_P}\left[T\left(\frac{dB}{dT}\right) - B\right] . \qquad (2.42)$$

It shows that at temperatures for which $B = T(dB/dT)$, the Joule–Thomson coefficient is zero. Such temperatures are called *Joule–Thomson inversion temperatures*, T_I. If we use the van der Waals approximation for $B(T)$, Eq. (1.22), we can find T_I as follows:

$$B(T_I) = T_I\left(\frac{dB}{dT}\right)_{T_I} \quad \text{(in general)}$$

or

$$b - \frac{a}{RT_I} = T_I\left(\frac{a}{RT_I^2}\right) \quad \text{(for the van der Waals virial expression).}$$

Solving for T_I, we find

$$T_I = 2\left(\frac{a}{Rb}\right) = 2T_B . \qquad (2.43)$$

The inversion temperature is predicted to be twice the Boyle temperature, in approximate agreement with experiment, and we should therefore expect inversion temperatures to be well above room temperature for most gases. Below T_I, gases *cool* when expanded, and that is the behavior of most real gases.

Typical values of μ_{JT} are collected in Table 2.1. A typical magnitude below the inversion temperature is a few tenths kelvin per atmosphere pressure drop. A large temperature drop requires a very large pressure drop, but one that can be realized.

TABLE 2.1 Joule–Thomson Coefficients, μ_{JT}/K atm^{-1}

Ar

T/°C	P/atm		
	1	20	200
−100	0.8605	0.8485	0.0395
0	0.4307	0.4080	0.1883
100	0.2413	0.2277	0.1255
200	0.1377	0.1280	0.0675
300	0.0643	0.0607	0.0276

N_2

T/°C	P/atm		
	1	20	200
−100	0.6490	0.5958	0.0587
0	0.2656	0.2494	0.0891
100	0.1292	0.1173	0.0419
200	0.0558	0.0472	0.0070
300	0.0140	0.0096	−0.0171

CO_2

T/°C	P/atm		
	1	100	200
−75	—	−0.0228	−0.0290
0	1.2900	0.0115	0.0045
100	0.6490	0.4320	0.2555
200	0.3770	0.2890	0.2455
300	0.2650	0.1700	0.1505

In fact, the major use of the Joule–Thomson effect is in gas liquefaction and refrigeration. Usually, the gas is subjected to several expansions, using the cold gas of the previous expansion to lower the initial temperature of the subsequent expansion until liquid forms. For H_2 and He, μ_{JT} is negative, and the gases *warm* on expansion, unless the temperature is very low. These gases must be pre-cooled from room temperature before they can be liquefied by expansion. For example, at 7 K, an expansion from only 10 atm to 1 atm will cool He below its liquefaction temperature of 4.2 K.

A Closer Look at the Joule–Thomson Effect

Equations (2.42) and (2.43) are valid for low-pressure, high-temperature conditions where non-ideality plays a small role. The inversion temperature demarcates the boundary between cooling and warming behavior. It represents another temperature at which attractive and repulsive forces are in some sense balanced, and the real gas behaves as if ideal. In general, however, real gases exhibit not just one inversion temperature, but a range of inversion temperatures that depend on the initial pressure.

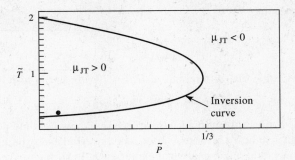

FIGURE 2.10 This is a graph of the Joule–Thomson inversion curve for the van der Waals equation of state in terms of the reduced variables of Eq. (1.25). Gases in states with $(\tilde{P}, \tilde{T})$ lying on this curve are in states for which $\mu_{JT} = 0$. The dot in the lower left corner locates the critical point. Behavior in this region is complicated by gas condensation phenomena.

To see this, we must use a complete, realistic equation of state in Eq. (2.41), such as the van der Waals equation. From Eq. (2.41), we see that $\mu_{JT} = 0$ whenever $(\partial V/\partial T)_P = V/T$. This condition can be applied to an equation of state to find the (T, P) coordinates of the inversion temperatures. The algebra is involved, but the result is worth the effort. One finds, using the reduced variables $\tilde{P}$ and $\tilde{T}$ of Eq. (1.25), that states for which $\mu_{JT} = 0$ lie on an *inversion curve* given by the equation

$$\tilde{P} = \left[3\left(\frac{\tilde{T}}{2}\right)^{1/2} - 1\right]\left[1 - \left(\frac{\tilde{T}}{2}\right)^{1/2}\right].$$

This equation is graphed in Figure 2.10. If the gas is expanded from a (P, T) state inside the inversion curve, μ_{JT} will be positive, and the gas will be cooled. An initial state outside the curve is characterized by $\mu_{JT} < 0$, and the gas will be warmed if the pressure drop is small. (If the pressure drop is very large, the gas may still cool, as was indicated in Figure 2.9.)

Recall that $\tilde{T} = T/T_B$. From Eq. (2.43), we found one inversion temperature to be $\tilde{T} = 2$. Figure 2.10 shows this to be the low-pressure limit of the *upper* inversion temperature. The *lower* inversion temperature at low pressure is $\tilde{T} = 2/9$. For all pressures below $\tilde{P} = 1/3$ there is an upper and a lower inversion temperature. (The dot on Figure 2.10 locates the critical point at $\tilde{P} = 1/27$. Condensation phenomena also complicate matters near the lower inversion temperature.)

There is one final comment to make. The porous plug wall that divides the apparatus plays a very important role.[7] It provides a rapid way of smoothing the pressure drop throughout the plug without leaving the gas in a turbulent state on the low-pressure side. If the plug were replaced by a thin plate with a small hole in it, the gas would jet into the receiving chamber with part of its enthalpy going into the bulk flow of the gas. With the plug, this flow never develops; instead, one maintains the desired uniform final pressure at all stages of the expansion.

Now that the Joule–Thomson effect has been described, we can return to the basic equations for dH and see its role. Eq. (2.38) becomes

$$\left(\frac{\partial H}{\partial P}\right)_T = -C_P\left(\frac{\partial T}{\partial P}\right)_H = -C_P\,\mu_{JT}$$

---

[7] Joule and Thomson used a silk handkerchief or a piece of meerschaum, a porous magnesium silicate, as barriers in their experiments.

and Eq. (2.39) becomes

$$\left(\frac{\partial H}{\partial T}\right)_V = C_P\left[1 - \mu_{JT}\left(\frac{\partial P}{\partial T}\right)_V\right].$$

We can also write the total differential of H, Eq. (2.36), in the form

$$dH = C_P\, dT - C_P\, \mu_{JT}\, dP\,. \tag{2.44}$$

To integrate this expression and find ΔH, we need to know not only how μ_{JT} varies with T and P, but also how C_P varies.[8] Often, one has tabulated experimental values for these quantities rather than analytical expressions. In such cases, one can fit the tabulated data to a simple empirical function that can be integrated, or approximate numerical integration methods can be used.

Finally we have to note that while the Joule–Thomson coefficient is very helpful in these sorts of calculations, it is limited to gases. We will see how to calculate ΔH over various paths for liquids and solids in the next few chapters, but we should not forget that the simplest way of calculating ΔH in general may often be to begin with the definition $dH = d(U + PV) = dU + P\, dV + V\, dP$ and integrate this equation term by term.

CHAPTER 2 SUMMARY

This chapter has introduced work, heat, the internal energy, and the First Law:

work: $w = \int_{\text{path}} \mathbf{F}(\mathbf{r}) \cdot d\mathbf{r}$ (in general),

$= -\int_{\text{path}} P_{\text{ext}}(V)\, dV$ (for PV work)

heat: $q = \int_{\text{path}} C_{\text{path}}\, dT$

internal energy and the First Law:
$dU = dq + dw = C\, dT - P_{\text{ext}}\, dV = dw_{\text{ad}}$

Specific results for an ideal gas include:

internal energy: $\Delta \overline{U}$(ideal gas) $= \overline{C}_V\, \Delta T = \frac{3}{2} R\, \Delta T$

$\left[\text{since } \left(\frac{\partial U}{\partial V}\right)_T = 0\right]$

heat capacities:

$\overline{C}_V$ (ideal gas) $= \frac{3}{2} R$,

$\overline{C}_P$ (ideal gas) $= \frac{5}{2} R$

reversible adiabatic path: $PV^{5/3} = P_i V_i^{5/3} = P_f V_f^{5/3}$

The second state function we introduced was the enthalpy:

enthalpy: $dH = dU + d(PV) = dq_P$

ideal gas enthalpy: $\Delta \overline{H} = \overline{C}_P\, \Delta T = \frac{5}{2} R\, \Delta T$

$\left[\text{since } \left(\frac{\partial H}{\partial P}\right)_T = 0\right]$

and its associated Joule–Thompson coefficient:

Joule–Thomson coefficient: $\mu_{JT} = \left(\frac{\partial T}{\partial P}\right)_H$

($= 0$ for an ideal gas)

total differential of enthalpy:
$dH = C_P\, dT - C_P\, \mu_{JT}\, dP$

As the systems we encounter become more complex, we will find additional paths to consider, but these will always involve the basic ideas of the path integrals introduced here.

The next chapter introduces entropy, the quantity that goes hand-in-hand with internal energy in any discussion of chemical systems at equilibrium. Since entropy is a less familiar quantity than energy, the next chapter devotes several sections to the description of entropy before introducing the Second Law of Thermodynamics and its physical consequences.

---

[8] The variation of C_P is discussed in Chapter 4.

FURTHER READING

Probably every text with *thermodynamics* somewhere in its title discusses material related to this chapter. It would be worth your while to scan your local library holdings to familiarize yourself with the wide range of such texts. Library of Congress catalog numbers around QC311, QD501, QD504, QD511 (for chemistry and physics applications), TJ265, TN673, TP155 (for engineering applications), and QH345 (for biological applications) are the best places to look for thermodynamics books. The preface will usually tell you the intended level of the book. A particularly readable elementary introduction to this and the following two chapters is *The Second Law*, by P. W. Atkins (Scientific American Library, distributed by W. H. Freeman, New York, 1984).

PRACTICE WITH EQUATIONS

2A What mass should be placed on a piston of area 100 cm^2 to produce a pressure component of 1.00 atm? (2.2)

ANSWER: 103 kg

2B A constant pressure of 1.80 atm compresses a gaseous system so that the volume change is -2.50 dm^3. What is the work, expressed in both dm^3 atm and J units? (2.5)

ANSWER: 4.50 dm^3 atm, 456 J

2C What is the work associated with a quasi-static isothermal compression of 5.00 mol of an ideal gas at 295 K from a volume of 120 dm^3 to 80.0 dm^3? (2.9)

ANSWER: 4.97 kJ

2D What is the internal energy change for a process involving the work in Problem 2B and a heat of -600 J? (2.17)

ANSWER: -134 J

2E By how much does the internal energy of 2.00 mol of an ideal gas, held at constant volume, change per K increase in temperature? (2.18), (2.22)

ANSWER: 24.9 J K^{-1}

2F An adiabatic process carried out on 1.25 mol of an ideal gas involves -850 J of work. What is the temperature change of the gas? (2.26)

ANSWER: -54.5 K

2G What is the final temperature of an ideal gas initially at 298 K that is subjected to a reversible, adiabatic expansion that doubles its volume? (2.28)

ANSWER: 188 K

2H A 1.50 mol sample of an ideal gas at 245 K and 0.75 atm is altered by some process to 295 K and 0.90 atm. What is the enthalpy change of the gas? (2.33)

ANSWER: 1.56 kJ

2I What is the molar enthalpy of Ar at 1.00 atm and 300 K per atm increase in pressure? (2.19), (2.38), Table 2.1

ANSWER: -1.34 J atm^{-1}

PROBLEMS

SECTION 2.1

2.1 Were you surprised to read in the introduction that a 10 oz milkshake has a greater metabolic energy content than the kinetic energy of a car? It is important to gain a sense of the magnitudes of chemically common energies. Listed below are several such quantities, expressed in the units in which they are commonly tabulated or measured. Using conversion factors from Appendix I, convert these to J units and "graph" them to display their relative magnitudes. The graph should be a line marked off logarithmically in powers of 10 spanning the range of your values with each value located along the line.

(a) metabolic energy of 1 tablespoon of peanut butter = 95 kcal (kilocalorie)
(b) ionization energy of one H atom = 13.6 eV (change to J mol^{-1} units)
(c) bond energy of F$_2$ = 37 kcal mol^{-1} (change to J mol^{-1} units)
(d) energy to melt 1 g of ice at 0 °C = 79.7 cal
(e) energy equivalent of 1 atomic mass unit (1 amu), expressed as $E = mc^2$
(f) decay energy of the radioactive ^{60}Co nucleus = 2.8 MeV
(g) kinetic energy of a 180 lb track star running 24 mph, expressed as $mv^2/2$

(h) kinetic energy of N_2 gas at room temperature = 3.7 kJ mol^{-1}

(i) energy needed to lift this book 1 m above a table = 4.3 cal

SECTION 2.2

2.2 Decide for the following processes, all carried out on an ideal gas, whether the associated work is positive, negative, or zero:

(a) the gas is left out in the sun, warms noticeably, and expands its container
(b) the gas is stirred by a paddle stirrer driven by an external motor
(c) the gas is held in a rigid container, placed in a refrigerator, and cooled
(d) the gas, contained in a balloon, is placed in an evacuated chamber and the balloon is popped.

2.3 The magnitude of the force between an electron and a proton a distance r apart is $(2.307 \times 10^{-28}$ N m$^2)/r^2$. Calculate the work associated with bringing these particles from $r = \infty$ to $r = 0.5 \times 10^{-10}$ m, roughly their average separation in a hydrogen atom.

2.4 A 200 kg brick is placed on a piston enclosing 5.00 mol of an ideal gas initially at 295 K and 1.00 atm as in Figure 2.1(a). The piston's area is 300 cm^2. Calculate the work associated with this isothermal process, and find the distance the piston moved.

2.5 Repeat the previous problem using the hard-sphere gas equation of state from Example 2.2.

SECTION 2.3

2.6 In Example 2.2, it is shown that the work for a one-step isothermal compression of an ideal gas depends only on the temperature of the gas. Suppose the process involved n steps instead of one, but was still isothermal, so that each step ended at an intermediate equilibrium state with volume V_j for step j such that $V_f < V_j < V_i$. Show that the work, while different from that for a one-step process, still depends on the temperature alone and not on the initial pressure.

2.7 Expand Problem 2.6 as follows. Imagine that the process divides $V_f - V_i$ into n equal volume decrements, find the work, then take the limit as $n \to \infty$ and recover the quasi-static work expression, Eq. (2.9).

2.8 Repeat Example 2.3 using the hard-sphere gas equation of state in Example 2.2.

2.9 An ideal gas at $(T, P, V) = (200$ K, 1 atm, 10 dm$^3)$ is suddenly and irreversibly compressed by a constant external pressure of 10 atm. It is then expanded quasi-statically back to its initial state. The entire process is isothermal. Calculate the total work for this process, both algebraically and graphically.

SECTION 2.4

2.10 One way to increase the internal energy of any system is to raise its temperature. Another is to throw it, giving it a kinetic-energy component. A third is to raise it in a gravitational field, increasing its potential-energy component. Consider a 1 cm^3 cube of pure gold. Find (a) the temperature increase and (b) the velocity needed to increase this cube's internal energy by an amount equal to that corresponding to raising the cube 1 km in the Earth's gravitational field. The mass density of gold is 19.3 g cm^{-3}, its molar heat capacity is 25 J mol^{-1} K^{-1}, and its molar atomic mass is 197 g mol^{-1}.

2.11 Can you invent a graphical method for calculating the heat associated with a path that is analogous to the P–V diagram graphical method for calculating work? The work calculation is based on $|dw| = P_{ext}\, dV$. Can you find a useful analogous expression for dq?

2.12 Find the heat associated with the process in Problems 2.4 and 2.5. What is the physical interpretation of the difference in the two heats?

SECTION 2.5

2.13 The molar heat capacity of liquid water is discussed in Chapter 4, where we find $\overline{C}_P = 75$ J mol^{-1} K^{-1}. Verify that, as quoted in Section 2.1, "water, when it falls through 778 feet, raises one degree Fahrenheit."

2.14 Use data in Example 2.5 and the previous problem to calculate the final temperature when a 50 g block of Cu at 75 °C is immersed in 50 mol H$_2$O at 25 °C.

2.15 The 50 g, 75 °C block of Cu from the previous problem is placed in a 10 L evacuated and thermally insulated container. The container is then rapidly filled to 1.00 atm with an ideal gas initially at 25 °C. What is the final temperature of the gas, and what is its internal energy change due to this temperature equilibration with the Cu?

2.16 Suppose the gas in the previous problem followed the van der Waals equation of state but still had the ideal gas $\overline{C}_V$ value. (Xenon is such a gas, as Example 2.7 discusses.) What would happen to the temperature of the equilibrated system in Problem 2.15 if the container's volume is increased?

2.17 Figure 2.7 shows how an adiabatic, reversible expansion is related to an isothermal, reversible expansion followed by an isochoric cooling. Construct the analogous diagram (to scale if you can, especially if you have access to a good plotting computer program) relating an adiabatic, reversible compression to:

(a) an isothermal, reversible path and an isochoric path,
(b) an isothermal, reversible path and an isobaric (constant pressure) path,

and

(c) an isochoric path and an isobaric path.

2.18 To what final volume should an ideal gas be expanded so that its temperature drops from 300 K to 225 K if it starts at 8 atm and 4 dm³ following:
(a) a reversible adiabatic expansion, or
(b) a one-step constant-pressure adiabatic expansion.
What is the lowest temperature a one-step constant-pressure adiabatic expansion can reach here?

SECTION 2.6

2.19 What is the difference between the molar enthalpy and the molar internal energy of liquid water at 1.00 atm? (*Hint:* How is molar volume related to mass density, and what is the density of water?) What can you conclude about the $\Delta \overline{H} - \Delta \overline{U}$ difference for condensed phases in contrast to that for gases at ordinary pressures and temperatures?

2.20 Follow these steps and derive the expression plotted in Figure 2.10 for the van der Waals Joule–Thomson inversion curve. First, from Eq. (2.41), we note that $\mu_{JT} = 0$ whenever $\overline{V} = T(\partial \overline{V}/\partial T)_P$. Show that this criterion is equivalent to

$$T\left(\frac{\partial P}{\partial T}\right)_V = -\overline{V}\left(\frac{\partial P}{\partial \overline{V}}\right)_T .$$

Next, use Eq. (1.25) as the equation of state and turn this new criterion into a quadratic equation in $\widetilde{V}$. Solve this equation, substitute the answer into Eq. (1.25), and show that your result can be simplified to

$$\widetilde{P} = \left[3\left(\frac{\widetilde{T}}{2}\right)^{1/2} - 1\right]\left[1 - \left(\frac{\widetilde{T}}{2}\right)^{1/2}\right] .$$

2.21 Use the identity

$$\left(\frac{\partial V}{\partial T}\right)_P = \frac{(\partial P/\partial T)_V}{(\partial P/\partial V)_T}$$

to find the exact expression for μ_{JT} for a virial equation of state through the second virial coefficient, and show that Eq. (2.42) is its low-pressure limit. (*Hint:* Don't forget, as you take derivatives, that B is a function of T!)

2.22 Does the temperature of a hard-sphere gas go up or down in a Joule–Thomson expansion? Use the abbreviated van der Waals equation of state, $P = RT/(\overline{V} - b)$, to find out, and estimate the magnitude of the effect for a pressure drop of 100 atm, an initial temperature of 300 K, and the H₂ b value: 26.61 cm³ mol⁻¹. Is it safe to open the valve on a high pressure (150 atm) cylinder of H₂ without a pressure regulator?

2.23 While it is true that $d(PV) = P\,dV + V\,dP$, it is *not* true that $\Delta(PV) = P\,\Delta V + V\,\Delta P$. What is the correct expression for $\Delta(PV)$, written in terms of P_i, V_i, ΔP, and ΔV?

2.24 A modern variation of Joule's free-expansion experiment employs two chambers separated by a valve, with one evacuated and of variable volume, but with the surrounding water bath replaced by an adiabatic enclosure. The gas is allowed to expand and its temperature change is measured. The experiment is repeated for various final volumes, in analogy to the varying final pressures of the Joule–Thomson experiment.

(a) Explain why this process takes place at constant U.
(b) One can define a *Joule coefficient* for this experiment as $\mu_J = (\partial T/\partial V)_U$. Explain how this coefficient would be determined from the experimental data.
(c) Derive a general expression for μ_J, and show that $\mu_J = -(a/\overline{V}^2)/C_V$ for a van der Waals gas.
(d) Find the temperature change when 1 mol of Ar at 300 K is expanded in this apparatus from $V_i = 1$ dm³ to $V_f = 2$ dm³, using the expression in part (c).

GENERAL PROBLEMS

2.25 A perfect spring exerts a force in direct proportion to its displacement from its rest position, but in a direction opposite to its displacement: force = $-$(constant) · (distance of displacement from rest), or $F = -kx$ where k is the spring's *force constant* and x is the displacement. Such a spring is connected at rest (i.e., under neither compression nor extension) in the right-hand chamber of the apparatus below. This side is evacuated, but an ideal gas at the initial state (P_i, V_i, T) is confined in the left chamber by a piston held in place by stops.

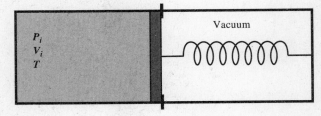

The spring is connected to the piston so that when the stops are removed, the gas expands to the state (P_f, V_f, T) and compresses the spring a distance x_0, as shown below. Note that the process is isothermal.

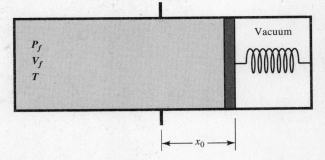

Calculate (algebraically) the work involved in this process, first taking the *gas* to be the system, then taking the *spring* to be the system. Show the path for the gas expansion on a P–V diagram, and verify that your calculations and the graphical area corresponding to w agree in magnitude.

2.26 A small cylindrical plug of mass m and cross-sectional area A is suspended at rest in the neck of a flask containing an ideal gas. The plug can slide without friction up and down the neck of the flask. If the plug is tapped lightly, it will oscillate with a frequency ν (cycles per second). Show that ν depends on the ratio $C_P/C_V = 5/3$ for the gas. (*Hints:* The motion is *fast*, and thus can be approximated as adiabatic and reversible. The oscillation can be treated as if the mass were connected to a spring of force constant k (see also the previous problem), for which $\nu = (2\pi)^{-1} (k/m)^{1/2}$; thus, the problem can be reduced to finding k in terms of properties of the gas, the path, and the plug's area. If V_0 is the volume of the gas with the plug at rest, and x is the amplitude of the plug's motion as it oscillates, then note that $xA/V_0 \ll 1$ always, since the amplitude of the motion is small. This inequality will allow you to write $(V_0 + xA)^{5/3} = V_0^{5/3} (1 + xA/V_0)^{5/3} \cong V_0^{5/3} (1 + 5xA/3V_0 + \ldots)$.)

2.27 The temperature variation of the second virial coefficient $B(T)$ for a van der Waals gas is discussed in Section 1.3 and graphed in Figure 1.6. Show that the Joule–Thomson inversion temperature condition, $B(T_I) = T_I (dB/dT)_{T_I}$, has the following graphical interpretation: plot $B(T)$, then draw a straight line from the origin with a positive slope such that this line is just tangent to the $B(T)$ curve, and prove that the point of tangency is at T_I.

Spontaneity, Equilibrium, and Entropy

THE last chapter described energy transfer in two general situations: irreversible processes (that occur naturally) and reversible processes (that are well-defined limiting cases of natural processes). The First Law can also treat a vast number of situations that fall into a third class: processes that observation and experience have shown to be impossible or unnatural. One of the very great powers of thermodynamics is its ability to distinguish between possible, natural (and hence irreversible) processes and impossible, unnatural processes. Closely related is thermodynamics' ability to predict the final equilibrium state of a natural process. The First Law alone lacks this ability. Hence, it is necessary to introduce the *Second Law of Thermodynamics* and consider a new state function, *entropy*, that can be used to distinguish between the possible and the unobserved.

A few examples will show how the First Law is insufficient. Once we have defined entropy and stated the Second Law, we will return to these examples and make quantitative statements about them.

Imagine two cubes of metal, one at a higher temperature than the other, but otherwise identical. Experience, but not the First Law, tells us that when the two cubes are brought into contact, they will reach a final equilibrium state in which both cubes have the same temperature. This is the natural, spontaneous outcome we expect, but the First Law would not be violated if energy were to flow from the colder cube to the hotter, ending with the two cubes even further apart in temperature.

3.1 A Statistical View of Entropy

3.2 A Thermodynamic View of Entropy

3.3 A Mathematical View of Entropy

3.4 The Second Law of Thermodynamics

3.5 General Properties of Entropy Changes

3.6 Consequences of Entropy as a State Function

Similarly, consider the Joule expansion: a gas in one container spontaneously flows into an evacuated adjoining container. In the final equilibrium state the gas fills both containers equally. Recall that this process involves no change in internal energy if the gas is ideal. The reverse process, however, does not occur: no one expects a gas that occupies two adjoining containers to rush spontaneously into only one of them leaving the other empty. Again, such a strange process would not violate the First Law, but it would violate common-sense expectations about what can and cannot occur in Nature.

The Second Law of Thermodynamics uses entropy to make "common sense" a quantitative concept. Many processes and contraptions are so complex that common sense offers little hope in understanding them. Many clever (and some not so clever, but exceedingly convincing) people have claimed to have invented contraptions that may seem possible but violate either the First or the Second Law. These imaginary contraptions are "perpetual motion" machines that seem to generate energy spontaneously (violating the First Law) or to produce work by an unnatural process (violating the Second Law). Such contraptions and their promoters have had, and continue to have, a colorful history. Around the year 1500, Leonardo da Vinci, who was centuries ahead of his contemporaries, wrote in one of his notebooks, "Oh speculators on perpetual motion! How many vain projects of the like character you have created! Go and be the companions of the searchers after gold!"[1]

Leonardo had a very finely tuned common sense. We lesser mortals need the Second Law and the concept of entropy. The next three sections present three introductory arguments that establish entropy. Each argument is superficially very different from the others, but together they show how entropy can be understood from different points of view. Since entropy is not as familiar a concept as energy, each view contributes toward the full picture.

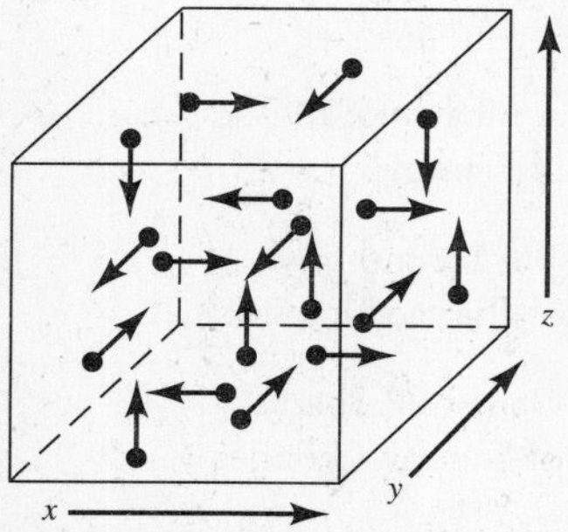

3.1 A STATISTICAL VIEW OF ENTROPY

Imagine a cubical box confining N molecules of an ideal gas. Let the box be oriented in a Cartesian axis system as shown in Figure 3.1. Suppose one-third of the molecules move back and forth along the x direction only, another $N/3$ along y, and the remaining $N/3$ along z. Suppose as well that the speeds of all the molecules are equal and that when they bounce from the walls they reverse direction but do not change speed. This microscopic description of an ideal gas will not violate any aspect of the First Law, but we know gases do not behave with such microscopic order. In reality, gas molecules go every which way, and experiments have shown that they move with a variety of speeds. Our curious gas is in a *microscopic configuration* (a *microstate*) that is not the *macroscopic equilibrium state* of reality.

The statistical view of entropy, discussed in Chapters 22 and 23 by the theory of statistical mechanics, shows that equilibrium means special configurations of microscopic motion. Equilibrium is a *unique* macroscopic state characterized by the *most probable microscopic configuration* the system can attain given the macroscopic constraints (pressure, volume, temperature, total energy, etc.) imposed by the sur-

FIGURE 3.1 An ideal gas with the microscopic configuration in which one-third of the molecules have velocities along the x direction, one-third along the y direction, and one-third along the z direction would not be inconsistent with the First Law. Moreover, the magnitudes of these velocities could all be equal. Experimentally, however, it is known that gas molecules have a variety of velocities and move in any possible direction.

[1] Leonardo didn't think highly of alchemists either!

roundings and boundary on the whole system. "Configuration" means not only a distribution in space (such as a gas filling a container uniformly), but also a distribution in energy (such as is caused by the variation in molecular velocities in a gas). While the energy of a molecule may vary from molecule to molecule, the sum of these individual microscopic energies must equal the total energy of the system.

Suppose N gas molecules are in a true equilibrium in a container of volume V at some temperature. If we consider only one molecule, then the probability W_1 that it is *somewhere* in the entire volume is clearly

$$W_1(V) = 1 \tag{3.1}$$

since probabilities vary between 0 and 1, and we know with certainty that the molecule is in the volume. The probability that the molecule is in one particular half of the volume is 0.5, and, in general, the probability that it is in any subvolume V' where $V' < V$ is

$$W_1(V') = \frac{V'}{V}. \tag{3.2}$$

Next we ask for the probability that two molecules are *simultaneously* in the subvolume V'. We do not expect these molecules to form a conspiracy against us and always travel together or always avoid one another. We expect each to behave as if the other were not there. Thus, the probability that both are in V' must be

$$W_2(V') = W_1(V') \cdot W_1(V') = \left(\frac{V'}{V}\right)^2. \tag{3.3}$$

Continuing this argument molecule by molecule, we find that the probability of finding all N molecules in V' is

$$W_N(V') = \left(\frac{V'}{V}\right)^N. \tag{3.4}$$

Typically, N is on the order of 10^{23}, so that $W_N(V')$ is very, very much less than 1 for any $V' < V$.

EXAMPLE 3.1

How many molecules need to be in a 1 cm³ volume so that the probability is less than 10^{-6}, one chance in a million, that all of them are simultaneously in a sub-volume of 0.999 cm³?

SOLUTION   We solve Eq. (3.4) for N:

$$N = \frac{\ln(W_N)}{\ln\left(\dfrac{V'}{V}\right)},$$

which for $W_N = 10^{-6}$ and $V'/V = 0.999$ yields $N = 13\ 800$. This is roughly the molecular density of the very best vacuum that can be obtained in the laboratory. Obviously, molecules do not "fill" an entire volume at any one instant. What we have in mind here is keeping our eye on a fixed sub-volume and asking if, at any one time, all N molecules are somewhere in that sub-volume. It is interesting to note that W_N is 0.25 (a one in four chance) for this N if $V'/V = 0.9999$, but falls rapidly below 10^{-6} to 6×10^{-61} as V'/V is increased to 0.99.

⟹ RELATED PROBLEMS 3.1, 3.2

And so we see that the unobserved spontaneous flight of a gas to one corner of its container is not strictly impossible, it is merely overwhelmingly improbable. The improbability is so great that the process is *never* observed. To show this, suppose we ask for the probability that all 6×10^{23} molecules in a mole of gas suddenly rush into a sub-volume $V' = V/10$. From Eq. (3.4), $W_N = 10^{-10^{23}}$. If we check the positions of each molecule from time to time, how long will we have to wait, on average, before this improbable configuration is observed? Suppose we could locate all 10^{23} molecules every 10^{-12} s. This is roughly the time required for a gas molecule to move a distance equal to its own diameter. We could thus check 10^{12} configurations each second. The age of the universe is on the order of 10^{10} years, or 10^{17} s; therefore, during the age of the universe we could observe only 10^{29} configurations. The chance that one of these configurations would be the one in $10^{10^{23}}$ sought is still vanishingly small.

Next we must see how to turn probabilities into a function of state. Entropy should be a monotonic function of V (one with no minima or maxima) so that it has a unique value at each volume. It should be defined for all equilibrium states, and the entropy of a system with several easily identified subsystems should be equal to the sum of the entropies of each subsystem. Entropy should be extensive.

Entropy is symbolized S. We cannot write $S = $ (constant) $\cdot W$, since probabilities are neither intensive nor extensive. Suppose we try

$$S = \text{(constant)} \cdot \ln W \ . \tag{3.5}$$

A logarithmic dependence on probability satisfies our requirements. If we double the number of particles: $N \rightarrow 2N$, we square the probability: $W \rightarrow W^2$, but we would double the entropy: $\ln W \rightarrow 2 \ln W$. Similarly, a system composed of subsystems with N_1 and N_2 particles would exhibit a probability $W_{N_1+N_2} = W_{N_1} W_{N_2}$ but an entropy $S_{12} = S_1 + S_2$, as desired. The volume dependence of entropy, incorporating Eq. (3.4) into Eq. (3.5), is logarithmic as well, and thus monotonic.

Equation (3.5) contains a constant of proportionality that we do not yet know, but we can use Eq. (3.5) to calculate the way entropy changes in a simple process: the isothermal Joule expansion. The gas is confined to an initial volume V_i and freely expands into a final volume V_f. We write

$$\Delta S = S_f - S_i = S(V_f) - S(V_i) = k(\ln W_f - \ln W_i)$$

where k is the constant in Eq. (3.5). Combining the difference in logarithms gives

$$\Delta S = k \ln \left(\frac{W_f}{W_i}\right) \ . \tag{3.6}$$

V_f and V_i play the roles of V' in Eq. (3.4), and V in that equation is arbitrary, since it will cancel in Eq. (3.6). We conclude that

$$\Delta S = k \ln \left(\frac{V_f}{V_i}\right)^N = Nk \ln \left(\frac{V_f}{V_i}\right) \ . \tag{3.7}$$

This is as far as we can take probability arguments without entering a full exposition of statistical mechanics. The next section gives a thermodynamic view of entropy that verifies Eq. (3.7) and identifies the constant k.

EXAMPLE 3.2

A container is divided into halves by a removable partition. One half contains gas A and the other contains gas B, both at the same temperature and pressure. We know that when the partition is removed, the gases will spontaneously mix. Show that this process involves a change in entropy.

SOLUTION Each gas doubles its original volume: $V_f = 2V_i$, and assuming each gas is ideal, the number of molecules of each is the same. From Eq. (3.7) we have

$$\Delta S_{tot} = \Delta S_A + \Delta S_B = 2Nk\ln(2) \ .$$

This little calculation has an importance much greater than its simplicity would indicate. We will find k to be a positive constant in the next section, making ΔS_{tot} positive. This will be the sign that a process can occur spontaneously. But note that *none* of the other thermodynamic variables of this system have changed. The volume of the entire system is the same, as is its pressure, temperature, and internal energy. We will return to mixing phenomena in Chapter 6 and derive this expression in a different way.

⟹ **RELATED PROBLEM 3.3**

3.2 A THERMODYNAMIC VIEW OF ENTROPY

In 1824, well before the First Law was established, a young French military engineer, Sadi Carnot, published an article titled "Réflexions Sur la Puissance Motrice du Feu et Sur les Machines Propres à Développer cette Puissance"—"Reflections on the Motive Power of Fire and on Machines Appropriate to Develop that Power"—in which he arrived at conclusions that can be taken as a basis for the Second Law. He did so starting from an incorrect theory of heat, and thus his conclusions are all the more remarkable. Much of Carnot's article is of little more than historical interest now, but one particular process remains of fundamental importance and is known as the *Carnot cycle*.

The Carnot cycle in modern terminology consists of four steps carried out on any system:

1. a reversible isothermal expansion at temperature T_1;
2. a reversible adiabatic expansion during which the temperature drops from T_1 to T_2;
3. a reversible isothermal compression at temperature T_2; and
4. a reversible adiabatic compression during which the temperature rises from T_2 back to T_1.

The cycle can begin with any step, but each step is performed so that the cycle is closed; the system is returned to its initial state.

Figure 3.2 shows the Carnot cycle as it applies to an ideal gas. Each step can be analyzed using the First Law methods of Chapter 2. Whether the system is an ideal gas or anything else, we can make the following general observations:

$\Delta U = 0$ so that $q_{cy} = -w_{cy}$ (since the process is cyclic)

$q_{cy} = q_1 + q_3$ (since steps 2 and 4 are adiabatic)

$q_1 > 0$ (since step 1 is an isothermal *expansion*)

$q_3 < 0$ (since step 3 is an isothermal *compression*)

$w_{cy} = w_1 + w_2 + w_3 + w_4$ (since work is associated with each step)

w_1 and w_2 are < 0 (since these are *expansion* steps)

w_3 and w_4 are > 0 (since these are *compression* steps)

Carnot then considered this cycle as an engine, a device for producing useful work when fueled by heat. This engine's flow of energy as work and heat to and from

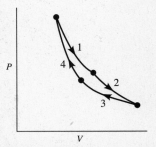

FIGURE 3.2 The Carnot cycle for an ideal gas (plotted in P–V coordinates) consists of the four connected paths described in the text. Paths 1 and 3 are isotherms (at different temperatures), and paths 2 and 4 are reversible adiabatic paths.

the system is diagrammed below. Energy flows from the high-temperature energy reservoir to the system, which converts part of the energy to useful work, returning the remainder of the energy to the low temperature energy reservoir.

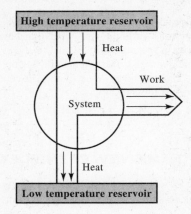

A practical measure of the usefulness of such an engine is the ratio of the work it produces to the energy flow from the higher temperature (such as is provided by burning fuel). We thus define the *efficiency* ε as

$$\varepsilon = \frac{\text{work done } by \text{ the system}}{\text{heat transferred to the system from the surroundings}}$$

and find

$$\varepsilon = \frac{-w_{cy}}{q_1} \quad \text{(note the minus sign!)}$$

$$= \frac{q_1 + q_3}{q_1} = 1 + \frac{q_3}{q_1} \tag{3.8}$$

Since $q_3/q_1 < 0$ (q_3 is negative), $\varepsilon < 1$.

These observations are completely general and rely on the First Law only. If we assume the cycle operates on an ideal gas, then we can compute ε in terms of the state variables of the gas. Problem 3.4 uses the methods of Chapter 2 to deduce the simple equation

$$\varepsilon = 1 - \frac{T_2}{T_1}. \tag{3.9}$$

The most remarkable aspect of the Carnot cycle is that Eq. (3.9) can be shown to hold for *any* system, not just the ideal gas. Moreover, ε is a function of any two empirical temperatures used to characterize the two isotherms of the Carnot cycle. Any physical property that varies reproducibly with a system's "degree of hotness" can be used to establish an empirical thermometer. We saw in the last chapter, in Eq. (2.13), how the PV product of an ideal gas can be used to establish a temperature scale, and the value 273.16 K was chosen for the triple point of water simply to make the Kelvin degree have the same magnitude as the Celsius degree (which was established first).

EXAMPLE 3.3

Carnot was interested in engines: put heat in and get work out. How should the efficiency be defined for:

(a) a *heat pump*, which takes an energy input as work from some source and as heat from a low temperature and outputs energy as heat at a high temperature (as in heating a room in winter)

(b) a *refrigerator,* which takes an input of work and removes energy as heat from a low-temperature source (such as a room being air conditioned, or the food in a refrigerator), depositing energy as heat at a high temperature (the great outdoors, or the kitchen).

SOLUTION Both of these devices are based on a cycle run "backwards" from that of an engine meant to "do work." A Carnot cycle can be run backwards from the way we described it, but many other types of cycles can be used as well. For a heat pump, the relevant measure of efficiency is

$$\varepsilon_{\text{heat pump}} = \frac{\text{(heat transferred to the higher temperature)}}{\text{(work done on the system)}}$$

$$= \frac{-q_1}{w_{cy}} \qquad (q_1 \text{ is negative})$$

$$= \frac{T_1}{T_1 - T_2} \qquad (T_1 > T_2)$$

while for the refrigerator it is

$$\varepsilon_{\text{refrigerator}} = \frac{\text{(heat transferred from the lower temperature)}}{\text{(work done on the system)}}$$

$$= \frac{q_3}{w_{cy}} \qquad (q_3 \text{ is positive})$$

$$= \frac{T_2}{T_1 - T_2} \qquad (T_1 > T_2)$$

These expressions can be appreciably greater than 1; thus, they are often called "coefficients of performance" rather than "efficiencies." Note too that a refrigerator trying to remove energy from a very cold source will require a *very* large energy input; if $T_2 = 0.1$ K and $T_1 = 300$ K, $\varepsilon \cong 1/3000$ and the work done on the system (the refrigerator) $\cong 3000$ times the heat transferred from the 0.1 K source.

➡ RELATED PROBLEMS 3.4, 3.8, 3.9, 3.10

Since Eq. (3.9) is general, a Carnot engine itself could be used as a thermometer simply by measuring ε with one isotherm at a reference temperature and the other at the temperature to be measured. This establishes a *thermodynamic temperature scale* equivalent to the ideal gas temperature scale.

Comparing Equations (3.8) and (3.9) shows

$$\frac{q_3}{q_1} = -\frac{T_2}{T_1}$$

or

$$\frac{q_3}{T_2} + \frac{q_1}{T_1} = 0 \ . \tag{3.10}$$

This is a key result that indicates we are on the way to discovering a state function; it appears that the ratio of heat to temperature vanishes when summed over a cyclic process,[2] just as a state function would. In fact, entropy *is* the state function if q/T is computed over a *reversible* cycle. If the cycle is in any way irreversible, Eq. (3.10) is not true.

---

[2] Since steps 2 and 4 are adiabatic, $q_2 = q_4 = 0$ so that Eq. (3.10) is the sum of all nonzero (heat/temperature) terms for the Carnot cycle.

But if $q_3/T_2 + q_1/T_1 \neq 0$ for an irreversible cycle, then perhaps this sum is *always* either positive or negative. If so, we have discovered a way to distinguish between natural and unnatural processes. We will return to this possibility once we have more fully defined entropy.

We can give entropy a formal thermodynamic definition in the differential form

$$dS = \frac{dq_{\text{rev}}}{T} . \qquad (3.11)$$

Recall that dq is not an exact differential in general. What we have found here is a special path (the reversible one) for which dq/T becomes an exact differential, much in the way we related dU to a special path for work in writing $dU = dw_{\text{ad}}$ or a special path for heat in writing $dU = dq_V$, except here the factor $1/T$ *must be included*.

We can use Eq. (3.11) to compute ΔS for the Joule expansion to see if it gives the same answer found in the last section by statistical arguments. The real Joule expansion is irreversible, of course, but *to use Eq. (3.11) we must imagine a reversible path connecting the initial and final states of the process*. This is a slight annoyance, forcing us to come up with a reversible path that may be very different from the real path.[3] But since entropy is a state function, this calculation will still give us the correct ΔS for the real, irreversible path.

Since the gas is ideal and the process is isothermal, $\Delta U = 0$. Consequently we imagine a reversible expansion from V_i to V_f so that

$$dq_{\text{rev}} = -dw_{\text{rev}} = -(-P\,dV) = \frac{nRT}{V}dV .$$

Thus

$$\Delta S = \int_{S_i}^{S_f} dS = \int_{\text{path}} \frac{dq_{\text{rev}}}{T} = nR \int_{V_i}^{V_f} \frac{dV}{V} = nR \ln\left(\frac{V_f}{V_i}\right) . \qquad (3.12)$$

This has the same dependence on V_i and V_f as the statistical expression, Eq. (3.7), and we can now identify the unknown constant in that expression as

$$k = \frac{nR}{N} = \frac{R}{N_A} \qquad (3.13)$$

where N_A is Avogadro's constant. The constant k is simply the universal gas constant on a per molecule basis rather than a per mole basis. It is called *Boltzmann's constant* (and we will write it k_B from now on) in honor of the man whose research led to Eq. (3.5), Ludwig Boltzmann.[4]

It is not obvious that reversible heat divided by temperature should have anything to do with the probability of a particular spatial and energetic configuration of molecules. As we calculate entropy changes in various situations, this connection will become clearer.

---

[3] We will soon derive expressions for dS that cover a variety of generic reversible paths so that we needn't think of each new process as a new challenge to our path-making creativity.

[4] Ironically, Boltzmann was not the first to write Eq. (3.5); Max Planck was. Planck also first introduced the constant k, but Boltzmann introduced the idea of a proportionality between entropy and the logarithm of probability, and this was the key conceptual step. Boltzmann also discovered the key expressions of the kinetic theory of gases at a time when atoms were not taken for granted. Some doubted his work, and his life ended tragically in suicide. The equation $S = k \log W$ is carved on his memorial in the Central Cemetery in Vienna.

Our last view of entropy is a more abstract one based on the mathematical theory of differentials. This view will eventually give us a very compact statement of the Second Law and will be of use in explaining the Third Law of Thermodynamics.

3.3 A MATHEMATICAL VIEW OF ENTROPY

In 1909, well after entropy was established, the Greek mathematical physicist Constantin Carathéodory published an analysis of entropy based on mathematical theorems of exact and inexact differentials. The full details of his analysis are complex, but the essence of them is worth our consideration. If we rewrite the differential form of the First Law, Eq. (2.16), as

$$dq = dU - dw ,$$

substitute Eq. (2.21a) for dU:

$$dq = \left(\frac{\partial U}{\partial V}\right)_T dV + \left(\frac{\partial U}{\partial T}\right)_V dT - dw ,$$

and finally specify a reversible path ($dw_{rev} = -P\, dV$), we arrive at

$$dq_{rev} = \left(\frac{\partial U}{\partial T}\right)_V dT + \left[\left(\frac{\partial U}{\partial V}\right)_T + P\right] dV . \tag{3.14}$$

This is, in general, an inexact differential because the coefficients of dT and dV fail the test for exactness. Carathéodory asked if Eq. (3.14) could be multiplied by some function that would turn it into an exact differential and thus into the differential of a state function.

The mathematical theory of equations such as Eq. (3.14) (known as Pfaff equations in the mathematical literature) makes certain statements about the existence of such functions. One statement is that such an equation that has only two independent variables (T and V) will *always* have an integrating factor. This is a mathematical result that is in no way dependent on thermodynamics. It therefore tells us nothing new.

If we extend Eq. (3.14) to more than two variables (such as in a system with two independent volumes linked by a boundary that conducts heat and ensures both volumes have the same temperature), the mathematics become more restrictive. Carathéodory was able to show two important connections between the abstract mathematics and the physical reality of thermodynamics:

1. the conditions under which dq_{rev} can be made exact by an integrating factor are precisely those that are allowed by the Second Law, and
2. the integrating function is $1/T$ where T is the thermodynamic temperature of Section 2.3.

The importance of this view of entropy comes from the mathematical conditions dq_{rev} must meet in order to possess an integrating function. We will state these conditions in the next section.

While developing his theory, Carathéodory was one of the first to point out the importance of an observation that seems so obvious as to be trivial. This observation is now called the Zeroth Law of Thermodynamics:

> If system A is in thermal equilibrium with system B, and if B is in thermal equilibrium with system C, then A and C must also be in thermal equilibrium.

This statement not only defines the condition for thermal equilibrium between separated systems (A and C), it also defines the role played by a thermometer (system B). Moreover, Carathéodory showed why thermometers could use any temperature scale (absolute, Celsius, Fahrenheit, etc.), but that once a particular thermometer is chosen, the temperature it measures becomes defined for any system.

A Closer Look at Temperature Scales

We can prove that the reciprocal of the ideal gas temperature is an acceptable integrating function and that the ideal gas thermometer has a thermodynamic basis. For simplicity, we give the proof for only two independent variables, but it can be generalized to any number.

The change in PV for an ideal gas is empirically proportional to any arbitrary temperature change as measured by any thermometer that distinguishes hot from cold. Thus we write

$$PV = \alpha(T' + \theta)$$

where α is some constant of proportionality for a given quantity of gas and θ is a constant temperature offset to make $PV = 0$ at absolute zero. The quantity T' establishes an arbitrary temperature scale. We will also need the empirical observation of the Joule experiment that $(\partial U/\partial V)_{T'} = 0$ for an ideal gas so that U is a function of T' alone.

Next we write Eq. (3.14) for an ideal gas:

$$dq_{\text{rev}} = dU + P\,dV = \frac{dU}{dT'}dT' + \frac{\alpha(T' + \theta)}{V}dV .$$

According to Carathéodory this equation must have an integrating function that we will write as $1/\tau$ so that

$$\frac{dq_{\text{rev}}}{\tau} = \frac{dU}{dT'}\frac{dT'}{\tau} + \frac{\alpha(T' + \theta)}{V\tau}dV$$

is an exact differential. Therefore the cross-derivative test must hold:

$$\frac{\partial}{\partial V}\left[\left(\frac{dU}{dT'}\right)\frac{1}{\tau}\right]_{T'} = \frac{\partial}{\partial T'}\left[\frac{\alpha(T' + \theta)}{V\tau}\right]_{V} .$$

The left-hand side must be zero, since the quantity in square brackets is a function of temperature only. Thus we have

$$0 = \frac{\partial}{\partial T'}\left[\frac{\alpha(T' + \theta)}{V\tau}\right]_{V} = \frac{d}{dT'}\left[\frac{(T' + \theta)}{\tau}\right] = \frac{1}{\tau} - \frac{(T' + \theta)}{\tau^2}\frac{d\tau}{dT'}$$

or

$$\frac{dT'}{(T' + \theta)} = \frac{d\tau}{\tau} .$$

Integrating this equation gives

$$\ln(T' + \theta) = \ln \tau + \text{constant}$$

or, exponentiating,

$$(T' + \theta) = C\tau .$$

Here, C is an arbitrary constant that determines the relative sizes of one degree between each temperature scale. If we let $C = 1$, $\theta = 0$, and $T' = T$ where T is the ideal gas temperature defined operationally by Eq. (2.13), we have completed the proof that the ideal gas temperature is equivalent to the thermodynamic temperature (the integrating denominator of Carathéodory). Finally, this choice for C and T establishes a value for α that is just nR.

The three views of entropy in these last sections are based on one or another of the many possible statements of the Second Law. Since we have not yet stated this law, we have been putting the cart before the horse. The next section states the Second Law in a variety of ways and shows how these statements are related to each view of entropy.

3.4 THE SECOND LAW OF THERMODYNAMICS

The Second Law is a summary of those processes that are shown by experience to be possible, natural processes. Thus there are many ways of stating this law, and most of them are in terms of processes that are impossible. One such statement was made by Clausius in 1850:

> No process is possible in which the *only* transfer of energy is as heat transferred from a colder to a hotter system.

This statement is in accord with our observation that two systems at differing temperatures would not spontaneously increase their temperature difference when brought in contact.

A second statement of this law was given by Thomson and later by Planck. It is more closely related to the Carnot analysis of entropy:

> No cyclic process can transfer a quantity of energy as heat from a reservoir of energy at one temperature and produce work without some fraction of this energy appearing as heat transferred to a colder reservoir.

This says the Carnot efficiency ε must always be less than 1; the work must always be less than the heat transferred from the higher temperature. The Clausius statement is, most fundamentally, that all natural processes are irreversible. Together with the Thomson–Planck statement we can deduce that not only is $\varepsilon < 1$, but also that all natural cycles have efficiencies less than the limiting efficiency of a Carnot cycle.

From the Carathéodory point of view, the Second Law becomes a statement of the mathematical properties of dq that must be satisfied in order for dq to possess an integrating function:

> From any arbitrary state of any system there are a finite number of states arbitrarily close to the initial state that cannot be reached by an adiabatic process, reversible or irreversible.

That this statement has anything in common with the other two is not immediately obvious. First we will show that it is true for a trivial situation, just to give the words a physical interpretation, and then we will establish a more complete connection.

Imagine an ideal gas in the state (P_i, V_i, T_i). If the gas is expanded *reversibly* and adiabatically to some arbitrary V_f, then we would find $T_f < T_i$, as discussed in Chapter 2. (See Eq. (2.28) and path I of Figure 2.7.) If the gas is expanded *freely*

FIGURE 3.3 A system in any arbitrary state (P_i, V_i, T_i) can be taken to any state with $V_f > V_i$ and any $T_f > T_f(\min)$ by a combination of adiabatic processes. If a reversible adiabatic expansion to $(P_f, V_f, T_f(\min))$ is followed by an adiabatic work process at constant volume, any $T_f > T_f(\min)$ can be reached, as indicated by the vertical line. This diagram is drawn for an ideal gas system, but the argument is general.

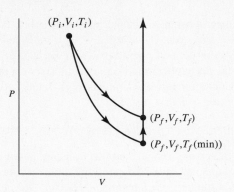

and adiabatically to the same V_f, we would find $T_f = T_i$ (the Joule experiment). If the gas is expanded *irreversibly* and adiabatically to V_f, the final temperature must lie *above* that for a reversible expansion, but *below* that for a free expansion.[5] We can, however, reach any temperature greater than T_i by following a free adiabatic expansion step with a constant-volume adiabatic work step (a stirrer, electrical heater, etc.) raising the temperature arbitrarily above T_i.

This argument is summarized in Figure 3.3. Since V_f is arbitrary, we can reach any state with $V > V_i$ and $T > T_f(\min)$ where $T_f(\min)$ represents the reversible adiabatic expansion T_f. All those states with $V > V_i$ and $T < T_f(\min)$ are represented by the Carathéodory statement. Similar arguments can be made about adiabatic compressions (see Problem 3.12).

We can relate this statement about inaccessibility to the Clausius and Thomson–Planck statements of the Second Law using Figure 3.4. There we locate states in a T–V coordinate system, beginning in state 1. A reversible adiabatic process takes us to state 2 at some lower temperature. We then follow an isothermal path that expands the system to state 3. Finally, we *hypothesize* an adiabatic path from state 3 back to state 1.

Since this is a cycle,

$$\Delta U = q_{cy} + w_{cy} = 0 \ .$$

Since the paths from 1 to 2 and from 3 to 1 were adiabatic

$$q_{1\text{-}2} = q_{3\text{-}1} = 0 \quad \text{and} \quad q_{cy} = q_{2\text{-}3} \ .$$

Therefore

$$q_{2\text{-}3} = -w_{cy} \ .$$

But $q_{2\text{-}3}$, a positive number, was energy transferred as heat at one temperature. By the Thomson–Planck statement, not all of this energy can be converted into work. Hence our hypothesis that path 3-1 was adiabatic must be incorrect. State 1 must be one of those inaccessible states that cannot be reached by an adiabatic path from state 3.

This argument has an important corollary: *No two reversible adiabatic paths can cross.* This is clear if we assume path 3-1 is reversible.[6]

FIGURE 3.4 If the path $1 \rightarrow 2$ is an adiabatic reversible path, and path $2 \rightarrow 3$ is an isothermal path, then path $3 \rightarrow 1$ cannot be adiabatic without violating the Thomson–Planck statement of the Second Law. We conclude that state 1 is one of the states that cannot be reached from state 3 by an adiabatic path.

---

[5]This must be true since work done *by* the system is a maximum for a reversible expansion. Irreversible expansions do less work—withdraw less energy—and hence lower the temperature less.

[6]There are paths passing through state 1 that are adiabatic but irreversible. Such paths connect state 1 to states with volumes *less* than that of state 2.

These three statements of the Second Law—Clausius, Thomson–Planck, and Carathéodory—allow us to define entropy and exclude certain imaginable processes as impossible. In the next two sections, we will see some general aspects of entropy changes and how they relate to equilibrium, and then we will see how entropy allows us to draw new thermodynamic conclusions about possible processes.

3.5 GENERAL PROPERTIES OF ENTROPY CHANGES

Consider a pendulum swinging back and forth, and contrast that to a Joule expansion in which gas in one container rushes into an empty container, filling both and ending there. The gas does not oscillate back and forth from container to container. Similarly, two metal cubes of different temperature equilibrate at one temperature when brought together; they do not oscillate in temperature. In this section, we will use these two situations to study what entropy changes have to say about natural, irreversible processes.

If the Joule expansion is done with a "gas" consisting of only one molecule, then an oscillatory description might be appropriate—we can easily keep track of the detailed motion of such a system on a *microscopic* scale. But when we have a mole or so of molecules we must abandon such a detailed picture. Instead we characterize the system by a few *macroscopic* variables: P, V, T, n, U, H, and now, S. It is this inability to have detailed microscopic information that is at the heart of understanding irreversible macroscopic processes that proceed in one direction and stop at a definite equilibrium state.

The entropy change in a Joule expansion was given by Eq. (3.12):

$$\Delta S = nR \ln \left(\frac{V_f}{V_i}\right) . \tag{3.12}$$

Since $V_f > V_i$, ΔS is positive. Had the gas filled only part of the volume, say some V' such that $V_i < V' < V_f$, then ΔS would still be positive but not *as* positive as it is when the gas takes advantage of the full volume. If the gas moved completely from one container to the other and stopped, ΔS could be either positive or negative (depending on the relative volumes of the two containers). But of all the imaginable processes, *the only one which is observed is that for which the entropy increases as much as possible* given the constraints of a fixed maximum final volume, a fixed total energy, and a fixed amount of gas.

Next, consider two identical metal cubes (call them a and b) at initial temperatures T_a and T_b where $T_a > T_b$. When brought into contact they spontaneously attain equal temperatures T. If the cubes each have a heat capacity C_P (which we will assume to be temperature independent), common sense tells us that T will be half way between T_a and T_b, a result we now verify from the First Law.

Cube a transfers energy as heat to cube b since $T_a > T_b$. Thus

$$q_a = \int_{\text{path}} dq_P = \int_{T_a}^{T} C_P \, dT = C_P (T - T_a) < 0$$

is the energy change for cube a, while

$$q_b = \int_{\text{path}} dq_P = \int_{T_b}^{T} C_P \, dT = C_P (T - T_b) > 0$$

is the energy change for b. The First Law says the energy lost by a ($-q_a$) is exactly equal to that gained by b (q_b):

$$-C_P(T - T_a) = C_P(T - T_b)$$

or

$$T = \frac{1}{2}(T_a + T_b) \tag{3.15}$$

as common sense told us.

The total entropy change is the sum of the entropy changes for each cube. For either cube we integrate dS:

$$\Delta S = \int_{S_i}^{S_f} dS = \int_{\text{path}} \frac{dq_{\text{rev}}}{T} . \tag{3.16}$$

It would appear that we need to figure out how to transfer heat *reversibly* to do this integral (remember—the real process is irreversible). But if dq happens to be equal to a state function, then the details of the path are not important, and such is the case here: $dq = dq_P = C_P\, dT = dH$. Even though the process is irreversible, dq will be the same as for a reversible heat transfer since the pressure is constant and thus dq equals dH. So we can write for cube a

$$\Delta S_a = \int \frac{dq_P}{T} = \int \frac{dH}{T} = \int_{T_a}^{T} \frac{C_P}{T} dT$$

which, since C_P is constant, is

$$\Delta S_a = C_P \ln\left(\frac{T}{T_a}\right) . \tag{3.17}$$

We can find ΔS_b similarly, add the two, and arrive at

$$\Delta S_{\text{tot}} = \Delta S_a + \Delta S_b = C_p \ln\left(\frac{T^2}{T_a T_b}\right) \tag{3.18}$$

where T is $(T_a + T_b)/2$, Eq. (3.15). It will always be true that $T^2/T_a T_b = (T_a + T_b)^2/4T_a T_b$ is greater than one and therefore ΔS_{tot} is positive. What's more, this ΔS_{tot} is *greater than that for any final cube temperature we could imagine* as long as the temperature is consistent with the First Law, as the Example below proves.

EXAMPLE 3.4

Prove that Eq. (3.18) is indeed the *maximum* ΔS_{tot} allowed by the conservation of energy.

SOLUTION Imagine that the cubes were brought into contact and equilibrated at final temperatures T'_a and T'_b that were not necessarily equal. The First Law would still require

$$C_P(T'_a - T_a) = -C_P(T'_b - T_b)$$

so that

$$T'_a = T_b + T_a - T'_b .$$

The total entropy change would be

$$\Delta S_{\text{tot}} = C_P \left[\ln\left(\frac{T'_a}{T_a}\right) + \ln\left(\frac{T'_b}{T_b}\right) \right]$$

$$= C_P \ln\left(\frac{T'_a T'_b}{T_a T_b}\right)$$

$$= C_P \ln\left[\frac{(T_a + T_b - T'_b)T'_b}{T_a T_b}\right]$$

which is an expression in the form $y(x) = \ln[(A - x)x/B]$ where A and B are constants. We want to find that value of x for which $y(x)$ is a maximum. The derivative of $y(x)$ is $dy/dx = (2x - A)/(x^2 - Ax)$, which equals zero when $x = A/2$. In terms of our variables, $x = T'_b$ and $A = T_a + T_b$. Thus, when $T'_b = (T_a + T_b)/2 = T'_a$, the entropy is a maximum, as we expected. (For completeness, we should double check that this value maximizes, rather than minimizes, ΔS_{tot} by plotting ΔS_{tot} as a function of T'_b or by looking at the second derivative of ΔS_{tot}. Neither of these are difficult and are left for you to explore.)

▶ RELATED PROBLEMS 3.13, 3.14, 3.24

Once again, the *only observed process is the one for which the entropy increased as much as possible* given the constraints of the system.

We could continue calculating entropy changes for other isolated systems undergoing spontaneous processes, but for each calculation, we would arrive at the same conclusion:

> The entropy change in any spontaneous process occurring in an isolated system is positive. Moreover, the process will continue until the entropy has increased the maximum amount allowed by the constraints imposed on the system.

This important result expresses the hope we held for entropy in Section 3.2. (See Eq. (3.10) and the discussion following it.) The total entropy change, ΔS_{tot}, is *always* positive. As an added bonus, we can find the end point to spontaneous processes, the final equilibrium state that maximizes ΔS_{tot}.

This statement concentrates on *isolated* systems, and we must now consider systems in contact with their surroundings. If we consider the system and surroundings together as one giant isolated super-system, then we can generalize:

> The sum of the entropy changes in the system and the surroundings can never be negative for any natural process.

Note that ΔS_{tot} could be zero, to allow for reversible adiabatic processes for which $dS = 0$ since $dq_{\text{rev}} = 0$.[7]

A variant of the two cubes process will show how the surroundings can be incorporated. Suppose we have only one cube at a cold temperature T_i that we bring into the room so that it warms to T_f, room temperature. The surroundings are the room, the room's air, you, etc. The entropy change of the cube is

$$\Delta S_{\text{cube}} = C_P \ln\left(\frac{T_f}{T_i}\right). \tag{3.19}$$

---

[7]For *irreversible* adiabatic processes, ΔS is not necessarily zero, even though dq is zero. Only for *reversible* adiabatic processes can we be sure that $dS = 0$.

To heat the cube, the surroundings supplied (irreversibly) energy as heat in the amount

$$q_{\text{cube}} = -q_{\text{surr}} = C_P(T_f - T_i) \ . \tag{3.20}$$

But an important point is this: the air, the room, you, etc., *stayed at essentially the same temperature*. The surroundings are so vast and energy rich that q_{surr} in Eq. (3.20) represents a negligible energy loss. This behavior defines a *reservoir* in thermodynamics. Energy transferred to or from a reservoir is effectively infinitesimal due to the reservoir's vast size.[8]

We can therefore write

$$\Delta S_{\text{surr}} = \frac{q_{\text{surr}}}{T_f} = C_P\left(\frac{T_i}{T_f} - 1\right) \tag{3.21}$$

because T_f is the *only* temperature appropriate to the surroundings. For the entire process

$$\Delta S_{\text{tot}} = C_P\left[\ln\left(\frac{T_f}{T_i}\right) + \frac{T_i}{T_f} - 1\right] , \tag{3.22}$$

which is *always positive* (or zero, if $T_i = T_f$).[9]

Note that

$$\Delta S_{\text{cube}} > 0 \quad \text{(the cube was heated)}$$

$$\Delta S_{\text{surr}} < 0 \quad \text{(energy left the surroundings as heat)}$$

and

$$\Delta S_{\text{tot}} > 0 \quad \text{(since the process was a natural one).}$$

Suppose the room had cooled the cube ($T_i > T_f$). Then we would find

$$\Delta S_{\text{cube}} < 0 \quad \text{(the cube cooled)}$$

$$\Delta S_{\text{surr}} > 0 \quad \text{(energy entered the surroundings as heat)}$$

but we would still find

$$\Delta S_{\text{tot}} > 0 \quad \text{(since the process was natural).}$$

If the surroundings are not as infinite as an ideal reservoir, then we are back to the two cubes problem: each cube acts as the other's surroundings. If the cubes are of different sizes, the heat capacities would be different, and the analysis leading to Eq. (3.15) would lead to

$$-C_P^a(T - T_a) = C_P^b(T - T_b) \ .$$

Solving for T gives

$$T = \frac{C_P^a T_a + C_P^b T_b}{C_P^a + C_P^b} \ . \tag{3.23}$$

---

[8]Reservoirs need not be infinite in extent, as we will see in Chapter 6. They need only have an infinite heat capacity. An ice water mixture behaves as a reservoir because heat can be input to it (melting a bit of ice) or extracted from it (freezing some water) without changing the temperature.

[9]Let $x = T_i/T_f$. Then Eq. (3.22) is just $y(x) = \ln(1/x) + x - 1 = -\ln(x) + x - 1$. This function is zero when $x = 1$ and positive for all $x > 0$, which means $\Delta S_{\text{tot}} > 0$ even if the cube had been warmer than the room ($T_f < T_i$).

3.5 GENERAL PROPERTIES OF ENTROPY CHANGES

If $C_P^a = C_P^b$, we recover Eq. (3.15); but if cube a is arbitrarily larger than b, then C_P^a is arbitrarily larger than C_P^b and T approaches T_a. The giant cube a became a reservoir.

EXAMPLE 3.5

One mole of He is confined at 1 atm and 300 K in a container connected by a valve to a second container holding 1 mole of Ar at 1 atm and 500 K. The valve is opened. What happens, and what is the entropy change?

SOLUTION From the discussion above, we know that each gas will fill the combined volumes and, since there are equal amounts of each gas with identical heat capacities, they will equilibrate at a final temperature of 400 K. The initial volumes are given by the ideal gas calculations:

$$V_{He} = \frac{(1.00 \text{ mol})(0.082 \text{ dm}^3 \text{ atm mol}^{-1} \text{ K}^{-1})(300 \text{ K})}{1.00 \text{ atm}} = 24.6 \text{ dm}^3$$

$$V_{Ar} = \frac{(1.00 \text{ mol})(0.082 \text{ dm}^3 \text{ atm mol}^{-1} \text{ K}^{-1})(500 \text{ K})}{1.00 \text{ atm}} = 41.0 \text{ dm}^3 \ .$$

The final state has 2.00 mol of gas at 400 K in a volume $V_f = 65.6 \text{ dm}^3$ at a pressure

$$P_f = \frac{(2.00 \text{ mol})(0.082 \text{ dm}^3 \text{ atm mol}^{-1} \text{ K}^{-1})(400 \text{ K})}{65.6 \text{ dm}^3} = 1.00 \text{ atm} \ ,$$

which is also P_i, a result that may not be obvious.

There are two components to ΔS: the volume change and the temperature change. We can imagine reaching the final state by applying the following steps to each gas:

(1) expand at constant temperature to V_f
(2) alter T to 400 K at constant volume.

For (1), we use Eq. (3.12), applied to each gas:

$$\Delta S_{He}^{(1)} = (1.00 \text{ mol})(8.31 \text{ J mol}^{-1} \text{ K}^{-1}) \left[\ln\left(\frac{65.6 \text{ dm}^3}{24.6 \text{ dm}^3}\right)\right] = 8.15 \text{ J K}^{-1}$$

$$\Delta S_{Ar}^{(1)} = (1.00 \text{ mol})(8.31 \text{ J mol}^{-1} \text{ K}^{-1}) \left[\ln\left(\frac{65.6 \text{ dm}^3}{41.0 \text{ dm}^3}\right)\right] = 3.91 \text{ J K}^{-1} \ .$$

For (2), we are changing T at *constant volume* instead of constant pressure; Eq. (3.17) does *not* apply. But, since $dq = C_V dT = dU$ here, we can follow the derivation of Eq. (3.17), substituting C_V for C_P so that

$$\Delta S_V = C_V \ln\left(\frac{T_f}{T_i}\right) \ .$$

Applying this to each gas gives (with $C_V = 3R/2 = 12.5 \text{ J K}^{-1}$):

$$\Delta S_{He}^{(2)} = (12.5 \text{ J K}^{-1})\left[\ln\left(\frac{400 \text{ K}}{300 \text{ K}}\right)\right] = 3.59 \text{ J K}^{-1}$$

$$\Delta S_{Ar}^{(2)} = (12.5 \text{ J K}^{-1})\left[\ln\left(\frac{400 \text{ K}}{500 \text{ K}}\right)\right] = -2.78 \text{ J K}^{-1} \ .$$

The total entropy change is just the sum of these four components:

$$\Delta S_{tot} = \Delta S_{He}^{(1)} + \Delta S_{Ar}^{(1)} + \Delta S_{He}^{(2)} + \Delta S_{Ar}^{(2)} = 12.9 \text{ J K}^{-1} \ .$$

There are several points worth noting here. First, the units of entropy are those of (energy)/(temperature), the same as for nR and heat capacity. Second, even though the Ar lost entropy in cooling, it lost less than the He gained by warming, in accord with our two-metal-cubes discussion. But the most important point is the way we chose a *hypothetical* path from the initial state to the final state and calculated ΔS contributions for each step along this path. You may find it helpful to recalculate ΔS_{tot} by a set of steps that first alters T at constant V and then alters V, or one that alters T at constant P and then V. Your answer for any path should still be $\Delta S_{tot} = 12.9$ J K^{-1}.

⟹ RELATED PROBLEMS 3.15, 3.16

This section has established the ground rules for simple entropy calculations and has shown how these calculations can indicate the natural spontaneous direction of any process and its equilibrium stopping point. In the next section we look at entropy from a more mathematical point of view and find how to calculate ΔS in more general situations.

3.6 CONSEQUENCES OF ENTROPY AS A STATE FUNCTION

We now know

$$dS = \frac{dq_{rev}}{T} \tag{3.11}$$

in general, and the First Law tells us

$$dq_{rev} = dU - dw_{rev} \,. \tag{3.24}$$

Since $-dw_{rev} = P\,dV$, we combine Equations (3.24) and (3.11) and find

$$dS = \frac{dU}{T} + \frac{P}{T}dV \tag{3.25}$$

or

$$dU = T\,dS - P\,dV \,. \tag{3.26}$$

Compare this to the general First Law expression:

$$dU = dq + dw \,, \tag{2.16}$$

which expresses a natural, path-dependent division of energy transfer into work and heat components. In contrast, Eq. (3.26) expresses a more fundamental division in terms of state variables. It is a compact statement of both the First and Second Laws known as the *master equation* or the *fundamental equation of thermodynamics*.

We can express S as some function of U and V for any closed system of one pure substance:

$$S = S(U, V)$$

so that

$$dS = \left(\frac{\partial S}{\partial U}\right)_V dU + \left(\frac{\partial S}{\partial V}\right)_U dV \tag{3.27}$$

and we can similarly express U as some function of S and V:

$$U = U(S, V)$$

3.6 CONSEQUENCES OF ENTROPY AS A STATE FUNCTION

so that

$$dU = \left(\frac{\partial U}{\partial S}\right)_V dS + \left(\frac{\partial U}{\partial V}\right)_S dV. \quad (3.28)$$

If we compare Eq. (3.25) to Eq. (3.27) and Eq. (3.26) to Eq. (3.28), we see that

$$\left(\frac{\partial S}{\partial U}\right)_V = \frac{1}{T} \quad \text{and} \quad \left(\frac{\partial S}{\partial V}\right)_U = \frac{P}{T}$$

and

$$\left(\frac{\partial U}{\partial S}\right)_V = T \quad \text{and} \quad \left(\frac{\partial U}{\partial V}\right)_S = -P.$$

The first equalities in each of these pairs of equations are equivalent, and we will usually write

$$\left(\frac{\partial U}{\partial S}\right)_V = T. \quad (3.29)$$

This rather remarkable equation relates the three variables that are most intimately a part of thermodynamics. It can be used in a microscopic statistical discussion of thermodynamics to define temperature, as we will see in the next chapter.

We can now apply the cross-derivative equality for exact differentials to Eq. (3.28). In particular

$$\frac{\partial}{\partial V}\left(\frac{\partial U}{\partial S}\right) = \frac{\partial}{\partial S}\left(\frac{\partial U}{\partial V}\right) \quad (3.30)$$

so that, using the expressions above for $(\partial U/\partial S)$ and $(\partial U/\partial V)$,

$$\left(\frac{\partial T}{\partial V}\right)_S = -\left(\frac{\partial P}{\partial S}\right)_V. \quad (3.31)$$

This type of cross-derivative expression is called a *Maxwell relation*. We will derive several others in a later chapter.

We can use the master equation to find:

(a) the general expression for $(\partial U/\partial V)_T$ alluded to in Section 2.5;
(b) a general expression for $C_P - C_V$;
(c) a general expression for $(\partial H/\partial P)_T$; and
(d) the expression for μ_{JT} from Eq. (2.41).

First, if we introduce Eq. (2.21a)

$$dU = \left(\frac{\partial U}{\partial T}\right)_V dT + \left(\frac{\partial U}{\partial V}\right)_T dV \quad (2.21a)$$

into Eq. (3.25), we find

$$dS = \frac{1}{T}\left(\frac{\partial U}{\partial T}\right)_V dT + \frac{1}{T}\left[\left(\frac{\partial U}{\partial V}\right)_T + P\right] dV. \quad (3.32)$$

Since this is an exact differential, the cross derivative equality holds:

$$\frac{\partial}{\partial V}\left[\frac{1}{T}\left(\frac{\partial U}{\partial T}\right)_V\right] = \frac{\partial}{\partial T}\left\{\frac{1}{T}\left[\left(\frac{\partial U}{\partial V}\right)_T + P\right]\right\}. \quad (3.33)$$

These derivatives (given a closer look below) yield

$$\left(\frac{\partial U}{\partial V}\right)_T = T\left(\frac{\partial P}{\partial T}\right)_V - P. \quad (3.34)$$

This equation allows us to find $(\partial U/\partial V)_T$ from any equation of state for any pure substance: gas, liquid, or solid.

A Closer Look at the Derivation of $(\partial U/\partial V)_T$

Equation (3.33) calls for some care. Glance back at Equations (3.28), (3.30), and (3.31), and note how the subscripts (indicating the variable held constant) disappeared in Eq. (3.30) and then reappeared in Eq. (3.31).

When we write $(\partial U/\partial S)_V$ in Eq. (3.29), we mean, "pick a volume and find out how U varies with S *at that volume*." When we write $\partial(\partial U/\partial S)/\partial V$ in Eq. (3.30), we mean, "evaluate $(\partial U/\partial S)$ at a chosen volume *and entropy* and see how $(\partial U/\partial S)$ varies from volume to volume *at that entropy*," which leads to $(\partial T/\partial V)_S$ in Eq. (3.31). The subscript S points out that entropy was held constant while $(\partial U/\partial S)$ was differentiated.

Now look at Eq. (3.33). On the left, the quantity in square brackets is to be evaluated at constant temperature. Hence $1/T$ is treated as a constant:

$$\frac{\partial}{\partial V}\left[\frac{1}{T}\left(\frac{\partial U}{\partial T}\right)_V\right] = \frac{1}{T}\left(\frac{\partial^2 U}{\partial V \partial T}\right).$$

The quantity in braces on the right-hand side of Eq. (3.33) is to be evaluated at constant volume, but since volume does not appear explicitly, this is no problem:

$$\frac{\partial}{\partial T}\left\{\frac{1}{T}\left[\left(\frac{\partial U}{\partial V}\right)_T + P\right]\right\} = -\frac{1}{T^2}\left[\left(\frac{\partial U}{\partial V}\right)_T + P\right] + \frac{1}{T}\left[\left(\frac{\partial^2 U}{\partial T \partial V}\right) + \left(\frac{\partial P}{\partial T}\right)_V\right].$$

Equating these and remembering that $(\partial^2 U/\partial V \partial T) = (\partial^2 U/\partial T \partial V)$ gives Eq. (3.34).

EXAMPLE 3.6

Under what conditions does U increase with V at constant T for a gas that follows the virial equation of state through the $B(T)/\overline{V}$ term?

SOLUTION We apply Eq. (3.34) to $P = RT/\overline{V} + RTB(T)/\overline{V}^2$:

$$\left(\frac{\partial P}{\partial T}\right)_V = \frac{R}{\overline{V}} + \frac{RB(T)}{\overline{V}^2} + \frac{RT}{\overline{V}^2}\left(\frac{dB(T)}{dT}\right)$$

so that

$$T\left(\frac{\partial P}{\partial T}\right)_V - P = \frac{RT}{\overline{V}^2}\left(\frac{dB(T)}{dT}\right) = \left(\frac{\partial U}{\partial V}\right)_T.$$

When $dB/dT = 0$ (as it does exactly for a hard-sphere gas at all T and approximately for a real gas for T well above the Boyle temperature), the internal energy is independent of volume: $(\partial U/\partial V)_T = 0$. Below, at, and for some range above the Boyle temperature, $dB/dT > 0$, and U increases with increasing V. Physically this means we have to input energy, increasing U, in order to pull the gas apart under the attractive intermolecular forces that dominate at low T.

⟹ RELATED PROBLEM 3.18

Equation (3.34) can be used to derive a general expression for $C_P - C_V$. We multiply Eq. (3.32) by T and recognize that $T\,dS = dq_{rev}$ and $(\partial U/\partial T)_V = C_V$, giving

$$T\,dS = dq_{rev} = C_V\,dT + \left[\left(\frac{\partial U}{\partial V}\right)_T + P\right]dV\;.$$

If we now divide by dT and hold P constant, we turn dq_{rev} into $(dq/dT)_P = C_P$ and arrive at

$$C_P - C_V = \left(\frac{\partial U}{\partial V}\right)_T \left(\frac{\partial V}{\partial T}\right)_P + P\left(\frac{\partial V}{\partial T}\right)_P\;. \qquad (3.35)$$

Both terms on the right of this expression have a physical interpretation that is worth pointing out before we simplify it further. The first term represents the energy of expansion against intermolecular forces ($(\partial U/\partial V)_T$ is the "internal pressure" due to these forces) per unit temperature rise. (It is zero for an ideal gas.) The second represents the work done *by* the system ($+P\,dV$) per unit temperature rise at constant pressure. Together they represent the extra energy involved when a system's volume changes due to temperature changes.

Substituting Eq. (3.34) for $(\partial U/\partial V)_T$ into Eq. (3.35) gives

$$C_P - C_V = T\left(\frac{\partial V}{\partial T}\right)_P \left(\frac{\partial P}{\partial T}\right)_V\;. \qquad (3.36)$$

We can express $(\partial V/\partial T)_P$ in terms of the *isobaric bulk thermal expansivity* α defined as

$$\alpha \equiv \frac{1}{V}\left(\frac{\partial V}{\partial T}\right)_P\;. \qquad (3.37)$$

For most substances, α is positive; most substances expand when the temperature is increased. (An important exception is water near its freezing point. Ice floats!)

This definition turns Eq. (3.36) into

$$C_P - C_V = TV\alpha\left(\frac{\partial P}{\partial T}\right)_V\;. \qquad (3.38)$$

We can replace $(\partial P/\partial T)_V$. If $dP = 0$ in Eq. (1.11):

$$dP = \left(\frac{\partial P}{\partial V}\right)_T dV + \left(\frac{\partial P}{\partial T}\right)_V dT = 0 \qquad (1.11)$$

a little rearrangement yields

$$\left(\frac{\partial P}{\partial T}\right)_V = -\left(\frac{\partial V}{\partial T}\right)_P \left(\frac{\partial P}{\partial V}\right)_T\;,$$

which equals

$$\left(\frac{\partial P}{\partial T}\right)_V = \frac{\alpha}{\kappa}$$

where κ is the isothermal bulk compressibility, Eq. (1.26). Substituting this into Eq. (3.38) gives

$$C_P - C_V = \frac{TV\alpha^2}{\kappa}\;. \qquad (3.39)$$

Since κ is always positive, C_P is always $\geq C_V$. (They are equal when $\alpha = 0$.) We will evaluate this expression for several substances in the next chapter.

To close this section, we will derive the expression for μ_{JT} given by Eq. (2.41) and an expression for $(\partial H/\partial P)_T$, a quantity that is to enthalpy what $(\partial U/\partial V)_T$ is to internal energy.

We start with the definition of μ_{JT}:

$$\mu_{JT} = \left(\frac{\partial T}{\partial P}\right)_H . \tag{2.40}$$

We write the total derivative of H from Eq. (2.36) as

$$dH = \left(\frac{\partial H}{\partial T}\right)_P dT + \left(\frac{\partial H}{\partial P}\right)_T dP \tag{2.36}$$

and set $dH = 0$ to find $(\partial T/\partial P)_H$ in the form

$$\left(\frac{\partial T}{\partial P}\right)_H = -\frac{\left(\frac{\partial H}{\partial P}\right)_T}{\left(\frac{\partial H}{\partial T}\right)_P} = -\frac{\left(\frac{\partial H}{\partial P}\right)_T}{C_P} . \tag{3.40}$$

Now we need an expression for $(\partial H/\partial P)_T$.[10] From the definition $H = U + PV$, we write

$$dH = dU + d(PV) = dU + P\,dV + V\,dP$$

or

$$dU = dH - P\,dV - V\,dP .$$

Substitution into the master equation in the form of Eq. (3.25) yields

$$dS = \frac{dH}{T} - \frac{V}{T}dP . \tag{3.41}$$

The expression for dH from Eq. (2.36) turns this into the analog of Eq. (3.32):

$$dS = \frac{1}{T}\left(\frac{\partial H}{\partial T}\right)_P dT + \frac{1}{T}\left[\left(\frac{\partial H}{\partial P}\right)_T - V\right] dP . \tag{3.42}$$

Taking cross derivatives of this expression as we did to find $(\partial U/\partial V)_T$ leads to

$$\left(\frac{\partial H}{\partial P}\right)_T = V - T\left(\frac{\partial V}{\partial T}\right)_P . \tag{3.43}$$

Substitution into Eq. (3.40) gives us Eq. (2.41):

$$\mu_{JT} = \frac{1}{C_P}\left[T\left(\frac{\partial V}{\partial T}\right)_P - V\right] . \tag{2.41}$$

EXAMPLE 3.7

It is easy to use Eq. (3.43) to show that H is independent of P for an ideal gas: $(\partial H/\partial P)_T = 0$. Show that U is also independent of P at constant temperature.

---

[10] Of course, Eq. (3.40) can be written as $(\partial H/\partial P)_T = -\mu_{JT} C_P$, which is an acceptable expression for $(\partial H/\partial P)_T$. Here we are deriving a fresh expression for $(\partial H/\partial P)_T$.

SOLUTION Applying Eq. (3.43) to the ideal gas verifies $(\partial H/\partial P)_T = 0$:

$$\left(\frac{\partial \overline{H}}{\partial P}\right)_T = \overline{V} - T\left(\frac{\partial \overline{V}}{\partial T}\right)_P = \overline{V} - T\frac{R}{P} = 0 \ .$$

Next, we note that since $H = U + PV$, or $U = H - PV$, we have

$$\left(\frac{\partial \overline{U}}{\partial P}\right)_T = \left(\frac{\partial \overline{H}}{\partial P}\right)_T - \left(\frac{\partial(P\overline{V})}{\partial P}\right)_T = 0 - \left(\frac{\partial(RT)}{\partial P}\right)_T = 0 \ .$$

This logic can also be used to prove that $(\partial H/\partial V)_T = 0$ for an ideal gas.

➠ RELATED PROBLEMS 3.19, 3.22, 3.23

CHAPTER 3 SUMMARY

This chapter has defined entropy from three points of view: statistical, thermodynamic, and mathematical. It has stated the Second Law in three different ways, calculated entropy changes for a variety of simple situations, and used mathematical properties of entropy to establish several useful relationships. We have seen that entropy is a greedy and aggressive property of matter when contrasted to the fair and even-handed nature of conservative energy.

In spite of all this scrutiny (or maybe because of it), you may have the feeling that entropy, like the national debt, is incomprehensible. Since entropy cannot be experienced by our senses the way energy can, it is an inherently more abstract quantity that takes more practice to adjust to.

For now, concentrate on the *physical consequences* of the Second Law rather than the mathematical tricks and details of this chapter. Most important is the definition $dS = dq_{rev}/T$, which is often the starting point for entropy calculations. Since S is a state function, any imaginable path from the initial to the final state can be used *as long as it is a reversible path or a path for which dq equals the differential of a state function*, such as $dq_V = dU$ and $dq_P = dH$.

We saw how to partition dS into system and surroundings components and used the Second Law criterion $dS_{tot} > 0$ to signal a natural process. We combined the path-dependent version of the First Law, $dU = dq + dw$, with $dS = dq_{rev}/T$ to arrive at a succinct relation among state variables of the system that combines aspects of the First and Second Laws:

Master Equation:
$$dU = T\,dS - P\,dV$$

Two other important results from this chapter are the pair of relations:

Volume derivative of internal energy:
$$\left(\frac{\partial U}{\partial V}\right)_T = T\left(\frac{\partial P}{\partial T}\right)_V - P$$

Pressure derivative of enthalpy:
$$\left(\frac{\partial H}{\partial P}\right)_T = V - T\left(\frac{\partial V}{\partial T}\right)_P$$

Together with heat capacity information and an equation of state, these expressions are all we need to compute ΔU and ΔH for any process involving a pure substance in one phase. We will see in following chapters how mixtures, phase transitions, and chemical reactions contribute to ΔU, ΔH, and ΔS.

Since entropy is the signpost for spontaneous processes, it will become the focus of attention in the next two chapters as we see its role in several new contexts. The next chapter begins by considering entropy in a very special state of affairs—the absolute zero of temperature.

FURTHER READING

There have been several books that discuss entropy and the Second Law (as well as, of course, the discussions in all general thermodynamics texts). Among these are *Understanding Energy: Energy, Entropy, and Thermodynamics for Everyman*, by R. Stephen Berry (World Scientific, Singapore, 1991); *Entropy and Energy Levels*, by R. P. H. Gasser and W. G. Richards (Rev. ed., Oxford University Press, New York, 1986); *The Second Law: An Introduction to Classical and Statistical Thermodynamics*, by Henry A. Bent (Oxford University Press, New York, 1965); *Engines, Energy, and Entropy*, by J. B. Fenn (W. H. Freeman, New York, 1982); and *The Second Law*, by P. W. Atkins (Scientific American Library, distributed by W. H. Freeman, New York, 1984). In addition, there are many books that discuss the application of entropy (or phenomena related to entropy) as they apply to such diverse areas as economics, information theory, and sociology.

PRACTICE WITH EQUATIONS

3A What is the probability that five coins thrown at random onto a chess board will all land with their centers over a black square? (3.4)

ANSWER: 1/32

3B An engine is to operate by rejecting wasted energy as heat into a river that has a mean temperature of 285 K. What must the minimum temperature of its energy source be if it is to produce work with an efficiency of 0.75? (3.9)

ANSWER: 1140 K

3C A cycle consisting of two isothermal steps (at 375 K and 890 K) and two adiabatic reversible steps inputs 14.5 kJ to the system as heat at 890 K and outputs 6.11 kJ as heat at 375 K. Are the isothermal steps reversible? (3.10)

ANSWER: Yes

3D An ideal gas at 1.50 atm, 40.0 dm^3, and 250 K is compressed isothermally to 35 dm^3. What is the entropy change of the gas? (3.12)

ANSWER: −3.25 J K^{-1}

3E A 1.00 mol sample of Hg at 40 °C is poured into 1.00 mol of Hg at 20 °C. If C_P(Hg) = 27.8 J K^{-1} mol^{-1}, what is the entropy change for this process? (3.18)

ANSWER: 30.3 mJ K^{-1}

3F What is the entropy change of the surroundings that cools 1.00 mol of Hg from 40 °C to 30 °C? (C_P(Hg) = 27.8 J K^{-1} mol^{-1}) (3.20), (3.21)

ANSWER: 0.917 J K^{-1}

3G An isothermal, isochoric process on a system at 298 K required an increase of 2.57 kJ to the system's internal energy. What was its entropy change? (3.29)

ANSWER: 8.62 J K^{-1}

3H What is the isobaric thermal expansivity of 3.00 mol of an ideal gas at 2.5 atm and 350 K? (3.37)

ANSWER: 2.86 × 10^{-3} K^{-1}

3I What is the isobaric rate of change of entropy with enthalpy for the gas of Problem 3H? (3.41)

ANSWER: 2.86 × 10^{-3} K^{-1}

PROBLEMS

SECTION 3.1

3.1 A popular state lottery configuration typically runs like this: pick six different integers from 1 to, say, 36. If the lottery drawing picks the same six numbers (in any order), you win. Which is more likely, winning this lottery or finding all 10^6 molecules in a gas sample simultaneously in a sub-volume 99.99% the size of the full gas container's volume?

3.2 Consider a sub-volume $V' = 0.99\,V$ in Eq. (3.4). For what value of N is $W_N(V') = 0$ to within the precision of your calculator?

3.3 One mole of an ideal gas doubles in volume at constant temperature. The entropy increase in this process is given by Eq. (3.7). What volume change is required for 2 mol of an ideal gas to exhibit the same entropy increase? for 0.5 mol?

SECTION 3.2

3.4 Prove Eq. (3.9), $\varepsilon = 1 - T_2/T_1$, for the Carnot cycle operating on an ideal gas. To do so, find expressions for q_1 and q_3 and verify your calculation by calculating w_{cy}, too.

3.5 One mol of an ideal gas at $T_1 = 300$ K and 10 atm is expanded reversibly and isothermally to twice its initial volume. It is then expanded reversibly and adiabatically to $T_2 = 200$ K. These are the first two steps in a Carnot cycle. Find the (P, V) coordinates of the end of the third step: a reversible isothermal compression at T_2 that can be followed by a reversible adiabatic compression back to the initial state.

3.6 Figure 3.2 shows the Carnot cycle for an ideal gas in (P, V) coordinates. Sketch the Carnot cycle (qualitatively) in temperature–entropy coordinates, labeling each path by the corresponding four step numbers in the text.

3.7 Take the Carnot cycle of Problem 3.5 and plot the cycle to scale in temperature–volume coordinates, labeling each path by the corresponding step numbers in the text.

3.8 It is recommended that home refrigerators maintain an internal temperature around 40 °F for safe food storage. Assuming that the typical refrigerator operates at 20% of the ideal coefficient of performance and that electricity costs 10¢/kW hr, what is the annual cost of operation in a 70 °F room if the door is opened ten times a day with each opening introducing an average "heat load" of 200 kJ? How much would it cost to keep the inside at 35 °F? to keep the inside at 35 °F if the room were kept at 65 °F?

3.9 Attaining very low temperatures is obviously difficult. Great care must be taken to eliminate "heat leaks," thermal connections to the warm surroundings, and the following calculation shows why. What is the minimum energy input needed to remove 1 J from a sample held by a refrigerator at 10^{-3} K? Assume the refrigerator is dumping its removed energy into the 300 K room. (See Example 3.3 for needed definitions.)

3.10 What physical interpretation can be given to the coefficients of performance of heat pumps or refrigerators if $T_1 = T_2$?

SECTION 3.3

3.11 We stated that $C_V = 3nR/2$ was a result for the ideal gas that comes from theories beyond thermodynamics, yet Carathéodory would seem to have given us a means of finding C_V. If we take T to be the integrating denominator for Eq. (3.14), recognize $(\partial U/\partial T)_V = C_V$, and take the empirical fact that $(\partial U/\partial V)_T = 0$ for an ideal gas, we have the exact differential

$$\frac{dq_{\text{rev}}}{T} = \frac{C_V}{T} dT + \frac{P}{T} dV \ .$$

Apply the cross-derivative test for exact differentials to this expression, and see if it can yield a value for C_V. Assume that we know the ideal gas equation of state, and assume C_V is constant.

SECTION 3.4

3.12 Carry out the arguments in the text relating to the Carathéodory statement of the Second Law and Figure 3.3 for the case of an adiabatic *compression*.

SECTION 3.5

3.13 Consider a gas in volume V_1 connected to an initially empty container of volume V_2. With the help of Eq. (3.12), show that ΔS is a maximum when $V_f = V_1 + V_2$ (i.e., for all $V_f < V_1 + V_2$, ΔS is less).

3.14 Show that $T_b' = (T_a + T_b)/2 = T_a'$ indeed *maximizes* ΔS_{tot} as claimed in Example 3.4.

3.15 Which involves the greater entropy change, isothermal doubling the volume of 1 mol of an ideal gas, or isobaric doubling its temperature? In considering your answer, consider, too, the volume change in the second process. Your answer should show you that entropy changes come from both configurational changes in the system (volume changes here) as well as energy changes (temperature changes here).

3.16 If an isolated stretched rubber band is released and snaps back to its relaxed state, what can be said about its entropy change?

3.17 Find the entropy change in Example 3.5 by choosing a different path of your devising.

SECTION 3.6

3.18 What is ΔU if 1.00 mol He(g) at 1.00 atm is doubled in volume at the Boyle temperature? What is ΔU if the gas is Xe? Use the van der Waals approximation for $B(T)$ and the expression for $(\partial U/\partial V)_T$ derived in Example 3.6. Are the two magnitudes for ΔU different in the way you expect for these gases?

3.19 Carry out the argument mentioned in Example 3.7 to show that $(\partial H/\partial V)_T = 0$ for an ideal gas.

3.20 Find the Joule–Thomson coefficient of NH_3 at 300 °C and 40 atm given $\overline{C}_P = 46$ J mol^{-1} K^{-1} and the V–T data below for $P = 40$ atm.

T/°C	$\overline{V}$/cm³ mol⁻¹
225	962
250	1017
275	1076
300	1136
325	1186

3.21 Apply both Equations (3.35) and (3.36) to, first, the ideal gas, and, second, the hard-sphere gas. (See Example 2.2 for the hard-sphere gas equation of state.)

GENERAL PROBLEMS

3.22 J. Willard Gibbs, about whom much more will be said in later chapters, was the first to show that the fundamental thermodynamic equation for an ideal gas could be written as

$$U = V^{-R/C_V} \exp(S/C_V)$$

where C_V and R are constants. Is this expression in agreement with Eq. (3.12)?

3.23 Use Gibbs' Equation from the previous problem to verify that $(\partial U/\partial V)_T = 0$ for an ideal gas. Go on to prove that this is true for any gas obeying the equation of state of the form $P f(V) = RT$ where $f(V)$ is any continuous function of volume.

3.24 If the two cubes discussed in Section 3.5 equilibrate freely and irreversibly, they reach the common equilibrium temperature $T = (T_a + T_b)/2$, the arithmetic average of their initial temperatures. Show that if they equilibrate *reversibly*, the final common temperature is the geometric mean of the initial temperatures: $T = \sqrt{T_a T_b}$. This will be discussed further in Chapter 5, but you know enough now to prove it.

CHAPTER 4

Heat Capacities, Absolute Zero, and the Third Law

4.1 The Variation of Entropy with Temperature

4.2 Life at Absolute Zero

4.3 The Meaning of Heat Capacity

4.4 A First Look at Phase Transitions

4.5 The Third Law and Absolute Entropy

4.6 The Calculation of Absolute Entropy

THE last chapter introduced entropy and its calculation in a few specific and simple situations. This chapter concentrates on temperature-dependent paths for which the important system parameter is the heat capacity. Heat capacities can be related to microscopic molecular motions (see Chapter 23), and they are central to an understanding of the interplay among entropy, temperature, and internal energy.

This chapter also introduces the last major statement of thermodynamics, the Third Law. This law gives entropy an *absolute* status so that a system at equilibrium has a definite total entropy. Of course, entropy differences will still have importance, but new benefits accrue as a result of the Third Law.

The first section will set the stage for these discussions by considering the general approach to entropy calculations along paths at constant volume or constant pressure, but changing temperature.

4.1 THE VARIATION OF ENTROPY WITH TEMPERATURE

The fundamental thermodynamic expression for entropy is Eq. (3.11):

$$dS = \frac{dq_{\text{rev}}}{T} . \tag{3.11}$$

Since the arguments for integrating this expression over constant volume and constant pressure paths are very similar, we will develop them in parallel. Recall from Chapter 2 that at constant volume there is no pressure–volume work: $-P_{\text{ext}}dV = 0$ if $dV = 0$. Thus, if we have no electrical, magnetic, or "shaft" work (such as a stirrer driven by a shaft passing through the boundary), the only way to alter the internal energy is via heat: $dU_V = dq_V$. At constant pressure, again neglecting all but P–V work, the analogous expression relates enthalpy and heat: $dH_P = dq_P$.

From the definitions of heat capacities,

$$C_V = \frac{dq_V}{dT} = \left(\frac{\partial U}{\partial T}\right)_V \quad \text{and} \quad C_P = \frac{dq_P}{dT} = \left(\frac{\partial H}{\partial T}\right)_P , \tag{4.1}$$

we can write

$$dS_V = \frac{C_V}{T}dT = \left(\frac{\partial S}{\partial T}\right)_V dT \quad \text{and} \quad dS_P = \frac{C_P}{T}dT = \left(\frac{\partial S}{\partial T}\right)_P dT . \tag{4.2}$$

Experiments (and statistical theories) show that heat capacities are functions of temperature. Thus

$$\Delta S_V = \int_{T_i}^{T_f} \frac{C_V(T)}{T}dT \quad \text{and} \quad \Delta S_P = \int_{T_i}^{T_f} \frac{C_P(T)}{T}dT \tag{4.3}$$

are the general expressions for the entropy changes that accompany temperature changes of a pure substance in a single phase.

We could end this chapter here. All we need say is that, from theory or experiment, we find $C(T)$ and use the appropriate form of Eq. (4.3) to find ΔS. With a functional form for $C(T)$, the problem is reduced to an exercise in evaluating integrals. Experiments give us graphs or tables of C_V or C_P at various temperatures, and we could use graphical or numerical integration methods. But heat capacities are so fundamental that we should consider them in greater detail.

4.2 LIFE AT ABSOLUTE ZERO

Heat capacities, entropies, and the absolute zero of temperature are intimately related. To establish their relationships, we will begin with a very commonplace thermodynamic system—a row of bricks. Careful analysis of this system, which is a model of a real molecular system, makes several subtle aspects of equilibrium clearer. We begin with a row of ten identical bricks next to a table, as shown in Figure 4.1(a).

We can increase the internal energy of this system if we move a number of bricks up onto the table. If we move n bricks onto the table, ΔU will be

$$\Delta U(n) = U(n) - U(n = 0) = nMgh = n\varepsilon \tag{4.4}$$

where $\varepsilon = Mgh$ is the internal energy increment (the work) associated with moving one brick of mass M onto the table.

Now note that there is only *one* way to have all ten bricks on the floor, but *ten* ways to *attain the thermodynamic state characterized by* $U = U(n = 1)$. Moving *any* of the ten bricks onto the table achieves this state. The state $U = U(n = 2)$, with *any* two bricks on the table, can be achieved in 45 distinct ways. There are

FIGURE 4.1(a) Ten identical bricks of mass M and a table of height h constitute a thermodynamic system. To describe the system, we must use macroscopic state variables, such as U, the internal energy. This energy, U, can be increased by doing work on the system: we raise one or more bricks up onto the table. **(b)** If two of the ten bricks are placed on the table, the internal energy increases by $2Mgh$. But *nowhere* in the thermodynamic description of this state can we specify *which* two bricks have been moved. There are 45 ways of raising two of the ten bricks and arriving at this thermodynamic state.

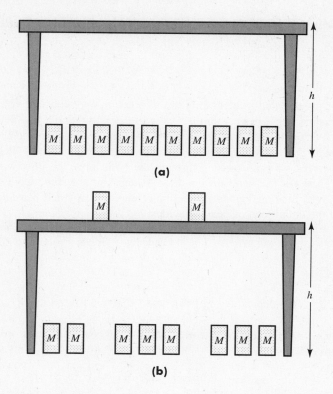

ten choices to move the first brick, then there are nine choices to move the second. But, as indicated in Figure 4.1(b), whether we move brick 3 followed by brick 7, for instance, or first move 7, then 3, we arrive at the same state. Thus the number of distinct ways of attaining $U(n = 2)$ is $(10 \times 9)/2 = 45$.

Continuing brick by brick, and generalizing to a system with a total of N bricks, we would find that the number of distinct ways of placing n bricks on the table is[1]

$$W(n) = \frac{N!}{n!\,(N-n)!} \tag{4.5}$$

where $N!$ (read "N factorial") $= N \cdot (N - 1) \cdot (N - 2) \cdot \ldots \cdot 2 \cdot 1$ and $0! = 1$ by definition. We should expect from symmetry, and Eq. (4.5) confirms, that $W(n) = W(N - n)$. For example, just as $W(0) = 1$, likewise $W(N) = 1$, since there is only one way to have all N bricks on the table. The quantity $W(n)$ increases from $W(0) = 1$ to a maximum value at $n = N/2$ (or at $n = (N \pm 1)/2$ if N is an odd number) and then decreases, symmetrically, back to $W(N) = 1$.

Next we bring entropy into the picture. We will call any specific arrangement of bricks a *microstate* of the system, as in "any of 10 microstates leads to $U(n = 1)$ if $N = 10$." The $W(n)$ function tells us how many microstates—detailed distributions of energy on a microscopic scale—lead to the same thermodynamic state. We turn numbers of microstates into entropy through the Boltzmann expression from Section 3.1:

$$S(n) = k_B \ln W(n) \,. \tag{4.6}$$

This is an important statistical interpretation of entropy. If one thermodynamic state (characterized *only* by *macroscopic* variables such as P, V, T, U, and S) can be

---

[1] $W(n)$ is the so-called n^{th} *binomial coefficient*, the coefficient of x^n in the expansion of $(1 + x)^N$.

achieved by a greater number of microscopic arrangements than another thermodynamic state, then the first state has the greater entropy.

Next, we leave the bricks behind, return to molecules, and introduce temperature into the discussion. But one subtle and vitally important distinction between molecules and bricks must be pointed out first. This distinction forms the molecular basis for attaining thermal equilibrium and will be treated in detail in Chapter 22.

Suppose we take the ten bricks in Figure 4.1(a) and divide them in half, five on the floor in front of one table and five on top of a second table. Considering each as separate subsystems, they each have the same entropy, according to Eq. (4.6) with $N = 5$ and $n = 0$ or 5, while one has an energy $U(n = 0) = 0$ and the other has $U(n = 5) = 5\varepsilon$. We now push the two tables side by side, simulating thermal contact. What happens? Absolutely nothing, of course. The bricks just sit there. We now have a system of ten bricks with a well-defined internal energy ($U = 0 + 5\varepsilon$, a conserved quantity!), but we *cannot* say the entropy of the combined system is appropriate for an equilibrium system. Since the bricks just sat there, we are *still certain* which five are on the floor and which are on the table.

If the bricks could communicate via "inter-brick forces" the way molecules communicate via intermolecular forces, then when we pushed the tables into "thermal contact," bricks would start jumping from the floor to the table or vice-versa, randomizing the *identity* of the bricks on the table, but *always* keeping five on the table and five on the floor to conserve the total energy. This microscopic motion, in which the attainment of thermal equilibrium involves the chaotic redistribution of energy among all the microstates of the system in a constantly changing fashion, is at the heart of molecular equilibria.

Thus we concentrate on a molecular system of N molecules (where N is very large), but we still imagine only two energies are possible for each molecule. Such a system is known as a *two-level system*. We can define thermodynamic variables and their changes according to

$$\Delta U(n) = U(n) - U(n = 0) = n\varepsilon \tag{4.4}$$

and

$$\Delta S(n) = S(n) - S(n = 0) = k_B \ln\left(\frac{N!}{n!\,(N-n)!}\right). \tag{4.7}$$

Suppose we plot $U(n)$ versus $S(n)$ at a variety of values for n as in Figure 4.2. The solid portion of the curve, which locates states with $n < N/2$, is of most interest. The dashed part is rather curious and is given a closer look later on.

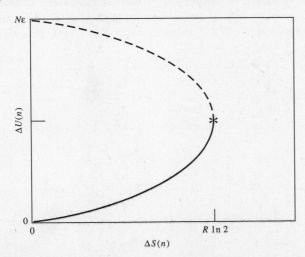

FIGURE 4.2 If the internal energy is plotted versus entropy for a system with only two microscopic energies, this curve results. The solid portion of the curve has a positive slope and corresponds to the states of positive absolute temperature. At the state noted by the asterisk, the temperature is infinite. In this state, $N/2$ molecules have extra energy, and the excess entropy is $Nk_B \ln 2$ which equals $R \ln 2$ for one mole of molecules.

The slope of this curve,

$$\left(\frac{\partial(\Delta U(n))}{\partial(\Delta S(n))}\right)_V = \left(\frac{\partial U(n)}{\partial S(n)}\right)_V, \qquad (4.8)$$

is, remarkably enough, the temperature of the system! Look back to Eq. (3.29). Following the solid curve in Figure 4.2, it is clear that for small n, the slope (and thus the temperature) is small, while at $n = N/2$, the slope and thus the temperature has become *infinite*. At this point, half the molecules have ε more energy than the other half. (Half the bricks are on the floor and half are on the table.)

With the chain rule for differentiation

$$\left(\frac{\partial U(n)}{\partial S(n)}\right)_V = \left(\frac{\partial U(n)}{\partial n}\right)_V \left(\frac{\partial n}{\partial S(n)}\right)_V = \left(\frac{\partial U(n)}{\partial n}\right)_V \left(\frac{\partial S(n)}{\partial n}\right)_V^{-1}$$

and a very good approximation for ln N! if N is very large,

Stirling's approximation: $\ln N! \cong N \ln N$,

we can find an expression for temperature as a function of n. We begin with

$$\left(\frac{\partial U(n)}{\partial n}\right)_V = \varepsilon . \qquad (4.9)$$

Next, using Stirling's approximation,

$$S(n) = k_B [\ln N! - \ln n! - \ln (N - n)!]$$
$$\cong k_B [N \ln N - n \ln n - (N - n) \ln (N - n)] ,$$

we find

$$\left(\frac{\partial S(n)}{\partial n}\right)_V = k_B \ln \left(\frac{N}{n} - 1\right) . \qquad (4.10)$$

Finally, we combine Equations (4.9) and (4.10) according to the chain rule to give

$$T(n) = \frac{\varepsilon}{k_B \ln \left(\dfrac{N}{n} - 1\right)} . \qquad (4.11)$$

Note that $T \to \infty$ as $n \to N/2$ and that $T \to 0$ as $n \to 0$. Figure 4.3 plots $\Delta U(n)$ versus the dimensionless temperature, $k_B T/\varepsilon$, to demonstrate these limits in terms of energy.

Concentrate on the $T \to 0$ limit first. At the absolute zero of temperature, not only is the internal energy at its lowest value, but so is the entropy. In fact, Eq. (4.8) suggests that $S = 0$ at $T = 0$, since $n = 0$ at $T = 0$. This will become one

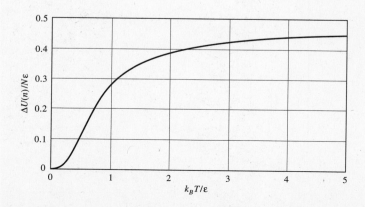

FIGURE 4.3 Over the positive temperature branch of Figure 4.2, the internal energy gradually approaches $N\varepsilon/2$ as T approaches infinity.

statement of the Third Law, but we will postpone a full discussion until Section 4.5. Nevertheless, the microscopic arrangement at absolute zero is clear: all the molecules are characterized by a common value of energy, the lowest possible. All the bricks are on the floor at absolute zero.

If we rearrange Eq. (4.11) a bit, we can write it as

$$\frac{n}{N-n} = \frac{\text{population of upper level}}{\text{population of lower level}} = e^{-\varepsilon/k_B T},$$

which is an expression known as the *Boltzmann distribution*. We will see in Chapter 22 that this result is general: $\exp[-(\text{energy difference})/k_B T]$ equals or is at least proportional to the equilibrium ratio of population between any two levels of different energy.

A Closer Look at the Two-Level System

You may be asking yourself two questions at this point. First, is there a real molecular system with only two microscopic energies per molecule? Second, what about the dashed part of the curve in Figure 4.2? The answer to the first question is a qualified yes. The system is one of nuclear magnets, the inherent magnetic moments of certain nuclei. Roughly speaking, just as it takes energy (work) to turn a bar magnet through 180° in an external magnetic field, so it takes energy to manipulate nuclear magnetic moments in external magnetic fields. These phenomena, which require quantum mechanics for their correct explanation, will be discussed in Chapters 15 and 21. The first experiments of this type were done by Pound and Purcell in 1950 using the Li nuclei in a LiF crystal. The nuclei communicate with one another very well, but they communicate with the rest of the molecular system poorly. Thus, since the nuclei are virtually isolated by an adiabatic wall of sorts (the poor communication with the rest of the crystal), the temperature in these experiments is one characterizing the nuclear magnetic system and not the entire molecular system.

If you guessed that the dashed part of Figure 4.2 corresponds to systems with *negative* absolute temperatures, since the slope is negative, you were correct. As crazy as it may sound at first, negative absolute temperatures *can* be attained, but only in molecular systems with a small number of possible energy states per molecule. What's more, negative temperature systems are *hotter* than positive ones, curiously enough. To see that this must be true, we need only state the Clausius form of the Second Law (energy flows from the hotter to the colder system if only heat transfers of energy are allowed) and recognize that negative temperature states have more energy than do positive temperature states. Any system at a negative temperature will transfer energy to any system at a positive temperature when the two are brought into contact.

As this analysis shows, it is not T that characterizes "degree of hotness," but rather $-1/T$, as the following diagram indicates:

coldest system	$\longrightarrow$			hottest system
$T = 0$	$\longrightarrow$	$(T = +\infty, T = -\infty)$	$\longrightarrow$	$T = 0$
(but just positive)				(but just negative)
$-1/T = -\infty$	$\longrightarrow$	$-1/T = 0$	$\longrightarrow$	$-1/T = +\infty$

Note that all negative temperature states have more molecules *with* microscopic energy ε than *without* ε ($n > N/2$). This type of microscopic arrangement is called a *population inversion* and is related to the arrangement required for masers and lasers to operate.

4.3 THE MEANING OF HEAT CAPACITY

The heat capacity of the two-level system is the slope of the curve in Figure 4.3. If we solve Eq. (4.11) for n,

$$n = N \frac{e^{-x}}{(1 + e^{-x})} \tag{4.12}$$

where $x = \varepsilon/k_B T$, then

$$U(T) = n\varepsilon = N\varepsilon \frac{e^{-x}}{(1 + e^{-x})}. \tag{4.13}$$

From this we can find

$$C_V(T) = \left(\frac{\partial U(T)}{\partial T}\right)_V = Nk_B x^2 \frac{e^{-x}}{(1 + e^{-x})^2}, \tag{4.14}$$

which is graphed in Figure 4.4(a) as C_V/Nk_B versus T (or rather $k_B T/\varepsilon = 1/x$).

Note how $C_V \to 0$ as $T \to 0$; zero heat capacity at $T = 0$ is a general phenomenon of Nature. In this simple system, we can see how this comes about on a molecular level. Figure 4.4(b) plots C_V/Nk_B versus n/N, the fraction of molecules in the upper energy level. Various points on the curve denote corresponding temperatures (again

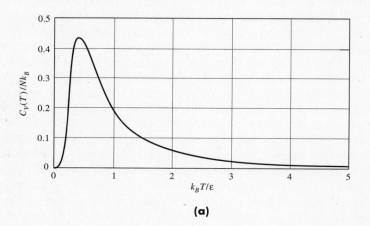

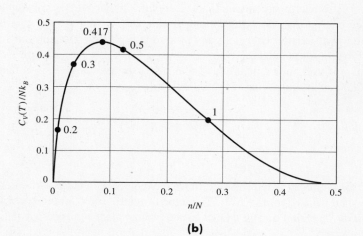

FIGURE 4.4(a) The temperature derivative of Figure 4.3 is the heat capacity $C_V(T)$. Note that $C_V(T)$ is large only over a short span of temperatures. At large T, the system's energy increases slowly with T, leading to the small value for C_V. **(b)** Another way to consider heat capacity is in terms of the number of molecules in excited states. This figure graphs C_V as a function of n/N, the fraction of molecules in the upper energy level of a two level system. Points on the line represent values of $k_B T/\varepsilon$.

scaled as $k_B T/\varepsilon$). The heat capacity is a maximum at $k_B T/\varepsilon = 0.417$ or $x = 1/0.417 \cong 2.4$, and Eq. (4.12) says $n/N \cong 0.083$ at this value of x. With only a bit more than 8% of the molecules excited, the heat capacity is as large as possible. Relatively large energy transfers into or out of the system have the smallest effect on T in this region. The heat capacity rapidly falls to zero at lower T, and it gradually approaches zero at higher T. Since $dT = dq_V/C_V$, small energy transfers have large temperature effects as $C_V \to 0$.

This high-temperature behavior is peculiar to systems with only a few molecular energies. Molecular systems generally have a near infinite number of possible energy values per molecule, and heat capacities generally continue to increase at high temperatures.

We will return to the molecular basis of heat capacities in Chapter 23 as part of our treatment of statistical mechanics. Here we summarize typical values of heat capacities for solids, liquids, and gases, and we use thermodynamics to relate C_V to C_P.

Solids

Solids have great variety in their macroscopic properties: hard and shiny metals, hard and brittle salt crystals, soft polymers, uniquely low-temperature states of matter that are less familiar, and so on. One compound may exist in several solid forms, such as grey, metallic arsenic and yellow, soft arsenic. In spite of this diversity, a small number of empirical rules were recognized long ago to predict C_V for solids. The first observations were generalized around 1819 by Dulong and Petit, who noted that C_V was nearly independent of T over wide temperature ranges. In modern terminology, the Dulong and Petit approximation is that

$$\overline{C}_V \text{(atomic solids)} \cong 3R \ . \tag{4.15}$$

Kopp made an empirical extension of this approximation to polyatomic solids in 1865. He suggested that polyatomic solid heat capacities were just the sum of the heat capacities of the solid's constituent atoms. His empirical atomic contributions varied from atom to atom and are given in Table 4.1.

EXAMPLE 4.1

Use the Kopp parameters to predict $\overline{C}_V$ for CuO, Cu$_2$O, and Fe$_2$O$_3$, and compare the predictions to the experimental values $5.08\,R$, $7.66\,R$, and $12.53\,R$, respectively.

TABLE 4.1 Kopp Parameters for Estimating $\overline{C}_V$

atom	$\overline{C}_V/R$
H	1.20
Be	1.50
B	1.36
C	0.90
N	1.50
O	2.01
F	2.53
Si	1.91
P	2.72
S	2.72
All others	3.02

SOLUTION The Kopp rules predict molar heat capacities for the two copper oxides to be

$$\overline{C}_V(\text{CuO}) = (3.02 + 2.01)\,R = 5.03\,R$$

$$\overline{C}_V(\text{Cu}_2\text{O}) = [2(3.02) + 2.01]\,R = 8.05\,R\ .$$

Kopp predicts $\overline{C}_V(\text{Fe}_2\text{O}_3) = 12.07\,R$. These calculations show the typical ranges of errors in the Kopp approximation, which was developed for ionic solids (halides, oxides, sulfides, metal hydrides, etc.) near room temperature. At elevated temperatures, ionic solid heat capacities are generally larger than at room temperature. For instance, at 1000 K, $\overline{C}_V(\text{CuO}) = 7.20\,R$, $\overline{C}_V(\text{Cu}_2\text{O}) = 10.33\,R$, and $\overline{C}_V(\text{Fe}_2\text{O}_3) = 18.12\,R$.

▶ **RELATED PROBLEM 4.5**

At very low temperatures, $\overline{C}_V$ decreases toward zero. The first theory to predict this behavior accurately (as well as the Dulong–Petit room temperature limit) was advanced by Debye in 1912 as a modification of Einstein's 1907 theory. Einstein made the first superb step in the right direction; Debye modified his approach toward a more realistic model. Both models are based on the idea that it is the vibration of atoms around their rest positions in solid crystals that provides the energy-absorbing motion leading to a non-zero heat capacity. For now, the important aspects of Debye's theory are the way it predicts $\overline{C}_V$ to approach zero as T approaches zero and the way it predicts $\overline{C}_V$ to vary from substance to substance at any one temperature.

Debye's theory introduces a parameter, Θ_D, the *Debye temperature*, for each substance and predicts a low temperature variation of $\overline{C}_V$ with T of the form

$$\frac{\overline{C}_V}{R} = \frac{12\pi^4}{5}\left(\frac{T}{\Theta_D}\right)^3 = \frac{233.8}{\Theta_D^3}T^3\ . \tag{4.16}$$

This result, known as the *Debye T-cubed law*, is the low temperature limit of the full theory.[2] Equation (4.16) is accurate to 1% or better for $T < \Theta_D/11.5$ and to 10% or better for $T < \Theta_D/8$.

To find Θ_D, one can graph measured values of $\overline{C}_V$ versus T^3 at successively lower temperatures. At some point, a straight line emerges as T^3 approaches zero; as Eq. (4.16) indicates, Θ_D is related to the slope of this line. Representative values of Θ_D are listed in Table 4.2 where we see that relatively soft metals, such as Pb and the heavier alkalis, have small values, while harder materials, especially diamond, have large values.

For *metals*, other phenomena can contribute to $\overline{C}_V$. The most important of these is made by the motion of conduction electrons in the normal metal. (The onset of superconductivity in some metals and ceramics is an exotic detail we will ignore.) This motion does not contribute to $\overline{C}_V$ in the same way as the nuclear motion contribution treated by Debye. This electronic heat capacity follows the simple expression

$$\frac{\overline{C}_V(\text{electrons})}{R} = \frac{T}{\Theta_{\text{el}}} \tag{4.17}$$

at all T where Θ_{el} is a constant that varies from metal to metal (Table 4.3). The total heat capacity of metals is the sum of the nuclear and electron contributions, which for sufficiently low T is

$$\frac{\overline{C}_V(\text{metals})}{R} = \frac{T}{\Theta_{\text{el}}} + \frac{12\pi^4}{5}\left(\frac{T}{\Theta_D}\right)^3\ .$$

---

[2]The full theory, for which Eq. (4.16) is only the low-temperature limit, also approaches $\overline{C}_V = 3R$ at high T. For $T = \Theta_D$, $\overline{C}_V = 2.86\,R$; for $T = 2\Theta_D$, $\overline{C}_V = 2.96\,R$.

TABLE 4.2 Debye Temperatures, Θ_D/K

Li	344.	KBr	177.
Na	156.	KCl	230.
K	91.1	NaCl	281.
Rb	55.5	CaF$_2$	474.
Cs	39.5	FeS$_2$	645.
Cu	343.	C(diamond)	1890.
Ag	226.	Si	645.
Au	162.	Ge	374.
		Sn	199.
Fe	470.	Pb	88.
Zn	317.		
Al	428.	Ne	75.
		Ar	92.
		Kr	72.
		Xe	64.

EXAMPLE 4.2

Since Θ_{el} values are typically several thousand K, it might seem that electrons do not contribute significantly to $\overline{C}_V$ except at very high temperatures. Explore this by calculating the electronic and Debye contributions to $\overline{C}_V$ for Ag at 300 K and at 3 K.

SOLUTION For Ag at room temperature, too hot for Eq. (4.16) to apply, $\overline{C}_V \cong 3R$, of which the electron contribution is only RT/Θ_{el} = 300 R/(12 900) $\cong$ 0.023 R. But at 3 K, even though $\overline{C}_V$(electron motion) has fallen two orders of magnitude to 2.4×10^{-4} R, the Debye equation shows that $\overline{C}_V$(nuclear motion) = 5.53×10^{-5} R. What was a < 1% contribution at 300 K has become a dominating 80% contribution to $\overline{C}_V$ at 3 K. Thus, electrons make major contributions to metal heat capacities at both very low and very high temperatures.

⟹ *RELATED PROBLEMS* 4.6, 4.7, 4.8

Liquids

Liquids are characterized by the close vicinity of molecules, as in a solid, but unlike a solid, there is molecular motion that destroys the extended order of a crystal. As a result of both the closeness and the motion, liquid heat capacities are generally larger than those of the corresponding solid at temperatures near the melting point, and larger than those of the gas at temperatures near the boiling point.

TABLE 4.3 Electronic Heat-Capacity Constants, Θ_{el}/K

Li	5 100.	Cu	12 000.
Na	6 025.	Ag	12 900.
K	4 000.	Au	11 400.
Rb	3 450.	Zn	13 000.
Cs	2 600.	Cd	12 100.
		Hg	4 650.
Al	6 160.		
Sn	4 670.	Fe	1 670.
Pb	2 790.	Co	1 760.
		Ni	1 180.

The most important case is H$_2$O. At the normal melting point, 273.15 K, $\overline{C}_V$(ice) = 4.69 R, a value close to the Kopp empirical value of 4.41 R. At the boiling point, experimental measurements give $\overline{C}_V$(H$_2$O(g)) = 3.098 R. But for liquid H$_2$O, the molar heat capacity is much larger and virtually constant at around 9 R over the range 273–373 K.

The heat capacity of H$_2$O(l) has also played a role in the definition of energy units. For many years, the calorie was defined as the energy required to raise the temperature of 1 g of water at 1 atm from 14.5 °C to 15.5 °C. This amount of energy is referred to as the "15° calorie" and is equal to 4.1855 J. The modern calorie is defined to be exactly 4.184 J, and this value is sometimes, but not always, called the "thermochemical calorie" to distinguish it from the 15° calorie. Several other slightly different calories have been used over the years, and for precise calculations with tabulated data, one must be aware of the convention used by the tabulators.

Values of $\overline{C}_P$ for other common liquids at 298 K span a wide range, but all are large, except for Hg(l), which has a value not much different from a metallic solid, 3.366 R. Other typical values are listed below. Note the large isotope effect between H$_2$O and D$_2$O and the general increasing trend with increasing molecular complexity. This trend is carried into the gas phase, as discussed below.

H$_2$O	Br$_2$	CH$_3$OH	D$_2$O	C$_2$H$_5$OH	CCl$_4$
9.055 R	9.103 R	9.814 R	10.145 R	13.41 R	15.85 R

C$_6$H$_6$	C$_6$H$_{12}$(cyclohexane)	C$_7$H$_{16}$(heptane)
16.37 R	18.82 R	26.98 R

Gases

The $\overline{C}_V$ values for gases exhibit not only an important temperature dependence, but also a dependence on molecular structure that is fully explained by statistical mechanics. As with liquids, $\overline{C}_V$ for polyatomic molecules generally increases the more "polyatomic" the molecule. For example, the heat capacities of gaseous straight-chain hydrocarbons increase regularly with the number of carbon atoms:

$$\overline{C}_V(\text{CH}_4, 1 \text{ atm}, 298 \text{ K}) = 3.30 \, R$$

$$\overline{C}_V(\text{C}_2\text{H}_6, 1 \text{ atm}, 298 \text{ K}) = 5.34 \, R$$

$$\overline{C}_V(\text{C}_3\text{H}_8, 1 \text{ atm}, 298 \text{ K}) = 7.89 \, R$$

$$\overline{C}_V(n\text{-C}_4\text{H}_{10}, 1 \text{ atm}, 298 \text{ K}) = 10.9 \, R \; .$$

We have been concentrating on $\overline{C}_V$, but most data are tabulated as $\overline{C}_P$ values. Recall that C_P is never less than C_V and that the two are related (for any phase, not just gases) via Eq. (3.35):

$$C_P - C_V = \left(\frac{\partial U}{\partial V}\right)_T \left(\frac{\partial V}{\partial T}\right)_P + P\left(\frac{\partial V}{\partial T}\right)_P \qquad (3.35)$$

or the equivalent Eq. (3.39):

$$C_P - C_V = \frac{TV\alpha^2}{\kappa} \; . \qquad (3.39)$$

For a real gas, we can use the van der Waals equation of state to see how intermolecular forces alter the ideal gas relation $C_P = C_V + nR$.

Substituting $(\partial U/\partial V)_T = a/\overline{V}^2$, Eq. (2.23), into Eq. (3.35) and using the equation of state for P, Eq. (1.20), we find

$$\overline{C}_P - \overline{C}_V = \frac{RT}{(\overline{V} - b)}\left(\frac{\partial \overline{V}}{\partial T}\right)_P.$$

Next we write the total differential of the van der Waals equation,

$$dP = \frac{R}{(\overline{V} - b)}dT + \left[\frac{2a}{\overline{V}^3} - \frac{RT}{(\overline{V} - b)^2}\right]d\overline{V},$$

set $dP = 0$, and rearrange to find

$$\left(\frac{\partial \overline{V}}{\partial T}\right)_P = \frac{R/(\overline{V} - b)}{[RT/(\overline{V} - b)^2] - (2a/\overline{V}^3)}.$$

This gives us the exact result

$$\overline{C}_P - \overline{C}_V = R\left[\frac{RT/(\overline{V} - b)^2}{[RT/(\overline{V} - b)^2] - (2a/\overline{V}^3)}\right]. \quad (4.18)$$

For most gas densities, $\overline{V} \gg b$, and Eq. (4.18) can be approximated as

$$\overline{C}_P - \overline{C}_V \cong R\left[\frac{1}{1 - (2a/RT\overline{V})}\right].$$

Typical values for a span 1 to 10 dm^6 atm mol^{-2} (see Table 1.2); thus, at $T \cong 300$ K and $\overline{V} \cong 24$ dm^3 mol^{-1}, $\overline{C}_P - \overline{C}_V$ is on the order of $1.003\,R$ to $1.03\,R$, which shows that the ideal gas expression $\overline{C}_P - \overline{C}_V = R$ is very accurate for real gases. For gas densities such that $\overline{V}$ is unusually small (as near the critical point) $\overline{C}_P$ can be significantly greater than $\overline{C}_V + R$.

EXAMPLE 4.3

Estimate $\overline{C}_P - \overline{C}_V$ for Xe at $T = 300$ K and $P = 55$ atm (just above the critical temperature and just below the critical pressure—see Table 1.3).

SOLUTION First, since the system is too close to the critical point to expect the ideal gas equation of state to apply, we find $\overline{V}$ from the van der Waals equation of state using the a and b parameters in Table 1.2. A numerical solution for $\overline{V}$ yields 269 cm^3 mol^{-1}. (The ideal gas value is 448 cm^3 mol^{-1}.) The exact van der Waals expression for $\overline{C}_P - \overline{C}_V$, Eq. (4.18), is appropriate because $\overline{V}$ is comparable to b (51.05 cm^3 mol^{-1}). We find $\overline{C}_P - \overline{C}_V = 5.94\,R$, a surprisingly large value. As an aside, the approximation for $\overline{V} \gg b$ gives $\overline{C}_P - \overline{C}_V = -3.75\,R$, which, being negative, is clearly in error!

➡ RELATED PROBLEM 4.10

Gaseous $\overline{C}_P$ values are tabulated as either measured values at a variety of temperatures or as coefficients of semi-empirical functions that have been fit to such values. These functions are "semi-empirical" in that theory suggests the *type* of fitting functions that should be best. Gaseous heat capacities are also usually tabulated for the low pressure, ideal gas limiting values; the pressure variation of $\overline{C}_P$ will be discussed at the end of this section.

The most common functions for $\overline{C}_P(T)$ are

$$\frac{\overline{C}_P(T)}{R} = a + bT + cT^2 + dT^3 \quad (4.19a)$$

or

$$\frac{\overline{C_P(T)}}{R} = a' + b'T + \frac{c'}{T^2}. \quad (4.19b)$$

Table 4.4 lists the parameters of Eq. (4.19a) for a number of gases, and Figure 4.5 graphs several of them.

A Closer Look at C_P and C_V for Gases

For an ideal gas, the ratio $C_P/C_V = 5/3$ appeared naturally in the discussion of the reversible adiabatic path, leading to Eq. (2.29):

$$P_i V_i^{5/3} = P_f V_f^{5/3} = P_{sys} V^{5/3} = P_{ext} V^{5/3} = \text{constant}. \quad (2.29)$$

If we were to repeat that discussion using a real gas, we would find a similar-looking result, but the exponent would not necessarily be 5/3. We could assume that the path was over states of the real gas that were low enough in pressure to allow the ideal equation of state to be a good approximation. But we could *not* assume the ideal gas heat capacities, no matter what the pressure, unless the gas were monatomic. The rare gases, of course, are monatomic at ordinary pressures and temperatures, and most metals vaporize as atoms. All other gases, and even some atomic metal vapors, have $\overline{C_V} > 3R/2$.

Anticipating a result from statistical mechanics, gaseous heat-capacity magnitudes can be classified into three categories, based on molecular structure:

Monatomic gases: $\quad\quad\quad\quad \overline{C_V} = \frac{3}{2}R$

TABLE 4.4 Heat-Capacity Function Coefficients

$\dfrac{\overline{C_P(T)}}{R} = a + bT + cT^2, \quad 300\,\text{K} \leq T \leq 1500\,\text{K}$

gas	a	$b/10^{-3}$ K^{-1}	$c/10^{-7}$ K^{-2}
H_2	3.496	-0.1006	2.419
N_2	3.245	0.7108	-0.406
O_2	3.067	1.637	-5.118
Cl_2	3.812	1.220	-4.856
Br_2	4.239	0.4901	-1.789
HCl	3.388	0.2176	1.860
CO_2	3.205	5.083	-17.13
H_2O	3.633	1.195	1.34
H_2S	3.213	2.870	-6.09
NH_3	3.114	3.969	-3.66
CH_4	1.701	9.080	-21.64
C_2H_2	3.689	6.352	-19.57
C_2H_4	1.424	14.39	-43.91
C_2H_6	1.131	19.22	-55.60
C_6H_6	-0.206	39.06	-133.0

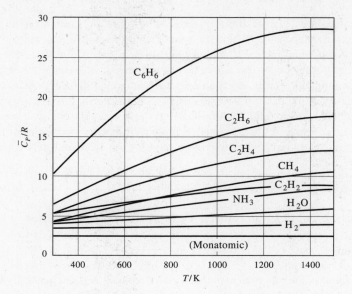

FIGURE 4.5 The heat capacity functions from Table 4.4 are graphed for several gases. Note how the larger molecules generally have larger heat capacities that vary most rapidly with temperature.

Linear polyatomics, composed of N atoms:

$$\overline{C}_V = \frac{5}{2} R \quad \text{at least, at low } T$$

increasing towards

$$\overline{C}_V = \left[\frac{5}{2} + (3N - 5)\right] R \quad \text{at high } T$$

Non-linear polyatomics, composed of N atoms:

$$\overline{C}_V = 3R \quad \text{at least, at low } T$$

increasing towards

$$\overline{C}_V = [3 + (3N - 6)] R \quad \text{at high } T.$$

For instance, acetylene, C_2H_2, is a linear molecule with $N = 4$ so that $\overline{C}_V(C_2H_2)$ should vary between $2.5 R$ and $9.5 R$. Experimental values for C_2H_2 increase from $3.50 R$ at 200 K to $7.91 R$ at 1500 K; this is a typical span within the allowed range for $\overline{C}_V$.

Since $\overline{C}_P = \overline{C}_V + R$ to a good approximation, the ratio $\gamma = \overline{C}_P/\overline{C}_V$ also falls into the same categories.

Monatomic gases:

$$\gamma = \frac{5}{3}$$

Linear polyatomics:

$$\underset{(\text{low } T)}{\frac{7}{5}} \geq \gamma \geq \underset{(\text{high } T)}{\frac{6N - 3}{6N - 5}}$$

Non-linear polyatomics:

$$\underset{(\text{low } T)}{\frac{4}{3}} \geq \gamma \geq \underset{(\text{high } T)}{\frac{3N - 2}{3N - 3}}.$$

For example, γ for acetylene ranges from 1.28 at 200 K *down* to 1.13 at 1500 K.

If γ could equal 1, the limiting ratio for a large polyatomic at high T, the reversible, adiabatic path, $PV^\gamma =$ constant, would be indistinguishable from the reversible, isothermal path, $PV =$ constant. (In fact, $\gamma > 1$ always.) Any gas with $\gamma < 5/3$ that is subjected to an adiabatic expansion (not necessarily reversible) will exhibit a temperature drop *less* than an ideal gas would. The reason is clear; the polyatomic molecule can tap internal storehouses of energy that are unavailable to monatomic real gases (and to the ideal gas) and use this energy to aid in the performance of expansion work. Internal molecular motion—rotations of the entire molecule and interatomic vibrations—account for the increased heat capacities of large molecules.

We can use Eq. (3.39) to compare $\overline{C}_P$ to $\overline{C}_V$ for condensed phases:

$$\overline{C}_P - \overline{C}_V = \frac{T\overline{V}\alpha^2}{\kappa}. \tag{3.39}$$

The coefficient of bulk thermal expansion, α, is zero at $T = 0$ and generally increases with temperature for both solids and liquids. Typical values are $\cong 10^{-5}$ K^{-1} for solids and $\cong 10^{-3}$ to 10^{-4} K^{-1} for liquids. The bulk isothermal compressibility, κ, is not very sensitive to temperature and is $\cong 10^{-6}$ atm^{-1} for most solid and liquid metals, but $\cong 10^{-4}$ atm^{-1} for molecular solids and liquids. The molar volume is related to the mass density, ρ, and the molecular weight, M, via $\overline{V} = M/\rho$.

With so many widely varying quantities, it is difficult to draw any general conclusions about $\overline{C}_P - \overline{C}_V$ for condensed phases. Instead, Table 4.5 collects the relevant data for five rather different materials: Ar(l), Fe(s), Hg(l), H$_2$O(l), and C$_6$H$_6$(l). As the last column shows, the $\overline{C}_P - \overline{C}_V$ values vary among themselves by nearly two orders of magnitude, although this difference is generally quite small for solids.

Finally, we consider how $\overline{C}_V$ varies from volume to volume and how $\overline{C}_P$ varies from pressure to pressure. If we differentiate $(\partial U/\partial V)_T$ with respect to T, we find

$$\frac{\partial}{\partial T}\left(\frac{\partial U}{\partial V}\right) = \frac{\partial^2 U}{\partial T \partial V} = \frac{\partial^2 U}{\partial V \partial T} = \frac{\partial}{\partial V}\left(\frac{\partial U}{\partial T}\right) = \left(\frac{\partial C_V}{\partial V}\right)_T$$

$\uparrow$ Expand the derivatives $\uparrow$ Swap order of differentiation $\uparrow$ Regroup the derivatives $\uparrow$ Since $C_V = (\partial U/\partial T)_V$

and, substituting Eq. (3.34) for $(\partial U/\partial V)_T$,

$$\frac{\partial}{\partial T}\left(\frac{\partial U}{\partial V}\right) = \frac{\partial}{\partial T}\left[T\left(\frac{\partial P}{\partial T}\right)_V - P\right] = T\left(\frac{\partial^2 P}{\partial T^2}\right)_V$$

TABLE 4.5 Data Relating to $C_P - C_V$

	T/K	$\overline{V}$/cm^3 mol^{-1}	α/K^{-1}	κ/atm^{-1}	$(\overline{C}_P - \overline{C}_V)/R$
Ar(l)	88.	28.74	4.51×10^{-3}	2.23×10^{-4}	2.81
Fe(s)	300.	7.1	3.51×10^{-5}	0.52×10^{-6}	0.0615
Hg(l)	273.	14.27	1.81×10^{-4}	3.83×10^{-6}	0.419
H$_2$O(l)	288.	18.00	2.1×10^{-4}	4.66×10^{-5}	0.061
C$_6$H$_6$(l)	298.	88.86	1.24×10^{-3}	9.21×10^{-5}	5.39

so that

$$\left(\frac{\partial C_V}{\partial V}\right)_T = T\left(\frac{\partial^2 P}{\partial T^2}\right)_V .\qquad(4.20)$$

We can find $(\partial C_P/\partial P)_T$ by a similar procedure starting from $(\partial H_P/\partial P)_T$:

$$\left(\frac{\partial C_P}{\partial P}\right)_T = -T\left(\frac{\partial^2 V}{\partial T^2}\right)_P .\qquad(4.21)$$

Equation (4.20) has implications for the van der Waals gas. Since this gas has $(\partial^2 P/\partial T^2)_V = 0$, C_V is independent of volume. But the van der Waals gas approaches the ideal gas at very large volumes. Therefore, we can conclude that C_V for a van der Waals gas should equal C_V for an ideal gas at *all* temperatures and volumes. Of course, when one models a *real* gas by the van der Waals *equation of state,* the real gas heat capacities must be used. All we have shown here is that the van der Waals equation of state does not predict a volume dependence to C_V.

This section has established the properties of C_V and C_P needed for entropy calculations. Before we can return to the entropy calculation itself and a discussion of the Third Law, we must consider the elementary properties of phase transitions.

4.4 A FIRST LOOK AT PHASE TRANSITIONS

Consider ice cubes floating in a glass of water. At a pressure of 1 atm, the temperature of the ice and the liquid is 272.15 K = 0 °C. Add a few more ice cubes, and, as long as both ice and liquid are present at equilibrium, the temperature does not change. Warm the glass with your hands, and some ice will melt, but the temperature is *still* 0 °C as long as both phases are present at equilibrium. But since such a warming *adds energy to the system in the form of heat ($dq \neq 0$) without changing the temperature ($dT = 0$)*, we must conclude that *the heat capacity ($C = dq/dT$) of a system containing two phases in equilibrium is infinite.*

Recall our definition of a reservoir at the end of Section 3.6. An ice water bath must be a reservoir at the temperature 273.15 K. Similarly, other solid–liquid or liquid–gas equilibria involving one pure compound can be used as reservoirs for whatever temperature is characteristic of the two phase equilibrium. This is put to practical use in a *two-phase calorimeter.* As its name implies, a calorimeter is a device for measuring the heat associated with any process. An ice water calorimeter is simply an equilibrium mixture of ice and water, thermally insulated from its surroundings, but in thermal contact with the system under study. For instance, we could place our gas cylinder-piston arrangement used to introduce work calculations (Figure 2.1) into an ice calorimeter and use the calorimeter to measure the heat associated with gas compressions under the isothermal condition $T = 273.15$ K.

Imagine transferring energy as heat to one mole of ice at 273.15 K and 1 atm until all the ice had melted, leaving one mole of water at 273.15 K and 1 atm. Since the process was isobaric ($P = 1$ atm = constant), we can write the path as

$$H_2O(s, 1 \text{ atm}, 273.15 \text{ K}) \rightarrow H_2O(l, 1 \text{ atm}, 273.15 \text{ K})$$

and the associated heat as

$$q = q_P = \Delta\overline{H} = \overline{H}(\text{water}) - \overline{H}(\text{ice}) .$$

The formal term for melting is *fusion,* and we call $\Delta\overline{H}$ for this type of path the *molar enthalpy change of fusion,* $\Delta\overline{H}_{\text{fus}}$. A pressure of 1 bar is the current standard reference pressure,[3] and we refer to enthalpy changes at 1 bar as standard enthalpy changes. Standard conditions are denoted by adding a superscript ○ (or sometimes ⊖) to the symbol. Experiments have found the *standard molar enthalpy change of fusion* for water to be $\Delta \overline{H}^{\circ}_{\text{fus}} = 6.008$ kJ mol^{-1}.

EXAMPLE 4.4

At 0 °C, the densities of ice and liquid water are

$$\rho(H_2O, s) = 0.9168 \text{ g cm}^{-3}$$

$$\rho(H_2O, l) = 0.9998 \text{ g cm}^{-3}.$$

The ice–water two phase calorimeter takes advantage of the difference in densities between ice and water to turn an energy transfer into a measurable change in the *volume* of the ice–water mixture. Relate this volume change to the corresponding energy transfer.

SOLUTION Since the molecular mass of H_2O is 18.02 g mol^{-1}, we can compute the molar volumes

$$\overline{V}(H_2O, s) = (18.02 \text{ g mol}^{-1})/(0.9168 \text{ g cm}^{-3}) = 19.66 \text{ cm}^3 \text{ mol}^{-1}$$

$$\overline{V}(H_2O, l) = (18.02 \text{ g mol}^{-1})/(0.9998 \text{ g cm}^{-3}) = 18.02 \text{ cm}^3 \text{ mol}^{-1}$$

or the molar volume change on fusion:

$$\Delta \overline{V}_{\text{fus}}(H_2O) = \overline{V}(H_2O, l) - \overline{V}(H_2O, s) = -1.63 \text{ cm}^3 \text{ mol}^{-1}.$$

This last value is the key; whenever one mole of ice melts, the volume of the equilibrium mixture decreases by 1.63 cm^3. But we also know that melting one mole of ice requires $(\Delta \overline{H}^{\circ}_{\text{fus}})(1 \text{ mol}) = 6008$ J. Thus, we can relate an observed volume change ΔV, either positive or negative, to the transferred energy:

$$q = \Delta V \left(\frac{6008 \text{ J}}{1.63 \text{ cm}^3} \right) = (\Delta V/\text{cm}^3) \times (3686 \text{ J}).$$

If ΔV is negative, q is negative, representing the energy transferred *from* the system *to* the ice calorimeter in the surroundings.

➠ RELATED PROBLEMS 4.13, 4.14

Similar arguments apply to boiling (vaporization), and one has *standard molar enthalpy changes of vaporization*, $\Delta \overline{H}^{\circ}_{\text{vap}}$. For water, $\Delta \overline{H}^{\circ}_{\text{vap}}(H_2O) = 40.66$ kJ mol^{-1} at the normal boiling point. The vaporization enthalpy $\Delta \overline{H}^{\circ}_{\text{vap}}$ is significantly greater than $\Delta \overline{H}^{\circ}_{\text{fus}}$. This difference can be traced to the unusually large intermolecular bonding energies in water. When ice melts, the orderly *pattern* of these bonds is disrupted, but the bonds *themselves* are not. Vaporization, however, breaks these intermolecular bonds (called *hydrogen bonds*), and $\Delta \overline{H}^{\circ}_{\text{vap}}$ is a measure of the energy needed to pull molecules one from another.

Some solids coexist at 1 atm in equilibrium with a gas phase rather than a liquid phase. An example is $CO_2(s)$, "dry ice." The solid → gas phase transition is called *sublimation*, and one can measure standard molar enthalpies of sublimation, $\Delta \overline{H}^{\circ}_{\text{sub}}$, in analogy with $\Delta \overline{H}^{\circ}_{\text{vap}}$ and $\Delta \overline{H}^{\circ}_{\text{fus}}$.

In addition to an enthalpy or energy change, phase transitions are also characterized by an entropy change. Since both phases are always in equilibrium during a

---

[3] By definition of both the atmosphere and bar units, 1 bar = 1/1.013 25 atm = 10^5 Pa.

phase transition at constant T and P, the heat associated with the transition is automatically a *reversible* heat: the system is never more than infinitesimally removed from equilibrium during the phase transition. If we use the subscript ϕ to represent a general phase transition,

$$\Delta S_\phi = \frac{q_P}{T_\phi} = \frac{\Delta H_\phi}{T_\phi}, \quad \text{the } \textit{entropy change for transition } \phi \ . \tag{4.22}$$

For water,

$$\Delta \overline{S}^\circ_{\text{vap}} = \frac{\Delta \overline{H}^\circ_{\text{vap}}}{T_{\text{vap}}} = 13.10\, R$$

and

$$\Delta \overline{S}^\circ_{\text{fus}} = \frac{\Delta \overline{H}^\circ_{\text{fus}}}{T_{\text{fus}}} = 2.645\, R \ .$$

In general, $\Delta \overline{S}^\circ_{\text{vap}} > \Delta \overline{S}^\circ_{\text{fus}}$, since solids and liquids have similar molecular order, but gases have considerably less.

The last type of phase transition we will consider now occurs between two distinct crystalline solid phases. For example, elemental sulfur exists in one atomic crystalline arrangement, known as a rhombic crystal phase, from $T = 0$ up to $T = 368.54$ K. At this temperature, sulfur exhibits a *solid–solid phase transition* to a different crystalline form known as the monoclinic phase. Calorimetric measurements have found

$$\Delta \overline{H}^\circ_\phi(\text{S(rhombic)} \to \text{S(monoclinic)}) = 401.7 \text{ mol}^{-1}$$

so that

$$\Delta \overline{S}^\circ_\phi = \frac{401.7 \text{ J mol}^{-1}}{368.54 \text{ K}} = 1.090 \text{ J mol}^{-1}\text{K}^{-1} = 0.1311\, R \ .$$

This small value for $\Delta \overline{S}^\circ_\phi$ indicates that the atomic arrangements in the two phases are not very different.[4]

4.5 THE THIRD LAW AND ABSOLUTE ENTROPY

A thermodynamic state is not easily characterized by a total energy, since only *differences,* rather than *absolute,* energies can be measured. For entropy, however, absolute quantities *can* be assigned to any state, and the Third Law shows the way. The Third Law is a summary of experimental observation, as are the other laws, but in a slightly different way. The Third Law, in one form, says something about systems at $T = 0$; in another form, it says that $T = 0$ is an unattainable state! As with the Second Law, the various ways of stating the Third Law all contribute to its full meaning. We begin with a statement most relevant to the concept of absolute entropy.

Think back to our pile of bricks system (two-level molecular system) introduced in Section 4.2. When all the bricks were on the floor or when all the two-level molecules were in their lowest energy state, our microscopic information about the

---

[4]These values are per mole of S atoms, but in both solid phases (and in the normal liquid phase), S exists as S_8 molecules. Thus, per mole of S_8 molecules, the values in the text should be multiplied by eight.

detailed configuration of the system was complete. In fact, if we set $n = 0$ in Eq. (4.5), we find

$$W(n = 0) = \frac{N!}{0!\,N!} = 1$$

and from Eq. (4.8),

$$\Delta S(n = 0) = k_B \ln W(n = 0) = 0 \;.$$

As we wrote Eq. (4.8), it was intended to represent a difference, $\Delta S(n) = S(n) - S(n = 0)$, and the zero in the above equation should not be such a surprise. But if we adopt $S = k_B \ln W$ as a general statement of entropy in terms of microscopic order, we are led directly to the conclusion that the state $n = 0$ is characterized by zero entropy. Remember, too, that this state is characterized by $T = 0$.

We can extend this idea to any molecular system in the following way. At $T = 0$, we can picture a perfect crystalline solid as one with maximum microscopic order. Every atom is in its place and atomic motion is at a minimum. (One of the curiosities of quantum mechanics is that atomic motion never ceases completely, even at $T = 0$.) Such an atomic configuration is microscopically simple, and we can specify the position and motion of each and every atom with a minimum of microscopic variables. Consequently, one statement of the Third Law is:[5]

> At absolute zero, the entropy of every perfectly crystalline equilibrium compound is zero.

A more precise way of saying the same thing is as follows:

> The entropy of any equilibrium system is zero for the state in which $(\partial U/\partial S)_V = 0$.

This statement allows us to include $He(l)$, which exists as a *liquid* at $T = 0$ and reduced pressures (including 1 atm). Thus the system need not be crystalline, but it does need to exhibit microscopic equilibrium in which every molecular compound is in its lowest energy state.

Another statement of the Third Law is in the spirit of Carathéodory:

> The isotherm $T = 0$ and the isentrope $S = 0$ are the same.

From this statement, another follows:

> It is impossible to attain absolute zero by any process with a finite number of steps.

To see that absolute zero is unattainable, we combine the Carathéodory statements of the Second and Third Laws. Experimentally, it has not been possible to attain bulk temperatures lower than about 10^{-5} K, and temperatures lower than 10^{-3} K can be attained only with difficulty.[6] (Nuclear magnetic systems have attained 10^{-7} K, but this is not a bulk, molecular temperature.) Thus, we could (in principle) make a reservoir for temperatures as low as, say, 10^{-5} K and cool other systems by equilibrating them with such a reservoir. To reach lower temperatures, we might use adiabatic processes, and to gain a maximum amount of cooling, each such path should be reversible. By the Second Law, no two reversible adiabatic paths cross. Thus, none of our cooling steps can be along the isentrope $S = 0$ nor can they cross over to this isentrope. We can move from isentrope to isentrope by constant

---

[5]This form of the Third Law was first suggested by Planck in 1913. Nernst had discussed a related but somewhat weaker statement in 1906, and it is he who is usually credited with introducing the Third Law.

[6]Methods of obtaining these temperatures are discussed in Chapters 6 and 9.

temperature processes (isotherms) that will allow us to lower the entropy and move onto a new isentrope. But by the Third Law, the isotherm $T = 0$ and the isentrope $S = 0$ are the same paths. Thus a non-zero temperature isotherm will not reach the isentrope $S = 0$ either. Therefore, no finite number of steps in any process, adiabatic or not, reversible or not, starting from any temperature, can reach a final temperature $T = 0$, even though the properties of matter are well-defined at this temperature.[7]

So we can't reach $T = 0$. But we can compute absolute entropies, tabulate them, and interpret their relative magnitudes for any one compound. Typical values of absolute entropies near room temperature span the range $\sim 5 < \overline{S}/R < \sim 50$. We can use these numbers and Eq. (4.6), the Boltzmann equation for entropy, to make a rather amazing statement about the molecular basis of entropy in equilibrium systems. Recall from Section 4.2 how W (in $S = k_B \ln W$) counted the number of microstates of the system—the number of ways the energy and configuration of the molecules in the system can be chosen to result in the given thermodynamic state. Since $R = N_A k_B$, we have, taking $\overline{S}/R \cong 20$,

W = # of microstates capable of giving the observed thermodynamic state
$$= e^{S/k_B} = e^{N_A \overline{S}/R} \sim \exp[(6 \times 10^{23})(20)] \cong 10^{5 \times 10^{24}}.$$

This incredibly huge number is at the heart of both the statistical basis for thermodynamics and the occurrence of spontaneous, irreversible processes.

In fact, this number is so big it points out how far from $T = 0$ *any* realizable temperature is. Suppose we cooled a crystal of NaCl to 10^{-5} K. The next section demonstrates how to turn the Debye heat capacity function into an absolute entropy. The result for NaCl with $\Theta_D = 281$ K is $\overline{S}/R = 3.5 \times 10^{-21}$ at $T = 10^{-5}$ K. This is a *very* small entropy, but it corresponds to $W \cong 10^{920}$! That is still a *very* long way from having all the bricks on the floor.

4.6 THE CALCULATION OF ABSOLUTE ENTROPIES

With the Third Law taking care of the value of entropy at $T = 0$ and various theories taking care of heat capacities near $T = 0$, we are all set to calculate absolute entropies by integrating Eq. (4.2):

$$dS_V = \frac{C_V(T)}{T} dT \quad \text{and} \quad dS_P = \frac{C_P(T)}{T} dT. \qquad (4.2)$$

Suppose we want to know the entropy change when a substance is taken at constant pressure from T_i to T_f. Then the integral of the second form of Eq. (4.2) becomes

$$\Delta S = S(T_f) - S(T_i) = \int_{T_i}^{T_f} \frac{C_P(T)}{T} dT.$$

This expression is incomplete if the substance undergoes phase transitions on the way from T_i to T_f. If a phase transition from, say, phase I to phase II is encountered at temperature T_ϕ, we alter the above expression to read

$$\Delta S = S(T_f) - S(T_i) = \int_{T_i}^{T_\phi} \frac{C_P^I(T)}{T} dT + \frac{\Delta H_\phi(\text{I} \to \text{II})}{T_\phi} + \int_{T_\phi}^{T_f} \frac{C_P^{II}(T)}{T} dT \qquad (4.23)$$

to include the phase transition entropy contribution.

---

[7]If you recall our argument that Nature cares more about $-1/T$ than about T, the state with $T = 0$ is more reasonably unattainable in that it corresponds to $-1/T \to -\infty$ and is thus a state "infinitely" far away from any state at a non-zero temperature.

EXAMPLE 4.5

Find the molar entropy change when $H_2O(l)$ at 1 atm and 300 K is taken to $H_2O(g)$ at 1 atm and 500 K. The $H_2O(g)$ heat-capacity function coefficients can be found in Table 4.4, while for the liquid

$$\frac{\overline{C_P}(H_2O(l))}{R} = 12.16 - 1.943 \times 10^{-2} T + 3.042 \times 10^{-5} T^2 .$$

SOLUTION In addition to the gas and liquid contributions, we expect a vaporization contribution at $T_\phi = 373.15$ K. From Table 4.4,

$$\frac{\overline{C_P}(H_2O(g))}{R} = 3.633 + 1.195 \times 10^{-3} T + 1.34 \times 10^{-7} T^2 ,$$

which can be substituted along with the liquid $\overline{C}_P$ function into Eq. (4.23) with $T_i = 300$ K, $T_\phi = 373.15$ K, and $T_f = 500$ K. The liquid contribution is

$$\frac{\Delta \overline{S}(l)}{R} = 12.16 \ln\left(\frac{373.15}{300}\right)$$

$$- 1.943 \times 10^{-2} (373.15 - 300)$$

$$+ 3.042 \times 10^{-5} \frac{[(373.15)^2 - (300)^2]}{2} = 1.98.$$

The vaporization contribution is

$$\frac{\Delta \overline{S}_{vap}}{R} = 13.10 ,$$

as noted in the text. The gas-phase contribution is

$$\frac{\Delta \overline{S}(g)}{R} = 3.633 \ln\left(\frac{500}{373.15}\right)$$

$$+ 1.195 \times 10^{-3} (500 - 373.15)$$

$$+ 1.34 \times 10^{-7} \frac{[(500)^2 - (373.15)^2]}{2} = 1.22 .$$

The total entropy change, which is largely due to the vaporization, is

$$\frac{\Delta \overline{S}}{R} = 1.98 + 13.10 + 1.22 = 16.30$$

$$= \frac{\overline{S}}{R}(H_2O, g, 500 \text{ K}, 1 \text{ atm}) - \frac{\overline{S}}{R}(H_2O, l, 300 \text{ K}, 1 \text{ atm}) .$$

If the explicit temperature dependence of the heat capacities had not been taken into account and only the 300 K and 500 K $\overline{C}_P$ values had been used as constants, we would have found an identical liquid phase contribution (to three-figure accuracy) and a gas phase contribution of 1.25 instead of 1.22. Thus, while the power-series expansion of $\overline{C}_P(T)$ would seem to be necessary only for the most accurate work or for very large temperature excursions, calculations with a power series are only slightly more difficult and should be used whenever the data are available.

➠ *RELATED PROBLEMS* 4.16, 4.17, 4.18

If we take $T_i = 0$ and include every phase transition entropy contribution, Eq. (4.23) yields the *absolute entropy* (also called the "Third Law entropy") of any pure compound at any temperature. To illustrate this calculation, we will use the classic 1929 data of Giauque and Johnston to find the absolute entropy of $O_2(g)$ at 300 K.

4.6 THE CALCULATION OF ABSOLUTE ENTROPIES

TABLE 4.6 Phase Transition Data for O_2

	T_ϕ/K	$(\Delta \overline{H}^\circ_\phi/R)/K$	$\Delta \overline{S}^\circ_\phi/R$
I → II	23.66	11.28	0.477
II → III	43.75	89.38	2.043
III → l	54.39	53.50	0.984
l → g	90.13	819.73	9.095

Solid O_2 exhibits three distinct crystalline phases, I, II, and III, with I the stable phase at lowest temperatures. Data for all O_2 phase transitions are given in Table 4.6, adapted from the original Giauque and Johnston paper.

The lowest temperature in their study was 12.97 K, where they measured $\overline{C}_P/R = 0.554$. For lower temperatures, we (as did they) assume that the Debye theory is applicable.[8] Writing Eq. (4.16) in the compact form

$$\frac{\overline{C}_P^\circ}{R} = aT^3$$

where $a = 12\pi^4/5\Theta_D^3$, we find the entropy contribution from $T = 0$ to $T = T'$ is

$$\frac{\overline{S}^\circ(T)}{R} = \int_0^{T'} \frac{aT^3}{T} dT = \int_0^{T'} aT^2 \, dT = \frac{aT'^3}{3} = \frac{\overline{C}_P^\circ(T')}{3R}.$$

Substituting O_2 data into this expression, we find

$$\frac{\overline{S}^\circ(12.97 \text{ K})}{R} = \frac{0.554}{3} = 0.185.$$

(Note that we write $\overline{S}^\circ$ rather than $\Delta \overline{S}^\circ$. The calculations are identical to those used for any entropy change, since it is a change which is being calculated, but the Δ symbol is omitted to emphasize that the result is an absolute entropy.)

From 12.97 K to the first phase transition, and between subsequent pairs of phase transitions, we will use the Giauque and Johnston tabulated C_P values in a graphical integration of Eq. (4.23). Figure 4.6 plots the integrand, $\overline{C}_P(T)/RT$, versus T for

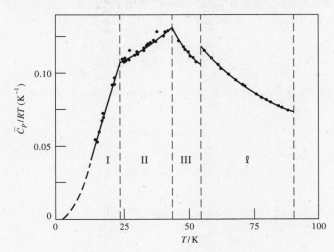

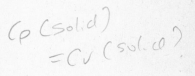

FIGURE 4.6 The absolute entropy of a substance at any temperature can be found from the area of a graph of $\overline{C}_P/T$ versus T. This figure plots data for O_2. The short dashed line is the Debye extrapolation to $T = 0$, and vertical dashed lines locate phase transition temperatures. Note that the large jump in $\overline{C}_P$ at the fusion temperature (III → l), indicative of the general trend that $C_P(l) > C_P(s)$. These data and those of Table 4.6 are from the classic paper, "The heat capacity of oxygen from 12°K to its boiling point and its heat of vaporization," by W. F. Giauque and H. L. Johnston *J. Am. Chem. Soc.* **51**, 2300 (1929).

---

[8] There is always a concern that another phase transition may be lurking at lower temperatures. Subsequent measurements on O_2 have shown that none exists.

each solid phase and for the liquid phase. The Debye extrapolation to $T = 0$ is shown by the dashed line, and each phase is noted over its range of stability.

We compute the various contributions to $\overline{S}°$ from $T = 0$ up to the vaporization temperature from the areas under each portion of the discontinuous curve in Figure 4.6:

0.185	(Debye extrapolation)
0.832	(I from 12.97 K to 23.66 K)
0.477	(I → II)
2.345	(II from 23.66 K to 43.75 K)
2.043	(II → III)
1.208	(III from 43.75 K to 54.39 K)
0.984	(III → l)
3.251	(l from 54.39 K to 90.13 K)
9.095	(l → g)
Total = 20.420 = $\overline{S}°(O_2(g), 90.13 \text{ K})/R$.	

To calculate the contribution as $O_2(g)$ is raised from 90.13 K to 300 K, we *cannot* use the power series coefficients of Table 4.4, since we are well outside their range of validity.[9] Instead we will use the simple approximation (which is very good here) that $C_P = C_V + R = 7R/2$, based on the discussion in Section 4.3. This approximation yields

$$\frac{\Delta \overline{S}°(g)}{R} = \int_{90.13 \text{ K}}^{300 \text{ K}} \frac{7}{2} \frac{dT}{T} = \frac{7}{2} \ln\left(\frac{300}{90.13}\right) = 4.209 .$$

The total absolute entropy at 300 K is

$$\overline{S}°(O_2(g), 300 \text{ K})/R = 20.420 + 4.209 = 24.629 .$$

Two corrections lead to a more accurate value. The first uses accurate gas heat-capacity data. The second depends on our intended use for the result, and the source of the more accurate gas data. Most gaseous data are tabulated by correcting real data to an ideal gas limit. This correction, which relies on a Maxwell relation we will meet in the next chapter, uses compressibility data to evaluate the isothermal entropy change from $P = 1$ atm to $P \to 0$. The Maxwell relation is

$$\left(\frac{\partial S}{\partial P}\right)_T = -\left(\frac{\partial V}{\partial T}\right)_P ,$$

and the path we follow to correct real data to ideal behavior takes the form, at constant T,

O_2(real, 1 atm)
↓
O_2(real, but behaving ideally, $P \to 0$)
↓
O_2(assumed ideal, $P \to 0$)
↓
O_2(ideal, 1 atm) .

---

[9] An integration based on that power series yields $\Delta \overline{S}°(g) = 4.011\ R$, which is too low because the series underestimates $\overline{C}_P/R$ for $T \ll 300$ K.

There is no entropy change in the second step. The combined entropy changes in the other two steps, with $(\partial V/\partial T)_P = R/P$ for the ideal gas step, is calculated from

$$\frac{\overline{S}°(\text{ideal, 1 atm})}{R} - \frac{\overline{S}°(\text{real, 1 atm})}{R} = \int_{P \to 0}^{1\,\text{atm}} \left[\left(\frac{\partial \overline{V}(\text{real})}{\partial T} \right)_P - \frac{R}{P} \right] dP \ .$$

For $O_2(g)$ at the boiling point, this correction is only $\overline{S}°/R = 0.09$.

CHAPTER 4 SUMMARY

We have now covered the Laws of Thermodynamics as they apply to pure compounds. This chapter introduced two major topics: absolute entropies and the variation of heat capacities from phase to phase, compound to compound, and state to state. Integration of Eq. (4.2) motivated us:

$$dS_V = \frac{C_V}{T} dT \quad \text{and} \quad dS_P = \frac{C_P}{T} dT$$

and we saw we had to consider the enthalpy and entropy effects of phase transitions as well:

$$\Delta S_\phi = \frac{q_P}{T_\phi} = \frac{\Delta H_\phi}{T_\phi}$$

We also discussed the basic magnitudes of important types of heat capacities:

$$\overline{C}_V(\text{solids}) \cong 3R \text{ per atom in the molecular formula}$$

$$\overline{C}_V(H_2O(l)) \cong \overline{C}_P(H_2O(l)) \cong 9R$$

$$\overline{C}_P(\text{gases}) \cong \overline{C}_V(\text{gases}) + R$$

along with some specific relations important to the Third Law calculation of absolute entropies:

Debye low temperature T-cubed law:
$$\frac{\overline{C}_V(\text{solids})}{R} = \frac{233.8}{\Theta_D^3} T^3$$

additional contribution for metals:
$$\frac{\overline{C}_V(\text{metal electrons})}{R} = \frac{T}{\Theta_{\text{el}}}$$

The next chapter puts the last three chapters together in a way that exposes one of the major powers of thermodynamics: the ability to predict a spontaneous process and to find the equilibrium point of complex processes.

FURTHER READING

For a comprehensive review of thermodynamic data sources, see E. F. G. Harrington, "Thermodynamics Quantities, Thermodynamic Data and Their Uses," in *Chemical Thermodynamics: Specialist Periodical Reports*, vol. 1, pp. 31–94, ed. by M. L. McGlashan (The Chemical Society, London, 1973). Another guide to the literature is *An Annotated Bibliography of Compiled Thermodynamic Data Sources for Biochemical and Aqueous Systems (1930 to 1975): Equilibrium, Enthalpy, Heat Capacity, and Entropy Data*, by George T. Armstrong and Robert N. Goldberg (Institute for Materials Research, National Bureau of Standards, Washington, 1976).

ACCESS TO DATA

D. R. Stull and G. C. Sinke, *Thermodynamic Properties of the Elements*, Advances in Chemistry Series, No. 18 (American Chemical Society, Washington, 1956).

D. D. Wagman, *et al.*, *Selected Values of Chemical Thermodynamic Properties*, National Bur. Stand. Tech. Note 270-3, 270-4, 270-5, 270-6, 270-7, and 270-8 (National Bureau of Standards, Washington, 1968–1981).

K. K. Kelley, *Contributions to the Data on Theoretical Metallurgy*, U.S. Bureau of Mines Bulletins. **476**(1949); **477**(1950); and **584**(1960).

F. D. Rossini *et al.*, *Selected Values of Physical and Thermodynamic Properties of Hydrocarbons and Related Compounds* (Carnegie Press, Pittsburgh, 1953).

CODATA Bulletin No. 28, *Recommended Key Values of Thermodynamics,* J. Chem. Thermo. **10** 903 (1978).

N. B. Vargaftik, *Tables on the Thermophysical Properties of Liquids and Gases*, 2nd ed. (John Wiley, New York, 1975).

F. D. Rossini, D. Wagman, W. Evans, S. Levine, and I. Jaffe, *Selected Values of Chemical Thermodynamic Properties*, National Bur. Stand. Circular 500 (National Bureau of Standards, Washington, 1952); periodic revisions as booklet supplements have appeared.

PRACTICE WITH EQUATIONS

4A What is the number of ways of fielding a baseball team of nine players from a roster of 24 players? (4.5)

ANSWER: 1 307 504

4B How many microstates are possible for a 1 ng sample of Ar(s) at 0.5 K, for which the molar entropy is only $1.25 \times 10^{-5} R$? (4.6)

ANSWER: $\sim 10^{8 \times 10^7}$

4C What is the energy separation ε of a two-level system that has 25% of its molecules in the upper level at $T = 145$ K? (4.11)

ANSWER: 2.2×10^{-21} J

4D The cerium ions in cerium ethyl sulfate behave as a two-level system with $\varepsilon/k_B = 6.7$ K. What contribution to $\overline{C_V}$ will these ions make at $T = 10$ K? (4.14)

ANSWER: 0.10 R

4E Which has the larger molar heat capacity, ZnO or ZnS? Table 4.1

ANSWER: ZnS

4F What is $\overline{C_V}$ for Pb at He(l) temperature, 4.2 K? (4.16), (4.17), Tables 4.2, 4.3

ANSWER: $2.69 \times 10^{-2} R$

4G By what factor does $\overline{C_P}(Br_2(g))$ exceed that of Ar at 500 K? (4.19a), Table 4.4

ANSWER: 1.78

4H The molar enthalpy of sublimation of CO_2 is 25.23 kJ mol^{-1} at the standard sublimation temperature, 194.6 K. What is the molar entropy of sublimation? (4.22)

ANSWER: 129.7 J mol^{-1} K^{-1}

4I What is the molar entropy change for C_2H_6(350 K, 1 atm) → C_2H_6(400 K, 1 atm)? (4.23), Table 4.4

ANSWER: 1.27 R

PROBLEMS

SECTION 4.1

4.1 Since entropy must be a finite number if it is to have absolute status, Eq. (4.3) sets some general restrictions on the temperature variability of heat capacities and the behavior of $\overline{C}$ at $T = 0$. Explore this by considering the three cases

(a) $\overline{C}$ = constant
(b) $\overline{C} = \alpha T$, α = a constant
(c) $\overline{C} = \beta/T$, β = a constant

with $T_i = 0$ K and T_f = some finite value.

SECTION 4.2

4.2 The Third Law has come under greater scrutiny and skepticism than the First or Second Laws. Part of the criticism is based around the argument that there might be "hidden" motions (degrees of freedom) associated with the incompletely understood structure of nuclei or atomic particles. Such motions, or so the argument goes, may not necessarily follow the rules for thermal equilibration that atoms and molecules follow. A more practical concern about the Third Law focuses on the many obvious "exceptions" to the Law that arise from incomplete thermal equilibration. For example, heteronuclear diatomics with chemically similar ends (such as CO) can form solids with randomly oriented directions, schematically something like

CO CO OC CO OC OC OC CO CO OC OC

instead of an energetically lower arrangement like

CO CO CO CO CO CO CO CO CO CO CO.

(While the molecules in CO(s) do not lie along a line, it is the idea of *configurational disorder* rather than structural detail that matters.) This disorder can be modeled as a two-level system: one level being the "correct," or energetically lowest, orientation and the other level, the "incorrect" one. As the solid freezes, assume that either level will be chosen by any one molecule with equal probability so that the populations of each level are equal. What "residual entropy" contribution will this configurational disorder make to the entropy at absolute zero? (The observed value for CO is 0.55 R.)

4.3 Here's a whimsical exercise that may or may not have validity. Consider the handedness of humans as a two-level system. Roughly 11% of the population is left handed. Assuming that handedness is determined by some unknown biochemical event and assuming that this event reaches equilibrium at conception (at 98.6 °F = 37 °C), what is the energy difference between the left- and right-handed outcomes of this event? (The answer is a chemically reasonable one, but the premise is certainly questionable.)

SECTION 4.3

4.4 Ferric methylammonium sulfate contains Fe^{3+} ions that form a *three-level* system at low temperatures.

Spectroscopic measurements can show the separations among these levels, but not necessarily the energy ordering. Given that the middle energy level is separated from the other two by $\varepsilon/k_B = 0.58$ K and $\varepsilon'/k_B = 1.05$ K, explain *qualitatively* how low-temperature heat capacity measurements could distinguish whether the middle level is 0.58 K or 1.05 K above the lowest level. The actual experiment is described in A. H. Cooke, H. Meyer, and W. P. Wolf, *Proc. Roy. Soc.* **A237**, 404 (1956).

4.5 Two samples, 1.00 mol each, one of crystalline B_2O_3 and the other of crystalline Al_2O_3, both at 300 K are subjected to the same heat in a particular process. Which compound has the higher final temperature? (Assume the heat is sufficient to raise the temperature less than 100 K.)

4.6 Equation (4.16) gives the low-temperature limit to the Debye expression for $\overline{C_V}$, valid when $T \ll \Theta_D$. A high temperature ($T > \Theta_D$) expression can be written as a power series:

$$\frac{\overline{C_V}}{3R} = 1 - \frac{1}{20}\left(\frac{\Theta_D}{T}\right)^2 + \frac{1}{560}\left(\frac{\Theta_D}{T}\right)^4 + \cdots$$

Compare the predictions of this expansion (and the electronic contribution, too!) to the following observed values at 298.15 K:

	Cu	Fe	Pb
$\overline{C_V}/R$	2.850	2.958	2.794

Your comparison should convince you that the simple rule $\overline{C_V} = 3R$ is about as accurate a predictor of reality as the simple Debye model at high temperatures.

4.7 W. H. Lien and N. E. Phillips (*Phys. Rev.* **133**, A1370 (1964)) measured the low-temperature heat capacities of K, Rb, and Cs. Some of their data for K are tabulated below. Use them to find Θ_D and Θ_{el} and compare your results to the values given in Tables 4.2 and 4.3. (*Hint*: How can you write $\overline{C}/R = T/\Theta_{el} + (12\pi^4/5)(T/\Theta_D)^3$ as the equation for a straight line? If you can't think of a way, look up the original paper.)

T/K	$\overline{C}/R \times 10^5$
0.3067	8.544
0.4578	14.53
0.6122	22.64
0.7236	30.20
0.8296	38.99
0.9334	49.11
1.013	58.92
1.101	70.77
1.180	82.68
1.238	92.56

4.8 If the heat capacity of a solid nonmetal is plotted versus T^3 at low T, a straight line results, according to the Debye expression. Against what power of T should the *entropy* of such a solid be plotted to yield a straight line? What about the *enthalpy*?

4.9 Guess values for $\overline{C_P}/R$ at 500 K for N_2, H_2O, NH_3, and C_2H_4 based on the limits discussed in the text, then calculate values based on data in Table 4.4 and compare.

4.10 What is $\overline{C_P} - \overline{C_V}$ for a hard-sphere gas? for a van der Waals gas at the critical point?

4.11 At what temperature does:

(a) Ag have the $\overline{C_V}$ value that Cu has at 4.2 K?
(b) H_2O have the $\overline{C_P}$ value that CO_2 has at 300 K?
(c) He have the $\overline{C_P}$ value that Ne has at 600 K?
(d) $H_2O(l)$ have $\overline{C_P} = \overline{C_V}$? (Look up the temperature dependence of the density of water and think about Eq. (3.39) if the answer here isn't apparent to you.)

SECTION 4.4

4.12 The enthalpy of vaporization of He(*l*), per cm³ of liquid, is about 3 J cm⁻³. Estimate the volume of He(*l*) needed to cool 1 g of Cu to He(*l*) temperature (4.2 K) from: (a) $H_2(l)$ temperature (20.3 K) and (b) $N_2(l)$ temperature (77.4 K). Your answer should convince you that cooling to very low temperatures is increasingly more efficient the lower the initial temperature. Hence, it is foolish to waste expensive He(*l*) to cool something from room temperature when much cheaper $N_2(l)$ can be used to prechill to 77 K.

4.13 A 100 g sample of BeO at 298 K is placed in a large amount of Hg at the fusion temperature of Hg, 234.28 K. The Hg contains both solid and liquid both before and after the BeO addition. Once the BeO has cooled to the Hg temperature, it is removed, and one finds that the Hg volume has increased by 1.37 cm³. Find $\Delta \overline{H}°_{fus}$ (Hg) from these data, given $\overline{C_P}$(BeO) = 2.76 R and, at 234.28 K, $\rho(Hg(s)) = 14.193$ g cm⁻³ and $\rho(Hg(l)) = 13.690$ g cm⁻³.

4.14 A 1.00 mol sample of ice and a 2.00 mol sample of water, both at 0 °C, are mixed in an adiabatic container. What is the mixture's volume? The 298 K BeO sample from the previous problem is added to the container, equilibrium is attained, and the sample is removed. Find the ice water's volume and composition (i.e., amounts of each phase) at the end of this experiment.

SECTION 4.6

4.15 The data below, taken from L. Pierce and E. L. Pace, *J. Chem. Phys.* **23**, 551 (1955), give heat-capacity and phase-transition data for NF_3. Use them to calculate the entropy of $NF_3(g)$ at the normal boiling point,

144.15 K. (The correction for nonideal gas behavior can be neglected. Pierce and Pace found it to be 0.04 R.)

solid-solid phase transition: $T_\phi = 56.52$ K

$\Delta \overline{H}_\phi = 361.8$ cal mol^{-1}

fusion transition: $T_{fus} = 66.37$ K

$\Delta \overline{H}_{fus} = 95.1$ cal mol^{-1}

vaporization transition: $T_{vap} = 144.15$ K

$\Delta \overline{H}_{vap} = 2769$ cal mol^{-1}

T/K	$\overline{C}_P$/cal mol^{-1} K^{-1}
12.67	2.098
24.61	6.596
36.71	9.526
44.8	10.95
54.44	12.79
56.52–66.37	16.55
70.1	17.27
85.69	16.99
89.75	16.90
109.62	16.68
124.72	16.82
142.21	17.40

4.16 Walther Nernst, who first discussed the Third Law and made seminal contributions to electrochemistry, also made low-temperature heat-capacity measurements such as the data below on benzene [(Z. Elektrochem. **17**, 265 (1911)].

T/°C	$\overline{C}_P/R$
−250.	1.57
−225.	3.56
−220.	4.87
−150.	6.67
−100.	8.91
−50.	11.7
0.	14.7

These data can be used in a graphical or numerical integration of Eq. (4.3) to yield the molar entropy change when benzene is raised from −250 °C to 0 °C. Note that one can either plot $\overline{C}_P/T$ versus T, as suggested by the way Eq. (4.3) is written, or, since $d(\ln T)/dT = 1/T$ so that $dT/T = d(\ln T)$, one can plot $\overline{C}_P$ versus $\ln T$. The area under either curve will be the entropy change. Apply both methods to these data and comment as to which produces the "simpler" curve to integrate graphically.

4.17 If "the entropy of every pure crystalline substance is zero at the absolute zero of temperature," then the entropies of different crystalline solid phases of any one substance should be zero at $T = 0$ as well. Check this corollary of the Third Law using heat-capacity data for S measured by E. D. Eastman and W. C. McGavock [J. Am. Chem. Soc. **59**. 145 (1937)]. As mentioned in the text, S(s) is stable below 368.54 K in the rhombic crystal form. The higher-temperature, monoclinic form, however, can be cooled rapidly through the transition temperature and studied as a metastable solid at low temperatures. The Eastman and McGavock data for rhombic S can be represented over the 15–370 K range by

$$\frac{\overline{C}_P^\circ}{R} = -0.08 + 2.17 \times 10^{-2}\, T - 6.2 \times 10^{-5}\, T^2 + 6.9 \times 10^{-8}\, T^3$$

while their monoclinic data are represented by

$$\frac{\overline{C}_P^\circ}{R} = 1.96 \times 10^{-2}\, T - 4.8 \times 10^{-5}\, T^2 + 4.7 \times 10^{-8}\, T^3.$$

(Note the absence of a constant term in the monoclinic expression. Such a term is statistically undetermined by a least-squares fit to the original data; for rhombic S, this term is small and just barely statistically determined—only one significant figure.) Find the molar entropy changes for each phase from 15 to 368.54 K by integration of $\overline{C}_P^\circ/RT$, and use the Debye relation to find $\overline{S}^\circ$ (15 K). Add these two contributions to obtain $\overline{S}^\circ(T_\phi)$ for each phase, subtract them, and compare your answer to the $\Delta \overline{S}_\phi$ value quoted in the text. Can you conclude that the Third Law applies to both phases of S? (The experimental uncertainty in $\overline{C}_P$ here leads to uncertainties in absolute entropies on the order of ±0.05 R.)

4.18 At the transition temperature, $T_\phi = 368.54$ K, $\Delta \overline{H}_{fus}^\circ/R = 48.31$ K for the S(rhombic) → S(monoclinic) transition. While the *entropies* of both phases are equal at $T = 0$ K (from the previous problem), the *enthalpies* are not, and $\Delta \overline{H}_{fus}^\circ \neq 0$. Use the heat capacity data from the previous problem to find $\Delta \overline{H}_{fus}^\circ$ at $T = 0$ K, and comment on the applicability of the relationship between $\Delta \overline{H}_{fus}^\circ$ and $\Delta \overline{S}_{fus}^\circ$, Eq. (4.22).

GENERAL PROBLEMS

4.19 The Debye heat capacity rises as T^3 at low temperatures. How does the heat capacity of a two-level system rise from $T = 0$? (Recall that this is the $T \ll \varepsilon/k_B$ limit, or the $x \to \infty$ limit in the notation of Eq. (4.14).)

4.20 For the two most important paths, we know that $\overline{C}_P/R = 5/2$ and $\overline{C}_V/R = 3/2$ for the ideal gas. Prove that the general heat capacity for an arbitrary reversible path is given by either of the following expressions:

$$\overline{C} = \left(\frac{\partial \overline{U}}{\partial T}\right)_{\overline{V}} + \left\{\left(\frac{\partial \overline{U}}{\partial \overline{V}}\right)_T + P\right\}\frac{d\overline{V}}{dT}$$

$$= \left(\frac{\partial \overline{H}}{\partial T}\right)_P + \left\{\left(\frac{\partial \overline{H}}{\partial P}\right)_T - \overline{V}\right\}\frac{dP}{dT}.$$

The details of the path are contained in the total derivatives $d\overline{V}/dT$ and dP/dT. What is $\overline{C}$ for an ideal gas along the paths:

(a) $\overline{V}T^{1/2}$ = constant?

(b) $\overline{V}T^2$ = constant?

(c) $\overline{V}T^{3/2}$ = constant?

What does a negative heat capacity mean? (You may find it helpful to plot these paths on a (V, T) diagram along with the simple isochoric path V = constant.) Where have you seen path (c) before?

4.21 In his book *Thermodynamics and Statistical Mechanics,* which is Volume V of his *Lectures on Theoretical Physics,* Arnold Sommerfeld quotes a 1938 note by Robert Emden published in *Nature,* **141,** 908 (1938) and titled, "Why do we have Winter Heating?" In it, Emden writes:

> The layman will answer: "To make the room warmer." The student of thermodynamics will perhaps so express it: "To impart the lacking (inner, thermal) energy." If so, then the layman's answer is right, the scientist's, wrong.

Emden then goes on to show in three simple equations that the energy per unit volume of an ideal gas is independent of the temperature at constant pressure. He concludes by writing:

> Then why do we have heating? ... Our conditions of existence require a determinate degree of temperature, and for the maintenance of this there is needed not an addition of energy but addition of entropy.

But in commenting on this, Sommerfeld shows a flaw in Emden's derivation. Sommerfeld concludes that the energy density, rather than staying constant, *decreases* with increasing temperature. Show that both conclusions are wrong by showing that $(\partial(U/V)/\partial T)_P$ for an ideal gas is essentially *indeterminant*. Recall that U is a function of only T for an ideal gas and that we may write $U(T) = U_0 + C_V T$ where U_0 is some (unknown and unknowable) reference energy. You should find that the algebraic sign of $(\partial(U/V)/\partial T)_P$ is opposite to the algebraic sign of U_0. Emden assumed $U_0 = 0$, and Sommerfeld assumed U_0 to be positive. It is, however, true that the ratio $\Delta U/\Delta V$ is independent of temperature for an isobaric temperature change in an ideal gas. Show this as well.

Stability, Free Energy, and Thermodynamic Potentials

5.1 Conditions for Stable Equilibrium

5.2 Free Energy

5.3 Multicomponent Systems and Chemical Potential

5.4 The Calculation of Chemical Potentials

5.5 A Summary of Thermodynamic Calculus

THE previous four chapters introduced nearly all the basic ideas and equations of thermodynamics. Now that the fundamental laws have been discussed, this and the following five chapters focus on important chemical applications of these laws. Here we will look at the mathematics behind the laws, introduce several convenient definitions that arise from them, and begin to discuss the variables that are necessary in the description of mixtures of compounds. We will find new ways of stating stability and spontaneity criteria in terms of new state functions, and we will begin to see how to calculate the most important of these functions, the Gibbs free energy.

5.1 CONDITIONS FOR STABLE EQUILIBRIUM

In Chapter 3, the direction for spontaneous change and the location of an equilibrium point were both described in terms of entropy. A spontaneous process continues until the total entropy has increased as much as any external constraints allow:

$$dS_{sys} > 0 \quad \text{(spontaneous process, isolated system)}$$

or

$$dS_{tot} = dS_{sys} + dS_{surr} > 0 \quad \text{(spontaneous process, nonisolated system)}.$$

The latter case is more usual, and the inequality can be written as

$$dS_{sys} > -dS_{surr}$$

or as

$$dS_{sys} > \frac{dq}{T_{surr}} \quad (5.1)$$

since $-dq/T_{surr} = dS_{surr}$. The surroundings are a reservoir of energy, and dq(entering the system) $= -dq$(leaving the surroundings).

Since $dq = dU - dw$,

$$dS_{sys} > \frac{dU_{sys}}{T_{surr}} + \frac{P_{ext}\,dV}{T_{surr}}$$

if only mechanical work is involved. This inequality can be written more compactly as

$$T_{surr}\,dS > dU + P_{ext}\,dV \quad \text{for a system moving spontaneously to a new equilibrium} \quad (5.2)$$

where, from now on, quantities without subscript labels will refer to system state variables. For an isolated system, $dU = dV = 0$, leading to $dS_{U,V} > 0$, as deduced above.

Equation (5.2) is the general expression from which the direction of spontaneity, the location of equilibrium, and the stability of the system at equilibrium can all be deduced. At equilibrium, $T_{surr} = T$ and $P_{ext} = P$, and the inequality becomes the equality of Eq. (3.26), the master equation:

$$T\,dS = dU + P\,dV \quad \text{at equilibrium.} \quad (3.26)$$

Equation (5.2) incorporates a variety of ways in which the system could be in contact with the surroundings. If the process constrains the system to a constant volume and a constant (system) entropy, $(dV = dS = 0)$, then

$$dU_{V,S} < 0. \quad (5.3)$$

Under these constraints, the internal energy will *decrease as much as possible*. Contrast this to the *different* constraints $dU = dV = 0$ that showed that the entropy will *increase* as much as possible: $dS_{U,V} > 0$.

The constraints of constant volume and entropy are not the most common for spontaneous processes. These are the constraints of systems that are isolated and closed from contact with their surroundings, or of processes that proceed adiabatically and reversibly at constant volume. Most systems of chemical interest are coupled to their surroundings by one of the following mechanisms:

- external forces (such as pressure or external fields)
- thermal contact (such as with a reservoir at constant T)
- material contact (such that the system exchanges matter with the surroundings)
- reactive contact (in which an exchange of matter influences chemical reactions within the system).

When we introduced enthalpy in Section 2.6, we did so primarily to give us an auxiliary state function that was convenient for constant pressure heat calculations. Now we can see a more fundamental reason for inventing enthalpy. Since $H = U + PV$, we have

$$dH = dU + d(PV) = dU + PdV + VdP$$

or

$$dU = dH - PdV - VdP \; .$$

If this expression is substituted into Eq. (5.2), we find

$$T_{\text{surr}}\, dS > dH - PdV - VdP + P_{\text{ext}}\, dV \; .$$

For constant pressure and constant entropy constraints ($dP = dS = 0$ and $P_{\text{ext}} = P$),

$$dH_{P,S} < 0 \; . \tag{5.4}$$

Under *these* constraints, the *enthalpy* (rather than the internal energy) decreases as much as possible during a spontaneous process.

Neither Eq. (5.3) nor Eq. (5.4) expresses the two most common constraints. Rather than constant entropy, it is experimentally more convenient and common to constrain a system to a constant temperature. At constant temperature and either constant volume or pressure, it is possible for a spontaneous process to occur during which neither U nor H decrease. For such constraints we invent two more auxiliary state functions.

Consider first $dT = dV = 0$. We want a function that *decreases* during a spontaneous process subjected to these constraints. The tendency of U to decrease and S to increase suggests we try the combination $U - TS$. (We have the factor of T for dimensional harmony; we cannot add energy and entropy directly.) If we define an auxiliary state function

$$A = U - TS \tag{5.5}$$

then

$$dA = dU - d(TS) = dU - TdS - SdT$$

or

$$dU = dA + TdS + SdT \; .$$

Substituting this expression into Eq. (5.2) gives

$$T_{\text{surr}}\, dS > dA + TdS + SdT + P_{\text{ext}}\, dV$$

and, when we apply the constraints $dT = dV = 0$ and $T = T_{\text{surr}}$ (*thermal* equilibrium with the surroundings, but not necessarily *total* equilibrium), we find

$$dA_{T,V} < 0 \; . \tag{5.6}$$

This new function is the *Helmholtz free energy*. This intriguing term, free energy, is explained in detail in the next section.

For systems constrained to a constant pressure and temperature, we define our final auxiliary state function as

$$G = H - TS \tag{5.7}$$

for which

$$dG = dH - d(TS) = dH - TdS - SdT$$
$$= dU + PdV + VdP - TdS - SdT$$

or

$$dU = dG - PdV - VdP + TdS + SdT \; .$$

Substituting into Eq. (5.2) gives

$$T_{\text{surr}}\, dS > dG - P\, dV - V\, dP + T\, dS + S\, dT + P_{\text{ext}}\, dV$$

and, when the system is in *thermal and mechanical equilibrium with the surroundings* ($dT = dP = 0$ and $T = T_{\text{surr}}$ and $P = P_{\text{ext}}$) but not necessarily at total equilibrium (with respect to matter exchange or ongoing spontaneous chemical reactions), we find

$$dG_{T,P} < 0 \ . \tag{5.8}$$

This function is the *Gibbs free energy*.[1]

To summarize, we have turned the general criterion of total entropy increase into four appropriately constrained statements that point to the direction of spontaneous change:

$$dU_{V,S} < 0 \tag{5.3}$$

$$dH_{P,S} < 0 \tag{5.4}$$

$$dA_{T,V} < 0 \tag{5.6}$$

$$dG_{T,P} < 0 \ . \tag{5.8}$$

We have *not* learned anything new. All we have done is use some algebra (formally called "Legendre transformations") to turn U, S, T, P, and V into other functions that are convenient to use. These functions, U, H, A, and G, are defined for *any* transformation due to *any* process. The *inequalities* listed above, however, are guaranteed to be true *only* under the appropriate constraints.

These are directions for spontaneous changes. Equilibrium is a stable condition, corresponding to one (or more) of these functions reaching some sort of minimum and unchanging value. In a gravitational field, things "roll down hill" until they reach the bottom and stop, at some local minimum of gravitational potential energy. In the same way, we can think of U, H, A, and G as *thermodynamic potential energy functions*. Just as it takes work to push something back up a hill in a gravitational potential, it takes work to move a system from an equilibrium state to another state of greater thermodynamic potential. In summary:

> Equilibrium is that thermodynamic state for which the appropriate thermodynamic potential, depending on constraints, is at a minimum value. Any displacement of the system from that constrained state requires an input of energy as work, a removal of one or more constraints, or a transfer of matter across the boundary.

A Closer Look at Stability

Consider the final state of two identical metal cubes (a and b) initially at different temperatures but allowed to equilibrate to the same temperature and pressure (as was done in Chapter 3). Suppose we now take these cubes, at equilibrium, and alter the constraints from those of constant U and P to, for instance, constant S and P, and ask if any new spontaneous process can happen once the entropy has been fixed. To do this, we imagine connecting the cubes, as in Figure 5.1, through some sort of engine that transfers energy from one cube to the other. These transfers can be in the form of heat or adiabatic work, as long as the entropy sum, $S_a + S_b$, stays the same. We might expect that this engine will have to transfer energy from (or to) the cubes to (or from) the surroundings in order to maintain $S_a + S_b$ constant.

---

[1] The Gibbs free energy has also been called the "free enthalpy" and symbolized F, especially in physics and older chemical literature.

FIGURE 5.1 Two identical cubes at the same temperature are connected to an energy transfer engine, but are otherwise isolated. The engine can transfer energy among the cubes and the surroundings, but the transfers are always such that the entropy $S_a + S_b$ is constant.

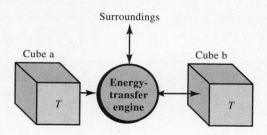

For simplicity, we assume that the cubes' temperatures may change, but that their volumes stay the same. What we must compute under these constraints is $dH_{S,P}$. If this is negative, then a spontaneous process will occur as a result of changing the constraints from $dU = dP = 0$ to $dS = dP = 0$. If $dH_{S,P} = 0$, then the process can occur, but the system remains in equilibrium; no work would be required for the process to occur. (The mechanical analogy here is sliding a mass across a horizontal table without friction.) If $dH_{S,P} > 0$, the process is not spontaneous; the system is stable under the new constraints.

To keep S constant, we must have

$$0 = dS = dS_a + dS_b = \frac{C_P}{T_a} dT_a + \frac{C_P}{T_b} dT_b .$$

This shows that the temperature changes of the two cubes are not independent:

$$dT_a = -\frac{T_a}{T_b} dT_b .$$

We also have

$$dH = dH_a + dH_b = C_P (dT_a + dT_b)$$

or, using the temperature relation above,

$$dH_{S,P} = C_P \left(1 - \frac{T_a}{T_b}\right) dT_b .$$

Initially, $T_a = T_b = T$. Suppose the temperature of cube b increases. Then $dT_b > 0$ and $T_a/T_b < 1$. These conditions make $dH_{S,P} > 0$. The same conclusion is reached if $dT < 0$ (and $T_a/T_b > 1$). Thus we can conclude that the system *is* stable, not only under constant U and P, but also under constant S and P constraints.

When a system in stable equilibrium requires work to move it to another state of equilibrium, it is clear that the reverse process, the spontaneous approach to equilibrium, can be characterized by the possible work that can be done by the system. The next section discovers what this work is and describes why the term "free energy" is used in connection with it.

5.2 FREE ENERGY

Free energy is a very descriptive term for a somewhat subtle concept. It is of considerable practical interest to be able to calculate the maximum amount of work that can be associated with any constant-temperature process, and free energy is related to this work.

We find a clue to this calculation if we write our main inequality, Eq. (5.2), to include all forms of work and to include equilibrium:

$$T_{surr}\, dS \geq dU - dw. \qquad (5.9)$$

The equals sign applies at equilibrium (and $T_{surr} = T$) and

$$dw = -P_{ext}\, dV + dw'.$$

Here we have divided work into two components: $-P_{ext}\, dV$, the work done in moving or otherwise deforming the boundaries of the system, and dw', which lumps together all other forms of work. If we rearrange Eq. (5.9) slightly:

$$\underbrace{T_{surr}\, dS - dU}_{\substack{\uparrow \\ \text{State-variable} \\ \text{dependent}}} \underbrace{\geq}_{\substack{\uparrow \\ \text{Process} \\ \text{dependent}}} \underbrace{P_{ext}\, dV - dw'}_{\substack{\uparrow \\ \text{Work done by} \\ \text{the system}}} \qquad (5.10)$$

we have what we need. The right-hand side is the work done *by* the system (a process variable). The left-hand side is state-variable dependent. The middle part, $\geq$, is process-dependent, in the sense that equality is established by a *reversible* process (for which $T_{surr} = T$ and $P_{ext} = P$). The inequality is established by *any* irreversible process. Thus, to *maximize* the work done *by* the system, we should establish the equality and consider reversible isothermal processes.

If the volume and entropy of the system are held constant,

$$-\Delta U_{S,V} = -w'_{S,V} = \text{maximum work done } by \text{ the system}. \qquad (5.11)$$

As the system falls down this thermodynamic potential, the released energy is available for useful work. But the maximum amount of this work is equal to ΔU *only* under the constraints of constant S, V, and reversibility.

If T and V had been constant, then we would conclude

$$T\Delta S_{V,T} - \Delta U_{V,T} = -w'_{V,T}$$

but, since $A = U - TS$, this is equivalent to

$$-\Delta A_{V,T} = -w'_{V,T}. \qquad (5.12)$$

To maintain constant T, energy is tied up in the system or wasted as useless transfers of energy as heat. (If S is constant, as above, and the process is reversible, then there is no heat: $dq_{rev} = TdS = 0$.)

It is easy to show that

$$-\Delta H_{S,P} = -w'_{S,P} \qquad (5.13)$$

and

$$-\Delta G_{T,P} = -w'_{T,P}. \qquad (5.14)$$

The minus signs in Equations (5.11)–(5.14) have been left in to emphasize that the motivation for this section was the discovery of the maximum work that could be done *by* a system under various conditions.

You might find it helpful at this point to review Section 2.3 on the variation of work with path. We showed how the work for the quasi-static isothermal expansion path of Figure 2.4(b) was greater (in magnitude) than for any other isothermal expansion path between the same two states. (Compare Figures 2.4(b) and 2.5.)

Those arguments were rather intuitive and were limited to mechanical work. Now we can discuss the same process in terms of the Helmholtz free energy. At constant temperature, and with only PdV work, we integrate

$$TdS - dU = PdV = -dw.$$

Since $dA = dU - TdS$, this equation is equivalent to

$$dA_T = dw = -PdV$$

or to

$$\Delta A_T = -\int_{V_i}^{V_f} P(V)\, dV = w\ . \tag{5.15}$$

Compare this equation to Eq. (2.9); they are the same, but the point of view is different. This equation shows how a free-energy function can be used to incorporate *both* internal energy *and* entropy effects into one compact statement.

A Closer Look at the Meaning of Work, w

Referring to Equations (5.11)–(5.14), you may find yourself asking, "How can a system constrained to have a constant volume and entropy produce *any* work?" In the figure below, we have a variant of the Joule apparatus of Figure 2.6. In this variant, the total system is divided into two volumes, V_I and V_{II}, by a frictionless sliding partition. The total volume, $V = V_I + V_{II}$, is fixed (there's one constraint) and the system is adiabatically isolated (there's *almost* the second constraint). If, instead of a free expansion, we withdraw the partition reversibly (there's the second constraint—a reversible, adiabatic process is isentropic), the system will perform work. This work, strictly speaking, falls into the w' category rather than the PV category, since an *internal* partition was moved. But practically speaking, we calculate w' by an integral over PdV, and the distinction is not very important. According to Eq. (5.11), the work will equal ΔU.

For constant T and V, we need only immerse the system in a constant temperature reservoir and remove its adiabatic blanket. Then $w' = \Delta A$.

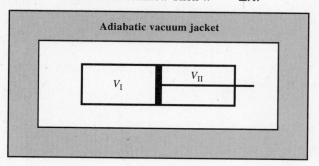

To perform work at constant S and P, we imagine a situation similar to that of Figure 5.1. If we fuel our Carnot engine by finite reservoirs at *different* temperatures, T_a and T_b, as in Figure 5.2, we can run the engine (producing work) until the temperatures are equal. This final temperature, T_f, will *not* be $(T_a + T_b)/2$, as it was in Section 3.6 (see Eq. (3.15)), since the system is *losing* energy as work. From Eq. (5.13), $w' = \Delta H$. To calculate ΔH, we need to find T_f. For constant entropy,

$$0 = dS = dS_a + dS_b = \frac{C_P}{T_a}dT_a + \frac{C_P}{T_b}dT_b$$

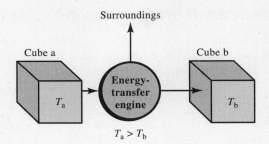

FIGURE 5.2 As in Figure 5.1, two identical cubes are connected to an engine. Here, the cubes are initially at *different* temperatures ($T_a > T_b$), and the engine reversibly transfers energy as heat from cube a to cube b and as work from the engine to the surroundings.

or

$$\frac{dT_a}{T_a} = -\frac{dT_b}{T_b}$$

which can be integrated (from T_a to T_f on the left, from T_b to T_f on the right), giving

$$\ln\left(\frac{T_f}{T_a}\right) = -\ln\left(\frac{T_f}{T_b}\right)$$

or, exponentiating both sides,

$$T_f = (T_a T_b)^{1/2} \ . \tag{5.16}$$

This is an interesting intermediate result. A reversible, adiabatic (isentropic) thermal equilibration should have a *lower* final temperature than a spontaneous, irreversible, thermal equilibration, since the Carnot engine exchanges energy between the surroundings and the system. The mathematics support this suspicion: $(T_a T_b)^{1/2}$ is always less than $(T_a + T_b)/2$.

The total ΔH is

$$\Delta H = \Delta H_a + \Delta H_b = C_P(T_f - T_a) + C_P(T_f - T_b) \ .$$

Using Eq. (5.16) for T_f, we find

$$\Delta H_{S,P} = C_P[2(T_a T_b)^{1/2} - T_a - T_b] = -C_P(T_a^{1/2} - T_b^{1/2})^2 \ .$$

This is a negative quantity and is equal in magnitude to the work done by the Carnot engine.

Finally, we come to the chemically important constraints of constant T and P. One system that produces work at constant T and P is very familiar to you: an ordinary chemical battery. If potentially reactive chemicals are held isolated (as in jars on the stockroom shelf) and then freely mixed (as in an open flask or beaker), any energy change brought about by the mixing and/or the resulting reaction is released (or consumed) as heat, rather than work. (Of course, the heat can be used to fuel a work-producing engine, but the reactive system itself does no work.) This is the chemical analog of the free Joule expansion or the spontaneous thermal equilibrations we have discussed in detail.

A chemical battery keeps the reactants largely isolated, but the reaction occurs by means of an external electrical circuit. These very important reactions are discussed in Chapter 10. Devices connected to the external circuit produce electrical work by means of the energy, the *free* energy, released by the reactive system as it undergoes a controlled fall down its Gibbs free-energy thermodynamic potential. The maximum work is $-\Delta G_{T,P}$, the quantity of central concern to the next five chapters. At chemical equilibrium, the battery has "run down" and "gone dead," at the bottom of its thermodynamic potential.

5.3 MULTICOMPONENT SYSTEMS AND CHEMICAL POTENTIAL

Imagine two systems, I and II, each at the same temperature and pressure. One *new* way we can allow these systems to interact is to allow them to *exchange molecules*. Simple examples of such interactions include:

System I		System II
liquid water	in contact with	water vapor
a container of hydrogen	in contact with	a container of deuterium
$Cu(s)$	in contact with	$Ag(s)$
oil	floating on	water

In the first example, we have a simple two-phase equilibrium in which each phase is a separate subsystem. The boundary between them is the open surface of the liquid. In the second example, the two isotopes of hydrogen spontaneously mix, as will Cu and Ag in the third example. In the fourth, the liquid–liquid interface is a boundary across which very little molecular transfer occurs. The oil and water stay relatively pure.

To discuss equilibrium in such situations, we turn first to the hydrogen isotope example. Since the process occurs at constant T and P, the appropriate thermodynamic potential is G. Since G is extensive,

$$G(\text{total}) = G_I + G_{II}$$

where G_I and G_{II} are the Gibbs free energies of each system at any point in the process. (We imagine the two containers connected by a valve, as in Figure 5.3, which we can close at any time to establish equilibrium in each container.) In addition to the T and P dependence of G, which is now of secondary importance, we should expect concentration dependencies:

$$G_I = G_I(n_1, n_2) \quad \text{and} \quad G_{II}(n_1, n_2)$$

where n_1 and n_2 are the numbers of moles of H_2 and D_2 in each container at any point. Since it is an experimental fact that the isotopes will spontaneously mix, we expect $G(\text{total})$ to *decrease*. We write

$$dG_I = \left(\frac{\partial G_I}{\partial n_1}\right)_{T,P,n_2} dn_1 + \left(\frac{\partial G_I}{\partial n_2}\right)_{T,P,n_1} dn_2 \quad (5.17)$$

and similarly for dG_{II}. Since the total amount of each isotope is constant, conservation of mass implies

$$dn_1(\text{container I}) = -dn_1(\text{container II})$$

and

$$dn_2(\text{container I}) = -dn_2(\text{container II}) \ .$$

We can combine these statements into the expression

$$dG = dG_I + dG_{II}$$

$$= \left[\left(\frac{\partial G_I}{\partial n_1}\right)_{T,P,n_2} - \left(\frac{\partial G_{II}}{\partial n_1}\right)_{T,P,n_2}\right] dn_1(\text{container I}) + \quad (5.18)$$

$$\left[\left(\frac{\partial G_I}{\partial n_2}\right)_{T,P,n_1} - \left(\frac{\partial G_{II}}{\partial n_2}\right)_{T,P,n_1}\right] dn_2(\text{container I}) \ .$$

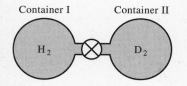

FIGURE 5.3 If a valve connecting a container of H_2 to a container of D_2 is opened, spontaneous mixing of the gases will begin and will continue until either the valve is closed or the chemical potential of each gas is uniform throughout both containers.

5.3 MULTICOMPONENT SYSTEMS AND CHEMICAL POTENTIAL

The process continues until G stops decreasing, that is, until $dG = 0$. This can happen in two ways. If $dn_1 = dn_2 = 0$, $dG = 0$. This happens when we close the connecting valve. With the valve open, the only way to reach equilibrium is to have both terms in square brackets in Eq. (5.18) vanish, which happens when

$$\left(\frac{\partial G_\mathrm{I}}{\partial n_1}\right)_{T,P,n_2} = \left(\frac{\partial G_\mathrm{II}}{\partial n_1}\right)_{T,P,n_2} \text{ and } \left(\frac{\partial G_\mathrm{I}}{\partial n_2}\right)_{T,P,n_1} = \left(\frac{\partial G_\mathrm{II}}{\partial n_2}\right)_{T,P,n_1}. \quad (5.19)$$

These partial derivatives are the key to equilibrium. Dimensionally, they are molar energies. For a one-component system, we would write

$$G(T, P, n) = n\overline{G}(T, P)$$

where

$$\overline{G}(T, P) \equiv \frac{G(T, P, n)}{n} = \left(\frac{\partial G}{\partial n}\right)_{T,P}. \quad (5.20)$$

The derivatives in Eq. (5.19) are the logical extensions of Eq. (5.20) to systems of more than one component. We call them *partial molar quantities*, and write

$$\overline{G}_1 = \left(\frac{\partial G}{\partial n_1}\right)_{T,P,n_2} \text{ and } \overline{G}_2 = \left(\frac{\partial G}{\partial n_2}\right)_{T,P,n_1} \quad (5.21)$$

where the subscript on $\overline{G}$ indicates the component of interest. Note that a partial molar derivative is evaluated *at constant T and P and with the number of moles of all components held fixed except for the one of interest.*

Thermodynamics, as you have no doubt noticed by now, has a propensity for highly decorated symbols. The partial molar Gibbs free energy plays such an important role that it is given a special symbol and a special name:

$$\overline{G}_i = \mu_i = \text{the } \textit{chemical potential} \text{ of component } i. \quad (5.22)$$

Though we do not yet know how to calculate or measure a chemical potential, we have shown that the spontaneous mixing of H_2 and D_2 will *continue until the chemical potential of either isotope in system I is equal to the chemical potential of that isotope in system II*. Just as the master equation, $dU = TdS - PdV$, indicated that T and P must be uniform to establish equilibrium in a one-component, one-phase system, we can now state that *uniform chemical potential for each component in all phases is an additional requirement for equilibrium in a many component and/or many phase system.*

There is nothing magical about hydrogen isotopes in this respect. We could repeat our argument for each example system and reach similar conclusions. Liquid water and water vapor in equilibrium have *equal chemical potentials*. The oil in the mostly oil phase has the same chemical potential as the oil in the mostly water phase in the last example.

Equation (5.22) defines the chemical potential as the partial molar Gibbs free energy. However, if we repeated the arguments leading to Eq. (5.22) using other systems of more than one component but constrained in different ways, we would find equations similar to Eq. (5.18). From each of these, we would identify a different expression for the chemical potential. Equation (5.23) summarizes all the equivalent definitions of the chemical potential of component i:

$$\mu_i = \left(\frac{\partial U}{\partial n_i}\right)_{S,V,n_j} = \left(\frac{\partial H}{\partial n_i}\right)_{S,P,n_j} = \left(\frac{\partial A}{\partial n_i}\right)_{V,T,n_j} = \left(\frac{\partial G}{\partial n_i}\right)_{T,P,n_j} = -T\left(\frac{\partial S}{\partial n_i}\right)_{U,V,n_j}. \quad (5.23)$$

where the subscript n_j is shorthand for "evaluate the derivative holding constant the amounts of all components except component i." Note that, by the definition of a partial molar quantity, μ_i is the partial molar derivative of G *only*, since the other derivatives do not specify constant T and P.

A Closer Look at the Uniformity of Chemical Potential

Imagine a complex system comprising c components all distributed throughout a total of ϕ different phases. All phases are assumed to be in either direct or indirect contact so that any one component has the opportunity to be found in any phase, but we assume the system is closed so that the total amount of each component is fixed.

The master equation for any one phase, say phase α, is

$$dU^\alpha = T^\alpha dS^\alpha - P^\alpha dV^\alpha + \sum_{i=1}^{c} \mu_i^\alpha dn_i^\alpha .$$

Since we are out to *prove* that the temperature, pressure, and chemical potential of component i are the same in each phase, we must assume that each phase *could* have unique temperatures, pressures, and individual chemical potential; thus, we *superscript* these variables by a phase label: $\alpha, \beta, ..., \phi$. The entropies, volumes, and compositions of each phase also require these labels, since these are the independent variables.

The total internal energy of the system is the sum of the internal energies of each phase:

$$\begin{aligned} dU &= dU^\alpha + dU^\beta + \cdots + dU^\delta + \cdots + dU^\phi \\ &= T^\alpha dS^\alpha - P^\alpha dV^\alpha + \mu_1^\alpha dn_1^\alpha + \cdots + \mu_c^\alpha dn_c^\alpha \\ &+ T^\beta dS^\beta - P^\beta dV^\beta + \mu_1^\beta dn_1^\beta + \cdots + \mu_c^\beta dn_c^\beta \\ &+ \cdots \\ &+ T^\delta dS^\delta - P^\delta dV^\delta + \mu_1^\delta dn_1^\delta + \cdots + \mu_c^\delta dn_c^\delta \\ &+ \cdots \\ &+ T^\phi dS^\phi - P^\phi dV^\phi + \mu_1^\phi dn_1^\phi + \cdots + \mu_c^\phi dn_c^\phi . \end{aligned}$$

At equilibrium, U is at a minimum *if* the system is constrained to constant entropy, volume, and total composition, which means

$$dU_{S, V, n_1, n_2, \ldots, n_c} = 0$$

if

$$\begin{aligned} dS &= dS^\alpha + dS^\beta + \cdots + dS^\delta + \cdots + dS^\phi = 0, \\ dV &= dV^\alpha + dV^\beta + \cdots + dV^\delta + \cdots + dV^\phi = 0, \text{ and} \\ dn_i &= dn_i^\alpha + dn_i^\beta + \cdots + dn_i^\delta + \cdots + dn_i^\phi = 0, \; i = 1, 2, \ldots, c . \end{aligned}$$

The full expression for dU contains $c + 2$ independent variables ($n_1, ..., n_c$ plus S and V) in each of the ϕ rows as written above, for a total of $\phi(c + 2)$ independent variables. The equilibrium constraining equations reduce the number of those variables that are truly independent. Since there are a total of $c + 2$ constraining equations above, we can rewrite the full expression for dU in terms of $c + 2$ *fewer* variables.

For instance, suppose we choose to solve the total entropy constraint for dS^α,

$$dS^\alpha = -dS^\beta - dS^\gamma - \cdots - dS^\phi,$$

and similarly for dV^α, dn_1^α, dn_2^α, etc. When we substitute these expressions into the full expression for dU, we find

$$dU_{S, V, n_1, \ldots, n_c} = 0$$
$$= (T^\beta - T^\alpha)\,dS^\beta - (P^\beta - P^\alpha)\,dV^\beta + (\mu_1^\beta - \mu_1^\alpha)\,dn_1^\beta + \cdots + (\mu_c^\beta - \mu_c^\alpha)\,dn_c^\beta + \cdots$$
$$+ (T^\phi - T^\alpha)\,dS^\phi - (P^\phi - P^\alpha)\,dV^\phi + (\mu_1^\phi - \mu_1^\alpha)\,dn_1^\phi + \cdots + (\mu_c^\phi - \mu_c^\alpha)\,dn_c^\phi .$$

Each variable (S^β, V^β, ..., n_c^ϕ) in this equation is truly independent. The only way for this large sum of terms to be zero and retain the independence of these variables is for *each* coefficient of each independent variable's differential to be zero:

$$T^\beta - T^\alpha = 0, \quad P^\beta - P^\alpha = 0, \quad \ldots, \quad \mu_c^\phi - \mu_c^\alpha = 0 .$$

This establishes the general criteria of equilibrium among many components in many phases:

1. uniform temperature: $T^\alpha = T^\beta = \cdots = T^\phi = T$
2. uniform pressure: $P^\alpha = P^\beta = \cdots = P^\phi = P$
3. uniform chemical potential of each component throughout the system:

$$\mu_1^\alpha = \mu_1^\beta = \cdots = \mu_1^\phi = \mu_1,$$
$$\mu_2^\alpha = \mu_2^\beta = \cdots = \mu_2^\phi = \mu_2,$$
etc. to
$$\mu_c^\alpha = \mu_c^\beta = \cdots = \mu_c^\phi = \mu_c.$$

We will return to these equations in Section 6.6 when we discuss the possible number of phases that can coexist in a system of many components.

5.4 THE CALCULATION OF CHEMICAL POTENTIALS

The last section showed how chemical potentials indicate equilibrium in many component (or many phase) systems. This section establishes ways of calculating them. We look first at general expressions for μ_i and then concentrate on specific expressions for ideal model systems as well as real systems.

If we start with the master equation for a one-component system and add $d(PV - TS)$ to both sides:

$$dU + d(PV - TS) = TdS - PdV + d(PV - TS)$$
$$= TdS - PdV + PdV + VdP - TdS - SdT$$
$$= VdP - SdT$$

we arrive at the total differential of G in terms of its natural variables, P and T:

$$d(U + PV - TS) = dG = VdP - SdT . \qquad (5.24)$$

As usual, we also write the mathematically formal total derivative of G as a function of P and T,

$$dG = \left(\frac{\partial G}{\partial P}\right)_T dP + \left(\frac{\partial G}{\partial T}\right)_P dT , \qquad (5.25)$$

and compare coefficients of dP and dT in Eq. (5.24) to those in Eq. (5.25):

$$\left(\frac{\partial G}{\partial P}\right)_T = V \quad \textbf{(always positive)} \qquad (5.26)$$

and

$$\left(\frac{\partial G}{\partial T}\right)_P = -S \quad \text{(always negative)} . \qquad (5.27)$$

These show that for a *one component system* G always *increases* with increasing P at constant T (Eq. (5.26), plus the fact that $V > 0$) and that G always *decreases* with increasing T at constant P (Eq. (5.27), plus the Third Law, which ensures $S \geq 0$).

For a multicomponent system, the chemical potential of a component follows the partial molar version of Eq. (5.24):

$$d\mu_i = \overline{V}_i \, dP - \overline{S}_i \, dT . \qquad (5.28)$$

We can also make connections between partial derivatives of μ_i and the coefficients $\overline{V}_i$ and $-\overline{S}_i$, as we did in arriving at Equations (5.26) and (5.27), but the signs of these derivatives are not so clearly defined. This is worth looking at in more detail.

The partial molar volume, $\overline{V}_i = (\partial V/\partial n_i)_{T, P, n_j}$, has an interesting physical interpretation that allows its experimental evaluation. In words, it means: "Take a system of fixed composition and measure its volume. Then add a very small amount of component i and measure the volume again, being sure to make both measurements at the same T and P. Then evaluate $\overline{V}_i$ by the ratio

$$\frac{V(\text{after addition}) - V(\text{before addition})}{(\text{\# of moles of } i \text{ added})}$$

and repeat with smaller and smaller additions of component i until this ratio, within experimental error, does not change." (Exactly the same type of experiments would lead to the partial molar entropy, etc. Just substitute, for instance, the word "entropy" into the preceding statements wherever "volume" appears.)

The interesting point is that the volume change can be either positive *or* negative! Some mixtures shrink noticeably when one component's amount is increased.[2] Microscopic rationalizations of this behavior come to mind easily. One can imagine one component interacting strongly with another, causing the other to cluster tightly around the first. (The phase—gas, liquid, or even solid—is irrelevant.) Just a pinch more of that one component could draw many otherwise disinterested molecules together in a more densely-packed configuration that would lower the volume. This is nonideal behavior in the sense that differences in intermolecular forces account for the unexpected volume change.

One other general expression, which has its origins in pure mathematics but makes sense on an intuitive level, is the relationship among G, μ_i, and n_i. (The same relationship also holds among V, $\overline{V}_i$, and n_i, or S, $\overline{S}_i$, and n_i, etc.) This expression, for a total of c components, is

$$G(\text{total}) = \mu_1 n_1 + \mu_2 n_2 + \cdots + \mu_c n_c = \sum_{i=1}^{c} \mu_i n_i . \qquad (5.29)$$

In differential form, it becomes

$$dG(\text{total}) = \sum_{i=1}^{c} (\mu_i \, dn_i + n_i \, d\mu_i) = \sum_{i=1}^{c} \mu_i \, dn_i + \sum_{i=1}^{c} n_i \, d\mu_i . \qquad (5.30)$$

---

[2]One well-studied example is $MgSO_4$ in water. Small, highly charged ions tend to form solutions with negative partial molar volumes, and $MgSO_4$ solutions shrink on addition of as much as 0.1 mol per liter of water.

Equation (5.30) can be written in an interesting way with particular meaning at equilibrium. We substitute Eq. (5.28) for $d\mu_i$

$$dG = \sum_{i=1}^{c} [\mu_i \, dn_i + n_i (\overline{V}_i \, dP - \overline{S}_i \, dT)] \;,$$

collect a few terms:

$$dG = \sum_{i=1}^{c} \mu_i \, dn_i + \left[\sum_{i=1}^{c} n_i \overline{V}_i\right] dP - \left[\sum_{i=1}^{c} n_i \overline{S}_i\right] dT \;,$$

and recognize that the quantities in square brackets are just V and S (in analogy with Eq. (5.29)) so that

$$dG = V \, dP - S \, dT + \sum_{i=1}^{c} \mu_i \, dn_i \;. \tag{5.31}$$

Equating the expression for dG in Eq. (5.30) to that in Eq. (5.31) yields

$$V \, dP - S \, dT - \sum_{i=1}^{c} n_i \, d\mu_i = 0 \;. \tag{5.32}$$

When we consider the impact of this expression on equilibrium, we reach two conclusions. The first is trivial: *at* equilibrium, P, T, and all μ_i are constant, so that $dP = dT = d\mu_i = 0$, making Eq. (5.32) true. But the second and more important conclusion is that Eq. (5.32) represents a *constraining relation* among P, T, and μ_i in the sense that evolution of a system from one equilibrium state to another *cannot* be made by *arbitrary* changes in P, T, and μ_i. In particular, if P and T are held constant, Eq. (5.32) simplifies to a constraint among the μ_i's only:

$$\sum_{i=1}^{c} n_i \, d\mu_i = 0 \;.$$

This is known as the Gibbs–Duhem expression, and we will find use for it in our discussion of mixtures in the next chapter.[3]

Now we look at the calculation of chemical potential in some simple systems, starting from Eq. (5.28) and integrating.

One-Component Ideal Gas—In General

With only one component, the chemical potential is just the ordinary molar Gibbs free energy: $\mu_i = \overline{G} = G/n$. We evaluate $\Delta\mu$ as

$$\Delta\mu = \int d\mu = \int_{P_i}^{P_f} \overline{V}(T_i, P) \, dP - \int_{T_i}^{T_f} \overline{S}(T, P_f) \, dT \tag{5.33}$$

for a "change P, then change T" path. The first integral is easy:

$$\int_{P_i}^{P_f} \overline{V}(T_i, P) \, dP = \int_{P_i}^{P_f} \frac{RT_i}{P} \, dP = RT_i \ln\left(\frac{P_f}{P_i}\right) \;.$$

The second integral is less easy. We lack an expression for the entropy of an ideal gas as a function of T and P. Thermodynamics gives this very important expression only to within an unknown constant:

$$\frac{\overline{S}(T, P)}{R} = \ln\left[\frac{T^{5/2}}{P} \, (\text{constant})\right] \;.$$

---

[3]Some authors call Eq. (5.32) the Gibbs–Duhem equation. The equation here is just the $dP = dT = 0$ result of Eq. (5.32).

The constant cancels in calculations of entropy *changes,* but we cannot ignore it here. The Third Law helps quite a bit; experimental absolute entropies of the rare gases at low pressure should give us this constant. Such experiments show that these entropies depend on the chemical identity of the rare gas. How do we work chemical identity into this "constant"? Around 1911, Sackur showed that it depends on *molecular mass* in the following way:

$$\frac{\overline{S}(T, P)}{R} = \ln\left[\frac{T^{5/2}M^{3/2}}{P} \text{(another constant)}\right] .$$

This final constant is truly universal and was first deduced by Tetrode in 1912. A correct derivation of this entire expression, known as the Sackur–Tetrode equation, requires quantum and statistical mechanics (Chapter 23). If we express T in K, P in atm, and M in g mol^{-1}, the Sackur–Tetrode equation is

$$\frac{\overline{S}(T, P)}{R} = \ln\left[\frac{T^{5/2}M^{3/2}}{P} (0.311\,968 \text{ g}^{-3/2} \text{ mol}^{3/2} \text{ K}^{-5/2} \text{ atm})\right] . \quad (5.34)$$

Using this in the second integral of Eq. (5.33) gives

$$-\int_{T_i}^{T_f} \overline{S}(T, P_f)\, dT = -R\left[T_f \ln\left(\frac{CM^{3/2} T_f^{5/2}}{P_f}\right) - T_i \ln\left(\frac{CM^{3/2} T_i^{5/2}}{P_f}\right)\right] - \frac{5}{2} R (T_i - T_f)$$

where C is the Sackur–Tetrode constant, $0.311\,968$ g$^{-3/2}$ mol$^{3/2}$ K$^{-5/2}$ atm.

Adding the two integrals and combining similar terms yields

$$\Delta\mu = RT_i \ln\left(\frac{CM^{3/2} T_i^{5/2}}{P_i}\right) - RT_f \ln\left(\frac{CM^{3/2} T_f^{5/2}}{P_f}\right) + \frac{5}{2} R (T_f - T_i) . \quad (5.35)$$

which equals $\Delta\overline{H} - \Delta(T\overline{S})$.

One-Component Ideal Gas—The Standard State

If we take an ideal gas at 1 bar and any temperature and change the pressure only, Eq. (5.35) simplifies to

$$\mu(T, P) = \mu°(T) + RT \ln(P/1 \text{ bar}) \quad (5.36)$$

where P is the final pressure and T is the uniform temperature throughout the process. $\mu°(T)$ is *the standard state ($P = 1$ bar) chemical potential at temperature T.* Since $\mu°(T)$ is an energy without absolute status, it is *not* calculable. Instead, it forms a reference point for measuring *changes* in μ at constant T.

The 1 bar standard pressure[4] is explicitly included in the logarithm term. Since one cannot evaluate the logarithm of a dimensioned quantity, P must be expressed in bar units, and the 1 bar denominator formally removes the dimensions. However, rather than write this 1 bar factor again and again, we will just write ln P, but you must remember that *all* such terms require that P is expressed in bar units.

One-Component Real Gas—Fugacity

The first integral leading to Eq. (5.35) was easy to do because we had the simple integrand $\overline{V} = RT/P$. For a nonideal gas, we could use a model equation of state for $\overline{V}(T, P)$ or resort to numerical integration of tabulated $\overline{V}$ data:

$$\mu(T, P) = \mu°(T) + \int_{1 \text{ bar}}^{P} \overline{V}(T, P)\, dP .$$

---

[4]The former standard pressure, 1 atm, is virtually the same as the 1 bar standard pressure (1 atm = 1.013 25 bar). For all but the most exacting work, either atm or bar units for P may be used.

5.4 THE CALCULATION OF CHEMICAL POTENTIALS

This approach, while perfectly correct, is awkward. The simplicity of Eq. (5.36) is very attractive, and, to preserve this simplicity, G. N. Lewis made a very clever suggestion in 1901. He defined a quantity that he called the *fugacity*, symbolized f.

You probably do not use this word in your day-to-day speech, but, unlike the words energy, entropy, and enthalpy, fugacity is an ordinary (if now obsolete) English word: it is the defining attribute of a fugitive! The driving force for spontaneous processes produced by chemical potential differences was described as an "escaping tendency" early in the development of chemical thermodynamics. Fugacity was a natural term at the time.

Lewis defined fugacity as

$$d\mu = \frac{RT}{f} df = RT\, d\ln f \quad \text{(at constant } T\text{)} . \tag{5.37}$$

Note that fugacity has the dimension of pressure and will be expressed in bar units (but see the previous footnote concerning bar and atm units). We integrate Eq. (5.37) from a very low pressure, P_0, to a final pressure P:

$$\mu(P, T) - \mu(P_0, T) = RT \ln f - RT \ln P_0 .$$

(At low enough pressures, all gases follow the ideal gas equation of state, and the fugacity becomes the actual pressure.) We add to this the chemical-potential difference for taking an ideal gas from P_0 to 1 bar from Eq. (5.36),

$$\mu(P_0, T) - \mu°(1 \text{ bar}, T) = RT \ln P_0 ,$$

and obtain

$$\mu(P, T) - \mu°(1 \text{ bar}, T) = RT \ln f . \tag{5.38}$$

Now *all* the nonideality is associated with f. The standard state, $\mu°$, is that of the real gas *behaving ideally at 1 bar*. This is hypothetical, but perfectly acceptable, since $\mu°$ does not appear in the final expression for any practical calculation.

Equation (5.38) is central to chemical thermodynamics. The monotonic relationship between chemical potential and fugacity means the equilibrium condition of equal potentials of any component in all phases can be replaced by a statement of *uniform fugacity*. (In multicomponent-multiphase systems, each *component* may have a unique fugacity, but each has that unique fugacity *in all phases*.)

To calculate the fugacity of a pure compound, we start with the equivalent expressions for $d\mu$ at constant T,

$$d\mu_T \quad = \quad \overline{V}\, dP \quad = \quad \frac{RT}{f} df,$$

$$\uparrow \qquad\qquad \uparrow$$

$$(\partial\mu/\partial P)_T\, dP \qquad \text{Definition of fugacity}$$

subtract $(RT/P)\, dP = RT\, d\ln P$ from both sides,

$$\left(\overline{V} - \frac{RT}{P}\right) dP = RT\,(d\ln f - d\ln P) ,$$

and integrate from $P = 0$ to the pressure of interest, P:

$$RT \ln\left(\frac{f}{P}\right) = \int_0^P \left(\overline{V} - \frac{RT}{P}\right) dP . \tag{5.39}$$

If we write the molar volume in terms of the compressibility factor, $\overline{V} = Z(P)RT/P$, Eq. (5.39) becomes

$$\ln\left(\frac{f}{P}\right) = \int_0^P \frac{Z(P) - 1}{P} dP . \tag{5.40}$$

A table of Z values at various pressures can be fit to a $Z(P)$ function or other numerical methods can be used to evaluate the integral. If P is so low that the virial expansion through the second virial coefficient is accurate, the integral can be approximated

$$\int_0^P \frac{Z-1}{P} dP \cong \int_0^P \frac{B(T)}{P\overline{V}} dP \qquad \text{(since } Z - 1 \cong B(T)/\overline{V}\text{)}$$

$$= \frac{B(T)}{RT} \int_0^P \frac{RT}{P\overline{V}} dP \qquad \text{(since } B \text{ is independent of } P\text{)}$$

$$= \frac{B(T)}{RT} \int_0^P \frac{1}{Z} dP \qquad \text{(since } Z = P\overline{V}/RT\text{)}$$

$$\cong \frac{B(T)}{RT} \int_0^P \frac{1}{1 + B(T)/\overline{V}} dP \qquad \text{(expanding } Z\text{)}$$

$$\cong \frac{B(T)}{RT} \int_0^P dP = \frac{B(T) P}{RT} \qquad \text{(assuming } 1 \gg B(T)/\overline{V}\text{)}$$

so that

$$\underset{\underset{\text{Fugacity}}{\uparrow}}{f} \cong \underset{\underset{\substack{\text{Real} \\ \text{pressure}}}{\uparrow}}{P} \underbrace{\exp\left[\frac{B(T) P}{RT}\right]}_{\underset{\substack{\text{Nonideal} \\ \text{correction}}}{\uparrow}}. \tag{5.41}$$

If $BP/RT \ll 1$, we can expand the exponential ($\exp(x) \cong 1 + x$, $x \ll 1$):

$$f \cong P\left(1 + \frac{B(T) P}{RT}\right) \cong P\left(1 + \frac{B(T)}{\overline{V}}\right) \cong PZ .$$

The simple approximation, $f = PZ$, connects the fugacity's deviation from the real pressure to the behavior of Z above and below the Boyle temperature:

$$f \overset{>}{\underset{<}{=}} P \quad \text{as} \quad T \overset{>}{\underset{<}{=}} T_B , \quad \text{the Boyle temperature.}$$

We can use the van der Waals equation of state for a more complete picture of fugacity. The van der Waals fugacity expression is derived in Problem 5.20, and in terms of the dimensionless variables of Eq. (1.25), the result is

$$\ln\left(\frac{f}{P}\right) = \ln\left[\frac{\tilde{T}}{\tilde{P}(\tilde{V} - 1)}\right] + \frac{1}{(\tilde{V} - 1)} - \frac{2}{\tilde{T}\tilde{V}} . \tag{5.42}$$

(Note: That's the real pressure, P, on the left!) This fugacity is graphed for several values of $\tilde{T}$ in Figure 5.4. Recall from Section 1.3 that the critical point is $\tilde{P} = 1/27$ and $\tilde{T} = 8/27$ (the lowest isotherm of the figure). Since the real critical P and T are often well above 1 atm and near or below room temperature (see Table 1.3), most of the graph covers quite high pressures. Note how $f \cong P$ at high temperatures, but how f is generally lower than P at lower temperatures.

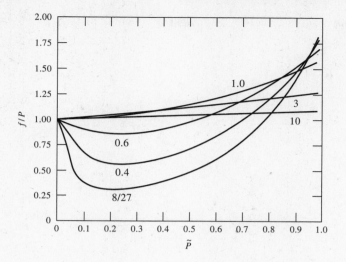

FIGURE 5.4 A graph of the van der Waals fugacity coefficient f/P versus $\tilde{P}$ at the $\tilde{T}$ values indicated shows increasing nonideality at high $\tilde{P}$ for all $\tilde{T}$ and at all $\tilde{P}$ for $\tilde{T}$ near the critical temperature of 8/27.

EXAMPLE 5.1

Consider $CO_2(g)$ at 320 K, only 16 K above the critical temperature. Compressibility data at this temperature can be represented by the following polynomial:

$$Z(P, 320\text{ K}) = 1.000 - (4.007 \times 10^{-3}\text{ atm}^{-1})P - (1.16 \times 10^{-5}\text{ atm}^{-2})P^2 \\ + (5.5 \times 10^{-8}\text{ atm}^{-3})P^3 - (1.65 \times 10^{-9}\text{ atm}^{-4})P^4 \ .$$

Graph f versus P using Eq. (5.40) with this polynomial to find $f(P)$, and compare your results to the approximation $f \cong PZ$ and the "ideal fugacity," $f = P$.

SOLUTION If we write the $Z(P)$ polynomial in general as $Z(P) = 1 + A_1P + A_2P^2 + A_3P^3 + A_4P^4$, the integral in Eq. (5.40) becomes

$$\int_0^P \frac{Z-1}{P}\,dP = \int_0^P (A_1 + A_2P + A_3P^3 + A_4P^3)\,dP \ ,$$

which can be integrated easily to give

$$f(P) = P \exp\left(A_1P + \frac{A_2P^2}{2} + \frac{A_3P^3}{3} + \frac{A_4P^4}{4}\right) \ .$$

This function is graphed in Figure 5.5 (solid line) along with the approximation $f \cong PZ$ (dashed line) and the (more difficult to calculate) van der Waals approximation (dot-dashed line), using parameters from Table 1.2. Note how the $f \cong PZ$ approximation grossly underestimates the true f at high pressures.

▶ *RELATED PROBLEMS* 5.14, 5.15

The ratio f/P is a convenient dimensionless measure of nonideality known as the *fugacity coefficient*:

$$\Phi = \frac{f}{P} \ . \tag{5.43}$$

In the next chapter we will define fugacity and fugacity coefficients for the components of a mixture. The most important use for fugacities is in explaining and predicting equilibrium among many components in many phases, the topic of Chapter 6, and in reacting mixtures, Chapter 7.

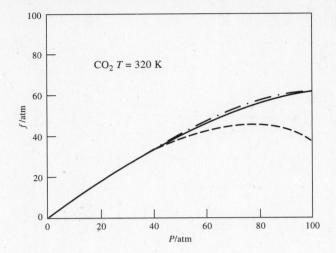

FIGURE 5.5 The experimental fugacity of CO_2 at 320 K (solid line) is compared to the approximation $f = PZ$ (dashed line) and the van der Waals approximation (dot-dash line). The ideal fugacity is the straight line $f = P$ (not shown).

We conclude this chapter with a summary of the major relations among the state functions that describe one phase, one component systems. Of the myriad equations we have encountered so far, you should memorize only a very few definitions and a very few mathematical techniques. Couple these to a physical understanding of paths, processes state variables, and process variables, and new equations can be derived as needed.

5.5 A SUMMARY OF THERMODYNAMIC CALCULUS

These are the key definitions and mathematics of thermodynamics, largely devoid of physical insight. Without the insight, you may know the words of thermodynamics, but you surely will not be able to carry the tune. Insight comes from practice. While these equations have appeared elsewhere, we list them together here with simple consecutive numbering to emphasize the logical progressions and relations among them.

We need

$$\text{First Law:} \quad dU = dq + dw \tag{1}$$

$$\text{Second Law:} \quad dS = \frac{dq_{rev}}{T} \tag{2}$$

and the definitions

$$dq_{path} = C_{path}\, dT \tag{3}$$

$$dw_{rev} = -P\, dV. \tag{4}$$

From these we substitute (4) and (2) into (1) to obtain the *master equation*

$$dU = T\,dS - P\,dV. \tag{5}$$

(The physical insight largely ends here. The rest is mathematics.)

Next come the definitions

$$H = U + PV \tag{6a}$$

$$A = U - TS \tag{7a}$$

$$G = H - TS \tag{8a}$$

5.5 A SUMMARY OF THERMODYNAMIC CALCULUS

or their differential equivalents, simplified by using (5),

$$dH = dU + PdV + VdP = TdS + VdP \tag{6b}$$

$$dA = dU - TdS - SdT = -PdV - SdT \tag{7b}$$

$$dG = dH - TdS - SdT = VdP - SdT. \tag{8b}$$

Now we use the properties of an exact, total differential in terms of the natural variables to derive

- first: the partial derivatives of state functions in terms of the natural variables, and
- second: the cross-derivative relations (Maxwell relations) found from differentiating the first expressions.

Internal Energy

$$U = U(S, V) \tag{9a}$$

thus

$$dU = \left(\frac{\partial U}{\partial S}\right)_V dS + \left(\frac{\partial U}{\partial V}\right)_S dV \tag{9b}$$

therefore, from (5), we have

first: $\quad \left(\dfrac{\partial U}{\partial S}\right)_V = T \quad \text{and} \quad \left(\dfrac{\partial U}{\partial V}\right)_S = -P$ \hfill (9c)

second: $\quad \dfrac{\partial}{\partial V}(T) = \dfrac{\partial}{\partial S}(-P)$

or

$$\left(\frac{\partial T}{\partial V}\right)_S = -\left(\frac{\partial P}{\partial S}\right)_V. \tag{9d}$$

Enthalpy

$$H = H(S, P) \tag{10a}$$

thus

$$dH = \left(\frac{\partial H}{\partial S}\right)_P dS + \left(\frac{\partial H}{\partial P}\right)_S dP \tag{10b}$$

therefore, from (6b) we have

first: $\quad \left(\dfrac{\partial H}{\partial S}\right)_P = T \quad \text{and} \quad \left(\dfrac{\partial H}{\partial P}\right)_S = V$ \hfill (10c)

second: $\quad \dfrac{\partial}{\partial P}(T) = \dfrac{\partial}{\partial S}(V)$

or

$$\left(\frac{\partial T}{\partial P}\right)_S = \left(\frac{\partial V}{\partial S}\right)_P. \tag{10d}$$

Helmholtz Free Energy

$$A = A(V, T) \tag{11a}$$

thus

$$dA = \left(\frac{\partial A}{\partial V}\right)_T dV + \left(\frac{\partial A}{\partial T}\right)_V dT \tag{11b}$$

therefore, from (7b), we have

first: $\left(\dfrac{\partial A}{\partial V}\right)_T = -P$ and $\left(\dfrac{\partial A}{\partial T}\right)_V = -S$ (11c)

second: $\dfrac{\partial}{\partial T}(-P) = \dfrac{\partial}{\partial V}(-S)$

or

$$\left(\dfrac{\partial P}{\partial T}\right)_V = \left(\dfrac{\partial S}{\partial V}\right)_T .\quad (11d)$$

Gibbs Free Energy

$$G = G(P, T) \quad (12a)$$

thus

$$dG = \left(\dfrac{\partial G}{\partial P}\right)_T dP + \left(\dfrac{\partial G}{\partial T}\right)_P dT \quad (12b)$$

therefore, from (8b), we have

first: $\left(\dfrac{\partial G}{\partial P}\right)_T = V$ and $\left(\dfrac{\partial G}{\partial T}\right)_P = -S$ (12c)

second: $\dfrac{\partial}{\partial T}(V) = \dfrac{\partial}{\partial P}(-S)$

or

$$\left(\dfrac{\partial V}{\partial T}\right)_P = -\left(\dfrac{\partial S}{\partial P}\right)_T . \quad (12d)$$

Now for a closing word of physical insight. While each thermodynamic potential has a unique set of natural variables, these potentials and changes in them are well-defined for *any* process. For example, one of the problems for this chapter considers the changes in A and G along the adiabatic paths of the Carnot cycle. These paths are characterized by changes in P, T, and V, yet ΔA and ΔG are well-defined along them. As a corollary, the Maxwell relations (9d, 10d, 11d, and 12d above) are valid for use in any reversible path integral or differential expression. In fact, that is their major value.

CHAPTER 5 SUMMARY

This chapter has introduced the state functions that point the way toward equilibrium (Equations (5.3), (5.4), (5.6), and (5.8)) under various constraints and signal the arrival at equilibrium (the same equations with < replaced by =). These functions do so by cleverly incorporating all three laws into thermodynamic potential functions. For chemical applications, the Gibbs free energy, $G = H - TS$, is the most valuable, especially the molar Gibbs free energy for a pure compound or the partial molar Gibbs free energy, the *chemical potential* μ_i, for mixtures, solutions, and multiphase equilibria. Carry from this chapter to the next this key idea:

If an isothermal, isobaric system can lower its chemical potential, a spontaneous process will occur to do just that. The process will continue until the chemical potential is as low as the external constraints on the system allow.

This chapter also introduced *fugacity* as a measure of chemical potential:

definition of fugacity: $d\mu = RT \, d \ln f$

gas fugacity:
$$f = P \exp\left[\dfrac{1}{RT} \int_0^P \left(\bar{V} - \dfrac{RT}{P}\right) dP\right]$$
$$= P \exp\left(\int_0^P \dfrac{Z-1}{P} dP\right)$$

low pressure gas fugacity:
$$f \cong P \exp\left[\frac{B(T)\,P}{RT}\right] \cong PZ$$

Fugacity measures chemical potential deviations from ideal behavior. The next chapter will use fugacity in the context of nonidealities beyond those of simple gases. In some situations, we will concentrate directly on chemical potential, but in others, fugacity will be more convenient. Remember that f and μ are measuring the same ideas, only in different ways.

FURTHER READING

Particularly relevant to this chapter is the classic text *Thermodynamics*, originally written by G. N. Lewis and M. Randall in 1923 and revised by K. S. Pitzer and L. Brewer (McGraw-Hill, New York, 1961). The original starts with a very well-known dedication:

> Let this book be dedicated to the chemists of the newer generation, who will not wish to reject all inferences from conjecture or surmise, but who will not care to speculate concerning that which may be surely known. The fascination of a growing science lies in the work of the pioneers at the very borderland of the unknown, but to reach this frontier one must pass over well travelled roads; of these one of the safest and surest is the broad highway of thermodynamics.

ACCESS TO DATA

In addition to the references at the end of Chapter 4, the following sources contain specific data useful for fugacity and free energy calculations.

J. Hilsenrath, *et al.*, *Vapor Pressures and Heats of Vaporization of Hydrocarbons and Related Compounds; Selected Values of Properties of Hydrocarbons and Related Compounds* (Thermod. Res. Center, Texas A&M Univ., College Station, Texas, 1971–72).

R. C. Reid and T. K. Sherwood, *The Properties of Gases and Liquids* (McGraw-Hill, New York, 1966).

A. N. Nesmeyanov, *Vapour Pressure of the Elements*, translated by J. I. Carasso (Pergamon Press, New York, 1963).

J. H. Keenan, F. G. Keyes, P. G.-Hill, and J. G. Moore, *Steam Tables: Thermodynamic Properties of Water Including Vapor, Liquid, and Solid Phases*, 2nd ed. (J. Wiley & Sons, New York, 1978).

PRACTICE WITH EQUATIONS

5A What is the absolute entropy of Hg(g) at 400 K and 1.00×10^{-3} atm, assuming ideal behavior? (5.34)

ANSWER: $\overline{S} = 21.89\,R$

5B At what pressure does He at 200 K have the same chemical potential as He at 300 K and 1.00 atm? (5.35)

ANSWER: 6.42×10^{-4} atm

5C At what pressure is the fugacity of He at 300 K 1% greater than the pressure? (5.41) and Table 1.1

ANSWER: 20.4 atm

5D What is the fugacity coefficient of a van der Waals gas at the critical point? (5.42), (5.43)

ANSWER: 0.695

PROBLEMS

SECTION 5.1

5.1 Derive Eq. (5.24), $dG = V\,dP - S\,dT$, starting from the definition $G = H - TS$. Then go on to derive differential expressions for the following two state functions that are largely of historical interest now. The first is called the Massieu function, J:

$$J = -\frac{U}{T} + S$$

and the second is called the Planck function, Φ:

$$\Phi = -\frac{H}{T} + S\,.$$

Find the total differentials of $J = J(T, V)$ and $\Phi = \Phi(T, P)$.

5.2 Shown below is a plot of the van der Waals isotherm $\tilde{T} = 7/27$. Show that the transition indicated, from $(\tilde{P} = 0.65/27, \tilde{V} = 6)$ to $(\tilde{P} = 0.65/27, \tilde{V} = 1.702)$, is spontaneous. What physical interpretation does this spontaneous transition have? Much more about this will be discussed in the next chapter.

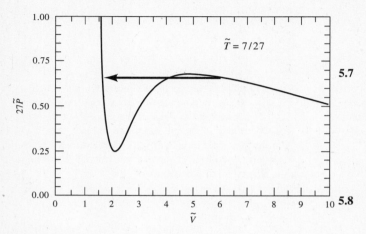

SECTION 5.2

5.3 Derive Equations (5.13) and (5.14), $-\Delta H_{S,P} = -w'_{S,P}$ and $-\Delta G_{T,P} = -w'_{T,P}$.

5.4 Consider the apparatus shown below. One side of an isothermal cell contains an amount of ideal gas A while the other side, of equal volume, contains the same amount of ideal gas B. The partition between the sides is removed and the gases spontaneously mix.

Ideal gas A	Ideal gas B
Pressure P	Pressure P
Volume V	Volume V
Temperature T	Temperature T

What is the maximum amount of work that could be obtained from this spontaneous process? Can you derive a process to extract this work? Note that since the process is isothermal and involves ideal gases, $\Delta U = \Delta H = 0$. Chapter 6 will have much more to say about spontaneous mixing phenomena.

SECTION 5.4

5.5 In Section 4.6, the absolute entropy of O_2 at 300 K was calculated to be $\overline{S}° = 24.629\ R$ at 1 atm. Tables of thermodynamic data report that N_2 has a Third Law entropy $\overline{S}° = 23.092\ R$. Calculate these quantities from the Sackur–Tetrode equation, Eq. (5.34). You will find rather poor agreement due to an extra contribution to the entropy from the internal motion characteristic of any diatomic or polyatomic molecule. (This disagreement is also due in part to the nature of the chemical bond in O_2, a topic discussed in Chapter 14.) In contrast, monatomic gases have $\overline{S}°$ values that agree well with the predictions of the Sackur–Tetrode equation. Verify this for the following gases and their associated Third Law entropies at 298.15 K: Kr ($\overline{S}° = 19.7\ R$), Xe ($\overline{S} = 20.4\ R$), and Ca ($\overline{S}° = 18.63\ R$).

5.6 Problem 3.5 asks for the missing $(P, \overline{V})$ coordinate of a particular Carnot cycle for an ideal gas. Let the gas be He and find ΔA and ΔG for each of the four steps in this cycle. (As a check, verify that the sum of your values for ΔA and ΔG add to zero, as they should for a cyclic process.)

5.7 Imagine pouring 10 L of water into a 20 L container half full of small ball bearings packed as densely as possible. Chapter 16 will show that such a packing has 74% of the packed volume occupied by the balls; 26% is open volume around the balls. Plot the total volume of the water–ball-bearing "solution" as a function of the amount of water added and deduce the partial molar volume of the water as a function of amount added.

5.8 The partial molar volume of NaCl(aq) at room temperature and pressure is 18.0 cm^3 mol^{-1} when [NaCl] = 0.50 M. How does this compare to the molar volume of NaCl(s) (density = 2.165 g cm^{-3})? to the molar volume of 0.5 mol of NaCl spread uniformly through a 1 L volume? Calculate the change in the volume of 1 L of this solution upon addition of 0.01 mol of NaCl(s) and contrast this to the volume of 0.01 mol NaCl(s). What conclusions can you make about the microscopic effects of this addition on the structure of the water solvent?

5.9 If the pressure on the initial solution in the previous problem is doubled from 1.00 atm to 2.00 atm, what effect does this have on the chemical potential of NaCl(aq)? What is $\Delta \mu$ for a similar pressure increase for pure water?

5.10 We have seen how curves of constant temperature (isotherms), pressure (isobars), volume (isochores), and entropy (reversible adiabats) look for an ideal gas. What do curves of constant chemical potential look like in $(P, \overline{V})$ coordinates? in (P, T) coordinates?

5.11 Derive an expression for the molar Helmholtz free energy change of an ideal gas going from state $(\overline{V}_i, T_i)$ to state $(\overline{V}_f, T_f)$. What do curves of constant molar Helmholtz free energy look like in $(P, \overline{V})$ coordinates? in $(\overline{V}, T)$ coordinates?

5.12 Problem 1.26 introduced the Berthelot equation of state, a simple modification of the van der Waals equation. In terms of its critical point coordinates (which that problem asks you to find), the Berthelot equation can be written

$$P = \frac{RT}{\overline{V}}\left[1 + \frac{9}{128}\frac{T_c}{P_c T}\left(1 - 6\frac{T_c^2}{T^2}\right)P\right].$$

Find the fugacity of such a gas. (*Hint*: Note that at

5.13 Use Eq. (5.40) to find the fugacity of a hard-sphere gas (the van der Waals gas with the a parameter set to zero). You should find

$$\frac{f}{P} = \exp\left(\frac{bP}{RT}\right).$$

Expand the exponential through the second term ($e^x \cong 1 + x$) and manipulate your result to show that

$$\frac{f}{P} \cong \frac{P\overline{V}}{RT} = \frac{P}{P_{ideal}}$$

where $P_{ideal} = RT/\overline{V}$ is the ideal pressure that the gas would exhibit at the observed molar volume. In other words, show that the real pressure is approximately the geometric mean of the ideal pressure and the fugacity (and thus lies between the two). Explore the range of validity of this approximation using the following data for O_2 at 0 °C.

P/atm	f/atm	$\frac{P}{P_{ideal}}$
50.0	48.0	0.961
100.	92.5	0.929
200.	174.	0.910
400.	338.	1.05
600.	540.	1.29

5.14 Problem 1.9 lists Z values for H_2 at 35 K and various $\overline{V}$ values. Use those data to find the fugacity of H_2 at the smallest tabulated $\overline{V}$, 2381 cm³ mol⁻¹.

5.15 The coefficients of Eq. (1.17), the compressibility factor as a power series in pressure, are: $A_1 = -5.14 \times 10^{-4}$ atm⁻¹, $A_2 = 4.375 \times 10^{-6}$ atm⁻², $A_3 = -4.348 \times 10^{-9}$ atm⁻³, and $A_4 = 1.56 \times 10^{-12}$ atm⁻⁴ for N_2 at 0 °C and pressures from 1 to 1000 atm. Compare f calculated from these precise data to the approximation $f \cong PZ$.

5.16 Derive expressions for the isothermal change in chemical potential due to a change in pressure from P_i to P_f for substances:
(a) that are incompressible.
(b) that have constant coefficients of isothermal bulk compressibility, κ, Eq. (1.26).

Use data from Table 4.5 to calculate the change in chemical potential of Fe(s) and benzene(l) under both assumptions (a) and (b) from $P_i = 1$ atm to $P_f = 1000$ atm.

5.17 The vapor pressure of Ar follows the empirical equation (which is given some theoretical justification in Chapter 6)

$$\log_{10}(P/\text{bar}) = A + \frac{B}{T} + C\log_{10} T + DT$$

with T in K and the parameters

$A = 25.807\ 9$ $B = -603.278\ 3$ K
$C = -10.664\ 47$ $D = 0.020\ 639\ 3$ K⁻¹.

Find the fugacity of Ar(l) in equilibrium with Ar(g) at 100.0 K. It is sufficient to use the second virial coefficient expression for f, Eq. (5.41), and data from Table 1.1. Next, find the fugacity of Ar(l) at 100.0 K and 105.0 bar (well above the 100 K vapor pressure, but also well below the 100 K solidification pressure of 684 bar) at which point the molar volume is 29.66 cm³ mol⁻¹. Assume Ar(l) is incompressible and simply integrate $\overline{V} dP = RT\ d(\ln f)$ from the vapor pressure to 105.0 bar. This calculation should illustrate to you that while the fugacity of a *gas* is closely related to its pressure, that is not necessarily the case for the fugacity of a *condensed* phase.

SECTION 5.5

5.18 Show that the general entropy change for a van der Waals gas with a constant heat capacity $\overline{C}_V = 3R/2$ taken from (T_i, V_i) to (T_f, V_f) is

$$\Delta S = nR\left[\frac{3}{2}\ln\left(\frac{T_f}{T_i}\right) + \ln\left(\frac{V_f - nb}{V_i - nb}\right)\right].$$

(*Hint*: Integrate $dS = (\partial S/\partial T)_V\ dT$ at constant V_i, then add to this the integral of $dS = (\partial S/\partial V)_T\ dV$ at constant T_f. You will need a Maxwell relation.)

GENERAL PROBLEMS

5.19 Gravitational potential functions are easy to visualize: a hill, an inclined plane, etc. But what about thermodynamic potential functions? Sketch qualitative graphs for an ideal gas system of:
(a) $\overline{U}$ versus $\overline{S}$ at constant $\overline{V}$ and $\overline{U}$ versus $\overline{V}$ at constant $\overline{S}$
(b) $\overline{H}$ versus $\overline{S}$ at constant P and $\overline{H}$ versus P at constant $\overline{S}$
(c) $\overline{A}$ versus T at constant $\overline{V}$ and $\overline{A}$ versus $\overline{V}$ at constant T
(d) $\overline{G}$ versus T at constant P and $\overline{G}$ versus P at constant T

Here's how to proceed. For example, to find $\overline{U}$ versus $\overline{S}$ at constant $\overline{V}$, we need to integrate $d\overline{U} = Td\overline{S}$ but under the constant volume constraint. Thus we need to know how T varies with $\overline{S}$ at constant $\overline{V}$. From the Sackur–Tetrode equation, $\overline{S} \propto \ln[(\text{constant})T^{5/3}/P]$. Use the equation of state and the constant $\overline{V}$ constraint to show that $T \propto \exp(2\overline{S}/3R)$ and thus $\int d\overline{U} \propto \int \exp(2\overline{S}/3R)\ d\overline{S}$. Similarly, to find $\overline{U}$ versus $\overline{V}$ at constant $\overline{S}$, integrate $d\overline{U} = -Pd\overline{V}$ remembering that constant $\overline{S}$ means the path is adiabatic and reversible: $P\overline{V}^{5/3} = $ constant.

5.20 The expression for the van der Waals gas fugacity is not difficult to derive. Start with the identity for a constant-temperature path:

$$dP = \left(\frac{\partial P}{\partial \overline{V}}\right)_T d\overline{V},$$

combine this with the differential relation between fugacity and pressure, $\overline{V} dP = RT d(\ln f)$, and subtract the ideal gas relationship $RT d(\ln P) = (RT/P) dP$ from both sides. Integrate from some pressure P_0, which will be allowed to go to zero at the end of the calculation, to some finite pressure of interest, P. The integral of $d(\ln (f/P))$ over these limits equals $\ln (f/P)$ since $f = P_0$ as $P_0 \to 0$. Evaluate $(\partial P/\partial \overline{V})_T$ from the full van der Waals equation of state, Eq. (1.20). You should obtain

$$\ln\left(\frac{f}{P}\right) = \left[-\frac{2a}{RT\overline{V}} + \frac{b}{\overline{V} - b} - \ln(\overline{V} - b)\right]_{\overline{V}_0}^{\overline{V}} - [\ln P]_{P_0}^{P}$$

where $\overline{V}_0$ is the molar volume at P_0 and $\overline{V}_0 \to \infty$ as $P_0 \to 0$. The first two terms on the right present no difficulty when evaluated at the integral limits and when $\overline{V}_0 \to \infty$. Combine the two logarithmic terms, however, before taking this limit, and note that $P_0(\overline{V}_0 - b) \to P_0 \overline{V}_0 = RT$. You should then be able to recover the expression in the text in terms of scaled coordinates.

CHAPTER 6

Phase Equilibria, Mixtures, and Solutions

WHY do phase transitions occur? Why does water, for instance, suddenly freeze into solid ice *at one sharp temperature?* On a molecular level, phase transitions seem magical. Suddenly, the regular order of a solid vanishes, not just here and there, but throughout a mole or more of molecules. Suddenly, an entire liquid drop can increase its molar volume by a factor of one thousand and vanish as a puff of gas.

The details of phase transitions are still areas of intense contemporary research, on both the thermodynamic and the molecular levels. The experiments and theories of this research have explained the main features of most phase transitions, but these are largely twentieth-century results. Had you been studying chemistry in the late 1820s, you might well have used a text entitled *The First Lines of Philosophical and Practical Chemistry as Applied to Medicine and the Arts Including the Recent Discoveries and Doctrines of the Science,* written by one J. S. Forsyth, "Surgeon, etc." and published in London by "Sustenance and Stretch, Percy Street" in 1828. It was an up-to-date text for its time and included an entire chapter on the (erroneous) caloric theory of heat, in which it is stated of caloric that "in the hands of chemists it is the most powerful agent they are acquainted with." The caloric theory also figured heavily in theories of phase transitions. Thus Forsyth wrote, "By the abstraction of caloric solid bodies rendered fluid by it are reproduced." Most interesting is a long footnote that opens a chapter on light:

> With regard to light, we would mention the singularly useful and hitherto unobserved effect of moonlight, in assisting the completion of certain

6.1	Why Do Phase Transitions Occur?
6.2	The Coexistence of Phases in Pure Substances
6.3	Composition Variables in Mixtures
6.4	Differences Between Solutions and Mixtures
6.5	Colligative Properties
6.6	Many Compounds, Many Phases—The Phase Rule

important natural phenomena. The crystallization of water, under the form of those light frosts which so much prevail during the early spring, and which are of such important service in assisting the operations of agriculture, by rendering the surface of the earth mellow, and better susceptible of the manure that is necessary to it, are greatly assisted, and in many cases entirely brought about by the intervention of moonlight. It is well known, that under certain circumstances, water will sink to the temperature of 22° before it freezes, or takes a form of crystals. Indeed it will invariably do so in the absence of any mechanical agitation, and in the absence of light. It is an unquestionable fact, but one which has not hitherto been observed generally, or attended to, that during that period of the year which has been alluded to, and indeed at other periods, before the moon rises on a still clear night, when the atmosphere is at a lower temperature than 32, the water remains in a liquid state, but immediately on the moon rising, and diffusing its light around, the water freezes, and performs the salutary offices required of it, without subjecting us to the severity of a low temperature.

A quaint explanation. How many explanations in this text might we suppose will be similarly quaint in a century-and-a-half's time?

6.1 WHY DO PHASE TRANSITIONS OCCUR?

The thermodynamic explanation (today) of this section's title question avoids moonlight and concentrates on the usual reason for any spontaneous process: an opportunity for the system to lower its chemical potential. On the one hand, thermodynamics predicts the conditions for phase transitions to arise. On the other hand, thermodynamics could not care less *that* phase transitions occur. This second point is important to establish.

Since thermodynamics has nothing to say about the *rate* at which phase transitions (or anything else) occur, we are free to describe *metastable phases* as if they were truly stable (as long as experiments establish that such phases exist for times comparable to or much longer than required observation times). Thus, Forsyth could comment on liquid water at 22 °F, a temperature where water is thermodynamically stable only when solid. Such super-cooled liquids are quite common, as are metastable phases of solids. Much of metallurgy is concerned with establishing metastable alloys with useful properties, including the property of remaining in the metastable state for many years.

The driving force that converts a metastable phase into a stable one is discussed at a microscopic level in Section 8.3. For now, we can assume, correctly, that the reason we do not routinely find liquid water at 22 °F is one from the realm of process *rates*, not process thermodynamics. This assumption allows us to compute thermodynamic quantities *for any phase at any temperature and pressure*. In practice, we may well find experimental data lacking outside the temperature and pressure range over which the phase is truly stable. In such cases, we resort to theoretical models to guide extrapolations of known experimental data.

We remarked in Section 1.3 that the van der Waals equation of state seemed capable of expressing the properties of a gas (large isothermal bulk compressibility) in certain states and the properties of a liquid (small compressibility and small molar volume) in other states. We can now show these features quantitatively and use them to indicate the thermodynamic origins of phase transitions.

Recall the arguments that led to Figure 5.6, the graph of the van der Waals fugacity coefficient. To turn fugacity coefficients into chemical potentials, it is convenient to write Eq. (5.48) in the form

$$\mu(P, T) = \mu°(1 \text{ bar}, T) + RT \ln\left(\frac{P_f}{P}\right)$$

so that

$$\mu(P, T) - \mu°(1 \text{ bar}, T) = RT \ln P + RT \ln \Phi .\quad (6.1)$$

Since $\mu°$ is a function of temperature only, we can subtract $RT \ln (a/b^2)$ from each side (with a and b the van der Waals parameters)

$$\mu(P, T) - \left[\mu° + RT \ln\left(\frac{a}{b^2}\right)\right] = RT \ln P + RT \ln \Phi - RT \ln\left(\frac{a}{b^2}\right)$$

$$= RT \ln\left(\frac{Pb^2}{a}\right) + RT \ln \Phi \quad (6.2)$$

$$= RT \ln \tilde{P}\Phi .$$

Call the left-hand side $\Delta\mu$, the isothermal pressure change in μ from the $\mu° + RT \ln (a/b^2)$ reference point, and we have a convenient expression for the van der Waals chemical potential in dimensionless form:

$$\frac{\Delta\mu}{RT} = \ln \tilde{P}\Phi .\quad (6.3)$$

Figures 6.1(a–c) plot Eq. (6.3) at temperatures above, at, and below the critical temperature. Refer back to Figure 1.7 (the van der Waals isotherms) as you look

FIGURE 6.1(a) Above the critical temperature, the chemical potential of a van der Waals gas is a smoothly increasing function of pressure. **(b)** At the critical temperature, μ increases smoothly through the critical pressure (the * on the figure). **(c)** For *every* temperature below the critical temperature, the van der Waals chemical potential exhibits a sharp-cornered loop as a function of pressure. The crossing point is indicative of a g–l phase transition.

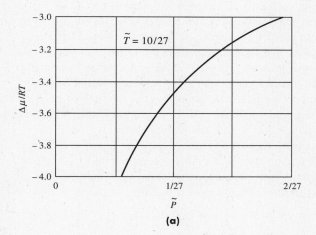

(a)

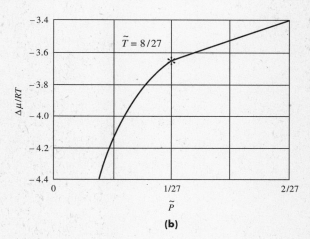

(b)

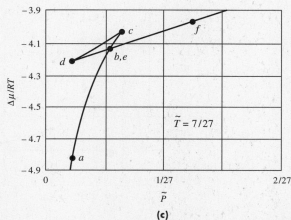

(c)

at these figures. Nothing spectacular happens in Figures 6.1(a) and 6.1(b), even at the critical point noted in the latter figure. But notice the peculiar shape of the curve in Figure 6.1(c), at $\tilde{T} = 7/27$. Several key points on that curve, lettered a–f, are also noted in Figure 6.2, which plots the $\tilde{T} = 7/27$ isotherm.

Now, with one eye on Figure 6.1(c) and the other on Figure 6.2, while your mind's eye visualizes a gas being compressed, follow the system as the volume is decreased from point a. From a to c, the pressure rises smoothly, as does the chemical potential. From c to d, the van der Waals equation is nonphysical; the isothermal bulk compressibility is negative (as discussed in Section 1.3). This nonphysical part of the isotherm is also a rather unique segment of the chemical potential curve. From d to f, the pressure again rises (rapidly) while the chemical potential rises gently.

What about points b and e, where the chemical potential curve crosses itself at one unique pressure? States between b and c are *metastable*, as are those between d and e. Recall that a system will do something spontaneously if, at constant T and P, the change will lower the system's chemical potential. Pick any pressure between, for example, points b and c. A point at the same pressure but on the curve from e to f is at a lower chemical potential. The system will make such a change spontaneously. (The volume will drop.) Points from b to a and on to lower pressures represent a *stable gas*. Points from e to f and on to higher pressures represent a *stable liquid*. The one pressure at point b represents the *equilibrium coexistence of a liquid phase and a gas phase at the vapor pressure of the substance*.

Picture a gas at point a subjected to an isothermal compression, as shown in Figure 6.3. When the system reaches point b, the first drop of liquid will form. What happens to the pressure when the system's volume is decreased from point b? If you answered immediately that the pressure stays the same, then you have a good, intuitive feeling for vapor pressure. As the volume is decreased from that at point b to that at point e, the equilibrium system's pressure is constant. More and more of the gas is condensing into a liquid. At point e, *all* of the molecules in the system will be in a liquid phase, and due to the relative incompressibility of liquids, a large pressure increase is required to reduce the volume further.

At the system's vapor pressure (between points b and e of Figure 6.2), a real equilibrium system has a horizontal isotherm (infinite compressibility). Maxwell was the first to prove that the vapor pressure of the van der Waals gas has an interesting geometrical interpretation. The horizontal isotherm segment, or *tie-line*

FIGURE 6.2 The coordinates a–f correspond to the same points on Figure 6.1(c). The pressure at points b and e is the *vapor pressure*.

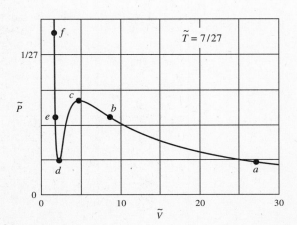

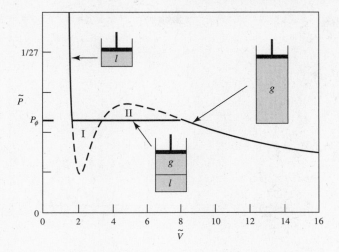

FIGURE 6.3 The real behavior of the system along the isotherm of Figure 6.2 is illustrated here. When the gas is compressed to P_ϕ, the vapor pressure, condensation begins and the pressure remains constant until all the gas is converted to liquid. The pressure P_ϕ is chosen to equalize the areas labeled I and II, bounded by the van der Waals isotherm and the P_ϕ isobar.

as it is called, is placed so that the area under it is exactly equal to the area under the van der Waals isotherm from points b and e. Calling the vapor pressure P_ϕ, these words imply

$$P_\phi(V_b - V_e) = \int_{V_e}^{V_b} P(V)dV \qquad (6.4)$$

where $P(V)$ is the van der Waals isotherm, Eq. (1.20). Since $PdV = d(PV) - VdP$ and since $P(V_b) = P(V_e) = P_\phi$, Eq. (6.4) can be rewritten

$$P_\phi(V_b - V_e) = \int_{V_e}^{V_b}[d(PV) - V\,dP] = P_\phi(V_b - V_e) - \int_{P(V_e)}^{P(V_b)} V(P)\,dP$$

so that we have proven

$$\int_{P(V_e)}^{P(V_b)} V(P)\,dP = 0 \ . \qquad (6.5)$$

The geometrical interpretation of this equation is that area I in Figure 6.3, between the tie-line and the isotherm, is equal to area II.

We will make further use of Figure 6.3 as we explore other aspects of phase transitions, but the main result for now is the thermodynamic reason for the condensation of the van der Waals gas at temperatures below the critical temperature. A spontaneous phase transition becomes possible *when the chemical potential of an existing phase exceeds that of another phase*. If the phases have equal chemical potentials, then *they can coexist in equilibrium in any proportion*. Thus, at point b, the first drop of liquid establishes the two phase equilibrium that lasts at constant T, P, and μ, until the last whiff of gas has condensed at point e.

Now we expand these ideas to a more general situation: a pure substance at any temperature under a constant pressure of 1 atm. For simplicity, we imagine the substance has only one solid phase, and single liquid and gas phases—we are below the critical pressure and we ignore the rare possibility that the substance can form a liquid crystal phase. We start in the solid phase and raise the temperature, encountering fusion at the normal melting point and vaporization at the normal boiling point. As a representative substance that fits these requirements, we choose zinc, which melts at 692.7 K and boils as a monatomic gas at 1181 K.

FIGURE 6.4 The chemical potential of Zn, plotted versus T at 1 atm, shows the origin of melting and boiling, spontaneous processes that lower zinc's Gibbs free energy. The dashed portions of each curve represent unstable regions for each type of phase.

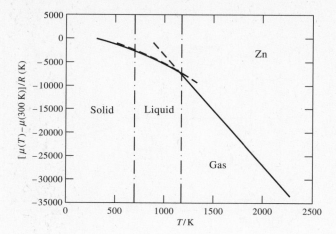

Figure 6.4 graphs $[\mu(T) - \mu(300\ \text{K})]/R$ for Zn versus T over the range 300–2300 K. The dashed lines are extensions of $\mu(T)$ for each phase. The slope of all these lines is negative, since $(\partial \mu/\partial T)_P = -\overline{S}$, and the slopes become *more* negative as one goes from solid to liquid to gas, since at each transition, the entropy jumps up by $\Delta \overline{S}_\phi$. Finally, note that the curves are in fact curved rather than straight! The second derivative of $\mu(T)$ governs curvature and is related to the *heat capacity,* since

$$\left(\frac{\partial^2 \mu}{\partial T^2}\right)_P = -\left(\frac{\partial \overline{S}}{\partial T}\right)_P = -\frac{\overline{C}_P(T)}{T}. \tag{6.6}$$

We conclude this section with a summary of the reasons for phase transitions as expressed by Figure 6.4.

- First, chemical potential curves exist for each phase, in principle at all temperatures and pressures.
- Second, whenever two such curves cross, a phase transition will occur spontaneously with the equilibrium system always following the lowest available curve.
- Third, at a phase transition, the chemical potentials of each phase are equal, but the slopes of the crossing curves are different, due to the finite entropy change.

EXAMPLE 6.1

Consider Fe at 1 atm. Along the way to melting at 1812 K, one finds two solid–solid phase transitions at 1183 K and 1673 K. Structural studies of Fe show that the low-temperature phase and the highest-temperature phase are microscopically the same. In other words, this phase disappears at 1183 K only to reappear at 1673 K. How can this happen?

SOLUTION If you sketch a few types of $\mu(T)$ curves that might explain this behavior, you will realize that the two curves (one for each unique solid phase) must cross each other twice. The only way to do this, with curves of negative slope everywhere, is shown in Figure 6.5 (a, b), which graphs the experimental $\mu(T)$ curves for Fe(s). Not only do the shapes of these curves explain the phenomenon, they also indicate that the heat capacity of the lowest- and highest-temperature phase must be *larger* than that of the intermediate-temperature phase. As Eq. (6.6) indicates, the larger C_P, the more negative is the second derivative of $\mu(T)$, which governs the curvature.

➠ *RELATED PROBLEMS* 6.1, 6.2

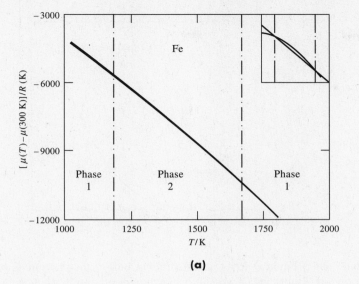

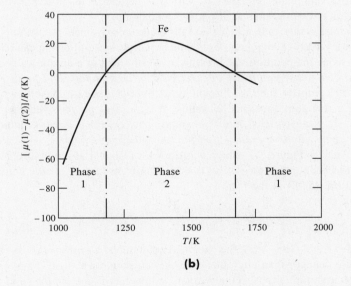

FIGURE 6.5(a) The chemical potentials of the two possible phases of solid iron are virtually identical. The slightest differences are enough to alter the relative stabilities of the two. A schematic picture of these curves (inset) shows that, due to the curvature differences between them, phase 1 must have the larger heat capacity. **(b)** The difference $\mu(\text{phase 1}) - \mu(\text{phase 2})$ for solid Fe, throughout the stable region for phase 2, is only a very small fraction of the absolute change in μ for either phase over the same region.

To summarize all possible one-phase stability regions and many-phase coexistence states, we plot the temperature and pressure values of the μ-curve crossing points as a *phase diagram*. This is the subject of the next section.

6.2 THE COEXISTENCE OF PHASES IN PURE SUBSTANCES

The simplest complete phase diagram we can construct now is that for the van der Waals gas. If we compute $\tilde{P}_\phi$, the reduced vapor pressure, at various $\tilde{T}$, using Eq. (6.4), we can plot the *liquid–vapor coexistence line,* as is done in Figure 6.6. Note how this line extends from the critical point down to the origin; the van der Waals gas liquifies, but does not solidify.

On a pure compound (P, T) phase diagram, *any one phase can exist in an area on the diagram.* As the figure indicates, the van der Waals "fluid" is a gas at high

FIGURE 6.6 The phase diagram for the van der Waals gas is a simple line of two-phase coexistence states terminated at the critical point.

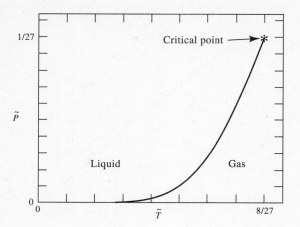

temperature and a liquid at low temperature, but only if the pressure is below the critical pressure. But whether we call it gas, liquid, or simply fluid, the important point is that one phase is stable at a variety of temperatures at any given pressure, or at a variety of pressures at any given temperature.

In contrast, a system in a state that falls on the coexistence line is more restricted. There are a variety of temperatures at which two-phase equilibrium can be observed, but at any one temperature, the pressure is constrained to a unique value, the vapor pressure. *Two-phase equilibria are confined to lines on the (P, T) phase diagram.*

We can complete the basic topology of pure-compound phase diagrams if we consider a real substance that solidifies. We can also prove a statement made in Section 1.1: water (or any pure compound) can coexist as a solid, liquid, and gas at *only one unique temperature and pressure,* the *triple point.*

Look back to Figure 6.4, which was drawn for Zn at 1 atm pressure. How would this figure look at other pressures? In other words, how does μ vary with pressure? We need Eq. (5.38) in the form

$$\left(\frac{\partial \mu}{\partial P}\right)_T = \overline{V} \ . \tag{6.7}$$

Since $\overline{V}(g) \gg \overline{V}(l) \sim \overline{V}(s)$, the chemical potential of a gas is much more sensitive to pressure changes than any other phase. Each point on the $\mu(T)$ curve for a gas, at each temperature, drops as the pressure is lowered. Points on the solid and liquid curves drop as well, but not as rapidly. At some pressure, all three curves will cross at one point. The temperature at this triple crossing is the *triple-point temperature* T_{tp} and the *only* pressure at which this can happen is the *triple-point pressure* P_{tp}. Table 6.1 lists several triple-point coordinates.

If the pressure is lowered below P_{tp}, the liquid $\mu(T)$ curve is *above* either the solid or the gas curves at all temperatures. The liquid phase is no longer stable, either alone or in coexistence with one or more phases. Instead, the sublimation transition, solid $\rightarrow$ gas, occurs at a unique temperature for each pressure below P_{tp}, and we can speak of the vapor pressure of the solid. For example, CO_2 (P_{tp} = 5.11 atm and T_{tp} = 216.6 K) sublimes with a vapor pressure of 1 atm at 195 K.

Figure 6.7(a, b) on page 150 summarizes, schematically, $\mu(T)$ curves at various pressures for *s–l–g* phases, with reference to the simple phase diagram for Ne. The phase diagram itself shows the three coexistence lines meeting at the triple point (0.426 atm, 24.56 K), the *l–g* line ending at the critical point (27.2 atm, 44.40 K),

6.2 THE COEXISTENCE OF PHASES IN PURE SUBSTANCES

TABLE 6.1 Triple Point Coordinates (s–l–g Equilibrium)

	T_{tp}/K	P_{tp}/atm
H_2O	273.160 0§	6.033×10^{-3}
^4He	2.186‡	5.04×10^{-2}
Ne	24.56	0.425 7
Ar	83.810	0.680 1
Kr	115.78	0.722 0
Xe	161.37	0.805 5
H_2	13.84	0.069 5
D_2	18.63	0.168
O_2	54.361	1.50×10^{-3}
N_2	63.18	0.124
CO_2	216.6	5.105
SO_2	197.68	1.653×10^{-3}
NH_3	195.40	5.996×10^{-2}

§By definition. The normal melting (fusion) temperature for pure water saturated with air at 1 atm is 0.0098 K lower.
‡Coordinates of the λ point (He I–He II–He(g)) equilibrium.

and the areas in which only one phase is possible. Note that the sublimation line is barely visible near the bottom of the plot with the linear pressure axis, Fig. 6.7(a).

As we saw with O_2 in Section 4.6, many solids exist in several crystal forms. The phase diagram for water, with several distinct solid phases, is shown in Figure 6.8 (shown on page 151). Note that there are several lines on this diagram corresponding to solid–solid two phase coexistence equilibria and several triple points involving three solid phases in coexistence (such as among ice VI, VII, and VIII) or coexistence among two solid phases and the liquid.

If we could express the two-phase coexistence lines as functions of T, we could predict quantities such as vapor pressures at various temperatures. It is possible to find such functions, and it is instructive to begin with an expression for the *slopes* of the coexistence lines.

Consider two phases, ϕ_1 and ϕ_2, in equilibrium:

$$\phi_1(T, P) \rightleftarrows \phi_2(T, P) . \tag{6.8}$$

How should we change T and P together to maintain this equilibrium? For infinitesimal changes, we have

$$\mu_1(T + dT, P + dP) = \mu_1(T, P) + d\mu_1$$

and

$$\mu_2(T + dT, P + dP) = \mu_2(T, P) + d\mu_2 .$$

To maintain equilibrium, we require $d\mu_1 = d\mu_2$, which, from Eq. (5.38), implies

$$-\overline{S}_1 \, dT + \overline{V}_1 \, dP = -\overline{S}_2 \, dT + \overline{V}_2 \, dP$$

or

$$(\overline{S}_2 - \overline{S}_1)dT = (\overline{V}_2 - \overline{V}_1)dP .$$

FIGURE 6.7(a) The phase diagram for Ne is representative of simple substances that have a single solid phase, a single triple point, and a critical point. **(b)** If the phase diagram in (a) is displayed with a logarithmic pressure axis, the solid phase area is more obvious. The qualitative behavior of $\mu(T)$ at pressures, from top to bottom, between the critical pressure and the triple point pressure, at the triple point pressure, and below the triple point pressure shows the crossing points for melting, vaporization, and sublimation. (Compare Figure 6.4.) The thick portion of each curve is the stable portion followed by a system at equilibrium.

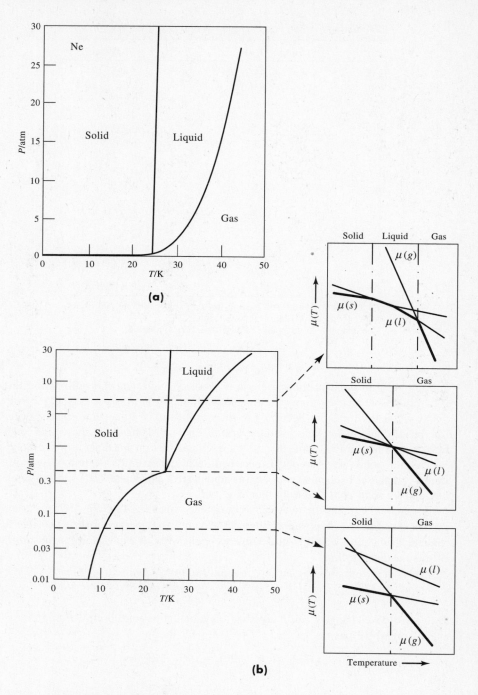

If we adopt the convention of writing the low-temperature phase as the "reactant" and the high-temperature phase as the "product" of an expression like Eq. (6.8), then we can define

$$\Delta \overline{S}_\phi \equiv \overline{S}_2 - \overline{S}_1 \tag{6.9}$$

and

$$\Delta \overline{V}_\phi \equiv \overline{V}_2 - \overline{V}_1 \ . \tag{6.10}$$

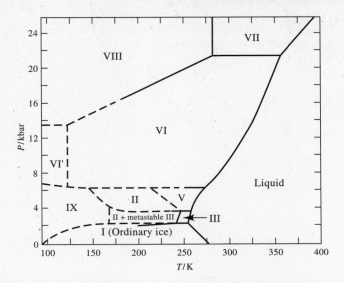

FIGURE 6.8 The phase diagram for water at high pressures exhibits the various solid phases of water and their coexistence lines (extrapolated with dashed lines in some cases from incomplete data). The entire gas-phase region has disappeared due to the pressure scale used in this figure. The solid phase I is ordinary ice.

These definitions give us a compact expression for the slope of a two-phase coexistence line, the *Clapeyron equation*

$$\left(\frac{dP}{dT}\right)_\phi = \left(\frac{\Delta \overline{S}_\phi}{\Delta \overline{V}_\phi}\right). \quad \text{(6.11a)}$$

The subscript ϕ on dP/dT reminds us that the expression is valid only for two-phase equilibrium.

Since $\Delta \overline{H}_\phi = T_\phi \, \Delta \overline{S}_\phi$, this equation can be rewritten[1] as

$$\left(\frac{dP}{dT}\right)_\phi = \left(\frac{\Delta \overline{H}_\phi}{T_\phi \, \Delta \overline{V}_\phi}\right). \quad \text{(6.11b)}$$

A Closer Look at the Clapeyron Equation

We have defined $\Delta \overline{S}_\phi$, etc., as if we had made an isobaric passage from phase 1, through 1-2 equilibrium, to phase 2 by raising the temperature. Suppose we had made an isothermal passage instead, or a passage through the two-phase line in any arbitrary direction on a diagram like Figure 6.8. Equation (6.11) ensures that the *ratio* $\Delta \overline{S}_\phi / \Delta \overline{V}_\phi$ is path-independent, but what about $\Delta \overline{S}_\phi$ or $\Delta \overline{V}_\phi$ alone? The only change these quantities can have with path is algebraic sign, depending on the direction of passage: $1 \to 2$ or $2 \to 1$.

What happens at the critical point, where $\Delta \overline{V}_\phi \to 0$? Does $(dP/dT)_\phi$ become infinite? A glance at Figure 6.6 would seem to say that it does not. If so, then $\Delta \overline{S}_\phi$ must also go to zero at the critical point, and it must do so *faster* than $\Delta \overline{V}_\phi$ vanishes. For the van der Waals gas, we can use Eq. (5.9) (with $T_f = T_i = T_\phi$ and

---

[1]There is an amusing historical anecdote associated with this form of the Clapeyron equation. It was once commonly written $\frac{dP}{dT} = \frac{1}{T}\frac{dq}{dV}$, and the factor $\frac{dq}{dV}$ was called the "isothermal heat of expansion," symbolized M. The $1/T$ factor was called the "Carnot function" (of Second-Law integrating-factor fame), symbolized C. Since q was measured in calories and dP/dT was a mechanical energy term, the conversion factor from calories to mechanical-energy units was needed as well. This was called the "mechanical equivalent of heat," symbolized J. Thus the equation read $\frac{dP}{dT} = JCM$ and thus, the eminent physicist James Clerk Maxwell used $\frac{dP}{dT}$ as his pen name!

$V_i = V_1$, $V_f = V_2$) to explore this behavior quantitatively. Figure 6.9(a) plots the van der Waals entropy of transition from the expression

$$\frac{\Delta \overline{S}_\phi}{R} = \ln\left(\frac{\tilde{V}_2 - 1}{\tilde{V}_1 - 1}\right)$$

while Figure 6.9(b) plots $\Delta \overline{V}_\phi / b$, using Eq. (6.4) to find $\tilde{V}_1$ and $\tilde{V}_2$ in both graphs. Finally, Figure 6.9(c) plots $(d\tilde{P}/d\tilde{T})_\phi$ in the vicinity of the critical point. Note that these arguments also imply that $\Delta \overline{H}_\phi$ vanishes at the critical point.

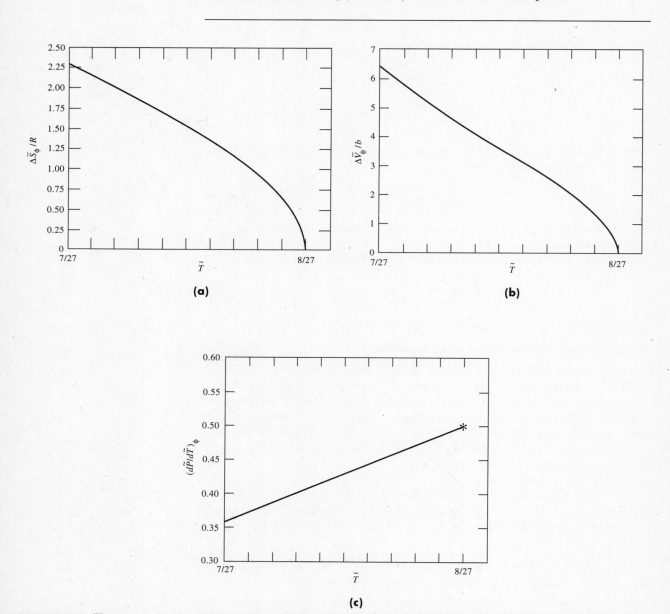

FIGURE 6.9(a) The entropy change of vaporization of a van der Waals gas rapidly falls to zero in the vicinity of the critical temperature. **(b)** The volume change on vaporization of a van der Waals gas also goes to zero at the critical temperature, as the two phases become indistinguishable. **(c)** Even though $\Delta \overline{V}_\phi$ and $\Delta \overline{S}_\phi$ vanish at the critical point (the * on the figure), the slope of the coexistence line $(dP/dT)_\phi$ remains finite.

The Clapeyron equation allows us to study a phase diagram and infer from the slopes of coexistence lines the algebraic signs of $\Delta \overline{S}_\phi$ and $\Delta \overline{V}_\phi$. Table 6.2 summarizes the possibilities.

Figures 6.10(a) and 6.10(b) show the phase diagrams for two absolutely unique substances, the helium isotopes, ^{4}He and ^{3}He. Both isotopes lack a s-l-g triple point and resist solidification at absolute zero. (Not shown in these diagrams are high-pressure solid–solid phase transitions, or remarkable liquid phases of ^{3}He that exist below ~3 mK.)

Helium-4 has a *liquid–liquid* phase transition, and the l–l coexistence line has a negative slope. He I(l), the high temperature form, is an ordinary (if very cold) liquid, but He II(l) is a "superfluid," exhibiting no viscosity; it leaks freely through the smallest hole and spontaneously flows as a very thin liquid film, even up container walls. The origin of this behavior can be traced to an unusual macroscopic manifestation of quantum-mechanical phenomena. But whatever the reason, the negative slope of the coexistence line implies that either $\Delta \overline{S}_\phi$ or $\Delta \overline{V}_\phi$ is negative. Experiments have established that $\Delta \overline{V}_\phi$ is negative, and that just as ice floats on liquid water, He II(l) floats on He I(l).

Now consider the s–l line below about 1 K. This line is virtually horizontal, and it becomes exactly horizontal at 0 K, since $\Delta \overline{S}_\phi$ must vanish at absolute zero according to the Third Law.

Contrast this with ^{3}He, which exists as a normal liquid above 3 mK. The $l \to s$ coexistence line below about 0.3 K has a negative slope, and again either $\Delta \overline{S}_\phi$ or $\Delta \overline{V}_\phi$ is negative, but here experiment establishes that $\Delta \overline{S}_\phi$ is negative. This is the only known case of a negative entropy (and thus a negative enthalpy) of fusion in a pure substance. In other words, for ^{3}He below 0.3 K and 32 atm, $\overline{S}(l) < \overline{S}(s)$. The surprising implication is that one must *heat this liquid to solidify it!* (And, of course, ^{3}He does obey the Third Law. The l–s line becomes horizontal as $T \to 0$.)

Returning to the full Clapeyron equation, how can we integrate it to find an expression for the coexistence line? In general, the following steps are called for.

1. Begin with $(dP/dT)_\phi = \Delta \overline{H}_\phi / T_\phi \Delta \overline{V}_\phi$, Eq. (6.11b).
2. Express $\Delta \overline{V}_\phi$ as $\overline{V}_2 - \overline{V}_1$, and use equations of state to find $\overline{V} = \overline{V}(T, P)$ for each phase.

TABLE 6.2 Clapeyron Equation Possibilities

Transition	$\Delta \overline{V}_\phi$	$\left(\dfrac{dP}{dT}\right)_\phi$
solid → gas	> 0 ($\cong \overline{V}(g)$)	> 0 (but T dependent)
liquid → gas	> 0 ($\cong \overline{V}(g)$)	> 0 (but T dependent)
solid → liquid	usually > 0 (exceptions: H$_2$O, Bi, Ga, Ge, Si)	usually > 0 (but very constant)
solid → solid	$\gtreqless 0$	$\gtreqless 0$ (but very constant)

FIGURE 6.10(a) The phase diagram for ^{4}He lacks an s–l–g triple point. The gas-phase area is too small (at too low a pressure) to be seen on this graph. Note that ^{4}He does not solidify, even at $T = 0$, unless the pressure is >25 atm. **(b)** The ^{3}He isotope has a very different phase diagram from the ^{4}He isotope. Note the negative slope of the s–l coexistence line below $T \sim 0.5$ K. One must heat ^{3}He(l) to solidify it!

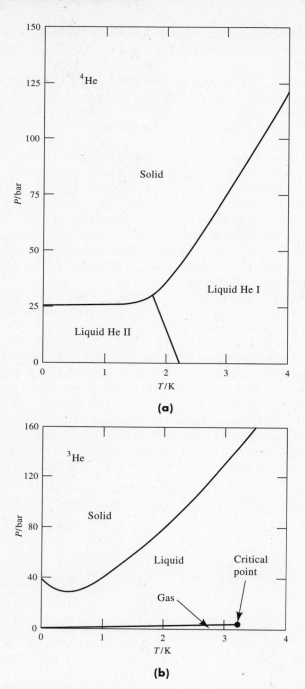

3. Express $\Delta \overline{H}_\phi$ as $\overline{H}_2 - \overline{H}_1$, and use Equations (5.4) and (5.5) to find $\overline{H} = \overline{H}(T, P)$ for each phase.
4. Integrate:

$$\int_{T_i}^{T_f} \left(\frac{dP}{dT} \right)_\phi dT = \int_{T_i}^{T_f} \frac{\Delta \overline{H}_\phi}{T_\phi \Delta \overline{V}_\phi} dT_\phi = \int_{P_i}^{P_f} dP \ .$$

These are not always easy steps to follow, but useful expressions can be obtained by approximating several steps, as derived below.

Vapor–Liquid or Vapor–Solid Coexistence Line (Vaporization and Sublimation)

If one phase is a gas far from the critical point, then $\overline{V}_g \gg \overline{V}$(condensed phase) and $\Delta\overline{V}_\phi \cong \overline{V}_g$. Moreover, if the ideal gas equation of state is sufficiently accurate, then we have simplified the Clapeyron equation to

$$\left(\frac{dP}{dT}\right)_\phi = \frac{\Delta\overline{H}_\phi}{T_\phi \Delta\overline{V}_\phi} \cong \underset{\underset{\Delta\overline{V}_\phi \cong \overline{V}_g}{\text{Since}}}{\frac{\Delta\overline{H}_\phi}{T_\phi \overline{V}_g}} = \underset{\underset{\overline{V}_g \cong RT_\phi/P}{\text{Since}}}{\frac{\Delta\overline{H}_\phi P}{RT_\phi^2}}.$$

If the temperature range of interest is not large, $\Delta\overline{H}_\phi$ is approximately constant, and we rearrange and integrate:

$$\int_{P_i}^{P_f}\left(\frac{dP}{P}\right) \cong \frac{\Delta\overline{H}_\phi}{R}\int_{T_i}^{T_f}\frac{dT}{T^2}$$

yielding a useful, approximate expression for vapor pressure:

$$P(T_f) = P(T_i)\exp\left[-\frac{\Delta\overline{H}_\phi}{R}\left(\frac{1}{T_f} - \frac{1}{T_i}\right)\right]. \qquad (6.12)$$

EXAMPLE 6.2

As we noted in Section 4.4, $\Delta\overline{H}_{\text{vap}}^\circ(H_2O) = 40.66$ kJ mol^{-1}. At the normal boiling point, 373.15 K (which we will take to be T_i), the vapor pressure is, by definition of a normal boiling point, 1 atm. These are sufficient data to allow us to predict the vapor pressure of water at other temperatures using Eq. (6.12). Compare its predictions to the following observed values that span the temperature range from the triple point to the critical point.

T_f/K	647.27	500.	400.	350.	300.	273.16
P_{obs}/atm	218.2	26.1	2.43	0.411	0.0349	0.00603

SOLUTION Direct calculation yields the following values and their percentage errors:

T_f/K	647.27	500.	400.	350.	300.	273.16
P_{calc}/atm	257.	27.8	2.41	0.420	0.0410	0.00825
$\dfrac{P_{\text{calc}} - P_{\text{obs}}}{P_{\text{obs}}}$	17.8 %	6.5 %	−0.8 %	2.2 %	17.5 %	36.8 %

At the extremes, far from the reference temperature, the agreement between calculated and observed values is not particularly good, but accurate predictions can be made over a range of at least 50 K from the reference temperature.

→ RELATED PROBLEMS 6.4, 6.6, 6.7, 6.9, 6.10

We assumed $\Delta\overline{H}_\phi$ (rather than $\Delta\overline{S}_\phi$) was constant in deriving Eq. (6.12). In fact, neither are, as Figure 6.11(a, b) illustrates for water from the triple point to the critical point (where both $\Delta\overline{H}_{\text{vap}}$ and $\Delta\overline{S}_{\text{vap}}$ are zero).

Solid–Liquid Coexistence Line (Fusion)

Since solids and liquids are very incompressible phases with similar molar volumes, we can assume $\Delta\overline{V}_{\text{fus}}$ is independent of T and P. Also, the molecular order of a solid

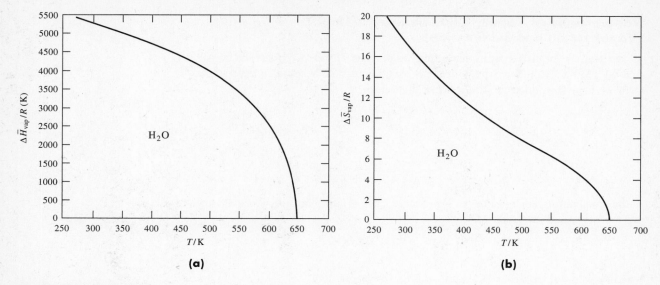

FIGURE 6.11(a) The enthalpy of vaporization of water is a slowly varying function of T except in the close vicinity of the critical temperature. These data span the full region from the triple point to the critical point, which, of course, includes the normal boiling temperature of 373.15 K. **(b)** Similarly, $\Delta \overline{S}_{vap}$ (H$_2$O) changes with T throughout the stability region of liquid water.

about to melt is relatively insensitive to T and P (a good assumption unless a solid–solid phase transition occurs) and similarly for a liquid (a less-good assumption in general), so that $\Delta \overline{S}_{fus}$ is fairly constant. This means the fusion coexistence curve (the melting curve) is approximately straight:

$$\left(\frac{dP}{dT}\right)_{fus} = \frac{\Delta \overline{S}_{fus}}{\Delta \overline{V}_{fus}} \cong \text{a constant} .$$

This slope is positive for most substances (and rather large typically—see Ne in Figure 6.7, for example), but as Table 6.2 points out, for water and a few other substances such as Bi, the solid is less dense than the liquid (frozen water and Bi both float in their liquids), and the slope is negative.

A more accurate, but semiempirical, expression for the fusion coexistence curve was suggested by Simon and Glatzel in 1929 and is now written in the form

$$\frac{P}{P_0} = \left(\frac{T}{T_0}\right)^a - 1 \tag{6.13}$$

where a and P_0 are empirical constants and T_0 is a reference temperature that is often taken to be at or near the triple-point temperature. For the rare gases Ar, Kr, and Xe, experimental (and highly accurate) data for the melting curve are satisfactorily fit by the equations

$$\frac{P}{2234 \text{ atm}} = \left(\frac{T}{83.2 \text{ K}}\right)^{1.5} - 1 \quad \text{(Ar)}$$

$$\frac{P}{3000 \text{ atm}} = \left(\frac{T}{116.1 \text{ K}}\right)^{1.4} - 1 \quad \text{(Kr)}$$

$$\frac{P}{3400 \text{ atm}} = \left(\frac{T}{161.5 \text{ K}}\right)^{1.31} - 1 . \quad \text{(Xe)}$$

If the Simon equation is differentiated and the result is carefully arranged and compared to the Clapeyron equation:

$$\underbrace{\frac{\Delta \overline{H}_{fus}}{T_{fus} \Delta \overline{V}_{fus}}}_{\text{Clapeyron}} = \left(\frac{dP}{dT}\right)_{fus} = \underbrace{\left(\frac{P_0 a}{T_{fus}}\right)\left(\frac{T_{fus}}{T_0}\right)^a}_{\text{Simon}}$$

we can identify $P_0 a$ (which has the dimensions of pressure) as $\Delta \overline{H}_{fus}/\Delta \overline{V}_{fus}$ and consider $(T_{fus}/T_0)^a$ an empirical "correction factor" for the temperature dependence of the melting curve's slope. (While this indicates the origins of the Simon equation, the constants in it are just "best fit" parameters that may not accurately reflect $\Delta \overline{H}_{fus}/\Delta \overline{V}_{fus}$ near T_0.)

Solid–Solid Coexistence Line

The coexistence line between two solid phases is often very straight, but with a slope that is difficult to predict *a priori*. Vertical lines are known, corresponding to two phases that differ in entropy (due to a reordering of atoms in the solid) but not in volume. Nearly horizontal lines are known for which the reverse is true, as are lines of virtually all intermediate slopes.

We conclude this important section with some general comments on phase transition enthalpies and entropies. For vaporization or sublimation, one can show (see Problem 6.38)

$$\frac{d(\Delta \overline{H}_\phi)}{dT} \cong \Delta \overline{C}_{P\phi}. \quad (\text{g–l or g–s only})$$

Since $\overline{C}_P(l) > \overline{C}_P(g)$ near T_{vap}, $\Delta \overline{C}_{P,vap} < 0$ so that $\Delta \overline{H}_{vap}$ generally decreases with increasing T.

The entropy of vaporization of many simple liquids in which the intermolecular forces are neither very strong nor very anisotropic (i.e., excluding H_2O, small alcohols and carboxylic acids, HF, and other hydrogen bonding species) is remarkably constant at the normal boiling point. This constancy is expressed by *Trouton's rule*:

$$\Delta \overline{S}^\circ_{vap} \cong 10.5\, R \quad \text{and} \quad \Delta \overline{S}^\circ_{sub} \cong 11.5\, R\,.$$

The accuracy and generality of this rule is improved if one considers vaporization not at constant pressure, but at temperatures for which the vapor densities are the same from compound to compound. This is the basis of *Hildebrand's rule*:

$$\Delta \overline{S}^\circ_{vap} \cong 11.1\, R \quad \text{when} \quad \overline{V}_{gas} = 22.414 \text{ L mol}^{-1}\,.$$

A similar rule due to Guggenheim states that

$$\Delta \overline{S}^\circ_{vap} \cong 9.0\, R \quad \text{when} \quad P_{vap} = P_c/50\,.$$

The value of 50 is arbitrary, but the idea is clearly related to that of corresponding states, as is Hildebrand's rule.

Trouton's rule allows a rapid (if sometimes reckless) estimation of vapor pressure given only the normal boiling temperature, T°_{vap}. From Eq. (6.12), we find

$$P(T)/\text{atm} \cong \exp\left[10.5\left(1 - \frac{T^\circ_{vap}}{T}\right)\right]\,. \tag{6.14}$$

For example, benzene has a normal boiling temperature of 353.25 K. Equation (6.14) predicts a vapor pressure of 0.155 atm at 300 K; the experimental value is 0.154 atm. But methane, which might be expected to follow Trouton's rule rather well, has $\Delta \overline{S}^\circ_{vap} = 8.83\, R$. Equation (6.14) would be in error for methane by a factor on the order of $\exp(10.5/8.83) \cong 3.3$.

6.3 COMPOSITION VARIABLES IN MIXTURES

Imagine a closed system in some two-phase equilibrium. How can we find the relative amounts of each phase in this system? At constant pressure, *all* the system is one phase below T_ϕ and *all* is the other phase above T_ϕ, but *at* T_ϕ, any fraction of the system could be in either phase. Call the phases 1 and 2, as in Eq. (6.8), and let n be the total amount (in moles) of the compound comprising the system. Then we can define the *mole fraction*

$$x(i) = \frac{n(i)}{n} = \frac{\text{amount in phase } i}{\text{total amount}} \qquad (6.15)$$

and recognize that mole fractions add to 1:

$$\sum_{\substack{\text{all}\\\text{phases}}} x(i) = 1 \; .$$

We further define the molar volume of the system (in the usual way) and relate it to the molar volume of each phase:

$$\overline{V} \equiv \frac{V}{n} = \sum_{\substack{\text{all}\\\text{phases}}} \overline{V}(i)\, x(i) \; . \qquad (6.16)$$

For two-phase equilibrium, Eq. (6.16) can be written in a very revealing form:

$$\overline{V} = \underbrace{\overline{V}[x(1) + x(2)]}_{\text{Since } x(1) + x(2) = 1} = \underbrace{\overline{V}(1)\, x(1) + \overline{V}(2)\, x(2)}_{\text{From Eq. (6.16)}},$$

which can be rearranged to

$$x(2)[\overline{V} - \overline{V}(2)] = x(1)[\overline{V}(1) - \overline{V}] \; . \qquad (6.17)$$

This equation is known as the *lever rule,* a name that is interpreted in Figure 6.12 using the system of Figures 6.2 and 6.3. We locate a fulcrum at $\tilde{V}(\text{system})$ and place it under the horizontal tie line, which we think of as the lever. On either end of this lever, we hang masses proportional to each mole fraction; Eq. (6.17) tells us the ratio these masses must have to balance the lever.

The absolute mole fraction of either phase is found from alternate versions of the lever rule:

$$x(1) = \frac{\overline{V} - \overline{V}(2)}{\overline{V}(1) - \overline{V}(2)} \qquad (6.18a)$$

$$x(2) = \frac{\overline{V} - \overline{V}(1)}{\overline{V}(2) - \overline{V}(1)} \; . \qquad (6.18b)$$

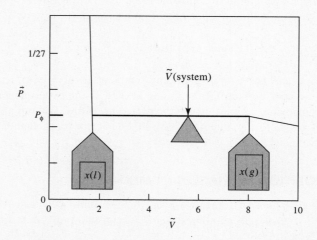

FIGURE 6.12 This figure illustrates the lever rule using the van der Waals isotherm of Figures 6.2 and 6.3. When the two-phase system has the molar volume $\tilde{V}(\text{system})$ noted by the arrow, the fraction in the gas phase $x(g)$ is represented by a weight that just balances a similar weight representative of the liquid fraction $x(l)$ with a fulcrum at the system's volume. The tie-line at the vapor pressure is the lever.

We will find further use for these expressions in two-phase equilibria between solutions as well as between pure compound phases.

We use the notation $x(\phi)$ for the mole fraction of a pure substance in phase ϕ when there is more than one phase, and the notation $x_i(\phi)$ for the mole fraction of compound i in phase ϕ when we have a multicomponent, multiphase system. But when do we have a *mixture* of compounds rather than a *solution* of compounds?

6.4 DIFFERENCES BETWEEN SOLUTIONS AND MIXTURES

Melt Bi and Cd, pour the liquid metals together above ~600 K, and a *liquid solution* results. We call this a solution because at a microscopic level, the atoms are randomly dispersed throughout the phase. Cool this solution below 413 K, and it all freezes. But if the solidification is carried out slowly enough to ensure equilibrium at every temperature, the solid will have macroscopic regions of pure Bi and pure Cd, as if we had taken the pure metals, powdered them, mixed the powder, and pressed the mixture together. We call this a two-phase *solid mixture* rather than a *solid solution*. In contrast, if we repeat this experiment with Cu and Ni, we find a solid that, under close scrutiny, has randomly dispersed atoms, just as the liquid solution. This is characteristic of a *solid solution*.

These terms take some getting used to, because we are compelled to use phrases like, "Cu(*l*) and Ni(*l*) spontaneously *mix* to form a homogeneous *solution*," as well as phrases like, "Cd(*s*) and Bi(*s*) form an inhomogeneous solid *mixture* when cooled from the liquid *solution*."

Why do some pairs of compounds form homogeneous solutions while others may spontaneously "unmix" from a solution, as do Cd and Bi when solidified? To answer this question, we need to understand the differences between homogeneous solutions and inhomogeneous mixtures.

Ideal Mixing and Partial Pressure

We define first an *ideal mixture*. Moreover, we consider an *ideal mixture of ideal gases*. On a molecular level, ideal gases are distinguished *only* by their molecular mass. They interact neither with their own kind nor with other species, ideal or not. Thermodynamically, an ideal gas is described by the simple equation of state $PV = nRT$, the Sakur–Tetrode expression for entropy, Eq. (5.44), and an internal energy that depends only on temperature.

A low-density monatomic gas is the closest we can come to this ideal. To follow the mixing of such gases, we could use two different isotopes of a rare gas, such as ^{83}Kr and ^{84}Kr. Each isotope can be distinguished by mass spectroscopy or by high resolution optical spectroscopy. We imagine, as in Figure 6.13(a), equal quantities of each isotope separated by a valve. Each container is adiabatically isolated and at the same T and P (i.e., all *intensive* variables are the same for each gas). We open the valve and periodically close it long enough to measure the equilibrium concentration of each isotope in each container. Eventually the amounts of each isotope in each container would be equal. We would also find that the temperature and pressure of each container had not changed, as in Figure 6.13(b).

Spontaneous mixing at constant T and P is best analyzed in terms of a decrease in G. At any point in the process, Eq. (5.39) will hold (whenever we close the valve to stop and measure concentrations):

$$G(\text{total}) = \mu_1 n_1 + \mu_2 n_2 .$$

Each μ is given by Eq. (5.48), and since the gases are ideal, we replace fugacity

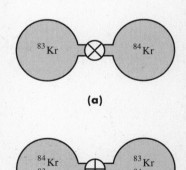

FIGURE 6.13 Isolated rare gas isotopes, (a), spontaneously mix to form an ideal mixture of ideal gases as they diffuse to a uniform composition in both containers, (b).

with pressure. But the appropriate pressure is not the total pressure P; rather, it is the subtle thing called the *partial pressure*. We write

$$P = P_1 + P_2 \quad \textit{(Dalton's law of partial pressures)} \quad (6.19)$$

where P_1 and P_2 are the partial pressures of components 1 and 2. It is the intensive nature of pressure that makes partial pressures somewhat subtle. Extensive quantities can be added and their sums readily given physical meaning. Not so for intensive properties, however. (Consider the density of a system of ice and water, for instance. The system density is clearly not the sum of the density of ice and the density of water!)

Equation (6.19) is so compelling that a physical interpretation of it has been around since about 1805, when Dalton first expressed it. Gibbs, in 1876, best stated Eq. (6.19) in words:

> It is in this sense that we should understand the law of Dalton, that every gas is as a vacuum to every other gas.

Think how reasonable this is. Two (or more) gases are mixed, and each thinks itself all alone in the entire volume of the gaseous solution.[2]

Since it is not at all obvious how one separates an intensive quantity into a sum of intensive quantities, each related to a unique compound, we must define partial pressure in a way that is at least useful (theoretically and practically) and measurable (experimentally), if not completely open to physical interpretation. The definition is based on composition variables and total pressure:

$$P_i = x_i P = \textit{partial pressure of component i} \ . \quad (6.20)$$

This definition is in accord with the spirit of Dalton, and when his spirit is violated by strong intermolecular interactions, it still has an operational flavor. We can always compute $x_i P$ and call this number "the partial pressure," even if its physical interpretation is a bit cloudy.

A Closer Look at Partial Pressure

The thermodynamic definition of pressure comes from the relations

$$\left(\frac{\partial U}{\partial V}\right)_{S,n} = \left(\frac{\partial A}{\partial V}\right)_{T,n} = -P \ .$$

There is no need to speak of molecules smashing into walls to define pressure. Since U and A are extensive, suppose we approach partial pressure by attacking these quantities and then performing the volume derivative. We concentrate on A, since the isothermal constraint is easier to imagine than the isentropic constraint on $(\partial U/\partial V)$.

At constant T and P, it will be true (compare Eq. (5.38)) that

$$A = \sum_i n_i \overline{A}_i = n \sum_i x_i \overline{A}_i$$

where $\overline{A}_i$ is the partial molar Helmholtz free energy of component i. We differentiate with respect to volume, remembering that each x is constant since the system is

---

[2]Note, however, that room is left for molecules of each gas to behave nonideally due to intermolecular forces among their own kind, but *inter-species* forces are tacitly ruled out. This is unreasonable unless like-species forces are also negligible.

closed (and we are assuming no chemical reactions are possible which can change x_i's), and find

$$\left(\frac{\partial A}{\partial V}\right)_{T,n} = \sum_i x_i\, n \left(\frac{\partial \overline{A_i}}{\partial V}\right)_{T,n}.$$

This suggests we should identify the partial pressure as

$$P_i = -x_i\, n \left(\frac{\partial \overline{A_i}}{\partial V}\right)_{T,n}$$

so that $P = \Sigma P_i$. One problem with this definition is that $(\partial \overline{A_i}/\partial V)$ has no simple operational interpretation. Moreover, the definition is a trivial identity for an ideal mixture of ideal gases. Consequently we define partial pressures by the simple expedient $P_i = x_i P$.

Now consider an apparatus sketched below, which would seem to provide a convenient way to measure partial pressures directly. In one side is pure H_2; in the other, we have H_2 and N_2. But instead of an impermeable wall between the two sides, we have a rigid foil of palladium metal, which has the very clever ability to be invisible to H_2.

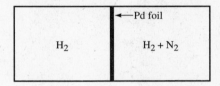

A material that allows one compound to pass through it while blocking others is said to be *semipermeable* to that compound. Unfortunately, only very few semipermeable substances are known. But if they were commonplace, then we could *use them to define a partial-pressure measurement*.

In the system at equilibrium above, the chemical potential of H_2 is the same on both sides of the foil. The *total* pressures on each side can be different, since the foil is fixed and rigid. But if we add a bit of H_2 to one side or the other, both pressures will change, while if we add a bit of N_2 to the mixture, the key experimental result is that the pressure in the pure H_2 chamber could change. This experiment would *not* necessarily find the H_2 chamber pressure $= P_{H_2} = x_{H_2} P =$ (H_2 mole fraction in the gaseous solution)×(solution-chamber total pressure).

It is convenient to define the changes in extensive quantities (such as G, V, S, etc.) that accompany mixing as

$$\Delta G_{\text{mix}} = G(x_1, x_2, \ldots) - \sum_i G(\text{pure, isolated species } i) \tag{6.21}$$

where $G(x_1, x_2, \ldots)$ is

$$G(x_1, x_2, \ldots) = n \sum_i x_i\, \mu_i(x_i) = n \sum_i x_i(\mu_i^\circ + RT \ln P_i). \tag{6.22}$$

We assume T and P are constant, along with n. Similarly,

$$G(\text{pure } i) = n_i\, \mu_i(x_i = 1) = n_i(\mu_i^\circ + RT \ln P). \tag{6.23}$$

In detail, Eq. (6.21) is

$$\Delta G_{\text{mix}} = n\left[\sum_i x_i(\mu_i^\circ + RT \ln P_i) - \sum_i x_i(\mu_i^\circ + RT \ln P)\right]$$

$$= n\left[\sum_i x_i(\mu_i^\circ + RT \ln P + RT \ln x_i) - \sum_i x_i(\mu_i^\circ + RT \ln P)\right],$$

which simplifies to

$$\Delta G_{\text{mix}}(x_1, x_2, \ldots) = nRT \sum_i x_i \ln x_i. \tag{6.24}$$

If only two distinguishable species are mixing, then $x_2 = 1 - x_1$, and Eq. (6.24) is

$$\frac{\Delta G_{\text{mix}}(x_1)}{nRT} = x_1 \ln x_1 + (1 - x_1) \ln (1 - x_1)$$

which is graphed in Figure 6.14(a).

We can quickly derive changes in other extensive quantities that accompany mixing:

$$\Delta S_{\text{mix}} = -\left(\frac{\partial \Delta G_{\text{mix}}}{\partial T}\right)_{x_i} = -nR \sum_i x_i \ln x_i \tag{6.25}$$

$$\Delta V_{\text{mix}} = \left(\frac{\partial \Delta G_{\text{mix}}}{\partial P}\right)_{x_i} = 0 \tag{6.26}$$

$$\Delta H_{\text{mix}} = \Delta G_{\text{mix}} + T\Delta S_{\text{mix}} = 0 \tag{6.27}$$

$$\Delta U_{\text{mix}} = \Delta H_{\text{mix}} - \Delta(PV)_{\text{mix}} = 0 \tag{6.28}$$

$$\Delta A_{\text{mix}} = \Delta U_{\text{mix}} - T\Delta S_{\text{mix}} = \Delta G_{\text{mix}}. \tag{6.29}$$

These results, Equations (6.24)–(6.29), are for ideal mixtures of ideal gases, and collectively they indicate that *spontaneous ideal mixing is due to an increase in entropy* (Eq. (6.25) has $x_i < 1$; so, $\ln x_i < 0$ and $\Delta S_{\text{mix}} > 0$), as is graphed in Figure 6.14(b) for a binary mixture. There is no volume change; there is no enthalpy or internal energy change. The system falls down the chemical potential well of Figure 6.14(a) as far as the total composition constraint of the mixture will allow.[3]

---

[3]There is a subtle aspect to mixing that Gibbs recognized in what has become known as "Gibbs' Paradox." Suppose in Figure (6.13) both containers held the same isotope. Then when the valve is opened, the gas would diffuse from container to container, but no macroscopic mixing would be measured and no entropy change would exist. The "paradox" is that ΔS_{mix} vanishes suddenly if components 1 and 2 are identical. This behavior is not paradoxical, in fact. Rather it points out the role of distinguishability in the calculation of entropy. One cannot continuously or infinitesimally alter a species' identity.

FIGURE 6.14(a) The Gibbs free energy change for an ideal mixture is everywhere negative, implying such mixing is always a spontaneous process. **(b)** The origin of spontaneous mixing in an ideal mixture is the entropy *increase* accompanying such mixing at all compositions.

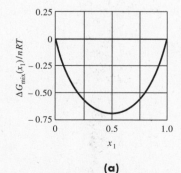
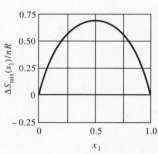

(a) (b)

Nonideal Mixing

We now turn to nonideal situations. First, we reintroduce fugacities. The simplest way, first used by Lewis and Randall to define a general ideal mixture (of real gases, or, as we shall see, of liquids), is to define *partial fugacities* in analogy with partial pressures. Their definition is called the *Lewis–Randall rule*:

$$f_i = x_i f(\text{pure } i, \text{ at the } T \text{ and total } P \text{ of the solution})$$

so that

$$\mu_i(T, x_i) = \underbrace{\mu°(\text{pure } i, T) + RT \ln f}_{\substack{\uparrow \\ \mu \text{ of pure} \\ \text{species } i}} + \underbrace{RT \ln x_i}_{\substack{\uparrow \\ \text{Correction for} \\ \text{solution composition}}}.$$

(with the first underbrace labeled "μ of species i in the solution" pointing to $\mu_i(T, x_i)$)

This definition relies on *Amagat's law,* which states

$$V = \sum_i n_i \overline{V}(\text{pure } i) \;.$$

Ideal gas mixtures follow Amagat's law exactly, and many real gas mixtures do so to a useful approximation.

To use the Lewis–Randall rule for solids or liquids, we need to look further at condensed phase fugacities. For a pure component l–g or s–g equilibrium, the fugacity of the liquid or solid equals that of the gas, and Chapter 5 discussed gas fugacities.

The fugacity of a pure condensed phase at any arbitrary[4] T and P comes from Eq. (5.49),

$$RT \ln \left(\frac{f}{P}\right) = \int_0^P \left(\overline{V} - \frac{RT}{P}\right) dP \;. \tag{5.49}$$

We break the integral into two parts, one going from zero pressure to P_ϕ, where P_ϕ is the l–g or s–g phase transition pressure at the temperature of interest, and another going from P_ϕ to P, the final pressure of interest:

$$RT \ln \left(\frac{f}{P}\right) = \int_0^{P_\phi} \left(\overline{V} - \frac{RT}{P}\right) dP + \int_{P_\phi}^P \left(\overline{V} - \frac{RT}{P}\right) dP \;.$$

The first integral gives the fugacity at the phase transition, $RT \ln (f_\phi/P_\phi)$, while the second integral is best written as a sum of its two terms:

$$RT \ln \left(\frac{f}{P}\right) = RT \ln \left(\frac{f_\phi}{P_\phi}\right) + \int_{P_\phi}^P \overline{V}(\text{condensed phase}) \, dP - RT \ln \left(\frac{P}{P_\phi}\right) \;.$$

Since condensed-phase molar volumes are fairly insensitive to pressure, the last integral is approximately $\overline{V}(P - P_\phi)$. With this approximation, we can divide both sides of the previous equation by RT and exponentiate both sides. This yields

$$f(\text{condensed phase}, T, P) = P_\phi \left(\frac{f_\phi}{P_\phi}\right) \exp \left[\frac{\overline{V}(P - P_\phi)}{RT}\right] \;. \tag{6.30}$$

The right-hand side could have factors of P_ϕ cancelled, but they are left in to emphasize the three parts of this expression. First, $f \cong P_\phi$. Next, f is corrected by

---

[4]Is negative pressure an impossible or meaningless concept? For example, the van der Waals isotherm for $\tilde{T} = 6.5/27$, shown in Figure 1.7, has a portion that extends below $\tilde{P} = 0$, but we now know that two-phase separation makes this portion unstable. However, it *is* possible to have a liquid *at negative pressure* in a metastable state; it is simply a liquid *under tension,* a stretched liquid.

the fugacity coefficient f_ϕ/P_ϕ to account for the nonideality of the vapor at the phase transition. Finally, f is corrected by the exponential term that takes the condensed phase from P_ϕ to the pressure of interest.[5]

EXAMPLE 6.3

The vapor pressure of Ar(s) at 75 K is 0.185 atm, and the solid's molar volume is 24.65 cm^3 mol^{-1}. What is the solubility of Ar(s) in He(g) if the solid is compressed isothermally to 100 atm by He(g)?

SOLUTION The solubility of a solid in a gas is the equilibrium gas-phase mole fraction of the solid, and this is the quantity we seek.

If we assume the Ar vapor pressure is unchanged by the high pressure He, this pressure is the Ar partial pressure in the gas-phase solution. We find $x_{Ar} = P_{Ar}/P = (0.185 \text{ atm})/(100 \text{ atm}) = 1.85 \times 10^{-3}$. This assumption neglects two important effects, however: the vapor pressure increase on compression and the nonideality of high-pressure gaseous mixtures.

Thus we argue as follows. At equilibrium, the fugacity of the solid argon equals the fugacity of Ar in the gas-phase solution. (We can safely assume that He is insoluble in solid Ar.) We apply Eq. (6.30) to find the fugacity of Ar(s) at 100 atm. The first factor, P_ϕ, is simply $P_{sub}(75 \text{ K}) = 0.185$ atm. The next factor requires f_ϕ. If we assume that the second virial coefficient is adequate to describe the nonideality of Ar(g) at this pressure, then from Table 1.1 (with the help of Figure 1.6) we extrapolate a value for $B(75 \text{ K})$ from those at higher T and find $B(75 \text{ K}) \cong -263$ cm^3 mol^{-1}. The accuracy of such an extrapolation is probably not very great, but we should worry only if f_ϕ is appreciably different from P_ϕ. From Eq. (5.41),

$$\frac{f_\phi}{P_\phi} \cong \exp\left[\frac{B(T)P_\phi}{RT}\right] = 0.992 .$$

As a check, we can compute the van der Waals fugacity as well. This yields $f_\phi/P_\phi = 1.007$. Both approximations are sufficiently close to 1 that we can set the pure saturated vapor nonideality correction f_ϕ/P_ϕ equal to 1. Finally, we compute the Poynting correction for the compression. With $\overline{V}(s) = 24.65$ cm^3 mol^{-1} and $P = 100$ atm,

$$\exp\left[\frac{\overline{V}(s)(P - P_\phi)}{RT}\right] = 1.49 .$$

This is not negligible. We put all three factors together and find

$$f(Ar(s), 75 \text{ K}, 100 \text{ atm}) = (0.185 \text{ atm})(1)(1.49) = 0.276 \text{ atm} .$$

This is also the fugacity of Ar(g) in the He/Ar mixture, and it is small enough that we can take it to be the Ar partial pressure as a first approximation. We now need to find that Ar mole fraction in the gas-phase solution for which Ar has this partial pressure. We solve $P_{Ar} = x_{Ar} P$ for $x_{Ar} = (0.276 \text{ atm})/(100 \text{ atm}) = 2.76 \times 10^{-3}$.

A more advanced calculation that is beyond the scope of our discussion here takes into account nonideal mixing in the high-pressure gaseous solution and gives $x_{Ar} = 3.53 \times 10^{-3}$.

Notice the effects of various assumptions on the calculation:

$x_{Ar} = 1.85 \times 10^{-3}$ **if both the compression effect and nonideal mixing are neglected**

$x_{Ar} = 2.76 \times 10^{-3}$ **if the compression effect is included and nonideal mixing is neglected**

$x_{Ar} = 3.53 \times 10^{-3}$ **if both compression and nonideal mixing effects are included**

---

[5]This latter factor is often called a *Poynting correction*; see also Problem 5.17.

Nonideal mixing effects can sometimes be much larger (orders of magnitude rather than factors of two or so) and often are the most difficult to calculate.

⟹ RELATED PROBLEMS 6.39, 6.41

Excess Functions and Activity

We will measure deviations from ideality with *excess functions*, such as the excess Gibbs free energy, defined as

$$G^E = G(\text{real}; T, P, x_i) - G(\text{ideal}; T, P, x_i) .$$

Excess functions can be either positive or negative, and one speaks of corresponding positive and negative deviations from ideality. These functions, especially G^E, lead to the concept of *activity*, which we now go into some detail to introduce.

The discussion so far has centered around mixtures of gases that form homogeneous single-phase gaseous solutions. It could equally well apply to liquid solutions or solid solutions, if such solutions obeyed equations such as Eq. (6.24) for an ideal mixture. Now we specifically introduce all possible phases and compositions.

The fugacity of a component of a solution need only represent the deviation from some reference state. Lewis recognized the practicality of reference states (standard states), and he took advantage of this by defining the *activity*. The activity of component i at a particular T, P, and composition is the ratio of the fugacity of i in such a system to the fugacity of i at the same temperature, but at a reference pressure $P°$ and a reference composition $x°$:

$$a_i(T, P, x_1, x_2, \ldots) = \frac{f_i(T, P, x_1, x_2, \ldots)}{f_i°(T, P°, x°)} . \tag{6.31}$$

Note that $x°$ is shorthand for $x_1°, x_2°, x_3°$, etc. The standard state need not be pure, but it often is. Note also that the activity is dimensionless. *As long as all species are referred to the same standard state, the equilibrium requirement of equal fugacities for any component in all phases can be replaced by a requirement of equal activities.*

Lewis further defined the *activity coefficient*, γ_i, in terms of the composition variable x_i as

$$\gamma_i = \frac{a_i}{x_i} . \tag{6.32}$$

Equations (6.31) and (6.32) together show that

$$f_i(T, P, x) = f_i°(T, P°, x°) \, a_i(T, P, x) = f_i° \gamma_i x_i .$$

This looks useful, but it is no more than an identity, and a rather vague one at that until the standard state (which defines both $f_i°$ and γ_i) is specified. To make it useful and to bring the excess Gibbs free energy back into the discussion, we define an *ideal solution*.[6]

Ideal Solution

In an ideal solution, the fugacity of every component is related to its mole fraction by a simple proportionality factor, K_i, representing $f_i° \gamma_i$:

$$f_i(\text{ideal}; T, P, x_i) = K_i(T, P) x_i . \tag{6.33}$$

---

[6]This is distinct from an *ideal mixture*. The two can be equivalent, but in general, we will reserve *ideal solution* for liquid (or perhaps solid) single-phase mixtures.

More than one kind of ideal solution is possible, depending on the meaning of K_i.

From the definition of an excess function,

$$\begin{aligned}\overline{G}_i^E &= \overline{G}_i(\text{real}) - \overline{G}_i(\text{ideal}) = \mu_i(\text{real}) - \mu_i(\text{ideal}) \\ &= \mu_i^\circ(\text{real}, T) + RT \ln f_i(\text{real}) - \mu_i^\circ(\text{ideal}, T) - RT \ln f_i(\text{ideal}) \\ &= RT \ln \frac{f_i(\text{real})}{f_i(\text{ideal})} \ . \end{aligned}$$

Since *all* the nonideal behavior in μ is in the fugacity, $\mu_i^\circ(\text{real}) = \mu_i^\circ(\text{ideal})$.

Now we use the ideal fugacity, Eq. (6.33), in the denominator:

$$\overline{G}_i^E = RT \ln \frac{f_i(\text{real})}{K_i x_i} \ . \tag{6.34}$$

If $K_i = f_i^\circ(T, P^\circ, x^\circ)$, the activity coefficient in an ideal solution is unity since by definition of γ_i

$$\gamma_i(\text{real or ideal}) = \frac{f_i(\text{real or ideal})}{f_i^\circ x_i}$$

and, for an ideal solution with $K_i = f_i^\circ$,

$$\gamma_i(\text{ideal}) = \frac{f_i(\text{ideal})}{K_i x_i} = \frac{f_i(\text{ideal})}{f_i(\text{ideal})} = 1 \ .$$

Substituting $f_i(\text{real}) = f_i^\circ \gamma_i(\text{real}) x_i$ in the numerator and the ideal-solution fugacity $f_i(\text{ideal}) = f_i^\circ x_i$ in the denominator of Eq. (6.34) yields

$$\overline{G}_i^E = RT \ln \frac{f_i^\circ \gamma_i(\text{real}) x_i}{f_i^\circ x_i} = RT \ln \gamma_i(\text{real}) \tag{6.35}$$

so that

$$G^E = \sum_i n_i \overline{G}_i^E = nRT \sum_i x_i \ln \gamma_i(\text{real}) \ . \tag{6.36}$$

The implications of this equation are explored in the remainder of this chapter and again in Chapters 7 and 10.

Phase Separation and Spontaneous Un-mixing

To close this section, we look at G^E in Eq. (6.36) in very general terms. We will see how to predict and understand how a one-phase liquid solution can spontaneously separate into two phases when the composition or temperature of the system is changed. We take the pure liquid of each component at the temperature and pressure of the mixture as standard states: $P^\circ = P$, $x^\circ =$ pure compound i, and we also consider only a binary solution. In such a case, G^E must go to zero as x_1 goes to zero (pure liquid 2) or as x_2 goes to zero (pure liquid 1). At intermediate compositions, the simplest empirical function that satisfies these requirements is

$$G^E = nAx_1x_2 \tag{6.37}$$

where A is an empirical constant that depends on T and the chemical identity of 1 and 2, but not on the solution's composition.

This assumption is in accord with experiment for binary mixtures of chemically similar species. Comparing Equations (6.36) and (6.37), it follows that

$$\ln \gamma_1 = \frac{A}{RT} x_2^2 \quad \text{and} \quad \ln \gamma_2 = \frac{A}{RT} x_1^2 \tag{6.38}$$

since

$$nRT(x_1 \ln \gamma_1 + x_2 \ln \gamma_2) = nRT\left(x_1 \frac{A}{RT}x_2^2 + x_2 \frac{A}{RT}x_1^2\right)$$
$$= nAx_1 x_2 (x_2 + x_1) = nAx_1 x_2 \, .$$

(Note that γ_1 depends on x_2 and vice versa.) Guggenheim called a solution that follows Eq. (6.37) a *simple solution,* and the expressions for γ_1 and γ_2 are referred to as Margules expressions.[7] While mathematically simple, a solution that follows Eq. (6.37) is capable of a very complex physical phenomenon—spontaneous separation into two phases of different composition.

Mixing to form a homogeneous, single-phase solution occurs spontaneously if the free energy of the mixture is lower than that of the isolated components *and* if no *other* spontaneous process can further lower the system's total free energy. Note that, for real solutions, the entropy change on mixing can be *negative,* as long as a large, negative (nonideal) enthalpy change ensures $\Delta G_{mix} = \Delta H_{mix} - T\Delta S_{mix} < 0$.

From the definition of excess functions, G(real) is

$$G(\text{real}) = G^E + G(\text{ideal}) \quad \text{(by definition of } G^E\text{)}$$
$$= G^E + \sum_i n_i \mu_i (\text{ideal}) \quad \text{(by definition of } G(\text{ideal})\text{)}$$
$$= G^E + n\sum_i x_i (\mu_i^\circ + RT \ln f_i) \quad \text{(by definition of } x_i \text{ and } \mu_i(\text{ideal})\text{)}$$
$$= G^E + n\sum_i x_i (\mu_i^\circ + RT \ln f_i^\circ x_i) \quad \text{(since } f_i(\text{ideal}) = f_i^\circ x_i\text{)}$$
$$= G^E + n\sum_i x_i (\mu_i^\circ + RT \ln f_i^\circ) + nRT \sum_i x_i \ln x_i \quad \text{(since } \ln f_i^\circ x_i = \ln f_i^\circ + \ln x_i\text{)}$$
$$= G^E + n\sum_i x_i \mu_i (\text{pure } i) + nRT \sum_i x_i \ln x_i \quad \text{(by definition of } \mu_i(\text{pure } i)\text{)}$$

(and we specify f_i° to refer to *pure* reference states). For a binary simple solution, this becomes

$$\overline{G}(\text{real}) = A\, x_1 x_2 + x_1 \mu_1(\text{pure}) + x_2 \mu_2(\text{pure}) + RT\,(x_1 \ln x_1 + x_2 \ln x_2)\, .$$

Since $\Delta G_{mix}(\text{real}) = G^E + \Delta G_{mix}(\text{ideal})$, the previous equation becomes

$$\frac{\Delta \overline{G}_{mix}(x_1)}{RT} = \frac{A}{RT} x_1 x_2 + x_1 \ln x_1 + x_2 \ln x_2$$
$$= \frac{A}{RT} x_1 (1 - x_1) + x_1 \ln x_1 + (1 - x_1) \ln (1 - x_1)\, .$$

This function is graphed in Figure 6.15(a) for various values of A/RT from 0 (the ideal case—a repeat of Figure 6.14(a)) to 6. (Experimental values of A are almost always positive.) Mixtures with $A/RT \geq 5$ (or A about 12 kJ mol^{-1} or greater at room temperature) are virtually immiscible at any composition, since homogeneous solution formation would result in an *increase* in free energy for the system. The $\Delta \overline{G}_{mix}$ is positive at virtually all compositions for large A.

Consider the curve for $A = 2.5RT$, shown in Figure 6.15(b). The $\Delta \overline{G}_{mix}$ is everywhere negative, but it has a local maximum at $x_1 = 0.5$ and local minima at $x_1 \cong 0.145$ and 0.855. If equal amounts of components 1 and 2 were to mix and form *one homogeneous solution,* the solution would have the free energy at point *a*. It would have a lower free energy than that of the unmixed components, and the homogeneous solution would seem to form spontaneously.

---

[7]Margules, in 1895, first suggested the empirical relation between $\ln \gamma_1$ and a power series in x_2. Equation (6.38) is the simplest such power series.

FIGURE 6.15(a) For a simple solution, the free energy of mixing depends on the value of the parameter A/RT, which labels each of these curves. For $A/RT > 2$, only limited composition ranges lead to a homogeneous, single-phase solution. Note that for $A/RT = 6$, $\Delta \overline{G}_{mix}/RT$ is positive for nearly all compositions, implying total immiscibility. **(b)** For a simple solution characterized by $A/RT = 2.5$, spontaneous immiscibility occurs over the total composition range $0.145 \leq x_1 \leq 0.855$. A system that tries to form a single homogeneous solution phase at point a will instead spontaneously lower its free energy to point b by forming two phases of composition x_1(phase 1) = 0.145 and x_1(phase 2) = 0.855.

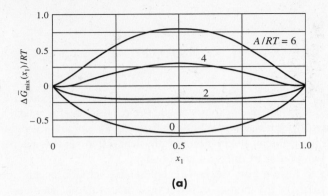

(a)

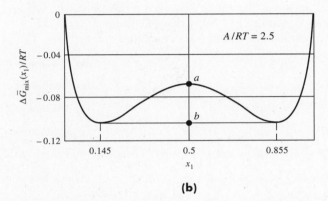

(b)

However, the mixture can do better, in a free energy sense, if it *spontaneously separates into two phases,* each of different composition. Of all possibilities, the one of lowest free energy has one phase characterized by $x_1 = 0.145$ and the other by $x_1 = 0.855$ (point b on the figure). Each of these phases is a homogeneous solution. With the lever rule, we can see that the amount of the *system* in each phase is the same, but one is rich in component 1 (the phase with $x_1 = 0.855$), while the other is rich in component 2 (with $x_2 = 1 - x_1 = 1 - 0.145 = 0.855$ as well).

Spontaneous segregation into two phases has lowered the total Gibbs free energy from that at point a to that at point b in Figure 6.15(b).

In general, nonideal $\Delta \overline{G}_{mix}(x_1)$ functions are not as symmetrical as those of Figure 6.15. But whenever a part of the function has a segment that is concave downward, that is, whenever

$$\left(\frac{\partial^2 \Delta \overline{G}_{mix}}{\partial x_1^2} \right)_{T,P,n} < 0 \quad ,$$

the system is *unstable* as one homogeneous phase, and it will spontaneously segregate into two homogeneous phases of different composition. The compositions of these phases are determined by drawing the (unique) straight line that is tangent to the $\Delta \overline{G}_{mix}$ curve at two points, such as the line containing point b in Figure 6.15(b). (The tangent line happens to be horizontal there; in general, it will not be if $\Delta \overline{G}_{mix}$ is less symmetric.)

For the simple solution, the behavior depends on the value of A/RT:

$$\frac{A}{RT} < 2 \quad \text{(one stable homogeneous phase)}$$

$$\frac{A}{RT} > 2 \quad \text{(two stable homogeneous phases)}$$

$$\frac{A}{RT} = 2 \quad \text{(something unique must happen!)} .$$

When $A/RT = 2$, the system is just at the verge of going one way or the other. This point defines the *critical solution temperature* (or the *consolute temperature*, as it is also known). It is a critical point for a mixture, just as the gas–liquid critical point for a pure substance makes two pure phases indistinguishably one phase.

It is instructive to look at the change in activity of component 1 in a simple solution. From Equations (6.32) and (6.38), we have

$$\ln a_1 = \ln \gamma_1 + \ln x_1 = \frac{A}{RT}(1 - x_1)^2 + \ln x_1 .$$

For $A/RT < 2$, a_1 increases with x_1 monotonically from $a_1 = 0$ at $x_1 = 0$ toward $a_1 = 1$ at $x_1 = 1$. Figure 6.16 plots the measured activity of methyl, ethyl, *n*-propyl, and *n*-butyl alcohol in water at various concentrations at 25 °C. Methyl, ethyl, and *n*-propyl alcohol are miscible with water at any composition. For these solutions, $A/RT < 2$. But *n*-butanol is *immiscible* with water over the dashed line in Figure 6.16 (roughly from $x_1 = 0.05$ to 0.5). A mixture of, for instance, 1 mol *n*-butanol with 3 mol water will lead to one phase that is nearly pure water in equilibrium with a second phase that is roughly equal amounts of water and alcohol.

It would be valuable to calculate A on a molecular basis, and many attempts to formulate molecular theories of mixing have been made with varying degrees of success. The problem is very difficult, especially for species as complex as alcohols and water, but even for species as simple as Ar and Kr, which, surprisingly, are

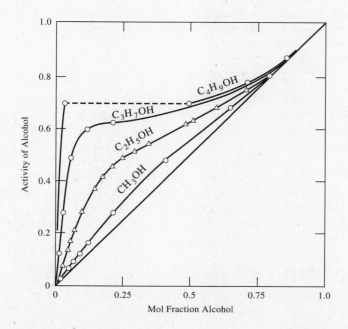

FIGURE 6.16 Simple alcohols form well-studied solutions with water. At 25 °C, methyl, ethyl, and propyl alcohols are miscible with water at all compositions; their activities rise smoothly without a horizontal point of inflection. But *n*-butyl alcohol exhibits a range of composition (the dashed line) that corresponds to two-phase immiscibility.

not completely miscible at all temperatures and pressures. Even helium isotopes are immiscible at certain compositions below about 0.87 K, the critical solution temperature.

EXAMPLE 6.4

The recovery of fission products from spent U fuel in fission reactors is a problem of considerable chemical complexity. Fission turns U into a mixture of elements with virtually every physical property imaginable. One suggestion has been to extract these products from molten U with molten Ag. At the suggested process temperature, ~1400 K, Ag and U are fairly immiscible. The activity coefficients for the Ag/U system at this temperature are

$$\ln \gamma_{Ag} = 2.83\, x_U^2 \quad \text{and} \quad \ln \gamma_U = 3.50\, x_{Ag}^2\ .$$

Calculate the solubility of Ag in the U-rich phase and of U in the Ag-rich phase.

SOLUTION First note from the form of the activity coefficient expressions that the Ag/U system *almost* follows the simple solution criteria, but doesn't quite, since the numerical factors, 2.83 and 3.50, are not equal. If we superscript variables pertaining to the U-rich phase with U (and similarly with Ag for the Ag-rich phase), then the criterion of uniform chemical potential can be written as one of uniform activity:

(1) uniform activity of U: $\quad a_U^U = x_U^U \gamma_U^U = a_U^{Ag} = x_U^{Ag} \gamma_U^{Ag}$

(2) uniform activity of Ag: $\quad a_{Ag}^U = x_{Ag}^U \gamma_{Ag}^U = a_{Ag}^{Ag} = x_{Ag}^{Ag} \gamma_{Ag}^{Ag}\ .$

We can relate the activity coefficients to the mole fractions, and in any phase, only one mole fraction is independent. Thus, (1) and (2) can be written as two coupled nonlinear equations in two unknowns, x_U^U and x_U^{Ag}:

(1) $\quad a_U^U = x_U^U\, e^{3.50(1 - x_U^U)^2} = a_U^{Ag} = x_U^{Ag}\, e^{3.50(1 - x_U^{Ag})^2}$

(2) $\quad a_{Ag}^U = (1 - x_U^U)\, e^{2.83(x_U^U)^2} = a_{Ag}^{Ag} = (1 - x_U^{Ag})\, e^{2.83(x_U^{Ag})^2}\ .$

These equations can be solved to sufficient accuracy by a simple iterative method. First, assume $x_U^U = 1.0$ and solve (1) for x_U^{Ag}, finding $x_U^{Ag} = 0.0396$. Use this value in (2) and solve (2) for an improved value of x_U^U, finding 0.9050. Return to (1) with this value, find an improved x_U^{Ag}, and continue until successive iterations yield sufficiently constant values. After three iterations, one finds, to $< 1\%$ accuracy, $x_U^{Ag} = 0.0362$ and $x_U^U = 0.9045$ (so that $x_{Ag}^U = 1 - 0.9045 = 0.0955$).

➠ RELATED PROBLEMS 6.18, 6.41

We have discussed and defined an ideal mixture of ideal gases, an ideal solution, and a simple solution. One can also define, through special cases of various excess functions, a *regular solution* (for which $S^E = 0$ at constant volume and temperature), a *van Laar solution* (for which S^E and V^E are both zero), an *athermal solution* (for which $\Delta H_{mix} = 0$), a *two-liquid solution* (which is especially useful in the case of He isotopes), and so on. Each of these definitions is applicable to special situations; athermal-solution theory has had good success with certain polymer solutions, for instance. Solutions of electrolytes are discussed in Chapter 10.

The next section discusses several predictions about solution properties that can be made from very simple assumptions.

6.5 COLLIGATIVE PROPERTIES

Several phenomena associated with mixtures and multiphase, multicomponent systems in general can be understood with simple models of the type introduced in the last section. Taken together, these phenomena are called *colligative properties,* a term that emphasizes their common theoretical similarity. (The word *colligative* comes from the Latin verb *colligare,* to bind together.)

The first phenomenon is the effect of the liquid (or, in principle, solid) solution's composition on the composition of the vapor in equilibrium with it. Here's a quick quiz to test your intuition and set the stage for our first discussion. We went to some effort to point out that water, ice, and water vapor can coexist at only one temperature (273.16 K) and pressure (0.00603 atm). So how can a glass of ice water exist at 1 atm pressure? Surely one would find $H_2O(g)$ above the glass, but the triple point pressure is way below 1 atm! The answer, of course, is that ice water in front of you in an open glass is a rather complex mixture. How does the external atmosphere alter the vapor pressure of water? How soluble is air in liquid water? In ice?

Vapor Pressures Over Solutions

Imagine a solution of two volatile components and consider the composition of the gas phase in equilibrium with it. To distinguish the liquid-phase composition from the gas-phase composition, we will use the symbols

$$y_i = \text{mole fraction of i in the } gas \text{ phase}$$
$$x_i = \text{mole fraction of i in the } liquid \text{ phase}.$$

For an ideal solution, we can choose the proportionality between $f_i(\text{solution})$ and x_i to be (see Eq. (6.33))

$K_i = f(i) =$ fugacity of pure i at the temperature of the solution and at the vapor pressure of pure i.

We will further assume the system is sufficiently ideal so that we can replace fugacities with pressures. These assumptions turn Eq. (6.33), $f_i = K_i x_i$, into the simple expression[8]

$$P_i \quad = \quad P(i) \quad \quad x_i \qquad \qquad (6.39)$$
$$\uparrow \qquad \qquad \uparrow \qquad \qquad \uparrow$$
Partial pressure of i — Vapor pressure — Mole fraction of i
above liquid solution — of *pure* liquid i — in liquid solution

known as *Raoult's law.* It is closely followed by solutions of chemically similar species, such as benzene–toluene. The total pressure[9] is the following function of solution composition:

$$P(x_1) = P_1 + P_2 = P(1) x_1 + P(2) x_2 = P(2) + [P(1) - P(2)] x_1 , \qquad (6.40)$$

but the composition of the gas phase may be very different from that of the liquid, as we can easily show.

By definition of the gas phase partial pressure, the gas phase composition is

$$y_1 = \frac{P_1}{P} .$$

---

[8]Don't confuse $P_i = P(i)x_i$ with the partial pressure expression $P_i = P y_i$. The pressure $P(i)$ is the vapor pressure of the pure condensed phase and x_i is the mole fraction of i in the condensed phase; P is the total pressure of the (mixed) system and y_i is the mole fraction of i in the gas phase.

[9]This pressure is the equilibrium pressure for a closed system containing only compounds 1 and 2. No open beakers under atmospheric pressure!

FIGURE 6.17 An ideal solution exhibits a two-phase *area* $(g + l)$ on an isothermal plot of pressure versus composition. In the all liquid area, the mole fraction coordinate refers to the composition of the single (liquid) phase; in the all gas area, it refers to the single (gas) phase. In the $(g + l)$ area, the mole fraction coordinate refers to the system total composition (the *isopleth*), or to the l or the g compositions at the intersections of the system isobar with the dew-point and bubble-point lines.

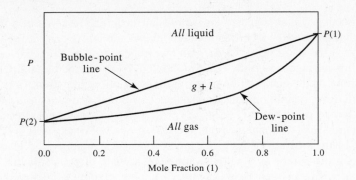

Substituting Eq. (6.39) for P_1 and Eq. (6.40) for P gives an expression for the gas phase composition in terms of the liquid phase composition:

$$y_1(x_1) = \frac{x_1 P(1)}{P(2) + [P(1) - P(2)] x_1} .$$

If we solve this equation for x_1 in terms of y_1 and substitute the result into Eq. (6.40), we obtain

$$P[x_1(y_1)] = P(y_1) = \frac{P(1)P(2)}{P(1) + [P(2) - P(1)]y_1} , \qquad \textbf{(6.41)}$$

which gives the total pressure as a function of the gas-phase composition.

If we graph Equations (6.40) and (6.41) versus a common system-composition variable, we obtain a *binary phase diagram* at one temperature, as is shown in Figure 6.17. In the area above the upper curve, the system is *entirely* liquid. The mole-fraction variable then describes the composition of the homogeneous liquid solution. In the area below the lower curve, the system is similarly *all* in the gas phase. Should the system's total composition and pressure land in the area *between* the two curves, the system is two phases (g and l) in equilibrium.

In contrast to a pure substance, a mixture vaporizes over a *range* of pressures rather than at one unique pressure, and the *compositions of the liquid and vapor phases are different*. In any two-phase area, we can apply the lever rule to determine not only the compositions but also the relative amounts of each phase.

Figure 6.18 describes what happens as a liquid solution of composition $x = 0.5$ is brought into the two-phase area by lowering the pressure. At point *a*, the gas phase has just begun to appear. The gas phase composition is $y_1(a)$; it is rich in the

FIGURE 6.18 As an ideal solution follows the isopleth $x = 0.5$ into the two-phase area, the gas-phase (y) and liquid-phase (x) compositions are determined from the isobar/two-phase-boundary intersection points.

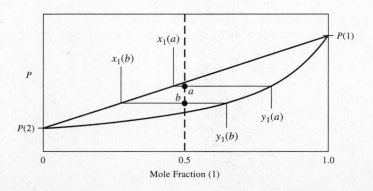

more volatile component. The remaining liquid is depleted in component 1, as is indicated by $x_1(a)$. The amounts of gas and liquid at this pressure are

$$\frac{n(g)}{n} = \frac{0.5 - x_1(a)}{y_1(a) - x_1(a)}$$

and

$$\frac{n(l)}{n} = \frac{y_1(a) - 0.5}{y_1(a) - x_1(a)}$$

in accord with the lever rule.

At point b, the liquid phase has nearly disappeared. The gas-phase composition, $y_1(b)$, is nearly at the value (0.5) it will have when the system is completely vaporized, and the liquid phase is very rich in the less-volatile component 2.

Diagrams such as Figure 6.17 (and related diagrams that consider temperature variations at constant pressure) are vital to understanding distillation. In the literature of distillation theory, the vertical line drawn at the system's total composition is known as an *isopleth*. The boundary between the two-phase region and the gas region is called the *dew-point* line (or surface, in P, T, x coordinates), while the two-phase/liquid-phase boundary is called the *bubble-point* line (or surface).

Henry's Law and the Ideal Dilute Solution

Raoult's law is rarely followed by two species at all compositions. It is, however, often a good approximation to the behavior of one compound when that compound is greatly in excess. This leads to the definition of an *ideal dilute solution,* and we can now use two terms you may have been expecting earlier in this chapter but which have been avoided. These are *solvent,* the compound in excess, and *solute,* the minor compound. In an ideal dilute solution, the *solvent* follows Raoult's law, while the *solute* obeys a semiempirical variation of Eq. (6.33):

$$P_2 = K_{2,1} x_2 \qquad (6.42)$$

known as *Henry's law*. The Henry's law constant, $K_{2,1}$, is a property of the *mixture*, not of the solute or solvent alone. The subscript 2,1 specifies *solute 2 in solvent 1*. The Henry's law constant acts as a hypothetical vapor pressure for the pure solute. It may be smaller or larger than the actual pure solute vapor pressure, but it is most often larger.

This definition, a direct descendant of Eq. (6.33), leads to convenient expressions for the chemical potential in terms of composition variables other than mole fraction. Since $\mu_i = \mu_i^\circ + RT \ln f_i$ in general, Eq. (6.33) implies

$$\mu_i(\text{ideal}) = \underbrace{\mu_i^\circ(\text{ideal}) + RT \ln K_i}_{} + RT \ln x_i$$

↑ ↑ ↑

Chemical potential of   Effective reference   Dependence of μ
i in an ideal solution   chemical potential   on composition

so that $\mu_i^\circ(\text{ideal}) + RT \ln K_i$ becomes a new standard-state reference chemical potential because this sum is, at any one pressure, a function of T alone. If mole fraction is not a convenient composition variable, but, say, molality[10] is, then we relate mole fraction to molality:

$$x_i = \frac{(10^{-3} \text{ kg g}^{-1}) M m_i}{1 + (10^{-3} \text{ kg g}^{-1}) M m}$$

---

[10]Moles of solute per *kg of solvent* is the *molality*. It is more convenient than molarity (moles of solute per *liter of solution*) in that it is temperature and pressure independent.

where M is the molar mass of the solvent (in g mol^{-1}), m_i is the molality of solute i, and m is the total molality of all solutes.

In the ideal dilute solution, $m \to 0$, and x_i approaches $10^{-3} M m_i$, which can also be written as

$$x_i = (10^{-3} \text{ kg g}^{-1}) \, M \, m° \left(\frac{m_i}{m°}\right)$$

where $m°$ is unit molality. The chemical potential becomes

$$\mu_i \;=\; \underbrace{\mu_i° \;+\; RT \ln K_i \;+\; RT \ln (10^{-3} \text{ kg g}^{-1} \, M \, m°)}_{} \;+\; RT \ln \left(\frac{m_i}{m°}\right)$$

↑   ↑   ↑

Chemical potential of i Effective reference   Dependence of μ
in ideal dilute solution   chemical potential   on composition

and again we have shifted the standard-state definition. As long as any one standard state is used throughout a calculation, we may take μ_i for a solute in an ideal dilute solution to vary as $RT \ln$(composition variable of our choice).

EXAMPLE 6.5

The Henry's law constant is a convenient measure of gaseous solubilities in liquids. For example,

$$K_{O_2, H_2O} = 4.3 \times 10^4 \text{ atm at } 25 \,°C \;.$$

The partial pressure of O_2 in ordinary air is 0.21 atm. What is the solubility of atmospheric oxygen in water?

SOLUTION For air in contact with water,

$$\frac{P_{O_2}}{K_{O_2, H_2O}} = x_{O_2} = \frac{n_{O_2}}{(n_{H_2O} + n_{O_2} + n_{N_2} + \cdots)} \cong \frac{n_{O_2}}{n_{H_2O}}$$

or

$$n_{O_2} = \frac{P_{O_2} n_{H_2O}}{K_{O_2, H_2O}} = \frac{0.21 \text{ atm}}{4.3 \times 10^4 \text{ atm}} n_{H_2O} \;.$$

If we choose 1 kg of H_2O as a reference amount, the *concentration* of O_2 is

$$\left(\frac{0.21 \text{ atm}}{4.3 \times 10^4 \text{ atm}}\right)\left(\frac{1000 \text{ g}}{18 \text{ g mol}^{-1}}\right) = 2.7 \times 10^{-4} \text{ mol } O_2 \text{ per kg solvent} \;.$$

This is the *molal* concentration.

Calculations such as this are important in designing physiologically acceptable artificial atmospheres, as in diving gear or space craft. A partial pressure of oxygen greater than ~0.5 atm is toxic. If ordinary air were used at a pressure great enough to repel sea water at a depth of 200 ft (61 m), which is $P = \rho_{H_2O} \, g \, h \cong (1 \text{ g cm}^{-3})(980 \text{ cm s}^{-2})(6100 \text{ cm}) \cong 6 \times 10^5 \text{ J m}^{-3} \cong 6$ atm, the partial pressure of O_2 would be at the lethal level of 1.5 atm. Instead of air, a mixture of nearly pure He with O_2 is used with an O_2 partial pressure of roughly 0.2 atm. Helium is used, as first suggested by Hildebrand, due to its very large (1.4 $\times 10^5$ atm) Henry's law constant with water (and, of equal importance, with human blood), which implies a very small solubility for He.

Note that not just any inert gas will do, however. Nitrogen has sufficient solubility to produce painful gas bubbles in the blood stream on decompression while Kr and Xe are anesthetics at sufficiently high partial pressure!

▶ RELATED PROBLEMS 6.19, 6.20, 6.42

Figure 6.19 graphs the binary phase diagram for a hypothetical mixture of two liquids that are assumed to have the following properties:

$P(1) = P(2) = P°$ (equal vapor pressures for the pure liquids)

$K_{1,2} = K_{2,1} = K$ (equal Henry's law constants)

$\dfrac{K}{P°} = 2$ (Henry's-law-constant-to-vapor-pressure ratio)

$P_i = Kx_i + (P° - K)x_i^2$ (partial-pressure-composition relation)

There are *two* areas in which two-phase behavior is found, and a liquid system with $x = 0.5$ would boil at a single pressure, as if it were pure. This phenomenon, a sharp boiling point for a mixture at a unique composition, is quite common. The unique composition is called the *azeotrope* composition, the existence of which has important consequences on the ability to purify mixtures by distillation. The liquid phase tends to achieve the azeotrope composition and stick there.

This hypothetical system, while qualitatively representative of many real systems, cannot be rigorously correct. We can prove that if one component follows Henry's law, then the other must follow Raoult's law. We start with Eq. (5.30) for a binary system at equilibrium, so that $dG(\text{total}) = 0$, closed, so that $dn_1 = dn_2 = 0$, and at constant T and P:

$$n_1 \, d\mu_1 + n_2 \, d\mu_2 = 0 \; .$$

A few steps transform this into

$$x_1 \left(\dfrac{\partial \ln f_1}{\partial x_1}\right)_{T,P} = x_2 \left(\dfrac{\partial \ln f_2}{\partial x_2}\right)_{T,P} .$$

Suppose component 1 follows Henry's law, so that $f_1 = Kx_1$. (We write K instead of $K_{2,1}$ in order to simplify notation here.) Then, since $(\partial \ln f/\partial x) = (1/f)(\partial f/\partial x)$, we have

$$\dfrac{x_1}{f_1}\left(\dfrac{\partial f_1}{\partial x_1}\right)_{T,P} = \dfrac{x_1}{f_1}K = \dfrac{x_1}{f_1}\left(\dfrac{f_1}{x_1}\right) = 1 \; .$$

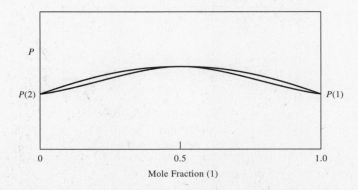

FIGURE 6.19 The interplay between Raoult's law (or ideal solution) and Henry's law (or ideal dilute solution) behavior can produce phase diagrams with *two* areas representing g–l equilibrium and a unique composition ($x = 0.5$ here) where the solution vaporizes at one sharp pressure. This behavior is characteristic of the *azeotrope* composition.

Thus

$$\frac{x_2}{f_2}\left(\frac{\partial f_2}{\partial x_2}\right)_{T,P} = 1$$

or, rearranging and integrating,

$$\int^{x_2}_{x_2=1} \frac{dx_2}{x_2} = \int^{f_2}_{f_2^\circ} \frac{df_2}{f_2}$$

so that

$$\ln x_2 = \ln\left(\frac{f_2}{f_2^\circ}\right)$$

or

$$f_2 = f_2^\circ x_2 ,$$

which is Raoult's law. (Note how the limits on the integrals were chosen. They are equivalent to x_1 going from 0 to $x_1 = 1 - x_2$.) The hypothetical mixture of Figure 6.19 follows Henry's law, since as $x_1 \to 0$, $P_1 \to Kx_1$, but as $x_1 \to 1$, $P_1 = Kx_1(1 - x_1) + P^\circ x_1^2 \to P^\circ x_1^2$, which is not Raoult's law.

Freezing Point Depression and Boiling Point Elevation

In this subsection, we will assume that

1. *the solute, 2, is involatile,*
2. *the solvent, 1, is volatile,* and
3. *the solute does not dissolve in the solid solvent.*

Thus the gas phase is *pure solvent, pure solid solvent* can be formed, and the chemical potential of the solvent in solution is always *less* than that of the pure liquid solvent. Applying these assumptions to a diagram like Figure 6.7 leads to the predictions of Figure 6.20. The dashed line is the chemical potential of the solvent *in solution*. Note that the freezing point is *lowered* and the boiling point is *raised*.

Before we compute the magnitudes of these effects, you should recognize that the phenomena are quite common. Salt (ordinary NaCl) added to water at one temperature extreme raises the boiling point to increase cooking temperatures and at the other extreme lowers the freezing point to melt winter snow and ice.

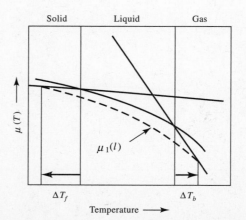

FIGURE 6.20 Compare this figure to Figures 6.4 and 6.7. If the solute forms a liquid solution *only* with solvent 1, then the chemical potential of the solvent in solution (the dashed line) will shift down in energy at all temperatures. This behavior causes a *decrease* in the freezing point (ΔT_f) and a *rise* in the boiling point (ΔT_b) of the solution compared to the pure solvent.

We want to find an expression for the solution freezing point T_{fus} in terms of the solution composition. First, we notice that the solution is in equilibrium with pure solid solvent at $T_{\text{fus}}(\text{solution}) < T_{\text{fus}}^{\circ}(\text{pure solvent})$ so that μ of solvent 1 in the solution equals μ of pure solid 1:

$$\mu_1(T_{\text{fus}}, P, x_1) = \mu(T_{\text{fus}}, P, \text{pure solid 1}) .$$

If the solution is ideal,

$$\mu_1(T_{\text{fus}}, P, x_1) = \mu_1^{\circ} + RT_{\text{fus}} \ln f_1^{\circ} x_1 .$$

Since $\mu_1^{\circ} + RT \ln f_1^{\circ} = \mu(\text{pure liquid 1})$ and $\Delta \overline{G}_{\text{fus}}^{\circ} = \mu(\text{pure liquid}) - \mu(\text{pure solid})$,

$$\ln x_1 = -\frac{\Delta \overline{G}_{\text{fus}}^{\circ}(\text{pure 1}, T_{\text{fus}})}{RT_{\text{fus}}} .$$

($\Delta \overline{G}_{\text{fus}}^{\circ} = 0$ at T_{fus}°, but *not* at T_{fus}.) Next we differentiate the expression above with respect to x_1,

$$\left(\frac{\partial \ln x_1}{\partial x_1}\right)_{T,P} = \frac{1}{x_1} = -\frac{1}{R}\left[\left(\frac{\partial (\Delta \overline{G}_{\text{fus}}^{\circ}/T)}{\partial T}\right)_P\right]_{T_{\text{fus}}} \left(\frac{\partial T}{\partial x_1}\right)_P$$

and use a general expression that follows from the definition of G:

$$\left(\frac{\partial (G/T)}{\partial T}\right)_P = \frac{1}{T}\left(\frac{\partial H}{\partial T}\right)_P - \frac{H}{T^2} - \left(\frac{\partial S}{\partial T}\right)_P = -\frac{H}{T^2} .$$

Therefore

$$\frac{1}{x_1} = \frac{\Delta \overline{H}_{\text{fus}}^{\circ}(\text{pure 1}, T_{\text{fus}} < T_{\text{fus}}^{\circ})}{RT_{\text{fus}}^2}\left(\frac{\partial T}{\partial x_1}\right)_P .$$

We rearrange this equation and integrate

$$\int_1^{x_1} \frac{dx_1}{x_1} = \int_{T_{\text{fus}}^{\circ}}^{T_{\text{fus}}} \frac{\Delta \overline{H}_{\text{fus}}^{\circ}}{RT_{\text{fus}}^2} dT_{\text{fus}} .$$

If we assume $\Delta \overline{H}_{\text{fus}}^{\circ}$ is independent of T, we carry out the integration and find

$$\ln x_1 = \frac{\Delta \overline{H}_{\text{fus}}^{\circ}}{R}\left(\frac{1}{T_{\text{fus}}^{\circ}} - \frac{1}{T_{\text{fus}}}\right) . \tag{6.43}$$

This is a perfectly useful expression, but is it helpful to write it in terms of variables more convenient to experiment. If we define the *freezing point depression*

$$\Delta T_{\text{fus}} = T_{\text{fus}}^{\circ}(\text{pure 1}) - T_{\text{fus}}(\text{solution})$$

and express the solute concentration as its molality, m_2, we have, from the integrands that lead to Eq. (6.43),

$$dT_{\text{fus}} = -d(\Delta T_{\text{fus}}) = \frac{RT_{\text{fus}}^2}{\Delta \overline{H}_{\text{fus}}^{\circ}} \frac{dx_1}{x_1} = \frac{RT_{\text{fus}}^2}{\Delta \overline{H}_{\text{fus}}^{\circ}} \frac{1}{x_1}\left(\frac{dx_1}{dm_2}\right) dm_2 .$$

In a solution containing 1 kg of solvent,

$$n_1 = \frac{1000}{M_1} \quad \text{and} \quad n_2 = m_2$$

where M_1 is the molar mass of the solvent in g mol^{-1}, so that

$$x_1 = \frac{n_1}{n_1 + n_2} = \frac{1}{1 + \dfrac{M_1 m_2}{1000}}$$

and from this one can show that

$$\frac{1}{x_1}\left(\frac{dx_1}{dm_2}\right) = -\frac{M_1 x_1}{1000}.$$

We therefore have the compact expression

$$\frac{d(\Delta T_{\text{fus}})}{dm_2} = \text{rate of freezing point lowering per unit molality increase of the solute}$$

$$= \frac{M_1 R T_{\text{fus}}^2 x_1}{1000 \Delta \overline{H}_{\text{fus}}^\circ (\text{pure solvent})}.$$

With the two good approximations, $x_1 \cong 1$ and $T_{\text{fus}} \cong T_{\text{fus}}^\circ$, this simplifies to

$$\frac{d(\Delta T_{\text{fus}})}{dm_2} \cong \frac{M_1 R T_{\text{fus}}^{\circ 2}}{1000 \Delta \overline{H}_{\text{fus}}^\circ} = K_f = \text{the } \textit{cryoscopic constant}. \tag{6.44}$$

The cryoscopic constant (dimensions: degrees K per mole of solute per kg of solvent) depends on *solvent properties only*. Moreover, only the *number of solute particles* (expressed by n_2) governs the freezing point depression, not their chemical identity. Thus, one millimole of NaCl dissolved in H_2O produces *two* millimoles of particles, $Na^+(aq)$ and $Cl^-(aq)$, $CaCl_2$ produces *three*, $Ca^{2+}(aq)$ and two $Cl^-(aq)$, etc.

We can use, for small m_2, the approximation

$$\Delta T_{\text{fus}} \cong \frac{d(\Delta T_{\text{fus}})}{dm_2} m_2 = K_f m_2 \tag{6.45}$$

to calculate freezing-point depressions. Table 6.3 collects K_f values for several substances, and this section's problems explore the various applications of freezing-point depression measurements.

For the boiling-point elevation, arguments similar to those that led to Eq. (6.43) give

$$\ln x_1 = \frac{\Delta \overline{H}_{\text{vap}}^\circ}{R}\left(\frac{1}{T_{\text{vap}}} - \frac{1}{T_{\text{vap}}^\circ}\right) \tag{6.46}$$

where T_{vap}° is the normal boiling point temperature of the pure solvent. If we also define the boiling point elevation

$$\Delta T_{\text{vap}} = T_{\text{vap}}(\text{solution}) - T_{\text{vap}}^\circ(\text{pure solvent})$$

we can derive, under assumptions similar to those leading to the freezing-point depression,

$$\Delta T_{\text{vap}} = \left(\frac{M_1 R T_{\text{vap}}^{\circ 2}}{1000 \Delta \overline{H}_{\text{vap}}^\circ}\right) m_2 = K_b m_2 \tag{6.47}$$

TABLE 6.3 Cryoscopic Constants

	T_{fus}°/K	$K_f/(\text{K kg mol}^{-1})$
H_2O	273.15	1.86
CH_3COOH	289.8	3.57
C_6H_6	278.6	5.07
$p\text{-}C_6H_4Cl_2$	325.9	7.11
camphor ($C_{10}H_{16}O$)	451.6	37.7
$BaCl_2$	1235.	108.

TABLE 6.4 Ebullioscopic Constants

	T_{vap}°/K	$K_b/(\text{K kg mol}^{-1})$
H_2O	373.15	0.512
CH_3OH	337.9	0.83
CH_3COOH	391.5	3.07
C_6H_6	353.4	2.53
C_6H_{12}	354.6	2.79

where K_b is the *ebullioscopic constant*, with the same dimensions as K_f. As with K_f, K_b depends only on properties of the solvent, and ΔT_{vap} is controlled by K_b and the *number* of solute particles, *not their chemical identity*. Table 6.4 collects K_b values for several substances.

Osmotic Pressure

We now consider another phenomenon that is a consequence of $\mu_1(\text{solution}) < \mu(\text{pure liquid solvent 1})$. Consider the experimental setup shown schematically in Figure 6.21. In side A we have pure solvent in equilibrium with its vapor at the normal vapor pressure. Side B contains the solution; again, by assumption, the vapor is pure solvent, at virtually the vapor pressure of side A, since the solution is dilute. The boundary between these two sides is rigid and transfers energy as heat, equalizing temperature, but more to the point here, it has a segment below the levels of both liquids that is *permeable to the solvent only*. Solute is confined to side B, but the boundary in the semipermeable region is invisible to the solvent.

Imagine the spontaneous process if the boundary is removed completely. Spontaneous mixing would occur, diluting the solution. The semipermeable membrane permits that same spontaneous process to occur, up to a point. Note that, due to gravity, the pressure at a depth below the surface of any liquid is increased due to the weight of the liquid above. Figure 6.21, drawn for a system at equilibrium, shows that the semipermeable membrane is at *different depths below the surfaces* of each side. This causes a pressure imbalance $P_B > P_A$ at the membrane. Since the system is at equilibrium, the chemical potential of the solvent is the *same* on both sides of the membrane at any one depth:

$$\mu(\text{pure 1}, P_A) = \mu_1(x_1, P_B) \ .$$

We define the *osmotic pressure* Π as

$$\Pi = \text{solution pressure} - \text{solvent pressure} = P_B - P_A$$

and find

$$\mu(\text{pure 1}, P_A) = \mu_1(x_1, P_A + \Pi)$$
$$= \mu(\text{pure 1}, P_A + \Pi) + RT \ln x_1$$
$$= \mu(\text{pure 1}, P_A) + \int_{P_A}^{P_A + \Pi} \overline{V}(\text{pure 1}) \, dP + RT \ln x_1$$

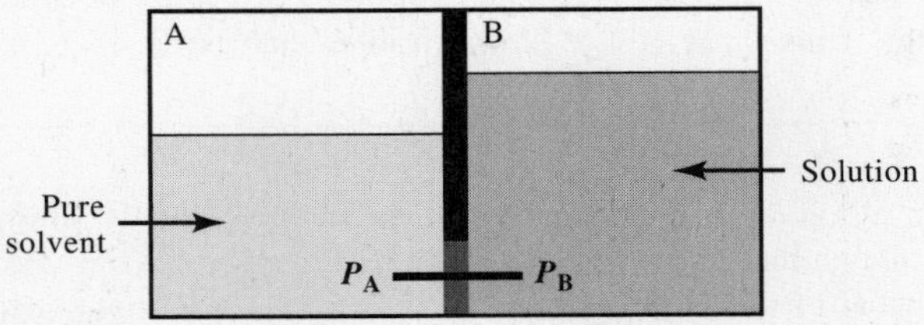

FIGURE 6.21 To demonstrate the effect of osmotic pressure, pure solvent (in side A) and a dilute solution (in side B) are separated by a barrier, the lower portion of which is semipermeable to the solvent. Solvent will spontaneously flow into the solution, causing the pressure mismatch $P_B > P_A$ at any horizontal plane below the solvent surface.

so that

$$-RT \ln x_1 = \int_{P_A}^{P_A + \Pi} \overline{V}(\text{pure 1}) \, dP \ .$$

If Π is not very great, $\overline{V}$ is constant, and

$$-\ln x_1 = \frac{\Pi \overline{V}}{RT} \ .$$

This can be further approximated if we first write

$$-\ln x_1 = \ln\left(\frac{1}{x_1}\right) = \ln\left(\frac{n_1 + n_2}{n_1}\right) = \ln\left(1 + \frac{n_2}{n_1}\right)$$

and, since $n_2/n_1 \ll 1$, expand the last expression as

$$\ln\left(1 + \frac{n_2}{n_1}\right) = \frac{n_2}{n_1} - \frac{1}{2}\left(\frac{n_2}{n_1}\right)^2 + \frac{1}{3}\left(\frac{n_2}{n_1}\right)^3 - \cdots$$

and retain only the first term, giving

$$-\ln x_1 = \frac{\Pi \overline{V}}{RT} \cong \frac{n_2}{n_1} \ ,$$

which, since $\overline{V} n_1 \cong V$, the total solution volume, becomes the somewhat surprising expression

$$\Pi V = n_2 RT \tag{6.48}$$

known as the *van't Hoff equation*. Shades of $PV = nRT$! After several approximations, we find that the osmotic pressure (a component of the total pressure) appears to be due to an ideal-gas-like effect of the solute! But we should not over-interpret this similarity. In the first place, osmotic pressure is a very general phenomenon that *all* dilute solutions, including those which vary from *any* definition of ideal behavior, exhibit. In the second place, Eq. (6.48) came from approximations that are very different from those for ideal gas behavior.

On the other hand, the similarity is not simply a matter of random chance. Osmotic pressure depends in this approximation on the number of solute particles, not their chemical identity. Several solutes would give a total osmotic pressure proportional to the total number of solute particles, as is the pressure in an ideal mixture of ideal gases. In a more detailed theory of osmotic pressure, one can include nonideal behavior in an expansion similar to the virial equation of state.

While semipermeable membranes may appear to be magical, they can be found in abundance in every living thing, since cell walls are semipermeable to certain compounds. For example, if intravenous injections are done with pure water, blood cell walls spontaneously allow the water to dilute their insides. The equilibrium osmotic pressure is so high in this case that the cell walls rupture. To avoid this unpleasant outcome, intravenous solutions are controlled in their solute concentration (using salts, nutrients, etc.) so that the solution establishes equilibrium with the cell's insides without appreciable transport of diluting material through the wall. Such a solution is said to be *isotonic*, a term that has made its way into the description of less life-threatening solutions such as shampoo. (Pure water is called *hypotonic*, and a solution that causes the cell to shrink, rather than burst, is *hypertonic*.)

EXAMPLE 6.6

Suppose we put a piston on the liquid surface of the solution (side B) in Figure 6.21 and increase the pressure on that side above the equilibrium pressure. We would then increase the chemical potential of the solvent in side B, creating disequilibrium, and pure solvent would flow into side A. This is the basis for reverse osmosis, a practical scheme for the purification of water.

Sea water (at around 280 K, a typical surface temperature worldwide) has a solute (salt) concentration of roughly 1 mol L^{-1}. What is the pressure needed to effect reverse osmosis, and how much work is needed to purify 1 mol of sea water by this process?

SOLUTION The equilibrium osmotic pressure of a water/sea-water system is

$$\Pi = \frac{n_2}{V}RT = \left(\frac{1 \text{ mol}}{1 \text{ L}}\right)\left(0.082 \frac{\text{L atm}}{\text{mol K}}\right)(280 \text{ K}) \cong 23 \text{ atm}.$$

Pressures above 23 atm induce reverse osmosis, and the energy we expend (as work at constant external pressure) to purify one mole of water is on the order of

$$w = -P_{\text{ext}} \Delta \overline{V} = -(23 \text{ atm})\left(-\frac{18 \text{ g mol}^{-1}}{1000 \text{ g L}^{-1}}\right) \cong 0.4 \text{ L atm mol}^{-1};$$

$\Delta \overline{V}$ is negative since we are compressing the solution. Note also that we have approximated the solution molar volume with that of pure water, because we need only an order of magnitude estimate.

In contrast, the energy required to purify one mole of water by distillation, $\Delta \overline{H}_{\text{vap}}^\circ$, is roughly 400 L atm mol^{-1}! This very great difference in energies is the impetus for much research and development of reverse osmosis equipment. The technological problems center around designing a suitable semipermeable membrane. (On the other hand, solar energy is free. Hence, solar distillation is also of interest to areas with abundant sunshine but a lack of fresh water.)

➡ **RELATED PROBLEMS** 6.31, 6.32

Osmotic pressure measurements are sensitive indicators of the molecular mass of the solvent for macromolecules, either natural or synthetic. For example, a small protein of molecular mass $\sim 10^4$ g mol^{-1} at a concentration of 300 mg per 10 cm^3 of water (or 3×10^{-3} molal) has a freezing point depression of only

$$\Delta T_{\text{fus}} = K_f m_2 = (1.86 \text{ K m}^{-1})(3 \times 10^{-3} \text{ m}) \cong 5.6 \times 10^{-3} \text{ K}$$

which is difficult to measure, certainly to any accuracy. The osmotic pressure (measured by bringing the solution in contact with pure water through a suitable membrane), however, has the readily measured value

$$\Pi = \frac{n_2}{V}RT = \left(\frac{3 \times 10^{-5} \text{ mol}}{10 \text{ cm}^3}\right)(82 \text{ cm}^3 \text{ atm mol}^{-1} \text{ K})(300 \text{ K}) \cong 7.4 \times 10^{-2} \text{ atm}.$$

The next section looks at the type of problem introduced at the beginning of this one—air, ice, liquid water, and water vapor all in equilibrium. Can the air dissolve in the water? In the ice? Can they exist in equilibrium at arbitrary pressures, temperatures, and compositions? These questions are answered in general by the *Phase Rule*, which restricts the types of equilibria that can be attained among many components in many phases.

6.6 MANY COMPOUNDS, MANY PHASES—THE PHASE RULE

We proved in Chapter 5 that the general conditions for equilibrium in a simple, closed system of ϕ phases and c components are

uniform temperature: $\quad T^\alpha = T^\beta = \ldots = T^\delta = \ldots = T^\phi = T$
uniform pressure: $\quad P^\alpha = P^\beta = \ldots = P^\delta = \ldots = P^\phi = P$
uniform chemical potential: $\quad \mu_i^\alpha = \mu_i^\beta = \ldots = \mu_i^\delta = \ldots = \mu_i^\phi = \mu_i, i = 1, 2, \ldots, c.$

In this section we will see how these conditions affect the number and compositions of phases that can exist in equilibrium.

We begin with a revealing quotation from the classic text *Thermodynamics,* written by G. N. Lewis and M. Randall (of the Lewis–Randall fugacity approximation for mixtures, among myriad other contributions to physical chemistry) and revised by K. S. Pitzer and L. Brewer, who wrote

> ... there is at every temperature some dissociation [of water] into hydrogen and oxygen and that at 25 °C the partial pressure of the hydrogen is 2.50×10^{-28} atm which is equivalent to the pressure exerted by a single molecule in a space of about 1 million liters. Yet this value has a precise significance and is certainly known within a few per cent.
>
> One of the most striking results of this character is obtained if we calculate the vapor pressure of tungsten at 100° from experiments at very high temperatures. The result, 10^{-105} atm, would mean that the concentration of tungsten vapor would be less than one molecule in a space equivalent to the known sidereal universe. Such a calculation need not alarm us. Allowing for the possibilities of experimental uncertainty, we may utilize such a calculated vapor pressure in our thermodynamic work with the same sense of security as we use the vapor pressure of water.

This quotation is at the heart of the two thermodynamic concepts "a component in every phase" and "an independent component." Thus, W(s) and W(g) are, thermodynamically, in equilibrium at 100° even though we would never find that one W atom predicted to be in the gas phase. (Thermodynamics cares more about intensive properties than extensive ones; the molar volume is more important than the total volume or total number of moles, for instance.) As a corollary, no two substances are completely immiscible, although the concentration of one may be so vanishingly small as to be below any possible means of detection. This is the meaning of "uniform chemical potential of any component in all phases."

Now consider the 2.50×10^{-28} atm partial pressure of H_2 that is in equilibrium with every sample of water at 25 °C. The H_2 must also be in the liquid water as well. We also know that water is a complicated liquid containing spontaneously formed ions, like $H^+(aq)$ and $OH^-(aq)$, and that these ions attract water molecules forming species such as H_3O^+, $H_5O_2^+$, etc., that have unique chemical properties. What has happened to the concept of a "pure substance"? This stuff we have treated as pure water all along and have used for such fundamental definitions as the size of an absolute degree is a rather complex mixture of water molecules, gaseous H_2 (and O_2), and a host of positive and negative ions!

Maybe we should not worry about these other species since their concentrations are so low. After all, you can readily calculate that the freezing point depression of water due to a solute at $\sim 2 \times 10^{-7}$ molal concentration (the H^+ and OH^- concentration sum in pure water) is $\sim 4 \times 10^{-7}$ K, a clearly insignificant amount.

There is a much more fundamental reason why we should not worry. This reason is the topic of the next chapter: *chemical equilibrium. Species that are in chemical equilibrium among themselves can not all be independent components.* In the case of water, all we need say is "water at 25 °C" and we immediately know the H^+ and OH^- concentrations, the H_2 partial pressure, and, in principle, the concentrations of hydrated ions such as $H_5O_2^+$. We know these through the use of *chemical reaction equilibrium constants,* a concept you doubtless encountered long ago in elementary chemistry courses. Recall expressions such as

$$H_2O(l) \rightleftarrows H^+(aq) + OH^-(aq) \quad K_{eq} = [H^+][OH^-] = 10^{-14}$$

for the autoionization equilibrium of water. Remember the further constraint imposed by total charge neutrality:

$$\text{total positive charge} = \text{total negative charge}$$

which ensures

$$[H^+] + [H_3O^+] + [H_5O_2^+] + \cdots = [OH^-] + [O_2H_3^-] + \cdots$$

Each of these constraints—chemical equilibrium or total charge neutrality—reduces by one the number of components that are truly independent. For example, ignore for the moment every species in water except H_2O, OH^-, and H^+. Charge neutrality reduces by one the number that are independent, as does the chemical equilibrium written above, leaving only one component truly independent. (By "independent," we mean "has a concentration variable—molarity, molality, mole fraction, etc.—that can be varied as an independent experimental variable, along with P and T.") Which of these three we choose to be independent is not important. Once we specify the concentration of any one of these three (and state that no other species are present), the concentrations of the other two are implicitly defined, and they are calculable via the equilibrium constant and charge neutrality.

A third type of constraint is often present as well. Consider the reaction

$$H_2O(l) \rightleftarrows H_2(g) + \frac{1}{2} O_2(g)$$

that leads to the very small amount of H_2 and O_2 as a spontaneous consequence of liquid water's existence. In the absence of *external sources* of H_2 and/or O_2, this reaction imposes the *stoichiometric constraint*

$$P_{H_2} = 2P_{O_2}$$

since every mole of H_2O that dissociates produces one mole of H_2 and one half mole of O_2 *simultaneously*. Thus the concentrations (as measured by partial pressures in a closed system) of H_2 and O_2 are related, and these two cannot both be independent species. (The equilibrium constraint among $H_2O(l)$ and the two gases further ensures that only one of the three is truly independent.)

Therefore, in any system containing several species in equilibrium, the number of those species that are truly independent is reduced by one for every chemical equilibrium expression connecting two or more species, by one for every stoichiometric constraint due to disproportionation reactions that connect the relative concentrations of a reaction's products, and by one for the charge neutrality constraint, if it applies.

EXAMPLE 6.7

One of the chemical coups of the 1960s was the synthesis of a number of stable compounds containing a rare gas atom, such as the xenon fluorides: XeF_2, XeF_4, and XeF_6. Over certain ranges of T and P, Xe and F_2 react to produce measurable amounts of all three fluorides, leading to a gaseous mixture of five species: Xe, F_2, XeF_2, XeF_4, and XeF_6. (The fluorides are volatile solids; we will concentrate on the gas-phase mixture.) How many of these are independent species?

SOLUTION Due to the three chemical equilibria

$$Xe + F_2 \rightleftarrows XeF_2$$

$$Xe + 2F_2 \rightleftarrows XeF_4$$

$$Xe + 3F_2 \rightleftarrows XeF_6$$

only 5 (total species) $-$ 3 (chemical equilibrium constraints) = 2 species are independent.

But what about equilibria such as

$$XeF_2 + F_2 \rightleftarrows XeF_4$$

$$XeF_4 + F_2 \rightleftarrows XeF_6$$

that are suggested by experiments that produce the higher fluorides from F_2 and a lower fluoride? Can they reduce the number of independent components further? No. These two equilibria are simply combinations of pairs of the first three. They therefore add nothing new.

➥ RELATED PROBLEMS 6.33, 6.34

What about mixtures of isotopes? Does a mixture of ^{12}CO and ^{13}CO, for example, contain one component, "carbon monoxide," or two? This is a fairly subtle question, since we mentioned earlier in this chapter that ^3He and ^4He isotopes are immiscible at low enough temperature. In general, however, compounds are not easily distinguished by their isotopic composition. Chemistry is largely insensitive to the slight changes in mass that distinguish most isotopes. (The H, He, and Li isotopes are the most important exceptions.) Thus, we take isotopic mixtures to be pure, one-component systems unless we have good chemical reasons to expect the macroscopic behavior of one isotope to be experimentally very different from another.

A similar situation arises in the case of optical isomers. A mixture of optical isomers is a simple binary mixture with the special added feature of chemical symmetry. Consider a D-L isomer-pair solution with $x_L = 0.25$. It would have the same thermodynamic properties as one with $x_D = 0.25$, except for the change in optical properties (which, after all, is the defining property for such isomers).

The preceding few paragraphs have put the "number of components" under control. Now we consider the number of independent intensive variables available to a system. Temperature is certainly one, but we have seen (refer to osmotic pressure, for example) that equilibrium can be established without requiring a homogeneously uniform pressure. Truly uniform pressure is possible only in gravitationless systems and is thus only an approximation here on Earth's surface. In Chapter 9, we will see that "uniform pressure" in a gravitational field is interpreted to mean uniformity of pressure in a thin slab of material oriented perpendicularly to the direction of gravity.

In other situations, other intensive variables may be important. Among these are

1. surface energies (the topic of Chapter 8);
2. external magnetic or electric fields (Section 9.3);
3. a linear stretch or push (deforming stresses on solids); and
4. the rate of rotation of the system (Section 9.2).

Surface energies are important for thin systems (thin films, liquid crystals, etc.) or for "homogenized" dispersions of two phases, one in the other (such as butterfat in milk). The importance of magnetic or electric fields clearly depends on the composition of the system. The properties of certain crystal phases and of certain polymer systems can be altered by a one-dimensional force (a stress) imposed in addition to a uniform pressure, and centrifugal fields can play a role analogous to gravity.

Let I symbolize the number of relevant external intensive variables beyond composition variables. Normally, $I = 2$, since T and P are the most common intensive variables, but I may be greater than two in certain situations, or it may be less if we stipulate a "constant P" or "constant T" constraint.

Now we can state the phase rule in general. We have a total of ϕ phases, each with c independent components, and each subject to I external intensive variables.

From the c components, we can use $c - 1$ mole fractions[11] as additional intensive composition variables that apply to each phase. Thus

$$c - 1 + I \text{ intensive variables define each phase}$$

or

$$\phi(c - 1) + I \text{ intensive variables define the system}.$$

(Note that each phase has the same T, P, or other external intensive variable, but, in general, each phase has a *different* composition.)

The uniformity of the chemical potential of any one component in all phases implies a total of

$$c(\phi - 1) \text{ phase equilibrium constraints}.$$

(This is the total number of equals signs in the chemical potential expressions $\mu_i^\alpha = \mu_i^\beta = \ldots = \mu_i^\delta = \ldots = \mu_i^\phi$, $i = 1, 2, \ldots, c$.) *Each* such constraint reduces the number of *independent* intensive variables by one, giving *the phase rule:*

> The number of truly independent intensive variables in a system composed of c independent components among ϕ phases subject to I external intensive variables is
>
> $$\phi(c - 1) + I - c(\phi - 1) = c - \phi + I.$$

The number $c - \phi + I$ is termed the *variance* of the system, or the number of *degrees of freedom* of the system. A few examples will illustrate the meaning of these terms.

Consider first a pure substance ($c = 1$) with only T and P as relevant external intensive variables ($I = 2$). Then $c - \phi + I = 3 - \phi$. If the system is all one phase, ($\phi = 1$), the number of degrees of freedom is $3 - 1 = 2$; we can vary T and P independently in this system. One phase regions are areas on the phase diagram. If the system has two phases in equilibrium ($\phi = 2$), the number of degrees of freedom is $3 - 2 = 1$. Here, we can vary *either T or P independently, but not both.* Two phase equilibrium is a line on the phase diagram. Three phases in equilibrium have zero degrees of freedom. This is the triple point, a unique value of T and P.

Now consider a binary system ($c = 2$), with T and P the relevant external intensive variables, but *constrained to constant pressure* so that $I = 1$ and $c - \phi + I = 3 - \phi$. If the system is a homogeneous solid or liquid solution, $\phi = 1$, and there are two degrees of freedom; we can independently vary T and the composition variable x_1. If the system is a homogeneous solid solution in equilibrium with a homogeneous liquid solution, then $\phi = 2$, and there is only one degree of freedom. We can vary T, for instance, but in so doing, the composition of *both* phases must change as well.

This binary system can form an invariant ($c - \phi + I = 0$) point of a type we have not yet discussed. This type of behavior will be described in the next section, which applies the phase rule to the general problem of phase diagrams, their construction and their interpretation.

---

[11] It is $c - 1$, because c mole fractions are connected through one constraining relationship: all c of the mole fractions add to 1. If we know $c - 1$ of them, we can find the last one, so only $c - 1$ are independent.

6.7 PHASE DIAGRAMS

We used chemical potential curves to introduce phase diagrams in Section 6.2 (recall Figure 6.7). Experimentally, phase diagrams are constructed by a very different approach. Imagine a sample of some pure liquid confined at constant pressure in a device that constantly monitors the temperature of the liquid while transferring energy as heat from (or to) the liquid at a constant, but slow, rate. If we graph the system's temperature as a function of time, not only will the phase transition temperatures become apparent, but we will also measure thermodynamic phase transition parameters such as ΔH_ϕ.

From the definitions

$$C_P = \frac{dq_P}{dT} \quad \text{and} \quad dH = dq_P$$

we can see that (t is time)

$$\frac{dT}{dt} = \frac{dq_P}{dt}\frac{1}{C_P(T)} = \frac{dH}{dt}\frac{1}{C_P(T)}$$

so that the *slopes* of the lines of temperature–time plots are inversely proportional to heat capacity, while the *duration* (in time) of any segment of such a plot is proportional to the enthalpy change of the system during that time period, since dH/dt is constant in such experiments.

Figure 6.22 graphs the result of such an experiment that starts with 1 mol of $H_2O(l)$ at 25 °C and ends with ice at -25 °C. The energy transfer rate in this figure is $dq/dt = -1$ J s^{-1}. It is apparent that $C_P(l) > C_P(s)$ (since the slope of the liquid cooling line is less than the solid's) and that these heat capacities are temperature independent (since the lines are straight over this temperature range). The phase transition occurs over the horizontal portion of the curve at 0 °C, and the length of this segment (~ 6000 s) times dq/dt is $-\Delta \overline{H}_{\text{fus}}$ (since fusion is the inverse transformation to $s \rightarrow l$).

Ordinary fusion and vaporization are the most common examples of *first-order phase transitions*. The *order* of a phase transition depends on μ and its derivatives at the transition temperature. In general, since

$$\left(\frac{\partial \mu}{\partial T}\right)_P = -\overline{S} \quad \text{and} \quad \left(\frac{\partial^2 \mu}{\partial T^2}\right)_P = -\left(\frac{\partial \overline{S}}{\partial T}\right)_P = -\frac{\overline{C}_P}{T},$$

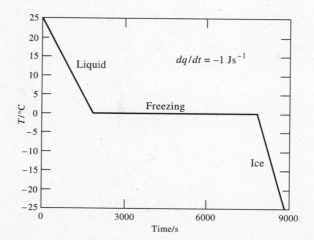

FIGURE 6.22 To measure heat capacities and phase transition enthalpies, a sample (1 mol of H_2O here) is subjected to a constant rate of energy withdrawal. At the phase transition temperature, a *temperature arrest* is observed until the phase transition is completed.

a first-order transition has

1. a continuous chemical potential ($\mu(\phi_1) = \mu(\phi_2)$)
2. a discontinuous entropy (due to a finite $\Delta \overline{S}_\phi$ jump)
3. a divergent heat capacity ($C_P \to \infty$ at T_ϕ).

A *second-order* phase transition has

1. a continuous chemical potential *and* entropy
2. a discontinuity in the heat capacity.

Second-order transitions are very rare. They correspond to equilibrium between two phases that differ not by μ, $\overline{S}$, $\overline{H}$, or $\overline{V}$, but by C_P, α, and κ, quantities that involve second derivatives of μ.

A number of phase transitions in condensed phases have the heat capacity behavior characteristic of a second-order transition, but they occur throughout a narrow range of temperature and generally involve a distinct entropy change. Many of these transitions feature a very rapid heat-capacity rise as the transition occurs, peaking to an experimental maximum at one temperature, then falling rapidly. An example is shown in Figure 6.23. (Theories suggest that C_P goes to infinity, but experiments never have the temperature precision to find a true infinity. A number of remarkable experiments have approached this problem with mK resolution, and the important result is more the temperature dependence of $C_P(T)$ near the predicted infinity rather than the infinity itself.)

The shape of the data in Figure 6.23 is used to name these types of transitions. They are called *lambda transitions,* since the data appear to draw the Greek letter λ. In the transition of Figure 6.23, the ammonium ion in NH$_4$Cl goes from a definite

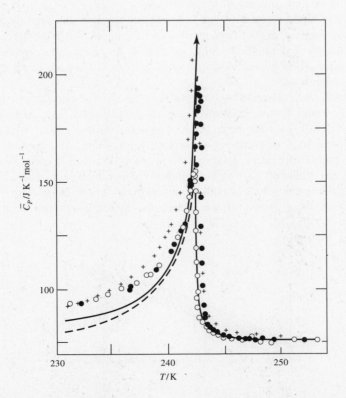

FIGURE 6.23 The heat capacity of NH$_4$Cl(s) at 1 atm shows the characteristic sharp rise near the order-disorder transition temperature. These data are from several different investigations.

orientational order with respect to Cl^- to a lack of orientational order. This type of *order–disorder* transition can occur in other ways. In the so-called β phase of brass (a Cu/Zn binary alloy), the composition can be in the range near $x_{Cu} = 0.5$. At low T (below ~740 K, where the alloy is said to be in the β′ phase), the atomic arrangement has each Cu atom surrounded by eight Zn atoms (at the corners of a cube) and vice versa. Above the 740 K transition temperature, this order is lost; the structure is the same, but the atoms are randomly distributed. (This system has certain similarities to the two-level system of Section 4.2. Recall the heat capacity curves of Figure 4.4.) As the transition temperature is approached, the degree of random disorder increases throughout the alloy. This is largely an entropy effect, and it points out how an entropy change can dominate a heat effect. When full disorder is achieved, the transition is abruptly over. The heat capacity rapidly falls back to a normal value on the high-temperature side of the transition temperature.

Other lambda transitions include the fluid-superfluid transition in He, various magnetic and electric property transitions, and the l–g transition at the critical point. Certain other transitions appear as bumps in otherwise smooth $C_P(T)$ curves, rather than as sharp λ spikes.

As we have already noted, a phase diagram denotes the stability boundaries between single-phase and multiphase equilibria. Cooling measurements or $C_P(T)$ measurements find these boundaries at various pressures and total system compositions. The remainder of this section discusses examples of some common binary phase diagrams.

Cu/Ni

At the beginning of Section 6.4, it was pointed out that Cu and Ni form homogeneous liquid and solid solutions at all compositions. From the point of view of the phase rule, $c = 2$, and, at constant P, $I = 1$, leaving $c - \phi + I = 3 - \phi$ degrees of freedom. The liquid phase thus has 2 degrees of freedom and appears as an area on the phase diagram, as does the one solid-solution phase. When the s–l equilibrium is established, $\phi = 2$ and the variance drops to 1. Any temperature that establishes s–l equilibrium fixes the compositions of *both* phases. They are different, but not at our disposal to vary.

The Cu/Ni phase diagram is shown in Figure 6.24. The upper area corresponds to a totally liquid system; the lower area, to totally solid. The area between is interpreted as follows. Imagine an alloy of any composition cooled from the liquid state, following the total composition isopleth down into this intermediate area, and stopping at one temperature. Examination of the system would show that some of the liquid has solidified so that s–l equilibrium is established. According to the phase rule, the compositions of both phases are fixed, but different from the system isopleth composition. In the two-phase area, the lever rule can find the amounts of

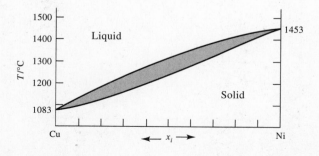

FIGURE 6.24 The Cu–Ni phase diagram has an *area* (shaded in the figure) corresponding to the two-phase equilibrium between a homogeneous liquid-solution phase and a homogeneous solid-solution phase. Outside this area, the system is *all* solid solution or *all* liquid solution.

each phase, as usual. The liquid composition is at the intersection of the upper curve (the *liquidus* curve) with the horizontal isotherm, and the solid composition is at the intersection of the lower curve (the *solidus* curve) and the isotherm.

Cd/Bi

The very different Cd/Bi phase diagram is shown in Figure 6.25. The liquid area represents a one-phase homogeneous liquid solution. The rest of the phase diagram represents various two-phase equilibria. The rectangular area below 144 °C is an inhomogeneous *mixture* of pure Cd and pure Bi solids. A homogeneous liquid solution of Cd and Bi spontaneously segregates into mixed chunks of pure solids below 144 °C. The Cd atoms find their own kind, as do the Bi atoms.

These atoms meet the requirements of our freezing-point depression discussion in Section 6.5. The solute (whichever has $x < 0.5$) is insoluble in the solid solvent, while the two are miscible in the liquid phase. Pure Cd melts at 321 °C and pure Bi at 271 °C. The almost-triangular regions on the phase diagram correspond to pure Cd(*s*) (in the left region) or pure Bi(*s*) (in the right region) in equilibrium with liquid solution. Their liquidus boundaries are sloped down towards the center of the diagram, corresponding to freezing point depressions of either solvent.

We can use Eq. (6.43) to find an expression for the liquidus line:

$$T_{\text{fus}}(x_1) = \frac{\Delta \overline{H}^\circ_{\text{fus}}}{R} \left(\frac{\Delta \overline{H}^\circ_{\text{fus}}}{RT^\circ_{\text{fus}}} - \ln x_1 \right)^{-1}.$$

For Cd $\Delta \overline{H}^\circ_{\text{fus}}/R = 730$ K while for Bi, $\Delta \overline{H}^\circ_{\text{fus}}/R = 1308$ K . Plotting this equation, first for Cd, then for Bi, lets us predict the T and x coordinates of the intersection point of the two liquidus lines. Experimentally, these coordinates are $T = 144$ °C and $x_{\text{Bi}} = 0.45$. Simple freezing-point depression theory gives (the details are left to a problem at this chapter's end) $T = 132$ °C and $x_{\text{Bi}} = 0.44$, which is very good agreement.

The experimental lowest freezing-point composition, $x_{\text{Bi}} = 0.45$, is a special point on the phase diagram known as the *eutectic point,* from the Greek for "easily melted." This mixture of Cd(*s*) and Bi(*s*) has the lowest possible melting point, the *eutectic temperature* (144 °C here), of any mixture. This mixture melts sharply at that temperature, as if it were a pure substance.

The phase rule has important things to say about eutectic behavior. In general, this system has (at constant P) $3 - \phi$ degrees of freedom. Solid Bi, solid Cd, and homogeneous liquid solution are all present. Thus $\phi = 3$, and the variance is zero. The temperature and compositions of all phases are fixed, as at the triple point of a pure compound.

Imagine melting a solid mixture with $x_{\text{Bi}} = 0.2$. At 144 °C, the first drop of liquid appears with the eutectic composition. Melting continues, *at this temperature,*

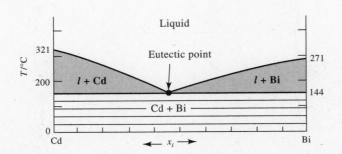

FIGURE 6.25 The Cd–Bi system is characterized by homogeneous solution formation in the liquid phase and heterogeneous mixing in the solid phase (below 144 °C). The areas labeled l + Cd and l + Bi correspond to homogeneous liquid solution in equilibrium with solid pure Cd or solid pure Bi, respectively.

until *all* the Bi(s) has melted into the eutectic liquid (along with enough Cd to maintain the liquid at the eutectic composition). When the last crystal of Bi(s) melts into the solution, the system becomes a two-phase system: Cd(s) and homogeneous liquid of the eutectic composition. Continued melting produces a temperature rise, Cd(s) continues to melt, the liquid composition follows the left-hand liquidus until it intercepts the $x_{Bi} = 0.2$ isopleth. At that temperature, the last Cd(s) crystal would melt, and we would have a one-phase liquid solution of composition $x_{Bi} = 0.2$.

Sn/Pb

The tin–lead system (common solder) has a phase diagram shown in Figure 6.26. It features a eutectic point, but also two new *one-phase* areas labeled α and β. Tin and lead exhibit *limited solid miscibility*. The α phase is a *homogeneous* solid solution of Pb in (mostly) Sn, and the β phase is a solid solution of Sn in (mostly) Pb. The degree of solid miscibility is a function of temperature, with miscibility dropping rapidly with decreasing T.

In the triangular-shaped regions (bordered by the liquidus line, the eutectic isotherm, and the α or β boundary, which is a solidus line), l–α or l–β two-phase equilibrium is found. The area below the eutectic isotherm and bounded by the α or β boundaries (called *solvus lines* below the eutectic temperature) is a two-phase area for the coexistence of α and β solid phases (as an inhomogeneous mixture of solids).

The inhomogeneity of immiscible solids formed from a homogeneous liquid solution is quite subtle. Figure 6.27 is a microphotograph of a solidified sample of a Pb/Sn eutectic solution. The alternating stripes are alternate α and β solid phases. On a truly atomic scale, these regions are vast. To the unaided eye, they are not discernible. (The dark regions are the β phase in the photograph.)

Nature can be far more diverse than these examples show. Other features, such as compound formation, multiple pure-compound phases, and various special points that are variants on the eutectic point are common. Regions of miscibility can come and go as T or x are varied, but all seemingly more complex phase diagrams must follow the dictates of the phase rule. For instance, there can never be five phases in equilibrium in a binary system with only T and P as relevant external intensive variables. (Four are possible: g–l–α–β perhaps, at a special T, P, and x_1.)

The extension to three component (ternary) and general multicomponent systems involves no new principles. The diagrams become rapidly more complex, and we will not discuss them here.

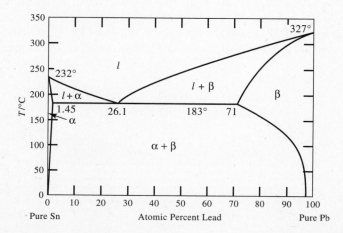

FIGURE 6.26 The Pb–Sn system exhibits areas (α and β) corresponding to *single phase* homogeneous solid solutions. The single-phase liquid-solution area is labeled l. The other areas are two-phase areas, labeled by the phases in equilibrium in each.

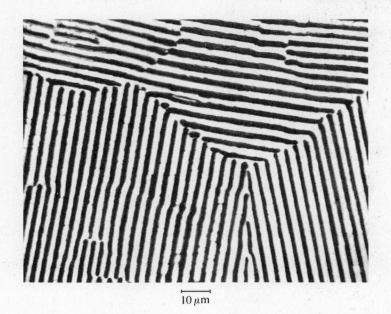

FIGURE 6.27 A microphotograph of a frozen sample of Pb–Sn at the eutectic composition shows stripes that are α and β phase regions, so intimately mixed that they appear only under high magnification (approximately 800 times here). The various directions of the stripes are due to microcrystallite formation characteristic of most solidification.

10 μm

CHAPTER 6 SUMMARY

This long chapter has introduced many important applications of thermodynamics to multicomponent, multiphase systems. The common themes have been the use of chemical potential changes to predict spontaneous processes and the attainment of chemical potential uniformity at equilibrium.

We started with pure-compound phase transitions, and derived the expression for the slope of two-phase equilibrium lines:

$$\text{Clapeyron equation:} \left(\frac{dP}{dT}\right)_\phi = \frac{\Delta \overline{S}_\phi}{\Delta \overline{V}_\phi} = \frac{\Delta \overline{H}_\phi}{T_\phi \Delta \overline{V}_\phi}$$

and we developed this into an expression for vapor pressure calculations:

$$\text{Vapor pressure:} \\ P(T_f) = P(T_i) \exp\left[-\frac{\Delta \overline{H}_\phi}{R}\left(\frac{1}{T_f} - \frac{1}{T_i}\right)\right]$$

We considered the composition and concentration variables of *mole fractions* and *molality* in describing mixtures and solutions. We defined partial pressure in terms of mole fractions:

$$\text{Partial pressure:} \ P_i = x_i P_{\text{tot}} = \frac{n_i}{n_{\text{tot}}} P_{\text{tot}}$$

and defined an *ideal mixture of ideal gases* that follows

$$\text{Ideal free energy of mixing:} \\ \Delta G_{\text{mix}} = nRT \sum_i x_i \ln x_i$$

as well as Dalton's and Amagat's laws.

We introduced *activity* through the use of standard fugacities:

$$\text{Definition of activity:} \ a_i = \frac{f_i}{f_i^\circ(T, P^\circ, x^\circ)}$$

and we defined an *ideal solution,* for which

$$\text{Ideal solution fugacity:} \ f_i(\text{ideal}) = K_i(T, P) \, x_i$$

Special cases were

$$\text{Raoult's law ideal solution:} \ P_i = P_i^\circ x_i$$

and

$$\text{Henry's law ideal dilute solution:} \ P_2 = K_{2,1} x_2$$

The excess Gibbs free energy was used to describe the nonideality in the

$$\text{Nonideal simple solution:} \\ G^E = G(\text{real}) - G(\text{ideal}) = nAx_1 x_2$$

that could exhibit spontaneous immiscibility.

We described the *colligative properties* of *vapor pressure lowering, freezing-point depression, boiling-point elevation,* and *osmotic pressure* in Section 6.5:

> Osmotic pressure:
> $$\Pi = \frac{n_2 RT}{V}$$
> Freezing point depression:
> $$\ln x_1 = \frac{\Delta \overline{H}^\circ_{fus}}{R}\left(\frac{1}{T^\circ_{fus}} - \frac{1}{T_{fus}}\right) \quad \text{or} \quad \Delta T_{fus} = K_f\, m_2$$
> Boiling point elevation:
> $$\ln x_1 = \frac{\Delta \overline{H}^\circ_{vap}}{R}\left(\frac{1}{T_{vap}} - \frac{1}{T^\circ_{vap}}\right) \quad \text{or} \quad \Delta T_{vap} = K_b\, m_2$$

Finally, we showed how the number of degrees of freedom could be calculated from the phase rule, $c - \phi + I$, and used to predict and/or explain the phenomena associated with those complex interactions displayed in phase diagrams for binary systems.

The next chapter describes how chemical reaction equilibrium fits into the scheme of thermodynamics. We will find that most of the formalism has already been worked out, but the new concepts that will dominate the chapter will be the chemical reaction equilibrium constant and the incorporation of a chemical reaction path into our repertoire of thermodynamic paths.

FURTHER READING

A good (and somewhat more advanced) discussion of much of this chapter's material can be found in the two books *Molecular Thermodynamics of Fluid-phase Equilibria*, by J. M. Prausnitz, R. N. Lichtenthaler, and E. G. Azevedo, 2nd ed., (Prentice-Hall, Englewood Cliffs, N.J., 1986) and *Regular and Related Solutions: the Solubility of Gases, Liquids, and Solids*, by J. H. Hildebrand, J. M. Prausnitz, and R. L. Scott, (Van Nostrand Reinhold Co., New York, 1970).

ACCESS TO DATA

Several standard references give phase equilibrium data. Data on pure compounds can be found in several of the references given in Chapters 4 and 5.

W. Hansen, *Constitution of Binary Alloys*, 2nd. ed., (McGraw-Hill, New York, 1958). Supplements appeared in 1965 and 1969.

E. M. Levin, H. F. McHurdie, and F. P. Hall, *Phase Diagrams for Ceramists*, (American Ceramic Study, Columbus, OH, 1956).

J. Timmermans, *The Physico-Chemical Constants of Binary Systems in Concentrated Solutions*, (4 volumes, Interscience, New York, 1959–60); *Physico-Chemical Constants of Pure Organic Compounds*, (Elsevier, New York, 1950, 1965).

Metals Handbook (Am. Soc. for Metals, Cleveland, OH, 1961–1990).

H. Stephen and T. Stephen, Ed., *Solubilities of Inorganic and Organic Compounds*, (4 volumes, Macmillan, New York, 1964).

Selected Data on Mixtures, (Thermo. Res. Center, Texas A&M Univ., College Station, TX, 1972).

PRACTICE WITH EQUATIONS

6A What is the rate of increase of bromine's vapor pressure with temperature if, at 298 K, $\Delta \overline{H}_{vap} = 31.01$ kJ mol^{-1} and $P_{vap} = 0.30$ atm? (6.11)

ANSWER: 1.5×10^{-3} atm K^{-1}

6B Use data from 6A to predict the normal boiling temperature of Br_2. (6.12)

ANSWER: 330 K

6C How well does Trouton's rule predict the 298 K vapor pressure of Br_2? (6.14)

ANSWER: prediction: 0.32 atm; observed: 0.30 atm

6D If 10.0 g of Br_2 are placed in a 1 L evacuated container at 298 K, what mole fraction of the Br_2 will be in the gas phase if $\rho(l) = 3.12$ g cm^{-3}? (6.18) and data in exercises above

ANSWER: 0.196

6E What is the 300 K free-energy change on mixing the major components of air: N_2, 78.08% by volume; O_2, 20.95%; Ar, 0.93%; and CO_2, 0.04%? (6.24)

ANSWER: -1.79 kJ mol^{-1}

6F When 1.00 mol each of O_2 and CO_2 are mixed at 1.00 atm and 298.2 K, the observed volume is 48.81 L. What would Amagat's Law predict?

ANSWER: 48.94 L

6G In an acetone–methanol mixture at 57.2 °C with $x_{acetone} = x_1 = 0.400$, the activity coefficients are $\gamma_1 = 1.248$

and $\gamma_2 = 1.113$. What is the molar excess Gibbs free energy of the mixture? (6.36)

ANSWER: $G^E/n = 420$ J mol^{-1}

6H How well does the acetone–methanol mixture described in 6G follow the expectations of a simple solution? (6.38)

ANSWER: A/RT(from γ_1) = 0.615; A/RT(from γ_2) = 0.669

6I Two liquids mix according to Raoult's law. At what liquid composition will they have an equilibrium pressure that is the arithmetic average of their pure vapor pressures? (6.40)

ANSWER: $x_1 = 0.5$

6J What is the gas-phase composition of the solution in 6I above? (6.41)

ANSWER: $y_1 = P(1)/[P(1) + P(2)]$

6K The solubility of Ar in H_2O is 5.15 mg Ar per 100 g H_2O at 25 °C and 1.00 atm Ar pressure. What is K_{Ar, H_2O}?

ANSWER: 4.31×10^4 atm

6L How many grams of NaCl should be added to 1.0 L of water to raise the boiling point 1 °C? (6.47) and Table 6.4

ANSWER: 115 g

6M If 5.0 g of glucose, $C_6H_{12}O_6$, is dissolved in 1.00 L of an aqueous solution at 300 K, what osmotic pressure would the solution exhibit? (6.48)

ANSWER: 0.68 atm

PROBLEMS

SECTION 6.1

6.1 How would a μ–T diagram like those in Figure 6.7(b) look for a system (a) above the critical pressure? (b) at the critical pressure? (c) if the substance formed a stable *liquid-crystal phase* between the liquid and solid phases?

6.2 Each of the following plots of μ versus T or P for a pure substance has at least one fault. What is wrong with each?

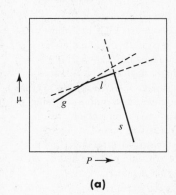
(a)

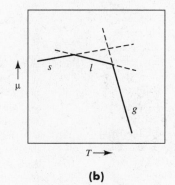
(b)

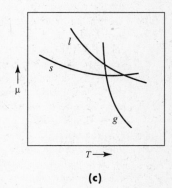
(c)

6.3 Construct a diagram similar to Figure 6.4 for Ar at 1 atm using Eq. (5.35) to find $\Delta\mu$ for the gas phase and the simple expression $\mu(T) = \mu°(T°) - \overline{S}(T - T°)$ for the liquid and solid phases. You will need to know $T°_{vap} = 87.293$ K, $T°_{fus} = 84.0$ K, $\overline{S}(l) = 6.31\,R$, and $\overline{S}(s) = 4.60\,R$. For the brief span of temperature between melting and boiling, assume that $\overline{S}(l)$ is constant, limit your graph to the temperature range 80–90 K, and similarly assume $\overline{S}(s)$ is constant.

SECTION 6.2

6.4 Derive a variant of Eq. (6.12) for the temperature variation of vapor pressure starting from Eq. (6.11a) and assuming $\Delta\overline{S}_\phi$ is temperature independent. Compare the predictions of your equation to that of Eq. (6.12) using the data in Example 6.2 for water. Which equation is preferable?

6.5 Plot the melting curve for Ar using the Simon and Glatzel expression in the text along with the vapor pressure curve ($\Delta\overline{H}_{vap}/R = 826.6$ K, $T°_{vap} = 87.293$ K). Do they intersect at the triple point? (See Table 6.1 for triple-point coordinates. The melting curve expression is designed for a very large pressure range with modest temperature accuracy. Consequently, limit your temperature span to only a few degrees on either side of the triple point temperature.)

6.6 The vapor pressure of $CO_2(s)$ is 0.0386 atm at 162.0 K and 0.395 atm at 184.0 K. Find the molar enthalpy of sublimation of CO_2, the standard sublimation temperature of CO_2 (the temperature at which the solid vapor pressure is 1 atm), and the slope of the solid–vapor coexistence curve, $(dP/dT)_\phi$, at this temperature.

6.7 Iodine boils normally at 183.0 °C, and its vapor pressure at 116.5 °C is 100 torr. What is the molar enthalpy of vaporization of iodine? If the molar enthalpy of fusion of iodine is 15.65 kJ mol^{-1}, and the vapor pressure of the *solid* is 1.00 torr at 38.7 °C, calculate the triple-point temperature and pressure.

6.8 If an ice skater weighs 155 lb (about 70 kg) and if the area of an ice skate blade is about 0.2 cm^2 (20 cm long and 0.1 mm wide), the skater exerts a pressure

$$\frac{(70\ \text{kg})(9.8\ \text{m s}^{-2})}{2\times 10^{-5}\ \text{m}^2} = 3.43\times 10^7\ \text{N m}^{-2} = 343\ \text{atm}$$

on the ice. Calculate the melting of ice at this pressure, given $\overline{V}(\text{water}) = 18.00\ \text{cm}^3\ \text{mol}^{-1}$, $\overline{V}(\text{ice}) = 19.65\ \text{cm}^3\ \text{mol}^{-1}$, and $\Delta \overline{H}^\circ_{\text{fus}}(\text{water}) = 6.03\ \text{kJ mol}^{-1}$.

6.9 Here you are, on top of Mt. McKinley, at 20 300 ft. You would like to know the atmospheric pressure, since you seem to have trouble breathing, but your manometer has fallen down an ice cave. Fortunately, your thermometer wasn't lost, and you use it to measure the boiling point of water as you boil a cup to make some tea. You find 75 °C. What is the atmospheric pressure?

6.10 There is an interesting paper in *Review of Scientific Instruments*, **57**, 2827 (1986) by J. E. Graebner of AT&T Bell Labs. In the paper, Graebner describes a very sensitive calorimeter that operates at low temperature and is designed to measure the amount of H_2 dissolved in solids. The molecular basis of this instrument requires some concepts from later chapters to understand, but what is of immediate interest is a graph in the paper of his raw data. The calorimeter is immersed in liquid He (at its boiling point), and data are collected over a 200-hour time period. Graebner needed to know the temperature of the He(l) during this time, which can fluctuate due to atmospheric pressure changes. His graph of one run of such measurements shows a huge drop in the He(l) temperature about 70 hours into the experiment. In the caption to this figure, the drop is explained as follows: "The sharp dip at 70 h is caused by a nearby hurricane"! The temperature drops from about 4.22 K to 4.16 K at this dip. What was the atmospheric pressure in Murray Hill, NJ, site of Bell Labs, when the hurricane went by ($\Delta \overline{H}^\circ_{\text{fus}}$ (He) = 84.0 J mol^{-1})?

SECTION 6.3

6.11 While mole fraction is often the most convenient theoretical measure of composition, it is rarely the most convenient experimentally. The practical concentration units of most utility are: percent by weight (g of species i per 100 g of solution), mass concentration (g of i per unit volume of solution), molarity (moles of i per L of solution), and molality (moles of solute i per kg of solvent). Find expressions relating mole fraction to each of these other concentration measures. An aqueous solution of 69.0 g of ethanol per L of solution has a density of 0.9862 g cm^3 at 20 °C. The density of pure water is 0.9982 g cm^3 at 20 °C. Express the concentration of ethanol in this solution in each of these units.

6.12 Derive Eq. (6.18), the versions of the lever rule that yield absolute mole fractions of either of two phases in equilibrium.

SECTION 6.4

6.13 Every chemist knows to "add acid to water with constant stirring" when diluting a concentrated acid in order to keep the solution from spewing boiling acid all over the place. Explain how *this one fact* is enough to prove that strong acids and water do not form ideal solutions.

6.14 Prove that $T = A/2R$ is the critical solution temperature for a simple solution.

6.15 For the benzene–cyclohexane system at 40 °C, $A/RT = 0.425$. By how much does the actual Gibbs free energy of mixing of a 50/50 solution of benzene and cyclohexane differ from the ideal value? If $A/RT = 0.455$ at 30 °C and 0.399 at 50 °C, what are the entropies and enthalpies of mixing this solution at 40 °C? (*Hint:* Look back to Equations (6.25)–(6.27).)

6.16 Construct a generic phase diagram for a simple solution in T, x_1 coordinates from the critical solution temperature, $T = A/2R$, to lower temperatures. Do this by finding the x_1 values for several temperatures at the first minimum in $\Delta \overline{G}_{\text{mix}}$, as was done for Figure 6.15(b), note that by symmetry the second x_1 minimum must occur at $1 - x_1$, and plot the data to produce the two-phase region boundary.

6.17 It is difficult to position a system exactly at its critical point; consequently, various extrapolation techniques are used to determine critical coordinates from multiphase equilibrium data away from the critical point. One such method, known as the "Cailletet–Mathias law of rectilinear diameter," states that the arithmetic mean of the densities of coexisting liquid and gaseous phases is a linear function of temperature:

$$\frac{\rho(g) + \rho(l)}{2} = aT + b\ .$$

Use this empirical law and the data below to find the critical temperature and volume of CO_2, and compare your results to the values in Table 1.3. First plot the data as ρ versus T to map out the two-phase boundary. Connect these points by a smooth line. Then fit the linear function above to the data, finding a and b. Plot the resulting line on the same graph, and read the critical temperature from the intersection of this line with the coexistence boundary line.

T/°C	ρ(l)/g cm^{-3}	ρ(g)/g cm^{-3}
30.0	0.6016	0.3380
25.0	0.7165	0.2375
20.0	0.7784	0.1910
15.0	0.8236	0.1594
10.0	0.8626	0.1350
5.0	0.8966	0.1154
0.0	0.9273	0.0993
−5.0	0.9556	0.0854

6.18 Gold and copper form a simple solution with $A/R = -2894$ K. Plot the activity of Au in Cu as a function of x_{Au} at $T = 1500$ K. Do you expect "unmixing" in this system, or are Au and Cu likely miscible in all proportions at 1500 K?

SECTION 6.5

6.19 If two optical isomers are mixed as liquids and follow Raoult's law, will the solution exhibit a range of vaporization pressures, or will any mixture vaporize at a single pressure? What can you say about the composition of the gas phase over the solution?

6.20 At 100 °C, the vapor pressures of hexane and octane are 2.416 atm and 0.466 atm. Solutions of these follow Raoult's law. One particular solution is mixed and held at 100 °C and 5.00 atm, where it is all in the liquid phase. The pressure is slowly reduced, and it is noted that the first gas bubble appears when the pressure has reached 1.00 atm. What was the solution's composition?

6.21 Construct a diagram analogous to Figure 6.19 for the case $K/P° = 0.5$. You should find a phase diagram that looks like Figure 6.19 upside down, a so-called *minimum boiling azeotrope*.

6.22 Problems 4.17 and 4.18 contain data on the S(rhombic) → S(monoclinic) phase transition.

(a) Find $\Delta\mu$ for the transformation

$$\text{S(rhombic, 298 K, 1 atm)} \rightarrow \text{S(monoclinic, 298 K, 1 atm).}$$

(b) At 298 K and 1 atm, a saturated solution of rhombic S in CCl$_4$ contains 8.4 g S per kg of solvent. Use this value and the $\Delta\mu$ value from part (a) to find the solubility of monoclinic S in CCl$_4$ at this T and P. Make use of the fact that the solutions are in the ideal dilute regime and take into account that dissolved S consists entirely of S$_8$ molecules (as do both solids, for that matter, but data for the solids are written as if they consisted of S atoms).

6.23 The compact freezing point expression, Eq. (6.45), came from the more exact Eq. (6.43) via several approximations. Test the validity of those approximations by computing the freezing point of a solution of 10.0 g of NaCl in 1.00 kg of H$_2$O using both equations.

6.24 Here are three related freezing point depression questions:

(a) Two grams of benzoic acid (C$_6$H$_5$COOH) dissolved in 25 g of benzene produce a freezing-point depression of 1.65 K. The cryoscopic constant for benzene is 5.07 K kg mol^{-1}. Find the molecular mass of benzoic acid from these data and interpret the answer.

(b) Two grams of LiF dissolved in 100 g of water produce a freezing-point depression of 2.9 K. What molecular mass do these data give for LiF, and why isn't the answer 26 g mol^{-1}?

(c) When 3.25 g of Se are dissolved in 225 g of benzene, the freezing-point depression is 0.117 K. What is the molecular formula of Se in this solvent?

6.25 There is a purification technique known as "freeze-pump-thaw" that is commonly used to remove impurities from gas samples. The entire sample is frozen, the frozen sample is subjected to a dynamic vacuum (i.e., it is "pumped on" with a vacuum pump), the sample is then isolated and thawed. This process is often repeated several times. In which of the three steps—freezing, pumping, or thawing—does the actual purification occur?

6.26 Salt, NaCl, is used effectively to lower water's freezing point and keep streets and sidewalks clear of winter ice. But any salt would work; NaCl happens to be cheap. Likewise, CaCl$_2$ is readily available (and used in some commercial de-icing formulations for sidewalks) and has an advantage over NaCl of providing three solute particles per mole of compound rather than two. If NaCl costs $x per kg, at what cost per kg does CaCl$_2$ become economically advantageous? The economic factor is degrees of freezing-point depression per unit mass of salt.

6.27 The melting points of substances that melt at very high temperatures are often difficult to measure precisely, even if one ignores the very great problem of finding a container in which to melt them! One problem is purity: what mole fraction of impurity would alter the fusion temperature of W ($\Delta\overline{H}°_{fus}(W)/R = 4237$ K, $T°_{fus} = 3660$ K) by 10 K? by 50 K?

6.28 The footnote to Table 6.1 states, "The normal melting (fusion) temperature for pure water saturated with air at 1 atm is 0.0098 K lower" than the triple-point temperature. Is all of this decrease due to a freezing-point depression from the dissolved air? (See exercise 6E for the composition of air, neglect Ar and CO_2, see Example 6.5 for the O_2 Henry's law constant, and use $K_{N_2,H_2O} = 8.04 \times 10^4$ atm.) How large a change is due to the pressure difference (1 atm versus P_{tp})?

6.29 Since vaporization temperatures are sensitive to pressure, so are ebullioscopic constants. Find a correction factor for $K_b(H_2O)$ in terms of the ambient pressure difference from 1 atm.

6.30 Derive an approximate expression for K_b for solvents that follow Trouton's rule. How well does this approximation predict the accurate values in Table 6.4?

6.31 A 0.31 M solution of sucrose ($C_{12}H_{22}O_{11}$) at human body temperature (37 °C) is isotonic to human blood serum. What osmotic pressure would red blood cells achieve (could they stand such a pressure without rupturing) if placed in pure water?

6.32 In practice, molecular masses of macromolecules are determined from osmotic pressure measurements by an extrapolation procedure to zero concentration. If we let c_2 represent the concentration of solute in mass per unit volume units and M_2 represent the solute molecular mass, the van't Hoff equation becomes $\Pi/c_2 = RT/M_2$. One measures Π at a variety of c_2 values, plots Π/c_2 versus c_2, and extrapolates to $c_2 = 0$. Then M_2 is found from

$$M_2 = \frac{RT}{\lim_{c_2 \to 0} \frac{\Pi}{c_2}}$$

Use this method to find the molecular mass of bovine serum albumin from the following data.

c_2/g L^{-1}	$\Pi/10^{-3}$ atm
8.95	3.30
17.69	6.67
27.28	11.0
56.20	25.4

SECTION 6.6

6.33 The compound $CaCO_3(s)$ exists in equilibrium with measurable amounts of $CaO(s)$ and $CO_2(g)$ at moderate to high temperatures. Suppose pure $CaCO_3(s)$ is placed in an evacuated container at low temperature and raised to some sufficiently high temperature so that an appreciable amount of $CO_2(g)$ is generated. In the context of the phase rule at this temperature, how many phases are present? How many independent components? How many degrees of freedom? Suppose a small amount of pure $CO_2(g)$ is added to the container at this temperature. How must the system respond? Repeat this problem replacing $CaCO_3(s)$ with $PCl_5(g)$, $CaO(s)$ with $PCl_3(g)$, and $CO_2(g)$ with $Cl_2(g)$.

6.34 Here are two questions related to the inherent discontinuity in structure between solid and fluid phases or between two solid phases. (a) Is there a solid–liquid critical point for a pure substance, or does the solid–liquid coexistence line extend up from the triple point forever? (b) Suppose there were two solid phases of a pure substance with a solid–solid coexistence line that, by chance, had just the right properties to intersect the triple point exactly so that solid 1, solid 2, liquid, and gas could coexist at once. What implications would this have on the phase rule?

SECTION 6.7

6.35 In 1855, Matthew Fontaine Maury wrote *The Physical Geography of the Sea and its Meteorology*, a book that was an immediate popular success and remained in print for 20 years. In it, Maury quotes sea-water temperatures from all over the world. The coldest is the arctic ice bath, which is brine (salty sea water) in equilibrium with icebergs. Maury gives 27.2 °F (−2.67 °C) as the freezing point for this brine. A partial phase diagram for the NaCl/H_2O system is shown below. Use it to answer the following questions. (Note that the composition variable is *percent by weight*, not mole fraction.)

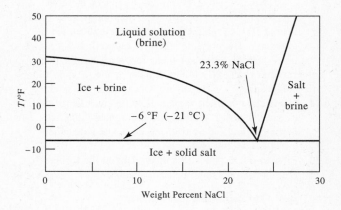

(a) Do arctic icebergs taste salty?
(b) What is the molality of NaCl in arctic sea water?
(c) Suppose in your wanderings among the arctic ice you encounter a puddle of brine in an iceberg with salt crystals at the bottom of the puddle. What can you say about the temperature and composition of the puddle?

(d) The diagram shows that a solution 23.3% by weight NaCl has the lowest freezing point (−21 °C). How well does this agree with the predictions of our freezing point depression expression?

6.36 The phase diagram for the Ag/Cu system is shown below.

(e) The Au/Si system deviates significantly from the predictions of Eq. (6.43) as parts (c) and (d) above will convince you. This is not always the case, however. Repeat part (d) for the Bi/Cd system and compare your results to Figure 6.25 ($T^\circ_{\text{fus}}(\text{Bi}) = 271.3$ °C, $T^\circ_{\text{fus}}(\text{Cd}) = 765$ °C, $\Delta \overline{H}^\circ_{\text{fus}}(\text{Bi})/R = 1308$ K, $\Delta \overline{H}^\circ_{\text{fus}}(\text{Cd})/R = 730$ K).

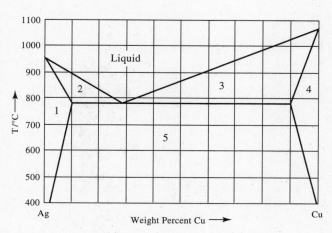

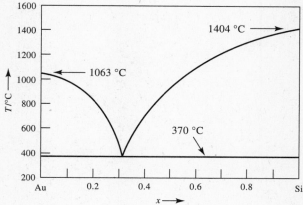

(a) Identify each numbered area by the phase or phases they represent.

(b) A sample 70% Cu (by weight) is liquified and mixed thoroughly at 1100 °C. The sample is then cooled to 900 °C. What is the *composition of each phase present* at this temperature?

(c) What are the relative amounts of each phase at this temperature?

(d) Sketch a cooling curve for the sample in (b) when brought from 1100 °C to 600 °C.

6.37 The Au/Si phase diagram is shown below. This system has importance in the semiconductor industry whenever Au contacts are deposited on Si circuit substrates.

(a) Identify the phase or phases associated with each area of the diagram.

(b) What would happen if one tried to melt gold in a silicon crucible?

(c) Assuming that Au and Si follow the freezing-point depression expression of Eq. (6.43), find $\Delta \overline{H}^\circ_{\text{fus}}$ for each element from the observed eutectic temperature and composition.

(d) The true values are $\Delta \overline{H}^\circ_{\text{fus}}(\text{Au})/R = 1524$ K and $\Delta \overline{H}^\circ_{\text{fus}}(\text{Si})/R = 4765$ K. Use these to construct a phase diagram based on Eq. (6.43) and locate the predicted eutectic temperature and composition.

GENERAL PROBLEMS

6.38 Show that the molar enthalpy of vaporization $\Delta \overline{H}^\circ_{\text{vap}}$ changes with temperature according to $d(\Delta \overline{H}^\circ_{\text{vap}})/dT = \Delta \overline{C}^\circ_{P,\text{vap}}$ where $\Delta \overline{C}^\circ_{P,\text{vap}}$ is the molar heat capacity difference between the two phases. Follow these steps. Consider $\Delta \overline{H}^\circ_{\text{vap}}$ as a function of T and P so that

$$d(\Delta \overline{H}^\circ_{\text{vap}}) = (\overline{C}_P(g) - \overline{C}_P(l))\, dT + \left[\left(\frac{\partial \overline{H}(g)}{\partial P} \right)_T - \left(\frac{\partial \overline{H}(l)}{\partial P} \right)_T \right] dP .$$

Use Eq. (3.43) to express the pressure derivatives, relate dT and dP by the Clapeyron equation, note that $\Delta \overline{V}_{\text{vap}} \cong \overline{V}(g)$, and invoke the ideal gas equation of state.

6.39 Verify that the Lewis–Randall ideal mixture also implies that the partial molar volume of any gas in the mixture is equal to the molar volume of the pure gas. This is enough to establish Amagat's Law rigorously.

6.40 The phase diagram for ^{3}He and ^{4}He solutions at 1 atm is shown below. Note that ^{3}He appears to retain roughly 6% solubility all the way to absolute zero, and that both isotopes and their mixtures remain liquids. In this case, the nonideality of the mixture that leads to phase separation can be traced to an excess enthalpy of mixing that has a quantum mechanical origin. Sketch the rough shape that ΔG_{mix} must have for this system at 0.5 K. What implications do stable solutions at absolute zero have on the Third Law?

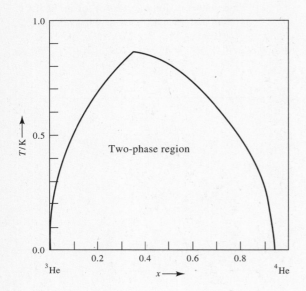

6.41 Compute the solubility of Cu(s) in Bi(l) at 800 K given $\overline{T}^\circ_{fus}$(Cu) = 1356 K, $\Delta \overline{H}^\circ_{fus}$(Cu)/R = 1585 K, and the activity coefficient of Cu in Bi solutions (relative to pure liquid Cu):

$$\ln \gamma_{Cu} = \frac{1240 \text{ K}}{T}(1 - x_{Cu})^2 .$$

Argue as follows: at equilibrium, the chemical potential of Cu(s) must equal that of Cu in solution (symbolized Cu(l, x_{Cu})). One can reach this equality by relating μ(Cu (s)) to μ(Cu(l)) at 800 K, then relating μ(Cu(l)) to μ(Cu(l, x_{Cu})). To make the first link, show that

$$\frac{\Delta \overline{G}^\circ_{fus}}{RT} = \frac{\Delta \overline{H}^\circ_{fus}}{R}\left(\frac{1}{T} - \frac{1}{T^\circ_{fus}}\right)$$

for the chemical potential change $s \to l$ at T (thus introducing $\Delta \overline{H}^\circ_{fus}/R$ and $\overline{T}^\circ_{fus}$ into the calculation). Next, for the transformation $l \to l$, x_{Cu} the change is just

$$\frac{\Delta \mu}{RT} = \ln a_{Cu} = \ln \gamma_{Cu} x_{Cu}$$

since the activity of the pure liquid is 1. Adding these two expressions yields an expression $\Delta \mu /RT$ for the $s \to l$, x_{Cu} transformation. At equilibrium, this equals zero. Solve for x_{Cu}; you should find x_{Cu} = 0.142 at 800 K.

6.42 In 1917, G. N. Lewis and H. Storch published the measurements below of the vapor pressure of Br_2 over room-temperature solutions of Br_2 in CCl_4. (760 torr = 1 atm)

$10^3 \, x_{Br_2}$	P_{Br_2}/torr
3.94	1.52
4.20	1.60
5.99	2.39
10.2	4.27
13.0	5.43
23.6	9.57
23.8	9.83
25.0	10.27

(a) Find the Henry's law constant K_{Br_2,CCl_4} from these data.

(b) They also measured the distribution of Br_2 between immiscible H_2O and CCl_4 phases in contact, finding that the ratio of the molality of Br_2 in H_2O to the mole fraction of Br_2 in CCl_4 was 0.371 g Br_2/kg H_2O. What is the partial pressure of Br_2 over a dilute Br_2/H_2O solution as a function of the Br_2 molality?

6.43 There are two theorems, known as the Gibbs–Konovalow theorems, that have particular relevance to the appearance of azeotrope (P, x) diagrams. The theorems state that if $dP/dx_i = 0$ along a dew-point line, then the liquid- and gas-phase compositions must be equal (i.e., $x_i = y_i$) and that if the dew point line has an extremum, the bubble point line must as well at the same point (i.e., $dP/dy_i = 0$). To prove these theorems, we write Eq. (5.32) at constant T for each phase in terms of mole fractions:

$$\overline{V}(l) \, dP = x_1 \, d\mu_1 + x_2 \, d\mu_2$$
$$\overline{V}(g) \, dP = y_1 \, d\mu_1 + y_2 \, d\mu_2.$$

Next, subtract the first of these from the second, divide by dx_i, and argue how the resulting expression ensures the first theorem: that if $dP/dx_i = 0$, $y_i = x_i$ and vice versa. Then divide by dy_i instead of dx_i and prove the second theorem.

6.44 Shown below is a generic phase diagram for species A and B that form a stoichiometric solid compound AB_2 with unusual melting properties. (The A = K, B = Na system closely approximates this phase diagram, as does the silica (SiO_2)/alumina (Al_2O_3) system.)

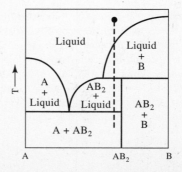

(a) At what mole fraction is the vertical line at AB_2 drawn?

(b) The compound AB_2 is said to melt *incongruently*. At the temperature corresponding to the upper end of the AB_2 vertical line, melting produces a mixture of a liquid-solution phase and solid B, rather than a liquid of AB_2 molecules. Sketch a cooling curve for a sample prepared at the dot in the one-phase homogeneous liquid area and cooled along the dashed line. What phase or phases would be observed along the way? Be sure to keep the phase rule in mind and accommodate any temperature arrests required by it.

Chemical Reactions and Chemical Equilibria

7.1 Chemical Reactions as a Thermodynamic Path

7.2 Energy and Entropy Changes in Chemical Reactions

7.3 Named Enthalpies and Enthalpy Diagrams

7.4 Chemical Reaction Equilibria

7.5 The Stability of Chemical Equilibria

7.6 Methods of Calculating Equilibrium Constants

IMAGINE having the magic ability to stop chemical reactions at your whim. You could mix explosively reactive chemicals as easily as you mix ethanol and water. You could start a reaction for a period of time and then stop it. You could mix the reactants and products in any proportion, allow the reaction to proceed naturally, and periodically analyze the reacting mixture. Eventually, you would find that the composition of the system stops changing. The system has spontaneously achieved *chemical reaction equilibrium*.

In this chapter, we use the path independence of thermodynamic state functions to simulate, on paper at least, this ability. We treat the conversion of reactants to products as another type of thermodynamic path, and we devise the tools to traverse such paths. Sections 7.2 and 7.3 discuss the very important energy, enthalpy, and entropy changes that accompany any reaction on the way to chemical equilibrium. The first section lays the foundation for the remainder of this chapter with the methods that describe the progress of a chemical reaction.

7.1 CHEMICAL REACTIONS AS A THERMODYNAMIC PATH

Our first task is to establish a consistent and complete notation for chemical reactions. Consider the reaction for the oxidation of ethanol to carbon dioxide and water:

$$C_2H_5OH(l) + 3O_2(g) \rightleftarrows 2CO_2(g) + 3H_2O(l) \ .$$

The reaction is balanced, in accord with conservation of mass requirements. We could multiply all of the stoichiometric coefficients by any factor and still have a balanced reaction. If the factor is negative, the roles of reactants and products are reversed. We can write a balanced reaction in *any* way that is convenient, but all calculations based on that reaction must be consistent with the way it is written. Next note that the phase of each species is specified. In general, we will use g, l, and s symbols to denote either pure or mixed phases, except we will reserve aq to denote a solute in an aqueous solution.

Note the use and implications of the double arrow symbol, $\rightleftarrows$. This will *always imply that we have the attainment of chemical equilibrium in mind*. In contrast, the single arrow, $\rightarrow$, will have a different meaning. For example,

$$C_2H_5OH(l) + 3O_2(g) \rightarrow 2CO_2(g) + 3H_2O(l)$$

means "take one mole of ethanol and three moles of oxygen and convert them *completely* to two moles of carbon dioxide and three moles of water." We need to distinguish *complete* conversion of reactants to products from attainment of chemical equilibrium for several reasons. First, at equilibrium, many reaction mixtures contain products *and* reactants in measurable amounts. (And *all* reactions at equilibrium can be considered to contain reactants and products in the same spirit as we considered tungsten vapor to be present over W(s) in Section 6.7.) Second, one often needs to consider processes that are chemical in nature, but not in the same sense as an ordinary reaction. An example is

$$NaCl(s) + 1000H_2O(l) \rightarrow Na^+(aq) + Cl^-(aq)$$

for the dissolution of one mole of NaCl in 1000 moles of water. In contrast,

$$NaCl(s) \rightleftarrows Na^+(aq) + Cl^-(aq)$$

describes the chemical equilibrium of a saturated aqueous NaCl solution. (Visualize a pile of NaCl(s) at the bottom of a beaker of water that is as salty as possible.)

Finally, for any reaction, we should specify any applicable thermodynamic constraints. Is the reaction isothermal? At what T? Is it isobaric or isochoric? Is it to be carried out under adiabatic conditions? Are any external fields present? This attention to detail is necessary in order to specify the initial and final states of a reaction, but how do we specify a *path* for a chemical reaction?

As with any process, we need to specify the initial state from which the path evolves. To do this, imagine a schematic reaction in which reactants A and B form products C and D according to

$$\alpha A + \beta B \rightleftarrows \gamma C + \delta D$$

with α, β, γ, and δ the appropriate stoichiometric coefficients. We imagine mixing, at a given T and P, arbitrary amounts n_A°, n_B°, etc., of A, B, C, and D. Reaction ensues, and at any later time, we find n_A, n_B, etc., moles of A, B, etc. These amounts are related to n_A°, n_B°, etc., and to the stoichiometric coefficients through the equalities

$$\frac{n_A^\circ - n_A}{\alpha} = \frac{n_B^\circ - n_B}{\beta} = \frac{n_C - n_C^\circ}{\gamma} = \frac{n_D - n_D^\circ}{\delta} = \xi \ . \tag{7.1}$$

This quantity, ξ, is a convenient measure of the reaction's progress. It is called the *degree of advancement* or the *reaction progress variable*. (Note that ξ is measured in units of moles.) The differential degree of advancement, $d\xi$, is

$$d\xi = -\frac{dn_A}{\alpha} = -\frac{dn_B}{\beta} = \frac{dn_C}{\gamma} = \frac{dn_D}{\delta} \ . \tag{7.2}$$

For $d\xi > 0$, dn_A and dn_B are negative, corresponding to the disappearance of reactants, while dn_C and dn_D are positive, as more of the products appear. Thus, $d\xi > 0$ corresponds to a reaction progressing from left to right as written.

The degree of advancement starts at zero, no matter what the initial composition, since $n_A = n_A^\circ$, etc., at the start. As reaction ensues, ξ can either increase (become positive) or decrease (become negative), according to the spontaneous direction of the reaction away from the initial composition and towards the equilibrium composition. The degree of advancement cannot increase or decrease without limit, however; its range is governed by the initial composition and the reaction stoichiometry.

The quantity ξ will become our measure of travel along a chemical reaction path. It is an extensive variable, and we can use it as an independent variable in thermodynamic functions, differentiate or integrate with respect to it, etc.

EXAMPLE 7.1

Initially 4 mol H_2, 2 mol N_2, and 1 mol NH_3 are mixed in preparation for studying the reaction

$$3H_2(g) + N_2(g) \rightleftarrows 2NH_3(g)$$

at elevated temperatures. What are the limits on ξ for this mixture?

SOLUTION The initial composition is

$$n_{H_2}^\circ = 4 \text{ mol} \qquad n_{N_2}^\circ = 2 \text{ mol} \qquad n_{NH_3}^\circ = 1 \text{ mol}$$

and $\alpha = 3$, $\beta = 1$, $\gamma = 2$, and $\delta = 0$. Imagine first all the ammonia reverting to H_2 and N_2 (i.e., $n_{NH_3} = 0$). Then, using Eq. (7.1),

$$\frac{4 - n_{H_2}}{3} = \frac{2 - n_{N_2}}{1} = \frac{0 - 1}{2} = -\frac{1}{2} \text{ mol} = \xi(n_{NH_3} = 0) \ .$$

The most negative ξ can be is therefore $-1/2$ mol ($n_{H_2} = 11/2$ mol and $n_{N_2} = 5/2$ mol when $\xi = -1/2$ mol).

If N_2 and H_2 react to form more NH_3, then ξ will be positive. The maximum ξ will be determined by the *limiting reagent:* does one deplete H_2 first, or N_2? Since every mole of N_2 that reacts consumes 3 mol H_2, we see that when $n_{H_2} = 0$, n_{N_2} will be 2/3 mol, and thus H_2 is the limiting reagent for this particular reactant mixture. The largest positive value for ξ will therefore be

$$\frac{4 - 0}{3} = \frac{4}{3} \text{ mol} = \xi(n_{H_2} = 0)$$

and $n_{NH_3} = 11/3$ mol is the maximum amount of NH_3 possible from this mixture since

$$n_{NH_3} = 2\xi + n_{NH_3}^\circ = \left[2\left(\frac{4}{3}\right) + 1\right] \text{ mol} = \frac{11}{3} \text{ mol} \ .$$

Suppose the reaction had been written

$$6H_2(g) + 2N_2(g) \rightleftarrows 4NH_3(g) \ .$$

What would be the limits on ξ for this reaction, given the same initial mixture? Since ξ is inversely proportional to stoichiometric coefficients, multiplying a balanced reaction by a given factor *divides* the ξ limits by the same factor. Thus $-1/4$ mol $\leq \xi \leq 2/3$ mol for the same initial mixture viewed in the context of this balanced reaction.

⟹ RELATED PROBLEMS 7.1, 7.2

EXAMPLE 7.2

A 0.50 mol sample of $H_2(g)$ is mixed with 1.00 mol $I_2(s)$ at a temperature so low that no reaction occurs. The mixture is then raised to a temperature high enough to vaporize the iodine and equilibrate the reaction $H_2(g) + I_2(g) \rightleftarrows 2HI(g)$. The total pressure is adjusted to 1.00 atm, and the partial pressure of HI is found to be 0.578 atm. What is the degree of advancement at equilibrium, and what are the equilibrium amounts of each compound?

SOLUTION From the reaction stoichiometry, we see that the total amount of gas must be 1.50 mol, since that is the total amount of material before reaction begins and the stoichiometry assures us that the amount of gas will not change as equilibrium is attained. We can use the definition of partial pressure to find the amount of HI: $P_{HI} = x_{HI} P_{tot} = (n_{HI}/1.5 \text{ mol})(1.00 \text{ atm}) = 0.578$ atm or $n_{HI} = 0.867$ mol. From the definition of ξ in Eq. (7.1) and the initial amounts of reagents,

$$\xi = 1.00 \text{ mol} - n_{I_2} = 0.50 \text{ mol} - n_{H_2} = \frac{n_{HI}}{2}$$

so $\xi = (0.867 \text{ mol})/2 = 0.434$ mol, $n_{I_2} = 0.566$ mol, and $n_{H_2} = 0.066$ mol.

⟹ RELATED PROBLEMS 7.3, 7.4

7.2 ENERGY AND ENTROPY CHANGES IN CHEMICAL REACTIONS

In this section, we will consider the energy, enthalpy, and entropy changes that accompany the transformation of stoichiometric amounts of reactants to products. We begin with entropy.

Consider the transformation

$$\frac{1}{2}H_2(g) + \frac{1}{2}F_2(g) \rightarrow HF(g)$$

at constant T and P. (Remember that $\rightarrow$ means *complete* conversion of stoichiometric amounts of reactants to products.) We start with the absolute total entropy of the isolated pure reactants:

$$S(\text{isolated reactants}) = \left(\frac{1}{2}\text{mol}\right)\overline{S}(H_2(g)) + \left(\frac{1}{2}\text{mol}\right)\overline{S}(F_2(g))$$

then mix them:

$$S(\text{mixed reactants}) = S(\text{isolated reactants}) + \Delta S_{mix}$$

(where ΔS_{mix} is given by Eq. (6.25) if the mixture is ideal), then carry the reaction to completion:

$$S(\text{isolated product}) = (1 \text{ mol}) \overline{S}(HF(g)) \ .$$

The total entropy change going from isolated products to isolated reactants consists of two parts. One corresponds to an entropy difference between the total entropy of isolated products and the total entropy of isolated reactants. We will call this part $\Delta_r \overline{S}$, the *molar entropy of reaction*, and write in general

$$\Delta_r \overline{S} = \sum_{\text{products}} \nu_i \overline{S}(\text{product } i) - \sum_{\text{reactants}} \nu_i \overline{S}(\text{reactant } i) \quad (7.3)$$

where ν_i is the stoichiometric coefficient (a dimensionless number) for species i.

Note that whenever stoichiometric amounts of reactants are converted completely to products (which is what both Δ_r and $\rightarrow$ will mean), ξ varies from 0 to 1 mol. Thus, $\Delta_r \overline{S}$ is a molar quantity since it corresponds to the entropy change *per unit degree of advancement* (i.e., per unit ξ).

The second part of the total entropy change is an entropy of mixing. We will leave the mixing term alone for now, but later it will be seen to play an extremely important role in establishing the composition of a reaction mixture at chemical equilibrium.

EXAMPLE 7.3

(1) At 500 K, the molar entropies of $H_2(g)$, $S_2(g)$, and $H_2S(g)$ are $\overline{S}(H_2(g)) = 17.51\,R$, $\overline{S}(S_2(g)) = 29.53\,R$, and $\overline{S}(H_2S(g)) = 26.94\,R$. What is $\Delta_r \overline{S}$ for the synthesis of H_2S from the reaction

$$H_2(g) + \frac{1}{2} S_2(g) \rightarrow H_2S(g) \ ?$$

(2) A mixture of 2 mol $H_2(g)$ and 4 mol $S_2(g)$, at 500 K, is allowed to reach chemical equilibrium. Analysis of the equilibrium mixture indicates that virtually all the H_2 has disappeared. What is the entropy change accompanying this process?

SOLUTION (1) From Eq. (7.3), we have

$$\begin{aligned}\Delta_r \overline{S} &= \overline{S}(H_2S) - \overline{S}(H_2) - 0.5\overline{S}(S_2) \\ &= 26.94\,R - 17.51\,R - 0.5(29.53\,R) = -5.34\,R \ .\end{aligned}$$

Note the negative sign. Entropy *decreased* due to a *net loss of gas-phase species* as the reaction converted 1.5 mol of reactant gases to only 1 mol of product gases.

(2) The final mixture must contain 2 mol $H_2S(g)$ and 3 mol $S_2(g)$, by stoichiometry. To reach this mixture, ξ has increased from zero to the equilibrium value $\xi_{eq} = 2$ mol since, for instance, $n^\circ_{H_2S} = 0$ and $n_{H_2S} = 2$ mol. Thus, *one* aspect of the total entropy change is due to reaction:

$$\xi_{eq}\, \Delta_r \overline{S} = (2 \text{ mol})(-5.34\,R) = -88.8 \text{ J K}^{-1} \ .$$

The *second* aspect is due to mixing. For the isothermal, isochoric mixing of 2 mol $H_2S(g)$ and 3 mol $S_2(g)$, we assume (safely) ideal behavior and calculate from Eq. (6.25)

$$\begin{aligned}\Delta S_{\text{mix}} &= -nR\,[x_{S_2} \ln x_{S_2} + x_{H_2S} \ln x_{H_2S}] \\ &= -(5 \text{ mol})R\left[\frac{3}{5}\ln\frac{3}{5} + \frac{2}{5}\ln\frac{2}{5}\right] \\ &= +28.0 \text{ J K}^{-1} \ .\end{aligned}$$

The total entropy change going from $2H_2(g)$ isolated from $4S_2(g)$ to the equilibrium mixture $3S_2(g) + 2H_2S(g) +$ insignificant $H_2(g)$ is the sum

$$\Delta S = \Delta_r S + \Delta S_{\text{mix}} = -60.8 \text{ J K}^{-1} \ .$$

➞ RELATED PROBLEMS 7.6, 7.7

Now we turn to energy changes. The origin of these changes is most important to establish early in the discussion. Consider the HF synthesis reaction. Imagine taking 0.5 mol of $H_2(g)$ and pulling apart all of the H_2 molecules into H atoms. A certain amount of energy must be transferred to the H_2 molecules to dissociate them. The same is true if we dissociate 0.5 mol of $F_2(g)$ into 1 mol of F atoms, but the energy required to dissociate F_2 is distinctly different (less, in fact) from that required to dissociate H_2. At this point, we could reassemble our atoms into 1 mol of $HF(g)$. In so doing, energy would be released. The amount released is, in this case, greater than the total amount required to dissociate 0.5 mol of H_2 and 0.5 mol F_2, but the important result is the *difference* between the total energy released and the total energy consumed.

Chemical reactions rarely, if ever, proceed by a *microscopic* mechanism such as this. Reactant molecules do not disassemble themselves into atoms and then reassemble themselves into product molecules, but since internal energy (or enthalpy, for that matter) is a state function, we are free to choose this path in our minds. This leads to the important concept of *bond energies,* which are different for every type of bond, to greater or lesser degrees, and which constitute the microscopic origin of energy changes in chemical reactions. The next section describes the magnitudes of bond energies and enthalpies in greater detail.

While conceptually very useful, bond energies can be devilishly difficult to measure experimentally. Spectroscopic methods, discussed in Chapter 19, are capable in principle of measuring bond energies to high accuracy, but only a handful of simple molecules have been studied in this way. This technique places too many special experimental demands on molecules to be of general use. Instead of bond energies, the experimentally simpler quantity to measure is the constant-pressure (or sometimes constant-volume) heat associated with a chemical reaction. In particular, at constant P, this heat represents an *enthalpy change* brought about by the reaction.

This heat can be measured by a number of types of calorimeters. But it is hopeless (and fortunately unnecessary) to expect calorimetric measurements to be made for every reaction. Instead, measurements of carefully selected reactions can be combined to establish tables of characteristic enthalpies for any one compound.

An example that illustrates this idea is the enthalpy change associated with

$$2CO(g) + O_2(g) \rightarrow 2CO_2(g) \ . \tag{7.4}$$

While assembling CO, O_2, and CO_2 from isolated C and O atoms is a hopeless task, the synthesis of these molecules from the stable forms of elemental carbon[1] and oxygen is experimentally tractable. These syntheses are

$$(1) \quad 2C(graphite) + O_2(g) \rightarrow 2CO(g) \tag{7.5a}$$

and

$$(2) \quad C(graphite) + O_2(g) \rightarrow CO_2(g) \tag{7.5b}$$

and calorimetric measurements yield enthalpy changes

$$2\overline{H}(CO) - 2\overline{H}(graphite) - \overline{H}(O_2) = \Delta\overline{H}(\text{reaction 1}) \tag{7.6a}$$

and

$$\overline{H}(CO_2) - \overline{H}(graphite) - \overline{H}(O_2) = \Delta\overline{H}(\text{reaction 2}) \tag{7.6b}$$

---

[1] Note that the stable form of carbon is graphite at ordinary temperatures and pressures, not diamond. Diamond should undergo a spontaneous phase transition to graphite, but the rate of this transition is so slow that diamond appears to be as stable as graphite or C_{60}.

Since enthalpy is a state function, we can combine reactions and their associated enthalpies algebraically. For reaction (7.4), the enthalpy change is

$$2\overline{H}(CO_2) - 2\overline{H}(CO) - \overline{H}(O_2) = 2\Delta\overline{H}(\text{reaction 2}) - \Delta\overline{H}(\text{reaction 1}) \ ,$$

since we can combine reactions (7.5a) and (7.5b) to yield (7.4), the net reaction of interest:

$$2CO(g) \rightarrow 2C(\text{graphite}) + O_2(g)$$
$$2C(\text{graphite}) + 2O_2(g) \rightarrow 2CO_2(g)$$
$$\overline{\phantom{2C(\text{graphite}) + 2O_2(g) \rightarrow 2CO_2(g)}}$$
$$2CO(g) + O_2(g) \rightarrow 2CO_2(g)$$

This scheme, synthesis from the stable form of the elements, is the experimental and conceptual backbone of reaction enthalpy calculations. The details yet to be discussed are the choices of reference enthalpies for elements (a choice of standard states) and the variation of reaction enthalpies with temperature and pressure. Note also that the enthalpy of mixing, which is zero only for an ideal mixture, has not entered the calculation.

The next section completes the discussion of reaction enthalpy changes and describes several generic types of reactions. These reactions are known by special names, and the magnitudes of the enthalpy changes brought about by them are important physical properties, many of which you will recognize from elementary chemistry courses.

7.3 NAMED ENTHALPIES AND ENTHALPY DIAGRAMS

The most important reaction enthalpy is that for the formation of a compound from its constituent elements. Since the elements themselves are immutable (except for spontaneous or induced nuclear reactions, which we will not consider), they form a natural set of reference species. If we choose standard states applicable to all elements, compounds, and mixtures, we can begin to tabulate reference enthalpies for any substance.

Standard Enthalpies of Formation

Here are the rules that define standard states:

1. We take, at any temperature, that phase of each element that is thermodynamically stable. (This is often called the *reference state* of the element when the standard pressure and temperature of interest are specified as well.)
2. If the stable phase is a gas, we take $P = 10^5$ Pa = 1 bar, but imagine that the gas follows the ideal gas equation of state. (Refer to Eq. (5.48) for our first use of this concept.) Formerly, the standard pressure was 1 atm = 101 325 Pa.
3. If the phase is solid or liquid, we also take 1 bar as the standard pressure.

The reactions we write for the formation of a compound should form only one mole, as in

$$C(\text{graphite}) + \frac{3}{2}H_2(g) + \frac{1}{2}I_2(s) \rightarrow CH_3I(\text{ideal gas}), \quad T = 298.15 \text{ K}, \quad P = 1 \text{ bar} \ ,$$

for the formation of methyl iodide at a commonly tabulated temperature, 25 °C = 298.15 K. Note the reference phases for the elements: graphite, gaseous H_2, solid I_2. Most elemental reference phases at room temperature are obvious, but a few may not be, such as Hg(l), Br$_2$(l), S(rhombic) (see Section 4.4 on the solid phases of sulfur), and P, which is sometimes referred to the metastable "white" phase rather than the more stable "red" phase, the latter being difficult to characterize.

The enthalpy change in this reaction is called the *standard molar enthalpy of formation* of CH$_3$I and is symbolized $\Delta_f \overline{H}°$(CH$_3$I(g), 298.15 K). These enthalpies are defined for every compound, neutral or ionized, stable or highly reactive. For instance, gas-phase sulfur can exist as atomic S, as diatomic S$_2$, or as octatomic S$_8$. Their standard molar enthalpies of formation at 25 °C are the reaction enthalpies for the following:

$$\begin{aligned} &\text{S(rhombic)} \rightarrow \text{S}(g) & \Delta_r \overline{H}° &= \Delta_f \overline{H}°(\text{S}(g)) = 278.805 \text{ kJ mol}^{-1} \\ &2\text{S(rhombic)} \rightarrow \text{S}_2(g) & \Delta_r \overline{H}° &= \Delta_f \overline{H}°(\text{S}_2(g)) = 128.37 \text{ kJ mol}^{-1} \\ &8\text{S(rhombic)} \rightarrow \text{S}_8(g) & \Delta_r \overline{H}° &= \Delta_f \overline{H}°(\text{S}_8(g)) = 101.9 \text{ kJ mol}^{-1} \end{aligned}$$

One usually finds enthalpies of formation (also called "heats of formation") tabulated in cal mol^{-1}, kcal mol^{-1}, kcal g^{-1} (a "specific enthalpy," per unit mass, rather than a "molar enthalpy"), or, in more recent tables, in kJ mol^{-1} units as we have here. Some values are reported as $\Delta_f \overline{H}°/R$ (units = K) in contemporary literature, but since enthalpies of formation are functions of T, and are so tabulated, it is also convenient to consider dimensionless $\Delta_f \overline{H}°/RT$ values.

As a corollary to our definition of elemental standard states (but *not* part of the definition itself), the *enthalpy of formation of any element in its standard state is zero, at any temperature.*

Enthalpy of formation values are not particularly informative by themselves. But $\Delta_f \overline{H}°$ values for all the products and reactants of a reaction can be combined to give the reaction enthalpy[2] from

$$\Delta_r \overline{H}° = \sum_{\text{products}} \nu_i \Delta_f \overline{H}°(\text{product } i) - \sum_{\text{reactants}} \nu_i \Delta_f \overline{H}°(\text{reactant } i) \ . \tag{7.7}$$

Note that this equation follows the usual "final state (products) − initial state (reactants)" prescription. For example, we can use the sulfur data above to calculate

$$8\text{S}(g) \rightarrow \text{S}_8(g) \qquad \Delta_r \overline{H}° = 101.9 - 8(278.805) = -2128.5 \text{ kJ mol}^{-1} \tag{7.8a}$$

$$\text{S}_8(g) \rightarrow 4\text{S}_2(g) \qquad \Delta_r \overline{H}° = 4(128.37) - 101.9 = +411.6 \text{ kJ mol}^{-1} \tag{7.8b}$$

$$\text{S}_2(g) \rightarrow 2\text{S}(g) \qquad \Delta_r \overline{H}° = 2(278.805) - 128.37 = +429.2 \text{ kJ mol}^{-1} \ . \tag{7.8c}$$

The standard molar reaction enthalpy is negative for the first of these. Enthalpy leaves the system if the reaction occurs at constant T and P. A *positive* value means enthalpy must enter the system to maintain the products at the same T and P of the reactants. In general, we call reactions for which $\Delta_r H$ is negative *exothermic*, and those for which $\Delta_r H$ is positive *endothermic*. The rare case for which $\Delta_r H = 0$ (to within experimental error) is termed *thermoneutral*.

If we string reactions (7.8a)–(7.8c) together sequentially and balance them:

$$8\text{S}(g) \rightarrow \text{S}_8(g) \rightarrow 4\text{S}_2(g) \rightarrow 8\text{S}(g) \ ,$$

we form a *reaction cycle* that ends where it begins. As with any state function, $\Delta_r H(\text{cycle}) = 0$. Here we have

$$\Delta_r \overline{H}°(\text{cycle})/\text{kJ mol}^{-1} = -2128.5 + 411.6 + 4(429.2) = -0.1$$

---

[2]The quantity $\Delta_r H$ is the "reaction enthalpy"; $\Delta_r \overline{H}$ is the "molar (per unit ξ) reaction enthalpy"; and $\Delta_r \overline{H}°$ is the "standard ($P = 1$ bar) molar reaction enthalpy."

which, to within the precision of the data, is zero, as expected. The advantage of reaction cycles is that enthalpy values for all but one of the steps in the cycle determines the value for the unknown step since all steps must add to zero.

EXAMPLE 7.4

An experimental study of the synthesis of $Ir_2S_3(s)$ from $Ir(s)$ and $S_2(g)$ [E. T. Chang and N. A. Gokcen *High Temp. Sci.* **4**, 432 (1972)] yielded $\Delta_r\overline{H} = -400.68$ kJ mol^{-1} at 298.15 K from a careful extrapolation of measurements made in the range 920–1110 K. Since $S_2(g)$ is not the standard state of sulfur at 298 K, this is not the standard molar enthalpy of formation of $Ir_2S_3(s)$. What is $\Delta_f\overline{H}°(Ir_2S_3(s), 298.15\ K)$?

SOLUTION The $\Delta_r\overline{H}$ value is for reaction (1):

(1) $2Ir(s) + \dfrac{3}{2}S_2(g) \rightarrow Ir_2S_3(s).$

The standard state for sulfur is the rhombic crystal form, and for

(2) $2S(rhombic) \rightarrow S_2(g)$

$\Delta_r\overline{H} = \Delta_f\overline{H}°(S_2(g)) = 128.37$ kJ mol^{-1}. We obtain the standard formation reaction if we combine reaction (1) with 3/2 reaction (2); therefore,

$$\Delta_f\overline{H}° = \Delta_r\overline{H}(1) + \frac{3}{2}\Delta_r\overline{H}(2)$$

$$= -400.68\text{ kJ mol}^{-1} + \left(\frac{3}{2}\right)(128.37\text{ kJ mol}^{-1})$$

$$= -208.13\text{ kJ mol}^{-1}.$$

⟹ RELATED PROBLEMS 7.10, 7.12

Before moving on, we discuss the effect of temperature and pressure on reaction enthalpies. Since we can write the reaction in terms of individual compound molar enthalpies as

$$\Delta_r\overline{H} = \sum_{\text{products}} \nu_i\overline{H}(\text{product }i) - \sum_{\text{reactants}} \nu_i\overline{H}(\text{reactant }i),$$

at any one T and P, variations in individual molar enthalpies cause $\Delta_r\overline{H}$ to vary at any other T and P. (Compare to Eq. (7.7), and note that $\overline{H}$ does *not* equal $\Delta_f\overline{H}°$.) Suppose we calculate $\Delta_r\overline{H}$ at one T and P from $\Delta_f\overline{H}°$ values and ask for $\Delta_r\overline{H}$ at a new temperature, T', and the same P. Formally, we can write

$$\Delta_r\overline{H}(T') = \Delta_r\overline{H}(T) + \int_T^{T'}\left(\frac{\partial \Delta_r\overline{H}}{\partial T}\right)_P dT$$

which becomes, with $(\partial \overline{H}/\partial T)_P = \overline{C}_P$,

$$\Delta_r\overline{H}(T') = \Delta_r\overline{H}(T) + \int_T^{T'}\Delta_r\overline{C}_P\, dT \qquad (7.9)$$

where

$$\Delta_r\overline{C}_P = \sum_{\text{products}}\nu_i\overline{C}_P(\text{product }i) - \sum_{\text{reactants}}\nu_i\overline{C}_P(\text{reactant }i).$$

For pressure variations at constant T, we write

$$\Delta_r \overline{H}(P') = \Delta_r \overline{H}(P) + \int_P^{P'} \left(\frac{\partial \Delta_r \overline{H}}{\partial P}\right)_T dP$$

and use Eq. (3.43)

$$\left(\frac{\partial \overline{H}}{\partial P}\right)_T = \overline{V} - T\left(\frac{\partial \overline{V}}{\partial T}\right)_P \qquad (3.43)$$

for each product and reactant in a sum of integrals analogous to Eq. (7.9).

If any product or reactant undergoes a phase transition along either of these paths, the appropriate $\Delta \overline{H}_\phi$ is added or subtracted for a product or reactant, respectively.

EXAMPLE 7.5

Given $\Delta_f \overline{H}°(Fe_2O_3(s), 298.15\ K) = -824.2\ kJ\ mol^{-1}$, estimate $\Delta_f \overline{H}°$ at 600 K.

SOLUTION We apply Eq. (7.9) to the standard formation reaction

$$2Fe(s) + \frac{3}{2}O_2(g) \rightarrow Fe_2O_3(s)$$

with $T = 298.15\ K$ and $T' = 600\ K$. For an estimate, we can approximate the relevant heat capacities:

$\overline{C}_P°(Fe(s)) = 3\,R$ **(Dulong-Petit value)**
$\overline{C}_P°(O_2(g)) = 3.9\,R$ **(Table 4.4 and Figure 4.5)**
$\overline{C}_P°(Fe_2O_3(s)) = 12\,R$ **(Kopp estimate)**

Estimates for the solids assume $C_P = C_V$, while the O_2 estimate uses a representative value at the intermediate temperature of 450 K. Likewise, it is safe to approximate 298.15 K = 300 K, leaving us with a very simple calculation:

$$\Delta_f \overline{H}°(600\ K) = \Delta_f \overline{H}°(300\ K) + \Delta T \left[\overline{C}_P°(Fe_2O_3) - 2\overline{C}_P°(Fe) - \left(\frac{3}{2}\right)\overline{C}_P°(O_2)\right]$$

$$= (-824.2\ kJ\ mol^{-1}) + (300\ K)\left[12 - 6 - \left(\frac{3}{2}\right)3.9\right]R$$

$$= -824.6\ kJ\ mol^{-1}.$$

The heat capacity correction is very small, and estimates of C_P usually yield an acceptable value, as they do here. A more accurate calculation that considers the correct, temperature dependent heat capacities gives a value about 6 kJ mol^{-1} higher than this estimate, but one rarely needs accuracy this great.

➡ **RELATED PROBLEMS** 7.13, 7.14

Bond Enthalpies and Energies

As pointed out in the previous section, the differences among chemical bond energies account for reaction enthalpies. The defining reaction for a bond enthalpy is, schematically,

$$AB(g) \rightarrow A(g) + B(g). \qquad (7.10)$$

Since bond enthalpies are intended to be isolated molecule properties, we specify the low pressure gaseous state. AB could represent a diatomic molecule:

$$HCl(g) \rightarrow H(g) + Cl(g)$$

or a polyatomic molecule broken into two fragments:

$$CF_3CH_3(g) \rightarrow CF_3(g) + CH_3(g) \quad \text{(break C—C)}$$

or

$$CF_3CH_3(g) \rightarrow CF_3CH_2(g) + H(g) \quad \text{(break C—H)}$$

but *not*

$$CF_3CH_3(g) \rightarrow CHF_3(g) + CH_2(g) \; ,$$

which represents a rearrangement as well as a bond cleavage.

As with any reaction enthalpy, bond enthalpies vary with temperature according to Eq. (7.9). At absolute zero, these enthalpies become true *bond dissociation energies*

$$\Delta_r \overline{H}(\text{bond enthalpy, 0 K}) = \mathcal{D}_0 = \textit{bond dissociation energy} \; . \tag{7.11}$$

The subscript in $\mathcal{D}_0$ represents the 0 K limit and, as will be discussed in Chapter 19, it also specifies a particular internal state for the molecule that is microscopically equivalent to this temperature limit.

Bond energies span an enormous range. Figure 7.1 is a histogram of over 500 diatomic molecule dissociation energies. The heights of the various histogram bins reflect, in part, the experimental accessibility of certain molecules, but they also reflect the natural grouping of bond energies around values characteristic of the *type* of bond involved.

For example, the bond between any two rare gas atoms is so weak that we do not normally classify it as a chemical bond. The bonds in F_2, O_2, and N_2 are covalent *single, double,* and *triple* bonds, respectively, and such bond energies regularly increase in this order. Ionic bonds (NaCl, BeO, etc.) also show periodic trends based on the degree of ionicity (i.e., Na^+Cl^- versus $Be^{2+}O^{2-}$) characteristic of the bond.

Average, representative bond enthalpies can be assigned to each atom–atom bond in each class of bond. These average values are only approximate, but they are quite useful in predicting reaction enthalpies when other data are lacking. Values that are worth committing to memory are given in Table 7.1 in a variety of units. More exact bond enthalpy values are given in Table 7.2 in $kJ \, mol^{-1}$ units.

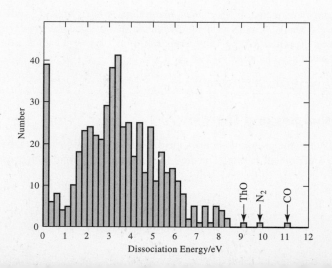

FIGURE 7.1 A histogram of the dissociation energies of gas-phase diatomic molecules shows the spread in bond energies due to different types of bonding. The peak near zero is for "non-bonded" species such as Ar_2. At high energies, one finds covalent double and triple bonds as well as certain ionic bonds.

7.3 NAMED ENTHALPIES AND ENTHALPY DIAGRAMS

TABLE 7.1 Representative Chemical Bond Energy Magnitudes

Benchmark: RT at 300 K:	~ 2.5 kJ mol^{-1}
	$\sim 1/40$ eV molecule^{-1}
Single bonds:	$\sim 270 \pm 100$ kJ mol^{-1}
	$\sim 3 \pm 1$ eV molecule^{-1}
Double bonds:	~ 570 kJ mol^{-1}
	~ 5 eV molecule^{-1}
Triple bonds:	~ 890 kJ mol^{-1}
	~ 10 eV molecule^{-1}
Ionic bonds:	$\sim 400-680$ kJ mol^{-1}
	$\sim 4-7$ eV molecule^{-1}

To see the origin of these values, consider plucking the hydrogens, one after the other, from methane at 25 °C:

$$CH_4(g) \rightarrow CH_3(g) + H(g) \quad \Delta_r\overline{H} = 432 \text{ kJ mol}^{-1}$$
$$CH_3(g) \rightarrow CH_2(g) + H(g) \quad \Delta_r\overline{H} = 469 \text{ kJ mol}^{-1}$$
$$CH_2(g) \rightarrow CH(g) + H(g) \quad \Delta_r\overline{H} = 422 \text{ kJ mol}^{-1}$$
$$CH(g) \rightarrow C(g) + H(g) \quad \Delta_r\overline{H} = 339 \text{ kJ mol}^{-1}$$

net: $\quad CH_4(g) \rightarrow C(g) + 4H(g) \quad \Delta_r\overline{H} = 1662$ kJ mol^{-1}

The average C—H bond enthalpy is the average of this last number: 1662/4 kJ mol^{-1} = 416 kJ mol^{-1}. The value in Table 7.2 is a similar average, but one calculated from a variety of hydrocarbons. Note that specific C—H bond enthalpies can vary by $\pm$ 20% or more from the average.

TABLE 7.2 Average Bond Enthalpies/kJ mol^{-1} at 25 °C

	Br	C	Cl	F	H	I	N	O	P	S	Si	
	Single bonds											
Br	193											Br
C	285	348										C
Cl	219	339	242									Cl
F	249	489	253	159								F
H	366	413	431	567	436							H
I	178	218	211	280	298	151						I
N	—	305	192	278	391	—	163					N
O	234	358	208	193	463	234	201	146				O
P	264	264	322	503	322	184	—	335	172			P
S	218	272	271	327	367	—	—	—	—	255		S
Si	325	285	397	586	318	234	—	451	—	293	176	Si
	Br	C	Cl	F	H	I	N	O	P	S	Si	
	Double bonds											
		C=C	614	C=N	615	C=O	745	C=S	536			
		N=N	418	N=O	607	O=O	498					
	Triple bonds											
		C≡C	839	C≡N	891	N≡N	945					

Source: G. H. Aylward and T. J. V. Findlay, *SI Chemical Data*, 2nd ed. (John Wiley, NY, 1971).

EXAMPLE 7.6

Precise enthalpy of formation data yield $\Delta_r \overline{H} = -92.22$ kJ mol^{-1} at 298 K for the ammonia synthesis reaction

$$N_2(g) + 3H_2(g) \rightarrow 2NH_3(g) \ .$$

How well does this value compare with an estimate based on bond enthalpies?

SOLUTION Using data from Table 7.2, we would first break the N_2 triple bond and the three H_2 single bonds, expending

$$945 \text{ kJ mol}^{-1} + (3)(436 \text{ kJ mol}^{-1}) = 2253 \text{ kJ mol}^{-1} \ ,$$

and then release the enthalpy of three N—H bonds in each of the two NH_3 product molecules:

$$-2[3(391 \text{ kJ mol}^{-1})] = -2346 \text{ kJ mol}^{-1} \ ,$$

for a net enthalpy change of $2253 - 2346 = -93$ kJ mol^{-1} which compares favorably with the precise value -92.22 kJ mol^{-1} (which is twice the enthalpy of formation of ammonia). Note, however, that the excellent agreement is largely fortuitous. A 0.25% change in $\mathcal{D}_0$(N—H) from 391 to 392 kJ mol^{-1} changes the calculated value by 6.5% to -99 kJ mol^{-1}! This is a common problem with procedures that require subtracting large numbers of roughly equal magnitude.

⮕ RELATED PROBLEMS 7.8, 7.9

Other Named Enthalpies

Certain classes of reactions are known by special names that reflect a particular type of chemical change. The more important of these are collected here.

Ionization Enthalpies. Two important reactions pertain to gas-phase ions. The first is the *ionization potential* (IP), the energy required to remove an electron from an atom or molecule, either neutral or previously ionized. The reaction is considered to take place at absolute zero and in the ideal gas phase. Examples are

$$H_2(g) \rightarrow H_2^+(g) + e^-(g)$$

or

$$C^+(g) \rightarrow C^{2+}(g) + e^-(g) \ .$$

Note that $\Delta_f \overline{H}°(e^-(g)) = 0$ by definition. The ionization enthalpy at a non-zero temperature is calculated from $\Delta_r \overline{H}(T=0) = $ IP and Eq. (7.9), with the electron given the ideal gas heat capacity.

The second special reaction is that for the removal of an electron from a gas-phase anion, as in

$$OH^-(g) \rightarrow OH(g) + e^-(g) \ .$$

The $T = 0$ energy change for this class of reactions is termed the *electron affinity* (EA). While the EA is just the IP of an anion, the term electron affinity is always used in discussing negative ions.

Ionization potentials are inherently positive (since ionization is an endothermic process), but certain atoms (the rare gases, alkaline earths, Zn, Cd, Hg, N, and a few others) and some molecules (such as N_2, CO, HF, and others) have *negative* electron affinities. For these species, the negative ion is not stable in the gas phase, for reasons often readily traced to elementary electronic structure and chemical-bonding consequences.

7.3 NAMED ENTHALPIES AND ENTHALPY DIAGRAMS

TABLE 7.3 Ionization Potentials/eV

H	13.598	H_2	15.427	F_2	15.7
He	24.586	O_2	12.063	Cl_2	11.48
Ne	21.564	N_2	15.576	Br_2	10.54
Ar	15.759	NO	9.25	I_2	9.28
Xe	12.130				
NH_3	10.15	CH_4	12.704	HF	15.77
O_3	12.3	C_2H_2	11.40	HCl	12.74
H_2O	12.60	C_2H_4	10.45	HBr	11.62
CO_2	13.769	C_2H_6	11.52	HI	10.38
N_2O	12.90	C_6H_6	9.25		
Li	5.390	Be	9.320		
Na	5.138	Mg	7.644		
K	4.339	Ca	6.111		
Rb	4.176	Sr	5.692		
Cs	3.893	Ba	5.210		

Electron affinities and ionization potentials are frequently tabulated in electron volt per molecule (eV) units where 1 eV molecule^{-1} = 1.602 18 × 10^{-19} J molecule^{-1} = 96.485 3 kJ mol^{-1}. (One eV is the kinetic energy acquired by a singly charged ion accelerated through an electrical potential difference of one volt.) Tables 7.3 and 7.4 give IP and EA values for representative atoms and molecules.

EXAMPLE 7.7

Is the charge transfer reaction

$$Na^+(g) + Cl^-(g) \rightarrow Na(g) + Cl(g)$$

endothermic or exothermic?

SOLUTION The charge is transferred from Cl$^-$ to Na$^+$, but we can imagine it happening in the following two steps:

(1) $Na^+(g) + Cl^-(g) \rightarrow Na^+(g) + Cl(g) + e^-(g)$
(2) $Na^+(g) + Cl(g) + e^-(g) \rightarrow Na(g) + Cl(g)$.

TABLE 7.4 Electron Affinities/eV

H$^-$	0.7542	OH$^-$	1.825
F$^-$	3.399	O_2^-	0.440
Cl$^-$	3.615	S_2^-	1.663
Br$^-$	3.364	NO$^-$	0.024
I$^-$	3.061	NH$^-$	0.38
O$^-$	1.4611		
		CH_3^-	0.08
SO_2^-	1.097	CH_3O^-	1.570
NH_2^-	0.771	$C_5H_5^-$	1.786
PH_2^-	1.271		
PH$^-$	1.028		
PO$^-$	1.092		

Step (1) is endothermic by EA(Cl$^-$) = 3.615 eV, while (2) is exothermic by $-$IP(Na) = $-$5.138 eV. The net reaction is thus exothermic by $-$5.138 + 3.615 = $-$1.523 eV (or $-$147.0 kJ mol^{-1}).

⇒ RELATED PROBLEM 7.15

Proton Affinity. The proton affinity is the energy change in a reaction such as

$$NH_4^+(g) \rightarrow NH_3(g) + H^+(g)$$

in which abstraction of H$^+$ yields a neutral molecule. Note that $\Delta_f \overline{H}°$(H$^+$(g), 298.15 K) = 1 536.31 kJ mol^{-1}.

Atomization Enthalpies. The atomization enthalpy, $\Delta \overline{H}_{at}$, corresponds to complete disassembly of the molecule into isolated atomic fragments, as in

$$CH_2N_2(g) \rightarrow C(g) + 2H(g) + 2N(g)$$

or, for a metal or an ionic solid,

$$Na(s) \rightarrow Na(g)$$

or

$$NaCl(s) \rightarrow Na(g) + Cl(g) \ .$$

Lattice Enthalpies. In addition to the atomization enthalpy, one speaks of other ways of disassembling a solid lattice. For any solid, we have the *sublimation enthalpy,* as in

$$CO_2(s) \rightarrow CO_2(g)$$

for molecular solids or

$$NaCl(s) \rightarrow NaCl(g)$$

for ionic solids that sublime as discrete molecules. One also speaks of the *crystal enthalpy* of an ionic solid, as in

$$NaCl(s) \rightarrow Na^+(g) + Cl^+(g) \ .$$

Enthalpies of Combustion. Also commonly called *heats of combustion,* enthalpies of combustion usually correspond to the conversion of a compound to simple oxides. With a compound containing only C, H, and perhaps O, an excess of O$_2$(g) is a co-reactant leading to CO$_2$(g) and H$_2$O(l) products. Compounds containing C, H, O, and N also produce N$_2$ and/or HNO$_3$ or HNO$_2$. An important case is the combustion of benzoic acid

$$C_6H_5COOH(s) + \frac{15}{2}O_2(g) \rightarrow 7CO_2(g) + 3H_2O(l) \ ,$$

which forms a reference reaction for the calibration of combustion calorimeters. The internal energy change (which is appropriate to a constant-volume combustion calorimeter) for this reaction is accurately known to be $\Delta_r \overline{U}° = -3\,228.29 \pm 0.24$ kJ per mole of benzoic acid.

Solution Enthalpies. A variety of enthalpy changes are associated with forming different types of solutions. The most important of these relate to aqueous solutions of electrolytes as discussed in detail in Chapter 10. Here we give only the definitions of the major processes used in describing solution formation. Recall from Chapter 6 that $\Delta \overline{H}_{\text{mix}}$ is an appropriate measure of the enthalpy change on solution formation, and that $\Delta \overline{H}_{\text{mix}} = 0$ for an ideal mixture of ideal gases, Eq. (6.27).

One distinguishes between the *integral enthalpy of solution,* which is the enthalpy change for the transformation

$$\text{solute} + \text{excess pure solvent} \to \text{solution}$$

and the *differential enthalpy of solution,* as in either

$$1 \text{ mole } \textit{solute} + \infty \text{ moles } \textit{solution} \to \text{solution}$$

or

$$1 \text{ mole } \textit{solvent} + \infty \text{ moles } \textit{solution} \to \text{solution} \ .$$

By "excess pure solvent," we mean "enough solvent *at least* to dissolve *all* the solute," and by "∞ moles solution" we mean "a very large excess of solution over solute (or solvent)." The integral enthalpy of solution depends on T, P, and the *final* solution composition, while the differential enthalpy (which is a partial molar enthalpy) depends on T, P, and the *initial* solution composition. If we have a solution containing n_2 moles of solute and to this we add dn_2 moles while measuring the heat, dq, associated with this addition, then

$$\frac{dq}{dn_2} = \text{differential enthalpy of solution at composition } n_2 \ .$$

The integral enthalpy of solution is the heat associated with its defining reaction or, equivalently,

$$\int_0^{n_2} \left(\frac{dq}{dn_2}\right)_{T,P} dn_2 = \text{integral enthalpy of solution} \ .$$

In the limit of infinite dilution, the differential and integral enthalpies of solution are identical.

Ions in solution present special problems, one of which is the necessary presence of ions of both charges in order to maintain net charge neutrality. Processes such as

$$\text{H}^+(g) \to \text{H}^+(aq)$$

cannot be realized experimentally, but theoretical studies of hydrated protons in the species (H_3O^+, H_5O_2^+, ..., $(\text{H}_2\text{O})_n\text{H}^+$) indicate this is a very exothermic process ($\Delta_r \overline{H} \cong -1000 \text{ kJ mol}^{-1}$). Consequently, relative *enthalpies of hydration* are tabulated based on the definition

$$\Delta_f \overline{H}^\circ(\text{H}^+(aq)) \equiv 0 \ .$$

These are relative molar enthalpies of formation for ions in solution. Several are listed in Table 7.5, where it can be seen that smaller ions have more negative values (compare the halogens) as do more highly charged ions (compare CO_3^{2-} to NO_3^-). Both small size and high charge favor a strong attractive interaction between the ion and water, an exothermic process.

Enthalpy-level Diagrams. A very convenient way to represent enthalpy changes in a sequence of reactions, especially a cycle, is to chart *enthalpy levels.* Figure 7.2 illustrates this, using reactions involving H and O. We (arbitrarily) assign a zero enthalpy value to the species $\text{H}_2(g) + \frac{1}{2}\text{O}_2(g)$.

TABLE 7.5 Aqueous Ion Standard Molar Enthalpies of Formation/kJ mol^{-1}

H$^+$	0 (by definition)		
Li$^+$	−278.49	F$^-$	−332.63
Na$^+$	−240.12	Cl$^-$	−167.159
K$^+$	−252.38	Br$^-$	−121.55
Rb$^+$	−251.17	I$^-$	−55.19
Cs$^+$	−258.28	I$_3^-$	−51
NH$_4^+$	−132.51	OH$^-$	−229.994
Be^{2+}	−383	HCO$_3^-$	−691.99
Al^{3+}	−524.7	CO$_3^{2-}$	−677.14
Ag$^+$	105.579	NO$_3^-$	−205.0
Cu$^+$	71.67	SO$_4^{2-}$	−909.27
Cu^{2+}	64.77		
Zn^{2+}	−153.89		

Step 1 is endothermic:

(1) $H_2(g) + \frac{1}{2}O_2(g) \rightarrow H_2(g) + O(g)$

$\Delta_r \overline{H}(1) = \Delta_f \overline{H}°(O(g)) = 249.1$ kJ per mole of $O(g)$

and we draw a horizontal line at 249.1 kJ mol^{-1} labeled $H_2(g) + O(g)$. (Note that $H_2(g)$ appears as both a reactant and a product in (1), but we carry it along in order to keep all species in mind at each enthalpy level.)

Step 2 is

(2) $H_2(g) + O(g) \rightarrow 2H(g) + O(g)$

$\Delta_r \overline{H}(2) = 2\Delta_f \overline{H}°(H(g)) = 436.0$ kJ mol^{-1}

and is also endothermic. The $2H(g) + O(g)$ level is therefore at $436.0 + 249.0 = 685.1$ kJ mol^{-1}.

Continuing,

(3) $2H(g) + O(g) \rightarrow H(g) + OH(g)$

$\Delta_r \overline{H}(3) = \Delta_f \overline{H}°(OH(g)) - \Delta_f \overline{H}°(H(g)) - \Delta_f \overline{H}°(O(g))$

$\quad = -428.$ kJ mol^{-1} (exothermic—a lower level)

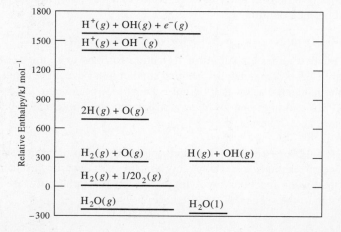

FIGURE 7.2 An enthalpy level diagram for various combination of H + H + O shows graphically how large the enthalpy differences are among various sets of reactants and products.

(4) $H(g) + OH(g) \rightarrow H^+(g) + OH(g) + e^-(g)$
$\Delta_r \overline{H}(4) = IP(H(g)) = 1318.\ \text{kJ mol}^{-1}$ **(up again)**

(5) $H^+(g) + OH(g) + e^-(g) \rightarrow H^+(g) + OH^-(g)$
$\Delta_r \overline{H}(5) = -EA(OH(g)) = -176.5\ \text{kJ mol}^{-1}$ **(down)**

(6) $H_2(g) + \frac{1}{2}O_2(g) \rightarrow H_2O(g)$
$\Delta_r \overline{H}(6) = \Delta_f \overline{H}^\circ(H_2O(g)) = -241.82\ \text{kJ mol}^{-1}$

(7) $H_2O(g) \rightarrow H_2O(l)$
$\Delta_r \overline{H}(7) = -\Delta \overline{H}_{vap}(H_2O) = -44.01\ \text{kJ mol}^{-1}$.

We can close this long cycle with a very unusual reaction:

(8) $H_2O(l) \rightarrow H^+(g) + OH^-(g)$

the enthalpy for which can be measured from the diagram or computed from $\Delta_r \overline{H}(1)$ through $\Delta_r \overline{H}(7)$:

$$\Delta_r \overline{H}(8) = 1684\ \text{kJ mol}^{-1}.$$

Enthalpy level diagrams summarize many calculations in a visually informative way. One can tell from the figure that, for instance,

$$H_2(g) + O(g) \rightarrow H(g) + OH(g)$$

is nearly thermoneutral since the enthalpy levels for the reactants and products are at nearly the same height, or that the phase transition enthalpy of step 7 is small compared to the enthalpy of formation of $H_2O(g)$, and so on.

7.4 CHEMICAL REACTION EQUILIBRIA

This section discusses what may well be termed the central problem of equilibrium thermochemistry: predicting the equilibrium composition of a mixture of reactive compounds. We will look at this problem in very general terms right from the start. There is no "ideal reaction" model that can be applied as a special case of general reactions, but there will be simplifying assumptions that we can make about the behavior of individual reactants or products, or about a mixture of these, that will clarify some aspects of the general discussion.

We start with the schematic reaction, at constant T and P,

$$\alpha A + \beta B \rightarrow \gamma C + \delta D$$

and imagine an initial state consisting of n_A°, n_B°, etc., moles of each compound isolated one from the next. At this point, the total Gibbs free energy is

$$G(\text{initial}) = n_A^\circ \mu(A) + n_B^\circ \mu(B) + n_C^\circ \mu(C) + n_D^\circ \mu(D) \quad (7.12)$$

where, for example,

$$\mu(A) = \mu(\text{pure A}) = \mu^\circ(T) + RT \ln f(A).$$

Next, we invoke the magic ability to mix these species without having them react. The free energy changes to

$$G(\text{poised to react}) = G(\text{initial}) + \Delta G_{mix} = n_A^\circ \mu_A + n_B^\circ \mu_B + n_C^\circ \mu_C + n_D^\circ \mu_D \quad (7.13)$$

where

$$\mu_A = \mu_A^\circ(T) + RT \ln(f_A/f_A^\circ).$$

Note that f_A is the fugacity of A in the mixture while $f(A)$, without the subscript, is the fugacity of pure A. Also f_A is a function of the mixture's composition, while f_A°, the standard-state fugacity, must be specified to determine the reference $\mu_A^\circ(T)$ value. Of course, partial pressures, activities, activity coefficients, and mole fractions, etc., can be brought into this expression in place of fugacities wherever warranted, but to keep the discussion general, we will focus on fugacities.

At this point, it is useful to introduce the reaction progress variable. Using Eq. (7.1), the definition of ξ, we can express G in terms of the system composition at any point in the reaction. Since, at *any* composition,

$$G = n_A \mu_A + n_B \mu_B + n_C \mu_C + n_D \mu_D$$

we have, with $n_A = n_A^\circ - \alpha\xi$, $n_B = n_B^\circ - \beta\xi$, $n_C = \gamma\xi - n_C^\circ$, and $n_D = \delta\xi - n_D^\circ$ from Eq. (7.1),

$$G(\xi) = n_A^\circ \mu_A + n_B^\circ \mu_B - n_C^\circ \mu_C - n_D^\circ \mu_D + \xi(\gamma\mu_C + \delta\mu_D - \alpha\mu_A - \beta\mu_B) \ . \quad \textbf{(7.14)}$$

This is the central expression for chemical reaction equilibria. It is a sum of two terms:

$$G(\xi) = G(\text{poised to react}) + G(\text{due to an amount } \xi \text{ of reaction}) \ .$$

Recall that $\xi = 0$ at any initial state, and that as a spontaneous reaction occurs, ξ will change in either direction. If the initial amounts of *reactants* exceed the equilibrium amount, ξ will increase from zero as reactants are consumed, but if the initial amounts of *products* exceed the equilibrium amount, ξ will decrease from zero to negative values.

The spontaneous direction of change of ξ will be governed by the requirement that G must decrease in a spontaneous process. *At some ξ, G will reach a minimum value and the reacting mixture will achieve chemical equilibrium.* The equilibrium degree of advancement will be denoted ξ_{eq}, and, from the general rule for finding a minimum,[3]

$$\text{at } \xi_{eq}, \left(\frac{\partial G}{\partial \xi}\right)_{T,P} = 0 \ . \quad \textbf{(7.15)}$$

Rather than differentiating Eq. (7.14) with respect to ξ, we will take a more general route to $(\partial G/\partial \xi)$. We begin with Eq. (5.31) with $dG = 0$:

$$VdP - SdT + \sum \mu_i \, dn_i = 0 \quad \textbf{(at equilibrium)} \quad \textbf{(5.42)}$$

specify constant T and P:

$$\sum \mu_i \, dn_i = 0 \quad \textbf{(at equilibrium, } dT = dP = 0\textbf{)}$$

and introduce $d\xi$ from Eq. (7.2):

$$\left(\sum_{\text{products}} \nu_i \mu_i - \sum_{\text{reactants}} \nu_i \mu_i \right) d\xi = 0 \ . \quad \textbf{(7.16)}$$

---

[3]Note that Nature will ensure the system reaches a true minimum, rather than a maximum. It is possible, though rare, for $(\partial G/\partial \xi)$ to be zero at more than one value of ξ. Some of these values will correspond to minima, and some to maxima. The minima would constitute a set of multiple "stationary points," as they are called, and the maxima would correspond to metastable equilibria. Multiple stationary points would correspond to several possible equilibria, a concept of current interest in regards to certain reactions that oscillate between one state and another.

Since ξ is an independent variable, Eq. (7.16) is satisfied only if

$$\left(\sum_{\text{products}} \nu_i \mu_i - \sum_{\text{reactants}} \nu_i \mu_i\right) = \left(\frac{\partial G}{\partial \xi}\right)_{T,P} = 0 , \quad (7.17)$$

which becomes our criterion for chemical equilibrium at constant T, P, and total amount of system.

Equation (7.17) is true only if each μ_i is evaluated at the equilibrium composition. Conversely, we can use the equality to *find* the equilibrium composition. To illustrate this most important point, we return to our schematic reaction and Eq. (7.14). The equilibrium criterion, Eq. (7.17), shows that the coefficient of ξ in Eq. (7.14) must vanish at equilibrium:

$$\gamma \mu_C + \delta \mu_D - \alpha \mu_A - \beta \mu_B = 0 .$$

Substituting $\mu_i = \mu_i^\circ + RT \ln(f_i/f_i^\circ)$ for each μ_i gives

$$\gamma \mu_C^\circ + \delta \mu_D^\circ - \alpha \mu_A^\circ - \beta \mu_B^\circ = -RT \ln\left[\frac{\left(\frac{f_C^{eq}}{f_C^\circ}\right)^\gamma \left(\frac{f_D^{eq}}{f_D^\circ}\right)^\delta}{\left(\frac{f_A^{eq}}{f_A^\circ}\right)^\alpha \left(\frac{f_B^{eq}}{f_B^\circ}\right)^\beta}\right] \quad (7.18)$$

where the superscript eq has been added to emphasize the equilibrium composition.

We will usually specify a *pure* standard state composition for chemical equilibrium calculations (i.e., $x^\circ = $ pure i in Eq. (6.31)), and a pressure of 1 bar. These choices ensure that the left hand side of Eq. (7.18) is a function of T only, and that it corresponds to the free energy of reaction change for the standard state transformation (note the single arrow!)

$$\alpha A + \beta B \rightarrow \gamma C + \delta D$$

with

$$\Delta_r \overline{G}^\circ = \gamma \mu_C^\circ + \delta \mu_D^\circ - \alpha \mu_A^\circ - \beta \mu_B^\circ . \quad (7.19)$$

With this definition for $\Delta_r \overline{G}^\circ$, Eq. (7.18) becomes

$$-\frac{\Delta_r \overline{G}^\circ}{RT} = \ln\left[\frac{\left(\frac{f_C^{eq}}{f_C^\circ}\right)^\gamma \left(\frac{f_D^{eq}}{f_D^\circ}\right)^\delta}{\left(\frac{f_A^{eq}}{f_A^\circ}\right)^\alpha \left(\frac{f_B^{eq}}{f_B^\circ}\right)^\beta}\right] \quad (7.20)$$

The argument of the logarithm is also a function of temperature only, and we denote it $K_{eq}(T)$, the *thermodynamic reaction equilibrium constant*. For a general reaction, we have

$$\Delta_r \overline{G}^\circ = \sum_{\text{products}} \nu_i \mu_i^\circ - \sum_{\text{reactants}} \nu_i \mu_i^\circ \quad (7.21)$$

and[4]

$$K_{eq} = \frac{\prod_{\text{products}} \left(\frac{f_i^{eq}}{f_i^\circ}\right)^{\nu_i}}{\prod_{\text{reactants}} \left(\frac{f_i^{eq}}{f_i^\circ}\right)^{\nu_i}} = \exp\left(-\frac{\Delta_r \overline{G}^\circ}{RT}\right) . \quad (7.22)$$

---

[4]The $\prod$ symbol may be unfamiliar to you; it is to multiplication what $\sum$ is to addition. Thus, just as $\sum_{i=1}^{3} x_i = x_1 + x_2 + x_3$, we can write $\prod_{i=1}^{3} x_i = x_1 x_2 x_3$.

We compute $\Delta_r \overline{G}^\circ$, find K_{eq}, then find the equilibrium composition. Practical methods for these steps are discussed in Section 7.6, but to close this section, we look at the equilibrium constant for several model systems.

Ideal Mixture of Reacting Ideal Gases

For this model system, $f_i^\circ = 1$ bar and $f_i = P_i = x_i P$ so that

$$K_{eq}(\text{ideal gases}) = \frac{\prod_{\text{products}} P_i^{\nu_i}}{\prod_{\text{reactants}} P_i^{\nu_i}}.$$

For example, the ammonia synthesis reaction

$$3H_2 + N_2 \rightarrow 2NH_3$$

at low pressure has equilibrium partial pressures related by

$$K_{eq} = \frac{P_{NH_3}^2}{P_{H_2}^3 P_{N_2}}.$$

The partial pressures must be expressed in bar units, but due to the implicit 1 bar standard state dividing each, K_{eq} is *always dimensionless*.

We can also express K_{eq} in terms of mole fractions x_i and the total pressure P:

$$K_{eq}(\text{ideal gases}) = \frac{\prod_{\text{products}} x_i^{\nu_i}}{\prod_{\text{reactants}} x_i^{\nu_i}} P^{\Delta \nu}$$

where

$$\Delta \nu = \sum_{\text{products}} \nu_i - \sum_{\text{reactants}} \nu_i.$$

For the ammonia synthesis, $\Delta \nu = 2 - (3 + 1) = -2$ so that

$$K_{eq} = \frac{x_{NH_3}^2}{x_{H_2}^3 x_{N_2}} P^{-2}.$$

From Amagat's law, which holds for any ideal mixture of ideal gases, we can use partial molar volumes to express K_{eq} in concentration units. We have

$$\overline{V}_i = \overline{V}(i) = \frac{V}{n_i} = \frac{1}{C_i}$$

with C_i = amount of i per unit volume. Since

$$P_i = \frac{n_i RT}{V} = C_i RT$$

the partial pressure expression for the ammonia synthesis becomes

$$K_{eq} = \frac{C_{NH_3}^2}{C_{H_2}^3 C_{N_2}} (RT)^{-2}$$

with R expressed in volume units consistent with those used for the concentrations and in pressure units of bar (such as dm^3 bar mol^{-1} K^{-1} or cm^3 bar mol^{-1} K^{-1}).

Real Reactive Mixtures

The equilibrium constant can be expressed in terms of activities:

$$a_i = \frac{f_i}{f_i^\circ} \tag{6.31}$$

or in terms of activity coefficients and mole fractions:

$$a_i = \gamma_i x_i \,. \tag{6.32}$$

We have

$$K_{eq} = \frac{\prod_{\text{products}} a_i^{\nu_i}}{\prod_{\text{reactants}} a_i^{\nu_i}} = \frac{\prod_{\text{products}} \gamma_i^{\nu_i} \prod_{\text{products}} x_i^{\nu_i}}{\prod_{\text{reactants}} \gamma_i^{\nu_i} \prod_{\text{reactants}} x_i^{\nu_i}} \,.$$

Several standard-state choices for activity simplify K_{eq}. Many of these are tacitly assumed in elementary chemistry courses, and the expressions for K_{eq} that result from them may look more familiar to you than our general expressions.

Pure Condensed Phases. If the pressure is the standard pressure, then $f_i = f_i^\circ$, and the activity is unity. This is why pure condensed-phase compounds disappear from elementary, approximate expressions for K_{eq}. If the pressure is not the standard pressure, then a Poynting correction from Eq. (6.30) determines the activity:

$$a(\text{pure } i, \text{condensed phase}, P \neq P^\circ) = \exp\left(\int_{P^\circ}^{P} \frac{\overline{V}(i)}{RT} dP\right) \,.$$

If P is not far from P° (and $\overline{V}$ is not huge, as it is for macromolecules, for instance), this activity is not far from unity. Most molar volumes (excluding macromolecules) fall in the range ~ 10 to ~ 200 cm^3 mol^{-1}, and a 1 bar pressure increase raises the activity an insignificant amount (to only 1.008 if $\overline{V} = 200$ cm^3 mol^{-1}).

Almost Pure Condensed Phases. The solvent in dilute solutions is nearly pure, and frequently both x_i(solvent) and γ_i(solvent) are unity to a good approximation. Of course, if the solvent does not appear in the net reaction under consideration, then it does not appear directly in K_{eq}.

Dilute Solutes. A convenient standard state for a dilute solute's activity is

$$\gamma(\text{solute}) \to 1 \quad \text{as} \quad x(\text{solvent}) \to 1 \,.$$

This means the solute behaves as an ideal solution ($\mu_i = \text{constant} + RT \ln x_i$; see Eq. (6.33)) in the limit of infinite dilution. This is the *ideal dilute solution* standard state, and it is assumed to hold at *all* compositions. It is the hypothetical behavior of a *pure* solute behaving as if it were infinitely dispersed in the solvent, an argument analogous to the ideal gas standard state used for fugacity. It retains all the nonideality away from the standard chemical potential.

Molality units are commonly used to express dilute solution composition. In terms of mole fractions,

$$m_i = x_i \frac{1000 \text{ g kg}^{-1}}{x_{\text{solvent}} M}$$

if M, the solvent molecular mass, is in g mol^{-1} units so that

$$a_i = \gamma_i x_i = \gamma_i m_i \frac{x_{\text{solvent}} M}{1000 \text{ g kg}^{-1}} \,.$$

A common convention assumes a unit molality standard state. One writes

$$\gamma_i^\circ = \gamma_i x_{\text{solvent}} \approx \gamma_i$$
$$a_i^\circ = \gamma_i^\circ m_i/(1 \text{ mol kg}^{-1})$$

and the factor $M(1\ \mathrm{mol\ kg^{-1}})/(1000\ \mathrm{g\ kg^{-1}})$ is absorbed into the reference chemical potential:

$$\mu_i(\text{dilute solute, unit-molality standard state}) = \mu_i^\circ(T) + RT \ln [M(1\ \mathrm{mol\ kg^{-1}})/(1000\ \mathrm{g\ kg^{-1}})]$$
$$+ RT \ln [\gamma_i^\circ m_i/(1\ \mathrm{mol\ kg^{-1}})] \quad (7.23)$$
$$= \mu_i^*(T) + RT \ln \gamma_i^\circ m_i$$

where, in the last expression, the $1\ \mathrm{mol\ kg^{-1}}$ standard-state concentration is tacitly assumed, and $\mu_i^*(T)$ represents the new standard state.

In the final section of this chapter (and again in Chapter 10), we will look at several typical calculations based on these various expressions for K_{eq}. The next section completes the general discussion by looking at the variation of chemical equilibrium with temperature, pressure, and composition.

7.5 THE STABILITY OF CHEMICAL EQUILIBRIA

The fundamental expression for K_{eq} is

$$\ln K_{eq} = -\frac{\Delta_r \overline{G}^\circ}{RT} \quad (7.24)$$

from Eq. (7.21). The equilibrium constant depends on T and P through the partial derivatives

$$\left(\frac{\partial \ln K_{eq}}{\partial T}\right)_P \quad \text{and} \quad \left(\frac{\partial \ln K_{eq}}{\partial P}\right)_T.$$

We look at each of these in turn to see how changes in T and P alter a state of chemical equilibrium.

Temperature Stability

The temperature derivative

$$\left(\frac{\partial \ln K_{eq}}{\partial T}\right)_P = -\frac{1}{R}\left(\frac{\partial (\Delta_r \overline{G}^\circ/T)}{\partial T}\right)_P$$

is simplified if we first use the identity

$$\frac{\Delta_r \overline{G}^\circ}{T} = \frac{\Delta_r \overline{H}^\circ}{T} - \Delta_r \overline{S}^\circ$$

then differentiate term by term on the right:

$$\left(\frac{\partial (\Delta_r \overline{H}^\circ/T)}{\partial T}\right)_P = \frac{1}{T}\left(\frac{\partial \Delta_r \overline{H}^\circ}{\partial T}\right)_P - \frac{\Delta_r \overline{H}^\circ}{T^2} = \frac{1}{T}\Delta_r \overline{C}_P^\circ - \frac{\Delta_r \overline{H}^\circ}{T^2}$$

and

$$\left(\frac{\partial \Delta_r \overline{S}^\circ}{\partial T}\right)_P = \frac{1}{T}\Delta_r \overline{C}_P^\circ .$$

so that

$$\left(\frac{\partial \ln K_{eq}}{\partial T}\right)_P = \frac{\Delta_r \overline{H}^\circ}{RT^2}$$

or, since $d(1/T) = -dT/T^2$,

$$\left(\frac{\partial \ln K_{eq}}{\partial (1/T)}\right)_P = -\frac{\Delta_r \overline{H}^\circ}{R} .$$

If $\Delta_r\overline{H}°$ is temperature-independent, a graph of $\ln K_{eq}$ versus reciprocal temperature is a straight line, and $\Delta_r\overline{H}°$ is computed from its slope. If the temperature range of the measurements is small, $\Delta_r\overline{H}°$ will be approximately constant, and the line will often appear to be straight.

A constant $\Delta_r\overline{H}°$ allows us to integrate the previous derivative expression between two temperatures:

$$\int_{\ln [K_{eq}(T_1)]}^{\ln [K_{eq}(T_2)]} d\ln K_{eq} = -\frac{\Delta_r\overline{H}°}{R} \int_{1/T_1}^{1/T_2} d\left(\frac{1}{T}\right)$$

so that

$$\ln\left(\frac{K_{eq}(T_2)}{K_{eq}(T_1)}\right) = -\frac{\Delta_r\overline{H}°}{R}\left(\frac{1}{T_2} - \frac{1}{T_1}\right)$$

or

$$K_{eq}(T_2) = K_{eq}(T_1) \exp\left[-\frac{\Delta_r\overline{H}°}{R}\left(\frac{1}{T_2} - \frac{1}{T_1}\right)\right]. \qquad (7.25)$$

This equation should look familiar to you; it is a generalization of Eq. (6.12), the integrated Clapeyron expression for the temperature dependence of vapor pressure. Vapor pressure is just the equilibrium constant for the reaction (condensed phase) $\rightleftarrows$ (vapor phase).

From Eq. (7.25), we can generalize the qualitative behavior of K_{eq} with T at constant P based on the sign of $\Delta_r\overline{H}°$:

exothermic ($\Delta_r\overline{H}° < 0$): K_{eq} *decreases* with increasing T, favoring reactants
endothermic ($\Delta_r\overline{H}° > 0$): K_{eq} *increases* with increasing T, favoring products.

Increasing K_{eq} favors the production of *products* while *decreasing* K_{eq} favors the production of *reactants*. Consequently, to produce more products at constant pressure in a closed system, one should *raise* the temperature of an endothermic reaction mixture but *lower* the temperature of an exothermic reaction mixture.

EXAMPLE 7.8

What value of $\Delta_r\overline{H}$ will double K_{eq} when T is raised from 300 K to 400 K? from 200 K to 300 K?

SOLUTION First, note that $\Delta_r\overline{H}$ must be *positive* (the reaction must be endothermic) in order for K_{eq} to increase with increasing T. If we assume $\Delta_r\overline{H}$ is independent of temperature over this range, we can use Eq. (7.25) with $K_{eq}(400\ \text{K})/K_{eq}(300\ \text{K}) = 2$:

$$\ln 2 = -\frac{\Delta_r\overline{H}}{R}\left(\frac{1}{400\ \text{K}} - \frac{1}{300\ \text{K}}\right)$$

which yields $\Delta_r\overline{H} = 6.92$ kJ mol^{-1}. This is a relatively small endothermicity; larger values increase K_{eq} by a larger factor. For the change from 200 K to 300 K, we find $\Delta_r\overline{H} = 3.46$ kJ mol^{-1}, one half the previous value.

↠ RELATED PROBLEMS 7.23, 7.24, 7.25, 7.27

Adiabatic Reactions

An important type of reaction condition that has great practical value is the *adiabatic reaction*. If an initial reaction mixture is allowed to achieve chemical equilibrium under constant pressure adiabatic conditions, then the final temperature will reflect the enthalpy change that accompanied the reaction's progress towards equilibrium.

The enthalpy released (or consumed) is $(\Delta_r \overline{H}°) \xi_{eq}$. Under adiabatic conditions, this enthalpy has no place to go (or had no external place to come from) except to (or from) the reaction mixture itself. The final temperature must therefore be given by

$$-(\Delta_r \overline{H}°)\xi_{eq} = \int_{T_i}^{T_f} C_P(T)\, dT$$

where

$C_P(T)$ = total heat capacity of the system at chemical equilibrium.

This can be seen by considering the path

initial mixture $(T_i, P) \to$ equilibrium mixture (T_i, P) ,

which involves the net enthalpy change $(\Delta_r \overline{H}) \xi_{eq}$, followed by

equilibrium mixture $(T_i, P) \xrightarrow{-q_{in}}$ equilibrium mixture (T_f, P)

with $-q_{in} = (\Delta_r \overline{H}°) \xi_{eq}$.

Reaction enthalpies that vary rapidly with temperature make adiabatic reaction temperatures somewhat more difficult to calculate, especially if the equilibrium composition is sensitive to temperature as well. But the idea remains the same: the effects of the reaction enthalpy are confined to the system.

EXAMPLE 7.9

Given $\Delta_f \overline{H}° = -241.818$ kJ mol^{-1} for H$_2$O(g) at 298.15 K, find the final temperature when 1.00 mol H$_2(g)$ and 0.50 mol O$_2(g)$ react adiabatically at constant pressure. Suppose air, rather than pure O$_2$, had been used to supply the 0.5 mol O$_2$. How would this change alter the final temperature?

SOLUTION The reaction in the absence of air is

$$H_2(g) + \frac{1}{2} O_2(g) \to H_2O(g) ,$$

a reaction that "goes to completion;" the equilibrium composition starting from stoichiometric amounts of H$_2$ and O$_2$ is virtually pure H$_2$O so that $\xi_{eq} = 1$ mol. If the reaction is maintained at 298 K, then, since H$_2(g)$ and O$_2(g)$ are elemental standard states, $\Delta_r \overline{H} = \Delta_f \overline{H}°(H_2O)$, and $\Delta_r H = \Delta_r \overline{H} \xi_{eq} = -241.818$ kJ is the enthalpy released by this exothermic reaction. Under adiabatic reaction conditions, this enthalpy is retained in the H$_2$O(g) product, raising its temperature:

$$-\Delta_r H = \int_{298.15 \text{ K}}^{T_f} C_P(T, H_2O(g))\, dT.$$

From Table 4.4 we find $C_P(T)$ for H$_2$O as a power series in T, and evaluate the integral:

$$-\Delta_r H = 241.818 \text{ kJ} = (1 \text{ mol})R[3.633(T_f - 298.15)$$

$$+ \frac{1}{2}(1.195 \times 10^{-3})(T_f^2 - 298.15^2)$$

$$+ \frac{1}{3}(1.34 \times 10^{-7})(T_f^3 - 298.15^3)].$$

An iterative solution gives $T_f = 4300$ K, a value that is likely inaccurate, since the temperature is far out of the range of validity of Table 4.4 and since subsequent chemistry (such as the dissociation of water) may be important at very high temperatures, but one that is certainly of the correct magnitude.

In air, since each 0.5 mol O_2 is accompanied by 1.88 mol N_2, the final reaction mixture is $H_2O(g) + 1.88\ N_2(g)$ with a total heat capacity

$$(1\ \text{mol})\ \overline{C}_P(H_2O) + (1.88\ \text{mol})\ \overline{C}_P(N_2)\ ,$$

a *larger* value than before, implying a *lower* flame temperature. (Hence, welding torches fueled by pure O_2 have a much higher flame temperature than those fueled by air alone.) A calculation of the final temperature here yields the impressively lower value $T_f = 2470$ K.

⏵ RELATED PROBLEM 7.29

Pressure Stability

The effect of pressure on an equilibrium constant is related to the role played by pressure in defining the standard state. Virtually all useful standard-state definitions specify a reference pressure. In such cases, since each $\mu_i^\circ(T)$ used to compute $\Delta_r \overline{G}^\circ$ is independent of pressure, K_{eq} is *independent of pressure* as well.

It is important to understand the implications of this statement in some detail. While K_{eq} may be the same at all pressures, the *equilibrium composition may vary considerably with pressure*. Consider the ammonia synthesis at low pressures again, for which

$$K_{eq} = \frac{x_{NH_3}^2}{x_{H_2}^3 x_{N_2}} P^{-2}\ .$$

If K_{eq} is a constant at any P, then the mole fraction ratio *must* change with pressure as its cofactor P^{-2} changes. This rapid dependence of ξ_{eq} on P is a direct consequence of any reaction that produces different amounts of gases than it consumes. The equilibrium composition of a reaction with $\Delta \nu = 0$ such as

$$H_2(g) + F_2(g) \rightleftarrows 2HF(g)$$

is *not* sensitive to pressure, except to the degree that fugacities vary with P in the real gas, high-pressure regime.

A related concern is the role of an inert gas. Imagine the ammonia synthesis reaction carried out in the presence of He. The mole fraction expressions are, for instance, $x_{N_2} = n_{N_2}/(n_{N_2} + n_{H_2} + n_{NH_3} + n_{He})$ and K_{eq} is

$$K_{eq} = \frac{n_{NH_3}^2 (n_{N_2} + n_{H_2} + n_{NH_3} + n_{He})^2}{n_{H_2}^3 n_{N_2}} P^{-2}$$

showing that the amount of inert gas directly influences the equilibrium composition. But again, the HF synthesis reaction is unaffected by an inert gas due to the equal numbers of moles of gaseous reactants and products. The equilibrium constant is still

$$K_{eq} = \frac{x_{HF}^2}{x_{H_2} x_{F_2}} = \frac{n_{HF}^2}{n_{H_2} n_{F_2}}\ .$$

Another example of the influence of pressure on equilibrium composition is the solubility of solids under pressure, a common problem in geochemistry or in oceanic chemistry near the ocean floor. Here the effect centers around

$$\left(\frac{\partial \ln f_i}{\partial P}\right)_T = \frac{\overline{V}_i}{RT}$$

so that

$$\Delta_r \overline{V} = \sum_{\text{products}} \nu_i \overline{V}_i - \sum_{\text{reactants}} \nu_i \overline{V}_i$$

governs the change in equilibrium composition with pressure for condensed phases.

Consider the effect of pressure on the solubility of solid A:

$$A(s) \rightleftarrows A(x_A, \text{solution}) \ .$$

In general, since $d\mu_i = \overline{V}_i \, dP$, we can write

$$\left(\frac{\partial(\mu_A - \mu(A))}{\partial P}\right)_T = \overline{V}_A - \overline{V}(A) = \Delta_r \overline{V} \ .$$

If the partial molar volume of A in solution is greater than the molar volume of solid A ($\Delta_r \overline{V}$ is positive), a pressure increase will raise the chemical potential of A in solution above that of the solid, precipitating A in order to lower the solution-phase chemical potential and regain equilibrium.

If the ideal-solution fugacity expression holds, (Eq. (6.33)), we have

$$f_A(x_A, T, P) = x_A f(A, l, T, P)$$

where $f(A, l, T, P)$ is the fugacity of *pure liquid A at the temperature and pressure of interest*. Since pure A is assumed to be a *solid* at (T, P), this reference fugacity is for the *supercooled liquid phase*. These are the conditions used in Section 6.5 to discuss freezing point depressions, and consequently Eq. (6.43) defines the *ideal solubility* of A. (What was called component 1 in that equation is A here. An ideally soluble compound has a partial molar volume in solution equal to the molar volume of its pure liquid phase.)

At equilibrium, the criterion of uniform fugacities for any component in all phases means $f_A(x_A, T, P) = f(A, T, P)$. Because the fugacities are equal, so must be the logarithms of the fugacities, or the changes in these logarithms with pressure and composition.[5] Thus

$$d \ln f_A(x_A, T, P) = d \ln f(A, T, P)$$

or, at constant T,

$$\left(\frac{\partial \ln f_A}{\partial P}\right)_{T, x_A} dP + \left(\frac{\partial \ln f_A}{\partial x_A}\right)_{T, P} dx_A = \left(\frac{\partial \ln f(A)}{\partial P}\right)_T dP$$

since $f(A)$ does not depend on composition. Using $(\partial \ln f/\partial P) = \overline{V}/RT$ and $(\partial \ln f_A/\partial x_A) = 1/x_A$,

$$\frac{dx_A}{x_A} = d \ln x_A = \left[\frac{\overline{V}(A(s)) - \overline{V}(A(l))}{RT}\right] dP = -\frac{\Delta \overline{V}_{\text{fus}}}{RT} dP \ .$$

To find the solubility at P_f when it is known at P_i we integrate the above expression (assuming that each $\overline{V}$ is independent of pressure) and find:

$$\int_{x_A(P_i)}^{x_A(P_f)} d \ln x_A = \ln\left[\frac{x_A(P_f)}{x_A(P_i)}\right] = -\int_{P_i}^{P_f} \frac{\Delta \overline{V}_{\text{fus}}}{RT} dP = -\frac{\Delta \overline{V}_{\text{fus}} \Delta P}{RT} \ .$$

This is in accord with our previous general argument: if the liquid-phase molar volume of A is larger than that of the solid, the solubility of A will decrease with increasing pressure.

This analysis also shows that we can take K_{eq} to equal x_A (which, when mole fraction is converted to concentration, yields a solubility equilibrium constant in the form one usually encounters in general chemistry courses). This choice causes

---

[5] Note that the argument to follow is virtually identical to that used in Section 6.2 to derive the Clapeyron equation, Eq. (6.11). There, P and T were varied under equilibrium constraints at constant composition while here, P and composition are varied at constant T.

K_{eq} to depend on pressure, but it is obviously a very convenient approach for practical solubility calculations, especially when the pressure is constant, as it usually is.

EXAMPLE 7.10

The oceans are a vast reservoir for carbon dioxide (in the form of various carbonates), and the concern about climatic effects of CO_2 from fossil fuel burning (the Greenhouse Effect) makes the role played by the oceans in the global CO_2 balance very important to understand. Calcium carbonate (which has two important mineralogical forms, calcite and aragonite) is not very soluble and can precipitate important amounts of carbonate. Predict the effect of the ocean's pressure on the solubility of $CaCO_3$ at great depths where $P \cong 1000$ bar and $T \cong 0$ °C given that at 1 bar and 25 °C the solubility of calcite is 1.4×10^{-4} molal and given the following molar and partial molar volumes:

$\overline{V}(CaCO_3(s)) = 36.8$ cm^3 mol^{-1} $\overline{V}_{CaCl_2}(aq) = 18.3$ cm^3 mol^{-1}

$\overline{V}_{Na_2CO_3}(aq) = -6.7$ cm^3 mol^{-1} $\overline{V}_{NaCl}(aq) = 19.6$ cm^3 mol^{-1} .

SOLUTION   Since $CaCO_3$ is so insoluble, reliable values of its partial molar volume are difficult to measure. Hence, we employ data on more soluble species (which are of relevance to sea water's typical composition) to turn the equilibrium of interest:

$$CaCO_3(s) \rightleftarrows CaCO_3(aq)$$

into an equilibrium for which data are readily available.

$$CaCO_3(s) + 2NaCl(aq) \rightleftarrows Na_2CO_3(aq) + CaCl_2(aq) \; .$$

For this equilibrium, we have

$$\Delta_r \overline{V} = [-6.7 + 18.3 - 36.8 - 2(19.6)] \text{ cm}^3 \text{ mol}^{-1} = -64.4 \text{ cm}^3 \text{ mol}^{-1}$$

which, being negative, tells us that a pressure increase will increase the solubility of $CaCO_3$. We can calculate the free energy imbalance that a 1000 bar pressure increase induces: it is $\Delta_r \overline{V} \Delta P \cong (-64.4$ cm^3 mol^{-1}) (1000 bar) $= -6.44$ kJ mol^{-1}, but we cannot so easily calculate the new solubility without expressions for $(\partial \ln f_i / \partial x_i)$ for the aqueous species. These are, of course, dissociated electrolytes with strong nonidealities. Fugacity expressions for them are discussed in Chapter 10.

➧ RELATED PROBLEMS 7.28, 7.40

Stability in Open Systems

Finally, we consider the effects of adding reactants or products (or both) to a reaction mixture that has achieved equilibrium. In general, the new mixture's composition will no longer be an equilibrium composition, and the system will spontaneously evolve toward a new equilibrium. The direction the new mixture takes to find equilibrium again (the algebraic sign of $d\xi$) depends on the type of material added (reactant or product) as well as the constraints invoked during the addition (added at constant P? adiabatically? at constant V?). Problem 7.54 discusses this further.

The response of a system, *any* system, to a perturbation that displaces it from equilibrium is a general problem for thermodynamic stability theory. We have looked at chemical equilibrium here, but the general principles were formulated in Chapter 5. The special ways of reattaining *chemical* equilibrium are often expressed by the Le Châtelier (or Le Châtelier–Braun) Principle: the response of a system to a perturbation away from equilibrium is one that restores equilibrium (which is no surprise).

Frequently the direction back to equilibrium (the sign of $d\xi$) can be deduced by elementary arguments. For example, consider the effect of temperature on the solubility of a compound for which the reaction enthalpy is endothermic. We add enthalpy to the equilibrium system

$$\text{solid solute} \rightleftarrows \text{solute in solution} ,$$

and the system responds by consuming at least part of this enthalpy as it dissolves more solid at the higher temperature. Or consider raising the activity of $Cl^-(aq)$ in an equilibrium mixture of $Ag^+(aq)$, $Cl^-(aq)$ and $AgCl(s)$ (by adding, for instance, a few drops of a sodium chloride solution). The response to this addition will be to lower the $Cl^-(aq)$ activity by precipitating more $AgCl(s)$ until the equilibrium expression

$$K_{eq} = a_{Ag^+} a_{Cl^-}$$

is once again satisfied.

The next section describes the practical calculation of equilibrium constants and the way one finds an equilibrium reaction mixture's composition. For chemists and chemical engineers, this sort of calculation is obviously a very important one to master. We will emphasize the use of real, tabulated data, and this section will be as much a guide to the use of these data as it is a description of equilibrium constants.

7.6 METHODS OF CALCULATING EQUILIBRIUM CONSTANTS

The key to an equilibrium constant is the standard Gibbs free-energy change for a reaction, $\Delta_r \overline{G}°$. Faced with a reaction of interest, one frequently must turn to compilations of thermodynamic data to find the relevant enthalpies, entropies, free energies, and heat capacities needed to compute K_{eq} under any given set of conditions.

One of the most extensive such compilations consists of the so-called JANAF Thermochemical Tables. Begun in 1959 by a group of researchers sponsored by the U.S. Department of Defense (JANAF stands for "Joint Army, Navy, Air Force"), the JANAF tables are continued, revised, and supplemented today, and they constitute part of the National Standard Reference Data System of the U.S. National Institute for Standards and Technology (the former National Bureau of Standards). Similar tabulations for special classes of compounds such as metallic alloys or hydrocarbons are published by several other groups.

Tables 7.6–7.9 are the JANAF tables for the simple substances $H^+(g)$, $H^-(g)$, $H(g)$, and $H_2(g)$. Each begins with the name, symbol, and physical state,[6] and each contains eight columns of data. In order from left to right, these are:

1. the temperature from 0 K to 6000 K including the reference temperature $T_r = 298.15$ K $= 25$ °C;
2. the constant pressure molar heat capacity at 0.1 MPa $= 1$ bar;
3. the absolute molar entropy at 1 bar;
4. the quantity $-[\overline{G}°(T) - \overline{H}°(T_r)]/T$, to which we will return;
5. the molar enthalpy increment, $\overline{H}°(T) - \overline{H}°(T_r)$;
6. the standard molar enthalpy of formation at each temperature;
7. the standard molar Gibbs free energy of formation at each temperature, $\Delta_f \overline{G}°$, discussed below; and
8. the dimensionless quantity $\log_{10} K_f$, which we also discuss below.

---

[6]While the most recent edition is in SI units and contemporary notation, older editions used the following units and notation. The standard pressure was 1 atm, the molecular mass was called GFW for "gram formula weight" in g mol^{-1} units, the energy unit was the calorie (or kilocalorie, kcal), and for entropy and heat capacity, the unit "gibbs" was sometimes used for "calorie per absolute degree." Also, some early data sheets used the symbol "F" rather than "G" for the Gibbs free energy.

Comments, references, and spectroscopic data appear with each table as well. These data are of importance to the statistical-mechanical calculation of thermodynamic properties and will not be of direct interest now.

First, compare the data sheets for $H^+(g)$ and $H^-(g)$, both treated as pure, ideal gases at 1 bar. This is the standard state, which is clearly hypothetical for these species. Note that the heat capacities of each are a constant $5R/2 = 20.786$ J K^{-1} mol^{-1}, as expected.

The entropies of each are computed from the Sackur–Tetrode equation, Eq. (5.44). The slight difference between $\overline{S}°(H^+)$ and $\overline{S}°(H^-)$ at any one T is a mass effect. The H^+ table is for the bare proton; the H^- mass is larger by twice $M_e = N_A m_e$, the electron molar mass.

Note next the enormous differences between the molar enthalpies of formation of H^+ and H^-. These differences are traced to the very endothermic ionization reaction

$$\tfrac{1}{2} H_2(g) \rightarrow H^+(g) + e^-(g) \tag{7.26}$$

(remember that $\Delta_f \overline{H}°(e^-(g)) = 0$), which yields $\Delta_f \overline{H}°(H^+(g))$, while the electron attachment reaction

$$\tfrac{1}{2} H_2(g) + e^-(g) \rightarrow H^-(g) \tag{7.27}$$

is considerably less endothermic due to the small electron affinity of atomic hydrogen (0.7542 eV = 72.77 kJ mol^{-1}). For Eq. (7.26) at $T = 0$ K we have

$$\Delta_r \overline{H} = \Delta_f \overline{H}°(H^+(g)) = \tfrac{1}{2} \mathcal{D}_0(H_2) + \text{IP}(H)$$

while for Eq. (7.27), we have

$$\Delta_r \overline{H} = \Delta_f \overline{H}°(H^-(g)) = \tfrac{1}{2} \mathcal{D}_0(H_2) - \text{EA}(H^-) \ .$$

Note that the H^+ values *increase* with increasing temperature, while those for H^- *decrease*. This is a direct consequence of Eq. (7.9) for the temperature dependence of $\Delta_r \overline{H}$; the heat capacity change for (7.26) is positive while that for (7.27) is negative.

Now compare the data sheet for $H(g)$ to these two. Again, the heat capacity is a uniform $5R/2$, but the entropy of H is uniformly greater than that for H^+ or H^-. This is not a mass effect, but a quantum-mechanical effect. In H atom, the one electron possesses a magnetic moment called spin (see Section 12.5) that can be in either of two states of essentially the same energy. This extra degree of freedom adds $R \ln 2 = 5.763$ J mol^{-1} K^{-1} to the Sackur–Tetrode entropy. (Recall our discussion of the two-level system in Section 4.1 and the $R \ln 2$ value in Fig. 4.2.) This degree of freedom is not available to H^+ (no electrons!) or to H^- (two electrons that are "spin paired" in a definite configuration). The proton in the H nucleus also has a spin, but its entropy (another $R \ln 2$ addition) is not included in these tables, since nuclear spin effects are normally inconsequential to chemical reactions (H_2 will be a curious exception).

Finally, compare the H_2 and H data sheets. Note the abundance of zeroes in the last three columns of the H_2 table, since $H_2(g)$ is the elemental reference state, and note the temperature variation of the heat capacity and the larger entropy of H_2 over H. The entropy of H_2 is a curious quantity, since there are *two types* of H_2 molecules. These are termed *ortho* and *para* (designations that have nothing in common with the use of these terms in organic nomenclature), and the distinction between them is based on nuclear spin differences. The reactivities and physical properties of o–H_2 and p–H_2 are virtually the same, except at low temperatures, but, as discussed on the H_2 data sheet, the entropy of H_2 includes a mixing term due to the presence of

TABLE 7.6 JANAF Table for H⁺

Enthalpy Reference Temperature = T_r = 298.15 K 	Standard State Pressure = $p°$ = 0.1 MPa

T/K	$C_p°$	$S°$	$-[G° - H°(T_r)]/T$	$H° - H°(T_r)$	$\Delta_f H°$	$\Delta_f G°$	Log K_f
	J K⁻¹mol⁻¹			kJ mol⁻¹			
0	0.	0.	INFINITE	−6.197	1528.085		
100	20.786	86.239	127.427	−4.119			
200	20.786	100.647	110.848	−2.040			
250	20.786	105.285	109.289	−1.001			
298.15	20.786	108.946	108.946	0.	1536.246	1516.990	−265.770
300	20.786	109.075	108.947	0.038	1536.297	1516.871	−264.111
350	20.786	112.279	109.200	1.078	1537.651	1513.526	−225.881
400	20.786	115.055	109.762	2.117	1539.001	1509.987	−197.184
450	20.786	117.503	110.489	3.156	1540.349	1506.280	−174.844
500	20.786	119.693	111.302	4.196	1541.697	1502.422	−156.957
600	20.786	123.483	113.026	6.274	1544.389	1494.314	−130.092
700	20.786	126.687	114.754	8.353	1547.078	1485.755	−110.868
800	20.786	129.463	116.423	10.431	1549.758	1476.812	−96.426
900	20.786	131.911	118.011	12.510	1552.428	1467.533	−85.173
1000	20.786	134.101	119.512	14.589	1555.084	1457.956	−76.156
1100	20.786	136.082	120.930	16.667	1557.721	1448.117	−68.765
1200	20.786	137.891	122.269	18.746	1560.339	1438.038	−62.596
1300	20.786	139.554	123.536	20.824	1562.936	1427.740	−57.367
1400	20.786	141.095	124.735	22.903	1565.511	1417.244	−52.878
1500	20.786	142.529	125.874	24.982	1568.065	1406.564	−48.981
1600	20.786	143.870	126.958	27.060	1570.596	1395.715	−45.565
1700	20.786	145.130	127.990	29.139	1573.107	1384.708	−42.547
1800	20.786	146.319	128.976	31.217	1575.597	1373.554	−39.859
1900	20.786	147.442	129.918	33.296	1578.068	1362.261	−37.451
2000	20.786	148.509	130.821	35.375	1580.520	1350.840	−35.280
2100	20.786	149.523	131.688	37.453	1582.954	1339.296	−33.313
2200	20.786	150.490	132.521	39.532	1585.372	1327.636	−31.522
2300	20.786	151.414	133.322	41.610	1587.774	1315.867	−29.884
2400	20.786	152.298	134.095	43.689	1590.161	1303.994	−28.381
2500	20.786	153.147	134.840	45.768	1592.533	1292.021	−26.995
2600	20.786	153.962	135.560	47.846	1594.891	1279.954	−25.715
2700	20.786	154.747	136.256	49.925	1597.236	1267.797	−24.527
2800	20.786	155.503	136.930	52.004	1599.569	1255.553	−23.423
2900	20.786	156.232	137.583	54.082	1601.889	1243.225	−22.393
3000	20.786	156.937	138.216	56.161	1604.198	1230.818	−21.430
3100	20.786	157.618	138.831	58.239	1606.495	1218.334	−20.529
3200	20.786	158.278	139.429	60.318	1608.781	1205.777	−19.682
3300	20.786	158.918	140.010	62.397	1611.057	1193.148	−18.886
3400	20.786	159.538	140.575	64.475	1613.322	1180.450	−18.135
3500	20.786	160.141	141.125	66.554	1615.576	1167.685	−17.427
3600	20.786	160.726	141.662	68.632	1617.821	1154.857	−16.757
3700	20.786	161.296	142.185	70.711	1620.056	1141.966	−16.122
3800	20.786	161.850	142.695	72.790	1622.281	1129.014	−15.519
3900	20.786	162.390	143.193	74.868	1624.497	1116.005	−14.947
4000	20.786	162.916	143.680	76.947	1626.703	1102.938	−14.403
4100	20.786	163.430	144.155	79.025	1628.899	1089.817	−13.884
4200	20.786	163.931	144.620	81.104	1631.087	1076.642	−13.390
4300	20.786	164.420	145.075	83.183	1633.265	1063.416	−12.918
4400	20.786	164.897	145.520	85.261	1635.435	1050.138	−12.467
4500	20.786	165.365	145.956	87.340	1637.596	1036.812	−12.035
4600	20.786	165.821	146.383	89.418	1639.748	1023.437	−11.621
4700	20.786	166.268	146.801	91.497	1641.891	1010.016	−11.225
4800	20.786	166.706	147.211	93.576	1644.027	996.549	−10.845
4900	20.786	167.135	147.613	95.654	1646.154	983.038	−10.479
5000	20.786	167.555	148.008	97.733	1648.274	969.484	−10.128
5100	20.786	167.966	148.395	99.811	1650.386	955.887	−9.790
5200	20.786	168.370	148.776	101.890	1652.491	942.249	−9.465
5300	20.786	168.766	149.149	103.969	1654.588	928.570	−9.152
5400	20.786	169.154	149.516	106.047	1656.680	914.852	−8.849
5500	20.786	169.536	149.877	108.126	1658.765	901.095	−8.558
5600	20.786	169.910	150.231	110.204	1660.845	887.301	−8.276
5700	20.786	170.278	150.579	112.283	1662.919	873.469	−8.004
5800	20.786	170.640	150.922	114.362	1664.988	859.601	−7.742
5900	20.786	170.995	151.259	116.440	1667.053	845.697	−7.487
6000	20.786	171.344	151.591	118.519	1669.113	831.758	−7.241

PREVIOUS: March 1977 (1 atm) 	CURRENT: March 1982 (1 bar)

7.6 METHODS OF CALCULATING EQUILIBRIUM CONSTANTS

HYDROGEN, ION (H$^+$)   IDEAL GAS   Mr = 1.00739

$S°(298.15\ \text{K}) = 108.946 \pm 0.02\ \text{JK}^{-1}\ \text{mol}^{-1}$

$\Delta_f H°(0\ \text{K}) = 1528.085 \pm 0.04\ \text{kJ mol}^{-1}$
$\Delta_f H°(298.15\ \text{K}) = [1536.246]\ \text{kJ mol}^{-1}$

Enthalpy of Formation

$\Delta_f H°(\text{H}^+, g, 0\ \text{K})$ is calculated from $\Delta_f H°(\text{H}, g, 0\ \text{K})$ (*1*) using the spectroscopic value of IP(H) = 109678.764 ± 0.005 cm^{-1} (1312.0498 ± 0.0001 kJ mol^{-1}) from Moore (*2*). The ionization limit is converted from cm^{-1} to kJ mol^{-1} using the factor, 1 cm^{-1} = 0.01196266 kJ cm^{-1}, which is derived from the 1973 CODATA fundamental constants (*3*). Rosenstock et al. (*4*) and Levin and Lias (*5*) have summarized additional ionization and appearance potential data.

$\Delta_f H°(\text{H}^+, g, 298.15\ \text{K})$ is calculated from $\Delta_f H°(\text{H}, g, 0\ \text{K})$ by using IP(H) with JANAF (*1*) enthalpies, $H°(0\ \text{K}) - H°(298.15\ \text{K})$, for H(g), H$^+$(g), and e$^-$(g). $\Delta_f H°(\text{H} + \text{H}^+ + e^-, 298.15\ \text{K})$ differs from a room temperature threshold energy due to inclusion of these enthalpies and to threshold effects discussed by Rosenstock et al. (*4*). $\Delta_f H°(298.15\ \text{K})$ should be changed by -6.197 kJ mol^{-1} if it is to be used in the ion convention that excludes the enthalpy of the electron.

Heat Capacity and Entropy

The thermodynamic functions of the proton gas are calculated using the recent CODATA fundamental constants (*2*) and assuming that the proton is an ideal monatomic gas. Since there is no electron associated with this species, there is only a translational contribution to the thermochemical function.

References
1. JANAF Thermochemical Tables: H(g), 3-31-82; e$^-$(g), 3-31-82.
2. C. E. Moore, U. S. Nat. Bur. Stand., NSRDS-NBS-3, Section 6, 1972.
3. E. R. Cohen and B. N. Taylor, J. Phys. Chem. Ref. Data *2*, 663 (1973).
4. H. M. Rosenstock, K. Draxl et al., J. Phys. Chem. Ref. Data *6*, Supp. 1 (1977).
5. R. D. Levin and S. G. Lias, U. S. Nat. Bur. Stand., NSRDS-NBS-71, 1982.

---

two distinguishable types of H$_2$. At temperatures above roughly 200 K, an equilibrium sample of H$_2$ is 75% *ortho*, and, from Eq. (6.25), the mixing entropy is

$$\Delta \overline{S}_{\text{mix}} = -R(0.75 \ln 0.75 + 0.25 \ln 0.25) = 4.245\ \text{J mol}^{-1}\ \text{K}^{-1}.$$

The value 216.035 kJ mol^{-1} for $\Delta_f \overline{H}°(\text{H}(g))$ at $T = 0$ K is one-half $\mathcal{D}_0(\text{H}_2)$. The $\Delta_f \overline{H}°(\text{H}(g))$ values increase with T up to 5600 K, then they decrease. The maximum value occurs when $2\overline{C}_P°(\text{H}) = \overline{C}_P°(\text{H}_2)$.

Now we turn to the columns of data relating to free energy and equilibrium constants. To be specific, we will find the equilibrium composition of the reaction mixture

$$\text{H}_2(g) \rightleftarrows 2\text{H}(g) \qquad P = 1\ \text{bar},\ T = 4000\ \text{K}. \tag{7.28}$$

The first step is to find $\Delta_r \overline{G}$ for this reaction using the H$_2$ and H data sheets. There are three ways to proceed.

First, we can use the general expression

$$\Delta_r \overline{G}°(T) = \Delta_r \overline{H}°(T) - T\Delta_r \overline{S}°(T). \tag{7.29}$$

We compute $\Delta_r \overline{H}°$ using Eq. (7.7):

$$\begin{aligned}\Delta_r \overline{H}° &= 2\Delta_f \overline{H}°(\text{H}) - \Delta_f \overline{H}°(\text{H}_2) \\ &= (2)(231.509\ \text{kJ mol}^{-1}) - 0\ \text{kJ mol}^{-1} \\ &= 463.018\ \text{kJ mol}^{-1}\end{aligned}$$

then $\Delta_r \overline{S}°$ using Eq. (7.3):

$$\begin{aligned}\Delta_r \overline{S}° &= 2\overline{S}°(\text{H}) - \overline{S}°(\text{H}_2) \\ &= (2)(168.686\ \text{J K}^{-1}\ \text{mol}^{-1}) - 213.848\ \text{J K}^{-1}\ \text{mol}^{-1} \\ &= -123.524\ \text{J K}^{-1}\ \text{mol}^{-1}\end{aligned}$$

and finally $\Delta_r \overline{G}°$ using Eq. (7.29):

$$\begin{aligned}\Delta_r \overline{G}° &= \Delta_r \overline{H}° - T\Delta_r \overline{S}° \\ &= 463.018\ \text{kJ mol}^{-1} - (4000\ \text{K})(0.123\,524\ \text{kJ K}^{-1}\ \text{mol}^{-1}) \\ &= -31.078\ \text{kJ mol}^{-1}.\end{aligned}$$

TABLE 7.7 JANAF Table for H⁻

Enthalpy Reference Temperature = T_r = 298.15 K
Standard State Pressure = $p°$ = 0.1 MPa

T/K	$C_p°$	$S°$	$-[G° - H°(T_r)]/T$	$H° - H°(T_r)$	$\Delta_f H°$	$\Delta_f G°$	Log K_f
		J K⁻¹mol⁻¹			kJ mol⁻¹		
0	0.	0.	INFINITE	−6.197	143.266		
100	20.786	86.253	127.440	−4.119			
200	20.786	100.661	110.861	−2.040			
250	20.786	105.299	109.302	−1.001			
298.15	20.786	108.960	108.960	0.	139.032	132.282	−23.175
300	20.786	109.089	108.960	0.038	139.006	132.240	−23.025
350	20.786	112.293	109.213	1.078	138.281	131.170	−19.576
400	20.786	115.068	109.776	2.117	137.553	130.203	−17.003
450	20.786	117.517	110.502	3.156	136.823	129.328	−15.012
500	20.786	119.707	111.315	4.196	136.091	128.535	−13.428
600	20.786	123.496	113.039	6.274	134.627	127.161	−11.070
700	20.786	126.701	114.768	8.353	133.158	126.033	−9.405
800	20.786	129.476	116.437	10.431	131.682	125.116	−8.169
900	20.786	131.924	118.024	12.510	130.194	124.384	−7.219
1000	20.786	134.114	119.526	14.589	128.692	123.819	−6.468
1100	10.786	136.096	120.943	16.667	127.173	123.405	−5.860
1200	20.786	137.904	122.283	18.746	125.634	123.130	−5.360
1300	20.786	139.568	123.549	20.824	124.073	122.985	−4.942
1400	20.786	141.108	124.749	22.903	122.491	122.960	−4.588
1500	20.786	142.542	125.888	24.982	120.887	123.050	−4.285
1600	20.786	143.884	126.971	27.060	119.262	123.247	−4.024
1700	20.786	145.144	128.004	29.139	117.615	123.546	−3.796
1800	20.786	146.332	128.989	31.217	115.948	123.943	−3.597
1900	20.786	147.456	129.932	33.296	114.262	124.433	−3.421
2000	20.786	148.522	130.835	35.375	112.557	125.012	−3.265
2100	20.786	149.536	131.701	37.453	110.834	125.677	−3.126
2200	20.786	150.503	132.534	39.532	109.094	126.425	−3.002
2300	20.786	151.427	133.336	41.610	107.339	127.252	−2.890
2400	20.786	152.312	134.108	43.689	105.568	128.156	−2.789
2500	20.786	153.160	134.853	45.768	103.783	129.133	−2.698
2600	20.786	153.976	135.573	47.846	101.984	130.183	−2.615
2700	20.786	154.760	136.269	49.925	100.172	131.302	−2.540
2800	20.786	155.516	136.943	52.004	98.348	132.488	−2.472
2900	20.786	156.245	137.596	54.082	96.511	133.740	−2.409
3000	20.786	156.950	138.230	56.161	94.662	135.055	−2.352
3100	20.786	157.632	138.845	58.239	92.802	136.432	−2.299
3200	20.786	158.292	139.442	60.318	90.931	137.869	−2.250
3300	20.786	158.931	140.023	62.397	89.049	139.365	−2.206
3400	20.786	159.552	140.589	64.475	87.157	140.918	−2.165
3500	20.786	160.154	141.139	66.554	85.255	142.527	−2.127
3600	20.786	160.740	141.675	68.632	83.342	144.190	−2.092
3700	20.786	161.309	142.198	70.711	81.420	145.907	−2.060
3800	20.786	161.864	142.709	72.790	79.488	147.676	−2.030
3900	20.786	162.404	143.207	74.868	77.546	149.496	−2.002
4000	20.786	162.930	143.693	76.947	75.595	151.365	−1.977
4100	20.786	163.443	144.169	79.025	73.635	153.284	−1.953
4200	20.786	163.944	144.634	81.104	71.665	155.250	−1.931
4300	20.786	164.433	145.088	83.183	69.686	157.264	−1.910
4400	20.786	164.911	145.534	85.261	67.699	159.323	−1.891
4500	20.786	165.378	145.969	87.340	65.702	161.428	−1.874
4600	20.786	165.835	146.396	89.418	63.697	163.577	−1.857
4700	20.786	166.282	146.815	91.497	61.684	165.770	−1.842
4800	20.786	166.720	147.225	93.576	59.662	168.007	−1.828
4900	20.786	167.148	147.627	95.654	57.632	170.285	−1.815
5000	20.786	167.568	148.022	97.733	55.594	172.604	−1.803
5100	20.786	167.980	148.409	99.811	53.549	174.965	−1.792
5200	20.786	168.383	148.789	101.890	51.497	177.366	−1.782
5300	20.786	168.779	149.163	103.969	49.437	179.806	−1.772
5400	20.786	169.168	149.530	106.047	47.372	182.285	−1.763
5500	20.786	169.549	149.890	108.126	45.300	184.802	−1.755
5600	20.786	169.924	150.245	110.204	43.222	187.358	−1.748
5700	20.786	170.292	150.593	112.283	41.139	189.950	−1.741
5800	20.786	170.653	150.936	114.362	39.051	192.579	−1.734
5900	20.786	171.009	151.273	116.440	36.958	195.244	−1.729
6000	20.786	171.358	151.605	118.519	34.862	197.944	−1.723

PREVIOUS: March 1977 (1 atm) CURRENT: March 1982 (1 bar)

| HYDROGEN, ION (H⁻) | IDEAL GAS | $M_r = 1.00849$ |

$EA(H, g) = 0.754209 \pm 0.000003$ eV
$S°(298.15 \text{ K}) = 108.960 \pm 0.017$ J K⁻¹ mol⁻¹

$\Delta_f H°(0 \text{ K}) = 143.27 \pm 0.02$ kJ mol⁻¹
$\Delta_f H°(298.15 \text{ K}) = [139.032]$ kJ mol⁻¹

Electronic Level and Quantum Weight

State	ϵ_1, cm⁻¹	g_1
¹S₀	0.0	1

Enthalpy of Formation

$\Delta_f H°(H^-, g, 0 K)$ is calculated from $\Delta_f H°(H, g, 0 K)$ (*1*) using the adopted electron affinity of $EA(H) = 0.754209 \pm 0.000003$ eV (72.7695 ± 0.0003 kJ mol⁻¹). This value, recommended by Hotop and Lineberger (*2*), is based on extensive Hylleras-type variational calculations on two electon systems (*3, 4, 5*). Additional information on H⁻(g) may be obtained in the critical discussions of Hotop and Lineberger (*2, 6*), Rosenstock et al. (*7*) and Massey (*8*). Experimentally, Dehmer and Chupka (*9*) have reported $EA(H) \geq 0.7540 \pm 0.0003$ eV.

$\Delta_f H°(H^-, g, 298.15 K)$ is obtained from $\Delta_f H°(H, g, 0 K)$ by using $EA(H)$ with JANAF (*1*) enthalpies, $H°(0 K) - H°(298.15 K)$, for H⁻(g), H(g), and e⁻(g). $\Delta_f H°(H^- + H + e^-, 298.15 K)$ differs from a room-temperature threshold energy due to inclusion of these enthalpies and to threshold effects discussed by Rosenstock et al (*7*). $\Delta_f H°(298.15 K)$ should be changed by $+6.197$ kJ mol⁻¹ if it is to be used in the ion convention that excludes the enthalpy of the electron.

Heat Capacity and Entropy

The ground state electronic configuration for H⁻(g) is given by Hotop and Lineberger (*2, 6*) and Rosenstock et al. (*7*). A comparison of the isoelectronic sequence—H⁻(g), He(g), Li⁺(g)—would suggest that stable electronic states may exist at 0.8 EA(H) or roughly 6400 cm⁻¹. This would greatly affect the entropy. However, Pekeris (*4*) states that he was unable to find any bound states. In addition, Seman and Branscomb (*10*) state that theoretical and semiempirical evidence suggests that atomic negative ions have very few if any excited states below the continuum. We assume no stable excited states exist.

References
1. JANAF Thermochemical Tables: H(g) and e⁻(g), 3-31-82.
2. H. Hotop and W. C. Lineberger, J. Phys. Chem. Ref. Data, to be published, 1985.
3. C. L. Pekeris, Phys. Rev. *112*, 1649 (1958).
4. C. L. Pekeris, Phys. Rev. *126*, 1470 (1962).
5. K. Aaskamar, Nucl. Instr. Meth. *90*, 263 (1970).
6. H. Hotop and W. C. Lineberger, J. Phys. Chem. Ref. Data *4*, 539 (1975).
7. H. M. Rosenstock, K. Draxel et al., J. Phys. Chem. Ref. Data *6*, Supp. 1 (1977).
8. H. S. W. Massey, "Negative Ions", 3rd ed., Cambridge University Press, Cambridge, 1976.
9. P. M. Dehmer and W. A. Chupka, Bull. Am. Phys. Soc. *20*, 729 (1975).
10. M. L. Seman and L. M. Branscomb, Phys. Rev. *125*, 1602 (1962).

Note how the large negative $T\Delta_r \overline{S}°$ term has dominated the positive $\Delta_r \overline{H}°$ term making $\Delta_r \overline{G}°$ negative.

A second method uses *standard Gibbs free energies of formation*, which are the values in column 7 of the tables. Defined analogously to enthalpies of formation, these quantities yield $\Delta_r \overline{G}°$ from an expression analogous to Eq. (7.7)

$$\Delta_r \overline{G}° = \sum_{\text{products}} \nu_i \Delta_f \overline{G}°(\text{product } i) - \sum_{\text{reactants}} \nu_i \Delta_f \overline{G}°(\text{reactant } i) . \qquad (7.30)$$

We find

$$\begin{aligned} \Delta_r \overline{G}° &= 2\Delta_f \overline{G}°(H) - \Delta_f \overline{G}°(H_2) \\ &= (2)(-15.541 \text{ kJ mol}^{-1}) - 0 \text{ kJ mol}^{-1} \\ &= -31.082 \text{ kJ mol}^{-1} , \end{aligned}$$

which is essentially the value calculated by the first method.

A third method uses the quantities in columns 4 and 6. Column 4 tabulates the *negative* of a quantity known variously as the *Giauque function*, the *Gibbs energy function* (Gef), or the *free energy function* (fef):

$$\text{fef} = \left[\frac{\overline{G}°(T) - \overline{H}°(T_r = 298.15 \text{ K})}{T} \right] . \qquad (7.31)$$

The fef is inherently negative, and $-$fef is tabulated to make all entries positive. The fef is a slowly varying function of temperature (compared to $\Delta_r \overline{G}°$ or enthalpy

TABLE 7.8 JANAF Table for H

Enthalpy Reference Temperature = t_r = 298.15 K Standard State Pressure = $p°$ = 0.1 MPa

T/K	$C_p°$	S°	$-[G° - H°(T_r)]/T$	$H° - H°(T_r)$	$\Delta_f H°$	$\Delta_f G°$	Log K_f
	J K^{-1}mol^{-1}			kJ mol^{-1}			
0	0.	0.	INFINITE	−6.197	216.035	216.035	INFINITE
100	20.786	92.009	133.197	−4.119	216.614	212.450	−110.972
200	20.786	106.417	116.618	−2.040	217.346	208.004	−54.325
250	20.786	111.055	115.059	−1.001	217.687	205.629	−42.964
298.15	20.786	114.716	114.716	0.	217.999	203.278	−35.613
300	20.786	114.845	114.717	0.038	218.011	203.186	−35.378
350	20.786	118.049	114.970	1.078	218.326	200.690	−29.951
400	20.786	120.825	115.532	2.117	218.637	198.150	−25.876
450	20.786	123.273	116.259	3.156	218.946	195.570	−22.701
500	20.786	125.463	117.072	4.196	219.254	192.957	−20.158
600	20.786	129.253	118.796	6.274	219.868	187.640	−16.335
700	20.786	132.457	120.524	8.353	220.478	182.220	−13.597
800	20.786	135.232	122.193	10.431	221.080	176.713	−11.538
900	20.786	137.681	123.781	12.510	221.671	171.132	−9.932
1000	20.786	139.871	125.282	14.589	222.248	165.485	−8.644
1100	20.786	141.852	126.700	16.667	222.807	159.782	−7.587
1200	20.786	143.660	128.039	18.746	223.346	154.028	−6.705
1300	20.786	145.324	129.305	20.824	223.865	148.230	−5.956
1400	20.786	146.865	130.505	22.903	224.361	142.394	−5.313
1500	20.786	148.299	131.644	24.982	224.836	136.522	−4.754
1600	20.786	149.640	132.728	27.060	225.289	130.620	−4.264
1700	20.786	150.900	133.760	29.139	225.721	124.689	−3.831
1800	20.786	152.088	134.745	31.217	226.132	118.734	−3.446
1900	20.786	153.212	135.688	33.296	226.525	112.757	−3.100
2000	20.786	154.278	136.591	35.375	226.898	106.760	−2.788
2100	20.786	155.293	137.458	37.453	227.254	100.744	−2.506
2200	20.786	156.260	138.291	39.532	227.593	94.712	−2.249
2300	20.786	157.184	139.092	41.610	227.916	88.664	−2.014
2400	20.786	158.068	139.864	43.689	228.224	82.603	−1.798
2500	20.786	158.917	140.610	45.768	228.518	76.530	−1.599
2600	20.786	159.732	141.330	47.846	228.798	70.444	−1.415
2700	20.786	160.516	142.026	49.925	229.064	64.349	−1.245
2800	20.786	161.272	142.700	52.004	229.318	58.243	−1.087
2900	20.786	162.002	143.353	54.082	229.560	52.129	−0.939
3000	20.786	162.706	143.986	56.161	229.790	46.007	−0.801
3100	20.786	163.388	144.601	58.239	230.008	39.877	−0.672
3200	20.786	164.048	145.199	60.318	230.216	33.741	−0.551
3300	20.786	164.688	145.780	62.397	230.413	27.598	−0.437
3400	20.786	165.308	146.345	64.475	230.599	21.449	−0.330
3500	20.786	165.911	146.895	66.554	230.776	15.295	−0.228
3600	20.786	166.496	147.432	68.632	230.942	9.136	−0.133
3700	20.786	167.066	147.955	70.711	231.098	2.973	−0.042
3800	20.786	167.620	148.465	72.790	231.244	−3.195	0.044
3900	20.786	168.160	148.963	74.868	231.381	−9.366	0.125
4000	20.786	168.686	149.450	76.947	231.509	−15.541	0.203
4100	20.786	169.200	149.925	79.025	231.627	−21.718	0.277
4200	20.786	169.700	150.390	81.104	231.736	−27.899	0.347
4300	20.786	170.190	150.845	83.183	231.836	−34.082	0.414
4400	20.786	170.667	151.290	85.261	231.927	−40.267	0.478
4500	20.786	171.135	151.726	87.340	232.009	−46.454	0.539
4600	20.786	171.591	152.153	89.418	232.082	−52.643	0.598
4700	20.786	172.038	152.571	91.497	232.147	−58.834	0.654
4800	20.786	172.476	152.981	93.576	232.204	−65.025	0.708
4900	20.786	172.905	153.383	95.654	232.253	−71.218	0.759
5000	20.786	173.325	153.778	97.733	232.294	−77.412	0.809
5100	20.786	173.736	154.165	99.811	232.327	−83.606	0.856
5200	20.786	174.140	154.546	101.890	232.353	−89.801	0.902
5300	20.786	174.536	154.919	103.969	232.373	−95.997	0.946
5400	20.786	174.924	155.286	106.047	232.386	−102.192	0.989
5500	20.786	175.306	155.646	108.126	232.392	−108.389	1.029
5600	20.786	175.680	156.001	110.204	232.393	−114.584	1.069
5700	20.786	176.048	156.349	112.283	232.389	−120.780	1.107
5800	20.786	176.410	156.692	114.362	232.379	−126.976	1.144
5900	20.786	176.765	157.029	116.440	232.365	−133.172	1.179
6000	20.786	177.114	157.361	118.519	232.348	−139.368	1.213

PREVIOUS: March 1977 (1 atm) CURRENT: March 1982 (1 bar)

HYDROGEN (H)   IDEAL GAS $A_r = 1.00794$

$S°(298.15 \text{ K}) = 114.716 \pm 0.017 \text{ J K}^{-1} \text{ mol}^{-1}$

$\Delta_f H°(0 \text{ K}) = 216.035 \pm 0.006 \text{ kJ mol}^{-1}$
$\Delta_f H°(298.15 \text{ K}) = 217.999 \pm 0.006 \text{ kJ mol}^{-1}$

Electronic Level and Quantum Weight

State	ϵ_1, cm^{-1}	g_1
$^2S_{1/2}$	0.00	2

Enthalpy of Formation

The enthalpy of formation is calculated from the dissociation energy, $D_0^0(H_2) = 36118.3 \pm 1$ cm^{-1} (432.071 $\pm$ 0.012 kJ mol^{-1}), from Herzberg (2) and auxiliary data from H$_2$ (3). The adopted value for the dissociation energy of hydrogen is the value recommended by CODATA (1).

Earlier experimental values for $D_0^0(H_2)$ were obtained by Herzberg and Monfils (36113.0 $\pm$ 0.3 cm^{-1}, 4) and Beutler (36116 $\pm$ 6 cm^{-1}, 5). Kolos and Wolniewicz (6) calculated the adiabatic dissociation energy of H$_2$, corrected for relativistic and radiative effects, to be 36117.4 cm^{-1}.

Heat Capacity and Entropy

The electronic levels for H(g) are given in the compilation by Moore (7). Our calculations indicate that the inclusion of levels through n = 12 has no effect on the thermodynamic functions to 6000 K. This is a result of the high energy of these levels; the first excited state lies at 82258 cm^{-1} above the ground state. Since the inclusion of these upper levels has no effect on the thermodynamic functions (to 6000 K) we list only the ground state. The reported uncertainty in S°(298.15 K) is due to uncertainties in the relative atomic mass and the fundamental constants. Extension of these calculations above 6000 K may require consideration of the higher excited states and utilization of proper fill and cut off procedures (8).

The thermal functions at 298.15 K differ from the CODATA recommendations for two reasons: a difference of 0.001 J K^{-1} mol^{-1} in the entropy due to the use of more current fundamental constants and a difference of 0.1094 J K^{-1} mol^{-1} in the entropy due to the use of a different standard state pressure.

References
1. J. D. Cox, chairman, ICSU-CODATA Task Group on Key Values for Thermodynamics, J. Chem. Thermodynamics *10*, 903 (1978).
2. G. Herzberg, J. Mol. Spectrosc. *33*, 147 (1970).
3. JANAF Thermochemical Tables: H$_2$(g), 3-31-77.
4. G. Herzberg and A. Monfils, J. Mol. Spectrosc. *5*, 482 (1960).
5. H. Beutler, Z. Phys. Chem. *B39*, 315 (1935).
6. W. Kolos and L. Wolniewicz, J. Chem. Phys. *49*, 404 (1968).
7. C. E. Moore, NSRDS-NBS 3, Section 6 (1972).
8. J. R. Downey, Jr., Dow Chemical Company, Thermal Research, to be published, 1977.
9. E. R. Cohen and B. Taylor, J. Phys. Chem. Ref. Data *2*, 663 (1973).

values themselves) so that interpolation at nontabulated temperatures has good accuracy. (Note that there is currently no agreed-upon symbol or name for this function.)

The fef is used as follows. We first compute $\Delta_r(\text{fef})$, remembering to add a minus sign to the tabulated values:

$$\Delta_r(\text{fef}) = (2)(-149.450 \text{ J K}^{-1} \text{ mol}^{-1}) - (-182.129 \text{ J K}^{-1} \text{ mol}^{-1})$$
$$= -116.771 \text{ J K}^{-1} \text{ mol}^{-1}$$

then, using $\Delta_f \overline{H}°(T_r)$, the *fifth* entries in column 6, we compute $\Delta_r \overline{H}°(T_r)$:

$$\Delta_r \overline{H}°(T_r) = (2)(217.999 \text{ kJ mol}^{-1}) - 0 \text{ kJ mol}^{-1}$$
$$= 435.998 \text{ kJ mol}^{-1}$$

and finally compute $\Delta_r \overline{G}°(4000 \text{ K})$ from

$$\Delta_r \overline{G}° = T(\Delta_r(\text{fef})) + \Delta_r \overline{H}°(298 \text{ K}) \qquad (7.32)$$

yielding

$$\Delta_r \overline{G}° = (4000 \text{ K})(-116.771 \text{ J K}^{-1} \text{ mol}^{-1}) + 435.998 \text{ kJ mol}^{-1}$$
$$= -31.086 \text{ kJ mol}^{-1}$$

in accord with the first and second methods.

Now we can use Eq. (7.24) to compute K_{eq}:

$$K_{eq} = \exp(-\Delta_r \overline{G}°/RT)$$
$$= 2.545\ 8 \qquad \text{from Eq. (7.29)}$$
$$= 2.546\ 1 \qquad \text{from Eq. (7.30)}$$
$$= 2.546\ 4 \qquad \text{from Eq. (7.32)}$$

TABLE 7.9 JANAF Table for H_2

Enthalpy Reference Temperature = T_r = 298.15 K
Standard State Pressure = $p°$ = 0.1 MPa

T/K	$C_P°$	$S°$	$-[G° - H°(T_r)]/T$	$H° - H°(T_r)$	$\Delta_f H°$	$\Delta_f G°$	Log K_f
	J K^{-1}mol^{-1}			kJ mol^{-1}			
0	0.	0.	INFINITE	−8.467	0.	0.	0.
100	28.154	100.727	155.408	−5.468	0.	0.	0.
200	27.447	119.412	133.284	−2.774	0.	0.	0.
250	28.344	125.640	131.152	−1.378	0.	0.	0.
298.15	28.836	130.680	130.680	0.	0.	0.	0.
300	28.849	130.858	130.680	0.053	0.	0.	0.
350	29.081	135.325	131.032	1.502	0.	0.	0.
400	29.181	139.216	131.817	2.959	0.	0.	0.
450	29.229	142.656	132.834	4.420	0.	0.	0.
500	29.260	145.737	133.973	5.882	0.	0.	0.
600	29.327	151.077	136.392	8.811	0.	0.	0.
700	29.441	155.606	138.822	11.749	0.	0.	0.
800	29.624	159.548	141.171	14.702	0.	0.	0.
900	29.881	163.051	143.411	17.676	0.	0.	0.
1000	30.205	166.216	145.536	20.680	0.	0.	0.
1100	30.581	169.112	147.549	23.719	0.	0.	0.
1200	30.992	171.790	149.459	26.797	0.	0.	0.
1300	31.423	174.288	151.274	29.918	0.	0.	0.
1400	31.861	176.633	153.003	33.082	0.	0.	0.
1500	32.298	178.846	154.652	36.290	0.	0.	0.
1600	32.725	180.944	156.231	39.541	0.	0.	0.
1700	33.139	182.940	157.743	42.835	0.	0.	0.
1800	33.537	184.846	159.197	46.169	0.	0.	0.
1900	33.917	186.669	160.595	49.541	0.	0.	0.
2000	34.280	188.418	161.943	52.951	0.	0.	0.
2100	34.624	190.099	163.244	56.397	0.	0.	0.
2200	34.952	191.718	164.501	59.876	0.	0.	0.
2300	35.263	193.278	165.719	63.387	0.	0.	0.
2400	35.559	194.785	166.899	66.928	0.	0.	0.
2500	35.842	196.243	168.044	70.498	0.	0.	0.
2600	36.111	197.654	169.155	74.096	0.	0.	0.
2700	36.370	199.021	170.236	77.720	0.	0.	0.
2800	36.618	200.349	171.288	81.369	0.	0.	0.
2900	36.856	201.638	172.313	85.043	0.	0.	0.
3000	37.087	202.891	173.311	88.740	0.	0.	0.
3100	37.311	204.111	174.285	92.460	0.	0.	0.
3200	37.528	205.299	175.236	96.202	0.	0.	0.
3300	37.740	206.457	176.164	99.966	0.	0.	0.
3400	37.946	207.587	177.072	103.750	0.	0.	0.
3500	38.149	208.690	177.960	107.555	0.	0.	0.
3600	38.348	209.767	178.828	111.380	0.	0.	0.
3700	38.544	210.821	179.679	115.224	0.	0.	0.
3800	38.738	211.851	180.512	119.089	0.	0.	0.
3900	38.928	212.860	181.328	122.972	0.	0.	0.
4000	39.116	213.848	182.129	126.874	0.	0.	0.
4100	39.301	214.816	182.915	130.795	0.	0.	0.
4200	39.484	215.765	183.686	134.734	0.	0.	0.
4300	39.665	216.696	184.442	138.692	0.	0.	0.
4400	39.842	217.610	185.186	142.667	0.	0.	0.
4500	40.017	218.508	185.916	146.660	0.	0.	0.
4600	40.188	219.389	186.635	150.670	0.	0.	0.
4700	40.355	220.255	187.341	154.698	0.	0.	0.
4800	40.518	221.106	188.035	158.741	0.	0.	0.
4900	40.676	221.943	188.719	162.801	0.	0.	0.
5000	40.829	222.767	189.392	166.876	0.	0.	0.
5100	40.976	223.577	190.054	170.967	0.	0.	0.
5200	41.117	224.374	190.706	175.071	0.	0.	0.
5300	41.252	225.158	191.349	179.190	0.	0.	0.
5400	41.379	225.931	191.982	183.322	0.	0.	0.
5500	41.498	226.691	192.606	187.465	0.	0.	0.
5600	41.609	227.440	193.222	191.621	0.	0.	0.
5700	41.712	228.177	193.829	195.787	0.	0.	0.
5800	41.806	228.903	194.427	199.963	0.	0.	0.
5900	41.890	229.619	195.017	204.148	0.	0.	0.
6000	41.965	230.323	195.600	208.341	0.	0.	0.

PREVIOUS: March 1961 (1 atm)
CURRENT: March 1977 (1 bar)

HYDROGEN (H$_2$) REFERENCE STATE – IDEAL GAS $M_r = 2.01588$
0 to 6000 K Ideal Gas

$D_0^0 = 432.071 \pm 0.012$ kJ mol^{-1}
$S°(298.15 \text{ K}) = 130.680 \pm 0.033$ J K^{-1} mol^{-1}

$\Delta_f H°(0 \text{ K}) = 0$ kJ mol^{-1}
$\Delta_f H°(298.15 \text{ K}) = 0$ kJ mol^{-1}

Vibrational and Rotational Levels (cm^{-1})

Direct Summation of Electronic Ground States

$E = G - G_0 + F = G - G_0 + BZ - DZ^2 + HZ^3 - LZ^4 + \cdots \approx G - G_0 + BZ - DZ^2 + H^2Z^3/(H+LZ)$,
where $Z = J(J+1)$, $Y = v + 1/2$, and we omit subscript v on G, F, B, D, H, and L

$G = 4403.566Y - 123.8573Y^2 + 1.87269Y^3 - 0.173514Y^4 + 9.93128 \times 10^{-3}Y^5 - 4.38015 \times 10^{-4}Y^6$

$B = 60.8904 - 3.16997Y + 0.155932Y^2 - 4.60094 \times 10^{-2}Y^3 + 8.72205 \times 10^{-3}Y^4 - 9.59207 \times 10^{-4}Y^5 + 5.31722 \times 10^{-5}Y^6 - 1.21393 \times 10^{-6}Y^7$

$D = 4.6573 \times 10^{-2} - 1.5085 \times 10^{-3}Y - 2.7385 \times 10^{-4}Y^2 + 1.0242 \times 10^{-4}Y^3 - 1.172 \times 10^{-5}Y^4 + 4.684 \times 10^{-7}Y^5$

$H = 5.224 \times 10^{-5} - 7.240 \times 10^{-6}Y + 9.619 \times 10^{-7}Y^2 - 4.838 \times 10^{-8}Y^3$

$L = 6.70 \times 10^{-8} - 1.426 \times 10^{-8}Y + 1.388 \times 10^{-9}Y^2$

$v_{max} = 14$, $J_{max} = 38 - 32 v/v_{max}$
Normalized statistical weights = 1/4 (even J) and 3/4 (odd J)

Ground State Configuration: $^1\Sigma_g^+$ $r_e = 0.7414$ Å

Enthalpy of Formation
Zero by definition. Refer to the monatomic hydrogen gas table for a discussion of the dissociation energy.

Heat Capacity and Entropy
These are calculated by direct summation over vibration-rotation energy levels of the electronic ground state. We performed the direct summation with an extended version of a program written by W. H. Evans and provided through cooperation of D. D. Wagman, both of the U. S. National Bureau of Standards. Contributions of excited states ($T_0 > 90000$ cm^{-1}) are negligible at 6000 K. Polynomials G, B, D, and H are our fits of data from Stoicheff (*1*), Herzberg and Howe (*2*) and Rank et al. (*3, 4*). We estimate polynomial L such that our approximation (*5, 6*) for the infinite series F yields high-J rotational levels in reasonable agreement with the theoretical values of Waech and Bernstein (*7*). Maximum deviations in our F values are about ±400 cm^{-1}; these occur near $1 \le v \le 4$ and $29 \le J \le 36$ which is far into the extrapolated region of F. Only about one-third of the vibration-rotation levels have been observed spectroscopically and the theoretical calculations (*7, 6*) provide the best available extrapolation to high values of J. Accuracy of the thermodynamic functions near 6000 K depends on this extrapolation and on the rotational cutoff procedure. We assume a linear approximation (*5*) for the limiting values (J_{max}) of rotational quantum number. Values in the J_{max} equation are estimated from theoretical calculations (*7, 6*). We omit the nuclear-spin contribution (R ℓn 4) to entropy and Gibbs energy function.

We adopt ortho-para "equilibrium" H$_2$ as the reference state at all temperatures. The previous JANAF reference state (*8*) referred to "normal" H$_2$ (75% ortho and 25% para). Our new reference state has significant changes in $C_P°$, $S°$ and $H°(T) - H°(298.15 \text{ K})$ at 100 K and slight changes at 200 K. Use of "equilibrium" H$_2$ as a reference state was proposed on the NBS H$_2$ table (*9*) which discusses three alternatives. Preferred alternatives are either "equilibrium" H$_2$ or "normal" H$_2$. "Normal" H$_2$ is the form always encountered except in low-temperature generation or catalytic ortho-para equilibrium. Use of "normal" H$_2$ involves a possible complication, depending on the choice of zero energy for ortho-H$_2$. If we chose the lowest allowed level (v = 0, J = 1) instead of v = 0, J = 0), then $H°(298.15 \text{ K}) - H°(0 \text{ K})$ would be 0.254 kcal mol^{-1} less for "normal" than for "equilibrium" H$_2$. This would change the difference between $\Delta_f H°(0 \text{ K})$ and $\Delta_f H°(298.15 \text{ K})$ for all species involving hydrogen (*8, 10*). No such change would occur if we chose the lowest level (v = 0, J = 0) as the energy zero for ortho-H$_2$. "Equilibrium" H$_2$ is the form which parallels most substances, i.e., those maintaining equilibrium among all rotational levels (*9*).

JANAF values and uncertainties at 298.15 K are the same as those selected by CODATA (*10*). Previous thermochemical tables based on direct-summation calculations of Woolley et al. (*6*) include that of Gurvich et al. (*11*) and that of NBS-JANAF (*9, 8*). Differences of the new JANAF values from the NBS table are greatest near 5300 K, reaching maximum of 0.005 cal K^{-1} mol^{-1} in $S°$ and 0.018 kcal mol^{-1} in $H°(T) - H°(298.15 \text{ K})$. Differences from Gurvich et al. at 6000 K are 0.006 cal K^{-1} mol^{-1} in $S°$ and 0.030 kcal mol^{-1} in $H°(T) - H°(298.15 \text{ K})$. Errors larger than these differences arise from uncertainty in extrapolation of the rotational levels.

References
1. B. P. Stoicheff, Can. J. Phys. *35*, 730 (1957).
2. G. Herzberg and L. L. Howe, J. Phys. *37*, 636 (1959).
3. J. V. Foltz, D. H. Rank, and T. A. Wiggins, J. Mol. Spectrosc. *21*, 203 (1966).
4. U. Fink, T. A. Wiggins and D. H. Rank, J. Mol. Spectrosc. *18*, 384 (1965).
5. G. A. Khachkuruzov, Opt. Spectrosc. *30*, 455 (1971).
6. H. W. Woolley, R. B. Scott, and F. G. Brickwedde, J. Res. Natl. Bur. Stand. *41*, 379 (1948).
7. T. G. Waech and R. B. Bernstein, J. Chem. Phys. *46*, 4905 (1967).
8. H. Prophet and D. R. Stull, ed., JANAF Thermochemical Tables, 2 ed., NSRDS-NBS 37, 1971.
9. S. Abramowitz et al., U. S. Natl. Bur. Stand. Rept. 10904, 239, July, 1972.
10. J. D. Cox, chairman, ICSU-CODATA Task Group on Key Values for Thermodynamics, J. Chem. Thermodynamics *4*, 331 (1972); *8*, 603 (1976).
11. L. V. Gurvich, G. A. Khachkuruzov et al., "Thermodynamic Properties of Individual Substances", 2nd ed., Vol. II, Nauka, Moscow, 1962.

We can compute K_{eq} in a fourth way from the values in column 8 labeled Log K_f, which is

$$\text{Log } K_f = \log_{10} K_f = -\Delta_f \overline{G}°/(RT \ln 10) = -\Delta_f \overline{G}°/(2.303\, RT). \quad (7.33)$$

We combine Log K_f values in the usual "products − reactants" fashion and find K_{eq} from

$$K_{eq} = 10^{\Delta_r(\text{Log } K_f)}. \quad (7.34)$$

Here $\Delta_r(\text{Log } K_f) = 2(0.203) - 0 = 0.406$ and

$$K_{eq} = 10^{0.406} = 2.546\,8. \quad \text{from Eq. (7.34)}$$

The slight variation in K_{eq} from method to method is common to the use of tabulated data and reflects both the experimental uncertainties of the data as well as their finite precision.

Now we can use the equilibrium constant to find the equilibrium composition of the reaction mixture, Eq. (7.28). We will assume that H_2 and H can be represented by an ideal mixture of ideal gases so that fugacities equal partial pressures (with P, the total pressure, fixed at 1 bar). We have, in general,

$$K_{eq} = \frac{P_H^2}{P_{H_2}} = \frac{x_H^2 P}{x_{H_2}} = \frac{n_H^2 P}{n_{H_2}(n_H + n_{H_2})}. \quad (7.35)$$

Since the initial composition is $n_{H_2}° = 1$ mol and $n_H° = 0$, we can introduce ξ using Eq. (7.1)

$$\xi = \frac{n_{H_2}° - n_{H_2}}{1} = \frac{n_H - n_H°}{2} = 1 - n_{H_2} = \frac{n_H}{2}$$

so that (note that ξ varies between 0 and 1 mol)

$$n_H = 2\xi \quad \text{and} \quad n_{H_2} = 1 - \xi.$$

Substituting these expressions into Eq. (7.35) gives

$$K_{eq} = \frac{4\xi_{eq}^2 P}{1 - \xi_{eq}^2} \quad (7.36)$$

or

$$\xi_{eq} = \left(\frac{4P}{K_{eq}} + 1\right)^{-1/2}. \quad (7.37)$$

With $P = 1$ bar and $K_{eq} = 2.546$, we calculate $\xi_{eq} = 0.624$ mol. At equilibrium, $n_{H_2} = 1$ mol $- \xi_{eq} = 0.376$ mol and $n_H = 2\xi_{eq} = 1.248$ mol.

It is instructive at this point to go back to Eq. (7.14) and follow the course of the total Gibbs free energy as ξ advances towards the equilibrium value. We write

$$G(\xi) = n_{H_2}\mu_{H_2} + n_H \mu_H$$
$$= n_{H_2}° \mu_{H_2} + n_H° \mu_H + \xi(2\mu_H - \mu_{H_2}).$$

Since $P = P°$, the standard pressure, the general expression for each μ simplifies to:

$$\mu_i = \mu_i°(T) + RT \ln(P_i/P°) = \mu_i°(T) + RT \ln x_i.$$

Substituting expressions like this for H and for H_2 into the $G(\xi)$ expression and writing x_i in terms of ξ gives, for $n_{H_2}° = 1$ mol and $n_H° = 0$,

$$\frac{\overline{G}(\xi) - \mu_{H_2}°}{RT} = \ln\left(\frac{1-\xi}{1+\xi}\right) + \xi\left[\frac{\Delta_r \overline{G}°}{RT} + \ln\left(\frac{4\xi^2}{1-\xi^2}\right)\right]\left(\frac{1}{1\text{ mol}}\right),$$

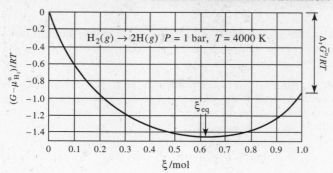

FIGURE 7.3 One mole of H_2, held at 1 bar and heated to 6000 K, follows this Gibbs free-energy curve as it approaches equilibrium. The equilibrium degree of advancement determines the composition of this reaction mixture, and it is found at the minimum point of the free-energy curve.

which describes the change in G away from the initial state of 1 mol H_2. This expression is plotted in Figure 7.3, where one can see that ξ_{eq} does locate the minimum of $\overline{G}(\xi)$ and that $\overline{G}(\xi = 1) - \overline{G}(\xi = 0) = \Delta_r \overline{G}°$.

These calculations have been based on 1 bar total pressure conditions. Had the pressure been lower, Eq. (7.37) shows that ξ_{eq} would have been larger, implying more complete dissociation of H_2. Figure 7.4 plots the variation in x_H with P for this reaction for a variety of equilibrium constants. As this figure shows, dissociation is favored by a combination of high temperatures (large K_{eq}) and low pressures. The effect of increasing temperature is understandable, since the reaction is endothermic. The pressure variation of the equilibrium *composition* at any one temperature is as predicted by the elementary Le Châtelier Principle. An increase in pressure evokes a response from the system that tries to decrease the pressure by reducing the number of gas-phase particles.

This survey of the JANAF tables has indicated the general use of tabulated data in equilibrium constant calculations. Other aspects of the JANAF tables are worth mentioning. Since H_2 does not liquify until 20 K at 1 bar, Table 7.9 is confined to the gas phase. For compounds with higher melting and boiling points, separate tables are available for each phase that extrapolate properties into temperature regions of phase instability, and one reference-state table is included that lists heat capacities, etc., for only the stable phase at each tabulated temperature.

EXAMPLE 7.11

A convenient way to make $F(g)$ is to exploit the small $F_2(g)$ bond energy. Raising the temperature of $F_2(g)$ makes thermal dissociation an increasingly efficient process, but the high reactivity of $F(g)$ limits the maximum temperature. In a Ni tube passivated by a surface layer of nickel fluoride, the maximum temperature is roughly 1000 K. Above this, the

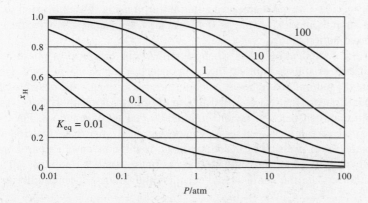

FIGURE 7.4 The mole fraction of hydrogen atoms varies rapidly with total pressure (note the logarithmic scale) at any of a wide variety of K_{eq} values (i.e., at any of a variety of temperatures).

protective layer sublimes rapidly, and the Ni beneath is rapidly attacked. An example of a typical calculation related to this sort of technique is the following: at what pressure is one-half the initial $F_2(g)$ dissociated at 1000 K?

SOLUTION This is formally the same problem as the H_2 dissociation discussed in the text. Imagine starting with 1 mol $F_2(g)$. (The initial amount cannot matter, since we are asked for only intensive quantities—a pressure and a dissociation *fraction*.) Then "one-half dissociated" means, at equilibrium,

$$n_{F_2} = \frac{1}{2} = 1 - \xi_{eq} \quad \text{or} \quad \xi_{eq} = \frac{1}{2}.$$

Thus, since $K_{eq} = 4\xi_{eq}^2 P/(1 - \xi_{eq}^2)$, $K_{eq} = 4P/3$. A consultation of JANAF or similar tables shows that, at 1000 K, $\Delta_f \overline{G}°(F(g))/RT = 2.323$, and thus $K_{eq} = \exp(-2.323) = 9.80 \times 10^{-2}$. The required pressure is $3K_{eq}/4 = 7.35 \times 10^{-2}$ bar.

➡ RELATED PROBLEMS 7.38, 7.46, 7.50

In general, calculating the equilibrium composition of a reaction mixture is often not as simple a task as it was for the hydrogen dissociation problem. One frequently is confronted with a polynomial expression for ξ_{eq}, or a transcendental equation. Such equations must be solved by trial and error successive approximation methods or by other numerical techniques best done by a suitable computer algorithm.

For example, the common and important problem of the water solubility of a simple electrolyte is often much more complicated than you may have been led to believe in elementary chemistry courses. Consider the solubility of $KBrO_3(s)$. At 25 °C, the equilibrium constant for the process

$$KBrO_3(s) \rightleftarrows K^+(aq) + BrO_3^-(aq)$$

is 0.072 if the dilute solution, unit molality standard state is used. As discussed in detail in Chapter 10,

$$K_{eq} = a_+ a_- = \gamma_+ \gamma_- m_+ m_- = 0.072$$

The elementary chemistry approximation sets the activity coefficients to 1 and predicts

$$m_+ = m_- K_{eq}^{1/2} = 0.27 \text{ mol kg}^{-1}.$$

In fact, the $\gamma_+ \gamma_-$ product is far from unity in this solution and is given by the transcendental expression

$$\ln \gamma_+ \gamma_- = \frac{-2.34 \, m^{1/2}}{(1 + m^{1/2})} - 0.184 \, m - 0.51 \, m^2$$

where $m = m_+ = m_-$. A numerical solution for m using this expression yields the equilibrium molalities $m_+ = m_- = 0.48$ mol kg^{-1}, a value nearly twice the elementary approximation value.

A further complication arises when more than one chemical or phase transition equilibrium is established among the components of a reacting mixture. One must then write the minimum number of reactions needed to specify the possible chemical or physical transformations and find the equilibrium composition that satisfies all the equilibrium constant expressions implied by that set of reactions. For example, if the partial pressure of a gaseous component of a gas-phase reaction mixture achieves the vapor pressure of that component, it will condense. Its partial pressure will be stuck at its vapor pressure and the gas-phase composition must adjust itself to take into account the simultaneous reaction and condensation equilibria.

But however complicated the system, the following steps will (eventually) lead to the equilibrium composition.

1. *Find the equilibrium constant.* This may involve the use of tabulated free-energy data, or it may involve experimentally determined equilibrium constants directly.
2. *Specify the initial composition.* This requires identifying the relevant independent species involved in the reaction, their phases, and their amounts.
3. *Express the equilibrium constant in terms of composition variables.* This requires one to recognize the correct composition variables based on the correct standard states and to use approximate expressions where warranted. It also requires a choice of model systems for nonideal behavior if measured fugacities or activity coefficients are not available. This is often the hardest step.
4. *Express the composition variables in terms of* ξ. This is simple algebraic substitution based on the definition of ξ in Eq. (7.1).
5. *Solve the equilibrium expression for* ξ_{eq}. This is often the most tedious step, which can be made less tedious by suitable computer methods.
6. *Calculate the equilibrium composition from* ξ_{eq} *and the initial composition.* Essentially the reverse of step 4, this should *not* be the last step. Rather, one should *always* proceed to step 7, which states:
7. *Check the final answers for physical validity.* Did you predict more of a compound at equilibrium than could possibly be present (i.e., did ξ_{eq} exceed the boundaries imposed by the initial composition)? If so, there is an error. Do any compositions violate a phase stability requirement (could a gas condense or a liquid completely volatilize)? Is the sign of ξ_{eq} what you expected from the value of K_{eq}, the initial composition, and the Le Châtelier Principle? If not, double check the calculation. It may be correct and your intuition may be at fault.

Other expressions for K_{eq} will be discussed in Chapter 10, but the principles established here will still be valid.

EXAMPLE 7.12

A 1.0 mol sample of HI, 0.50 mol I_2, and 0.10 mol H_2 are mixed in a variable-volume reactor at room temperature. The temperature is then raised to 150 °C and the volume is adjusted to give a total pressure of 4.0 bar. What is the equilibrium composition in the reactor? At 150 °C, the vapor pressure of $I_2(l)$ is 0.392 bar, and the equilibrium constant for $H_2(g) + I_2(g) \rightleftarrows 2HI(g)$ is $K_{eq} = 207$.

SOLUTION Assuming first that all species remain in the gas phase, the equilibrium composition is relatively easy to find. We have

$$n_{HI} = n_{HI}^\circ + 2\xi = 1.0 \text{ mol} + 2\xi$$
$$n_{H_2} = n_{H_2}^\circ - \xi = 0.10 \text{ mol} - \xi$$
$$n_{I_2} = n_{I_2}^\circ - \xi = 0.50 \text{ mol} - \xi$$

$$K_{eq} = \frac{P_{HI}^2}{P_{H_2} P_{I_2}} = \frac{x_{HI}^2}{x_{H_2} x_{I_2}} = \frac{n_{HI}^2}{n_{H_2} n_{I_2}} = \frac{(1.0 \text{ mol} + 2\xi_{eq})^2}{(0.10 \text{ mol} - \xi_{eq})(0.50 \text{ mol} - \xi_{eq})} = 207 \; ,$$

which can be solved to yield $\xi_{eq} = 0.084$ mol, from which the equilibrium n_i's can be easily computed. In particular, $n_{I_2} = 0.416$ mol so that $x_{I_2} = 0.416$ mol/1.60 mol $= 0.26$, and we would predict $P_{I_2} = x_{I_2} P = (0.26)(4.0 \text{ bar}) = 1.04$ bar. But this exceeds the vapor pressure of I_2 at 150 °C, which is only 0.392 bar. Thus, our assumption that all species remain gaseous at equilibrium is incorrect. Some I_2 will liquify, and the gas-phase partial pressure of I_2 will be a constant 0.392 bar. (We must also assume that HI and H_2 are insoluble in $I_2(l)$.)

We therefore modify our calculation to take this $g-l$ equilibrium into account. We note that

$$P_{tot} = P_{H_2} + P_{I_2} + P_{HI} = P_{H_2} + 0.392 \text{ bar} + P_{HI} = 4.0 \text{ bar}$$

so that P_{HI} can be expressed in terms of P_{H_2} giving us the equilibrium constant expression

$$K_{eq} = \frac{P_{HI}^2}{P_{H_2} P_{I_2}} = \frac{(4.0 \text{ bar} - 0.392 \text{ bar} - P_{H_2})^2}{P_{H_2}(0.392 \text{ bar})} = 207 \;,$$

which has the solution $P_{H_2} = 0.148$ bar. We find $P_{HI} = (4.0 - 0.392 - 0.148)$ bar $= 3.460$ bar, and the gas-phase mole fractions are

$$x_{HI} = 3.460 \text{ bar}/4.0 \text{ bar} = 0.865$$
$$x_{H_2} = 0.148 \text{ bar}/4.0 \text{ bar} = 0.037$$
$$x_{I_2} = 0.392 \text{ bar}/4.0 \text{ bar} = 0.098 \;.$$

We can complete our description of the equilibrium system with a simple accounting of atoms. We know we started with 1.20 mol of H atoms (from 1.0 mol HI and 0.10 mol H_2) so that

$$1.20 \text{ mol} = n_{HI} + 2n_{H_2} = n_{H_2}\left(\frac{x_{HI}}{x_{H_2}} + 2\right)$$

which yields $n_{H_2} = 0.047$ mol and $n_{HI} = [1.20 - 2(0.047)]$ mol $= 1.106$ mol. The amount of gas phase I_2 can be found from

$$n_{I_2}(g) = n_{HI}\left(\frac{x_{I_2}}{x_{HI}}\right) = 0.125 \text{ mol}$$

and the amount of liquid I_2 from

$$2.00 \text{ mol total I} = 2n_{I_2}(g) + 2n_{I_2}(l) + n_{HI}$$

so that $n_{I_2}(l) = 0.322$ mol.

⇒ **RELATED PROBLEMS** 7.41, 7.50, 7.51

CHAPTER 7 SUMMARY

The thermodynamics of chemical equilibrium involves two major themes. The first centers around the "Δ_r" type of calculation that involves the change in a property when *stoichiometric amounts of unmixed reactants* (the initial state) are converted to *stoichiometric amounts of unmixed products* (the final state). The relevant expressions are Equations (7.3), (7.7), (7.21), and (7.30). The second major theme is the change in a thermodynamic function (most fundamentally G) as a reactive mixture evolves towards equilibrium. The path variable is ξ, the degree of advancement, as defined by Eq. (7.1).

The demonstration that $G(\xi)$ is a minimum at ξ_{eq}, the equilibrium degree of advancement, led to the introduction of a standard Gibbs free-energy change for reaction, $\Delta_r \overline{G}°$, and its relation to composition through the equilibrium constant. Equations (7.20) and (7.22) are central to this chapter.

The roles played by various standard states and by various model systems were seen to give simplified versions of the equilibrium constant expression in terms of composition variables, leading to expressions for ξ_{eq} in terms of K_{eq}, initial amounts of species, and the system intensive variables T and P. The solution of such expressions for ξ_{eq} is the goal of practical chemical equilibrium calculations.

This chapter has incorporated ideas from every preceding chapter. In a sense, each of those chapters has been building towards this one, since chemical reactions inherently involve a variety of temperatures, pressures, mixtures, phases, and nonidealities.

Before the advent of modern thermodynamics, years were spent in exploring various reaction conditions experimentally, often with incorrect theories or no guidance at all, in attempts to optimize the yield of a desired product, or to synthesize a new compound. Today, the theory is well established, and progress is most often limited by a lack of key experimentally measured properties. Thermodynamic research is far from dead, however. For example, the theoretical and experimental study of highly nonideal systems constitutes a major contemporary research area. The financial impact of a quick, accurate estimate instead of an expensive, time-consuming measurement provides part of the impetus for this research, but the inherent surprises of nature and the challenge of trying to understand complex phenomena will ensure the vitality of the field for years to come.

ACCESS TO DATA

In addition to those sources listed at the end of Chapters 4 and 6, the following compilations of data are especially important to chemical equilibrium calculations.

M. W. Chase, Jr., et al., Eds. *JANAF Thermochemical Tables,* 3rd ed., vol. 14, supplement 1 to the *Journal of Physical and Chemical Reference Data* (American Chemical Society, Washington, DC, 1985). These are the tables described in the text.

D. D. Wagman, et al., *The NBS Tables of Chemical Thermodynamic Properties,* vol. 11, supplement 2 to the *Journal of Physical and Chemical Reference Data,* (American Chemical Society, Washington, DC, 1983). Subtitled *Selected Values for Inorganic and C_1 and C_2 Organic Substances in SI Units,* this most recent compilation contains enthalpy, entropy, heat-capacity, and free-energy data at 0 and 298.15 K, as well as sample calculations illustrating the use of these data.

S. G. Lias, et al., *Gas-Phase Ion and Neutral Thermochemistry,* vol. 17, supplement 1 to the *Journal of Physical and Chemical Reference Data,* (American Chemical Society, Washington, DC, 1988). This compilation contains ionization energies, proton affinities, and electron affinities as well as enthalpies of formation on many ions and their related neutral species. This is often the best place to find enthalpy data on free radicals and other molecular fragments.

G. H. Aylward and T. J. V. Findlay, *SI Chemical Data,* 2nd ed. (J. Wiley and Sons, Australasia Ltd., Sydney, 1971). This is a handy paperback containing more than thermochemical data.

I. Barin and O. Knacke, *Thermochemical Properties of Inorganic Substances,* (Springer-Verlag, Berlin, 1973). Supplementary to the JANAF tables, this work contains data on many oxides, sulfides, carbides, halides, etc. The temperature range is not as large as for the JANAF tables, but functional forms (as a power series in T) for heat capacities are given, which can save one the effort of interpolation. Another bonus is the inclusion of vapor-pressure data.

R. C. Weast, Ed., *Handbook of Chemistry and Physics,* (CRC Press, Boca Raton, FL). This is the venerable "CRC Handbook," 2300+ pages of data on all aspects of physical science. Every student should spend an hour or more becoming familiar with the diverse amount of invaluable and arcane data in this book. As new editions appear, choices are made to include new material and omit old material, making old editions sometimes valuable sources of obscure information that has fallen from favor in newer editions.

PRACTICE WITH EQUATIONS

7A If 0.100 mol each of LiH and BF_3 are mixed in preparation for a synthesis of diborane according to $6LiH + 8BF_3 \rightleftarrows 6LiBF_4 + B_2H_6$, what are the limits for ξ? (7.1)

ANSWER: $0 \le \xi \le 0.0125$ mol

7B A 1.00 mol sample of pure $NO_2(g)$ is brought from a high temperature and low pressure to 298 K and 1.00 bar to establish the equilibrium $2NO_2(g) \rightleftarrows N_2O_4(g)$. The equilibrium degree of advancement of the reaction is found to be 0.103 mol. How much NO_2 is present at equilibrium? (7.1)

ANSWER: 0.794 mol

7C If the standard molar entropies of NO_2 and N_2O_4 are $\overline{S}° = 28.86\ R$ and $36.58\ R$, respectively, what is the entropy change (including the entropy of mixing) due to the establishment of the equilibrium in 7B above? (7.3), (6.25)

ANSWER: -15.44 J K^{-1}

7D The NO_2 units in N_2O_4 are linked by a N—N bond. If $\Delta_f \overline{H}°(NO_2) = 33.18$ kJ mol^{-1} K^{-1} and $\Delta_f \overline{H}°(N_2O_4) = 9.16$ kJ mol^{-1} K^{-1} at 298 K, what is the N—N bond enthalpy? (7.7), (7.10), (7.11)

ANSWER: 57.20 kJ mol^{-1}

7E The NO_2 dimerization reaction is exothermic, since the major chemical change is the formation of a new bond. If $\overline{C}_P°(NO_2) = 37.20$ J mol^{-1} K^{-1} and $\overline{C}_P°(N_2O_4) = 77.28$ J mol^{-1} K^{-1}, does the reaction become more or less exothermic as the temperature is raised above 298 K? (7.9)

ANSWER: less ($\Delta_r \overline{C}_P° > 0$)

7F Estimate $\Delta_r \overline{H}°$ for the 1,1 dibromination of ethylene: $C_2H_4(g) + Br_2(g) \rightarrow CH_2BrCH_2Br(g)$ using average bond energies. The accurate value at 298 K is -121.5 kJ mol^{-1}. Table 7.1

ANSWER: -111 kJ mol^{-1}

7G The NO$_2$ dimerization (2NO$_2$(g) $\rightleftarrows$ N$_2$O$_4$(g)) equilibrium constant calculated from tabulated $\Delta_f G°$ values is 0.146 at 298.15 K. Does the $\Delta_r G°$ value this number implies agree with that calculated from $\Delta_r G° = \Delta_r \overline{H}° - T\Delta_r \overline{S}°$? (See 7D for enthalpy data and 7C for entropy data; values within ~1% should be considered to "agree.") (7.24), (7.29)

ANSWER: $\Delta_r \overline{H}° - T\Delta_r \overline{S}° = -4.80$ kJ mol^{-1};
$-RT \ln K_{eq} = -4.77$ kJ mol^{-1}

7H What is the NO$_2$ dimerization equilibrium constant at 320 K? (See 7D for enthalpy data and 7G for K_{eq} at 298.15 K.) (7.25)

ANSWER: 0.0302

PROBLEMS

SECTION 7.1

7.1 The following reactions are all of the "$\alpha A + \beta B \rightarrow \gamma C + \delta D$" type. Calculate the amounts of each species present when $\xi = 0.20$ mol given the initial amounts of each tabulated below.

(1) $2C_8H_{18}(l) + 25O_2(g) \rightarrow 16CO_2(g) + 18H_2O(l)$
(2) $4NH_3(g) + 3O_2(g) \rightarrow 2N_2(g) + 6H_2O(l)$
(3) $2S_2O_3^{2-}(aq) + I_3^-(aq) \rightarrow 3I^-(aq) + S_4O_6^{2-}(aq)$
(4) $Ca_3P_2(s) + 6H_2O(l) \rightarrow 2PH_3(g) + 3Ca(OH)_2(aq)$

Reaction	$n_A°$/mol	$n_B°$/mol	$n_C°$/mol	$n_D°$/mol
(1)	1.5	14.	2.3	0.0
(2)	4.0	3.0	2.0	6.0
(3)	1.0	1.0	0.0	0.0
(4)	0.5	3.0	1.6	0.0

7.2 For the reaction $2N_2O_5(g) \rightleftarrows 4NO_2(g) + O_2(g)$ by how many moles must the amounts of each substance change to advance ξ 1 mol?

7.3 One mole of CO and 1 mol Cl$_2$ are prepared for a study of the synthesis of phosgene, COCl$_2$, according to CO + Cl$_2$ $\rightleftarrows$ COCl$_2$. They are mixed and held at a high temperature until equilibrium is attained. It is found that the total pressure is 1.00 atm and the partial pressure of COCl$_2$ is 0.171 atm. What is the equilibrium degree of advancement, and how many moles of each compound are found at this equilibrium? (Hint: Express the number of moles of each compound and the total number of moles in terms of ξ, and use the ideal gas equation of state for the total pressure as well as for the individual partial pressures.)

7.4 A mixture of 1.00 mol CO$_2$(g) and 0.500 mol CO(g) is raised to a temperature that produces graphite and establishes the equilibrium C(graphite) + CO$_2$(g) $\rightleftarrows$ 2CO(g). What are the limits on ξ? The equilibrium ξ is found to be $-1/9$ mol. What are the equilibrium amounts of each compound?

7.5 The algebra of the degree of advancement variable ξ is worth exploring further.

(a) Show that the mole fraction of species i with stoichiometric coefficient ν_i is given by

$$x_i = \frac{x_i°}{1 + \left(\dfrac{\xi \Delta \nu}{n°}\right)} \pm \frac{\nu_i \xi}{n° + \xi \Delta \nu}$$

where the + sign holds for products and the − sign for reactants and where

$$x_i° \equiv \frac{n_i°}{n°},\quad n° \equiv \sum_{\text{all } i} n_i°,\quad \text{and}\quad \Delta \nu \equiv \sum_{\text{products}} \nu_i - \sum_{\text{reactants}} \nu_i .$$

(b) Define an *intensive* (and dimensionless) reaction progress variable λ by $\xi = n° \lambda$. Show that

$$x_i = \frac{x_i° \pm \nu_i \lambda}{1 + \Delta \nu \lambda} .$$

SECTION 7.2

7.6 Arrange the following reactions in order of increasing $\Delta_r \overline{S}°$, justifying your arrangement by simple arguments.

(1) $HgO(s) \rightarrow Hg(l) + \tfrac{1}{2}O_2(g)$
(2) $6H_2(g) + P_4(s) \rightarrow 4PH_3(g)$
(3) $NH_2COONH_4(s) \rightarrow 2NH_3(g) + CO_2(g)$
(4) $2TiBr_3(s) \rightarrow TiBr_2(s) + TiBr_4(g)$

7.7 A large positive $\Delta_r \overline{S}°$ can often be the deciding factor in a reaction's ability to produce products in abundance. With this in mind, consider the generic reaction between a solid A and a gas B$_x$ (where x is usually either 1 or 2) to produce a gaseous product:

$$\alpha A(s) + \beta B_x(g) \rightarrow A_\alpha B_{\beta x}(g) .$$

What restrictions (if any) must be imposed on α, β, and/or x to cause $\Delta_r \overline{S}°$ to be positive? For a specific case, consider A to be Pt and B to be O. Explain why, at temperatures and pressures such that O$_2$(g) dominates O(g), one would not expect the Pt$_\alpha$O$_{\beta x}$(g) products to contain three or more atoms of O, but perhaps any number of Pt atoms. (Experiments have determined that $PtO_2(g)$ is the major product of this reaction.)

SECTION 7.3

7.8 In the gas phase, sulfur vapor at the normal boiling point (717.824 K at 1.00 atm) is a mixture of molecular species, S_n, $n = 8, \ldots, 1$ with S_8, S_7, and S_6 predominating. All eight species are fairly well characterized, making the choice of a gas-phase reference species not very obvious. Consequently, tabulated data on gaseous S frequently (and arbitrarily) use 0.5 $S_2(g)$ as the reference species. Given the $\Delta_f \overline{H}°$ data below at 1000 K (based on this reference), calculate the sulfur to sulfur bond enthalpies for the processes $S_n(g) \rightarrow S_{n-1}(g) + S(g)$ for $n = 8, \ldots, 2$. Why is the value for $n = 2$ anomalous? What would you recommend for an average sulfur-to-sulfur bond enthalpy?

$S_n(g)$	$\Delta_f \overline{H}°/\text{kJ mol}^{-1}$
1	215.834
2	0
3	−50.809
4	−106.741
5	−202.924
6	−270.383
7	−318.800
8	−392.026

7.9 In the course of your work on the atmospheric ozone problem, you find you need the molar enthalpy of reaction at 298 K for the ozone-destruction reaction

$$O_3(g) + O(g) \rightarrow 2O_2(g) \ .$$

You know $\Delta_r \overline{H}°$ for two reactions that form a catalytic cycle for ozone destruction:

$$O_3(g) + NO(g) \rightarrow NO_2(g) + O_2(g)$$
$$\Delta_r \overline{H}° = -199.77 \text{ kJ mol}^{-1}$$
$$O(g) + NO_2(g) \rightarrow NO(g) + O_2(g)$$
$$\Delta_r \overline{H}° = -192.1 \text{ kJ mol}^{-1}$$

and you have Table 7.2 at hand. Find $\Delta_r \overline{H}°$ for the reaction of interest first from a combination of the catalytic reactions and then from an average bond enthalpy calculation. For the average bond enthalpy calculation, first disassemble ozone into atoms, then form the two O_2 molecules. Your two answers will be in disturbingly large disagreement. Can you explain why?

7.10 The enthalpy of formation of $LiAlH_4(s)$ at 298 K was measured by calorimetry in 1964 [L. G. Fasolino, *J. Chem. Eng. Data* **9**, 68 (1964)]. In three different experiments, the heat associated with the reaction at constant volume of $Li(s)$, $Al(s)$, and $LiAlH_4(s)$ with excess 4 M $HCl(aq)$ was measured. While several runs were made for each solid, the table below gives representative data from the original paper. Write the balanced reaction for each run, calculate $\Delta_r \overline{H}°$ for each (remembering that the raw data yield $\Delta_r \overline{U}°$ because the calorimeter was at constant volume), then combine the three reactions appropriately and calculate $\Delta_f \overline{H}°$ ($LiAlH_4(s)$).

Reactant	Mass/g	Heat/cal
$Al(s)$	0.2707	1295.8
$Li(s)$	0.1395	1358.4
$LiAlH_4(s)$	0.3367	1492.1

7.11 Your organic chemist friend has a supply of one of the isomers of hexane, but he isn't sure which it is. All the lab's spectrometers are broken; so, he brings a sample to you hoping that your calorimeter can be of help. You are given a 1.356 g sample of $C_6H_{14}(l)$ to burn in excess oxygen in your constant-volume adiabatic calorimeter. Careful calibration with benzoic acid has told you that the calorimeter's heat capacity is 12.630 J K^{-1}, but your thermometer has a precision of only 0.002 K. Can you do the analysis? Enthalpy of formation data follow:

	$\Delta_f \overline{H}°/\text{kJ mol}^{-1}$
$H_2O(l)$	−285.83
$CO_2(g)$	−393.51
n-hexane	−198.66
2-methylpentane	−204.65
3-methylpentane	−202.39
2,2-dimethylbutane	−212.66
2,3-dimethylbutane	−206.15

7.12 When the amino acid tyrosine

HO—⟨benzene ring⟩—$CH_2CH(NH_2)COOH$

is burned in excess oxygen, the products are $CO_2(g)$, $H_2O(l)$, and $HNO_3(aq)$. A 1.0472 g sample of tyrosine is found to liberate 25.879 kJ in a calorimeter (when corrected to constant pressure and standard temperature conditions). What is the standard molar enthalpy of combustion of tyrosine? Given the following enthalpies of formation:

$$\Delta_f \overline{H}°(CO_2(g)) = -393.51 \text{ kJ mol}^{-1}$$
$$\Delta_f \overline{H}°(H_2O(l)) = -285.83 \text{ kJ mol}^{-1}$$
$$\Delta_f \overline{H}°(HNO_3(aq)) = -202.86 \text{ kJ mol}^{-1}$$

find $\Delta_f \overline{H}°$ for tyrosine.

7.13 Given $\Delta_f \overline{H}°(H_2O(g), 298.15 \text{ K}) = -241.818$ kJ mol^{-1} and the $\overline{C}_P/R$ data of Table 4.4, find $\Delta_r \overline{H}°$ at 1500 K for the reaction

$$2H_2(g) + O_2(g) \rightarrow 2H_2O(g) \ .$$

Next, repeat the calculation using just the 298.15 K values for $\overline{C}_P$. How large are the errors that this approximation introduces?

7.14 For each of the following reactions, predict whether $\Delta_r \overline{H}$ would increase or decrease with increasing T. Briefly justify your predictions, but do not consult any tables of data (other than information in Chapter 4, if need be).

(a) $\quad 2H_2O_2(l) \rightleftarrows 2H_2O(l) + O_2(g)$
(b) $\quad 2Mg(s) + O_2(g) \rightleftarrows 2MgO(s)$
(c) $\quad N_2(g) + 3H_2(g) \rightleftarrows 2NH_3(g)$
(d) $\quad Ba(OH)_2(s) \rightleftarrows BaO(s) + H_2O(l)$

7.15 An important reaction cycle, known as the *Born–Haber cycle*, was applied to alkali halides to find quantities such as electron affinities at a time when such quantities were very difficult to obtain experimentally. The cycle runs as follows, using NaCl as an example:

```
           I
NaCl(s) ─────────→ Na⁺(g) + Cl⁻(g)
   ↑                      │
IV │                      │ II
   │         III          ↓
Na(s) + ½Cl₂(g) ←───── Na(g) + Cl(g)
```

Step I can also be considered to be the sum of the following three processes:

sublimation:
$$NaCl(s) \to NaCl(g) \quad \text{(Ia)}$$
dissociation:
$$NaCl(g) \to Na(g) + Cl(g) \quad \text{(Ib)}$$
charge transfer:
$$Na(g) + Cl(g) \to Na^+(g) + Cl^-(g) \quad (-II)$$

Using data from Tables 7.2–7.4 along with $\Delta_f \overline{H}°(NaCl(s)) = -411.153$ kJ mol^{-1} and $\Delta_r \overline{H}°(Ib) = 410$ kJ mol^{-1}, construct an enthalpy-level diagram for these reactions, and find from it the sublimation enthalpy of NaCl(s). Place the system $Na(s) + \frac{1}{2}Cl_2(g)$ at the arbitrary zero of enthalpy in your diagram.

SECTION 7.4

7.16 The enzyme *alanine racemase* interconverts L-alinine and D-alinine:

$$\begin{array}{c} CO_2^- \\ | \\ ^+H_3N-C-H \\ | \\ CH_3 \end{array} \quad \underset{\text{racemase}}{\overset{\text{alanine}}{\rightleftarrows}} \quad \begin{array}{c} CO_2^- \\ | \\ H-C-NH_3^+ \\ | \\ CH_3 \end{array}$$

starting from either isomer. This is a thermoneutral reaction with an equilibrium constant of unity (and thus $\Delta_r \overline{S}° = 0$ as well). What, then, spontaneously drives the reaction toward a 50/50 equilibrium mixture?

7.17 Use the enthalpy of formation and absolute entropy data below to predict which of the following compounds would have a thermodynamic tendency to form spontaneously starting from stoichiometric amounts of their elements, each in their standard states at 298 K: NO(g), N$_2$O(g), NO$_2$(g), NH$_3$(g), HN$_3$(g), SO$_2$(g), SO$_3$(g).

	$\Delta_f \overline{H}°$/kJ mol^{-1}	$\overline{S}°$/J K^{-1} mol^{-1}
H$_2$(g)	0	130.684
N$_2$(g)	0	191.61
O$_2$(g)	0	205.138
S(s, rhombic)	0	31.80
NO(g)	90.25	210.76
N$_2$O(g)	82.05	291.85
NO$_2$(g)	33.18	240.06
NH$_3$(g)	−46.11	192.45
HN$_3$(g)	294.1	238.97
SO$_2$(g)	−296.83	248.22
SO$_3$(g)	−395.72	256.76

7.18 Write the equilibrium constant expression for the following reactions in terms of (a) partial pressures, (b) mole fractions, (c) amounts in moles, and (d) amount per unit volume concentrations.

(1) $CH_4(g) + H_2O(g) \rightleftarrows CO(g) + 3H_2(g)$
(2) $2NO_2(g) + F_2(g) \rightleftarrows 2NO_2F(g)$
(3) $2HO_2(g) \rightleftarrows H_2O_2(g) + O_2(g)$
(4) $ClONO_2(g) \rightleftarrows ClO(g) + NO_2(g)$

7.19 It is stated in the text that certain gas-phase anions such as Ne$^-$, CO$^-$, HF$^-$, etc., are not stable because these species have negative electron affinities. How can an enthalpy argument, such as this, be used to specify a stability criterion that must in fact be based on a free-energy argument?

7.20 Consider two reactions with equal $\Delta_r \overline{H}°$ values. One produces two moles of gaseous products per mole of gaseous reactants and the other produces only one mole of gaseous products per mole of gaseous reactants. Which reaction will have the larger equilibrium constant?

7.21 Limitations on our ability to detect trace amounts of many compounds often determine the practical meaning of the phrase "a reaction that goes to completion" in spite of what thermodynamics predicts for equilibrium concentrations. One of the more sensitive analytical methods for certain gas-phase species is laser induced fluorescence, which is explained more fully in Chapter 19. This technique can detect, for example, OH(g) concentrations as low as 10^6 molecules cm^3. If one has H$_2$O(g) at 10^{-2} bar, what is the smallest K_{eq} that could be measured for the reaction

$$H_2O(g) \rightleftarrows H(g) + OH(g) \ ?$$

SECTION 7.5

7.22 Since NO_2 is brown and N_2O_4 is colorless, the degree of NO_2 dimerization can be measured by visible-light spectrophotometry. If a colorless inert gas is added to a constant-volume cell containing NO_2 and N_2O_4 in equilibrium, will the brownish color darken, lighten, or stay the same? Why?

7.23 When XeF_2 was first synthesized in the early 1960s, there was a great rush to characterize its physical properties. Near room temperature XeF_2 is a fairly volatile solid, and absorption spectrophotometry could measure the concentration of $XeF_2(g)$ via the quantity known as the *optical density*, which is directly proportional to the XeF_2 pressure. In 1963, the first such measurements found the optical density to be 10 at 294 K and 1.2 at 267 K in a particular experiment. What is the molar enthalpy of sublimation of XeF_2?

7.24 The reduction of stannous sulfide by hydrogen

$$SnS(s) + H_2(g) \rightarrow Sn(l) + H_2S(g)$$

can be studied at high temperature from the ratios of the gaseous species' concentrations at equilibrium. The table below gives experimental data from such measurements in terms of volume ratios at various temperatures. Use these data to find the reaction enthalpy.

T/K	806.	855.	904.	952.
$10^4 \, (V_{H_2S}/V_{H_2})$	10.6	16.1	27.6	46.2

7.25 A curious example of isomerization that has features superficially similar to dimerization occurs when a single strand of an oligonucleotide (a series of amino acids bound in a chain by phosphodiester links) has self-complementary ends. This means one end has a string of amino acids (such as adenine, A) that "match" and hydrogen bond strongly to a complementary string (such as uracil, U) on the other end. Between the ends are a series of "spacer" amino acids (such as cytosine, C). The strand can thus be represented schematically as, for example,

AAAAAACCCCCCUUUUUU .

This strand can form a so-called base-paired hairpin loop and establish the equilibrium

AAAAAACCCCCCUUUUUU $\rightleftarrows$

```
      C—C
     /    \
    C      U—U—U—U—U—U
    |      · · · · · ·
    C      A—A—A—A—A—A
     \    /
      C—C
```

The equilibrium constant for this reaction is 0.86 at 25 °C and 2.44 at 37 °C. Before you calculate $\Delta_r\overline{H}°$ and $\Delta_r\overline{S}°$ at 37 °C from these data, predict their algebraic signs. Is your intuition supported by your calculation? [Adapted from *Physical Chemistry*, 2nd ed. by I. Tinoco, Jr., K. Sauer, and J. C. Wang (Prentice-Hall, Englewood Cliffs, NJ, 1985).]

7.26 How does $\Delta_r\overline{H}$ vary with pressure for: (a) a reaction among ideal gases? (b) a reaction among gases that follow the virial equation of state in the form of Eq. (1.17) with only the A_1 term?

7.27 Is the autoionization of water

$$H_2O(l) \rightarrow H^+(aq) + OH^-(aq)$$

increased or decreased by increasing the temperature $(\Delta_f\overline{H}°(H_2O(l)) = -285.83$ kJ mol^{-1})?

7.28 Predict the effect of the following on the equilibrium composition of each reaction:

(a) increasing the pressure of

$$H_2O(l) + CO(g) \rightleftarrows CO_2(g) + H_2(g)$$

(b) increasing the temperature of

$$CH_4(g) + 2F_2(g) \rightleftarrows CF_4(g) + 2H_2(g)$$

(c) adding He at constant T and P to

$$(CH_3)_3CCH_2Cl(g) \rightleftarrows (CH_3)_2C=CHCH_3(g) + HCl(g)$$

(d) adding He at constant T and V to reaction (c).

7.29 In the course of your work for NASA, you hit upon the invention shown below. It is an adiabatic container with three compartments: one that holds about 1 cup (~0.25 L) of water (or broth, soup, etc.), one that holds $Na(s)$, and one that holds $Cl_2(g)$. The partition between the latter two is removed to initiate the reaction $Na(s) + \frac{1}{2}Cl_2(g) \rightarrow NaCl(s)$, which, being exothermic, heats the water and provides salt for seasoning at the same time! If the water starts at around 70 °F (21 °C) and should be heated to 200 °F (93 °C), what masses of $Na(s)$ and $Cl_2(g)$ should be used? How much salt will be produced, given that at equilibrium, the reaction mixture is virtually pure $NaCl(s)$? For $NaCl(s)$, $\Delta_f\overline{H}° = -411.153$ kJ mol^{-1}, and $\overline{C}_P°(H_2O(l)) = 75.291$ J K^{-1} mol^{-1}. Assume the heat capacity of the container walls is negligible, but consider heating the salt product: $\overline{C}_P°(NaCl(s)) = 50.50$ J K^{-1} mol^{-1}.

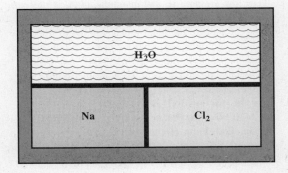

7.30 The heat capacity of a reacting mixture contains a component due to the reaction enthalpy. To see how this comes about, imagining a simple gas-phase isomerization (such as $CH_3CN \rightleftarrows CH_3NC$, a well-studied example), which we can take symbolically to be $A \rightleftarrows B$. Imagine that the reaction is endothermic so that K_{eq} increases with increasing temperature and that we start with $n_A^\circ = 1$ mol, $n_B^\circ = 0$, and T low enough to allow negligible B at equilibrium. Now we raise T until B appears in an appreciable amount. At this point, the total enthalpy of the system is

$$H(T) = n_A \overline{H}_A(T) + n_B \overline{H}_B(T)$$

where n_A and n_B are related to K_{eq}. Show first that

$$n_A = \frac{1}{K_{eq}(T) + 1} \quad \text{and} \quad n_B = \frac{K_{eq}(T)}{K_{eq}(T) + 1}.$$

Then use $C_P = (\partial H/\partial T)_P$ and the following variant of Eq. (7.25):

$$\left(\frac{\partial K_{eq}}{\partial T}\right)_P = K_{eq}\frac{\Delta_r \overline{H}^\circ}{RT^2}$$

to show that the reacting system's heat capacity is given by

$$C_P = \left(\frac{\partial H}{\partial T}\right)_P = \frac{\overline{C}_P^\circ(B) K_{eq} + \overline{C}_P^\circ(A)}{K_{eq} + 1} + \frac{(\Delta_r \overline{H}^\circ)^2 K_{eq}}{RT^2 (K_{eq} + 1)^2}.$$

Verify that this expression has the correct limits as $K_{eq} \to 0$ and as $K_{eq} \to \infty$.

SECTION 7.6

7.31 Many gases dimerize to an appreciable extent (alkali metals, alkali halides, hydrogen-bonding species, and NO_2 being among the best studied). The generic dimerization reaction $2A \rightleftarrows A_2$ indicates that the equilibrium amount of gas is tied to a chemical reaction that will cause the gas to behave nonideally. Find the equation of state for such a gas assuming A and A_2 behave individually as ideal species and form an ideal solution, and compare the isotherms of this equation to those of an ideal gas. To simplify the problem, imagine starting the system at 300 K with 1.00 mol A at very low pressure so that insignificant A_2 is present. (Why will the amount of A_2 decrease as P is lowered?) Compare the ideal molar volume to the molar volume of this gas at 1 bar for dimerization equilibrium constants in the range 0.1–0.001. Are P–V–T measurements likely to be accurate measures of the dimerization K_{eq}?

7.32 For the reaction $2NO_2(g) \rightleftarrows N_2O_4(g)$ at 298.15 K, $K_{eq} = 0.146$. (See Practice with Equations 7G.) If 0.240 g of an equilibrium mixture of this system is placed in a 150 mL container at 298.15 K, what will the pressure be? This calculation is a generalization of the previous one in that you will need to consider an arbitrary initial amount of gas. As a check, note that your answer should be *less* than the pressure 0.240 g pure NO_2 would exhibit in the same container.

7.33 The mass-three isotopic variant of ordinary molecular hydrogen, HD(g), is unusual in that it has a positive $\Delta_f \overline{H}^\circ$ (0.318 kJ mol^{-1} at 298.15 K) but a negative $\Delta_f \overline{G}^\circ$ (-1.464 kJ mol^{-1} at 298.15 K). For most substances both quantities have the same algebraic sign. Explain why HD(g) has this unusual feature.

7.34 Consider the generic disproportionation reaction $2AB \rightleftarrows A_2 + B_2$. Show that if one starts with 1 mol AB and allows the reaction to come to equilibrium at some T where the equilibrium constant is K_{eq}, the fractional disproportionation, defined as $n_{AB}^{eq}/1$ mol is given by $1/(2\sqrt{K_{eq}} + 1)$. Next, consider the special case of disproportionation due to isotope scrambling, as in $2\,^{37}Cl^{35}Cl \rightleftarrows \,^{37}Cl_2 + \,^{35}Cl_2$. Such reactions can be followed by spectroscopic methods that are sensitive to isotopic composition, and they are very nearly, but not exactly, thermoneutral. One might argue that $\Delta_r \overline{S}^\circ$ should also be zero; then $\Delta_r \overline{G}^\circ$ would be zero, K_{eq} would be 1 at all temperatures, the fractional disproportionation would be 1/3, and an initial 1 mol AB would equilibrate to 1/3 mol each of AB, A_2, and B_2. This is wrong. Instead, one might argue as follows: 1 mol AB constitutes 1 mol A and 1 mol B. Plucking at random from these atoms two at a time should give an AB pair twice as often as an AA or BB pair (since plucking A followed by B or B followed by A both yield AB). Then, the fractional disproportionation would be 1/2, and an initial 1 mol AB would equilibrate to 1/2 mol AB and 1/4 mol each of A_2 and B_2. This is correct. To what K_{eq} does this correspond? The K_{eq} you will find is non-unity because of an entropy effect that (as is explained in Chapter 23) distinguishes heteronuclear from homonuclear diatomics. If $\Delta_r \overline{G}^\circ = -T\Delta_r \overline{S}^\circ = -RT \ln K_{eq}$ and if $\Delta_r \overline{S}^\circ = R \ln W_r$, as in Eq. (3.5), then $K_{eq} = W_r$ where W_r is a "reaction probability ratio" of the form

$$W_r = \frac{W_{A_2} W_{B_2}}{W_{AB}^2}.$$

What must the ratio W_{A_2}/W_{AB} be?

7.35 More can be learned from disproportionation reactions than what was discussed in the previous problem. Consider, for example, the six interhalogens, ClF, BrF, etc., through IBr. How stable are they to disproportionation? Use the bond energy data in Table 7.2 to estimate $\Delta_r \overline{H}^\circ$ for each interhalogen disproportionation reaction. Given that entropy is a small factor (as determined in the previous problem), estimate the disproportionation equilibrium constants for each interhalogen. Which is most susceptible to disproportionation? Can you give a simple reason for the periodic trend in your calculations based on elementary chemical-bonding arguments?

7.36 From around the thirteenth century and on for several centuries, alchemical writings were strongly influenced by the legendary writings of one Hermes Trismegistos (Hermes the "Thrice-Great") who is supposed to have written thirteen alchemical precepts on an Emerald Tablet. These oracular precepts (the seventh reads, "Separate the earth from the fire, the subtle from the gross, acting prudently and with judgement.") were thought to give the key to discovering the Philosopher's Stone and the secret to transmutation of elements. Suppose such a Stone could be found (and in a certain sense, nuclear chemistry provides it). How would the facile transmutation of elements affect the JANAF tables?

7.37 In the text's discussion of the JANAF table for H(g), it was pointed out that $\Delta_f \overline{H}°(H(g))$ increases to a maximum at 5600 K and that this maximum corresponds to the point where $2\overline{C}_P°(H) = \overline{C}_P°(H_2)$. Prove that this must be so.

7.38 Compute K_{eq} for the reaction

$$2H \rightleftarrows H^+ + H^-$$

at 6000 K using data from the JANAF tables. Use each of the four methods discussed in the text: Equations (7.29), (7.30), (7.32), and (7.34).

7.39 At 298 K, the molar entropies of $O_2(g)$ and $O_3(g)$ are 205.0 J K^{-1} mol^{-1} and 237.7 J K^{-1} mol^{-1} respectively. For $O_3(g)$, $\Delta_f \overline{H}° = 142.2$ kJ mol^{-1}. What is $\Delta_f \overline{G}°$ for ozone?

7.40 Superman could squeeze a lump of coal until it transformed into diamond. Since coal is a poorly characterized material, consider instead the transformation

$$C(graphite) \rightarrow C(diamond) \; .$$

Calculate the pressure at which this transformation reaches equilibrium at 298.15 K given $\Delta_f \overline{G}°(diamond) = 2.900$ kJ mol^{-1}, $\rho(graphite) = 2.266$ g cm^{-3}, and $\rho(diamond) = 3.52$ g cm^{-3}. What would happen if the densities were reversed?

7.41 Formation data, such as free energies of formation, are assumed to hold for the substance at 1 bar pressure. For example, $\Delta_f \overline{G}°(H_2O(g)) = -228.572$ kJ mol^{-1} at 298.15 K even though the gas phase is not stable at this temperature and 1 bar pressure. Considering $\Delta_f \overline{G}°(H_2O(g))$ to be just another free energy, calculate its change as $H_2O(g)$ is expanded from 1 bar to 3.17×10^{-2} bar (the H$_2$O vapor pressure at 298 K), then condensed to liquid, and finally compressed as a liquid to 1 bar (with a molar volume $\overline{V}(l) = 18$ cm^3 mol^{-1}). Your final answer is $\Delta_f \overline{G}°(H_2O(l))$, the tabulated value for which is -237.129 kJ mol^{-1}. Which step in this process dominates the other two?

7.42 The previous problem used a simple path to relate free energies of formation data for two phases in equilibrium. Another way of looking at that calculation is to consider the phase transition to be a reaction in equilibrium. Given that the room-temperature vapor pressure of methanol is 0.176 bar and $\Delta_f \overline{G}°(CH_3OH(g))$ is -161.96 kJ mol^{-1}, find $\Delta_f \overline{G}°(CH_3OH(l))$ by using K_{eq} for

$$CH_3OH(l) \rightleftarrows CH_3OH(g) \; .$$

7.43 At 1100 K, the free energies of formation of BaO(s) and BaO$_2$(s) are nearly equal:

$$\Delta_f \overline{G}°(BaO(s)) = -452 \text{ kJ mol}^{-1}$$

$$\Delta_f \overline{G}°(BaO_2(s)) = -454 \text{ kJ mol}^{-1} \; .$$

Predict what will happen when, in separate experiments, first BaO(s) and then BaO$_2$(s) is heated in air to 1100 K. (Ba is a liquid at 1100 K.)

7.44 Example 7.10 mentioned the two crystallographic forms of CaCO$_3$(s): *calcite* and *aragonite*. For T in the range 298–450 K and P in the range ≤ 5000 bar, molar free energy of formation data for these can be expressed as

$$\Delta_f \overline{G}/R = -145\,160 \text{ K} + 31.50\, T$$
$$+ (0.451 \text{ K bar}^{-1})\, P \quad calcite$$

$$\Delta_f \overline{G}/R = -145\,185 \text{ K} + 31.975\, T$$
$$+ (0.416 \text{ K bar}^{-1})\, P \quad aragonite \; .$$

Use these to draw the two-phase equilibrium line for the *calcite* $\rightleftarrows$ *aragonite* phase transition in (P, T) coordinates over the 300–400 K range. Label the areas on either side of the line by the phase that is stable in that area.

7.45 From the data given later in this problem, you can easily calculate that $\Delta_r \overline{G}°$ at 298 K is 11.1 kJ mol^{-1} for the reaction

$$2BaCl_2 \cdot H_2O(s) \rightarrow BaCl_2 \cdot 2H_2O(s) + BaCl_2(s)$$

but how should this number be interpreted? Turning it into an equilibrium constant is easy enough:

$$K_{eq} = \exp\left(-\frac{\Delta_r \overline{G}°}{RT}\right) = 1.14 \times 10^{-2} = \frac{a_{BaCl_2 \cdot 2H_2O}\, a_{BaCl_2}}{a^2_{BaCl_2 \cdot H_2O}}$$

but if the activities of pure solids are supposed to equal unity, what does this expression mean? To explore this, consider placing three dishes in an evacuated chamber. In one dish, place 0.001 mol BaCl$_2$(s); in another, 0.001 mol BaCl$_2 \cdot 2$H$_2$O(s); and in the third, 0.002 mol BaCl$_2 \cdot$H$_2$O(s). Equilibrium among these hydrates can now occur via vapor transport of H$_2$O(g) according to the equilibria

$$BaCl_2 \cdot H_2O(s) \rightleftarrows BaCl_2(s) + H_2O(g)$$

and

$$BaCl_2 \cdot 2H_2O(s) \rightleftharpoons BaCl_2 \cdot H_2O(s) + H_2O(g) \ .$$

Find the equilibrium composition of the dishes if the volume of the container is (a) 1000 L, (b) 10 L, and (c) 0.1 L. Relevant $\Delta_f \overline{G}°$/kJ mol^{-1} data are given below.

$BaCl_2(s)$	$BaCl_2 \cdot H_2O(s)$	$BaCl_2 \cdot 2H_2O(s)$	$H_2O(g)$
-810.9	-1059	-1296	-228.572

How should $\Delta_r \overline{G}°$ for the first reaction (among all solids) be interpreted?

7.46 The vast majority of the Universe consists of interstellar regions that are populated almost exclusively by H(g) at a density of around 1 H atom per cm^3 and a temperature of $\sim$10 K. These data imply the very low pressure of $\sim$1.4 $\times$ 10^{-21} bar. Thus a possible explanation for the relative lack of H$_2$ is simply that low pressure favors dissociation. What does the 2H(g) $\rightleftharpoons$ H$_2$(g) equilibrium constant at $\sim$10 K have to say about this explanation? Is this reaction at equilibrium in the interstellar medium?

7.47 Mixing plays an important role in the establishment of chemical equilibrium. To illustrate this point, redraw Figure 7.3 for the H$_2$ $\rightleftharpoons$ 2H equilibrium in the absence of mixing. Assume that one has two containers, one for each species, and that the initial state ($\xi = 0$) has the H$_2$ container holding 1 mol H$_2$ while the H container is empty. At $\xi = 1$ mol, the H container holds 2 mol H and the H$_2$ container is empty. Imagine adjusting the container volumes so as to keep the total pressure of both a constant. What state has the lowest G in this scenario?

7.48 In the course of your research on the recovery of gold from metallic ores, you find you need to know the room-temperature $\Delta_f \overline{G}°$ values for Au$_2$O$_3$(s) and Fe$_2$O$_3$(s). You go to the library, look them up, and write them on a scrap of paper, having forgotten to carry your research notebook with you for the first and only time in your life. When you return to your lab, you discover that the scrap of paper has only the two numbers -742 kJ mol^{-1} and 126 kJ mol^{-1} written on it without an indication of which is which. How can a little chemical knowledge make it clear which value goes with which compound?

7.49 What is the activity of CO$_2$(aq) if the partial pressure of CO$_2$(g) over the solution is 1.00 bar, given $\Delta_f \overline{G}°$(CO$_2$(g)) $= -394.359$ kJ mol^{-1} and $\Delta_f \overline{G}°$(CO$_2$(aq)) $= -385.98$ kJ mol^{-1}? (This system, and the hydrolysis of CO$_2$ to carbonic acid, is discussed further in Problem 10.14.)

7.50 No discussion of gas-phase equilibria would be complete without mention of the classic work on the synthesis of ammonia from N$_2$(g) and H$_2$(g) (the Haber process) done in the 1920s by Larson and Dodge. Thermodynamically, the ammonia yield is favored by high pressures (fewer moles of gaseous products than reactants) and low temperatures (the reaction is exothermic), but at temperatures too low, the rate of reaction is impractically small. In one study by Larson [*J. Am. Chem. Soc.* **46**, 367 (1924)] an initial mixture that was 75.00% by volume H$_2$(g) and 25.00% N$_2$(g) was equilibrated at 450 °C and 300 atm, conditions under which nonidealities are important. The equilibrium mixture contained 35.82% NH$_3$(g) by volume. Writing the reaction as

$$\tfrac{1}{2} N_2(g) + \tfrac{3}{2} H_2(g) \rightleftharpoons NH_3(g)$$

implies the equilibrium constant expression

$$K_{eq} = \frac{f_{NH_3}}{f_{H_2}^{1.5} f_{N_2}^{0.5}} = \frac{P_{NH_3}}{P_{H_2}^{1.5} P_{N_2}^{0.5}} \frac{\Phi_{NH_3}}{\Phi_{H_2}^{1.5} \Phi_{N_2}^{0.5}}$$

where each Φ is a fugacity coefficient as in Eq. (5.52). Factoring K_{eq} this way allows you to apply our usual formalism to the partial pressure expression (finding ξ_{eq} from the data given above—if $n^\circ_{N_2} = 1.000$ mol, you should find $\xi_{eq} = 1.055$ mol) and to calculate the Φ expression independently. Use Eq. (5.41) with $B_{H_2} = 16.$ cm^3 mol^{-1}, $B_{N_2} = 24.4$ cm^3 mol^{-1}, and $B_{NH_3} = -31.$ cm^3 mol^{-1}. From Haber's original data at lower pressure [*Z. Elektrochem.* **20**, 597 (1914)], a value $K_{eq} = 5.9 \times 10^{-3}$ can be inferred. How well does your value agree? How important is the fugacity coefficient correction?

7.51 Equilibrium constant data at a variety of temperatures can often be expressed most directly and compactly in terms of the reaction free-energy change. For T in the range 270–350 K, $P \leq 1$ bar, and molalities ≤ 0.1 mol/kg, the following expressions for free energy hold for their corresponding processes where I$_2$(CCl$_4$) represents iodine dissolved in CCl$_4$:

$$I_2(s) \rightleftharpoons I_2(aq) \quad \Delta_r \overline{G}/R = 2720 \text{ K} - 2.54\,T + T \ln m_{I_2}$$

$$I_2(s) \rightleftharpoons I_2(CCl_4) \quad \Delta_r \overline{G}/R = 3020 \text{ K} - 5.66\,T + T \ln m_{I_2}$$

$$I_2(s) \rightleftharpoons I_2(g) \quad \Delta_r \overline{G}/R = 7510 \text{ K} - 17.37\,T + T \ln P_{I_2} \ .$$

Use these expressions to find:

(a) the molality of an aqueous solution saturated by I$_2$(s) at 300 K and at 350 K,

(b) the molality of I$_2$(aq) if a solution of 0.010 mol I$_2$ per kg CCl$_4$ is equilibrated with a pure water phase at 300 and 350 K, and

(c) the vapor pressure of I$_2$ that would be in equilibrium with such solutions.

GENERAL PROBLEMS

7.52 Chapter 6 emphasized the equality of the chemical potential of any species in all phases of a system. For closed reactive systems, this is especially important as phases may come and go during the attainment of equilibrium. (See Example 7.12.) Prove that if a new phase appears, all species must have a non-zero concentration in that phase; that is, if $n_i = 0$ for some species i in some phase, then n_{tot}, the total number of moles in that phase, must also be zero. Argue as follows. We can write in general for a nonideal solution that

$$\mu_i = \mu_i^\circ + RT \ln \gamma_i + RT \ln x_i$$
$$= \mu_i^\circ + RT \ln \gamma_i + RT \ln \frac{n_i}{n_{tot}}$$

and by definition,

$$\left(\frac{\partial G}{\partial n_i}\right)_{T, P, n_j} = \mu_i .$$

Assuming that γ_i is finite, use these two expressions to argue that if $n_i = 0$ and $n_{tot} \neq 0$, G would be lowered by an infinitesimal addition of i to the phase in question. Thus the only equilibrium possibilities are $n_i = n_{tot} = 0$ (the phase is missing) or *both* n_i and n_{tot} are non-zero (the phase is present and all species are at least infinitesimally present in it).

7.53 The Le Châtelier Principle can be given a simple theoretical proof for chemical reactions in a closed system by following these steps. First, we write the total differential of G as

$$dG = -S\,dT + V\,dP + \left(\frac{\partial G}{\partial \xi}\right)_{T, P} d\xi .$$

We introduce $(\partial G/\partial \xi)_{T, P}$ from Eq. (7.17):

$$dG = -S\,dT + V\,dp + \left(\sum_{\text{products}} \nu_i \mu_i - \sum_{\text{reactants}} \nu_i \mu_i\right) d\xi$$

where ν_i is the stoichiometric coefficient of species i. The total differential of $(\partial G/\partial \xi)_{T, P}$ can also be written as

$$d\left(\frac{\partial G}{\partial \xi}\right)_{T, P} = -\left(\frac{\partial S}{\partial \xi}\right)_{T, P} dT + \left(\frac{\partial V}{\partial \xi}\right)_{T, P} dP + \left(\frac{\partial^2 G}{\partial \xi^2}\right)_{T, P} d\xi .$$

At equilibrium, this must be zero, since $(\partial G/\partial \xi)_{T, P}$ is zero at equilibrium. Use this to find expressions for $(\partial \xi/\partial T)_P$ and $(\partial \xi/\partial P)_T$. Then recall that G is at a minimum with respect to its intensive variables at any stable equilibrium so that $(\partial^2 G/\partial \xi^2)_{T, P} > 0$. With this in mind, interpret your final expressions to show how they imply that an increase in T at constant P causes a reaction to proceed in the direction that transfers heat into the system while an increase in P at constant T causes the reaction to proceed in a direction that decreases the volume.

7.54 Suppose we have the reaction $3H_2 + N_2 \rightleftarrows 2NH_3$ at equilibrium, and we add more N_2 to this mixture. Will reattainment of equilibrium involve consumption or production of N_2? If N_2 is added at constant T and V, the partial pressure of nitrogen increases since $P_{N_2} = n_{N_2} RT/V$ while the partial pressures of H_2 and NH_3 are unchanged. More ammonia must be produced in order to raise P_{NH_3} and lower P_{N_2} (and P_{H_2} as well) to regain the partial-pressure equilibrium-constant expression, $K_{eq} = P_{NH_3}^2/P_{H_2}^3 P_{N_2}$. This is the elementary expectation of a Le Châtelier Principle argument. If we add N_2 at constant T and P, the paths back to equilibrium are slightly more complex. Since $P_i = x_i P$, the focus is on *mole fraction* change rather than *amount* change. We still expect P_{N_2} to decrease, but at constant P, this need *not* happen by consuming more N_2! Write

$$\left(\frac{\partial P_{N_2}}{\partial \xi}\right)_{T, P} = \left(\frac{\partial (x_{N_2} P)}{\partial \xi}\right)_{T, P} = P\left(\frac{\partial x_{N_2}}{\partial \xi}\right)_{T, P}$$

and show that since $x_{N_2} = n_{N_2}/n$, and since both n_{N_2} and n change as equilibrium is reattained,

$$P\left(\frac{\partial (n_{N_2}/n)}{\partial \xi}\right)_{T, P} = \frac{P}{n} [2x_{N_2} - 1]$$

since $n = n_{N_2} + n_{H_2} + n_{NH_3}$, $(\partial n_{N_2}/\partial \xi) = -1$, and $(\partial n/\partial \xi) = -1 - 3 + 2 = -2$. Since we want to decrease the partial pressure of nitrogen, we must have

$$dP_{N_2} = \left(\frac{\partial P_{N_2}}{\partial \xi}\right)_{T, P} d\xi < 0 .$$

Explain how this condition predicts that N_2 is consumed if the N_2 mole fraction is < 0.5, but if $x_{N_2} > 0.5$, *more* N_2 will be produced as the system returns to equilibrium. A more complete discussion of the reattainment of equilibrium can be found in articles by J. deHeer in *J. Chem. Educ.* **34**, 375 (1957); **35**, 133 (1958), on which this discussion was based.

7.55 Consider a simple isomerization reaction of the $A \rightarrow B$ type, or any similar reaction with the same number of moles of products as reactants. If you were told that K_{eq} for such a reaction was, say, 10^9, you would immediately recognize that the equilibrium reaction mixture was essentially pure B. But what if you were told instead that $\Delta_r \overline{G}^\circ$ was, say, -9 kJ mol^{-1}? Could you so easily connect $\Delta_r \overline{G}^\circ$ to equilibrium composition? To help you do so, consider $A \rightarrow B$ with $n_A^\circ = 1$ mol and $n_B^\circ = 0$. Plot ξ_{eq} as a function of $\Delta_r \overline{G}^\circ$ over the range -30 kJ mol^{-1} ≤ $\Delta_r \overline{G}°$ ≤ 30 kJ mol^{-1} for $T = 300, 700$, and 1500 K. For what $\Delta_r \overline{G}°$ values is ξ_{eq} ≥ 0.90 mol at each T?

7.56 When more than one reaction is involved in establishing chemical equilibrium, the calculation of the equilibrium composition can become considerably more involved. Consider, as a simple example, the equilibrium among $C(g)$, $C_2(g)$, and $C_3(g)$ at high temperatures. (The experiment can be done in a graphite furnace or in the atmosphere of carbon-rich stars where this type of equilibrium is a sensitive indicator of stellar temperatures.) We can take the following two reactions to describe the system:

(A) $\qquad 2C \rightleftarrows C_2$
(B) $\qquad C + C_2 \rightleftarrows C_3$.

(Note that these are not unique choices. Reaction (B) could also be written as, for example, $3C \rightleftarrows C_3$.) Each reaction will have its own degree of advancement variable (call them ξ_A and ξ_B). Assume, for simplicity, that the initial composition is $n_C° = 1$ mol and $n_{C_2}° = n_{C_3}° = 0$ and that $P = 1$ bar. Show that the equilibrium constant expressions for this case are

$$K_A = \frac{(\xi_A - \xi_B)(1 - \xi_A - \xi_B)}{(1 - 2\xi_A - \xi_B)^2}$$

and

$$K_B = \frac{\xi_B(1 - \xi_A - \xi_B)}{(1 - 2\xi_A - \xi_B)(\xi_A - \xi_B)} .$$

These are two nonlinear equations in ξ_A and ξ_B and are thus not so simple to solve. One graphical technique for their solution starts by writing each expression above as functions of the form $\xi_A = f(\xi_B)$. One graphs these two functions and reads the solution (ξ_A^{eq}, ξ_B^{eq}) from the intersection of the two curves. Since the K_A and K_B expressions are quadratic, this is tedious but not impossibly so. Should you care to try such a solution, the numerical values at 4500 K are $K_A = 2.87$ and $K_B = 6.87$. You should find $\xi_A^{eq} = 0.326$ mol and $\xi_B^{eq} = 0.216$ mol. Given these values (whether you verify them or not) find the equilibrium amounts of C, C_2, and C_3 and check that $n_C + 2n_{C_2} + 3n_{C_3} = 1.00$ mol, as it should.

7.57 The vapor phase of S was discussed in Problem 7.8. Here's another aspect of that system. At temperatures from the normal boiling point up to ~1000 K, the octamer, S_8, is present in appreciable amounts at pressures near 1 bar. One of its dissociation paths can be written as the reaction $S_8 \rightleftarrows 4S_2$. Imaging starting with 1 mol S_8 and asking what ξ_{eq} would be for this reaction. (In reality, this and a host of other dissociations are coupled, making the true equilibrium composition somewhat more complicated to find.) Show first that the equation for ξ_{eq} in the ideal mixture of ideal gases approximation is

$$K_{eq} = \frac{(4\xi_{eq})^4 P^3}{(1 - \xi_{eq})(1 + 3\xi_{eq})} .$$

This is a quartic equation in ξ_{eq}, and its solutions are not simple to find. One method writes the equation above as a polynomial in ξ that is graphed: $y = f(\xi)$. The value for which $f(\xi) = K_{eq}$ is ξ_{eq}. Try this for $P = 1$ bar and $K_{eq} = 580$, the 1000 K value. What is the physically allowable range of ξ for the problem as we have stated it? If you have access to a computer that can plot polynomial functions, explore the solutions as a function of P and K_{eq}. Have your program plot $f(\xi)$ over the allowed range of ξ, and scale the y axis between $+2K_{eq}$ and $-2K_{eq}$. Let K_{eq} range from 0.1 to 1000 by powers of 10 at $P = 1$ bar, plotting all curves on the same graph. Then pick one K_{eq} value and make a similar plot varying P. A quartic polynomial has potentially four solutions that, in this case, would correspond to multiple equilibria. Do you find any evidence for more than one equilibrium composition here?

CHAPTER 8

Thermodynamics of Surfaces and Interfaces

IN previous chapters, the boundaries between phases defined the physical extent of a phase, but they did not contribute to thermodynamics. In this chapter, we concentrate on these boundaries, the real surfaces and interfaces between phases, and see to what extent they govern properties of a system.

The practical implications of interfacial phenomena are enormous. Detergents, catalysts, electrochemical electrodes, lubricants, adhesives, colloids, emulsions, living-cell membranes, materials' strengths: all depend critically on the explicit properties of the narrow region that defines an interface. On a molecular level, it is easy to understand the gross origin of these properties. Inside a phase, an atom or molecule is surrounded by a uniform environment that is homogeneous over a region large compared to the range of intermolecular forces. A surface or interfacial molecule finds a different environment on opposite sides of the interface. The distance over which the environment changes is on the order of the range of intermolecular forces, several nanometres or so, but the "thickness" of an interface is an elusive quantity. On a solid surface in equilibrium with a gas, the surface layer of molecules clearly forms part of the interface, but this layer may be coated with an adsorbed layer of atoms from the gas phase with a different composition from the bulk gas, and molecules in the solid layers beneath the surface slowly, layer by layer, blend into those of the bulk.

8.1 The Energy Contributions of Interfaces

8.2 Small Drops and Bubbles

8.3 The Nucleation of Phases

8.4 Equilibrium Forms Among Interfaces

In some situations, the interface can grow to macroscopic widths. In two-phase fluid equilibrium (liquid–vapor or liquid–liquid) near a critical point, the system is changing to a single phase. As the critical point is approached (typically when within ~1 K of the critical temperature), the interface grows rapidly. This growth leads to macroscopic regions with wild and rapid fluctuations in density (and other properties) that produce a dynamic light scattering known as *critical opalescence*. The interface dominates the system.

But in most situations, the number of surface molecules (a generic term we will use to represent those molecules in the interfacial region) is such a small fraction of the total that the previous chapters were safe in ignoring surface effects. Consider an atomic metal in which the atom-to-atom spacing is 3 Å, a typical distance, and the atomic packing is a simple cubic array. Each atom is at the center of a cube of edge length 3 Å. The volume of a mole of such atoms is $(6 \times 10^{23})(3 \times 10^{-8} \text{ cm})^3 \approx 16 \text{ cm}^3$. Such a sample could be shaped into a cube ~2.5 cm on an edge. On each of the faces of this cube, we would find $(2.5 \text{ cm}/3 \times 10^{-8} \text{ cm})^2 \approx 7 \times 10^{15}$ atoms, or about 6×10^{16} surface atoms on all eight faces. Thus, the fraction of surface to bulk atoms is on the order of 6×10^{16} surface atoms$/6 \times 10^{23}$ bulk atoms $\cong 10^{-7}$. The thermodynamic properties (free energy, entropy, etc.) of these surface atoms would have to be roughly one million times or more larger per atom than those of the bulk atoms to make a noticeable, surface specific contribution. This is highly unlikely; therefore, the role played by surface molecules must be fairly subtle. They can be dominant only when the total number of surface atoms is an appreciable fraction (such as 10^{-1} rather than 10^{-7}) of the total number of system atoms (as in small drops of fluid) or when surface effects govern the geometry of the system (as in governing the equilibrium shape of the system).

8.1 THE ENERGY CONTRIBUTIONS OF INTERFACES

As with so much of thermodynamics, we begin with the work of J. Willard Gibbs, who delineated virtually everything thermodynamic over a century ago. Consider a system of two phases α and β in contact. Prior to this chapter, we would write the total system's Gibbs free energy as $G = \mu(\alpha)n(\alpha) + \mu(\beta)n(\beta)$. To incorporate the interface, we add a new term:

$$G = \mu(\alpha)n(\alpha) + \mu(\beta)n(\beta) + \gamma_{\alpha\beta}A \tag{8.1}$$

where $\gamma_{\alpha\beta}$ is the *Gibbs free energy per unit area of the interface* and A is the interfacial area.[1] Note that $\gamma_{\alpha\beta}$ depends on the nature of both phases; it will depend on T and P also, as well as on the crystallographic geometry of the interface in the case of solids. As with any Gibbs free energy, we can write

$$\gamma_{\alpha\beta} = \epsilon_{\alpha\beta} - T\eta_{\alpha\beta} + Pv_{\alpha\beta} \tag{8.2}$$

(in analogy with $G = U - TS + PV$) where $\epsilon_{\alpha\beta}$ is the surface energy per unit area, $\eta_{\alpha\beta}$ is the surface entropy per unit area, and $v_{\alpha\beta}$ is a curious quantity called the surface volume per unit area. It is associated with the change in atomic density

---

[1] Don't confuse $\gamma_{\alpha\beta}$, the surface free energy, with γ_i, the activity coefficient, or γ, the heat capacity ratio, C_P/C_V!

(or rather the reciprocal of the density) in the vicinity of the interface. For instance, near the surface of a solid, the layer-to-layer spacing may depart significantly from the bulk value. In analogy with Eq. (5.36), $v_{\alpha\beta} = (\partial\gamma_{\alpha\beta}/\partial P)_T$, but the $Pv_{\alpha\beta}$ term is very small (except near critical points) and may be almost always ignored.

In analogy with Eq. (5.37), $\eta_{\alpha\beta} = -(\partial\gamma_{\alpha\beta}/\partial T)_P$ for surface entropy. This entropy can be positive or negative, but is most often positive, and usually $\epsilon_{\alpha\beta} > \gamma_{\alpha\beta}$.

To find $\gamma_{\alpha\beta}$, we use a simple experiment. Consider first γ_{sg}, the solid–gas surface free energy. Take a perfect crystal of the solid of interest, grab each end, and pull until the crystal breaks into two pieces. The surface free energy created this way is the work associated with breaking, per unit area:

$$\gamma_{sg} = \int_0^\infty \frac{F(x)\,dx}{A} \qquad (8.3)$$

where $F(x)$ is the force applied along the x direction (the pulling direction) and A is the cross-sectional area of the crystal perpendicular to this direction. While conceptually simple, this experiment has a tacit requirement that is often very difficult to satisfy: the crystal must be atomically perfect throughout (a so-called *single crystal*).

The force that fractures most manufactured materials is less than the force that fractures a single crystal. (Have you ever seen a solid break so cleanly that two perfectly flat surfaces were exposed?) Consequently, most fracture strengths are related to γ_{ss}, a solid-crystallite–solid-crystallite surface free energy, in which two improperly oriented crystals form a solid–solid interface. Nevertheless, Eq. (8.3) is the starting point for theoretical calculations of surface free energies based on models of the interatomic binding forces in a single crystal.

A related simple experiment illustrates another aspect of $\gamma_{\alpha\beta}$ that is used to measure liquid–vapor surface free energy. In Figure 8.1, we have a simple U-shaped frame with a sliding bar enclosing a film of the liquid of interest. (Obviously, soap films come to mind here, but the argument is general for any liquid.) If the sliding bar is free to move without friction along the edges of the U, we can pull on it and increase the film's area. If, in the coordinate system of the figure, we move the bar a distance dx, a new area $dA = 2L\,dx$ (there are two sides to the film!) will be created at the expense of energy in the form of work ($F\,dx$). The surface free energy is

$$\gamma_{lg} = \frac{F\,dx}{2L\,dx} = \frac{F}{2L}\ . \qquad (8.4)$$

If we release the bar, the film will spontaneously pull it back, since γ is positive. (Lowering the area lowers G; hence, in the absence of a constraining force, surface areas tend to decrease spontaneously.)

In this experiment, and in the discussion of liquid surfaces in general, the surface free energy is often called the *surface tension*. There is a slight difference between this experiment and the previous one, however. The surface free energy corresponds to the work done in *forming* a new surface in the former experiment, while the surface tension is related here to the work done in *stretching* the surface. As long as the surface structure produced by stretching is not deformed (the experiment is done slowly enough to allow the stretched surface to relax), the experiments are equivalent. For simple liquids, this will always be the case, but for very viscous macromolecular fluids, relaxation may be very slow, and even solids can be stretched.

An additional method for measuring $\gamma_{\alpha\beta}$ will be discussed in Section 8.4. Now we turn to the magnitude of $\gamma_{\alpha\beta}$ for various systems. Table 8.1 collects solid–liquid, solid–vapor, liquid–vapor, and a few liquid–liquid surface free energy values. Note

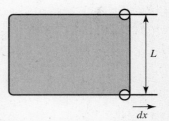

FIGURE 8.1 A simple device for measuring the surface free energy of liquid films suspends the film in a frame with one moveable side. The force required to displace this side by a small distance dx is related to the film's surface free energy.

TABLE 8.1 Interfacial Free Energies/mJ m^{-2}

	γ_{sl}	γ_{sg}	γ_{lg}
H$_2$	—	—	2.88
He	—	—	0.308
Ne	—	17.2	15.1
Ar	—	41.1	13.34
Kr	—	53.3	8.34
Xe	—	60.7	18.46
S	—	—	60.9
Hg	24.4	—	485.
Pb	33.3	—	460.
Ag	126.	1140.	900.
Au	132.	1400.	1128.
Cu	177.	1650.	1350.
Ni	255.	1900.	1725.
Co	234.	1950.	1870.
W	—	2900.	2400.
NaCl	—	310.	190.
MgO	—	1300.	—
LiF	—	370.	—
Si	—	1240.	—
benzene	—	—	28.88
ethanol	—	—	22.75
n-octane	—	—	21.8
H$_2$O	32.1	—	71.99
N$_2$	—	—	8.75
Br$_2$	—	—	41.5

	liquid/water interfaces
Hg/H$_2$O	375.
n-hexane/H$_2$O	51.1
n-octane/H$_2$O	50.8
benzene/H$_2$O	35.
n-butanol/H$_2$O	1.8

Note: As is obvious, these data represent a variety of temperatures. Since γ depends on temperature, they should be considered representative values only. Specific values at any one temperature can be found in data references.

that they range in magnitude from ~ 1 mJ m^{-2} to ~ 1 J m^{-2}. Next, compare the sl, sg, and lg values for the simple atomic metals. Note that $\gamma_{sl} \ll \gamma_{lg} < \gamma_{sg}$, since solid and liquid surfaces are atomically similar (in atomic density per unit area), while the high values for condensed-phase–vapor-phase interfaces reflect the atomically naked condensed-phase surface. (The vapor pressure of these metals, near the melting points where these data were measured, is very low. Thus the solid and liquid surface atoms are exposed to near vacua.) The values for metals are larger than for ionic solids, which in turn are larger than for simple liquids, including the rare gases.

Finally, note the liquid–liquid surface free energies, especially the very small value for the water/n-butanol interface. Recall from Figure 6.16 that n-butanol is "just barely" immiscible in water; hence, the interfacial free energy is quite small, and would decrease to zero at the critical solution temperature when the two fluids become miscible.

The temperature dependence of γ leads to the surface entropy through $\eta = -(\partial \gamma/\partial T)_P$. Unfortunately, very few surface free energies have been measured over a wide range of temperatures, and information on η is scarce.[2] Certain empirical forms for $\gamma(T)$ have been suggested, especially for the liquid–vapor interface. The most common of these is $\gamma(T) = \gamma_0 (1 - T/T_c)^a$ where T_c is the critical temperature, γ_0 is a reference surface free energy, and a is an empirical exponent $\cong 1.2$. (See Problem 8.15.) For the sl interface, very little is known, while values of $\eta \sim 10^{-3}$ J m^{-2} K^{-1} have been measured for a few sg metal interfaces.

8.2 SMALL DROPS AND BUBBLES

Since a sphere is the shape that has a minimum area for a fixed volume, it is obvious why small drops and bubbles are spherical in the absence of external forces (such as gravity or the force of air that rushes past a falling raindrop). These small particles, if not solid, have naturally curved surfaces.[3]

A general element of a curved surface is described by two radii of curvature. These radii are equal for a spherical surface, but in the general case of an arbitrary shape, they are unequal and vary from point to point on the surface. As indicated in Figure 8.2, the radii are found by passing two perpendicular planes through the surface such that the line of intersection of these planes contains the normal to the surface at the point in question. The radii of curvature are those for circles lying in each plane and just tangent to the surface at the desired point. We will call these radii r_1 and r_2, and we will state a theorem, based on mechanical stability, for their equilibrium value if a pressure difference ΔP exists across an interface with surface free energy γ. This theorem, proven by Thomas Young in 1805 and by Pierre Laplace in 1806, states

$$\Delta P = \gamma \left(\frac{1}{r_1} + \frac{1}{r_2} \right) . \qquad (8.5)$$

For a sphere, $r_1 = r_2 = r$, and we can prove this theorem very easily. We distinguish between drops (one interface) and film bubbles (two interfaces—one inside and one outside). When the radius of a sphere is altered by dr, its surface area, $4\pi r^2$, changes by $8\pi r dr$. The differential work done against a pressure change ΔP is

$$dw = \text{(force per unit area)} \times \text{(area)} \times \text{(distance)} = \Delta P \, 4\pi r^2 \, dr \, ,$$

which must equal the change in surface free energy. For a drop,

$$d(\text{surface energy}) = 8\pi r \gamma \, dr \, ,$$

but for a film bubble with two interfaces, inside and outside,

$$d(\text{surface energy}) = 8\pi r (2\gamma) \, dr \, .$$

Equating these yields

$$\Delta P = 2\gamma/r \quad \text{or} \quad \Delta P = 4\gamma/r \qquad (8.6)$$

Liquid drop in vapor or   Film bubble
vapor bubble in liquid

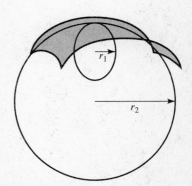

FIGURE 8.2 Any point on a curved plane surface can be characterized by two radii of curvature. In general, these are different, but they become equal for a sphere.

---

[2] Note too that a surface heat capacity at constant area, $C_A = (\partial \epsilon/\partial T)_A$, can be readily defined, but is difficult to measure!

[3] For a view of cubic bubbles the reader is referred to the pink elephants scene in the movie, *Dumbo*.

which is Eq. (8.5) with $r_1 = r_2 = r$ (and the film bubble versus drop distinction is taken into account).

In these equations, $\Delta P = P(\text{inside}) - P(\text{outside})$, and $P(\text{inside}) > P(\text{outside})$. Moreover, this difference increases (from zero for $r = \infty$, a flat surface) as r decreases. Typical values for γ show that ΔP is not very large unless the drop is very small. For example, ΔP for a water drop of radius 1 mm is $2(72.75 \times 10^{-3} \text{ J m}^{-2})/10^{-3}$ m = 150 J m^{-3} = 0.0015 bar, but a droplet of radius 10^{-5} mm has $\Delta P \sim 150$ bar! (Such a droplet has about 140 000 molecules and is "macroscopic" in size.)

The most important consequence of the pressure difference for small drops is its effect on the vapor pressure of the drop. Equation (6.30) gives the effect of pressure on the fugacity of a condensed phase. We focus on the exponential factor (the Poynting correction):

$$(\text{drop vapor pressure}) = (\text{bulk vapor pressure}) \exp\left[\frac{\overline{V}(\Delta P)}{RT}\right] \quad (8.7)$$

and substitute Eq. (8.6) for ΔP:

$$(\text{drop vapor pressure}) = (\text{bulk vapor pressure}) \exp\left[\frac{2\gamma \overline{V}}{rRT}\right]. \quad (8.8)$$

For bulk water at 25 °C, the vapor pressure is 0.031 bar, but the internal pressure of a 10^{-5} mm radius drop is, as we found, 150 bar. This pressure increase, due *entirely* to the inward pull of the outer skin of molecules, raises the vapor pressure of the drop to

$$(0.031 \text{ bar}) \exp\left[\frac{2(72.75 \times 10^{-3} \text{ J m}^{-2})(1.8 \times 10^{-5} \text{ m}^3 \text{ mol}^{-1})}{(10^{-8} \text{ m})(8.314 \text{ J mol}^{-1} \text{ K}^{-1})(298 \text{ K})}\right] = 0.034 \text{ bar}.$$

Thus, bulk water is in equilibrium with water vapor at a pressure of 0.031 bar, but the water vapor pressure must be raised to 0.034 bar to attain equilibrium with the small drops.[4] These situations are contrasted in Figure 8.3, which points out these pressures.

Equation (8.8) was first derived by Kelvin. Not long after, the Wilson cloud chamber was invented in 1896 as a means of viewing charged products of radioactive decay or X-ray irradiation. In the cloud chamber, ions form nucleation sites for the growth of visible-sized drops from a super saturated vapor. J. J. Thompson showed that a drop carrying a charge q followed a variant of Eq. (8.8). If P_0 is the bulk vapor pressure and P is the drop vapor pressure, then Thompson found

$$\ln \frac{P}{P_0} = \left[\left(\frac{2\gamma}{r}\right) - \left(\frac{q^2}{32\pi^2 \epsilon_0 r^4}\right)\right] \frac{\overline{V}}{RT},$$

which is the Kelvin equation with an extra term accounting for the charge. This extra term reduces the tendency of a small drop to evaporate in the vapor at the normal bulk vapor pressure.

With $q = e = 1.6 \times 10^{-9}$ C (the elementary charge) and $\epsilon_0 = 8.854 \times 10^{-12}$ J^{-1} C^2 m^{-1} (the permittivity of vacuum), a 10^{-6} mm radius charged drop of water has a vapor pressure of 0.083 bar, while an uncharged drop the same size has a vapor pressure of 0.089 bar. This difference is not large, but by careful control of the gas pressure in the chamber, the charged droplets will grow to visible sizes preferentially, allowing the tracks of ions to be seen.

---

[4]Hence small drops tend to evaporate at equilibrium pressures appropriate to the bulk phase.

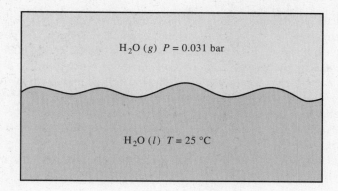
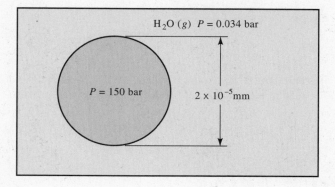

FIGURE 8.3 Water vapor in equilibrium with bulk liquid exhibits a pressure of 0.031 bar (top), but a drop 2×10^{-5} mm in diameter is in equilibrium with vapor at 0.034 bar. The internal pressure of this drop is 150 bar.

8.3 THE NUCLEATION OF PHASES

Here we look at conditions for nucleation and growth of a new phase. We will concentrate on freezing, since many of the measurements of metal γ_{sl} values have come from the study of homogeneous nucleation, but condensation from a vapor phase follows a similar sequence of events with obvious changes in notation (*lg* or *sg* instead of *sl* parameters, etc.).

Recall Figure 6.7 and the arguments related to it. At the fusion temperature the chemical potentials of the solid and liquid phases are equal, but there is a discontinuity in entropy and enthalpy. What concerns us here is the passage of the system through fusion. A pure liquid has to contend with the *sl* surface free energy (a positive contribution) as the initially small regions of solid try to form. We therefore imagine the system is at some temperature T' slightly below the normal fusion point (T_f°). At T_f°, $\Delta \overline{G}_f = 0$, but at T', *after* freezing has taken place, $\Delta \overline{G}_f(T') = \mu_s(T') - \mu_l(T') < 0$.

We imagine that a small solid sphere of radius r has formed by chance in the liquid as a result of random molecular motion. Forming this sphere changes the total free energy of the molecules that comprise it by two readily identifiable amounts, a surface term and a bulk term:

$$\Delta G_{l \to s} = 4\pi r^2 \gamma_{sl} + \frac{4}{3}\pi r^3 \left(\frac{\rho_s}{M}\right) \Delta \overline{G}_f(T') \quad (8.9)$$

$\uparrow$ Free energy of interface

$\uparrow$ Bulk free energy for forming solid sphere from liquid where ρ_s is the solid density, M is the molecular mass, and $\Delta \overline{G}_f(T')$ is negative. As the sphere grows from $r = 0$, the first term dominates, and $\Delta G_{l \to s}$ is positive. For large enough r, however, the second term dominates, and $\Delta G_{l \to s}$ becomes negative.

The free energy change $\Delta G_{l \to s}$ has two key values as r increases (see Figure 8.4). The first is a critical nucleation radius r^* that locates the maximum in $\Delta G_{l \to s}$. Any spheres of radius $r < r^*$ will spontaneously remelt. Any with $r > r^*$ can lower $\Delta G_{l \to s}$ by growing. In particular, any that reach the second key value r_0 (where $\Delta G_{l \to s} = 0$ again) can rapidly lower the free energy and grow to solidify the entire system.

Differentiation of Eq. (8.9) with respect to r leads to the position of the maximum:

$$r^* = -\frac{2\gamma_{sl} M}{\Delta \overline{G}_f(T') \rho_s} \qquad (8.10)$$

and further algebra shows

$$r_0 = \frac{3}{2} r^* . \qquad (8.11)$$

If $T_f^\circ - T'$ is small (which we require it to be), we can expand $\mu(T)$ in a Taylor's series about the fusion temperature and retain only the first two terms:

$$\mu(T') = \mu(T_f^\circ) + \left(\frac{\partial \mu(T_f^\circ)}{\partial T}\right)_P (T' - T_f^\circ)$$
$$= \mu(T_f^\circ) - \overline{S}(T_f^\circ)(T' - T_f^\circ)$$

and find, since $\mu_s(T_f^\circ) = \mu_l(T_f^\circ)$,

$$\Delta \overline{G}_f(T') = -\overline{S}_s(T' - T_f^\circ) + \overline{S}_l(T' - T_f^\circ) = \Delta \overline{S}_f(T' - T_f^\circ) = \Delta \overline{H}_f \frac{\Delta T}{T_f^\circ}$$

where $\Delta T = T' - T_f^\circ < 0$ is the degree of supercooling below the normal freezing temperature. Substituting this expression for $\Delta \overline{G}_f(T')$ into Eq. (8.10) gives a compact expression for the critical nucleation radius:

$$r^* = -\left(\frac{2\gamma_{sl} M}{\rho_s \Delta \overline{H}_f}\right) \frac{T_f^\circ}{\Delta T} . \qquad (8.12)$$

The first factor is of roughly atomic dimensions. For example, data for gold give $r^* = -(2.1 \text{ Å})(T_{\text{fus}}^\circ/\Delta T)$. For gold, $T_{\text{fus}}^\circ = 1336$ K; supercooling by one degree ($\Delta T = -1$ K) indicates a critical nucleation radius ~ 2800 Å, which would be a fairly large sphere of gold containing roughly 10^9 atoms. Anything smaller would tend to remelt.

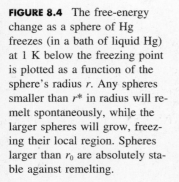

FIGURE 8.4 The free-energy change as a sphere of Hg freezes (in a bath of liquid Hg) at 1 K below the freezing point is plotted as a function of the sphere's radius r. Any spheres smaller than r^* in radius will remelt spontaneously, while the larger spheres will grow, freezing their local region. Spheres larger than r_0 are absolutely stable against remelting.

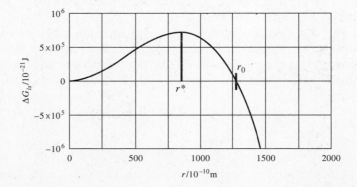

This sensitivity of r^* to ΔT is one reason the melting points of metals are difficult to measure with accuracy. A related problem is the release of the enthalpy of fusion to the liquid, resulting in a local temperature rise that could easily offset the intended supercooling. While a larger supercooling will lower r^*, liquid gold can be supercooled by as much as 230 K.

We have developed here a theory of *homogeneous nucleation:* the solid forms spontaneously in the midst of pure liquid. In most cases, *heterogeneous nucleation* due to impurities (such as the ions in a cloud chamber or the usual things one considers as impurities) or to nucleation at imperfections on the container's wall dominate the growth of a new, lower temperature phase.

Our theory is also applicable to condensation from a vapor, as in atmospheric cloud formation. We will have more to say about this in the next chapter where we will discuss the natural cooling mechanism—adiabatic expansion—that accompanies a parcel of air as it rises in the atmosphere.

We close this section with a retrospective comment on the "theory" of freezing based on moonlight that was quoted in the introduction to Chapter 6. We no longer believe moonlight intervenes directly in the freezing of water, but the phenomenon is real. Forsyth quotes a supercooling of water to 22 °F ($\Delta T = -5.5$ K), which is not unreasonable. Clearly, the crystallization of light frosts on a meadow are not likely to be homogeneous in nature. One suggestion that has been made to explain the correlation between crystallization and moonrise is that moonlight is a signal for small meadow bugs to arise and scurry about, doing whatever meadow bugs do, but also providing the mechanical agitation known to enhance the crystallization of supercooled liquids!

8.4 EQUILIBRIUM FORCES AMONG INTERFACES

Droplets of mist are spherical; snowflakes are any of a myriad of fantastic shapes. Raindrops sit like beads on a waxed car roof, but spread as a thin sheet on clean glass. These various phenomena are all related to the interplay among surface free energies. The spherical shape of an unperturbed liquid drop is easy to understand as a result of the spontaneous minimization of surface free energy. But how to explain snowflakes! The six-fold symmetry of snowflakes reflects the six-fold symmetric packing of water molecules in ice. The filigree details that complete one's mental picture of a snowflake results from kinetic factors as well as surface factors. Slight temperature changes can alter the rate and geometry of snowflake growth, and a full explanation is beyond the scope of our discussion here.

What would seem to be more tractable than the snowflake is the rationale behind the structure of many ionic and metallic crystals, such as those shown in Figure 8.5. These crystals include the simple cubes of table salt and many other natural crystals that have aroused curiosity and admiration for centuries. Rather than develop the most detailed theory of crystal geometry, which is based on topics treated in detail in Chapter 16, we will turn instead to the first systematic treatment of crystallography proposed with remarkable insight by R. J. Haüy in 1784. His ideas were published in 1801 in his *Traité de Minéralogie,* which was decades ahead of its time. We will use one of his drawings, Figure 8.6, to illustrate the modern theory.

Haüy proposed that crystal facets could be understood by the packing of identical building blocks, as illustrated in the figure. He was almost right, but a consideration of surface energy is also necessary.

Consider a flat (atomically flat) surface such as the plane EOO'E' of the figure. For a simple cubic symmetry atomic packing, this plane has the least amount of

FIGURE 8.5 The regular geometric shapes of many simple crystals (sodium chlorate on the left and ethylene diamine tartrate on the right) is governed by surface free energy minimization criteria, but often kinetic or other factors play important roles, such as in the six-fold symmetric snowflake.

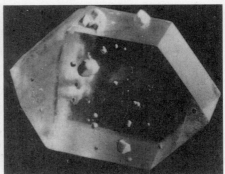

FIGURE 8.6 This is Figure 13, Plate II from Vol. 5 of *Traité de Minéralogie,* by R. J. Haüy, published in Paris in 1801. It shows how various crystal faces could have different surface free energies as a result of differing degrees of atomic exposure.

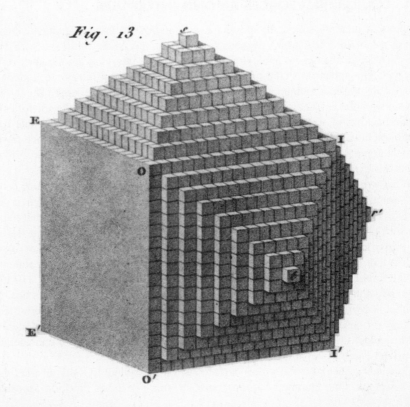

exposed atomic area of any that can be formed by cutting the crystal. It therefore has the lowest surface free energy. In contrast, any of the remaining planes in the figure (the pyramidal faces) that are at 45° to the flat planes have a maximum amount of exposed atomic area and the highest surface free energy of any plane. Those cut at intermediate angles to the flat plane will have a series of terraces and steps and will exhibit intermediate surface free energies. At any finite temperature, the surface entropy term will tend to make surfaces cut at slightly different angles vary smoothly in free energy. The entropy term will arise from small surface defects, such as occasional extra or missing atoms on a terrace or step.

We can summarize these intuitive conclusions in a polar plot in which the radial distance from the origin represents the magnitude of γ and the angle represents the direction of a line drawn perpendicular to the plane in question. Such a plot is shown as the heavy line in Figure 8.7. (This is actually a two-dimensional slice through a three-dimensional surface of values for planes cut at any angle). It is a plot that reflects the cubic symmetry of atomic parking, and, of most interest, it can be used to find the *equilibrium shape* of the crystal.

Intuition tells us that, for this simple crystal, the equilibrium shape should be a perfect cube. A theorem derived by Wulff in 1901 uses the γ plot to reach the same conclusion. One draws a radial line from the origin to the γ curve followed by a line (or a plane, in three dimensions) perpendicular to the radial line at its intersection with the γ curve. When done at all radial directions, the innermost area (volume) formed by those planes is the equilibrium shape, as indicated by the thin lines in the figure.

This is a particularly simple crystal. As with most general theorems, Wulff's is most useful in less obvious situations. Figure 8.8 shows a γ plot for a more complex crystal structure along with the equilibrium shape predicted by Wulff's construction. The origin of crystal facets is made particularly clear by this type of analysis.

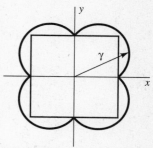

FIGURE 8.7 A Wulff plot for a simple, cubic crystal determines the crystal's shape (thin line) by plotting the surface free energy (thick line) in a polar coordinate system. The radius gives the magnitude of γ while the angle (or angles in three dimensions) locates a line perpendicular to any imagined crystal plane surface. Those planes of lowest γ will determine the crystal's shape.

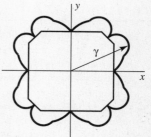

FIGURE 8.8 A more complex but still basically cubic crystal may develop *facets* due to several planes of singularly low surface free energy, as this Wulff plot demonstrates.

Wetting and Capillarity

We turn now to a different sort of equilibrium geometry: a liquid in equilibrium with its vapor and in contact with a solid surface. A drop of water on glass or a column of mercury in a thermometer are two classic examples of this situation, and the general theory of capillarity and capillary action start with this problem in equilibrium theory.

We focus on a point somewhere along a line that includes the intersection of all three phases, as shown in Figure 8.9. Since the gas and liquid phases are freely mobile, they will adjust their relative shapes until our focal point comes to rest with

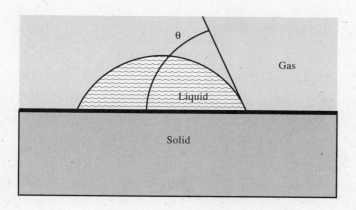

FIGURE 8.9 The equilibrium shape of a liquid on a flat solid surface depends on the three interfacial free energies and leads to a single tangential angle θ between the liquid–gas interface and the liquid–solid interface.

no net force on it. The rigidity of the solid simplifies the analysis somewhat; since it is immobile, we can treat the solid surface as a plane, but it will be under tension at equilibrium. The surface free energy is best thought of as a surface tension here, a force pulling at the focal point along the direction of each interface. For Figure 8.9, this equilibrium of forces is expressed by

$$\gamma_{sg} = \gamma_{sl} + \gamma_{lg} \cos\theta \tag{8.13}$$

where θ is the limiting angle made by the liquid–vapor interface with the solid. If $\gamma_{lg} > |\gamma_{sg} - \gamma_{sl}|$, then no value for θ can satisfy Eq. (8.13). Either the liquid phase wets the solid phase completely ($\gamma_{sg} > \gamma_{sl} + \gamma_{lg}$) or the vapor "wets" the surface ($\gamma_{sl} > \gamma_{sg} + \gamma_{lg}$), causing the liquid to detach from the solid or sit as a ball on the solid.

Eq. (8.13) was first deduced by Thomas Young in 1805, but consequences of it are still being discovered. At the critical point, γ_{lg} and $\gamma_{sg} - \gamma_{sl}$ both vanish, and the contact angle becomes indeterminate. It was not until 1977 that Cahn showed that just below the critical point one of the fluid phases will wet the solid phase completely, excluding contact between the solid and the other fluid phase.

There are several experimental techniques for measuring interfacial free energies based on Eq. (8.13), or variants of it that apply to three fluid phases in equilibrium. Some of these are discussed in the problems for this chapter. Care must be taken that the fluid phases are pure (or of known composition) and that the solid surface is clean in order to obtain reproducible results.

CHAPTER 8 SUMMARY

The surface free energy contribution to the total free energy, $\gamma_{\alpha\beta} \, dA$, is the new thermodynamic variable of this chapter. The surface free energy per unit area, γ, depends on the composition and physical state of both bulk phases that form the interface, and γ is inherently a positive quantity. (If it were not, the interface would spontaneously grow and consume the bulk completely!) In general, one expects $\gamma_{sg} > \gamma_{lg} \gg \gamma_{sl}$ based on atomic views of these interfaces.

Surface contributions to the total free energy are important only for small amounts of a phase: small drops, bubbles, or grains of solid. Due to the inward force a surface exerts on a small particle, the pressure is greater inside the particle than outside, and this pressure difference is directly proportional to γ and inversely proportional to the particle's radius:

$$P(\text{inside}) - P(\text{outside}) = \frac{2\gamma}{r} \text{ (for drops, grains, and bubbles)}, = \frac{4\gamma}{r}\text{(for film bubbles)}$$

This pressure difference affects the vapor pressure of a condensed phase drop (or grain):

$$(\text{drop vapor pressure}) = (\text{bulk vapor pressure}) \exp\left(\frac{2\gamma \overline{V}}{rRT}\right)$$

The formation of a new phase is also controlled by surface phenomena. Homogeneous nucleation theory predicts the size a new, low-temperature phase must reach in order to grow to truly macroscopic size:

$$\text{Critical nucleation radius} \quad r^* = -\left(\frac{2\gamma_{sl}M}{\rho_s \Delta \overline{H}_f}\right)\frac{T_f^\circ}{T' - T_f^\circ}$$

but most nucleation processes commonly encountered are governed (or at least initiated) by heterogeneous nucleation phenomena.

The shape an interface assumes at equilibrium is also governed by a free energy interplay. Not only condensed phase shapes but also capillarity phenomena fall into this category:

$$\text{Equilibrium balance of capillarity forces} \quad \gamma_{sg} = \gamma_{sl} + \gamma_{lg} \cos\theta$$

These phenomena are among the most widely applied to the experimental measurement of surface free energies.

FURTHER READING

A more through discussion of surface phenomena including discussions of many important experimental techniques can be found in *Physical Chemistry of Surfaces*, 4th ed., by A. W. Adamson (John Wiley & Sons, New York, 1982). In addition, the references at the end of Chapter 16 include surface science texts and monographs that go beyond this chapter's discussions.

ACCESS TO DATA

A very thorough compilation of liquid surface tensions can be found in J. J. Jasper, *J. Phys. Chem. Ref. Data*, **1**, 841 (1972). A critical survey of water's surface tension can be found in N. B. Vargaftik, *et al.*, *J. Phys. Chem. Ref. Data*, **12**, 817 (1983).

PRACTICE WITH EQUATIONS

8A How much greater is the surface contribution to the free energy of a cube compared to that of a sphere of the same volume? (8.1)

ANSWER: $6/[3^{2/3}(4\pi)^{1/3}] = 1.24$

8B How much work must be performed to reversibly increase the area of a water film by 10 cm^2? (8.1), (8.4), Table 8.1

ANSWER: 72.75 µJ

8C What is the internal pressure of a drop of Hg(l) 0.1 mm in diameter? (8.6), Table 8.1

ANSWER: 0.188 bar

8D At what temperatures will a droplet of water 50 Å in radius have a tendency to evaporate rather than grow when in water vapor at 1 atm, 373.15 K ($\Delta \overline{H}_{vap}$ = 40.656 kJ mol^{-1}, ρ_l = 958.36 kg m^{-3}, and γ_{lg} = 61.06 mJ m^{-2} at 373 K)? (8.12)

ANSWER: T > 367 K

8E Over the range $T = T_f = 660.37$ °C to ~1300 °C, γ_{lg}(Al)/mJ m^{-2} = 865 − 0.14(T/°C − 660). At 1200 °C, the contact angle of Al(l) on Al$_2$O$_3$(s) is 60°. What is $\gamma_{sg} - \gamma_{sl}$? (8.13)

ANSWER: 395 mJ m^{-2}

PROBLEMS

SECTION 8.1

8.1 The fraction of molecules that are on the surface of a system is usually negligibly small, but this fraction can be appreciable for small drops or crystals. Consider a small cluster composed of spherical molecules 2r in diameter. How does the fraction of molecules on the surface vary with the radius R of such a cluster? The diagram below shows a schematic of one such cluster to aid your thinking. Neglect the "open" volume between the packed molecules, assume $r << R$, and graph the fraction as a function of R/r for $10 \leq R/r \leq 50$. (Such spheres hold ≅ 125 000 molecules.)

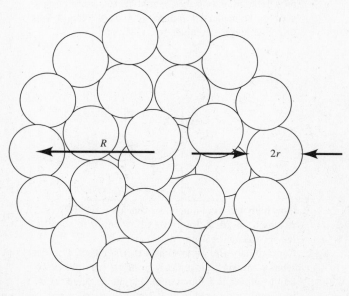

8.2 Two liquids with surface free energies γ_1 and γ_2 (assume $\gamma_2 > \gamma_1$) are placed as films in the device shown below, which is similar to that shown in Figure 8.1.

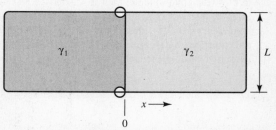

A wire frame of width L is fitted with a sliding bar initially held fixed at $x = 0$ in the center of the frame so that the initial areas of both films are equal. The bar is released, and the unequal γ's cause it to move. What is its final equilibrium position?

8.3 What is the minimum energy needed to turn a 1.0 L sample of bulk water at 20 °C and 1.0 bar into a mist of 0.010 mm radius droplets? If this mist were to condense adiabatically back to a 1.0 L bulk, how much would the temperature rise ($\overline{C}_P^\circ/R = 9.06$ for water)?

8.4 Consider two immiscible liquids α and β in contact. Write an expression for the work per unit area required to pull these liquids apart at the interface in terms of the three relevant surface free energies $\gamma_{\alpha\beta}$, $\gamma_{\alpha g}$, and $\gamma_{\beta g}$. Compute the magnitude of this work for the case of a benzene–water interface using data from Table 8.1. As was mentioned in the text, if $\gamma_{\alpha\beta} = 0$ the liquids will mix. What would it mean if $\gamma_{\alpha\beta} > \gamma_{\alpha g} + \gamma_{\beta g}$?

SECTION 8.2

8.5 What is the maximum depth below the surface of a glass of champagne at which bubbles of radius 0.1 mm can form? (Assume champagne has the surface free energy and density of water. The formation of champagne bubbles is actually a more complicated phenomenon than this problem suggests.)

8.6 If we write the differential of Eq. (8.1) in the form

$$dG = -S\,dT + V\,dP + \gamma\,dA$$

and consider $G = G(T, P, A)$, new Maxwell relations can be derived along the lines used in Section 5.6. First, derive

$$\left(\frac{\partial V}{\partial A}\right)_{T,P} = \left(\frac{\partial \gamma}{\partial P}\right)_{A,T},$$

then apply this relation to a spherical drop, and show that it leads directly to the Laplace equation, Eq. (8.6). Then derive

$$\left(\frac{\partial \gamma}{\partial T}\right)_{A,P} = -\left(\frac{\partial S}{\partial A}\right)_{T,P} \equiv \eta,$$

which is the fundamental expression for surface entropy.

8.7 Shown below is a tube equipped with three valves, and on either end of the tube are soap bubbles with different radii. With valve 1 closed, valve 2 was opened to blow the small bubble, then it was closed and the large bubble was blown through valve 3. What will happen if valve 1 is opened so that the two bubbles are connected by an open tube? Does the large bubble get larger or does the small bubble get larger?

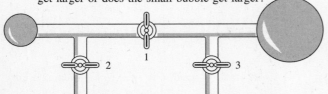

SECTION 8.3

8.8 Gibbs was the first to show that the change in free energy on formation of a sphere of a new phase having the critical radius r^* is equal to one-third the surface free energy of such a sphere. Prove that this is so.

8.9 The molar enthalpy of fusion of mercury is 2.292 kJ mol^{-1}. If Hg(g) is supercooled 20 K below its normal boiling point (from 630 K to 610 K), and if a liquid sphere of radius r_0 as in Eq. (8.11) forms adiabatically from the vapor at this temperature, what will be the sphere's final temperature? Is the sphere likely to survive and grow? (At this temperature, the density of Hg(l) is 12.74 g cm^{-3}, $\gamma_{lg} = 420$ mJ m^{-2}, and $\overline{C}_P^\circ/R = 3.36$.)

SECTION 8.4

8.10 One way to measure surface tensions is to measure the rise (or, in some cases, depression) of a column of fluid in a narrow capillary tube. In terms of the geometry shown below for capillary rise, the upward pull of the liquid on the inside walls of the capillary will continue until the downward gravitational force on the raised column of liquid equals that upward pull. The downward gravitational force is

$$2\pi g(\rho - \rho_0) \int_0^r (h + y)\,x\,dx$$

where $y(x)$ is the curve describing the shape of the meniscus, h is the height of the bottom of the meniscus above the bulk surface, ρ is the liquid density, and ρ_0 is the density of the surrounding gas. The upward force is $2\pi r \gamma \cos\alpha$ where α is the limiting contact angle ($\gamma \cos\alpha$ is the vertical component of the force acting around the circumference, $2\pi r$). Derive the simplest relation among these variables by assuming that $y(x)$ can be neglected and that $\alpha = 0$ (the liquid wets the inner surface completely).

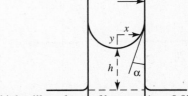

How high will a column of benzene ($\rho = 0.8765$ g cm^{-3}) rise in air at 20 °C ($\rho_0 = 0.0012$ g cm^{-3}) in a capillary of radius 0.15 mm?

8.11 Can capillary rise be used as a thermometer? Use the relation among h, r, ρ and γ from the previous problem to plot h as a function of T from 20 °C to 90 °C if water is the liquid, $r = 0.2$ mm, ρ_0 is neglected, and the temperature variations of ρ and γ are given by the approximate relations

$$\rho/g\ cm^{-3} = 1.002 - 1.2 \times 10^{-4}\ (T/°C) \\ - 3.2 \times 10^{-6}\ (T/°C)^2$$

$$\gamma/mJ\ m^{-2} = 75.83 - 0.1477\ (T/°C).$$

8.12 Following the logic of Problem 8.10, derive an expression for the capillary rise between two parallel flat plates spaced a distance ℓ as shown below and dipped into a fluid of surface free energy γ. Imagine the plates to be much longer than ℓ so that "end effects" can be neglected. (*Hint:* Consider the capillary rise per unit length along the plates.) For a given fluid, is the rise greater for these plates or for a capillary tube of diameter ℓ?

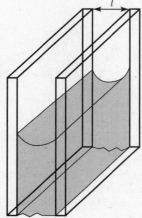

8.13 Consider a crystal growing from its melt as long, slender needles. Call the plane perpendicular to the long (needle) axis the x-y plane. If $\gamma_{s\ell}$ is 165 mJ m^{-2} for crystal faces perpendicular to the x and y axes but 150 mJ m^{-2} for crystal faces at 45° to these axes, predict the cross-sectional shape of the needles.

GENERAL PROBLEMS

8.14 Imagine the device in Figure 8.1 turned so that the sliding bar is horizontal at the bottom of the frame with the liquid film tending to pull the bar up against gravity. What mass can be hung from the bar to counteract this pull and achieve equilibrium? What is the magnitude of the mass (per unit length of the bar) if the liquid is water? if it is ethanol?

8.15 In 1945, Guggenheim wrote an interesting paper on "The Principle of Corresponding States" [E. A. Guggenheim, *J. Chem. Phys.* **13**, 253 (1945)], a principle discussed in Chapter 1. In it, he remarked how data for the equilibrium temperature dependence of the densities of liquid and gas phases of Ar were very well fit by the empirical formulae

$$\frac{\rho_l}{\rho_c} = 1 + \frac{3}{4}\left(1 - \frac{T}{T_c}\right) + \frac{7}{4}\left(1 - \frac{T}{T_c}\right)^{1/3}$$

and

$$\frac{\rho_g}{\rho_c} = 1 + \frac{3}{4}\left(1 - \frac{T}{T_c}\right) - \frac{7}{4}\left(1 - \frac{T}{T_c}\right)^{1/3}$$

where ρ_c is the critical density and T_c is the critical temperature. If these two expressions are added, one obtains the "law of rectilinear diameter" that is discussed in Problem 6.17. If they are subtracted, a "new" expression is obtained. Guggenheim then went on to discuss the temperature dependence of surface tension, noting that Katayama had found the proportionality

$$\gamma y^{-2/3} \propto \left(1 - \frac{T}{T_c}\right)$$

where

$$y = \frac{1}{v_l} - \frac{1}{v_c} = \frac{1}{v_c}\frac{(\rho_l - \rho_g)}{\rho_c}$$

fit data for a number of organic substances quite well. Use these expressions to derive Guggenheim's final expression for the temperature variation of γ:

$$\gamma = \gamma_0\left(1 - \frac{T}{T_c}\right)^{11/9}$$

where γ_0 is a constant. How well does this expression work for water, and what value of γ_0 would you recommend? Data for water are given below, and $T_c(H_2O) = 647.3$ K.

$T/°C$	$\gamma/mJ\ m^{-2}$
360.	1.90
320.	9.81
280.	18.94
240.	28.42
200.	37.69
160.	46.58
120.	54.96
80.	62.67
40.	69.60
0.01	75.64

8.16 Continuing from the previous problem, what is the surface entropy of water at its normal boiling point? For Ar, $\gamma_0 = 36.0$ mJ m^2, $T_{vap} = 87.45$ K, and $T_c = 150.8$ K. How does η_{lg} for Ar at T_{vap} compare to that for water? Is the trend that which you would expect for a comparison between a hydrogen-bonding fluid and an inert-gas fluid?

8.17 How does particle size influence a solid's solubility? Imagine that the solid forms an ideal solution so that $\mu_i = \mu_i° + RT \ln x_i$ describes the chemical potential of solute i in solution at mole fraction x_i. Imagine the solid (in equilibrium with the saturated solution!) to consist of N small spheres of radius r and surface free energy γ. If the solubility of i when the solid consists of a single large particle of negligible surface area is x_i°, show that the solubility of the small particles is given by

$$x_i = x_i^\circ \exp\left(\frac{3M\gamma}{\rho rRT}\right)$$

where M is the solid's molecular mass and ρ is its density. (*Hint:* Write the chemical potential of the solid as $\mu^\circ + \gamma \overline{A}$ where $\overline{A}$ is the molar surface area.) By what factor is the solubility enhanced at 300 K if $\gamma \sim 500$ mJ m^{-2}, $M \sim 200$ g mol^{-1}, $\rho \sim 2.5$ g cm^3, and $r \sim 10^{-4}$ mm? A common practice in gravimetric analysis is to "digest" a freshly precipitated solid such as BaSO$_4$ by holding the solution at elevated temperatures for some time during which the average particle size is found to grow. Explain how this happens.

8.18 When a small drop of one liquid rests on the surface of another, the cross-sectional shape (at its most ideal) is defined by two contact angles as shown below. Derive the equilibrium relation among the various γ's and angles analogous to Eq. (8.13) that describes this equilibrium.

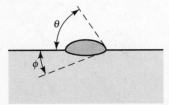

If a small solid sphere of radius r rests on the liquid, the geometry can be described as below.

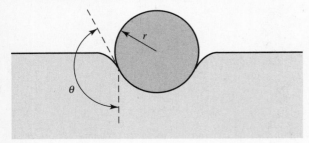

The angle θ here is measured from the vertical to the sphere's tangent line at the ls contact. If the weight of the sphere is W, how is W related to θ and surface tensions at equilibrium? In the 1941 *Report of the Annual Meeting of the New Jersey Mosquito Extermination Association,* **28,** 19 (1941), M. A. Manzelli reported that soapy water can eradicate mosquitoes at soap concentrations well below the level of toxicity, around 0.1 – 0.25% soap by weight. Explain this observation.

8.19 Consider a spherical container filled with a liquid and its vapor. In a gravitational field ($g \neq 0$), we know the liquid will lie at the bottom of the sphere with a flat surface. If gravity is removed ($g = 0$), the shape the system can take will be one of these three general forms:

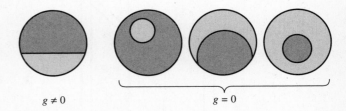

The liquid will be detached from the sphere and surrounded by vapor (left), both liquid and vapor will be in contact with the sphere (center), or the vapor will be surrounded by liquid (right). If we define a free energy change $\Delta G = G(g = 0) - G(g \neq 0)$, use Eq. (8.1) to show that

$$\frac{\Delta G}{\gamma_{lg}} = \Delta A_{lg} - \Delta A_{sl} \cos \theta$$

where ΔA_{lg} is the difference in the liquid–gas interfacial areas, ΔA_{sl} is the solid–liquid interfacial area difference, and θ is the contact angle with the container defined so that $\theta = 0$ if the liquid surrounds the gas and $\theta = \pi$ if the gas surrounds the liquid. This expression means that the weightless shape of the system is governed by the fraction of the container filled with liquid (which governs ΔA_{sl}) as well as the chemical nature of the container and its contents (which governs $\cos \theta$).

CHAPTER 9

Gravitational, Magnetic, and Electric Fields

EXTERNAL force fields are a fact of life on Earth that we have more or less ignored in previous chapters. We did invoke gravity to introduce work in Chapter 2, and we have tacitly imagined gravity holding beakers onto tables and solutions in beakers, but we have yet to introduce a gravitational field into, for instance, a free-energy expression.

In addition to gravity, we consider here the thermodynamic consequences of static electric and magnetic fields. These fields interact in strong and important ways with molecules on a microscopic level, but we will discuss here primarily the macroscopic consequences of these fields. The microscopic view is discussed in Chapters 15 and 21.

The first section treats gravitational fields and some important consequences such fields have on our lives: planetary atmosphere composition, temperature, and pressure profiles, and the equilibrium distribution of elements in the Earth's crust. The second section discusses centrifugation, while electric and magnetic field effects are discussed in the third section.

Each section will follow the same logic: identify how the field interacts with the system, see how the field's energy influences the free energy of the system, apply the usual stability criteria of equilibrium, and explore the consequences.

9.1   Equilibrium in Gravitational Fields

9.2   Centrifugal Fields

9.3   Magnetic and Electric Field Energies

9.1 EQUILIBRIUM IN GRAVITATIONAL FIELDS

A gravitational force exists between any two masses, m_1 and m_2, according to an inverse-square distance expression

$$F = \frac{G m_1 m_2}{r^2} \qquad (9.1)$$

where r is the separation between the centers of mass of objects 1 and 2 and G is the universal gravitational constant. The magnitude of G ($\sim 6.673 \times 10^{-11}$ N m^2 kg^{-2}) is so small that the force is usually of no importance unless at least one of the masses is huge. If m_2 is the mass of the Earth and r is the distance from the surface to the center of mass of the Earth, we rewrite Eq. (9.1) as

$$F = m_1 \left(\frac{G m_2}{r^2} \right) = m_1 g \qquad (9.2)$$

where g is the local acceleration of gravity. Near sea level, $g \cong 980$ cm s^{-2}. On the surface of Venus, $g \cong 882$ cm s^{-2}, while $g \cong 392$ cm s^{-2} on Mars, and so on, from planet to planet. Equation (9.2) shows that g varies with altitude above the reference surface (sea level on Earth), but most altitudes of interest are small compared to the Earth's radius, and g is considered constant in this section.

Moving mass 1 along the r direction changes its gravitational potential energy. This change is (compare Eq. (2.1))

$$\Delta \phi_G = \phi_G(r_2) - \phi_G(r_1) = \int_{r_1}^{r_2} m_1 g \, dr = m_1 g (r_2 - r_1) \; . \qquad (9.3)$$

This equation brings gravity into thermodynamics. The work associated with this displacement is transferred to the system and alters its internal energy.

The criteria for equilibrium in a gravitational field are still those discussed in Chapter 5, especially in Section 5.2. At equilibrium, entropy under the constraints of constant volume and internal energy is a maximum, and any perturbation of the system away from equilibrium (produced by the isolated system itself, not by external intervention) lowers the entropy. In the absence of gravity, these conditions produce the uniformity of T, P, and μ_i that have characterized equilibrium states in the preceding chapters. While we will not prove it, one can show that *uniform T and uniform* $\mu_i + \overline{\phi}_G(r)$ hold for equilibrium states in the presence of gravity. As common experience indicates, pressure is no longer uniform in a gravitational field, and it will be one of our tasks in this section to predict how pressure might vary with r.

Since the pressure varies, so will the density of the system or its molar volume. Consider a tall cylinder of gas in the absence of gravity, but at equilibrium. The density is everywhere uniformly

$$\rho = \text{mass density} = \frac{1}{V} \sum_i M_i n_i$$

where M_i is the molar mass of component i. Apply a gravitational field along the long axis of the cylinder. One expects the density to be greater at the "bottom" of the cylinder (at regions of lower gravitational potential energy) than at the "top." But as we travel up and down the cylinder (along the r direction), we must find $\mu_i + \overline{\phi}_G(r)$ to be the same for each component in the system:

$$\left(\frac{\partial}{\partial r} [\mu_i(T) + \overline{\phi}_G(r)] \right)_T = 0 \qquad \text{for all } r \; . \qquad (9.4)$$

9.1 EQUILIBRIUM IN GRAVITATIONAL FIELDS

There are two terms in this equation, $(\partial \mu_i/\partial r)_T$ and $(\partial \overline{\phi}_G/\partial r)_T$. The first can be written

$$\left(\frac{\partial \mu_i}{\partial r}\right)_T = \left(\frac{\partial \mu_i}{\partial P}\right)_T \left(\frac{\partial P}{\partial r}\right)_T = \overline{V}_i \left(\frac{\partial P}{\partial r}\right)_T$$

(using Eq. (6.7) to introduce $\overline{V}_i$) while the second is

$$\left(\frac{\partial \overline{\phi}_G(r)}{\partial r}\right)_T = M_i g = \frac{m_i}{n_i} g \ .$$

Combining these shows what Eq. (9.4) really says:

$$\left(\frac{\partial P(r)}{\partial r}\right)_T = -\frac{M_i}{\overline{V}_i} g = -\rho_i(r) g \ . \tag{9.5}$$

(The mass density now depends on r through the r dependence of $\overline{V}_i$.) This equation expresses the equilibrium between the weight of a section of the system and its buoyancy.

Suppose the system is a one-component ideal gas. Then

$$P(r) = \frac{RT}{\overline{V}(r)} = \frac{\rho(r) RT}{M}$$

and Eq. (9.5) becomes

$$\left(\frac{\partial P(r)}{\partial r}\right) = -\frac{M P(r) g}{RT} \ ,$$

which can be rearranged to

$$\frac{dP(r)}{P(r)} = -\frac{Mg}{RT} dr$$

and integrated from $r = 0$ to $r = h$ yielding

$$P(h) = P(0) \, e^{-Mgh/RT} \tag{9.6a}$$

for the pressure variation, or

$$\rho(h) = \rho(0) \, e^{-Mgh/RT} \tag{9.6b}$$

for the density variation with altitude h, at constant temperature.

EXAMPLE 9.1

What is the fractional change in density from top to bottom of a column of $N_2(g)$ 100 m tall (a) at 300 K and (b) at the normal boiling point of N_2, 77 K?

SOLUTION Note first that

$$\frac{g}{R} = \frac{(980 \text{ cm s}^{-2})(1 \text{ m}/100 \text{ cm})}{8.314 \text{ J K}^{-1} \text{ mol}^{-1}} = 1.179 \text{ K mol kg}^{-1} \text{ m}^{-1}$$

or, with $M = 2.8 \times 10^{-2}$ kg mol^{-1} for N_2,

$$\frac{Mg}{R} = 3.30 \times 10^{-2} \text{ K m}^{-1} \ .$$

Thus

$$\frac{\Delta \rho}{\rho} = \frac{\rho(100 \text{ m}) - \rho(0)}{\rho(0)} = \exp\left[-\frac{Mgh}{RT}\right] - 1$$

and since Mg/R is small, whenever the ratio h/T is ~ 1 m K^{-1} or less (as it is here), we can expand the exponential and write the approximate expression

$$\frac{\Delta \rho}{\rho} = \exp\left[-\frac{Mgh}{RT}\right] - 1 = \left[1 - \frac{Mgh}{RT} + \ldots\right] - 1 \cong -\frac{Mgh}{RT} .$$

At 300 K, this is

$$\frac{\Delta \rho}{\rho} = -1.1 \times 10^{-2} \quad \text{(hence density variations over normal distances in gases are quite small)}$$

and at 77 K

$$\frac{\Delta \rho}{\rho} = -4.3 \times 10^{-2} .$$

It is instructive to return to the Clapeyron equation, Eq. (6.11), and explore how, for instance, the measured value of the boiling point of N_2 might change if the experiment were performed under various pressures characteristic of a first floor lab, a second floor lab, one on a coast versus one in Denver, etc., and to compare these pressure changes to normal weather induced changes in pressure.

➡ **RELATED PROBLEMS** 9.1, 9.5

On a grand scale, Eq. (9.6) could be a model for an *isothermal planetary atmosphere*. For a multicomponent (ideal gas) atmosphere, it generalizes to

$$P(h) = P(0) \sum_i x_i(0) e^{-M_i g h/RT} \tag{9.7}$$

where $x_i(r)$ is the mole fraction of component i at altitude r. Moreover, the composition would vary with altitude according to

$$x_i(h) = x_i(0) \frac{P(0)}{P(h)} e^{-M_i g h/RT} . \tag{9.8}$$

This equation predicts, as expected, that heavier species are most concentrated near the surface. Implicit in this discussion is ideal mixing of the gaseous components, and a very different conclusion results if this criterion is not valid.

Consider, for example, the distribution of elements in a planetary crust. If elements mixed ideally, we would find lots of Li, C, Al, etc. near the surface and very little U, Th, etc. Of course, elements do react and mix nonideally. (It is clear that the Earth's crust is not in equilibrium or even uniformly mixed; yet, locally, equilibrium is often closely approximated.) To understand the elemental distribution, we must return to the fundamental expression, Eq. (9.4).

We would need to consider the variation of μ_i with height, differentiating terms such as Eq. (5.48), in this form:

$$\left(\frac{\partial \mu_i}{\partial h}\right)_T = \left(\frac{\partial(\mu_i^\circ + RT \ln f_i)}{\partial h}\right)_{T, n_j} = RT \left(\frac{\partial \ln f_i}{\partial x_i}\right)_{T, n_j} \left(\frac{dx_i}{dh}\right) .$$

A sum of such terms, one for each *other* element, would govern the distribution of any *one* element (along with the buoyancy term), resulting in a series of coupled equations. This topic remains a very complex problem, yet with simplifications and reasonable estimates for unknown quantities, Brewer first showed in 1951 that this theory could closely approximate the known distributions of major elements and could explain why uranium, in spite of its mass, *decreases* in concentration with depth. This topic remains an area of continuing interest for many practical applications in geochemistry.

9.1 EQUILIBRIUM IN GRAVITATIONAL FIELDS

Returning to the atmosphere, real atmospheres are neither particularly isothermal nor at equilibrium. Winds blow; rains and snows fall; the barometer goes up and down. A model somewhat more accurate than isothermal can be made if we turn off the horizontal winds and dry the atmosphere. The major air flow (due to polar cold and equatorial heat) is vertical, and a parcel of air will spontaneously rise from an equatorial surface. Gas near atmospheric pressure is not a good medium for conduction of heat, and each parcel of air is fairly well insulated (adiabatically) from each neighboring parcel. Thus, a good approximation is that each parcel of air rises, expands, and cools *adiabatically* and *reversibly*.

We return to the differential form of Eq. (9.6)

$$\frac{dP(r)}{P(r)} = -\frac{Mg}{RT} dr$$

where M is the average molar mass for air (about 29 g mol^{-1}). We can introduce the reversible adiabatic path through Eq. (2.29), $PV^{5/3} = $ const., which was derived for an ideal gas with $\overline{C}_V/R = 3/2$. Recall that the exponent 5/3 equals $\overline{C}_P/\overline{C}_V$ (symbolized γ). With good accuracy, a real gas follows $PV^\gamma = $ constant if the temperature change resulting from the process is not large. Of most use is the version of this equation that relates T and P:

$$\frac{T}{P^{(\gamma-1)/\gamma}} = \text{constant} .$$

The differential form of this equation,

$$\frac{dT}{T} - \left(\frac{\gamma-1}{\gamma}\right)\frac{dP}{P} = 0 ,$$

can incorporate the differential form of Eq. (9.6) and yield

$$\frac{dT}{T} + \left(\frac{\gamma-1}{\gamma}\right)\frac{Mg}{RT} dr = 0$$

or

$$\frac{dT}{dr} = -\left(\frac{\gamma-1}{\gamma}\right)\frac{Mg}{R} . \tag{9.9}$$

This equation shows that the temperature drops with increasing altitude (dT/dr is negative), and it predicts $dT/dr = -9.9$ K km^{-1} for air (for which $\gamma = 1.4$). In fact, the air temperature does drop with increasing altitude, up to around 10 km (the top of the troposphere), but at an observed average rate of -6.5 K km^{-1}. The disagreement is largely due to neglect of water-vapor condensation, which releases the enthalpy of vaporization, heating the atmosphere slightly. (See Problem 9.6.) From 10 to 50 km, (the stratosphere) the temperature slowly rises from ~220 to ~270 K, only to fall back to ~180 K by 85 km (the mesopause). Solar energy input on the atmosphere above 85 km (the thermosphere) causes a temperature increase from a variety of photochemical and photophysical processes. Our simple picture of an adiabatic atmosphere is incomplete, but the tropospheric dT/dr (called the *adiabatic lapse rate* in meteorology) is reasonably close to the observed value, even though Eq. (9.9) predicts an upper edge to the atmosphere (at about 30 km) characterized by $T = 0$ K!

To find how pressure varies with altitude in our adiabatic model, we write the adiabatic-path expression in the form

$$T(r) = T(0)\left[\frac{P(r)}{P(0)}\right]^{(\gamma-1)/\gamma}$$

and substitute this into the differential form of Eq. (9.6),

$$\frac{dP}{P(r)} = -\frac{Mg}{R}\left[\frac{P(r)}{P(0)}\right]^{(\gamma-1)/\gamma}\frac{dr}{T(0)}$$

which, when rearranged and integrated from $r = 0$ to $r = h$, yields

$$P(h) = P(0)\left[1 - \left(\frac{\gamma-1}{\gamma}\right)\frac{Mgh}{RT(0)}\right]^{\gamma/(\gamma-1)}. \qquad (9.10)$$

Figure 9.1 compares this equation to the isothermal atmosphere (Eq. (9.6a)) for Earth ($g = 980$ cm s^{-2}, $M = 29$ g mol^{-1}, $T(0) = 300$ K, $\gamma = 1.4$, $P(0) = 1$ atm), Venus ($g = 882$ cm s^{-2}, $M = 44$ g mol^{-1} (CO_2), $T(0) = 750$ K, $\gamma = 1.2$, $P(0) = 91$ atm), and Mars ($g = 392$ cm s^{-2}, $M = 44$ g mol^{-1}, $T(0) = 250$ K, $\gamma = 1.3$, $P(0) = 5 \times 10^{-3}$ atm) over a 10 km altitude range. Note how the two atmospheric models yield very similar pressure profiles, even though the temperature is constant in one and varying appreciably in the other.

These are the major practical results of gravitational thermodynamics. One we have not discussed is gravity's lack of influence on chemical equilibrium. In any gravitational field, K_{eq} is still a constant and the only influence of gravity is through the *equilibrium composition* changes that a pressure change might bring about.

9.2 CENTRIFUGAL FIELDS

A centrifuge produces a radial force on a system much like that of gravity. Instead of the radially inward force of gravity on Earth's surface, a centrifugal force is directed radially outward; instead of a constant acceleration, g, a centrifugal acceleration depends on the system's angular velocity and distance from the rotation axis.

The gravitational free-energy contribution

$$d\phi_G = Mg\, dr \qquad (9.11a)$$

becomes the centrifugal free-energy contribution

$$d\phi_C = -M\omega^2 r\, dr \qquad (9.11b)$$

with a change in sign (due to a change in direction of force) and a swap of $\omega^2 r$ for g. The centrifugal acceleration is $\omega^2 r$ where ω is the angular velocity (measured in *radians per second* rather than cycles per sound; one cycle equals 2π radians). In analogy with Eq. (9.5), we have

$$\left(\frac{\partial P(r)}{\partial r}\right)_T = \rho(r)\,\omega^2 r \qquad (9.12)$$

for the density–pressure relationship.

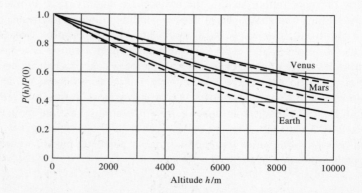

FIGURE 9.1 Predictions of an isothermal (solid line) and an adiabatic (dashed line) model of the atmospheres of, from top to bottom, Venus, Mars, and Earth. Data for these calculations are given in the text.

Centrifugal separation of multicomponent mixtures (gaseous isotopes, macromolecular mixtures, etc.) is the major practical application of centrifugation, and we will want to understand the limitations of the method. At (isothermal) equilibrium (with the centrifuge still spinning away at a constant ω), the chemical potential of any component must be uniform at all radii:

$$d\mu_i = \left(\frac{\partial \mu_i}{\partial r}\right)_{P, x_i} dr + \left(\frac{\partial \mu_i}{\partial P}\right)_{r, x_i} dP + \left(\frac{\partial \mu_i}{\partial x_i}\right)_{r, P} dx_i = 0 \ . \tag{9.13}$$

The first term is just $-M_i \omega^2 r \, dr$ from Eq. (9.11b). The second is

$$\left(\frac{\partial \mu_i}{\partial P}\right)_{r, x_i} dP = \underbrace{\overline{V}_i}_{(\partial \mu_i/\partial P)} \underbrace{\rho \omega^2 r \, dr}_{\substack{dP \text{ from} \\ \text{Eq. (9.12)}}}$$

The third term varies with the system at hand and how we model it.

If the system is an ideal solution for which Eq. (6.33) holds:

$$\mu_i = \mu_i^\circ + RT \ln f_i = \mu_i^\circ + RT \ln K_i + RT \ln x_i$$

where K_i is a simple constant of proportionality, the third term in Eq. (9.13) is

$$\left(\frac{\partial \mu_i}{\partial x_i}\right) dx_i = RT \frac{dx_i}{x_i} = RT \, d \ln x_i \ ,$$

and we have

$$(\overline{V}_i \rho - M_i) \omega^2 r \, dr + RT \, d \ln x_i = 0 \ .$$

Since $2r \, dr = d(r^2)$, we can write this equation as

$$\frac{d \ln x_i}{d(r^2)} = \frac{(M_i - \overline{V}_i \rho) \omega^2}{2RT} \ . \tag{9.14}$$

If an experiment can measure ρ and $\overline{V}_i$ independently, then this equation shows how centrifugation can determine molecular masses. Such an experiment is called a *sedimentation equilibrium* experiment and has been especially valuable in the study of macromolecules. One measures the radial *concentration profile* of the molecule of interest, and Eq. (9.14) says that a graph of $\ln x_i(r)$ versus r^2 will be a straight line of slope $(M_i - \overline{V}_i \rho) \omega^2 / 2RT$.

An independent measurement of $\overline{V}_i$ would seem to require prior knowledge of the molecular mass. In practice, the *specific volume* v_i (volume per unit mass) of the solute and the density of the solution are measured independently. The product $\overline{V}_i \rho$ becomes $M_i v_i \rho$, and the slope is $M_i \omega^2 (1 - v_i \rho)/2RT$.

Modern centrifuges can achieve enormous centrifugal accelerations. Rotor angular speeds approaching 100 000 rpm can be achieved, corresponding to $\omega \cong 10^4$ s^{-1}. At a 5 cm radius, the acceleration is on the order of $(10^4 \text{ s}^{-1})^2$ (5 cm) = 5 × 10^8 cm s^{-2} or $\omega^2 r/g \cong 500\,000$. At these enormous forces, the solute would equilibrate as a thin layer along the outermost edge of the centrifuge cell; consequently, somewhat lower rotational speeds (2000 to ~20 000 rpm) are used to spread the concentration gradient along a more useful distance. In a clever variant of this method, a solution of the macromolecule of interest and a heavy species such as CsCl or sucrose is centrifuged. The salt or sugar forms a *density gradient*[1] along the cell, and the macromolecule moves to a narrow region of this gradient (called the *isopycnic point* from the Greek *pyknos* meaning dense) in which $1 - v_i\rho = 0$. On the high-density side of this region, buoyant forces push the macromolecule back up the tube, while centrifugal forces push it down on the low-density side. Separation of macromolecular mixtures can be accomplished in this way.

EXAMPLE 9.2

The protein lysozyme (chicken egg white) has a specific volume of 0.688 cm³ g⁻¹. When subjected to equilibrium centrifugation at 19.5 °C and 7500 rpm in a solution of density 0.998 g cm⁻³, concentrations C were observed at the radial distances noted below. What is the molecular mass of lysozyme?

r/cm	5.24	5.73	6.11
C(arb)	3.68	4.97	6.40

SOLUTION A least-squares fit of $\ln C$ versus r^2 for these data yield a straight line of slope 0.0567 cm⁻², as is shown in the graph below. Note that we do not need an absolute value for concentrations, only relative values in arbitrary units.

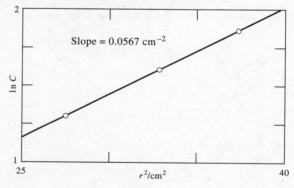

This slope is

$$0.0567 \text{ cm}^{-2} = \frac{M_i \omega^2 (1 - v_i \rho)}{2RT}$$

and to find M_i, we express ω as

$$\omega = (2\pi \text{ radians rev}^{-1})(7500 \text{ rev min}^{-1})/(60 \text{ s min}^{-1}) = 785 \text{ s}^{-1}$$

and find

$$M_i = \frac{2 \,(8.31 \text{ J K}^{-1} \text{mol}^{-1}) \,(292.65 \text{ K}) \,(0.0567 \text{ cm}^{-2}) \,(100 \text{ cm m}^{-1})^2}{(785 \text{ s}^{-1})^2 [1 - (0.688 \text{ cm}^3 \text{ g}^{-1})(0.998 \text{ g cm}^{-3})]}$$
$$= 14.3 \text{ kg mol}^{-1} = 14\,300 \text{ g mol}^{-1},$$

which makes lysozyme a fairly small protein.

➭ RELATED PROBLEMS 9.7, 9.8, 9.9

---

[1] Note that ρ has represented the density of the uncentrifuged solution, a constant during centrifugation. In contrast, ρ is a function of r in the density-gradient technique due to the concentration gradient of the heavy salt or sugar solute. In this method, $[1 - v_i\rho(r)]$ is proportional to the buoyant force of minor solute i replacing solution with density ρ at point r.

For gaseous centrifugation we use an ideal-mixture-of-ideal-gases model and focus on the partial pressure of component i to write Eq. (9.13) in the form

$$d\mu_i = \left(\frac{\partial \mu_i}{\partial r}\right)_{P_i} dr + \left(\frac{\partial \mu_i}{\partial P_i}\right)_r dP_i = 0 \ .$$

Since $\mu_i = \mu_i^\circ + RT \ln P_i$ in this model, we have

$$d\mu_i = -M_i \omega^2 r \, dr + RT \, d \ln P_i = 0$$

or

$$\frac{d \ln P_i}{d(r^2)} = \frac{M_i \omega^2}{2RT} \ . \tag{9.15}$$

Gaseous centrifugation usually has separation of components of known mass as its goal. Thus, we will integrate Eq. (9.15) from $r = r_1$ to $r = r_2$ to obtain

$$P_i(r_2) = P_i(r_1) \exp\left[\frac{M_i \omega^2}{2RT}(r_2^2 - r_1^2)\right] \ . \tag{9.16a}$$

Now suppose we have a mixture of two gases. These gases could be obviously different, such as $N_2 + O_2$, or subtly different, such as ^{3}He + ^{4}He, or ^{235}UF$_6$ + ^{238}UF$_6$ (fluorine is monoisotopic). At any one r, the total pressure is the same for both components, even though P will vary with r, so that

$$\frac{P_1(r)}{P_2(r)} = \frac{x_1(r) P}{x_2(r) P} = \frac{x_1(r)}{x_2(r)}$$

and, from Eq. (9.16a),

$$\frac{x_1(r_2)}{x_2(r_2)} = \frac{x_1(r_1)}{x_2(r_1)} \exp\left[\frac{(M_1 - M_2)\omega^2}{2RT}(r_2^2 - r_1^2)\right] \ . \tag{9.16b}$$

This expression shows that the separation ratio x_1/x_2 varies exponentially with the *absolute* mass difference $(M_1 - M_2)$ rather than a *relative* mass difference. This means centrifugal separation of ^{235}UF$_6$ from ^{238}UF$_6$ is more efficient than ^{3}He–^{4}He separation due to the greater absolute mass difference.

In practice, of course, He isotopes are much simpler to separate by distillation than by centrifugation, and isotopes enriched centrifugally must go through several enrichment steps to attain usefully high separation ratios. We can see this if we look at the magnitude of the exponential term above. If $M_1 - M_2 = 3$ g mol^{-1}, $\omega = 1000$ s^{-1}, and $(r_2^2 - r_1^2) = 10^{-2}$ m^2, the exponential term is only ~ 1.006.

9.3 MAGNETIC AND ELECTRIC FIELD ENERGIES

Static magnetic and electric fields can be easily established in the laboratory over quite wide magnitudes. In this section, we discuss the macroscopic phenomenological consequences of these fields; the microscopic effects are discussed in Chapters 15 and 21.

The phenomena in this section both increase our general understanding of matter and lead to many important practical devices. While there is not room to discuss such devices in detail, many of the expressions we will derive point to the principles on which such devices are based. We begin with external electric fields, progressing from Coulomb's force law, through a new free-energy contribution, to the phenomena of electrostriction and piezoelectricity that lead to devices such as microphones and crystal oscillators.

Electric Field Phenomena

Two charges Q_1 and Q_2 (measured in units of coulombs, C) separated in vacuum by a distance r exhibit a force

$$F = \frac{Q_1 Q_2}{4\pi\epsilon_0 r^2} \tag{9.17}$$

where the universal constant ϵ_0, the *permittivity of vacuum,* is given by the exact expression

$$\epsilon_0 = \frac{10^7 \, C^2 \, N^{-1} \, m^{-2}}{4\pi \, (c/m \, s^{-1})^2} = \frac{10^7 \, C^2 \, N^{-1} \, m^{-2}}{4\pi \, (299\,792\,458)^2} \cong 8.854 \times 10^{-12} \, C^2 \, N^{-1} \, m^{-2} \, ,$$

where c is the speed of light.

The units for ϵ_0 can be expressed in various ways:

$$1 \, C^2 \, N^{-1} \, m^{-2} = 1 \, A^2 \, s^4 \, kg^{-1} \, m^3 = 1 \, F \, m^{-1}$$

where 1 F (one farad, named for Michael Faraday) = $1 \, C \, V^{-1}$ and 1 V (one volt) = $1 \, J \, C^{-1}$. The farad is a unit of electrical capacitance, while the volt is a unit of electric potential (and the electric *potential* is a *potential energy per unit charge*).

It is convenient to consider *the electric field,* the force per unit charge experienced by an infinitesimal test charge in the vicinity of one (or more) finite charges. The electric field is a vector quantity (as is the force itself), and will be denoted by **E** (and measured equivalently in either $N \, C^{-1}$ or $V \, m^{-1}$ units). A finite charge Q_1 placed at a point where the electric field is **E** will experience a force

$$\boldsymbol{F} = Q_1 \boldsymbol{E} \, .$$

Throughout most of this section, we will imagine **E** to be a *uniform, constant field,* denoted $\boldsymbol{E_0}$. Such a field can be realized in the laboratory by a parallel plate capacitor with plate areas large compared to the square of the plate separation so that the distortions of the field due to the edges of the plates are small. If one plate is connected to an electrical potential source that is different from that supplying the other plate, an electrical potential difference (in volts)

$$\Delta\phi_{el} = \phi_{el}^{(1)} - \phi_{el}^{(2)}$$

will exist between the plates, and the electric field magnitude in vacuum will be

$$E_0 = \frac{\Delta\phi_{el}}{d}$$

where d is the plate separation. The vector direction of $\boldsymbol{E_0}$ is from the plate connected to the more positive potential to the plate connected to the more negative potential as in Figure 9.2(a).

Our next step will be to introduce matter into the volume between the plates. As Figure 9.2(a) shows, a parallel plate capacitor has an excess *surface charge density* $+\sigma$ (in $C \, m^{-2}$) on one plate and $-\sigma$ on the other. When matter is placed in such a capacitor, its electrons are displaced toward the positive surface charge of one plate while the nuclei are displaced toward the negative surface charge of the other. In electrical conductors (metals, solutions containing ions, or ionized gases) the magnitude of the displacement is macroscopic; charges move more or less freely. In contrast, insulators lack mobile charges, and the displacements are slight distortions of the charge distribution away from their field-free average position. We will limit this section to the behavior of such insulators.

9.3 MAGNETIC AND ELECTRIC FIELD ENERGIES

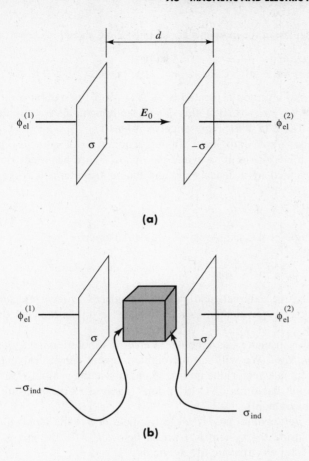

FIGURE 9.2 A parallel plate capacitor (a) with one plate at electrical potential $\phi_e^{(1)}$ and the other at $\phi_e^{(2)}$ establishes a uniform electric field E_0 as a result of the charge densities σ and $-\sigma$ established on the plates. For plate separation d, $E_0 = \Delta\phi_e/d$. Introducing an insulator into the field, (b), induces surface charge densities σ_{ind} and $-\sigma_{ind}$ on the insulator.

Imagine, then, filling the capacitor's volume with an insulator (gas, liquid, or solid). The external field E_0 will cause a positive-negative charge displacement throughout the insulator, leading to an *induced surface charge* σ_{ind} as is shown in Figure 9.2(b). It is observed experimentally that the magnitude of σ_{ind} depends not only on E_0 but also on the kind of matter, its temperature, etc. Our goal is thus to relate the usual thermodynamic variables and the external-field strength to one another through quantities descriptive of the system's composition. We can expect density, molecular mass, etc., to play a role, but we must also introduce new parameters to express the electrical properties of matter.

To do this, we start by recalling the definition of a *polar* molecule and the distinctions between polar and nonpolar substances. Most molecules are polar. Molecules that, by symmetry, have coincident centers of charge and centers of mass are nonpolar. Examples include all homonuclear diatomics, linear symmetric molecules such as CO_2, C_2H_2, and C_3O_2, and compounds such as CH_4, C_2H_4, C_6H_6, SF_6, and so on.

The *dipole moment* measures the degree of polarity. Two charges, $+Q$ and $-Q$, separated by a distance d, constitute the simplest arrangement of charge that exhibits a dipole moment p given by

$$p = Qd. \qquad (9.18)$$

The dipole moment is a vector quantity pointing from the negative to the positive charge.[2] Molecular dipole moments are tabulated in C m and in debye units (symbol: D) where

$$1 \text{ D} \cong 3.335\ 64 \times 10^{-30} \text{ C m} \quad \text{or} \quad 10^{-30} \text{ C m} = 0.299\ 792\ 458 \text{ D exactly.}$$

(An electron and a proton separated by 1 Å = 10^{-10} m exhibit a dipole moment of 1.60×10^{-29} C m = 4.80 D; this is a typical magnitude for a molecular dipole moment. See Table 15.1 for representative values.)

Note that the sample in Figure 9.2(b) has acquired a *macroscopic dipole moment*. One face has a net charge $\sigma_{ind}A$ while the opposite face has a net charge $-\sigma_{ind}A$. These are separated by a distance, d, and thus constitute a macroscopic induced moment

$$P_{ind} = (\sigma_{ind}A)\,d \ .$$

Since the volume of the sample is Ad, we can also write

$$\sigma_{ind} = \frac{\sigma_{ind}Ad}{Ad} = \frac{P_{ind}}{V} = \text{dipole moment density} \ . \tag{9.19}$$

In other words, the external field can be considered to have induced a surface-charge density or, equivalently, to have induced a dipole-moment density in the system.

Whenever an insulator has a non-zero electric dipole moment, the insulator is said to be *polarized*. We will call the dipole-moment density (a vector, since the induced moment is a vector) the *polarization* and use the symbol **P** rather than σ_{ind}. (The polarization has units of C m^{-2}; don't confuse P, the pressure, and P, the polarization magnitude!)

Due to the *polarization field* (the electric field due to the induced charges), the electric field inside the system is *reduced* below E_0.[3] In the empty capacitor, a simple result from electrostatic theory yields

$$E_0 = \frac{\sigma}{\epsilon_0} \quad \text{or} \quad \Delta\phi_{el} = \frac{\sigma d}{\epsilon_0} \ .$$

This suggests that the polarization field **P** yields an internal field **E** of magnitude

$$E = E_0 - \frac{P}{\epsilon_0} = \frac{\sigma - P}{\epsilon_0} \ . \tag{9.20}$$

For most simple substances (gases, liquids, and isotropic solids) at low fields, a simple proportionality holds between **P** and **E**:

$$\boldsymbol{P} = \chi_e \epsilon_0 \boldsymbol{E} \tag{9.21}$$

where χ_e is the (dimensionless) *dielectric susceptibility*, a quantity that varies from substance to substance, temperature to temperature, etc.

While χ_e gives a simple, phenomenological connection between **P** and **E**, it is more common to discuss and tabulate the related quantity ϵ_r, the *relative permittivity* or *dielectric constant*

$$\epsilon_r = \chi_e + 1 \ . \tag{9.22}$$

---

[2]Often chemists, especially in discussing dipole moments in complex molecules, write an arrow pointing from + to −, as in H⟶Cl. Actually, the correct dipole-moment vector is H⟵Cl.

[3]One of the properties of a conductor is that the electric field in the conductor is always zero; in an insulator, the field is reduced, but not to zero.

This trivial equation is not just a matter of convenience or historical convention. There are fundamental reasons for the quantity $\chi_e + 1$ to appear in dielectric-theory equations, and they all revolve around an auxiliary field first deduced by Maxwell in his unification of electromagnetic theory. Called the *displacement field D*, it is

$$\boldsymbol{D} = \epsilon_0 \boldsymbol{E} + \boldsymbol{P} . \tag{9.23}$$

Combining Equations (9.21) and (9.23) yields

$$\boldsymbol{D} = \epsilon_0 \boldsymbol{E} + \chi_e \epsilon_0 \boldsymbol{E} = (\chi_e + 1)\epsilon_0 \boldsymbol{E} = \epsilon_r \epsilon_0 \boldsymbol{E} , \tag{9.24}$$

which shows that the dielectric constant is a scale factor between $\boldsymbol{E}$ and $\boldsymbol{D}$.[4]

Table 9.1 collects representative values of ϵ_r for various substances. Note that gases at ordinary pressures have dielectric constants only slightly greater than unity, while condensed phases, especially polar liquids, can have appreciably larger values. We will derive a microscopic expression for ϵ_r for gases later in this section.

We now have an expression relating $\boldsymbol{P}$ to the internal electric field $\boldsymbol{E}$, Eq. (9.21), or to the displacement field $\boldsymbol{D}$, Eq. (9.23). But to return to Figure 9.2, how do we relate $\boldsymbol{P}$ to $\boldsymbol{E_0}$, the experimental field in the empty capacitor? To make this connection, we will give the results of a simple model system for which the relevant electrostatic-field calculation can be made exactly. As in Figure 9.3, consider a spherical dielectric with dielectric constant ϵ_r and radius r imbedded in a different dielectric with dielectric constant ϵ_r' (which could be a vacuum with $\epsilon_r' = 1$). Outside the sphere we establish an electric field that would be uniformly equal to E_0 if $\epsilon_r = \epsilon_r'$, and ask how this field is modified inside the sphere if $\epsilon_r \neq \epsilon_r'$. The answer is

$$E = \text{field in the sphere} = \frac{3\epsilon_r'}{\epsilon_r + 2\epsilon_r'} E_0 , \tag{9.25a}$$

TABLE 9.1 Dielectric Constants (Relative Permittivities)

H_2	1.000 24	H_2O(25 °C)*	78.30
HBr	1.003 1		
HCl	1.004 6	hexane	1.91
CO_2	1.000 92	benzene	2.274
H_2O(383 K)	1.012 6	cyclohexane	2.015
		CCl_4	2.228
CH_3OH	32.63		
C_2H_5OH	24.30	NH_3(288 K)	17.8
n-propanol	19.5	NH_3(239 K)	22.0
n-butanol	18.0		
acetone	21.5	HF(l, 273 K)	83.6
		H_2O_2(l, 273 K)	91.
S(s)	4.0	H_2SO_4(l, 298 K)	101.
NaCl(s)	6.12	HCN(l, 298 K)	107.
LiF(s)	8.9	formamide(l, 298 K)	111.
TlBr(s)	29.8		
polyethylene	2.25–2.3		

*Note: The following empirical equation gives ϵ_r for $H_2O(l)$ at 1 atm from 273.15 to 373.15 K:

$\epsilon_r = 295.9 - 1.2291\,(T/K) + 2.095 \times 10^{-3}\,(T/K)^2 - 1.410 \times 10^{-6}\,(T/K)^3$

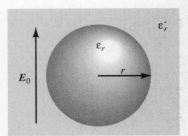

FIGURE 9.3 A simple model problem considers a uniform sphere with dielectric constant ϵ_r imbedded in a region of dielectric constant ϵ_r'.

---

[4]Equation (9.24) is one of the so-called *constitutive equations* of electromagnetic theory in that it depends on the constitution of the system through ϵ_r.

which, for a sphere in vacuum, reduces to

$$E = \frac{3}{\epsilon_r + 2} E_0 . \tag{9.25b}$$

Since $\epsilon_r > 1$ always, the field inside the sphere is always *less* than the external field in vacuum. From Eq. (9.24) we can write immediately

$$D = \frac{3\epsilon_r \epsilon_0}{\epsilon_r + 2} E_0 > \epsilon_0 E_0$$

and from Equations (9.23) and (9.24)

$$P = D - \epsilon_0 E = (\epsilon_r - 1)\epsilon_0 E = \left(\frac{\epsilon_r - 1}{\epsilon_r + 2}\right) 3\epsilon_0 E_0 .$$

The total induced dipole moment of the sphere is (see Eq. (9.19))

$$p = VP = \frac{4}{3}\pi r^3 P = 4\pi r^3 \epsilon_0 \left(\frac{\epsilon_r - 1}{\epsilon_r + 2}\right) E_0 . \tag{9.26}$$

This is a special case of a general relationship between electric field and induced moment. In general, for field strengths not too large,[5] one finds experimentally that

$$p = \alpha E_0 \tag{9.27}$$

where α (units: C V^{-1} m^{-2}, C^2 m^2 J^{-1}, or perhaps best, as will become apparent, J m^2 V^{-2}) is called the *polarizability*. Comparing Eq. (9.27) to Eq. (9.26) shows that, for a spherical dielectric,

$$\alpha = 4\pi\epsilon_0 \left(\frac{\epsilon_r - 1}{\epsilon_r + 2}\right) r^3 .$$

Note that the ratio $\alpha/4\pi\epsilon_0$ has the dimensions of *volume* and is often called the *polarization volume*.[6] One can also speak of the polarizability of an individual atom or molecule, and this important molecular quantity is discussed further in Chapter 15.

EXAMPLE 9.3

What is the total dipole moment induced in a 1.0 mm radius drop of water in an electric field of 10^5 V m^{-1} (as can be easily created by applying a 1000 V potential difference to two parallel plates spaced by 1 cm)?

SOLUTION By direct application of Eq. (9.26) with $\epsilon_r = 78.4$, we find

$$p = 4\pi\epsilon_0 (10^{-3} \text{ m})^3 \left(\frac{78.4 - 1}{78.4 + 2}\right) (10^5 \text{ V m}^{-1}) = 1.07 \times 10^{-14} \text{ C m} ,$$

which, by itself, is not a very illuminating number. But if we compute p on a *per molecule* basis, we find (with 1.4×10^{20} molecules of water in the drop $= N$)

$$p/N = 7.64 \times 10^{-35} \text{ C m} = 2.29 \times 10^{-5} \text{ D} ,$$

---

[5] Since the electric field holding atoms together is on the order of $e/(4\pi\epsilon_0 r^2)$ with $r \sim 10^{-10}$ m so that $E \sim 10^{11}$ V m^{-1}, very large external fields can overcome atomic internal fields, stripping electrons from atoms, and making the dielectric a conducting, ionized plasma. Here, we are concerned with fields well below this limit.

[6] Older tabulations of molecular data frequently list polarization volumes, but call them simply polarizations. One can tell which is which simply by the units.

which should be compared to the permanent dipole moment of one water molecule, which is 1.85 D, a considerably larger value.

▶ RELATED PROBLEMS 9.10, 9.11

We have now introduced the central expressions for dielectrics in external electric fields. The major parameter is the dielectric constant ϵ_r that can be measured with the parallel-plate capacitor apparatus shown schematically in Figure 9.2. The capacitance of such a capacitor with plate area, A, and plate separation, d, filled with a dielectric is

$$C = \frac{\epsilon_r \epsilon_0 A}{d} .$$

The ratio of the capacitance with the dielectric present to that with the dielectric absent is the dielectric constant if a homogeneous dielectric fills the volume between the capacitor plates.

Next, we calculate the energy change brought about by the field. The main approach to the energy change in electrostatic systems is a path integral for the work of assembling the charged system. Since electrical potentials are state functions, this work is path independent. When two charges Q_1 and Q_2 are placed in a uniform dielectric, the force between them is reduced by the dielectric:

$$F_{\text{diel}} = \frac{F_{\text{vac}}}{\epsilon_r} = \left(\frac{Q_1 Q_2}{4\pi\epsilon_0 r^2}\right) \frac{1}{\epsilon_r} .$$

Since ϵ_r can be $\cong 100$, the reduction can be very significant.

The work required to bring the charges through the dielectric from infinite separation to a finite separation is

$$w_{\text{el}} = -\int_\infty^r F(r)\, dr = \frac{Q_1 Q_2}{4\pi\epsilon_0 \epsilon_r r} = Q_1 \left(\frac{Q_2}{4\pi\epsilon_0 \epsilon_r r}\right) = Q_1 \phi_{\text{el}}(r) \tag{9.28}$$

where $\phi_{\text{el}}(r)$ is the electric potential at r due to charge Q_2. (Note that we could have just as well singled out Q_1 as the source of electric potential for charge Q_2.)

Next, we assemble a set of n charges, one at a time, so that the net electric potential is a sum of terms:

$$\phi_{\text{el}} = \sum_n \phi_n(r_n) = \sum_n \frac{Q_n}{4\pi\epsilon_0 \epsilon_r r_n}$$

where r_n locates charge n. These charges could be dispersed throughout a dielectric, or they could be attached to the plates of our capacitor, or both; the argument is quite general.

We next ask for the energy associated with the introduction of an electric dipole into this potential. Imagine a dipole of finite size: charge $+q$ at position r_+ and $-q$ at r_- with a separation d. The work required to bring these charges into the system is, from Eq. (9.28),

$$w_{\text{el}} = q\phi_{\text{el}}(r_+) - q\phi_{\text{el}}(r_-) = q[\phi_{\text{el}}(r_+) - \phi_{\text{el}}(r_-)] .$$

While this is exact, a more useful expression considers the electrical properties at a single point, the "center of the dipole." We reduce the finite dipole to an idealized *point dipole*. We imagine $q \to \infty$ and $d \to 0$ in such a way that the dipole moment, $p = qd$, remains finite and constant. To introduce the vector nature of p, we write

$$p = qd \quad \text{with} \quad r_+ = r_- + d \ .$$

After some algebra and limit-taking, the exact expression for w_{el} above becomes the following approximate expression, correct through first derivatives of the electric potential:

$$w_{el} = qd \cdot \left[i \left(\frac{\partial \phi_{el}}{\partial x} \right) + j \left(\frac{\partial \phi_{el}}{\partial y} \right) + k \left(\frac{\partial \phi_{el}}{\partial z} \right) \right] = qd \cdot \nabla \phi_{el}(r) = -p \cdot E(r) \quad (9.29a)$$

where $i, j,$ and k are unit vectors in the $x, y,$ and z directions, respectively, and the ∇ symbol (the *vector gradient operator*) is short-hand for the vector sum of partial derivatives in the first expression. The final equality comes from the general relationship between an electric potential function and its associated electric field:

$$E(r) = -\nabla \phi_{el}(r) \ .$$

From the definition of the vector dot-product, Eq. (9.29a) can also be written in terms of the magnitudes p and E:

$$w_{el} = -p \cdot E = -pE \cos\theta \quad (9.29b)$$

where θ is the angle between p and E. As Figure 9.4 illustrates, the energy of a point dipole in an external field is a function of relative direction, and thus there are energetically favorable and unfavorable orientations. Moreover, an external field will tend to align a free dipole along the preferred direction, but more about this will be said below.

Equation (9.29) leads us directly to a general expression for the total derivative of the chemical potential of component i of the dielectric, since our macroscopic induced dipole moment points in the direction of E:

$$d\mu_i = -\overline{S} \, dT + \overline{V} \, dP - \overline{p}_i \, dE \quad (9.30)$$

where $\overline{p}_i = (\partial p_i/\partial n_i)_{T, P, E, n_j}$, the partial molar macroscopic dipole moment. (Note that P in Eq. (9.30) is the pressure, not the polarization!) Since

$$p_i = VP_i = V\chi_{e,i}\epsilon_0 E = V(\epsilon_{r,i} - 1)\epsilon_0 E = V\epsilon_{r,i}\epsilon_0 E - V\epsilon_0 E$$

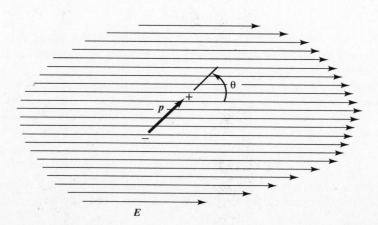

FIGURE 9.4 A point dipole p in an electric field E has an energy of interaction with the field given by $-pE \cos\theta$ where θ is the angle between the dipole and field vectors.

we have

$$\bar{p}_i = \left(\frac{\partial p_i}{\partial n_i}\right)_{T,P,E,n_j} = V\left(\frac{\partial \epsilon_{r,i}}{\partial n_i}\right)_{T,P,E,n_j} \epsilon_0 E \qquad (9.31)$$

so that

$$d\mu_i = -\bar{S}\, dT + \bar{V}\, dP - V\left(\frac{\partial \epsilon_{r,i}}{\partial n_i}\right)_{T,P,E,n_j} \epsilon_0 E\, dE\ . \qquad (9.32)$$

An interesting Maxwell relation comes immediately from this expression:

$$\left(\frac{\partial \bar{V}}{\partial E}\right) = \left(\frac{\partial^2 \mu_i}{\partial E \partial P}\right) = \left(\frac{\partial^2 \mu_i}{\partial P \partial E}\right) = \frac{\partial}{\partial P}\left[-V\left(\frac{\partial \epsilon_{r,i}}{\partial n_i}\right)\epsilon_0 E\right]$$

so that

$$d\bar{V} = \frac{\partial}{\partial P}\left[-V\left(\frac{\partial \epsilon_{r,i}}{\partial n_i}\right)\epsilon_0 E\right] dE\ . \qquad (9.33)$$

This equation states how the system's volume might change with the field. It is found experimentally that *V decreases for a condensed phase as E increases*. This is the phenomenon known as *electrostriction*. It is at least part of the reason that some aqueous solutions of electrolytes can shrink as the electrolyte dissolves.

A related phenomenon is *piezoelectricity,* a change in polarization at constant field with changing pressure. This phenomenon was first observed by the Curie brothers in 1880 when they detected a surface charge on quartz as they deformed it. These two phenomena work together in a number of practical devices ranging from microphones to piezoelectric positioning devices capable of controlling very slight (~ 1 μm and up) displacements.

One can also find changes in polarization with temperature (a *pyroelectric effect*) and the inverse, a temperature change with electric field (an adiabatic *electrocaloric effect*). These are entropy-related effects. Consider a polarized dielectric, adiabatically isolated, but at equilibrium. Compared to an identical unpolarized dielectric, the order imposed by the electric field *lowers* the entropy of a polarized system. If the field is adiabatically and reversibly lowered, the system must spontaneously *lower* its temperature to maintain the entropy constant. We will return to this phenomenon (*adiabatic depolarization,* as it's known) when we discuss the related magnetic effect in the next subsection.

An expression for μ_i that is more useful than Eq. (9.32) is one that shows how ϵ_r varies with n and P. We start with a gaseous sample of nonpolar molecules. If the field is E, each molecule has an induced dipole moment αE giving the system a net polarization $P = N_A n\alpha E/V$ so that $\chi_e = N_A n\alpha/V\epsilon_0$ or

$$\epsilon_r = 1 + \frac{N_A n\alpha}{V\epsilon_0} = 1 + \frac{N_A \alpha P}{RT\epsilon_0} \qquad (9.34a)$$

using the ideal gas equation of state. The second term on the right is

$$\frac{N_A n\alpha}{V\epsilon_0} = \frac{4\pi N_A \alpha}{\bar{V}\, 4\pi\epsilon_0} = 4\pi\, \frac{(N_A \alpha/4\pi\epsilon_0)}{\bar{V}} = 4\pi \times \frac{\text{(molar polarizability volume of molecules)}}{\text{(molar volume occupied by gas)}}\ .$$

Since the molecular polarizability volume is on the order of the molecular volume, this term is very similar in magnitude to the hard-sphere second virial coefficient term $B/\bar{V} = 4(\text{hard-sphere molar volume})/(\text{gas molar volume}) = 4(N_A v/\bar{V})$ of Equations (1.18) and (1.19). Thus $\epsilon_r \cong 1$, increasing linearly with P.

As the gas density increases, the field at each molecule is no longer simply the external field E, but E is augmented by the net field from its polarized neighbors. A more accurate expression that takes this into account for higher densities is the Clausius-Mossotti equation:[7]

$$\frac{N_A n \alpha}{V \epsilon_0} = (\epsilon_r - 1)\left(\frac{3}{\epsilon_r + 2}\right). \tag{9.34b}$$

Note that the second factor on the right approaches 1 as ϵ_r approaches 1, as it does for a low density gas, correctly recovering Eq. (9.34a).

If the molecular molar mass is M and the mass density is ρ, then Eq. (9.34b) can be written

$$\frac{\epsilon_r - 1}{\epsilon_r + 2} = \left(\frac{\alpha}{3\epsilon_0}\right)\left(\frac{\rho N_A}{M}\right) = C\rho$$

where C is the collection of (molecule-dependent) constants $\alpha N_A/3\epsilon_0 M$. Solving for ϵ_r yields

$$\epsilon_r = \frac{2C\rho + 1}{1 - C\rho}, \tag{9.34c}$$

which not only shows that ϵ_r increases with increasing density, but also predicts ϵ_r would be *infinite* if $\rho = 1/C$! The trend is correct, but the infinity is of no concern; the density $1/C$ is huge. (For example, $1/C$ for Ar is 9.75 g cm^{-3} while the density of solid Ar is 1.623 g cm^{-3} at the triple point. Truly heroic pressures, unrealizable in any current laboratory, would be required to compress Ar to a density six times the triple point value!) Figure 9.5 plots ϵ_r versus density for H$_2$ gas at 100 °C over the pressure range 0–1500 bar both as measured experimentally and as predicted by Eq. (9.34c); the agreement is excellent, and $\alpha = 0.90 \times 10^{-40}$ C^2 m N^{-1}.

If the molecules have a permanent dipole moment, the Clausius–Mossotti equation must be modified to include the extra polarization brought about by the tendency of the dipoles to align with the external field. This is not at all easy to do. The

---

[7]This expression is derived along the following lines. From Equations (9.21) and (9.22), we start with $(\epsilon_r - 1)\epsilon_0 E = P = N_A n \alpha E_i/V$ where $E_i (\neq E)$ is the field magnitude at any molecule, and Eq. (9.27) has been used to express the induced molecular dipole moment. Lorentz showed that $E_i = (\epsilon_r + 2)E/3$ in a spherical cavity surrounding any molecule as in Eq. (9.25b). Putting all this together yields the Clausius–Mossotti equation.

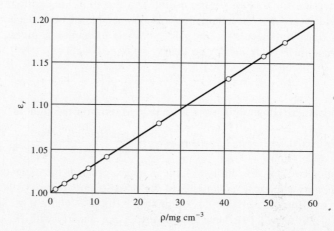

FIGURE 9.5 The measured dielectric constant of H$_2$ agrees well with the Clausius–Mossotti prediction (solid line) over a density range corresponding to pressures of roughly 0–1500 bar.

dipole orientation must compete with random, reorienting, molecular collisions (in gases and liquids at least), and the effect is temperature dependent. An original expression derived by Debye was shown to be slightly in error by Onsager, who replaced Debye's expression by a more complex one, yet one that was still only an approximation. (Debye assumed that the internal field acting on a polar molecule was the same Lorentz field mentioned in connection with Clausius–Mossotti expression, but Onsager showed that this was incorrect.) Nevertheless, the Debye expression is simple and reasonably accurate for two important cases: gases at low-to-moderate pressure and dilute solutions of polar molecules in nonpolar solvents. The equation is

$$\frac{\epsilon_r - 1}{\epsilon_r + 2} = \frac{N_A n}{3V\epsilon_0}\left(\alpha + \frac{p^2 N_A}{3RT}\right), \quad (9.35)$$

which is the Clausius–Mossotti equation with α replaced by $\alpha + p^2 N_A/3RT$ where p is the *molecular* dipole moment.

If we measure ϵ_r at various densities and use the Clausius–Mossotti equation for a nonpolar molecule, we can find α, while both α and p can be found for polar molecules if we measure ϵ_r as a function of density or temperature.

We can use Eq. (9.35) to find a more useful expression for the chemical potential of a low-pressure ideal gas in an electric field. At low pressure, the Clausius–Mossotti correction factor $3/(\epsilon_r + 2)$ can be set to unity, so that

$$\epsilon_r = 1 + \frac{N_A n}{V\epsilon_0}\left(\alpha + \frac{p^2 N_A}{3RT}\right)$$

and

$$\left(\frac{\partial \epsilon_r}{\partial n}\right) = \frac{N_A}{V\epsilon_0}\left(\alpha + \frac{p^2 N_A}{3RT}\right).$$

Eq. (9.32) becomes

$$d\mu_i = -\overline{S}\, dT + \overline{V}\, dP - EN_A\left(\alpha + \frac{p^2 N_A}{3RT}\right)dE. \quad (9.36)$$

If we integrate this at constant T and P from $E = 0$ to E, we find[8]

$$\Delta\mu_i = -\frac{1}{2}E^2 N_A\left(\alpha + \frac{p^2 N_A}{3RT}\right). \quad (9.37)$$

It is worthwhile to emphasize that this energy change is very small, as the following Example demonstrates. And note the sign of $\Delta\mu_i$: as E increases $\Delta\mu_i$ becomes more negative. This means not only that the energy of a dielectric is lowered in an external field, but also that an electric field *gradient* will exert a *force* on the system. The system will spontaneously be attracted to the region of highest field, since that region lowers the free energy the most. While this sounds like an intriguing phenomenon that might have practical applications, the force is unfortunately too weak to be of much use on macroscopic systems. This force can influence the motion of isolated gas-phase molecules, however, and it forms the basis of a technique for measuring molecular polarizations, as discussed in Chapter 15. (See also Problems 9.26 and 9.27.)

---

[8] This expression also shows why J m² V⁻² units are so useful for polarizability: (energy/J) ∝ (polarizability/J m² V⁻²) × (electric field/V m⁻¹)².

EXAMPLE 9.4

What is the chemical-potential change of methane when brought into an electric field of 10^5 V m^{-1}? The polarizability of CH_4 is (see Table 15.2) 1.17×10^{-39} J m^2 V^{-2}, and there is no molecular dipole moment.

SOLUTION From Eq. (9.37) we calculate

$$\Delta \overline{G}_{el} = -\frac{1}{2} N_A \alpha E^2$$

$$= -(0.5)(6.022 \times 10^{23} \text{ mol}^{-1})(1.17 \times 10^{-39} \text{ J m}^2 \text{ V}^{-2})(10^5 \text{ V m}^{-1})^2$$

$$= -3.52 \times 10^{-6} \text{ J mol}^{-1}.$$

We can see just how small this is by comparing it to the chemical-potential change the sample would experience if its temperature is increased by only 10^{-3} K. Using

$$\Delta \mu \cong \left(\frac{\partial \mu}{\partial T}\right) \Delta T = -\overline{S} \Delta T$$

with $\overline{S} = 22.4\, R$ for methane at room temperature gives

$$\Delta \mu \cong -(22.4\, R)(10^{-3} \text{ K}) = -0.186 \text{ J mol}^{-1}.$$

This is about 5×10^4 times the change brought about by the field!

➡ **RELATED PROBLEMS** 9.26, 9.27

A Closer Look at Polarization

You may have noticed that not much has been said about the polarization of solids. In general, solids must be discussed on the one hand with slightly more sophisticated mathematics and on the other, with radically different theories.

First, consider the solid phase of a nonspherical molecule. The induced dipole moment in such a molecule depends on the *relative orientation of the molecule* in the polarizing field. This is because such a molecule is more polarizable in some directions than in others simply because it lacks a highly symmetrical shape, as discussed more fully in Chapter 15. In fact, such a molecule placed in a random direction in an external field exhibits an induced dipole moment pointing in a direction different from the external field direction. In a gas, all random orientations are present, and an average polarizability is observed. But in a single-crystal solid, a definite macroscopic orientation can be maintained, and the induced macroscopic polarization will, in general, point in a different direction from the external field.

Mathematically, one replaces the simple numbers χ_e (or ϵ_r or α) by second-rank *tensors*, which are nine-element quantities, each element saying, in effect, "this gives the amount of polarization in direction i due to a component of the field in direction j" with i and j both ranging over x, y, and z Cartesian directions. We replace Eq. (9.21), for instance, by the three equations

$$P_x = (\chi_{xx} E_x + \chi_{xy} E_y - \chi_{xz} E_z)\epsilon_0$$
$$P_y = (\chi_{yx} E_x + \chi_{yy} E_y + \chi_{yz} E_z)\epsilon_0$$
$$P_z = (\chi_{zx} E_x + \chi_{zy} E_y + \chi_{zz} E_z)\epsilon_0$$

and similarly for Equations (9.24) and (9.27).

Another interesting aspect of solid polarization is the phenomenon of *ferroelectric polarization*. A single-crystal ferroelectric solid has a *macroscopic polarization even in the absence of an electric field.* One can suspend a ferroelectric crystal from a thread, dangle the crystal in an electric field, and watch the crystal swing into alignment with the field, just as a compass needle swings into alignment with a magnetic field.

None of the simple relations of Equations (9.21), (9.24), or (9.27) hold for ferroelectrics. Their dielectric constants are huge (several hundred to several thousand), strongly dependent on external-field strength, and sensitive to temperature. In fact, as T is increased, a ferroelectric will undergo a *phase transition* to a normal dielectric solid (or else it melts, in a few cases). The transition temperature is called the *Curie temperature,* and the transition occurs when there is sufficient thermal disorder to disrupt the ferroelectric order.

It would seem that an easy way to make a ferroelectric crystal would be to solidify a crystal of polar molecules in such a way as to align all the molecular moments. However, virtually no polar molecules are so accommodating (a notable exception is $(NH_2)_2CS$). Instead, most ferroelectrics work by a subtle asymmetric shift of discrete ions in the solid. Barium titanate ($BaTiO_3$), a well-studied ferroelectric with a Curie temperature of 393 K, gains its polarization by a shift of Ba^{2+} and Ti^{4+} ions in one direction and O^{2-} ions in the opposite direction from their non-ferroelectric symmetric locations as shown below for the crystal phase stable around room temperature. (This diagram is a projection through the crystal. The atoms do not all lie in the same plane.) The interesting aspect of this shift is that it is stable below the Curie temperature in the absence of an external field.

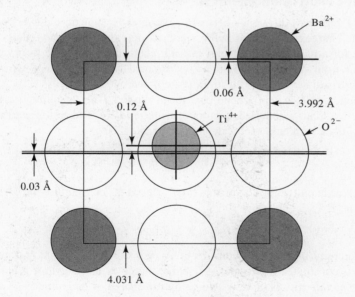

Less common is a phase in which the shift of rows of ions *alternates* in direction, leading to no net residual polarization, but to a phase change detectable in the temperature dependence of ϵ_r. This type of phase is called *antiferroelectric* and is observed, for example, in some zirconates as well as in low-temperature phases of some molecular hydrogen halides.

We also remark that all ferroelectrics (or *electrets,* as they are also called) are piezoelectric, but that the converse need not be true. Quartz is the classic example of a piezoelectric material that is not ferroelectric.

If the electric field imposed on a sample is not static, the frequency of the field's variation is an important parameter. The most common nonstatic electric field is provided by light, and the most common measure of the dielectric properties of matter at light frequencies is the *index of refraction n*. When light travels through matter, its speed is reduced below its value in vacuum, and the index of refraction is

$$n = \frac{\text{speed in vacuum}}{\text{speed in matter}} \geq 1 \;.$$

Electromagnetic theory shows that $n^2 = \epsilon_r$, and both ϵ_r and n depend on the frequency of light. Consequently index of refraction information can be related to polarizability. At visible light frequencies ($\sim 5.09 \times 10^{14}$ Hz for the commonly used yellow light of a sodium lamp), any permanent molecular dipole moment will have no effect, since the field is varying faster than the molecular framework can respond.[9]

Refraction data are often expressed in terms of the *molar refraction* $\overline{R}$, which comes from Eq. (9.35). We ignore the p^2 term and write ϵ_r as n^2:

$$\overline{R} = \frac{n^2 - 1}{n^2 + 2}\overline{V} = \frac{n^2 - 1}{n^2 + 2}\frac{M}{\rho} = \frac{N_A \alpha}{3\epsilon_0} \;. \tag{9.38}$$

Measurements of $\overline{R}$ are particularly useful in dilute solutions of a polar molecule in a nonpolar solvent, as Problem 9.18 explains.

Magnetic Field Phenomena

The most important difference between electric and magnetic fields is the lack of "magnetic charge." In spite of many searches, no experiment has found definitive evidence for a magnetic monopole. Instead, magnetic fields arise from charges in motion or from inherent *magnetic dipoles,* such as those that accompany many elementary particles like the electron, proton, and neutron.

Magnetic force centers around charges in motion. The magnitude of the force between two long, parallel conductors of negligible cross-section, separated a distance, d, in vacuum, and each carrying a current, I, is

$$F = \frac{2\mu_0}{4\pi}\frac{I^2 l}{d} \tag{9.39}$$

where l is the length of the conductors ($l \gg d$), and the constant μ_0, the *permeability of vacuum,* is[10]

$$\mu_0 = 4\pi \times 10^{-7}\,\text{N A}^{-2} \cong 1.257 \times 10^{-6}\,\text{N A}^{-2}$$

with the current measured in amperes (1 A = 1 C s^{-1}). A charge Q moving with velocity v constitutes a current, and if the charge is moving in a region with a *magnetic flux density* $\boldsymbol{B}$ there is a force on the charge of magnitude

$$F = Qv B \sin \theta$$

where θ is the angle between the velocity vector and the magnetic flux density vector. (The direction of the force is perpendicular to the directions of both $\boldsymbol{v}$ and

---

[9]This effect can be simulated by placing a magnetic stir-bar (representing the polar molecule) on a magnetic stirring-motor drive. As the motor speed is increased slowly from zero (simulating an increase in field frequency), the spin-bar at first follows the field and spins synchronously with it. But at high motor speed, the spin-bar tends to quiver slightly about one fixed direction, as its own inertia and friction inhibit it from keeping up with the high-frequency driving field.

[10]Note that $\mu_0 \epsilon_0 = c^{-2}$, where c is the speed of light in vacuum.

B.) While we will not use this force equation explicitly, it is introduced to emphasize that **B** (also called the magnetic induction vector) plays a fundamental role. The vector **B** is *not* the magnetic *field* vector; that will appear in the next paragraph. The magnetic flux density is measured in tesla units where $1\text{ T} = 1\text{ V s m}^{-2} = 1\text{ J A}^{-1}\text{ m}^{-2}$. One tesla is rather large; the Earth's magnetic flux density at the surface is $\sim 50\ \mu$T, while only strong laboratory electromagnets or superconducting solenoids can produce B values in the 1–10 T range.

The magnetic *field* vector is symbolized **H**. An infinitely long solenoid, constructed by winding n turns of wire per meter of length, produces a uniform magnetic field when current I is passed through the wire. The field is aligned along the solenoid axis, and has the magnitude (in vacuum)

$$H = In. \tag{9.40}$$

The magnetic field, H, has SI units of A m^{-1}.

Just as we placed matter in a parallel-plate capacitor, we now consider what happens when matter is placed in the solenoid. In analogy with Eq. (9.21), we find

$$M = \chi H \tag{9.41}$$

where **M** is the *magnetization* and χ is the (dimensionless) *volume magnetic susceptibility* (and we have assumed the matter to be isotropic, so that **M** and **H** are collinear). Although dimensionless, χ is a susceptibility per unit volume because, as we will show, the magnetization **M** is a magnetic dipole moment per unit volume, in analogy with the polarization. One can also write

$$B = \mu_0(M + H) = \mu_0(1 + \chi)H = \mu_0 \mu_r H \tag{9.42}$$

where the first equality defines the relation[11] among **B, H,** and **M** and where $\mu_r = 1 + \chi$ is the (dimensionless) *relative permeability*. While we focused on ϵ_r (with $\epsilon_r = 1 + \chi_e$) for electric phenomena, it will be χ, not μ_r, that will be of most interest here. The reasons are two-fold: first, χ is very small for most substances (and thus $\mu_r \cong 1$), but second, and more importantly, χ can be positive *or* negative (and thus μ_r can be > 1 or < 1, while $\epsilon_r \geq 1$ always).

Since χ can have either sign, we see from Eq. (9.41) that **M** will point in the opposite direction from **H** if $\chi < 0$. Substances with negative χ are called *diamagnetic* (and comprise most compounds). A diamagnetic χ is independent of H and T over wide ranges of these parameters. Substances with positive χ are further classified according to χ's magnitude and behavior with **H**. Positive χ compounds all have temperature- and field-dependent susceptibilities. Those with relatively small χ and relatively weak H dependence are termed *paramagnetic,* while huge χ values with strong H dependencies characterize *ferromagnetic* materials, such as Fe, Co, and Ni.

Table 9.2 collects a number of representative susceptibility values. Before discussing them, it is worthwhile to relate these values to the cgs unit values one finds in most older tabulations. These usually list the so-called "molar susceptibility," often symbolized χ_M. The relationship is

$$\chi(\text{SI}) = 4\pi \chi_M(\text{cgs})\ \rho(\text{in g cm}^{-3})/M(\text{in g mol}^{-1})$$

---

[11]Like Eq. (9.24), Eq. (9.42) is another constitutive equation of electromagnetic theory.

TABLE 9.2 Magnetic Susceptibilities

Gases: $\chi/10^{-9}$ (25 °C, 1 bar, except where noted)

He	−0.96	CO	−5.04	O_2(293 K)	1803.
Ne	−3.46	N_2	−6.17	NO(293 K)	763.6
Ar	−10.1	CO_2	−10.8	NO(204 K)	1424.
Kr	−14.8	N_2O	−10.1	NO(147 K)	2423.
Xe	−22.5	NH_3	−9.25	NO_2(408 K)	56.3
H_2	−2.05				

Liquids: $\chi/10^{-6}$ (25 °C, 1 bar, except where noted)

Hg	−28.5	CCl_4	−8.68
CH_3OH	−6.66	CBr_4	−12.1
C_2H_5OH	−7.23	CI_4	−14.2
C_6H_6	−7.68	Br_2	−13.7
CS_2	−8.55	H_2O	−9.054
C_2H_6(173 K)	−6.42	D_2O	−8.82
C_2H_4(171 K)	−3.04	O_2(90.1 K)	3447.
		NO(117.6 K)	62.

Solids: $\chi/10^{-6}$ (25 °C, 1 bar)

Li	12.9	LiF	−11.3
Na	8.70	NaCl	−14.3
K	5.75	CsI	−18.0
Rb	3.69		
Cs	5.15	Zn	−15.6
		Cu	−9.72
Al	20.7	Au	−34.5
Pt	278.	Si	−4.0
Cr	313.	C(diamond)	−21.6
Eu	14 800.	C(graphite)	−14.5
		P(red)	−19.4
$NiCl_2 \cdot 6H_2O$	403.	P(black)	−29.1
$FeCl_2 \cdot 4H_2O$	1550.		
$FeCl_3 \cdot 6H_2O$	1210.		

where ρ is the density and M is the molecular molar mass, expressed in the units shown. Also sometimes tabulated is a specific susceptibility, $\chi/(4\pi\rho)$, and a cgs-unit volume susceptibility, $\chi/4\pi$, where χ in all these relations is the SI volume susceptibility.

Now turn to Table 9.2. Note that the density factor causes χ(condensed phases) $\cong$ 10^3 χ(gases). Next, look at the uniformity of the negative values for diamagnetic compounds. There is a trend comparing CCl_4, CBr_4, and CI_4, and a trend in the rare gases, but in general the values follow no simply discernable pattern and are all rather uniformly the same magnitude (within any one type of phase). The microscopic theory of diamagnetism states that the susceptibility is proportional to the square of the average radius of electrons around their atomic nuclei. Since atoms are fairly closely the same size, so are diamagnetic susceptibilities, but since Xe is larger than He, so is its χ, and similarly for I in CI_4 and Cl in CCl_4. (Note that it is *atomic* size, not molecular size, that matters, except for molecules with bonds of a type allowing electrons to roam over several atoms, a phenomenon called *delocalization*.)

Finally, note the positive χ compounds (which are all paramagnetic in this table). About the only common paramagnetic nonmetallic compounds are O_2, NO_2, and NO. Elemental metallic paramagnets range from the modest (alkali metals and Al) to the robust (Eu, typical of many lanthanides), while the ferromagnetic pure elements Fe and Ni form paramagnetic compounds and, in turn, the paramagnetic alkali metals form diamagnetic alkali halides. This rich (and intertwined) chemistry is discussed on a microscopic level in Chapters 15 and 21.

The experimental measurement of χ is not quite as simple as the capacitance-ratio method for mesuring ϵ_r. We could measure an *inductance* change in the solenoid and relate that to χ, but the change would be so small for diamagnets and most paramagnets that this is not a viable technique. The usual technique involves a magnetic field *gradient*. A paramagnet (or a ferromagnet) is drawn towards regions of high field, while a diamagnet is drawn towards regions of low field. These forces can be balanced by gravity and measured as an apparent weight change via the *Gouy balance,* which suspends a sample from an ordinary balance into a field gradient.

Now we turn to **M,** the magnetization. Just as the polarization was an electric dipole-moment density, so the magnetization is a *magnetic dipole-moment density.* An elementary magnetic dipole is formed by a loop of current flowing in a plane and enclosing an area (of any shape). If the current is I and the area is A, then

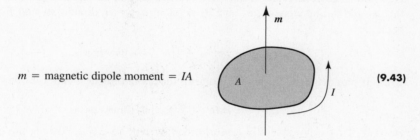

$$m = \text{magnetic dipole moment} = IA \qquad (9.43)$$

and we take the limit $A \to 0$, $I \to \infty$, $IA = $ constant to obtain an ideal point magnetic dipole. Note that

$$M(\text{in A m}^{-1}) = \frac{m(\text{in A m}^2)}{V(\text{in m}^3)}$$

in analogy with Eq. (9.19). The magnetization is a magnetic dipole-moment density.

We now seek the energy expression analogous to Eq. (9.29). As with electric fields, magnetostatic equations depend on the shape of the sample, and a spherical sample yields the simplest expressions. A spherical sample, introduced into a (vacuum) field $\boldsymbol{H_0}$ exhibits a uniform *internal* field $\boldsymbol{H}$ given by

$$\boldsymbol{H} = \boldsymbol{H_0} - \frac{1}{3}\boldsymbol{M} \qquad (9.44a)$$

or, in terms of the magnetic flux density,

$$\boldsymbol{B} = \boldsymbol{B_0} + \frac{2}{3}\mu_0 \boldsymbol{M}, \qquad (9.44b)$$

which follows from $\boldsymbol{B} = \mu_0(\boldsymbol{H} + \boldsymbol{M})$. Outside the sphere, the contributions to $\boldsymbol{H}$ and $\boldsymbol{B}$ from the sphere are somewhat complex but of no direct interest and will be ignored.

The work associated with changing B_0 by dB_0 is

$$dw = -\boldsymbol{m} \cdot d\boldsymbol{B_0} \qquad (9.45)$$

in analogy with Eq. (9.29). Equations (9.44) and (9.45) do not depend on the relationship between M (or m) and H (or B). If Equations (9.41) and (9.42) hold, which they do for diamagnets and paramagnets, but *not* for ferromagnets, then B/H is constant at constant temperature (as is χ); thus M/B_0 is also constant, and we can integrate Eq. (9.45):

$$w_{\text{mag}} = -\int_0^{B_0} \boldsymbol{m} \cdot d\boldsymbol{B_0} = -\int_0^{B_0} V\boldsymbol{M} \cdot d\boldsymbol{B_0} = -\frac{\boldsymbol{M}}{B_0} \cdot \int_0^{B_0} VB_0 \, dB_0 \, .$$

Now, if we assume V is independent of B_0 (which is a good approximation, although a small *magnetostriction* effect is known), we obtain

$$w_{\text{mag}} = -\frac{1}{2} V\boldsymbol{M} \cdot \boldsymbol{B_0} \qquad (9.46)$$

and note the units: w_{mag}(J), M(A m^{-1}), V(m^3), B_0(T = J A^{-1} m^{-2}). This immediately gives the magnetic free-energy contribution to the entire sample:

$$\Delta G_{\text{mag}} = -\frac{1}{2} \boldsymbol{m} \cdot \boldsymbol{B_0} \, . \qquad (9.47)$$

Note that the vector relationship has been maintained throughout the derivation; if the sample is diamagnetic, $\boldsymbol{m}$ and $\boldsymbol{B_0}$ point in *opposite* directions, and Eq. (9.47) becomes

$$\Delta G_{\text{mag}} = +\frac{1}{2} m B_0 \qquad \uparrow\boldsymbol{m} \quad \downarrow\boldsymbol{B_0} \text{ diamagnetic}$$

while if paramagnetic, $\boldsymbol{m}$ and $\boldsymbol{B_0}$ point in the *same* direction and

$$\Delta G_{\text{mag}} = -\frac{1}{2} m B_0 \qquad \uparrow\boldsymbol{m} \quad \uparrow\boldsymbol{B_0} \text{ paramagnetic} \, .$$

And so, if B_0 is not spatially uniform, neither will be ΔG_{mag}. This is the origin of the force that pulls paramagnets into a strong field but repels diamagnets.

We now turn to the temperature dependence of the magnetization of paramagnets. In $FeCl_3 \cdot 6H_2O$, only the Fe^{3+} ions are paramagnetic. The other atoms are diamagnetic (and even the Fe^{3+} ions have a diamagnetic component). The origin of paramagnetism in these ions can be traced to the inherent magnetic dipole moment of the electron, known as the *electron spin*. This moment is cancelled by electron spin pairing in diamagnetic compounds, but molecules with an odd number of electrons (such as NO and NO_2) or with unpaired spins (such as many Fe^{2+} compounds and O_2) are inherently paramagnetic. Just as thermal equilibrium produces a randomizing motion that counteracts the polarization alignment of permanent *electric* dipoles in an electric field, that motion also counteracts the aligning force of an external magnetic field on inherent *magnetic* dipoles.

The natural measure[12] of spin magnetic moment is the Bohr magneton:

$$\mu_B = 9.274\,015\,4 \times 10^{-24} \text{ A m}^2 \text{ (or J T}^{-1}, \text{ which is the same thing)},$$

and the critical value for the comparison of magnetic flux density magnitudes and temperatures is the ratio

$$\frac{N_A \mu_B}{R} = 0.6717 \text{ K T}^{-1} \, .$$

---

[12]The Bohr magneton is expressed in terms of fundamental constants in Section 15.2.

If $T \gg N_A\mu_B B_0/R$, the thermal motion controls the magnetization, but if the inequality is reversed, B_0 controls (and saturates) the magnetization. Since B_0 in the 1–10 T range is large, thermal motion dominates except at very low temperatures.

When thermal motion controls the magnetization, the molar magnetic moment due to spin paramagnetism is[13]

$$M\overline{V} = \overline{m} = \frac{4S(S+1)(N_A\mu_B)^2}{3RT}B_0 \quad (9.48)$$

where S is the integral or half-integral *spin quantum number*. At low-temperature–high-field conditions, this expression is replaced by the saturation value

$$\overline{m} = 2SN_A\mu_B \ . \quad (9.49)$$

The quantum number S varies from compound to compound over the range 1/2 to 7/2, and thus $S(S+1)$ is in the range 0.75 to 15.75; it is a constant for each paramagnetic compound (or paramagnetic phase of each compound, to be precise). Equation (9.48) can be rewritten as

$$M = \left[\frac{4S(S+1)(N_A\mu_B)^2}{3\overline{V}RT}\right]B_0 \quad (9.50a)$$

which, with Equations (9.41) and (9.42), becomes

$$M = \left[\frac{4S(S+1)(N_A\mu_B)^2}{3\overline{V}RT}\right]\mu_0(1+\chi)H_0 \cong \left(\frac{4S(S+1)(N_A\mu_B)^2\mu_0}{3\overline{V}RT}\right)H_0 \quad (9.50b)$$

(since $1 + \chi \cong 1$) so that

$$\chi = \left(\frac{4S(S+1)(N_A\mu_B)^2\mu_0}{3\overline{V}RT}\right) \quad \text{low-field, high-}T\text{, spin paramagnet} \ . \quad (9.50c)$$

Long before the quantum details were understood, this inverse-temperature relationship was known as *Curie's Law*. (Note that $\overline{V}$ also depends on T, due to thermal expansion, but the dependence is weak, and one often factors $\overline{V}$ from χ and reports a "specific susceptibility," as noted earlier.)

EXAMPLE 9.5

What is the spin quantum number S for O_2?

SOLUTION From Table 9.2, the gaseous magnetic susceptibility at 293 K and 1 bar is 1.803×10^{-6}. Using Eq. (9.50c) and the ideal gas expression for $\overline{V}$

$$\chi = \left(\frac{4S(S+1)(N_A\mu_B)^2\mu_0}{3\overline{V}RT}\right) = \left(\frac{4S(S+1)(N_A\mu_B)^2\mu_0 P}{3(RT)^2}\right)$$

we solve for S, finding $S = 1.016$. But since S must be integral or half-integral, we deduce that $S = 1$. This is an inherent molecular property of O_2, and it must hold for $O_2(l)$ as well, but if we use $S = 1$ and the density 1.1409 g cm³ of $O_2(l)$ at 90.1 K (so that $\overline{V} = 2.80 \times 10^{-5}$ m³ mol^{-1}), Eq. (9.50c) predicts $\chi = 4.98 \times 10^{-3}$ while Table 9.2 lists 3.447×10^{-3}. This disagreement is *not* due to mathematical approximations but to a *physical* one leading to Eq. (9.50), which is tacitly based on an assumption that the spin magnetic moments of

---

[13] The factor of 4 is actually $(2.002\ 319\ 304\ 422)^2 \cong 4.009$, the square of the free electron g factor, explained in Chapter 15. See Eq. (15.16).

each molecule are spatially far apart and do not interact with each other. This is true in low pressure $O_2(g)$, but not true at the densities of $O_2(l)$.

⇒ RELATED PROBLEMS 9.20, 9.21

You may have noticed that we do not seem to have introduced a "magnetizability" in analogy with the "polarizability" of electric phenomena. In fact we have, but not in so many words. The always present diamagnetic part of the total magnetization plays the role of a magnetizability, since diamagnetism is a purely induced effect.

Equations (9.48)–(9.50) do not tell the full story of paramagnetism. Unpaired electrons can create a magnetic moment through their orbital motion in an atom or molecule in addition to the inherent spin moment, although the spin moment is often dominant.

A Closer Look at Magnetization

We have said relatively little about ferromagnetism (and nothing as yet about *ferrimagnetism* and *antiferromagnetism*). Nor have we spoken of *superconductivity* with a total lack of resistance to electron current flow and a *perfect diamagnetic behavior,* which means a magnetization that causes B to be exactly zero inside the sample. This is known as the *Meissner effect*. The normal-to-superconducting phase transition is a so-called *second order* transition, with $\Delta H_\phi = T_\phi \Delta S_\phi = 0$, $\Delta V_\phi = 0$, and $\Delta C_{P,\phi} =$ finite (if there is no applied external field). The transition temperature to superconductivity is *lowered* in an external magnetic field. Above a critical field, the material remains a normal conductor even at 0 K. (See Problems 9.28 and 9.29.)

Transition to a superconducting phase is only one of four fates a paramagnetic substance can suffer at reduced temperatures. The remaining three are transitions to the ferromagnetic, antiferromagnetic, or ferrimagnetic states. Stated in reverse, ordinary "permanent" magnets (magnetized ferromagnetic iron, for instance) will revert to a paramagnetic phase above a certain phase-transition temperature, known as the *Curie temperature*. For Fe, Co, and Ni, this temperature is 1043, 1400, and 631 K, respectively.

A ferromagnetic phase can exhibit a permanent magnetization even in the absence of an external field. It does so by a macroscopic alignment of spins into regions known as *domains*. In contrast, an antiferromagnetic material (such as NiO, FeF_2, MnO, MnO_2, and CrSb), has *alternating* ordered directions of spins. No net magnetism results, but these phases have susceptibilities that are large, positive, and *increasing* functions of temperature, up to the transition temperature to a paramagnetic phase. This transition temperature is called the *Néel temperature* and ranges from lows on the order of a few K (24 K for $FeCl_2$; 2.1 K for $MnBr_2 \cdot 4H_2O$) to highs on the order of 10^3 K (723 K for CrSb; 953 K for Fe_2O_3). The ordering of spins in alternating directions in parallel layers also implies a highly anisotropic susceptibility.

Finally we mention ferrimagnets, such as Fe_3O_4 (magnetite), or $Y_3Fe_5O_{12}$ (called yttrium iron garnet or YIG for short). These exhibit a permanent magnetization, although typically somewhat less than that of a ferromagnet. In magnetite, which should be described as $FeO \cdot Fe_2O_3$, the Fe^{3+} ions' spins are in an antiparallel array and cancel, while the Fe^{2+} spins are aligned to yield the net magnetization. Since ferrimagnets (often called simply *ferrites*) are often electrical insulators, they constitute an important class of compounds in the electronics industry whenever the dual properties of strong magnetization and nonconductivity are called for. Even a species of bacteria use magnetite in the form of a dozen or so small cubical crystals about 5×10^{-5} mm across, arrayed in a chain, to orient themselves in the earth's magnetic field, and very recently, magnetite has been found in human brain cells.

We close this section with a discussion of the relation among magnetic susceptibility, temperature, magnetic field, and entropy. A common general expression for the temperature dependence of χ for a paramagnet is the *Curie–Weiss* expression

$$\chi = A + \frac{B}{T + \Delta}$$

where A is the diamagnetic component and B and Δ are constants that generalize the simple Curie expression, Eq. (9.50c). The constant Δ (with units of temperature) can be either positive or negative, and it attempts to lump together the behavior of other sources of paramagnetism aside from the spin part, as well as the behavior of spin–spin interactions. The Third Law requires $(d\chi/dT) \to 0$ as $T \to 0$, and neither the Curie nor the Curie–Weiss Laws allow such behavior. Thus, all paramagnets must undergo a phase transition to another magnetic phase at low enough temperature. Some paramagnets, notably hydrated salts of certain lanthanides, follow Curie's Law to temperatures well below 1 K. These compounds are particularly important to the technique of *adiabatic demagnetization,* first predicted (independently[14]) by Giauque and Debye. Such compounds serve as *refrigerants* in this technique. Currently, $Ce_2Mg_3(NO_3)_{12} \cdot 24H_2O$ (abbreviated, by physicists, as CMN) and its deuterated variety as well as the mixed cerrous-lanthanous salt of typical stoichiometry $Ce_{0.10}La_{1.90}Mg_3(NO_3)_{12} \cdot 24H_2O$ (CLMN) are routinely used to reach millikelvin temperatures by adiabatic demagnetization. In these compounds, the Ce^{3+} ions are paramagnetic, and Curie's Law is followed to temperatures ~ 2 mK, although mechanisms other than just electron spin contribute. In pure CMN single crystals, χ is very anisotropic, but the Ce^{3+} ions are separated by ~ 8 Å. This reduces spin–spin interactions, an idea behind the synthesis of CLMN that dilutes Ce^{3+} even further by diamagnetic La^{3+} ion substitution.

A historically important compound for adiabatic demagnetization is $Gd_2(SO_4)_3 \cdot 8H_2O$ in which the Gd^{3+} ions are spin-paramagnetic. Giauque showed that χ for this compound followed the full theoretical expression from which Equations (9.48) and (9.49) were obtained as limiting values. (See Problem 9.21.) It is now known that this substance loses paramagnetism and thus its usefulness as a refrigerant at 0.182 K, which is like a blast furnace to an ice cube if the goal is attaining $T \sim 1$ mK.

Giauque also showed how an adiabatic decrease of external field changes the paramagnet's temperature. If the field is not too large, we can use Eq. (9.48) in Eq. (9.47) to write

$$\Delta \overline{G}_{\text{mag}}(\text{paramagnet}) = -\frac{4S(S+1)(N_A\mu_B)^2}{6R}\frac{B_0^2}{T}$$

---

[14] Giauque's paper, "A thermodynamic treatment of certain magnetic effects. A proposed method of producing temperatures considerably below 1° absolute." *J. Am. Chem. Soc.* **49,** 1864 (1927), and Debye's paper, *Annalen der Physik* **81,** 1154 (1926), appeared almost simultaneously, but Giauque's paper included a footnote in which he referenced Debye's but was also careful to point out that he had "discussed the idea with colleagues since 1924" and that his Berkeley colleague, W. M. Latimer, had discussed it at a California American Chemical Society meeting on April 9, 1926. The first successful experiment, however, was not realized until 1933. Giauque won the Chemistry Nobel Prize for this line of research in 1949; Debye won it for his work with polar molecules (and for X-ray and electron diffraction of gases) in 1936.

and use $(\partial \Delta \overline{G}/\partial T) = -\Delta \overline{S}$ to find ($\overline{S}$ is molar entropy and S is spin quantum number!)

$$\Delta \overline{S}_{\text{mag}}(\text{paramagnet}) = -\frac{4S(S+1)(N_A\mu_B)^2}{6R}\left(\frac{B_0}{T}\right)^2. \tag{9.51}$$

This is the entropy change associated with an *isothermal magnetization*; it is *negative*, corresponding to the *ordering* of the spin system brought about by the field.

Now imagine isolating the magnetized sample adiabatically and reducing the field back to zero. Under adiabatic (and reversible) conditions, $\Delta \overline{S} = 0$, and Eq. (9.51) shows that if B_0 decreases, *T must also decrease* to hold the entropy constant.

Another way to think about the origin of adiabatic demagnetization is as follows. In the absence of the field, the entropy would decrease if the temperature is lowered. Thus, we could associate molar entropy $\overline{S}_1$ with temperature T_1 and $\overline{S}_2$ with T_2 where we will suppose $T_1 > T_2$. Now we hold the system at T_1 and apply a field that lowers the entropy to $\overline{S}_2$. Then, under adiabatic demagnetization, the system stays at $\overline{S}_2$ and has no choice but to find itself at T_2 when B_0 is reduced to 0. Were it to do otherwise the entropy would not be the path-independent, continuous, single-valued function of temperature that we know it must be. Similar arguments hold for adiabatic *depolarization*.

To reach temperatures below 1 mK, adiabatic demagnetization not of electron spins but of *nuclear* spins is used. Temperatures in the μK range and below can be attained in this way, and the nuclear magnetic moments of Cu form the most commonly used refrigerant. Nuclear magnetic moments are ~2000 times smaller than electronic moments (see Section 15.2), which implies that the initial value of B/T must be ~2000 times larger than that for electronic demagnetization to achieve the same degree of entropy reduction. Initial Cu temperatures ~10 mK and $B_i \sim$ 5 T are used. The μK to nK temperatures attained this way correspond to the nuclear spin system only, however, since heat transfer from the nuclear spins to the metal lattice is slow. (And so, ultimately, is transfer to the sample of experimental interest, which is often ^{3}He, the substance of most diverse physical chemistry at these temperatures. See Section 6.2 and Figure 6.10.)

From Equations (4.16) and (4.17), we can see why a conducting metal is used as the refrigerant. Heat transfer from the nuclear spins to the metal lattice is mediated by electron spins. In an insulator, the T^3 behavior of the heat capacity (Eq. (4.16)) means that the lattice heat capacity is nearly zero at mK temperatures. In contrast, the conduction electrons of a metal (with C_V proportional to T; see Eq. (4.17) and Example 4.2) carry the heat capacity and mediate the cooling.

CHAPTER 9 SUMMARY

The major expressions of each external-field discussion are:

Gravity:

isothermal atmosphere: $P(h) = P(0)\, e^{-Mgh/RT}$

adiabatic atmosphere:
$$P(h) = P(0)\left[1 - \left(\frac{\gamma - 1}{\gamma}\right)\frac{Mgh}{RT(0)}\right]^{\gamma/(\gamma-1)}$$

Centrifuge:

sedimentation: $\dfrac{d \ln x_i}{d(r^2)} = \dfrac{M_i \omega^2 (1 - v_i \rho)}{2RT}$

gas centrifugation:
$$P_i(r_2) = P_i(r_1) \exp\left[\frac{M_i \omega^2}{2RT}(r_2^2 - r_1^2)\right].$$

Electric Field:

induced dipole moment: $\boldsymbol{p} = \alpha\, \boldsymbol{E_0}$

Clausius–Mossotti relation:

$(\epsilon_r - 1)\left(\dfrac{3}{\epsilon_r + 2}\right) = \dfrac{N_A n \alpha}{V \epsilon_0}$

Debye–Clausius–Mossotti relation:

$$\frac{\epsilon_r - 1}{\epsilon_r + 2} = \frac{N_A n}{3V\epsilon_0}\left(\alpha + \frac{p^2 N_A}{3RT}\right)$$

Magnetic Field:

molar paramagnetic moment:

$$M\overline{V} = \overline{m} = \frac{4S(S+1)(N_A\mu_B)^2}{3RT} B_0$$

adiabatic demagnetization entropy:

$$\Delta\overline{S}_{mag} = -\frac{4S(S+1)(N_A\mu_B)^2}{6R}\left(\frac{B_0}{T}\right)^2$$

The important free-energy equations are:

gravity: $d\overline{\phi}_G = Mg\,dr$ (9.3)

centrifugation: $d\overline{\phi}_C = -M\omega^2 r\,dr$ (9.11)

electric field: $d\mu_i = -\overline{p}_i\,dE$ (9.30)

magnetic field: $d\mu_i = -\overline{m}_i \cdot dB_0$ (9.45)

Remember also the following parallels between related quantities:

Gravity	Centrifuge
g	$-\omega^2 r$

Electric	Magnetic
χ_e	χ
ϵ_r	μ_r
p	m
P	M
E_0	B_0

You should also remember that the magnitude of the energy or entropy effects brought about by these external fields are usually significantly smaller than those caused by ordinary temperature changes or chemical reactions.

FURTHER READING

Three seminal and now classic original research papers on thermodynamics in electric and magnetic fields are: E. A. Guggenheim, "On Magnetic and Electrostatic Energy," *Proc. Roy. Soc. London* **115A**, 49 (1936); E. A. Guggenheim, "The Thermodynamics of Magnetization," *Proc. Roy. Soc. London* **115A**, 70 (1936); and E. A. Guggenheim, "The Thermodynamics of the Electric Field with Special Reference to Chemical Equilibrium," *J. Phys. Chem.* **41**, 597 (1937). While these papers are over fifty years old and are couched in the unit systems of the day, they are still quite readable and can fill in and expand on the discussion of this chapter.

An excellent current text on electromagnetism is *Electricity and Magnetism*, 2nd edition, by Edward M. Purcell (McGraw-Hill, New York, 1985). In particular, Chapter 10, *Electric Fields in Matter*, and Chapter 11, *Magnetic Fields in Matter*, have material relevant to the discussion here.

Access to Data. The important material parameters introduced in this chapter are the dielectric constant or relative permittivity ϵ_r, and the magnetic susceptibility χ. Tables of these quantities may be found in the *Chemical Rubber Company Handbook of Chemistry and Physics*, mentioned at the end of Chapter 7.

PRACTICE WITH EQUATIONS

9A The largest moon of Saturn is Titan, with a mass of 1.35×10^{23} kg and a radius of 2575 km. What is the acceleration of gravity on the surface of Titan? (9.2)

ANSWER: 1.36 m s^{-2}

9B What would be the atmospheric pressure in Denver (1 mile above sea level) on a warm day (70 °F) if the sea level pressure was 1.00 atm? (9.6a)

ANSWER: 0.83 atm

9C The largest volcanic structure known is Olympus Mons on Mars, which is 29 km high and 500 km across. (On Earth, Mauna Loa on Hawaii holds the record at 9 km high and 200 km across at the ocean floor.) Predict the pressure and temperature on top of Olympus Mons. (9.10) and data in text following Eq. (9.10)

ANSWER: $P = 1.5 \times 10^{-4}$ atm; $T = 111$ K

9D The Earth, with a mean radius of 6371 km, rotates once per day (actually, once per 86 164.10 s, about 4 min short of a full day). What centrifugal acceleration does this rotation produce at the Earth's surface, expressed as a fraction of g? (9.11)

ANSWER: 3.45×10^{-3}

9E Mixed ^{20}Ne–^{22}Ne and ^{235}UF$_6$–^{238}UF$_6$ gases are placed in a gas centrifuge. By what factor will the separation ratio for UF$_6$ exceed that for Ne? (9.16b)

ANSWER: 4.48

9F By what factor does the electrical repulsive force between two electrons exceed their attractive gravitational force? (9.1), (9.17)

ANSWER: $F_{el}/F_{grav} = 4.16 \times 10^{42}$

9G If $\alpha(\text{Ar}) = 1.85 \times 10^{-40}$ J m² V⁻², what is the induced dipole moment of an Ar atom in a field $E = 10^5$ V m⁻¹? What separation of $Q_+ = 18e$ and $Q_- = -18e$ would have the same dipole moment? (9.27), (9.18)

ANSWER: $p = 1.85 \times 10^{-35}$ C m; $d = 6.41 \times 10^{-18}$ m

9H How much work is required to turn 1.0 mol of molecules with $p = 10^{-29}$ C m from the energetically most favorable to the least favorable orientation in a field of 10^5 V m⁻¹? How does this molar energy compare to the product RT for $T \cong 300$ K? (9.29b)

ANSWER: $w_{el} = 1.2$ J mol⁻¹; $RT \cong 2500$ J mol⁻¹

9I If $\alpha(\text{H}_2\text{O}) = 1.65 \times 10^{-40}$ J m² V⁻² and $p(\text{H}_2\text{O}) = 6.17 \times 10^{-30}$ C m, over what temperature range will the permanent dipole contribution to $\epsilon_r(\text{H}_2\text{O})$ dominate the polarizability contribution? (9.35)

ANSWER: $T \leq 5570$ K

9J Five thousand turns per meter of (insulated) wire are wound around a bar of Eu. What current must flow through the wire to produce a magnetic flux density of 0.200 T in the solenoid? How much current would be needed if the Eu bar is removed? (9.40), (9.42), Table 9.2

ANSWER: $I_{Eu} = 31.4$ A; $I_{vac} = 31.8$ A

9K For GdCl_3 at 293 K, $\chi = 5800 \times 10^{-6}$, $\rho = 4.52$ g cm⁻³, and $M = 263.609$ g mol⁻¹. The paramagnetism is due to spin only. What is S? (9.50c)

ANSWER: $S = 7/2$

PROBLEMS

SECTION 9.1

9.1 Combine the expression for the temperature dependence of vapor pressure, Eq. (6.12), with the isothermal atmosphere expression, Eq. (9.6a), to show that the boiling point of a liquid at height h is

$$T_b = T_b^\circ \left(1 + \frac{Mgh}{\Delta \overline{H}_{vap}^\circ}\right)^{-1}$$

where T_b° is the normal boiling point (i.e., the temperature at which the vapor pressure equals an external pressure of 1 atm), M is the average molecular mass of the atmosphere, and $\Delta \overline{H}_{vap}^\circ$ is the molar enthalpy of vaporization of the liquid. For water near 300 K, $\Delta \overline{H}_{vap}^\circ = 44.0$ kJ mol⁻¹, and for air, $M \cong 29$ g mol⁻¹. If the bottom of the World Trade Towers in New York is at 1.00 atm and 300 K, what is the boiling point of water at the top, 411 m higher, assuming the atmosphere at the top is also at 300 K?

9.2 The equilibrium distribution of small particles in a gravitational field can be used to find a value for Avogadro's constant, N_A. We recognize that $M = N_A \rho_i V$ where ρ_i is the particle's density and V is its volume and make a further correction to Eq. (9.6b) to account for the buoyancy of the particle in a medium of density ρ_0:

$$e^{-Mgh/RT} \quad \text{becomes} \quad e^{-N_A(\rho_i - \rho_0)Vgh/RT}.$$

Westgren [*Z. anorg. Chem.* **94**, 193 (1916)] prepared an aqueous suspension (a colloidal *sol*) of gold particles of uniform size (radius = 6.25×10^{-8} m) by reduction of AuCl_3 solutions. His observations of the average number of particles found in adjacent layers 1.11×10^{-3} cm apart is given below. Find N_A, given $\rho_{Au} = 19.281$ g cm⁻³, $\rho_0 = 0.999$ g cm⁻³, and $T = 16.7$ °C.

Layer	1	2	3	4	5	6	7	8
ρ	959.4	601.2	328.9	219.8	140.0	67.6	40.4	25.0

9.3 When a spacecraft views a planet's surface, it naturally looks through the entire planet's atmosphere. Use the isothermal model to find an expression for the *integrated column density,* the total amount of gas per unit cross-sectional area contained in a column extending upward from the surface. If the planet is Earth, how tall a column of gas *at standard conditions* (1 bar, 298.15 K) would you need to contain all this gas? This value is called the "equivalent thickness" of the atmosphere. What if the planet were Mars?

9.4 The surface temperature of Venus is 750 K, but at the base of the lowest layer of clouds, $T = 493$ K. Assuming an adiabatic model, find the height of the base of the clouds. If you use data from the text, you will find a height somewhat larger than the observed height of 29 km. Suggest reasons for the discrepancy.

9.5 The cover headline of the November 13, 1979, issue of *National Enquirer* states "Top Russian Scientists Discover UFO Base on Saturn Moon." It seems that someone in the (then) Soviet Union took a few UFO sighting trajectories and back-calculated them to converge on Titan, Saturn's largest moon. Titan is interesting for other reasons as well. Voyager spacecraft data have shown that Titan's atmosphere is largely N_2 with $(P, T) = (1.0$ mbar, 167 K) at 190 km and (6.2×10^{-9} mbar, 165 K) at 1265 km. While the atmospheric temperature is 94 K at the surface, falling to ~70 K at

~45 km, the temperature rises to ~165 K at and above ~170 km. Is the observed pressure drop from 190 to 1265 km in agreement with the isothermal atmosphere model? Apply Eq. (9.6a) directly to these data (with P(0) = 1.0 mbar, h = 1265 km − 190 km, and the g value from Practice Problem 9A), and note that your answer is ~2 × 10^{-5} times lower than it should be! The reason Eq. (9.6a) fails is simple: Titan is a small body, and the atmosphere is very extended; thus, g is not a constant over the distances involved. This is easy to fix. Simply return to the differential form of Eq. (9.6a):

$$\frac{dP(r)}{P(r)} = -\frac{Mg}{RT}dr$$

and incorporate the explicit dependence of g on r from Eq. (9.2). Integrate the result from r_1 to r_2, and recompute the pressure drop, remembering that r is now the distance from the moon's center. (See 9A for necessary data.) Your answer should be in much better agreement with observation.

9.6 There is an alternate way to derive the adiabatic lapse rate, Eq. (9.9). First show that the molar enthalpy analog to the molar internal-energy equality $d\overline{U} = \overline{C}_V dT = -P\, d\overline{V}$ for an adiabatic reversible process can be written for an ideal gas as

$$d\overline{H} = \overline{C}_P\, dT = \overline{V}\, dP = \frac{M\, dP}{\rho}.$$

Combine this with Eq. (9.5) to arrive at

$$\frac{dT}{dr} = -\frac{Mg}{\overline{C}_P}.$$

Show that since $\overline{C}_V = \overline{C}_P - R$, this expression and Eq. (9.9) are identical. Find the numerical value for dT/dr at 300 K for a terrestrial atmosphere 78.084% N_2, 20.9476% O_2, 0.934% Ar, and 0.0314% CO_2. (This is the composition of the four major components of the "1976 U.S. Standard Atmosphere," an idealized representation of the lower atmosphere often used in meteorological models.) Table 4.4 contains data for calculating $\overline{C}_P$.

SECTION 9.2

9.7 T. Svedberg, the inventor of the ultracentrifuge, reported the following observations on the sedimentation equilibrium of the CO–hemoglobin complex in 1926 [Z. physik. Chem. **121**, 65 (1926)]. The solution used had 1.0 g hemoglobin per 100 mL of solution; T = 293.3 K; ω = 912 s^{-1}; ρ = 0.9988 g mL^{-1} = mean density of solutions ranging in concentration from 0.56–1.22% by mass; v_2 = 0.749 mL g^{-1} = specific volume of hemoglobin complex; and the table below gives measured concentrations C (as percent by mass) at the stated radii. Find the molecular mass of the complex.

r/cm	4.61	4.56	4.51	4.46	4.41	4.36
C	1.220	1.061	0.930	0.832	0.732	0.639

9.8 Suppose you suspect that the bovine serum albumin protein has a molecular mass in the 50 000–75 000 g mol^{-1} range, and you know it has a specific volume of 0.734 cm^3 g^{-1}. You would like to do a sedimentation equilibrium experiment in aqueous solution (ρ = 0.998 g cm^{-3}) in your new ultracentrifuge to measure this number more precisely. This machine has a maximum rotor speed of 50 000 rev min^{-1} and provisions for measuring concentrations over a radial distance from 6.0 to 7.2 cm. It can control the temperature over the 15–30 °C range. Pick experimental conditions that will give a reasonable concentration gradient (say, a factor of 10 change) over the central 1 cm of the apparatus.

9.9 In a density gradient sedimentation experiment, the density in the vicinity of the isopycnic point r_0 can be accurately written as a Taylor's series expansion

$$\rho(r) = \rho(r_0) + \rho'(r_0)(r - r_0) = v_i^{-1} + \rho'(r_0)(r - r_0)$$

where $\rho'(r_0) \equiv (d\rho/dr)_{r_0}$ is the density gradient at r_0 and v_i is the solute's specific volume. Substitute this into the solute concentration-gradient expression

$$\frac{d \ln x_i}{dr} = \frac{M\omega^2 r_0(1 - v_i\rho)}{RT}$$

and show that the predicted spread in concentration around r_0 is in the form of a Gaussian function:

$$x_i(r) = A \exp\left(-\frac{B(r - r_0)^2}{2}\right)$$

where A and B are constants. The density gradient can be expressed as

$$\frac{d\rho}{dr} = \frac{\omega^2 r}{\beta_0}$$

where β_0 is a density-dependent constant characteristic of the substance used to establish the density gradient. Calculate the full width at half maximum of the sedimentation band that a macromolecule of $M = 10^6$ g mol^{-1} would have if $v_i = 0.74$ cm^3 g^{-1}, ω = 2000 s^{-1}, $r_0 = 6.5$ cm, T = 290 K, and $\beta_0 = 1.41 \times 10^9$ cm^5 g^{-1} s^{-2}, the appropriate value for a CsCl gradient under these conditions.

SECTION 9.3

9.10 If the molecular dipole moment of water is 6.17 × 10^{-30} C m, what fraction of the molecules in the water drop of Example 9.3 would need to be aligned to produce the drop's 1.07×10^{-14} C m macroscopic moment? Could all of these molecules be found on the surface of the drop? (See Problem 8.1 as well.)

9.11 How large must a drop of hexane be to have the same induced dipole moment as the water drop of Example 9.3?

9.12 If a gas follows the ideal equation of state, the Clausius–Mossotti equation, Eq. (9.34), can be written as a linear function of P:

$$\frac{(\epsilon_r - 1)}{(\epsilon_r + 2)} = \frac{N_A \alpha}{3\epsilon_0 RT} P \; .$$

Test this expression on the data below for air and CO_2, finding a value for α for each. For CO_2, the agreement will be particularly bad, due to the high pressures involved and the concomitant nonideal behavior. Does the version of the Clausius–Mossotti equation written in terms of density agree better with the CO_2 data?

Air: T = 292 K

P/atm	1.0	20.	40.	60.	80.	100.
ϵ_r	1.00054	1.0108	1.0218	1.0333	1.0439	1.0548

CO_2: T = 373 K

P/atm	98.7	292.	470.	691.	958.
ϵ_r	1.1041	1.3895	1.4900	1.5570	1.6097
ρ/g cm^{-3}	0.192	0.657	0.805	0.899	0.975

9.13 One of the earliest precise measurements of the gaseous hydrogen halide dipole moments was reported in 1924 [C. T. Zahn, *Phys. Rev.* **4**, 400 (1924)] at a time when Debye's expression, Eq. (9.35), was twelve years old, but our understanding of polar molecules was in its infancy. Zahn's original data for HCl are given below. Write Eq. (9.35) in the form

$$\left(\frac{\epsilon_r - 1}{\epsilon_r + 2}\right) T = \frac{N_A P \alpha}{3\epsilon_0 R} + \frac{p^2 N_A^2 P}{9R^2 \epsilon_0} \frac{1}{T}$$

so that a plot of the left-hand side is a linear function of $1/T$, and find values for the molecular dipole moment p and the molecular polarizability α. The data were corrected to a pressure of 1.00 atm by Zahn. How does your value for p compare to the current value, 1.1085 D?

T/K	201.4	260.1	294.2	359.2	433.9	503.9	588.8
$10^3(\epsilon_r - 1)$	7.452	4.716	3.792	2.672	1.948	1.526	1.182

9.14 When an electrical inductor and a capacitor are connected in series as indicated schematically by

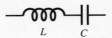

charge can flow from one to the other at a constant frequency ν (in "cycles per second," or Hz) given by

$$\nu = \frac{\omega}{2\pi} = (2\pi \sqrt{LC})^{-1}$$

where L is the inductance (measured in henry units, where 1 H = 1 V s^2 C^{-1}) and C is the capacitance (measured in farad units, where 1 F = 1 C V^{-1}). The periodic voltage developed across either component can be detected and its frequency measured to a precision of 1 part in 10^8 without great difficulty. Suppose the inductor has a value of 10^{-4} H and the capacitor is a pair of parallel plates 10 cm^2 in area, spaced by 0.1 mm. What is the frequency ν_{vac} of this circuit if the capacitor is evacuated? Now suppose the capacitor is filled with a gas at some T and P. Can this circuit measure T? P? the gas identity? For $P < \sim 1$ atm, Eq. (9.34a) gives ϵ_r to sufficient accuracy. If $\alpha(H_2)$ = 9.112×10^{-41} J m^2 V^{-2}, find the circuit's frequency at 0.01 atm, 0.10 atm, and 1.00 atm for T = 300 K. Repeat these calculations for D_2, for which α = 9.001 $\times 10^{-41}$ J m^2 V^{-2}. How well must T be controlled? Show that

$$\frac{1}{\nu_{vac}} \left(\frac{\partial \nu}{\partial T}\right) \cong \frac{N_A \alpha P}{RT^2 \epsilon_0}$$

and, taking $d\nu/\nu_{vac} = 10^{-8}$, find the order of magnitude of dT, the allowable temperature variation.

9.15 When two capacitors are connected in series, as in the schematic

the equivalent capacitance of the pair is given by

$$\frac{1}{C_{eq}} = \frac{1}{C_1} + \frac{1}{C_2} \; .$$

If a single capacitor is filled with two different dielectrics of thicknesses a and b as shown below

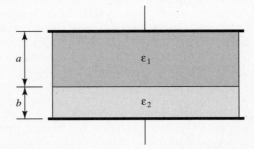

the effect is that of one capacitor of spacing a in series with one of spacing b. Let the plate areas be A and let ϵ_1 and ϵ_2 be the dielectric constants as shown. Show that the capacitance of this system is given by

$$C = \frac{\epsilon_0 A}{a/\epsilon_1 + b/\epsilon_2} \,.$$

Now suppose substances 1 and 2 are nonpolar gases (with polarizabilities α_1 and α_2) at the same T and P (which we will take to be low enough to have Eq. (9.34a) apply). Suppose further that the partition between the gases is removed and the gases mix. Write an expression for the capacitance in terms of ϵ_r for a binary gas mixture, and explain why C changes when the gases mix. (An expression for ϵ_r for a low-pressure binary gas mixture can be derived easily by extending the arguments behind Eq. (9.34a) in the text.)

9.16 Aluminum foil and waxed paper are commonly available in rolls of total area 200 ft² and 100 ft², respectively. If the dielectric constant of waxed paper is ~2.2 and its thickness is ~0.002 in., what capacitance would result if the Al-foil roll was cut in half and the waxed paper roll was sandwiched between the Al half-rolls? Your answer should be on the order of a few μF, a common magnitude required in many electronic circuits. Modern capacitors of this magnitude are considerably more compact, however, with a typical size on the order of 1 cm × 1 cm × 2 mm. If these devices consisted of two parallel plates spaced by a high-dielectric-constant material such as a solid solution of BaTiO$_3$/20 mol % BaZrO$_3$ ($\epsilon_r \sim 12\,500$ near room temperature), would they have a capacitance in the μF range? If not, how might such capacitors be made?

9.17 The diagram of BaTiO$_3$ in the text shows the shifts of each ion from the "perfect" positions in the crystal. Use these shifts to find the permanent dipole moment of BaTiO$_3$ given that the crystal can be thought of as an array of cubes (unit cells) each containing one Ba^{2+}, one Ti^{4+}, and three O^{2-} ions. (See Chapter 16 for a complete discussion of the unit cell concept.) Note that the positive ions shift "up" and the anions shift "down" from their perfect positions and that the dipole moment in such a case is a simple sum of terms of the form

$$p = \sum(\text{ion charge})(\text{amount of shift})$$

where the algebraic sign of both charge and shift must be taken into account. Given that the unit cell's volume is 3.992 Å × 3.992 Å × 4.013 Å compute the crystal's polarization and compare your answer to the observed value 0.26 C m^{-2}. The agreement will be close, but not exact. What other physical effects do you think need to be considered to improve the agreement?

9.18 To extract the dipole moment of a dilute dipolar solute in a polar solvent from static-field dielectric-constant measurements, it is helpful to begin with the following expression for the molar polarizability of the solution:

$$\overline{P} = x_1\overline{P}_1 + x_2\overline{P}_2 = \frac{\epsilon_r - 1}{\epsilon_r + 2} \frac{x_1 M_1 + x_2 M_2}{\rho} \,.$$

Component 1 is the solvent, and 2 is the dilute solution, as usual, and ρ and ϵ_r refer to the solution. At low concentrations, both ϵ_r and ρ can be taken to be linear functions of x_2:

$$\epsilon_r = \epsilon_1 + \epsilon' x_2 \quad \text{and} \quad \rho = \rho_1 + \rho' x_2 \,.$$

Substitution of these functions into that for $\overline{P}$ and taking the limit as $x_2 \to 0$ (to eliminate effects due to solute–solute interactions) leads to

$$\overline{P}_2^\circ = \left[\left(\frac{\epsilon_1 - 1}{\epsilon_1 + 2}\right)\frac{1}{\rho_1}\right]\left(M_2 - \frac{M_1\rho'}{\rho_1}\right) + \frac{3 M_1 \epsilon'}{(\epsilon_1 + 2)^2 \rho_1} \,.$$

Since the molar polarization of the solute is a sum of a polarizability and a dipolar contribution: $\overline{P}_2 = \overline{P}_{2,\alpha} + \overline{P}_{2,p}$, the molar refraction of the pure solute ($\overline{R}_2$ from Eq. (9.38)) can be used to find the dipolar contribution and thus the solute dipole moment p_2:

$$\overline{P}_{2,p}^\circ = \overline{P}_2^\circ - \overline{P}_{2,\alpha}^\circ = \overline{P}_2^\circ - \overline{R}_2 = \frac{p_2^2 N_A^2}{9\epsilon_0 RT} \,.$$

Use the data below and this formalism to find the dipole moment of cyclohexanol from ϵ_r and ρ measurements of dilute solutions of cyclohexanol in benzene. Note that you will need to fit $\rho(x_2)$ and $\epsilon_r(x_2)$ to the linear functions above to find ϵ_1, ϵ', ρ_1, and ρ'.

$T = 20\,°C$, $n_2 = 1.4641$, $\rho_2 = 962.4$ kg m^{-3}

$x_2/10^{-2}$	0	0.669	1.116	2.437	3.120	5.338	8.122
ρ/kg m^{-3}	879.27	879.64	879.91	880.75	881.17	882.73	884.67
ϵ_r	2.2836	2.3186	2.3409	2.4066	2.4402	2.5492	2.6921

9.19 In a Gouy balance measurement, a small cylindrical sample of material of cross-sectional area A is lowered into an inhomogeneous magnetic field. The inhomogeneity is along the direction of gravity (the z direction), and the resulting magnetic force is measured by a balance. The force per unit volume of the sample is

$$f_z = \chi \mu_0 H \frac{dH}{dz} \,.$$

Show that the work done in moving volume dV of the sample from a position where $H = H_i$ to one where $H = H_f$ is

$$dw_{\text{mag}} = \frac{1}{2}\chi\mu_0(H_f^2 - H_i^2)\,dV$$

and from this show that if the sample is brought from a region of zero field to one where the field is H that the net force is

$$F_z = \frac{1}{2}\chi\mu_0 H^2 A \ .$$

If a cylinder of Al with $A = 1$ cm^2 and height = 1 cm is lowered into a region where $H = 6 \times 10^5$ A m^{-1}, what is its weight change? How does this compare to the field-free weight of the sample? (The density of Al is 2.6989 g cm^{-3}.) What if the sample is NaCl, density = 2.165 g cm^{-3}?

9.20 Since $O_2(l)$ is paramagnetic, it is attracted towards regions of increasing magnetic field; since this force has the direction of the field gradient, it can be directed at will. In particular, a field gradient pointing upwards could perhaps be used as a pump for $O_2(l)$ that would have no moving parts. Explore this possibility as follows. The force in the z direction on a volume V of material with magnetic susceptibility χ placed in a region with magnetic flux density B and gradient dB/dz is given by

$$F_z = \frac{\chi V}{\mu_0} B \frac{dB}{dz} \ .$$

Imagine a field which is zero at $z = 0$ and increases linearly with z according to $B(z) = (dB/dz)\, z$. The material is placed at $z = 0$ and rises against gravity to $z = z_0$. The work involved by the field is just the integral of F_z over z from 0 to z_0. This must equal the gravitational work, mgz_0 where m is the mass of volume V and $g = 9.8$ m s^{-2} is the acceleration of gravity. Find z_0 for a disk of $O_2(l)$ 1.0 cm in diameter and 1.0 mm thick, oriented with the thin edge parallel to gravity. The density of $O_2(l)$ is 0.85 g cm^{-3}, and take $dB/dz = 10$ T m^{-1}. Do you think this is a viable method for pumping liquid oxygen up large distances?

9.21 In the article following Giauque's first discussion of adiabatic demagnetization [W. F. Giauque J. Am. Chem. Soc. **49,** 1870 (1927)], he took data on $Gd_2(SO_4)_3 \cdot 8H_2O$ magnetization measured by Woltjer and Kamerlingh-Onnes and showed that it followed the full quantum-mechanical expression on which Equations (9.48) and (9.49) are based. Woltjer and Kamerlingh-Onnes had used an expression derived in 1905 from classical mechanics by Langevin, which happens to lead to the same high-temperature–low-field expression as Eq. (9.48), but differs systematically from observation at low-temperature–high-field conditions where quantum details become important. Use the data below, adapted from Giauque's paper, to find both the spin quantum number S of Gd^{3+} and the range of validity of Eq. (9.48). Note that the experimental magnetizations are expressed as the dimensionless ratio of M divided by the saturation value, M_{sat}. The experimental temperatures and magnetic-flux densities are listed for each data point, but recall that paramagnetic magnetization is most naturally expressed in terms of the ratio of B to temperature. (The last row of data, B/T, means "field B measured in tesla units"; it is *not* the field to temperature ratio!) Plot M/M_{sat} versus the field to temperature ratio along with the line predicted by Equations (9.48) and (9.49) for various choices of S, remembering that S must be integral or half-integral.

M/M_{sat}	0.124	0.245	0.441	0.609	0.772	0.207	0.268	0.335	0.905	0.940
T/K	4.20	4.20	4.20	4.20	2.30	1.41	1.41	1.41	1.41	1.31
B/T	0.260	0.506	0.997	1.600	1.321	0.142	0.187	0.231	1.555	1.735

9.22 Show that, since the magnetization remains constant for a substance that follows Curie's Law, the word "demagnetization" is somewhat misleading in describing an "adiabatic demagnetization" experiment.

9.23 One of the important experimental parameters in the design of a low-temperature adiabatic demagnetization experiment is the heat associated with the initial magnetization of the refrigerant salt. One such salt for which our electron-spin expressions provide good approximations to observed behavior is "iron alum," ferric ammonium sulfate, $Fe_2(SO_4)_3(NH_4)_2SO_4 \cdot 24H_2O$ in which the Fe^{2+} ions have a spin quantum number $S = 5/2$. Using Eq. (9.51) to express the molar entropy and Eq. (4.2) to relate entropy to heat capacity in the form

$$\overline{C}_B = T\left(\frac{\partial \overline{S}}{\partial T}\right)_B$$

where $\overline{C}_B$ is the molar heat capacity at constant B, the heat associated with magnetization at T_i can be expressed in the form of an integral of $\overline{C}_B$ over temperature at constant B:

$$\overline{Q} = \int_\infty^{T_i} \overline{C}_B \, dT \ .$$

Explain why the limits of this integral are chosen as they are, why in fact the integral does represent the heat associated with magnetization, and why $\overline{Q}$ is negative. Then evaluate Q for a 100 g sample of iron alum using $T_i = 1.1$ K and $B_0 = 2.0$ T. (Note that the molar quantities here are *per mole of paramagnetic ions*.) What volume of He(l) would this Q evaporate given $\Delta H_{vap}(He(l)) = 2.6$ J mL^{-1}?

GENERAL PROBLEMS

9.24 Problem 9.6 can be expanded to allow the adiabatic atmosphere to be wet instead of dry. If x_w represents the mole fraction of saturated water vapor in air, then the enthalpy released on condensation of this water is $-\Delta \overline{H}^\circ_{vap} \, dx_w$. Work this term into the enthalpy expression in Problem 9.6 to show that

$$\left(\frac{dT}{dr}\right)_{wet} = \frac{\left(\dfrac{dT}{dr}\right)_{dry}}{1 + \dfrac{\Delta \overline{H}^\circ_{vap}}{\overline{C}_P}\left(\dfrac{dx_w}{dT}\right)}.$$

At 25 °C, $\Delta \overline{H}^\circ_{vap}(H_2O) = 44.016$ kJ mol^{-1}, and the vapor pressure of water is tabulated below.

$T/°C$	25.0	26.0	27.0
P/mbar	31.672	33.609	35.649

Use these data to find an approximate value for (dx_w/dT), and calculate the wet adiabatic lapse rate for such a completely water-saturated atmosphere. Since the air is neither completely dry nor wet, the observed average value of dT/dr, -6.5 K km^{-1}, is between the dry and wet limits.

9.25 While the technical details of gas centrifuges used for uranium isotope separation are not available to the public, certain aspects and limitations can be calculated from simple thermodynamic considerations. First, the uranium carrier is UF$_6(g)$, but the reactivity and vapor pressure of UF$_6$ set some immediate limits. At 25 °C, UF$_6(s)$ sublimes with a vapor pressure of 150 torr; thus, the pressure at the outer wall of the centrifuge cannot be much more than 100 torr in order to inhibit solidification at ordinary temperatures. Rotor material strengths are expressed in terms of the maximum peripheral speed v they can maintain where $v = \omega r$. The strongest materials for rotor construction are carbon-fiber–resin composites and the nylon–resin composite known as Kevlar. These have maximum v values around 10^3 m s^{-1}. Rather than the separation ratio defined in the text in Eq. (9.16b), the engineering measure often used is the so-called separation factor defined by

$$q_0 = \frac{(x_1/x_2)_0}{(x_1/x_2)_v}$$

where $x_1 =$ mole fraction of ^{235}UF$_6$, $x_2 =$ mole fraction of ^{238}UF$_6$, the 0 subscript refers to the stagnant gas with natural isotopic abundance, and the v subscript refers to a point in the centrifuge with peripheral speed v. Show that

$$q_0 = \exp\left[\frac{(M_2 - M_1)v^2}{2RT}\right]$$

and plot or tabulate q_0 for $T = 310$ K, $M_2 - M_1 = 3$ g mol^{-1}, and v ranging up to 1000 m s^{-1}. You will find that q_0 is a fairly slowly rising function of v. In contrast, the ratio of pressures between the outer wall of the centrifuge and its axis is a very rapidly rising function of v. (The axis is a legitimate place to find the gas, since these centrifuge rotors are usually hollow cylinders spun by external shafts.) Find an expression for this ratio and plot or tabulate it over the same range of v. What would you select as operating conditions for such a centrifuge in light of these calculations?

9.26 Can we make an antigravity machine using electrical fields, at least for gases? Equation (9.6a) gives the pressure ratio with increasing height for an isothermal ideal gas. Use Eq. (9.36) to find the corresponding expression for the pressure ratio between two points with electric field strengths E_1 and E_2. You should find that the pressure increases as E increases. Combine your expression with Eq. (9.6a) to give an equation representing a balance between gravitational and electrical forces that will make pressure independent of height. (You will, of course, find that the electric field cannot be uniform. How should it vary with h?) Can such a device work over useful distances? Suppose first that the gas is N$_2$ at 300 K, for which $\alpha = 1.97 \times 10^{-40}$ J m^2 V^{-2}. Nitrogen gas around 1 bar can sustain an electric field $\sim 10^7$ V m^{-1} before suffering electrical breakdown. Over what height can an electric field counteract gravity? Repeat the calculation for steam at 373 K, using $\alpha = 1.65 \times 10^{-40}$ J m^2 V^{-2} and $p = 6.17 \times 10^{-30}$ C m. Given that the fair-weather vertical electric field of the earth is about 130 V m^{-1}, decreasing exponentially with height, are you surprised that this field does not have to be taken into account when discussing atmospheric pressure variations?

9.27 The central phenomenon of the previous problem is the attraction of a gas toward regions of highest electric field, a gaseous electrostriction phenomenon. This was put to use by O. E. Frivold [*Physik. Zeitschr.* **22**, 603 (1921)] to measure the dipole moment of SO$_2$. His apparatus was a capacitor in the shape of two concentric cylinders for which the electric field is not uniform. In this paper, he derives the following expression (written here in SI units) for the relative volume contraction of a dipolar gas:

$$\left|\frac{\Delta \overline{V}}{\overline{V}_0}\right| \cong \frac{N_A}{2}\left(\alpha + \frac{p^2 N_A}{3RT}\right)\frac{E^2}{RT}$$

where $\overline{V}_0$ is the molar volume in the absence of the field E. Derive this expression, noting that the approximation comes from recognizing that

$$\frac{N_A}{2}\left(\alpha + \frac{p^2 N_A}{3RT}\right)E^2 \ll RT$$

for any realizable field at ordinary temperatures. How large would $|\overline{\Delta V}/\overline{V}_0|$ be for SO_2, for which $p = 5.22 \times 10^{-30}$ C m (Frivold's measurements gave $p = 1.83$ D $= 6.10 \times 10^{-30}$ C m) and $\alpha = 4.72 \times 10^{-40}$ J m^2 V^{-2}, in a field $E = 10^6$ V m^{-1} at 298. K? Frivold measured volume changes $\sim 4 \times 10^{-4}$ cm^3 in a total volume ~ 2500 cm^3 under similar conditions.

9.28 Superconductors are said to be "type I" if they completely exhibit the Meissner effect, expulsion of the **B** field from their interiors. ("Type II" superconductors expel the field incompletely.) Above a certain critical applied field (typically ≤ 0.1 T), the material undergoes a superconducting to normal-conducting phase transition. The value of the applied field is a function of temperature, which we now seek. If a superconducting sample is placed in a solenoid generating field **B,** then the work per unit volume associated with expelling the field from the sample is the same as the work needed to create a field $-$**B** by a second solenoid in the absence of the sample. Electromagnetic theory gives this work as $B^2/2\mu_0$. The superconducting sample must therefore perform this work to expel the field, and thus it raises its free energy by this amount. The normal conducting phase has a free energy that is essentially independent of **B,** since it is usually a normal diamagnetic or weakly paramagnetic phase. Put all this together in a qualitative plot of free energy versus **B** at any T below the normal-to-superconducting T_ϕ and explain why type I superconductors with relatively high T_ϕ values at zero field also have relatively high critical-field values at $T = 0$.

9.29 Continuing from the previous problem, how would a phase diagram of the normal and superconducting phases look on a plot of B versus T? In other words, how does the coexistence line look in such a plot? Define the field at any $T < T_\phi$ that induces the superconducting-to-normal transition as B_c, the critical field. Find an expression for the difference in entropies of the two phases in terms of B_c and its temperature derivative, dB_c/dT_ϕ. What does the Third Law have to say about dB_c/dT_ϕ as $T \to 0$? When a field $> B_c$ is applied to a superconductor, what is the heat associated with the transition to normal behavior? Given that dB_c/dT_ϕ at the onset of superconductivity in zero field is finite, what can you say about the heat associated with the normal-to-superconducting transition at $B = 0$?

CHAPTER 10

Ionic Solutions and Electrochemistry

THE topic of this chapter was largely responsible for making physical chemistry a distinct subject around the turn of the century. While the study of ionic solutions (electrolytes) began roughly a century ago, it continues today, largely because of the immense practical side of the topic, but also because ionic solutions are so nonideal (in the Chapter 6 sense) that satisfactory theories, especially at high concentrations, are not yet available to predict *a priori* the practical consequences of technological importance.

The bold hypothesis of Arrhenius that aqueous solutions of electrolytes consisted of free, charged species (ions) in large abundance marked in 1884 the beginning of a correct understanding of ionic solutions. His idea was not immediately accepted. Ions had been proposed as minor species that would explain electrical conductivity, but to suggest that they were major species in solution was a radical departure from contemporary thought. Electrical conduction was a key phenomenon. Consider ordinary sodium chloride, for example. Both solid and gas phases are poor conductors of electricity, but liquid NaCl conducts electricity quite well, roughly as well as a concentrated aqueous NaCl solution.

We begin this chapter with the ionic solution parameters needed to describe such solutions. One of the very first requirements we will face is that of net charge neutrality. One simply cannot create a solution of just negative ions; a counter ion must always be present.

10.1 Aqueous Solutions of Ions—Extreme Nonideality

10.2 Chemical Reactions That Transfer Electrons

10.3 The Electrochemical Potential

10.4 Electrochemical Devices

10.5 Using Electrochemical Data

The second section will concentrate briefly on those chemical reactions that are mediated by electron transfer in condensed phases. Such reactions are not only the basis for practical electrochemical devices discussed in the following two sections, but they also play vital roles in many biochemical reactions and in technologically important reactions such as corrosion.

The final section discusses the importance of electrochemical data to all of chemical thermodynamics. In that section, we will see how an electrochemical cell operates as a free-energy meter, allowing us to measure the equilibrium constant of a reaction without ever allowing the reaction to occur.

10.1 AQUEOUS SOLUTIONS OF IONS—EXTREME NONIDEALITY

You will recall from elementary chemistry that NaCl(s) is a regular array of Na and Cl ions, and that the crystal enthalpy of the solid is overwhelmingly dominated by the electrostatic attraction between these ions. From Coulomb's law, written for charges in any medium as

$$F = \frac{Q_1 Q_2}{4\pi\epsilon_0 \epsilon_r r^2} \tag{10.1}$$

we see that the force between ions 1 and 2 depends on three factors: the ions' charges, Q_1 and Q_2, the relative permittivity or dielectric constant of the medium, ϵ_r, and the ions' separation, r. When NaCl dissolves in water, the charges do not change, but ϵ_r increases from 1 (the vacuum value) to 78 (the value for H_2O at 25 °C), lowering the force appreciably.

Of course, the ions do move apart on dissolution, and we can easily estimate the magnitude of the motion. The density of NaCl(s) is about 2.2 g cm^{-3}, which, along with the molecular mass, implies a molar density of 3.8×10^{-2} mol cm^{-3} or 38 mol L^{-1}, a concentration of 38 M. Suppose we make a 1 M aqueous solution of NaCl. If the ions spread apart isotropically, the increase over the ions' solid spacing would be a factor of $(38 \text{ M}/1 \text{ M})^{1/3} = 3.4$. The interionic force would be reduced by a combination of permittivity and expansion factors equal to $(78)(3.4)^2 = 900$.[1]

While this roughly three-order-of-magnitude decrease in interionic force is important, it is not enough to quench nonideality completely. In fact, we will see that even mM solutions are still measurably nonideal, while M solutions are virtually beyond the scope of current theories.

Several other factors contribute to the complexity of electrolyte solutions. First, ions do not enter solutions and retain their regular, equal spacing. Ions of opposite charge tend to cluster together spontaneously, if only momentarily. As they dance around under the influence of solvent collisions and interionic forces, they fluctuate from cluster to cluster, forming transient associated ions. For instance, $BaCl_2$ in water can form an appreciable concentration of $BaCl^+(aq)$ species that, for some period of time, are not far enough apart to be considered discrete Ba^{2+} and Cl^- ions. The coming and going of associated ions is a very dynamic process, controlled by temperature, diffusion rates, and chemical interactions. Even the associated ions' identity is somewhat whimsical, in that such clusters require an arbitrary minimum separation distance for their definition.

---

[1] If the solution were more dilute, the factor would of course be larger. The infinite dilution limit will be an important extrapolation point for us.

But if an associated ion is deemed to exist, it reduces both charge and number of particles. One Ba^{2+} and one Cl^- (with the effect of three elementary charges) are reduced to one $BaCl^+$ (with the effect of only one charge). Moreover, the association of solvent around any ion has the effect of increasing solute concentration as free solvent becomes tied to the solute.

In addition to these weak associations, true chemical associations are often important. For example, AgI is a very insoluble compound. Electrochemical techniques can measure the concentrations of Ag^+ and I^- in a saturated AgI solution, and in pure water, they are 6.7×10^{-9} M. But undissociated molecular AgI is also present at a concentration of 6.0×10^{-9} M, and minor amounts of AgI_2^- and Ag_2I^+ are also present. Dissolve AgI in 1 M KI, in which the solubility is a factor of 10^6 greater than for pure water, and one finds appreciable amounts of I^-, AgI_2^-, AgI_3^{2-}, AgI_4^{3-}, $Ag_2I_6^{4-}$, $Ag_3I_8^{5-}$, and even larger species, with hardly any free Ag^+. Repeat the experiment with a 1 M $AgNO_3$ solvent, and one finds Ag^+, Ag_2I^+, Ag_3I^{2+} and Ag_4I^{3+} and virtually no I^-. These discrete, stable, complex ions make chemistry a rich subject, but they play havoc with any simple quantitative theory by introducing many species and many simultaneous reaction equilibria.

We will limit our discussion, for the most part, to those simple electrolyte solutions (and concentration ranges) that do not have such complex chemistry. This limitation does not overlook any major physical phenomena. It is simply a choice of model solutions that exempts us from worrying about the number of independent species. (Recall the discussion in Section 6.6 on the choice of independent components in the phase rule.)

From this preamble we now turn to the task of defining the chemical potential of aqueous ions. Since positive and negative ions always coexist in solution, we will need to define several *mean ionic* quantities that can be measured. In the definitions below, the general expression is given on the left, while the specific expression for an example electrolyte, $BaCl_2$, is given on the right.

We start with a general electrolyte $C_{\nu_+}A_{\nu_-}$ that dissociates into the cation C^{z+} carrying charge $Q_+ = z_+e$ (a positive number) and the anion A^{z-} carrying charge $Q_- = z_-e$ (a negative number).[2]

$$C_{\nu_+}A_{\nu_-} \to \nu_+ C^{z+} + \nu_- A^{z-} \quad BaCl_2 \to Ba^{2+} + 2Cl^- \quad (10.2)$$
$$\nu_+ = 1, \nu_- = 2, z_+ = 2, z_- = -1$$

We define a stoichiometric total number of particles

$$\nu = \nu_+ + \nu_- \quad \quad \nu = 3 \quad (10.3)$$

and stoichiometric molalities of each ion, based on the analytic molality of the solid electrolyte, m_i

$$m_+ = \nu_+ m_i, \quad m_- = \nu_- m_i, \quad m_+ = m_i \quad m_- = 2m_i \;. \quad (10.4)$$

Next, we define the *ionic strength* of the solution, I. Introduced empirically by Lewis, the ionic strength will emerge naturally in the dilute solution theory discussed later in this section.

$$I = \frac{1}{2}\sum_{j=1}^{\text{all ions}} m_j z_j^2 \quad \quad I = \frac{1}{2}(4m_+ + m_-) = 3m_i \quad (10.5)$$

---

[2]Note the use of the single arrow, indicating complete dissolution. Saturation (solubility) equilibria will be discussed later in this section.

EXAMPLE 10.1

What analytical concentration of the following electrolytes would have the same ionic strength as a 0.01 mol kg^{-1} solution of KCl?

(a) Ag_2SO_4
(b) $K_3Fe(CN)_6$
(c) $K_4Fe(CN)_6$
(d) $Al_2(SO_4)_3$

SOLUTION   For a 1:1 electrolyte like KCl, $m_+ = m_- = m_i$ so that

$$I = \frac{1}{2}(m_+ + m_-) = m_i .$$

Thus, for the other compounds, we want to find that m_i for which $I = 0.01$ mol kg^{-1}.

(a) Ag_2SO_4: $m_+ = 2m_i$ $m_- = m_i$
$z_+ = +1$ $z_- = -2$

$$I = \frac{1}{2}(m_+ + 4m_-) = 3m_i \qquad m_i = \frac{0.01 \text{ mol kg}^{-1}}{3} = 3.33 \text{ mmol kg}^{-1}$$

(b) $K_3Fe(CN)_6$: $m_+ = 3m_i$ $m_- = m_i$
$z_+ = +1$ $z_- = -3$

$$I = \frac{1}{2}(m_+ + 9m_-) = 6m_i \qquad m_i = \frac{0.01 \text{ mol kg}^{-1}}{6} = 1.67 \text{ mmol kg}^{-1}$$

(c) $K_4Fe(CN)_6$: $m_+ = 4m_i$ $m_- = m_i$
$z_+ = +1$ $z_- = -4$

$$I = \frac{1}{2}(m_+ + 16m_-) = 10m_i \qquad m_i = \frac{0.01 \text{ mol kg}^{-1}}{10} = 1.00 \text{ mmol kg}^{-1}$$

(d) $Al_2(SO_4)_3$: $m_+ = 2m_i$ $m_- = 3m_i$
$z_+ = +3$ $z_- = -2$

$$I = \frac{1}{2}(9m_+ + 4m_-) = 15m_i \qquad m_i = \frac{0.01 \text{ mol kg}^{-1}}{15} = 0.67 \text{ mmol kg}^{-1}$$

➠ **RELATED PROBLEMS**   10.1, 10.2, 10.3

Finally, we define the *mean ionic molality* $m_\pm$ and the *mean ionic activity coefficient* $\gamma_\pm$ based on individual ionic molalities, activity coefficients, and stoichiometric factors:

$$m_\pm = [m_+^{\nu_+} m_-^{\nu_-}]^{1/\nu} = m_i(\nu_+^{\nu_+} \nu_-^{\nu_-})^{1/\nu} \qquad m_\pm = [(m_i)(2m_i)^2]^{1/3} = 4^{1/3} m_i \quad \textbf{(10.6)}$$

and

$$\gamma_\pm = [\gamma_+^{\nu_+} \gamma_-^{\nu_-}]^{1/\nu} \qquad\qquad \gamma_\pm = [\gamma_{Ba^{2+}}^2 \gamma_{Cl^-}]^{1/3} . \quad \textbf{(10.7)}$$

The chemical potential of the dissociated electrolyte can now be written very compactly in terms of these quantities. We write first, in the usual way, (see Eq. (7.23))

$$\mu_i(aq) = \mu_i^\circ + RT \ln a_i \quad \textbf{(10.8)}$$

where a_i = activity of electrolyte i = $\gamma_i m_i$ (and the standard state $m_i^\circ = 1$ mol kg^{-1} is the basis against which m_i is measured). Next we recognize that the equilibrium *in solution* (even if $C_{\nu_+}A_{\nu_-}(aq)$ is present in a vanishingly small concentration)

$$C_{\nu_+}A_{\nu_-}(aq) \rightleftarrows \nu_+ C^{z+}(aq) + \nu_- A^{z-}(aq)$$

implies the chemical potential equality

$$\mu_i(aq) = \nu_+\mu_+(aq) + \nu_-\mu_-(aq) \qquad (10.9)$$

and that we can write

$$\mu_+(aq) = \mu_+^\circ + RT \ln a_+ \quad \text{and} \quad \mu_-(aq) = \mu_-^\circ + RT \ln a_-$$

even if a_+ and a_- are not individually measurable. Therefore

$$\mu_i(aq) = \nu_+\mu_+^\circ + \nu_-\mu_-^\circ + RT \ln [\gamma_+^{\nu_+} \gamma_-^{\nu_-} m_+^{\nu_+} m_-^{\nu_-}] = \mu_i^\circ + RT \ln (\gamma_\pm m_\pm)^\nu \qquad (10.10)$$

where $\gamma_\pm \to 1$ as $m_\pm \to 0$. The problem is now centered on $\gamma_\pm$, since ν is given by stoichiometry, $m_\pm$ is under experimental control, and, since only changes in μ_i are measured, we do not need μ_i°'s numerical value. Formally,

$$\mu_i^\circ = \lim_{m_\pm \to 0} [\mu_i - RT \ln (\gamma_\pm m_\pm)^\nu]$$

by our standard state definition.

Now we ask: how can $\gamma_\pm$ be measured, how can it be calculated by itself or accurately extrapolated from limited measurements, and, once we have it, what do we do with it?

First, the measurement question. Electrochemical cell measurements will be discussed in the last section of this chapter, but several other equilibrium methods based on phenomena discussed in previous chapters exist as well. These include:

(a) solubility measurements, relying on chemical equilibrium between the solid solute and its ions in solution,
(b) colligative property measurements, such as solvent vapor pressure, freezing-point depression, and osmotic-pressure measurements,
(c) solute vapor pressure measurements,
(d) measurements of solute distribution between two immiscible solvents, and
(e) ultracentrifuge solute sedimentation methods.

Consider first the solubility equilibrium. The reaction is

$$C_{\nu_+}A_{\nu_-}(s) \rightleftarrows \nu_+ C^{z+}(aq) + \nu_- A^{z-}(aq) \ .$$

Since

$$\mu(\text{solid } C_{\nu_+}A_{\nu_-}) = \mu(C_{\nu_+}A_{\nu_-} \text{ in solution}) = \mu(\text{ions in solution})$$

and since, by Eq. (10.10),

$$\mu(\text{ions}) = \nu_+\mu_+^\circ + \nu_-\mu_-^\circ + RT \ln (\gamma_\pm m_\pm)^\nu$$

we have

$$-\mu(\text{solid}) + [\nu_+\mu_+^\circ + \nu_-\mu_-^\circ] = -RT \ln (\gamma_\pm m_\pm)^\nu$$

which, since $\mu(\text{solid}) = \mu^\circ(\text{solid})$ if the pressure is standard, corresponds to Eq. (7.22) in the form

$$\Delta_r \overline{G}^\circ = -RT \ln K_{eq}$$

with $K_{eq} = (\gamma_\pm m_\pm)^\nu$. Thus, if K_{eq} is known (from, say, previously measured $\Delta_f \overline{G}^\circ$ values) an equilibrium measurement of $m_\pm$ yields $\gamma_\pm$, but only at that one equilibrium concentration.

Colligative-property measurements are based on the theories derived in Chapter 6, except the general, nonideal, chemical potentials for both solvent and solute (in terms of activities) must be used. Consider, for example, the change of solvent (water) vapor pressure due to dissolution of some involatile solute (the electrolyte). Formally, the solution vapor pressure yields a fugacity (given equation of state data; see Equations (5.49)–(5.51)), that, when divided by the pure solvent fugacity (Eq. (6.31) with $x° =$ pure solvent) yields the activity *of the solvent*. In practice, the vapor-pressure ratio itself is often an accurate measure of the activity, since the pressures are small at moderate temperatures and the approximation $f_i = P_i$ is valid. To find the *solute* activity, we must make many measurements of the solvent activity at various concentrations and integrate[3] the isobaric, isothermal version of the Gibbs–Duhem equation, Eq. (5.42):

$$n_s \, d\mu_s + n_i \, d\mu_i = 0$$

where the s subscript corresponds to the solvent and i corresponds to the solute.

Solute vapor pressure measurements require, of course, a volatile solute. The hydrogen halides are the classic examples of volatile electrolytes, and HCl is the prototype. The vapor pressure of HCl(l) varies from 41.6 atm at 293 K to 64.5 atm at 313 K. If HCl formed an ideal aqueous solution, Raoult's law would hold, and a 10 m solution (for which $x_{HCl} = 0.18$) would have an HCl partial pressure around (42 atm)(0.18) = 7.5 atm. In fact, such a solution has $P_{HCl} = 5.4 \times 10^{-3}$ atm, decreasing rapidly if the solution is diluted, as Figure 10.1 shows.

To extract activity coefficients from these data, we use the chemical potential equality

$$\mu(HCl(g), \text{ over solution}) = \mu(HCl(aq), \text{ in solution})$$

or

$$\mu°(g) + RT \ln P_{HCl} = \mu°(H^+(aq)) + \mu°(Cl^-(aq)) + RT \ln (\gamma_\pm m_\pm)^2 ,$$

which rearranges to

$$RT \ln \left[\frac{P_{HCl}}{(\gamma_\pm m_\pm)^2} \right] = \Delta\mu° = \text{a constant at constant } T \text{ and } P$$

---

[3]There are technical difficulties encountered in this integration, all of which can be treated by various tricks; details are given in the references at this chapter's end.

FIGURE 10.1 The experimental vapor pressure of HCl over an aqueous solution is plotted at various HCl concentrations. An ideal-solution vapor pressure would be tens of atmospheres instead of tenths or less. Below 5 mol kg^{-1}, the pressure is less than 10^{-4} atm.

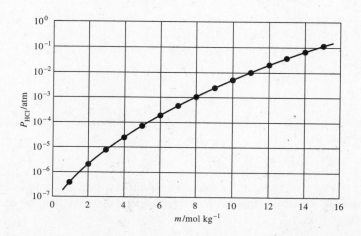

where $\Delta\mu° = \mu°(H^+) + \mu°(Cl^-) - \mu°(g)$. If the left-hand side is a constant, then so must be the quantity in square brackets. Call this constant K (and recognize that we have derived an equilibrium constant):

$$K = \frac{P_{HCl}}{(\gamma_{\pm} m_{\pm})^2} \ . \tag{10.11}$$

Since $\gamma_{\pm} \to 1$ as $m_{\pm} \to 0$, a limiting value of $P_{HCl}/m_{\pm}^2$ as $m_{\pm} \to 0$ yields K. One could then compute $\gamma_{\pm}$ from Eq. (10.11) at finite $m_{\pm}$, but unfortunately, the extrapolation is difficult to do reliably, since P_{HCl} becomes unmeasurable long before $\gamma_{\pm}$ approaches unity.

EXAMPLE 10.2

Laser spectroscopic techniques can detect as few as 10^8 molecules of HCl(g) per cm³. If $\gamma_{\pm}(HCl(aq, 10\ m)) = 10.5$ and $\gamma_{\pm}(HCl(aq, 0.01\ m)) = 0.904$, can HCl(g) be detected above a 0.01 m solution of HCl(aq)?

SOLUTION The text quotes $P_{HCl} = 5.4 \times 10^{-3}$ atm over a 10 m solution. Therefore, Eq. (10.11) becomes, with $m_{\pm} = m_i$

$$K = \frac{P_{HCl}}{(\gamma_{\pm}\, m_{\pm})^2} = \frac{5.4 \times 10^{-3}}{[(10.5)(10.0)]^2} = 4.9 \times 10^{-7}$$

so that at 0.01 m, we have

$$P_{HCl} = K(\gamma_{\pm}\, m_{\pm})^2 = (4.9 \times 10^{-7})[(0.904)(0.01)]^2 = 4.0 \times 10^{-11}\ \text{atm}\ .$$

This pressure corresponds to a molecular number density of

$$\frac{N}{V} = \frac{nN_A}{V} = \frac{PN_A}{RT}$$
$$= \frac{(4.0 \times 10^{-11}\ \text{atm})(1.01 \times 10^5\ \text{Pa atm}^{-1})(6.02 \times 10^{23}\ \text{mol}^{-1})}{(8.31\ \text{J mol}^{-1}\ \text{K}^{-1})(298\ \text{K})}$$
$$= 9.8 \times 10^{14}\ \text{m}^{-3} = 9.8 \times 10^8\ \text{cm}^{-3}\ ,$$

which is just above the limits of detection.

➡ **RELATED PROBLEMS** 10.4, 10.5

Solute distribution measurements between two fluid phases is an extension of the solute vapor-pressure method, since the vapor phase certainly qualifies as a fluid phase. This method is not often used, since it is difficult to find two immiscible solvents that will both support measurable amounts of an electrolyte solute. The ultracentrifuge method relies on Eq. (9.14) to describe the concentration gradient established by centrifugation. The major use of centrifugal electrolytes is not as an activity measurement technique, but rather as a means of establishing a strong density gradient, as discussed in Chapter 9.

The results of these various measurements can be summarized by plotting $\gamma_{\pm}$ versus $m_{\pm}$ for a variety of electrolytes as in Figures 10.2 and 10.3. Note that $\gamma_{\pm}$ starts at 1 at $m_{\pm} = 0$ and *decreases,* at least initially, and that the high-concentration behavior seems to have little rhyme or reason. General trends in $\gamma_{\pm}$ at high concentration have been explained by three phenomena, any of which complicate theory. The very low values are indicative of extensive stable complex-ion formation, while the moderately low values are indicative of transient ion association. The very high values are usually indicative of ion hydration, which is the formation of appreciable

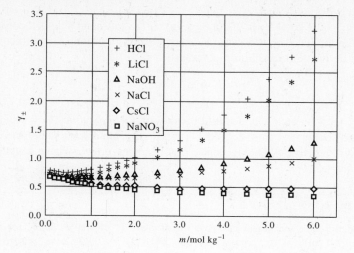

FIGURE 10.2 The mean ionic activity coefficients $\gamma_\pm$ for a number of simple 1:1 electrolytes is plotted as a function of electrolyte molality.

amounts of stable species in which several water molecules move with the ion, such as $Na(H_2O)_4^+$. This association increases the effective molality of the electrolyte as it ties up an appreciable amount of solvent.

It appears that if any theoretical headway can be made, it will likely start from low concentrations where the chemical description of the solution is simplest. Such a theory was first presented by Debye and Hückel in 1923.

Why did nearly 40 years pass from the first suggestion that dissociation into ions was extensive until a firm physical theory was advanced? Was physical chemistry, in its infancy, incapable of a correct explanation? In part, yes. It was a problem that has been aptly called one "which only physicists could solve and in whose solution only chemists were interested." There were false starts; futile attempts to fit uncertain data by, frankly, absurd empirical functions; an uncommon timidity by two giants of physical chemistry, G. N. Lewis and A. A. Noyes; a period of polemics, excitement, and activity (called the "swear-word episode" by G. Scatchard) that centered around a series of papers in 1918 by a twenty-four-year-old lecturer at the University of Calcutta, J. Chandra Ghosh; and, in the middle of it all, in 1908, a short paper by Niels Bjerrum (taken as a personal affront by Arrhenius),

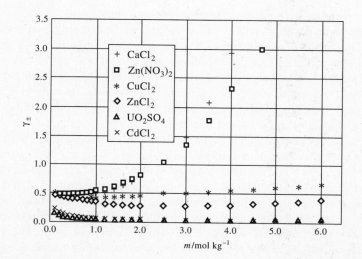

FIGURE 10.3 The $\gamma_\pm$ values for several 1:2 salts and the 2:2 salt UO_2SO_4 are plotted at various molalities. Note the much wider range of values compared to the 1:1 salt data in Figure 10.2, especially for $CaCl_2$ and $Zn(NO_3)_2$, which rise to 11.1 and 6.38, respectively, at 6 M, (both off scale) while $\gamma_\pm$ for $CdCl_2$ and UO_2SO_4 is roughly 10^2 times smaller.

who proposed that certain electrolytes could be *completely* dissociated into ions.

A full description of the Debye–Hückel theory is given a closer look below. Here, we outline the major assumptions and present the result as it applies at lowest concentrations, the so-called *Debye–Hückel limiting law*. The assumptions are:

(1) the solute is completely dissociated (Bjerrum's suggestion in action),
(2) the solvent acts as a continuous fluid with relative permittivity ϵ_r,
(3) the interionic forces compete with thermal jiggling by the solvent to determine the equilibrium spatial distribution of ions,
(4) the ion concentration is presumed to be low, and
(5) the nonideality is solely due to the free-energy component that results from the electrostatic interaction of ions in their final equilibrium configuration.

With these ideas and assumptions about the nature of electrolyte solutions, the limiting-law expression for $\gamma_\pm$ is (remember: $z_+ > 0$, $z_- < 0$)

$$\ln \gamma_\pm = \frac{e^3(2N_A\rho_s)^{1/2}}{8\pi}\left(\frac{N_A}{\epsilon_0\epsilon_r RT}\right)^{3/2} z_+ z_- I^{1/2} \qquad (10.12)$$

where ρ_s is the solvent density (kg m^{-3}) and I is the ionic strength, Eq. (10.5). For water as a solvent at 25.0 °C, $\rho_s = 997.044\,9$ kg m^{-3} and $\epsilon_r = 78.30$, so that the constants evaluate to

$$\ln \gamma_\pm = 1.174\, z_+ z_- (I/\text{mol kg}^{-1})^{1/2} < 0 \quad (\gamma_\pm < 1)\ . \qquad (10.13)$$

Note that the chemical identity of the electrolyte does *not* enter the equation; only the ionic charges matter. Equation (10.13) predicts the same $\gamma_\pm$ for, say, HCl, LiF, CsI, NH$_4$NO$_3$, at any one total molality (m_i), since $I = m_i$, $z_+ = 1$, and $z_- = -1$ for all these compounds.

Equation (10.13) is compared to experimental data for HCl in Figure 10.4. Notable disagreement is found for ionic strengths around 0.07 mol kg^{-1} and higher, and this is typical behavior. For electrolytes of higher charge per ion, such as CuSO$_4$, the range of agreement is often less. But the low end of the concentration scale, <0.001 mol kg^{-1}, is correctly predicted, and the limiting law has been successfully verified in solvents other than water at varying temperatures and with varying densities and relative permittivities. So, in spite of its limited range of validity (Eq. (10.12) can never predict $\gamma_\pm > 1$, for instance), the Debye–Hückel theory must be classified as a success.

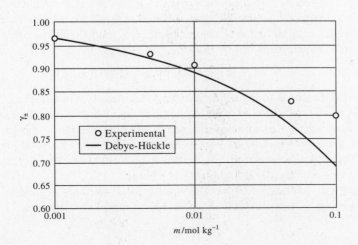

FIGURE 10.4 Measured $\gamma_\pm$ values for HCl are compared to the Debye–Hückle limiting-law expression $\exp(-1.174\, m^{1/2})$, the solid line. Note how deviations are significant for concentrations as low as 0.01 mol kg^{-1}.

EXAMPLE 10.3

How does the mean ionic activity coefficient for a 1.00 mmol kg^{-1} solution of Ca(NO$_3$)$_2$ in water at 298 K compare to that in liquid ammonia at 239 K? For NH$_3$(l), $\epsilon_r = 22.0$ (Table 9.1) and $\rho_s = 683.2$ kg m^{-3}.

SOLUTION Since $m_+ = m_i$, $m_- = 2m_i$, $z_+ = +2$, and $z_- = -1$, $I = 3m_i = 3.00 \times 10^{-3}$ mol kg^{-1} for both solutions. For the aqueous solution, we can use Eq. (10.13) directly and find

$$\gamma_\pm = e^{1.174\, z_+ z_- (I/\text{mol kg}^{-1})^{1/2}} = e^{1.174(2)(-1)(3.00 \times 10^{-3})^{1/2}} = 0.879 \ .$$

For the liquid ammonia solution, we need to evaluate Eq. (10.12) directly:

$$\ln \gamma_\pm = \frac{e^3 (2 N_A \rho_s)^{1/2}}{8\pi} \left(\frac{N_A}{\epsilon_0 \epsilon_r RT} \right)^{3/2} z_+ z_- I^{1/2} = 9.109\, z_+ z_- (I/\text{mol kg}^{-1})^{1/2} \ .$$

Therefore, in liquid ammonia

$$\gamma_\pm = e^{9.109\, z_+ z_- (I/\text{mol kg}^{-1})^{1/2}} = e^{9.109(2)(-1)(3.00 \times 10^{-3})^{1/2}} = 0.368$$

a significantly smaller (less ideal) value. We can explore the factors that contribute to this decrease if we write the proportionalities between $\ln \gamma_\pm$ and the three factors that have changed: T, ϵ_r, and ρ_s:

$$\ln \gamma_\pm \propto \rho_s^{1/2}\, \epsilon_r^{-3/2}\, T^{-3/2}$$

so that

$$\frac{\ln \gamma_\pm(aq)}{\ln \gamma_\pm(\text{NH}_3)} = \left(\frac{997.0}{683.2} \right)^{1/2} \left(\frac{78.3}{22.0} \right)^{-3/2} \left(\frac{298}{239} \right)^{-3/2} = (1.21)(0.149)(0.718) = 0.129 \ .$$

By far, the largest effect comes from the drop in ϵ_r; the density and temperature effects largely counteract each other.

➠ **RELATED PROBLEMS** 10.8, 10.9

A Closer Look at the Debye–Hückel Theory

A derivation of Eq. (10.12) is not very difficult, and it is worth following for two reasons. First, the assumptions stated in the text can be turned into equations, one at a time, and second, the derivation gives us a glimpse at a statistical mechanics approach to thermodynamics. We start with ions of type i at a volumetric concentration c_i/mol L^{-1} and, since this is a molecular theory, immediately turn these concentrations into number densities, $\overline{N}_i$/ions m^{-3}, via

$$\overline{N}_i^\circ = N_A\, \overline{n}_i \qquad \text{and} \qquad \overline{n}_i = c_i (1000 \text{ L m}^{-3}) \tag{10.14}$$

where $\overline{N}_i^\circ$ corresponds to the solution *average* ion density. On a microscopic scale, $\overline{N}_i$ would equal $\overline{N}_i^\circ$ if there were no interionic forces. We thus hypothesize a microscopic $\overline{N}_i(r)$ that is a function of distance from some arbitrarily chosen reference ion, which we will assume to be a positive ion just for sake of argument. The nonideality will all come from the deviation of $\overline{N}_i(r)$ from the spatially uniform $\overline{N}_i^\circ$ value that would exist if there were no charges.

Around our reference ion, we also specify $\rho(r)$, the *charge* density function, computed simply from the ion charges and their spatial distributions:

$$\rho(r) = \sum Q_i\, \overline{N}_i(r) \tag{10.15}$$

10.1 AQUEOUS SOLUTIONS OF IONS—EXTREME NONIDEALITY

but we still do not know $\overline{N}_i(r)$. The connection between $\overline{N}_i^\circ$ and $\overline{N}_i(r)$ is based on Boltzmann's statistical-mechanical result for the competition between thermal energy (which tries to randomize ion positions) and electrostatic energy (which tries to cluster negative ions around positive ions, and vice versa). The Boltzmann expression is very simple:

$$\overline{N}_i(r) = \overline{N}_i^\circ \exp\left(-\frac{Q_i \psi(r)}{k_B T}\right) \tag{10.16}$$

where $k_B = R/N_A$, Boltzmann's constant, and $\psi(r)$ is the electrostatic potential at distance r from the reference ion. Recall (see the discussion in Chapter 9) that $Q\psi$, charge times potential, is potential energy. The argument of the exponential is the negative of the ratio of the electrical potential energy to the thermal energy.

Now we can calculate $\overline{N}_i(r)$, once we find $\psi(r)$, which comes from classical electrostatics.[4] The relation between $\psi(r)$ and $\rho(r)$ for a spherically symmetric distribution around the reference ion comes from the following form of the general electrostatic expression known as *Poisson's equation*

$$\frac{1}{r}\frac{d^2[r\psi(r)]}{dr^2} = -\frac{\rho(r)}{\epsilon} \tag{10.17}$$

where $\epsilon = \epsilon_0 \epsilon_r$, the solvent's permittivity.

A glance back through Equations (10.16) and (10.15) shows that this will be a difficult differential equation to understand and solve without some simplifications. The first simplification expands Eq. (10.16) through the first two terms,

$$\overline{N}_i(r) = \overline{N}_i^\circ \left[1 - \frac{Q_i \psi(r)}{k_B T}\right]$$

and simplifies the expression for $\rho(r)$ to

$$\rho(r) = \sum_i Q_i \overline{N}_i^\circ \left[1 - \frac{Q_i \psi(r)}{k_B T}\right] = -\frac{\psi(r)}{k_B T}\sum_i Q_i^2 \overline{N}_i^\circ .$$

The last equality follows from $\sum Q_i \overline{N}_i^\circ = 0$ because the solution has equal numbers of positive and negative charges. (Note that this is where the ionic strength, which is related to $\sum Q^2 N^\circ$, enters the theory.)

Now we have simplified[5] Eq. (10.17) to

$$\frac{d^2[r\psi(r)]}{dr^2} = r\psi(r)\left(\frac{\sum Q_i^2 \overline{N}_i^\circ}{\epsilon k_B T}\right) = \frac{r\psi(r)}{r_D^2} \tag{10.18}$$

where we have abbreviated the quantity in parentheses on the right by the reciprocal of the square of a naturally occurring distance r_D, called the *Debye length*. Equation (10.18) can be readily solved, but before we do so, it is worth taking a moment to explore the magnitude of r_D to see if our expansion of the exponential is justified. For a 1:1 salt (NaCl, HCl, etc., with $z_+ = -z_- = 1$),

$$r_D = \frac{3.04 \text{ Å}}{(c/\text{mol L}^{-1})^{1/2}}$$

---

[4] Note that, even though the ions are mobile and are jumping around the solution, we take a time-average view of the charge distribution to allow us to compute a continuous charge density $\rho(r)$ and a continuous electrostatic potential $\psi(r)$. An analogy is an aerial picture of an auto race taken with a very long exposure time. The picture will show that most of the cars stayed on the track, went down the middle of the straightaways, etc., but it won't tell us who won the race or where any one car was at any time.

[5] The technical term is *linearized*, since we turned the exponential dependence of $\overline{N}_i(r)$ on ψ into a linear one.

for water at 25 °C. For $c = 10^{-3}$ mol L^{-1}, $r_D = 98.2$ Å, and for $c = 0.1$ mol L^{-1}, $r_D = 9.62$ Å. Since r_D will turn out to measure the distance of maximum ion density around the reference ion, these are reassuring values. They are not so large that interionic forces would be trivially small, nor are they smaller than the radius of a typical central reference ion ($\cong 1$–3 Å).

For the exponential expansion to be valid, $Q\psi/k_B T$ should be small compared to unity. If we take $\psi \cong e/4\pi\epsilon r_D$, $r_D \cong 10$ Å, $Q \cong e$, and $T \cong 300$ K, we find $Q\psi/k_B T \cong 0.7$, which is less than unity, but just barely. If $r_D \cong 100$ Å (the mmol L^{-1} value), this ratio drops by a factor of 10 to a more comfortably small value. Here we have found an approximation in the theory that becomes marginally valid at those concentrations where the final theoretical expression and experiment begin to disagree.

Now to solve Eq. (10.18). The general solution is of the form

$$\psi(r) = A\frac{\exp(-r/r_D)}{r} + B\frac{\exp(+r/r_D)}{r}$$

where A and B are constants that are determined by the physical constraints on the problem. The constant B must be zero, since the $\exp(+r/r_D)$ factor would cause the potential to increase arbitrarily as $r \to \infty$. To find A, we assume that other ions can approach the reference ion up to, but no closer than, a distance $r = a$, where a is a distance on the order of an ionic radius. For $r \leq a$, $\psi(r)$ is the sum of the reference-ion potential, $Q°/4\pi\epsilon r$, and some average potential ϕ due to ions outside the $r = a$ boundary. (We write the charge on the central ion as $Q°$.) Thus

$$\psi(r) = \frac{Q°}{4\pi\epsilon r} + \phi \qquad r \leq a \tag{10.19}$$

and, at $r = a$, $\psi(r)$ and the electric field, $-d\psi/dr$, must be continuous. With the general expression $\psi = A \exp(-r/r_D)/r$ and the specific expression above for $r \leq a$, the two continuity equations let us find A and ϕ:

$$A = \frac{Q° \exp(a/r_D)}{4\pi\epsilon(1 + a/r_D)}$$

and

$$\phi = -\frac{Q°}{4\pi\epsilon r_D(1 + a/r_D)}.$$

The origin of ϕ, which will be our focal point in all of ψ, is worth exploring further. Given A, we can now write $\psi(r)$ as

$$\psi(r) = \frac{Q°}{4\pi\epsilon(1 + a/r_D)}\frac{\exp[-(r-a)/r_D]}{r}. \tag{10.20}$$

If $r - a < r_D$ (i.e., we are close to the reference ion in a dilute solution), we can expand the exponential, keeping two terms

$$\psi(r) = \frac{Q°}{4\pi\epsilon(1 + a/r_D)}\left(\frac{1}{r}\right)\left(1 - \frac{r-a}{r_D}\right),$$

which can be simplified to

$$\psi(r) = \frac{Q°}{4\pi\epsilon r} - \frac{Q°}{4\pi\epsilon r_D(1 + a/r_D)}.$$

This is just the assumed form (Eq. (10.19)) for $\psi(r \leq a)$, with ϕ a natural piece of the full potential. (Note that Eq. (10.19) will not satisfy Eq. (10.18), which is only valid for $r > a$, since, by hypothesis, ions approach no closer than a.)

We now have the potential, which gives us the charge and number density distributions. What remains is to relate these to a free energy and thus to an activity coefficient. As discussed in Chapter 5, any Gibbs free-energy change is equivalent to the maximum work done by the system at constant T and P (see Eq. (5.24)). Here, we use that result in reverse. We will perform reversible, isothermal, isobaric work on the system and equate that work to ΔG. We imagine our reference ion is *uncharged* initially, but in a solution described by the number densities $\overline{N}_i$ (rather than $\overline{N}_i^\circ$). In effect, we have *all* the ions (since any could be the representative one) in their final, real positions, but with their charges set to zero.

We bring the charge from zero to the final value, Q°. This involves electrical work w_{el}:

$$w_{el} = \int_0^{Q^\circ} \psi(r)\, dQ$$

but what do we choose for r? It would seem that $r = 0$ would be logical, but both our expressions for $\psi(r)$ (for $r \leq a$, Eq. (10.19), or for $r > a$, Eq. (10.20)) diverge at $r = 0$.

In fact, $r = 0$ *is* the correct location; the way out of the divergence dilemma is to recognize that we do not need *all* of w_{el} to compute the nonideality due to interionic forces. Look at Eq. (10.19). The first term is the potential of the reference ion itself. The work done in creating this ion, all by itself, is *never* available to us in any real process. Only the interaction-energy part of the work can be recovered, and this part is due to the second term, ϕ. (The first part is part of μ°.)

Thus, an integral of $\phi(r)$ gives the nonideality part of μ directly. For a mole of positive reference ions

$$RT \ln \gamma_+ = N_A \int_0^{Q_+} \phi\, dQ^\circ = -N_A \int_0^{Q_+} \frac{Q^\circ\, dQ^\circ}{4\pi\epsilon r_D(1 + a/r_D)}$$
$$= -\frac{N_A Q_+^2}{8\pi\epsilon r_D(1 + a/r_D)}.\qquad(10.21)$$

This is, in fact, a better expression than the limiting-law result, Eq. (10.12). Although it applies to the positive ions only (and thus cannot be compared to experiment), the negative-ion expression is identical (just change γ_+ to γ_- and Q_+ to Q_-), and we find $\gamma_\pm$ from Eq. (10.7) in the form

$$\nu \ln \gamma_\pm = \nu_+ \ln \gamma_+ + \nu_- \ln \gamma_-.$$

The limiting-law expression comes from replacing $(1 + a/r_D)$ by 1, since $r_D \gg a$ as the concentration approaches zero. It is left as an exercise to recover Eq. (10.12), the limiting law in terms of molality-based ionic strength, from Equations (10.21), (10.7), and the definition of r_D from Eq. (10.18).

The expression for $\ln \gamma_\pm$ based on Eq. (10.21) contains the parameter a in addition to the fundamental constants and macroscopic solvent properties of the limiting law. Since a is a microscopic radius that is difficult to predict or measure, it is often taken as a free parameter and used to improve the fit of Eq. (10.21) to experimental data. But since, in general, we should have an "a_+" and "a_-" to represent the different sizes of positive and negative ions, the single radius that is required to derive an expression for $\ln \gamma_\pm$ has very little physical significance.

Several semiempirical expressions have been advanced to improve the range of agreement between theory and experiment. Some of these expressions have been very successful, especially in their ability to explain $\gamma_\pm$ values above 1 on the basis of ion hydration. The parameters in the best of these theories have physical interpretations (such as the ionic radius or the number of solvent molecules attached to the solvated ion), and their values have provided valuable insight into the microscopic structure of ionic solutions. Of particular importance is the result that the H^+ ion in solution does not have the size of a bare proton. It is hydrated in a form probably best represented as $H_3O^+(H_2O)_n$ with n on the order of 5–7.

Discussions of these improved forms of the basic Debye–Hückel limiting law are given in this chapter's references. We list here only a few of the more successful. The first, often called the *extended Debye–Hückel law,* follows from the derivation given a closer look above. For water at 25 °C, it is

$$\ln \gamma_\pm = 1.174 \, z_+ z_- \left[\frac{(I/\text{mol kg}^{-1})^{1/2}}{1 + (I/\text{mol kg}^{-1})^{1/2}} \right],$$

the limiting law with the addition of an $I^{1/2}$ term in the denominator. This term slows down the drop in $\ln \gamma_\pm$ with increasing I and extends the agreement with experiment to $I \cong 0.05$ mol kg^{-1} for many electrolytes.

Adding the positive term βI, where β is an empirical parameter, produces a minimum in $\ln \gamma_\pm$. A value for $\beta \cong 0.35$ kg mol^{-1} gives estimates of $\gamma_\pm$ good to $\cong 10\%$ for I as large as several tenths mol kg^{-1}. For example, the experimental value for HCl(*aq*, $I = 0.1$ mol kg^{-1}) is $\gamma_\pm = 0.796$. The limiting law predicts 0.690; the extended law predicts 0.754; the extended law with 0.35 I added predicts 0.781, which is in error by only 2%. Figure 10.5 compares this modified extended law to experimental data for several electrolytes of varying charge types. The improvement over the simple limiting law is evident.

We have now discussed the measurement of and theoretical basis for $\gamma_\pm$. To put $\gamma_\pm$ to good use, we start with some simple applications. Of course, a good theory not only gives us an enhanced chemical and physical understanding of a set of phenomena, but it also allows us to predict the results of an experiment in advance. Thus, the theory can be used to predict solubilities of electrolytes, their colligative properties, etc. We will explore solubilities in some detail.

Solubility-product calculations are probably familiar to you from elementary chemistry courses. The simple approximation that replaces activities by concentra-

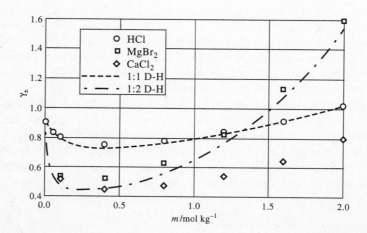

FIGURE 10.5 The extended Debye–Hückle expression, with 0.35 $I^{1/2}$ added to the basic term for $\ln \gamma_\pm$, is compared to measured values for HCl—where the agreement is good—and to two 1:2 salts, CaCl$_2$ and MgBr$_2$—where the agreement is less satisfactory. Better agreement results if the factor 0.35 is allowed to vary freely, taking on a unique value for each electrolyte.

tions in solubility equilibrium-constant expressions is often seriously in error. The correct expression for a solubility equilibrium constant is

$$K_{eq} = (\gamma_{\pm} m_{\pm})^{\nu} . \tag{10.22}$$

The elementary approximation sets $\gamma_{\pm} = 1$ and approximates molalities with volumetric concentrations (molarities). The second approximation is often acceptable, while the first can be not only grossly in error, but it can also fail to predict other phenomena.

The following few paragraphs explore solubility calculations in the form of an example based on the single experimental observation that one can dissolve no more than 0.220 g of $Ba(IO_3)_2$ in 1.00 L of pure water at 25 °C. The molar mass of $Ba(IO_3)_2$ is 487.14 g mol^{-1}; thus, the molar solubility is

$$\frac{0.220 \text{ g L}^{-1}}{487.14 \text{ g mol}^{-1}} = 4.52 \times 10^{-4} \text{ M} .$$

At these concentrations, it is a good approximation to take molalities equal to molarities. The ionic strength, based on the equilibrium

$$Ba(IO_3)_2(s) \rightleftarrows Ba^{2+}(aq) + 2IO_3^-(aq)$$

is, with m_i = solubility molality,

$$I = \frac{1}{2} [m_+ (2)^2 + m_- (-1)^2] = 3m_i = 1.35 \times 10^{-3} \text{ mol kg}^{-1}$$

since $m_+ = m_i$ and $m_- = 2m_i$. Likewise

$$m_{\pm} = 4^{1/3} m_i = 7.17 \times 10^{-4} \text{ mol kg}^{-1}$$

and, from the limiting law,

$$\ln \gamma_{\pm} = 1.174 \, (+2)(-1)(1.35 \times 10^{-3})^{1/2} = -8.64 \times 10^{-2}$$

or $\gamma_{\pm} = 0.917$. Therefore, with $\nu = 3$,

$$K_{eq} = [(7.17 \times 10^{-4})(0.917)]^3 = 2.84 \times 10^{-10} .$$

The elementary approximation $\gamma_{\pm} = 1$ predicts $K_{eq} = 4(4.52 \times 10^{-4})^3 = 3.69 \times 10^{-10}$.

Armed with K_{eq}, we can approach a variety of other situations. For instance: what is the solubility of $Ba(IO_3)_2$ in a 0.01 mol kg^{-1} solution of $MgCl_2$? The Mg^{2+} and Cl^- ions do not *chemically* influence the solubility, but they raise the ionic strength and *physically* influence the solubility. For a first approximation, we will assume that $MgCl_2$ is the only source of ions and I = 0.03 mol kg^{-1}. The limiting law predicts $\gamma_{\pm}$ (for $Ba(IO_3)_2$ in a solution of this ionic strength!) to be

$$\gamma_{\pm} = e^{1.174(2)(-1)(0.03)^{1/2}} = 0.666 .$$

This value is probably inaccurate, since we are outside the range of likely validity of the limiting law, but it is a good first approximation. The mean $Ba(IO_3)_2$ molality is thus

$$m_{\pm} = \frac{K_{eq}^{1/3}}{\gamma_{\pm}} = \frac{(2.84 \times 10^{-10})^{1/3}}{0.666} = 9.87 \times 10^{-4} \text{ mol kg}^{-1}$$

and the solubility m_i, which is also m_+ by stoichiometry, is

$$m_{Ba^{2+}} = m_+ = \frac{m_{\pm}}{4^{1/3}} = 6.22 \times 10^{-4} \text{ mol kg}^{-1} .$$

If the extended Debye–Hückel law is used, a more accurate value of 5.91×10^{-4} mol kg^{-1} is calculated for the solubility. Note that this is a significantly *greater* solubility than the pure water value, 4.52×10^{-4} mol kg^{-1}. Such a solubility increase is not predicted by an elementary solubility equilibrium theory.

The physical origin of enhanced solubility by a chemically inert electrolyte is easy to understand. The inert ions provide a stabilizing ionic atmosphere for Ba^{2+} and IO$_3^-$ (they lower the chemical potential of these ions below the pure-water value), and thus more Ba^{2+} and IO$_3^-$ ions have to enter the solution to create the $\nu_+\mu_+ + \nu_-\mu_-$ value (see Eq. (10.10)) that matchs μ of the solid and attains chemical equilibrium.

Other solubility phenomena, such as the common-ion effect, are readily treated by similar calculations, as are acid and base dissociation equilibrium calculations. Several of these are explored in the problems at this chapter's end and in Example 10.4.

EXAMPLE 10.4

The equilibrium constant for the aqueous dissociation of acetic acid (HOAc) to acetate ion (OAc$^-$):

$$\text{HOAc}(aq) \rightleftarrows \text{H}^+(aq) + \text{OAc}^-(aq)$$

is $K_{eq} = 1.75 \times 10^{-5}$ at 298 K. Find the molality of H$^+$ and the pH, defined as pH = $-\log_{10}(a_{H^+})$, for the solutions:

(a) 0.010 mol kg^{-1} total acetate (in both forms, HOAc and OAc$^-$) in pure water
(b) 0.010 mol kg^{-1} total acetate in a 0.020 mol kg^{-1} NaCl(aq) solution.

SOLUTION The equilibrium constant expression is

$$K_{eq} = \frac{a_{H^+}\, a_{OAc^-}}{a_{HOAc}} = \frac{m_{H^+}\, m_{OAc^-}\, \gamma_{\pm}^2}{m_{HOAc}} = \frac{m_{H^+}\, m_{OAc^-}\, e^{-1.174(2)I^{1/2}}}{m_{HOAc}}$$

where we have made the good approximation that $a_{HOAc} = m_{HOAc}$, since HOAc is uncharged, and $I = m_{H^+} = m_{OAc^-}$ in the Debye–Hückel expression for $\gamma_{\pm}$. Since molality is directly proportional to number of moles, we can apply Chapter 7 methodology and introduce the degree of advancement ξ via

$$\xi = m_{HOAc}^\circ - m_{HOAc} = m_{H^+} = m_{OAc^-} = 0.01 - m_{HOAc}\;.$$

For solution (a), we solve the equilibrium expression

$$K_{eq} = 1.75 \times 10^{-5} = \frac{\xi^2\, e^{-1.174(2)\sqrt{\xi}}}{(0.01 - \xi)},$$

which is suited to an iterative solution if written in the form

$$\xi_{n+1} = \sqrt{1.75 \times 10^{-5}\,(0.01 - \xi_n)}\; e^{1.174(2)\xi_n^{1/2}}$$

and the iterations are started with the guess $\xi_0 = 0$. After three iterations, we find

$$\xi = 4.194 \times 10^{-4} \text{ mol kg}^{-1} = m_{H^+}$$

$$a_{H^+} = m_{H^+}\, \gamma_+ = m_{H^+}\, e^{-1.174\sqrt{m_{H^+}}} = 4.095 \times 10^{-4}$$

$$\text{pH} = -\log_{10}(a_{H^+}) = 3.387\;.$$

For solution (b), the chemically inert NaCl alters $\gamma_{\pm}$ via the ionic strength. The iterative expression becomes

$$\xi_{n+1} = \sqrt{1.75 \times 10^{-5}\,(0.01 - \xi_n)}\; e^{1.174(2)(0.02 + \xi_n)^{1/2}}$$

and we find
$$\xi = 4.828 \times 10^{-4} \text{ mol kg}^{-1} = m_{H^+}$$
$$a_{H^+} = m_{H^+} \gamma_+ = m_{H^+} \, e^{-1.174\sqrt{(0.02 + m_{H^+})}} = 4.081 \times 10^{-4}$$
$$\text{pH} = -\log_{10}(a_{H^+}) = 3.389 \ .$$

Note that while m_{H^+} increased 15% in the NaCl solution, the pH is essentially unchanged due to a similar decrease in activity coefficient.

➥ RELATED PROBLEMS 10.4, 10.11

10.2 CHEMICAL REACTIONS THAT TRANSFER ELECTRONS

This section reviews a topic from elementary chemistry: oxidation–reduction or "redox" reactions. You probably recall, and perhaps have done, the experiment in which a strip of Zn metal is placed in a solution of Cu^{2+}. Immediately, the Zn begins to dissolve and the characteristic blue color of $Cu(H_2O)_6^{2+}$ fades as $Cu(s)$ appears. The spontaneous reaction is

$$Zn(s) + Cu^{2+}(aq) \rightleftarrows Zn^{2+}(aq) + Cu(s)$$

and we say[6] the zinc was *oxidized* (lost electrons) while the Cu^{2+} ions were *reduced* (gained electrons) as the reaction proceeded from left to right.

Suppose we imagined the Zn/Cu^{2+} reaction proceeding by the following path:

$$Zn(s) + Cu^{2+}(aq)$$
$$\downarrow \text{(1) vaporize Zn}$$
$$Zn(g) + Cu^{2+}(aq)$$
$$\downarrow \text{(2) move } Cu^{2+} \text{ to gas phase}$$
$$Zn(g) + Cu^{2+}(g)$$
$$\downarrow \text{(3) doubly ionize Zn}$$
$$Zn^{2+}(g) + Cu^{2+}(g) + 2e^-(g)$$
$$\downarrow \text{(4) attach electrons to } Cu^{2+}$$
$$Zn^{2+}(g) + Cu(g)$$
$$\downarrow \text{(5) put } Zn^{2+} \text{ ions in aqueous phase}$$
$$Zn^{2+}(aq) + Cu(g)$$
$$\downarrow \text{(6) condense gaseous Cu}$$
$$Zn^{2+}(aq) + Cu(s)$$

Steps (3) and (4) together represent a *charge-transfer reaction* (as gas-phase redox reactions of this type are traditionally called). The thermodynamics of such reactions are readily discussed in terms of ionization potentials, ideal gas entropies, etc., as was done in Chapter 7. Similarly, steps (1) and (8) are simple sublimations. The

---

[6]You may have once been taught the time-honored mnemonic "Leo the lion goes ger" for "*Lose Electrons—Oxidation . . . Gain Electrons—Reduction*," which has kept oxidation and reduction straight in the minds of many students over many years. Redox chemistry and electrochemistry have many traditional definitions, and several other mnemonics have arisen over the years to help keep these definitions straight in contemporary minds. We owe the "negative" charge on an electron to Benjamin Franklin; in fact, the electron's charge is only "opposite" or "different in the only possible way" from that of the proton (neither of which were discovered in Franklin's lifetime, of course). It is fortunate that Franklin did not say "right handed" or "upside down" instead of "negative" when describing the charge, since mathematical theories of electrostatics are obviously suited to "positive/negative" differences. But the choice was arbitrary, and like many other aspects of science, a need for some agreed-to set of nomenclature arose before the phenomena were fully understood. Other conventions and definitions will appear as we progress toward electrochemical cell behavior.

TABLE 10.1 Dimensionless Standard Molar Entropies ($\overline{S}°/R$) of Aqueous Ions

H^+	0.000 by definition				
Li^+	1.71	OH^-	-1.293	NH_4^+	13.57
Na^+	7.25	F^-	-1.66	HCO_3^-	11.4
K^+	12.3	Cl^-	6.64	CO_3^{2-}	-6.39
Rb^+	14.4	Br^-	9.71	CH_3COOH^-	10.5
Cs^+	16.0	I^-	13.15	Al^{3+}	-37.7
Mg^{2+}	-14.2	I_3^-	28.7	Sn^{2+}	-3.0
Ca^{2+}	-6.64	CN^-	14.2	Pb^{2+}	2.6
Sr^{2+}	-4.7	NO_3^-	17.6	Zn^{2+}	-12.81
Ba^{2+}	1.5	SO_4^{2-}	2.42	Cd^{2+}	-7.35
Cu^{2+}	-11.9	HSO_4^-	15.9	Fe^{2+}	-16.56
Ag^+	8.89	Co^{2+}	-13.6	Fe^{3+}	-37.99

difficult steps to carry out experimentally are (2) and (5). We cannot pull one ion out of solution (step (2)), nor can we inject gas-phase ions of one charge into solution (step (5)).

Consequently, practical calculations on aqueous ions use *relative* values for $\overline{S}°$, $\Delta_f\overline{H}°$, $\Delta_f\overline{G}°$, and even $\overline{C}_P°$ for individual ions. The relative enthalpies of hydration, based on $\Delta_f\overline{H}°(H^+(aq)) = 0$, were discussed in Section 7.3 and tabulated in Table 7.5. In Tables 10.1 and 10.2, relative aqueous entropies and free energies of formation, respectively, are listed for several ions, again based on the $H^+(aq)$ reference[7] values

$$\overline{S}°(H^+(aq)) = 0 \quad \text{and} \quad \Delta_f\overline{G}°(H^+(aq)) = 0 .$$

Look first at the entropy values. Note that increasing size at constant ion charge (the alkali metal ions, for instance) increases the relative entropy, while increasing the charge at (roughly) constant size (compare Fe^{2+} to Fe^{3+}) decreases the entropy.

Now turn to the relative molar free energies of formation in Table 10.2. As with most energy of formation data, the individual values are less telling than the

---

[7] We also take $\Delta_f\overline{G}°(e^-) = 0$.

TABLE 10.2 Dimensionless Standard Molar Free Energies of Formation ($\Delta_f\overline{G}°/RT$) of Aqueous Ions at $T = 298.15$ K

H^+	0.000 by definition				
Li^+	-118.5	OH^-	-63.4312	NH_4^+	-32.07
Na^+	-105.6	F^-	-111.5	HCO_3^-	-236.8
K^+	-113.8	Cl^-	-52.91	CO_3^{2-}	-213.0
Rb^+	-114.2	Br^-	-41.47	CH_3COOH^-	-150.25
Cs^+	-119.5	I^-	-20.84	Al^{3+}	$-194.$
Mg^{2+}	-184.0	I_3^-	-20.78	Sn^{2+}	-10.6
Ca^{2+}	-223.1	CN^-	69.54	Pb^{2+}	-9.81
Sr^{2+}	-224.8	NO_3^-	-44.61	Zn^{2+}	-59.38
Ba^{2+}	-226.21	SO_4^{2-}	-300.34	Cd^{2+}	-31.36
Cu^{2+}	26.21	HSO_4^-	-304.93	Fe^{2+}	-34.26
Ag^+	31.11	Co^{2+}	-22.11	Fe^{3+}	-4.27

combinations that predict $\Delta_r\overline{G}°$ or K_{eq} values. For example, using the Zn/Cu^{2+} reaction, we calculate

$$\frac{\Delta_r\overline{G}°}{RT} = -\ln K_{eq} \quad \text{or} \quad K_{eq} = e^{-[(-59.38)-(26.21)]} = 1.48 \times 10^{37} \;.$$

This huge value is typical for many redox reactions. (The more telling observation is that a redox K_{eq} is often many orders of magnitude larger *or* smaller than unity.) Such reactions "go to completion," with vanishing amounts of the limiting reagent present at equilibrium.

EXAMPLE 10.5

Zinc powder is added to a dilute solution of CuSO$_4$ and ZnSO$_4$ and allowed to come to equilibrium. Analysis of the solution shows that [Zn^{2+}] = 4.26×10^{-4} mol kg^{-1}. What is [Cu^{2+}]?

SOLUTION This is a simple question to answer, but it is worth looking at in some detail to illustrate the way mean ionic quantities enter equilibrium expressions. We can write immediately (since the activities of the pure metals are unity)

$$K_{eq} = 1.48 \times 10^{37} = \frac{a_{Zn^{2+}}}{a_{Cu^{2+}}} = \frac{\gamma_+(Zn^{2+})\, m_+(Zn^{2+})}{\gamma_+(Cu^{2+})\, m_+(Cu^{2+})}$$

and ignore the common SO$_4^{2-}$ ion, since its activity would cancel from the equilibrium activity-ratio expression. Moreover, the activity coefficients need *not* be calculated since they are *equal* and cancel. If we include the sulfate counter ion and consider $\gamma_\pm$ from the Debye–Hückel expression, *since both cations are in the same solution at the same ionic strength with the same ionic charge and counter ion,* both have the same mean ionic activity coefficient. This need not be true at high concentrations where chemical distinctions among ions become important, but it is true at low concentrations where only the physical effects of the Debye–Hückel theory govern $\gamma_\pm$. The elementary expression

$$[Cu^{2+}] = \frac{[Zn^{2+}]}{K_{eq}} = \frac{4.26 \times 10^{-4} \text{ mol kg}^{-1}}{1.48 \times 10^{37}} = 2.88 \times 10^{-41} \text{ mol kg}^{-1}$$

is the correct one. The magnitude of the answer should impress you; this concentration is equivalent to 1.73×10^{-17} *ions* per kg of solution. This number is so small that a probabilistic interpretation is warranted. If 1 kg of such a solution had been sitting around throughout the Earth's history ($\cong 4.5 \times 10^9$ y = 1.4×10^{17} s) one and only one Cu^{2+} ion would appear in the solution for a total time of roughly one second!

➤ RELATED PROBLEMS 10.13, 10.14

The large free-energy change[8] of redox reactions implies a large amount of energy potentially available for useful isobaric, isothermal work. This is the key to the practical utility of redox reactions. Consider the path

$$\text{Zn}(s) + \text{Cu}^{2+}(aq)$$
$$\downarrow$$
$$\text{Zn}^{2+}(aq) + \text{Cu}^{2+}(aq) + 2e^-(\text{somewhere})$$
$$\downarrow$$
$$\text{Zn}^{2+}(aq) + \text{Cu}(s)$$

---

[8]The $\Delta_r\overline{G}°/RT$ value $(-59.38) - (26.21) = -85.59$ at 298.15 K corresponds to $(-85.59)(8.314\ 51$ J mol^{-1} K^{-1})(298.15 K) = -212.2 kJ mol^{-1}.

Here e^-(somewhere) indicates that the *transferred electrons are the species that could be used to generate work* if we can grab them during the course of the reaction. This leads directly to the electrochemical cell, a valuable practical device (in many forms) as well as a valuable research tool. The next section discusses how the redox free energy can be related to an electrical potential-energy change.

10.3 THE ELECTROCHEMICAL POTENTIAL

Imagine dipping a piece of a pure metallic element constructed with an unnatural isotope distribution (a piece of cobalt with some known fraction of the radioactive ^{60}Co isotope comes to mind) into a solution of the metal's ions in the natural isotope distribution. After some time, the isotope distribution in solution will be altered (^{60}Co^{2+}(aq) would be found, for example). This illustrates a dynamic aspect to equilibrium, an equilibrium we would write

$$\text{Co}^{2+}(aq) + 2e^- \rightleftarrows \text{Co}(s)$$

for the cobalt example. The solubility equilibria we have talked about are also dynamic in nature on a molecular level (like all chemical equilibria). We could write reactions such as AB$(s) \rightleftarrows$ AB(aq) and understand the equilibrium to mean equality of chemical potential: $\mu(\text{AB}(s)) = \mu(\text{AB}(aq))$. How do we incorporate the ionization of the typical redox reaction, or the chemical potential of the electrons?

With any charged species in any phase, the electrical potential of that phase directly alters the species' chemical potential. This follows from the definition of electrical work as the integral that raises the charge from zero to its final value in a region of potential ϕ or, equivalently, that brings the final charge from an infinite distance, where $\phi = 0$, to the final position at potential ϕ. The work per mole of charge Q_i is $Q_i N_A \phi$, and this adds directly to the chemical potential.

To simplify the nomenclature, we write $Q_i = z_i e$, and introduce *Faraday's constant, F,* the charge magnitude of 1 mol of elementary charges:

$$F = e\, N_A = 96\,485.309 \text{ C mol}^{-1} .$$

The total chemical potential (which we will call the *electrochemical potential, μ^e*) is the sum of μ_i and $Q_i N_A \phi = z_i F \phi$:

$$\mu^e = \mu_i + z_i F \phi . \qquad (10.23)$$

Note that a positive charge in a region of positive potential or a negative charge in a negative potential will have $\mu^e > \mu_i$ and will thus have an enhanced tendency to leave that region, while charges of sign opposite that of the potential have an enhanced tendency to remain in the region.

Armed with Eq. (10.23), we could return to the analysis of chemical equilibrium in Section 7.4 and replace the ordinary μ by μ^e. We would, for our example Co redox reaction, find that equilibrium is given according to Eq. (7.17) by

$$\left(\sum_{\text{products}} \nu_i \mu_i^e - \sum_{\text{reactants}} \nu_i \mu_i^e \right) = \mu^e(\text{Co}(s)) - \mu^e(\text{Co}^{2+}(aq)) - 2\mu^e(e^-) = 0 . \quad (10.24)$$

Note that since $z(\text{Co}(s)) = 0$, $\mu^e(\text{Co}(s)) = \mu(\text{Co}(s))$, but note too that the ions are in solution at potential $\phi(aq)$ while the electrons are in the metal at perhaps a different potential $\phi(s)$. An electrical potential *difference* is what a voltmeter measures. We should find a voltage $\Delta\phi$ between the solid metal and the ionic solution given from Eq. (10.24) by these steps:

$$\mu(\text{Co}(s)) = \mu(\text{Co}^{2+}(aq)) + 2F\phi(aq) + 2\mu(e^-) - 2F\phi(s)$$

or

$$\Delta\phi = \phi(s) - \phi(aq) = -\frac{1}{2F}[\mu(Co^{2+}(aq)) + 2\mu(e^-) - \mu(Co(s))] \ . \quad (10.25)$$

The potential $\phi(s)$ is uniform throughout the solid and $\phi(aq)$ is uniform throughout the solution since both these phases are electrical conductors at equilibrium. If ϕ were not uniform, then electric fields would exist and the free charges would experience forces and start to move; this motion would dissipate energy (generate heat) and would be indicative of an irreversible, non-equilibrium process.

The practical problem with Eq. (10.25) is that this $\Delta\phi$ *cannot,* in fact, be measured by a voltmeter. Here are some of the reasons. The first, which is not very serious, centers around our inability to construct the ideal voltmeter, which would require zero current in order to measure voltage. (Any non-zero current would necessarily flow through the system as well, destroying the uniform equilibrium.) Modern research voltmeters can measure a 1 V potential difference with a current $\sim 10^{-15}$ A (which is 10^{-15} C s^{-1}/1.6×10^{-19} C electron^{-1} = 6.25×10^3 electrons s^{-1}) flowing through the voltmeter. Common hobbyist or routine test voltmeters, by contrast, require currents $\sim 5 \times 10^{-5}$ A V^{-1}.

The second difficulty arises because the metal wires leading from the voltmeter will introduce their own $\Delta\phi$'s at each connection to the system. Suppose these wires were made of gold, which is a good choice due to the chemical inertness of gold. If we touched these wires together, the voltmeter would sensibly read zero volts. If we touched one wire to one end of the Co(s) strip and the other wire to the other end, again the reading would be zero, but *only because two equal but opposite* $\Delta\phi$'s, Au|Co and Co|Au, were introduced into the circuit.[9] So imagine the sequence of connections:

$$Au|Co(s)|Co^{2+}(aq)|Au \ .$$

We want $\Delta\phi$ for the middle interface, but the voltmeter must record the algebraic sum of $\Delta\phi$'s for all three interfaces. It would be useless to try and make the wires out of cobalt, since the arrangement Co(s)|Co^{2+}(aq)|Co(s) would yield a zero voltage just as the Au|Co(s)|Au arrangement would.

While $\Delta\phi$ in Eq. (10.25) is not readily measurable, the equation itself indicates how we can control it. The most obvious way is to alter $\mu(Co^{2+}(aq))$ through the solution's concentration (and thus the ion's activity). Changing the temperature is another way, since μ depends on T. We pick up these ideas in the next section where we construct electrochemical cells, step by step, and finally construct devices with measurable, controllable, and interpretable potential differences.

10.4 ELECTROCHEMICAL DEVICES

The easiest way to see how an electrochemical cell is constructed is to use a simple process to establish a Gibbs free-energy difference through which electrons can move, producing useful work. Imagine two solutions of CdCl$_2$, one at 2 mol kg^{-1} and the other at 1 mol kg^{-1}, separated by a removable partition, as in Figure 10.6a. If both solutions have the same volume and we remove the partition, mixing occurs as indicated by

$$CdCl_2(2\ m) + CdCl_2(1\ m) \rightarrow 2CdCl_2(1.5\ m) \ .$$

---

[9]The notation Au|Co means "Au in contact with Co." The | symbol will represent a simple interface between two phases in contact.

FIGURE 10.6 In (a), it is clear that removing the partition leads to spontaneous mixing. Similarly, in (b), the membrane permeable to Cd^{2+} and Cl^- will permit mixing. In (c), Ag/AgCl electrodes and the Cd/Hg amalgam allow mixing via electron transfer through an external circuit between the Ag electrodes.

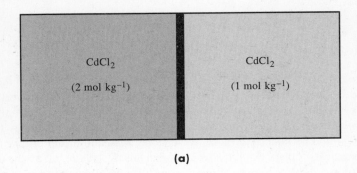

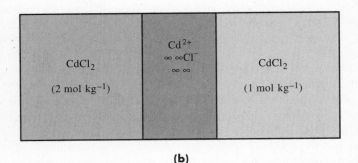

This is a *spontaneous* process characterized by a negative ΔG. If we replaced the partition with a barrier semipermeable to Cd^{2+} and Cl^- ions, the same spontaneous process would occur (Figure 10.6b). Neither of these experiments allow us to grab ΔG and produce useful work, however.

Suppose we had a barrier that was semipermeable to Cd^{2+} ions only, and in addition we had devices in each solution that could act as either a source or a sponge for Cl^-, as needed. Then we could transfer Cd^{2+} from the high- to the low-concentration cell (from left to right in the figure) while supplying Cl^- separately from the device in the right-hand solution and soaking up the excess Cl^- with the left side's device. This would quickly build up an excess negative charge on the left and a positive charge on the right. We connect these devices through an external electric circuit that includes a light bulb, motor, or whatever, and allow the excess charges to be neutralized by an electron flow through this circuit. Voilà. We have used the ΔG of spontaneous mixing to perform external work. We have constructed an electrochemical cell, a battery.

The Cd^{2+} permeable barrier could be made from a solution of Cd in liquid Hg. Such alloys with mercury are called *amalgams*. At the amalgam surface, the reversible and rapid redox reaction

$$Cd^{2+}(aq) + 2e^- \rightleftarrows Cd(Hg)$$

takes place, and the cadmium atoms are sufficiently mobile in the liquid mercury solution to defeat the concentration gradient that ion motion into the amalgam (on the left) and out of the amalgam (on the right) tries to establish. The Cl^- source/sponge device is easy to construct from, for instance, a silver wire packed tightly into a paste of $AgCl(s)$ in the solution. Here the redox reaction is

$$AgCl(s) + e^- \rightleftarrows Ag(s) + Cl^-(aq) .$$

This cell is diagrammed in Figure 10.6(c), and the electrical chain of interfaces, left to right, is written

$$Ag(s)|AgCl(s)|CdCl_2(aq, 2 \text{ m})|Cd(Hg)|CdCl_2(aq, 1 \text{ m})|AgCl(s)|Ag(s) .$$

If we assemble this cell, but do not connect the two Ag wires through an external circuit, the ions move instantaneously in the direction toward equilibrium, charges of opposite sign build on opposite wires, and migration stops once the charge imbalances across each interface have established the proper electrochemical potentials. If we then connected a voltmeter between the Ag wires, we would measure a constant voltage. Experiment would verify what we are about to learn how to calculate: the voltage is proportional to the $CdCl_2$ activity difference between the two solutions. If we connect the negative terminal of the voltmeter to the Ag wire in the more concentrated solution and the positive terminal to the less concentrated, the voltmeter will read a positive voltage. Note that the negative terminal is connected to the Ag electrode with the excess negative charge due to the oxidation reaction

$$Ag(s) + Cl^-(aq) \rightarrow AgCl(s) + e^-$$

and, historically, the electrode at which oxidation occurs is called the *anode*. (Reduction occurs at the *cathode,* as the other electrode is called.)

We have constructed a *concentration cell* as our prototypical electrochemical device. We have also acquired some extra nomenclature, and another mnemonic will help us keep the terms straight. If we arrange the cell in front of us with the negative terminal of the voltmeter connected to the left electrode and if the voltmeter reads a positive voltage, then the following pairs of terms, *each pair written in alphabetical order,* match:

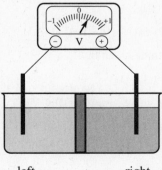

left	right	(convention giving positive voltage)
negative	positive	(voltmeter connections)
anode	cathode	(electrode names)
oxidation	reduction	(electrode reactions).

10 IONIC SOLUTIONS AND ELECTROCHEMISTRY

This is purely coincidental, but it is very easy to remember.[10]

Next we follow the electrochemical potentials from interface to interface through our cell. Our goal is an expression for the cell voltage, ΔE_{cell}. (The symbol E stands for "electromotive force" or "emf," which is just electrical potential difference.) The voltmeter measures *the electrochemical potential difference of the two electrodes*

$$\Delta E_{cell} = -\frac{1}{F}[\mu^e(e^-, R) - \mu^e(e^-, L)]$$
$$= -\frac{1}{F}[\mu(e^-, R) - F\phi(R) - \mu(e^-, L) + F\phi(L)] \qquad (10.26)$$
$$= \phi(R) - \phi(L)$$

where L, R denotes the left and right electrodes, and where $\mu(e^-, R) = \mu(e^-, L)$ since the electrodes are chemically identical.

The following simplified notation will prove helpful:

- on either side of the cell: $\quad \mu(e) = \mu(e^-, \text{in pure Ag})$
- in the amalgam: $\quad \mu(e, \text{Hg}) = \mu(e^-, \text{in Cd/Hg solution})$
  $\mu(\text{Cd}) = \mu(\text{Cd(Hg)})$
  $\phi(\text{Hg}) = \phi(\text{Cd(Hg)})$
- in the 1 molal solution: $\quad \phi(1) = \phi(\text{Cl}^-, 1\text{ m}) = \phi(\text{Cd}^{2+}, 1\text{ m})$
  $\mu(\text{Cl}^-(1)) = \mu(\text{Cl}^-(aq, 1\text{ m}))$
  $\mu(\text{Cd}^{2+}(1)) = \mu(\text{Cd}^{2+}(aq, 1\text{ m}))$

and similarly for the 2 m solution.

The first interface, moving from left to right, involves the anode reaction

$$\text{Ag}(L) + \text{Cl}^-(2) \rightleftarrows \text{AgCl}(L) + e^-(L)$$

and the electrochemical equilibrium expressed by

$$\mu^e(\text{Ag}) + \mu^e(\text{Cl}^-(2)) = \mu^e(\text{AgCl}) + \mu^e(e^-, L)$$

becomes, using Eq. (10.23),

$$\mu(\text{Ag}) + \mu(\text{Cl}^-(2)) - F\phi(2) = \mu(e) - F\phi(L) + \mu(\text{AgCl}) . \qquad (10.27a)$$

Next comes the 2 m solution|amalgam interface:

$$\text{Cd}^{2+}(2) + 2e^-(\text{Hg}) \rightleftarrows \text{Cd(Hg)}$$

$$\mu(\text{Cd}^{2+}(2)) + 2F\phi(2) + \mu(e, \text{Hg}) - F\phi(\text{Hg}) = \mu(\text{Cd}) \qquad (10.27b)$$

followed by the amalgam|1 m solution interface:

$$\text{Cd(Hg)} \rightleftarrows \text{Cd}^{2+}(1) + 2e^-(\text{Hg})$$

$$\mu(\text{Cd}) = \mu(\text{Cd}^{2+}(1)) + 2F\phi(1) + \mu(e, \text{Hg}) - F\phi(\text{Hg}) \qquad (10.27c)$$

and finally the right solution|paste|electrode assembly:

$$e^-(R) + \text{AgCl}(R) \rightleftarrows \text{Ag}(R) + \text{Cl}^-(1)$$

$$\mu(e) - F\phi(R) + \mu(\text{AgCl}) = \mu(\text{Ag}) + \mu(\text{Cl}^-(1)) - F\phi(1) . \qquad (10.27d)$$

Adding Equations (10.27b) and (10.27c), we find

$$\mu(\text{Cd}^{2+}(1)) + 2F\phi(1) = \mu(\text{Cd}^{2+}(2)) + 2F\phi(2) . \qquad (10.28)$$

---

[10] Unfortunately, it does not hold for *electrolysis cells*, in which an external voltage source (a "bigger battery") is connected to the cell to drive the cell chemistry in the non-spontaneous direction.

We solve Eq. (10.27a) for $F\phi(2)$ and Eq. (10.27d) for $F\phi(1)$. Substitution into Eq. (10.28) gives, after much cancellation,

$$\mu(Cd^{2+}(1)) + 2\mu(Cl^-(1)) + 2F\phi(R) = \mu(Cd^{2+}(2)) + 2\mu(Cl^-(2)) + 2F\phi(L) \ . \quad (10.29)$$

Solving for $\phi(R) - \phi(L) = \Delta E_{cell}$ yields

$$\Delta E_{cell} = -\frac{1}{2F}[\mu(CdCl_2(1)) - \mu(CdCl_2(2))] \quad (10.30)$$

where $\mu(CdCl_2) = \mu(Cd^{2+}) + 2\mu(Cl^-)$.

Note what Eq. (10.30) tells us: this complex set of interfacial reactions has reduced to a ΔE_{cell} governed only by the tendency to equilibrate the spontaneous process of interest:

$$CdCl_2(aq, 2\ m) \rightleftarrows CdCl_2(aq, 1\ m)$$

for which $\Delta_r \overline{G}$ is the chemical potential difference in Eq. (10.30). This expression can be further simplified, since

$$\mu(CdCl_2) = \mu°(Cd^{2+}) + 2\mu°(Cl^-) + RT \ln (\gamma_\pm m_\pm)^3 \ . \quad (10.31)$$

The standard chemical potentials cancel when Eq. (10.31) is substituted in Eq. (10.30), leaving only

$$\Delta E_{cell} = -\frac{RT}{2F} \ln \left[\frac{\gamma_\pm m_\pm(1)}{\gamma_\pm m_\pm(2)}\right]^3 \ . \quad (10.32)$$

Since $m_\pm = 4^{1/3} m_i$ for $CdCl_2$ (see Eq. (10.6)), and since measurements of $\gamma_\pm$ for $CdCl_2$ solutions give $\gamma_\pm(1\ m) = 0.0664$ and $\gamma_\pm(2\ m) = 0.0439$ at 298 K, we can calculate

$$\Delta E_{cell} = -\frac{(8.314\ 51\ J\ mol\ K^{-1})(298\ K)}{2\ (96\ 485\ C\ mol^{-1})} \ln \left(\frac{(0.0664)(1)}{(0.0439)(2)}\right)^3 = +0.0108\ V \ .$$

The sign is positive, as expected since the process is spontaneous, but the magnitude[11] is not large enough to expect concentration cells to have great practical utility as batteries.

Equation (10.32) is our first view of the *Nernst equation*. Before deriving the full equation, here is what we have so far:

At 298.15 K the constants in Eq. (10.32) are

$$\frac{RT}{F} = 0.025\ 693\ V = 25.693\ mV \ .$$

The minus sign comes from $z(e^-) = -1$.

The factor of 2 in the denominator counts *the number of electrons involved in the net reaction*. This factor is discussed in detail below.

The argument of the logarithm factor is a ratio of activities that has the *structure* of an equilibrium constant, but it is computed with *actual cell activities* instead of equilibrium net reaction activities.

If the activities of the species in the cell satisfy K_{eq} for the net reaction, $\Delta E_{cell} = 0$, implying a "run-down battery," no free-energy gradient, and chemical equilibrium.

---

[11]Note that if we had set $\gamma_\pm = 1$, the calculated ΔE_{cell} would have been in error by a factor of 2.5; $\gamma_\pm$ is very small due to the complex ions $CdCl_3^-$ and $CdCl^+$.

To derive the complete and general Nernst equation, we need to consider an electrochemical cell in which a spontaneous chemical reaction, rather than just the physical process of mixing, governs the voltage. The cell we will use is shown in Figure 10.7 and can be represented schematically as

$$Pt(s), H_2(1\ atm)|HCl(0.1\ m)|AgCl(s)|Ag(s)|Pt(s)\ .$$

Note that we have been careful to begin and end the cell with electrodes of the same metal (Pt) to assure that our voltmeter will not be biased by a dissimilar metal interfacial potential, but note also that Pt will not enter into the chemistry of this cell. On the left, the Pt serves only as a source or sponge for electrons in the reaction

$$H^+(aq) + e^- \rightarrow \tfrac{1}{2}H_2(g)$$

(which could go in either direction for all we know now), and thus it allows us to introduce a gaseous reagent into electrochemical reactions.[12] On the right, the Ag|Pt junction will not alter the cell chemistry, but, since all metal–metal interfaces are permeable to electrons, it forms a convenient way to make proper voltmeter connections to the cell. We have previously discussed the reaction on the right:

$$AgCl(s) + e^- \rightarrow Ag(s) + Cl^-(aq)\ .$$

If we follow $\Delta \phi$ from interface to interface, we will arrive at an expression for ΔE_{cell} that, at one point in the derivation, will look like this

$$\begin{aligned}-F[\phi(R) - \phi(L)] &= -F\,\Delta E_{cell}\\ &= \mu^\circ(Ag^+) + \mu^\circ(H^+) + \mu^\circ(Cl^-)\\ &\quad - \tfrac{1}{2}\mu^\circ(H_2) - \mu^\circ(AgCl) + RT\ln\left(\frac{a_\pm^2}{f^{1/2}}\right)\end{aligned} \quad (10.33)$$

where $a_\pm$ is the mean activity of 0.1 m HCl(aq), and f is the fugacity of H_2 at 1.0 atm pressure. The terms on the right-hand side should strike you as having the same *form* as the $\Delta_r \overline{G}$ expression for the net reaction

$$AgCl(s) + \tfrac{1}{2}H_2(g) \rightarrow Ag(s) + H^+(aq) + Cl^-(aq)\ .$$

But while the stoichiometric sum and difference of μ°'s is clearly $\Delta_r\overline{G}^\circ$, the

---

[12]When H_2 contacts Pt (often in the form of a fine, porous coating of electrodeposited Pt on Pt wire), it dissociates into H atoms bound to the Pt surface. These atoms are more reactive than H_2 and can be oxidized, entering solution as H^+(aq).

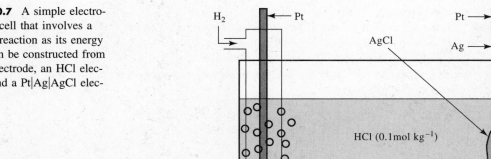

FIGURE 10.7 A simple electrochemical cell that involves a chemical reaction as its energy source can be constructed from a Pt|H_2 electrode, an HCl electrolyte, and a Pt|Ag|AgCl electrode.

$RT \ln (\ )$ term involves *actual cell activities and fugacities rather than equilibrium values*. Dividing Eq. (10.33) through by $-F$ gives

$$\Delta E_{\text{cell}} = -\frac{\Delta_r \overline{G}^\circ}{F} - \frac{RT}{F} \ln \left(\frac{a_\pm^2}{f^{1/2}} \right)$$

and defining $\Delta E^\circ = -\Delta_r \overline{G}^\circ / F$ gives the Nernst equation for the cell of Figure 10.7:

$$\Delta E_{\text{cell}} = \Delta E^\circ - \frac{RT}{F} \ln \left(\frac{a_\pm^2}{f^{1/2}} \right) . \qquad (10.34)$$

If we construct the cell so that $a_\pm = 1$ and $f = 1$ atm, then $\Delta E_{\text{cell}} = \Delta E^\circ$, and thus ΔE° is directly measurable. For this cell, $\Delta E^\circ = 0.2223$ V at 298 K. (Much more will be said of ΔE° in the next section.)

For 0.100 molal HCl, $\gamma_\pm = 0.7964$, so that

$$a_\pm^2 = [(0.7964)(0.100)]^2 = 6.343 \times 10^{-3}$$

and if $P_{H_2} = 1$ atm, $f = 1$ atm to good accuracy. Thus we calculate

$$\Delta E_{\text{cell}} = 0.2223 \text{ V} - 0.025\ 693 \ln (6.343 \times 10^{-3}) = 0.3523 \text{ V} .$$

Since ΔE_{cell} is positive, the reaction

$$\text{AgCl}(s) + \frac{1}{2}\text{H}_2(g, 1 \text{ atm}) \to \text{Ag}(s) + \text{HCl}(aq, 0.1 \text{ m})$$

would proceed spontaneously in the direction indicated. As we mentioned early in this chapter, ΔE_{cell} can give this information *without* allowing the reaction to occur. It is in this sense that *a cell voltmeter is a free-energy meter*.

Before going on to the details of Nernst equation applications in the next section, we mention briefly one other type of interface that you may have encountered in elementary discussions of electrochemical cells. This is the *salt bridge*. A salt bridge is an electrolyte solution, usually saturated KCl, confined to a tube by plugs of a conductive paste such as agar. It forms an electrolyte "wire" for connecting two otherwise isolated electrolyte solutions. While conceptually simple, electrolyte interfaces ("liquid junctions") introduce the equivalent of a small-voltage battery internal to the cell. Due to the unequal rate at which anions and cations move in opposite directions across the liquid junction, a potential difference builds up across the junction, even when no net current is flowing through the equilibrium cell. This potential difference builds to a constant value as soon as the liquid junction interfaces are established when the cell is constructed. Its magnitude can be as large as 10–20 mV, which is very large compared to the accuracy ($\cong 0.01$ mV) with which cell voltages can be measured. The saturated KCl salt bridge has a liquid junction potential difference of only 1 mV or so due to the nearly equal rates of motion of $K^+(aq)$ and $Cl^-(aq)$.

10.5 USING ELECTROCHEMICAL DATA

For the general net reaction of an electrochemical cell held poised to react:

$$\alpha A + \beta B \to \gamma C + \delta D$$

the Nernst equation is

$$\Delta E_{\text{cell}} = \Delta E^\circ - \frac{RT}{nF} \ln \left(\frac{a_C^\gamma a_D^\delta}{a_A^\alpha a_B^\beta} \right) \qquad (10.35)$$

where the a's are the actual activities of the reactants and products.

The last section pointed out how $\Delta E°$ could be measured (at least in principle for any cell) if we arrange $a_C^\gamma a_D^\delta / a_A^\alpha a_B^\beta = 1$ so that the logarithm term is zero. A table of *standard reduction potentials* for *reduction half-reactions* can be made, however, making this step unnecessary. The electrochemical potential difference for a reaction such as

$$Co^{2+}(aq) + 2e^- \rightarrow Co(s) \tag{10.36}$$

cannot be measured directly. But since cells *always* measure the difference between the potential of a reduction half-reaction (like Eq. (10.36)) and an oxidation half-reaction, such as

$$Ag(s) \rightarrow Ag^+(aq) + e^- \tag{10.37}$$

we need only choose *one* reduction half-reaction, assign it an arbitrary potential of zero, and measure other half-reaction potentials relative to this standard.

The standard half-reaction must be readily constructed, behave reproducibly in a variety of cells, and have a chemistry sufficiently versatile to warrant its joining with a wide variety of other half-reactions. Just as $H^+(aq)$ is the reference substance of aqueous thermodynamic functions, the reference reduction half-reaction is

$$H^+(aq, a = 1) + e^- \rightarrow H_2(g, f = 1 \text{ bar}) \tag{10.38}$$

and the corresponding cell electrode, the *standard hydrogen electrode,* is

$$Pt, H_2(f = 1 \text{ bar})|H^+(a = 1)$$

(as used in Figure 10.7).

We assign the standard hydrogen electrode a standard reduction potential $E_{red}°$ of zero, conduct experiments with a variety of cell partners, and build a table of relative $E_{red}°$ values. Measurements at a variety of temperatures away from the standard 298.15 K value allow us to represent $E_{red}°(T)$ as a power series of the form

$$E_{red}°(T) = E_{red}° + E_{red}°{}'(T - 298.15 \text{ K}) + \frac{1}{2}E_{red}°{}''(T - 298.15 \text{ K})^2 \tag{10.39}$$

where

$$E_{red}°{}' = \left(\frac{dE_{red}°(T)}{dT}\right)_{P, T = 298.15 \text{ K}} \quad \text{and} \quad E_{red}°{}'' = \left(\frac{d^2 E_{red}°(T)}{dT^2}\right)_{P, T = 298.15 \text{ K}}. \tag{10.40}$$

Note that we have properly indicated isobaric conditions in these derivatives, but the pressure variation of $E_{red}°$ is slight, and we will not explore it. These derivatives are more than power-series expansion coefficients; they contain thermodynamic information by themselves, as we shall see.

Table 10.3 lists $E_{red}°$, $E_{red}°{}'$, and $E_{red}°{}''$ values for many important half-reactions. As you look through this table, note that there is a general division between acidic and basic solution half-reactions (and moreover one can certainly do electrochemistry in nonaqueous media, such as alcohols or liquid ammonia). These values are for $T = 298.15$ K, and they implicitly assume all species to be in their unit-activity standard states (pure solids, unit fugacity for gases, and unit activity for solutes).

This table clearly contains a wealth of information, since N half-reactions can be combined into $N(N - 1)/2$ full net reactions. For example, consider combining the half-reactions of Eq. (10.36) and Eq. (10.37) stoichiometrically in the following sense:

$$\begin{array}{r} Co^{2+} + 2e^- \rightarrow Co \\ 2Ag \rightarrow 2Ag^+ + 2e^- \\ \hline Co^{2+} + 2Ag \rightarrow Co + 2Ag^+ \end{array} \tag{10.41}$$

TABLE 10.3 Standard Reduction Potentials and Their Temperature Derivatives† for Aqueous Half-reactions at $T = 298.15$ K

Half-reaction	$E°_{red}$ (V)	$\dfrac{dE°_{red}}{dT}$† (mV K^{-1})	$\dfrac{d^2E°_{red}}{dT^2}$† ($\mu$V K^{-2})
Acidic solutions			
$F_2(g) + 2H^+ + 2e^- \rightarrow 2HF$	3.06	−0.60	—
$F_2(g) + 2e^- \rightarrow 2F^-$	2.87	−1.830	−5.339
$OH + H^+ + e^- \rightarrow H_2O$	2.85	−1.855	1.078
$O(g) + 2H^+ + 2e^- \rightarrow H_2O$	2.422	−1.15	0.427
$Ag^{2+} + e^- \rightarrow Ag^+$	1.980	—	—
$Co^{3+} + e^- \rightarrow Co^{2+}$	1.808	—	—
$XeO_3 + 6H^+ + 6e^- \rightarrow Xe + 3H_2O$	1.8	—	—
$H_2O_2 + 2H^+ + 2e^- \rightarrow 2H_2O$	1.776	−0.658	—
$MnO_4^- + 4H^+ + 3e^- \rightarrow MnO_2 + 2H_2O$	1.695	−0.666	—
$PbO_2 + SO_4^{2-} + 4H^+ + 2e^- \rightarrow PbSO_4 + 2H_2O$	1.682	0.326	2.516
$Ce^{4+} + e^- \rightarrow Ce^{3+}$	1.61	—	—
$MnO_4^- + 8H^+ + 5e^- \rightarrow Mn^{2+} + 4H_2O$	1.51	−0.66	—
$Mn^3 + e^- \rightarrow Mn^{2+}$	1.51	1.23	—
$PbO_2 + 4H^+ + 2e^- \rightarrow Pb^{2+} + 2H_2O$	1.455	−0.238	—
$Cl_2 + 2e^- \rightarrow 2Cl^-$	1.359 5	−1.260	−5.454
$Cr_2O_7^{2-} + 14H^+ + 6e^- \rightarrow 2Cr^{3+} + 7H_2O$	1.33	−1.263	—
$MnO_2 + 4H^+ + 2e^- \rightarrow Mn^{2+} + 2H_2O$	1.23	−0.661	—
$O_2 + 4H^+ + 4e^- \rightarrow 2H_2O$	1.229	−0.846	0.552
$Br_2 + 2e^- \rightarrow 2Br^-$	1.087	−0.478	—
$V(OH)_4^+ + 2H^+ + e^- \rightarrow VO^{2+} + 3H_2O$	1.00	—	—
$2Hg^{2+} + 2e^- \rightarrow Hg_2^{2+}$	0.911 0	—	—
$Ag^+ + e^- \rightarrow Ag$	0.799 1	−1.000	−0.378
$Hg_2^{2+} + 2e^- \rightarrow 2Hg(l)$	0.796 0	—	—
$Fe^{3+} + e^- \rightarrow Fe^{2+}$	0.771	1.188	—
$C_2H_2(g) + 2H^+ + 2e^- \rightarrow C_2H_4(g)$	0.731	−0.580	−0.508
$I_3^- + 2e^- \rightarrow 3I^-$	0.536	−0.214	—
$I_2(s) + 2e^- \rightarrow 2I^-$	0.535 5	−0.148	−5.965
$Cu^+ + e^- \rightarrow Cu$	0.521	−0.058	—
$C_2H_4(g) + 2H^+ + 2e^- \rightarrow C_2H_6(g)$	0.52	−0.625	−0.343
$VO^{2+} + 2H^+ + e^- \rightarrow V^{3+} + H_2O$	0.361	—	—
$Cu^{2+} + 2e^- \rightarrow Cu$	0.337	0.008	—
$Hg_2Cl_2(s) + 2e^- \rightarrow 2Hg(l) + 2Cl^-$	0.267 6	−0.317	−5.664
Saturated Calomel Electrode*	0.241	—	—
$AgCl(s) + e^- \rightarrow Ag + Cl^-$	0.222 3	−0.658	−5.744
$Cu^{2+} + e^- \rightarrow Cu^+$	0.153	0.073	—
$C(graphite) + 4H^+ + 4e^- \rightarrow CH_4(g)$	0.131 6	−0.209	−0.266
$2H^+ + 2e^- \rightarrow H_2(g)$	**0.000 0**	**0.000**	**0.000**
$2D^+ + 2e^- \rightarrow D_2(g)$	−0.003 4	—	—
$Pb^{2+} + 2e^- \rightarrow Pb$	−0.126	−0.451	—
$Ni^{2+} + 2e^- \rightarrow Ni$	−0.250	0.06	—
$V^{3+} + e^- \rightarrow V^{2+}$	−0.255	—	—
$Co^{2+} + 2e^- \rightarrow Co$	−0.277	0.06	—
$Cd^{2+} + 2e^- \rightarrow Cd(Hg)$	−0.351 6	−0.250	—
$PbSO_4 + 2e^- \rightarrow Pb + SO_4^{2-}$	−0.358 8	−1.015	−1.555
$Cd^2 + 2e^- \rightarrow Cd$	−0.402 9	−0.093	—
$Cr^{3+} + e^- \rightarrow Cr^{2+}$	−0.408	—	—
$Fe^{2+} + 2e^- \rightarrow Fe$	−0.440 2	0.052	—
$Cr^{3+} + 3e^- \rightarrow Cr$	−0.744	0.468	—

TABLE 10.3 *Continued*

Half-reaction	$E°_{red}$ (V)	$\dfrac{dE°_{red}{}^\dagger}{dT}$ (mV K^{-1})	$\dfrac{d^2E°_{red}{}^\dagger}{dT^2}$ (μV K^{-2})
Acidic solutions			
$Zn^{2+} + 2e^- \rightarrow Zn(Hg)$	$-0.762\,7$	0.100	0.62
$Zn^{2+} + 2e^- \rightarrow Zn$	$-0.762\,8$	0.091	—
$Mn^{2+} + 2e^- \rightarrow Mn$	-1.180	-0.08	—
$Al^{3+} + 3e^- \rightarrow Al$	-1.662	0.504	—
$Mg^{2+} + 2e^- \rightarrow Mg$	-2.363	0.103	—
$Na^+ + e^- \rightarrow Na$	-2.714	-0.772	—
$Ca^{2+} + 2e^- \rightarrow Ca$	-2.866	-0.175	—
$Sr^{2+} + 2e^- \rightarrow Sr$	-2.888	-0.191	—
$Ba^{2+} + 2e^- \rightarrow Ba$	-2.906	-0.395	—
$Cs^+ + e^- \rightarrow Cs$	-2.923	-1.197	—
$Rb^+ + e^- \rightarrow Rb$	-2.925	-1.245	—
$K^+ + e^- \rightarrow K$	-2.925	-1.080	—
$Li^+ + e^- \rightarrow Li$	-3.045	-0.534	—
Basic solutions			
$OH(g) + e^- \rightarrow OH^-(aq)$	2.02	-2.689	-6.194
$ClO^- + H_2O + 2e^- \rightarrow Cl^- + 2OH^-$	0.89	-1.079	—
$MnO_4^- + 2H_2O + 3e^- \rightarrow MnO_2 + 4OH^-$	0.588	-1.778	—
$O_2 + 2H_2O + 4e^- \rightarrow 4OH^-$	0.401	-1.680	-6.719
$S + 2e^- \rightarrow S^{2-}$	-0.447	-0.93	—
$PbCO_3 + 2e^- \rightarrow Pb + CO_3^{2-}$	-0.509	-1.294	—
$Cd(NH_3)_4^{2+} + 2e^- \rightarrow Cd + 4NH_3(aq)$	-0.613	—	—
$2H_2O + 2e^- \rightarrow H_2(g) + 2OH^-$	$-0.828\,06$	$-0.834\,2$	-7.272
$Se + 2e^- \rightarrow Se^{2-}$	-0.92	-0.89	—
$PbS + 2e^- \rightarrow Pb + S^{2-}$	-0.93	-0.90	—
$Te + 2e^- \rightarrow Te^{2-}$	-1.143	—	—
$Ca(OH)_2 + 2e^- \rightarrow Ca + 2OH^-$	-3.02	-0.965	—

Notes: Unless noted to the contrary, all species are assumed to be in an aqueous phase, all pure metals are as solids in contact with an aqueous solution, and all species are at unit activity.
†These temperature derivatives are based on a convention that the reduction potential of the standard hydrogen electrode is temperature invariant. In fact, this electrode has a temperature derivative at 298.15 K of 0.871 mV K^{-1}. See Example 10.8 to see how these temperature derivatives are incorporated in thermodynamic calculations.
*The Saturated Calomel Electrode (SCE) is the half-reaction $Hg_2Cl_2(s) + 2e^- \rightarrow 2Hg(l) + 2Cl^-$ operated with a saturated KCl electrolyte. The $E°_{red}$ quoted, 0.241 V, incorporates the effects of the ionic strength of this electrolyte.

From Table 10.3, we find, to a consistent three-significant-figure accuracy

$$Co^{2+} + 2e^- \rightarrow Co \quad E°_{red} = -0.277 \text{ V}$$

$$Ag^+ + e^- \rightarrow Ag \quad E°_{red} = 0.799 \text{ V}$$

so that the standard cell potential for the net reaction would be (remember: "right minus left" with reduction presumed to be on the right and oxidation on the left)

$$\begin{aligned}\Delta E° &= E°_{red}(\text{reduction}) - E°_{red}(\text{oxidation}) \\ &= -0.277 \text{ V} - 0.799 \text{ V} \\ &= -1.076 \text{ V}\,.\end{aligned}$$

Thus, a cell with these electrodes and unit activities of the ions in solution, connected to a voltmeter according to our convention, would exhibit a *negative* voltage. This means that Eq. (10.41) as written is *spontaneous in the reverse direction*.

One of the major uses of tables such as Table 10.3 is, therefore, to predict spontaneous reaction directions at a glance. In a table of reduction reactions arranged with the most positive at the top, the spontaneously reacting combination of reductants and oxidants is found by a diagonal relationship such as that indicated schematically in Figure 10.8. For instance, $F_2(g)$ and $Li(s)$ react spontaneously, a fact that should not surprise you, while $F^-(aq)$ and $Li^+(aq)$ do not spontaneously yield the elements in their standard states. (Remembering this fact helps you remember Figure 10.8.) It is not so obvious that dipping Ag in a Co^{2+} solution yields only wet silver, while a piece of Co will dissolve in a solution of Ag^+, yielding $Ag(s)$. This is the first great benefit of reduction-potential tables.

For many combinations, the reduction potentials are quite close, and variations away from standard concentrations might alter the sign of ΔE_{cell} (*via* the Nernst expression). This is explored in the following example.

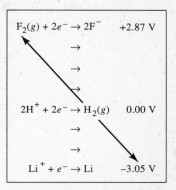

FIGURE 10.8 Spontaneous reactions can be predicted from tables of standard reduction potentials arranged as shown. *Reactants* in one half-reaction, such as F_2, will react spontaneously with the *products* of a half-reaction *below* it, such as Li.

EXAMPLE 10.6

For the following cell at 298 K (where || represents a salt bridge)

$$Co|Co^{2+}(a = 1)| |Ni^{2+}(a = ?)|Ni$$

what range of Ni^{2+} activity leads to a positive ΔE_{cell}?

SOLUTION The question asks for those activities for which

$$Co + Ni^{2+}(a) \rightarrow Co^{2+}(a = 1) + Ni$$

is spontaneous as written. We first find

$$\Delta E° = E°_{red}(R) - E°_{red}(L) = E°_{red}(Ni^{2+}) - E°_{red}(Co^{2+}) = -0.250 - (-0.277) = 0.027 \text{ V}$$

from Table 10.3. The Nernst equation for unit Co^{2+} activity is, with $n = 2$ (n does *not* figure in the $\Delta E°$ calculation!),

$$\Delta E_{cell} = \Delta E° - \frac{RT}{nF} \ln \left(\frac{a_{Co^{2+}}}{a_{Ni^{2+}}}\right)$$

$$= 0.027 \text{ V} + \frac{0.025\ 693 \text{ V}}{2} \ln a_{Ni^{2+}}$$

$$= 0.027 \text{ V} + (0.013 \text{ V}) \ln a_{Ni^{2+}} .$$

For Ni^{2+} activities greater than $\exp(-0.027/0.013) = 0.125$, ΔE_{cell} is positive, and the net cell reaction would proceed spontaneously as written.

➠ **RELATED PROBLEMS** 10.19, 10.20, 10.21, 10.22

As we pointed out earlier, if our voltmeter reads zero when connected to a cell, that would mean that the net cell reaction had been constructed with all species at the *activities of chemical equilibrium*. The argument of the logarithm term in the Nernst equation would thus be a true *equilibrium constant*. Therefore, if $\Delta E_{cell} = 0$ in Eq. (10.35),

$$\Delta E° = \frac{RT}{nF} \ln K_{eq} \tag{10.42}$$

or, rewriting this in accord with Eq. (7.20),

$$\ln K_{eq} = -\frac{\Delta_r \overline{G}°}{RT} = \frac{nF\Delta E°}{RT}$$

we see that

$$\Delta_r \overline{G}^\circ = -nF\Delta E^\circ .\qquad(10.43)$$

Again using Eq. (10.41) as an example, we can calculate

$$\ln K_{eq} = \frac{(2)(-1.076\text{ V})}{(0.025\ 693\text{ V})} = -83.76$$

for the Co/Ag reaction, or

$$K_{eq} = 4.21 \times 10^{-37} .$$

This value is so small that ordinary methods of running the reaction to equilibrium and then analyzing the equilibrium mixture would be hopeless. Similarly, we find

$$\Delta_r \overline{G}^\circ = -(2)(96\ 485\text{ C mol}^{-1})(-1.076\text{ V}) = 207.6\text{ kJ mol}^{-1} .$$

Essentially the same value comes from the $\Delta_f \overline{G}^\circ(aq)$ data of Table 10.2:

$$\Delta_r \overline{G}^\circ = RT[2(31.11) - (-22.11)] = 209.1\text{ kJ mol}^{-1} .$$

We have now shown how free-energy and standard-potential information are interrelated. This is the second benefit of standard reduction potential data. Not only can we predict the spontaneous direction of a hypothetical net reaction at its initial composition, but we can also predict its equilibrium composition from K_{eq} and the procedures of Chapter 7.

EXAMPLE 10.7

When Fe(s) is placed in a solution of ferric ion, $Fe^{3+}(aq)$, the following reaction takes place spontaneously and rapidly:

$$2Fe^{3+}(aq) + Fe(s) \to 3Fe^{2+}(aq)$$

so that the reduction potential for the half-reaction

$$Fe^{3+} + 3e^- \to Fe$$

cannot be measured directly (and hence is not among those listed in Table 10.3). Nevertheless, we can compute E°_{red} for this reaction from $\Delta_f \overline{G}^\circ$ data in Table 10.2 or by combining other half-reactions listed in Table 10.3. Use both methods to compute a E°_{red} value for the ferric to Fe(s) half-reaction.

SOLUTION From Table 10.2, $\Delta_f \overline{G}^\circ/RT = -4.27$ for $Fe^{3+}(aq)$, while $\Delta_f \overline{G}^\circ = 0$ for both Fe(s) and e^-. For $Fe^{3+} + 3e^- \to Fe$, we have

$$E^\circ_{red} = -\frac{\Delta_r \overline{G}^\circ}{3F} = -\frac{[0 - (-4.27)]RT}{3F} = -36.6\text{ mV} .$$

Alternatively, we find from Table 10.3

(a) $Fe^{3+} + e^- \to Fe^{2+}$ $E^\circ_{red}(a) = 0.771\text{ V}$
(b) $Fe^{2+} + 2e^- \to Fe$ $E^\circ_{red}(b) = -0.4402\text{ V} .$

It is clear that these reactions sum to the half-reaction of interest, but we *cannot* simply add the two E°_{red} values themselves. Instead, we add nE°_{red} values for each, since these are proportional to $\Delta_r \overline{G}^\circ$ values, and divide the sum by n for the reaction of interest. Thus,

$$E^\circ_{red}(Fe^{3+}/Fe) = \frac{E^\circ_{red}(a) + 2E^\circ_{red}(b)}{3} = -36.5\text{ mV}$$

in agreement with the first calculation.

➠ *RELATED PROBLEMS* 10.21, 10.22, 10.23, 10.24

Next we discuss the variation of E_{red}° with temperature and the thermodynamic meanings of $E_{red}^{\circ\prime}$ and $E_{red}^{\circ\prime\prime}$. Since $(\partial\mu/\partial T)_P = -\overline{S}$,

$$\left(\frac{d(\Delta_r\overline{G}^\circ)}{dT}\right)_P = -nF\left(\frac{d(\Delta E^\circ)}{dT}\right)_P = -\Delta_r\overline{S}^\circ$$

so that

$$\Delta_r\overline{S}^\circ = nF\left(\frac{d(\Delta E^\circ)}{dT}\right)_P. \tag{10.44}$$

Since $\overline{C}_P = T(\partial\overline{S}/\partial T)_P$, we also find

$$\Delta_r\overline{C}_P^\circ = nFT\left(\frac{d^2(\Delta E^\circ)}{dT^2}\right)_P. \tag{10.45}$$

Finally we can find $\Delta_r\overline{H}^\circ$ from

$$\Delta_r\overline{H}^\circ = \Delta_r\overline{G}^\circ + T\Delta_r\overline{S}^\circ = -n\Delta E^\circ + nFT\left(\frac{d(\Delta E^\circ)}{dT}\right)_P. \tag{10.46}$$

All the basic thermochemical quantities for a reaction are recoverable from ΔE° and its temperature dependence.

EXAMPLE 10.8

Find the thermodynamic properties $\Delta_f\overline{G}^\circ$, $\Delta_f\overline{H}^\circ$, $\overline{S}^\circ$, and $\overline{C}_P^\circ$ for $Ag^+(aq)$ using only data in Table 10.3 and the following information:

- for Ag(s): $\overline{S}^\circ/R = 5.118$ $\overline{C}_P^\circ/R = 3.049$
- for $H_2(g)$: $\overline{S}^\circ/R = 15.7176$ $\overline{C}_P^\circ/R = 3.4667$.

SOLUTION For $\Delta_f\overline{G}^\circ$, we can turn easily and directly to the half-reaction

$$Ag^+(aq) + e^- \rightarrow Ag(s)$$

and find from Table 10.3 that $E_{red}^\circ = 0.7991$ V, $dE_{red}^\circ/dT = -1.000$ mV K^{-1}, and $d^2E_{red}^\circ/dT^2 = -0.378$ μV K^{-2}. From Eq. (10.43), applied to a half-reaction,

$$\Delta_r\overline{G}^\circ = -nFE_{red}^\circ = -F(0.7991 \text{ V}) = -77.10 \text{ kJ mol}^{-1},$$

which also must be $-\Delta_f\overline{G}^\circ(Ag^+(aq))$, since $\Delta_f\overline{G}^\circ(Ag(s)) = 0$ as does $\Delta_f\overline{G}^\circ(e^-)$. We can find ΔH from $G = H - TS$ once we find $\Delta_r\overline{S}^\circ$. Entropy is related to the temperature derivative of E_{red}°, as Eq. (10.44) indicates. But since E_{red}° is based on combinations of a given half-reaction with the standard hydrogen electrode, and since that half-reaction contains $H_2(g)$, which has a non-zero entropy at non-zero temperatures, we need to account for this entropy. The values in Table 10.3 for temperature derivatives have incorporated this entropy contribution (see the footnote to Table 10.3) so that

$$\Delta_r\overline{S}^\circ = nF\left(\frac{dE_{red}^\circ}{dT}\right) = F(-1.000 \times 10^{-3} \text{ V K}^{-1}) = -96.49 \text{ J mol}^{-1}\text{ K}^{-1}$$

for the reaction

$$Ag^+(aq) + \frac{1}{2}H_2(g) \rightarrow Ag(s) + H^+(aq)$$

and we find

$$\overline{S}^\circ(Ag^+(aq)) = \overline{S}^\circ(Ag(s)) + \overline{S}^\circ(H^+(aq)) - \frac{1}{2}\overline{S}^\circ(H_2(g)) - \Delta_r\overline{S}^\circ$$
$$= 5.118\,R + 0 - 0.5\,(15.7176\,R) - (-96.49 \text{ J mol}^{-1}\text{ K}^{-1})$$
$$= 73.70 \text{ J mol}^{-1}\text{ K}^{-1} = 8.864\,R$$

in good agreement with the value $\overline{S}°/R = 8.89$ in Table 10.1. We can now find $\Delta_f \overline{H}°(Ag^+(aq))$ from

$$\Delta_f \overline{H}°(Ag^+) = -\Delta_r \overline{H}° = -\Delta_r \overline{G}° - T\Delta_r \overline{S}°$$
$$= 77.10 \text{ kJ mol}^{-1} - (298.15 \text{ K})(-96.49 \text{ J mol}^{-1}\text{ K}^{-1})$$
$$= 105.9 \text{ kJ mol}^{-1}$$

since $\Delta_r \overline{G}°$ for the half-reaction equals $\Delta_r \overline{G}°$ for the full reaction with H_2 and H^+. This is in good agreement with the value in Table 7.5, 105.579 kJ mol^{-1}. We find $\overline{C}_P°(Ag^+(aq))$ in a manner similar to the entropy calculation, beginning with Eq. (10.45):

$$\Delta_r \overline{C}_P° = nFT\left(\frac{d^2 E°_{\text{red}}}{dT^2}\right) = F(298.15 \text{ K})(-0.378 \times 10^{-6} \text{ V K}^{-2})$$
$$= -10.9 \text{ J mol}^{-1} \text{ K}^{-1}$$

so that

$$\overline{C}_P°(Ag^+(aq)) = \overline{C}_P°(Ag(s)) + \overline{C}_P°(H^+(aq)) - \frac{1}{2}\overline{C}_P°(H_2(g)) - \Delta_r \overline{C}_P°$$
$$= 3.049\,R + 0 - 0.5\,(3.466\,7\,R) - (-10.9 \text{ J mol}^{-1}\text{ K}^{-1})$$
$$= 21.8 \text{ J mol}^{-1}\text{ K}^{-1} = 2.63\,R.$$

In summary, the temperature derivative data in Table 10.3 incorporate the H^+/H_2 half-reaction so that $\Delta_r \overline{S}°$ and $\Delta_r \overline{C}_P°$ for the net reaction between any tabulated half-reaction and the H^+/H_2 half-reaction can be calculated immediately. The calculation of individual ion $\overline{S}°$ and $\overline{C}_P°$ values, however, require data beyond $\Delta_r \overline{S}°$ and $\Delta_r \overline{C}_P°$ alone.

⟹ **RELATED PROBLEMS** 10.27, 10.28

We close this section with a discussion of perhaps the most pervasive electrochemical measurement, that of a pH meter.[13] To define pH as "what a well-calibrated pH meter reads" is not far off the mark. You may recall the definition pH = $-\log_{10}$ [H^+], which is often *close* to what a pH meter reads, but we know now that pH = $-\log_{10}(a_{H^+})$ would make more thermodynamic sense. (See Example 10.4.) Due to variations of a_{H^+} with ionic strength, to the replacement of standard hydrogen electrodes by *glass* electrodes (about which more will be said), and to the inherent uncertainty of liquid junction potentials, the practical pH scale ("what the meter reads") is designed around several standard solutions (standard buffers) that are chosen and assigned pH values so that the meter reads within ±0.01 to ±0.02 pH units of $-\log_{10}(a_{H^+})$ over a wide pH and a reasonable ionic-strength range. A common reference solution for pH is an aqueous solution of potassium hydrogen phthalate at exactly 0.05 mol kg^{-1}, which at 298.15 K has pH = 4.008.

The glass pH electrode is constructed with a very thin ($\cong$50 μm) membrane of a particular glass formulation containing mobile Na^+ ions. The glass acts like a solid electrolyte, separating the unknown solution from the solution (most often $HCl(aq)$ in contact with a Ag|AgCl paste on a Pt wire) sealed into the inner part of the electrode. These electrodes are stored in water, so that a surface exchange of Na^+ for H^+ ions takes place. Similarly, an equilibrium involving H^+ occurs in any unknown solution, as well as on the inner surface of the membrane in contact with the sealed HCl solution. The pH meter responds to the electrochemical potential difference across the membrane.

---

[13]The abbreviation pH comes originally from the French *puissance d'Hydrogène*, "power of hydrogen," but it may equivalently be thought of as the "potential of hydrogen."

Altering the chemical composition of the glass and the nature of the sealed inner solution and electrode leads to a variety of other *ion-specific* electrodes. These have great value in analysis, especially for species such as Na^+, NO_3^-, NH_4^+, F^-, etc., that are difficult to analyze by any but electrochemical means.

CHAPTER 10 SUMMARY

We started this chapter with a number of new terms and definitions specific to electrolyte solutions:

$$\text{Ionic strength: } I = \frac{1}{2} \sum_{j=1}^{\text{all ions}} m_j z_j^2$$

$$\text{Mean ionic molality:}$$
$$m_\pm = [(m_+)^{\nu_+} (m_-)^{\nu_-}]^{1/\nu} = m_i(\nu_+^{\nu_+} \nu_-^{\nu_-})^{1/\nu}$$

$$\text{Mean ionic activity coefficient: } \gamma_\pm = (\gamma_+^{\nu_+} \gamma_-^{\nu_-})^{1/\nu}$$

$$\text{Ionic solution chemical potential:}$$
$$\mu_i(aq) = \mu_i^\circ + RT \ln (\gamma_\pm m_\pm)^\nu$$

We found an expression for the mean ionic activity coefficient valid at low concentrations:

$$\text{Debye–Hückel limiting law:}$$
$$\gamma_\pm = \exp[1.174\, z_+ z_-(I/\text{mol kg}^{-1})^{1/2}] \text{ (water at 25.0 °C)}$$

and saw how this expression or extensions of it allowed $\gamma_\pm$ to be calculated.

Next, we argued how redox reactions could be coupled so that the transferred electrons could release the reaction's free energy as useful work. The initial measure of this free energy was the cell voltage ΔE_{cell}. This is related to the cell chemistry via the

$$\text{Nernst equation:}$$
$$\Delta E_{\text{cell}} = \Delta E^\circ - \frac{RT}{nF} \ln \left(\frac{\text{product activities}}{\text{reactant activities}} \right)$$

with the actual, not equilibrium, activities controlling the cell voltage.

We introduced standard reduction potentials, E_{red}°, so that $\Delta E^\circ = E_{\text{red}}^\circ$ (reduction half-reaction) $- E_{\text{red}}^\circ$ (oxidation half-reaction) and found the

$$\text{Standard molar free-energy change of reaction:}$$
$$\Delta_r \overline{G}^\circ = -nF\Delta E^\circ = -RT \ln K_{\text{eq}}$$

along with the expressions for $\Delta_r \overline{S}^\circ$, $\Delta_r \overline{H}^\circ$, and $\Delta_r \overline{C}_P^\circ$ in Equations (10.44)–(10.46).

FURTHER READING

Modern Electrochemistry, by J. O'M. Bockris and A. K. N. Reddy (Plenum, N.Y., 1970) and *Electroanalytical Chemistry,* volumes 1 and 2, edited by A. J. Bard (Edward Arnold, London, 1966) are two modern references. The second covers much more than this chapter's topics and deals with aspects of electrochemistry for analytical purposes. Most texts on quantitative analytical chemistry typically cover much more electrochemistry and ionic solution phenomena than is covered here. Look under Library of Congress catalog numbers QD101.2, QD450.2, and QD553 in your library. Several of the data sources below also cover topics from this chapter.

ACCESS TO DATA

Aside from the usual handbook tables of data, several works specific to electrochemistry and ionic solutions are available. *The Physical Chemistry of Electrolyte Solutions,* 3rd ed., by H. S. Harned and B. B. Owen, (Reinhold, N.Y., 1958) contains extensive tables of activity coefficients, etc., along with a full discussion (as of 1958) of electrolyte solutions. *Electrolyte Solutions,* 2nd revised ed., by R. A. Robinson and R. H. Stokes, (Butterworth, London, 1965) is a second source of information similar to the previous reference.

Determination of pH: *Theory and Practice,* 2nd ed., by R. G. Bates, (Wiley, New York, 1973) covers just what its title says. The pH values of standard buffers can be found in R. G. Bates, *J. Res. Nat. Bur. Stand.* **66A,** 179 (1962) and B. R. Staples and R. G. Bates, *J. Res. Nat. Bur. Stand.* **73A,** 37 (1969).

The Oxidation States of the Elements and Their Potentials in Aqueous Solutions, 2nd ed., by W. M. Latimer (Prentice Hall, N.Y., 1952) tabulates standard *oxidation* potentials, E_{ox}°

PRACTICE WITH EQUATIONS

10A What is $m_\pm$ for a solution made by adding 0.015 mol $Al_2(SO_4)_3$ to 0.100 kg of H_2O? (10.6)

ANSWER: 0.383 mol kg^{-1}

10B The measured $\gamma_\pm$ values for 0.10 mol kg^{-1} solutions of KCl and NaCl are 0.769 and 0.778, respectively, while for 0.01 mol kg^{-1}, they are 0.901 and 0.904. What values does the Debye–Hückel limiting law predict? (10.13)

ANSWER: 0.690 for 0.10 m and 0.889 for 0.01 m

10C What are the values calculated in 10B if the extended Debye–Hückel law is used?

ANSWER: 0.754 for 0.10 m and 0.899 for 0.01 m

10D What is the solubility equilibrium constant for $BaSO_4(s)$ if its solubility is 0.246 mg in 0.100 L H_2O at 298 K? (10.22), (10.13)

ANSWER: 1.05×10^{-10}

10E In 10D, the solubility equilibrium constant for $BaSO_4$ was found to be 1.05×10^{-10}. Repeat this calculation using $\Delta_f \overline{G}°$ values, given $\Delta_f \overline{G}°(BaSO_4(s))/RT = -549.5$. Table 10.2, (7.20)

ANSWER: 1.08×10^{-10}

10F In the gas phase, the dissociation of water into the ion pair $H^+ + OH^-$ would have a positive $\Delta_r \overline{S}°$. What is $\Delta_r \overline{S}°$ for $H_2O(l) \to H^+(aq) + OH^-(aq)$? Table 10.1, $\overline{S}°(H_2O(l))/R = 8.408$

ANSWER: $-9.701\, R$ (Note that $\Delta_r \overline{S}°$ is negative for the solution. Why?)

10G One of the first equilibrium constants any chemistry student learns is that for the dissociative autoionization of liquid water: $K_{eq} \cong 10^{-14}$ at 25 °C. Compute this number from $\Delta_f \overline{G}°$ data. Table 10.2, $\Delta_f \overline{G}°(H_2O(l))/RT = -95.6562$.

ANSWER: 1.011×10^{-14}

10H A cell such as in Figure 10.7 contains HCl(aq) at 0.001 mol kg^{-1} and $H_2(g)$ at an unknown pressure, all at 298 K. The observed ΔE_{cell} is 0.5548 V. What is the H_2 pressure? (10.13), (10.34)

ANSWER: 0.15 atm

10I What would ΔE_{cell} be for the cell in 10H if $T = 0$ °C? (10.13), (10.34), (10.38), Table 10.3

ANSWER: 0.5416 V

10J Bars of Cd(s) are dipped into aqueous solutions of the following, each at unit activity: H^+, Zn^{2+}, Ag^+, Fe^{3+}, Co^{2+}, and Al^{3+}. With which of them will there be a spontaneous reaction? Table 10.3

ANSWER: H^+, Ag^+, Fe^{3+}, and Co^{2+}

10K What is the equilibrium constant at 298.15 K for the reaction $Ni^{2+} + Co \to Ni + Co^{2+}$? Table 10.3, (10.42)

ANSWER: 8.2

10L If the equilibrium constant for $2Cl_2 + 2H_2O \rightleftarrows O_2 + 4Cl^- + 4H^+$ is $K_{eq} = 1.122 \times 10^9$ at 298.15 K, what is $\Delta E°$ for a cell designed around this reaction? (10.42)

ANSWER: 0.1338 V

10M If $E°_{red}$ for $ClO_3^- + 6H^+ + 6e^- \to Cl^- + 3H_2O$ is 1.45 V, what is $E°_{red}$ for $ClO_3^- + 6H^+ + 5e^- \to \frac{1}{2}Cl_2 + 3H_2O$? Table 10.3, (10.43)

ANSWER: 1.47 V

PROBLEMS

SECTION 10.1

10.1 Fill in the table below relating ionic strength I to the molality m of the electrolyte for the various electrolyte-ion stoichiometries shown. (For example, Na_2SO_4 is a 2:1 electrolyte since 1 mol Na_2SO_4 produces 2 mol Na^+ and 1 mol SO_4^{2-}.)

Stoichiometry	Molality	I
1:1	m	?
2:2	m	?
3:1	m	?
4:3	m	?

10.2 What is the ionic strength of a 0.020 mol kg^{-1} solution of ZnSO$_4$? of a 0.045 mol kg^{-1} solution of AlCl$_3$? of a solution of both at these concentrations?

10.3 The compound TlN$_3$ is a moderately insoluble electrolyte as measured by its solubility product equilibrium constant:

$$\text{TlN}_3(s) \rightleftarrows \text{Tl}^+(aq) + \text{N}_3^-(aq) \qquad K_{eq} = 2.2 \times 10^{-4}.$$

Find an expression for the ionic strength of a saturated TlN$_3$ solution in terms of K_{sp}. Repeat this problem for Cr(OH)$_3$ for which K_{eq} is 1.6×10^{-30}.

10.4 The activity coefficients of H$^+$ and OH$^-$ in a KCl solution at 0.10 mol kg^{-1} ionic strength are 0.83 and 0.76, respectively. Do these two ions, H$^+$ and OH$^-$, have equal concentrations? equal activities? What is the pH of this solution? (The definition of pH, discussed further in Section 10.4 and Example 10.4, is pH = $-\log_{10} a_{\text{H}^+}$.)

10.5 One can dissolve 0.13 g of LiF in pure water at 25 °C, and the solubility-product equilibrium constant for LiF at this temperature is 1.7×10^{-3}. Find the mean ionic activity coefficient for a saturated LiF solution. What are the activity coefficients of the individual ions if they are assumed to be equal? If the activity coefficient for Li$^+$ is in fact $\gamma_+ = 0.836$, what is the activity coefficient for F$^-$, γ_-?

10.6 We wrote

$$\mu_i^\circ = \lim_{m_\pm \to 0} [\mu_i - RT \ln (\gamma_\pm m_\pm)^\nu]$$

as the formal expression for μ_i° in the dilute-solution limit. Why would the following expression, where m_i is the electrolyte molality, be incorrect?

$$\mu_i^\circ = \lim_{m_i \to 0} [\mu_i - RT \ln (\gamma_\pm^\nu m_i)]$$

(*Hint:* What are the consequences of electrolyte dissociation on the two expressions? Is dissociation treated on an equal footing by them?)

10.7 The Debye–Hückel theory most naturally leads to an expression for the activity coefficient of a single ion in the limiting-law form

$$\ln \gamma_i = -A z_i^2 I^{1/2}$$

with $i = +$ or $-$, but the experimentally relevant quantity is $\gamma_\pm$, given by Eq. (10.13) with $A = 1.174$ at 25 °C. Derive Eq. (10.13) from the single-ion expression above, starting with the definitions

$$\gamma_\pm = (\gamma_+^{\nu_+} \gamma_-^{\nu_-})^{1/\nu}, \qquad \nu = \nu_+ + \nu_-$$

and using the electroneutrality condition $\nu_+ z_+ + \nu_- z_- = 0$.

10.8 The closer look at the Debye–Hückel theory leads to Eq. (10.21) for $\ln \gamma_\pm$. Neglect the term a/r_D in that expression (which is the limiting-law approximation), and use the definition of r_D in Eq. (10.18) to arrive at the collection of physical constants appearing in Eq. (10.12). (Note that Eq. (10.12) applies to $\gamma_\pm$; see Problem 10.7 to relate $\gamma_\pm$ to γ_+ and γ_-.) Along the way, you will discover another approximation implicit in Eq. (10.12): the replacement of a molarity by a molality.

10.9 From the Debye–Hückel limiting law, Eq. (10.12), $\ln \gamma_\pm$ for any one solvent varies with temperature in three ways: through the temperature variations of ρ_s and ϵ_r as well as through the $T^{-3/2}$ factor itself. Compute $\gamma_\pm$ for HCl at 0.002 mol kg^{-1} at 5, 25, and 60 °C given the following corresponding values for ρ_s: 999.963 8, 997.044 9, and 983.198 9 kg m^{-3}. The temperature dependence of ϵ_r can be found in Table 9.1. The corresponding measured values of $\gamma_\pm$ are 0.9539, 0.9521, and 0.9491. How well does the limiting law do at temperatures away from 25 °C?

10.10 Practice problems 10B and 10C ask you to compare the Debye–Hückel limiting- and extended-law predictions for $\gamma_\pm$ to measured values for KCl and NaCl at 0.10 and 0.01 mol kg^{-1} concentrations. The text mentions the following variation of the extended law

$$\ln \gamma_\pm = 1.174 \, z_+ z_- \left[\frac{(I/\text{mol kg}^{-1})^{1/2}}{1 + (I/\text{mol kg}^{-1})^{1/2}} \right] + \beta I$$

where β is an empirical parameter. Use the measured $\gamma_\pm$ values at 0.01 mol kg^{-1} to find a value for β for each electrolyte, then use those β values to predict $\gamma_\pm$ for each at 0.10 mol kg^{-1}. Is the agreement markedly improved over the simpler, parameter-free extended-law expression? Can you conclude that a parameter such as β can be usefully and reliably determined from a single measurement?

10.11 Problem 10.5 discusses the solubility of LiF in pure water. What is the solubility of LiF in a solution of 0.003 mol kg^{-1} NaF? Note that the enhanced solubility is as if the LiF had been dissolved in a pure solvent with a dielectric constant different from that of water. What would the dielectric constant of pure water have to be to give LiF the solubility it has in this NaF solution?

SECTION 10.2

10.12 We can relate partial molar entropy and free energy through the usual expression $\partial \mu / \partial T = -\overline{S}$. If we apply this to an electrolyte in the dilute regime with the Debye–Hückel limiting law expression for $\gamma_\pm$, we can gain insight into the entropy contribution from ionic nonideality. Write, for a 1:1 electrolyte,

$$\mu = \mu^\circ + RT \ln a_\pm^2 = \mu^\circ + RT \ln m_\pm^2 + RT \ln \gamma_\pm^2$$

and consider the entropy contribution from the last term, which represents all the nonideality:

$$\Delta \overline{S} = -\frac{\partial (RT \ln \gamma_\pm^2)}{\partial T}.$$

The limiting law expression for $\ln \gamma_\pm$ is given by Eq. (10.12). Assume ρ_s and ϵ_r are independent of T (which they aren't, but the assumption will simplify your work without changing any conclusions) and find an expression for $\Delta \overline{S}$. As important as its magnitude is its algebraic sign. Can you give this sign a physical interpretation?

10.13 The following table gives measured values for $\gamma_\pm$ for $H_2SO_4(aq)$ at various molalities. Compare these $\gamma_\pm$ values to the Debye–Hückel limiting-law predictions. You should find that the calculated values are significantly greater, even at the lowest concentration. What could account for this disagreement? The extended Debye–Hückel law won't help, since it *increases* $\gamma_\pm$ over the limiting-law value. The answer must be chemical in nature: use data in Table 10.2 to find the equilibrium constant for the second ionization step, $HSO_4^- \to H^+ + SO_4^{2-}$. Your answer should be $\sim 10^{-2}$, which indicates that the second ionization step isn't complete. Many apparently anomalous properties of sulfuric acid can be explained by this observation.

$m/\text{mol kg}^{-1}$	$\gamma_\pm(\text{obs})$
0.0005	0.885
0.001	0.830
0.002	0.757
0.005	0.639
0.010	0.544

10.14 The commonly tabulated molar free energy of formation of carbonic acid, H_2CO_3, is -623.16 kJ mol^{-1}. This number and data in Table 10.2 can be used to find the equilibrium constant for the first ionization of carbonic acid: $H_2CO_3 \rightleftarrows HCO_3^- + H^+$. Calculate this number. You should find $K_{eq} \cong 4 \times 10^{-7}$, a surprisingly small number for a carboxylic acid. (For example, K_{eq} is 1.80×10^{-4} for formic acid and 1.75×10^{-5} for acetic acid.) The reason can be traced to the meaning of "$H_2CO_3(aq)$" in thermodynamic tables. It represents all dissolved CO_2 (i.e., $H_2CO_3(aq)$ + $CO_2(aq)$), but in fact the hydrolysis of aqueous CO_2 is incomplete:

$$H_2O(l) + CO_2(aq) \rightleftarrows H_2CO_3(aq) \qquad K_{eq} \cong 1.3 \times 10^{-3}$$

Thermodynamic tables assume this reaction has an equilibrium constant of 1. Given $\Delta_f \overline{G}^\circ = -237.18$ kJ mol^{-1} for water, find first $\Delta_f \overline{G}^\circ$ for $CO_2(aq)$ as it is tabulated under this assumption. Then take into account the real equilibrium constant for hydrolysis to find the real first ionization constant for $H_2CO_3(aq)$. (*Hint:* Take advantage of the fact that the hydrolysis equilibrium constant is small so that the concentration sum $H_2CO_3(aq) + CO_2(aq)$ is approximately equal to the concentration of $CO_2(aq)$ alone.) Finally, use this corrected equilibrium constant to find more honest $\Delta_f \overline{G}^\circ$ values for $H_2CO_3(aq)$ and $CO_2(aq)$. (The tables are not actually dishonest, they are set up this way for convenience. One usually needs information on the net effects of $H_2CO_3(aq) + CO_2(aq)$ in carbonic acid solutions so that their individuality does not matter.)

SECTION 10.3

10.15 If an electrochemical cell was set up with 1.00 L of solution to represent the reaction

$$2Ag^+(aq) + Fe(s) \to 2Ag(s) + Fe^{2+}(aq)$$

with, initially, $[Ag^+] = 0.01$ M, $[Fe^{2+}] = 0$, and sufficient $Fe(s)$ and $Ag(s)$ to represent stoichiometric excesses were the reaction to proceed in either direction, how long would it take for the reaction to achieve equilibrium if the cell were connected to a device that drew a constant 10^{-9} A current? For this reaction, K_{eq} is about 8×10^{41}. What does your answer tell you about the rate of advancement of a cell reaction when connected to a voltmeter that draws very little current?

SECTION 10.4

10.16 Sometimes material incompatibilities pose serious problems to the construction of an electrochemical cell designed to study a particular net reaction. Consider, for example, the following cell, devised by G. Singh and P. A. Rock [*J. Chem. Phys.* **57,** 5556 (1972)]:

$$^7Li(s)|^7LiBr(PC)|^7Li(Hg)|^7LiCl(aq)|Hg_2Cl_2(s)|$$
$$Hg(l)|Hg_2Cl_2(s)|^6LiCl(aq)|^6Li(Hg)|^6LiBr(PC)|^6Li(s)$$

where PC is the solvent propylene carbonate. This sequence of ten interfaces is cleverly designed to overcome the spontaneous reaction between Li and water. What is the net reaction represented by this cell? You may expect this net reaction to have $K_{eq} = 1$ and thus to cause the cell to have $\Delta E^\circ = 0$. In fact, Singh and Rock found $\Delta E^\circ = 1.16$ mV at 296.65 K. What is K_{eq}? Statistical thermodynamics (Chapter 23) explains why $K_{eq} \neq 1$.

10.17 Can $\gamma_\pm$ have any value? Consider the ΔE_{cell} expression for the Cd concentration cell given by Eq. (10.32). Suppose $\gamma_\pm(2\ m) = 0.0332$ instead of 0.0439. Suppose it was even lower. What would be the physical consequences? Are these possible?

10.18 The cell

$$Pt, H_2(g)|HCl(aq)|AgCl(s)|Ag(s)$$

as diagrammed in Figure 10.7 and discussed in the text has been studied over a wide range of T, H_2 pressure, and HCl concentration. The data below represent ΔE_{cell} at varying HCl molalities for $P_{H_2} = 1$ atm and $T = 298.15$ K:

m/mol kg^{-1}	0.005 0	0.010	0.020	0.050	0.100
ΔE_{cell}/V	0.498 38	0.464 11	0.430 17	0.385 76	0.352 28

An appropriate extrapolation of these data to $m = 0$ will yield $\Delta E°$, but how should such an extrapolation be made? First, show that the Nernst equation for this cell at these conditions can be written as

$$\Delta E° = \Delta E_{cell} + 0.051\,385\,\ln \gamma_\pm m \ .$$

Then use the extended Debye–Hückel law, $\ln \gamma_\pm = -1.174\,m^{1/2}/(1 + m^{1/2})$ here, to evaluate the right-hand side of the expression above at each m. Plot these values versus m and extrapolate (via a linear least squares fit, if you can) to $m = 0$ to obtain $\Delta E°$. How well does your $\Delta E°$ agree with that in the text?

SECTION 10.5

10.19 The Cd(Hg) solution played a major role in the discussion of the Cd^{2+} concentration cell in the text. Use data in Table 10.3 to find the activity of Cd in a saturated amalgam. *Big hint:* Consider the equilibrium constant for the reaction $Cd(s) \rightleftarrows Cd(Hg)$.

10.20 Which ion, Tl^+ or Tl^{3+}, is stable in aqueous solution in the presence of $Tl(s)$? You will need to know

$$Tl^+ + e^- \rightarrow Tl \qquad E°_{red} = -0.3363\ V$$
$$Tl^{3+} + 2e^- \rightarrow Tl^+ \qquad E°_{red} = 1.25\ V\ .$$

10.21 Aqueous vanadium chemistry is quite rich due to the variety of oxidation states that are stable under various conditions. Find $E°_{red}$ for the following half-reaction of the vanadyl ion, VO^{2+}

$$VO^{2+} + 2H^+ + 2e^- \rightarrow V^{2+} + H_2O$$

from data in Table 10.3. Then go on to find equilibrium constants for the net reactions

$$VO^{2+} + 2H^+ + V^{2+} \rightleftarrows 2V^{3+} + H_2O$$
$$V(OH)_4^+ + V^{3+} \rightleftarrows 2VO^{2+} + 2H_2O\ .$$

10.22 Calculate the equilibrium constant for the disproportionation reaction of $Cu^+(aq)$ in acidic solution using Table 10.3:

$$2Cu^+(aq) \rightarrow Cu(s) + Cu^{2+}(aq)\ .$$

What can you conclude about the stability of $Cu^+(aq)$? Can the potential of the $Cu^+ + e^- \rightarrow Cu$ half-reaction actually be measured? Given

$$Cu^{2+} + Cl^- + e^- \rightarrow CuCl \qquad E°_{red} = 0.538\ V$$

what can you conclude about the stability of $CuCl(aq)$?

10.23 Using data in Table 10.3, find the solubility equilibrium constant for $PbSO_4(s)$. Note that there are two ways to proceed: combine the half-reactions $PbSO_4(s) + 2e^- \rightarrow Pb + SO_4^{2-}$ and $Pb^{2+} + 2e^- \rightarrow Pb$ or combine $PbO_2 + SO_4^{2-} + 4H^+ + 2e^- \rightarrow PbSO_4 + 2H_2O$ and $PbO_2 + 4H^+ + 2e^- \rightarrow Pb^{2+} + 2H_2O$. Try both methods to see how closely they agree. (The literature value is 1.82×10^{-8}.) Next, having shown that $PbSO_4$ is fairly insoluble, consider the cell

$$Pb(s)|PbSO_4(s)|CuSO_4(aq)|Cu(s)$$

with $CuSO_4$ at a concentration of 0.010 mol kg^{-1}. Such a cell has $\Delta E_{cell} = 0.5554$ V at 298.15 K. What is $\gamma_\pm$ for $CuSO_4$ at 0.010 mol kg^{-1}? If the Cu is alloyed with Pd and all else in the cell is kept constant, ΔE_{cell} changes to 0.5643 V. What is the activity of Cu in the alloy?

10.24 Find $E°_{red}$ for

$$PbO_2(s) + 4H^+(aq) + 4e^- \rightarrow Pb(s) + 2H_2O(l)$$

from data in Table 10.3, then go on to find the equilibrium constant and $\Delta_r\overline{G}°$ for

$$PbO_2(s) + Pb(s) + 4H^+(aq) \rightleftarrows 2Pb^{2+}(aq) + 2H_2O(l)\ .$$

Finally, compute $\Delta_f\overline{G}°$ for $PbO_2(s)$ from both reactions. A value for $\Delta_f\overline{G}°(Pb^{2+}(aq))$ can be found in Table 10.2; $\Delta_f\overline{G}°(H_2O(l))$ is quoted in Practice problem 10G.

10.25 The energy released when O_2 and H_2 combine to form water is quite large:

$$H_2(g) + \tfrac{1}{2}O_2(g) \rightarrow H_2O(l) \qquad \Delta_r\overline{G}° = -237.129\ \text{kJ mol}^{-1}$$

yet these gases can be mixed without reacting at ordinary T and P. Add a spark, however, and the reaction is explosively dramatic. How can the energy release of this reaction be released in a controlled fashion? The device that does so is called, generically, a fuel cell. Such a cell can operate in either acidic or alkaline media. Find the corresponding half-reactions for each case, and verify that they combine to yield a $\Delta E°$ in accord with the $\Delta_r\overline{G}°$ quoted for the net reaction. How many electrons are involved in the net redox reaction? A green plant performs somewhat the reverse reaction, producing $O_2(g)$ from H_2O and $CO_2(g)$ in its photosynthetic apparatus.

10.26 We first met the two crystal forms of $CaCO_3(s)$, calcite and aragonite, in Example 7.10 and Problem 7.44. Use data quoted in Problem 7.44 to find ΔE_{cell} for the following cell at 300 K and 1 bar pressure:

$$Pb(s)|PbCO_3(s)|CaCO_3(\text{calcite})|CaCl_2(aq, 0.1\ m)|$$
$$Hg_2Cl_2(s)|Hg(l)|Hg_2Cl_2(s)|CaCl_2(aq, 0.1\ m)|$$
$$CaCO_3(\text{aragonite})|PbCO_3(s)|Pb(s)\ .$$

10.27 Tabulated below are ΔE_{cell} values for the cell

$$Pt(s)|H_2(g, f = 1 \text{ bar})|HCl(aq, m)|Hg_2Cl_2(s)|Hg(l)|Pt(s)$$

at 298.15 K as a function of the cell molality, m. What is the net reaction represented by this cell? What is its equilibrium constant? (Use Table 10.3.) What values for $\gamma_\pm(HCl)$ do these data represent? Graph them as $\ln \gamma_\pm$ versus $I^{1/2}$ (which is just $m^{1/2}$ here). How well do they follow the Debye–Hückel limiting law, which says they should fall on a straight line on such a plot?

m/mol kg^{-1}	0.001 608	0.005 040	0.013 968	0.037 690	0.075 081	0.119 304
ΔE_{cell}/V	0.600 80	0.543 66	0.493 39	0.445 16	0.411 87	0.389 48

10.28 If we have an analytic expression for $\Delta E°$ as a function of T, we can not only find $\Delta_r \overline{G}°$ immediately from Eq. (10.43) and $\Delta_r \overline{S}°$ from Eq. (10.44), but we can find $\Delta_r \overline{H}°$ as a function of T from either Eq. (10.46) or, returning to Eq. (7.25), as the single derivative

$$\Delta_r \overline{H}° = -nF \left(\frac{\partial(\Delta E°/T)}{\partial(1/T)} \right).$$

Derive this expression, then use the following power series in T for the cell

$$Pt|H_2(g)|HCl(aq)|AgCl(s)|Ag(s)$$

[from R. G. Bates and V. E. Bower, *J. Res. Nat. Bur. Stand.* **53**, 283 (1954)] to find $\Delta_r \overline{H}°$ at 298.15 K for the corresponding net reaction

$$AgCl(s) + \tfrac{1}{2}H_2(g) \rightarrow Ag(s) + HCl(aq).$$

$$\frac{\Delta E°}{V} = -5.574\,16 \times 10^{-3} + 2.696\,65 \times 10^{-3} \frac{T}{K}$$
$$- 8.229\,85 \times 10^{-6} \left(\frac{T}{K}\right)^2 + 5.869 \times 10^{-9} \left(\frac{T}{K}\right)^3$$

(*Hint:* Before attacking the power series with your calculator, you might find it useful to show first that if

$$y = a_0 + a_1 x + a_2 x^2 + a_3 x^3$$

then

$$\left(\frac{\partial(y/x)}{\partial(1/x)}\right) = a_0 - a_2 x^2 - 2a_3 x^3 .)$$

10.29 The power series of Eq. (10.39) can be used (one per half-reaction) to write a power series for $\Delta E°$. Do so, and then rewrite the result in terms of the thermodynamic state functions of the corresponding net cell reaction: $\Delta_r \overline{G}°$, $\Delta_r \overline{S}°$, and $\Delta_r \overline{C}_P°$.

10.30 Since many biochemical reactions are redox reactions, reduction potentials are important measures of energy flow in biochemical pathways. But since most such reactions occur in media at or near physiological pH, standard reduction potentials, based on $a_{H^+} = 1$ (or pH = 0) are somewhat inconvenient. Consequently, a biochemical reduction-potential scale has been adopted that is based on $a_{H^+} = 10^{-7}$, representing neutral pH, a more appropriate condition for a living cell than pH = 0. For example, the important half-reaction

$$O_2 + 4H^+ + 4e^- \rightarrow 2H_2O \qquad E°_{red} = 1.229 \text{ V}$$

has, from the Nernst equation,

$$E_{\frac{1}{2}rxn} = E°_{red} - \frac{RT}{4F} \ln \frac{a_{H_2O}^2}{a_{O_2} a_{H^+}^4}$$
$$= E°_{red} + \frac{RT}{F} \ln 10^{-7} - \frac{RT}{4F} \ln \frac{a_{H_2O}^2}{a_{O_2}}$$
$$= E°'_{red} - \frac{RT}{4F} \ln \frac{a_{H_2O}^2}{a_{O_2}}$$

where the biochemical reduction potential $E°'_{red}$ is

$$E°'_{red} = E°_{red} + \frac{RT}{F} \ln 10^{-7}$$
$$= 1.229 \text{ V} + (2.569 \times 10^{-2} \text{ V})(-16.118)$$
$$= 0.815 \text{ V}.$$

Find the standard free-energy change that accompanies the reaction between the coenzyme NADH, nicotinamide adenine dinucleotide:

(a coenzyme derived from the water-soluble vitamin B$_3$, nicotinic acid, found in whole grains and liver—R is an adenosine diphosphate ribose moiety) and O$_2$ according to

$$NADH + H^+ + \tfrac{1}{2}O_2 \rightarrow NAD^+ + H_2O$$

where NAD$^+$ is NADH less a hydride at the 4 position on the nicotinamide ring. You will need to know

$$NAD^+ + H^+ + 2e^- \rightarrow NADH \qquad E°'_{red} = -0.320 \text{ V}.$$

The net reaction, by the way, takes place through a chain of electron transfer reactions involving four enzymes, producing three adenosine triphosphates (ATP) from the released energy in the inner mitochondrial membranes of your respiratory system.

10.31 There is more to the biochemical reduction potential $E_{red}^{\circ\prime}$ than was covered in the previous problem. If the half-reaction contains a polyfunctional acid or base, then not only do we set $a_{H^+} = 10^{-7}$, but we write the Nernst equation in terms of the total analytic activity (which is closely equal to the concentration, for most situations) of the neutral form of the compound. For example, the half-reaction between dehydroascorbic acid and the diprotic acid ascorbic acid (vitamin C) is

[structures of dehydroascorbic acid $+ 2H^+ + 2e^- \rightarrow$ ascorbic acid]

with $E_{red}^{\circ} = 0.390$ V, which we will abbreviate D + $2H^+ + 2e^- \rightarrow H_2A$. The Nernst equation is

$$E_{\frac{1}{2}\text{rxn}} = E_{red}^{\circ} - \frac{RT}{2F} \ln \frac{a_{H_2A}}{a_D\, a_{H^+}^2}$$

and while $a_{H_2A} \cong [H_2A]$ and $a_D \cong [D]$, due to the pH dependence of $[H_2A]$, the total analytic concentration of H_2A in all forms:

$$c_{H_2A} = [H_2A] + [HA^-] + [A^{2-}]$$

is a function of pH and the successive acid-dissociation equilibrium constants. For ascorbic acid, these are

$$H_2A \rightleftharpoons H^+ + HA^- \qquad K_{eq1} = 7.94 \times 10^{-5}$$
$$HA^- \rightleftharpoons H^+ + A^{2-} \qquad K_{eq2} = 1.62 \times 10^{-12}\ .$$

Show that if we write the Nernst equation in terms of the biochemical reduction potential and analytic total concentrations as

$$E_{\frac{1}{2}\text{rxn}} = E_{red}^{\circ\prime} - \frac{RT}{2F} \ln \frac{c_{H_2A}}{[D]}$$

then

$$E_{red}^{\circ\prime} = E_{red}^{\circ} - \frac{RT}{2F} \ln \frac{1}{[H^+]^2 + [H^+]K_{eq1} + K_{eq1}K_{eq2}}\ .$$

Evaluate $E_{red}^{\circ\prime}$ for this half-reaction.

GENERAL PROBLEMS

10.32 R. E. Powell and W. M. Latimer [*J. Chem. Phys.* **19**, 1139 (1951)] found that the entropies of monatomic aqueous ions was given to remarkable accuracy by the semiempirical expression

$$\frac{\overline{S}^{\circ}}{R} = \frac{3}{2} \ln \left(\frac{M}{\text{g mol}^{-1}}\right) - \frac{135.8\,|z|}{(r/\text{Å} + x)^2} + 18.62$$

where M is the atomic mass, z is the ion's charge magnitude, r is the ion's *crystal radius,* and x is 2.00 for positive ions and 1.00 for negative ions. (Note the $3/2 \ln M$ term; it also appears in the Sackur–Tetrode expression for the entropy of a monatomic gas, Eq. (5.44). We will later see the molecular origin of this term. The $|z|/r^2$ term was a surprise, since Born had predicted a $|z|^2/r$ dependence!) Check the accuracy of this expression against Table 10.1 for the alkali ions (r/Å, Li$^+$–Cs$^+$, = 0.60, 0.95, 1.33, 1.48, 1.69), and then predict $\overline{S}^{\circ}/R$ for Au$^+$, La^{3+}, Se^{2-}, and BH$_4^-$, for which r/Å = 1.37, 1.15, 1.98, and 2.03, respectively. For BH$_4^-$, you should add $2.9\,R$ to the Powell and Latimer expression to account for the entropy due to the rotational motion of the ion. (See W. H. Stockmayer, D. W. Rice, and C. C. Stephenson, *J. Am. Chem. Soc.* **77,** 1980 (1955) for more on the thermodynamic properties of the BH$_4^-$ ion.)

10.33 Rusting of Fe is a complex redox system with tremendous economical implications. While "rust" is mainly Fe$_2$O$_3$ and Fe(OH)$_4$, other species such as FeOOH and amorphous mixtures of varying composition have been detected by X-ray crystallography and other spectroscopies. To simplify the problem, we will consider here only Fe$_2$O$_3$. If we already have Fe in an acidic solution as Fe^{2+}, the following net reaction occurs:

$$4\text{Fe}^{2+}(aq) + O_2(g) + (4 + 2x)H_2O$$
$$\rightarrow 2\text{Fe}_2O_3 \cdot xH_2O(s) + 8H^+(aq)$$

where x represents a variable hydration stoichiometry. But if oxygenated water is in contact with Fe, the free-energy drop that ultimately drives this net reaction comes from the half-reactions

$$\text{Fe}^{2+}(aq) + 2e^- \rightarrow \text{Fe}(s)$$
$$O_2(g) + 4H^+(aq) + 4e^- \rightarrow 2H_2O(l)\ .$$

Armed with this knowledge, we can answer the following question: Why does a drop of salty water on Fe lead to corrosion in the middle of the drop and rust deposits on the drop's periphery? To help you answer this latter question, consider that O_2 will more readily diffuse to the Fe through the thin solution at the drop edge, while the thick middle of the drop will be more rapidly depleted of O_2. Consider as well the spontaneous process represented by the concentration cell shown on next page:

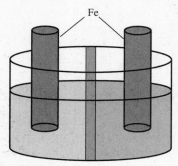

In this cell, the container has a porous barrier that prevents rapid mixing and pure Fe rods on each side. On the right, the acidic solution is heavily oxygenated; on the left, the solution is less oxygenated but otherwise identical. If the Fe rods are connected by an external wire, which will dissolve?

10.34 A *potentiometric titration* has several interesting aspects that are explored in this and the next two problems. The experimental apparatus is typically something like that shown below for the titration of Fe^{2+} by Ce^{4+}:

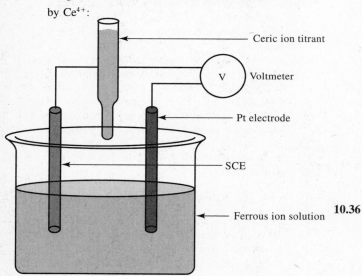

The ceric ion titrant at a known concentration is added from a buret to the solution of unknown ferrous ion concentration. The titration reaction is

$$Ce^{4+}(aq) + Fe^{2+}(aq) \rightleftarrows Ce^{3+}(aq) + Fe^{3+}(aq).$$

Show first that this reaction "goes to completion" by calculating its K_{eq}. To be somewhat more realistic, use the following "formal reduction potentials" E_f° that apply in 1 M $HClO_4$, a common analytical medium. These potentials account for the deviations from unit activity (the standard-potential condition) in the perchloric acid medium.

$$Fe^{3+}(aq) + e^- \rightarrow Fe^{2+}(aq) \qquad E_f^\circ = 0.767 \text{ V}$$
$$Ce^{4+}(aq) + e^- \rightarrow Ce^{3+}(aq) \qquad E_f^\circ = 1.70 \text{ V}$$

Since all species in the net titration reaction are in the same pot, it is always in equilibrium during the titration. The Pt electrode and the Saturated Calomel Electrode (SCE) do not influence the titration reaction. They only *measure* changes in activities, not *cause* those changes. At the SCE, the half-reaction is

$$Hg_2Cl_2(s) + 2e^- \rightarrow 2Hg(l) + 2Cl^-(aq)$$

while at the Pt electrode, both the iron and the cerium half-reactions come to equilibrium. Thus, there are *two* net reactions that govern ΔE_{cell}. Find them, and show that either gives the same ΔE_{cell}, since the titration reaction is always in equilibrium.

10.35 Continuing the potentiometric titration from the previous problem, consider the potential at the equivalence point, at which the amount of Ce^{4+} added to the solution stoichiometrically matches the amount of Fe^{2+} originally present according to the net titration reaction. Show that, in general, for titration half-reactions involving only aqueous ions (Ox represents the oxidized form; Rd represents the reduced form):

$$Ox_1(aq) + n_1 e^- \rightarrow Rd_1 \qquad \text{titrant 1}$$
$$Ox_2(aq) + n_2 e^- \rightarrow Rd_2 \qquad \text{analyte 2}$$

the cell potential at the equivalence point is given in terms of the formal reduction potentials of each by

$$\Delta E_{cell}(eq) = \frac{n_1 E_{f,1}^\circ + n_2 E_{f,2}^\circ}{n_1 + n_2}.$$

You may find it helpful to write the general net titration reaction based on the half-reactions above in order to properly account for stoichiometry. Does the equivalence-point potential change if the net titration reaction does not go to completion?

10.36 Now we turn to the titration curve for the potentiometric titration begun two problems above. To simplify the algebra without sacrificing the chemistry, assume that the ceric ion solution is so concentrated that dilution of the analyte solution as it is added can be neglected. Also assume that we can set activity coefficients to unity and consider only concentrations. Prior to the equivalence point, excess Fe^{2+} remains in solution, and since the titration goes to completion, the amount of Fe^{3+} produced is simple to find. Thus it is convenient to consider the net *cell* reaction to involve the SCE and the Fe half-reaction (which is equilibrated at the Pt electrode). Find an expression for the cell voltage in terms of the amount of Ce^{4+} added prior to the equivalence point. Beyond the equivalence point, we know the cerous/ceric ion concentrations most easily, and thus it is convenient to consider the Ce half-reaction instead of the Fe half-reaction. Do so, and find an expression for the cell voltage in this region.

CHAPTER 11

The Need for Quantum Mechanics

THIS chapter is an introduction to the major topic of the next several chapters—the microscopic structure of atoms and molecules. Quantum theory, as we will use it, is a product of the first third of this century. Refinements in the practice of quantum theoretical calculations have continued since then, as have developments in fundamental theory. The refinements will interest us to the extent that they have made calculations of chemical interest more accurate and more widely applicable to complex chemical systems, but developments in fundamental theory since roughly the late 1920's have been largely of secondary importance to chemistry. True quantum chemistry, the application of that theory to significant chemical problems, had a much more recent birth, coincident with the rapid development of high speed, large memory computers.

The size and energy realm of molecular interactions is so far removed from our usual macroscopic view of the world that most quantum mechanical phenomena went unnoticed until around the turn of the century, when experiments in atomic physics first began and grew with increasing sophistication. Many of these phenomena are not at all intuitive in the way classical, macroscopic phenomena are. Certain motions that are classical impossibilities become possible, while other motions that classical mechanics allows become prohibited. Moreover, important new phenomena have no classical analogy at all.

We will not review the sequence of experiments that lead to the complete theory. Instead, we start with a section pointing out situa-

11.1 Basic Quantum Mechanical Phenomena

11.2 What to Expect from Quantum Mechanics

11.3 How to Think Quantum Mechanically

tions, some with historical importance, for which a quantum explanation is required. In the second section, we lay the ground rules for the theory and discuss how one uses it in a very general way. Finally, we close with a section on certain quantum consequences that are worth remembering. These will be the recurring patterns of quantum mechanical results, and you will be aided in your quantum intuition if you learn to recognize and anticipate them.

11.1 BASIC QUANTUM MECHANICAL PHENOMENA

The centerpiece of chemistry is the Periodic Table of the hundred-plus elements. Mendeleev recognized the organization of this table and Ramsay discovered a whole family of elements (the rare gases) without quantum theory, but the structure and extent of the table is inexplicable until quantum ideas are used. Take atomic fluorine, add just one more proton to the nucleus and one more electron to make neon, and chemical properties change from violent to placid.

One of the clues to quantum theory came from the study of light interacting with atoms. A key observation was that the high-temperature vapor of each element emitted a unique set of discrete light frequencies. For all electromagnetic radiation moving through a vacuum

$$\underset{\text{Frequency}}{\nu} \quad \underset{\text{Wavelength}}{\lambda} = \underset{\text{Speed of light}}{c} \qquad (11.1)$$

where ν is the frequency of alternation of the electromagnetic field (expressed in cycles per second, or hertz, abbreviated Hz), λ is the wavelength, and c is the speed of light.[1] We divide the electromagnetic spectrum into traditionally named regions according to frequency or wavelength:

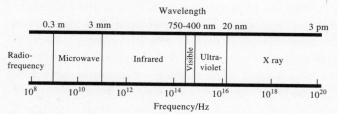

This division is explained fully in Chapter 18.

As the ability to detect light spread from the narrow range of visible frequencies up through the ultraviolet and X-ray frequencies and down through the infrared and microwave, more and more atomic emissions were found.

The first interesting observation was the discrete nature of these frequencies. No matter what one tried (varying temperature, external field strengths, etc.) they remained virtually fixed.[2] Typically hundreds of atomic emissions could be measured for any one atom, and they at first made no sense. Creative numerology was applied in an attempt to categorize, at least, and perhaps understand their pattern. For the

---

[1]The speed of light is a defined physical constant equal to 299 792 458 m s^{-1} exactly (or approximately 3.0×10^8 m s^{-1}).

[2]External electric and magnetic fields had a subtle effect that we will return to in Chapters 15 and 21.

most part, these attempts failed. But when relationships were found, they were strikingly simple. For H, a series of emissions were observed in groups throughout the infrared, through the visible, and on into the ultraviolet. In the visible, Balmer, in 1885, analyzed a set (called "lines" by spectroscopists due to their appearance as such on the photographic plates of early spectrographs) with wavelengths that followed the simple formula

$$\lambda = (364.70 \text{ nm}) \frac{n^2}{n^2 - 4} \quad (11.2)$$

where n was an integer numbering each line, but starting with $n = 3$ (for which $\lambda = 656$ nm, at the red end of the visible spectrum).

Other sets of lines followed variations on this equation. The most general equation is

$$\lambda = (91.176 \text{ nm}) \frac{n^2 m^2}{n^2 - m^2} \quad (11.3)$$

where n and m are both integers with $n > m$, and m is unique to each set of lines. Balmer's set has $m = 2$, and Lyman found the ultraviolet set with $m = 1$. The sets with $m = 3$, 4, and 5, moving out through the infrared, were found by Paschen, Brackett, and Pfund, respectively, and these names for each set remain in use. A schematic spectrum of these five sets is shown in Figure 11.1 which locates the wavelengths of the first six lines (those with $n = m + 1$ through $m + 6$) of each set. Note how each is converging to a short-wavelength limit as n increases.

EXAMPLE 11.1

One H atom emission line is in the infrared at $\lambda = 2\,625.9$ nm. What are the n and m values of Eq. (11.3) for this transition, and to what named series does it belong?

SOLUTION This would seem to be an insoluble problem of one equation in two unknowns. But n and m must be integers with $n > m$, and a solution is possible. We write Eq. (11.3) in the form

$$\frac{1}{m^2} = \frac{1}{n^2} + \frac{91.176}{\lambda/\text{nm}} = \frac{1}{n^2} + \frac{91.176}{2\,625.9}$$

and note that the right-hand side (RHS) is largest when $n = m + 1$ (the smallest n allowed) and smallest (at $91.176/2\,625.9 = 0.034\,72$) as $n \to \infty$. Thus we begin a search with $m = 1$ and $n = m + 1$, incrementing m. For $m = 1$, 2, and 3, $1/m^2$ is always larger than the

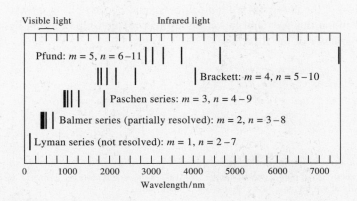

FIGURE 11.1 The H atom emission, from Eq. (11.3) with m and n as shown, spans the spectrum from the ultraviolet through the visible and through and beyond the infrared. The predictions are in excellent agreement with observation, and emissions not resolved on the scale of this figure are easily resolved experimentally.

RHS. For $m \geq 6$, the opposite is true. Thus, $m = 4$ or 5, and trial values of n will quickly (a computer spreadsheet program is a great help here) show that $m = 4$, $n = 6$ satisfies Eq. (11.3). The line must therefore be the second line ($n = m + 2$) in the Brackett ($m = 4$) series.

▶ RELATED PROBLEMS 11.1, 11.2

For some other atoms, notably the alkali metals, some sets of lines were found that followed an expression similar to Eq. (11.3), but with two notable differences. First, the constant wavelength factor varied from element to element, decreasing from Li to Cs. Second, m and n in Eq. (11.3) were no longer exactly integers in the best fit to the alkali spectra.

For several years, those empirical expressions were no more than inspired numerology; they reproduced the spectroscopic data but explained nothing. They were important advances nevertheless, because you should recognize just how strikingly nonclassical and nonintuitive line spectra really are. Think about the color changes a macroscopic piece of metal undergoes as its temperature is increased. Turn on a stove burner, and it is first notably warm, but dark. As its temperature increases, its color passes from dull, faint red through brighter reds into red-orange. If its temperature is raised above that of household necessity, the orange becomes yellow and eventually white at temperatures characteristic of incandescent filaments, around 2500 K or so. This same sequence would hold *no matter what metal* was used to construct the burner. *There is an inherent difference between the emission spectra of macroscopic solids and individual atoms.*

The study of light emitted by incandescent solids had been going on at the same time as atomic emission studies, and it provided another unexpected quantum clue. In contrast to atomic line spectra, incandescent solids emit a *continuous range of wavelengths*. The intensity of each incandescent wavelength depends on the temperature of the solid and very slightly on the material itself. This led to the idea of a perfect incandescent emitter called a "blackbody."[3] Life inside a perfect blackbody would be very strange. The light would be completely isotropic: you could not discern the shape of the blackbody's walls around you, nor could you tell how far you were from any wall!

Experiments deduced the shape of the *blackbody curve*—the curve of intensity versus emitted wavelength, at any temperature—and classical theory could not explain its shape. The breakthrough, and in many ways the birth of quantum theory, came from Planck in 1901. We will explore the blackbody curve in detail in Chapters 18 and 23; for now, we quote the famous result Planck was forced to introduce to explain this curve.

Classically, the energy of a light source should depend on its *intensity*. Planck had to abandon this idea and introduce a direct relationship between light energy and its *frequency* of the form

$$E = h \nu \quad (11.4a)$$

where E is Energy, h is Planck's constant, and ν is Frequency in Hz.

---

[3] A blackbody is not only a perfect emitter, but it is also a perfect absorber. Hence, it would appear black—totally absorbing at all wavelengths—at absolute zero.

where h is a universal constant. Its modern value is $6.626\,075\,5 \times 10^{-34}$ J Hz^{-1}.[4] We will frequently use the related expression

$$E = \hbar\omega \tag{11.4b}$$

where $\hbar$ (read "h-bar") is

$$\hbar = \frac{h}{2\pi} \cong 1.055 \times 10^{-34} \text{ J s}$$

and ω is the frequency in radians per second units.

This new constant and its interpretation were quickly applied to atomic spectra. If each frequency of light corresponds to a unique energy, and if each atom could emit (or absorb) only a discrete and unique number of frequencies, then it follows that atoms might well exist with only *discrete and unique energies*. We say "might well" at this point, because only light–matter interactions have been discussed. That *matter–matter* interactions could exhibit the same sort of discrete energy structure was provided by a series of experiments begun by Franck and Hertz in 1914.

In these experiments, an atomic gas such as Hg was bombarded with electrons of known kinetic energy. The apparatus allowed Franck and Hertz to control this energy, analyze for any changes in energy after electron–atom collisions, and observe any atomic emissions. They observed the following sequence of events as the electron energy was increased from zero. First, there was no electron energy loss or emission from the atoms until the energy reached a threshold. For Hg, this was 7.83×10^{-19} J per electron, corresponding to the energy acquired by an electron moving through an electrical potential difference of

$$\frac{7.83 \times 10^{-19} \text{ J electron}^{-1}}{1.602 \times 10^{-19} \text{ C electron}^{-1}} = 4.89 \text{ V} .$$

A convenient energy unit for these types of experiments is the *electron-volt,* eV, where 1 eV $\cong 1.602 \times 10^{-19}$ J. Thus, the threshold energy for Hg is 4.89 eV. At, and just above, this threshold, only a single emission line was observed (at $\lambda = 394$ nm), and the electrons were found to have lost an amount of energy equal to $h\nu = hc/\lambda$ with $\lambda = 394$ nm, as predicted by the Planck relation.

When they increased the electron energy, other emissions appeared, and with each, a corresponding energy loss was observed in the electron beam. For some thresholds, more than one line appeared, but the corresponding electron energy loss generally equalled $h\nu$ for the line with the greatest frequency.

EXAMPLE 11.2

When atomic oxygen is bombarded by electrons of 5.00 eV energy, emissions with wavelengths around 297, 558, and 637 nm are observed. What do these emissions tell us about the energy levels of O?

SOLUTION First, we turn the wavelengths into frequencies, since frequencies are directly comparable to energies and are additive in the way energies are. We find 297 nm corresponds to 1008 THz; 588 nm, to 537 THz; and 637 nm, to 471 THz. (1 THz = 10^{12} Hz, and 1000 THz corresponds to 4.136 eV.) The first frequency is the sum of the second and third, suggesting that only three energy levels are found below 5 eV, and that we are observing

---

[4]Note the use of Hz (hertz) as the frequency unit. We will always distinguish a frequency in cycles per second (Hz) by the symbol ν and a frequency in radians per second (s^{-1}) by ω where $\omega = 2\pi\nu$. No agreed-to unit for radians per second exists, but a strong case has been made for introducing the "Avis" unit, which of course stands for "Angular velocity in inverse seconds."

transitions from the upper-most to the middle and lowest levels, plus a transition from the middle to the lowest, as indicated below:

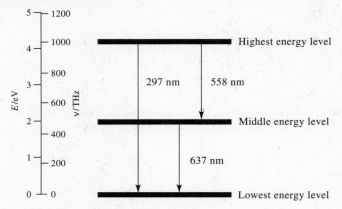

We will look at these levels in more detail in Chapter 13.

▶ *RELATED PROBLEMS* 11.4, 11.5, 11.6

Taken together, these phenomena—discrete-frequency absorption or emission of light by atoms, the Planck blackbody explanation, and the discrete energy transfer in particle–particle collisions—all point to a unique and discrete set of allowed energies for each atom. Classical mechanics, which was constantly at a loss to explain atomic structure, had no way to accommodate anything but a continuously variable energy. Moreover, while classical physics might arrange a stable assembly of electrons and protons, its stability would be destroyed whenever the charges accelerate, as they would if they orbited the atom's center of mass. A charged particle radiates light when accelerated, and a classical atom would emit a *continuous spectrum* of frequencies. As it emits, it loses energy, and the classical atom would collapse to a point in a time on the order of 10^{-8} s.

Another unexpected clue with wide-ranging implications came from another experiment involving the interaction among light, matter, and electrons: the *photoelectric effect*. In the photoelectric experiment, light of one frequency illuminates a solid surface in vacuum, and one measures the electron stream from that surface as a function of the light's intensity and frequency. The intensity and energy of these *photoelectrons*, as they are called, are of prime importance. One finds that no photoelectrons are emitted from the surface unless the frequency is above a threshold value ν_0, and that once emitted, the electrons have a kinetic energy that does not exceed $h(\nu - \nu_0)$ where ν is the illuminating light's frequency. The *number* of photoelectrons, however, depends on the light's *intensity*. Classically, the *energy* of light depends on its *intensity*, not its frequency. This is contrary to the photoelectron experiments in that no matter how intense the light, *no* photoelectrons are seen if the light frequency is below a threshold.

Einstein first used the Planck relationship to explain the photoelectric effect in a 1905 paper that was the basis for his Nobel Prize award in 1921.[5] The subtle conclusion Einstein drew was that light could transfer energy in a way closely

---

[5]Einstein is, of course, more popularly famous for his theories of relativity, and this work, also begun in a 1905 publication, was certainly worthy of a Nobel Prize. Several political and quite frankly anti-Semitic international factors conspired to force the Nobel committee to award the prize for this less controversial photoelectric effect work.

parallel to the way a *particle* transfers energy. Today, we call light particles *photons* (a name suggested by G. N. Lewis). Photons in the photoelectric effect, as do electrons in the Franck–Hertz experiment, produce energy thresholds and discrete energy transfers.

Prior to this, theories of light had settled comfortably into a *wave description*. Light shows diffraction and interference phenomena, all of which are explained by classical wave theory. But if light, traditionally described by wave-like phenomenon, is a particle, then *perhaps matter,* the classical source of particle behavior, *can exhibit wave effects*. This would mean that a stream of particles could be diffracted and interfere in a wave-like fashion. This simple but striking suspicion of a *dualism* in nature (wave and particle behavior everywhere) was first made by Louis de Broglie in 1923.

De Broglie's argument won him a Nobel Prize as well, and, in essence, it runs as follows. Light has an energy $E = h\nu$ and, from Einstein's special relativity theory, a momentum $p = E/c$. Thus, light has a momentum–wavelength relationship

$$p = \frac{E}{c} = \frac{h\nu}{c} = \frac{h}{\lambda}. \tag{11.5}$$

De Broglie brazenly proposed that the same relationship would hold for matter. He was not widely believed until experiments confirmed his proposal.

With $p = mv$, the classical mass–velocity product, matter should behave as a wave of wavelength

$$\underset{\substack{\uparrow \\ \text{Particle} \\ \text{wavelength}}}{\lambda} = \frac{h}{\underset{\uparrow}{p}} = \frac{h}{mv}. \tag{11.6}$$

$$\underset{\substack{\text{Planck's constant} \\ \text{Momentum}}}{}$$

Could this wavelength be observed experimentally? And if it could, why did matter waves escape notice until this century?

Macroscopic momenta typically range in magnitude from 10^{-9} kg m s^{-1} (a 1 mg mass moving 1 mm s^{-1}) to perhaps 10^3 kg m s^{-1} (a 100 kg mass moving 10 m s^{-1}). These momenta imply de Broglie wavelengths in the range 6×10^{-25} to 6×10^{-37} m. These are so small that no wave-like properties could be observed from any macroscopic matter interactions.

A *microscopic* mass, however, can have a de Broglie wavelength of microscopic size. An electron with a 1 eV kinetic energy has a de Broglie wavelength[6]

$$\lambda = \frac{h}{p} = \frac{h}{(2m_e E)^{1/2}}$$
$$= \frac{6.6 \times 10^{-34} \text{ J s}}{[(2)(9.1 \times 10^{-31} \text{ kg})(1 \text{ eV})(1.6 \times 10^{-19} \text{ J eV}^{-1})]^{1/2}}$$
$$= 1.2 \times 10^{-9} \text{ m}.$$

This length is comparable to molecular sizes. Similarly, a He atom at room temperature has an average speed around 1000 m s^{-1} and a de Broglie wavelength

$$\lambda = \frac{h}{p} = \frac{h}{mv} = \frac{hN_A}{Mv}$$
$$= \frac{(6.6 \times 10^{-34} \text{ J s})(6.0 \times 10^{23} \text{ mol}^{-1})}{(4 \text{ g mol}^{-1})(10^{-3} \text{ kg g}^{-1})(10^3 \text{ m s}^{-1})}$$
$$= 9.9 \times 10^{-11} \text{ m}$$

that is also comparable to molecular sizes.

---

[6] Recall that kinetic energy is $E = mv^2/2 = p^2/2m$ so that $p = (2mE)^{1/2}$.

Thus, if de Broglie matter-wave effects are to be seen, they will surely involve atomic-sized particles interacting among themselves rather than macroscopic-sized particles. Confirmation of Eq. (11.6) came in 1927, when Davisson and Germer reported that (by chance, rather than by deliberate design) they had observed the *diffraction* of electrons by the regularly-spaced atoms at the surface of a crystal of Ni. The electron energy was such that the de Broglie wavelength closely matched the atomic spacing of the crystal, just as closely-spaced lines cut in glass diffract light of wavelength comparable to the line spacing.

EXAMPLE 11.3

In the 1913 Bohr theory of atomic hydrogen, the electron was constrained to move around the proton in circular orbits only when the condition

$$m_e v r = n\hbar, \qquad n = 1, 2, 3, 4, \ldots$$

was met. Here, m_e is the electron mass; v, its velocity; r, the orbit radius; and n is a non-zero positive integer. What de Broglie wavelengths does this expression represent?

SOLUTION Writing the Bohr expression in the form of Eq. (11.6) gives

$$\lambda = \frac{h}{m_e v} = \frac{hr}{n\hbar} = \frac{2\pi r}{n}$$

or $n\lambda = 2\pi r$. But $2\pi r$ is just the *circumference of the circular orbit,* and de Broglie would say that the Bohr atom's orbits must have an integral number of de Broglie wavelengths wrapped about them. This is consistent with the view that only *standing waves* are stable. A non-integral number of wavelengths would not match amplitude over the 2π periodicity of a circle and would be unstable due to destructive interference. The diagrams below show this schematically. On the right, a wave with $n = 12$ is wrapped smoothly around a circle; on the left, $n = 12.1$, and the mismatch is obvious.

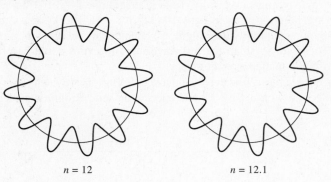

$n = 12$ $\qquad\qquad n = 12.1$

➥ *RELATED PROBLEMS* 11.8, 11.9, 11.10, 11.11

Quantum effects are confined to the realm of the microscopic, and it is the magnitude of Planck's constant that makes this so. No one has a clear clue *why* this fundamental constant has the magnitude it has; maybe one day we will understand it on yet more fundamental grounds. But for now, we accept this small magnitude, perhaps with thanks that it is not thirty orders of magnitude larger, so that we macroscopic beings do not diffract and interfere with one another in quantum-like ways.

The phenomena we have briefly surveyed here point to the simple new relations of quantum theory, $E = h\nu$ and $\lambda = h/p$, and to the discrete nature of atomic energy transfers. Several other phenomena played important roles in the development and acceptance of quantum theory, but we will move on to the next section and begin to set out the ground rules for the full theory.

11.2 WHAT TO EXPECT FROM QUANTUM MECHANICS

How does one construct a theory of physical phenomena? What requirements must be met in order to say, "This body of equations, hypotheses, postulates, etc., constitutes a viable theory"? The Balmer expression, Eq. (11.2), is *predictive* in that it says, "given an integer, this equation will yield a possible wavelength emitted by H atoms," but it does not explain *why* that wavelength would be emitted, or what wavelengths any other atom might emit. It lacks both an *understandable basis* and *generality*. The Balmer expression and its variant, Eq. (11.3), do not constitute theories of the H atom.

Recall the basis for thermodynamics: three fundamental laws that are postulated to hold universally. They are the starting point of the theory, and they are remarkably few in number, yet they do not (yet) spring from a more fundamental source. So it will be with quantum theory.

One way of inventing a theory is to write down a set of postulates, derive consequences from them, and check with Nature to see if the consequences are observed as predicted. Without enormous creative instinct, genius, and grasp of natural facts, this approach can very quickly make one look like a crackpot, but from the hands and minds of the talented and insightful, valid new theories are born. It is thus remarkable that two different versions of quantum theory were developed in nearly complete form at nearly the same time by Erwin Schrödinger and Werner Heisenberg, in 1926 and 1925, respectively. The two versions were quickly shown to be equivalent, but we will base most of our discussion here on Schrödinger's version.

When a new theory breaks into the open, advances move rapidly. Chemists and physicists of the time were ready to exploit the theory, having pushed earlier theories to their limits in the preceding years. Fortunately, the mathematical basis of Schrödinger's theory was well known. It had much in common with other areas of mathematical physics, ranging from hydrodynamics to electromagnetic-wave theory, areas which were familiar to the physics community at large.

A few crucial tests were passed quickly. Paul Ehrenfest showed in 1927 that quantum theory could be reduced to classical mechanics (Newton's second law, in particular), if the limit $h \to 0$ was correctly applied to the theory. This was vindication of the *Correspondence Principle* first set forth in Bohr's seminal 1913 theory of the H atom. If quantum theory is universal, it must contain classical theory as a subset, or else some hidden dividing line between the microscopic and the macroscopic must exist in Nature. Max Born and Schrödinger showed that Schrödinger's and Heisenberg's mathematics, superficially so different, were physically equivalent, and Born gave Schrödinger's theory greater physical interpretation. Numerous simple physical situations were explored, many to be discussed in the next section, and approximation methods were worked out to treat more complex situations. Paul Dirac expanded the theory into the relativistic velocity range, with important new consequences. The theory was in essential agreement with experiment at every step, and within a decade, it was generally accepted and widely applied.

In quantum theory, one speaks of a *system* much as one does in thermodynamics. One inquires, via the theory, as to the *observable properties* of the system. Among these might be the system's position, or energy, or angular momentum, or ability to emit light, or undergo a chemical change. In classical mechanics, masses, forces, and initial conditions define a *trajectory*. Kick a stationary ball off a cliff with a particular force a thousand times over, and it will move through space in exactly the same way and land on the same readily calculated spot each time.

Classical mechanics is said to be *deterministic* in this respect. Not so with quantum mechanics, to the initial dismay and profound concern of many. Quantum results are often (but not always) in the form of *probability distributions*. A quantum outcome is typically phrased, "most likely this, but quite probably any of these other outcomes." This takes some getting used to, since it is a fundamental departure from classical mechanics and is directly related to the wave-like nature of matter.

Next we turn to those postulates pulled from creative air by theorists and introduce much of the mathematics of Schrödinger's form of quantum theory. These postulates (sometimes stated in different order or broken into a greater number by some authors) together with the Planck expression $E = h\nu$ and the de Broglie expression $p = h/\lambda$ will be the basis for our work in future chapters.

Postulate I. All that can be known about a system is contained in a mathematical function of the coordinates of all the particles in the system and of time. This function is called the system's *wavefunction* or *state function*.

We will symbolize the wavefunction here (and rather often in general) by Ψ and write

$$\Psi = \underbrace{\Psi(x_1, y_1, z_1, x_2, y_2, z_2, \ldots, t)}_{} \qquad (11.7)$$

$\uparrow$ Wavefunction $\qquad\qquad \uparrow$ Function of position and time

where (x_1, y_1, z_1) are the Cartesian coordinates of particle 1 of the system, and so on for particle 2, etc., and t is the time. The wavefunction must also satisfy certain other mathematical requirements to be physically acceptable:

(1) Ψ must be continuous, finite, and single valued,
(2) Ψ may, in general, be a function of complex numbers,
(3) Ψ must be quadratically integrable, and
(4) all partial first derivatives of Ψ must be continuous except possibly at a finite number of points.

In detail, these requirements mean:

(1) *continuous:* a graph of Ψ versus any one coordinate must be a smooth curve with no discontinuous jumps. Thus, ⌢⌣ is acceptable, but ⌢⌣⌢ is not due to the discontinuous jump in the graph.
finite: at all coordinate values, including infinity, Ψ must remain finite. At infinity, or at finite extremes of the coordinates if they are physically limited, Ψ must equal zero.
single-valued: at any set of coordinate values, Ψ can have one and only one value. Thus $\sin x$ is potentially a wavefunction, but $\sin^{-1} x$, which is multivalued, is not.
(2) *complex:* (not to be confused with complicated!) Ψ may be a complex number of the form $f + ig$ where f and g are real and $i = (-1)^{1/2}$, the unit imaginary number. The *complex conjugate of* Ψ, written Ψ^*, is just $f - ig$. One replaces i by $-i$ everywhere when forming a complex conjugate.

(3) *quadratically integrable:* the integral of $\Psi^*\Psi$ over all space coordinates must be a finite, real number. We will introduce the abbreviation

$$dx_1\, dy_1\, dz_1\, dx_2\, dy_2\, dz_2 \ldots = d\tau$$

and write

$$\int_{\substack{\text{all}\\\text{space}}} \Psi^*\Psi\, d\tau = \text{finite, real number}$$

and, in particular, note that we can scale Ψ in such a way that it is said to be *normalized,* which means

$$\int_{\substack{\text{all}\\\text{space}}} \Psi^*\Psi\, d\tau = 1 \; . \tag{11.8}$$

(4) *continuous first derivatives:* while $\Psi(\tau)$ is continuous, we will allow, at a finite number of points, $\partial\Psi/\partial x$ to change suddenly at such points. (Here, x is any of the coordinates in the set constituting τ.) These discontinuities will generally occur only in simple model problems that approximate natural situations.

EXAMPLE 11.4

Which of the following functions are acceptable as wavefunctions by Postulate I? In each, x is the only variable; the other symbols represent arbitrary real non-zero constants.

(a) $\Psi(x) = N \ln (x - a)$
(b) $\Psi(x) = \begin{cases} Nx(x - a) & 0 \leq x \leq a \\ 0 & x < 0 \text{ or } x > a \end{cases}$
(c) $\Psi(x) = N \cos(ax) e^{bx^2}$
(d) $\Psi(x) = N (x^2 - a) e^{ibx + cx^2}$

SOLUTION Note that each function has a constant factor N. This is typical of all wavefunctions with N representing a scaling factor that is usually adjusted to satisfy the normalization integral, Eq. (11.8). Now for the individual choices:

(a) as $x \to a$, $\Psi \to -\infty$, and as $x \to \infty$, $\Psi \to +\infty$, neither of which is allowed. Ψ must remain finite. Thus, unless we know $a < x < \infty$, this function is not acceptable.
(b) Here there are two regions over which Ψ takes different forms: x between 0 and a and x *not* between 0 and a. At the boundaries of these regions, Ψ is continuous: $\Psi(x = 0) = \Psi(x = a) = 0$. From 0 to a, Ψ is finite, continuous, and quadratically integrable. At $x = 0$ and $x = a$, $d\Psi/dx$ is discontinuous as these points are approached from opposite directions, but Postulate I allows a finite number of slope discontinuities. This function is acceptable.
(c) The only question with this function is its behavior at $x = \pm\infty$ (assuming x ranges this far). If b is a real constant, it must be < 0 in order to keep Ψ finite at $x = \pm\infty$; if x is restricted to a finite region, then b is not restricted.
(d) As above, the constant c must be negative (if purely real) to keep Ψ finite at $x = \pm\infty$. If purely real, b can be anything, since $e^{ibx} = \cos(bx) + i \sin(bx)$. We will see this factor in the first wavefunction of the next chapter. The binomial factor $x^2 - a$ is acceptable as long as c is real and negative. What if c and b are both purely imaginary numbers?

⟹ RELATED PROBLEM 11.12

Postulate II. The wavefunction itself has the following physical interpretation: $\Psi^*(\tau, t)\Psi(\tau, t)\, d\tau$ is proportional to the *probability* that the system will be found with coordinates in the range $d\tau$ about $\tau = (x_1, y_1, z_1, x_2, y_2, z_2, \ldots)$ at time t.

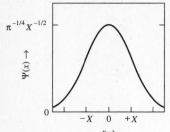

FIGURE 11.2 This is the wavefunction for the lowest energy state of a one-dimensional harmonic oscillator. Note that ψ has the dimension of $1/(\text{length})^{-1/2}$.

Max Born recognized this interpretation after Schrödinger had first set out his theory. We can expand the postulate in these ways:

(1) $\int_{\tau_1}^{\tau_2} \Psi^*\Psi \, d\tau$ is proportional to the probability of finding the system between τ_1 and τ_2,

(2) if Ψ is normalized, then this integral *is* the probability, since, if normalized,

$$\int_{\substack{\text{all}\\\text{space}}} \Psi^*\Psi \, d\tau = 1$$

which means the system is definitely *somewhere,* at any time, and

(3) the *dimensions* of Ψ are those that make the normalization integral, Eq. (11.8), dimensionless.

At this point it helps to pull a valid wavefunction out of the next chapter and explore it. We will use one of the wavefunctions that describes the motion of a particle in a one-dimensional harmonic oscillator, such as a mass vibrating at the end of a massless spring attached to a wall:

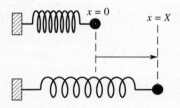

The spatial part of the wavefunction is

$$\Psi(x) = \frac{1}{\sqrt{X\sqrt{\pi}}} \exp(-x^2/2X^2) \qquad (11.9)$$

for motion along the x axis with $x = 0$ the oscillator's classical rest point. Classically, the mass vibrates symmetrically about $x = 0$. The constant X is the maximum excursion from $x = 0$, the so-called *classical turning point*.

This wavefunction is graphed in Figure 11.2, where the smooth nature of the curve shows that it is continuous. Note that it is also single-valued and finite, and that Ψ has dimensions of $(\text{length})^{-1/2}$. We graph $\Psi^*\Psi$ (or just Ψ^2 here, since Ψ is real) in Figure 11.3. Since Ψ^2 is a maximum at $x = 0$, this must be the *most probable position* of the mass. Similarly, the *area* under Ψ^2 between any two points is the probability of finding the mass between those points. Since Ψ^2 is symmetric about $x = 0$,

$$\int_{-\infty}^{0} \Psi^2(x) \, dx = \int_{0}^{\infty} \Psi^2(x) \, dx = \frac{1}{X\sqrt{\pi}} \int_{0}^{\infty} e^{-x^2/X^2} \, dx = \frac{1}{2}$$

and thus[7] the mass is as likely to be found to the right of $x = 0$ ($0 \leq x \leq \infty$) as it is to be found to the left ($-\infty \leq x \leq 0$), as expected. Both regions have probabilities of 1/2. The expression $\Psi^*\Psi$ is thus a *probability density:* a probability per unit dimensionality.

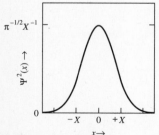

FIGURE 11.3 The square of a wavefunction is a probability distribution. This distribution for the wavefunction of Figure 11.2 peaks at $x = 0$, which is therefore the most probable position of the oscillator. Since ψ is normalized, the total area under this curve equals 1.

---

[7]We use here $\int_{0}^{\infty} \exp(-z^2) \, dz = \sqrt{\pi}/2$. Don't be alarmed if integrals crop up that are not familiar to you. Concentrate on the result of the integration now, and take heart that we will frequently be able to evaluate integrals "by intuition" using a combination of Postulate requirements and physical arguments.

EXAMPLE 11.5

Another wavefunction for the harmonic oscillator can be written

$$\psi(q) = \frac{(2q^3 - 3q)e^{-q^2/2}}{\sqrt{3\sqrt{\pi}}}$$

where q, explained in the next chapter, is a scaled, dimensionless coordinate that varies from $-\infty$ to ∞. What is the most probable value for q for this state?

SOLUTION There are several ways to answer this question. First, note that ψ is a product of a simple cubic polynomial, $(2q^3 - 3q)$, and an exponential in q^2, $e^{-q^2/2}$. The exponential causes ψ to vanish at $q = \pm\infty$, while the polynomial gives ψ its structure. A cubic polynomial can have no more than three real roots, or, in quantum language, this ψ will have no more than three *nodes*, points where $\psi = 0$. If we graph ψ:

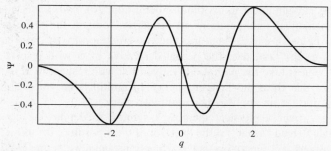

we see the nodes at $q = 0$ and at $q \cong \pm 1.2$. But we want *most probable* positions, not nodes, and thus we should focus on the maximum around $q = 2$. Since it is ψ^2 rather than ψ that gauges probability, we should graph ψ^2 as well:

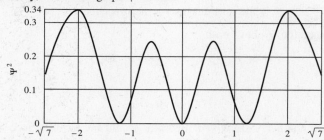

This graph shows that $q \cong \pm 2$ locates the two symmetric and equivalent maxima or most probable positions.

We can also approach this problem analytically. All maxima will be given by the roots of $d\psi^2/dq = 0$, but before attacking what promises to be a tedious calculus and algebra exercise, it is always a good idea to stop, think, and look for simplifications. Note that, in general, we seek solutions to the equation

$$\frac{d\psi^2}{dq} = \frac{d\psi^2}{d\psi}\frac{d\psi}{dq} = 2\psi\frac{d\psi}{dq} = 0 \ .$$

This may be written $\psi(d\psi/dq) = 0$, which is equivalent to $\psi = 0$ and/or $d\psi/dq = 0$. But $\psi = 0$ is just the equation for the nodes, which we know are not the most probable points. (In fact, they are the *least* probable points!) Thus we need to solve $d\psi/dq = 0$. *The most probable positions will always be extrema of the wavefunction, even though the square of the wavefunction gauges probability in general.* For this ψ,

$$\frac{d\psi}{dq} = 0 = 2q^4 - 9q^2 + 3$$

with roots corresponding to absolute maxima at

$$q = \pm \sqrt{\frac{9 + \sqrt{57}}{4}} = \pm 2.034 \ .$$

By the way, a classical oscillator in this state is most probably found at $q = \pm\sqrt{7} = 2.646$, the extreme q values of the ψ^2 graph above.

➡ *RELATED PROBLEMS* 11.13, 11.31

Postulate III. Each physically observable quantity is represented mathematically by a unique *operator* that acts on the wavefunction in a prescribed way. Physical operators must be linear and Hermitian.

Again, the mathematical terms require definition. An *operator* is a set of rules for the mathematical manipulations to be carried out on the wavefunction. An operator can be as simple as "multiply the wavefunction by x," or it can be more complicated, as in "first, take the partial derivative of ψ with respect to y, then multiply by x, next take the partial derivative of ψ with respect to x, then multiply by y, subtract this from the first product, then multiply the whole thing by $i\hbar$," which is the rule for an angular momentum operator.

We place ^ over a physical quantity's symbol to denote the corresponding operator. Thus $\hat{x}$ (read "x-hat") is the operator of the x position. It follows the simple rule "multiply by x." Operators act on functions, and we write $\hat{O}\psi$ to indicate operator $\hat{O}$ operating on function ψ. The operation just gives a new function (that may or may not be another *wavefunction*, but it is a function nonetheless):

$$(\text{Operator})(\text{function}) = \text{new function} \ .$$

A *linear* operator $\hat{O}$ obeys both

$$\hat{O}(\psi_1 + \psi_2 + \ldots) = \hat{O}\psi_1 + \hat{O}\psi_2 + \ldots$$

and

$$\hat{O}(c\psi) = c\hat{O}\psi$$

where c is any number. A *nonlinear* operator would be, for instance, "square the function" or "take the square root of the function," while linear operators, the kind of interest here, are limited to "multiply by something" or "take the derivative" sorts of operations.

An operator is said to be *Hermitian* if, for two functions ψ_1 and ψ_2

$$\underset{\underset{\substack{\text{Conjugate} \\ \text{of function 1}}}{\uparrow}}{\int \psi_1^*} \quad \underset{\underset{\substack{\text{Result of} \\ \text{operating on} \\ \text{function 2}}}{\uparrow}}{(\hat{O}\psi_2)} \quad d\tau \quad = \quad \underset{\underset{\text{Function 2}}{\uparrow}}{\int \psi_2} \quad \underset{\underset{\substack{\text{Conjugate of} \\ \text{result of} \\ \text{operating on 1}}}{\uparrow}}{(\hat{O}\psi_1)^*} \, d\tau$$

We will find that operators corresponding to physically observable quantities must be Hermitian.

A particularly important aspect of operators is that they must be applied in a stated order. When we write $\hat{A}\hat{B}\psi$ we mean "do operation $\hat{B}$ on ψ, producing a new function, and *then* do operation $\hat{A}$ on the new function" so that $\hat{A}\hat{B}\psi = \hat{A}\psi' = \psi''$ where $\psi' = \hat{B}\psi$. Operators do not operate on other operators!

We will see that profound consequences follow when two operators fail to *commute*, which means that the order of applying the operators is important. If operators $\hat{A}$ and $\hat{B}$ do *not* commute, then

$$\hat{A}\hat{B}\psi \neq \hat{B}\hat{A}\psi .$$

We will write the *commutator* of $\hat{A}$ and $\hat{B}$ as

$$[\hat{A}, \hat{B}] = \hat{A}\hat{B} - \hat{B}\hat{A}$$

and recognize that the commutator is itself an operator:

$$[\hat{A}, \hat{B}]\psi = (\hat{A}\hat{B} - \hat{B}\hat{A})\psi = \hat{A}\hat{B}\psi - \hat{B}\hat{A}\psi .$$

If $\hat{A}$ and $\hat{B}$ do not commute, this expression is not zero.

One major use of operators will be evident once we state the next postulate.

Postulate IV. If the result of $\hat{A}\psi$ is to reproduce ψ multiplied by a real constant, then that constant is the value of the physical quantity represented by $\hat{A}$ that would be measured in a system represented by ψ.

Here is another physical interpretation of ψ along with a rule for pulling information about the state of the system from ψ. We call the equation

$$\hat{A}\psi = \alpha\psi , \qquad (11.10)$$

where α is a number, an *eigenvalue relationship* (also called a *characteristic-value relationship* in English, but the German prefix *eigen* has stuck). We call α the *eigenvalue* and say that ψ is an *eigenfunction* of $\hat{A}$. Note that elementary rules of algebra do not apply to Eq. (11.10)! We *cannot* cancel the ψ's on both sides of Eq. (11.10) and write $\hat{A} = \alpha$; this statement is *meaningless*.

Since only real numbers can represent physical measurements, it is certainly reasonable to expect α to be a real number. But it is *forced* to be real by that part of Postulate III that requires physically meaningful operators to be Hermitian. *Hermitian operators have real eigenvalues* as we can now prove.

Let ϕ_n be a normalized eigenfunction of operator $\hat{A}$ with eigenvalue a_n so that

$$\underset{\text{Operator}}{\hat{A}} \quad \underset{\text{Eigenfunction}}{\phi_n} \quad = \quad \underset{\text{Eigenvalue}}{a_n} \quad \underset{\substack{\text{Same}\\\text{eigenfunction}}}{\phi_n} .$$

Next, multiply both sides from the left by ϕ_n^*, the complex conjugate of ϕ_n, and integrate over all coordinates. Call this Integral I, yielding

$$\underbrace{\int \phi_n^* \hat{A} \phi_n \, d\tau}_{\text{Integral I}} = \int \phi_n^* a_n \phi_n \, d\tau$$

$$= a_n \int \phi_n^* \phi_n \, d\tau \quad \text{(since } a_n \text{ is just a constant)}$$

$$= a_n \quad \text{(since } \phi_n \text{ is normalized).}$$

Next, take the complex conjugate of the eigenvalue relationship, $(\hat{A}\phi_n)^* = a_n^* \phi_n^*$, multiply both sides of this from the right by ϕ_n, and integrate. Call this Integral II, yielding

$$\underbrace{\int (\hat{A}\phi_n)^* \phi_n \, d\tau}_{\text{Integral II}} = \int a_n^* \phi_n^* \phi_n \, d\tau$$

$$= a_n^* \int \phi_n^* \phi_n \, d\tau$$

$$= a_n^* .$$

But since $\hat{A}$ is Hermitian,

$$\underbrace{\int \phi_n^* \hat{A} \phi_n \, d\tau}_{\text{Integral I}} = \underbrace{\int \phi_n (\hat{A}\phi_n)^* \, d\tau}_{\substack{\text{Definition of} \\ \text{Hermitian}}} = \underbrace{\int (\hat{A}\phi_n)^* \phi_n \, d\tau}_{\text{Integral II}}$$

and it follows that

$$a_n = a_n^* \ .$$

A number can equal its complex conjugate only if the number is purely real. Thus, a_n must be real, and eigenvalues of Hermitian operators are always real numbers.

EXAMPLE 11.6

Are both of the following operators potentially acceptable by Postulate III?

(a) $\hat{A} = \dfrac{d^2}{dx^2}$

(b) $\hat{B} = -i \dfrac{d^2}{dx^2}$

Is $\psi(x) = N e^{i\alpha x}$ (N and α both real constants) an eigenfunction of either operator? If so, what is/are the eigenvalue(s)?

SOLUTION We first check linearity of each operator:

Does $\hat{A}(\psi_1 + \psi_2) = \dfrac{d^2}{dx^2}(\psi_1 + \psi_2) = \dfrac{d^2\psi_1}{dx^2} + \dfrac{d^2\psi_2}{dx^2} = \hat{A}\psi_1 + \hat{A}\psi_2$? Yes.

Does $\hat{A}(c\psi) = \dfrac{d^2}{dx^2}(c\psi) = c\left(\dfrac{d^2\psi}{dx^2}\right) = c(\hat{A}\psi)$? Yes.

Does $\hat{B}(\psi_1 + \psi_2) = -i\dfrac{d^2}{dx^2}(\psi_1 + \psi_2) = -i\dfrac{d^2\psi_1}{dx^2} - i\dfrac{d^2\psi_2}{dx^2} = \hat{B}\psi_1 + \hat{B}\psi_2$? Yes.

Does $\hat{B}(c\psi) = -i\dfrac{d^2}{dx^2}(c\psi) = -ic\left(\dfrac{d^2\psi}{dx^2}\right) = c(\hat{B}\psi)$? Yes.

Thus, both operators are linear. Are they both Hermitian?

Does $\int \psi_1^*(\hat{A}\psi_2) \, dx = \int \psi_1^*\left(\dfrac{d^2\psi_2}{dx^2}\right) dx = \int \psi_2(\hat{A}\psi_1)^* \, dx = \int \psi_2 \left(\dfrac{d^2\psi_1}{dx^2}\right)^* dx$?

This is less obvious than the linearity conditions, but if we use the formula for integration by parts: $\int u \, dv = uv - \int v \, du$, we can decide. We start with

$$\int \psi_1^* \left(\dfrac{d^2\psi_2}{dx^2}\right) dx = \int \psi_1^* \dfrac{d}{dx}\left(\dfrac{d\psi_2}{dx}\right) dx = \int \psi_1^* \, d\left(\dfrac{d\psi_2}{dx}\right)$$

and apply integration by parts to the last integral with $u = \psi_1^*$ and $v = d\psi_2/dx$:

$$\int \psi_1^* \, d\left(\dfrac{d\psi_2}{dx}\right) = \psi_1^* \dfrac{d\psi_2}{dx} - \int \dfrac{d\psi_2}{dx} \, d\psi_1^* = \psi_1^* \dfrac{d\psi_2}{dx} - \int \dfrac{d\psi_2}{dx} \dfrac{d\psi_1^*}{dx} \, dx \ .$$

The term $\psi_1^* \, d\psi_2/dx$ is to be evaluated at the limits of x, since these are definite integrals even though we have written them without limits for simplicity. Since ψ must vanish at the limits, this term is zero. Therefore

$$\int \psi_1^* \left(\dfrac{d^2\psi_2}{dx^2}\right) dx = - \int \dfrac{d\psi_2}{dx} \dfrac{d\psi_1^*}{dx} \, dx \ .$$

Similarly we can show

$$\int \psi_2 \left(\frac{d^2\psi_1}{dx^2}\right)^* dx = -\int \frac{d\psi_2}{dx}\frac{d\psi_1^*}{dx} dx$$

and thus the first operator is Hermitian. Carry through the same argument on the second operator and you will see that the factor i makes this operator *non-Hermitian*. We now look for eigenvalues:

$$\hat{A}\psi(x) = \hat{A}(N e^{i\alpha x}) = N\frac{d^2 e^{i\alpha x}}{dx^2} = -\alpha^2 N e^{i\alpha x} = -\alpha^2 \psi(x),$$

which says ψ is an eigenfunction of $\hat{A}$ with eigenvalue $-\alpha^2$. Similarly,

$$\hat{B}\psi(x) = \hat{B}(N e^{i\alpha x}) = -iN\frac{d^2 e^{i\alpha x}}{dx^2} = i\alpha^2 N e^{i\alpha x} = i\alpha^2 \psi(x),$$

which says ψ is an eigenfunction of $\hat{B}$ with eigenvalue $i\alpha^2$. But since $\hat{B}$ is not Hermitian, *this eigenvalue is not purely real*, and the operator $-i\, d^2/dx^2$ (or this operator times any real constant) will not interest us.

▶ **RELATED PROBLEMS** 11.14, 11.15

We will explore an eigenvalue relationship with our test wavefunction once we state one more postulate. This one tells us how to find ψ in the first place.

Postulate V. To every physical system there corresponds a unique operator representing the total energy of the system. This operator is symbolized $\hat{H}$, and the observable physical states of the system are described by those wavefunctions that satisfy the equation

$$\hat{H}\psi = i\hbar \frac{\partial \psi}{\partial t}. \tag{11.11}$$

This is the heart of Schrödinger's form of quantum theory. Equation (11.11) is known as the *time dependent Schrödinger equation*. The operator for total energy, $\hat{H}$, is called the *Hamiltonian operator*, in honor of William Hamilton, who rewrote Newton's classical mechanics in a form that centered around the total energy. Evidently, all we have to do is write down $\hat{H}$ (which, we will see, is frequently easy to do) and then solve an operator equation to find ψ (a step that is often difficult or impossible to do exactly).

Note the explicit appearance of time in Eq. (11.11). One important physical class of systems are those that are isolated and static. They should have a probability density that is time independent. Recall that $\psi^*\psi$ tells where particles are likely to be found in a system. In a static system, this should not depend on time at all, so that ψ can depend on time only in a way that disappears in $\psi^*\psi$.

This suggests we can write ψ in what is called a *separable* form:

$$\underset{\substack{\uparrow \\ \text{Total} \\ \text{wavefunction}}}{\psi(\tau, t)} = \underset{\substack{\uparrow \\ \text{Spatial} \\ \text{part}}}{\psi(\tau)} \quad \underset{\substack{\uparrow \\ \text{Temporal} \\ \text{part}}}{\phi(t)}$$

where $\phi(t)$, the time-dependent part, obeys $\phi^*\phi = 1$ for all t. We substitute this into Eq. (11.11) (and note that $\hat{H}$ cannot involve time in an isolated system, since the total energy is a fixed, time-invariant quantity, just as it is in thermodynamics) and find

$$\hat{H}[\psi(\tau)\phi(t)] = i\hbar \frac{\partial(\psi(\tau)\phi(t))}{\partial t}$$

or, since $\hat{H}[\psi\phi] = \phi\hat{H}\psi$,

$$\phi(t)\hat{H}\psi(\tau) = \psi(\tau)i\hbar\frac{\partial\phi(t)}{\partial t} .$$

This can be rearranged to

$$\frac{\hat{H}\psi(\tau)}{\psi(\tau)} = \frac{i\hbar}{\phi(t)}\frac{\partial\phi(t)}{\partial t} ,$$

which has spatial coordinates only on the left and the time coordinate only on the right. For these two expressions to be equal for arbitrary positions or times, they must individually equal the same constant, which we will call E. Thus,

$$\hat{H}\psi(\tau) = E\psi(\tau) \tag{11.12}$$

and

$$i\hbar\frac{\partial\phi(t)}{\partial t} = E\phi(t) . \tag{11.13}$$

Note that Eq. (11.12) is an eigenvalue relationship; $\hat{H}$ is a Hermitian operator, and E, its eigenvalue, is a real constant. Equation (11.12) is the *time independent Schrödinger equation,* and E is the *total energy of the isolated system.*

Equation (11.13) has the general solution

$$\phi(t) = C e^{-iEt/\hbar} \tag{11.14}$$

where C is an arbitrary constant. Thus we can write

$$\psi(\tau, t) = \psi(\tau) e^{-iEt/\hbar}$$

with C absorbed into $\psi(\tau)$.

When this separation of ψ is possible, we say ψ is a *stationary state*. Our harmonic oscillator wavefunction, Eq. (11.9), is such a state (written without the uninteresting $\phi(t)$ factor), and we should now find its energy.

The Hamiltonian is always the sum of all the kinetic energy operators, $\hat{T}$, and all the potential energy operators, $\hat{V}$:

$$\hat{H} = \hat{T} + \hat{V} . \tag{11.15}$$

$\hat{H}$ ↑ Total Hamiltonian $\hat{T}$ ↑ Kinetic energy $\hat{V}$ ↑ Potential energy

For a one-dimensional harmonic oscillator with a mass m, we will see later that

$$\hat{T} = -\frac{\hbar^2}{2m}\frac{d^2}{dx^2}$$

and

$$\hat{V} = E\left(\frac{x}{X}\right)^2$$

where E is the total energy of the oscillator.

We seek E in Eq. (11.12), which will look like this:

$$\hat{H}\psi(x) = \left[-\frac{\hbar^2}{2m}\frac{d^2}{dx^2} + E\left(\frac{x}{X}\right)^2\right]\psi(x) = E\psi(x) .$$

The operator algebra for $\hat{H}\psi(x)$ yields, first, the kinetic energy part:

$$-\frac{\hbar^2}{2m}\frac{d^2}{dx^2}\psi(x) = -\frac{\hbar^2}{2m}\frac{1}{\sqrt{X}\sqrt{\pi}}\frac{d^2}{dx^2}(e^{-x^2/2X^2})$$

$$= -\frac{\hbar^2}{2m}\frac{1}{\sqrt{X}\sqrt{\pi}}(e^{-x^2/2X^2})\left(-1 + \frac{x^2}{X^2}\right)\left(\frac{1}{X^2}\right)$$

and, trivially, the potential energy part:

$$E\left(\frac{x}{X}\right)^2 \psi(x) = \frac{1}{\sqrt{X}\sqrt{\pi}} E\left(\frac{x}{X}\right)^2 e^{-x^2/2X^2} .$$

Adding these, equating the sum to $E\psi(x)$, and cancelling common factors yields

$$\frac{\hbar^2}{2mX^2}\left(1 - \frac{x^2}{X^2}\right) + E\left(\frac{x}{X}\right)^2 = E ,$$

which we solve for E to give

$$E = \frac{\hbar^2}{2mX^2} .$$

This is the total energy of the *eigenstate* (to use the eigen prefix in another common way) of the harmonic oscillator represented by $\psi(x)$. Note how this energy eigenvalue is cleverly coded into ψ and is pulled out of ψ by the operator algebra:

$$\hat{H}\psi(x) = \frac{\hbar^2}{2mX^2}\psi(x) = E\psi(x) .$$

We can now specify $\phi(t)$ and write $\psi(x, t)$:

$$\psi(x, t) = \psi(x) e^{-i\hbar t/2mX^2} .$$

EXAMPLE 11.7

The Hamiltonian operator for a mass m constrained to move on a circle of radius R is

$$\hat{H} = -\frac{\hbar^2}{2mR^2}\frac{d^2}{d\theta^2}$$

where θ locates the angular position of the mass on the circle. If such a system's wavefunction is

$$\psi(\theta) = (2\pi)^{-1/2} e^{i2\theta}$$

what is its total energy?

SOLUTION We operate on ψ with the Hamiltonian operator and fish the energy eigenvalue out of the result:

$$\hat{H}\psi(\theta) = -\frac{\hbar^2}{2mR^2}\frac{d^2\psi(\theta)}{d\theta^2} = -\frac{\hbar^2}{2mR^2}\frac{-4e^{i2\theta}}{\sqrt{2\pi}} = \frac{4\hbar^2}{2mR^2}\frac{e^{i2\theta}}{\sqrt{2\pi}} = \frac{4\hbar^2}{2mR^2}\psi(\theta)$$

so that $E = 4\hbar^2/2mR^2$.

➠ *RELATED PROBLEMS* 11.17, 11.18, 11.19

Suppose we seek just the potential energy of the oscillator. While the *total* energy is fixed, classical harmonic motion tells us that the potential energy is constantly changing as the mass oscillates. Will quantum mechanics say the same thing? We ask if the operator equation

$$\hat{V}\psi(x) = V\psi(x)$$

has a *constant* eigenvalue V. Evidently it cannot, since $\hat{V}\psi(x)$ is just

$$\hat{V}\psi(x) = E\left(\frac{x}{X}\right)^2 \psi(x)$$

and $E(x/X)^2$ is not a constant, since x is a variable.

Since $\psi(x)$ is not an eigenfunction for $\hat{V}$, repeated measurements of potential energy on identical systems will vary from measurement to measurement. But if we make many measurements, we could average them and predict something about future measurements. The next postulate says how to calculate this average.

Postulate VI. The average measurement of a physical observable represented by the operator $\hat{A}$ is $\int \psi^*\hat{A}\psi \, d\tau / \int \psi^*\psi \, d\tau$ or, for a normalized wavefunction, simply $\int \psi^*\hat{A}\psi \, d\tau$.

We call this average the *expectation value* and symbolize it $\langle \hat{A} \rangle$. If ψ is an eigenfunction of $\hat{A}$ with eigenvalue α, then

$$\langle \hat{A} \rangle = \underset{\underset{\text{Definition of } \langle \hat{A} \rangle}{\uparrow}}{\int \psi^*\hat{A}\psi \, d\tau} = \underset{\underset{\text{Since } \hat{A}\psi = \alpha\psi}{\uparrow}}{\int \psi^* \alpha\psi \, d\tau} = \underset{\underset{\substack{\text{Since } \alpha \text{ is} \\ \text{a constant}}}{\uparrow}}{\alpha \int \psi^*\psi \, d\tau} = \underset{\underset{\substack{\text{Since } \psi \text{ is} \\ \text{normalized}}}{\uparrow}}{\alpha} .$$

We measure α *each time*. The total energy of our oscillator falls into this category.

In contrast, our oscillator state is *not* an eigenstate of the potential energy operator, $\hat{V}$. We expect repeated measurements of potential energy to vary from measurement to measurement, but if we make many, they will have an average value

$$\langle \hat{V} \rangle = \int_{-\infty}^{\infty} \psi^*\hat{V}\psi \, dx = \frac{1}{X\sqrt{\pi}} \int_{-\infty}^{\infty} e^{-x^2/2X^2} E\left(\frac{x^2}{X^2}\right) e^{-x^2/2X^2} \, dx$$

$$= \left(\frac{E}{X^3\sqrt{\pi}}\right) \int_{-\infty}^{\infty} x^2 \, e^{-x^2/X^2} \, dx = \left(\frac{E}{X^3\sqrt{\pi}}\right) \left(\frac{X^3\sqrt{\pi}}{2}\right) = \frac{E}{2} .$$

The average potential energy of the harmonic oscillator is half the total energy.

We can easily prove $E = \langle \hat{T} \rangle + \langle \hat{V} \rangle$. First, $\langle \hat{H} \rangle = E$, since E is the eigenvalue of $\hat{H}$. Also $\langle \hat{H} \rangle = \langle \hat{T} + \hat{V} \rangle = \langle \hat{T} \rangle + \langle \hat{V} \rangle$ since finding the expectation value is a linear operation. Thus

$$E = \langle \hat{T} \rangle + \langle \hat{V} \rangle . \tag{11.16}$$

While Eq. (11.16) is true *in general*, the result that $\langle \hat{V} \rangle = \langle \hat{T} \rangle = E/2$ is specific to a harmonic oscillator. This is predicted by the *Virial Theorem*, which we will discuss later and which holds for both quantum and classical mechanics.

At the beginning of this section, we mentioned Paul Ehrenfest's reconciliation of quantum mechanics and Newtonian mechanics in 1927. Ehrenfest proved two relations. The first was

$$m\frac{d}{dt}\langle \hat{x} \rangle = \langle \hat{p}_x \rangle$$

where $\langle \hat{p}_x \rangle$ is the expectation value of the x component of linear momentum (recall the classical $m(dx/dt) = mv_x = \hat{p}_x$), and the second was

$$\frac{d}{dt}\langle \hat{p}_x \rangle = -\left\langle \frac{\partial \hat{V}}{\partial x} \right\rangle ,$$

which is Newton's second law: the time derivative of the linear momentum is the negative of the potential energy gradient. A common special case of this more general statement is $F = ma$. The classical force in the x direction is $-\partial V/\partial x$.

EXAMPLE 11.8

Show that $\langle \hat{p}_x \rangle$ does not depend on time for the harmonic-oscillator ground state.

SOLUTION Turning English into calculus, we want to show that $d\langle\hat{p}_x\rangle/dt = 0$. By Ehrenfest's theorem, this is equivalent to showing that $\langle d\hat{V}/dx\rangle = 0$. Since $\hat{V} = E(x^2/X^2)$, $d\hat{V}/dx = 2Ex/X^2$ and $\langle d\hat{V}/dx\rangle$ is thus

$$\left\langle \frac{d\hat{V}}{dx} \right\rangle = \int_{-\infty}^{\infty} \psi^*(x) \left(\frac{d\hat{V}}{dx}\right) \psi(x)\, dx = \frac{2E}{X^3\sqrt{\pi}} \int_{-\infty}^{\infty} x\, e^{-x^2/X^2}\, dx \; .$$

This integral must be zero, since the limits of the integral are symmetric about the origin, and the integrand is an odd function of the variable of integration.[8] Thus $\langle d\hat{V}/dx\rangle = d\langle\hat{p}_x\rangle/dt = 0$ and $\langle\hat{p}_x\rangle$ does not depend on time.

RELATED PROBLEMS 11.33, 11.34

We have only one more postulate left to consider. It says something about how many possible eigenstates a system may have and something about the way a system can exist in a state that is *not* an eigenstate of the Hamiltonian.

Postulate VII. The eigenfunctions of a given Hamiltonian form a complete set of orthogonal, normalizable functions.

We have already encountered the term *normalizable*. Imagine we have found all the wavefunctions for a given system. (There may be an infinite number of them, as we will see!) The nth function is normalized if

$$\underbrace{\int \psi_n^* \psi_n\, d\tau}_{\text{Same system, same wavefunctions}} = 1$$

but the nth and m^{th} functions, $n \neq m$, are said to be *orthogonal* if

$$\underbrace{\int \psi_n^* \psi_m\, d\tau}_{\substack{\text{Same system,}\\ \textit{different } \text{wavefunctions}}} = \underbrace{\int \psi_m^* \psi_n\, d\tau}_{\substack{\text{Same system,}\\ \textit{different } \text{wavefunctions}}} = 0 \; . \qquad (11.17)$$

All these functions are said to form a *complete set* if *any* physical state of the system can be represented by a general *linear combination* of the eigenfunctions of the set. This means the general state ψ can be written in terms of the eigenfunctions ψ_n, $n = 1, 2, 3, \ldots,$ as

$$\underset{\substack{\uparrow\\ \text{General}\\ \text{state of}\\ \text{system}}}{\psi} = \sum_n \underset{\substack{\uparrow\\ \text{Constant}\\ \text{coefficient}}}{c_n} \underset{\substack{\uparrow\\ \text{Specific}\\ \text{eigenfunction}}}{\psi_n} \qquad (11.18)$$

where the coefficients c_n are, in general, complex number constants. If the ψ_n's are both orthogonal and normalized ("orthonormal"), then ψ is normalized if

$$\sum_n c_n^* c_n = 1 \; . \qquad (11.19)$$

---

[8] A function is *even* if $f(x) = f(-x)$; it is *odd* if $f(x) = -f(-x)$. For example, $\cos(x)$, x^2, $\exp(x^2)$, etc. are even functions, while $\sin(x)$, x, x^3, etc. are odd. The product of an even and an odd function is an odd function, while (even) $\times$ (even) = (odd) $\times$ (odd) = even. Symmetric integrals of odd functions are always zero. Some functions, such as e^x, are neither odd nor even: $e^x \neq e^{-x} \neq -e^{-x}$.

If the system is in the nth eigenstate, then $\psi = \psi_n$, and all the c's are zero except for c_n, which is trivially equal to 1.

Suppose the system is in a state not described by a stationary eigenstate. Then any measurement of a physical observable represented by an operator for which the ψ_n's are eigenfunctions must yield one eigenvalue as a result, but which one and with what probability? We can never predict *which* one *a priori*, but we can easily show that the *probability* of finding the nth eigenvalue is $c_n^* c_n$. Let $\hat{A}$ be the operator and a_j the jth eigenvalue. Then

$$\langle \hat{A} \rangle = \int \psi^* \hat{A} \psi \, d\tau = \int \left(\sum_i c_i^* \psi_i^* \right) \hat{A} \left(\sum_j c_j \psi_j \right) d\tau$$
$$= \sum_i \sum_j c_i^* c_j \int \psi_i^* \hat{A} \psi_j \, d\tau = \sum_i \sum_j c_i^* c_j \int \psi_i^* a_j \psi_j \, d\tau$$
$$= \sum_i \sum_j c_i^* c_j a_j \int \psi_i^* \psi_j \, d\tau \; .$$

But, since the complete set of ψ_n's is orthonormal, all the integrals with $i \neq j$ vanish, and all those with $i = j$ are unity. The double sum reduces to a single sum:

$$\langle \hat{A} \rangle = \sum_i c_i^* c_i a_i \; . \tag{11.20}$$

This, together with Eq. (11.19), lets us call $c_i^* c_i$ the probability of measuring eigenvalue a_i. Note that, in general, $\langle \hat{A} \rangle$ will *not* equal any of the a_i's but that *every measurement must yield one of the a_i's*. The eigenvalues are the observable outcomes; the expectation values are averages of many, many observations.[9]

When the system has a wavefunction such as ψ in Eq. (11.18), we say it is in a *linear superposition* of eigenstates. (Linear, because ψ involves only first powers of eigenfunctions; superposition, because ψ is a simple sum of varying fractions of these component eigenfunctions.) Often, even though there may be an infinite number of eigenfunctions in the complete set for any system, only a few will sensibly superpose to yield ψ, and the number may be as small as two. For instance, an atom irradiated with light of the proper frequency has a certain probability of absorbing the light and ending in a new state of higher energy. Only two states are directly involved in this process (call their eigenfunctions ψ_1 and ψ_2) with the system evolving from ψ_1 initially to some final superposition state $c_1 \psi_1 + c_2 \psi_2$ where $c_2^* c_2$ is the probability that the transition is successful.

EXAMPLE 11.9

Repeated measurements of some property of a system are found to give either of only two values. What must the system's wavefunction be?

SOLUTION Let the operator for the property be $\hat{A}$. The system must be in a two-state mixture of eigenfunctions, since measurements yielded only two values. Thus we can write

$$\psi = c_1 \psi_1 + c_2 \psi_2 \; .$$

Call the two observed values (which must be eigenvalues of $\hat{A}$) a_1 and a_2. Averaging all the observations yields the observed expectation value $\langle \hat{A} \rangle$. But we also know that

$$\langle \hat{A} \rangle = c_1^* c_1 a_1 + c_2^* c_2 a_2 \quad \text{and} \quad c_1^* c_1 + c_2^* c_2 = 1 \; .$$

---

[9] A commonplace analogy helps here. The *average* age, $\langle \text{age} \rangle$, of all n of your friends, $\langle \text{age} \rangle = \frac{1}{n} \sum_{i=1}^{n} \text{age}_i$, is not necessarily equal to the age of any one of them, age_i.

We thus have two equations in two unknowns: $c_1^* c_1$ and $c_2^* c_2$. We cannot find the c's more precisely, only their squares, but the unknown part of the c's amounts to at most a pure imaginary-number factor that is unimportant. Solving for $c_1^* c_1$ and $c_2^* c_2$ and taking the positive square roots of the answers yields

$$\psi = \sqrt{\frac{a_2 - \langle \hat{A} \rangle}{a_2 - a_1}} \psi_1 + \sqrt{\frac{\langle \hat{A} \rangle - a_1}{a_2 - a_1}} \psi_2 \ .$$

⟹ RELATED PROBLEMS 11.21, 11.22

We have now set forth seven postulates and introduced several new terms. Review the postulates, concentrating on their main ideas:

I. All we can know is embodied in a system's wavefunction.
II. The probability of the system's configuration is given by $\psi^* \psi$.
III. Observable quantities are represented by operators.
IV. Eigenvalues give the results of measurements of corresponding operators.
V. The Schrödinger equation yields the wavefunctions.
VI. Expectation values represent the average of many measurements.
VII. The wavefunctions are sufficient to describe any state, eigenstate or not, of the system.

You should also understand the new terms and definitions. For now, much of this is necessarily rather abstract, but as you see more examples and work more problems on your own, the mathematics will become more familiar and transparent, and the physical interpretation of quantum phenomena will make more sense. We close this chapter with a section pointing out some general consequences of our postulates and the striking physical significance of them.

11.3 HOW TO THINK QUANTUM MECHANICALLY

For certain problems, it is the structure (the shape, symmetry, and extent) of the wavefunction that is of primary concern. For others, it is the pattern of eigenvalues of a particular operator. There are several general aspects to these structures and patterns, and learning to recognize them helps to hone your quantum intuition.

The first is perhaps the most famous; it is the Uncertainty Principle first recognized by Heisenberg. This principle has a clear physical interpretation, and it is often discussed in terms of the effects the *act of making a measurement* has on the outcome of the measurement itself. But it is a natural, if unexpected, *consequence of the postulates themselves,* and thus is an inherent part of quantum theory.

The Uncertainty Principle

Postulate VI explains how to calculate an expectation value:

$$\langle \hat{A} \rangle = \int \psi^* \hat{A} \psi \, d\tau \ . \tag{11.21}$$

Equally important is the *spread* of measurements about this average. Are 90% of the measurements within, say, 1% of $\langle \hat{A} \rangle$, or are they within 10%? Is percentage difference the best way to measure the spread?

One of the most common measures of the spread about a mean is the *root mean squared* (abbreviated *rms*) deviation, which is the square root of the averaged squared deviation of the measurements from their mean. The rms deviation ΔA is

$$(\Delta A)^2 = \langle (\hat{A} - \langle \hat{A} \rangle)^2 \rangle = \int \psi^* (\hat{A} - \langle \hat{A} \rangle)^2 \psi \, d\tau \ . \tag{11.22}$$

(Remember that $\langle \hat{A} \rangle$ in Eq. (11.22) is just a number.) The Uncertainty Principle states that if *any* two operators, $\hat{A}$ and $\hat{B}$ obey the commutation relation[10]

$$[\hat{A}, \hat{B}] = \frac{\hbar}{i} \tag{11.23}$$

then

$$\Delta A \, \Delta B \geq \frac{\hbar}{2} \tag{11.24}$$

must hold for simultaneous measurements of the quantities represented by $\hat{A}$ and $\hat{B}$.

If there were no operators that satisfied Eq. (11.23), then the Uncertainty Principle would be a mathematical curiosity only. But a number of important pairs of physical operators follow this commutation relation. To find which pairs these might be, it is helpful to do a little dimensional analysis. The dimensions of the product $\hat{A}\hat{B}$ (or $\hat{B}\hat{A}$) must be the same as the dimensions of Planck's constant: energy times time, or some equivalent product of quantities. In classical mechanics, the integral of energy with respect to time is called the *action*.[11]

There are several products of physical quantities that have the dimensions of action. First, of course, is the energy–time product, but we will return to this pair at the end of the discussion. Another product is the linear displacement and momentum pair, as is angular displacement and angular momentum. We will apply the operators for these quantities to many systems in the next chapter, but we can introduce them now to see the way they follow the dictates of the Uncertainty Principle. For linear displacement (or simply position), the operator is "multiply by the position coordinate." Thus

$$\hat{x} = x, \quad \hat{y} = y, \quad \hat{z} = z \,. \tag{11.25}$$

Similarly, the operator for the angular orientation of a system with respect to a fixed coordinate system is "multiply by the angle."

For linear momentum, the operators involve differentiation. Each Cartesian component is[12]

$$\hat{p}_x = \frac{\hbar}{i} \frac{\partial}{\partial x}, \quad \hat{p}_y = \frac{\hbar}{i} \frac{\partial}{\partial y}, \quad \hat{p}_z = \frac{\hbar}{i} \frac{\partial}{\partial z} \,. \tag{11.26}$$

If we restrict the system's rotation to a plane, so that only one angle specifies its orientation, the angular momentum operator $\hat{J}$ is similar to a linear-momentum component operator:

$$\hat{J} = \frac{\hbar}{i} \frac{d}{d\theta}$$

where θ is the angle.

Now we construct commutators for these operator pairs starting with linear momentum and displacement. It helps to imagine a commutator acting on some

---

[10]In general, if $[\hat{A}, \hat{B}] = \frac{\hat{C}}{i}$, then the most general form of the Uncertainty Principle is $(\Delta A)^2 (\Delta B)^2 \geq \frac{1}{4} \langle [\hat{A}, \hat{B}] \rangle^2$.

[11]Thus, h is often called the "quantum of action." We will not go into the importance of action here, but note only that it is a legitimate concept in mechanics.

[12]Note that $\frac{\hbar}{i} = \left(\frac{i}{i}\right) \frac{\hbar}{i} = -i\hbar$, and one often finds the operators written $-i\hbar \partial/\partial x$.

function that need not be specified in detail. Let $\phi(x, y, z)$ be any acceptable wavefunction. Then, for example,

$$[\hat{p}_x, \hat{x}] \phi(x, y, z) = (\hat{p}_x\hat{x} - \hat{x}\hat{p}_x)\phi = \hat{p}_x\hat{x}\phi - \hat{x}\hat{p}_x\phi$$
$$= \left(\frac{\hbar}{i}\frac{\partial}{\partial x}\right)x\phi - x\left(\frac{\hbar}{i}\frac{\partial}{\partial x}\right)\phi = \frac{\hbar}{i}\phi + x\left(\frac{\hbar}{i}\frac{\partial \phi}{\partial x}\right) - x\left(\frac{\hbar}{i}\frac{\partial \phi}{\partial x}\right)$$
$$= \frac{\hbar}{i}\phi,$$

which shows that the operator $[\hat{p}_x, \hat{x}]$ operating on a function yields $\hbar/i$ times that same function. Thus, displacement in one direction and the linear-momentum component along that direction are a pair of observables that follows the general relations Equations (11.23) and (11.24).[13]

What about, say, $\hat{x}$ and $\hat{p}_y$? Following the same argument as above shows

$$[\hat{p}_y, \hat{x}] \phi(x, y, z) = \left(\frac{\hbar}{i}\frac{\partial}{\partial y}\right)x\phi - x\left(\frac{\hbar}{i}\frac{\partial}{\partial y}\right)\phi = x\left(\frac{\hbar}{i}\frac{\partial \phi}{\partial y}\right) - x\left(\frac{\hbar}{i}\frac{\partial \phi}{\partial y}\right) = 0$$

so that spreads in displacements and linear momentum components along nonidentical directions do *not* satisfy the Uncertainty Principle inequality. You can verify that $[\hat{J}, \hat{\theta}] = \hbar/i$ so that angular momentum and angular position about the rotation axis are an Uncertainty Principle pair.

Here, then, are two pairs of important physical observables with operators that satisfy the Uncertainty Principle inequality. How do we interpret this inequality? Think first about the rotating system pictured in Figure 11.4(a). Suppose this dumb-

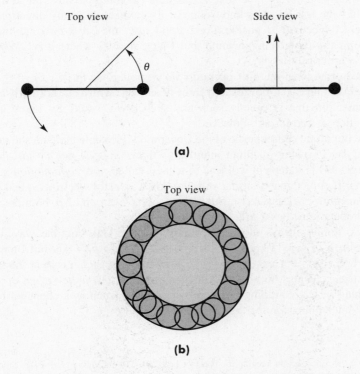

FIGURE 11.4 In classical mechanics, (a), a dumbbell constrained to rotate in a plane is located by a single angle θ that varies with time in a way governed by the angular momentum, the vector **J** oriented perpendicular to the plane of rotation. A quantum description of the same motion, shown in (b), also uses **J**, but it cannot specify θ. The motion is a blur with all values of θ equally likely at any instant.

---

[13]The sharp-eyed reader may note that, while $[\hat{p}_x, \hat{x}] = \hbar/i$, $[\hat{x}, \hat{p}_x] = -\hbar/i$, which is not exactly the form of Eq. (11.23). The sign change does not alter the Uncertainty Principle outcome of Eq. (11.24).

bell is in an eigenstate of the angular momentum. Then every measurement will yield the magnitude of the angular momentum exactly, since, for state $\psi(\theta)$,

$$\hat{J}\psi(\theta) = j\psi(\theta)$$

where j is the eigenvalue. Thus $\langle \hat{J} \rangle$ is also j, and ΔJ, the rms deviation of many measurements, is zero, since[14]

$$\begin{aligned}(\Delta J)^2 &= \int \psi^*(\theta) (\hat{J} - \langle \hat{J} \rangle)^2 \psi(\theta) \, d\theta \\ &= \int \psi^* (\hat{J}\hat{J} - \hat{J}\langle \hat{J} \rangle - \langle \hat{J} \rangle \hat{J} + \langle \hat{J} \rangle^2) \psi \, d\theta \\ &= \int \psi^* \hat{J}\hat{J} \psi \, d\theta - 2j \int \psi^* \hat{J} \psi \, d\theta + j^2 = j^2 - 2j^2 + j^2 = 0 \ .\end{aligned}$$

The Uncertainty inequality says $\Delta J \Delta \theta \geq \hbar/2$, so if $\Delta J = 0$, then $\Delta \theta$ must be *indeterminate*. If we measure the dumbbell's angular momentum and angular orientation *simultaneously*, we can find j exactly, but we will have *no* idea what θ was at the time we did the measurement. Classically, of course, we could determine both simultaneously to arbitrary precision by, for instance, making a movie of the dumbbell and analyzing the frame-by-frame motion. A quantum movie camera would show a homogeneous, 360° blur in each frame, as in Figure 11.4(b).

Since $\Delta p_x \Delta x \geq \hbar/2$ for linear motion, we can simultaneously measure the displacement from some origin and the momentum along that direction only with tradeoffs in precision. The better we measure x, the worse our simultaneous knowledge of p_x.[15] Again, classical mechanics allows us to say, "The cat was running due east with such and such a momentum when it passed me," specifying position and momentum to arbitrary precision. A quantum cat could pass me, but at that instant, I would have less and less knowledge of its momentum in my direction the more precisely I specified its position, or I could say, "the cat is now moving due east with a precise easterly momentum, but I have no idea where it is." When $\Delta p_x = 0$, Δx is arbitrarily large.

Since orthogonal pairs of operators *do* commute, as in $[\hat{p}_y, \hat{x}] = 0$, we could say the quantum cat is "now exactly three blocks east with a due northerly component of momentum that is exactly such and such." We would not know its easterly momentum at that instant, however.

One important consequence of the Uncertainty Principle that you should incorporate in your growing quantum intuition is that *things do not sit still in quantum mechanics*. This statement will lead to what is called *zero-point motion* in problems to come. If something is sitting still, you know exactly where it is, and you know its momentum is exactly zero. But we cannot have Δx and Δp_x both be zero, since their product must be $\geq \hbar/2$.

Thus, things always move. Or almost always. One possible way out of this restriction is to say, "I know it is sitting still ($\Delta p_x = p_x = 0$), but I have no idea where it is ($-\infty \leq x \leq \infty$, $\Delta x = \infty$)." We will meet such a case at the start of the next chapter. Things do not sit still *when confined to a finite region of space*. Put something in any confining potential energy field (such as a chemical bond), and Δx becomes finite, forcing Δp_x to be non-zero.

---

[14]Note the care taken in keeping the operator order clear in expanding $(\hat{J} - \langle \hat{J} \rangle)^2$. Since $\langle \hat{J} \rangle$ is just a number, the care was not really necessary here, but you should keep in mind that care is often necessary.

[15]Hence the origin of the bumper sticker, "Heisenberg may have been here," which could be expanded to add, "and I'm not sure where he was headed."

EXAMPLE 11.10

In the next chapter, we will show that the lowest energy state of a particle of mass m moving freely back and forth in a box of length L ($0 \leq x \leq L$) has the wavefunction

$$\psi(x) = \left(\frac{2}{L}\right)^{1/2} \sin\left(\frac{\pi x}{L}\right).$$

Find the $\Delta x\, \Delta p_x$ uncertainty product for this state, and verify that it does not violate the Uncertainty Principle.

SOLUTION To find Δx and Δp_x, we first find $\langle \hat{x} \rangle$ and $\langle \hat{p}_x \rangle$:

$$\langle \hat{x} \rangle = \int \psi^*(x)\hat{x}\psi(x)\,dx = \left(\frac{2}{L}\right)\int_0^L x\sin^2\left(\frac{\pi x}{L}\right)dx = \left(\frac{2}{L}\right)\left(\frac{L^2}{4}\right) = \frac{L}{2}$$

(so that the particle is in the middle of the box, on average) and

$$\langle \hat{p}_x \rangle = \frac{\hbar}{i}\left(\frac{2}{L}\right)\int_0^L \sin\left(\frac{\pi x}{L}\right)\frac{d}{dx}\sin\left(\frac{\pi x}{L}\right)dx = \frac{2\pi\hbar}{iL^2}\int_0^L \sin\left(\frac{\pi x}{L}\right)\cos\left(\frac{\pi x}{L}\right)dx = 0$$

(so that on average the momentum is zero; the particle is moving to the left as often as it is moving to the right). Next we find $(\Delta x)^2$ from Eq. (11.22):

$$(\Delta x)^2 = \langle(\hat{x} - \langle\hat{x}\rangle)^2\rangle = \left\langle\left(x - \frac{L}{2}\right)^2\right\rangle = \left\langle x^2 - 2x\frac{L}{2} + \frac{L^2}{4}\right\rangle$$

$$= \langle x^2 \rangle - 2\langle x\rangle\frac{L}{2} + \frac{L^2}{4} = \langle x^2\rangle - \frac{L^2}{4}$$

and with

$$\langle x^2 \rangle = \left(\frac{2}{L}\right)\int_0^L x^2\sin^2\left(\frac{\pi x}{L}\right)dx = \frac{L^2}{4}\left(\frac{4}{3} - \frac{2}{\pi^2}\right)$$

we have

$$(\Delta x)^2 = \frac{L^2}{4\pi^2}\left(\frac{\pi^2}{3} - 2\right).$$

Similarly, we can find[16]

$$(\Delta p_x)^2 = \langle(\hat{p}_x - \langle\hat{p}_x\rangle)^2\rangle = \langle\hat{p}_x^2\rangle = \left\langle -\hbar^2\frac{d^2}{dx^2}\right\rangle = \frac{2\hbar^2\pi^2}{L^3}\int_0^L \sin^2\left(\frac{\pi x}{L}\right)dx = \frac{\hbar^2\pi^2}{L^2}.$$

so that

$$\Delta x\, \Delta p_x = \left(\frac{\hbar\pi}{L}\right)\left[\frac{L}{2\pi}\left(\frac{\pi^2}{3} - 2\right)^{1/2}\right] = \frac{\hbar}{2}\left(\frac{\pi^2}{3} - 2\right)^{1/2} = 1.136\frac{\hbar}{2} > \frac{\hbar}{2}$$

in accord with the Uncertainty Principle.

⟹ **RELATED PROBLEMS** 11.23, 11.24, 11.32

---

[16] Note the way the square of an operator is written: $\hat{p}_x^2 = \hat{p}_x\hat{p}_x = [(\hbar/i)(\partial/\partial x)][(\hbar/i)(\partial/\partial x)] = -\hbar^2(\partial^2/\partial x^2)$.

What about energy and time? Can we say

$$\Delta E \, \Delta t \geq \frac{\hbar}{2} \;?$$

Yes, and no. This relation has physical meaning, and it impacts spectroscopy, as we will see in a later chapter. But, no, it does not come from the Postulates in the same way $\Delta x \Delta p_x$ does. There are no truly time-dependent forces in nature, no true time-dependent Hamiltonian operators.[17] Rather, time is a *parameter,* and we should interpret Δt here as a *time interval* during which we observe the system's energy. After all, the heart of the Uncertainty Principle is the result of a pair of measurements made *simultaneously in time.*

Discrete Energy Levels

We now ask how our postulates can predict that some systems must have discrete energies only. How does energy quantization arise from the wavefunctions that solve Schrödinger's equation?

For simplicity, we will imagine a system with only one direction of motion, *a one-dimensional (1-D) system.* (These systems are explored in detail at the beginning of the next chapter.) We have already encountered a 1-D kinetic-energy operator for the harmonic oscillator. In general, such an operator has the form

$$\hat{T} = -\frac{\hbar^2}{2} \frac{\text{(second derivative with respect to the dimension)}}{\text{(a characteristic positive parameter)}}$$

where the parameter is related to the dimension in the following way:

Dimension	Parameter	System
net displacement (x)	total mass (m)	Uniform 1-D translation
relative displacement ($\|x_1 - x_2\|$)	reduced mass (μ) $\mu = m_1 m_2/(m_1 + m_2)$	Relative two-body motion
angular displacement (θ)	moment of inertia (I) $I = \mu R^2$	Two-body rotation

The reduced mass and moment of inertia are explained in the next chapter, but for now, only the general *form* of $\hat{T}$ is important. It is a "take the second derivative" operator.

The parameter in $\hat{T}$ distinguishes one class of system from another, but $\hat{V}$, the potential energy operator, distinguishes *specific* systems of any one class. Consider the class of systems for one-dimensional linear motion. When we say "harmonic

---

[17]Suppose we could invent a time operator $\hat{t}$ that satisfied a commutation relation with the energy operator $\hat{H}$ such that $[\hat{H}, \hat{t}] = \hbar/i$. Such an operator would allow a time–energy Uncertainty relation, but Pauli showed in 1933 that such an operator *cannot exist if the energy has a minimum value,* as it always must. Time is not a dynamical variable with a corresponding operator.

oscillator," we mean $\hat{V} = Ex^2/X^2$. When we say "free particle," we mean $\hat{V} = $ a constant for all x.

Potential energy functions depend only on *powers of displacements* (including negative powers, and the power zero, as in a free particle). The operator $\hat{V}$ does not involve derivative operators, but it may involve transcendental functions of x (such as sin x) that are equivalent to an infinite power series in x.

This property of $\hat{V}$ leads to a helpful visualization. We can graph $\hat{V}$ as a function of x, while we cannot graph the $\hat{T}$ operator. In fact, numerical solutions for ψ carried out on a computer need only a *digitized* graph of $\hat{V}$, such as a table of $\hat{V}$ at closely spaced x values.

Imagine mass m moving under the influence of the potential-energy function $V(x)$ graphed in Figure 11.5. This could be the potential energy an atom feels in a chemical bond or as it moves in a solid or across a surface.

Suppose the total energy, E in the figure, is an allowed eigenvalue for the time independent Schrödinger equation

$$\hat{H}\psi(x) = \left(-\frac{\hbar^2}{2m}\frac{d^2}{dx^2} + V(x)\right)\psi(x) = E\psi(x) \ .$$

If this is rearranged to

$$\frac{d^2\psi(x)}{dx^2} = \frac{2m}{\hbar^2}[V(x) - E]\psi(x) \ ,$$

we can deduce something about the shape of the wavefunction. If $x = x_A$, then $[V(x_A) - E]$ is negative since $E > V(x_A)$. If $\psi(x_A)$ is positive at this point, the equation above tells us that $d^2\psi/dx^2$ is negative, and vice versa. Since the second derivative of a function is related to its *curvature*,[18] a negative ψ has a positive curvature, and vice versa.[19] This is a characteristic property of an *oscillatory* function, such as sine or cosine, among others. Thus, *if $E > V(x)$, $\psi(x)$ is always curving toward the $\psi = 0$ axis*.

Regions in which $E > V$ are *classically allowed*. Since the classical kinetic energy ($T = E - V$) must be *positive*, classical motion is confined exclusively to

---

[18]The correct definition of the curvature κ of a function $f(x)$ is $\kappa = \dfrac{d^2f/dx^2}{[1 + (df/dx)^2]^{3/2}}$. The second derivative alone equals κ only when the first derivative equals zero, as at a local minimum or maximum in f. However, we will say a function shaped like this: ⌢ has negative curvature and one shaped like this: ⌣ has positive curvature.

[19]Note that since $\psi^*\psi$ has a physical interpretation while ψ alone does not, multiplying ψ by -1 yields a wavefunction with the same physical significance as ψ itself.

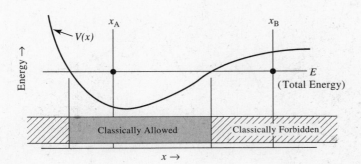

FIGURE 11.5 For an arbitrary potential energy function $V(x)$, classical motion at any one total energy E is constrained to those regions in which $V(x) \leq E$. Quantum mechanical motion, in contrast, can explore the classically forbidden regions with $V(x) > E$.

such regions. For $x = x_B$ in the figure, $E < V(x_B)$, or $E - V$ is *negative*. Such a region is *classically forbidden;* the classical system has insufficient total energy to reach x_B. But if we can find a $\psi(x)$ that is valid by our postulates at x_B, then the system *will* have some probability of being found at that point.

If $E < V$, $d^2\psi/dx^2$ must have the *same* sign as ψ: if ψ is positive, the *slope* of ψ is increasing with increasing x. This limits the general shape ψ can have that satisfy our postulates. The wavefunction must be finite at some x_B and decrease monotonically to zero as $x \to \infty$ (if the classically forbidden region extends that far) *or* finite and decrease to zero as $x \to -\infty$ (if the mass is in the left-hand classically forbidden region in Figure 11.5) *or* be some other smooth, finite, *nonoscillatory* function if the mass is in a classically forbidden region of finite extent.

The most important of these three situations are the first two. The wavefunction must decrease to zero as $x \to \pm\infty$ through a classically forbidden region extending to $\pm\infty$. If ψ remained finite at infinity, the normalization integral would not be finite, as it must to ensure finite probabilities.

Of course, ψ might simply be zero *everywhere* in *all* classically forbidden regions. This would satisfy our postulates and agree with classical mechanics. We will find, however, that ψ is finite through all naturally occurring classically forbidden regions. *The system can be in regions that are impossible to reach classically.* This behavior is called *tunneling,* a term that suggests the mass burrowing a tunnel into a potential energy that is impenetrable classically. Whether or not we can *find* the particle in such a region with a negative kinetic energy (and whatever that may mean) are questions we postpone for now.

In Figure 11.6, we sketch a plausible shape to $\psi(x)$ for the potential energy in Figure 11.5. Focus on the classically allowed region, where $\psi(x)$ is oscillatory. The farther E is above $V(x)$, the greater the kinetic energy and, thus, the greater the momentum. From de Broglie's relationship, the greater the momentum, the *smaller* the de Broglie wavelength. This leads to another generalization about the shape of ψ: *the oscillations in ψ become more closely spaced as the kinetic energy increases.*

Note how the number of oscillations is just right in the classically allowed region to tie it smoothly into the decaying part of ψ in the tunneling regions. Suppose we raise E by a small amount. Then, the kinetic energy in the allowed region would be everywhere greater by that amount, and the oscillations would be slightly closer together. This would be acceptable in the allowed region, but the oscillations might no longer have the correct magnitude and shape at the allowed-forbidden interface to join smoothly onto a decaying portion of ψ.

If we view the Schrödinger equation as simply a differential equation in which the energy is just a constant, we can find a *mathematically* acceptable ψ for *any E.* The argument above, however, indicates that a mathematically acceptable ψ may

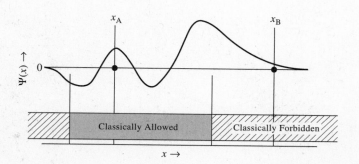

FIGURE 11.6 A schematic wavefunction for the potential of Figure 11.5 has characteristic oscillations in the classically allowed regions and monotonically decaying behavior in the classically forbidden regions.

not be *physically* acceptable. This is illustrated in Figure 11.7. The central panel is a repeat of Figure 11.2 for the harmonic oscillator wavefunction, Eq. (11.9). The surrounding panels plot the functions that satisfy the Schrödinger equation for various energies above and below the physically allowed energy, E. In the classically allowed region, shaded in each panel, the shape of ψ is fairly constant, but even the slightest variation in E has dramatic effects on ψ in the tunneling regions. For any wrong E, ψ rapidly diverges toward $\pm\infty$ as $x \to \pm\infty$.

We have plausibly argued and demonstrated an important result that can be proven rigorously. *For a system with a total energy such that the classically allowed region is of finite extent, only certain discretely spaced total energies are possible.*

The *number* of discrete energies might be infinite in some systems, but the number is, at most, *denumerably* infinite, which means we can always count them, even if we need every last one of the infinite number of integers. This is a crude way to see the origin of *quantum numbers* that appear naturally in many solutions of Schrödinger's equation. We can even see how to approach this counting if we return to Figure 11.6. That figure shows three values for x (other than $x = \pm\infty$) for which $\psi(x) = 0$. Such points are called *nodes*. The allowed wavefunction for the next *higher* energy would have *four* nodes, and that for the next *lower* energy would have only *two* nodes. And the next lower would have one node, and the next lower than that would have no nodes, and *there we must stop. The wavefunction for the lowest allowed total energy has no nodes.* Our harmonic oscillator wavefunction in Figure 11.7 is of this type. The lowest allowed total energy state is called the *ground state,* and the ground state wavefunction is uniformly all one algebraic sign. (Whether all positive or all negative is irrelevant, remember.)

As we raise the energy from the ground state value, we encounter the set of *excited states,* each having one more node than the next lowest. This general principle

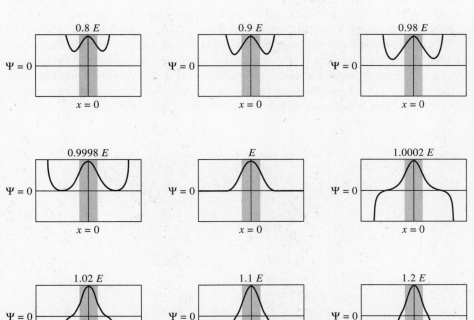

FIGURE 11.7 While the Schrödinger equation can be solved in a mathematical sense for any energy, not all such solutions are acceptable on physical grounds. Here the lowest energy state of a harmonic oscillator is given the value E, and the Schrödinger equation *mathematical* solutions for energies close to E are shown. Within the classically allowed regions (shaded), all the solutions are quite similar, but as x approaches $\pm\infty$, only the physically allowed solution stays finite, as it must.

is very important to remember, since node counting and the spatial form of the nodes in three-dimensional systems have important chemical consequences.

Continuous Energy Levels

Discrete energies appear for so-called *bound states,* a particle confined to a classically allowed region of finite extent. What happens if that region is extended? What if it has an infinite extent?

The general result, which we will not prove here, is that *increasing the extent of the bound motion decreases the separation between successive discrete energies.* If the extent is infinite, *the spacing between successive allowed energies is infinitesimal.* In this sense, at least, we recover the classical notion of a continuously variable energy.

The potential-energy function for some systems may have both discrete and continuous energy levels. One such system is the chemical bond, for which the potential-energy function has the general shape shown in Figure 11.8 (or Figure 1.5, where we first encounterd it). For total energies below the asymptotic value of V at $x = +\infty$, the system is confined to a finite region with discrete energies. Above this asymptote, the motion is unbound for arbitrarily large positive x (the bond is "broken"), and the energies are continuous.

The Correspondence Principle

All the nonclassical behavior that distinguishes quantum systems must disappear in some appropriate limit in which we recover the familiar world of classical mechanics. Finding and verifying this limit (or these limits, in fact, since there is more than one way to recover classical mechanics) is our next task. Such a limit is at the heart of the *Correspondence Principle,* first used successfully and very cleverly by Bohr in his 1912 solution for the energies of the H atom.

One almost trivial limit is to imagine that $h \to 0$. Note that the generalized 1-D kinetic-energy operator has the factor $\hbar^2/$(a characteristic parameter) so that the effects of shrinking $\hbar$ could be equally well realized by increasing the value of the characteristic parameter. For simple 1-D displacement, the parameter is the system mass. Thus, *for identical $V(x)$'s, the greater m is, the more classical the system.* Stated in reverse, as m decreases, quantum effects are more important. Thus, the motion of an H atom in a chemical reaction is inherently more dependent on quantum mechanics than is the motion of a much heavier Br atom, for instance.

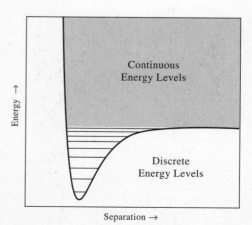

FIGURE 11.8 A potential-energy function for which motion is classically bounded for some total energies but unbounded for others has discrete energy levels for the bound motion and a continuous range of energy levels for the unbound motions.

But what does "more classical" really mean here? It means that successive energy levels are more closely spaced in a bound potential region for systems with greater mass (or greater characteristic parameter, in general). This also means that the ground-state energy is lower, and that the *zero-point energy* (which is the alternate name for the ground-state energy in bound-state problems) is also lower.

Finally, a third approach to classical mechanics, and this is the one Bohr used, considers the large quantum-number limit. For every system with only bound states *the spacing between successive energy levels is an ever decreasing fraction of the total energy in the limit of large quantum numbers*. This is automatically satisfied in continuous energy-level regions, but even for systems without a continuous region, the larger the quantum number, the more classically behaved is the system.

Degeneracy

Quantum theory uses the word *degenerate* in a way very different from its use in ordinary speech. In quantum theory, the term arises when *two or more different wavefunctions have exactly the same energy*. Such wavefunctions are said to be *degenerate*. How this arises is intimately related to the *symmetry* of the system, and in particular to the symmetry of the Hamiltonian operator itself.

Here is a simple example. Imagine an otherwise free particle confined to a perfectly cubical box. There is nothing in Nature that says, "This way is the x (or y or z) axis." We impose coordinates on a system to suit ourselves, not because the system has an inherently fixed set of coordinates. Thus, if you look at the box one way, then lie on your side or stand on your head and look at it, the box and the particle in it could not care less. The three directions parallel to the cube faces are completely independent and, *due to the symmetry of the system*, they are *equivalent*. The particle has a certain wavefunction and energy for motion back and forth along the direction we choose to call x. An equal energy must exist for motion along y and along z. Each such motion has a *unique wavefunction* (one would be a function of x; another, of y; and the third, of z), but all three have *identical energies*. That energy is *triply* (or three-fold) *degenerate*.

In general, the more highly symmetric a system, the more likely are degeneracies to arise, due to two or more equivalent motions. Therefore, the detailed classification of the symmetry of a system is a topic of importance unto itself. We will see symmetry's great role in chemical structure, reactivity, and spectroscopy in subsequent chapters.

CHAPTER 11 SUMMARY

This chapter has had some philosophy, some very abstract mathematics, and some arguments that may have stretched your intuition. We introduced Planck's constant and the simple expression for the energy of a photon:

$$E = h\nu = \hbar\omega$$

↑ ↑ ↑ ↑ ↑
Photon energy | Planck's constant | Photon frequency in Hz | $\frac{h}{2\pi}$ | Photon frequency in s^{-1}

We tied together wave and particle behavior for both light and matter with de Broglie's expression:

$$p = \frac{h}{\lambda}$$

↑ ↑
Momentum (of light or matter) | $\frac{\text{Planck's constant}}{\text{Light or matter wavelength}}$

We next set forth the postulates that define the ground rules of quantum mechanics. At the heart of these postulates is the wavefunction, a function of the coordinates of the system and of time, that is less useful by itself than when acted on in a variety of ways:

(1) $\psi^*\psi$ yields the probability of finding the system in

any given coordinate configuration: Probability = $\int_{\tau_1}^{\tau_2} \psi^*\psi \, d\tau$.

(2) Operator algebra pulls from ψ the value of a physically observable quantity characteristic of the state of the system, the eigenvalue: $\hat{A}\psi_n = a_n\psi_n$.

(3) The expectation or average value of the observable represented by $\hat{A}$ is the integral $\int \psi^*\hat{A}\psi \, d\tau = \langle\hat{A}\rangle$.

We stated conditions a valid wavefunction must obey: continuous, finite, single-valued, perhaps a complex-number function, and quadratically integrable.

Operators also had restrictions: linear and Hermitian. Hermitian operators have real-number eigenvalues and thus represent physical observables. Any two operators may or may not commute. The total energy of a system is represented by the Hamiltonian operator, $\hat{H}$, and the observable energies and static wavefunctions of the system solve the time-independent Schrödinger equation: $\hat{H}\psi = E\psi$. All of the (possibly infinite in number) solutions to the Schrödinger equation, ψ_n, together form a complete set of functions capable of representing any physical state of the system. Generally we will normalize each wavefunction and say the complete set is orthonormal: $\int \psi_n^* \psi_m \, d\tau = 1$ if $n = m$, $\int \psi_n^* \psi_m \, d\tau = 0$ if $n \neq m$.

From these postulates, a series of general consequences follows naturally. The first is the Uncertainty Principle that limits the precision of our simultaneous knowledge of certain pairs of physical observables: if $[\hat{A}, \hat{B}] = \hbar/i$, then $\Delta A \, \Delta B \geq \hbar/2$ where $\Delta A = \langle(\hat{A} - \langle\hat{A}\rangle)^2\rangle^{1/2}$.

The second set of generalities relates to the spatial structure of ψ and the separation in energies between two systems with differing wavefunctions. Systems with classically allowed regions of finite extent have some wavefunctions representing discrete energies differing by finite amounts. The state of lowest energy always represents some non-zero kinetic energy and has a wavefunction that is everywhere of the same algebraic sign (has no nodes), while successively higher energy states have successively one more node. Energies of unbound systems behave more classically in that they are continuously variable.

A quantum system in classically forbidden regions (the tunnelling phenomenon) is strikingly bizarre. It is more than a curiosity; tunnelling behavior is integral to many chemical systems.

Quantum and classical mechanics can be connected in several limits: $h \to 0$, large masses, and large quantum numbers. It is gratifying that such limits exist, and for many chemical problems, a mix of quantum and classical language is a very great simplifying tool.

Finally, we pointed out the role symmetry plays in the energy-level pattern of a system. Degenerate motions are tied to symmetry, and we will see in later chapters important ways to alter or destroy symmetry, thereby altering the energy pattern.

The next chapter applies quantum theory to simple *model systems* that are often close approximations to real systems. They are also easily solved: ψ can be found, played with, and readily interpreted. An understanding of these models is the next step toward a more detailed application of quantum theory to chemistry.

FURTHER READING

Any text with the title *Quantum Mechanics* or some variant is likely to cover the material of this chapter. Differentiating among the readable and the obtuse of these books is another matter usually settled by a scan of the Table of Contents and a few selected chapters. The historical background of quantum theory is covered in several books, such as *From X-Rays to Quarks*, by Emilio Segrè (W. H. Freeman, New York, 1980) and *The Strange Story of the Quantum*, by B. Hoffmann (Penguin, London, 1959). Somewhat more advanced discussions are in *The Conceptual Development of Quantum Mechanics*, by Max Jammer (McGraw-Hill, New York, 1966).

PRACTICE WITH EQUATIONS

11A What is the frequency of the $n = 3$ line in the Balmer series for H? (11.1), (11.2)

ANSWER: 4.567×10^{14} Hz

11B What is the smallest H emission wavelength predicted by Eq. (11.3)?

ANSWER: 91.176 nm ($n = \infty$, $m = 1$)

11C What are the energies of the light in 11A and 11B? Express them in both J and eV units. (11.4)

ANSWER: (A) 3.026×10^{-19} J = 1.888 eV;
(B) 2.179×10^{-18} J = 13.598 eV

11D An FM radio station broadcasts at a nominal frequency of 91.5 MHz. What is the energy of one photon of this frequency? If the station's power is 50 000 W, how many moles of photons per second is it emitting? (11.4)

ANSWER: 6.06×10^{-26} J; 1.37×10^6 mol s^{-1}

11E Your eye is most sensitive to green light with a wavelength $\cong$ 550 nm. What energy (in eV) must an electron have to exhibit a de Broglie wavelength of 550 nm?

What energy (also in eV) does the green photon have? (11.1), (11.4a), (11.6)

ANSWER: electron: 4.97×10^{-6} eV; photon: 2.25 eV

11F An H atom is struck by light of energy 15.6 eV and thereby ionized. What is the de Broglie wavelength of the ejected electron? 11C above, (11.6)

ANSWER: 8.67×10^{-10} m

11G What N normalizes the wavefunction $N x(x - L)$ if $0 \leq x \leq L$? (11.8)

ANSWER: $N = \sqrt{30/l^5}$

11H What is the eigenvalue of $\psi = (2\pi)^{-1/2} e^{-i4\phi}$ with respect to the operator $\hat{A} = (\hbar/i) d/d\phi$? (11.10)

ANSWER: $4\hbar$

11I The Hamiltonian for the system of 11H above is $\hat{H} = -(\hbar^2/2I)(d^2/d\phi^2)$ where I is a constant. What is the energy of the state with the wavefunction of 11H? (11.12)

ANSWER: $4^2\hbar^2/2I = 8\hbar^2/I$

11J Another wavefunction for the system of 11H above is $\psi = (2\pi)^{-1/2} e^{-i3\phi}$. Is this wavefunction orthogonal to that of 11H (ϕ varies from 0 to 2π)? (11.17)

ANSWER: Yes, since $\int_0^{2\pi} \psi_J^* \psi_H \, d\phi = (1/2\pi) \int_0^{2\pi} e^{i3\phi} e^{-i4\phi} \, d\phi = 0$

11K If the system of 11H–J is found in the state $\psi = (2\pi)^{-1/2} [(\sqrt{3})^{-1/2} e^{-i3\phi} + c\, e^{-i4\phi}]$, what value(s) for the constant c normalizes ψ? (11.18)

ANSWER: $c = \pm (2/3)^{1/2}$

11L What is $\langle \hat{A} \rangle$ for the $\hat{A}$ operator of 11H and the ψ of 11K? (11.20)

ANSWER: $(11/3)\hbar$

11M What is ΔA, the rms deviation of $\hat{A}$, for the situation of 11L? (11.22)

ANSWER: $(\sqrt{2}/3)\hbar$

11N If $\hat{B} = \phi$ and $\hat{A}$ is given in 11H, then $[\hat{A}, \hat{B}] = \hbar/i$. What is the smallest allowed ΔB for the ΔA of 11M? (11.23), (11.24)

ANSWER: $3/(2\sqrt{2})$

PROBLEMS

SECTION 11.1

11.1 Show that Eq. (11.3) can be written as

$$E = 2.1787 \times 10^{-18} \, \text{J} \left(\frac{1}{m^2} - \frac{1}{n^2} \right)$$

where E is the energy of the emitted photon. This expression will be derived in the next chapter.

11.2 For the one-electron ion He$^+$, there is an expression for emission wavelengths analogous to Eq. (11.3) for H:

$$\lambda = (22.794 \, \text{nm}) \left(\frac{n^2 m^2}{n^2 - m^2} \right) \quad m = 1, 2, 3, \ldots, \; n > m.$$

Certain lines emitted by He$^+$ happen to have the same wavelength as certain lines emitted by H. For example, the line from He$^+$ corresponding to $n = 4$ and $m = 2$ in the expression above has the same wavelength as a H line predicted from Eq. (11.3). What n and m values in Eq. (11.3) yield this wavelength? Can you find a general relation between the n and m values of these two wavelength expressions that will yield these common emissions?

11.3 A common research laser is the Nd:YAG laser. The acronym YAG stands for yttrium aluminum garnet, $Y_3Al_2O_{15}$, and in Nd:YAG, some of the Y^{3+} ions are replaced by Nd^{3+} ions, which are the active lasing species. This later emits radiation at 1064 nm in the infrared. What is the energy of each laser photon? If the laser is operated in a pulsed mode, it emits intense pulses that last only 10 ns but have a total energy that can easily be 0.5 J. How many photons are in each pulse? Note, by the way, that the peak power of this laser is an impressive $(0.5 \, \text{J}/10 \times 10^{-9} \, \text{s}) = 50$ MW. If the laser emits these pulses at a rate of only 10 per second, what is its average power? How long does it take, at this rate, for the laser to emit a mole of 1064 nm photons?

11.4 When the radioactive ^{134}Cs nucleus decays, it emits a neutrino and a high-energy electron, leaving behind a ^{134}Ba nucleus in a variety of excited states. These spontaneously emit light in the form of high-energy photons called γ rays, and analysis of these photons is a sensitive analytical probe for the presence of ^{134}Cs. One level is at an energy 604.6 keV above the ground state. What is the wavelength and frequency of the photon emitted as this level relaxes to the ground state?

11.5 The calcium ion, Ca$^+$, is a relatively abundant species in stellar atmospheres. Emission from this ion can be used to deduce information about stellar atmospheric temperatures and compositions. The ion has two closely spaced excited energy levels that are $5.004\,17 \times 10^{-19}$ and $5.048\,44 \times 10^{-19}$ J above the ground state. What wavelengths are observed in emission from each of these two excited energy levels to the ground state? What wavelength would result from emission *between* these two levels? Specify the wavelength category (i.e., visible, infrared, microwave, etc.) of each transition.

11.6 Radioastronomy can detect atomic C from emission lines originating from very low-energy excited levels of the atom. It is known that there are two excited levels above the ground state, and that emissions are observed from the middle level to the ground state and from the highest level to the middle level. Moreover, the middle level is closer in energy to the ground state than it is to the highest level. If the emissions occur at wavelengths of 0.6097 mm and 0.3690 mm, find the energies (in J units) of the two excited energy levels.

11.7 In a photoelectric effect experiment, the energy corresponding to the threshold frequency ν_0 for photoelectron emission is called the *work function*, ϕ. Of all the elements, Cs has the smallest work function. In an experiment with a clean Cs surface, a tunable laser was used to generate photoelectrons, and the maximum photoelectron energy was measured at various laser wavelengths:

λ/nm	electron energy/eV
650	0.097
600	0.256
550	0.444
500	0.670

Use these data to find a value for Planck's constant and the Cs work function.

11.8 One of the largest most nearly spherical molecules to be studied in the gas phase is buckminsterfullerene, C_{60}, which has the symmetry of a soccer ball with a diameter around 9 Å or so. A typical magnitude for the gas-phase velocity of C_{60} vaporized at ~500 K is ~130 m s^{-1}. What is the de Broglie wavelength of such a molecule? In contrast, He(g) at its normal boiling point (4.2 K) has an average velocity somewhat greater, ~160 m s^{-1}. What is this atom's de Broglie wavelength?

11.9 It is possible with microlithography techniques used in manufacturing integrated circuits to prepare a grid of lines spaced by 0.3 μm. These lines would constitute a grating from which particles could be diffracted if they had a de Broglie wavelength on the order of the line spacing. It is easy to make an atomic beam of any rare gas in which the atoms are all moving close to the speed $\sqrt{5RT/M}$ where M is the molar mass and T is the temperature of the gas reservoir. What must T be in order to see diffraction of He? What energy (in eV) must *electrons* have to diffract from this grating? Is either experiment possible?

11.10 High-energy neutrons can be "thermalized" to ordinary temperatures and used in diffraction experiments. They have an advantage over X rays in that (see Chapter 16 for details) X rays do not scatter effectively from H atoms while neutrons do. If the characteristic speed of thermal neutrons is $v = (3k_BT/m_n)^{1/2}$ where k_B is Boltzmann's constant (R/N_A) and m_n is the neutron mass, what temperature yields neutrons with a 1 Å de Broglie wavelength?

11.11 Davisson and Germer's original paper [*Physical Review* **30**, 705 (1927)] describing, in the words of the title, "Diffraction of electrons by a crystal of nickel," is a model of careful science and careful interpretation. It is also a model of honesty, opening with the sentence, "The investigation reported in this paper was begun as the result of an accident which occurred in this laboratory in April 1925." The "accident" had turned their original Ni sample from a polycrystalline mess into a crystal with large surface regions of well-defined atomic regularity. This regularity was essential to the observation of diffraction. On the Ni surface they used, the atoms have the arrangement shown below.

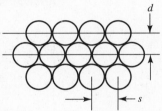

They are spaced center-to-center by a distance $s = 2.48$ Å in a hexagonal array, but it is *lines* of atoms, spaced by a distance d, that cause diffraction. As explained in greater detail in Chapter 16, diffraction in their experiment is observed as strong electron scattering at an angle θ such that $\lambda = d \sin \theta$ where λ is the electron de Broglie wavelength. Find d from the geometry of the sketch above, and compare the predicted angle from this relation to two of Davisson and Germer's observations: 54 eV electrons scatter at $\theta = 50°$ and 65 eV electrons scatter at 44°. You might find it helpful to derive first this "working formula" for electron de Broglie wavelengths: $\lambda/\text{Å} = [150.4/(E/\text{eV})]^{1/2}$. The wavelength, λ, is in Å when the electron energy, E, is in eV units.

SECTION 11.2

11.12 Which of the following are acceptable possible wavefunctions? In each, N and a are constants. For those that are acceptable, find a value for N that normalizes the function. Assume $-\infty \leq x \leq \infty$.

(a) $\psi(x) = N e^{-a|x|}$, $|x|$ = absolute value of x

(b) $\psi(x) = N |x|$

(c) $\psi(x) = N\cos(ax)\, e^{-x^2/2}$

(d) $\psi(x) = N e^{+a|x|}$

11.13 In the next chapter, we will find the normalized wavefunction $\psi(x) = (2)^{1/2} \sin(\pi x)$ where $0 \le x \le 1$. (ψ is 0 outside this range.) Find the probability that the system is: (a) anywhere in the interval $0 \le x \le 1/2$; (b) anywhere in either the interval $0 \le x \le 1/4$ *or* the interval $3/4 \le x \le 1$; (c) anywhere in the interval $1/4 \le x \le 3/4$. Think before you integrate! With some thought, this problem can be done by evaluating at most one integral! And by the way, where is the system most likely to be found?

11.14 Show that the operators "take the square root of the function," "take the reciprocal of the function," and "square the function" are not linear operators. Show that "integrate the function over all space" and "multiply the function by i and take the first derivative" are linear operators. Is the operator "add a constant to the function" linear or not?

11.15 If operator $\hat{A}$ and $\hat{B}$ are Hermitian, are the following operators also Hermitian?

(a) $\hat{C} = \hat{A}\hat{B}$

(b) $\hat{D} = \hat{A} + \hat{B}$

(c) $\hat{E} = \hat{A} + i\hat{B},\ i = \sqrt{-1}$

11.16 The Schrödinger equation for the H atom can be written in certain circumstances as

$$2\frac{dR}{dr} + r^2 \frac{d^2R}{dr^2} = \left(\frac{r}{a_0^2} - \frac{2}{a_0}\right) R$$

where $R = R(r)$ is that part of the wavefunction that depends on the electron–nucleus separation r and a_0 is a constant. Show that the function $R(r) = e^{-\alpha r}$ where α is a constant that satisfies this equation. Find the relation between α and a_0.

11.17 A system with the Hamiltonian operator $\hat{H} = -(\hbar^2/2m)(d^2/dx^2)$ is in a state with wavefunction $\psi(x) = (2/L)^{1/2} \sin(4\pi x/L)$ where L is a constant. What is the total energy of this state? What can you say about the potential energy of this system?

11.18 The time-dependent part of the wavefunction for a particular state of the hydrogen atom is

$$\phi(t) = e^{i\omega_0 t/4}$$

where ω_0 is a constant. What is this state's energy?

11.19 Classically, the angular momentum of a mass m rotating in a plane a fixed distance R about a point (as in Example 11.7) is $J = mvR$ where v is the mass's speed. Imagine a 1 g mass rotating at $R = 1$ cm such that its wavefunction is $\psi(\theta) = (2\pi)^{-1/2} e^{-i15\theta}$. Use this wavefunction and the angular momentum operator $\hat{J} = (\hbar/i)(d/d\theta)$ to find the number of revolutions the mass makes around the point every billion years. Now suppose the mass rotates once per second. Find its wavefunction, noting that the only change in the wavefunction will be the numerical constant in the exponential.

11.20 Suppose functions ψ_i and ψ_j are eigenfunctions of the Hermitian operator $\hat{A}$ with different eigenvalues a_i and a_j. Show that ψ_i and ψ_j must be orthogonal: $\int \psi_i^* \psi_j\, d\tau = 0$.

11.21 It is often the case that only two eigenstates of a system are important in a particular context. One example is absorption or emission of light. Such a *two-level system*, as it's called, helped introduce entropy in Chapter 4. If the two eigenfunctions for this system are ψ_1 and ψ_2, show how any general state of the system can be described by the expression

$$\psi = \cos\gamma\, \psi_1 + \sin\gamma\, \psi_2$$

where γ is called the *mixing angle* since it alone describes the "mixture" of eigenstates 1 and 2 that make up ψ.

11.22 The linear combination expression for ψ of Eq. (11.18) tells us how to construct an arbitrarily general ψ given a complete set of eigenfunctions ψ_n. But suppose we know ψ and want to find the c_n's that were used to make it. Prove that

$$\int \psi_n^* \psi\, d\tau = c_n.$$

For a specific example, the harmonic oscillator wavefunction

$$\psi = \frac{Nq^2}{\pi^{1/4}} e^{-q^2/2}$$

where N is a normalization constant is a linear combination of the two harmonic oscillator eigenfunctions

$$\psi_0 = \frac{1}{\pi^{1/4}} e^{-q^2/2} \quad \text{and} \quad \psi_2 = \frac{(2q^2 - 1)}{\sqrt{2}\,\pi^{1/4}} e^{-q^2/2}$$

that is, $\psi = c_0 \psi_0 + c_2 \psi_2$. Find N, c_0, and c_2. *Hints:* take advantage of $c_0^2 + c_2^2 = 1$. Consider the definite integral $\int_{-\infty}^{\infty} e^{-\alpha q^2}\, dq = \sqrt{\pi/\alpha}$ where $\alpha > 0$. Repeated differentiation of both sides of this equality with respect to α will generate all the integrals you will need for this problem. On the other hand, this problem can be worked without doing any integrals! Can you see how?

SECTION 11.3

11.23 It is often helpful to have alternate expressions for any one quantity, since in some contexts, one expression may be simpler to evaluate than another. Thus, show that the rms deviation of an operator $\hat{A}$ can be calculated by the expression $(\Delta \hat{A})^2 = \langle \hat{A}^2 \rangle - \langle \hat{A} \rangle^2$ as well as by Eq. (11.22).

11.24 Prove that if a set of functions ψ_i are eigenfunctions of two different operators $\hat{A}$ and $\hat{B}$ (i.e., $\hat{A}\psi_i = a_i \psi_i$ and $\hat{B}\psi_i = b_i \psi_i$ where a_i and b_i are corresponding eigenvalues) then operators $\hat{A}$ and $\hat{B}$ must commute. The reverse of this statement, that two commuting operators have simultaneously observable eigenvalues, is the important corollary to the Uncertainty Principle.

11.25 We stated that the position operator is simply $\hat{x} = x$ and the corresponding momentum operator is $\hat{p}_x = (\hbar/i)(\partial/\partial x)$. These are not unique choices, however. If we choose the momentum operator to be $\hat{p}_x = p_x$, (i.e., simply multiply by the x component of momentum) what is the $\hat{x}$ operator if we still require $[\hat{x}, \hat{p}_x] = \hbar/i$?

11.26 The figure below plots a one-dimensional potential-energy function (heavy line):

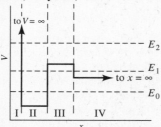

The potential is constant, but at different values, in each of four regions: in (I), it is infinite, in (II)–(IV) it is finite at different values. Region (IV) extends to $x = \infty$. Also shown are three total energies, E_0–E_2, and E_0 is the *lowest allowed energy* for the system. For each energy in each region (I)–(IV), state which of the following characteristics of the wavefunction is/are applicable:

(a) has no nodes.

(b) is zero.

(c) does not oscillate.

(d) oscillates with greatest wavelength for this total energy.

(e) oscillates with least wavelength for this energy.

(f) oscillates with an intermediate wavelength for this energy.

11.27 The diagram below plots the potential energy function V that can be found for certain chemical bonds as a function of the bond length r. It is characterized by a zero of potential energy at large r, a rapidly rising potential energy at very small r, and a minimum of depth D with a slight local maximum of height U at intermediate r. (Compare to Figures 11.8 and 1.5.)

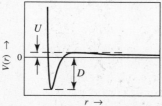

Discuss the allowed energies (discrete or continuous) and the qualitative wavefunctions (localized or delocalized) that would be expected for this potential-energy function.

GENERAL PROBLEMS

11.28 It is very useful to carry around in one's head the photon wavelength equivalents of energies characteristic of various chemical processes. Make a table of the wavelengths (order of magnitude and maybe one significant figure will do) characteristic of the following chemically interesting energies (see Chapter 7 for tables of these quantities): a chemical bond energy (3 eV), an ionization energy (10 eV), an electron affinity (1.5 eV), and "room temperature" energy (1/40 eV). Keep a copy of the table handy, and memorize it. As we discover other characteristic molecular energies, add them to the table.

11.29 The potential energy function known as the *Kratzer potential* has been suggested as a model potential to describe the bonding between two atoms. It can be written in its simplest form as

$$V(x) = \left(1 - \frac{1}{1+x}\right)^2$$

where x measures the deviation of the bond length from its equilibrium value ($x = 0$). The energy is scaled in units such that the energy difference between $x = 0$ and $x = \infty$ (the dissociated bond) is 1, and the potential looks like this when graphed:

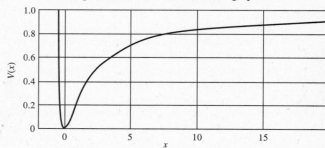

The Kratzer potential is of interest because it has more or less the shape a bond potential energy function should have, and moreover, it is a function for which the Schrödinger equation can be solved analytically. The bound state energy eigenvalues are

$$E_v = 1 - \frac{1}{v + \frac{7}{4}}$$

in these energy units where v, the quantum number, is 0, 1, 2, . . . (a) Find E_0, E_1, E_2, and E_{10}, and locate these energies on a graph of $V(x)$. Pay attention to the classical turning points (those values of x for which $V(x) = E_v$) so that your sketch resembles the discrete portion of Figure 11.8. (b) The value of v for which $E_v = 1$ gives the quantum number of the bound state of highest energy. It also represents the number of bound states (minus 1, since v starts at $v = 0$). Does the Kratzer potential have a finite or an infinite number of bound states?

11.30 If we define a variable $\xi = x/X$ and ignore the normalization constant, our example harmonic-oscillator wavefunction takes the simple form $\psi(\xi) = e^{-\xi^2}$. This is the correct shape for the lowest energy harmonic-oscillator wavefunction, but suppose we were trying to invent quantum mechanics knowing only the de Broglie hypothesis. We might argue as follows. First, a wave of constant wavelength λ has the shape $y = \cos(2\pi x/\lambda)$. The de Broglie hypothesis says $\lambda = h/p$, and $T = p^2/2m = E - V$ so that $p = \sqrt{2m(E - V)}$. For the harmonic oscillator, $V = E\xi^2$ so that we can express p as a function of ξ and thus express λ as a function of ξ as well. Find $\lambda(\xi)$, remembering that the total energy of our oscillator is $E = \hbar^2/2mX^2$, and substitute the result into $y = \cos(2\pi x/\lambda)$. This $y(\xi)$ might be a suitable wavefunction based on the idea of a "local" de Broglie wavelength that varies with position according to the dictates of simple classical mechanics. Plot this $y(\xi)$ along with $\psi(\xi)$, and comment as to how well our theory works.

11.31 What is the probability that the system described by the wavefunction in Example 11.5 will be found somewhere between the most probable classical positions, $q = \pm\sqrt{7}$? The easy part to this problem is writing the integral for this probability based on Postulate II; the hard part is evaluating the integral, since it cannot be done analytically. A graphical approximation to the integral may suffice, and the graph of ψ^2 shown in that example is drawn with this problem in mind. Alternately, the integral can be done by a numerical approximate method. If you have access to a computer capable of such a calculation, try it and compare your result to the answer: probability $\cong 0.9145$.

11.32 Let there be a complete set of eigenfunctions ϕ_i for the operator $\hat{A}$ such that $\hat{A}\phi_i = a_i\phi_i$, and write in accord with Postulate VII and Eq. (11.18)

$$\psi = \sum_i c_i\phi_i .$$

Show that the rms uncertainty of measuring A on system ψ is given by

$$(\Delta A)^2 = \sum_i |c_i|^2(a_i - \langle \hat{A} \rangle)^2 .$$

11.33 Discrete energy levels come from systems confined to a finite region of space. They are said to be "bound" to that region, and one thus speaks of "bound states." For bound states, $\langle \hat{p} \rangle$, the average momentum, is zero: the system is as likely to be moving in one direction as it is to be moving in the opposite direction. To prove this, let $\phi(x)$ represent the normalized, real, one-dimensional bound state wavefunction. The operator is $\hat{p} = (\hbar/i) \, d/dx$, and thus we want to show

$$\langle \hat{p} \rangle = \int \phi^* \left(\frac{\hbar}{i}\frac{d}{dx}\right) \phi \, dx = 0 .$$

Use the fact that ϕ is real and must vanish at the boundaries of the system (which are the limits of the integral) to complete the proof.

11.34 Evaluate $\langle \hat{x}^2 \rangle$ for the harmonic oscillator wavefunction of Eq. (11.9) first by using the definition of an expectation value (Eq. (11.21)) and then more simply by invoking the harmonic oscillator Virial Theorem.

CHAPTER 12

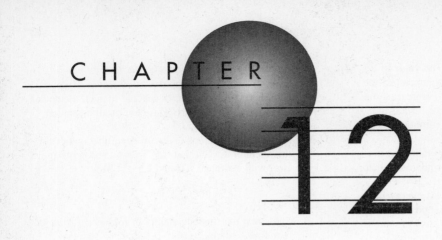

Quantum Mechanical Model Systems

12.1 The One-Dimensional Free Particle

12.2 The One-Dimensional Trapped Particle

12.3 Multidimensions

12.4 Angular Momentum and the Hydrogen Atom

12.5 "It Might as Well Be Spin"

IN this chapter, we solve Schrödinger's equation for a variety of model systems. Each has a simple potential-energy function, and each is important for three reasons. First, we can find exact wavefunctions and energies for each. Second, each in turn adds a bit more complexity, until at the end of the chapter we have the wavefunctions and energies of the hydrogen atom. Finally, these systems model real chemical systems. They exhibit the rudimentary aspects of atom–atom scattering, chemical bonding, atomic and molecular internal motions, molecular energy transfer, and other chemical phenomena.

Thus, this chapter is a training ground for the last chapter's abstract theory. All of the systems here are of such general applicability to so many scenarios that their key results are worth memorizing. These are gathered in the summary at the chapter's end.

12.1 THE ONE-DIMENSIONAL FREE PARTICLE

We begin with a particle of mass m moving in free space. The potential energy is everywhere a constant (which we can take to be zero) and, in accord with Newton's classical description, the particle moves with a constant linear momentum. We align the x axis along the direction of this momentum and seek the allowed wavefunctions and energies.

No particle is truly free in this sense, but consider a gas at low density. Compared to the size of a gas molecule, any macroscopic container is nearly infinitely larger. And, as the density is lowered, the chances that any one molecule will strike another before it strikes and rebounds from the container's wall also decreases. At low enough pressure, the gas behaves as a collection of free particles, unaware of any interactions except those with the nearly infinitely distant walls. This is the ideal gas limit, and the free particle is not only the starting point for a quantum discussion of an ideal gas, but it is also the starting point for any gas-phase scattering process in which molecule–molecule collisions (so-called *binary* collisions) are rare events.

The de Broglie relation, $\lambda = h/p = h/mv$, suggests that the wavefunction waves with a periodicity in space equal to λ. When x is advanced by λ, ψ should go through one cycle. Thus, we guess that ψ has the form

$$\psi \propto \sin\left(\frac{2\pi x}{\lambda}\right)$$

and test this guess in the Schrödinger equation.

The Hamiltonian operator is all kinetic energy:

$$\hat{H} = \hat{T} = \frac{\hat{p}_x^2}{2m}. \quad (12.1)$$

This is a direct transcription from classical mechanics where $T = p^2/2m = mv^2/2$, the familiar kinetic energy expression.

From Eq. (11.26), $\hat{p}_x = (\hbar/i)d/dx$, and the Hamiltonian operator is

$$\hat{H} = -\frac{\hbar^2}{2m}\frac{d^2}{dx^2}. \quad (12.2)$$

Next we test ψ in the Schrödinger equation,

$$\hat{H}\psi = -\frac{\hbar^2}{2m}\frac{d^2}{dx^2}\psi = E\psi,$$

which can be rearranged to

$$\frac{d^2\psi}{dx^2} = -\frac{2mE}{\hbar^2}\psi. \quad (12.3)$$

Note that ψ could be multiplied by any constant, and that constant would cancel from Eq. (12.3). The constant is related to normalization and will be ignored for now.

Substituting $\psi = \sin(2\pi x/\lambda)$ into Eq. (12.3) yields

$$-\left(\frac{2\pi}{\lambda}\right)^2 \sin\left(\frac{2\pi x}{\lambda}\right) = -\frac{2mE}{\hbar^2}\sin\left(\frac{2\pi x}{\lambda}\right),$$

which we solve for E:

$$E = \frac{(2\pi\hbar)^2}{2m\lambda^2} = \frac{h^2}{2m\lambda^2}. \quad (12.4)$$

Our guessed wavefunction survived the test. Operating with $\hat{H}$ gives ψ back again multiplied only by constants. If Eq. (12.4) for E makes sense, we can have more confidence in the wavefunction. Since $\lambda = h/p$,

$$E = \frac{h^2}{2m\lambda^2} = \frac{h^2 p^2}{2mh^2} = \frac{p^2}{2m},$$

which is the expected result since the total energy of a free particle is its kinetic energy.

The choice $\psi \propto \sin(2\pi x/\lambda)$ is not unique. An equally valid wavefunction is $\psi \propto \cos(2\pi x/\lambda)$ corresponding to a displacement of the x axis origin an arbitrary number of half wavelengths from our first choice. Thus, the most general solution is an arbitrary linear combination of sine and cosine functions, and, in particular, the linear combination

$$\cos\left(\frac{2\pi x}{\lambda}\right) + i \sin\left(\frac{2\pi x}{\lambda}\right) = e^{i2\pi x/\lambda}$$

is a solution, as is its complex conjugate, $e^{-i2\pi x/\lambda}$.

It is convenient at this point to introduce the *wavevector*, k, which is the linear momentum in units of $\hbar$:

$$\underset{\text{Wavevector}}{k} = \underset{\frac{\text{Momentum}}{\hbar}}{\frac{p}{\hbar}} \quad \text{or} \quad \hbar k = p. \tag{12.5}$$

(If the vector linear momentum **p** is divided by $\hbar$, a vector quantity **k** results.) With this definition, the relation between k and λ is[1]

$$k = \frac{2\pi}{\lambda}$$

so that our more general wavefunction is

$$\psi \propto e^{ikx} + e^{-ikx} \tag{12.6}$$

and the energy $p^2/2m$ is

$$E = \frac{\hbar^2 k^2}{2m}. \tag{12.7}$$

We can pull one more physical interpretation from Eq. (12.6) before we consider probability and normalization. Why is Eq. (12.6) the sum of two pieces? If we operate on the first piece with just the momentum operator,

$$\hat{p}_x e^{ikx} = \frac{\hbar}{i} \frac{d}{dx} e^{ikx} = \hbar k\, e^{ikx}$$

we find an eigenvalue, $\hbar k$, which is just the momentum. Or, more precisely, it is the momentum for a particle moving towards the positive end of the x axis, towards $+\infty$. If we operate on the second piece,

$$\hat{p}_x e^{-ikx} = \frac{\hbar}{i} \frac{d}{dx} e^{-ikx} = -\hbar k\, e^{-ikx},$$

---

[1] Note that k has units of inverse length.

we find the eigenvalue $-\hbar k$, which is the momentum for a particle moving the other way, towards $x = -\infty$. Thus, the most general wavefunction specifies the *magnitude of the momentum*, but *allows the direction of the momentum to be in either sense*.

This is as it should be. The free particle moves without reference to any inherent direction, and what we choose to call the $+x$ direction could as easily and correctly be called the $-x$ direction by, for instance, someone facing us. Motion to our right would be motion to that person's left. However, if we specified an initial source of particles, by saying, for example, that a source far to our left sent particles by us moving left to right, then this constraint would make the particle less free, in that only e^{+ikx} would be valid, since the system would have a definite direction to its momentum.

If the most general free particle is moving with equal probability to the right or left, then a series of measurements of momentum for such particles should give $+\hbar k$ as often as $-\hbar k$, for an average of zero: $\langle \hat{p}_x \rangle = 0$. On the other hand, if we knew the particle *had* to be moving from left to right, then $\langle \hat{p}_x \rangle = +\hbar k$, since $\psi \propto e^{+ikx}$. This has implications on the Uncertainty Principle statement for a free particle. If we get exactly $+\hbar k$ every time we measure p_x then clearly Δp_x is zero. There is no spread to the measurements.

If Δp_x is zero, then Δx must be arbitrarily large; a certainty in the value of p_x implies a complete lack of certainty in position. You may now, rightly, ask how we could have measured p_x for a particle we could never find! The answer is in the initial condition to the system. We deliberately set up a source of particles at $-\infty$, of exact momentum moving left to right, towards $+\infty$. But in so doing we gave up our ability to locate the particle. Such a particle has an equal (and infinitesimal) probability of being anywhere on the x axis.

To find the particle, we must give up our ability to specify p_x exactly. Suppose our source emits particles with a distribution in momenta, still left to right, but not all at the same $\hbar k$ magnitude. For example, suppose one fourth of the particles moved with slightly less momentum, one fourth with slightly more, and one half with exactly $\hbar k$. Let $\delta k \ll k$ be the slight difference. Then the wavefunction would be

$$\psi \propto \left(\frac{1}{4}\right)^{1/2} e^{i(k-\delta k)x} + \left(\frac{1}{2}\right)^{1/2} e^{ikx} + \left(\frac{1}{4}\right)^{1/2} e^{i(k+\delta k)x} \quad (12.8)$$

in accord with Postulate VII, which tells us, for example, that the probability of finding momentum $\hbar(k + \delta k)$ is $\left(\frac{1}{4}\right)^{1/2}\left(\frac{1}{4}\right)^{1/2} = \frac{1}{4}$, as we stipulated. The average momentum is just

$$\langle \hat{p}_x \rangle = \frac{1}{4}\hbar(k - \delta k) + \frac{1}{2}\hbar k + \frac{1}{4}\hbar(k + \delta k) = \hbar k$$

as expected, but Δp_x is no longer zero. Rather, it is

$$\begin{aligned}
\Delta p_x &= \langle (\hat{p}_x - \langle \hat{p}_x \rangle)^2 \rangle^{1/2} = \langle (\hat{p}_x - \hbar k)^2 \rangle^{1/2} &&\text{(by definition)}\\
&= \langle (\hat{p}_x^2 - 2\hat{p}_x \hbar k + \hbar^2 k^2) \rangle^{1/2} &&\text{(expanding the square)}\\
&= [\langle \hat{p}_x^2 \rangle - 2\hbar k \langle \hat{p}_x \rangle + \hbar^2 k^2]^{1/2} &&\text{(averaging each term)}\\
&= [\langle \hat{p}_x^2 \rangle - \hbar^2 k^2]^{1/2} &&\text{(collecting terms)}\\
&= [2m\langle E \rangle - \hbar^2 k^2]^{1/2} &&\left(\text{since } \langle E \rangle = \frac{\langle p^2 \rangle}{2m}\right)
\end{aligned}$$

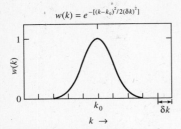

FIGURE 12.1 This function of wavevector k represents a controlled uncertainty in momentum (governed by δk) about some nominal value (k_0).

and, since $\langle E \rangle$ is one fourth the energy of the $k - \delta k$ particles plus one half the energy of the k particles plus one fourth that of the $k + \delta k$ particles, this becomes, with some algebra,

$$\Delta p_x = \frac{1}{\sqrt{2}} \hbar \delta k \ .$$

Since $\Delta x \Delta p_x \geq \hbar/2$, the particle can be located with a positional uncertainty Δx no less than $1/\sqrt{2}\delta k$.

A more realistic situation would have the momentum vary *continuously* about the most probable value with some finite spread. In analogy with the sum in Eq. (12.8) we would have

$$\psi(x) \propto \int_{-\infty}^{+\infty} w(k) e^{-ikx} \, dk \tag{12.9}$$

where $w(k)$ is a *weighting function*, in that $|w(k)|^2$ is proportional to the probability of finding momentum $\hbar k$. The effect of this smooth, continuous smear of momenta about some average value is startling. To see this effect, we use the *Gaussian* weighting function

$$w(k) = \exp\left(-\frac{(k - k_0)^2}{2(\delta k)^2}\right) \tag{12.10}$$

plotted in Figure 12.1. It peaks at $k = k_0$ and has a width related to δk. We take $\delta k = 0.1 \, k_0$ to represent a small spread in momentum about the nominal $\hbar k_0$ value.

Without $w(k)$, $\psi(x)$ would be a simple oscillatory function, $e^{ik_0 x}$, extending from $-\infty$ to $+\infty$. In Figure 12.2(a), we plot the real part of such a function for $k_0 = 10^{10}$ m^{-1}. (This is the wavevector for an electron with a kinetic energy of 3.8 eV, or of a typical gaseous He atom near the normal boiling point of He, 4.2 K.) This is the *exact momentum, unknown position* wavefunction's shape.

Next, we plot the real part of the wavefunction for Eq. (12.9) with the Gaussian weighting function of Eq. (12.10). The integral in Eq. (12.9), aside from multiplicative constants, is

$$\psi(x) \propto e^{ik_0 x} e^{-x^2(\delta k)^2/2} \ ,$$

the real part of which is plotted in Figure 12.2(b). The startling aspect of momentum uncertainty, when Figures 12.2(a) and (b) are compared, is that the particle has become *localized*. The wavefunction is appreciable in magnitude over a finite region in space; we can begin to visualize the spatial extent of a particle.

This localization is another manifestation of the Uncertainty Principle. We say the infinite wave of Figure 12.2(a) has become a *wavepacket* due to momentum uncertainty. Two final aspects of wave packet behavior deserve at least brief mention. One is the time dependence of the packet, due to the $\exp(-iEt/\hbar)$ time dependence of the wavefunction in general. One can show that the peak of the wave packet moves with a velocity $\hbar k_0/m$, which is just the expected mean velocity, but, unexpectedly, the wave packet *spreads* (becomes wider) in time.

EXAMPLE 12.1

How well does the probability distribution of the wavefunction in Eq. (12.8) express particle localization? To be specific, take $k = 10 \, \delta k$.

SOLUTION The probability distribution is proportional to $\psi^* \psi$. To simplify the math, define $\xi = kx$. With $k = 10 \, \delta k$, Eq. (12.8) becomes

$$\psi(\xi) = \frac{1}{\sqrt{2}} e^{i\xi} + \frac{1}{2}(e^{i 9\xi/10} + e^{i 11\xi/10})$$

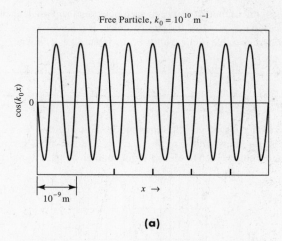

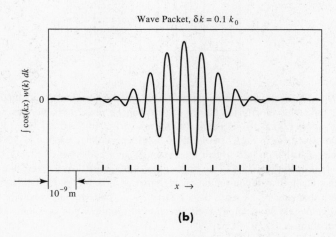

FIGURE 12.2 For a free particle of definite wavevector, (a), the wavefunction oscillates uniformly throughout space. However, a wavevector uncertainty, (b), *localizes* the particle, generating a *wavepacket*. The uncertainty function of Figure 12.1 has been used here.

and the probability distribution is (after some algebra using $e^{ix} + e^{-ix} = 2\cos x$)

$$\text{Probability} \propto 1 + \frac{\cos(\xi/5)}{2} + \frac{\cos(\xi/10)}{\sqrt{2}},$$

which looks like this:

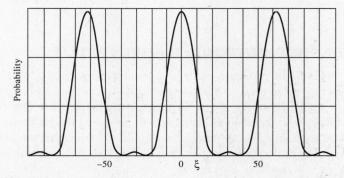

In contrast to the Gaussian weighted wave packet, this wave is spatially periodic. The particle would be most likely found at a series of equally spaced positions (the maxima in the probability distribution), but around each such position, the particle is only slightly localized. Since we found $\Delta x \geq 1/\sqrt{2}\delta k$ for this system, and since $\Delta x = \Delta \xi / k$, we have $\Delta \xi \geq 10/\sqrt{2} = 7.07$, comparable to the widths of the intense peaks in the probability distribution.

➠ *RELATED PROBLEM* 12.1

EXAMPLE 12.2

A beam of He atoms has the velocity distribution shown below. How well can any one atom be located? Compare this number to the "size" of a He atom, as given, for example, by the high-temperature limit to the He second virial coefficient. (See Table 1.1 and Example 1.8.)

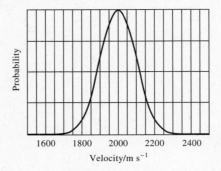

SOLUTION This distribution is an approximate Gaussian weighting function for momentum. The most probable velocity is 2000 m s^{-1} so that the most probable wavevector is $k_0 = mv/\hbar = (4 \text{ amu})(1.66 \times 10^{-27} \text{ kg amu}^{-1})(2000 \text{ m s}^{-1})/(1.054 \times 10^{-34} \text{ J s}) = 1.26 \times 10^{11} \text{ m}^{-1}$. A Gaussian written as in Eq. (12.10) falls to half its peak value when $k = k_0 \pm \sqrt{2 \ln(2)} \, \delta k$. The velocity distribution falls to half its maximum for speeds about 120 m s^{-1} on either side of the most probable velocity. This implies $\delta k \cong 6.3 \times 10^9$ m^{-1}. Thus, we can locate an atom in this beam to within a region on the order of $1/\delta k \cong 1.6 \times 10^{-10}$ m = 1.6 Å. The hard-sphere diameter of He (Example 1.8) is 2.2 Å, and we can conclude that we could locate individual He atoms along their direction of motion while simultaneously measuring their momentum.

➠ *RELATED PROBLEM* 12.2

The second aspect is the normalization condition for a free particle. If the range of motion is truly infinite in extent, then the probability of finding the particle at any point must always be infinitesimal, even if a wave packet has been created. To consider normalization in a more practical and realistic setting, we move on to the next section, in which we consider situations that confine the particle to a finite region.

12.2 THE ONE-DIMENSIONAL TRAPPED PARTICLE

The first way we will confine a particle is perhaps the simplest. We construct a potential-energy function that is zero over some distance L and arbitrarily large on and beyond either end of this range. Experimentally, such a function could be a closed container with the walls of the container providing the confining force, or, if the particle were charged, an arrangement of electric fields that trap the particle. This model problem is called the "particle-in-a-box." Here, our box is one-dimensional, but we will see three-dimensional boxes in the next section.

Figure 12.3(a) describes the 1-D box and its coordinate system. Inside the box, from $x = 0$ to $x = L$, the particle is free, just as in the previous section. Outside

12.2 THE ONE-DIMENSIONAL TRAPPED PARTICLE

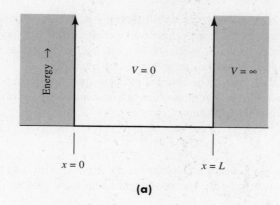

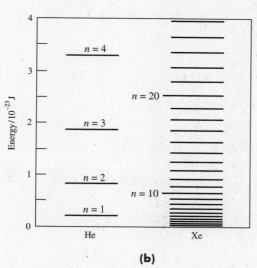

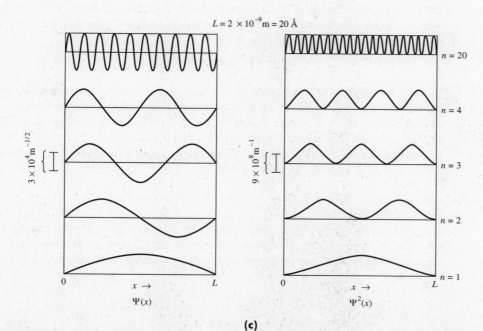

FIGURE 12.3(a) A model potential-energy function shown here is infinite over all space except over the region of length L where it is constant and finite. **(b)** For a 1-D particle-in-a-box of length 20 Å, the allowed energies of He and Xe, contrasted here, show the effect that increasing mass has on lowering the energy of any one level of a given quantum number n. **(c)** While the mass alters the energy of a particle-in-a-box, the wavefunctions, shown here with their corresponding probability distributions, depend only on the length of the box L and the quantum number n.

the box the potential energy is infinite, and the particle can never be found in these regions. A particle can never tunnel into a region of infinite potential energy. The wave function, ψ, is zero outside the box and oscillatory inside the box, and, since the particle is confined, the energy levels are quantized.

By Postulate I, ψ must be continuous. Thus, just inside the box near $x = 0$ and $x = L$, ψ must approach zero in order to join continuously with $\psi = 0$ outside the box. The oscillatory functions

$$\psi(x) = N \sin\left(\frac{n\pi x}{L}\right), \qquad n = 1, 2, 3, \ldots \qquad (12.11)$$

where N is a normalization constant satisfy these criteria, and must, therefore, be acceptable wavefunctions. The normalization constant can be found easily:

$$1 = \int_0^L \psi^2(x)\, dx = N^2 \int_0^L \sin^2\left(\frac{n\pi x}{L}\right) dx = N^2 \frac{L}{2}$$

or

$$N = \left(\frac{2}{L}\right)^{1/2}$$

so that the full wavefunction is

$$\psi(x) = \begin{cases} 0 & x < 0 \quad \text{or} \quad x > L \\ \left(\dfrac{2}{L}\right)^{1/2} \sin\left(\dfrac{n\pi x}{L}\right) & 0 \le x \le L \end{cases}.$$

The discrete values for the energy follow from associating $\sin(n\pi x/L)$ with $\sin kx$ for a free particle:

$$k = \frac{n\pi}{L}, \qquad n = 1, 2, 3 \ldots$$

$$E = \frac{p^2}{2m} = \frac{\hbar^2 k^2}{2m} = \frac{\hbar^2 \pi^2 n^2}{2mL^2}. \qquad (12.12)$$

The dimensionless integer n is our first example of a *quantum number*. This integer indexes the wavefunctions and their corresponding energies. Note that the energy varies *quadratically* with n,

$$E_n = n^2 E_1 \text{ where } E_1 = \frac{\pi^2 \hbar^2}{2mL^2},$$

and that we speak of the first, second, etc., *energy level* associated with quantum number $n = 1, 2$, etc. The lowest allowed energy is non-zero; since this energy is all kinetic, this result is also in accord with our expectation that "things don't sit still" in quantum mechanics. The energy E_1 is the *zero-point energy*.

Can we recover classical mechanics? Since the spacing between energy levels, the difference $E_{n+1} - E_n$, is *growing* with n:

$$E_{n+1} - E_n = \frac{\pi^2 \hbar^2}{2mL^2}[(n+1)^2 - n^2] = (2n+1)E_1,$$

it might seem that the system is becoming more obviously quantum mechanical in the limit of large quantum numbers, rather than more classical. But the correct measure is the *relative* rate of energy-level spacing with respect to the total energy, which is

$$\frac{E_{n+1} - E_n}{E_n} = \frac{(2n+1)}{n^2}$$

and which *does* go to zero as $n \to \infty$. This is the classical limit: an energy-level spacing that is a vanishingly small fraction of the total energy as the quantum number increases.

Note too that as the mass increases, E_n decreases. This is also in accord with classical mechanics. Figure 12.3(b) compares the first few allowed energies for He in a box 20 Å long (such as in the cavity of a molecular sieve) to those for Xe in the same size box. The mass ratio $m_{He}/m_{Xe} = 4/131.3 = 0.0305$ gives Xe dozens of allowed levels in the energy range of the first few He levels.

We can also alter the spacing if we keep m fixed and change the size of the box. As L increases, E_n and the spacing between levels decrease. These simple proportionalities:

$$E_n \propto n^2, \qquad E_n \propto \frac{1}{m}, \qquad \text{and} \qquad E_n \propto \frac{1}{L}$$

are important to remember.

Finally, we graph a few ψ_n's for He in a 20 Å box, along with corresponding ψ_n^2 probability functions, in Figure 12.3(c). Note how ψ_n has one node more than ψ_{n-1}, as we expect. To find the particle, we look for maxima in ψ_n^2. For $n = 1$, the particle is most likely found exactly in the middle of the box, but for $n = 2$ (or 4, 6, 8, etc.) the particle is *never* found *exactly* in the middle! For even values of n, $x = L/2$ is always a node. The average position for any n, however, is

$$\langle \hat{x} \rangle = \left(\frac{2}{L}\right) \int_0^L \sin\left(\frac{n\pi x}{L}\right) x \sin\left(\frac{n\pi x}{L}\right) dx = \left(\frac{2}{L}\right) \frac{L^2}{4} = \frac{L}{2},$$

which is the middle of the box. Moreover, as n increases, the spacing from maximum to maximum decreases. This means that all points are becoming more and more equally probable, which is also a classical limit.

Suppose we now alter the box somewhat. Suppose the walls have a *finite* potential energy. Just as one can put a steel ball in a paper box and expect the ball to burst through the box if the ball is given enough energy, we should expect our quantum particle to be able to escape this 1-D box. Put another way, this model problem also corresponds to a truly free particle that suddenly encounters a potential energy change *down* in energy at $x = 0$ as it moves from $x = -\infty$ to $x = +\infty$. Put yet another way, the only important parameters in this model are the *magnitude of the potential-energy change and the distance over which it changes*.

Here we expect wavefunctions of two types: those with total energy less than that needed to escape the well from $x = 0$ to L and those with total energy sufficient to roam from $x = -\infty$ to $+\infty$. The first type has discrete energies while the second is similar to continuous-energy free-particle states.

Exact expressions for ψ cannot be found as easily here as for the infinite wall box problem, but this model is worth exploring in some detail, at least qualitatively, since it models much of the phenomenology of molecular scattering, particularly certain aspects of reactive collisions. A free molecule suddenly encounters another for which it has an affinity (to speak loosely). The two together are capable of a lower potential energy than when they are separated. What can happen? Can they stick together? Can they pass without notice of one another? Can they rebound? These are the sorts of classical expectations we might have, and their quantum counterparts are worth exploring.

First, if the particle has enough energy to be free as it approaches the well, then it will *not* fall into the well or become localized in the region $x = 0$ to L but will eventually pass from one side to the other. To be trapped in the well, the particle

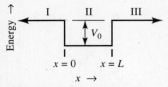

FIGURE 12.4 A simple variation on the particle-in-a-box potential-energy function has a constant, *finite* potential energy over all space except over the region of length L where it is less by the amount V_0. The solutions break naturally into the three spatial regions, I, II, and III, as shown.

must lose energy. We will not consider the ways energy could be lost here but will focus only on the motion for a constant energy.

The quantum solution breaks the problem into three spatial regions (see Figure 12.4):

- region I: $x < 0$ free particle, kinetic energy E
- region II: $0 \leq x \leq L$ free particle, kinetic energy $E + V_0$
- region III: $x > L$ free particle, kinetic energy E.

In regions I and III, k is just

$$k_\text{I} = k_\text{III} = \frac{p}{\hbar} = \frac{(2mE)^{1/2}}{\hbar}$$

while in II we have

$$k_\text{II} = \frac{[2m(E + V_0)]^{1/2}}{\hbar}.$$

But in any region, ψ is of the general form of a free particle, Eq. (12.6). We join regions again via Postulate I by requiring continuity of ψ *and* $d\psi/dx$ at $x = 0$ and $x = L$. (We need not require $d\psi/dx$ to be continuous in the particle in a box model since the particle would never be found outside the box.)

If we write the wavefunction for region I as

$$\psi_\text{I} = \underset{\substack{\uparrow \\ \text{Particle} \\ \text{moving} \rightarrow}}{e^{ik_\text{I}x}} + \underset{\substack{\uparrow \\ \text{Particle} \\ \text{moving} \leftarrow}}{B\, e^{-ik_\text{I}x}} \qquad (12.13a)$$

where B is to be determined, we can use $|B|^2$ to measure the probability *per unit incident particle moving to the right* that a reflection (a particle moving towards $x = -\infty$) occurs. This probability is *zero* in classical mechanics! Classically, the particle moves faster over the well, but it is never reflected. Whether or not $|B|^2$ is zero quantum mechanically remains to be seen.

We write, similarly,

$$\psi_\text{II} = C\, e^{ik_\text{II}x} + D\, e^{-ik_\text{II}x} \qquad (12.13b)$$

$$\psi_\text{III} = E\, e^{ik_\text{I}x} \qquad (12.13c)$$

where our choice for ψ_III means[2] that any particle that clears the well goes to $+\infty$ and never returns. There is no motion toward $-\infty$ in region III. Thus, $|E|^2$ is the probability of *transmission* across the well, and, by conservation of particles (they must all reflect or be transmitted),

$$|E|^2 + |B|^2 = 1.$$

To find B, C, D, and E, we solve the four simultaneous continuity equations:

$$\psi_\text{I}(x = 0) = \psi_\text{II}(x = 0)$$

$$\left(\frac{d\psi_\text{I}}{dx}\right)_{x=0} = \left(\frac{d\psi_\text{II}}{dx}\right)_{x=0}$$

$$\psi_\text{II}(x = L) = \psi_\text{III}(x = L)$$

$$\left(\frac{d\psi_\text{II}}{dx}\right)_{x=L} = \left(\frac{d\psi_\text{III}}{dx}\right)_{x=L}$$

---

[2]The coefficient E is *not* the energy, of course. The context should keep the two symbols straight in your mind.

These become, in order

$$1 + B = C + D$$
$$ik_I - ik_I B = ik_{II} C - ik_{II} D$$
$$Ce^{ik_{II}L} + De^{-ik_{II}L} = Ee^{ik_I L}$$
$$ik_{II} Ce^{ik_{II}L} - ik_{II} De^{-ik_{II}L} = ik_I Ee^{ik_I L}.$$

After some algebra, we could find expressions for B, C, D, and E. When the measurable quantities derived from them are graphed, some unexpected behavior arises.

Consider $|E|^2$, the probability for transmission across the potential well. We plot $|E|^2$ in Figure 12.5 for the system of a hydrogen atom traversing a step down of magnitude $V_0 = 300$ kJ mol^{-1}, which is a typical bond energy magnitude. We take the zero of energy to be that of the potential energy outside the well and graph $|E|^2$ as a function of T/V_0, the ratio of the atom's kinetic energy outside the well to the well depth. We take $L = 2.0$ Å, a typical atomic interaction length. Note first the large T/V_0 limit. At great enough T, V_0 is insignificant, and the well has no effect. The quantity $|E|^2$ is constant (at unity), just as it is classically for any energy.

At low T/V_0, however, the structure in $|E|^2$ is striking. For certain energies, the transmission is peaked at 1, but other nearby energies have reduced transmissions. In general, a transmission probability that is modulated with varying energy is said to be characteristic of *resonance* behavior. When not strongly transmitted, the particle must be reflected with enhanced probability, since $|B|^2 + |E|^2 = 1$.

The sharply rising and falling peaks characterizing the resonances in Figure 12.5 are qualitatively found in several types of experiments. Real potential-energy profiles are smoother than our potential step, and as a result, resonances are not as pronounced in reality as our model calculation predicts. One of the first observations of scattering resonances came from the study of electron transmission through rare gases. In these measurements, the scattering of low-energy electrons (~1–10 eV) by rare gases shows prominent resonances (the *Ramsauer–Townsend effect*) at characteristic energies for each rare gas. More recently, a more complicated resonance has been found in the reactive scattering of F and H_2, in which the probability of finding the HF product in a particular quantum state rises and falls as the F + H_2 collision energy is varied.

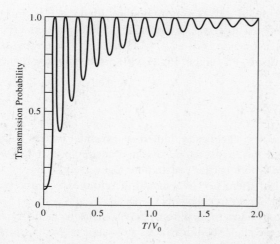

FIGURE 12.5 For an H atom traversing a potential-energy well of magnitude 300 kJ mol^{-1} and width 2 Å, the transmission probability across the well as a function of the atom's kinetic energy T shows pronounced oscillations called *resonances*.

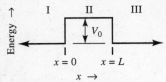

FIGURE 12.6 Another simple model potential has a constant potential-energy barrier of height V_0 and length L in an otherwise uniform potential energy.

We can look at another model potential directly related to that of Figure 12.4 by making the potential change *up* in energy instead of down. This model (Figure 12.6) can be solved just as the last model was for any total energy that exceeds the potential barrier height. As before, and again in contrast with classical mechanics, a particle with total energy greater than V_0 has a certain probability of reflection from the barrier, and resonance behavior is observed.

The more interesting case here is for an energy in Region I less than V_0. Classically, the particle bounces back from the barrier. Quantum mechanically, there is some probability of *transmission* through the barrier due to tunneling. To solve this model problem, we use ψ_I and ψ_III as before, Equations (12.13a) and (12.13c), but ψ_II is different. We write

$$k_\mathrm{II} = \frac{[2m(E - V_0)]^{1/2}}{\hbar}$$

but, since $E < V_0$, k_II involves the square root of a negative number, and hence is a pure imaginary number. Thus we define

$$\kappa_\mathrm{II} = \frac{k_\mathrm{II}}{i} = \frac{[2m(V_0 - E)]^{1/2}}{\hbar}$$

so that

$$\psi_\mathrm{II} = C\,e^{\kappa_\mathrm{II} x} + D\,e^{-\kappa_\mathrm{II} x}$$

and (compare Eq. (12.13b)) the tunneling wavefunction is exponential rather than oscillatory.

Expressions for B, C, D, and E are again obtained from continuity conditions. The final expression for $|E|^2$ can be written in terms of $\epsilon \equiv E/V_0$, the ratio of the total energy to the potential height, as

$$|E|^2 = \left(\frac{\sinh^2 \kappa L}{4\epsilon(1-\epsilon)} + 1\right)^{-1} \tag{12.14}$$

where sinh is the hyperbolic sine function

$$\sinh \kappa L \equiv \frac{e^{\kappa L} - e^{-\kappa L}}{2}$$

and

$$\kappa L = \frac{(2mV_0)^{1/2} L}{\hbar}(1-\epsilon)^{1/2}\ .$$

Two features of Eq. (12.14) are important. First, $|E|^2$ is, in general, non-zero for any ϵ, in contrast to classical mechanics, which predicts $|E|^2 = 0$ for all ϵ. Second, $|E|^2$ is not unity even at $\epsilon = 1$, and $|E|^2$ decreases monotonically for smaller ϵ. Thus, the $\epsilon \to 1$ limit of Eq. (12.14) is the *maximum* tunneling probability:

$$\lim_{\epsilon \to 1} |E|^2 = \left(\frac{1}{1+\alpha}\right)$$

where $\alpha = mV_0 L^2/2\hbar^2$.

If atomic masses, chemically important potential-barrier heights, and barrier widths are such that $\alpha \gg 1$, the maximum tunneling probability will be small, and we chemists can ignore barrier penetration. Conversely, if $\alpha \ll 1$, then tunneling will pervade all of chemistry. We should therefore explore the typical range of α values we might expect to encounter.

Potential-energy barriers arise from many sources in chemistry. Most commonly, barriers arise as molecular conformations are changed. We expect, for instance, that the barrier to cis-trans isomerization in an alkene is large, since this corresponds to rotation about a C—C double bond and such bonds are very rigid. On the other hand, rotation about the C—O bond in methyl alcohol should be less restricted, but there should still be small barriers every 120° in this rotation as the alcoholic H eclipses the methyl hydrogens. Similarly, chemical reactions that have temperature-dependent reaction rate constants usually exhibit this behavior because of a real potential-energy barrier that exists when, typically, the reactants are close together and an old bond is breaking while a new bond is forming.

These barriers have heights on the order of 5–200 kJ mol^{-1} and widths on the order of 1 Å. Per kJ mol^{-1} of barrier height and per amu of mass, for $L = 1$ Å, the parameter α is

$$\alpha = 1.24 \, (m/\text{amu})(V_0/\text{kj mol}^{-1}) \; .$$

Thus, if $m = 1$ amu and $V_0 = 5$ kJ mol^{-1}, we find $\alpha \cong 6$, which indicates a maximum tunneling probability close to 14%, which is not negligible. But if $m = 35$ amu (Cl) and $V_0 = 100$ kJ mol^{-1}, the value drops to less than 0.03%. Chemistry thus finds itself virtually free of barrier tunneling, except for the chemically important cases of hydrogen (and its isotopes) involved with low, narrow barriers.

EXAMPLE 12.3

A smoother potential energy barrier that is more realistic than a square step is $V(x) = V_0/\cosh^2(x/a)$ where V_0 and a are constants and cosh is the hyperbolic cosine function. The function looks like this:

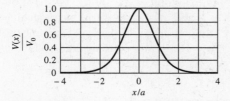

and Schrödinger's equation can be solved for it. For the reaction H + H$_2$ → H$_2$ + H, a simple reaction that will be discussed in detail in later chapters, as the H approaches H$_2$ collinearly, the potential energy changes in this general way. The peak V corresponds to the configuration H—H—H with adjacent atoms about 0.93 Å apart, V_0 is about 0.43 eV, and a is about 0.5 Å. The (difficult) solution to this problem leads to an expression for the barrier transmission probability P of the form

$$P = \frac{\sinh^2(28\sqrt{E/\text{eV}})}{\sinh^2(28\sqrt{E/\text{eV}}) + 2.08 \times 10^{15}}$$

where sinh is the hyperbolic sine function and the collision energy E is in eV units. How high does the energy have to be in order that 10% of the collisions react via tunneling?

SOLUTION If we plot the transmission probability as a function of E:

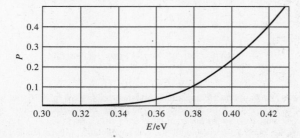

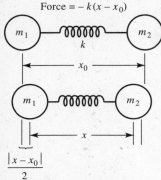

FIGURE 12.7 The linear harmonic-oscillator potential-energy function is equivalent to two masses connected by a massless spring. The force on the masses depends on the characteristic spring constant k and on the displacement of the spring from its rest position x_0.

we see that the transmission is negligibly small unless the collision energy is $\cong 0.32$ eV or greater. It is less than 10% for all $E < \cong 0.38$ eV. Consequently, we can assume tunneling is not a major pathway for this reaction except at very high collision energies. For example, at room temperature, a typical collision energy is well below 0.1 eV.

▸ *RELATED PROBLEMS* 12.3, 12.4, 12.5

To close this section, we turn to the *harmonic potential*. Ever since Hooke wrote "Ut tensio sic vis" ("As the tension, so the force")[3] in 1678, the harmonic oscillator has played a central role as a model for nearly all periodic, bound motions (excluding orbiting circular motion). Before looking at the quantum mechanics of harmonic motion, we should review the classical description.

Hooke's law involves a single constant of proportionality k, the *force constant*, relating force to displacement away from the spring's equilibrium position. If the rest length of the spring is (see Figure 12.7) x_0, Hooke says

$$\underset{\text{Force}}{F} = \underset{\substack{-\text{Force} \\ \text{constant}}}{-k} \underset{\substack{\text{Displacement} \\ \text{from rest}}}{(x - x_0)} .$$

We go from Hooke to Newton via $F = ma = m(d^2x/dt^2)$ and find the classical *trajectory* of the spring system to be

$$x(t) = X \cos \omega t + x_0$$

where X is the maximum extension of the spring. We have pulled the spring from length x_0 to length $x_0 + X$ and, at $t = 0$, we let go, watching the system vibrate. The circular frequency of vibration, ω, is related to the number of cycles of vibration per second ν by $\omega = 2\pi\nu$. The spring constant enters via

$$\omega = \sqrt{\frac{k}{\mu}} \qquad (12.15)$$

where μ is the *reduced mass* of the system

$$\frac{1}{\mu} = \frac{1}{m_1} + \frac{1}{m_2} \quad \text{or} \quad \mu = \frac{m_1 m_2}{m_1 + m_2}$$

where m_1 and m_2 are the masses on either end of the (massless) spring. (Note that if $m_1 \gg m_2$, as when we tie one end of the spring to a wall, then μ becomes simply m_2.)

The total energy of the system depends on the maximum extension:

$$E = -\int_{x_0}^{x_0 + X} F(x)\, dx = k \int_{x_0}^{x_0 + X} (x - x_0)\, dx = \frac{1}{2} k X^2 \qquad (12.16)$$

and the energy oscillates between all potential (at the *classical turning points*, $x = x_0 \pm X$) and all kinetic (at $x = x_0$):

$$E = T + V = \frac{1}{2} \mu \left(\frac{dx}{dt}\right)^2 + \frac{1}{2} k (x - x_0)^2$$
$$= \frac{1}{2} \mu X^2 \omega^2 \sin^2 \omega t + \frac{1}{2} k X^2 \cos^2 \omega t = \frac{1}{2} k X^2 .$$

---

[3] Hooke was clever, but cautious. He apparently first deduced his law in 1676, but waited until 1678 to publish it, and then only in the form of the cryptic anagram "ceiiinosssttuu." Two years later, in *De potentia restitutiva*, he unscrambled the anagram into "ut tensio sic uis" (i.e., "vis"), writing "the power of any spring is in the same proportion with the tension thereof," where Hooke, rather brilliantly, generalized "spring" to mean any "springy body."

File away into your intuition the fact that a more energetic oscillator *moves over a greater distance* but oscillates *at the same frequency* as a less energetic one with the same μ and k. Equations (12.15) and (12.16) say just that, but it is somehow easy to associate (erroneously) an increased energy in any one spring with an increased oscillation frequency.

At the turning points the oscillator is momentarily stationary, while at $x = x_0$ it is moving with greatest velocity. If we sampled the oscillator's position from time to time, at random, we could plot the probability of finding a particular extension or compression. It is inversely proportional to the oscillator's velocity, and consequently, the most probable classical position is at either turning point.

We will want to compare this to the quantum solution that we now seek. The time-independent Schrödinger equation has the kinetic energy operator

$$\hat{T} = -\frac{\hbar^2}{2\mu}\frac{d^2}{dx^2}$$

and the potential energy operator (with $x_0 = 0$)

$$\hat{V} = \frac{1}{2}kx^2 .$$

The full equation for bound, discrete energy eigenvalues E and wavefunctions $\psi(x)$ is

$$-\frac{\hbar^2}{2\mu}\frac{d^2\psi}{dx^2} + \frac{1}{2}kx^2\psi = E\psi . \tag{12.17}$$

This equation can be readily written into a form that has well-known solutions; in fact, this was one of the first problems solved by Schrödinger, who was able to exploit the extensive study of this and related second-order differential equations that was the hallmark of mathematical physics around the turn of the century. Rather than turning immediately to such solutions, we will first guess a solution, based on what we learned from previous models.

We expect nonclassical tunneling behavior as the quantum oscillator extends or compresses beyond the classical turning points. In these regions, we consider an analogy to the barrier-tunneling problem. The tunneling wavefunction was $\exp(\pm\kappa L)$ where κ depended on the square root of the difference between the potential energy and the total energy. Here, the potential energy varies with the square of the distance, $V \propto x^2$, so that $\kappa \propto V^{1/2}$ suggests $\kappa \propto x$. We also have $L \propto x$, so that $\kappa L \propto x^2$. This suggests that functions of the form $\exp(\pm\alpha^2 x^2)$ where α is a constant might be suitable. Of the two choices for sign, we can reject the positive one on the basis that the tunneling regions extend to infinity, and $\exp(+\alpha^2 x^2)$ would diverge at large $|x|$.

Thus, we try $\psi \propto \exp(-\alpha^2 x^2)$, again ignoring the normalization constant for now. Eq. (12.17) becomes

$$-\frac{\hbar^2}{2\mu}\frac{d^2 e^{-\alpha^2 x^2}}{dx^2} + \frac{1}{2}kx^2 e^{-\alpha^2 x^2} = E e^{-\alpha^2 x^2} .$$

Differentiating and then cancelling the common factor $\exp(-\alpha^2 x^2)$ leaves us with

$$-\frac{\hbar^2}{2\mu}(-2\alpha^2 + 4x^2\alpha^4) + \frac{1}{2}kx^2 = E$$

or

$$\frac{\hbar^2 \alpha^2}{\mu} + x^2\left(\frac{1}{2}k - \frac{2\alpha^4 \hbar^2}{\mu}\right) = E .$$

The terms multiplying x^2 must vanish in order for the left-hand side of this equation to equal a constant energy. This gives us a value for α:

$$\frac{1}{2}k - \frac{2\alpha^4\hbar^2}{\mu} = 0$$

or

$$\alpha^2 = \frac{1}{2}\frac{(k\mu)^{1/2}}{\hbar} \;.$$

We can now find *the energy eigenvalue corresponding to our guessed wavefunction*:

$$\frac{\hbar^2\alpha^2}{\mu} = \frac{1}{2}\hbar\left(\frac{k}{\mu}\right)^{1/2} = \frac{1}{2}\hbar\omega = E \;.$$

Thus

$$\psi \propto \exp\left(-\frac{1}{2}\frac{(k\mu)^{1/2}}{\hbar}x^2\right) \qquad \text{exhibits} \qquad E = \frac{1}{2}\hbar\omega \;.$$

Moreover, since this wavefunction has no nodes, it must be the ground-state wavefunction, and $E = \hbar\omega/2$ must be the *zero-point energy*. This is the wavefunction we used for practice in Chapter 11, Eq. (11.9).

It is not difficult from here to guess the excited-state wavefunctions. We first introduce a quantum number v to label each function. The first excited state will have one node; this can be expressed most easily by multiplying our ground-state wavefunction, which we now call ψ_0, by x, placing a single node at $x = 0$ where symmetry indicates it must be. Thus, $\psi_1 \propto x\psi_0$. Similarly, a quadratic polynomial of the form $(ax^2 + c)$ multiplying ψ_0 would, with proper choice of a and c, give us two nodes symmetrically placed about $x = 0$. Each excited-state wavefunction takes the form

$$\psi_v = f_v(x)\psi_0$$

where $f_v(x)$ is a polynomial with all odd or all even powers of x, with v the highest power.

To find these polynomials, we will first write the Schrödinger equation with dimensionless variables to simplify the algebra and notation. We define a dimensionless length q

$$q = \sqrt{2}\alpha x = \left(\frac{(k\mu)^{1/2}}{\hbar}\right)^{1/2} x = \left(\frac{k}{\hbar\omega}\right)^{1/2} x$$

and a dimensionless energy ϵ_v corresponding to the eigenvalue E_v

$$\epsilon_v = \frac{E_v}{E_0} = \frac{2E_v}{\hbar\omega} \;.$$

In these variables, our ground state wavefunction is $e^{-q^2/2}$, and its energy is $\epsilon_v = 1$. The Schrödinger equation is now

$$-\frac{d^2}{dq^2}\psi_v(q) + q^2\psi_v(q) = \epsilon_v\psi_v(q)$$

or

$$\left(q^2 - \frac{d^2}{dq^2}\right)\psi_v = \epsilon_v\psi_v$$

or, better yet, for reasons that will become apparent,

$$\left(q + \frac{d}{dq}\right)\left(q - \frac{d}{dq}\right)\psi_v = (\epsilon_v + 1)\psi_v \;.$$

(The right-hand side changes from ϵ_v to $\epsilon_v + 1$ due to the operator algebra: $(q + d/dq)(q - d/dq) = (q^2 - d^2/dq^2 + 1)$, as you should verify.)

This equation introduces two linear operators $(q + d/dq)$ and $(q - d/dq)$ that are worth exploring by themselves. Consider first the action of $(q + d/dq)$ on the ground-state wavefunction $\psi_0 = N_0 \exp(-q^2/2)$:

$$\begin{aligned}\left(q + \frac{d}{dq}\right)\psi_0 &= N_0\left(q + \frac{d}{dq}\right)e^{-q^2/2} \\ &= N_0\left(q\, e^{-q^2/2} + \frac{d}{dq}e^{-q^2/2}\right) \\ &= N_0(q\, e^{-q^2/2} - q\, e^{-q^2/2}) \\ &= 0\ ,\end{aligned}$$

which is a surprise. The operator $(q + d/dq)$ has annihilated ψ_0! In fact, this operator is called the *annihilation operator*, and could be turned around to give us ψ_0, had we not already found it, by solving

$$\left(q + \frac{d}{dq}\right)\psi_0 = 0\ .$$

Next we turn to $(q - d/dq)$ and compute

$$\left(q - \frac{d}{dq}\right)\psi_0 = N_0\left(q - \frac{d}{dq}\right)e^{-q^2/2} = N_0\, 2q\, e^{-q^2/2}\ ,$$

which (since $q \propto x$) is a function with the nodal properties we expect for ψ_1. Continuing, with $\psi_1 = N_0 2q \exp(-q^2/2)$,

$$\left(q - \frac{d}{dq}\right)\psi_1 = N_0\left(q - \frac{d}{dq}\right)2q\, e^{-q^2/2} = N_0(4q^2 - 2)e^{-q^2/2}\ ,$$

which looks like our expected ψ_2. The factor $(4q^2 - 2)$ is the $(ax^2 + c)$ quadratic we expected. And we could go on forever, applying $(q - d/dq)$ again and again, creating new functions that are at least proportional to the wavefunction with the next highest energy. Thus, $(q - d/dq)$ is called the *creation operator*. Given any one eigenfunction, it creates the wavefunction[4] of next highest energy.

One can find the correct normalization constants along the way and show that, in general,

$$\left(q - \frac{d}{dq}\right)\psi_v = [2(v + 1)]^{1/2}\psi_{v+1} \qquad v = 0, 1, 2, \ldots$$

$$\left(q + \frac{d}{dq}\right)\psi_v = (2v)^{1/2}\psi_{v-1} \qquad v = 1, 2, 3, \ldots$$

These operator equations give us the energy eigenvalues ϵ_v if we return to the Schrödinger equation:

$$\begin{aligned}\left(q + \frac{d}{dq}\right)\left[\left(q - \frac{d}{dq}\right)\psi_v\right] &= \left(q + \frac{d}{dq}\right)[2(v+1)]^{1/2}\psi_{v+1} \\ &= [2(v+1)]^{1/2}\left[\left(q + \frac{d}{dq}\right)\psi_{v+1}\right] \\ &= [2(v+1)]^{1/2}[2(v+1)]^{1/2}\psi_v \\ &= 2(v+1)\psi_v = (\epsilon_v + 1)\psi_v\end{aligned}$$

---

[4] Note that neither operator leads to an eigenvalue equation. The wavefunctions are *not* eigenfunctions of these operators.

or

$$\epsilon_v = 2v + 1, \quad v = 0, 1, 2, 3, \ldots$$

so that

$$E_v = \frac{1}{2}\hbar\omega\epsilon_v = \hbar\omega\left(v + \frac{1}{2}\right), \quad v = 0, 1, 2, 3, \ldots \quad \text{(12.18)}$$

The energy levels of the harmonic oscillator start with the zero-point value, $\hbar\omega/2$, and continue through an infinite number of discrete energies, each $\hbar\omega$ higher than the previous.

The natural unit of energy is $\hbar\omega$, the *energy quantum* for the oscillator, and the natural unit of length for state v is its classical turning point X_v where

$$\frac{1}{2}kX_v^2 = E_v \,,$$

which, with a little algebra, shows that

$$q = \sqrt{\frac{k}{\hbar\omega}}\,x = \sqrt{\frac{2E_v}{\hbar\omega}}\frac{x}{X_v} = \sqrt{\epsilon_v}\frac{x}{X_v} = \sqrt{2v+1}\,\frac{x}{X_v}\,.$$

Also, since

$$-\alpha^2 x^2 = -q^2/2 = -\frac{\frac{1}{2}kx^2}{\hbar\omega}\,,$$

the argument of the exponential part of each wavefunction, $-\alpha^2 x^2$, is minus the ratio of the potential energy to the energy quantum. These simple expressions make the otherwise complicated and opaque collection of $\hbar$, ω, k, μ symbols in ψ_v somewhat easier to remember.

EXAMPLE 12.4

Normalize the ground-state wavefunction and show that it is the same function we used in Chapter 11, Eq. (11.9):

$$\Psi(x) = \frac{1}{\sqrt{X\sqrt{\pi}}}\exp(-x^2/2X^2)\,.$$

SOLUTION We have $\psi_0(q) = N_0 e^{-q^2/2}$, and we seek N_0 from the normalization integral

$$\int_{-\infty}^{\infty} \psi(q)\psi(q)\,dq = N_0^2 \int_{-\infty}^{\infty} e^{-q^2}\,dq = 1\,.$$

Since we want the wavefunction in terms of the real distance variable x rather than in terms of q, we transform variables. Since $q = \sqrt{2v+1}\,(x/X_v)$, $dq = (\sqrt{2v+1}/X_v)\,dx = dx/X$ here ($v = 0$). Thus

$$N_0^2 \int_{-\infty}^{\infty} e^{-q^2}\,dq = \frac{N_0^2}{X} \int_{-\infty}^{\infty} e^{-(x/X)^2}\,dx = 1$$

The integral is known: $\int_{-\infty}^{\infty} e^{-(x/X)^2}\,dx = X\sqrt{\pi}$ so that the normalization constant is

$$N_0 = \frac{1}{\pi^{1/4}}\,.$$

We recognize that the normalization integral in terms of x is just $\int \Psi(x)\Psi(x)\,dx$, which, together with the value for N_0, recovers Eq. (11.9). Note that the transformation from q to

x involved $dq = (\sqrt{\epsilon_v}/X_v)\,dx$. The variable q and wavefunctions written in terms of it are dimensionless, but wavefunctions written in terms of the real distance x have the dimension (length)$^{-1/2}$.

⇒ RELATED PROBLEMS 12.10, 12.12

The series of polynomials produced by repetitive operation with $(q - d/dq)$ are known as *Hermite polynomials* $H_v(q)$. The full wavefunction is

$$\psi_v(q) = N_v H_v(q) \exp(-q^2/2) \tag{12.19}$$

where a separate normalization constant N_v has been retained away from the polynomials, as is conventionally done. It is

$$N_v = \frac{1}{\sqrt{2^v v!\,\pi^{1/2}}}. \tag{12.20}$$

The Hermite polynomials can be found from well-known relations among them, starting with the customary choice $H_0(q) = 1$. The first relation is the *generating function*

$$H_v(q) = (-1)^v e^{+q^2} \frac{d^v e^{-q^2}}{dq^v},$$

which produces any polynomial after v differentiations. (Note how this is reminiscent of our creation operator method.)

The first few Hermite polynomials are gathered in Table 12.1. They satisfy the differential equation

$$\frac{d^2 H_v}{dq^2} - 2q \frac{dH_v}{dq} + 2v H_v = 0$$

and the *recurrence relation*

$$H_{v+1} = 2q H_v - 2v H_{v-1}$$

that is especially useful for computer generation of any polynomial given the first two.

Several complete wavefunctions and their squares are shown in Figure 12.8. In each wavefunction graph, the classical range of motion is limited by the turning points $\pm X_v$ in

$$\frac{1}{2} k X_v^2 = \hbar\omega \left(v + \frac{1}{2}\right)$$

or, in reduced units, turning points $\pm Q_v$ where

$$Q_v^2 = 2v + 1.$$

TABLE 12.1 Hermite Polynomials, $H_v(q)$

$H_0(q) = 1$
$H_1(q) = 2q$
$H_2(q) = 4q^2 - 2$
$H_3(q) = 8q^3 - 12q$
$H_4(q) = 16q^4 - 48q^2 + 12$
$H_5(q) = 32q^5 - 160q^3 + 120q$
$H_6(q) = 64q^6 - 480q^4 + 720q^2 - 120$
$H_{v+1} = 2q H_v - 2v H_{v-1}$

FIGURE 12.8 The $v = 1, 2, 3,$ and 100 wavefunctions of the harmonic oscillator show the characteristic symmetry and nodal pattern of the Hermite polynomials. The corresponding ψ^2 for $v = 1, 2,$ and 3 are also shown, along with the classical turning points $\pm Q$. (The wavefunction for $v = 0$ and its square are shown in Figures 11.2 and 11.3). The dashed lines are probability distributions predicted from classical mechanics.

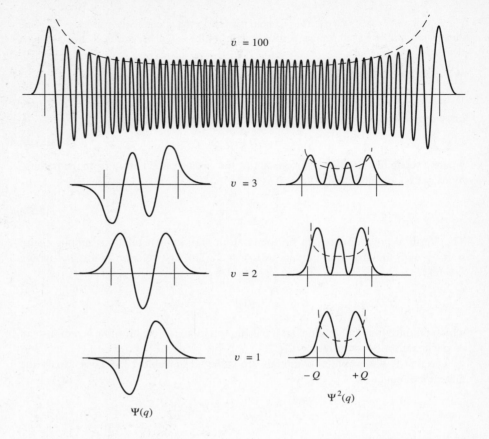

These show clearly the importance of tunneling outside the classical region, especially for small v. The graphs of ψ_v^2 also show, as dashed lines, the classical probability distributions mentioned earlier in this section. These distributions diverge at the turning points where classically the oscillator is instantaneously at rest. Contrast the classical and quantal behavior for the ground state: the classical oscillator is *least* likely to be found at $x = 0$, while the quantal oscillator is *most* likely to be found there. At large v, even though ψ_v^2 is oscillating madly, the local average magnitude of ψ_v^2 is more closely approximating the classical distribution, as the Correspondence Principle demands.

We will have many opportunities to return to the harmonic oscillator, but we close this section now with some general comments about its importance to chemical problems. Whenever a group of atoms interact, whether it be to form a stable molecule, a transient intermediate, a solid, a liquid, or whatever, the exact form of the interatomic potential energy function is always difficult to describe with precision and often difficult to discuss even qualitatively. But whatever this function of many variables might be, the regions of most importance are those that correspond to local minima of some type. These are the locally stable potential wells that correspond to chemically important structures. For simplicity, let us suppose that we are interested in motion in only one direction through this multidimensional function. We write the function as $V(R)$ with R the direction of interest. At some $R = R_0$, V has a local minimum. We expand $V(R)$ in a Taylor's series about this minimum:

$$V(R) \cong V(R_0) + \left(\frac{dV}{dR}\right)_{R_0} (R - R_0) + \frac{1}{2}\left(\frac{d^2V}{dR^2}\right)_{R_0} (R - R_0)^2 + \cdots$$

The first term is a constant, and we can set our zero of energy to make $V(R_0) = 0$. The second term contains the first derivative of V evaluated at R_0; this factor is zero since R_0 locates a minimum of $V(R)$. The third term remains, and, if R_0 represents a minimum, then $(d^2V/dR^2)_{R_0} \equiv V''(R_0) > 0$, and we can write

$$V(R) \cong \tfrac{1}{2}V''(R_0) \underbrace{(R - R_0)^2}_{} = \underbrace{\tfrac{1}{2}k(R - R_0)^2}_{},$$

↑ Potential near a local minimum ↑ Second derivative at the minimum ↑ (Displacement from the minimum)² ↑ Harmonic oscillator potential energy

which shows that $V''(R_0)$ acts as a force constant. Thus, unless V'' vanishes for some higher symmetry reason, we can approximate any sort of motion around a local potential-energy minimum as harmonic. The next few terms of the expansion,

$$\tfrac{1}{6}V'''(R_0)(R - R_0)^3, \quad \tfrac{1}{24}V''''(R_0)(R - R_0)^4, \quad \text{etc.}$$

represent *anharmonic corrections,* as they are called, to the basic harmonic approximation. We will see how these corrections can be incorporated, at least approximately, when we discuss molecular bonding and spectroscopy.

12.3 MULTIDIMENSIONS

There are other 1-D models with exact solutions, but those of the last two sections are the most important. We now consider them in more than one dimension and along the way, prepare for the next section's discussion of the hydrogen atom.

The 1-D particle-in-a-box was really a particle-on-a-*line,* since the particle was confined to one linear direction. Now we consider the true particle-in-a-box, a 3-D potential-energy field with sides of arbitrary lengths. We align a Cartesian axis system (x, y, z) parallel to the edges of the box with the origin in one corner (see Figure 12.9) and specify

$$\hat{V}(x, y, z) = 0 \begin{cases} 0 \leq x \leq L_x \\ 0 \leq y \leq L_y \\ 0 \leq z \leq L_z \end{cases}$$

$$\hat{V}(x, y, z) = \infty \text{ elsewhere}.$$

We place one particle of mass m in the box and seek wavefunctions $\psi = \psi(x, y, z)$.

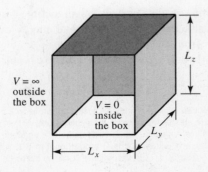

FIGURE 12.9 A 3-D particle-in-a-box model has a region of constant potential energy surrounded by a region of infinite potential energy as shown.

The kinetic-energy operator is a sum of terms for the three independent directions, as in classical mechanics where $T = (p_x^2 + p_y^2 + p_z^2)/2m$. Thus,

$$\hat{T} = -\frac{\hbar^2}{2m}\left[\frac{\partial^2}{\partial x^2} + \frac{\partial^2}{\partial y^2} + \frac{\partial^2}{\partial z^2}\right] = -\frac{\hbar^2}{2m}\nabla^2 \qquad (12.21)$$

where ∇^2 (read "del squared") is a standard shorthand for the sum of partial derivatives $\partial^2/\partial x^2 + \partial^2/\partial y^2 + \partial^2/\partial z^2$. It is also called the *Laplacian operator*.

Schrödinger's equation, inside the box where $\psi \neq 0$, is

$$\hat{H}\psi(x, y, z) = \hat{T}\psi = -\frac{\hbar^2}{2m}\nabla^2\psi = E\psi \qquad (12.22)$$

since $\hat{V}$ is zero. This was again a well-studied type[5] of linear second-order partial differential equation by Schrödinger's time.

To solve Eq. (12.22), we use a technique known as *separation of variables,* first introduced in the last chapter to separate time and distance coordinates. The Hamiltonian is a sum of operators

$$\hat{H} = \hat{H}_x + \hat{H}_y + \hat{H}_z$$

where, for example,

$$\hat{H}_x = -\frac{\hbar^2}{2m}\frac{\partial^2}{\partial x^2} \ .$$

Any one of these operators alters $\psi(x, y, z)$ through the single variable in that operator. Consider solutions $\psi_x(x)$ to the equation

$$\hat{H}_x\psi_x(x) = -\frac{\hbar^2}{2m}\frac{d^2\psi_x}{dx^2} = E_x\psi_x \ .$$

This is just Schrödinger's equation for the 1-D box, solved by Eq. (12.11). We can write $\psi_y(y)$ and $\psi_z(z)$ and the corresponding energies E_y and E_z in the same way.

We return to Eq. (12.22) for the 3-D box, try the solution

$$\psi(x, y, z) = \psi_x(x)\psi_y(y)\psi_z(z) \ , \qquad (12.23)$$

and find

$$\begin{aligned}\hat{H}\psi(x, y, z) &= (\hat{H}_x + \hat{H}_y + \hat{H}_z)\psi \\ &= (\hat{H}_x + \hat{H}_y + \hat{H}_z)\psi_x\psi_y\psi_z \\ &= \psi_y\psi_z(\hat{H}_x\psi_x) + \psi_x\psi_z(\hat{H}_y\psi_y) + \psi_x\psi_y(\hat{H}_z\psi_z) \\ &= \psi_y\psi_z(E_x\psi_x) + \psi_x\psi_z(E_y\psi_y) + \psi_x\psi_y(E_z\psi_z) \\ &= (E_x + E_y + E_z)\psi_x\psi_y\psi_z \\ &= (E_x + E_y + E_z)\psi = E\psi \ .\end{aligned}$$

This shows that not only is the product wavefunction $\psi_x\psi_y\psi_z$ a solution for the 3-D box problem, but also that the particle's total energy can be decomposed into a sum of contributions from each direction.

And so we find that when we solved the 1-D box problem, we also solved the

---

[5]In fact, Schrödinger's equation, along with Laplace's equation in electrostatics and Helmholtz's equation for wave motion, are all special cases of the general form $\nabla^2 f + k^2 f = 0$. If $k = 0$, we have Laplace's equation; if k is a real constant, we have Helmholtz's equation; and, if k^2 is a function of coordinates, we have Schrödinger's equation with k^2 proportional to $V(x, y, z) - E$. Here, since we care only for solutions in the box where $V = 0$, Schrödinger's equation becomes a Helmholtz wave equation.

3-D (and, for that matter, the 2-D, 4-D, etc.) box problem. Our specific solution $\psi(x, y, z) = \psi_x\psi_y\psi_z$ yields wavefunctions for any state of the particle-in-a-3-D-box.

Separation of variables is a powerful and general technique. *Whenever the total Hamiltonian can be written as a sum of terms, each depending on a set of coordinates unique to that term, then the total wavefunction will be a product of functions, each depending on the corresponding set of coordinates.*

A Closer Look at Separation of Variables

Both the mathematical and physical consequences of the separation of variables method are worth exploring. Using a 2-D box Schrödinger equation as an example, the usual derivation of separability runs like this. We have to solve

$$\frac{\partial^2 \psi(x, y)}{\partial x^2} + \frac{\partial^2 \psi(x, y)}{\partial y^2} + \epsilon\psi(x, y) = 0$$

where $\epsilon = -2mE/\hbar^2$. Assume $\psi = \psi_x\psi_y$, and divide the equation by ψ:

$$\frac{1}{\psi_x}\frac{\partial^2 \psi_x}{\partial x^2} + \frac{1}{\psi_y}\frac{\partial^2 \psi_y}{\partial^2 y} + \epsilon = 0 \; .$$

The first term is a function of x alone; the second, of y alone; and the third is just a constant. For the equation to be satisfied for arbitrary values of x and y, *each term must be a constant.* Hence, one writes

$$\frac{1}{\psi_x}\frac{\partial^2 \psi_x}{\partial x^2} = \epsilon_x \quad \text{or} \quad \frac{\partial^2 \psi_x}{\partial x^2} = \epsilon_x \psi_x$$

and similarly for ψ_y. The mathematical dilemma (and it is a minor one) centers around dividing by ψ before knowing what ψ might be. At the very least, ψ might have nodes, and to claim that the divided equation holds, a priori, for all x and y is brash unless we know the behavior of $(d^2\psi_x/dx^2)/\psi_x$ at a nodal point. The way out is to anticipate the result (which is that ψ_x and $d^2\psi_x/dx^2$ differ by only a multiplicative constant at all x), plunge ahead, and be rewarded by solutions that satisfy the Postulates as well as the primary differential equation for ψ.

Physically, the relevance of separability includes not only the decomposition of the eigenvalue into a sum, but also the decomposition of probability density into independent factors *for all times*. The motions along separable directions are, by definition, completely decoupled from one another, and only some *external force with a nonseparable Hamiltonian representation* can transfer motion in one direction into motion in another.

The 1-D energies give a 3-D energy

$$E = E_x + E_y + E_z = \frac{\hbar^2 \pi^2}{2m}\left(\frac{n_x^2}{L_x^2} + \frac{n_y^2}{L_y^2} + \frac{n_z^2}{L_z^2}\right) \tag{12.24}$$

where n_x, n_y, and n_z are the three independent quantum numbers needed to specify the state. Since each n can be any integer starting with 1, the particle *always* has some motion (zero-point motion again) *in each direction*.

If the lengths of the box are all arbitrary in size, it is not likely (or at least not obvious) that any one set of n's yields the same energy as any other set. But suppose the box is a cube of common length L. Then Eq. (12.24) is

$$E = \frac{\hbar^2 \pi^2}{2mL^2}(n_x^2 + n_y^2 + n_z^2)$$

and the symmetry of the cube (or, more fundamentally, the symmetry of the potential-energy operator) causes many different sets of n's to have the same energy. For instance, the sets $(n_x, n_y, n_z) = (1, 1, 5), (1, 5, 1), (5, 1, 1)$, and $(3, 3, 3)$ all have sums of squares equal to 27, and the sets $(1, 5, 7), (1, 7, 5), (5, 1, 7), (5, 7, 1)$, $(5, 5, 5), (7, 1, 5)$ and $(7, 5, 1)$ all have sums of squares equal to 75. Each member of these sets represents a *different quantum-mechanical state of the system*, but each such state in any one set has the same energy.

Recall from Chapter 11 that when two or more states have the same energy, the states are said to be *degenerate*, and one would say that the states $(1, 1, 5)$ etc. are *four-fold degenerate*. Here the equivalence of x, y, and z ensures that states $(1, 1, 5), (1, 5, 1)$, and $(5, 1, 1)$ are degenerate, but the fact that state $(3, 3, 3)$ is degenerate with these three is less obvious.

Armed with the separation of variables technique, we can solve the multidimensional free-particle problem or the multidimensional harmonic oscillator at once. Details of these systems are left to this chapter's problems.

The potential functions for these systems are very simple when written in Cartesian coordinates. There are many other orthogonal coordinate systems that are useful in various scenarios, but among them, the most important is the spherical polar coordinate system.

Spherical polar coordinates, shown in relation to Cartesian coordinates in Figure 12.10, contain one radial distance, r, and two angles, θ (called the polar angle) and ϕ (called the azimuthal angle). They are related to x, y, and z via

$$x = r \sin \theta \cos \phi, \quad y = r \sin \theta \sin \phi, \quad \text{and} \quad z = r \cos \theta$$

or, inversely,

$$r = (x^2 + y^2 + z^2)^{1/2}, \quad \theta = \cos^{-1}(z/r), \quad \text{and} \quad \phi = \tan^{-1}(y/x)$$

with ranges

$$0 \leq r \leq \infty, \quad 0 \leq \theta \leq \pi, \quad \text{and} \quad 0 \leq \phi \leq 2\pi$$

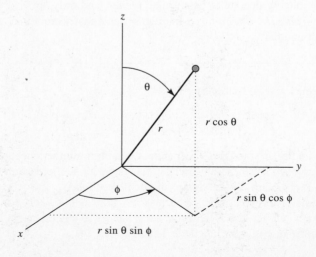

FIGURE 12.10 Spherical polar coordinates locate a point by the distance r and the angles θ and ϕ which are related to the Cartesian x, y, and z coordinates as shown.

and a differential volume element

$$dx\,dy\,dz = r^2 \sin\theta\,dr\,d\theta\,d\phi\ .$$

We will also need the Laplacian operator ∇^2 in spherical polar coordinates. Converting ∇^2 from Cartesian to polar coordinates is an exercise somewhat like writing, "I will not chew gum in class" one hundred times on the blackboard. It is tedious and sometimes necessary, but not that instructive. Thus, we will skip the tedium and write the result:

$$\nabla^2 = \frac{1}{r^2}\frac{\partial}{\partial r}\left(r^2\frac{\partial}{\partial r}\right) + \frac{1}{r^2}\left[\frac{1}{\sin\theta}\frac{\partial}{\partial\theta}\left(\sin\theta\frac{\partial}{\partial\theta}\right) + \frac{1}{\sin^2\theta}\frac{\partial^2}{\partial\phi^2}\right]\ .$$

When are spherical polar coordinates called for? The most obvious use is in those model problems (those potential-energy operators) with spherical symmetry. Some examples are the H atom (coming up in the next section), the problem of two rare-gas atoms colliding (at a certain level of approximation), or two atomic ions colliding under the influence of Coulomb's law. Less obvious perhaps is the usefulness of spherical polar coordinates to the general discussion of angular momentum, also a topic for the next section.

Spherically symmetric potential-energy operators depend only on r and have Schrödinger equations of the general type

$$\hat{H}\psi = -\frac{\hbar^2}{2m}\nabla^2\psi + \hat{V}(r)\psi = E\psi\ .$$

They are ready candidates for the spherical polar Laplacian operator and separation of variables. We write

$$\psi(r,\theta,\phi) = R(r)\,\Theta(\theta)\,\Phi(\phi)$$

and find

$$\nabla^2\psi = \Theta\Phi\left[\frac{1}{r^2}\frac{d}{dr}\left(r^2\frac{dR}{dr}\right)\right] + \frac{R\Phi}{r^2}\left[\frac{1}{\sin\theta}\frac{d}{d\theta}\left(\sin\theta\frac{d\Theta}{d\theta}\right)\right] + \frac{R\Theta}{r^2\sin^2\theta}\left(\frac{d^2\Phi}{d\phi^2}\right)$$

where the partial derivatives can be written as ordinary derivatives since R, Θ, and Φ are functions of only one variable. Separation seems possible, but the details depend on the explicit form of $\hat{V}$.

Nevertheless we can state that *for all $\hat{V} = \hat{V}(r)$, the Θ and Φ functions are the same.* This is true because we can separate the Schrödinger equation into two terms, one containing r alone and one containing θ and ϕ alone. We write

$$-\frac{\hbar^2}{2mr^2}\left[\frac{\partial}{\partial r}\left(r^2\frac{\partial\psi}{\partial r}\right) + \frac{1}{\sin\theta}\frac{\partial}{\partial\phi}\left(\sin\theta\frac{\partial\psi}{\partial\theta}\right) + \frac{1}{\sin^2\theta}\left(\frac{\partial^2\psi}{\partial\phi^2}\right)\right] + V(r)\psi = E\psi\ ,$$

multiplying through by r^2, and rearrange, finding

$$\left[-\frac{\hbar^2}{2m}\frac{\partial}{\partial r}r^2\left(\frac{\partial\psi}{\partial r}\right) + r^2V(r)\psi - r^2E\psi\right] - \frac{\hbar^2}{2m}\left[\frac{1}{\sin\theta}\frac{\partial}{\partial\theta}\left(\sin\theta\frac{\partial\psi}{\partial\theta}\right) + \frac{1}{\sin^2\theta}\frac{\partial^2\psi}{\partial\phi^2}\right] = 0\ .$$

This has an r part (the collection of terms in the first square brackets) and a θ, ϕ part (the second collection of terms).

EXAMPLE 12.5

Consider a He atom stuck inside a buckminsterfullerene molecule, C_{60}. (This is denoted He@C_{60}.) To a good approximation, this is a particle-in-a-spherical-box system. What are the radial wavefunctions and energies for He moving in such a sphere?

SOLUTION The C_{60} molecule has a C nuclear framework diameter 7.04 Å, and if we assign C a 1.85 Å radius and He a 1.22 Å radius as suggested by the structures of their solids, our spherical box allows the He atom to move through a maximum radius $r_m = [(7.04/2) - 1.85 - 1.22]$ Å $\cong 0.45$ Å, as shown below:

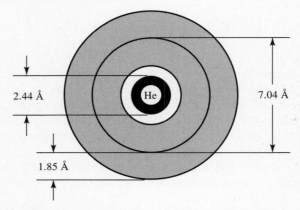

The potential energy function is thus $V(r) = 0$, $0 \leq r \leq 0.45$ Å; $V(r) = \infty$, $r > 0.45$ Å, and we seek radial wavefunctions $R(r)$ and energies inside the box ($r \leq r_m$) that satisfy the r part of the full Schrödinger equation:

$$-\frac{\hbar^2}{2m}\frac{1}{r^2}\frac{d}{dr}\left(r^2\frac{dR(r)}{dr}\right) = E\, R(r) \;.$$

We can introduce the wavevector k through

$$k^2 = \frac{2mE}{\hbar^2} \;,$$

which leads to

$$\frac{1}{r^2}\frac{d}{dr}\left(r^2\frac{dR}{dr}\right) + k^2 R = 0 \;.$$

This differential equation has solutions of the form $R(r) = (N/kr) \sin kr$. We find the allowed values for k by requiring $R(r_m) = 0$ (just as in the 1-D box, but note that $R(0) \neq 0$!). This gives us the energies and wavevectors

$$E_n = \frac{\hbar^2\pi^2}{2mr_m^2}n^2 \;, \quad k = \frac{n\pi}{r_m} \;, \quad n = 1, 2, 3, \ldots$$

in analogy to the 1-D box problem. The normalization constant is found from (note that the spherical polar differential for r is $r^2\, dr$!)

$$\int_0^{r_m} R^2(r)\, r^2\, dr = N^2 \int_0^{r_m} \left(\frac{\sin kr}{kr}\right)^2 r^2\, dr = \frac{N^2}{k^2}\int_0^{r_m} \sin^2 kr\, dr = 1 \;,$$

which yields

$$N = \frac{\sqrt{2}\, n\pi}{r_m^{3/2}} \quad \text{and} \quad \frac{N}{k} = \left(\frac{2}{r_m}\right)^{1/2}$$

so that the final wavefunction is

$$R(r) = \left(\frac{2}{r_m}\right)^{1/2} \frac{\sin(n\pi r/r_m)}{r}.$$

With $m = 4$ amu and $r_m = 0.45$ Å, the ground state wavefunction looks like this:

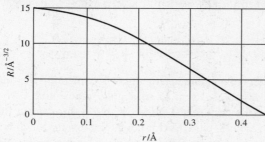

Note that the atom is most probably found in the middle of the cavity. In addition to these states, there are others in which the He atom has non-zero angular momentum about the center of the spherical box. This is discussed in the next section.

⇒ *RELATED PROBLEMS* 12.19, 12.20

In the next section, we will consider the θ, φ part first, then specify $V(r)$ for the hydrogen atom. Once again, differential equations with well-studied solutions will arise, and we will approach them much as we did that for the harmonic oscillator.

12.4 ANGULAR MOMENTUM AND THE HYDROGEN ATOM

Recall the classical origins of angular momentum: something with linear momentum goes *around* or *by* a fixed point. Like linear momentum, angular momentum is a vector quantity. Given the (vector) linear momentum **p** of a particle located by the (vector) **r** with respect to some coordinate origin, the angular momentum, vector **L** (*not* to be confused with our use of *L* as a potential box length earlier!) is

L	=	**r**	×	**p**
↑		↑		↑
Angular momentum vector		Radius vector to particle from arbitrary origin	Vector cross-product	Linear momentum vector

where the × symbolizes the vector cross-product. In Cartesian coordinates, the components of **L**, given by the rule for forming a cross-product, are

$$L_x = yp_z - zp_y, \quad L_y = zp_x - xp_z, \quad \text{and} \quad L_z = xp_y - yp_x, \quad (12.25)$$

or, in terms of the magnitudes of the three vectors and the angle α between **r** and **p**,

$$L = rp \sin \alpha. \quad (12.26)$$

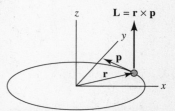

FIGURE 12.11 The angular momentum vector of a particle with vector linear momentum **p** located by vector **r** from the origin is given by the vector cross product **r** × **p**. This vector is perpendicular to the plane containing **r** and **p**.

As Figure 12.11 demonstrates, **L** is perpendicular to the plane containing **r** and **p**, and, as Equations (12.25) and especially (12.26) show, if **r** and **p** are collinear (α = 0), **L** vanishes. Something thrown straight at you has no angular momentum with respect to you.

The dimensions of angular momentum are the same as those of Planck's constant: (energy) (time) = (distance) (momentum). If you spin a 0.1 kg mass around your head on a 1 m string at one revolution per second, (covering the circumferential distance 2π m each second)

$$r = 1 \text{ m}, \quad p = mv = (0.1 \text{ kg})(2\pi \text{ m s}^{-1}), \quad \alpha = 90°,$$

and thus

$$L = rp \sin\alpha = (2\pi/10) \text{ kg m}^2 \text{ s}^{-1} \cong 0.63 \text{ J s}.$$

This is about 33 orders of magnitude larger than $\hbar$, explaining why quantum phenomena do not enter classical descriptions of angular momentum.

If you spin the mass such that it starts on your right, goes in front of you, then to your left, and on around, the **L** vector points up; reverse the direction of rotation, and the direction of **L** is down. The mass has a kinetic energy $p^2/2m$, and, since $\alpha = 90°$ so that $L = rp$, we can write

$$T = \frac{p^2}{2m} = \frac{r^2 p^2}{2mr^2} = \frac{L^2}{2mr^2}. \tag{12.27}$$

The product mr^2 in the denominator (or some similar product of a mass and a distance squared) occurs frequently in discussions of rotating systems and is called the *moment of inertia, I*:

$$I = mr^2. \tag{12.28}$$

We can also write the mass velocity as $v = r\omega$ where ω is the rotational angular velocity so that

$$T = \frac{1}{2} mv^2 = \frac{1}{2} mr^2 \omega^2 = \frac{1}{2} I\omega^2 \tag{12.29}$$

which further implies

$$\underbrace{L}_{\text{Angular momentum}} = \underbrace{I}_{\text{Moment of inertia}} \underbrace{\omega}_{\text{Angular velocity}}. \tag{12.30}$$

This summarizes the important classical equations for angular momentum (Equations (12.25)–(12.30)), at least for the simple case of circular orbital motion of a single mass about a point. We will expand these equations for more complicated types of rotational motion as the need arises, but for now, we have the framework for a quantum mechanical discussion.

The first step is straightforward. We transcribe the classical components of angular momentum in Eq. (12.25) into operators. We have

$$\hat{L}_x = \frac{\hbar}{i}\left(y\frac{\partial}{\partial z} - z\frac{\partial}{\partial y}\right)$$

$$\hat{L}_y = \frac{\hbar}{i}\left(z\frac{\partial}{\partial x} - x\frac{\partial}{\partial z}\right) \tag{12.31}$$

$$\hat{L}_z = \frac{\hbar}{i}\left(x\frac{\partial}{\partial y} - y\frac{\partial}{\partial x}\right)$$

in Cartesian coordinates, or, in spherical polar coordinates,

$$\hat{L}_x = \frac{\hbar}{i}\left(-\sin\phi\frac{\partial}{\partial\theta} + \cot\theta\cos\phi\frac{\partial}{\partial\phi}\right) \quad (12.32a)$$

$$\hat{L}_y = \frac{\hbar}{i}\left(\cos\phi\frac{\partial}{\partial\theta} - \cot\theta\sin\phi\frac{\partial}{\partial\phi}\right) \quad (12.32b)$$

$$\hat{L}_z = \frac{\hbar}{i}\left(\frac{\partial}{\partial\phi}\right). \quad (12.32c)$$

The classical expression for L^2 is $L_x^2 + L_y^2 + L_z^2$, and we can transcribe this sum into an operator $\hat{L}^2$ for the square of the magnitude of the total angular momentum. More algebra leads to

$$\hat{L}^2 = (\hat{L}_x\hat{L}_x + \hat{L}_y\hat{L}_y + \hat{L}_z\hat{L}_z)$$
$$= -\hbar^2\left[\frac{1}{\sin\theta}\frac{\partial}{\partial\theta}\left(\sin\theta\frac{\partial}{\partial\theta}\right) + \frac{1}{\sin^2\theta}\frac{\partial^2}{\partial\phi^2}\right]. \quad (12.33)$$

The operator in square brackets is just the angular part of ∇^2. By itself, it is called the *Legendrian operator*.

Commutators tell us those features of angular momentum that are simultaneously observable. Consider, for instance,

$$[\hat{L}_x, \hat{L}_y] = \hat{L}_x\hat{L}_y - \hat{L}_y\hat{L}_x.$$

Equation (12.31) can be used to show that

$$[\hat{L}_x, \hat{L}_y] = -\hbar^2\left(y\frac{\partial}{\partial x} - x\frac{\partial}{\partial y}\right) = i\hbar\hat{L}_z. \quad (12.34a)$$

The other component commutators must be cyclic permutations[6] of x, y, and z in Equation (12.34a):

$$[\hat{L}_z, \hat{L}_x] = i\hbar\hat{L}_y \quad (12.34b)$$

$$[\hat{L}_y, \hat{L}_z] = i\hbar\hat{L}_x. \quad (12.34c)$$

Since all of these commutators are non-zero, *we cannot measure two (or more) components of the angular momentum vector simultaneously*. The vector **L** cannot be located exactly in a 3-D system.

While we cannot find **L** exactly, we can find its magnitude and something about its direction. We can evaluate $[\hat{L}^2, \hat{L}_z]$ using Equations (12.32c) and (12.33). We find

$$[\hat{L}^2, \hat{L}_z] = 0, \quad (12.35a)$$

and it must also be true that

$$[\hat{L}^2, \hat{L}_x] = 0 \quad (12.35b)$$

and

$$[\hat{L}^2, \hat{L}_y] = 0 \quad (12.35c)$$

since the x, y, and z directions are equivalent.

---

[6]The cyclic permutations of three things such as (x, y, z) are $(x, y, z) \rightarrow (y, z, x) \rightarrow (z, x, y)$ and can be visualized by rotating the following diagram like a Ferris wheel in 120° steps:

Since $\hat{L}^2$ commutes with any component, *we can simultaneously measure the magnitude of* **L** *(the square root of the eigenvalue of $\hat{L}^2$) and the magnitude of any one component of* **L**.

We will find the eigenvalues of $\hat{L}_z$ first. Imagine a mass orbiting a point source of a spherically symmetric potential. If we fix the orbiting radius, we have a *particle-on-a-sphere*. Calling the radius R and setting $V(R) = 0$ reduces the Schrödinger equation to

$$\hat{H}\psi(r, \theta, \phi) = \hat{H}\psi(R, \theta, \phi) = \frac{\hat{L}^2}{2mR^2}\psi = E\psi .$$

(Note the correspondence to the classical Eq. (12.27).) Since r is fixed at R, we can write $\psi = \Theta(\theta)\Phi(\phi)$. For the z component of **L**, we expect an eigenvalue equation of the form

$$\hat{L}_z\psi = m\hbar\psi$$

where m is some real number (*not* the mass!) that measures L_z in multiples of the quantum of angular momentum, $\hbar$. Using Eq. (12.32c), we have

$$\hat{L}_z\psi = \frac{\hbar}{i}\frac{\partial}{\partial\phi}\psi = \frac{\hbar}{i}\Theta(\theta)\frac{d\Phi(\phi)}{d\phi} = \Theta(\theta)\,m\hbar\Phi(\phi)$$

or simply

$$\frac{d\Phi(\phi)}{\Phi(\phi)} = im\,d\phi ,$$

which can be integrated immediately to give

$$\Phi(\phi) = N\,e^{im\phi} \tag{12.36}$$

where N is a normalization constant. Wavefunctions must be single-valued, and, since ϕ varies from 0 to 2π, the angles ϕ, $\phi + 2\pi$, $\phi + 4\pi$, etc., represent the same point in space. Thus we must require[7]

$$\Phi(\phi) = \Phi(\phi + 2\pi n) \quad n = \text{any integer (positive or negative)} .$$

This requirement can be met only if

$$m = 0, \pm 1, \pm 2, \ldots \tag{12.37}$$

since $e^{im\phi} = \cos m\phi + i\sin m\phi$ has a 2π periodicity. Thus, any component of **L** is quantized in integral steps of size $\hbar$.

Our particle is free to roam the surface of the sphere, and we might expect a connection with free linear motion. If we make $\theta = $ constant, then the particle-on-a-sphere becomes a *particle-on-a-ring* problem, and $\Phi(\phi)$ is the entire wavefunction. Positive (or negative) m values correspond to rotation in one direction (or the opposite direction) around the ring. Note the correspondences:

Free linear motion	Free circular motion
$\psi \propto e^{\pm ip_z z/\hbar} \propto e^{\pm ikz}$	$\psi \propto e^{\pm iL_z\phi/\hbar} \propto e^{\pm im\phi}$

since $p_z = k_z\hbar$ and $L_z = m\hbar$.

---

[7]This is called imposing *periodic boundary conditions* on the wavefunction.

The normalization constant is found easily, due to the periodicity of ϕ:

$$1 = N^2 \int_0^{2\pi} \Phi^*\Phi \, d\phi = N^2 \int_0^{2\pi} d\phi = 2\pi N^2$$

so that

$$N = \frac{1}{\sqrt{2\pi}} \ .$$

EXAMPLE 12.6

To a first approximation, the two least strongly bound valence electrons of benzene are bound to a ring of 6 carbon atoms with a radius 1.40 Å:

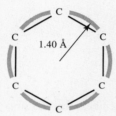

What energies are allowed under the rather severe approximations that the electrons do not interact with each other and that they feel a constant potential energy around the ring?

SOLUTION For a particle-on-a-ring problem, which is what we have here, the z component of the angular momentum (z is perpendicular to the plane of the ring) is the total angular momentum, and the Schrödinger equation is

$$\frac{\hat{L}^2}{2m_e R^2} \Phi(\phi) = \frac{\hat{L}_z \hat{L}_z}{2m_e R^2} \Phi(\phi) = E \, \Phi(\phi)$$

where m_e is the electron mass, R is the ring radius, and $\Phi(\phi) = e^{im\phi}/\sqrt{2\pi}$. Since we know $\hat{L}_z \Phi(\phi) = m\hbar \Phi(\phi)$, we can find E:

$$E = \frac{m^2 \hbar^2}{2m_e R^2} \ , \quad m = 0, \pm 1, \pm 2, \ldots$$

Note that $E = 0$ is possible (if $m = 0$), unlike the particle-in-a-box energy. This does not violate the Uncertainty Principle, however, since the wavefunction for $m = 0$ is just $1/\sqrt{2\pi}$ = a constant for all ϕ. We have no knowledge of the angular position of the electron in this state. The numerical energy is

$$E = m^2 \, (3.11 \times 10^{-19} \text{ J}) = m^2 \, (1.94 \text{ eV}) \ .$$

Although some severe approximations have been made here, we nevertheless find a result that is in rough order of magnitude agreement with observations: the excited states of molecules that correspond to electron excitations are typically spaced by energies of a few eV.

⇒ **RELATED PROBLEMS** 12.21, 12.22

Going back to the particle-on-a-sphere problem, we now seek the magnitude of **L**. We can do this without finding the eigenfunction $\Theta(\theta)$ if we manipulate operator expressions carefully. First, since $\hat{L}_x$, $\hat{L}_y$, $\hat{L}_z$, and $\hat{L}^2$ are all Hermitian operators with real eigenvalues, all *squares* of their eigenvalues must be positive. This means we can assume that

$$\hat{L}^2 \psi = L^2 \hbar^2 \psi$$

where $L^2\hbar^2$ is the eigenvalue of $\hat{L}^2$, with the real number L^2 yet to be determined.

Next note that

$$(\hat{L}^2 - \hat{L}_z^2)\psi = \hat{L}^2\psi - \hat{L}_z^2\psi = (L^2\hbar^2 - m^2\hbar^2)\psi$$

is equivalent to

$$(\hat{L}^2 - \hat{L}_z^2)\psi = (\hat{L}_x^2 + \hat{L}_y^2 + \hat{L}_z^2 - \hat{L}_z^2)\psi = (\hat{L}_x^2 + \hat{L}_y^2)\psi \ .$$

Since eigenvalues of $\hat{L}_x^2$ and $\hat{L}_y^2$ are positive (or perhaps zero, at least), we can conclude that

$$L^2\hbar^2 - m^2\hbar^2 \geq 0$$

or

$$L^2\hbar^2 \geq m^2\hbar^2 \ ,$$

which makes sense if one remembers that the square of one component of a vector ($m^2\hbar^2$) cannot exceed the sum of the squares of all three components ($L^2\hbar^2$).

Thus, *for any given* **L**, *m is bounded*. It is clear that if $|\mathbf{L}| = 0$, then $m = 0$ is the only possibility; if there is no angular momentum, there is no component. What we have shown is that for any given $|\mathbf{L}|$, there will be some maximum m, call it m^+, and some corresponding minimum, m^-, with $m^+ = -m^-$, since the inequality above involves m^2.

Next, we note that $L \neq m^+$, for if it did, we would know the magnitude of **L** (it would be $L\hbar = m^+\hbar$), and one component ($m^+\hbar$), which is fine, except that the component would then be the entire vector, and the other two components would be zero. This is not possible, since we cannot know two (or more) components exactly and simultaneously. Thus, L must be a more complex quantity than a simple integer. Consequently, we write $L = L(l)$ where l is some indexing quantum number for the full structure of L.

Next, we introduce two operators, similar to the operators $(q \pm d/dq)$ used in the harmonic oscillator problem. These are

$$\hat{L}_+ = \hat{L}_x + i\hat{L}_y \quad \text{and} \quad \hat{L}_- = \hat{L}_x - i\hat{L}_y \ , \tag{12.38}$$

which are called *raising and lowering operators*, respectively. They have the following commutation relations:

$$[\hat{L}_z, \hat{L}_+] = \hbar\hat{L}_+ \quad \text{but} \quad [\hat{L}_z, \hat{L}_-] = -\hbar\hat{L}_-$$
$$[\hat{L}_+, \hat{L}_-] = 2\hbar\hat{L}_z$$
$$[\hat{L}_\pm, \hat{L}^2] = 0 \quad (\hat{L}_\pm \text{ is either } \hat{L}_+ \text{ or } \hat{L}_-) \ .$$

Note especially the last of these. Since $\hat{L}_+$ and $\hat{L}_-$ commute with $\hat{L}^2$, neither operator, operating on an eigenfunction of $\hat{L}^2$, can produce a new function with a different $\hat{L}^2$ eigenvalue. Thus $\hat{L}_+$ and $\hat{L}_-$ can change the state without changing the magnitude of the angular momentum. The only way this can happen is if these operators *reorient* the angular momentum, changing components of **L** without changing $|\mathbf{L}|$.

If ψ_m is the eigenfunction for which $\hat{L}_z\psi_m = m\hbar\psi_m$, to what value of m does the eigenfunction $\hat{L}_+\psi_m$ correspond? To find out, we consider the action of $\hat{L}_z$ on $\hat{L}_+\psi_m$:

$$\begin{aligned}
\hat{L}_z(\hat{L}_+\psi_m) &= \hat{L}_z\hat{L}_+\psi_m \\
&= (\hat{L}_+\hat{L}_z + \hat{L}_z\hat{L}_+ - \hat{L}_+\hat{L}_z)\psi_m &&\text{(adding and subtracting } \hat{L}_+\hat{L}_z\text{)} \\
&= (\hat{L}_+\hat{L}_z + [\hat{L}_z, \hat{L}_+])\psi_m &&\text{(recognizing the commutator)} \\
&= (\hat{L}_+\hat{L}_z + \hbar\hat{L}_+)\psi_m &&\text{(evaluating the commutator)} \\
&= \hat{L}_+\hat{L}_z\psi_m + \hbar\hat{L}_+\psi_m &&\text{(distributing the product)} \\
&= m\hbar\hat{L}_+\psi_m + \hbar\hat{L}_+\psi_m &&\text{(using } \hat{L}_z\psi_m = m\hbar\psi_m\text{)} \\
&= (m + 1)\hbar(\hat{L}_+\psi_m) &&\text{(collecting common factors)} ,
\end{aligned}$$

which shows that the function $(\hat{L}_+\psi_m)$ is an eigenfunction of $\hat{L}_z$ with a component $\hbar$ larger than ψ_m. The operator $\hat{L}_+$ *raises* the z component one unit while keeping the magnitude fixed. Similarly, one can show that $\hat{L}_-$ *lowers* the component one unit.

Now recall how useful the harmonic oscillator annihilation operator was. Operating on the lowest energy eigenfunction, it gave zero. Similarly, since m is bounded, we cannot raise m forever. We must have

$$\hat{L}_+\psi_{m^+} = 0 .$$

Moreover, (operate by $\hat{L}_-$ on both sides of this equation)

$$\hat{L}_-\hat{L}_+\psi_{m^+} = 0 ,$$

and a little operator algebra shows that

$$\hat{L}_-\hat{L}_+ = \hat{L}_x^2 + \hat{L}_y^2 + i[\hat{L}_x, \hat{L}_y] = \hat{L}^2 - \hat{L}_z^2 - \hbar\hat{L}_z$$

so that we can also write

$$(\hat{L}^2 - \hat{L}_z^2 - \hbar\hat{L}_z)\psi_{m^+} = 0 \quad \text{or} \quad \hat{L}^2\psi_{m^+} = (\hat{L}_z^2 + \hbar\hat{L}_z)\psi_{m^+} .$$

The eigenvalue expression for $\hat{L}_z$ gives

$$\begin{aligned}
\hat{L}^2\psi_{m^+} &= \hat{L}_z^2\psi_{m^+} + \hbar\hat{L}_z\psi_{m^+} \\
&= (m^+)^2\hbar^2\psi_{m^+} + m^+\hbar^2\psi_{m^+} \\
&= m^+(m^+ + 1)\hbar^2\psi_{m^+} .
\end{aligned}$$

Comparing this result to our assumption that $\hat{L}^2\psi = L^2\hbar^2\psi$ shows that

$$L^2 = m^+(m^+ + 1) .$$

Since $\hat{L}_+$ and $\hat{L}_-$ do not alter the magnitude of **L,** we can identify l as m^+ and write

$$\hat{L}^2\psi_{l,m} = l(l + 1)\hbar^2\psi_{l,m} \quad (12.39a)$$

$$\hat{L}_z\psi_{l,m} = m\hbar\psi_{l,m} . \quad (12.39b)$$

We already argued that $m^+ = -m^-$, and this can be proven rigorously. One can also rigorously establish that $l = m^+ = -m^-$ and that l is not bounded, since there is no inherent upper bound to an angular momentum magnitude. Taken all together, we have

$$m = l, \; l - 1, \; l - 2, \; \ldots, -l \quad (12.40)$$

If m varies symmetrically about zero in integer steps, then l must be either integral or half-integral:

$$l = 0, \frac{1}{2}, 1, \frac{3}{2}, 2, \ldots, \quad (12.41)$$

which is a result that turns out to be more profound and general than we might have expected, since, as the last section in this chapter shows, the half-integral values are reserved by Nature for *intrinsic particle angular momenta,* so-called *spin*. For orbital motion, which derives from $\mathbf{L} = \mathbf{r} \times \mathbf{p}$, *only integral values of l are allowed*.

For the rest of this section, we will restrict ourselves to orbital motion and thus to integral l. The $\Theta(\theta)$ functions and the energy eigenvalues remain to be found. The latter come immediately. We write our particle-on-a-sphere Schrödinger equation again, but now index ψ with the two quantum numbers l and m:

$$\frac{\hat{L}^2}{2mR^2} \psi_{l,m} = E \psi_{l,m} = \frac{l(l+1)\hbar^2}{2mR^2} \psi_{l,m} \quad (12.42)$$

or simply

$$E = \frac{l(l+1)\hbar^2}{2mR^2} = \frac{l(l+1)\hbar^2}{2I}.$$

Note that E depends on l, but not on m. For any l, there are $2l + 1$ possible values of m:

$l = 0$	$m = 0$	$2l + 1 = 1$
$l = 1$	$m = +1, 0, -1$	$2l + 1 = 3$
$l = 2$	$m = +2, +1, 0, -1, -2$	$2l + 1 = 5$
etc.		

Thus, each energy state is $(2l + 1)$-fold degenerate, since each allowed orientation of $\mathbf{L}$ has the same energy.

To find $\Theta(\theta)$, given that we know $\Phi(\phi)$, we can solve the equation for $m = m^+ = l$ using the raising operator relation

$$\hat{L}_+ \psi_{l,l} = \hat{L}_+ \Theta_{l,l}(\theta) \, \Phi_l(\phi) = 0$$

and a spherical polar coordinate representation for $\hat{L}_+$. This is left as an exercise, and the result is

$$\Theta_{l,l}(\theta) = N_{l,l} \sin^l \theta$$

where $N_{l,l}$ is a normalization constant and where we have anticipated the result that Θ will in general depend on both l and m. Given $\psi_{l,l}$, we can find $\psi_{l,l-1}$ and so on down to $\psi_{l,-l}$ by repeated use of $\hat{L}_-$. This sequence of operations produces a series of functions of θ known as *Legendre polynomials*.

The product $\Theta_{l,m}\Phi_m$ is written $Y_{l,m}(\theta, \phi)$, the *spherical harmonic functions*. These functions are the angular factors of *any* wavefunction involving a spherically symmetric potential function and are thus quite important. A number of them are collected in Table 12.2.

All of this work with the properties of angular momentum has been building toward the hydrogen atom. We still have to solve the radial part of the H atom Schrödinger equation, and we need some way to picture angular-momentum wavefunctions. Since these pictures are so basic to all of contemporary chemical thought, we will postpone them until we can put them into the context of an atom.

12.4 ANGULAR MOMENTUM AND THE HYDROGEN ATOM

TABLE 12.2 Spherical Harmonic Functions, $Y_{l,m}(\theta, \phi)$

$Y_{l,m}(\theta, \phi) = \Theta_{l,m}(\theta)\,\Phi_m(\phi)$ $\quad\Phi_m(\phi) = \dfrac{1}{\sqrt{2\pi}}\,e^{im\phi}$

l	m	$\Theta_{l,m}(\theta)$
0	0	$(1/2)^{1/2}$
1	0	$(3/2)^{1/2}\cos\theta$
1	± 1	$(3/4)^{1/2}\sin\theta$
2	0	$(5/8)^{1/2}(3\cos^2\theta - 1)$
2	± 1	$(15/4)^{1/2}\sin\theta\cos\theta$
2	± 2	$(15/16)^{1/2}\sin^2\theta$
3	0	$(63/8)^{1/2}\left(\tfrac{5}{3}\cos^3\theta - \cos\theta\right)$
3	± 1	$(21/32)^{1/2}(5\cos^2\theta - 1)\sin\theta$
3	± 2	$(105/16)^{1/2}\sin^2\theta\cos\theta$
3	± 3	$(35/32)^{1/2}\sin^3\theta$

Note: $\Theta_{l,l} \propto \sin^l\theta$. See this chapter's Summary for a general expression for $Y_{l,m}$.

EXAMPLE 12.7

The HI bond length is about 1.6 Å. What rotational energies can this molecule have?

SOLUTION We will see how to answer this question in detail in Chapter 19, but here we can note that the great mass difference between H and I causes HI to rotate much as we picture our moon rotating about the Earth. The I atom is approximately stationary while the H orbits 1.6 Å away:

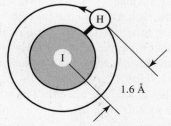

Eq. (12.42) is the energy expression we need. Using $R = 1.6$ Å and $m = 1$ amu, we have

$$E = \frac{l(l+1)\hbar^2}{2mR^2}$$

$$= l(l+1)\,\frac{(1.055 \times 10^{-34}\,\text{J s})^2}{2\,(1.66 \times 10^{-27}\,\text{kg})\,(1.6 \times 10^{-10}\,\text{m})^2}$$

$$= l(l+1)\,1.3 \times 10^{-22}\,\text{J}$$

where $l = 0, 1, 2, \ldots$. Two aspects of this exercise deserve emphasis. The first is that the energies of a system rotating in 3-D vary like $l(l+1)$(constant) while a system confined to 2-D has $E = m^2$(constant). The second point is the small magnitude of the energy constant here: $1.3 \times 10^{-22}\,\text{J} = 8.2 \times 10^{-4}$ eV. An electron "in orbit" will have energy levels spaced by amounts on the order of a few eV, but the greater mass of the orbiting H atom here makes the system more classical, lowering the energy level spacing by orders of magnitude.

→ **RELATED PROBLEMS** 12.29, 12.30

Probably no aspect of the general public's picture of an atom is more firmly rooted than that of electrons whizzing in orbits around a nucleus. This is dead wrong, a false popularization that got its start from Bohr's seminal theory of atomic hydrogen at the dawn of quantum theory. If the electron is going *around* the nucleus, then it *must* have orbital angular momentum. But we have found perfectly good wavefunctions for $l = 0$ that represent zero angular momentum. So if the electron is not whizzing *around* the nucleus, what is it doing?

The answer to this and related questions about the structure of atoms comes from considering the relative motion of the electron and proton, which is governed by the Coulombic potential-energy operator

$$\hat{V} = -\frac{e^2}{4\pi\epsilon_0 r} \tag{12.43}$$

where ϵ_0 is the permittivity of vacuum (a universal constant), e is the elementary charge, and r is the electron–proton distance.

The H atom solutions can be easily adapted to *any* atom or ion containing only two charged particles: H, He$^+$, Li^{2+}, etc., as well as so-called *exotic atoms*, such as *positronium* (symbolized Ps), which is composed of a normal electron and an antielectron (a *positron*) with the mass of an electron but an opposite charge. For the one-electron ions, such as He$^+$, etc., the Coulombic operator is Z times larger than that for H, where Z is the atomic number. We will not carry Z explicitly through derivations, but we will write the final results including Z as if we had.

Relative motion is governed by the reduced mass μ that we first encountered in the harmonic oscillator:

$$\mu = \frac{m_p m_e}{m_p + m_e} \cong m_e$$

since $m_p \gg m_e$. This inequality ($m_p/m_e \cong 1836$) has profound and recurring implications for chemistry. For atoms, it means that the atomic center of mass is located quite close to the center of the nucleus. The H center of mass is $m_e/(m_p + m_e) \cong 1/1837^{\text{th}}$ of the way from the proton toward the electron, a distance that, on average, will turn out to be about 36 times the characteristic radius of the proton (which is about 1.2×10^{-15} m). Thus, while the center of mass is much closer to the proton than the electron, both particles are appreciable distances from the center of mass in terms of their individual sizes.

The ground state of H has a nodeless wavefunction without angular momentum. The Schrödinger equation for this state is

$$-\frac{\hbar^2}{2\mu}\nabla^2\psi(r, \theta, \phi) - \frac{e^2}{4\pi\epsilon_0 r}\psi(r, \theta, \phi) = E\psi(r, \theta, \phi)$$

with ∇^2 expressed in spherical polar coordinates. Separating variables and setting the Θ and Φ functions to $Y_{0,0}(\theta, \phi)$ yields the equation for the radial function $R(r)$:

$$-\frac{\hbar^2}{2\mu}\left[\frac{1}{r^2}\frac{d}{dr}\left(r^2\frac{dR}{dr}\right)\right] - \frac{e^2}{4\pi\epsilon_0 r}R(r) = ER(r) \ . \tag{12.44}$$

This is just a one-dimensional problem; in fact, it resembles a simple trapped-particle problem if we look at it properly. First, the zero of energy corresponds to infinite separation ($r = \infty$), or, physically, to an ionized atom: H$^+$ and e^- infinitely separated. All *positive* energy solutions correspond to H$^+$ and e^- with finite kinetic energy at $r = \infty$. These are the unbound, continuum solutions. *Negative* energy solutions correspond to discrete energies with H$^+$ and e^- bound in a finite region of space as something we can recognize as an H atom.

For states with zero angular momentum, the H$^+$ and e^- move along a line connecting their centers. Classically, we would say they approach, accelerate due to mutual attraction, *pass through each other at $r = 0$*, then decelerate to a classical turning point, pause instantaneously, then reverse directions, approach, etc., *oscillating along a line*.

Atoms are supposed to be roundish in shape, not linear. And so they are; our quantum discussion of angular momentum told us that for $L = 0$ *we have no knowledge of the direction in space along which this linear oscillation occurs.* This argument is reviewed pictorially in Figure 12.12. Each attempt to find the direction in space along which the H$^+$ and e^- are oscillating will vary randomly over all possibilities. The average is spherically symmetric, and zero angular momentum atoms are spherical. (But the electron clearly does *not* orbit the nucleus in the classical sense!)

Next we find the ground-state wavefunction and its corresponding energy from Eq. (12.44). We begin by looking for a natural length with which we can scale r. For the harmonic oscillator, the classical turning point proved useful, and we try this choice again here. At the classical turning point, r_0, the total energy equals the potential energy:

$$E = -\frac{e^2}{4\pi\epsilon_0 r_0}.$$

Angular momentum = $\mathbf{r} \times \mathbf{p} = 0$

Classical Linear Motion

Quantum Mechanical Orientational Averaging

FIGURE 12.12 For a hydrogen atom in a state of zero angular momentum, the electron–proton (e^-–H$^+$) relative motion is confined to a line. Orientational uncertainty of this line yields a spherical distribution of e^-–H$^+$ separations on average.

Let $\rho = r/r_0$ so that $dr = r_0 d\rho$. Then Eq. (12.44) becomes, after canceling extra factors of r_0,

$$\frac{\hbar^2}{2\mu r_0}\left[\frac{1}{\rho^2}\frac{d}{d\rho}\left(\rho^2\frac{dR}{d\rho}\right)\right] + \frac{e^2}{4\pi\epsilon_0}\left[\frac{1}{\rho} - 1\right]R = 0$$

or, defining $\alpha^2 = 2\mu r_0 e^2/4\pi\epsilon_0 \hbar^2$ and performing the differentiation in the first term,

$$\frac{d^2R}{d\rho^2} + \frac{2}{\rho}\frac{dR}{d\rho} + \alpha^2\left[\frac{1}{\rho} - 1\right]R = 0.$$

If we go into the tunneling region ($r \gg r_0$ or $\rho \gg 1$), we expect the wavefunction to approach zero smoothly. In this limit, the terms containing $1/\rho$ above become smaller and smaller, and the equation approaches the asymptotic form

$$\rho \to \infty: \quad \frac{d^2R}{d\rho^2} = \alpha^2 R$$

with solutions $R(\rho) = $ (constant) $\exp(\pm\alpha\rho)$. Since $\rho \geq 0$ always, we reject the $+$ sign to keep R finite at $r = \infty$. We try $R = \exp(-\alpha\rho)$ in the full equation, neglecting the normalization constant for now:

$$\frac{d^2[\exp(-\alpha\rho)]}{d\rho^2} + \frac{2}{\rho}\frac{d[\exp(-\alpha\rho)]}{d\rho} + \alpha^2\left[\frac{1}{\rho} - 1\right]\exp(-\alpha\rho) = 0,$$

which becomes

$$\alpha^2 - \frac{2\alpha}{\rho} + \frac{\alpha^2}{\rho} - \alpha^2 = \frac{1}{\rho}(\alpha^2 - 2\alpha) = 0.$$

For this equation to hold for all ρ, the term multiplying $1/\rho$ must be zero:

$$\alpha^2 - 2\alpha = 0 \quad \text{or} \quad \alpha^2 = 4$$

so that, with the definition of α^2,

$$\frac{2\mu r_0 e^2}{4\pi\epsilon_0 \hbar^2} = 4$$

or

$$r_0 = 2\left(\frac{4\pi\epsilon_0 \hbar^2}{\mu e^2}\right) = 2a_0$$

where we define

$$a_0 = \frac{4\pi\epsilon_0 \hbar^2}{\mu e^2} \cong 52.918 \text{ pm} \cong \frac{1}{2} \text{ Å} \qquad (12.45)$$

the *Bohr radius*[8], which appeared naturally in Bohr's early theory of atomic hydrogen. (There has probably never been a wrong theory as useful as Bohr's theory of hydrogen.)

We now have shown that

$$\psi_0(r, \theta, \phi) = (\text{constant}) \exp(-2r/r_0)$$
$$= (\text{constant}) \exp(-r/a_0)$$

is the ground state wavefunction, and we have a numerical value for the ground-state energy:

$$E_1 = -\frac{e^2}{4\pi\epsilon_0 r_0} = -\frac{\mu e^4}{2(4\pi\epsilon_0)^2 \hbar^2} = -\frac{e^2}{2(4\pi\epsilon_0)a_0} \qquad (12.46a)$$
$$= -2.1787 \times 10^{-18} \text{ J} = -13.598 \text{ eV} = -109\,678 \text{ cm}^{-1}$$

where the cm^{-1} energy unit is common to many discussions of atomic and molecular systems. It is defined by

$$(\text{Energy/cm}^{-1}) = \frac{(\text{Energy/J})}{(hc/\text{J cm})} \cong \frac{(\text{Energy/J})}{1.9865 \times 10^{-23}} \,.$$

Note that E_1 is the negative of the ionization energy of H.

The normalization constant (call it N_0) is found from

$$1 = \iiint \psi_0^*(r, \theta, \phi)\, \psi_0(r, \theta, \phi)\, dx\, dy\, dz$$
$$= N_0^2 \iiint e^{-2r/a_0}\, r^2 \sin\theta\, dr\, d\theta\, d\phi$$
$$= 4\pi N_0^2 \int_0^\infty e^{-2r/a_0}\, r^2\, dr$$
$$= 4\pi N_0^2 \left[\frac{4}{(2/a_0)^3}\right]$$

so that[9]

$$N_0 = \left(\frac{1}{\pi a_0^3}\right)^{1/2} \,.$$

---

[8] Actually, a_0 is defined in terms of m_e rather than μ; the a_0 of Eq. (12.45) is a generalization of the expression Bohr first found for the specific case of H atom. Problems 12U and 12.33 consider the effect of varying nuclear masses.

[9] That last integral is not obvious. Starting from $\int_0^\infty e^{-\beta r} dr = \beta^{-1}$ and differentiating both sides with respect to β yields the general integral (after n such differentiations)

$$\int_0^\infty r^n e^{-\beta r}\, dr = \frac{n!}{\beta^{n+1}} \,.$$

We used the $n = 2$ form of this general result. See Problem 11.22 for another use of this trick.

These results are specific to H. For the general one-electron ion of nuclear change Z, one finds[10]

$$E_1 = -\frac{Z^2 e^2}{2(4\pi\epsilon_0)a_0} \quad \text{(12.46b)}$$

and

$$\psi_0 = \left(\frac{Z^3}{\pi a_0^3}\right)^{1/2} e^{-Zr/a_0} . \quad \text{(12.47)}$$

There are, of course, *excited* states of spherical symmetry, and each one successively higher in energy has one more node than the preceding. In spherically symmetric 3-D systems, these nodes are *spherical surfaces,* spheres at various r for which $\psi = 0$. Just as the Hermite polynomials appeared in the harmonic oscillator, a set of polynomials in r appear as a factor in the H atom wavefunction. Similarly, we index each wavefunction by a quantum number, n, the *principal quantum number,* with

$$n = 1, 2, 3, 4 \ldots \quad \text{(12.48)}$$

Since the state with principal quantum number n has $n - 1$ nodes, the spherical states of hydrogen have wavefunctions of the general form

$$\underset{\substack{\uparrow \\ \text{Spherically} \\ \text{symmetric} \\ \text{wavefunction}}}{\Psi_n(r)} = \underset{\substack{\uparrow \\ \text{Normalization} \\ \text{constant}}}{N_n} \underset{\substack{\uparrow \\ \text{Polynomial in } n \\ \text{with } n-1 \text{ nodes}}}{\left(\sum_{i=0}^{n-1} c_i r^i\right)} \underset{\substack{\uparrow \\ \text{Common} \\ \text{exponential} \\ \text{factor}}}{\exp(-\alpha_n r)}$$

where the c_i's must be chosen to satisfy Eq. (12.44), the radial Schrödinger equation, and where the coefficient α_n depends on n. Substitution of $\Psi_n(r)$ into Eq. (12.44) leads to a differential equation with solutions known as *associated Laguerre polynomials.* They are derivatives of functions called simply Laguerre polynomials, and hence they are indexed by two integers to denote the polynomial and the degree of differentiation. One integer is n, and the other is l, the total orbital angular-momentum quantum number.

Solutions to the full Schrödinger equation involve associated Laguerre polynomials as well. Therefore, we go directly to the full equation and the complete set of H atom wavefunctions. We introduce angular factors into Eq. (12.44):

$$-\frac{\hbar^2}{2\mu}\nabla^2 \psi(r, \theta, \phi) - \frac{e^2}{4\pi\epsilon_0 r}\psi(r, \theta, \phi) = E\psi(r, \theta, \phi)$$

with ∇^2 the full polar coordinate Laplacian operator. In detail, this is

$$-\frac{\hbar^2}{2\mu}\left[\frac{1}{r^2}\frac{\partial}{\partial r}\left(r^2\frac{\partial \psi}{\partial r}\right)\right] - \frac{\hbar^2}{2\mu r^2}\left[\frac{1}{\sin\theta}\frac{\partial}{\partial \theta}\left(\sin\theta \frac{\partial \psi}{\partial \theta}\right) + \frac{1}{\sin^2\theta}\frac{\partial^2 \psi}{\partial \phi^2}\right] - \frac{e^2}{4\pi\epsilon_0 r}\psi = E\psi ,$$

$$\text{(12.49)}$$

which looks formidable until we remember that *the angular part has already been solved.* Using Equations (12.33) and (12.42), the second term is

$$\frac{\hat{L}^2}{2\mu r^2}\psi(r, \theta, \phi) = \frac{l(l+1)\hbar^2}{2\mu r^2}\psi(r, \theta, \phi)$$

---

[10]In this and all expressions for species other than H, one must also correct the Bohr radius a_0 for the change in nuclear mass. This is a small correction; see Practice Problem 12U.

so that Eq. (12.49) simplifies to

$$-\frac{\hbar^2}{2\mu r^2}\left[\frac{\partial}{\partial r}\left(r^2\frac{\partial \Psi}{\partial r}\right)\right] + \left[\frac{l(l+1)\hbar^2}{2\mu r^2} - \frac{e^2}{4\pi\epsilon_0 r}\right]\Psi = E\Psi .\qquad(12.50)$$

Note that *this is just a one-dimensional equation again.* Since $\Psi = R\Theta\Phi$, we can immediately cancel the $\Theta\Phi$ factors (the spherical harmonic polynomials $Y_{l,m}$) and reduce Eq. (12.50) to an equation for $R(r)$ alone.

The sum $l(l+1)\hbar^2/2\mu r^2 + V(r)$ is known as the *effective potential*, and $l(l+1)\hbar^2/2\mu r^2$ alone is the *centrifugal potential*. It expresses the centrifugal (radially outward) fictitious force felt by the system in a non-zero angular momentum state. For each value of l, we have a *different effective potential*, and thus a new problem to solve. Fortunately, the associated Laguerre polynomials handle all l values, but it is helpful to think of the effective potentials classically before jumping to the quantum solution. Figure 12.13 plots effective potentials for several l values, including the $l = 0$ spherically symmetric case we have been discussing. Since the centrifugal potential is positive and varies as r^{-2}, it dominates the Coulomb potential at small r. Every $l \ne 0$ effective potential diverges to $+\infty$ as $r \to 0$, but the $l = 0$ potential diverges to $-\infty$. Classically, $l \ne 0$ implies orbiting, going around instead of through the origin, and we can speak of *inner and outer classical radial turning points* for such states. Quantum mechanically, we should expect tunneling toward $r = 0$ for $l \ne 0$, *but, unlike the $l = 0$ case, Ψ is zero at $r = 0$.*

The full solutions to the general radial Schrödinger equation, Eq. (12.50), contain four factors that we list here for reference. These complicated expressions will make sense once we graph a few of them and discuss their general properties. The general radial wavefunction is:

$$R_{nl}(r) = N_{nl}\left[L_{n+l}^{2l+1}\left(\frac{2r}{na_0}\right)\right]\left[\exp\left(-\frac{r}{na_0}\right)\right]\left[\left(\frac{2r}{na_0}\right)^l\right] .$$

In turn, the four factors in R are the *normalization constant*

$$N_{nl} = -\left[\frac{(n-l-1)!}{2n[(n+l)!]^3}\right]^{1/2}\left(\frac{2}{na_0}\right)^{3/2},$$

the *associated Laguerre polynomials*, defined by

$$L_{n+l}^{2l+1}\left(\frac{2r}{na_0}\right) = \sum_{k=0}^{n-l-1}(-1)^{k+1}\frac{[(n+l)!]^2}{(n-l-1-k)!(2l+1+k)!k!}\left(\frac{2r}{na_0}\right)^k,$$

the *exponential factor* $\exp(-r/na_0)$, and a *radial factor* $(2r/na_0)^l$ that assures

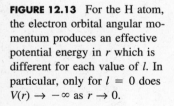

FIGURE 12.13 For the H atom, the electron orbital angular momentum produces an effective potential energy in r which is different for each value of l. In particular, only for $l = 0$ does $V(r) \to -\infty$ as $r \to 0$.

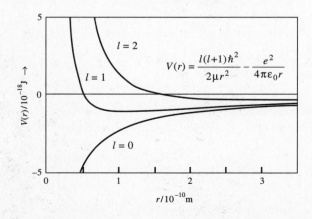

12.4 ANGULAR MOMENTUM AND THE HYDROGEN ATOM

TABLE 12.3 Hydrogenic Atom Wavefunction Radial Factors, $R_{nl}(r)$

n	l	$R_{nl}(r)$
1	0	$2\alpha e^{-\rho}$
2	0	$(8)^{-1/2}\alpha(2-\rho)e^{-\rho/2}$
2	1	$(24)^{-1/2}\alpha\rho e^{-\rho/2}$
3	0	$\dfrac{2\alpha}{81\sqrt{3}}(27-18\rho+2\rho^2)e^{-\rho/3}$
3	1	$\dfrac{4\alpha}{81\sqrt{6}}(6\rho-\rho^2)e^{-\rho/3}$
3	2	$\dfrac{4\alpha}{81\sqrt{30}}\rho^2 e^{-\rho/3}$

$\alpha = (Z/a_0)^{3/2}$ $\qquad \rho = Zr/a_0$

$R(r=0) = 0$ for $l \neq 0$. (The normalization constant is negative due to the conventional definition of L_{n+l}^{2l+1}.) For a one-electron ion of nuclear charge Z, we change the ratio r/na_0 to Zr/na_0 in each factor and multiply N_{nl} by $Z^{3/2}$. The first few R_{nl} functions are given in Table 12.3.

These functions are normalized in the sense that

$$\int_0^\infty R_{nl}^2(r)\, r^2\, dr = 1 \; .$$

Note that the integrand includes r^2 from the spherical polar volume element $r^2 \sin\theta\, dr\, d\theta\, d\phi$.

For each effective potential (for each l) an infinite number of n values are allowed, but with the restriction $n \geq l + 1$. Thus, for $l = 0$, $n = 1, 2, 3, \ldots$, while for $l = 1, n = 2, 3, 4, \ldots$, and so on. This restriction is usually stated with an emphasis on n rather than l:

$$n = 1, 2, 3, \ldots \tag{12.48}$$

$$l = 0, 1, 2, \ldots, n - 1 \; , \tag{12.51}$$

which shows that l is bounded for any n. We can include the m quantum number for the angular momentum component from Eq. (12.40):

$$m = l, l - 1, l - 2, \ldots, -l \; . \tag{12.40}$$

All three quantum numbers are needed to specify H wavefunctions. We write

$$\Psi_{nlm}(r, \theta, \phi) = R_{nl}(r)\, \Theta_{lm}(\theta)\, \Phi_m(\phi) = R_{nl}(r)\, Y_{l,m}(\theta, \phi) \; .$$

The first few complete wavefunctions are given in Table 12.4.

In elementary discussions of atoms in general chemistry, you probably encountered these quantum numbers and the alternate nomenclature used in connection with them. Buried somewhere in the prequantum history of spectroscopy are the terms *sharp, principal, diffuse,* and *fundamental* that were used to classify various series of atomic transitions. These series were ultimately found to be related to transitions with common l values, and the first letter of each word was associated

TABLE 12.4 Hydrogenic Atom Wavefunctions, $\Psi_{nlm}(r, \theta, \phi)$

n	l	m	$\Psi_{nlm}(r, \theta, \phi)$
1	0	0	β_1
2	0	0	$\dfrac{\beta_2}{4\sqrt{2}}(2 - \rho)$
2	1	0	$\dfrac{\beta_2}{4\sqrt{2}}\rho \cos\theta$
2	1	± 1	$\dfrac{\beta_2}{8}\rho \sin\theta\, e^{\pm i\phi}$
3	0	0	$\dfrac{\beta_3}{81\sqrt{3}}(27 - 18\rho + 2\rho^2)$
3	1	0	$\dfrac{\sqrt{2}\,\beta_3}{81}(6\rho - \rho^2)\cos\theta$
3	1	± 1	$\dfrac{\beta_3}{81}(6\rho - \rho^2)\sin\theta\, e^{\pm i\phi}$
3	2	0	$\dfrac{\beta_3}{81\sqrt{6}}\rho^2(3\cos^2\theta - 1)$
3	2	± 1	$\dfrac{\beta_3}{81}\rho^2 \sin\theta \cos\theta\, e^{\pm i\phi}$
3	2	± 2	$\dfrac{\beta_3}{2\cdot 81}\rho^2 \sin^2\theta\, e^{\pm 2i\phi}$

$\beta_n = (1/\sqrt{\pi})(Z/a_0)^{3/2}\, e^{-\rho/n}$ $\rho = Zr/a_0$

with its corresponding value. To this day, we use the notation

$$\begin{array}{c} l = 0 \ 1 \ 2 \ 3 \ 4 \ 5 \ 6 \ \ldots \\ s \ p \ d \ f \ g \ h \ i \ \ldots \end{array}$$

where, from f on, we proceed alphabetically, skipping j (which is reserved to represent angular momentum in general). One rarely needs to go past $l = 8$ in *any* chemical problem, with the vast majority of chemistry confined to s, p, d, and f orbital angular momenta. Thus, we speak of a *1s state*, meaning $n = 1$ and $l = 0$, or a *3d state* ($n = 3$, $l = 2$), etc.

Finally, we need a general energy expression. The Bohr theory, which agreed with observed H spectra, predicted that energy varied with $1/n^2$, implying that l and m (which did not appear in Bohr's original theory) do not govern the energy. The Schrödinger theory agrees with this, finding

$$E_n = \frac{E_1}{n^2} \leq 0 \qquad (12.52)$$

with, from Eq. (12.46b),

$$E_1 = \text{the ground-state } (n = 1) \text{ energy} = -\frac{Z^2 e^2}{2(4\pi\epsilon_0)a_0}. \qquad (12.46b)$$

Since any n value has n different l values associated with it, and each l has

$(2l + 1)$ different m values, the nth energy level is (at least, since another factor appears in the next section) n^2-fold degenerate, since one can show that

$$\sum_{l=0}^{n-1} (2l + 1) = n^2 .$$

We have now found the energies and wavefunctions for the *bound* states of all one-electron atoms. The *unbound* states have wavefunctions that are mathematically more difficult to treat, especially to normalize (recall the similar difficulty for the free-particle wavefunctions), and we will not discuss them here. Instead, we concentrate on the bound states and ask, "Where is the electron?" The answer will show us what the associated Laguerre and spherical harmonic functions look like.

The answer to this question is equivalent to the outcome of many repeated measurements of the electron's position and tells us all we can hope to know about the *size* and *shape* of the atom. We can make a few general observations immediately. We know that any s state (one with $l = 0$) is spherically symmetric and contains $n - 1$ spherical nodal surfaces. The $l \neq 0$ functions also have nodes, but in the form of one or more *planes* or other nonspherical surfaces. The lth angular function contains l such nodal planes. Since all states of a given n are degenerate, the *total* number of nodal spheres and planes must equal $n - 1$, no matter what the value of l. Thus, a state characterized by quantum numbers (n, l) has

$n - 1$ *total nodal surfaces* of which $n - l - 1$ *are spherical.*

Note that m does *not* govern the *number* of nodes; it will, however, govern their *shape and orientation in space.*

The *size* of the atom is the other quantity of interest. One measure of size is $\langle \hat{r} \rangle$, the average electron–nuclear separation, which is computed in the usual way:

$$\langle \hat{r} \rangle = \int_{\substack{\text{all} \\ \text{space}}} \Psi^*_{nlm} r \, \Psi_{nlm} \, d\tau .$$

From the general expression for Ψ_{nlm}, one can show that this integral is

$$\langle \hat{r} \rangle = \frac{a_0}{2Z} [3n^2 - l(l + 1)] .$$

Note that m does not enter the equation, that $\langle \hat{r} \rangle = 3a_0/2$ for the wavefunction of the 1s ground state, and that $\langle \hat{r} \rangle$ increases as n^2. The effect of l (for any given n) is relatively slight ($\langle \hat{r} \rangle$ decreases from $24a_0/Z$ to $18a_0/Z$ for the 4s to 4f wavefunctions).

EXAMPLE 12.8

How does the probability of finding the electron a distance r from the nucleus vary with r for the 1s, 2s, and 3s wavefunctions?

SOLUTION This question requires some careful interpretation. Every s wavefunction is spherically symmetric; the angular part is just $Y_{0,0}(\theta, \phi) = (\sqrt{4\pi})^{-1}$. If we sit on the nucleus and look out to larger r, we see the same thing in all directions. If we walk away from the nucleus along a radial straight line, we could imagine probing for the electron as we go. This experiment would find a probability equal to R^2_{nl} per unit length along the radial path.

An alternate and somewhat more useful measure of the radial variation of probability is the *radial distribution function* based on the following idea. We again start at the nucleus and move out radially to some position r, but instead of measuring the *local* probability at r along this radius, we probe around the *entire spherical surface of radius r*. We always look back and forth along a small radial increment dr, but instead of probing just a length

dr along the radius, we probe *throughout* a thin spherical shell of volume $4\pi r^2\, dr$. What we would find is the quantity $4\pi r^2 R_{nl}^2$ per unit radius. This is called the *radial distribution function*. It is plotted below (as the dimensionless quantity $4\pi r^2 a_0 R_{nl}^2$ versus the dimensionless length r/a_0) for the 1s, 2s, and 3s wavefunctions. Also shown are $\langle \hat{r} \rangle$ and the distances over which the electron would move classically. The radial distribution function (which is valid for any wavefunction, not just the spherically symmetric ones) clearly shows the radial nodes and the radial localization of the electron to predominant regions of space that are quite rapidly moving to larger r as n increases.

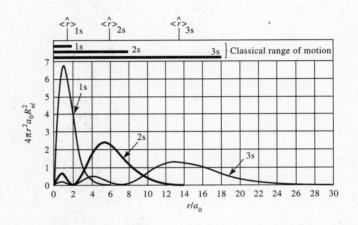

⟹ **RELATED PROBLEMS** 12.36, 12.37, 12.38

To picture the overall size, shape, and orientation of the atom, imagine the following. You make repeated measurements of the instantaneous position of the electron relative to the nucleus, localizing the electron in some small volume. To keep track of these measurements, you construct a 3-D grid of cubes, each cube representing a small volume in space. Whenever you make a measurement, you place a marker of some kind in the corresponding cube. After many such measurements, you stand back, look at the grid, and use the markers to picture the atom as best as it can be pictured.

This measurement cannot be done easily, as you might imagine, but the mathematical calculation that simulates it is not so difficult. We know to interpret

$\Psi^*_{nlm} \Psi_{nlm}\, d\tau$ = probability of being in volume element $d\tau$ centered at (r, θ, ϕ).

Thus, we compute these probabilities over a 3-D grid, and display them by markers (a varying number of dots, an intensity or color change on a video display, etc.) that are coded according to the probability value. Figures 12.14(a–d) show such calculations for the 1s, 2s, 3s, and 2p($m = 0$) wavefunctions of H.

These figures, which are drawn to the same scale, show a plane of points through the nucleus. For the s wavefunctions, the orientation of the plane is immaterial, due to spherical symmetry. For the $n = 2$, $l = 1$, $m = 0$ figure, the plane contains the z axis that is oriented along the vertical symmetry line of the two globs of dots.

The spherical nodes of the s wavefunctions are very obvious in these figures, as is the increase in size with increasing n. The p wavefunction has a single nodal plane (the x-y plane), and the shape is obviously nonspherical. The figure shows

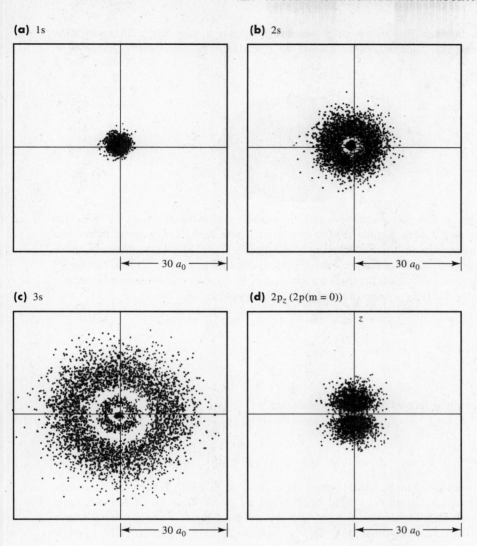

FIGURE 12.14 (a) Repeated measurements (5000 in these figures) of the electron's position in the 1s state yield a distribution as shown. The "size" of the atom is ill-defined, but the shape is spherical. (b) As in (a), but for the 2s orbital. The distribution is shown as a slice through the nucleus, exposing the radial node where the dot density goes to zero. (c) As in (b), but for the 3s state. Note the spherical shape, the two spherical nodes, and the increase in size on comparing 1s to 2s to 3s. (d) For the $2p_z$ orbital, the shape is nonspherical. Note the one nodal plane (the x-y plane) characteristic of p orbitals.

the $m = 0$ version of the 2p wavefunctions, which, from Table 12.4, depends on (r, θ, ϕ) as follows:

$$\psi_{210} \propto r\, e^{-r/2a_0} \cos\theta$$

Since $z = r\cos\theta$, we could write this as

$$\psi_{210} \propto z\, e^{-r/2a_0},$$

which says $\psi_{210} = 0$ whenever $z = 0$ (i.e., in the x-y plane), as the figure shows. For this reason, the $2p(m = 0)$ wavefunction is called $2p_z$.

What about the $2p(m = \pm 1)$ wavefunctions? Since (again from Table 12.4)

$$\psi_{21\pm 1} \propto r\, e^{-r/2a_0} \sin\theta\, e^{\pm i\phi}$$

we see that the probability density is again cylindrically symmetric about the z axis (since the ϕ dependence goes away in $\psi^*\psi$). The nodal surface for the probability is the line $x = y = 0$ (the z axis, since $\psi^*\psi \propto r^2 \sin^2\theta \propto x^2 + y^2$). The probability plot has the shape of a donut around the z axis for *both* functions:

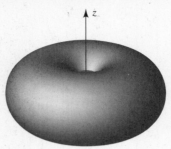

The $2p_z$ function is attractive since it is purely real, while the $m = \pm 1$ functions have the complex number factors $\exp(\pm i\phi)$. These three functions are the correct eigenfunctions for the z component of the atom's orbital angular momentum, in the sense that $\hat{L}_z\psi_{nlm} = m\hbar\psi_{nlm}$, but two functions other than $2p(m = \pm 1)$ are commonly used by chemists. Like $2p_z$ these are purely real functions, but they are *not* eigenfunctions of $\hat{L}_z$. They are the (normalized) linear combinations

$$\psi_+ = \frac{1}{\sqrt{2}}[\psi_{211} + \psi_{21-1}]$$

and

$$\psi_- = \frac{i}{\sqrt{2}}[\psi_{211} - \psi_{21-1}] \ .$$

The angular factors of these functions are

$$\psi_+ \propto \sin\theta\,[e^{i\phi} + e^{-i\phi}] = 2\sin\theta\cos\phi$$

and

$$\psi_- \propto -i\sin\theta[e^{i\phi} - e^{-i\phi}] = 2\sin\theta\sin\phi \ .$$

The wavefunction ψ_+ has the nodal plane $\cos\phi = 0$, which is the y-z plane, while ψ_- is zero when $\sin\phi = 0$, or when in the x-z plane. Thus, ψ_+ is zero whenever $x = 0$, and ψ_- is zero whenever $y = 0$. We call these linear combination wavefunctions *hybrids,* and, in analogy to $2p_z$, these are the $2p_x$ and $2p_y$ (for ψ_+ and ψ_-) hybrid wavefunctions. In summary, the 2p wavefunctions most often used by chemists are

$$2p_z = \psi_{210}$$

$$2p_x = \frac{1}{\sqrt{2}}[\psi_{211} + \psi_{21-1}]$$

$$2p_y = -\frac{i}{\sqrt{2}}[\psi_{211} - \psi_{21-1}] \ .$$

A picture of $2p_x$ and $2p_y$ probability distributions looks like that for $2p_z$ (Figure 12.14(d)) except the symmetry axis of the lobes is changed from z to x or y. No matter what value we pick for n, we can make p_x and p_y hybrids by the preceding prescription.

For states of higher orbital angular momentum a similar construction of hybrids yields wavefunctions that are purely real and are commonly used by chemists to visualize the directional properties that govern bonding geometries and the like. Table 12.5 summarizes the hybrids used for p, d, and f functions in terms of the fundamental ψ_{nlm} functions. Just as p_x is subscripted by x to denote the nodal plane ($x = 0$), subscripts are used on the d and f functions to indicate the nodal shapes and orientations. For example, d_{xy} is zero whenever $xy = 0$, that is, when $x = 0$

TABLE 12.5 Hybrid H-atom Wavefunctions

$n \geq 2$

$$np_z = \Psi_{n10}$$
$$np_x = \frac{1}{\sqrt{2}}(\Psi_{n11} + \Psi_{n1-1}) \qquad np_y = \frac{-i}{\sqrt{2}}(\Psi_{n11} - \Psi_{n1-1})$$

$n \geq 3$

$$nd_{3z^2-r^2} = \Psi_{n20} \text{ (often called } nd_{z^2})$$

$$nd_{xz} = \frac{1}{\sqrt{2}}(\Psi_{n21} + \Psi_{n2-1}) \qquad nd_{yz} = \frac{-i}{\sqrt{2}}(\Psi_{n21} - \Psi_{n2-1})$$

$$nd_{x^2-y^2} = \frac{1}{\sqrt{2}}(\Psi_{n22} + \Psi_{n2-2}) \qquad nd_{xy} = \frac{-i}{\sqrt{2}}(\Psi_{n22} - \Psi_{n2-2})$$

$n \geq 4$

$$nf_{z(5z^2-3r^2)} = \Psi_{n30} \text{ (often called } nf_{z^3})$$

$$nf_{x(5z^2-r^2)} = \frac{1}{\sqrt{2}}(\Psi_{n31} + \Psi_{n3-1}) \qquad nf_{y(5z^2-r^2)} = \frac{-i}{\sqrt{2}}(\Psi_{n31} - \Psi_{n3-1})$$

$$nf_{z(x^2-y^2)} = \frac{1}{\sqrt{2}}(\Psi_{n32} + \Psi_{n3-2}) \qquad nf_{xyz} = \frac{-i}{\sqrt{2}}(\Psi_{n32} - \Psi_{n3-2})$$

$$nf_{x(x^2-3y^2)} = \frac{1}{\sqrt{2}}(\Psi_{n33} + \Psi_{n3-3}) \qquad nf_{y(y^2-3x^2)} = \frac{-i}{\sqrt{2}}(\Psi_{n33} - \Psi_{n3-3})$$

Note: The ns ($n \geq 1$) wavefunctions are the Ψ_{n00} wavefunctions.

(the *y-z* plane) *or* when $y = 0$ (the *x-z* plane). For a complicated function like $f_{z(5z^2-3r^2)}$, saying that the nodal surfaces are given by $z(5z^2 - 3r^2) = 0$ and *visualizing* them are obviously very different matters! Thus, to guide your visualization, Figures 12.15 and 12.16 show probability distributions for the 3d and 4f functions. For most of these, three orthogonal views are necessary, but the $m = 0$ functions still (always) have cylindrical symmetry.

Note how the p functions have one distinct nodal plane, the d's have two, and the f's have three, as expected: l is the total number of nodal planes. The $d_{3z^2-r^2}$ and $f_{z(5z^2-3r^2)}$ functions have two nodal surfaces that are *cones* with the z axis as the cone axis. These two functions are frequently called simply d_{z^2} and f_{z^3}.

We will end this very important section on the hydrogen atom here, but there are many more aspects of the solutions to be explored in later chapters. We have yet to go deeply into the H atom spectrum, or to begin a discussion of the other hundred-plus elements, or to start chemistry by making H_2! Before we can do any of these, we need to introduce the last remaining simple quantum system, the intrinsic angular momentum of the elementary particles.

12.5 "IT MIGHT AS WELL BE SPIN"

This section's title is the title of an article published in the July, 1976, issue of *Physics Today* celebrating the fiftieth anniversary of the discovery of the intrinsic angular momentum known as spin. The author, Samuel A. Goudsmit, was the codiscoverer along with George E. Uhlenbeck, who has a companion paper in the same issue.

436 12 QUANTUM MECHANICAL MODEL SYSTEMS

FIGURE 12.15 Views through a 3-dimensional array of 5000 measurements of the electron's position in each of the 3d H-atom orbitals.

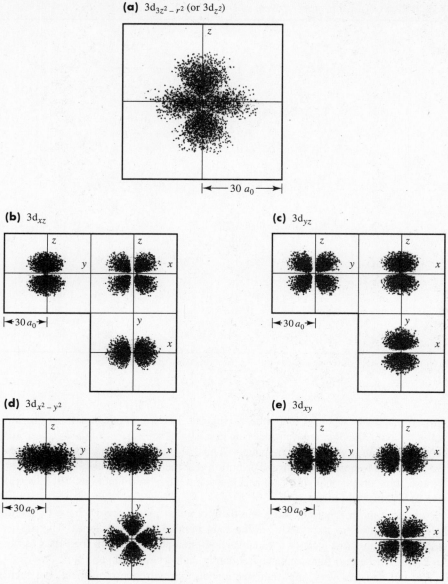

Goudsmit and Uhlenbeck were students in Holland in 1925, working on the forefront of atomic spectroscopy, trying to make some sense of the emission spectra of atoms larger than H. They had Bohr's theory and the Bohr–Sommerfeld extended theory, but not Schrödinger's theory, to guide them. The spectra made sense, up to a point. Goudsmit and Uhlenbeck were, in Goudsmit's continuation of his title's analogy, "starry-eyed and vaguely discontented." Something was missing, and being deeply involved with the details of atomic spectroscopy while also being free to consult other leading physicists of their time—Pauli, Ehrenfest (their mentor), Bohr, etc.—they were able to find that something. It was an additional property of the electron. Not only do electrons have mass and charge, but they (and, as found later, protons and neutrons) have a built-in, non-zero angular momentum.

The term used for this property, *spin,* was chosen in analogy to a classical situation. If a charged sphere is spinning, it will have a magnetic moment. It was the electron's magnetic moment that altered the atomic spectra, and hence the

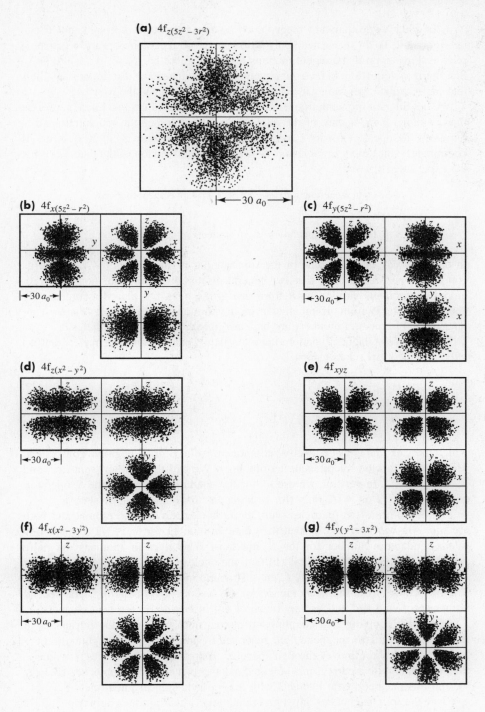

FIGURE 12.16 Views through a 3-dimensional array of 5000 measurements of the electron's position in each of the 4f H-atom orbitals.

angular momentum was inferred. But the picture of a spinning electron cannot be taken literally. In fact, H. A. Lorentz, the Dutch master of physics in the 1920s, quickly pointed out that for an electron to have the angular momentum postulated, its surface would have a speed about ten times that of light! So while the name is only suggestive, the phenomenon is real. It was not until Dirac produced a relativistic version of quantum theory that spin emerged naturally from postulates and became better understood.

A nonrelativistic quantum theory such as ours cannot *predict* spin, but it can incorporate it. To do so, we need only introduce the correct operators and eigenvalue relationships, even if the eigenfunctions for spin cannot be found by a solution to Schrödinger's equation. In fact, our eigenfunctions will be pure abstractions, symbols that indicate the eigenvalues as explicitly as the eigenvalues themselves.

We use angular momentum operators to describe spin functions. Let $\hat{S}^2$ represent the operator for the square of the total spin angular momentum magnitude and $\hat{S}_z$ represent the operator for its z component. (We cannot write $\hat{S}^2$ or $\hat{S}_z$ in terms of coordinates, since there *are no spin coordinates* in our version of the theory!) What the experiment showed is

$$\hat{S}^2 \text{ (spin eigenfunction)} = \frac{1}{2}\left(\frac{1}{2} + 1\right)\hbar^2 \text{ (spin eigenfunction)}$$

$$\hat{S}_z \text{ (spin eigenfunction)} = \pm\frac{1}{2}\hbar \text{ (spin eigenfunction)} .$$

In other words, the magnitude of the spin angular momentum is $(\sqrt{3}/2)\hbar$ and its component is $\pm\hbar/2$. Recall that our general discussion of angular momentum required components varying by $\hbar$ from one to the next, but placed symmetrically about zero. The H atom orbital angular momentum quantum numbers (l and m) are integers; spin quantum numbers are half-integral for elementary particles.[11]

The electron spin eigenfunctions are traditionally called α and β. They are defined *only* by the formal relationships

$$\hat{S}^2\alpha = \frac{1}{2}\left(\frac{1}{2} + 1\right)\hbar^2\alpha \qquad \hat{S}^2\beta = \frac{1}{2}\left(\frac{1}{2} + 1\right)\hbar^2\beta \qquad \textbf{(12.53a)}$$

$$\hat{S}_z\alpha = +\frac{\hbar}{2}\alpha \qquad \hat{S}_z\beta = -\frac{\hbar}{2}\beta , \qquad \textbf{(12.53b)}$$

which indicate that α has a positive component while β has a negative component. This has led to more picturesque terminology. We call the state α "spin up," and β, "spin down." In general, we use m_s for the quantum number of the $\hat{S}_z$ operator, with $m_s = +1/2$ or $-1/2$ only (for a single electron, or proton, or neutron).

The H atom has two particles with spin, the electron and the proton, and both are *spin one-half particles,* in that the quantum number for $\hat{S}^2$ is 1/2 (the analogy to the l quantum number for the $\hat{L}^2$ operator). The electron spin has the more pervasive influence on chemistry. (In fact, the nuclear spin is usually symbolized I instead of S; hence we write $\hat{I}^2$, $\hat{I}_z$, etc.) This importance stems from two sources. The first comes from the phenomenon known as *spin–orbit coupling.* The second is more profound and will occupy much of our discussion in the next chapters.

Spin–orbit coupling is a coupling—a mutual interaction—of the intrinsic-spin magnetic field and the magnetic field generated by non-zero orbital angular momentum. As the electron moves through space in a non-s state, it produces a magnetic field (in analogy to a current in a solenoid electromagnet, for example; see Chapter 9), and this magnetic field interacts with the intrinsic-spin magnetic field.

Since the spin magnet has only two orientations, it will produce only two slightly different interactions with the orbital motion field, in analogy to holding one permanent magnet in the field of another in either of two orientations: aligned, or "anti-aligned." These interactions slightly alter the H atom energies we found in the last section, where

---

[11]Electrons, protons, and neutrons have spin angular momentum component quantum numbers of $\pm 1/2$, but *nuclei composed of more than a single nucleon* can have *net* spin angular momentum components that are integral. For example, the D nucleus (proton + neutron) can have nuclear spin angular momentum components of $+1$, 0, or -1.

spin was ignored. It splits the three degenerate 2p energy levels, for example, into two levels separated by about 4 parts per million of the $n = 1$ to $n = 2$ energy separation. For larger values of l, the effect is even smaller, but still observable.

The proton spin also interacts with the electron spin, but it does so only *very* slightly, since the wavefunction for the electron is spread out over a volume many times that of the proton. But when it can[12] occur, *spin–spin interaction* is observable. The H 1s state is split by this interaction into a pair of energy levels, separated by only 0.0474 cm^{-1} (two million times less than the H-atom ionization energy, for comparison). These levels correspond to parallel and antiparallel nuclear and electron spins, to a rough approximation. The energy separation corresponds, via $\Delta E = h\nu = hc/\lambda$, to a 21.1 cm photon wavelength in the radio frequency region. These photons are observed in emission from H atoms in interstellar space, where the atoms spontaneously change from the higher to the lower energy level.

For the most part, spin effects on the energy levels of H are not chemically important. (They are important for larger atoms, however, as will be seen in the next chapter.) But if we neglect the way spin interactions *alter* H atom energy-level degeneracies, we have to include the spin quantum number m_s as part of that degeneracy. Since m_s can take either of two values, *each energy level is $2n^2$-fold degenerate*, instead of n^2-fold degenerate, in this approximation.

Figure 12.17 diagrams the H atom energy-level splittings at successively better approximation for the $n = 1$, 2, and 3 states. The final tiny splitting in the $n = 2$ state requires a theory that was developed in the 1940s, called quantum electrodynamics (QED). It is relativistic in origin and requires a uniform quantum description of both the atom and the electromagnetic field.

The most profound consequence of spin, however, is a topic for the start of the next chapter. It is embodied in the Pauli Exclusion Principle which governs the allowed quantum numbers of a system of several indistinguishable particles such as the identical electrons in a many-electron atom. This Principle is a postulate, in that Nature exhibits its consequences, but we do not understand it on a more fundamental basis any more than we understand why energy is conserved.

---

[12]For $l \neq 0$ states, the electron is never found at the nucleus, since $\psi(r = 0) = 0$ for such states.

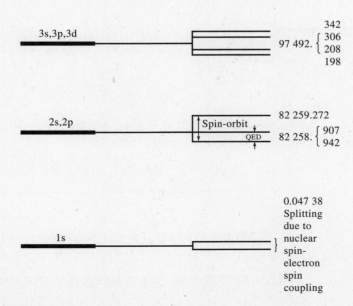

FIGURE 12.17 The degeneracy of the *nl* levels of H is only an approximate one. When spin interactions are considered, the degeneracy is lost. The experimental energies (in cm^{-1} units with the ground state at 0 cm^{-1}) are shown, along with the physical origins of the very small splittings discussed in the text.

CHAPTER 12 SUMMARY

The important results of this chapter are the various wavefunctions and eigenvalues for the half-dozen or so model potential-energy functions we have discussed, collected here for easy reference. The most important relations are outlined and are the ones you should concentrate on learning and understanding.

1-D free particle

$V(x) = $ a constant, from region to region

$\Psi(x) \propto \exp(\pm ikx)$, classically allowed region ($E > V$)
$\quad\quad\propto \exp(\pm \kappa x)$, classically forbidden region ($E < V$)

$$k = \frac{\sqrt{2m(E-V)}}{\hbar}, \quad \kappa = \frac{\sqrt{2m(V-E)}}{\hbar}$$

$$\boxed{E = \frac{\hbar^2}{2m}k^2}$$

1-D particle-in-a-box

$V(x) = 0, \ 0 \leq x \leq L; \ V(x) = \infty$ elsewhere

$$\boxed{\Psi(x) = \left(\frac{2}{L}\right)^{1/2} \sin\left(\frac{n\pi x}{L}\right), \ n = 1, 2, 3, \ldots; \ 0 \leq x \leq L \\ n = 0, \ x < 0 \text{ or } x > L}$$

$$\boxed{E = \frac{\hbar^2 \pi^2}{2mL^2} n^2}$$

1-D harmonic oscillator

$V(x) = \frac{1}{2} kx^2, \quad \omega = \sqrt{\frac{k}{\mu}} = 2\pi \nu$

$\Psi(x) = N_v H_v(q) e^{-q^2/2}, \quad q = \sqrt{\frac{k}{\hbar\omega}} x$

$N_v = \left[\sqrt{\frac{k}{\pi\hbar\omega}} \frac{1}{2^v v!}\right]^{1/2}$

$H_0(q) = 1, \quad H_1(q) = 2q,$

$H^{v+1}(q) = 2qH_v(q) - 2vH_{v-1}(q)$

$$\boxed{E = \hbar\omega\left(v + \frac{1}{2}\right) = h\nu\left(v + \frac{1}{2}\right), \quad v = 0, 1, 2, 3, \ldots}$$

Particle in 3-D rectangular box

$V(x) = 0, \ 0 \leq x \leq L_x, \ 0 \leq y \leq L_y, \ 0 \leq z \leq L_z$
$\quad\quad = \infty$ elsewhere

$\Psi(x, y, z) = \psi_x(x)\,\psi_y(y)\,\psi_z(z)$

$\psi_x(x) = \left(\frac{2}{L_x}\right)^{1/2} \sin\left(\frac{n_x \pi x}{L_x}\right), \ n_x = 1, 2, 3, \ldots, \ 0 \leq x \leq L_x$
$\quad\quad = 0$ elsewhere

(and similarly for ψ_y and ψ_z)

$$\boxed{E = \frac{\hbar^2 \pi^2}{2m}\left(\frac{n_x^2}{L_x^2} + \frac{n_y^2}{L_y^2} + \frac{n_z^2}{L_z^2}\right)}$$

3-D orbital angular momentum
angular solutions to *any* potential of the form $V = V(r)$

$\Psi_{lm}(\theta, \phi) = Y_{lm}(\theta, \phi)$

$Y_{lm}(\theta, \phi) = \left[\frac{(2l+1)}{4\pi} \frac{(l-|m|)!}{(l+|m|)!}\right]^{1/2} P_l^{|m|}(\cos\theta) e^{im\phi}$

$P_l^{|m|}(x) = (1-x^2)^{|m|/2} \frac{d^{|m|} P_l(x)}{dx^{|m|}}$

$P_0(x) = 1, \quad P_1(x) = x,$
$P_{l+1}(x) = [(2l+1)xP_l(x) - lP_{l-1}(x)]/(l+1)$

$$\boxed{\text{eigenvalues of } \hat{L}^2 = l(l+1)\hbar^2, \quad l = 0, 1, 2, \ldots}$$

$$\boxed{\text{eigenvalues of } \hat{L}_z = m\hbar, \quad m = -l, -l+1, \ldots, +l}$$

One-electron atom of atomic number Z

$V(r) = -\frac{Ze^2}{4\pi\epsilon_0 r}, \quad a_0 = \frac{4\pi\epsilon_0 \hbar^2}{\mu e^2} \cong 0.53 \text{ Å}$

$\Psi_{nlm}(r, \theta, \phi) = R_{nl}(r) Y_{lm}(\theta, \phi)$

$R_{nl}(r) = N_{nl}\left[L_{n+l}^{2l+1}\left(\frac{2r}{na_0}\right)\right]\left[\exp\left(-\frac{r}{na_0}\right)\right]\left[\left(\frac{2r}{na_0}\right)^l\right]$

$N_{nl} = -\left\{\frac{(n-l-1)!}{2n[(n+l)!]^3}\right\}^{1/2} \left(\frac{2}{na_0}\right)^{3/2}$

$L_{n+l}^{2l+1}\left(\frac{2r}{na_0}\right)$
$\quad = \sum_{k=0}^{n-l-1} (-1)^{k+1} \frac{[(n+l)!]^2}{(n-l-1-k)!(2l+1+k)!\,k!}\left(\frac{2r}{na_0}\right)^k$

$$\boxed{E_n = -\frac{Z^2 e^2}{2(4\pi\epsilon_0)a_0}\frac{1}{n^2}, \quad n = 1, 2, 3, \ldots}$$

$$\boxed{\begin{array}{l} l = 0, 1, 2, \ldots, n-1; \\ m = -l, -l+1, \ldots, 0, \ldots, l-1, l \end{array}}$$

Electron, proton, and neutron (individual) spin

$$\hat{S}^2\alpha = \frac{3}{4}\hbar^2\alpha \qquad \hat{S}^2\beta = \frac{3}{4}\hbar^2\beta$$

$$\hat{S}_z\alpha = +\frac{1}{2}\hbar^2\alpha \qquad \hat{S}_z\beta = -\frac{1}{2}\hbar^2\beta$$

This summary is a reference point. It is *not* a tabulation of formulas to memorize beyond the key results boxed above, but a few trends and proportionalities are worth committing to memory along with the math:

1-D particle in a box

- unevenly spaced energy levels
- energy proportional to n^2, to $\frac{1}{L^2}$, and to $\frac{1}{m}$
- sinusoidal wavefunction with $n-1$ nodes (not counting ends of the box) for state n

1-D harmonic oscillator

- evenly spaced (by $\hbar\omega$) energy levels
- wavefunction with v nodes for state v

3-D box

- symmetry (equal side lengths) induces degeneracy

3-D orbital angular momentum

- angular momentum magnitude $\sqrt{l(l+1)}\hbar$
- angular momentum axial component equally spaced (by $\hbar$)
- spherical symmetry for $l = 0$
- cylindrical symmetry about z axis for $m = 0$

One-electron atom

- energy: $E_n \cong (-13.6 \text{ eV})(Z^2/n^2)$
- $n-1$ total nodes, l of which are *not* spherical in shape
- degeneracy = $2n^2$ (including spin degeneracy)
- n governs size and energy, l governs shape and size, and m governs orientation and shape

Spin

- intrinsic angular momentum
- electron spin axial components $+\frac{1}{2}\hbar$ and $-\frac{1}{2}\hbar$

Finally there are pictures you should remember as well. Among these are the wavefunction pictures from this chapter, but another is the diagram below that plots on the same energy scale the energy levels of an electron in a 1-D box, a harmonic oscillator, and an H atom. The parameters of each model are shown below the corresponding diagram. The *pattern* of energy levels in each system is their most important feature.

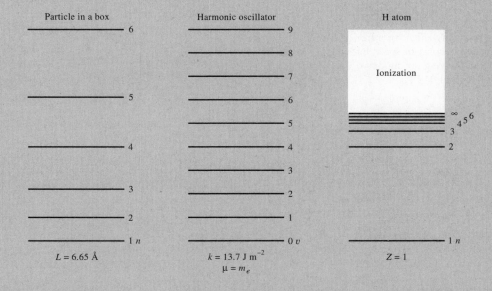

FURTHER READING

There are several classic texts that discuss model problems in either greater mathematical detail or from a different point of view than this chapter's. The earliest of these is *Introduction to Quantum Mechanics With Applications to Chemistry*, by Linus Pauling and E. Bright Wilson, Jr., (McGraw-Hill, New York, 1935). Many other model problems are worked in more or less detail, and the genesis of quantum chemistry is discussed in this book that appeared only nine years after Schrödinger's seminal work in the first half of 1926.

After another nine years, the book *Quantum Chemistry*, by Henry Eyring, John Walter, and George E. Kimball (Wiley, New York, 1944) appeared. The preface begins, "In so far as quantum mechanics is correct, chemical questions are problems in applied mathematics. In spite of this, chemistry, because of its complexity, will not cease to be in large measure an experimental science..." Written originally as a graduate level text, much of this book is at the level of this chapter. Undergraduates did not study quantum chemistry in any depth in 1944. It is a particularly interesting book for the earliest application of quantum mechanics to reaction rate theory, Eyring's specialty.

More recent texts of note include *Atoms and Molecules*, by Martin Karplus and Richard N. Porter, (Benjamin-Cummings, New York, 1970), *Molecular Quantum Mechanics*, by Peter W. Atkins, (Oxford University Press, New York, 1983), *Quantum Chemistry*, by Ira N. Levine, (Allyn & Bacon, Boston, 1983), and *Quantum Chemistry*, by Donald A. McQuarrie (University Science Books, Mill Valley, CA, 1983). These are specialized texts at more or less the level of this one.

PRACTICE WITH EQUATIONS

12A What is the wavevector of an electron moving 6×10^5 m s^{-1} (about 1 eV)? of a He atom moving at the same speed (about 7.3 keV)? of a He atom moving 1100 m s^{-1} (a typical room temperature He(g) speed)? (12.5)

ANSWER: $k_{e^-} = 5.2 \times 10^9$ m^{-1}; fast $k_{He} = 3.8 \times 10^{13}$ m^{-1}; slow $k_{He} = 6.9 \times 10^{10}$ m^{-1}

12B What is the energy of a free H atom with a 10^{10} m^{-1} wavevector? (12.7)

ANSWER: 3.4×10^{-21} J $\cong 0.02$ eV

12C What is the wavefunction for a free particle moving in the x direction at constant speed but with twice the probability of moving towards $x = +\infty$ as towards $x = -\infty$? (12.6), (12.8)

ANSWER: $\psi \propto \sqrt{\frac{2}{3}} e^{ikx} + \sqrt{\frac{1}{3}} e^{-ikx}$

12D How long should a 1-D potential box be to give an electron a 1 eV zero-point energy? (12.12)

ANSWER: 6.1×10^{-10} m = 6.1 Å

12E How would the energy in 12D above change if the box were doubled in length? (12.12)

ANSWER: 0.25 eV

12F An H$_2$ molecule moving in a 10 Å 1-D box in quantum state $n = 3$ has energy E. A D$_2$ molecule moving in a 5 Å 1-D box in quantum state $n = 4$ has energy E'. How are E and E' related? (12.12)

ANSWER: $E \propto 3^2/(2 \cdot 10^2)$; $E' \propto 4^2/(4 \cdot 5^2)$; $E/E' = 9/32$

12G A force of 1 N (about a 100 g mass hanging in gravity) extends a spring 1 cm. What is the spring's force constant? Hooke's Law

ANSWER: 100 N m^{-1}

12H If both ends of an H$_2$ molecule could be grabbed and pulled apart with a force of 5.7×10^{-10} N, the H$_2$ bond length would increase by 0.01 Å. What is the bond's force constant? Hooke's Law

ANSWER: 570 N m^{-1}, comparable to ordinary spring force constants

12I What is the ground-state energy for a harmonic oscillator vibrating at a frequency of 1 cycle per second? 10^{15} cycles per second? (12.18)

ANSWER: 3.31×10^{-34} J; 3.31×10^{-19} J

12J Where are the nodes in the $v = 2$ harmonic-oscillator wavefunction? (12.19)

ANSWER: $q = \pm 1/\sqrt{2}$

12K What is the next highest Hermite polynomial not listed in Table 12.1?

ANSWER: $H_7(q) = 128q^7 - 1344q^5 + 3360q^3 - 1680q$

12L What is the kinetic energy operator for a particle of mass m confined to a plane? (12.21)

ANSWER: $\hat{T} = -(\hbar^2/2m)(\partial^2/\partial x^2 + \partial^2/\partial y^2)$

12M What is the ground-state energy for a particle in a 3-D box with sides L, $\sqrt{2}L$, and $2L$? (12.24)

ANSWER: $7\hbar^2\pi^2/8mL^2$

12N For a particle-in-a-cube, how many states have the same energy as the state $(n_x, n_y, n_z) = (3,2,1)$? (12.24)

ANSWER: six (total)

12O What are the possible spherical polar coordinates of a point that is one unit distant from the x, y, and z axes?

ANSWER: $r = \sqrt{3}$, $\theta = 54.74°$ or $125.26°$, $\phi = 45°, 135°, 225°,$ or $315°$

12P What are the components of the angular momentum vector about the origin for a 2.0 kg particle located at $x = 0.3$ m ($y = z = 0$) moving along the $+z$ direction with a momentum of 2.5 kg m s^{-1}? (12.25)

ANSWER: $L_x = 0$, $L_y = -0.75$ kg m^2 s^{-1}, $L_z = 0$

12Q If the particle in 12P above is in fact orbiting the origin at a constant radius, what is its angular velocity ω and its moment of inertia? (12.28), (12.30)

ANSWER: $\omega = -4.17$ s^{-1}, $I = 0.18$ kg m^2

12R What is the value of this commutator of commutators: $[[\hat{L}_x, \hat{L}_y], [\hat{L}_y, \hat{L}_z]]$? (12.34)

ANSWER: $-i\hbar^3 \hat{L}_y$

12S What is the magnitude of the angular momentum for the state with the angular wavefunction $\psi_{4,2}$? (12.39)

ANSWER: $2\sqrt{5}\hbar$

12T What is the largest z component of angular momentum for a state with a total angular momentum equal to that of the state in 12S above? (12.39)

ANSWER: $4\hbar$

12U The ^{4}He nuclear mass is 4.002 6 u, and the ^{3}He nuclear mass is 3.016 0 u. What are the ionization energies in cm^{-1} units of ^{4}He$^+$ and ^{3}He$^+$? (12.46), Footnote 10

ANSWER: ^{4}He$^+$: 438 889.1 cm^{-1}; ^{3}He$^+$: 438 869.4 cm^{-1}

12V How many spherical nodes and planar nodes does a 6h wavefunction have?

ANSWER: no spherical nodes; five planar nodes

PROBLEMS

SECTION 12.1

12.1 A source of particles at $x = -\infty$ fires them towards $x = \infty$ so that half the particles have twice the momentum of the other half. The entire x axis is at a constant potential energy. What is the wavefunction for this system? What does the probability distribution look like? Are the particles in any way localized?

12.2 Use the Gaussian weighting function of Eq. (12.10) in the wave-packet integral of Eq. (12.9) and derive the Gaussian wave-packet expression given in the text.

12.3 This and the next two problems consider aspects of the simple 1-D potential step shown below:

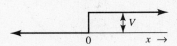

The potential is constant on both sides of $x = 0$, but at different values; for $x > 0$ it is an amount V greater than for $x < 0$. Consider first a source of particles at $x = -\infty$ moving toward $x = \infty$ with a total energy E such that classically the particles would pass over the potential step (slowing down) and continue on to $x = \infty$. Take the wavefunction to be

$$\psi_<(x) = e^{ik_< x} + B e^{-ik_< x}, \quad x < 0$$
$$\psi_>(x) = C e^{ik_> x} + D e^{-ik_> x}, \quad x > 0$$

where $k_<$ and $k_>$ are the wavevectors for $x < 0$ and $x > 0$, respectively.

(a) One of the coefficients, C or D, must be zero by the conditions of the problem. Which is it and why?

(b) Find an expression for B and the non-zero coefficient C or D by requiring $\psi_<(0) = \psi_>(0)$ and $(d\psi_</dx)_{x=0} = (d\psi_>/dx)_{x=0}$.

(c) The product B^*B is the probability that the particle is reflected from the potential step. It is zero classically. Verify that it goes to zero quantum mechanically as the energy is increased.

12.4 Continuing from the previous problem, consider lowering the energy to a value that cannot surmount the barrier. Classically, no particles are found in the region $x > 0$, and all particles are reflected. The wavefunction for $x < 0$ takes the same form as in the previous problem, but for $x > 0$, the most general form is

$$\psi_>(x) = C e^{\kappa_> x} + D e^{-\kappa_> x}, \quad x > 0$$

where $\kappa_>$ is the (real number) tunneling wavevector.

(a) Again, either C or D must be zero. Which is it and why?

(b) Use continuity expressions as before to find expressions for B and the non-zero coefficient C or D.

(c) Show that B^*B, the reflection probability, equals 1, as in classical mechanics.

(d) Unlike classical mechanics, one of the coefficients C or D is non-zero, and the particle can be found in the region $x > 0$. What total energy maximizes the probability for barrier penetration?

12.5 Now work the following variant of the past two problems without much guidance. Consider the same potential function, but now place a source of particles at $x = \infty$ moving towards $x = -\infty$. Find the probability that the particles will be reflected at $x = 0$ back to their source. This is again a classical impossibility. Find the appropriate limit to the quantum probability that recovers this classical expectation.

SECTION 12.2

12.6 Suppose an electron is confined to a 1-D box that is so short in length that the zero-point energy of the electron equals the energy equivalent of its mass, $m_e c^2$. Find the length of such a box (algebraically and numerically). Your answer will be of the same order of magnitude as the quantity called the *Compton wavelength of the electron*, $h/m_e c$.

12.7 The probability distributions for particle-in-a-box states show maxima that are more closely spaced as n increases. (See Figure 12.3(c).) For some n, (call it n^*), the uncertainty in the particle's location (measured by Δx) is going to be larger than the spacing between adjacent maxima in ψ^2. This is one criterion for the onset of classical behavior. Find n^*, noting that the maxima are spaced by L/n. To help you find Δx, use the expression mentioned in Problem 11.23, and use the result

$$\langle \hat{x}^2 \rangle = L^2 \left(\frac{1}{3} - \frac{1}{2n^2\pi^2} \right).$$

Before you begin the calculation, take a guess at the magnitude of the n^*. Is it on the order of 1? of 10? of many powers of 10? You may be surprised at how small n^* turns out to be.

12.8 The classical probability distribution for a harmonic oscillator (shown in Figure 12.8) is easy to derive. Let $P(x) \, dx$ be the probability that the oscillator is found in the region x to $x + dx$. This probability must be proportional to the time dt spent in the interval dx. In fact, it is just the ratio of dt to the time $\tau_{1/2}$ spent traversing one half-cycle of motion where $\tau_{1/2} = 1/(2\nu) = \pi/\omega$. Classically, $x(t) = X \cos \omega t$ for an oscillator with maximum displacement X from its rest length x_0. Use the identity $dt = dx/(dx/dt)$ to show that

$$P(x) \, dx = \frac{dx}{\pi(X^2 - x^2)^{1/2}}.$$

Verify the normalization requirement that

$$\int_{-X}^{+X} P(x) \, dx = 1.$$

12.9 Consider the Schrödinger equation for the harmonic oscillator. What does it say about the shape of the wavefunction (the derivatives of ψ) at the *classical turning points* of the motion? Rewrite Eq. (12.17) as

$$-\frac{\hbar^2}{2\mu} \frac{d^2\psi_v}{dx^2} + \left(\frac{1}{2} kx^2 - E_v \right) \psi_v = 0$$

and consider the definition of the classical turning point for state v: $E_v = kX_v^2/2$. Are your conclusions born out in Figure 12.8? Can you draw a general conclusion about the behavior of ψ at the classical turning point of any bound potential?

12.10 The harmonic oscillator ground-state wavefunction of Eq. (11.9) is written in terms of x and the classical turning point, $X = X_0$. Write the wavefunction for the first and second excited states also in terms of x and each state's classical turning point, X_1 and X_2.

12.11 Bonded atoms are connected by a potential function that is closely harmonic for small stretches and compressions of the bond about its equilibrium length. An interesting question is the range of vibrational motion about this equilibrium. Express this range as the ratio of the ground state's classical turning point X_0 to the equilibrium bond length for the diatomics H_2 and I_2. These species have, respectively, about the largest and smallest vibrational motion ranges. You will need to know the harmonic force constants k, equilibrium bond lengths R_e, and effective oscillator masses m:

	k/N m^{-1}	R_e/Å	m/kg
H_2	575.17	0.741 44	8.368×10^{-28}
I_2	172.0	2.666	1.053×10^{-25}

The effective mass is the reduced mass which, for a homonuclear diatomic, is just half the mass of the atom.

12.12 Consider the potential below, which is a hybrid of a 1-D particle-in-a-box and a harmonic oscillator:

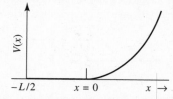

For $x > 0$, $V(x)$ is harmonic with force constant k; for $-L/2 < x < 0$, $V(x) = 0$; and for $x < -L/2$, $V(x) = \infty$. Imagine a particle of mass m in this potential. What condition(s) must hold among m, k, and L in order that the ground-state wavefunction have its maximum at $x = 0$?

12.13 The position operator for a harmonic oscillator can be written in terms of the creation and annihilation operators:

$$\hat{q} = \frac{1}{2} \left[\left(q - \frac{d}{dq} \right) + \left(q + \frac{d}{dq} \right) \right].$$

Use this expression to show that the integrals

$$q_{nm} = \int_{-\infty}^{\infty} \psi_n \hat{q} \psi_m \, dq$$

are zero unless $n = m \pm 1$, and find expressions for $q_{n, n+1}$ and $q_{n, n-1}$.

12.14 The q_{nm} integral defined in the previous problem is useful in understanding certain molecular spectroscopy questions. If we think of n and m each varying from 0 on up, we could arrange all the possible values of q_{nm} in the form of a square array (of infinite extent) called the *matrix representation* of the $\hat{q}$ operator:

$$\mathbf{q} = \begin{pmatrix} q_{00} & q_{01} & q_{02} & \cdots \\ q_{10} & q_{11} & q_{12} & \cdots \\ q_{20} & q_{21} & q_{22} & \cdots \\ \vdots & & & \end{pmatrix} = \begin{pmatrix} 0 & q_{01} & 0 & \cdots \\ q_{10} & 0 & q_{12} & \cdots \\ 0 & q_{21} & 0 & \cdots \\ \vdots & & & \end{pmatrix}$$

where the first matrix is the general one and the second matrix shows the pattern of zero and non-zero values that the previous problem predicts. Now suppose we want values for higher powers of $\hat{q}$ than the first. To find $\hat{q}^j$, all we have to do is multiply the $\mathbf{q}$ matrix by itself j times. Use this fact to find (a) the *general pattern of zero and non-zero values* for q_{nm}^2 and (b) the explicit value for q_{00}^2. (See also Problem 11.34 for another way to find this latter quantity.) Recall the definition of matrix multiplication: If $\mathbf{A} = \mathbf{BC}$ where matrix $\mathbf{A}$ has elements a_{nm} and similarly for $\mathbf{B}$ and $\mathbf{C}$, then

$$a_{nm} = \sum_{i=1}^{N} b_{ni} c_{im}$$

where N is the number of columns in matrix $\mathbf{B}$ or rows in matrix $\mathbf{C}$, which must be the same in order to define matrix multiplication. $N = \infty$ here, but that shouldn't scare you!

12.15 The Hermite polynomial recurrence relation is easy to prove. Differentiate the generating function

$$H_v(q) = (-1)^v e^{q^2} \frac{d^v e^{-q^2}}{dq^v}$$

and show that

$$\frac{dH_v}{dq} = 2qH_v - H_{v+1} \ .$$

Next, differentiate this expression and find an expression for d^2H_v/dq^2 in terms of H_v, H_{v+1}, and H_{v+2}. Substitute these expressions into Hermite's differential equation

$$\frac{d^2H_v}{dq^2} - 2q\frac{dH_v}{dq} + 2v H_v = 0$$

and show that the result can be written as the recurrence relation

$$H_{v+1} = 2q H_v - 2v H_{v-1} \ .$$

Finally, prove that

$$\frac{dH_v}{dq} = 2v H_{v-1} \ .$$

12.16 We will encounter the *Lennard–Jones potential function* later in the book in discussions of intermolecular forces. It contains two parameters that can be chosen to be the well depth D and the position of the potential minimum, x_e:

$$V(x) = D\left[1 - \left(\frac{x_e}{x}\right)^6\right]^2 \ .$$

(See Fig. 1.5 for a picture of this type of potential.) Expand this potential in a Taylor's series about x_e, and find an expression for its effective harmonic force constant in terms of D, x_e, and simple numerical constants.

SECTION 12.3

12.17 What is the Hamiltonian operator for a particle moving freely in 3-D space? Neglecting normalization, what is the most general wavefunction and total energy expression for such a particle? How many quantum numbers are needed to describe its state? What values can these quantum numbers assume?

12.18 Consider a particle of mass m confined to a square 2-D potential box with sides of length L. What are the energies and degeneracies of all states with a total energy $E \leq 9\hbar^2\pi/mL^2$? Where is the particle most likely to be found in the states $(n_x, n_y) = (1, 1)$? $(3, 1)$? $(1, 3)$? $(2, 2)$?

12.19 An isotropic 3-D harmonic oscillator is one for which the force constants in x, y, and z are all equal. What are the energies and degeneracies of the first four lowest energy states of such a system? Note that only one parameter, $\omega = \sqrt{k/m}$, is needed to specify the motion of one particle in this system.

12.20 The energy of a C—C bond in C_{60} is on the order of 200 kJ mol^{-1}, comparable to graphite. Consider the endohedral He@C_{60} discussed in Example 12.5. In what quantum state does the He atom have sufficient energy to break out of its C_{60} cage?

SECTION 12.4

12.21 Consider a particle confined to move in the (x, y) plane. Its angular momentum vector is necessarily perpendicular to the plane, since only the L_z component is non-zero. (See Eq. (12.25); z and p_z are both zero.) In Cartesian coordinates, $\hat{L}_z$ is given by Eq. (12.31), but plane polar coordinates (r, θ) are often better suited for 2-D systems with central potential fields (i.e., $V = V(r)$). These coordinates are related to Cartesian coordinates through $x = r \cos \theta$ and $y = r \sin \theta$ so that

$$r = (x^2 + y^2)^{1/2} \quad \text{and} \quad \theta = \tan^{-1}\left(\frac{y}{x}\right).$$

The operator $\hat{L}_z$ can be written in terms of these coordinates if we use the chain rule for partial differentiation of an arbitrary function f:

and
$$\left(\frac{\partial f}{\partial x}\right)_y = \left(\frac{\partial f}{\partial r}\right)_\theta \left(\frac{\partial r}{\partial x}\right)_y + \left(\frac{\partial f}{\partial \theta}\right)_r \left(\frac{\partial \theta}{\partial x}\right)_y$$

$$\left(\frac{\partial f}{\partial y}\right)_x = \left(\frac{\partial f}{\partial r}\right)_\theta \left(\frac{\partial r}{\partial y}\right)_x + \left(\frac{\partial f}{\partial \theta}\right)_r \left(\frac{\partial \theta}{\partial y}\right)_x.$$

Use these expressions to show that $\hat{L}_z = (\hbar/i)(\partial/\partial\theta)$.

12.22 The Laplacian operator in two Cartesian dimensions is simply

$$\nabla^2 = \frac{\partial^2}{\partial x^2} + \frac{\partial^2}{\partial y^2}.$$

In plane polar coordinates (r, θ) (defined in the previous problem), ∇^2 is

$$\nabla^2 = \frac{1}{r}\frac{\partial}{\partial r}\left(r\frac{\partial}{\partial r}\right) + \frac{1}{r^2}\frac{\partial^2}{\partial \theta^2}.$$

Derive this expression from the Cartesian expression using the chain rule relations of the previous problem and their extension to second derivatives. Show first that

$$\left(\frac{\partial F}{\partial x}\right)_y = \cos\theta \left(\frac{\partial F}{\partial r}\right)_\theta - \frac{\sin\theta}{r}\left(\frac{\partial F}{\partial \theta}\right)_r$$

$$\left(\frac{\partial F}{\partial y}\right)_x = \sin\theta \left(\frac{\partial F}{\partial r}\right)_\theta + \frac{\cos\theta}{r}\left(\frac{\partial F}{\partial \theta}\right)_r$$

for any function $F = F(x,y) = F(r,\theta)$. Next, evaluate

$$\frac{\partial^2 F}{\partial x^2} = \frac{\partial}{\partial x}\left(\frac{\partial F}{\partial x}\right)_y = \left(\frac{\partial r}{\partial x}\right)_y\left[\frac{\partial}{\partial r}\left(\frac{\partial F}{\partial x}\right)_y\right] + \left(\frac{\partial \theta}{\partial x}\right)_y\left[\frac{\partial}{\partial \theta}\left(\frac{\partial F}{\partial x}\right)_y\right]$$

$$\frac{\partial^2 F}{\partial y^2} = \frac{\partial}{\partial y}\left(\frac{\partial F}{\partial y}\right)_x = \left(\frac{\partial r}{\partial y}\right)_x\left[\frac{\partial}{\partial r}\left(\frac{\partial F}{\partial y}\right)_x\right] + \left(\frac{\partial \theta}{\partial y}\right)_x\left[\frac{\partial}{\partial \theta}\left(\frac{\partial F}{\partial y}\right)_x\right]$$

and combine these to complete the proof.

12.23 Equation (12.33) states that $\hat{L}^2 = \hat{L}_x\hat{L}_x + \hat{L}_y\hat{L}_y + \hat{L}_z\hat{L}_z$, an expression that has its origin in the vector expression for angular momentum: $\mathbf{L} = L_x\mathbf{x} + L_y\mathbf{y} + L_z\mathbf{z}$ where $\mathbf{x}$, $\mathbf{y}$, and $\mathbf{z}$ are unit vectors. Classically, $|L|^2 = \mathbf{L}\cdot\mathbf{L} \equiv L_x^2 + L_y^2 + L_z^2$ where the $\cdot$ indicates a vector dot product, and the quantum expression is just a direct transcription. Equation (12.33) gives $\hat{L}^2$ in terms of spherical polar coordinates. Derive this expression, and also derive an expression for $\hat{L}^2$ in Cartesian coordinates.

12.24 Prove the following commutator expressions:
(a) $[\hat{L}_x, \hat{L}_y] = i\hbar\hat{L}_z$
(b) $[\hat{L}^2, \hat{L}_z] = 0$
(c) $[\hat{L}_z, \hat{L}_+] = \hbar\hat{L}_+$
(d) $[\hat{L}_+, \hat{L}^2] = 0$

Do the operators $\hat{L}_+$ and $\hat{L}_-$ commute with each other?

12.25 The proof that $\hat{L}_+$ raises the z component of angular momentum one unit is given in the text. Repeat that argument for the $\hat{L}_-$ operator, and show that it lowers the component one unit.

12.26 Consider an angular momentum state with $l = 1$ and $m = 1$. Think of the classical vector $\mathbf{L}$ that is analogous to this situation. It has a magnitude $|\mathbf{L}| = \sqrt{l(l+1)}\hbar = \sqrt{2}\hbar$ and a z component $m\hbar = \hbar$. The quantum result cannot specify the x and y components of this vector exactly, but if we recognize that $\hat{L}_x^2 + \hat{L}_y^2 = \hat{L}^2 - \hat{L}_z^2$, we can find the magnitude of the projection of the $\mathbf{L}$ vector onto the x, y plane, given by the square root of the eigenvalue of $\hat{L}^2 - \hat{L}_z^2$. Find this projection, and show that the closest classical analog to the quantum result is a vector $\mathbf{L}$ lying somewhere on a cone that has its apex at the origin, the z axis as the cone's symmetry axis, and a 45° angle between $\mathbf{L}$ and the z axis. Next, find a general expression for the cone angle in the case $l = m$ and show that this angle goes to zero as $l \to \infty$. As the cone angle goes to zero, $\mathbf{L}$ becomes more nearly coincident with the z axis, again recovering classical mechanics in the sense that states with large l and $l = m$ have an angular momentum vector that can be specified exactly in all components: two are zero and the third is $|\mathbf{L}|$.

12.27 Write $\hat{L}_+$ in spherical polar coordinates, and show that if $l = m$, the function $\Theta(\theta) = \sin^l \theta$ satisfies

$$\hat{L}_+ \psi_{l,l} = \hat{L}_+ \Theta_{l,l}(\theta)\Phi_l(\phi) = 0.$$

Next, set $l = 2$, write $\hat{L}_-$ in spherical polar coordinates, operate with it on $(\sin^2\theta)\Phi_2(\phi)$, and show that the result is at least proportional to $\psi_{2,1}$.

12.28 There is a marvelous theorem about spherical harmonic functions known as *Unsöld's Theorem*. It states that

$$\sum_{m=-l}^{l} |Y_{lm}(\theta,\phi)|^2 = \sum_{m=-l}^{l} Y_{lm}^* Y_{lm} = \text{a constant}.$$

This will be important in understanding why certain atoms are spherical (such as the rare gases) and others are not (such as the halogens). Verify this theorem for the particular cases of $l = 1$ and 2.

12.29 Suppose the H atom consisted of an electron a fixed distance a_0 from the (massive) proton but otherwise free to orbit in 3-D. What would its allowed energies be? (*Hint:* Note the resemblance of this scenario to that of Example 12.7.)

12.30 Bohr's theory of the H atom runs like this. First, we imagine the electron to orbit a stationary proton at a radial distance a. The classical condition for a stable orbit is applied: we equate the attractive force $e^2/(4\pi\epsilon_0 a^2)$ between the particles to the centrifugal force on the electron $m_e a \omega^2$. Next, we require the electron's angular momentum to be an integral multiple of $\hbar$: $m_e a^2 \omega = n\hbar$. Finally, we invoke the Virial Theorem result that the total energy is the negative of the kinetic energy: $E = -T = -m_e a^2 \omega^2/2$. Show that these conditions make $a = n^2 a_0$ and lead to the energy expression of Eq. (12.52).

12.31 The scattering of high-energy electrons from protons shows that the positive charge in the proton is localized to a region of radius $\sim 1.2 \times 10^{-15}$ m that we may take as the effective radius of the proton. If we magnify things so that this radius is 1 mm and imagine scaling a H atom in the 1s state by the same factor, on average how far is the center of mass from the proton? the electron (as measured by $\langle \hat{r} \rangle$)?

12.32 Use our results on the hydrogen atom to express the wavelength constants in Balmer's Eq. (11.2) and the more general Eq. (11.3) in terms of fundamental constants.

12.33 How different in size and ionization potential are the three isotopes of hydrogen, ^{1}H = H (ordinary hydrogen, sometimes called *protium*), ^{2}H = D (deuterium), and ^{3}H = T (tritium)? In atomic mass units, the nuclear masses are: H = 1.007 825 u, D = 2.014 102 u, and T = 3.016 050 u.

12.34 Deduce H-atom quantum numbers from the following descriptions of different wavefunctions:

(a) spherically symmetric with three spherical nodes.
(b) three nodal planes: the *xy, xz,* and *yz* planes.
(c) one spherical node and one nodal plane, the *xy* plane.
(d) two spherical nodes and two conical nodal planes.
(e) cylindrically symmetric about the *z* axis with one nodal plane.

12.35 A common representation of the associated Legendre functions is as a polar plot with coordinates (r, θ). In such a plot, a point on the plotted curve is placed a distance r from the origin at an angle θ such that r equals the value of the function at θ. Make polar plots of the θ dependent parts of the $2p_z$ and $3d_{z^2}$ wavefunctions, and compare your plots to the probability density pictures in Figures 12.14(d) and 12.15(a).

12.36 Since the H-atom wavefunctions extend to $r = \infty$ (although with rapidly decreasing magnitude), the notion of *atomic size* is ill-defined. Two measures are a_0 and $\langle \hat{r} \rangle$, of course, but another might be that radius that encloses some specified probability. Find the radius for the 1s state within which the H atom will be found with a probability of 0.90. Compare this number to a_0 and $\langle \hat{r} \rangle$ for this state.

12.37 Example 12.8 introduced the radial distribution function. Find the explicit form of the radial distribution function for the 1s and $2p_z$ states, and compare the distance at which they are maximum to the corresponding $\langle \hat{r} \rangle$ values of each state. Similarly, find the 2s radial distribution function, and find the point (other than $r = 0$ and $r = \infty$) where this function is zero. This is the location of the state's spherical node.

12.38 Consider a U atom stripped of all but one electron: U^{91+}. What is this ion's ionization potential (in both eV units and as a multiple of the H-atom IP)? What is $\langle \hat{r} \rangle$ for the 1s state of this ion (in both Å units and as a multiple of a_0)? For some principal quantum number n^*, U^{91+} will have $\langle \hat{r} \rangle_{n^*s} \geq \langle \hat{r} \rangle_{1s}$ for H. What is the smallest n^* can be? What would a plot of $\langle \hat{r} \rangle_{1s}$ versus ionization potential look like for all the one-electron atoms from H through U^{91+}?

12.39 The Earth (mass $\cong 6 \times 10^{24}$ kg) rotates about the sun (at an average distance $r \cong 1.5 \times 10^{11}$ m) once per year (1 year $\cong 10\pi$ million seconds). What is the Earth's angular momentum with respect to the sun in units of $\hbar$? What is the smallest principal quantum number n that allows a hydrogen atom to have this angular momentum? How large is such an atom? Use $\langle \hat{r} \rangle$ to express this size.

12.40 Show that the $2p_x$ and $2p_y$ hybrid orbitals are *not* eigenfunctions of the $\hat{L}_z$ operator.

12.41 Write explicit forms for the $3p_x$ and $3p_y$ hybrid wavefunctions, and show that they have the same nodal planes as the $2p_x$ and $2p_y$ hybrids.

12.42 Use Tables 12.4 and 12.5 to write explicit forms for the $3d_{xy}$, $3d_{xz}$, and $3d_{yz}$ hybrid wavefunctions. Show that each has nodal planes appropriate to the hybrid's name, that is, $3d_{xy}$ is zero everywhere $xy = 0$ and similarly for the other two hybrids.

SECTION 12.5

12.43 Equation (12.53) focuses on the *z* component of the spin, but as always, this axis choice is arbitrary. Since *x, y,* and *z* are equivalent independent directions, each must have a spin component that contributes equally to the magnitude of the total spin:

$$\langle \hat{S}^2 \rangle = \langle \hat{S}_x^2 \rangle + \langle \hat{S}_y^2 \rangle + \langle \hat{S}_z^2 \rangle = 3 \langle \hat{S}_z^2 \rangle \ .$$

Show that this expression and Eq. (12.53b) leads to Eq. (12.53a) for either spin eigenfunction α or β.

GENERAL PROBLEMS

12.44 It is often true that insight to the behavior of chemical systems starts with an understanding of the potential-energy function for a particular kind of motion. Listed below are several simple 1-D chemical motions and their associated coordinates. For each, sketch the qualitative shape you would expect the potential energy to take as this coordinate is varied. Be sure to note the allowed range of the coordinate, any unique points such as equilibrium minima, and be sure your sketches display any symmetry elements dictated by the system.

(a) The bending angle θ in water:

(b) The distance between two electrons, r_{12}:

(c) The isomerization angle θ between methyl cyanide and the isocyanide:

(d) The C—O bond rotation angle θ in methanol:

12.45 Sometimes symmetry can be exploited to your advantage but only after it's uncovered. Consider for example a solid surface composed of atoms A and B arranged in a square pattern with the atoms alternating places on the squares' corners. (Alkali halides do this.) Now imagine we are interested in the adsorption of a small atom on this surface. The surface interaction with this atom can be such that the atom is bound to the center of a square, but the different sizes of atoms A and B create a potential well that at first glance seems to have an awkward symmetry. The diagram below plots the two dimensional potential function $V(x, y)$ as equipotential contours centered on the square superimposed on the outlines of the A and B atoms at the square's corners.

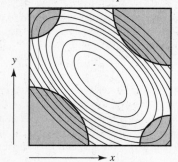

In these coordinates, the potential function has the form

$$V(x, y) = k_0(5x^2 + 5y^2 + 6xy)$$

where k_0 is a constant. The contour map has a symmetry that suggests we should be able to separate the potential energy function into two independent pieces, but it also suggests that (x, y) are not the symmetry coordinates, as does the term in xy that prevents separation of variables. No physical property of any system can depend on how we define our artificial coordinates. Consequently, show that the new coordinates x' and y' defined by

$$x' = \frac{1}{\sqrt{2}}(x + y) \quad \text{and} \quad y' = \frac{1}{\sqrt{2}}(y - x)$$

allows V to be written in a separable form, and then find an expression for the energies of an atom of mass m bound by this potential.

12.46 Positronium (symbol: Ps) is a stable (if only for times on the order of 0.1 ns) atom composed of an ordinary electron and its antiparticle, a positron. The positron, e^+, has the same mass as an electron, but an opposite charge. Positronium dies when the two particles annihilate each other, releasing two photons with a combined energy equal to twice the rest-mass energy equivalent of the electron: $2m_e c^2$. (There is a longer-lived (0.1 μs) version of Ps that must decay via three-photon emission. We'll save this atom for the next chapter.) How should one interpret the e^- and e^+ motion in Ps classically? Show that the ionization potential of Ps is half that of H. What is the representative size of ground-state Ps, that is, what is the Bohr radius for Ps? What is the longest wavelength of light that ground-state Ps can absorb? One-tenth nanosecond is not a very long time. Classically, how many "classical orbits" do the electron and positron make about their mutual center of mass in the ground state during 0.1 ns? (By the way, the ^{22}Na isotope with a half-life of 2.58 y spontaneously emits positrons and is a common laboratory source of them.)

12.47 An important characteristic of quantum systems is the *density of states,* the number of quantum states per unit energy at any given energy. For example, the harmonic oscillator states are equally spaced in energy by an amount $\hbar\omega$. Thus we say the density of states is $1/\hbar\omega$, independent of energy; there is one state every $\hbar\omega$ at all energies. What is the density of states for a particle in a 1-D box? a particle-on-a-ring?

12.48 The Virial Theorem in classical mechanics states that the time derivative of the time average of the quantity $\mathbf{r}\cdot\mathbf{p} = xp_x + yp_y + zp_z$ is zero for a system with any periodic motion:

$$\frac{d}{dt}\langle\mathbf{r}\cdot\mathbf{p}\rangle = 0 \; .$$

The quantum analog to this can be written (after some work) as

$$2 \langle \hat{T} \rangle = \left\langle x \frac{\partial V}{\partial x} + y \frac{\partial V}{\partial y} + z \frac{\partial V}{\partial z} \right\rangle,$$

which, for a spherically symmetric potential function $V(r)$, becomes simply

$$2 \langle \hat{T} \rangle = \left\langle r \frac{\partial V}{\partial r} \right\rangle$$

where $\langle \hat{T} \rangle$ is the expectation value of the kinetic-energy operator. Note that the harmonic oscillator and the H-atom problem are both examples of systems for which $V(r) = $ (constant) r^n where $n = 2$ for the oscillator and $n = -1$ for the atom. Find general relations among $\langle \hat{T} \rangle$, $\langle \hat{V} \rangle$, n, and E, the total energy, for the general $V(r) \propto r^n$ potential and for these two specific systems.

12.49 One measure of the size of an H atom in any given state is $\langle \hat{r} \rangle$, but another is the variation in classical radial turning points for various n, l quantum numbers. Define a dimensionless radius ρ in terms of the Bohr radius: $\rho = r/a_0$, and show that the expression for the inner and outer turning points can be written as

$$\rho^2 - 2n^2\rho + l(l + 1)n^2 = 0.$$

Do this by equating the effective potential to the total energy. Solve this quadratic, and compare the outer turning points for the 4s–4f states to the $\langle \hat{r} \rangle$ values for those states.

12.50 Very simple arguments can yield insight to the H atom, as the Bohr theory (Problem 12.30) shows. Here's an even simpler approach. Consider the electron in the H atom to be bound to the proton so that it is confined to a region of space of characteristic dimension R. Take this size to be the positional uncertainty of the electron, and take the momentum uncertainty to have the same order of magnitude as its uncertainty so that we have $p \sim \Delta p \sim \hbar/R$. The kinetic energy is then $T = p^2/2m = \hbar^2/2mR^2$. This energy represents a radially outward motion, and we can define an outward force via $F_{\text{out}} = -dT/dR$. The inward force on the electron, the binding force, is just Coulombic: $F_{\text{in}} = -dV/dR = -e^2/(4\pi\epsilon_0)R^2$. At equilibrium, $F_{\text{out}} + F_{\text{in}} = 0$. Use this relation to find an expression for R, and compare it to the Bohr radius. Then use the H-atom Virial Theorem $E = -T = V/2$ to find an expression for the total energy E, and compare it to the exact ground-state energy.

CHAPTER 13

The Structure of Atoms

13.1 Basic Ideas—A Review of the Periodic Table

13.2 The Problem of Too Many Particles

13.3 An Interlude on Approximation Methods

13.4 The Wavefunctions of Atoms

13.5 The Description of Atomic Energy Values

THE last chapter showed, in some detail, how quantum theory could describe the hydrogen atom. Since the Hamiltonian for any atom is a kinetic-energy term for each electron and the nucleus plus potential-energy terms for each Coulombic attraction or repulsion, we know how to write the Schrödinger equation for any atom. But for any system of three or more particles, it is impossible to find an exact wavefunction or an exact energy. Consequently we develop in this chapter a series of approximation methods that give wavefunctions and energies sufficiently accurate for chemists.

We begin with a review of the basic structure of the Periodic Table. We then discuss how even a He atom has too many particles, and we begin to see ways to use approximations to approach otherwise intractable problems. Two major kinds of approximations that have applications beyond the arena of multielectron atoms are described in some detail, and then these are applied to atomic structure questions. Finally, we categorize the interactions that govern atoms' energies by a systematic nomenclature that will greatly aid our subsequent discussions of molecular bonding and structure.

13.1 BASIC IDEAS—A REVIEW OF THE PERIODIC TABLE

We start with a review of the basic layout of the Periodic Table to refresh your mind on some elementary ideas of atomic structure. First, recall the elementary shell and subshell picture of an atom's electronic structure. These structures can be deduced empirically by the periodic variation of chemical properties with atomic number. You probably remember the so-called "building-up principle" (also known by the German *aufbau prinzip*) that gives a scheme for adding electrons, one by one, into subshells and shells, leading to *electron configurations* such as $1s^2$ for He or $1s^2 2s^2 2p^6 3s^2 3p^6 4s^1$ for K. You probably also remember that there is something important about "electron pairs," two electrons with "paired spins" but otherwise equal quantum numbers. Our task in this chapter is to put these elementary descriptions on a firm basis and to point out the limitations of these descriptions.

The fundamental reason for electron pairing is discussed at the beginning of the next section, but we can understand the shell and subshell structure on the basis of the H atom. The principle shells represent groups of electrons with similar energies, and we use a nomenclature (1, 2, 3, etc.) borrowed directly from the H-atom principal quantum number.

Subshells have common shell numbers, but are labeled with the analog of the H atom's l quantum number coded by the letters s, p, d, and f. We also understand why an s subshell holds two electrons, a p holds six, and so on from the H atom's m quantum number. For the s subshell, we have only $m = 0$, and we say that He ($1s^2$) represents one electron with quantum numbers $(n, l, m, m_s) = (1, 0, 0, +1/2)$ and another with $(n, l, m, m_s) = (1, 0, 0, -1/2)$. Similarly, the six electrons in a p subshell correspond to three pairs, each pair more or less associated with one of the three m values ($+1, 0, -1$) allowed for $l = 1$.

Figure 13.1 shows a schematic Periodic Table indicating each block of elements for which the chemistry is dominated by a particular subshell. One can derive an enormous amount of chemistry just from this diagram and a few elementary ideas about molecular bonding. (General chemistry courses do just that, of course.) A more detailed description of atomic structure may seem to be of secondary importance, rather like learning French only to have someone point out the differences between accents in Paris and Québec. A good knowledge of basic French will take you a long way in either city, just as a good understanding of general chemistry will take you far in this science. But when you move into one city (or adopt the science professionally), the finer points, the details, become daily necessities.

For example, consider atomic oxygen with the electron configuration $1s^2 2s^2 2p^4$. One detail missing from general chemistry discussions is that this configuration actually corresponds to *five distinctly different* energy states. The three that are

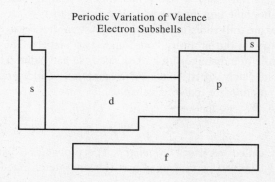

FIGURE 13.1 The Periodic Table breaks naturally into blocks of elements, the chemistry of which is largely governed by the valence electron subshells occupied in each block.

lowest in energy are nearly degenerate, while the fourth is higher in energy by about 2 eV (a chemically significant amount), and the fifth is about 2 eV higher yet. Moreover, each of the states—all somehow related to that $1s^2 2s^2 2p^4$ configuration—has unique chemistry. The lowest three, for instance, are virtually nonreactive towards H_2 while the fourth reacts with H_2 on nearly every encounter.

Oxygen is not unique in this regard. Electron configurations of atoms are only the starting points for an understanding of atomic energy states and bonding and reactivity differences, not only among different elements, but also among different eigenstates of any one element. A full understanding of electron configurations is a central goal of this chapter. The next section begins to explore configurations starting with the special properties and difficulties inherent in systems of many particles.

13.2 THE PROBLEM OF TOO MANY PARTICLES

All the model systems we solved in the last chapter had only a single particle. The harmonic oscillator and the hydrogen atom started as two-body problems but were quickly reduced to one-body problems via the use of the relative separation coordinate and the reduced mass of the system. New and unexpected phenomena arise when the system has more than two interacting particles, as do all the elements past H. This section ends with the He-atom Schrödinger equation, but we begin with a simpler situation: two noninteracting identical particles in a 1-D potential box.

As usual, imagine the box length is L and the particles' mass is m. Let one have energy $E_1 = \hbar^2 \pi^2 / 2mL^2$ (or quantum number $n_1 = 1$) and the other have energy $E_2 = 4\hbar^2 \pi^2 / 2mL^2$ (or $n_2 = 2$). By assumption, the particles do not interact so that the wavefunctions for each are

$$\psi_1(x_1) = \left(\frac{2}{L}\right)^{1/2} \sin\left(\frac{\pi x_1}{L}\right) \quad \text{and} \quad \psi_2(x_2) = \left(\frac{2}{L}\right)^{1/2} \sin\left(\frac{2\pi x_2}{L}\right)$$

where x_1 and x_2 are the coordinates for each particle. From the separation of variables theorem, the total wavefunction is

$$\Psi_{\text{total}}(x_1, x_2) = \psi_1(x_1)\psi_2(x_2) \; , \qquad (13.1)$$

which satisfies the system's Schrödinger equation and represents the energy $E_{\text{total}} = E_1 + E_2$.

But now consider

$$\Psi_{\text{total}}^2(x_1, x_2)\, dx_1 dx_2 = \psi_1^2(x_1)\psi_2^2(x_2)\, dx_1 dx_2$$
$$= \text{probability of finding particle } \textit{one} \text{ at } x_1 \text{ to } x_1 + dx_1$$
$$\textit{and at the same time} \text{ finding particle } \textit{two} \text{ at } x_2 \text{ to } x_2 + dx_2 \; .$$

This, in general, is *not* equal to the probability of finding particle *two* at x_1 to $x_1 + dx_1$ and at the same time finding particle *one* at x_2 to $x_2 + dx_2$. But this cannot be! If the particles are truly indistinguishable, the system's total wavefunction *cannot depend in any way* on our arbitrary "one" and "two" labels. These labels, in Eq. (13.1), play the role of a distinguishing mark, as if we had painted one particle blue and the other gold. Such labels violate the assumption of indistinguishability, and the electrons in an atom are indistinguishable.

We must write Ψ_{total} so that

$$\Psi_{\text{total}}^2(x_1, x_2) = \Psi_{\text{total}}^2(x_2, x_1) \; . \qquad (13.2)$$

This property is called *exchange symmetry*. The probability distribution cannot change when the labels of two identical particles are interchanged. We can define an exchange operator (also called a *permutation operator*), $\hat{P}_{nm}$, which exchanges the labels we give to particles n and m:

$$\hat{P}_{nm} \Psi(r_1, r_2, \ldots, r_n, \ldots, r_m, \ldots) = \Psi(r_1, r_2, \ldots, r_m, \ldots, r_n, \ldots) . \quad (13.3a)$$
$$\uparrow \qquad \uparrow$$
$$\text{Was } r_n \quad \text{Was } r_m$$

From Eq. (13.2), the following eigenvalue equation follows:

$$\hat{P}_{nm} \Psi(r_1, r_2, \ldots, r_n, \ldots, r_m, \ldots) = (\pm 1) \Psi(r_1, r_2, \ldots, r_n, \ldots, r_m, \ldots) . \quad (13.3b)$$

The eigenvalue of the exchange operator can be either $+1$ (in which case the wavefunction is said to be *symmetric with respect to exchange*) or -1 (in which case Ψ is *antisymmetric with respect to exchange*).

Our initial Ψ_{total}, Eq. (13.1), is *not* an eigenfunction of $\hat{P}_{12}$. But the two linear combinations

$$\Psi_{\pm}(x_1, x_2) = \frac{1}{\sqrt{2}} [\psi_1(x_1)\psi_2(x_2) \pm \psi_1(x_2)\psi_2(x_1)] \quad (13.4)$$

are a pair of (normalized) wavefunctions Ψ_+ and Ψ_- that *are* eigenvalues of $\hat{P}_{12}$, since

$$\hat{P}_{12}\Psi_+ = \frac{1}{\sqrt{2}} [\psi_1(x_2)\psi_2(x_1) + \psi_1(x_1)\psi_2(x_2)] = +1\Psi_+$$

and

$$\hat{P}_{12}\Psi_- = \frac{1}{\sqrt{2}} [\psi_1(x_2)\psi_2(x_1) - \psi_1(x_1)\psi_2(x_2)] = -1\Psi_- .$$

Thus, Ψ_+ is symmetric and Ψ_- is antisymmetric.

It would seem that we have found a curious type[1] of degeneracy. The wavefunctions Ψ_+ and Ψ_- both represent total energies $E_1 + E_2$, although Ψ_+^2 and Ψ_-^2 are different. Which eigenstate is observed? Either? Both? The answer, amazingly enough, depends on the *spin of the particles* and forms the basis of the *Pauli Exclusion Principle*. Nature only allows wavefunctions that are *antisymmetric with respect to exchange of any two identical particles with half-integral spin quantum numbers* and *symmetric with respect to exchange of any two identical particles with integral (or zero) spin*. Those particles with half-integral spin (electrons, protons, neutrons, and the nuclei of atoms with an odd total number of nucleons, such as ^{3}He, ^{7}Li, etc.) are called *fermions* in honor of Enrico Fermi, while those with integral spin (photons,[2] α-particles or ^{4}He nuclei, ^{14}N nuclei, ^{16}O nuclei, etc.) are called *bosons* in honor of S. N. Bose.

EXAMPLE 13.1

The statement was made in the text that Ψ_+^2 and Ψ_-^2 for the two particles-in-a-box state were different. In what way are they different?

---

[1]The astute reader will have noticed a parallel between this problem and the one particle in a 2-D box problem; just change x_1 into x and x_2 into y. Exchange symmetry, however, is an independent attribute of many-particle systems, beyond whatever spatial symmetries the potential energy operator may have.

[2]Photons are inherently relativistic particles, and the spin of a photon is usually expressed by its so-called *helicity state,* which is related to the photon's state of circular polarization. The projection of the photon's state of inherent angular momentum along its direction of propagation can be only $\pm 1\hbar$, a result that requires a relativistic theory to explain.

SOLUTION We could approach this question algebraically, but a good way to explore wavefunctions of two variables is through contour plots. Drawing these plots requires an algebraic expression for the wavefunctions, of course, but with practice, contour plots can be sketched at least qualitatively for many wavefunctions. Here are the contour plots for our Ψ_+^2 and Ψ_-^2:

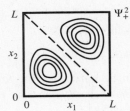 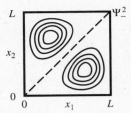

Note how the symmetries of each plot (the dashed contours are nodes) treat each position variable, x_1 and x_2, equally, as required by exchange symmetry. If particles 1 and 2 are indistinguishable, no wavefunction for the system can show a bias in favor of either particle. These pictures and the linear combinations that make up Ψ_+^2 and Ψ_-^2 might remind you of the symmetries of the p_x and p_y hybrids. The symmetries are similar, but the hybrids are constructed with different physical requirements in mind.

⟹ *RELATED PROBLEMS* 13.4, 13.5

The Pauli Principle is essentially another postulate, verified by experiment, but not understood on more fundamental grounds. Its consequences are profound. To jump the gun a bit, the Pauli Principle prohibits the $1s^3$ electron configuration for Li, for example, (although it is not yet clear to us how $1s^3$ represents an invalid wavefunction). It causes ^{4}He and ^{3}He to behave very differently, especially at low temperatures (see Figure 6.10 for phase diagrams), since ^{4}He nuclei are bosons while ^{3}He nuclei are fermions. It plays a major role in the phenomena of superconductivity, which we will not discuss, and in molecular spectroscopy, which we will.

To return to our box again, if the particles have integral spin, then only Ψ_+ is observed, since it is symmetric. A more interesting case, which will lead us directly into the He atom, considers two noninteracting spin 1/2 particles, both with energy E_1, in the same box.

In this case, Ψ_{total} must include both the spatial and the spin coordinates of the two particles. We can construct four total wavefunctions that satisfy exchange symmetry, since the spatial part is trivial:

$$\Psi_1 = \psi_1(x_1)\psi_1(x_2)\alpha(1)\alpha(2) \quad \textbf{(13.5a)}$$

$$\Psi_2 = \psi_1(x_1)\psi_1(x_2)\beta(1)\beta(2) \quad \textbf{(13.5b)}$$

$$\Psi_3 = \frac{1}{\sqrt{2}}\psi_1(x_1)\psi_1(x_2)[\alpha(1)\beta(2) + \alpha(2)\beta(1)] \quad \textbf{(13.5c)}$$

$$\Psi_4 = \frac{1}{\sqrt{2}}\psi_1(x_1)\psi_1(x_2)[\alpha(1)\beta(2) - \alpha(2)\beta(1)] \ . \quad \textbf{(13.5d)}$$

Of these four possibilities, *only* Ψ_4, which is the *only one that is antisymmetric with respect to exchange,* will satisfy the Pauli Principle. Notice the subtle difference among these wavefunctions. The spatial parts are identical, but the spin parts vary. We could call Ψ_1 the "both spins up" function and Ψ_2 the "both spins down" function, while *both* Ψ_3 and Ψ_4 seem to indicate spin-pairing.

The spin parts of Ψ_1–Ψ_4 are eigenfunctions of the total spin magnitude operator. For a two-particle system $\hat{S}^2$ is not $\hat{S}^2(1) + \hat{S}^2(2)$, but we will not go into its form here. However, $\hat{S}_z$ is $\hat{S}_z(1) + \hat{S}_z(2)$. The eigenvalues for $\hat{S}_z$ can thus be found immediately:

$$\hat{S}_z \Psi_1 = +1\hbar \Psi_1$$
$$\hat{S}_z \Psi_2 = -1\hbar \Psi_2$$
$$\hat{S}_z \Psi_3 = 0(\hbar \Psi_3)$$
$$\hat{S}_z \Psi_4 = 0(\hbar \Psi_4)$$

while one can show that

$$\hat{S}^2 \Psi_i = 1(1 + 1)\hbar^2 \Psi_i, \ i = 1, 2, \text{ or } 3$$

and

$$\hat{S}^2 \Psi_4 = 0(0 + 1)\hbar^2 \Psi_4$$

where the zero quantum numbers have been shown explicitly to help you recall the general form of angular momentum eigenvalue expressions.

The wavefunctions Ψ_1–Ψ_3 correspond to the three allowed components ($m_s = +1, 0, -1$) of a total spin $S = 1$ system, while Ψ_4 corresponds to a total spin $S = 0$ state. *This* is the "paired spins" state.

We can now approach He. We assume that the nucleus (with $Z = 2$) is sufficiently massive to be considered stationary. We locate the electrons relative to the nucleus and relative to each other via the vectors $\mathbf{r}_1$, $\mathbf{r}_2$, and $\mathbf{r}_{12}$ shown in Figure 13.2. The potential-energy operator is just a sum of Coulombic terms:

$$\hat{V}(\mathbf{r}_1, \mathbf{r}_2) = \frac{1}{4\pi\epsilon_0}\left[\frac{e^2}{|\mathbf{r}_1 - \mathbf{r}_2|} - \frac{2e^2}{r_1} - \frac{2e^2}{r_2}\right] = \frac{1}{4\pi\epsilon_0}\left[\frac{e^2}{r_{12}} - \frac{2e^2}{r_1} - \frac{2e^2}{r_2}\right] \quad (13.6)$$

where $r_{12} = |\mathbf{r}_1 - \mathbf{r}_2|$. The total Hamiltonian is

$$\hat{H} = -\frac{\hbar^2}{2m}\nabla_1^2 - \frac{\hbar^2}{2m}\nabla_2^2 + \hat{V}(\mathbf{r}_1, \mathbf{r}_2) \quad (13.7)$$

where m is the electron mass and ∇_i^2 is the Laplacian operator for electron i.

Without the term in $\hat{V}$ that depends on r_{12}, we could solve the Schrödinger equation immediately, and we will do so in the next paragraph. With this term, we cannot solve the equation exactly. The full equation is neither separable nor is it amenable to solution by any of the methods in the previous chapter. It is not a question of being clever enough to find a solution; an analytic solution simply does not exist.

If the term $e^2/4\pi\epsilon_0 r_{12}$ is ignored, the Hamiltonian is

$$\hat{H} = -\frac{\hbar^2}{2m}\nabla_1^2 - \frac{2e^2}{(4\pi\epsilon_0)r_1} - \frac{\hbar^2}{2m}\nabla_2^2 - \frac{2e^2}{(4\pi\epsilon_0)r_2},$$

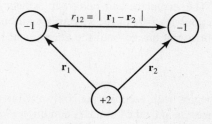

FIGURE 13.2 The three natural lengths for the He Hamiltonian are the two nuclear-electron distances r_1 and r_2 and the electron–electron distance r_{12}.

which is the sum of two Hamiltonians for the He$^+$ one-electron system:

$$\hat{H}_{He^+} = -\frac{\hbar^2}{2m}\nabla^2 - \frac{2e^2}{(4\pi\epsilon_0)r} .$$

We know how to solve the Schrödinger equation for this Hamiltonian, and, in analogy with the two identical but noninteracting spin 1/2 particles in a box, we write

$$\Psi(r_1, r_2) = \psi_{1s}(r_1)\psi_{1s}(r_2)\left\{\frac{1}{\sqrt{2}}[\alpha(1)\beta(2) - \alpha(2)\beta(1)]\right\} \quad (13.8)$$

for the ground state where ψ_{1s} is the 1s function (for $Z = 2$) and we have been careful to choose an antisymmetric total wavefunction. The energy of this system is twice the energy of the He$^+$ ion from Eq. (12.46b):

$$E = 2\left[-\frac{4e^2}{2(4\pi\epsilon_0)a_0}\right] = -108.78 \text{ eV} .$$

The assumption that the r_{12} term is unimportant, called the *independent electron approximation,* leads to this energy. The experimental number to which we should compare this is (the negative of) the sum of the first and second ionization energies of He, which is

$$E_{expt} = -(IP_1 + IP_2) = -(24.6 + 54.4) \text{ eV} = -79.0 \text{ eV} .$$

This is very far from -108.78 eV, and it shows that our neglected term *is* important.

We should not be surprised that the independent electron approximation gives an energy significantly *lower* than experiment. The r_{12} term in $\hat{V}$ represents a repulsive interaction and leads to a positive energy contribution. We can guess its rough magnitude if we guess that r_{12}, on average, is on the order of a_0 so that $e^2/4\pi\epsilon_0 r_{12} \cong e^2/4\pi\epsilon_0 a_0 = 2 \times$ (ionization energy of H) $= 2(13.598 \text{ eV}) = 27.2$ eV. The fact that -108.78 eV $+ 27.2$ eV $= -81.6$ eV is very close to the experimental -79.0 eV is fortuitous, since we pulled the length a_0 more or less out of the air, but it does indicate a correct order of magnitude and trend.

We need to introduce better approximation methods to deal with the r_{12} term. The next section does this, and the methods (called variation and perturbation theories) have very general applicability. We leave the He atom at this point and return to it and other atoms once these methods have been introduced.

13.3 AN INTERLUDE ON APPROXIMATION METHODS

There are several methods for obtaining approximate wavefunctions, energies, and other eigenvalues for systems that cannot be solved exactly. This section examines the two with the greatest applicability to chemical problems. The first, called the variation method, uses an inexact wavefunction that has one or more initially undetermined parameters. The method shows how to choose these parameters so that the expectation value of the total energy is as close to the exact energy as the inexact wavefunction can come. The second method, perturbation theory, uses wavefunctions that are exact eigenfunctions for part of the system Hamiltonian and then gives a series of successive approximations to both the energy and the wavefunction for the full Hamiltonian.

Variation Theory

The variation method begins with a choice for the wavefunction that, at the very least, satisfies the symmetry, nodal, and boundary conditions we expect for the exact wavefunction. The method is most useful for approximating ground-state energies, and we will limit our discussion to this case.

The basis for the variation method is the theorem, which we will prove, that the quantity

$$\langle E \rangle = \frac{\int \phi^* \hat{H} \phi \, d\tau}{\int \phi^* \phi \, d\tau} \qquad (13.9)$$

can never be less than the true ground-state energy, E_0, where ϕ is some function that depends on all the coordinates of the system at hand and has the general shape we expect for the true ground-state wavefunction. The denominator is unity if we have previously normalized ϕ, and if ϕ is the true ground state wavefunction, $\langle E \rangle = E_0$. The theorem says that an arbitrary ϕ can approach the exact E_0 "from above;" $\langle E \rangle$ is never less than the exact energy.

The true wavefunctions do, of course, exist. We simply cannot write them in terms of a finite sum of analytic functions, as we can for, say, the H-atom or harmonic oscillator systems. So, even though we do not know them explicitly, we can represent the exact wavefunctions symbolically as the complete set ψ_n where n is just a running index. Moreover, we can express our guessed function in terms of the ψ_n's as an expansion of the form

$$\phi = \sum_i a_i \psi_i \ .$$

If we substitute this expansion into Eq. (13.9), and use the fact that the ψ_n's are orthonormal, we find

$$\langle E \rangle = \sum_i \sum_j a_i^* a_j \int \psi_i^* \hat{H} \psi_j \, d\tau = \sum_i a_i^* a_i E_i$$

where E_i represents the ith exact energy. If we subtract the ground state energy E_0 from both sides, we find

$$\langle E \rangle - E_0 = \sum_i a_i^* a_i E_i - E_0 = \sum_i a_i^* a_i (E_i - E_0)$$

where the last step follows from $\sum_i a_i^* a_i = 1$. But, since E_0 is the lowest of the E_i values (so that each $(E_i - E_0)$ factor is ≥ 0), and since $a_i^* a_i$ is always positive or zero, it must be true that

$$\langle E \rangle - E_0 \geq 0$$

or, as we set out to prove,

$$\langle E \rangle \geq E_0 \ . \qquad (13.10)$$

We now go through a few sample calculations using the variation method. First, suppose we wanted to find the ground state particle-in-a-box energy. We know the exact answer and can verify the variational theorem quantitatively. The wavefunction should be zero at $x = 0$ and L, and it should have no nodes. One simple function that satisfies these criteria is

$$\phi = N x (L - x) \ . \qquad (13.11)$$

We compute, with $\hat{H} = -\hbar^2 (d^2/dx^2)/2m$,

$$\langle E \rangle = \frac{\int \phi^* \hat{H} \phi \, d\tau}{\int \phi^* \phi \, d\tau} = \frac{\frac{\hbar^2}{m} \int_0^L x(L-x) \, dx}{\int_0^L x^2 (L-x)^2 \, dx} = \frac{5\hbar^2}{mL^2} \ .$$

The exact answer is $E_0 = \hbar^2\pi^2/2mL^2$, and, since $\pi^2/2 = 4.935$, our guessed function gives an energy that is high by only $(\langle E \rangle - E_0)/\langle E \rangle = (5 - 4.935)/5 = 0.013$ or 1.3%, which is very good.

We can improve this if we make the guessed wavefunction more flexible. The somewhat more general function

$$\phi' = N[x(L - x) + \alpha x^2(L - x)^2]$$

where α is a constant to be determined, also satisfies the boundary conditions. Using it in Eq. (13.9) gives

$$\langle E \rangle = \frac{3\hbar^2}{mL^2}\left(\frac{2\alpha^2 L^4 + 14\alpha L^2 + 35}{\alpha^2 L^4 + 9\alpha L^2 + 21}\right),$$

which depends on the parameter α. We choose α so that $\langle E \rangle$ is as small as possible. We compute $d\langle E \rangle/d\alpha$, set this equal to zero:

$$\frac{d\langle E \rangle}{d\alpha} = 0 = 4\alpha^2 L^4 + 14\alpha L^2 - 21$$

and solve for α:

$$\alpha = \frac{-7 \pm (133)^{1/2}}{4L^2}.$$

Substituting these two values for α into the expression for $\langle E \rangle$ gives the two energies

$$\langle E \rangle = \frac{\hbar^2}{mL^2}(51.065\,1) \quad \text{or} \quad \frac{\hbar^2}{mL^2}(4.934\,87).$$

We choose the latter, since it gives the lower energy, and compare 4.934 87 to the exact coefficient, $\pi^2/2 = 4.934\,80$. The error is less than 0.002%.

This exercise shows two important features of the variational method. First, it shows how the energy estimate can be improved with a more flexible wavefunction. If we add a term varying as $[x(L - x)]^4$, for instance, to our last guessed function, we could improve our energy estimate even more. Second, we now see where the term "variational" comes from. Our second trial wavefunction had a parameter α that we were free to vary, and we used the α that minimized $\langle E \rangle$.

These are the two general steps in a variational calculation: pick a flexible function (with perhaps many parameters) that satisfies the symmetry, nodal, and boundary conditions of the problem, then choose values for those parameters such that they minimize $\langle E \rangle$. With these steps, we can compute a ground-state energy that is as close to the exact value as our ability (and patience) in writing a flexible function permits. The major drawback is that we have no assurance that the guessed wavefunction accurately represents anything but the energy. It may give bad estimates of other physical quantities.

EXAMPLE 13.2

How well does the trial wavefunction $\phi(x) = N(xL - x^2 + \beta x^3)$, where β is a variational parameter, reproduce the ground-state energy of the particle-in-a-box system?

SOLUTION We know the answer if $\beta = 0$, since this gives the first trial wavefunction we explored in the text. The question here asks if a cubic contribution improves the simplest trial wavefunction. The normalization constant N cancels from the variational energy expression, Eq. (13.9). The two integrals in that expression are

$$\int_0^L \phi \hat{H} \phi \, dx = -\frac{\hbar^2}{2m}\int_0^L \phi \frac{d^2\phi}{dx^2} dx = -\frac{\hbar^2 N^2}{2m}\left(\frac{6}{5}L^5\beta^2 - \frac{1}{3}L^3\right)$$

and

$$\int_0^L \phi^2(x)\,dx = N^2\left(\frac{1}{7}L^7\beta^2 + \frac{1}{15}L^6\beta + \frac{1}{30}L^5\right).$$

so that, after some algebraic simplification,

$$\langle E \rangle = -\frac{7\hbar^2}{2mL^2}\left(\frac{18L^2\beta^2 - 5}{30L^2\beta^2 + 14L\beta + 7}\right).$$

We want to choose β so that this expression is a minimum. We find, after more algebra,

$$\frac{d\langle E \rangle}{d\beta} = 0 \quad \text{implies} \quad 126L^2\beta^2 + 276L\beta + 35 = 0,$$

which has the roots

$$\beta L = -\frac{23}{21} \pm \frac{1}{2}\sqrt{\frac{542}{147}} = -2.055\,3 \quad \text{and} \quad -0.135\,15$$

corresponding to the energies

$$\langle E \rangle = 5.781\,4\,\frac{\hbar^2}{mL^2} \quad \text{and} \quad -4.737\,9\,\frac{\hbar^2}{mL^2}.$$

The second is nonphysical and is rejected. The energy must be positive. But note the first value. It is *higher* than the energy predicted by $\beta = 0$ (the example worked in the text) by $(5.781\,4 - 5)\hbar^2/mL^2 = (0.781\,4)\hbar^2/mL^2$. Adding an extra term has made the wavefunction and the energy *worse!*

Why is this? The answer is revealed if we look at the wavefunction corresponding to this β:

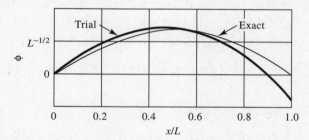

The variational wavefunction (normalized) is shown as the heavy line, while the exact wavefunction is shown as the light line. Note how the βx^3 term has destroyed the exact symmetry: the node that should be at $x = L$ has moved to $x \cong 0.9\,L$. The moral here is that a variational wavefunction should have not only suitable parameters to vary but also the appropriate symmetry dictated by the potential energy.

➡ RELATED PROBLEMS 13.9, 13.10

One other aspect of the variation method is worth looking at in some detail. This is the common and quite useful method of constructing our guessed function as a linear combination of N functions, none of which is internally parameterized.[3] If we write

$$\phi = \sum_{i=1}^{N} c_i \phi_i = c_1\phi_1 + c_2\phi_2 + \cdots + c_N\phi_N \qquad (13.12)$$

---

[3]This method is often called the *Rayleigh–Ritz variation method*. Lord Rayleigh devised variational methods for classical mechanics, and Ritz (in 1909) introduced a linear combination of functions in a variational study of differential equations that, some years later, turned out to be applicable to quantum mechanics.

then the coefficients c_i are the parameters we vary to minimize $\langle E \rangle$. If we substitute Eq. (13.12) into Eq. (13.9), we find

$$\langle E \rangle = \frac{\sum_i \sum_j c_i^* c_j H_{ij}}{\sum_i \sum_j c_i^* c_j S_{ij}} \qquad (13.13)$$

where

$$H_{ij} = \int \phi_i^* \hat{H} \phi_j \, d\tau \quad \text{and} \quad S_{ij} = \int \phi_i^* \phi_j \, d\tau \, .$$

These integrals are just sets of numbers indexed by i and $j = 1, 2, \ldots, N$. They can all be represented as the *square matrices* **H** and **S** where

$$\mathbf{H} = \begin{pmatrix} H_{11} & H_{12} & \cdots \\ H_{21} & H_{22} & \cdots \\ \vdots & \vdots & \ddots \end{pmatrix}$$

and

$$\mathbf{S} = \begin{pmatrix} S_{11} & S_{12} & \cdots \\ S_{21} & S_{22} & \cdots \\ \vdots & \vdots & \ddots \end{pmatrix} .$$

Each entry in each matrix represents the corresponding definite integral, and thus integrals of this type are often called *matrix elements*. One calls H_{ij}, for example, the "i-jth matrix element of the Hamiltonian."

The next step is to find that set of coefficients c_i that minimizes $\langle E \rangle$ in Eq. (13.13). To demonstrate how this is done, we will assume the particularly simple $\phi = c_1 \phi_1 + c_2 \phi_2$ for which Eq. (13.13) can be written (with the c's taken to be real)

$$\langle E \rangle (c_1^2 S_{11} + 2 c_1 c_2 S_{12} + c_2^2 S_{22}) = c_1^2 H_{11} + 2 c_1 c_2 H_{12} + c_2^2 H_{22} \, .$$

If we define $S = c_1^2 S_{11} + 2 c_1 c_2 S_{12} + c_2^2 S_{22}$ and similarly define H, we have

$$\langle E \rangle S = H \, .$$

We differentiate this with respect to c_1 and c_2:

$$\frac{\partial \langle E \rangle}{\partial c_1} S + \langle E \rangle \frac{\partial S}{\partial c_1} = \frac{\partial H}{\partial c_1} \quad \text{and} \quad \frac{\partial \langle E \rangle}{\partial c_2} S + \langle E \rangle \frac{\partial S}{\partial c_2} = \frac{\partial H}{\partial c_2}$$

and set each $\partial \langle E \rangle / \partial c_i$ derivative to zero. Expanding $\partial S / \partial c_i$ and $\partial H / \partial c_i$ gives the two equations

$$c_1(H_{11} - \langle E \rangle S_{11}) + c_2(H_{12} - \langle E \rangle S_{12}) = 0$$
$$c_1(H_{12} - \langle E \rangle S_{12}) + c_2(H_{22} - \langle E \rangle S_{22}) = 0 \, .$$

The theory of linear simultaneous equations, such as these are, says that non-zero values for c_1 and c_2 exist only if the determinant[4] formed from the coefficients of c_1 and c_2 vanishes:

$$\begin{vmatrix} H_{11} - \langle E \rangle S_{11} & H_{12} - \langle E \rangle S_{12} \\ H_{12} - \langle E \rangle S_{12} & H_{22} - \langle E \rangle S_{22} \end{vmatrix} = 0 \, .$$

Expanding the determinant yields a quadratic expression for $\langle E \rangle$:

$$(H_{11} - \langle E \rangle S_{11})(H_{22} - \langle E \rangle S_{22}) - (H_{12} - \langle E \rangle S_{12})^2 = 0 \, .$$

---

[4] Recall that a two-by-two determinant is evaluated by the rule $\begin{vmatrix} a & b \\ c & d \end{vmatrix} = ad - bc$.

Solving this quadratic for $\langle E \rangle$ yields the desired minimum energy; it is the lesser of the two solutions. We could stop with the energy, or we could

1. substitute it into the simultaneous equations for c_1 and c_2,
2. find c_1 in terms of c_2,
3. use this expression to write ϕ in terms of c_2 alone,
4. normalize ϕ to find c_2, and[5]
5. have the corresponding wavefunction as well as the energy.

EXAMPLE 13.3

Suppose we write a trial wavefunction for the particle-in-a-box in the form $\phi = c_1\phi_1 + c_2\phi_2$ where $\phi_1 = x(L - x)$ and $\phi_2 = x^2(L - x)^2$. Find ϕ, c_1, and c_2 such that $\langle E \rangle$ is minimized.

SOLUTION This problem has been (partially) solved in the text, since we can write $\phi = c_1\phi_1 + c_2\phi_2$ as $\phi = N(\phi_1 + \alpha\phi_2)$ where $c_1 = N$ and $c_2 = N\alpha$. The optimum energy was found in the text by minimizing just the single parameter α; the wavefunction requires both parameters, but given one, the other is determined by normalization. This is true of all two-parameter variational wavefunctions of the form $\phi = c_1\phi_1 + c_2\phi_2$, but we will solve this example in a general way to illustrate the method of varying more than one parameter. We first find the integrals that make up the **H** and **S** matrices. They are (note that $H_{12} = H_{21}$ and $S_{12} = S_{21}$ since the wavefunctions are real number functions):

$$\mathbf{H} = \begin{pmatrix} H_{11} & H_{12} \\ H_{12} & H_{22} \end{pmatrix} = \frac{L^3\hbar^2}{m}\begin{pmatrix} \frac{1}{6} & \frac{L^2}{30} \\ \frac{L^2}{30} & \frac{L^5}{105} \end{pmatrix}, \quad \mathbf{S} = \begin{pmatrix} S_{11} & S_{12} \\ S_{12} & S_{22} \end{pmatrix} = L^5\begin{pmatrix} \frac{1}{30} & \frac{L^2}{140} \\ \frac{L^2}{140} & \frac{L^4}{630} \end{pmatrix}$$

The determinant equation is a quadratic in $\langle E \rangle$ that has the same roots found in the text:

$$\langle E \rangle = \frac{\hbar^2}{mL^2}(28 \pm 2\sqrt{133}) = 4.934\,87\,\frac{\hbar^2}{mL^2} \quad \text{and} \quad 51.065\,1\,\frac{\hbar^2}{mL^2}\,.$$

We substitute the first value into the pair of linear equations for c_1 and c_2:

$$c_1(H_{11} - \langle E \rangle S_{11}) + c_2(H_{12} - \langle E \rangle S_{12}) = 0$$

$$c_1(H_{12} - \langle E \rangle S_{12}) + c_2(H_{22} - \langle E \rangle S_{22}) = 0$$

and find, after some algebra,

$$(2\sqrt{133} - 23)c_1 + \frac{L^2}{7}(3\sqrt{133} - 35)c_2 = 0$$

$$(3\sqrt{133} - 35)c_1 + \frac{2L^2}{3}(\sqrt{133} - 11)c_2 = 0\,.$$

We cannot find both c_1 and c_2 from these two equations, but we can solve either for, say, c_1 in terms of c_2:

$$c_1 = \frac{L^2}{7}\frac{(35 - 3\sqrt{133})}{2\sqrt{133} - 23}c_2$$

then use normalization:

$$\int_0^L \phi^2\,dx = \frac{1}{30}L^5\,c_1^2 + \frac{1}{70}L^7c_2c_1 + \frac{1}{630}L^9c_2^2 = 1$$

---

[5]The simultaneous linear equations alone cannot be solved for both c_1 and c_2, since the equations both equal zero—they are said to be *homogeneous* linear equations—and such equations do not have unique solutions by themselves.

to find a second relation between c_1 and c_2. The final exact expressions for c_1 and c_2 are numerically complicated, but they evaluate to

$$c_1 = \frac{4.404\ 0}{L^{5/2}} \quad \text{and} \quad c_2 = \frac{4.990\ 3}{L^{9/2}}.$$

The final, normalized wavefunction is so close to the exact function that we can see the differences between them only if we plot ϕ − (exact wavefunction):

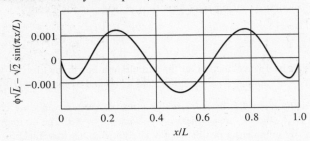

The maximum difference is $\sim 10^{-3}$ times the exact function. You may be wondering why one would want to find an approximate wavefunction when an exact wavefunction is available. For a simple system with simple wavefunctions, one rarely does. But for more complicated wavefunctions, an accurate approximation by simpler functions might result in a significant increase in computation speed in highly involved computer calculations. Computers evaluate simple polynomials much faster than transcendental functions (sin, cos, exp, etc.).

➔ **RELATED PROBLEMS** 13.36, 13.37

Perturbation Theory

In contrast to variation theory, where the full Hamiltonian is used with an approximate wavefunction, the perturbation technique starts with exact wavefunctions for an approximate Hamiltonian.[6] We apply perturbation theory to those systems that are almost soluble in the sense that the full Hamiltonian includes only one or two terms that make it insoluble. We write $\hat{H}$ as

$$\underset{\underset{\text{Hamiltonian}}{\text{Full}}}{\hat{H}} = \underset{\underset{\text{Hamiltonian}}{\text{Unperturbed}}}{\hat{H}_0} + \underset{\underset{\text{term}}{\text{Perturbation}}}{\hat{H}'} \quad \textbf{(13.14)}$$

where $\hat{H}_0$ represents the Hamiltonian for a soluble system and the term or terms that constitute $\hat{H}'$ are called the perturbation.

A "soluble Hamiltonian" is one for which we know the eigenfunctions and eigenenergies:

$$\hat{H}_0 \psi_n^{(0)} = E_n^{(0)} \psi_n^{(0)}. \quad \textbf{(13.15)}$$

We use the superscript (0) on ψ and E to indicate that these are the unperturbed *zeroth-order solutions* to the full Hamiltonian; that is, they solve the full problem if we neglect the perturbation completely.

Perturbation theory is a systematic procedure for using $\hat{H}'$, $E_n^{(0)}$, and $\psi_n^{(0)}$ to generate a series of corrected (and, we hope, improved) energies and wavefunctions. Each correction constitutes a successively higher order of approximation. We consider

---

[6]We have Schrödinger to thank for perturbation theory. He devised it in his third paper on quantum theory in the spring of 1926.

here only the case where the zeroth-order solutions are nondegenerate, and we describe the corrections through second order. The second-order corrections, especially to the energy, are worth looking at since several important perturbations have first-order corrections that are zero due to the symmetry of $\psi_n^{(0)}$ and $\hat{H}'$.

The first-order theory (which is given a Closer Look derivation) states

$$E_n^{(1)} = \int \psi_n^{(0)*} \hat{H}' \psi_n^{(0)} d\tau = H'_{nn}. \tag{13.16}$$

The first-order correction to the energy is just the expectation value of the perturbation calculated using the zeroth-order wavefunctions.

For example, consider the case where $\hat{H}_0$ is the 1-D harmonic-oscillator Hamiltonian and the perturbation is a sum of cubic and quartic terms:

$$\hat{H} = \underbrace{-\frac{\hbar^2}{2m}\frac{d^2}{dx^2} + \frac{k}{2}x^2}_{\hat{H}_0} + \underbrace{ax^3 + bx^4}_{\hat{H}'}$$

where a and b are constants. The $\psi_n^{(0)}$'s are just the harmonic oscillator wavefunctions, Eq. (12.19). According to Eq. (13.16),

$$E_0^{(1)} = \hat{H}'_{00} = \int_{-\infty}^{\infty} \psi_0^{(0)*}(ax^3 + bx^4)\psi_0^{(0)} dx.$$

The integral over the ax^3 term in $\hat{H}'$ is zero by symmetry, since $\psi_0^{(0)}$ is symmetric about $x = 0$ (an even function of x) while ax^3 is antisymmetric (an odd function). The bx^4 integral, however, is not zero:

$$\int_{-\infty}^{\infty} \psi_0^{(0)*} bx^4 \psi_0^{(0)} dx = 2b\left(\frac{\alpha^2}{\pi}\right)^{1/2} \int_{-\infty}^{\infty} e^{-\alpha^2 x^2} x^4 dx = \frac{3b}{4\alpha^4}$$

where $\alpha^2 = k/\hbar\omega$. To first order, the ground-state energy of an oscillator with $\hat{V}(x) = kx^2/2 + ax^3 + bx^4$ is

$$E_0 = E_0^{(0)} + E_0^{(1)} = \frac{1}{2}\hbar\omega + \frac{3}{4}b\left(\frac{\hbar\omega}{k}\right)^2.$$

The first-order correction to the wavefunction is

$$\psi_n^{(1)} = \sum_{m \neq n} \frac{H'_{mn}}{E_n^{(0)} - E_m^{(0)}} \psi_m^{(0)}. \tag{13.17}$$

This is, in general, an infinite sum (but note that the summation extends over all states m *except* the nth one). But consider the ground state correction, $\psi_0^{(1)}$. If the integrals H'_{m0} are not increasing in size with m (and they often do not), then the coefficient multiplying each $\psi_m^{(0)}$ is getting smaller and smaller in magnitude with increasing m since $|E_0^{(0)} - E_m^{(0)}|$ is getting larger. Thus, one can approximate the infinite sum by ignoring all those states that contribute small amounts (by whatever criterion seems appropriate) to the sum.

Note how the correction to ψ for state n involves, in principle, *all* the other states. It may be that many of the H'_{mn} integrals vanish by symmetry, but whether or not that is the case, we say the perturbation has *mixed* varying amounts of other states into the nth state under consideration. If $H'_{mn} \neq 0$, we say the perturbation has *connected* states m and n.

The second order correction to the energy is also an infinite sum:

$$E_n^{(2)} = \sum_{m \neq n} \frac{H'_{nm} H'_{mn}}{E_n^{(0)} - E_m^{(0)}}. \tag{13.18}$$

This sum is also often truncated, and two special cases are worth looking at in detail. The first is the ground-state correction, $n = 0$. In this case, all the denominators, $E_n^{(0)} - E_m^{(0)}$, are *negative*, since $E_m^{(0)} > E_0^{(0)}$ for all $m \neq 0$, while the numerators, $H'_{0m} H'_{m0} = |H'_{0m}|^2$, are always *positive*. Therefore, *the second-order energy correction to the ground state is always negative.*

To see how this comes about, consider the second special case of interest. Suppose only two zeroth-order states contribute significantly to the correction. This may be because only two states exist in the unperturbed solution or, more commonly, because the energy separation between the ground and first excited states is much smaller than the energy separation between either of these states and all others. For a large separation, the denominator $E_n^{(0)} - E_m^{(0)}$ is large, and the sum in Eq. (13.18) is approximated by ignoring all such terms. The correction to the ground and first excited states are

$$E_0^{(2)} = \frac{|H'_{01}|^2}{E_0^{(0)} - E_1^{(0)}} < 0 \quad \text{and} \quad E_1^{(2)} = \frac{|H'_{01}|^2}{E_1^{(0)} - E_0^{(0)}} = -E_0^{(2)} > 0 \;.$$

The ground-state energy is lowered the same amount that the excited-state energy is raised. We say the two states connected by the perturbation are *repelled* from one another. Figure 13.3 illustrates this repulsion for the case where the first-order correction is zero. Note how the magnitude of the corrections *increases* if the zeroth-order energies are close together.

These two common examples show how a perturbation can connect zeroth-order states and push them around in energy. Connected states that are close together will have the greatest mutual repulsion; hence, the ground state, with no states below it, will always be "pushed down" when connected (in second order) to one or more excited states.

If we return to our perturbed harmonic oscillator example, with $\hat{H}' = ax^3 + bx^4$, we can evaluate H'_{nm} integrals and find the second order correction to $E_0^{(0)}$. Only a few are non-zero:

$$H'_{01} = \frac{3a}{2\sqrt{2}}\left(\frac{\hbar\omega}{k}\right)^{3/2} \quad \text{(from the } ax^3 \text{ term)}$$

$$H'_{02} = \frac{3b}{\sqrt{2}}\left(\frac{\hbar\omega}{k}\right)^{2} \quad \text{(from the } bx^4 \text{ term)}$$

$$H'_{03} = \frac{3a}{2\sqrt{3}}\left(\frac{\hbar\omega}{k}\right)^{3/2} \quad \text{(from the } ax^3 \text{ term)}$$

$$H'_{04} = \frac{3b}{\sqrt{6}}\left(\frac{\hbar\omega}{k}\right)^{2} \quad \text{(from the } bx^4 \text{ term)} \;.$$

FIGURE 13.3 A two-level system, subjected to a perturbation with a leading second-order term, will show a characteristic "repulsion" of the unperturbed levels away from their common average energy.

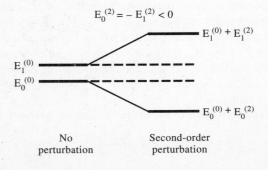

Note how the contributions to H'_{0m} alternate from the cubic to the quartic perturbation, reflecting the differing symmetries of these terms. We find an energy correction, with $E_0^{(0)} - E_m^{(0)} = -m\hbar\omega$,

$$E_0^{(2)} = \sum_{m=1}^{4} \frac{|H'_{0m}|^2}{-m\hbar\omega} = -\frac{11}{8}\frac{a^2}{\hbar\omega}\left(\frac{\hbar\omega}{k}\right)^3 - \frac{21}{8}\frac{b^2}{\hbar\omega}\left(\frac{\hbar\omega}{k}\right)^4$$

or, a final estimate of E_0, through second order,

$$E_0 = \underbrace{\frac{1}{2}\hbar\omega}_{E_0^{(0)}} + \underbrace{\frac{3}{4}b\left(\frac{\hbar\omega}{k}\right)^2}_{E_0^{(1)}} \underbrace{- \frac{11}{8}\frac{a^2}{\hbar\omega}\left(\frac{\hbar\omega}{k}\right)^3 - \frac{21}{8}\frac{b^2}{\hbar\omega}\left(\frac{\hbar\omega}{k}\right)^4}_{E_0^{(2)}}.$$

In order for this expansion to be valid, a and b must have magnitudes such that the correction terms, as a sum, are small compared to $\hbar\omega/2$. That is, after all, the operational definition of a perturbation.

EXAMPLE 13.4

What is the first-order corrected ground-state wavefunction for a harmonic oscillator perturbed by a cubic term, ax^3?

SOLUTION As discussed in the text, the only perturbation Hamiltonian matrix elements that are non-zero for a cubic perturbation to the ground state are H'_{01} and H'_{03}. Eq. (13.17) becomes

$$\psi_0^{(1)} = \frac{H'_{01}}{-\hbar\omega}\psi_1^{(0)} + \frac{H'_{03}}{-3\hbar\omega}\psi_3^{(0)} = -\frac{a}{\hbar\omega}\left(\frac{\hbar\omega}{k}\right)^{3/2}\left(\frac{3}{2\sqrt{2}}\psi_1^{(0)} + \frac{1}{2\sqrt{3}}\psi_3^{(0)}\right)$$

where we have used $E_0^{(0)} - E_m^{(0)} = -m\hbar\omega$ and $\psi_m^{(0)}$, $m = 1$ and 3, are given by Eq. (12.19), the unperturbed harmonic-oscillator wavefunctions. The final wavefunction, correct through first order, is $\psi_0 = \psi_0^{(0)} + \psi_0^{(1)}$:

$$\psi_0 = \left(\frac{k}{\hbar\omega\pi}\right)^{1/4} e^{-kx^2/2\hbar\omega}\left[1 - \left(\frac{a}{\hbar\omega}\right)\left(\frac{x^3}{3} - \frac{\hbar\omega}{k}x\right)\right].$$

Suppose $k = 570$ N m^{-1}, the value for the H—H bond's potential energy function (see Practice Problem 12H), and suppose we choose a so that $E_0^{(0)}$ is lowered by about 3%. This is in the direction of, but a bit more than, typical anharmonic effects in chemical bonds. If we take a mass appropriate to H$_2$, the unperturbed ground-state energy is 4.37×10^{-20} J, and with $a = -4.5 \times 10^{12}$ J m^{-3}, we can generate the following graph:

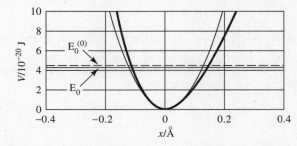

where the perturbed potential is the heavy line, the unperturbed potential is the light line, and the perturbed and unperturbed ground-state energies are the solid and dashed lines, respectively. Note how the perturbed potential rises *more* steeply for $x < 0$ (bond compression) and *less* steeply for $x > 0$ (bond extension) in comparison to the pure harmonic potential. This is typical of chemical-bond potential-energy functions. Finally, we can graph the perturbed and unperturbed wavefunctions:

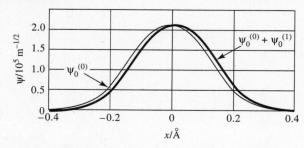

Note how the perturbed wavefunction has shifted to the right, and note that it has one glaring defect: a node around $x = -0.3$ Å. True ground-state wavefunctions never have nodes, remember, but perturbation theory often gives approximate wavefunctions with less than perfect nodal properties.

➡ RELATED PROBLEM 13.12

The second-order correction to the wavefunction is a complicated (and little used) expression that we will not consider here. Also, for the case where the soluble system is degenerate, a more general form of perturbation theory must be applied. Details of the general technique and higher order corrections can be found in this chapter's references.

A Closer Look at Perturbation Theory

A derivation of Equations (13.16)–(13.18) illustrates the power of using a set of orthonormal functions (the $\psi_n^{(0)}$'s) as well as a useful mathematical trick. First, here is the trick.

Suppose you are in a quiet room of people who are studying when suddenly the urge strikes you to listen to some music. This music will be a decided perturbation to the others. The degree of the perturbation will depend on the volume level you choose for the music—volume = 0, no perturbation; volume = full blast, total perturbation. But no matter what the volume, the nature of the perturbation (Chopin, Ice T, Willie Nelson, whatever) and the people it perturbs are unchanged.

The mathematical equivalent of a volume knob is a dimensionless parameter λ that varies from 0 to 1. We write

$$\hat{H} = \hat{H}_0 + \lambda \hat{H}'$$

with the intention of setting $\lambda = 1$ (perturbation fully on) at the conclusion of the derivation.

Since $\hat{H}$ now depends on λ, so will the true wavefunctions and energies. We can expand these quantities in a power series in λ as

$$\psi_n = \psi_n^{(0)} + \lambda \psi_n^{(1)} + \lambda^2 \psi_n^{(2)} + \cdots$$

and

$$E_n = E_n^{(0)} + \lambda E_n^{(1)} + \lambda^2 E_n^{(2)} + \cdots$$

which shows how each order of correction is associated with a particular power of λ. Thus, λ also serves as a bookkeeping device.

We substitute these expansions into the complete Schrödinger equation,

$$\hat{H}\psi_n = (\hat{H}_0 + \lambda \hat{H}')\psi_n = E_n \psi_n$$

and find

$$(\hat{H}_0 + \lambda \hat{H}')(\psi_n^{(0)} + \lambda \psi_n^{(1)} + \lambda^2 \psi_n^{(2)} + \cdots)$$
$$= (E_n^{(0)} + \lambda E_n^{(1)} + \lambda^2 E_n^{(2)} + \cdots)(\psi_n^{(0)} + \lambda \psi_n^{(1)} + \lambda^2 \psi_n^{(2)} + \cdots) \ .$$

Next, we expand the products and group them in a form containing common powers of λ. We will have a power series in λ that looks like this:

$$(\ldots) + \lambda(\ldots) + \lambda^2(\ldots) + \cdots = 0 \ .$$

Each collection of terms in parentheses must be individually equal to zero, since λ is an independent parameter. The first such collection of terms is just

$$\hat{H}_0 \psi_n^{(0)} - E_n^{(0)} \psi_n^{(0)} = 0 \ ,$$

which is the zeroth-order unperturbed equation we assume we have already solved.

The collection of terms multiplying λ is

$$\hat{H}' \psi_n^{(0)} + \hat{H}_0 \psi_n^{(1)} - E_n^{(0)} \psi_n^{(1)} - E_n^{(1)} \psi_n^{(0)} = 0 \ ,$$

which contains two unknown quantities, the first-order energy correction, $E_n^{(1)}$ and the corresponding wavefunction correction, $\psi_n^{(1)}$. We expand $\psi_n^{(1)}$ in the set of known functions $\psi_n^{(0)}$:

$$\psi_n^{(1)} = \sum_m c_{nm} \psi_m^{(0)}$$

(which, at first glance, seems to make things worse, since we now have an infinite number of unknowns, the c_{nm}'s) and substitute this expansion into the first-order equation, yielding

$$(\hat{H}' - E_n^{(1)})\psi_n^{(0)} + (\hat{H}_0 - E_n^{(0)})\sum_m c_{nm} \psi_m^{(0)} = 0 \ .$$

We now invoke the power of orthonormality. We multiply this equation (from the left) by $\psi_n^{(0)*}$ and integrate over all space:

$$H'_{nn} - E_n^{(1)} + \sum_m c_{nm} \int \psi_n^{(0)*} \hat{H}_0 \psi_m^{(0)} \, d\tau - E_n^{(0)} \sum_m c_{nm} \int \psi_n^{(0)*} \psi_m^{(0)} \, d\tau = 0 \ .$$

The integrals in the first sum are

$$E_m^{(0)} \int \psi_n^{(0)*} \psi_m^{(0)} \, d\tau = \begin{cases} 0 & n \neq m \\ E_n^{(0)} & n = m \end{cases}$$

while the integrals in the second sum are similarly zero if $n \neq m$ and 1 if $n = m$, in both cases due to orthonormality. Thus most of the terms in the sums are zero, and we have

$$H'_{nn} - E_n^{(1)} + c_{nn} E_n^{(0)} - c_{nn} E_n^{(0)} = 0$$

or simply

$$E_n^{(1)} = H'_{nn} = \int \psi_n^{(0)*} \hat{H}' \psi_n^{(0)} \, d\tau$$

which is Eq. (13.16).

Now we return to the first-order equation and multiply by $\psi_k^{(0)*}$, $k \neq n$, (again from the left) and integrate. This will give us

$$H'_{kn} - E_n^{(1)} \int \psi_k^{(0)*} \psi_n^{(0)} d\tau + \sum_m c_{nm} \int \psi_k^{(0)*} \hat{H}_0 \psi_m^{(0)} d\tau$$
$$- E_n^{(0)} \sum_m c_{nm} \int \psi_k^{(0)*} \psi_m^{(0)} d\tau = 0$$

and we again invoke orthonormality. The only terms that survive are

$$H'_{kn} + c_{nk} E_k^{(0)} - c_{nk} E_n^{(0)} = 0$$

or

$$c_{nk} = \frac{H'_{kn}}{E_n^{(0)} - E_k^{(0)}}$$

so that

$$\psi_n^{(1)} = \sum_m c_{nm} \psi_m^{(0)} = \sum_{m \neq n} \frac{H'_{mn}}{E_n^{(0)} - E_m^{(0)}} \psi_m^{(0)},$$

which is Eq. (13.17). (What happened to the $m = n$ term that appears in the first sum, but not in the second? See Problem 13.40!)

The second-order energy and wavefunction corrections (and the third, and so on) are found by successive application of these methods to, in turn, the various collection of terms that multiplied λ^2, λ^3, etc., in the original expansion of the full Schrödinger equation.

Finally, you should not let the parameter λ confuse you. It is merely a device by which we can conveniently collect various orders of the perturbation solution and, at the end, set $\lambda = 1$.

These two approximation methods form the basis of attack on all problems of atomic and molecular structure, with various elaborations built in as called for. We can now return to the He atom problem and go beyond H in the Periodic Table.

13.4 THE WAVEFUNCTIONS OF ATOMS

We left He in Section 13.2 at the independent electron approximation, where we neglected the r_{12} term in the potential energy. As a next step, we apply perturbation theory to this term. Our zeroth-order wavefunction is just the spatial part of Eq. (13.8):

$$\psi_0^{(0)} = \Psi(r_1, r_2) = \psi_{1s}(r_1)\psi_{1s}(r_2) \ .$$

The perturbation is

$$\hat{H}' = \frac{e^2}{4\pi\epsilon_0} \frac{1}{r_{12}} = \frac{e^2}{4\pi\epsilon_0} \frac{1}{|\mathbf{r}_1 - \mathbf{r}_2|}$$

and so, recognizing that ψ_{1s} is a real function,

$$E_0^{(1)} = H'_{00} = \iint \psi_{1s}^2(r_1)\psi_{1s}^2(r_2)\left(\frac{e^2}{4\pi\epsilon_0}\frac{1}{|\mathbf{r}_1 - \mathbf{r}_2|}\right) d\tau_1 \, d\tau_2$$

where $d\tau_1 = r_1^2 \sin\theta_1 dr_1 d\theta_1 d\phi_1$ and similarly for $d\tau_2$. This is not a trivial integral to do, but it can be evaluated exactly:

$$E_0^{(1)} = \frac{5}{8}\left(\frac{Ze^2}{(4\pi\epsilon_0)a_0}\right) = \frac{5}{4}\left(\frac{e^2}{(4\pi\epsilon_0)a_0}\right) = \frac{5}{4}(27.20 \text{ eV}) = 34.00 \text{ eV} \ . \quad (13.19)$$

We argued at the end of Section 13.2 that the independent-electron approximation was likely to be in error by something on this order of magnitude, and it is encouraging to see that $E_0^{(1)}$ is both positive and appreciable. The total energy is now

$$E_0 = E_0^{(0)} + E_0^{(1)} = -108.78 \text{ eV} + 34.00 \text{ eV} = -74.78 \text{ eV} ,$$

which is getting closer to, but now above, the experimental energy of -79.0 eV.

A Closer Look at the "r_{12} Integral"

Every so often, one runs across a mathematical expression that seems so fantastically unlikely or amazing, one can only marvel that anyone had the audacity or good fortune to uncover it. The integral that leads to Eq. (13.19) involves such an expression. The wavefunctions depend on the radial distances of each electron, the integral itself is over the *vector* locations of each, and the perturbation depends on the inverse of the electron–electron separation:

$$\frac{1}{r_{12}} = \frac{1}{|\mathbf{r}_1 - \mathbf{r}_2|} .$$

It is this insidious factor that makes the integral seemingly intractable until one is shown the following identity:

$$\frac{1}{r_{12}} = \frac{1}{|\mathbf{r}_1 - \mathbf{r}_2|} = \begin{cases} \dfrac{4\pi}{r_1} \displaystyle\sum_{l,m} \frac{1}{2l+1} \left(\frac{r_2}{r_1}\right)^l Y^*_{lm}(\theta_1, \phi_1) Y_{lm}(\theta_2, \phi_2), & r_1 > r_2 \\ \dfrac{4\pi}{r_2} \displaystyle\sum_{l,m} \frac{1}{2l+1} \left(\frac{r_1}{r_2}\right)^l Y^*_{lm}(\theta_1, \phi_1) Y_{lm}(\theta_2, \phi_2), & r_2 > r_1 \end{cases}$$

The angles θ_i and ϕ_i locate the vectors $\mathbf{r}_i$. While the sums over the spherical harmonic products are infinite, the wavefunctions are spherically symmetric, and all terms in the sums vanish when we integrate over angles except those for $l = m = 0$! The perturbation integral becomes

$$E_0^{(1)} = \frac{4e^2}{(4\pi\epsilon_0)\pi}\left(\frac{Z}{a_0}\right)^6 \int_0^\infty e^{-2Zr_2/a_0}\left[\frac{1}{r_1}\int_0^{r_1} e^{-2Zr_2/a_0} r_2^2\, dr_2 + \int_{r_1}^\infty e^{-2Zr_2/a_0} r_2\, dr_2\right] r_1^2\, dr_1 ,$$

which contains known integrals that lead to the expression given in the text. It is also interesting to note that this integral can be evaluated by a classical electrostatic energy argument (for spherically symmetric systems), and it was first attacked by Unsöld in 1927 using such an argument.

If we were able to calculate the second-order correction to $E_0^{(0)}$, we would surely improve our estimate, since the second-order correction is guaranteed to be negative. But the calculations for higher-order corrections past the first are more and more complex. We can do better, with less effort, by a variational method.

First, we need a parameter to vary. Looking at ψ_{1s} for nuclear charge Z:

$$\psi_{1s}(r) = \left(\frac{Z^3}{\pi a_0^3}\right)^{1/2} \exp(-Zr/a_0) ,$$

Z/a_0 seems to be the only built-in parameter. We can focus on just Z, and give this parameter a physical meaning, in the following way. The real Z for He is, of course, 2. But consider the effect one electron has on the *total effective charge* seen by the other electron. If that first electron is buried in the nucleus, the effective atomic number (call it Z_e) is $+1$. If it is completely absent, $Z_e = 2$. In fact, the electron is somewhere between $r = 0$ and ∞, spending most of its time near the nucleus, so that $1 < Z_e < 2$ would seem to be a likely range.

This is the concept called *shielding* or *screening;* any one electron, on average, senses a nuclear atomic number that is somewhat reduced from Z by the counteracting effect of the other electrons. For He, we can get a rough value for Z_e if we consider He as a one-electron atom of atomic number Z_e and the experimental first ionization energy:

$$\text{first ionization energy of He} = 24.6 \text{ eV} = Z_e^2\left(\frac{e^2}{2(4\pi\epsilon_0)a_0}\right) = Z_e^2(13.598 \text{ eV})$$

so that $Z_e = 1.34$. This is not a very useful number, since it is just another way of expressing the ionization energy, which itself has to be measured.

We can find a more appropriate Z_e from a variational calculation. We compute

$$\langle E \rangle = \iint \Psi(r_1, r_2; Z_e) \hat{H} \, \Psi(r_1, r_2; Z_e) \, d\tau_1 \, d\tau_2$$

where $\hat{H}$ is the full Hamiltonian and the trial wavefunction is just the zeroth-order function, Eq. (13.8), with Z replaced by the variational parameter, Z_e. The full Hamiltonian requires some care in this case. To conserve total charge, when electron 1 feels a nuclear charge eZ_e, electron 2 must feel a charge $e(2 - Z_e)$, and, moreover, the total Hamiltonian must be invariant to electron exchange. We write

$$\hat{H} = \underbrace{-\frac{\hbar^2}{2m}\nabla_1^2 - \frac{Z_e e^2}{(4\pi\epsilon_0)r_1}}_{\substack{\text{One-}e^-\text{ atom of} \\ \text{nuclear charge } Z_e}} \underbrace{- \frac{(2 - Z_e)e^2}{(4\pi\epsilon_0)r_1}}_{\substack{\text{Residual } e^- \\ \text{attraction}}} \underbrace{- \frac{\hbar^2}{2m}\nabla_2^2 - \frac{Z_e e^2}{(4\pi\epsilon_0)r_2}}_{\substack{\text{One-}e^-\text{ atom of} \\ \text{nuclear charge } Z_e}} \underbrace{- \frac{(2 - Z_e)e^2}{(4\pi\epsilon_0)r_2}}_{\substack{\text{Residual } e^- \\ \text{attraction}}} \underbrace{+ \frac{e^2}{(4\pi\epsilon_0)r_{12}}}_{\substack{e^-\text{-}e^- \\ \text{repulsion}}}.$$

The first and second terms represent a hydrogenic atom of nuclear charge Z_e as do the fourth and fifth. The last term is the electron-interaction term as we have already used it. The third and sixth terms are new, but easily evaluated.

Term by term, the contributions to $\langle E \rangle$ are (see Problem 13.18 for an easy derivation)

$$\langle E \rangle = \frac{e^2}{(4\pi\epsilon_0)a_0}\left[\frac{Z_e^2}{2} - Z_e^2 + (Z_e^2 - 2Z_e) + \frac{Z_e^2}{2} - Z_e^2 + (Z_e^2 - 2Z_e) + \frac{5}{8}Z_e\right]$$

$$= \frac{e^2}{(4\pi\epsilon_0)a_0}\left(Z_e^2 - \frac{27}{8}Z_e\right).$$

We minimize $\langle E \rangle$ with respect to Z_e and find

$$Z_e = \frac{27}{16} = 1.687\,5 \qquad (13.20a)$$

$$\langle E \rangle = -\frac{729}{256}\frac{e^2}{(4\pi\epsilon_0)a_0} = -2.847\,7(27.20 \text{ eV}) = -77.46 \text{ eV}, \qquad (13.20b)$$

which compares favorably with the experimental -79.0 eV.

More extensive variational calculations have been done using trial wavefunctions with more parameters to vary. The champion variational calculation of the He ground state was performed by C. L. Pekeris in 1959 using a function with 1078 parameters (!) that gave essentially the exact experimental energy. Similarly, a perturbation theory calculation through thirteenth order (!) by C. W. Scheer and R. E. Knight in 1963 also gave essentially exact agreement.

To use the variational method for atoms with more than two electrons, we need to devise a systematic way to generate trial wavefunctions, complete with spin factors and, of course, antisymmetric with respect to electron exchange. In these trial functions, we also have to choose a spatial function for any one primitive electron state (i.e., for any given (n, l, m) quantum number set in the desired configuration). These spatial functions will have the variational parameter (or parameters) built into them, but their shape is up to us to choose. Wise choices, of course, follow the general shape, symmetry, and orientation expected from the H-atom solutions.

The systematic method and the first useful, general spatial variational functions were both devised by John C. Slater, one of the pioneers in the application of quantum theory to atomic structure. We will use an abbreviated symbol for the spatial and spin wavefunctions of an individual electron in any given configuration, writing, for electron i,

$$\psi_{nlm}(r_i)\alpha(i) \equiv nl_m\alpha(i)$$

so that the wavefunction for electron 1 in spin state α with $n = 1$, $l = 0$, $m = 0$ is abbreviated $1s\alpha(1)$ (omitting the m subscript, which is redundant for s functions). Slater noticed that the requirement of antisymmetry with respect to exchange was automatically expressed by the mathematical properties of a determinant. The columns of the determinant are labeled by the primitive wavefunctions we choose to represent the state, while the rows are labeled by the electron number, 1 to N, for an N-electron atom.

For example, consider the Li atom. The ground state electron configuration is $1s^2 2s^1$. According to Slater's prescription, we construct the determinant

One primitive wavefunction per column:

Electron 1 → $\begin{vmatrix} 1s\alpha(1) & 1s\beta(1) & 2s\alpha(1) \\ 1s\alpha(2) & 1s\beta(2) & 2s\alpha(2) \\ 1s\alpha(3) & 1s\beta(3) & 2s\alpha(3) \end{vmatrix}$
Electron 2 →
Electron 3 →

that, when expanded, is a correctly antisymmetrized total wavefunction. (See Problem 13.24 for the expansion rule for a 3×3 determinant.) Notice that the last column could be written with $2s\beta$ functions, instead of $2s\alpha$, but if we wrote the second column with $1s\alpha$ instead of $1s\beta$, the determinant would have two identical columns. *A determinant with two identical columns is always zero.* Thus, in the sense that the Pauli Exclusion Principle requires all electrons to have a unique set of quantum numbers, the Slater determinant automatically kills the wavefunction (i.e., evaluates to zero) if we try to violate this principle.

The determinant above is not normalized. If the individual primitive wavefunctions ϕ_i are separately normalized, then we multiply the determinant by $(N!)^{-1/2}$ to yield a total, normalized, antisymmetrized trial wavefunction:

$$\Psi_{\text{total}} = \frac{1}{\sqrt{N!}} \begin{vmatrix} \phi_1(1) & \phi_2(1) & \phi_3(1) & \cdots & \phi_N(1) \\ \phi_1(2) & \phi_2(2) & \phi_3(2) & \cdots & \phi_N(2) \\ \vdots & \vdots & \vdots & \ddots & \vdots \\ \phi_1(N) & \phi_2(N) & \phi_3(N) & \cdots & \phi_N(N) \end{vmatrix} \quad (13.21)$$

A common abbreviation is used for this determinant that specifies only the principal diagonal of the full Slater determinant:

$$\Psi_{\text{total}} = ||\,\phi_1(1)\phi_2(2)\ldots\phi_N(N)\,||$$

where the $||\ldots||$ symbol implies the $(N!)^{-1/2}$ normalization constant as well as the full determinant. It is a highly symbolic notation for a very complex function.

Slater also suggested a set of approximate individual-electron functions. These can be used in a Slater determinant to give an approximate total function that need not be further refined by the variational method. Slater used two parameters: an effective atomic number Z_e and an effective principal quantum number n_e to govern the function's size and energy. Two parameters are needed to express the experimental observation that, for example, the $1s^22s^1$ and $1s^22p^1$ configurations for Li have different energies since the 2p electron interacts with the $1s^2$-shielded nuclear charge differently from a 2s electron.

These functions, called *Slater atomic orbitals* or *Slater-type orbitals* (STOs), are written as a close approximation to the H-atom functions:

$$\psi_{nlm} = N\, Y_{lm}(\theta, \phi)\, r^{(n_e-1)} \exp(-Z_e r/n_e a_0) \tag{13.22}$$

where N is a normalization constant and Y_{lm} is a spherical harmonic function. The original values Slater suggested for Z_e and n_e have been superceded by improved values based on more detailed modern calculations. Table 13.1 gives Z_e values while the effective principal quantum number is related to the "true" number as follows:

n	1	2	3	4	5	6
n_e	1	2	3	3.7	4.0	4.2

For more elaborate uses of the total wavefunction, such as in molecular bonding, the STOs suffer from one drawback: they are mathematically somewhat intractable since the integrals needed in perturbation or variational calculations using them often cannot be written in closed form. Functions of the type $\exp(-\alpha r^2)$, so-called *Gaussian-type orbitals*, GTOs, are more amenable to complex integrals than are the STO functions. (See also Problem 13.22.) Any slight advantage in computational efficiency is greatly magnified in contemporary calculations that may easily involve over 10^6 integrals. Nevertheless, caution is called for, since the STO and GTO have notably different shapes. The graph below compares the essence of the STO (e^{-x})

TABLE 13.1 Selected Slater Orbital Parameters Z_e

	H							He
1s	1.0000							1.6875
	Li	Be	B	C	N	O	F	Ne
1s	2.6906	3.6843	4.6795	5.6727	6.6651	7.6579	8.6501	9.6421
2s	1.2792	1.9120	2.5762	3.2166	3.8474	4.4916	5.1276	5.7584
2p			2.4214	3.1358	3.8340	4.4532	5.1000	5.7584
	Na	Mg	Al	Si	P	S	Cl	Ar
1s	10.6259	11.6089	12.5910	13.5745	14.5578	15.5409	16.5239	17.5075
2s	6.5714	7.3920	8.2136	9.0200	9.8250	10.6288	11.4304	12.2304
2p	6.8018	7.8258	8.9634	9.9450	10.9612	11.9770	12.9932	14.0082
3s	2.5074	3.3075	4.1172	4.9032	5.6418	6.3669	7.0683	7.7568
3p			4.0656	4.2852	4.8864	5.4819	6.1161	6.7641

to a Gaussian with α chosen to fit e^{-x} in a least-squares sense ($\alpha \cong 0.967$):

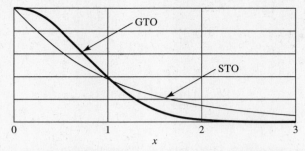

To see how accurate calculations of atomic wavefunctions progress beyond this simple approximation, it is helpful to consider first the excited states of He. This analysis leads to definitions of intra-atomic electron-interaction integrals that appear in the more general method, and it shows us how a single configuration is insufficient to specify an atomic energy.

The first excited state of He is based on the configuration $1s^12s^1$. Four distinct Slater determinants can be written based on different spin states:

$$\Psi_a = ||\, 1s\alpha(1) \quad 2s\alpha(2)\, ||$$

$$\Psi_b = ||\, 1s\beta(1) \quad 2s\beta(2)\, ||$$

$$\Psi_c = ||\, 1s\alpha(1) \quad 2s\beta(2)\, ||$$

$$\Psi_d = ||\, 1s\beta(1) \quad 2s\alpha(2)\, ||\ .$$

To see how these four functions lead to states with different energies, it is helpful to consider the following linear combinations of them:

$$\Psi_1 = \Psi_a = \frac{1}{\sqrt{2}} [1s(1)2s(2) - 1s(2)2s(1)]\alpha(1)\alpha(2) \qquad \textbf{(13.23a)}$$

$$\Psi_2 = \Psi_b = \frac{1}{\sqrt{2}} [1s(1)2s(2) - 1s(2)2s(1)]\beta(1)\beta(2) \qquad \textbf{(13.23b)}$$

$$\Psi_3 = \frac{1}{\sqrt{2}} (\Psi_c + \Psi_d)$$
$$= \frac{1}{\sqrt{2}} [1s(1)2s(2) - 1s(2)2s(1)] \frac{1}{\sqrt{2}} [\alpha(1)\beta(2) + \alpha(2)\beta(1)] \qquad \textbf{(13.23c)}$$

$$\Psi_4 = \frac{1}{\sqrt{2}} (\Psi_c - \Psi_d)$$
$$= \frac{1}{\sqrt{2}} [1s(1)2s(2) + 1s(2)2s(1)] \frac{1}{\sqrt{2}} [\alpha(1)\beta(2) - \alpha(2)\beta(1)]\ . \qquad \textbf{(13.23d)}$$

Notice that Ψ_4 has a *symmetric* spatial part (with respect to electron exchange), while the others are spatially *antisymmetric*. Thus, in contrast to the wavefunctions written in Eq. (13.5) that had the same spin factors, *each* of these four functions satisfies the Pauli Principle (while only that of Eq. (13.5d), or better, Eq. (13.8), could exist). Each of these four functions represents the $1s^12s^1$ configuration.

In a perturbation calculation, with $\hat{H}'$ = the r_{12} term and $\psi^{(0)}$ = any of these four wavefunctions, only the spatial parts of the functions contribute to $E^{(1)}$. Since the spatial parts of Ψ_1 through Ψ_3 are the same, they have the same energy (at this level of approximation) while Ψ_4 might well represent a different energy. We need to carry out the calculation to see.

The first-order energy correction breaks into two integrals:

$$E^{(1)}_1 = E^{(1)}_2 = E^{(1)}_3 = J - K$$

and

$$E^{(1)}_4 = J + K$$

where

$$J = \frac{e^2}{(4\pi\epsilon_0)} \iint [1s(1)]^2 [2s(2)]^2 \frac{1}{r_{12}} d\tau_1\, d\tau_2 \qquad (13.24)$$

and

$$K = \frac{e^2}{(4\pi\epsilon_0)} \iint [1s(1)2s(1)] \frac{1}{r_{12}} [1s(2)2s(2)]\, d\tau_1\, d\tau_2 \ . \qquad (13.25)$$

These integrals are known as the *Coulomb integral* (J) and the *exchange integral* (K). The interpretation of J is straightforward. The integrand is

$$\frac{e^2}{(4\pi\epsilon_0)} \iint \begin{bmatrix}\text{probability for}\\ \text{electron in 1s}\end{bmatrix} \begin{bmatrix}\text{probability for}\\ \text{electron in 2s}\end{bmatrix} \begin{bmatrix}\dfrac{1}{\text{electrons'}\\ \text{separation}}\end{bmatrix} d\tau_1\, d\tau_2$$

$$= \iint \begin{bmatrix}\text{charge density}\\ \text{for first electron}\end{bmatrix} \begin{bmatrix}\text{charge density}\\ \text{for second electron}\end{bmatrix} \begin{bmatrix}\dfrac{1}{\text{charge}\\ \text{separation}}\end{bmatrix} d\tau_1\, d\tau_2 \ ,$$

which is just the classical electrostatic interaction energy between unit charge distributed in space like $(1s)^2$ and unit charge distributed like $(2s)^2$. Since like charges repel, J is inherently positive.

The K integral is less easily interpreted, although an interesting interpretation is zngiven a closer look below. The K integral is, however, inherently positive as well, and we can conclude that

$$E_1 = E_2 = E_3 = E_{1s} + E_{2s} + J - K$$

and

$$E_4 = E_{1s} + E_{2s} + J + K$$

so that E_4 is greater than $E_{1, 2, \text{or } 3}$ by $2K$. The single configuration $1s^1 2s^1$ leads to four states, three with one energy and the fourth with a higher energy.

The sum of one-electron atom energies for nuclear charge $Z = 2$ is $E_{1s} + E_{2s}$. This is the zeroth-order energy:

$$E^{(0)} = E_{1s} + E_{2s} = -\frac{e^2}{2(4\pi\epsilon_0)a_0}\left(\frac{1}{1^2} + \frac{1}{2^2}\right)Z^2$$

$$= -(13.598 \text{ eV})\left(\frac{5}{4}\right)(4) = -67.99 \text{ eV} \ .$$

The Coulomb integral is

$$J = \frac{5}{27}\left(\frac{e^2}{2(4\pi\epsilon_0)a_0}\right)Z^2 = \frac{20}{27}(13.598 \text{ eV}) = 10.07 \text{ eV}$$

and the exchange integral is

$$K = \frac{64}{(27)^2}\left(\frac{e^2}{2(4\pi\epsilon_0)a_0}\right)Z^2 = \frac{64}{(27)^2}(13.598 \text{ eV}) = 1.194 \text{ eV}$$

so that the predicted energies are

$$E_1 = E_2 = E_3 = -59.11 \text{ eV}$$

and

$$E_4 = -56.73 \text{ eV} .$$

These two states are found experimentally to lie 19.8 and 20.6 eV above the ground state. The experimental separation is 0.8 eV, which is not in stunning agreement with the theoretical separation, $2K = 2.4$ eV. Since the ionization energy of ground state He is 24.6 eV, and that of He^+ is 54.4 eV, the experimental energies of these two states, relative to an energy of zero for $He^{2+} + e^- + e^-$, are

$$(-54.4 - 24.6 + 19.8) \text{ eV} = -59.2 \text{ eV}$$

(which compares favorably to the predicted -59.11 eV) and

$$(-54.4 - 24.6 + 20.6) \text{ eV} = -58.4 \text{ eV}$$

(which is significantly lower than the predicted -56.73 eV).

But the lower level *is* found experimentally to be three-fold degenerate (a *triplet* state) and the upper is nondegenerate (a *singlet* state). Since we only used first-order perturbation theory, and since our perturbation was a large one (J is nearly 15% of $E^{(0)}$), we should not be surprised that the predicted energies are not in striking agreement with experiment. However, the trends, splitting, and degeneracies are given correctly. Figure 13.4 summarizes this calculation with a pictorial energy-level diagram.

A Closer Look at the Exchange Integral

The integral we have symbolized K, Eq. (13.25), is called the exchange integral for more reasons than simply the fact that electron 2 is "exchanged" for electron 1 in the functions on either side of the r_{12} factor. In fact, this integral can be loosely interpreted as measuring the rate that electrons 1 and 2 exchange orbital positions. The interpretation is very loose, since it ignores the spin part of the wavefunctions and

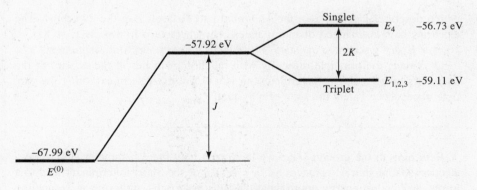

FIGURE 13.4 For the 1s2s He configuration, electron–electron repulsion raises and splits the independent-electron approximate energy $E^{(0)}$.

requires one to imagine a way of distinguishing the electrons, but the development of the argument is a useful exercise. It will give us our first opportunity to consider the explicit time dependence of the wavefunction itself.

Suppose we had a state of He that was prepared at time $t = 0$ with electron 1 in the 1s orbital and electron 2 in the 2s orbital. We might do this in an experiment in which $He^+(1s)$ captures an electron that finds its way into the 2s orbital at time $t = 0$. We would write

$$\Psi(t = 0) = 1s(1)2s(2) = \frac{1}{\sqrt{2}}(\psi_1 + \psi_4)$$

where ψ_1 and ψ_4 are the spatial parts of Ψ_1 and Ψ_4 in Equations (13.23c) and (13.23d). We assume that ψ_1 and ψ_4 are stationary states for the atom with energies that differ by $2K$, as above, while the state Ψ is *not* a stationary state, since

$$\hat{H}\Psi = \hat{H}\left[\frac{1}{\sqrt{2}}(\psi_1 + \psi_4)\right] = \frac{1}{\sqrt{2}}[\hat{H}\psi_1 + \hat{H}\psi_4]$$
$$= \frac{1}{\sqrt{2}}[(E^{(0)} + J - K)\psi_1 + (E^{(0)} + J + K)\psi_4]$$
$$\neq \text{(an energy eigenvalue)} \times \left[\frac{1}{\sqrt{2}}(\psi_1 + \psi_4)\right].$$

The explicit time dependencies of ψ_1 and ψ_4 according to Eq. (11.3) are

$$\phi_1(r_1, r_2, t) = \psi_1(r_1, r_2) \exp[-i(E^{(0)} + J - K)t/\hbar]$$

and

$$\phi_4(r_1, r_2, t) = \psi_4(r_1, r_2) \exp[-i(E^{(0)} + J + K)t/\hbar]$$

so that, with $e^{ix} = \cos x + i \sin x$, we can arrive at the time-dependent form of Ψ,

$$\Psi(r_1, r_2, t) = \left[1s(1)2s(2) \cos\left(\frac{Kt}{\hbar}\right) + i\, 1s(2)2s(1) \sin\left(\frac{Kt}{\hbar}\right)\right] \exp[-i(E^{(0)} + J)t/\hbar].$$

This reduces at $t = 0$ to our initial assumption about Ψ. But after a time $\tau = \pi\hbar/2K$,

$$\Psi(r_1, r_2, \tau) = 1s(2)2s(1)\{i \exp[-i(E^{(0)} + J)\tau/\hbar]\}$$

which describes electron *2* in the 1s orbital and electron *1* in the 2s orbital. The electrons have exchanged quantum states. The characteristic time is $\tau = 8.6 \times 10^{-16}$ s. Every τ seconds, the state Ψ oscillates between electron assignments.

Of course, the principle difficulty with this argument lies in the neglect of the spin parts of Ψ_1 and Ψ_4. Nevertheless, it is a useful exercise in exploring the way time dependence enters the state of a system.

Returning to the ground state, we have seen how more parameters in the one-electron spatial-spin wavefunctions, or *orbitals,* of the Slater determinant lead to a lower, and better, energy approximation. This approach—write the wavefunction as a product of one-electron orbitals—is called the *Hartree–Fock,* or HF, approach. (We have been using this method without explicitly calling it anything special.) The variation method, when applied to Hartree–Fock wavefunctions, leads to a series of equations that are solved iteratively by what is called the *self-consistent field* or SCF, method. Taken together, one speaks of the HF-SCF method.

HF-SCF for a general $2N$-electron atom with closed shells starts with the Hamiltonian

$$\hat{H} = \hat{H}_1 + \hat{H}_2$$

where

$$\hat{H}_1 = -\frac{\hbar^2}{2m}\sum_i \nabla_i^2 - \frac{Ze^2}{4\pi\epsilon_0}\sum_i \frac{1}{r_i} \equiv \sum_i \hat{h}_i$$

is the sum of each electron's individual interaction with a nucleus of charge Z and where

$$\hat{H}_2 = -\frac{1}{2}\left(\frac{e^2}{4\pi\epsilon_0}\right)\sum_i\sum_{j\neq i}\frac{1}{r_{ij}}$$

is the electron–electron repulsion sum (with a factor of 1/2 in front so that identical pairs of interactions are not counted twice). If Ψ is a Slater determinant, Eq. (13.21), the variational energy can be shown to be

$$\langle E \rangle = 2\sum_i^N \langle \hat{h}_i \rangle + \sum_{i=1}^N \sum_{j=1}^N (2J_{ij} - K_{ij}) \qquad (13.26)$$

where

$$\langle \hat{h}_i \rangle = \int \phi_i^* \hat{h}_i \phi_i \, d\tau$$

is the expectation value of $\hat{h}_i$ for the one-electron orbital ϕ_i and where

$$J_{ij} = \iint \phi_i^*(1)\phi_j^*(2)\left[\frac{e^2}{(4\pi\epsilon_0)r_{12}}\right]\phi_i(1)\phi_j(2)\, d\tau_1\, d\tau_2$$

is a generalized Coulomb integral and

$$K_{ij} = \iint \phi_i^*(1)\phi_j^*(2)\left[\frac{e^2}{(4\pi\epsilon_0)r_{12}}\right]\phi_i(2)\phi_j(1)\, d\tau_1\, d\tau_2$$

is a generalized exchange integral.

The variation method requires us to find those ϕ's that minimize $\langle E \rangle$. No matter how we parameterize ϕ, this minimization yields a set of coupled equations, one per orbital, of the general form

$$\hat{F}_i \phi_i = \epsilon_i \phi_i, \quad i = 1, 2, \ldots, N \qquad (13.27)$$

where $\hat{F}_i$ is the *Fock operator*:

$$\hat{F}_i = \hat{h}_i + \sum_j (2\hat{J}_j - \hat{K}_j)$$

and ϵ_i is interpreted as the *orbital energy* of orbital i. The Coulomb and exchange operators are

$$\hat{J}_j \phi_i(1) = \left\{\int \phi_j^*(2)\left[\frac{e^2}{(4\pi\epsilon_0)r_{ij}}\right]\phi_j(2)\, d\tau_2\right\}\phi_i(1)$$

(r_{ij} here is the separation between the electron in orbital ϕ_j, in the integral, and the electron in the orbital i on which $\hat{J}_j$ operates), and

$$\hat{K}_j \phi_i(1) = \phi_j(1)\int \phi_j^*(2)\left[\frac{e^2}{(4\pi\epsilon_0)r_{ij}}\right]\phi_i(2)\, d\tau_2 \, .$$

The coupled equations of Eq. (13.27) are difficult to solve since orbitals in every equation appear as orbital products and/or integrals over them. The "self-consistent" name of the theory comes from the approach used to solve these equations. We first assume a set of ϕ_i's (such as general Slater-type orbitals) and use them to generate a set of Fock operators. These, in turn, are used via Eq. (13.27) to generate a new set of orbitals (and orbital energies), which give new Fock operators, which give new orbitals, and so on, iteratively producing orbitals and their energies until the energies from one iteration to the next converge to some desired tolerance.

The converged energies are said to be "at the Hartree–Fock limit" and are the result of a self-consistent solution. A simple example, the He ground state, points out the steps and allows us some physical interpretation. We start with the Slater determinant

$$\Psi = || \phi_1(1) \quad \phi_2(2) || = || 1s\alpha(1) \quad 1s\beta(2) ||.$$

The first Fock operator is

$$\hat{F}_1 = \hat{h}_1 + \hat{J}(1)$$

where

$$\hat{J}(1) = \int \phi_2^* \left[\frac{e^2}{(4\pi\epsilon_0)r_{12}} \right] \phi_2 \, d\tau_2 ,$$

which is the average repulsion electron 1 feels from electron 2 when 2 is spread over the region $|\phi_2|^2$. (The only other term in the Fock operator is $\hat{J}$ since $2J - K = J$ for the He ground state.)

We solve the Fock equation

$$\hat{F}_1 \phi_1'(1) = \epsilon \phi_1'(1)$$

variationally to get an improved wavefunction $\phi_1'(1)$ and a first estimate of the 1s orbital energy ϵ. We then use ϕ' to generate a new $\hat{F}$ and repeat this step. For He(1s^2) the whole procedure collapses to our familiar total energy expression

$$\langle E \rangle = 2E_{1s} + J_{1s,1s} ,$$

which we would have used to begin a direct variation calculation. For more complex atoms, the procedure is just an elaboration of these steps through the various spatial orbitals in the Slater determinant.

For an atom with some electrons unpaired (as in our He excited state, 1s^{1}2s^1), one generally needs a linear combination of Slater determinants for the total wavefunction. This increases the mathematical work (which is all done by computer) but does not significantly alter the plan of attack.

There are two major problems with the HF-SCF method, quite apart from the problem of picking suitable ϕ's in the first place. One is that the orbital energies are independent of the number of electrons present in the atom (or ion). We could apply HF theory to the F atom, for example, and find ϵ_{1s}, ϵ_{2s}, and ϵ_{2p} energies. These would be the same for F$^-$ or F$^+$, and HF-SCF theory predicts that the ϵ_i's are just the ionization energies[7] of the various orbitals.

In fact, the electrons *relax* or adjust to the loss of one of their number, and orbital energies are only *approximately* additive to give correct total energies. This defect

---

[7]This result is known as *Koopmans' theorem;* note where the apostrophe is. Tjalling C. Koopmans proved the theorem, then switched fields and went on to share (with Leonid V. Kantorovich) the 1975 Nobel Prize in economics!

of the theory can be traced to averaging in the Coulomb and exchange operators. They treat the electrons as independent particles, but their motion must be *correlated*. One electron knows (or feels) the motion of its spin-coupled partner. There are more advanced theoretical methods for treating the energy contribution due to electron correlation.

The second defect of HF-SCF is its assumption of a single configuration. The orbitals that lead to one configuration must be only an approximation to the true spatial-spin wavefunctions, and a set of orbitals for some other configuration may have the same spatial-spin symmetry as the original configuration. This means a trial wavefunction of the type

$$\Psi = c_1\psi(\text{major configuration}) + c_2\psi(\text{alternate configuration})$$

where c_1 and c_2 are variable coefficients could (and should) be used. The coefficients show how much of any one configuration contributes to the total wavefunction for any one energy state of the atom. They are thus often called *mixing coefficients*, and one says "state Ψ is $|c_1|^2 \times 100\%$ of the major configuration and $|c_2|^2 \times 100\%$ of the alternate." More than two configurations can lead to Ψ of course, and the general *configuration interaction (CI) approach* (also known as *MCSCF* for Multi-Configuration SCF) uses as many configurations of the appropriate symmetry as one's computer budget and computer memory limitations allow.

Thus, the penultimate technique for finding atomic energies and wavefunctions is the HF-SCF-CI method with electron correlation, which has been widely applied across the Periodic Table, leading to tables of energies and wavefunction parameters, such as those for the STO's of Table 13.1. For elements of large nuclear charge (around Hg and beyond), however, the SCF-CI technique cannot hope to be accurate. Increasingly large nuclear charges imply increasingly large electron velocities, large enough to require a *relativistic* theory. There are several well-known chemical facts, such as the importance of both +1 and +2 oxidation states of Hg, and even the gold color of Au, that can be explained only by a relativistic theory. Important contemporary efforts are underway to apply such theory to heavy elements.

This section has made several important points about many-electron atoms. The first is that even approximate wavefunctions must satisfy the Pauli Exclusion Principle. The second, as our look at the excited states of He showed, is that an electron configuration can lead to *more* than one distinct energy, depending on the spin assignments of the configuration. Finally, we described a systematic approximation method, variational HF-SCF-CI, for calculating atomic wavefunctions and energies. The next section concentrates on just the energy levels of atoms and describes a systematic and compact notation for them.

13.5 THE DESCRIPTION OF ATOMIC ENERGY VALUES

Our atomic Hamiltonians have been sums of obvious terms: electron kinetic energy and charge-charge potential energy terms. Less obvious terms need to be included, however, and we begin this section discussing one of these: the coupling between the intrinsic spin angular momentum of an electron and its orbital angular momentum (if any) in the atom. This is *spin-orbit coupling* first mentioned in Section 12.5.

We have two possible sources of angular momentum (neglecting nuclear spin), but any atom has only one total angular momentum, of course. Let **J** be the total angular momentum vector. Due to spin-orbit coupling, *every value of **J** could correspond to a different energy and thus to a different atomic state*. Thus, we must learn how angular momenta—spin and orbital—add.

We can write

$$\mathbf{J} = \mathbf{L} + \mathbf{S} \quad (13.28a)$$

where $\mathbf{L}$ is the total orbital angular momentum vector and $\mathbf{S}$ is the total spin angular momentum vector. But we could also write

$$\mathbf{J} = \sum_i \mathbf{j}_i \quad (13.28b)$$

where $\mathbf{j}_i$ = the total angular momentum vector of electron i = $\mathbf{l}_i + \mathbf{s}_i$, the sum of the spin and orbital angular momentum vectors of electron i. Since

$$\mathbf{L} = \sum_i \mathbf{l}_i \quad \text{and} \quad \mathbf{S} = \sum_i \mathbf{s}_i$$

either approach gives the same $\mathbf{J}$.

These are the two *coupling schemes* commonly used to find $\mathbf{J}$. They are called *L–S coupling* (Eq. (13.28a)) and *j–j coupling*[8] (Eq. (13.28b)), respectively. (The L–S coupling scheme is also known as *Russell–Saunders coupling*.)

If we neglect this coupling, as we have until now, we know that

$$\hat{L}^2 \Psi = L(L + 1)\hbar^2 \Psi$$

and

$$\hat{S}^2 \Psi = S(S + 1)\hbar^2 \Psi$$

since Ψ is simultaneously an eigenfunction of both $\hat{L}^2$ and $\hat{S}^2$. In the presence of spin-orbit coupling, these relations are no longer valid; instead, we have only

$$\hat{J}^2 \Psi = J(J + 1)\hbar^2 \Psi$$

and we say J is "a good quantum number" (i.e., valid in an eigenvalue expression) while L and S have become "bad quantum numbers." The "goodness" and "badness" are rigorously true attributes, but even a bad quantum number can serve as a useful label, as we will see.

The spin-orbit operator has the general form, for electron i,

$$\hat{H}_{so} = h_{so} \hat{\mathbf{l}}_i \cdot \hat{\mathbf{s}}_i \quad (13.29)$$

where h_{so} is the *spin-orbit coupling constant*. The theory of angular momentum coupling, which we will not explore in detail here, describes the addition of two quantized angular momenta. The quantum numbers[9] that can result from adding $\mathbf{J}_1$ and $\mathbf{J}_2$ ($\mathbf{J}_{12} = \mathbf{J}_1 + \mathbf{J}_2$) are the numbers J_{12} where

$$J_{12} = J_1 + J_2, J_1 + J_2 - 1, \ldots, |J_1 - J_2|. \quad (13.30)$$

The maximum J_{12} is the sum of J_1 and J_2, the minimum is the absolute value of the difference between J_1 and J_2, and every value in between spaced one unit apart is also possible. With this rule, which permits an arbitrary number of angular momenta to be combined by adding one after the other as in $\mathbf{J}_1 + \mathbf{J}_2 = \mathbf{J}_{12}$, then $\mathbf{J}_{12} + \mathbf{J}_3 = \mathbf{J}_{123}$, etc., we can see how $\hat{H}_{so}$ has an effect on an atomic electron configuration's energy.

---

[8] Science fiction buffs may have encountered an author named J. J. Coupling. This is the pseudonym of John R. Pierce, who has contributed to acoustics research, satellite communications, synthetic music, and other areas at the engineering/physics interface. He obviously learned atomic physics as well at some point in his distinguished career!

[9] Be careful of the notation: $\mathbf{J}$ is the angular momentum vector; $\hat{\mathbf{J}}$ is the corresponding operator; $|\mathbf{J}|$ is the eigenvalue for the magnitude of $\mathbf{J}$; and J is the corresponding quantum number. Thus we write $\hat{\mathbf{J}} \cdot \hat{\mathbf{J}} \Psi = \hat{J}^2 \Psi = |\mathbf{J}|^2 \Psi = J(J + 1)\hbar^2 \Psi$.

13.5 THE DESCRIPTION OF ATOMIC ENERGY VALUES

We take H as a simple case. It is observed that h_{so} is small compared to the total energy. (After all, if it were large, we would have considered it sooner in our discussion!) Thus, we treat it as a perturbation. From the definition of **J** in Eq. (13.28a), and the coupling rule of Eq. (13.30), we have

$$J = j = l + s, l + s - 1, \ldots, |l - s|$$

so that for a d electron ($l = 2$) with $s = 1/2$, we have

$$j = 2 + \frac{1}{2}, 2 + \frac{1}{2} - 1, \ldots, \left|2 - \frac{1}{2}\right|$$

$$= \frac{5}{2} \text{ or } \frac{3}{2} \text{ since } 2 + \frac{1}{2} - 1 = \left|2 - \frac{1}{2}\right| \text{ here}.$$

Moreover, since $\hat{\mathbf{l}}$ and $\hat{\mathbf{s}}$ commute,

$$\hat{\mathbf{j}} \cdot \hat{\mathbf{j}} = (\hat{\mathbf{l}} + \hat{\mathbf{s}}) \cdot (\hat{\mathbf{l}} + \hat{\mathbf{s}}) = \hat{\mathbf{l}} \cdot \hat{\mathbf{l}} + 2\hat{\mathbf{l}} \cdot \hat{\mathbf{s}} + \hat{\mathbf{s}} \cdot \hat{\mathbf{s}}$$

or

$$\hat{\mathbf{l}} \cdot \hat{\mathbf{s}} = \frac{1}{2}(\hat{\mathbf{j}}^2 - \hat{\mathbf{l}}^2 - \hat{\mathbf{s}}^2)$$

so that the first-order perturbation theory correction[10] from $\hat{H}_{so}$ is

$$\begin{aligned} E_{so}^{(1)} &= \int \psi_{nls}^* \hat{H}_{so} \psi_{nls} \, d\tau \\ &= \frac{h_{so}}{2} \int \psi_{nls}^* (\hat{\mathbf{j}}^2 - \hat{\mathbf{l}}^2 - \hat{\mathbf{s}}^2) \psi_{nls} \, d\tau \\ &= \frac{h_{so}}{2} [j(j+1)\hbar^2 - l(l+1)\hbar^2 - s(s+1)\hbar^2] \int \psi_{nls}^* \psi_{nls} \, d\tau \\ &= \frac{\hbar^2 h_{so}}{2} [j(j+1) - l(l+1) - s(s+1)]. \end{aligned} \quad (13.31)$$

For the d electron, there are two possible j values and thus two different perturbation corrections:

$$E_{so}^{(1)}(j = 5/2) = \frac{\hbar^2 h_{so}}{2}\left[\frac{5}{2}\left(\frac{7}{2}\right) - 2(3) - \frac{1}{2}\left(\frac{3}{2}\right)\right] = \hbar^2 h_{so}$$

and

$$E_{so}^{(1)}(j = 3/2) = \frac{\hbar^2 h_{so}}{2}\left[\frac{3}{2}\left(\frac{5}{2}\right) - 2(3) - \frac{1}{2}\left(\frac{3}{2}\right)\right] = -\frac{3}{2}\hbar^2 h_{so}.$$

Thus, $\hat{H}_{so}$ splits the d energy level into *two* levels of differing energy. What about other l values? Since $l = 0$ (s states) implies no orbital part to **j**, the perturbation is zero ($j = s$). For every other l value, there will be two possible j's ($l + 1/2$ and $l - 1/2$) for H, and *each* $l \neq 0$ level is split into two states. The splitting is l and n dependent (see the previous footnote) and quite small. It is only 0.37 cm^{-1} for the 2p level and decreases roughly as $1/(n^3 l^3)$.

What about the m and m_s quantum numbers? Recall that they do not enter into the zeroth-order H-atom energy expression, but they do constitute the $2n^2$ degeneracy of that expression. For j, we speak of an m_j quantum number such that the possible m_j values are

$$m_j = -j, -j + 1, \ldots, +j$$

---

[10]The coupling constant also depends on the l quantum number (and the n quantum number—see Chapter 21). Here, we assume we know its value for the state under consideration.

as usual for an angular momentum component. Thus, for the H-atom d electron,

$$\text{for } j = \frac{5}{2}: \quad m_j = \frac{5}{2}, \frac{3}{2}, \frac{1}{2}, -\frac{1}{2}, -\frac{3}{2}, -\frac{5}{2}$$

$$\text{for } j = \frac{3}{2}: \quad m_j = \frac{3}{2}, \frac{1}{2}, -\frac{1}{2}, -\frac{3}{2}.$$

Note that ten values of m_j exist in all. This is consistent with the $(2l + 1) = 5$ degeneracy of the d level and the $(2s + 1) = 2$ spin degeneracy. *But we cannot in general associate unique (m, m_s) values with any one m_j.* It is true that the largest m_j, 5/2, is equal to the largest m (+2) plus the largest m_s (+1/2), but we cannot uniquely associate any one $m_j = 1/2$ value, say, with a pair of (m, m_s) values. All we can do, and it is a very valuable thing to do, is to check that we have "conserved the total numbers of states;" that is, we have a total number of m_j values equal to the product of the number of m and m_s values.

Our first classification of atomic energy levels begins with J values. We start with an L–S coupling scheme and later point out why the j–j coupling scheme is more appropriate for certain heavy atoms. We assume a single configuration wavefunction, and although we neglect configuration interaction, our result will show us how to construct a CI wavefunction.

Consider first the He ground state. Since each $l_i = 0$, $L = 0$; since each $s_i = 1/2$, S should be either 1 or 0, and thus J should be either 1 or 0. Now we count states: $J = 1$ implies $m_J = +1, 0$, or -1 for three states, and $J = 0$ is the single $m_J = 0$ state, for a total of four. But we know the $1s^2$ ground state configuration wavefunction is a *single, nondegenerate wavefunction,* Eq. (13.8)!

What went wrong? Why did we predict *four* states from the total J degeneracy, but only *one* state from looking at the $1s^2$ wavefunction? The answer comes from the Pauli Exclusion Principle. If $S = 1$, m_s could be 1, or each m_{s_i} would *have to be* +1/2. That would violate the Pauli Principle, and thus the three $S = J = 1$ states *do not exist* for the $1s^2$ configuration.

But for the $1s^1 2s^1$ configuration, $S = J = 1$ *is* allowed, and the four m_J states agree in number with the four Slater determinant wavefunctions, Eq. (13.23a–d), that we discussed at length. Straightforward angular momentum coupling will never undercount states, but it may *overcount* them if electrons of the same (n, l) quantum numbers are involved. (Such electrons are said to be *equivalent*.)

Can we associate a given (J, m_J) with a particular wavefunction? To a useful approximation, yes. Since we used first-order perturbation theory to find energy corrections to zeroth-order energies, we should find first-order corrected wavefunctions and see how $\hat{H}_{so}$ couples zeroth-order states. We would need to develop degenerate-state perturbation theory to do this correctly, and thus we will stick with the zeroth-order states. Consider the He $1s^1 2s^1$ configuration again. We found two distinct energies, one for wavefunctions from Eq. (13.23a, b, and c) and a higher one for the wavefunction from Eq. (13.23d). Consider Eq. (13.23a) first. It has the $\alpha(1)\alpha(2)$ spin part, suggesting $m_s = +1$, while $m_s = -1$ seems to go with Eq. (13.23b), $\beta(1)\beta(2)$. That Eq. (13.23c) and Eq. (13.23d) both represent $m_s = 0$ is less obvious, as we have discussed, but, at the zeroth-level of approximation, we can associate

Eq. (13.23a) $\to L = 0, S = 1, J = 1, m_J = 1$

Eq. (13.23b) $\to L = 0, S = 1, J = 1, m_J = 0$

Eq. (13.23c) $\to L = 0, S = 1, J = 1, m_J = -1$

Eq. (13.23d) $\to L = 0, S = 0, J = 0, m_J = 0$.

Since $L = 0$ (or $S = J$ for each J), the spin–orbit interaction is zero. We have just reclassified, but not altered, these states' energies. The lower level remains three-fold degenerate and the upper remains non-degenerate.

But now consider the $1s^1 2p^1$ configuration. Here, $l_1 = 0$, $l_2 = 1$, and $s_1 = s_2 = 1/2$. Thus, $L = 1$ and $S = 1$ or 0. Combining $L = 1$ and $S = 1$ gives $J = 1 + 1$, $1 + 1 - 1$, and $|1 - 1|$ or $2, 1,$ and 0, while combining $L = 1$ and $S = 0$ gives $J = 1$. The total number of states (total number of m_J values) is

$$\underbrace{[2(2) + 1] + [2(1) + 1] + [2(0) + 1]}_{\substack{2J + 1 \text{ for } J = 2, 1, \text{ and } 0 \\ \text{from } L = S = 1}} + \underbrace{[2(1) + 1]}_{\substack{2J + 1 \text{ for } J = 1 \\ \text{from } L = 1, S = 0}} = 12 \; .$$

This is the number of Slater determinants we could write for the $1s^1 2p^1$ configuration: four spin combinations for each of the three m values associated with the 2p orbital.

If we take *any one* of the three sets of four $1s^1 2p^1$ Slater determinants through the HF-SCF method, we find two energies, just as for the $1s^1 2s^1$ configuration. The lower energy states are $3 \times 3 = 9$-fold degenerate overall, and the upper is $1 \times 3 = 3$-fold degenerate overall. Figure 13.5 shows the splitting of the twelve independent-electron $1s^1 2p^1$ states into these two energies.

Next, we include spin-orbit coupling, which, since $J \neq 0$, will not vanish for this configuration. We have four different (J, L, S) combinations that give four different energies from Eq. (13.31):

$$(J, L, S) = (2, 1, 1) \rightarrow E_{so}^{(1)} = \hbar^2 h_{so}$$

$$(J, L, S) = (1, 1, 1) \rightarrow E_{so}^{(1)} = -\hbar^2 h_{so}$$

$$(J, L, S) = (0, 1, 1) \rightarrow E_{so}^{(1)} = -2\hbar^2 h_{so}$$

$$(J, L, S) = (1, 1, 0) \rightarrow E_{so}^{(1)} = 0 \; .$$

Figure 13.5 shows these energies as a splitting of the lower $S = 1$ level of the HF-SCF approximation. When we compare them to experimentally determined energies and quantum numbers, the pattern of levels—three very closely spaced and one significantly higher energy—agrees with our theory, but the observed J's for the lowest three levels do not. A complete theory would have to include effects we have neglected, such as CI, electron-spin–electron-spin interaction, and the coupling of the spin of the 1s electron to the orbital angular momentum of the 2p electron (the so-called *spin–other-orbit interaction*). All of these are small effects, but no one of them is dominant.

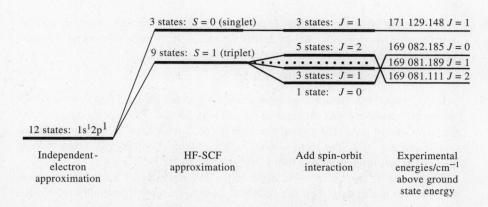

FIGURE 13.5 For the 1s2p He configuration, spin–orbit interactions (not shown to scale here) are important along with other CI and spin related effects that lead to the final experimental energies shown.

Our major conclusion, nevertheless, is that J is a legitimate and necessary label for atomic energy levels. We have gone through a series of steps to reach this conclusion, as summarized below:

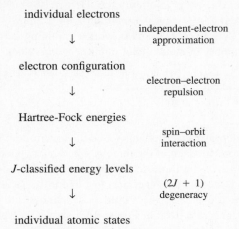

A systematic nomenclature, called *term symbols,* is used to label individual states. (The experimental equivalent of Hartree-Fock energies, the levels without J classification and spin-orbit splitting, were called *terms* by early atomic spectroscopists.)

The term symbol itself displays J, L, and S values in a compact, but somewhat cryptic, way. For L, capital letters analogous to s, p, d, etc., are used to form the main part of the symbol. Thus, $L = 0$ is an S term, which is confusing, since we have already used S to represent the total spin quantum number. With a little practice, the distinction will be second nature to you. For example, the $1s^12p^1$ level with $L = 1$ uses P at the heart of the symbol. The J value is written as a numerical subscript to the letter representing L. Thus, $L = 1$ and $J = 2$ is represented P_2. Finally, the S quantum number is converted to the spin degeneracy $2S + 1$ and written as a leading superscript, so that $L = 1$, $J = 2$, and $S = 1$ is $^3P_2$ (read "triplet P two"). (The $2S + 1$ degeneracy is read as a *multiplet:* singlet, doublet, triplet, quartet, quintet, sextet, etc.) In summary, the term symbol is written

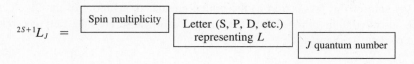

A complete state specification lists the configuration, the term symbol, and the m_J value. Our four levels for the $1s^12p^1$ configuration, in order of increasing observed energy, are (ignoring m_J)

$$1s^12p^1\ {}^3P_2, \quad 1s^12p^1\ {}^3P_1, \quad 1s^12p^1\ {}^3P_0, \quad \text{and} \quad 1s^12p^1\ {}^1P_1\ .$$

It may seem only natural that there should be "three triplets" and "one singlet," but we can easily see that the number of levels is not always equal to the multiplet number. Ground state H is a good example. Here, $S = 1/2$ (so $2S + 1 = 2$), while $L = 0$ and $J = 1/2$. The term symbol is $^2S_{1/2}$ ("doublet S one half"), but there is only *one* level, in spite of the "doublet" name of the term symbol.

The power of the term symbol approach is that we can pick a configuration (an easy step for a chemist), then derive all the possible term symbols allowed by that configuration, and finally, with the help of a few empirical rules, predict their energy ordering and their rough energy spacing. The hard part is to derive the term symbols when the configuration contains equivalent electrons (as in the $1s^2$ He example earlier). But once this is done for one generic type of configuration (p^2 for example), it is done for all: $2p^2$, $3p^2$, etc., and we can tabulate commonly occurring configurations and their associated terms once and for all. This is done in Table 13.2 for several important configurations, and Problem 13.30 takes you through the derivation of one of them.

To use these tables, you should remember that *only electrons in unfilled subshells contribute to the total L, S, and J values.* Filled subshells ($1s^2$, $2s^2$, $2p^6$, etc.) have $L = S = 0$. Moreover, *if a subshell can hold n electrons, but has only m, the term symbols for such a configuration is the same as for the complementary configuration with $n - m$ electrons.* Thus, p^1 and p^5 configurations have the same term symbols, as do p^2 and p^4 or d^2 and d^8, etc.

For example, consider the term symbols for the ground state configurations of C ($1s^2 2s^2 2p^2$) and O ($1s^2 2s^2 2p^4$). We consider only the p electrons, since the 1s and 2s subshells are filled, and, since p^2 and p^4 configurations have equivalent term symbols, we consult Table 13.2 and find, for both atoms, the possibilities (including *J*)

$$^1S_0 \quad ^1D_2 \quad ^3P_2 \quad ^3P_1 \quad ^3P_0$$

To put these states in order of increasing energy, we use three semiempirical rules first discussed by Hund and known as *Hund's rules*. These state, for any one configuration,

1. Levels of greatest spin multiplicity lie lowest:

    thus $(^1S_0, {}^1D_2) > ({}^3P_2, {}^3P_1, {}^3P_0)$ in energy.

2. Levels of equal multiplicity have the greatest *L* lowest:

    thus $^1S_0 > {}^1D_2 > ({}^3P_2, {}^3P_1, {}^2P_0)$

TABLE 13.2 Term Symbols for Various Electron Configurations

s	^2S	s^2, p^6, d^{10}	^1S		
sp	$^{1,3}P$	sd	$^{1,3}D$	pd	$^{1,3}P,D,F$
pp	$^{1,3}S,P,D$	dd	$^{1,3}S,P,D,F,G$		
p, p^5	^2P				
p^2, p^4	$^1S,D; {}^3P$				
p^3	$^2P,D; {}^4S$				
d, d^9	^2D				
d^2, d^8	$^1S,D,G; {}^3P,F$				
d^3, d^7	$^2P,D,D,F,G,H; {}^4P,F$				
d^4, d^6	$^1S,S,D,D,F,G,G,I; {}^3P,P,D,F,F,G,H; {}^5D$				
d^5	$^2S,P,D,D,D,F,F,G,G,H,I; {}^4P,D,F,G; {}^6S$				

Notes: (1) The notation pp means two *inequivalent* p electrons (as in 2p3p) while p^2 denotes two *equivalent* p electrons (as in $2p^2$).

(2) The notation $^{1,3}P,D,F$ means six possible terms: ^1P, ^1D, ^1F, ^3P, ^3D, ^3F.

(3) Certain d electron configurations have one (or more) term symbols possible in two (or more) ways, e.g., d^5 has three possible and distinguishable ^2D terms, two ^2F, etc.

3. Levels of equal L and S have:

minimum J lowest if a single open subshell is *less* than half full,

(thus $^1S_0 > {}^1D_2 > {}^3P_2 > {}^3P_1 > {}^3P_0$ for C)

but *maximum* J lowest if a single open subshell is *more* than half full,

(thus $^1S_0 > {}^1D_2 > {}^3P_0 > {}^3P_1 > {}^3P_2$ for O)

and (usually) maximum J is lowest if exactly half full.

EXAMPLE 13.5

What are the term symbols for the ground state configuration of the halogens? Which term symbol represents the true atomic ground state?

SOLUTION The halogens have the ns^2np^5 valence configuration. The s subshell is full and can be ignored. According to Table 13.2, five equivalent p electrons form only ^2P terms. The superscript 2 implies $2S + 1 = 2$ or $S = 1/2$, while the P represents $L = 2$. Consequently, $J = 3/2 = L + S$ or $1/2 = L - S$ only. The full term symbols are $^2P_{1/2}$ and $^2P_{3/2}$ ("doublet P one half" and "doublet P three halves"). The unfilled p subshell is more than half full. According to Hund's rule (3), the higher J term, $^2P_{3/2}$, is the true ground state. It is interesting to note the periodic increase in the energy difference between these two (lowest) energy states of the halogens:

atom	F	Cl	Br	I
$[E(^2P_{1/2}) - E(^2P_{3/2})]/\text{cm}^{-1}$	404	881	3685	7603

The systematic increase is due to the increasing importance of spin–orbit coupling as one goes down a column of the Periodic Table. The different chemistries of these two halogen states has been an area of great contemporary research interest.

➠ *RELATED PROBLEMS* 13.26, 13.27, 13.32, 13.33

Hund's rules are most reliable for ground-state configurations of elements in the first five or so rows of the Periodic Table, but excited state configurations can lead to terms that are ordered in energy in violation of one or more rules. For detailed atomic-state information (observed energies, degeneracies, configurations, etc.) one turns to tabulated data, the most important of which is *Atomic Energy Levels*, compiled by Charlotte E. Moore. (A complete reference is given at the end of the chapter.)

Tables 13.3 and 13.4 reproduce parts of the C and O data from this compilation (which includes atomic positive ions as well as neutrals). Look first at the C data. The notation "C I" (not "configuration interaction" here!) is commonly used by astronomers and astrophysicists to represent ordinary neutral atomic C. In this notation, C^+ becomes C II, C^{2+} is C III, etc. The columns are

1. *Edlén*   term symbol notation used by the Swedish spectroscopist B. Edlén, who contributed much to the spectroscopy of atoms and atomic ions
2. *Config.*   the configuration leading to the observed levels
3. *Desig.*   the abbreviated configuration and term-symbol notation as we have used it
4. *J*   the J value of the level

13.5 THE DESCRIPTION OF ATOMIC ENERGY VALUES

TABLE 13.3 C Atom Energy Levels and Term Symbol Designations

	C I						C I				
Edlén	Config.	Desig.	J	Level	Interval	Edlén	Config.	Desig.	J	Level	Interval
$2p\ ^3P_0$	$2s^2\ 2p^2$	$2p^2\ ^3P$	0	0.0		$3d\ ^1D_2$	$2s^2\ 2p(^2P°)\ 3d$	$3d\ ^1D°$	2	77680.5	
$^3P_1$			1	16.4	16.4						
$^3P_2$			2	43.5	27.1	$4s\ ^3P_0$	$2s^2\ 2p(^2P°)\ 4s$	$4s\ ^3P°$	0	78105.23	
						$^3P_1$			1	78117.06	11.83
$2p\ ^1D_2$	$2s^2\ 2p^2$	$2p^2\ ^1D$	2	10193.70		$^3P_2$			2	78148.36	31.30
$2p^1\ S_0$	$2s^2\ 2p^2$	$2p^2\ ^1S$	0	21648.4		$3d\ ^3F_2$	$2s^2\ 2p(^2P°)\ 3d$	$3d\ ^3F°$	2	78199.34	
						$^3F_3$			3	78215.82	16.48
	$2s\ 2p^3$	$2p^3\ ^5S°$	2	33735.3		$^3F_4$			4	78250.22	34.40
$3s\ ^3P_0$	$2s^2\ 2p(^2P°)\ 3s$	$3s\ ^3P°$	0	60333.80		$3d\ ^3D_1$	$2s^2\ 2p(^2P°)\ 3d$	$3d\ ^3D°$	1	78300.8	
$^3P_1$			1	60353.00	19.20	$^3D_2$			2	78307	6
$^3P_2$			2	60393.52	40.52	$^3D_3$			3	78316	9
$3s\ ^3P_1$	$2s^2\ 2p(^2P°)\ 3s$	$3s\ ^1P°$	1	61982.20		$4s\ ^1P_1$	$2s^2\ 2p(^2P°)\ 4s$	$4s\ ^1P°$	1	78338	
$2p'\ ^3D_3$	$2s\ 2p^3$	$2p^3\ ^3D°$	3	64088.56		$3d\ ^1F_3$	$2s^2\ 2p(^2P°)\ 3d$	$3d\ ^1F°$	3	78531	
$^3D_2$			2	64093.19	−4.63						
$^3D_1$			1	64092.01	1.18	$3d\ ^1P_1$	$2s^2\ 2p(^2P°)\ 3d$	$3d\ ^1P°$	1	78727.91	
$3p\ ^1P_1$	$2s^2\ 2p(^2P°)\ 3p$	$3p\ ^1P$	1	68858		$3d\ ^3P_2$	$2s^2\ 2p(^2P°)\ 3d$	$3d\ ^3P°$	2	79311.10	
						$^3P_1$			1	79319.06	−7.96
$3p\ ^3D_1$	$2s^2\ 2p(^2P°)\ 3p$	$3p\ ^3D$	1	69689.79					0	79323.32	−4.26
$^3D_2$			2	69710.99	21.20						
$^3D_3$			3	69744.40	33.41	$4p\ ^3D_1$	$2s^2\ 2p(^2P°)\ 4p$	$4p\ ^3D$	1	80173.29	
						$^3D_2$			2	80192.49	19.20
$3p\ ^3S_1$	$2s^2\ 2p(^2P°)\ 3p$	$3p\ ^3S$	1	70744.26		$^3D_3$			3	80222.74	30.25
$3p\ ^3P_0$	$2s^2\ 2p(^2P°)\ 3p$	$3p\ ^3P$	0	71352.81		$4p\ ^1P_1$	$2s^2\ 2p(^2P°)\ 4p$	$4p\ ^1P$	1	80563.57	
$^3P_1$			1	71365.23	12.42						
$^3P_2$			2	71385.70	20.47	$4p\ ^3S_1$	$2s^2\ 2p(^2P°)\ 4p$	$4p\ ^3S$	1	81105.70	
$3p\ ^1D_2$	$2s^2\ 2p(^2P°)\ 3p$	$3p\ ^1D$	2	72611.06		$4p\ ^3P_0$	$2s^2\ 2p(^2P°)\ 4p$	$4p\ ^3P$	0	81311.52	
						$^3P_1$			1	81326.33	14.81
$3p\ ^1S_0$	$2s^2\ 2p(^2P°)\ 3p$	$3p\ ^1S$	0	73976.23		$^3P_2$			2	81344.48	18.15
$2p'\ ^3P$	$2s\ 2p^3$	$2p^3\ ^3P°$	2, 1, 0	75256.3		$4p\ ^1D_2$	$2s^2\ 2p(^2P°)\ 4p$	$4p\ ^1D$	2	81770.36	
						$4p\ ^1S_0$	$2s^2\ 2p(^2P°)\ 4p$	$4p\ ^1S$	0	82252.31	

5. *Level*   the energy, in cm^{-1} units, relative to the ground-level energy of zero
6. *Interval*   the energy separation between adjacent levels of any one term, a positive number if the lower J is lower in energy (a *normal* interval) and negative otherwise (an *inverted* interval).

Look at the first five energy entries. These are our five levels from the ground state configuration. Note how the spin–orbit splitting in the ^3P term is slight compared

TABLE 13.4 O Atom Energy Levels and Term Symbol Designations

	O I					O I			
Config.	Desig.	J	Level	Interval	Config.	Desig.	J	Level	Interval
$2s^2\,2p^4$	$2p^4\,{}^3P$	2	0.0	−158.5	$2s^2\,2p^3({}^4S^\circ)4s$	$4s\,{}^3S^\circ$	1	96225.5	
		1	158.5	−68.0					
		0	226.5		$2s^2\,2p^3({}^4S^\circ)3d$	$3d\,{}^5D^\circ$	4	97420.24	
							3, 2	97420.37	−0.13
$2s^2\,2p^4$	$2p^4\,{}^1D$	2	15867.7				2, 1, 0	97420.50	−0.13
$2s^2\,2p^4$	$2p^4\,{}^1S$	0	33792.4		$2s^2\,2p^3({}^4S^\circ)3d$	$3d\,{}^3D^\circ$	3, 2, 1	97488.14	
$2s^2\,2p^3({}^4S^\circ)3s$	$3s\,{}^5S^\circ$	2	73767.81		$2s^2\,2p^3({}^4S^\circ)4p$	$4p\,{}^5P$	1	99092.64	0.67
							2	99093.31	1.21
$2s^2\,2p^3({}^4S^\circ)3s$	$3s\,{}^3S^\circ$	1	76794.69				3	99094.52	
$2s^2\,2p^3({}^4S^\circ)3p$	$3p\,{}^5P$	1	86625.35	2.02	$2s^2\,2p^3({}^4S^\circ)4p$	$4p\,{}^3P$	2, 1, 0	99680.4	
		2	86627.37	3.67	$2s^2\,2p^3({}^2D^\circ)3s$	$3s'\,{}^3D^\circ$	3	101135.04	−12.17
		3	86631.04				2	101147.21	−7.89
$2s^2\,2p^3({}^4S^\circ)3p$	$3p\,{}^3P$	2	88630.84	0.54			1	101155.10	
		1	88630.30	−0.70	$2s^2\,2p^3({}^4S^\circ)5s$	$5s\,{}^5S^\circ$	2	102116.21	
		0	88631.00						
$2s^2\,2p^3({}^4S^\circ)4s$	$4s\,{}^3S^\circ$	2	95476.43		$2s^2\,2p^3({}^4S^\circ)5s$	$5s\,{}^3S^\circ$	1	102411.65	

to the ${}^3P$–${}^1D$ or ${}^1D$–${}^1S$ separations. Now turn to the O data in Table 13.4. The columns are labeled as for C (except Edlén does not get a column all to himself here). Note how the intervals in the first ${}^3P$ level are negative numbers, indicating an inverted J ordering (highest J lowest in energy), as Hund's rule (3) predicts. Again, the spin–orbit splitting is small (but notably larger than for C). These are the states of O mentioned in Section 13.1. The singlet levels have significantly different chemistry from the triplet levels, even though all five are nominally "$1s^22s^22p^4$ atomic oxygen."

There are mountains of information in these compilations. We will return to them when we discuss atomic spectroscopy in depth, and explain why some level energies are in italics and what the superscript "o" means on the term designations. For now, notice how terms from one configuration can be interspersed among those from another (see the $2s2p^3$ and $2s^22p3s$ terms of C, for example). Excited levels are generally entangled this way. Notice how some J values are entangled for one term (as in the $2p^3\,{}^3D_J$ levels of C, which go in the order $J = 3$, $J = 1$, $J = 2$). And finally, remember that these tables are incomplete. There are predicted terms of every element that have not been observed experimentally. In some cases, the energies of these terms lie *above* the first ionization energy. Such states will *autoionize*—spontaneously eject an electron and leave behind the positive ion. Examples are the ${}^1D$ and ${}^1P$ terms of the C $2s2p^3$ configuration, which lie about 7000 cm^{-1} and 29 000 cm^{-1} above the C ionization limit. These states exist for short times (usually $\sim 10^{-14}$ s) before they ionize, and by the $\Delta E \Delta t \cong \hbar$ relationship, they have uncertain energies. (A $\Delta t \sim 10^{-14}$ s implies a $\Delta E \sim 500$ cm^{-1}.) Note too how levels from the $2s2p^3$ configuration are spread out over about 90 000 cm^{-1} in energy, from the $2p^3\,{}^5S_2$ level at 33 735.2 cm^{-1} to the $2p^3\,{}^1P$ level at 119 878 cm^{-1}!

Other states of C that are completely unknown (and presumably autoionizing) include all from the $2p^4$ configuration. This is a so-called *doubly excited* configuration: $2s^2 2p^2 \rightarrow 2p^4$ involves the promotion of two electrons. These are quite commonly autoionizing; all the doubly excited states of He are such, for example.

We close this chapter on atoms first with a final comment on configuration interaction. Configuration interaction accounts, by and large, for the failures of Hund's rules in excited configurations and for the higgledy-piggledy way that excited state levels are intertwined. In deciding which configurations interact in a CI calculation, term symbols are invaluable. For any type of term (^3P, for example) only configurations that can generate that term (1s2p, 1s3p, etc. for ^3P) will interact. This tells us how to pick the interacting configurations. Once they are chosen, the CI calculation will automatically give energy level values for *all* the levels with that term symbol for *all* the configurations in the calculation. These configurations will mix and push the non-CI energies around, so that saying 1s2p ^3P, for instance, really means "a ^3P level dominated by the $1s^1 2p^1$ configuration."

Next, we point out what happened to j–j coupling. Suppose the spin–orbit interaction is not small compared to the electrostatic electron–electron repulsion interaction. Then, each electron's spin and orbital motion are strongly coupled to form **j** and the **j**'s couple to form **J**. This can be seen as we go down a column (a group) in the Periodic Table. Spin–orbit interaction becomes progressively more important for heavier elements. This can be seen in Figure 13.6, which plots the np^4 configuration energies for the O–Po group. This figure shows clearly how the pattern of three closely-spaced levels, then two widely spaced, as seen in O, gradually shifts to a pattern of five more equally spaced levels in Po. For O, L–S coupling is physically more in line with the relative energy contributions of electrostatic and spin–orbit effects, while for Po, j–j coupling is more appropriate.

Finally, we should mention something about the shape of many-electron atoms. Chapter 12 discussed the shapes of various H-atom states, and many of those ideas carry over to other atoms. In fact, a relation among spherical harmonics known as Unsöld's theorem (see Problem 12.28) tells us what atomic states are spherically symmetric. This theorem states, in effect, that only atoms in configurations with completely filled or half-filled shells or subshells can have spherically symmetric

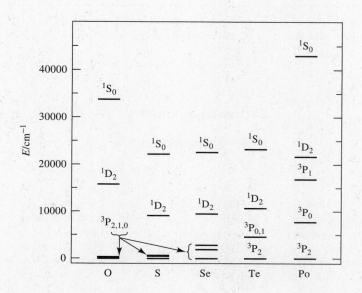

FIGURE 13.6 The five energy levels of the $ns^2 np^4$ configuration of the O–Po group show the increasing importance of spin–orbit coupling as the elements become heavier. This coupling increases the $J = 2, 1,$ and 0 levels' separation in the ^3P term and indicates the natural switch from L–S to j–j coupling as the atomic number increases. Note as well the irregularity of the J components of ^3P for Po.

states. Consequently, *only atoms in states with S term symbols ($L = 0$) are spherical.* This observation has consequences for chemical bonding. Nonspherical atoms interact with each other in different and more numerous ways than do spherical atoms.

This chapter opened with a Periodic Table blocked off by dominant valence-electron l quantum number. It ends with a much more informative Periodic Table, Figure 13.7, which gives the term symbol for the ground state of the elements (where this is known with any certainty). This Table, coupled with valence configurations, is the starting point for an understanding of molecular bonding, molecular structure, and chemical reactivity. The next chapter carries the story into the realm of molecular bonding.

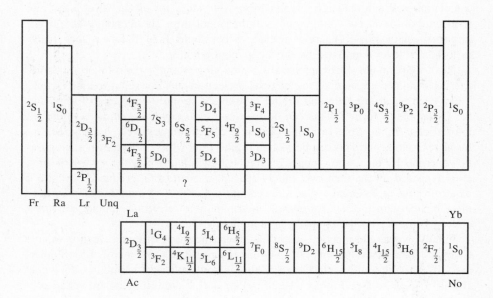

FIGURE 13.7 The ground-state term symbols of the elements show periodic variations. Compare this figure to Figure 13.1, and note how each block in that figure ends here with the closed-shell $^1S_0$ term symbol. Note also how spin multiplicities *alternate* from odd to even across any row, as the number of electrons varies from even to odd. The term symbols for the elements 105–109 (denoted ?) are unknown, and those for the final few actinides are presumed to follow their lanthanide counterpart.

CHAPTER 13 SUMMARY

This chapter has added to both our quantum mechanics repertoire and our chemical insight. From a pure quantum point of view, we have

1. seen how the Pauli Exclusion Principle governs many particle wavefunctions,
2. developed two approximation methods: perturbation and variation theory,
3. applied approximation methods to atomic systems, leading to the HF-SCF-CI theory,
4. explored the spin–orbit coupling operator, and
5. learned how to add angular momenta.

From a pure chemical point of view, we have

1. found the correct interpretation of paired spins,
2. used elementary electron configurations as a starting point for more correct atomic state descriptions,
3. learned how to construct term symbols from configurations, and

4. seen how various terms are ordered in energy and the physical effects that govern their spacing.

The key new expressions in this chapter are summarized below:

Exchange Symmetry

For identical particles n and m:

$$\hat{P}_{nm}\Psi(\ldots, r_n, \ldots, r_m, \ldots) = \Psi(\ldots, r_m, \ldots, r_n, \ldots)$$

$$= \begin{cases} +\Psi(\ldots, r_n, \ldots, r_m, \ldots), \\ \quad n \text{ and } m \text{ are bosons} \\ \quad \text{(integral spin)} \\ -\Psi(\ldots, r_n, \ldots, r_m, \ldots), \\ \quad n \text{ and } m \text{ are fermions} \\ \quad \text{(half-integral spin)} \end{cases}$$

Variation Theory

For trial wavefunction ϕ containing variable parameters α_i, the best approximation to the true ground state energy E_0 is

$$\langle E \rangle = \frac{\int \phi^* \hat{H} \phi \, d\tau}{\int \phi^* \phi \, d\tau} \geq E_0$$

where the parameters are chosen so that

$$\frac{\partial \langle E \rangle}{\partial \alpha_i} = 0$$

Perturbation Theory

For a total Hamiltonian $\hat{H}$ with a soluble part $\hat{H}_0$ and a perturbation $\hat{H}'$:

Zeroth-order solutions: $\hat{H}_0 \psi_n^{(0)} = E_0^{(0)} \psi_n^{(0)}$ (presumed known, states n nondegenerate)

First-order solutions:

$$E_n = E_n^{(0)} + E_n^{(1)} = E_0^{(0)} + H'_{nn}$$

$$E_n^{(1)} = H'_{nn} = \int \psi_n^{(0)*} \hat{H}' \psi_n^{(0)} \, d\tau$$

$$\psi_n = \psi_n^{(0)} + \psi_n^{(1)}$$

$$\psi_n^{(1)} = \sum_{m \neq n} \frac{H'_{mn}}{E_n^{(0)} - E_m^{(0)}} \psi_m^{(0)}$$

Second-order energies:

$$E_n^{(2)} = \sum_{m \neq n} \frac{H'_{nm} H'_{mn}}{E_n^{(0)} - E_m^{(0)}}$$

Slater Determinant Wavefunctions

For an N electron atom approximated by N one-electron normalized orbitals:

$$\Psi_{\text{total}} = \frac{1}{\sqrt{N!}} \begin{vmatrix} 1s\alpha(1) & 1s\beta(1) & \cdots \\ 1s\alpha(2) & 1s\beta(2) & \cdots \\ \vdots & \vdots & \end{vmatrix}$$

$$= ||\, 1s\alpha(1) \; 1s\beta(2) \; \cdots \,||$$

Addition of Angular Momenta

For two angular momenta $\mathbf{J}_1$ and $\mathbf{J}_2$ added to produce $\mathbf{J}_{12}$, the quantum numbers possible for the sum are

$$J_{12} = J_1 + J_2, \, J_1 + J_2 - 1, \ldots, |J_1 - J_2|$$

Spin–Orbit Coupling

For orbital angular momentum $\mathbf{L}$ interacting with spin angular momentum $\mathbf{S}$, both coupled to form a total angular momentum $\mathbf{J}$ and interacting with a spin–orbit coupling constant h_{so}, the first-order perturbation theory energy contribution is

$$E_{so}^{(1)} = \frac{\hbar^2 h_{so}}{2} [J(J+1) - L(L+1) - S(S+1)]$$

FURTHER READING

Any of the references mentioned at the end of Chapter 12 can be consulted for further information on perturbation and variation theories.

An interesting historical narrative and guide to key research papers is *Quantum Chemistry: The development of ab initio methods in molecular electronic structure theory,* by Henry F. Schaefer III, (Clarendon Press, Oxford, 1984). Schaefer has chosen about 150 research papers that represent landmarks in the history of quantum chemistry, and each gets a short and insightful commentary. (The papers themselves are not reprinted in this very readable—especially after the next chapter!—140 page volume.) The book shows what constitutes *real* quantum chemistry, a concept Schaefer (a *real* quantum chemist) states succinctly in an anonymous quote in his preface: "running a few molecular orbital calculations doesn't make one an electronic structure theorist any more than sleeping in one's garage makes him/her an automobile."

An excellent general reference source for atomic data is the book *Reference Data on Atoms, Molecules, and Ions,* by A. A. Radzig and B. M. Smirnov, (Springer-Verlag, Berlin, 1985). The standard reference for atomic energy levels, as discussed in the text, is *Atomic Energy Levels,* by C. E. Moore, National Standard Reference Data System-NBS 35 (in three volumes, Superintendent of Documents, US GPO, Washington, 1971). In three volumes spanning the s, p, and d block elements (H–V in Vol. I, Cr–Nb in Vol. II, and Mo–La, Hf–Ac in Vol. III), one finds not only neutral atoms but also positive atomic ions. The lanthanides are covered in a separate volume, *Atomic Energy Levels, the Rare-Earth Elements,* by W. C. Martin, R. Zalubas, and L. Hagen, National Standard Reference Data System-NBS-60 (Superintendent of Documents, US GPO, Washington, 1978). Data on a number of individual elements have been published as separate articles in *J. Phys. Chem. Ref. Data* as new data demand.

PRACTICE WITH EQUATIONS

13A What is the valence electron configuration of the alkali metal elements? the alkaline earths? the halogens? Periodic Table

ANSWER: ns^1; ns^2; ns^2np^5

13B What elements are characterized by the valence electron configuration $nd^{10}(n + 1)s^2$? Periodic Table

ANSWER: Zn, Cd, Hg

13C What is the first excited configuration of Mg? of Ne? of C? Periodic Table

ANSWER: $1s^22s^22p^63s^13p^1$; $1s^22s^22p^53s^1$; $1s^22s^22p^13s^1$

13D What is the atomic number and electron configuration of the (currently undiscovered) rare gas element below Rn? Periodic Table

ANSWER: $Z = 118$; $1s^22s^22p^63s^23p^63d^{10}4s^24p^64d^{10}5s^2$ $5p^64f^{14}5d^{10}6s^26p^65f^{14}6d^{10}7s^27p^6$

13E If $\psi_1(x_1) = N_1 e^{-ix_1}$ and $\psi_2(x_2) = N_2 e^{+ix_2}$, what is the spatial wavefunction $\Psi(x_1, x_2)$ that is symmetric with respect to exchange of particles 1 and 2? (13.2), (13.3), (13.4)

ANSWER: $\Psi(x_1, x_2) = \dfrac{N_1 N_2}{\sqrt{2}}[e^{i(x_2 - x_1)} + e^{i(x_1 - x_2)}]$

13F What is the eigenvalue for the operator $\hat{P}_{12}$ operating on the wavefunction $\Psi(x_1, x_2) = N(e^{-\alpha x_1^2}e^{-\beta x_2^2} - e^{-\alpha x_2^2}e^{-\beta x_1^2})$? (13.3)

ANSWER: -1

13G What is the potential energy operator for the two-electron ion C^{4+}? (13.6)

ANSWER: $\hat{V}(r_1, r_2) = \dfrac{1}{4\pi\epsilon_0}\left[\dfrac{e^2}{r_{12}} - \dfrac{6e^2}{r_1} - \dfrac{6e^2}{r_2}\right]$

13H What is the variational energy for a particle of mass m in a 1-D box of length L for the trial wavefunction $\phi = Nx^2(L - x)^2$? (13.9)

ANSWER: $6\hbar^2/mL^2$

13I What is the perturbation term in the Hamiltonian $\hat{H} = -(3.35 \times 10^{-42}\text{ J m}^2)(d^2/dx^2) + (200\text{ J m}^{-2})x^2 + (1.5 \times 10^{-23}\text{ J})\cos[x/(0.45\text{ Å})]$? (13.14)

ANSWER: the last term: 3.35×10^{-42} J m^2 is $\hbar^2/2m$ for $m = 1$ amu

13J A two-level system is connected by a perturbation that has matrix elements $H'_{00} = H'_{11} = 0$ and $H'_{01} = H'_{10} = 0.1\epsilon$ where $\epsilon = E_1^{(0)} - E_0^{(0)}$. What is the energy separation between the two states from second-order perturbation theory? (13.18), Figure 13.3

ANSWER: 1.02ϵ

13K What is Ψ_{total} for the Slater determinant $||\,1s\alpha(1)\ 3s\alpha(2)\,||$?

ANSWER: $\Psi_{\text{total}} = (\sqrt{2})^{-1}[1s(1)3s(2) - 1s(2)3s(1)]\alpha(1)\alpha(2)$

13L What is the Slater wavefunction for the O atom $2p_z$ orbital? (13.22), Table 13.1, Table 12.2

ANSWER: $n_e = 2$; $Z_e = 4.4531$;

$\psi_{210} = N\left(\dfrac{3}{4\pi}\right)^{1/2}\cos\theta\ r\exp(-2.2266r/a_0)$

13M What are the possible quantum numbers for the angular momentum that results from adding an angular momentum of magnitude $|\mathbf{J}_1| = \sqrt{20}\hbar$ to one of magnitude $|\mathbf{J}_2| = \sqrt{6}\hbar$? (13.30)

ANSWER: $J_{12} = 6, 5, 4, 3$, or 2 (note that $J_1 = 4$ and $J_2 = 2$)

13N What are the S, L, and J quantum numbers for a state with term symbol $^4D_{5/2}$?

ANSWER: $S = 3/2$, $L = 2$, $J = 5/2$

13O What is the ground-state term symbol for Al, valence configuration $3s^23p^1$? Table 13.2

ANSWER: $^2P_{1/2}$

13P What is the ground-state term symbol for Ni, valence configuration $3d^84s^2$? Table 13.2, Hund's rules

ANSWER: $^3F_4$

13Q What is the energy difference between the $^2P_{1/2}$ and $^2P_{3/2}$ states of an atomic configuration that has a spin–orbit coupling constant $h_{so} = 9.0 \times 10^{45}$ J^{-1} s^{-2}? (13.29)

ANSWER: 1.5×10^{-22} J (or 7.6 cm^{-1})

PROBLEMS

SECTION 13.1

13.1 Write the ground-state electron configurations of the elements Sc–Zn that would be expected on a regular filling of the nd subshell that characterizes that section of the Periodic Table. Then look up the actual ground state configurations and explain any differences.

13.2 If "super-heavy" artificial elements are ever synthesized that are so large they require g ($l = 4$) electrons in their configurations, a new element block will have to be created for them in the Periodic Table. How many columns across will this block be?

SECTION 13.2

13.3 Apply the operators $\hat{P}_{12}$, $\hat{P}_{13}$, and $\hat{P}_{23}$ to the simple function $y(x_1, x_2, x_3) = x_1 + x_2 - x_3$. Is y an eigenfunction of any of these operators? If so, what is/are the eigenvalue(s)?

13.4 Construct both a symmetric and an antisymmetric function with respect to exchange of indices 1 and 2 from the two functions $y_1(x_1) = \sin x_1$ and $y_2(x_2) = \cos x_2$.

13.5 Write the correctly symmetrized wavefunction for two (spinless, boson) noninteracting He atoms placed in a box of length L. Let one atom be in the state $n = 1$ and the other in $n = 2$. Next, repeat this problem for two (spin 1/2, fermion) noninteracting neutrons.

13.6 Practice Problem 13E has some interesting implications if explored further. Let the system be the particle-on-a-ring (Example 12.6) with two identical noninteracting particles, and let particle 1 be in the state with quantum number $m = +1$ and particle 2 be in $m = -1$. Thus

$$\psi_1(\phi_1) = \frac{1}{\sqrt{2\pi}} e^{i\phi_1} \quad \text{and} \quad \psi_{-1}(\phi_2) = \frac{1}{\sqrt{2\pi}} e^{-i\phi_2}.$$

Show that the total spatial wavefunction for the system can be written in terms of only *one* variable, the relative angular separation between the particles, $\Delta\phi = \phi_1 - \phi_2$, and show that the total spatial wavefunction can be written as either

$$\Psi(\Delta\phi) = \frac{\sqrt{2}}{2\pi} \cos \Delta\phi \quad \text{or} \quad \Psi(\Delta\phi) = \frac{i\sqrt{2}}{2\pi} \sin \Delta\phi$$

depending on the spin type of the particles. What about the case with both particles in the state $m = 0$?

13.7 Write an expression for the ground state energy of a two-electron atom of arbitrary atomic number Z in the independent-electron model. What rough order of magnitude correction to the energy will the term involving r_{12} in the Hamiltonian make for such an atom?

SECTION 13.3

13.8 If we use the exact wavefunction in a variational calculation, we should get the exact energy. Show that this is the case for the ground state of a 1-D harmonic oscillator by using the trial wavefunction $\phi(x) = N \exp(-\beta x^2)$ where β is the parameter to vary. Show that the best β gives the exact wavefunction.

13.9 As particle-in-a-box trial wavefunctions go, the choice $\phi(x) = N \sin^2(\pi x/L)$ would seem to have great merit. It has no nodes except at $x = 0$ and L, and, unlike the exact wavefunction (note that ϕ is essentially the square of the exact wavefunction), it also has $d\phi/dx = 0$ at $x = 0$ and L so that ϕ and its first derivative are both continuous everywhere. What energy does this trial function predict? How does it compare to the exact ground-state energy? The integrals

$$\int_0^L \sin^4\left(\frac{\pi x}{L}\right) dx = \frac{3L}{8} \quad \text{and} \quad \int_0^L \sin^2\left(\frac{\pi x}{L}\right) dx = \frac{L}{2}$$

will perhaps prove useful.

13.10 Sometimes a potential function has a minimum that, when expanded in a Taylor series, has a vanishing harmonic and cubic term and a leading quartic term. Use the trial function $\phi(x) = N \exp(-\alpha x^2)$ and vary α to find the best variational ground-state energy for the pure quartic oscillator, $V(x) = bx^4$.

13.11 Consider the two modified 1-D particle-in-a-box potentials below:

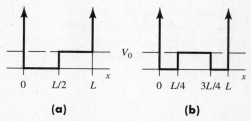

(a) (b)

In both, a potential step of height V_0 extends over half the box. In (a), the step covers the right half of the box; in (b), it covers the middle. Suppose V_0 is less than the zero-point energy of the system so that you may treat it as a perturbation to the normal particle-in-a-box potential. Find the first-order energy corrections in each case, and give a physical explanation for any difference you find in the total energy E_n for the two cases. (*Hints:* For (a), you should not have to do any integrals if you think about the location of the perturbation and the symmetry of the unperturbed wavefunctions. For (b), you may need the integral

$$\int_b^c \sin^2(ax) dx = \frac{1}{4a}[\sin(2ab) - \sin(2ac)] + \frac{c-b}{2}.$$

You should find that the two potentials have the same energies as $n \to \infty$. Why is this?)

13.12 What is the first-order corrected wavefunction for a harmonic oscillator perturbed by only a quartic term, bx^4? Follow Example 13.4 closely.

13.13 Consider a two-level system connected by a perturbation that has only a second-order energy correction. Let $\Delta E^{(0)} = E_1^{(0)} - E_0^{(0)}$, the unperturbed energy difference, and let $\Delta E = E_1 - E_0$, the perturbed energy difference. Derive an expression relating ΔE to $\Delta E^{(0)}$ and H'_{01}, the perturbation matrix element. How large does H'_{01} have to be in order to make ΔE twice $\Delta E^{(0)}$?

13.14 This and the next few problems consider a perturbation due to a uniform electric field F_0 in the x direction acting on a particle of charge q and mass m. We will show in Chapter 15 that the perturbation term in this case is $\hat{H}' = -qF_0 x$. Consider here the effect of this perturbation on the energies of the particle-in-a-box system. Find the first-order energy correction for all states and show that it is a constant. (*Hint:* Look closely at the H'_{nn} integral, and you should be able to evaluate it by a simple physical argument once you recognize that $\langle \hat{H}' \rangle = -qF_0 \langle x \rangle = H'_{nn}$.) This problem has an exact solution, by the way. See T. C. Dymski *Am. J. Phys.* **36** 54 (1968) and J. N. Churchill and F. O. Arntz *ibid.* **37** 693 (1969).

13.15 Now imagine placing the charged particle from the previous problem in a harmonic-oscillator potential with an external field. The perturbation is the same, but its effect is different. Show that the first-order energy correction must be zero for all states. (Invoke symmetry, not integral tables!) Since $E_0^{(1)} = 0$, you must consider the second-order energy correction. You will need the general matrix elements H'_{mn}, and the result of Problem 12.13 gives them to you:

$$x_{mn} = \int_{-\infty}^{\infty} \psi_m^{(0)} x \psi_n^{(0)} dx = \begin{cases} \dfrac{1}{\alpha}\sqrt{\dfrac{m+1}{2}}, & n = m+1 \\ \dfrac{1}{\alpha}\sqrt{\dfrac{m}{2}}, & n = m-1 \end{cases}$$

where $\alpha^2 = k/\hbar\omega$. We can never measure absolute state energies, only differences between them. With this in mind, is this perturbation observable?

13.16 The final question of the previous problem might have caused you to wonder what happens if perturbation theory is extended to higher orders. In fact, that problem can be easily solved exactly, and it turns out that the exact energy *is* the second-order perturbation energy. We write the perturbed potential function $V(x) = kx^2/2 - qF_0 x$ in the form

$$V(x) = \left(\frac{1}{2}kx^2 - qF_0 x + E'\right) - E' = a(x - x')^2 - E'$$

where we have first added and subtracted a constant E' and then expressed the potential as a simple quadratic (a and x' are constants) potential minus E'. Find a, x', and E' in terms of k, q, and F_0, and write down the exact energy by inspection. You should find the same answer as in the previous problem.

13.17 Here, we consider the charged particle of the last few problems to be constrained to move on a circle of radius R in the (x, y) plane. (See Example 12.6 for the unperturbed energy.) The electric field in this case produces a perturbation $\hat{H}' = -qRF_0 \cos \phi$. Find a general expression for the perturbation matrix elements H'_{mn} and evaluate the energy through second-order perturbation theory. (*Hint:* Use $\cos\phi = (e^{i\phi} + e^{-i\phi})/2$.)

SECTION 13.4

13.18 The variational energy expression appearing just before Eq. (13.20) can be easily derived using the hydrogenic atom energy expression, the H-atom Virial Theorem, and the r_{12} integral, Eq. (13.19). Look at the seven terms in the Hamiltonian that leads to $\langle E \rangle$. The first two and the fourth and fifth terms' contributions follow from the hydrogenic atom energy expression, Eq. (12.46b). The last contributes the r_{12} integral amount. Write the third and sixth as

$$-\frac{(2 - Z_e)e^2}{(4\pi\epsilon_0)r} = -\frac{2 - Z_e}{Z_e} \frac{Z_e e^2}{(4\pi\epsilon_0)r}$$

and apply the Virial Theorem (see Problem 12.48) to them to complete the derivation.

13.19 Use the variational screening calculation leading up to Eq. (13.20) to see if this level of theory can predict the simplest two-electron atom, H$^-$, to be stable. Table 7.4 gives this ion's ionization potential; it is just the electron affinity of H, 0.7542 eV.

13.20 Compare the Slater orbital for the 3s orbital of H to the exact 3s orbital. What is the most striking difference between the two? How would Figure 12.14c change if the STO were used?

13.21 Explain the general trend of the Slater Z_e parameters in Table 13.1 using general arguments about screening and H orbital shapes and sizes. If the screening was complete, what would the various Z_e parameters be for each of the orbitals of, say, Al? Do the tabulated values show that the 3p orbitals are more or less well screened from the full nuclear charge than the 3d orbitals?

13.22 Gaussian orbitals were mentioned in the text as useful approximations to Slater orbitals for atoms. Apply the variational method to the trial Gaussian wavefunction $\phi(r) = N \exp(-\alpha r^2)$ to find the best value for the parameter α and the corresponding energy for the ground state of H. Since ϕ is a function of r alone, you can ignore all angular factors so that the integration volume element is $r^2 dr$. The variation-energy integrals can be evaluated with the help of

$$\int_0^\infty x e^{-ax^2} dx = \frac{1}{2a}, \quad \int_0^\infty x^2 e^{-ax^2} dx = \frac{1}{4}\sqrt{\frac{\pi}{a^3}}, \quad \text{and}$$

$$\int_0^\infty x^4 e^{-ax^2} dx = \frac{3}{8a^2}\sqrt{\frac{\pi}{a}}$$

and the Hamiltonian is just (see Eq. (12.44))

$$\hat{H} = -\frac{\hbar^2}{2\mu}\left[\frac{1}{r^2}\frac{d}{dr}\left(r^2\frac{d}{dr}\right)\right] - \frac{e^2}{(4\pi\epsilon_0)r}.$$

You should find $\alpha = 8/(9\pi a_0^2)$ and $\langle E \rangle = (8/3\pi)E_1$. The text plots an STO along with a GTO that was least-square fit to the STO. How sensible is this type of fit?

13.23 Write the twelve Slater determinants that describe the states derived from the He $1s^1 2p^1$ configuration. How many Slater determinants can be written for the $1s^1 3d^1$ configuration?

13.24 The rule for expanding a 3×3 determinant is

$$\begin{vmatrix} a & b & c \\ d & e & f \\ g & h & i \end{vmatrix} = aei + bfg + cdh - gec - afh - bdi .$$

Use this to show explicitly that a Slater determinant for Li based on the configuration $1s^3$ must be zero.

13.25 Show that $K = J$ for the He ground state configuration, as claimed in the text. How are J_{ii} and K_{ii} in Eq. (13.26) related?

SECTION 13.5

13.26 Example 13.5 tabulates the intervals between the $^2P_{3/2}$ and $^2P_{1/2}$ levels of the halogens. Calculate the magnitude of the spin–orbit coupling constant h_{so} for each element.

13.27 Expand Table 13.2's entry for the d^2, d^8 configuration to include all the possible J values for each term symbol in the table. Predict the lowest energy term for each configuration, then use Figure 13.7 to find two elements that have these configurations in their ground state and verify your predictions.

13.28 The first excited-state configuration of the heavier alkaline earths is responsible for the states that emit the brilliant colors often used in fireworks displays. What is this configuration, what term symbols are derived from it, and what would you expect to be the energy ordering of the states represented by these symbols?

13.29 If you compare Figures 13.1 and 13.7, you will notice that half-way across each l block in Figure 13.1 (the groups H–Fr, N–Bi, Mn–Re, and Eu–Am), there is a column of S_J term symbols with increasing spin multiplicity: 2, 4, 6, and 8. Explain why these elements have ground state S terms and why the spin multiplicity has the values shown.

13.30 The possible term symbols for two inequivalent p electrons (pp in the notation of Table 13.2) are easy to derive: $l_1 = 1$ and $l_2 = 1$ so $L = 2, 1,$ or 0 (S, P or D terms), and $s_1 = 1/2$ and $s_2 = 1/2$ so $S = 1$ or 0 (singlet or triplet terms). Finding the possible terms for two (or more) *equivalent* electrons is less easy. Here is one approach. Make a table with three rows, one for each of the three possible m values (-1, 0, or $+1$) either electron could have. Enter "spin arrows" ↑ or ↓ to represent individual m_s assignments ($+1/2$ or $-1/2$), column by column in your table, until all possible entries that do not violate the Pauli Principle have been made. For example, your table might start with these four columns:

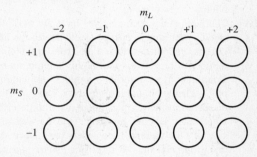

You should find 15 unique columns in the table without violating the Pauli Principle. For each column, add the two m values to obtain a corresponding m_L and likewise add the m_s values represented by ↑ and ↓ to obtain a corresponding m_S. For example, the four columns above represent $m_L = 2, 1, 0,$ and 1, respectively, and $m_S = 0, 1, 1,$ and 0, respectively. You should find $-2 \leq m_L \leq +2$ and $-1 \leq m_S \leq +1$. The term symbols are hidden in these numbers, and they can be revealed by a device known as a Slater diagram. Arrange a series of circles (as Slater did originally) in a grid five across (for the m_L values) and three down (for m_S):

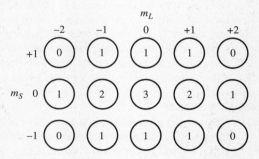

Next, fill in each circle with the *number of times the corresponding (m_L, m_S) values appeared in your original table*. The Slater diagram should look like this at the end:

	m_L				
	-2	-1	0	$+1$	$+2$
$+1$	0	1	1	1	0
m_S 0	1	2	3	2	1
-1	0	1	1	1	0

Note that all the numbers add to 15, the total number of columns in the original table, and note that the entries are symmetric about the center of the diagram. The largest number in all the circles (3) is the number of term symbols represented by the diagram. Any one term symbol has a *unique pattern* in the diagram, and the full diagram is just the sum of the individual patterns. For example, a ^1S term represents $L = S = 0$, so that $m_L = m_S = 0$ only. The ^1S pattern is thus a single circle at the origin of the diagram. Deduce the patterns for the two other terms represented in the full diagram, ^1D and ^3P.

13.31 Use the method of the preceding problem to verify the Table 13.2 term symbols for three equivalent p electrons, p^3.

13.32 The ground state of Ce has the valence configuration $4f^1 5d^1 6s^2$. What are the possible term symbols that can come from this configuration?

13.33 Write the term symbols for the ground states of the *stable* negative ions (singly charged only) of the atoms from H through Ne. Do the same for the singly-charged positive ions of these atoms.

13.34 The *Atomic Energy Levels* tables for Ti I list energies for all but one of the terms possible for the $3d^2 4s^2$ ground-state configuration. The missing term, as yet unobserved, is the $^1S_0$. Rounded to the nearest cm^{-1}, the tabulated energies are at 0, 170, 387, 7255, 8437, 8492, 8602, and 12 118 cm^{-1}. Use Table 13.2 and Hund's rules to assign complete term symbols (including J) to these energies.

13.35 The lanthanides and actinides have the most complicated set of atomic energy levels due to the large angular momentum of their f electrons and their large number of closely spaced, highly interacting orbitals. Consequently, a number of angular momentum coupling schemes in addition to the *LS* and *jj* schemes discussed in the text have been used to describe their levels. One scheme, however, is just a simple extension of *LS* coupling. It considers the f electrons as one group and the other valence electrons as another group, assigns ordinary *LS* term symbols to each, then couples the two spin and orbital angular momenta of these groups to arrive at final term symbols. For example, in Eu there is a set of levels that derive from the $4f^7 6s^1 6p^1$ configuration in which the f electrons are coupled as an ^8S term and the sp electrons are coupled as a ^3P term (written $4f^7(^8S)6s^1 6p^1(^3P)$). Couple the spin and orbital angular momenta of these two terms to arrive at the final nine (including J distinctions) term symbols for these states.

GENERAL PROBLEMS

13.36 Consider the perturbed particle-in-a-box potential of Problem 13.11(a) in which the step of height V_0 is between $L/2$ and L. Use a linear variation trial function of the form

$$\phi(x) = c_1 \psi_1(x) + c_2 \psi_2(x)$$

where $\psi_1(x)$ is the exact, normalized, *unperturbed* ground-state wavefunction (see Eq. (12.11)) and $\psi_2(x)$ is the corresponding first excited state wavefunction, and find the best energy. To make the algebra somewhat simpler, let $V_0 = E_1$, the unperturbed ground-state energy. Be sure to take advantage of the orthonormality of ψ_1 and ψ_2, and note that these are also eigenfunctions of the unperturbed Hamiltonian (so that all energies can be expressed in terms of E_1).

The only integral you should need is

$$\frac{2}{L}\int_{L/2}^{L} \sin\left(\frac{\pi x}{L}\right) \sin\left(\frac{2\pi x}{L}\right) dx = -\frac{4}{3\pi}.$$

13.37 Now consider the wavefunction for the previous problem. Classically, one would expect the particle to spend more time (be more likely found) in the region from $L/2$ to L because the step up in potential energy slows it down in this region. Does the trial wavefunction support this contention? (Along these lines, can you see why it is smart to mix a bit of the *first* excited-state wavefunction into the unperturbed ground-state wavefunction? Think about the desired shape of the true wavefunction: bigger on the right than on the left, and think about the way the trial wavefunction can achieve this.) Use the best energy from the previous problem (it's about $1.44 E_1$ in case you didn't work that problem) to find c_1 and c_2, and remember that $c_1^2 + c_2^2 = 1$ since ψ_1 and ψ_2 are normalized. At some point, you will have to make a choice about the algebraic signs of c_1 and c_2. Be sure your choice leads to the physically correct wavefunction.

13.38 Repeat the derivation of Eq. (13.20) for a two electron atom of arbitrary nuclear charge Z and show that, in general,

$$Z_e = Z - \frac{5}{16},$$

$$\langle E \rangle = E_1 \left(2Z^2 - \frac{5}{4}Z + \frac{25}{128} \right),$$

and the first ionization energy is

$$IP = -\langle E \rangle - Z^2 E_1 = -E_1\left(Z^2 - \frac{5}{4}Z + \frac{25}{128} \right)$$

where

$$E_1 = -\frac{e^2}{2(4\pi\epsilon_0)a_0}$$

is the ground state energy of H atom ($Z = 1$), -13.595 eV. Experimental first ionization energies for several two-electron atoms are tabulated below. How well does the general expression predict these values? Does the agreement improve or worsen as Z increases?

Species	H$^-$	He	Li$^+$	Be^{2+}	Be^{3+}	C^{4+}	N^{5+}	O^{6+}	F^{7+}
IP/eV	0.7542	24.596	75.619 3	153.85	259.298	391.986	551.925	739.114	953.6

13.39 Problem 12.46 introduced the positronium atom, Ps, and mentioned that a longer lived state has to decay (self-annihilate) via a three-photon emission. The short-lived state has the electron and positron spins *singlet* coupled. It can decay via two-photon emission and still conserve angular momentum, since singlet ground-state Ps has a net angular momentum of zero and two photons (each of intrinsic angular momentum $1\hbar$—see Footnote 2 in this chapter) can have their individual angular momenta opposed and add to zero, as required. Now explain why *triplet* spin-coupled Ps *must* decay by emitting at least three photons. Why won't one photon do? (*Hint:* Remember that photons have linear momentum as well as angular momentum.)

13.40 The Closer Look at perturbation theory obtained the expression

$$\psi_n^{(1)} = \sum_m c_{nm} \psi_m^{(0)} = \sum_{m \neq n} \frac{H'_{mn}}{E_n^{(0)} - E_m^{(0)}} \psi_m^{(0)}$$

in its derivation of Eq. (13.17) and asked what happened to the $m = n$ term that appears in the first sum, but not in the second. Here's what happened to it. We write the true wavefunction through first-order in the form

$$\psi_n = \psi_n^{(0)} + \lambda \psi_n^{(1)} = \psi_n^{(0)} + \lambda \sum_{m \neq n} c_{nm} \psi_m^{(0)} + \lambda c_{nn} \psi_n^{(0)}$$

where the mystery c_{nn} term has been explicitly pulled out of the sum. Next, require ψ_n to be normalized, and write the normalization integral $\int \psi_n^* \psi_n \, d\tau$. Invoke the orthonormal properties of the zeroth-order wavefunctions $\psi_n^{(0)}$ and show that

$$\int \psi_n^* \psi_n \, d\tau = 1 + 2\lambda \, c_{nn} \ .$$

Finally, explain how this expression requires c_{nn} to be zero.

CHAPTER 14

The Origins of Chemical Bonding

14.1 The Simplest Molecules: H_2^+ and H_2

14.2 Approximate Descriptions of Covalent Bonds

14.3 Named Bonds and the Concept of Delocalization

14.4 The Forces That Shape Molecules

14.5 Taking Molecules Apart

THE last chapter described atoms in some detail. This chapter's central topic is molecules. We could discuss the quantum theory of bonding and the interpretation of molecular wavefunctions in depth, but one of the strengths of chemistry as a science is its ability to work successfully with simplified models of complex phenomena, and the chemical bond is among these.

Consider the subtle change in energy when two atoms bond. This change, the bond energy, is typically of the same magnitude (1–5 eV) as the difference between an atom's ground and first excited energy level. An accurate description of bonding requires very high accuracy wavefunctions, since a 1 eV error in a calculated atomic energy can mean an appreciable error in a chemical bond energy. In contrast, 1 eV is a small fraction of an atom's total energy.

Molecules also have a glut of coordinates. Each electron and each nucleus is a separate particle; the wavefunction for a polyatomic molecule is a function of $4 \times$ (number of electrons and nuclei) spatial and spin coordinates. For even a simple molecule of chemical interest, H_2O, this is a function of 52 variables! Such a function is obviously more complicated than the chemist's picture of water as two H atoms singly-bonded, covalently, to oxygen with a bond angle of about 105°, yet this simple picture is often all one needs.

Note the terms in the last sentence—*singly-bonded, covalently, bond angle*, common words chemists use routinely to describe bonding. They represent effects that are undoubtedly present in an exact molecular wavefunction, but they may also be difficult to extract from it. One of our goals here is to put the chemist's daily vocabulary on a fairly precise theoretical and observational footing.

Thus, while the last chapter discussed the computational basis for atomic quantum calculations in some detail, this chapter will *not* teach you how to construct molecular wavefunctions in general, to arbitrary accuracy, according to any of several approximation methods, or how to interpret the results. That task is easily the topic of one (or more) separate texts and courses. We will see a few of the approaches used with simple molecules, starting with H_2^+ and H_2 in the next section. This will introduce the language and some of the methodology needed to understand more complex molecules, but we will quickly pass from mathematics to symbols and pictures representative of theory.

14.1 THE SIMPLEST MOLECULES: H_2^+ and H_2

The H_2^+ molecule is very reminiscent of He. There are only three particles involved, all interacting through Coulomb's law. But why is it that our mind's picture of He is so very different from that for H_2^+? Figure 14.1 contrasts the similar interactions and different pictures; the interactions are essentially the same, but He "looks like an atom" while H_2^+ "looks like a molecule." The heart of chemical thought and intuition pictures nuclei fixed in space, defining molecular structure, while electrons fuzz and blur around them.

These contrasting views—fixed nuclei and blurred electrons—are the result of what is arguably the most important approximation in all of chemistry, first described in detail by Max Born and J. Robert Oppenheimer and now known as the *Born–Oppenheimer approximation*. The entire approximation and the way we view molec-

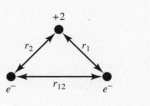

He Atom

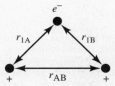

H_2^+ Molecule

FIGURE 14.1 While the electrostatic interactions in He and H_2^+ are very similar (left), our mental pictures of He and H_2^+ are very different (right). This is a consequence of the differences in dynamics of light electrons and heavy nuclei.

ular structure are due to the dynamical consequences of the proton/electron mass ratio.[1] Our way of thinking about chemistry would be vastly different if this ratio had been closer to 1 instead of 1836.

It is worthwhile to jump ahead and use results we will later derive in detail to illustrate the meaning of the Born–Oppenheimer approximation. In brief, it states that electron speeds in molecules are so much greater than nuclear speeds that we can consider the nuclei to be fixed in space and the electrons to move in the electrostatic potential-energy field they generate. To incorporate nuclear motion, we move the nuclei from one fixed configuration to another, solving Schrödinger's equation for the electron motion at each configuration.

A typical electron speed in a molecule is on the same order of magnitude as the speed of the 1s electron in ground-state H. This speed is called the *atomic unit of velocity*, in analogy to calling the Bohr radius the atomic unit of length. It has the value

$$v_0 = \frac{e^2}{(4\pi\epsilon_0)\hbar} = 2.187\,690\,6 \times 10^6 \text{ ms}^{-1}\,.$$

The time it takes an electron to move one Bohr radius (the *atomic unit of time*) is

$$t_0 = \frac{a_0}{v_0} = \frac{4\pi\epsilon_0\hbar^2/m_e e^2}{e^2/4\pi\epsilon_0\hbar} = \frac{(4\pi\epsilon_0)^2\hbar^3}{m_e e^4} = 2.418\,885 \times 10^{-17} \text{ s}\,.$$

Nuclear motion, within the Born–Oppenheimer approximation, comes from rotation of the molecule and relative vibrational motion of the nuclei. The vibrational speed is usually the larger of the two, and we can use it to gauge the relative nuclear/electron speed ratio, in comparison to v_0 and t_0 above. To a good approximation, nuclear vibrational motion is like that of a harmonic oscillator with a unique force constant for each type of intramolecular vibration. Typical force constants fall in the range 100–1000 N m^{-1}. For example, the force constant in F_2 is 472 N m^{-1} so that the vibrational frequency, in Hz, is

$$\nu = \frac{1}{2\pi}\left(\frac{k}{\mu}\right)^{1/2}$$

$$= \frac{(472 \text{ N m}^{-1})^{1/2}}{2\pi[(9.5 \text{ g mol}^{-1})(10^{-3} \text{ kg g}^{-1})/6.02 \times 10^{23} \text{ mol}^{-1}]^{1/2}}$$

$$= 2.75 \times 10^{13} \text{ Hz}$$

or the time scale for one vibrational cycle, t_{vib}, is

$$t_{\text{vib}} = 1/\nu = 3.63 \times 10^{-14} \text{ s}\,.$$

In atomic units of time, this is

$$\frac{t_{\text{vib}}}{t_0} = \frac{3.63 \times 10^{-14} \text{ s}}{2.42 \times 10^{-17} \text{ s}} = 1500\,.$$

This ratio is the Born–Oppenheimer approximation in action. Electron motion is several hundred times more rapid than nuclear motion. The electrons are quite literally like the cartoon image of a swarm of gnats chasing someone through a field, keeping up with ease even if the person is running.

---

[1]The 1986 Universal Physical Constants adjusted value for the proton to electron mass ratio is $m_p/m_e = 1\,836.152\,8$. It is an easy number to remember if one remembers $6\pi^5 = 1\,836.12$, which gives the ratio to five or so significant figures. At one point, the experimental value for this ratio was $1\,836.12 \pm 0.05$, and when the coincidental agreement of this number and $6\pi^5$ was noticed, perhaps the shortest research paper ever—two sentences in length—was the result. See F. Lenz, *Phys. Rev.* **82** 554 (1951).

Now we apply the Born–Oppenheimer approximation to the Schrödinger equation for a molecule. Variants of this approximation exist, since, for instance, treating the nuclei as static point charges in space is rather non-quantal—we would know both the position and momentum of the nuclei exactly, in violation of the Uncertainty Principle—and the approximation itself can be improved by minor corrections that are most important for molecules such as H_2^+ and H_2 with light, high-velocity nuclei. But by and large, the basic Born–Oppenheimer approximation yields results in satisfactory agreement with experiment.

We will use a simplified notation and system of units: natural atomic units of energy, mass, charge, etc. These units are discussed in the Appendix, and Table 14.1 collects basic information on many of them. They have two primary advantages. First, they simplify the appearance of the equations of atomic and molecular quantum theory. For example, the H-atom Hamiltonian, Eq. (12.49), is

$$\hat{H} = -\frac{1}{2}\nabla^2 - \frac{1}{r} \quad (14.1)$$

in atomic units. All the e's, m's, $\hbar$'s, and $(4\pi\epsilon_0)$'s are gone. The ground-state energy is simply

$$E_{1s} = -\frac{1}{2} \text{ au} .$$

The atomic unit of energy is often called one Hartree, equal to twice the ionization energy of H.

EXAMPLE 14.1

What are the following characteristic energies expressed in atomic units?

(a) The total electron binding energy of He.
(b) A typical single bond energy.

TABLE 14.1 SI equivalents for selected atomic units

Quantity	Atomic unit		SI equivalent
Length	$a_0 = \dfrac{4\pi\epsilon_0 \hbar^2}{m_e e^2}$	= Bohr radius	$5.291\ 772\ 49 \times 10^{-11}$ m
Mass	m_e	= electron rest mass	$9.109\ 389\ 7 \times 10^{-31}$ kg
Charge	e	= elementary charge	$1.602\ 177\ 33 \times 10^{-19}$ C
Time	$\dfrac{a_0^2 m_e}{\hbar}$	= 1st Bohr orbit period	$2.418\ 884\ 3 \times 10^{-17}$ s
Energy	$\dfrac{e^2}{4\pi\epsilon_0 a_0}$	$= -2E_{1s}(H) = 1$ hartree	$4.359\ 748\ 2 \times 10^{-18}$ J (2 625.500 kJ mol^{-1})
Speed	$\dfrac{e^2}{4\pi\epsilon_0 \hbar}$	$= e^-$ speed in 1st Bohr orbit	$2.187\ 691\ 4 \times 10^6$ m s^{-1}
Angular momentum	$\hbar$	= Planck's constant/2π	$1.054\ 572\ 66 \times 10^{-34}$ J s

(c) The fusion energy of 1 mol of ice at 0 °C.
(d) The vaporization energy of 1 mol of water at 0 °C.

SOLUTION This is a simple conversion problem, but the answers help make one aware of the magnitude of the atomic unit of energy.

(a) From Chapter 13, 79.0 eV are required to remove both electrons from He. In atomic units, this is 79.0 eV/(27.2 eV au^{-1}) = 2.90 au. This is a characteristic magnitude for the total binding energy of an atom's valence electrons, 1–10 au.
(b) From Table 7.1, a characteristic single bond energy is 3 eV or 3 eV/(27.2 eV au^{-1}) = 0.11 au. Thus, chemical bonds represent energies of a few tenths of an au.
(c) From Example 4.4, the molar enthalpy of fusion of water is 6007 J mol^{-1} or [(6007 J mol^{-1})/(6.022 × 10^{23} mol^{-1})]/(4.36 × 10^{-18} J au^{-1}) = 0.0023 au so that physical transformations involve only thousandths of an au.
(d) Again from Section 4.4, the molar enthalpy of vaporization of water is 40.7 kJ mol^{-1} or [(40 700 J mol^{-1})/(6.022 × 10^{23} mol^{-1})]/(4.36 × 10^{-18} J au^{-1}) = 0.0155 au so that an unusually strong intermolecular attraction can be in the hundredths of an au range, but still roughly an order of magnitude or more smaller than a true chemical bond energy.

⟹ **RELATED PROBLEMS** 14.1, 14.2

The second advantage of atomic units is that a result quoted in terms of them is *independent* of uncertainties in fundamental constants. The 1s energy will *always* be $-1/2$ au, but its equivalent amount in joule units, or whatever, is subject to (minor) changes as $\hbar$, e, etc., are determined to greater and greater precision.

The original derivation of the Born–Oppenheimer approximation began with the complete Hamiltonian written in a way that allowed certain terms to be expanded as a power series in the small quantity m_e/m_p, the electron–proton mass ratio. This derivation included those small effects that the basic approximation neglects. We will use a simplified derivation, but this is acceptable since the corrections are rarely needed.

The Hamiltonian for a general polyatomic molecule with N nuclei and n electrons contains, as usual, kinetic energy and Coulombic potential-energy terms. In atomic units, it is

$$\hat{H} = -\frac{1}{2}\sum_{i=1}^{n} \nabla_i^2 - \frac{1}{2}\sum_{j=1}^{N} \frac{m_e}{m_j} \nabla_j^2 \\ -\sum_{i=1}^{n}\sum_{j=1}^{N} \frac{Z_j}{r_{ij}} + \frac{1}{2}\sum_{i=1}^{n}\sum_{\substack{i'=1 \\ i'\neq i}}^{n} \frac{1}{r_{ii'}} + \frac{1}{2}\sum_{j=1}^{N}\sum_{\substack{j'=1 \\ j'\neq j}}^{N} \frac{Z_j Z_{j'}}{r_{jj'}} \quad (14.2)$$

In order, these terms are

$-\dfrac{1}{2}\sum_{i=1}^{n}\nabla_i^2$ the kinetic energy of the electrons (indexed by i)

$-\dfrac{1}{2}\sum_{j=1}^{N}\dfrac{m_e}{m_j}\nabla_j^2$ the kinetic energy of the nuclei (indexed by j), each of mass m_j

$-\sum_{i=1}^{n}\sum_{j=1}^{N}\dfrac{Z_j}{r_{ij}}$ the attractive potential energy between electron i and nucleus j (of charge magnitude Z_j)

$+\dfrac{1}{2}\sum_{i=1}^{n}\sum_{\substack{i'=1 \\ i'\neq i}}^{n}\dfrac{1}{r_{ii'}}$ the repulsive potential energy between electron i and electron i'

$+\dfrac{1}{2}\sum_{j=1}^{N}\sum_{\substack{j'=1 \\ j'\neq j}}^{N}\dfrac{Z_j Z_{j'}}{r_{jj'}}$ the repulsive potential energy between nucleus j and nucleus j' .

In the second term, the Born–Oppenheimer parameter m_e/m_j appears naturally when $\hat{H}$ is written in atomic units. In the last two terms, the factors of 1/2 ensure that identical pairs of interactions (such as r_{12} and r_{21}) are not counted twice.

It is sobering that the majority of chemistry is contained in Eq. (14.2) and the wavefunctions and energies it implies. We specify n and the N values of Z_j to define any molecule, but we do *not a priori associate any number of these electrons with any one nucleus*. This most general Hamiltonian is not obviously a sum of atomic Hamiltonians, as is proper, given electron indistinguishability.

Two steps lead to the Born–Oppenheimer approximation. The first sets the nuclear kinetic energy term (the second term in $\hat{H}$) to zero (as if $m_e/m_j = 0$) and, correspondingly, sets the nuclear–nuclear repulsion term (the last term in $\hat{H}$) equal to a constant. We fix the nuclei in some configuration and solve for the electron motion in the electrostatic field of these nuclei.

This step alters the way we view the wavefunction. For the full Hamiltonian, we write

$$\hat{H}\Psi(r_i, r_j) = E\Psi(r_i, r_j)$$

where $\Psi(r_i, r_j)$ means that the wavefunction depends *explicitly* on all the electron and nuclear coordinates. After the first step in the B–O method, we write

$$\hat{H}_{BO}\Psi(r_i; r_j) = E(r_j)\Psi(r_i; r_j) \quad (14.3)$$

where the semicolon separating r_i and r_j in Ψ emphasizes that the nuclear positions r_j are now *parameters*, as does the notation $E(r_j)$.

The second step in a B–O calculation uses $E(r_j)$, the electronic energy as a function of nuclear position, to describe nuclear motion. The energy function $E(r_j)$ is the potential energy operator in a multidimensional, many particle (the nuclei) problem. Note the complementarity in these two steps. In the first, the electron motion is solved for many different nuclear configurations, and the second step uses the energies from the first step to solve for the nuclear motion.

In this chapter, we will concentrate primarily on the first step, the total electronic energy plus nuclear–nuclear repulsive potential energy. The nuclear motion is discussed in detail in Chapter 19.

The B–O electronic Hamiltonian for the simplest molecule, H_2^+, is

$$\hat{H}_{BO} = -\frac{1}{2}\nabla_1^2 - \frac{1}{r_{1A}} - \frac{1}{r_{1B}} + \frac{1}{R} \quad (14.4)$$

where the nuclei are called A and B and are set at a fixed separation R. The wavefunction is an explicit function of r_1 and a parametric function of R: $\Psi = \Psi(r_1; R)$.

The Schrödinger equation for H_2^+ is simplified in a coordinate system known as *confocal elliptical coordinates*. In this system (remember, R is a constant!), we define two distance variables, ξ and η, with

$$\xi = \frac{r_{1A} + r_{1B}}{R} \quad (1 \leq \xi \leq \infty)$$

$$\eta = \frac{r_{1A} - r_{1B}}{R} \quad (-1 \leq \eta \leq +1)$$

and one angle, ϕ, the azimuthal angle (0 to 2π) about the axis passing through the nuclei. Figure 14.2 illustrates these coordinates. In these coordinates, the B–O Schrödinger equation is separable, and the wavefunction and electron energy can be found at any R to very high accuracy.

FIGURE 14.2 Confocal elliptical coordinates ξ and η are useful in describing the location of one particle relative to two other fixed particles. A cylindrically symmetric angle ϕ about the fixed particle axis completes the 3-D location.

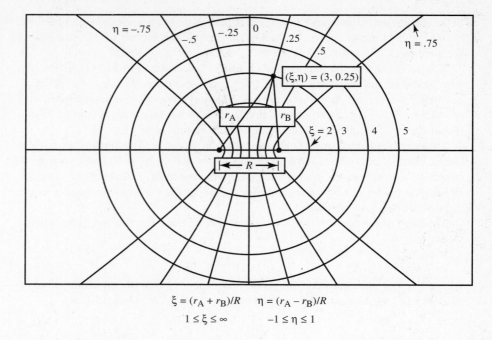

$$\xi = (r_A + r_B)/R \quad \eta = (r_A - r_B)/R$$
$$1 \leq \xi \leq \infty \quad -1 \leq \eta \leq 1$$

The wavefunctions are restricted, of course, to this one molecule, but their general properties appear in all bonds and are worth study. Theory, in agreement with experiment, shows that the average H–H nuclear separation in the H_2^+ ground state is almost exactly 2 au, or about 1.06 Å. We can fix R at 2 au, as a representative value, and look at the sequence of electron wavefunctions calculated for that nuclear separation.

Figure 14.3 shows the first few wavefunctions schematically. These pictures represent those regions in space where $|\Psi|$ is appreciable along with a shading for the algebraic sign of Ψ. Look first at the function labeled $\sigma_g 1s$, the nodeless electronic ground state. The wavefunction Ψ is uniformly one algebraic sign in all regions of space. The next, $\sigma_u^* 1s$, has one node: a plane perpendicular to the internuclear axis. The $\pi_u 2p$ also has one nodal plane, but it is a plane *containing* the internuclear axis.

The nodal planes are not the only symmetry elements present, however. Three important *symmetry operations* can be defined:

1. reflection in a plane midway between the nuclei and perpendicular to the nuclear axis,
2. reflection in a plane containing the nuclear axis, and
3. inversion through a point midway along the nuclear axis.

Each of the operations produces a new wavefunction that is either identical to the original or different only by a change in sign. In fact, each can be treated formally as a quantum mechanical operator with an eigenvalue of $+1$ or -1. Operation (1) is traditionally called[2] $\hat{\sigma}_h$; operation (2) is called $\hat{\sigma}_v$ (or sometimes $\hat{\sigma}_d$); operation (3) is $\hat{i}$.

Table 14.2 gives the eigenvalues for each of these operators acting on each of the wavefunctions of Figure 14.3. Note that the functions with a g subscript have $\hat{i}\Psi = (+1)\Psi$ while those with a u subscript have $\hat{i}\Psi = (-1)\Psi$. These subscripts

---

[2]The σ in $\hat{\sigma}_h$ or $\hat{\sigma}_v$ has nothing to do with the σ in $\sigma_g 1s$, etc.! The notation is explained in the following paragraphs.

TABLE 14.2 Symmetry operator eigenvalues for H_2^+ MOs

Molecular orbital	$\hat{i}$	Operator $\hat{\sigma}_h$	$\hat{\sigma}_v$
$\sigma_g 1s$	+1	+1	+1
$\sigma_u^* 1s$	−1	−1	+1
$\sigma_g 2s$	+1	+1	+1
$\sigma_u^* 2s$	−1	−1	+1
$\pi_u 2p$	−1	+1	−1
$\pi_g^* 2p$	+1	−1	−1
$\sigma_u^* 2p$	−1	−1	+1
$\sigma_g 2p$	+1	+1	+1

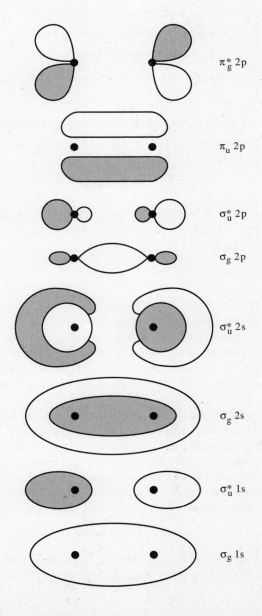

FIGURE 14.3 The eight lowest-energy one electron MOs of H_2^+ have the qualitative shapes and symmetry labels shown here.

stand for the German words *gerade* (even) and *ungerade* (odd). Also note that those with -1 eigenvalues for $\hat{\sigma}_h$ have an asterisk superscript, as in σ_u^*1s. Finally, those wavefunctions labeled σ-something have $+1$ eigenvalues for $\hat{\sigma}_v$, while those labeled π-something have -1 eigenvalues.

This last association, between the effect of $\hat{\sigma}_v$ and the σ, π label, is a false clue, however. The σ, π notation is the beginning of a set of labels that run σ, π, δ, and ϕ, the Greek parallels of s, p, d, and f. They connote *electron angular momentum* information. Unlike the spherically symmetric atomic problem, a diatomic molecule has a *unique direction* (in the B–O approximation)—the internuclear axis. The internuclear electric field establishes a quantization direction for the electron's orbital angular momentum. The labels give the quantum number for *the component of angular momentum along this axis:*

$$\sigma \rightarrow \text{no angular momentum}$$
$$\pi \rightarrow \text{angular momentum component} = \pm\hbar$$
$$\delta \rightarrow \text{angular momentum component} = \pm 2\hbar$$
$$\phi \rightarrow \text{angular momentum component} = \pm 3\hbar$$

Higher values exist in principle, but they are of no importance to chemistry.

A Closer Look at Molecular Electronic Angular Momentum

In the confocal elliptical separation of the H_2^+ Hamiltonian, the equation for ϕ, the angle about the R axis, is

$$\frac{d^2\Phi(\phi)}{d\phi^2} = -m^2\Phi(\phi)$$

where $\Phi(\phi)$ is that factor in Ψ depending on ϕ. This is exactly the same problem we solved in Chapter 12 as the "particle-on-a-ring" problem. There, we solved $d\Phi/d\phi = im\Phi$ and found

$$\Phi(\phi) = \frac{1}{\sqrt{2\pi}} e^{-im\phi}, \quad m = 0, \pm 1, \pm 2, \ldots,$$

which is also the function that solves $d^2\Phi/d\phi^2 = -m^2\Phi$.

The m quantum number can take either sign, corresponding classically to rotation in either of the two senses about the R axis, and thus each state with $m \neq 0$ is doubly degenerate. Also, in analogy to the 2p atomic wavefunctions with the purely real $2p_z$ function and the complex $2p(m = \pm 1)$ functions, we have the option for forming *hybrid molecular electronic wavefunctions* in analogy to the $2p_x$ and $2p_y$ atomic hybrids.

Effects such as spin–orbit coupling appear in molecules as well, but in a much more complex form, not only because of a molecule's inherent lack of spherical symmetry, but also because there is now one extra type of angular momentum—molecular rotation—with which the electronic angular momentum can couple. These couplings are usually unimportant in most molecules in their lowest electronic energy states, but they are important for many excited states and for many states in general of molecular fragments, or *free radicals,* such as OH, CH_3, and the like.

"Chemical bonding" means that the electrons in a molecule are doing something to keep the nuclei from flying apart under their repulsive interaction, to speak in very general terms. For H_2^+, with only one electron, intuition suggests that electron localization between the nuclei leads to bonding. This is largely, but only partly, correct. Bonding involves not only changes in *electrostatic potential energy* (putting a negative charge between two positive charges) but also changes in *electron kinetic energy*. The *total energy* must be lowered to lead to a stable bond. We will see how these two effects can be separated in our discussion of H_2 and again in Section 14.3 when we discuss electron delocalization.

For H_2^+, however, the simple idea—stick the electron between the nuclei—is good enough. The $\sigma_g 1s$ electron state (or *molecular orbital*, MO) does lead to a stable H_2^+ as alluded to above. We say this MO is *bonding*. The $\sigma_u^* 1s$ orbital, with a nodal plane just in the middle of the bonding region, does *not* promote atomic binding. If we hold H_2^+ at the 2 au H–H separation with the electron in the $\sigma_u^* 1s$ orbital and let go, the nuclei spontaneously fly apart. In contrast to the $\sigma_g 1s$ orbital, $\sigma_u^* 1s$ is *antibonding*. The asterisk will be our general symbol for an antibonding molecular orbital.

Two other important internuclear separations are also worth characterizing. These are the *separated atom* (SA, $R = \infty$) and *united atom* (UA, $R = 0$) limits. First, we look at $R = \infty$.

Imagine pulling H_2^+ apart. (This sort of operation will occupy all of Section 14.5; it is at the center of chemical reaction language.) We grab one nucleus in each hand and pull. Which hand has H and which has H^+? Our discussion of exchange symmetry and identical particle behavior in the last chapter suggests that, on average, each hand has a 50:50 chance of ending up with the H atom instead of the bare proton, H^+. *A priori*, we cannot assign the electron uniquely to one nucleus.

We do expect, however, that the H atom is found in the 1s state if we start from the $\sigma_g 1s$ molecular electronic state. (This, in fact, is the origin of the 1s part of the $\sigma_g 1s$ designation.) Thus, our $R = \infty$ wavefunction looks like either nucleus A or B has a 1s electron associated with it, and therefore, our $\sigma_g 1s$ orbital should reflect this symmetry. We can now perform a simple approximate calculation of $\sigma_g 1s$ as a function of R based on this idea.

We start, in a variational approximation, with the trial wavefunction

$$\Psi(r_{1A}, r_{1B}; R) = c_A(1s_A) + c_B(1s_B) \tag{14.5}$$

where $1s_A$ and $1s_B$ are ordinary H atom 1s orbitals centered about nuclei A and B, respectively. These functions are fixed, and neither depends on R. The R dependence is contained in the coefficients c_A and c_B, and these are the parameters of the variational calculation.

With this wavefunction and the H_2^+ Hamiltonian of Eq. (14.4), we compute, at any R,

$$\langle E(R) \rangle = \frac{\int \Psi^* \hat{H}_{BO} \Psi d\tau}{\int \Psi^* \Psi d\tau}$$

and minimize this quantity with respect to c_A and c_B. Using matrix element notation introduced in Chapter 13, this procedure leads to two coupled equations for c_A and c_B, the simultaneous solution of which involves the *secular determinant* equation

$$\begin{vmatrix} H_{AA} - S_{AA}\langle E \rangle & H_{AB} - S_{AB}\langle E \rangle \\ H_{BA} - S_{BA}\langle E \rangle & H_{BB} - S_{BB}\langle E \rangle \end{vmatrix} = 0 \; .$$

The matrix elements are the *Hamiltonian integrals*

$$H_{ij} = \int (1s_i) \hat{H}_{BO} (1s_j) d\tau, \qquad i, j = A \text{ or } B$$

and the *overlap integrals*

$$S_{ij} = \int (1s_i)(1s_j)d\tau, \quad i, j = \text{A or B}$$

with the properties

$$S_{AB} = S_{BA} = S = e^{-R}\left(1 + R + \frac{R^2}{3}\right)$$

$$S_{AA} = S_{BB} = 1 .$$

Note that $S \to 0$ as $R \to \infty$ (no overlap) and $S \to 1$ as $R \to 0$ (full overlap). The Hamiltonian integrals are

$$H_{AA} = H_{BB} = H = -\frac{1}{2} + \frac{1}{R} + J$$

$$H_{AB} = H_{BA} = H' = -\frac{1}{2}S + \frac{S}{R} + K$$

where each $-\frac{1}{2}$ represents the H-atom 1s energy in atomic units, $1/R$ represents the nuclear repulsion energy, and J and K are variants of the Coulomb and exchange integrals we needed for atoms:

$$J = \int (1s_A)\frac{1}{R}(1s_A)d\tau_1 = e^{-2R}\left(1 + \frac{1}{R}\right) - \frac{1}{R}$$

$$= \text{Electrostatic energy of proton a distance } R$$
$$\text{from an electron with charge density } (1s)^2$$

$$K = -\int (1s_A)\frac{1}{r_{1A}}(1s_B)d\tau_1 = -e^{-R}(1 + R) .$$

Note that K involves, as does H', an integration with respect to *two* nuclear positions (A and B, separated by R). Such integrals are called *two center integrals*.

We expand the secular determinant in terms of H, H', S, and $\langle E \rangle$ and solve for $\langle E \rangle$. Two solutions are found:

$$\langle E(R) \rangle = \frac{H \pm H'}{1 \pm S} \tag{14.6}$$

and, going back to Ψ, two relations between c_A and c_B are found: $c_A = \pm c_B$, leading to two wavefunctions, that, when normalized, are

$$\Psi_g = \frac{(1s_A + 1s_B)}{[2(1 + S)]^{1/2}} \tag{14.7a}$$

and

$$\Psi_u = \frac{(1s_A - 1s_B)}{[2(1 - S)]^{1/2}} . \tag{14.7b}$$

The g and u subscripts remind us of symmetry—symmetric and antisymmetric, respectively, with respect to inversion.

We can also write $\langle E(R) \rangle$ in an illuminating way using the results for H, H', and S. Each $\langle E(R) \rangle$ is the sum of three terms:

$$\langle E \rangle_g = -\frac{1}{2} + \frac{1}{R} + \frac{J + K}{1 + S} \tag{14.8a}$$

↑ Electronic ground state ↑ 1s H-atom energy ↑ Nuclear repulsion ↑ Chemical bond term

$$\langle E \rangle_u = -\frac{1}{2} + \frac{1}{R} + \frac{J - K}{1 - S} \tag{14.8b}$$

↑ Electronic excited state ↑ 1s H-atom energy ↑ Nuclear repulsion ↑ Nonbonding interaction

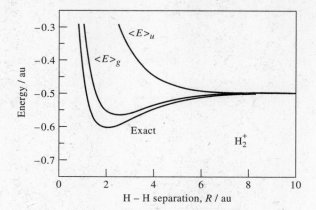

FIGURE 14.4 The approximate B–O total-energy curves for the lowest two electronic states of H_2^+ show a bound state $\langle E \rangle_g$ and a repulsive state $\langle E \rangle_u$. The exact bound-state curve is shown for comparison.

The first term is the asymptotic ($R \to \infty$) energy of a 1s H atom, $-1/2$ au; the second is the nuclear repulsive potential energy; the third, with the subtle sign difference, accounts for the chemical bond. In the ground state, this term overcomes nuclear repulsion and leads to bonding. In the excited state, it does not, and no bond is formed.

Equations (14.8a, b) are our first explicit expressions for *Born–Oppenheimer potential-energy functions*. These functions describe bonding—its existence, extent, and structure—and they serve as the potential-energy fields within which the nuclei vibrate. Figure 14.4 graphs $\langle E \rangle_g$ and $\langle E \rangle_u$ as functions of R along with the essentially exact curve for the g electronic state.

The approximate curves miss the exact ones, but not by much. (The exact u curve is quite close to the approximate $\langle E \rangle_u$ curve and is not shown.) Note that the bonding energy for the g state is a fairly small fraction of the total energy, as we discussed earlier, and that the u state leads to *no* bonding since the $\langle E \rangle_u$ curve has no binding potential well. Finally, note that the curves merge (toward $E = -1/2$) as $R \to \infty$. Put another way, a proton and an H atom have only a 50:50 chance of *approaching* along the g curve and interacting in a bonding instead of an antibonding way.

Many of these general observations can be applied to any molecule. Some electronic states will be bonding; some will not. Some will become degenerate at large R. Some will bind with deep potential wells; some, with shallow wells.

The approximation technique is easily generalized. It is termed the *linear combination of atomic orbitals to give molecular orbitals* or *LCAO-MO* method. Moreover, the same techniques we used to improve atomic structure calculations improve molecular structure calculations. We can have more variational parameters, and/or we can include more types of atomic functions. We can use SCF-HF methods, and we can incorporate configuration interaction and electron correlation effects.

Now we turn to the other R limit of interest, $R \to 0$, or the *united atom limit*. Figure 14.5 shows the approximate $\sigma_g 1s$ orbital as a contour diagram at three values of R: 20, 2, and 0.2 au. At 20 au, the separated atom limit is quite obvious; Ψ_g looks like isolated 1s atomic functions, peaked at each nucleus. At 2 au, Ψ_g looks like the qualitative picture of $\sigma_g 1s$ in Figure 14.3, with electron density localized between the nuclei. At 0.2 au, Ψ_g looks atomic-like again, with near spherical symmetry about the closely spaced nuclei. As $R \to 0$, this similarity becomes exact,

FIGURE 14.5 Contour diagrams of the H_2^+ Ψ_g wavefunction at various H–H separations R clearly show the UA, MO, and SA limits as, respectively, a uniform, nearly spherical function, a diatomic with a bonding region, and two separated atoms.

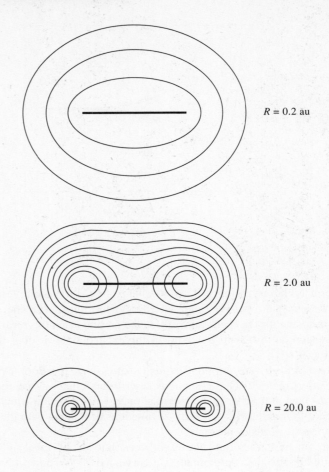

and takes on the appearance of a 1s orbital for He.[3] We say the $\sigma_g 1s$ *correlates* to the 1s orbital in the united atom limit.

Figure 14.6 shows the same sequence for the Ψ_u orbital ($\sigma_u^* 1s$). At $R = 20$ au, the atoms are again well separated; Ψ_u looks like two 1s atoms with wavefunctions of different sign. The nodal plane is always there and becomes a plane through the united atom nucleus at $R = 0$. The wavefunction Ψ_u correlates to the $2p_z$ orbital (taking R to lie along the z axis) in the united atom limit.

The other molecular orbitals of Figure 14.3 can be followed toward $R = 0$, preserving all the nodal planes and correlating to familiar atomic orbitals. They can also be followed toward $R \to \infty$ to find their atomic orbital parentage. At each limit, and at R values characteristic of bonding, a different MO nomenclature is appropriate. The nomenclature we have used so far (that of Figure 14.3) is most appropriate for the separated atom limit. It denotes the symmetry of the MO and the separated atom AO parentage. In the united atom limit, we write the UA orbital designation first, followed by its MO parentage:

(MO symmetry)(SA parent) → numbered MO symbol → (UA orbital)(MO parent)
     ↑   ↑   ↑
     SA   MO   UA

---

[3]But only the appearance. If you look at Ψ_g's mathematical form, as $R \to 0$, S approaches 1, so that Ψ_g approaches $(1/\pi)^{1/2} e^{-r}$, the H 1s function, not the He 1s function. Had we included an effective nuclear charge parameter in Ψ_g, we would have been able to recover the He 1s function at $R = 0$.

14.1 THE SIMPLEST MOLECULES: H_2^+ AND H_2

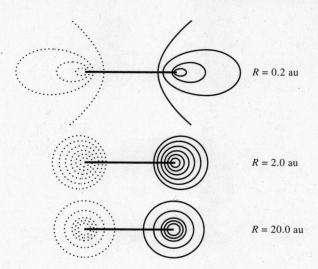

FIGURE 14.6 For the H_2^+ Ψ_u wavefunction, the SA limit shows separate atoms, the MO region shows the antibonding nodal plane, and the UA limit shows a 2p shape at small R. Dotted contours have the opposite algebraic sign from solid contours.

Thus, $\sigma_g 1s$ is called $1s\sigma_g$ in the UA limit, which seems redundant until we realize that some UA limits have AO forms very different from the original separated atom parentage.

Figure 14.7 is an elaboration of Figure 14.3 showing the UA, diatomic, and SA symmetries of the first eight H_2^+ MOs. For the R region characteristic of bonding, the MO symmetry itself is of greatest importance, and we simply number each orbital of a given symmetry in order of increasing energy. We write the ground state correlations $\sigma_g 1s \rightarrow 1\sigma_g \rightarrow 1s\sigma_g$ with $1\sigma_g$ indicating that this is the first (lowest in energy) orbital with σ_g symmetry, while we write, for example, $\sigma_u^* 2p_z \rightarrow 3\sigma_u^* \rightarrow 4f_{z(5z^2-3r^2)}\sigma_u^*$ for the seventh excited-state orbital correlation. The 3 in $3\sigma_u^*$ indicates that this is the third orbital of σ_u^* symmetry.

A more quantitative way to picture these correlations locates the SA and UA orbital energies, one on either side of a graph that has R as the abscissa, and connects them with lines indicating the MO energy as a function of R. This is done in Figure 14.8. Note the following general features of these correlations:

(1) The UA nl level gives rise to l distinct MOs; all odd l have g symmetry, and all even l have u symmetry.

(2) Each pair of nl SA orbitals combine to form $2(2l + 1)$ MOs:

$(l = 0)$ s $\rightarrow \sigma_g, \sigma_u^*$

$(l = 1)$ p $\rightarrow \sigma_g, \sigma_u^*, \pi_g^*, \pi_u$ (the π's are doubly degenerate)

$(l = 2)$ d $\rightarrow \sigma_g, \sigma_u^*, \pi_g^*, \pi_u, \delta_g, \delta_u^*$ (the π's and δ's are doubly degenerate and not shown).

We can now discuss H_2. We begin much as we began with He, taking account of electron spin and the Pauli Exclusion Principle. We start with our two lowest energy MOs, $\Psi_g = 1\sigma_g$ and $\Psi_u = 1\sigma_u^*$ from H_2^+, and consider the spatial and spin orbital combinations we can make from them. The possibilities are (numbering the electrons 1 and 2):

$$\Psi_1 = \Psi_g(1)\Psi_g(2)\frac{1}{\sqrt{2}}[\alpha(1)\beta(2) - \beta(1)\alpha(2)] \quad (14.9a)$$

$$\Psi_2 = \Psi_u(1)\Psi_u(2)\frac{1}{\sqrt{2}}[\alpha(1)\beta(2) - \beta(1)\alpha(2)] \quad (14.9b)$$

FIGURE 14.7 UA (left)—MO (center)—SA (right) symmetry correlations.

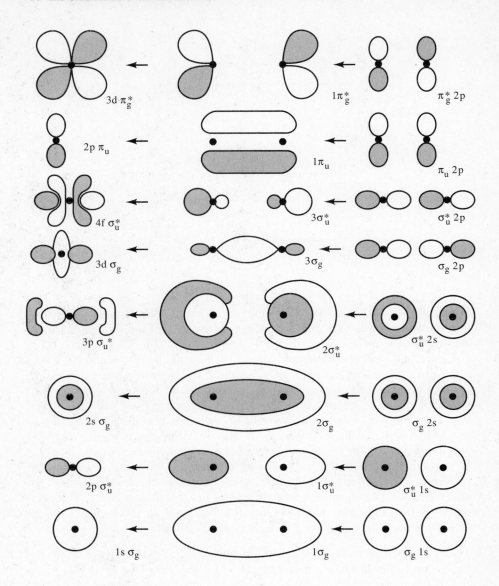

$$\Psi_3 = \frac{1}{\sqrt{2}} [\Psi_g(1)\Psi_u(2) + \Psi_u(1)\Psi_g(2)] \frac{1}{\sqrt{2}} [\alpha(1)\beta(2) - \beta(1)\alpha(2)] \quad \textbf{(14.9c)}$$

$$\Psi_{4a} = \frac{1}{\sqrt{2}} [\Psi_g(1)\Psi_u(2) - \Psi_u(1)\Psi_g(2)]\alpha(1)\alpha(2) \quad \textbf{(14.9d)}$$

$$\Psi_{4b} = \frac{1}{\sqrt{2}} [\Psi_g(1)\Psi_u(2) - \Psi_u(1)\Psi_g(2)]\beta(1)\beta(2) \quad \textbf{(14.9e)}$$

$$\Psi_{4c} = \frac{1}{\sqrt{2}} [\Psi_g(1)\Psi_u(2) - \Psi_u(1)\Psi_g(2)] \frac{1}{\sqrt{2}} [\alpha(1)\beta(2) + \beta(1)\alpha(2)] \quad \textbf{(14.9f)}$$

The first three, Ψ_1–Ψ_3, are *singlet* states, corresponding to *paired* spin functions, while the last three, Ψ_{4a}–Ψ_{4c}, have identical spatial parts (and thus identical B–O energies), but different *unpaired* spin parts. We will lump this *triplet* of degenerate states into a single spatial part, Ψ_4. (Each of these six functions could be found

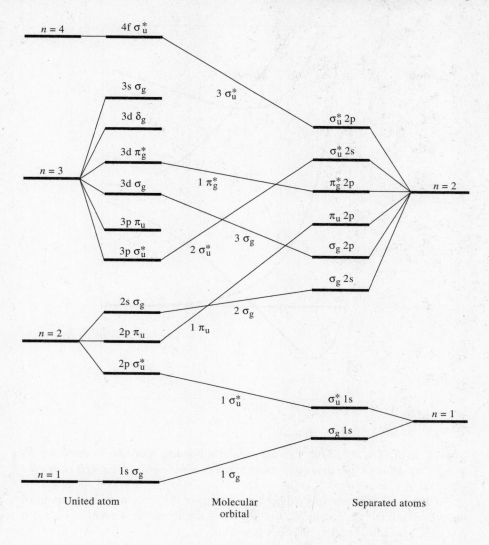

FIGURE 14.8 A correlation diagram from the SA to the UA limit connects orbitals of equivalent symmetry. Orbitals such as $3s\sigma_g$ and the remaining $n = 4$ UA orbitals correlate to higher energy SA limits not shown here.

from *molecular* Slater determinants, with Ψ_g and Ψ_u for spatial functions. Compare with Eq. (13.21).)

We can associate *molecular-orbital electron configurations* with these functions just as we did for atoms. One configuration can lead to more than one state, and, in Chapter 19, we will introduce *molecular electronic-state term symbols*. Looking first at the configurations, we see that

- Ψ_1 has the configuration $1\sigma_g^2$,
- Ψ_2 has the configuration $1\sigma_u^{*2}$, and
- Ψ_3 and Ψ_4 have the configuration $1\sigma_g 1\sigma_u^*$.

To see what bonding, if any, these functions predict, we compute $\langle E(R) \rangle$ for each. Note that there is *no* variational minimization to perform; that was done, once and for all, when we constructed Ψ_g and Ψ_u for H_2^+.

Figure 14.9 plots all four $\langle E(R) \rangle$ curves. You should recognize that there is a significant problem with curves 1–3. Only the curve for Ψ_4's energy approaches the expected *asymptotic* $(R \to \infty)$ *dissociation limit energy* of -1 au corresponding to two isolated H atoms. It is, however, an unbound state with a repulsive internuclear

FIGURE 14.9 Of the four simple MO wavefunctions for H_2, only Ψ_4 has a potential energy curve with the proper dissociation limit $\langle E(R = \infty) \rangle = -1$ au.

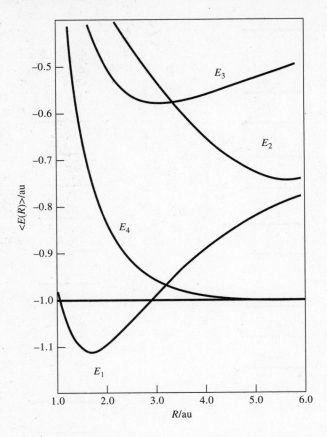

force at all R. The curve for Ψ_1 *does* exhibit a bonding well, as do those for Ψ_2 and Ψ_3 but *all of these curves dissociate towards some energy limit well above the expected -1 au value.*

To understand this unexpected behavior, we write the wavefunctions in terms of the *atomic* orbitals that define Ψ_g and Ψ_u. Using the abbreviated notation $1s_A(1) = A(1)$, etc., we find, omitting the spin functions,

$$\Psi_1 = \frac{1}{2(1+S)} \{[A(1)A(2) + B(1)B(2)] + [A(1)B(2) + B(1)A(2)]\}$$

$$\Psi_2 = \frac{1}{2(1-S)} \{[A(1)A(2) + B(1)B(2)] - [A(1)B(2) + B(1)A(2)]\}$$

$$\Psi_3 = \frac{1}{[2(1+S)(1-S)]^{1/2}} [A(1)A(2) - B(1)B(2)]$$

and

$$\Psi_4 = \frac{-1}{[2(1+S)(1-S)]^{1/2}} [A(1)B(2) - B(1)A(2)] .$$

Note how two physically distinct electron–nucleus assignments appear. The first is the linear combination pair

$$A(1)B(2) \pm B(1)A(2) = \phi_{cov}^{\pm} \quad (14.10a)$$
$$\uparrow \qquad \uparrow$$
Electrons 1 and 2
on *different* nuclei

which we give a "cov" subscript to indicate *covalent*. These are the exchange-symmetrized combinations of atomic orbitals that describe a covalent bond—one electron associated with one nucleus and the second electron associated with the other nucleus. Note that Ψ_4 dissociates properly to two H atoms and is purely covalent in character.

The second distinct assignment is

$$A(1)A(2) \pm B(1)B(2) = \phi_{\text{ion}}^{\pm} \quad (14.10b)$$

$$\underset{\substack{\uparrow \quad \quad \uparrow \\ \text{Electrons 1 and 2} \\ \text{on } same \text{ nuclei}}}{}$$

where the "ion" subscript indicates an *ionic bond*. These combinations have both electrons associated with the *same* nucleus. They correspond to H$^-$ + H$^+$ at dissociation, an *ion pair* state. Note that Ψ_3 is purely ionic while Ψ_1 and Ψ_2 are equally ionic and covalent. The energy of the ion pair state (at $R = \infty$) is greater than the energy of the atomic pair state (H + H) since the ionization energy of H (1/2 au or 13.6 eV) is much greater than the electron affinity of H (about 0.0555 au or 0.754 eV).[4]

Thus, states Ψ_1 and Ψ_2 dissociate to a separated atom limit that is equally ionic-pair and ground-state atomic-pair in character, and, even in the bonding region, these states correspond to equal covalent and ionic contributions to the molecular wavefunction. This is, given the high energy of the ion pair states and the known physical and chemical properties of H$_2$, an unlikely state of affairs, especially for Ψ_1, the electronic ground state.

Does this result mean that our MO approach is inherently flawed? Not at all. But our implementation of it was too simplistic. We need to allow the covalent and ionic character of the wavefunctions to vary with R. We can benefit from a *configuration interaction* calculation right from the start. We write, concentrating on the ground state,

$$\Psi_1' = N[c_{\text{cov}}(R)\phi_{\text{cov}}^{+} + c_{\text{ion}}(R)\phi_{\text{ion}}^{+}] \quad (14.11)$$

where the c's are variational functions of R. At $R = \infty$, we expect c_{ion} to be zero, since this choice minimizes the energy. At smaller R, in the bonding region, chemical intuition says that c_{ion} should still be small, but perhaps not zero. We can calculate the *percentage ionic character* ($|c_{\text{ion}}|^2 \times 100$) of the ground state at any R.

An alternate approach is suggested by Eq. (14.11). It is not likely to be as accurate, but it is simpler. Suppose we drop the ionic term completely and try a purely covalent function, ϕ_{cov}^{+}, so that $c_{\text{cov}} = 1$ and $c_{\text{ion}} = 0$ at all R. This is called the *valence bond* (VB) approach. Our trial wavefunction has only valence electrons (a trivial statement for H$_2$, but for larger molecules, not so trivial), and these are placed in AOs from which the covalent bond evolves. Thus, for the ground state, again omitting spin,

$$\Psi_{\text{VB}}^{+} = N\phi_{\text{cov}}^{+} = \frac{1}{[2(1+S)]^{1/2}}[1s_A(1)1s_B(2) + 1s_A(2)1s_B(1)] \ . \quad (14.12)$$

This function gives $\langle E(R) \rangle_{\text{VB}}$ in terms of Coulomb, exchange, and overlap integrals.

---

[4]It is a simple and worthwhile calculation to use these numbers to show that the ion pair energy lies between the energies of the excited atom pairs H($n = 1$) + H($n = 4$) and H($n = 1$) + H($n = 5$). See Problem 14.15.

Our Ψ_4 MO is also Ψ_{VB}^-, since it is purely ϕ_{cov}^- in character. Both VB functions have the correct dissociation limit energy: Ψ_{VB}^+ is the bonding ground state and Ψ_{VB}^- is the repulsive (triplet spin character) first excited state. Since $\Psi_4 = \Psi_{VB}^-$, perhaps there is a relationship between the VB and MO methods in general. If there is *not*, we are likely doing something fundamentally wrong, since both approaches seem to have firm physical bases.

The MO approach first takes a given nuclear arrangement and finds all the electron states. Next, one counts the total number of electrons and assigns the molecule an electron MO configuration. This is analogous to the SCF-HF atomic orbital calculation.

The VB approach says instead that one knows *a priori* (from even the most elementary bonding theories, such as Lewis electron dot structures) the electron assignments to chemical bonds in the target molecule. Given these bond locations, we construct a trial wavefunction based on the atomic valence orbitals that must contribute to them.

Equation (14.11), the covalent and ionic mix, shows us how MO and VB theories are equivalent at this level of approximation. This equation is not only an MO-CI type of function, it is also a VB function since the ionic term could be considered a VB starting point. For H_2, chemical intuition says that covalency outweighs ionic bonding, but for LiH, for example, we might well start with a purely ionic VB function, since LiH has the physical properties of an ionic molecule: high melting point, low vapor pressure, significant dipole moment, etc.

The two major experimental observables of any ground electronic state are the bond dissociation energy $\mathcal{D}_e$ and the equilibrium bond length R_e. The bond length locates the minimum in $\langle E(R) \rangle$, and $\mathcal{D}_e$ is the energy difference $\langle E(R = \infty) \rangle - \langle E(R = R_e) \rangle$. For H_2, both of these quantities are known experimentally to very high accuracy. They are compared to our best trial wavefunction predictions (Eq. (14.11)) below:

	Experiment	Simple Theory
R_e	0.7414 4 Å	0.880 Å
	(1.401 1 au)	(1.66 au)
$\mathcal{D}_e$	4.748 33 eV	3.230 eV
	(0.174 597 au)	(0.118 8 au)

While not spectacular, the agreement is at least encouraging. The simple theory also predicts H_2 is 2.4% ionic at R_e.[5]

More functions and more variational parameters improve the agreement, as does consideration of electron correlation. All of these effects can be, and have been, accounted for, so that theoretical B–O wavefunctions and electronic-state potential-energy curves now exist for H_2 that agree quite well with observation.

Before we turn to a general discussion of other molecules, it is worthwhile to review some of the subtle effects we have found here that govern bonding. Figure 14.10 plots the B–O potential curves for the two lowest-energy electronic states of H_2. Both dissociate to the atom pair limit $H(1s) + H(1s)$. The singlet, bound state corresponds to paired electron spins, while the triplet, unbound state corresponds to unpaired spins. Thus, spin configurations can play a determining role in bonding.

---

[5]Of course, H_2 does not have a dipole moment, but it still can have an ionic component to its bond, as Eq. (14.11) indicates. Even a "pure" ionic H_2 would not have a dipole moment since the nuclei are equivalent. It would be as much H^-H^+ as it would be H^+H^-.

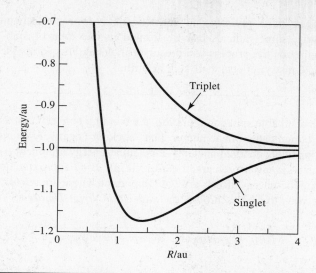

FIGURE 14.10 The interaction of two 1s state H atoms can follow either the repulsive, nonbonding triplet curve or the attractive, bonding singlet curve. These curves are based on essentially exact quantum chemical calculations.

A Closer Look at Spin Effects on Bonding

Suppose we had a bottle of ground state H atoms, with all their electron spins aligned. As the atoms collide, they do so under the influence of the antibonding, repulsive B–O potential curve. Thus, they bounce from one another and never bond. Now, suppose we could cause the spin directions to randomize. One out of four such random spin collisions between two atoms will now be under the influence of the bonding curve. The 1:4 ratio is just the ratio of one singlet state to a total of four (one singlet plus three triplet) possible electronic states dissociating to H(1s) + H(1s). Eventually (and in fact very quickly) all the H atoms would form H_2. In the process, the system would release the bonding energy of H_2. The energy released amounts to -436 kJ per mol of H_2 formed.

This is a substantial exothermicity. Since the relative atomic mass of H is 1 amu, -436 kJ accompany the recombination of every gram of H. In contrast, the chemical energy released for common fuels such as gasoline, coal, wood, natural gas, etc., falls in the range -20 to -50 kJ per gram of fuel burned in pure O_2. Can a bottle of H atoms actually be made and used as an energy source?

This idea has received some theoretical and experimental study. *Spin-polarized hydrogen*, as a collection of H atoms with parallel spins is known, can be trapped for short times in a magnetic field. (These atoms could be cooled all the way to absolute zero, and they would still be in the gas phase. This is the only system known that remains gaseous at absolute zero.) The idea is to extend these times and release the energy as needed simply by removing the field.

Unfortunately, this is not as easy as it sounds, and it does not really sound that easy to begin with! First, we need a source of H atoms. Dissociating H_2 would certainly do the trick, but, of course, we would have to input at least as much energy (and in practice, considerably more) than we can ultimately recover. Next, the bottle must be made of some material that will not catalyze (by spin-flipping) the recombination. Probably no such material exists.

Research, however, has verified that the effect exists. Spin-polarized H can be held for short times and will release the expected amount of energy, but a practical energy source based on this idea is currently unlikely.

We will see that atomic and molecular fragments in general often have the same spin constraints on bonding that H_2 has. Several ingenious experiments have been performed using these radicals and magnetic fields to control chemical reaction rates. Similarly, some radicals can be formed in one type of spin state or the other, depending on the process and precursor molecule that forms them, and their subsequent chemistry reflects their spin at birth.

Even with a suitable spin configuration, the bonding between two atoms remains a somewhat complicated phenomenon. Just "sticking negatively charged electrons in between positively charged nuclei"—a potential-energy argument—is not a good enough description, as we remarked earlier. In the B–O approximation, potential-energy and kinetic-energy effects are *both* important. These are related, in a surprisingly simple way, to $\langle E(R) \rangle$ through the *molecular Virial Theorem*.

If we let

$$\langle T(R) \rangle = \text{average total } \textit{electron kinetic energy} \text{ at } R$$

and

$$\langle V(R) \rangle = \text{average total } \textit{electron–electron, electron–nuclear, and} \\ \textit{nuclear–nuclear Coulombic potential energy} \text{ at } R,$$

then the molecular Virial Theorem states

$$\langle E(R) \rangle = \langle T(R) \rangle + \langle V(R) \rangle \tag{14.13a}$$

$$\langle T(R) \rangle = -\langle E(R) \rangle - R \frac{d\langle E(R) \rangle}{dR} \tag{14.13b}$$

and

$$\langle V(R) \rangle = 2\langle E(R) \rangle + R \frac{d\langle E(R) \rangle}{dR}. \tag{14.13c}$$

Thus, given *any* $\langle E(R) \rangle$ curve, whether from theory or experiment, we can decompose it into $\langle T \rangle$ and $\langle V \rangle$ contributions. Figure 14.11 shows the H_2 bonding curve (from Figure 14.10) along with the $\langle T \rangle$ and $\langle V \rangle$ curves derived from it.[6]

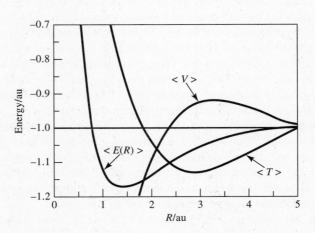

FIGURE 14.11 The average potential- and kinetic-energy curves (shifted to have a common large R limit) are compared here to the B–O total-energy curve (the singlet curve of Figure 14.10). The physical interpretation of these curves is discussed in the text.

---

[6]These curves have been displaced vertically so that they approach the same energy at large R as the $\langle E(R) \rangle$ curve. The real $R \to \infty$ limits are $\langle E(R) \rangle \to -1$, $\langle T(R) \rangle \to +1$, and $\langle V(R) \rangle \to -2$. See also Problems 14.7 and 14.8.

First, follow $\langle V(R) \rangle$ in from large R. Initially, $\langle V \rangle$ rises as the electrons are pulled away from the nuclei, increasing their Coulombic energy long before reaching the true bonding region near R_e. When $R \cong R_e$, however, $\langle V \rangle$ has fallen as the electrons congregate between the closely spaced nuclei. This is the "stick negative charges into the bonding region" effect. At considerably smaller R, $\langle V \rangle$ reaches a minimum nearly four times lower than $\langle E(R) \rangle$, the effects of internuclear repulsion begin to dominate, and $\langle V \rangle$ finally increases as $R \to 0$ (not shown in the figure). If there were no kinetic-energy effects, the $\langle V \rangle$ curve would describe H_2, and it would be a very different molecule. It would have about half the real bond length and four times the real dissociation energy.

Now consider $\langle T(R) \rangle$. It drops, as R decreases, faster than $\langle V \rangle$ increases, leading to a real attractive force at large R. This drop is again related to motion of the electrons away from the nuclei. Recall that for a particle-in-a-box, which has only a kinetic-energy component, making the box bigger causes the kinetic energy to decrease. At large R, the electrons begin to move away from the local region of their parent atoms (their box gets bigger) toward the bonding region. But in the R region characteristic of bonding, $\langle T \rangle$ has begun to rise steeply. Again, a box analogy is appropriate: the electrons are confined to the bonding region between the nuclei, which is a *smaller* volume than was available to them at larger R. The kinetic-energy component $\langle T \rangle$ reaches a minimum at $R > R_e$ and is rising rapidly at $R = R_e$.

We leave H_2 at this point and, by and large, we leave much of our theoretical framework behind as well. Except for a few very severe approximations that are nevertheless useful, the mathematical theory of molecules in general is a complex subject best viewed at first in terms of its output rather than its input. Thus, we will concentrate on graphical and pictorial representations of wavefunctions and energies rather than on their mathematical form.

14.2 APPROXIMATE DESCRIPTIONS OF COVALENT BONDS

The MO picture of bonding has the distinct advantage of providing a uniform framework for the discussion of bonding in any of the possible nuclear configurations of families of molecules. The language of homonuclear diatomic MOs is the same whether we are discussing H_2 or N_2 or Cs_2 or whatever. In general, all linear molecules (H_2, CO_2, N_2O, C_2H_2, HCCN, etc.) fall into a family, with two subfamilies distinguished by the presence (as in H_2, CO_2, and C_2H_2) or absence (as in N_2O and HCCN) of a center of symmetry. Next could come bent triatomics, further classified by symmetry (compare the symmetry of H_2O to that of HNO, for example), and then perhaps tetrahedral symmetry molecules, and so on.

Notice how nuclear configuration symmetry—the general classes of shapes of molecules—plays a natural role in the classification of molecules. The study of symmetry is such an important task that a major branch of mathematics, *group theory*, has been applied to molecular problems. We will not delve into group theory in detail in this text, but we will frequently point out some of its results and use some of its notation. In fact, we already have on several occasions, most recently in our classification of H_2^+ MOs.

We start with homonuclear diatomics. The MO nomenclature is that of Figure 14.7. We specify the number of electrons in the molecule and the order in which these MOs occur in increasing energy. The first step is easy, but the second is devilishly subtle, not only for homonuclear diatomics, but for molecules in general, especially those of high symmetry.

To see why (and how) the MO order can vary, it is important to recognize two

FIGURE 14.12 In a photoelectron spectrometer, ionizing light from a gas discharge lamp photoionizes a target gas. Ejected photoelectrons are energy analyzed by a device that passes electrons of a selected kinetic energy onto an electron detector. Electrons of an improper energy (for example, too high, as in the figure) are absorbed by the analyzer and thus rejected.

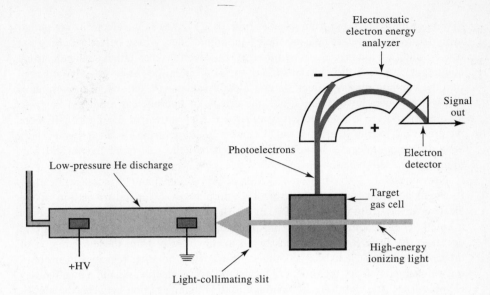

basic facts. First, molecular orbitals for electrons are not just mathematical artifacts. They have observable and assignable properties, such as energies. Second, these orbitals can be approximated by suitable combinations of atomic orbitals, but without detailed calculations, these LCAO approximations cannot, in many cases, be easily ordered in energy.

Molecular orbital energies come most strikingly from the powerful experimental technique known as *photoelectron spectroscopy*—the photoelectric effect of Chapter 11 applied to individual molecules. In this method, a photon of known energy photoionizes the target (an atom, a molecule, or even a macroscopic surface, as in the usual photoelectric effect experiment), and one of several types of electron kinetic energy analyzers measures the energy of the released electron. In its most elaborate form, the electron's energy is measured as a function of its recoil angle from the molecule (and perhaps also with respect to the incident light's polarization direction), but the simplest experiment, diagrammed in Figure 14.12, measures the electron energy at a fixed angle such as 90° from the photon beam. Several monochromatic photon sources are used, ranging from high photon energy atomic-emission sources (most often H, He, and Ne) to higher energy X-ray sources to monochromatized synchrotron radiation sources of virtually any photon energy.

Figure 14.13 shows a photoelectron spectrum of N_2. This experiment used 21.22

FIGURE 14.13 A photoelectron spectrum of N_2 displays the energy structure of the N_2 molecular orbitals as three groups of features, each corresponding to photoelectron emission from a different valence MO.

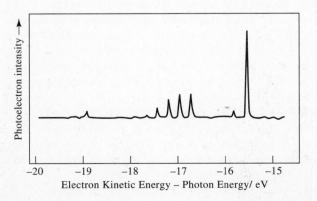

eV photons. Any electron with an orbital energy less than -21.22 eV could not be photoionized. The spectrum is plotted as a function of the orbital energy of the detected photoelectron. Note that the signals are in three well separated regions. The most intense signal is at an orbital energy of -15.57 eV. These electrons come from

$$21.22 \text{ eV photon} + N_2 \rightarrow N_2^+ + 5.65 \text{ eV measured electron kinetic energy}$$

from which the orbital energy is calculated:

$$\text{orbital energy} = \text{measured electron kinetic energy} - \text{photon energy}$$
$$= 5.65 \text{ eV} - 21.22 \text{ eV} = -15.57 \text{ eV} \;.$$

There is a small signal near -15.8 eV that corresponds to removing an electron from the same MO as the large feature, but leaving the N_2^+ product with excess internal energy (in the form of nuclear vibrational energy). Thus, photoelectron spectroscopy also yields important information about the molecular ion product. The next group of signals, from about -16.7 eV to -18 eV, corresponds to removing an electron from a second MO, again with varying amounts of N_2^+ vibrational excitation from peak to peak. The feature at -16.69 eV represents the second MO energy, free of N_2^+ vibrational-energy contributions. Finally, the small feature at -18.75 eV corresponds to a third MO's energy.

We will not explain the peaks' relative intensities, but we will say more about the reason the central clump shows several peaks while the other two are dominated by single peaks. Our main task is to understand the nature of the three MOs that lead to this spectrum. No electrons are found with orbital energies greater than (less negative than) -15.57 eV. This must correspond, therefore, to the energy of the *highest* (energy) *occupied molecular orbital,* the *HOMO*. There are electrons with orbital energies lower than -21.22 eV that could not be detected by this experiment. An X-ray or high photon-energy synchrotron source could be (and has been) used to observe the more tightly bound electrons. They have orbital energies of -37.9 and -410.0 eV.

We can begin to predict which MO corresponds to which peak by simple electron-counting and a simple node-counting, overlap, and degeneracy argument based on Figure 14.3 or 14.7. The N_2 molecule has a total of 14 electrons, and we populate MOs with pairs of electrons. The lowest energy MO is the nodeless $1\sigma_g$, with $1\sigma_u^*$ next. The two N 1s AOs dominate an LCAO–MO approximation to these MOs. These two *core* atomic orbitals (as opposed to the *valence* orbitals that produce the higher energy MOs) are so tightly localized about the N nuclei that their overlap is negligible. As a result, the $1\sigma_g$ and $1\sigma_u^*$ MOs are essentially degenerate,[7] leading to a single photoelectron feature at -410.0 eV. Considerably higher in energy are the $2\sigma_g$ and $2\sigma_u^*$ MOs, dominated by N 2s AOs. These are four of the necessary seven N_2 MOs. The other three are the $3\sigma_g$ (from the $\sigma_g 2p_z$ SA limit) and the pair of doubly degenerate $1\pi_u$ MOs. These originate from the 2p AOs, and should have similar energies, not far different from the $2\sigma_u^*$ energy. On the one hand, $3\sigma_g$ has more AO overlap (electron density) along the internuclear axis than the $1\pi_u$ orbitals, but, on the other hand, $3\sigma_g$ has nodal planes through the nuclei, implying confined electrons with high kinetic energy.

We conclude that the three features in the photoelectron spectrum correspond to the $2\sigma_u^*$, $3\sigma_g$, and $1\pi_u$ MOs, with the peak at -18.75 eV probably corresponding

---

[7]The analogy to our H_2^+ paradigm is the degeneracy of $1\sigma_g$ and $1\sigma_u^*$ at large R. For N_2, the large nuclear charge so greatly shrinks the 1s orbital that even chemical-bonding distances are "large" as far as this orbital is concerned.

to the $2\sigma_u^*$ MO. However, only a detailed calculation can reveal that $3\sigma_g$ is the HOMO (corresponding to an orbital energy of -15.57 eV), and $1\pi_u$ is the orbital of second-least binding energy (at -16.69 eV).

We can write these three ionization processes (with the pair of degenerate π orbitals called simply $1\pi_u$ with a maximum capacity of four electrons)

$$N_2(1\sigma_g^2\, 1\sigma_u^{*2}\, 2\sigma_g^2\, 2\sigma_u^{*2}\, 1\pi_u^4\, 3\sigma_g^2)$$

(1) $\rightarrow N_2^+\,(\ldots 3\sigma_g^1)$ at -15.57 eV

(2) $\rightarrow N_2^+\,(\ldots 1\pi_u^3\, 3\sigma_g^2)$ at -16.69 eV

(3) $\rightarrow N_2^+\,(\ldots 2\sigma_u^{*1}\, 1\pi_u^4\, 3\sigma_g^2)$ at -18.75 eV.

We have not only deduced the MO configuration for N_2, but, as a bonus, we have the *MO electron configurations for the ground and first two excited electronic states of* N_2^+.

Process (2) leads experimentally to extensive N_2^+ vibration. An explanation based on MO configurations is not hard to find. First, recall the simple triple-bond picture of N_2 (:N:::N:). These three bonds correspond to three pairs of electrons, one in the $3\sigma_g$ HOMO and one in each of the two $1\pi_u$ MOs. Process (1) removes an electron from the HOMO, and very little N_2^+ vibration is observed. This indicates very similar bonding in N_2 and N_2^+ in these electronic states; remove one $3\sigma_g$ electron, and the resulting B–O potential-energy curve that binds and governs the N_2^+ nuclear motion is virtually the same as that for N_2. This lack of profound change (actually, it is a lack of bond *length* change that is important here[8]) indicates that the HOMO electrons in $3\sigma_g$ are relatively *nonbonding* in nature; other experiments have shown that the bond energy of N_2^+ in this electronic state is only slightly less than that of N_2 (about 8.71 eV versus 9.76 eV).

The extensive vibrational structure of process (2) indicates that the N_2^+ electronic state that comes from removing one $1\pi_u$ electron is very different from the N_2 ground state. The N_2^+ bond is considerably longer in this state due to the important bonding character of the $1\pi_u$ MO. Finally, process (3), which lacks vibrational structure, also indicates removing a nonbonding electron. In fact, the $2\sigma_u^*$ MO is *antibonding* in character, and other types of spectroscopy on N_2^+ show that while this electronic state has a decidedly smaller bond energy, it has a "tighter" bond as measured by the N_2^+ vibrational motion in this state.

These photoelectron data describe the N_2 ground electronic state electron configuration and the first three (ground and two excited) electronic states of N_2^+. Other experiments, most importantly optical absorption and emission spectroscopy, reveal the nature of neutral N_2 excited electronic-state configurations. From Figure 14.7 and our ground-state configuration assignments, we can predict that excitations such as

$$N_2(\ldots 1\pi_u^4\, 3\sigma_g^2) \rightarrow N_2^*(\ldots 1\pi_u^4\, 3\sigma_g^1\, 1\pi_g^{*1})$$

or

$$N_2(\ldots 1\pi_u^4\, 3\sigma_g^2) \rightarrow N_2^*(\ldots 1\pi_u^4\, 3\sigma_g^1\, 3\sigma_u^{*1})$$

(the * on N_2^* simply indicates an excited electronic state) should all be possible, and electronic states derived from these configurations are known.

---

[8]This will be proven in Chapter 19. Photoelectron spectroscopy cannot give accurate bond *energies* without auxiliary information from other types of experiments.

EXAMPLE 14.2

The ionization potentials of C and O atoms are 11.26 eV and 13.62 eV, respectively. The photoelectron spectrum of CO shows four distinct features at low energy: 14.5, 17.2, 20.1, and 38.3 eV. What do these data say about the MOs of CO?

SOLUTION CO is isovalent to N_2, and we can draw on our analysis of the N_2 photoelectron spectrum for guidance. Since CO lacks a center of symmetry, the g and u MO labels cannot be used, and we simply number the MOs of either type, σ and π, in order of increasing energy. The two lowest σ orbitals, 1σ and 2σ, are O and C atomic core MOs. (Since the greater nuclear charge of O places the energy of its 1s orbital well below that for C, 1σ is assigned to O(1s).) Next, we expect the 2s AOs to mix and form two σ MOs, 3σ and 4σ. (In SA notation, these are $2s\sigma$ and $2s\sigma^*$, respectively.) The three 2p AOs on each atom mix to form six MOs: the next two σ's, 5σ and 6σ ($2p\sigma$ and $2p\sigma^*$ in SA notation), and two doubly degenerate π MOs, 1π and 2π ($2p\pi$ and $2p\pi^*$ in SA notation). We have 14 electrons to pair in these orbitals to form the CO ground-state configuration. In analogy to N_2, we write this as $1\sigma^2 2\sigma^2 3\sigma^2 4\sigma^2 1\pi^4 5\sigma^2$. The photoelectron data suggest we assign the outer four orbitals the ionization energies $E_{5\sigma} = 14.5$ eV, $E_{1\pi} = 17.2$ eV, $E_{4\sigma} = 20.1$ eV, and $E_{5\sigma} = 38.3$ eV. Since the IP of C is less than the IP of O, we can display the energies of the various valence AOs and MOs in the following diagram:

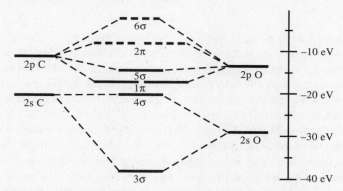

The unfilled 2π and 6σ orbital energies (dashed lines) are not located by these data and are only qualitatively positioned—even their relative order is not obvious. The 2s AO energies are likewise estimated from atomic excitation energies (see Tables 13.3 and 13.4). We can use this diagram to analyze the simple Lewis dot structure :C:::O: and decide which MOs form the triple bond. The 1π orbitals are obviously two of the three, but which is the third, 4σ or 5σ? The question does not have an obvious answer, even with detailed MO calculations as an aid. The 3σ and 4σ orbitals form a bonding and antibonding pair, which argues for 5σ as the "third bond." But the AO energy differences between C and O make all the valence σ orbitals asymmetric so that 3σ is best viewed as the pair of dots on the outside of O in the Lewis structure (a so-called *lone pair*) while 5σ, the HOMO, is the pair outside C. That makes 4σ the "third bond," and this agrees with the coordination chemistry of CO in transition metal carbonyls, but, as you should now recognize, it is often impossible to make a firm association between MOs and more elementary bonding descriptions.

➡ **RELATED PROBLEMS** 14.9, 14.10

This detailed walk through the MO configurations of the N_2–N_2^+ system can be carried out for every homonuclear diatomic, and Table 14.3 gives the experimentally determined (and theoretically assigned) MO configurations for the homonuclear diatomics H_2–Ne_2, along with the experimental bond lengths, bond dissociation energies, and the elementary *bond orders*. These are defined as

bond order = (# of electron pairs in bonding MOs) − (# of electron pairs in antibonding MOs) .

TABLE 14.3 Ground electronic-state molecular-orbital configurations, bond lengths, bond-dissociation energies, and bond orders for the homonuclear diatomics H_2–Ne_2.

Species	Configuration	Bond length (in Å)	Bond energy (in eV)	Bond order
H_2	$1\sigma_g^2$	0.741 44	4.478 1	1
He_2	$1\sigma_g^2 1\sigma_u^{*2}$	very long†	$\sim 10^{-7}$	0
Li_2	$1\sigma_g^2 1\sigma_u^{*2} 2\sigma_g^2$	2.672 9	1.05	1
Be_2	$1\sigma_g^2 1\sigma_u^{*2} 2\sigma_g^2 2\sigma_u^{*2}$	2.465	0.1	0
B_2	$1\sigma_g^2 1\sigma_u^{*2} 2\sigma_g^2 2\sigma_u^{*2} 1\pi_u^2$	1.590	3.02	1
C_2	$1\sigma_g^2 1\sigma_u^{*2} 2\sigma_g^2 2\sigma_u^{*2} 1\pi_u^4$	1.242 5	6.21	2
N_2	$1\sigma_g^2 1\sigma_u^{*2} 2\sigma_g^2 2\sigma_u^{*2} 1\pi_u^4 3\sigma_g^2$	1.097 7	9.759	3
O_2	$1\sigma_g^2 1\sigma_u^{*2} 2\sigma_g^2 2\sigma_u^{*2} 3\sigma_g^2 1\pi_u^4 1\pi_g^{*2}$	1.207 5	5.115	2
F_2	$1\sigma_g^2 1\sigma_u^{*2} 2\sigma_g^2 2\sigma_u^{*2} 3\sigma_g^2 1\pi_u^4 1\pi_g^{*4}$	1.411 9	1.602	1
Ne_2	$1\sigma_g^2 1\sigma_u^{*2} 2\sigma_g^2 2\sigma_u^{*2} 3\sigma_g^2 1\pi_u^4 1\pi_g^{*4} 3\sigma_u^{*2}$	3.03	0.003 68	0

†The He_2 bond energy and average bond length are uncertain. The nuclear motion has only one bound quantum state in its ground electronic state. The average bond length has been estimated to be $\cong 55$ Å!

Thus, $He_2(1\sigma_g^2 1\sigma_u^{*2})$, $Be_2(1\sigma_g^2 1\sigma_u^{*2} 2\sigma_g^2 2\sigma_u^{*2})$, and $Ne_2(1\sigma_g^2 \ldots 3\sigma_u^{*2})$, each with equal numbers of bonding and antibonding electrons, have bond orders of zero and no formal bond. Dihelium is a most remarkable molecule. The first conclusive report of its existence came in 1993. Its binding well is so shallow that one and only one bound state for nuclear motion in $^4He_2$ exists, and that state sits almost exactly at the dissociation limit. The bond is so fragile that the lighter isotopes, either $^3He_2$ or $^3He^4He$, are not bound at all. The tiny binding in these formally unbound molecules can be understood at an elementary level as a consequence of configuration interaction. If we treat the CI part of the electron energy as a perturbation to the non-CI ground-state energy, we know to expect an energy *lowering*.[9] The small admixture of *excited* MO configurations into the nominally zero bond-order MO configuration is enough to stabilize He_2, Be_2, and Ne_2. Be_2 is more strongly bound than Ne_2 since the first excited configuration in Be_2 that can interact with the ground configuration is derived from 2p AOs on Be (not far in energy from the 2s AOs), while in Ne, the interaction requires 3s AOs (very far in energy, due to the large $n = 2$ to $n = 3$ energy gap). Configurations close in energy generally have the greatest influence.

Other features of Table 14.3 are important to note. First is the change in MO ordering going from N_2 to O_2, F_2, and Ne_2. For these latter three, $1\pi_g^*$ is *higher* in energy than $3\sigma_u^*$. This subtle switch in energy ordering is due to changes in the atomic orbital energies (and to interactions among them) as the nuclear charge is increased. The second feature to note is the correlation among bond order and bond length/energy values. *Higher bond orders imply shorter bonds of greater bond dissociation energy.* Finally, note B_2 and O_2, with doubly occupied $1\pi_u$ and $1\pi_g^*$ HOMOs, respectively. Since the π orbitals are doubly degenerate, there is a choice to be made here: do we put both electrons, spin-paired, in one π MO, or do we put one electron in each of the degenerate pair, and if we do the latter, do we spin-pair them or not? The answer comes from a molecular version of the atomic Hund's rules. We put one electron in each, and we do *not* spin-pair them. Thus, B_2 and O_2

---

[9] Recall the discussion of the second-order perturbation energy term, Eq. (13.18), which is always *negative* for the ground state. The first-order term happens to vanish for homonuclear diatomics.

have the unusual magnetic properties associated with unpaired spin magnetic moments. For example, liquid O_2, poured between the poles of a permanent magnet, congregates in the strong field, but liquid N_2 (all spins paired) does not.

EXAMPLE 14.3

Predict the MO configurations of O_2^+ and O_2^-, and use them to predict the relative bond energies of O_2^+, O_2, and O_2^-. Do the following ionization potentials, electron affinities, and bond energy support your predictions? IP(O) = 13.617 eV, IP(O_2) = 12.063 eV, EA(O) = 1.4611 eV, EA(O_2) = 0.440 eV, $\mathcal{D}_e(O_2)$ = 5.115 eV.

SOLUTION From Table 14.3, we see that the electron removed to form the cation or added to form the anion will come from or go into the antibonding $1\pi_g^*$ MO of O_2. Taking one electron from an antibonding MO raises the bond order by 0.5, and adding one lowers the bond order 0.5. The configurations and bond orders are:

$$O_2^+: 1\sigma_g^2 \, 1\sigma_u^{*2} \, 2\sigma_g^2 \, 2\sigma_u^{*2} \, 3\sigma_g^2 \, 1\pi_u^4 \, 1\pi_g^{*1}, \quad BO = 2.5$$

$$O_2: 1\sigma_g^2 \, 1\sigma_u^{*2} \, 2\sigma_g^2 \, 2\sigma_u^{*2} \, 3\sigma_g^2 \, 1\pi_u^4 \, 1\pi_g^{*2}, \quad BO = 2$$

$$O_2^-: 1\sigma_g^2 \, 1\sigma_u^{*2} \, 2\sigma_g^2 \, 2\sigma_u^{*2} \, 3\sigma_g^2 \, 1\pi_u^4 \, 1\pi_g^{*3}, \quad BO = 1.5 \; .$$

This predicts the bond energies to fall in the order $O_2^+ > O_2 > O_2^-$. From the IP, EA, and $\mathcal{D}_e$ data, we can construct the following energy level diagram:

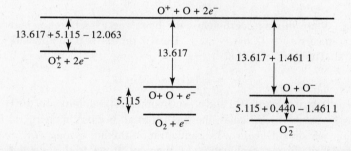

The O_2^+ bond energy is (13.617 + 5.115 − 12.063) eV = 6.669 eV, ~1.5 eV greater than $\mathcal{D}_e(O_2)$, while the O_2^- bond energy is (5.115 + 0.440 − 1.461 1) eV = 4.094 eV, ~1 eV less than $\mathcal{D}_e(O_2)$, as predicted.

➡ *RELATED PROBLEMS* 14.11, 14.12, 14.14

So far, we have considered only about a dozen of the several million molecules for which we would like to understand chemical bonding. Quantum chemistry would be hopeless if we had to consider every molecule from scratch. Fortunately, this is not the case, as our dozen molecules have begun to show. There is a manageably small number of *physically distinct types of bonding* that account for all molecular bonding. Here, we have seen the simplest type of *covalent* bonding between homonuclear atoms, so-called *homopolar* bonding. In the next section, we look at several of the more important types of interactions that, based on simple physical models, have distinct features and characteristic properties. From these (and a few more obscure types we will not consider), all of molecular structure can be discussed in a systematic and coherent way.

14.3 NAMED BONDS AND THE CONCEPT OF DELOCALIZATION

Heteropolar interactions between atoms of heteronuclear diatomics are probably the next simplest after the purely covalent interactions of homonuclear diatomics.

Recall the *ionic bond* in a molecule such as LiF. The dominant interaction is the Coulombic attraction between Li^+ and F^- caused by an electron transfer from Li to F, and the hallmark of the heteropolar bond is its dipole moment. The next chapter will discuss dipole moments in detail. Here, we recall that a dipole moment is, most simply, a consequence of an imbalance of the electronic charge distribution about the center of mass of the molecule.

To see how MO theory describes this imbalance, we will consider the simplest stable neutral ionic molecule, LiH. Our MO nomenclature must change for heteropolar bonding, since two symmetry elements, the inversion center (the *g/u* label) and the reflection plane perpendicular to the nuclear axis, are missing. We still have rotational (cylindrical) symmetry about the axis so that the σ, π, etc. notation is still valid. Consequently, we simply number MOs in order of increasing energy: 1σ, 2σ, 3σ, etc. (See also Example 14.2.)

The ground electronic state configuration of LiH is $1\sigma^2 2\sigma^2$, which says very little by itself. We know 1σ is nodeless, and 2σ has one nodal surface (not a flat plane, due to lack of symmetry), but only when we look at the *dominant AO contributions* to these MOs can we make physical statements about charge migration, etc. Figure 14.14 diagrams the H 1s AO energy, the Li 1s and 2s AO energies, and (qualitatively) the lowest three LiH MO energies. Consider first the 1σ MO. If we write this in the LCAO form

$$1\sigma = c_H 1s(H) + c_{Li} 1s(Li)$$

where the *c*'s are mixing coefficients, the huge energy difference between 1s(H) and 1s(Li) makes $c_H \cong 0$ and $c_{Li} \cong 1$. The 1σ orbital is predominantly the Li 1s core orbital. The next LCAO is

$$2\sigma = c'_H 1s(H) - c'_{Li} 1s(Li) \; ,$$

which, to no surprise, has $c'_H \cong 1$ and $c'_{Li} \cong 0$, and 2σ is predominantly the H 1s orbital.

But, as Figure 14.14 indicates, the 2s(Li) orbital is close in energy to the 1s(H). As a result, we should *begin* with configuration mixing among *all three* of these orbitals and write linear combinations of them all. If we do this, we find 1σ is still strongly dominated by 1s(Li), but 2σ and 3σ are both mixtures of 2s(Li) and 1s(H). These mixings are indicated in the figure by the lines connecting AOs to the central MOs. The 2σ MO is, however, mostly 1s(H) while 3σ is mostly 2s(Li).

Now we can categorize the $1\sigma^2 2\sigma^2$ molecular configuration. The $1\sigma^2$ electrons are core-localized on Li; they constitute Li^+ in the ionic bond. The $2\sigma^2$ electrons are localized on H; they constitute H^-.

Several factors, all related to configuration mixing, conspire to make the bond less than purely ionic, however. The 2s(Li) contribution to 2σ keeps the Li valence

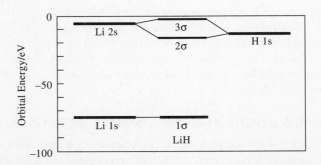

FIGURE 14.14 In forming the ionic bond of LiH, the AOs of each atom mix as shown to form the lowest-energy s MOs.

electron partly on Li, as does configuration interaction mixing other configurations ($1\sigma^2 2\sigma 3\sigma$, etc.) into the basic $1\sigma^2 2\sigma^2$ configuration. A recent detailed MO calculation of LiH ground and excited electronic states concluded that the ground state is 88% $1\sigma^2 2\sigma^2$, 5.3% $1\sigma^2 2\sigma 3\sigma$, and smaller fractions of several other excited configurations.

The last section in this chapter looks at molecular dissociation in more detail, but we can make an important observation here. Our LiH discussion has centered about the *electronic structure of the molecule in the vicinity of the observed bond length,* 1.595 7 Å. Here, the Li^+H^- configuration $1\sigma^2 2\sigma^2$ dominates. At large R, we should find the *neutral* atoms, Li + H, lowest in energy. And, in fact, we do. The same calculation that had 88% $1\sigma^2 2\sigma^2$ at R_e found only 0.5% of this configuration contributing at $R = 12$ au (6.35 Å) where $1\sigma^2 2\sigma 3\sigma$ was the dominant configuration. This is the configuration that most closely approximates $Li(1s^2 2s) + H(1s)$.

Ionic bonding is a consequence of the energy mismatch between the atomic orbitals forming a heteropolar bond. The *degree* of ionicity is a function of both the degree of mismatch and the nature (symmetry, size, and orientation) of the dominant AOs that interact to form each MO. Thus, atoms that are chemically similar in these respects form predominantly covalent, but slightly ionic, bonds (such as CO, NO, etc.), while great differences lead to enhanced ionic contributions (as in the hydrogen halides, metal hydrides, and all "salt" molecules, such as LiF). The elementary concept of *electronegativity,* a dimensionless number assigned to each element, has been used to predict, at least qualitatively, the degree of ionic bonding.

Several different scales of electronegativities have been proposed, and Table 14.4 lists a current set. All try to apportion the bonding between covalent and ionic contributions and to predict the *direction* of charge flow. In a heteropolar bond, the more electronegative element is relatively more negative than its partner. The greater the electronegativity difference, the greater the ionic character of the bond. But despite several empirical attempts to correlate electronegativities with measurable properties such as dipole moments, no fully satisfactory expressions have been

TABLE 14.4 Electronegativity values based on the scale of A. L. Allred and E. G. Rochow *J. Inorg. Nucl. Chem.* **5,** 264 (1958)

H 2.20																	He —
Li 0.97	Be 1.47											B 2.01	C 2.50	N 3.07	O 3.50	F 4.10	Ne —
Na 1.01	Mg 1.23											Al 1.47	Si 1.74	P 2.06	S 2.44	Cl 2.83	Ar —
K 0.91	Ca 1.04	Sc 1.20	Ti 1.32	V 1.45	Cr 1.56	Mn 1.60	Fe 1.64	Co 1.70	Ni 1.75	Cu 1.75	Zn 1.66	Ga 1.82	Ge 2.02	As 2.20	Se 2.48	Br 2.74	Kr —
Rb 0.89	Sr 0.99	Y 1.11	Zr 1.22	Nb 1.23	Mo 1.30	Tc 1.36	Ru 1.42	Rh 1.45	Pd 1.35	Ag 1.42	Cd 1.46	In 1.49	Sn 1.72	Sb 1.82	Te 2.01	I 2.21	Xe —
Cs 0.86	Ba 0.97	Lu 1.14	Hf 1.23	Ta 1.33	W 1.40	Re 1.46	Os 1.52	Ir 1.55	Pt 1.44	Au 1.42	Hg 1.44	Tl 1.44	Pb 1.55	Bi 1.67	Po 1.76	At 1.96	Rn —
Fr 0.86	Ra 0.97	Lr (1.2)	Unq	Unp	Unh	Uns	Uno	Une									

La 1.08	Ce 1.06	Pr 1.07	Nd 1.07	Pm 1.07	Sm 1.07	Eu 1.01	Gd 1.11	Tb 1.10	Dy 1.10	Ho 1.10	Er 1.11	Tm 1.11	Yb 1.06
Ac 1.00	Th 1.11	Pa 1.14	U 1.22	Np 1.22	Pu 1.22	Am (1.2)	Cm (1.2)	Bk (1.2)	Cf (1.2)	Es (1.2)	Fm (1.2)	Md (1.2)	No (1.2)

found or should be expected. Electronegativity is more a qualitative than quantitative concept.

Covalent and ionic are terms best applied to the local bonding between two atoms. In a polyatomic molecule, we may speak of one bond being ionic and another covalent (as in NaOH, for example: O—H is largely covalent, Na$^+$—OH$^-$ is largely ionic), but additional terminology is useful to discuss aspects of polyatomic bonding that are not satisfactorily covered by localized bonding models. These can all be classified under the effects of *charge delocalization,* and we discuss several prototypical cases in the remainder of this section.

The simplest polyatomic molecule is H_3^+. It is a known, stable species in gaseous hydrogen discharges, and spectroscopic experiments have confirmed its long predicted structure: an equilateral triangle.[10] How can we describe bonding among three nuclei given only two electrons? A naive guess might come up with the equilateral structure just on the basis of the "symmetric fairness" this nuclear arrangement presents: all three nuclei are equivalent and have an equal shot at electron attraction. This explanation is not that far off the mark, but it cannot be complete once we are told that adding one more electron (to make H_3) produces a *linear* molecule (which happens to be unstable, but less unstable than the triangular structure).

We can use a simple LCAO-MO method, as derived by Hückel (also of Debye–Hückel electrolyte theory fame), to understand the bonding in H_3^+, and, as we will see, to understand the bonding in many other molecules as well. We start with the LCAO built from 1s AOs on each of the three nuclei (and thus we expect to find three MOs). We write

$$\psi = c_1\phi_1 + c_2\phi_2 + c_3\phi_3 \tag{14.14}$$

where ϕ_i is the 1s orbital centered on nucleus i and c_i is a mixing coefficient. Hückel used this type of function in a variational calculation based on an independent electron Hamiltonian. He, as the first of several approximations, assumed that electron–electron interaction energies were less important than the primary electrostatic interactions of the Coulomb and exchange integrals we have already encountered.

The variational method leads to a set of equations, one for each atom:

$$c_1(H_{11} - ES_{11}) + c_2(H_{12} - ES_{12}) + c_3(H_{13} - ES_{13}) = 0$$

$$c_1(H_{21} - ES_{21}) + c_2(H_{22} - ES_{22}) + c_3(H_{23} - ES_{23}) = 0$$

$$c_1(H_{31} - ES_{31}) + c_2(H_{32} - ES_{32}) + c_3(H_{33} - ES_{33}) = 0$$

where H_{ii} (the diagonal elements of the **H** matrix) are Coulomb integrals, $H_{ij}, i \neq j$ (off the diagonal) are exchange integrals, (or, in the usual terminology of Hückel theory, *resonance* integrals), and S_{ij} are overlap integrals.

At the time of Hückel's work, computers did not exist, and direct calculation of the c's and E's of such a problem was a formidable task, especially for larger molecules. Dramatic approximations were called for. Hückel made three:

1. All off-diagonal overlap integrals $S_{ij}, i \neq j$, are assumed to be zero. The diagonal ($i = j$) overlaps are taken to be unity.
2. All diagonal Coulomb integrals H_{ii} are assumed to be equal (and are given the common symbol α).

---

[10]Actually, H_3^+ undergoes a complex internal motion that averages to equilateral. This motion, due to what is called a *Jahn–Teller distortion,* has its origin in the degenerate electronic ground state of the molecule.

3. All off-diagonal resonance integrals H_{ij}, $i \neq j$, are assumed to be
    (i) *zero* if atoms i and j are not bonded neighbors
    (ii) equal to a common value β if atoms i and j are neighbors.

These are about the most drastic assumptions that one can make and still retain some semblance of reality. Assumption (1) is sensible for diagonal overlaps, since it corresponds to a normalized set of AOs. For off-diagonal overlaps, it is severe, since it is now known that neighboring atoms can have S integrals in the range 0.1–0.3. Assumption (2) makes sense if all the atoms are the same *and* are expected by symmetry and chemical reactivity to have identical (or at least very similar) charges. Assumption (3) is difficult to evaluate *a priori,* but it tries to express the R dependence of the important resonant exchange interaction. Recall that resonance integrals (compare Eq. (13.25) or Equations (14.8a, b) and the discussion of them) are *negative,* and thus $\beta < 0$.

With these assumptions, the variational problem is reduced to solving

$$c_1(\alpha - E) + c_2\beta + c_3\beta = 0 \tag{14.15a}$$

$$c_1\beta + c_2(\alpha - E) + c_3\beta = 0 \tag{14.15b}$$

$$c_1\beta + c_2\beta + c_3(\alpha - E) = 0 . \tag{14.15c}$$

The energy eigenvalues are solutions to the *secular determinant* (a cubic polynomial here) equation

$$\begin{vmatrix} \alpha - E & \beta & \beta \\ \beta & \alpha - E & \beta \\ \beta & \beta & \alpha - E \end{vmatrix} = 0 .$$

We substitute each eigenvalue in turn into Equations (14.15a, b, c) to find the c's associated with each energy. If we divide the secular determinant equation by β and define the dimensionless quantity $x = (\alpha - E)/\beta$, the equation becomes

$$\begin{vmatrix} x & 1 & 1 \\ 1 & x & 1 \\ 1 & 1 & x \end{vmatrix} = 0 \tag{14.16}$$

which, when expanded, is the cubic equation $x^3 - 3x + 2 = 0$ with the roots $x = 1, 1, -2$. Thus, the energies are, in increasing order (remember that $\beta < 0$),

$$E' = \alpha + 2\beta \text{ (lowest)}$$

$$E'' = \alpha - \beta \quad \text{and} \quad E''' = \alpha - \beta \text{ (degenerate pair)} .$$

To find the c's, we first substitute $E' = \alpha + 2\beta$ into Equations (14.15a, b, c). This leads to an indeterminant set of equations, however, as discussed in Section 13.3. We can take one such equation, say that from Eq. (14.15a),

$$-2c_1' + c_2' + c_3' = 0$$

and the normalization condition,

$$c_1'^2 + c_2'^2 + c_3'^2 = 1 \tag{14.17}$$

and find that the choice $c_1' = c_2' = c_3' = 1/\sqrt{3}$ is satisfactory. Thus, the lowest energy MO is the nodeless function

$$\psi' = \frac{1}{\sqrt{3}}(\phi_1 + \phi_2 + \phi_3) \tag{14.18a}$$

with an energy $E' = \alpha + 2\beta$.

For the degenerate pair of energies, $\alpha - \beta$, an infinite number of c_i choices can be found.[11] Two that have appealing symmetry properties (we are looking for one node) are

$$\psi'' = \frac{1}{\sqrt{2}}(\phi_1 - \phi_2) \quad (14.18b)$$

and

$$\psi''' = \frac{1}{\sqrt{6}}(\phi_1 + \phi_2 - 2\phi_3) \ . \quad (14.18c)$$

These three MOs are shown in Figure 14.15. Note that ψ', the lowest energy MO, extends over the entire molecule. In H_3^+, it is doubly occupied, and we see how delocalization works here; electron density is spread symmetrically over the entire nuclear framework that is chemically bonded.

The nodal properties of ψ'' and ψ''' are also interesting. In ψ'', $c_3'' = 0$ so that a nodal plane cuts through nucleus 3 and bisects the 1-2 axis, since $c_1'' = -c_2''$. In ψ''', the plane is rotated 90°.

We cannot use the σ, π, etc., symmetry labels for triangular H_3^+ MOs (or for any nonlinear nuclear framework). Instead, other symmetry symbols are used. These were first introduced by Mulliken as he was inventing much of MO theory. The rules for these symbols are:

1. If the MO is nondegenerate, use the letters *a* or *b* (as determined below); if doubly degenerate, use *e*; if triply degenerate, use *t* (in older literature, *f* is used).

FIGURE 14.15 Contour diagrams of the three lowest-energy MOs of triangular H_3^+ show symmetry features common to any X_3 equilateral triangle system. The algebraic signs of the MOs are distinguished by differing shadings.

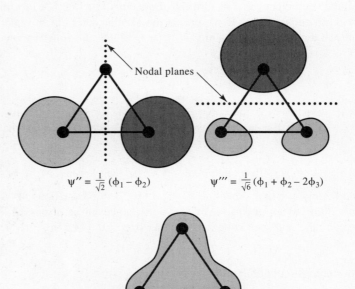

---

[11] This should not upset you. It is just a manifestation of the fact that any linear combination of solutions to Schrödinger's equation is also a solution. It is a common outcome for degenerate wavefunctions.

14.3 NAMED BONDS AND THE CONCEPT OF DELOCALIZATION

2. Look for an axis of rotational symmetry in the molecule. (Often, many can be found, such as a three-fold axis piercing the H_3^+ triangular plane, or a two-fold axis passing through one nucleus and bisecting the opposite bond.) Find the "largest-fold" such axis.[12] If the MO is nondegenerate and *symmetric* (does not change sign) with respect to one symmetry increment about this axis, label it *a;* if antisymmetric, label it *b*.
3. Use a subscript 1 or 2 on *a* or *b only* to denote, respectively, symmetric or antisymmetric with respect to a two-fold rotation about any axis perpendicular to the one used in (2). If no such symmetry exists, consider reflection in a plane *containing* that axis, and assign 1 and 2 if such a symmetry plane exists. Otherwise, the 1 and 2 subscripts will not apply.
4. Use the *g* and *u* in symbols as usual to indicate symmetric or antisymmetric behavior with respect to inversion, if this symmetry operation exists.
5. Add a superscript prime or double prime to all letters to indicate symmetric or antisymmetric behavior with respect to reflection in a plane *perpendicular* to the major axis used in (2). If no such symmetry plane exists, these superscripts do not apply.

For triangular H_3^+, the symmetry elements are shown in Figure 14.16. Using this figure with the MOs of Figure 14.15 and the rules above, we see that

ψ' has a_1' symmetry
 a, since symmetric with respect to the three-fold rotation axis,

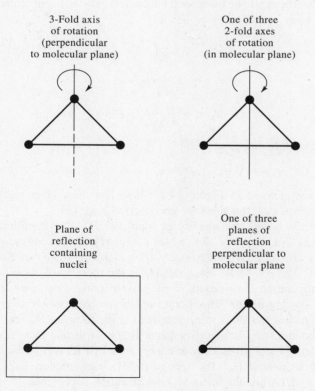

FIGURE 14.16 The symmetry elements of equilateral triangle H_3^+ include the rotation axes and reflection planes shown. The major rotation axis is the 3-fold symmetry axis perpendicular to the molecular plane.

---

[12] An *n*-fold axis of symmetry gives you what you started with after a rotation of $2\pi/n$ about that axis. The sign of the wavefunction is ignored when searching for symmetry axes and reflection planes.

1, since symmetric with respect to the two-fold axis, and
', since symmetric with respect to reflection in the molecular plane.
ψ'' and ψ''' both have e' symmetry
e, since doubly degenerate, and, again,
', since symmetric with respect to reflection in the molecular plane.

Thus, just as we wrote H_2 ($1\sigma_g^2$) to indicate the MO configuration for the ground state of H_2, we write H_3^+ ($1a_1'^2$).

It is helpful in picturing MO shapes to keep these nomenclature rules in mind, but, on a first encounter with them, you may want to ignore their meaning, view them as simply labels, and concentrate on the overall bonding or antibonding nodal properties of a given set of MOs.

We have found the MO structure for triangular H_3^+, but we have yet to prove it is a lower energy structure than the linear one. The Hückel secular determinant for linear H_3^+ (or, for that matter, for linear A_3 where A is any atom with one valence AO of importance) is

$$\begin{vmatrix} x & 1 & 0 \\ 1 & x & 1 \\ 0 & 1 & x \end{vmatrix} = 0 \qquad (14.19)$$

which has zeroes in the 1,3 and 3,1 corners, indicating the lack of direct bonding between atoms 1 and 3. The cubic equation here is $x^3 - 2x = 0$, and the solutions are $x = -\sqrt{2}, 0, +\sqrt{2}$, or energies $\alpha + \sqrt{2}\beta$, α, and $\alpha - \sqrt{2}\beta$. The LCAO-MO coefficients can also be found, and we can return to the linear molecule symmetry labels. We find

$$\psi' = 1\sigma_g = \frac{1}{\sqrt{2}}\left(\phi_1 + \frac{1}{\sqrt{2}}\phi_2 + \phi_3\right)$$

$$\psi'' = 1\sigma_u^* = \frac{1}{\sqrt{2}}(\phi_1 - \phi_3)$$

$$\psi''' = 2\sigma_g = \frac{1}{\sqrt{2}}\left(\phi_1 - \frac{1}{\sqrt{2}}\phi_2 + \phi_3\right)$$

with corresponding energies

$$E' = \alpha + \sqrt{2}\beta, \qquad E'' = \alpha, \quad \text{and} \quad E''' = \alpha - \sqrt{2}\beta \ .$$

The MOs are pictured in Figure 14.17; note how they resemble the first three particle-in-a-box wavefunctions we encountered long ago.

Now we can assess the stability of triangular H_3^+. The configuration $1a_1'^2$ has a total energy $2(\alpha + 2\beta) = 2\alpha + 4\beta$. Linear H_3^+ in the configuration $1\sigma_g^2$ has a total energy $2(\alpha + \sqrt{2}\beta) = 2\alpha + 2\sqrt{2}\beta$, which is *greater* than $2\alpha + 4\beta$ (since $\beta < 0$) by $1.17\,|\beta|$. Thus, triangular H_3^+ is the more stable (lower total energy) nuclear configuration *as long as the H—H bond lengths are the same in the triangular and linear configurations*. This point, which we have tacitly overlooked, is, of course, the rest of the B–O story. But even if we compare H_3^+ total energies at a variety of bond lengths to those for linear H_3^+, we will find that the equilibrium R for triangular H_3^+ still gives a lower energy than that for *any* linear configuration.

What about neutral H_3? The triangular MO configuration $1a_1'^2 1e'$ has a total energy $3\alpha + 3\beta$, while the linear MO configuration $1\sigma_g^2 1\sigma_u^*$ has $3\alpha + 2\sqrt{2}\beta$. This predicts linear H_3 to be *higher* in energy than triangular H_3 by $0.171\,|\beta|$, but a full B–O calculation, including all the details ignored by Hückel, shows, in accord with experiment, that *no* configuration of H_3 is stable, but that a linear configuration is

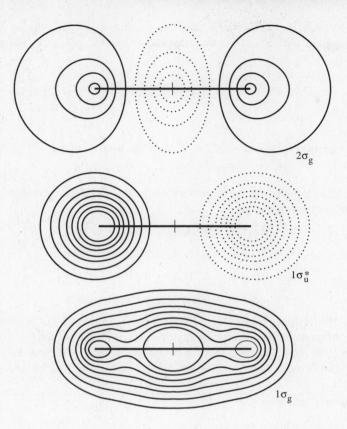

FIGURE 14.17 Contour diagrams of the three lowest-energy linear A_3 MOs are shown here. The solid and dotted contours denote regions of opposite algebraic sign.

less unstable than any others (at internuclear distances comparable to chemical bond lengths).

The electron delocalization in Figures 14.15 and 14.17 has important and pervasive consequences for bonding. The analogy with lowering particle-in-a-box energies as the box is lengthened cannot be overemphasized. *Conjugated polyenes*, hydrocarbons of alternating single and double carbon–carbon bonds are dramatic examples of delocalization stability. (See Problem 14.36 for a particularly simple model.) We focus on the p electrons of carbon, and, in particular, those *not* used in forming the covalent C—C bond that is locally σ in symmetry. These electrons form the bonds that are locally of π symmetry (with a nodal plane containing the carbon–carbon internuclear axis).

We start with ethylene and the Hückel approximation. Each carbon contributes one p AO to the C—C π bond (that, of course, is the "second" bond of the ethylene C=C double bond). The secular determinant is simply

$$\begin{vmatrix} x & 1 \\ 1 & x \end{vmatrix} = 0$$

(as, in fact, it is for H_2, Li_2, etc., using s AOs—see Problem 14.16) with energy solutions $\alpha + \beta$ and $\alpha - \beta$. The π-type MOs resemble the $1\pi_u$ and $1\pi_g^*$ MOs of Figure 14.7. With two electrons in the lower energy MO, we predict a π-bond energy contribution of $2\alpha + 2\beta$.

Now consider 1,3-butadiene. This molecule is rather like two ethylenes stuck together end to end: . We inquire if the butadiene π-bond energy is different from $2(2\alpha + 2\beta)$, the π-bond energy of two isolated ethylenes, to see if conjugation delocalization has added an appreciable stabilization. The Hückel secular equation is

$$\begin{vmatrix} x & 1 & 0 & 0 \\ 1 & x & 1 & 0 \\ 0 & 1 & x & 1 \\ 0 & 0 & 1 & x \end{vmatrix} = 0$$

a quartic equation ($x^4 - 3x^2 + 1 = 0$) with the four roots

$$x = \pm[(3 \pm \sqrt{5})/2]^{1/2} = \pm 1.618, \pm 0.618 \ .$$

The four MOs (labeled by the symmetry elements of the trans configuration) are, in order of increasing energy,

$$1a_u = 0.371\,\phi_1 + 0.600\,\phi_2 + 0.600\,\phi_3 + 0.371\,\phi_4$$

$$1b_g = 0.600\,\phi_1 + 0.371\,\phi_2 - 0.371\,\phi_3 - 0.600\,\phi_4$$

$$2a_u = 0.600\,\phi_1 - 0.371\,\phi_2 - 0.371\,\phi_3 + 0.600\,\phi_4$$

$$2b_g = 0.371\,\phi_1 - 0.600\,\phi_2 + 0.600\,\phi_3 - 0.371\,\phi_4 \ .$$

Figure 14.18 illustrates these MOs. The MO configuration $1a_u^2\,1b_g^2$ for the ground-state π configuration has a total energy $2(\alpha + 1.618\,\beta) + 2(\alpha + 0.618\,\beta) = 4\alpha + 4.472\,\beta$. Two ethylenes have a π energy of $4\alpha + 4\beta$, and the difference, $0.472\,|\beta|$, in favor of butadiene, is the *delocalization stabilization energy*.

EXAMPLE 14.4

Is cyclobutadiene stabilized by π-electron delocalization?

SOLUTION Cyclobutadiene, ☐, is also related to two separated ethylenes, and we can again use the reference energy $4\alpha + 4\beta$ to decide the question. The Hückel determinant differs from the butadiene determinant only in the 1, 4 and 4, 1 corners, since cyclobutadiene is cyclic:

$$\begin{vmatrix} x & 1 & 0 & 1 \\ 1 & x & 1 & 0 \\ 0 & 1 & x & 1 \\ 1 & 0 & 1 & x \end{vmatrix} = 0 \ .$$

A 4×4 determinant can be expanded in terms of 3×3 subdeterminants (called *minors* or *cofactors* in the language of linear algebra):

$$\begin{vmatrix} a_{11} & a_{12} & a_{13} & a_{14} \\ a_{21} & a_{22} & a_{23} & a_{24} \\ a_{31} & a_{32} & a_{33} & a_{34} \\ a_{41} & a_{42} & a_{43} & a_{44} \end{vmatrix} = a_{11}\begin{vmatrix} a_{22} & a_{23} & a_{24} \\ a_{32} & a_{33} & a_{34} \\ a_{42} & a_{43} & a_{44} \end{vmatrix} - a_{21}\begin{vmatrix} a_{12} & a_{13} & a_{14} \\ a_{32} & a_{33} & a_{34} \\ a_{42} & a_{43} & a_{44} \end{vmatrix}$$

$$+ a_{31}\begin{vmatrix} a_{12} & a_{13} & a_{14} \\ a_{22} & a_{23} & a_{24} \\ a_{42} & a_{43} & a_{44} \end{vmatrix} - a_{41}\begin{vmatrix} a_{12} & a_{13} & a_{14} \\ a_{22} & a_{23} & a_{24} \\ a_{32} & a_{33} & a_{34} \end{vmatrix}$$

Note how each 3×3 determinant is multiplied by an element in the first column of the 4×4, and each 3×3 contains the elements of the 4×4 *not* in the same row or column as the single element multiplying it. Any row or column in the 4×4 can be selected as the elements multiplying the 3×3 cofactors. It is therefore a computational help to pick a row or column with as many zero elements as possible, but here, all rows and columns have only one zero.

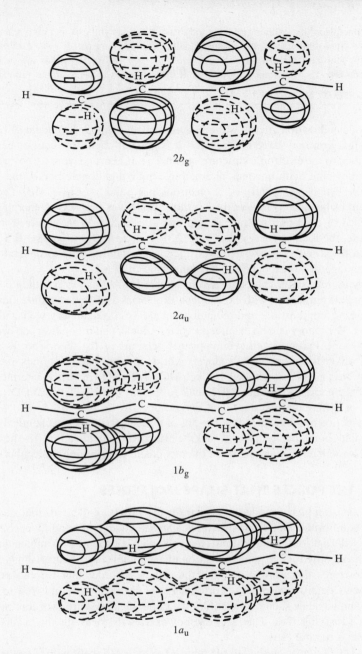

FIGURE 14.18 Lowest-energy π-type MOs for 1,3-butadiene.

Expansion of our determinant yields the quartic polynomial

$$x^4 - 4x^2 = 0 \quad \text{or} \quad (x+2)(x-2)x^2 = 0$$

so that the energy roots are $x = 0, 0, +2$, and -2 or, in order of increasing energy, $E = \alpha + 2\beta, \alpha, \alpha,$ and, $\alpha - 2\beta$. We have one π electron per C, for a total of four, and the first two go into the lowest energy MO for an energy contribution $2(\alpha + 2\beta)$. The next two enter the degenerate pair of MOs at $E = \alpha$. Hückel theory has nothing to say about spins, and the exact assignment does not alter the Hückel energy. A Hund's rule argument would place these electrons in a triplet configuration, one in each orbital. But whatever the spin assignments, the total Hückel energy is $2(\alpha + 2\beta) + 2\alpha = 4\alpha + 4\beta$, exactly the same as two isolated ethylenes. Consequently, we can predict that cyclobutadiene is not stabilized by π delocalization, a conclusion that explains its long and difficult initial synthesis and high reactivity. The definitive structural study of cyclobutadiene [S. Masamune, F. A. Souto-Bachiller, T. Machiguchi, and J. E. Bertie *J. Am. Chem. Soc.* **100** 4889 (1978)] has the succinct title, "Cyclobutadiene is Not Square." Other work indicates that it has a singlet ground state.

> RELATED PROBLEMS 14.17, 14.18, 14.22, 14.37

This delocalization energy is the MO equivalent of the VB concept of *resonance energy* that you may have encountered in a general chemistry course, particularly with regard to the electronic structure of benzene. If benzene were a true "cyclohexatriene" molecule, with unequal, alternating, single and double bonds, the π energy would be $3(\alpha + \beta)$ bonds $\times$ 2 electrons per bond $= 6\alpha + 6\beta$. The Hückel treatment of benzene (Problem 14.22), however, gives one MO at energy $\alpha + 2\beta$, a pair at $\alpha + \beta$, a pair at $\alpha - \beta$, and one at $\alpha - 2\beta$. With six electrons to pair into these, the total π energy is $2(\alpha + 2\beta) + 4(\alpha + \beta) = 6\alpha + 8\beta$. Compared to "cyclohexatriene," benzene has an increased stability (due to delocalization in MO language or to resonance in VB language) of $2|\beta|$.

Many more systems can be treated by simple Hückel theory, yielding predictions in reasonable agreement with experiment. Properties as diverse as ionization energy, enthalpies of combustion, and reduction potentials correlate very well with Hückel energies. The theory can be extended to heteroatoms, and σ bonding can be incorporated as well. Parametrized semiempirical versions of the theory are also in use, and an advertising executive's dream-world of acronyms describes the various elaborations: EHMO (extended Hückel MO), MINDO (modified intermediate neglect of differential overlap), MINDO/2 (a variant of MINDO), CNDO (complete neglect of differential overlap), and on and on.

This section has considered the origins of three main themes in chemical bonding: the covalent bond, the ionic bond, and the effects of delocalization. The next section considers the interplay among these interactions in determining molecular structure.

14.4 THE FORCES THAT SHAPE MOLECULES

One of the triumphs of modern chemistry is the variety of methods that successfully predict equilibrium molecular structures. Our knowledge of bonding is so advanced that it is the curiously anomalous (for example, H_2 bound as a molecular adduct rather than a dihydride in some transition metal complexes) or surprisingly beautiful (C_{60}) structures that attract attention these days in order that we can pounce on unexpected details and quirks. Fortunately for those of us employed to do such studies and advance such explanations, the curiosities and surprises continue to turn up at a satisfying rate. This section, however, will focus on the mainstream of molecular structure.

Why is CO_2 linear, with C in the middle? Why is OF_2 bent with O in the middle? These are the first kind of questions we need to address. Of course, the glib answer is, "because those structures minimize the total molecular energy." That statement is true, but it is not predictive. What we need is an overview of energy changes with changes in molecular structure. The first detailed overview was Walsh's in the early 1950s. *Walsh's rules* for triatomic molecular geometries are the result. As you read about these rules below, keep in mind their generic nature. They apply with reasonable accuracy to neutral molecules, closed shell or free radical, as well as to molecular ions, positive or negative, and once a prototype electron configuration has been chosen, its geometry is often applicable to every *isovalent analog*. *Isovalent* implies an identical MO symmetry and population of *valence* electrons, as in the isovalent sequences CF_4, CCl_4, . . . , SiF_4, $SiCl_4$, etc. or CO_2, N_2O, NCO^-, NO_2^+, etc.

14.4 THE FORCES THAT SHAPE MOLECULES

We begin with triatomic dihydrides, first in a linear and then in a bent geometry. We will consider the rough shapes and (exact) symmetries of the valence MOs that can be formed from the 1s AOs of each H and the ns and np valence AOs of the central atom, symbolized A, as in AH_2. To be specific, we will use an MO numbering scheme based on first-row A atoms, so that the 2s and 2p orbitals will be important, but we will quickly generalize so that isovalent sequences (such as NH_2, PH_2 or OH_2, SH_2, etc.) can be included.

In Figure 14.19(a), we look at the seven MOs that can be made from the seven AOs of importance in a symmetric, linear, HAH arrangement (on A: 1s, 2s, $2p_x$, $2p_y$, $2p_z$; on each H: 1s). The major AO contributions to each MO are shown shaded, while minor contributions are outlined only, and all AO sizes are schematic. The same sorts of combinations for a bent geometry are shown in Figure 14.19(b). In this figure, the a and b symmetry symbols refer to the two-fold symmetry axis

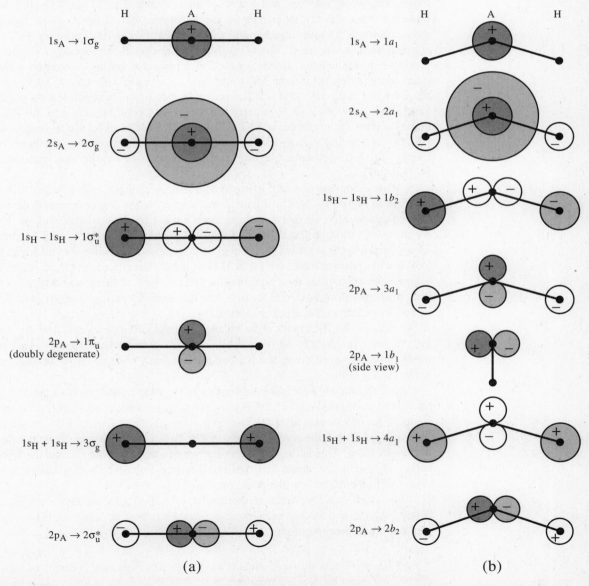

FIGURE 14.19 (a) Atomic parentage for AH_2 linear geometry MOs. (b) Atomic parentage for AH_2 bent geometry MOs.

bisecting the bond angle, and the 1 and 2 subscripts refer to reflection symmetry in a plane that bisects the bond angle.

We will want to see how the energies of these orbitals change as the molecule is bent away from linearity, but first note a few general features. First, the linear molecule's doubly degenerate $1\pi_u$ MO becomes two nondegenerate MOs ($3a_1$ and $1b_1$) when bent. Next, note how symmetry changes on bending can alter the amount of the secondary AO contributions to the MO. For example, the $2p_A$ orbitals do not contribute to the linear $3\sigma_g$ MO (by symmetry), but, on bending, a $2p_A$ orbital in the plane of the molecule can contribute to the $4a_1$ MO.

Now, one by one, we can see what happens to each MO's energy if we consider the changes in nodal structure and degree of AO overlap that occur on bending. The $(1\sigma_g-1a_1)$ MO is so predominantly A-atom core in nature that its energy is virtually unchanged. The $(2\sigma_g-2a_1)$ MO, on bending, brings the H-atom 1s orbitals closer, enhancing overlap, and lowering the orbital energy. The $(1\sigma_u^*-1b_2)$ MO, however, *rises* sharply in energy as the H atoms, with a nodal plane between them, are brought closer together and the favorable 1s–2p–1s overlap in the linear configuration is worsened. The $(1\pi_u-3a_1)$ MO energy is appreciably *lowered* for just the opposite reasons. Bending produces a favorable 1s–2p–1s overlap without intervening nodes. (Different $2p_A$ orbitals are involved in $1b_2$ and $3a_1$. Refer to the figures.) The $(1\pi_u-1b_1)$ MO that lies on either side of the bent molecule's plane is least affected by bending. Symmetry precludes H 1s contributions, and this MO is largely a pure $2p_A$ nonbonding MO. The $(3\sigma_g-4a_1)$ MO should rise in energy, as a pair of antibonding nodes develops on mixing $2p_A$ character into the bent MO. Finally, $(2\sigma_u^*-2b_2)$ should rise due to the disruption of overlap and unfavorable H–H nodal plane.

A *Walsh diagram* is a plot of these MO energy changes as a function of bond angle, as in Figure 14.20. This figure is qualitative, but the degree of energy changes are semiquantitative in that they have been guided by detailed calculations. Now we can move through the Periodic Table and populate these MOs with electrons, seeing what orbitals are filled and what geometry the corresponding dihydrides prefer.

We will consider BeH_2, CH_2, and H_2O in detail. Beryllium dihydride, with four valence electrons,[13] has the configuration $1\sigma_g^2 2\sigma_g^2 1\sigma_u^{*2}$ if linear and $1a_1^2 2a_1^2 1b_2^2$ if bent. The controlling MO here is $(1\sigma_u^*-1b_2)$ that strongly prefers linearity, and BeH_2 is, in fact, linear in this electronic state.

Next, consider H_2O, with eight valence electrons. If linear, it is $1\sigma_g^2 2\sigma_g^2 1\sigma_u^{*2} 1\pi_u^4$; if bent, $1a_1^2 2a_1^2 1b_2^2 3a_1^2 1b_1^2$. While $(1\sigma_u^*-1b_2)$ wants to be linear, $(1\pi_u-3a_1)$ wants to be bent even more, and the bending tendency of $(2\sigma_g-2a_1)$ aids this trend. Thus, H_2O is bent.

Both BeH_2 and H_2O are stable molecules due to their closed ground-state molecular orbital configurations. But CH_2, a fascinating transient species, has a more complex set of electron assignments.

If linear, CH_2 has the configuration $1\sigma_g^2 2\sigma_g^2 1\sigma_u^{*2} 1\pi_u^2$. But where do we put those two π electrons? This is the same question we encountered with O_2, and the lowest energy answer is the same: they go into different $1\pi_u$ MOs with unpaired spins. Linear CH_2 should be a triplet-spin species.

If bent, CH_2 could be assigned the singlet, all paired spin, configuration $1a_1^2 2a_1^2 1b_2^2 3a_1^2$, which, like H_2O, (since $1b_1$ does not do much of anything on H_2O) would be a favorable geometry.

---

[13]We will ignore the A atom's $1s^2$ electrons, which reside rather harmlessly in the $1\sigma_g-1a_1$ MO. As with most of chemistry, only the valence electrons are important.

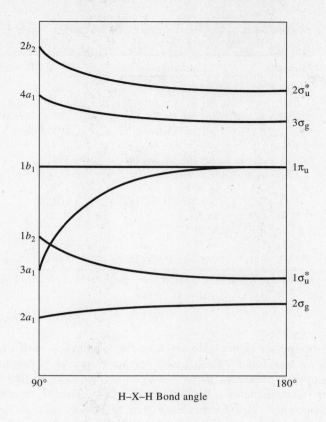

FIGURE 14.20 The Walsh orbital correlation diagram for the dihydrides, plotted as a function of the H—X—H bond angle.

It would seem that, to test these two choices, all we need to do is determine CH_2's magnetic properties. If magnetic, then it is a *linear triplet;* if nonmagnetic, a *bent singlet.* The CH_2 radical can be made readily from compounds such as CH_2N_2 and CH_2CO and has been studied for roughly a century, but only in the last decade or so have definitive answers from both very sophisticated experiments and theoretical calculations settled the question.

The current view of CH_2 has its ground electronic state a *bent triplet* derived from the configuration $1a_1^2 \, 2a_1^2 \, 1b_2^2 \, 3a_1^1 \, 1b_1^1$. The bond angle is now accurately known to be $133.84° \pm 0.05°$ and the equilibrium C—H bond length is $1.074\,8 \pm 0.000\,4$ Å. The precision of these numbers indicates the detail of information available on this molecule: CH_2 is among the best understood transient polyatomic species. The reason a simple Walsh argument is misleading (and led to twenty years of debate and confusion) is the close energetic proximity of an excited bent singlet electronic state.[14] Moreover, the tendencies for $(1\sigma_u^*$–$1b_2)$ to be linear and $(1\pi_u$–$3a_1)$ to be bent are qualitative ones in Walsh terminology, and only calculations can decide which is dominant, especially given the other energy effects (electron kinetic energy in particular) that are almost impossible to estimate with inspired guesses.

Table 14.5 summarizes configuration and geometry data for a variety of dihydrides, some in excited as well as ground states. Particularly striking is the decrease in bond angle as one goes down a column of the Periodic Table. But, except for

---

[14]The lowest-energy singlet state is a mixture of the MO configurations $1a_1^2 \, 2a_1^2 \, 1b_2^2 \, 3a_1^2$ and $1a_1^2 \, 2a_1^2 \, 1b_2^2 \, 1b_1^2$ and is known to lie only 0.390 eV above the triplet. It has a bond angle of 105°.

TABLE 14.5 Triatomic dihydride electron configurations and geometries

Molecule	Configuration	H—X—H angle
BeH_2	$1\sigma_g^2\, 2\sigma_g^2\, 1\sigma_u^{*2}$	180°
BH_2	$1a_1^2\, 2a_1^2\, 1b_2^2\, 3a_1^1$	131°
CH_2	$1a_1^2\, 2a_1^2\, 1b_2^2\, 3a_1^1\, 1b_1^1$	133.8° (triplet)
	$1a_1^2\, 2a_1^2\, 1b_2^2\, 3a_1^2$	105° (singlet)
NH_2	$1a_1^2\, 2a_1^2\, 1b_2^2\, 3a_1^2\, 1b_1^1$	103.4°
	$1a_1^2\, 2a_1^2\, 1b_2^2\, 3a_1^1\, 1b_1^2$	144°
PH_2	$a_1^2\, b_1^1$	91.5°
	$a_1^1\, b_1^2$	123.1°
H_2O	$1a_1^2\, 2a_1^2\, 1b_2^2\, 3a_1^2\, 1b_1^2$	105.2°
H_2S	$a_1^2\, b_1^2$	92.2°
H_2Se	$a_1^2\, b_1^2$	91°
H_2Te	$a_1^2\, b_1^2$	90.3°
FH_2^+	$1a_1^2\, 2a_1^2\, 1b_2^2\, 3a_1^2\, 1b_1^2$	113.9°

Note: For PH_2, H_2S, H_2Se, and H_2Te, only representative valence MO labels are shown.

BeH_2, all the species in the table are bent. For dihydrides, Walsh's rules do not seem to be very profound. Walsh's interpretation, however, is profound.

For nonhydride triatomics, the Walsh arguments are more useful, since there are many more AO combinations. We will take CO_2 (linear) and O_3 (bent) as prototypes for MO shapes and relative energies. Figure 14.21 does this, and gives the MO energies of an Extended Hückel calculation, which is basic Hückel theory extended to include dissimilar atoms and AOs other than those of one valence type. Table 14.6 gives the primary AO contributions to these MOs.

Look first at CO_2. The lowest four MOs are all σ type, ordered by nodal properties. Next come π bonding, π nonbonding, and π antibonding MOs, the latter unoccupied in CO_2. Finally, the (unoccupied) four and five node σ orbitals round out the picture. Note in particular the nonbonding[15] nature of the $1\pi_g$ HOMO.

Next, compare the CO_2 MOs to those of O_3 and try to anticipate the Walsh correlation diagram. Bending once again disrupts the π-orbital degeneracy, as it did for the dihydrides. Ozone has two more electrons than CO_2, and, if O_3 had been linear, those electrons would enter the $2\pi_u$ antibonding MO. But by bending, one of the $2\pi_u$ MOs becomes the stabilized $6a_1$ MO. In fact, the HOMO of O_3 is a nonbonding MO ($1a_2$) derived from one of the $1\pi_u$ MOs in the linear symmetry.

Figure 14.22 gives the full Walsh diagram correlating MOs with bond angle for symmetric AB_2 triatomics. (Here, as in Figure 14.21, the 1s core MOs, $1a_1$, $2a_1$, and $1b_2$, have been omitted.) Energy changes of the π-type MOs with decreasing bond angle are again the most dramatic. On the basis of these changes, we deduce Walsh's rules for non-hydride triatomics:

1. up to 16 valence electrons—linear
   (CO_2 and N_2O are 16 e^- examples)

---

[15]And note that we will drop the * superscript for these molecules. With three atoms contributing p-type AOs, we get three MOs—bonding, nonbonding, and antibonding—of u, g, and u symmetry in a centro-symmetric triatomic. The * is usually reserved to denote true antibonding character, which would be somewhat misleading here if applied to the $1\pi_g$ MO.

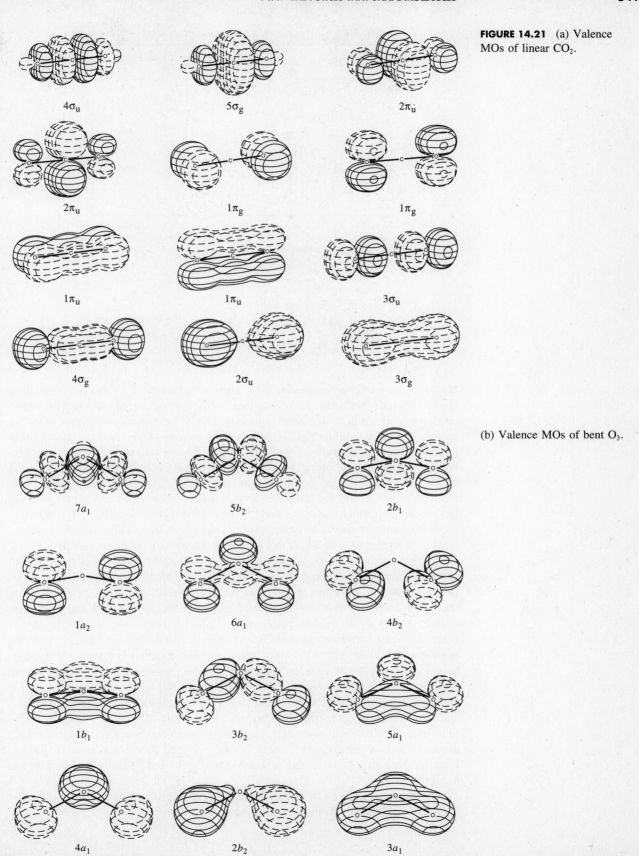

FIGURE 14.21 (a) Valence MOs of linear CO_2. (b) Valence MOs of bent O_3.

TABLE 14.6 Primary AO contributions to linear AB_2 MOs

MO	AO parentage on atom:		
	A	B	A
$4\sigma_u^*$	$+np_z$	$+n'p_z$	$+np_z$
$5\sigma_g$	$+np_z$		$-np_z$
$2\pi_u$	$+np_{x,y}$	$-n'p_{x,y}$	$+np_{x,y}$
$1\pi_g^*$	$+np_{x,y}$		$-np_{x,y}$
$1\pi_u$	$+np_{x,y}$	$+n'p_{x,y}$	$+np_{x,y}$
$3\sigma_u^*$	$+np_z$	$-n'p_z$	$+np_z$
$4\sigma_g$	$+ns$	$-n's$	$+ns$
$2\sigma_u^*$	$+ns$		$-ns$
$3\sigma_g$	$+ns$	$+n's$	$+ns$

Note: The A—B—A axis is the z direction. Atom A has valence electrons with principle quantum number n, and atom B has valence electrons with principle quantum number n'.

2. 17–20 valence electrons—bent
   (NO_2, 17 e^-, 134° bond angle; O_3, 18 e^-, 117°; ClO_2, 19 e^-, 117°; OF_2, 20 e^-, 103°)
3. 22 electrons—linear
   (I_3^- and other trihalide anions, KrF_2, XeF_2).

No 21 valence electron molecules are known with structural certainty. Trihalogens (F_3, etc.) would be in this class, but they are not sufficiently stable to be readily produced and studied. They are probably highly bent. Similarly, photoelectron spectra of rare gas difluorides suggest that the *cations* (21 e^- XeF_2^+, for example) are bent. The rare gas trimers (Ne_3, Ar_3, etc.) are 24 e^- equilateral triangular species, but they are bound by forces very different from those of ordinary chemical bonds.

The 22 e^- species are particularly interesting. We can understand why C is in the middle of CO_2 on the basis of any of a number of elementary bonding ideas, but which atom should be in the middle in the known species $ClIBr^-$? The way the formula is written gives away the answer: iodine is in the middle. But why? If we look back to the CO_2 MO pictures in Figure 14.20 and count up eleven MOs to accommodate all 22 electrons, we find $5\sigma_g$ is the HOMO and $4\sigma_u$ is the LUMO, *L*owest (energy) *U*noccupied *M*olecular *O*rbital. Note the large amount of electron density on the central atom in $5\sigma_g$. This is a highly antibonding MO, and anything the molecule can do to lower the amount of charge on this center is stabilizing. In $ClIBr^-$, I goes in the middle because *it is the least electronegative of the three*. Both Cl and Br withdraw electron density from the central I. Likewise, F—Xe—F is more stable than Xe—F—F, and, perhaps most remarkably of all, the weakly bound molecule ArClF is linear with atomic arrangement Ar—Cl—F rather than Ar—F—Cl.

Walsh arguments can be extended to polyatomics with more than three atoms and other possible geometries, predicting, for example, BF_3 to be planar and NF_3 to be pyramidal. We will not present these arguments here, but rather turn to another aspect of polyatomic geometry, one that is grounded in valence-bond language, localized to a given bond, and probably familiar to you. This is the concept of *hybrid orbitals*.

We first encountered hybridization with H-atom AOs. There, we made new p orbitals (p_x and p_y) from linear combinations of $p(m = +1)$ and $p(m = -1)$

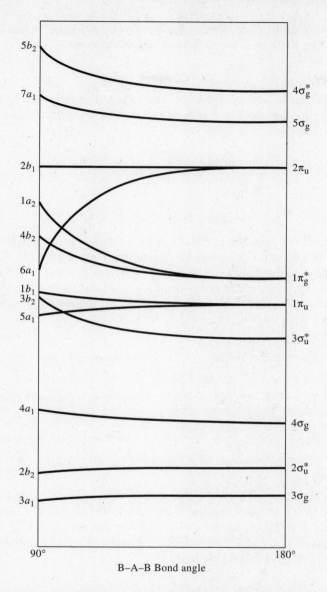

FIGURE 14.22 The Walsh orbital correlation diagram for an AB$_2$ triatomic, plotted as a function of the B—A—B bond angle.

orbitals. The hybrids were both real, highly directional functions. We seek the same goals here, especially directionality.

Consider linear BeH$_2$, $1\sigma_g^2 \, 2\sigma_g^2 \, 1\sigma_u^{*2}$, ignore the core $1\sigma_g$ MO, and note from Figure 14.19(a) how $2\sigma_g$ and $1\sigma_u^*$ involve the H 1s AOs and the Be 2s and 2p$_z$ AOs, with 2p$_z$ oriented along the bond. Now, instead of pure atomic AOs on Be, consider the hybrid linear combinations $(2s \pm 2p_z)/\sqrt{2}$. These *sp hybrid orbitals* point along the H—Be—H bond direction, and, in VB terminology, each is poised to overlap with one H 1s orbital, forming the σ *sp single bonds* of BeH$_2$.

Similarly, *sp^2 hybrids* explain the trigonal planar molecular geometry of BF$_3$. They are combinations of B 2s, 2p$_x$, and 2p$_y$ orbitals (with z perpendicular to the molecular plane), and the C 2s and all three 2p orbitals can combine to form *sp^3 hybrids* that explain the tetrahedral molecular geometry of CH$_4$.

We take advantage of symmetry and orthonormality to make a hybrid point in a desired direction. Consider, for example, the sp^2 hybrids made from 2s, 2p$_x$, and

$2p_y$. We can imagine them lying in the x-y plane, 120° from each other, with one pointing along the x axis. It will have no $2p_y$ component:

$$\phi_1 = c_{11}\, 2s + c_{12}\, 2p_x$$

with coefficients c_{11} and c_{12} yet to be determined. The other two hybrids are not directed along either the x or y axis, and so they have contributions from both $2p_x$ and $2p_y$:

$$\phi_2 = c_{21}\, 2s + c_{22}\, 2p_x + c_{23}\, 2p_y$$

and

$$\phi_3 = c_{31}\, 2s + c_{32}\, 2p_x + c_{33}\, 2p_y\,.$$

The 2s orbital must contribute equally to each hybrid since it is spherical and the three hybrids are equivalent except for orientation. This means $c_{11} = c_{21} = c_{31} = c_1$. The directions and symmetry of ϕ_2 and ϕ_3 (once ϕ_1 is fixed to point along $+x$) make $c_{22} = c_{32} = c_2$ (equal x components for each) and $c_{23} = -c_{33} = c_3$ (opposite y components). We can find c_1, c_2, and c_3 if we require ϕ_1, ϕ_2, and ϕ_3 to be orthonormal. This leads to the complete hybrids

$$\phi_1 = \frac{1}{\sqrt{3}} 2s + \frac{\sqrt{2}}{\sqrt{3}} 2p_x \qquad \text{(points along } +x \text{ axis)}$$

$$\phi_2 = \frac{1}{\sqrt{3}} 2s - \frac{1}{\sqrt{6}} 2p_x + \frac{1}{\sqrt{2}} 2p_y \qquad \text{(120° from } +x \text{ axis on } +y \text{ side)}$$

$$\phi_3 = \frac{1}{\sqrt{3}} 2s - \frac{1}{\sqrt{6}} 2p_x - \frac{1}{\sqrt{2}} 2p_y \qquad \text{(120° from } +x \text{ axis on } -y \text{ side)}$$

illustrated in Figure 14.23.

We find the four sp^3 hybrids in a similar way. Imagine the hybrids pointing toward the four numbered corners of the cube in Figure 14.24.[16] Each hybrid has the same 2s contribution, as did the sp^2 hybrids, and the relative 2p contributions are the signs of x, y, and z in each numbered corner. Normalization determines the overall multiplicative factor. We find

$$\phi_1 = \frac{1}{2}(2s + 2p_x + 2p_y + 2p_z)$$

$$\phi_2 = \frac{1}{2}(2s - 2p_x - 2p_y + 2p_z)$$

$$\phi_3 = \frac{1}{2}(2s - 2p_x + 2p_y - 2p_z)$$

$$\phi_4 = \frac{1}{2}(2s + 2p_x - 2p_y - 2p_z)\,.$$

---

[16]This picture also gives the simplest way to find the tetrahedral bond angle, θ_t. Imagine four unit vectors $\mathbf{e}_1$–$\mathbf{e}_4$ pointing from the center of the cube toward each numbered corner in Figure 14.24. Their vector sum must be zero: $\mathbf{e}_1 + \mathbf{e}_2 + \mathbf{e}_3 + \mathbf{e}_4 = 0$. Take the vector dot product of this equation with any one unit vector: $\mathbf{e}_1 \cdot (\mathbf{e}_1 + \mathbf{e}_2 + \mathbf{e}_3 + \mathbf{e}_4) = 1 + \cos\theta_t + \cos\theta_t + \cos\theta_t = 1 + 3\cos\theta_t = 0$ or $\theta_t = \cos^{-1}(-1/3) = 109.47°$.

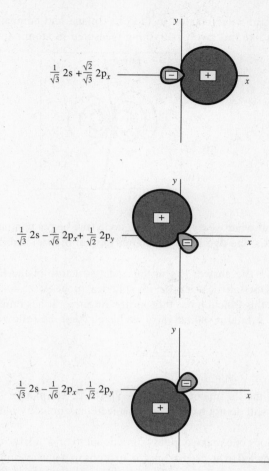

FIGURE 14.23 An orthonormal set of sp² hybrid orbitals can be made from the usual 2s, 2p$_x$, and 2p$_y$ orbitals. This particular choice depends on the x and y directions being defined with respect to the hybrids as shown.

A Closer Look at Hybrid Orbitals

Figure 14.23 gives the impression of pronounced directionality to the sp² hybrids, but it is important to recognize that those pictures are only polar coordinate plots of the angular part of the wavefunction in one plane and they may not tell the entire story. Consider the sp hybrids. A polar plot of the hybrid that is largest along the $+z$ axis is just a plot of $r = |(1 + \sqrt{3}\cos\theta)|/\sqrt{2}$, the spherical harmonic parts of the hybrid. It looks like this:

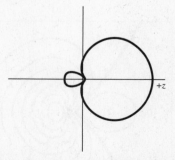

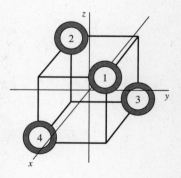

1: $+x, +y, +z$
2: $-x, -y, +z$
3: $-x, +y, -z$
4: $+x, -y, -z$

FIGURE 14.24 The four tetrahedrally directed sp³ hybrids can be chosen to point into the four octants of Cartesian space shown in the diagram. The signs of x, y, and z in each octant give the signs of the p$_x$, p$_y$, and p$_z$ orbitals that contribute to each hybrid.

The large lobe has the opposite algebraic sign from the small one.

Now suppose we turn back to Chapter 12 and fish out the full H-atom 2s and 2p$_z$ wavefunctions and write this sp hybrid in its entirety. We can make a contour plot of the hybrid wavefunction in the (x, z) plane and compare it to the picture above. It looks like this (with everything measured in atomic units):

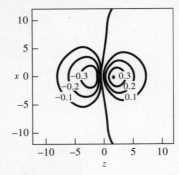

This is not at all what we expected! The node (the heavy contour) is on the wrong side of the z axis, the two lobes are roughly equal in size, and the larger one is on the $-z$ side!

Why is this? The answer lies in the nodal structure of the H 2s wavefunction. Most of this function lies *outside* its spherical node at $r = 2$ au, and the sign convention for this function has this portion negative. It has reinforced the negative lobe of the $2p_z$ orbital. Qualitatively, we have added the following functions:

If we subtract the 2s function: $(-2s + 2p_z)/\sqrt{2}$, the picture becomes its mirror image, but we still do not have the pronounced directionality indicated by the polar coordinate plot.

How then does one make a highly directional hybrid? First, remember that these hybrids are intended to be used as approximations to real wavefunctions in numerical electronic-structure calculations. They must be streamlined for computational efficiency, but we are free to choose their shapes to satisfy our prejudices about the nature of real bonds. Consequently, hybrids used in calculations have s and p components with simple radial parts, such as Slater orbitals or Gaussian orbitals. (See Problem 13.20.) For example, the normalized O atom sp hybrid using Slater orbitals is (in atomic units)

$$\phi_{sp}^{STO} = 1.740\,9\,r\,e^{-2.245\,8r} + 2.951\,3\,z\,e^{-2.226\,6r}$$

and looks like this:

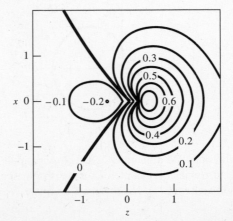

This hybrid looks much more like the polar plot, but only because it was *designed to do so*. See Problems 14.34 and 14.35 for more ways to make hybrids of our choosing.

Hybrid orbitals are not limited to s and p AO combinations. In particular, d orbital contributions are useful for third row and higher compounds. For example, six equivalent d^2sp^3 hybrids formed from two d orbitals and all four s and p's point towards the corners of a regular octahedron. Molecules such as SF_6, $Cr(CO)_6$, UF_6, etc., have this symmetry about the central atom. Table 14.7 summarizes the important hybrids that use s, p, and d AOs, their geometries, and molecular examples.

With d orbital participation, bonding can be extended to include δ *bonds*, with *two* nodal planes containing the bond axis. These are formed from d_{xy} AOs on each atom (see Figure 12.15 for this orbital's shape), if z is the bond axis. *Quadruple bonding* is thus possible and participates in metal–metal bonding in species such as $Re_2Cl_2^{2-}$ that have unusually short metal–metal bond lengths.

Another advantage of hybrid orbital language is the natural way it incorporates nonbonding pairs of electrons. For example, ammonia has a triangular pyramidal structure. If we imagine sp^3 hybrids around N, then one of the five valence electrons of N is in each of three hybrids (forming three NH σ bonds) and the remaining two are paired in the fourth. This nonbonding pair is called a *lone pair*. Thus, we can say the *electron-pair geometry* about the N is *tetrahedral*, the *atomic geometry* is *trigonal pyramidal*, there are three equivalent σ bonding pairs, and there is one lone pair. Table 14.8 expands Table 14.7 considerably and makes the distinction between electron-pair geometry and atomic geometry. We can work our way through complex polyatomics using this language for the local bonding of each atom and build up a complete structural picture.

Still to come in later chapters are other aspects of bonding and structure. We have not discussed bonding in the solid state, for example. Why are solid Ar, NaCl, and Mg so very different in chemical and physical properties? Why should Ar (or any closed-shell atom or molecule) liquify and solidify in the first place? Why does Mg conduct an electric current while solid Ar does not? We will answer these questions eventually, but now we close our discussion of molecular bonding by considering molecular dissociation.

14.5 TAKING MOLECULES APART

Our reference energy for an *n*-electron atom defines $E = 0$ when the nucleus and all *n* electrons are infinitely far apart. Similarly, a convenient reference energy for molecular bonding in an *N*-atom polyatomic has all *N* atoms infinitely far apart *and in their ground electronic states*. For LiH, we found that $Li^+ + H^-$ was the dominant

TABLE 14.7 Hybrid AO geometries

Hybrid	Geometry	Molecular example
sp	linear	BeH_2
sp^2	trigonal planar	BF_3
sp^3	tetrahedral	CH_4
dsp^2	square planar	$PdBr_4^{2-}$
dsp^3	trigonal bipyramidal	PF_5
d^2sp^3	octahedral	SF_6

14 THE ORIGINS OF CHEMICAL BONDING

TABLE 14.8 Electron pair and molecular geometries for various hybridization schemes

Number of Electron Pairs	Electron Pair Geometry	Molecular Geometry	Example	
2	Linear	Linear	H—Be—H	Beryllium dihydride
3	Trigonal Planar	Trigonal Planar		Boron trifluoride
3	Trigonal Planar	Bent		Methylene (singlet)
4	Tetrahedral	Tetrahedral		Methane
4	Tetrahedral	Trigonal Pyramidal		Ammonia
4	Tetrahedral	Bent		Water
4	Tetrahedral	Linear		Hydrogen fluoride
5	Trigonal Bipyramidal	Trigonal Bipyramidal		Phosphorus pentafluoride
5	Trigonal Bipyramidal	See-saw		Sulfur tetrafluoride
5	Trigonal Bipyramidal	T-shaped		Chlorine trifluoride
5	Trigonal Bipyramidal	Linear		Xenon difluoride
6	Octahedral	Octahedral		Sulfur hexafluoride
6	Octahedral	Square-based Pyramidal		Chlorine pentafluoride
6	Octahedral	Square Planar		Xenon tetrafluoride
7	Pentagonal Bipyramidal	Pentagonal Bipyramidal		Iodine heptafluoride
7	Pentagonal Bipyramidal	Distorted Octahedral		Xenon hexafluoride
8	Square Anti-prism	Square Anti-prism		Xenon octafluoride
9	Tri-capped Trigonal Prism	Tri-capped Trigonal Prism		Rhenium nonahydride

configuration at bonding distances while Li + H (nonionic) dominated at large R. To start this section, we return to this molecule and look at its dissociation in detail.

If we define the energy of the atomic ground states, Li ($1s^2 2s\ ^2S_{1/2}$) + H ($1s\ ^2S_{1/2}$) to be zero, the energy of the *separated ions,* Li$^+$ ($1s^2\ ^1S_0$) + H$^-$ ($1s^2\ ^1S_0$), is *higher* by

$$\text{IP(Li)} - \text{EA(H)} = 5.390 \text{ eV} - 0.754 \text{ eV} = 4.636 \text{ eV} .$$

This is the net energy associated with moving the Li 2s electron to the H 1s orbital when the atoms are infinitely far apart. Suppose we start Li$^+$ and H$^-$ at $R = \infty$ and bring them together. Coulombic attraction makes their mutual potential energy at any finite R lower than that at $R = \infty$.

It is easy to calculate that two singly charged ions of opposite sign lower their potential energy 4.636 eV when brought from infinity to $R = 3.11$ Å.[17] If we make the drastic assumption that *neutral* Li and H have *no* covalent *or* ionic interaction when brought together, the Li + H B–O potential energy is constant for all R. For $R > 3.11$ Å this constant energy is less than the Li$^+$ + H$^-$ Coulombic energy, but for all $R \leq 3.11$ Å, the Coulombic energy is lower.

Figure 14.25 plots the Coulombic potential energy and the no-interaction energy (the horizontal line at 0 eV) as well as the true ground-state singlet B–O potential curve and the true triplet curve. (Just as H ($^2S_{1/2}$) + H ($^2S_{1/2}$) leads to singlet and triplet H$_2$ electronic states, so does Li ($^2S_{1/2}$) + H ($^2S_{1/2}$).) A *curve-crossing* occurs between the pure ionic state and the no-interaction state at 3.11 Å. The valence electron of Li can jump to the H atom at this R, turning on the Coulombic attraction that leads to binding.

But, as the figure shows and as we saw in the last section, LiH has a *gradual* increase in ionic character as R is decreased. There is no abrupt turn-on of ionic attraction, but the true singlet curve in Figure 14.25 does show that a significant increase in attraction begins near 3.11 Å.

What happens if we bring Li$^+$ and H$^-$ together? The Coulombic energy of Figure 14.25 cannot continue as shown to small R, since if the neutral atoms can develop

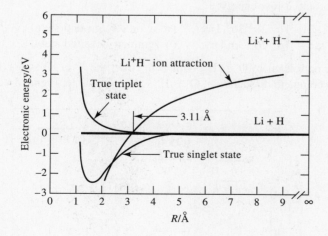

FIGURE 14.25 The ion pair (Li$^+$ + H$^-$) attractive Coulombic interaction is energetically favorable only over short separations. As Li and H approach, they interact as either a nonbonding triplet or a bonding singlet, but the fictitious crossing point at 3.11 Å is a useful measure of the distance at which a strong ionic bonding component starts.

---

[17]Here is one easy way. Remember that the ground-state H-atom energy is -13.6 eV, that the Virial Theorem for H atom says $2E = \langle V \rangle = \text{constant} \times \langle 1/r \rangle$, and that $\langle 1/r \rangle$, the average reciprocal charge separation, is $1/(0.529 \text{ Å})$, the reciprocal of the Bohr radius, for the 1s state. Thus, the constant is $2(-13.6 \text{ eV})(0.529 \text{ Å}) = -14.4$ eV Å, and the Coulombic energy of a singly charged ion pair is -4.636 eV at $R = -14.4$ eV Å$/-4.636$ eV $= 3.11$ Å.

ionic character, the ions can develop covalent character. The curve crossing at 3.11 Å does *not* in fact occur. It is said to be *avoided*.

Covalent/ionic mixing is a strong perturbation. Mixing between different electronic configurations can keep curves from crossing, a result first derived by Wigner and Witmer. They stated a *non-crossing rule,* the essence of which is that potential curves for two electronic states cannot cross if the two states have the same total electron spin and the same total MO symmetry.

Both Li$^+$ (1s^2 $^1S_0$) + H$^-$ (1s^2 $^1S_0$) and Li (1s^{2}2s $^2S_{1/2}$) + H (1s $^2S_{1/2}$) combine to form a σ HOMO. The ion pair combination must be a singlet spin species, and the neutral pair can form a singlet. The non-crossing rule keeps the ionic and the singlet neutral potential curves from crossing.

Polyatomic molecule dissociation is fascinating and complicated simply because there are so many ways to take a polyatomic, even a triatomic, apart. For example, consider H_2O dissociation. The energy zero is, as usual, ground-state atoms infinitely separated.[18] Only the ground state (1s ^2S) of H is important, but both the ^3P ground state and ^1D first excited state of O are low enough in energy to be vital to the story. You may wish to review Figure 13.6 and Table 13.4 to refresh your memory on O energy levels. We will neglect spin–orbit splitting and lump the $^3P_{2,1,0}$ levels into a single ^3P level.

The ground electronic state configuration of H_2O is $1a_1^2\,2a_1^2\,1b_2^2\,3a_1^2\,1b_1^2$. Three excited states are important, and the configurations of all four (valence electrons only) are:

- $H_2O(1)$ $1b_2^2\,3a_1^2\,1b_1^2$ (singlet ground state)
- $H_2O(2)$ $1b_2^2\,3a_1^2\,1b_1\,4a_1$ (triplet, unbound state)
- $H_2O(3)$ $1b_2^2\,3a_1^2\,1b_1\,4a_1$ (singlet, unbound state)
- $H_2O(4)$ $1b_2^2\,3a_1\,1b_1^2\,4a_1$ (singlet).

Note that the electron excitation $1b_1^2 \rightarrow 1b_1 4a_1$ leads to states that are unbound; H_2O spontaneously falls apart in these states. Our question here is, "Falls apart into what?"

One possible dissociation path leads to H + OH, and OH has two electronic states important at low energy:

- OH(1) $1\sigma^2\,2\sigma^2\,3\sigma^2\,1\pi^3$ (π symmetry ground state)
- OH(2) $1\sigma^2\,2\sigma^2\,3\sigma\,1\pi^4$ (σ symmetry excited state)

Another path leads to O + H_2, with O in either the ^3P or ^1D states, and, finally, the path to complete dissociation, O + H + H, is possible. Figure 14.26 gives the

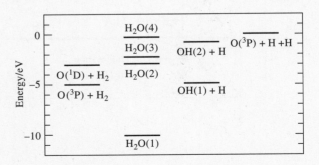

FIGURE 14.26 If we set the energy of O(^3P) + H + H arbitrarily to zero, we can easily locate the energies of the other electronic states important to H_2O. The important electronic structure question is how these electronic states are connected.

---

[18] For most atomic-molecular interactions, a separation of only 10 to 20 Å is as good as an infinite separation. As Figure 14.25 demonstrates, ion–ion interactions are an exception. The Coulombic force is said to be "very long range."

experimental energies of these dissociation products. Each line corresponds to the lowest energy B–O nuclear configuration, except for states (2)–(4) of H_2O, where the ground state H_2O geometry is assumed. (Compare this molecular diagram to the thermodynamic bulk equilibrium diagram—the enthalpy level diagram—of Figure 7.2. In that figure, $H_2O(g)$ corresponds to $H_2O(1)$ here, $H_2(g) + O(g)$ corresponds to $O(^3P) + H_2$, $H(g) + OH(g)$ corresponds to $OH(1) + H$, and $2H(g) + O(g)$ corresponds to $O(^3P) + H + H$.)

Electronic structure theory can indicate, at least qualitatively, how the energy varies as we go around this diagram, and whether or not two levels are connected. As a first step, notice that $H_2O(1)$ is a singlet state while $O(^3P) + H_2$ must be a *triplet* combination, since $O(^3P)$ is a triplet and H_2 is a singlet. Thus, we *cannot* directly dissociate the ground state of H_2O into ground state $O + H_2$.

Figure 14.27 makes this point. We pull the O away from H_2O in its isosceles triangle equilibrium geometry, keeping it equidistant from each H, but allowing the H's to move closer together as the O recedes. We end with $O + H_2$, and the figure shows that the electronic state of O depends on the electronic state of H_2O from which we begin. In particular, the three singlet states of H_2O—(1), (3) and (4)— all yield excited state $O(^1D)$. Put another way, ground state O will *not* react with H_2 to yield ground state H_2O *in this nuclear geometric symmetry*. The singlet and triplet curves cross.

Suppose we first straightened H_2O into a linear H—O—H geometry and then pulled off one H. What states of $OH + H$ would we have? Figure 14.28 gives the answer. Straighten ground-state H_2O, and the energy goes up. (Or else H_2O would have been linear to begin with! Recall the dihydride Walsh diagram.) States (3) and (4) lead to a degenerate singlet level (with π symmetry) that is lower in energy when collinear. (Thus, these states do not want to be bent in the first place.) Triplet state (2) leads to a slightly lower-energy π triplet state. On dissociation, both π's correlate to $OH(1)$ (of π symmetry) $+ H$ while the lower-energy singlet σ state correlates to excited state $OH(2) + H$. Here we conclude that ground state $OH + H$, approaching collinearly, will also *not* yield ground-state H_2O. Each crossing is allowed: singlet crosses triplet, σ crosses π.

Suppose we try a third dissociation path. We grab one H and pull it away directly along its H—O bond direction. This lowers the molecule's symmetry compared to

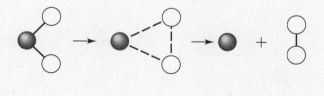

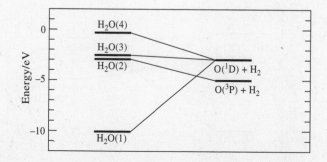

FIGURE 14.27 The lowest energy electronic state of H_2O and the two highest states connect to an *excited* state (^1D) of O when H_2O is dissociated with the symmetry shown.

FIGURE 14.28 If H_2O is bent to a linear configuration (center) and then dissociated to OH + H, ground state H_2O connects to an excited state of OH.

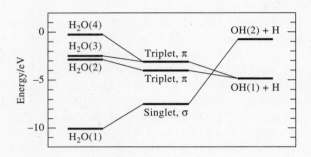

either case above. The only symmetry element left is the plane of the molecule. States (1) and (4) are both *symmetric* with respect to reflection in this plane, while (2) and (3) are *antisymmetric*. Figure 14.29 shows the correlation, but *now the crossing of curves coming from states (1) and (4) is avoided*. The dashed lines in the figure indicate the correlations had the non-crossing rule not existed. But it does, and at last we can see a path from ground state H_2O to ground-state dissociation products, OH(1) + H.

A price has been paid, however. The avoided crossing, as is often the case, produces an *energy barrier* on the way to dissociation. The curve leading from H_2O(1) goes *up* in energy to a *barrier maximum* before falling to the final energy of OH(1) + H.

Chapters 26–28 discuss chemical reactions and explain how *correlation diagrams* such as these can be used to understand differences in reactivity on a quantum state basis. The diagrams here explain why excited state O(^{1}D) reacts with H_2 (to form OH + H) on every collision, but only one in 10^7 O(^{3}P)–H_2 collisions is reactive (at ordinary temperatures).

FIGURE 14.29 An asymmetric dissociation of H_2O lowers the symmetry of the electronic states leading to OH + H. As a result, the correlations between $H_2O(1) \rightarrow OH(2) + H$ and $H_2O(4) \rightarrow OH(1) + H$ are no longer allowed to cross. The resulting avoided crossing allows $H_2O(1)$ to correlate to OH(1) + H, but at the expense of a potential-energy barrier peaking at the avoided crossing.

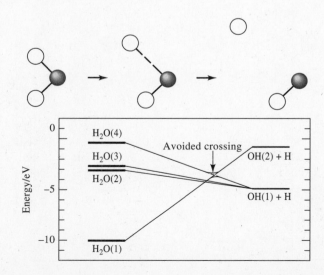

Figures 14.27–14.29 are molecular electronic *state* correlation diagrams. We could have used molecular electronic *orbital* correlation diagrams to reach the same conclusions. (Figure 14.7 is an elementary orbital-correlation diagram, as is the Walsh diagram of Figure 14.20.) Spectroscopy and thermodynamics give us experimental information on electronic-state energies and molecular geometries, but, lacking these data, we can often guess orbital correlations, at least qualitatively. But however it is done, there is as much chemistry to be learned taking a molecule apart as there is studying the intact molecule.

CHAPTER 14 SUMMARY

The main theme of this chapter has been the Born–Oppenheimer approximation, which allows us to discuss electronic structure and molecular geometry in the everyday language of chemistry. It takes us from the general molecular Hamiltonian of Eq. (14.2) to the simpler Hamiltonian such as Eq. (14.4). The simplest molecule, H_2^+, gave us our first view of MOs and their symmetries (Figure 14.3). The LCAO-MO approximation (such as Eq. (14.5)) lets us describe a molecule as a connected set of atoms, and we view any one bond in terms of its B–O potential-energy function (Eq. (14.8a, b) and Figure 14.4).

As soon as we focus on the atoms in molecules and allow the bond length to vary, we find that the United Atom and Separated Atom limits have valuable information (Figures 14.7 and 14.8). Just as we assigned electrons to atomic-orbital configurations, we assigned them to molecular-orbital configurations, for both the ground state and a series of excited states (Eq. (14.9a–f) for H_2 and Table 14.3 for homonuclear diatomics). The reality of MOs and measures of their energies came from photoelectron spectroscopy (Figures 14.12 and 14.13), and the molecular Virial Theorem (Eq. (14.13a–c)) helped us understand the various energy effects that contribute to bonding. Elementary names for bonds, such as covalent, ionic, hybrid, delocalized, and their classification into bonding, nonbonding, and antibonding categories, have simple pictorial interpretations and, at a first approximation, simple mathematical forms.

While you may not carry away from this chapter all the numerical details of the Hückel theory that led to the pictures of 1,3-butadiene in Figure 14.1(a), you should certainly remember the shapes, symmetries, and energy ordering of those orbitals. Likewise, you should remember the general rules governing molecular structure, such as the Walsh diagrams of Figures 14.20 and 14.22 and the hybrid orbital summary of Tables 14.6 and 14.7.

Just as learning the multiplication tables for the integers 0–9 allows you to multiply an infinity of numbers, the basic rules of electronic structure allow you to predict the geometries of an infinite number of molecules. Generalizations of these rules allow you to take a molecule, rearrange it or dissociate it, and predict chemical consequences.

FURTHER READING

Pictures of MOs and general ideas about their energies are very valuable tools for any chemist. Two excellent picture books are *The Organic Chemist's Book of Orbitals,* by W. L. Jorgensen and L. Salem, (Academic Press, NY, 1973) and *A Pictorial Approach to Molecular Structure and Reactivity,* by R. F. Huot, Jr., W. J. Pietro, and W. J. Hehre, (Wiley & Sons, NY, 1984). The first is an MO coloring book: over 200 pages of MO diagrams generated from several theoretical methods on over 100 basic molecules. Figures 14.21(a–b) are from this book. The second has over 1500 photographs of computer generated MO surfaces for a similarly wide array of molecules.

An excellent and very readable introduction to the roles symmetry and group theory play in bonding is *Symmetry and Structure,* by S. F. A. Kettle, (Wiley, NY, 1985). For a wider view of the bonding ideas in this chapter, see *The Chemical Bond,* by J. N. Murrell, S. F. A. Kettle, and J. M. Tedder, 2nd ed., (Wiley, NY, 1985). Hückel MO theory is very well explained and explored in *Molecular Orbital Theory for Organic Chemists,* by A. Streitwieser, Jr., (Wiley, NY, 1961). It is far fresher than its thirtysomething age might suggest. *Molecular Photoelectron Spectroscopy,* by D. W. Turner, C. Baker, A. D. Baker, and C. R. Brundle, (Wiley-Interscience, London, 1970), is a monograph on the (prelaser and synchrotron radiation sources) method of photoelectron spectroscopy with extensive chapters on individual molecules.

Finally, no reference list on bonding can be complete without mentioning *The Nature of the Chemical Bond,* by Linus Pauling, 3rd ed., (Cornell University Press, Ithaca, NY, 1960). Pauling used pictures, insightfully correlated data, the simplest of physical models, and his encyclopedic knowledge of chemistry to cover structural chemistry at a level and scope that will probably never be matched again by anyone in under 650 pages.

PRACTICE WITH EQUATIONS

14A What is the Hamiltonian for the He atom in atomic units? (14.1)

ANSWER: $-\frac{1}{2}(\nabla_1^2 + \nabla_2^2) - \frac{2}{r_1} - \frac{2}{r_2} + \frac{1}{r_{12}}$

14B How many individual terms are in the full Hamiltonian for H_2O? (14.2)

ANSWER: $N = 3$; $n = 10$; total of 91 terms

14C Write a triplet spin wavefunction for H_2 based on the MO configuration $1\sigma_g 2\sigma_g$. (14.9)

ANSWER: $\Psi = \frac{1}{\sqrt{2}}[1\sigma_g(1)2\sigma_g(2) - 2\sigma_g(1)1\sigma_g(2)] \times$ (spin part of Eq. (14.9d, e, or f))

14D How are $\langle E(R)\rangle$, $\langle T(R)\rangle$, and $\langle V(R)\rangle$ related at $R = \infty$? at $R = R_e$? (14.13)

ANSWER: $d\langle E(R)\rangle/dR = 0$ at both points; $\langle T\rangle = -\langle E\rangle$ and $\langle V\rangle = 2\langle E\rangle$

14E What is the MO configuration for the lowest energy state of He_2 that does not have a zero bond order? Table 14.3

ANSWER: $1\sigma_g^2 1\sigma_u^{*1} 2\sigma_g^1$

14F What is the electron configuration and bond order of the carbide ion, C_2^{2-}? Table 14.3

ANSWER: $1\sigma_g^2 1\sigma_u^{*2} 2\sigma_g^2 2\sigma_u^{*2} 1\pi_u^4 3\sigma_g^2$; bond order = 3

14G Should the F_2^+ ion have a larger bond energy than F_2? Table 14.3

ANSWER: Yes. An antibonding electron is removed. The bond order is 1.5.

14H What metal hydride bond should have the largest ionic character? Table 14.4

ANSWER: CsH

14I What is the Hückel theory total energy of linear H_3^-? (14.18)

ANSWER: $4\alpha + 2\sqrt{2}\beta$

14J Would you expect NO_2^- to be linear or bent? What about NO_2^+? Figure 14.22

ANSWER: NO_2^- is bent (18 e^-); NO_2^+ is linear (16 e^-)

PROBLEMS

SECTION 14.1

14.1 Write the Hamiltonians for the model systems: (a) a proton in a 1-D box of length $L = 10a_0$, (b) an electron in a harmonic oscillator, and (c) an H atom orbiting in a plane at a distance $20a_0$ from the origin, all in atomic units. Be careful to define all symbols you use.

14.2 Write the B–O Hamiltonian for the two-electron molecular ion HeH^+ in atomic units. Give each term a physical interpretation; that is, kinetic energy of electron 1, electron–electron repulsion, etc.

14.3 Show that the H_2^+ Schrödinger equation is separable in confocal elliptical coordinates. The Laplacian in these coordinates is (remember R is a constant)

$$\nabla^2 = \frac{4}{R^2(\xi^2 - \eta^2)}\left\{\frac{\partial}{\partial\xi}\left[(\xi^2-1)\frac{\partial}{\partial\xi}\right] + \frac{\partial}{\partial\eta}\left[(1-\eta^2)\frac{\partial}{\partial\eta}\right] + \frac{\xi^2-\eta^2}{(\xi^2-1)(1-\eta^2)}\frac{\partial^2}{\partial\phi^2}\right\}$$

and you should write the total wavefunction as $\psi(\xi, \eta, \phi) = \Xi(\xi)H(\eta)\Phi(\phi)$. (H is a "capital eta ($\eta$)," and Ξ is a "capital xi (ξ).")

14.4 Evaluate the H_2^+ overlap integral $S = \int 1s_A \, 1s_B \, d\tau$ using elliptical coordinates. The volume element is $d\tau = (R^3/8)(\xi^2 - \eta^2)d\xi \, d\eta \, d\phi$, and the 1s orbital in atomic units is simply $e^{-r}/\sqrt{\pi}$. Note the limits on ξ and η from Figure 14.2 (and $0 \le \phi \le 2\pi$). The integrals

$$\int e^{ax}\,dx = \frac{e^{ax}}{a} \quad \text{and} \quad \int x^2 e^{ax}\,dx = e^{ax}\left(\frac{x^2}{a} - \frac{2x}{a^2} + \frac{2}{a^3}\right)$$

will prove useful.

14.5 Pictured below are several wavefunctions with different shadings distinguishing different algebraic signs. Nuclei are located by black dots, and the squares or rectangles denote reflection planes of symmetry as indicated. Give the eigenvalues for each of the symme-

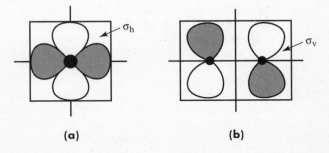

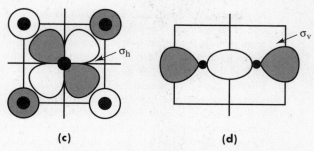

(c)   **(d)**

try operators $\hat{\sigma}_h$, $\hat{\sigma}_v$, and $\hat{i}$ operating on each wavefunction.

14.6 Plot the g and u wavefunctions of H_2^+ along the internuclear axis for $R = 2$ au, the equilibrium ground-state bond length. Ignore normalization constants, and take each H atom 1s orbital to be simply e^{-r} where r is the distance away from the nucleus. For the g wavefunction, what is the ratio of the wavefunction at the bond midpoint to its value at either nucleus? What is this ratio for a point $R/2$ beyond either nucleus? What do these numbers say about electron localization in the bonding region?

14.7 The molecular Virial Theorem can be applied to the H_2^+ ground state $\langle E(R) \rangle_g$ curve, Eq. (14.8a), yielding the graph below in which $\langle V \rangle$ (the heavy line) has been raised 1/2 au and $\langle T \rangle$ (the dashed line) has been lowered 1 au so that all three curves approach a common asymptote at large R.

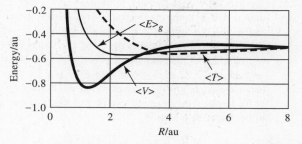

Note that $\langle V \rangle$ has a minimum $\cong -1.34$ au (when the curve is corrected for the energy shift in plotting) at $R \cong 1.2$ au. Compare this potential energy to the total electrostatic potential energy of an electron halfway between two protons spaced 1.2 au. You should find an energy significantly lower than -1.34 au. Now repeat the calculation with the protons spaced 1.2 au, but with the electron equidistant from each proton by an amount r. Express the total electrostatic energy in terms of r, and find r so that the electrostatic energy equals -1.34 au. What do your answers tell you about the localization of charge between the nuclei?

14.8 Nonbonding (purely repulsive) potential energy curves are often simply approximated by a single exponential function of internuclear distance. For the $\langle E(R) \rangle_u$ curve of H_2^+, Eq. (14.8b), a least-squares fit of the exact curve to such a function over the range $R = 1$–10 au gives $\langle E(R) \rangle_u = -0.5 + 2.887 \exp(-1.050 R)$. Calculate $\langle V(R) \rangle$ and $\langle T(R) \rangle$ using the molecular Virial Theorem and this approximate $\langle E(R) \rangle_u$ and compare your results to the exact curves plotted below. (In this plot, $\langle V \rangle$ (the heavy line) has been raised 1/2 au and $\langle T \rangle$ (the dashed line) has been lowered 1 au so that all three curves approach a common asymptote at large R, as in Problem 14.7.)

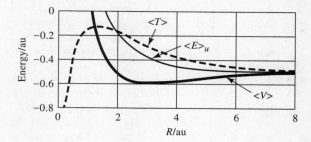

SECTION 14.2

14.9 The photoelectron spectra of the rare gases Ar, Kr, and Xe taken with 21.22 eV photon energy light show two peaks each with electron energies 5.46 and 5.28 eV for Ar, 7.22 and 6.56 eV for Kr, and 9.09 and 7.78 eV for Xe. Explain why two peaks are seen, and assign atomic-ion term symbols to each. (*Hint:* Think about Example 13.5.)

14.10 The photoelectron spectrum of HF with 21.22 eV radiation shows signals at electron energies of 5.18 eV and 2.13 eV. The higher energy peak is accompanied by three successively smaller peaks at lower electron energy, spaced by about 0.35 eV from each other, while the lower-energy peak is quite broad ($\sim$1 eV) and highly structured. Assign the two major peaks to appropriate MO ionizations, and suggest explanations for the structure associated with each.

14.11 Predict the ground-state electron configurations, bond orders, and relative bond energies of NO^+, NO, and NO^-. Use the following ionization potentials, electron affinities, and bond energy to check your bond-energy prediction. IP(N) = 14.534 eV, IP(O) = 13.617 eV, IP(NO) = 9.25 eV, EA(N) = $\leq$ 0 eV, EA(O) = 1.461 eV, EA(NO) = 0.024 eV, $\mathscr{D}_e$(NO) = 6.496 eV. (Why doesn't N have a positive electron affinity?)

14.12 What is the ground state MO configuration for HeBe? What bond order would you assign this molecule? What configuration would you assign to its first excited state? Is the first excited MO likely to be localized on one atom, and if so, which one?

14.13 What are the ground electronic state MO configurations of CF, CH, CH^+, and CN^-? Do any of these have unpaired electrons?

14.14 What would be the bond order of the dication He_2^{2+}? Would you expect it to be stable in its ground state? Be as quantitative as you can with your answer.

SECTION 14.3

14.15 Graph potential energy curves as in Figure 14.25 for the $H^+ + H^-$ system along with $H(1s) + H(ns)$, $n = 1$–5. Draw the neutral curves as simple horizontal lines, and locate the curve crossing points between the ion pair curve and the neutral potential lines. You should find that the ion-pair asymptote falls between the $n = 4$ and $n = 5$ neutral energies. Is the reaction $H^+ + H^- \rightarrow H(2s) + H(2s)$ exothermic or endothermic?

14.16 Apply Hückel MO theory to H_2 with 1s AOs on each atom, derive the corresponding wavefunctions for the two Hückel energies, and compare them to the various wavefunctions for H_2^+ and H_2 of Section 14.1. To which of those expressions does the Hückel theory come closest?

14.17 The π electron MOs of the linear allyl radical, C_3H_5, and its ions, $C_3H_5^+$ and $C_3H_5^-$, can be easily treated by Hückel theory. Write the Hückel secular determinant, find the predicted energies, and determine the total π energies and electron configurations of each species.

14.18 Continuing from the previous problem, you should have found MO energies $\alpha + \sqrt{2}\beta$, α, and $\alpha - \sqrt{2}\beta$ for the allyl system. Use these energies to find the wavefunctions for each MO in terms of the three p orbitals (call them p_a, p_b, and p_c) on the three C atoms that contribute one electron each to the molecular π framework. Sketch each MO, and compare your sketches to Figure 14.17.

14.19 Which of the following is the Hückel secular determinant for trimethylene methane, $C(CH_2)_3$? What simple Lewis electron-dot structures could one draw for this compound? The Hückel prediction is that it is a "diradical" with unpaired (triplet arranged) spins on different carbons. Is this in accord with a Lewis picture?

(a) $\begin{vmatrix} x & 1 & 1 & 1 \\ 1 & x & 1 & 1 \\ 1 & 1 & x & 1 \\ 1 & 1 & 1 & x \end{vmatrix}$ (b) $\begin{vmatrix} x & 1 & 1 & 1 \\ 1 & x & 0 & 0 \\ 1 & 0 & x & 0 \\ 1 & 0 & 0 & x \end{vmatrix}$

(c) $\begin{vmatrix} x & 0 & 1 & 1 \\ 0 & x & 0 & 1 \\ 1 & 0 & x & 0 \\ 1 & 1 & 0 & x \end{vmatrix}$ (d) $\begin{vmatrix} x & 1 & 0 & 1 \\ 1 & x & 0 & 0 \\ 0 & 0 & x & 1 \\ 1 & 0 & 1 & x \end{vmatrix}$

14.20 Hückel theory would be rather boring if it could not be extended to molecules with more than C and H atoms. There have been several suggestions as to how this can be done, and this problem considers one of the simplest. We take C to be the reference atom with its associated α value, and any heteroatom X has its α value adjusted by adding some fraction of β to the C value: $\alpha_X = \alpha_C + f\beta$ where f is the fraction (a positive number for atoms more electronegative than C) and β is the C value (or more precisely, the C in benzene value, since benzene is a convenient reference for C). Find the π MO energies for the $H_2C{=}X$ system with the simple assumptions $\beta_{CX} = \beta_{CC}$ and $f = 1$. Compare these energies to those for ethylene.

14.21 The previous problem should have given you the π MO energies $\alpha_C + (1 \pm \sqrt{5})\beta_{CC}/2$ for $H_2C{=}X$. Use these energies and the secular equations for the wavefunction coefficients:

$$c_1(\alpha - E) + c_2\beta = 0$$
$$c_1\beta + c_2(\alpha + \beta - E) = 0$$

along with the normalization condition $c_1^2 + c_2^2 = 1$ to find c_1 and c_2 for each MO. You should reach the important conclusion that in the *bonding* MO, the electron density is more localized on the *more* electronegative atom (X = atom 2 here), but in the excited, antibonding MO (look for the node!) the electron density is localized *away from* the more electronegative atom.

14.22 The Hückel π MO energies for benzene and other monocyclic systems can be found very quickly once one is told the following. If the regular geometric ring for the cyclic system is inscribed inside a circle of radius 2β so that one vertex points down, and if a vertical energy scale is drawn with $E = \alpha$ at the center of the circle, the π energies fall at the locations of the vertices. For example, cyclobutadiene (assumed to be a square) is drawn

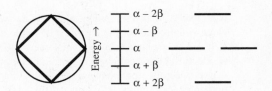

with the construction on the left and the corresponding energy levels on the right: one at $E = \alpha + 2\beta$, a degenerate pair at $E = \alpha$, and one at $E = \alpha - 2\beta$, as in Example 14.4. Use this construction to find the energies of the six π MOs of benzene. Do you find benzene stabilized by $2|\beta|$ as discussed in the text? The nodal patterns of these MOs are also instructive. The lowest is node-free, of course, while the highest has three nodal planes perpendicular to the molecular plane. These can be shown qualitatively as

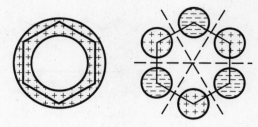

What would you draw for the other four MOs' nodal patterns?

SECTION 14.4

14.23 Use the Walsh diagram of Figure 14.20 to explain why BH_2 is bent in its ground state but linear in its first excited state.

14.24 If the bond angle for the dihydrides in the Walsh diagram of Figure 14.20 is reduced to 0°, the triatomic is folded into a diatomic. What happens to the $1b_2$ and $1b_1$ orbitals in this limit?

14.25 What does the periodic trend in bond angles for the species H_2O–H_2Te tell you about the nature of the bonding MOs in these dihydrides? Can you suggest a reason for this trend?

14.26 Give an additional example of a triatomic molecule (neutral or ion) with, respectively, 16, 17, 18, 19, 20, and 22 valence electrons beyond those mentioned in the text. Predict the bond angles of each as best you can.

14.27 Refer back to Chapter 12's discussion of H atom orbitals, and decide which p and d AOs should be mixed with an s AO to form each of the hybrids in Table 14.7 that involves d orbitals. Be sure to specify a coordinate system with reference to each molecular example in the table.

14.28 What sort of hybrid orbitals should be used to describe the collinear C—Zn bonds in dimethyl zinc, $Zn(CH_3)_2$? Recall that Zn has the valence electron configuration $3d^{10}4s^2$.

14.29 Explain why the first excited electronic state of CO_2 is not linear. What do you think happens if one tries to stick an extra electron on CO_2? Does your answer make sense in light of the experimental observation that CO_2 has a negative electron affinity?

SECTION 14.5

14.30 The simplest bound heteronuclear diatomic is HeH^+. Define a zero of energy corresponding to He^{2+} + H^+ + $2e^-$ infinitely far apart. Locate the energies of the pairs $He(1s^2\ ^1S_0) + H^+$, $He^{2+} + H^-(1s^2\ ^1S_0)$, $He^+(1s\ ^2S_{1/2}) + H(1s\ ^2S_{1/2})$, $He^*(1s2s\ ^1S_0$, 20.6 eV above the He ground state) $+ H^+$, and $He^*(1s2s\ ^3S_1$, 19.8 eV above the He ground state) $+ H^+$. (Chapter 13 has further energy information on these species.) If ground state HeH^+ is pulled apart, which pair do you get? What is the dissociation limit of the first excited state of HeH^+? How would you describe the MOs in both of these states?

14.31 The H_2O electronic states called (2) and (3) in the text (based on the $1b_1 4a_1$ configuration) are known to be unbound. What would you predict their geometries to be if they *were* bound: linear, more bent than the ground state, less bent, or bent to roughly the same angle?

14.32 When ground state H_2O is straightened into linearity, as in Figure 14.28, what MO configuration does it have?

GENERAL PROBLEMS

14.33 Adolph Butenandt shared the 1939 Nobel Prize for Chemistry for his work on steroidal sex hormones, and he spent much of his career on the silkworm moth attractant pheromone, bombykol. Butenandt needed the glands of 500 000 female moths to extract only 6.4 mg of bombykol from which he determined its structure:

$$CH_3CH_2CH_2\underset{H}{\overset{H}{\underset{|}{C}}}{=}\underset{H}{\overset{H}{\underset{|}{C}}}{-}\underset{H}{\overset{H}{\underset{|}{C}}}{=}\underset{|}{\overset{H}{C}}-CH_2CH_2CH_2CH_2CH_2CH_2CH_2CH_2OH$$

Other isomers of bombykol elicit physiological responses in male silkworm moths, but only at concentrations 10^9 to 10^{13} times higher. What does simple Hückel MO theory have to say about the π electron MOs of bombykol? Would you expect these MOs to include the HOMO of bombykol?

14.34 Hybrid orbitals can be constructed to point in arbitrary directions. Suppose one wanted to make hybrids from s, p_x, and p_y such that two of them made an arbitrary angle θ. The directionality comes from the p orbitals, and the s orbital can be used to make the hybrids orthogonal. In fact, we can consider the p_x and p_y orbitals as if they were unit vectors along the x and y axes, and our hybrids as unit vectors with the directional properties of interest. If we call the three hybrids α, β, and γ, with θ the angle between α and β, we can use the diagram below to deduce the p components of each (except for orthonormality constants):

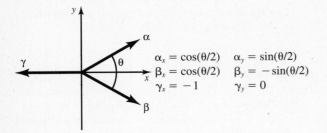

Consequently, the three hybrids must be

$$\alpha = a_1 s + a_2(p_x \cos(\theta/2) + p_y \sin(\theta/2))$$

$$\beta = a_1 s + a_2(p_x \cos(\theta/2) - p_y \sin(\theta/2))$$

$$\gamma = a_3 s - a_4 p_x$$

where a_1, a_2, a_3, and a_4 ($a_4 > 0$) are determined by orthonormality. Show that orthonormality of α and β (s, p_x, and p_y are individually orthonormal) leads to

$$a_1^2 = \frac{\cos\theta}{\cos\theta - 1} \quad \text{and} \quad a_2^2 = \frac{1}{1 - \cos\theta}.$$

We can interpret a_1^2 as the amount of *s character* in these hybrids. What is the s character of a pure sp^2 hybrid? What is it for hybrids that have the H_2O bond angle, $\theta = 105.2°$?

14.35 Continuing the previous problem, we can find a_3 and a_4 for γ from normality: $a_3^2 + a_4^2 = 1$, and the condition $2a_1^2 + a_3^2 = 1$ ensures we have used "all" the s orbital in the three hybrids. Find an expression for a_3^2 in terms of θ, and compute the s character of this third hybrid (presumably not used in bonding) for water.

14.36 A simple model for the π electrons of linear conjugated polyenes is based on the 1-D particle-in-a-box model. We assume all C—C bond lengths are equal (to $R = 1.4$ Å) and that the π electrons from N double bonds are confined to a line that extends from $R/2$ ahead of the first double-bonded C to $R/2$ beyond the last double-bonded C in the chain:

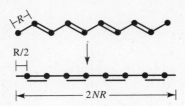

This is the length of the 1-D box into which we place one π electron per C atom, spin-paired in each energy level. What is the HOMO energy in this model? What is the excitation energy from the HOMO to the LUMO? As simple as this model is, it does a remarkably good job in predicting this excitation energy as a function of polyene chain length.

14.37 We can make cyclobutadiene not square in a Hückel calculation (see Example 14.4) if we introduce two different resonance integrals β_1 and β_2:

Find the π MO energies for this system. Define $x = (\alpha - E)/\beta_2$ and $b = \beta_1/\beta_2$, and you should find the secular determinant polynomial is $x^4 - 2(b^2 + 1)x^2 + (b^2 - 1)^2 = 0$. Verify that "square" cyclobutadiene is recovered if $b = 1$. What does $b = 0$ correspond to physically? Both β's are negative, of course, but which would you expect to be more negative? Do you predict nonsquare cyclobutadiene to be a singlet or a triplet?

14.38 Problems 12.46 and 13.39 introduced singlet and triplet positronium atoms, Ps, respectively. Here we consider Ps chemistry. When Ps and H bind to form positronium hydride, how would this species be described? Is the B–O approximation of any use? How would one picture PsH in the context of Figure 14.1? It is known that PsH is stable to the dissociation PsH → Ps + H by about 1.1 eV. Make a quick energy-level diagram with a zero at $H^+ + 2e^- + e^+$ infinitely far apart, and locate the species Ps + H^+ + e^-, $e^- + e^+$ + H, Ps + H, e^+ + H^-, and PsH relative to this zero. Is PsH stable against the dissociation PsH → e^+ + H^-?

CHAPTER 15

Electric and Magnetic Properties of Molecules

CHAPTER 9 described some of the macroscopic effects of electric and magnetic fields on bulk matter. Here, we concentrate on their effects on the molecular level, and in Chapter 21, we will continue into the realm of molecular spectroscopies in external fields.

Chapter 9 considered fields generated by some external source. Here, we will also consider electric or magnetic fields that are *not* generated externally. These are the fields generated by and carried around with individual molecules. One molecule's field is the external field to its neighbors, and since *any* molecule can generate an electric field and many can generate magnetic fields of importance, we will also consider the effects of external fields on individual molecules. Sections 15.2 and 15.3 cover this topic, but we will begin in the first section with the electric fields of individual molecules.

These fields have several sources that are familiar to you: the net charge of an ion or the charge distribution in a molecule with a permanent dipole moment. Since *all* molecular electric fields are due to the molecule's electrons and nuclei, we start with a very general picture of molecular charge distributions. We will find a systematic way to relate its net electric field to the total charge, dipole moment, and other properties of the general distribution.

15.1 Basic Properties of Molecular Charge Distributions

15.2 Special Consequences of Unpaired Electrons

15.3 The Motion of Molecules in External Fields

15.1 BASIC PROPERTIES OF MOLECULAR CHARGE DISTRIBUTIONS

If molecules were collections of static, classical point charges, the description of molecular electric fields would be simple. Classical electrostatics tells us that the electric field from a collection of point charges is simply the vector sum at any point of the electric fields of the individual charges. This is the principle of *superposition*. For real molecules, however, the charges are not static points, and we speak of the *charge distribution* in molecules *via* the molecular wavefunction.

We should use exact molecular wavefunctions, of course, but they are elusive. The charge distribution can be discussed in the Born–Oppenheimer framework, using whatever approximate electronic wavefunction we can concoct, treating the nuclei as static point charges. Thus, for an *n*-electron molecule, the *electron* charge distribution is given in terms of the total electronic wavefunction, $\psi(r_1, r_2, \ldots, r_n)$, as

$$\rho_{el}(r) = -ne \int \psi^* \psi \, dr_2 \, dr_3 \ldots dr_n \qquad (15.1)$$

Note that $\psi^*\psi$ is integrated over all but *one* electron coordinate. This integral by itself is the probability distribution for finding any one electron at r no matter what the positions of the others. We multiply this by n because the electrons are indistinguishable, and we multiply by $-e$, the electron charge, to turn the probability distribution into a charge distribution. The total charge distribution ρ is just ρ_{el} plus point charges $Z_i e$ at each nucleus of atomic number Z_i.

Since we so rarely have the ψ's of Eq. (15.1), this expression is of less practical value than one might hope. Instead, we picture ρ much in the way we picture molecular shape. Only the simplest properties of these pictures are important.

To see how these pictures can do the job, consider the following analogy. Someone is walking towards you from a distance. At first all you can discern is that it is someone and not a dog, bear, etc. As the person approaches, you begin to recognize features—male or female, then perhaps a familiar face, and finally, up close, you can make out details. So it is with the electric field of a molecule. At very large distances, all molecules look alike. At closer distances, a set of features can be discerned, one by one. First comes net charge: is it a neutral molecule or a positive or a negative ion? Next come the grossest features of total charge symmetry: does the molecule have a dipole moment? Then come finer details until the molecular charge distribution is in full view.

This way of thinking, which takes the observer to a point much farther away from the molecule than the dimensions of the molecule itself, is applicable to many situations. Its mathematical basis is called the *multipole expansion* of the electric potential. The electric potential[1] $\phi(r)$ due to a charge distribution $\rho(r')$ at a point r such that $r > r'$ is

$$\phi(r) = \int_{\substack{\text{all} \\ \text{space}}} \frac{\rho(r') \, dr'}{4\pi\epsilon_0 |r - r'|} \, .$$

The details of the multipole expansion are given a closer look below, but the general scheme breaks down as follows. We write the potential as a sum of physically distinct terms:

$$\phi(r) = \text{monopole term (the net charge)}$$
$$+ \text{ dipole term}$$
$$+ \text{ quadrupole term}$$
$$+ \text{ octupole term}$$
$$+ \text{ higher-order terms .}$$

---

[1] The electric *potential* is *not* to be confused with the electric potential *energy*. The potential is the potential energy *per unit charge*.

15.1 BASIC PROPERTIES OF MOLECULAR CHARGE DISTRIBUTIONS

For the charge distribution of a *linear molecule,* these terms are (through the quadrupole term, which is as far as we will go in any detail[2])

$$\phi(r) = \frac{q}{(4\pi\epsilon_0)r} + \frac{p\cos\theta}{(4\pi\epsilon_0)r^2} + \frac{\Theta(3\cos^2\theta - 1)}{2(4\pi\epsilon_0)r^3} + \ldots \quad (15.2)$$

↑ ↑ ↑ ↑ ↑
Total electric Monopole (net Dipole Quadrupole Higher-order
potential charge) potential potential potential terms

where

$$q = \int \rho(r')\, d\mathbf{r}' = \text{the } \textit{net charge} \quad (15.3a)$$

$$p = \int \rho(r')\, r'\cos\theta\, d\mathbf{r}' = \text{the } \textit{dipole moment} \quad (15.3b)$$

$$\Theta = \int \rho(r')\, r'^2\left(\frac{3\cos^2\theta - 1}{2}\right) d\mathbf{r}' = \text{the } \textit{quadrupole moment} \quad (15.3c)$$

and θ is the angle between the vector pointing to the point at **r**, far from the molecule, and the line (the z axis) passing through the collinear nuclei: In each integral, the coordinate origin is at the molecule's center of mass. These expressions can also be written in Cartesian coordinates: $z' = r'\cos\theta$ and $d\mathbf{r}' = dx'\, dy'\, dz' = r'^2\sin\theta\, dr'\, d\theta\, d\phi$.

The net charge is simply a number, but, in general, p is a vector quantity with three independent components. For a linear molecule, only the component p_z along the bond axis is non-zero. The general quadrupole moment has *nine* components ($\Theta_{xx}, \Theta_{xy}, \Theta_{xz}, \ldots, \Theta_{zz}$), but only five of these are independent, at most. Methods for calculating Θ in general can be found in this chapter's references.

Figure 15.1 shows simple point charge distributions that (a) have a dipole moment and (b) a quadrupole moment but no dipole moment. For (b), both the linear and the square arrays have only one quadrupole moment component. The linear array is similar to the charge distribution in CO_2. (Note how it looks like two counterposed dipoles: O=C=O .) The square array is similar to the shape of alkali halide dimers (Na_2Cl_2, for example), which are not square but have the same alternating + − charge distribution with no dipole moment.

EXAMPLE 15.1

A simple MO calculation for LiH at the experimental bond length of 1.59 Å predicts net charges of ±0.53 e at the atomic centers. What are the dipole and quadrupole moments of this charge distribution?

SOLUTION For linear arrays of point charges, Equations (15.3b,c) are

$$p = \sum_i q_i z_i \quad \text{and} \quad \Theta = \sum_i q_i z_i^2$$

and we locate the coordinate origin at the center of mass, defined so that $m_{Li}\, z_{Li} + m_H\, z_H = 0$. We draw the molecule as

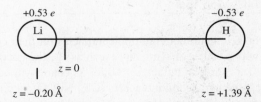

---

[2]But note the spellings: quad**ru**pole, not quadrapole, and oct**u**pole, not octapole.

FIGURE 15.1 Figure (a) is an elementary charge distribution with a dipole moment but no net charge while (b) shows two elementary charge distributions with quadrupole moments but no dipole moment or net charge. In each, R is the $+$ to $-$ charge separation.

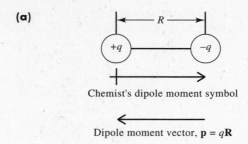

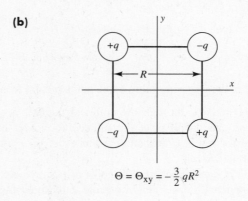

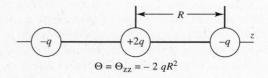

and, since $z_H - z_{Li} = 1.59$ Å, we find, for the ^{7}Li isotope, $z_H = 1.39$ Å and $z_{Li} = -0.20$ Å as center of mass coordinates.

The dipole moment is

$$p = (0.53\ e) \times (-0.20\ \text{Å}) + (-0.53\ e) \times (1.39\ \text{Å}) = -0.53\ e \times 1.59\ \text{Å}$$

(which shows that, for a neutral diatomic, $p = $ (excess atomic charge)$\times$(bond length) gives the dipole moment magnitude), or $p = -13.5 \times 10^{-30}$ C m, in reasonable agreement with the measured magnitude, 19.62×10^{-30} C m.

The quadrupole moment is

$$\Theta = (0.53\ e) \times (-0.20\ \text{Å})^2 + (-0.53\ e) \times (1.39\ \text{Å})^2$$
$$= -16.1 \times 10^{-40}\ \text{C m}^2$$

This moment has not been accurately measured for LiH, but it has been estimated at $(-16 \pm 1) \times 10^{-40}$ C m^2, in good agreement.

➠ RELATED PROBLEMS 15.4, 15.8

Symmetry can make one or more multipole moments (or their components) zero. Linear molecules with a center of symmetry (such as H_2, CO_2, acetylene, etc.) have no dipole moment, for example. In fact, $p = 0$ for all centrosymmetric species (SF_6, $Cr(CO)_6$, XeF_4, etc.) and tetrahedral molecules (CF_4, $Ni(CO)_4$, etc.). The quadrupole moment is non-zero *except* for tetrahedral or octahedral symmetries. The first nonvanishing multipole moment for CF_4 is the *octupole* moment, while even this term is zero for octahedral SF_6 (which has a non-zero *hexadecapole* moment). Finally, a spherical atom can have only a net charge.

In SI units, these moments are expressed in C m^n units with $n = 0, 1, 2, \ldots$, for net charge, dipole, quadrupole, ... moments. Dipole moments have until recently been expressed almost exclusively in the non-SI unit named for Peter Debye. One debye[3] (1 D) equals 10^{-18} esu cm, or about 3.336×10^{-30} C m. Dipole moments are always tabulated as positive numbers, but, when used to calculate a potential (as in Eq. (15.2)) or a related quantity, the vector nature of **p** (as shown in Figure 15.1(a)) must be taken into account.

Quadrupole moments can have either algebraic sign. For example, the linear array of Figure 15.1(b) has $\Theta = 2er^2$ if the charges are $+--+$ instead of $-++-$. The first array should remind you of H_2, and, in fact, $\Theta(H_2)$ is positive, while $\Theta(CO_2)$ is negative due to the more electropositive central C atom.

Table 15.1 collects dipole and quadrupole moments for these and other molecules. Note the dipole entry for HD. This tiny moment is due to a breakdown of the Born–Oppenheimer approximation. The electrons cannot exactly keep up with the nuclear vibrational motion, and the asymmetry of this motion about the center of mass produces the moment. Also note the one molecular ion entry: ArH^+. This ion is isoelectronic to HCl and has an appreciable bond energy and a substantial dipole

TABLE 15.1 Molecular gas phase dipole and quadrupole moments

\multicolumn{6}{c}{Dipole moments†/10^{-30} C m}
LiH	19.62	HD (+HD−)	0.001 95	CO (−CO+)	0.374
LiF	21.11	HF	6.069	SO	5.17
LiCl	23.75	DF	6.071	CS	6.53
LiBr	24.24	HCl	3.697	OCS	2.385 3
LiI	24.79	HBr	2.76	SO_2	5.22
NaCl	30.03	HI	1.49	NO	0.517
KCl	34.25	HCN	9.94	NO_2	1.054
RbCl	35.06	OH	5.74	N_2O	0.536 4
CsCl	34.76	H_2O	6.284	O_3	1.77
BaO	23.53	H_2S	3.25	ClF (+ClF−)	2.94
BaS	36.27	H_2Se	0.80	BrF	4.30
LiNa	1.5	NH_3	4.920	ClF_3	1.86
LiK	11.5	PH_3	1.914	SF_2	3.50
LiRb	13.3	H_2CO	7.77	SF_4	5.93
NaRb	10.3				
NaCs	15.8			ArH^+	10. ± 2.
\multicolumn{6}{c}{Quadrupole moments/10^{-40} C m^2}
$H(2p_z)$	−53.8	HF	7.9	CO	−8.3
$H(3d_{z^2})$	−161.5	HCl	12.7	CO_2	−14.3
H_2^+	6.87	HBr	13.	OCS	−3.0
H_2	2.17	HI	20.	O_2	−1.3
Li_2	46.0			N_2	−4.7
				N_2O	−12.2

†10^{-30} C m ≅ 0.300 D

---

[3]The relation between D and C m units is perhaps better remembered this way: dipole moments are on the order of 10^{-30} C m, and 10^{-30} C m = 0.299 792 458 D exactly, or 0.3 D approximately. Yes, those are the digits of the speed of light! 1 esu = $10/(c/\text{cm s}^{-1})$ C.

moment, but it is very difficult to measure dipole moments for ions. This ion was the first, in 1987.

EXAMPLE 15.2

Table 15.1 states that the H atom in the $2p_z$ state has a quadrupole moment $\Theta = -53.8 \times 10^{-40}$ C m². If we make a simple point charge model of this state:

how far from the nucleus are the (fractional) point charges; that is, what is R?

SOLUTION This is another way of seeing that atoms not in s states are not spherical, and it is also a simple application of the idea that cleverly placed point charges of appropriate magnitudes can approximate more complicated continuous charge distributions. Here, we imagine the electron split in half so that $-0.5e$ is on either side of the nucleus. This is the simplest point charge distribution that has the properties of the true distribution: no net charge or dipole moment, but a negative quadrupole moment. Comparing the diagram above to Figure 15.1, we see that $q = 0.5e$ so that $\Theta = -eR^2$, or

$$R = \sqrt{\frac{-\Theta}{e}} = \sqrt{\frac{53.8 \times 10^{-40} \text{ C m}^2}{1.602 \times 10^{-19} \text{ C}}} = 1.83 \times 10^{-10} \text{ m}.$$

Problem 15.7 shows that $R = 2\sqrt{3}a_0 = 3.464\ a_0$. This number can be compared to $\langle \hat{r} \rangle$ for the $2p_z$ state, which Chapter 12 shows is $5a_0$.

➞ *RELATED PROBLEM* 15.7

At this point, it is interesting to return to quantum mechanics and reconsider the general symmetry requirements for dipole moments. For an isolated molecule in the absence of all external fields, we know that the Hamiltonian depends on electron and nuclear coordinates in a simple way. The kinetic energy operator is a "take a second derivative" operator, and the Coulombic potential-energy operators depend only on *relative* rather than absolute positions. Thus, neither term is affected by the *coordinate inversion operator*[4] $\hat{\imath}$ that, for every particle, turns (x, y, z) into $(-x, -y, -z)$. We can say that $\hat{\imath}$ and $\hat{H}$ commute, or, equivalently, that $\hat{H}$ is invariant to $\hat{\imath}$ and the energy of a state is not changed by $\hat{\imath}$. Thus we can find wavefunctions that are simultaneously eigenfunctions of $\hat{H}$ and $\hat{\imath}$. Obviously, $\hat{\imath}\cdot\hat{\imath}$ (doing $\hat{\imath}$ twice) gives us what we began with, so that

$$\hat{\imath}^2\psi = \psi$$

and, if ψ is an eigenfunction of $\hat{\imath}$, we must have

$$\hat{\imath}\psi = (\pm 1)\psi$$

so that the eigenvalues of $\hat{\imath}$ are $+1$ or -1. An eigenfunction with a $+1$ eigenvalue is said to have *even parity*, and -1 implies *odd parity*. For example, the H-atom wavefunctions with even l have even parity and those with odd l have odd parity.

EXAMPLE 15.3

Prove that H-atom wavefunctions with even l have even parity and those with odd l have odd parity.

---

[4]This is the same inversion operator we first used in Chapter 14. See Table 14.2.

SOLUTION We write the H-atom wavefunctions from Chapter 12 as

$$\psi_{nlm}(r, \theta, \phi) = R_{nl}(r)\,\Theta_{lm}(\cos\theta)\,e^{im\phi}$$

with Θ_{lm} written as a function of $\cos\theta$ for reasons that will become clear. The inversion operator changes spherical polar coordinates in the following way:

$$r \to r, \quad \theta \to \pi - \theta, \quad \phi \to \pi + \phi .$$

The $R(r)$ function is unchanged, and the ϕ function becomes

$$e^{im(\pi + \phi)} = e^{im\pi}e^{im\phi} = (-1)^m e^{im\phi}$$

since $e^{i\pi} = -1$. The wavefunction Θ_{lm} has the properties

$$\Theta_{lm}(\cos(\pi - \theta)) = \Theta_{lm}(-\cos\theta) = (-1)^{(l-m)}\,\Theta_{lm}(\cos\theta)$$

so that the net result is

$$\hat{\imath}\psi_{nlm}(r, \theta, \phi) = \psi_{nlm}(r, \pi - \theta, \pi + \phi) = (-1)^l\,\psi_{nlm}(r, \theta, \phi)$$

and states with l even (s, d, etc.) have even parity: $(-1)^l = 1$, while odd l states (p, f, etc.) have odd parity: $(-1)^l = -1$. We called even parity MOs "g" and odd parity MOs "u" in Chapter 14.

⇒ **RELATED PROBLEMS** 15.16, 15C

The multipole moments can be written as operators as well. The *dipole-moment operator* is a vector operator with components

$$\hat{p}_x = \sum_i q_i x_i, \quad \hat{p}_y = \sum_i q_i y_i, \quad \text{and} \quad \hat{p}_z = \sum_i q_i z_i$$

summed over all charges i, including nuclei. The dipole-moment operator has *odd* parity, since the inversion operator changes $x_i \to -x_i$, etc.

The expectation value of the dipole-moment operator is the observable dipole moment. For a wavefunction Ψ of definite parity we write

$$\langle \hat{p} \rangle = \int \Psi^* \hat{p}\, \Psi\, d\tau = \int \hat{p}\, \Psi^* \Psi\, d\tau$$

and recognize that the product $\Psi^*\Psi$ has *even* parity (no matter what the parity of Ψ) so that the full integrand, $\hat{p}\,\Psi^*\Psi$ has overall *odd* parity. The integration extends symmetrically about the origin to infinity, and the integral is *exactly zero by symmetry*. Thus, $\langle \hat{p} \rangle = 0$ for systems of definite parity.[5]

A Closer Look at Parity

Following his discussion of parity, A. D. Buckingham (an acknowledged contemporary successor to Peter Debye in this area) wrote (see *Further Reading*), "The conclusions of the previous paragraph require that a diatomic molecule . . . like HCl does not possess a dipole moment in any of its stationary states. However, we have it on high authority (Debye, 1929) that the dipole moment of HCl is approximately 1 D." How can this apparent contradiction be resolved, and what does it mean to "have definite parity"?

---

[5]This argument turns out to be incomplete for molecules with a molecular rotational angular-momentum component along one molecular symmetry axis. (It takes a three-fold rotational symmetry axis to do this.) Such molecular states are degenerate and have to be treated differently.

The key phrase in our discussion of parity is that we *can* find wavefunctions that are simultaneously eigenfunctions of $\hat{H}$ and $\hat{\imath}$, but writing, "$\hat{\imath}^2 \psi = \psi$; therefore, $\hat{\imath}\psi = \pm\psi$," begs the question. It may well be that $\hat{\imath}\psi = \psi'$ with $\psi' \neq \pm\psi$, but we would still have $\hat{\imath}^2\psi = \hat{\imath}\psi' = \psi$. For example, let ψ be the wavefunction for HCl (with the coordinate origin at H) so that ψ and $\hat{\imath}\psi = \psi'$ look something like this:

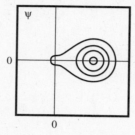 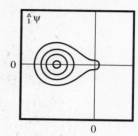

Neither of these wavefunctions "have definite parity," and both have the same energy and non-zero dipole moment. However, we can make wavefunctions with the same energy *and* with definite parity *via* the linear combinations

$$\psi^{\pm} = \frac{1}{\sqrt{2}}(\psi \pm \hat{\imath}\psi) = \frac{1}{\sqrt{2}}(\psi \pm \psi')$$

that look like this:

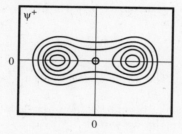 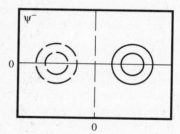

Neither of these would have a dipole moment, but then neither looks like HCl! This type of construction, however, is useful in understanding *chiral* molecules and how such left-handed and right-handed molecules are related to each other.

Finally, when we measure or calculate dipole moments, we do so with respect to a coordinate system *fixed to and rotating with* the molecule, and the Uncertainty Principle prohibits us from orienting this axis in space (see also Fig. 12.12). Averaging over all orientations in a laboratory fixed coordinate system causes $\langle p \rangle = 0$.

An important adjunct to the total molecular dipole moment is the *localized bond moment*. Here, we use the measured moment for one molecule to estimate the moments for a homologous series of molecules. Consider the dipole moments of monochlorobenzene, *o*-dichlorobenzene, and *m*-dichlorobenzene in Figure 15.2. The dipole moment of monochlorobenzene is 5.64×10^{-30} C m (or 1.69 D), and electronic-structure arguments indicate that, to a first approximation, this moment is due to the C—Cl bond. Thus, we assign this bond a local moment of 1.69 D.

In the two polar dichloro isomers, we estimate total moments with a simple *vector* sum of the C—Cl local moments. The *ortho* isomer's dipole moment vector bisects the two C—Cl bond directions with a magnitude equal to that of the vector sum of two C—Cl moments 60° apart. The calculation, shown in the figure, predicts

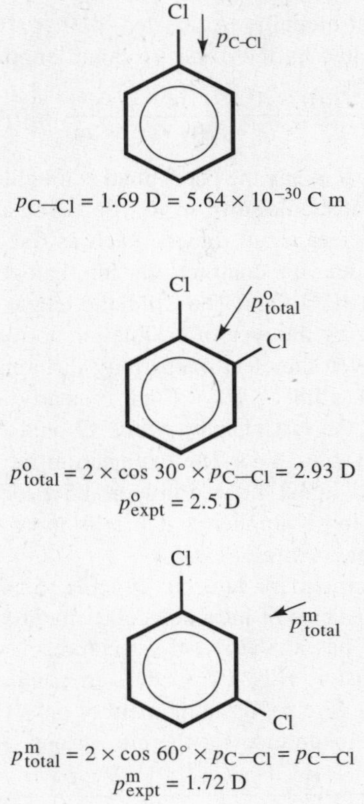

15.1 BASIC PROPERTIES OF MOLECULAR CHARGE DISTRIBUTIONS

FIGURE 15.2 Local bond dipole moments can be used to estimate molecular moments, as is illustrated here for the two polar dichlorobenzenes.

$p_{C-Cl} = 1.69\ D = 5.64 \times 10^{-30}\ C\ m$

$p^o_{total} = 2 \times \cos 30° \times p_{C-Cl} = 2.93\ D$
$p^o_{expt} = 2.5\ D$

$p^m_{total} = 2 \times \cos 60° \times p_{C-Cl} = p_{C-Cl}$
$p^m_{expt} = 1.72\ D$

2.93 D, in pretty good agreement with the observed 2.5 D moment. Likewise, the *meta* isomer has a predicted moment that exactly equals the monochloro isomer, and again, the observed and predicted values are in close agreement.

The predictions of a local-moment calculation can do more than estimate an unknown moment. They can also help explain total moments when the two disagree sharply. For example, the observed moments of fluorobenzene and iodobenzene are quite closely the same: 1.60 D and 1.70 D, respectively, predicting a moment for *p*-iodofluorobenzene around 0.1 D or 3.3 D (depending on whether one adds or subtracts the local moments), but the observed value is 0.89 D, indicating an electron redistribution not assumed by the local-moment model.

A second use is in the study of molecular conformation changes. The local moment description attaches little dipole vectors to each heteropolar bond. As the entire molecule folds or twists or rotates into varying configurations, the vector net moment changes, often by significant amounts. An observed total moment can be compared to a set of predicted moments for varying configurations, and even if a unique match does not appear, often one can at least rule out many possible conformations.

The magnitudes of net molecular dipole moments involve three factors. The most complicated is the molecular-geometry factor we have just discussed, but in simpler molecules, p can profitably be interpreted in terms of the other two factors: the *amount* of charge transferred and the *distance* over which it is transferred. For example, the NaCl dipole moment is $30.0 \times 10^{-30}\ C\ m$, which is the moment (see Figure 15.1) of an electron and a proton separated by 1.87 Å. The bond length in NaCl is 2.36 Å, a comparable but larger distance. A better way to compare the molecule and the simplest model is to say that 30.0×10^{-30} C m corresponds to a *fractional* charge separated by the *observed* bond length:

$$\frac{(30.0 \times 10^{-30} \text{ C m})}{(1.602 \times 10^{-19} \text{ C}/e^-)(2.36 \times 10^{-10} \text{ m})} = 0.794 \ e^- .$$

It is this calculation that lets us say the NaCl bond is roughly 80% ionic in character.

This is no surprise. What should surprise you, however, are the tabulated moments for some of the heteronuclear alkali dimers, such as CsNa, etc. They seem to be ridiculously large! Consider, for contrast, the interhalogens, such as ClF with a dipole moment of 2.94×10^{-30} C m. The ClF bond length is 1.63 Å, and we would say the bond is 11% ionic by the type of calculation above. This percentage seems chemically acceptable, given the electronegativity differences between Cl and F.

But the CsNa moment, 15.8×10^{-30} C m, is nearly as large as that for LiH, our ionic prototype from the last chapter, while Cs and Na have nearly identical electronegativities. (See Table 14.4.) The explanation (at least in part) lies in the unusually long CsNa bond, about 4.0 Å, implying 25% ionic character. This is still a somewhat surprisingly large number, but it tells more of the story of Cs—Na bonding than the dipole moment alone can.

Similarly the large moment for aqueous glycine, 52.4×10^{-30} C m, can be explained by noting the distance of intramolecular charge transfer. In dilute neutral aqueous solution, glycine has a *zwitterionic* structure in which there is a genuine large distance charge transfer: $H_3N^+CH_2COO^-$. In contrast, nitroethane, with the same empirical formula as glycine, has a moment of only 10.8×10^{-30} C m, which is a typical local moment magnitude for the nitro group: $H_3CCH_2NO_2$.

A Closer Look at the Multipole Expansion

The most general expression for the electric potential at a point **r** from a charge distribution $\rho(\mathbf{r}')$ is

$$\phi(\mathbf{r}) = \int \frac{\rho(\mathbf{r}')d\mathbf{r}'}{4\pi\varepsilon_0|\mathbf{r} - \mathbf{r}'|} .$$

Equation (15.2) follows from this when ρ is localized in space and the point at **r** is far away.

You may have recognized that the angular factors in Equations (15.3 a–c) are the H atom's s, p_z, and d_{z^2} angular wavefunctions. This is not by chance. It arises from that amazing equality we first saw in a closer look at the "r_{12} integral" in Chapter 13:

$$\frac{1}{|\mathbf{r} - \mathbf{r}'|} = \sum_{l=0}^{\infty} \sum_{m=-l}^{l} \frac{4\pi}{2l + 1} \frac{r_<^l}{r_>^{l+1}} Y_{lm}(\theta', \phi')Y_{lm}^*(\theta, \phi)$$

where $r_<$ is either r or r', whichever has the lesser magnitude, $r_>$ is whichever has the greater magnitude, Y and Y^* are the spherical harmonic function and its complex conjugate, and (θ, ϕ) and (θ', ϕ') are the spherical polar angles of **r** and **r**'.

Substitution of this powerful expression into the general expression for $\phi(\mathbf{r})$ leads to *angle averaged* charge distributions

$$\rho_{lm}(r') = \iint \rho(r', \theta', \phi')P_{lm}(\cos\theta)e^{im\phi'} \sin\theta' \ d\theta' \ d\phi'$$

where $P_{lm}(\cos\theta)$ is the associated Legendre polynomial that makes up part of the spherical harmonics. We can also define *moments* of these distributions as

$$Q_{lm} = \int \rho_{lm}(r')r'^{l+2} \ dr'$$

and finally write in a compact way

$$\phi(\mathbf{r}) = \sum_{l=0}^{\infty} \sum_{m=-l}^{l} \frac{(l-|m|)!}{(l+|m|)!} \frac{Q_{lm}}{r^{l+1}} P_{lm}(\cos\theta) e^{-im\phi} .$$

In Eq. (15.2) (for a linear molecule, with the z axis taken along the molecular axis), we can associate

$$q = Q_{00}, \quad p = p_z = Q_{10}, \quad \text{and} \quad \Theta = \Theta_{zz} = 2Q_{20}$$

and, in general,

$$p_x = \frac{1}{2}[Q_{11} + Q_{1-1}], \quad p_y = \frac{1}{2i}[Q_{11} - Q_{1-1}]$$

with similar expressions for the remaining components of Θ.

The multipole analysis gives the electric field (or potential) generated by a particular molecular charge distribution. Another basic property of such a distribution is its response to an *external* electric field. This field could be generated by a laboratory power supply connected to a set of conducting plates or it could be generated by a neighboring molecule, as we remarked in the introduction. We concentrate now on that part of the response known as the *molecular polarizability* and seek an expression for the change in molecular energy any such field causes.

Polarizability was first discussed in Chapter 9 for macroscopic systems. Here, we are interested in the polarizability of a single atom or molecule. Imagine placing a spherical atom in an electric field and measuring its charge distribution. Then take that distribution and calculate its multipole moments. The dipole moment would not be zero. Repeating this experiment for a variety of electric fields (all smaller than the internal fields of the atom) would show that the *induced dipole moment* is a linear function of the external field:

$$p_{\text{ind}} = \alpha E \tag{15.4}$$

where p_{ind} is the induced moment, E is the field magnitude, and α is the *atomic polarizability*.

A classical picture of the origin of p_{ind} is shown in Figure 15.3 for a spherical atom in a uniform external electric field. We will discuss a quantum mechanical view shortly. As the figure shows, the atomic charge distribution reacts to the field by shifting charge in the directions of the forces felt by them. (Compare Figure 15.3 to Figure 9.2.)

EXAMPLE 15.4

Table 15.1 mentions the molecular ion ArH^+. Its bond length is 1.280 Å. If we picture ArH^+ as a bare proton 1.28 Å from Ar, what is the dipole moment induced in Ar by the electric field of the proton? The polarizability of Ar is 1.826×10^{-40} C m^2 V^{-1}.

SOLUTION The electric field a distance R from a proton is $e/(4\pi\epsilon_0 R^2)$ so that the induced moment is (Eq. (15.4))

$$p_{\text{ind}} = \frac{\alpha e}{4\pi\epsilon_0 R^2} = \frac{(1.826 \times 10^{-40})(1.602 \times 10^{-19})}{4\pi(8.854 \times 10^{-12})(1.280 \times 10^{-10})^2} \text{ C m} = 16.0 \times 10^{-30} \text{ C m} ,$$

a value 1.6 times greater than the observed total moment (Table 15.1). This molecule's dipole moment is looked at further in Problem 15.10.

➡ *RELATED PROBLEMS* 15.10, 15.23, 15.24

FIGURE 15.3 When a spherical atom (top) is placed in a uniform electric field (bottom), a charge distortion occurs, leading to an induced dipole moment p_{ind}. The proportionality constant between p_{ind} and the field E is the polarizability α.

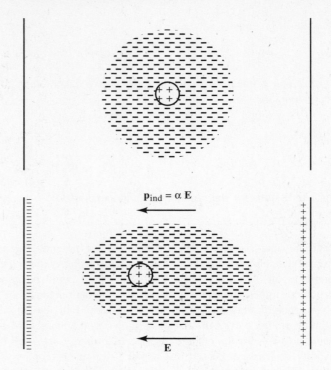

For nonspherical molecules, p_{ind} has different values for different orientations of the field with respect to the molecule. Consider, for example, H_2 in a typical laboratory field of 10^5 V m^{-1}. If the H_2 is aligned parallel to this field, the observed p_{ind} is 1.14×10^{-34} C m. (Note how small this is compared to *permanent* dipole moments, which are typically 10^4 times larger.) Orient H_2 *perpendicular* to the field, and p_{ind} drops to 0.794×10^{-34} C m. The electron charge is more easily shifted *parallel* to the bond axis than perpendicular to it.

Suppose the H_2 is oriented at a 45° axis to the field direction, as shown and contrasted to the first two cases in Figure 15.4. In this case, the *magnitude* of p_{ind} has the intermediate value 0.905×10^{-34} C m, but the *vector* $\mathbf{p}_{ind}$ is *no longer parallel* to $\mathbf{E}$. Instead, it makes an angle of 34.8° to the molecular axis due to the greater parallel component of the molecular polarizability.

Molecular polarizability is a more complicated quantity than a single number can express. For a linear molecule (or any molecule with a three-fold or higher axis of rotational symmetry, such as CH_3F, benzene, etc.) α has both a parallel and a perpendicular component, $\alpha_\parallel$ and $\alpha_\perp$. If z is the molecular symmetry axis, the components of p_{ind} are

$$p_x = \alpha_\perp E_x, \quad p_y = \alpha_\perp E_y, \quad \text{and} \quad p_z = \alpha_\parallel E_z. \tag{15.5}$$

For H_2, $\alpha_\parallel = 11.4 \times 10^{-41}$ C m^2 V^{-1} and $\alpha_\perp = 7.94 \times 10^{-41}$ C m^2 V^{-1}.

One often finds tabulated the *average polarizability*

$$\overline{\alpha} = \frac{1}{3}(\alpha_\parallel + 2\alpha_\perp) \tag{15.6}$$

and the *polarizability anisotropy*, $\alpha_\parallel - \alpha_\perp$. These are more useful than $\alpha_\parallel$ and $\alpha_\perp$ alone, in that $\overline{\alpha}$ gives the average induced moment for an ensemble of randomly oriented molecules while $\alpha_\parallel - \alpha_\perp$ measures the relative stiffness of the electronic

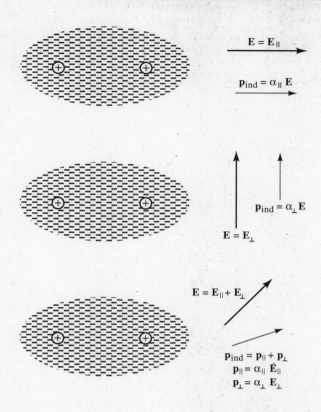

FIGURE 15.4 A linear molecule, such as H_2, is polarizable to different degrees in directions parallel and perpendicular to the bond direction. As a result, the induced dipole moment may not point along the field direction, as the lower panel shows.

charge in the parallel and perpendicular directions. Another measure of this stiffness is the dimensionless *polarizability anisotropy ratio* η:

$$\eta = \frac{\alpha_\parallel - \alpha_\perp}{\bar{\alpha}} . \qquad (15.7)$$

The units of α, C m² V⁻¹, are somewhat awkward and can be improved. Classical electrostatics shows that the polarizability of a perfectly conducting sphere of radius r is $4\pi\epsilon_0 r^3$. This suggests defining *polarizability volumes* α' as

$$\alpha' = \frac{\alpha}{4\pi\epsilon_0} \qquad (15.8)$$

with SI units of m³. Older literature often quotes α' in Å³ units (1 Å³ = 10^{-30} m³) and refers to these as "the polarizability."

Table 15.2 collects several α' values for spherical atoms (and tetrahedral molecules for which only one number specifies α'). Glance through the values for the rare gases, and note that α' is roughly the same magnitude as the atomic volume, increasing with atomic number. (The cube of the Bohr radius, for comparison, is 0.148×10^{-30} m³.) In fact, if one computes the ratio $\alpha'/(4\pi r^3/3)$ for the rare gases where r is the atomic radius,[6] one finds a smooth trend:

gas	He	Ne	Ar	Kr	Xe
$\alpha'/(4\pi r^3/3)$	0.061	0.067	0.107	0.123	0.141

---

[6] The atomic radius for the rare gases is half the nearest neighbor distance in the solid.

TABLE 15.2 Polarizability volumes α' for atoms, atomic ions, and tetrahedral molecules

			Polarizability volume/10^{-30} m^3			
		H	0.666 831	Be^{2+}	0.007 65
H$^-$	13.8	He	0.204 956	Li$^+$	0.028 6
F$^-$	0.759	Ne	0.395 6	Na$^+$	0.255
Cl$^-$	2.974	Ar	1.641 1	K$^+$	1.201
Br$^-$	4.130	Kr	2.484 4	Rb$^+$	1.797
I$^-$	6.199	Xe	4.044	Cs$^+$	3.137
Li	24.3	He*	46.6	CH$_4$	2.593
Na	23.6	Ar*	48.4	CF$_4$	3.838
K	43.4	O^{2-}	2.8	CCl$_4$	10.5
Rb	47.3	S^{2-}	5.9		
Cs	59.6	Se^{2-}	6.5		

Notes: The uncertainty in the neutral alkali atom values is typically $\pm 2 \times 10^{-30}$ m^3. The ion values are derived from optical measurements of alkali halide crystals. The two excited states listed, He*($1s^12s^1$ $^3S_1$) and Ar*($3p^54s^1$ $^3P_2$), are metastable triplet states that can be studied experimentally. For tetrahedral molecules, a single value of α' describes the polarizability in any direction. For H, $\alpha' = 9a_0^3/2$. In general, $\alpha = 4\pi\epsilon_0\alpha'$.

Compare also the values for an isoelectronic series such as H$^-$, He, and Li$^+$ or Cl$^-$, Ar, and K$^+$, and note the marked decrease in α' as the nuclear charge increases for a constant number of electrons. This shows that polarizability also measures how tightly the electrons are bound.

Molecular values of $\bar{\alpha}'$, $\alpha_{\parallel}' - \alpha_{\perp}'$, and η are collected in Table 15.3. The negative $\alpha_{\parallel}' - \alpha_{\perp}'$ and η values for cyclopropane and benzene mean that these molecules are more polarizable *in* the plane of the C atoms than they are perpendicular to it. (The $\parallel$ axis for symmetric nonlinear molecules is the molecular axis of greatest rotational symmetry.) A small η implies highly isotropic polarizability ($\eta = 0$ for spheres, tetrahedra, and octahedra) while a large *magnitude* (η ranges from -1.5 to $+3$) implies high anisotropy. The molecules C$_2$H$_2$, CO$_2$, CS$_2$, and N$_2$O, for example, are much more polarizable *along* the molecular axis than perpendicular to it. The explanation lies in the extended π bonding in these molecules. The π system is like a tube of charge density around the nuclei. The tube can be most readily distorted *along* its axis.

The Clausius–Mossotti equation, Eq. (9.34a), and the Debye extension for polar molecules, Eq. (9.35), described how to find α from relative permittivity measurements. We will see other techniques for measuring α in this chapter's final section. Now we look briefly at how α is calculated, and in the process we will obtain an interesting interpretation of the origin of polarizability.

Imagine the external field is small compared to the internal fields of the molecule so that perturbation theory can be used. We assume we know the exact wavefunctions for the molecule in a molecule-fixed coordinate system, and we ask what happens to its ground state. We also assume the molecule is linear and has a nondegenerate ground state.

Of the several experimental methods for measuring α, each relies at some point on knowing the energy of the molecule in the field. This will be our point of attack. Since E is a natural symbol for both energy and electric field, we will use E for energy and $\boldsymbol{F}$ for the electric field vector (or F for its magnitude) in what follows.

TABLE 15.3 Molecular average polarizability volumes $\bar{\alpha}'$, polarizability volume anisotropies $\alpha'_\parallel - \alpha'_\perp$, and polarizability anisotropy ratios η

Molecule	$\bar{\alpha}'/10^{-30}$ m³	$(\alpha'_\parallel - \alpha'_\perp)/10^{-30}$ m³	η
H_2	0.802	0.314	0.392
D_2	0.792	0.299	0.378
N_2	1.74	0.70	0.40
CO	1.95	0.53	0.27
O_2	1.581	1.10	0.696
Cl_2	4.61	2.98	0.646
HF	2.46	0.72	0.29
HCl	2.63	0.74	0.28
HBr	3.61	0.910	0.252
HI	5.44	1.690	0.311
CO_2	2.91	2.10	0.722
CS_2	8.74	9.60	1.098
OCS	5.71	4.18	0.732
N_2O	3.03	2.79	0.921
NH_3	2.26	0.288	0.127
C_2H_2	3.33	2.69	0.808
C_3H_6	5.66	−0.81	−0.14
C_6H_6	10.4	−5.68	−0.546
CH_3Cl	4.72	1.55	0.328
$CHCl_3$	8.56	−2.68	−0.313

Note: The $\parallel$ axis is the molecular axis for linear molecules or the axis of greatest rotational symmetry for nonlinear molecules; that is, the axis perpendicular to the plane of the C atoms in C_3H_6 (cyclopropane) and benzene.

For simplicity, let F be constant along the laboratory z axis. (Remember the molecule-fixed axes are *rotating* with respect to this direction.)

The perturbation Hamiltonian for a weak, constant field interacting with a neutral molecule is

$$\hat{H}' = -\left(\sum_i q_i \mathbf{r}_i\right) \cdot \mathbf{F} \tag{15.9}$$

where the sum is over all the charges in the molecule ($q_i = -e$ for an electron, $= Z_i e$ for the ith nucleus). This sum is the total dipole-moment operator:

$$\hat{p} = \sum_i q_i \mathbf{r}_i . \tag{15.10}$$

For a linear molecule, $\hat{p}$ has only a single non-zero component, and Eq. (15.9) reduces to

$$\hat{H}' = -\hat{p} F \cos\theta \tag{15.11}$$

where θ is the angle between the electric field vector and the dipole-moment vector.

Now we expand $E(F)$, the energy as a function of field strength, in a Taylor's series about the known zero-field ground-state energy E_0:

$$E(F) = E_0 + \left(\frac{\partial E}{\partial F}\right)_0 F + \frac{1}{2}\left(\frac{\partial^2 E}{\partial F^2}\right)_0 F^2 + \frac{1}{6}\left(\frac{\partial^3 E}{\partial F^3}\right)_0 F^3 + \cdots$$

Note that $\hat{p}\cos\theta$ is the lab's z-axis component of the dipole-moment operator and that

$$\left(\frac{\partial \hat{H}'}{\partial F}\right) = -\hat{p}\cos\theta = -\hat{p}_z \ .$$

There is a theorem known as the *Hellmann–Feynman* theorem that relates this derivative to $(\partial E/\partial F)$. The theorem states simply

$$\frac{\partial E}{\partial F} = \left\langle\frac{\partial \hat{H}}{\partial F}\right\rangle \qquad (15.12)$$

where $\hat{H}$ is the total Hamiltonian and $\langle\ \rangle$ represents an expectation value. Thus, since $\hat{H}'$ is the only term in $\hat{H}$ that depends on F,

$$\frac{\partial E(F)}{\partial F} = -\langle\hat{p}_z\rangle \ . \qquad (15.13)$$

We substitute the Taylor expansion of $E(F)$ into this expression and obtain

$$\langle\hat{p}_z\rangle = -\left(\frac{\partial E(F)}{\partial F}\right) = -\left(\frac{\partial E}{\partial F}\right)_0 - \left(\frac{\partial^2 E}{\partial F^2}\right)_0 F - \frac{1}{2}\left(\frac{\partial^3 E}{\partial F^3}\right)_0 F^2 - \cdots \qquad (15.14)$$

This is exact. We have only required the field to be small. Now we apply perturbation theory to $\hat{H}'$ and seek the first two corrections to the ground-state energy

$$E_0(F) = E_0^{(0)} + E_0^{(1)} + E_0^{(2)} \ .$$

We know $E_0^{(0)} = E_0$, and Equations (13.16) and (13.18) tell us

$$E_0^{(1)} = H_{00}' = -\int \psi_0^* (\hat{p}\, F \cos\theta)\, \psi_0\, d\tau$$

and

$$E_0^{(2)} = \sum_{m\neq 0} \frac{H_{0m}'\, H_{m0}'}{E_0 - E_m}$$

or, pulling the field dependencies from the H' integrals and using $p_z = p\cos\theta$,

$$E_0^{(1)} = -F p_{z,00} \qquad (15.15a)$$

$$E_0^{(2)} = F^2 \sum_{m\neq 0} \frac{|p_{z,m0}|^2}{E_0 - E_m} \ . \qquad (15.15b)$$

Comparing these terms to the Taylor's expansion of $E(F)$ lets us associate terms with the same power of F, and we can define

$$p_{z,00} = -\left(\frac{\partial E}{\partial F}\right)_0 = \text{the permanent dipole moment}$$

$$-2\sum_{m\neq 0}\frac{|p_{z,m0}|^2}{E_0 - E_m} = -\left(\frac{\partial^2 E}{\partial F^2}\right)_0 = \text{the polarizability} \ .$$

These are operative definitions. In particular, the polarizability here is $\alpha_\parallel$, the component parallel to the molecular axis. Note that we carried the $E(F)$ Taylor's expansion through third order, but stopped perturbation theory at second order. Had we continued perturbation theory, we would have found an expression for $(\partial^3 E/\partial F^3)$, a *hyperpolarizability* that is important at high electric fields such as those generated by light pulses from intense lasers or nearby ions.

15.1 BASIC PROPERTIES OF MOLECULAR CHARGE DISTRIBUTIONS

We can also go back to the Hellmann–Feynman expression and write a perturbation expansion for $\langle \hat{p}_z \rangle$, the expectation value of the dipole moment in the field, as

$$\langle \hat{p}_z \rangle \;=\; p_{z,00} \;-\; 2\sum_{m \neq 0} \frac{|p_{z,m0}|^2}{E_0 - E_m} F \;+\; \cdots ,$$

$\uparrow$ Total dipole moment in field F $\uparrow$ Permanent moment $\uparrow$ Field induced moment $(+\alpha F)$ $\uparrow$ Higher-order induction terms

which shows that the expectation value for the dipole moment is the permanent moment plus a polarization term linear in field strength, plus higher (and smaller) terms. Note the minus sign and recall that the second-order perturbation sum for the ground state is itself always negative: $|p_{z,m0}|^2$ is positive, but $E_0 - E_m$ is negative for all $m \neq 0$. Thus, α is inherently positive for the ground state.

We can make an illuminating approximation to this sum. First, we approximate all the energy difference denominators by a constant, average amount. This is reasonable for molecules in which the first excited-state energy is an appreciable fraction of the ionization energy, as is often the case. Thus $E_1, E_2, \ldots, E_+$ where E_+ represents the ionization energy are all assumed to be similar in magnitude, and we write

$$\alpha = -2\sum_{m \neq 0} \frac{|p_{z,m0}|^2}{E_0 - E_m} \cong \frac{2}{\Delta E}\sum_{m \neq 0} |p_{z,m0}|^2$$

where $\Delta E \equiv$ (the representative average excited-state energy) $- E_0 > 0$. Next, we extend the sum to include $m = 0$ by adding and then subtracting $|p_{z,00}|^2$:

$$\alpha = \frac{2}{\Delta E}\left[\left(\sum_{m=0}^{\infty} |p_{z,m0}|^2\right) - |p_{z,00}|^2\right].$$

The sum, which in detail is

$$\sum_{m=0}^{\infty} \int \psi_0^* \hat{p}_z \psi_m \, d\tau \int \psi_m^* \hat{p}_z \psi_0 \, d\tau ,$$

can be shown[7] to equal the *single* term

$$\int \psi_0^* \hat{p}_z^2 \psi_0 \, d\tau .$$

Thus, our approximate expression for polarizability looks like

$$\alpha \cong \frac{2}{\Delta E}(\langle \hat{p}_{z,00}^2 \rangle - \langle \hat{p}_{z,00} \rangle^2)$$

where the terms in parentheses represent the *mean square deviation of p_z from its mean value*.

These terms are called the dipole moment *fluctuation*. At any instant, the charges in the molecule (whether it has a permanent moment or not) are likely to be arranged in space so that the charge distribution yields an instantaneous non-zero dipole moment vector. An instant later, the charges have moved, and this vector has changed magnitude and direction. The ordinary permanent moment is the time average, in effect, of this flickering, fluctuating instantaneous moment. What we see here is that the polarizability is proportional to the mean square deviation of this flicker about the average dipole-moment magnitude.

---

[7] Heisenberg's version of quantum mechanics makes this next step crystal clear. It is somewhat more difficult to show via Schrödinger's formulation. Hence, we merely state the result. (See also Problem 12.14. Think of the integrals $p_{z,m0}$ as entries in a matrix of all $p_{z,mn}$ values.)

A Closer Look at the Hellmann–Feynman Theorem

This theorem, in general, is easy to prove. The idea is to find an expression for the derivative of the total energy with respect to *any* parameter appearing in the system's Hamiltonian. Here, we used the electric field strength as the parameter. Other possibilities are, for example, the nuclear charge number Z or the reduced mass of an oscillator, etc.

We consider any state with the exact wavefunction ψ that we assume is normalized. We assume the Hamiltonian contains a parameter ξ. Thus, the total energy E is a function of ξ:

$$E(\xi) = \int \psi^* \hat{H} \psi \, d\tau ,$$

and we take the total derivative of this equation with respect to ξ. We find

$$\frac{\partial E}{\partial \xi} = \int \left(\frac{\partial \psi^*}{\partial \xi}\right) \hat{H} \psi \, d\tau + \int \psi^* \left(\frac{\partial \hat{H}}{\partial \xi}\right) \psi \, d\tau + \int \psi^* \hat{H} \left(\frac{\partial \psi}{\partial \xi}\right) d\tau .$$

Since $E\psi = \hat{H}\psi$,

$$\int \left(\frac{\partial \psi^*}{\partial \xi}\right) \hat{H} \psi \, d\tau = E \int \left(\frac{\partial \psi^*}{\partial \xi}\right) \psi \, d\tau$$

and, since $\hat{H}$ is Hermitian,

$$\int \psi^* \hat{H} \left(\frac{\partial \psi}{\partial \xi}\right) d\tau = \int \left(\frac{\partial \psi}{\partial \xi}\right) (\hat{H}\psi)^* d\tau = E \int \left(\frac{\partial \psi}{\partial \xi}\right) \psi^* d\tau .$$

Thus

$$\frac{\partial E}{\partial \xi} = E \left[\int \left(\frac{\partial \psi^*}{\partial \xi}\right) \psi \, d\tau + \int \left(\frac{\partial \psi}{\partial \xi}\right) \psi^* d\tau \right] + \left\langle \left(\frac{\partial \hat{H}}{\partial \xi}\right) \right\rangle .$$

The terms in square brackets are equal to $d(\int \psi^*\psi \, d\tau)/d\xi$, which, since $\int \psi^*\psi \, d\tau = 1$, is zero. Thus we arrive at the theorem:

$$\frac{\partial E}{\partial \xi} = \left\langle \left(\frac{\partial \hat{H}}{\partial \xi}\right) \right\rangle .$$

It is left as an exercise (see Problem 15.26) to verify this theorem for hydrogenic atoms with Z as the parameter. You may also appreciate knowing that Richard Feynman proved this theorem in 1939 as part of his undergraduate senior thesis!

15.2 SPECIAL CONSEQUENCES OF UNPAIRED ELECTRONS

Chapter 9 described magnetic field effects on bulk matter. Categories of behavior—diamagnetic, paramagnetic, and ferromagnetic—were discussed by and large without recourse to their microscopic origins. We will discuss only a fraction of these origins here, since a full microscopic discussion of diamagnetism alone would take many pages. Thus, we will settle for an abbreviated treatment focusing on the effects of most interest to chemists: those due to unpaired electron spins.

The spin of a free electron produces a *magnetic dipole moment* **m** where

$$\mathbf{m} = g_e \left(\frac{-e}{2m_e}\right) \mathbf{S} = -\frac{g_e \mu_B \mathbf{S}}{\hbar} . \tag{15.16}$$

15.2 SPECIAL CONSEQUENCES OF UNPAIRED ELECTRONS

In this expression,

$$g_e = \text{\textit{free electron g factor}} = 2.002\,319\,304\,386$$
$$\mu_B = \frac{e\hbar}{2m_e} = \text{\textit{the Bohr magneton}} = 9.274\,015\,4 \times 10^{-24}\,\text{J T}^{-1}$$

and

$$\mathbf{S} = \text{\textit{the spin angular momentum}}.$$

In turn, these have the following origins.

First, g_e is a pure number which is *almost* equal to 2. That g_e is not *exactly* 2 is a relativistic result. Relativistic theory predicts

$$g_e = 2\left(1 + \frac{\alpha}{2\pi} - 11.892\frac{\alpha^2}{4\pi^2} + \cdots\right)$$

where α (*not* the polarizability!) is one of the most interesting quantities in physical science, the *fine structure constant*:

$$\alpha = \frac{e^2}{(4\pi\epsilon_0)\hbar c} = 7.297\,35 \times 10^{-3} \cong \frac{1}{137}. \tag{15.17}$$

This is a *dimensionless* quantity composed of fundamental physical constants from seemingly everywhere: electromagnetism (e and ϵ_0), quantum mechanics ($\hbar$), and relativity (c).

The second part, μ_B, is the classical coefficient between magnetic moment and angular momentum, multiplied by $\hbar$. Classically, a circular loop of current i produced by charge q moving with speed v around a circle of radius r, $i = qv/2\pi r$, has a magnetic moment equal to i times the circle's area, πr^2:

$$m_\text{cl} = \underbrace{\left(\frac{qv}{2\pi r}\right)}_{\substack{\uparrow \\ \text{Current in loop} \\ \text{of radius } r}} \underbrace{(\pi r^2)}_{\substack{\uparrow \\ \text{Loop area}}} = \frac{qvr}{2} = \frac{qpr}{2m_e} = \frac{q}{2m_e}L$$

$\uparrow$ Classical moment

where m_e = the electron mass and $L = pr$ = the charge's angular momentum magnitude. For an electron, $q = -e$, and we write the classical moment as

$$m_\text{cl} = -\frac{e}{2m_e}L.$$

The quantum expression for the spin moment, Eq. (15.16), is g_e times larger. Note the last factor, $\mathbf{S}$, the angular momentum. Some authors make $\mathbf{S}$ dimensionless by absorbing $\hbar$ from it into μ_B. We will keep $\mathbf{S}$ with the usual dimensions for angular momentum.

The magnitude of μ_B largely governs the magnitude of $\mathbf{m}$ and the energy changes an external magnetic field can cause. Its value, 9.274×10^{-24} J T^{-1}, is equal to 0.467 cm^{-1} T^{-1} or 5.79×10^{-5} eV T^{-1}. Recall that 1 T (one tesla) is moderately large for laboratory magnetic fields; thus, spin magnetic-energy changes are quite small.

For a field $\mathbf{B}$, the Hamiltonian term is

$$\hat{H}'_s = -\mathbf{m}\cdot\mathbf{B} = \frac{g_e\mu_B}{\hbar}\hat{\mathbf{S}}\cdot\mathbf{B}, \tag{15.18}$$

and, choosing a uniform field along the z axis, the eigenvalues for this operator are simply $+g_e\mu_B B m_s$ where m_s, the spin component quantum number, is $\pm 1/2$.

578   15 ELECTRIC AND MAGNETIC PROPERTIES OF MOLECULES

We rarely deal with free electrons in chemical systems. A careful choice of systems, however, lets us start with the "almost" free electron and then move to more complicated and typical systems. Consider ^{4}He$^+$ in its ground state. The ^{4}He nucleus is spinless, and the single electron has no orbital angular momentum. Therefore, Eq. (15.18) applies, and the energy of ^{4}He$^+$(1s) in a magnetic field splits into *two* values according to m_s. The magnetic field lifts the spin degeneracy of this state. The state with $m_s = -1/2$ is lower in energy than that with $m_s = +1/2$, and both energies are linear functions of the field magnitude, as is shown in Figure 15.5.

Now consider atomic H, again in the 1s state, but with a *nuclear* spin. The spin of the proton has the same $\pm 1/2$ quantum numbers as the electron, but nuclear magnetic moments have two important differences from their electron counterparts. The nuclear magnetic moment is

$$\mathbf{m}_N = \frac{g_N \mu_N \mathbf{I}}{\hbar} \qquad (15.19)$$

where $\mathbf{I}$ is the nuclear spin angular momentum, μ_N is the *nuclear* magneton $+e\hbar/2m_p = \mu_B(m_e/m_p) = 5.050\ 786\ 6 \times 10^{-27}$ J T^{-1} = 3.152×10^{-8} eV T^{-1} (m_p = proton rest mass), and g_N is the *nuclear g factor*. Unlike g_e, g_N varies considerably in magnitude and even in algebraic sign from one nucleus to another.[8] This variability comes from details of nuclear structure, and g_N is not readily calculable. For a proton, $g_N = 5.585\ 7$.

With the Hamiltonian $\hat{H}'_N = -\mathbf{m}_N \cdot \mathbf{B}$, the eigenvalues are $-g_N \mu_N B m_I$. The energy expression for H in a magnetic field now has two terms, electron and nuclear spin, but there is a third: the electron–nuclear spin–spin coupling term. Called the *hyperfine* interaction term,[9] it can be approximated[10] in high fields as

$$\hat{H}'_{hf} = \frac{h a_N \mathbf{S} \cdot \mathbf{I}}{\hbar^2} \qquad (15.20)$$

where a_N (units of Hz) is the *hyperfine coupling constant*. The chemical importance of a_N is that it depends on the magnitude of the *electronic wavefunction at the*

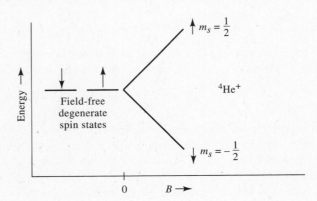

FIGURE 15.5 The energy of the one-electron ion ^{4}He$^+$ varies linearly with B. The charge depends on the spin projection quantum number m_s as shown.

---

[8]Note that the electron spin magnetic moment, Eq. (15.16), contains a minus sign, reflecting the sign of the electron charge and the constancy of g_e, while the nuclear expression, Eq. (15.19), has the algebraic sign of g_N.

[9]"Fine structure" describes the electron spin–orbit interaction (such as the J-dependent splitting of energy levels in atoms) while the smaller "hyperfine" splitting describes nuclear-spin related splittings.

[10]The approximation here neglects the magnetic-dipole–magnetic-dipole interaction between the two spins (which is zero in an s state) and assumes the spins are interacting with the applied field and not with each other. This is called the high-field or *Paschen–Bach* limit, and the only two good quantum numbers are m_I and m_S. See Problems 15.20 and 15.27.

nucleus. This interaction can thus probe the s character of wavefunctions. Including it in the full Hamiltonian introduces a coupling that destroys the individual integrity of **S** and **I**. Much as spin–orbit coupling in atoms produced **J** from **S** and **L**, we consider a total angular-momentum vector **F** = **S** + **I** (+**L** to be complete, but **L** = 0 for an s state) and corresponding quantum number F.

Since $\hat{H}'_{hf}$ is small, we can use perturbation theory to calculate its effects. For H(1s), a_N = 1.420 4 GHz, or ha_N = 5.85 × 10^{-6} eV. Although the coupling term is small compared to typical magnitudes of either spin effect alone, hyperfine splitting exists *even in the absence of an external magnetic field*. This splitting leads to two energy levels (see Figure 12.17) in the field-free 1s state. (The lower level has F = 0, and the other has F = 1.) A transition between these levels (an emission) allows radio frequency telescopes to see H atoms in interstellar space. For other atoms, a detailed theory of the hyperfine splitting (see Problem 15.27) shows that it depends on both I and g_N so that experiments can measure these important nuclear parameters.

A first order perturbation calculation of $\hat{H}'_{hf}$'s effect on the ground state of H yields

$$E^{(1)}_{hf} = ha_N m_s m_I$$

and the total energy expression

$$E = E^{(0)} + E^{(1)}_{hf} = g_e \mu_B B\, m_s - g_N \mu_N B\, m_I + ha_N m_s m_I \qquad (15.21)$$

where $E^{(0)}$ is the field-dependent energy (often called the *Zeeman energy*; the *Zeeman effect* is a generic term for magnetic-field–molecular interactions).

Of these three terms, the first (the electron spin Zeeman term) is by far the largest in typical laboratory magnetic fields. Since each quantum number can be ±1/2, there are four possible energies:

m_s	m_I						
+1/2	+1/2	: $E = +g_e \mu_B B/2$	−	$g_N \mu_N B/2$	+	1/4 ha_N
+1/2	−1/2	: $E = +g_e \mu_B B/2$	+	$g_N \mu_N B/2$	−	1/4 ha_N
−1/2	+1/2	: $E = -g_e \mu_B B/2$	−	$g_N \mu_N B/2$	−	1/4 ha_N
−1/2	−1/2	: $E = -g_e \mu_B B/2$	+	$g_N \mu_N B/2$	+	1/4 ha_N

Figure 15.6 shows these energies, term by term, for one value of magnetic field.

It should be no surprise that molecular systems with electron orbital angular momenta and molecular rotational angular momenta have Zeeman energy expressions much more complex than this one. For example, the NO_2 molecule, with one unpaired electron spin and a non-zero N nuclear spin, has an energy-level structure in a magnetic field that permits >10^6 different spectroscopic transitions! The important case of two unpaired spins (a triplet electronic state) adds the possibility of electron spin–spin coupling as well.

In spite of this complexity, some headway can be made in understanding and interpreting the energy-level scheme of a molecule like NO_2. For example, since the molecule is bent, the unpaired electron can produce a different magnetic moment in each of the three unique molecular directions: perpendicular to the molecular plane, along the O—N—O bond bisector, and perpendicular to this bisector, but in the plane. These directions lead to three different effective g_e values (since the electron is, after all, no longer free). In NO_2, these values are close to the free-electron value, and, along with measured hyperfine coupling constants, they can be interpreted to show that the unpaired electron is in a molecular orbital formed largely

FIGURE 15.6 In contrast to ^{4}He$^+$ in Figure 15.5, atomic H has a more complex behavior in a magnetic field. Not only electron spin, but also nuclear spin and spin–spin interactions need to be considered.

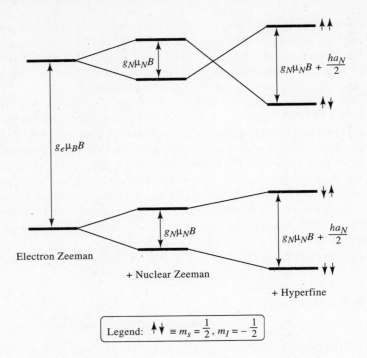

from N and O 2p orbitals in the molecular plane, with some localization towards the O atoms.

We will return to these measurements in Chapter 21 where we discuss spectroscopies that rely on magnetic fields. There we will see that useful chemical information can often be garnered from a choice of Hamiltonian that focuses on the big effects, parameterizes them, and ignores the smaller details.

15.3 THE MOTION OF MOLECULES IN EXTERNAL FIELDS

The last two sections concentrated on the electric and magnetic field *energies* of a molecule. Here, we take the results of those sections and see how the energy expressions can be used to measure some of the molecular parameters, such as the polarizability or magnetic moment.

Consider first the polarizability of an uncharged atom. How can we exploit the change in that atom's energy in an external electric field? We know, for the ground state, that

$$E = E_0 - \frac{1}{2}\alpha F^2 \qquad (15.22)$$

when the field magnitude is F and the zero field energy is E_0. This shows us that an external field *lowers* the ground-state energy.

Suppose we construct a spatially nonuniform electric field. This can be done in many ways (and in fact, a good *uniform* field is harder to construct). One simple arrangement makes one electrode a plane and the other a knife edge, as shown in Figure 15.7. In the vicinity of the edge, the field is quite large compared to its value near the plane so that a field *gradient* is produced away from the knife edge.

Now suppose we place this field in a vacuum and direct a beam of atoms between the electrodes but parallel to the knife edge. The spatially nonuniform field produces

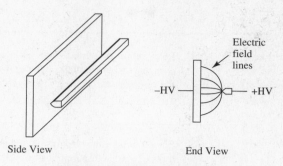

FIGURE 15.7 One technique for measuring polarizabilities involves measuring an atom or molecule's trajectory through an inhomogeneous electric field that deflects the particles an amount proportional to α. The device shown here can produce such a field.

a spatially nonuniform distribution of energy. In particular, let the x direction be perpendicular to the flat electrode so that $(\partial F/\partial x) \neq 0$. From Eq. (15.22) we have

$$\frac{\partial E}{\partial x} = \left(\frac{\partial E}{\partial F}\right)\left(\frac{\partial F}{\partial x}\right) = -\alpha F \left(\frac{\partial F}{\partial x}\right) .$$

The derivative of energy with distance is the negative of the *force* in that direction:

$$\text{force along } x = -\left(\frac{\partial E}{\partial x}\right) = \alpha F \left(\frac{\partial F}{\partial x}\right) .$$

Thus, a beam of atoms is deflected in an inhomogeneous field *towards* the high-field region. If we know the initial position of the beam as it enters the field, its initial velocity, and the field distribution $F(x)$, it is a simple matter of kinematics to calculate the beam's position and velocity at the end of the field. Measuring these quantities leads to α. This technique was first applied by Scheffers and Stark in 1934 to Li, K, and Cs, and electric field effects are called *Stark effects* in general.

Similarly, an inhomogeneous *magnetic* field produces a force proportional to the magnetic moment. This is the basis of the classic Stern–Gerlach experiment that first displayed angular momentum quantization via the discrete deflections of a Ag atom beam.

In both experiments, the magnitudes of the forces and deflections are quite small. Field lengths of about 0.5 m are typical, as are atomic or molecular speeds on the order of 300 m s^{-1}. Thus, the molecule is in the field for a millisecond or two. A typical electric field is $\sim 10^6$ V m^{-1}, and a typical gradient is $\sim 10^9$ V m^{-2}. With a typical polarizability volume of $\sim 10^{-29}$ m^3, the force is on the order of

$$(4\pi\epsilon_0)(10^{-29} \text{ m}^3)(10^6 \text{ V m}^{-1})(10^9 \text{ V m}^{-2}) \cong 10^{-24} \text{ N} .$$

For something weighing 100 g mol^{-1}, the deflection is $\sim 10^{-2}$ mm over a 0.5 m distance. This is on the borderline of observability. Consequently, these experiments are best suited to slowly moving species of large polarizability moving through long, intense, and highly inhomogeneous fields. The heavier alkali metals fit these criteria, as do the highly polarizable metastable excited electronic states of rare gases mentioned in Table 15.2.

CHAPTER 15 SUMMARY

In this chapter, we have looked at the overall charge distribution in a molecule. The total charge, the dipole moment, the quadrupole moment, and the polarizability often provide the best descriptions of the charge distribution, given that a more detailed map is often either impossible to obtain or too complex to interpret.

From this discussion, you should carry away the following major points:

(a) The dipole moment represents two effects: an amount of charge and its location:

$$\mathbf{p} = \int \rho_{el}(r)\, \mathbf{r}\, d\mathbf{r} + e \sum_{nuclei} Z_i \mathbf{r}_i$$

(b) Dipole magnitudes and molecular geometry data reflect molecular symmetry and ionic character information:

$$\% \text{ ionic character} = \frac{100\, p_{bond}}{e \times \text{bond length}}$$

(c) Quadrupole moments typically arise from charge distributions based on the elementary pictures of Figure 15.1 and are the leading multipole moments for linear molecules with centers of symmetry:

$$+ = +\ \text{has}\ \Theta > 0,\quad -\,{}^+_+\,- \ \text{has}\ \Theta < 0$$

(d) Polarizabilities increase with molecular size and/or with increasing numbers of electrons. They measure the response stiffness of charges to external fields:

$$p_{ind} = \alpha E$$

(e) Polarizability anisotropies complement MO pictures of charge distributions by adding a *dynamical* difference to the stiffness of these distributions in varying directions.

The second section outlined the elementary interactions of an unpaired electron and a non-zero nuclear spin with a magnetic field. These interactions contain chemical information about the atomic orbital character of the unpaired electron and about its degree of interaction with the nuclear spin:

$$E = g_e \mu_B m_s B - g_N \mu_N m_I B + h a_N m_s m_I$$

Finally, we saw how field gradients can produce a force on molecules acting through such parameters as polarizability and magnetic moment. These forces can alter molecular trajectories, and they provide a valuable technique for measuring such parameters:

$$\text{force} = \alpha F \left(\frac{\partial F}{\partial x} \right)$$

FURTHER READING

While largely of historical interest today, *Polar Molecules,* by Peter Debye, (Chemical Catalog Co., 1929, reprinted by Dover Publications, NY) is an interesting overview of the field by its master. The quote in the text concerning the parity of HCl is taken from *Electric Moments of Molecules,* by A. D. Buckingham, in *Physical Chemistry, An Advanced Treatise,* D. Henderson, Ed., (Academic Press, NY, vol IV, p. 349, 1970). This is an excellent article on the multipole expansion as applied to molecular charge distributions. It includes a brief table of dipole and quadrupole moments as well as references to more extensive tabulations. Standard handbooks also tabulate dipole moments and polarizabilities. Further details on spin–magnetic-field interactions can be found in *Introduction to Magnetic Resonance,* by A. Carrington and A. D. McLachlan, (Harper & Row, NY, 1967) and *Principles of Magnetic Resonance,* 2nd ed., by C. P. Slichter, (Springer, Berlin, 1978).

PRACTICE WITH EQUATIONS

15A What is the electron charge distribution in a 1s H atom? (15.1), Table 12.4

ANSWER: $\rho_{el}(r) = -e\psi_{1s}^2 = -\dfrac{e}{\pi a_0^3} e^{-2r/a_0}$

15B What is the electric potential 100 Å from an electron? 100 Å from a dipole $p = 10^{-30}$ C m and along the line of the dipole? (15.2)

ANSWER: -0.144 J C^{-1}; $\pm 8.99 \times 10^{-5}$ J C^{-1}

15C What is the parity of the vth harmonic oscillator state? Fig. 12.8

ANSWER: $(-1)^v$

15D At what distance from an electron does an atom of polarizability $\alpha = 10^{-40}$ C m^2 V^{-1} have an induced dipole moment of 10^{-30} C m? (15.4)

ANSWER: 3.79×10^{-10} m (3.79 Å)

15E What is the energy change associated with changing the spin state of a free electron in a 10 T field? (15.18)

ANSWER: 1.85×10^{-22} J (1.16×10^{-3} eV)

15F The magnetic moment of the spin 1/2 ^{13}C nucleus is 6.144×10^{-27} J T^{-1}. What is the ^{13}C g_N value? (15.19)

ANSWER: 1.405

15G The magnetic moment of the ^{7}Li nucleus is 4.203 94 μ_N and $g_N = 2.170\ 7$. What is the spin quantum number S for ^{7}Li? (15.19)

ANSWER: 3/2

15H How much does the energy of a Na atom change when placed between two large parallel metal plates 1 mm apart that have a 10^4 V potential difference? (15.22)

ANSWER: $F = 10^7$ V m^{-1};
$\Delta E = E - E_0 = -1.31 \times 10^{-25}$ J (-8.2×10^{-7} eV)

PROBLEMS

SECTION 15.1

15.1 The dipole moment of toluene is 0.36 D, and that for fluorobenzene is 1.60 D. Predict the dipole moments of o-fluorotoluene and m-fluorotoluene if the moment of p-fluorotoluene is 2.00 D. (The observed values are: o-, 1.37 D; m-, 1.86 D.) Why do you need to know the dipole moment of p-fluorotoluene?

15.2 If we write a diatomic's wavefunction in the VB form of Eq. (14.11), $\psi = c_{cov}\phi_{cov} + c_{ion}\phi_{ion}$, we can call $100 \times |c_{ion}|^2$ the percentage ionic character of the bond. We can also consider $100p/eR_e$ the percentage ionic character so that a list of dipole moments p and bond lengths R_e allow us to calculate $|c_{ion}|$. Carry out this calculation for the species tabulated below.

Molecule	BaO	CO	NO	OH
$p/10^{-30}$ C m	23.53	0.374	0.517	5.74
$R_e/10^{-10}$ m	1.940	1.128	1.151	0.971

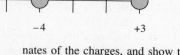

(a)

(b)

1 unit   +1   −2   −4   +3

Next, repeat both calculations ignoring the −2 charge (so that the net charge is zero). This should give you the same dipole moment for both choices of origin.

15.5 The result of the previous problem is easy to prove in general. Suppose we have N charges q_i arranged along the z axis at points z_i. Then the dipole moment is

$$p = \sum_i^N q_i z_i .$$

Now shift the origin by z_0 so that the new locations are at $z' = z - z_0$. Write p in terms of the z' coordi-

Is BaO well described by the ionic formula Ba^{2+}O^{2-}? How should one calculate $|c_{ion}|$ if BaO is very nearly Ba^{2+}O^{2-}?

15.3 A strong laboratory electric field is $\sim 10^7$ V m^{-1}, usually generated by a potential difference on the order of 10^4 V applied across a distance ~ 1 mm. How far away from an electron or proton does one have to go to get the field to drop to this magnitude? What does this tell you about the magnitude of fields inside molecules versus those we apply to them externally?

15.4 For systems with a net charge (such as molecular ions), the dipole moment depends on the choice of coordinate origin. (In general, the nth moment depends on this choice if any of the lower moments are non-zero.) Show that this is true by calculating the dipole moment of the following array of charges assuming first that the origin is at point (a) and then at point (b). Each charge is a multiple of a unit charge and each distance is a multiple of a unit distance as shown.

nates of the charges, and show that p is different only if the net charge (Σq_i) is non-zero.

15.6 An important molecular quantity is the *dipole-moment function*, the variation of p with bond length. What is this function for a molecular ion like ArH$^+$ as the bond length increases to large distances? Qualitatively, what does this function look like for a molecule like LiH? Refer to Figure 14.25 and the discussion of it.

15.7 The quadrupole moment of H in the 2p$_z$ state, -53.8×10^{-40} C m^2, is not difficult to calculate. The charge distribution is $-e\psi^2_{2p_z}$ plus a point $+e$ charge at the nucleus that we can forget about since the quadrupole moment integral (Eq. (15.3c)) includes a factor of r^2 that gives everything at the origin zero weight. The wavefunction is in Table 12.4, and you should show that

$$\rho(r) = -\frac{e}{32\pi a_0^5} r^2 e^{-r/a_0} \cos^2\theta \ .$$

The quadrupole moment is Eq. (15.3c) (which works for any cylindrically symmetric charge distribution, not just those of linear molecules):

$$\Theta = \int \rho(r) r^2 \left(\frac{3\cos^2\theta - 1}{2} \right) r^2 \sin\theta \, dr \, d\theta \, d\phi \ .$$

The following integrals will help, and you should find a final answer $\Theta = -12 e a_0^2$.

$$\int_0^\infty x^n e^{-x} dx = n! \quad \int_0^\pi \cos^2\theta \sin\theta \, d\theta = \frac{2}{3}$$

$$\int_0^\pi \cos^4\theta \sin\theta \, d\theta = \frac{2}{5}$$

15.8 Given the net charge, dipole and quadrupole moments, and bond lengths of a linear triatomic, one can assign effective point charges to each atom. Do this for OCS. To make the arithmetic easier, divide the tabulated p and Θ values by e, express lengths in Å units, and thus find point charges q_O, q_C, and q_S expressed as fractions of e. The atomic positions in Å relative to the OCS center of mass at $x = 0$ are shown below. Interpret your answer qualitatively in terms of electronegativity differences.

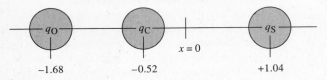

15.9 Which species in each of the following pairs would you expect to have the larger average polarizability?
 (a) HgI_2 or $HgCl_2$ (b) 1-butyne or 1,3-butadiene
 (c) CHF_3 or CH_3F (d) SF_6 or SeF_6
 (e) Cs^+ or Ba^{2+} (f) Xe or Te^{2-}

15.10 Example 15.4 found an induced dipole moment in the Ar atom of ArH$^+$ equal to 16.0×10^{-30} C m, about 1.6 times the measured magnitude. Induction is only part of the story of the total dipole moment, however. Calculate the dipole moment due simply to a proton located $(40/41) \times (1.280$ Å$)$ from the molecular center of mass, pay attention to its vector direction and the direction of the induced moment, add these two vector components of the total dipole moment, and see if the agreement is improved. Can we expect a simple sum of these two basic sources of dipole moment to produce accurate molecular dipole moments? (The factor 40/41 is the ratio of the Ar mass to the ArH mass and measures the bond length fraction of the H$^+$ center of mass coordinate.)

15.11 This and the following two problems consider the polarizability of a very simple system, first classically and then quantum mechanically. The system is a charge $+e$ connected by a harmonic spring to a charge $-e$:

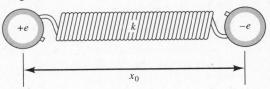

We imagine the Coulombic attraction between the charges has compressed the spring to an equilibrium length x_0, and we apply a uniform electric field F along the spring direction in an orientation that pulls the charges apart to a separation x. We assume the extension, $x - x_0$, is small enough that the system still responds harmonically with a force constant k (a combination of the real spring constant and the Coulombic attraction force). In the absence of the field, $p = ex_0$. With the field, the spring expands to $x > x_0$. Write an expression for the total force on either of the charges, (remember: field is force per unit charge), set this expression to zero to indicate equilibrium, and then manipulate it into a form you can identify as $p_{\text{ind}} = p_{F\ne 0} - p_{F=0} = \alpha F$, and finally identify α.

15.12 Here, we work the previous problem quantum mechanically. We need to find

$$\alpha = -2 \sum_{m \ne 0} \frac{|p_{x,m0}|^2}{E_0 - E_m}$$

for a harmonic oscillator with $\hat{p}_x = ex$. (We consider here only the ground state, but it is easy to show that the answer is the same for all states.) The only matrix element that is non-zero is (see Problems 12.13 and 12.14)

$$p_{x,10} = e\left(\frac{\hbar}{2\mu\omega} \right)^{1/2}$$

where μ is the reduced mass of the charges. You should find α is the same here as in the classical calculation.

15.13 The previous two problems suggest a source of polarizability unique to ionic molecules. The question here is the magnitude of the effect. Consider LiF to be 100% ionic. (Based on its dipole and bond length, it is really about 84% ionic.) For small excursions about the equilibrium bond length, the nuclei vibrate as if they were connected by a harmonic spring of force constant 248.22 N m^{-1}. What contribution does this make to the total polarizability of LiF, which is $\alpha' = 10.8 \times 10^{-30}$ m^3?

15.14 What is the dipole moment—magnitude and direction—induced in an N_2O molecule with its molecular axis oriented (a) at 0°, (b) at 30°, (c) at 60°, and (d) at 90° with respect to an electric field of 10^4 V m^{-1}? Repeat the calculation for the much more isotropically polarizable H_2 molecule. You might find it helpful to derive expressions for $\alpha_\parallel$ and $\alpha_\perp$ in terms of $\overline{\alpha}$ and $\alpha_\parallel - \alpha_\perp$ and to write a general expression for p_{ind} in terms of them and the molecule–field angle.

15.15 For a 1s H atom, the dipole fluctuation expression for the polarizability reduces to

$$\alpha = \frac{2\langle \hat{p}^2_{z,00}\rangle}{\Delta E}$$

since the atom has no permanent moment. If we write $p^2 = p_x^2 + p_y^2 + p_z^2$, recognize that the spherical symmetry of the atom makes $p_x^2 = p_y^2 = p_z^2$, and write $p^2 = e^2 r^2$, the polarizability expression becomes

$$\alpha = \frac{2e^2 \langle r^2\rangle_{1s}}{3\Delta E}.$$

Given that the exact expression for α is $9(4\pi\varepsilon_0)a_0^3/2$ and $\langle r^2\rangle_{1s} = 3a_0^2$, find ΔE and express it in terms of the H-atom ionization energy. If we write $\Delta E = E' - E_1$ where E_1 is the energy of the H-atom ground state, what state of H has an energy closest to E'?

15.16 Show that the inversion operator $\hat{i}$ commutes with the total angular momentum operator $\hat{L}^2$ and with the z component operator $\hat{L}_z$.

SECTION 15.2

15.17 Suppose an electron is a charge $-e$ running around a circle of radius r at the speed of light. Classically, it would have a magnetic dipole moment of magnitude $ecr/2$. Equate this to the true expression, Eq. (15.16), and find r. How does this length compare to another measure of the "size" of an electron, the Compton wavelength $h/m_e c$? Now repeat this calculation for a proton, for which nuclear scattering experiments deduce a radius $\sim 10^{-15}$ m.

15.18 Calculate the energies of the four levels of Figure 15.6 for the case of a hydrogen atom in a 2 T magnetic field. It helps if you first calculate the magnitude of each of the three terms (the two Zeeman terms and the hyperfine term) that contribute to each energy. If you were to draw these energy levels to scale on a piece of paper so that the distance between the highest- and lowest-energy levels is 10 inches, what would be the distance between the two levels closest in energy?

15.19 The deuterium atom has a nuclear spin $I = 1$ so that three m_I values, -1, 0, and $+1$, are possible. How would Figure 15.6 look for the deuterium atom? Label each final energy level with spin component labels such as α_e and β_e for the electron ($m_s = +1/2$, $-1/2$) and α_D, β_D, and γ_D for the deuteron ($m_I = +1, 0, -1$). You will also need to know that g_N for D is 0.857 4. By the way, the hyperfine coupling constant for D is about 1.86 times that for H.

15.20 If the field **B** is large, the H-atom electron-spin–nuclear-spin interaction is smaller than the interaction of either with the field. Thus, to a first approximation, both spins are independently aligned with the field (which we take to be in the z direction) and their residual hyperfine interaction, Eq. (15.20), can be treated by first-order perturbation theory. Derive the first-order energy correction in the text remembering that

$$\hat{S}\cdot\hat{I} = \hat{S}_x\hat{I}_x + \hat{S}_y\hat{I}_y + \hat{S}_z\hat{I}_z$$

and that our assumption of a strong field means the total spin wavefunction can be written as a product of an electron-spin wavefunction and a nuclear-spin wavefunction each of which is an eigenfunction of the corresponding spin's z component operator. (The field is so high that the spins are uncoupled to zeroth-order, and this product of wavefunctions is appropriate.)

15.21 For one-electron atoms in their ground state, the dominant term in the theoretical expression for a_N (a few very small corrections are known) is

$$a_N = \frac{g_e \mu_B g_N \mu_N \mu_0}{3\pi\hbar}\left(\frac{m_N}{m_N + m_e}\right)^3 \psi_{1s}^2(r=0)$$

where μ_0 is the permittivity of vacuum, $4\pi \times 10^{-7}$ T^2 J^{-1} m^3, and m_N is the atom's nuclear mass. Note that the wavefunction is evaluated at $r = 0$; that is, at the nucleus. Consequently this term is called the *Fermi contact term*. Fermi predicted it in 1930. Calculate a_N for H, compare your answer to the value given in the text, then calculate it for ^{3}He$^+$ for which $g_N = -4.255\ 2$ (and $I = 1/2$, by the way). What are the consequences of the sign change for ^{3}He$^+$? What is the biggest factor influencing the change in a_N's magnitude? (See Table 12.4 for the 1s wavefunction.)

SECTION 15.3

15.22 Let's try to invent an isotope separation machine for Ar isotopes. The largest part, 99.6%, of Ar is ^{40}Ar, but 0.337% is ^{36}Ar and 0.063% is ^{38}Ar. Our plan will be to use tunable laser radiation to excite a minor isotope to the excited Ar* (3p^{5}4s^1 ^{3}P$_2$) state, taking advantage of the slight isotope shifts in atomic energy levels. (This is actually the hardest part of the scheme, as Chapter 18 will make clear.) We imagine doing this to a beam of Ar atoms that is then directed into a region of nonuniform electric field so that we can take advantage of the polarizability differences between Ar and Ar* to deflect Ar* out of the beam. If the Ar beam has a speed of 500 m s^{-1} and the field has an average magnitude of 10^6 V m^{-1} and a gradient of 10^9 V m^{-2}, how long does the field have to be to produce a 1 mm deflection of ^{36}Ar* atoms? In the course of answering this question, you will need to calculate the acceleration of Ar* due to the field gradient. How does this acceleration compare to the acceleration due to gravity, 9.8 m s^{-2}?

GENERAL PROBLEMS

15.23 In his classic monograph, *Polar Molecules*, Debye gives a simple calculation of the H$_2$O bond angle based on a purely electrostatic model of the molecule. He considers 100% ionic bonds and assumes the H$^+$ ions are located a radial distance a from the O^{2-} ion as shown below.

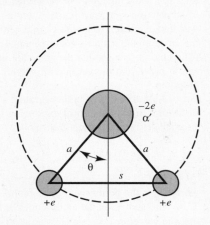

If the O^{2-} ion's polarizability volume α' is taken into account, the potential energy is

$$4\pi\epsilon_0 V(\theta) = \frac{e^2}{s} - \frac{4e^2}{a} - \frac{\alpha'}{2}\left(\frac{2e}{a^2}\cos\theta\right)^2$$

where the first two terms are the H$^+$–H$^+$ repulsion and the two H$^+$–O^{2-} attractions while the third is (see Eq. (15.22)) the polarization energy of O^{2-} in the net electric field $2e\cos\theta/a^2$ of the two protons. Note that $s = 2a\sin\theta$, substitute this into the expression above, and differentiate with respect to θ. Set the derivative to zero, and show that $\cos\theta = 0$ (i.e., a linear molecule) and $\sin^3\theta = a^3/8\alpha'$ locate extrema in $V(\theta)$. For expected values of a and α', the first is a local maximum in V, and thus the linear configuration is unstable. The second is a minimum. What bond angle does it predict? Assume $a = 0.958$ Å, the known O—H bond length. The known bond angle is about 104.5°, or $\theta = 52.25°$.

15.24 Debye goes on to consider water's dipole moment in the electrostatic model of the previous problem. Write an expression for p in terms of its two components: that due to the three ions and that due to the polarization of O^{2-} by the field of the protons. What moment does your expression give using the known bond angle and bond length?

15.25 Dipole moments are almost always measured in terms of the *square* of the moment, and thus the *direction* of the moment is somewhat elusive. For many molecules, such as HF, chemical bonding intuition makes the direction clear: +HF−. Note that Table 15.1 points out the direction of the moment in ClF: +ClF−, as intuition (guided by electronegativity) might well suggest. However, an accurate measurement in 1972 of the magnitude, 0.8881 D, was followed the next year by a difficult experiment that was sensitive to the direction and found 2.1 ± 1.4 D in the surprising sense −ClF+. Another year passed with a high-quality theoretical calculation yielding 0.839 D in the expected sense +ClF−. The matter was settled the following year by a very clever experiment [K. C. Janda, W. Klemperer, and S. E. Novick *J. Chem. Phys.* **64**, 2698 (1976); *ibid.* **65**, 5115 (1976)]. They made the weakly bound molecular complex HFClF, determined its structure and dipole moment magnitude precisely, and argued as follows. The structure has an interesting "antihydrogen bond" in that the Cl of ClF is weakly bound to the F of HF, rather than the H of HF to the ClF as in a normal hydrogen bond:

To a good approximation, the F—Cl—F axis is linear, and the H—F bond is 55° from this axis. The experiment found that the dipole moment component along the F—Cl—F axis is 2.313 D. Assume this moment is the sum of the ClF dipole and the *component* of the HF dipole along the F—Cl—F axis, and compute this sum for the two choices of ClF dipole direction, taking the HF dipole to be in the direction +HF−, which is not in doubt. Which choice gives the better agreement? The agreement can be improved, and was in the original paper, by considering polarization effects as well.

15.26 It is easy to verify the Hellmann–Feynman theorem in the form $\partial E/\partial Z = \langle \partial \hat{H}/\partial Z \rangle$ for hydrogenic atoms of nuclear charge Z. Assume that the nucleus is infinitely heavy so that the Hamiltonian in atomic units is

$$\hat{H} = \hat{T} + \hat{V} = -\frac{1}{2}\nabla_e^2 - \frac{Z}{r}$$

and the energy is $E = -Z^2/2n^2$, and take advantage of the Virial Theorem: $E = \langle \hat{V} \rangle/2$.

15.27 Figure 12.17 shows that the 2s–2p levels of H are split into three levels, very closely spaced, due to spin–orbit and quantum electrodynamics effects. From the top down in energy, these levels have the term symbols $^2P_{3/2}$, $^2S_{1/2}$, and $^2P_{1/2}$. We can add hyperfine splitting to this figure (as was already done for the 1s level) and finish off the H-atom energy-level scheme, at least through the $n = 2$ energies. We start by defining **F** for these levels. Here, **F** = **I** + **J**. What possible F quantum numbers can be associated with each of the three term symbols? A detailed theory of the hyperfine interaction (one that exposes the things that make up our rather phenomenological coupling constant a_N) shows that the hyperfine splittings ΔE_{hf} of a general term of H with quantum numbers n, L, and J follow

$$\Delta E_{hf} = E_{hf} \frac{I + 1/2}{n^3(2L + 1)(J + 1)}, J \leq I, \text{ or}$$

$$E_{hf} \frac{J(J + 1/2)}{n^3(2L + 1)(J + 1)}, J \geq I$$

where E_{hf} is a constant. Calculate the hyperfine splittings of each of these three $n = 2$ terms given the 1s $^2S_{1/2}$ splitting $\Delta E_{hf} = 4.738 \times 10^{-2}$ cm^{-1}.

15.28 As has become tradition over the past three chapters, we end with a question about positronium. The hyperfine splitting in the ground state of ordinary H is due to an interaction between the spin magnetic moments of the electron and proton and amounts to ~0.047 cm^{-1}. In Ps, the corresponding splitting (which is now between what we would simply call the singlet and triplet states) is much larger: ~6.78 cm^{-1}. What *one* difference between Ps and H accounts for this much larger splitting?

CHAPTER 16

The Solid State

16.1 Periodicity and Symmetry in Solids

16.2 Bonding Types in Solid Phases

16.3 Crystallography: Measuring Solid Structures

16.4 Amorphous Solids and Crystal Defects

16.5 Solid Surface Structures

SOLIDS are a most pervasive form of matter. At ordinary conditions, essentially all compounds are solids. Major segments of the economy are based on understanding, controlling, and exploiting narrowly-defined types of solids—metals, plastics, semiconductors, glasses, etc. The physical chemistry of the solid state takes many lines of approach. Thermodynamic properties of solids from simple parameters to complex phase diagrams pervade metallurgy. Molecular engineering of physical and chemical properties pervade polymer science. Electrical properties of solids are ultimately questions of valence electron energies, and optical property design, whether for thermally insulating windows or high purity optical fibers for long distance communication, requires a combination of thermodynamic, kinetic, and molecular structure decisions.

On top of these considerations, or rather surrounding them, is the topic of solid surfaces. The surface atoms of a solid cleaved in vacuum are a vast layer of raw bonds, chemically unsatisfied and often capable of important catalytic chemistry. The past half century has seen surface science grow into a mature discipline of its own.

Each of these focal points of solid state physical chemistry, whether the focus is metallurgy, polymers, electronics, glasses, surfaces, or a host of others, is a topic (or, in fact, a career) unto itself. Chapter 8 described the thermodynamic properties of surfaces and interfaces. In this chapter, we introduce some of the language, concepts, and techniques common to all solid state discussions.

We begin here with a microscopic view of the hallmark of much of the solid state—atomic or molecular periodicity in space. The second section discusses the bonding that leads to these periodicities and structures. Section 16.3 describes solid structure measurements via crystallographic spectroscopy. Then, we leave the world of crystalline regularity and describe the importance of imperfections in amorphous solids and defective crystals. Finally, we discuss surfaces on the atomic level.

16.1 PERIODICITY AND SYMMETRY IN SOLIDS

The macroscopic symmetry of natural crystals is a beautifully striking consequence of the microscopic order of solids. At the microscopic level, the simplest solids are the rare gases and metals, and we begin with them.

The Rare Gases

The rare gas solids, except for He, have the same atomic packing symmetry. It is chemically sensible to expect these atoms to pack like stacks of oranges often are in a supermarket. If you play with about two dozen spheres, you will quickly discover a *close-packed scheme* for symmetrically stacking and packing spheres. Odds are 50-50 that the first such scheme you discover is the one Ne–Xe use, while the *only* other close-packed scheme (with periodicity and symmetry) is used by He.

The hallmark of the crystalline solid is not only atomically local symmetry of one kind or another, but also *long-range periodicity*. This means that one portion of the solid can represent the entire solid. This portion can be cloned again and again and used to build the entire crystal by simple *translation* of that portion along unique directions in the crystal. This representative portion is called the *unit cell*. There is an infinite number of ways to choose a unit cell, but particularly simple and instructive choices can be made.

A related concept is the *Bravais lattice*. Bravais, in 1848, sixty-five years before the first crystal structures were measured by X-ray diffraction, discussed how only 14 types of lattices could exist. The unit cell and the primitive lattice are different but related concepts. We will generally focus on the unit cell, but it is worth using Bravais lattices to establish some terminology.

Figure 16.1 shows these lattices, classified into seven crystal systems. The systems are defined in Table 16.1 in terms of the conventional unit cell side lengths, *a*, *b*, and *c*, and angles, α, β, and γ. Notice how the presence or absence of special lattice points in highly symmetric positions distinguishes different *lattices* of the same *system*.

For example, consider the cubic system. The *simple cubic (sc) lattice* (or primitive cubic, as it is also called) has lattice points at the corners of a cube. These points could be located by Cartesian coordinates such as $(0, 0, 0)$, (a, a, a), $(a, 0, 0)$, etc., where *a* is the cube edge length. The *body-centered cubic* (bcc) lattice has all these points plus one at $(a/2, a/2, a/2)$ at the center of the cube. The *face-centered cubic* (fcc) lattice has the points of sc plus those at $(a/2, a/2, 0)$, $(a/2, 0, a/2)$, $(a/2, a/2, a)$, etc., the centers of the cube faces.

The power of crystal symmetry lies in the small amount of information needed to specify the location of *every atom in the crystal*. The rare gases Ne–Xe all form fcc crystals. For Ar, for example, $a = 5.43$ Å. This number and the fcc label are all we need to locate any atom. We can place one atom at the coordinate origin

FIGURE 16.1 There are 14 distinct Bravais lattices that fill three-dimensional space. These are classified into seven crystal systems. See also Table 16.1.

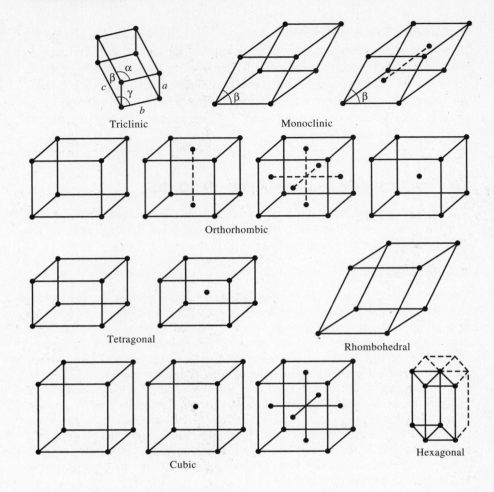

TABLE 16.1 The Seven Crystal Systems

System	Unit-cell dimensions	Lattice types
Triclinic	$a \neq b \neq c \neq a$ $\alpha \neq \beta \neq \gamma \neq \alpha$, none = 90°	Primitive
Monoclinic	$a \neq b \neq c \neq a$ $\alpha = \gamma = 90° \neq \beta$	Primitive, Base centered
Orthorhombic	$a \neq b \neq c \neq a$ $\alpha = \beta = \gamma = 90°$	Primitive, Base centered, Face centered, Body centered
Tetragonal	$a = b \neq c$ $\alpha = \beta = \gamma = 90°$	Primitive, Body centered
Cubic	$a = b = c$ $\alpha = \beta = \gamma = 90°$	Primitive (sc), Face centered (fcc), Body centered (bcc)
Trigonal	$a = b = c$ $\alpha = \beta = \gamma < 120°, \neq 90°$	Primitive rhombohedron
Hexagonal	$a = b \neq c$ $\alpha = \beta = 90°, \gamma = 120°$	Primitive

and locate a corner atom at (5.43 Å, 0, 0) or a face-centered atom at ((5.43/2) Å, (5.43/2) Å, 0), and so on.

Figure 16.2 shows atoms at the fcc lattice positions. Figure 16.2(a) shows the full atom at each point, while Figure 16.2(b) shows the *sliced portions of the atoms that contribute to the unit cell*. Since each face is common to two unit cells, only a fraction of any one atom belongs to one cell. We see from Figure 16.2(b) that the number of atoms in the fcc unit cell is (1/8 atom per corner) × (8 corners) + (1/2 atom per face) × (6 faces) = 4 atoms.

If the atoms are touching spheres, a convenient geometric idea but not necessarily a physically realistic one, we can compute the *packing fraction,* the fraction of the unit cell's volume occupied by spheres. Since a face diagonal is $\sqrt{2}a$ long, or, in terms of the sphere radius r, $4r$ long, we compute

$$\text{fcc packing fraction} = \frac{\text{atomic sphere volume in unit cell}}{\text{unit cell volume}}$$
$$= \frac{(4 \text{ atoms})(4\pi r^3/3 \text{ per atom})}{a^3}$$

but, since $\sqrt{2}a = 4r$ or $a^3 = 16\sqrt{2}r^3$, this is

$$\text{fcc packing fraction} = \frac{(4)(4\pi r^3/3)}{16\sqrt{2}r^3} = \frac{\pi}{3\sqrt{2}} = 0.7405 \ .$$

It is easy to repeat this calculation for the other cubic lattices, and one finds the packing fraction is $\pi/6 = 0.5236$ for sc and $\sqrt{3}\pi/8 = 0.6802$ for bcc. The fcc lattice is called *closest packed* since it has the greatest packing fraction, and the fcc lattice is often called *cubic closest packed* (ccp) for this reason.

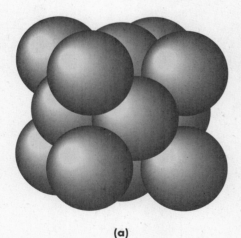

(a)

(b)

FIGURE 16.2 The face centered cubic unit cell has the sphere packing shown in (a). In (b), only the fractions of each sphere that contribute to the cubic unit cell are shown—one eighth of each corner atom and one half of each face atom—for a total of four atoms per unit cell.

It is not chemically surprising that the rare gases condense into closest packed lattices, since we expect them, among all elements, to behave most like spheres under nearly all conditions. What is surprising is that He does *not* form the fcc lattice of the other rare gases.

When ^{4}He condenses, it forms a *hexagonal* lattice structure.[1] Figure 16.3 shows the hexagonal atomic packing and the unit-cell slice. While a first glance might lead you to think that cubic and hexagonal packing are very different, they are in fact closely related. The hexagonal packing fraction is also the closest packing fraction, 0.7405, and we call the structure hcp for *hexagonal closest packed*. Figure 16.4 shows how fcc and hcp lattices are related. Slice off a corner of the fcc lattice, and you find a close-packed plane with the six-fold symmetry that gives hcp its name.

To see how these two closest packed lattices (and they are the *only* two) differ, we introduce the idea of *nearest neighbors*. In both fcc and hcp, any one sphere is touching a total of *12* other spheres. These are the nearest neighbors, each with a center $2r$ from the center of the reference sphere. The subtle difference between fcc and hcp

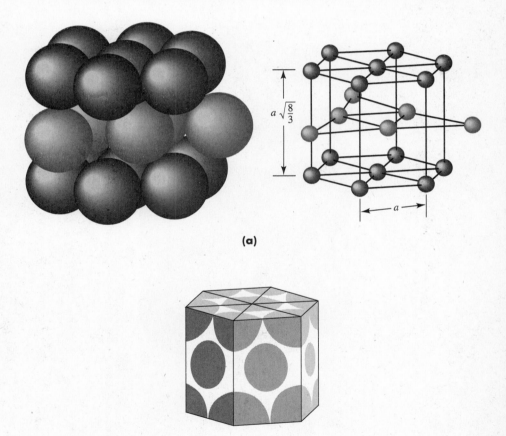

FIGURE 16.3 The hexagonal closest packed (hcp) unit cell (shown as packed spheres in (a) and as fractions of spheres in the unit cell in (b)) has alternating layers of close-packed atoms.

---

[1]The He phase diagrams are shown on Figure 6.10. At 0 K, ^{4}He is hexagonal above 25 bar, but ^{3}He solidifies as bcc at 34 bar and transforms to a hexagonal phase at 105 bar. The differences are due to the different zero-point energies of the two structures for the two isotopes. Both isotopes can form fcc crystals at $P > \sim 10^3$ bar and $T > \sim 18$ K.

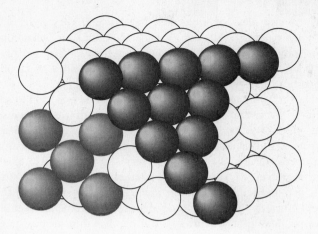

FIGURE 16.4 The fcc lattice has close-packed planes that are visible if the fcc unit-cell array has a few corner atoms removed. The darkly shaded atoms lie in an exposed close-packed plane, while the lightly shaded atoms form a face of the fcc unit cell shown in Figure 16.2.

is shown in Figure 16.5. In each lattice, we take one *close-packed plane* of seven atoms (the central and six nearest neighbor atoms) as a reference plane. The three hcp neighbors above this plane *eclipse* the three neighbors below it, while in fcc, the three above and below are *not* eclipsed. One says hcp has close-packed planes stacked in an ABABAB ... order, but fcc has such planes stacked ABCABC ... .

It would be fascinating to know the structure of the Ar_{13} molecule, or of the He_{13} molecule. There is great current interest in small atomic and molecular clusters.

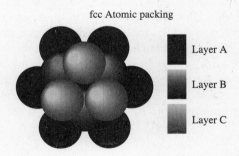

FIGURE 16.5 The difference between fcc and hcp packing lies in the way close-packed planes are layered. In fcc, three layers repeat throughout the crystal, while hcp has only two.

They represent a vast class of molecules about which very little is known. We understand Ar and Ar_2 very well, and we understand bulk Ar (what could be called Ar_N with $N \cong$ Avogadro's constant) very well, but between Ar_2 and Ar_N (and freely replace Ar by any other element) lies a series of molecules of largely unknown chemical properties.

For Ar or He, the 13-atom cluster might reasonably be expected to adopt a close-packed structure. But would it be hcp or fcc layered? If we assume that any one atom interacts *only* with its nearest neighbor, then both hcp and fcc 13-atom clusters have the *same energy*. This assumption of *pairwise interactions* is reasonable, since we know the interaction in Ar_2 is not only very slight but is also localized over a short distance. However, when we include interactions between one atom and *more distant* atoms, hcp and fcc lattices differ due to the layer stacking difference.

Both have 12 nearest neighbors (at a distance $2r$ from the central atom) and 6 next-nearest neighbors (at $\sqrt{8}r$ away). But hcp has third-nearest neighbors only $4\sqrt{2/3}\ r = 3.266\ r$ away, while these are $2\sqrt{3}\ r = 3.464\ r$ away in fcc. This would seem to favor hcp over fcc, but hcp has only *two* third-nearest neighbors to fcc's *24*. Nevertheless, this difference, which continues in a varying pattern over farther neighbors, leads to the prediction that atoms bound by *only* pair-wise interactions are marginally (by about 0.01% of the total binding energy) more stable in the hcp form.

Thus, Ne–Xe, which should be among the simplest atomic solids, pose a puzzle. Why do they form fcc crystals? Several suggestions have been advanced, and the answer must involve several of them. First, the assumption of pairwise additive interactions is an approximation, and three-body mutual interactions have been calculated. Second, Ne–Xe have excited electron configurations involving p, d, etc., electrons at much lower energies than He. Configuration interaction, which is highly symmetry dependent, may allow these directional excited state interactions to stabilize the fcc structures. And finally, the observed fcc lattice may be the result of crystal growth from small crystal nuclei for which the surface free energy favors the fcc structure that, as the nuclei grow, becomes irreversibly locked in place.

For whatever reasons, the fcc/hcp balance is very tenuous. Helium–4 prefers hcp for zero-point energy reasons, but as little as one atomic percent N_2 impurity in Ar will force the hcp structure onto the Ar lattice surrounding the N_2 impurity.

Metallic Elements

We postpone a discussion of *why* metals have the structures they do and begin with some structural observations. We start with the alkali metals, which all form bcc lattices at ordinary temperature and pressure.[2] The bcc unit cell and its two atoms are shown in Figure 16.6.

It would be satisfying if the metallic elements packed into lattices that, at worst, varied from column to column across the Periodic Table, but such is not the case. Table 16.2 is a Periodic Table of crystal structures. There is great variability only one column past the alkali metals. Be and Mg are hcp, Ca and Sr are fcc, and Ba and Ra are bcc. On top of this variability from element to element, most elements undergo solid–solid phase transitions from one crystal lattice to another as T and P are varied. For example,[3] Fe at 1 atm is bcc up to 1183 K, but fcc from 1183 K

---

[2] All structures mentioned in this section are those found at ordinary temperature and pressure unless stated otherwise.

[3] See Figure 6.5 for a thermodynamic discussion of these Fe phase changes.

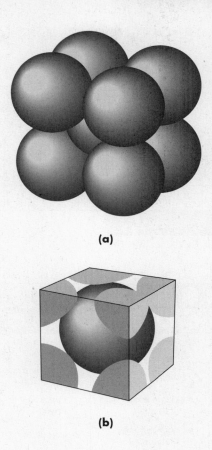

FIGURE 16.6 The body centered cubic (bcc) unit cell (shown as packed spheres in (a) and as fractions of spheres that contribute to the unit cell in (b)) has two atoms per cell. In (b), one eighth of each atom at a corner and all of the central atom (lightly shaded for clarity) contribute.

to 1673 K, then it is bcc again until the melting point at 1812 K. Cerium has a distinct phase transition from an fcc lattice of one lattice parameter to an fcc lattice of *different size,* as if the atoms themselves suddenly changed size! This transition has been tracked in temperature and pressure to a very rare solid–solid critical point (near 550 K and 17.5 kbar) where the lattice parameter difference vanishes. (See Problem 16.30 for more on this transition.)

Curiously, the sc lattice is not used by pure metals (except for Po) at ordinary pressures. This lattice has too small a packing fraction, and the octahedral nearest-neighbor geometry (six nearest neighbors) is not an optimum symmetry for metallic bonding, as the next section will discuss. Thus, while Cr is perfectly content to coordinate six CO ligands in an octahedral fashion in $Cr(CO)_6$, $Cr(s)$ is bcc with eight nearest neighbor Cr atoms.

Not all metallic elements form a cubic or hexagonal lattice, in fact. The so-called gray form of Sn forms a lattice *isomorphic* (i.e., of the same structural symmetry) with that of diamond. We discuss this structure in the next section. The actinides have the most bizarre packing. Uranium packs in an orthorhombic unit cell at low temperatures (below 938 K) in which atoms form zig-zag corrugated sheets stacked in an ABAB... fashion with a highly anisotropic thermal expansion. The distance between the sheets decreases while the sheet area increases. From 938 K to 1048 K, U forms a very hard phase with a complex tetragonal unit cell that contains 30 atoms. Above 1050 K, it has a simple bcc phase.

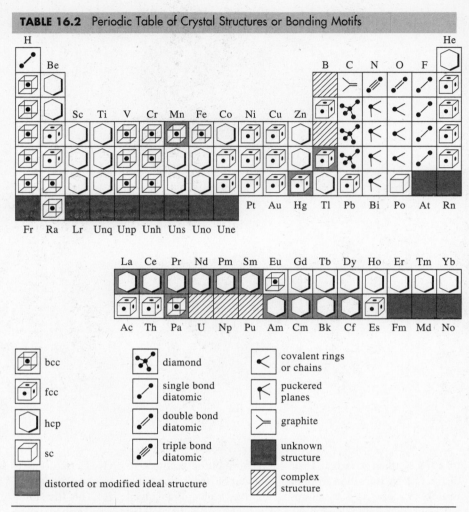

TABLE 16.2 Periodic Table of Crystal Structures or Bonding Motifs

Structures are the most stable form at room temperature and pressure except for the rare gases.

Abrupt changes in thermal expansion can signal phase transitions. Figure 16.7 shows six distinct phases (of the total of eight that are known) in the thermal[4] expansion of Pu, owner of one of the most complicated elemental phase diagrams.

It is clear that an understanding of metal-atom packing will require some careful chemical thought. Simple isotropic sphere packing will not do. Thus, we move on in the next section to the forces that bind atoms and molecules in solids.

16.2 BONDING TYPES IN SOLID PHASES

In this section, we give an overview of the major classes of solid bonding. We want to understand why each class has the obvious physical properties that lead to an almost intuitive classification scheme. We ask questions such as these: Why do

---

[4]This experiment is particularly simple to do with Pu. Its natural radioactivity provides the energy to heat the sample! Of course, this radioactivity and Pu's toxicity largely offset the self-heating simplicity of the experiment.

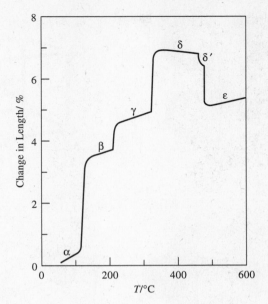

FIGURE 16.7 Since solid–solid phase transitions often result in significant structural reorganization, abrupt changes in size can accompany them. This phenomenon is illustrated here by a graph of the thermal expansion of Pu versus temperature. The six regions α–ε correspond to six distinct phases with transition temperatures at the discontinuities of the curve. [From *The Nature of Metals,* by B. A. Rogers (MIT Press, Cambridge, MA, 1964.)]

metals conduct electricity? Why do ionic solids (salts) melt at such high temperatures (as do metals) yet fail (as solids) to conduct electricity? Why do solid N_2, CO_2, O_2, etc., melt at such low temperatures, while C(*s*) melts somewhere around 4500 K?

Ionic Solids

The previous section discussed the rare-gas solids as close-packed spheres. Sphere packing is also a natural starting point for simple ionic solids (alkali halides, alkaline earth oxides, etc.) since the ions are isoelectronic to the rare gases. Two major differences come to mind immediately, however. The first is the *net charge* on the spheres, which is obviously a major point. The second is the *size difference* of the spheres.

The size effect can be very important. The two important alkali halide crystal lattices are fcc and sc. To describe an entire crystal with more than one kind of atom, we have to specify more than just the lattice. We must also specify *which atom is at any one lattice point and where the other atoms are relative to this one.* For example, CsCl forms the sc lattice. We place the anion (Cl$^-$) on a lattice point and locate the cation (Cs$^+$) at the center of the lattice cube. Thus, Cl$^-$ is at (0, 0, 0) and Cs$^+$ is at (*a*/2, *a*/2, *a*/2) where *a* is the lattice constant. The result (Figure 16.8(a)) looks very "body-centered-cubic," but the lattice is *not* bcc, since *lattice points must be equivalent.*[5]

In the CsCl structure, we say the cations have occupied all the *cubic sites* of the anions' sc lattice. (A cubic site has eight nearest neighbors at each corner of a cube.) In contrast, NaCl forms an fcc lattice of anions with cations located *a*/2 along fcc unit cell edges (Figure 16.8(b)). The result looks very "simple cubic," but, just as CsCl is not bcc, NaCl is not sc, but fcc. In NaCl, we say the cations occupy all the *octahedral sites* (six nearest neighbors at each corner of an octahedron) of the fcc lattice of the anions.

---

[5] It is, however, equally correct to say that the *cations* are on the lattice points and that the *anions* are in the center of the lattice cubes. There is in general no unique way to assign a crystal geometry to a lattice, but the lattice itself will always be unique.

FIGURE 16.8 The CsCl crystal structure (a) has one ion coordinated by *eight* nearest neighbor ions of opposite charge. The sphere packing (left) is contrasted with a lattice picture (right) to emphasize the eightfold coordination. The NaCl crystal structure (b) has any one ion coordinated in an octahedral fashion by *six* nearest-neighbor ions of opposite charge. The lattice picture (right) shows the ions on interpenetrating fcc lattices.

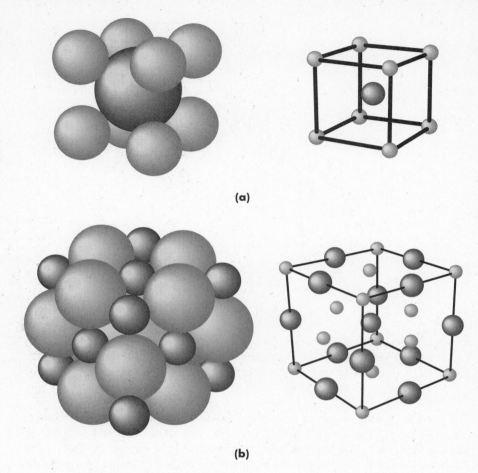

The Cl^- ions are presumably about the same size in CsCl and NaCl, but Cs^+ is considerably larger than Na^+. Simple geometry shows that an sc lattice of spheres of radius r has a cubic site that can accommodate a sphere of radius $r' \leq (\sqrt{3} - 1)\,r = 0.7321\,r$ and still allow the sc lattice spheres to touch:

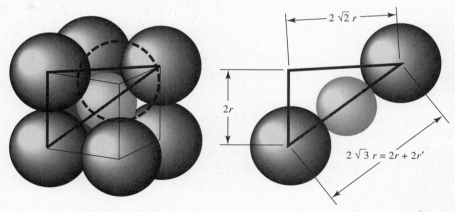

Pack spheres of radius r in an fcc lattice, and you find that the octahedral sites (located at $(a/2, 0, 0)$, $(0, a/2, 0)$, $(0, 0, a/2)$, etc. on the unit cell edge midpoints plus one in the center of the cell) can hold a sphere of radius $r' \leq (\sqrt{2} - 1)\,r =$

0.4142 r. There are also fcc *tetrahedral sites* (eight in all, at ($a/4$, $a/4$, $a/4$), etc.) in which four lattice atoms surround a void that can hold a sphere of radius $r' \leq (\sqrt{3}/\sqrt{2} - 1) r = 0.2247\ r$. In K_2O, the O^{2-} anions occupy the fcc lattice points and the K^+ cations occupy all the tetrahedral sites. (There are four lattice atoms in the fcc unit cell, and eight tetrahedral sites; hence, the stoichiometry is satisfied.) In CaF_2, the mineral *fluorite,* the larger Ca^{2+} cations are on the fcc lattice and the smaller F^- anions are in the tetrahedral sites.

EXAMPLE 16.1

There is a rare ionic crystal structure in which the ionic radii are so different that the smaller ions occupy a trigonal planar interstitial site. In B_2O_3, the small B^{3+} ions are coplanar with three "touching" O^{2-} ions. Find the largest value for the ionic radius ratio r'/r so that an ion of radius r' can nestle among three touching ions of radius r.

SOLUTION This problem illustrates the general approach to limiting radius ratio questions. The geometry here is shown below. (A carefully drawn diagram is always necessary in questions of sphere packing!)

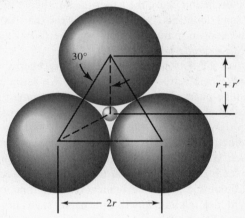

The sum of the radii, $r + r'$, forms two sides of an isosceles triangle, and the third side is $2r$ long. From the law of sines,

$$\frac{\sin 120°}{2r} = \frac{\sin 30°}{r + r'} \quad \text{or} \quad \frac{\sqrt{3}}{2r} = \frac{1}{r + r'},$$

which can be solved to give

$$\frac{r'}{r} = \frac{2 - \sqrt{3}}{\sqrt{3}} = 0.1547\ .$$

➡ RELATED PROBLEMS 16.10, 16.11, 16.12

You might expect that an ion of one charge would find it energetically advantageous to pack as *many* ions of the opposite charge around itself as *closely* as possible. But which factor, number of neighbors or distance to them, is more important? Consider two singly charged ions with a radius ratio $r'/r = 0.2247$, the maximum value for tetrahedral coordination. In a tetrahedral site, the attractive Coulombic energy is proportional to

$$V_{\text{tet}} \propto 4 \text{ neighbors}/(0.2247 \text{ relative distance units away})$$
$$\cong 17.8 \text{ relative energy units.}$$

Put this same small sphere in an octahedral site, and

$$V_{oct} \propto 6 \text{ neighbors}/(0.4142 \text{ relative distance units})$$
$$\cong 14.5 \text{ relative energy units}$$

or, in a cubic site,

$$V_{cub} \propto 8 \text{ neighbors}/(0.7321 \text{ relative distance units})$$
$$\cong 10.9 \text{ relative energy units.}$$

Thus, a small ion favors the closer distance of the tetrahedral site in spite of its lower coordination number.

These simple size/packing arguments, based purely on radius ratios, go a surprisingly long way toward understanding ionic solid crystal structures, but this cannot be the entire story, since there is also a *repulsive* interaction among ions of the same charge. We need to consider the full Coulombic energy of the crystal.

Crystal enthalpies, which are $\Delta \overline{H}$ for $MX(cr) \rightarrow M^-(g) + X^-(g)$ at 0 K, are accurately calculated with a model that assumes 100% ionic bonding in the solid. The last two chapters pointed out several ways of showing that molecular, gas phase NaCl, LiH, etc., have only *partial* ionic character, whether measured by the ionic component of the wavefunction or by the experimental dipole moment. Thus, it is somewhat surprising (and doubtless somewhat fortuitous) that 100% ionic character works. This should be kept in mind as we develop the following bonding model, known as the *Born–Mayer ionic crystal model*.

We start with a pair of ions and seek the binding energy as a function of internuclear separation. The electrostatic part is easy. We assume pure Coulombic attraction between unlike charges, and we will further assume that there are only two types of ion, each of unit charge magnitude. This gives an energy term $-e^2/4\pi\epsilon_0 R$. Between *like* charges we have the repulsive term $+e^2/4\pi\epsilon_0 R$.

The repulsive interaction that must exist even between unlike charged ions is more difficult to describe. Look back at Figure 14.25, describing the bonding in LiH. Solid structures focus on the region near the binding function's minimum, and we have chosen a Coulombic attractive branch for this function. The repulsive interaction alone should look qualitatively like the triplet state curve in Figure 14.25. No simple function readily predicts this curve on a sound theoretical basis, but a simple exponential function (see also Problem 14.8)

$$U_{rep}(R) = A\, e^{-R/\rho} \qquad (16.1)$$

is in reasonable agreement with a variety of measured repulsive interactions. It has two parameters that roughly measure the strength (A) and range (ρ) of the repulsive interaction.

Consider NaCl. The Na–Na repulsive forces (aside from the Coulombic term) should be different from the Cl–Cl or the Na–Cl repulsive forces. Thus, we should have to consider *six* repulsive interaction parameters: A_{Na-Na}, A_{Cl-Cl}, A_{Na-Cl}, ρ_{Na-Na}, ρ_{Cl-Cl}, and ρ_{Na-Cl}. These are more than we can expect to obtain from a few measurable properties, such as the crystal binding energy. Thus, we assume a single A and a single ρ for all non-Coulombic repulsions.

The potential energy of the ith ion in this model is a sum of pair potentials

$$U_i = \sum_{j \neq i} U_{ij} \qquad (16.2)$$

where

$$U_{ij}(R) \quad = \quad Ae^{-R/\rho} \quad \pm \quad \frac{e^2}{(4\pi\epsilon_0)R} . \qquad (16.3)$$

$\uparrow$ Pair potential $\quad$ $\uparrow$ Exponential repulsion $\quad$ $\uparrow$ Coulombic potential

We use the $+$ sign if ions i and j have the same charge and the $-$ sign if they do not.

The molar crystal energy $\overline{U}_{cr}$ is

$$\overline{U}_{cr} = N_A U_i = N_A \sum_{j \neq i} U_{ij} ,$$

but to calculate $\overline{U}_{cr}$ from this, we need two independent measurements that can give us A and ρ. We will choose the *nearest neighbor separation* (from X-ray diffraction) and the *isothermal bulk compressibility* κ first introduced in Chapter 1:

$$\kappa \equiv -\frac{1}{V}\left(\frac{\partial V}{\partial P}\right)_T . \qquad (1.26)$$

Using Eq. (3.34):

$$\left(\frac{\partial U}{\partial V}\right)_T = T\left(\frac{\partial P}{\partial T}\right)_V - P \qquad (3.34)$$

we have, at $T = 0$,

$$\left(\frac{\partial U}{\partial V}\right)_{T=0} = -P . \qquad (16.4)$$

Differentiating Eq. (16.4) with respect to V yields

$$\left(\frac{\partial^2 U}{\partial V^2}\right)_{T=0} = -\left(\frac{\partial P}{\partial V}\right)_T = \frac{1}{V\kappa} \qquad (16.5)$$

so that we can relate $\overline{U}_{cr}$ (and thus A and ρ) to κ.

Now we turn to the lattice sum in Eq. (16.2). Since the exponential repulsion term falls rapidly with increasing distance, we set it to zero for all but the interactions of our central ion with its nearest neighbors. This approximation further justifies our choice of a single pair of A and ρ parameters.

We will measure all distances in terms of the nearest-neighbor separation R_0, writing $R = rR_0$ with r dimensionless. With N nearest neighbors around our central ion, (6 for NaCl and 8 for CsCl), $\overline{U}_{cr}$ becomes

$$\overline{U}_{cr} = N_A \sum_{j \neq i} U_{ij} = N_A \left[NA\, e^{-R_0/\rho} - \frac{e^2}{(4\pi\epsilon_0)R_0} \sum_{j \neq i} \frac{(\pm 1)}{r_{ij}} \right]$$

where the remaining sum is just *a number that depends only on the crystal structure*. It is called the *Madelung constant* M for the crystal:

$$M = \sum_{j \neq i} \frac{(\pm 1)}{r_{ij}} . \qquad (16.6)$$

For the NaCl structure, $M = 1.747\,558$, while $M = 1.762\,670$ for the CsCl structure. Thus we can write

$$\overline{U}_{cr} = N_A \left[NA\, e^{-R_0/\rho} - \frac{Me^2}{(4\pi\epsilon_0)R_0} \right] . \qquad (16.7a)$$

Since R_0 by definition minimizes $\overline{U}_{cr}$:

$$\left(\frac{\partial \overline{U}_{cr}}{\partial R}\right)_{R=R_0} = 0 = N_A \left[-\frac{NA}{\rho} e^{-R_0/\rho} + \frac{Me^2}{(4\pi\epsilon_0)R_0^2} \right]$$

or

$$NA\, e^{-R_0/\rho} = \frac{Me^2}{(4\pi\epsilon_0)R_0}\left[\frac{\rho}{R_0}\right],$$

we can write

$$\overline{U}_{cr} = \frac{N_A Me^2}{(4\pi\epsilon_0)R_0}\left[\frac{\rho}{R_0} - 1\right]. \tag{16.7b}$$

To relate this to κ, we must find $(\partial^2 U/\partial V^2)$. First, note from Figure 16.8(b) that N_A molecules of NaCl have a molar volume $\overline{V} = 2N_A R_0^3$, since there are four molecules in the volume $(2R_0)^3 = 8R_0^3$. Thus

$$\left(\frac{\partial \overline{V}}{\partial R}\right)_{R_0} = 6N_A R_0^2 \quad \text{and} \quad \left(\frac{\partial \overline{U}_{cr}}{\partial \overline{V}}\right) = \left(\frac{\partial \overline{U}_{cr}}{\partial R}\right)\left(\frac{\partial R}{\partial \overline{V}}\right)$$

so that

$$\left(\frac{\partial^2 \overline{U}_{cr}}{\partial \overline{V}^2}\right) = \left(\frac{\partial^2 \overline{U}_{cr}}{\partial R_0^2}\right)\left(\frac{dR}{d\overline{V}}\right)^2 + \left(\frac{\partial \overline{U}_{cr}}{\partial R}\right)\left(\frac{d^2 R}{d\overline{V}^2}\right) = \frac{(\partial^2 \overline{U}_{cr}/\partial R_0^2)}{(6N_A R_0^2)^2} \quad \text{at } R = R_0$$

since $(\partial \overline{U}_{cr}/\partial R)_{R_0} = 0$. From Eq. (16.7a),

$$\left(\frac{\partial^2 \overline{U}_{cr}}{\partial R_0^2}\right) = N_A\left[\frac{NA}{\rho^2}e^{-R_0/\rho} - 2\frac{Me^2}{(4\pi\epsilon_0)R_0^3}\right].$$

The equation $(\partial \overline{U}_{cr}/\partial R)_{R_0} = 0$, evaluated for Eq. (16.7a), gives an expression for $e^{-R_0/\rho}$, which simplifies the equation above to

$$\left(\frac{\partial^2 \overline{U}_{cr}}{\partial R_0^2}\right) = \frac{N_A Me^2}{(4\pi\epsilon_0)R_0^3}\left(\frac{R_0}{\rho} - 2\right)$$

so that

$$\left(\frac{\partial^2 \overline{U}_{cr}}{\partial \overline{V}^2}\right) = \frac{Me^2}{36(4\pi\epsilon_0)N_A R_0^7}\left(\frac{R_0}{\rho} - 2\right).$$

Eq. (16.5) says

$$\kappa = \left[\overline{V}\left(\frac{\partial^2 \overline{U}_{cr}}{\partial \overline{V}^2}\right)\right]^{-1}.$$

We know $\overline{V} = 2N_A R_0^3$, and our expression above for $(\partial^2 \overline{U}_{cr}/\partial \overline{V}^2)$ finally yields an expression for R_0/ρ in terms of κ:

$$\frac{R_0}{\rho} = 2 + \frac{18(4\pi\epsilon_0)R_0^4}{\kappa Me^2}. \tag{16.8}$$

This expression can go directly in Eq. (16.7b), giving us an expression for $\overline{U}_{cr}$ in terms of independently measurable quantities, R_0, κ, and the crystal symmetry (which governs M). We can compare $\overline{U}_{cr}$ to experimental $\Delta \overline{H}_{cr}$ values extrapolated to $T = 0$.

For NaCl, $R_0 = 2.820$ Å, $\kappa = 4.17 \times 10^{-11}$ m^3 J^{-1}, and $M = 1.747\,558$. We find $\overline{U}_{cr} = -763$ kJ mol^{-1}. (The minus sign means that the crystal is 763 kJ mol^{-1} lower in energy than one mole of Na$^+(g)$ and one mole of Cl$^-(g)$, all infinitely far apart.) The experimental value is -775.3 kJ mol^{-1} in remarkably good agreement.

EXAMPLE 16.2

What does the Born–Mayer model predict for the internuclear distance of a free NaCl molecule?

SOLUTION If we assume the Born–Mayer potential function, Eq. (16.3) with a minus sign, holds for the free molecule as well as for the bulk crystal, then NaCl will have an equilibrium bond length at the minimum of $\overline{U}_{ij}(R)$. This bond length, R_e, will *not* equal R_0, however. We find ρ from the discussion in the text: $R_0 = 2.820$ Å, $R_0/\rho = 8.771$ (from Eq. (16.8)), and thus $\rho = 0.3215$ Å. To find A, the coefficient of the exponential repulsion term, we solve

$$\left(\frac{\partial \overline{U}_{cr}}{\partial R}\right)_{R=R_0} = 0 = N_A\left[-\frac{NA}{\rho}e^{-R_0/\rho} - \frac{Me^2}{(4\pi\epsilon_0)R_0^2}\right]$$

for A and find $A = 175.05$ aJ. Algebraic minimization of Eq. (16.3) is not possible, but a numeric solution for the minimum yields $R_e = 2.289$ Å. The observed bond length is 2.360 9 Å, in reasonable agreement for such a simple model potential function. A graph of the full function is shown below. Note how the Coulombic attraction is appreciable at $R \gg R_e$. This is why we could not limit ion–ion interactions to just nearest neighbors.

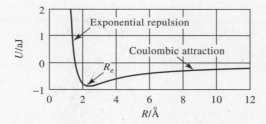

It is interesting to compare this graph to one of Eq. (16.7a) with R_0, the nearest neighbor distance, as the variable. See also Problem 16.20 for more on the difference between molecular bond lengths and solid nearest-neighbor distances.

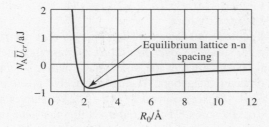

➡ **RELATED PROBLEMS** 16.18, 16.19, 16.20

What have we left out of the model that might either improve the agreement with experiment or perhaps totally destroy it? The model assumes the ions are stationary at $T = 0$; thus, $\overline{U}_{cr}$ should be *increased* to account for zero-point energy. Using $(\partial^2 U/\partial R^2)$ as a force constant and a mean mass m to represent both Na and Cl motions, the harmonic oscillation frequency about R_0 is $\omega = [(\partial^2 U/\partial R^2)/m]^{1/2}$, corresponding to a molar zero-point energy $N_A\hbar\omega/2 \cong 1.6$ kJ mol^{-1}. The second neglected effect is the polarizability of the ions. Table 15.2 lists them, but anything only a few Å from a point charge feels an *enormous* electric field, and a simple

$-\alpha E^2/2$ polarization energy may not be correct. But, on the other hand, symmetry comes to the rescue. At, for instance, the Cl$^-$ *nucleus,* the electric field from the six Na$^+$ neighbors is zero by symmetry, so we expect the polarization energy to be small. A final correction, also negative and small, comes from the attractive forces that are present whether or not the ions are charged. These are the forces that bind the rare gas solids, and they contribute about -4 kJ mol^{-1}.

Thus, by and large, the physical effects ignored by the ionic model together contribute a small, negative total amount to $\overline{U}_{cr}$. In general, the simple model predicts $\overline{U}_{cr}$ to be too large by ~ 10 kJ mol^{-1}. This agreement is cause to cheer, since we will soon find that the binding energies in metals, diamond, etc., cannot be found as accurately by any simple theory.

A Closer Look at the Madelung Constant

We defined the Madelung constant as a sum over a lattice:

$$M = \sum_{j \neq i} \frac{(\pm 1)}{r_{ij}} \tag{16.6}$$

where r_{ij} is the distance from target ion i to any other ion j, measured in terms of the nearest-neighbor distance, and the $+$ sign goes with *unlike* charges here. We then quoted values for M for various lattices without pointing out just how difficult M is to calculate. We look at that calculation here.

Suppose we had only a linear array of equally spaced, alternating charges:

$$\cdots + - + - + - + - + - \cdots$$

We pick a $+$ ion as the target and compute the Madelung sum (with $r_{ij} = 1, 2, 3,$ etc.) as (note how the $\pm$ sign choice is made)

$$M = 2\left(\frac{1}{1} - \frac{1}{2} + \frac{1}{3} - \frac{1}{4} + \cdots\right).$$

The expansion $\ln(1 + x) = x - x^2/2 + x^3/3 - x^4/4 + \cdots$ with $x = 1$ shows $M = 2 \ln 2 = 1.386\ 3$ exactly here.

There is no such closed sum for alternating charges on a simple cubic lattice. So we try brute force, summing over nearest neighbors, then next nearest, etc. Starting again with a positive target ion in the NaCl lattice, we find six negative ions at $r_{ij} = 1$, the nearest neighbors. Then come twelve positives at $r_{ij} = \sqrt{2}$, eight negatives at $r_{ij} = \sqrt{3}$, six positives at $r_{ij} = 2$, 24 negatives at $r_{ij} = \sqrt{5}$, 24 positives at $r_{ij} = \sqrt{6}$, and so on. The sum, through the first nine nearest-neighbor shells, is

$$\begin{aligned}M &= \frac{6}{1} - \frac{12}{\sqrt{2}} + \frac{8}{\sqrt{3}} - \frac{6}{2} + \frac{24}{\sqrt{5}} - \frac{24}{\sqrt{6}} + \frac{12}{\sqrt{8}} - \frac{30}{3} + \frac{24}{\sqrt{10}} - \cdots \\ &= 6 - 8.485 + 4.619 - 3 + 10.733 - 9.798 + 4.23 - 10 + 7.589 - \cdots \\ &= 1.900\ 8\ .\end{aligned}$$

The sum is clearly failing to converge. None of the terms is small, and they are guaranteed to alternate in sign.

But these days, the brute force approach has a powerful ally—the high-speed digital computer. The expression for this lattice's Madelung constant is very simple to program directly on a computer. Most simply, we go left, right, up, down, back, and forth N ions from our target ion, looping for larger and larger N until we have converged at an answer of sufficient accuracy. The computer time is increasing rapidly, like N^3, but we try anyway. We find the following:

N	# of ions included in sum[6]	Madelung constant
2	124	1.516 6
3	342	1.912 5
4	728	1.619 3
5	1 330	1.852 5
6	2 197	1.658 7
7	3 374	1.824 5
8	4 912	1.679 6
9	6 858	1.808 3
10	9 261	1.692 6
11	12 166	1.797 8
12	15 624	1.701 4

So much for brute force. Even with >15 000 terms in the sum, we have barely two significant figures in the result, $M \cong 1.7$.

An elegant solution is called for. The first approach that found the accurate $M = 1.747\ 558$ value broke the unit charges into fractional charges (much as we sliced atoms into fractions in the unit cells of Figures 16.2(b) and 16.3(b)) and summed over regions with zero net charge. This sum converges much more rapidly. Details of the calculation and Madelung constants for many other ionic lattices can be found in this chapter's references, and Problem 16.52 applies this approach to a planar array of charges.

Metals

When chemists and physicists speak to each other about common topics of interest, it can seem that the physicists never took a course in chemistry and the chemists never learned any physics. Consider metal bonding. To the physicist, the metal's chemical identity has secondary importance. What matters most are the *conduction electrons*. These are, in the *free electron gas* model, free to roam inside the metal. To the chemist, the chemical identity is most important. The number of valence electrons; the nature and accessibility of excited electronic states; hybrid orbitals of various shapes: in short, all the usual concerns for molecular bonding fit into a chemical picture of metals.

Of course, physicists and chemists do share ideas and points of view. In the *band theory* of solids they find a common ground. We will start with simple physical and chemical pictures of metal bonding and end with an abbreviated look at band theory.

Free Electron Model of Metals

Since the electrons that conduct electric current are known to move rather freely through the metal, it makes some sense to approximate their energies and wavefunctions as if they were truly free, noninteracting particles confined to the metal's volume. This is a severe approximation: how can electrons, with unit negative

---
[6]This is $(2N + 1)^3 - 1$. The -1 is due to the target ion at $r_{ij} = 0$, which is skipped.

charge, packed closely together, not interact? The answer is, of course, that they must interact. Our model is in error in the same way that the independent electron model of He was in Chapter 13.

Imagine a 1 mol cube of Na. Even the physicists would agree that each Na atom contributes only one electron to the pool of conduction electrons, not all 11. The other ten remain localized around the Na nuclei. Thus we ask for the energies of one mole of independent electrons confined to a cubic box. We use the quantum numbers (n_x, n_y, n_z) for the 3-D particle-in-a-box states, and we seek first the ground state of this many-particle system.

We are ignoring electron charge, but we cannot ignore electron spin. We assign two electrons, spin-paired, to the lowest energy orbital, $(n_x, n_y, n_z) = (1, 1, 1)$, two to the next lowest—and then remember that for a cubic box, this level is three-fold degenerate and can hold six electrons—and so on, until all 6×10^{23} electrons are assigned. We ask the seemingly very difficult question, "What is the energy of the *last* pair of electrons?"

This question is easier to answer than it sounds if we introduce a very good approximation. Let (N_x, N_y, N_z) be the quantum numbers of the last pair. The energy of this state is (Eq. (12.23) with $L_x = L_y = L_z = L$, the cube's length):

$$E(N_x, N_y, N_z) = \frac{\hbar^2 \pi^2}{2mL^2}(N_x^2 + N_y^2 + N_z^2)$$
$$= (\text{constant})(N_x^2 + N_y^2 + N_z^2)$$
$$= (\text{constant}) R^2$$

where we have written $(N_x^2 + N_y^2 + N_z^2)$ as R^2 in the last step to point out that $(N_x^2 + N_y^2 + N_z^2)$ is *the square of a radius in a Cartesian quantum-number coordinate system*. We imagine a 3-D lattice of equally spaced points with the axes labeled n_x, n_y, and n_z, as in Figure 16.9. Each lattice point corresponds to a unique set of quantum numbers (and each n must be a positive integer).

Each lattice point locates a unique quantum state, and each state holds two spin-paired electrons. *Unit volume in this quantum-number space must therefore contain two electrons,* and minimum energy means minimum R^2, which in turn means those unit volumes located within the positive quantum-number octant of a sphere of

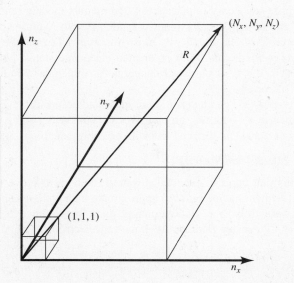

FIGURE 16.9 For a 3-D particle-in-a-box free-electron model of metals, the three quantum numbers (n_x, n_y, n_z) can be thought of in terms of a 3-D Cartesian "quantum number space." Unit volume in this space holds two spin-paired electrons. The radius R locates the highest energy state (N_x, N_y, N_z), and the energy is proportional to R^2.

radius R will be occupied. Our mole of spin-paired electrons thus occupy $N_A/2$ unit volumes. We find R from

(volume of octant of sphere) = (total volume of $N_A/2$ unit volumes)

$$\frac{1}{8}\left(\frac{4\pi R^3}{3}\right) = \frac{N_A}{2}$$

or $R = (3N_A/\pi)^{1/3}$ so that the energy[7] of the most energetic electrons is

$$E_F = \frac{\hbar^2 \pi^2}{2mL^2}\left(\frac{3N_A}{\pi}\right)^{2/3}$$

where the subscript F stands for Fermi. This highest energy is called the *Fermi energy*. In terms of the molar volume of the solid $\overline{V}$, it is

$$E_F = \frac{\hbar^2 \pi^2}{2m}\left(\frac{3N_A}{\pi \overline{V}}\right)^{2/3}. \tag{16.9}$$

EXAMPLE 16.3

Sodium has a bcc lattice with a lattice constant $a = 4.290\,6$ Å. What is the conduction-electron concentration in Na, and what is the Na Fermi energy?

SOLUTION The unit cell volume is $a^3 = 7.899 \times 10^{-29}$ m^3. Each unit cell in the bcc lattice holds two atoms, and each atom of Na contributes one conduction electron. Thus, the conduction-electron concentration is (2 electrons)/(7.899×10^{-29} m^3) = 2.532×10^{28} electrons m^{-3}. This is equal to $N_A/\overline{V}$, so the Fermi energy according to this model is

$$E_F = \frac{\hbar^2 \pi^2}{2m}\left(\frac{(3)(2.532 \times 10^{28} \text{ m}^{-3})}{\pi}\right)^{2/3} = 5.037 \times 10^{-19} \text{ J} = 3.144 \text{ eV}.$$

This is a typical magnitude—a few eV—for the Fermi energy of metals.

➡ *RELATED PROBLEMS* 16.21, 16.24, 16.25, 16.53, 16.54

The Engel–Brewer Model

How would a chemist describe Na(s)? Metal conduction electrons are the ultimate example of delocalization. The perfect metal crystal is a giant molecule, and we could write giant Slater determinant—vast LCAO-MO wavefunctions. Qualitatively, we know what to expect. Figure 16.10 shows the progression of the energy level scheme from Na$_2$ to Na(s). The solid has N_A MOs produced from the N_A AOs (3s, in the case of Na), and exactly half of them are occupied in the ground state of the solid at $T = 0$. The unoccupied orbitals are a continuum of states infinitesimally close by in energy, poised to accept electrons at $T \neq 0$ (or even at $T = 0$ in the presence of an electric field that induces conduction).

There seems to be a problem with this chemist's viewpoint if we pursue it across the Periodic Table. Consider Mg(s) instead of Na(s). Now there are two valence conduction electrons per atom, and Figure 16.10 would have *all* the MOs doubly occupied. There would be as many occupied orbitals *above* the separated atom limit as below it, and it would seem that Mg(s) should have a very small binding energy. This is the same argument used to explain why He$_2$ and He(s) have such a slight binding energy. The bond order is zero. While Mg(s) is a strongly bound solid, the

---

[7]The constant N_A is a dimensionless number here. It is formally (1 mol)$N_A = 6.022 \times 10^{23}$.

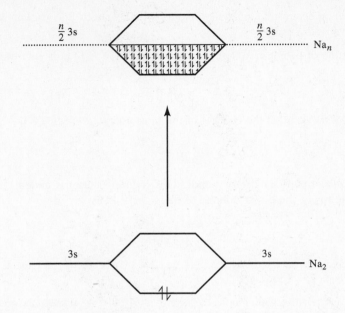

FIGURE 16.10 An LCAO-MO picture of metallic bonding is illustrated here for the Na case. As the number of atoms that bond increases from 2 towards the number in a solid, the bonding orbitals fill with paired electrons up to the energy of the separated atoms' 3s orbitals.

dimer, Mg_2, is very weakly bound. Like He_2, (or Be_2—see Table 14.3), Mg_2 has a formal bond order of zero.

How, then, are we to explain the strong bonding in Mg(s) on chemical terms? One of the most useful chemically based models of metal bonding was first advanced by the Danish scientist Niels Engel in the 1940s (based on earlier ideas of Hume–Rothery), only to be forgotten until the late 1960s, when Engel's ideas were unearthed and greatly extended by the American physical chemist Leo Brewer. The Engel–Brewer model, as it has come to be known, begins with the total binding energy (the enthalpy of atomization, to be correct) of elemental metals. The atomization energy is the energy required at zero temperature and pressure to separate the most stable form of the element into ground-state atoms. The model attempts to correlate simple bonding ideas with these energies and, along the way, it introduces an empirical metal structure correlation as well.

Ordinary covalent molecular-bond energies fall into three major classes, based on the bond multiplicity: triple bonds (N≡N, —C≡C—, etc.) have a larger bond energy than double bonds (O=O, $\rangle$C=C$\langle$, etc.), which have a larger bond energy than single bonds (F—F, —C—C—, etc.). Within each class, the bond energies are about the same, except that bonds between elements farther down the Periodic Table tend to have smaller bond energies (such as I_2 versus Br_2 versus Cl_2). (See Tables 7.1 and 7.2 for numerical data.)

If we look at the atomization energies in Table 16.3, we can see these trends carried over into the metals. Note how values for single covalent bonds cluster around 80–150 kJ mol^{-1} (except for the unusual stability of H_2 at 216 kJ mol^{-1}), values for elements with two covalent bonds (O_2, S, Se, etc.) range around 150–250 kJ mol^{-1}, values for three covalent bonds (N_2, As, etc.) range around 250–350 kJ mol^{-1}, and values for four covalent bonds (C, Si, etc.) are around 300 kJ mol^{-1} and up.

TABLE 16.3 Periodic Table of Atomization Energies/kJ mol^{-1}

H 216																	He 0.06
Li 158	Be 320											B 566	C 711	N 474	O 251	F 81	Ne 1.9
Na 108	Mg 145											Al 327	Si 451	P 331	S 276	Cl 135	Ar 7.7
K 90	Ca 178	Sc 376	Ti 467	V 511	Cr 395	Mn 282	Fe 413	Co 427	Ni 428	Cu 336	Zn 130	Ga 271	Ge 372	As 302	Se 206	Br 118	Kr 11.1
Rb 82	Sr 164	Y 424	Zr 607	Nb 718	Mo 656	Tc 688	Ru 650	Rh 552	Pd 376	Ag 284	Cd 112	In 243	Sn 301	Sb 264	Te 196	I 107	Xe 15.9
Cs 78	Ba 183	Lu 428	Hf 619	Ta 782	W 848	Re 775	Os 788	Ir 668	Pt 564	Au 368	Hg 60.3	Tl 182	Pb 196	Bi 210	Po 145	At (93)	Rn (19)
Fr (75)	Ra 144	Lr (305)	Unq	Unp	Unh	Uns	Uno	Une									

La 431	Ce 423	Pr 357	Nd 328	Pm (310)	Sm 206	Eu 175	Gd 400	Tb 391	Dy 294	Ho 303	Er 318	Tm 233	Yb 153
Ac 410	Th 598	Pa 569	U 531	Np 465	Pu 344	Am 284	Cm 387	Bk 310	Cf 196	Es 133	Fm 143	Md (113)	No (108)

The process is: most stable form of element at 0 K → ground-state gaseous atom.

The alkali metal values fall into the single-bond range, as we might have expected. The values also decrease going down the group, paralleling the trend of the halogens (except for anomalous F_2). Likewise, the alkaline earth metals fall in the two covalent-bond range, while B and Al parallel N and P. A histogram of all the metal atomization energies shows clusters around values roughly characteristic of these ranges:

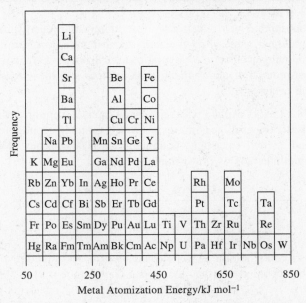

The alkali metals, alkaline earth metals, and Al all use s and/or p electrons to bind. If we abandon the s^2 atomic configuration and adopt sp *hybridization,* we can explain the stability of, for example, Mg. Magnesium can form these hybrids only by paying the promotion energy price of an $s^2 \rightarrow$ sp configuration change, but the gain in bonding energy more than offsets this price. (This is exactly the same argument used in discussing C-atom hybridization in CH_4, for example. The $2s^2 2p^2$ ground state configuration of C is imagined to be promoted to a $2s2p_x2p_y2p_z$ configuration, and the sp^3 hybrids are formed from this energetic configuration. The gain in bonding energy from the hybrids is greater than the loss in energy associated with forming the promoted configuration.)

The Engel–Brewer model goes beyond the s-p electron contribution to bonding to include the d electrons important to transition elements. For example, contrast V and Ge. Counting s, p, and d electrons, V has five and Ge has four (which are all s and p) and, consequently, V has a notably higher stability than Ge. But even Ti, just to the left of V with four s-p-d electrons, has a higher stability than Ge. Not only do d electrons contribute to bonding, but they do so very effectively.

Metal crystal structures can be worked into the story in an empirical way. Consider Na, Mg, and Al, which form bcc, hcp, and fcc lattices and which have one, two, and three s-p bonding electrons, respectively. If we adopt the point of view that *structure* is determined by s-p bonding, no matter how many d electrons are involved, then we can understand why, for instance, Cu(s) is fcc. Ground state atomic Cu has the valence configuration $3d^{10}4s$. On this basis, Cu(s) should be bcc (one valence electron) with an atomization energy ~ 100 kJ mol^{-1}. The observed atomization energy, however, is 336 kJ mol^{-1}, almost identical to that for Al and P, indicating three electron bonding. Thus, Cu adopts a $3d^8 4s^1 4p^2$ configuration, forms hybrid bonds, and adopts a tightly bound fcc lattice.

From these and many other comparisons, Engel arrived at a set of empirical rules based on the average number of valence electrons per atom.

1. A configuration with paired s electrons will not be used in metallic bonding.
2. Crystal structure correlates with the number of s-p electrons used in bonding: one = bcc, two = hcp; three = fcc; four = diamond (a new structure discussed below).
3. The contribution of d electrons to the atomization energy increases with atomic number, while the s-p electron contribution decreases.
4. The predominant electron configuration used in bonding is that which requires the least promotion energy in the gaseous atom to achieve reasonable metal bonding energies.

This final point helps explain why one element may have more than one crystal structure. Consider the alkaline earth metals with a ground s^2 configuration (useless by rule 1.) and excited sp and sd configurations. The gaseous atoms have the following promotion energies (in kJ mol^{-1}):

	Mg	Ca	Sr	Ba
$s^2 \to sp$	335	230	218	184
$s^2 \to sd$	586	251	234	121

For Mg, the sd configuration (which would yield a bcc structure by rule 2.) is considerably higher in energy than the sp-hcp configuration. Thus, Mg forms hcp crystals exclusively. For Ba, the sd-bcc configuration is considerably lower than sp-hcp, and Ba adopts the bcc configuration exclusively below ~ 50 kbar.

We would not predict an fcc structure for the alkaline earths, but both Ca and Sr form these structures at room temperature, transforming to bcc at 721 K and ~ 880 K, respectively. An fcc phase could come from configurations such as p^5dsp (three unpaired s-p electrons but four bonding electrons in all) that are of unknown gas-phase energy. The increase in bonding energy from four bonding electrons could offset the high promotion energy ($p^6s^2 \to p^5dsp$).

These ideas can be extended to account for the stability of alloys and to predict phase diagrams. The arguments are strikingly chemical in nature. For example, the NaGa alloy at around 50% atomic composition has a very high melting temperature, far above that of Na or Ga alone. (Galium melts in your hand, at 303 K!) The Ga atoms in the alloy form a diamond lattice, indicating four s-p bonding electrons. Galium has the configuration $[Ar]3d^{10}4s^24p^1$. If Ga takes an electron from Na, it can form four bonding hybrids: $4s^24p^2$ hybridizes to $Ga^-(sp^3)$. This leads to the observed crystal structure, in line with Engel's rule (2.), and the electron transfer increases the alloy's stability. It is an ionic alloy, Na^+Ga^-, with a covalent network of Ga—Ga bonds.

Covalent Solids and Band Theory

We will discuss band theory first for solids that are clearly covalently bonded. This theory is applicable to all solids, however, and our choice of the word "covalent" for a class of solids distinct from "metals" is a bit forced, since, as we have seen, metals can be profitably discussed in terms of covalent bonding.

Consider the C–Pb group. Diamond is an excellent electric insulator. Silicon and germanium are notable *semiconductors*. They have electrical conductivities much better than diamond, but much worse than the obvious metals Sn and Pb. Can we understand these conduction trends, along with the crystal structures and binding energies of these elements, in a uniform way?

We start with Pb and the Engel rules. Lead forms an fcc lattice that indicates three s-p bonding electrons. The atomic ground state valence configuration of Pb (and its group companions) is s^2p^2, but an fcc lattice suggests[8] an excited sp^2d configuration that yields three s-p electrons (determining structure) and one d electron (especially effective in bonding, since Pb is in the last row of the group—rule 3).

Now we move up one to Sn, in particular to the phase called gray Sn that is stable below 292 K. Its lattice is isomorphic to that of diamond, which, by rule 1., requires four s-p electrons. We take the s^2p^2 ground state to an sp^3 configuration and form the tetrahedrally directed bonds of the diamond structure (Figure 16.11). The same argument holds for Ge, Si, and diamond itself.

For the electrical conduction question, we modify the free electron model (at least qualitatively) to conform to atomic energy level ideas. The free electron model allows electrons to have *any* energy above E_F without recognizing that some remnants of atomic quantum numbers should exist. For Na, the conduction electrons should, at some point, look like they came from a 3s orbital. Expanding on Figure 16.10, we form Na(s) from LCAO's of *all* the atomic orbitals of Na, not just the 3s valence orbitals. As the atoms approach, the larger orbitals (of higher n) overlap first, forming *bands* of allowed energies. This is shown in Figure 16.12.

The *energy width* of a band (the height of a shaded region in Figure 16.12) at any R increases as R decreases. At the lattice spacing of Na(s), the 2p band has not yet formed. These core electrons are still atomically localized. But the 3s band is fully formed, wide in energy, and overlapping the 3p, 4s, 3d, etc. bands. This overlap of wide bands is simulated in the free electron model by the continuous range of energies above E_F.

---

[8] In addition, relativistic effects, important in the heaviest elements, destabilize the sp^3 diamond configuration.

FIGURE 16.11 The diamond crystal structure has C atoms on an fcc lattice (lightly shaded spheres) plus four additional C toms (darkly shaded spheres).

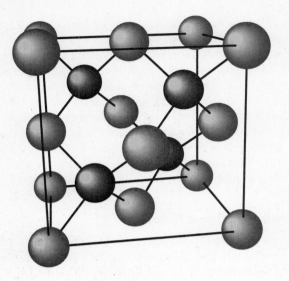

For a solid to exhibit ordinary electrical conduction, it must have either a partially full band or a series of overlapping bands that give the conduction electrons a pool of available energy levels. This is the case for both Pb and Sn that have many low-energy orbitals, all of which are large (have large n values) and overlap considerably at the ordinary lattice spacing.

For Ge, Si, and C(diamond), bands form, but they do not overlap, and the occupied bands are filled. An electron must acquire an appreciable energy increment called the *band gap energy* to occupy the next available energy levels. For semiconductors,

FIGURE 16.12 As Na atoms in a solid are brought closer together, their atomic orbital energies interact and spread into bands. At the observed solid spacing, 3.67 Å, the 2p orbitals remain localized on the atoms, but the 3s, 3p, etc. orbitals have formed a delocalized band, indicated by the shaded areas on the diagram.

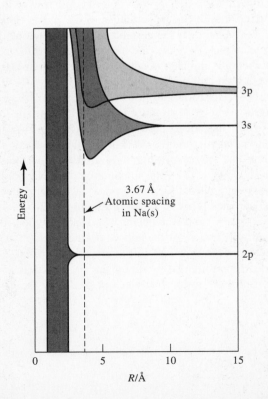

it is a moderate amount (0.72 eV for Ge, 1.14 eV for Si),[9] but for insulators (which are really just very lousy semiconductors) it is large (5.33 eV for diamond). Large band gaps shut down electrical conduction at ordinary temperatures.

Figure 16.12 suggests that band overlap, and thus electrical conduction, could be controlled by varying interatomic spacing. If the atoms in Na(s) were pulled far enough apart, Na would cease to conduct. After all, Na(g) is an insulator! Turning this argument around, *decreasing* atomic spacing in an insulator should induce conductivity. For most insulators, the atomic repulsion is so great (and the insulator band gap is so great) that pressures unknown on Earth would be required. But at least two insulating elemental solids, hydrogen and xenon, should have insulator → metal transitions at attainable laboratory pressures. There is experimental evidence that the 5p-5d band gap in Xe is overcome at around 1.3–1.5 Mbar, and Xe becomes an electrical conductor (but, interestingly, it is still transparent). Similarly, there is evidence for metallic hydrogen at $T \leq 100$ K and $P > 1.5$ Mbar, conditions that might well exist deep in hydrogen-rich Saturn and Jupiter.

The traditional definition of covalent solid applies to those elements, like C, Si, and Ge, that are insulators or semiconductors with an extended lattice of covalently linked atoms. Other covalent elemental solids include S, Se, and Te that form large rings (S) or long spiral chains of atoms (Se and Te), and the P, As, Sb, Bi group that forms puckered sheets of covalently bonded atoms.

The graphite structure of C is thermodynamically more stable than the diamond structure, and it will lead us into our final solid-bonding type. Graphite forms stacked layers of hexagonal C rings (Figure 16.13). The rings have C—C distances of 1.42 Å (compare to 1.40 Å in benzene) and form vast, delocalized π-electron bonding networks in the planar layer.

The layers are separated by a much larger distance, 3.35 Å. While the rings are strongly bound as covalent nets, the planes are only weakly bound together. It is this weak plane-to-plane attraction that gives graphite its lubricating ability and its easy direction of cleavage parallel to the planes. Moreover, graphite is a reasonably good conductor of electricity if the electric field that supports the current lies *in* the π-bonded plane of atoms, but it is a poor conductor perpendicular to these planes, indicating again the poor electron communication between planes.[10]

Molecular Crystals

The planes in graphite are (admittedly gigantic) covalent molecules held together by the sort of forces that bind solid benzene, or, for that matter, solid N_2, O_2, CO_2, F_2, polyethylene, etc. We use the term *molecular crystal* to describe those solids in which discrete molecules with closely spaced, chemically bound atoms pack under relatively weaker attractive forces. In a sense, we have come full circle back to the rare-gas solids, since the forces that bind those solids must exist in other molecular solids as well.

This connection is particularly clear in the recently discovered solid forms of carbon called the *fullerenes* of which C_{60}, buckminsterfullerene, is the major form. Buckminsterfullerene is an exceptionally elegant and beautiful molecule by itself (Figure 16.14(a)). The C atoms lie at the corners of connected hexagons and pentagons that together make the shape and pattern of a soccer football. All the C atoms

---

[9] And even gray Sn has a band gap, but of only ~0.08 eV magnitude. Such a slight gap can permit appreciable conduction at ordinary temperatures. These materials are called *intrinsic semiconductors*.

[10] An ordinary sample of graphite is polycrystalline, with planes oriented in all directions. Such a sample has an electrical conductivity intermediate to that in and perpendicular to the atomic planes of a perfect single crystal of graphite.

FIGURE 16.13 The graphite crystal structure (shown in a perspective view of a space-filling model in (a)) features sheets of hexagonal C atom rings, stacked ABABAB... with an offset indicated by the bold vertical line in (b) that connects eclipsed atoms.

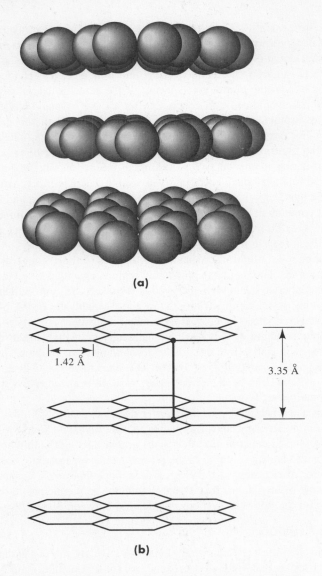

are equivalent. Each can be considered to be sp^2 hybridized with one π bond per C. These are delocalized over the molecule, adding to its stability. To a first approximation, C_{60} is a big—7.1 Å diameter—electron-rich sphere, a giant rare-gas atom. These nearly spherical molecules pack (Figure 16.14(b)) in an fcc lattice, and at ordinary temperatures, each molecule rotates fairly freely around its center at each lattice site. The nearest neighbor spacing, 10.0 Å, translates to a 2.9 Å gap between nearest C atoms on adjacent C_{60} molecules. This is less than the interplanar spacing in graphite (3.35 Å), but graphite and $C_{60}(s)$ ("fullerite") have nearly identical *linear* compressibilities (defined as $-d$(lattice constant)$/dP \times 1/$(lattice constant)) for the single fcc lattice constant of fullerite and the interplanar lattice constant of graphite; both are $\cong 2.3 \times 10^{-11}$ Pa^{-1}.

Another important interaction in certain molecular solids is *hydrogen bonding*, with ice the most important example by far. Water's molar volume increase on freezing under normal pressure is due to the formation of a regular lattice of H_2O molecules oriented so that the H of one molecule is closely along the line connecting

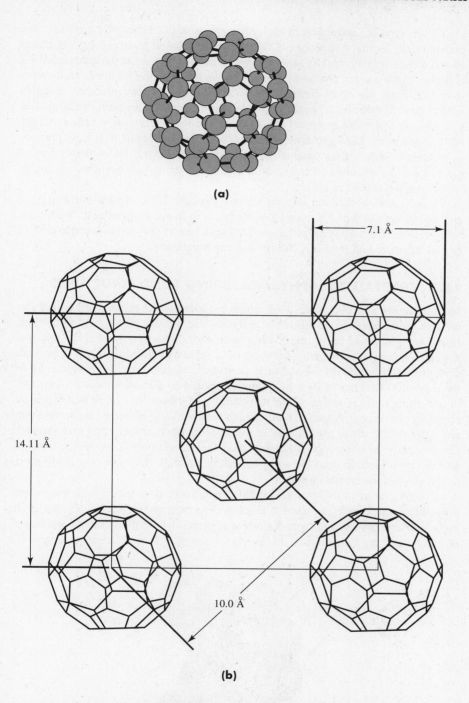

FIGURE 16.14 Solid C_{60} packs the nearly spherical C_{60} molecules (a) in an fcc unit cell, one face of which is shown in (b). At room temperature, the molecules are nearly freely rotating about their centers.

its covalent O partner and the O of a nearest-neighbor molecule forming a hydrogen bond from one molecule to the next.

Alcohols, carboxylic acids, and many biochemical macromolecules have structures and binding energies governed by hydrogen bonds. A typical hydrogen-bond strength is in the range 20–60 kJ mol^{-1}, just below the single covalent bond range, but appreciably larger than the strength of the interactions that bond Ar_2, $(CO_2)_2$, or other simple molecular dimers.

A third type of interaction in molecular solids borders the truly covalent. The classic examples are the halogens Cl_2–I_2 that form stacked planes of zig-zag chains of molecules (Figure 16.15). As the figure shows, the nominally nonbonded I–I distances in the zig-zag chains are appreciably shorter than the I–I distances between the planes, and this short contact can be interpreted in terms of iodine doing its best to form a delocalized bonding path along the chain. This path, weak as it is, gives $I_2(s)$ a noticeable electrical conductivity and its silvery cast in reflected light. At 0.17 Mbar, the band gap is closed and I_2 becomes a diatomic metal. As pressure is increased further, four other phase transitions are found, all corresponding to gradual loss of diatomic identity, until, at 0.55 Mbar, iodine becomes an atomic metal with a normal fcc lattice.

We have now worked our way through the Periodic Table of solid-bonding types and picked up many of the main ideas for bonding in solid compounds, from ionic to molecular, from alloy to polymer. The next section returns to the question of crystal structure and how it is determined experimentally.

16.3 CRYSTALLOGRAPHY: MEASURING SOLID STRUCTURES

Many butterfly and bird wings look iridescent because the regular array of scales or veins on these wings are spaced by distances that diffract visible light from them. If human eyes could see in the X-ray region of the spectrum (and if the sun was hot enough to emit appreciable numbers of X rays, which fortunately it is not), all crystalline solids would look similarly iridescent. Visible light wavelengths are in the range 400–700 nm while solid atoms are spaced by distances more on the order of 0.1–1.0 nm. X-ray radiation has wavelengths of this order, but, as de Broglie so elegantly pointed out, beams of particles can behave as waves of definite wavelength and exhibit diffraction phenomena as well, as Davisson and Germer first showed. Consequently, modern crystallography encompasses not only X-ray diffraction but also neutron, electron, and He atom diffraction as well, the latter two being most useful for surface crystallography.

Crystallography at its best can provide much more than lattice symmetries and unit-cell dimensions. It can provide electron density maps, molecular geometries in molecular crystals ranging from diatomics to proteins with thousands of atoms per molecule, and an indication of the average thermal motion of atoms in a crystal.

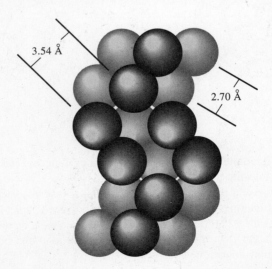

FIGURE 16.15 The I_2 molecular crystal structure shows zig-zag chains of molecules stacked in layers. Here, molecules in one layer are shaded differently from those in the layer below. The I_2–I_2 distance between closest atoms in the chains (3.54 Å) is notably less than the layer spacing (3.64 Å) but considerably greater than the bonded atom distance (2.70 Å).

Ever since Röntgen discovered X rays and won the first Nobel Prize for physics in 1901, crystallography has been a continuing basis for this award: 1914, von Laue, X-ray diffraction; 1915, William and Lawrence Bragg, X-ray analysis of crystal structures; 1937, Davisson and Thomson, electron diffraction; and 1986, in part to Binning and Rohrer for the scanning tunneling microscope (described in Section 16.5). In chemistry, the 1936 prize to Debye mentioned his work in gas-phase diffraction. Several other chemistry prizes have been awarded for structural work based on crystallographic research: 1962, Kendrew and Perutz; 1964, Hodgkin; 1969, Barton and Hassel; and the 1985 prize went to Hauptman and Karle in recognition of their work in solving the "phase problem" discussed later in this section.

Max von Laue's work explained the physical origin of X-ray diffraction patterns and described at least the rudiments of an analysis to yield structural information. To understand X-ray crystallography at the simplest level, we need to understand something of the nature of X-ray sources, of how X-ray diffraction patterns are generated and what they look like, and finally how X rays interact with matter to produce these patterns.

Traditional X-ray sources, from the time of Röntgen's discovery, are no more than a metal target that is bombarded in vacuum with electrons having kinetic energies in the tens to hundreds of kiloelectron-volt range. When these electrons strike the target, their energy is released in the metal as they undergo collisions with the metal's atoms and ultimately are slowed to energies near zero.

For X-ray production, the important energy transfer step is one in which the incident electron strikes an atom and transfers to it enough energy to remove one of the atom's *core* electrons. For a Cu target, a common X-ray source, this process is

$$e^- + Cu(1s^22s^22p^6\ldots) \rightarrow Cu^+(1s^12s^22p^6) + 2e^-$$

where one of the product electrons is the ejected core electron and the other is simply the incident electron now moving with somewhat less energy than it had originally. The Cu^+ left behind is highly energized; it is in an excited configuration as a result of the 1s electron loss.

One way this ion can spontaneously lose its excitation energy is the photon emission process

$$Cu^+(1s^12s^22p^6\ldots) \rightarrow Cu^+(1s^22s^22p^5\ldots) + photon$$

where the product ion is still in an excited configuration and will relax further, but for our purposes, the job has been done. The photon is in the X-ray region.

EXAMPLE 16.4

The $Cu^+(1s^12s^22p^6\ldots) \rightarrow Cu^+(1s^22s^22p^5\ldots)$ X ray has a wavelength $\lambda = 1.540\,6$ Å. What Cu-atom energy change is necessary to produce a photon with this wavelength?

SOLUTION Starting from $E = h\nu = hc/\lambda$, with a few conversion factors we find a useful relationship between photon wavelength in Å and photon energy in eV:

$$E = \frac{hc}{\lambda} = \frac{(6.626\,075\,5 \times 10^{-34}\text{ J s})(299\,792\,458\text{ m s}^{-1})}{\lambda(10^{-10}\text{ m Å}^{-1})(1.602\,177\,33 \times 10^{-19}\text{ J eV}^{-1})} = \frac{12\,398.456}{\lambda}\text{ eV Å}.$$

For the Cu X-ray wavelength, $E = 8.048$ keV. This is roughly 1000 times the energy associated with excitations of valence electrons and is a typical magnitude for X rays.

→ RELATED PROBLEM 16.31

Thus, each kind of metal target in a source of this type produces a set of discrete X rays with wavelengths varying from element to element. It has recently become possible to use a *continuously tunable* X-ray source, the *synchrotron accelerator source*. Synchrotron radiation is emitted when charged particles are accelerated in a circular track (the synchrotron) and, as do all accelerated charged particles, emit radiation. The radiation's spectrum is continuous and can have appreciable intensity in the X-ray region if the particle energy is high enough. Diffraction gratings disperse this continuous band of radiation and tune the output to a desired X ray. In addition, the light is rapidly pulsed (since the charged particles travel in bunches around the synchrotron) and polarized, two features difficult or impossible to obtain with conventional sources.[11]

The X-ray diffraction pattern itself is produced by directing the X-ray beam, collimated to a narrow pencil of radiation, onto the sample. Most of the radiation passes directly through the sample (as in X-ray radiography used for medical diagnosis), but some is scattered in well-defined directions away from the initial direction. This pattern of scattered radiation, both its direction and intensity, contains the diffraction information. It is recorded by either photographic film or an X-ray detector than can be scanned around the sample, usually under computer control.

The nature of the diffraction pattern depends strongly on the physical state of the sample. A macroscopic single crystal sample will yield a series of diffraction "spots" (see Figure 16.16a), and the location of these spots change as the sample is reoriented with respect to the incident X-ray beam. In contrast, a polycrystalline or powdered sample exposes all possible crystal orientations to the beam at once and yields a very different looking pattern (Figure 16.16b). This pattern has rotational symmetry about the incident beam direction and is called a *powder pattern*.

Now we turn to von Laue's theory. Given an X-ray source of wavelength λ and a diffraction pattern, how can we determine crystal and molecular structure? We look at a simple 1-D array of atoms first, and then generalize to the 3-D case.

Figure 16.17 shows a linear array of objects capable of scattering X rays. Obviously, we have atoms in mind, but all that matters now is that they scatter X rays. (In reality, X rays scatter only from atomic electrons. Nuclei are too small and heavy.) The periodic spacing is a, and the incident beam makes an angle θ with the array. The angle of observation is θ', and we ask if the diffracted intensity is large in this direction. If it is, the relative phase of the incident beam must be carried over to the diffracted beam. If the diffracted beams from each scatterer are not all in phase, destructive interference will reduce the diffraction intensity.

On the way in, the ray striking scatterer $(n + 1)$ has travelled a distance $a \cos \theta$ farther than that ray striking scatterer n. On the way out, the ray leaving scatterer n travels a distance $a \cos\theta'$ farther than that leaving scatterer $(n + 1)$. To retain phase and produce intense diffraction, the (in − out) path-length difference of the two rays must be an integral multiple of the wavelength λ. This is the simplest von Laue diffraction condition:

$$a \cos \theta \;-\; a \cos \theta' \;=\; a(\cos \theta - \cos \theta') \;=\; n\lambda \qquad (16.10)$$

$\uparrow$ Inward path-length difference $\qquad$ $\uparrow$ Outward path-length difference $\qquad$ $\uparrow$ Net path-length difference $\qquad$ $\uparrow$ Integral number (n) of wavelengths

where n is an integer.

---

[11] A conventional X-ray tube fits comfortably in your hand. A synchrotron fits more or less comfortably in a large building and costs $\sim 10^2$ times more than a conventional diffractometer.

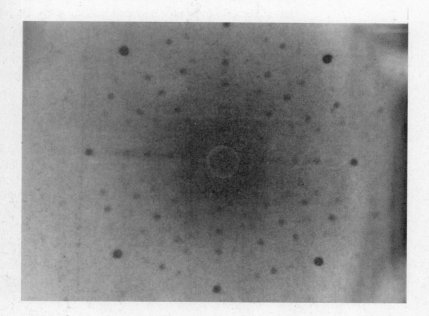

FIGURE 16.16 (a) A single crystal diffraction pattern (shown here for bcc W) shows diffraction spots that can be analyzed in detail for atomic positions in the unit cell. [From *Laué Atlas*, by E. Preuss, B. Krahl-Urban, and R. Butz (John Wiley & Sons, NY, 1974.)] (b) An X-ray powder pattern (shown here, for NaCl, as a slice of film that recorded the cylindrically symmetric diffraction intensity) has less information than an oriented single crystal pattern. Courtesy Professor R. H. Soderberg.

FIGURE 16.17 Diffraction by a line of scatterers, equally spaced by an amount a, relates the incident angle of the radiation θ to the angle of diffraction intensity θ′ through $a(\cos\theta - \cos\theta') = n\lambda$ where n is an integer and λ is the radiation's wavelength. If d is the spacing of hypothetical mirror planes in a crystal, then $2d\sin\phi = n\lambda$ where φ is defined in the figure.

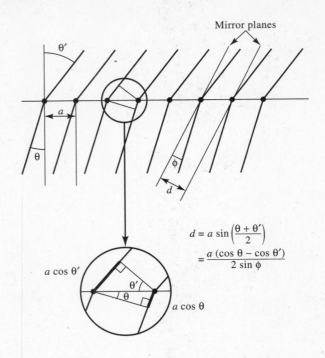

W. L. Bragg introduced an alternate way of looking at diffraction. He noticed that intense diffraction could be viewed as if the X rays had *reflected* from an imaginary mirror (or a set of parallel mirrors) in the array. Specular reflection (angle of incidence = angle of reflection = $\pi - \phi$; see Figure 16.17) from these imaginary mirrors must satisfy the Bragg condition, which is equivalent to the von Laue condition:

$$2d \sin \phi = n\lambda \qquad (16.11)$$

where d = mirror plane spacing = $a \sin[(\theta + \theta')/2]$. Due to Bragg's analogy, crystallographers call intense diffraction spots "reflections," much as molecular spectroscopists call transitions between energy levels "lines." The "lines" are so called because they appeared that way on early photographic plate spectrographs, but the term is now used for all sorts of spectroscopies and detectors. Likewise, an X-ray "reflection" is a term for a more complicated phenomenon.

In three dimensions, the Bragg diffraction condition is the simpler to introduce. To do so, we need an expression for the distance d between the hypothetical mirror planes, and to do this, we introduce the *Miller indices* of a crystal structure.

We start with a simple orthorhombic unit-cell lattice. The sides all intersect at right angles, but the edge lengths, a, b, and c, are all different: Miller indices use a cunning code to identify all the atoms that lie on a given plane.

For example, consider a plane containing the atoms on one face of the unit cell (and a zillion other atoms as well, but we need only locate those in the unit cell). They are at $(x, y, z) = (a, 0, 0)$, $(a, b, 0)$, (a, b, c), and $(a, 0, c)$ in the face shown in Figure 16.18. This plane satisfies the equation $x = a$. It is awkward to measure lengths in terms of *real* distances when the unit-cell dimensions are natural scaling lengths. We define dimensionless distances $(\tilde{x}, \tilde{y}, \tilde{z})$ such that

$$\tilde{x} = \frac{x}{a}, \quad \tilde{y} = \frac{y}{b}, \quad \tilde{z} = \frac{z}{c}$$

and the equation for this plane is $\tilde{x} = 1$.

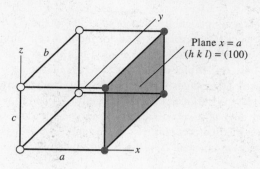

FIGURE 16.18 In the orthorhombic ($a \neq b \neq c$, all angles 90°) unit cell shown here, the shaded atoms all lie in the plane described by the equation $x = a$. This plane is called the (1 0 0) plane, since the x intercept of the plane is at $x = a$ and the y and z intercepts are at infinity.

An equation of the form

$$h\tilde{x} + k\tilde{y} + l\tilde{z} = 1 \qquad (16.12)$$

represents *any* plane in the crystal. Our plane has $h = 1$, $k = l = 0$. These are the *reciprocals* of the $\tilde{x}$, $\tilde{y}$, and $\tilde{z}$ *intercepts* of the plane. Since our plane does not intercept the $\tilde{y}$ or $\tilde{z}$ axis (or "intercepts them at ∞"), $k = l = 0$ (since "$1/\infty = 0$"). The set of three numbers, written $(h\ k\ l)$ without commas, are called the *Miller indices of the plane*. Our plane is the (1 0 0) plane. If the plane has a *negative* axis intercept, the notation uses a bar over the index. Thus, the (0 $\bar{1}$ 0) plane is parallel to the ($\tilde{x}$, $\tilde{z}$) plane and intercepts $\tilde{y}$ at -1 (or, equivalently, intercepts y at $-b$).

Figure 16.19 shows other important planes. The (1 1 0), (2 2 0), etc., planes are parallel to the $\tilde{z}$ axis while the (1 1 1) plane slices through the body of the unit cell with unit intercepts on each axis.

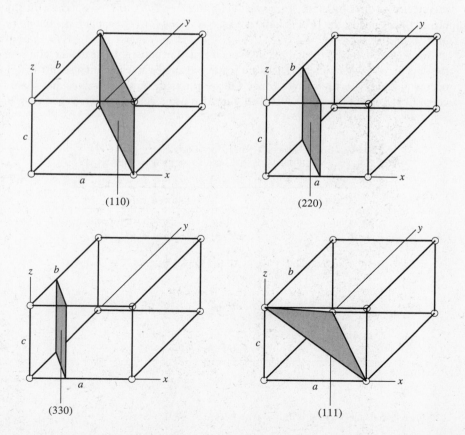

FIGURE 16.19 Other planes, identified by their Miller indices, can be found in the orthorhombic unit cell of Figure 16.18. The crystallographically equivalent (1 1 0), (2 2 0), and (3 3 0) planes along with the (1 1 1) plane are shown here.

If the unit cell is simple cubic ($a = b = c$), the planes (1 0 0), (0 1 0), (0 0 1), ($\bar{1}$ 0 0), (0 $\bar{1}$ 0), and (0 0 $\bar{1}$), the six faces of the unit cell, are *crystallographically equivalent* by symmetry. We designate a *set* of equivalent planes by curly brackets: {1 0 0}, using only one triplet of indices to represent the set.

There are many advantages to the Miller index notation. (We will meet them again when we classify surfaces.) For the crystallographer, they code for interplanar distances. From now on, we will consider only cubic lattices (sc, fcc, bcc, and diamond) and one lattice constant. For these lattices, the distance between planes in the {$h\,k\,l$} set (the constant d in the Bragg equation) is

$$d_{hkl}^2 = \frac{a^2}{h^2 + k^2 + l^2} \qquad (16.13)$$

where d_{hkl} is the interplanar distance for the set.[12] For example, the interplanar distance of the {1 1 0} set is $d_{110} = a/\sqrt{2}$.

Now we can incorporate Miller indices into the Bragg equation, Eq. (16.11). We write this first as $\sin \phi = n\lambda/2d$, square both sides, set $n = 1$ (as explained below), and use Eq. (16.13) for d^2:

$$\sin^2 \phi = \frac{\lambda^2}{4a^2}(h^2 + k^2 + l^2) \,. \qquad (16.14)$$

In the original equation, the integer n was related to the integral number of wavelength's of path difference between rays involving adjacent scatters. If $n = 1$, we say the diffraction is "first order"; if $n = 2$, "second order"; and so on. A second order (1 1 0) reflection appears in Eq. (16.14) as $2^2(1^2 + 1^2 + 0^2) = 8$, which is identical to a first order (2 2 0) reflection: $1^2(2^2 + 2^2 + 0^2) = 8$. Thus, in Eq. (16.14), we specify the ($h\,k\,l$) indices *in terms of the lowest common denominator of h, k, and l.*

We are almost ready to analyze an X-ray diffraction pattern. In particular, consider the data for a powder pattern. These are scattering angles at which intense diffraction occurs, a table of ϕ values. We convert these to a table of $\sin^2 \phi$ values that should equal $(\lambda^2/4a^2)(h^2 + k^2 + l^2)$. We then guess (or use a common divisor of $\sin^2 \phi$ to

TABLE 16.4 Allowed ($h\,k\,l$) Plane Indices for X-Ray Reflections in Cubic Crystals with All Atoms Alike

Crystal	Allowed values
Simple cubic	All ($h\,k\,l$) except those for which ($h^2 + k^2 + l^2$) = 7, 15, 23, ...
Body-centered cubic	All ($h\,k\,l$) such that $h + k + l$ is *even*
Face-centered cubic	h, k, and l *all* even *or* h, k, and l *all* odd
Diamond	$h + k + l$ is odd *or* $h + k + l$ is an *even* multiple of 2

---

[12] In an orthorhombic crystal, $1/d_{hkl}^2 = h^2/a^2 + k^2/b^2 + l^2/c^2$.

help us guess) values of (h k l) that can predict $\sin^2 \phi$ to within a common divisor. That divisor is ($\lambda^2/4a^2$). We know λ and thus compute a.

But how do we tell if the crystal is sc, fcc, bcc, or diamond? The answer comes from finding which reflections are *missing*. Each of these crystal structures has certain values of (h k l) that may represent a real plane of atoms in the crystal but *a plane from which reflections cannot occur*. The allowed values of (h k l) are summarized in Table 16.4. Thus, while *one* assigned diffraction angle is sufficient to find a, *many* assignments are needed to pinpoint the full crystal structure.

EXAMPLE 16.5

A Cu X-ray source ($\lambda = 1.540$ Å) produced a powder pattern in W with reflections at $\phi = 20.1°, 29.1°, 36.6°, 43.5°, 50.3°, 57.4°$, and $65.5°$. Assuming W has one of the cubic structures, find the lattice and lattice constant.

SOLUTION The solution can be followed through the table below:

ϕ	20.1°	29.1°	36.6°	43.5°	50.3°	57.4°	65.5°
$\sin^2 \phi$	0.118	0.237	0.355	0.474	0.592	0.710	0.828
$\dfrac{\sin^2 \phi}{\sin^2(20.1)}$	1.000	2.003	3.010	4.012	5.012	6.009	7.011
$\dfrac{2\sin^2 \phi}{\sin^2(20.1)}$	2.000	4.005	6.020	8.024	10.025	12.019	14.022
index	(1 1 0)	(2 0 0)	(2 1 1)	(2 2 0)	(3 1 0)	(2 2 2)	(3 2 1)
$h + k + l$	2	2	4	4	4	6	6
$\lambda^2/4a^2$	0.059 05	0.059 13	0.059 25	0.059 23	0.059 20	0.059 14	0.059 14
a/Å	3.169	3.167	3.163	3.164	3.165	3.166	3.166

The second row tabulates $\sin^2 \phi$, and the third starts our search for indexing (h k l) values with ratios of $\sin^2 \phi/\sin^2(20.1°)$. These are all closely integral, which is encouraging, except the last entry is $\cong 7$, and no set of integers (h k l) can make $(h^2 + k^2 + l^2) = 7$. The next row simply doubles the previous row. Each of these values (rounded to integers) is indexed to an (h k l) value in the following row (and note that (1 1 0) is equivalent to (1 0 1) and (0 1 1), etc.) For example, the last index is (3 2 1) since $(3^2 + 2^2 + 1^2) = 14 \cong 14.022$. The sixth row shows that $h + k + l$ is an even number for each reflection. Table 16.4 shows that this is the hallmark of bcc reflections. The following row is $\sin^2 \phi/(h^2 + k^2 + l^2) = \lambda^2/4a^2$, a constant for each angle, and finally, we calculate a row of a values. We would report the average, $a = 3.166$ Å. Note that without the reflection at 65.5°, we could have been misled into thinking the structure was sc. The remaining reflections could be indexed (1 0 0), (1 1 0), (1 1 1), (2 0 0), (2 1 0), (2 1 1), and we would find $a = (3.166$ Å$)/\sqrt{2} = 2.239$ Å. A measurement of the sample's density would settle the matter.

⇨ *RELATED PROBLEMS* 16.38, 16.39

The powder pattern, coupled with an analysis of allowed and forbidden reflections, allows one to find the unit-cell geometry. But to find detailed atomic structures for molecules packed in the unit cell, true single-crystal diffraction is used. Different elements scatter X rays with different efficiencies, called *atomic scattering factors* f. They are defined in such a way that f is a function of $\sin \phi/\lambda$ and equal to the number of electrons in the scatterer at $\phi = 0$, as is shown in Table 16.5.

The fact that f decreases at *any* ϕ as atomic number decreases means that light atoms, H in particular, scatter poorly and are difficult to locate. As a result, neutrons, which scatter from nuclei, are used to produce diffraction patterns sensitive to H-atom positions.

TABLE 16.5 Atomic Scattering Factors for Selected Elements

Atom	\sin \phi/(\lambda/Å)					
	0.0	0.20	0.40	0.60	0.80	1.0
H	1	0.48	0.13	0.04	0.02	0.01
He	2	1.45	0.74	0.36	0.18	0.10
B	5	2.71	1.69	1.41	1.15	0.90
C	6	3.58	1.95	1.54	1.32	1.11
N	7	4.60	2.40	1.70	1.44	1.26
O	8	5.63	3.01	1.94	1.57	1.37
Ne	10	7.82	4.62	2.79	1.98	1.61
Na	11	8.34	5.47	3.40	2.31	1.78
Na$^+$	10	8.39	5.51	3.42	2.31	1.79
S	16	11.21	7.83	6.31	4.82	3.56
Cl	17	12.00	8.07	6.64	5.27	4.00
Cl$^-$	18	12.20	8.03	6.64	5.27	4.00
Ar	18	12.93	8.54	6.86	5.61	4.43
Kr	36	28.53	21.34	16.54	12.57	9.66
Xe	54	43.7	33.1	25.9	20.9	17.2

Not only do different elements scatter with varying efficiencies (which means a variation in the scattered wave's *amplitude*), they will scatter the X ray with a *phase* that depends on the element's position in the unit cell. All we can observe is the collective result of the varying amplitude and phase combinations when an X-ray wave scatters from all the atoms in the unit cell. In particular, if F represents the collective effect of f's of varying amplitude and phase, what we see is proportional to $|F|^2$. Intensities of waves are proportional to squares of amplitudes.[13]

The complex function $f_i e^{i\Phi_i}$ where Φ_i is the phase angle represents the amplitude and phase of the wave scattered from atom i (at unit-cell position $(\tilde{x}_i, \tilde{y}_i, \tilde{z}_i)$). For a cubic crystal,

$$\Phi_i = 2\pi(h\tilde{x}_i + k\tilde{y}_i + l\tilde{z}_i)$$

where $(h\ k\ l)$ denotes the diffracting plane. What we observe is

$$|F|^2 = \left(\sum_i f_i e^{i\Phi_i}\right)^2. \tag{16.15}$$

For example, a bcc crystal has two atoms per unit cell, at $(\tilde{x}_i, \tilde{y}_i, \tilde{z}_i) = (0, 0, 0)$ and $(\frac{1}{2}, \frac{1}{2}, \frac{1}{2})$. Equation (16.15) becomes

$$|F|^2 = [f_1 + f_2 e^{\pi i(h+k+l)}]^2 = \{f_1 + f_2\cos[\pi(h + k + l)]\}^2$$

since $e^{ix} = \cos x + i \sin x$ and $\sin[\pi(h + k + l)] = 0$ for all integral $(h\ k\ l)$. If atoms 1 and 2 are the same, $f_1 = f_2 = f$, and if $(h + k + l)$ is an even number, $|F|^2 = (f_1 + f_2)^2 = 4f^2$. But if $(h + k + l)$ is odd, $|F|^2 = (f_1 - f_2)^2 = 0$. Analyses such as these lead to the rules in Table 16.4. If atoms 1 and 2 are different (as in CsCl, for example), then if $(h + k + l)$ is odd, $|F|^2 = (f_1 - f_2)^2 \neq 0$, but it may be *small* if atoms 1 and 2 have similar f values.

---

[13]The parallel with ψ, the quantum mechanical wave function amplitude, and $|\psi|^2$, the observable probability distribution, may have occurred to you.

We have glossed over the origin of the f values themselves. Without going through a full derivation, f is proportional to the charge density $\rho_{el}(r)$ of the scatterer.[14] But of more chemical importance, $\rho_{el}(r)$ is proportional to $|\psi(r)|^2$, and thus X-ray techniques can reveal full electron distributions as well as nuclear positions.

We now focus on the full charge distribution in the unit cell. We know this distribution will have maxima in the vicinity of atomic nuclei, but we should be able to recover ρ_{el} at any point in the unit cell. We write (still for a cubic crystal, but more general expressions are known) an analog of Eq. (16.15):

$$F \propto \int_{\substack{\text{unit}\\\text{cell}}} \rho_{el}(\tilde{r}) e^{2\pi i (h\tilde{x} + k\tilde{y} + l\tilde{z})} \, d\tilde{r} \ .$$

To find ρ_{el}, we invert this integral equation by a technique known as Fourier transformation. This is, in itself, not difficult, but a major difficulty is that we need F to find ρ_{el} while what we measure is $|F|^2$. Given $|F|^2$, F is determined only to within an unknown phase factor.

This is the *phase problem* in structure determination that was mentioned in this section's opening. Since the phase appears to be unknowable, any technique that can find it would certainly seem to be worthy of a Nobel Prize, as the 1985 prize in chemistry to Hauptman and Karle acknowledged. Several approaches to this problem have been suggested. The current method of most widespread use (called the *direct method*) is an algorithm for systematically improving estimates of $\rho_{el}(r)$ through various relations that effectively constrain or limit the phases to certain values or narrow ranges of values. As many bits of information as one can think of (e.g., all the symmetry properties of the crystal, the fact that ρ_{el} is everywhere the same algebraic sign, etc.) are brought to bear. The process is now so well refined that one can buy a precision computer controlled diffractometer that automatically positions and scans both the crystal[15] and the detector, collecting data, analyzing them by programs that can routinely locate ~50 atoms in a molecule, and finally plotting publication quality views of the molecular structure and/or the electron charge distribution. Standardized formats for such data have been established, leading to computer data bases of structures that have a growing number of uses.

16.4 AMORPHOUS SOLIDS AND CRYSTAL DEFECTS

A macroscopic crystal of ~10^{22} identical molecules all correctly positioned at the proper locations in some space lattice is a practical impossibility. Through heroic efforts, some compounds can be purified to the point that they are 99.999 999 9% pure: impurities are at the sub-parts-per-billion level. That still implies ~6×10^{14} foreign molecules per mole. Through further heroic efforts, these pure compounds can be "grown" into macroscopic crystals that otherwise approach lattice perfection.

But, by and large, solids are dirty, imperfect phases. Often, the dirt—the impurities—have value, as in the carefully added dopants of semiconductor electronics or carefully formulated metal alloys. The difference between structural steel and the much stronger steel of piano wire is largely due to a variation in C content between roughly 0 and 1% by weight. Mechanical failure in virtually all materials occurs at stresses far below the theoretical maximum strength. This maximum corresponds to the energy required to break bonds in a pure, perfect crystal. The observed strength is due to fracture of the much weaker interactions at a variety of defects in real samples.

---

[14] Recall Eq. (15.1). Since $\rho_{el}(r) \propto n$, the number of electrons, f has a maximum value of n.

[15] Which need not be large: crystals of ~0.1 mm sides will suffice.

Defects are classified into several families based on their geometry. There are *point defects* (Figure 16.20(a)) in which atoms are missing, are of the wrong type, or are in the wrong place. These yield important electronic and optical properties without seriously altering mechanical properties.

Line defects (Figure 16.20(b)) arise from packing faults of several types. *Grain boundaries* locate the touching-planes of microcrystals. *Dislocations* are due to extra planes of atoms stuck part way into the crystal, or to a twist or spiral slip of plane-to-plane registration (in what are called *screw dislocations*). To see how easily these defects arise, throw a large number of marbles into a box or grab a handful of drinking straws. In neither case are you likely to find perfect close-packing.

Defects can be treated as if they were physical entities. They have entropies and energies associated with them, and they can be tracked as they diffuse (and sometimes collide with and annihilate one another). Their study and control is a major topic of materials science.

At the extreme of defects we find the amorphous solid, which is a solid lacking any long range structural order. The most common amorphous solid is ordinary

FIGURE 16.20 (a) All real crystals are imperfect, and a number of common *point defects* can be expected. Note that ionic crystals are subject to additional types of defects that relate to charge as well as chemical identity. (The *F* center is also called a *color center* from the German, *farbe,* meaning color. Such centers can absorb visible or near-IR light.)

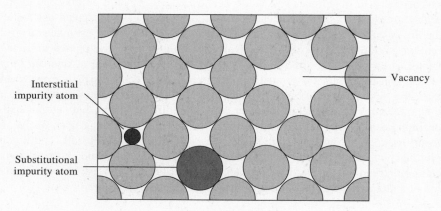

Simple solid point defects

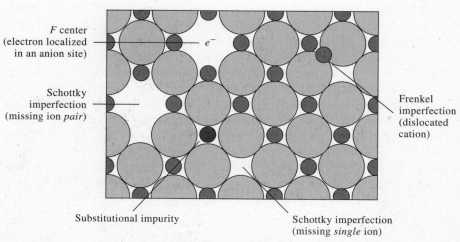

Ionic solid point defects

(a)

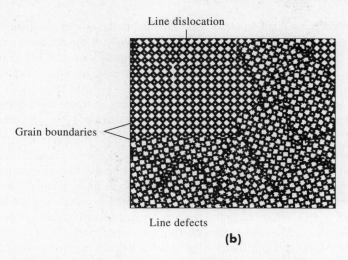

FIGURE 16.20 (b) In addition to point defects, *line defects* can cause crystal imperfection. The packing of crystal grains leads to *grain boundaries*, in which crystal lattices mismatch. Within a grain, a *dislocation* can occur when an extra row of atoms tries to slip into an otherwise regular lattice.

glass. (In spite of the common name, "crystal" glass is not very crystalline.) Amorphism does imply mechanical failure in and of itself, however. For instance, ordinary glass fractures because of defects of larger than microscopic size. These are typically surface cracks (called Griffith flaws): large, built-in points of strain that, with a little help, can be relieved by propagating a crack, leading to brittle fracture. It is a quite dramatic demonstration to dip a microscopic slide in an alcoholic HF solution (etching away the outer layer with the largest flaws) and observe the unbelievably large angle through which the etched slide can be bent before finally breaking.

The term *glass* can be applied to a class of noncrystalline solids, including ordinary glass (much as we use "salt" for ionic compounds in general and edible NaCl in particular). Among the elements, only S and Se form amorphous, glassy solids as long spiral chains of covalently bonded atoms. When the crystalline form of these elements is melted, the chains that pack in the crystal become mobile, shortened somewhat, and entangled. Rapid cooling freezes the tangles in place, forming a brittle, amorphous solid—a glass.

Similarly, many important polymeric materials are amorphous. Crystalline polyvinylidene chloride becomes amorphous when copolymerized with vinyl chloride ($CH_2=CHCl$) to yield the less brittle material used as a transparent food-wrap. But even amorphous polymers become brittle, stiff, and hard at sufficiently low temperatures.

The transition from pliant to brittle glass behavior in polymers is noted by measuring, for example, the coefficient of bulk thermal expansion. At low enough temperatures, the molar volume of the brittle glassy state is a slight function of temperature, but at a certain temperature, called the *glass transition temperature,* the thermal expansion coefficient increases notably. Above this temperature, the molecular chains in the glass (or at least long segments of them) gain considerable extra motion. The material expands, and other physical properties—viscosity, thermal conductivity, heat capacity, etc.—undergo notable changes. The glass transition temperature is typically between one-half to two-thirds the normal melting temperature of the polymer.

16.5 SOLID SURFACE STRUCTURES

This section will touch (one is tempted to say, "only the surface") just the basic structural terminology that is the language of the larger topic of surface science.

Techniques for surface-structure characterization have been available for roughly thirty years or less, but they are largely responsible for the great activity in surface science during that time. X-ray diffraction is generally of little use to the study of surfaces. X rays penetrate too deeply into the sample to diffract from the surface alone.

Just as 3-D solids have only 14 Bravais lattices, surfaces have only five corresponding "nets" or 2-D lattices: square (☐), rectangular (▭), centered rectangular (▭•), hexagonal (◊), and "oblique" (▱). For pure, isolated solids, one might expect that the surface structure is just like some plane of atoms that can be found in the bulk crystal. By and large, this is true, but two important phenomena make surface crystallography an important topic in its own right. The first is the study of *adsorbate* structures on a crystal surface. For instance, when a monolayer or less of a gas is adsorbed onto a regular crystalline surface, do the gas molecules form a lattice based on that of the underlying surface, do they form no regular lattice, or do they form a lattice that has no relation to the underlying surface structure?

The second phenomenon pertains to the pure crystal surface itself. Since a surface is an exposed plane of atoms rather than a surrounded plane in the bulk, the surface atom distances vary slightly from distances in the bulk. In some cases, the crystal surface may take on an entirely different symmetry. It may *reconstruct* itself into a new pattern.

Figure 16.21 shows top and side views of the atomic packing of several typical crystal surfaces for pure atomic solids. These faces are named by their Miller indices and by the cubic unit cell to which they belong. The top plane of atoms is generally pulled slightly into the crystal since there are no bonding interactions on the vacuum side of the surface.

A reconstructed surface, the Ir fcc (1 1 0) plane, is shown in Figure 16.22. It is called a "(2 × 1) missing row" reconstruction to emphasize that, in one direction, atoms are found one every *two* idealized positions (the missing row is between the positions) while in the orthogonal direction, atoms are found at every idealized position.

We know these patterns primarily from *Low Energy Electron Diffraction,* or LEED (commonly pronounced as one word, rhyming with "need"). For an electron with a kinetic energy E (in eV), the de Broglie wavelength (in Å) is approximately $\lambda/\text{Å} = [150/(E/\text{eV})]^{1/2}$. (See Problem 11.11.) For diffraction, we need λ values comparable to atom spacings, or $\lambda \sim 0.5$–5 Å or so. This corresponds to electron energies in the range 600–6 eV, which happen to be quite easily attained experimentally. Moreover, electrons in this range penetrate a metal surface, on average, to a depth of only 5–20 Å before scattering. Lower energy electrons penetrate more deeply ($\sim$100 Å or more), as do higher energy electrons. We thus have a fortuitous match of easily obtained electron energies, yielding appropriate de Broglie wavelengths, in a range which is most sensitive to the surface atoms. The only drawback is that the sample needs to be electrically conducting. A salt crystal, for instance, would quickly attain a static electric charge from the electron beam. This charge would deflect the electrons that follow.

The LEED experiments (and most surface science experiments in general) are conducted in vacuum chambers capable of very low pressures, typically on the order of or lower than 10^{-11} atm. At ordinary vacuum levels (10^{-9} atm is quite easy to attain), the residual gas bombards the surface so frequently that a monolayer of adsorbed gas can form within a second or so. Thus, extra effort is required to attain ultrahigh vacuum, to clean the surface, and only then to begin the experiment.

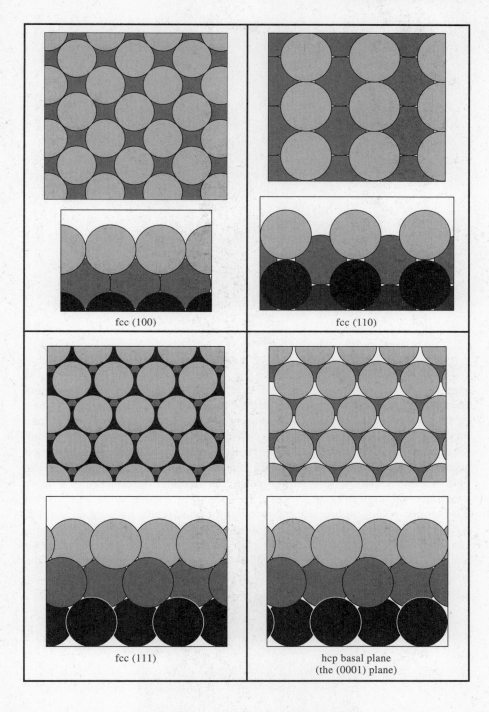

FIGURE 16.21 Top and side views of common simple surface structures.

Even at 10^{-11} atm ambient pressure, one has only a few hours' time before a monolayer accumulates.

The LEED diffraction pattern is typically recorded by directing the electron beam normal to the surface and observing the back-scattered diffraction intensities as spots on a fluorescent screen. Figure 16.23 shows the LEED pattern resulting from 51 eV electrons striking the close-packed (1 1 1) surface plane of Pt. The hexagonal

FIGURE 16.22 Iridium surfaces are said to *reconstruct* in that they have a structure related to, but different from, any plane internal to the bulk. This Ir surface has a missing row of atoms on the surface compared to the patterns expected for a (1 1 0) plane.

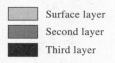
Surface layer
Second layer
Third layer

symmetry is quite evident. As with X rays, more detailed structural information comes from a combined analysis of both the diffraction *positions* and their *intensities*.

Due to the unique chemical environment of the surface, an alloy surface may have a *different composition* from that of the bulk. Surface thermodynamics, using ideas from regular solution theory, can predict these composition changes with good accuracy in many cases. One finds that the top layer in a binary alloy may be enriched in one element to a greater extent, the second layer enriched to a lesser extent, and the third may be depleted in this element, relative to the bulk, so that a *composition wave* is observed as one leaves the bulk, heading toward the surface.

Surface compositions are measured most often by a technique known as *Auger electron spectroscopy*. The Auger (a French name, pronounced, more or less, "Oh-zhay'") process begins with high energy (10^3 eV or more) electron impact on the

FIGURE 16.23 The LEED pattern from a Pt (1 1 1) surface (shown here for a 51 eV electron beam energy) displays the hexagonal symmetry of the surface. [From *Chemistry in Two Dimensions: Surfaces,* by G. A. Somorjai (Cornell University Press, Ithaca, NY, 1981.)]

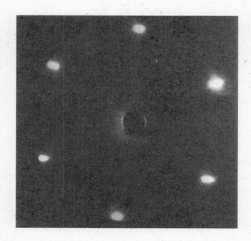

16.5 SOLID SURFACE STRUCTURES

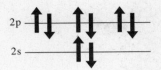

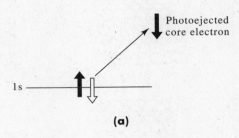

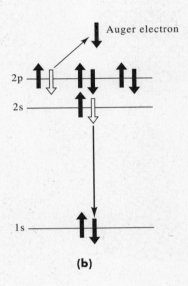

FIGURE 16.24 The surface analysis method known as Auger electron spectroscopy begins—in (a)—with the photoejection of a core electron due to bombardment by photons (X rays) or electrons of 1–5 keV energy. This leaves behind an ion in an excited state that can relax by the Auger process, (b), in which the core electron vacancy is filled by a higher energy electron while a valence electron gains energy and is ejected. It is the characteristic energy of this second ejected electron that forms the basis for chemical analysis.

metal surface, as if we had an X-ray source tube. These electrons can eject an inner, core electron (Figure 16.24(a)) from an atom at or near the surface. As an alternate to X-ray emission, this excited atom can undergo a *two-electron* process (the Auger effect, Figure 16.24(b)) in which one electron drops in energy while a second (which need not be from the same atom) gains energy in excess of its binding energy and escapes from the surface. The analytical information is contained in the kinetic energy of the ejected electrons. The energies of these electrons depend on the elemental identity *and* the chemical environment of the atom. Moreover, alternating Auger spectroscopy with high-energy atomic ion bombardment (keV energy Ar^+ is commonly used), one can slowly remove layer after layer of surface atoms and obtain a composition depth profile.

A final aspect of surface structure worth discussing is the ability to prepare, characterize, and study surfaces with high Miller indices. Such surfaces (see Figure 16.25) are composed of *steps* of atomic layers with step terraces of varying width and step edges of varying orientation and regularity. The steps themselves can have *kinks,* as they are called, as can be seen in the fcc (14 11 10), (10 8 7) and (13 11 9)

FIGURE 16.25 When an fcc lattice is cut along a plane with high Miller indices, the resulting surface is characterized by steps and terraces of varying sizes with kinks along the steps. [From *Chemistry in Two Dimensions: Surfaces,* by G. A. Somorjai (Cornell University Press, Ithaca, NY, 1981.)]

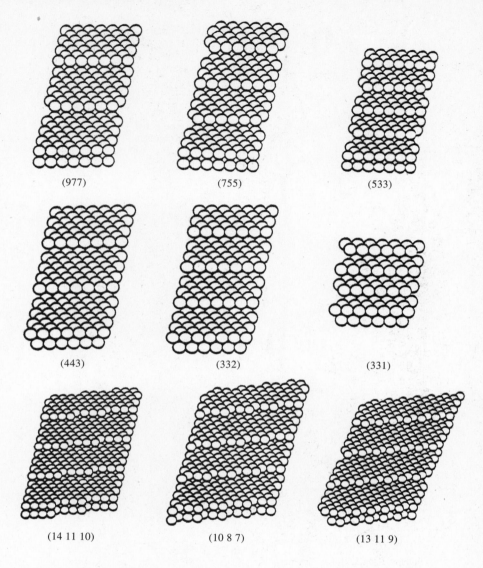

surfaces of Figure 16.25. The atoms on the corners of a kink are highly exposed (or, equivalently, they have a very low number of nearest neighbors). They are chemically unsaturated and are likely sites for crystal growth or for unique chemical activity. Such surfaces clearly give the best hope for systematic study of the random surfaces exposed in uncontrolled conditions such as those commonly encountered in industrial catalytic processes.

The most recent tool of surface crystallography, *scanning tunneling microscopy* (STM), has made surfaces visible in stunning detail.[16] The STM technique uses a very sharply pointed metal tip that is brought to within roughly an atomic diameter of the surface under study. At such a distance, the electron clouds of the surface and the probe tip interact, and an electrical current flows via an electron tunneling mechanism through the surface-tip gap when an electrical potential is applied across

---

[16] A related instrument is the atomic force microscope (AFM).

the gap. The magnitude of the current is a strong function of the tip to surface distance so that circuitry controlling the 3-D position of the tip can scan the tip over the surface at a constant tunneling current value. What is remarkable is that this positional control can be maintained at subnanometer levels, scanning the surface topography with atomic resolution. A striking example of this resolution is shown in Figure 16.26, an STM picture of the terraces and kinks of the Si (100) 2 $\times$ 1 reconstructed surface. Each step is one atom high, and one can clearly see that the reconstructed rows alternate direction from layer to layer and that the kink density is very different from terrace edge to terrace edge. One can also see other imperfections: isolated atoms sitting on terrace planes and isolated vacancies one layer deep on terraces.

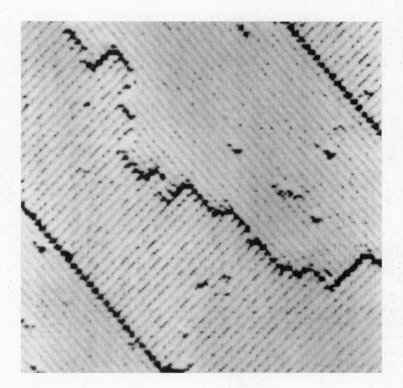

FIGURE 16.26 A scanning tunneling micrograph of a stepped surface of Si shows terraces, kinks, terrace vacancies, and reconstructed terraces in alternating layers.

CHAPTER 16 SUMMARY

Solids are characterized by long-range order and crystalline symmetry (or by lack of both for amorphous solids). This chapter has discussed the elementary symmetries of common crystalline solids and focused first on the properties of the *unit cell:* coordination numbers, packing fractions, symmetries, dimensions, and interstitial sites. Some of these properties are summarized below for the cubic symmetry lattices:

Lattice	sc	bcc	fcc	hcp	diamond
Unit cell		Fig. 16.6b	Fig. 16.2b	Fig. 16.3b	Fig. 16.11
Atoms in unit cell	1	2	4	6	8
Packing fraction	0.5236	0.6802	0.7405	0.7405	0.5101
Number of nearest neighbors	6	8	12	12	4

Solid bonding was discussed in terms of several models based on the dominant interactions among the solid's constituents: ionic bonding, free-electron metals, the Engel–Brewer view of metal bonding, the molecular packing of rare gases, CO_2, N_2, etc., and the covalent bonding of C, Si, etc., P, As, etc., and S, Se, etc. Each of these models has features that most directly account for macroscopic solid properties, such as the high melting point of ionic solids (due to strong Coulombic binding), the electrical conductivity of metals (due to the free-electron motion at the Fermi energy), the low melting point of molecular solids (due to the generally weak attractive forces between stable molecules), and the hardness of diamond or the cleavage planes of graphite (due to the covalent bond network differences between these solids).

We saw how crystallographic diffraction reveals crystal symmetry and atomic identity, and how the Bragg diffraction condition could be generalized to more complicated situations, ultimately arriving at a measure of the detailed electron density in the solid. Diffraction was also seen to be useful in deducing solid surface geometries due to a fortuitous match of electron energies, penetrating powers, and de Broglie wavelengths brought forth in LEED measurements.

The theme of symmetry and its exploitation, so important for solids, is largely lost when we consider liquids in the next chapter. The liquid state gains its unique features primarily by giving up any claim to symmetry.

FURTHER READING

If you like reading short stories in bed at night, the book *Phase Diagrams of the Elements,* by David A. Young (University of California Press, Berkeley, CA, 1991) is for you. You can turn to any chapter and spend a few minutes reading about the often fascinating phases and structures of a favorite element. Excellent general chapters also review experimental and theoretical techniques with emphasis on high-pressure phases.

Structural data on elements and numerous compounds are collected in *Crystal Structures,* 2nd ed., by R. W. G. Wyckoff, (Wiley-Interscience, NY, 1963). More extensive discussions of many of this chapter's topics are in *Introduction to Solid State Physics,* by C. Kittel, (Wiley, NY, 1976). The Engel–Brewer model is described in "The Role and Significance of Empirical and Semiempirical Correlations," by L. Brewer in *Structure and Bonding in Crystals,* M. O'Keeffe and A. Navrotsky, Ed., vol. I, p. 155, (Academic Press, NY, 1981). This volume also has an interesting article, "Some Aspects of the Ionic Model of Crystals," by M. O'Keeffe, (*ibid.,* p. 299), that goes beyond the ionic model discussion here.

X-ray crystallography is covered in many texts and monographs. M. M. Woolfson's *An Introduction to X-ray Crystallography* (Cambridge University Press, Cambridge, UK, 1970) and *Structure Determination by X-ray Crystallography,* by M. F. C. Ladd and R. A. Palmer, (Plenum, NY, 1985) are two good sources.

Surfaces are covered in depth in *Surface Science, an Introduction,* by J. B. Hudson, (Butterworth-Heinemann, Boston, 1992), *Chemistry in Two Dimensions: Surfaces,* by G. A. Somorjai, (Cornell University Press, Ithaca, NY, 1981), *Principles of Surface Chemistry,* by G. A. Somorjai, (Prentice-Hall, Englewood Cliffs, NJ, 1972), and *Physical Chemistry of Surfaces,* by A. W. Adamson, 4th ed., (Wiley-Interscience, NY, 1982).

PRACTICE WITH EQUATIONS

16A A 1.00 L cube is carefully filled with small marbles packed in an fcc fashion. How much water can be poured into the cube?   Figure 16.2

ANSWER: 0.26 L

16B The CsCl lattice constant is 4.11 Å. What is the Cs-Cl distance?   Figure 16.8a

ANSWER: 3.56 Å

16C The NaCl lattice constant is 5.63 Å. What is the Na–Cl distance?   Figure 16.8b

ANSWER: 2.82 Å

16D How does the compressibility vary as the exponential repulsion range parameter ρ decreases for a crystal of constant lattice parameter?   (16.8)

ANSWER: $\kappa \to 0$ as $\rho \to 0$

16E What is the speed of an electron in Na with the Fermi energy? (16.9), Example 16.3

ANSWER: $v = (2E_F/m)^{1/2} = 1.05 \times 10^6$ m s^{-1}

16F The valence configuration of atomic V is $3d^34s^2$. What configuration does solid V use? Tables 16.2, 16.3, Engel's rules

ANSWER: $3d^44s^1$

16G X rays of 0.70 Å wavelength impinge at an angle $\theta = 20°$ on a row of scatterers spaced 2.5 Å apart. At what angle does the first-order diffracted beam appear? (16.10)

ANSWER: 48.7°

16H In a cubic lattice, which set of planes are more closely spaced, {1 1 1} or {2 0 0}? (16.13)

ANSWER: $d_{111} = a\sqrt{3}$; $d_{200} = a/2$

16I What is the Bragg diffraction angle for the (2 0 0) reflection of fcc Au ($a = 4.078\,2$ Å) recorded with Cu X-rays ($\lambda = 1.54$ Å)? (16.14)

ANSWER: 22.2°

16J To what wavelength should you set your synchrotron light source so that the (2 0 0) reflection from a powder sample with $a = 4.46$ Å emerges at 30°? (16.14)

ANSWER: $\lambda = a/2 = 2.23$ Å

PROBLEMS

SECTION 16.1

16.1 Derive the packing fractions $\pi/6$ for the sc lattice and $\sqrt{3}\pi/8$ for the bcc lattice that are quoted in the text.

16.2 Prove that the ratio of the hcp unit cell's basal plane edge length a to the cell's height (traditionally called c) is $c/a = \sqrt{8/3}$, as shown in Figure 16.3(a). How well do the following data on hcp forms of elements agree with this perfect-packing prediction?

Element	a/Å	c/Å
Be	2.27	3.59
Cd	2.97	5.61
Co	2.51	4.07
Gd	3.62	5.75
He	3.57	5.83
Mg	3.20	5.20
Ti	2.95	4.73
Zn	2.66	4.94

Calculate c/a for each, along with a percent deviation of each from the perfect-packing value. Are you surprised by the elements that best agree with perfect packing? Do you see any periodic trends among those that do not? Explaining the deviations is more difficult than observing them!

16.3 The densest element is Os, $\rho = 22.950$ g cm^{-3}. Osmium forms an hcp lattice with $a = 2.734\,1$ Å. What is c, the hexagonal unit-cell height? (See the previous problem for a discussion of a and c.)

16.4 The central atom in a bcc unit cell has eight nearest neighbors, as Figure 16.6 shows. If the bcc packed spheres have radius r, what is the volume of the unit cell? How many second-nearest neighbors are there? How far away are they from the central atom?

16.5 The edge length of the Po unit cell (the only sc lattice element) is 3.36 Å. What is its density?

16.6 From 1183 to 1673 K, Fe forms an fcc lattice with a density 7.648 g cm^{-3}. What is the unit-cell edge length? Above 1673 K, Fe(s) converts to a bcc structure. Predict the density of this phase. In general, when a close-packed structure transforms to a bcc structure, do you expect the sample to expand or contract in size? What assumption must you make in these calculations?

16.7 Gold, density 1.932 g cm^{-3}, packs in one of the cubic lattices fcc or bcc. The unit cell's edge length is 4.078 33 Å. Which lattice does Au use?

16.8 The two sketches below show the 12 nearest neighbors around any one atom in the fcc and hcp lattices. Which sketch goes with which lattice? How can you tell?

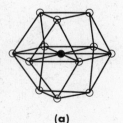 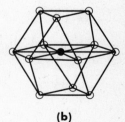

(a)   (b)

16.9 There is no unique way to assign a unit cell. The fcc unit cell of Figure 16.2 (shown in heavy outline on the next page) is the traditional and visually simple one. Shown in light outline is an alternate unit cell, the *primitive* unit cell. What is the ratio of the primitive cell's volume to the traditional cell's volume? How many atoms are in the primitive cell?

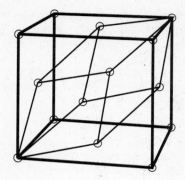

SECTION 16.2

16.10 There are many biochemical and macromolecular systems in which the packing of long cylindrical molecules is important. Calculate the packing fraction for long cylinders, stacked like drinking straws in a box, for each of these two packing arrangements:

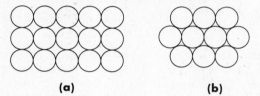

What is the largest radius ratio for packing cylinders of unequal radii in each of these two arrangements with the smaller cylinders in interstitial sites?

16.11 Derive the radius ratios $r'/r = (\sqrt{2} - 1)$ for the fcc octahedral site and $r'/r = (\sqrt{3}/\sqrt{2} - 1)$ for the fcc tetrahedral site. What are the geometries of the interstitial sites (and how many of any one kind are there) in the hcp unit cell?

16.12 There are two interstitial sites in the bcc lattice. One is $(0, \frac{1}{2}, \frac{1}{2})$ and the other is $(0, \frac{1}{2}, \frac{1}{4})$ as shown below. Find the radius ratio for both sites. Which is larger?

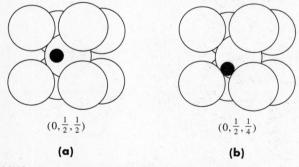

16.13 Draw, to scale, the plane in the CsCl crystal in which four Cl⁻ ions "touch" the central Cs⁺ ion and the centers of all five ions are coplanar. Remember that in the CsCl crystal, one edge of the rectangle connecting the Cl⁻ ions' centers is $\sqrt{2}$ times longer than the other in this plane. Use the following ionic radii: $r_{Cl^-} = 1.70$ Å; $r_{Cs^+} = 1.84$ Å. Do the Cl⁻ ions touch each other in your drawing?

16.14 Use the Born–Meyer potential model to calculate the molar crystal energy of LiF for which $R_0 = 2.014$ Å, $\kappa = 1.49 \times 10^{-11}$ m³ J⁻¹, and $M = 1.747\,558$. The experimental value is -1032.6 kJ mol⁻¹, the most negative of all alkali halide crystals. (Can you explain why?)

16.15 Solve Eq. (16.8) for κ, and use the result to explain why κ is smallest for LiF and largest for RbI among all the alkali halides that form the NaCl crystal lattice. *Hint:* For all the NaCl-type alkali halides, $R_0/\rho \cong 9 \pm 1$ while R_0 ranges from 2.014 Å (LiF) to 3.671 Å (RbI).

16.16 An alternate to the exponential repulsive potential of Eq. (16.1) is the *inverse power repulsion* $U_{rep}(R) = B/R^m$ where B and m are parameters with m usually in the range 8–20. Use this repulsive term in place of the exponential term for $U_{ij}(R)$ in Eq. (16.3) and show that the analog of Eq. (16.7b) is

$$\overline{U}_{cr} = \frac{-N_A Me^2}{4\pi\epsilon_0 R_0}\left(1 - \frac{1}{m}\right).$$

Is the assumption that the inverse-power repulsion acts only between the central ion and its nearest neighbors necessary here? (This is the form of the repulsive energy first used by Born to discuss alkali halide packing.)

16.17 Use the molar crystal-energy function of the previous problem to find an expression for the parameter m in terms of the isothermal bulk compressibility κ. Find a value for m for NaCl and then find $\overline{U}_{cr}$, following the calculation in the text that used the Born–Meyer potential with an exponential repulsion.

16.18 In Example 16.2, the Born–Meyer potential was used to find a bond length for NaCl(g). Use this same potential to find the NaCl(g) bond energy, and compare your value to the experimental value of 4.23 eV.

16.19 The bond lengths of free alkali halide molecules (such as NaCl, 2.360 9 Å) are *shorter* than the nearest-neighbor distances in the crystalline solids (such as Na–Cl, 2.820 Å). Why? In contrast, the bond lengths of free rare gas dimers are almost exactly equal to the nearest-neighbor distances in the solid rare gases. Is this observation in accord with your explanation for the alkali halides?

16.20 Suppose a linear chain of alternating + and − charges interact through a Coulombic and a (nearest-neighbor only) repulsive potential of the form A/R^{12}. What is the equilibrium nearest-neighbor distance? How different is it from the equilibrium separation of a + − ionic diatomic with the same interatomic potential?

16.21 The alkali metals all form bcc lattices at room temperature. Write the free-electron model Fermi-energy expression in terms of the bcc lattice constant (and any other aspect of bcc packing that is relevant), and use the result to predict the trend in Fermi energy from Li to Cs.

16.22 The chemical view of electrons at the Fermi energy would have them in the highest occupied molecular orbital of the bulk metal. Such an orbital would have the largest number of nodes, a number on the order of the number of atoms in the solid. Show that the free-electron model is qualitatively in agreement with this by writing an expression for the de Broglie wavelength of an electron at the Fermi energy and showing that this wavelength is of the same order of magnitude as the atomic spacing.

16.23 Consider a one-dimensional metal, such as has been proposed for certain organometallic compounds that conduct well in only one crystallographic direction. Use a 1-D particle-in-a-box model to find an expression for the Fermi energy of N electrons confined to a line of length L. Find also the total energy of these N electrons by direct summation from $E = 0$ to $E = E_F$. You will need the summation formula

$$\sum_{n=1}^{k} n^2 = \frac{1}{6} k (2k^2 + 3k + 1) \cong \frac{1}{3} k^3 \text{ for large } k \ .$$

16.24 A metal's conduction electrons' contribution to $\overline{C}_V$ is given by Eq. (4.17), $\overline{C}_V = RT/\Theta_{el}$ where R is the universal gas constant. In the free electron model (from a statistical mechanics calculation), $\Theta_{el} = 2E_F N_A/\pi^2 R$. Compute Θ_{el} for Na and compare your answer to the experimental value in Table 4.3. The agreement will not be spectacular, but it is remarkable that such a simple model comes as close as it does.

16.25 The expression for the free electron model's heat capacity $\overline{C}_V$ from the previous problem predicts $\overline{C}_V$ should vary with one elemental property, the molar volume, in a simple way. Use the experimental data in Table 4.3 and the molar volumes below to check this prediction for the alkali metals. Plot Θ_{el} versus $(1/\overline{V})^{2/3}$. How good is the correlation?

	Li	Na	K	Rb	Cs
$\overline{V}$/m³ mol⁻¹	1.300×10^{-5}	2.368×10^{-5}	4.536×10^{-5}	5.579×10^{-5}	7.096×10^{-5}

16.26 The free-atom ground-state valence configurations of Pd, Ag, and Cd are $4d^{10}$, $4d^{10}5s^1$, and $4d^{10}5s^2$, respectively. Palladium is fcc, silver is fcc, and cadmium is hcp. Use these facts and the atomization energies of Table 16.3 to decide the number of electrons used in bonding and the bonding configurations of these elements.

16.27 The C—C bond distance in diamond is accurately known to be 1.544 52 Å at 18 °C. Calculate the density diamond would have if the C atoms were close-packed in an fcc or hcp lattice, using the ideal packing fraction for these lattices. The observed density of diamond is 3.516 g cm⁻³, about half the value you will calculate assuming closest packing. Explain the discrepancy on the basis of the diamond lattice shown in Figure 16.11.

16.28 Imagine the diamond lattice shown in Figure 16.11 to be made of equal-radii spheres packed so that they touch along covalent bond directions. Taking the 18 atoms in that figure to describe the diamond unit cell, find the number of atoms contained *in* the unit cell and the packing fraction. In almost all problems of geometry around a tetrahedral center, it helps to view the tetrahedral geometry centered in a cube as shown below:

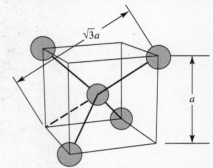

This diagram makes it clear that the bond length is one-half the body diagonal of the cube in which the tetrahedral shape is inscribed.

16.29 You can see how a band of orbitals of finite energy width is formed in a solid if you consider a simple one dimensional model in the Hückel approximation introduced in Section 14.3. One can show that the Hückel orbital energies of N identical orbitals equally spaced on a line is (with N an even number)

$$E_n = \alpha + 2\beta \cos \frac{n\pi}{N+1}, n = 1, 2, \ldots, N \ .$$

Find the difference between E_N, the highest energy orbital, and E_1, the lowest of the band, in the limit $N \rightarrow \infty$. How much larger is this band width than the $E_2 - E_1$ difference for an $N = 2$ Hückel system like ethylene? Now let $N = 10$. Calculate all ten energies (and note that the energies are symmetrically distributed about $E = \alpha$ so that you need calculate only five numbers), and plot them in an energy-level diagram. What does your diagram say about the density of levels across the band? Where is the density highest? lowest?

16.30 Atomic Ce has the configuration $4f^15d^16s^2$, and an explanation of the origin of Ce's curious isostructural phase transition (a sudden fcc unit-cell volume collapse from $a = 5.16$ Å to 4.85 Å) has focused on this configuration. The earliest explanation, which further study shows not to be the full story, starts with the conduction band, essentially the 5d and 6s orbitals, having three electrons in the expanded phase. Collapse is blamed on promotion of the localized 4f electron into this band. Why would this promotion be expected to induce collapse?

SECTION 16.3

16.31 In addition to the $\lambda = 1.54$ Å X ray discussed in Example 16.4, Cu also emits one with $\lambda = 13.357$ Å. Find this photon's energy and suggest the excitation that leads to it.

16.32 The diagrams below are variants of Figure 16.19. What are the Miller indices for the planes shown? (Each cuts an axis at 1, 1/2, 1/3, or ∞ times the unit cell edge length.)

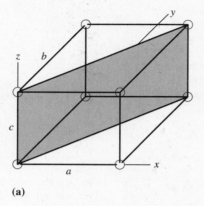

(a)

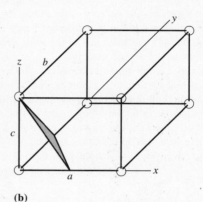

(b)

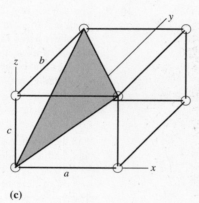

(c)

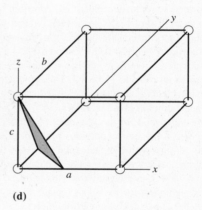

(d)

16.33 Show that the spacing between planes with Miller indices $(h\ k\ 0)$ is

$$\frac{1}{d_{hk0}^2} = \frac{h^2}{a^2} + \frac{k^2}{b^2}$$

for the unit cell shown below.

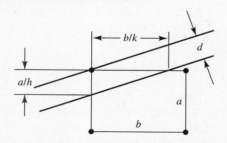

Remember that the edges a and b are perpendicular, look for places to use the law of sines for a triangle [sin(angle)/(opposite side) = constant], and remember that $\sin \theta_1 = \cos \theta_2$ if $\theta_1 + \theta_2 = 90°$.

16.34 One of the most remarkable recent uses of precision X-ray measurements was in the determination of Avogadro's constant. A group at the (then) U.S. National Bureau of Standards (now the National Institute for Science and Technology), over a period of several years, combined a precision lattice plane measurement of Si with a precision Si density measurement to refine the precision of N_A. (Why do you think Si was used?) This group found the distance between the (2 2 0) planes in Si to be $d_{220} = 0.192\ 017\ 07$ nm at 25 °C. They also measured $\overline{V} = 12.059\ 027\ 4$ cm^3 mol^{-1}. Use these data (and the fact that Si has the diamond lattice) to compute N_A. These values were obtained around 1980; the 1986 adjusted value is $6.022\ 136\ 7 \times 10^{23}$ mol^{-1}.

16.35 Why are the simple cubic $(h\ k\ l)$ values for which $h^2 + k^2 + l^2 = 7, 15, 23$, etc. excluded in Table 16.4?

What is the next number in the series 7, 15, 23, . . . ? (*Big hint:* h, k, and l are integers.) What is Eq. (16.15) for a simple cubic lattice with only one unit-cell atom?

16.36 Derive the rules in Table 16.4 for the allowed values of h, k, and l in an fcc lattice with all atoms alike. This lattice can be taken to have its four atoms in the unit cell at coordinates $(0, 0, 0)$, $(\frac{1}{2}, 0, \frac{1}{2})$, $(\frac{1}{2}, \frac{1}{2}, 0)$, and $(0, \frac{1}{2}, \frac{1}{2})$. How do the rules change if there are two different atoms, A and B, in the unit cell with A at positions $(0, 0, 0)$ and $(\frac{1}{2}, 0, \frac{1}{2})$ and B at the other two positions?

16.37 How many sc powder pattern reflections can be seen from a crystal with a lattice parameter four times the X-ray wavelength? To see more reflections, should one use a larger or a smaller wavelength?

16.38 Suppose that the first reflection in an X-ray powder diffraction comes at an angle ϕ_1. Write expressions for the cubic lattice constant in terms of λ and ϕ_1 for each of the four cubic lattices, sc, fcc, bcc, and diamond. Order the values from largest to smallest, and give the reflection plane index for each.

16.39 A powder pattern for V taken with 1.54 Å (Cu source) X rays shows lines at the following angles from the X-ray direction: 21.1°, 30.7°, 38.6°, 46.1°, 62.0°, 53.7°, and 72.5°. Show that these reflections are consistent with a bcc lattice, and find the lattice constant.

16.40 Solid C_{60}, fullerite, has been shown by NMR spectroscopy at room temperature to consist of essentially freely rotating molecules centered on an fcc lattice. Freely rotating molecules in an X-ray pattern would be indistinguishable from another packing situation that would be physically different. What other crystal packing characteristic would give the same X-ray diffraction as freely rotating C_{60} molecules?

16.41 Both KCl and KBr pack in the NaCl fcc lattice type, but X-ray powder patterns of KCl look remarkably like those of a sc lattice while KBr powder patterns clearly show all the fcc allowed reflections. Why is this? (*Hint:* Think isoelectronically and refer to Table 16.5.)

SECTION 16.4

16.42 One solid has a large number of Schottky-pair defects. Another has a large number of Frenkel defects. Which has the greater density difference in comparison with the perfect solids? In which direction is the difference?

16.43 What sort of defects would be introduced if a small amount of $CaCl_2$ is added to $NaCl(s)$ as it crystallizes? The ions Na^+ and Ca^{2+} have about the same size.

16.44 Stoichiometric FeO cannot be prepared. Iron-deficient compounds in the range $Fe_{0.91}O$ to $Fe_{0.95}O$ are the best one can do. These compounds form an ionic lattice isomorphous to NaCl. Suggest the types of defects that must exist in the nonstoichiometric compounds. Pay particular attention to the requirement of electroneutrality.

16.45 It takes energy to create a Schottky-pair defect in an ionic solid, 195 kJ mol^{-1} for NaCl(s). A statistical mechanics argument shows that the number of defects in a crystal of N ion pairs at temperature T is $N \exp(-E_s/2RT)$ where E_s is the energy to create one mole of defects, as quoted above. The density of NaCl is 2.17 g cm^{-3}. Calculate the number of ion pairs in a 1 cm^3 sample and the number of Schottky defects per cm^3 at room temperature and at 1064 K, 10 K below the melting temperature.

SECTION 16.5

16.46 The nearest-neighbor distance in fcc Ag is 2.88 Å. Calculate the surface density of atoms (atoms cm^{-2}) for the (1 1 1), (1 1 0), and (1 0 0) surfaces of Ag.

16.47 Assign each of the surface morphologies (top layer of atoms only) in Figure 16.21 to a primitive surface net: square, rectangular, centered rectangular, hexagonal, or oblique. Draw the surface atom arrangements and assign nets to the (1 0 0) and (1 1 0) bcc surfaces as well.

16.48 Imagine you have a single fcc crystal of Pt in the form of a cube with (1 0 0) faces. Describe how you would cut this cube to expose a (3 3 1) stepped face as shown in Figure 16.25.

16.49 The lattice constant of Ir (fcc) is $a = 3.838\,9$ Å. What is the distance between the surface layer of rows in the reconstructed Ir (1 1 0) surface of Figure 16.22? Draw a side view of this surface showing the surface rows as peaks and the third layer atoms as valleys. How deep are the valleys?

GENERAL PROBLEMS

16.50 He(s) in the hcp form has lattice constants $a = 3.531$ Å and $c = 5.693$ Å. Since an hcp lattice of hard spheres would have a = hard-sphere diameter (see Figure 16.3), it would seem to be reasonable to assign He an effective diameter of 3.531 Å. How well does this value agree with the hard-sphere gas estimate of the radius using Eq. (1.19) and equating B_{hs} to the high temperature limit $B \cong 12$ cm^3 mol^{-1} quoted in Table 1.1? You will find a to be substantially larger than the gas-phase hard-sphere diameter. Can you suggest a reason for the disagreement? (*Hint:* Helium is very light and very weakly bound in the solid. What are the quantum consequences of a light mass in the lowest bound state of a shallow potential well?) Repeat the calculation for fcc Xe with lattice constant 6.131 7 Å and $B_{hs} = b$(van der Waals) = 51.05 cm^3 mol^{-1} (Table 1.2). You will again find the crystal radius to be larger than the hard sphere radius (but by a much smaller factor). Roughly speaking, the crystal radius is sensitive to the location of the interatomic potential minimum while the gas-phase hard sphere radius is sensitive to the location of the interatomic potential's repulsive wall.

16.51 The close-packed planes in the hcp structure are stacked <u>AB</u>AB... and those in fcc are stacked <u>ABC</u>ABC... as shown in Figure 16.5. (The repeating units are underlined.) The lanthanides from La–Pm adopt a variant of hexagonal packing with layers ordered <u>ABAC</u>ABAC... while Sm adopts the more complex <u>ABABCBCAC</u>ABABCBCAC... order. If we represent hcp by the sketch

draw the corresponding sketch for fcc and the early lanthanides. Start and end your sketch with close-packed A layers, top and bottom, as in the sketch above. (Try the Sm arrangement only if you have a full clean sheet of paper and draw exceptionally well!)

16.52 Calculate the Madelung constant for an array of alternating $+$ and $-$ charges on a square planar lattice of unit edge length. A portion of this lattice is shown below:

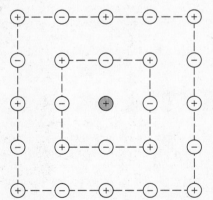

Center the Madelung sum on the shaded positive charge, and add successive contributions from ever larger square areas such as the two outlined above. In each square, partition the unit charges into fractional charges inside and outside the square. For example, the innermost square has four corner $+$ charges that each contribute $+1/4$ to the inside of the square and $+3/4$ to the outside along with four edge $-$ charges that contribute $-1/2$ inside and $-1/2$ outside. The net charge enclosed by this square

is thus $+1$ (the central ion) $+ 4 \times (+1/4) + 4 \times (-1/2) = 0$. The Madelung sum will converge rapidly if summed over regions of zero net charge. The first square's contribution is (note the sign convention that makes M positive)

$$M \cong -\left[\frac{(+1)\left(-\frac{1}{2}\right)(4)}{1} + \frac{(+1)\left(\frac{1}{4}\right)(4)}{\sqrt{2}}\right] = 1.292\,893$$

$$= -\left[\frac{\begin{pmatrix}\text{central-}\\\text{ion}\\\text{charge}\end{pmatrix}\begin{pmatrix}\text{fractional}\\\text{edge}\\\text{charge}\end{pmatrix}\begin{pmatrix}\text{number}\\\text{of edge}\\\text{charges}\end{pmatrix}}{\text{center-to-edge distance}}\right.$$

$$\left.+ \frac{\begin{pmatrix}\text{central-}\\\text{ion}\\\text{charge}\end{pmatrix}\begin{pmatrix}\text{fractional}\\\text{corner}\\\text{charge}\end{pmatrix}\begin{pmatrix}\text{number}\\\text{of corner}\\\text{charges}\end{pmatrix}}{\text{center-to-corner distance}}\right]$$

Continue the sum over the next two squares. Be sure to include the fractional charges outside the first square as part of the second square's contribution, and so on. You should find $M \cong 1.292\,89 + 0.313\,98 + 0.003\,65 = 1.610\,52$. Additional terms converge to $1.615\,54$.

16.53 An important quantity in the free electron model is the total molar energy of the electrons at absolute zero, a quantity we will denote E_0. Three steps are needed to derive it. First, show that for a monovalent metal $N(\epsilon)$, the *number of free-electron states* with energy $\leq \epsilon$, is given by

$$N(\epsilon) = \frac{\pi}{3}\left(\frac{2m\epsilon}{\hbar^2\pi^2}\right)^{3/2} V$$

where $V = L^3$ = the system's volume. Second, show that the *density of states* $\rho(\epsilon)\,d\epsilon$, the number of states with energy between ϵ and $\epsilon + d\epsilon$, is

$$\rho(\epsilon)\,d\epsilon = \frac{\pi}{2}\left(\frac{2m}{\hbar^2\pi^2}\right)^{3/2} V\epsilon^{1/2}\,d\epsilon\ .$$

At this point, we can derive the molar Fermi-energy expression directly from

$$N_A = \int_0^{E_F} \rho(\epsilon)\,d\epsilon\ ,$$

which you can verify recovers Eq. (16.9). Finally, find E_0 from

$$E_0 = \int_0^{E_F} \rho(\epsilon)\epsilon\,d\epsilon$$

and show that your answer can be written as $E_0 =$

$3N_A E_F/5$, which can be considered to be the *zero-point energy* of the ensemble of free electrons.

16.54 We can use Eq. (3.34) to compute an impressive number: the internal pressure of the free electrons in a metal. If we recognize that $\overline{U} = E_0$ (found in the previous problem) at $T = 0$, we have simply $P = -(\partial E_0/\partial \overline{V})$. Show (easily) that

$$P = \frac{2}{3}\frac{E_0}{\overline{V}}$$

and then evaluate P for Na (in both Pa and atm units). The magnitude of the number should impress you, but what is more impressive, recall, is that this number comes from a model which *ignores* electron–electron repulsion, yet it gives a value of the correct order of magnitude when compared to experiment. Thus, the enormous pressure you calculate is *not* due to free electrons pushing away from each other. What physical reason can you give for this pressure?

CHAPTER 17

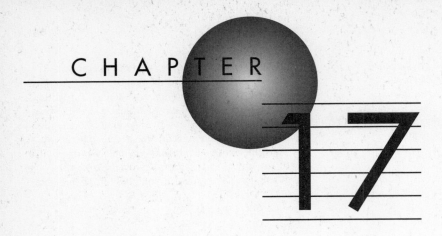

The Liquid State

17.1 Why Liquids Are Messy
17.2 The Connections to Solids and Gases
17.3 Simple Liquids
17.4 Ordinary Liquids—Water
17.5 Liquid Crystals

THE majority of chemistry involves liquids in one way or another. Reactions are commonly run in liquid solutions, and solution properties are commonly used for the separation and/or analysis of products. Of the 50 chemicals produced in greatest mass by the U.S. chemical industry, typically 20 are liquids with sulfuric acid leading the list. About 5×10^{13} g of H_2SO_4 are manufactured annually, a staggering number,[1] but one that pales in comparison to the approximately 1.45×10^{24} g of water in the oceans.

It is clearly important to understand liquids. But, for reasons discussed in this chapter's first section, liquids are inherently more difficult to characterize and model than either gases or solids. We will concentrate here on liquid structure, but since many of the results depend on statistical mechanics, a topic for later in this book, we will not see the basis of these results. Once we learn how to think about liquids on a microscopic level in a general way, however, the results will be plausible and sensible.

After the introductory section, the following section discusses an important measure of liquid structure, the *radial distribution function*. This function puts qualitative impressions of the differences among gas, liquid, and solid structures on a quantitative basis. Next, the properties and structures of simple real liquids and some hypothetical

---

[1]But comparable to the roughly 5×10^{13} g of soft drinks manufactured each year!

models are discussed, followed by a section on more everyday liquids that focuses on water, one of the most complex liquids known. Finally, a closing section discusses the curious state of matter that may well be found in your digital watch's or calculator's display—the liquid crystal.

17.1 WHY LIQUIDS ARE MESSY

There is nothing inherently intractable about bulk properties of liquids. We can (and do throughout this book) discuss the thermodynamic properties of liquids and liquid solutions with the same ease as we do those for solids and gases. If we roam through the (T, P) domain of pure liquids below the critical point, we find that it occupies a large area on any phase diagram. If we follow gas–liquid phase boundaries, we find that the vapor pressure curves of most liquids are remarkably similar.

Liquids often have bulk properties that are unique to that phase: solubility properties, electrical conductivity of liquid salts (something solid and gaseous salts do very poorly), vast changes in viscosity with temperature, etc.

So what is it that makes a microscopic description of liquids difficult? Think first about the simplest model of a gas, the ideal gas. All the molecules are noninteracting points. There is no potential energy, only kinetic energy. Temperature and pressure are the important parameters. Structure enters only through the volume of the container, controlling entropy from the Sackur–Tetrode equation, Eq. (5.44). Molecular identity (beyond mass) allows us to distinguish one gas from another only through "imperfections," the deviations from $PV = nRT$ behavior described by virial coefficients or van der Waals parameters.

Contrast this model with the ideal crystalline solid of the previous chapter. Atoms or molecules of definite size and shape are arrayed immobile on a symmetric, periodic lattice. Our simplest solid models are all *potential* energy, no kinetic energy. Temperature and pressure play secondary roles to the packing structure, which is itself determined through molecular interactions.

These two extremes, gas and crystal, have such simple first descriptions as a result of density and order differences. Gaseous molecules are spaced by distances large compared to molecular size, and their motion is chaotic. Crystalline molecules are packed in regular arrays and they "touch." In contrast, although they have densities comparable to solids, liquids have a general lack of structure beyond the first few nearest neighbors; liquid molecules are in constant motion, diffusing, ebbing, and flowing through their surroundings, always touching several neighbors.

Thus, liquids are difficult to describe if only because they demand a structural, potential-energy-based description, like crystals, but one that lacks the simplicity of long range order, while they also require a dynamic, motion-based description, like gases, but one that insists on molecular collisions. Consequently, a unique point of view is called for. The next section introduces this by means of a measure of structure, the pair distribution function.

17.2 THE CONNECTIONS TO SOLIDS AND GASES

The previous section mentioned the high entropy of gases and the low entropy of solids. Thus, we will begin to place the structure of liquids on some scale relative to gases and liquids by considering a liquid's entropy. We will consider benzene's entropy change on vaporization as a specific example. Here is our plan of attack. The Sackur–Tetrode equation, Eq. (5.44), gives the molar entropy of an ideal gas in any state. The perfect crystal has $\overline{S} = 0$ at $T = 0$. If we compare $\Delta\overline{S}_{vap}$ to the absolute gaseous molar entropy, we can find if the liquid's entropy is more "gaseous" in nature ($\overline{S}(g) \gg \Delta\overline{S}_{vap}$) or more "crystalline" in nature ($\overline{S}(g) \sim \Delta\overline{S}_{vap}$).

The data we need are the molar enthalpy of vaporization at the normal boiling point:

$$\Delta\overline{H}_{vap} = 30\,780 \text{ J mol}^{-1} \text{ at } T_{vap} = 353.26 \text{ K}, P_{vap} = 1 \text{ atm}$$

and the molecular mass, 78.11 g mol^{-1}. The Sackur–Tetrode equation says the absolute entropy of benzene vapor at the normal boiling point is $\overline{S} = 20.04\,R$. This is the entropy change for the process

$$C_6H_6(\text{crystal}, T = 0) \rightarrow C_6H_6(\text{ideal gas}, T = T_{vap}, P = 1 \text{ atm}) \ .$$

Since $\Delta\overline{S}_{vap} = \Delta\overline{H}_{vap}/T_{vap}$, we find $\Delta\overline{S}_{vap} = 10.48\,R$. (This is the Trouton's constant for benzene, as you may recall from Chapter 6.) Thus, for

$$C_6H_6(\text{liquid}, T = T_{vap}, P = 1 \text{ atm}) \rightarrow C_6H_6(\text{vapor}, T = T_{vap}, P = 1 \text{ atm})$$

the molar entropy change is 10.48 R. The liquid entropy at the boiling point must be $20.04\,R - 10.48\,R = 9.56\,R$.

Thus liquid benzene has about half the entropy of the gas:

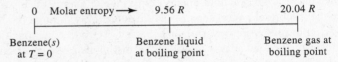

It is neither fully crystalline nor fully gaseous, at least in so far as these limits are measured by entropy.

EXAMPLE 17.1

Is this entropy ranking of liquid benzene fair? How does the picture change if the molar enthalpy of fusion, 9.951 kJ mol^{-1} at the melting point, 278.7 K, and the liquid heat capacity, $\overline{C}_p = 135.6$ J mol^{-1} K^{-1}, are considered?

SOLUTION The entropy change on melting is just $\Delta\overline{H}_{fus}/T_{fus} = 35.7 \text{ mol}^{-1} \text{ K}^{-1} = 4.29\,R$, which means the solid at the melting point is 4.29 R lower in entropy than the liquid. The liquid has $\overline{S} = 9.56\,R$ at the boiling point, and on going from melting to boiling, the liquid gains $\Delta\overline{S} = \overline{C}_p \ln(T_{vap}/T_{fus}) = 3.35\,R$; so, at the melting point, the liquid has $\overline{S} = (9.56 - 3.35)R = 6.21\,R$, and the solid has $\overline{S} = (6.21 - 4.29)R = 1.92\,R$. Thus, melting roughly triples the molar entropy, and vaporization roughly doubles it:

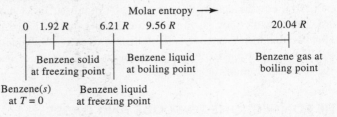

➠ RELATED PROBLEM 17.3

How else can we place liquid structure in contrast with those of gases and solids? One way is through what are called *molecular configurational distribution functions*. These functions describe the location and arrangement of molecules in any phase.

The simplest is the local *particle density distribution* $n^{(1)}$. It is defined so that

$$n^{(1)}(x, y, z)\, dx\, dy\, dz = \text{number of molecules at position } (x, y, z) \text{ in the volume } dx\, dy\, dz\ . \quad (17.1)$$

In any isotropic phase (gas or normal liquid), $n^{(1)}$ is the constant particle density, independent of position:

$$n^{(1)} = \frac{N}{V} \text{ for an isotropic fluid} = \text{the particle density}\ . \quad (17.2)$$

The second configuration distribution function is much more useful. Called the *pair distribution function* $n^{(2)}$, it is defined so that $n^{(2)}(\mathbf{r}_1, \mathbf{r}_2)\, dx_1\, dy_1\, dz_1\, dx_2\, dy_2\, dz_2$ is the probability that two molecular centers (any two) are simultaneously at positions $\mathbf{r}_1$ and $\mathbf{r}_2$ in volumes $dx_1\, dy_1\, dz_1$ and $dx_2\, dy_2\, dz_2$ around them. In an ideal gas,

$$n^{(2)} = \frac{N(N-1)}{V^2} \cong \left(\frac{N}{V}\right)^2 = (n^{(1)})^2\ . \quad (17.3)$$

The pair distribution function can be used to calculate the *configurational energy* U_c. This is the total intermolecular potential energy, a complicated function of the positions and orientations of all N molecules: $U_c(\mathbf{r}_1, \mathbf{r}_2, \ldots, \mathbf{r}_N)$. If we assume that molecules interact only as *pairs* through so-called *pair potentials*, U_c can be written in a simple way.

Here is the idea behind pair potentials. Suppose molecules 1, 2, and 3 are brought from positions infinitely far apart (where they do not interact) to some closely spaced configuration. If we assume each *i–j* interaction is independent of the location of molecule *k*, then the change in U_c when the molecules are brought together is the sum of the 1–2, 2–3, and 1–3 interactions only: If the interaction between molecules *i* and *j* depends on the location of molecule *k*: then the pair approximation is invalid and three-body (or in general, *many-body*) interactions have to be considered explicitly.

If U_{ij} is the pair potential between molecules *i* and *j* (which we take to be a function of the *i–j* separation only, ignoring any dependence U_{ij} might have on the relative *orientation* of *i* and *j*), then U_c is the sum

$$U_c(\mathbf{r}_1, \mathbf{r}_2, \ldots, \mathbf{r}_N) = \sum_{i=1}^{N-1} \sum_{j=i+1}^{N} U_{ij}(r_{ij})\ . \quad (17.4)$$

For example, with $N = 3$, $U_c = U_{12} + U_{13} + U_{23}$, as alluded to in the previous paragraph. For charged particles, the Coulombic potential that we have used in previous chapters many times is an example of a pair potential, and we will discuss a few others later in this chapter.

If the fluid has only one component, then U_{ij} is the same function for all *i* and *j*, and it can be shown that

$$U_c = \frac{4\pi V}{2} \int_0^\infty U_{ij}(r) n^{(2)}(r) r^2 dr$$

where the factor $4\pi V/2$ (V is the volume) has been left in this form to emphasize its origins: 4π comes from a spherical-coordinates angular integration, and a factor of 1/2 ensures each *i–j* interaction is counted only once, rather than twice (as in *i* interacting with *j* and then *j* with *i*, which would be an erroneous doubling of the interaction). This expression shows how $n^{(2)}(r)$ acts as a distance-dependent weighting factor for interactions. The expression $4\pi r^2 n^{(2)}(r)\,dr$ counts the fraction of the pairs that are at a separation between r and $r + dr$.

It is convenient to segregate the $(N/V)^2$ part of $n^{(2)}$ as a separate factor and introduce a distance-dependent, density-*independent* factor called the *radial distribution function* $g(r)$ defined (for an isotropic system) as

$$n^{(2)}(r) = \left(\frac{N}{V}\right)^2 g(r) . \tag{17.5}$$

$Ng(r)/V$ is the average particle density at a distance r from an arbitrarily chosen reference molecule. With this definition, U_c becomes

$$U_c = \frac{2\pi N^2}{V} \int_0^\infty U_{ij}(r)\, g(r) r^2 dr .$$

This is a key expression. Define U_{ij} based on the nature of the molecules, and, given $g(r)$, a host of *macroscopic* quantities can be calculated. For example, the equation of state in this approximation is

$$P\overline{V} = RT - \frac{2\pi N_A^2}{3\overline{V}} \int_0^\infty r^3 \left(\frac{dU_{ij}}{dr}\right) g(r)\, dr \tag{17.6}$$

where $\overline{V}$ is the molar volume and N_A is Avogadro's constant. This expression is the basis for a virial expansion of the equation of state as discussed in Chapter 1. (The quantity $r(dU/dr)$ is called "the virial" in classical mechanics and plays a key role in the Virial Theorem. See Problem 12.48.)

EXAMPLE 17.2

Given a $g(r)$, how can one find the number of molecules, on average, within a radius r^* of a reference molecule?

SOLUTION This is an important use of $g(r)$, since it lets us find the average number of nearest neighbors if we make r on the order of a molecular diameter. This is a number directly comparable to a solid-state structure, and it will let us see how ordered or close-packed a fluid is. The key is the interpretation that $Ng(r)/V$ is the average density of molecules at a point r from the reference molecule. We simply integrate this expression over spherical polar coordinates from $r = 0$ to r^*. Since $g(r)$ is independent of angle, the angular integrations yield 4π, and we have

$$\text{\# molecules at } r^* \text{ or closer} = \frac{4\pi N}{V} \int_0^{r^*} g(r) r^2\, dr .$$

We can relate the particle density, N/V, to the mass density ρ or to a gaseous equation of state.

➡ RELATED PROBLEMS 17.4, 17.10

For much of the rest of this chapter, $g(r)$ and its interpretation, calculation, and experimental measurement will be the focus of the discussion. We start this discussion with a few general properties:

1. $g(r) \to 1$ for large values of r
2. $g(r) = 0$ means that no pair of molecules have a separation r
3. $g(r) = 1$ at all r for an ideal gas; all separations are equivalent and the fluid has no structure

4. $g(r) > 1$ means the probability of finding a pair separated by r is greater than that of the random distribution of an ideal gas
5. at low density, $g(r)$ approaches $\exp(-U_{ij}(r)/k_B T)$ where $k_B = R/N_A$ is Boltzmann's constant.

The third property, $g(r) = 1$ for an ideal gas, follows from our comments about $n^{(2)}(r)$ and the definition of $g(r)$ in Eq. (17.5). The fifth property we will not prove. But note that these two, along with the limit expressed in 1. are the only clues we have so far to a general expression for $g(r)$ at any T, $\overline{V}$, and U_{ij}. In fact, this general expression is by and large the central, difficult, and sometimes elusive quantity in the theory of liquids.

We can begin to understand more about $g(r)$ if we invoke some simple model potential functions and explore the low density limit. The two simplest U_{ij} choices, $U_{ij} = 0$ (ideal gas) and $U_{ij} = \infty$, $r < r_0$, $U_{ij} = 0$, $r \geq r_0$ (hard-sphere gas) yield the $g(r)$ functions shown in Figure 17.1(a) and (b). They follow trivially from properties 1.–3.: $g(r) = 0$ for all r less than the hard-sphere diameter, r_0. (The hard-sphere second virial coefficient in Eq. (1.19) also follows from Eq. (17.6).)

A typical U_{ij} has the general shape of the Ar–Ar interatomic potential function in Figure 1.5. A simple approximation to this shape is the *square well potential*

$$U_{ij}(r) = \begin{cases} \infty & r < r_0 \\ -\epsilon & r_0 \geq r \geq r_1 \\ 0 & r > r_1 \end{cases}$$

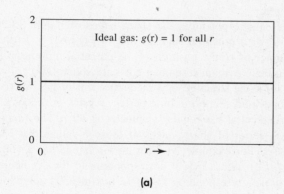

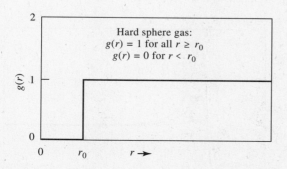

FIGURE 17.1 (a) The radial distribution function $g(r)$ for an ideal gas (a system of noninteracting points) is 1 at all r. All separations are equally likely. (b) For a hard-sphere gas, $g(r) = 0$ for $r < r_0$, the hard-sphere diameter. No pairs of molecules can be closer than r_0.

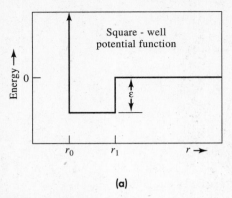
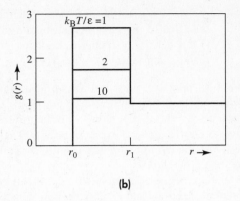

FIGURE 17.2 For the square well potential, (a), the radial distribution function, (b), is greater than 1 for distances corresponding to the well region. The lower the temperature, the greater is $g(r)$, reflecting the greater chance of finding pairs with separations from r_0 to r_1.

plotted in Figure 17.2(a). From property 5., $g(r)$ at low density is

$$g(r) = \exp(-U_{ij}/k_B T) = \begin{cases} 0 & r < r_0 \\ \exp(\epsilon/k_B T) & r_0 \geq r \geq r_1 \\ 1 & r > r_1 \end{cases},$$

which is plotted in Figure 17.2(b) for three temperatures: high ($T = 10\epsilon/k_B$), medium ($2\epsilon/k_B$), and low (ϵ/k_B). Note how a peak appears in $g(r)$ throughout the binding region of the potential well. This peak corresponds to greater than random odds of finding i–j pairs at these distances. Note also how the magnitude of the peak decreases with increasing T. As T increases, the average total energy of any pair of molecules becomes larger and larger with respect to ϵ, and the square-well fluid approaches the hard-sphere fluid.

For a smooth, realistic potential, the sharp edges of Figure 17.2(b) round, as Figure 17.3 shows for a potential such as that of Figure 1.5. In this figure, r is measured in multiples of r_e, the position of the potential minimum. Note how the highest temperature curve again resembles the hard sphere $g(r)$.

These figures show the structure of a moderate density gas. We could also find $g(r)$ for a crystalline solid. For an infinite perfect crystal, $g(r)$ does not approach 1 at large r, since the crystal has regular symmetry at all r. The function $g(r)$ for a perfect crystal is a series of sharp spikes at those r values that locate neighbor coordination spheres. For example, $g(r)$ for an fcc lattice has a spike at $r = a/2$ (for lattice constant a) with a height corresponding to the six nearest neighbors, then one at $r = a/\sqrt{2}$ corresponding to the 12 next-nearest neighbors, and so on.

Contrast this picture of $g(r)$—a series of spikes separated by gaps—to those for the moderate density gas—a single rounded peak at short distances only. In between we find the liquid $g(r)$, as the next section shows.

FIGURE 17.3 The radial distribution function for a realistic potential resembles a "rounded" version of that for a square well potential.

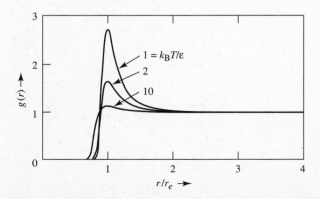

17.3 SIMPLE LIQUIDS

The simplest liquid exists only in computer simulations. It is the hard-sphere fluid, for which $U_{ij}(r)$ was discussed in the previous section. This fluid is not approached in Nature, but its study provides a base of data against which new theories may compete and real fluids may be compared. Moreover, the hard-sphere fluid can be studied in other than three dimensions, leading to the 2-D hard-disk fluid and the 1-D hard-rod fluid. The equation of state of the latter is known exactly. For rods of individual length l confined to a line of length L,

$$P\overline{L} = RT \left(\frac{1}{1 - N_A l/\overline{L}} \right) \qquad (17.7)$$

where $\overline{L} = L/n$, the "molar distance," the 1-D analog of $\overline{V}$. The hard-rod fluid does not have a phase transition. It is "fluid" at all $(P, \overline{L}, T)$ points.

For two and three dimensions, computer simulations of two types have shown how hard-sphere fluids behave and, in fact, undergo a phase transition to a solid phase. The first type of simulation is called *molecular dynamics,* a technique Isaac Newton might have used had there been high-speed computers in the seventeenth century, since it consists of solving Newton's classical equations of motion for N particles. The details of the method—keeping track of time, position, and velocity, and relating these to temperature, pressure, and volume—are somewhat involved, but basically one specifies U_{ij}, finds the force from $-dU_{ij}/dr$, and then applies $F = ma$ over and over.

Since particle trajectories are a natural outcome of this calculation, one can make movies of molecular motion. A multiple exposure of several frames from such a movie is shown in Figure 17.4 for a 2-D hard-disk system, taken from Alder and Wainwright's pioneering work with this technique. This picture shows a segregation of the system into solid and liquid phases. Note the regular hexagonal packing of the solid islands surrounded by the tangled, irregular squiggles of liquid trajectories.

The second simulation method is more from the world of Wayne Newton[2] than that of Sir Isaac. It is called the *Monte Carlo* technique, in that one rolls the dice

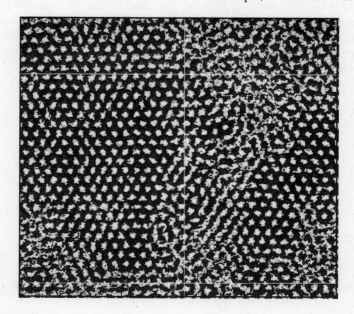

FIGURE 17.4 The motion of several hundred hard disks at a density corresponding to solid–fluid coexistence shows localized crystals surrounded by fluid regions. [From "Phase Transitions in Elastic Disks," by B. J. Alder and T. E. Wainright *Physical Review* **127**, 359 (1962).]

---

[2] Mr. Wayne Newton made his fortune with a Las Vegas nightclub act.

FIGURE 17.5 The radial distribution function for a hard sphere liquid at high densities ($Nd^3/V = 0.90$ here) shows considerable structure, indicative of some local ordering. [Data from "Monte Carlo Values for the Radial Distribution Function of a System of Fluid Hard Spheres," by J. A. Barker and D. Henderson *Molecular Physics* **21**, 187, (1971).]

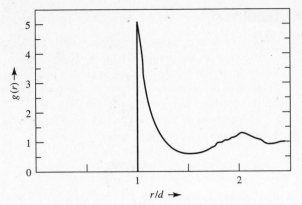

of a computer's random number generator to study the liquid. The idea is that a particular configuration of molecules has a probability of existing in real life that is readily calculable. It is proportional to $\exp(-U_c/k_BT)$, the *configurational Boltzmann factor*. (Note that U_c, not U_{ij}, appears here.) Thus, the computer generates configuration after configuration at random,[3] but rejects those that are unlikely according to the Boltzmann factor. Once many (typically 10^5 to $>10^6$) acceptable configurations have been generated, thermodynamic averages can be computed.

Figure 17.5 shows a radial distribution function for a dense hard-sphere fluid calculated by a Monte Carlo method. In contrast to the low density $g(r)$ of Figure 17.1(b), this calculation shows pronounced structure in $g(r)$, indicating local structure around any one sphere in the fluid.

A Closer Look at Computer Simulations

Computer simulations, either molecular dynamics or Monte Carlo, have to contend with a paradox: they claim to represent the behavior of a mole of molecules, but they have a practical limit of no more than a few hundred molecules. We look here at one way these methods make the most of a small number of simulated molecules.

Suppose we want to simulate $g(r)$ for a very simple system, a 2-D ideal gas. The positions of ideal gas molecules are completely random; so, we assign (x, y) coordinates with a random-number generator. For 50 molecules, the initial configuration might look like this:

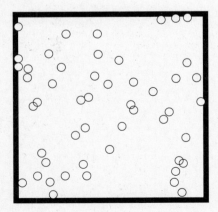

---

[3]Not really at random. Truly random samples of configurations would almost all have overlapping molecules and thus be nonphysical. A technique called *importance sampling* is used to improve the odds.

While the molecules are drawn as disks, they are in fact points, and since they have no inherent size to act as a length scale, we can consider the square above to have unit area. Next, the computer goes from molecule to molecule and calculates all the intermolecular distances r_{ij}. We make a histogram of these distances, counting the number of times we find values in a range r to $r + \Delta r$ where Δr, the histogram bin width, is a small interval, say 0.01. This histogram is *almost* $g(r)$. Since the annular area represented by a bin, $2\pi r \Delta r$, is increasing with r, we divide each bin's count by the average r it represents to correct for this increasing area. This corrected histogram is directly proportional to $g(r)$, and for our 50 points, it looks like this:

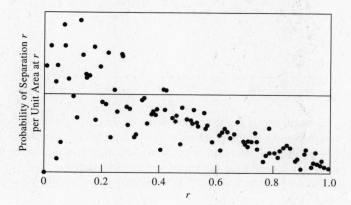

It is very noisy, since only 50 molecules went into the simulation, but most troubling is the distribution's fall to zero at large r. For an ideal gas (in any number of dimensions) $g(r)$ should be 1 at all r. What's wrong with the simulation?

We have found a problem that is common to all simulations. There are so few molecules that the *fraction near a boundary of the system* is far greater than a real system would have. To avoid these edge (or surface, in 3-D) effects, simulations use *periodic boundary conditions* to replicate the system around its edges, providing a source of molecules surrounding the set under study. For us, we instruct the computer to surround our square with eight replicas that provide a surrounding source of molecules:

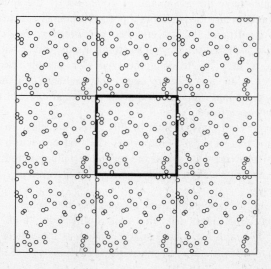

Now we repeat the distance measurements, scanning over the original, central 50 particles, but extending the measurements over *all* 450 molecules in the simulation. We again construct a weighted histogram out to $r = 1$ (still inside the central square) to represent $g(r)$, and this time we find what we expect:

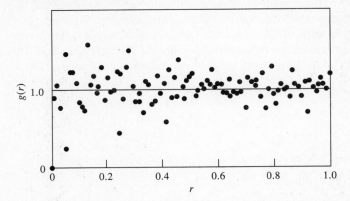

The distribution is still noisy, but the "noise level" is decreasing with r. This reflects the statistics of the problem: there are more random points at large r than at small. In fact, the first point is at zero simply because our initial random distribution had *no* pair of molecules, even in the replicated set, so close that they fell into the first histogram bin from $r = 0$ to $r = 0.01$. (The closest pair was 0.016 units apart in this simulation.)

We could smooth the noise in $g(r)$ by repeated averaging over many random initial configurations. This is the essence of the Monte Carlo method. For any property that depends on the motion of the system in a molecular dynamics calculation, periodic boundary conditions also come to the rescue. A molecule in the initial set that, say, moves out the right side of the initial area is replaced from the left by its replicated twin. Periodic boundary conditions thus supply a surrounding field of molecules that can provide either a static force or a dynamic source of particles that makes a simulation approach the behavior of bulk matter using a minimum of particles.

These techniques have the advantage of generality in the sense that computer programs readily accept any intermolecular potential-energy function and can bridge the gap between experiment and analytic theory. Moreover, they can be adapted to molecules with internal structure as well as atomic fluids. But as with most computer simulation methods, they fail to offer the type of ready physical insight that an analytic theory, even an approximate one, can provide. Such theories are under continual development and refinement, and an important concept to many of them is the *potential of mean force*.

To see the origin and consequence of this force, we can use a modified low-density expression for $g(r)$ that incorporates the first correction to the ideal gas value, $g(r) = 1$. This expression yields the second virial coefficient, and thus cannot be expected to be accurate at high density, but it will illustrate a physical consequence of increased density.

Before we write this new expression for $g(r)$ for hard spheres, we will describe a convenient measure of the density that makes clear how "high" is high density. The natural unit of length for a hard-sphere system is the sphere diameter d, and

d^3 is a natural unit of volume. Likewise Nd^3/V is a natural (dimensionless) measure of density. When $Nd^3/V = 1$, the hard-sphere density is that of a simple cubic solid. The maximum Nd^3/V is that of a close-packed solid, either hcp or fcc, and is equal to $\sqrt{2} = 1.41$. Simulations of the 3-D hard-sphere fluid have shown that at $Nd^3/V \cong 0.94$, the fluid begins to solidify. From ~ 0.94 to ~ 1.04, the system is in two-phase (solid/fluid) equilibrium, while the interval from ~ 1.04 to $\sqrt{2}$ corresponds to a solid.

With these values in mind, consider a modified low-density expression for $g(r)$

$$g(r) = \begin{cases} 0 & r \leq d \\ 1 + \dfrac{4\pi}{3}\left(\dfrac{Nd^3}{V}\right)\left[1 - \dfrac{3}{4}\left(\dfrac{r}{d}\right) + \dfrac{1}{16}\left(\dfrac{r}{d}\right)^3\right] & d < r \leq 2d \\ 1 & r \geq 2d \end{cases} \quad (17.8)$$

This is different from the function of Figure 17.1(b) only over $d < r < 2d$. It is plotted in Figure 17.6 for $Nd^3/V = 0.1$, safely away from the high-density limit.

We can reach an amazing conclusion if we compare Figure 17.6 to the $T = 10\epsilon/k_B$ curve for the square well potential, Figure 17.2(b), and the $T = 10\epsilon/k_B$ curve for the smooth well potential in Figure 17.3. All three have $g(r) > 1$ over a short distance beyond the repulsive wall characteristic of each potential. For the lowest density situations, we argued that $g(r) > 1$ because the *attractive* portions of the pair potential allowed two molecules to stay in each other's vicinity with greater-than-random odds. Figure 17.6 shows this same increase *even though the pair potential on which it is based, hard spheres, has no attractive portion*. The increase in density has brought about an *apparent attraction* between the hard spheres.

This apparent attraction can be traced to an unsymmetrical interaction any two closely spaced spheres have with the remaining $N - 2$ spheres, as shown in Figure 17.7. Spheres 1 and 2 shield each other from collisions with surrounding spheres in such a way that those around 1 (but "to its left" in the diagram) tend to collide with 1 in a way that directs it towards 2, and vice versa.

Recall the second virial coefficient $B(T)$ for the van der Waals gas, Eq. (1.22),

$$B(T) = b - \dfrac{a}{RT}. \quad (1.22)$$

The excluded volume parameter is b in the van der Waals equation while a is an attraction parameter. Increasing a lowers the system's pressure, since $PV = RT(1 + B(T)/V)$, as does decreasing b. This effect can be seen in Figure 17.7 as well. When spheres 1 and 2 are separated by more than $2d$, they exclude a total

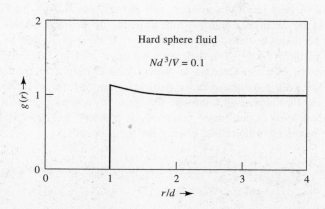

FIGURE 17.6 At moderate to high density, a hard-sphere fluid can exhibit a $g(r) > 1$ region, even though there are no pair attractive forces between hard spheres. The origin of this behavior can be traced to the dynamics of the many collisions that occur at these densities.

FIGURE 17.7 At the high densities of a liquid, a pair of neighboring hard-sphere molecules will remain near each other with greater than random probability. The dynamic effects of collisions with surrounding molecules helps to keep them together, as does the reduced excluded volume (dashed circles around 1 and 2) of two close neighbors.

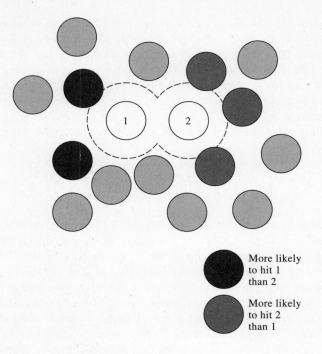

volume $2(4\pi d^3/3)$. At separations closer than $2d$, together they exclude a volume *smaller* than $2(4\pi d^3/3)$, indicated by the dashed lines around the 1–2 pair in Figure 17.7. It is as if the hard-sphere diameter itself decreased, which is equivalent to a decrease in the van der Waals excluded volume parameter.

The pressure is lowered whether we focus on size or attraction effects, but $g(r)$ shows that the structure is still that of spheres of size d moving in such a way that they linger in close pairs with greater than random odds, suggesting that we should focus on the dynamic motion of the entire fluid rather than that of isolated pairs.

The paths of colliding spheres have conspired to constrain the positions of 1 and 2 in much the same way as they would be constrained had a *real* attractive force existed between them. This is the physical origin of the hump in $g(r)$ at moderate densities for $r \sim d$. Since $g(r)$ represents an *average* of all such effective forces throughout the fluid, the term *potential of mean force* is applied to describe *all* the interactions, real and effective, that exist between pairs of molecules.

The following argument shows how to incorporate the potential of mean force into $g(r)$. The lowest density expression, $g(r) = \exp(-U_{ij}/k_B T)$, is simple and direct. We can find $g(r)$ at higher density from a variety of techniques such as computer simulation or moderate density approximations.[4] If these $g(r)$'s can be attributed to a potential of mean force $W(r)$, then we must have[5] $g(r) = \exp[-W(r)/k_B T]$.

The potential of mean force has a direct thermodynamic interpretation: $W(r)$ is the reversible work required to move particles 1 and 2 from an infinite separation (in the liquid of infinite extent), through the liquid, by any path, to a final separation r. Since this process is performed reversibly at constant N, V, and T, $W(r) = \Delta A$, the change in the Helmholtz free energy for the process.

---

[4]The general expression that leads to Eq. (17.8) for hard spheres is much more difficult to evaluate for any other pair potential.
[5]This can be proven rigorously.

We seem to have transferred the computation burden from calculating $g(r)$ to calculating $W(r)$, which is a similarly complicated function: W depends not only on r, but also on N/V and T. The hope is that W depends *slowly* on N/V and T and perhaps can be approximated by an expansion in them. Experimental $g(r)$ functions (typically obtained by X-ray and/or neutron scattering) do show that $g(r)$ is fairly insensitive to T at constant density.

For example, the modified hard-sphere $g(r)$ expression, Eq. (17.8), is

$$g(r) = 1 + x(r)$$

over the range $d \leq r \leq 2d$ where $x(r)$ is the distance-dependent term. Thus, if

$$g(r) = \exp[-W(r)/k_B T] \qquad (17.9)$$

then

$$-W(r)/k_B T = \ln[g(r)] = \ln[1 + x(r)] \cong x(r)$$

since $\ln(1 + x) \cong x$ for $x \ll 1$, so that

$$W(r) \cong -x(r) k_B T, \qquad d \leq r \leq 2d$$

is the potential of mean force between hard spheres at moderately low densities.

For the simplest real liquids, the rare gases (with the exception of quantum-dominated lightweight He), computer simulations based on accurate experimental pair potentials, the potential of mean force approach, and more sophisticated extensions yield structural, dynamic, and equilibrium thermodynamic properties in essentially exact agreement with experiment.

17.4 ORDINARY LIQUIDS—WATER

For non-monatomic liquids, we have to interpret $g(r)$ differently. Consider $N_2(l)$, a well-studied diatomic liquid. Here, $g(r)$ still picks one *atom* and measures the distribution of other atoms about it, but in N_2, every N carries along with it a chemically-bonded N partner always about 1.09 Å away. Thus, $g(r)$ for $N_2(l)$ has a sharp peak at 1.09 Å, followed by other, broader humps at larger distances corresponding to nonbonded neighboring N atoms. Since each nonbonded neighbor also has a bonded partner, each broad hump has a width on the order of 1.09 Å. This chemical bond length causes a loss of resolution in $g(r)$. Within one or two coordination distances, all structure in $g(r)$ is blurred.

Now consider water. Here, there are several kinds of $g(r)$ functions: O—O distributions, O—H distributions, and H—H distributions. Since X-ray scattering is insensitive to H atoms, the O—O distribution is a logical focal point. (For substances with two or more distinct atoms per molecule, both of which scatter X rays well, distinguishing one from another can be difficult.) But the greatest differences between molecular fluids such as water and those such as $N_2(l)$ arises from the *strong, directional, attractive intermolecular forces* in water: the hydrogen bonds.

Directional forces are always important in fluids of polar molecules. The pair potential is not a simple function of separation, but it must include orientation coordinates as well. In water, these orientational forces are so strong and directional that water is termed an *associated fluid*. Any one water molecule is strongly associated with its neighbors by an attractive potential well of depth greater than the average thermal energy of the liquid. The consequences of this strong, direction interaction are many.

First, water has an unusually high boiling point compared to other covalently bonded triatomics (or diatomics, for that matter), as well as a high heat capacity.

The quantity $\overline{C}_V(l) - \overline{C}_V(g)$ is often called the *configurational* or *residual* molar heat capacity. For Ar, N_2, and CH_4 at the normal boiling point of each, $\overline{C}_V(l) - \overline{C}_V(g)$ is about $0.9\,R$, $1.2\,R$, and $1\,R$, respectively; for water, it is $5.1\,R$. This extra heat capacity can be traced to the relatively open structure of $H_2O(l)$ and its temperature dependence.

Second, the structure itself is unusual. The O—O radial distribution function shows a peak in the vicinity of 3.0 Å and a second peak, much smaller, at around 4.5 Å. If this $g(r)$ is integrated from $r = 0$ to $r \sim 3.5$ Å, the position of the minimum between the first two peaks, one finds (see Example 17.1)

$$\text{number of waters in first coordination shell} = \frac{4\pi N}{V} \int_0^{3.5\,\text{Å}} g(r) r^2 \, dr \cong 4\;.$$

In contrast, the corresponding calculation for Ar yields $\cong 12$, the expected "close packed" number, and even $N_2(l)$ has a coordination number $\cong 12$.

The number 4 indicates tetrahedral coordination, as the nearly perfectly tetrahedral bond angle in H_2O suggests. Ice is an infinite, ordered array of hydrogen bonded, tetrahedrally coordinated water. In the liquid, the ordering is not perfect, the structure collapses slightly, and the liquid density is greater than that of ice.

This tetrahedral structure is, however, a very open one, even in the liquid. It would collapse if it were not held in place by strong hydrogen bonds. It does at least partially collapse around many solutes, not only those with small cations, such as $MgSO_4$ with a negative partial molar volume, but also locally for virtually any dissociating ionic solute.

For nonassociated molecules, even though they are nonspherical, the liquid structure is largely determined by the *repulsive* part of the intermolecular potential. It is this repulsive part that gives us our usual view of a molecule's size and shape. Bond lengths and angles position nuclei, but the intermolecular potential's repulsive wall defines the periphery of the molecule. Figure 17.8 shows a likely configuration of N_2 molecules in the liquid phase, drawn with the N_2 nuclei spaced by the correct bond length, 1.09 Å, and with molecular boundaries drawn as 1.25 Å spheres centered at each nucleus. These boundaries may be thought of as impenetrable surfaces, a "hard molecules" variant to the hard-sphere potential. It is the packing of these hard molecules that largely governs the structure of nonassociated liquids.

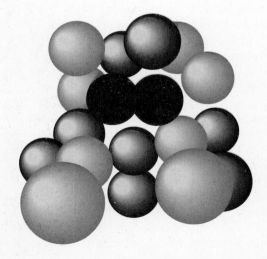

FIGURE 17.8 Packing in a liquid (such as $N_2(l)$ shown here) has roughly the same density and number of nearest neighbors as a solid phase. What is lacking in a liquid is the very long-range order of a solid.

Thus, it has been suggested that the structure of liquid benzene is easily simulated by pouring a bowl of the donut-shaped cereal, Cheerios. The hard Cheerios have the important symmetry elements of the hard molecule surface of benzene.

Our look at liquid structures in this chapter closes with the next section, which describes what is sometimes called a *mesophase*. This is the *liquid crystal* phase, which sounds like a contradiction in terms, an oxymoron in a class with postal express and similar aspects of contemporary life.

17.5 LIQUID CRYSTALS

While liquid crystals have been carried on our wrists, in the displays of digital watches, since only the mid-1970s, they have been known and studied since the 1880s. Liquid crystals are called *mesophases* or *mesomorphic phases* from the Greek for "intermediate form." They are "intermediate" in two ways: they not only appear between the solid and ordinary liquid as temperature is increased, but they also have structural features of both solids and liquids.

Liquid crystals have most in common with the macroscopic behavior of ordinary liquids. They flow freely and assume the shape of their container. Phase transitions from one pure liquid phase to another are extremely rare. The normal fluid to superfluid transition in liquid He is one notable case, and the liquid crystal to ordinary liquid transition is another.

Liquid crystals warrant being called crystals because they display some type of long-range order. If *three*-dimensional translational order is not allowed for a liquid, *one*-dimensional translational order is, and one type of liquid crystal has this order. Another type has *orientational* order with a preferential molecular *alignment*.

Liquid crystals are classified into three structural types based on these two broad classes of order:

1. *nematic:* translationally disordered but orientationally ordered
2. *smectic:* translationally ordered in one dimension (yielding a system of layers of two-dimensional liquids with a well-defined layer spacing) and orientationally ordered in the layers (at least)
3. *cholesteric:* orientationally ordered like a nematic, but with a helical twist to the direction of orientation. (This phase is also called *twisted nematic*.)

The name nematic is derived from the Greek word for thread. This is less a statement about the shape of the molecules in a nematic liquid than a description of the long, thread-like patterns often formed by liquid crystalline packing defects. But the molecules of nematic liquid crystals (and of all types of liquid crystals to some extent) are themselves commonly "rod-shaped" if not truly "thread-like." For example, *p*-azoxyanisole (PAA) forms a well-studied nematic phase. Figure 17.9 shows three orthogonal views of PAA, each outlined by a rectangle to demonstrate the rod-like shape of the molecule.

A schematic view of nematic PAA, which forms when the solid melts at 392.7 K and disappears at the nematic–ordinary liquid-phase transition[6] temperature of 408 K, is shown in Figure 17.10 as a packing of the solid rectangular boxes in Figure 17.9. The positional disorder and orientational order are both evident in this picture. The molecular centers of mass are positioned randomly, but the long axes

---

[6]Like any phase transition, there are phase transition enthalpy and entropy changes at these temperatures. For transitions from one liquid crystal phase to another, the molar enthalpy change is usually very small, on the order of 0.4 kJ mol^{-1}.

FIGURE 17.9 The PAA molecule, *p*-azoxyanisole, shown here in a space-filling model without the hydrogens, is approximately a long rectangular rod. This is a common shape for molecules that form a liquid crystal phase.

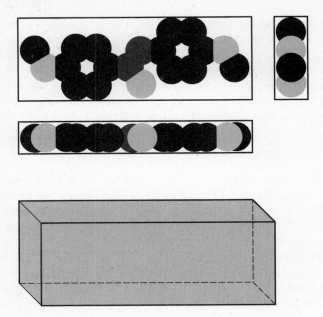

tend to point in one direction.[7] Around the long axis, there is disorder. The aromatic rings, for instance, are lying in random planes about the long axis.

A single number, an *order parameter,* can describe this alignment. If we draw a line along the long axis of *each* molecule, measure the angle this line makes with any arbitrary external axis, and average our results, we will get an angle that represents the general alignment with respect to the external axis. The order parameter measures the *spread* in molecular axis tilts from this average orientation direction.

If we try to measure this spread by just the average of the tilt angles, we will obviously get zero, since this is the technique we used to find the average orientational

FIGURE 17.10 A schematic picture of a nematic liquid crystal phase of PAA shows that there is a preferential direction of alignment of the long molecular axes. This ordering persists throughout the phase.

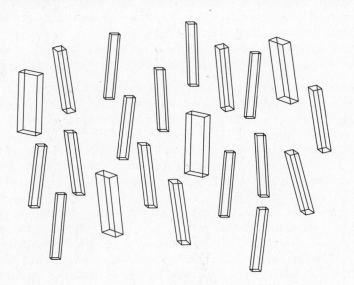

---

[7] And note that "up and down" play no role in this figure. Gravity plays no role, and you can rotate the picture any way you like and obtain a valid view of the nematic phase.

direction in the first place. Similarly, an average of cos θ (θ is the tilt angle) gives zero. The quantity that is most often used as an order parameter is

$$S = \frac{1}{2} \langle 3 \cos^2 \theta - 1 \rangle \qquad (17.10)$$

where $\langle \rangle$ represents averaging. If all the molecules are perfectly aligned (θ = 0), $S = 1$. Typically, S varies from around 0.7 to around 0.4 (θ around 25° to 40°) as the temperature is raised in a nematic phase. Direct measurement of S would seem to be a formidable task, but it can be measured by NMR (nuclear magnetic resonance) spectroscopy.

The smectic phase comes in a number of varieties. All are characterized by *layers* of molecules (this is the 1-D ordering) and distinguished into subclasses called smectic *A, B,* and *C*. In smectic *A,* there is no translational order within the layers, but there is orientational order such that the average orientation direction is perpendicular to the layers. In smectic *C,* the average orientation is tilted away from this perpendicular, and in smectic *B,* not only is there orientation order, but also some degree of translational order (most commonly hexagonal packing) within the planes. Figure 17.11 shows a schematic of a smectic *A* phase.

Some molecules exhibit not only a variety of smectic phases, but also smectic-nematic transitions. As expected, the order of these phases with increasing temperatures is in the direction of increasing disorder: solid → smectic *B* → smectic *A* → nematic → ordinary liquid. The name smectic is derived from the Greek word for soapy, and smectic layering is thought to be due to the same sort of forces that operate in soaps and detergents. A smectic molecule often has the sort of central core of PAA, with any of a variety of groups linking the aromatic rings, but with long, aliphatic chains in the *para* positions. These aliphatic chains prefer to be in the immediate vicinity of other, similarly nonpolar chains, and this produces layering. Similar interactions operate not only in soaps, but also in biological membranes *(lipid bilayers)* and in clustered agglomerates of molecules such as micelles.

The final major category, cholesteric, is formed by a variety of esters of cholesterol. (But not cholesterol itself. See Figure 17.12; the HO—group of cholesterol is thought to form hydrogen bonds that preclude a liquid crystal phase.) These are not particularly "rod-like" molecules, but a number of rod-like, noncholesterol-derivative molecules form a cholesteric phase. What all these molecules have in

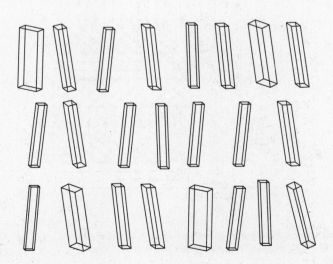

FIGURE 17.11 The smectic type of liquid crystal phase consists of layers of molecules with a definite layer spacing. Within a layer, the molecules are more or less aligned, and thus smectics are more ordered than nematics. Shown here is the variant called smectic A.

FIGURE 17.12 While cholesterol itself (R = OH) does not form a liquid crystal phase, many ester and thioester derivatives of cholesterol do, and the phase is called *cholesteric* in general.

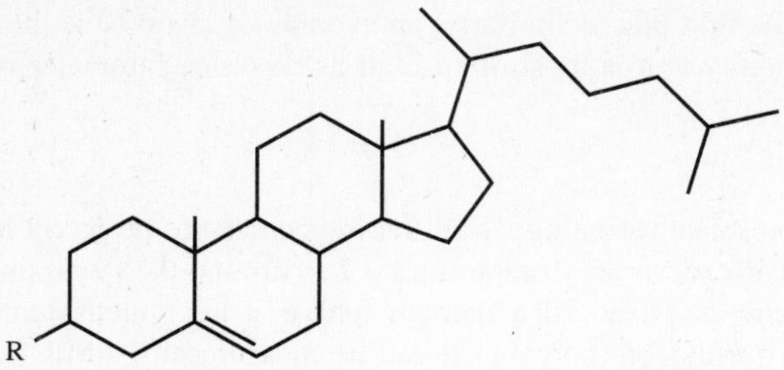

R = OH cholesterol (not liquid crystalline)

R = OOC(CH$_2$)$_n$H (*n*-alkanoates)

common is not shape, but *chirality*. A racemic mixture of right- and left-handed isomers will not form a cholesteric phase, but a pure isomer will.

The cholesteric phase is a subtle and clever variant of the nematic. Locally, within a few molecular diameters, the structure is nematic with a well-defined average orientation direction. But wander away from a local spot in the cholesteric and you find that the average orientation direction is slowly spiraling, as does the vector pointing from the cylinder axis of a spring to the spring itself. Figure 17.13 demonstrates this spiral. The average orientation direction eventually points back in the original direction, then continues its spiral. The periodicity of the spiral (called the spiral's *pitch*) is often 200 nm or more and very temperature sensitive.

This brings us to the first practical application of liquid crystals. Cholesteric phases with a pitch distance on the order of a visible light wavelength reflect that wavelength strongly and appear highly colored.[8] Any physical parameter—pressure, temperature, external field, etc.—that changes the pitch length changes the cholesteric

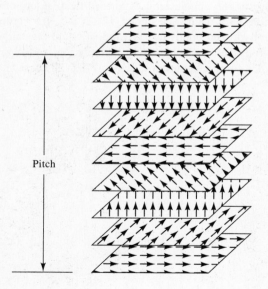

FIGURE 17.13 A cholesteric phase is characterized by a helical twist to the orientation direction within any plane of molecules. The pitch of the helix is typically tens to hundreds of molecular thicknesses in length.

---

[8]There is a direct analogy to X-ray scattering here. One period of the spiral covers a long distance so that Bragg scattering requires a wavelength longer than X ray.

phase's color. Temperature happens to be particularly effective for certain cholesterics, and simple temperature sensors are based on this fact. In medical applications, a strip of various cholesteric molecular liquid crystals, each masked so that it displays the midpoint of its temperature range, is used as a quick, local, skin thermometer. Alternately, a cholesteric liquid crystal can be painted directly on the skin to highlight the localized temperature changes caused by tumors or infection. Similar paints highlight overheating components on electronic circuit boards.

We close this section with a description of liquid crystal displays used in watches, calculators, and portable computers. These are "black and white," typically, but color liquid crystal displays for large, thin, color television screens are continually evolving. Such displays were claimed to be "only two years away" for roughly the past twenty years, but small displays are cheaply and routinely fabricated now.

One common black and white display uses a liquid crystal to alter the plane of polarized light as it enters the liquid crystal, strikes a reflective back surface, and passes through the liquid crystal again. The liquid crystal is sandwiched between ordinary polarizing filters, and an electric field is applied to the liquid crystal to alter its polarization properties. This is called "electrically controlled birefringence" and works with nematic phases.

Two conditions must be met in these devices. The first involves the "bread" of the sandwich, which must have a transparent but electrically conducting surface *that can preferentially orient the nematic in the absence of an applied voltage.* The transparent, conducting material is most often a thin layer of SnO_2 and/or In_2O_3 (known as "ITO," indium tin oxide). The glass substrate is polished in one direction, the surface is coated, and the nematic cooperates by aligning along the polishing direction. If the top and bottom substrates are polished at 90° to each other, the nematic becomes "twisted" through the liquid crystal phase. (See Figure 17.14(a).) This twist rotates the polarization of transmitted light by 90° so that crossed polarizers make the liquid crystal appear transparent.

When an electric field is applied, the nematic molecules experience a torque as the field tends to align them parallel to its direction. This alignment happens through-

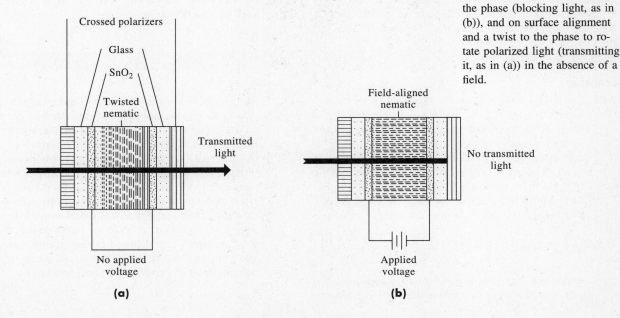

FIGURE 17.14 Liquid crystal displays using a nematic phase rely on an electric field to align the phase (blocking light, as in (b)), and on surface alignment and a twist to the phase to rotate polarized light (transmitting it, as in (a)) in the absence of a field.

out the phase when the field has reached a minimum threshold value. At and above this value (typically a few volts across a nematic layer around 10 μm thick), the orientation is virtually complete, and the polarizers are effective at blocking reflected light. The device looks "black." (See Figure 17.14(b).) The switching time is important, particularly for computer and TV displays, and can be made to fall in the ms range.

Many other variations on this scheme exist, and all are based on using an electric field to alter either the optical properties of the liquid crystal itself or of molecules added as solutes to the liquid crystal. In fact, liquid crystals can be used as nonisotropic solvents in many types of experiments that can benefit from one dimensional ordering.

CHAPTER 17 SUMMARY

The important structural principles of liquids discussed in this chapter are embodied in two major themes: the packing of molecules based on shape in all but associated liquids, and the description of order via the radial distribution function, $g(r)$. It is the interpretation of $g(r)$ and the phenomena that give it structure that have occupied most of our attention.

We saw how to find $g(r)$ in the very low density fluid and how to anticipate the shape of $g(r)$ in the dense fluid when we could not directly calculate it. It is controlled by the intermolecular potential energy, and given this energy and $g(r)$, we can find thermodynamic and structural features of the liquid, such as its equation of state and its nearest neighbor coordination number.

Simple liquids are now very well understood while ionic fluids, associated fluids, macromolecular fluids, and liquid crystals remain important areas for further research.

FURTHER READING

Two books on liquids, both going far beyond the topics of this chapter, are *Properties of Liquids and Solutions*, by J. N. Murrell and E. A. Boucher (Wiley-Interscience, NY, 1982) and *Computer Simulation of Liquids*, by M. P. Allen and D. Tildesley, (Clarendon Press, Oxford, 1986). At a more advanced level, but still very readable, the review article "What is 'liquid'? Understanding states of matter," by J. A. Barker and D. Henderson *Rev. Mod. Phys.* **48**, 587, (1976) covers simulation methods, analytic theories, and a number of model intermolecular potentials for simple systems.

The classic work on liquid crystals is *The Physics of Liquid Crystals*, by P. G. DeGennes (Clarendon Press, Oxford, 1974), the 1991 Nobel laureate in physics, awarded in part for his work on liquid crystals.

PROBLEMS

SECTION 17.1

17.1 Consider Ar at its triple point. (See Table 6.1.) Calculate the average volume available per atom in each phase, solid, liquid, and gas. For the solid, the lattice parameter is $a = 5.4659$ Å. For the liquid, assume it is also an fcc "crystal" with the experimental density 1.4146 g cm^{-3}. (Remember that there are four atoms in the fcc unit cell.) Take the cube root of each volume as an estimate of the average spacing R, then use the simple Ar–Ar potential-energy function

$$V(R) = (1.98 \times 10^{-21} \text{ J}) \left[\left(\frac{3.75 \text{ Å}}{R}\right)^{12} - 2\left(\frac{3.75 \text{ Å}}{R}\right)^{6} \right]$$

to estimate the average potential energy between a pair of molecules in each phase.

17.2 The energy constant in the potential-energy expression in the previous problem measures the energy required to pull two Ar atoms apart from their equilibrium position. The molar enthalpy of vaporization (6.51 kJ mol^{-1}) measures the energy to pull a mole of liquid atoms apart, and the molar enthalpy of sublimation (7.70 kJ mol^{-1}) does the same for the solid. If we model these two condensed phases by pair interactions, these enthalpy values (when converted from a per mole to a per molecule basis) should be a small multiple of the diatomic binding energy. Calculate these multiples for each process, and interpret them in terms of the solid and liquid structures.

SECTION 17.2

17.3 Repeat the text's calculation on benzene's entropy for the following substances. Find the liquid's entropy at the boiling point as a fraction of the gas entropy at T_{vap}. Can you suggest reasons for the variations among your answers?

	$\Delta \bar{H}_{vap}$/J mol^{-1}	T_{vap}/K	M/g mol^{-1}
Ar	6 510	87.29	40.0
F$_2$	3 160	85.0	38.0
Na	98 010	1 156.	23.0
H$_2$O	40 656	373.15	18.0
CH$_3$OH	35 270	337.2	32.0

17.4 Since, for an ideal gas, the number density $N/V = PN_A/RT$, and since ideal gas molecules are uniformly and randomly distributed in their volume, the number of ideal gas molecules within a distance r of any one molecule is the volume $4\pi r^3/3$ times the number density, or simply $4\pi Nr^3/3V$. Derive this quantity starting from the radial distribution function expression

$$\frac{4\pi N}{V} \int_0^r g(r) \, r^2 \, dr \ .$$

17.5 Why does the $T = 10\epsilon/k_B$ curve in Figure 17.3 fall to zero at r values significantly smaller than the $T = \epsilon/k_B$ curve or the $T = 2\epsilon/k_B$ curve? (If you can answer Problem 1.12(b), you can answer this one!)

17.6 Consider 1 mol of Ar(g) in equilibrium with Ar(l) at the normal boiling point ($T = 87.29$ K, $P = 1$ atm = 101 325 Pa). That gas had to overcome the interatomic cohesion of the liquid (expressed by $\Delta \bar{H}_{vap} = 6.53$ kJ mol^{-1}), but at this low temperature and moderate pressure, we might expect that there is still some residual cohesive energy in the gas. We will estimate that here using a square well approximation to the intermolecular potential in a calculation of the configurational energy, U_c. For the potential, take $\epsilon = 1.98 \times 10^{-21}$ J, $r_0 = 3.34$ Å, and $r_1 = 4.77$ Å. Show that the configurational energy can be written

$$U_c = -\frac{PN_A^2 \epsilon}{RT} \exp(N_A\epsilon/RT) \int_{r_0}^{r_1} r^2 \, dr$$

since either $g(r) = 0$ or $U_{ij}(r) = 0$ outside the integration limits r_0 to r_1. Evaluate this expression (check units carefully!), and compare your answer to the molar vaporization enthalpy.

SECTION 17.3

17.7 Write the equation of state for the 1-D hard-rod system, Eq. (17.7), in the form of a compressibility factor expression in analogy with Eq. (1.18) ($\bar{L}$ plays the role of $\bar{V}$), and find the second virial coefficient. While you're at it, find *all* the virial coefficients!

(*Hint:* A virial expansion is valid when the system is at fairly low density, which means $\bar{L} \gg N_A l$ here.)

17.8 The expression for the second virial coefficient of a hard-sphere gas given in Eq. (1.19), $B_{hs} = 4N_A v$, lets us write the hard-sphere compressibility factor expression as $P\bar{V}/RT = 1 + 4\eta$ where $\eta = N_A v/\bar{V}$. An analytic theory of the hard-sphere liquid predicts (in close agreement with computer simulations up to rather high densities) that $P\bar{V}/RT = (1 + \eta + \eta^2)/(1 - \eta)^3$. Since η is proportional to density, note that both expressions approach $P\bar{V}/RT = 1$ at zero density, as they should. By how much do the two expressions differ at $\eta = 0.05, 0.1$, and 0.5? Are these appreciably high densities? For Ar, if the solid atomic diameter is used to calculate v, $N_A v$ is 6.65×10^{-5} m^3 mol^{-1}. How does the density $\eta = 0.1$ compare to the density of Ar(g) at 300 K and 1 atm pressure?

17.9 If we write the potential of mean force expression $g(r) = \exp(-W/k_BT)$ in the form $W = -k_BT \ln g(r)$, we get something reminiscent of the free-energy/equilibrium-constant expression $\Delta \bar{G}^0 = -RT \ln K_{eq}$ with $g(r)$ playing the role of K_{eq}. For what equilibrium "reaction" (net physical process) is $g(r)$ acting as the "equilibrium constant"?

SECTION 17.4

17.10 The experimental $g(r)$ for Ar(l) at 85 K from $r = 0$ past the first maximum to the first minimum in g (i.e., through the first coordination sphere) is shown below:

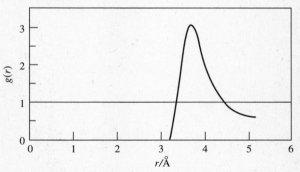

It was measured by a very accurate neutron scattering experiment using the ^{36}Ar isotope that scatters neutrons much better than the ordinary, abundant ^{40}Ar isotope. (See J. L. Yarnell, M. J. Katz, R. G. Wenzel, and S. H. Koenig *Phys. Rev. A* **7**, 2130, (1973).) If these data are numerically integrated, one finds $4\pi \int g(r) r^2 dr = 567$ Å^3. The liquid density is 1.409 6 g cm^{-3}. Find the number of nearest neighbors in this liquid. Are you surprised?

SECTION 17.5

17.11 Sketch a phase diagram in T, P coordinates for a pure compound that forms a single liquid crystal phase in addition to ordinary solid, liquid, and vapor phases. You might want to look back at Section 6.1 and think about Figure 6.7 and Problem 6.1.

17.12 The diagrams below show nematic phases with order parameters $S = 0.92, 0.72,$ and 0.33. Which order parameter goes with which picture?

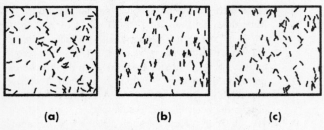

(a)   **(b)**   **(c)**

17.13 Predict what might happen if a nematic phase of PAA is in contact with the reconstructed Ir surface shown in Figure 16.22.

GENERAL PROBLEM

17.14 Chapter 16 showed that a crystalline close-packed solid of hard spheres (either fcc or hcp) has a packing fraction of $\pi/2\sqrt{3} = 0.7405$ so that 74.05% of the volume of the solid is occupied by the spheres. In contrast, if you dump a large number of hard spheres into a bag, shake them up a bit, and squeeze the bag around them, you will find that the packing is much more random, and no more than about 63.7% of the bag's volume is occupied by the spheres. This is called *random close-packing*, and the packing fraction cannot be calculated but must be measured by experiments very much like that just described. (See "The density of random close packing of spheres," by G. D. Scott and D. M. Kilgour *J. Phys. D* **2**, 863, (1969).) If a mole of hard spheres of diameter d undergoes a maximum density liquid (i.e., random close-packed) to close-packed crystalline-solid phase transition, what will be the molar volume change (in terms of d^3)? How does your answer compare to the observation that the densities of liquid and solid Ar are 1.451 5 g cm^{-3} and 1.622 g cm^{-3}, respectively, at the triple point?

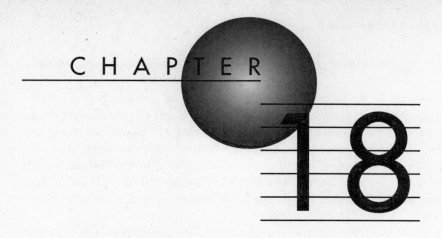

CHAPTER 18

Molecules and Radiation

THIS chapter is a broad introduction to molecular spectroscopies of all types and atomic spectroscopy in particular. We will see the major phenomena, experimental techniques, and terminology of matter–radiation interactions. The following three chapters focus on specific molecular spectroscopies where we will see how our most precise and detailed information about molecular structure is measured.

Light—its explanation and its effects—has occupied scientific thought throughout recorded history. Most speculations were misguided. Sometime in the first century A.D., Hero of Alexandria speculated, more or less correctly, that the angle of reflection from a mirror equaled the angle of incidence so that the light travelled the *minimum* path from source to observer. From then until the sixteenth and seventeenth centuries, speculations on light passed through the Dark Ages, revived only in the Enlightenment by experiments such as those Isaac Newton performed with two prisms, first dispersing sunlight into a rainbow of color, then recombining the colors into white light.

In this period and on through the 1800s, light was a topic of serious investigation and widespread interest by Natural Philosophers of the times. Goethe wrote an interesting treatise on the theory of color; René Descartes calculated, by hand, the paths of dozens of light rays through raindrops to explain the rainbow; Newton founded the field of "Optiks"; telescopes and microscopes flourished; Benjamin Franklin invented bifocal glasses; impressionist painters discussed

18.1 What Is Electromagnetic Radiation?

18.2 How Spectroscopies Work—Absorption and Emission

18.3 Atomic Spectroscopy and Selection Rules

18.4 Too Much Energy—Photoionization and Photodissociation

in great detail theories of color and pigments; and Bunsen began spectroscopy in a systematic way.

By the start of the twentieth century, Maxwell had unified all of electricity and magnetism with light, and the theory of light as an electromagnetic wave phenomenon was secure. Then came Planck's theory of blackbody radiation, Einstein's theory of photoelectric effect, the birth of quantum theory, and with it, the birth of the photon, a particle/wave of light. Quantum mechanics advanced in the middle of the twentieth century to include light in the theory called quantum electrodynamics, which has been further refined to include forces beyond the classical electromagnetic.

We do not need the most current quantum theories to understand the important light–matter interactions relevant to chemistry. Our simpler quantum mechanics can use Maxwell's wave language in the equations, modified by Einstein's photon language in the text. Consequently, the first section discusses certain basic features of electromagnetic radiation. The second section gives a broad overview of light–matter interactions, followed by a section on atomic spectroscopy. The final section discusses those spectroscopies that lead to significant and long-lived chemical changes: photoionization and photodissociation.

18.1 WHAT IS ELECTROMAGNETIC RADIATION?

We start with a number of simple descriptions of electromagnetic radiation. These include ways of specifying the radiation's energy, direction, and electric field strengths as well as the more familiar descriptors of "color," such as wavelength. Section 9.3 introduced the terminology of *static* electric and magnetic fields. Here, the fields are time dependent.

In vacuum, radiation moves in a constant direction with an invariant speed $c \cong 3 \times 10^8$ m s^{-1}. Any one photon has a frequency ν and a wavelength λ such that

$$c = \lambda \nu \qquad (18.1)$$

and a photon energy

$$E = h\nu \ . \qquad (18.2)$$

(These expressions first appeared in Chapter 11.) Traditional names are given to various frequency/wavelength/energy regions of the full electromagnetic spectrum. These regions have somewhat ill-defined boundaries, but their names are useful in specifying the rough order of magnitude of the radiation's parameters. Spectroscopy also associates a particular type of molecular process with each region. Table 18.1 lists these regions and includes a very useful measure of radiation energy, the *wavenumber*. Since $E \propto \nu$ and $\nu \propto 1/\lambda$, E is proportional to $1/\lambda$. The wavenumber is simply the reciprocal of the wavelength. It counts the number of waves per unit length, almost always expressed in cm^{-1} units.[1]

The quantity that "waves" in an electromagnetic field is the field strength. There are two such fields, electric and magnetic, directed at right angles to each other and

---

[1] Purists would correctly read (for example) 10 cm^{-1} as "the wavenumber is ten per centimeter" or "ten reciprocal centimeters," but common usage turns "cm^{-1}" into a unit called "the wavenumber," so that 10 cm^{-1} is often read "ten wavenumbers."

TABLE 18.1 Regions of Electromagnetic Radiation

Region name	Frequency/Hz ν	Wavelength c/ν	Wavenumber/cm^{-1} ν/c	Photon energy/J $h\nu$
	0	∞	0	0
Radiofrequency				
	10^9	0.3 m	0.03	7×10^{-25}
Microwave				
	10^{11}	3 mm	3	7×10^{-23}
Far infrared (IR)				
	10^{13}	30 μm	300	7×10^{-21}
Mid IR				
	1.5×10^{14}	2 μm	5000	10^{-19}
Near IR				
	4×10^{14}	750 mm	13 000	3×10^{-19}
Visible				
	1.5×10^{15}	400 nm	25 000	5×10^{-19}
Ultraviolet (UV)				
	1.5×10^{15}	200 nm	50 000	10^{-18}
Vacuum UV				
	1.5×10^{16}	20 nm	5×10^5	10^{-17}
X ray				
	10^{20}	3 pm	3×10^9	7×10^{-14}
Gamma ray				
	∞	0	∞	∞

Note: Entries are rounded to one or two representative significant figures. Boundaries between regions are somewhat arbitrary.

to the direction of propagation. If the z direction is the direction of propagation, then the electric field vector (the only one we will be concerned with) is

$$\mathbf{E} = \mathbf{x}E_x + \mathbf{y}E_y \quad (18.3)$$

where **x** and **y** are unit vectors. (See Figure 18.1.) The amplitudes E_x and E_y for any one frequency are

$$E_x(z, t) = E_x^\circ \sin(kz - \omega t + \delta_x)$$
$$E_y(z, t) = E_y^\circ \sin(kz - \omega t + \delta_y) \quad (18.4)$$

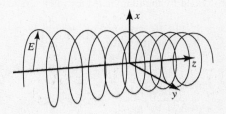

FIGURE 18.1 The intensity of a radiation field propagating along the z axis can be characterized by the electric field vector **E**. This vector has components along both the x and y directions in the general case of light that is not plane polarized.

where

$$k = \frac{2\pi}{\lambda} = \frac{2\pi\nu}{c} = \frac{\omega}{c} \qquad (18.5)$$

and δ_x, δ_y are arbitrary constants. These expressions hold for radiation in vacuum. In an isotropic medium with a relative electric permittivity (a dielectric constant) ϵ_r and a relative magnetic permeability μ_r, the speed of propagation is slowed from c to $c/(\epsilon_r\mu_r)^{1/2}$. (See Section 9.3 for representative values of ϵ_r and μ_r.) This causes k to change from ω/c to $(\epsilon_r\mu_r)^{1/2}\omega/c$. Vacuum is characterized by $\epsilon_r = \mu_r = 1$.

If $E_x^\circ = E_y^\circ$ and $\delta_x = \delta_y$ (or $\delta_x = \delta_y + \pi$), the electric field oscillates in one plane and the radiation is said to be *plane polarized*. If $E_y^\circ = 0$, then the radiation is confined to the zx plane and is *x polarized*, while $E_x^\circ = 0$ produces y-polarized radiation. A more general polarization is *elliptical*:

$$\mathbf{E}\pm = \mathbf{x}E_x^\circ \cos(kx - \omega t) \pm \mathbf{y}E_y^\circ \sin(kx - \omega t) \ .$$

The $+$ sign leads to *right elliptically polarized light*; the $-$ sign, to left. If you look along the z axis back towards the source, "right" corresponds to a counterclockwise rotation of **E**, while "left" corresponds to a clockwise rotation. *Circularly polarized light* is just the special case of elliptical polarization with $E_x^\circ = E_y^\circ$.

The energy in the radiation field can be specified by either the number of photons at each frequency or by the electric field intensity. A *monochromatic* radiation field, produced by electric circuits in the radiofrequency and microwave regions and by lasers in the infrared, visible, and ultraviolet regions, has an *energy density* ρ with

$$\rho = \frac{1}{2}\epsilon_0(E^\circ)^2 \qquad (18.6)$$

where ϵ_0 is the permittivity of free space and E° is the maximum electric field amplitude. The energy density has units of J m^{-3}, energy per unit volume, which is equivalent to pressure, and the pressure of a radiation field can be measured, although it is normally insignificant.

The flux of radiant energy, the *radiation intensity,* is a measure of brightness. It is

$$I = \text{energy per unit area per unit time} = \rho c \ . \qquad (18.7)$$

The *power* of the radiation field, measured in watts (1 W = 1 J s^{-1}), is the product of I and the total area illuminated.

EXAMPLE 18.1

Consider a He/Ne laser with a 1 mW power. This is quite a common laser, often used for demonstration purposes. It emits red light of wavelength 632.8 nm. If this laser is focused to a 1 mm radius spot, what is the electric field amplitude at the spot and how many photons strike it per second?

SOLUTION The power divided by the illuminated area is the radiation intensity:

$$I = \frac{10^{-3}\,\text{W}}{\pi \times 10^{-6}\,\text{m}^2} = 318\,\text{W m}^{-2} \ .$$

The energy density is

$$\rho = \frac{I}{c} = \frac{318\,\text{W m}^{-2}}{3 \times 10^8\,\text{m s}^{-1}} = 1.06 \times 10^{-6}\,\text{J m}^{-3} \ .$$

(This is a pressure $\sim 10^{-11}$ atm.) The field amplitude is

$$E° = \left(\frac{2\rho}{\epsilon_0}\right)^{1/2} = \left(\frac{2.12 \times 10^{-6}}{10^7/4\pi c^2}\right)^{1/2} = 489 \text{ V m}^{-1}.$$

This is not a very large electric field. It is comparable to the field between the terminals of a common 9 V electronics battery.

Since one 632.8 nm photon has an energy

$$E = h\nu = \frac{hc}{\lambda} = \frac{(6.6 \times 10^{-34} \text{ J s})(3 \times 10^8 \text{ m s}^{-1})}{6.328 \times 10^{-7} \text{ m}} = 3.1 \times 10^{-19} \text{ J}$$

the radiation intensity is

$$I = \frac{318 \text{ J m}^{-2} \text{ s}^{-1}}{3.1 \times 10^{-19} \text{ J photon}^{-1}} \cong 10^{21} \text{ photons m}^{-2} \text{ s}^{-1}$$

so that $(10^{21}$ photons m^{-2} s$^{-1})(\pi \times 10^{-6}$ m$^2) = 3.2 \times 10^{15}$ photons strike the illuminated spot each second.

⮕ *RELATED PROBLEMS* 18.1, 18.2, 18.3

While the radiation energy density and electric field in the example above are not particularly impressive, it is important to realize that these are large values compared to ordinary light sources. Lasers are uniquely bright. To show this, we move on to the next section and consider ordinary radiation sources.

18.2 HOW SPECTROSCOPIES WORK—ABSORPTION AND EMISSION

Section 11.1 discussed the role of blackbody radiation in the development of quantum theory. The blackbody is an idealized thing in which the interior walls of a cavity at some finite temperature are in thermal equilibrium with the radiation inside the cavity. This radiation field spreads over a wide range of wavelengths, and the theoretical explanation of this distribution was Planck's major achievement. In an ideal blackbody, the walls absorb and emit with unit probability at all frequencies.

An experimental approximation to a blackbody starts with a cavity that has a small aperture to the outside. The cavity walls must be as absorbing (as "black") as possible at all wavelengths. The light that escapes from the aperture is therefore in equilibrium with the cavity walls, and any light that strikes the aperture from outside passes into the cavity and is absorbed. Thus, the best experimental blackbody is *the aperture of the cavity,* not the cavity itself. It is the area of the aperture that appears to absorb all wavelengths striking it from outside the cavity and to emit light to the outside with a distribution characteristic of an equilibrium temperature.

Planck's expression for the energy density of blackbody radiation at temperature T in the frequency interval ν to $\nu + d\nu$ is

$$\rho(\nu; T) \, d\nu = \frac{8\pi h \nu^3}{c^3} \frac{1}{e^{h\nu/k_B T} - 1} \, d\nu \qquad (18.8)$$

where k_B is Boltzmann's constant. Figure 18.2 plots $\rho(\nu; T)$ at several temperatures. Note how ρ peaks at greater frequencies as T is increased.

If we define $x = h\nu/k_B T$ so that $d\nu = k_B T \, dx/h$, Eq. (18.8) is

$$\rho(x; T) \, dx = \frac{8\pi h}{c^3} \left(\frac{k_B T}{h}\right)^4 \frac{x^3}{e^x - 1} \, dx = (\text{constant}) T^4 f(x) \, dx \ .$$

FIGURE 18.2 The blackbody energy density expression is a strong function of both frequency and source temperature. Note how high the temperature must be before there is appreciable emission in the visible. (The ordinate scale is logarithmic.)

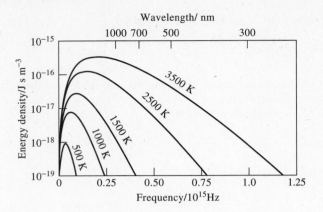

The function $f(x) = x^3/(e^x - 1)$, which gives ρ its shape, has a maximum at $x \cong 2.822$. At any one temperature, the corresponding wavenumber is $(2.822/1.438)T \cong 2T$ cm^{-1} (for T in K, using $hc/(100$ cm m$^{-1})k_B = 1.438$ K cm). Thus, the sun, which closely approximates a 6000 K blackbody, has an energy density distribution peaked at $\sim 12\,000$ cm^{-1}. This, as Table 18.1 shows, is just at the near IR fringe of the visible. Equation (18.8) can also be converted into a *photon number density* per unit frequency interval if we divide by $h\nu$: $\rho_\phi(\nu; T) = \rho(\nu; T)/h\nu$.

Integrating Eq. (18.8) over all frequencies gives an expression for the total radiation energy density at any temperature. The integral is known exactly, $\int [x^3/(e^x - 1)]\, dx = \pi^4/15$, so that

$$\rho(T) = \frac{8\pi^5}{15} \frac{(k_B T)^4}{(hc)^3} = \eta T^4$$

where $\eta = 8\pi^5 k_B^4/15 h^3 c^3 = 7.566 \times 10^{-16}$ J m^{-3} K^{-4}. (This is closely related to the *Stephan–Boltzmann constant* σ, where $\sigma = \eta c/4$. Boltzmann used thermodynamics to show that $\rho \propto T^4$.)

Integration over a finite range of frequencies gives the energy density over that range alone. Multiplication by c gives the radiation intensity I from Eq. (18.7). A few more steps, discussed in Problem 18.6, let us calculate that a blackbody at 2000 K positioned 1 m from a surface has about 6.8×10^{11} photons per second striking a 1 mm radius spot over the frequency region of the He/Ne laser in Example 18.1.

This modest laser is thus about 4700 times brighter than a fairly hot blackbody *over the frequency range of the laser*. This is an important point, since the laser concentrates all its light over a very narrow range of frequencies while the blackbody radiates energy (at the rate $I = \sigma T^4 = (5.67 \times 10^{-8}$ W m^{-2} K$^{-4})(2000$ K$)^4 \cong 90$ W per cm^2 of its surface) *over the entire electromagnetic spectrum.*

A Closer Look at Blackbody Radiation

What color is the sun? This question can be answered by at least two approaches: look up in the sky on a sunny day, or consider the blackbody radiation distribution law. We will take the latter, less poetic, approach here.

The distribution of sunlight radiation closely follows a blackbody distribution for $T \cong 6000$ K, as mentioned earlier, and the distribution of Eq. (18.8) peaks at $\cong 12\,000$ cm^{-1} for this temperature. This suggests that the sun is predominantly "near IR colored," which is about as unpoetic a description as one can imagine.

But this answer depends on *our choice for the meaning of the radiation distribution function.*

The 12 000 cm^{-1} answer is based on assigning "the color" to *the most probable frequency* (and thus photon energy) in the distribution *written in terms of frequency*. Suppose instead we did the following experiment. We counted the *number of photons* in a series of narrow, but equal, frequency intervals rather than counting the radiant energy. As mentioned in the text, the *photon density* distribution is related to the *energy density* distribution by

$$\rho_\phi(x; T)\, dx = \frac{\rho(x; T)}{h\nu}\, dx = \frac{\frac{8\pi h}{c^3}\left(\frac{k_B T}{h}\right)^4 \frac{x^3}{e^x - 1}}{x k_B T}\, dx = \frac{8\pi}{c^3}\left(\frac{k_B T}{h}\right)^3 \frac{x^2}{e^x - 1}\, dx$$

where $x = h\nu/k_B T$. This distribution depends on x like $x^2/(e^x - 1)$, while $\rho(x; T)$ varied like $x^3/(e^x - 1)$. The different power of x causes the peak of the ρ_ϕ distribution to shift from $x \cong 2.82$ to $x \cong 1.59$. In this experiment, we would say that the *most probable photon* measured by an instrument with a *constant photon frequency bandpass* has a frequency $\nu = 1.59 k_B T/h$, or a wavenumber $(1.59\, T)/1.438 \cong 6600$ cm^{-1}, near the mid IR/near IR boundary.

Suppose instead our instrument was sensitive to energy density, but instead of a constant *frequency* bandpass it had a constant *wavelength* bandpass. Since $\lambda = c/\nu$, $d\nu = -c/\lambda^2\, d\lambda$ (the $-$ sign tells us only that λ decreases as ν increases and can be ignored) and Eq. (18.8) becomes

$$\rho^\lambda(\lambda; T)\, d\lambda = \frac{8\pi hc}{\lambda^5} \frac{1}{e^{hc/\lambda k_B T} - 1}\, d\lambda\ .$$

Likewise, an instrument that counted photons with constant wavelength bandpass measures the distribution

$$\rho^\lambda_\phi(\lambda; T)\, d\lambda = \frac{\rho^\lambda(\lambda; T)}{hc/\lambda}\, d\lambda = \frac{8\pi}{\lambda^4} \frac{1}{e^{hc/\lambda k_B T} - 1}\, d\lambda\ .$$

The distribution ρ^λ peaks at $x = h\nu/k_B T = hc/\lambda k_B T = 4.96$ or at a wavenumber $\cong 20\,700$ cm^{-1}, which is blue-green light, while the distribution ρ^λ_ϕ peaks at $x = 3.92$, or a wavenumber $\cong 16\,400$ cm^{-1}, which is visible red light.

Thus we see that the most probable "color" of a blackbody depends dramatically on the type of measurement (photon counting or radiation energy) as well as the type of instrument (constant frequency or constant wavelength bandpass) used experimentally. These four distributions are plotted below as functions of x.

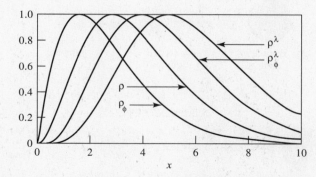

It may seem to you that the simple question "What color is the sun?" should not have such a complicated and apparently ambiguous answer. After all, we've seen the sun, and it's obviously yellowish, but we reach that conclusion from observations with a very unique light detector, the human eye. This exercise points out the importance of interpreting distribution functions with care more than it answers the original question!

On a microscopic level, three fundamental matter–radiation interactions sustain the thermal equilibrium of a blackbody. These interactions apply to isolated atoms and molecules as well, but in preparation for the next section, we will focus on photon–atom interactions. The three interactions are (see also Figure 18.3):

1. Absorption:
$$\text{atom} + \text{photon} \rightarrow \text{energized atom}$$

2. Spontaneous Emission:
$$\text{energized atom} \rightarrow \text{de-energized atom} + \text{photon}$$

3. Stimulated Emission:
$$\text{energized atom} + \text{photon} \rightarrow \text{de-energized atom} + \text{two identical photons}$$

Note that "spontaneous absorption" does not exist, but that absorption and stimulated emission are both photon mediated. Stimulated emission gives the words laser and maser two of their letters: "**l**ight (or **m**icrowave) **a**mplification by **s**timulated **e**mission of **r**adiation," and the two identical photons (they are said to be *coherent*) that result provide the amplification mechanism.

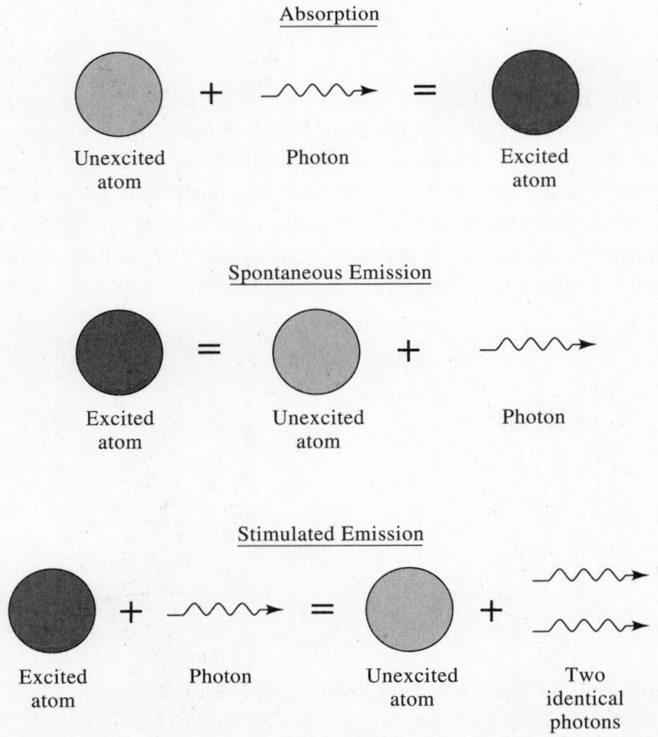

FIGURE 18.3 These are the three fundamental types of matter–radiation interactions: absorption, spontaneous emission, and stimulated emission.

18.2 HOW SPECTROSCOPIES WORK—ABSORPTION AND EMISSION

Thermal equilibrium is a dynamic situation, as the last third of this book will show in some detail. In 1917, Einstein used this dynamic aspect to introduce three quantities that measure the probabilities of each of these interactions. Here is his argument. First, consider *any* two energy levels of an atom that can interact with light, levels 1 and 2 with energies $E_2 > E_1$. (One says such levels are *connected* by radiative interactions.) In a sample of many such atoms at equilibrium, the ratio of the number in the upper state (state 2) to the number in the lower state (state 1) is given by a relationship Boltzmann first derived (and which we derive in Chapter 22):

$$\frac{N_2}{N_1} = \frac{g_2}{g_1} e^{-(E_2 - E_1)/k_B T} = \frac{g_2}{g_1} e^{-h\nu_{12}/k_B T} \quad (18.9)$$

where the g's are the degeneracies of the states and $\nu_{12} = (E_2 - E_1)/h$ is the frequency of light with a photon energy equal to the energy difference of the two levels.

Absorption increases the number of atoms in level 2, while stimulated emission and spontaneous emission decrease this number. Einstein said, quite simply, that at equilibrium this number is not changing with time:

$$\frac{dN_2}{dt} = \text{absorption rate} - \text{stimulated emission rate} - \text{spontaneous emission rate} = 0 \quad (18.10)$$

and, again simply but with great insight, that the absorption rate is proportional to the radiation density at ν_{12} times the number of atoms in state 1 times a proportionality constant:

$$\text{absorption rate} = N_1 B_{12} \rho(\nu_{12}) \quad (18.11)$$

where B_{12} is called the *absorption coefficient*. Similarly, the rate of both kinds of $2 \rightarrow 1$ transitions was assumed to be

$$\text{stimulated emission rate} + \text{spontaneous emission rate} = N_2(B_{21} \rho(\nu_{12}) + A_{21}) \quad (18.12)$$

where B_{21} is called the *stimulated emission coefficient* and A_{21} is the *spontaneous emission rate coefficient*. From Eq. (18.10), we can write

$$\frac{N_2}{N_1} = \frac{B_{12} \rho(\nu_{12})}{B_{21} \rho(\nu_{12}) + A_{21}}. \quad (18.13)$$

Next, we equate this expression for N_2/N_1 with that in Eq. (18.9) and solve for $\rho(\nu_{12})$:

$$\rho(\nu_{12}) = \frac{A_{21} \left(\frac{g_2}{g_1}\right) \exp(-h\nu_{12}/k_B T)}{B_{12} - B_{21} \left(\frac{g_2}{g_1}\right) \exp(-h\nu_{12}/k_B T)}.$$

This expression for $\rho(\nu)$ must equal the Planck blackbody expression, Eq. (18.8), since the radiation and the atoms are assumed to be in thermal equilibrium with each other. A little algebra shows that

$$B_{12} = \left(\frac{g_2}{g_1}\right) B_{21} \quad (18.14a)$$

and

$$A_{21} = \left(\frac{8\pi h \nu_{12}^3}{c^3}\right) B_{21}. \quad (18.14b)$$

An important conclusion is that A_{21} cannot be zero. Spontaneous emission is necessary to maintain equilibrium in a non-zero radiation field.

We will explore the magnitudes and properties of A_{21}, B_{12}, and B_{21} further in the next section, but to close this section, we discuss these interactions in the context of a more familiar view of spectroscopy.

You may have encountered the Beer–Lambert law: the intensity of radiation exiting from an absorbing sample, $I(\nu)$, is related to the incident radiation intensity, $I_0(\nu)$, according to

$$I(\nu) = I_0(\nu) \exp(-abc) \qquad (18.15)$$

$$\underset{\text{Transmitted intensity}}{\uparrow} \qquad \underset{\text{Incident intensity}}{\uparrow} \qquad \underset{\text{Attenuation factor}}{\uparrow}$$

where a is the *absorbance coefficient* of the sample at frequency ν, b is the sample's *breadth* (or optical path length, but breadth serves as a mnemonic), and c is the *concentration* of absorbing molecules. Figure 18.4 illustrates this equation.

You might guess that the phenomenological absorbance coefficient is somehow related to B_{12}, the Einstein absorption coefficient, and it is. Stimulated emission and spontaneous emission do not seem to be involved in the Beer–Lambert law, and there are two reasons why. First, stimulated emission requires that state 2 is populated *in advance of photon arrival*. The Boltzmann expression, Eq. (18.9), generally keeps this from happening in an *equilibrium* sample. For example, consider absorption at a frequency corresponding to 20 000 cm^{-1} (blue light). Eq. (18.9) says, at room temperature with $(g_2/g_1) = 1$,

$$N_2 \cong N_1 \exp\left[-\frac{(2 \times 10^4 \text{ cm}^{-1})(1.438 \text{ K cm})}{300 \text{ K}}\right] \cong e^{-96} N_1 \cong 2 \times 10^{-42} N_1 \ .$$

Thus, at equilibrium, state 2 is unpopulated, and stimulated emission is of no importance for room-temperature samples with excited-state energies > 200 cm^{-1} or so. ($N_2/N_1 \sim e^{-1}$ for a 200 cm^{-1} excited state at room temperature.)

Spontaneous emission is also usually negligible. As in Figure 18.4, the probing light beam is collimated to a tightly defined direction. Spontaneously emitted photons move in random directions from the sample so that very few of them arrive at the spectrometer's detector. Also, spontaneous emission may not occur due to other de-excitation processes. The energy brought in by a photon need not leave by another photon. Various *quenching* mechanisms can occur to de-excite the atom or molecule before the photon can be emitted. These processes can be as trivial as collision of the molecule with the walls of the apparatus, which often results in de-excitation with unit probability, or as subtle as an internal re-distribution of the energy within the molecule, or as interesting as a photochemical reaction. We will have much more to say about these processes in later chapters, but we turn now to the specific case of atomic spectroscopy.

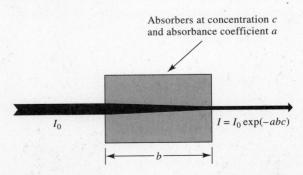

FIGURE 18.4 The Beer–Lambert law predicts the radiation intensity I that emerges from a sample struck by intensity I_0. The absorption coefficient of the sample, its concentration, and the optical path length all enter the attenuation factor.

18.3 ATOMIC SPECTROSCOPY AND SELECTION RULES

Section 13.5 described atomic energy levels and the quantum numbers summarized in atomic-state term symbols. Here, we use those symbols as a guide for finding states that are radiatively connected. We will answer questions such as this: Given a table of atomic energy levels (such as Table 13.3 for C), what are the photon energies a ground state atom (such as C $2s^2 2p^2$ $^3P_0$) can absorb?

We assume we know enough about the atomic states to specify the term symbols if not actual wavefunctions. This will give us a series of yes or no answers to our question, at the least, while the harder question—What is B_{12}?—depends on the entire wavefunction. Further, we will assume that the light intensity is small enough to allow the matter–radiation interaction to be treated by perturbation theory.

The type of perturbation theory needed here is, in contrast to that of Chapter 13, explicitly time dependent. We will not derive time-dependent perturbation theory in detail but will instead outline its main ideas.

Imagine some time-dependent part of the total Hamiltonian is the perturbation. Call this term $\hat{H}'(t)$ and, as usual, assume a set of exact wavefunctions ψ_k^0 that satisfy the total Hamiltonian *minus* the perturbation term. We seek the time-dependent wavefunction $\Psi(x, t)$ for the system in the presence of $\hat{H}'(t)$. We expand $\Psi(x, t)$ in the complete set of ψ_k^0 functions, as suggested by Eq. (11.18) and Postulate VII, and we explicitly include time dependence:

$$\Psi(x, t) = \sum_k c_k(t)\, e^{-iE_k^0 t/\hbar}\, \psi_k^0(x)\ . \qquad (18.16)$$

The $c_k(t)$ coefficients tell us the probability that the perturbation has taken the system from one state to another. Suppose $\hat{H}'(t) = 0$ for all times earlier than $t = 0$. (In other words, we switch on the light at $t = 0$.) At earlier times, the system is in some particular state n so that $c_n(t < 0) = 1$ and $c_k(t < 0) = 0$, $k \neq n$. At some time T, we switch the light off and expect to find $c_n(T) < 1$ (since the system has perhaps left state n) while certain c_k's are now non-zero. Perturbation theory says that

$$\Psi(x, t > T) = \sum_k c_k(T)\, e^{-iE_k^0 t/\hbar}\, \psi_k^0(x)$$

where now $c_k(T)$ are just time-independent *constants* such that

$|c_k(T)|^2$ = probability of finding the system in state k with energy E_k^0 at any time $t > T$.

We can picture these equations for a system that starts with only the ground state populated and ends with the second excited state also populated:

<pre>
       □ c₃ = 0   □ c₃ = 0

       □ c₂ = 0   ▨ c₂ > 0
                  ↑
       □ c₁ = 0   hν on   □ c₁ = 0
                  for time T
       ■ c₀ = 1   ▨ c₀ < 1
        t = 0   t = T
</pre>

Theory also gives the following expression for the $c_k(T)$'s

$$c_k(T) = c_k(0) - \frac{i}{\hbar} \int_0^T e^{i\omega_{kn} t}\, H'_{kn}(t)\, dt \qquad (18.17)$$

where

$$\omega_{kn} = \frac{E_k^0 - E_n^0}{\hbar} = \frac{\text{energy difference between connected states}}{\hbar}$$

and

$$H'_{kn}(t) = \int \psi_k^{0*} \hat{H}'(t) \psi_n^0 \, d\tau \, .$$

Looking back through these expressions, we see that the one for $H'_{kn}(t)$ is key. We have encountered such integrals many times, calling them *matrix elements*, and we see here that *if this perturbation matrix element equals zero, there is no transition between states n and k*. Thus, our goal is to find those conditions that make $H'_{kn}(t)$ zero or non-zero. These conditions will be restrictions on changes in quantum numbers called *selection rules*.

To evaluate $H'_{kn}(t)$, we need to specify the matter–radiation interaction Hamiltonian $\hat{H}'(t)$. We will not write the full expression for $\hat{H}'(t)$ but will instead concentrate on its leading and dominant term, a time-dependent version of the *static* electric field–molecule interaction we introduced in Chapter 15, where we first defined the dipole moment operator

$$\hat{\mathbf{p}} = \sum_i q_i \mathbf{r}_i \qquad (15.10)$$

in terms of the charges q_i at positions $\mathbf{r}_i$. It is the interaction between $\mathbf{p}$ and the radiation's electric field that accounts for most of spectroscopy:

$$\hat{H}'(t) = -\hat{\mathbf{p}} \cdot \mathbf{E}(t) \, . \qquad (18.18)$$

This is called *the electric dipole approximation*.

We can simplify Eq. (18.18) if the light is linearly polarized along the x direction and moving along the z direction so that (see Eq. (18.4)) $E(t) = E_x^\circ \cos(kz - \omega t)$. Since most spectroscopic wavelengths are longer than ~ 100 nm and thus much larger than the dimensions of the molecule, $kz = 2\pi z/\lambda$ will be small over the region where the wavefunction is large (i.e., over the size of the molecule). This means we can approximate the field inside the integral as

$$E(t) \cong E_x^\circ \cos(-\omega t) = E_x^\circ \cos \omega t \, .$$

This is called the *long wavelength approximation*, and it is equivalent to writing

$$\hat{H}'(t) = -p_x E_x^\circ \cos \omega t \qquad (18.19)$$

from the start.

The key matrix element is now

$$H'_{kn}(t) = -E_x^\circ \cos \omega t \int \psi_k^{0*} \hat{p}_x \psi_n^0 \, d\tau \, .$$

We will write

$$p_{x,kn} = \int \psi_k^{0*} \hat{p}_x \psi_n^0 \, d\tau \qquad (18.20)$$

for a general matrix element of the vector dipole moment operator.

We showed in Chapter 15 (just past Eq. (15.15)) that Eq. (18.20) with $k = n$ is the *permanent dipole moment* of the state. For transitions, we want $k \neq n$, and $p_{x,kn}$ is called the *transition dipole moment*.

When time-dependent perturbation theory is carried out with Eq. (18.19), one finds that B_{12}, the Einstein absorption coefficient, is

$$B_{12} = \frac{|p_{12}|^2}{6\epsilon_0 \hbar^2}, \qquad (18.21)$$

which shows that the absorption rate is directly proportional to the square of the transition dipole moment. A typical atomic p_{12} is about 1.3×10^{-29} C m, about the same as a typical molecular permanent dipole moment, so that $B_{12} \cong 4 \times 10^{20}$ m^3 J^{-1} s^{-2}. This is not a very interesting number[2] by itself, but when used in Eq. (18.14b) for the *spontaneous* emission rate coefficient, we find

$$A_{21} \text{ (atoms, allowed transition)} \cong 3 \times 10^{-37} \, \nu_{12}^3 \text{ s}^{-1}$$

or, for a typical value of ν_{12} in the visible, say 5×10^{14} Hz, $A_{21} \cong 3 \times 10^7$ s^{-1}.

An equally interesting number is the reciprocal of A_{21}, called the *radiative lifetime*. Our representative A_{21} corresponds to a radiative lifetime

$$\tau_{\text{rad}} = \frac{1}{A_{21}} \cong 3 \times 10^{-8} \text{ s} \cong 30 \text{ ns} .$$

The simplest physical interpretation of τ_{rad} is the same as that for radioactive decay processes or other *first-order rate processes*, a term we will describe in detail in our discussion of chemical kinetics later in this book. If we take a large number of ground-state atoms and irradiate them with a pulse of light of duration much less than τ_{rad}, we create a certain number N_2° in the excited state almost instantaneously. The radiative lifetime tells us the number remaining in this state as time passes:

$$N_2(t) = N_2^\circ \, e^{-t/\tau_{\text{rad}}} . \qquad (18.22)$$

A more profound interpretation of spontaneous emission is given a Closer Look below.

A Closer Look at Spontaneous Emissions

An absorption event at ν_{12} is often followed by spontaneous emission at a host of frequencies including, but not limited to, ν_{12}. This can come about by the following type of energy-level structure:

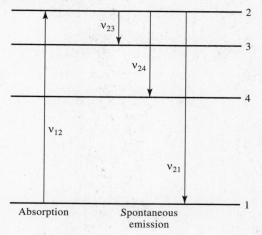

---

[2]Note that the units of B_{12}, m^3 J^{-1} s^{-2}, could be simplified to m kg^{-1}, but since Eq. (18.11) shows that B_{12} times an energy density per unit frequency (units: J s m^{-3}) equals a rate (units: s^{-1}), the m^3 J^{-1} s^{-2} set of units are more instructive.

Here, absorption from state 1 to state 2 is followed by emission from state 2 not only back to state 1, but also to intermediate states such as 3 and 4. Each spontaneous emission has a unique A value, and the radiative lifetime of state 2 is

$$\tau_{rad} = \frac{1}{A_{21} + A_{23} + A_{24}}$$

to account correctly for the three different rates.

But, whatever the value for τ_{rad}, *once we recognize that radiative processes can occur spontaneously, no excited state can be said to be "stationary."* When one state spontaneously reverts to another, it is said to have *decayed*. Our postulated mechanism for specifying energy with arbitrary precision was that the state was stationary: it could exist in isolation forever. Now we see that matter–radiation interactions give finite lifetimes to excited states and *therefore introduce finite uncertainties to these states' energies.* An uncertainty in the excited state energy implies a *range* of frequencies of light that can be absorbed in producing that state, and thus the absorption profile (the *lineshape function,* as it is called) is *not* infinitely narrow at $\nu = \nu_{12}$, but has some finite width.

In 1930, Weisskopf and Wigner showed that the probability of absorption follows what is called a *Lorentzian line shape*:

$$P(1 \rightarrow 2) \propto [(\nu - \nu_{12})^2 + b^2]^{-1}$$

where

$$b = (\tau_1^{-1} + \tau_2^{-1})/4\pi$$

with τ_1 and τ_2 the radiative lifetimes of states 1 and 2. (Of course, $\tau_1^{-1} = 0$ if state 1 is the ground state.) The quantity b is the half-width at half-height of the lineshape, as is shown in Figure 18.5.

Suppose state 1 is the ground state. Then the uncertainty in the transition frequency is all due to the excited state lifetime, and

$$\Delta \nu = \frac{\Delta E_2}{h} \cong b = \frac{1}{4\pi\tau_2}$$

or

$$\Delta E_2 \tau_2 = \frac{1}{2}\hbar$$

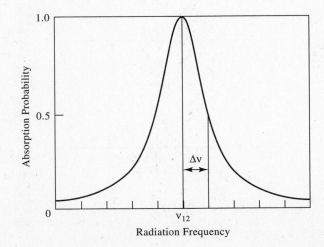

FIGURE 18.5 An isolated, stationary system will undergo a radiative transition with a lineshape given by the Lorentzian function of frequency. The transition's half-width at half height, $\Delta \nu$, is a measure of the radiative lifetimes of the two states in the transition.

where ΔE_2 is the uncertainty in state 2's energy. This is exactly in the form of an Uncertainty Principle expression (see Eq. (11.24)), but it is obtained by a different line of reasoning. Unlike conjugate dynamic variables, such as position and linear momentum, time plays the role of a parameter. The more time a system has to "find itself," the more definite becomes its energy.

EXAMPLE 18.2

Consider an electron in a 1-D particle-in-a-box potential of length $L = 10$ Å. What is the radiative lifetime of the $n = 2$ state of this system?

SOLUTION We first calculate p_{12} for the $n = 2$ to $n = 1$ transition. The wavefunctions are given by Eq. (12.11):

$$\psi_1(x) = \left(\frac{2}{L}\right)^{1/2} \sin\left(\frac{\pi x}{L}\right) \quad \text{and} \quad \psi_2(x) = \left(\frac{2}{L}\right)^{1/2} \sin\left(\frac{2\pi x}{L}\right)$$

and $\hat{p} = -ex$ so that

$$p_{12} = \int \psi_1 \hat{p} \psi_2 \, dx = -\frac{2e}{L} \int_0^L \sin\left(\frac{\pi x}{L}\right) x \sin\left(\frac{2\pi x}{L}\right) dx .$$

With a change of variable $y = \pi x/L$ and the identity $\sin 2y = 2 \sin y \cos y$, the integral can be found in integral tables:

$$p_{12} = -\frac{4e}{L} \frac{L^2}{\pi^2} \int_0^\pi y \cos y \sin^2 y \, dy = \left(-\frac{4e}{L}\frac{L^2}{\pi^2}\right)\left(-\frac{4}{9}\right) = \frac{16eL}{9\pi^2} .$$

Since this integral is non-zero, the transition is allowed with high probability. We also need ν_{12} from the particle-in-a-box energy expression, Eq. (12.12):

$$\nu_{12} = \frac{E_2 - E_1}{h} = \frac{\pi^2 \hbar^2}{2m_e hL^2}(2^2 - 1^2) = \frac{3\pi^2 \hbar^2}{2m_e hL^2} .$$

We use Eq. (18.21) with p_{12} to find B_{12} and Eq. (18.14b) to find A_{12}:

$$B_{12} = \frac{p_{12}^2}{6\epsilon_0 \hbar^2} \quad \text{and thus} \quad A_{12} = \frac{8\pi h \nu_{12}^3}{c^3} \frac{p_{12}^2}{6\epsilon_0 \hbar^2} .$$

Numerically, $\nu_{12} = 2.728 \times 10^{14}$ Hz (or $\lambda = 1099$ nm, in the near IR), $p_{12} = 2.886 \times 10^{-29}$ C m, $B_{12} = 1.410 \times 10^{21}$ m kg^{-1}, $A_{12} = 1.768 \times 10^7$ s^{-1}, and $\tau_{rad} = 56.6$ ns. Algebraically, we have, after some substitutions

$$\tau_{rad} = \frac{9}{8} \frac{c^3 m_e^3 \pi \epsilon_0 L^4}{e^2 h^2} ,$$

which shows explicitly how the lifetime of this transition depends on the parameters (L, m_e, and e) of the system.

⇒ **RELATED PROBLEMS** 18.9, 18.10, 18.11, 18.12

Next, we discuss *selection rules,* those constraints on the quantum numbers of the states involved in an allowed transition. If we dig back into our collection of atomic wavefunctions in spherical polar coordinates and write the dipole operator also in these coordinates, we can find the conditions that make transition dipole moments non-zero. This approach gives the right answer, but it lacks physical insight. Thus, we will simply state the results and then discuss them to glean some physical interpretations.

Dipole selection rules allow transitions to other states that meet the following conditions on atomic quantum numbers:

$$\Delta L = -1, 0, \text{ or } +1$$
$$\Delta S = 0$$

so that

$$\Delta J = -1, 0, \text{ or } +1$$

but excluding transitions of the type

$$L = 0 \rightarrow L = 0 \quad \text{and} \quad J = 0 \rightarrow J = 0 \ .$$

These are remarkably simple rules. They all focus on angular momentum changes, and we take this as a starting point for a discussion of their physical origin.

First, $\Delta S = 0$, implying a *conservation of spin multiplicity*. The origin of this rule can be traced to our perturbation term, $\hat{H}'(t)$. Since $\hat{H}'(t)$ does not mention spin, the total spin should not change.

Next, the $\Delta L = 0, \pm 1$ rule puts a boundary on the amount of electron orbital angular momentum change that can accompany photon absorption or emission. This boundary, $|\Delta L| = 1$, reflects the angular momentum of a photon and the requirement that angular momentum is conserved. On an individual electron basis, in fact, the selection rule $\Delta l = \pm 1$ is strict and $\Delta l = 0$ cannot happen. This rule holds for H atom and for the majority of transitions in alkali metals, which are dominated by a single valence-electron interaction.

The $\Delta l = \pm 1$ rule also reflects atomic wavefunction symmetry properties. One-electron wavefunctions with even l values have *even parity*, and those with odd l values have *odd parity* (see Example 15.3). *Dipole-allowed transitions must change parity*. This is known as *Laporte's rule*. Thus, in atomic hydrogen, 1s → 4p is allowed (even to odd parity, $\Delta l = 1$), while 1s → 4s is not ($\Delta l = 0$), nor is 1s → 4f (which changes parity, but has a forbidden $\Delta l = 3$).

In atomic energy level tables, such as Tables 13.3 and 13.4, a right superscript ○ codes for odd parity. In Table 13.3 for C, the ground state is $2p^4$ $^3P_2$, *without* the superscript, meaning this state has overall *even* parity. Farther down the table is 3s $^5S_2^\circ$, which has the configuration $2p^3 3s$ and overall *odd* parity, as the superscript indicates. For many-electron atoms, *states derived from configurations with an odd number of electrons of odd l value (p, f, etc.) have odd parity*. The others are even. (Odd-parity state energies are italicized in these tables.)

The $\Delta J = 0, \pm 1, J = 0 \not\rightarrow J = 0$, rule comes from spin–orbit coupling. Since spin–orbit interactions are greatest for the heaviest elements, this selection rule can dominate as atomic number increases and L and S lose their identities as good quantum numbers.

Now we can predict atomic absorption and emission spectra. As an example, the lowest seven energy levels of O are listed below, taken from Table 13.4.

	Designation		Energy/cm^{-1}
(1)	$2p^4$	$^3P_2$	0.0
(2)		$^3P_1$	158.5
(3)		$^3P_0$	226.5
(4)		$^1D_2$	15 867.7
(5)		$^1S_0$	33 792.4
(6)	$2p^3 3s$	$^5S_2^\circ$	73 767.81
(7)	$2p^3 3s$	$^3S_1^\circ$	76 794.69

What is the first excited state connected to the ground state? State (2) would seem to be acceptable on the basis of $\Delta S = 0$ (both are triplets), $\Delta L = 0$ (both are P terms) and $\Delta J = -1$ ($J = 2$ going to $J = 1$), but states (1) and (2) are both *even parity,* and Laporte's rule prohibits the transition. The first odd parity state is (6), but the $\Delta S = 0$ rule goes against the (1)–(6) (triplet to quintet) transition. State (7), however, satisfies all the selection rules.

Of all the possible transitions among these seven lowest states (and there are 21 in all), only three of them, (1)–(7), (2)–(7), and (3)–(7), are robust, fully dipole-allowed transitions. They and their experimental Einstein A coefficients are shown below:

Transition	A_{21}/s^{-1}
(1) ← (7)	2.1×10^8
(2) ← (7)	1.3×10^8
(3) ← (7)	0.4×10^8

If all the other transitions are "forbidden" to occur, what type of experiment could determine their energies to such precision? The answer is again atomic spectroscopy. *No transition is absolutely forbidden to occur.* Beyond the electric-dipole approximation are higher approximations to the true radiation–matter interaction Hamiltonian, and these approximations introduce new and somewhat less restrictive selection rules, but they also carry with them much smaller transition probabilities. The next two higher approximations beyond the electric dipole are the magnetic dipole (focusing on light's magnetic field) and the electric quadrupole (reminiscent of the multipole expansion hierarchy discussed in Chapter 15).

Intermediate between pure electric dipole transitions for *L–S* coupled electrons and higher order transitions come those that relax the ΔL and ΔS selection rules but maintain the ΔJ rules. For example, both the (1)–(6) and (2)–(6) transitions violate $\Delta S = 0$, but satisfy $\Delta J = 0$ and $\Delta J = \pm 1$ rules, respectively. These are called *intercombination* transitions since they connect two states of different spin multiplicity. While they do occur, especially for heavier elements, they have unusually small transition moments or, measured another way, small A coefficients:

Transition	A_{21}/s^{-1}
(1) ← (6)	1.3×10^3
(2) ← (6)	3.8×10^2

Magnetic dipole and electric quadrupole allowed transitions are typically a few more orders of magnitude less likely to occur. One classic example is the transition (5) → (4), an emission at ~558 nm that is seen naturally in the green light of the aurora borealis. This transition has $\Delta S = 0$, $\Delta L = \Delta J = 2$, which is allowed by electric quadrupole selection rules, and has $A_{21} = 1.34$ s^{-1}.

Magnetic dipole selection rules can account for so-called *intramultiplet transitions* in which J changes *within* a given L, S, and configuration. Transitions such as (1)–(2) and (2)–(3) fall in this category, and they have very small spontaneous emission rates: 9×10^{-5} s^{-1} and 1.7×10^{-5} s^{-1}, respectively. These small numbers are partly due to the relative forbiddenness of the transitions and partly due to the close spacing of the energy levels, since $A_{21} \propto \nu_{12}^3 \propto$ (energy spacing)^3.

Finally, we mention another higher-order process that is important in He. Recall from Chapter 13 that the $1s^1 2s^1$ configuration yields $^1S_0$ and $^3S_1$ terms while the $1s^2$ ground state is $^1S_0$. The triplet to singlet transition could use the $\Delta J = 1$ rule (but with small probability, since He is so light and spin–orbit coupling is therefore small), but the $1s^1 2s^1\ ^1S_0$ excited state appears to have no chance to connect to the ground state because of the $L = 0 \not\to L = 0$ and $J = 0 \not\to J = 0$ rules. Both states do have exceedingly long radiative lifetimes. In general, such atomic states (and states (4) and (5) of O also fall in this category) are called *metastable*. They are energy-rich and sufficiently long-lived to be chemically unique species. Thus, ^3P ground-state oxygen and ^1D metastable oxygen are chemically very different atoms, as Figures 14.26–14.29 showed.

The $1s^1 2s^1\ ^1S_0$ state has a fascinating way of getting around selection rules. It emits two photons at once! As long as the photons have frequencies ν_1 and ν_2 such that $h(\nu_1 + \nu_2)$ equals the energy difference between the $1s^1 2s^1\ ^1S_0$ excited state and the $1s^2\ ^1S_0$ ground state, energy is conserved, and angular momentum is conserved if two photons of opposite angular momentum are emitted. Instead of emitting a very narrow range of frequencies, this state emits a broad, continuous range. We will see molecular examples of multiphoton processes in later chapters, since the intensity of laser sources has made multiphoton absorptions of various types fairly easy to study and exploit.

EXAMPLE 18.3

Use Table 13.3 to predict the main radiative decay paths of a C atom in the $2s^2 2p^1 3p^1\ ^1P_1$ level at 68 858 cm^{-1}.

SOLUTION The spin selection rule, $\Delta S = 0$, immediately limits the choices to another singlet state of lower energy. Laporte's rule further limits the choice to an odd state. The only singlet odd state of lower energy is $2s^2 2p3s\ ^1P_1^\circ$ at 61 982.20 cm^{-1}. The transition $2s^2 2p3p\ ^1P_1 \to 2s^2 2p3s\ ^1P_1^\circ$ is allowed, since $\Delta S = 0$, $\Delta L = 0$, $\Delta J = 0$, but $J \neq 0$, and the parity selection rule is satisfied.

➞ *RELATED PROBLEMS* 18.17, 18.19, 18.20, 18.21, 18.22

18.4 TOO MUCH ENERGY—PHOTOIONIZATION AND PHOTODISSOCIATION

While photon absorption can alter chemical behavior and reactivity, sometimes dramatically, as in the ^3P versus ^1D oxygen-atom example, the absorber retains its basic chemical identity. Any energy loss process can return it to its original state. But there are a variety of radiation-induced changes that can *irreversibly alter chemical identity*. We discuss two of these, photoionization and photodissociation, in this section.

We have mentioned or hinted at these processes in earlier chapters, and we will return to photodissociation in particular in subsequent chapters. Photoionization was discussed in terms of molecular photoelectron spectroscopy in Section 14.2, and at the end of Chapter 13, we discussed atomic states that have a total energy greater than the ionization energy and autoionize. In Chapter 11, we hinted at photodissociation (see Figure 11.8) with the potential energy function that binds two atoms and has two energy regions, one with the discrete levels of an intact molecule and one with the continuous levels of a broken bond.

Photoionization

We start with atomic photoionization and discuss how atomic spectroscopy can measure ionization energies. Once again, atomic oxygen is our example. The $2p^4$

$^3P_2$ to $2p^33s^1\ ^3S_1^\circ$ transition (what we called (1)–(7) in the previous section) is fully dipole allowed. There must also be states of the type $2p^3ns^1\ ^3S_1^\circ$ where $n = 4, 5, 6, \ldots$, and transitions to these states from the ground state must also be allowed. Just as in atomic H, this series of one-electron states with successively larger principle quantum number must converge to an ionization limit. Thus, the absorption spectrum of O from the ground state must contain, among other allowed transitions, a *progression* of transitions that approaches a maximum energy as n increases. Various theoretical expressions guide the extrapolation of this progression toward the $n = \infty$ ionization limit, which is about 109 836.7 cm^{-1} for O.

This is a simple idea, but it assumes O^+ is formed at the ionization limit in its ground state:[3] $2p^3\ ^4S_{3/2}^\circ$. The $2p^3$ configuration can also lead to $^2D^\circ$ and $^2P^\circ$ terms, (see Table 13.2), and Table 13.4 lists $^3D^\circ$ states based on the $2p^33s$ configuration with the $^2D^\circ$ ion core around 101 100 cm^{-1}. Not shown in that Table are $^3P^\circ$ states with a $^2P^\circ$ core and a $2s^22p^3(^2P^\circ)3s$ configuration that lie around 113 920 cm^{-1}, about 4000 cm^{-1} *above* the ionization limit. These and others of the $2s^22p^3(^2D^\circ)ns$ and $2s^22p^3(^2P^\circ)ns$ progression are autoionizing.

There are two types of experiments that can detect autoionization. The more direct of the two monitors the photoelectric current[4] as the radiation's energy is increased. The photocurrent starts at the first ionization limit, and as higher-energy autoionizing states are reached, there are features in the photocurrent that signal a new ionization channel.

The second experiment is based on the fast, irreversible, nonradiative nature of autoionization. The photoelectron is emitted

$$A + h\nu \to A^{**} \xrightarrow{\tau_{ion}} A^+ + e^-$$

on a time scale τ_{ion} shorter than spontaneous emission

$$A + h\nu \to A^{**} \xrightarrow{\tau_{rad}} A^* + h\nu'$$

where A^{**} is the autoionizing state. Since $\tau_{ion} < \tau_{rad}$, the absorption linewidth of the A to A^{**} transition is notably broadened.

Photodissociation

Molecular photodissociation is a somewhat surprising phenomenon. Many molecules can absorb a photon that has an energy greater than a bond energy without dissociating. The excited molecule often relaxes by spontaneous emission, which is not surprising until you realize that it spends a rather long time deciding to do so. Molecular radiative lifetimes in this energy region are typically in the 100 ns to 10 µs range, which is a very long time when measured in units of a molecular heartbeat, the typical vibrational time of bonded atoms. This time is usually in the 10^{-14} to 10^{-13} s range, so that some 10^6 to 10^9 vibrations take place before spontaneous emission. Thus, total energy input alone is not the sole deciding factor for a photodissociation event.

An absorbed photon with an energy greater than a bond energy generally causes the molecule to *change its electronic structure,* and *bond energies change as molecular structure changes.* Sometimes the change is trivially small, but sometimes it is dramatic. Two atoms that are bonded in the ground state may be completely non-

---

[3] Note how this "ion core" is coded into the "Config." column of Table 13.4.

[4] Ions and electrons can be detected in high vacuum with 100% efficiency. Photon detectors, however, are typically much less efficient, often by a factor >10.

bonded in an excited state. If these two states are connected by radiative selection rules, the molecule undergoes what is called *direct photodissociation*.

Halogen atom bond photodissociation often follows this route. For example, HI has a bond energy of 3.05 eV, corresponding to the energy of a 407 nm photon. Such a photon is *not* absorbed by HI, however. The absorption spectrum begins at ~280 nm and extends *continuously* to ~180 nm. The molecule simply *cannot absorb* light with energy in the vicinity of the bond energy. Radiative transition energies generally have *no* direct correspondence to bond energies. Note also that the absorption is *continuous,* since the final state, H + I, has a continuous range of energies.

The lowest energy dissociation path for ground-state HI is to ground-state atoms: H (1s $^2S_{1/2}$) + I (2p^5 $^2P^\circ_{3/2}$). The higher energy spin–orbit state of iodine (2p^5 $^2P^\circ_{1/2}$) is 7603 cm^{-1} above the ground state (see Example 13.5). Consider irradiation with photons in the middle of the HI absorption band, around 230 nm or 43 500 cm^{-1}. We need 24 600 cm^{-1} (3.05 eV) to break the HI bond into ground state atoms, leaving 18 900 cm^{-1} of excess energy. This is more than enough to produce the I $^2P_{1/2}$ electronically excited state, but will it be produced? We need to look at the HI electronic-state potential-energy curves to decide.

Figure 18.6 shows the HI electronic-state curves of importance to photodissociation. The ground state is a singlet, and an excited repulsive[5] singlet state also leads to the ground-state dissociation limit. The excited atomic-pair limit connects to an excited repulsive triplet molecular state. Just as in atomic spectroscopy, molecular absorption tends to preserve spin multiplicity, and we can assign the onset of absorption to the singlet–singlet transition. At this onset, we expect the ground state I atom products ($^2P_{3/2}$) exclusively, as has been confirmed by experiment. But, also in parallel with the atomic case, molecules containing heavy atoms can conserve the molecular equivalent of the *J* quantum number while violating spin multiplicity conservation. Thus, at higher photon energies, the excited triplet state has some absorption probability, leading to excited I atom ($^2P_{1/2}$). This has also been observed experimentally.

Variations on Figure 18.6 can also lead to photodissociation. Some molecules

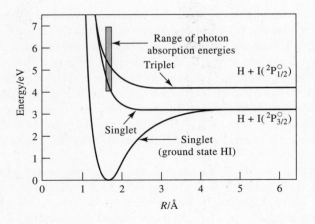

FIGURE 18.6 The continuous absorption spectrum of HI is due to the unbound (photodissociated) nature of the final states of the transition. Two such states, a singlet and a triplet, are important in the ultraviolet region of the spectrum.

---

[5]There is also a triplet repulsive state connected to the ground-state atoms, but it can be ignored for our purposes here.

can be excited to a state that is bound, but crossed, or mixed, with another state that is repulsive:

$$AB + h\nu \rightarrow AB^* \text{ (bound)} \rightarrow AB^{**} \text{ (repulsive)} \rightarrow A + B .$$

This process features a *nonradiative transition* (the $AB^* \rightarrow AB^{**}$ step) and is called *predissociation*, a funny term that is meant to imply that dissociation is preordained to occur as a result of a process following absorption. We will have much more to say about nonradiative transitions in Chapter 20.

Some photodissociation events occur with chemical-bond rearrangement. For example, the process

$$CH_4 + h\nu \rightarrow CH_2 + H_2$$

has been observed, and so-called 1,2 eliminations of the type

$$H_2C=CHF + h\nu \rightarrow HCCH + HF$$

have been studied intensely since they can produce HF in states suitable for operation as a *chemical laser*, about which we will have more to say in later chapters. In both of these processes, two bonds are broken and at least one new bond is formed. It is rare, but not unknown, for more than two bonds to break as a result of single photon absorption. For example, the process

$$Fe(CO)_5 + h\nu \rightarrow Fe + 5CO$$

has been observed in the vacuum UV.

In the photodissociation processes discussed so far, absorption has changed the molecule's electronic state. With the advent of intense laser sources in the infrared, however, a new type of photodissociation process that does *not* change electronic state has been discovered and studied in detail. Called *infrared multiphoton dissociation*, several photons are rapidly absorbed, any one of which has significantly lower energy than a bond energy. The best studied example is

$$SF_6 + nh\nu \rightarrow SF_5 + F .$$

Here, each photon has an energy ~940 cm^{-1} (provided by a CO$_2$ laser), but about 33 000 cm^{-1} are required to break the F$_5$S—F bond. The number of absorbed photons n is about $33\,000/940 \cong 35$ *at least*. During the brief (≤ 100 ns), intense laser pulse, all 35-plus photons are absorbed. The energy accumulates in the molecule, photon by photon, because the absorption rate is greater than the rate of any process that can remove energy. Spontaneous emission in the infrared is relatively slow, due to the ν^3 factor in A_{21} (Eq. (18.14b)), and at the gas pressures used in these experiments, the time between molecule collisions is also long compared to the laser pulse duration. Once the molecule has absorbed a sufficient number of photons to reach the bond-energy threshold, dissociation can occur. The rate of dissociation is small at threshold so that there is typically time to absorb an excess dozen or so photons, but the rate increases very rapidly as the total molecular energy is increased.

CHAPTER 18 SUMMARY

This chapter has discussed four major ideas:

1. light as a source of electromagnetic energy:

   general radiation energy density:
   $$\rho = \frac{1}{2}\epsilon_0(E^\circ)^2 = \frac{I}{c}$$

   blackbody energy density distribution:
   $$\rho(\nu; T)d\nu = \frac{8\pi h\nu^3}{c^3}\frac{1}{e^{h\nu/k_BT}-1}d\nu$$

2. the three matter–radiation interactions (absorption, stimulated emission, and spontaneous emission):

   relation between absorption and stimulated emission:
   $$B_{12} = \left(\frac{g_2}{g_1}\right)B_{21}$$

   relation between absorption and spontaneous emission:
   $$A_{21} = \left(\frac{8\pi h\nu_{12}^3}{c^3}\right)\left(\frac{g_2}{g_1}\right)B_{12}$$

   relation between absorption and transition dipole moment:
   $$B_{12} = \frac{|p_{12}|^2}{6\epsilon_0\hbar^2}$$

   relation between spontaneous emission and radiative lifetime:
   $$\tau_{\text{rad}} = \frac{1}{A_{21}}, \qquad N_2(t) = N_2^0\, e^{-t/\tau_{\text{rad}}}$$

3. atomic spectroscopy selection rules:

   angular momentum selection rules:
   $\Delta L = 0, \pm 1$ (but $\Delta l = \pm 1$ and $L = 0 \leftrightarrow L = 0$)
   $\Delta S = 0$
   $\Delta J = 0, \pm 1$ (but $J = 0 \leftrightarrow J = 0$)

   parity selection rules:
   odd ↔ even but
   odd ↮ odd and even ↮ even

4. two photochemical processes with long-lasting chemical consequences: photoionization and photodissociation.

The central molecular quantity in this chapter is the transition dipole-moment matrix element, which measures the ability of a molecule or atom to undergo a particular quantum-state change as a result of a radiative event:

transition dipole moment:
$$\mathbf{p}_{12} = \int \psi_1^{0*}\, \hat{\mathbf{p}}\, \psi_2^0\, d\tau$$

The next chapter focuses on molecular spectroscopy and its use as a high precision probe of molecular structure. For atoms, the electronic state's energy, term symbol, and configuration are the focal points of spectroscopy. These same quantities apply to molecules, but as a bonus, molecular spectroscopy also yields interatomic distances, bond angles, bond force constants, and a number of probes of electron distributions that, taken together, define our mental pictures of molecules.

FURTHER READING

The tables of atomic energy levels described in Chapter 13 are obviously a rich source of data for atomic spectroscopy. A related set of tables is *Atomic Transition Probabilities,* by W. L. Wiese, M. W. Smith, and B. M. Glennon (U.S. Department of Commerce, NSRDS-NBS, 2 vol., 1966–69), a critically evaluated tabulation of information about Einstein A coefficients and related quantities. Problems 18.11 and 18.12 take you through a calculation of these tables' first entry.

Wavelengths and Transition Probabilities for Atoms and Atomic Ions, by J. Reader, C. H. Corliss, W. L. Wiese, and G. A. Martin (NSRDS-NBS 68, 1980), is a similar compilation, both in the National Standard Reference Data Series.

A classic, readable, and still useful account of atomic spectroscopy is *Atomic Spectra and Atomic Structure,* by Gerhard Herzberg (Dover, New York, 1944). This book covers many aspects of atomic spectra that are not discussed here.

PRACTICE WITH EQUATIONS

18A A common commercial research laser produces a radiation energy density 1.06×10^{-3} J m^{-3}. What is its electric field amplitude? (18.6)

ANSWER: 15.5 kV m^{-1}

18B The energy density in 18A is attained if the laser illuminates a 2 mm radius spot. What is its power? (18.7)

ANSWER: 4.0 W

18C The average Einstein A_{21} coefficient for the $n = 2$ to $n = 1$ transitions in atomic H is 4.699×10^8 s^{-1}. What is the Einstein absorption coefficient B_{12} for this transition? (18.14), (12.52)

ANSWER: $B_{12} = 2.025 \times 10^{20}$ m^3 J^{-1} s^{-2}; ($g_2 = 8$, $g_1 = 2$, $\nu_{12} = 2.467 \times 10^{15}$ s^{-1})

18D What is the average transition dipole moment for the transition in 18C? (18.21)

ANSWER: 1.094×10^{-29} C m

18E If 10^{12} $n = 2$ H atoms are created at $t = 0$, how many remain 10 ns later? (18.22)

ANSWER: 9.1×10^9

18F Deduce the parity of the following configurations: (a) $3d^14s^1$, (b) $2s^22p^5$, (c) $3s^23p^2$, (d) $3d^{10}4f^1$. (basis for Laporte's rule)

ANSWER: even: (a), (c)

18G Which of the following transitions are allowed in atomic H: (a) $7s \rightarrow 4s$, (b) $7s \rightarrow 9p$, (c) $7f \rightarrow 9p$, (d) $125s \rightarrow 124p$, (e) $5d \rightarrow 6p$. (atomic selection rules)

ANSWER: allowed: (b), (d), (e)

PROBLEMS

SECTION 18.1

18.1 It is easy to build microwave radiation sources that have powers in the range of a few watts. What power should such a source have to produce a flux of 1 mole of photons cm^{-2} s^{-1} when operating at $\nu = 10$ GHz?

18.2 Your eye has a maximum sensitivity for green light around 555 nm and can detect a radiation intensity as small as 2×10^{-16} W cm^{-2}. How many photons of 555 nm light would strike your eye each second at this intensity? (You may want to look in a mirror and estimate the area of your eye's pupil.)

18.3 Pulsed lasers can have seemingly outrageous specifications at first glance. For instance, commercial pulsed CO_2 lasers can have power specifications of many MW. Obviously, these lasers have to be pulsed, since a continuous MW power requirement would tax most power stations. This specification thus refers to the power of a single laser pulse of duration ~5 ns. How many photons (of wavenumber ~1000 cm^{-1} for this laser) are in a 10 MW pulse? If the laser fires ten of these pulses each second, what is its average power? Another useful measure of the intensity of a pulsed source is the *fluence* of the pulse, the energy per unit area. What is the fluence of this laser pulse if it is focused to a 1 mm^2 spot?

SECTION 18.2

18.4 The long-wavelength limiting expression of the Planck blackbody distribution, Eq. (18.8), is called the Rayleigh–Jeans equation. It was derived from classical physics, and its failure to explain the experimental distribution was an early sign that classical physics was incomplete. Derive this expression by considering Eq. (18.8) in the classical limit $h\nu \ll k_BT$. (Why is this the classical limit?) As a check, you should find $\rho \, d\nu \propto \nu^2 \, d\nu$, and your answer should not contain h. After all, it was derived (along different lines) before Planck discovered h!

18.5 Commercial infrared spectrometers use a "glowbar" as their light source. These are coiled conductive ceramics through which current is passed heating them until they "glow" like a stove-top burner. What is the theoretical minimum power needed to operate a glowbar with a 10 cm^2 area at 1500 K?

18.6 This problem fills in the missing steps in the text's comparison between a He/Ne laser and a blackbody. The laser radiation is narrowly centered around 632.8 nm or $\nu_0 = 4.738 \times 10^{14}$ Hz, but with a frequency spread $\Delta\nu$ around ν_0 that is typically 3×10^8 Hz. The energy density of the blackbody over $\Delta\nu$ can be approximated by $\int\rho(\nu)d\nu \cong \rho(\nu_0)\Delta\nu$ since $\Delta\nu$ is so small. Divide this by c to get the intensity, multiply the intensity by the surface area of a 1 m radius sphere, divide by the energy of one 632.8 nm photon, and you have the number of photons illuminating the interior of a 1 m sphere with the blackbody at the center. Multiply this number by the ratio of the area of a 1 mm radius spot to the surface area of the sphere, and you have the number of photons per second striking the spot, 6.8×10^{11} photons s^{-1}.

18.7 We can turn the previous problem around a bit and find another very impressive measure of the brightness difference between a blackbody and a laser. The laser of Example 18.1 generated 3.2×10^{15} photons s^{-1} at the illuminated spot. If we increase the temperature of the blackbody, its radiant intensity in the laser's bandwidth will increase. What blackbody temperature is needed to match the laser's photon flux at the spot? (You will have to use successive approximations to find this temperature. Start with a guess around 7500 K.) Your answer will be higher than is experimentally possible, but remember that this laser is neither remarkably bright nor narrow in frequency spread compared to research lasers that can be several thousand

18.8 In 1905, R. W. Wood, one of the pioneers of spectroscopy, first observed what he termed *resonance radiation*. If you sprinkle some NaCl into a flame, you will see the characteristic yellow light of Na emission. Wood was the first to observe that Na atoms could absorb this yellow light and re-emit it in all directions. He vaporized Na in an evacuated tube and illuminated it with the yellow light from a gas flame spiked with a NaCl solution's mist. He observed the yellow light emerging from the tube at the point where the illumination entered. If he raised the Na vapor pressure, the cone of fluorescence shrank back toward the wall of the tube to the point of illumination. Explain this observation. Why does increasing the Na vapor pressure decrease the length of the resonant fluorescence region?

SECTION 18.3

18.9 Problems 15.4 and 15.5 point out that the permanent dipole moment of a system with a net charge depends on the origin of the system. Example 18.2 used $x = 0$ as an origin for an electron in a box extending from $x = 0$ to L, and this is a charged system. The center of mass is the point from which a dipole moment of a molecular ion is measured, and for this system, the center of mass is at $L/2$. Show that a shift of coordinate origin for the dipole moment operator (so that it is $-e(x - L/2)$ instead of $-ex$) does *not* change the results of that Example.

18.10 Follow Example 18.2 and calculate the spontaneous emission rate for an electron in a box of length L going from state $n = 3$ to $n = 2$. You will need the integral

$$\int_0^\pi y \sin(2y) \sin(3y) \, dy = -\frac{24}{25}.$$

Can state $n = 3$ radiate to state $n = 1$? (*Hint:* Problem 18.9 shows that the transition dipole moment is independent of the location of the origin. Put the origin at the midpoint of the box, and sketch the $n = 1$ and $n = 3$ wavefunctions along with the dipole moment function, $-ex$, and look carefully at the symmetry of their product, the transition moment integrand.)

18.11 This and the following problem consider the lowest energy absorption allowed for ground state H. First, show that only light polarized along the z direction can cause a transition from 1s to $2p_z$. Use symmetry arguments about the transition dipole-moment integral to show this, and then evaluate the integral. You will need the 1s and $2p_z$ wavefunctions from Chapter 12, you should write the dipole moment operator, $-ez$, as $-er \cos \theta$, and you will need the integrals

$$\int_0^\pi \cos^2 \theta \sin \theta \, d\theta = \frac{2}{3} \quad \text{and} \quad \int_0^\infty x^4 e^{-x} \, dx = 24.$$

At some point, it will be convenient to make the variable switch $x = 3r/2a_0$.

18.12 In the previous problem, you should have found that the transition dipole moment for the $1s \rightarrow 2p_z$ H transition is 6.315×10^{-30} C m. Use this value to find the Einstein spontaneous emission coefficient A_{12} for this transition. Your answer should be 6.258×10^8 s^{-1}. Problem 18C states, "The average Einstein A_{21} coefficient for the $n = 2$ to $n = 1$ transitions in atomic H is 4.699×10^8 s^{-1}." This number is the first entry in the tabulation *Atomic Transition Probabilities* and it is three-fourths what you calculate here. Can you explain the factor 3/4? Pay attention to the wording of Problem 18C, and think about the degeneracies of $n = 1$ and $n = 2$ and the role of light polarization.

18.13 A low pressure sample of Ne is placed in the beam line of a synchrotron, and the synchrotron light is tuned to a strong vacuum UV absorption at 73.59 nm. This light source is pulsed with a pulse duration of a few ps. The arrival of a pulse starts a fast Ne fluorescence detector that measures the atomic fluorescence intensity as a function of time. The experiment is repeated for many pulses to average this signal. At the end of the experiment, the fluorescence decay curve is plotted:

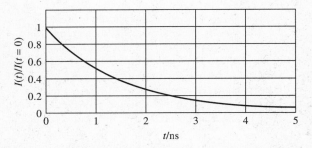

What is the radiative lifetime of this excited state? What is its spontaneous emission rate coefficient?

18.14 The Einstein A coefficient for the hyperfine transition $F = 1 \rightarrow F = 0$ in ground-state H (see Section 15.2) is 2.87×10^{-15} s^{-1}. What is the radiative lifetime of the $F = 1$ level in units of years? This transition has been observed in emission from interstellar H atoms. What does τ_{rad} tell you about the rate at which collisions can relax the $F = 1$ state?

18.15 For each of the following types of transitions, assume $p_{12} \cong 1.6 \times 10^{-29}$ C m (i.e., unit charges separated by 1 Å), and calculate the radiative lifetime of each. The wide range of your answers has important implications for experiments in each frequency region.

(a) an X-ray atomic transition ($\Delta E \cong 2 \times 10^5$ cm^{-1})
(b) a visible transition ($\Delta E \cong 2 \times 10^4$ cm^{-1})
(c) an IR transition ($\Delta E \cong 10^3$ cm^{-1})
(d) a microwave transition ($\Delta E \cong 1$ cm^{-1})

18.16 It is possible to produce laser light pulses that are incredibly brief, on the order of 10^{-14} s = 10 fs (femtoseconds). How many cycles of light of nominal wavelength 600 nm are in a 10 fs pulse? How long (in space) is this pulse? How precisely can you specify this pulse's frequency? Express this in terms of $\Delta \nu / \nu_0$ where $\Delta \nu$ is the frequency uncertainty and ν_0 is the nominal frequency. In contrast, CW (*Continuous Wave*) lasers that are not pulsed can attain a $\Delta \nu / \nu_0$ precision $\sim 10^{-10}$ or less.

18.17 The lowest five states of Cd, in order of increasing energy, are $5s^2$ $^1S_0$ (the ground state), 5s5p $^3P_0$, 5s5p $^3P_1$, 5s5p $^3P_2$, and 5s5p $^1P_0$. Assign each state's parity, and decide which excited state or states have an allowed transition from the ground state. Cadmium is a fairly heavy atom, and we might expect spin–orbit coupling to make S and L less reliable quantum numbers. If only changes in J govern Cd's selection rules, how does your answer change?

18.18 Laporte's rule is sometimes stated, "Transitions that do not change atomic configuration are forbidden." Is this a valid, complete statement? Why or why not?

18.19 The *Atomic Energy Levels* compilation discussed in Chapter 13 lists the following first eight entries for N:

State	Energy/cm^{-1}	State	Energy/cm^{-1}
$2p^3$ $^4S^\circ_{3/2}$	0.0	$2p^3$ $^2D^\circ_{5/2}$	19 223.0
$2p^3$ $^2D^\circ_{3/2}$	19 231.0	$2p^3$ $^2P^\circ_{3/2}$	28 840.0
$2p^3$ $^2P^\circ_{1/2}$	28 840.0	3s $^4P_{1/2}$	83 285.5
3s $^4P_{3/2}$	83 319.3	3s $^4P_{5/2}$	83 366.0

What is the longest wavelength dipole allowed absorption of ground state N? What would the emission spectrum of the $2p^3$ $^2P^\circ_{3/2}$ state look like? the 3s $^4P_{5/2}$ state?

18.20 The green color of some fireworks is due to an allowed emission in Ba that ends in its ground state. The transition wavelength is 553.7 nm. Suggest the configuration and term symbol of the excited state in this transition.

18.21 In the course of your work for the EPA, you are asked to devise a trace analysis method for the toxic metal thallium. You decide to try atomic spectroscopy and convince the agency to buy a tunable laser for your work. Budget restrictions, however, limit you to a laser that can only generate wavelengths longer than 350 nm. You look up Tl in a standard table of energy levels and find that the lowest few levels and their energies are: $6s^26p$ $^2P^\circ_{1/2}$, 0 cm^{-1}; $6s^26p$ $^2P^\circ_{3/2}$, 7793 cm^{-1}; $6s^27s$ $^2S_{1/2}$, 26 478 cm^{-1}; $6s^27p$ $^2P^\circ_{1/2}$, 34 160 cm^{-1}; and $6s^27p$ $^2P^\circ_{3/2}$, 35 161 cm^{-1}. Can you still do the experiment? Where would you tune your laser?

18.22 Consider transitions in atomic C that change the configuration from $2s^22p^2$ to $2s^12p^3$. Transitions of this type are called *inner shell* transitions since the excited electron is not one of the outermost. Work out the term symbols (ignore J) for the $2s^12p^3$ configuration (some are in Table 13.3, but you should find six in all). Which of these term symbols represent states that can be accessed by allowed transitions from the $2s^22p^2$ ^3P ground-state term?

18.23 The radiative lifetime of H(2s $^2S_{1/2}$) is 0.125 s while that of H(2p $^2P_{1/2}$), at essentially the same energy, is 1.60 ns, about 10^8 times less. Explain this difference and suggest a radiative mechanism for the 2s $^2S_{1/2}$ state.

SECTION 18.4

18.24 Ground state Ca has a very broad absorption centered at 148 nm. Its width, in wavenumber units, is about 500 cm^{-1}. Use this width to calculate the lifetime of the excited state and suggest a reason why it is so short. (Table 7.3 has a clue.)

18.25 Autoionization ideas can be applied to excited negative ions falling apart to the neutral atom and an electron. The ground state of He$^-$, $1s^22s^1$, is very unstable, as expected, but excited He$^-$ states with the term symbol ^4P have unusually long (~ 10 μs) autoionization lifetimes and can be observed. What is the lowest energy electron configuration that can lead to a ^4P term? What state of He would be a good candidate for an experiment designed to attach an electron to He and produce this anion configuration?

18.26 Fluorine gas, F_2, is very pale yellow because it has a very weak continuous absorption that starts in the blue part of the spectrum and extends (and increases in intensity) into the UV. It results in F_2 dissociation. The absorption peaks around 300 nm, and the F_2 bond energy is 1.56 eV. If F_2 is irradiated at 300 nm, how much excess kinetic energy will the two F atoms have? What wavelength should you use to give the atoms 2 eV excess kinetic energy? (Assume the F atoms are formed in their ground state.)

18.27 Hydrogen bromide has a ground-state dissociation energy of 3.75 eV. If HBr is irradiated at 185 nm, it photodissociates, and the H-atom fragments are found to have a kinetic energy of 2.96 eV. At 248 nm, this drops to 1.25 eV. Since H is so much lighter than Br, these kinetic energies accurately reflect the total excess energy of the photodissociation. Is the Br atom left in its ground state or its first excited state in either of these processes? See Example 13.5 for data on Br's lowest two states.

GENERAL PROBLEMS

18.28 Experiments that use more than one laser are quite common now, since two photons can populate states that are inaccessible by one photon. For example, the ground state of Cu is $^2S_{1/2}$ and the first excited states are $^2D_{3/2}$ and $^2D_{1/2}$ at 11 203 and 13 245 cm^{-1}, respectively. Since an S → D transition is forbidden ($\Delta L = 2$), the D states cannot be reached by one photon from the ground state. The next excited states, however, are $^2P^°_{1/2}$ and $^2P^°_{3/2}$ at 30 535 and 30 784 cm^{-1}, respectively. Suppose you have one laser that can be widely tuned around 325 nm and another that can be tuned anywhere from 500 to 600 nm. Explain how these two lasers could populate either one of the D states from the ground state, and give explicit tuning wavelengths for each laser. (*Hint:* This technique is more commonly used for molecules where it is known as *Stimulated Emission Pumping*.)

18.29 Photons do not interact with each other (except very rarely at huge photon energies). Thus, blackbody radiation should behave like an ideal gas, and it does, but with two major differences from a particle ideal gas. First, the photon gas always moves at the speed of light, and relativistic effects are the whole story of its kinematics. Second, photons can be created and destroyed at will. There is no "closed system" for photons. Nevertheless, the thermodynamics of a photon gas are very simple to derive. The total internal energy in volume V is just $V\rho(T) = \eta V T^4 = U$. It can be shown that the free energy is $G = -U/3$. Find the photon-gas entropy from $S = -(\partial G/\partial T)$ and show that an adiabatic compression of this gas follows the path VT^3 = constant (contrast to Eq. (2.28)). Then find its pressure from $P = -(\partial G/\partial V)$ and show that its equation of state is $PV = U/3$.

CHAPTER 19

Molecular Spectroscopy of Small Free Molecules

CHAPTER 16 discussed how X-ray diffraction measures molecular structure. In this chapter, we see how to measure molecular structure in ways very different from diffraction. Diffraction techniques require a large number of molecules, packed with regularity in a crystal. In contrast, *single* molecules, free in the gas phase, are our targets in this chapter, and *discrete* photon absorption and emission events will carry all the information.

The chapter's title includes the adjective "small" as well as "free." What constitutes a "small" molecule? Surely H_2O is small and DNA is not, but is benzene small? Is NO_2? And why make a size distinction at all? These questions will be addressed as the chapter progresses. It is true that the very high precision and often exquisite sensitivity of modern molecular spectroscopic techniques allow us to measure, say, bond lengths to very high precision, even for elusive species such as the hydroxyl radical, OH. But it is also true that we *cannot* measure the O—H bond length of any hydroxyl group stuck somewhere in DNA to anywhere near the same precision, nor do we currently have a need to. This is one of the great strengths of chemistry—the ability to transfer information from one molecule to another with a near certainty that the transfer will be correct. We study OH, H_2O, CH_3OH, and a few other small molecules in detail, and, *voilà*, we can make a picture of DNA that is useful and valid.

In the introduction to Chapter 6, we quoted from the 1828 chemistry text of J. S. Forsyth, who, when speaking of "caloric," said that

19.1 The Classification of Spectroscopies by Molecular Motions

19.2 Small Molecule Energy Levels

19.3 Molecular Structure as Deduced by Spectroscopy

19.4 Potential Energy Surfaces

19.5 Line Shapes and Spectral Congestion—What's Observable?

"in the hands of chemists it is the most powerful agent they are acquainted with." It is fair to say that molecular spectroscopy and its extensions into various analytical tools have replaced "caloric" and become "the most powerful agent" of today's chemists. To put this agent in your hands, we will see first how molecular motions naturally group themselves into different types with characteristic energies. Then, we will use familiar model systems from Chapter 12 to write explicit energy expressions in terms of molecular parameters and quantum numbers. The third section will show how the parameters can be measured and related to molecular structure, while the fourth extends this idea to discuss the entire intramolecular force field, or potential-energy surface. Finally, we confront the reality facing most of modern spectroscopy: the problem of too many observable transitions in many systems.

19.1 THE CLASSIFICATION OF SPECTROSCOPIES BY MOLECULAR MOTIONS

Just as we did in Chapter 12 for our model quantum mechanical systems, we start a study of molecular spectroscopy with a classical analogy. The Born–Oppenheimer approximation allows us to discuss the motions of heavy nuclei in a different context from those of the light and rapid electrons. We assume that the electrons' motion has been treated with the electronic structure techniques of Chapter 14.

Picture H$_2$O, for example, as three spheres, one 16 times the mass of the other two, all connected by a collection of springs. These springs are arbitrary; they simply keep the masses from flying apart if the whole "molecule" is tossed around. After all, the only reason we picture H$_2$O like this

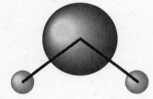

instead of any of the following

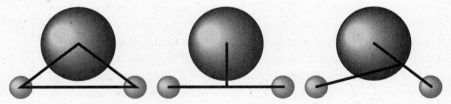

is because we have been taught to do so on the basis of the very successful theories of electronic structure in Chapter 14. But, if we are talking about the relative motion of the nuclei in general terms, any of these four pictures will do.

What we seek, classically, is the position in space of all the atoms as a function of time. Suppose there are N atoms in the molecule, each located by x, y, and z coordinates, so that there are $3N$ coordinates in all. The *center of mass* of the molecule has coordinates X, Y, and Z such that

$$\sum_{i=1}^{N} m_i(x_i - X) = 0, \quad \sum_{i=1}^{N} m_i(y_i - Y) = 0, \quad \sum_{i=1}^{N} m_i(z_i - Z) = 0 \quad (19.1)$$

where m_i is the mass of atom i. We follow the center of mass of the molecule and measure the atoms' motion relative to it. Thus, of the $3N$ total coordinates, three of them follow the entire molecule as if it were a single particle of mass $M = \sum m_i$.

EXAMPLE 19.1

The equilibrium H—O—H bond angle in water is 104.5°, and the equilibrium O—H bond length is 0.958 Å. Where is the center of mass of water?

SOLUTION The symmetry of H_2O can be used to advantage if we place the molecule in the (x, y) plane (so that all atomic z coordinates are zero) with the O atom at the origin and the x axis bisecting the H—O—H bond angle. This makes $Y = Z = 0$, and X is found from Eq. (19.1):

$$X = \frac{\sum_{i=1}^{3} m_i x_i}{M} = \frac{2 m_H (0.958 \text{ Å}) \cos(104.5°/2)}{2 m_H + m_O} = 0.0656 \text{ Å} \ .$$

Since O is so much heavier than the two H atoms, the center of mass is almost centered on O. The displacement towards the hydrogens is only 0.0656 Å:

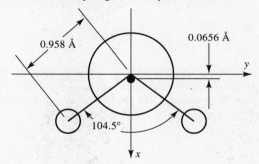

➡ **RELATED PROBLEM** 19.1

This leaves us with $3N - 3$ coordinates. As our model flies about, it will in general tumble also, and classical mechanics tells us how to treat this *rotational* motion. We place a coordinate system $(\bar{x}, \bar{y}, \bar{z})$ at the center of mass and choose an axis orientation so that

$$\sum_{i=1}^{N} m_i \bar{x}_i \bar{y}_i = 0 \quad \text{and} \quad \sum_{i=1}^{N} m_i (\bar{x}_i^2 + \bar{y}_i^2) \neq 0 \ . \tag{19.2}$$

This axis system is called the *principal inertial axis system*, and the symbols (a, b, c) are conventionally used in place of $(\bar{x}, \bar{y}, \bar{z})$. The rotation of the entire molecule follows the rotation of these three axes.

Three angular coordinates are needed to locate the (a, b, c) axis system in space relative to the molecule. We are now down to $3N - 3 - 3 = 3N - 6$ coordinates, except for the special case of *linear molecules*. The non-zero sums of Eq. (19.2) are the three *principal moments of inertia*:

$$I_a = \sum_{i=1}^{N} m_i (b_i^2 + c_i^2)$$

$$I_b = \sum_{i=1}^{N} m_i (a_i^2 + c_i^2) \tag{19.3}$$

$$I_c = \sum_{i=1}^{N} m_i (a_i^2 + b_i^2)$$

For a linear molecule, one of the principal axes will always lie along the atomic axis. By convention, this is called the a axis. Since $b_i = c_i = 0$ for all atoms in a line along the a axis, $I_a = 0$ and $I_b = I_c$. Two angles locate a linear molecule's orientation, the θ, ϕ coordinates of a spherical polar coordinate system, so that $3N - 5$ coordinates instead of $3N - 6$ remain for a linear molecule once the center of mass and rotation angle coordinates are subtracted.

EXAMPLE 19.2

What are the principal moments of inertia for water?

SOLUTION We place the (a, b, c) origin at the center of mass found in Example 19.1 and again orient the axes so as to reflect the molecule's symmetry. By convention, the axes are named so that $I_a < I_b < I_c$, an assignment that can often be made only after the three moments have been found! Here we will use the correct assignment from the start. The O atom is at $(a, b, c) = (0, -0.065\ 6\ \text{Å}, 0)$ while the H atoms are at $(\pm \sin(104.5°/2) \times (0.958\ \text{Å})$, $\cos(104.5°/2) \times (0.958\ \text{Å}) - 0.0656\ \text{Å}, 0) = (\pm 0.757\ \text{Å}, 0.521\ \text{Å}, 0)$. The moments follow directly from Eq. (19.3):

$$I_a = 2m_H(0.521\ \text{Å})^2 + m_O(-0.065\ 6\ \text{Å})^2 = 0.616\ \text{amu Å}^2$$

$$I_b = 2m_H(0.757\ \text{Å})^2 = 1.157\ \text{amu Å}^2$$

$$I_c = 2m_H[(0.521\ \text{Å})^2 + (0.757\ \text{Å})^2] + m_O(-0.065\ 6\ \text{Å})^2 = 1.772\ \text{amu Å}^2\ ,$$

(using 1 amu Å^2 = 1.660 54 × 10^{-47} kg m^2). Symmetry guided alignment of the (a, b, c) axes here, but for nonsymmetrical molecules, it is more difficult to find the unique orientation that makes the first kind of sums in Eq. (19.2) zero. Methods for doing this are discussed in this chapter's references.

➡ **RELATED PROBLEMS** 19.1, 19.2

The remaining $3N - 6$ or $3N - 5$ coordinates all correspond to true intramolecular, interatomic distances. As these coordinates evolve in time, they usually oscillate, corresponding to internal vibrations. We therefore conclude that a free molecule with N atoms has

- three degrees of overall translational freedom,
- two (if linear) or three (if not) rotational degrees of freedom, and
- $3N - 5$ (if linear) or $3N - 6$ (if not) internal vibrational degrees of freedom.

On top of this (and we now return to the quantum world) are electronic degrees of freedom (of arbitrary number) and spin degrees of freedom (both nuclear and electronic). The overall translation will not interest us here, and spin is discussed in Chapters 15 and 21. That leaves us with *electronic, vibrational, and rotational motions*.

We start with a rough idea of the energies characteristic of these motions. Suppose we have a homonuclear diatomic of atomic mass M and bond length R. The order-of-magnitude energy of an electron bound in this molecule is the electron-in-a-box energy. To within factors of order unity, the spacing of electronic energies is (compare Eq. (12.12))

$$\Delta E_{el} \sim \frac{\hbar^2}{m_e R^2}$$

where m_e is the electron mass.

For vibration, we picture the bond as a harmonic spring. Doubling the bond length dissociates or nearly dissociates most bonds, and since bond energies are comparable to electronic-state energy differences, stretching the bond to twice its equilibrium length takes an energy comparable to the electronic energy level spacing. Again ignoring factors near unity, the harmonic potential-energy increase is on the order of $k(\Delta R)^2 \sim kR^2$ if $\Delta R = R$ where k is the spring force constant. This energy is on the order of ΔE_{el} so that we can write

$$k \sim \frac{\Delta E_{el}}{R^2} \sim \frac{\hbar^2}{m_e R^4}.$$

For a harmonic oscillator (see Equations (12.15) and (12.18))

$$\Delta E_{vib} \sim \hbar \left(\frac{k}{\mu}\right)^{1/2} \sim \hbar \left(\frac{k}{M}\right)^{1/2} \sim \frac{\hbar^2}{R^2}\left(\frac{1}{m_e M}\right)^{1/2} \sim \left(\frac{m_e}{M}\right)^{1/2} \Delta E_{el}.$$

The typical zero-point vibration amplitude ΔR_{vib} is such that

$$k\Delta R_{vib}^2 \sim \hbar \left(\frac{k}{M}\right)^{1/2}$$

or, substituting from above for k,

$$\Delta R_{vib} \sim \left(\frac{m_e}{M}\right)^{1/4} R.$$

Finally, rotational energies are approximately spaced by 3-D rigid-rotor energies. (See Example 12.7 and Eq. (12.42).) We have the order of magnitude relation

$$\Delta E_{rot} \sim \frac{\hbar^2}{MR^2} \sim \left(\frac{m_e}{M}\right) \Delta E_{el}.$$

If we compare these expressions, we see that the following rough proportionalities hold:

$$\Delta E_{el} : \Delta E_{vib} : \Delta E_{rot} :: 1 : (m_e/M)^{1/2} : (m_e/M) \quad \textbf{(19.4)}$$

and that zero point vibrational excursions are on the order of

$$\Delta R_{vib} \sim (m_e/M)^{1/4} R. \quad \textbf{(19.5)}$$

All of these relations are controlled by that magic ratio at the heart of the Born–Oppenheimer approximation, the electron-mass/nuclear-mass ratio.

We can test these relations with H_2 and I_2, a light and a heavy diatomic. For each ΔE, we take an average value for the first few electronic states. We should not expect fantastic agreement; after all, those many $\sim$ signs represent ignoring factors that could be on the order of 0.1–10 or so, and our basic models are pretty crude pictures of a diatomic bond. Nevertheless, the numbers are convincing:

	H_2	I_2
ΔE_{el}	100 000 cm^{-1}	15 000 cm^{-1}
ΔE_{vib}	2500 cm^{-1}	120 cm^{-1}
ΔE_{rot}	30 cm^{-1}	0.03 cm^{-1}
(m_e/M)	5.5×10^{-4}	4.3×10^{-6}
$\Delta E_{rot}/\Delta E_{el}$	3×10^{-4}	2×10^{-6}
$(m_e/M)^{1/2}$	0.023	2.1×10^{-3}
$\Delta E_{vib}/\Delta E_{el}$	0.025	8×10^{-3}

What is most remarkable about the numbers above is not so much the agreement between simple theory and observation but rather the numbers' *magnitudes*. These tell us that *rotational, vibrational, and electronic transition energies occur in isolated regions of the spectrum* separated by factors of 100 to 1000 in photon energy. The spectroscopy of pure rotational transitions occurs in the ~0.03–30 cm^{-1} region, or through the microwave and into the far infrared. (See Table 18.1.) Vibrational spectroscopy occurs in the mid infrared (~300–5000 cm^{-1}) and, just as for the majority of atomic spectroscopy, molecular electronic spectroscopy occurs in the visible through vacuum UV regions (~10^4–10^5 cm^{-1}).

Spectroscopy involving changes in *rotational* energy give us our most detailed *structural* information, but not without some effort and not in every molecule. The major problem is this: we can find values for, at most, the three moments of inertia, but the general polyatomic has many more than three bond lengths and angles. We do not seem to have enough information to find all the quantities of interest. If we can synthesize (or have occurring naturally) *isotopically variant molecules,* then we can measure spectra in which the masses are different (in known ways) but the bond lengths are the same (or very nearly so, as we shall see).

Vibrational spectra can be interpreted in terms of the force field—the strengths of the springs—that binds the atoms together as a molecule. By themselves, vibrational transition energies carry no *rigid* structural information. For polyatomic molecules, the total number of observed vibrational transitions can carry structural information due to *molecular symmetry* effects. We know to expect $3N - 6$ (or $3N - 5$) types of vibrations, but symmetry may cause transitions among one or more of these to be forbidden. They will simply not appear in our spectra. For example, the high symmetry of benzene allows only seven of the $3N - 6 = 30$ possible vibrations to absorb individually.

An additional matter–radiation interaction, *light scattering,* can probe beyond the absorption/emission spectroscopy we have so far discussed. Light can be nearly instantaneously absorbed and re-emitted with only a change in the light's direction of travel. *Elastic* or *Rayleigh scattering* (which accounts for the blue color of the sky) is not too exciting in itself, since the molecule does not exchange energy with the photon. More interesting is the rare event in which the photon and molecule *do* exchange energy. This is called *inelastic* or *Raman scattering,* and it is the *change in energy* of the photon that measures a molecular transition energy. Raman spectroscopy has its own set of symmetry-related selection rules. Continuing the benzene example, another 12 of the 30 vibrational transitions can be observed by Raman spectroscopy, and these are different from the seven allowed in IR absorption. But note that $12 + 7 \neq 30$, so that in benzene there are still 11 vibrations that infrared and Raman spectroscopy cannot observe directly. These are called *silent* vibrations.

Electronic spectroscopy, when done with sufficient resolution to observe the vibrational and rotational transitions that accompany an electronic transition, can provide information on the structures of molecules in electronically excited states. It also plays a role in elucidating the ground-state structure. A simple example is N_2. We will see that N_2 has *no* dipole-allowed rotational *or* vibrational transitions, although it does have a Raman spectrum. But emission from an excited electronic state of N_2 to the ground state can carry an enormous amount of ground-state information.

19.2 SMALL MOLECULE ENERGY LEVELS

Our next task is to derive expressions for molecular energy levels and the quantum numbers and molecular parameters that govern them. We start with diatomic molecules, which can be discussed in detail without enormous effort, and then we will generalize to polyatomic molecules.

Diatomic Molecules

In Chapter 14 we discussed electronic states in terms of MO configurations. We follow the same course used for atoms and introduce *molecular electronic-state term symbols* as adjuncts to MO configurations. As with atoms, one configuration can lead to many states. We will not go into great detail on the derivation of a molecular term symbol from a configuration, but you will recognize many parallels with atomic term symbols.

First, molecular spin multiplicity appears as a leading superscript. But, unlike an atom, in a diatomic only the *component* of the total spin angular momentum along the internuclear axis is relevant. We use a quantum number Σ (a symbol that appears in a totally different context with a totally different meaning in the next paragraph—beware!) for this component. For spin S, Σ varies from S to $-S$ in unit steps. If $S = 0$, the $2S + 1$ multiplicity is 1, and we have a singlet state, which we know to expect from configurations with all electrons paired. In general, S can be integral or half-integral, leading to singlet, doublet, triplet, etc. states.

Electron *orbital* angular momentum likewise has meaning only in terms of its axial component, Λ (capital Greek lambda, in analogy with the atomic L quantum number). The value of Λ can be integral only (including zero), and it is coded into the center of the term symbol just as L is for atoms. Instead of S, P, D, etc., capital Greek Σ, Π, Δ, etc., letters are used. Thus, if $\Lambda = 0$, we have a Σ state (beware!— here is the different context for the symbol Σ mentioned above); if $\Lambda = 1$, we have a Π state; and so on. All states except Σ states are doubly degenerate.

The analog of atomic L-S coupling adds the components Σ and Λ to produce a total component magnitude called Ω (capital Greek omega):

$$\Omega = |\Sigma + \Lambda| . \tag{19.6}$$

If $\Lambda = 0$ (a common case), Σ is not defined, since it is the magnetic field of $\Lambda \neq 0$ electrons that defines the components of Σ, and Ω is not needed. If $\Lambda \neq 0$, there are $2S + 1$ values for Σ, and thus $2S + 1$ for Ω as well.

Just as j-j coupling is appropriate for atoms as they become heavier, Ω can become the dominant quantum number for heavy diatomics. While Ω would seem to be the direct analog of J and should appear in term symbols, the $\Sigma + \Lambda$ sum (which can be negative!) rather than Ω is the subscript to the Λ-coded letter. (See Problem 19.6.)

So far, our term symbol looks very much like that for an atom:

$$^{2S+1}\Lambda_{\Sigma + \Lambda} =$$

- Spin multiplicity
- Letter (Σ, Π, Δ, etc.) representing Λ
- Spin and orbital angular momentum component sum

Two other features of diatomic electronic wavefunctions have to be considered as well. The first is the *g* or *u* inversion symmetry label for the total electronic wavefunction of *homonuclear* diatomics. The *g* or *u* is added to the right of $\Sigma + \Lambda$ as an additional subscript to Λ. The second applies only to $\Lambda = 0$ states (Σ states) and expresses the effect of reflecting the wavefunction in a plane containing the internuclear axis. This operation is a little mysterious, since there is an *infinite number* of such planes, and the wavefunction is difficult to visualize. But the effect of the operation is simple; it either leaves the wavefunction unchanged (a +1 eigenvalue) or else it changes its sign (a −1 eigenvalue). Consequently, Σ states can be either + or −, and this is denoted as a following superscript: Σ^+ or Σ^-.

The vast majority of stable diatomic molecules have an even number of electrons, all spin-paired in closed MOs in their ground electronic states. These states are all

$^1\Sigma_g^+$ (read "singlet sigma g plus") if homonuclear or $^1\Sigma^+$ if heteronuclear. Note that the $\Sigma + \Lambda$ subscript is left off for $^1\Sigma$ states, since $\Omega = 0$ is obvious. These diatomics are the analogs of the $^1S_0$ atomic ground states of the rare gases.

A Closer Look at + and − States

We go into some of the finer points of the $+, -$ symmetry label here. This symmetry is related to electronic orbital angular momentum. If $\hat{L}_z$ is the operator for the z component (along the internuclear axis) of an electron's orbital angular momentum, then for any one MO, Φ_i, the eigenvalue relation

$$\hat{L}_z \Phi_i = \lambda \hbar \Phi_i$$

will hold where $|\lambda| = 0, 1, 2, \ldots$ for $\sigma, \pi, \delta, \ldots$ MOs. The λ quantum number is just the individual electron's contribution to Λ.

If ϕ is the cylindrical polar angle about the z axis, $\hat{L}_z = (\hbar/i)(\partial/\partial\phi)$, and (see Eq. (12.36)) the ϕ dependence of Φ_i is $\exp(i\lambda\phi)$. Let $\hat{R}_z$ be the operator for reflection in any plane perpendicular to the z axis. This operator turns the ϕ coordinate of a point into $-\phi$, so that $\hat{R}_z \Phi_i(\phi) = \Phi_i(-\phi)$. If the result of $\hat{R}_z$ operating on Ψ, the *total* electronic wavefunction, is $\hat{R}_z \Psi = \Psi$, we have a + state; if $\hat{R}_z \Psi = -\Psi$, we have a − state.

Now for the hard part: Ψ is, at the very least, an antisymmetric combination of spatial-spin functions. Thus, $\hat{R}_z$ will operate on a number of sums of products of one-electron MOs ϕ_i. If ϕ_i is a σ MO, $\lambda = 0$ and $\hat{R}_z \sigma = +\sigma$. For any $\lambda \neq 0$, we have doubly degenerate pairs of MOs corresponding to $+|\lambda|$ and $-|\lambda|$. If one of these is $\exp(i\lambda\phi)$, then, since $\hat{R}_z \exp(i\lambda\phi) = \exp(-i\lambda\phi)$, $\hat{R}_z$ turns one of the pair into the other, as if we had replaced λ with $-\lambda$ instead of replacing ϕ with $-\phi$.

At this point, we can see some general rules:

1. If $\Lambda \neq 0$, the $+/-$ symbol is *not needed,* since the + and − states are degenerate.
2. If, outside a closed core of MOs, there is a single σ electron, the state will always be Σ^+, since $\hat{R}_z \sigma = +\sigma$ (and $\hat{R}_z$ (closed core MO) = + (closed core MO)).
3. A Σ^- state, therefore, must have an *MO configuration with non-σ MOs less than fully occupied.*

From Table 14.3, we find that B_2 and O_2, of all the first row homonuclear diatomics, fulfill rule 3. B_2 has $1\pi_u^2$ and O_2 has $1\pi_g^{*2}$ occupancy beyond closed core subshells. For either of these, we can ignore the u, g difference for the moment and concentrate on the doubly degenerate 1π MO pair. We call one of these (the one with $\lambda = +1$) $1\pi^+$ and the other ($\lambda = -1$) $1\pi^-$. We can assign two electrons and their spins to these MOs in four ways. This gives us *three* electronic states (since $^1\Delta$ is doubly degenerate):

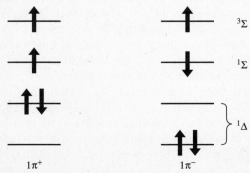

The first two are Σ states since $\lambda = +1$ for $1\pi^+$ and -1 for $1\pi^-$, making $\Lambda = 1 - 1 = 0$. The $^1\Delta$ state does not need $+$, $-$ notation ($\Lambda \neq 0$), but what about the $^3\Sigma$ and $^1\Sigma$ states?

If the *spin* part of Ψ is a triplet and thus symmetric with respect to exchange (compare Equations (13.23a–d)), the spatial part must be antisymmetric and look like this somewhere in its expansion:

$$\Psi(^3\Sigma) \propto [1\pi^+(1)1\pi^-(2) - 1\pi^+(2)1\pi^-(1)] \ .$$

Since $\hat{R}_z(1\pi^+) = 1\pi^-$ and vice versa, $\hat{R}_z\Psi(^3\Sigma) = -\Psi(^3\Sigma)$ and the state is a $-$ state, $^3\Sigma_g^-$, in fact, for both B_2 and O_2. The $^1\Sigma$ state, however, has an antisymmetric (singlet) spin part and thus a symmetric spatial part:

$$\Psi(^1\Sigma) \propto [1\pi^+(1)1\pi^-(2) + 1\pi^+(2)1\pi^-(1)] \ .$$

This means $\hat{R}_z\Psi(^1\Sigma) = +\Psi(^1\Sigma)$, a $+$ state, $^1\Sigma^+$ (also g, for both B_2 and O_2, from the rules $g \times g = u \times u = g$, but $g \times u = u \times g = u$).

The two most important exceptions to the rule that stable diatomics are $^1\Sigma_g^+$ or $^1\Sigma^+$ are NO and O_2. These have the MO configurations

NO: $1\sigma^2 1\sigma^{*2} 2\sigma^2 2\sigma^{*2} 3\sigma^2 1\pi^4 1\pi^{*1}$

O_2: $1\sigma_g^2 1\sigma_u^{*2} 2\sigma_g^2 2\sigma_u^{*2} 3\sigma_g^2 1\pi_u^4 1\pi_g^{*2}$

both of which have closed MOs up to the outer $1\pi^*$ or $1\pi_g^*$ orbital. These unfilled MOs determine the electronic state term symbol. For NO, the state must be a doublet (one unpaired electron spin), and it must be a Π state (since the electron is in a π MO). There can be two Ω values, however, corresponding to $\Sigma = \pm 1/2$ and $\Lambda = 1$ giving $\Omega = 3/2$ if these components add or $\Omega = 1/2$ if they subtract. These two *molecular* spin–orbit states are different in energy, with $^2\Pi_{1/2}$ lower[1] (by about 120 cm^{-1}) than $^2\Pi_{3/2}$.

In O_2, the $1\pi_g^{*2}$ configuration leads to three different molecular states: $^3\Sigma_g^-$, $^1\Delta_g$, and $^1\Sigma_g^+$. Of these, paralleling the atomic Hund's rules, the triplet is lowest, and of the singlets, Δ is lower. (Note the similarity to the ground-configuration states of O: ^3P, ^1D, and ^1S. See Table 13.4.)

EXAMPLE 19.3

What are the expected term symbols for the ground electronic states of NO$^+$ and NO$^-$?

SOLUTION These ions have the ground state configurations

NO$^+$: $1\sigma^2 1\sigma^{*2} 2\sigma^2 2\sigma^{*2} 3\sigma^2 1\pi^4$

NO$^-$: $1\sigma^2 1\sigma^{*2} 2\sigma^2 2\sigma^{*2} 3\sigma^2 1\pi^4 1\pi^{*2}$.

For NO$^+$, all the MOs are fully occupied, and the term symbol is $^1\Sigma^+$. The molecule is isoelectronic to CO. Similarly, NO$^-$ is isoelectronic to O_2, and the $1\pi^{*2}$ electrons govern the term symbol. The ground state is $^3\Sigma^-$. Note that neither term symbol carries a g or u subscript since these are heteronuclear molecules.

➠ *RELATED PROBLEMS* 19.5, 19.6

---

[1] Whenever the state with the smaller of two possible Ω components is lower in energy, the states are said to be *regular*, and often the shorthand $^2\Pi_r$ is used to denote both states. When the converse is true, the states are called *irregular*, and denoted, for example, $^2\Pi_i$, as is the case in the OH radical's ground state, with $\Omega = 3/2$ lower than $\Omega = 1/2$ by about 126 cm^{-1}.

Excited electron configurations can lead to a variety of states, as in the atomic case, and a molecule can have a number of, say, $^1\Sigma_g^+$ states. Even (maybe especially) practicing spectroscopists find it tedious to write the entire symbol, so that a further shorthand has come into use. One calls the ground state X, the first excited state *of the same spin multiplicity* A, followed by B, C, etc. The lowest state of *alternate* spin multiplicity is called a, followed by b, etc. Thus, for O_2 we write X $^3\Sigma_g^-$, a $^1\Delta_g$, and b $^1\Sigma_g^+$ in detail but often speak of "the a to X transition," for example. There is an excited $^3\Sigma_u^+$ state in O_2 called the A $^3\Sigma_u^+$ state, but a somewhat lower triplet state, $^3\Delta_u$, was discovered later, and it is called A′ $^3\Delta_u$. Thus, historical accidents of discovery or reassignment can make even the shorthand sometimes ambiguous.

We now turn to nuclear motion. We will assume that the electronic state is $^1\Sigma^+$ (or $^1\Sigma_g^+$) and that the potential-energy function that governs the nuclear motion is known. For a diatomic with nuclei A and B, bound by potential function $V(R)$, the B–O Hamiltonian for nuclear motion is

$$\hat{H}_{nuc} = -\frac{\hbar^2}{2m_A}\nabla_A^2 - \frac{\hbar^2}{2m_B}\nabla_B^2 + V(R) . \qquad (19.7)$$

$\uparrow$ Nuclear motion Hamiltonian $\quad\uparrow$ Kinetic energy of nucleus A $\quad\uparrow$ Kinetic energy of nucleus B $\quad\uparrow$ Internuclear potential energy function

This has six coordinates (the (x, y, z) coordinates of each nucleus), which we know to be three too many, since we expect only one vibration coordinate (the internuclear distance R) and two rotation coordinates (θ and ϕ). When the center of mass coordinates are eliminated, we have the Hamiltonian for the relative motion of the nuclei in the center of mass reference frame:

$$\hat{H}_{nuc} = -\frac{\hbar^2}{2\mu}\nabla_R^2 + V(R) \qquad (19.8)$$

where μ is the reduced mass

$$\frac{1}{\mu} = \frac{1}{m_A} + \frac{1}{m_B} \quad \text{or} \quad \mu = \frac{m_A m_B}{(m_A + m_B)} . \qquad (19.9)$$

Since V depends only on the internuclear distance R and not on the angular coordinates, the nuclear motion wavefunction, Ψ_N, has an R part and an angular part such that

$$\Psi_N(R, \theta, \phi) = \psi(R) Y_{J,M}(\theta, \phi) \qquad (19.10)$$

where $Y_{J,M}(\theta, \phi)$ is the *spherical harmonic* function tabulated in Table 12.2. The *nuclear rotation* angular momentum quantum number is J, and the quantum number for its projection on some space-fixed axis is M. As usual, $J = 0, 1, 2, \ldots$, so that the molecular rotational angular momentum has the magnitude $[J(J + 1)]^{1/2}\hbar$ (compare Eq. (12.39a)).

Eliminating the angular coordinates in Eq. (19.8) leaves the *radial Schrödinger equation* (RSE) for a diatomic molecule:

$$-\frac{\hbar^2}{2\mu R^2}\left[\frac{d}{dR}\left(R^2 \frac{d\psi}{dR}\right)\right] + \left[\frac{J(J+1)\hbar^2}{2\mu R^2} + V(R)\right]\psi = E\psi . \qquad (19.11)$$

Compare this to Eq. (12.50), the radial equation for the H atom:

$$-\frac{\hbar^2}{2\mu r^2}\left[\frac{\partial}{\partial r}\left(r^2 \frac{\partial \Psi}{\partial r}\right)\right] + \left[\frac{l(l+1)\hbar^2}{2\mu r^2} - \frac{e^2}{4\pi\epsilon_0 r}\right]\Psi = E\Psi . \qquad (12.50)$$

They are virtually identical.

If we define $\chi(R) \equiv R\psi(R)$, we turn Eq. (19.11) into a simpler equation for $\chi(R)$:

$$-\frac{\hbar^2}{2\mu}\frac{d^2\chi(R)}{dR^2} + \left[\frac{J(J+1)\hbar^2}{2\mu R^2} + V(R)\right]\chi(R) = E\chi(R) \ . \qquad (19.12)$$

Each J value gives a different equation, since J is a *parameter* in the effective potential energy, the sum of terms in square brackets in Eq. (19.12). This is the same idea that led to Figure 12.13 for atomic H.

We can expand $V(R)$ as discussed at the end of Section 12.2. Equation (19.5) shows that zero-point vibration is confined to the vicinity of R_e, the equilibrium bond length: $|R - R_e|/R_e \sim (m_e/\mu)^{1/4} = 0.13$ for H_2, 0.038 for I_2, to pick two extremes. We define the relative extension or compression of the bond

$$\rho = R - R_e \qquad (19.13)$$

so that ρ/R_e is small.

We expand $V(R)$ about R_e:

$$V(R) \cong V(R_e) + V'(R_e)\rho + \frac{1}{2}V''(R_e)\rho^2 + \cdots \qquad (19.14)$$

where

$V(R_e) \equiv V_e =$ a constant,

$V'(R_e) \equiv (dV/dR)_{R=R_e} = 0$ since R_e locates a minimum in $V(R)$, and

$V''(R_e) \equiv (d^2V/dR^2)_{R=R_e} = k$, the *bond harmonic force constant* .

If the higher derivatives of V are small, we can ignore them and write

$$V(R) \cong V_e + \frac{1}{2}k\rho^2 \ . \qquad (19.15)$$

Since $dR = d\rho$, we can work ρ and Eq. (19.15) into the RSE:

$$-\frac{\hbar^2}{2\mu}\frac{d^2\chi(\rho)}{d\rho^2} + \left[\frac{J(J+1)\hbar^2}{2\mu(\rho+R_e)^2} + V_e + \frac{1}{2}k\rho^2\right]\chi(\rho) = E\chi(\rho) \ .$$

We approximate the centrifugal-potential term with

$$\frac{1}{(\rho+R_e)^2} = \frac{1}{R_e^2}\left[\frac{1}{(1+\rho/R_e)^2}\right] \cong \frac{1}{R_e^2}\left[1 - 2\left(\frac{\rho}{R_e}\right) + 3\left(\frac{\rho}{R_e}\right)^2 - \cdots\right] \qquad (19.16)$$

and, since $\rho/R_e < 1$, we keep only the first term for now, giving us a twice-approximated RSE:

$$-\frac{\hbar^2}{2\mu}\frac{d^2\chi}{d\rho^2} + \left\{\left[\frac{J(J+1)\hbar^2}{2\mu R_e^2} + V_e - E\right] + \frac{1}{2}k\rho^2\right\}\chi = 0 \ . \qquad (19.17)$$

The terms in square brackets are all constants for any one J, and we define the *vibrational energy* E_{vib} as

$$E_{vib} = E - V_e - \frac{J(J+1)\hbar^2}{2\mu R_e^2} \ . \qquad (19.18)$$

Each of the terms in E_{vib} has a clear physical interpretation:

$E =$ the *total molecular energy* (except for translational energy)

$V_e =$ the *total electronic energy plus nuclear repulsion potential energy* at R_e

$\dfrac{J(J+1)\hbar^2}{2\mu R_e^2} =$ the *rotational energy of a rigid diatomic* with quantum number J .

Rearranging Eq. (19.18) shows the total energy partitioned according to type of motion:

$$E \quad = \quad V_e \quad + \quad E_{\text{vib}} \quad + \quad \frac{J(J+1)\hbar^2}{2\mu R_e^2} \quad (19.19)$$

$\uparrow$ Total internal energy $\quad$ $\uparrow$ B–O electronic energy $\quad$ $\uparrow$ Vibrational energy $\quad$ $\uparrow$ Rotational energy

The RSE now looks familiar:

$$-\frac{\hbar^2}{2\mu}\frac{d^2\chi}{d\rho^2} + \frac{1}{2}k\rho^2\chi = E_{\text{vib}}\chi \ . \quad (19.20)$$

This is the Schrödinger equation for a harmonic oscillator, just like Eq. (12.17). We can immediately write down the energy eigenvalues, (from Eq. (12.18)), but before we do, we should think more about our approximations.

We have made what is called the *rigid rotor, harmonic oscillator approximation,* and these two terms are contradictory. If the rotor is rigid, how can it oscillate? If the oscillations are small, as they are for vibrations near the bottom of the potential well, the rotor bond length is always close to R_e (i.e., *almost* rigid).

Also, the true radial wavefunction is not $\chi(\rho)$, but $\psi(R) = \chi(R)/R$, and the range of R is restricted to $0 \le R \le \infty$, or $-R_e \le \rho \le \infty$, which is different from the harmonic oscillator coordinate range. We can safely imagine ρ to vary from $-\infty$ to ∞, however, since $\chi(\rho)$ will approach zero very rapidly as $\rho \to -R_e$, on the inner tunneling region of the true potential.

Thus, this approximation is justified, and it implies the following quantum numbers:

$J =$ nuclear rotational quantum number,[2]

$v =$ vibrational quantum number, and

$n =$ an electronic state quantum number that specifies the electronic state represented by $V(R)$,

with standard energy expressions for each:

$$E_{\text{rot}}(J) = \frac{J(J+1)\hbar^2}{2\mu R_e^2}, \qquad J = 0, 1, 2, \ldots \quad (19.21)$$

$$E_{\text{vib}}(v) = \hbar\left(\frac{k}{\mu}\right)^{1/2}\left(v + \frac{1}{2}\right), \qquad v = 0, 1, 2, \ldots \quad (19.22)$$

and

$$E_{\text{el}}(n) = V_e \text{ for state } n \quad (19.23)$$

so that

$$E = E_{\text{el}} + E_{\text{vib}} + E_{\text{rot}} \ . \quad (19.24)$$

Since the total energy is a *sum* of terms, the total molecular wavefunction is a *product* of an electronic, a vibrational, and a rotational wavefunction.

---

[2] When there are other sources of angular momentum in a molecule (electron spin, orbital electronic, nuclear spin), the symbol N is used for the nuclear rotational quantum number. The symbol J is always reserved for the total angular momentum excluding nuclear spin.

EXAMPLE 19.4

Example 12.7 considered a simplified model of rotation by HI. Improve on it using the accurate equilibrium bond length $R_e = 1.6092$ Å, and find the *rotational quantum number for the ground vibrational state that has the same energy as the $J = 0$ state of the first excited vibrational state*. Use the experimental harmonic force constant 314 N m^{-1}.

SOLUTION Example 12.7 assumed the H atom orbited a stationary I atom, but here we will (correctly) use the HI reduced mass. Since $m_H \cong 1$ amu and $m_I \cong 127$ amu, $\mu \cong 127/128$ amu = 1.66×10^{-27} kg $\cong 1$ amu (Eq. (19.9)), and the approximation in Example 12.7 was very good. Equation (19.22) gives us the energy of the $v = 1, J = 0$ level. If we measure energy from the $v = 0, J = 0$ level, $v = 1, J = 0$ is higher by

$$E_{vib}(1) - E_{vib}(0) = \hbar \left(\frac{k}{\mu}\right)^{1/2} = (1.05457 \times 10^{-34} \text{ J s})\left(\frac{314 \text{ N m}^{-1}}{1.66 \times 10^{-27} \text{ kg}}\right)^{1/2}$$
$$= 4.59 \times 10^{-20} \text{ J} \ .$$

Equation (19.21) gives the energies of states with $v = 0$ and $J = 0, 1, 2, \ldots$:

$$E_{rot}(J) = J(J+1)\frac{\hbar^2}{2\mu R_e^2} = J(J+1)\frac{(1.05457 \times 10^{-34} \text{ J s})^2}{2(1.66 \times 10^{-27} \text{ kg})(1.6092 \times 10^{-10} \text{ m})^2}$$
$$= J(J+1)(1.29 \times 10^{-22} \text{ J}) \ .$$

Now we have to find a J (call it J^*) such that $E_{rot}(J^*) \cong 4.59 \times 10^{-20}$ J. We find $E_{rot}(18) = 4.42 \times 10^{-20}$ J and $E_{rot}(19) = 4.92 \times 10^{-20}$ J. Since J is integral, it was unlikely that the $v = 1, J = 0$ energy and the $v = 0, J = J^*$ energy would be equal, (an *accidental degeneracy*, a very rare thing). The calculation shows that HI has about 19 rotational levels ($J = 0 \rightarrow J = 18$) in the energy region between $v = 0$ and $v = 1$. This is shown below in a stack of energy levels for various J's in each vibrational state.

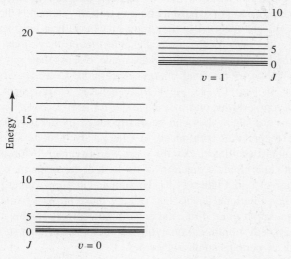

→ **RELATED PROBLEM** 19.7

Now we return to the exact RSE and consider the approximations that lead to the rigid rotor, harmonic-oscillator energy expressions. We do so for three reasons. First, the approximation must fail for large vibrational quantum numbers, since the true $V(R)$ curve looks nothing like a harmonic potential as $R \rightarrow \infty$. Figure 19.1 demonstrates this with a plot of the exact H_2 ground state $V(R)$ and its harmonic oscillator approximation.

The second reason to look beyond the harmonic approximation is related to the energy differences among the three terms of Eq. (19.24). We assume we know E_{el}

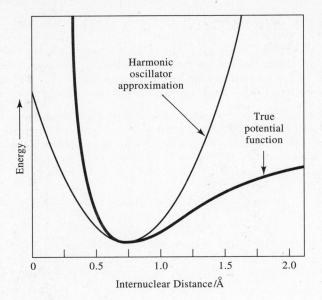

FIGURE 19.1 The true H_2 ground-state potential-energy function and the harmonic-oscillator approximation to it are compared here. The comparison is discussed in the text.

exactly, or at least to great precision, either from theory or experiment. But since $E_{vib} \gg E_{rot}$, if errors in E_{vib} are at all appreciable, it is silly to bother with E_{rot} unless we were interested only in changes in rotational energy for constant E_{el} and E_{vib}.

The third reason has two parts. Improved approximations are fairly easy to make and result in an energy expression that is compact and full of physical insight. Moreover, the high precision of experimental spectroscopy demands them, and they are valuable not only for the *energy* they represent (usually a very small fraction of the total energy) but also for the molecular information they alone can yield.

We will describe how the improvements are made, discuss their physical origins, and very quickly move to the final energy expression. Consider first the terms represented by ... in Eq. (19.14), the $V(R)$ expansion. Taken together, they are called *anharmonic corrections,* and they have the form of a power series in ρ. As long as these terms are small, they can be incorporated by perturbation theory. In fact, perturbation theory was introduced in Chapter 13 with terms of this type perturbing a harmonic oscillator, exactly what we have here. Moreover, we can guess the effect of anharmonic corrections if we look at the shape of $V(R)$ in Figure 19.1 (and, qualitatively, in Figure 11.8). Harmonic vibrational energy levels are equally spaced; the vibrational levels of $V(R)$ are *more closely spaced the closer we are to the dissociation limit.* Thus, the anharmonic corrections must be *negative* and they must vary with vibrational quantum number v faster than $(v + 1/2)$; that is, like $(v + 1/2)^2$ or perhaps some higher power.

Now consider the centrifugal-potential approximation that led to the rigid rotor, Eq. (19.16). The next two terms in that expansion describe two physical phenomena. The first

$$-\frac{J(J+1)\hbar^2}{2\mu R_e^2}\left(\frac{2\rho}{R_e}\right) = -\frac{J(J+1)\hbar^2}{2\mu R_e^2}\left[\frac{2(R-R_e)}{R_e}\right]$$

expresses the way rotational energy is *lowered* as the bond length is *increased* beyond R_e. This is called *centrifugal distortion.* A rotor held together by a spring of finite force constant experiences a centrifugal force that lengthens the bond as its angular momentum is increased.

The second phenomenon comes from the next term in the expansion of Eq. (19.16):

$$\frac{J(J+1)\hbar^2}{2\mu R_e^2}\left[3\left(\frac{\rho}{R_e}\right)^2\right] = \frac{1}{2}\left[\frac{3J(J+1)\hbar^2}{\mu R_e^4}\right]\rho^2.$$

This looks, for constant J, *just like a harmonic potential term*, $k'\rho^2/2$, with a force constant $k' = 3J(J+1)\hbar^2/\mu R_e^4$. This means the *total* force constant, $k + k'$, depends on J. This is *vibration-rotation coupling*. Somewhere in our improved expression for E there will be terms that depend on v and J *simultaneously*. The neat partition of energy into a sum of terms, each uniquely associated with one type of motion, as in Eq. (19.24), is invalid if we include this term in the Hamiltonian.

These three physical phenomena—anharmonicity, centrifugal distortion, and vibration-rotation coupling—lead to a series of corrections to the rigid rotor, harmonic oscillator energy expression. We will now discuss the corrected energy expression in the language of molecular spectroscopists. Previous energy expressions in this section have been in SI units with energy in joules. The energy unit of molecular spectroscopy is the wavenumber, in cm^{-1}. *The energy expressions we write now will contain molecular constants in cm^{-1} units and dimensionless quantum numbers only.*

The most general expression for a diatomic's vibrational and rotational energies in a $^1\Sigma$ electronic state, called the *Dunham expansion,* is a sum of terms each with $(v + 1/2)$ and/or $J(J + 1)$ raised to a different power. It contains molecular constants Y_{ij} (*not* to be confused with the spherical harmonic functions, $Y_{J,M}$!) that are related to R_e, μ, derivatives of the potential, and so forth. The subscript i is the power of $(v + 1/2)$, and j is the power of $J(J + 1)$. The first few terms show us which are new and which are familiar, but in disguise:

$$
\begin{aligned}
E_{\text{vib,rot}}(v, J) = \; & Y_{00} && \text{(a new term, } not \text{ the zero-point energy)} \\
& + Y_{10}\left(v + \tfrac{1}{2}\right) && \text{(the harmonic oscillator)} \\
& + Y_{01}J(J + 1) && \text{(the rigid rotor)} \\
& + Y_{20}\left(v + \tfrac{1}{2}\right)^2 && \text{(the first anharmonic term, } Y_{20} \text{ negative)} \\
& + Y_{02}[J(J + 1)]^2 && \text{(centrifugal distortion, } Y_{02} \text{ negative)} \quad (19.25) \\
& + Y_{11}\left(v + \tfrac{1}{2}\right)[J(J + 1)] && \text{(vibration-rotation coupling)} \\
& + Y_{30}\left(v + \tfrac{1}{2}\right)^3 && \text{(the next anharmonic term)} \\
& + \cdots && \text{(and so on.)}
\end{aligned}
$$

The constant Y_{00} is very small. It represents an extra contribution to the zero-point energy of the diatomic's vibration:

$$E_{\text{vib,rot}}(v = 0, J = 0) = \sum_i Y_{i0}\left(\frac{1}{2}\right)^i = Y_{00} + \frac{Y_{10}}{2} + \frac{Y_{20}}{4} + \cdots.$$

Since spectroscopy measures only energy *differences*, Y_{00} does not affect transition energies in any one electronic state.

The Dunham expression is compact, but it came too late (1932) to overthrow the original notation. Diatomic spectroscopic constants are usually referred to by these original symbols, which are only marginally systematic. For instance, the harmonic oscillator constant is called ω_e, the "e" standing for "equilibrium value," and originally meant $\hbar(k/\mu)^{1/2}/[(100 \text{ cm m}^{-1})hc]$ (to give it cm^{-1} units).[3] Table 19.1 is a dictionary of these constants, their method of entering $E_{\text{vib,rot}}$, and their common name. Since Y_{20}, Y_{01}, and Y_{11} are almost always negative, $\omega_e x_e$, D_e, and α_e appear in the energy expression with a negative sign so that positive numbers are tabulated for these constants. (Note that $\omega_e x_e$, $\omega_e y_e$, and on to $\omega_e z_e$, $\omega_e a_e$, etc., where warranted, are each *single* symbols for *unique numbers*; that is, $\omega_e x_e$ is *not* ω_e times some x_e!)

Often, the traditional symbols are used in expressions for the *vibrational energy in the absence of rotation*, called $G(v)$, where

$$G(v) = \omega_e\left(v + \tfrac{1}{2}\right) - \omega_e x_e\left(v + \tfrac{1}{2}\right)^2 + \omega_e y_e\left(v + \tfrac{1}{2}\right)^3 + \cdots \quad (19.26)$$

where the terms are: Vibrational energy; Major harmonic contribution; 1st anharmonic correction; 2nd anharmonic correction.

and the *rotational energy in any one vibrational state* v, called $F_v(J)$, where

$$F_v(J) = B_v J(J+1) - D_v[J(J+1)]^2 + \cdots \quad (19.27)$$

where the terms are: Rotational energy in vibrational state v; Rigid rotor energy in vibrational state v; Centrifugal distortion correction.

with v subscripts instead of e, so that one speaks of B_0, B_1, etc., for vibrational state $v = 0, 1,$ etc. The effective rotational constants and centrifugal distortion constants for state v are related to the equilibrium constants through

$$B_v = B_e - \alpha_e\left(v + \tfrac{1}{2}\right) + \cdots \quad (19.28)$$

where the terms are: Rotational constant for vibrational state v; Equilibrium rotational constant; 1st vibration-rotation correction.

and

$$D_v = D_e + \beta_e\left(v + \tfrac{1}{2}\right) + \cdots \quad (19.29)$$

where the terms are: Centrifugal distortion for vibrational state v; Equilibrium centrifugal distortion constant; Correction for vibrational state v.

TABLE 19.1 Glossary of major diatomic spectroscopic constants

Dunham symbol	Traditional symbol	Physical name	Coefficient of
Y_{10}	ω_e	harmonic constant	$\left(v + \tfrac{1}{2}\right)$
Y_{20}	$-\omega_e x_e$	1st anharmonic correction	$\left(v + \tfrac{1}{2}\right)^2$
Y_{30}	$-\omega_e y_e$	2nd anharmonic correction	$\left(v + \tfrac{1}{2}\right)^3$
Y_{01}	B_e	rotational constant	$J(J+1)$
Y_{02}	$-D_e$	centrifugal distortion	$[J(J+1)]^2$
Y_{11}	$-\alpha_e$	vibration-rotation interaction	$\left(v + \tfrac{1}{2}\right)[J(J+1)]$

---

[3] The factor (100 cm m^{-1}) reminds us that the speed of light is most naturally expressed in m s^{-1} units while wavenumbers are usually in cm^{-1} units. This factor handles the length conversion.

where β_e is usually quite small and can be neglected. Since

$$E_{\text{vib,rot}}(v, J) = G(v) + F_v(J) ,$$

Equations (19.26)–(19.29) are just selected collections of terms from the Dunham expansion.

There are several important references for diatomic molecule spectroscopic constants. One of the best and most recent is *Constants of Diatomic Molecules*, by K. P. Huber and G. Herzberg. Published in 1979, this is the fourth volume of Herzberg's definitive treatise, *Molecular Spectra and Molecular Structure*. (See Further Reading for the complete reference.) Table 19.2 collects the most important constants for several diatomics.

EXAMPLE 19.5

A CO molecule in the quantum state $v = 2$, $J = 1$ collides with an H_2 molecule in the $v = 0$, $J = 0$ state. Can the collision excite H_2 to the first vibrationally excited state? Ignore the translational energy of the collision.

SOLUTION The energy needed to excite H_2 to $v = 1$, $J = 0$ from $v = 0$, $J = 0$ is

$$\Delta E(v = 0 \to 1) = G(1) - G(0) = \omega_e - 2\omega_e x_e = 4\,158.54 \text{ cm}^{-1}$$

using the constants in Table 19.2. The maximum CO internal energy that can be transferred is the energy difference between $v = 2$, $J = 1$ and $v = 0$, $J = 0$. Again using the constants in Table 19.2, we first use Equation (19.28) and (19.27) to find

$$B_2 = B_e - 2.5\alpha_e = 1.887\,6 \text{ cm}^{-1}$$

$$F_2(1) = 2B_2 - 4D_e = 3.775\,1 \text{ cm}^{-1} .$$

This is the rotational energy in the CO. The vibrational energy is $G(2) - G(0) = 2\omega_e - 6\omega_e x_e = 4\,260.26 \text{ cm}^{-1}$, far more than the rotational energy. This alone is enough to excite H_2 to the first excited vibrational level.

➥ **RELATED PROBLEM 19.7**

Table 19.3 reproduces the Huber–Herzberg entry[4] for Na_2. It begins with the reduced mass, $\mu = 11.49\ldots$ amu, calculated from precise nuclear masses. Next is the dissociation energy *from the zero-point, rotationless level*, $\mathcal{D}_0^0 = 0.720$ eV, and the ionization potential I.P. = 4.90 eV. A few other notes come next, followed by individual state data to the extent they are known, and a series of footnotes (and references, not shown here).

The ground state, X $^1\Sigma_g^+$, is at the bottom of the table. The second column, T_e, is the energy of the equilibrium point of $V(R)$ for each state with $V(R_e,$ ground state) $\equiv 0$. The constant T_e is related to V_e in Eq. (19.23) through:

$$T_e(\text{state } n) = V_e(n) - V_e(\text{ground state}) .$$

Next are columns of ω_e and $\omega_e x_e$ constants (in cm^{-1}). For the X state, footnote u tells us that the Dunham expansion has been extended through the $(v + 1/2)^6$ term, an unusually high order. Next come B_e and α_e followed by D_e, the centrifugal distortion constant, and r_e (our R_e), the equilibrium bond length, in Å. The "Observed

---

[4]This molecule was chosen because it is well known, but not *too* well known. The entry for H_2 and its isotopes runs over 36 pages, and that for N_2 covers 14 pages (and 47 electronic states)!

TABLE 19.2 Spectroscopic constants of $^1\Sigma$ diatomic molecule ground states (in cm^{-1} and Å units)*

Molecule	ω_e	$\omega_e x_e$	B_e	D_e	α_e	R_e
Hydrogen						
H_2	4 401.213	121.336	60.853 0	4.71(−2)	3.062 2	0.741 44
HD	3 813.15	91.67	45.655	2.605(−2)	1.986	†
D_2	3 115.50	61.82	30.444	1.141(−2)	1.078 6	†
T_2	2 546.47	41.23	20.335		0.588 7	†
Alkali metals						
$^7Li_2$	351.43	2.610	0.6726	9.87(−6)	7.04(−3)	2.672 9
Na_2	159.125	0.725 5	0.1547	5.81(−7)	8.736(−4)	3.078 9
K_2	92.021	0.282 9	0.05674	8.63(−8)	1.65(−4)	3.905 1
Rb_2	57.31	0.105				
Cs_2	42.022	0.082 3	0.0127	4.64(−9)	2.64(−5)	4.47
Halogens						
F_2	916.64	11.236	0.8902	3.3(−6)	1.385(−2)	1.411 9
$^{35}Cl_2$	559.72	2.675	0.2440	1.86(−7)	1.49(−3)	1.987 9
$^{79}Br_2$	325.321	1.077 4	0.0821	2.092(−8)	3.187(−4)	2.281 1
I_2	214.502	0.614 7	0.03737	4.25(−9)	1.138(−4)	2.666 3
Interhalogens						
^{35}ClF	786.15	6.161	0.5165	8.77(−7)	4.358(−3)	1.628 3
^{79}BrF	670.75	4.054	0.3558	4.01(−7)	2.612(−3)	1.758 9
IF	610.24	3.123	0.2797	2.37(−7)	1.873(−3)	1.909 8
$^{79}Br^{35}Cl$	444.276	1.843	0.1525	7.183(−8)	7.697(4)	2.136 1
$I^{35}Cl$	384.293	1.501	0.1142	4.03(−8)	5.354(−4)	2.320 9
$I^{79}Br$	268.64	0.814	0.0568 3	1.02(−8)	1.969(−4)	2.469 0
Hydrogen halides						
HF	4 138.32	89.88	20.955 7	2.151(−3)	0.798	0.916 8
$H^{35}Cl$	2 990.946	52.819	10.593 4	5.319(−4)	0.307 2	1.274 6
$H^{81}Br$	2 648.975	45.218	8.464 9	3.458(−4)	0.233 3	1.414 4
HI	2 309.014	39.644	6.426 4	2.069(−4)	0.168 9	1.609 2
Multiple bonds						
C_2	1 854.71	13.34	1.819 8	6.92(−6)	1.765(−2)	1.242 5
CO	2 169.814	13.288	1.931 3	6.1215(−6)	1.75(−2)	1.128 3
CS	1 285.08	6.46	0.820 0	1.43(−6)	5.922(−3)	1.534 9
N_2	2 358.57	14.324	1.998 2	5.76(−6)	1.732(−2)	1.097 7
P_2	780.77	2.835	0.303 6	1.88(−7)	1.49(−3)	1.893 4
Alkaline earth oxides						
BeO	1 487.32	11.830	1.651 0	8.20(−8)	1.90(−2)	1.330 9
MgO	785.06	5.18	0.574 3	1.22(−6)	5.0(−3)	1.749 0
CaO	732.11	4.81	0.444 5	6.58(−7)	3.38(−3)	1.822 1
SrO	653.49	3.96	0.338 0	3.6(−7)	2.19(−3)	1.919 8
BaO	669.76	2.028	0.312 6	2.724(−7)	1.392(−3)	1.939 7
Alkali fluorides						
^7LiF	910.34	7.929	1.345 3	1.175(−7)	2.028 7(−2)	1.563 9
NaF	536	3.4	0.436 9	1.16(−6)	4.559(−3)	1.925 9
KF	428	2.4	0.279 9	4.83(−7)	2.335(−3)	2.171 5
RbF	376	1.9	0.210 7	2.68(−7)	1.523(−3)	2.270 3
CsF	352.56	1.615	0.184 4	2.017(−7)	1.176(−3)	2.345 3

Data from *Molecular Spectra and Molecular Structure IV. Constants of Diatomic Molecules*, by K. P. Huber and G. Herzberg (van Nostrand Reinhold, NY, 1979).
*The notation 1.23(−4) means 1.23 × 10^{-4}.
†The isotopes of hydrogen have the same R_e, the B–O potential minimum bond length, as does H_2.

TABLE 19.3 Spectroscopic constants for Na$_2$

State		T_e	ω_e	$\omega_e x_e$	B_e	α_e	D_e (10^{-7} cm^{-1})	r_e (Å)	Observed transitions Design.	ν_{00}		References
^{23}Na$_2$			$\mu = 11.4948852$		$D_0^0 = 0.720$ eV^a		I.P. $= 4.90$ eV^b					JUN 1977 A
		Diffuse bands of Na$_2$ van der Waals molecules close to the lines of the principal series of Na. (5)(7) Several fragments of other UV emission and absorption band systems.^{c}										
E	($^1\Pi_u$)	35557.0	106.2	H 0.6$_5$					E←X,	R 35530.6	H	(15)(19)
D	$^1\Pi_u$	33486.8	111.3^d	H 0.4$_8$	e				D↔X,	R 33462.9	H	(13)(15)(17) (19)
	$^1\Sigma_g^+$	(33000)	Fragment observed in two-photon excited Na$_2$ fluorescence.									(37)
C	$^1\Pi_u$	29382	119.33^f	H 0.53	e				C↔X,	R 29362	H	(12)(14)(17)
B	$^1\Pi_u$	20320.02	124.090	Z 0.6999^g	0.12527$^h_7$	0.000723$^i_7$	3.24$^j_8$	3.4228	B^k↔X,^{l}	R 20302.49^m	Z	(1)(6)(21) (30)
A	$^1\Sigma_u^+$	14680.58	117.323	Z 0.3576^n	0.11078$^{op}_4$	0.000548$^q_8$	3.88$^r_2$	3.6384	A^s↔X,	E 14659.80	Z	(2)(8)(31) (36)(40)
a	$^3\Pi$	<14680	(145.)^t		(0.140)^t							
X	$^1\Sigma_g^+$	0	159.124$_5$	Z 0.7254$^u_5$	0.154707^h	0.0008736^v	5.81$^w_1$	3.0788$_7$	Mol. beam magn. reson.^{x}			

Na$_2$: ^{a}From $D_e^0 = 5890 \pm 70$ cm^{-1} based on the RKR potential curve for the ground state (21)(30). The thermochemical value of (3), obtained by a molecular beam technique, is 0.73$_2$ eV.

^{b}From photoionization (20)(23). A similar value is obtained by extrapolation of the Rydberg series B, C, D, E (17)(19)(26).

^{c}Molecular absorption cross sections 27000 – 62500 cm^{-1} (20).

^{d}Vibrational constants from (15).

^{e}(17) report the following rotational constants for
D: $B_e = 0.1185$, $\alpha_e = 0.001$;
C: $B_e = 0.1281_5$, $\alpha_e = 0.0008_4$.
Considerably different constants, however, are quoted by Richards in (25):
D: $B_e = 0.1152$, $\alpha_e = 0.00110$;
C: $B_e = 0.1185$, $\alpha_e = 0.00096$.

^{f}Vibrational constants from (14) (except T_e which has been recalculated). (17), without details, give $T_e = 29393$, $\omega_e = 117.3$, $\omega_e x_e = 0.5_5$, while Richards (25) quotes 29384.8, 119.53, and 0.782, respectively.

^{g} $-0.00495 (v+\frac{1}{2})^3 - 0.000153 (v+\frac{1}{2})^4 + 7.01 \times 10^{-6} (v+\frac{1}{2})^5 - 1.804 \times 10^{-7} (v+\frac{1}{2})^6$; from the laser-induced fluorescence spectrum, including levels with $v' \leq 29$ (30). This state has a potential hump of ~ 550 cm^{-1} (0.069 eV); see (30), also (27). The non-appearance of levels with $v' > 26$ in the magnetic rotation spectrum may be due to weak predissociation; see (11).

^{h}RKR potential functions (30).

^{i} $-3.15_9 \times 10^{-5} (v+\frac{1}{2})^2 + 1.040 \times 10^{-6} (v+\frac{1}{2})^3 - 2.92_0 \times 10^{-8} (v+\frac{1}{2})^4$. The constants are for P and R lines, B(R, P) − B(Q) = $+0.0000128$ (30).

^{j} $\beta_e = +4.75 \times 10^{-8}, \ldots; H_e = -2.14_5 \times 10^{-11}$, and higher order constants (30).

^{k}Radiative lifetimes of 24 different levels (v', J') have been measured with an accuracy of $\sim$1% by means of a delayed coincidence single-photon counting technique (34); the observed lifetimes [see also (22)(24)] vary from 7.0 to 7.5 ns and have been used to determine the electronic transition moment and its variation with r. The results are in good agreement with (28) and with the *ab initio* calculations of (39). See, however, (27)(29).

^{l}Franck–Condon factors, dependence on rotation (34).

^{m}The band origin given here does not include $-B\Lambda^2$.

^{n} $+5.167 \times 10^{-6} (v+\frac{1}{2})^3 + 9.277 \times 10^{-6} (v+\frac{1}{2})^4 - 1.456 \times 10^{-7} (v+\frac{1}{2})^5$; from (40), see also (31). The observations extend to $v' = 44$ (36)(40).

^{o}Rotational perturbations in $v = 0$ and 1 are caused by levels belonging to the three components of the lower-lying a $^3\Pi$ state (31). Similar perturbations affect the higher vibrational levels and are responsible for the appearance of an A-X magnetic rotation spectrum (8)(9).

^{p}RKR potential function (40).

^{q} $+1.625 \times 10^{-8} (v+\frac{1}{2})^2 + 3.165 \times 10^{-8} (v+\frac{1}{2})^3 - 9.205 \times 10^{-10} (v+\frac{1}{2})^4$; from (40), see also (31).

^{r} $H_e = +1.129 \times 10^{-12}$ (40), see also (31).

^{s}Radiative lifetimes (35) are nearly constant for $1 \leq v \leq 25$, $\tau = 12.5 \pm 0.5$ ns, in very good agreement with theory (39).

^{t}Constants estimated from the perturbations in A $^1\Sigma_u^+$ (9), see ^{o}. (33) predict $T_e = 13500$, $\omega_e = 160$, $B_e = 0.154$.

^{u} $-0.00109_5 (v+\frac{1}{2})^3 - 4.72 \times 10^{-5} (v+\frac{1}{2})^4 + 3.2_1 \times 10^{-7} (v+\frac{1}{2})^5 - 7.5_3 \times 10^{-9} (v+\frac{1}{2})^6$; the analysis of the B→X system includes levels with $v'' \leq 46$ (30).

^{v} $-3.14_6 \times 10^{-5} (v+\frac{1}{2})^2 - 2.40_0 \times 10^{-7} (v+\frac{1}{2})^3 + 4.8_4 \times 10^{-9} (v+\frac{1}{2})^4 - 8.7_3 \times 10^{-11} (v+\frac{1}{2})^5$, from (30).

^{w} $\beta_e = +3.59 \times 10^{-9}$; $H_e = +1.92 \times 10^{-12}$, also higher order constants (30).

^{x} $g_J = (+)0.0389_2 \mu_N$ (18). From the nuclear resonance spectrum (16) determined eqQ, but much improved hyperfine coupling constants eqQ and c [spin-rot. const., see also (38)] in the $v = 0$, $J = 28$ level have recently been obtained by (32) using a laser-fluorescence molecular-beam-resonance technique.

Transitions" columns designate ("Design.") observed transitions in shorthand; for example, A ↔ X, meaning both absorption from state X to state A and emission from A to X have been observed.

The adjoining column, ν_{00}, lists the wavenumber of the transition from the zero-point vibrational ($v = 0$), rotationless ($J = 0$) level of one state to $v = 0$, $J = 0$ of the other. We can calculate ν_{00} for the $A \leftarrow X$ transition from the constants in the body of the table and compare to the tabulated value, 14 659.80 cm^{-1}. The calculation is shown schematically in Figure 19.2, which uses the common notation that a *single* prime (as in T'_e) refers to the state of *higher* energy and a *double* prime (T''_e) refers to the state of *lower* energy. Thus, ν_{00} is

$$\nu_{00} = T'_e + G'(v' = 0) - [T''_e + G''(v'' = 0)]$$
$$= (T'_e - T''_e) + \frac{\omega'_e - \omega''_e}{2} - \frac{\omega_e x'_e - \omega_e x''_e}{4}$$
$$= \left(14\,680.58 + \frac{117.323 - 159.125}{2} - \frac{0.3576 - 0.7255}{4}\right) \text{cm}^{-1} \quad (19.30)$$
$$= 14\,659.77 \text{ cm}^{-1}$$

in excellent agreement with the tabulated value. The higher anharmonic constants, $\omega_e y_e$, etc., which are lurking in the table's footnotes, can be ignored since they contribute very little to $G(v)$ for $v = 0$.

We can also use these constants to illustrate other effects of anharmonicity. Figure 19.3 compares $G(v)$ for $v = 0$–10 in the ground state to the harmonic oscillator approximation. Anharmonicity causes the increasing mismatch between the real levels and the equally spaced harmonic levels as v increases. We can also use $G(v)$ to estimate the *total number* of bound vibrational levels. The dissociation energy is $\mathcal{D}^0_0 = 0.720$ eV $\cong 5800$ cm^{-1}, and since

$$\mathcal{D}^0_0 \cong G(v_{\max}) - G(v = 0)$$

where $v_{\max}$ is the largest vibrational quantum number that can exist in the bound portion of the potential, we can find $v_{\max}$. We first calculate $G(v = 0) = 79.38$ cm^{-1} so that

$$G(v_{\max}) = \mathcal{D}^0_0 + G(v = 0) \cong 5800 \text{ cm}^{-1} + 79 \text{ cm}^{-1} \cong 5880 \text{ cm}^{-1}.$$

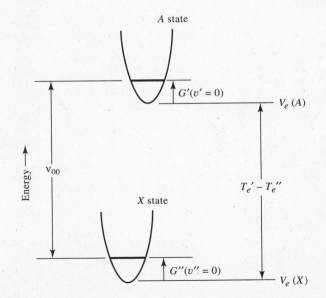

FIGURE 19.2 The relationships among the tabulated quantities ν_{00} and T_e and the $G(v = 0)$ vibrational energy (calculated from vibrational constants) is illustrated. In general, quantities with the *e* subscript refer to the equilibrium point of the interatomic potential-energy surface, while 0 subscripts refer to zero-point vibrational-level quantities.

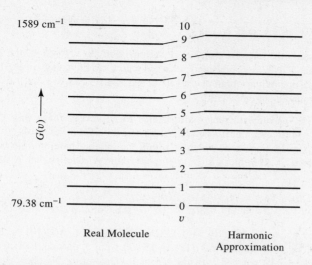

FIGURE 19.3 The true vibrational levels for Na_2 ($v = 0$–10), on the left, show the effects of anharmonicity when compared to the equally spaced harmonic-oscillator levels on the right.

Here, the higher anharmonic corrections to $G(v)$ (found in footnote u of the table) *are* important, since we want $G(v)$ at large v where anharmonicity is very important. We write

$$G(v)/\text{cm}^{-1} = 159.124\,5(v + 1/2) - 0.725\,47(v + 1/2)^2 + \text{(the terms in footnote } u\text{)} .$$

A few trial guesses for v quickly converges on $G(v = 54) = 5\,880.8$ cm^{-1}, which tells us that $v = 0$ through $v_{\max} = 54$ (or thereabouts) account for all the vibrational levels. Footnote u tells us that levels up to $v'' = 46$ in the X state have, in fact, been observed.

Figure 19.4 shows the first 21 rotational levels of the $v = 0$ and 1 vibrational states of the ground state. (Note how the words in that last sentence express in English the mathematics of Eq. (19.19)—the total energy is a sum of rotational, vibrational, and electronic energies.) Even though the rotational level spacing *increases* with J, rotational levels are much more closely spaced than vibrational levels. The state $v = 0$, $J = 20$ is only midway between $v = 0$, $J = 0$ and $v = 1$, $J = 0$.

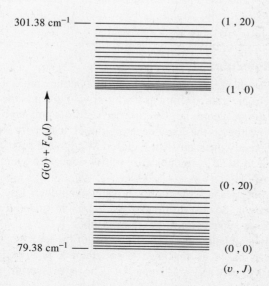

FIGURE 19.4 The typically larger vibrational spacing, compared to the rotational spacing, is displayed here for Na_2. Only the lowest 21 rotational levels ($J = 0$–20) are shown for each vibrational state, and the lowest energy is the zero-point energy.

EXAMPLE 19.6

You want to study the chemistry of Na_2 in various vibrational levels of the $A\,^1\Sigma_u^+$ excited state. Your tunable dye laser is currently charged with a laser dye that covers the wavelength region 630–685 nm. What vibrational levels of the A state can you reach from the $v = 0$ level of the ground electronic state?

SOLUTION The wavelength range corresponds to an energy range (use the conversion expression $E = (10^7\ cm^{-1}\ nm)/\lambda$) 15 873–14 599 cm^{-1}. Table 19.3 lists $\nu_{00} = 14\,659.80$ cm^{-1} for the A ← X transition, which is within the range. Therefore, the laser can pump molecules from $v'' = 0$ in the X state to $v' = 0$ in the A state. Higher A state vibrational levels are

$$\Delta G(v')/cm^{-1} = [G_A(v') - G_A(0)]/cm^{-1}$$
$$= 117.323(v' + 1/2) - 0.3576(v' + 1/2)^2 - 58.572$$

above this. For $v' = 10$, $\Delta G_A(10) = 1\,133.9\ cm^{-1}$, or $\nu_{00} + \Delta G_A(10) = 15\,793.7\ cm^{-1}$, also within the range, but $\nu_{00} + \Delta G_A(11) = 15\,903.1\ cm^{-1}$, outside the range. Thus, states with $v' = 0$–10 can be reached *from the zero-point level* of the X state:

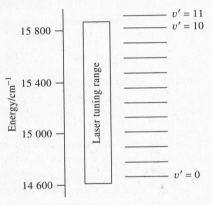

However, one cannot easily make Na_2 at temperatures so low that only the zero-point level is occupied. Thus, absorption from *excited vibrational levels* of the X state to even higher levels of the A state is also possible. Finally, a source of photons of the right energy is not enough to ensure we can cause a particular transition. The molecule must also have a large transition probability.

➡ **RELATED PROBLEMS** 19.9, 19.10

Polyatomic Molecules

As we did with diatomics, we will approach the polyatomic molecule energy expression from its electronic component and work our way towards the rotational. Term symbols for polyatomic electronic states are derived from polyatomic MO configurations along the same lines as the diatomic term symbol. In fact, for linear polyatomics, the derivation is exactly the same. Thus, closed MO ground state configurations for HCCH and CO_2 lead to $^1\Sigma_g^+$ ground states, or, for noncentrosymmetric species like HCN and OCS, to $^1\Sigma^+$.

For nonlinear polyatomics, the Λ and Ω quantum numbers lose meaning, as does Σ. Spin multiplicity is, however, often a "good-enough" quantum number that it is retained as a leading superscript to the term symbol.

A symmetry-related letter is used at the center of the term symbol, replacing the Λ code letter. It is based on the lower-case letters *a, b, e,* and *t* introduced for polyatomic MO symmetries in Chapter 14. The term symbol analogous to the diatomic $^1\Sigma^+$ for closed MO polyatomic ground states is ^1A, often decorated by other labels appropriate to the nuclear symmetry of the molecule: $^1A_1$, $^1A'$, $^1A_{1g}$. The subscript 1 has nothing to do with angular momentum, in contrast to the $\Omega = 1$ subscript that can appear in diatomic term symbols. Also, the $+$, $-$ notation no longer applies to nonlinear polyatomics.

These symbols carry somewhat less information than those of diatomics. Often, they are replaced with a set of generic symbols. Since most ground states are singlets, they are often called simply S_0, and excited singlet states are, in order of increasing energy, S_1, S_2, etc. The lowest energy triplet spin state is called T_1, followed by T_2, etc.

We will use H_2O and CO_2 as examples for the polyatomic vibrational energy expression. A mental picture of a diatomic's single vibration is obvious; not so for a polyatomic. Consider first H_2O. We could perhaps imagine first one O—H bond vibrating and then the other, for two of the three vibrations, while the H—H distance oscillates for the third, as in Figure 19.5(a). This picture predicts that the two O—H bond vibrational frequencies would be *identical* while the third (H—H) would be different. Similarly, for CO_2 (Figure 19.5(b)), we could imagine two degenerate O—C vibrations and *two* O—O vibrations. There are two O—O vibrations since the molecule must *bend* if the O—O distance is to change while keeping the O—C lengths the same, and there are two unique bending directions. If z is the O—C—O axis, the molecule can bend along either the x or the y direction equivalently.

These motions are called *local mode* vibrations, and they do correspond to motions that are possible in the molecule. They are not, however, what we might call "eigenmotions," which means they do not appear naturally when we solve for the nuclear wavefunction and vibrational energy under the influence of a B–O potential-energy function. Instead, the "eigenmotions" (more commonly called *normal modes of vibration*) reflect the symmetry of the potential-energy function directly.

For water, the three normal modes are shown in Figure 19.6(a). They each have a *distinctly different frequency*. They are numbered 1, 2, and 3 by convention and are called the *symmetric stretch,* the *bend,* and the *antisymmetric stretch,* respectively. Each normal mode has an associated harmonic vibration constant, and for H_2O these are:

$\omega_1 = 3\,657.05$ cm^{-1} **(symmetric stretch)**

$\omega_2 = 1\,594.78$ cm^{-1} **(bend)**

$\omega_3 = 3\,755.79$ cm^{-1} **(antisymmetric stretch)** .

Carbon dioxide (Figure 19.6(b)) has similar normal modes, except, as the local-mode picture predicts, the bending vibration is *doubly degenerate*. The observed harmonic constants are

$\omega_1 = 1\,354.31$ cm^{-1} **(symmetric stretch)**

$\omega_2 = 672.95$ cm^{-1} **(bend, doubly degenerate)**

$\omega_3 = 2\,396.32$ cm^{-1} **(antisymmetric stretch)** .

In both molecules the bending constant is much smaller than the stretch constants and, of the two stretching constants, the antisymmetric one is larger. This is typical for triatomics.

Each mode has unique harmonic (and anharmonic) constants in the vibrational energy expression, and there are also constants representing a coupling between

FIGURE 19.5 (a) The vibrational motion of H_2O might seem to be best described in terms of local atom–atom bond stretches. This approach, however, is not appropriate for the description of most molecular vibrations in small molecules. (b) As in (a), one could imagine local vibrations of bonds in CO_2, but again, these are not the most appropriate nuclear motions for vibrational states in CO_2.

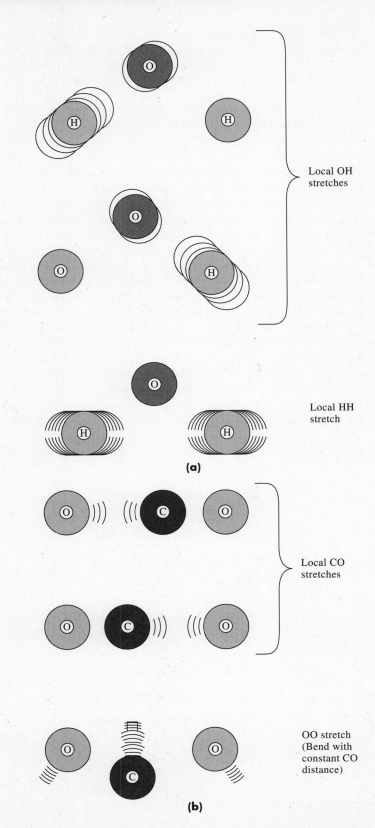

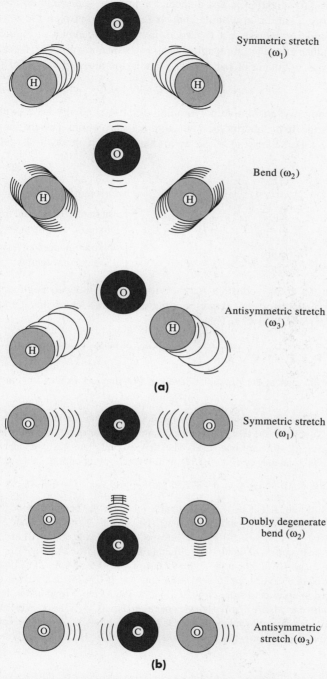

FIGURE 19.6 (a) In contrast to Figure 19.5(a), the normal modes of H_2O show symmetry in their motion which reflects the symmetry of the molecule's nuclear framework. (b) The normal modes of CO_2 also reflect the molecule's symmetry. Since the molecule is linear at its equilibrium configuration, the bending mode is doubly degenerate. In addition to the bend in the plane of the figure shown here, the molecule can bend perpendicularly to this plane at the same frequency.

normal modes. One final energy term is not so obvious, however. It corresponds to an *effective rotational angular momentum* that can appear only in *linear molecules* as a result of a doubly degenerate bending mode. A simple analogy shows how this arises. Recall from Chapter 18 how the general expression for the electric field of radiation could express both linear and circular polarizations. (See the discussion following Eq. (18.4) and Figure 18.1.) Carbon dioxide, for example, can bend in the x direction or the y direction, just as the field can be x polarized or y polarized.

Since circular polarization is just a special amplitude and phase combination of x, y polarization, a similar amplitude and phase relation between the x, y bends will induce a *rotational motion in CO_2 about the z axis*, as shown in Figure 19.7. This vibration-induced angular momentum has its own quantum number, ℓ, and magnitude $\ell\hbar$. If CO_2 has v_2 quanta of bending excitation, the allowed values of ℓ are

$$\ell = v_2, v_2 - 2, \ldots, 1 \text{ or } 0 .$$

We can now write a simple polyatomic vibrational energy level expression. We will stop at the first anharmonic and coupling corrections (where the constant is traditionally called x instead of ωx as in diatomics). If mode i has degeneracy d_i, we have

$$\begin{aligned} G(v_1, v_2, \ldots) = &\sum_i \omega_i \left(v_i + \frac{d_i}{2}\right) &&\Leftarrow \text{harmonic normal modes} \\ &+ \sum_i \sum_{j \geq i} x_{ij}\left(v_i + \frac{d_i}{2}\right)\left(v_j + \frac{d_j}{2}\right) &&\Leftarrow \begin{cases} \text{anharmonic } (i=j) \text{ and mode} \\ \text{ to mode coupling } (i \neq j) \end{cases} \\ &+ \sum_i \sum_{j \geq i} g_{ij} \ell_i \ell_j &&\Leftarrow \begin{cases} \text{vibration-induced angular} \\ \text{ momentum energies} \end{cases} \end{aligned} \quad (19.31)$$

where g_{ij} is the energy constant for the vibrationally induced angular momentum. It is non-zero only for degenerate bending normal modes of linear molecules and equals -0.97 cm^{-1} for CO_2.

EXAMPLE 19.7

What are the energies of the vibrational levels of CO_2 that are within 2500 cm^{-1} of the zero-point level?

SOLUTION We will use only the harmonic constants (and the single g_{ij} constant, $g_{22} = -0.97$ cm^{-1}) quoted in the text. We use the quantum numbers v_1, v_2, v_3, and ℓ in the compact notation (v_1, v_2^ℓ, v_3) for any one level. The first excited state has one quantum of bend: $(0, 1^1, 0)$ and is at an energy (Eq. (19.31) with the zero-point energy subtracted)

$$G(0, 1^1, 0) = \omega_2 + g_{22} = 672.85 \text{ cm}^{-1} - 0.97 \text{ cm}^{-1} = 671.88 \text{ cm}^{-1} .$$

Note that ℓ must equal 1 if $v_2 = 1$. The next levels have the bending mode doubly excited ($v_2 = 2, \ell = 0, 2$): $(0, 2^0, 0)$ and $(0, 2^2, 0)$ at $2(672.85 \text{ cm}^{-1}) = 1\,345.7 \text{ cm}^{-1}$ and $2(672.85 \text{ cm}^{-1}) - 4(0.97 \text{ cm}^{-1}) = 1\,341.82 \text{ cm}^{-1}$, respectively. Next is $(1, 0^0, 0)$ at 1 354.31 cm^{-1} followed by three quanta of bend: $(0, 3^1, 0)$ and $(0, 3^3, 0)$ at $3(672.85 \text{ cm}^{-1}) - 0.97 \text{ cm}^{-1} = 2\,017.58$ cm^{-1} and $3(672.85 \text{ cm}^{-1}) - 9(0.97 \text{ cm}^{-1}) = 2\,009.82$ cm^{-1}. The first *combination level*, $(1, 1^1, 0)$ at 1 354.31 cm^{-1} + 671.88 cm^{-1} = 2 026.19 cm^{-1}, is next. Finally, the singly excited antisymmetric stretch $(0, 0^0, 1)$ at 2 396.32 cm^{-1} rounds out the picture.

Some of these levels are involved in the operation of the CO_2 IR laser. While anharmonicity changes the ordering and location of these levels—see Problem 19.34—we can use them to understand qualitatively how this laser works. A mixture of He, CO_2, and N_2 is subjected to

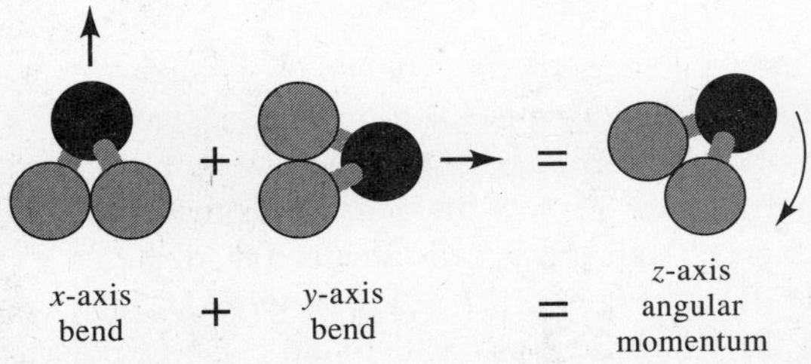

FIGURE 19.7 The doubly degenerate bend in CO_2 (or in other linear molecules) can lead to an angular momentum about the molecular axis.

an electrical discharge, and electron collisions populate the $(0, 0^0, 1)$ level, in part by direct excitation, but also by collisional energy transfer from $v = 1$ excited N_2:

$$N_2(v = 1) + CO_2(0, 0^0, 0) \rightarrow N_2(v = 0) + CO_2(0, 0^0, 1) \ .$$

This nitrogen level is only 19 cm^{-1} below the $(0, 0^0, 1)$ level of CO_2, and the energy transfer is very efficient. The lower energy $(1, 0^0, 0)$ and $(0, 2^0, 0)$ levels are not highly populated in the discharge, and a population inversion is created between them and the $(0, 0^0, 1)$ level. Lasing action between the $(0, 0^0, 1)$ level and these lower two (at wavelengths around 10.6 µm and 9.6 µm) is the result:

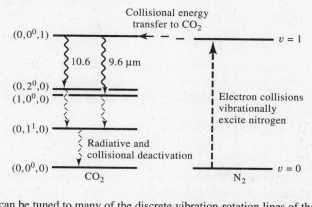

The laser can be tuned to many of the discrete vibration-rotation lines of these transitions, and isotopically enriched CO_2 is used to produce still more lines.

▶ RELATED PROBLEMS 19.11, 19.34, 19.35

Polyatomic rotational-energy expressions introduce new phenomena and terminology. The three principal rotation axes are labeled by convention so that (see also Eq. (19.3)) the moments of inertia are ordered

$$I_c \geq I_b \geq I_a \ . \tag{19.32}$$

For any

linear molecule $\quad I_c = I_b \quad$ and $\quad I_a = 0 \ .$

If two moments are equal and the third is non-zero, we have a *symmetric top* molecule. There are two kinds of these:

prolate symmetric top $\quad I_c = I_b > I_a$

oblate symmetric top $\quad I_a = I_b > I_c \ .$

Examples of prolate tops include CH_3F, $CH_3C{\equiv}CH$, and a baseball bat; they tend to have most of their mass spread *along* a high-symmetry axis. Examples of oblate tops include CF_3H, benzene, and a discus; they tend to have most of their mass spread *in planes perpendicular* to a high-symmetry axis.

If all three moments are equal, the molecule is a *spherical top*:

spherical top $\quad I_c = I_b = I_a \ .$

Examples include tetrahedral molecules (CH_4, CF_4, etc.), octahedral molecules (SF_6, $Cr(CO)_6$, etc.), C_{60}, and a sphere. The most common situation, however, is the *asymmetric top* molecule:

asymmetric top $\quad I_c > I_b > I_a \ .$

The rotational energy expression for a linear polyatomic has a single rotational constant (called B_e, as for diatomics):

$$B_e = \left(\frac{\hbar^2}{2I_b}\right) / (100 \text{ cm m}^{-1})hc \ . \tag{19.33a}$$

For nonlinear molecules, we define two other rotational constants similarly:

$$A_e = \left(\frac{\hbar^2}{2I_a}\right) / (100 \text{ cm m}^{-1})hc \tag{19.33b}$$

$$C_e = \left(\frac{\hbar^2}{2I_c}\right) / (100 \text{ cm m}^{-1})hc \ . \tag{19.33c}$$

The subscript e means the *equilibrium* atomic positions are used to calculate the moments of inertia. Since atoms are always vibrating, the *vibrationally averaged* moments actually govern the molecule's rotational energy. Consequently, the subscript e is often omitted.

From Eq. (19.32), these constants are ordered

$$A \geq B \geq C \ . \tag{19.34}$$

They can be cleverly combined into a single, dimensionless number, called *Ray's asymmetry parameter,* that scales asymmetric tops between prolate and oblate symmetric top limits:

$$\kappa = \frac{2B - A - C}{A - C} \ . \tag{19.35}$$

A prolate top ($B = C$) has $\kappa = -1$, an oblate top ($A = B$) has $\kappa = +1$, and all asymmetric tops fall in between. Tops with $|\kappa| \cong 1$ are called *near* (prolate or oblate) symmetric tops. For example, in H_2O (an asymmetric top in spite of its considerable symmetry), $A = 27.877$ cm^{-1}, $B = 14.512$ cm^{-1}, and $C = 9.285$ cm^{-1} so that $\kappa = -0.4377$, on the prolate side of $\kappa = 0$, but not "near" the $\kappa = -1$ limit. In contrast, H_2CO, with $A = 9.405\ 3$ cm^{-1}, $B = 1.295\ 4$ cm^{-1}, $C = 1.134\ 3$ cm^{-1}, is near prolate: $\kappa = -0.9610$.

The rigid-rotor energy expression for nonlinear symmetric tops comes from the classical energy expression (compare Equations (12.27)–(12.29))

$$E_{\text{rot}} = \frac{1}{2}(I_a\omega_a^2 + I_b\omega_b^2 + I_c\omega_c^2) = \frac{1}{2}\left(\frac{J_a^2}{I_a} + \frac{J_b^2}{I_b} + \frac{J_c^2}{I_c}\right) \tag{19.36}$$

where ω_i is the *angular velocity* about axis i and $J_i = I_i\omega_i$ is the angular momentum component along axis i. Since quantum mechanics causes us to consider $J^2 = J_a^2 + J_b^2 + J_c^2$ and any one component, we must manipulate this classical expression before transcribing it into quantum terms.

For a prolate symmetric top ($I_b = I_c$), we add and subtract $J_a^2/2I_b + J_c^2/2I_b$ to Eq. (19.36) and rearrange the result:

$$\begin{aligned}E_{\text{rot}} &= \frac{1}{2}\left(\frac{J_a^2}{I_a} + \frac{J_b^2}{I_b} + \frac{J_c^2}{I_c} + \frac{J_a^2}{I_b} - \frac{J_a^2}{I_b} + \frac{J_c^2}{I_b} - \frac{J_c^2}{I_b}\right) \\ &= \frac{1}{2}\left(\frac{J^2}{I_b} + \frac{J_a^2}{I_a} + \frac{J_c^2}{I_c} - \frac{J_a^2}{I_b} - \frac{J_c^2}{I_b}\right) \\ &= \frac{1}{2}\left(\frac{J^2}{I_b} + J_a^2\left(\frac{1}{I_a} - \frac{1}{I_b}\right)\right)\end{aligned}$$

where we have used $I_b = I_c$ to reach the last expression. The operator for J^2 has the familiar J quantum number and $\sqrt{J(J+1)}\hbar$ eigenvalues. The component of $\mathbf{J}$ along a *molecule-fixed* direction, the principal symmetry axis, is J_a. Its corresponding quantum number is K, with $-J \leq K \leq J$ and an eigenvalue $K\hbar$. We can treat the oblate top in a similar way and find the same two quantum numbers in the energy expression.

If we express the energy in cm^{-1} units and introduce J, K and the rotational constants into Eq. (19.36), we find the symmetric top rotational energy expressions (neglecting centrifugal distortion):

$$\text{prolate top:} \quad E_{\text{rot}}(J, K) = BJ(J+1) + (A-B)K^2 \quad (19.37a)$$

and

$$\text{oblate top:} \quad E_{\text{rot}}(J, K) = BJ(J+1) + (C-B)K^2 \ . \quad (19.37b)$$

These expressions yield energy level stacks with qualitatively different appearances. For the prolate top, $A > B$ and $(A - B)K^2$ is *positive*; for the oblate top, $B > C$ and $(C - B)K^2$ is *negative*. Thus, for constant J, E_{rot} *increases with increasing K for a prolate top*, but *decreases for an oblate top*. This is shown in Figure 19.8(a) for prolate CF$_3$I and in Figure 19.8(b) for oblate CF$_3$H.

In these figures, K ranges from 0 to J. Since K enters the energy expression as K^2, only the *magnitude* of K matters. Classically, $+K$ and $-K$ correspond to degenerate rotations in opposite senses about the top's principle symmetry axis.

FIGURE 19.8 (a) The rotational energy levels for the prolate symmetric top CF$_3$I form the pattern shown here when sorted according to J and K quantum numbers. (b) The J, K sorted rotational energy levels of the oblate symmetric top CF$_3$H form this pattern. To see the difference between this pattern and that in (a), trace the evolution of the $J = 10$ level with K in both figures. Here, $J = 10$ decreases in energy as K increases. It does the opposite for a prolate top.

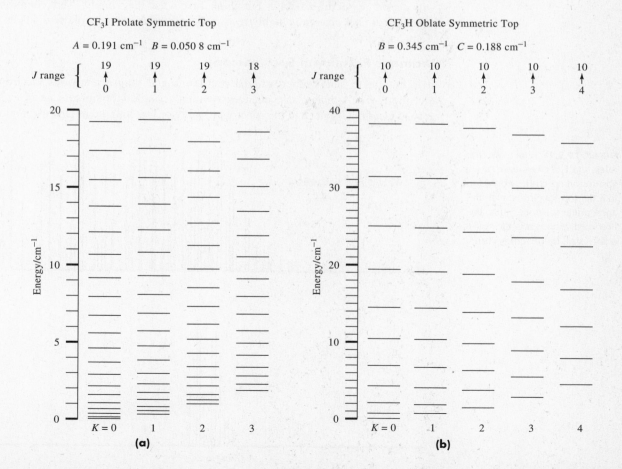

For the rigid spherical top, $A = B = C$, and the rotational energy expression is as simple as that for a rigid linear rotor:

$$\text{spherical top} \qquad E_{\text{rot}}(J) = BJ(J + 1) \ . \qquad (19.37\text{c})$$

Logically, we should now write an expression for E_{rot} for the asymmetric top, but there is no simple, general expression. The K quantum number is no longer "good," J alone is not sufficient to express the complexity of motion of an asymmetric rotor, and numerical methods are often used to find energies.

This section has presented the central energy terms in any discussion of molecular spectroscopy. Spectroscopy measures energy *differences,* and the next section gets to the heart of spectroscopy: selection rules and *observable* ΔE's.

19.3 MOLECULAR STRUCTURE AS DEDUCED BY SPECTROSCOPY

Selection rules might seem like an unfortunate fact of Nature since they limit what we can observe. In fact, they are a blessing. Without them, molecular spectra would be vastly more complicated. As an example, Figure 19.9 plots the IR absorption spectrum of HF first (top spectrum) with no selection rules (any ΔE could be observed) and then (lower spectrum) as it really appears.

The top spectrum goes on and on to higher frequency, as higher and higher energy (v, J) states are accessed. The true spectrum is only a handful of lines derived from the vibration-rotation selection rules $\Delta v = v' - v'' = 1$ and $\Delta J = J' - J'' = \pm 1$ that hold for any $^1\Sigma^+$ electronic state of any heteronuclear diatomic. We will derive these rules below, but think first what they imply. The vibrational quantum number change is highly restricted, and the $\Delta J = \pm 1$ rule reminds us of atomic spectroscopy.

Diatomic Rotational Spectroscopy

It is helpful to start with rotational transitions and progress towards electronic transitions. The simplest case changes J within a single vibrational state of a $^1\Sigma^+$ ground electronic state, and the spectra fall more or less into the microwave region.

FIGURE 19.9 Without selection rules, the HF vibration-rotation absorption spectrum would look like the top spectrum; with the appropriate selection rules, the observed spectrum is the considerably simpler lower spectrum.

Without selection rules

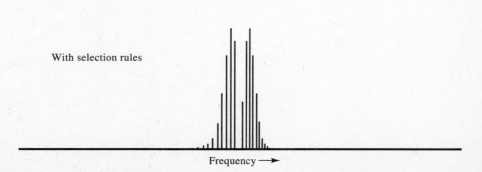

With selection rules

Frequency →

19.3 MOLECULAR STRUCTURE AS DEDUCED BY SPECTROSCOPY

As with atoms, selection rules appear when we explore the conditions that make the dipole transition moment non-zero. Recall Eq. (18.20):

$$\mathbf{p}_{kn} = \int \psi_k^* \hat{\mathbf{p}} \, \psi_n \, d\tau \, . \tag{18.20}$$

Since $\hat{\mathbf{p}}$, the dipole moment vector operator, depends on both nuclear and electron coordinates, it is amenable to a B–O approximation. If we average $|\hat{\mathbf{p}}|$ over the *electronic* wavefunction as a function of R, we have the *permanent dipole-moment function*, $p(R)$:

$$p(R) = \int \psi_{el}^*(R) |\hat{\mathbf{p}}| \psi_{el}(R) \, d\tau_{el} \, . \tag{19.38}$$

Since $p(R) = 0$ for all R for homonuclear diatomics, *homonuclear diatomics can have no pure rotation spectrum*. Thus, H_2, N_2, F_2, Cl_2, etc., which are all $^1\Sigma_g^+$ molecules, are transparent[5] in the microwave region.

The average of $p(R)$ over the vth vibrational wavefunction is the permanent dipole moment of the vth vibrational state

$$p_v = \langle p(R) \rangle_v = \int_0^\infty \psi_{vib}^* \, p(R) \psi_{vib} \, dR \, . \tag{19.39}$$

This quantity enters Eq. (18.20) along with the rotational wavefunctions. The total transition moment expression is thus

$$p_{kn} = \int \psi_{J',M'}^* \underbrace{\left[\int \psi_{vib}^* \underbrace{\left(\int \psi_{el}^* \hat{p} \psi_{el} \, d\tau_{el} \right)}_{p(R), \text{ the dipole moment function}} \psi_{vib} \, dR \right]}_{p_v, \text{ the vibrationally averaged dipole moment}} \psi_{J'',M''} \, d\Omega = \int \psi_{J',M'}^* p_v \psi_{J'',M''} \, d\Omega$$

where $d\Omega = \sin\theta \, d\theta \, d\phi$. Since $\psi_{J,M}$ is a spherical harmonic function, $Y_{J,M}(\theta, \phi)$, and $\mathbf{p}_v$ is a dipole moment vector (directed along the bond), we can invoke the parity selection rules on J and M in exactly the same way as we did in Chapter 18 for atoms. The analog of the atomic $\Delta l = \pm 1$ rule is

$$\Delta J = J' - J'' = \pm 1 \quad \text{and} \quad \Delta M = M' - M'' = 0, \pm 1 \tag{19.40}$$

where M is the quantum number for the component of $\mathbf{J}$ along a space-fixed axis. (Such an axis is relevant only in the presence of an external electric or magnetic field, discussed in Chapter 21.)

Before we apply Eq. (19.40) to an energy level expression and predict the spectrum, we should consider the roles of absorption and spontaneous emission in molecular spectroscopy. Consider a simple polar gas such as CO at equilibrium. We irradiate this sample with light that matches the $v = 0$, $J'' = 0$ to $J' = 1$ energy difference and satisfies the $\Delta J = \pm 1$ selection rule. There are molecules in the sample in *both* rotational states at equilibrium, and absorption changes $J = 0$ molecules into $J = 1$, but stimulated emission changes $J = 1$ molecules into $J = 0$. Which effect will dominate? Is there a net *absorption* or a net *emission*?

We will see expressions for the fraction of molecules in a given quantum state at equilibrium in later chapters. They show that the net result is *absorption* in a sample *at thermal equilibrium*, no matter what pair of adjacent J states we choose. One of the great achievements of molecular science in the second half of this century is the development of many clever ways to generate and sustain *nonequilibrium quantum state distributions*. If such a distribution has sufficiently greater population in state $J + 1$ over that in state J, a *nonequilibrium population inversion* is said to

---

[5]O_2 has a $^3\Sigma_g^-$ ground state that can absorb in the microwave region via magnetic dipole selection rules. These transitions do not change vibrational or rotational states, but rather reorient the electronic spin.

exist. (The thermodynamics of such situations was discussed in Chapter 4 for a two-level system.) In this case, spectroscopy at the J to $J + 1$ transition frequency is dominated by stimulated emission. This is the principle of the *maser*. A maser (or a laser) takes energy stored in a nonequilibrium molecular distribution and transfers it to a stimulated radiation field.

Now we can predict the ordinary, equilibrium, rotational spectrum under the absorption selection rule $\Delta J = J' - J'' = +1$. The rotational energy expression is $F_v(J)$, Eq. (19.27), so that the transition energy for vibrational state v is

$$
\begin{aligned}
\Delta E &= F_v(J') - F_v(J'') & \text{(in general)} \\
&= F_v(J'' + 1) - F_v(J'') & \text{(using the selection rule)} \\
&= 2B_v(J'' + 1) - 4D_v(J'' + 1)^3 + \cdots & \text{(expanding each } F_v \text{)} \\
&\cong 2B_v(J'' + 1) & \text{(for a rigid rotor)}
\end{aligned}
\quad (19.41)
$$

where $\cdots$ stands for the minor higher-order corrections that we will ignore and J'' is the rotational quantum number of the lower energy (initial) level.

Suppose we record the microwave spectrum of a sample of CO which has only a single isotopic form ($^{12}C^{16}O$) of any abundance and such a great vibrational-energy quantum that every molecule is in its ground vibrational state at room temperature. The spectrum is a series of nearly equally spaced transitions, as Figure 19.10(a) shows. If there had been no centrifugal distortion, a very small effect, the transitions would be equally spaced by $2B_0$, the rigid rotor spectrum.

EXAMPLE 19.8

A sample of CO at room temperature is placed in a spectrometer that can cover only the frequency range 400–700 GHz. Three transitions are seen, at 461.040, 576.267, and 691.473 GHz. What transitions are these?

SOLUTION Spectra do not roll off a spectrometer neatly annotated as in Figure 19.10(a), and this example considers the first task that follows acquisition of a spectrum: its *assignment* to quantum number changes. These transitions are purely rotational, and our task is to find the J'' and J' quantum numbers of each. First, note that the lines are almost equally spaced: $576.267 - 461.040 = 115.227$ GHz and $691.473 - 576.267 = 115.206$ GHz. From Eq. (19.41), this spacing, $\cong 115.2$ GHz or $(115.2 \times 10^9 \text{ Hz})/(c \times 100 \text{ cm m}^{-1}) = 3.843 \text{ cm}^{-1}$, equals $2B_0$. (Since only $v = 0$ is appreciably populated at room temperature, the rotational constant is called B_0.) Thus, $B_0 = 3.843/2 = 1.922 \text{ cm}^{-1}$. The transitions themselves, again from Eq. (19.41), appear at $\Delta E = 2B_0(J'' + 1) = (3.843 \text{ cm}^{-1})(J'' + 1)$. In cm^{-1} units, the observed transitions are at $\Delta E = 15.379, 19.222$, and 23.065 cm^{-1}. We find J'', the quantum number of the initial state of the absorption, from $J'' = (\Delta E/3.843 \text{ cm}^{-1}) - 1$, or $J'' = 3$, 4, and 5, and the observed transitions are $J = 3 \to 4$, $4 \to 5$, and $5 \to 6$, as Figure 19.10(a) indicates.

➡ RELATED PROBLEMS 19.14, 19.15, 19.16

Figure 19.10(b) shows another advantage of the high resolution of microwave spectroscopy. It is the $J = 3 \to 4$ spectrum for a variety of CO isotopomers. Each species is distinctly isolated. Isotopic substitution changes not only the reduced mass of the molecule, but also the region of $V(R)$ sampled by the molecule's vibrational wavefunction, since changing μ changes the zero-point energy. This slightly changes R_0, the bond length averaged over $\psi_{vib}(v = 0)$, so that analysis of many isotopes can be used to find[6] the equilibrium bond length, R_e.

---

[6]Since rotational constants depend on $1/R^2$, what is actually measured is $\langle 1/R^2 \rangle_v$, the average of $1/R^2$ over the vth vibrational wavefunction. Since $\langle R \rangle \neq \langle 1/R^2 \rangle^{-1/2}$, one does not actually measure the average R, but the inequality is close enough to an equality that one can say "rotational constants measure bond lengths."

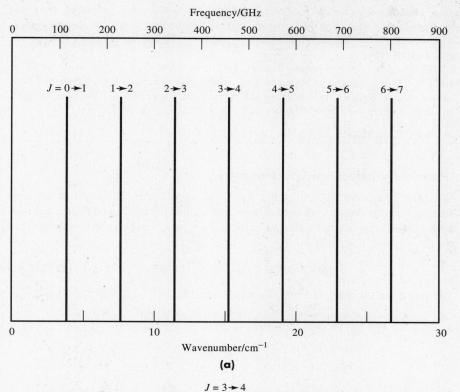

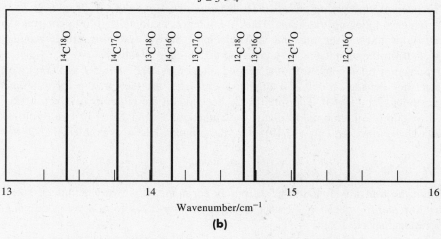

FIGURE 19.10 (a) The pure rotational spectrum of CO in the microwave region is a simple series of very nearly equally spaced transitions that are assigned according to the selection rule $\Delta J = +1$. All of these transitions are in the $v = 0$ vibrational state. (b) One advantage of microwave spectroscopy is its inherently high resolution. The transitions of various CO isotopomers, shown here for the $J = 3$ to 4 transition, are at different wavenumbers due to the reduced mass dependence of the moment of inertia. Although the differences are small, they are easily resolved.

EXAMPLE 19.9

What is the CO bond length, based on the spectrum from the previous Example?

SOLUTION Structural information comes from the relations among spectroscopic constants and $V(R)$. In this case, the important relation is

$$B_0 = \frac{\hbar^2}{2\mu R_0^2} \frac{1}{(100 \text{ cm m}^{-1})hc}.$$

The reduced mass (for the common $^{12}C^{16}O$ isotope) is $\mu = (12)(15.995)/(12 + 15.995)$ amu $= 6.856\ 2$ amu $= 1.138\ 5 \times 10^{-26}$ kg. Note that the masses are *not* the common "atomic weights" of the Periodic Table, which are 12.011 for C and 15.999 4 for O, but are instead the masses of the specific isotopes at hand. This distinction is particularly important for elements that have several isotopes of appreciable abundance, such as Br, atomic weight = 79.904 amu, with ^{79}Br (50.69% abundant) weighing 78.918 amu and ^{81}Br (49.31% abundant) weighing 80.916 amu.

The B_0 value of the previous Example, 1.922 cm^{-1}, and the reduced mass give $R_0 =$ 1.131 Å. This number can be improved if centrifugal distortion (the reason the spectral lines are not exactly equally spaced) is taken into account, and the current R_e value for CO, based on numerous spectra of many isotopes, is 1.128 3 Å. (See Table 19.2.)

⟹ RELATED PROBLEMS 19.13, 19.14, 19.23

Diatomic Vibrational Spectroscopy

We again imagine a $^1\Sigma^+$ molecular state, and we consider the transition moment starting with the dipole moment function, $p(R)$. Since the initial and final states now have *different* vibrational quantum numbers, we cannot average this function, but we can expand it about $R = R_e$:

$$p(R) = p(R_e) + (dp/dR)_{R_e}(R - R_e) + \cdots \quad (19.42)$$

The transition moment, Eq. (18.20), now looks like ($p' = (dp/dR)_{R_e}$)

$$\begin{aligned} p_{kn} &= \int \psi^*_{v',J'} \, p(R) \psi_{v'',J''} \, d\tau_{\text{nuc}} \\ &= \int \psi^*_{v',J'} [p(R_e) + p'(R_e)(R - R_e) + \cdots] \psi_{v'',J''} \, d\tau_{\text{nuc}} \\ &\cong p(R_e) \int \psi^*_{v',J'} \psi_{v'',J''} \, d\tau_{\text{nuc}} + p'(R_e) \int \psi^*_{v',J'} (R - R_e) \psi_{v'',J''} \, d\tau_{\text{nuc}} \ . \end{aligned}$$

The integral multiplying $p(R_e)$ is zero by orthogonality unless $v' = v''$ and $J' = J''$, which implies no transition. Thus, the integral multiplying $p'(R_e)$ controls the transition, and we see that the molecule must have a *changing* dipole moment ($p' \neq 0$) in the vicinity of R_e, a criterion that is easy to satisfy.

Selection rules come from the second integral as well. The ΔJ and ΔM rules, since they depend on only the angular coordinates, are the same as for rotational transitions, Eq. (19.40). The rule for Δv depends on the vibrational wavefunction. If it is purely a harmonic oscillator wavefunction, a strict $\Delta v = \pm 1$ rule follows since these functions (Eq. (12.19)) have the property (see also Problem 12.13) that

$$\int \psi^*_{v'} (R - R_e) \psi_{v''} \, dR = 0 \quad \text{unless} \quad v' - v'' = \pm 1 \ .$$

While the true wavefunction is not purely harmonic, it is overwhelmingly so for the lowest few vibrational levels, and the $\Delta v = \pm 1$ selection rule dominates vibrational spectroscopy.

The minor anharmonic component of ψ_{vib}, which increases with increasing v, allows transitions with larger values of Δv. The strongly allowed $\Delta v = \pm 1$ transitions are called *fundamental* transitions, while the higher Δv transitions are called *overtones*. The intensity of successively *higher* energy overtones is successively *weaker* by a factor ~0.1 to 0.01. In summary, the selection rules for vibrational transitions are

$$\begin{aligned} \Delta v &= \pm 1, \text{ primarily} \quad (+1 \text{ is absorption}, -1 \text{ is emission}) \\ &= \pm 2, \pm 3, \text{ etc., with diminished intensity} \ . \end{aligned} \quad (19.43)$$

The $\Delta J = \pm 1$ rule means a rotational state change accompanies a vibrational state change. We do not find a single line in the $\Delta v = +1$ absorption spectrum, but a series of them, as Figure 19.9 (lower panel) showed. The transitions naturally fall into two *branches* traditionally called[7]

P branch	(Q branch)	R branch
$\Delta J = -1$	$(\Delta J = 0)$	$\Delta J = +1$

where the $\Delta J = 0$ Q branch is not allowed for $^1\Sigma^+$ states, but can occur in electronic states with $\Lambda \neq 0$. Transitions are denoted, for example, R(J'') so that transition R(3) means J starts at 3 and increases by 1. (Note that there is an R(0) transition, but *no* P(0) transition, since P(0) implies starting at $J = 0$ and decreasing J by 1, which is impossible.)

Figure 19.11 shows the HF spectrum again with an accompanying annotated energy level diagram. The intensity variations from line to line primarily reflect the

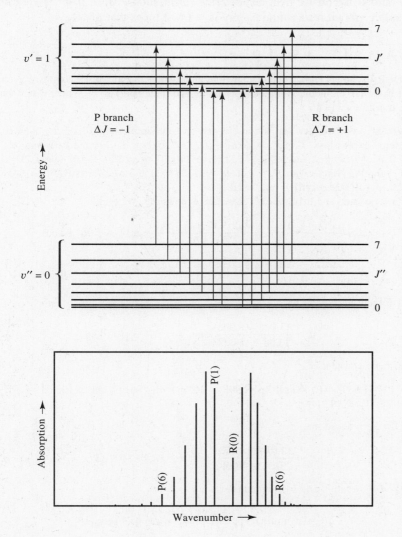

FIGURE 19.11 The IR absorption spectrum of HF around 4000 cm^{-1} (lower panel) can be assigned to a series of $\Delta J = J' - J'' = +1$ (R branch) and $\Delta J = -1$ (P branch) transitions for the vibrational excitation $\Delta v = 1, v'' = 0 \rightarrow v' = 1$.

---

[7]Note that the branch names, P, Q, R, increase in alphabetical order just as the corresponding ΔJ values, $-1, 0, +1$, increase in numerical order.

populations of the initial rotational state of each transition, since any one transition intensity is directly proportional to the number of absorbing molecules in the initial state.

The primary new piece of structural information from IR spectra is the vibrational constant (or constants). The ω_e constant yields the harmonic force constant k, a characteristic of $V(R)$. We find k from Eq. (19.22) written for the units of ω_e:

$$(100 \text{ cm m}^{-1})hc\omega_e = \hbar(k/\mu)^{1/2} \quad \text{or} \quad k = \mu[2\pi c\omega_e(100 \text{ cm m}^{-1})]^2 .$$

Isotopic substitution alters the location of an IR transition; k stays the same, but μ changes, changing ω_e. For two isotopomers with reduced masses μ and μ^*,

$$\omega_e^* = \omega_e \left(\frac{\mu}{\mu^*}\right)^{1/2} \quad \text{and} \quad B_e^* = B_e \frac{\mu}{\mu^*} .$$

This ratio is largest for hydride/deuteride substitutions. Since μ/μ^* is very nearly 1/2, a DX vibration transition energy is $\cong 1/\sqrt{2}$ times that of HX.

EXAMPLE 19.10

The HF R(0) and P(1) fundamental transitions are at 4 003.35 and 3 919.92 cm^{-1}, respectively. What is the HF harmonic vibration constant ω_e and force constant k? Where would we find these transitions in DF?

SOLUTION With only two transitions, we must use the simplest rigid-rotor harmonic oscillator energy expressions: $E(v, J) = \omega_e(v + 1/2) + BJ(J + 1)$. For the R(0) line, $v'' = 0$, $J'' = 0, v' = 1, J' = 1$ and $\Delta E_{R(0)} = E(1, 1) - E(0, 0) = \omega_e(3/2) + 2B - \omega_e(1/2) = \omega_e + 2B$. For the P(1) line, $v'' = 0, J'' = 1, v' = 1, J' = 0$ and $\Delta E_{P(1)} = E(1, 0) - E(0, 1) = \omega_e(3/2) - [\omega_e(1/2) + 2B] = \omega_e - 2B$.

If we first add and then subtract these ΔE expressions, we find

$$\omega_e = \frac{\Delta E_{R(0)} + \Delta E_{P(1)}}{2} = \frac{4\,003.35 + 3\,919.92}{2} \text{ cm}^{-1} = 3\,961.64 \text{ cm}^{-1}$$

$$B = \frac{\Delta E_{R(0)} - \Delta E_{P(1)}}{4} = \frac{4\,003.35 - 3\,919.92}{4} \text{ cm}^{-1} = 20.86 \text{ cm}^{-1} .$$

The HF reduced mass is

$$\mu = \frac{(1.008)(18.998)}{1.008 + 18.998} \text{ amu} = 1.589 \times 10^{-27} \text{ kg}$$

so the force constant is

$$k = (1.589 \times 10^{-27} \text{ kg})[(2\pi)(2.998 \times 10^8 \text{ m s}^{-1})(3\,961.64 \text{ cm}^{-1})(100 \text{ cm m}^{-1})]^2$$
$$= 885 \text{ N m}^{-1} .$$

For DF, the reduced mass is

$$\mu^* = \frac{(2.014)(18.998)}{2.014 + 18.998} \text{ amu} = 3.024 \times 10^{-27} \text{ kg} ,$$

and the DF vibration and rotation constants are

$$\omega_e^* = 3\,961.64 \left(\frac{1.589}{3.024}\right)^{1/2} \text{ cm}^{-1} = 2\,871.75 \text{ cm}^{-1}$$

$$B^* = 20.86 \frac{1.589}{3.024} \text{ cm}^{-1} = 10.96 \text{ cm}^{-1} .$$

The R(0) and P(1) lines appear at

$$\Delta E_{R(0)} = \omega_e^* + 2B^* = 2\,893.67 \text{ cm}^{-1}$$

$$\Delta E_{P(1)} = \omega_e^* - 2B^* = 2\,849.83 \text{ cm}^{-1} \, .$$

⟹ RELATED PROBLEMS 19.16, 19.17, 19.19

Note that diatomics with a pure rotational spectrum ($p \neq 0$) also have an IR vibration-rotation spectrum ($dp/dR \neq 0$). We have not yet discussed a way to observe homonuclear diatomics ($p = dp/dR = 0$), but *Raman spectroscopy* allows a way. We hinted in Section 19.1 that the Raman effect would come to the rescue when we introduced it and pointed out that it has a unique set of selection rules.

Raman spectroscopy operates not on a permanent dipole moment but on an *induced* moment. The radiation's electric field induces a dipole moment via the molecular polarizability (see Eq. (15.5)). If the polarizability of a linear molecule is written in terms of its components parallel and perpendicular to the internuclear axis, $\alpha_\parallel$ and $\alpha_\perp$, then $\alpha_\parallel - \alpha_\perp$, the *anisotropy,* allows a *pure rotational* Raman spectrum with selection rules $\Delta J = 0, \pm 2$ (with $\Delta J = 0$ uninteresting elastic or *Rayleigh* scattering). Derivatives of α's components with respect to R allow a vibration-rotation Raman spectrum with selection rules $\Delta v = \pm 1$, $\Delta J = 0, \pm 2$.

The $\Delta J = 0, \pm 2$ (instead of ± 1) selection rule is a consequence of the two-photon nature of the Raman process. One photon strikes the molecule and a second photon leaves it, not by the time-delayed process of spontaneous emission, but by instantaneous scattering. Raman vibration-rotation spectral branches are designated O and S for $\Delta J = -2$ and $+2$, respectively, building on the P, Q, R alphabetic order. A scattered photon of *less* energy than the incident photon is said to be "Stokes scattered," while the opposite, a scattered photon of *greater* energy, is said to be "Anti-Stokes" scattered.

The laser has been the saving grace of Raman spectroscopy, which would not have been nearly so useful a technique otherwise. The reason is simply that a Raman scattering event is depressingly rare. The majority of scattered photons (often by a factor of 10^9 or more) are *elastically* scattered. Only those rare inelastic events that exchange energy with the molecule carry molecular structure information. Prior to the laser, an intense Hg lamp was used (with suitable filters to isolate strong atomic emission lines) as the radiation source. While this worked, (after all, Chandrasekhara Raman won the Nobel Prize in 1930 for his discovery of the effect, 30 years before the word laser was coined), the Hg lamp has neither the intensity, monochromaticity, nor intensity stability of a modern laser.

Diatomic Electronic Spectroscopy

We start with a general expression for the dipole transition moment and derive not only selection rules but also an additional consequence of the B–O approximation. We write the dipole moment vector operator as a sum of its electron and nuclear charge terms:

$$\hat{\mathbf{p}} = \hat{\mathbf{p}}_{el} + \hat{\mathbf{p}}_{nuc}$$

and the molecular wavefunction as the B–O product of the electronic and vibrational wavefunctions, ignoring rotation for now:

$$\Psi = \psi_{el}\psi_{vib}$$

so that

$$\begin{aligned}\mathbf{p}_{kn} &= \int (\psi'_{el}\psi'_{vib})^*(\hat{\mathbf{p}}_{el} + \hat{\mathbf{p}}_{nuc})(\psi''_{el}\psi''_{vib})\,d\tau \\ &= \int \psi'^*_{el}\,\hat{\mathbf{p}}_{el}\,\psi''_{el}\,d\tau_{el}\int \psi'^*_{vib}\,\psi''_{vib}\,dR + \int \psi'^*_{el}\,\psi''_{el}\,d\tau_{el}\int \psi'^*_{vib}\,\hat{\mathbf{p}}_{nuc}\,\psi''_{vib}\,dR\ .\end{aligned}$$

The second term is zero since the electronic wavefunctions are orthogonal, but the first term is *not* necessarily zero. The vibrational wavefunctions are *not* orthogonal, since they represent vibrations in *different electronic states*. For electronic transitions, $\mathbf{p}_{kn}$ is the product of two factors. The first is called the *electronic transition moment* $\mathbf{M}(R)$:

$$\mathbf{M}(R) = \int \psi'^*_{el}\,\hat{\mathbf{p}}_{el}\,\psi''_{el}\,d\tau_{el}\ . \qquad (19.44)$$

It governs selection rules for the quantum numbers of electronic states.

The second factor is fascinating. Called the *Franck–Condon overlap integral*, it modulates the transition moment as v'' and v' change. Since the transition probability is proportional to $|\mathbf{p}_{kn}|^2$, the *square* of this integral is known as the *Franck–Condon factor*:

$$g_{v',v''} = \left[\int \psi'^*_{vib}\,\psi''_{vib}\,dR\right]^2\ . \qquad (19.45)$$

The Franck–Condon factor exposes the B–O approximation at work in electronic spectroscopy. When a molecule absorbs radiation that alters the electronic state, the electrons reconfigure themselves so quickly that the nuclei are essentially stationary during the transition. The nuclei, which were in a definite vibrational level before the transition, suddenly find themselves under the influence of the forces of a new electronic state. The Franck–Condon overlap integral measures the probability that the molecule can start in one vibrational level of one electronic state and end in another level of another state via this mechanism.

Figure 19.12 shows schematic potential energy curves for two electronic states

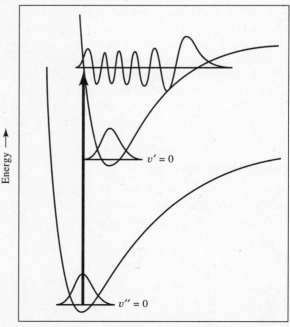

FIGURE 19.12 The Franck–Condon principle governs the levels of the excited electronic state that can be accessed from a lower state level. Access is facilitated when the two vibrational wavefunctions have appreciable overlap.

that have characteristics common to diatomics: the bond length of the ground state is *less* than that of the excited state, while the ground state dissociation energy is *greater* than that of the excited state.[8] The diagram assumes the molecule is initially in the lowest vibrational level, $v'' = 0$. The excited state has a set of vibrational levels, and the wavefunctions for $v' = 0$ and for a higher level (count the nodes to see which one!) are shown in the figure.

Suppose we wanted to induce a transition from $v'' = 0$ to the higher v' level. For the transition to be probable, the v' and the v'' vibrational wavefunctions must have appreciable magnitude over some common range of internuclear separation. This makes $g_{v',v''}$ appreciable. Since the $v'' = 0$ wavefunction peaks near R_e'', the ground-state equilibrium separation, we want the v' wavefunction to be appreciable in the vicinity of R_e'', as the diagram shows it to be.

In contrast, the $v'' = 0$ to $v' = 0$ transition would be *much* less probable, since $R_e' > R_e''$ and the $v' = 0$ wavefunction is very small near R_e''. In the $v'' = 0$ level, the nuclei simply fail to reach the bond length characteristic of the excited state $v' = 0$ level. The Franck–Condon factor is very small.

In the spirit of the B–O approximation, electronic transitions are called *vertical*. The vertical arrow in Figure 19.12 expresses the idea that R does not change during the time of the transition. This idea is also known as the *Franck–Condon principle*. A vertical electronic transition need not end at a bound vibrational level, however. Section 18.4's discussion of photodissociation can be continued here with reference to HI and Figure 18.6. A vertical transition from R_e'' terminates on the repulsive, nonbound portion of the excited singlet state. If R happened to be greater than R_e'' at the instant of the transition,[9] the vertical line to the excited state curve would be shorter. The low energy, long wavelength end of the photodissociation absorption band selects those ground state molecules with *extended* bond lengths. Similarly, *compressed* bond lengths contribute to *higher* energy transitions in the absorption.

There is no strict selection rule on vibrational quantum number changes in electronic transitions, only a Franck–Condon modulation of transition probability. The electronic state quantum number selection rules, however, have the final say. We will not derive them nor even state them all; they can be found in many of this chapter's references. The majority of diatomic electronic transitions are covered by the rules

$$\Delta \Lambda = 0, \pm 1 \qquad (19.46a)$$

so that $\Sigma \to \Sigma$, $\Sigma \to \Pi$, $\Pi \to \Sigma$, $\Pi \to \Pi$, etc., transitions are allowed,

$$\Delta S = 0 \qquad (19.46b)$$

so that singlet $\to$ singlet, triplet $\to$ triplet, etc., transitions are allowed, and

$$\text{for } \Sigma \text{ to } \Sigma \text{ transitions, } + \leftrightarrow +, \; - \leftrightarrow -, \text{ but } + \not\leftrightarrow -, \qquad (19.46c)$$

while for homonuclear diatomics,

$$g \leftrightarrow u, \text{ but } g \not\leftrightarrow g \text{ and } u \not\leftrightarrow u . \qquad (19.46d)$$

These are the rules for strongly allowed transitions, but transitions that violate the first two are often observed, although weakly.

---

[8] It is rare, therefore, to find an excited state that is "longer and stronger" than the ground state, but one is lurking in this chapter! See Problem 19.9.

[9] This is very "semiclassical" language. We are talking about the nuclear position as if it were classical in nature while keeping the electronic transition otherwise quantum-like.

Rotational selection rules are also more extensive than we can discuss here. In the most important cases,

$$\Delta J = \pm 1 \quad (\Sigma \text{ to } \Sigma \text{ electronic transitions, P and R branches})$$

or

$$\Delta J = 0, \pm 1 \quad (\text{for any transition involving } \Lambda \neq 0) \ . \quad (19.46\text{e})$$

Thus, for a $^1\Sigma_g^+$ to $^1\Sigma_u^+$ transition (such as the Na_2 X to A state transition in Example 19.6), there are P and R rotational branches to *each* vibrational band of the electronic transition, while a $^1\Sigma_g^+$ to $^1\Pi_u$ transition (X to B in Na_2) has P, Q, and R branches.

With everything changing in a molecular electronic transition, electronic spectra can be very complex. It is common to find several thousand transitions in one electronic spectrum of a typical diatomic.

Spontaneous emission is especially useful for electronic transitions. For pure rotational or for vibration-rotation transitions, the radiative lifetimes are generally too long to be of use. The excited molecule has collided with the apparatus walls, typically, before emitting. Radiative lifetimes of excited electronic states, in contrast, often fall in the 10^{-4} to 10^{-8} s range, which is sufficiently brief.

One type of spontaneous emission experiment is particularly useful. Called *laser induced fluorescence*, LIF, it uses the experimental arrangement diagrammed in Figure 19.13. In one type of experiment, a narrow frequency tunable laser is directed into the gaseous sample. As the scanned laser frequency hits an allowed molecular transition, excited states are produced and spontaneously emit, or *fluoresce*,[10] in all directions. A sensitive light detector, such as a photomultiplier, placed perpendicular to the laser direction records the spontaneous emission intensity as a function of the laser frequency.

This experiment has much in common with an ordinary absorption experiment, but two differences are important. The first is an increase in sensitivity. This comes about from the change in detection scheme in LIF versus absorption. Direct absorption measures a decrease in intensity from a high value to, in the case of weak absorptions, a slightly lower value. The inherent "noise" of any light source and detector limits the measurable decrease and hence limits the minimum detectable absorption. In contrast, LIF uses a detector that emits *no* signal unless an LIF event has caused even a few photons per second to strike it. Modern photomultipliers can readily count individual photons at rates of only a few per second, even if the laser is blasting away at 10^{15} photons per second or more. (Great care must be taken, of course, to keep the laser photons from scattering into the photomultiplier. This can be the hardest part of the experiment!)

The second difference between LIF and absorption is more subtle. Absorbance records a signal when a photon has been absorbed, but LIF requires that one photon is absorbed and a second is emitted. Nonradiative events *after* absorption can preclude emission. For example, the HI photodissociation spectrum is observed by absorption spectroscopy, but *not* by LIF, since the HI molecule dissociates before it can emit (and the atomic fragments do not emit).

A second variant of the LIF experiment tunes the laser to a transition and holds it there. A device that can analyze the emitted photons' wavelengths is inserted

---

[10]"Fluorescence" is a generic term for spontaneous emission. It is also sometimes limited to states of very brief (≤ 10 μs or so) radiative lifetime with the term "phosphorescence" reserved for radiative lifetimes in the ms or longer range. Any lifetime less than about 10 ms or so appears to the eye to be instantaneous, while longer lifetimes, the "phosphorescent" states, appear to have emission that lingers after the excitation is extinguished. These terms are considered further in Chapter 20.

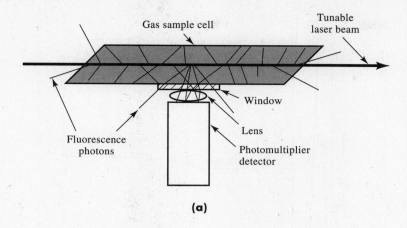

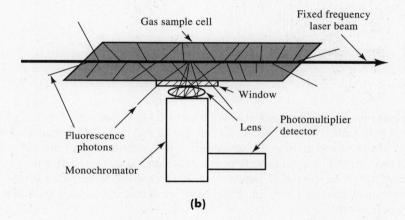

FIGURE 19.13 Laser induced fluorescence spectra can be recorded in two different ways: in (a), a tunable laser is used to excite the sample, and that fraction of the fluorescent photons that strike a detector are recorded versus laser frequency; in (b), the laser frequency is fixed to a particular molecular transition, and the fluorescent photon's frequencies are analyzed by a monochromator or other type of light frequency analyzer.

between the sample and the photomultiplier, and a *dispersed fluorescence spectrum* is recorded. (The device can be as simple as a set of filters or as complex as a diffraction grating monochromator or an interferometer.) This type of spectrum is very simple. It begins with the molecule in a single known electronic, vibrational, rotational state, fluorescing under the control of selection rules to lower states. The rotational structure of the emission spectrum is reduced to a simple P, R (or P, Q, R) pair (or triplet) of lines for each final vibrational state that is Franck–Condon connected to the prepared excited-state level. Figure 19.14 demonstrates how the Franck–Condon principle works in this experiment.

Polyatomic Spectra

We will not discuss polyatomic spectroscopy and spectroscopic structural determination in great detail. The next chapter considers polyatomic spectra from other points of view, and further details can be found in the chapter's references.

For rotational spectra, we still require a permanent dipole moment. Thus, N_2O (linear, polar) has a pure rotational spectrum while CO_2 (linear, nonpolar) does not. For nonlinear polyatomics, any total dipole moment can be decomposed into components along the directions of the principal moments of inertia. For symmetric tops, the permanent moment lies completely along the top's symmetry axis.

Selection rules for a symmetric top are $\Delta J = \pm 1$ and $\Delta K = 0$. (The dipole moment is along the symmetry axis, but a change in K would require a

FIGURE 19.14 If one disperses the spontaneous fluorescence in a LIF experiment, the emission is often a series of transitions to vibrationally excited levels of the ground state. The Franck–Condon principle governs which transitions are observable. Here, $R_e' > R_e''$, which causes many such transitions to be observable.

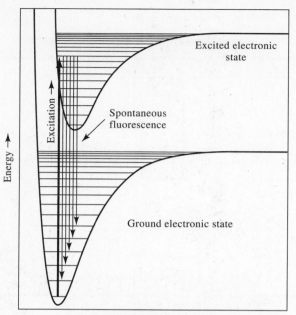

component perpendicular to this axis. Without such a component for the electromagnetic field to grab, it cannot change K.) For asymmetric tops, the rules are complex. Spherical tops would seem to be incapable of a pure rotational spectrum, since they lack a permanent dipole moment. In their equilibrium structure, spherical tops (CH_4, SF_6, SiH_4, etc.) are nonpolar, but as they rotate, centrifugal distortion lowers the symmetry and induces a very small dipole moment and thus a very weak absorption spectrum.

Polyatomic vibration-rotation spectra are strongly governed by molecular symmetry. Each normal mode has to be examined. If the nuclear motion of a particular mode changes the total dipole moment, then that mode will absorb with a $\Delta v = 1$ selection rule. For example, the three normal modes of H_2O in Figure 19.6(a) are all infrared active, but only two of the three modes of CO_2 in Figure 19.6(b) are active. The inactive CO_2 mode is the symmetric stretch, which has $\mathbf{p} = 0$ at all times. For Raman spectra, the molecular polarizability must change with normal mode motion. Group theory can be used to classify normal mode motions and determine if a mode is Raman or IR active. (Or both, but for centrosymmetric molecules like CO_2, benzene, etc., an *exclusion rule* exists: starting from the ground vibrational state, IR active modes are Raman inactive and vice versa.)

Overtones of diminished intensity can be observed as in diatomics. Moreover, when symmetry allows, *combination* transitions can be observed in which two or more modes are excited together. These transitions occur close to the sum of the transition energies of the modes involved.

High resolution polyatomic spectra (in all wavelength regions) can be assigned and analyzed much as those for diatomics, and in favorable cases, complete structural and bond force-constant information can be deduced. But molecular spectroscopy has turned increasingly from the collection of molecular constants and structural data to an analysis aimed at uncovering as much as possible about interatomic forces. This detailed picture of chemical bonding is embodied in the Born–Oppenheimer potential energy function, the topic of the next section.

19.4 POTENTIAL ENERGY SURFACES

If ab initio theory can calculate a B–O potential energy function for the motion of a molecule's nuclei, and if the bound eigenstates of this function can be found, then, given selection rules, molecular spectra can be predicted. In many cases, this can be done today to useful accuracy. Such predictions are valuable guides to the experimental search for the spectra of elusive species. But can the reverse be done? Can a spectrum be converted to tables of energy levels and can these levels be turned into a potential function? Can a B–O potential energy function be *measured* by experimental spectroscopy?

This question is known as the *inversion problem;* the path from potential surface to spectrum is relatively straightforward, but the inverse is much less so. In this section, we see how some initial attacks on the inversion problem are made. The first two steps—attaining and assigning spectra to yield energy-level expressions— are the easy steps, and we assume they have been carried out. Now we are confronted with an energy-level expression (such as the Dunham expression, Eq. (19.25), for a diatomic) and its associated molecular constants.

We will start with the simplest case: a diatomic in a $^1\Sigma^+$ state. We seek the function $V(R)$ that enters the B–O Hamiltonian of Eq. (19.8). We know that $V(R)$ has the general shape of, say, the true H_2 $^1\Sigma_g^+$ $V(R)$ in Figure 19.1. Many simple functions have more or less this shape, and well over 50 of them have been used at one time or another. The idea is to choose one with several parameters and vary the parameters until the function closely reproduces the observed bound levels.

To do this, it is a great advantage if the observable spectroscopic constants (ω_e, B_e, etc.) can be written as analytic functions of the parameters. We will look at only one function in detail, but it is very important because its parameters *can* be written in terms of observable constants. It turns out to be only marginally accurate (by spectroscopy's high standards), but it is very instructional and almost invaluable when data that would warrant a more accurate function are lacking.

The function is called the *Morse (oscillator) function:*

$$V(R) = \mathcal{D}_e(1 - e^{-\beta(R - R_e)})^2 = \mathcal{D}_e(1 - e^{-\beta\rho})^2 \qquad (19.47)$$

introduced by Morse in 1929. It has a minimum number of parameters: three. We need one to express the bond length, R_e, one to express the dissociation energy, $\mathcal{D}_e$, and one to express the harmonic force constant, $k \equiv V''(R_e)$. The parameter β is related to k via

$$\left(\frac{d^2V}{dR^2}\right)_{R_e} = k = 2\beta^2\mathcal{D}_e$$

so that

$$\beta = \left(\frac{k}{2\mathcal{D}_e}\right)^{1/2}.$$

We will not discuss the Morse vibrational wavefunction,[11] but we will spend some time looking at its simple energy-level expression. The vibration energy has a *single* anharmonic constant, $\omega_e x_e$, in addition to the harmonic term:

$$G(v) = \omega_e\left(v + \frac{1}{2}\right) - \omega_e x_e\left(v + \frac{1}{2}\right)^2$$

---

[11]The Schrödinger equation for this potential can be solved *exactly* for $J = 0$ (and to a high degree of accuracy for any J). Were the mathematics of this solution simpler, it would have appeared in Chapter 12 with our other model systems.

where, with ω_e in cm^{-1} units, $\mathcal{D}_e$ in J, and β in m^{-1},

$$\omega_e = \frac{\beta}{2\pi c(100\text{ cm m}^{-1})}\left(\frac{2\mathcal{D}_e}{\mu}\right)^{1/2} \tag{19.48a}$$

and

$$\omega_e x_e = \frac{\omega_e^2}{4\mathcal{D}_e}[hc(100\text{ cm m}^{-1})] \ . \tag{19.48b}$$

(The molecular identity enters as a fourth parameter through μ, the reduced mass.)

The rotational energy expression through its three most important constants is:

$$F_v(J) = B_e J(J+1) - D_e[J(J+1)]^2 - \alpha_e J(J+1)\left(v + \frac{1}{2}\right)$$

(where D_e is the centrifugal distortion constant) and one finds

$$B_e = \frac{\hbar^2}{2\mu R_e^2}\left(\frac{1}{hc(100\text{ cm m}^{-1})}\right) \tag{19.48c}$$

(which is just Eq. (19.33a)), and (do not confuse D_e, the centrifugal distortion constant, with $\mathcal{D}_e$, the dissociation energy)

$$D_e = \frac{4B_e^3}{\omega_e^2} = \text{Morse centrifugal distortion constant} \tag{19.48d}$$

while

$$\alpha_e = \frac{6}{\omega_e}[(B_e^3 \omega_e x_e)^{1/2} - B_e^2] \ . \tag{19.48e}$$

Note how measured constants give potential parameters. First, we find ω_e and $\omega_e x_e$ from the spectrum to find the dissociation energy, $\mathcal{D}_e$, from Eq. (19.48b), then use μ, $\mathcal{D}_e$, and ω_e to find β from Eq. (19.48a), and finally find R_e from B_e as usual. Measured and calculated (from Equations (19.48d,e)) values of D_e and α_e can be compared to see how "Morse-like" the molecule is. For example, the ground state of H_2 has the following observed spectroscopic constants

$$\omega_e = 4\,401.213\text{ cm}^{-1} \qquad B_e = 60.853\,0\text{ cm}^{-1}$$
$$\omega_e x_e = 121.336\text{ cm}^{-1} \qquad D_e = 4.71 \times 10^{-2}\text{ cm}^{-1}$$
$$\alpha_e = 3.062\,2\text{ cm}^{-1} \ .$$

These constants give a Morse $\mathcal{D}_e = 7.928 \times 10^{-19}$ J while the actual $\mathcal{D}_e$ (measured by a variety of other experiments) is 7.607×10^{-19} J. The value for β is 1.904 Å^{-1}, and R_e is 0.741 4 Å. As a check on consistency, Eq. (19.48d) predicts $D_e = 4.653 \times 10^{-2}$ cm^{-1} (off by 1.2% from the observed) and Eq. (19.48e) predicts $\alpha_e = 2.08$ cm^{-1} (off by 32%). If the exact and Morse potentials for H_2 are compared, one finds the agreement is surprisingly good at all R, especially since only three parameters determine the Morse potential. It is somewhat unexpected and astounding that the essence of the H_2 bond can be represented in such detail by only three numbers.

A Closer Look at the Morse Potential

In Chapter 13, we used cubic and quartic perturbations to a harmonic oscillator to introduce perturbation theory, and Example 13.4 looked closely at the cubic perturbation. It is instructive to take apart the Morse function (expand it about $R = R_e$), and consider its quadratic and cubic parts. First, however, we strip the function to its essence. If we define a dimensionless length, $x = \beta\rho = \beta(R - R_e)$, and measure energy in units of the dissociation energy, we have $v(x) = V(R)/\mathcal{D}_e = (1 - e^{-x})^2$. At $x = -1$ (i.e., $R = 0$), $v(-1) = (1 - e)^2 = 2.95$. This is nonphysical. As $R \to 0$, the potential energy should increase rapidly to a very high value, not to a small finite value. Thus, the Morse function does not represent the repulsive "wall" of an interatomic potential very well, but it does a much better job representing the binding well region.

An expansion of $v(x)$ about $x = 0$ (i.e., $R = R_e$) gives $v(x) \cong x^2 - x^3 + 7x^4/12 - x^5/4 \cdots$. The graph below compares the full potential (heavy line) to the quadratic (harmonic) approximation (medium line) and the quadratic + cubic approximation (light line):

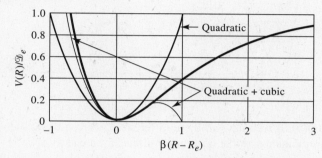

Compare this figure to the plot of the cubic perturbation to a harmonic oscillator in Example 13.4. The $-x^3$ term in the expansion of $v(x)$ shows an origin of anharmonicity, but at large R, the cubic term goes to $-\infty$ rather than to a finite asymptotic dissociation limit.

The Morse harmonic force constant is $k = 2\beta^2\mathcal{D}_e$. Our expansion of $v(x)$ gives us the next few terms: $V(R) = k\rho^2/2 - \mathcal{D}_e\beta^3\rho^3 + 7\mathcal{D}_e\beta^4\rho^4/12 \cdots$ where $\rho = R - R_e$. In Chapter 13, we wrote cubic and quartic perturbations in a general way: $a\rho^3$ and $b\rho^4$. What we have found here are expressions for a and b specific to the Morse function: $a = -\mathcal{D}_e\beta^3$ and $b = 7\mathcal{D}_e\beta^4/12$.

While a perturbation expansion of the Morse function is not very useful, since accurate energy expressions are known for the full function, the exercise is instructive. Perturbation methods are often applied to more complicated potential functions as a useful alternate to numerical methods.

If we write Eq. (19.48b) in the form

$$\mathcal{D}_e = \frac{\omega_e^2}{4\omega_e x_e}[hc(100 \text{ cm m}^{-1})] \qquad (19.49)$$

we get an expression for the bond energy in terms of vibrational parameters, and Problem 19.27 considers an additional expression based on ω_e, B_e, and α_e. Equation (19.49) can be derived by a very physical argument that shows another experimental approach to $\mathcal{D}_e$.

Figure 19.15 shows the H_2 Morse potential and its bound vibrational levels (for $J = 0$). The function $G(v)$ gives the energy of these levels (referenced to an energy zero at the potential well's bottom), and we can tabulate a series of energy differences,

FIGURE 19.15 The rotationless Morse potential for the H$_2$ ground state has 18 bound vibrational levels, compared to 15 in the true molecule. Nevertheless, the Morse approximation is often the simplest model function with any claim to accuracy.

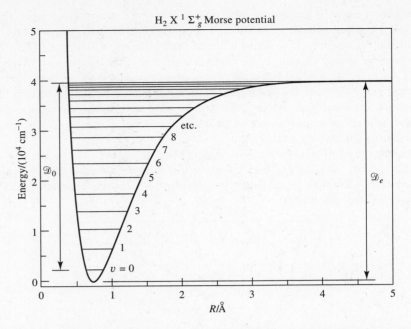

$G(v + 1) - G(v)$. If we knew *all* these differences, we could add them and get a very close approximation to the observable dissociation energy $\mathcal{D}_0$:

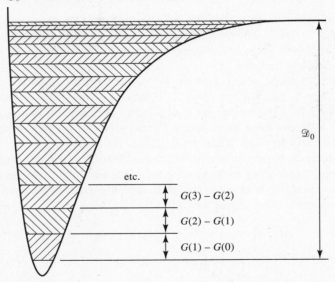

Our only error would be the tiny energy difference from the last bound vibrational level to the true atomic dissociation limit.

In practice, we usually measure only a few of these energy differences for the lowest few vibrational levels. Birge and Sponer suggested in 1926 how such differences could be extrapolated to yield an estimate of the bond energy; their method was later shown to be exact for the Morse oscillator. The Birge–Sponer extrapolation follows these steps. First, a running index, n, is defined with $n = v + 1/2$. Then the differences $\Delta G(n) = G(n + 1/2) - G(n - 1/2) = G(v + 1) - G(v)$ are plotted versus n. (Note that $n = 1/2, 3/2, 5/2$, etc.) Birge and Sponer noticed that such values invariably fell close to a straight line, at least for the lowest few n

values. They drew the best such line and extrapolated it to the abscissa. (The line has a negative slope since anharmonicity ensures $\Delta G(n + 1) < \Delta G(n)$.) The *area under that line is a close approximation to the dissociation energy.*

We can carry out this calculation exactly with the Morse $G(v)$. If we define n_D to be that value of n for which $\Delta G(n_D) = 0$ (i.e., the x intercept of the line), we find

$$n_D = \frac{\omega_e - \omega_e x_e}{2\omega_e x_e} .$$

The area under the $\Delta G(n)$ line is

$$\text{Area} = \int_0^{n_D} \Delta G(n)\, dn = \frac{\omega_e^2}{4\omega_e x_e} - \left(\frac{\omega_e}{2} - \frac{\omega_e x_e}{4}\right),$$

which we can recognize as

$$\text{Area} = \mathcal{D}_e - G(v = 0) \quad (\text{all in cm}^{-1} \text{ units})$$

or

$$\mathcal{D}_e = \text{Area} + \text{zero-point energy} .$$

The area itself, $\mathcal{D}_e - G(0)$, is the observable dissociation energy, $\mathcal{D}_0$, the energy required to pull the molecule apart from the zero-point level.

Given only ω_e and $\omega_e x_e$, we can calculate how many bound vibrational levels the corresponding Morse oscillator should have. Let v_D be the pseudo (non-integral) vibrational quantum number for the (fictitious) vibrational level just at the dissociation limit. Then

$$G(v_D) = \mathcal{D}_e = \frac{\omega_e^2}{4\omega_e x_e}$$

or

$$v_D = \frac{\omega_e - \omega_e x_e}{2\omega_e x_e} .$$

The true quantum number of the last level is the largest integer not larger than v_D. For H$_2$, $v_D = 17.64$, so that the Morse representation of H$_2$ has eighteen bound levels ($v = 0$ through 17). The *true* H$_2$ potential has only fifteen ($v = 0$ through 14). This sort of discrepancy causes the Birge–Sponer extrapolation to be in error, as Figure 19.16 shows. This figure plots the experimental $\Delta G(n)$ values as points and the Morse $\Delta G(n)$ line. The shaded triangular area is the Morse–Birge–Sponer

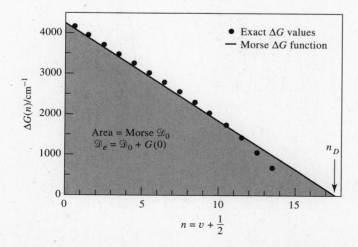

FIGURE 19.16 The Birge–Sponer extrapolation, which plots the energy difference between successive vibrational levels versus $v + 1/2$, yields the dissociation energy $\mathcal{D}_0$ as the area under the plot. For the Morse function, the extrapolation is a straight line, and the area is the shaded triangular portion of the figure.

approximation, while the area under a smooth line drawn through the true values is a more accurate approximation of the dissociation energy.

The inversion problem for polyatomics is much more difficult. For a diatomic, we can plot $V(R)$ in two dimensions. A function of many more variables for a polyatomic, V is a *multidimensional surface* that is often difficult to visualize. Consequently, a polyatomic V comes in bits and pieces. For example, in H_2O, we could use a Morse function to represent each O—H bond and a *bending potential function* for the H—O—H angle. This is a local mode view of V. The normal mode vibrational constants give us harmonic bending force constants, accurate for small amplitude vibrations of the more complicated normal mode coordinates.

We will leave further discussion of polyatomic potential energy surfaces for later chapters and our discussion of chemical reaction kinetics and dynamics. In the next section, we consider a final aspect of polyatomic spectroscopy: the great number of observable transitions.

19.5 LINE SHAPES AND SPECTRAL CONGESTION—WHAT'S OBSERVABLE?

With effort, and in limited spectral regions, today's best spectral *resolution,* the ability to distinguish two different transitions at virtually the same transition energy, can exceed one part in 10^9. More typically (and much less expensively), resolutions in the range of 10^3 to 10^5 are routinely available. In this section, we explore questions of observability in molecular spectroscopy, and experimental resolution is only part of the answer. The rest of the answer depends on the molecule and its physical state.

An experimental example indicates how we approach the question, "What's observable?" In the vicinity of 2000 cm^{-1}, the OCS molecule has extensive vibration-rotation transitions that are completely understood. The normal mode is a stretch in which the O═C bond is mostly involved. Figure 19.17(a) shows a 5 cm^{-1} portion

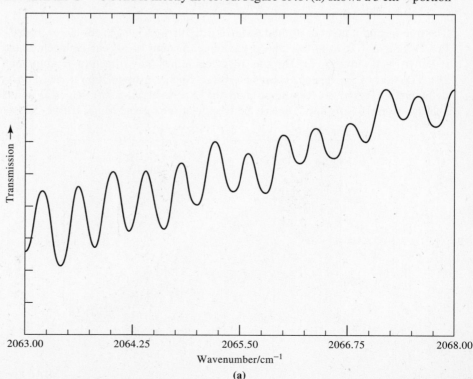

FIGURE 19.17 A portion of the OCS IR spectrum at (a) 0.4 cm^{-1} resolution, (b) 0.04 cm^{-1} resolution, and (c) 0.004 cm^{-1} resolution.

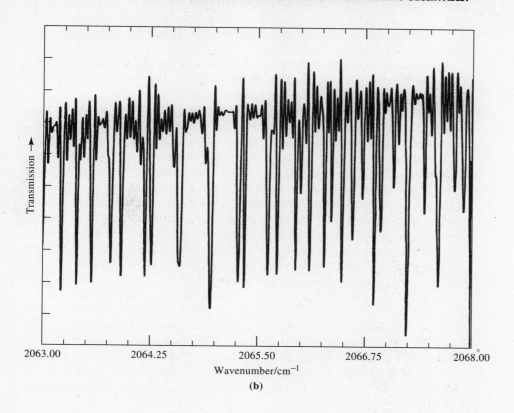

(b)

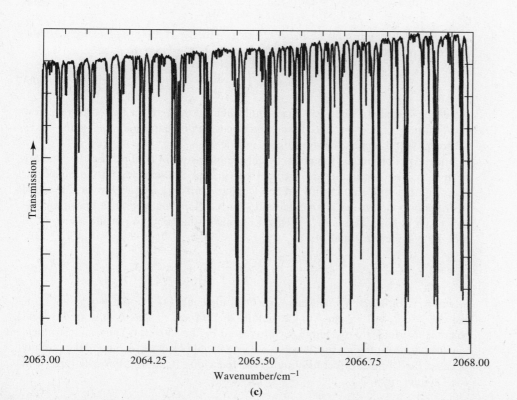

(c)

of the spectrum at a resolution of 0.4 cm^{-1}, which is as good or better than the resolution of most routine instruments.

Some obvious absorption features can be seen. At ten times higher resolution (Figure 19.17(b)), more structure can be seen. Finally, at 100 times higher resolution (Figure 19.17(c)), dozens of discrete features appear.

What if we increase the resolution another factor of ten? These spectra were recorded on an interferometric spectrometer (an "FTIR"—*F*ourier *T*ransform *I*nfra-*R*ed—spectrometer) for which significantly higher resolution is generally not possible, but IR laser spectra can be obtained at higher resolution. They look *essentially the same* (i.e., have the same *width* to each feature) as the spectrum in Figure 19.17(c). Something on the *molecular* level has limited resolution.

Chapter 18 pointed out that the radiative lifetime of an excited state automatically introduces an uncertainty in the state's energy and a corresponding finite width to spectroscopic transitions. A transition's frequency width, $\Delta \nu$, is on the order of $1/\tau_{rad}$ for the upper state, and for an IR transition with $\tau_{rad} \sim 1$ ms, $\Delta \nu \sim 10^3$ Hz or only $\sim 3 \times 10^{-8}$ cm^{-1}, far less than the observed linewidths of Figure 19.17(c). Something else must be governing this linewidth.

The major phenomenon that limits the linewidth here is called *Doppler broadening*. The acoustical Doppler effect is familiar to you; it accounts for the rise in pitch of, say, an ambulance siren as it moves *towards* you, and a fall in pitch as it moves *away*. The optical Doppler effect is similar. It is a consequence of special relativity. Suppose a molecule at rest can absorb a photon of frequency ν_0. Now suppose that the molecule moves *away from* the light source at some velocity v. That motion causes a *Doppler shift* in the transition frequency from ν_0 to the *higher* frequency ν_+ given by

$$\nu_+ = \nu_0 \left(\frac{1 + \beta}{1 - \beta} \right)^{1/2} \tag{19.50a}$$

where $\beta = v/c$, the ratio of the molecular velocity to the speed of light. If the molecule moved *towards* the source at velocity v, the *lower* frequency

$$\nu_- = \nu_0 \left(\frac{1 - \beta}{1 + \beta} \right)^{1/2} \tag{19.50b}$$

would be resonant. Since gas-phase molecular speeds are on the order of a few hundred m s^{-1}, $\beta \ll 1$, and these expressions can be expanded, neglecting terms of order β^2 and higher, yielding $\nu_\pm = \nu_0(1 \pm \beta)$.

In an equilibrium gas sample, molecules are moving in all directions at a variety of speeds. The distribution of speeds (see Chapter 24) depends on the molecular mass and the temperature only. When Doppler shifts are averaged over this speed distribution, one arrives at an expression for the *Doppler width* of a transition: the full width at half maximum, FWHM, in terms of the nominal transition energy E_0 and the parameters T and M:

$$\Delta E = \frac{E_0}{c} \left(\frac{8 k_B T \ln 2}{M} \right)^{1/2} \cong 7.16 \times 10^{-7} E_0 \left(\frac{T/K}{M/\text{g mol}^{-1}} \right)^{1/2} \tag{19.51}$$

where k_B is Boltzmann's constant. The OCS molecule ($M = 60$ g mol^{-1}) at 300 K has $\Delta E \cong 3.2 \times 10^{-3}$ cm^{-1} in the vicinity of 2000 cm^{-1}. This is only slightly less than the instrumental resolution used to record Figure 19.17(c). Thus, an instrument capable of much higher resolution would be unable to observe more detail simply because the Doppler effect has limited the linewidth. (One says the spectrum is "at the Doppler limit.")

The Doppler effect can be beaten. One way is to arrange for all the molecules to travel in one direction and the light to travel at a right angle to this direction. This reduces the Doppler shift to a factor on the order of β^2, which is negligible, and it is surprisingly easy to realize. We make a *molecular beam* of the molecules under study. We expand a gas from high pressure (~0.1–10 atm, typically) through a small hole or slit (of diameter or slit width ~0.1–1 mm) into an evacuated chamber. As the gas streams through the hole, intermolecular collisions force the molecular velocities from their random spread in magnitude and direction in the high pressure reservoir into a beam characterized by a unique direction and a very narrow spread in speeds. The molecular density in a beam is not great, but it is often sufficient for direct absorption and generally more than sufficient for laser induced fluorescence.

A Closer Look at the Doppler Effect

There is another technique that can beat the Doppler effect. It uses a static sample, not a beam, but requires a high power, narrow frequency spread, tunable radiation source such as a microwave source or a laser. The technique was described by Willis Lamb in 1964 and first observed in 1969 by C. Costain. The radiation passes through the sample in two parallel and coincident directions, using either a special waveguide for microwaves or a mirror for lasers, as Figure 19.18 indicates.

Suppose the radiation frequency, ν_+, is greater than the nominal transition frequency, ν_0, but still within the Doppler linewidth. From Eq. (19.50a), radiation

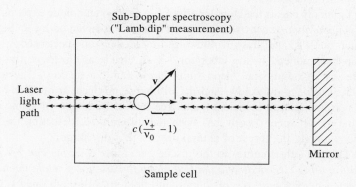

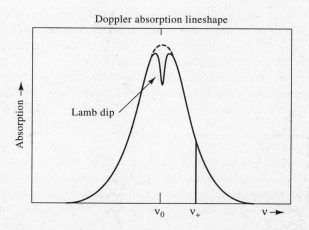

FIGURE 19.18 One way to increase spectral resolution is to exploit the Doppler shift instead of being limited by it. A *Lamb dip* experiment schematically diagrammed here and explained in the text does just this.

moving to the right is absorbed by molecules with a velocity component to the right of magnitude

$$v \cong c \left(\frac{\nu_+}{\nu_0} - 1 \right).$$

When the radiation returns to the left, it is absorbed by a *different* batch of molecules, those with a velocity component to the *left* with the same magnitude v.

Suppose the radiation frequency is exactly ν_0. As it moves through the sample to the right, it is absorbed by molecules with no velocity component to either the right or left. Now, *if the radiation intensity is great enough,* it *saturates* the transition on its first pass through the sample. *Saturation* means the initial and final states' populations have been made *equal*. (A weak light source typically places only a small fraction of the absorbers in the excited state. Light absorption can never create a population inversion, with more excited states than initial states, since stimulated emission and absorption compete equally. The best light can do is equalize the populations.)

On the return trip, which happens instantaneously compared to the time it takes a molecule to leave an excited state, the radiation would be absorbed by those same molecules *if any still had the correct velocity and initial quantum state.* But the first saturating pass made the sample *transparent* at ν_0. As a result, the magnitude of the net absorption at ν_0 is notably reduced, producing what is called a *Lamb dip* at the center of the Doppler profile. The width of this dip is the width of the frequency spread of the radiation source, which is less than the Doppler width, as Figure 19.18 also shows.

The Lamb dip cheats the Doppler effect, but another technique exploits it. Termed *velocity modulation spectroscopy,* this technique was invented by R. Saykally and coworkers in 1983 and has revolutionized molecular ion spectroscopy. Molecular ions can be produced in the laboratory in an electrical discharge through a low pressure gas. (Commercial "neon signs" are a familiar analog.) The ion density, however, is always quite low, and the discharge produces a mixture of ions, hot neutrals (from electron collisions), and free radical fragments of stable molecules. The bane of ion spectroscopy has been the difficulty of detecting the minor ion transitions among the forest of neutral spectra in a discharge.

The Doppler effect comes to the rescue here since *the ions can be accelerated by an alternating electrical field applied along the discharge while the neutrals are not affected by this field.* As this alternating field goes through one cycle, the ions are forced to oscillate along the discharge axis with velocities that can exceed thermal velocities. A high resolution laser is directed along the axis as well, and the laser detector signal is processed to reject any laser intensity changes that are *not at the alternating field frequency* (or a simple multiple of that frequency). Thus, the detector sees only absorptions due to the Doppler-modulated ions, and moreover, the signal processing can be arranged to distinguish between positive and negative ions. At any one instant in the field's oscillation, positive ions are Doppler shifted in the opposite sense from negative ions.

Inherent linewidth is only part of the observability problem. Another is called *spectral congestion*. A molecule may have more transitions than *any* technique can resolve. Spectral congestion can be greatly reduced in certain cases, and the molecular beam technique does it. As a beneficial consequence of the narrow velocity spread of a beam, there is also dramatic internal *cooling* of beam molecules. Rotationally, only the lowest few *J* states are populated, as if the gas was at equilibrium at perhaps 1–10 K, while vibrational states are populated as if the gas was at perhaps 10–200 K, depending on the vibration's frequency. For polyatomics in particular, the ability to cool the molecule while keeping it in an isolated gaseous state eliminates many thermally populated initial states and dramatically lowers spectral congestion.

Thus, many experimental techniques improve spectral observability, but one aspect of observability remains. For some molecules, the Pauli Exclusion Principle plays a role. Some molecular transitions cannot be observed simply because *the molecule cannot exist in certain seemingly innocuous quantum states*. In these cases, molecular symmetry and *nuclear spin* conspire in an unexpected way.

Consider a homonuclear diatomic in a $^1\Sigma_g^+$ electronic state with half-integral (fermion) nuclear spin quantum numbers *I*, such as H_2 with $I = 1/2$. The total wavefunction (including *all* spins) must change sign if the nuclei are relabeled, just as we saw for the He atom electrons in Chapter 13. If we call the spins A and B and let $\hat{R}_{nuc}$ be the operator that relabels nuclei (by writing B for A and A for B), we require

$$\hat{R}_{nuc}\Psi_{tot}(A, B) = \Psi_{tot}(B, A) = -\Psi_{tot}(A, B) \ . \quad (19.52)$$

$\hat{R}_{nuc}$ is equivalent to the following operations diagrammed in Figure 19.19.

1. Rotate a space-fixed coordinate system 180° about an axis perpendicular to the nuclear axis.
2. Invert the molecule-fixed coordinates of the electrons through the origin.
3. Reflect the electron coordinates in a plane containing the nuclei and the rotation axis of step 1.
4. Exchange the nuclear spins.

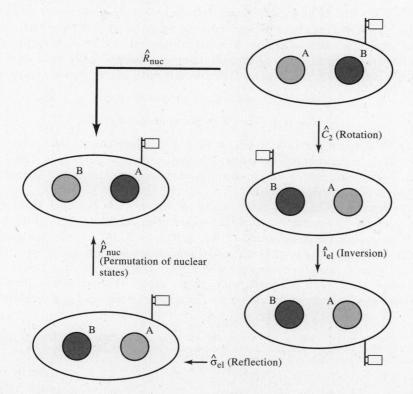

FIGURE 19.19 The operation of relabeling the nuclei in a homonuclear diatomic is equivalent to the sequence of rotation, inversion, reflection, and permutation operations shown here. [Adapted from *Molecular Quantum Mechanics,* by P. W. Atkins (Oxford University Press, Oxford, 1983).]

As the figure shows, these operations together relabel the nuclei. Symbolically

$$\hat{R}_{nuc} = \underset{4.}{\underset{\uparrow}{\hat{P}_{nuc}}} \quad \underset{3.}{\underset{\uparrow}{\hat{\sigma}_{el}}} \quad \underset{2.}{\underset{\uparrow}{\hat{i}_{el}}} \quad \underset{1.}{\underset{\uparrow}{\hat{C}_{2}}}$$

If we write

$$\Psi_{tot} = \psi_{el}\psi_{vib}\psi_{rot}\psi_{nuc}$$

we see that ψ_{vib} is not changed by any of these operators (none of them depend on the internuclear separation) while

$$\hat{i}_{el}\psi_{el} = +\psi_{el} \quad \text{for } g \text{ states}$$
$$= -\psi_{el} \quad \text{for } u \text{ states}$$

and

$$\hat{\sigma}_{el}\psi_{el} = +\psi_{el} \quad \text{for } + \text{ states}$$
$$= -\psi_{el} \quad \text{for } - \text{ states}$$

and, though less obviously,

$$\hat{C}_{2}\psi_{rot} = (-1)^{J}\psi_{rot}$$

where J is the angular momentum (of everything *except* nuclear spin) quantum number. The nuclear spin exchange operator has the eigenvalues

$$\hat{P}_{nuc}\psi_{nuc} = +\psi_{nuc} \quad \text{for triplet-paired nuclear spins}$$
$$= -\psi_{nuc} \quad \text{for singlet-paired nuclear spins}.$$

Putting all this together for a $^1\Sigma_g^+$ state yields

$$\hat{R}_{nuc}\Psi_{tot} = -\Psi_{tot} = (-1)^{J}\psi_{el}\psi_{vib}\psi_{rot}(\hat{P}_{nuc}\psi_{nuc}),$$

which shows that *the molecule's rotational state is coupled to its nuclear spin state*. States with *even J must* have nuclear spins that are *singlet*-paired. Once two nuclear spins settle into one spin-pairing, it is difficult to convert them to the other. One can prepare and isolate for long times H_2 samples with, say, only even J rotational states. It is as if two different types of H_2 molecules exist, and each has its own name:[12]

ortho-H_2: odd J, triplet spin-paired nuclei

para-H_2: even J, singlet spin-paired nuclei.

The triplet-spin states (ortho-H_2) are three-fold degenerate and thus three times more likely to occur at true ortho-para equilibrium than the para states. This leads to an *intensity alternation* in, say, the H_2 Raman spectrum. Transitions alternate in roughly a 3 to 1 intensity ratio.

For *spinless* nuclei in $^1\Sigma_g^+$ states (and there are very few molecules that satisfy these criteria, $^{40}Ar_2$ being one well-studied example), the boson form of the Pauli Principle applies. For boson nuclei, $\hat{R}_{nuc}\Psi_{tot}(A, B) = +\Psi_{tot}(A, B)$, and, since spinless nuclei can form *only* a singlet state, *odd-J rotational levels are not allowed*. Thus, ground state Ar_2 *simply cannot exist* with $J = 1, 3, 5, \ldots$.

Ground state $^{16}O_2$ is a $^3\Sigma_g^-$ state with spinless nuclei. The state is a $-$ state, and the *total* angular momentum is the sum of molecular rotation and spin components.

---

[12]These names have nothing in common with the ortho-, meta-, para- nomenclature of organic chemistry. It may help you remember them if you remember that "ortho" and "odd" both start with "o" so that *ortho*-H_2 has *odd J*: 1, 3, 5, ...

We use N to represent the molecular rotation quantum number here, since, by convention, $\mathbf{J}$ = the total angular momentum except for nuclear spin = $\mathbf{N} + \mathbf{S}$ where $\mathbf{S}$ is the electron spin angular momentum. The nuclei are spinless bosons, and we have, including a (-1) factor from the $-$ nature of the state,

$$\hat{R}_{\text{nuc}} \Psi_{\text{tot}} = (-1)^{N+1} \Psi_{\text{tot}} = \Psi_{\text{tot}}$$

so that N must be *odd*. Ground-state $^{16}O_2$ is doomed to rotate! There are *no* $N = 0, 2, 4, \ldots$ levels.

These arguments can be carried on to other molecules with symmetrically located equivalent nuclei, such as acetylene ($^1H^{12}C^{12}C^1H$), ammonia, planar BF_3, methane, etc. Nuclear spin degeneracies (or, more spectacularly, the prohibition of certain levels as in $^{40}Ar_2$ and $^{16}O_2$) play an important role in the appearance of the spectrum. Figure 19.20 shows a vibration-rotation spectrum of acetylene in which the intensity alternation from line to line is very obvious.

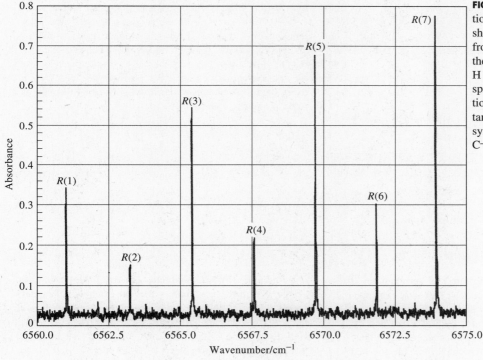

FIGURE 19.20 This high-resolution spectrum of acetylene shows a 3:1 intensity alternation from one line to the next due to the $I = 1/2$ nuclear spins of the H atoms. The ^{12}C nuclei are spinless. The vibrational excitation shown here involves simultaneous excitation of both the symmetric and antisymmetric C—H stretching normal modes.

CHAPTER 19 SUMMARY

This has been a long chapter, but an important one, since it is through molecular spectroscopy that we can measure the energies and structures of individual molecules in greatest detail. There have been several themes:

1. the classification of molecular motions—rotational, vibrational, and electronic—and the energy expressions for each,
2. the selection rules that permit radiation to change quantum states,
3. the relationships among spectroscopic observables, molecular structure parameters, and potential energy functions, and
4. the experimental techniques that measure spectra of isolated molecules.

The Born–Oppenheimer approximation and the rigid-rotor, harmonic-oscillator approximation gave us our first, simplest expressions for a molecular wavefunction and its associated energy:

$$\Psi_{tot} = \psi_{el}\psi_{vib}\psi_{rot}, \quad E_{tot} = E_{el} + E_{vib} + E_{rot} \quad (19.24)$$

with

$$E_{vib}(v) = \hbar\left(\frac{k}{\mu}\right)^{1/2}\left(v + \frac{1}{2}\right), v = 0, 1, 2, \ldots \quad (19.22)$$

$$E_{rot}(J) = \frac{\hbar^2}{2\mu R_e^2}J(J+1), \quad J = 0, 1, 2, \ldots \quad (19.21)$$

These expressions were improved, leading to the diatomic spectroscopic constants of Table 19.1 and the energy expression they enter:

$$E_{vib, rot}(v, J) = G(v) + F_v(J)$$
$$G(v) = \omega_e\left(v + \frac{1}{2}\right) - \omega_e x_e\left(v + \frac{1}{2}\right)^2 + \cdots$$
$$F_v(J) = B_e J(J+1) - \alpha_e J(J+1)\left(v + \frac{1}{2}\right)$$
$$- D_e[J(J+1)]^2 + \cdots$$

We indicated how these constants apply to polyatomic molecules with several normal modes of vibration and as many as three distinct rotational constants.

Selection rules showed us that

(a) rotational spectroscopy requires a permanent dipole moment,
(b) vibrational spectroscopy requires a *changing* dipole moment,
(c) electronic spectroscopy follows close parallels to atomic spectroscopy,
(d) Raman spectroscopy focuses on molecular polarizability, and
(e) the Franck–Condon principle modulates transition probabilities in electronic transitions.

From rotational constants we derive basic bond-length information; from vibrational spectroscopy we derive bond-stiffness information. With the help of empirical model potential-energy functions, we can estimate bond strengths, and we can do this for each observable electronic state.

The next chapter carries the story of spectroscopy into the realm of molecules for which detailed, resolved spectra of isolated molecules is either impossible or so difficult as to be no longer worthwhile.

FURTHER READING

There are four classic works on molecular spectroscopy that still have relevance decades later. The first is *Microwave Spectroscopy,* by C. H. Townes and A. L. Schawlow (McGraw-Hill, NY, 1955, reprinted by Dover) written by two of the leaders of the field. (Townes shared the Nobel Prize for inventing the maser.) The next three are the first three volumes of the series *Molecular Spectra and Molecular Structure* written by Nobel laureate G. Herzberg, the dominant figure in molecular spectroscopy this century. The three volumes are individually titled, *Spectra of Diatomic Molecules,* 2nd ed. (Van Nostrand Reinhold, NY, 1950), *Infrared and Raman Spectra* (Van Nostrand Reinhold, NY, 1945), and *Electronic Spectra of Polyatomic Molecules* (Van Nostrand Reinhold, NY, 1966). All four of these books go well beyond the material of this chapter.

Several more recent books on molecular spectroscopy are also worth noting. *Molecular Spectroscopy,* by J. D. Graybeal (McGraw-Hill, NY, 1988) is a graduate level text that also goes beyond the material in this chapter. *Molecules and Radiation,* by J. I. Steinfeld, 2nd ed. (MIT Press, Cambridge, MA, 1985) has many excellent examples of modern spectroscopic techniques and experiments. *Quantum Chemistry and Molecular Spectroscopy,* by C. L. Dykstra (Prentice Hall, Englewood Cliffs, NJ, 1992) is at about the same level as this chapter (and Chapter 14). *Modern Spectroscopy,* by J. M. Hollas, 2nd ed. (J. Wiley, New York, 1992) is similar to Steinfeld's book in its coverage of modern methods of spectroscopy.

The fourth volume of Herzberg's series, *Constants of Diatomic Molecules,* with K. P. Huber (Van Nostrand Reinhold, NY, 1979) was discussed in the text as an excellent source of data for diatomics. For polyatomics, some data are tabulated in Herzberg's third volume, and the article "Molecular Structures of Gas-Phase Polyatomic Molecules Determined by Spectroscopic Methods," by M. D. Harmony, *et al. J. Phys. Chem. Ref. Data* **8,** 619, (1979) lists ground-state structural parameters for many molecules along with references to the original literature.

PRACTICE WITH EQUATIONS

19A The HF bond length is 0.9168 Å. How far is the center of mass from the H atom? (19.1)

ANSWER: 0.8706 Å

19B What is the moment of inertia of HF? (19.3)

ANSWER: 0.8045 amu Å^2 = 1.336 × 10^{-47} kg m^2

19C The S—F bond length in SF$_6$ is 1.56 Å. What are its moments of inertia? (19.3)

ANSWER: $I_a = I_b = I_c$ = 185 amu Å^2 = 3.07 × 10^{-45} kg m^2

19D What values for Ω are possible for a $^3\Pi$ molecular term symbol? (19.6)

ANSWER: Ω = 0, 1, 2 (S = 1; Λ = 1)

19E What is the approximate reduced mass (in amu) of H$_2$? of HD? of D$_2$? (19.9)

ANSWER: H$_2$: ≅0.5 amu; HD: ≅0.67 amu; D$_2$: ≅1 amu

19F How much does the energy of HF change if it is excited from J = 0 to J = 1? (19.21), 19A, 19B

ANSWER: 8.32 × 10^{-22} J = 41.9 cm^{-1}

19G The harmonic force constant for the HF bond is 965 N m^{-1}. How much does the energy of HF change if it is excited from v = 0 to v = 1? (19.22)

ANSWER: 8.22 × 10^{-20} J = 4138 cm^{-1}

19H Repeat Problem 19G including the anharmonic corrections $\omega_e x_e$ = 89.9 cm^{-1} and $\omega_e y_e$ = 0.90 cm^{-1}. (19.26)

ANSWER: 3961 cm^{-1} = 7.87 × 10^{-20} J

19I Repeat Problem 19F including vibration-rotation coupling and centrifugal distortion: α_e = 0.798 cm^{-1}, D_e = 2.151 × 10^{-3} cm^{-1}. (19.27), (19.28), (19.29)

ANSWER: 8.16 × 10^{-22} J = 41.1 cm^{-1}

19J The HF molecule has a bound excited state for which T_e = 84 777 cm^{-1}, ω_e = 1159 cm^{-1}, and $\omega_e x_e$ = 18 cm^{-1}. What is ν_{00} between this state and the ground state? (19.30), 19G, 19H

ANSWER: 83 305 cm^{-1} = 1.65 × 10^{-18} J

19K The rotational constants for O$_3$ are 0.4453, 3.553, and 0.3948 cm^{-1}. What kind of top is O$_3$? (19.34), (19.35)

ANSWER: nearly prolate asymmetric top (κ = −0.968)

19L What is the approximate energy of the first rotationally excited state of ozone? (19.37), 19K

ANSWER: 0.89 cm^{-1} (assume prolate top, J = 1, K = 0)

19M The lowest three singlet states of O$_2$ are the a $^1\Delta_g$, b $^1\Sigma_g^+$, and c $^1\Sigma_u^-$ states. What transitions are fully allowed among these states? (19.46)

ANSWER: none

19N The dissociation energy of HF is $\mathcal{D}_e$ = 6.122 eV. What is the β parameter for a Morse potential representation of HF? (19.47), 19G

ANSWER: 2.218 Å^{-1}

19O An H atom is flying toward the Earth from an exploded star with a speed 1 × 10^6 m s^{-1}. Where would an astronomer find this atom's n = 1 to n = 2 transition? (19.50), (12.52)

ANSWER: 82 029 cm^{-1} (rather than 82 303 cm^{-1} if the atom was at rest)

19P What is the Doppler width of a transition in the fundamental vibration-rotation spectrum of HF at 300 K? (19.51), 19H

ANSWER: ~1.1 × 10^{-2} cm^{-1} (based on E_0 ≅ 4000 cm^{-1})

PROBLEMS

SECTION 19.1

19.1 Find the center of mass and moments of inertia for D$_2$O and H$_2^{18}$O assuming the bond lengths and bond angle are the same as in ordinary H$_2^{16}$O.

19.2 Note in Example 19.2 that, to within their rounded precision, $I_a + I_b = I_c$. Prove that for any planar molecule this equality is exact. Experimentally, the degree to which measured moments of inertia satisfy this equality is a good measure of planarity. The observed moments of inertia for N$_3$H are 41.78, 42.61, and 0.8286 amu Å^2, and those for HFCO are 5.544, 42.97, and 48.51 amu Å^2. Are either of these molecules planar?

19.3 How many normal modes of vibration do each of the following molecules have?

(a) H$_2$C=C=CH$_2$ (allene)

(b) O=C=C=C=O (carbon suboxide)

(c) HFClF (see Problem 15.25)

(d) H$_2$C=C=O (ketene)

19.4 Test Eq. (19.5) on Li_2 and Cs_2. Assume the atoms are bound by a harmonic potential with force constants 25.25 N m^{-1} and 6.91 N m^{-1}, respectively. Find the classical turning points for the zero-point motion (the zero-point energies are 3.478×10^{-21} J and 4.170×10^{-22} J, respectively), take ΔR_{vib} to be the difference between the classical turning points, and take $R = R_e = 2.6729$ Å and 4.47 Å, respectively.

SECTION 19.2

19.5 What are the ground and first excited state configurations and term symbols for the following: H_2^+, He_2^+, HeH, HeH$^+$, and Li_2^+? For each configuration, give all the possible term symbols whenever more than one exists.

19.6 The reason $\Sigma + \Lambda$ is used instead of Ω as a subscript to a diatomic term symbol can be understood if we consider a $^4\Pi$ state and the diatomic equivalent of the atomic spin–orbit energy expression, Eq. (13.29). First, find the subscripts and Ω values for a $^4\Pi$ state. Next, the diatomic equivalent to Eq. (13.29) is an energy term $A\Sigma\Lambda$ where A is a *molecular* coupling constant. Use this expression to explain why $\Sigma + \Lambda$ has to be used to distinguish all possible energies instead of Ω.

19.7 Example 19.5 showed that CO in the $v = 2$, $J = 1$ state could excite H_2 from $v = 0$, $J = 0$ to $v = 1$. What is the largest J state of $v = 1$ H_2 that could be reached in this collision? You may be worrying about conserving angular momentum in this problem, but collisions have varying amounts of angular momentum depending on the details of the collision geometry (head-on, glancing, etc.).

19.8 The rotational constant for the first excited singlet state of H_2, B $^1\Sigma_u^+$, is almost exactly one-third as large as that for the ground state. What is R_e for the B state if R_e for the ground state is 0.741 Å?

19.9 Use data in Table 19.3 to calculate the dissociation energy of the Na_2 A $^1\Sigma_g^+$ excited state. You will need to know that this state dissociates to one Na in its ground state and one in the first excited $^2P_{1/2}$ state, 16 956.183 cm^{-1} above the ground state. (Atomic emission from this state and its spin–orbit $^2P_{3/2}$ partner 17.2 cm^{-1} higher account for the yellow color of Na flames. Emission from the two excited states are known as the Na "D lines"; see also Problem 18.8.)

19.10 Footnote k of Table 19.3 tells us that the B $^1\Pi_u$ excited state of Na_2 has a radiative lifetime of about 7 ns. Calculate the number of rotations and vibrations Na_2 would make (classically) in 7 ns if it was in the $J = 1$ and $v = 1$ level of the B state. Is 7 ns a long time in the life of a molecule?

19.11 The six normal mode harmonic frequencies of H_2O_2 are (with the conventional numbering) $\omega_1 = 2869$ cm^{-1}, $\omega_2 = 1435$ cm^{-1}, $\omega_3 = 1408$ cm^{-1}, $\omega_4 = 870$ cm^{-1}, $\omega_5 = 3417$ cm^{-1}, and $\omega_6 = 1370$ cm^{-1}. How many vibrational states are there within 3000 cm^{-1} of the zero-point level? You may find it helpful to use the standard notation for polyatomic vibrational states $(v_1, v_2, v_3, v_4, v_5, v_6)$ where each v is a normal mode quantum number. In this notation, the zero-point state is $(0, 0, 0, 0, 0, 0)$, the state at 2869 cm^{-1} is $(1, 0, 0, 0, 0, 0)$, the one at $(2869 + 870)$ cm^{-1} is $(1, 0, 0, 1, 0, 0)$, etc.

19.12 Come up with three ordinary objects that are spherical tops, three that are prolate symmetric tops, and three that are oblate symmetric tops. Minor details of the object can be ignored, such as the seams on a baseball so that a baseball can be called spherical. Then come up with three molecules not discussed in this chapter for each type of top.

19.13 The equilibrium C—H bond length in acetylene is 1.061 Å, and that for the C≡C bond is 1.203 Å. What is acetylene's moment of inertia? What is its equilibrium rotational constant? Figure 19.20 shows part of a vibration-rotation spectrum of C_2H_2. Analysis of this spectrum shows that the energy difference between the states $J = 0$ and $J = 1$ in the excited vibrational state is 2.3275 cm^{-1}. What is the moment of inertia in this vibrational state? Assuming the C≡C bond length does not change, what is the C—H bond length?

SECTION 19.3

19.14 There are vast clouds of fascinating molecules spread here and there throughout the interstellar regions of the Universe. Optical and radioastronomy techniques have identified dozens of molecules, and a recent observation found the ^{13}CO isotopomer in a cloud called IRc2 in the vicinity of the familiar constellation Orion. The observed transition was an emission from the $J = 6$ to the $J = 5$ rotational level of the ground vibrational and electronic state. Calculate the transition energy using data in Table 19.2, the proportionalities $B_e \propto \mu^{-1}$ and $D_e \propto \mu^{-2}$, and the accurate nuclear masses $m_{^{12}C} = 12$ amu, $m_{^{13}C} = 13.00335$ amu, and $m_{^{16}O} = 15.99491$ amu. Find the transition energy in cm^{-1} and the transition frequency in GHz.

19.15 The microwave spectrum of isotopically enriched ^{6}LiF vapor at high temperature shows six prominent transitions of varying intensity:

Frequency/GHz	Relative Intensity
448.491 07	1.00
441.386 83	0.25
358.856 19	0.78
353.172 23	0.19
269.179 18	0.56
264.915 79	0.14

What are the quantum numbers of the initial and final state of each transition? (Table 19.2 lists vibration/rotation constants for the other Li isotope, ^{7}LiF, but remember, R_e is the same for both isotopes.)

19.16 The $v = 0 \to 7$ overtone vibration-rotation spectrum of HI was measured in 1977 [P. Niay, P. Bernage, C. Coquant, and H. Bocquet *J. Mol. Spectrosc.* **68**, 329 (1977)]. Eighteen transitions were reported, and three of them are R(0) = 13 891.493 cm^{-1}, R(1) = 13 899.315 cm^{-1}, and P(1) = 13 868.273 cm^{-1}. Use the difference between *two* of these transition energies (study Figure 19.13 to decide which two) to find the energy difference between two different rotational states of the excited vibrational state. Then use the rigid-rotor energy expression to find the rotational constant B_7 for the $v = 7$ state and the corresponding bond length. Compare this length to R_e in Table 19.2, and interpret the (appreciable) difference.

19.17 Your IR spectrometer can scan only a 400 cm^{-1} region at a time. Where would you center a scan if you were looking for the *first overtone* ($v = 0$ to $v = 2$) transition in HCl? in DCl? Your sample of HCl contains both Cl isotopes in natural abundance. Which isotopomer, H^{35}Cl or H^{37}Cl, will have the lower frequency spectrum?

19.18 The graph below shows the dipole moment functions $p(R)$ for HF and CO over the R range for which these functions are accurately known from experiment. Use these curves to explain why the permanent dipole moment of HF is 16 times larger than that for CO, but the transition dipole moments for the $v = 0 \to 1$ transitions in both molecules are about equal. You will also need the R_e values in Table 19.2.

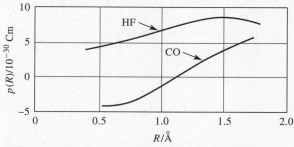

19.19 If a rigid-rotor, harmonic oscillator diatomic in a $^1\Sigma^+$ electronic state could have a Q branch in its vibration-rotation absorption spectrum, where would it be found, and what would it look like (i.e., how many different lines would make up this branch)? Which would be closer to the hypothetical Q(0) line, the R(0) line or the P(1) line?

19.20 Refer back to Chapter 14 and the discussion of the photoelectron spectrum of N$_2$ (see Figure 14.13). That discussion deduced the ground and first two excited configurations of N$_2^+$. What term symbols come from these configurations? Now that you know about Franck–Condon factors and the Franck–Condon principle, explain the photoelectron spectrum's structure with a sketch of the potential energy curves of N$_2$ and the ion states.

19.21 Use the data in Table 19.3 to find the wavelengths of the emissions from $v = 22$, 23, and 24 of the A state of Na$_2$ to $v = 0$, 1, and 2 of the X state. (Use only the harmonic and first anharmonic vibrational constants for each state.) Sketch this spectrum as a "stick spectrum" with straight vertical lines along a wavelength axis covering the range of emission. Include the atomic emission lines $^2P_{1/2} \to {}^2S_{1/2}$ and $^2P_{3/2} \to {}^2S_{1/2}$ at 589.76 and 589.16 nm. What molecular transitions are closest to these atomic transitions?

19.22 Your fluorocarbon chemist friend has a sample cylinder labeled only "Difluoroethylene" without any indication of the isomeric form. She asks you to help her determine the isomer, and you place some of the material in your microwave spectrometer. Careful searches around regions where you might expect to see transitions yield none. What isomer (or isomers) does she have?

19.23 Show that one moment of inertia for a rigid planar AX$_3$ symmetric top molecule like BF$_3$ is twice the other. The smaller rotational constant for ^{11}BF$_3$ is 0.173 cm^{-1}. What would this constant be for ^{10}BF$_3$? The rotational constant for diatomic ^{11}BF is B_e = 1.516 cm^{-1}. Which has the longer B—F bond, the diatomic or BF$_3$?

19.24 Use Eq. (19.37b) and the rotational constants in Figure 19.8(b) along with selection rules to predict the transition energies for $J = 0 \to 1$ and $J = 1 \to 2$ in CF$_3$H. You will find that Eq. (19.37b) predicts only two lines, since K will not appear in the transition-energy expression. In reality, centrifugal distortion changes this picture. One adds the terms $-D_J J^2(J + 1)^2 - D_{JK} J(J + 1)K^2 - D_K K^4$ to Eq. (19.37b) where D_J, D_{JK}, and D_K are centrifugal distortion constants. Now how many lines are seen in these two transitions? (You do not need the values of the D's here. Just derive a general expression for the transition energy and decide how many lines are seen.) Can the spectrum determine all three centrifugal distortion constants?

19.25 Acetylene has two doubly degenerate bending normal modes called the *trans bend* and the *cis bend*. The atomic motion in one plane is shown below:

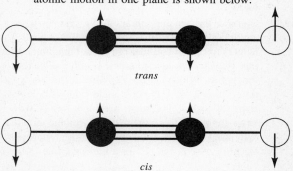

One of these modes is infrared active and the other is Raman active. Which is which?

SECTION 19.4

19.26 Use the data in Table 19.3 for the ground state of Na_2 to find: (a) the Morse function β parameter, (b) the Morse dissociation energy, and (c) the Morse values for the centrifugal distortion constant D_e and the vibration-rotation constant α_e. How well do these latter two constants agree with the tabulated values?

19.27 Equation (19.49) gives a way of estimating bond dissociation energies from ω_e and $\omega_e x_e$. While anharmonicity reflects the shape of $V(R)$ and leads to more closely spaced vibrational levels as v increases, the vibration-rotation constant α_e also reflects the shape of $V(R)$ by measuring the increase in average bond length with increasing v. Show that this constant also governs the dissociation energy of a Morse potential function by solving Eq. (19.48e) for $\omega_e x_e$ and substituting the result in Eq. (19.49). Your answer should contain only rigid-rotor, harmonic oscillator constants in addition to α_e. Use data from Table 19.2 to check your expression and the original Eq. (19.49) expression against the following observed dissociation energies: F_2, 13 370 cm^{-1}; CO, 90 544 cm^{-1}; HF, 49 390 cm^{-1}; and H_2, 38 298 cm^{-1}. Would you recommend one expression over the other?

19.28 The electronic absorption spectrum of the Ar dimer's ground state was analyzed in detail in 1976 [E. A. Colbourn and A. E. Douglas *J. Chem. Phys.* **65**, 1741 (1976)]. The analysis yielded, among other things, energies of the lowest six vibrational levels of the weakly bound ground state. The energies of the $v = 1$–5 levels above the $v = 0$ level were found to be 25.740, 46.149, 61.755, 72.661, and 79.441 cm^{-1}. Use the Birge–Sponer method to estimate the Ar_2 dissociation energy, $\mathcal{D}_0$. Note that this method does not require you to know the zero-point energy, and note that when corrections we have not discussed are used, one finds $\mathcal{D}_0 = 84.75$ cm^{-1}. Your answer will be smaller. What would you predict for v_D, the last bound vibrational state quantum number? Finally, estimate the zero-point energy and suggest a value for $\mathcal{D}_e$.

19.29 Consider the bending vibration of a symmetric triatomic such as CO_2. If we keep the C—O bond lengths fixed and consider only small amplitude bending, we could write a harmonic bending potential function as $V(\theta) = k_b(\Delta\theta)^2/2$ where $\Delta\theta$ is the deviation of the O—C—O bond angle from its equilibrium 180° value. For larger amplitude bends, we know to correct this $V(\theta)$ with anharmonic terms. Why can there be no anharmonic term proportional to $(\Delta\theta)^n$, $n = 3, 5, \ldots$, for a molecule like CO_2?

SECTION 19.5

19.30 In Problem 19.14, you should have found that the $J = 6 \rightarrow 5$ transition in ^{13}CO occurs at about 664.01 GHz. If you tuned your radiotelescope to that frequency and pointed it toward the cloud where this transition was first observed, you would see nothing since the cloud is moving away from us at about 11 km s^{-1}. Where should you tune your telescope? Another interesting aspect of this observation is the relatively high J value of the upper level. It implies that the cloud is fairly warm (as interstellar temperatures go), and a temperature ~100 K has been assigned to this cloud. What is the Doppler width of the transition?

19.31 The acetylene spectrum of Figure 19.20 was recorded at room temperature with an FTIR at a resolution of ~0.004 cm^{-1}. Is this spectrum "at the Doppler limit"?

19.32 One can show that a homonuclear diatomic with nuclear spin I has $(I + 1)(2I + 1)$ symmetric and $I(2I + 1)$ antisymmetric nuclear-spin wavefunctions. The ratio of these numbers leads to the 3:1 intensity alternation in the spectrum of Figure 19.20. Predict the intensity alternation from one line to the next in the rotational Raman spectra of $^{14}N_2$ ($I = 1$), $^{15}N_2$ ($I = 1/2$), and $^{14}N^{15}N$.

GENERAL PROBLEMS

19.33 Franck–Condon factors are usually difficult to calculate, but this problem considers a simplified model that is soluble. We imagine a diatomic with two electronic states that are identically harmonic except for the location of their minima. We imagine one has a minimum displaced a distance δ from the other. We will consider only the $v = 0$ levels of each state so that the vibrational wavefunctions are just the zero-point harmonic-oscillator functions of Eq. (12.19):

$$\psi'_{vib} = \frac{1}{\pi^{1/4}} e^{-q_1^2/2} = \frac{1}{\pi^{1/4}} \exp\left(-\frac{k}{2\hbar\omega}(R - R_e)^2\right)$$

$$\psi''_{vib} = \frac{1}{\pi^{1/4}} e^{-q_2^2/2} = \frac{1}{\pi^{1/4}} \exp\left(-\frac{k}{2\hbar\omega}(R - R_e - \delta)^2\right)$$

where R_e is the bond length of state 1 and $R_e + \delta$ is the bond length of state 2. Show that the integral for the Franck–Condon factor in Eq. (19.45) can be written

$$\int_{-\infty}^{\infty} \psi'_{vib} \psi''_{vib} \, dR = \frac{e^{-k\delta^2/2\hbar\omega}}{\pi^{1/2}} \int_{-\infty}^{\infty} \exp(-q_1^2 + \sqrt{k/\hbar\omega}\, \delta\, q_1)\, dq_1 \,.$$

This integral is known and yields the Franck–Condon factor

$$g_{0,0} = \exp\left(-\frac{k}{\hbar\omega}\frac{\delta^2}{2}\right)\,.$$

Note that this expression equals 1 for $\delta = 0$ (equal R_e's) and that it depends only on the relative shift between the states (δ), not on their absolute positions. Typical values for $k/\hbar\omega$ fall in the range 200–800 Å^{-2}. Calculate $g_{0,0}$ for $\delta = 0.01$ Å, 0.05 Å, 0.1 Å, and 0.5 Å for $k/\hbar\omega = 200$ Å^{-2} and 800 Å^{-2}. The numbers should convince you of the importance of the Franck–Condon factor in controlling one's access to vibrational levels of excited electronic states.

19.34 Follow Example 19.7 and find the energies of all the vibrational states of CO_2 at or below the $(0, 2^0, 0)$ level, this time including anharmonicity. The anharmonic constants for CO_2, all in cm^{-1}, are: $x_{11} = -2.93$, $x_{22} = 1.35$, $x_{33} = -12.47$, $x_{12} = -4.61$, $x_{13} = -19.82$, and $x_{23} = -12.31$. You should find that the states $(1, 0^0, 0)$, $(0, 2^0, 0)$, and $(0, 2^2, 0)$ are quite close, within about 6 cm^{-1} of each other. Your calculation accurately locates the $(0, 2^2, 0)$ state, but not the other two. An interaction between them due to anharmonicity and known as *Fermi resonance* couples and shifts them. In addition, the wavefunctions of the two interacting states is a roughly 50:50 mixture of the two unperturbed states' wavefunctions so that transitions from the ground state to both coupled states are allowed with roughly equal probability (in Raman spectra). A second-order perturbation theory of Fermi resonance predicts that the interacting levels (call them a and b) move from their unperturbed energies E_a and E_b to energies $\{E_a + E_b \pm [4W_e^2 + (E_a - E_b)^2]^{1/2}\}/2$ where W_e is another molecular energy constant. Apply this theory to the $(1, 0^0, 0)$ and $(0, 2^0, 0)$ levels of CO_2, for which $W_e = -52.84$ cm^{-1}, and compare your energies to the observed transition energies from the ground state: 1 285.41 and 1 388.19 cm^{-1}.

19.35 The graph below is a contour plot of the potential energy function for CO_2 drawn with the molecule held linear. The origin is at the equilibrium CO bond lengths, 1.160 Å, and the axes represent compression or expansion of the two C—O bonds away from equilibrium. The contours are spaced in 1000 cm^{-1} intervals.

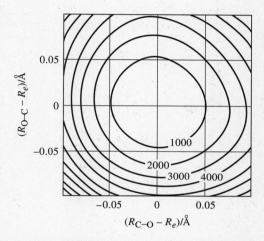

Make a sketch of this graph and indicate on it the *two straight lines* that represent the symmetric and antisymmetric normal mode motions. The zero-point energy of CO_2 is 2532 cm^{-1}. What are the C—O bond lengths at the classical turning points of the zero-point motion in, first, the symmetric normal mode and second, the antisymmetric mode?

CHAPTER 20

Molecular Spectroscopy of Large Molecules

20.1 Why a Triatomic May Be a Large Molecule

20.2 Electronic Structure in Large Molecules

20.3 Radiationless Transitions—Where Did the Energy Go?

20.4 Atomic Chaos at Large Energies

20.5 Spectroscopy in Condensed Media

THE majority of the molecular spectra chemists encounter in their day-to-day work do not fall into the "small, free molecule" category of the last chapter. Nuclear magnetic resonance (NMR) spectra, discussed in the next chapter, are probably the most common and useful spectra to chemists in general, and even most routine IR and UV spectra (microwave spectroscopy never achieved "routine" status) are of polyatomic molecules in a condensed phase. It is just as well that large, complicated polyatomics are generally too involatile for gas-phase study. Their spectra would be so congested with unresolved transitions that the methods of Chapter 19 would be inapplicable. With the aid of some of the tricks described in Section 19.5, important large molecules have been studied in detail as free molecules, but such experiments are the exception.

For large molecules, we must learn what a coarser view of molecular structure can teach. This view has important chemical justification, particularly for qualitative analysis. For example, IR spectra can tell the trained observer the "functional groups" in a molecule and quite a lot about its structure as a result.

This chapter is considerably more qualitative than the last. When chemists talk about a spectrum, we often do so in terms that relay the most information with the least effort: What gross features appeared in the spectrum? How intense were they? Were they in unexpected positions? Did expected characteristic features appear?

The first section describes the very nebulous term "large molecule." The next section says more about the gross features of large molecule electronic structure, adding to Chapters 14 and 19. The third section brings in new phenomena as it discusses radiationless transitions in more detail, while the fourth takes a side-trip into the quantum states characterized by internal energies that are an appreciable fraction of a chemical bond energy. Finally, Section 20.5 summarizes the common features of spectra taken in condensed media.

20.1 WHY A TRIATOMIC MAY BE A LARGE MOLECULE

Figure 20.1 shows an expanded 1 cm^{-1} portion of the OCS gas-phase spectrum from Figure 19.17(c). It contains about 45 distinct transitions with 35 or so clearly resolved. This morsel of the OCS spectrum includes: rotationally resolved lines (all in the R branch) of (1) ground vibrational state OCS excited to the predominantly O=C stretching normal mode, (2) excited bending states going to states with both stretch and bend excited, and (3) excited C=S normal-mode molecules going to states with both O=C and C=S normal modes excited. Most of the transitions are due to the two major isotopes of S, ^{32}S and ^{34}S, present in natural abundance. This is only 1 cm^{-1} in a spectrum that, at room temperature, spans over 160 cm^{-1} and contains over 2800 observable transitions of these three types.

The term "large molecule" refers not so much to the *size* of the molecule, but more to the *complexity of its spectrum*. Increased size brings spectral complexity for most molecules, but size is not the only factor. A complex spectrum has many transitions crowded closely together; many transitions imply many closely spaced

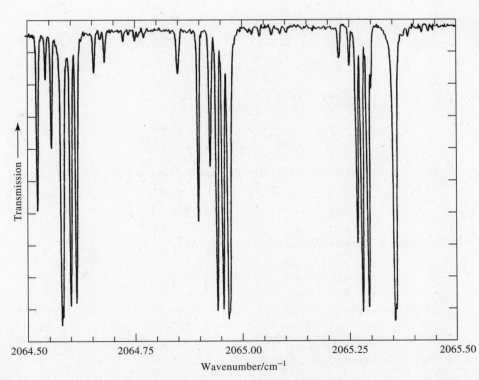

FIGURE 20.1 A 1 cm^{-1} portion of the OCS spectrum of Figure 19.17(c).

energy levels. Thus, more than anything, it is the *close spacing of energy levels* that gives a molecule its spectroscopic largeness.

There is a conceptually simple way of measuring energy level spacing. Called the *density of states function*, $\rho(E)$, it is the *number of quantum states per unit energy* at any total energy. For example, the energy levels of the harmonic oscillator are equally spaced. Every ω_e cm^{-1} you find one state. Thus, the density of states is "one per ω_e cm^{-1}," or

$$\rho(E) = 1/\omega_e \; . \tag{20.1}$$

For a rigid rotor, the states are not evenly spaced (see Figure 19.4), so that the density states would seem to be *greatest* for energies near low J, where levels are close, but rigid-rotor levels are *degenerate*. Each energy level corresponds to $(2J + 1)$ distinct states. Its density of states is the product of the degeneracy of each level, $(2J + 1)$, and the number of levels per unit energy, (dJ/dE), so that, with $E(J) = B_e J(J + 1)$,

$$\begin{aligned}\rho(E) &= (2J + 1)\left(\frac{dJ}{dE}\right) = (2J + 1)\left(\frac{dE}{dJ}\right)^{-1} \\ &= (2J + 1)[B_e(2J + 1)]^{-1} = \frac{1}{B_e} \; .\end{aligned} \tag{20.2}$$

The OCS rotational constant is 0.203 cm^{-1} so that the rotational density of states is about 5 per cm^{-1}. Multiply this number by the number of vibrational levels populated at room temperature (three), by the number of observable isotopomers (two, at least), and include another factor of two for the doubly degenerate bending vibrations, and you reach the right order of magnitude for the number of transitions seen in the 1 cm^{-1} slice of the OCS spectrum in Figure 20.1.

EXAMPLE 20.1

What is the density of states for a 1-D particle in a box?

SOLUTION The energy is (Eq. (12.12)):

$$E = \frac{\hbar^2 \pi^2 n^2}{2mL^2} \; ,$$

the levels are not degenerate, and the density of states is

$$\rho(E) = \left(\frac{dn}{dE}\right) = \left(\frac{dE}{dn}\right)^{-1} = \left[n\left(\frac{\hbar^2 \pi^2}{mL^2}\right)\right]^{-1} \; .$$

Usually, we express ρ as a function of energy rather than quantum number. We solve the energy expression for n:

$$n = \left(\frac{2mL^2 E}{\hbar^2 \pi^2}\right)^{1/2}$$

and substitute this in the density of states expression:

$$\rho(E) = \left[\left(\frac{2mL^2 E}{\hbar^2 \pi^2}\right)^{1/2}\left(\frac{\hbar^2 \pi^2}{mL^2}\right)\right]^{-1} = \frac{L}{\hbar \pi}\sqrt{\frac{m}{2E}} \; .$$

Note that increasing the box length or the mass increases ρ at any one energy, but here ρ *decreases* as E increases, in contrast to the harmonic oscillator or rigid rotor. The 1-D box does not have a degeneracy factor to offset the increasing spacing between levels as E increases.

➡ RELATED PROBLEMS 20.1, 20.2, 20.3

Generally, the density of electronic states is quite small. Electronic states are typically separated by large energies, and they are characterized by small degeneracies, but this need not be the case. A striking example that has received much experimental and theoretical study is diatomic Ni. When two Ni atoms (in the ground-state $3d^84s^2$ configuration) are joined, a total of 100 different electronic states of Ni_2 result! (Contrast this to ground-state H + H, which can combine in only *two* H_2 electronic states.)

Polyatomics typically have several of the characteristics of a large density of states—small vibrational constants, small rotational constants, or highly degenerate open-shell electronic configurations—and, since the total density of states is the product of those for each degree of freedom, they have significantly greater state densities than do diatomics.

This forces us to consider the spectroscopy of polyatomics in a different light. A density of 10^4–10^6 states per cm^{-1} is common for molecules of moderate size, and such a large number of states may never be resolved, assigned, and interpreted in the way the techniques of Chapter 19 allowed for small molecules. In Section 20.4, we will see that such very dense spectra are not completely without interesting information, but for now, we concentrate on the qualitative language used to describe polyatomic spectroscopy.

20.2 ELECTRONIC STRUCTURE IN LARGE MOLECULES

It is a great help to think of a molecule in terms of pieces with a certain individuality. Organic chemistry, with its functional groups, does this routinely and successfully. For electronic spectroscopy, the relevant pieces of a large molecule are those that can be said to "carry" the transition; that is, to be the more or less localized source of the electronic structure change in a transition. Such carriers are called *chromophores,* and there are many different types: isolated or conjugated double-bonded carbons, carbonyls, lone pairs, aromatic rings, and transition-metal atom centers, to name a few of the most common.

A characteristic MO change can also classify electronic transitions. Consider, for example, benzene. Ground state benzene has four π electrons in the HOMO: a doubly degenerate set called e_{1g} with the configuration e_{1g}^4. The first excited states come from moving one π electron to the LUMO, which has π^* character and is called e_{2u}. The transition is $e_{1g}^4 \rightarrow e_{1g}^3 e_{2u}^1$ in terms of configurations, but it is called a $\pi \rightarrow \pi^*$ *transition* (read "pi to pi star") in general.

The excited $e_{1g}^3 e_{2u}^1$ configuration leads to six distinct molecular electronic states, B_{1u}, B_{2u}, and E_{1u}, each in singlet and triplet spin multiplicities. The ground-state configuration is a unique singlet electronic state, $^1A_{1g}$ (read "singlet a one g"), and we can ignore the excited triplets by the spin selection rule. Just as in diatomics, the $g \rightleftarrows u$ rule holds where applicable, so transitions from $^1A_{1g}$ to $^1B_{1u}$, $^1B_{2u}$ or $^1E_{1u}$ would all seem to be allowed. Other polyatomic selection rules, however, allow transitions to only the $^1E_{1u}$ state, and a strong transition to this state is observed around 180 nm.

A second, but weaker, transition is observed around 260 nm. The excited state in this transition is the $^1B_{2u}$ state, and a breakdown of the Born–Oppenheimer approximation allows the transition. Since it is an approximation to write $\psi = \psi_{el}\psi_{vib}$, any part of the *true* wavefunction (the only one the molecule knows about, after all) that mixes vibrational and electronic motions (through a mechanism called *vibronic interaction*) can cause B–O selection rules to fail.

These are interesting details, common to many large molecules, but in both transitions, the chromophore is the aromatic ring of π electrons; to a good approximation, the C—H and C—C σ bonding electrons are unchanged in either transition. This is the central, chemical origin of the benzene spectrum, and it carries over to all single aromatic rings. All amino acids that contain a benzene ring chromophore have a characteristic "signature" absorption around 260 nm, even if they are part of a protein many thousands of atoms big.

In this section, we will survey the major types of generic MO changes in electronic transitions without discussing the details of vibronic interactions or specific electronic-state selection rules. The first type is common to polyatomics with only single bonds (σ bonds) and no lone pairs of electrons such as saturated hydrocarbons, silanes, etc. The LUMO in such molecules can be either of two distinct kinds, and both lie at high energies so that spectra occur typically in the UV or vacuum UV. The first kind of LUMO (which dominates the first excited electronic state) is a localized σ^* orbital, and the transition is termed $\sigma \rightarrow \sigma^*$. The second kind is a localized *atomic* orbital that has predominantly the character of a single atom in an excited state of increased principal quantum number n. For example, the MO may be dominated by a N atom with $n = 3$ (or 4, 5, etc.). States dominated by this MO are called *Rydberg states,* and the transition to them is called a *Rydberg transition.* A series of Rydberg transitions to successively higher n values appears very atomic-like and converges to an ionization limit. Often, the σ^* and Rydberg levels are closely spaced in energy, especially for hydrocarbons, and one cannot readily assign the nature of the excited state or states responsible for a broad absorption feature.

The next generic type of transition involves a localized excitation of an electron in a nonbonding MO, typically a lone-pair electron. The excitation moves this electron to a σ^* LUMO, and the transition is called n (for "nonbonding") $\rightarrow \sigma^*$. Halogenated saturated hydrocarbons, saturated amines, water, and alcohols enter this category. The $n \rightarrow \sigma^*$ transition generally falls at a lower energy than the $\sigma \rightarrow \sigma^*$ transition, but still well in the deep UV. As a consequence, H_2O, CH_3OH, and saturated hydrocarbons are convenient solvents for spectroscopy in the ordinary UV since they are transparent at wavelengths ≥ 200 nm.

When unsaturated bonds are present, such as $>C=C<$ or $>C=O$, π and π* MOs enter the picture. We have discussed the $\pi \rightarrow \pi^*$ transition in benzene, and such transitions dominate the lowest energy transitions of noncyclic alkenes as well. In an homologous series of conjugated alkenes, the $\pi \rightarrow \pi^*$ transition falls regularly as the alkene length increases. As discussed in Problem 14.36, this phenomenon can be approximated by a simple particle-in-a-box model. The first $\pi \rightarrow \pi^*$ transition in ethylene lies at ~170 nm, deep in the UV, but increased conjugation lowers the $\pi \rightarrow \pi^*$ transition into the visible region. This is the defining transition of dye molecules and the photoactive parts of the molecules of human vision.

EXAMPLE 20.2

The β-carotene molecule in its all trans configuration:

has an intense electronic absorption that peaks around 440 nm, accounting for its yellow-orange color (as in carrots, mangos, etc.). What is the energy separation and nature of the MOs responsible for this transition?

SOLUTION The energy between the HOMO and LUMO must be on the order of the electronic transition peak: 440 nm corresponds to $(10^7 \text{ nm cm}^{-1})/(440 \text{ nm}) = 22\,700 \text{ cm}^{-1}$. The HOMO is a filled π orbital, and the transition is $\pi \to \pi^*$, promoting one π electron to the lowest energy π^* MO. The intensity of the transition (i.e., the large magnitude of the electronic transition dipole moment) can be traced in part to the large spatial extent of the MOs involved. Moving charge through a large distance in a transition means a large transition dipole moment.

➡ *RELATED PROBLEMS* 20.5, 20.6, 20.31

If a lone-pair MO is occupied in the ground state of an unsaturated molecule, as in formaldehyde and more complex aldehydes and ketones, the lowest energy transition can become a lone-pair to π antibonding transition, an $n \to \pi^*$ transition.

In summary, most covalently bonded polyatomics have lowest energy electronic spectra dominated by one of the following types of transitions:

$\sigma \to \sigma^*$:	saturated hydrocarbons
$n \to \pi^*$:	saturated, but with a lone-pair MO
σ or $n \to$ Rydberg:	saturated molecules
$\pi \to \pi^*$:	unsaturated or aromatic molecules
$n \to \pi^*$:	unsaturated, with a lone-pair MO.

Figure 20.2 summarizes these transitions with a generic MO diagram that illustrates the relative ordering and energy spacing of the orbitals. Of course, this diagram is not applicable to any one molecule unless and until the ground-state MO configuration and the unoccupied or nonexistent MOs are specified.

In the past twenty or so years, the areas of organometallic chemistry and metal coordination chemistry have grown so rapidly that we must include them in our discussion. To do so, we need to consider the role of a transition metal atom as a chromophore.

The electronic structure and the electronic spectra of transition metal compounds are often dominated by the metal's d orbitals. The vivid and varying colors of metal ions in aqueous solution are well-known examples of such transitions. In solution, ions are not free, but are tightly coordinated to, typically, six water molecules so that the polyatomic that causes the transition is $M(H_2O)_6^{n+}$ and not M^{n+} alone.

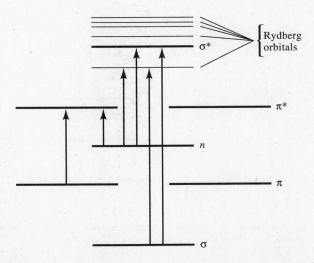

FIGURE 20.2 The various generic types of large molecule electronic transitions can be classified according to the molecular orbitals involved in the transition.

Consequently, the spectra are not as simple as those of an isolated M^{n+} ion in the gas phase.

The simplest model that gives most of the spectral details correctly is called *crystal field theory*. At its simplest, it is an electrostatic theory for the interaction between a central ion and a set of point charges (simulating the coordinating ligands) symmetrically arrayed about the ion. For hexaaqua complexes, six negative charges are arranged at the corners of an octahedron about the central positive ion.[1]

The assembly's stability is attributed to the negative ligand–positive ion attraction, of course, but the *spectrum* reflects changes in the ion's d orbital energies and degeneracy. Picture first an ion free in space with one electron in each of the five d orbitals. Then enclose this ion in a spherical shell of uniform negative charge (totalling $-6e$). As the radius of this shell is decreased from infinity to some value of atomic dimensions, the shell-charge–ion-orbital-charge repulsion raises the total energy of the system. Work must be done to decrease the shell's radius.

Next rearrange the shell of charge into six symmetrically localized equal charges. The energies of the d orbitals change as a result of the rearrangement. These d orbitals have two different types of symmetries with respect to the octahedral ligands:

directed towards ligands: d_{z^2} and $d_{x^2-y^2}$
directed between ligands: d_{xy}, d_{yz}, and d_{zx} .

Figure 20.3 diagrams the differences between these two groups for the representative $d_{x^2-y^2}$ and d_{xy} orbitals. (You may want to review Figure 12.15 if you have forgotten the shapes of the d orbitals.) Those orbitals that point directly *towards* a ligand have *increased* their energy further, while those that point *between* ligands *lower* their energy slightly. The octahedral crystal field has *split* the five-fold degeneracy of the free d orbitals.

Figure 20.4 is an energy level diagram for the d orbitals as they evolve in this theory. The classical model is written in energy units that are the product of two naturally occurring parameters, symbolized D (the strength of the electrostatic potential energy) and q (the value of a radial overlap integral) in the original theory. In these units, the octahedral field of the ligands splits the d orbitals into two groups separated by a total energy of $10Dq$. (In inorganic chemistry texts and literature, $10Dq$ is often symbolized Δ or Δ_o, the "o" subscript standing for "octahedral.") The three "between ligand" orbitals are lowered $4Dq$ while the two "towards ligand" orbitals are raised $6Dq$. The two orbital groups are renamed e_g for the doubly degenerate pair and t_{2g} for the lower energy group of triply degenerate orbitals.

FIGURE 20.3 In crystal field theory, the energies of the central atom's d orbitals break into two categories depending on the relative orientation of the orbitals and the octahedrally positioned ligands, pictured as negative charges in the model.

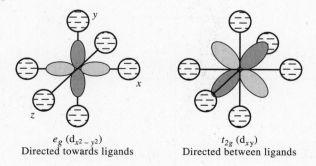

e_g ($d_{x^2-y^2}$)
Directed towards ligands

t_{2g} (d_{xy})
Directed between ligands

---

[1]This theory was first advanced by Hans Bethe to explain the spectra of impurity species that could fit into holes with octahedral symmetry in simple ionic crystals. Hence the name.

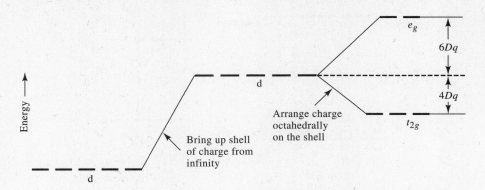

FIGURE 20.4 As the ligand charges are brought up to the central ion and arranged into an octahedral pattern, the d orbital energies split into a high-energy pair (called e_g) and a lower energy triplet of orbitals (called t_{2g}).

In the hexaaqua transition-metal ion complexes of the first transition row (Sc–Zn), we fill the d orbitals according to the number of d electrons in the ion. Thus, in solution, Sc^{3+} has *no* d electrons,[2] and these levels are empty. The $Sc(H_2O)_6^{3+}$ ion is colorless since the HOMO is a 3p atomic orbital that is far removed in energy from the LUMO of the complex. Similarly, Zn^{2+} in solution (configuration: $[Ar]4s^2 3d^{10}$) has all five of these orbitals filled and is similarly colorless.

In between Sc^{3+} and Zn^{2+} are the colored transition-metal ions with *d-d transitions* in which an electron is promoted from a t_{2g} orbital to an e_g orbital. For example, "one-electron" Ti^{3+} ($[Ar]3d^1$) has the MO transition $t_{2g}^1 \to e_g^1$. The water molecule electrons are not involved directly, even though $Ti(H_2O)_6^{3+}$ is an intact polyatomic species in solution.

Before we discuss the ground-state MO configurations (and electron spin multiplicities) of the important aqua complexes, two additional features of this MO scheme need discussion.

First, consider Cr^{2+} in solution with four d electrons. The first three are assigned one to each t_{2g} orbital, with parallel spins, according to Hund's rule. Now where does the fourth electron go? Placing it spin-paired into one of the now half-occupied t_{2g} orbitals costs a certain amount of spin-pairing energy. Placing it into an empty e_g orbital does not cost spin-pairing energy, but it does cost $10Dq$'s worth of excitation energy. *If $10Dq$ is small enough,* the electron goes into an e_g orbital and we have a *quintet* spin multiplicity for the $t_{2g}^3 e_g^1$ configuration. If not, the electron will spin-pair in a t_{2g} orbital and we have a *triplet* spin multiplicity for the t_{2g}^4 configuration.

Figure 20.5 shows these two possibilities. The "small $10Dq$" case leads to a *high-spin* ground state while the "large $10Dq$" case leads to a *low-spin* (smaller spin multiplicity) state. The spin multiplicity enters dipole selection rules and governs the probability of the electronic transition and thus the intensity of the color.

This allows us to understand why Mn^{2+} species (five d electrons) are a very pale (pink) color in solution: Mn^{2+} forms a high-spin aqua complex with configuration $t_{2g}^3 e_g^2$. By Hund's rules, the five electrons are all aligned in the highest spin configuration in the ground state, a *sextet* spin state. Any excitation from t_{2g} to e_g *must* pair two spins and therefore *must* yield a state of *lower* spin multiplicity: a quartet state. The ground state is the *only* sextet state around, and thus any excitation must violate the spin selection rule, which implies a small transition probability and a pale color.

---

[2]As is also true in the gas phase. Scandium has the ground state $[Ar]3d4s^2$ $^2D_{3/2}$, and Sc^{3+} is $^1S_0$, isoelectronic to Ar.

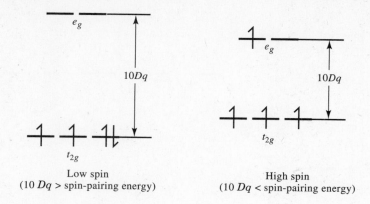

FIGURE 20.5 For systems such as the 4 e^- (d^4) Cr^{2+} ion shown here there are alternate electron configuration assignments possible, depending on the e_g-t_{2g} orbital energy separation. The two configurations differ in total spin multiplicity.

Table 20.1 collects data on the octahedral aqua complexes of the first-row transition-metal ions. Note how the multiplicities smoothly increase and then decrease across the row. This table gives only the coarsest view of the electronic spectra of the species. Intramolecular distortions from perfect octahedral coordination, called *Jahn–Teller distortions*,[3] play a role as do a number of excited states of varying spin multiplicity. (See Problems 20.15 and 20.16.)

The crystal field calculation can be performed for several other ligand geometries. For tetrahedral coordination, the d orbitals again split, but the doubly degenerate pair (called simply *e* orbitals, *g/u* symmetry being absent) are *lower* than the triply degenerate set (called simply t_2 orbitals), and the separation is four-ninths that of octahedral coordination. Two ligand (linear), four ligand (square planar), and eight ligand (cubic) coordination can also be treated. (An octahedron is the "sum" of a square planar and a linear array, suitably oriented.)

Crystal field theory fails to take into account the chemical nature of ligands in a complex, and for many molecules, there is clear evidence that the ligands are doing more than just sitting around as point sources of negative charge, which is

TABLE 20.1 Octahedral aqua complexes of the first transition row ions

Ion	# of d electrons	MO config.† (ground state)	Spin multiplicity	$10Dq/10^3$ cm^{-1} (observed)
Sc^{3+}	0	—	singlet	—
Ti^{3+}	1	t_{2g}^1	doublet	20.3
V^{3+}	2	t_{2g}^2	triplet	18.0
Cr^{3+}	3	t_{2g}^3	quartet	17.6
Cr^{2+}	4	$t_{2g}^3 e_g^1$	quintet	14.0
Mn^{2+}	5	$t_{2g}^3 e_g^2$	sextet	7.5
Fe^{3+}	5	$t_{2g}^3 e_g^2$	sextet	14.0
Fe^{2+}	6	$t_{2g}^4 e_g^2$	quintet	10.0
Co^{2+}	7	$t_{2g}^5 e_g^2$	quartet	10.0
Ni^{2+}	8	$t_{2g}^6 e_g^2$	triplet	8.6
Cu^{2+}	9	$t_{2g}^6 e_g^3$	doublet	13.0
Zn^{2+}	10	$t_{2g}^6 e_g^4$	singlet	—

†Jahn–Teller distortions that change MO symmetries and degeneracies have been neglected here.

---

[3] We first met this distortion in Chapter 14 when discussing why H_3^+ was not a perfect equilateral triangle.

what crystal field theory would have us believe. Thus, an elaborated theory, *ligand field theory,* has been developed to incorporate ligand orbitals directly into the electronic structure.

We will illustrate the central ideas of ligand field theory for one class of compounds, the stable *metal carbonyls*. The theory points to a total of nine important MOs, six bonding and three nonbonding, of which five are directly related to our crystal field, d-orbital MOs. Thus, *18 electrons* becomes an important stability landmark, one which fills these nine MOs and is characteristic of the stable metal carbonyls. Table 20.2 summarizes the known stable neutral metal carbonyls of the first transition row in terms of these valence electrons.

Note how the last column alternates between 17 and 18. The neutral compound $V(CO)_6$ is less stable than its anion, the 18 electron $V(CO)_6^-$, while the 17 electron fragments $Mn(CO)_5$ and $Co(CO)_4$ *couple via metal–metal σ bonds* to form the dimetal compounds $Mn_2(CO)_{10}$ and $Co_2(CO)_8$. These latter two compounds have chemistries that are based on this 18 electron propensity. Thus $Mn_2(CO)_{10}$ reacts with I_2:

$$Mn_2(CO)_{10} + I_2 \rightarrow 2Mn(CO)_5I ,$$

and $Co_2(CO)_8$ undergoes reactions of the types

$$Co_2(CO)_8 + H_2 \rightarrow 2Co(CO)_4H$$

$$Co_2(CO)_8 + 2NO \rightarrow 2Co(CO)_3NO + 2CO$$

all of which yield 18 electron products.

The metal–ligand bond is an interesting combination of MO interactions with readily observable consequences. The σ-type lone pair on the carbon end of the ligand, ●$C\equiv O$, donates charge to the metal, while the metal d electrons *back-donate* charge to the π^* antibonding orbital of CO:

metal d to ligand π^* back-donation

This latter interaction adds antibonding character to the CO bond, lowering its bond order and its IR stretching frequency. The free CO vibration is at 2143 cm^{-1}, while this vibration is at 2090, 2000, and 1860 cm^{-1} in $Mn(CO)_6^+$, $Cr(CO)_6$, and $V(CO)_6^-$, respectively. The overall decrease in this frequency is due to π^* character, and the specific trend in this isoelectronic series can be traced to the two valence-bond resonances we can write for the metal–CO interaction:

$$M^- - C \equiv O^+ \leftrightarrow M = C = O .$$

TABLE 20.2 Characteristic metal carbonyls				
Compound	CO ligands per metal atom	# of ligating CO electrons	Metal atom electrons	Total electrons per metal atom
$V(CO)_6$	6	12	5	17
$Cr(CO)_6$	6	12	6	18
$Mn_2(CO)_{10}$	5	10	7	17
$Fe(CO)_5$	5	10	8	18
$Co_2(CO)_8$	4	8	9	17
$Ni(CO)_4$	4	8	10	18

The greater the positive charge on the metal, the greater the σ donation from CO and the more the first structure dominates. Thus, $Mn(CO)_6^+$ has the highest CO frequency of this series. In contrast, the negative charge on $V(CO)_6^-$ provides greater π* donation, enhancing the second resonance structure, and lowering the CO frequency the most.

A new type of generic transition dominates the electronic spectra of metal carbonyls. Since the metal's d orbital MOs are all occupied in the ground state, the LUMO represents a flow of charge from the metal to the ligands. Transitions that bring this about are called *charge-transfer* transitions and typically end with the charge in the π* orbitals of the CO ligands.[4]

The next section describes the sequence of physical events that can follow the electronic excitation of a large molecule and lead to important photochemical processes based on spectroscopic selection rules and molecular energy-level interactions. These processes form the basis for photochemistry: the ability to effect a permanent chemical change with light.

20.3 RADIATIONLESS TRANSITIONS—WHERE DID THE ENERGY GO?

In Chapters 18 and 19, we went to some length to discuss the complementary nature of photon absorption and emission. We look further into these phenomena in this section. Our key parameter is the time required for spontaneous emission to occur after photon absorption. For any isolated excited state, the amount of spontaneous emission from that state falls exponentially with time. This exponential decay is governed by a single constant, the radiative lifetime τ_{rad}, which is different for each excited state. The following two important proportionalities (see Equations (18.14b) and (18.21)) control τ_{rad}:

$$\tau_{rad} \propto \frac{(\text{emission wavelength})^3}{(\text{transition dipole moment})^2}.$$

For large molecules, spontaneous emission is influenced by several *intra*molecular effects that are a direct consequence of a large density of states. For example, benzene can be excited to a single rotational-vibronic level of the lowest allowed π → π* transition we discussed before, and after excitation, the molecules emit with $\tau_{rad} \cong 26$ ns (for benzene dissolved in *n*-hexane at room temperature), as expected for an allowed transition. But not all the photons that go into a sample of benzene molecules come out on this time scale. If one keeps watching, benzene continues to dribble out photons on a much longer time scale (many seconds if the temperature is low enough). These delayed photons have a longer wavelength than the prompt emission photons, but only by a factor of about 2. This alone cannot account for the increased radiative lifetime, since the wavelength factor changes τ_{rad} by a factor on the order of $2^3 = 8$. We must conclude that the molecule has spontaneously turned itself into a state of greatly reduced transition moment.

For historical reasons, the expected *prompt* emission is termed *fluorescence* while the *slow* (large τ_{rad}) emission is called *phosphorescence*.[5] Phosphorescence is a

---

[4]Charge-transfer interactions are of greater importance than we have indicated here, in that they also account for the stability of the *ground* states of many important types of molecular complexes.

[5]This term has an interesting history. Around 1603, an Italian alchemist, Vicenzo Cascariolo, stumbled across a synthesis of a powder that glowed after exposure to sunlight. When widely described in 1640, the powder was termed "Litheosphoros," from the Greek for "stoney light bearer." In 1669, a new chemical element was discovered that "glowed in the dark." It was named phosphorus. Thus, the phenomenon preceded the element.

familiar phenomenon in Nature. The glow of fireflies and certain bacteria is sustained in time by a slow emission from a molecule prepared in an excited state by a biochemical reaction. The flicker-free display of a video screen is likewise due to the slow emission of the aptly named "phosphors" coating the inside of the display tube and excited by an electron beam.

Molecular phosphorescence takes us beyond the Born–Oppenheimer approximation. The B–O electronic states of most stable large molecules begin with a singlet ground state (that we denote S_0). The first excited state is typically a triplet (T_1), and the *second* excited state is another singlet (S_1). Still in B–O language, the initial absorption event connects a single vibration-rotation level in S_0 to a single level in S_1, according to dipole selection rules and Franck–Condon factors. Since the zero-point level of T_1 is lower than that of S_1, the attained S_1 level has approximately the energy of very many levels of T_1, since the T_1 density of states increases as we move up in energy from its zero-point level. This scheme is diagrammed in Figure 20.6.

If the B–O approximation was exact, the molecule could only fluoresce rapidly back to S_0. Neglected terms in the full Hamiltonian, especially spin–orbit coupling and various vibronic terms, mix singlet and triplet character so that *no* one level is purely one spin type or the other. Thus, the real molecule is excited to a mixed level, no matter how narrow the excitation source's frequency spread.

Suppose the molecule was large but not too large, so that the T_1 density of states was perhaps only two or three times that of the S_1 state. As Figure 20.7 shows for this case, all these levels have a certain natural energy spread, and excitation into the broad, nominally S_1, level could also excite one or more T_1 levels due to singlet-triplet mixing. As time passes after the absorption, the molecule's wavefunction oscillates between predominantly singlet and predominantly triplet character. The frequency of this oscillation is on the order of (energy level spacing)/$\hbar$, which can correspond to a time period comparable to the fast S_1 radiative lifetime.

This oscillation in spin character has an amusing and observable consequence. As the molecules oscillate from singlet to triplet to singlet, etc., so does the radiative transition moment. The molecules blink on and off as they relax. This phenomenon

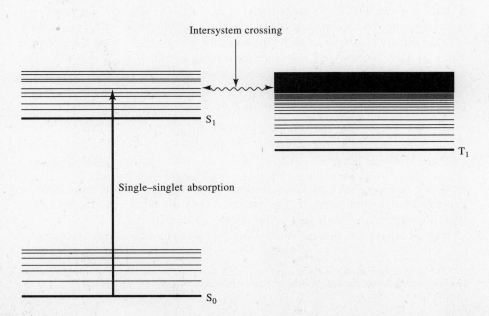

FIGURE 20.6 Excitation to a B–O level of S_1 is often accompanied by a coupling to the lowest triplet state (T_1) through processes (such as spin–orbit coupling) that are neglected in the B–O approximation.

FIGURE 20.7 Any one level in S_1 generally has a natural width greater than isoenergetic levels of T_1 due to the smaller radiative lifetime of the singlet state. Such a level can thus mix strongly with more than one triplet level.

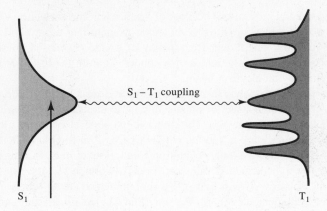

is called "quantum beating" in analogy with the "beating" of sound frequencies that one hears if two tuning forks of closely spaced natural frequencies are struck together. Quantum beats were first observed in a free molecule in biacetyl[6] ($CH_3CO\ COCH_3$), and quantum beats have since been observed in other molecules from acetylene on up in size.

If the T_1 density of states is large enough so that dozens to hundreds of T_1 levels are very close to the S_1 level that was initially excited, the singlet level becomes merely a doorway for absorption of energy into a vast maze of triplet levels. The molecule simply gets lost in the triplet levels very quickly and never returns. It can radiate only via a triplet to singlet transition moment. This is phosphorescence. Singlet-singlet absorption followed by rapid singlet-triplet interconversion is called *intersystem crossing*:

$$S_0 \xrightarrow{\text{Allowed absorption}} S_1 \xrightarrow{\text{Spin-orbit coupling}} T_1$$

If intersystem crossing cannot occur because of a low T_1 density of states, the S_1 level can mix with high energy levels of S_0 itself. The vibronic coupling terms ignored by the B–O approximation mix levels of S_1 with those of S_0, particularly if the S_0 density of states is high. In this case, the energy again enters the molecule via an S_1 door, but now gets lost in the labyrinth high vibrational levels of S_0. The only radiative processes available to high levels of S_0 are vibration-to-vibration transitions with radiative lifetimes on the order of seconds. This process, which pumps a molecule up in energy in the S_0 state via the sequence

$$S_0 \text{ (low vibrational state)} \xrightarrow{\text{Allowed absorption}} S_1 \xrightarrow{\text{Vibronic coupling}} S_0 \text{ (high vibrational state)}$$

as shown in Figure 20.8, is called *internal conversion*. In internal conversion, a photon goes into the molecule, but one does not come back out. The energy is typically dispersed by molecular collisions or dissociation of S_0.

Internal conversion, intersystem crossing, and photodissociation are all examples

---

[6]The molecules were prepared in a supersonic molecular beam to cool the vibration-rotation state distribution in S_0 before the absorption and to lower the Doppler spread of the transition.

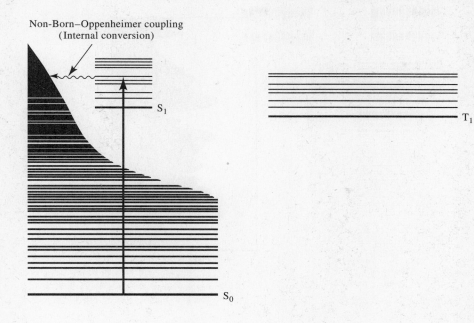

FIGURE 20.8 Vibronic coupling, also ignored by the B–O approximation, can lead to *internal conversion,* in which a level in S_1 is coupled to a high vibrational level of the S_0 ground state.

of *radiationless transitions*. An experimental observable that is often a clue to the occurrence of a radiationless transition is the *fluorescence quantum yield* Φ_F:

$$\Phi_F = \frac{(\text{\# of photons emitted})}{(\text{\# of photons absorbed})}. \qquad (20.3)$$

A measurement of Φ_F follows this sequence. First, a pulse of light is flashed into the sample (in a few ns, with a pulsed laser), and the absorption is measured. Next, fluorescent (and phosphorescent) photons are collected and counted, usually by observing the sample through a known solid angle, since the sample will fluoresce in all directions. If the emission is integrated in time, then the rapid fluorescence can be measured separately from the long time phosphorescence. (If the emission is recorded as a *function* of time, information such as quantum beats and radiative lifetimes can be measured as well.) Once the total emission is corrected for the observation solid angle (and other instrumental response factors), Φ_F can be calculated.

Contemporary experiments of this type give great insight into radiationless transitions. For example, radiative lifetime measurements of individual vibration-rotation levels of the S_1 state of formaldehyde, a fairly small and simple polyatomic, vary wildly from level to level. If S_1 had been pure in the B–O sense, these lifetimes would vary smoothly and slightly. The wild variations in τ_{rad} are due to wild variations in the density of high vibrational levels of S_0 that mix with S_1. These S_0 levels form what has been called a "lumpy continuum," as is diagrammed in Figure 20.9.

Thus, radiationless transitions are not only a fact of life, they can also be exploited to gain molecular information that is otherwise difficult to obtain. In the next section, we continue this theme as we look at techniques that explore the energy levels of a polyatomic with abnormally high amounts of vibrational energy.

20.4 ATOMIC CHAOS AT LARGE ENERGIES

Since a typical chemical bond energy is on the order of tens of thousands of cm^{-1} while vibrational energy quanta in a polyatomic are only a few hundred to maybe

FIGURE 20.9 An experiment that measured the lifetimes of single vibration-rotation levels of the S_1 state in formaldehyde showed how the density of high energy levels of the S_0 ground state varied in a wild, irregular, and "lumpy" fashion. These levels are connected by internal conversion (Figure 20.8).

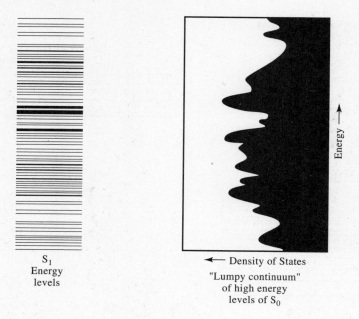

3000 cm^{-1}, ordinary vibrational spectroscopy can access only the lowest few vibrational levels of a molecule and explore only a small fraction of all the bound levels of a polyatomic. Those levels that lie near the dissociation limit are important to understand for several reasons. They are of fundamental interest on their own; they are levels that must be encountered when a molecule undergoes an internal rearrangement; and they are necessarily involved in chemical reactions as old bonds break and new ones are formed.

The traditional method for observing such levels is electronic transition emission spectroscopy. This method is useful for small molecules and has provided information on high energy vibrational levels for many important diatomics and triatomics. But for large molecules, there may be too many levels for ordinary emission methods to resolve. Several new experimental methods have been devised to probe them.

The first two methods use direct absorption from ground vibrational levels to high levels. They exploit the small transition probability for $\Delta v > 1$ *overtone* transitions. For the usual $\Delta v = 1$ fundamental transition of an IR active normal mode, the phenomenological absorption coefficient (the quantity a in the Beer-Lambert law, Eq. (18.15)) is $\sim 10^{-20}$–10^{-18} m^2. A gas at a pressure $\sim 10^{-3}$ atm in a cell ~ 10 cm long absorbs $>50\%$ of the incident light under these conditions.

For overtone transitions ($\Delta v = 2, 3, 4$, etc.) the absorption coefficient drops by a factor on the order of 10^2 for each successive overtone. Thus, the $\Delta v = 3$ overtone typically absorbs less than 1% of the incident light in the same sample that absorbs over half in the fundamental transition. To observe such weak transitions in a conventional absorption experiment we have three choices: increase the sample concentration, increase the optical path length, or increase both. The most dramatic way to increase concentration is to liquify the sample. For example, liquid benzene, although it appears "colorless," has weak absorptions in the visible corresponding to $\Delta v = 5$ and 6 overtone transitions of C—H stretching modes. These absorptions can be observed with an ordinary spectrophotometer set to the most sensitive absorption scale. Similarly, the blue color seen in white light transmission through long paths of pure water is due to O—H vibrational overtones absorbing in the near IR and red regions.

A 10 cm length of liquid benzene (density 0.88 g cm^{-3}) exposes (10 cm)$N_A\rho/M$ = 6.8×10^{22} molecules per cm^2 of illuminated area. The room-temperature vapor pressure of benzene is 0.10 bar, which gives a molecular density of 2.5×10^{18} cm^{-3}. Thus, to gather 6.8×10^{22} molecules of *gaseous* benzene per cm^2 of area, the path length must be

$$l = \frac{6.8 \times 10^{22} \text{ cm}^{-2}}{2.5 \times 10^{18} \text{ cm}^{-3}} = 2.7 \times 10^4 \text{ cm} = 270 \text{ m} .$$

This great length would seem to put long-path gas-phase measurements out of reach, but cleverly placed mirrors around a sample cell of modest length (~2 m or so) reflect the light back and forth through the gas many, many times, increasing the optical path length without increasing the physical sample length. Research multipass cells of this type with optical path lengths of 100–1000 m have been used to observe such weak transitions in gases at ordinary pressures.[7]

Overtone absorptions have played an important role in extraterrestrial spectroscopy as well. For example, the atmosphere of Jupiter—very dense and very big—contains sufficient NH$_3$ to observe and track across the planet N—H overtone transitions in the red region of the spectrum. (Overtone absorptions do not, however, account for the colors of planetary atmospheres.)

If long paths and high molecular densities cannot be used, an alternate approach is to design a new type of spectrometer of fantastic sensitivity. Amazing as it may seem, one such spectrometer was first invented over a century ago by Alexander Graham Bell, who was more interested in others of his inventions. Rediscovered and refined over the past thirty years or so, his device is called a *spectrophone* and relies on the *optoacoustic effect*.

This effect is the conversion of absorbed radiation energy into a pressure increase in a constant-volume sample cell. The sample (which can be a gas or an absorbing solid or liquid in contact with a transparent gas) is confined in a closed cell equipped with a sensitive microphone. The incident radiation is rapidly chopped on and off, typically by a rotating toothed wheel (a "chopper") that modulates the radiation intensity at 10^2–10^4 Hz or by a pulsed laser. (See Figure 20.10.) When the light is

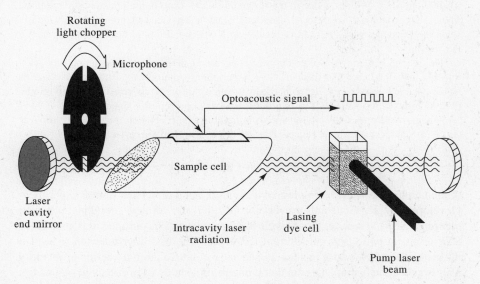

FIGURE 20.10 An optoacoustic apparatus (a *spectrophone*) typically consists of a constant-volume sample cell equipped with a sensitive microphone. The sample is irradiated with an intense, tunable radiation source, such as that from a dye laser. The radiation is modulated by a chopper to facilitate detection and amplification of the microphone signal.

---

[7]Likewise, one can detect the fundamental transitions of ordinary absorption strength for trace amounts of gases such as air pollutants in "natural" abundance through such long paths.

on and absorbed, the energy is equilibrated by molecular collisions (if fluorescence is negligible) and produces a local temperature rise along the light path. This periodic temperature rise leads to a periodic pressure fluctuation, which is a sound wave. The microphone picks up this sound and, on resonance, one can "hear" the sample hum at the chopping frequency. Tune the radiation off resonance, and the sample is quiet. The greater the absorption, the louder the hum, as recorded by the sensitive, tuned amplifiers that listen to the microphone's signal.

The spectrophone is surprisingly sensitive. If the cell is placed inside the radiation cavity of a tunable laser, where the radiation power can be $\sim 10^2$ W or so, highly forbidden transitions such as weak overtones can be recorded with precision limited only by the laser radiation's linewidth or by Doppler broadening.

A laser induced fluorescence spectrum's intensity is proportional to the rate at which excited states fluoresce; the optoacoustic spectrum's intensity is proportional to the rate at which excited states *fail* to fluoresce but instead lose energy through molecular collisions. Thus, LIF and optoacoustic spectra are sensitive to different fates of the absorbed energy.

EXAMPLE 20.3

The reddish-brown gas, NO_2, has an absorption spectrum that rises slowly from around 700 nm to a maximum near 410 nm and then falls smoothly as the wavelength is decreased to 300 nm and below. In contrast, the optoacoustic spectrum follows the absorption spectrum to about 420 nm where it drops rather abruptly to about two-thirds its peak value. At smaller wavelengths, the spectrum is flat to about 350 nm where it again joins the trend of the optical absorption. Explain the difference between these spectra in terms of the thermodynamics of the following reactions (all among ground-state species):

$$NO_2(^2A_1) \rightarrow NO(^2\Pi) + O(^3P) \qquad \Delta_r\overline{H}° = 306.24 \text{ kJ mol}^{-1}$$

$$NO_2(^2A_1) + O(^3P) \rightarrow NO(^2\Pi) + O_2(^3\Sigma_g^-) \qquad \Delta_r\overline{H}° = -192.10 \text{ kJ mol}^{-1}$$

SOLUTION The first reaction, the endothermic dissociation of NO_2, has a thermodynamic threshold of 306.24 kJ mol^{-1} = 25 600 cm^{-1} or a wavelength threshold of $(10^7 \text{ nm cm}^{-1})/(25\,600 \text{ cm}^{-1}) = 390$ nm, close to the observed sudden drop in the optoacoustic spectrum. Thus, the drop is due to the onset of dissociation, and the optoacoustic pressure rise is reduced since the incident energy is used to break one N—O bond and is not returned as heat to the sample.

The second reaction, however, describes the fate of the O atom dissociation product. It releases 192.10 kJ mol^{-1} as heat that will contribute to the optoacoustic signal. Thus the signal does not drop to zero at the photodissociation threshold but instead drops to roughly (192 kJ mol^{-1})/(306 kJ mol^{-1}) $\cong$ two-thirds the peak optoacoustic signal.

Note how the optical absorption spectrum faithfully reflects the *absorbed energy directly* while the optoacoustic signal reflects the *fate* of the absorbed energy.

⇒ RELATED PROBLEMS 20.24, 20.25

A final technique for reaching high overtones is not a direct absorption experiment. Developed by R. Field and coworkers in 1981 and termed *stimulated emission pumping* (SEP), it reaches these levels through a detour. As outlined in Figure 20.11, SEP relies on two lasers of different frequency. The first laser connects a level of the ground electronic state to a level in an excited state of high transition probability. This is called the "pump" laser in that it pumps molecules up in energy in preparation for the effects of the second laser. If the second laser beam is missing, the pumped molecules would simply fluoresce, and this fluorescence is monitored as a key phase

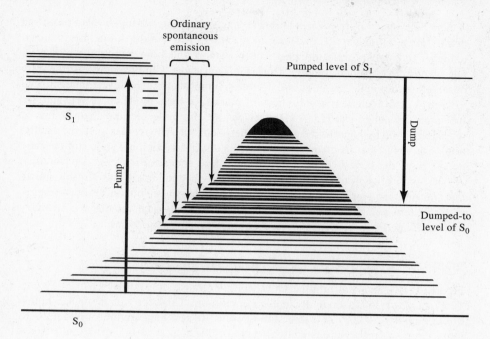

FIGURE 20.11 A stimulated emission pumping (SEP) experiment carries a molecule from a low-lying to a high-lying level of the ground state via a two photon (pump and dump) absorption–stimulated-emission process.

of the overall detection scheme. Now we turn on the second laser and tune it in the vicinity of expected transitions between the pumped level and high levels of the ground state. When we hit such a transition, the second laser causes the pumped molecules to end up in a high level of the ground state not by absorption but by stimulated emission *down* in energy. This second laser is termed the "dump" laser since it dumps molecules from the excited state level to the desired ground state level. Thus, the molecule is taken to a high ground state level via a detour through an excited state.

The dump transition is monitored by its effect on fluorescence. The dump transitions at least partially decrease this fluorescence, and by recording the fluorescence modulations as the dump laser is tuned, one maps the vast number of high overtone transitions that this two laser process can reach, consistent with the usual selection rules and Franck–Condon factors.

These new types of experiments—long path absorption, optoacoustic spectroscopy, and SEP—have revealed surprising aspects to the levels that lie near a dissociation limit. The first two methods pick out those levels that have the largest transition moments with zero-point levels. In small- to medium-sized hydrocarbons, which have been studied extensively, a series of overtone transitions from the $\Delta v = 1$ fundamental to $\Delta v = 6$ or higher have been found based around IR active normal modes that involve *C—H stretching motion*. If the center frequencies of these transitions are analyzed, overtone to overtone, they are found to be represented accurately by a simple Morse function energy expression: $G(v) = \omega_e(v + 1/2) - \omega_e x_e(v + 1/2)^2$. A single anharmonic constant is needed.

This diatomic-like behavior is reinforced by experiments on partially deuterated benzenes. The spectra of C_6D_5H in the high overtone region looks very much like that of C_6H_6, as if the C—H bond were acting as an IR chromophore. This has led to a view of these transitions as *local-mode* transitions rather than normal-mode transitions.

The SEP spectra, however, can probe many more high-lying levels due to the detour taken through an excited state in reaching them. As the dump laser is tuned to higher and higher levels, the spectra become more and more complex, as expected, but unexpected features emerge from the confusion. Clumps of transitions are observed with regular spacings and uniform clump widths. The levels that can lead to such spectra include not only those of normal or local mode character, but also those that are called *chaotic*. A classical picture of a molecule in a chaotic state of vibration would look just as the name implies—chaotic. The regular, periodic oscillation of atoms is lost. A quantum picture of a chaotic vibrational state shows many nodes, since the level is highly excited, but these nodes form no regular pattern.

The study of such unusual levels is still in its beginnings. It has already shown that the simpler, comfortable picture of regular molecular motion at low energies, near the zero-point level, cannot be extended to levels near dissociation in many molecules, and a more dynamic view of these levels is called for.

20.5 SPECTROSCOPY IN CONDENSED MEDIA

Figure 20.12 shows our familiar OCS vibration transition, shown in part in Figures 19.17 and 20.1, but taken under unusual conditions. This spectrum is of OCS highly dilute (about 1:50 000) in a thin slab of solid Ar at a temperature of about 20 K. Only three sharp transitions can be found in this region of the spectrum. (One is obscured on the scale of the figure by the more intense transition near it.) Each transition is due to a different isotope of S in natural abundance, shifted in frequency by simple mass effects.

The technique of recording spectra in high dilution in a cryogenic, transparent solid is called *matrix isolation* spectroscopy. The molecule of interest is mixed at a dilution of 1:100 to $1:10^5$ in the gas phase with an inert, transparent gas such as a rare gas or N_2. This mixture is sprayed through a vacuum onto a transparent window cooled to typically 4–30 K. As the gas hits the window, it rapidly freezes forming a thin (μm thickness) solid in which the solute molecules are isolated from each other (by dilution) in the solid solvent matrix.

FIGURE 20.12 When OCS is frozen at high dilution in solid Ar at very low temperatures, the IR spectrum shown in Figures 19.20 and 20.1 for OCS(*g*) is dramatically simplified. The major simplification is due to the loss of OCS rotational motion from OCS frozen in a cage of solid Ar.

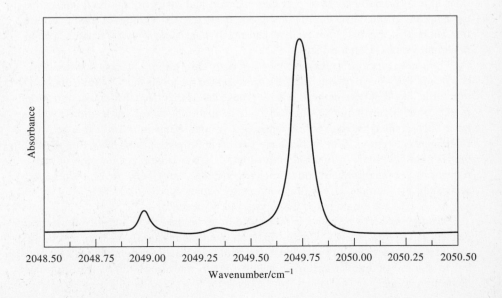

What is striking about Figure 20.12 is its obvious simplicity—three narrow transitions—when compared to the thousands of transitions found in the same region for a gas-phase spectrum. Two effects account for this simplicity. The first is the low temperature of the sample, which ensures that all the OCS molecules are in the zero-point vibrational level of each normal mode. The second comes from the rigid packing of the solid matrix. Each OCS is caged by Ar atoms close-packed around it, and OCS simply cannot rotate in this cage. As a result, the rotational structure of the gas phase spectrum is lost.

It is this second effect—the loss or hindrance of rotational motion—that is the most important in understanding spectra in condensed phases. While most such spectra are not recorded under the unusual conditions of matrix isolation, even liquid solution spectra at room temperature are subject to a solvent cage that holds large solute molecules in place. Especially in the IR, large molecule spectra in condensed phases are reduced to a series of a few fairly narrow transitions.

When rotational structure is quenched, other effects govern the frequency width of an IR transition. The most important is called *inhomogeneous broadening*. Even in a solvent cage, each solute molecule (especially in a liquid solution) is in a slightly different environment as the solute and solvent jostle around one another. If all solutes were identically caged, their vibrational transition energies would be identical (but shifted from the free molecule value). Since each cage is slightly different, the observed spectrum is a broadened smear over typically 0.1–10 cm^{-1}.

Configurational changes in a solute can also alter its spectrum from that of the free molecule. These changes can range from solvent-induced folding or coiling of chain-like solutes to specific solute–solvent interactions such as hydrogen bonding. Internal solute motions such as methyl group rotations may not be quenched by the solvent cage and may also contribute to broadening.

All of these effects cause mid-IR solution spectra to exhibit sets of transitions characteristic of the IR-active functional groups in the molecule. Viewed together, these sets form a unique spectral signature for the molecule. Qualitative IR analysis is based on the presence or absence (and the detailed shape, in some cases) of these characteristic transitions, which can be broadly classified into the spectral regions (with boundaries that are only approximate) shown in Table 20.3.

Large molecule electronic spectra in condensed phases (in the visible and UV regions) are often broad and featureless. Characteristic chromophore transitions can lead to maxima in identifiable positions, such as the aromatic ring $\pi \to \pi^*$ transition discussed earlier, but these spectra are often less useful for qualitative analysis than are IR spectra. Moreover, solvent–solute interactions, especially hydrogen bonding (of two types: solute to solvent or, in nonhydrogen bonding solvents, solute to solute), can dramatically alter the spectrum's appearance.

Due to the restrictions of the Franck–Condon principle, the electronic *fluorescence* spectrum often is *red-shifted* from the absorption spectrum, but similarly broad. As Figure 20.13 shows, the emission spectrum often looks closely like a mirror image of the absorption spectrum.

A Closer Look at Tunable Dye Lasers

Continuous large molecule electronic absorption and emission spectra along with the red-shift of emission compared to absorption has been put to great practical use in the tunable dye laser. In this laser, a dye is dissolved in a solvent such as an alcohol at a concentration on the order of 10^{-3} mol L^{-1}. A laser dye must have the following properties: a large absorption probability, a large quantum yield for

TABLE 20.3 Characteristic infrared spectral features

4000–2500 cm^{-1}   H-atom stretching:

O—H: 3700–3600 cm^{-1}
N—H: 3400–3300 cm^{-1}
C—H: 3000–2850 cm^{-1} (aliphatic)
      3100–3000 cm^{-1} (aromatic, olefinic)
      ~3300 cm^{-1} (acetylenic)
S—H: ~2500 cm^{-1}

2500–2000 cm^{-1}   Alkyne, allene, and nitrile stretching:

C≡C: 2300–2050 cm^{-1}
C≡N: 2300–2200 cm^{-1}
C=C=C: ~1900 cm^{-1}

2000–1600 cm^{-1}   Double bond and carbonyl ligand stretches:

C=O: ~1700 cm^{-1} (ketones, aldehydes, etc.)
C≡O: 2000–1800 cm^{-1} (carbonyl ligands)
C=C: 1700–1600 cm^{-1} (olefinic)
C=N: ~1650 cm^{-1}

1600–700 cm^{-1}   Single bond bends and nonhydride stretches

below ~1000 cm^{-1}   Skeletal bends, ring bends, torsions, out of plane bends of olefinic and aromatic C—H bonds.

fluorescence, a resistance to photochemical degradation, and a particular type of energy level and Franck–Condon factor scheme.

A fixed-frequency laser or an intense flash lamp pumps dye molecules from S_0 to S_1 (usually by a $\pi \rightarrow \pi^*$ transition in the visible). The Franck–Condon factors have to be such that S_1 is produced in vibrationally excited levels. Collisions with the solvent relax the vibrational excitation on a ps time scale, far faster than the 1–10 ns S_1 to S_0 radiative lifetime. Fluorescence occurs from the lowest vibrational levels of S_1 down to S_0, again governed by Franck–Condon factors that leave the

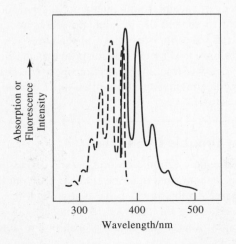

FIGURE 20.13 Often, the fluorescence spectrum of a large molecule appears as an approximate mirror image of the absorption spectrum, as is illustrated here for anthracene in ethanol solution. The solid line is the fluorescence spectrum, and the dashed line is the absorption spectrum.

molecule vibrationally excited in S_0. Normally these levels of S_0 are not populated. Thus the sequence

S_0 (no vibrational excitation)

↓ pump laser

S_1 (vibrationally excited)

↓ solvent collisions

S_1 (no vibrational excitation)

produces a population inversion between the lower levels of S_1 and the vibrationally excited upper levels of S_0. The sequence continues as:

S_1 (no vibrational excitation)

↓ dye laser emission

S_0 (vibrationally excited)

↓ solvent collisions

S_0 (no vibrational excitation)

where the last collisional relaxation step is again very fast.

A radiation tuning element (a prism, grating, or other device) in the dye-laser optical cavity stimulates dye emission over a very narrow frequency band within the continuous emission spectrum of the dye. Repositioning the tuning element tunes the laser frequency, and the spectrum from the near IR to the near UV can be covered with a series of different dye solutions.

Other effects, such as intersystem crossing, photochemical reactions, and self-absorption of the fluorescence, enter the story to govern the overall efficiency and tunability range of any one dye. These processes are now so well understood that newly synthesized dyes can be evaluated quickly and in detail.

CHAPTER 20 SUMMARY

This chapter has discussed large molecule spectra in terms of three major themes: the complexity of such spectra due to a large density of states, the localized MOs of chromophores, and the role of nonradiative transitions—intersystem crossing, internal conversion, and photodissociation—after photon absorption.

Large molecule electronic state energy spacings and spin characteristics generally follow a simple pattern, such as that of Figure 20.6. Within these electronic states, we identified vibrational levels at high energy that had spectroscopic characteristics somewhat more complicated than those at low energy in the same state. For the ground electronic state, we discussed three techniques for placing the molecule in such high levels: direct overtone absorption, stimulated emission pumping, and internal conversion from an excited electronic state.

In condensed phase spectra, we remarked how the rotational motion of the gas phase was quenched but transitions were broadened by solvent interactions with the solute. These interactions altered both electronic and vibrational transitions. We also categorized regions of the IR by the characteristic vibrations found in common organic molecule functional groups.

FURTHER READING

Many of the references of Chapter 19 are also relevant to this chapter. In addition, R. S. Drago's *Physical Methods for Chemists,* 2nd ed. (Saunders, New York, 1992) covers chemical spectroscopy (of virtually all types) from the point of view of the inorganic or organic chemist. It also covers group theory and its applications to spectra. This latter topic and its role in bonding and spectroscopy are covered in *Chemical Applications of Group Theory,* by F. A. Cotton,

3rd ed. (Wiley-Interscience, New York, 1990), the standard text in this area. These topics are also covered in *Symmetry and Spectroscopy,* by D. C. Harris and M. D. Bertolucci (Oxford University Press, New York, 1978).

Additional insight into crystal field theory and ligand field theory can be found in most good inorganic chemistry texts such as *Inorganic Chemistry,* by J. E. Huheey, 4th ed. (Harper-Collins College Publishers, New York, 1993).

PRACTICE WITH EQUATIONS

20A What is the ratio of the vibrational density of states for HI and DI at low energy (well below the bond dissociation energy)? (20.1)

ANSWER: $\rho_{HI}^{vib}/\rho_{DI}^{vib} = (\mu_{HI}/\mu_{DI})^{1/2} \cong \sqrt{0.5}$

20B What is the ratio of the rotational density of states for HI and DI, both in their lowest vibrational state? (20.2)

ANSWER: $\rho_{HI}^{rot}/\rho_{DI}^{rot} = \mu_{HI}/\mu_{DI} \cong 0.5$

20C What is the ratio of the rotational density of states to the vibrational density of states for HI? (20.1), (20.2), Table 19.2 (See also Example 19.4.)

ANSWER: $\rho_{HI}^{rot}/\rho_{HI}^{vib} = \omega_e/B_e = 2309/6.426 \cong 360$

20D What is the color of $Cr^{2+}(aq)$ solutions? Table 20.1

ANSWER: blue (10Dq corresponds to a red absorption)

20E What is the expected molecular formula for the simplest stable carbonyls of Tc and W? Table 20.2

ANSWER: $Tc_2(CO)_{10}$ and $W(CO)_6$

20F How would you use IR spectroscopy to distinguish between allene, $H_2C=C=CH_2$, and methyl acetylene, $H_3C-C\equiv C-H$, dissolved in *n*-hexane? Table 20.3

ANSWER: allene, sharp feature near 1900 cm^{-1}; methyl acetylene, feature near 2100 cm^{-1}

PROBLEMS

SECTION 20.1

20.1 What is the density of states for the electronic states of atomic hydrogen? What happens to this density in the vicinity of the ionization limit? (Don't forget the total degeneracy of each level, including electron spin.)

20.2 If we make a table of the energies of every state from the ground state on up and plot the *total number of states* at or below E as a function of E, the resulting plot will generally increase monotonically with E at a rate that varies from system to system. This rate, the slope of the plot, is the density of states function. (See also Problem 16.53 for another use of this idea.) Show that this is true for a rigid rotor with rotational constant $B_e = 0.02$ cm^{-1}. Tabulate the energies of every level from $J = 0$ through $J = 10$, the $2J + 1$ degeneracies of each level, and the total number of states (the sum of the current level's degeneracy and those of all lower energy levels) at each energy. Plot the number of states versus the corresponding energies. The plot should be quite linear. Find its slope and compare it to $1/B_e$, the density of states expression derived in the text.

20.3 A simple expression for the density of vibrational states of a polyatomic (one that neglects anharmonicity) is

$$\rho(E) = \frac{E^{n-1}}{(n-1)! \prod_{i=1}^{n} \omega_i}$$

where E is the total energy above the zero-point level, ω_i is the harmonic vibrational constant of the ith normal mode, and n is the total number of modes. Use this expression to calculate $\rho(E)$ for CO, CO_2, and H_2CO at $E = 10\,000$ cm^{-1} and 30 000 cm^{-1}. For CO, $\omega = 2170$ cm^{-1}; for CO_2, the constants are 2349, 667 (use it twice! doubly degenerate bend!), and 1388 cm^{-1}; for H_2CO, they are 1167, 1251, 1501, 1746, 2766, and 2843 cm^{-1}. You may be surprised to find that CO_2 has a larger density of states at 10 000 cm^{-1} than does formaldehyde with two more vibrational modes. Why is this? At what energy does formaldehyde's ρ overtake CO_2's?

SECTION 20.2

20.4 The first singlet-singlet electronic transition in ethylene starts near 40 000 cm^{-1}. The carbon–carbon stretching frequency in the ground state is 1623 cm^{-1}, but only 850 cm^{-1} in the excited state. Use the generic electronic structure change of this transition to explain the vibrational frequency drop.

20.5 Use the Hückel MO description of 1,3-butadiene in Chapter 14 (see especially Figure 14.18) to predict the type and number of electronic transitions this molecule should exhibit. Consider only singlet states.

20.6 A simplified MO description of the valence electrons in formaldehyde places two of the four highest energy electrons spin-paired in the (single) π C—O MO just below a doubly occupied nonbonding MO (an O lone-pair). Singlet-singlet electronic transitions from this ground state place the origins of the $n \rightarrow \pi^*$ transition at 28 188 cm^{-1}, the $n \rightarrow \sigma^*$ transition at 57 133 cm^{-1}, and the $\pi \rightarrow \pi^*$ transition at 64 264 cm^{-1}. While these numbers reflect energy differences between electronic *states*, they can be a good guide to the energy separations between MOs. Construct an MO diagram similar to Figure 20.2 for the π, n, π^* and σ^* MOs using these transition energies to deduce the MO energy separations. Where would you predict a $\pi \rightarrow \sigma^*$ transition?

20.7 The molecular geometry of Rydberg states often closely resemble the geometry of the molecule's positive ion. Why is this?

20.8 Qualitatively, how would you expect the d orbital degeneracies to be broken for (a) linear coordination by two equivalent ligands and (b) square planar coordination by four equivalent ligands? Which d orbitals remain in degenerate groups in each case?

20.9 What would you predict for the ground-state MO configuration of the metal d electrons in $FeF_6^{3-}(aq)$? How does knowledge that the ion is colorless aid your MO assignments?

20.10 The ground state of gas phase Co^{3+} is [Ar]$3d^6$ $^5D_4$. How many unpaired electrons does this ion have? The $Co(NH_3)_6^{3+}$ molecular ion in solution is diamagnetic and orange-yellow colored, but the CoF_6^{3-} ion is paramagnetic and blue. Explain both the magnetic properties and the color changes in terms of the occupation of the crystal field MOs t_{2g} and e_g.

20.11 The compound ReF_6 has an absorbance maximum at 32 500 cm^{-1}. What electron excitation accounts for this transition? The ground state of Re is [Xe]$4f^{14}5d^56s^2$ $^6S_{5/2}$, and you should consider the Re—F bonds to be 100% ionic.

20.12 Ruby is Al_2O_3 with Cr^{3+} substitutional impurities that account for the characteristic color. Experiments at very high pressure used the shift in frequency of this absorption as a pressure meter: one places a small ruby crystal in the sample under study (typically compressed between the flats of two highly polished diamonds) and measures the ruby spectrum changes. These are converted to pressure via a standard calibration curve. If the Cr^{3+} ion is in an octahedral coordination site, which way does the absorption change under increased pressure, to higher or lower frequencies?

20.13 Hemoglobin contains four Fe^{2+} ions, each pentacoordinated by the protein with the sixth coordination site used in O_2 transport. Red, oxygenated arterial blood is a low-spin species, but blue-red, deoxygenated venous blood is a high-spin species. Explain how the spin changes are consistent with the color changes.

20.14 Table 13.2 shows that a d^5 atomic electron configuration leads to a total of 16 distinguishable atomic electronic states: one sextet, four quartets, and 11 doublets. Similarly, the $t_{2g}^3 e_g^2$ configuration of d^5 $Mn(H_2O)_6^{2+}$ leads to a large number of *molecular* electronic states, 21 in all, but again only one sextet, the ground state. Draw MO diagrams for the t_{2g} and e_g orbitals as in Figure 20.5 and show one possible *quartet* spin assignment for $t_{2g}^3 e_g^2$ and one possible *doublet* spin assignment. What spin multiplicities are possible for the configurations t_{2g}^5 and $t_{2g}^1 e_g^4$?

20.15 The ground-state configuration of $Ti(H_2O)_6^{3+}$ has a single electron in a t_{2g} orbital. This orbital is degenerate, and the Jahn–Teller theorem says that a molecule will distort to remove such a degeneracy and lower its energy whenever possible. Show that this is possible for $Ti(H_2O)_6^{3+}$. Consider the distortion shown below in which two ligands move closer to the central ion:

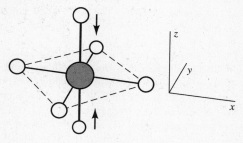

How will this motion change the relative energies of the five d orbitals? You should be able to show that one of the t_{2g} orbitals (which one?) drops in energy, the other two stay degenerate, and the e_g orbitals split as well. The distortion and splitting is small ($<10Dq$), and you should be able to show that three d-d transitions are possible, one of very low energy (in the near IR) and two in the visible. Draw an MO diagram showing the relative locations of the orbitals in this case, and indicate the three possible transitions.

20.16 Laporte's rule for atomic transitions stated $\Delta l = \pm 1$ in order to change parity. For a d-d transition, $\Delta l = 0$, and such a transition would seem to be forbidden. (Another way to see the same thing is to note that transitions among the t_{2g} and e_g orbitals violate the $g \rightarrow u$ selection rule.) Show how a normal mode of vibration in an octahedral complex can lower the symmetry so that g and u are no longer applicable symmetry labels and the $g \rightarrow u$ selection rule thus no longer applies. Focus on two normal modes (and by the way, how many normal modes would an octahedral complex like MX_6^{n+} have in all?) that involve motion of the central atom and two axial ligands. Use the CO_2 stretching normal modes discussed in Chapter 19 as a model, and decide which of these two modes of the complex break the inversion symmetry and allow a d-d transition. This is a simple example of vibronic coupling.

20.17 The spectra below show Ti^{3+} and Yb^{3+} absorptions in dilute acidic solutions.

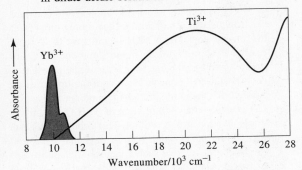

Explain the differences in these spectra in terms of the gas phase ions' ground states: Ti^{3+}, $[Ar]3d^1$ $^2D_{3/2}$, and Yb^{3+}, $[Xe]4f^{13}$ $^2F_{7/2}$, and the different sizes of the 3d and 4f orbitals. (*Hint:* The first excited state of gas-phase Yb^{3+} is $[Xe]4f^{13}$ $^2F_{5/2}$, 10 214 cm^{-1} above the ground state.)

20.18 The IR stretching frequencies of the CO ligands in $Ni(CO)_4$ and $Fe(CO)_4^{2-}$ are at 2060 cm^{-1} and 1790 cm^{-1}, respectively. Where would you expect to find the CO frequency for $Co(CO)_4^-$? What accounts for the trend in these frequencies?

20.19 If V atoms and excess CO are cocondensed with a rare gas in a matrix isolation experiment, the IR spectrum of the product can be interpreted in terms of a $V_2(CO)_{12}$ species that is consistent with the *bridged carbonyl* structure

$$(OC)_5V \overset{\overset{O}{\underset{|||}{C}}}{\underset{\underset{|||}{\underset{O}{C}}}{}} V(CO)_5$$

Is the 18 electron configuration satisfied around each V? What might account for the stability of this species?

20.20 The permanganate ion, MnO_4^-, has an intense purple color and the chromate ion, CrO_4^{2-}, is intensely orange. The intensity of the color rules out a d-d transition (nominally forbidden; see Problem 20.16) as the source of these transitions, but they can be understood as charge-transfer transitions. What is the direction of the (negative) charge transfer in these transitions, and what aspect of charge transfer helps to make the transition dipole moment so large? Recall that the transition moment is the product of two effects: an amount of charge transferred times a transfer distance. (*Hint:* The formal oxidation number for Mn in MnO_4^-, for example, is +7, and the transition changes this number. Is it more likely to increase or decrease?)

SECTION 20.3

20.21 The origin of the formaldehyde S_1 state is 28 188 cm^{-1} above the zero-point level of S_0, and the T_1 origin is 25 194 cm^{-1} up. Use the expression and data in Problem 20.3 to estimate the density of states of S_0 at the S_1 origin. Not all of the vibrational constants of T_1 are well-known; so, use the S_0 constants to predict the T_1 density of states at the S_1 origin. Do you expect the actual T_1 density of states to be larger or smaller than this estimate?

20.22 The radiative lifetime of a vibronic level of D_2CO 68 cm^{-1} above the zero-point level of S_1 is 4.5 μs, that of a level 2400 cm^{-1} up is 0.55 μs, while one 4154 cm^{-1} up has $\tau_{rad} = 0.053$ μs. What accounts for the decrease in these fluorescence lifetimes?

SECTION 20.4

20.23 Neglecting anharmonicity, what would be the first overtone to fall in the visible portion of the spectrum for a hydrocarbon with a local mode C—H vibrational frequency of 3000 cm^{-1}? What about NH_3 with an N—H frequency near 3400 cm^{-1}? Would a 20 cm^{-1} anharmonicity change either answer?

20.24 The NO_2 spectrum described in Example 20.3 was recorded under conditions (~10 torr) that inhibited NO_2 fluorescence and forced the optical excitation energy to be released through collisions as heat. What would the optoacoustic spectrum have looked like if the pressure has been so low that the dominant energy-loss mechanism for *undissociated* NO_2 was fluorescence? Be sure to include the spectral region with wavelengths shorter than the dissociation threshold in your answer.

20.25 Air pollutants in trace quantities can be detected optoacoustically. Suppose an experiment designed to detect trace atmospheric NO via its vibrational spectrum around 1875 cm^{-1} uses a 1 W laser chopped at 3000 Hz and demonstrates a sensitivity such that an absorption of 10^{-5} of the incident light produces a useful optoacoustic signal in a 15 cm^3 cell with a total pressure of 500 torr at room temperature. What would be the pressure fluctuation in this experiment? Assume air is pure N_2 with $\overline{C}_V = 20.81$ J mol^{-1} K^{-1}. Would the signal be stronger or weaker if the NO was diluted in He instead of air?

20.26 Acetylene has been studied extensively by the SEP technique. The pump laser connects the (linear) ground electronic state to an excited state in which the equilibrium geometry is trans bent. (See Problem 19.25.) The dump laser then returns the molecule to highly excited vibrational levels of the ground electronic state. Invoke Franck–Condon arguments to show that the final levels will be those that cannot be easily reached (in practice, not at all reached) by direct overtone absorption from the ground vibrational state.

SECTION 20.5

20.27 You are planning a matrix-isolation experiment that will look for the elusive, high-energy isomer of formaldehyde, the hydroxycarbene molecule, HCOH. You plan to photolyze H_2CO frozen in Xe and search for the IR spectrum of HCOH. (It is known that formaldehyde can be photodissociated into H + HCO in the gas phase. The idea here is to effect a recombination of these fragments into HCOH that will be stabilized in the frozen matrix.) What characteristic changes in the IR spectrum of formaldehyde should you expect? Would it help to study deuterated formaldehyde, D_2CO, as well?

20.28 The IR spectrum of the compound with the empirical formula C_4H_5N shows, in addition to features characteristic of aliphatic and olefinic C—H stretches, sharp features at 2260 and 1647 cm^{-1}. At lower wavenumbers, several skeletal C—C and methylene features appear. What can you deduce about the structure of this compound?

20.29 A sample of 0.01 M ethanol dissolved in CCl_4 has an IR spectrum in the 2500–4000 cm^{-1} region with sharp absorption peaks at 2950, 2835 and 3640 cm^{-1}. If the concentration is raised to 1.0 M, the 2950 and 2835 cm^{-1} peaks are unaffected, the 3640 cm^{-1} peak is greatly reduced in intensity, and a very broad, intense feature appears centered around 3350 cm^{-1} and ~300 cm^{-1} wide. Assign the 0.01 M concentration features, and account for the spectral changes at the higher concentration.

GENERAL PROBLEMS

20.30 Does an octahedral crystal field reduce the three-fold degeneracy of p orbitals? What about f orbitals? Such a field does split the f orbital degeneracy into a low-energy triply degenerate group, a medium-energy triply degenerate group, and a highest energy single orbital. The f orbitals as pictured in Figure 12.16 and described in Table 12.5 are not the best set of linear combinations of Ψ_{nlm} H-atom wavefunctions for this situation. A better set uses the function f_{xyz} (unchanged from Table 12.5) that looks like this:

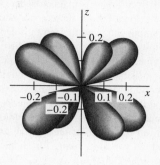

the set of three functions called $f_{x^3} = -[\sqrt{6}\,f_{x(5z^2-r^2)} - \sqrt{10}\,f_{x(x^2-3y^2)}]/4$, $f_{y^3} = -[\sqrt{6}\,f_{y(5z^2-r^2)} + \sqrt{10}\,f_{y(y^2-3x^2)}]/4$, and f_{z^3} (unchanged from Table 12.5), all of which look alike except for the axis, x, y, or z, about which they are cylindrically symmetric:

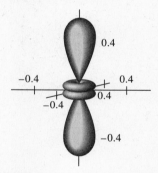

and the set of three called $f_{x(z^2-y^2)} = [\sqrt{10}\,f_{x(5z^2-r^2)} + \sqrt{6}\,f_{x(x^2-3y^2)}]/4$, $f_{y(z^2-x^2)} = [\sqrt{10}\,f_{y(5z^2-r^2)} - \sqrt{6}\,f_{y(y^2-3x^2)}]/4$, and $f_{z(x^2-y^2)}$ (unchanged from Table 12.5) that look like this:

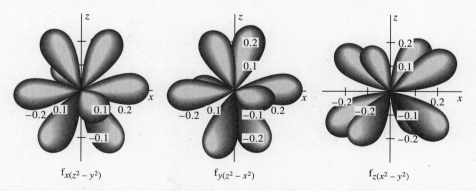

$f_{x(z^2-y^2)}$ $f_{y(z^2-x^2)}$ $f_{z(x^2-y^2)}$

Use Figure 20.3 and symmetry arguments as a guide, and assign these f orbitals to each energetically distinct group in an octahedral field.

20.31 Apply the simple conjugated polyene model of Problem 14.36 to the following situation. Saturn's principle moon, Titan, has an atmosphere that is largely N_2 (see Problem 9.4), but several larger components such as CH_4, C_3H_8, HCN, C_2N_2, HC_3N, and C_2H_2

have been observed by IR spectroscopy. The atmospheric chemistry is thought to be dominated by methane photodissociation from the sun along with nitrogen dissociation and ionization from high-energy electrons from Saturn's magnetosphere. Chemistry among the fragments of these processes lead to the larger components, and they in turn drift to the colder, lower regions of the atmosphere, condense, and polymerize further. Titan has an orange color, and it is possible that polymerized conjugated hydrocarbons and nitriles account for this color. What would be the length (in terms of number of C atoms) of a conjugated polyene that would look orange? Remember: an orange color means blue and green are absorbed, so you are looking for an absorption in the region of 500 ± 100 nm or so. Could a mixture of polyenes of different lengths look orange, or do you predict a single species absorbing in this region?

20.32 So-called *binary recombination* in which two molecular fragments collide and stick to form an intact molecule is impossible without some mechanism to remove the kinetic energy of the colliding fragments. If they have enough energy to come together, they have enough to fly apart again unless they lose some energy either by photon emission (in what is then called *radiative recombination*) or by collision with a third molecule (*ternary recombination*). Two atoms rarely collide and stay in each other's vicinity long enough for either to happen (unless the pressure is sufficiently high to promote ternary recombination), but polyatomic molecular fragments can have "stickier" collisions and linger in proximity for a much longer time. How do phenomena such as internal conversion and intersystem crossing help explain this observation?

CHAPTER 21

Spectroscopy in Magnetic and Electric Fields

CHAPTER 9 (in Section 9.3) considered the *macroscopic* energy changes produced by external electric and magnetic fields, Chapter 15 considered them on a *microscopic* level, and here we use them in spectroscopic experiments. We will find that precision spectroscopy in external fields measures new microscopic quantities: nuclear spin magnitudes, molecular dipole moments, features of electronic wavefunctions, new molecular structure information, and even internal molecular dynamics information.

We divide the chapter into two sections, since there are two broad types of external field spectroscopies. The first section covers interactions with the entire molecule or atom. Spectroscopy in a magnetic field is called *Zeeman spectroscopy*, named for the Dutch physicist Pieter Zeeman. In 1896, he found that a magnetic field split atomic emission lines by an amount proportional to the strength of the field. His Dutch colleague, Hendrik A. Lorentz, advanced a theory that explained many of these splittings, and the two of them shared the second Nobel Prize for physics in 1902. In 1908, the American astronomer George E. Hale used this effect to explain the splitting of atomic lines in sunspot spectra, the first time a magnetic field was observed and measured outside Earth. The effects Lorentz failed to explain, called "anomalous Zeeman effects," required Goudsmit and Uhlenbeck's discovery of electron spin, discussed in Section 12.5, for their explanation.

21.1 The Zeeman and Stark Effects

21.2 Spin Resonance Spectroscopies

Spectroscopy in an electric field is called *Stark spectroscopy,* named after the German physicist, Johannes Stark. In 1913, just before World War I and in the same year as Bohr's H-atom theory, Stark found that an electric field split visible H-atom emission lines, and he won the 1919 Nobel Prize for physics. In the 1930s, he conducted the first experiments on the deflection of atoms in an inhomogeneous electric field, discussed in Section 15.3. Stark had demonstrated considerable experimental brilliance since the turn of the century, when he was working in areas relevant to Einstein's ideas on light quanta. Lorentz considered Stark and Einstein equal proponents of these ideas, which were the foundations of photochemistry, but, while Stark initially admired Einstein, he increasingly felt that Einstein had not sufficiently acknowledged Stark's contribution. During World War I, Stark grew ever more nationalistic while Einstein espoused internationalism and pacifism. As Hitler rose to power, Stark became openly anti-Semitic and nationalistic. He renounced much of modern physics and embraced what became known as "Aryan physics," a curious blend of classical physics and wrong physics. Stark and several other scientists exerted considerable influence on German science during this time, but outside Germany, whatever fame he once had was gone, the effect he discovered was called simply the "electrochromic effect," and he was regarded as an enemy of science.[1]

The second section considers spectroscopies that couple magnetic fields directly to nuclear or electron spins. These techniques—nuclear magnetic resonance (NMR) and electron spin resonance (ESR)—typically operate in the rf and microwave frequency ranges. Electronic circuits manipulate and control these frequencies, and detectors respond directly to them. This level of control is exploited in modern instruments in many very clever ways, culminating in magnetic resonance imaging (MRI) machines that produce NMR spectral pictures for medical diagnosis. NMR is probably the single most important chemical spectroscopy. It is used routinely by synthetic chemists, biophysical chemists, analytical chemists, and polymer chemists to provide structural, dynamic, and chemical-bonding information that is unavailable from any other technique.

21.1 THE ZEEMAN AND STARK EFFECTS

An external magnetic or electric field grabs whatever part of an atom or molecule it can: a magnetic dipole moment, an electric dipole moment, individual charges (leading to polarizability), or higher order moments. Chapter 15 introduced the most important ways these interactions change atomic or molecular energies, and spectroscopy measures, as always, the net energy difference between two levels connected by a radiative transition. Our job here is to explore the magnitude of these energy changes (they are in general quite small), the molecular parameters that govern them (they are important ones), and the changes in spectroscopic selection rules brought on by the field (they are relatively minor). We start with magnetic fields and the Zeeman effect on atoms.

---

[1]The story of Stark and German physics in the first half of the 20th century is told in *Scientists under Hitler: Politics and the Physics Community in the Third Reich,* by A. D. Beyerchen (Yale University Press, New Haven, 1977).

21.1 THE ZEEMAN AND STARK EFFECTS

The Zeeman interaction energy Hamiltonian, $\hat{H}'_Z$, for something with a magnetic dipole moment vector **m** (measured in A m² or, equivalently, C m² s⁻¹ units) placed in a magnetic flux density field **B** (measured in tesla units, T) is given by Eq. (15.18):

$$\hat{H}'_Z = -\mathbf{m}\cdot\mathbf{B} , \qquad (15.18)$$

and we know from Section 15.2 that an unpaired electron has a magnetic moment given by Eq. (15.16):

$$\mathbf{m}_S = -\frac{g_e\mu_B \mathbf{S}}{\hbar} \qquad (15.16)$$

where **S** is the spin angular momentum, g_e is the free-electron g factor ($g_e \cong 2$), and μ_B is the Bohr magneton, $e\hbar/2m_e$. Unpaired spin is not the only possible source of an atomic magnetic moment, however. *Orbital* angular momentum produces a moment as well, and for singlet states with all electron spins paired, this is the only moment possible (assuming spinless nuclei). Consequently, $^1S_0$ states with spinless nuclei have *no* interaction with a magnetic field, except for a weak diamagnetic interaction that varies like B^2 and will be ignored here.

If $L \neq 0$, the magnetic moment m_L follows the prescription of classical mechanics for a current I looping around an area A. (See Eq. (9.43) as well.) The current is $-ev/2\pi r$ for an electron moving with speed v around a circle of radius r, and $A = \pi r^2$ so that the magnitude of the moment is

$$m_L = IA = -\frac{ev}{2\pi r}\pi r^2 = -\frac{evr}{2} .$$

Since such an electron's orbital angular momentum has the magnitude $|\mathbf{L}| = m_e v r$, we can write

$$m_L = -\frac{evr}{2} = -\frac{e}{2m_e}m_e v r = -\frac{e|\mathbf{L}|}{2m_e} = -\frac{e\hbar|\mathbf{L}|}{2m_e\hbar} = -\frac{\mu_B|\mathbf{L}|}{\hbar} . \qquad (21.1)$$

(Recall the quantum expression $|\mathbf{L}| = \sqrt{L(L+1)}\hbar$ for quantum number L.)

If z is the field direction, the Hamiltonian is

$$\hat{H}'_Z = -\mathbf{m}_L\cdot\mathbf{B} = -m_{L,z}B = \frac{\mu_B}{\hbar}L_z B , \qquad (21.2)$$

and we can use first-order perturbation theory to find the corresponding energy corrections. From Eq. (13.16), we have

$$E_M^{(1)} = \int \psi_n^{(0)*}\hat{H}'_Z\psi_n^{(0)}d\tau = \frac{\mu_B B}{\hbar}\int \psi_n^{(0)*}\hat{L}_z\psi_n^{(0)}d\tau = \frac{\mu_B B}{\hbar}M\hbar = M\mu_B B \qquad (21.3)$$

where M is the quantum number for the component of **L** along the z axis. Note that if $L = 0$ (an S state), $M = 0$ only, and there is no energy change. But even if $L \neq 0$, the $M = 0$ component of such a state is unchanged by the field.

Now consider the specific case of He in the excited 1s2p $^1P_1$ state. It fluoresces readily to the 1s² $^1S_0$ ground state, and in zero magnetic field, the single transition is in the vacuum ultraviolet at 171 129.148 cm⁻¹. In a magnetic field, $^1P_1$ splits into three states, and the ground state is not affected, as illustrated in Figure 21.1.

The selection rules for L, S, and J in the Zeeman spectrum are the same as for the field-free case: $\Delta L = 0, \pm 1$, $\Delta S = 0$, $\Delta J = 0, \pm 1$, but excluding $L = 0 \to L = 0$ and $J = 0 \to J = 0$. The M selection rule is $\Delta M = 0, \pm 1$, but excluding $M = 0 \to M = 0$ if $\Delta J = 0$. Since the field imposes a direction on the system, radiation *polarization* enters the story. For $\Delta M = 0$ transitions, the light is polarized

FIGURE 21.1 The Zeeman effect on the excited 1s2p $^1P_1^o$ state of He splits the M degeneracy into three levels. The $M = +1$ state moves up by $+\mu_B B$, the $M = 0$ state stays put, and the $M = -1$ state moves down by $-\mu_B B$. Since $\mu_B = 0.467$ cm^{-1} T^{-1}, the emission splits into three closely spaced transitions, at 171 129.615, 171 129.148, and 171 128.681 cm^{-1} for a 1.0 T field (shaded region).

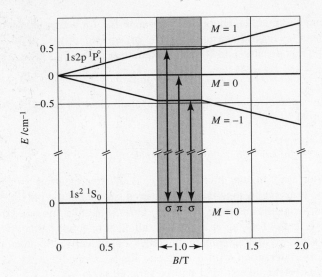

parallel to the field and is said to be π *polarized*. Transitions with $\Delta M = \pm 1$ lead to light polarized *perpendicular* to the field when viewed perpendicular to the field or to *left and right circularly polarized light* when viewed along the field. These transitions are said to be σ *polarized*, as noted in Figure 21.1.[2]

If the state is not a singlet, the splitting pattern is no longer so simple, leading to the *anomalous Zeeman effect*. Its explanation is traced to the "anomalous" magnetic moment of an electron spin. The electron g factor shows that an electron has about twice the magnetic moment expected classically, as a comparison of Equations (15.16) and (21.1) show. If we define the total angular momentum (except for nuclear spin) as $\mathbf{J} = \mathbf{L} + \mathbf{S}$, then the perturbation Hamiltonian for weak fields is

$$\hat{H}_Z' = \frac{g_J \mu_B}{\hbar} \hat{J}_z B \qquad (21.4)$$

if the field is along the z axis. Here, g_J is the *Landé g factor*, which the Closer Look below derives:

$$g_J = 1 + \left[\frac{J(J+1) + S(S+1) - L(L+1)}{2J(J+1)} \right] (g_e - 1) \ . \qquad (21.5)$$

The selection and polarization rules are the same, but the splittings depend on all three quantum numbers, J, L, and S. Perturbation theory gives an energy expression analogous to Eq. (21.3):

$$E_{M_J}^{(1)} = M_J g_J \mu_B B \ . \qquad (21.6)$$

EXAMPLE 21.1

What are the energies in a 0.1 T field of the Na states based on the field-free states 3p $^2P_{1/2}$ and 3p $^2P_{3/2}$?

SOLUTION The $^2P_{1/2}$ state has $S = 1/2$, $L = 1$, and $J = 1/2$ so that $g_J = 2/3$ from Eq. (21.5), taking $g_e \cong 2$. This state splits into the two states $M_J = +1/2$ and $-1/2$ (since

---

[2] Here, π and σ come from the German for parallel and perpendicular: *parallel* and *senkrecht*. This notation preceded the use of σ and π in molecular orbital theory.

M_J varies from $-J$ to $+J$ in unit steps) with energy shifts from the field-free energy (which is 16 956.183 cm^{-1}) equal to

$$E^{(1)}_{+1/2} = M_J g_J \mu_B B = +\frac{1}{2}\frac{2}{3}(0.467 \text{ cm}^{-1} \text{ T}^{-1})(0.1 \text{ T}) = 0.016 \text{ cm}^{-1}$$

and

$$E^{(1)}_{-1/2} = M_J g_J \mu_B B = -\frac{1}{2}\frac{2}{3}(0.467 \text{ cm}^{-1} \text{ T}^{-1})(0.1 \text{ T}) = -0.016 \text{ cm}^{-1} .$$

Similarly, the $^2P_{3/2}$ state, normally at 16 973.379 cm^{-1}, splits into *four* states: $M_J = -3/2, -1/2, +1/2,$ and $+3/2$. Here, $g_J = 4/3$, and the energies are

$$E^{(1)}_{+3/2} = M_J g_J \mu_B B = +\frac{3}{2}\frac{4}{3}(0.467 \text{ cm}^{-1} \text{ T}^{-1})(0.1 \text{ T}) = 0.093 \text{ cm}^{-1}$$

$$E^{(1)}_{+1/2} = M_J g_J \mu_B B = +\frac{1}{2}\frac{4}{3}(0.467 \text{ cm}^{-1} \text{ T}^{-1})(0.1 \text{ T}) = 0.031 \text{ cm}^{-1}$$

$$E^{(1)}_{-1/2} = M_J g_J \mu_B B = -\frac{1}{2}\frac{4}{3}(0.467 \text{ cm}^{-1} \text{ T}^{-1})(0.1 \text{ T}) = -0.031 \text{ cm}^{-1}$$

$$E^{(1)}_{-3/2} = M_J g_J \mu_B B = -\frac{3}{2}\frac{4}{3}(0.467 \text{ cm}^{-1} \text{ T}^{-1})(0.1 \text{ T}) = -0.093 \text{ cm}^{-1} .$$

Note that the splittings are still symmetric about the field-free energies, but they are notably different in the two states.

➽ *RELATED PROBLEMS* 21.2, 21.3, 21.4

A Closer Look at the Landé *g* Factor

The expression in Eq. (21.5) for the Landé *g* factor is not difficult to derive, and the derivation we use here gives us an opportunity to explore the coupling of angular momenta in greater depth. Classical mechanics and some elementary trigonometry suffice, but Problem 21.27 considers also a quantum mechanical derivation. Consider, for example, a $^2P_{3/2}$ atomic state (with quantum numbers $S = 1/2$, $L = 1$, and $J = 3/2$) that has no nuclear spin. In Figure 21.2(a), we align the z axis along the **B** field's direction. The **S** and **L** angular momentum vectors couple (add) to produce **J**, and we imagine a state with $M_J = 3/2$, giving the z component vector **J**$_z$ along the field with magnitude $J_z = M_J \hbar = 3\hbar/2$. This coupling (L–S coupling, in the language of Section 13.5) in classical language causes vectors **L** and **S** to *precess* around **J**. Figure 21.2(a) pictures these vectors at one instant in their precession. The **J** vector, in turn, precesses about **B**, dragging **L** and **S** with it.

In Figure 21.2(b), we use Equations (15.16) and (21.1) to find the magnetic moment vectors **m**$_S$ and **m**$_L$ corresponding to **S** and **L**. These add to produce **m**, the net magnetic moment, but note that while **m**$_S$ and **m**$_L$ are antiparallel to **S** and **L**, *m is not antiparallel to J* since the free-electron *g* factor, g_e, does not equal 1. This net moment, **m**, in turn precesses around **J** (and **J** is precessing around **B**, but at a slower rate for weak fields). The precession of **m** around **J** means that only the *component of m along the direction of J* matters. The other components average to zero. If we call the component of **m** along **J** the *effective moment*, **m**$_{\text{eff}}$, we can find its length using the drawing in Figure 21.2(c). Note that the length of **m**$_{\text{eff}}$ has two readily identified pieces: the component of **m**$_S$ along the direction of **J** and the component of **m**$_L$ along the direction of **J**. If we define θ_{JS} to be the angle between **J** and **S**, as shown in the figure, we see that the first piece of **m**$_{\text{eff}}$ has the length $|$**m**$_S| \cos \theta_{JS}$. We also define θ_{JL} as the angle between **J** and **L**, and we see that the

FIGURE 21.2 The sequence of vectors shown in (a)–(d) lead to a classical derivation of the Landé g factor. Each step is explained in the text.

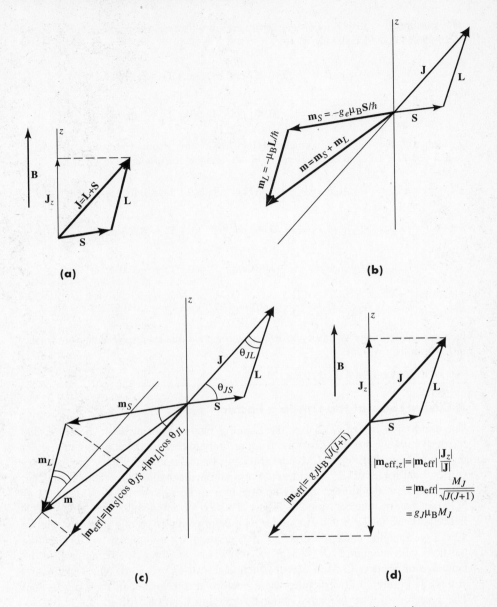

second piece of $\mathbf{m}_{\text{eff}}$ has the length $|\mathbf{m}_L| \cos \theta_{JL}$ so that $m_{\text{eff}} = |\mathbf{m}_{\text{eff}}| = |\mathbf{m}_S| \cos \theta_{JS} + |\mathbf{m}_L| \cos \theta_{JL}$.

To simplify notation, we will use J^* to represent $J(J+1)$ and similarly for S^* and L^* so that we write the length of $\mathbf{J}$ as $\sqrt{J^*}\hbar$, for example. We find the angles from the cosine law for triangles.[3] Applying this law to find θ_{JS}, we have (cancelling common factors of $\hbar^2$)

$$L^* = S^* + J^* - 2\sqrt{L^*}\sqrt{S^*} \cos \theta_{JS}$$

and for θ_{JL}, we have

$$S^* = L^* + J^* - 2\sqrt{L^*}\sqrt{J^*} \cos \theta_{JL} .$$

---

[3]For a triangle of sides a, b, and c with angle θ between sides b and c, this law states $a^2 = b^2 + c^2 - 2bc \cos \theta$.

If we solve each of these for their cos θ factors and substitute into the expression for m_{eff}, we can write (using $m_S = g_e \mu_B \sqrt{S^*}$ and $m_L = \mu_B \sqrt{L^*}$)

$$\frac{m_{\text{eff}}}{\mu_B} = g_e \sqrt{S^*} \frac{J^* + S^* - L^*}{2\sqrt{J^*}\sqrt{S^*}} + \sqrt{L^*} \frac{J^* + L^* - S^*}{2\sqrt{J^*}\sqrt{L^*}}.$$

Next, we simplify this expression with some careful algebra:

$$\begin{aligned}
\frac{m_{\text{eff}}}{\mu_B} &= \sqrt{J^*}\left(g_e \frac{J^* + S^* - L^*}{2J^*} + \frac{J^* + L^* - S^*}{2J^*}\right) \\
&= \sqrt{J^*}\left[\frac{g_e J^* + J^* + S^*(g_e - 1) - L^*(g_e - 1)}{2J^*}\right] \\
&= \sqrt{J^*}\left[\frac{2J^* + J^*(g_e - 1) + S^*(g_e - 1) - L^*(g_e - 1)}{2J^*}\right] \\
&= \sqrt{J^*}\left[1 + \frac{J^* + S^* - L^*}{2J^*}(g_e - 1)\right] = g_J \sqrt{J(J+1)}
\end{aligned}$$

where we define the final quantity in square brackets as the Landé g factor, g_J.

Finally, in Figure 21.2(d), we find $m_{\text{eff},z}$, the magnitude of the component of $\mathbf{m}_{\text{eff}}$ along the z axis. This component governs the energy, since the Hamiltonian has the form $-\mathbf{m}_{\text{eff}} \cdot \mathbf{B} = m_{\text{eff},z} B$. Note from the figure that simple geometry tells us $m_{\text{eff},z}/m_{\text{eff}} = |\mathbf{J}_z|/|\mathbf{J}|$ so that $m_{\text{eff},z} = m_{\text{eff}} |\mathbf{J}_z|/|\mathbf{J}| = m_{\text{eff}} M_J \hbar / \sqrt{J(J+1)}\hbar = M_J g_J \mu_B$, and the energy is $M_J g_J \mu_B B$, which is Eq. (21.6).

Spin–orbit coupling makes $\mathbf{J} = \mathbf{L} + \mathbf{S}$ a valid sum and makes J the "good quantum number" in the absence of a field. If the field is large enough, where "large enough" means the field interaction energy is larger than or comparable to the spin–orbit interaction energy, the orbital and spin magnetic moments decouple from each other. This is called the *Paschen–Back effect*. Momenta $\mathbf{L}$ and $\mathbf{S}$ behave independently, L and S become "good quantum numbers," and their components follow the selection rules $\Delta M = 0, \pm 1$ and $\Delta M_S = 0$. This (see Problem 21.6) leads to a spectrum that looks very much like the normal Zeeman effect pattern. A small residual spin–orbit interaction further splits the lines into closely spaced components.

The Zeeman effect on molecules is considerably more complicated than for atoms. In addition to orbital, electron spin, and nuclear spin moments, molecules have moments induced by molecular rotation. Molecular g values are considerably more complicated to calculate and interpret than the atomic Landé g factor. Nevertheless, the molecular Zeeman effect has been put to good use in the technique known as *laser magnetic resonance* (LMR) spectroscopy. In an LMR experiment, a magnetic field is used to *tune* a molecular transition into resonance with a fixed-frequency laser. Tunable lasers in the far infrared region are difficult and expensive devices. (A few have been made that mix fixed-frequency laser radiation with tunable microwave radiation, and synchrotron sources hold some promise.) Many polyatomic gases lase in the far IR in an electric discharge. These lasers operate between rotational transitions in the gas, and a grating placed inside the laser cavity allows one to select a single, narrow emission. Thus, by changing gases and isotopes, a large number of fairly closely spaced laser frequencies are available. It is still unlikely that any one of these frequencies is coincident with another molecule's transition.

The LMR spectrometer places the gas sample of interest (and molecules, free radicals, atoms, and molecular ions have all been studied this way) in a tunable magnetic field and also *inside the far-IR laser cavity*. If the sample is not resonant with the laser frequency, the laser glows happily. When the magnetic field tunes the sample transition into resonance with the laser frequency, the laser output drops measurably. Lasers are *very* sensitive to anything that absorbs the light inside their cavity. Thus, the experiment takes advantage of the high resolution of laser spectroscopy, the sensitivity of intracavity experiments, and the Zeeman effect to reach new transitions in interesting species.

In contrast, Stark spectroscopy is widely used for molecules but less so for atoms. The Stark Hamiltonian is given by Equations (15.9) and (15.10), which can be written in a form similar to Eq. (21.2):

$$\hat{H}'_S = -\mathbf{p}\cdot\mathbf{F} \tag{21.7}$$

where $\mathbf{p}$ is the dipole moment vector and $\mathbf{F}$ is the electric field vector. For the H atom, which has sets of nearly degenerate states differing by L (such as 2s and 2p, 3s, 3p and 3d, etc.), there is a *linear* Stark effect (see Problem 21.31). The energy splitting is proportional to the first power of the electric field magnitude F. This allowed Stark to discover the effect in the first place, and the H-atom Stark effect problem was the first application of quantum perturbation theory. In molecules, a first-order effect is found for certain states of symmetric tops. The degeneracy here is expressed by the K quantum number (see Equations (19.37a) and (19.37b)), which represents rotation around the symmetry axis. Since the dipole moment of a symmetric top must lie along the top's symmetry axis, $K \neq 0$ states have a component of the moment parallel to the total angular momentum, and this leads to a first-order interaction. See Problem 21.30 for more details.

Most atoms and molecules do not have degenerate states, however, and for them, the Stark effect is second order. For an atom, the interaction is through the polarizability, as in Eq. (15.22): $E^{(2)} = -\alpha F^2/2$, and this is exploited in the deflection experiments discussed in Section 15.3. The quadratic dependence can be thought of in the following way. First, the field polarizes the atom, creating an induced moment proportional to the field, since $p = \alpha F$ for polarization α. The field then interacts with this moment to alter the energy, $E \propto pF = \alpha F^2$, leading to a quadratic effect. This argument would seem to imply that a molecule with a *permanent* dipole moment should have a first-order Stark effect, but such is not the case. The reason can be traced to arguments similar to those in Section 15.1 based on the parity of the molecular wavefunction. Rotational wavefunctions have definite parity, and the first-order perturbation integral over the dipole moment vanishes. In contrast, an induced moment is tied more to the *field* than to the *atom or molecule*. It does not rotate with the molecule in the same way a permanent moment does.

Nevertheless, the second-order effect exists and is quite useful. For a $^1\Sigma$ linear molecule, the only case we will look at in any detail here, the second-order Stark energy expression is

$$\begin{aligned} E^{(2)}_{J,M_J} &= -\frac{1}{6}\frac{p^2 F^2}{B_e}\frac{1}{[(100\text{ cm m}^{-1})hc]^2} & \text{for } J = 0 \\ &= \frac{p^2 F^2}{2B_e J(J+1)}\frac{J(J+1) - 3M_J^2}{(2J-1)(2J+3)}\frac{1}{[(100\text{ cm m}^{-1})hc]^2} & \text{for } J \neq 0 \end{aligned} \tag{21.8}$$

where the energy is in cm^{-1} units (due to the $(100\text{ cm m}^{-1})hc$ factors), B_e is the rotational constant in cm^{-1}, and J and M_J are the usual rotational quantum number and its projection. Note that M_J enters *squared* so that the second-order Stark energy is insensitive to the algebraic sign of M_J. Note also that the dipole moment, p, enters squared as well. This means that second-order Stark spectroscopy can determine dipole moment *magnitudes* quite well, but not their direction in the molecule.

This effect is put to use in rotational spectroscopy in three important ways. First, laser Stark spectroscopy, the direct analog of laser magnetic spectroscopy, uses an electric field to tune a molecule into resonance with a fixed-frequency laser. Second, an alternating electric field (of the "off-on, square-wave" type, rather than the sinusoidal type) is often used to modulate microwave transitions and increase signal processing sensitivity in microwave spectroscopy.

EXAMPLE 21.2

You are designing a Stark spectrometer to detect CO at room temperature. What electric field do you need to shift the $(J, M_J) = (0,0) \rightarrow (1,0)$ transition by twice its normal Doppler width?

SOLUTION We find $B_e = 1.9313$ cm^{-1} from Table 19.2 and calculate the (approximate) position of the field-free $J = 0 \rightarrow 1$ transition to be $2B_e = 3.86$ cm^{-1}. The Doppler width is given by Eq. (19.51) with $M = 28$ g mol^{-1} and $T = 300$ K:

$$\Delta E = (7.16 \times 10^{-7})(3.86 \text{ cm}^{-1})\left(\frac{300}{28}\right)^{1/2} \cong 9 \times 10^{-6} \text{ cm}^{-1}.$$

From Eq. (21.8), we see that the Stark effect causes the (0,0) state to drop in energy, and the (1,0) state to rise, since

$$E_S^{(2)}(0,0) = -\frac{1}{6}\frac{p^2 F^2}{B_e}\frac{1}{[(100 \text{ cm m}^{-1})hc]^2}$$

$$E_S^{(2)}(1,0) = \frac{p^2 F^2}{2B_e J(J+1)}\frac{J(J+1) - 3M_J^2}{(2J-1)(2J+3)}\frac{1}{[(100 \text{ cm m}^{-1})hc]^2} = \frac{p^2 F^2}{10 B_e}\frac{1}{[(100 \text{ cm m}^{-1})hc]^2}.$$

The difference between these energies, $E_S^{(2)}(1,0) - E_S^{(2)}(0,0)$, is the Stark shift from the field-free transition:

$$E_S^{(2)}(1,0) - E_S^{(2)}(0,0) = \frac{p^2 F^2}{B_e}\frac{1}{[(100 \text{ cm m}^{-1})hc]^2}\left(\frac{1}{10} + \frac{1}{6}\right).$$

We want the field, F, to make this shift about twice the Doppler width, or about 1.8×10^{-5} cm^{-1}. The CO dipole moment is in Table 15.1, 0.374×10^{-30} C m, and we solve

$$1.8 \times 10^{-5} \text{ cm}^{-1} = \frac{p^2 F^2}{B_e}\frac{1}{[(100 \text{ cm m}^{-1})hc]^2}\left(\frac{1}{10} + \frac{1}{6}\right)$$

for F, finding $F \cong 6 \times 10^5$ V m^{-1}. If we have a 6000 V potential difference between two parallel plates 1 cm apart, we have this field. Carbon monoxide's small dipole moment and fairly large rotational constant force us to use this fairly large electric field.

➡ **RELATED PROBLEMS** 21.7, 21.9

Finally, the technique known as molecular beam electric resonance (MBER) spectroscopy uses the Stark effect not only to induce splittings (and, as we will see, to introduce new types of transitions), but also to push the molecular beam around with *inhomogeneous* electric fields. The OCS molecule has been heavily studied by this method, and we will use it as a model.

Figure 21.3 graphs Eq. (21.8) for the $J = 0$, $M_J = 0$ and $J = 1$, $M_J = \pm 1$ and 0 states of OCS using the dipole moment in Table 15.1 and the OCS rotational constant $B_e = 0.2028$ cm^{-1}. Note that the $J = 0$ and $J = 1$ states at zero field are separated by about $2B_e = 0.4056$ cm^{-1}, the usual field-free microwave transition energy. The Stark shifts and splittings are many times smaller. Consequently, high

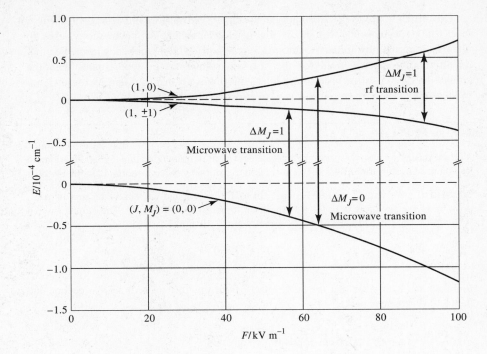

FIGURE 21.3 The Stark effect on the lowest two rotational levels of OCS changes and splits the field-free energies. Microwave ($\Delta J \neq 0$) transitions and rf transitions ($\Delta J = 0, \Delta M_J \neq 0$) are both allowed. Note that this Stark effect does not depend on the *sign* of M_J, only its magnitude.

resolution is called for, and the collision-free and Doppler-free environment of a molecular beam coupled to the inherent frequency precision of microwave and radiofrequency sources give MBER spectroscopy a resolution that can approach 1 kHz.

A new type of transition is indicated in Figure 21.3. The $J = 1$, $M_J = 0 \leftrightarrow J = 1$, $M_J = \pm 1$ transition is no longer degenerate in a non-zero field, and the transition is allowed. It occurs in the radiofrequency (rf) region of the spectrum, typically around 10 kHz to 100 MHz, depending on the molecule and the field. If the rotational constant is known from ordinary field-free rotational spectra, analysis of this transition yields precise dipole moments.[4]

The molecular beam in an MBER spectrometer is so dilute that too little radiation is absorbed to be measured directly. Consequently, the transition is detected in a very clever indirect way. First, the beam is passed through a strong inhomogeneous field designed (according to the ideas discussed in Section 15.3) to act like a focusing lens for the beam's molecules. The properties of this "lens" depends on the molecule's quantum state: those for which the energy *decreases* with increasing field (such as (0,0) and (1,±1) in Figure 21.3) are *defocused* and pushed away from the center axis of the beam and thrown out of the experiment. Those for which the energy *increases* are *focused* toward the axis. This first field becomes a *quantum state selector* that we can adjust to focus predominantly one state. Beyond it, the beam molecules are enhanced in one quantum state, (1,0), for example. The beam next enters a region of highly uniform electric field in which transitions are induced.[5]

---

[4] Accurate work includes not only polarizability effects on the energy, which we have neglected, but also higher order Stark terms (fourth-order comes next). The transition energy is followed as a function of electric field, and the full energy expression is used to extract dipole and polarizability information.

[5] The uniform electric field is not always necessary here, and experiments using radiation beyond microwave frequencies, up through visible, have been performed as well.

The beam then enters a second inhomogeneous field similar to the first that acts as a *quantum state analyzer*. In one common type of experiment, this field is set to refocus the state initially selected onto the entrance aperture of a mass spectrometer. If no transition is induced, the signal from the mass spectrometer beam detector set to the molecule of interest does not change as the radiation frequency is swept. If a transition occurs, however, the final state of the transition is one that the analyzing field rejects, and the detector signal drops. Thus, a spectrum is recorded by monitoring the *beam* intensity while scanning the *radiation frequency*. As transitions are encountered, the flux of molecules entering the mass spectrometer changes, signaling a transition.

Note from Figure 21.3 that an experiment designed to select the (1,0) state initially and induce transitions to, for example, $(1, \pm 1)$ in the rf region, operates through *stimulated emissions* rather than absorptions. Molecules emit photons *into* the radiation field rather than absorb them. This is put to use in a *maser* in which an inhomogeneous electric field selects molecules in a beam that are in an excited rotational state and passes only them into a tuned microwave cavity. The distribution of quantum states entering the cavity is thus no longer at equilibrium; it has a *population inversion* between the selected state and a lower lying state. A weak microwave source stimulates emission from the selected excited state, and the transition adds photons to the radiation field, amplifying it.

The MBER spectrometer has been used extensively to study small weakly bound molecular clusters known as *van der Waals molecules*. These can be formed in supersonic molecular beams in the natural course of the beam's production (see Chapter 24 for more details), and the mass spectrometer–microwave spectrometer combination inherent in the method allows one to measure precise structural information on these molecules, even when the beam is a mixture of several species. For example, one of the first van der Waals molecules studied this way was ArHCl, the weakly bound adduct of an Ar atom and an HCl molecule. One mixes Ar and HCl gases, expands the mixture through a small hole into the spectrometer vacuum, and species such as ArHCl, $(HCl)_2$, Ar_2, etc., are observed in the beam's mass spectrum. Dozens of molecules have been studied this way, including many rare gas–diatomic adducts like ArHCl, hydrogen bounded dimers like $(HF)_2$, $(H_2O)_2$, and $(NH_3)_2$, and polyatomic clusters like $(CO_2)(NH_3)$, $(HCCH)(NH_3)$, and many others. (Problem 15.25 considers the (HF)(ClF) dimer studied by this method.)

The Zeeman effect can be used instead of the Stark effect in the molecular beam *magnetic* resonance (MBMR) spectrometer. Inhomogeneous magnetic fields act as state selectors, and spectroscopy takes place in a homogeneous field. One of the most important applications of MBMR spectroscopy is the generation of precise time (or frequency) standards. Since 1967, the second has been defined as "the duration of 9 192 631 770 periods of the radiation corresponding to the transition between the two hyperfine levels of the ground state of the ^{133}Cs atom," and this transition is at the heart of the "Atomic Clock" frequency standard given a closer look below. Note the emphasis on counting "periods of the radiation." Integral counting is an *exact* measurement, and the passage of time is recorded by counting these periods directly from an agreed-upon time origin.

A Closer Look at Atomic Time Standards

The ^{133}Cs isotope (100% natural abundance) has a nuclear spin quantum number $I = 7/2$ and a 6s $^2P_{1/2}$ ground electronic state with $J = 1/2$: **I** and **J** add to produce the total atomic angular momentum $\mathbf{F} = \mathbf{J} + \mathbf{I}$ with possible quantum numbers $F = 4$ and 3. The $F = 4$ state has the higher energy of the two, about 0.3 cm^{-1} (or a frequency exactly 9 192 631 770 Hz) above the $F = 3$ ground state. In a weak magnetic field, these states split into nine and seven M_F components, respectively.

The Cs beam frequency standard maintained by the US National Institute of Standards and Technology (NIST) is an MBMR spectrometer equipped with a highly directional atomic Cs beam generated from an oven held at about 100 °C. The oven is charged with about 4 g of Cs, an amount that lasts through several years of operation, since the vapor pressure of Cs(ℓ) at 100 °C is only 6×10^{-7} atm. The first inhomogeneous field selects the $F = 4$, $M_F = 0$ state. These atoms then pass through a field of only 6×10^{-6} T, but one that is uniform to better than 1%. The clock transition (to the $F = 3$, $M_F = 0$ state) is stimulated in this field. Both states in this transition vary in energy with a weak magnetic field only very slightly,[6] but this small field is needed to split other M_F components out of the way, leaving the clock transition quite close to its field-free value. Microwave radiation at the 9.2 GHz transition frequency induces the transition, and a following inhomogeneous field state-analyzes the beam and allows only those atoms that have undergone the transition to reach the detector. The detector signal is then used in an electronic feedback circuit to stabilize the microwave source, and frequency division circuits typically shift this frequency down to the range of a quartz oscillator frequency. A separate quartz oscillator is controlled by the down-shifted frequency and supplies the time reference.

Numerous small effects ranging from temperature control through quantum effects beyond our discussion are known and compensated to the point that the clock is stable to about 10 ns per day, or 1 part in 10^{13}. Several other quantum sources that have the potential of 1 part in 10^{14} or more stability are under investigation, and it is likely that the Cs clock will be replaced as these new sources are developed.

We close this section with a comment on some effects that the precision of Zeeman or Stark spectroscopy makes visible, but that are beyond our scope to consider in detail. Several of these depend on very small energy shifts that result from interactions controlled by chemically interesting quantities. One is called *nuclear quadrupole coupling*. Nuclei with spins $I \geq 1$ (such as D, N, Cl, Br, and many others) have a charge distribution that has a non-zero quadrupole moment: the nuclear charge distribution looks like a cigar or a pancake rather than a sphere. This moment interacts with the *gradient* of the electronic charge at the nucleus, leading to a small energy shift that depends on the molecular angular momentum and the magnitude of the gradient. Transitions among such levels probe this gradient, which, in turn, is proportional to the s and p character of the electronic wavefunction. The nuclear quadrupole moments themselves are measured by atomic resonance experiments, as are nuclear spin magnitudes themselves. These spins interact with each other, with the rotational motion of the molecule, and with any unpaired electron spins, but the most important spectroscopic role for spins places them at the center of the stage in spin resonance spectroscopies, the topic of the next section.

---

[6]Our discussion of the Zeeman effect has them independent of field, but higher-order perturbation theory introduces a small B^2 dependence to their energies.

21.2 SPIN RESONANCE SPECTROSCOPIES

Unpaired spins, aligned in external magnetic fields and split into different energy levels, form an exquisite probe of molecular geometry in spite of the fact that the energy splittings are trivially small. In a single spin system there are only two levels, and their energy separation is so small compared to the typical temperature of the sample that, in the language of the two-level system as we used it in Chapter 4, Figure 4.1, there are nearly "as many bricks on the table as on the floor"—the upper and lower energy states have nearly the same population. If they were exactly equal, there could be no net absorption or emission—no spectroscopy. If the population difference is increased, the sensitivity of the spectroscopy is increased. For spin spectroscopies, this means larger magnetic fields, and most research NMR spectrometers use high-field superconducting magnets for this reason. Increased energy level splittings mean increased spectroscopic frequencies, of course, and the past few decades have seen NMR frequencies rise roughly an order of magnitude from around 60 MHz to 600 MHz. These are not cheap machines—they cost roughly 0.1¢ per Hz, which sounds cheap until one multiplies by a factor on the order of $3-6 \times 10^8$ Hz. It is not unusual for any research spectrometer of a specialized type to cost somewhere in the $\$10^5$ to $\$10^6$ range, but the NMR spectrometer, along with magnetic resonance imaging (MRI) machines used for noninvasive medical diagnosis, are the only ones with such wide use to support a large commercial market.

Unpaired electron spins, found in free radicals, paramagnetic ions, or triplet state molecules, form the working media for electron spin resonance, ESR, (also called electron paramagnetic resonance, EPR, and electron magnetic resonance, EMR). Since species with unpaired spins are relatively rare (compared to species with nonzero nuclear spin), ESR does not have the wide applicability of NMR. It is, however, an excellent probe of electronic structure. For example, it can pick out a paramagnetic ion buried in a large metalloprotein and tell one something about the electronic configuration of the ion and the structure of the protein in the ion's vicinity.

Since NMR is the more general spin resonance spectroscopy, we begin with it. Many of the ideas we develop for NMR hold for ESR as well.

Nuclear Magnetic Resonance

Nuclear magnetic resonance spectroscopy has several differences from the optical spectroscopies we have discussed in Chapters 18–20. Most of these differences are bad news—they lower the sensitivity of the method—but the advantages of operating in the radiofrequency or microwave regions where modern electronic circuits can shape, manipulate, and detect such radiation with ease and precision help overcome the disadvantages. First, as mentioned earlier, the small energy difference between levels involved in a transition limits sensitivity since any spectroscopic signal is proportional to the population difference between the levels.

Second, NMR transitions occur through a different mechanism than optical transitions. Only magnetic fields can grab and flip spin magnetic moments. Thus, it is the magnetic field of the radiation rather than the electric field that induces the transition, leading to a *magnetic dipole transition moment,* which is inherently smaller than the electric dipole moments of optical transitions.

Third, spontaneous emission plays no role; there is no "NMR fluorescence." The transition moment and the photon energy are so small that the radiative lifetime is effectively infinite. This means that the spin system has a tough time maintaining equilibrium. It is more strongly coupled to the radiation field than to its surroundings. In optical spectroscopy, collisions and/or spontaneous emission usually relax the excited state at a rate greater than the rate radiation populates it. In an NMR experiment, the relaxation mechanism analogous to gas-phase collisions is called *spin–lattice relaxation*. The spins are weakly coupled to each other through their own magnetic fields and those of electrons, and as these fields fluctuate due to molecular motion, they induce spin transitions that try to maintain an equilibrium distribution. Since this relaxation mechanism is weak, it is also slow. Consequently, a strong radiation field can *saturate* the transition, which means it can make the populations of the upper and lower energy levels equal. At that point, absorption and stimulated emission occur at equal rates, and the transition becomes invisible.

Finally, NMR needs to be an inherently high-resolution method to expose the most useful information in the spectrum. This means that the radiation field must have a very narrow bandwidth (easy to do) and that the magnetic field must have a very high spatial homogeneity and stability (less easy to do).

With these caveats setting the stage for the general NMR experiment, we turn now to some of the important quantum details. We will not go into the use of NMR for structure determination in any detail—most good organic chemistry textbooks cover this topic—but we will explore the major effects that govern all high-resolution spectra. The first of these is the way the nuclear g factor, g_N, works to place spectra of different nuclei in distinct spectroscopic regions. Recall from Eq. (15.19) that the nuclear magnetic moment depends on the product of g_N and $\mathbf{I}$, the nuclear spin angular momentum:

$$\mathbf{m}_N = \frac{g_N \mu_N \mathbf{I}}{\hbar} \quad (15.19)$$

where μ_N is the *nuclear* magneton, which is considerably smaller than the Bohr magneton: $\mu_N = e\hbar/2m_p = \mu_B(m_e/m_p) \cong \mu_B/1836$. Each isotope has a characteristic $\mathbf{I}$, and for $\mathbf{I} \neq 0$, a characteristic g_N. Table 21.1 collects these quantities (expressing $\mathbf{I}$ in terms of its corresponding quantum number I) for several important nuclei. Note that g_N has no obvious pattern from nucleus to nucleus—the value for ^{2}H (deuterium) is not equal to the sum or difference of the proton and neutron values, for instance—and g_N can be positive or negative.

TABLE 21.1 NMR Properties of Selected Nuclei

Nucleus	% abundance	I	g_N
^1n	—	1/2	−3.826 0
^{1}H	99.985	1/2	5.585 7
^{2}H	0.015	1	0.857 4
^{3}H	—	1/2	5.957 5
^{3}He	1.38×10^{-4}	1/2	−4.255 2
^{6}Li	7.42	1	0.821 91
^{7}Li	92.58	3/2	2.170 7
^{10}B	19.9	3	0.600 2
^{11}B	80.1	3/2	1.792
^{13}C	1.10	1/2	1.404 7
^{14}N	99.635	1	0.403 6
^{15}N	0.365	1/2	−0.566 1
^{17}O	0.037	5/2	−0.757 5
^{19}F	100.	1/2	5.256 7
^{31}P	100.	1/2	2.263 5
^{35}Cl	75.53	3/2	0.547 9
^{37}Cl	24.47	3/2	0.456 1

A single, isolated nucleus (which any nucleus closely approximates, even one in a molecule buried among a mole or so of neighbors) interacts with the field through the nuclear Zeeman Hamiltonian $\hat{H}'_N = -\mathbf{m}_N \cdot \mathbf{B} = -(g_N\mu_N/\hbar)\hat{\mathbf{I}} \cdot \mathbf{B}$, which leads to the energy $-g_N\mu_N M_I B$ for a field aligned along the z axis. Here, M_I is the quantum number for the z component of the spin: $M_I = +I, +I-1, \ldots -I$, as usual. The NMR selection rules in this case are $\Delta M_I = \pm 1$ so that the transition frequency, $\nu_{\text{NMR}} = E/h$, equals $|g_N|\mu_N B/h$ for a field of magnitude B. There are thus two obvious ways an NMR experiment could find this transition: scan the radiation frequency at constant field, or scan the field at constant frequency. In either case, the transition is induced when the equality

$$\nu_{\text{NMR}} = |g_N|\mu_N B/h \qquad (21.9)$$

is reached, which shows that ν_{NMR}/B is a constant for any one nucleus. There are arguments that favor either type of scan, but the fixed-frequency–scanned-field experiment dominated the early years of NMR and introduced a conventional way of displaying spectra that we discuss later. Modern spectrometers use neither method, relying instead on a *Fourier transform* method we also discuss later.

Figure 21.4 shows the fields needed to resonate the nuclei in Table 21.1 at $\nu_{\text{NMR}} = 100$ MHz and the frequencies needed for a 7 T field. *All* of these nuclei are so far apart that designing a spectrometer that could cover them all at high resolution in one experiment is not done. Consequently, one tunes an instrument for one particular nucleus at a time.

EXAMPLE 21.3

You are designing an NMR spectrometer to study boron hydrides, and you have a 5.0 T superconducting magnet. What frequencies will you need to study ^{1}H, ^{2}D, ^{10}B, and ^{11}B spectra with this magnet?

SOLUTION For each nucleus, the resonance frequencies will be quite close to the bare nucleus values given by Eq. (21.9) with $B = 5.0$ T and g_N values from Table 21.1. The common factor in these calculations is $\mu_N B/h = 38.11$ MHz:

$$\nu_{\text{NMR}}(^1\text{H}) = \frac{(5.585\ 7)\ \mu_N B}{h} = 212.9\ \text{MHz}$$

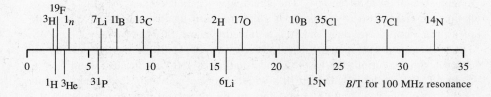

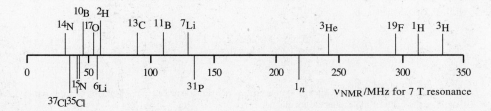

FIGURE 21.4 The magnetic fields that induce NMR transitions at 100 MHz (top panel) and the frequencies at a constant 7 T field (bottom panel) are well separated from one magnetic nucleus to another, due to the variability of the nuclear g factor, g_N.

$$\nu_{\text{NMR}}(^2\text{H}) = \frac{(0.857\ 4)\ \mu_N B}{h} = 32.68 \text{ MHz}$$

$$\nu_{\text{NMR}}(^{10}\text{B}) = \frac{(0.600\ 2)\ \mu_N B}{h} = 22.88 \text{ MHz}$$

$$\nu_{\text{NMR}}(^{11}\text{B}) = \frac{(1.792)\ \mu_N B}{h} = 68.30 \text{ MHz} \ .$$

This range of frequencies, 20–200 MHz, is larger than a single spectrometer detector circuit can cover with high sensitivity. Detection circuits are generally tuned to a fairly narrow range of frequencies.

⟹ *RELATED PROBLEMS* 21.10, 21.11

Once one picks a nucleus for study, ^{1}H for example, the next effect on the spectrum is known as *shielding*. The idea behind shielding is quite simple: the external magnetic field interacts with the entire molecule, of course, and the magnetic response of the molecule to the field generates a small, local field at the nucleus. This field depends on the electronic environment of any one nucleus and thus on the nature of the chemical bond to it. The field is often pointed in a direction that opposes the applied field so that the nucleus finds itself in a total field slightly less than the applied field.[7] If we define a *shielding constant* σ_i for nucleus i, we can write the local field B_i at that nucleus as

$$B_i = B(1 - \sigma_i) \ . \tag{21.10}$$

For H_2, the shielding constant is 2.66×10^{-5}, a typical value for protons. For heavier atoms (more electrons!), σ increases considerably. For the free atoms, it is 4.7×10^{-4} for ^{19}F, 5.6×10^{-3} for spin-$\frac{1}{2}$ ^{129}Xe, and about 10^{-2} for spin-$\frac{1}{2}$ ^{207}Pb. Positive values for σ reflect a net *diamagnetic shielding*, but even diamagnetic substances can produce a net *local paramagnetic shielding* at any one nucleus for which $\sigma < 0$.

For whatever the reason, shielding causes each distinct nucleus (or set of nuclei that are equivalent by symmetry, as we are about to see) to resonate at a unique position in the spectrum, since B_i rather than B enters Eq. (21.9). This is the origin of the *chemical shift:* the fingerprint pattern of distinct resonances for each nucleus in a unique environment in a molecule. Since spectrometers may operate at different field/frequency ranges, it is convenient to define a chemical shift in a way that is independent of these differences. The shielding constant itself does this, but a simpler definition makes use of a *chemical shift standard* against which all others are measured. For protons, the standard is tetramethylsilane, $Si(CH_3)_4$, abbreviated TMS. The predominant isotopes of Si and C have zero spin, all the methyl protons are equivalent in solution due to the rapid rotation of the CH_3 groups about Si—C bonds, TMS is soluble and inert in a wide variety of solvents, it is cheap, and its proton spectrum is a single line that appears (in a constant field spectrometer) at a higher frequency than almost all other proton resonances. Thus we define the chemical shift of proton i, δ_i, as

$$\delta_i = 10^6 \frac{\nu_i - \nu_{\text{TMS}}}{\nu_{\text{TMS}}} = 10^6 \frac{\sigma_{\text{TMS}} - \sigma_i}{1 - \sigma_{\text{TMS}}} \cong 10^6 (\sigma_{\text{TMS}} - \sigma_i) \tag{21.11}$$

---

[7]See the discussion following Eq. (9.47) concerning diamagnetic and paramagnetic interactions with a field. The majority of molecules are diamagnetic.

where we have used the proportionality $\nu_i \propto (1 - \sigma_i)$ and multiplied by the arbitrary but convenient factor 10^6 so that chemical shifts are expressed as *parts per million (ppm) shifts from TMS*. With this definition, proton chemical shifts lie in the range 0–12 or so, and ^{13}C chemical shifts, also based on a TMS standard, span the range 0–220 or so.

EXAMPLE 21.4

The reference compound for ^{17}O NMR is $H_2^{17}O$, and the ^{17}O signal from the P=O oxygen in phosphates has a chemical shift of 75 ppm. In a machine operating at $B = 2.36$ T, what is the frequency difference between the reference and the P=O resonances?

SOLUTION From Eq. (21.9) and Table 21.1, we find the operating frequency is around

$$\nu_{\text{NMR}}(^{17}\text{O}) = \frac{0.7575 \, \mu_N B}{h} = 13.6 \text{ MHz}$$

and from Eq. (21.11), we can write

$$\nu_{\text{P=O}} - \nu_{\text{ref}} = \delta_i \nu_{\text{ref}} \, 10^{-6} = (75)(13.6 \text{ Hz}) = 1020 \text{ Hz}$$

where ν_{ref} represents the resonance frequency of the $H_2^{17}O$ reference signal. Note that we did not need to know the resonance frequency of *either* resonance to high precision to calculate the frequency *difference* between them. We can calculate that difference to a given accuracy as long as we know both the frequency and the chemical shift to the same accuracy.

RELATED PROBLEM 21.11

Note that writing Eq. (21.9) in the form

$$\nu_i = g_N \mu_N B_i / h = g_N \mu_N B (1 - \sigma_i)/h = [g_N(1 - \sigma_i)] \mu_N B / h = g_{\text{eff}} \mu_N B / h$$

shows that we can also think of the chemical shift as giving each chemically distinct nucleus an effective g factor. This leads us to consider how nuclear spins that are distinguishable for whatever reason—different shielding or different g_N—interact with each other.

The obvious interaction between two nuclear magnetic moments is usually not the important one. Each magnetic moment has its own magnetic field, and the fields of two moments interact through what is called *direct dipolar coupling*, just as two macroscopic bar magnets interact. The magnitude of the interaction energy for any one instantaneous orientation of two spins depends on the distance between them (varying like $1/r^3$) as well as their relative orientations, but it can be appreciably larger than the energy associated with the chemical shift. Since a variety of distances and orientations are possible, direct dipolar interaction has the potential to smear a transition over a wide range, blurring and obscuring the interesting chemical shift information. In fact, this happens in an *oriented* sample, such as a crystalline solid and, to some extent in samples dissolved in liquid crystal solvents. Consequently, solid-state NMR requires quantum spectroscopic tricks, discussed in some of this chapter's references, to overcome dipolar broadening. Fortunately, in an *isotropic* sample, such as a liquid or gas, the rapid tumbling of molecules averages the direct dipolar coupling to zero.

While direct interaction is usually negligible, an *indirect spin–spin interaction* is not and leads to an important observable effect on the spectrum. This coupling is mediated by the electrons between two nuclei. One nuclear spin moment interacts with all electron spin moments, electron orbital angular momentum moments, and the electron spin density at the nucleus. Two different nuclear spins couple in a second-order perturbation term for the energy that connects one nucleus to one electron and another nucleus to the same electron, summed over all electrons. This leads to a term in the NMR Hamiltonian that can be written in a simple way:

$$\hat{H}'_{12} = \frac{hJ}{\hbar^2}\hat{\mathbf{I}}_1 \cdot \hat{\mathbf{I}}_2 \quad (21.12)$$

where $\hat{\mathbf{I}}_1$ and $\hat{\mathbf{I}}_2$ are the spin angular momentum operators for any two nuclear spins 1 and 2 and J is the *indirect spin–spin coupling constant*. This constant contains all the coupling terms in a complicated second-order perturbation expression, and it is traditionally written in the Hamiltonian so that it has frequency units, as we have done. Its magnitude can be measured directly from an NMR spectrum, and it depends on not only the kinds of nuclei involved but also their positions in any one molecule.

Note that we have seen this kind of Hamiltonian term two times before: in Eq. (13.29), for spin–orbit coupling in atoms, and in Eq. (15.20), the hyperfine interaction Hamiltonian. All three have the form (constant) × (angular momentum one)·(angular momentum two). We can apply this Hamiltonian to a simple molecule, H_2, and take advantage of the methodology developed in Chapter 13 to evaluate its effect. If we have H_2 in the absence of a field, the NMR Zeeman Hamiltonian is zero, and $\hat{H}'_{12}$ is the only new term. It provides the mechanism that couples the two ^1H spins, and we can write $\mathbf{I} = \mathbf{I}_1 + \mathbf{I}_2$ for the total spin angular momentum. Each spin alone could be either α ("spin up") or β ("spin down"), and the two are indistinguishable. *This is exactly the same problem as the coupling of the two electron spins in He atom* discussed in Chapter 13. We can lift the four total spin wavefunctions directly from Equations (13.5a)–(13.5d):

$$\phi_1 = \alpha(1)\alpha(2)$$

$$\phi_2 = \beta(1)\beta(2)$$

$$\phi_3 = \frac{1}{\sqrt{2}}[\alpha(1)\beta(2) + \beta(1)\alpha(2)] \quad (21.13)$$

$$\phi_4 = \frac{1}{\sqrt{2}}[\alpha(1)\beta(2) - \beta(1)\alpha(2)] \; .$$

We can also find the energies of these four wavefunctions if we adapt the first-order perturbation theory formalism behind Eq. (13.31) for spin–orbit splitting. Using the identity $\hat{\mathbf{I}}_1 \cdot \hat{\mathbf{I}}_2 = (\hat{\mathbf{I}}^2 - \hat{\mathbf{I}}_1^2 - \hat{\mathbf{I}}_2^2)/2$, the analog of Eq. (13.31) is

$$\begin{aligned} E^{(1)}_{12} &= \int \phi_i^* \hat{H}'_{12}\, \phi_i\, d\tau = \frac{hJ}{2\hbar^2} \int \phi_i^*(\hat{\mathbf{I}}^2 - \hat{\mathbf{I}}_1^2 - \hat{\mathbf{I}}_2^2)\phi_i\, d\tau \\ &= \frac{hJ}{2}[I(I+1) - I_1(I_1+1) - I_2(I_2+1)] \int \phi_i^* \phi_i\, d\tau \quad (21.14) \\ &= \frac{hJ}{2}[I(I+1) - I_1(I_1+1) - I_2(I_2+1)] \; . \end{aligned}$$

Here, we have $I_1 = I_2 = 1/2$, and $I = I_1 + I_2$ and $I_1 - I_2$, or $I = 1$ and 0. Substituting these quantum numbers into Eq. (21.14), we find two energies

$$E^{(1)}_{12} = \frac{hJ}{2}\left[1(2) - \frac{1}{2}\left(\frac{3}{2}\right) - \frac{1}{2}\left(\frac{3}{2}\right)\right] = \frac{1}{4}hJ \quad (\text{for } I = 1)$$

$$E^{(1)}_{12} = \frac{hJ}{2}\left[0(1) - \frac{1}{2}\left(\frac{3}{2}\right) - \frac{1}{2}\left(\frac{3}{2}\right)\right] = -\frac{3}{4}hJ \quad (\text{for } I = 0)$$

just as we did for spin–orbit coupling in H atom, and, just as we found an electron-spin triplet and singlet for the electrons of He, we find a nuclear-spin triplet ($I = 1$) and singlet ($I = 0$) for the nuclear spin states of H_2.

EXAMPLE 21.5

The coupling constant in $^1H^{19}F$ is 521 Hz. What total nuclear spin quantum numbers are possible, how many spin states are possible in zero field, and what are their relative energies?

SOLUTION Since the ^{19}F spin is also 1/2, HF and H_2 have quite similar spin–spin interactions. The I quantum number can be either 1 or 0, and each leads to a distinct state, as in H_2. They are again separated by $hJ = h(521 \text{ Hz}) = 3.45 \times 10^{-31} \text{ J} = 1.74 \times 10^{-8} \text{ cm}^{-1}$. Note that the individual I quantum numbers *alone* determine the number of states, and they together with J determine the energies of the states. The g_N value differences among nuclei do not enter the picture at zero field.

➠ RELATED PROBLEMS 21.12, 21.13

The H_2 coupling constant, typical of all proton coupling constants, is quite tiny: $J = 276$ Hz or about 10^{-8} cm^{-1}.[8] The next question is whether or not we can observe any effect of this splitting in any spectrum of H_2. Alas, we cannot, even in the presence of a magnetic field. The value for J quoted above is a calculated value. An additional NMR selection rule here mimics optical rules: $\Delta I = 0$, and the singlet-triplet transition is not allowed.[9] (Problems 21.15 and 21.16 discuss this result further.) In the presence of a field, the NMR Zeeman energies $-g_N\mu_N M_I B$ add to the two spin–spin energies. Figure 21.5 plots the energies of the four states (in frequency units) as a function of B (for very small fields to keep everything on

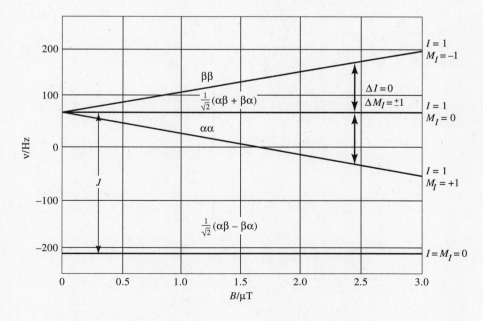

FIGURE 21.5 The Zeeman effect on the nuclei of H_2 splits the $I = 1$ triplet state. The NMR selection rules allow transitions only between adjacent levels of the $I = 1$ manifold, a single line. Contrast this figure to Figure 21.1. The Zeeman effects are similar in overall effect, but different selection rules lead to quite different spectra.

---

[8] We have now covered energy effects in H_2 in this book from a high at its ionization potential, about 10^5 cm^{-1}, down to this spin–spin coupling, a total of 13 orders of magnitude in energy!

[9] It is possible to use high radiation power and induce *multiphoton* NMR transitions in which the spin systems simultaneously absorb more than one photon. These transitions have different selection rules.

scale), and shows the two *degenerate* allowed transitions, both with $I = 1$, but one with $M_I = -1 \leftrightarrow M_I = 0$ and the other with $M_I = 0 \leftrightarrow M_I = +1$.

Consequently, the indirect spin–spin coupling has no effect, and the NMR spectrum of H_2 is a single line. This is disappointing, since J carries valuable information, but it is also a *general* result: *the NMR spectrum of any molecule with only one kind of magnetic nucleus, all of which are in symmetrically equivalent positions, consists of a single transition.* This is true of $^1H_2$, $^{13}C_{60}$, $^1H^{12}C^{12}C^1H$, $^{12}C_6^1H_6$, $^{12}C^1H_4$, $^{19}F_2$, etc., and it explains why TMS, the proton reference molecule, has a one-line proton NMR spectrum. All the methyl protons are equivalent.

We can measure the coupling constant if we have *inequivalent* magnetic nuclei. The simplest way to do this is to change one proton in H_2 into something else, as in HF, HCl, or, the case we will consider next, HD ($^1H^2H$). Theory behind the J constant indicates that, for identical electronic environments, J should scale with g_N. Using g_N values in Table 21.1, we predict $J_{HD} = J_{HH} g_N(D)/g_N(H) = 42.37$ Hz, and the measured value is 42.94 Hz. Table 21.1 also indicates that the D nucleus (the deuteron) has a spin of 1, rather than 1/2, which means the proton + deuteron spin coupling at zero field again leads to two energy levels, but with total nuclear spin quantum numbers $I = 1 \pm 1/2$, or 1/2 and 3/2. From Eq. (21.15), we find the $I = 3/2$ level has the energy $hJ_{HD}/2$ and the $I = 1/2$ level has the energy $-hJ_{HD}$ in zero field.

If we put HD in a non-zero field, the $I = 1/2$ level splits into two states, $M_I = \pm 1/2$, and the $I = 3/2$ level splits into four: $M_I = \pm 3/2$ and $\pm 1/2$. At very low fields, where the Zeeman energy is comparable to the spin–spin interaction energy, four of this total of six states are mixed. If we represent the M_I values for the proton by α and β ($+1/2$ and $-1/2$) as usual, and we represent those for the deuteron by $\uparrow$ for $M_I = +1$, $\rightarrow$ for 0, and $\downarrow$ for -1, the two states that do not mix and thus follow a pure Zeeman linear energy dependence on B are $\phi_1 = \alpha\uparrow$, with both spins as "up as possible," and $\phi_2 = \beta\downarrow$, both as "down as possible." These two states are the analogs of $\alpha(1)\alpha(2)$ and $\beta(1)\beta(2)$ in H_2.

Unlike H_2, however, the other states mix to *varying degrees depending on the magnitude of the field.* At extremes of the field strength, they are:

$$\left. \begin{aligned} \phi_3 &= \sqrt{\tfrac{2}{3}}\alpha\rightarrow + \sqrt{\tfrac{1}{3}}\beta\uparrow \\ \phi_4 &= \sqrt{\tfrac{1}{3}}\alpha\rightarrow - \sqrt{\tfrac{2}{3}}\beta\uparrow \\ \phi_5 &= \sqrt{\tfrac{1}{3}}\alpha\downarrow + \sqrt{\tfrac{2}{3}}\beta\rightarrow \\ \phi_6 &= \sqrt{\tfrac{2}{3}}\alpha\downarrow - \sqrt{\tfrac{1}{3}}\beta\rightarrow \end{aligned} \right\} \text{at zero field}$$

and

$$\left. \begin{aligned} \phi_3 &= \beta\uparrow \\ \phi_4 &= \alpha\rightarrow \\ \phi_5 &= \beta\rightarrow \\ \phi_6 &= \alpha\downarrow \end{aligned} \right\} \text{at large field}$$

so that all six states are simple combinations of M_I values for the proton and deuteron at high field. This makes the energies of all six states vary linearly with B, and since J_{HD} is so small, "high field" is not very high at all.

Figure 21.6(a) plots these energies at very low field (but not so low as to preclude an experiment—see Problem 21.17) and indicates the seven allowed transitions at a field of 1.5 µT, which is about 30 times smaller than the Earth's surface magnetic field. As the figure indicates, each transition is at a unique frequency. Figure 21.6(b) extends the diagram to fields 100 times larger, large enough to make the Zeeman energy much greater than the spin–spin interaction energy. Note that states 2, 5, and 3 have the same energy ordering here as in Figure 21.6(a), but states 1, 4, and 6 are in a *different* order, due to strong mixing at low fields. At large fields, the six states are sorted in energy in a simple order according to M_I values for the proton and deuteron.

Figure 21.6(b) suggests that the three allowed proton transitions are at the same frequency and the four allowed deuteron transitions are also at the same frequency (but one different from the proton's, of course). In fact, the three proton transitions are *each* at slightly different frequencies, and the four deuteron transitions fall at *two* slightly different frequencies. Spin–spin interaction causes these differences,

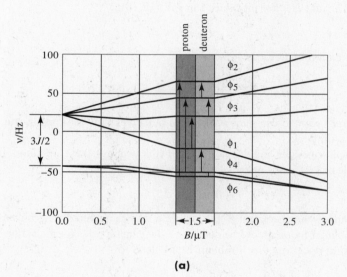

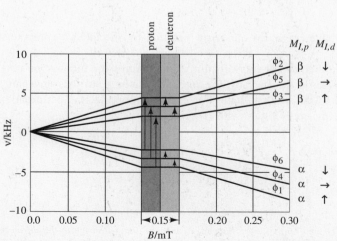

FIGURE 21.6 The Zeeman effect on the HD nuclei is nonlinear at low magnetic fields, shown in (a), but linear at higher fields, shown in (b). The spin–spin coupling between the proton and deuteron leads to a triplet of lines in the proton resonance region, and a doublet of lines in the deuteron resonance region. At high fields, the six states are labeled by the M_I values of each nucleus. At low fields, these labels are not valid due to mixing among four of the six states.

just as a residual spin–orbit interaction causes a slight splitting in the Paschen–Back limit of optical Zeeman spectra.[10]

We can use a first-order perturbation calculation to find the size of these splittings at high fields if we take a zeroth-order picture of independent spins with energies that depend only on the Zeeman effect of each spin:

$$E^{(0)} = \underset{\underset{\text{Proton Zeeman energy}}{\uparrow}}{-g_{N,p}\mu_N B M_{I,p}} - \underset{\underset{\text{Deuteron Zeeman energy}}{\uparrow}}{g_{N,d}\mu_N B M_{I,d}} \qquad (21.15)$$

We use the spin–spin Hamiltonian from Eq. (21.12) as the perturbation and write the spin operator dot product first as

$$\hat{H}'_{12} = \frac{hJ_{HD}}{\hbar^2}\hat{\mathbf{I}}_p \cdot \hat{\mathbf{I}}_d = \frac{hJ_{HD}}{\hbar^2}(\hat{I}_{x,p}\hat{I}_{x,d} + \hat{I}_{y,p}\hat{I}_{y,d} + \hat{I}_{z,p}\hat{I}_{z,d}) \ .$$

Since the operators for the x and y components of angular momentum are difficult to deal with, we convert them into the angular momentum *raising and lowering* operators, $\hat{I}_+$ and $\hat{I}_-$, first introduced in Eq. (12.38):

$$\hat{I}_\pm = \hat{I}_x \pm i\hat{I}_y \ .$$

The perturbation operator becomes

$$\hat{H}'_{12} = \frac{hJ_{HD}}{\hbar^2}\left[\frac{1}{2}(\hat{I}_{+,p}\hat{I}_{-,d} + \hat{I}_{-,p}\hat{I}_{+,d}) + \hat{I}_{z,p}\hat{I}_{z,d}\right] \ , \qquad (21.16)$$

which is very easy to use once we remind ourselves how the raising and lowering operators work.

For a spin one-half angular momentum (the proton), these operators have very simple rules:

$$\hat{I}_{+,p}\alpha = 0 \quad \hat{I}_{+,p}\beta = \hbar\alpha \quad \hat{I}_{-,p}\alpha = \hbar\beta \quad \hat{I}_{-,p}\beta = 0$$

since α cannot be raised, but β can, while α can be lowered, but β cannot. For a spin one angular momentum (the deuteron), the general rules for the action of raising and lowering operators (see Problem 21.14) lead to

$$\begin{array}{lll} \hat{I}_{+,d}\uparrow = 0 & \hat{I}_{+,d}\rightarrow = \sqrt{2}\hbar\uparrow & \hat{I}_{+,d}\downarrow = \sqrt{2}\hbar\rightarrow \\ \hat{I}_{-,d}\uparrow = \sqrt{2}\hbar\rightarrow & \hat{I}_{-,d}\rightarrow = \sqrt{2}\hbar\downarrow & \hat{I}_{-,d}\downarrow = 0 \end{array}$$

where we have continued our arrow notation for the three I_z component quantum numbers.

Finding the first-order perturbation energy expression for any one state takes two steps: operate on the state by $\hat{H}'_{12}$, then see if the result has any part that is not orthogonal to the original state. For example, consider the perturbation to $\phi_1 = \alpha\uparrow$. Operating on this state with $\hat{H}'_{12}$ gives

$$\begin{aligned}\hat{H}'_{12}\phi_1 &= \frac{hJ_{HD}}{\hbar^2}\left[\frac{1}{2}(\hat{I}_{+,p}\hat{I}_{-,d} + \hat{I}_{-,p}\hat{I}_{+,d}) + \hat{I}_{z,p}\hat{I}_{z,d}\right]\alpha\uparrow \\ &= \frac{hJ_{HD}}{\hbar^2}\left(\frac{\hat{I}_{+,p}\hat{I}_{-,d}}{2}\alpha\uparrow + \frac{\hat{I}_{-,p}\hat{I}_{+,d}}{2}\alpha\uparrow + \hat{I}_{z,p}\hat{I}_{z,d}\alpha\uparrow\right)\end{aligned}$$

---

[10] In fact, Figure 21.6(b) is at the "Paschen–Back limit" for nuclear spins—the two inequivalent spins are essentially decoupled from each other—but NMR literature usually does not refer to this limit by the name "Paschen–Back."

$$= \frac{hJ_{HD}}{\hbar^2}\left[\frac{(\hat{I}_{+,p}\alpha)(\hat{I}_{-,d}\uparrow)}{2} + \frac{(\hat{I}_{-,p}\alpha)(\hat{I}_{+,d}\uparrow)}{2} + (\hat{I}_{z,p}\alpha)(\hat{I}_{z,d}\uparrow)\right]$$

$$= \frac{hJ_{HD}}{\hbar^2}\left[\frac{(0)(\sqrt{2}\hbar\rightarrow)}{2} + \frac{(\hbar\beta)(0)}{2} + \left(\frac{1}{2}\hbar\,\alpha\right)(\hbar\uparrow)\right]$$

$$= \frac{hJ_{HD}}{2}\alpha\uparrow .$$

Continuing through the other five states gives (as you should verify for practice)

$$\hat{H}'_{12}\phi_2 = \hat{H}'_{s-s}\beta\downarrow = \frac{hJ_{HD}}{2}\beta\downarrow$$

$$\hat{H}'_{12}\phi_3 = \hat{H}'_{s-s}\beta\uparrow = hJ_{HD}\left(\frac{1}{\sqrt{2}}\alpha\rightarrow - \frac{1}{2}\beta\uparrow\right)$$

$$\hat{H}'_{12}\phi_4 = \hat{H}'_{s-s}\alpha\rightarrow = \frac{hJ_{HD}}{\sqrt{2}}\beta\uparrow$$

$$\hat{H}'_{12}\phi_5 = \hat{H}'_{s-s}\beta\rightarrow = \frac{hJ_{HD}}{\sqrt{2}}\alpha\downarrow$$

$$\hat{H}'_{12}\phi_6 = \hat{H}'_{s-s}\alpha\downarrow = hJ_{HD}\left(\frac{1}{\sqrt{2}}\beta\rightarrow - \frac{1}{2}\alpha\downarrow\right) .$$

The energy perturbations, $E_i^{(1)} = \int \phi_i \hat{H}'_{12}\phi_i d\tau$, are just the pieces of these expressions, if any, that are not orthogonal to the original state:

$$E_1^{(1)} = \frac{hJ_{HD}}{2}$$

$$E_2^{(1)} = \frac{hJ_{HD}}{2}$$

$$E_3^{(1)} = -\frac{hJ_{HD}}{2}$$

$$E_4^{(1)} = 0$$

$$E_5^{(1)} = 0$$

$$E_6^{(1)} = -\frac{hJ_{HD}}{2} .$$

Now we can add these perturbations to the zeroth-order, purely Zeeman energies for each state. This is done in the energy level diagram of Figure 21.7. States 1 and 2 go up by $hJ_{HD}/2$, states 4 and 5 do not change, and states 3 and 6 drop by $hJ_{HD}/2$. Consider first the effect on the high-energy, proton spin-flip transitions. Without spin–spin interaction, these three transitions (6–2, 4–5, and 1–3) have the same energy, $h\nu^{(0)}$. *With* the interaction, the 6–2 transition energy increases by hJ_{HD}, the 4–5 transition stays the same, and 1–3 decreases by hJ_{HD}, leading to *a triplet of transitions, spaced in frequency by J_{HD} from each other.*

Similarly, the four deuteron transitions, which are all degenerate without spin–spin interaction, become a *doublet spaced by J_{HD}*. If we measured the HD spectrum at a fixed frequency of 60 MHz, scanning the magnetic field, we would find the spectra shown in Figures 21.8(a),(b).[11]

---

[11] These spectra were calculated assuming no shielding.

FIGURE 21.7 If the combined effects of the Zeeman and spin–spin energies are taken into account in an energy-level diagram at constant field, we can see how the proton transitions are a triplet spaced in frequency by the coupling constant, J_{HD}. Similarly, the deuteron doublet is found to have the same spacing.

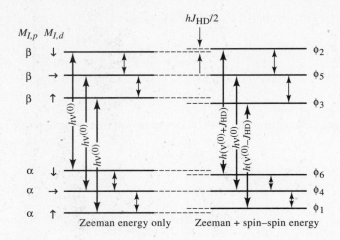

FIGURE 21.8 The HD NMR spectrum in the proton region, (a), and the deuteron region, (b), shows the splitting pattern predicted by the energy level diagram. These spectra assume the field is scanned at a constant 60 MHz irradiation frequency. Note the very short span of magnetic field covered in each spectrum.

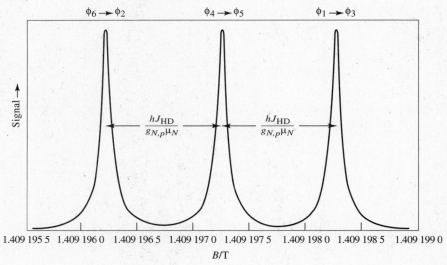

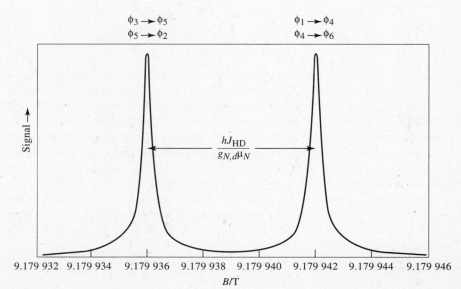

We can now predict the effects of spin–spin interactions on other systems of inequivalent nuclei. Consider first a molecule just with two inequivalent $I = 1/2$ nuclei,[12] such as $^1H^{19}F$ or $^1H^{12}C^{16}O^{16}O^1H$. If there is no spin–spin interaction, the spectrum has only two lines, separated by the combined effects of different g_N values and shielding constants (in a case like HF) or by shielding differences alone (as in formic acid). Including spin–spin interaction (which is always there, although it may be vanishingly small if the spins are far apart in the molecule) causes *each* line to split into a doublet, separated in frequency at constant field by J. This is exactly analogous to the splitting of the D resonance in HD by the two spin states of H.

If the two kinds of inequivalent nuclei differ only by shielding constants and if the spin–spin constant is large, doublets still result, spaced in frequency by J, but the intensity of the components of each doublet may be unequal. The relevant parameter is the ratio of J to the difference between the center of the doublets (the difference controlled by shielding). If this ratio is very small (less than about 0.1), the doublets have equal intensity. If it is very large (because J is large and the shielding difference is small), the closest component of the doublet patterns have increased intensity, as the schematic spectra below indicate:

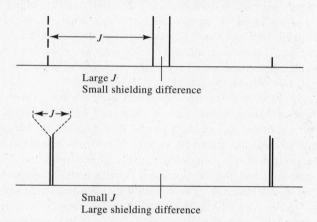

Large J
Small shielding difference

Small J
Large shielding difference

Next, consider two sets of inequivalent $I = 1/2$ nuclei such as the protons in $CH_3C\equiv CH$. The three methyl protons are equivalent, but have different shielding constants from the terminal proton. The coupling constant in such a molecule is quite small, since the protons are well separated, but the π electrons in the $C\equiv C$ triple bond help transmit the coupling, and $J \cong 2-6$ Hz for this and similar molecules. The methyl resonance splits into a doublet, due to the two spin states of the terminal proton, but this proton feels all possible spin states of the three methyl protons. There are two states for each of these three protons, or $2^3 = 8$ states in all, but they are not all distinct. There is only one way to have all methyl spins "up" ($\alpha\alpha\alpha$) and one way to have them "down" ($\beta\beta\beta$), but if only one is down, the three possible states are equivalent ($\alpha\alpha\beta$, $\alpha\beta\alpha$ and $\beta\alpha\alpha$). The same is true if one is up ($\alpha\beta\beta$, $\beta\alpha\beta$, and $\beta\beta\alpha$). The methyl protons have *four* unique interactions with the terminal

---

[12] Such species with only two inequivalent protons and no other magnetic nuclei are remarkably rare. Besides formic acid, the highly strained lactone , and the highly unstable hydroxyacetylene, $HOC\equiv CH$ are two possibilities. Can you think of others?

proton. Its resonance splits into a *quartet* (still spaced in frequency by J), but the intensities of these four transitions are unequal. They follow a 1:3:3:1 ratio, reflecting the statistical weights of the four unique methyl proton states.

Continuing this argument for other sets of coupled $I = 1/2$ nuclei shows that a spin (or a set of equivalent spins) coupled to N different equivalent nuclei is split into $N + 1$ components with an intensity pattern reflecting the statistical weights of the coupling spin states:

N	Intensity Pattern	
(number of equivalent coupled spins)	(of the spin or spin(s) coupled to)	
1	1:1	(doublet)
2	1:2:1	(triplet)
3	1:3:3:1	(quartet)
4	1:4:6:4:1	(quintet)

Coupling is important to observe, but it can be also a nuisance, since it spreads transition intensity over many lines, reducing sensitivity and crowding spectra. This is particularly annoying in ^{13}C spectra. Since ^{13}C has a natural abundance of only 1.1%, it is unlikely that any small to medium-sized molecule not enriched in ^{13}C has more than one ^{13}C nucleus at all, and unlikely that *any* molecule has two closely spaced ^{13}C nuclei. But even one ^{13}C typically interacts with one or more protons through a J on the order of 25–200 Hz.

Coupling can be eliminated by a quantum trick known as *decoupling*. Two radiation sources are used, one narrow in frequency tuned to the ^{13}C region, and one so intense and broad in frequency spread that it causes *all* the protons in the sample to undergo α–β spin flips very rapidly. The ^{13}C nucleus sees a time-average field from the protons that is independent of the proton spin state. This quenches the ^{13}C–^{1}H coupling, and the ^{13}C spectrum is reduced to a single line for each inequivalent C in the molecule. The intensity of each line is proportional to the number of ^{13}C nuclei contributing to each, and the spectrum effectively counts the number and types of C atoms in the molecule. Decoupling can be used in general to eliminate the effects of spin–spin coupling in any spectrum. Here, *broad-band* decoupling is used to squelch the effect of all of one kind of nuclei, but *narrow-band* experiments decouple two inequivalent protons, for example.

EXAMPLE 21.6

The proton-decoupled ^{13}C NMR spectrum of one of the xylenes has three lines, another xylene has four, and the third has five lines. Which spectrum goes with which isomer?

SOLUTION Since the number of lines in a ^{13}C decoupled spectrum counts the number of unique C atoms, symmetry plays the leading role here. The *ortho* isomer has four unique kinds of carbons and the four-line spectrum:

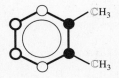

The *meta* isomer has five kinds of carbon and a five-line spectrum, while the *para* isomer has three kinds and thus three lines:

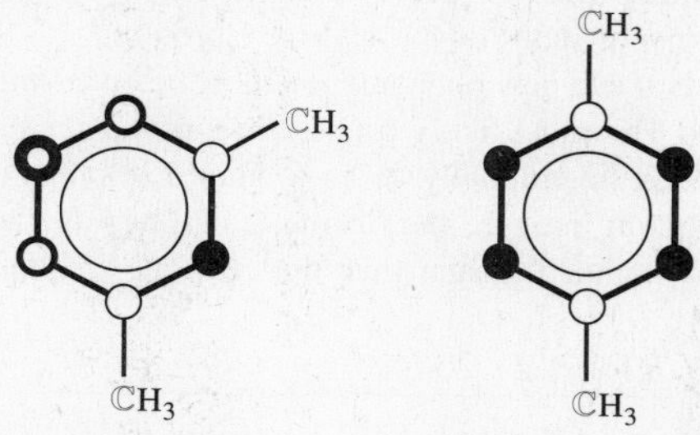

Spins appear equivalent in a spectrum because symmetry requires them to be (as in this example), because they move fast enough to average to one transition, or because their shielding differences or coupling, while different in reality, are so small that the differences are not resolved.

➡ *RELATED PROBLEMS* 21.12, 21.18, 21.20

Sometimes, the molecule can produce its own variant of decoupling due to its own internal dynamics. The tumbling motion of liquid and gas molecules decouples internal resonances of one molecule from intermolecular coupling with its neighbors. Taking this idea one step further, if an *intramolecular* motion is fast enough (and the rotation of a methyl group is one we have already considered), any stereochemical splittings a static configuration might have can be averaged away—decoupled—by the motion. Many interesting intramolecular motions have time scales that are temperature dependent over ordinary temperature ranges. For example, the ring inversion motion in cyclohexane is very rapid at room temperature, and the spectrum is a single line: all the axial and equatorial protons interconvert rapidly and become equivalent. At low temperature, however, the inversion motion slows so that the six axial protons are distinguishable from the six equatorial protons, the two sets couple, and a complex spectrum results. Similarly, the bonding of an ethylene ligand in some organometallic compounds has a fixed orientation at low temperature, but at higher T, the ethylene rotates like a propeller at a rate fast enough to average orientational inequivalences. The NMR spectrum as a function of T picks up this change, and dynamic and thermodynamic information can be measured.

The rate at which an intermolecular motion must occur to produce averaging, known as the *NMR time scale*, has physical origins that we will not cover here in detail. Roughly speaking, however, intermolecular interconversions that occur faster than about 10^6 s^{-1} always lead to averaging, while interconversions slower than about 10^2 s^{-1} do not.

EXAMPLE 21.7

At a temperature around 200 K, the NMR of methanol shows the expected pattern of resonances, but around 300 K, the spectrum has only two lines. What is the expected pattern, and why does it change with temperature?

SOLUTION Even at 200 K, we expect the methyl group to rotate freely about the C—O bond (or, equivalently, the C—O—H bond angle is not locked into any one configuration with respect to the methyl protons) so that the methyl protons form a set of three equivalent spins. Similarly, the hydroxyl proton is in a different environment from the methyl protons and has a different shielding constant. We expect the hydroxyl signal to appear as a 1:3:3:1 quartet, as did the terminal proton on $CH_3C{\equiv}CH$, and the methyl resonance to appear as a doublet, split by the two spin states of the hydroxyl proton.

At higher temperature, some new phenomenon must be occurring fast enough to decouple the spin–spin couplings. The most likely process is *proton exchange* among methanols (or with water, whether added intentionally or as a natural impurity). If the time a proton is bound to the oxygen is short enough, the distinct couplings average to a single resonance. The exchange frequency is much greater than the coupling frequency.

⟹ RELATED PROBLEMS 21.19, 21.20

There are many, many more quantum tricks that can be exploited in NMR spectroscopy to simplify spectra and track bonding connectivity and molecular conformation in molecules as complex as small proteins with several hundred atoms. These tricks are beyond the scope of our discussion here, and we close with a general outline of the experimental method that revolutionized NMR spectroscopy: Fourier transform NMR, or FT-NMR. This is the method used by all high-frequency spectrometers today, and without it, data acquisition times (and many quantum tricks) would be intractable.

The idea behind the Fourier transform method is fairly simple to grasp. Consider a grand piano. It has 88 strings, each of which resonates at a unique central frequency, although harmonic coupling (such as between the lowest C note and the one an octave above it) provides the "spectral richness" to the sound of a piano. Suppose you wanted to record the acoustic spectrum of a piano. One way, the analog of the way we have discussed NMR so far, would be to point at the strings a loudspeaker connected to an audio sweep oscillator. A microphone could detect the sympathetic resonances (the absorbed acoustic energy) of the piano strings as a function of the oscillator's frequency, but this takes time to do, especially if the string resonances are weak. One might need to sweep the oscillator slowly and repetitively through its range, averaging the microphone signal over all sweeps.

A faster way to record a piano spectrum (and, on the face of it, a much more fun way) would be to keep the microphone in place, but instead of the loudspeaker, whack *all* the strings at once with a big stick. To the ear, and to the microphone, the sound would be noise, although a trained ear could distinguish a piano from a harpsichord. Nevertheless, we record the microphone signal *as a function of time after the whack*. This signal, call if $f(t)$, contains all the sounds all the strings can make, mixed in a complicated way. If one takes the *Fourier transform* of this signal (a technique given a Closer Look below), the *entire* spectrum of the piano can be recovered. The Fourier transform of a signal recorded in time is a function of *reciprocal time*, or *frequency*, which is just what we want. Modern FT-NMR spectrometers give all the spins (at least all of them in one region, such as all protons) a sharp whack from a *pulse* of radiation that, like the stick, induces all the spin resonances in one broad region. The spectrometer then listens to the scream of these whacked spins as they relax in time—the details of the relaxation and its role in generating a signal are beyond our scope—and the scream of whack after whack is recorded, averaged with the other screams, and stored in a computer. The computer then Fourier transforms this averaged signal into the spectrum. Transformation is fast (due to a computer algorithm known as the *Fast Fourier Transform* or FFT), and time is saved since we have replaced a slow sweep by a fast whack-and-listen.

A Closer Look at Fourier Transforms

The Fourier transform, introduced by Jean Fourier in the early 1800s in the context of his study of heat transfer, finds uses in many areas of science. A Fourier transform method can be used to advantage in NMR, visible, IR, microwave, and mass spectroscopies. The details of each spectroscopic method vary, of course, and we will look here only at the transform itself and the reasons for trying to find ways to exploit it.

Consider some function of time, $f(t)$. Its Fourier transform, F, is a function of frequency, and it is defined as

$$F(\nu) = \int_{-\infty}^{\infty} f(t)\, e^{2\pi i \nu t}\, dt = \int_{-\infty}^{\infty} f(t)\, [\cos(2\pi \nu t) + i \sin(2\pi \nu t)]\, dt$$

where $i = \sqrt{-1}$ so that $F(\nu)$ is, in general, a complex number function.

Imagine that some time in the remote past, at time $t = -\infty$ in fact, a single-frequency radiation source at frequency ν_0 was started, and imagine that it stays on forever, until $t = +\infty$. In other words,

$$f(t) = \sin(2\pi\nu_0 t + \delta), \quad -\infty \leq t \leq +\infty$$

where δ is some arbitrary phase factor that sets the $t = 0$ point in time. The Fourier transform of this function[13] is a single spike, infinitesimally narrow and arbitrarily tall, but of unit area, at the frequency ν_0. This tells us to interpret $F(\nu)$ as the *distribution of frequencies* or the *spectrum* of the source evolving in time as $f(t)$.

Now imagine that this source is off for all times *except* for the time period covering one cycle, centered around $t = 0$:

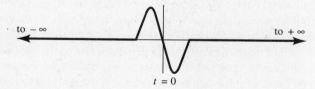

or, mathematically,

$$f(t) = \begin{cases} 0 & t < -\dfrac{1}{2\nu_0} \\ -\sin(2\pi\nu_0 t) & -\dfrac{1}{2\nu_0} \leq t \leq \dfrac{1}{2\nu_0} \\ 0 & t > \dfrac{1}{2\nu_0}. \end{cases}$$

It might seem that we could still specify the frequency of this radiation exactly—we have one full cycle to measure over time—but when we compute $F(\nu)$, we find otherwise. The transform is easy to do:

$$\begin{aligned} F(\nu) &= -\int_{-1/2\nu_0}^{1/2\nu_0} \sin(2\pi\nu_0 t)[\cos(2\pi\nu t) + i \sin(2\pi\nu t)]\, dt \\ &= \frac{i\nu_0 \sin(\pi\nu/\nu_0)}{\pi(\nu^2 - \nu_0^2)}, \end{aligned}$$

---

[13]Proving this by direct substitution into the definition of the transform is not at all simple, but we will see that it is true in the following paragraphs.

which is a purely imaginary function, since the integral over the $\cos(2\pi\nu t)$ term is zero, but that is not a worry: the observable energy in the radiation field is proportional to the square of the magnitude of F, which is F^*F:

$$F^*F = \left(\frac{-i\nu_0 \sin(\pi\nu/\nu_0)}{\pi(\nu^2 - \nu_0^2)}\right)\left(\frac{i\nu_0 \sin(\pi\nu/\nu_0)}{\pi(\nu^2 - \nu_0^2)}\right) = \frac{\nu_0^2 \sin^2(\pi\nu/\nu_0)}{\pi^2(\nu^2 - \nu_0^2)^2}.$$

This is *not* sharply peaked at ν_0. A very short pulse of light, even if during the pulse the radiation is a pure sine wave, represents a *spread* of frequencies. This is similar to (and is based on mathematics quite like) our discussion of wavepackets and the Uncertainty Principle in Chapter 12. Here, the time over which the radiation is present is small so that its frequency spread is large.

Suppose we added some integral number of cycles, n, before and after our one cycle of radiation so that the pulse had a total of $2n + 1$ cycles. The Fourier transform of this signal is

$$F(\nu) = \frac{i\nu_0 \sin\left(\frac{(2n + 1)\pi\nu}{\nu_0}\right)}{(\nu^2 - \nu_0^2)\pi},$$

and if we plot F^*F normalized by $(2n + 1)^2$ for several n values, as is done in Figure 21.9, we can see how more cycles—a longer pulse time—more closely defines the frequency.

Consequently, excitation with a radiation pulse, even if the radiation source is very narrow in frequency, leads to an appreciable spread in excitation frequency. For example, a 10 μs pulse of 100 MHz radiation contains $(10^{-5}\text{ s}) \times (10^8 \text{ cycles s}^{-1}) = 1000$ cycles of radiation, and F for this pulse shows a spread of energy about 10^5 Hz wide, centered on 100 MHz. This is sufficiently broad to resonate the full range of chemical shifts for one nuclear spin in almost any compound. Following the pulse, the excited spins slowly attain equilibrium through relaxation processes, and the component of the net magnetization that is perpendicular to the applied static field, which is the component detected, oscillates at the resonance frequency (or at a superposition of them, if more than one resonance is excited) while decaying exponentially as equilibrium is attained. This signal is called a *free induction decay*, and its Fourier transform is the spectrum.

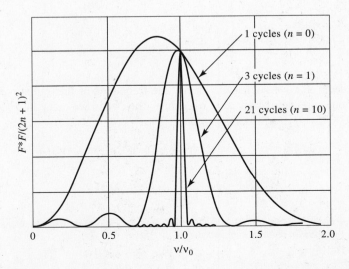

FIGURE 21.9 A single cycle of a sine wave at frequency ν_0 has a Fourier transform that is spread over a considerable range of frequencies. As more and more cycles are added, the frequency spectrum of the radiation becomes more closely defined at ν_0.

If more than one spin type is resonating, $f(t)$ is a sum of oscillatory functions for each unique resonance frequency. The transform of a sum of functions is a sum of transforms, one for each function, so that each resonance peak in the spectrum can be picked out of $f(t)$, which by itself looks very complicated. The exponential decay gives $f(t)$ a finite duration, and as we saw above, the *duration* of $f(t)$ governs the *width* of peaks in $F(\nu)$. Consequently, the peak width contains dynamic information on spin relaxation processes. The spectra in Figure 21.8 were drawn with typical line *shapes* for this relaxation, but their widths were exaggerated for the illustration.

The FT technique acquires a signal much more rapidly than any scan of either the field strength or the radiation frequency since a pulse excites *all* resonances at once (the "whack" mentioned above) and detects the signal from all of them very quickly. This can be repeated pulse after pulse, accumulating and averaging the signal to reduce noise, and the Fourier transform of the accumulated signal is performed once at the end of the experiment. This is a particularly important advantage for rare nuclei, such as ^{13}C in natural abundance, but it is a general advantage enjoyed by all Fourier transform spectroscopies.

Finally, the FT method in NMR allows one to use pulse sequences to manipulate spins in new ways—the quantum tricks mentioned in the text. These are beyond our scope here, but this chapter's references cover many of them.

Electron Spin Resonance

There are many parallels between NMR and ESR spectroscopies, but three important differences are worth attention at the start. The first is the scarcity of species with unpaired electron spins (compared to the abundance of species with magnetic nuclei). This makes ESR a somewhat less general technique than NMR. The second difference is related to the difference between the Bohr magneton, μ_B, and the nuclear magneton, μ_N. Since $\mu_B = (m_p/m_e)\mu_N$, unpaired spin states change energy with applied magnetic field much more rapidly than do nuclear spin states. Consequently, ESR transition frequencies for typical laboratory magnetic fields are in the microwave region (on the order of GHz) rather than the radiofrequency (MHz) region. Finally, there is only one kind of electron (in contrast to many different nuclear g_N factors and spin quantum numbers) so that one machine setup covers all ESR signals. Otherwise, the story is very much the same. One can do Fourier transform ESR, multiquantum ESR, ESR imaging, and a host of other quantum tricks that have analogs in NMR.

The ESR resonance condition, the analog of Eq. (21.9) for NMR, is

$$\nu_{ESR} = g\mu_B B_{loc}/h \qquad (21.17)$$

where g is a molecular variant of g_e, the free-electron g factor, and B_{loc} is the local field at the electron. Both of these quantities warrant further discussion, but first note the magnitude of ν_{ESR}. The molecular g value is typically close to 2, and if B is 1 T, ν_{ESR} is almost exactly 28 GHz. Microwave "optics," the sources, waveguides, and cavities used to generate, direct, and detect microwave radiation, must have physical sizes that are fairly tightly tied to the wavelength of the radiation, and traditional "bands" of microwave radiation are defined for various wavelength ranges. One of these, called the *X-band,* covers 8–12.4 GHz (or λ = 3.7–2.4 cm), and since optics in this region are readily available, many spectrometers operate around 10 GHz, and the magnetic field is around 1/3 T. Sensitivity is enhanced if the optics are tightly tuned to one frequency, and non-FT ESR spectrometers use a fixed irradiation frequency and a scanned magnetic field.

The molecular g value is the analog of the effective NMR g value we defined in terms of a shielding constant. We define an ESR shielding constant similarly and write $g = g_e(1 - \sigma_{ESR})$. Organic free radicals have g generally close to 2.002 5, while inorganic g values fall in the range 1.0–2.1 with transition metal complexes spread over a larger range. Values as high as 9 or so have been measured.

EXAMPLE 21.8

The g value for the ClO$_2$ radical trapped in a KClO$_4$ crystal is 2.008 8 for a field along the principle symmetry axis of the molecule. What field is needed to observe its ESR spectrum with a 9.367 GHz microwave source?

SOLUTION We solve Eq. (21.17) for B and find

$$B = \frac{h\nu_{ESR}}{g\mu_B} = \frac{(6.626 \times 10^{-34} \text{ J Hz}^{-1})(9.367 \times 10^9 \text{ Hz})}{(2.008\ 8)(9.274 \times 10^{-24} \text{ J T}^{-1})} = 0.3332 \text{ T}\ .$$

Note that the g value is specified for a particular orientation of the molecule with relation to the field. In solid samples, molecules of low symmetry show considerable shielding *anisotropy*. The electrons respond to the external field in different ways depending on the orientation. This means spectra of oriented samples can be quite sensitive to the relative field direction.

➡ **RELATED PROBLEMS** 21.21, 21.22

The B field is the "local" field, as in NMR, but note that shielding is associated with g rather than with the field, as we do in NMR. This is partly historical tradition, but it also expresses our inability to "localize" an electron to the degree we "localize" a nucleus in space. But when an electron can spend some time in the vicinity of a nucleus (typically because of a large s character in the wavefunction), the electron and nucleus can interact if the nucleus has a non-zero spin. The interaction is expressed in terms of a *hyperfine coupling constant*. Chapter 15 discussed hyperfine coupling in H atom in some depth—see Eq. (15.20). Here, we incorporate the coupling in a more phenomenological way as is often done in ESR spectroscopy. We write the local field at the molecule in terms of a *hyperfine coupling constant*, a (units of tesla), so that

$$B_{loc} = B + aM_I \tag{21.18}$$

where M_I is any of the possible component quantum numbers for nuclear spin I. This leads to a splitting of the ESR signal in analogy to indirect spin–spin coupling in NMR. For example, the benzene anion, C$_6$H$_6^-$, has a delocalized, unpaired π electron that roams around the ring and feels the six proton spins.[14] The hyperfine constant is 0.375 mT, and for six protons, M_I ranges from $+3$ (all spins up) down in unit steps to -3 (all spins down). This leads to a *seven* line spectrum with a 1:6:15:20:15:6:1 intensity pattern.

An ESR signal in a swept-field machine is detected in a way that has the effect of recording the *first derivative* of the signal with respect to the field. Instead of simple peaks, transitions appear as sharp up and down excursions that cross the zero axis at the line center. Figure 21.10 shows the C$_6$H$_6^-$ spectrum as it would appear in an instrument designed to have the center resonance fall exactly at 0.3 T. The spacing between each peak is equal to a, and the intensity pattern is quite obvious.

---

[14] An electron in a pure π MO has no density at any nucleus. In an aromatic ion, the unpaired π electron spin senses a neighboring proton spin indirectly through the other electrons.

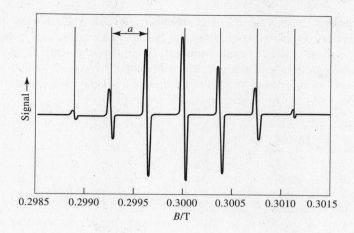

FIGURE 21.10 The benzene anion ESR spectrum, shown in the first-derivative representation typical of ESR spectrometers, is a septet of lines due to the coupling of the unpaired electron spin with the ensemble of six equivalent proton spins. The constant spacing from line to line equals the hyperfine coupling constant, a.

Rather simple radicals or radical ions can have amazingly complex ESR spectra. For example, there are three unique protons on anthracene anion:

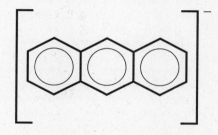

There are two of one kind (para in the central ring), and four of each of the other two kinds. Each kind has its own hyperfine coupling constant. The first pair splits the signal into a 1:2:1 triplet; the next four split *each* line in this triplet into a 1:4:6:4:1 quintet (we are now up to $3 \times 5 = 15$ lines); and the final four further split each of these lines into 1:4:6:4:1 quintets, for a grand total of $3 \times 5 \times 5 = 75$ lines. Sorting out all these lines takes a blend of patience, insight, and high resolution.

EXAMPLE 21.9

The methyl radical ESR hyperfine constant is 2.2 mT. How many lines are seen in the spectrum? What if the radical is CD_3?

SOLUTION First, each line is spaced by 2.2 mT from the next, since there is only one coupling constant (one kind of non-zero nuclear spin). The protons form a set of three equivalent spins, and thus the spectrum is a 1:3:3:1 quartet. If the radical is totally deuterated, the spectrum changes in two ways: the coupling constant changes, and the number and intensity pattern of the lines changes. For a deuteron, $I = 1$. Each splits the signal into an equal intensity triplet, and we might expect $3^3 = 27$ lines, but symmetry causes many of them to coincide. There is only one way to have $M_I = +3$ and one for $M_I = -3$. Using our spin arrows for each deuteron's M_I as we did for NMR, these are ↑↑↑ and ↓↓↓. We have three choices (↑↑→, ↑→↑, and →↑↑) for $M_I = +2$ and three (↓↓→, ↓→↓, →↓↓) for -2. For $M_I = \pm 1$, there are six (such as ↑→→ and ↓↑↑), and there are seven ways of reaching $M_I = 0$. Thus the spectrum is a 1:3:6:7:6:3:1 septet.

▸ **RELATED PROBLEMS** 21.23, 21.24, 21.25, 21.26

In the gas phase, ESR of small free radicals (OH, SH, ClO, etc.) and triplet excited states (O_2 is the classic example) have been measured and interpreted. The spectra of these species are quite complex, since there are potentially several sources of angular momenta: electron spin, electron orbital, molecular rotation, and nuclear spin. Every term gets its own kind of g factor, and coupling among them runs amok. Hundreds of lines can be seen in O_2, for example.

CHAPTER 21 SUMMARY

Spectroscopy in external fields is inherently a high-precision measurement. Laboratory fields simply do not change molecular energies by large amounts, but the way these changes vary with quantum numbers expose new molecular parameters that are reflected in the spectra. In magnetic fields, the energy change has a generic look to it no matter what the scenario: energy = (field) × (g factor) × (magneton) × (projection quantum number). We discussed four important types of g factors: g_e for the free electron, g_N for any one nucleus, g_J for electrons in atoms, and the effective molecular g for ESR. Those relevant to electrons require the Bohr magneton, μ_B, and nuclei require the much smaller nuclear magneton, μ_N. The projection quantum number is $\pm 1/2$ for a free electron (spin α or β), M_I for a nucleus with spin quantum number I, and M_J for an atomic electron with total angular momentum quantum number J.

Another major theme in this chapter is the *coupling* of angular momenta to produce new states and new transitions. In NMR and ESR, spin–spin couplings (nucleus to nucleus in NMR and electron to nucleus in ESR) lead to line splittings and characteristic intensity patterns that depend on the number and I value of the interacting spins. Symmetry and internal motion play a large role: methyl protons are equivalent by rapid methyl rotation, and symmetrically placed spins are always equivalent.

Stark molecular spectroscopy focuses on the molecular dipole moment. An external electric field couples to the dipole moment and splits M_J rotational-state degeneracy. Radiofrequency transitions between M_J states with the same J become possible. Several techniques such as Stark modulation of rotational transitions, masers, and MBER spectroscopy exploit electric fields to advantage. Zeeman spectroscopy is also put to use in atomic clocks, laser magnetic resonance, and magnetic molecular beam resonance techniques.

This chapter ends our discussion of molecular spectroscopies. In future chapters, we look more at the dynamic aspects of molecular motion rather than the properties of a more or less static, isolated molecule. Nevertheless, the ideas behind molecular spectroscopies and the information they reveal play continuing important roles in what is to come.

FURTHER READING

Many of the references in Chapter 19 have sections on molecular spectroscopy in external fields. Also highly recommended is *Microwave Spectroscopy of Free Radicals,* by Alan Carrington (Academic Press, New York, 1974). This book covers techniques and results on small gas-phase molecules with a minimum of mathematical tedium and a maximum of physical insight.

A good place to turn for more depth on modern NMR methods is *Nuclear Magnetic Resonance Spectroscopy,* by Robin K. Harris (Longman, London, 1986). The book *Physical Methods for Chemists* 2nd ed., by Russell S. Drago (Saunders, New York, 1992), covers both NMR and ESR in some depth with particular emphasis on inorganic systems. Both of these books have excellent tables of nuclear spin properties as well as extensive references to other texts and to the original literature.

PRACTICE WITH EQUATIONS

21A An $S = 0$ state of an atom has a magnetic moment $m_L = -1.144$ cm^{-1} T^{-1}. What is this state's L quantum number? (21.1)

ANSWER: 2

21B The Be $2s^2\ ^1S_0 \rightarrow 2s2p\ ^1P_1^\circ$ transition occurs at 42 565.3 cm^{-1} in the absence of a magnetic field. What field splits this line into three lines spaced 1.0 cm^{-1} apart? (21.3)

ANSWER: 2.14 T

21C What is the Landé g factor for singlet states? (21.5)

ANSWER: 1

21D What is the Landé g factor for S states? (21.5)

ANSWER: g_e

21E What is the Landé g factor for the 3p $^4D_{1/2}^\circ$ excited state of N? (21.5)

ANSWER: 0 (approximating $g_e = 2$)

21F How much does the energy of the lowest state of HF change in a 10^5 V m^{-1} electric field? (21.8), Tables 15.1, 19.2

ANSWER: -7.424×10^{-6} cm^{-1}

21G What rf frequency would induce the $J = 1$, $M_J = 0 \to J = 1$, $M_J = 1$ transition in HF in a 10^5 V m^{-1} electric field? (21.8), Tables 15.1, 19.2

ANSWER: 200 kHz

21H In a fixed magnetic field NMR that records proton spectra at 500 MHz, what frequency resonates ^{19}F nuclei? (21.9), Table 21.1

ANSWER: 470.5 MHz

21I The ^{1}H–^{19}F coupling constant in HC≡CF is $J_{HF} = 27$ Hz. What is the zero-field energy difference between the $I = 1$ and $I = 0$ states of HC≡CF? (21.14)

ANSWER: 1.8×10^{-32} J

21J What is the result of operating on the two proton spin state $\alpha\beta$ with $\hat{H}'_{12}$ and a coupling constant J_{12}? (21.12), (21.16)

ANSWER: $hJ_{12}\left(\frac{1}{2}\beta\alpha - \frac{1}{4}\alpha\beta\right)$

21K How many lines appear in the high resolution proton NMR spectrum of CH$_3$CH$_2$COH, assuming that only protons on *adjacent* carbons have indirect spin–spin coupling?

ANSWER: 14 (CH$_3$ triplet; CH$_2$ doublet of quartets; COH triplet)

21L The stable free radical 1,1-diphenyl-2-picrylhydrazyl (DPPH) is sometimes used to calibrate ESR spectrometers. Its g value is 2.003 6. What is the field at the center of its spectrum in an instrument operated at 9.268 GHz? (21.17)

ANSWER: 0.3305 T

PROBLEMS

SECTION 21.1

21.1 The NO$_2$ absorption spectrum at room temperature is so rich and complicated that an entire book exists listing and discussing it [*Spectral Atlas of Nitrogen Dioxide*, by D. K. Hsu, D. L. Monts, and R. N. Zare, (Academic Press, NY, 1978)]. You would like to study the photochemical effects various NO$_2$ excitations might have, but your research granting agency cut your request for a tunable laser from your grant funds. You do, however, have an assortment of atomic emission lamps left over from another project and a large electromagnet abandoned by a colleague down the hall. You note a fairly prominent absorption feature at 18 060.563 cm^{-1} in the NO$_2$ atlas, and you remember that the strongest emission line of Ba is near this. A quick trip to the *Atomic Energy Levels* tables referenced in Chapter 13 shows you that the 6s^2 ^{1}S$_0$ ← 6s6p ^{1}P$_1^\circ$ emission, the strongest in the Ba spectrum, is quite close: 18 060.264 cm^{-1}. You decide to place the Ba emission lamp in the electromagnet field, Zeeman-split this line, and tune one component into resonance with the NO$_2$ absorption. What field should you use? Which component of the Ba Zeeman spectrum will tune into resonance with NO$_2$?

21.2 Example 21.1 considered the Zeeman effect on the 3p excited states of Na. Compute the energies of the 3s ^{2}S$_{1/2}$ ground-state Zeeman components in the same 0.1 T field, and use those energies and the results of Example 21.1 to predict the emission spectrum from these states. Unlike the normal Zeeman effect spectrum of three lines, you should find that one of these excited-state emission lines becomes four lines and the other becomes six lines. (At high resolution, further splitting of these lines is observed due to the $I = 3/2$ nuclear spin of Na.)

21.3 It is possible to trap and observe atomic ions (even one at a time!) in a clever combination of magnetic and electric fields. Imagine you are studying the Ti$^+$ 3d^{2}4s ^{4}F$_{3/2}$ → 3d^{2}4p ^{4}D$^\circ_{1/2}$ transition (from the ground state) in such a trap. What is the Zeeman effect on this transition? Would you expect the trap's magnetic field (which is not very strong) to shift or split this transition a great deal?

21.4 Consider a normal Zeeman pattern from a ^{1}P$_1$ → ^{1}S$_0$ transition. Let λ_0 represent the emission wavelength in zero field. Show that the wavelength difference between the lowest and highest wavelength lines of the Zeeman spectrum, $\Delta\lambda$, is given by the approximate expression $\Delta\lambda \cong \text{(constant)}\lambda_0^2 B$, and find the value of the constant. You will need the approximation $(1 + x)^{-1} \cong 1 - x$ for $x \ll 1$. This expression is used in astrophysics for a quick measure of stellar magnetic fields when star spectra show resolved Zeeman patterns. What is $\Delta\lambda$ for the Ca ^{1}P$_1$ → ^{1}S$_0$ transition at $\lambda_0 = 422.79$ nm (a common feature in stellar emissions) if the field is 0.60 T?

21.5 The solar magnetic field in sunspot magnetic storms is not strong enough to reach the Paschen–Back limit for any element except Li. The Li 2p ^{2}P$_{1/2}$ to 2p ^{2}P$_{3/2}$ spin–orbit splitting is 0.34 cm^{-1}, the smallest among the alkali metals. Use this number and a simple, order-of-magnitude argument based on the meaning of the Paschen–Back limit to estimate the solar magnetic field.

21.6 Continuing from the previous problem, Paschen and Back used the Li 2p ^{2}P → 2s ^{2}S emissions to discover the effect named for them. They used a 4.3 T field, which the previous problem should tell you is well into the Paschen–Back limit. Calculate their spectrum. Assume $g_e = 2$, neglect spin–orbit splitting between ^{2}P$_{1/2}$ and ^{2}P$_{3/2}$, (i.e., neglect J and imagine a single ^{2}P upper state), and take the field-free transition as 14 904 cm^{-1}. In this limit, the splittings due to **L** and **S** are independent and add to produce the total splitting of each level. You should find a triplet of lines (remember the selection rules!) spread over 4 cm^{-1}, centered on the field-free transition.

21.7 Figure 21.3 considers the Stark effect for only the $J = 0$ and 1 levels of OCS. Calculate the Stark splittings for the $J = 2$ states of OCS in a 60 kV m^{-1} field. Note that, since the effect at low field is a second-order perturbation effect, the ground state, (0,0), drops in energy with increasing field, and the (1,0) state is "repelled" by it (see Figure 13.3) and rises. The (1,±1) states drop in Figure 21.3. Do you find that (2,±1) is repelled by (1,±1) and rises? Does (2,±2), the first state with $M_J = \pm 2$, drop?

21.8 Excited electronic state Stark effects are, of course, controlled by the excited state's dipole moment, which may be significantly different from the ground-state moment. For example, the CO X $^1\Sigma^+$ ground-state moment is rather small, 0.374×10^{-30} C m, but the electronic excited state C $^1\Sigma^+$, about 11.4 eV above the ground state, has the significantly larger moment 15×10^{-30} C m. Both states have nearly the same bond lengths, however, and thus nearly equal rotational constants: $B_e(X) = 1.931$ cm^{-1} and $B_e(C) = 1.953$ cm^{-1}. Calculate the Stark shift of the X($v = 0$, $J = 0$) → C($v = 0, J = 0$) transition in a 10^6 V m^{-1} field. In the absence of the field, this transition occurs around 91 919.15 cm^{-1}, or a wavelength of 108.79 nm, in the vacuum ultraviolet, a region where high-resolution lasers are scarce. Do you think ordinary spectroscopy with a grating spectrometer would have the resolution to observe this shift?

21.9 Where would you find the CO(J, M_J) = (1,0) → (1,±1) transition in the experiment described in Example 21.2?

SECTION 21.2

21.10 Magnetic resonance not only needs a magnetic field to exist, it can also be used to measure one in the device known as the *proton magnetometer,* which uses the NMR signal of protons in water or a simple hydrocarbon to measure the field magnitude. Archeologists use proton magnetometers to detect small variations in the Earth's field due to buried artifacts from ancient towns, and the geomagnetic field is constantly monitored at over 100 observatories around the world. The Earth's surface field varies over the range 20–70 μT, and observatories try to attain a resolution ~1 nT. Derive the following calibration expression for a proton magnetometer relating the field and resonance frequency: B/nT = 23.487 (ν/Hz).

21.11 A 250 MHz ^{1}H spectrometer operates in a field of 5.875 0 T. What frequency resonates ^{13}C in this field? The proton-decoupled ^{13}C spectrum of 2-propanol shows two peaks with chemical shifts of 63.4 ppm and 24.7 ppm from TMS. What is the frequency difference between these peaks in this spectrometer?

21.12 Tetramethylsilane (TMS) is the reference compound for both ^{1}H and ^{13}C NMR. What is the ^{13}C spectrum of TMS first with, and then without, proton decoupling? Assume natural abundance ^{13}C, and use the TMS C–H coupling constant $J_{CH} = 118$ Hz. Would TMS be so useful if ^{13}C was 50% abundant instead of 1.1% abundant?

21.13 Consider a general molecule with two non-zero nuclear spins, I_1 and I_2 with $I_1 \geq I_2$. Use Eq. (21.14) to find expressions for the energies of the zero-field coupled spin states. (*Hint:* Write $I = I_1 + I_2 - n$, $n = 0, 1, 2, \ldots$ in general, and remember that n stops when $I = I_1 - I_2$.) Find the number of states and their energies for the ^{17}O^{1}H$^-$ and ^{17}O^{2}H$^-$ hydroxyl ions in terms of the coupling constant J for each.

21.14 The general angular momentum raising and lowering operators, $\hat{L}_+$ and $\hat{L}_-$, were introduced in Eq. (12.38) and used again in this chapter. Here, we prove some further properties of these operators. Their general action is expressed by the operator equation $\hat{L}_\pm \Psi_{L,m} = c_\pm \Psi_{L,m\pm1}$ where $\Psi_{L,m}$ and $\Psi_{L,m\pm1}$ are normalized eigenfunctions with angular momentum quantum numbers L and m, as in Chapter 12, and $c_\pm$ is a proportionality constant that we seek. First, use commutator expressions in Chapter 12 to prove the operator identity $\hat{L}_- \hat{L}_+ = \hat{L}^2 - \hat{L}_z(\hat{L}_z + 1)$. One can show that

$$|c_+|^2 = \int \Psi_{L,m}^* \hat{L}_- \hat{L}_+ \Psi_{L,m} \, d\tau \ .$$

Use the operator identity to show that if we choose c_+ to be real, we have

$$c_+ = \hbar[L(L+1) - m(m+1)]^{1/2} \ .$$

One can also show that $c_- = \hbar[L(L+1) - m(m-1)]^{1/2}$. Apply these expressions to the case $L = I = 1$ (a deuteron) and $m = M_I = -1, 0$, and $+1$ to prove the six operator equations for $\hat{I}_\pm \Psi_{1,M_I}$ used in the text.

21.15 NMR transitions are induced by components of the radiation's magnetic field that are perpendicular to the applied static field. If the applied field is in the z direction, we can take the radiation field to be in the x direction. The transition operator is proportional to the x component of the total nuclear spin, $\hat{I}_{x,T}$, and for H$_2$, the case we consider here, $\hat{I}_{x,T} = \hat{I}_{x,1} + \hat{I}_{x,2}$, the sum of the individual proton spin x components. Thus, *selection rules* are proportional to the effects of the $\hat{I}_{x,T}$ operator. Show first that the raising and lowering operators can be introduced to give the equality $2\hat{I}_{x,T} = \hat{I}_{+,1} + \hat{I}_{+,2} + \hat{I}_{-,1} + \hat{I}_{-,2}$. Find the result of operating with $\hat{I}_{+,1} + \hat{I}_{+,2} + \hat{I}_{-,1} + \hat{I}_{-,2}$ on the four primitive spin functions $\alpha\alpha$, $\alpha\beta$, $\beta\alpha$, and $\beta\beta$. The following problem continues this line of thought to derive the selection rules.

21.16 Armed with the results of the previous problem, it is easy to derive the selection rules for the H$_2$ NMR spectrum. We have four spin states, ϕ_1–ϕ_4 in Eq. (21.13), and we need the *transition moment matrix*, the array that shows us which transitions are forbidden (zero matrix element) and which are allowed (non-zero). Here, this is a 4×4 array. If we define the transition operator as $\hat{H}' = \hat{I}_{+,1} + \hat{I}_{+,2} + \hat{I}_{-,1} + \hat{I}_{-,2}$, then its matrix (the elements of which are $H_{ij} = \int \phi_i \hat{H}' \phi_j \, d\tau$) has only 10 unique elements since $H_{ij} = H_{ji}$: the diagonal elements H_{ii} and either off-diagonal corner. The diagonal elements do not represent transitions, since the state does not change, and this reduces the number of interesting elements to the six unique off-diagonal elements H_{12}–H_{34}. Evaluate these, and show that the non-zero elements imply the selection rules $\Delta I = 0$ and $\Delta M_I = \pm 1$.

21.17 Figure 21.6(a) shows the HD nuclear spin energy levels as a function of B for very low fields, whereas most NMR is performed at fields 10^6–10^7 times larger. NMR spectra of HD (as the liquid at 20.4 K) have been taken at these low fields, as reported by H. Benoit and P. Piejus in *Compt. Rend.* **265B**, 101 (1967). They used a *prepolarization* method to enhance the signal; the liquid flowed first through a strong field (about 0.8 T) and then into the weak field (scanned from 0 to 3 µT) in which transitions were measured. The prepolarizing strong field created a larger population difference between states involved in transitions, and this difference was largely preserved in the short time it took the liquid to flow from the strong to the weak field. In one of their spectra, the irradiation frequency was only 54 Hz. Calculate the ratio of the equilibrium population at 20.4 K for two levels separated by $\epsilon = h\nu = h(54 \text{ Hz})$ using the two-level system Boltzmann distribution expression, Eq. (4.11) rearranged in the form (population of upper level)/(population of lower level) $= e^{-h\nu/k_BT}$. If your calculator reads 1 when you compute this number, consider the approximation $e^{-x} = 1 - x$, $x \ll 1$. At 54 Hz, one transition ($\phi_1 \to \phi_3$ in Figure 21.6(a)) appears at about 1.9 µT. Estimate the frequency of this transition in a 0.8 T field, and repeat the population ratio calculation for this energy level spacing. By what factor does the prepolarization trick enhance the signal?

21.18 Example 21.6 considered the ^{13}C NMR spectra of the xylenes. What would their proton spectra look like?

21.19 Predict the proton spectrum of propionic acid, CH$_3$CH$_2$COOH, assuming (correctly) that the acidic proton is too far from the other protons to couple to them, but including indirect coupling among the other protons. Carboxylic acid protons in general have very large chemical shifts—about 11.7 ppm here—while the methyl protons appear around 1.2 ppm and the methylene protons are around 2.3 ppm.

21.20 Consider the 1,4-dihalobenzenes. The halogens all have non-zero nuclear spins that couple to protons in varying degrees, but we will ignore that coupling here. Predict the proton NMR of these compounds in general, and then focus on the 1-halo-4-iodobenzenes. As the halogen on the 1 position is varied from F through I, the spectrum has a characteristic change in chemical shifts. Can you predict this change and suggest a reason for it?

21.21 Radiation in the microwave region known as the *K-band* can be used to observe the ESR of organic radicals in spectrometers with magnetic fields around 1.3 T. What is the representative frequency of K-band radiation? What is its typical wavelength in mm? What field is needed for an inorganic radical with $g = 4.68$?

21.22 The ESR spectrum of peroxylamine disulfonate, ON(SO$_3$)$_2^{2-}$, is a simple triplet. In an instrument operating at 9.487 GHz, the lines occur at 339.3 mT, 340.6 mT, and 341.9 mT. Find g and a from these data, and predict the relative intensities of the three lines. (Both sulfur and oxygen have spinless nuclei.)

21.23 How many lines are in the ESR spectra of complexes of Cu^{2+} (nuclear spin 3/2)? What are their relative intensities? How about ^{51}V^{2+} complexes with $I = 7/2$? (Assume the active electron interacts only with the metal nucleus no matter what the complex.)

21.24 Expand on the solution to Example 21.9 by writing the spin arrow diagrams for the six ways of having $M_I = +1$, the six for -1, and the seven for $M_I = 0$.

21.25 The benzene anion ESR spectrum has seven lines, but the benzyl anion, C$_6$H$_5$CH$_2^-$, has a 1:2:1 triplet spectrum. What accounts for the difference between these spectra?

21.26 How many lines would you expect in a fully resolved ESR spectrum of the naphthalene anion, C$_{10}$H$_8^-$? Experimentally, it is found that this spectrum appears only at low concentration. At higher concentrations, the hyperfine structure broadens, the lines blur together, and the splitting ultimately disappears. What could account for this behavior?

GENERAL PROBLEMS

21.27 The quantum mechanical derivation of the Landé g factor follows the steps outlined in this problem. First, the magnetic moment due to $\mathbf{L}$ and $\mathbf{S}$ is $\mathbf{m} = -\mu_B(\mathbf{L} + g_e\mathbf{S})/\hbar$, so that the first-order perturbation energy is $E^{(1)} = \mu_B\langle\mathbf{L} + g_e\mathbf{S}\rangle\cdot\mathbf{B}/\hbar$ where $\langle\rangle$ represents the expectation value of $\mathbf{L} + g_e\mathbf{S}$ for a state specified by quantum numbers L, S, J, and M_J (J and M_J are the only quantum numbers that appear in eigenvalues for this state, however). Show first that if the field is in the z direction, this expression can be written

$$E^{(1)} = \mu_B\left[M_J + \frac{(g_e - 1)\langle S_z\rangle}{\hbar}\right]B$$

since $\mathbf{J} = \mathbf{L} + \mathbf{S}$. Now we need $\langle S_z\rangle$, which is *not* equal to $M_S\hbar$ since M_S is no longer a good quantum number. Since $\mathbf{J}$ is the conserved angular momentum, $\langle\mathbf{S}\rangle$, the average value of $\mathbf{S}$, must lie along the direction of $\mathbf{J}$ so that $\langle\mathbf{S}\rangle = \alpha\mathbf{J}$ where α is a constant. Thus, $\mathbf{J}\cdot\langle\mathbf{S}\rangle = \alpha\mathbf{J}\cdot\mathbf{J} = \alpha J(J+1)\hbar^2$ since $\mathbf{J}\cdot\mathbf{J} = |\mathbf{J}|^2 = J(J+1)\hbar^2$ so that $\alpha = \mathbf{J}\cdot\langle\mathbf{S}\rangle/J(J+1)\hbar^2$. Similarly, $\langle S_z\rangle = \alpha J_z = \alpha M_J\hbar$. Now we need $\mathbf{J}\cdot\langle\mathbf{S}\rangle$. We note that $\mathbf{J}\cdot\langle\mathbf{S}\rangle = \langle\mathbf{J}\cdot\mathbf{S}\rangle$ (since $\mathbf{J}$ is a conserved quantity) $= \mathbf{J}\cdot\mathbf{S}$ (since the dot product is just a number, not another operator). Now we need $\mathbf{J}\cdot\mathbf{S}$. Consider any two angular momenta $\mathbf{L}_1$ and $\mathbf{L}_2$, and define $\mathbf{L} = \mathbf{L}_1 + \mathbf{L}_2$. Then $\mathbf{L}\cdot\mathbf{L} = \mathbf{L}^2 = \mathbf{L}_1^2 + 2\mathbf{L}_1\cdot\mathbf{L}_2 + \mathbf{L}_2^2$ or $\mathbf{L}_1\cdot\mathbf{L}_2 = (\mathbf{L}^2 - \mathbf{L}_1^2 - \mathbf{L}_2^2)/2 = [L(L+1) - L_1(L_1+1) - L_2(L_2+1)]\hbar^2/2$. Show that this expression leads to $\mathbf{J}\cdot\mathbf{S} = [J(J+1) - L(L+1) + S(S+1)]\hbar^2/2$. Put all of this together to find the Landé g factor, g_J, given in Eq. (21.5).

21.28 Practice Problem 21E shows that the Landé g factor for a $^4D_{1/2}$ state is essentially zero. Construct, to scale, a diagram like Figure 21.2 for this state, assume $g_e = 2$ exactly, and show that $\mathbf{m}$ is perpendicular to $\mathbf{J}$ so that $\mathbf{m}_{eff} = 0$. First, locate the $\mathbf{J}$ vector from its length and z component. The hard part is locating $\mathbf{L}$ and $\mathbf{S}$, since all you know are their lengths and that they add to $\mathbf{J}$, but the cosine law for triangles is helpful here.

21.29 In a recent experiment with an atomic Cs beam, B. P. Masterson, C. Tanner, H. Patrick, and C. E. Wieman [*Phys. Rev. A* **47**, 2139 (1993)] used polarized laser light and magnetic fields to produce a Cs atom beam in which 95% of the atoms were in the 6s $^2S_{1/2}$ $F = 3$ state, a so-called *spin-polarized beam*. The lasers connected the 6s $^2S_{1/2} \to$ 6p $^2P_{3/2}$ levels around 852.1 nm, in the near IR, and in fact were lasers similar to those found in commercial compact disk players. To check the quantum state purity of their beam, they recorded the microwave spectrum in the vicinity of the Cs Atomic Clock frequency, near 9.19 GHz. They used a 0.7 mT static magnetic field, and the microwave field was polarized parallel to the static field. They saw seven strong transitions. What are the F and M_F quantum numbers associated with these seven transitions?

21.30 The Stark energy expression for a symmetric rotor is linear in the field strength. A rough, but correct, derivation of this expression starts with the component of $\mathbf{p}$ along the direction of $\mathbf{J}$. Since $\mathbf{p}$ and $\mathbf{K}$ are collinear and the length of $\mathbf{J}$ is $\sqrt{J(J+1)}\hbar$ while the length of $\mathbf{K}$ is $K\hbar$, this component is $pK/\sqrt{J(J+1)}$. The interaction energy is $-(pK/\sqrt{J(J+1)})$ times the component of the field in the direction of this component of $\mathbf{p}$. The angle θ between the field $\mathbf{F}$ and the component is the same as the angle between $\mathbf{J}$ and $\mathbf{F}$. Since $\mathbf{J}$ has a component $M_J\hbar$ along the field direction, $\cos\theta = M_J/\sqrt{J(J+1)}$. Thus, $E^{(1)} = -\mathbf{p}\cdot\mathbf{F} = -[pK/\sqrt{J(J+1)}]F[M_J/\sqrt{J(J+1)}] = -pFM_JK/[J(J+1)]$. Note that this expression does not depend on rotational constants so that a Stark experiment can determine the dipole moment directly. Also, symmetric rotors with very small dipole moments have readily measured Stark shifts in moderate fields. Consider CH_3D with a small dipole moment along the C—D bond direction. In an MBER study of this molecule [S. C. Wofsy, J. S. Muenter, and W. Klemperer, *J. Chem. Phys.* **53**, 4005 (1970)], a 2.0×10^5 V m^{-1} field produced transitions around 2.8 MHz (more than one due to nuclear spin effects) between the $M_J = 1$ and 0 components of $J = 1$, $K = 1$. Find the CH_3D dipole moment, and compare it to the values for "ordinary" moments in Table 15.1.

21.31 The degenerate nature of many states of atomic hydrogen lead to a linear Stark effect since, for example, the 2s state and 2p$_z$ states can produce a nonzero first-order perturbation energy. If the field is along the z direction, the perturbation Hamiltonian is $-\mathbf{p}\cdot\mathbf{F} = -ezF$, and one can show that the first-order energy is $E^{(1)} = \pm 3ea_0F$ where a_0 is the Bohr radius. These degenerate states mix, one goes up in energy $3ea_0F$, and the other goes down the same amount for a total splitting of $6ea_0F$. Calculate the size of the splitting in a 10^5 V m^{-1} field. The mixed states can be written $\Psi_\pm = (2s \pm 2p_z)/\sqrt{2}$ where 2s and 2p$_z$ are the unperturbed H-atom wavefunctions. Sketch these wavefunctions in relation to the field, and decide which corresponds to the higher energy state and which to the lower energy state. Note that these states are sp hybrids, as discussed in Chapter 14. Before you make your sketches, however, review the Closer Look at Hybrid Orbitals in Chapter 14.

21.32 The power spectrum of a pulse, F^*F, should not depend on our arbitrary location of the zero point of time. Show this explicitly for the "one-cycle pulse" discussed in the text written in the form $f(t) = 0$, $t < 0$ and $t > 1/\nu_0$, $f(t) = \sin(2\pi\nu_0 t)$, $0 \le t \le 1/\nu_0$. First, compute the Fourier transform of $f(t)$ in two parts, real and imaginary. Call them Re(F) and Im(F). Why was the real part zero in the text but not here? Now note that $F^*F = [\text{Re}(F) - i\,\text{Im}(F)][\text{Re}(F) + i\,\text{Im}(F)] = [\text{Re}(F)]^2 + [\text{Im}(F)]^2$. Use this to find F^*F for this pulse, and show that it is the same as that in the text for a one-cycle pulse centered at $t = 0$.

21.33 In the course of your study of chlorofluorocarbon radicals, you find that one radical derived from CF_3CCl_3 has an ESR spectrum that consists of four patterns of seven lines each (a "quartet of septets"). The ^{19}F and ^{35,37}Cl spin quantum numbers are 1/2 and 3/2, respectively. Which radical did you observe?

CHAPTER 22

The Dynamic Nature of Equilibrium

22.1 The Meaning of Distribution Functions

22.2 The Statistical Necessity of Equilibrium

22.3 The Role of Intermolecular Forces in Maintaining Equilibrium

MACROSCOPIC equilibrium seems placid, static, uniform, dull. We cannot see the fast-paced, dynamic, textured, exciting microscopic action that sustains macroscopic calm picosecond after picosecond. As equilibrium is approached, we often see macroscopic dynamics: turbulence, pressure fluctuations, energy flows, chemical changes. These macroscopic dynamics are also backed by incredible microscopic activity, and we will look in Chapter 25 at some of them. The general microscopic theory of nonequilibrium situations, however, is not yet understood in all its detail, and aspects of macroscopic disequilibrium are still being discovered. (Consider one classic example—weather—and the relative uncertainty of predictions only a few days into the future.)

Equilibrium has one great advantage. The same equilibrium state can be approached from many nonequilibrium states. No matter how reacting species are mixed, K_{eq} and the initial amounts of compounds determine the equilibrium composition. An equilibrium state requires far fewer macroscopic variables in its description than does any nonequilibrium state leading to it. The system not only falls to the bottom of a chemical potential well at equilibrium, but it also sheds much of its complexity along the way.

This is true for the microscopic picture of equilibrium as well. It is not only hopeless to try to retain microscopic information about each molecule, it is also unnecessary. Given a mole of gas at equilibrium, we cannot specify, nor do we need to, the quantum state of each molecule from instant to instant. Instead, we exploit the indistinguishability of identical particles and speak of the *distribution of quantum states among the mole of molecules.*

Distributions are at the center of a microscopic picture of equilibrium, and the basic properties of these distributions are the topics of this and the following chapter. Here, we will describe distribution functions in general, the role various intermolecular forces play in establishing and maintaining them, and the reasons why they work in the first place. We will see that the very large numbers of molecules in a macroscopic system are both a curse and a blessing. The number is so large that we readily abandon hope of following all molecular motion in detail.[1] Consider 10^{23} $H_2(g)$ molecules, for example. Locating just their centers of mass requires three classical coordinates (x, y, and z) for each molecule (or a spatial wavefunction of 3×10^{23} coordinate variables). Their kinetic energies are another 10^{23} variables, or 3×10^{23} momentum components, and each has vibrational, rotational, electronic, and nuclear spin-state degrees of freedom as well. This totals over 10^{24} variables, and they may all be changing on a picosecond time scale. They cannot all be followed.

Large numbers are a blessing in this sense—we quickly realize they are beyond our grasp, but a curse in that we understand an individual H_2 molecule and the forces between two H_2 molecules extremely well—the temptation to try and conquer a mole of them, molecule by molecule, is compelling. If only we had the time and the vast reams of paper needed to write down the $>10^{24}$ variables! Moreover, in between one molecule and one mole are clusters of a few to perhaps a few thousand molecules (of anything, not just hydrogen) that can be made and studied. These small clusters represent an intermediate form of matter, neither atomic nor bulk, and they have fascinating and valuable properties. Understanding them will require a mix of microscopic and macroscopic ideas that is not completely understood today.

22.1 THE MEANING OF DISTRIBUTION FUNCTIONS

Section 4.2 introduced the two-level system with an analogy of a table and a set of bricks. (See Figure 4.1.) It was pointed out there that the thermodynamic (equilibrium) state of this system requires us to say how many bricks are on the table and how many are on the floor, but it did not matter which bricks were where, only the number of them. We called a specific arrangement of bricks (such as bricks 3 and 7 on the table as in Figure 4.1(b)) a *microstate* of the system. We also showed how to derive the distribution of bricks (molecules) between the floor and the table (the *ground quantum state* and the *excited quantum state*), and we wrote Eq. (4.11) as

$$\frac{n}{N-n} = \frac{\text{population of excited state}}{\text{population of ground state}} = e^{-\epsilon/k_B T} \qquad (22.1)$$

---

[1] As Charles Kittel put it in his book, *Elementary Statistical Physics* (John Wiley, New York, 1958), "It is often difficult to contemplate solving 10^{23} equations of motion."

known as the *Boltzmann distribution.* We can also write this equation as (see also Eq. (4.12))

$$\frac{n}{N} = \frac{e^{-\epsilon/k_B T}}{1 + e^{-\epsilon/k_B T}} = \text{fraction of molecules in the excited state}. \quad (22.2)$$

This is our first example of a *distribution function,* and we will use it to explore some of the general properties of such functions. We will also learn how to construct such functions in general.

Consider the quantities in Eq. (22.2): T, the macroscopic equilibrium temperature; ϵ, the microscopic quantum energy level spacing; k_B, Boltzmann's constant, a universal constant relating temperature to energy; N, the number of molecules in the macroscopic system; and n, the number of those molecules with excitation energy ϵ. This mixture of macroscopic and microscopic variables is a characteristic of distribution functions.

The function tells us several things about this system that are true in general for more complicated systems.

At $T = 0$, all the molecules are in the ground state: $n/N = 0$. (This is the microscopic picture of the Third Law of thermodynamics.)

At $T = \infty$, molecules are distributed equally among all quantum states: $n/N = \frac{1}{2}$ for the two-level system at infinite T.

At finite, non-zero temperatures, the fraction of molecules that are energetically excited is governed by the ratio of the excitation energy to $k_B T$. The larger is ϵ at any one T, the smaller (in an exponential way, $e^{-\epsilon/k_B T}$) is the number of excited molecules.

As a corollary to the last observation, we see that *at equilibrium there will be a negligible number of molecules in excited states with energies ϵ such that $\epsilon \gg k_B T$.*

EXAMPLE 22.1

Consider the ground ($v = 0$) and first excited ($v = 1$) vibrational states of a diatomic molecule as a two-level system. Let n_0 be the number of molecules with $v = 0$ (the ground-state population) and n_1 be the first excited-state population. What is the ratio n_1/n_0 for a sample of H_2 at 300 K? What about I_2? What about both at 4000 K?

SOLUTION The vibrational energies are spaced by a characteristic amount, unique to each molecule. To a very good approximation (the harmonic approximation—see Chapter 19), $\epsilon = hc\omega_e$, where ω_e is the harmonic vibrational constant. Since ω_e is tabulated in cm^{-1} units, the unit conversion factor $hc(100 \text{ cm m}^{-1})/k_B \cong 1.439$ cm K is convenient to compute at the start. From Table 19.2, $\omega_e(H_2) = 4\,401.213$ cm^{-1} and $\omega_e(I_2) = 214.502$ cm^{-1}. If we let $x = \epsilon/k_B T = \omega_e(100hc/k_B)/T$,

$$x = \frac{(4\,401.213 \text{ cm}^{-1})(1.439 \text{ cm K})}{300 \text{ K}} = 21.11 \quad H_2 \text{ at 300 K}$$

$$x = \frac{(4\,401.213 \text{ cm}^{-1})(1.439 \text{ cm K})}{4000 \text{ K}} = 1.583 \quad H_2 \text{ at 4000 K}$$

$$x = \frac{(214.502 \text{ cm}^{-1})(1.439 \text{ cm K})}{300 \text{ K}} = 1.029 \quad I_2 \text{ at 300 K}$$

$$x = \frac{(214.502 \text{ cm}^{-1})(1.439 \text{ cm K})}{4000 \text{ K}} = 0.0772 \quad I_2 \text{ at 4000 K}$$

we see that x for H_2 at 4000 K and x for I_2 at 300 K are similar. These systems have similar distributions between $v = 0$ and $v = 1$. The large x for H_2 at 300 K tells us to expect a very small fraction in $v = 1$, and the small x for I_2 at 4000 K tells us to expect nearly equal $v = 0$ and $v = 1$ populations. The population ratios from Eq. (22.1), e^{-x}, bear this out:

$$\frac{n_1}{n_0} = 6.79 \times 10^{-10} \quad H_2 \text{ at 300 K}$$

$$\frac{n_1}{n_0} = 0.205 \quad H_2 \text{ at 4000 K}$$

$$\frac{n_1}{n_0} = 0.357 \quad I_2 \text{ at 300 K}$$

$$\frac{n_1}{n_0} = 0.926 \quad I_2 \text{ at 4000 K}$$

The availability of higher energy states ($v = 2, 3$, etc.) means some of the molecules are in states other than $v = 0$ and 1, especially for the high-temperature I_2 situation.

⟹ RELATED PROBLEM 22.1

Since very few systems qualify as two-level systems, we need to learn how distribution functions are constructed in general. There are several questions. Must we invoke quantum mechanics, or will a classical description suffice (and perhaps be simpler)? If quantum mechanics are called for, do we treat systems with discrete energy levels differently from those with continuously variable energies? How is the Pauli Exclusion Principle taken into account?

In many ways, the quantum description is simpler than the classical description (and even the classical description requires Planck's constant and is "semiclassical"). If we have continuous energy levels (or nearly so, as for the quantized translational motion of a gas in a macroscopic container in which the discrete levels are very closely spaced), we will have a continuous distribution function. This means that we will speak of the "fraction of molecules with energies between E and $E + dE$," for example, rather than the fraction with one certain energy.

Most interesting is the role of the Pauli Exclusion Principle. The molecules in a macroscopic system are not only individually subject to quantum mechanics, but the entire macroscopic sample is as well, of course. It is a *very* many body system, with a wavefunction of $> 10^{24}$ variables per mole of system molecules as mentioned in the chapter's introduction. The Pauli Principle says this wavefunction must be antisymmetric with respect to interchange of any two half-integral spin particles and symmetric with respect to interchange of integral spin particles. All half-integral spin particles must have a unique set of quantum numbers, but this restriction does not hold for integral spin particles. The most important systems[2] for which this distinction is vital are the helium isotopes, ^{3}He and ^{4}He. Nuclei in ^{4}He are spinless, and an arbitrarily large number of them can occupy any one energy level of the macroscopic system. In contrast, ^{3}He nuclei have a nuclear spin quantum number of $\frac{1}{2}$. *Every* atom in a macroscopic sample of ^{3}He has a unique set of quantum numbers. This simple difference accounts for the profoundly different behaviors of the He(l) isotopes. (See the phase diagrams in Figure 6.10.)

These two classes of particles follow different distribution functions. For *half-integral spin* species (called *fermions*), we will speak of *Fermi–Dirac* (FD) distributions. *Integral spin* species (called *bosons*) are said to follow *Bose–Einstein* (BE)

---

[2]Our discussion of the free-electron model of metals in Chapter 16 also required us to use the Pauli Principle to distribute a mole of electrons among a large number of energy levels.

distributions. While this distinction can be important, there is a common physical situation for which the distinction disappears, and an ordinary analogy helps explain it. Imagine you and a friend are somehow placed at random in rooms in the largest building you can imagine. (Each room represents a unique quantum state, and you are a molecule.) If both of you follow FD rules, you are not allowed to be in the same room, but if you adhere to BE rules, this placement is allowed. But if you are both placed at random, and if the number of rooms is huge, the odds that you both land in the same room are quite small. Even as you wander from room to room, there is only a small chance you both will enter the same room at the same time.

This is the analog of what we will call a *dilute system:* many more, *very* many more, possible sets of quantum numbers are energetically allowed in comparison to the molecules in the system. Consider Ar(g) in a 1 m cube. The translational motion of the atoms follows the allowed energies of the 3-D particle-in-a-box quantum mechanical problem (see Eq. (12.24)). Each atom's quantum state is specified by a set of three quantum numbers, (n_x, n_y, n_z), each an integer greater than zero. Given this set of numbers, the energy of any one atom in that state is

$$E = \frac{\hbar^2 \pi^2}{2mL^2}(n_x^2 + n_y^2 + n_z^2)$$

where $\hbar$ is Planck's constant divided by 2π, m is the atom's mass, and L is the length of the cube edge, 1 m. Numerical values for these constants yield

$$E = (8.3 \times 10^{-43} \text{ J})(n_x^2 + n_y^2 + n_z^2) \ .$$

We will later prove that the average energy of an atom in this system is on the order of magnitude of $k_B T$, which is 4.1×10^{-21} J at 300 K. Consequently, the average quantum number is very large:

$$n_x^2 + n_y^2 + n_z^2 \cong \frac{4.1 \times 10^{-21} \text{ J}}{8.3 \times 10^{-43} \text{ J}} = 5 \times 10^{21}$$

or, assuming each n is of comparable magnitude, $3n^2 = 5 \times 10^{21}$ and $n \cong 4 \times 10^{10}$. The number of quantum states that we can assign to these atoms is therefore the number of different sets of three integers (n_x, n_y, n_z) with each n independently ranging from 1 to $\sim 4 \times 10^{10}$. We can estimate this number of states if we imagine cubes of unit volume stacked into a huge cube 4×10^{10} small cubes along each side. Each cube represents one quantum state, and the huge cube holds roughly as many small cubes as there are quantum states available to all the Ar atoms.[3] This number, $(4 \times 10^{10})^3 = 6.4 \times 10^{31}$, is many orders of magnitude larger than the number of atoms. (About 2.4×10^{25} Ar atoms fit in 1 m³ at room temperature and pressure.) Consequently, this is a dilute system, and the dictates of the Pauli Principle never have an opportunity to come into play. Each atom is very likely to be in a unique quantum state.

Finally, we will need to consider the roles of *degeneracy,* (two or more quantum states with the same energy, as in the states $(n_x, n_y, n_z) = (2, 1, 1)$ or $(1, 2, 1)$ or $(1, 1, 2)$ for the particle-in-a-box: three different states with identical energies), and *particle distinguishability*. Distinguishability may depend on our level of detail (are isotopic differences important?), or it may be imposed by the system. For example, the fixed positions in space make molecules in a crystal distinguishable from each other, even if the molecules are identical and become indistinguishable if the crystal is melted or vaporized.

---

[3]Again, see the free-electron model of metals in Chapter 16 for a more precise way to estimate this number. A rough order of magnitude is all we need here.

It is also important to recognize that there may be many different ways to assign energies to molecules so that the total energy is some fixed amount. In our new terminology, there are *many distributions that lead to the same thermodynamic energy.* An earlier example of this is shown in Figure 3.1; an ideal gas could have one third of its molecules moving in each of the *x, y,* and *z* directions, all with the same speed, as long as their total kinetic energy added to the thermodynamic internal energy. Nature, however, invented entropy to prevent this from happening. Entropy plays the role of arbitrator among *possible* distributions to produce the *one* observed distribution. It does so in a very reasonable way: it chooses the *most probable distribution.*

We specify a distribution when we list the number of particles in each quantum state; we say state 0 (the ground state) has n_0 particles, state 1 has n_1, etc. These *state populations* must add to the total number of particles in the system:

$$N = \sum_{i=0} n_i \tag{22.3}$$

and the total energy of the particles must equal the fixed, specified energy of the system:

$$E = \sum_{i=0} n_i \epsilon_i \tag{22.4}$$

where ϵ_i is the energy of the ith quantum level, and both sums extend over all quantum levels. These two equations, knowledge of the level energies ϵ_i and their degeneracies, and the quantum rules (FD or BE) followed by the system, allow us to specify which distributions are *possible*. The relative *probabilities* of these states tell us which distribution is observed, but assigning these probabilities requires one additional assumption, central to all statistical arguments. This is the assumption of *equal a priori probabilities.* We will call a specific way of realizing a particular distribution a *microstate,* and we must assume that *all microstates are equally probable.* The probability of a distribution is proportional to the number of ways of realizing it, and this in turn is simply a count of microstates that lead to the distribution. We can distinguish among distributions experimentally, but we cannot distinguish microstates. We will see, however, that the most probable distribution has an overwhelmingly larger number of microstates leading to it than does any other distribution.

Consider, for example, a system of three particles ($N = 3$) with a total energy $E = 6\epsilon$ where ϵ is a constant. Imagine that each particle has nondegenerate energies $\epsilon_i = i\epsilon$, $i = 0, 1, 2, \ldots$ (as in the harmonic oscillator with the zero-point energy level[4] taken to be the zero of energy—see Eq. (12.18) and let $i = (v + \frac{1}{2})$ and $\epsilon = \hbar\omega$). Suppose first that the particles are *distinguishable*. They are neither FD nor BE particles, and we can label them A, B, and C. One way we can distribute the total energy among them gives A all the energy, 6ϵ, and B and C zero energy. This distribution has $n_6 = 1$, n_5 through $n_1 = 0$, and $n_0 = 2$. (Clearly no level with $i > 6$ can be populated. We lack the total energy to do so.) This is one microstate leading to the distribution we could abbreviate $(n_6, n_5, n_4, n_3, n_2, n_1, n_0) = (1, 0, 0, 0, 0, 0, 2)$, but there are two others. If particle B had 6ϵ and A and C had zero,

---

[4]The English term "zero-point energy" comes from the German "Nullpunktsenergie," and while "Nullpunkt" can be translated as "zero point," the German word was devised to represent the residual energy that remains at the "Nullpunkt" of temperature—at absolute zero. Consequently, the term has its origins as much in statistical mechanics as in quantum mechanics and could as well have been translated "absolute zero energy."

FIGURE 22.1 If three distinguishable particles with identical, equally spaced energy levels have a total energy 6ϵ where ϵ is the energy level spacing, there are seven possible *distributions* of energy among the particles.

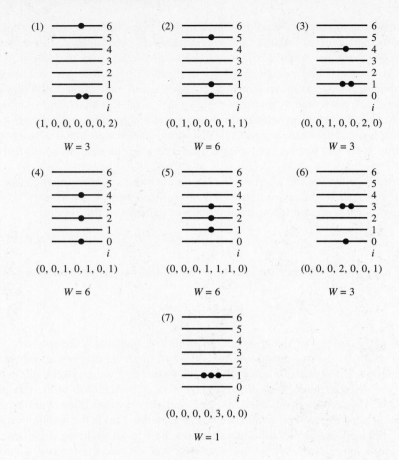

or if particle C had 6ϵ and A and B had zero, we would have the same distribution. Therefore, we assign this distribution a relative probability or *weight* $W = 3$.

There are many other microstates and distributions that will satisfy this system, however. Figure 22.1 summarizes them, and we see that seven distinct distributions of varying weights are possible. The weights can be calculated by a simple, general formula. If we think of the particles as marbles and the energy levels as boxes, the number of ways to select a marble to put in one box—any box—is N; the number of ways to select the second is $(N - 1)$; and so on until the last marble is chosen. The total number is $N(N - 1)(N - 2) \cdots 1 = N!$, the number of permutations of N objects. For $N = 3$, $N! = 6$: (A, B, C), (A, C, B), (B, A, C), (B, C, A), (C, B, A), and (C, A, B). Now we look in each box and recognize that the same *distribution* is attained no matter what order marbles are placed in the box. If a box has 3 marbles, there are 3! ways that box could have been filled; if it has n_i marbles, there are $n_i!$ ways. Thus, we divide the raw count of permutations, $N!$, by the permutations of each box, $n_i!$, yielding the final weight of the distribution:[5]

$$W = \frac{N!}{n_0! \, n_1! \cdots} = \frac{N!}{\prod_i n_i!} \quad \text{(distinguishable particles)} \, . \tag{22.5}$$

---

[5]Note that this is the same expression we obtained in Chapter 4, Eq. (4.5), for the two-level system. There, $n_1 = n$, and $n_0 = N - n$.

For example, the first distribution, (1, 0, 0, 0, 0, 0, 2), has $W = 3!/(1! \, 0! \, 0! \, 0! \, 0! \, 0! \, 2!) = 6/2 = 3$ (and recall that $0! = 1$) and number (7), (0, 0, 0, 0, 3, 0, 0), has $W = 3!/(0! \, 0! \, 0! \, 0! \, 3! \, 0! \, 0!) = 1$.

EXAMPLE 22.2

A system of 10 distinguishable particles that follow the energy level scheme of Figure 22.1 is adjusted to have $E = 8\epsilon$. Three possible distributions for this system are: $(n_4, n_3, n_2, n_1, n_0) = (2, 0, 0, 0, 8)$, $(0, 1, 1, 3, 5)$, and $(1, 1, 0, 1, 7)$. Which of these is most probable?

SOLUTION The probabilities are directly proportional to the statistical weights given by Eq. (22.5):

$$(2, 0, 0, 0, 8): \quad W = \frac{10!}{2! \, 0! \, 0! \, 0! \, 8!} = 45$$

$$(0, 1, 1, 3, 5): \quad W = \frac{10!}{0! \, 1! \, 1! \, 3! \, 5!} = 5040$$

$$(1, 1, 0, 1, 7): \quad W = \frac{10!}{1! \, 1! \, 0! \, 1! \, 7!} = 720 \; .$$

The second distribution, $(0, 1, 1, 3, 5)$, is considerably more probable than the other two. If we make simple bar graphs of these distributions:

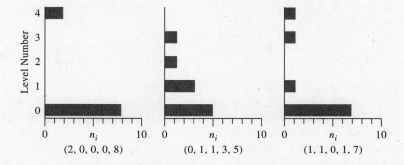

we see that the most probable distribution has smoothly decreasing populations as we go up in energy. This will be a common feature of equilibrium distributions in general.

➠ RELATED PROBLEMS 22.2, 22.3, 22.4

Now we can change the problem slightly and explore the ramifications of quantum rules. Note that this is not a dilute system; there are several possible distributions in Figure 22.1 that have levels with two or more particles. Suppose the particles are indistinguishable and follow FD rules with the Pauli Exclusion Principle. Immediately, all the distributions in Figure 22.1 with more than singly occupied levels (numbers (1), (3), (6), and (7)) are not possible. The FD rules allow nondegenerate levels to have one or zero particles only. Moreover, the allowed distributions now have weights of *one,* since the particles are indistinguishable.[6]

---

[6]In the language of Chapter 13, there is only *one* antisymmetric total system wavefunction for these allowed distributions. For example, distribution (2) represents the Slater determinant (see Eq. (13.21))

$$(3!)^{-1/2} \begin{vmatrix} \phi_5(A) & \phi_1(A) & \phi_0(A) \\ \phi_5(B) & \phi_1(B) & \phi_0(B) \\ \phi_5(C) & \phi_1(C) & \phi_0(C) \end{vmatrix} \; .$$

Suppose instead that the particles are indistinguishable but follow BE rules. In this case, every distribution in Figure 22.1 is allowed, since BE rules allow any number of particles to have the same quantum numbers, but, just as for FD particles, each distribution has a weight of one.

This simple example has one feature in common with more complicated systems. No matter what the energy-level and degeneracy pattern, it will always be true that the number of FD distributions for any N and E will be less than the number of BE distributions. Moreover, both types of distributions have limits that we can relate to our counts for distinguishable particles. For them, *each* microstate has a unique quantum state, a unique wavefunction. The total number of such quantum states in Figure 22.1 is the sum of the weights of all the distributions, $3 + 6 + 3 + 6 + 6 + 3 + 1 = 28$. For the FD distributions, numbers (2), (4), and (5), each corresponds to $6 = 3! = N!$ distinguishable particle quantum states, but some distinguishable states are not FD allowed. Consequently,

$$\binom{\text{the number of distributions}}{\text{of distinguishable particles}} \geq N! \binom{\text{the number of distributions of}}{\text{indistinguishable FD particles}}. \quad \textbf{(22.6a)}$$

Similarly, each BE distribution corresponds to one state, but the number of distinguishable particle distributions must be *less* than $N!$ times the number of BE distributions, since each distinguishable distribution has a weight less than or equal to $N!$ and every distinguishable distribution is an allowed BE distribution. The BE analog of Eq. (22.6a) is

$$\binom{\text{the number of distributions}}{\text{of distinguishable particles}} \leq N! \binom{\text{the number of distributions of}}{\text{indistinguishable BE particles}}. \quad \textbf{(22.6b)}$$

Most interesting is the condition that establishes the equalities in Eq. (22.6). This is simply the *dilute limit:* many more states than particles so that single occupancy is highly probable. We have already argued that the higher energy levels are populated only when the temperature is proportionately high, and here we see that the dilute limit is also a high-temperature limit. If the thermal energy, gauged roughly by $k_B T$, is higher than the characteristic spacing of energy levels *and* if those levels extend in energy up to and beyond the energy $k_B T$, then we can have many more levels than particles and ignore the FD/BE distinction.

Since the dilute limit is the usual case of chemical interest, we will (after giving them a Closer Look below) leave the details of FD/BE distinctions behind. We can closely approximate the number of distributions that lead to any one E for N particles in a given energy level scheme if we use the distinguishable-particle counting rules and then divide by $N!$. Since $N!$ is the number of permutations of N things, this division is the simplest way to account for indistinguishability.

In general, Eq. (22.5) lacks one finishing touch before we can use it. It does not, as it stands, allow for arbitrary degeneracy for any level. This is easy to correct, however. If level i with energy ϵ_i has degeneracy g_i (so that g_i quantum states correspond to that energy level), then each particle placed in level i has g_i choices. This is true for each of the n_i particles that enter this level so that a factor $g_i^{n_i}$ comes into play. For example, consider a level with $g_i = 3$ and $n_i = 2$. There are three places to place the first particle, and the second can be added to each of these in three ways for a total of $3^2 = 9 = g_i^{n_i}$, as illustrated in Figure 22.2. Equation (22.5) becomes

$$W = N! \prod_i \frac{g_i^{n_i}}{n_i!} \quad \text{(distinguishable particles with degeneracy)} \quad \textbf{(22.7a)}$$

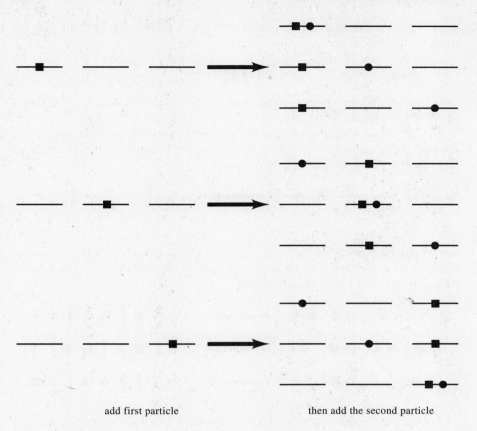

FIGURE 22.2 If n_i particles are placed in an energy level that has a degeneracy g_i, the total number of possible placements is $g_i^{n_i}$, as illustrated here for the case $g_i = 3$ and $n_i = 2$.

or, for the dilute limit,

$$W = \prod_i \frac{g_i^{n_i}}{n_i!} \text{ (indistinguishable dilute particles with degeneracy)} . \quad \text{(22.7b)}$$

These expressions were deduced before quantum theory was known. Systems that follow these rules are said to follow *Maxwell–Boltzmann* (MB) statistics[7] in honor of two prequantum scientists who first derived many of the results we will see later on.

A Closer Look at FD and BE Statistics

The individual weights for levels that follow FD or BE statistics are not difficult to find, and once we do, we can show that the dilute limit is intermediate between them. We imagine any one level has an arbitrary degeneracy g_i from the start. For FD particles, no more than one can enter each state so that $n_i \leq g_i$. If the particles were distinguishable, they could be assigned in g_i ways for the first, $g_i - 1$ for the second, etc., for a total of

$$g_i(g_i - 1)(g_i - 2) \cdots (g_i - n_i + 1) = \frac{g_i!}{(g_i - n_i)!}$$

---

[7] The particles that follow these rules are also sometimes called "boltzons" in analogy with the terms "fermions" and "bosons" for FD and BE particles.

ways. The particles are indistinguishable, however, and we correct for this with the usual factor $1/n_i!$ so that

$$W_{\text{FD}} = \prod_i \frac{g_i!}{n_i!\,(g_i - n_i)!}.$$

(The example in the text has nondegenerate levels so that $W_{\text{FD}} = 1$ for each allowed distribution.)

For BE particles, there is no limit to the number that enters any one state. If we think of the g_i states in level i as a series of $g_i - 1$ walls separating groups of particles from state to state, as in this example:

then any permutation of the $n_i + g_i - 1$ particles plus walls will be an allowed microstate for level i. For example, if $n_i = 6$, three of the allowed assignments of BE particles to the four states above are:

There are $(n_i + g_i - 1)!$ permutations of particles + walls if the particles and walls are distinguishable, but since they are not, we again divide by $n_i!$ to correct for particle indistinguishability and also by $(g_i - 1)!$ to correct for wall indistinguishability and find

$$W_{\text{BE}} = \prod_i \frac{(n_i + g_i - 1)!}{n_i!\,(g_i - 1)!}.$$

Consider any one factor in the product for W_{FD}:

$$\frac{g_i!}{n_i!\,(g_i - n_i)!} = \frac{g_i(g_i - 1)(g_i - 2)\cdots(g_i - n_i + 1)}{n_i!}.$$

Each factor in the numerator of this expression is less than or equal to g_i, and there are n_i such factors. Thus, the numerator is $\leq g_i^{n_i}$. For W_{BE},

$$\frac{(n_i + g_i - 1)!}{n_i!\,(g_i - 1)!} = \frac{(n_i + g_i - 1)(n_i + g_i - 2)\cdots(n_i + g_i - n_i)}{n_i!}$$

and each of the n_i factors in the numerator is greater than or equal to g_i so that the entire BE numerator is $\geq g_i^{n_i}$. These are quantitative statements of Eq. (22.6), and they further justify Eq. (22.7b).

We now seek the most probable distribution, that set of n_i's that maximizes W in Eq. (22.7b). We must do so keeping N (the sum of the n_i's—see Eq. (22.3)) constant and keeping the total energy (Eq. (22.4)) constant. Since N is constant, the distinction between Eq. (22.7a) and (22.7b) is unimportant here, but we will use Eq. (22.7b). Each n_i is large for any level populated with a measurable number of molecules (see, however, Problem 22.1), and so is each g_i, but for somewhat subtle reasons. Most model quantum mechanical problems, such as those in Chapter 12, yield, at most, rather modest level degeneracies for any states of practical interest. Certainly, none of them has degeneracies on the order of 10^{10} to 10^{20}, as we will assume here.[8]

For gases, our discussion earlier in this section indicated that the spacing from one level to the next was a very small fraction of the total energy. We can group many levels that lie within a small interval of energy into one vastly degenerate level without loss of generality. The small interval can be based on any reasonable criterion, such as the momentum uncertainty dictated by the Heisenberg Uncertainty Principle. For dense gases or condensed phases, we can assume what is called "weak interaction" among the particles of the system. For example, $N_2(l)$ is still a collection of N_2 molecules, but the individual molecular energy levels are perturbed slightly from their dilute-gas energies due to close molecular contact and interaction. These interactions blur molecular energies in comparison to the isolated molecule energies, but not so much as to change basic chemical identity. Levels spread over narrow groups that we can take to be degenerate. (Crystals, as we will see, attain degeneracies through a "strong interaction.")

If we maximize $\ln W$ instead of W, we will arrive at the same most probable distribution, but with less effort. We can use a very good approximation for the factorial of a large number, *Stirling's approximation*. An abbreviated version was used in Chapter 4, but here we use a somewhat more accurate version:[9]

$$\ln N! \cong N \ln N - N . \qquad (22.8)$$

With this approximation, Eq. (22.7b) becomes

$$\ln W = \sum_i \ln\left(\frac{g_i^{n_i}}{n_i!}\right) = \sum_i (n_i \ln g_i - \ln n_i!)$$
$$= \sum_i (n_i \ln g_i - n_i \ln n_i + n_i) . \qquad (22.9)$$

Since each n_i is very large compared to 1, we can treat each as a continuous variable and begin to maximize $\ln W$ in the usual way (remembering that the g_i's are constants):

$$\text{at maximum } W: d\ln W = 0 = \sum_i \left(\ln g_i - \ln n_i - \frac{n_i}{n_i} + 1\right) dn_i = \sum_i \left(\ln \frac{g_i}{n_i}\right) dn_i \qquad (22.10)$$

but we must have this true under the two constraints of Equations (22.3) and (22.4).

---

[8] One possible exception might be the hydrogen atom for which the principle quantum number n has no upper bound and the degeneracy of level n is $2n^2$. Only rarely does one encounter an H atom with n greater than roughly 10, however.

[9] The full Stirling's approximation is remarkable. It contains a little bit of everything, as if arrived at by consensus among a committee of mathematicians while the spokesperson for the trigonometric functions was out of the room. It is $N! \cong \sqrt{2\pi N} N^N e^{-N}$, and its derivation can be found in many of the references at this chapter's end. James Stirling first published it in his book *Methodus Differentialis* in 1730, and it was well known to mathematicians of his time. The form we use can be justified as follows: $\ln N! = \sum_{i=1}^{N} \ln i \cong \int_1^N \ln x \, dx = N \ln N - N + 1 \cong N \ln N - N$. The full Stirling's approximation for $\ln N!$ is $\ln (\sqrt{2\pi N})$ larger, a usually negligible amount for very large N.

The general method for maximizing a function of many variables that are constrained among themselves is known as the *Lagrange method of undetermined multipliers*. One writes the constraining equations in differential form:

$$\sum_i dn_i = 0 \quad \text{and} \quad \sum_i \epsilon_i\, dn_i = 0$$

↑ From Eq. (22.3) with $dN = 0$ ↑ From Eq. (22.4) with $dE = d\epsilon_i = 0$

multiplies each by arbitrary constants, the undetermined multipliers, and adds them to the central equation, Eq. (22.10).[10] We will use $-\alpha$ and $-\beta$ as the undetermined multipliers and write

$$d\ln W = 0 = \sum_i \ln\frac{g_i}{n_i} dn_i - \alpha \sum_i dn_i - \beta \sum_i \epsilon_i dn_i = \sum_i \left[\ln\frac{g_i}{n_i} - \alpha - \beta\epsilon_i\right] dn_i . \quad (22.11)$$

In this expression, the coefficients of each dn_i must be zero in order for the sum to be zero since the undetermined multipliers now account for the constraints among the n_i's. Therefore

$$\ln\frac{g_i}{n_i} - \alpha - \beta\epsilon_i = 0$$

or

$$n_i = g_i e^{-\alpha} e^{-\beta\epsilon_i} . \quad (22.12)$$

This equation expresses the most probable distribution, the n_i's, in terms of α and β, which we do not yet know. We can find α immediately, however, if we use the first constraining equation, Eq. (22.3):

$$N = \sum_i n_i = \sum_i g_i e^{-\alpha} e^{-\beta\epsilon_i} = e^{-\alpha} \sum_i g_i e^{-\beta\epsilon_i}$$

↑ Eq. (22.3) ↑ Using Eq. (22.12) ↑ Since $e^{-\alpha}$ is a constant

so that

$$e^{-\alpha} = \frac{N}{\sum_i g_i e^{-\beta\epsilon_i}} = \frac{N}{q}$$

where q is called the *molecular partition function*:

$$q = \sum_i g_i e^{-\beta\epsilon_i} . \quad (22.13)$$

We can now write Eq. (22.12) as

$$n_i = N\frac{g_i e^{-\beta\epsilon_i}}{q} , \quad (22.14a)$$

which shows that the *fraction* of molecules in level i, n_i/N, falls off exponentially with increasing level energy, since q and β are both constants (and β is a positive number, as we are about to see). Note that the undetermined multiplier $-\alpha$ has been swallowed into Eq. (22.14a) through the definition of q: $-\alpha = \ln(N/q)$.

---

[10] Roughly speaking, each constraining relation among the independent variables denies independence to one of them. For instance, if we require three otherwise independent variables to add to 7, then we can always find the third once we specify the first two. The third is no longer independent of the others. The undetermined multipliers (which become determined at the end of the method), one per constraint, make up for this loss of independent variables.

22.1 THE MEANING OF DISTRIBUTION FUNCTIONS

You may have guessed that $\beta = 1/k_BT$, and it does. Chapter 3 related Boltzmann's assumption about W and entropy, $S = (\text{constant}) \ln W$, (Eq. (3.5)), to a thermodynamic expression for entropy (Eq. (3.12)) to show that the constant in Eq. (3.5) is $R/N_A = k_B$. This was used in Chapter 4 to derive Eq. (22.2) for the two-level system, which is in the form of Eq. (22.14a). The sum for the partition function extends over the only two levels, $\epsilon_0 = 0$ and $\epsilon_1 = \epsilon$, neither of which is degenerate, and $q = 1 + e^{-\beta\epsilon}$, the denominator of Eq. (22.2) if $\beta = 1/k_BT$. A connection to thermodynamics provides a physical interpretation for β, and our working expression for the equilibrium distribution contains that key thermodynamic variable, the absolute temperature, in a prominent place:

$$n_i = N \frac{g_i e^{-\epsilon_i/k_BT}}{\sum_i g_i e^{-\epsilon_i/k_BT}} = N \frac{g_i e^{-\epsilon_i/k_BT}}{q}. \tag{22.14b}$$

This is the general form of the equilibrium *Boltzmann distribution function*. Equation (22.14b) gives the fractional population, n_i/N, in any one level. Equation (22.14b) written for level i divided by the same equation written for another level, j, yields a useful expression for the *ratio of populations between any two levels*:

$$\frac{n_i}{n_j} = \frac{g_i}{g_j} e^{-(\epsilon_i - \epsilon_j)/k_BT}. \tag{22.14c}$$

EXAMPLE 22.3

The lowest four energy levels of atomic C (see Table 13.3) have the energies and degeneracies:

i	$\epsilon_i/\text{cm}^{-1}$	g_i
0	0.0	1
1	16.4	3
2	43.5	5
3	10 193.70	5

Find the fractional population of each level for C atoms in a stellar atmosphere at 6000 K.

SOLUTION We can use the conversion factor from Example 22.1, 1.439 cm K, to convert the level energies from cm^{-1} to temperature equivalent units: 0 K, 23.6 K, 62.6 K, and 14 670 K. Next, we compute the partition function from Eq. (22.13):

$$q = 1 e^{-0/6000} + 3 e^{-23.6/6000} + 5 e^{-62.6/6000} + 5 e^{-14\,670/6000}$$
$$= 1 + 2.99 + 4.95 + 0.43$$
$$= 9.37$$

but since the sum for q is over *all* levels, we should consider whether or not levels higher than these add significantly to q. Table 13.3 shows that the fifth level is singly degenerate ($J = 0$, and $g = 2J + 1 = 1$) and at an energy roughly twice the fourth level's. Consequently, it will contribute roughly $(0.43/5)^2 \cong 0.007$ to q. It and higher levels can be ignored in the sum for q at this T. The level population fractions come from the Boltzmann distribution, Eq. (22.14):

$$\frac{n_0}{N} = \frac{1 e^{-0/6000}}{q} = \frac{1}{9.37} = 0.106$$

$$\frac{n_1}{N} = \frac{3 e^{-23.6/6000}}{q} = \frac{2.99}{9.37} = 0.319$$

$$\frac{n_2}{N} = \frac{5 e^{-62.6/6000}}{q} = \frac{4.95}{9.37} = 0.528$$

$$\frac{n_3}{N} = \frac{5 e^{-14\,670/6000}}{q} = \frac{0.43}{9.37} = 0.046.$$

These fractions verify several general aspects of distributions. First, the fractions add to 0.999; only 0.1% of the atoms are in energy levels higher than these at 6000 K. Second, the energies of the lowest three levels are small fractions of k_BT at 6000 K. Consequently, each *state* in these levels have roughly equal populations. *A state population is a level population divided by the level degeneracy:* $0.106/1 \cong 0.319/3 \cong 0.528/5$. The fourth level, however, is considerably higher than k_BT and is very slightly populated, only 4.6%.

➡ RELATED PROBLEMS 22.9, 22.10

The next section takes us from distributions towards thermodynamics. Consider what lies ahead. Once we know the allowed energies and degeneracies of a system from quantum mechanics, we can pick a temperature and use the Boltzmann distribution to describe the various states populated at equilibrium. This is something thermodynamics cannot do, but we will soon see that the partition function is the route to more than distributions. It is an entry into *all* equilibrium thermodynamic properties.

22.2 THE STATISTICAL NECESSITY OF EQUILIBRIUM

We have claimed that the Boltzmann distribution is the most probable one, but how probable is "most probable"? Experiment shows that equilibrium distributions do not change with time, dancing around the most probable one. Does our theory support this point of view? We have used the words "overwhelmingly large number" in the previous section. Can we prove this claim, and can we understand why it is true?

Consider the system in Example 22.3: C atoms at 6000 K with only four levels appreciably occupied. We can imagine altering the equilibrium distribution calculated there in such a way that we keep the total energy and number of atoms fixed. We can then use Eq. (22.9) to contrast the weight of the equilibrium distribution, W, to the weight of an altered distribution, W'. Suppose we move a few atoms from the third energy level down to the second and simultaneously move a few more up in energy from the first to the second in such a way that the total energy is constant. The energy spacing from the first to the second and from the second to the third level is not equal, but if we move about six atoms down for every ten we move up, we will keep the energy closely constant, since

$$\left.\begin{array}{l}43.5 \underline{\qquad}\\ 16.4 \underline{\qquad}\\ 0 \underline{\qquad}\end{array}\right\} \begin{array}{l}43.5 - 16.4\\ 16.4 - 0\end{array} \qquad \frac{\epsilon_1 - \epsilon_0}{\epsilon_2 - \epsilon_1} = \frac{(16.4 - 0)\text{ cm}^{-1}}{(43.5 - 16.4)\text{ cm}^{-1}} = 0.605 \cong \frac{6}{10}.$$

Suppose we have only 1000 atoms in the system, and suppose we move ten atoms from $i = 0$ to $i = 1$ and 6 from $i = 2$ to $i = 1$. Equation (22.9) will tell us that the equilibrium distribution weight is greater than this altered distribution's weight, but not by much: $W/W' \cong 7$. That is not a very "overwhelming" factor. If we increase N to 10^9 and alter the distribution at roughly the part per million level, taking 1000 atoms up to the first level and 600 down from the second, we find $W/W' \cong 1495$, a larger factor, but still not unfathomably large.

One thousand atoms or even 10^9 do not constitute what we usually think of as a macroscopic sample, however. (A billion C atoms weigh only 2×10^{-17} kg!) Suppose we have $N = 10^{18}$, about 20 µg of C atoms, and suppose we move one millionth of them, 10^{12}, up and 6×10^{11} down. Then Eq. (22.9) will tell us that $W/W' \cong 10^{3 \times 10^{10}}$. *That* is an overwhelming number! Even if we move only 10^6 atoms up and 600,000 down (surely they won't be noticed among the 10^{18} total), we find $W/W' \cong 10^{3167}$. Even a part-per-billion deviation from the most probable distribution is so improbable that it simply will not be observed in any system of macroscopic size.

It is often stated that the spontaneous flight of all the air molecules in a room to one corner is not impossible, only highly improbable.[11] This is entropy at work, selecting the most probable from the possible, and doing so with apparent ease, since the most probable distribution is so obviously more probable than any other. Entropy's agent in all of this is the variety of *intermolecular forces* that communicate energy distribution information among molecules. The next section looks at this most important aspect of equilibrium and its attainment, but in this section, we begin to tie statistical mechanics to thermodynamics. This will allow us to peer back through entropy and internal energy, through the partition function, to the quantized energy levels of a system's component molecules.

The average energy per molecule, $\langle \epsilon \rangle$, for N molecules with total energy E is simply E/N, the ratio of our two constraining equations, Equation (22.3) and (22.4). We can write this ratio in terms of the Boltzmann distribution:

$$\langle \epsilon \rangle = \frac{E}{N} = \frac{\sum_i \epsilon_i n_i}{N} = \sum_i \epsilon_i \left(\frac{n_i}{N}\right)$$
$$= \sum_i \epsilon_i \left(\frac{g_i e^{-\beta \epsilon_i}}{q}\right) = \frac{1}{q}\sum_i \epsilon_i g_i e^{-\beta \epsilon_i} \qquad (22.15)$$

and, since

$$-\frac{\partial q}{\partial \beta} = \sum_i \epsilon_i g_i e^{-\beta \epsilon_i} ,$$

we can arrive at very compact expressions for the average molecular energy:

$$\langle \epsilon \rangle = -\frac{1}{q}\frac{\partial q}{\partial \beta} = -\frac{k_B}{q}\frac{\partial q}{\partial(1/T)} = \frac{k_B T^2}{q}\frac{\partial q}{\partial T} = k_B T^2 \frac{\partial \ln q}{\partial T} \qquad (22.16)$$

because $dq/q = d \ln q$ and $d\beta = (1/k_B) d(1/T) = -dT/k_B T^2$. We are free to use whichever of these expressions is most convenient.

We can explore Eq. (22.16) with the electronic levels of carbon atoms from Example 22.3. First, consider how q itself varies with temperature. At $T = 0$, q always equals g_0, the degeneracy of the ground level, or $q = 1$ here. As T increases, each excited level is populated in turn, and the term in q corresponding to any one excited level, $g_i e^{-\beta \epsilon_i}$, approaches g_i as $T \gg \epsilon_i/k_B$. If we stay below temperatures that would significantly populate levels above the highest one mentioned in Example 22.3, ϵ_3 at 10 193 cm^{-1}, q will start at 1 at $T = 0$ and increase toward $\sum_{i=0}^{3} g_i = 1 + 3 + 5 + 5 = 14$ as T increases toward, say, 6000 K, the temperature of the example.

Figure 22.3 plots this partition function,

$$q = 1 + 3 e^{-23.6 \text{ K}/T} + 5 e^{-62.6 \text{ K}/T} + 5 e^{-14\,670 \text{ K}/T} ,$$

over three temperature ranges: 0–60 K, 0–600 K, and 0–6000 K. Note how q starts at *and stays near* 1 until T approaches the equivalent excitation temperature of the first excited level, $T = \epsilon_1/k_B = 23.6$ K. From there until $T \gg \epsilon_2/k_B$ but $\ll \epsilon_3/k_B$

---

[11]Or, scarier still, consider the spontaneous flight of all the molecules in your body to one corner of the room. Perhaps it is microscopic control over equilibrium distributions that enables Superman to fly!

FIGURE 22.3 The partition function for the electronic states of atomic C increases with increasing temperature, reflecting the number of states that are significantly populated at any one temperature.

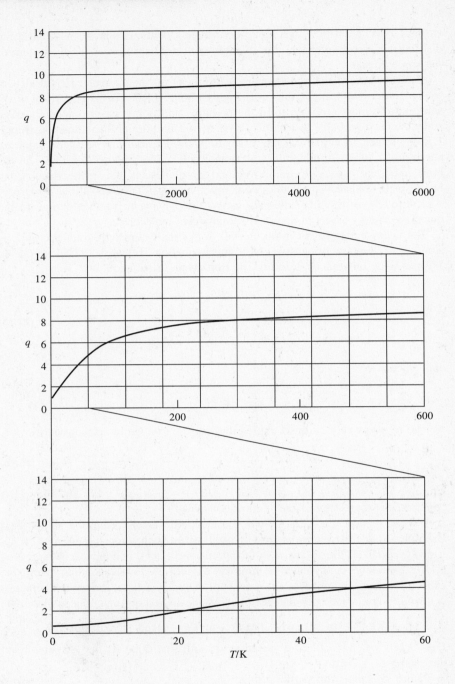

(from roughly 500 K to a few thousand K), $q \cong 9$, the sum of the degeneracies of the lowest three levels. As T approaches 6000 K (and somewhat beyond), q slowly rises. Before it reaches the limit of 14 we predicted above, however, other excited levels would be populated, and q would continue to rise beyond 14 at very high temperatures. Nevertheless, this analysis shows us an important general conclusion. *The molecular partition function is a rough measure of the number of appreciably populated states at any one T.* When the molecular energy level scheme has wide

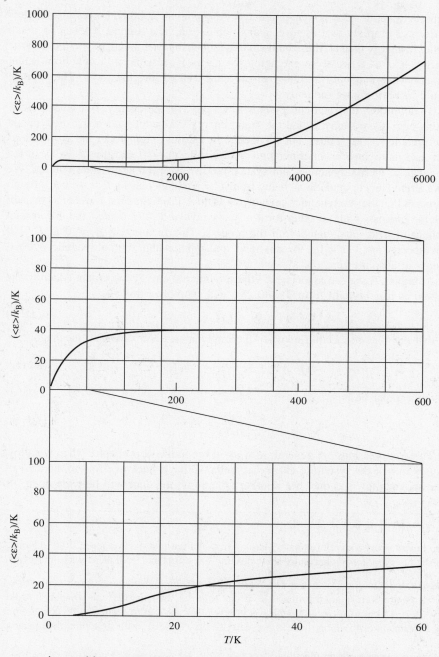

FIGURE 22.4 The average electronic energy of atomic C as a function of T rises in stages, reflecting the number of energy levels that are appreciably populated at any one temperature.

gaps, q is roughly constant over the temperature range in which the lower levels are "saturated" with population but the higher levels are relatively unpopulated.

We can use Eq. (22.16) to graph $\langle \epsilon \rangle$ as well, in Figure 22.4. The average electronic energy per atom stays near zero until the first excited level begins to be populated, at around 6 K.[12] The energy increases smoothly as the lower two excited levels gain population, but from roughly 100 K to about 1800 K, $\langle \epsilon \rangle$ is very constant. The

---

[12]Presumably these atoms are not sitting still at these temperatures! We are neglecting any translational (kinetic energy) contribution to $\langle \epsilon \rangle$. That will come in the next chapter.

highest excited level is not yet populated. Above about 2000 K, however $\langle \epsilon \rangle$ begins to rise very rapidly as this level gains population. The highest level is so much higher in energy than all those below it that it contributes dramatically to the average energy. It is as if we were averaging the ages of everyone in a new-born infant nursery when suddenly a 2000-year-old man walked into the room. The average would increase rather suddenly.

By definition, the total energy is N times the average energy. But what exactly is this quantity we have called the "total energy"? Does it have a thermodynamic interpretation? Yes, it does, but we need to be careful to remember the assumptions that lead us to such an interpretation. At $T = 0$, $\langle \epsilon \rangle = 0$ and thus $E = 0$. In the language of thermodynamics, our system has always been at constant volume. We have altered level populations, rising from $T = 0$ and increasing E as we do, through a process that thermodynamics categorizes as *heat*. From the First Law for a constant volume process (with no other form of work involved), $dU = dq_V$, the differential amount of heat[13] associated with the process. This means that $E = 0$ at $T = 0$ corresponds to $U(T = 0)$, the unknown and unknowable thermodynamic internal energy at absolute zero.

At higher T, we can associate E with the internal energy increment added to the system as it is brought from $T = 0$. We will symbolize this as

$$\Delta_0 U = U(T \neq 0) - U(T = 0) = E = N \langle \epsilon \rangle \tag{22.17}$$

so that our first connection between thermodynamics and statistical mechanics is

$$\Delta_0 U = -\frac{Nk_B}{q} \frac{\partial q}{\partial (1/T)} = Nk_B T^2 \frac{\partial \ln q}{\partial T} \tag{22.18a}$$

or, on a molar basis,

$$\Delta_0 \overline{U} = -\frac{R}{q} \frac{\partial q}{\partial (1/T)} = RT^2 \frac{\partial \ln q}{\partial T} . \tag{22.18b}$$

These expressions are general, as we will see in the next chapter when we allow the volume of the system to change as well. We have stuck one foot in the door of thermodynamics, and once we relate q to entropy, the door will be wide open.

EXAMPLE 22.4

The lowest energy level of atomic I has $g_0 = 4$ and the first excited level, 7 603.15 cm^{-1} higher, has $g_1 = 2$. For what temperature does $\Delta_0 \overline{U} = 4.0$ kJ mol^{-1} for these electronic levels?

SOLUTION One way to answer this question assumes no higher energy levels are involved at the required temperature so that $\Delta_0 \overline{U}$ is due *solely* to the energy of the population of the first excited level: $\Delta_0 \overline{U} = \epsilon_1 N_A$ (fraction in level 1) $= \epsilon_1 N_A (n_1/N)$ where n_1/N comes from the Boltzmann distribution. Instead, we will use Eq. (22.18b) to illustrate the more general method.

We write the partition function:

$$q = 4 + 2 e^{-(7\,603.15\text{ cm}^{-1})(1.439\text{ cm K})/T} = 4 + 2 e^{-(10\,939.2\text{ K})/T} .$$

differentiate it with respect to $1/T$:

$$\frac{\partial q}{\partial (1/T)} = -2(10\,939.2\text{ K}) e^{-(10\,939.2\text{ K})/T} ,$$

---

[13]Pay attention to the context, and the dual use of q to stand for both heat and molecular partition function should not confuse you.

write Eq. (22.18b) as

$$\Delta_0 \overline{U} = -\frac{R}{q}\frac{\partial q}{\partial (1/T)} = 2R\,(10\,939.2\text{ K})\,\frac{e^{-(10\,939.2\,\text{K})/T}}{4 + 2\,e^{-(10\,939.2\,\text{K})/T}} = 4.0\text{ kJ mol}^{-1}\ ,$$

and solve this equation (iteratively, guessing T, is fastest). We find $T = 4585$ K.

Note that this is the same expression as $\epsilon_1 N_A(n_1/N)$ since $\epsilon_1 N_A = R(\epsilon_1/k_B)$ and $(n_1/N) = 2e^{-(10\,939.2\,\text{K})/T}/q$.

⟹ **RELATED PROBLEMS** 22.13, 22.14

Think how thermodynamics would work in Example 22.4. One would use the molar heat capacity $\overline{C}_V$ and integrate both sides of Eq. (2.22):

$$\Delta_0 \overline{U} = \int_0^T \left(\frac{\partial \overline{U}}{\partial T}\right)_V dT = \int_0^T \overline{C}_V\,dT\ .$$

We are only one differentiation away from a statistical expression for the constant volume molar heat capacity, since

$$\overline{C}_V = \left(\frac{\partial \overline{U}(T)}{\partial T}\right)_V = \left(\frac{\partial [\overline{U}(T) - \overline{U}(T=0)]}{\partial T}\right)_V = \left(\frac{\partial \Delta_0 \overline{U}}{\partial T}\right)_V .$$

    ↑   ↑   ↑
  Using the   Since $\overline{U}(T=0)$ Using the
 definition of $\overline{C}_V$ is a constant   definition of $\Delta_0 \overline{U}$

Therefore,

$$\overline{C}_V = N_A \left(\frac{\partial \langle \epsilon \rangle}{\partial T}\right)_V = -R\beta^2 \left(\frac{\partial \langle \epsilon \rangle}{\partial \beta}\right)_V . \quad (22.19)$$

Figure 22.5 graphs the heat capacity contribution of the atomic C electronic levels of Example 22.3 using $\langle \epsilon \rangle$ from Figure 22.4. At low T, where only a few levels are populated, (and these are approaching saturation), the heat capacity curve resembles that for any system with a finite number of levels, as we saw earlier in Figure 4.4 for the two-level system. As these levels saturate at relatively low T, C_V approaches zero and stays there until around 2000 K and above. The rise at $T > 2000$ K is due to the increasing population of very high energy levels.

We relate entropy to the partition function through the Boltzmann hypothesis, $S = k_B \ln W$. The large N, dilute system approximation for $\ln W$ from Eq. (22.9) gives

$$S = k_B \ln W = k_B \sum_i (n_i \ln g_i - n_i \ln n_i + n_i)\ ,$$

which, since $N = \sum_i n_i$, can be written

$$S = k_B \left(N + \sum_i n_i \ln \frac{g_i}{n_i}\right) .$$

We use the Boltzmann distribution to express g_i/n_i:

$$\frac{g_i}{n_i} = \frac{q}{N\,e^{-\epsilon_i/k_B T}}$$

so that

$$S = k_B \left[N + \sum_i n_i \ln \left(\frac{q}{N} e^{\epsilon_i/k_B T} \right) \right]$$

$$= k_B \left(N + \sum_i n_i \ln \frac{q}{N} + \sum_i \frac{n_i \epsilon_i}{k_B T} \right)$$

$$= k_B \left(N + \ln \frac{q}{N} \sum_i n_i + \frac{1}{k_B T} \sum_i n_i \epsilon_i \right)$$

$$= N k_B \left(1 + \ln \frac{q}{N} \right) + \frac{\Delta_0 U}{T}.$$

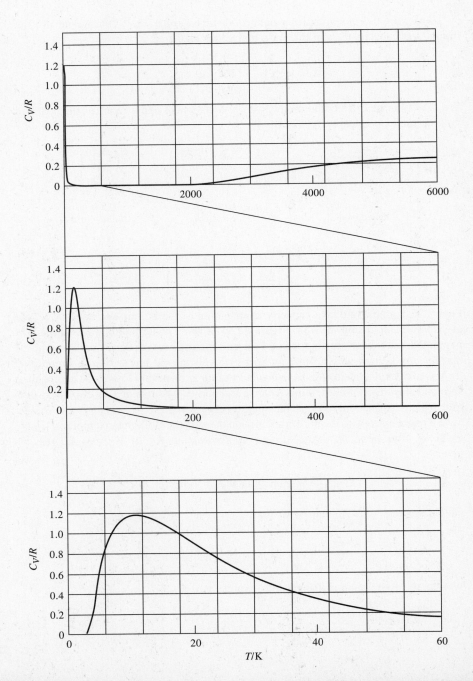

FIGURE 22.5 The electronic contribution to the heat capacity of atomic C passes through maxima and minima as a function of temperature, reflecting the degree of population saturation of energy levels at any one temperature.

We can write the first term as $k_B (N - N \ln N + \ln q^N)$, or, since $\ln N! \cong N \ln N - N$, it is also $k_B \ln (q^N/N!)$ so that

$$S = k_B \ln \frac{q^N}{N!} + \frac{\Delta_0 U}{T} \quad \text{(indistinguishable particles, dilute system)} . \quad (22.20)$$

Notice the ratio $q^N/N!$ in Eq. (22.20). The $N!$ in the denominator is the correction factor that turns the weight for distinguishable particles in the dilute limit into that for indistinguishable particles, as in Equations (22.7a) and (22.7b). Had we written $\ln W$ in Eq. (22.9) for distinguishable particles (using W from Eq. (22.7a)), $N!$ would not have appeared in Eq. (22.20). The entropy for a given distribution of distinguishable particles is larger than that for indistinguishable particles in the same distribution.

This makes sense if we consider an analogous situation. Imagine a bridge hand of 13 cards dealt from, first, an ordinary deck of cards and then from a deck that has no rank markings (A, 2, 3, ..., Q, K) only suit markings (♣, ♦, ♥, ♠). An ordinary deck has every card distinguishable, but the other deck is a mixture of four kinds of otherwise indistinguishable cards. The order in which cards are dealt does not matter in either case, the suits play the role of energy levels, and the dealt hand is a distribution. There are many more ways of dealing, for example, the distribution 5♠, 4♥, 3♦, and 1♣ with an ordinary deck than with the altered deck, and the vitality of the game relies on these distinctions. The statistical weight of any one hand is generally much greater for a hand—a distribution—dealt from an ordinary deck than from the altered deck. The greater the statistical weight, the greater the entropy for any one distribution.

The factor q^N also has a physical interpretation. For simplicity, imagine the molecules have only two levels so that

$$q^N = (g_0 + g_1 e^{-\beta \epsilon})^N$$
$$= g_0^N + N g_0^{N-1} g_1 e^{-\beta \epsilon} + \frac{N(N-1)}{2!} g_0^{N-2} g_1^2 e^{-2\beta \epsilon} + \cdots + g_1^N e^{-N\beta \epsilon}$$

from the binomial expansion rule.[14] The energies in the exponentials of each term represent, term by term, the *total system energy* for $n_0 = N$, $n_1 = 0$ (the first term, $\epsilon_0 = 0$), $n_0 = N - 1$, $n_1 = 1$ (the second term, $\epsilon_1 = \epsilon$), $n_0 = N - 2$, $n_1 = 2$ (the third term), ..., $n_0 = 0$, $n_1 = N$ (the last term). The factors in front of each exponential represent the *total degeneracy of the system level with the energy corresponding to that term*. Molecular energies add to the system energy, but molecular-level degeneracies *multiply* to the degeneracy of a system total energy. Moreover, the factors N, $N(N - 1)/2!$, etc., represent *additional* degeneracy factors for distinguishable molecules. For example, if $N = 2$ and the molecules are distinguishable (A and B), the system energy $E = \epsilon$ can be achieved with $n_0 = 1$ and $n_1 = 1$, but either level could hold molecule A. If level 0 can be formed from any of g_0 states and level 1 from g_1, the total number of states that lead to $E = \epsilon$ is $2 g_0 g_1 = N g_0^{N-1} g_1$.

Consequently, we can write q^N in terms of *energy states of the entire system*. This is called (for reasons explained in the next section) the *canonical partition function*

$$Q = \sum_j e^{-\beta E_j} \quad (22.21)$$

---

[14]The binomial expansion is $(x + y)^N = x^N + N x^{N-1} y + \frac{N(N-1)}{2!} x^{N-2} y^2 + \frac{N(N-1)(N-2)}{3!} x^{N-3} y^3 + \cdots + y^N$.

where E_j is the energy of the jth possible energy *state* (not *level*) of the system. We could write Q in terms of system levels and their degeneracies, but we will generally mention degeneracies only for molecular energy levels and q. We will find that the canonical partition function gives us the most direct route to all thermodynamic functions.

Equation (22.21) is a general definition of the canonical partition function, but it equals q^N *only if the system is composed of distinguishable particles.* For indistinguishable particles, the relation between q and Q is less simple except for our old friend the dilute system, for which the factor $1/N!$ corrects for indistinguishability:

$$Q = \begin{cases} q^N & \text{distinguishable particles} \\ \dfrac{q^N}{N!} & \text{indistinguishable particles, dilute system} \end{cases} \quad (22.22)$$

We can now write the entropy in a general way, valid for any system, dilute or not, distinguishable or not, as

$$S = k_\text{B} \ln Q + \frac{\Delta_0 U}{T}. \quad (22.23)$$

This expression for entropy and Eq. (22.18) for internal energy are the foundations of statistical thermodynamics. All the might of classical thermodynamics can be brought to bear on them, leading to statistical expressions for enthalpy, pressure (equations of state!), free energy, etc. We will work our way through these fairly quickly.

EXAMPLE 22.5

The absolute entropy of 1 mol of He(g) at 400 K and 1.00×10^{-3} atm is 21.89 R. (See Problem 5A.) What is the canonical partition function for this system? What is the molecular partition function?

SOLUTION If we solve Eq. (22.23) for Q:

$$Q = \exp\left(\frac{S}{k_\text{B}} - \frac{\Delta_0 U}{k_\text{B} T}\right)$$

and recall that $\Delta_0 U$ for 1 mol of an ideal gas is $3RT/2$ (since $\overline{C}_V = 3R/2$), we have

$$Q = \exp\left(\frac{S}{k_\text{B}} - \frac{3RT}{2k_\text{B} T}\right)$$
$$= \exp\left(\frac{21.89\,R}{k_\text{B}} - \frac{3R}{2k_\text{B}}\right)$$
$$= \exp(20.39\,N_\text{A}) \cong 10^{5.3 \times 10^{24}}$$

a gigantic number, but one known very precisely. It represents a measure of the number of quantum states available to the entire system.

Since an ideal gas is a dilute system, $Q = q^N/N!$ and $N = N_\text{A}$ so that $q = N_\text{A}!^{1/N_\text{A}} Q^{1/N_\text{A}}$; $Q^{1/N_\text{A}} = e^{20.39}$, and $\ln(N_\text{A}!^{1/N_\text{A}}) \cong N_\text{A}^{-1}(N_\text{A} \ln N_\text{A} - N_\text{A}) = \ln N_\text{A} - 1 = 53.76$. Thus, $q = e^{53.76} e^{20.36} = e^{74.12} = 1.55 \times 10^{32}$. This is a measure of the number of states available to any one molecule.

➤ RELATED PROBLEMS 22.16, 22.17

The first auxiliary thermodynamic state function we consider is the Helmholtz free energy, $A = U - TS$ (Eq. (5.5)). Using Eq. (22.23) for S and defining $\Delta_0 A = A(T) - A(T = 0)$, we have

$$\Delta_0 A = \Delta_0 U - TS = \Delta_0 U - T\left(k_B \ln Q + \frac{\Delta_0 U}{T}\right) = -k_B T \ln Q \ . \quad (22.24)$$

From here, the expression (see Section 5.5) $P = -(\partial A/\partial V)_T$ gives us the pressure:

$$P = -\left(\frac{\partial A}{\partial V}\right)_T = k_B T \left(\frac{\partial \ln Q}{\partial V}\right)_T \ . \quad (22.25)$$

We will see in the next chapter how Q depends on volume. The enthalpy, $H = U + PV$, is simply

$$\Delta_0 H = \Delta_0 U + PV = k_B T^2 \left(\frac{\partial \ln Q}{\partial T}\right)_V + k_B T \left(\frac{\partial \ln Q}{\partial V}\right)_T V$$

$$= k_B T \left[T\left(\frac{\partial \ln Q}{\partial T}\right)_V + V\left(\frac{\partial \ln Q}{\partial V}\right)_T \right] \quad (22.26)$$

where we have expressed $\Delta_0 U$ in terms of Q rather than N and q:

$$\Delta_0 U = k_B T^2 \left(\frac{\partial \ln Q}{\partial T}\right)_V \ . \quad (22.27)$$

The Gibbs free energy, $G = H - TS = A + PV$, and the partial molar Gibbs free energy, the chemical potential μ, are central to chemical thermodynamics. With Eq. (22.24) for A and Eq. (22.25) for P,

$$\Delta_0 G = \Delta_0 A + PV = -k_B T \left[\ln Q - V\left(\frac{\partial \ln Q}{\partial V}\right)_T \right] \ . \quad (22.28)$$

Finally, we have several ways to find the chemical potential, since μ is related to a derivative of any of U, H, A, G, or S (Eq. (5.23)). For a one-component system, the easiest route to μ is

$$\Delta_0 \mu = \left(\frac{\partial \Delta_0 A}{\partial n}\right)_{V,T} = \left(\frac{\partial N}{\partial n}\right)\left(\frac{\partial \Delta_0 A}{\partial N}\right)_{V,T} = -RT\left(\frac{\partial \ln Q}{\partial N}\right)_{V,T} \quad (22.29)$$

where the last equality follows from $N = nN_A$ for n moles of particles and $R = k_B N_A$. If $Q = q^N/N!$, as it does for a dilute gas,

$$\frac{\partial \ln Q}{\partial N} = \frac{\partial \ln q^N/N!}{\partial N} = \frac{\partial}{\partial N}(N \ln q - N \ln N + N) = \ln \frac{q}{N} + \frac{N}{q}\frac{\partial q}{\partial N} = \ln \frac{q}{N}$$

(since q, a *single molecule* quantity, is independent of N) and

$$\Delta_0 \mu = -RT \ln \frac{q}{N} \quad \text{(indistinguishable particles, dilute system)} \ . \quad (22.30)$$

EXAMPLE 22.6

The volume of an ideal gas sample is halved at constant temperature. How does this process change the molecular partition function for the gas?

SOLUTION Thermodynamics says (Eq. (5.36)) $\mu = \mu° + RT \ln P$, and statistical mechanics says (Eq. (22.30)) $\Delta_0 \mu = -RT \ln(q/N)$. If T (and N) are constant, then changes in μ will be due to changes in ln P in the first expression and to $-\ln q = \ln (1/q)$ in the second. Consequently, q is inversely proportional to P for an ideal gas. Doubling P halves q. Thermodynamics can reveal many proportionalities between macroscopic variables and partition functions, and it is important to verify these proportionalities whenever an expression for a partition function is derived, as we do in the next chapter.

⇒ RELATED PROBLEMS 22.20, 22.21

We now have expressions that relate microscopic energies and degeneracies to macroscopic thermodynamic functions. Once we derive expressions for partition functions in various physical model systems we will be able to relate microscopic quantum mechanics to macroscopic phenomena: chemical equilibria, phase transitions, mixing phenomena, etc. Statistical mechanics does not end here, however. There is more insight into the nature of equilibrium states to be learned, and the next section discusses some of the more important implications of the statistical point of view.

22.3 THE ROLE OF INTERMOLECULAR FORCES IN MAINTAINING EQUILIBRIUM

The last two sections have discussed and derived the working formulas of statistical mechanics. The next chapter feeds these formulas various energy-level expressions and puts them to work in important chemical settings. This section goes deeper into the foundations of statistical mechanics and the curious things it has to say about equilibrium.

We stressed that statistical arguments are required for macroscopic systems because there are too many dynamical variables to follow in time, but suppose we could follow them. We could average the motion of the particles over a sufficiently long time and calculate the average system energy, internal state distribution, and so on. For example, suppose we had a system with only one particle: a mass oscillating at constant frequency, confined to move along a line. We could record the motion of this mass with, say, a movie or video camera, capturing its position every Δt seconds, frame after frame. The movie might look like Figure 22.6. We

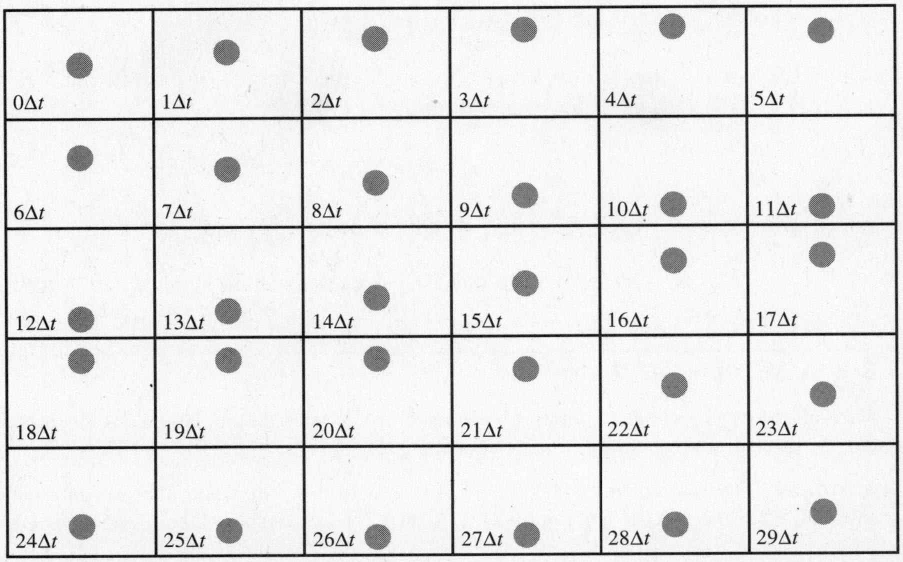

FIGURE 22.6 If a harmonic oscillator is filmed, successive frames in the movie plot the oscillator's *trajectory*. Since frames are equally spaced in time, we could use the movie to find the position and momentum of the oscillator as a function of time.

could measure the position of the particle in each frame, and since we know the motion from frame to frame in sequence, we could measure the velocity as, for example, $\Delta z/\Delta t$. From velocity and mass, we get momentum and kinetic energy. Analysis of the acceleration of the particle gives us forces, potential energy, and ultimately the total energy. Extension of this idea to a 3-D movie of N particles with Δt very small would give us a complete classical picture of the system from which we could deduce averages and thermodynamic behavior.

If we graphed the position and momentum of the oscillating mass from Figure 22.6 from instant to instant in a (z, p_z) coordinate system, we would have a view of the system in what is called *phase space*. Figure 22.7 shows this for simple harmonic oscillation. The system rides round and round an ellipse, between classical turning points along the z axis and between minimum and maximum momenta along the p_z axis. There is a general theorem in classical mechanics that is trivially satisfied in this system. Called the *Poincaré recursion theorem*, it states that *any* classical system enclosed in a finite region of (real) space will start its motion at some point in phase space and eventually return arbitrarily close to that starting point. The time it takes to do so is called the *recurrence time*. Our oscillator exactly passes through the same points on the ellipse each oscillation, and the recurrence time is the oscillation period.

Now imagine that we cut apart the frames in the movie of Figure 22.6, dump them in a box, mix them, draw them at random, and place them side by side. We would have no more than a collage of snapshots, Figure 22.8. Suppose we had a room full of identical oscillators, all swinging at random, and we took one snapshot of each at random times. It seems reasonable to expect a collage of these snapshots, one per system but taken from a collection of identical systems, would contain the same information as our scrambled movie in Figure 22.8.

This very clever idea—replacing the time average of one system by an instantaneous average over many identical systems—is at the heart of statistical mechanics as formulated by J. Willard Gibbs. While Boltzmann was busy in Europe devising the combinatorial statistical mechanics more or less as we have used it, Gibbs was at Yale University devising a broader theory founded more in thermodynamics than mechanics. Boltzmann had many detractors, some who we are about to meet and many who did not believe in the existence of atoms. Gibbs, however, chose to publish his work in a rather obscure place, the *Transactions of the Connecticut Academy of Science*, in the years 1876 and 1878. While he did circulate copies of these papers among many prominent scientists of the time, they were largely unknown until Ostwald, a strong opponent of both Boltzmann and the reality of atoms,

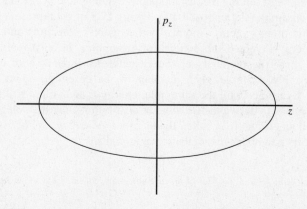

FIGURE 22.7 A graph of the momentum versus the position of the harmonic oscillator in Figure 22.6 is an ellipse. This type of representation of the system's trajectory is called the *phase-space* representation.

FIGURE 22.8 In contrast to the movie of Figure 22.6, we could take a large number of identical harmonic oscillators, give them all the same energy, and take *one picture of each oscillator at random times*. A collage of these pictures of this *ensemble* of oscillators is hypothesized to have the same dynamic information as the movie of Figure 22.6.

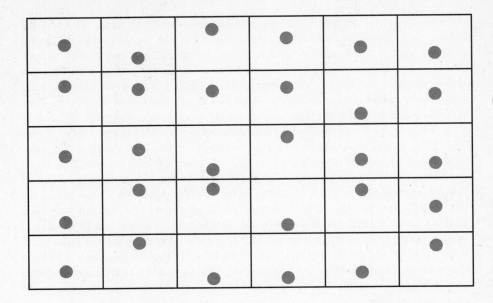

published a German translation in 1902 under the remarkably bland, understated title *Thermodynamische Studien,* "Thermodynamic Studies." In the same year, Yale University Press published Gibbs' *Elementary Principles in Statistical Mechanics.* Gibbs died the following year, and Boltzmann, who wrote, "I am conscious of being only an individual struggling weakly against the stream of time," died by his own hand in 1906. It is remarkable that statistical mechanics managed to survive such a difficult start.

Gibbs introduced the idea of an *ensemble*—a set of identical replicas of a system—into physics.[15] He postulated that the average of the instantaneous properties of all the (very many) members of the ensemble would be equivalent to the time averaged properties of any one member. The replicas of the system that together make the ensemble have *uncorrelated histories.* They are not related to each other through any dynamical laws (and quantum mechanics later imposed that the *phases* of the systems' wavefunctions must also be randomized). Created at random times in the distant past and allowed to evolve to equilibrium states before we inspect them, the members of an ensemble are roughly equivalent to our snapshots in Figure 22.8. A phase-space picture of an ensemble is a collection of points, one per member of the ensemble, representing the instantaneous dynamic and positional state of each system in the ensemble. For our oscillator, Figure 22.9 might represent the ensemble view (but many more systems—points on the figure—constitute an ensemble).

Gibbs not only invented the concept of ensembles, he also realized that it is beneficial to allow the members of the ensemble to interact with each other in different ways. For example, each member of the ensemble might be a system composed of N particles with the same total energy E and volume V. These are the constraints we used to derive the Boltzmann distribution and the molecular partition function. Such an ensemble is called the *microcanonical ensemble.* Each system is isolated from its neighbors by adiabatic, rigid walls.

---

[15]Einstein also came up with the idea of an ensemble, independently, but somewhat later.

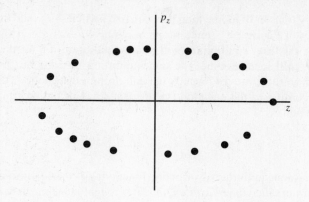

FIGURE 22.9 A phase space picture of the ensemble of oscillators photographed in Figure 22.8 is a series of points representing the instantaneous position and momentum of one oscillator in the ensemble at the time each picture was taken.

The *canonical ensemble* has N and V fixed, but the walls between members allow energy transfer as heat. Each member is at the same temperature (rather than at the same total energy). All the other members of the ensemble act as a heat bath to any one member, since they are all at equilibrium. Analysis of this ensemble leads to the canonical partition function as we have described it.

If the walls allow particle as well as energy transfer, we have the *grand canonical ensemble*. It is characterized by constant V, T, and *chemical potential* μ for each member of the ensemble. There are other types of ensembles that can be devised for various other constraints, such as constant P, T, and μ, but the key idea behind them all is Gibbs' analysis of statistical averages over all members of the ensemble.[16] The results we have derived in the previous two sections have tacitly relied on microcanonical and canonical ensemble concepts, but we will not go further into this idea here. Several of the references at the end of this chapter can be consulted for more information.

Gibbs was less concerned than Boltzmann with the question of equilibrium attainment and stability, at least on the microscopic level. Boltzmann, however, stressed the critical role *molecular collisions* play in establishing and maintaining equilibrium. Here is the key (and to this day, still somewhat vexing) question. If we take one system and, through Boltzmann's eye, follow in time the trajectory it takes through phase space and then contrast this line to the series of points in phase space that Gibbs' ensemble snapshots show, *would Boltzmann's trajectory run through all of Gibbs' points?*

Consider again the curious state of a gas shown in Figure 3.1: noninteracting molecules that move in three equal groups back and forth along straight lines perpendicular to the walls of their cubical container. Experience would indicate that Boltzmann's trajectory would never run through the phase space points represented by this state of a gas, since no one has ever observed such a state. But might it possibly be one of the members of Gibbs' microcanonical ensemble? It could certainly be constructed to have the same E, N, and V as an "ordinary" equilibrium sample of a gas.

To Boltzmann's mind, the *lack of interactions* among the particles of this system is so unphysical that Figure 3.1 can be dismissed. Collisions—which mean, at the

---

[16]In case you are wondering what is so "grand" about the "grand canonical ensemble," rest assured that nothing really is. The term comes from Gibbs who also called our canonical ensemble the "petit canonical ensemble."

very least, any transfer of momentum between two particles—would quickly destroy the simplicity of Figure 3.1's microscopic motion. We will see in Chapter 25 that a system not far from equilibrium will attain equilibrium on a time scale that is governed by collisional events. The nonequilibrium distribution function, which must depend on time, moves towards the equilibrium distribution function, which does not, according to the equation (valid for a small deviation from the equilibrium distribution)

$$\frac{df^*}{dt} = -\frac{f^* - f}{\tau} \qquad (22.31)$$

where f^* is the nonequilibrium distribution function, f is the equilibrium distribution function, and τ is called the *relaxation time*. It is roughly the time between collisions. This equation can be rearranged and integrated to give

$$f_t^* - f = (f_{t=0}^* - f)e^{-t/\tau}$$

or

$$f_t^* = f + \Delta f e^{-t/\tau} \qquad (22.32)$$

where Δf is the distribution function's initial perturbation away from equilibrium. If the perturbation involves only gas velocities, as in Figure 3.1, then τ is very short (on the order of ns at ordinary pressures) and the perturbation dies very quickly.

Buried in the nonequilibrium distribution function is the very complicated description of molecular collisions and the various momentum and energy transfers they cause. Boltzmann showed that a particular function of the nonequilibrium distribution function—basically an integral over all position and momentum coordinates of the quantity $f \ln f$—must *always decrease in time towards the equilibrium value*. Boltzmann called this function H, and his result is known as *Boltzmann's H theorem*.[17]

This remarkable theorem is unexpected for several reasons. First, it seems to give us a way to take microscopic mechanics and give them a *direction in time*, just as entropy does. This is unexpected because classical mechanics (Newton's equations of motion)—and quantum mechanics, for that matter—are *symmetrical* in time. If the movie in Figure 22.6 is run backwards, we would see a perfectly allowable sequence of states of the system. If a movie is made of 10^{23} molecules of gas initially confined to one corner of a box and allowed to move to uniform equilibrium throughout the box and if the movie is played backwards, the gas would seem to spontaneously rush into the corner *from an equilibrium state*. Experience says this does not happen, of course.

It is also unexpected from a phase-space point of view. The Poincaré recursion theorem states that the system *must* return arbitrarily close to any initial state, and this would also seem to contradict the H theorem. The perturbation would seemingly have to grow back again (H would increase) at some later time. Boltzmann was attacked along both these lines (although his attackers, primarily Loschmidt and Zermelo, generally raised objections to the *result* of the theorem without suggesting where its derivation might be in error), and fully satisfactory interpretations of his theorem were not advanced in his lifetime. Boltzmann did, however, correctly point out the answer to the Poincaré paradox. Poincaré says a time, the recurrence time,

---

[17]When Boltzmann wrote H, he had in mind the upper case Greek letter eta, H, in analogy with the lower case η that was commonly used to represent entropy. Thus, his theorem should really be called the "capital eta theorem," but the distinction between the Roman H and the Greek H has long been lost.

must pass before the system's phase-space trajectory returns[18] arbitrarily close to any previously visited point, but there is no simple statement that can be made about the *duration* of this recurrence time. Boltzmann showed that any molecular system composed of a macroscopic number of particles had a recurrence time that was not only longer than the age of the Universe, but *considerably* so, perhaps $10^{10^{20}}$ years!

Consequently, the statistical averages over an ensemble or over time on one system are stable and meaningful thermodynamic representations of systems of large particles. This equivalence is known as the *ergodic hypothesis*, and a large body of scientific literature is devoted to the conditions under which this hypothesis can be expected to hold.[19] The equilibrium average state is overwhelmingly most probable, as we have seen, but if we can compute averages, we can also compute other *moments of the distribution* such as deviations from the average. These deviations represent *fluctuations*, and fluctuations are not only observable but also necessary to maintain equilibrium. (The correct interpretation of Boltzmann's *H* theorem, for example, shows that *H* decreases in time in an average sense, but that it also fluctuates slightly about its equilibrium average. The fluctuations are very unlikely to grow, however, and the equilibrium state is said to be stable.)

Consider the canonical ensemble in which V, N, and T are constant in all members of the ensemble. Any one member might have an energy, pressure, entropy, etc., different from its neighbors, and ensemble theory allows us to find the average fluctuations in these quantities. For example, the mean squared fluctuation in the energy is

$$\sigma_E^2 = \overline{(E - \overline{E})^2} = \overline{E^2} - \overline{E}^2$$

where the bars over symbols represent ensemble averages. The average energy itself, $\overline{E}$, is

$$\overline{E} = \sum_j E_j P_j = \sum_j \frac{E_j e^{-E_j/k_B T}}{Q} = k_B T^2 \left(\frac{\partial \ln Q}{\partial T}\right)_{N,V},$$

which is our $\Delta_0 U$ of Eq. (22.27). Here, P_j is the probability that member j of the ensemble has a system energy E_j. This probability is the Boltzmann distribution factor $e^{-E_j/k_B T}/Q$. Similarly, the average of the square of the energy is

$$\overline{E^2} = \sum_j E_j^2 P_j = \sum_j \frac{E_j^2 e^{-E_j/k_B T}}{Q}$$

and the mean squared fluctuation becomes (see Problem 22.23)

$$\sigma_E^2 = k_B T^2 \left(\frac{\partial \overline{E}}{\partial T}\right)_{N,V} = k_B T^2 C_V \tag{22.33}$$

where C_V is the constant volume heat capacity.

In comparison to the average energy itself, this fluctuation is very small. Consider an ideal gas, for example. The average molar energy is $3RT/2$, and the root mean

---

[18] The uniqueness of the solutions to dynamical equations of motion ensure that a phase space trajectory will not cross itself.

[19] A technical point: the strict ergodic hypothesis would allow Boltzmann's trajectory to pass through a phase-space point representing Fig. 3.1, and it is known that this will not happen starting from an equilibrium state. A modification of the hypothesis, known as the *quasi-ergodic hypothesis*, is a more reasonable statement of the connection between time and ensemble averages, but the distinction between the two hypotheses has very little practical implications, although it has influenced many theories of irreversible processes.

squared fluctuation, σ_E, is $(k_B \overline{C}_V)^{1/2} T = (3R^2/2N_A)^{1/2} T = (3/2N_A)^{1/2} RT$. The ratio $\sigma_E/\overline{E}$ is a measure of the relative magnitude of the average fluctuation: $\sigma_E/\overline{E} = (2/3N_A)^{1/2} \cong N_A^{-1/2}$. This small number, $N_A^{-1/2} = 1.3 \times 10^{-12}$, shows that energy fluctuations are very small on average (and fluctuations much larger than average are *very* improbable—see Problem 22.24) so that the system energy at constant temperature appears to be a stable, invariant quantity.

CHAPTER 22 SUMMARY

This chapter has shown how thermodynamic equilibrium is the consequence of large numbers of molecules interacting among themselves, distributing energy among their quantum states in a way that leads to only one observable outcome. Equilibrium is the overwhelmingly most probable of all possible distributions. The central quantity that connects microscopic to macroscopic is the partition function. Much as a quantum mechanical wavefunction cleverly codes all that can be known about a quantum system, the partition function (in one or more of its varieties) codes the connection between microscopic energies and macroscopic states.

The key expressions of this chapter begin with the combinatorial expressions for the weights of distributions in various types of systems:

distinguishable particles, nondegenerate levels
$$W = N! \prod_i \frac{1}{n_i!}$$

distinguishable particles, degenerate levels
$$W = N! \prod_i \frac{g_i^{n_i}}{n_i!}$$

indistinguishable particles, dilute system, degenerate levels
$$W = \prod_i \frac{g_i^{n_i}}{n_i!}$$

The set of n_i's that maximizes W subject to the constraints of fixed E, V, and N is the most probable distribution:

Boltzmann distribution function
$$n_i = N \frac{g_i e^{-\epsilon_i/k_B T}}{q}$$

and it depends on all the molecular energy levels through

the molecular partition function
$$q = \sum_i g_i e^{-\epsilon_i/k_B T}$$

The system-based, rather than molecule-based, partition function is called

the canonical partition function
$$Q = \sum_j e^{-E_j/k_B T} = \begin{cases} q^N & \text{distinguishable particles} \\ q^N & \text{indistinguishable particles,} \\ \frac{q^N}{N!} & \text{dilute system} \end{cases}$$

where the sum is over states j of the entire system.

Statistical thermodynamics expressions are based on

the internal energy $\quad \Delta_0 U = k_B T^2 \left(\frac{\partial \ln Q}{\partial T} \right)_V$

and

the absolute entropy $\quad S = k_B \ln Q + \frac{\Delta_0 U}{T}$

From these basic expressions, thermodynamic relations can be used to find $\Delta_0 A$, P, $\Delta_0 H$, $\Delta_0 G$, and $\Delta_0 \mu$. Equations (22.24)–(22.29), in terms of Q and derivatives of it.

FURTHER READING

Several texts and monographs discuss the combinatorial mathematics of distribution functions in some detail. The very readable *Elements of Statistical Thermodynamics,* 2nd ed., by L. K. Nash (Addison-Wesley, Reading, MA, 1974) is a good place to start, as is *Elementary Statistical Mechanics,* by N. O. Smith (Plenum, New York, 1982). The text *Molecular Thermodynamics,* by R. E. Dickerson (Benjamin, New York, 1969) is written at a level comparable to this chapter.

Several advanced texts can be consulted for information that we have not covered here. *Statistical Mechanics,* by D. A. McQuarrie (Harper & Row, New York, 1976) would be a good place to turn after reading this chapter, as would *An Introduction to Statistical Thermodynamics,* by T. L. Hill (Dover, New York, 1986). *Introduction to Statistical Mechanics,* by D. Chandler (Oxford University Press, Oxford, 1987); *A Course in Statistical Mechanics,* by H. L. Friedman (Prentice-Hall, Englewood Cliffs, NJ, 1985); and *Statistical Mechanics and Kinetic Theory,* by C. Hecht (W. H. Freeman, New York, 1990) cover several contemporary areas of statistical mechanics in depth. A classic treatise that goes far beyond this text is *The Principles of Statistical Mechanics,* by R. C. Tolman (Dover, New York, 1979).

PRACTICE WITH EQUATIONS

22A If a two-level system has $\epsilon = 1.8 \times 10^{-20}$ J, at what temperature will the ground-state population be 10 times the excited-state population? (22.1)

ANSWER: 566 K

22B How many ways can ten people be seated at two tables, one that seats six and one that seats four, ignoring arrangement differences around either table? (22.5)

ANSWER: 210 different ways

22C How many five-card poker hands can be dealt from a deck of 52 cards? (22.5)

ANSWER: 2 598 960

22D What is the smallest integer N such that $N! > 10^{1000}$? (22.8)

ANSWER: 450

22E A doubly degenerate level is 4.6×10^{-21} J above a triply degenerate level in a system at 300 K. What is the ratio of populations between the two levels? (22.14c)

ANSWER: 0.22

22F If a lower level is nondegenerate and 10 cm^{-1} below a triply degenerate level, at what temperature will the two levels have equal populations? (22.14c)

ANSWER: 13.1 K

22G The molecular partition function for 1 mol of a gaseous system is $q = (1.5 \times 10^{27}\ \text{K}^{-3/2})T^{3/2}$. What is $\langle\epsilon\rangle$ at 280 K? (22.16)

ANSWER: $3k_BT/2 = 5.8 \times 10^{-21}$ J

22H What is the molar internal energy of the system in 22G above? (22.18b)

ANSWER: $3RT/2 = 3.5$ kJ mol^{-1}

22I What is the molar entropy of the system in 22G? (22.20)

ANSWER: 18.8 R

22J What is the canonical partition function for the system in 22G? (22.23), 22H, 22I

ANSWER: $\ln Q = 17.3\ N_A$

22K What is $\Delta_0 A$ for the system in 22G? (22.24), 22J

ANSWER: $-17.3\ RT$ (1 mol) = -40.3 kJ

22L What is $\Delta_0 \mu$ for the system in 22G? (22.30)

ANSWER: -37.9 kJ mol^{-1}

22M The root mean squared fluctuation of the total energy for the system in 22G is what fraction of the internal energy? (22.33), 22H

ANSWER: 1.05×10^{-12}

PROBLEMS

SECTION 22.1

22.1 An atom that is in an excited electronic state in the midst of hordes of ground-state atoms has the chemical properties of a totally different species. Can one such atom be found in a macroscopic, equilibrium sample? Consider Na(g). The first excited state (3p ^{2}P$^o_{1/2}$) is 16 956.183 cm^{-1} above the ground state. Imagine a 1-mm diameter laser that passes through a 1-cm length cell of Na(g). The laser could be tuned to a transition from this excited state to a higher one, and laser induced fluorescence from the higher state would signal population of the first excited state. Suggest experimental conditions that would ensure about 1 excited-state atom in the volume interrogated by the laser. Treat the ground and first excited states as a two-level system, and use the following Na vapor pressure data: $P = 10$ torr at 546 K, $P = 100$ torr at 700 K. Recall that the ideal gas equation of state can be written as $PV = Nk_BT$.

22.2 Find all the distributions of four distinguishable particles with a total energy 4ϵ for the energy-level scheme of Figure 22.1. What is the weight of each distribution?

22.3 Suppose the levels of Figure 22.1 are each doubly degenerate. Find all the distributions and their weights such that $E = 6\epsilon$ and $N = 3$, as in the text. Does this degeneracy change the relative probabilities of different distributions from the nondegenerate scheme of Figure 22.1?

22.4 For the two-level system, the weight of a given distribution can be written uniquely in terms of N and $n = n_1$ since $n_0 = N - n_1 = N - n$. Show that for a three-level system, the weight of a distribution can be written in terms of N, the total energy E, the level energies (ϵ_0, ϵ_1, and ϵ_2), and the population n of any one level.

22.5 Depending on the dynamic range of your calculator, there will be some maximum integer N for which $N!$ will not overflow your calculator. Find this N for your calculator. If it has a factorial function built in, see if you can figure out the algorithm it uses to compute factorials. Use a value that will not overflow, such as 30!, which is exactly 265 252 859 812 191 058 636 308 480 000 000. Does your calculator use either the abbreviated Stirling's approximation $N! = N^N e^{-N}$ or the full approximation $N! = \sqrt{2\pi N}\, N^N e^{-N}$?

22.6 For small N, Stirling's approximation is not very accurate, as the previous problem may have demonstrated. An improved approximation was derived by R. K. Pathria [*Am. J. Phys.* **53**, 81 (1985)]:

$$N! \cong \sqrt{2\pi}\,(N^*)^{N^*} e^{-N^*} e^{-1/24N^*},\ N^* = N + 1/2\ .$$

Compare this approximation for $N = 30$ to the Stirling approximation $N! = \sqrt{2\pi N}\, N^N e^{-N}$ and the exact value in the previous problem. For what systems might one want to consider the Pathria approximation over Stirling's?

22.7 Expressions for the FD and BE weights, W_{FD} and W_{BE}, are given in the text. Use Stirling's approximation along the lines that lead to Eq. (22.10) to show that

$$d \ln W = \sum_i \left(\ln \frac{g_i + cn_i}{n_i}\right) dn_i$$

where $c = -1$ for W_{FD} and $c = +1$ for W_{BE}. Note that setting $c = 0$ leads to the MB expression, Eq. (22.10).

22.8 Incorporate undetermined multipliers in the expression for $d \ln W$ in the previous problem and show that FD and BE distributions can be written

$$n_i = \frac{g_i}{e^\alpha e^{\beta \epsilon_i} \pm 1}$$

with the + sign for FD and the − sign for BE statistics.

22.9 Repeat Example 22.3 for O atom, using data in Table 13.4. Note that the degeneracy of a level with atomic quantum number J is $2J + 1$.

22.10 The energies and degeneracies of the lowest two levels of atomic iodine are $\epsilon_0 = 0$ cm^{-1}, $g_0 = 4$, and $\epsilon_1 = 7\,603.15$ cm^{-1}, $g_1 = 2$. Higher levels are sufficiently above these that they can be ignored in this problem. A high-temperature experiment finds 1% of the I atoms are in the excited energy level. What was the temperature of the experiment?

22.11 Many quantum mechanical systems have energy-level expressions for which the lowest allowed energy, the zero-point energy, is not zero. The derivation of the Maxwell–Boltzmann distribution in the text assumed the zero-point energy was zero. Does a zero-point energy alter this distribution? Imagine each energy level in a system is shifted by a zero-point energy ϵ' that is simply added to each ϵ_i. Show that this change does not alter the distribution and, as a consequence, the zero-point energy can never be accessed nor can it influence observable quantities.

SECTION 22.2

22.12 Show that $W/W' \cong 7$ as mentioned in the text for a system of 1000 C atoms with W the equilibrium distribution and W' the altered distribution in which 10 atoms are moved from level 0 to level 1 and six are moved from 2 to 1. Use Eq. (22.9), and also use the approximation $\ln(n + a) \cong \ln(n) + a/n$ for $a \ll n$.

22.13 Repeat Example 22.4 for the other halogens. Each has the same degeneracy pattern as I, but the excited level lies at 404 cm^{-1}, 881 cm^{-1}, and 3685 cm^{-1} for F, Cl, and Br, respectively.

22.14 The third excited level of iodine has $g_2 = 6$, $\epsilon_2 = 54\,633.46$ cm^{-1}. For what range of temperatures will this level contribute less than 1% to q? Was Example 22.4 safe in ignoring it? (See also Problem 22.10.)

22.15 Show that Eq. (22.19) can be written

$$\overline{C}_V = \frac{R}{qT^2}\left[\left(\frac{\partial^2 q}{\partial (1/T)^2}\right)_V - \frac{1}{q}\left(\frac{\partial q}{\partial (1/T)}\right)_V^2\right]\ .$$

Use this expression and the partition function for the nondegenerate two-level system, $q = 1 + e^{-\epsilon/k_B T}$, to verify Eq. (4.14).

22.16 The quantity $\ln Q$ appears in several statistical thermodynamics expressions. Show that it can be written $N \ln (qe/N)$ for a dilute, indistinguishable particle system. Use the value for q in Example 22.5 to find $\Delta_0 A$ for that system using this form of $\ln Q$, and compare your result to $\Delta_0 A = -k_B T \ln Q$ using Q from that Example. (*Hint:* Write N_A as $e^{\ln N_A}$.)

22.17 Find the electronic contribution to the molar entropy of the C atoms in Example 22.3 at 6000 K. The translational contribution is 24.3 R at 1 atm. Which term dominates?

22.18 The notation $\Delta_0 A = A(T) - A(T = 0)$, for example, is mostly a formality, since any measurable change in a thermodynamic quantity such as A is written $\Delta A = \Delta_0 A_f - \Delta_0 A_i = [A_f - A(T = 0)] - [A_i - A(T = 0)] = A_f - A_i$ as usual for a process from initial state i to final state f. Show that $\Delta_0 A$ is itself indeterminant since it depends on an energy zero choice for the entire system. Write Eq. (22.21) as $Q = e^{-E_0/k_B T} + e^{-E_1/k_B T} + \cdots$ where E_0 and E_1 are the energies of the lowest and first excited states of the system with $E_1 > E_0$, and show that as $T \to 0$, $\Delta_0 A \to E_0$, which is the arbitrary energy zero of the system.

22.19 Show that the most probable distribution's weight, W, for a system of indistinguishable particles in a dilute system can be written $W = q^N e^{\Delta_0 U/k_B T}/N!$

22.20 Show that a molecular partition function of the form $q = \alpha T^n$, $n > 0$ and α constant, has an internal energy that is a linear function of T and thus a temperature-independent constant-volume heat capacity. Can n be < 0 for any partition function?

22.21 The heat capacity contribution from the conduction electrons of a metal is a linear function of T (see Eq. (4.17)), and consequently, their internal energy contribution depends on T^2. How does q depend on T for this system? (*Hint:* Integrate Eq. (22.18b).)

SECTION 22.3

22.22 What is the phase space trajectory for a particle of mass m bouncing freely between walls at $\pm x$ with velocity $\pm v$?

22.23 Complete the derivation of Eq. (22.33) along these lines: start with the definitions

$$\bar{E} = \sum_j E_j P_j = \frac{1}{Q} \sum_j E_j e^{-\beta E_j} = -\frac{1}{Q}\frac{\partial Q}{\partial \beta}$$

$$\overline{E^2} = \sum_j E_j^2 P_j = \frac{1}{Q} \sum_j E_j^2 e^{-\beta E_j}$$

where $\beta = 1/k_B T$. Then note that

$$\frac{1}{Q}\sum_j E_j^2 e^{-\beta E_j} = -\frac{1}{Q}\frac{\partial}{\partial \beta}\sum_j E_j e^{-\beta E_j} = -\frac{1}{Q}\frac{\partial}{\partial \beta}(\bar{E}Q)$$

so that

$$\overline{E^2} - \bar{E}^2 = -\frac{1}{Q}\frac{\partial}{\partial \beta}(\bar{E}Q) - \left(\frac{1}{Q}\frac{\partial Q}{\partial \beta}\right)^2.$$

Show that this expression reduces to Eq. (22.33).

22.24 The distribution of total energies among members of a canonical ensemble at temperature T follows a Gaussian distribution centered around the most probable, equilibrium energy $\bar{E}$. If $P(E)$ is the probability that a member has energy E,

$$P(E) = \frac{1}{(2\pi k_B T^2 C_V)^{1/2}} \exp\left[-\frac{(E-\bar{E})^2}{2k_B T^2 C_V}\right].$$

What is the probability that 1 mol of an ideal gas will have an energy that is a part per million different from $\bar{E}$? How different from $\bar{E}$ is the energy that has a 1% chance of observation?

22.25 Generally, fluctuations are so small that they are not observed. However, the expression for the relative fluctuation of the density is

$$\frac{\sigma_\rho}{\rho} = \left(\frac{k_B T \kappa}{V}\right)^{1/2}$$

where κ is the isothermal bulk compressibility, Eq. (1.26). Use this expression to explain why the density fluctuations are macroscopic and easily observed in a fluid at its critical point.

GENERAL PROBLEMS

22.26 The artificial ^{22}Na isotope is a convenient laboratory source of positrons, the antiparticles with a charge opposite that of electrons, but otherwise identical to them. As they spontaneously decay, 90% of the time ^{22}Na nuclei emit a positron and leave behind a ^{22}Ne nucleus in an excited state 1.276 MeV higher than the ground state. This excited state emits a high energy photon (a "γ ray") almost instantaneously as its way of relaxing to its ground state. The γ ray is used to note the time of a positron emission event. How hot would the sample have to be so that 10^{-9} of the ^{22}Ne nuclei were *thermally* excited to this γ-ray emitting state? Are false γ-ray emissions due to thermal excitation likely to be a concern to experimenters? (Assume the nuclear levels have equal degeneracies.)

22.27 How many ways are there of dealing a bridge hand with the suit distribution 5♠, 4♥, 3♦, and 1♣? How many ways are there if one uses the unranked deck discussed in the text? The *order* in which the cards are dealt does not matter in either case.

22.28 When a solid melts or a liquid boils, its partition function changes. Show that vaporization increases Q by a factor approximately equal to e^N. Follow these steps. Use Eq. (22.23) to write an expression for $\Delta S_{vap} = S(g) - S(l)$. Substitute $\Delta_0 H - \Delta_0(PV)$ for $\Delta_0 U$, and invoke Eq. (4.22), $\Delta S_{vap} = \Delta H_{vap}/T_{vap}$. Next, use the approximation $\Delta V_{vap} = V(g)$ and the ideal gas equation of state (as was used to integrate the Clapeyron equation leading to Eq. (6.12)) to complete the problem.

22.29 Somewhere along the way to the final answer in the previous problem you should have derived an expression that can be written $Q_1/Q_2 = \exp(P\Delta V_\phi/k_B T)$ for the phase transition $\phi_1 \to \phi_2$. Evaluate this ratio for the ice → water transition given the densities $\rho(\text{ice}) = 0.917$ kg dm^3 and $\rho(\text{water}) = 1.000$ kg dm^3. Assume 1 mol of water is melted and compare your answer to the vaporization expression for 1 mol, $Q_1/Q_2 = e^{N_A}$, from the previous problem. Is the difference in the direction you expected?

22.30 Schrödinger's name is strongly associated with quantum mechanics, but he was a physicist of exceptional breadth. In 1944, he gave a series of lectures on statistical mechanics (reprinted as *Statistical Thermodynamics* (Dover, New York, 1989)) that were as much a series of critiques of the foundations of the theory as they were a course of instruction. In his discussion of fluctuations, he considers the following system:

> If you enclose a fluid with its saturated vapour above it in a cylinder, closed by a piston, loaded with a weight to balance the vapour pressure—the piston gliding frictionlessly within the cylinder—and put it in a heat bath, then you may include the piston and the weight in what you call the system and no "external" work is done, even if the piston moves.

What is this system's heat capacity? What does this say about fluctuations among members of an ensemble of such systems?

CHAPTER 23

The Molecular Basis of Equilibrium

23.1 Partition Functions—How Many States Contribute?

23.2 From Microscopic Individuality to Macroscopic Average

23.3 Statistical Thermochemistry

THE previous chapter showed how statistical mechanics bridges the gap between a quantum picture of individual molecules and a thermodynamic picture of a mole of them at equilibrium. This chapter builds the framework of that bridge, constructing many partition functions based on quantum mechanical energy-level expressions from Chapters 12–20. The previous chapter showed how to do this for simple systems such as open-shell atoms that have only a few important energy levels. The first section here uses other basic energy level schemes—the particle-in-a-box, the harmonic oscillator, the rigid rotor, etc.—to build quickly a repertoire of partition functions.

The second section uses these partition functions as tools to reach thermodynamic expressions in terms of microscopic molecular parameters—molecular masses, force constants, bond lengths, etc. The third section takes us into the heart of chemical thermodynamics from the statistical point of view. Statistical thermodynamic expressions are combined to yield the reaction equilibrium constant. We will see which microscopic parameters control the position of chemical equilibria and which fail to have much influence.

This is a very important chapter. It signals the end of our study of static, equilibrium situations, for the most part. The remainder of the book largely treats *dynamic* processes in chemistry, but it is important to remember, as the last chapter stressed, that even equilibrium is dynamic on a microscopic scale.

Many of the thermodynamic expressions of Chapters 1–10 are revisited here. We will see in this chapter the microscopic basis for the ideal gas heat capacity, Eq. (2.18), the Debye T-cubed law, Eq. (4.16), the power series expansions for molecular gas heat capacities, Eq. (4.19), the Sackur-Tetrode equation, Eq. (5.34), and the Giauque free energy function, Eq. (7.31), for example. It is therefore fair to ask why we did not start this book (or this entire science of physical chemistry, for that matter) with the material of this and the previous chapters. If we can express equilibrium quantities in terms of microscopic parameters, why do we need spectroscopy? Why not use the vast amount of thermodynamic data to deduce bond lengths, bond angles, electronic excited-state energies, etc., from the expressions of statistical thermodynamics? We will see in this chapter one important case where thermodynamic data gave important microscopic, quantum mechanical information—the heat capacity of molecular hydrogen—but for the most part, thermodynamic data have nowhere near the precision of spectroscopic data. The averaging inherent in a "statistical" expression is too course-grained to allow us to sort out the many fine details that come from individual molecular study. Moreover, most thermodynamic measurements are so far removed from an ideal limit—ideal gas, ideal mixture, etc.—that quite sophisticated statistical theories are required to reproduce the nonideal effects of intermolecular interactions. These nonidealities will be largely ignored in this chapter. We will see a statistical expression for the second virial coefficient (in the *next* chapter), but we will not see general expressions for arbitrary types of nonideal behavior. Many of these are elusive and the subject of contemporary research.

23.1 PARTITION FUNCTIONS—HOW MANY STATES CONTRIBUTE?

Our general approach in this section reaches back to the energy-level and degeneracy expressions of Chapters 12–20 for various model systems, couples them to the notion that, to a very good approximation, the total energy of an isolated molecule is the sum of independent parts each associated with one such model, and uses this sum in Eq. (22.13) for the molecular partition function. If the ith level's energy can be written as a sum of independent contributions ϵ_j each indexed by a quantum number n_j:

$$\epsilon_i = \sum_j \epsilon_j(n_j) \qquad (23.1a)$$

with a total degeneracy that is a product of independent degeneracies:

$$g_i = \prod_j g_j(n_j) \qquad (23.1b)$$

then the molecular partition function can be written as a *product* of independent partition functions, each associated with a term in the energy expression:

$$q = \prod_j q_j . \qquad (23.2)$$

For example, consider a gaseous closed-shell diatomic molecule with only a single electronic state of importance at the temperature of interest. Its total energy, to a very good approximation, is the sum of its translational, vibrational, and rotational energies:

$$\epsilon_i = \epsilon_{tr} + \epsilon_{vib} + \epsilon_{rot} .$$

The translational energy is that of a 3-D particle-in-a-box model with independent quantum numbers (n_x, n_y, n_z) (Eq. (12.24)). The vibrational energies follow, again to a good approximation, the harmonic-oscillator model with quantum number v and no degeneracy (Eq. (19.22)), and the rotational energies follow a rigid-rotor model with quantum number J and degeneracies $2J + 1$ (Eq. (19.21)). Consequently, the partition function is the product

$$q = q_{tr}\, q_{vib}\, q_{rot} .$$

EXAMPLE 23.1

Show explicitly that the translational molecular partition function for a structureless particle at low density (i.e., a rare gas atom at low temperature) can be written as a product of partition functions, one for each independent direction of translational motion.

SOLUTION The translational energies of independent particles confined to some rectangular volume V are given by the particle-in-a-box expression, Eq. (12.24):

$$\epsilon_i = \epsilon_x + \epsilon_y + \epsilon_z = \frac{a n_x^2}{L_x^2} + \frac{a n_y^2}{L_y^2} + \frac{a n_z^2}{L_z^2}$$

where $a = \hbar^2 \pi^2/2m$ and each L is the length of the corresponding side of the box: $V = L_x L_y L_z$. Each direction is independent of the others, and each direction's associated energy is nondegenerate ($g_i = 1$). The molecular partition function, Eq. (22.13), is

$$\begin{aligned}
q &= \sum_i g_i\, e^{-\beta \epsilon_i} \\
&= \sum_{n_x, n_y, n_z} e^{-\beta(\epsilon_x + \epsilon_y + \epsilon_z)} \\
&= \sum_{n_x, n_y, n_z} e^{-\beta \epsilon_x} e^{-\beta \epsilon_y} e^{-\beta \epsilon_z} \\
&= \left(\sum_{n_x=1}^{\infty} e^{-\beta \epsilon_x}\right) \left(\sum_{n_y=1}^{\infty} e^{-\beta \epsilon_y}\right) \left(\sum_{n_z=1}^{\infty} e^{-\beta \epsilon_z}\right) = q_x\, q_y\, q_z
\end{aligned}$$

and it is worth pointing out that q factors into independent pieces no matter what the dimensionality of the region confining the particle. A 2-D partition function for a He atom loosely bound to a graphite surface but considered free to move over the surface, for example, could be written $q = q_x q_y$. Note how i, the abstract summation index representing a simple count of states in the first expression, is turned into all the quantum numbers of the system, n_x, n_y, and n_z, in the second expression.

↠ **RELATED PROBLEM** 23.1

We can now go through simple model quantum mechanical systems, pulling their energy level and degeneracy expressions from past chapters, deriving a partition function for each, and piecing together those partition functions that represent the energy terms of a particular molecular system.

Translational Motion

The quantum mechanical model for translational motion is the particle-in-a-box. This model works for dilute, low-density systems where intermolecular interactions can be neglected. This, of course, is the *ideal gas limit*, and it is important to recognize that the particle-in-a-box, that simplest quantum system, is about to give us (in the next section) that simplest equation of state, $PV = nRT$!

Example 23.1 shows that the 3-D translational partition function is a product of the three independent 1-D partition functions for each Cartesian direction. Thus, we begin with the 1-D particle-in-a-box energy level expression, Eq. (12.12):[1]

$$\epsilon_n = \frac{\hbar^2 \pi^2 n^2}{2mL^2}, \qquad n = 1, 2, 3, \ldots$$

Each level is singly degenerate, the particles have mass m, and the box length is L.

It would be handy to have a closed, single expression for the molecular partition function sum

$$q_{tr} = \sum_{n=1}^{\infty} e^{-\beta \epsilon_n} = \sum_{n=1}^{\infty} \exp\left(-\frac{\beta \hbar^2 \pi^2 n^2}{2mL^2}\right), \qquad \text{(1-D)} \qquad (23.3)$$

but unlike some infinite sums, this one does not have a closed form. However, the spacing between energy levels is *very* small for any microscopic m and macroscopic L, as discussed in Chapter 22. Consequently, successive terms in the infinite sum vary so slightly from one to the next that they are essentially a continuous function of n. As Chapter 22 also pointed out, the typical translational quantum number for a gas at ordinary temperatures is very large. Consequently, we can turn the infinite sum into a continuous integral over the infinite number[2] of quantum numbers (extending the integral from $n = 0$ to ∞, rather than $n = 1$ to ∞) without making any measurable error:

$$q_{tr} = \int_0^\infty \exp\left(-\frac{\beta \hbar^2 \pi^2 n^2}{2mL^2}\right) dn = \left(\frac{mk_B T}{2\pi\hbar^2}\right)^{1/2} L \qquad \text{(1-D)} \qquad (23.4)$$

where we have used the known definite integral $\int_0^\infty \exp(-a^2 n^2)\, dn = \sqrt{\pi}/2a$ and the definition $\beta = k_B T$.

We can immediately use the result from Example 23.1 to write the 3-D translational partition function:

$$q_{tr} = q_{tr,x} q_{tr,y} q_{tr,z} = \left(\frac{mk_B T}{2\pi\hbar^2}\right)^{3/2} L_x L_y L_z = \left(\frac{mk_B T}{2\pi\hbar^2}\right)^{3/2} V \qquad \text{(3-D)} \qquad (23.5)$$

where V is the total volume of the system.

Because partition functions are dimensionless, the factor $(mk_B T/2\pi\hbar^2)^{1/2}$ must have the units of 1/(length). The numerator, $(mk_B T)^{1/2}$, has units of momentum (since $k_B T$ is an energy and $p \sim (mE)^{1/2}$), and the denominator, $(2\pi\hbar^2)^{1/2}$, is on the order of h. Thus, this factor looks like p/h, which (Eq. (11.6)) is the reciprocal of the *de Broglie wavelength*. We therefore define the *thermal de Broglie wavelength* Λ as

$$\Lambda = \left(\frac{2\pi\hbar^2}{mk_B T}\right)^{1/2} \qquad (23.6)$$

---

[1] We are considering a rectangular box here, but it can be shown that the final thermodynamic expression is independent of the shape of the container.

[2] The "infinite quantum number" limit may bother you. It would take an impressively strong box to hold a particle with infinite energy! This limit is more a mathematical convenience than a physical reality. The Boltzmann distribution assures us, through its exponential dependence on $-\epsilon$, that states of arbitrarily high energy have vanishingly small populations.

so that the 3-D translational partition function is $q_{tr} = V/\Lambda^3$. Since Λ is microscopic and L (or V) is macroscopic, the partition function is huge. *Very* many states are accessible, and the system is dilute (in the Chapter 22 sense): $q_{tr} \gg N$. If we write this inequality as $V/\Lambda^3 \gg N$ and rearrange it to $(V/N)^{1/3} \gg \Lambda$, we see that in a translationally dilute system the characteristic distance between molecules, $(V/N)^{1/3}$, is greater than the thermal de Broglie wavelength.

EXAMPLE 23.2

Contrast q_{tr} for He(g) at its boiling point, 4.2 K, to q_{tr} for N_2(g) at 300 K and 1 atm. Assume 1 mol of each so that $V = \overline{V}$, the molar volume.

SOLUTION The molar volumes come from the ideal gas equation of state: $\overline{V} = RT/(1 \text{ atm}) = 3.45 \times 10^{-4}$ m^3 mol^{-1} for He, 2.46×10^{-2} m^3 mol^{-1} for N_2. The thermal de Broglie wavelengths from Eq. (23.6) are 4.26×10^{-10} m for He and 1.90×10^{-11} m for N_2, both microscopic, as expected.

The partition functions are huge:

$$q_{tr}(\text{He}) = \frac{V}{\Lambda^3} = \frac{3.45 \times 10^{-4} \text{ m}^3}{(4.26 \times 10^{-10} \text{ m})^3} = 4.46 \times 10^{24}$$

$$q_{tr}(N_2) = \frac{V}{\Lambda^3} = \frac{2.46 \times 10^{-2} \text{ m}^3}{(1.90 \times 10^{-11} \text{ m})^3} = 3.56 \times 10^{30}$$

even for the very cold, light, and therefore more quantum-like He. Recall the interpretation of these *molecular* partition functions: they represent a measure of the number of states available *per molecule* so that each of the mole of molecules has more available states than there are molecules. This is the statistical dilute limit.

➥ **RELATED PROBLEMS** 23.2, 23.3, 23.4

A Closer Look at the Translational Partition Function

Since we argued that the individual terms in the infinite sum for q_{tr} formed a near continuum of values so that the sum could be written as an integral over the quantum number, Eq. (23.4), we should also be able to write an expression for q_{tr} in terms of an integral over the *energy* directly. In such an integral, the Boltzmann exponential factor, $e^{-\beta\epsilon}$, would have to be weighted by the degeneracy of the system as a function of energy. This degeneracy function is the *density of states* function, $\rho(\epsilon)$, the number of states at energy ϵ in the interval ϵ to $\epsilon + d\epsilon$. From this point of view, the partition function is written

$$q_{tr} = \int_0^\infty \rho(\epsilon) \, e^{-\beta\epsilon} \, d\epsilon \ .$$

The density of states function for a 3-D particle-in-a-box system is easy to find. (In fact, Problem 16.53, which discusses the free-electron model of metals, derives this density of states, or rather *twice* the density of states we want here, since the free-electron model considers the two electron-spin states associated with each translational state.) It is given by

$$\rho(\epsilon) \, d\epsilon = \frac{\pi}{4} \left(\frac{2m}{\hbar^2 \pi^2} \right)^{3/2} V \epsilon^{1/2} \, d\epsilon$$

so that

$$q_{tr} = \frac{\pi}{4} \left(\frac{2m}{\hbar^2 \pi^2} \right)^{3/2} V \int_0^\infty \epsilon^{1/2} \, e^{-\beta\epsilon} \, d\epsilon \ .$$

The integral is known: $\int_0^\infty \epsilon^{1/2} e^{-\beta\epsilon} d\epsilon = \pi^{1/2}\beta^{-3/2}/2$, and the partition function calculated this way agrees with Eq. (23.5), as it must:

$$q_{\text{tr}} = \frac{\pi}{4}\left(\frac{2m}{\hbar^2\pi^2}\right)^{3/2} V \frac{\sqrt{\pi}}{2}(k_BT)^{3/2} = \left(\frac{mk_BT}{2\pi\hbar^2}\right)^{3/2} V\ .$$

This point of view, an integral over a density of states weighted by $e^{-\beta\epsilon}$, has important mathematical roots. For a general function $F(x)$, the integral

$$f(s) = \int_0^\infty F(x) e^{-sx} dx$$

is called the *Laplace transform* of F, often written symbolically $\mathcal{L}(F(x))$. When applied to partition functions, x is ϵ; F is the density of states function ρ; s is $\beta = 1/k_BT$; and f is q, the partition function, a function of β and thus of T. This formalism is more useful when run backwards, however. The *inverse Laplace transform of the partition function is the density of states:* $\mathcal{L}^{-1}(q(\beta)) = \rho(\epsilon)$. One calculates the partition function for a complicated system numerically, by direct summation of $g_i e^{-\beta\epsilon_i}$ over energy levels measured by spectroscopy, repeats this calculation for a number of temperatures, fits the resulting table of $q(\beta)$ values to a power series in β, then uses standard tables of inverse Laplace transforms for each term in the power series to find the density of states as a power series in energy. This often leads to a more compact and useful expression for the density of states than a direct count of the number of states per unit energy interval, especially where the system has a large number of irregularly spaced energies, as do most atoms.

Vibrational Motion

The harmonic oscillator has the simplest vibrational motion, and we used this model for intramolecular vibrations in Chapter 19. Extrapolating to a crystal, the lattice vibrations of a solid can also be approximated as harmonic, and we will use our results here to discuss both single-molecule and solid-lattice vibrations in the next section.

A single harmonic oscillator has nondegenerate energy levels indexed by a quantum number v such that the energies are given by Eq. (12.18)[3]

$$E_v = \hbar\omega\left(v + \frac{1}{2}\right),\quad v = 0, 1, 2, 3, \ldots$$

where ω is the harmonic oscillation frequency. The partition function is

$$q_{\text{vib}} = \sum_{v=0}^\infty e^{-\beta E_v} = \sum_{v=0}^\infty e^{-\hbar\omega\beta(v+1/2)} = e^{-\hbar\omega\beta/2}\sum_{v=0}^\infty e^{-\hbar\omega\beta v}$$

where the zero-point energy contribution, $e^{-\hbar\omega\beta/2}$, has been factored from each term in the sum in the last expression. The final sum has a simple closed form:

$$\sum_{v=0}^\infty e^{-\hbar\omega\beta v} = \sum_{v=0}^\infty (e^{-\hbar\omega\beta})^v = \sum_{v=0}^\infty x^v = \frac{1}{1-x} = \frac{1}{1-e^{-\hbar\omega\beta}}$$

so that the partition function is

$$q_{\text{vib}} = \frac{e^{-\hbar\omega/2k_BT}}{1 - e^{-\hbar\omega/k_BT}}\ . \tag{23.7}$$

---

[3] We will use E instead of ϵ as the symbol for energy here to avoid confusion with the dimensionless energy ϵ_v used in Chapter 12.

While we have kept the zero-point energy contribution in this expression (so that the zero of energy is at the bottom of the harmonic potential energy function rather than at the lowest allowed energy level), it is important to remember that no observable thermodynamic quantities depend on the choice of the zero of energy (see Problem 22.11). Including the zero-point energy here does introduce one slight annoyance. It causes $q_{vib} \to 0$ as $T \to 0$. All our other partition functions are based on an energy zero at the lowest allowed level and approach the ground-state degeneracy as $T \to 0$.

Equation (23.7) is the whole story for the vibrations of a diatomic. For a polyatomic, *each mode* of vibration has a partition function of this form, and the total q_{vib} is the product of functions for each mode. If one or more modes is n-fold degenerate, as in the doubly degenerate bending modes of CO_2 (see Figure 19.6(b)), the partition function for that mode is raised to the nth power in the expression for the total vibrational partition function.

EXAMPLE 23.3

How does q_{vib} depend on T at high T? How high is "high T" if we are representing the vibrations of H_2? of I_2?

SOLUTION High T means $T \gg \hbar\omega/k_B$, or, defining $x = \hbar\omega/k_B T$, high T is $x \ll 1$. If we use the expansion $e^{-x} = 1 - x \ldots$, the partition function for small x is

$$q_{vib} = \frac{e^{-x/2}}{1 - e^{-x}} \cong \frac{1 - x/2 \cdots}{1 - (1 - x \cdots)} \cong \frac{1}{x}, \quad x \ll 1$$

or

$$q_{vib} \cong \frac{k_B T}{\hbar\omega}, \quad T \gg \frac{\hbar\omega}{k_B}.$$

The partition function should increase linearly with T at high T.

Data for the vibrational energies of H_2 and I_2 were used in Example 22.1. For H_2, $\hbar\omega/k_B = 6333$ K, and for I_2, $\hbar\omega/k_B = 309$ K. Consequently, vibration in I_2 is at the high-T limit near and above room temperature, but H_2 is far from this limit except at extraordinarily high temperatures. A reduced graph of q_{vib} as a function of $k_B T/\hbar\omega$ shows the onset of the linear, high-T behavior:

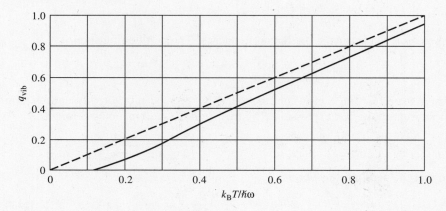

The dashed line is the high-temperature limiting expression, and the solid line is the full expression. Note that q_{vib} is a fairly linear function of T, and even at $T = \hbar\omega/k_B$, the simple high-temperature limiting expression is a good approximation.

⮕ RELATED PROBLEMS 23.6, 23.7

If the vibrational motion is appreciably *anharmonic,* the partition function sum cannot, in general, be written in a closed form. Instead, one must calculate the sum directly to whatever precision is desired using the anharmonic vibrational energies. For molecular vibrations, a vibrational energy expression such as Eq. (19.26) for diatomics can be used. The anharmonic correction is rarely large, however. For example, $q_{\text{vib}}(I_2)$ in the harmonic approximation at 300 K is 0.9305. Including anharmonicity changes this to 0.9360. (The partition function increases slightly because the anharmonicity of molecular vibration *lowers* excited vibrational energies below their harmonic positions.)

Electronic Motion

For most stable species, the atom or molecule has a closed-shell or MO electron configuration with a nondegenerate electronic ground state. The first excited electronic state is typically so high above the ground state that it is never significantly populated at ordinary temperatures. Consequently, for the vast majority of species,

$$q_{\text{el}} = 1 \quad \text{(closed-shell species)} \ . \tag{23.8}$$

For open-shell species, which typically have one or more low-energy excited states (see atomic C in Example 22.3, for instance), the electronic partition function has to be calculated by a direct sum over those levels and degeneracies that contribute significantly to the partition function sum:

$$q_{\text{el}} = g_0 + g_1 e^{-\beta\epsilon_1} + g_2 e^{-\beta\epsilon_2} + \cdots \quad \text{(open-shell species)} \tag{23.9}$$

where the ground electronic state has degeneracy g_0 and energy $\epsilon_0 = 0$, etc.

We can write a very accurate energy-level expression for one-electron atoms in terms of the principal quantum number n (Eq. (12.52)), and if we introduce this expression (and the $2n^2$ degeneracy of each level) into the partition function sum, we find an amusing paradox at first glance. Since the level energies approach a constant value as n increases (the ionization energy), each term in the sum is finite, but there is an infinite number of such terms ($n = 1, 2, \ldots, \infty$) so that, at *any* non-zero temperature, q_{el} would appear to be *infinite!* This cannot be, and the paradox is discussed further in Problem 23.10.

Nuclear Spin Motion

Nuclear excited states are so very far above the ground state that no chemically interesting temperature can be large enough to excite them. Consequently, nuclear spin *degeneracy* for the ground state is the whole story. If the nuclear spin quantum number is I, the degeneracy is $2I + 1$ and

$$q_{\text{nuc}} = 2I + 1 \ . \tag{23.10}$$

Normally, the nuclear spin state affects no thermodynamic quantity (except the entropy, and it is usually ignored there as well). However, nuclear spin does influence molecular rotational states through the Pauli Exclusion Principle (see the end of Section 19.5), and we must consider nuclear spins again when we consider rotation.

Rotational Motion

Rotational motion in the high-temperature, classical limit is the simplest to treat, and we begin with a linear rotor. The 3-D rotational energy levels of a rigid linear molecule are, from Eq. (19.21),

$$\epsilon_{\text{rot}} = \frac{J(J+1)\hbar^2}{2I}, \quad J = 0, 1, 2, \ldots$$

where I is the moment of inertia ($I = \mu R_e^2$ for a diatomic), and each level is $2J + 1$ fold degenerate. The rotational molecular partition function is thus

$$q_{\text{rot}} = \sum_{J=0}^{\infty} (2J + 1)e^{-\beta \epsilon_{\text{rot}}} = \sum_{J=0}^{\infty} (2J + 1)e^{-\beta J(J + 1)\hbar^2/2I}.$$

Most moments of inertia are sufficiently small to make $T \gg \hbar^2/2Ik_B$ at most temperatures of interest; for example, $\hbar^2/2Ik_B = 0.164$ K for IC1, 2.78 K for CO, 0.292 K for OCS, etc. Diatomic hydrides are an important exception that we will consider later. For most molecular gases, therefore, any temperature of interest is at the high-temperature limit. Many rotational levels will be populated, and, just as we did for translational motion, we can convert the partition function sum into an integral over J. If we define a new variable $z = J(J + 1)$ so that $dz = (2J + 1) \, dJ$, we can write

$$q_{\text{rot}} = \int_0^{\infty} (2J + 1)e^{-\beta J(J + 1)\hbar^2/2I} \, dJ = \int_0^{\infty} e^{-z\beta \hbar^2/2I} \, dz = \frac{2Ik_BT}{\hbar^2} \quad (23.11)$$

an expression reminiscent of the high-temperature vibrational partition function expression derived in Example 23.3. Both are linear in T, but while the high-temperature limit for vibration is not usually encountered, it is the usual case for molecular gas rotational motion.

Equation (23.11) is correct for *heteronuclear diatomics and other linear molecules lacking a center of symmetry*. For centrosymmetric linear molecules, we have overcounted the number of classical states. Just as we divided by a factor $N!$ to account for indistinguishability in the dilute limit in Chapter 22, we need to divide Eq. (23.11) by 2 to account for the two-fold rotational symmetry of centrosymmetric linear molecules. Everything is indistinguishable under a 360° rotation,[4] but centrosymmetric linear molecules are also indistinguishable under a 180° rotation about an axis perpendicular to the molecular axis:

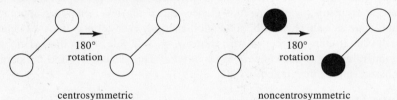

centrosymmetric   noncentrosymmetric

Consequently, we introduce a *symmetry number* σ so that $\sigma = 1$ for noncentrosymmetric linear molecules and $\sigma = 2$ for centrosymmetric linear molecules. Thus, we can write

$$q_{\text{rot}} = \frac{2Ik_BT}{\sigma \hbar^2} \quad \text{(linear molecule, high-temperature limit)} \quad (23.12)$$

in general.

Nonlinear molecules are described by three moments of inertia, I_a, I_b, and I_c (see Eq. (19.3)). Their high-temperature limiting partition function is difficult to derive (see references at the end of Chapter 22 for details), but the final expression is

$$\begin{aligned} q_{\text{rot}} &= \frac{\pi^{1/2}}{\sigma} \left(\frac{2I_a k_B T}{\hbar^2}\right)^{1/2} \left(\frac{2I_b k_B T}{\hbar^2}\right)^{1/2} \left(\frac{2I_c k_B T}{\hbar^2}\right)^{1/2} \\ &= \frac{(\pi I_a I_b I_c)^{1/2}}{\sigma} \left(\frac{2k_B T}{\hbar^2}\right)^{3/2} \quad \text{(nonlinear molecule, high } T\text{)} \end{aligned} \quad (23.13)$$

---

[4] Almost everything is indistinguishable under a 360° rotation. The wavefunctions of free spin one-half elementary particles (electrons, neutrons, etc.) are 4π, rather than 2π, rotationally invariant.

where the meaning of σ for nonlinear molecules is based on the same idea as for linear molecules: it is the *number of indistinguishable orientations that can be produced by rotations less than or equal to 360° about any axis through the center of mass of the molecule*. For example, rotating H_2O through 180° or 360° about an axis bisecting the H—O—H bond angle gives an orientation indistinguishable from the original. These are the only such rotations, and $\sigma = 2$. For HDO, $\sigma = 1$, since H and D are distinguishable. For CH_4, $\sigma = 12$: each of the four C—H bonds represents a three-fold symmetry axis about which three successive 120° indistinguishable rotations are possible. Table 23.1 summarizes various σ values for different molecular symmetry classes.

The rotational partition function for nonlinear molecules with three-fold or higher symmetry (symmetric or spherical tops, in the language of Chapter 19) is also given correctly by Eq. (23.13). For example, a spherical top (such as CH_4 or SF_6) has $I_a = I_b = I_c$, and the product $I_a I_b I_c$ in Eq. (23.12) is simply I^3 where I is the common moment of inertia.

Just as anharmonicity usually has a small effect on q_{vib}, centrifugal distortion and other nonrigid-rotor corrections to the level energies usually have a minor effect on q_{rot}. Hindered internal rotations, such as in the methyl groups' motion about the C—C bond in ethane, pose special problems. At sufficiently high temperature, they are free internal rotations, but at low T, they are better described as torsional oscillations. Some of the references at the end of this chapter discuss this problem, and it represents one of the few instances for which macroscopic measurements (typically, the temperature dependence of the heat capacity) were used to deduce a microscopic quantity (the height of the potential energy barrier to hindered rotation) that spectroscopy could not directly measure.

Finally, we turn to those molecules, mainly diatomic hydrides, for which the moment of inertia is sufficiently small that the high-temperature limit approximation can be questioned. For example, the moment of inertia for HF makes $\hbar^2/2Ik_B = 30.15$ K. The vapor pressure of HF (which solidifies at 190 K) is $\leq 10^{-3}$ atm for $T \leq 200$ K, so that most practical situations are at the high-temperature limit for HF rotation. Figure 23.1 makes this point with a comparison of the high-temperature approximation, Eq. (23.12) (which is simply $q = T/(30.15\text{ K})$), and the numerically evaluated exact partition function for $T \leq 200$ K. The exact function quickly becomes quite linear, but it is appreciably larger than the approximation over this range of temperatures.

The most important small moment of inertia hydrides are, of course, H_2 and its isotopic variants and molecular ions. For H_2, $\hbar^2/2Ik_B = 87.55$ K. Even well above room temperature, only a few rotational levels are populated, and an exact evaluation of the partition function sum is called for. For HD, this sum is straightforward, but

TABLE 23.1 Rotational Partition Function Symmetry Numbers σ for Various Molecular Symmetries

Representative Molecule	σ
C_{60}	60
SF_6	24
CH_4, C_6H_6	12
C_2H_4, B_2H_6, BrF_5	4
NH_3	3
H_2O	2

FIGURE 23.1 The exact rigid rotor partition function for the HF moment of inertia (solid line) quickly becomes a linear function of T. In comparison to the high-T approximation to q_{rot} (dashed line), the exact function is slightly greater, but the error is an ever decreasing fraction of the exact q_{rot} as T is increased, making the approximation useful at all practical temperatures.

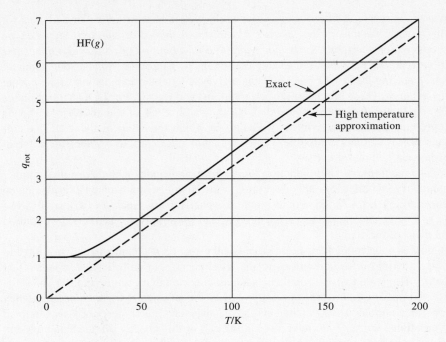

for H_2 or D_2, the coupling between nuclear spin states and rotational states must be considered. This important macroscopic manifestation of quantum mechanics is given a closer look below. Historically, the seemingly anomalous heat capacity of $H_2(g)$ was ultimately explained through quantum statistical mechanics and represented one of the first clear examples of nonclassical behavior in a macroscopic system.

EXAMPLE 23.4

How do the high-temperature rotational partition functions differ among the isotopomers H_2, HD, and D_2?

SOLUTION In the rigid-rotor approximation, the bond length is the same for all isotopomers of any molecule. Since $q_{rot} \sim I/\sigma = \mu R_e^2/\sigma$, only changes in the reduced mass μ and the symmetry number σ affect q_{rot}. Since (Eq. (19.19)) $\mu = m_A m_B/(m_A + m_B)$ for diatomic AB, we can make a short table of μ/σ ratios:

Isotopomer	$\dfrac{m_A m_B}{\sigma(m_A + m_B)} = \dfrac{\mu}{\sigma}$
H_2	$\dfrac{1 \cdot 1}{2(1 + 1)} = \dfrac{1}{4}$
HD	$\dfrac{1 \cdot 2}{1(1 + 2)} = \dfrac{2}{3}$
D_2	$\dfrac{2 \cdot 2}{2(2 + 2)} = \dfrac{1}{2}$

In spite of its intermediate moment of inertia, HD has the largest high-T rotational partition function of the three. At 1000 K, $q_{rot} = 5.71$ for H_2, 15.22 for HD, and 11.42 for D_2.

▸ RELATED PROBLEMS 23.11, 23.12

A Closer Look at the Symmetry Number

The symmetry number σ used in the rotational partition function was introduced to account for the classical indistinguishability of various rigid molecular orientations. This number can be traced to the subtle quantum interplay among the rotational wavefunction, the nuclear spin state, and the Pauli Exclusion Principle. This interplay was first discussed at the end of Section 19.5, and you may find it useful to review that material before reading this section.

We will use H_2 as an example of the quantum origin of σ, but with suitable elaboration, the argument holds for polyatomics as well. To summarize the discussion in Section 19.5, the $I = \frac{1}{2}$ nuclear spin quantum number of H leads to four nuclear spin wavefunctions for H_2 of two types: one that is antisymmetric with respect to nuclear exchange and three that are symmetric. The Pauli Exclusion Principle forces rotational states with *even* quantum number J to have only the antisymmetric nuclear spin state (para-H_2), while *odd* J states have any of the three symmetric nuclear spin states (ortho-H_2). Consequently, the *nuclear spin states and the rotational states are coupled,* and we can no longer write separate q_{rot} and q_{nuc}. Instead, we have to consider a combined $q_{rot,\,nuc}$ which looks like this:

$$q_{rot,\,nuc} = 1 \sum_{J \text{ even}} (2J + 1)\, e^{-\beta \epsilon_{rot}} + 3 \sum_{J \text{ odd}} (2J + 1)\, e^{-\beta \epsilon_{rot}}$$

where the factors 1 and 3 represent the nuclear spin-state degeneracies. Because collisions among H_2 molecules are not very effective at converting between the ortho and para forms, rapid equilibrium between them is possible only in the presence of a catalyst such as graphite or a paramagnetic substance. Failure to recognize this fact caused considerable grief in the early days of low temperature H_2 study, and gaseous H_2 acts like a mixture of two different substances, ortho and para.

Now consider the high-temperature limits of the sums in $q_{rot,\,nuc}$. At high enough T, the two should be equal, since all quantum distinctions blur at high enough T. Moreover, each sum should approach one half the classical sum (neglecting σ) over all rotational states, since effectively half the states are present in each. The graph below shows this behavior. The dotted line is one half the classical sum; the light solid line is the sum over even J (para-H_2); and the heavy line is the sum over odd J (ortho-H_2). The three lines merge even at fairly low T.

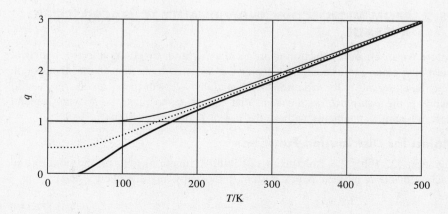

Thus, each of the two sums above *approach the classical value including σ at high T*. The classical, uncoupled, high-T expression for the combined nuclear and rotational partition functions is (using Equations (23.2), (23.10), and (23.12))

$$q_{\text{rot}} q_{\text{nucA}} q_{\text{nucB}} = \frac{2I k_B T}{\sigma \hbar^2} (2I_A + 1)(2I_B + 1)$$

$$= \frac{2I k_B T}{2\hbar^2} (2)(2) = \frac{4 I k_B T}{\hbar^2}$$

while the quantum, coupled, high-T expression is

$$q_{\text{rot, nuc}} = 1 \left(\frac{1}{2} \frac{2 I k_B T}{\hbar^2} \right) + 3 \left(\frac{1}{2} \frac{2 I k_B T}{\hbar^2} \right) = \frac{4 I k_B T}{\hbar^2} .$$
$$\uparrow \qquad\qquad \uparrow$$
$$\text{para sum} \qquad \text{ortho sum}$$

The two final expressions are identical, which shows how the symmetry number σ cleverly accounts for the quantum mechanical coupling between nuclear spins and rotational states.

Spectroscopic Quantities

The most common tabulations of molecular structure parameters are not in terms of moments of inertia or vibrational frequencies in s^{-1} units. Instead, as Chapters 13, 19, and 20 discussed for atoms and molecules, spectroscopic energies and the parameters of molecular energy level expressions are usually tabulated in cm^{-1} units. Similarly, molecular masses are more familiar in molar mass units, g mol^{-1}, than in single molecule mass units, kg. Table 23.2 collects several partition function expressions from this section and writes them in a practical form in terms of these more common units. Recall from Chapter 22 that the conversion factor $hc(100 \text{ cm m}^{-1})/k_B \cong 1.439$ cm K is commonly needed for these calculations. In Table 23.2, this factor is symbolized k'. Many of the examples in Chapter 22 showed how to use this factor for electronic state partition function calculations.

Armed with the collection of partition function expressions from this section, we can forge ahead in the next into a similar collection of thermodynamic quantities. Along the way, we will gain further insight into the molecular workings of thermodynamics.

23.2 FROM MICROSCOPIC INDIVIDUALITY TO MACROSCOPIC AVERAGE

Before we insert our collection of molecular partition functions into the statistical thermodynamic expressions of Chapter 22, we will look first at the distribution functions they imply for vibrations and rotations. We will then prove a very useful theorem, the *equipartition theorem*, that allows us to make some very general thermodynamic statements without the use of partition functions.

Molecular Distribution Functions

Equation (22.14b), the Boltzmann distribution function, tells us the equilibrium fraction of the N molecules in a system that occupy any one energy level:

$$\frac{n_i}{N} = \frac{g_i \, e^{-\epsilon_i / k_B T}}{q} . \qquad \textbf{(22.14b)}$$

TABLE 23.2 Practical Expressions for Molecular Partition Functions

Translation

M = molar mass/g mol^{-1} V = volume/m^3 L = length/m

1-D: $q_{\text{tr}} = 5.7280 \times 10^8 \, (MT)^{1/2} \, L$
3-D: $q_{\text{tr}} = 1.8793 \times 10^{26} \, (MT)^{3/2} \, V$

Vibration

ω_e = harmonic vibrational constant/cm^{-1}

$q_{\text{vib}} = \dfrac{e^{-k'\omega_e/2T}}{1 - e^{-k'\omega_e/T}}$ (per vibrational mode, including zero-point energy)

$q_{\text{vib}} = \dfrac{1}{1 - e^{-k'\omega_e/T}}$ (per vibrational mode, excluding zero-point energy)

Rotation

A_e, B_e, C_e = equilibrium rotational constant/cm^{-1} σ = symmetry number (Table 23.1)

linear: $q_{\text{rot}} = \dfrac{T}{\sigma k' B_e}$

nonlinear: $q_{\text{rot}} = \dfrac{\sqrt{\pi}}{\sigma} \left(\dfrac{T}{k'A_e}\right)^{1/2} \left(\dfrac{T}{k'B_e}\right)^{1/2} \left(\dfrac{T}{k'C_e}\right)^{1/2} \cong \dfrac{1.027}{\sigma} \dfrac{T^{3/2}}{(A_e B_e C_e)^{1/2}}$

Notes: $k' = (100 \text{ cm m}^{-1}) hc/k_B = 1.438\,769$ cm K, T = absolute temperature/K

We will save translational distribution functions for the next chapter, which also discusses them from the point of view of the kinetic theory of gases. Similarly, Chapter 22 used electronic state distribution functions extensively, and we will not consider them further here.

The vibrational distribution function for a single, nondegenerate mode is

$$\frac{n_v}{N} = \frac{e^{-\hbar\omega(v+\frac{1}{2})/k_B T}}{q_{\text{vib}}}$$
$$= \frac{e^{-\hbar\omega v/k_B T}\, e^{-\hbar\omega/2k_B T}\,(1 - e^{-\hbar\omega/k_B T})}{e^{-\hbar\omega/2k_B T}} \qquad (23.14)$$
$$= e^{-\hbar\omega v/k_B T}\,(1 - e^{-\hbar\omega/k_B T}) \,.$$

Note that the zero-point energy, $\hbar\omega/2$, does not appear in the final expression. Populations drop exponentially from one level to the next higher, as Figure 23.2 shows for the gas-phase halogens at room temperature. The vibrational frequencies decrease regularly from F_2 to I_2 (see Table 19.2), and thus the population of excited vibrational levels increases in the same order. (Compare the greater detail of this calculation to our first, simplest distribution calculation in Example 22.1.)

Various spectroscopic techniques can measure the relative populations of vibrational levels. These measurements, through Eq. (23.14), constitute, in effect, a *thermometer*. If the populations follow Eq. (23.14), the system must be at equilibrium, and its temperature can be deduced.[5] If the populations do not follow Eq.

---

[5] Or at least the *vibrational modes* of the system must be at equilibrium. We will see later that it is possible to create systems that are characterized by different apparent temperatures for each degree of freedom—translational, rotational, vibrational, and electronic.

FIGURE 23.2 At 300 K, the population fractions of various vibrational states of gaseous halogens fall by a constant factor (note the logarithmic axis for population) from one state to the next higher. These equilibrium distributions constitute a measurement of the *vibrational temperature* of the system.

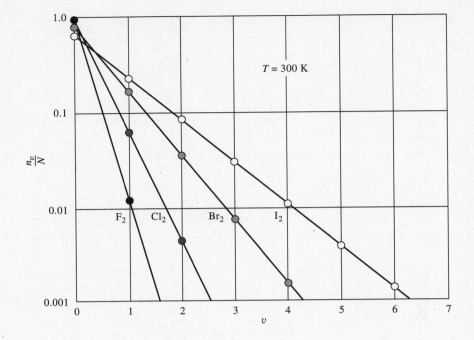

(23.14), the system is not at equilibrium. This is the situation in a *chemical laser*, such as the HF laser. The very exothermic reaction $F + H_2 \rightarrow HF + H$ produces HF vibrational levels that are populated far from equilibrium, so much so that populations *increase* with increasing v (for the lowest few v, at least). This is the *population inversion* characteristic of laser action.

EXAMPLE 23.5

The three vibrational constants of $Cl_2O(g)$ are 680 cm^{-1}, 330 cm^{-1}, and 973 cm^{-1}. Each vibration is nondegenerate. What fraction of the molecules at 300 K has the second vibration singly excited? What fraction has *both* the first *and* the second vibrations singly excited? What fraction has only the third vibration singly excited?

SOLUTION Numbering the frequencies 1, 2, 3 in the order given above, the fraction with $v_2 = 1$, $v_1 = v_3 =$ *anything* is given by Eq. (23.14) alone:

$$\frac{n_1}{N} = e^{-(1.439 \text{ cm K})(330 \text{ cm}^{-1})/300 \text{ K}}(1 - e^{-(1.439 \text{ cm K})(330 \text{ cm}^{-1})/300 \text{ K}}) = 0.163 \ .$$

In contrast, the fraction with $v_2 = 1$, $v_1 = v_3 = 0$ depends on the *full* vibrational partition function, $q_{vib} = q_1 q_2 q_3$. If we measure energy from the zero-point level ($v_1 = v_2 = v_3 = 0$), we have

$$q_{vib}^{-1} = \left(1 - e^{-\frac{1.439 \cdot 680}{300}}\right)\left(1 - e^{-\frac{1.439 \cdot 330}{300}}\right)\left(1 - e^{-\frac{1.439 \cdot 973}{300}}\right)$$
$$= (0.962)(0.795)(0.991) = 0.757$$

so that the fraction with $v_2 = 1$, $v_1 = v_3 = 0$ (the (0, 1, 0) level) is

$$\frac{n_{(0, 1, 0)}}{N} = 0.757 e^{-(1.439 \text{ cm K})(330 \text{ cm}^{-1})/300 \text{ K}} = 0.155 \ .$$

The fraction with $v_1 = v_2 = 1$, $v_3 = 0$ (i.e., (1, 1, 0)) is

$$\frac{n_{(1,1,0)}}{N} = 0.757 e^{-(1.439 \text{ cm K})[(680+330) \text{ cm}^{-1}]/300 \text{ K}} = 5.96 \times 10^{-3}$$

where the *total vibrational energy*, $(680 + 330)$ cm^{-1}, is used in the Boltzmann factor. Similarly, the fraction with $v_1 = v_2 = 0$, $v_3 = 1$ (i.e., (0, 0, 1)) is

$$\frac{n_{(0,0,1)}}{N} = 0.757 e^{-(1.439 \text{ cm K})(973 \text{ cm}^{-1})/300 \text{ K}} = 7.12 \times 10^{-3}$$

which is nearly the same as the (1, 1, 0) population simply because these two states have nearly equal total energies: 973 cm^{-1} ≅ 680 + 330 cm^{-1}.

⇒ RELATED PROBLEMS 23.15, 23.16

The rotational distribution function for polyatomics depends on the type of rotor (linear, spherical, symmetric, or asymmetric) through both the energy-level expression itself and the partition function. Since rotational energy-level expressions for nonlinear molecules are complicated or, in the case of asymmetric tops, impossible to write in general (see Eq. (19.37)), we will limit our discussion to heteronuclear linear molecules. The extension to nonlinear molecules and to homonuclear diatomics molecules involves consideration of nuclear spin degeneracies, along the lines of the Closer Look at the symmetry number from the previous section.

If the moment of inertia is small enough and/or the temperature is high enough, the high-temperature rotational partition function, Eq. (23.12), is accurate, and the distribution function is

$$\frac{n_J}{N} = \frac{(2J + 1) e^{-J(J+1)\hbar^2/2Ik_BT}}{q_{\text{rot}}}$$
$$= \frac{\hbar^2(2J + 1) e^{-J(J+1)\hbar^2/2Ik_BT}}{2Ik_BT} \quad (23.15)$$
$$= \frac{k'B_e(2J + 1) e^{-J(J+1)k'B_e/T}}{T}$$

where B_e is the equilibrium rotational constant (in cm^{-1}) and k' is defined in Table 23.2. Figure 23.3 graphs this distribution for several interhalogens at 300 K. Note

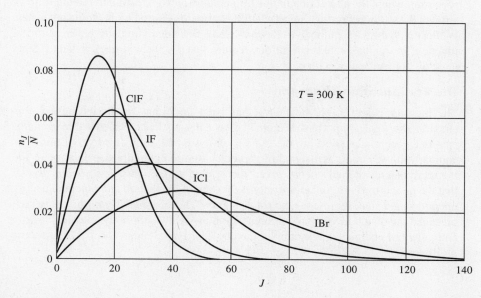

FIGURE 23.3 The population fractions of the rotational levels of gaseous interhalogens (shown here for 300 K) peak at a most probable J value, J_{mp}, that reflects the competition between the growing $(2J + 1)$ degeneracy of any one level and the decreasing Boltzmann factor. Typically very many levels have significant population, a characteristic of the high-T limit.

that quite high rotational quantum numbers are significantly populated, as expected in the high-temperature limit. No single J level has more than about 3% of the population for IBr, for example. Just as measured vibrational state distributions can serve as a thermometer, measured rotational state distributions can probe the degree of disequilibrium among rotational states and deduce the temperature if equilibrium is found.

Unlike the vibrational distribution, rotational distributions peak at a non-zero quantum number. This is caused by a competition between the increasing degeneracy, $2J + 1$, of successive levels and the exponentially decreasing Boltzmann factor. The position of the maximum in the distribution locates the most probable J value. Differentiating the distribution function with respect to J locates this most probable J. If we write $(\partial(n_J/N)/\partial J)_{J=J_{mp}} = 0$, then J_{mp}, the most probable J, is the integer closest to

$$J_{mp} = \left(\frac{T}{2k'B_e}\right)^{1/2} - \frac{1}{2} \cong \left(\frac{T}{2k'B_e}\right)^{1/2}. \qquad (23.16)$$

For the interhalogens at 300 K, J_{mp} = 14 (ClF), 19 (IF), 30 (ICl), and 42 (IBr), as Figure 23.3 indicates.

This rotational state distribution accounts, in part, for the appearance of rotationally resolved absorption spectra. The probability of absorption is proportional to the population of the lower energy state, and the equilibrium distribution of population varies with J according to Eq. (23.15) or its variant for molecules other than heteronuclear diatomics. Figure 19.9 shows the result as it appears in the vibration-rotation absorption spectrum of HF. One can almost read the sample's temperature from the spectrum, assigning the transitions to J values for the initial state, measuring which transition is most intense, and using its J value in Eq. (23.16) to find T. This is not exact, however, since even if every rotational state had equal populations, the observed intensities would vary due to an inherent dependence of transition probability on J.

For homonuclear diatomics and other symmetrical molecules, the rotational state distribution can alternate population from one J to another due to nuclear spin symmetry effects. This can be seen in Figure 19.20, the C_2H_2 vibration-rotation spectrum, where the alternation in transition intensity from feature to feature follows a rough 3:1 ratio, reflecting the population alternation caused by H atom spin states. While the symmetry number σ is very handy for calculating the entire partition function, individual state distributions require that we include nuclear spin effects in detail for molecules with $\sigma \neq 1$.

The Equipartition Theorem

The high-temperature, classical limit has been important for us throughout this chapter. There is an important theorem in classical statistical mechanics that strengthens and generalizes some of the conclusions we can draw from this limit. The equipartition theorem, as it is known, predicts the temperature dependence of the thermodynamic internal energy given the system's *classical Hamiltonian*. Recall that the quantum Hamiltonian operator is the sum of the kinetic and potential energy operators and thus represents the total energy. The classical Hamiltonian (which preceded the quantum version, of course) is likewise the sum of the classical kinetic and potential energies. For example, the classical Hamiltonian for a 1-D harmonic oscillator is

$$H = T + V = \frac{p^2}{2m} + \frac{1}{2}kx^2$$

where T is the kinetic energy (*not* the temperature!), V is the potential energy, and p is the momentum. The kinetic energy is written in terms of p rather than the perhaps more familiar velocity expression $mv^2/2$ for reasons that will become clear.

The classical 3-D molecular partition function for a molecule with N atoms, which we will not derive in detail, is

$$q_{cl} = \frac{1}{h^{3N}} \int \cdots \int e^{-\beta H} \, dp^{3N} \, dx^{3N} \ . \tag{23.17}$$

This is a $6N$-dimensional integral over momenta p and coordinates x: dp^{3N} stands for $dp_{x1} \, dp_{y1} \, dp_{z1} \cdots dp_{xN} \, dp_{yN} \, dp_{zN}$, and dx^{3N} is a similar shorthand for $dx_1 \, dy_1 \, dz_1 \cdots dx_N \, dy_N \, dz_N$. The factor $1/h^{3N}$ shows that this "classical" partition function in fact relies on quantum mechanics. The integrals carry units of (momentum·length)^{3N}, and the momentum·length product in classical mechanics is known as the *action*; h is the "quantum of action." Further justification for this factor is given in several of the references at the end of the chapter.

The multiple integral is the classical equivalent of the quantum "sum over states of $e^{-\beta \epsilon}$," just as we passed from a sum to an integral to derive q_{tr}. The classical Hamiltonian for a single particle's free 1-D translation over a distance L is just $H = $ kinetic energy $= p^2/2m$, so that Eq. (23.17) is

$$q_{cl} = \frac{1}{h} \int_{x=0}^{L} \int_{p=-\infty}^{\infty} e^{-p^2/2mk_B T} \, dp \, dx \ .$$

The integral over x equals L, and the integral over p is

$$\int_{-\infty}^{\infty} e^{-p^2/2mk_B T} \, dp = 2 \int_{0}^{\infty} e^{-a^2 p^2} \, dp \quad \left\{ \begin{array}{l} \text{the integrand is symmetric about} \\ x = 0, \text{ and we define } a^2 = 1/2mk_B T \end{array} \right.$$

$$= 2 \left(\frac{\sqrt{\pi}}{2a} \right) \quad \left\{ \text{since } \int_{0}^{\infty} e^{-a^2 p^2} \, dp = \sqrt{\pi}/2a \right.$$

$$= (2\pi m k_B T)^{1/2}$$

so that q_{cl} is

$$q_{cl} = \frac{L(2\pi m k_B T)^{1/2}}{h} = \left(\frac{m k_B T}{2\pi \hbar^2} \right)^{1/2} L$$

in agreement with Eq. (23.4).

The average molecular energy does *not* depend on h, however. It is given by the Boltzmann-weighted average of the Hamiltonian:

$$\langle \epsilon \rangle = \frac{\iint H \, e^{-\beta H} \, dp^{3N} \, dx^{3N}}{\iint e^{-\beta H} \, dp^{3N} \, dx^{3N}} \tag{23.18}$$

so that (compare Eq. (22.17)) the internal molar energy in the classical limit is

$$\Delta_0 \overline{U} = N_A \langle \epsilon \rangle \ .$$

Many common classical Hamiltonians contain quadratic terms ap^2 and/or bx^2 where a and b are constants or functions of the other momenta and coordinates. (The 1-D harmonic oscillator Hamiltonian mentioned above is one such, as is that for the 1-D free particle.) Since $\langle \epsilon \rangle$ breaks into a sum of averages over the separate terms of H, *any term in H that has this quadratic form*, whether in a momentum component or a positional coordinate, *contributes $k_B T/2$ to $\langle \epsilon \rangle$*. This is the *equipartition theorem*. Problem 23.22 takes you through its proof.

We can apply this theorem immediately to several high-temperature situations. The 3-D translational Hamiltonian has three quadratic terms: $(p_x^2 + p_y^2 + p_z^2)/2m$. Thus, translation contributes $3k_BT/2$ per molecule to $\langle\epsilon\rangle$, or $3RT/2$ to $\Delta_0\overline{U}$. Since $C_V = (\partial U/\partial T)_V$, the 3-D translational contribution to the constant-volume molar heat capacity is $3R/2$. This is the monatomic ideal gas $\overline{C}_V$ that we introduced long ago (Eq. (2.18)), and here we see its origin as well as the reason thermodynamics alone cannot predict it. It is based on a particular model of microscopic motion.

Similarly, rotational motion is in the high-temperature limit for almost all gas phase molecules. The classical Hamiltonian for rigid-body rotation has two (linear) or three (nonlinear) quadratic momentum terms (*angular* momenta, but momenta nevertheless). Consequently, free rotation contributes RT (linear) or $3RT/2$ (nonlinear) to $\Delta_0\overline{U}$ and R or $3R/2$ to $\overline{C}_V$.

The harmonic vibrational Hamiltonian has one quadratic momentum and one quadratic coordinate term. Each vibration in a molecule therefore contributes RT to $\Delta_0\overline{U}$ or R to $\overline{C}_V$, but only at the classical limit for that mode. Many molecular vibrations do not reach this limit at ordinary temperatures, however.

We can now understand the limits on $\overline{C}_V$ for gases that were stated in Chapter 4. A monatomic gas has $\overline{C}_V = 3R/2$ always (neglecting any *electronic* contribution to $\overline{C}_V$), since it only has translational motion. A linear molecule (except diatomic hydrides at low T, for which rotation may not be at the high-T limit) has $\overline{C}_V = 3R/2 + R = 5R/2$ at low T, where translation and rotation contribute fully, increasing toward $\overline{C}_V = [5/2 + (3N - 5)]R$ at high temperatures where the $3N - 5$ vibrational modes contribute fully as well. A nonlinear molecule similarly starts with $\overline{C}_V = 3R/2 + 3R/2 = 3R$ and increases toward $[3 + (3N - 6)]R$ as T increases. Often, these high-temperature limits are not reached in an equilibrium sample due to dissociation or other reactions of the stable low-temperature molecule. Moreover, the empirical expressions for heat capacities as a function of T (such as that used in Table 4.4 for $\overline{C}_P$) are guided by temperature expansions of the statistical expressions for the heat capacity.

Statistical Thermodynamics

The equipartition theorem has taken much of the work out of the conversion from molecular partition functions to basic thermodynamic quantities. We still have the low-temperature expression for vibrational contributions and all expressions for entropy to consider. We start with vibrations.

If we define $x = \hbar\omega/k_B$, Eq. (23.7) for q_{vib} is

$$q_{\text{vib}} = \frac{e^{-x/2T}}{1 - e^{-x/T}}$$

and Eq. (22.18b) for $\Delta_0\overline{U}$ gives

$$RT^2 \frac{\partial \ln q_{\text{vib}}}{\partial T} = R\left(\frac{x}{2} + \frac{x}{e^{x/T} - 1}\right).$$

The first term is just the zero-point energy, a constant we can ignore, so that

$$\Delta_0\overline{U}_{\text{vib}} = \frac{N_A\hbar\omega}{e^{\hbar\omega/k_BT} - 1}. \tag{23.19}$$

As $T \to 0$, $e^{\hbar\omega/k_BT} \to \infty$, and there is no vibrational contribution to the energy, as expected. At high T, $(e^{\hbar\omega/k_BT} - 1) \to \hbar\omega/k_BT$, and $\Delta_0\overline{U}_{\text{vib}} \to N_Ak_BT = RT$, as predicted by equipartition.

Differentiating Eq. (23.19) with respect to T gives us the vibrational contribution to the heat capacity:

$$\overline{C}_{V,\text{vib}} = R\left(\frac{\hbar\omega}{k_B T}\right)^2 \frac{e^{\hbar\omega/k_B T}}{(e^{\hbar\omega/k_B T} - 1)^2}, \quad (23.20)$$

which approaches R at high T, again in accord with equipartition. We can now look at the full temperature dependence of a molecular gas heat capacity. Figure 23.4 compares Eq. (23.20) (plus $5R/2$ for translation and rotation) with $\overline{C}_V$ for F_2 as given in the JANAF tables discussed in Chapter 7.[6] The agreement is quite good up to ~600 K where two phenomena—anharmonicity and dissociation—cause increasing disagreement. Neither of these phenomena is included in Eq. (23.20). Anharmonicity requires, typically, a direct sum over states calculation of q_vib, and dissociation is discussed in the next section.

Figure 23.5 applies Eq. (23.20) to acetylene, again comparing to JANAF tabulated values. Acetylene is linear with 7 vibrational modes. Two are doubly degenerate bending modes (see Problem 19.25) with harmonic vibrational constants 612.87 and 730.33 cm^{-1}. The three remaining modes are a symmetric and an antisymmetric C—H stretch (3372.85 and 3288.39 cm^{-1}) and a C≡C stretch (1974.32 cm^{-1}), all much higher in frequency than the bends. The heavy solid line in the figure uses the full heat capacity expression:

$$\overline{C}_V = \left(\frac{5R}{2}, \text{translation and rotation}\right) + \sum \overline{C}_{V,\text{vib}}(\text{stretches}) + 2\sum \overline{C}_{V,\text{vib}}(\text{bends})$$

while the light solid line neglects the degeneracy factor of 2 for the bends, and the dashed line neglects the stretches. It is clear that the degenerate bends contribute the most up to $T \cong 500$ K and that their degeneracy is vital. The dashed line shows these modes reaching the high-T limit ($4R$ above the translation + rotation amount)

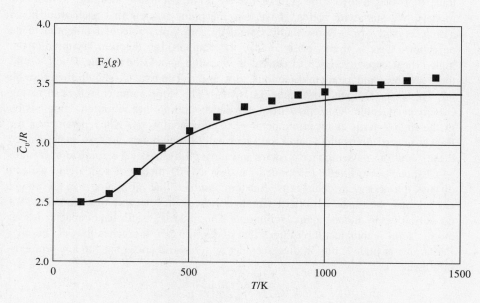

FIGURE 23.4 The heat capacity of F_2 calculated from statistical mechanics (solid line) using the equipartition expressions for translation and rotation and Eq. (23.20) for vibration (harmonic approximation) is in good agreement with tabulated thermodynamic values (points). The disagreement at highest temperatures can be traced to F_2 dissociation and anharmonic vibration.

---

[6]JANAF tables list $\overline{C}_P$, but since data for gases are for the ideal gas reference state, Eq. (2.35), $\overline{C}_V = \overline{C}_P - R$, can be used to find $\overline{C}_V$.

FIGURE 23.5 A heat capacity calculation for C_2H_2 similar to that for F_2 in Figure 23.4 (heavy solid line) shows good agreement to tabulated values (points). The two low-frequency, degenerate bending vibrations make major contributions as is shown by the light solid line, which neglects the degeneracy factor, and above around 800 K, the stretching vibrations contribute significantly as well, as shown by the dashed line, which neglects them completely.

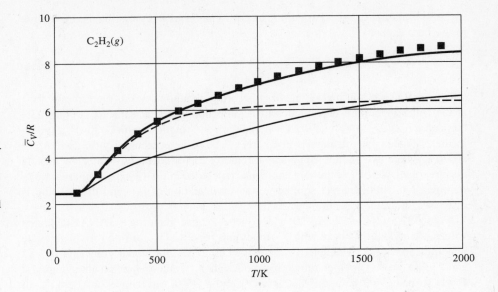

above $\sim$1500 K. The slight disagreement between the full calculation and the tabulated values at the highest temperatures can be traced to anharmonicity.

Before we present a statistical mechanical derivation of the ideal gas equation of state, we look at two other important applications of the vibrational partition function: the heat capacity of solids and the blackbody distribution law. It is somewhat surprising and certainly worth noticing that several seemingly unrelated physical situations have very common basic origins. Blackbody radiation, solid heat capacities, metal conduction, and the ideal gas equation of state, among others, all trace their origin to one or both of the harmonic oscillator and particle-in-a-box models. We start with solids. If we take the point of view that each atom in an atomic crystal is a 3-D oscillator, bound by some intermolecular potential to its neighbors, we can immediately invoke the equipartition theorem and predict the high-T heat capacity. Each direction of vibration should contribute R to $\overline{C}_V$, and there are three independent directions, x, y, and z. The total $\overline{C}_V$ should therefore be $3R$. This is the Dulong–Petit value, Eq. (4.15). The Kopp parameters for estimating $\overline{C}_V$ (given in Table 4.1) follow from our understanding that vibrations may not be in the high-T limit at room temperature. Consequently, the Kopp parameters for light atoms, or atoms that tend to form unusually strong bonds, are less than $3R$ since these vibrations have high frequencies and are not fully classical at room temperature.

Chapter 4 also briefly mentioned Einstein's 1907 theory of solids, and we can discuss this theory in detail now. Einstein assumed that all N atoms in an atomic crystal had the *same* harmonic vibrational frequency ω and thus constituted a collection of $3N$ independent oscillators.[7] The molar Einstein heat capacity for an atomic solid is thus just three times that of Eq. (23.20). It reaches the $3R$ Dulong–Petit value at high T, but, in disagreement with experiment, it has the low temperature behavior

$$\overline{C}_V \to 3R \left(\frac{\hbar\omega}{k_B T}\right)^2 e^{-\hbar\omega/k_B T}$$

while experiment showed that $\overline{C}_V$ varied like T^3 at low T.

---

[7]Moreover, these oscillators are *distinguishable*, since each sits at a fixed point in space and moves along fixed directions.

Debye introduced a correction to Einstein's theory that produced agreement at both temperature limits. The essence of this correction is based on a view of an atomic solid as a huge polyatomic molecule. Just as ordinary polyatomics, even if made of the same atom, have a *distribution* of vibrational frequencies, the normal mode frequencies, so does a crystal. Debye replaced Einstein's assumption of a single frequency with a simple distribution of frequencies. It would take us too far afield to derive this distribution in detail, but some features of it are worth pointing out. First, Debye assumed that the $3N - 6 \cong 3N$ normal modes of the crystal formed a continuous distribution of frequencies. Such a distribution must have a maximum frequency, since they must have a minimum *wavelength* comparable to the interatomic spacing in the crystal. This highest frequency normal mode, for a linear array of atoms, looks like an antisymmetric stretch:

We call this highest frequency the *Debye frequency* ω_D. The full distribution function, which acts like a degeneracy function for frequency, is

$$g(\omega)\, d\omega = \frac{9N}{\omega_D^3} \omega^2\, d\omega, \qquad 0 \le \omega \le \omega_D$$

and thus the heat capacity expression is not a sum of $3N$ oscillator terms of the same frequency, as in Einstein's theory, but an *integral* over frequency weighted by $g(\omega)$:

$$C_V = k_B \int_0^{\omega_D} \frac{(\hbar\omega/k_B T)^2\, e^{\hbar\omega/k_B T}}{(e^{\hbar\omega/k_B T} - 1)^2} g(\omega)\, d\omega$$

$$= 9Nk_B \left(\frac{k_B T}{\hbar\omega_D}\right)^3 \int_0^{x_D} \frac{x^4 e^x}{(e^x - 1)^2}\, dx, \qquad x_D = \frac{\hbar\omega_D}{k_B T}.$$

The integral must be evaluated numerically, except at low temperature where it is acceptable to replace the upper limit by ∞. In this limit, the integral equals $4\pi^4/15$, and we recover Eq. (4.16):

$$\frac{\overline{C}_V}{R} = \frac{12\pi^4}{5}\left(\frac{k_B T}{\hbar\omega_D}\right)^3 = \frac{12\pi^4}{5}\left(\frac{T}{\Theta_D}\right)^3$$

where $\Theta_D = \hbar\omega_D/k_B$, the *Debye temperature,* tabulated for several solids in Table 4.2. This expression has the correct low-temperature behavior, $C_V \propto T^3$, and the full expression approaches the Dulong–Petit value at high T. (See Problem 23.23.)

EXAMPLE 23.6

Contrast the Debye frequencies of the solids C(diamond) through Pb to the vibrational frequencies of the diatomic halogens.

SOLUTION The Debye temperatures in Table 4.2 can be converted to equivalent Debye frequencies in cm^{-1} units if they are divided by the conversion factor 1.439 cm K. These values can be compared directly to the halogen diatomic ω_e values in Table 19.2. We find

ω_D/cm^{-1} C: 1286   Si: 448   Ge: 260   Sn: 138   Pb: 61
ω_e/cm^{-1} F$_2$: 917   Cl$_2$: 560   Br$_2$: 325   I$_2$: 215

The trends are the same, decreasing down the Periodic column, but the wider range of magnitudes for the Debye frequencies reflect the considerable differences in bonding as one moves from light, strongly covalent-bonded diamond down to heavy, delocalized metallic-bonded Pb. (See Chapter 16 for a discussion of solid bonding.) The diatomic halogen bonds all have quite similar electronic origins, and the main factor in their ω_e value differences is simply the reduced mass differences. Just as F_2 is far from the vibrational high-T limit near room temperature, the heat capacities of diamond, Si, and Ge are far from the Dulong–Petit values.

⟹ RELATED PROBLEMS 23.24, 23.25, 23.33, 23.34

The Einstein and Debye expressions are quite similar functions of T, except at low T. Each is a function of a single parameter: the unique frequency ω in the Einstein expression, and ω_D in the Debye expression. As Figure 23.6(a) shows, these two parameters can be chosen so that the two functions nearly coincide. Figure 23.6(b) magnifies the low-T regions where the disagreement is most pronounced and compares the full Debye expression to the T^3 limiting expression.

The *electronic* contribution to the heat capacity of normally conducting metals (see Eq. (4.17) and Table 4.3) can also be derived from statistical mechanics using the free electron gas model of Chapter 16. Details of the derivation are given in several of this chapter's references.

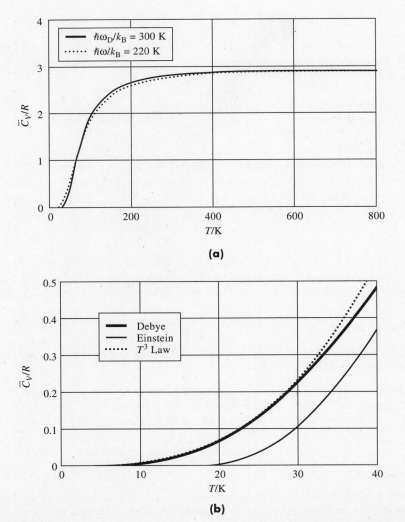

FIGURE 23.6 The Debye and Einstein heat capacity expressions are superficially quite similar at high T, (a), but at low T, (b), the Einstein expression falls too rapidly while the Debye expression recovers the experimentally observed T^3 behavior.

23.2 FROM MICROSCOPIC INDIVIDUALITY TO MACROSCOPIC AVERAGE

Statistical mechanics can also derive the Planck blackbody distribution law, discussed in Chapters 11 and 18 (Eq. (18.8)), although we must remember that Planck did not derive it this way.[8] The statistical view of blackbody radiation considers a collection of photons as an ideal gas (a relativistic gas, but an ideal one in the sense that photons do not interact with each other) in equilibrium with its material surroundings.[9] The photon frequencies (and thus their energies through $\epsilon = h\nu = \hbar\omega$) have no inherent upper limit, in contrast to the crystal vibrational frequencies of the Debye solid. Each photon energy state is doubly degenerate due to the two polarization directions of light,[10] and any one state is a *standing wave* analogous to the standing de Broglie waves in the quantum 3-D particle-in-a-box model's wavefunctions (see Figure 12.3(b)). Since the wavelength of light is generally always much smaller than a characteristic length of the cavity with which it is in equilibrium, the standing waves form a continuum of states. Since we know how to treat harmonic oscillators (i.e., diatomic molecules) at thermal equilibrium, and since it does not matter how they came to be in equilibrium, we could imagine them reaching equilibrium through radiative interactions alone. Turning this argument around, we could as easily imagine the radiation field reaching equilibrium through its interaction with equilibrated oscillators. We will see how these ideas show up in the mathematics of the problem.

The total energy of the photon gas is

$$E = \sum_\omega \hbar\omega n_\omega$$

where n_ω is the number of photons of energy $\hbar\omega$. Any one set of n_ω's constitutes a possible state of the photon gas, and since photons can come and go, any n_ω can be $0, 1, 2, \ldots$, without restriction. Consequently, we can write the canonical partition function for the gas, Eq. (22.21), as

$$\begin{aligned} Q &= \sum_j e^{-\beta E_j} & \text{(by definition of } Q\text{)} \\ &= \sum_j e^{-\beta \sum_{\omega=0}^{\infty} \hbar\omega n_\omega} & \text{(by definition of } E_j\text{)} \\ &= \prod_{\omega=0}^{\infty} \left(\sum_{n=0}^{\infty} e^{-\beta \hbar \omega n} \right) & \begin{array}{l}\text{(expanding the exponential sum,} \\ \text{allowing for all values of } n, \text{ and} \\ \text{writing the exponential sum as a} \\ \text{product of sums of exponentials)}\end{array} \\ &= \prod_{\omega=0}^{\infty} \frac{1}{1 - e^{-\beta \hbar \omega}} & \begin{array}{l}\text{(writing the sum over } n \text{ in} \\ \text{closed form)}\end{array} \end{aligned}$$

and use Eq. (22.27) to find the thermodynamic internal energy:

$$\Delta_0 U = k_B T^2 \frac{\partial \ln Q}{\partial T} = \sum_\omega \frac{\hbar\omega \, e^{-\hbar\omega/k_B T}}{1 - e^{-\hbar\omega/k_B T}}$$

since $\ln Q = -\sum \ln(1 - e^{-\beta\hbar\omega})$. This sum becomes an integral over ω if we weight the integral by the number of states at ω between ω and $\omega + d\omega$. This

---

[8] In his Nobel Prize address in 1920, Planck modestly described his distribution function as "an interpolation formula which resulted from a lucky guess." Arnold Sommerfeld's book, *Lectures on Theoretical Physics: Thermodynamics and Statistical Mechanics*, vol. 5 (Academic Press, New York, 1964) derives this function along the lines Planck first used. Planck did not use a statistical derivation until the fourth edition of his text on radiation theory, in 1921. He did, however, call it "an exceedingly simple derivation" in that edition.

[9] Since photons come and go through absorption and emission events, the number of them in the system is not constant. Only T and V, not N, count as thermodynamic variables for radiation.

[10] These polarizations correspond to a photon's two angular momentum components, $\pm\hbar$, along the direction of propagation.

degeneracy weight, which is derived from arguments similar to the density of translational states given a Closer Look in Section 23.1, is

$$g(\omega)\, d\omega = \frac{V\omega^2}{\pi^2 c^3} d\omega \ .$$

Consequently, the energy *density* is

$$\frac{\Delta_0 U}{V} = \int_0^\infty \frac{\hbar \omega^3\, e^{-\hbar\omega/k_B T}}{\pi^2 c^3\, (1 - e^{-\hbar\omega/k_B T})} d\omega = \int_0^\infty \frac{\hbar \omega^3}{\pi^2 c^3\, (e^{\hbar\omega/k_B T} - 1)} d\omega \ .$$

The integrand is the energy density per unit frequency, $\rho(\omega; T)$, which is the Planck blackbody distribution, Eq. (18.8) written in terms of ω rather than ν.

Note the third line leading to the final expression for Q. It is a product of terms, each of which is a harmonic oscillator partition function, $\Sigma\, e^{-\beta \hbar \omega n}$, and Q is thus a familiar product of partition functions for individual, independent modes of oscillation of the radiation field. Each mode is harmonic, at its own frequency, and the radiation field has components of the particle-in-a-box model—standing waves—and the harmonic oscillator model.

Next, we turn to the statistical mechanical derivation of the ideal gas equation of state. Equation (22.25) expresses the pressure P in terms of the canonical partition function Q, and, since a one-component ideal gas is a collection of indistinguishable particles in the dilute limit, Eq. (22.22) relates Q to q:

$$P = k_B T \left(\frac{\partial \ln Q}{\partial V}\right)_T = k_B T \left(\frac{\partial \ln (q^N/N!)}{\partial V}\right)_T = nRT \left(\frac{\partial \ln (q/N!)}{\partial V}\right)_T = nRT \left(\frac{\partial \ln q}{\partial V}\right)_T . \quad (23.21)$$

$\uparrow$ $\uparrow$ $\uparrow$ $\uparrow$

Eq. (22.25)   Using Eq. (22.22)   Since $nR = Nk_B$ Since $\partial \ln N!/\partial V = 0$

Here, q is a product of all the individual terms representative of the motions available to the molecules: $q_{tr} q_{el} q_{rot} \Pi q_{vib}$ where the product includes all the vibrational modes. But *no matter what the molecule*, the *only* term that depends on V in any way is q_{tr}. Equation (23.5) shows that $q_{tr} \propto V$, and thus $P = nRT(\partial \ln V/\partial V) = nRT/V$, the familiar ideal gas equation of state.

It is worthwhile to peer back through the steps that bring us to $PV = nRT$. The internal molecular partition functions are based on coordinates centered on the molecule, and as long as the container is large enough to ensure the internal energy levels are not perturbed from those of a completely isolated molecule, V will not affect any internal partition function. The translational partition function, however, follows the center of mass of the molecule through the coordinate system fixed to the container. In the classical partition function expression for q_{tr}, the factor V pops out simply because the classical Hamiltonian had no terms that depended on these container-fixed coordinates—no intermolecular potential energy terms. Integration over the three center-of-mass coordinates of the molecule gives the factor of V directly. As soon as we introduce intermolecular forces, this integral picks up a fairly complicated integrand, $\exp[-(\text{intermolecular potential energy})/k_B T]$. We will confront this in the next chapter.

We now consider statistical expressions for entropy. Equation (22.23) gives the absolute entropy in terms of the canonical partition function and the internal energy. For one mole of an ideal gas, with $Q = q_{tr}^{N_A}/N_A!$ and $\Delta_0 \overline{U} = 3RT/2$, the molar entropy is

$$\overline{S} = k_B \ln Q + \frac{\Delta_0 \overline{U}}{T} = k_B \ln q_{tr}^{N_A} - k_B \ln N_A! + \frac{3}{2} R \ .$$

The first term is $R \ln q_{tr}$, and the second is, with Stirling's approximation, $-k_B (N_A \ln N_A - N_A) = R \ln (1/N_A) + R$. If we write $R = R \ln e$ and $3R/2 = 3(R \ln e)/2$ in the second and third terms, we can combine everything into one term:

$$\bar{S}_{tr} = R \left[\ln q_{tr} + \ln\left(\frac{1}{N_A}\right) + \ln e + \frac{3}{2} \ln e \right] = R \ln \left[\left(\frac{mk_B T}{2\pi \hbar^2}\right)^{3/2} \frac{\bar{V} e^{5/2}}{N_A} \right] \quad (23.22a)$$

where we have used Eq. (23.5) for q_{tr}. This is the *Sackur–Tetrode equation* from Chapter 5, Eq. (5.34). That equation was written in terms of the molar mass M and the pressure P, but $M/\text{g mol}^{-1} = (m/\text{kg})(1000/\text{g kg}^{-1})(N_A/\text{mol}^{-1})$ and $P/\text{atm} = (R/\text{J mol}^{-1} \text{ K}^{-1})(T/\text{K})/(\bar{V}/\text{m}^3)(101\ 325/\text{J m}^{-3} \text{ atm}^{-1})$. Thus, we can also write the Sackur–Tetrode equation as

$$\bar{S}_{tr} = R \ln \left\{ \left[\left(\frac{k_B}{(1000\ N_A)\ 2\pi \hbar^2}\right)^{3/2} \frac{R e^{5/2}}{101\ 325\ N_A} \right] \frac{(T/\text{K})^{5/2}\ (M/\text{g mol}^{-1})^{3/2}}{P/\text{atm}} \right\} \quad (23.22b)$$

where the constants in square brackets are the Sackur–Tetrode constant $C = 0.311\ 968/\text{g}^{-3/2}\ \text{mol}^{3/2}\ \text{K}^{-5/2}$ atm as used in Chapter 5.

EXAMPLE 23.7

The JANAF Tables list the standard molar entropy of $F_2(g)$ at 300 K as $\bar{S}° = 202.983$ J K^{-1} mol^{-1}. How much of this entropy is due to internal motion (rotation and vibration) of F_2?

SOLUTION Each type of motion contributes to the total entropy. Writing $\bar{S}° = \bar{S}°_{tr} + \bar{S}°_{int}$, the internal contribution is $\bar{S}°_{int} = \bar{S}° - \bar{S}°_{tr}$ where the Sackur–Tetrode expression gives the translational component, $\bar{S}°_{tr}$. For F_2, $M = 37.996\ 806$ g mol^{-1}, and the JANAF standard state pressure is $P = 1$ *bar* rather than 1 *atm*. Equation (23.22b) is

$$\bar{S}°_{tr} = R \ln \left[0.311\ 968\ \frac{(300)^{5/2}(37.996\ 806)^{3/2}}{1.013\ 25} \right] = 154.132 \text{ J K}^{-1} \text{ mol}^{-1}$$

or about 76% of the total entropy. Note that this is roughly the total entropy of Ar ($M = 40$ g mol^{-1}) in the same state. We are about to see how to calculate the internal contributions directly.

➡ *RELATED PROBLEMS* 23.27, 23.28, 23.29, 23.30

We add the relevant internal entropy components directly to the translational component for gases. When we write $Q = q^N/N!$, and further write $q = q_{tr} q_{int}$, we associate the $N!$ factor with q_{tr} only, since $N!$ accounts for indistinguishability in the dilute limit, and it is the translational motion that produces indistinguishability.[11] For a gas, $q_{int} = q_{nuc} q_{rot} q_{vib} q_{el}$, and each of these factors has its own entropy contribution.

At temperatures low enough to keep excited electronic state populations negligible, $q_{el} = g_0$, the ground-state degeneracy. Since this degeneracy is often 1, the electronic degree of freedom often makes no entropy contribution. Likewise, the nuclear spin contribution, using Eq. (23.10), is independent of T, but it is almost always ignored since chemical reactions and most physical processes do not change spin states. One important exception, of course, is equilibrium in an external magnetic field where spin entropy can play an important role in phenomena such as adiabatic demagnetization, discussed in Chapter 9.

---

[11]For crystals, there is no translational (or rotational) term, nor is there a factor of $N!$ since atoms at crystal lattice positions are distinguishable. See Example 23.8 and Problem 23.35.

Since rotational motion is almost always in the high-T limit, we can use Eq. (23.12) or (23.13) for q_{rot} and the equipartition value, RT or $3RT/2$, for $\Delta_0\overline{U}$ in Eq. (22.23). We will consider only the linear molecule case (Eq. (23.12) and $\Delta_0\overline{U} = RT$) in detail, where we have

$$\overline{S}_{rot} = k_B \ln q_{rot}^{N_A} + \frac{\Delta_0\overline{U}}{T} = R \ln \left(\frac{2Ik_BT}{\sigma\hbar^2}\right) + R = R \ln \left(\frac{2Ik_BTe}{\sigma\hbar^2}\right). \quad (23.23)$$

We can write the factors in parentheses in more convenient terms using Table 23.2:

$$\overline{S}_{rot} = R \ln \left(\frac{Te}{\sigma k'B_e}\right) = R \ln \left(1.889\,311\,\frac{T/K}{\sigma B_e/\text{cm}^{-1}}\right)$$

where $1.899\,311 = e/k'$. Note that this entropy depends only on the symmetry number, the equilibrium rotational constant, and the temperature. For F_2 in the state discussed in Example 23.7, ($\sigma = 2$, $B_e = 0.8902$ cm^{-1}, $T = 300$ K), $\overline{S}_{rot} = 47.918$ J K^{-1} mol^{-1}. Translation and rotation account for 202.050 J K^{-1} mol^{-1}, nearly all of the 202.983 J K^{-1} mol^{-1} total.

The vibrational entropy is usually quite small, as in the F_2 case, since usually very few vibrational levels are populated at ordinary temperatures. We can find the vibrational entropy for any one mode using Eq. (23.7) for q_{vib} and Eq. (23.19) for $\Delta_0\overline{U}_{vib}$. If we omit the zero-point term, $e^{-\hbar\omega/2k_BT}$, in q_{vib}, which cannot lead to anything observable, we find

$$\overline{S}_{vib} = k_B \ln q_{vib}^{N_A} + \frac{\Delta_0\overline{U}_{vib}}{T} = -R \ln (1 - e^{-\hbar\omega/k_BT}) + \frac{R\hbar\omega}{k_BT(e^{\hbar\omega/k_BT} - 1)}, \quad (23.24)$$

which is not a very transparent expression. At low T, $\overline{S}_{vib} \to 0$, as only one state is populated. At high T, we can expand the exponentials ($e^x \cong 1 + x$) as we did in Example 23.3 and find

$$\overline{S}_{vib} = R \ln \left(\frac{k_BT}{\hbar\omega}\right) + N_Ak_B = R \ln \left(\frac{k_BTe}{\hbar\omega}\right).$$

For F_2 at 300 K, Eq. (23.24) gives (using the harmonic constant in Table 19.2 and the usual conversion factor for constants in cm^{-1} units) $\overline{S}_{vib} = 0.559$ J K^{-1} mol^{-1}. Adding the translation and rotation entropies gives a total of 202.609 J K^{-1} mol^{-1}, 0.374 J K^{-1} less than the tabulated amount. This difference can be traced to our use of the harmonic oscillator/rigid-rotor approximations for q_{vib} and q_{rot}. The tabulated values include anharmonicity, centrifugal distortion, and vibration-rotation coupling effects. Nevertheless, our simplest approximations account for 99.8% of the tabulated value.

Figure 23.7 shows a similar calculation for C_2H_2, again using JANAF data and the molecular constants used in Figure 23.5 (plus M and the rotational constant). The heavy line is the total molar entropy, and the light lines show the contributions of various degrees of freedom: translation, rotation, the two doubly degenerate low-frequency bends, and the three high-frequency stretches. The translational component is clearly the largest, but rotations and the low-frequency bends play appreciable roles, especially at higher T.

EXAMPLE 23.8

The low-temperature limiting expression for the Debye solid's entropy was used in Section 4.6 to calculate absolute entropies from experimental heat capacities. What is the low-temperature molecular partition function that gives this molar entropy, $\overline{S} = \overline{C}_V/3$, where $\overline{C}_V = 12\pi^4RT^3/5\Theta_D^3$?

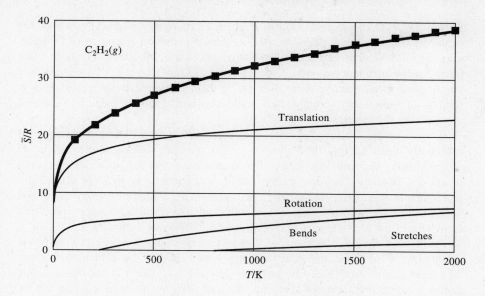

FIGURE 23.7 The statistical entropy expression for C_2H_2 (heavy line) agrees well with tabulated thermodynamic entropies (points). The statistical expression can be dissected into the various contributions shown as light lines.

SOLUTION We start with Eq. (22.23) for entropy, use the distinguishable particle version of Eq. (22.22) for Q, since crystal lattice sites are distinguishable, and use Eq. (22.27) for $\Delta_0 \overline{U}$:

$$\overline{S} = k_B \ln Q + \frac{\Delta_0 \overline{U}}{T} \quad \text{(Eq. (22.23))}$$

$$= k_B \ln q^{N_A} + k_B T \left(\frac{\partial \ln q^{N_A}}{\partial T} \right)_V \quad \text{(Eq.'s (22.22) and (22.27))}$$

$$= R \ln q + RT \left(\frac{\partial \ln q}{\partial T} \right)_V \quad \text{(Since } R = N_A k_B\text{)}$$

If we write $\overline{S}/R = aT^3$ where $a = 12\pi^4/15\Theta_D^3$, and let $y = \ln q$ and $x = T$, the final expression above simplifies to

$$ax^3 = y + x\left(\frac{\partial y}{\partial x}\right).$$

If we try the solution $y = bx^3$, this differential equation becomes $ax^3 = bx^3 + x(3bx^2) = 4bx^3$ or $b = a/4$, and thus $y = ax^3/4$ or $\ln q = aT^3/4 = 12\pi^4 T^3/60\Theta_D^3$. This is an example of a situation where it is more useful to start with a thermodynamic expression and work backwards, in a sense, toward the statistical expression.

➡ *RELATED PROBLEMS* 23.33, 23.35

There is one other statistical source of entropy that takes us back to our simplest model, the two-level system. Called *residual entropy,* this contribution is most often a nonequilibrium effect. It arises when, typically, a solid forms with configurational disorder that persists to $T = 0$. (See Problem 4.2 for an early discussion of this idea.) The true equilibrium solid at $T = 0$ has an ordered molecular arrangement that has absolute lowest energy status and zero entropy. It is common, however, to find molecular solids in which the molecules have distinguishable but chemically similar "ends" (such as CO, NO, N_2O, etc.) arranged in a more or less random fashion that persists at very low T. Once a random configuration forms, the energy barrier to reordering into the most ordered configuration is so high (imagine trying to turn one N_2O molecule around in a lattice of its tightly packed neighbors, for example) that it never happens and the random distribution is said to be "frozen" into the solid. Similarly, when isotopic distinctions are important, the slight zero-point energy differences among, for example, $^{35}Cl_2$, $^{37}Cl_2$, and $^{35}Cl^{37}Cl$ plus the homonuclear/heteronuclear distinction among them would mean that a solid random mixture of isotopically scrambled Cl_2 molecules would not likely be in a state of absolute lowest energy.

If we consider the simplest case: molecules with two different orientations of essentially equal energy, then when N such molecules solidify, there will be 2^N different ways the solid can form.[12] Boltzmann's original statistical interpretation of entropy predicts a residual entropy at $T = 0$ equal to $k_B \ln 2^N$ or $R \ln 2$ per mole. This is approximately the difference between the statistical entropy calculated *without* considering configurational disorder and the absolute entropy calculated from Third Law heat capacity and phase transition data as discussed in Chapter 4.

Finally, we turn to the roots of thermodynamics and consider the statistical interpretations of work and heat. Imagine 1 mol of Ar in a cubical box about 29 cm on a side. At 298 K, the gas pressure is 1 bar. The 3-D particle-in-a-box energy level expression applies here, and the total partition function is q_{tr}, Eq. (23.5). We can use the Boltzmann distribution, $e^{-\epsilon_i/k_B T}/q_{tr}$, to calculate the population fraction, n_i/N_A, of any translational state. Suppose we keep the box fixed in size and raise the temperature to 398 K. The energy level expression does not change with temperature, but the population fractions do, since both q_{tr} and the Boltzmann factor depend on T. The population of some states goes down, and that of others goes up. Thermodynamics tells us that the internal energy of the gas increases an amount $\Delta U = q_V = C_V \Delta T = 3R\Delta T/2$. Statistics finds ΔU through the expression

$$\Delta U = \Delta \left[\sum (\text{state energy}) \cdot (\text{state population}) \right] .$$

Since the state energies did not change, we see that *heat is most directly linked to changes in state populations.*

Now suppose we change the internal energy through a reversible adiabatic work process, $\Delta U = w_{ad}$. We thermally insulate the box and reversibly slide one wall in a bit, doing work and raising U. The methods of Chapter 2 tell us that moving one wall in about 10.3 cm does enough work to change U by the same amount as the $\Delta T = 100$ K constant volume heat process we just considered. Again, the temperature goes up to 398 K and state populations change, but *unlike the constant volume heat process,* the work process has *changed the state energies* as well, since particle-in-a-box energies depend on the size of the box. In the initial state, the volume was a cube, and the symmetry of the cube causes many states to be degenerate. Pushing in one side of a cube lowers this symmetry and breaks some of the degeneracies.

Figure 23.8 contrasts graphically these heat and work processes. The point in the upper left corner represents the quantum state energy of quantum numbers $n_x = n_y = n_z = 10^{10}$ in the initial thermodynamic state of the system. The heat process has kept the quantum state energy fixed, but lowered the state population. The reversible adiabatic work process has changed both the population and the state energy. Follow the curve connecting quantum states for the final state of the heated system and you will see that some populations fall, but some rise. Also follow the curve connecting initial quantum states and note that it is the same curve connecting

---

[12]If the molecules lock into only one orientation, due to highly directional intermolecular forces, then there is only one way ($1^N = 1$) to form the solid, and o^N ways in general for o different orientations. The *orientation number o* plays a role similar to the symmetry number σ for rotational states.

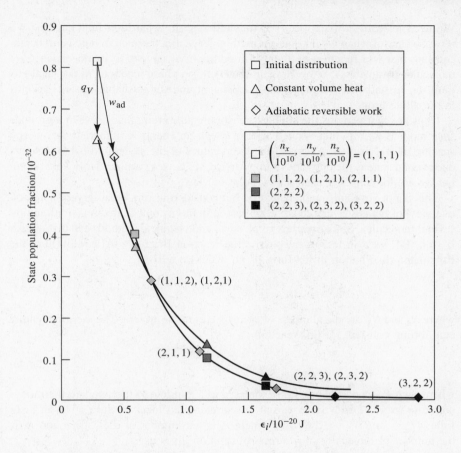

FIGURE 23.8 The population fractions of translational states of a gas change in clear ways if the gas is heated or subjected to reversible adiabatic compression. Heating at constant V alters populations without changing state energies. Work alters state energies and, in general, state populations as well.

the quantum states of the system after the work process. This is due to the nature of the thermodynamic path. Note as well how the compression has broken the degeneracy of some of the states in the initial cube. (The figure assumes the wall perpendicular to the x direction was pushed in.)

We have now considered statistical expressions for energy, entropy and the process variables of thermodynamics, heat and work. Statistical thermodynamic expressions for enthalpy and free energies follow directly from the expressions in Chapter 22 and the partition functions we have derived here. The most important of these is the chemical potential since it will give us a statistical expression for a chemical reaction equilibrium constant, the central topic of the next section.

23.3 STATISTICAL THERMOCHEMISTRY

Thermochemical equilibrium revolves around chemical potentials, and Chapter 22 gave us statistical expressions for the chemical potential in general, Eq. (22.29), and for the specific case of dilute, indistinguishable particles, Eq. (22.30). We can use these expressions to find the standard free energy change for reaction, Eq. (7.19):

$$\Delta_r \overline{G}^\circ = \gamma \mu_C^\circ + \delta \mu_D^\circ - \alpha \mu_A^\circ - \beta \mu_B^\circ \quad \text{for} \quad \alpha A + \beta B \rightarrow \gamma C + \delta D \quad \textbf{(7.19)}$$

and the reaction's thermodynamic equilibrium constant, Eq. (7.22):

$$K_{eq} = \exp\left(-\frac{\Delta_r \overline{G}^\circ}{RT}\right). \quad \textbf{(7.22)}$$

We need to learn how to specify a standard state in a partition function, and we must be careful when we do that our method leads to a thermodynamic equilibrium constant that is a function of T only (in addition, of course, to its dependence on molecular parameters). We will again focus on gas-phase reactions in the ideal gas limit, or, in statistical language, on independent, indistinguishable gaseous species in the dilute limit.

Consider first perhaps the simplest reaction equilibrium, that between a ground-state atom B and its first excited state B* with an energy ϵ higher. If the energy spacing of B's excited states and the temperature of the system are such that this one excited state is the only one of importance, we have a familiar two-level system, but we are looking at it from a new point of view.

This is not very profound chemistry, but considering it in detail gives us a good overview of the issues that must be addressed in more complicated reactions among several molecules. Since a variety of physical and chemical methods can distinguish B from B*, we can legitimately write the reaction B $\rightleftarrows$ B*. At equilibrium, the Boltzmann distribution in the form of Eq. (22.14c) tells us

$$\frac{N_{B^*}}{N_B} = \frac{g_{B^*}}{g_B} e^{-\epsilon/k_B T} \tag{23.25}$$

where N_B and N_{B^*} are the numbers of each species in the system. The thermodynamic equilibrium constant for this reaction is

$$K_{eq} = \frac{P_{B^*}}{P_B} \tag{23.26}$$

where each P is a partial pressure expressed as a multiple of the standard thermodynamic pressure, 1 bar. Since partial pressures for an ideal mixture of ideal gases follow $P_i = n_i RT/V = N_i RT/N_A V$ where n_i is the number of moles of i and N_i is the number of molecules of i, direct substitution gives us

$$K_{eq} = \frac{n_{B^*} RT/V}{n_B RT/V} = \frac{N_{B^*} RT/N_A V}{N_B RT/N_A V} = \frac{N_{B^*}}{N_B} ,$$

which is identical to Eq. (23.25).

To go from statistics to thermodynamics, we first write molecular partition functions for B and B*. Each is the product $q_{tr} q_{el}$, and we can introduce the standard pressure into the translational part. Since q_{tr} is proportional to the volume, Eq. (23.5), and since the gases are ideal, we can express a standard volume $V°$ as

$$V_i° = \frac{n_i RT}{P°} = \frac{N_i RT}{N_A P°}$$

where $P°$ is the standard pressure.

The electronic part is trickier. Since we are distinguishing B and B*, the ground state belongs to B and the excited state belongs to B*, but *since the two interconvert by a reaction that can rapidly establish equilibrium between them,*[13] we must express their energies *on a common scale*. Consequently, we pick the energy of B as the energy zero and assign B* an energy ϵ, as if ϵ was a zero-point energy for B* that we cannot ignore. With this choice, the electronic partition functions are

$$q_{el}(B) = g_B \quad \text{and} \quad q_{el}(B^*) = g_{B^*} e^{-\epsilon/k_B T} .$$

---

[13]"Rapidly" implies a rate that is appreciable on some time scale of interest. Rates are the topic of Chapters 26 and 27, but the idea here is that B and B* cannot exist in isolation at the temperature of the system. If we have one, we have the other, in a time period short enough to keep us from getting bored waiting for equilibrium.

The total partition functions are (with a superscript ○ to indicate the standard state)

$$q^\circ(B) = \frac{V^\circ}{\Lambda^3} g_B = \frac{N_B RT}{N_A P^\circ \Lambda^3} g_B \quad \text{and} \quad q^\circ(B^*) = \frac{V^\circ}{\Lambda^3} g_{B^*} = \frac{N_{B^*} RT}{N_A P^\circ \Lambda^3} g_{B^*} e^{-\epsilon/k_B T}$$

where we have exploited the fact that, since B and B* are at the same T, and have the same mass, they have the same thermal deBroglie wavelengths Λ.

Next we use Eq. (22.30) for the chemical potentials of each:

$$\Delta_0 \mu^\circ(B) = -RT \ln \frac{q^\circ(B)}{N_B} = -RT \ln\left(\frac{RT}{N_A P^\circ \Lambda^3} g_B\right)$$

$$\Delta_0 \mu^\circ(B^*) = -RT \ln \frac{q^\circ(B^*)}{N_{B^*}} = -RT \ln\left(\frac{RT}{N_A P^\circ \Lambda^3} g_{B^*} e^{-\epsilon/k_B T}\right)$$

and Eq. (7.19) for $\Delta_r \overline{G}^\circ$:

$$\Delta_r \overline{G}^\circ = \Delta_0 \mu^\circ(B^*) - \Delta_0 \mu^\circ(B)$$

so that Eq. (7.22) can be written

$$\ln K_{eq} = -\frac{\Delta_r \overline{G}^\circ}{RT} = -\frac{\Delta_0 \mu^\circ(B^*) - \Delta_0 \mu^\circ(B)}{RT}. \quad (23.27)$$

Substituting the partition function expressions for the chemical potentials gives

$$\ln K_{eq} = \ln\left(\frac{RT}{N_A P^\circ \Lambda^3} g_{B^*} e^{-\epsilon/k_B T}\right) - \ln\left(\frac{RT}{N_A P^\circ \Lambda^3} g_B\right)$$

$$= \ln\left(\frac{g_{B^*} e^{-\epsilon/k_B T}}{g_B}\right),$$

which shows that

$$K_{eq} = \frac{g_{B^*} e^{-\epsilon/k_B T}}{g_B} \quad (23.28)$$

in agreement with Eq. (23.25).

Had we chosen B* to have zero energy instead of B, we would have written $q_{el}(B^*) = g_{B^*}$ and $q_{el}(B) = g_B e^{+\epsilon/k_B T}$. This would have given us the same expression for K_{eq}, again establishing that the state we choose to place at zero energy is arbitrary, as long as the entire system is referenced to that choice.

We have now shown that the equilibrium constant can be approached from the thermodynamic *or* the statistical mechanic point of view. Only two steps were new: expressing the partition function in terms of a standard pressure and relating product and reactant species to the same energy zero. The first step makes the partition functions, and thus the equilibrium constant derived from them, *functions of T alone*, as thermodynamic equilibrium constants are supposed to be.

The second step introduces basic thermochemistry into the story. If we write Eq. (23.27) in terms of standard state partition functions:

$$\ln K_{eq} = \ln \frac{q_{B^*}^\circ/N_{B^*}}{q_B^\circ/N_B}$$

and differentiate this with respect to T, we find

$$\frac{\partial \ln K_{eq}}{\partial T} = \frac{\partial \ln q_{B^*}^\circ/N_{B^*}}{\partial T} - \frac{\partial \ln q_B^\circ/N_B}{\partial T}$$

$$= \frac{\partial \ln q_{B^*}^\circ}{\partial T} - \frac{\partial \ln q_B^\circ}{\partial T}$$

$$= \frac{1}{N_A}\left(\frac{\partial \ln Q_{B^*}^\circ}{\partial T} - \frac{\partial \ln Q_B^\circ}{\partial T}\right)$$

$$= \frac{1}{N_A}\left(\frac{\Delta_0 \overline{U}^\circ(B^*)}{k_B T^2} - \frac{\Delta_0 \overline{U}^\circ(B)}{k_B T^2}\right) = \frac{\Delta_r \overline{U}^\circ}{RT^2}.$$

If we differentiate $\ln K_{eq}$ from Eq. (23.28), we find

$$\frac{\partial \ln K_{eq}}{\partial T} = \frac{\partial \ln (g_{B^*} e^{-\epsilon/k_B T} g_B^{-1})}{\partial T}$$

$$= \frac{\partial (\ln g_{B^*} - \epsilon/k_B T - \ln g_B)}{\partial T}$$

$$= -\frac{\epsilon}{k_B}\frac{\partial (1/T)}{\partial T} = \frac{\epsilon}{k_B T^2} = \frac{N_A \epsilon}{RT^2},$$

which shows that $N_A \epsilon = \Delta_r \overline{U}^\circ$ for this reaction. Chapter 7 showed that $(\partial \ln K_{eq}/\partial T)_P = \Delta_r \overline{H}^\circ / RT^2$, which is consistent with this result, since for our ideal gas reaction, we have $\Delta_r \overline{H}^\circ = \Delta_r \overline{U}^\circ + \Delta_r(P^\circ V) = \Delta_r \overline{U}^\circ + RT(\Delta_r n) = \Delta_r \overline{U}^\circ$ because $\Delta_r n = 0$.

We can now make the first of two important deductive leaps. First, suppose B and B* were *isomers of the same basic molecule,* such as HCN and HNC, instead of two states of the same atom. What would change? The standard-state partition functions would include *internal vibrational and rotational degrees of freedom* in addition to electronic, and we would still need a single zero of energy, but *nothing else would change.* Second, suppose B is a molecule that dissociates into two fragments C and D. What would change here? Again, the standard-state partition function for B would include all internal motion as well as translational, but q° for the products C + D would be the *product of partition functions for C and D:* $q^\circ = q_C^\circ q_D^\circ$.

We can see that this must be so if we consider the nature of the states of the products of a chemical reaction from the thermodynamic point of view. Recall from Chapter 7 that we distinguished between chemical reactions written with a single arrow, →, and a double arrow, ⇌. For the double arrow, we had in mind equilibration among reactants and products. The single arrow was reserved to remind us of the meaning of "Δ_r": complete stoichiometric conversion of isolated reactants to stoichiometric amounts of isolated products. Since our statistical expression for K_{eq} relies on $\Delta_r \overline{G}^\circ$, our partition function for products (or reactants, in the general case of more than one reactant species) must refer to the states of stoichiometric amounts of isolated products. Again, all states must be referenced to a common energy zero.

For example, suppose C and D had only two nondegenerate states each, ϵ_{C1} and ϵ_{C2} for C and ϵ_{D1} and ϵ_{D2} for D. The system C + D could then exist in only four energy states: $\epsilon_1 = \epsilon_{C1} + \epsilon_{D1}$, $\epsilon_2 = \epsilon_{C1} + \epsilon_{D2}$, $\epsilon_3 = \epsilon_{C2} + \epsilon_{D1}$, and $\epsilon_4 = \epsilon_{C2} + \epsilon_{D2}$, and the partition function would be

$$\begin{aligned}q &= e^{-\beta\epsilon_1} + e^{-\beta\epsilon_2} + e^{-\beta\epsilon_3} + e^{-\beta\epsilon_4}\\ &= e^{-\beta(\epsilon_{C1}+\epsilon_{D1})} + e^{-\beta(\epsilon_{C1}+\epsilon_{D2})} + e^{-\beta(\epsilon_{C2}+\epsilon_{D1})} + e^{-\beta(\epsilon_{C2}+\epsilon_{D2})}\\ &= e^{-\beta\epsilon_{C1}}e^{-\beta\epsilon_{D1}} + e^{-\beta\epsilon_{C1}}e^{-\beta\epsilon_{D2}} + e^{-\beta\epsilon_{C2}}e^{-\beta\epsilon_{D1}} + e^{-\beta\epsilon_{C2}}e^{-\beta\epsilon_{D2}}\\ &= (e^{-\beta\epsilon_{C1}} + e^{-\beta\epsilon_{C2}})(e^{-\beta\epsilon_{D1}} + e^{-\beta\epsilon_{D2}}) = q_C q_D\end{aligned}$$

or simply the *product of the partition functions for C and D alone.* Note how "isolation of C and D" is expressed here. The species C and D do not have to be in different bottles, but they do have to be sufficiently dilute in each other that C–D interactions can be neglected so that each retains its own energy levels, unperturbed by the other. Also, any reference energy ϵ_0 common to ϵ_1–ϵ_4 factors *once* from q as a single term, $e^{-\beta\epsilon_0}$, so that $q = q_C' q_D' e^{-\beta\epsilon_0}$ where q_C' and q_D' are expressed in terms of energies measured from ϵ_0.

EXAMPLE 23.9

In high-temperature stellar atmospheres, atomic hydrogen is appreciably ionized. Write the statistical expression for K_{eq} of the ionization reaction $H \rightarrow H^+ + e^-$ and show that $\partial \ln K_{eq}/\partial T = \Delta_r \overline{H}^\circ/RT^2$.

SOLUTION In the language of the text, H is B, H^+ is C, and e^- is D. If we choose the H ground state (1s $^2S_{1/2}$) as the energy zero, then the ground state energies of the $H^+ + e^-$ products are higher by an amount ϵ_0 equal to the ionization energy of H. The standard-state partition function for H is a product of translational, electronic, and nuclear spin parts:

$$\frac{g_H^\circ}{N_H} = \frac{q_{tr}^\circ}{N_H} q_{el} q_{nuc} = \left[\left(\frac{m_H k_B T}{2\pi \hbar^2} \right)^{3/2} \frac{RT}{N_A P^\circ} \right] (2)(2)$$

where we have assumed only the ground electronic state ($g_0 = 2$) is appreciably populated. The H^+ and e^- functions are similar: q_{H^+} has a nuclear spin degeneracy of 2 and q_{e^-} has an electron spin degeneracy of 2. The product of these functions is

$$\frac{g_{H^+}^\circ}{N_{H^+}} \frac{g_{e^-}^\circ}{N_{e^-}} = \left[\left(\frac{m_{H^+} k_B T}{2\pi \hbar^2} \right)^{3/2} \frac{RT}{N_A P^\circ} \right] \left[\left(\frac{m_e k_B T}{2\pi \hbar^2} \right)^{3/2} \frac{RT}{N_A P^\circ} \right] (2)(2) \ ,$$

and we also have a single $e^{-\epsilon_0/k_B T}$ factor. Many terms cancel in the equilibrium constant expression (assuming H and H^+ have the same mass):

$$K_{eq} = \frac{(g_{H^+}^\circ/N_{H^+})(g_{e^-}^\circ/N_{e^-})}{g_H^\circ/N_H} e^{-\epsilon_0/k_B T} = \left[\left(\frac{m_e k_B}{2\pi \hbar^2} \right)^{3/2} \frac{R}{N_A P^\circ} \right] T^{5/2} e^{-\epsilon_0/k_B T} \ .$$

If we differentiate $\ln K_{eq}$ with respect to T, we find

$$\frac{\partial \ln K_{eq}}{\partial T} = \frac{\partial \left(\frac{5}{2} \ln T - \frac{\epsilon_0 N_A}{RT} \right)}{\partial T} = \frac{1}{RT^2} \left(\frac{5}{2} RT + \epsilon_0 N_A \right) \ .$$

The second term in parentheses is $\Delta_r \overline{U}^\circ(T = 0)$, and the first is the enthalpy difference between the reactant and products over the temperature range $T = 0$ to the temperature of the system, according to Eq. (7.9):

$$\Delta_r \overline{H}^\circ(T) = \Delta_r \overline{H}^\circ(T = 0) + \int_0^T \Delta_r \overline{C}_P^\circ \, dT = \Delta_r \overline{U}^\circ(T = 0) + \frac{5}{2} RT \ .$$

⇒ **RELATED PROBLEMS** 23.37, 23.38

We now have all the information we need to write a general statistical mechanical expression for the equilibrium constant of any gas-phase reaction. We need standard-state partition functions for each reactant and product, and we need to choose a reference energy for all species. The first step is easier, since we can write the partition functions in terms of isolated molecules referenced to their individual ground states as we did in Section 23.1, but corrected to standard pressure conditions. The second step often requires information from experimental thermochemistry.

Consider, for example, the reaction

$$H_2(g) + \frac{1}{2} O_2(g) \rightarrow H(g) + OH(g)$$

for which the enthalpies of the reactants and products are shown on an enthalpy-level diagram in Figure 7.2. If, as is done in that figure, we place the reactants at the zero of enthalpy, the product enthalpies are 257.1 kJ mol^{-1} higher. Since the statistical reference energy ϵ_0 used in K_{eq} is an *absolute zero energy* difference, not a *finite T enthalpy* difference, this number must be corrected. This is often a small, easily made correction for small molecules. Alternatively, ϵ_0 can be related to bond dissociation energies $\mathcal{D}_0$.

In this example, we can write

$$H_2(g) + \frac{1}{2}O_2(g) \rightarrow 2H(g) + O(g) \rightarrow H(g) + OH(g) \ .$$

The first step requires an energy $\mathcal{D}_0(H_2) + \frac{1}{2}\mathcal{D}_0(O_2)$, and the second releases $\mathcal{D}_0(OH)$. If we look up these dissociation energies, we find

$$\epsilon_0 = \mathcal{D}_0(H_2) + \frac{1}{2}\mathcal{D}_0(O_2) - \mathcal{D}_0(OH) = 255.1 \text{ kJ mol}^{-1}$$

as illustrated in Figure 23.9. This figure also shows various combinations of vibra-

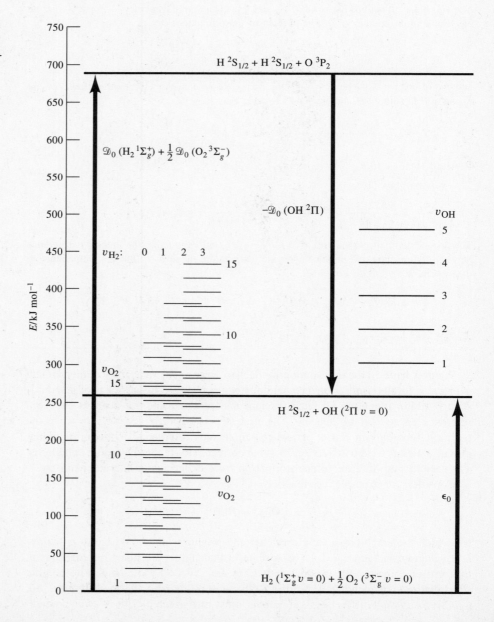

FIGURE 23.9 In a reacting mixture, all the energy levels of all the reactants and products are interconnected on a common energy scale. Bond energies are the most direct way of interconnecting these energies.

tional energies added to the ground electronic-state energies of the reactants and products to emphasize how all the energy levels of the system are related.

We can compare ϵ_0 to a corrected thermochemical value starting with $\Delta_r \overline{H}°(298.15 \text{ K}) = 257.1 \text{ kJ mol}^{-1}$ from Figure 7.2. Since $\Delta_r \overline{H}° = \Delta_r \overline{U}° + \Delta_r n(RT)$ and $\Delta_r n = 0.5$ here, $\Delta_r \overline{U}°(25 \text{ °C}) = 257.1 \text{ kJ mol}^{-1} - (0.5) R (298.15 \text{ K}) = 255.9 \text{ kJ mol}^{-1}$. We correct this to absolute zero by subtracting $T\Delta_r \overline{C}_V° = (298.15 \text{ K})(0.25 R) = 0.62 \text{ kJ mol}^{-1}$ so that $\Delta_r \overline{U}°(T = 0) = \epsilon_0 = 255.3 \text{ kJ mol}^{-1}$, in good agreement with the value based on dissociation energies.[14]

Dissociation energies are not as well known for polyatomics as they are for diatomics. When they are unavailable, ϵ_0 has to be calculated from experimental thermodynamic data corrected from $\Delta_r \overline{H}°$ at some finite T to $\Delta_r \overline{U}°$ at $T = 0$. If we always choose the ground states of the reactants as the reference energy, then $\Delta_r \overline{U}°(T = 0) = \epsilon_0$ always, and the algebraic sign of $\Delta_r \overline{U}°$ will correctly reflect the exo- or endothermicity of the reaction. For an endothermic reaction, such as in our example, the products' ground states lie above the reactants in energy, and ϵ_0 is positive. Consequently, the Boltzmann factor $e^{-\epsilon_0/k_B T}$ is < 1. For an exothermic reaction, $\epsilon_0 < 0$ and $e^{-\epsilon_0/k_B T} > 1$. This part of our expression for K_{eq} shows how absolute energy differences between reactants and products control equilibria.

Our general K_{eq} expression for the reaction $\alpha A + \beta B \rightarrow \gamma C + \delta D$ is

$$K_{eq} = \frac{(q_C°/N_C)^\gamma (q_D°/N_D)^\delta}{(q_A°/N_A)^\alpha (q_B°/N_B)^\beta} e^{-\epsilon_0/k_B T} \qquad (23.29)$$

where ϵ_0 is $\Delta_r \overline{U}°$ at absolute zero. We put this expression to work in the following examples in order to expose further connections between thermodynamics and statistical mechanics.

EXAMPLE 23.10

What is the equilibrium constant for the isotope scrambling reaction

$$2\,^{37}\text{Cl}^{35}\text{Cl} \rightarrow \,^{37}\text{Cl}_2 + \,^{35}\text{Cl}_2$$

and what factor dominates its value?

SOLUTION Problem 7.34 considers this reaction from a thermodynamic point of view, but even there, statistical arguments are needed to complete the story. The simplest approximation to K_{eq} assumes the reaction is thermoneutral ($\epsilon_0 = 0$), the masses are all equal, the rotational constants are all equal, and the vibrational constants are as well. The only fragments of any partition functions that are not cancelled from K_{eq} are the *symmetry numbers* σ in rotational partition functions for each molecule: $\sigma = 1$ for $^{37}\text{Cl}^{35}\text{Cl}$ and 2 for $^{37}\text{Cl}_2$ and $^{35}\text{Cl}_2$. Since rotational partition functions are inversely proportional to σ (Eq. (23.12)), K_{eq} is approximately

$$K_{eq} = \frac{\sigma_{37,35}^2}{\sigma_{37,37}\,\sigma_{35,35}} = \frac{1}{4}\;.$$

This is purely an entropy effect favoring the scrambled isotopomer. The various approximations we have made are not difficult to remove, and they do not affect the conclusion or the value for K_{eq} appreciably, as explored in Problem 23.41.

➠ RELATED PROBLEMS 23.41, 23.43, 23.44

---

[14]We find $\Delta_r \overline{C}_V°$ for this reaction using the approximation that translation and rotation, but not vibration, is at the high-T limit for each species at 298.15 K. This makes $\overline{C}_V° = 5R/2$ for H_2, $5R/4$ for $\frac{1}{2} O_2$, $3R/2$ for H, and $5R/2$ for OH. Consequently $\Delta_r \overline{C}_V° = R(3/2 + 5/2 - 5/2 - 5/4) = R/4$.

EXAMPLE 23.11

Compare the statistical expression for K_{eq} at 4000 K for the reaction

$$H_2(g) \to 2H(g)$$

to the value 2.546 calculated from thermodynamic data in Chapter 7.

SOLUTION We first collect standard-state partition function expressions. The atomic H expression has only translation and a doubly degenerate electronic ground state contribution:

$$q_H^\circ/N_H = q_{tr}\, q_{el} = \left[\left(\frac{m_H k_B T}{2\pi\hbar^2}\right)^{3/2} \frac{RT}{N_A P^\circ}\right] (2) \qquad (2)$$

while that for H_2 has translational, rotational (at the high-T limit—4000 K is high T for any rotation!), vibrational (perhaps *not* at the high-T limit), and a nondegenerate electronic contribution:

$$q_{H_2}^\circ/N_{H_2} = q_{tr}\, q_{rot}\, q_{vib}\, q_{el}$$

$$= \left[\left(\frac{2m_H k_B T}{2\pi\hbar^2}\right)^{3/2} \frac{RT}{N_A P^\circ}\right]\left(\frac{T}{2k' B_e}\right)\left(\frac{1}{1 - e^{-k'\omega_e/T}}\right) \qquad (1)$$

where k' is our cm^{-1} to K conversion factor. Note that the *vibrational partition function does not include the zero-point energy* as it did in Eq. (23.7). This is because *we place the zero of energy at the zero-point level of the reactant,* and this choice makes $\epsilon_0 = \mathcal{D}_0(H_2)$, the dissociation energy of H_2 from the zero-point level. This value is discussed in Table 7.8, the H atom JANAF Table. Since that table was published, improved spectroscopy on H_2 has shown that the theoretical value, 36 177.4 cm^{-1}, is very accurate, and we will use that value for ϵ_0. Assembling these pieces into K_{eq} gives, after cancellation of many common factors,

$$K_{eq} = \frac{Rk' B_e}{N_A P^\circ}\left(\frac{k_B m_H T}{\pi\hbar^2}\right)^{3/2}\left(1 - e^{-k'\omega_e/T}\right) e^{-k'\epsilon_0/T}$$

and we find spectroscopic constants in Table 19.2: $\omega_e = 4\,401.213$ cm^{-1} and $B_e = 60.853$ cm^{-1}. These values, along with other standard constants ($P^\circ = 1$ bar $= 10^5$ Pa), give $K_{eq} = 2.981$, in reasonable agreement with the JANAF value that is based on a direct sum partition function over the vibration/rotation states of H_2.

➡ *RELATED PROBLEMS* 23.45, 23.46, 23.47

EXAMPLE 23.12

Chapter 7 calculated that 1 mol $H_2(g)$ raised to 4000 K and 1 bar pressure dissociated to an equilibrium mixture of 0.376 mol H_2 + 1.248 mol H. What amount of the H_2 in this mixture is in an excited vibrational state?

SOLUTION The fraction in an excited state is just $1 -$ (the fraction in the ground state) $= 1 - n_0/N$, and Eq. (23.14) says $n_0/N = 1 - e^{-k'\omega_e/T} = 0.795$ for H_2 at 4000 K. Thus, only ~20% of the H_2 is vibrationally excited, and we could continue this calculation to show that only ~4% are in $v > 1$. In other words, the *overwhelming majority* of H_2 molecules have internal energies that are a *small fraction* of $\mathcal{D}_0$, the dissociation energy. (Note that the H_2 molecules would have essentially this distribution even if they could not dissociate.) How, then, can such a *large* fraction of the system be dissociated into atoms if most of the molecules have relatively little internal energy? The answer revolves around entropy. The products, $2H(g)$, have a much larger density of translational states than the $H_2(g)$ reactant has, and even though the atoms must pay a large energy price to reach these states, once they do, they are lost in them.

➡ *RELATED PROBLEMS* 23.44, 23.47

EXAMPLE 23.13

Given the dissociation energies $\mathcal{D}_0$ for I_2, Cl_2, and ICl: 12 440 cm^{-1}, 19 997 cm^{-1}, and 17 366 cm^{-1}, respectively, and data from Table 19.2, calculate K_{eq} at 300 K for the disproportionation reaction

$$2ICl(g) \rightleftarrows I_2(g) + Cl_2(g) \; .$$

SOLUTION We assign the ground state of $2ICl(g)$ as the energy zero and find ϵ_0 from $2ICl \rightarrow 2I + 2Cl \rightarrow I_2 + Cl_2$:

$$\epsilon_0 = 2(17\,366 \text{ cm}^{-1}) - 12\,440 \text{ cm}^{-1} - 19\,997 \text{ cm}^{-1} = 2295 \text{ cm}^{-1}$$

so that $e^{-k'\epsilon_0/T} = \exp[-(1.439 \text{ cm K})(2295 \text{ cm}^{-1})/(300 \text{ K})] = 1.659 \times 10^{-5}$. This reaction is discussed qualitatively in Problem 7.35 from a thermodynamic viewpoint. Using average bond enthalpies in that problem to find $\Delta_r \overline{H}°$ suggested $K_{eq} \cong e^{-\Delta_r \overline{H}°/RT} = 8.93 \times 10^{-6}$, quite close to the statistical energy factor alone.

The standard-state partition function expressions contain many common factors that cancel in K_{eq}, since $\Delta_r n = 0$ here. Those that survive are

$$\left(\frac{m_{I_2} m_{Cl_2}}{m_{ICl}^2}\right)^{3/2} \left(\frac{\sigma_{ICl}^2 B_{ICl}^2}{\sigma_{I_2} B_{I_2} \sigma_{Cl_2} B_{Cl_2}}\right) \frac{(1 - e^{-k'\omega_{ICl}/T})^2}{(1 - e^{-k'\omega_{I_2}/T})(1 - e^{-k'\omega_{Cl_2}/T})}$$

that, factor by factor, represent translation, rotation, and vibration, since the electronic ground states are nondegenerate. Numerically, these factors are $(0.564)(0.358)(1.183) = 0.239$, so that $K_{eq} = (0.239)(1.659 \times 10^{-5}) = 3.965 \times 10^{-6}$, within a factor of two of the bond enthalpy estimate.

→ **RELATED PROBLEMS** 23.36, 23.39, 23.48

Examples 23.10 and 23.13 contrast the two types of thermodynamic factors that can control chemical equilibria: entropy and enthalpy. As Problems 7.34 and 7.35 discuss in more detail, K_{eq} in Example 23.13 is strongly governed by the reaction *enthalpy* rather than the reaction *entropy,* and the statistical analysis agrees: the entropy component is, essentially, the 0.239 factor from the partition functions (note that only the vibrational part depends on T) while the reaction energy part, $e^{-k'\epsilon_0/T}$, is most of the story. Problem 23.39 contrasts the role of these two factors, and Problem 23.43 considers another way entropy can control equilibria.

To close this chapter, we return to the JANAF Tables in Chapter 7, Tables 7.6–7.9, and focus first on the commentary associated with the data listings. Table 7.6 for H^+ points out that only the translational contribution matters, that is, the nuclear spin is ignored. This is common practice in thermochemical tabulations. *All* the data in that table are calculated from statistical expressions in this and the previous chapter using the ionization energy of H. The data in Table 7.7 for H^- uses the electron affinity of H in a similar way. Table 7.8 for H discusses $\mathcal{D}_0(H_2)$ and the unimportance of the excited electronic states of H for $T < 6000$ K. Note as well the increased entropy of H compared to H^+ or H^- due to the $R \ln 2$ electronic degeneracy entropy of H. Table 7.9 for H_2 has a discussion of the treatment of ortho- and para-H_2 nuclear spin statistics as well as an elaborate ground electronic state (the only one of importance below 6000 K) bound energy-level expression. This expression accounts for anharmonicity, centrifugal distortion, and vibration-rotation coupling in some detail, and it allows, through direct summation, a quite accurate calculation of the coupled vibration-rotation partition function.

We can also see the utility of the Giauque free energy function introduced in Eq. (7.31):

$$\text{fef} = \left[\frac{\overline{G}°(T) - \overline{H}°(T_r = 298.15 \text{ K})}{T} \right] \quad (7.31)$$

and included (as its negative) in the JANAF tables. Consider, for example, H^+ in Table 7.6 with only a simple translational component. Using Eq. (22.30) for $\overline{G}°(T)$ (since the chemical potential is the molar Gibbs free energy) with Eq. (23.5) for q_{tr}, along with $\overline{H}°(T_r = 298.15 \text{ K}) = 5RT_r/2$, we have

$$\text{fef} = \frac{-RT \ln q_{tr}/N_A}{T} - \frac{5RT_r/2}{T} = -R \ln \left[\left(\frac{m_p k_B T}{2\pi\hbar^2} \right)^{3/2} \frac{RT}{N_A P°} \right] - \frac{5RT_r}{2T}$$

where m_p is the proton mass. Division by T has slowed the temperature dependence of the free energy or enthalpy alone, making table interpolation of the fef more accurate.

The expression above will reproduce all the fef entries in Table 7.6, but for more complicated molecules, the statistical fef expression is either much more complicated (as for a polyatomic gas) or impossible to write with any accuracy (as for a polyatomic liquid) so that experimentally measured thermodynamic properties are still the rule rather than the exception. Nevertheless, statistical thermodynamics is an impressive branch of physical chemistry. It brings virtually all aspects of the science to focus on a central problem: the equilibrium composition of a reacting mixture.

CHAPTER 23 SUMMARY

We have seen here the relations between the microscopic and macroscopic worlds in greater detail than anywhere else in this book. We use simple quantum mechanical model systems to generate model partition functions:

$$q_{tr} = \left(\frac{mk_B T}{2\pi\hbar^2} \right)^{3/2} V = \frac{V}{\Lambda^3}$$

$$q_{rot} = \begin{cases} \dfrac{2Ik_B T}{\sigma\hbar^2} & \text{linear molecule, high } T \\ \dfrac{(\pi I_a I_b I_c)^{1/2}}{\sigma} \left(\dfrac{2k_B T}{\hbar^2} \right)^{3/2} & \text{nonlinear molecule, high } T \end{cases}$$

$$q_{vib} = \frac{e^{-\hbar\omega/2k_B T}}{1 - e^{-\hbar\omega/k_B T}} \quad \text{per vibrational mode}$$

$$q_{nuc} = 2I + 1 \quad \text{for nuclear spin quantum number } I$$

invoke a theorem from classical statistical mechanics:

equipartition: every quadratic term in the classical Hamiltonian contributes $\frac{1}{2}k_B T$ to the average molecular energy

and we can calculate thermodynamic quantities such as heat capacities, entropies, and chemical equilibrium constants as well as internal-state distribution functions.

We found that vibrational-state populations fall exponentially from one state to the next higher:

$$\frac{N_v}{N} = e^{-\hbar\omega v/k_B T}(1 - e^{-\hbar\omega/k_B T})$$

$$\frac{N_{v+1}}{N_v} = e^{-\hbar\omega/k_B T}$$

while rotational states typically have a population distribution that peaks at $J = J_{mp} > 0$:

$$\frac{N_J}{N} = \frac{k' B_e (2J + 1) e^{-J(J+1)k'B_e/T}}{T}$$

$$J_{mp} \cong \left(\frac{T}{2k'B_e} \right)^{1/2} - \frac{1}{2}, \quad k' = 1.439 \text{ cm K}$$

We saw how to view a reacting mixture as a system with a common energy origin and thus an interrelated set of energy levels. This view showed us how an equilibrium constant was just another kind of distribution function calculated from partition functions:

$$K_{eq} = \frac{(q_C^\circ/N_C)^\gamma (q_D^\circ/N_D)^\delta}{(q_A^\circ/N_A)^\alpha (q_B^\circ/N_B)^\beta} e^{-\epsilon_0/k_B T} \quad \text{for the reaction} \quad \alpha A + \beta B \rightarrow \gamma C + \delta D$$

The next chapter prepares us for the study of chemical kinetics and dynamics as it considers distributions of velocities and speeds in gases and some of the consequences of these distributions.

FURTHER READING

All of the references for Chapter 22 contain material useful here as well. In particular, *Elements of Statistical Thermodynamics*, 2nd ed., by L. K. Nash (Addison-Wesley, Reading, MA, 1974) and *Elementary Statistical Mechanics*, by N. O. Smith (Plenum, New York, 1982) have extensive additional problems and questions.

PRACTICE WITH EQUATIONS

23A What is the ratio of q_{tr} for N_2 at 250 K in a 0.5 m³ volume to q_{tr} for CO_2 at 500 K in a 0.3 m³ volume? (23.5)

ANSWER: 0.299

23B At what temperature does Xe in a 1 m³ volume have the same translational partition function that He at 6000 K has in the same volume? (23.5)

ANSWER: 183 K

23C What are the thermal de Broglie wavelengths of He and Xe in Problem 23B above? (23.6)

ANSWER: $\Lambda = 1.126 \times 10^{-11}$ m for both

23D The vibrational frequency of CsI is $\omega = 2.245 \times 10^{13}$ s⁻¹ ($\omega_e = 119.178$ cm⁻¹), and that for LiF is 1.715×10^{14} s⁻¹ (910.34 cm⁻¹). What is q_{vib} for each of these at 1000 K? (23.7)

ANSWER: CsI: 5.825; LiF: 0.7114

23E Which has the larger q_{el} at 700 K, Rb(g) or Sr(g)? (23.9)

ANSWER: Rb ($g_0 = 2$ for Rb; $g_0 = 0$ for Sr)

23F The bond length in CsI is 3.315 Å and in LiF, it is 1.564 Å. What are the rotational partition functions of each at 1000 K? (23.11)

ANSWER: CsI: 2.941×10^4; LiF: 512.7

23G Which has the largest q_{rot} at 400 K, $^{79}Br_2$, $^{81}Br_2$, or $^{81}Br^{79}Br$ (23.12)

ANSWER: $^{81}Br^{79}Br$

23H The three moments of inertia of O_3 are 7.877×10^{-47} kg m², 6.287×10^{-46} kg m², and 7.086×10^{-46} kg m². What is the rotational partition function for O_3 at 300 K? (23.13), Table 23.1

ANSWER: 3375

23I What are the symmetry numbers σ for $H_2C=C=CH_2$ (allene), $Ni(CO)_4$, PF_3, H_2CO, and CF_2Cl_2? Table 23.1

ANSWER: 2, 12, 3, 2, 2

23J What is the internal partition function, $q_{int} = q_{el}q_{vib}q_{rot}$, for $^{14}N_2$ at 300 K? Table 19.2, Table 23.2

ANSWER: 52.18 excluding zero-point energy

23K Spectroscopy of a high-temperature sample of CsI finds 18% of the molecules in the ground vibrational state. What is the gas temperature? (23.14), 23D

ANSWER: 864K

23L An infrared spectrum of a sample of ICl(g) shows that $J = 45$ is the most probable J value. What is q_{rot} for this sample? (23.11), (23.16)

ANSWER: 4050 ($q_{rot} \cong 2J_{mp}^2$)

23M How close to the equipartition limit is the vibrational contribution to $\Delta_0 \overline{U}$ for CsI and LiF at 1000 K? (23.19), 23D

ANSWER: CsI: 91.7%; LiF: 48.4%

23N What is the translational molar entropy in J mol⁻¹ K⁻¹ units for a gas at 1 atm and 298.15 K in terms of its molar mass M? (23.22)

ANSWER: $108.75 + 12.472 \ln (M/\text{g mol}^{-1})$

23O What is the ratio of the translational entropy to the rotational entropy for the CsI sample in 23F? (23.23), 23F, 23N

ANSWER: 1.90

PROBLEMS

SECTION 23.1

23.1 The center of mass of a 1-D harmonic oscillator is constrained to move in a 1-D box so that its total energy is the sum of its internal vibrational energy and its bulk translational energy through the box. Write its total molecular partition function sum in terms of its allowed energies for translation and internal vibration. How would your answer change if an external field applied to the box made the oscillator's harmonic frequency change with position in the box?

23.2 Is the dilute system inequality $V/\Lambda^3 \gg N$ in q_{tr} ever violated? Example 23.2 considered ^{4}He at its boiling point, and V/Λ^3 decreases with decreasing mass and temperature. Consequently, repeat the calculation for ^{3}He(g) and H_2(g) at their normal boiling points: 3.19 K and 20.27 K, respectively.

23.3 Molecular ions are frequently studied in the gas phase in gas discharges through cold gases. The free electrons produced in such a discharge move under the influence of the electric field that sustains the discharge, and consequently they are not a very good model for an equilibrium gas. But if the electrons were behaving ideally, would the dilute limit for their translational partition function be attained? Take 77 K for a temperature (since the discharges are often cooled in $N_2(l)$ baths), and consider 10^{16} electrons in 1.0 m^3, a typical discharge density.

23.4 We have said that the molecular partition function represents a rough measure of the number of states available to a molecular system at any one temperature. At what T is q_{tr} for N_2(g) equal to 6×10^{23}, a "mole of states," if the gas is confined to a volume 1 mm^3 at 1 atm pressure? How many moles of gas are in this volume? Is this still a dilute system?

23.5 The translational partition function for a one-component ideal gas is proportional to $T^{3/2}V$. Any thermodynamic process that follows a path $T^{3/2}V =$ constant cannot, therefore, change q_{tr}. What thermodynamic process follows this partition function conserving path? *Hint:* The answer is one of the simple paths considered in Chapter 2.

23.6 If we consider $k_B T/\hbar\omega = 2$ sufficient to say vibrational motion is in the high-T limit, find the high-temperature limit for the diatomic halogens listed in Table 19.2.

23.7 The dimer (HF)$_2$ is sufficiently stable to be seen in HF(g) at room temperature. It is a nonlinear, four-atom molecule and thus has six normal modes of vibration. Two of these are high-frequency stretching motions involving the two H—F strong chemical bonds. Are these vibrations in the high-temperature limit at 300 K? Do you expect the other four vibrational modes to be in this limit?

23.8 The text quotes a value 0.9360 for $q_{vib}(I_2)$ at 300 K if vibrational anharmonicity is considered. Use the harmonic constant ω_e and the anharmonicity constant $\omega_e x_e$ in Table 19.2 to verify this value. Recall that the energy expression for an anharmonic oscillator in this approximation is (see Eq. (19.26)) $\epsilon_v = \omega_e(v + 1/2) - \omega_e x_e(v + 1/2)^2$ where ϵ_v is expressed in the units of ω_e and $\omega_e x_e$, which is cm^{-1} in Table 19.2.

23.9 Consider a quantum mechanical system in which the energy levels are equally spaced, as in the harmonic oscillator, but *unlike the harmonic oscillator,* the vth level is v-fold degenerate: $\epsilon_v = v\epsilon$, $g_v = v$, $v = 1, 2, 3, \ldots$. Find a closed expression for the molecular partition function for this system. You will need to know that

$$1 + 2x + 3x^2 + \cdots = \frac{1}{(1-x)^2}, \quad x < 1 .$$

Next, find the average energy $\langle \epsilon \rangle$ of this system defined as

$$\langle \epsilon \rangle = \frac{\sum_{v=1}^{\infty} N_v \epsilon_v}{N} .$$

For this part, you will need to know that

$$1 + 4x + 9x^2 + \cdots = \frac{1+x}{(1-x)^3}, \quad x < 1 .$$

23.10 We can write the energy-level expression for atomic hydrogen as $\epsilon_n = R(1 - 1/n^2)$ where R is the Rydberg constant (the ionization potential of H) and n is the principal quantum number, $n = 1, 2, 3, \ldots, \infty$. Each level is $2n^2$-fold degenerate, and the electronic partition function is

$$q_{el} = \sum_{n=1}^{\infty} 2n^2 e^{-\epsilon_n/k_B T} .$$

At 1000 K, for instance, the first few terms in the sum are

$$q_{el} = 2 + 2.99 \times 10^{-51} + 2.02 \times 10^{-60} + 1.67 \times 10^{-63} + 7.46 \times 10^{-65} ,$$

which seem to indicate convergence. While the 100th term is 5.48×10^{-65}, the 1000th is 5.39×10^{-63}, 100 times larger than the 100th, and there are an infinite number of terms yet to come! The $n = 10^{35}$ term is about 54, for instance. It and the infinite remaining terms are larger than the first term. Consequently, we must conclude $q_{el}(H) = \infty$, at any temperature. If this were true, the probability of finding a hydrogen atom would be zero, in conflict with experience,

23.11 Find the rotational partition function for the simplest molecular ion, H_2^+, in the high-temperature limit given its bond length 1.057 Å. What temperature range corresponds to "high T" for this species? Is it likely that H_2^+ would be found below these temperatures?

23.12 How would the rotational partition functions for $^{12}C^{16}O^{18}O$, $^{12}C^{18}O_2$, and $^{13}C^{16}O_2$ differ from that for $^{12}C^{16}O_2$? Give arguments that rank q_{rot} for each in relative magnitude with respect to $^{12}C^{16}O_2$, but do not try to calculate each q_{rot}.

23.13 One can imagine that some linear molecules loosely adsorbed to a solid might rotate in two dimensions like pinwheels spinning over the surface. Consequently, the partition function for a planar rotor is of interest. The planar rotor has only one angular coordinate, and in Chapter 12, it was discussed as the particle-on-a-ring problem (see Example 12.6). The energies are $\epsilon_m = m^2\hbar^2/2I$ where I is the rotor's moment of inertia. The quantum number $m = 0, \pm 1, \pm 2, \ldots$ so that every level except the first is doubly degenerate. Find the partition function for this system using the high-temperature approximation that allows the sum to be turned into an integral over m.

23.14 C_{60} is a very spherical molecule. As shown in Figure 16.14, the C atoms are spread over a sphere of radius $r = 3.55$ Å. Due to this high symmetry, the C_{60} moment of inertia is very easy to find. Since it is a spherical top, the three moments of inertia, I_a, I_b, and I_c, are equal. From Eq. (19.3), we can write this unique moment of inertia I in three equal ways:

$$I = m\sum_{i=1}^{60}(x_i^2 + y_i^2) = m\sum_{i=1}^{60}(y_i^2 + z_i^2) = m\sum_{i=1}^{60}(x_i^2 + z_i^2)$$

where m is the C atom mass and (x_i, y_i, z_i) are the coordinates of atom i measured from the center of the molecule. Show that adding these three expressions gives

$$I = \frac{2}{3} 60\, mr^2$$

and use this moment of inertia to find the rotational partition function for $C_{60}(g)$ at 400 K.

SECTION 23.2

23.15 In the course of your research on high-temperature fluorides, you decide to study the infrared spectrum of vibrationally excited LiF(g). You construct a cell 10 cm long that can attain $T = 1700$ K (where the vapor pressure of LiF is 0.145 atm), and you plan to use your tunable IR laser (with a 4 mm² beam cross-sectional area). You estimate that you can detect direct absorption by a single rotational/vibrational level if 3×10^{14} or more molecules in that level are in the volume illuminated by your laser beam. From data in Table 19.2, you calculate that the most probable J at 1800 K is ~21. Can you detect absorption by molecules with $v = 1$, $J = 21$? What about $v = 2$, $J = 21$? $v = 2$, $J = 1$?

23.16 You make several improvements to the apparatus described in the previous problem and increase the sensitivity and the temperature range of the cell to the point that many excited vibrational levels are observed in the spectrum. You record the spectrum at one temperature, and calculate from it the fractional populations in each vibrational level. To your surprise, you find that by chance you have chosen a temperature such that half the molecules are in $v = 0$, one-fourth are in $v = 1$, one-eighth in $v = 2$, etc. What was T?

23.17 Consider a gas of pure para-H_2 (only $J =$ even integer is allowed) at 30 K, about 10 K above the normal boiling point. What fraction of these molecules has $J = 0$? What fraction has $J = 2$? The rotational constant is in Table 19.2.

23.18 Derive Eq. (23.16) for the most probable J value of a heteronuclear diatomic gas and verify the J_{mp} values for the interhalogens quoted in the text in reference to Figure 23.3.

23.19 Do the peaks in the rotational distributions of Figure 23.3 mean that a thermal population inversion between, say, $J = 0$ and $J = J_{mp}$ can be attained? Why or why not?

23.20 We can use Eq. (23.14) to find the fraction of a sample of diatomic molecules in the ground vibrational state. What is the expression for the fraction in *all vibrational states except the ground state*? How is this fraction related to the ratio of the first excited-state population to the ground state's population? Can you explain why these two expressions are related in the way you have found them to be?

23.21 Show that the classical partition function applied to a 1-D harmonic oscillator's classical Hamiltonian, $H = p^2/2\mu + kx^2/2$, gives the high-T limit of the quantum vibrational partition function derived in Example 23.3. Here, μ is the oscillator's reduced mass and k is its force constant.

23.22 We can prove the equipartition theorem along these lines. Suppose the classical Hamiltonian has a momentum term ap_1^2 where a is a constant so that $H = ap_1^2 + H'$ where H' is all the other terms. The classical partition function, Eq. (23.17), is then

$$q_{cl} = \frac{1}{h^{3N}} \iint e^{-\beta a p_1^2 - \beta H'}\, dp_1\, dp^{3N-1}\, dx^{3N} = q_{p_1} q_{cl'}\,.$$

Evaluate the q_{p_1} factor explicitly, and use the relation $\langle \epsilon \rangle = -(1/q_{cl})(\partial q_{cl}/\partial \beta)$ to show that q_{p_1} contributes $k_BT/2$ to $\langle \epsilon \rangle$. Is it obvious to you that a coordinate term bx^2 would have the same contribution?

23.23 The high-temperature limit of the Debye solid heat capacity must be the classical DuLong–Petit value, $\overline{C}_V = 3R$. Show that it is, starting from the general expression:

$$\overline{C}_V = 9R\left(\frac{k_BT}{\hbar\omega_D}\right)^3 \int_0^{x_D} \frac{x^4 e^x}{(e^x-1)^2}\,dx, \quad x_D = \frac{\hbar\omega_D}{k_BT}.$$

High temperature means small x. Use this limit to expand the exponentials in the integral and simplify it.

23.24 The graph below is a universal plot of the Debye heat capacity expression (see Problem 23.23 for the full expression):

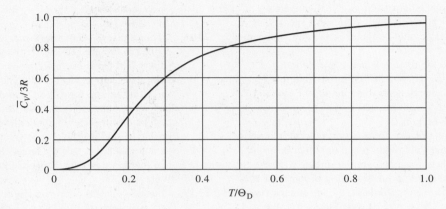

Use this graph and data from Table 4.2 to predict the heat capacities of Cu, Ag, and Au at 100 K, 200 K and 300 K.

23.25 For some solids such as high molecular weight polyethylene (a solid of stacked and entangled chains of chemically bonded CH_2 groups) and graphite (see Figure 16.13 for a view of graphite—parallel layers of hexagonal C atom rings), the heat capacity passes through a region proportional to T^2 or T before reaching the T^3 Debye form at very low temperatures. Can you point to the assumptions of the Debye theory that accounts for this behavior? Problem 23.33 is relevant to the situation here.

23.26 We can write the Sackur–Tetrode equation, Eq. (23.22a), in terms of an arbitrary number of molecules N if we change R into Nk_B and N_A into N in that equation. Derive the one- and two-dimensional versions of this equation for N molecules confined to a line of length L or an area A, respectively. Evaluate the 1-D, 2-D, and 3-D expressions for Xe at 300 K. Let $L = 1$ m, $A = 1$ m^2, and $V = 1$ m^3, and assume the particle densities are the same in each system taking $N = N_A^{n/3}$ where n is the dimensionality of the system. Your answers should convince you how strongly entropy decreases as a system is constrained to lower and lower dimensions.

23.27 What is the difference between the rotational contribution to the entropy of DF to that for HF at a temperature T high enough to place rotations in the high-T limit? Is this difference changed appreciably if we contrast DI and HI at the same T?

23.28 The molecules CO_2 (C=O=C) and N_2O (N=N=O) have the same mass and virtually the same rotational constant. They both have nondegenerate ground electronic states and virtually the same vibrational frequencies. Yet one has a notably larger standard molar entropy as a gas at any one temperature. Which has the greater entropy and why?

23.29 The standard molar entropy of $N_2(g)$ at 298.15 K is 191.61 J mol^{-1} K^{-1}. What is the N_2 bond length? Does your answer agree with the value in Table 19.2?

23.30 Derive an expression for the rotational contribution to the molar entropy of a symmetric top, and find $\overline{S}^\circ_{rot}$ for NH_3 given the rotational constants $B = 9.4443$ cm^{-1} and $C = 6.196$ cm^{-1}. The tabulated total standard molar entropy of NH_3 at 298.15 K is 192.45 J mol^{-1} K^{-1}. How much of this total entropy is due to vibrational contributions?

23.31 The following three numbers are standard molar entropies at 500 K taken from the JANAF Tables: 180.007 J mol^{-1} K^{-1}, 193.388 J mol^{-1} K^{-1}, and 191.533 J mol^{-1} K^{-1}. One of these is for I(g), one is for I$^+(g)$, and one is for I$^-(g)$. Which is which? (*Hint:* At 500 K, only the ground electronic state of each species contributes. What are the ground states of these species?)

23.32 The electronic partition function has to be calculated by a direct sum over levels, Eq. (23.9), at any one temperature. Electronic contributions to thermodynamic functions, however, often depend on temperature *derivatives* of q. The sums

$$q' = \sum_i g_i x_i e^{-x_i} \quad \text{and} \quad q'' = \sum_i g_i x_i^2 e^{-x_i}$$

where $x_i = \epsilon_i/k_B T$ can be calculated easily along with q, and thermodynamic functions can be written in terms of q, q', and q''. For example, $\Delta_0 \overline{U}_{el} = RTq'/q$. Derive this expression and find similar expressions for $\overline{S}_{el}$ and $\overline{C}_{V,el}$. For Ga at 298.15 K, only the lowest two energy levels contribute: $\epsilon_0 = 0$, $g_0 = 2$, $\epsilon_1 = 826.24$ cm^{-1}, and $g_1 = 4$. Use the expressions you derive to find $\overline{S}°$ and $\overline{C}_V°$ at 298.15 K (including translation!), and compare with the tabulated values: $\overline{S}° = 169.06$ J K^{-1} mol^{-1} and $\overline{C}_V° = 25.36$ J K^{-1} mol^{-1}.

23.33 What is the partition function for a solid of a diatomic like F$_2$? Would the Debye T^3 law still hold? (*Hint:* How do you think ω_D and ω_e compare for F$_2$?)

23.34 Use data in Table 4.2 to calculate the Debye frequencies ω_D (in cm^{-1} units) for the alkali metals and compare them to the diatomic ω_e values in Table 19.2. One might argue that the bonding environment in the solid is similar to that in the diatomic so that the force on an atom in either situation has about the same harmonic force constant. If this is true, then the difference in frequencies can be traced to a difference in the effective oscillator mass: the reduced mass for the diatomic versus the atomic mass in the solid. What frequency ratio ω_e/ω_D does this simple model predict? How well is it born out by the experimental data?

23.35 Some solids such as RbTi(SO$_4$)$_2 \cdot$12H$_2$O have an unpaired electron isolated on a metal atom (on the Ti^{3+} ion here), forming a system of N independent, isolated, and distinguishable electron spins. In an external magnetic induction field B, each spin can have one of only two energies, $\epsilon_1 = +mB$ or $\epsilon_2 = -mB$ where m is the electron magnetic moment, Eq. (15.16). Thus the molecular partition function for any one spin is $q = e^{-\epsilon_1\beta} + e^{-\epsilon_2\beta} = e^{-mB\beta} + e^{+mB\beta} = 2\cosh(mB\beta)$ where $\beta = 1/k_B T$ and cosh is the hyperbolic cosine function, $\cosh(x) = (e^{-x} + e^{+x})/2$. Find an expression for the temperature dependence of the net magnetic moment of the system, $M = m(N_1 - N_2)$, where N_i is the number of spins with energy ϵ_i. Find as well an expression for the average energy of any one spin, $\langle\epsilon\rangle$, first from the statistical thermodynamic expression $\langle\epsilon\rangle = \Delta_0\overline{U}/N_A$ and then from the expression $\langle\epsilon\rangle = (N_1\epsilon_1 + N_2\epsilon_2)/N$. The two expressions for $\langle\epsilon\rangle$ should be the same, of course, and can be written in terms of $\tanh(mB\beta)$ where tanh is the hyperbolic tangent function, $\tanh(x) = (e^x - e^{-x})/(e^x + e^{-x})$. What is the high-temperature limit of $\langle\epsilon\rangle$, and what is the physical reason for this limit?

SECTION 23.3

23.36 Imagine taking 1 mol HCN(g) from 300 K to 1000 K. Calculate the number of moles of the HNC isomer that you would find at this temperature. For the isomerization reaction HCN $\rightleftarrows$ HNC, take $\epsilon_0 = 1.0 \times 10^{-19}$ J, a value from quantum chemistry theory since no experimental value is known, and use the rotational constants $B_{HCN} = 1.478$ cm^{-1} and $B_{HNC} = 1.512$ cm^{-1}. Only the doubly degenerate bending vibration of each species has a low enough frequency to be important at this temperature. For HCN, it is $\omega_e = 713.46$ cm^{-1}. For HNC, the only vibrational data come from a low-temperature matrix isolation study that first isolated HNC and found $\omega_e = 535$ cm^{-1}. Both molecules have nondegenerate electronic ground states. Your answer should convince you why HNC is such an elusive molecule.

23.37 Continuing from Example 23.9, evaluate K_{eq} for the H(g) $\rightleftarrows$ H$^+$(g) + e^-(g) self-ionization reaction. Recall that ϵ_0 is 109 678 cm^{-1}, the ionization potential of H(g), from Eq. (12.46). Find the temperature for which $K_{eq} = 1$. (You can find this temperature iteratively. Take 16 000 K as your first guess.) Show that q_{el}(H) = 2 is a good approximation even at this very high T.

23.38 Contrast the thermal ionization equilibrium of H(g) considered in the previous problem to the ionization of Cs(g). (The Cs ionization potential is listed in Table 7.3.) What one physical property governs the ratio of the ionization equilibrium constants for these two systems, and what is this ratio at 3000 K?

23.39 The linear molecule ClXeF has been seen only at cryogenic temperatures, and only its IR spectrum is known. It is an interesting molecule since, unlike XeF$_2$ or XeCl$_2$, it is polar, and its dipole moment would reveal something new about Xe bonding. Estimate the equilibrium constant for the (not very promising, as you will see) synthetic route Xe(g) + ClF(g) $\rightarrow$ ClXeF(g). Data on ClF are in Table 19.2 and Problem 23.48. Very little is known about ClXeF. Its two stretching vibrations are 480 and 316 cm^{-1}, and the doubly degenerate bending vibration is below 200 cm^{-1}, but probably above 100 cm^{-1}. The bond lengths are unknown, but reasonable guesses put the rotational constant around 0.05 cm^{-1}. The Xe—F bond energy in XeF$_2$ and XeF$_4$ is around 130 kJ mol^{-1}, and the Xe—Cl bond energy is unknown, but is probably close to the Xe—O bond energy, $\sim$80 kJ mol^{-1}. Use this information to estimate K_{eq} at 600 K. You should find a depressingly small value. What one factor makes a reaction such as XeF$_2$ + NOCl $\rightarrow$ ClXeF + NOF potentially more favorable?

23.40 The ozone destruction reaction, O$_3$(g) + O(g) $\rightarrow$ 2O$_2$(g), is important in atmospheric chemistry. Problem 7.9 considers its reaction enthalpy change, and we consider its equilibrium constant at 298.15 K here. From thermochemistry, $\Delta_r\overline{H}°$(298 K) = -391.847 kJ mol^{-1}, and since $\Delta_r n = 0$, this is $\Delta_r\overline{U}°$(298 K) as well. Correct this value to 0 K, yielding ϵ_0, assuming only translation and rotation contribute to each species' heat capacity. Find the translational partition function contribution to K_{eq} first. Next, use the rotational constants $B_e(O_2) = 1.4456$ cm^{-1}, $A_e(O_3) = 3.5538$ cm^{-1}, $B_e(O_3) = 0.4453$ cm^{-1}, and $C_e(O_3) = 0.3948$ cm^{-1} to find the rotational partition function contribution. For both O_2 and O_3, $\sigma = 2$. The O_2 vibrational constant is 1580 cm^{-1}, and for O_3, there are three: 1110, 705, and 1043 cm^{-1}. Use them to find the vibrational contribution. The electronic ground state of O_2 has $g_0 = 3$, the O_3 ground state is nondegenerate, and the excited states of both can be ignored. Using the method of Problem 23.32 and data in Table 13.4, one finds $q_{el}(O(g), 298 \text{ K}) = 6.732$. Use this information to find the electronic contribution. Put all the pieces together to form K_{eq}, and compare your answer to the value calculated from JANAF tables: $K_{eq}(298.15 \text{ K}) = 1.54 \times 10^{69}$.

23.41 Take into account the small isotope effects on ϵ_0, rotational constants, vibrational constants, and masses, and find an improved value for the 300 K equilibrium constant in Example 23.10. Table 19.2 lists spectroscopic constants for $^{35}Cl_2$, and using the relationships $\omega_e \propto \mu^{-1/2}$ and $B_e \propto \mu^{-1}$ where μ is the reduced mass, you can find these constants for the other isotopomers. Note as well that ϵ_0 is *not* zero due to slight zero-point energy differences between the reactants and products. Does any one factor dominate the others?

23.42 Consider the reaction in Example 23.13 at a temperature sufficiently high so that the vibrational partition functions are at the high-T limit (see Example 23.3), but electronically excited states can be ignored. Show that the partition function ratio in K_{eq} is independent of T in this limit. Find $\Delta_r \overline{S}°$ and $\Delta_r \overline{H}°$ using $\Delta_r \overline{G}° = \Delta_r \overline{H}° + T \Delta_r \overline{S}°$.

23.43 Consider the equilibrium in the gas phase between meta and para disubstituted benzenes, such as the xylenes or the dichlorobenzenes. The equilibrium constant for meta $\rightleftarrows$ para over a wide temperature range is ~0.4 for the xylenes and ~0.7 for the dichlorobenzenes. What structural feature or features of these isomers dominates the statistical expression for these equilibrium constants?

23.44 At 1000 K, the equilibrium constant for the isomerization reaction propene $\rightleftarrows$ cyclopropane is 4.65×10^{-4}. Discuss in turn the various statistical factors that go into this number (i.e., $e^{-\epsilon_0/k_BT}$, the ratio of q_{tr} factors, of q_{vib} factors, and of q_{rot} factors), and give simple arguments based on molecular structure differences between propene and cyclopropane that cause each factor to favor one molecule over the other (or to play little or no role in K_{eq}).

23.45 Example 23.11 considers H_2 dissociation at 4000 K. A slight disagreement was found between the JANAF value $K_{eq} = 2.546$ and the simplest statistical value $K_{eq} = 2.981$. Repeat the statistical calculation for $T = 800$ K and compare it to the JANAF value, $K_{eq} = 8.39 \times 10^{-24}$. Why is the agreement better at 800 K than at 4000 K?

23.46 The helium dimer, He_2, is a quite remarkable molecule. Its interatomic potential is so shallow and the ^{4}He mass is so small that the zero-point level, $v = 0, J = 0$, is the *only* bound level in the ground electronic state, and that lone level sits so close to the dissociation limit that $\mathcal{D}_0 \cong 0$. With these facts in mind, calculate the equilibrium constant for the dimer association reaction $2He(g) \rightleftarrows He_2(g)$ at 4.2 K and at 300 K.

23.47 The recent research in carbon clusters such as C_{60} has turned up long tubes of C that look like a cylindrical tube of rolled up hexagonal graphite planes capped at each end by half a spherical fullerene. These nanotubes, as they are called, could hold atoms in a line inside them. Imagine stuffing some He atoms into such a tube. Would this improve the odds of finding He_2 based on the information and calculation of the previous problem? Can you estimate an equilibrium constant for $2He \rightleftarrows He_2$ in this 1-D system? Let the nanotube be 100 μm long, and put 10^4 atoms in it. Since these are real atoms, any one He is confined to a length $\cong$ 100 μm/10^4 = 100 Å. Imagine the system is at 4.2 K as well.

23.48 Find the equilibrium constant for the halogen exchange reaction $ClF + IBr \rightleftarrows ICl + BrF$ at 300 K. Use data from Table 19.2 and the following $\mathcal{D}_0$ values: 21 110 cm^{-1} for ClF, 14 660 cm^{-1} for IBr, 17 366 cm^{-1} for ICl, and 20 551 cm^{-1} for BrF.

23.49 Pick any temperature (except 0 K or 298.15 K) listed in Table 7.8, the JANAF table for H(g), and calculate the tabulated quantities $\overline{C}_P°, \overline{S}°, -[\overline{G}° - \overline{H}°(T_r)]/T$, and $\overline{H}° - \overline{H}°(T_r)$ at that temperature. For best agreement, use the accurate mass of H, 1.007 94 g mol^{-1}, and remember that, unlike H^+ discussed in the text, H has a doubly degenerate electronic ground state.

GENERAL PROBLEMS

23.50 Apply the argument given in the text to show how the symmetry number σ predicts the correct q_{rot} for a homonuclear diatomic with *spinless* nuclei, such as Ar_2.

23.51 We have used the integral $I = \int_0^\infty e^{-ax^2} dx$ and variants of it several times in this chapter, and we will again in the next. It is easy to derive their values, following these steps. Since

$$I^2 = \int_0^\infty e^{-ax^2} dx \int_0^\infty e^{-ax^2} dx = \int_0^\infty e^{-ax^2} dx \int_0^\infty e^{-ay^2} dy$$

$$= \int_0^\infty \int_0^\infty e^{-a(x^2+y^2)} dx dy$$

we can change from Cartesian coordinates (x, y) to plane polar coordinates (r, θ) where $x^2 + y^2 = r^2$ and $dx\,dy = r\,dr\,d\theta$ and write

$$I^2 = \int_0^{\pi/2} \int_0^{\infty} e^{-ar^2} r\,dr\,d\theta \ .$$

Let $z = r^2$ and evaluate this double integral to show that $I = \sqrt{\pi/4a}$. Next, show that repeated differentiations of I with respect to a gives the series of integrals $\int_0^{\infty} x^n e^{-ax^2}\,dx$, n = even integer, and evaluate these integrals for $n = 2$ and 4. Then evaluate $\int_0^{\infty} x\,e^{-ax^2}\,dx$ and show how repeated differentiations give $\int_0^{\infty} x^n e^{-ax^2}\,dx$, n = odd integer, and evaluate this for $n = 3$. Finally, evaluate all these integrals over the range $-\infty$ to ∞ instead of 0 to ∞.

23.52 A *Schottky defect* in a solid is one in which an interior crystal site is vacant due to motion of an atom to the surface. (See Figure 16.20(a).) This motion requires an energy ϵ_S per atom to accomplish, but creating a vacancy introduces entropy (in a two-level system sense—occupied versus unoccupied sites). Consequently, the equilibrium number of defects minimizes the free energy of the solid, which we can write

$$A(\text{defect solid}) - A(\text{perfect solid}) = \Delta A(N_S) = N_S \epsilon_S - k_B T \ln W$$

where N_S is the number of defects and W is the combinatorial factor for a two-level system, $N!/[N_S!\,(N - N_S)!]$. The first term is the internal energy, and the second is $-T$ times the entropy. Find the value for N_S that minimizes $\Delta A(N_S)$ (with $N_S \ll N$, the total number of sites in the crystal) and show that $N_S = N\,e^{-\epsilon_S/k_B T}$. (Problem 16.45 first used this expression for an ion *pair* defect that leads to a factor of 2 in the denominator of the exponential's argument.) We can estimate ϵ_S for a simple solid such as Ar(s) if we consider a surface atom to have half the binding energy of an interior atom. This assumption makes $\epsilon_S \cong \Delta \overline{U}^{\circ}_{\text{sub}}/2N_A$ where $\Delta \overline{U}^{\circ}_{\text{sub}}$ is the molar energy of sublimation. For Ar(s) around its melting temperature, 83.8 K, $\Delta \overline{U}^{\circ}_{\text{sub}} = 7.04$ kJ mol^{-1}. What fraction of sites are Schottky defects at 80 K?

23.53 We derived the harmonic oscillator vibrational partition function as if an infinite number of levels was available to the system. When the temperature is low, this is an excellent assumption, but for some situations, it makes no sense. Consider, for example, a weakly bound diatomic like Ar$_2$, which has only a few bound vibrational levels, all at quite low energy. For a molecule like this, either an exact sum or a *truncated* harmonic oscillator approximation is called for. Suppose we consider only levels $v = 0$ through n of a harmonic oscillator. Write, with $x = \hbar \omega / k_B T$ and $e^{-x/2}$ the zero-point energy factor,

$$q_{\text{vib}} = e^{-x/2} \sum_{v=0}^{n} e^{-vx} = e^{-x/2}\left(\sum_{v=0}^{\infty} e^{-vx} - \sum_{v'=n+1}^{\infty} e^{-v'x}\right)$$

and show that this leads to

$$q_{\text{vib}} = \frac{e^{-x/2}(1 - e^{-(n+1)x})}{1 - e^{-x}} \ .$$

An exact sum over the known vibrational levels of Ar$_2$ at 300 K gives $q_{\text{vib}} = 5.71$. Compare this value to the full harmonic approximation, Eq. (23.7), and the truncated harmonic approximation derived above. Use $\omega_e = 25.74$ cm^{-1}, the experimental energy difference between the ground and first excited vibrational levels of Ar$_2$, and let $n = 7$, since levels $v = 0-7$ are known to be bound. It is perhaps surprising that the truncated approximation works as well as it does, since, as the figure below shows, the real energy level pattern and that of the truncated oscillator are quite different! Can you explain why this approximation works? (*Hint:* What is the high-temperature limit of the truncated oscillator partition function? What is the high-temperature limit of the exact q_{vib}?)

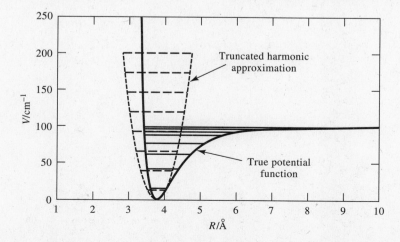

23.54 We can derive the ideal gas equation of state from a statistical argument that anticipates some of the ideas that appear in the next chapter. We imagine a monatomic gas in a rectangular container of volume $V = L_x L_y L_z$. The energy levels of the gas depend on the edge lengths L so that a small change in, say, L_x changes a level energy an amount $d\epsilon_i = (\partial \epsilon_i / \partial L_x) dL_x$. This allows us to identify $-(\partial \epsilon_i / \partial L_x)$ as a force acting in the x direction, and the statistically average force in that direction is

$$\langle F_x \rangle = - \frac{\sum_i \left(\frac{\partial \epsilon_i}{\partial L_x} \right) e^{-\epsilon_i / k_B T}}{q_{tr}}.$$

where q_{tr} is the translational partition function. Show that this expression is equivalent to

$$\langle F_x \rangle = \frac{k_B T}{q_{tr}} \left(\frac{\partial q_{tr}}{\partial L_x} \right)_T$$

and then use Eq. (23.5) to show that the average force for a mole of atoms, $N_A \langle F_x \rangle$, leads to $P\overline{V} = RT$. Remember that pressure is force per unit area!

CHAPTER 24

The Kinetic Theory of Gases

THIS chapter's title seems somewhat redundant today. How else can a gas be a gas except to have its molecules "be kinetic," constantly in motion, mostly flying past each other, occasionally colliding, and rapidly exploring every corner of any volume confining it? Today, this picture seems obvious, but for centuries, it was not universally accepted that molecules existed, let alone that they moved in a gas in the way we are about to discover. The atomic theory of matter has deep historical foundations, of course, but the word "gas" is fairly recent. It was coined by the Flemish chemist Johannes Baptista van Helmont, who died in 1644. Van Helmont based his new word on the Greek word "chaos," which seems today to have been a very apt choice.[1] Alas, he probably did not use "chaos" as we typically do today. What he had in mind was the word as used by one Theophrastus Philipus Aureolus Bombastus von Hohenheim, a German physician and chemist who called himself Paracelsus. He was quite influential in his time (he died in 1541), but he called "chaos" that phase through which gnomes moved (as we and birds do through air and fish do through water). To van Helmont, "gas" was an ultrararified form of water.

Robert Hooke may have been the father of the kinetic theory of gases when he proposed in 1678 that collisions of gas atoms with

24.1   Speeds and Velocities of Gases

24.2   The Dynamic Nature of Pressure

24.3   Intermolecular Forces and Nonideal Gases

---

[1] He also coined the now defunct word "blas" to represent the "flatus of the stars" that supposedly caused seasonal changes in weather!

their container's walls accounted for what he called the *elasticity* of a gas, and Daniel Bernoulli used this idea to derive Boyle's Law ($P \propto 1/V$) in 1738.

If Hooke was the father, the theory certainly took its time maturing to birth, since the current version had its beginnings roughly 200 years later, around 1860. When J. J. Waterston read a paper before the Royal Society in 1846 that first suggested that the mean kinetic energy of a gas was proportional to its temperature, the paper was rejected. It was not published until 1892 when Lord Rayleigh ran across it in the Society's files. (Waterston's paper was also the first to discuss the escape of light gases from planetary atmospheres, a phenomenon discussed in Problem 24.17.) Using only classical mechanics, of course, scientists such as Boltzmann, van der Waals, Maxwell, Clausius, and others from the genesis of classical thermodynamics incorporated kinetic ideas about heat to develop a theory based on atoms but made manifest by macroscopic variables such as temperature and pressure. At this point in the book, you are much better armed than they were, if only because you have covered many of the topics that followed classic kinetic theory: quantum and statistical mechanics in particular. Consequently, we will not follow the historic development of the theory very closely, but we will derive and use many of the earliest results. Their experimental confirmation—some of which extended well into the twentieth century—constitutes a landmark in science. One could no longer doubt the existence of atoms and molecules.

We start with the equilibrium distribution functions for translational motion—kinetic energy, velocity, and speed—and see how modern experimental methods can manipulate them. In the second section, we consider some of the dynamic variables of gas motion such as the number of collisions any one molecule has with the walls of the container each second. These ideas will take us into the world of chemical reaction dynamics in Chapter 28. The final section returns to Chapter 1 and its discussion of nonideal gases. In this section we explore the molecular basis for gas nonideality through the various types of forces that one kind of molecule can exert on another.

24.1 SPEEDS AND VELOCITIES OF GASES

Suppose 1 mol of Ar is at equilibrium at 300 K. From the equipartition theorem, we know that the total translational energy of this gas is $3nRT/2 = 3.741$ kJ. We also know something about the *distribution* of energy among the molecules, but not very much. That distribution is the focus of this section. We know to expect a smooth distribution, in accord with other equilibrium distributions, and finding it is not difficult, but it is worth considering some ridiculous distributions first. For example, if all the Ar atoms were still except one, that one would carry the total kinetic energy, and it would have a speed 3.36×10^{14} m s^{-1}, about 10^6 times the speed of light, an impossibility. At the other extreme, suppose all the atoms had the same velocity, again in accord with equipartition $\langle \epsilon \rangle = 3k_\text{B}T/2 = mv^2/2$:

$$v = \sqrt{\frac{3k_\text{B}T}{m}} = 432.8 \text{ m s}^{-1} \ .$$

This is a reasonable velocity, and we will see that it is one of the characteristic velocities of the full distribution.

While there is no known way to still all but one of a mole of atoms, there is a way to create the second nonequilibrium distribution in which all the atoms (or molecules) are moving at essentially the same speed. This is an experimentally interesting and important distribution, and we will look at ways of creating such customized, nonequilibrium distributions.

Chapter 23's discussion of partition functions and distributions shows us how to find the population fraction, $f(\epsilon) = N_\epsilon/N$, for the number N_ϵ of molecules with kinetic energy ϵ in a system of N molecules. This fraction is

$$f(\epsilon) = \frac{N_\epsilon}{N} = \frac{e^{-\epsilon/k_BT}}{q_{tr}} = \frac{e^{-\epsilon/k_BT}}{V(mk_BT/2\pi\hbar^2)^{3/2}} \tag{24.1}$$

where the energy is the particle-in-a-box expression from Chapter 12:

$$\epsilon = \frac{\hbar^2\pi^2}{2m}\left(\frac{n_x^2}{L_x^2} + \frac{n_y^2}{L_y^2} + \frac{n_z^2}{L_z^2}\right) \tag{12.24}$$

and the volume $V = L_xL_yL_z$. This distribution is a function of the three quantum numbers, n_x, n_y, and n_z, and is perfectly correct as it stands. In fact, it was used to construct Figure 23.8. It is not, however, very useful. It is simply too detailed. Translational energy levels are so closely spaced that quantum state distinctions are not very important.

A simpler and more useful distribution takes advantage of this close energy-level spacing. We pass from discrete states to a *continuum* of states and from a discrete distribution to a continuous distribution defined such that $f(\epsilon)\, d\epsilon$ equals the *fraction of molecules with kinetic energy between ϵ and $\epsilon + d\epsilon$*. We need the *translational density of states*, $\rho(\epsilon)\, d\epsilon$, the number of states with energy ϵ to $\epsilon + d\epsilon$, which was discussed along with the translational partition function in Section 23.1:

$$\rho(\epsilon)\, d\epsilon = \frac{\pi}{4}\left(\frac{2m}{\hbar^2\pi^2}\right)^{3/2} V\, \epsilon^{1/2}\, d\epsilon \tag{24.2}$$

so that the *continuous translational kinetic energy distribution function* is

$$f(\epsilon)\, d\epsilon = \frac{\rho(\epsilon)e^{-\epsilon/k_BT}}{q_{tr}}\, d\epsilon = 2\pi\left(\frac{1}{\pi k_BT}\right)^{3/2} \epsilon^{1/2}\, e^{-\epsilon/k_BT}\, d\epsilon\ . \tag{24.3}$$

Note that this distribution depends *only on the gas temperature*. The molecular mass does not matter, nor does any other molecular parameter. Figure 24.1 illustrates

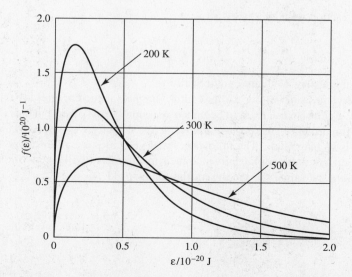

FIGURE 24.1 The equilibrium kinetic-energy distribution function for a gas depends on temperature only, not on the gas composition. As T is increased, the distribution is more uniformly spread over a greater energy range.

this distribution at three temperatures. At low T, the distribution is highly peaked—most molecules have similar energies—but at high T, the distribution is spread over a large energy range. A wide range of energies have comparable probabilities. The area under each curve, $\int_0^\infty f(\epsilon)\, d\epsilon$, is unity, and the fraction of molecules with kinetic energy between ϵ_1 and ϵ_2 is $\int_{\epsilon_1}^{\epsilon_2} f(\epsilon)\, d\epsilon$.

EXAMPLE 24.1

Show that the kinetic energy distribution function reproduces the equipartition average energy, $\langle \epsilon \rangle = 3k_B T/2$.

SOLUTION The average of any quantity distributed over a continuous range of values is the integral of that quantity weighted by the distribution. Here we have

$$\langle \epsilon \rangle = \int_0^\infty \epsilon f(\epsilon)\, d\epsilon = 2\pi \left(\frac{1}{\pi k_B T}\right)^{3/2} \int_0^\infty \epsilon^{3/2} e^{-\epsilon/k_B T}\, d\epsilon$$

and with the change of variable $\epsilon = x^2$, so that $d\epsilon = 2x\, dx$, and writing $1/k_B T = \beta$,

$$\langle \epsilon \rangle = 4\pi \left(\frac{\beta}{\pi}\right)^{3/2} \int_0^\infty x^4 e^{-\beta x^2}\, dx = 4\pi \left(\frac{\beta}{\pi}\right)^{3/2} \left(\frac{1}{4}\frac{3}{2} \pi^{1/2} \beta^{-5/2}\right)$$

since (see Problem 23.51) $\int_0^\infty x^4 e^{-\beta x^2}\, dx = \frac{1}{4}\frac{3}{2} \pi^{1/2} \beta^{-5/2}$. Simplifying gives us

$$\langle \epsilon \rangle = \frac{3}{2} \beta^{-1} = \frac{3}{2} k_B T \ .$$

⇒ **RELATED PROBLEMS** 24.1, 24.2, 24.3, 24.4, 24.5

It is also interesting to note that the average energy, $\langle \epsilon \rangle = 3k_B T/2$, is three times the *most probable energy*, $\epsilon_{mp} = k_B T/2$, the energy corresponding to the peak in the distribution. (See Problem 24.1.) The fraction with energy $\leq \epsilon_{mp}$ is

$$\int_0^{k_B T/2} f(\epsilon)\, d\epsilon = 2\pi \left(\frac{1}{\pi k_B T}\right)^{3/2} \int_0^{k_B T/2} \epsilon^{1/2} e^{-\epsilon/k_B T}\, d\epsilon = 0.1987$$

and the fraction with energy $\leq \langle \epsilon \rangle$ is

$$\int_0^{3k_B T/2} f(\epsilon)\, d\epsilon = 2\pi \left(\frac{1}{\pi k_B T}\right)^{3/2} \int_0^{3k_B T/2} \epsilon^{1/2} e^{-\epsilon/k_B T}\, d\epsilon = 0.6084 \ .$$

(Both integrals are evaluated numerically.)

These values support our earlier claims that gas kinetic energies are on the order of $k_B T$. We can also use them to begin a discussion of two other dynamical variables that are important to gases, *speed and velocity*. Velocity **v** is a vector quantity, and speed v is its magnitude. The following set of velocities are all different, since they point in different directions, but since their magnitudes (the lengths of the velocity vectors) are the same, they represent the same speed:

same speed, different velocities

The kinetic energy depends only on the speed, but it can be written in terms of the vector *dot product* of the velocity with itself:

$$\frac{1}{2}mv^2 = \frac{1}{2}m\mathbf{v} \cdot \mathbf{v} = \frac{1}{2}m(v_x^2 + v_y^2 + v_z^2),$$

which also shows how the speed can be written in terms of the Cartesian components of the velocity: $v = (v_x^2 + v_y^2 + v_z^2)^{1/2}$. If we average ϵ, we are also averaging the *square* of the speed:

$$\langle \epsilon \rangle = \left\langle \frac{1}{2}mv^2 \right\rangle = \frac{1}{2}m\langle v^2 \rangle.$$

Since equipartition gives us $\langle \epsilon \rangle = 3k_BT/2$, we can immediately find the *root mean square* (rms) speed, the square root of the average of the square of the speed:

$$v_{rms} = \langle v^2 \rangle^{1/2} = \sqrt{\frac{2\langle \epsilon \rangle}{m}} = \sqrt{\frac{3k_BT}{m}}. \tag{24.4}$$

For the rare gases at 300 K, v_{rms} ranges from 1367 m s^{-1} for He down to 239 m s^{-1} for Xe.

We can easily turn the kinetic energy distribution into a speed distribution. Since $\epsilon = mv^2/2$, $d\epsilon = mv\,dv$ and $\epsilon^{1/2} = m^{1/2}v/\sqrt{2}$ so that Eq. (24.3) becomes the *Maxwell–Boltzmann speed distribution*

$$f(v)\,dv = 4\pi \left(\frac{m}{2\pi k_BT}\right)^{3/2} v^2 e^{-mv^2/2k_BT}\,dv. \tag{24.5}$$

The probability distribution $f(v)$ gives the population fraction of molecules with speeds between v and $v + dv$. Figure 24.2 plots $f(v)$ for the rare gases at 300 K. In addition to the rms speed, several other speeds can be used to characterize this distribution. The *most probable* speed, v_{mp}, is the speed that maximizes $f(v)$; the *mean* or *average speed*, $\langle v \rangle$, is $\int v f(v)\,dv$; the *median speed*, v_{med}, is that speed for which half the molecules have $v \leq v_{\text{med}}$. It divides the area under $f(v)$ in half. Problems 24.6 and 24.7 derive expressions for v_{mp} and $\langle v \rangle$, while v_{med} is that speed for which

$$\int_0^{v_{\text{med}}} f(v)\,dv = \int_{v_{\text{med}}}^{\infty} f(v)\,dv = \frac{1}{2}$$

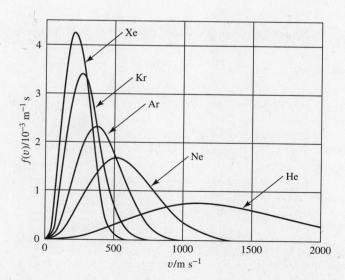

FIGURE 24.2 The Maxwell–Boltzmann equilibrium speed distribution is a function of the gas temperature (300 K here) and the gas composition. Heavy Xe's distribution is concentrated over a short speed range, while light He's distribution is spread over a very wide range.

and it has to be found by numerical methods. Finally, thermodynamics gives the following expression for the *speed of sound* in an ideal gas:

$$v_s = \sqrt{\frac{\gamma k_B T}{m}} \qquad (24.6)$$

where γ is the heat capacity ratio, C_P/C_V. For a monatomic gas, $\gamma = 5/3$, independent of temperature as long as excited electronic states are not populated. For polyatomic gases, $\gamma < 5/3$ decreasing toward 1 with increasing temperature. While the speed of a sound wave is not a molecular speed, it is interesting to contrast it to molecular speeds such as v_{mp} that are characteristic of $f(v)$. We will also consider the speed of sound in relation to the speed at which energy, matter, and momentum are transported through gases in the next chapter.

Figure 24.3 locates these various speeds in relation to the speed distribution for Ar at 300 K, and Table 24.1 lists expressions for them. In Figure 24.3, the speed axis is scaled from 0 to $\sim 2\langle v \rangle$, and it is clear from the distribution function that

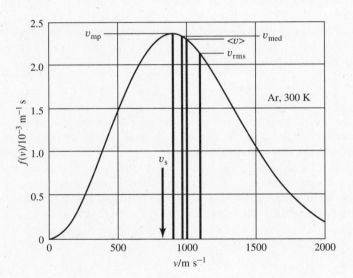

FIGURE 24.3 We can identify several characteristic speeds for a Maxwell–Boltzmann distribution: v_{mp}, the most probable speed; v_{med}, the median speed; $\langle v \rangle$, the average speed; and v_{rms}, the root mean square speed. While not a true characteristic of the distribution, v_s, the speed of sound, has roughly the same magnitude.

TABLE 24.1 Characteristic Speeds of Equilibrium Gas Speed Distributions

Most probable speed:	$v_{mp} = \sqrt{\dfrac{2k_B T}{m}}$	$= 128.95 \text{ m s}^{-1} \sqrt{\dfrac{T/K}{M/\text{g mol}^{-1}}}$
Median speed:	$v_{med} = 1.538\,2\sqrt{\dfrac{k_B T}{m}}$	$= 140.26 \text{ m s}^{-1} \sqrt{\dfrac{T/K}{M/\text{g mol}^{-1}}}$
Average speed:	$\langle v \rangle = \sqrt{\dfrac{8k_B T}{\pi m}}$	$= 145.51 \text{ m s}^{-1} \sqrt{\dfrac{T/K}{M/\text{g mol}^{-1}}}$
Rms speed:	$v_{rms} = \langle v^2 \rangle^{1/2} = \sqrt{\dfrac{3k_B T}{m}}$	$= 157.94 \text{ m s}^{-1} \sqrt{\dfrac{T/K}{M/\text{g mol}^{-1}}}$
Speed of sound:	$v_s = \sqrt{\dfrac{\gamma k_B T}{m}}$	$= 91.184 \text{ m s}^{-1} \sqrt{\dfrac{\gamma(T/K)}{M/\text{g mol}^{-1}}}$

T = temperature in K, m = molecular mass in kg, M = molar mass in g mol^{-1}.
γ = heat capacity ratio C_P/C_V. $\gamma = 5/3$ for monatomic gases, $\cong 7/5$ for diatomic gases.

very few molecules have speeds greater than twice $\langle v \rangle$. A rigorous calculation shows that only 1.7% are moving faster than $2\langle v \rangle$.

EXAMPLE 24.2

One can buy a "sonar yardstick" that sends a sound pulse toward a wall or other large object and measures the time it takes for the reflected pulse to return. This time is then displayed as a distance from the yardstick to the target object. What time resolution should it have to measure a 10 m distance with <1 cm error? If the yardstick is calibrated at 25 °C, what error will it have at −10 °C?

SOLUTION This is an example of a *time-of-flight* technique used here to turn a known speed and a measured time into a distance. We consider time-of-flight methods for measuring molecular speeds later in this chapter. If the target is a distance x_0 away from the sonar pulse source, the round-trip pulse transit time is $\tau = 2x_0/v_s$. For air, $\gamma = 7/5 = 1.4$, the value for any diatomic at temperatures low enough to keep vibrations and electronic excitations negligible, and $m = 4.8 \times 10^{-26}$ kg (~29 g mol^{-1}) so that $v_s = 346$ m s^{-1} at 25 °C. For $x_0 = 10$ m, $\tau = 57.81$ ms. Since $d\tau = (2/v_s)\, dx_0$, a 1 cm uncertainty ($dx_0 = 0.01$ m) implies a time resolution $d\tau = 58$ µs. Commercial devices have time resolutions typically an order of magnitude better than this.

If we let τ_1 and $v_s(1)$ represent the time and speed at the calibration temperature, 25 °C, and τ_2 the time at −10 °C, then for fixed x_0, $\tau_1/\tau_2 = \sqrt{T_2/T_1}$ so that the distance reported at T_1 is $x_0(1) = \tau_1 v_s(1)/2$ and the distance reported at T_2 is $x_0(2) = \tau_2 v_s(1)/2 = x_0(1)\sqrt{T_2/T_1}$ (since the device has memorized the calibration speed $v_s(1)$). The fractional error at −10 °C is thus

$$\frac{x_0(2) - x_0(1)}{x_0(1)} = 1 - \sqrt{\frac{T_1}{T_2}} = 1 - \sqrt{\frac{298.15 \text{ K}}{263.15 \text{ K}}} = -0.064 ,$$

so that the device should have a temperature calibration control if used at extreme temperatures.

➟ RELATED PROBLEMS 24.9, 24.10

We can turn our speed distribution function, Eq. (24.5), into a *velocity* distribution function. First, equipartition and the relation between speed and velocity components, $v^2 = (v_x^2 + v_y^2 + v_z^2)$, along with the equivalence of x, y, and z in an equilibrium gas (that is not strongly influenced by an external field such as gravity) tell us that

$$\langle v_x^2 \rangle = \langle v_y^2 \rangle = \langle v_z^2 \rangle = \frac{1}{3} \langle v^2 \rangle = \frac{v_{\text{rms}}^2}{3} = \frac{k_B T}{m} .$$

Second, if the gas is at rest macroscopically (i.e., its container is not moving with respect to us), the average velocity in any direction $\langle v_x \rangle$, etc., must be zero—there are as many molecules moving to the left, say, with speed v as there are moving to the right with this speed. The velocity distribution must be symmetric about $\mathbf{v} = 0$.

Classical statistical mechanics can give the velocity distribution directly (see Problem 24.11), but we can also deduce it from the speed distribution. If we rearrange Eq. (24.5) a bit, we can identify the velocity distribution, $f(v_x)f(v_y)f(v_z)\, dv_x\, dv_y\, dv_z$, buried in it:

$$\underbrace{\left[\left(\frac{m}{2\pi k_B T}\right)^{1/2} e^{-mv_x^2/2k_B T}\right]}_{f(v_x)} \underbrace{\left[\left(\frac{m}{2\pi k_B T}\right)^{1/2} e^{-mv_y^2/2k_B T}\right]}_{f(v_y)} \underbrace{\left[\left(\frac{m}{2\pi k_B T}\right)^{1/2} e^{-mv_z^2/2k_B T}\right]}_{f(v_z)} \underbrace{(4\pi\, v^2\, dv)}_{\substack{dv_x\, dv_y\, dv_z \\ \text{in polar coordinates}}}$$

The key is to note that the differential volume element in 3-D velocity space, $dv_x\, dv_y\, dv_z$, becomes $v^2 \sin\theta\, d\theta\, d\phi\, dv$ in spherical polar speed coordinates, and since speed is independent of direction, integration over polar angles θ and ϕ gives the factor of 4π:

$$\int_0^{2\pi} d\phi \int_0^{\pi} \sin\theta\, d\theta = 4\pi\ .$$

Consequently, any one velocity component distribution function, $f(v_x)\, dv_x$ for example, has the form

$$f(v_x)\, dv_x = \left(\frac{m}{2\pi k_B T}\right)^{1/2} e^{-mv_x^2/2k_B T}\, dv_x\ . \tag{24.7}$$

Velocity components can be positive or negative, and this distribution peaks at and is symmetric about $v_x = 0$. The average velocity component, $\langle v_x \rangle$, is thus zero, and it is easy to show that equipartition is satisfied so that $\langle v_x^2 \rangle = k_B T/m$.

These speed and velocity distributions describe equilibrium, and the next section describes experiments that have verified them, but for many modern research settings, they are a nuisance. They do not define molecular speed very precisely, and for modern spectroscopic and kinetic research, this is not acceptable. Fortunately, it is amazingly simple to decrease the spread in speeds and generate a *nonequilibrium* distribution that is very useful.

Imagine grabbing a handful of gas at equilibrium, cooling it to ~1 K *without condensing it,* and throwing the gas at several hundred m s^{-1} through a vacuum. The distribution in speeds would be dramatically narrowed through the temperature reduction, and if you rode along with the handful of gas as it flew through the vacuum, you would say the temperature was 1 K. In the lab, you would see the gas move past you in one direction (call it the x direction) with a bulk velocity V, superimposed on the 1 K spread in v_x. This seems like a very difficult distribution to arrange experimentally, but the *supersonic molecular beam* (also called the *free jet* or *nozzle* beam) technique does almost exactly this. Diagrammed in Figure 24.4, the supersonic beam has no moving parts (except for those in the vacuum pumps!)

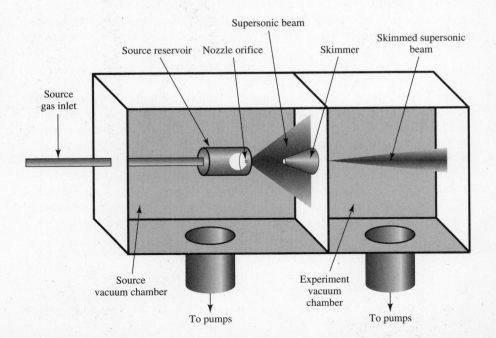

FIGURE 24.4 A supersonic molecular beam source typically consists of a source reservoir containing the beam gas and a small orifice, the nozzle, through which the gas expands into a source vacuum chamber. A skimmer selects the central portion of the beam for use in an adjacent experiment vacuum chamber.

An equilibrium gas at some temperature is confined in a source reservoir at a pressure typically in the 0.1 to 10 atm range. The container has a small hole (with a diameter typically in the 100–1000 μm range) through which the gas freely expands into the vacuum. *This alone does the job,* but often, the center region of the expanding gas is selected by a *skimmer* that allows a fraction of the expanding gas to enter a separate vacuum chamber in which the experiment of interest is carried out.[2]

A simple analogy describes how the nozzle beam works on a microscopic level. Picture a subway train arriving at a popular station at rush hour. Very many people (the gas molecules) rush from the train (the delivery line) onto the station platform (the source reservoir). They jostle each other in the crowd, rearranging packages, newspapers, etc., as they head to the exit (they are at a reasonably high temperature, but move under the influence of a pressure gradient). The only way out of the station is through a turnstile (the nozzle orifice) that is quite narrow. They must exit *at the same speed,* and they adjust their relative speeds (through more or less polite or violent collisions with each other) to do so. Once they exit the turnstiles (leave the source reservoir), they enter the outside world (the experimental vacuum chamber) with a remarkably constant speed, still somewhat fast (it is rush hour, after all), but in a state of considerably less agitation and considerably greater order than when they first left the train.

Molecules behave much the same way. While they do not go through the nozzle one at a time, the gas density is high enough that *intermolecular collisions dominate their rate of exit*. This happens very quickly so that the expansion is adiabatic. The enthalpy of the gas is converted during the expansion into a very uniform *flow velocity V* of the gas in the downstream direction. The *beam* temperature is a measure of the *local* spread of speeds *around* the flow velocity, but the *source* temperature determines the maximum value for V through enthalpy conversion:

$$\overline{C}_P (T_0 - T) = \frac{1}{2} MV^2$$

↑   ↑
Molar enthalpy lost cooling   Bulk flow molar
from source temperature T_0 kinetic energy
to beam temperature T

where M is the molar mass. Since T, the beam temperature, is easily very much less than the source temperature T_0, the maximum flow velocity is (for a monatomic gas with $\overline{C}_P = 5R/2$)

$$V = \sqrt{\frac{5k_B T_0}{m}} = 203.89 \text{ m s}^{-1} \sqrt{\frac{T_0/\text{K}}{M/\text{g mol}^{-1}}}, \quad (24.8)$$

which is not exceptionally larger than characteristic speeds of the equilibrium distribution.

EXAMPLE 24.3

What beam source conditions can produce a He flow velocity of 4500 m s^{-1}?

SOLUTION We solve Eq. (24.8) for T_0 and find

$$T_0 = \frac{mV^2}{5k_B} = \frac{MV^2}{5R} = \frac{(4 \times 10^{-3} \text{ kg mol}^{-1})(4500 \text{ m s}^{-1})^2}{5 (8.31 \text{ J mol}^{-1} \text{ K}^{-1})} = 1950 \text{ K},$$

---

[2]Separating the vacuum systems for the source and experimental chambers—a technique known as *differential pumping*—allows the vacuum pumps in the source chamber to bear the major pumping load, keeping the experimental chamber at lower pressure. It is an axiom of all molecular beam research that all the noncondensable gas that enters a vacuum system has to be pumped away. Differential pumping shares this job among many pumps.

which is a very high, but experimentally attainable source temperature. The source reservoir is typically a W or Ta tube connected across a high-current power supply. The (small) electrical resistance of the metal tube heats it and the gas, and temperatures in the vicinity of 2000 K can be reached. It is interesting to note that at 300 K, the fraction of He atoms with speeds ≥ 4500 m s^{-1} is $\int_{4500}^{\infty} f(v)\, dv \cong 3.7 \times 10^{-7}$, while in the beam, *all* the molecules have roughly this speed.

▶ **RELATED PROBLEMS** 24.12, 24.13, 24.16

The exceptional difference between the equilibrium and the supersonic speed distributions is the *narrowness* of the latter, and this is controlled by T, the beam temperature.[3] Along the beam direction (call it x), the velocity distribution is the distribution of Eq. (24.7) modified to account for the flow velocity:

$$f(v_x)\, dv_x = \left(\frac{m}{2\pi k_B T}\right)^{1/2} e^{-m(v_x - V)^2/2k_B T}\, dv_x\ . \qquad (24.9)$$

Supersonic flows are characterized by a dimensionless number, the *Mach number*:

$$\text{Mach number} = \frac{\text{mean flow velocity}}{\text{local speed of sound}} = \frac{V}{\sqrt{\gamma k_B T/m}}\ . \qquad (24.10)$$

Large Mach numbers mean small local beam temperatures. Calculating the beam temperature from a given set of experimental conditions (the nature of the gas, its source pressure and temperature, the nozzle diameter, and the downstream pressure all enter the calculation) is a difficult problem in hydrodynamics that we will not consider, but experiment and theory agree quite well.

For example, consider an Ar expansion from $P_0 = 1$ atm, $T_0 = 300$ K through a 200 μm diameter nozzle into a high vacuum. From Eq. (24.6), the flow velocity of this beam is 589 m s^{-1}, which can be compared to the characteristic speeds of the equilibrium gas in Figure 24.3. The beam temperature is found to be about 0.36 K, and the Mach number is 50. Figure 24.5 plots the equilibrium speed distribution of Figure 24.3 along with the beam velocity distribution along the beam direction (Eq. (24.9)). The widespread equilibrium distribution is compressed into a narrow spike centered around V.

As mentioned in Chapter 19, other benefits come from the supersonic expansion. Diatomic and polyatomic molecules have *internal temperatures* cooled as well, although not to the degree that translational temperatures are. Each degree of freedom is characterized by its own temperature (although highly nonequilibrium distributions that cannot be characterized by a temperature are also possible). Typically, T_{tr} is lowest, T_{rot} is 2 to 10 times larger than T_{tr}, and T_{vib} is 20 to 100 times larger than T_{tr}. This has profound spectroscopic advantages, since the population of excited vibration/rotation states is greatly lowered, simplifying the spectra. In addition, as the gas expands from a region where many multiple-body collisions occur at high pressure to densities so low that virtually no collisions occur, *condensation* into loosely bound dimers and higher clusters can occur.[4] These can be studied on their

---

[3] There are, in fact, two temperatures characteristic of a supersonic beam. What we call T here is more properly called $T_\parallel$, the temperature of the velocity distribution parallel to the flow direction. The velocity distribution perpendicular to this direction is characterized by a slightly different temperature, $T_\perp$. Expressions for either of these are beyond the scope of this book, but can be found in this chapter's Further Reading references.

[4] The density must ultimately fall so low that collisions stop, or else the gas would turn into 10 K chunks of solid falling out of the nozzle!

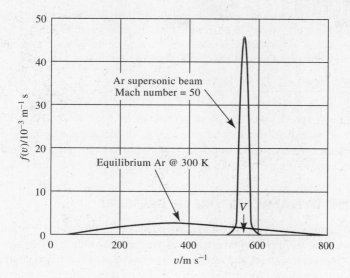

FIGURE 24.5 The speed distribution of a supersonic beam is characterized by a bulk flow velocity V and a measure of the local spread in velocity, the Mach number. The higher the Mach number, the narrower the velocity distribution. In contrast to an equilibrium distribution, the beam molecules are all moving quite closely at the same velocity, V.

own as bound, stable, isolated species in the beam. If the gas does not stick to itself very well (i.e., if it is He), very high Mach numbers (>100) can be attained, leading to beam temperatures ~1 mK, and velocity distributions so narrow that such He beams have found use in surface diffraction studies. These beams have very monochromatic de Broglie wavelengths.

Finally, the technique of *seeding* can be used to control *kinetic energies* in a dramatic way. If a small (typically <5%) amount of a heavy molecule is added to a light gas such as He, expansion of the mixture causes the heavy molecules to reach nearly the same speed as the light carrier gas. Just as a swarm of mosquitoes can chase you in the direction and at the speed of the swarm, the many collisions of the He atoms with the heavy molecule force it to attain the He flow velocity. Since the flow velocity is governed by the light gas, the heavy gas is accelerated to an exceptionally high kinetic energy. If m_H and m_L are the heavy and light species' masses, the heavy particle kinetic energy is

$$\frac{1}{2} m_H V^2 = \frac{1}{2} m_H \left(\frac{5 k_B T_0}{m_L} \right) = \frac{5}{2} k_B T_0 \left(\frac{m_H}{m_L} \right) . \qquad (24.11)$$

Compared to the equilibrium average kinetic energy, $3 k_B T_0 / 2$, this kinetic energy can be an order of magnitude or more larger.

EXAMPLE 24.4

Suppose Xe is mixed into the He in Example 24.3 at the 1% level. What kinetic energy will the Xe atoms have? Express the answer in eV units.

SOLUTION From Eq. (24.11), with $T_0 = 1950$ K,

$$\frac{1}{2} m_{Xe} V^2 = \frac{5}{2} k_B T_0 \left(\frac{m_{Xe}}{m_{He}} \right) = \frac{5 (1.38 \times 10^{-23} \text{ J K}^{-1})(1950 \text{ K})}{2} \left(\frac{131}{4} \right) = 2.2 \times 10^{-18} \text{ J}$$

or, converting to eV units, $(1.6 \times 10^{-19} \text{ J eV}^{-1})$, the Xe kinetic energy is

$$\frac{2.2 \times 10^{-18} \text{ J}}{1.6 \times 10^{-19} \text{ J ev}^{-1}} = 13.7 \text{ eV} ,$$

which is slightly more than the ionization energy of H! Put another way, the fraction of Xe atoms, even at 1950 K, with a kinetic energy this large or larger *at equilibrium* is ~3 × 10^{-35}. One mole of Xe at 1950 K would have one (and only one) atom at a time with this kinetic energy, for a total time of only ~0.5 ms each year!

⇒ *RELATED PROBLEMS* 24.14, 24.15

The energy, speed, and velocity distributions we have discussed in this chapter are based on statistical pictures of gases. A more mechanical picture—molecules colliding with container walls and with each other—yields further insight into the inner workings of a gas. We take up that line of thought in the next section.

24.2 THE DYNAMIC NATURE OF PRESSURE

Thermodynamics says (see Eq. (22.25)) pressure is the negative of the partial derivative of the Helmholtz free energy with respect to volume at constant temperature. This is a very abstract and sterile description of pressure when contrasted with the dynamic picture of pressure as a force per unit area, especially when that force is due to gas molecules bombarding a container's walls. We can calculate the pressure from this dynamic picture and show that it also predicts $PV = nRT$ for an ideal gas.

We start with

$$P = \frac{\text{force}}{\text{unit area}} = \frac{dp/dt}{\text{unit area}}$$

where $F = dp/dt$, the time derivative of momentum, is a generalization of $F = ma$, the familiar force equals mass times acceleration expression. We can also write the pressure in terms of molecular collisions with the wall and the rate of these collisions (the number per unit time):

$$P = (\text{momentum change per wall collision}) \frac{(\text{number of wall collisions})}{(\text{unit wall area})(\text{unit time})}.$$

In any gas at densities such that the average molecular separation is greater than the range of intermolecular forces, molecules fly freely toward the wall, collide with it, and rebound freely from it. The details of this collision cannot matter, since thermal equilibrium could not be maintained if the velocity distribution of the molecules heading toward the wall was different from that of molecules rebounding from it. Consequently, we consider the simplest collision, known as an *elastic* collision. In an elastic collision, a line perpendicular to the wall lies in the same plane as the molecule's incoming and outgoing trajectory. If the surface perpendicular is the x direction, the collision looks like this:

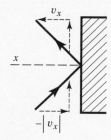

24.2 THE DYNAMIC NATURE OF PRESSURE

The x component of the velocity (or momentum) changes direction in an elastic collision, so the change in momentum per collision, (the first piece of the pressure expression), is

$$m|v_x| \quad - \quad (-m|v_x|) \quad = \quad 2m|v_x| \ .$$

↑ Momentum component away from the wall after the wall collision

↑ Momentum component towards the wall before the wall collision

↑ Net change in momentum due to the collision

We could specify the distribution of x velocity components from Eq. (24.7), but with the help of the equipartition theorem, we can arrive at $PV = nRT$ *without knowing this distribution in detail*. If there are $N = nN_A$ molecules in a volume V with an x-component velocity distribution $f(v_x)$,

$$\frac{Nf(v_x)\,dv_x}{V} = \frac{\text{number of molecules with velocities } v_x \text{ to } v_x + dv_x}{\text{unit volume}}.$$

Now let τ be an arbitrary period of time and A be an area on the wall. Every molecule in a volume $(v_x\tau)A$ extending from the wall into the gas that has velocity component v_x directed towards the wall will strike it in time τ since a molecule $v_x\tau$ away from the wall will just hit the wall at time τ while any closer to the wall will hit it sooner (assuming the density is so low that no molecule–molecule collisions occur during τ). (See Figure 24.6.) There are $(v_x\tau)A(N/V)f(v_x)\,dv_x$ such molecules in this volume. Thus the number that strike *unit* area of the wall in *unit* time is

$$\frac{(v_x\tau)A(N/V)f(v_x)\,dv_x}{\tau A} = \frac{v_x N f(v_x)\,dv_x}{V}.$$

This is the second piece of the pressure expression so that the total pressure, averaging over all velocities directed toward the wall, is

$$P = \int_0^\infty (2mv_x)\left(\frac{v_x N f(v_x)}{V}\right)dv_x = \left(\frac{N}{V}\right)m\int_{-\infty}^\infty v_x^2 f(v_x)\,dv_x = \left(\frac{N}{V}\right)m\langle v_x^2\rangle \ . \quad \textbf{(24.12)}$$

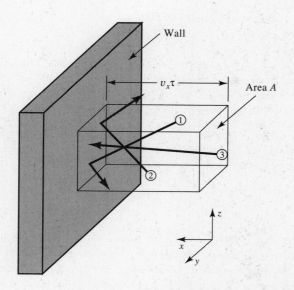

FIGURE 24.6 Molecules in a volume $Av_x\tau$ with velocity v_x toward a wall will all hit the wall in time τ.

From the equipartition theorem, $m\langle v_x^2\rangle/2 = k_B T/2$ or $m\langle v_x^2\rangle = k_B T$ so that

$$P = \left(\frac{N}{V}\right) m \langle v_x^2\rangle = \frac{Nk_B T}{V} = \frac{nN_A k_B T}{V} = \frac{nRT}{V}.$$

We have arrived at $PV = nRT$ through yet another route. We needed only elementary mechanics—the momentum change on an elastic collision of a particle with a wall—and the general equipartition theorem.

We can also use Figure 24.6 to calculate the *wall collision rate*, J_x, the number of molecules striking unit area in unit time. This is the average over v_x of $v_x N f(v_x)/V$ (for a wall perpendicular to the x axis):

$$J_x = \frac{N}{V}\int_0^\infty v_x f(v_x)\, dv_x = \frac{N}{V}\left(\frac{m}{2\pi k_B T}\right)^{1/2}\int_0^\infty v_x\, e^{-mv_x^2/2k_B T}\, dv_x = \frac{N}{V}\left(\frac{k_B T}{2\pi m}\right)^{1/2} \quad (24.13a)$$

where here we *do* need the full expression for $f(v_x)$, Eq. (24.9). While Eq. (24.13a) is a perfectly acceptable expression, we can write it in two other illuminating ways. First, since $P = Nk_B T/V$,

$$J_x = \frac{P}{(2\pi m k_B T)^{1/2}} \quad (24.13b)$$

and also, since the mean speed $\langle v\rangle = (8k_B T/\pi m)^{1/2}$, we have

$$J_x = \frac{N}{V}\left(\frac{k_B T}{2\pi m}\right)^{1/2} = \frac{1}{4}\frac{N}{V}\left(\frac{8k_B T}{\pi m}\right)^{1/2} = \frac{1}{4}\frac{N}{V}\langle v\rangle. \quad (24.13c)$$

At ordinary temperatures and pressures, J_x is huge. Since a typical $\langle v\rangle$ is ~ 500 m s^{-1} and a typical molecular density is $N/V = P/k_B T \cong 2.4 \times 10^{25}$ m^{-3}, (for 1 atm and 300 K), J_x is on the order of 3×10^{27} m^{-2} s^{-1}.

EXAMPLE 24.5

Imagine a microphone so sensitive that it could hear the tap of a single molecule. What air pressure (at room temperature) would be required to reduce the collision rate with the microphone to the audible frequency 10^3 Hz?

SOLUTION We write Eq. (24.13b) in the form

$$P = AJ_x(2\pi m k_B T)^{1/2}$$

where A is the area of the microphone. Since a large microphone is struck at a greater rate than a small one, all we require here is that the product $AJ_x = 1000$ s^{-1}. For air, $M \cong 29$ g mol^{-1} or $m \cong 4.8 \times 10^{-26}$ kg. For $T = 300$ K, we find $P = 3.5 \times 10^{-20}$ Pa $= 3.5 \times 10^{-25}$ atm. This corresponds to $N/V \cong 8.5$ molecules m^{-3}, far below the best attainable laboratory vacuum.

It is also interesting to note that, could the experiment be done, the microphone would *not* "hum" at 1000 Hz as a struck tuning fork would. Molecules strike at random, uncorrelated times, and the sound would be a roar of "static" that had 1 kHz as a significant but not unique tone.

→ RELATED PROBLEMS 24.18, 24.19, 24.20

While we have written and derived J_x as a rate of collisions with a real container wall, we could also consider J_x as the *flux of molecules* (the number per unit area per unit time) *passing in one direction through any imaginary plane inside the gas*. Of course, there are as many molecules per second going through such a plane in one direction as in the opposite direction (otherwise the gas would not be in equilibrium—a wind would blow through the gas). The next chapter will consider nonequilibrium states in terms of fluxes such as J_x. Also, we could place the gas container in a vacuum and consider the collision rate with a *hole in the container wall*. This would give us the rate of escape of gas, an idea that is behind molecular-beam generation.

Suppose the container has a small hole of area A. The rate at which molecules escape through the hole and into the vacuum is AJ_x, but *only if the density of the gas and the size of the hole have the correct relationship*. If the density is too high, the gas escapes as a supersonic beam, and the rate of escape is *not* AJ_x. Implicit in our derivation of the collision rate is the idea of independent, isolated molecule–wall collisions. At high densities, molecules collide *with each other as they are escaping through the hole*. (Recall the importance of intermolecular collisions in our discussion of the supersonic beam.) The escape rate in this regime is a complicated problem of hydrodynamic flow.[5]

We can find the proper relationship between hole area and gas density so that AJ_x is the escape rate if we consider another characteristic dynamic property of a gas: the *mean free path*, which is the average distance a molecule travels between collisions with other molecules. As soon as we introduce molecular collisions, we are leaving behind the ideal gas model. Point molecules never collide with each other. The simplest nonideal gas is one we have seen before: the hard-sphere gas. We imagine each particle is a sphere of radius r and diameter $d = 2r$. Whenever two spheres' centers are a distance d apart, a collision takes place:

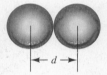

It is convenient to define the hard-sphere *collision cross section*

$$\sigma = \pi d^2 \qquad (24.14)$$

since this is the effective target area of any one sphere:

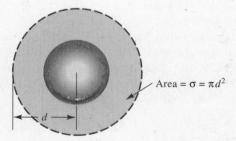

We now seek an expression for the mean free path λ in terms of σ.

The main idea is not difficult. If the other spheres are stationary, one moving sphere encounters one other sphere when the first moves a distance λ. As it does so, it generates a collision volume $\sigma\lambda$, and since it found one collision partner in

---

[5]Consider the very high density limit—a liquid—and you will recognize that AJ_x cannot be the escape rate at all densities.

that volume (by definition of λ), it must be true that $\sigma\lambda = V/N$, since V/N is the volume of the container available on average per sphere. This says $\lambda = V/N\sigma$, which is almost correct. It is not exact because the other spheres are moving as well, and it is not the absolute speed of a sphere, but the *relative speed of two spheres* that governs the collision rate.

The relative *speed* of two objects moving through space depends on the *velocities* of each. In general, the relative *velocity* is the *vector difference* of the two particle velocities, and the relative speed is the magnitude of the relative velocity, **g**:

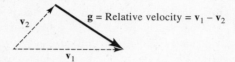

We will have much more to say about the relative velocity of a two-particle collision in Chapter 28. Note for now that $\mathbf{v}_1$ and $\mathbf{v}_2$ are distributed over a range of equilibrium values (Eq. (24.7) is the distribution for each component of **v**) and that **g** follows the same distribution *with one important difference:* instead of the particle mass m governing the distribution, the *reduced mass* μ takes its place. This is the same reduced mass that governs the dynamics of a two-body harmonic oscillator (Eq. (19.9)), and for two identical particles, $\mu = m/2$. Consequently, the *mean relative speed* $\langle g \rangle$ is somewhat larger than the mean speed $\langle v \rangle$:

$$\langle g \rangle = \left(\frac{8k_B T}{\pi \mu}\right)^{1/2} = \sqrt{2}\left(\frac{8k_B T}{\pi m}\right)^{1/2} = \sqrt{2}\,\langle v \rangle \;. \tag{24.15}$$

Thus, the collision volume $\sigma\lambda$ is swept out, on average, not on the time scale of $\langle v \rangle$ but of $\langle g \rangle$. Since $\langle g \rangle$ is *greater* than $\langle v \rangle$, the volume is swept out in a *shorter* time in direct proportion to $\langle g \rangle/\langle v \rangle = \sqrt{2}$. This shortens the mean free path by this factor, so that

$$\lambda = \frac{V}{\sqrt{2}\,N\sigma} = \frac{k_B T}{\sqrt{2}\,P\sigma} \tag{24.16}$$

where we have assumed that the ideal gas expression $V/N = k_B T/P$ is still valid.

It is interesting to contrast the magnitude of the mean free path to the *average distance between molecules*. For example, 1 mol of Ar at 1 atm and 300 K occupies $V = 2.46 \times 10^{-2}$ m³. On average, these atoms are $(V/N_A)^{1/3} = 34.4$ Å apart. We can estimate the hard-sphere radius for Ar from Figure 1.5, the graph of the true Ar–Ar intermolecular potential energy. "Hard sphere" means rapidly rising (infinitely rapidly, in fact) potential energy, and for Ar, this happens at an Ar–Ar separation ~3.6 Å. This is d, the effective hard-sphere diameter. The collision cross section $\sigma = \pi d^2 = 4.1 \times 10^{-19}$ m², and the mean free path is

$$\lambda = \frac{\overline{V}}{\sqrt{2}\,N_A \sigma} = \frac{2.46 \times 10^{-2}\text{ m}^3\text{ mol}^{-1}}{\sqrt{2}\,N_A\,(4.1 \times 10^{-19}\text{ m}^2)} \cong 7.0 \times 10^{-8}\text{ m} = 700\text{ Å}$$

about twenty times the average separation.

The mean free path is significantly larger than the average separation simply because the atoms are small compared to that separation. They can pass by each other as they move in random directions over random positions and collide less frequently than their average spacing might indicate. As the density increases, the average spacing decreases, but the mean free path decreases even faster. When the two are comparable, the gas is obviously nonideal, and our derivation of the mean free path becomes suspect. Problem 24.26 considers this limit, and Figure 24.7 illustrates the mean free path for 1 mol of Ar at 10 atm and 300 K. The hard spheres are drawn to scale, as is the collision volume, $\sigma\lambda$. In this state, the average spacing is 16 Å and $\lambda = 70$ Å.

We can now return to the gas escape rate through a small hole. This process is called *effusion* whenever *the mean free path is much larger than the characteristic size of the hole*. If the diameter of the hole is D, the ratio λ/D is called the *Knudsen number*, Kn. Whenever Kn $\gg$ 1, molecules randomly hit the hole's open area and pass through it without striking each other on the way out.[6] The hole area is $A = \pi D^2/4$, and the rate molecules are lost from the container is

$$-\frac{dN}{dt} = AJ_x = \frac{\pi D^2}{4} \frac{P}{(2\pi m k_B T)^{1/2}}.$$ (24.17)

Written this way, we can see how the effusion rate can be used to measure *vapor pressures*.[7] We place the material of interest (solid or liquid) in a container with a small hole (called a *Knudsen cell*) and place the cell on a sensitive balance in a vacuum chamber. The balance records the *mass loss* through vapor effusion out of the hole as a function of cell temperature, and converting the mass loss rate into particle number loss lets us calculate the vapor pressure. Conversely, if we know the vapor pressure, the mass loss rate lets us deduce something about the vapor *composition*.

EXAMPLE 24.6

At 763 K, the vapor pressure of Sb is 1.75×10^{-6} atm. Antimony in a Knudsen cell with a 700 μm diameter hole loses mass at the rate -2.37×10^{-7} g s^{-1}. What is the principle species in Sb vapor at this temperature?

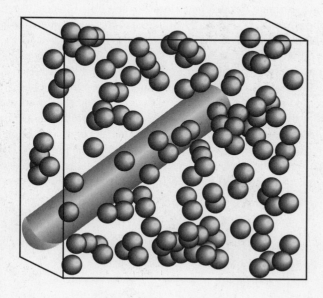

FIGURE 24.7 While gas molecules at any one instant (shaded spheres) can be reasonably close to each other, their relative motion can lead to significantly longer distances between collisions (the mean free path, represented by the shaded cylinder). The radius of the mean free path collision cylinder equals the molecular diameter.

---

[6]The container wall around the hole must also be very thin so that collisions inside the hole or with the inner walls of the hole can also be neglected.

[7]The $m^{-1/2}$ dependence of the effusion rate on mass is known as *Graham's law*. It forms the basis of gaseous isotope separation schemes that rely on differential effusion rates.

SOLUTION The mass loss rate is just m times the particle loss rate where m is the *mass of a gas molecule*. If, as is the case here, we know the particles are composed of one or more of the same atoms, dividing m by the atom mass determines the gas molecular formula. We multiply Eq. (24.17) by m and rearrange it to read

$$m^{1/2} = -\frac{d(mN)}{dt}\frac{4\,(2\pi k_B T)^{1/2}}{P\pi D^2}.$$

We substitute the experimental data in this expression (after converting to SI units: $D = 7.00 \times 10^{-4}$ m, $P = 0.177$ Pa, and $d(mN)/dt = -2.37 \times 10^{-10}$ kg s^{-1}) and find $m = 7.98 \times 10^{-25}$ kg. The mass of one Sb atom is 2.02×10^{-25} kg; so, the species in the gas phase has $(7.98 \times 10^{-25}$ kg$)/(2.02 \times 10^{-25}$ kg$) = 3.95 \cong 4$ Sb atoms: Sb_4. The mean free path at this pressure is $\sim 13\,000$ μm, much larger than D (Kn $\cong$ 18), so that the condition for effusive flow is satisfied.

▶ **RELATED PROBLEMS** 24.23, 24.24, 24.25

The gas that leaves a Knudsen cell through effusion is not at equilibrium. The faster molecules in the cell vapor strike the hole more often than the slower molecules. That means the speed distribution of the effusing gas is *not* the equilibrium speed distribution, Eq. (24.5), but v times larger. Once this is recognized, we can use effusion to *measure and verify* the Maxwell–Boltzmann speed distribution. Several clever experiments have been designed to do this. One that relies on the *free-fall of the effusing molecules due to gravity* is discussed in Problem 24.28, but most such experiments rely on a *time-of-flight* measurement. The Knudsen cell is placed a long distance from a particle detector, and a device that periodically interrupts the straight line molecular path from the cell to the detector is placed close to the cell a known distance L from the detector. The device is often a rotating disk (called a "chopper") with several slots cut in its periphery that acts as an on/off shutter, but one can also arrange to open and close the cell hole directly in what is called the "pulsed beam source." Figure 24.8 diagrams this experiment. When an open segment of the chopper wheel passes between the cell and detector, a brief pulse of gas passes through at what we can call time $t = 0$. This pulse contains molecules moving with a variety of speeds, and as it moves toward the detector, it spreads in space as the faster molecules in the pulse outrun the slower ones. The detector signal is recorded as a function of time past $t = 0$, and the speed distribution is calculated from the arrival times τ, since for any one speed $v = L/\tau$.

The detailed shape of the time-of-flight signal depends on the nature of the detector. For example, a mass spectrometer detector that uses electron impact to convert neutral molecules to easily detected positive ions is sensitive to the time a molecule spends in its electron beam. Slower molecules have a greater ionization

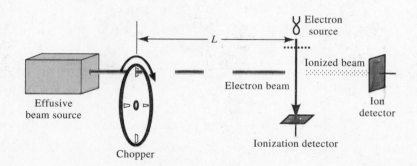

FIGURE 24.8 In a time-of-flight experiment, a device such as the chopper shown here periodically interrupts or otherwise pulses a molecular beam. A particle detector such as the electron impact ionizer shown here generates a signal proportional to the number of molecules that reach a known distance L from the chopper as a function of time.

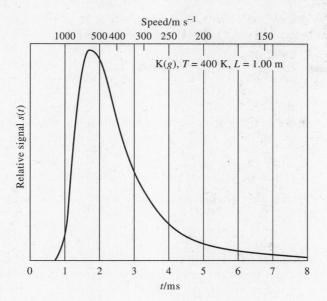

FIGURE 24.9 A time-of-flight measurement provides experimental verification of the Maxwell velocity distribution, as illustrated here for a K atom beam effusing from a 400 K source 1.00 m from a detector. Note that short flight times correspond to large velocities and vice versa.

probability than do faster molecules. This factor of v counteracts the same factor in the escape rate so that an ionization detector's response follows the equilibrium speed distribution directly. A detector such as this is said to be a particle *number-density detector*. Other types of detectors respond directly to the *particle flux* or the beam's *energy flux*. Figure 24.9 shows the time-of-flight signal for K atoms effusing from a source at 400 K over a flight path of 1.00 m as detected by a number-density detector.

A Closer Look at an Effusive Molecular Beam

The instantaneous particle density in a beam is related to its flux and speed through density = flux/speed. The flux of molecules in the beam with *speeds* between v and $v + dv$ is related to the expression used to derive Eq. (24.13), the wall collision flux. That derivation was in terms of the x component of the *velocity*, $v_x N f(v_x)/V$. Here, we picture the full vector velocity $\mathbf{v}$ in terms of its magnitude (the speed v) and the two spherical polar angles that orient $\mathbf{v}$ in space. First, $v_x = v \cos\theta$ where θ is the angle between $\mathbf{v}$ and the x direction (the direction of the flux). Second, the velocity vector is pointing in a *differential element of solid angle* $d\omega = \sin\theta\, d\theta\, d\phi$ where ϕ is the polar angle (0 to 2π) around the x direction:

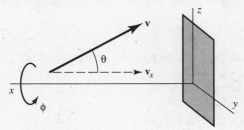

Consequently, the differential beam flux for velocity $\mathbf{v}$ is

$$dJ(\mathbf{v}) = \frac{N}{V}(v\cos\theta)f(v)\,dv\,\sin\theta\,d\theta\,d\phi \ .$$

We average this over all solid angles such that **v** is headed toward the wall ($0 \leq \phi \leq 2\pi$, $0 \leq \theta \leq \pi/2$) and arrive at the flux for speed v:

$$dJ(v) = \frac{N}{V} v f(v) \, dv \, \frac{\int_0^{\pi/2} \cos\theta \sin\theta \, d\theta \int_0^{2\pi} d\phi}{\int_0^{\pi} \sin\theta \, d\theta \int_0^{2\pi} d\phi} = \frac{N}{V} v f(v) \, dv \, \frac{\int_0^{\pi/2} \cos\theta \sin\theta \, d\theta \int_0^{2\pi} d\phi}{4\pi}$$

where the integrals in the denominator represent the solid angle of an entire sphere, 4π, and since $\int_0^{2\pi} d\phi = 2\pi$ and $\int_0^{\pi/2} \cos\theta \sin\theta \, d\theta = \frac{1}{2}$, we arrive at

$$dJ(v) = \frac{1}{4} \frac{N}{V} v f(v) \, dv \, .$$

The *density* of molecules with speed v in an effusive beam is therefore

$$dN(v) = \frac{dJ(v)}{v} = \frac{1}{4} \frac{N}{V} f(v) \, dv$$

where $f(v)$ is the Maxwell–Boltzmann speed distribution, Eq. (24.5).

The time-of-flight signal $s(t)$ for molecules with speed v is proportional to $dN(v) = (dN/dt) \, dt = s(t) \, dt$, and the arrival time is L/v. Since $dv = -L/t^2 \, dt$ (and the minus sign can be ignored—it simply reminds us that slower molecules arrive at greater times), we have the following expression for a density detector's response as a function of time:

$$s(t) \, dt = \frac{1}{4} \frac{N}{V} f(L/t) \, dv = \frac{1}{4} \frac{N}{V} \frac{L}{t^2} f(L/t) \, dt$$

$$= \frac{1}{4} \frac{N}{V} \frac{L}{t^2} \left[4\pi \left(\frac{m}{2\pi k_B T} \right)^{3/2} \left(\frac{L^2}{t^2} \right) e^{-mL^2/2t^2 k_B T} \right] dt$$

$$= \frac{\pi N L^3}{V t^4} \left(\frac{m}{2\pi k_B T} \right)^{3/2} e^{-mL^2/2t^2 k_B T} \, dt \, ,$$

which is the function shown in Figure 24.9. For a detector that responds directly to particle flux, the signal is proportional to t^{-5} rather than t^{-4}, and for a detector that responds to the particle *energy*, the signal is proportional to t^{-7}. A *bolometer*, which is a sensitive calorimeter often operated at liquid He temperature, responds to an incident energy flux through a change in electrical resistance caused by the incident beam's energy.

Analysis of experimental data takes into account the solid angle of the beam source viewed by the detector (which, since L is usually quite large compared to the entrance to the detector, can be a very small solid angle) as well as the finite open time of the chopper's slots and the finite response time of the detector.

Static, equilibrium gases also display their velocity distributions spectroscopically through the velocity's effect on a transition lineshape. This is the *Doppler broadening mechanism* discussed in Chapter 19. (See Eq. (19.51) and Problem 24.31.) Since the Maxwell–Boltzmann distribution is no longer in doubt, Doppler spectroscopy is more often used now to measure unknown nonequilibrium velocity distributions.

In later chapters, we will return to many of the ideas in this section. In particular, it is not difficult to go from the idea of a wall collision rate and a mean free path to a rate of *molecule–molecule* collisions. This will interest us when we consider gas-phase chemical reactions, since the collision of two molecules is the minimum requirement for a chemical reaction between them. In the next chapter, we will consider nonequilibrium situations for which velocity distribution functions and mean free paths play a central role. To close this chapter, we turn now to the macroscopic consequences of intermolecular interactions and explore the ways nonideality in gases reflect aspects of molecular collisions.

24.3 INTERMOLECULAR FORCES AND NONIDEAL GASES

Chapter 1 discussed the experimental evidence for the breakdown of the ideal gas equation of state at high densities as well as the modified equations of state that, with varying degrees of success, reproduce observed nonidealities. The first important nonideal equation of state is the virial expansion, Equations (1.17) and (1.18), especially (for reasons we will see here) in the form of a density or inverse molar volume expansion of the compressibility factor Z:

$$Z(T, \overline{V}) = 1 + \frac{B(T)}{\overline{V}} + \frac{C(T)}{\overline{V}^2} + \cdots = 1 + \frac{B(T)}{N_A}\rho + \frac{C(T)}{N_A^2}\rho^2 + \cdots \quad \text{(1.18)}$$

where $B(T)$ is the second virial coefficient, $C(T)$ is the third virial coefficient, $\overline{V} = V/n$ is the molar volume, and $\rho = N/V$ is the particle density. The second important equation of state is van der Waals', Eq. (1.20):

$$P = \frac{RT}{\overline{V} - b} - \frac{a}{\overline{V}^2}$$

with two parameters, a and b, that roughly measure the effects of intermolecular attractions and repulsions, respectively.

What we seek here are ways to calculate and understand the nonideal parameters in these equations of state ($B(T)$, etc.) in terms of molecular properties and realistic intermolecular forces. Our first goal is met through classical statistical mechanics. We will focus on the second virial coefficient, $B(T)$, and arrive at a simple expression for it in terms of the *intermolecular potential energy function*. Once we have this expression, we will spend some time considering the nature of these potential energy functions.

On the one hand, if we know the potential energy, we can calculate the gas compressibility in nonideal states, but on the other hand, perhaps we can also *invert* the nonideal equation of state into the potential energy function. This is an appealing prospect—measuring the detailed interactions between two molecules from the PVT behavior of a mole of them—but just as a measurement of the heat capacity as a function of temperature does not give molecular information as accurately and detailed as direct molecular spectroscopy, the intermolecular potentials deduced from bulk measurements will be fairly crude in comparison to more direct molecular measurements. These more direct methods—molecular scattering and molecular cluster spectroscopy—have largely replaced the inversion of bulk data in modern investigations.

We begin with the classical expression for the molecular partition function introduced in Chapter 23 in our discussion of the equipartition theorem. It was written for a molecule containing N atoms as

$$q_{cl} = \frac{1}{h^{3N}} \int \cdots \int e^{-\beta H} dp^{3N} dx^{3N} \ . \quad \text{(23.17)}$$

Here, $\beta = 1/k_B T$, and H is the classical molecular Hamiltonian, the sum of the kinetic and internal potential energies of the molecule. We can generalize this expression and write the *canonical* classical partition function for a system of N *nonbonded atoms* as

$$Q_{cl} = \frac{1}{N! h^{3N}} \int \cdots \int e^{-\beta H} \, dp^{3N} \, dx^{3N} \tag{24.18}$$

where the factor $N!$ is the usual classical, dilute system correction factor for N indistinguishable particles. If the atomic mass is m, we have

$$H = \frac{1}{2m} \sum_{i=1}^{N} (p_{x_i}^2 + p_{y_i}^2 + p_{z_i}^2) + U_N(x_1, y_1, \ldots, z_N) \tag{24.19}$$

where $U_N(x_1, y_1, \ldots, z_N)$ is the N-body potential energy of the system.[8] If the gas is molecular instead of atomic, H would also have various kinetic and potential energy terms representing internal molecular energies. These terms were covered in Chapter 23, and they will not concern us here since *internal molecular energy does not affect an equation of state*.[9]

As we showed in Chapter 23, integration over the momenta is easy and leads to

$$Q_{cl} = \frac{1}{N!} \left(\frac{m k_B T}{2 \pi \hbar^2} \right)^{3N/2} Z_N \tag{24.20}$$

where Z_N is called the *N body configuration integral*:

$$Z_N = \int \cdots \int e^{-U_N/k_B T} \, dx_1 dy_1 \ldots dz_N \,. \tag{24.21}$$

If there is no intermolecular potential energy ($U_N = 0$), this integral is simply

$$Z_N = \int \cdots \int dx_1 dy_1 \ldots dz_N = \left(\iiint dx_1 dy_1 dz_1 \right)^N = V^N$$

and Q_{cl} has the form $(f(T)V)^N$ where $f(T)$ is a function of T alone. This always leads to the ideal gas equation of state, no matter what $f(T)$ is. In Eq. (24.20), $f(T)$ contains only a translational kinetic-energy contribution. If the gas is molecular, then $f(T)$ contains internal-energy contributions as well, but these are still only functions of T.

Clearly, then, Z_N for $U_N \neq 0$ contains all the nonideal behavior of a gas. The connection between Z_N and the various virial coefficients is complicated to derive, and we will just state the main results. The theory focuses on the effects of a single atom, of two atoms, of three, $\ldots$, in that a series of $N = 1, 2, 3, \ldots$ configuration integrals emerge. For $N = 1$, there is no U_1, and Z_1 is just the volume of the system:

$$Z_1 = \iiint dx_1 dy_1 dz_1 = V \,. \tag{24.22a}$$

For $N = 2$, U_2 is the *two-body interaction* that in general depends on the distance between the two molecules and their orientation in space relative to each other. If only the distance between them matters, Z_2 is

$$Z_2 = \int \cdots \int e^{-U_2/k_B T} \, dx_1 dy_1 dz_1 dx_2 dy_2 dz_2 \,. \tag{24.22b}$$

The expressions for higher N are similar.

---

[8] To avoid confusion with the symbol for volume, we will use U to represent the potential energy rather than V.

[9] Exceptions to this statement include internal energies great enough to dissociate the molecule or otherwise radically alter its chemical structure. We will assume, as is usually the case, that we can avoid these severe conditions.

One finds that the Nth virial coefficient involves only Z_1 through Z_N. This is a particularly interesting result, since it says the *second* virial coefficient is governed by *the interaction of only two molecules at a time*. The third virial coefficient involves the mutual interaction of two *and* three molecules, and so on. The expression for the second virial coefficient is

$$B(T) = -\frac{N_A}{2V}(Z_2 - Z_1^2) \qquad (24.23a)$$

or, using Equations (24.22a) and (24.22b),

$$B(T) = -\frac{N_A}{2V}\int \cdots \int (e^{-U_2/k_B T} - 1)\, dx_1 dy_1 dz_1 dx_2 dy_2 dz_2 \ . \qquad (24.23b)$$

The vectors $\mathbf{r}_1 = (x_1, y_1, z_1)$ and $\mathbf{r}_2 = (x_2, y_2, z_2)$ locate the centers of mass of molecules 1 and 2 so that $\mathbf{R} = \mathbf{r}_2 - \mathbf{r}_1$ is the vector that locates molecule 2 relative to molecule 1. It is reasonable to assume that U_2 depends only on *the magnitude of this separation, R,* and not on the actual location of the molecules in the volume. The integrals in Eq. (24.23b) can then be transformed into integrals over $\mathbf{r}_1$ (which U_2 does not depend on) and integrals over $\mathbf{R}$. The $\mathbf{r}_1$ integration gives just a factor of V, and the $\mathbf{R}$ integral can be written in spherical polar coordinates:

$$B(T) = -\frac{N_A}{2V} \overbrace{\iiint dx_1 dy_1 dz_1}^{V} \int_0^\infty (e^{-U_2(R)/k_B T} - 1) R^2\, dR \overbrace{\int_0^\pi \sin\theta\, d\theta \int_0^{2\pi} d\phi}^{4\pi} \qquad (24.23c)$$

$$= 2\pi N_A \int_0^\infty (1 - e^{-U_2(R)/k_B T}) R^2\, dR$$

where the final expression is the one we will use.

Several general comments are in order before we turn to the nature of $U_2(R)$. First, the limits on the remaining integral are mathematically convenient rather than physically rigorous. Two molecules are never at $R = 0$, but as $R \to 0$, U_2 is so large compared to $k_B T$ (so that $e^{-U_2/k_B T}$ approaches zero) that we can safely extend the limit down to zero. Similarly, $R = \infty$ is physically artificial, but since U_2 rapidly approaches zero for R greater than a few tens of Å (so that $e^{-U_2/k_B T} - 1$ approaches zero), we can extend the upper limit to ∞. Finally, note that $R = \infty$ locates the energy zero. For *intra*molecular potential energies, we often locate the energy zero at the potential *minimum*. (See, for example, the Morse potential in Figure 19.15.) For *inter*molecular potentials, *infinite separation of the interacting molecules* is the traditional configuration for zero energy.

The Intermolecular Potential Energy

The types of forces between molecules have been the source of contemplation on and off throughout history since the early Greeks suggested that atoms had interlocking hooks in solids. Not until the late 1700s, however, was anything much more than idle speculation put forth. The discussion that began to bloom in the late 1700s and early 1800s centered around explanations of capillary phenomena. Why did water creep up a capillary tube? Was gravity at work between atoms? Was the force purely repulsive? If so, how did atoms condense into liquids and solids? Since capillary rise did not depend on the thickness of the capillary tube walls, isn't it reasonable to assume that these forces, whatever they are, act over only very small distances? Scientists such as Laplace and Gauss were involved in these early musings and theories, but not until Maxwell approached gas behavior from the point of view of intermolecular forces in 1868 did anything close to our current view begin to take shape.

Maxwell deduced a curious type of force law, one which we know to be incorrect now but one that was in agreement with gas behavior known at the time. Maxwell's force was purely repulsive. Boltzmann followed with arguments showing that Maxwell's force was only one of many kinds of forces that would predict known behavior, and in particular, attractive forces alone could do the job. Boltzmann, who called molecules "very complicated individuals," preferred attractive forces (to account for gas condensation), and in 1871, Tyndall wrote, "Atoms are endowed with powers of mutual attraction." Sutherland used the Joule–Thomson experiment to make another convincing argument in favor of attractive forces, and van der Waals considered attraction and repulsion in his equation of state. He also considered capillarity in a later publication.

Progress towards the detailed origin of intermolecular forces began early in the twentieth century. For some time, Debye made considerable progress assuming that *all* molecules had a permanent dipole moment. When it became clear that this was wrong, Debye simply moved up the multipole expansion and blamed everything on quadrupole moments (discussed in Chapter 15). By the mid 1920s, quantum mechanics began to show the way, however, and our current view emerged.

We will first approach intermolecular forces from two general categories—repulsion and attraction—and then further categorize the origins of various types of each. We start with repulsion.

It is the repulsive part of the intermolecular potential that gives molecules their *shape* in the usual chemist's meaning of the word. Consider your mental picture of, for example, PCl_5. It may be as simple as an abstract trigonal bipyramidal shape, showing connectivity and geometry, aspects of the *intra*molecular chemical forces, or as sophisticated as a scaled *space-filling* model:

The space-filling model is our simplest attempt to visualize an intermolecular potential energy function, since the only way we can measure the "size" of an atom or molecule is through its interaction with another atom or molecule. This is the idea we used to pick an effective hard-sphere diameter for Ar in our discussion of the mean free path (Figure 24.7). Similarly, the "shape" of a molecule can be expressed through the *anisotropy* of the intermolecular interaction of the molecule with a spherical probe such as He or Ar.

The hard-sphere potential function is zero for all $R > d$, the hard-sphere diameter (or for all $R > (R_1 + R_2)$ for two hard spheres of radii R_1 and R_2), and infinite for all $R \leq d$. This represents impenetrable spheres, but real molecules are compressible. The Born model for ionic solids discussed in Chapter 16 relied on compressibility to determine the parameters of one of the most common models for the real repulsive interaction, the *exponential repulsion*:

$$U(R) = A\, e^{-R/\rho} . \qquad (16.1)$$

This function reasonably approximates quantum chemical calculations of repulsions (see Problem 14.8), and it also appears in the Morse potential function used for chemically bonding interactions (see Eq. (19.47)). The two parameters, A and ρ, are chosen to reproduce whatever experimental data are available.

Another common representation of repulsive interactions is the *inverse-power repulsion*:

$$U(R) = \frac{B}{R^m} \qquad (24.24)$$

where m is usually chosen in the range 8–20 with 12 a particularly popular choice (for reasons we will see below). (See also Problems 16.16 and 16.17.) Neither repulsive model potential has a simple physical basis that can guide the choice of parameters. The exponential function is generally more realistic, but most macroscopic phenomena are not particularly sensitive to the detailed shape of the repulsion beyond its general range at low energy (which is why the hard-sphere model works at all well and why the exponential repulsion, which is finite at $R = 0$ rather than infinite, also works well).

In contrast to repulsions, intermolecular attractive interactions have a firm basis in simple theories. For example, if the two molecules have a net charge, the dominant interaction between them is the Coulomb potential (which, if the ions have the same charge, is also the only repulsive interaction with a simple theoretical basis). Since this potential varies like R^{-1}, it is said to be *very long range*. It continues to exert its influence at distances far beyond those of any other force, as the infinite number of bound states of atomic H or the significant nonideality of even very dilute electrolyte solutions indicate. Consequently, ionized gases—*plasmas*—do not follow a virial expansion or any other simple equation of state.

For molecules that do not both have a net charge, the intermolecular interaction dies much more rapidly as R increases than it does for the pure Coulombic interaction. The *types* of interactions among these species fall into one of three general categories: *electrostatic potential energies,* such as that between the two dipole moments of polar molecules, *induced potential energies,* such as between an ion and the polarizable charge distribution of a molecule, and what are called *dispersion energies* that act between the polarizability of one molecule and that of another.

Electrostatic interactions depend on the magnitude of the various permanent electric moments of each molecule (net charge, dipole moment, etc.) as well as their relative orientation. For example, it is clear that the interaction between two dipole moments is different if the dipole moment vectors are aligned: $\rightarrow\rightarrow$ or anti-aligned: $\rightarrow\leftarrow$ or parallel: $\uparrow\uparrow$ or anti-parallel: $\uparrow\downarrow$. Likewise, an ion–dipole interaction depends on orientation as well: $+\rightarrow$ and $+\leftarrow$, for example, are different, but $+\downarrow$ and $+\uparrow$ are the same. Moreover, the $+\leftarrow$ interaction is *repulsive* (recall that a dipole vector points from the negative to the positive end of an elementary dipole—see Figure 15.1) but the $+\rightarrow$ interaction is *attractive*.

In addition to orientation differences, each kind of electrostatic interaction has a characteristic *distance dependence*. If we rank electrostatic moments such that the net charge is the zeroth moment, the dipole is the first moment, etc., the interaction potential between the nth moment of one molecule and the mth moment of another varies like $1/R^{(n+m+1)}$. The strength of an interaction *decreases* ever more rapidly with R as $n + m + 1$ increases.

For many situations, the detailed orientational dependence is important, but for an isotropic gas, we can average over orientations. This average, however, must be weighted according to the energy differences of each orientation using a Boltzmann factor, $e^{-U_2/k_\text{B}T}$. Even without this averaging, the expressions for high-order electrostatic interactions are quite complicated in general. We will consider in detail only the simplest charge and dipole interactions.

FIGURE 24.10 The relative position of two dipole moment vectors is described by a separation R and three angles, as defined here. Each vector lies in its own shaded plane, and the dihedral angle, the angle between the planes, is ϕ.

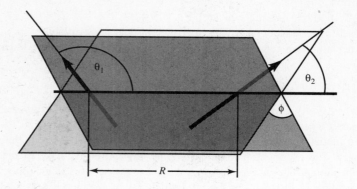

TABLE 24.2 Electrostatic Intermolecular Potential Energy Terms

Interaction	Orientation dependent	Thermally averaged
Charge–charge	$\dfrac{q_1 q_2}{(4\pi\epsilon_0)R}$	$\dfrac{q_1 q_2}{(4\pi\epsilon_0)R}$
Charge–dipole	$-\dfrac{q_1 p_2}{(4\pi\epsilon_0)R^2} \cos\theta_2$	$-\dfrac{q_1^2 p_2^2}{(4\pi\epsilon_0)3k_B T R^4}$
Dipole–dipole	$-\dfrac{p_1 p_2}{(4\pi\epsilon_0)R^3}(2\cos\theta_1 \cos\theta_2 - \sin\theta_1 \sin\theta_2 \cos\phi)$	$-\dfrac{2 p_1^2 p_2^2}{(4\pi\epsilon_0)3k_B T R^6}$

Note: q = net charge, p = permanent dipole moment. Coordinates are defined in Fig. 24.10.

Figure 24.10 shows the coordinate system used to specify a particular orientation of two dipolar molecules in terms of their dipole moment vectors, and Table 24.2 gives the various electrostatic potential energy[10] terms first in detail and then thermally averaged over Boltzmann-weighted orientations. Note how thermal averaging *squares* the R dependence of each term that depends on orientation, effectively *weakening* the interaction. The averaged potentials (except the charge–charge potential, which has no orientation dependence) are all *attractive* and *weaken as T increases*. As $T \to \infty$, they all vanish, since the molecular rotational energy is too great (the classical rotation speed is too fast) to notice the orientational dependence itself. Without the Boltzmann factor favoring attractive orientations, the averages are all zero.

EXAMPLE 24.7

Contrast the thermally averaged attractive electrostatic energies between two molecules of opposite net charge, $q = +e$ and $-e$, and equal dipole moment, $p = 4 \times 10^{-30}$ C m, to the average ideal gas kinetic energy, $3k_B T/2$ for $T = 300$ K and $R = 25$ Å.

SOLUTION The average ideal gas kinetic energy, $3k_B T/2$, is 6.2×10^{-21} J. The charge–charge attraction is

$$U_{q,q} = -\frac{e^2}{(4\pi\epsilon_0)R} = -9.2 \times 10^{-20} \text{ J}$$

---

[10]The charge–multipole moment potential energies are q, the charge on one molecule, times $\phi(R)$, the electric potential of the other, given in Eq. (15.2).

about 15 times $3k_BT/2$, which shows why a plasma cannot be considered an ideal gas. Charge–charge interactions are overwhelming. The thermally averaged charge–dipole interaction for this typical dipole moment is

$$U_{q,p} = -\frac{e^2 p^2}{(4\pi\epsilon_0)\, 3k_B T R^4} = -7.6 \times 10^{-33}\ \text{J}$$

far less than the kinetic energy, and the dipole–dipole interaction energy is smaller still:

$$U_{q,p} = -\frac{2p^4}{(4\pi\epsilon_0)\, 3k_B T R^6} = -1.5 \times 10^{-36}\ \text{J}\ .$$

25 Å is a characteristic distance in a gas at ordinary pressures, and while the plasma is dominated by charge–charge interactions, a neutral dipolar gas with a typical dipole moment is largely unaware of electrostatic interactions. The following example contrasts this situation to the close quarters of a molecular cluster.

⟹ RELATED PROBLEMS 24.33, 24.37

EXAMPLE 24.8

Estimate the magnitude of the attractive electrostatic energy between two neutral molecules of equal dipole moment, 4×10^{-30} C m, when the molecules are 3 Å apart in a specific orientation, as they might be in a bound molecular dimer.

SOLUTION In contrast to Example 24.7, we imagine a specific, close orientation of the two dipoles. The angular dependence in Table 24.2 is fairly complicated, but a little thought shows that an attractive interaction is maximized if the dipoles lie in the same plane ($\phi = 0$) and are favorably aligned: either →→ or ↑↓. The first configuration has $\theta_1 = \theta_2 = 0$, and the angular factors in parentheses in Table 24.2 equal 2. For the second configuration, $\theta_1 = 90°$ and $\theta_2 = -90°$, and the angular factors equal 1. Thus, the →→ orientation is lower in energy by a factor of two. At $R = 3$ Å, this orientation has an energy

$$U_{p,p} = -\frac{2p^2}{(4\pi\epsilon_0) R^3} = -1.1 \times 10^{-20}\ \text{J}\ ,$$

which is greater in magnitude than $3k_BT/2$ at 300 K. This interaction could stabilize a bound dimer against dissociation from the collisional kinetic energy of the average intermolecular collision.

⟹ RELATED PROBLEMS 24.34, 24.35, 24.36, 24.38, 24.39, 24.55

The second general class of attractive interactions are *induction forces*. Place any molecule in an external electric field, and the molecule's charge distribution responds in a way that induces a dipole moment whether or not the molecule has a permanent moment, as discussed in detail in Chapter 15. The induced moment per unit electric field is called the *polarizability* α (Eq. (15.4)). The *polarizability volume* α' is also often tabulated (see Tables 15.2 and 15.3) where $\alpha' = \alpha/4\pi\epsilon_0$. Induction forces are interactions between *permanent* moments on one molecule and the *induced moments* (and thus the polarizability) of another. These interactions in general depend on orientation, but we will consider only a spherical "molecule" with a single average polarizability interacting with either a point charge or a point dipole. Table 24.3 lists the form of these interactions, which are *all* attractive. Note also that averaging does *not* change the distance dependence of the interaction, and temperature does not enter the story here. An induced dipole moment follows the moment of the inducer no matter how fast the inducer is moving or rotating. The induced moment is *highly correlated* with the motion of the inducer.

TABLE 24.3 Inductive Intermolecular Potential Energy Terms

Interaction	Orientation dependent	Thermally averaged
Charge–polarizability	$-\dfrac{q_1^2 \alpha_2'}{(4\pi\epsilon_0)\, 2R^4}$	$-\dfrac{q_1^2 \alpha_2'}{(4\pi\epsilon_0)\, 2R^4}$
Dipole–polarizability	$-\dfrac{p_1^2 \alpha_2'}{(4\pi\epsilon_0)\, 2R^6}(3\cos^2\theta_1 + 1)$	$-\dfrac{p_1^2 \alpha_2'}{(4\pi\epsilon_0)\, R^6}$

Note: q = net charge, p = permanent dipole moment, α' = polarizability volume. Coordinates are defined in Fig. 24.10.

EXAMPLE 24.9

Repeat Example 24.7 for induction interactions assuming a typical α' for each molecule, 2×10^{-30} m^3. Consider only the charge–polarizability and dipole–polarizability interactions listed in Table 24.3.

SOLUTION The charge–polarizability interaction is the same whether or not the interaction is thermally averaged, and it is independent of the sign of the ion's charge:

$$U_{q,\alpha} = -\frac{q_1^2 \alpha_2'}{(4\pi\epsilon_0)\, 2R^4} = -\frac{(1.6 \times 10^{-19}\, \text{C})^2\, (2 \times 10^{-30}\, \text{m}^3)}{(1.113 \times 10^{-10}\, \text{C}^2\, \text{m}^{-1}\, \text{J}^{-1})\, 2\, (2.5 \times 10^{-9}\, \text{m})^4}$$
$$= -5.9 \times 10^{-24}\, \text{J}\,,$$

which is less than $3k_B T/2$, but significantly larger than the thermally averaged, charge–*permanent* dipole interaction in Example 24.7. The averaged dipole–polarizability interaction is

$$U_{p,\alpha} = -\frac{p_1^2 \alpha_2'}{(4\pi\epsilon_0)\, R^6} = -\frac{(4.0 \times 10^{-30}\, \text{C m})^2\, (2 \times 10^{-30}\, \text{m}^3)}{(1.113 \times 10^{-10}\, \text{C}^2\, \text{m}^{-1}\, \text{J}^{-1})\, (2.5 \times 10^{-9}\, \text{m})^6}$$
$$= -1.2 \times 10^{-27}\, \text{J}$$

about 5000 times weaker, but significantly greater than the averaged permanent-dipole–permanent-dipole interaction energy at this temperature and gas density (represented by the average distance $R = 25$ Å).

➥ *RELATED PROBLEMS* 24.34, 24.35, 24.38, 24.39

EXAMPLE 24.10

The charge–polarizability interaction can provide reasonable binding energies in otherwise nonchemically bonded molecules. Estimate the binding energy of the CaAr$^+$ molecular ion given its bond length, 2.8 Å.

SOLUTION The binding energy (which is the dissociation energy $\mathcal{D}_e$) is, to a first crude approximation, the magnitude of the charge–polarizability interaction between a Ca ion of charge $+e$ and an Ar atom of polarizability volume (see Table 15.2) 1.64×10^{-30} m^3:

$$\mathcal{D}_e = -U_{q,\alpha} = \frac{e^2 \alpha'}{(4\pi\epsilon_0)\, 2R^4} = \frac{(1.6 \times 10^{-19}\, \text{C})^2\, (1.64 \times 10^{-30}\, \text{m}^3)}{(1.113 \times 10^{-10}\, \text{C}^2\, \text{m}^{-1}\, \text{J}^{-1})\, 2\, (2.8 \times 10^{-10}\, \text{m})^4}$$
$$= 3.1 \times 10^{-20}\, \text{J}\,,$$

which is appreciably larger than the thermal energy. The experimental value is about two-thirds this value, since the repulsive interaction has been neglected. This is an appreciable interaction nevertheless. It is about an order of magnitude larger than the Ar$_2$ binding energy and about an order of magnitude smaller than a typical (single) chemical bond energy.

⟹ RELATED PROBLEMS 24.34, 24.35

The third type of attractive interaction, called variously the *dispersion* energy, the *London* energy (after its contemporary proponent), or simply the *van der Waals* energy (after its "discoverer," in spirit if not in detail) exists between *any* two molecules. It is a mutual interaction between the polarizable charge distributions of the two molecules, and it is always attractive for ground state molecules. Chapter 15 discussed polarizability from the point of view of perturbation theory and arrived at the expression

$$\alpha \cong \frac{2}{\Delta E} (\langle p^2 \rangle - \langle p \rangle^2)$$

where the terms in parentheses represent the *fluctuation* of the dipole moment due to the dynamic flow of charge around a molecule. At any instant, the charge distribution can lead to an instantaneous dipole moment in any atom or molecule. For example, the H atom at any one instant has the electron some distance and direction away from the proton, forming an instantaneous moment. Over time, the electron motion averages this moment to zero, and atoms have no permanent dipole moment, of course, but the *instantaneous moment of one atom (or molecule) generates an electric field that can induce a dipole moment in a neighboring polarizable atom (or molecule)*.

Consequently, another name for the strongest type of dispersion energy is the *induced-dipole–induced-dipole* energy. The instantaneous moment on one molecule induces a moment on another, which induces a change in the moment of the first, which . . . , and so on, back and forth. Higher charge distribution moments (quadrupole, etc.) enter the story as well, and each type has a unique R dependence. The leading (i.e., most attractive at any one distance) term is the induced-dipole–induced-dipole term, and it varies like $1/R^6$, just like the thermally averaged permanent dipole–dipole interaction. The other terms (see Problem 24.40) vary like $1/R^8$, $1/R^{10}$, etc. We will consider in detail only the leading term, which we write

$$U_{\text{disp}} = -\frac{C_6}{R^6} \qquad (24.25)$$

where C_6 depends on polarizabilities at least.

A full expression for C_6 is quite complicated, even in the approximate form given by second-order perturbation theory. An interesting highly simplified and stylized model, first proposed by Drude and known by his name, considers an atom (or molecule—the approximations are so severe that the distinction is irrelevant) to be only two particles of equal magnitude but opposite charge bound by a *harmonic oscillator* potential of known force constant rather than by Coulomb's law. Drude showed that the interaction energy between atom 1 with harmonic frequency ω_1 and atom 2 with ω_2, in the limit where the interaction energy is smaller than the natural energy scale of the oscillators, $\hbar\omega_1$ or $\hbar\omega_2$, is

$$U_{\text{disp}} = -\frac{3}{2}\left(\frac{1}{\hbar\omega_1} + \frac{1}{\hbar\omega_2}\right)^{-1} \frac{\alpha'_1 \alpha'_2}{R^6} = -\frac{3}{2}\left(\frac{\hbar\omega_1\omega_2}{\omega_1 + \omega_2}\right) \frac{\alpha'_1 \alpha'_2}{R^6}$$

where α' is the polarizability volume.

The Drude model of an atom seems perhaps hopelessly screwy at first glance, but it is capable of giving reasonable magnitudes and parameter dependencies, and it has a distinguished history in the theory of metals. When London approached dispersion forces with quantum mechanics, he found the following approximate expression

$$U_{\text{disp}} = -\frac{3}{2}\left(\frac{1}{E_1} + \frac{1}{E_2}\right)^{-1}\frac{\alpha'_1\alpha'_2}{R^6} = -\frac{3}{2}\left(\frac{E_1 E_2}{E_1 + E_2}\right)\frac{\alpha'_1\alpha'_2}{R^6}, \qquad (24.26)$$

which has the same form as the Drude expression.[11] Here, E_1 and E_2 are "representative excitation energies" for the molecule or atom, and either first ionization energies or energies of the first excited state that has a strongly allowed transition from the ground state is often used for them. Since ionization energies for most common molecules (see Table 7.3) are in the range 9–13 eV, and since polarizability volumes for small molecules (Tables 15.2 and 15.3) are on the order of 3–8 Å³, this term is on the order of $-(7 \text{ to } 10 \text{ eV}) (3 \text{ to } 8 \text{ Å}^3)^2/R^6 \sim -(50 \text{ to } 500 \text{ eV})/(R/\text{Å})^6$. For an internuclear distance of 4 Å or longer, characteristic of nonbonding interactions, the dispersion energy is on the order of hundredths to tenths of an electron volt (or ~ 100–1000 cm^{-1}), which is one to two orders of magnitude weaker than a chemical bond, but comparable to the cohesive energy of a molecular solid.

EXAMPLE 24.11

The Ne$_2$ intermolecular potential has a well depth of 29.4 cm^{-1}, and that for He$_2$ is 7.5 cm^{-1}. Both potentials have minima quite close to 3 Å. What would you predict the HeNe well depth to be?

SOLUTION The attractive force that holds these molecules together is the dispersion force. We should first check to see how well the dispersion energy alone accounts for the homonuclear dimers' binding energies at 3 Å. Using ionization energies from Table 7.3 and polarizability volumes from Table 15.2 (and with the conversion factor 1 eV = 8 065.54 cm^{-1}), we find, using Eq. (24.26),

$$\dot{U}_{\text{disp}} = -\frac{3}{4}I_1\frac{\alpha'^2}{R^6} = -\frac{3}{4}(21.564 \text{ eV})(8\,065.54 \text{ cm}^{-1}\text{ eV}^{-1})\frac{(0.3956 \text{ Å}^3)^2}{(3 \text{ Å})^6}$$
$$= -28.0 \text{ cm}^{-1}$$

for Ne$_2$, and for He$_2$,

$$U_{\text{disp}} = -\frac{3}{4}I_1\frac{\alpha'^2}{R^6} = -\frac{3}{4}(24.586 \text{ eV})(8\,065.54 \text{ cm}^{-1}\text{ eV}^{-1})\frac{(0.205 \text{ Å}^3)^2}{(3 \text{ Å})^6}$$
$$= -8.6 \text{ cm}^{-1}$$

both of which are quite close to the magnitudes of the observed binding energies. Thus, assuming a 3 Å distance for HeNe, we predict the interaction energy

$$U_{\text{disp}} = -\frac{3}{2}\left(\frac{I_1 I_2}{I_1 + I_2}\right)\frac{\alpha'_1\alpha'_2}{R^6}$$
$$= -\frac{3}{2}\left(\frac{21.564 \cdot 24.586}{21.564 + 24.586} \text{ eV}\right)(8\,065.54 \text{ cm}^{-1}\text{ eV}^{-1})\frac{(0.3956 \text{ Å}^3)(0.205 \text{ Å}^3)}{(3 \text{ Å})^6}$$
$$= -15.5 \text{ cm}^{-1},$$

which is in excellent agreement with the experimental bond length (3.00 Å) and well depth (15.3 cm^{-1}). As we are about to discover, however, this agreement is in part due to the small

---

[11] The term "dispersion" comes from the phenomenon that first attracted London's attention, the way the index of refraction varies with light frequency, a function called the "dispersion curve" for a substance.

but comparable polarizabilities of He and Ne that ensure the dominance of the $1/R^6$ leading term in the dispersion energy expansion.

RELATED PROBLEMS 24.41, 24.43, 24.53

A good intermolecular potential energy function, like a good novel, has a beginning, a middle, and an end, connected seamlessly one to the next. The beginning, at large R, is one or more of the electrostatic, induction, and dispersion terms, chosen according to the nature of the two interacting molecules. The end, at small R, is a repulsive term, chosen from experience or mathematical convenience. The middle represents the R region characteristic of the *potential energy minimum* (or, on occasion, *minima* if the function has an orientational dependence). The two dominant characteristics of the middle are the *location* and *magnitude* of its lowest energy point. For a simple potential that does not depend on relative orientations, the location is R_e, the equilibrium intermolecular bond length, and the magnitude of the potential's depth at R_e is $\mathcal{D}_e$, the classical intermolecular bond energy. We will now consider several simple model intermolecular potential-energy functions with these characteristics.

The hard-sphere potential lacks both R_e and $\mathcal{D}_e$; it has the hard-sphere diameter as its only parameter. The *square-well potential* is a crude model that adds an attractive energy region to the hard sphere. It has a hard-sphere repulsion for all $R \leq d_1$, but for $d_1 < R \leq d_2$, it has a constant potential energy at $-\mathcal{D}_e$, returning to zero potential energy for all $R > d_2$:

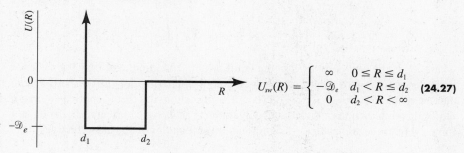

$$U_{sw}(R) = \begin{cases} \infty & 0 \leq R \leq d_1 \\ -\mathcal{D}_e & d_1 < R \leq d_2 \\ 0 & d_2 < R < \infty \end{cases} \quad (24.27)$$

This function models both "size" and "binding" as simply as possible.

We can easily smooth the sharp edges of the square well, and the way we do depends on the properties of the interacting molecules. For example, the attraction between a closed-shell positive ion and a rare gas (as in Na$^+$ + Ar) has no electrostatic term, a charge–polarizability induction term, and an induced-dipole–induced-dipole dispersion term. The induction term dominates the dispersion term, since the induction term is a longer range interaction (varies as $1/R^4$ rather than $1/R^6$), and Na$^+$ has a small polarizability. (See Problem 24.34.) Consequently, a good model for a nonbonding ion–molecule interaction is

$$U(R) = A\,e^{-R/\rho} - \frac{q_1^2\,\alpha_2'}{(4\pi\epsilon_0)\,2R^4}. \quad (24.28)$$

This type of model potential is called in general an *exponential-n* (exp-*n*) potential where here $n = 4$; it is an exponential repulsion term coupled to an attractive term that varies like $1/R^4$.

The most common interactions are between neutral molecules. If the molecules are both nonpolar, then the $1/R^6$ dispersion term dominates the attraction. If one is polar, the dipole–polarizability induction term also contributes to the $1/R^6$ attraction. If both are polar and we care only about the averaged electrostatic dipole–dipole attraction, we have another term that varies like $1/R^6$. Thus, for *many* common situations, $1/R^6$ is the most important attractive term, and an *exp-6* model potential is appropriate:

$$U(R) = A\,e^{-R/\rho} - \frac{C}{R^6}. \tag{24.29}$$

Here, the C parameter is usually chosen as a third freely variable parameter along with A and ρ rather than a fixed constant based on a particular model's prediction, such as Eq. (24.25). If it is fixed in advance at the C_6 value from Eq. (24.26), for example, the A and ρ parameters are too highly constrained to give a good overall model potential.

An alternate model for nonbonding interactions combines the $1/R^6$ attraction with an inverse-power repulsion:

$$U(R) = \frac{A}{R^m} - \frac{C}{R^6}. \tag{24.30}$$

By far the most common choice for m is 12, and this choice is known as the (6-12) or *Lennard-Jones* potential.[12] It can be written in several equivalent ways (see also Problem 12.16):

$$U(R) = \frac{A}{R^{12}} - \frac{C}{R^6} = \mathcal{D}_e\left[1 - \left(\frac{R_e}{R}\right)^6\right]^2 - \mathcal{D}_e$$
$$= \mathcal{D}_e\left[\left(\frac{R_e}{R}\right)^{12} - 2\left(\frac{R_e}{R}\right)^6\right] = 4\mathcal{D}_e\left[\left(\frac{\sigma}{R}\right)^{12} - \left(\frac{\sigma}{R}\right)^6\right] \tag{24.31}$$

where $\mathcal{D}_e$ is the well depth, R_e is the equilibrium separation, and σ is the distance such that $U(\sigma) = 0$. This distance is a rough approximation to the hard-sphere diameter, and $R_e = 2^{1/6}\,\sigma \cong 1.122\sigma$. This function has been applied to many molecular interactions, including polyatomics that have orientation-dependent forces. For these molecules, it represents an orientation-averaged effective interaction.

Unlike the exp-6 potential, the Lennard-Jones potential has only two parameters. These can always be related to R_e and $\mathcal{D}_e$, and Table 24.4 lists R_e and $\mathcal{D}_e$ for all the rare gas dimers, based on a variety of bulk, spectroscopic, and quantum chemical studies using elaborate multiparameter model potentials. Figure 24.11 compares the Lennard-Jones potential for Ar_2 (using parameters from Table 24.4) with a highly accurate eight parameter model potential. The two are in good agreement over the R range shown, even though the repulsive term in the accurate potential is a modified exponential, $AR^2\,e^{-R/\rho}$, rather than A/R^{12}. The accurate potential also includes $1/R^8$ and $1/R^{10}$ attractive dispersion energy terms with accurate C_6, C_8, and C_{10} parameters. As the figure shows, the C_6/R^6 dispersion approximation of Eq. (24.26) *alone* does not fall fast enough at distances just past R_e, but the (6-12) potential can mimic the full dispersion attraction fairly well.

There are several lessons hiding in the data of Table 24.4. Consider first the homonuclear rare gas interactions. The Lennard-Jones potentials for these species are plotted in Figure 24.12. The systematic increase in bond length (atomic radius)

---

[12] Kathleen Lennard married the British scientist John E. Jones in 1925, and they adopted a combined last name, Lennard-Jones, a practice that was much less common then than now.

24.3 INTERMOLECULAR FORCES AND NONIDEAL GASES

TABLE 24.4 Rare Gas Dimer Bond Lengths and Binding Energies

Equilibrium bond lengths R_e/Å					
	He	Ne	Ar	Kr	Xe
He	2.97				
Ne	3.00	3.08			
Ar	3.47	3.43	3.76		
Kr	3.67	3.58	3.88	4.01	
Xe	3.95	3.75	4.06	4.18	4.36

Binding energies $\mathcal{D}_e$/cm^{-1}					
	He	Ne	Ar	Kr	Xe
He	7.5				
Ne	15.3	29.4			
Ar	20.4	48.7	99.5		
Kr	21.0	50.7	116.4	138.9	
Xe	19.5	52.1	130.6	160.6	196.2

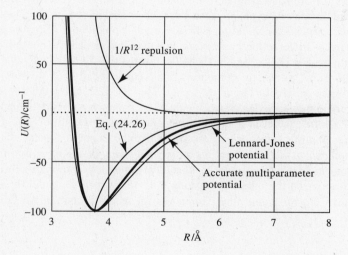

FIGURE 24.11 A highly accurate multiparameter Ar$_2$ interatomic potential-energy function (the heavy line) is fairly well reproduced by a simple Lennard-Jones function. Equation (24.26), the R^{-6} dispersion term alone, does not follow either model potential well. The R^{-12} repulsion term of the Lennard-Jones model is also shown for comparison.

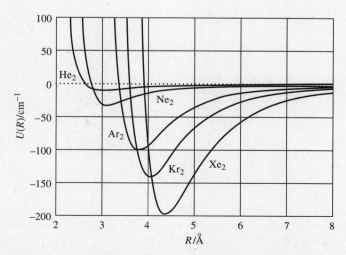

FIGURE 24.12 The homonuclear rare gas interatomic potential energy functions follow readily predictable trends in binding energy (well depth) and equilibrium separations (bond lengths).

from He$_2$ to Xe$_2$ is expected from elementary ideas of atomic size, and the corresponding increase in well depth follows from the increase in polarizability from He to Xe.

Something unexpected is found, however, if we look at a series of dimers with one common atom. Figure 24.13 plots the potentials of the Ne species HeNe–NeXe. The R_e increase is as expected, but the binding energies *stop increasing at NeAr*. They seem to be *governed by the less polarizable atom, Ne*. The NeAr, NeKr, and NeXe dimers have roughly the same binding energy even though the polarizability of Xe is about 2.5 times that of Ar.

Polarizability does indeed govern the strength of attraction at *long range* (i.e., $R > \sim 1.5\ R_e$), but it does *not* govern the net binding energy. As Ne approaches Xe, for example, the energy falls *until Ne's electrons encounter the outermost Xe electrons*. The nonbonding interaction between these two sets of electrons causes the *kinetic energy* of the Xe electrons to increase rapidly (in analogy to shortening the box of a particle-in-a-box). This energy increase keeps the dispersion attraction at bay and cuts short the binding energy. (This argument follows from, and uses the language of, the molecular Virial Theorem discussed at the end of Section 14.1.)

The principle of *corresponding states,* first discussed in Section 1.4, is exact for pure gases that interact only through a two-parameter pair potential. For gas mixtures, however, there are at least three (for a binary mixture) different interactions: those between molecules of type 1, those between molecules of type 2, and the 1–2 interactions. So-called *combining rules* are used to predict the 1–2 parameters from the 1–1 and 2–2 interaction parameters. For the binding energy, one often assumes the combining rule $\mathcal{D}_{12} = (\mathcal{D}_{11}\mathcal{D}_{22})^{1/2}$ holds. From Table 24.4, $\mathcal{D}_{Ne_2} = 29.4$ cm^{-1} and $\mathcal{D}_{Xe_2} = 196.2$ cm^{-1} so that the combining rule predicts $\mathcal{D}_{NeXe} = (29.4 \cdot 196.2)^{1/2}$ cm^{-1} = 69.9 cm^{-1}, considerably larger than the observed 52.1 cm^{-1} binding energy. A distance combining rule, $R_{12} = (R_{11} + R_{22})/2$, is somewhat more reliable; it predicts $R_e = 3.72$ Å for NeXe, in good agreement with the observed 3.75 Å.

Another important use for pair potentials like the Lennard-Jones is the construction of *molecule–molecule* potentials based on a *sum of atom–atom potentials*. Each atom on one molecule has its own set of (6–12) potentials representing its interaction with each atom on the other. For example, if we wanted to model the interaction of Ar with CO$_2$, we would find parameters for an Ar–C and an Ar–O interaction and write

$$U_{Ar-CO_2} = U_{Ar-O_1} + U_{Ar-C} + U_{Ar-O_2}$$

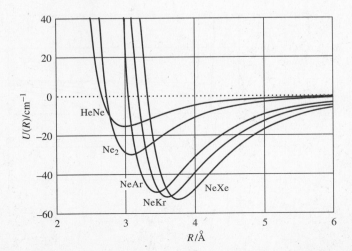

FIGURE 24.13 In contrast to the homonuclear rare gas interactions, the heteronuclear interactions *saturate* binding energy in spite of an increased polarizability in the series Ne—Ar, —Kr, —Xe.

where each atom–atom potential depends on the distance from Ar to a unique atom in CO_2. This approach to complicated molecule–molecule interactions can be successful, but it has built into it a structural bias: the equilibrium point of the potential has the two interacting molecules "touching" as much as possible so as to maximize the individual atom–atom binding energies. For Ar–CO_2, this leads to an equilibrium geometry in which the Ar atom lies along a line perpendicular to the CO_2 bond direction and passing through the C atom:

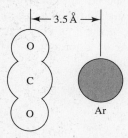

which is the experimentally observed structure. The Ar is 3.5 Å from the C atom.

Since CO_2 is cylindrically symmetric about its bond axis, the Ar–CO_2 potential energy must have this symmetry as well. We can fix CO_2 in place and slide the Ar atom around in a plane, calculating the potential energy as we do, and build a table of potential energy values at various x, y coordinates in the plane. Since only two coordinates describe the entire potential energy surface (if we keep the atoms in CO_2 still), we can use a *potential energy contour plot* to represent the full *potential energy surface*, as is shown in Figure 24.14. In this figure, U_{Ar-CO_2} was calculated from standard atom–atom Lennard-Jones parameters (see Problem 24.45). The contour values are labeled in cm^{-1} units, and shaded overlapping spheres with standard hard-sphere radii represent the CO_2 molecule. The closely spaced contours nearest

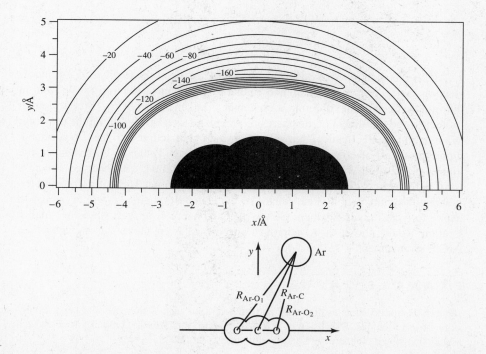

FIGURE 24.14 A common model for polyatomic interactions assumes atom–atom pair potentials, illustrated here for the interaction of Ar with CO_2. The energy is shown here as a contour map in the Ar–CO_2 coordinates. The minimum energy locates the equilibrium geometry of the ArCO_2 molecule.

these spheres locate the rapidly rising repulsive wall of the potential function. They represent the "size and shape" of CO_2 as measured by an Ar atom. The equilibrium geometry of the $ArCO_2$ molecule has the Ar sitting at the bottom of the well most closely outlined by the -160 cm^{-1} contour, 3.5 Å from the C with the Ar—C bond perpendicular to the O—C—O direction. The shape of the -160 cm^{-1} contour tells us that Ar motion *along* the O—C—O direction is easier (has a lower vibrational force constant) than motion *perpendicular* to that direction. The contour is narrow in the perpendicular direction and broad in the parallel direction.

While an atom–atom representation is easy to construct, it does not always give the right answer. A classic case where it fails is the Ar—ClF interaction. An atom–atom potential snuggles the Ar as close as possible to the Cl and F atoms and predicts the equilibrium structure

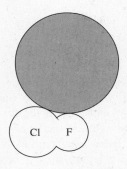

but the observed structure is linear:

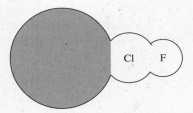

ArClF was mentioned in Section 14.4 as an example of the family of 22-valence-electron molecules such as XeF_2 and I_3^- that place the most electropositive atom in the middle of a linear triatomic arrangement, and the short Ar—Cl bond length (note how the traditional hard-spheres of Ar and Cl overlap slightly in the drawing) is a clue to a more chemical, but still quite weak, interaction. ClF has a respectable dipole moment, of course, (comparable to HBr—see Table 15.1), and since the dipole–polarizability interaction favors a linear orientation, the observed structure is in agreement with the wishes of this attractive force. However, the dipole–polarizability interaction alone cannot be the entire story, since it would be just as happy to have the arrangement ClFAr (perhaps happier, in fact, since the smaller F atom would allow Ar to approach the dipole even closer) as the observed ArClF arrangement.

EXAMPLE 24.12

Heteromolecular clusters can be made and studied using supersonic molecular beam expansions of gas mixtures. Expansion of a Xe/Ar mixture, for example, produces species such as Xe_2Ar. Estimate the structure and binding energy of this molecule.

SOLUTION We can approximate the interaction among these three atoms with a sum of atom–atom Lennard-Jones pair potentials, each with parameters taken from Table 24.4:

$$U_{Xe_2Ar} = U_{Xe_2}(R_1) + U_{XeAr}(R_2) + U_{XeAr}(R_3)$$

where R_1, R_2, and R_3 are interatomic distances. Since these distances form a closed triangle, they are not all independent, but the symmetry of U_{Xe_2Ar} tells us that the molecule will be triangular with $R_2 = R_3$. Since the three terms in U_{Xe_2Ar} are otherwise independent, the minimum energy configuration of the cluster must minimize each term. Consequently, we predict the cluster to have an isosceles triangle geometry with bond lengths equal to those in the isolated dimers Xe_2 and ArXe:

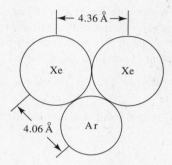

Similarly, the predicted binding energy (the energy for the process $Xe_2Ar \rightarrow 2Xe + Ar$) is the sum of the three pair potential well depths: $\mathcal{D}_e = 196.2 \text{ cm}^{-1} + 2(130.6 \text{ cm}^{-1}) = 457.4 \text{ cm}^{-1}$.

RELATED PROBLEMS 24.44, 24.45

Quite elaborate atom–atom plus electrostatic potential functions are used to predict minimum energy geometries of interacting, very large biochemicals. Drug design, for instance, increasingly relies on predictions of the equilibrium binding geometry of a drug with a target biomolecule, and conformations of proteins and biopolymers such as DNA and RNA are also studied this way. In biological systems in particular and aqueous media in general, the *hydrogen bond* can play a dominant role. Hydrogen bond energies fall between chemical bond energies and other non-bonded interaction energies. For example, $(HF)_2$, which has the equilibrium structure

deduced from an elaborate orientation dependent potential used to interpret microwave and infrared spectra, has a hydrogen bond dissociation energy $\mathcal{D}_e \cong 1650 \text{ cm}^{-1}$, and $\mathcal{D}_e$ for $(HCl)_2$ with a similar structure is 820 cm^{-1}. These are roughly an order of magnitude larger than rare gas dimer binding energies and roughly an order of magnitude smaller than single chemical-bond energies.

It is commonly assumed that weak bonding between two chemically bound molecules does not significantly alter *intramolecular* bonds. Hydrogen bonding shows the limits of this assumption. For example, the fundamental HF vibrational frequency is 3 961.422 9 cm^{-1}. In ArHF (a weak hydrogen bond to the Ar), the HF vibration drops slightly to 3 951.768 1 cm^{-1}, about a 10 cm^{-1} or 0.25% change. In (HF)$_2$, however, there are two distinct HF frequencies, one for the "inner" HF, the one that contributes an H to the hydrogen bond, and one for the "outer" HF, which has its H atom uninvolved in the hydrogen bond. The inner HF's fundamental frequency is 3 868.313 0 cm^{-1}, nearly a 100 cm^{-1} drop from free HF, and the outer HF's frequency is 3 930.903 0 cm^{-1}, only a 30.5 cm^{-1} drop.

The Molecular Description of Nonideal Gases

We are now armed with a reasonable stockroom of bits, pieces, and full models of intermolecular potential energy functions. We will continue to assemble these in various ways throughout the rest of this book, but our first job here takes us back to the second virial coefficient expression, Eq. (24.23c):

$$B(T) = 2\pi N_A \int_0^\infty (1 - e^{-U_2(R)/k_B T}) R^2 \, dR \ . \tag{24.23c}$$

For the hard sphere gas, U_2 is zero for $R \geq d$ and infinite for $R < d$, and the second virial coefficient is

$$B_{hs} = 2\pi N_A \int_0^d R^2 \, dR = \frac{2\pi}{3} N_A d^3 = 4 N_A v \tag{24.32}$$

where v is the hard-sphere volume. B_{hs} is independent of temperature. (See also the discussion leading to Eq. (1.19) and Problem 1.13.) If the gas interacts through an inverse power law repulsion, A/R^m, the $B(T)$ integral can be evaluated in terms of known functions, but one of them must be evaluated numerically. One finds for $m = 12$ that

$$B_{\text{rep}}(T) = 2.566\ 5\ N_A \left(\frac{A}{k_B T}\right)^{1/4} , \tag{24.33}$$

which is a *slowly decreasing function of T*. The greater average collision energy of higher temperature gases explains this behavior. Higher collision energy means deeper penetration of any repulsive potential, as if the molecule was a hard sphere with a diameter that decreased with increasing T. A gas with a weak attraction approaches this behavior at high T since the binding energy is a small fraction of the average collision energy.

For both of these purely repulsive potentials, $B > 0$ always. Attraction alone is not a very realistic model, since the "excluded volume" of the repulsive part of the potential cannot be ignored at any temperature. Thus, we move to the square well potential, Eq. (24.27), which predicts (see Problem 24.46)

$$B_{sw}(T) = B_{hs}[1 - (\lambda^3 - 1)(e^{\mathcal{D}_e/k_B T} - 1)] \tag{24.34}$$

where $\lambda = d_2/d_1$ and B_{hs} is Eq. (24.32) with $d = d_1$. At high T, $B_{sw} \to B_{hs}$ as the potential well plays a negligible role. At low T, $B_{sw} < 0$, as is observed experimentally, and it is easy to find the *Boyle temperature*, T_B, at which $B(T_B) = 0$.

EXAMPLE 24.13

The Ar Boyle temperature is 410 K. If the Ar–Ar interaction is modeled as a square well with the experimental Ar$_2$ binding energy and a repulsive wall (d_1 in the square well potential) equal to the Lennard-Jones σ value, $\sigma = R_e/2^{1/6}$, how wide is the potential's well?

SOLUTION At the Boyle temperature, $B(T_B) = 0$. If we set Eq. (24.34) equal to zero, we see immediately that *the repulsive wall does not determine* T_B:

$$B(T_B) = 0 = B_{hs}[1 - (\lambda^3 - 1)(e^{\mathscr{D}_e/k_BT_B} - 1)]$$

since B_{hs} can be cancelled from the equation. Solving for $\lambda = d_2/d_1$ gives

$$\lambda = (1 - e^{-\mathscr{D}_e/k_BT_B})^{-1/3}.$$

The $\mathscr{D}_e$ value in Table 24.4, 99.5 cm^{-1}, gives $\lambda = 1.502$. Since $R_e = 3.76$ Å, $\sigma = d_1 = 3.76$ Å$/2^{1/6} = 3.35$ Å, and thus $d_2 = \lambda d_1 = 5.03$ Å. This is a very reasonable value, as we can see if we compare the full Lennard-Jones potential to this square well potential:

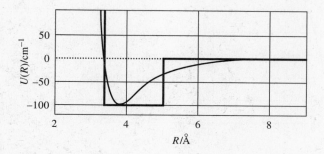

▶ **RELATED PROBLEMS** 24.46, 24.47, 24.48, 24.49

The full Lennard-Jones potential leads to an expression for $B(T)$ that cannot be integrated in closed form. If we define a dimensionless set of variables $x = R_e/R$ and $T^* = k_BT/\mathscr{D}_e$ and also define B_{hs} in terms of σ: $B_{hs} = (2\pi N_A/3)\sigma^3$, we can write Eq. (24.23c) as

$$B_{LJ}(T) = 3B_{hs}\int_0^\infty \left\{1 - \exp\left[\frac{4}{T^*}(x^{-12} - x^{-6})\right]\right\} x^2\, dx,$$

which is shown in Figure 24.15 along with experimental values for Ar, scaled using the accurate R_e and $\mathscr{D}_e$ values in Table 24.4. (Compare also Figure 1.6, the $B(T)$ expression based on the van der Waals equation of state.)

The qualitative shape of B_{LJ} is in agreement with experiment, and the disagreement between B_{LJ} and the experimental points for Ar (note the logarithmic temperature scale) is easily explained. If we had *only* second virial coefficient data on Ar, we could fit them quite accurately to B_{LJ} if we allow R_e and $\mathscr{D}_e$ to vary freely. The Boyle temperature[13] for B_{LJ} is 3.417 9 $\mathscr{D}_e/k_B$, or, since $T_B = 410$ K for Ar, $\mathscr{D}_e = 3.417\ 9 \cdot (410 \text{ K})/(1.438\ 8 \text{ cm K}) = (120.0 \text{ K})/(1.438\ 8 \text{ cm K}) = 83.37$ cm^{-1}. The experimental B for Ar at 200 K is -47.4 cm^3 mol^{-1} (see Table 1.1) so that $T^* = 200.0 \text{ K}/120.0 \text{ K} = 1.667$. At this T^*, $B_{LJ}/B_{hs} = -0.964\ 53$ so that $B_{hs} = -47.4/-0.964\ 53 = 49.15$ or

$$\sigma = \left[\left(\frac{B_{\text{expt}}}{B_{LJ}/B_{hs}}\right)\left(\frac{3}{2\pi N_A}\right)\right]^{1/3} = 3.39 \text{ Å}$$

so that $R_e = 2^{1/6}\sigma = 3.81$ Å. This is larger than the R_e value in Table 24.4, and the $\mathscr{D}_e$ value deduced from T_B is smaller than the tabulated value. These disagreements reflect the limitations of a simple two-parameter potential function applied to only

---

[13] Note that the Lennard-Jones Boyle temperature does not depend on σ.

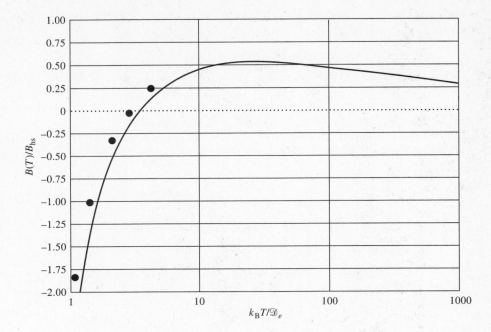

FIGURE 24.15 The line represents the predicted second virial coefficient as a function of temperature for the Lennard-Jones model potential. (Note the logarithmic, scaled temperature axis.) It can accurately reproduce experimental $B(T)$ data, but only for Lennard-Jones parameters that are somewhat different from accurately measured bond lengths and binding energies, as illustrated by the points for Ar experimental data that are located by accurate parameters in this figure.

one kind of experimental measurement. It produces parameters that are close to the truth, but only a multiparameter potential function applied to a variety of experimental data can yield accurate bond length and binding energy values (along, of course, with a full potential that can predict new experimental results accurately).

One can use orientation-dependent potentials in a numerical calculation of $B(T)$ for polyatomics, but one finds that $B(T)$ is remarkably insensitive to orientation-dependent forces. Most of our information about the orientation dependence of intermolecular potentials comes from spectroscopic studies of bound dimers and larger clusters rather than from bulk property measurements. Also, many other bulk properties (in principle, *all* other properties!) of gases are used to deduce intermolecular potential function information. For example, Joule–Thompson coefficients can be related to $B(T)$ (see Eq. (2.42)), and in the next chapter, we meet various nonequilibrium *transport properties* of gases that have also played an important role in discovering and refining intermolecular potentials.

CHAPTER 24 SUMMARY

This chapter has three main themes: the equilibrium distribution functions for the dynamic variables of a gas, the intermolecular forces that produce and maintain them, and the nonideal behavior caused by those forces. The three key distribution functions describe kinetic energy, speed, and velocity:

$$\text{speed: } f(v)\, dv = 4\pi \left(\frac{m}{2\pi k_B T}\right)^{3/2} v^2\, e^{-mv/2k_B T}\, dv$$

$$\text{velocity: } f(v_x)\, dv_x = \left(\frac{m}{2\pi k_B T}\right)^{1/2} e^{-mv_x^2/2k_B T}\, dv_x$$

$$\text{kinetic energy: } f(\epsilon)\, d\epsilon = 2\pi \left(\frac{1}{\pi k_B T}\right)^{3/2} \epsilon^{1/2}\, e^{-\epsilon/k_B T}\, d\epsilon$$

and each distribution has quantities, such as $\langle \epsilon \rangle = 3k_B T/2$ and the speeds in Table 24.1, that are characteristic of the full distribution.

We also saw how to transform the static equilibrium distribution into a nonequilibrium molecular-beam distribution through either a supersonic expansion:

$$\text{supersonic beam flow velocity: } V = \sqrt{\frac{2\overline{C}_P T_0}{M}}$$

$$\text{supersonic beam Mach number: } \frac{\text{flow velocity}}{\text{local speed of sound}} = \frac{V}{\sqrt{\gamma k_B T/m}}$$

or an effusive beam:

$$\text{effusive beam escape rate: } -\frac{dN}{dt} = AJ_x = \frac{AP}{\sqrt{2\pi m k_B T}}$$

The mean free path related molecular size to the average distance between collisions:

$$\text{mean free path: } \lambda = \frac{V}{\sqrt{2}\, N\pi d^2}$$

and served as a clue to effusive versus supersonic flow.

We classified intermolecular forces into two broad categories: repulsive, for which there was little simple theoretical guidance, and attractive, with a variety of simple physical origins: electrostatic (Table 24.2), inductive (Table 24.3), and dispersive (dominated by a $-C_6/R^6$ term). These pieces were assembled into several intermolecular potential models, such as the hard sphere, the square well, and the Lennard-Jones potentials, each with a number of parameters.

Finally, these potential models were used to predict non-ideal gas behavior through the second virial coefficient expression:

$$\text{general second virial coefficient:}$$
$$B(T) = 2\pi N_A \int_0^\infty \left(1 - e^{-U_2(R)/k_B T}\right) R^2\, dR$$

leading to simple model expressions:

$$\text{hard-sphere second virial coefficient: } B_{hs} = \frac{2\pi}{3} N_A d^3$$

$$\text{square-well second virial coefficient:}$$
$$B_{sw}(T) = B_{hs}\left[1 - (\lambda^3 - 1)(e^{\mathcal{D}_e/k_B T} - 1)\right]$$

The next chapter takes us farther from equilibrium. We will see how to construct distribution functions for simple nonequilibrium situations, derive from them the direction taken by a system as it spontaneously moves toward equilibrium, and explore the role played by intermolecular forces in these situations.

FURTHER READING

There are several classic texts on the kinetic theory of gases: *Kinetic Theory of Gases,* by R. D. Present (McGraw-Hill, New York, 1958); *An Introduction to the Kinetic Theory of Gases,* by J. H. Jeans (Cambridge University Press, Cambridge, 1940); and *Kinetic Theory of Gases,* by E. H. Kennard (McGraw-Hill, New York, 1938) all cover the material in this chapter (and much more). The practical side of the kinetic theory of gases is covered in *Scientific Foundations of Vacuum Technique,* (2nd ed., by S. Dushman, Wiley, New York, 1962).

Although it is over forty years old, the classic (and monumental—1249 pages) *The Molecular Theory of Gases and Liquids,* by J. O. Hirschfelder, C. F. Curtiss, and R. B. Bird (Wiley, New York, 1954) covers many of the topics of not only this chapter, but also statistical mechanics as applied to gases and liquids. It has extensive discussions of intermolecular forces, and summarizes virtually all the classic methods of inverting experimental data to intermolecular potential energy information. Modern methods (primarily molecular beam scattering and molecular cluster spectroscopy) have largely replaced the classic methods for simple systems, but the modern methods often rely on guidance from the classic.

Modern molecular beam methods are discussed in great depth and from a practical point of view in the two volume *Atomic and Molecular Beam Methods,* edited by G. Scoles (Oxford University Press, New York, 1988 (vol. 1), 1992 (vol. 2)). Volume 1 covers basic techniques of the type introduced in this chapter.

Intermolecular forces are discussed in several books. *Intermolecular Forces: Their Origin and Determination,* by G. C. Maitland (Oxford University Press, New York, 1981) is a recent example. *The Virial Equation of State,* by E. A. Mason and T. H. Spurling (Pergamon Press, Oxford, 1969) discusses intermolecular forces from the point of view of the virial expansion, and *The Virial Coefficients of Gases and Mixtures: a Critical Compilation,* by J. H. Dymond and E. B. Smith (Clarendon Press, Oxford, 1980) contains extensive tables of critically analyzed data.

PRACTICE WITH EQUATIONS

24A At what temperature does $N_2(g)$ have the same v_{rms} that $SF_6(g)$ has at 400 K? (24.4)

ANSWER: 76.7 K, roughly the N_2 boiling point

24B What is the most probable speed of the O_2 molecules you are breathing now? of the Ar atoms? Assume your air is at 25 °C. Table 24.2

ANSWER: O_2: 393.6 m s^{-1}; Ar: 352.3 m s^{-1}

24C How much greater is the speed of sound in He than in N_2, both at 300 K? (24.6), Table 24.1

ANSWER: 2.87

24D A spectroscopic experiment finds the most probable speed of a species in Na vapor at 800 K to be 537.9 m s^{-1}. What was the species? Table 24.1

ANSWER: Na_2

24E What is the maximum bulk flow velocity of a supersonic molecular beam of Ar expanded from 25 °C? How much larger is this than the equilibrium most probable speed? (24.8), 24B

ANSWER: V = 557.0 m s^{-1}; $V/v_{mp} = \sqrt{5/2}$ = 1.58

24F If the Mach number of the beam in 24E is 20, what is the local beam temperature? (24.10), 24E

ANSWER: 2.24 K

24G If a small amount of SF_6 is seeded in the supersonic Ar beam of 24E, what is its kinetic energy in the beam? How much larger is this than $3k_BT/2$? (24.11)

ANSWER: 3.76×10^{-20} J; 6.1 times larger

24H How many molecules strike a 1.00 cm^2 single-crystal W surface during a 1.0 h experiment if the crystal is in an ultrahigh vacuum apparatus at 300 K and 5×10^{-11} torr pressure of N_2? (24.13)

ANSWER: 6.90×10^{13}

24I On the surface of Saturn's largest moon, Titan, the predominantly N_2 atmosphere is at 94.5 K and 1.5 bar pressure. What is the N_2 mean free path if the N_2 effective hard-sphere diameter is 3.7 Å? (24.16)

ANSWER: 143 Å

24J How many molecules escape each second from a Knudsen cell with a 500 μm diameter hole containing $I_2(s)$ at 200 K, at which the vapor pressure is 10^{-6} torr? (24.17)

ANSWER: 3.06×10^{11} s^{-1}

24K What is the charge–dipole attractive energy between Na^+ and H_2O in its most favorable orientation at a distance of 4.0 Å? Tables 15.1 and 24.2

ANSWER: -5.66×10^{-20} J

24L What is the maximum dipole–dipole attractive energy between two H_2O molecules 4.0 Å apart? Tables 15.1 and 24.2

ANSWER: -1.11×10^{-20} J

24M What is the maximum dipole–polarizability attractive energy between H_2O and Xe 3.9 Å apart? Tables 15.1, 15.2, and 24.3

ANSWER: -8.16×10^{-22} J

24N What is the thermally averaged dipole–polarizability energy for H_2O and Xe 3.9 Å apart? Tables 15.1, 15.2, and 24.3

ANSWER: -4.08×10^{-22} J (half that of 24M)

24O What is the London dispersion energy between two CH_4 molecules 4.0 Å apart, using the CH_4 ionization potential to estimate the C_6 constant? (24.26), Tables 7.3 and 15.2

ANSWER: -2.51×10^{-21} J = -0.0136 eV

24P An early Ar–Ar exp-6 potential had parameters $A = 8.51 \times 10^7$ cm^{-1}, ρ = 0.273 Å, and $C = 5.14 \times 10^5$ cm^{-1} Å^6. What is the equilibrium bond length and binding energy of this potential? (24.29)

ANSWER: R_e = 3.82 Å; $\mathcal{D}_e$ = 94.1 cm^{-1}

24Q If the Lennard-Jones potential is written $U(R) = A/R^{12} - C/R^6$, what are A and C for the Ar–Ar interaction? (24.31), Table 24.4

ANSWER: $A = 7.94 \times 10^8$ cm^{-1} Å^{12}; $C = 5.62 \times 10^5$ cm^{-1} Å^6

24R If the Lennard-Jones σ parameter is considered a hard-sphere diameter, what are the hard-sphere radii of Ar and Kr based on Ar_2 and Kr_2 potentials? (24.31), Table 24.4

ANSWER: Ar: 1.67 Å; Kr: 1.79 Å

24S What error is made if the binding energy combining rule $\mathcal{D}_{12} = (\mathcal{D}_{11}\mathcal{D}_{22})^{1/2}$ is used to predict the He–Xe binding energy from accurate He_2 and Xe_2 binding energies? Table 24.1

ANSWER: Predicted: 38.4 cm^{-1}; Observed: 19.5 cm^{-1}

24T What is the high-temperature limit of the Kr second virial coefficient in the hard-sphere approximation? (24.32), 24R

ANSWER: 57.5 cm^3 mol^{-1}

24U What A parameter for the A/R^{12} purely repulsive potential reproduces the experimental second virial coefficient for He at 600 K? (24.33), Table 1.1

ANSWER: 1.90×10^{-137} J m^{12} = 1.90×10^{-17} J Å^{12} = 9.58×10^5 cm^{-1} Å^{12}

24V For N_2 at 200 K, $B(T) = -35.2$ cm^3 mol^{-1}. Assuming a square-well interaction with $d_1 = 3.28$ Å and $d_2 = 5.18$ Å, what is the N_2–N_2 binding energy? (24.34)

ANSWER: 1.31×10^{-21} J = 66.1 cm^{-1}

24W The Xe second virial coefficient at 225 °C is -39.1 cm^3 mol^{-1}. The Lennard-Jones potential based on the known $\mathcal{D}_e$ of Xe_2 predicts $B_{LJ}/B_{hs} = -0.8495$ at this temperature. What Lennard-Jones σ do these data imply? (24.35)

ANSWER: 3.31 Å

PROBLEMS

SECTION 24.1

24.1 Show that the translational kinetic energy distribution function predicts that the most probable translation energy, ϵ_{mp}, equals $k_B T/2$. Also show that this distribution is normalized: the area under $f(\epsilon)$ equals 1. What expression would you use to find the *median* translational energy such that half the molecules have ϵ less than the median? The expression cannot be evaluated analytically, but if you are adept at numerical integration, you can show that the median energy is $\cong 1.183\, k_B T$.

24.2 It is useful to memorize certain benchmark energies, such as the bond energies in Table 7.1. These numbers are useful even if you know them to only one or two significant figures, since order-of-magnitude estimates are often revealing. With this in mind, what is the characteristic average gas kinetic energy at (a) room temperature, (b) liquid nitrogen temperature, and (c) $T = 1$ K (from which you can estimate the energy at any T) in both cm^{-1} units and kJ mol^{-1} units? The Cs_2 bond energy is rather small, only about 3180 cm^{-1}. At what temperature is the average translational energy in Cs vapor equal to this bond energy? Would you expect Cs_2 to survive collisional dissociation in the gas phase at the normal Cs boiling point, 944 K?

24.3 Use the 1-D particle-in-a-box density of states function

$$\rho(\epsilon)\, d\epsilon = \frac{L}{\pi}\left(\frac{m}{2\hbar^2}\right)^{1/2} \epsilon^{-1/2}\, d\epsilon$$

to find the 1-D translational energy distribution function. You should find

$$f_{1\text{-D}}(\epsilon)\, d\epsilon = \left(\frac{\beta}{\pi}\right)^{1/2} \epsilon^{-1/2}\, e^{-\beta\epsilon}\, d\epsilon\ .$$

Show that this function is normalized and predicts an average energy in accord with the equipartition theorem. What is the most probable energy?

24.4 While we cannot integrate the translational energy distribution function analytically between two arbitrary energies ϵ_1 and ϵ_2, if the energy region is small (so that $(\epsilon_1 - \epsilon_2)/\langle\epsilon\rangle \ll 1$), we can approximate the integral:

$$\int_{\epsilon_1}^{\epsilon_2} f(\epsilon)\, d\epsilon \cong f(\epsilon)\,(\epsilon_2 - \epsilon_1)$$

whenever we need to know the fraction of molecules in a small energy range. With this in mind, estimate the fraction of Cs_2 molecules at 1000 K that have a translational energy within one Cs_2 vibrational quantum (42 cm^{-1}) of the most probable translational energy.

24.5 It is often useful to transform an equation into a more universal form by a judicious choice of a new dimensionless variable. Show, for example, that the transformation $x = \epsilon/k_B T$ transforms Eq. (24.3) into

$$f(x)\, dx = \frac{2}{\sqrt{\pi}} x^{1/2}\, e^{-x}\, dx\ .$$

The integral of this function from $x = 0$ to $x = x_m$ can be calculated numerically and tabulated for any x_m, and a short table of this kind is shown below. Use it to find (a) the fraction of molecules with kinetic energies between ϵ_{mp} and $\langle\epsilon\rangle$, (b) the fraction of molecules with $\epsilon \geq \langle\epsilon\rangle$, and (c) the fraction with $\epsilon > 2.5 \times 10^{-20}$ J at 600 K.

x_m	$\int_0^{x_m} f(x)\, dx$
0.5	0.1987
1.0	0.4276
1.5	0.6084
2.0	0.7385
3.0	0.8884
4.0	0.9540
5.0	0.9814

24.6 Differentiate Eq. (24.5) with respect to v, and use your result to derive the expression for v_{mp} given in Table 24.1. Calculate v_{mp} for N_2 at its normal boiling point, 77.35 K, and for W at its T_{vap}, 5828 K.

24.7 The average speed is defined as $\langle v \rangle = \int_0^\infty v\, f(v)\, dv$. Use Eq. (24.5) with this definition to derive the expression for $\langle v \rangle$ in Table 24.1. What temperature gives Ar an average speed equal to $\langle v \rangle$ for He at room temperature?

24.8 (a) Use the information in Problem 24.5 to find the fraction of gas molecules with speeds $\geq v_{mp}$. (b) A molecule with the most probable kinetic energy has a speed v' such that $mv'^2/2 = \epsilon_{mp} = k_B T/2$ or $v' =$

$(k_B T/m)^{1/2}$. Find an expression for v' in terms of T and M (the molar mass) similar to those in Table 24.1. Is v' always smaller than, larger than, or intermediate to the other characteristic speeds in Table 24.1?

24.9 Practice Problem 24I mentions the atmospheric conditions on the surface of Titan. What is the speed of sound on Titan, and how does it compare to the speed of sound on Earth? What is the speed of sound on the surface of Mars, where (see Chapter 9) $T = 250$ K and the atmosphere is CO_2 with $\gamma = 1.3$?

24.10 Can the speed of sound ever be larger than the most probable gas speed? Which would you predict to have the greater speed of sound, N_2 or C_2H_4, if both were at 1000 K?

24.11 The classical equipartition theorem tells us that $\langle mv_x^2/2 \rangle = k_B T/2$ where v_x is the x velocity component. Since space is isotropic (all directions are the same) in a gas at equilibrium, and since we live in a 3-D world, the three unique directions, x, y, and z, must have the same velocity distributions. Since we know the full kinetic energy distribution can be written $f(\epsilon) = f(v_x^2 + v_y^2 + v_z^2)$ because $\epsilon = mv^2/2 = (m/2)(v_x^2 + v_y^2 + v_z^2)$, it must also be true that $f(v_x^2 + v_y^2 + v_z^2) = f(v_x)f(v_y)f(v_z)$ where $f(v_x)$, etc., are velocity component distribution functions. This last equality is only satisfied if $f(v_x)$ is an exponential function of v_x^2, that is, if $f(v_x) = A \exp(\pm B v_x^2)$. The $+$ sign is physically unrealistic as $v_x \to \infty$ and is rejected. Use the facts that the integral of $f(v_x)$ over the range $-\infty > v_x > \infty$ must equal 1 and that the integral of $v_x^2 f(v_x)$ can be related to the equipartition theorem to find A and B and verify Eq. (24.7).

24.12 A supersonic beam of He is found to have a root mean squared velocity spread about its mean flow velocity and along its flow direction of 204 m s^{-1}. What is the beam's translational temperature?

24.13 You would like to study solid C_{60} surfaces by atomic diffraction methods to see if you can detect thermal motion of the C_{60} molecules. (C_{60} freely rotates in the bulk solid at high temperatures, but does not at low temperatures.) You decide to use a supersonic beam of either He or Ne, and your calculations indicate that you need an atomic de Broglie wavelength around 1 Å. Which gas can you use, and what should the beam's source temperature be?

24.14 This problem and the next explore ways of making supersonic beams of *slowly* moving species. The first way is obvious: lower the source temperature. The limitation is the need for sufficient vapor pressure to ensure a strong expansion and a high Mach number. If we consider the rare gases Ar–Xe, increasing mass will lower V, but also raise the boiling point and thus raise the practical lowest limit to T_0. Given the boiling points Xe: 165 K; Kr: 119.8 K; Ar: 87.3 K, find the maximum flow velocity for each gas at T_{vap} of each.

24.15 A clever way to slow large molecules in supersonic beams is called "inverse seeding" in which a *light* species is dilute in a *heavy* species and the light species assumes the flow velocity of the heavy carrier. Suppose 1% or so He is seeded into Xe and the mixture is expanded at T_{vap}(Xe) (given in the previous problem). The He will have roughly the same flow velocity as the Xe. What source temperature T_0 would be required to give a pure He beam the velocity it has in this inverse seeded beam?

24.16 What is the local beam temperature in a N_2 supersonic beam expanded from 300 K to a Mach number of 20?

24.17 If a molecule of mass m is brought from ∞ to the surface of the Earth, its gravitational potential energy changes from 0 at ∞ to $-mgr_e$ (see Eq. (9.3)) where r_e is the radius of the Earth, 6378 km, and g is the acceleration of gravity at the surface, 9.807 m s^{-2}. To escape gravity, the molecule must acquire sufficient kinetic energy to overcome this drop in potential energy. In a real atmosphere, molecules diffuse or are blown by winds to a height (~400 km or so on Earth) where the atmosphere's density is so low they can escape to ∞ without suffering collisions. This mechanism is known as *Jeans escape*, since it was first discussed in detail in Jeans' 1916 book, "The Dynamical Theory of Gases." Find the terrestrial escape velocity, v_{esc}, and show that it is independent of molecular mass. (You can neglect the 400 km trip up the atmosphere and use just r_e in your calculation.) At the high altitudes of free escape (called the *exobase*), the gas temperature is quite high, ~1000 K, due to a number of atmospheric energy input mechanisms. Calculate the ratio $f(v_{esc})/f(\langle v_x^2 \rangle^{1/2})$ for both O_2 and He under these conditions based on the velocity distribution of Eq. (24.7). Is it likely that any O_2 has been lost from the atmosphere due to this mechanism?

SECTION 24.2

24.18 Consider a cube 1.00 cm on each side containing Xe at 300 K and 1.00 bar pressure. How many Xe-wall collisions are there per second in this cube? (See also Problem 24.52.)

24.19 Imagine that the gas in the previous problem is an equimolar He/Xe mixture instead of pure Xe. How many He atoms strike the wall per Xe wall collision? How does this ratio change if the temperature is doubled? if the pressure is doubled?

24.20 A beam of metal shot, each weighing 0.1 g, strikes a balance pan at a rate of 10 s^{-1} with a velocity of 2.5 m s^{-1} and at an angle of 15° to the pan's normal. Assuming the shot rebounds elastically from the pan (and thus silently, since sould would indicate an inelastic event), what would the balance read? Suppose the beam was a stream of 0.1 g clay balls that crashed onto the balance pan and stuck there rather than rebounding. What would the balance read in this case?

24.21 A common "rule of thumb" used by vacuum engineers states that the room temperature mean free path of air in cm units is approximately 5 divided by the pressure in μm of Hg units ("microns" of pressure where 1 μm Hg = 10^{-3} torr). What collision cross section does this rule imply?

24.22 If the background pressure in a molecular beam vacuum apparatus is too high, the beam is scattered and attenuated from collisions with the gas. How low should the pressure be in order to assure a 1.0 m mean free path for Ar at room temperature? Express the pressure in both Pa and torr units. A mechanical vacuum pump that uses a low-vapor-pressure oil can routinely achieve $\sim 10^{-2}$ torr pressures. Are such pumps alone sufficient to create the vacuum needed here? What mean free path does this pressure represent for Ar?

24.23 The data below are experimental rates of W evaporation at various temperatures, measured as the mass loss rate in vacuum per unit area. Use them to find the vapor pressure of W at each T.

T/K	2600	3000	3400
$\left(\dfrac{1}{A}\dfrac{dm}{dt}\right)$/g cm^{-2} s^{-1}	8.41×10^{-9}	9.95×10^{-7}	3.47×10^{-5}

24.24 If two containers of a gas at different temperatures are connected by a large hole, the final state of the system is established when the pressures in both containers are equal. If the hole is small (diameter d << mean free path), then a phenomenon known as *thermal transpiration* determines the final state: the net flux of molecules through the hole is zero. Show that this condition means the pressures are *unequal* but related through $P_1/P_2 = (T_1/T_2)^{1/2}$. If one container is at liquid N$_2$ temperature (77 K), the other is at 300 K, and the gas has a sufficiently low pressure at either temperature to ensure $\lambda >> d$, what is the true pressure in the cold container if a pressure gauge in the hot container reads 2.5×10^{-5} torr?

24.25 You placed some ZrO$_2$(s) in your Ta Knudsen cell so that you could measure the vapor pressure of ZrO$_2$ at $T \sim$ 2000 K from a mass loss measurement. Unfortunately, your mass spectrometer was broken at the time so that you could not monitor the composition of the material effusing from the cell. Only later did you discover that the reaction Ta(s) + ZrO$_2$(s) $\rightarrow$ TaO(g) + ZrO(g) occurs at these temperatures. What was the true pressure of TaO and ZrO in the cell if P_2 represents the pressure calculated on the assumption that the gas was pure ZrO$_2$?

24.26 At what pressure does the average spacing between Ar(g) atoms at 300 K equal the mean free path? Is the gas significantly nonideal at this pressure?

24.27 If we write Eq. (24.17) in the form

$$-\frac{dN}{dt} = N\frac{A\langle v \rangle}{4V}$$

we can rearrange it to $-dN/N =$ (constant) dt and integrate to find the number of molecules at time t left in a Knudsen cell if the cell's hole is opened at $t = 0$. (This expression is invalid, of course, if the effusing molecules are maintained at a constant pressure inside the cell due to evaporation from a condensed phase.) Consider N$_2$ in a 100 cm^3 cell effusing through a 500 μm diameter hole at 300 K. Find the time required for half the gas to escape.

24.28 If we construct an effusive molecular beam apparatus that is big enough and have heavy enough atoms in the beam, we will observe the atoms fall under the influence of gravity. Such an experiment was first performed by one of the pioneers of molecular beam methods, Otto Stern, and his colleagues in 1947 [*Phys. Rev.* **71**, 238 (1947)] using Cs atoms. They detected the Cs beam with a 0.02 mm diameter hot W wire (a *Langmuir–Taylor* detector) that ionizes every Cs atom that strikes it and ejects the ion, forming a readily measured current. If the source temperature in such an experiment is 450 K, how far does a Cs atom fall over a 1.00 m distance if it is moving horizontally at the most probable speed of the gas in the oven? What reasons can you suggest for their choice of Cs over any other element?

24.29 In the course of your study of CdTe semiconductor surfaces, you decide to investigate *laser ablation* in which an intense laser beam is focused onto the surface, rapidly heating a small area and evaporating some of the material. You use a pulsed laser and a mass spectrometer located $L = 8.53$ cm from the surface so that you can measure the composition and velocity distribution of the ejected material. You find that the time-of-flight signal of Te$_2$ is consistent with a Maxwell–Boltzmann distribution, and the signal peaks 316 μs after the laser pulse (which lasts only $\sim$10 ns). Since your mass spectrometer is a density detector, the time-of-flight signal should follow

$$s(t) = \text{(constant)}\, t^{-4}\, e^{-mL^2/2t^2 k_B T}$$

where (constant) represents a collection of constants. If we define t_{max} as the time of the signal maximum, show that

$$s(t) \propto t^{-4}\, e^{-2t_{max}^2/t^2}$$

where t_{max} is related to L, m, T, etc. Use this relation and the data quoted above to find the ejected Te$_2$(g) temperature.

24.30 You have designed an experiment to study the chemistry of 3p excited Na atoms in either the $^2P_{1/2}$ state at 16 956.2 cm^{-1} or the $^2P_{3/2}$ state at 16 973.4 cm^{-1}. You will start with ground-state atoms seeded in a He supersonic beam such that their beam flow velocity is 2880 m s^{-1}. You will excite the atoms with a high-resolution tunable laser, and you can either cross the laser beam at a right angle to the molecular beam or you can direct the laser along the beam's direction, but opposing it. Calculate the laser frequencies needed to excite each state in each experimental configuration: transverse and counter-propagating. (You may want to review the discussion of the Doppler shift and Eq. (19.50) in Chapter 19.)

24.31 The essence of the Doppler broadening mechanism is not hard to expose. We assume that the molecule at rest has such a long radiative lifetime that its *natural* transition lineshape is infinitely sharp: it absorbs or emits only one frequency, ν_0. In the gas phase at equilibrium, the molecule has a distribution of velocities v_x given by Eq. (24.7), and we imagine the light travels along the x direction as well. The Doppler effect causes a molecule with any velocity v_x (which can be positive or negative) to absorb at $\nu_D = \nu_0(1 + v_x/c)$ where c is the speed of light. Write this in the form $v_x = c(\nu_D - \nu_0)(1/\nu_0)$ so that $dv_x = (c/\nu_0)\, d\nu_D$, and show that Eq. (24.7) becomes

$$f(\nu_D)\, d\nu_D = \frac{c}{\nu_0}\left(\frac{m}{2\pi k_B T}\right)^{1/2} \exp(-mc^2 v_D^2 / 2k_B T \nu_0^2)\, d\nu_D \; .$$

Show that this function reproduces the Doppler full width at half maximum expression of Eq. (19.51). A transition with a significant natural linewidth has a lineshape (known as a *Voigt lineshape*) that is slightly different from the Gaussian shape derived here.

24.32 In a photodissociation experiment with HI in a molecular beam (see Figure 18.6), high-resolution laser induced fluorescence (LIF) is used to probe the velocity distribution of the H atom photofragment using the 1s → 2p transition with a rest frequency ~82 259 cm^{-1}. The observed LIF spectrum is shown below.

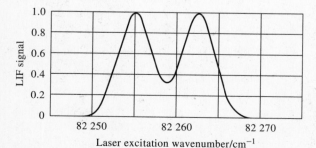

Why are there two peaks? What are they telling you about the H atom velocities?

SECTION 24.3

24.33 Categorize the various types of long-range attractive interactions that would best describe the intermolecular potential for each of the following pairs of species. Arrange them in order of decreasing interaction strength if there is more than one type.

(a) CO and N_2 (b) K^+ and F^-
(c) HCl and HF   (d) Ar and CF_4
(e) H^+ and He   (f) H_2 and HCN

24.34 A very simple model potential has been found to describe the BeAr$^+$ molecular ion's electronic states quite well. It has an exponential repulsion added to an *effective* charge–polarizability attractive term:

$$U(R) = A\, e^{-R/\rho} - \frac{Z^2 e^2 \alpha'}{(4\pi\epsilon_0)2R^4}$$

where Ze is the effective Be$^+$ charge felt by Ar at the potential minimum and α' is the Ar polarizability volume. Values for the three parameters A, ρ and Z can be determined from the experimental bond length R_e, dissociation energy $\mathcal{D}_e = -U(R_e)$, and vibrational force constant $k_e = (d^2U/dR^2)_{R=R_e}$. The ground state has Be$^+$ 2s ^2S interacting with Ar in a σ fashion, while the first excited state has Be$^+$ 2p ^2P interacting in a π fashion with the Be$^+$ 2p orbital perpendicular to the internuclear axis. It is found that $Z = 1.29$ for the ground state and 1.88 for the excited state. Why are both of these numbers greater than 1, and why is the excited-state Z significantly larger than the ground-state Z? (*Hint:* The ground state bond length is 2.085 5 Å, but the excited state bond length is 1.934 8 Å.)

24.35 Ionic molecules such as alkali halides (MX) have the ion-pair interaction dominating their attraction at distances near R_e, but other interactions play a significant role. In the *Rittner model*, charge–polarizability terms are also included (the charge of ion M$^+$ acting on the polarizability of ion X$^-$ and vice versa). Calculate the charge–charge and the two charge–polarizability energies for NaCl at its equilibrium bond length, 2.360 8 Å (α' values are in Table 15.2). The Rittner model includes an exponential repulsion, $A\, e^{-R/\rho}$, with $A = 5.285 \times 10^{-16}$ J and $\rho = 0.295$ Å for NaCl. Add this energy at R_e to the three attractive terms, and compare the sum to the experimental binding energy with respect to separated ions, -9.255×10^{-19} J.

24.36 The Rittner model in the previous problem also has something to say about the dipole moment of an ionic molecule. The pure Coulombic dipole moment for an alkali halide is $p_C = eR_e$, of course, but the induced dipoles alter this value. Calculate the dipole moments induced in each ion ($p_{ind} = \alpha' e/R_e$), consider their directions carefully, and combine them with the Coulombic moment to predict the net moment. How well does your value compare to the experimental value, 30.03×10^{-30} C m?

24.37 At its normal boiling point, 1686 K, NaCl(g) is a fairly dense gas of fairly large molecular dipole moment (quoted in the previous problem). What is the thermally averaged dipole–dipole interaction energy in this gas at the average intermolecular spacing? Is this a significant energy?

24.38 Consider adsorption of HCl(g) on the surface of NaCl(s). The salt surface looks like an array of point charges, and HCl is a polarizable dipole. Which of the attractive energy terms, charge–dipole or charge–polarizability, dominates if HCl is 5.0 Å from one surface charge and most favorably aligned? The HCl dipole moment is 3.697×10^{-30} C m, and the HCl polarizability volume (along the bond direction) is 3.12×10^{-30} m^3. At what distance are the two interactions equally strong? Is HCl likely to be found at this distance from either ion, Na$^+$ or Cl$^-$? (Consider that HCl has a 1.275 Å bond length, and R measures the distance between an ion and the HCl center of mass, located essentially at the Cl atom. Consider as well the traditional ionic radii of Na$^+$ and Cl$^-$: 0.97 Å and 1.81 Å, respectively.)

24.39 Calculate the dipole–dipole interaction energy for (HF)$_2$ based on the structure shown in the text. The molecule is planar, the HF dipole moment is 6.069×10^{-30} C m, and the angles used in Table 24.2 are defined in Figure 24.10. Is this energy sufficient to account for the observed dimer binding energy, 1650 cm^{-1}?

24.40 The accurate Ar–Ar C_6 coefficient is $3.065\ 16 \times 10^5$ cm^{-1} Å^6. How well does the approximate London formula for C_6 agree with this value? The next term in the London dispersion interaction expansion can be written $-C_8/R^8$ where

$$C_8 \cong \frac{45}{8} \frac{\alpha_1' \alpha_2' (4\pi\epsilon_0)}{e^2} \left(\frac{\alpha_1' E_1}{2E_1 + E_2} + \frac{\alpha_2' E_2}{E_1 + 2E_2} \right) E_1 E_2$$

with E_1 and E_2 representative excitation energies that can be taken to be ionization energies. Note that this expression simplifies considerably if species 1 and 2 are the same. With this in mind, calculate C_8 for Ar–Ar, and contrast the C_8 term's energy to that for the C_6 term at $R = 3.76$ Å, the Ar$_2$ bond length. (The accurate C_8 coefficient, by the way, is 2.382×10^6 cm^{-1} Å^8, quite close to what you should find here.)

24.41 Repeat the analysis of Example 24.11 for the species He$_2$, Xe$_2$ and HeXe, and compare your results to the observed HeXe bond length and binding energy in Table 24.4.

24.42 Relate the A and C parameters of the Lennard-Jones potential written simply $U(R) = A/R^{12} - C/R^6$ to the more common $\mathcal{D}_e$ and σ parameters. (See Eq. (24.31).) Derive the relation $R_e = 2^{1/6} \sigma$. How much lower would the potential energy be at $R = R_e$ if the Lennard-Jones potential had no $1/R^{12}$ repulsion term? Suppose the potential was written $U(R) = A/R^{10} - C/R^6$. How would R_e and σ be related?

24.43 The Ar$_2$ binding energy is ~ 100 cm^{-1}, and the Na$_2$ binding energy is ~ 5900 cm^{-1}. Sodium is considerably more polarizable than Ar (Table 15.2), but the binding energy in NaAr is only ~ 50 cm^{-1}. Suggest a reason (or reasons) why NaAr is less strongly bound than Ar$_2$.

24.44 Predict the structure and total binding energy of Ar$_4$ based on atom–atom pair potentials. The ions K$^+$ and Cl$^-$ are isoelectronic to Ar. Would you expect (KCl)$_2$ to have the same structure as Ar$_4$? If not, what structure would you expect and why?

24.45 Consider the Ar–CO interaction as a sum of two Lennard-Jones atom–atom pair potentials, one between Ar and C and one Ar and O. Let R_e(Ar–O) = 3.41 Å, R_e(Ar–C) = 3.81 Å, $\mathcal{D}_e$(Ar–O) = 82 cm^{-1}, and $\mathcal{D}_e$(Ar–C) = 43 cm^{-1}. Find the ArCO equilibrium geometry and the Ar–CO binding energy. Assume CO has its normal bond length, 1.128 3 Å. What is the Ar binding energy in the two *collinear configurations* Ar—C≡O and Ar—O≡C?

24.46 Show that Eq. (24.23c) breaks into three easily evaluated integrals for the case $U_2 = U_{sw}$, the square-well potential of Eq. (24.27), and derive the expression for $B_{sw}(T)$ in Eq. (24.34). How does B_{sw} change at constant T and d_1 if (a) $\mathcal{D}_e$ is increased at constant d_2—the well gets deeper—or (b) d_2 is increased at constant $\mathcal{D}_e$—the well gets wider?

24.47 For the square-well potential, the quantity $-\mathcal{D}_e(d_2 - d_1)$ is called the *well capacity,* since it equals the area bounded by the potential below the $U = 0$ axis. Show that the well capacity for the Lennard-Jones potential is

$$\int_\sigma^\infty 4\mathcal{D}_e \left[\left(\frac{\sigma}{R} \right)^{12} - \left(\frac{\sigma}{R} \right)^6 \right] dR = -\frac{24}{55} \mathcal{D}_e \sigma \ .$$

Equate this to the square well capacity and explain why $\lambda \cong 1.5$ is a reasonable value, as found in Example 24.13.

24.48 Methane is sufficiently spherical that a Lennard-Jones potential should accurately represent its second virial coefficient. The Boyle temperature is 509 K (Table 1.2), and several experimental $B(T)$ values are:

T/K	200	300	400	600
B/cm^3 mol^{-1}	-105	-42	-16	8.5

Use these data and the table below of Lennard-Jones B_{LJ}/B_{hs} values at selected reduced temperatures, $T^* = k_BT/\mathcal{D}_e$, to estimate Lennard-Jones potential parameters for CH_4.

$k_BT/\mathcal{D}_e$	B_{LJ}/B_{hs}
1.00	−2.538
1.25	−1.704
1.50	−1.201
1.75	−0.8659
2.00	−0.6276
2.50	−0.3126
3.00	−0.1152
4.00	−0.1154
5.00	−0.2433

24.49 For the O_2–O_2 interaction, Lennard-Jones parameters $\mathcal{D}_e = 82$ cm^{-1} and $R_e = 3.96$ Å have been deduced from a variety of experiments. How well do these parameters reproduce the O_2 Boyle temperature, 405 K? Predict B at 200 K, and compare to the experimental value −49 cm^3 mol^{-1}. The table in the previous problem is useful here as well.

GENERAL PROBLEMS

24.50 The fraction of molecules with velocities v_x between two arbitrary limits v_a and v_b can be expressed in terms of a well-known function called the *error function*. If we let $x^2 = mv_x^2/2k_BT$ in Eq. (24.7), we can write

$$\text{fraction between } v_a \text{ and } v_b = \int_{v_a}^{v_b} f(v_x)\, dv_x = \frac{1}{\sqrt{\pi}} \int_{x_a}^{x_b} e^{-x^2} dx.$$

The error function, erf(x), is defined as

$$\text{erf}(x_a) = \frac{2}{\sqrt{\pi}} \int_0^{x_a} e^{-x^2} dx$$

and, while the integral cannot be expressed in a closed analytical form, it can be evaluated numerically and tabulated. Use the table below to find the fraction of molecules with v_x between (a) 0 and $(2k_BT/m)^{1/2}$; (b) $(2k_BT/m)^{1/2}$ and ∞; (c) $(k_BT/2m)^{1/2}$ and $(2k_BT/m)^{1/2}$; and (d) $-(2k_BT/m)^{1/2}$ and $(2k_BT/m)^{1/2}$.

x	0	0.25	0.50	1.00	1.50	∞
erf(x)	0	0.2763	0.5205	0.8427	0.9661	1

24.51 The previous problem considered the velocity fraction over an arbitrary range. Here we consider the fraction with a *speed* between arbitrary limits. Use the identity

$$\int_0^{x_a} x^2 e^{-x^2} dx = \frac{\sqrt{\pi}}{4} \text{erf}(x_a) - \frac{x_a}{2} e^{-x_a^2}$$

where erf is the error function defined in Problem 24.50 to express $\int_{v_a}^{v_b} f(v)\, dv$ in terms of the dimensionless variable $x = (mv^2/2k_BT)^{1/2}$ where $f(v)$ is the speed distribution of Eq. (24.5). Check your expression by showing that it correctly predicts $\int_0^{v_{med}} f(v)\, dv = 0.5$ where v_{med} is the median speed. You will need to know that $x_{med} = 1.0877$ and erf(x_{med}) = 0.87599. What is the fraction of molecules with speeds greater than the most probable speed?

24.52 Imagine plunging the cube in Problem 24.18 into liquid He at 4.2 K, a temperature so low that every Xe atom that strikes the surface sticks to it, forming a Xe(s) film of negligible vapor pressure. Assuming the cube walls fall instantly from 300 K to 4.2 K but the gas remains at 300 K, how long does it take for the pressure in the sphere to drop to zero? Note that once the pressure is so low that the mean free path is ∼1 cm, each atom left in the gas freezes nearly instantaneously since it has an unobstructed path to the walls. Note also the similarity to Problem 24.27.

24.53 It was pointed out in the text that the simple combining rule for binding energies, $\mathcal{D}_{12} = (\mathcal{D}_{11}\mathcal{D}_{22})^{1/2}$, often fails if one species is much less polarizable than the other. An alternate combining rule that has some theoretical justification is $\mathcal{D}_{12} = 2\mathcal{D}_{11}\mathcal{D}_{22}/(\mathcal{D}_{11} + \mathcal{D}_{22})$. Calculate the heteronuclear He–rare gas binding energies using both rules and the homonuclear binding energies in Table 24.4, and compare the rules' predictions to the experimental values in that table. Is one rule appreciably better than the other?

24.54 The second virial coefficient expresses more than the nonideal equation of state, since an equation of state is an opening to more general thermodynamic functions. Consider, for example, the low-pressure Joule–Thomson coefficient expression in Eq. (2.42), $\mu_{JT} = [T(dB/dT) - B]/\overline{C}_P$. Use this to derive an expression for μ_{JT} for a square-well potential, and compare its predictions to the data in Table 2.1 for Ar at 1 atm. Use the square-well potential parameters derived in Example 24.13. You should find reasonable, but not spectacular, agreement.

24.55 The nonideal behavior of polar gases such as NH_3 has been discussed in terms of the *Stockmayer potential*, a sum of a Lennard-Jones model potential and a dipole–dipole electrostatic potential energy term. Graph this function for the most favorable orientation of two ammonia molecules, using the NH_3 dipole moment in Table 15.1 and the Lennard-Jones parameters $\mathcal{D}_e = 222.4$ cm^{-1} and $R_e = 2.92$ Å. Cover the range 2 Å < R < 8 Å, and plot just the Lennard-Jones potential as well. Do the Lennard-Jones parameters have a simple interpretation here? Which term, Lennard-Jones or dipole–dipole, dominates at large distances?

CHAPTER 25

Nonequilibrium Dynamics

IN this chapter, we take two somewhat different views of simple nonequilibrium situations. Each is characteristic of situations more complicated than we will see here, and each is reminiscent of our two descriptions—microscopic and macroscopic—of equilibrium situations. In the first and third sections, we take a microscopic point of view. We consider distribution functions and molecular collisions and see how collisions force a system towards equilibrium. The second section takes a macroscopic view with *phenomenological* expressions that tell us, in differential equation form, how *gradients* of macroscopic variables evolve in time as the system approaches equilibrium. These gradients—spatial variations in density, temperature, velocity, and electrical potential—describe a nonequilibrium state.

Gradients decay as an isolated system approaches equilibrium. Regions with, for example, a temperature above the equilibrium value fall in temperature, and as they do, energy is *transported* to regions initially below the equilibrium temperature. It is the transport of energy, matter, electric charge, and linear momentum that interests us here. We will see the microscopic origins of thermal conductivity (energy transport), diffusion (matter transport), conductivity (charge transport), and viscosity (linear momentum transport). These terms have familiar connotations in everyday language: silver is a better conductor of heat and electricity than plastic; motor oil is more viscous than water; and a drop of ink diffuses throughout a glass of water. Our concern here is the molecular mechanism behind the everyday experience.

25.1 The Nonequilibrium Distribution Function

25.2 Phenomenological Transport—Forces and Fluxes

25.3 Microscopic Transport Coefficients

25.4 Transport in Condensed Media

25.1 THE NONEQUILIBRIUM DISTRIBUTION FUNCTION

Consider a gas confined to the center of a larger container. The confining walls are removed, the gas expands into the full volume, and at some later time, the molecules are distributed at equilibrium throughout the entire volume. How did they get there? Is there a distribution function for the gas that is valid a nanosecond or two after the gas begins expansion, but before final equilibrium is attained? What aspects of molecular dynamics might govern such a function? Is it unique, or can many paths lead from one equilibrium state to another?

Collisions play a central role in nonequilibrium states, but they are much more difficult to incorporate than for equilibria. Simplicius, the great 6th century Greek commentator on Aristotle, had the right idea when he wrote:

> The atoms move in the void and catching each other up jostle together, and some recoil in any direction that may chance, and others become entangled with one another in various degrees according to the symmetry of their shapes and sizes and positions and order, and they remain together, and thus the coming into being of composite things is effected.

This one cryptic sentence encompasses the spirit of all nonequilibrium transformations, reactive and nonreactive. In this chapter, we consider only nonreactive situations, but we will see how to express the ideas "catching each other up" and "jostle together" in modern language.

If we go back to the equilibrium velocity distribution functions in Chapter 24, we can begin to see how they must change in nonequilibrium situations. First, the equilibrium distribution function Eq. (24.7), depends only on velocity, *not* on time or position. A nonequilibrium distribution depends in general on *velocity, time, and position,* since the distribution changes in time and space as the system moves toward a new equilibrium state.

A nonequilibrium distribution function f^* is defined such that

$$f^*(v_x, v_y, v_z, x, y, z, t)\, dv_x\, dv_y\, dv_z\, dx\, dy\, dz \tag{25.1}$$

is the *probability that a molecule at time t in the volume dx dy dz located at point (x, y, z) has velocity components in the range (dv_x, dv_y, dv_z) about (v_x, v_y, v_z)*. We are particularly interested in the time evolution of f^*, something an equilibrium distribution does not have. Note that f^* not only depends on time *explicitly,* it also depends on time through *coordinates and velocities that depend on time.* Consequently, the *total* time derivative of f^* is

$$\frac{df^*}{dt} = \frac{\partial f^*}{\partial x}\frac{dx}{dt} + \frac{\partial f^*}{\partial y}\frac{dy}{dt} + \frac{\partial f^*}{\partial z}\frac{dz}{dt} + \frac{\partial f^*}{\partial v_x}\frac{dv_x}{dt} + \frac{\partial f^*}{\partial v_y}\frac{dv_y}{dt} + \frac{\partial f^*}{\partial v_z}\frac{dv_z}{dt} + \frac{\partial f^*}{\partial t}, \tag{25.2a}$$

an expression that looks more formidable than it is. Consider the first three terms. The time derivatives of position, such as $dx/dt,$ are velocity components, and the $\partial f^*/\partial x$ factors that multiply them are *spatial gradients* of f^*. We can write these three terms together in a vector shorthand notation:

$$\frac{\partial f^*}{\partial x}\frac{dx}{dt} + \frac{\partial f^*}{\partial y}\frac{dy}{dt} + \frac{\partial f^*}{\partial z}\frac{dz}{dt} \equiv \nabla f^* \cdot \mathbf{v}$$

where ∇ is the vector gradient operator, $\mathbf{v}$ is the velocity vector, and the dot represents a vector dot product.[1] Similarly, the next three terms contain time derivatives of

---

[1] The vector gradient operator, ∇, operating on an arbitrary function, $w(x,y,z)$, is defined as $\nabla w \equiv \hat{x}(\partial w/\partial x) + \hat{y}(\partial w/\partial y) + \hat{z}(\partial w/\partial z)$ where $\hat{x}, \hat{y},$ and $\hat{z}$ are unit vectors in the $x, y,$ and z directions.

velocity components, which are components of the *acceleration* vector, **a**, or, since $\mathbf{F}_{ext} = m\mathbf{a}$, they are components of the total *external force* divided by mass. These terms together are

$$\frac{\partial f^*}{\partial v_x}\frac{dv_x}{dt} + \frac{\partial f^*}{\partial v_y}\frac{dv_y}{dt} + \frac{\partial f^*}{\partial v_z}\frac{dv_z}{dt} = \nabla_v f^* \cdot \frac{\mathbf{F}_{ext}}{m}$$

where ∇_v is the vector gradient operator for velocity. Thus, we can write df^*/dt compactly as

$$\frac{df^*}{dt} = \nabla f^* \cdot \mathbf{v} + \nabla_v f^* \cdot \frac{\mathbf{F}_{ext}}{m} + \frac{\partial f^*}{\partial t}. \qquad (25.2b)$$

EXAMPLE 25.1

Show that $df^*/dt = 0$ for the case $f^* = f$, the equilibrium distribution function of a gas.

SOLUTION If the gas is at equilibrium, any external forces that act on it cannot depend on time, and they must be derived from a static external potential energy, $U_{ext}(\mathbf{r})$, such that $\mathbf{F}_{ext} = -\nabla U_{ext}$. Since the equilibrium distribution is a function of only molecular energy ϵ, as Eq. (24.3) shows, and since $\epsilon = mv^2/2 + U_{ext}(\mathbf{r})$ is time-independent, the partial derivative $\partial f/\partial t = (\partial f/\partial \epsilon)(\partial \epsilon/\partial t)$ equals zero. Any non-zero spatial gradient of f, ∇f, must be due to a non-zero gradient in U_{ext}. We can write

$$\nabla f \cdot \mathbf{v} = \frac{\partial f}{\partial \epsilon} \nabla \epsilon \cdot \mathbf{v} = \frac{\partial f}{\partial \epsilon} \nabla U_{ext} \cdot \mathbf{v} = -\frac{\partial f}{\partial \epsilon} \mathbf{F}_{ext} \cdot \mathbf{v} .$$

Similarly, the velocity gradient of f, $\nabla_v f$, is due to the $mv^2/2$ kinetic energy component of ϵ, and we can write

$$\nabla_v f \cdot \frac{\mathbf{F}_{ext}}{m} = \frac{\partial f}{\partial \epsilon} \nabla_v \epsilon \cdot \frac{\mathbf{F}_{ext}}{m} = \frac{\partial f}{\partial \epsilon}(m\mathbf{v}) \cdot \frac{\mathbf{F}_{ext}}{m} = \frac{\partial f}{\partial \epsilon} \mathbf{v} \cdot \mathbf{F}_{ext}$$

where we have used

$$\nabla_v \epsilon = \nabla_v \left[\frac{1}{2}m(v_x^2 + v_y^2 + v_z^2)\right] = m\mathbf{v} .$$

Thus, the two gradient terms cancel each other: $\nabla_v f \cdot \mathbf{F}_{ext}/m = -\nabla f \cdot \mathbf{v}$, the explicit time dependence derivative is zero: $\partial f/\partial t = 0$, and therefore the total derivative is also zero: $df/dt = 0$.

▶ *RELATED PROBLEMS* 25.1, 25.2

Each of the three terms on the right of Eq. (25.2b) has a physical interpretation. The $\partial f^*/\partial t$ term tells us whether or not the system is in a *steady state:*

$$\frac{\partial f^*}{\partial t} = 0 \quad \text{in a steady state or at equilibrium.}$$

In a steady state, the properties of the system *at any one point* do not change in time even though the system may not be at equilibrium. For example, a metal bar with one end in an ice-water bath and the other in a boiling water bath has a temperature gradient along the bar, but as long as one end is maintained at 0 °C and the other at 100 °C, the gradient is maintained, and the temperature at any point along the bar is constant in time. Steady-state conditions are often easy to establish experimentally.

The middle term on the right of Eq. (25.2b) is zero whenever there are no external forces on the system. This is also easy to arrange experimentally (or to ignore, as we will with gravity). Examples of processes driven by an external force include electrical conduction (which we will consider) and equilibrium in a centrifuge (which we will not).

The first term generally contains most of the action. Quantities that are transported in the direction of spatial gradients of f^* have $\nabla f^* \neq 0$. We can usually arrange the gradient along only one direction so that $\nabla f^* \cdot \mathbf{v} = (\partial f^*/\partial z)\, v_z$. For example, the heated metal bar mentioned above has a temperature gradient along only one direction that we can take to be the z direction.

Equation (25.2) is a general, abstract expression for the way f^* changes. We now seek the molecular basis for these changes, and in so doing, we will find a microscopic expression for df^* that, together with Eq. (25.2), leads to a differential equation for f^* that we can solve.

The distribution function f^* is proportional to a probability, but we generally care more about *molecular concentrations* or *number densities*. Since nonequilibrium situations often have nonuniform number densities, the ratio of the total number of particles, N, to the total volume, V, is no longer sufficient. We introduce a function $n(x,y,z,t)$ that gives us the local, instantaneous number density. We combine f^* and n and write[2]

$$dN = n(x,y,z,t)\, f^*(v_x,v_y,v_z,x,y,z,t)\, d^3\mathbf{v}\, d^3\mathbf{r}$$
$$= F^*(v_x,v_y,v_z,x,y,z,t)\, d^3\mathbf{v}\, d^3\mathbf{r}$$

where dN is the *number of molecules at time t* in the six-dimensional volume element of velocity and position coordinates $d^3\mathbf{v}\, d^3\mathbf{r} = dv_x\, dv_y\, dv_z\, dx\, dy\, dz$ and F^* is *the number-density distribution function*, $F^* = nf^*$.

During a time interval dt, these molecules move due to the combined effects of their velocities and any accelerations external forces produce on them. Molecules at $\mathbf{r}$ move to $\mathbf{r}' = \mathbf{r} + \mathbf{v}\, dt$, and their velocities change from $\mathbf{v}$ to $\mathbf{v}' + \mathbf{a}\, dt$. The number in $d^3\mathbf{v}\, d^3\mathbf{r}$ at $\mathbf{v}'$ and $\mathbf{r}'$ changes to

$$dN' = F^*(\mathbf{v}', \mathbf{r}', t+dt)\, d^3\mathbf{v}\, d^3\mathbf{r}\ .$$

Collisions cause some of the molecules initially at $\mathbf{v}$ and $\mathbf{r}$ to leave the element $d^3\mathbf{v}\, d^3\mathbf{r}$, while others initially in another element will enter this one. At equilibrium, $dN = dN'$, but away from equilibrium, this equality may not hold. Thus we write

$$dN' - dN = dN_{\text{coll}} \tag{25.3}$$

where dN_{coll} is the net change due to collisions during time dt.

We can write two important expressions for dN_{coll}. The first comes from the definitions of dN and dN':

$$dN_{\text{coll}} = [F^*(\mathbf{v}',\mathbf{r}',t+dt) - F^*(\mathbf{v},\mathbf{r},t)]\, d^3\mathbf{v}\, d^3\mathbf{r}\ .$$

The terms in square brackets equal dF^*, the total derivative of F^*. Since $dF^* = (dF^*/dt)dt$, we have

$$dN_{\text{coll}} = \frac{dF^*}{dt} dt\, d^3\mathbf{v}\, d^3\mathbf{r}\ ,$$

and in analogy with Eq. (25.2b), we can write dF^*/dt explicitly:

$$dN_{\text{coll}} = \left(\nabla F^* \cdot \mathbf{v} + \nabla_v F^* \cdot \frac{\mathbf{F}_{\text{ext}}}{m} + \frac{\partial F^*}{\partial t}\right) dt\, d^3\mathbf{v}\, d^3\mathbf{r}\ . \tag{25.4}$$

This is half of our final differential equation for F^*.

---

[2] We will *not* use n to represent a number of moles in this chapter, and we will always write $\mathbf{F}_{\text{ext}}$ to represent the external force vector and F or F^* to represent a number-density distribution function.

The second half considers collisions explicitly. Since dN_{coll} represents a net change of molecules scattered into and out of $d^3\mathbf{v}\,d^3\mathbf{r}$ in time dt, we consider the *rate* of this change, dN_{coll}/dt, and write

$$\frac{dN_{coll}}{dt} = \text{(time rate of change of } F^*) \, d^3\mathbf{v}\,d^3\mathbf{r} \ .$$

Finding an expression based on collisions for the quantity written above as (time rate of change of F^*) is difficult. Boltzmann first did so and found that a multiple integral over collision conditions that take molecules into and out of $d^3\mathbf{v}\,d^3\mathbf{r}$ was required. The result, known as the *Boltzmann transport equation,* equates this integral over F^* to the total derivative of F^* in Eq. (25.4). It is thus quite a difficult equation to solve.

We will use an approximation that leads to informative approximate solutions. In particular, we imagine that the system is close to equilibrium so that $F^* \cong F$ where F is the *equilibrium* number density distribution function, Nf/V. In this case, F^* should settle smoothly and quickly toward F if the system is isolated.

Consider, for example, Ar(g) at equilibrium at 300 K. Suppose we took about half the molecules with $v_x \cong +250$ m s^{-1} and turned them around, creating the nonequilibrium distribution f^* shown in Figure 25.1. This distribution is Eq. (24.7), except for two spikes at ± 250 m s^{-1} where we have altered things. As f^* relaxes to equilibrium, the spikes shrink into the smooth curve. At $v_x = -250$ m s^{-1}, $f^* > f$, or $f^* - f > 0$. For this spike to fall, df^*/dt must be negative. Similarly, at $v_x = +250$ m s^{-1}, $f^* - f < 0$, and df^*/dt must be positive so that the spike rises as equilibrium is attained. These arguments show that df^*/dt (and thus dF^*/dt or dN_{coll}/dt) must have the *opposite* sign from $f^* - f$ (or $F^* - F$).

If F^* is not far from F, it is reasonable to assume (and it can be shown rigorously) that dN_{coll}/dt is directly proportional to the difference between F^* and F (i.e., to $F^* - F$ and not $(F^* - F)^2$, or some other power). When this is true, one says the nonequilibrium system is in the *linear response regime,* and we write the missing factor for dN_{coll}/dt as

$$\text{(time rate of change of } F^*) \propto -(F^* - F) \ .$$

This is only a proportionality, since it lacks a reference to the time scale on which F^* changes. Since collisions provide the relaxation mechanism, it is reasonable to

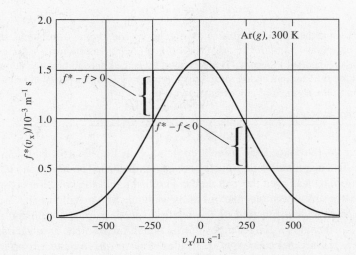

FIGURE 25.1 In this simple nonequilibrium distribution function, f^*, the equilibrium distribution, f, has been altered at only two velocities. About half the Ar atoms with $v_x = +250$ m s^{-1} have been turned around.

take τ_{coll}, the average time between collisions for molecules with velocity $\mathbf{v}$, as this time scale and write

$$\frac{dN_{coll}}{dt} = -\frac{F^* - F}{\tau_{coll}} d^3\mathbf{v}\, d^3\mathbf{r} \ . \tag{25.5}$$

Equation (25.5) is the other half of our differential equation for F^*. Combining it with Eq. (25.4) and cancelling the differentials gives us the *linearized Boltzmann transport equation*:

$$\nabla F^* \cdot \mathbf{v} + \nabla_v F^* \cdot \frac{\mathbf{F}_{ext}}{m} + \frac{\partial F^*}{\partial t} = -\frac{F^* - F}{\tau_{coll}} \ . \tag{25.6}$$

This is a differential equation for F^*, and we will solve it for specific situations in Section 25.3. To close this section, we look further at the collision time, τ_{coll}.

EXAMPLE 25.2

Consider a gas that has a spatially uniform density, is not subjected to any external forces, but has a nonequilibrium velocity distribution, such as the system in Figure 25.1. How does such a system's distribution function change in time?

SOLUTION If the gas density is spatially uniform, then $\nabla F^* = 0$. Similarly, if there are no external forces, the second term on the left-hand side of Eq. (25.6) is also zero, and the transport equation is

$$\frac{\partial F^*}{\partial t} = -\frac{F^* - F}{\tau_{coll}}$$

which we can rearrange and integrate:

$$\int_{F_0^*}^{F_t^*} \frac{\partial F^*}{F^* - F} = -\frac{1}{\tau_{coll}} \int_0^t dt$$

yielding

$$\ln\left(\frac{F_t^* - F}{F_0^* - F}\right) = -\frac{t}{\tau_{coll}}$$

or

$$F_t^* = F + (F_0^* - F)\, e^{-t/\tau_{coll}} \ ,$$

which shows that the initial perturbation away from the equilibrium distribution, $(F_0^* - F)$, decays exponentially until $F^* = F$. (This is Eq. (22.32), mentioned in the context of Boltzmann's H theorem.) Since τ_{coll} is a very short time, the nonequilibrium distribution does not last a long time, or, put another way, it is difficult to *maintain* a nonequilibrium velocity distribution in the absence of any external force.

⟹ **RELATED PROBLEM 25.3**

Chapter 24 introduced the mean free path, λ, in Eq. (24.16), and defined λ as the mean distance a molecule travels between collisions. The hard-sphere collision cross section, σ, was used there to define a collision, but the collision cross section is defined for any intermolecular force. In general, as we will see in Chapter 28, the cross section is a function of the relative speed, g, of the collision partners. Slower collisions allow longer times during which intermolecular attractions, even weak ones, can act and cause two molecules' trajectories to deflect from a straight line, which is the hallmark of a collision. Faster collisions are less aware of intermolecular attractions and generally have reduced collision cross sections.

This suggests we can define a velocity-specific free path—the average distance a molecule with a *specific* velocity travels between collisions—and relate it to our velocity-specific collision time, τ_{coll}. The relationship is very simple:

$$\underset{\substack{\uparrow \\ \text{Free path of a molecule} \\ \text{with velocity } \mathbf{v}}}{\lambda(\mathbf{v})} = \underset{\substack{\uparrow \\ \text{Speed corresponding} \\ \text{to velocity } \mathbf{v}}}{v} \underset{\substack{\uparrow \\ \text{Time between collisions for} \\ \text{a molecule with velocity } \mathbf{v}}}{\tau_{\text{coll}}(\mathbf{v})} \quad (25.7)$$

since a molecule that just had a collision and leaves it with velocity $\mathbf{v}$ will move with speed $v = |\mathbf{v}|$ for a time $\tau_{\text{coll}}(\mathbf{v})$, covering a distance $v\tau_{\text{coll}}$ before it has its next collision. The expression for λ in Chapter 24 is the average of this quantity over all velocities. Here, we consider that average in terms of the collision time.

If τ_{coll} is the time between collisions, its reciprocal, τ_{coll}^{-1}, is the number of collisions per unit time, or the *collision rate*. Consider molecules with velocity $\mathbf{v}_1$ colliding with molecules of velocity $\mathbf{v}$ and thus with relative collision speed $g = |\mathbf{v}_1 - \mathbf{v}|$. The *flux* of the $\mathbf{v}_1$ molecules toward their collision partners is gn where n is the number density of target molecules, as we can see if we choose the right reference frame. If we ride along with a molecule having velocity $\mathbf{v}_1$, we see the targets rushing at us with speed g. There are n of these per unit volume, and every second, gn of them cross unit area towards us, which is a flux. The product of flux and cross section is the collision rate:

$$\underset{\substack{\uparrow \\ \text{Collision rate for} \\ \text{molecules with} \\ \text{velocity } \mathbf{v}_1}}{\tau_{\text{coll}}^{-1}(\mathbf{v}_1)} = \underset{\substack{\uparrow \\ \text{Relative collision} \\ \text{speed, } g = |\mathbf{v}_1 - \mathbf{v}|}}{g} \underset{\substack{\uparrow \\ \text{Number density} \\ \text{of target molecules} \\ \text{with velocity } \mathbf{v}}}{n} \underset{\substack{\uparrow \\ \text{Collision} \\ \text{cross section} \\ \text{for speed } g}}{\sigma(g)} . \quad (25.8)$$

Figure 25.2 shows how the collision cross section represents the fraction of a unit area that causes collisions as a flux of molecules passes through it. In time dt, all

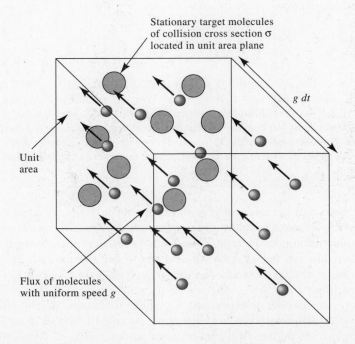

FIGURE 25.2 All the molecules within a distance $g\,dt$ of the unit area containing target molecules where g is their speed relative to the targets pass through the area or strike a target molecule.

molecules a distance $g\,dt$ or less from the unit area and moving with speed g towards it will pass through unless they collide with a target molecule in that area.

If the product $g\sigma(g)$ is averaged over all velocities (weighting the average with the velocity distribution function), we find an expression for $\langle \tau_{\text{coll}}^{-1} \rangle \equiv z_1$, the *mean collision rate for one molecule of type 1 in a gas of these molecules*:

$$z_1 \;=\; \int g\,\sigma(g)\,F^*(\mathbf{v})\,d^3\mathbf{v} \;=\; n\,\langle g\sigma \rangle, \qquad (25.9)$$

$$\underset{\text{Mean collision rate}}{\uparrow} \qquad \underset{\text{Weighted average of } g\sigma}{\uparrow} \qquad \underset{\text{Number density}}{\uparrow} \qquad \underset{\text{Mean value of } g\sigma}{\uparrow}$$

which is a constant for an equilibrium system. If the system is not at equilibrium, n can vary in time and/or space.

We can see that this expression leads to the mean free path expression in Eq. (24.16) if we substitute it into Eq. (25.7) and average over equilibrium conditions:

$$\lambda \;=\; \int \lambda(\mathbf{v})\,f(\mathbf{v})\,d^3\mathbf{v} \;=\; \langle v\,\tau_{\text{coll}} \rangle \;=\; \frac{\langle v \rangle}{n\langle g\sigma \rangle}.$$

If σ is independent of velocity (as it is for hard spheres), $\langle g\sigma \rangle = \sigma\langle g \rangle$. If the molecules are all the same, $\langle g \rangle = \sqrt{2}\langle v \rangle$ from Eq. (24.15). At equilibrium, $n = N/V$. Altogether, these conditions lead to $\lambda = \langle v \rangle V/N\sqrt{2}\langle v \rangle\sigma = V/\sqrt{2}N\sigma$, which is Eq. (24.16).

EXAMPLE 25.3

What is the mean collision rate for 1.00 mol of Ar(g) at equilibrium at 300 K and 1.00 atm pressure? Which increases the collision rate more: doubling the pressure at constant temperature, or doubling the temperature at constant pressure?

SOLUTION We will assume a hard-sphere cross section for Ar based on the Ar_2 bond length in Table 24.4 and the Lennard-Jones potential relation: hard-sphere diameter $d = R_e/2^{1/6}$. This gives $d_{Ar} = 3.35$ Å and $\sigma = 3.53 \times 10^{-19}$ m². Since the gas is pure Ar at equilibrium, $\langle g \rangle = \sqrt{2}\langle v \rangle$, and from Table 24.1, $\langle v \rangle = (8k_B T/\pi m)^{1/2}$. Also, $n = N/V = P/k_B T$. Equation (25.9) is

$$z_1 \;=\; \frac{P}{k_B T}\sqrt{2}\left(\frac{8k_B T}{\pi m}\right)^{1/2}\sigma \;=\; P\sigma\left(\frac{16}{\pi k_B T m}\right)^{1/2},$$

and with $P = 1.00$ atm $= 101\,325$ Pa, $T = 300$ K, and $m = 6.63 \times 10^{-26}$ kg, the mass of one Ar atom, we find $z_1 = 4.86 \times 10^9$ s^{-1}. We also see that doubling the pressure at constant T doubles the collision rate (because it doubles the density without changing the velocity distribution), but doubling the temperature at constant pressure *lowers* the collision rate since the system has to expand to keep P constant if T is increased. Even though the mean speed increases with increasing T, the density falls faster and the collision rate is less.

▸ **RELATED PROBLEMS** 25.4, 25.7, 25.8

The mean collision rate will play an important role when we consider the rates of gas-phase chemical reactions in later chapters. If a reaction can occur on every collision, collision rates indicate upper limits to reaction rates. Some reactions do proceed at this rate, but most do not, and one of our goals in later chapters is an understanding of the various reasons they do not. Problem 25.6 considers this type of comparison.

Finally, we consider collisions between unlike molecules and the total collision rate per unit volume in any gas. In a binary gas mixture (call the gases 1 and 2), we have to distinguish among three different collision cross sections, σ_{11} for 1–1 collisions,

σ_{22} for 2–2 collisions, and σ_{12} for 1–2 collisions. If 1 and 2 are hard spheres with diameters d_1 and d_2, then $\sigma_{ii} = \pi d_i^2$, as usual, and $\sigma_{12} = \pi(d_1 + d_2)^2/4 = \pi(r_1 + r_2)^2$ where r_i is the radius of sphere i. The mean collision rate of molecules of type 1 with molecules of type 2, which we write z_{12}, is

$$z_{12} = n_2 \langle g \rangle \sigma_{12} = n_2 \left(\frac{8k_B T}{\pi \mu_{12}}\right)^{1/2} \sigma_{12} \qquad (25.10)$$

where μ_{12} is the reduced mass of molecules 1 and 2. Note that this rate depends on the density of type 2 molecules, the *targets*, which makes sense if one considers a system with $n_1 \gg n_2$. If 2 is the minor species, the $2 \to 1$ collision rate will be large, reflecting the higher density of 1 while the $1 \to 2$ collision rate will be small. Type 1 molecules would have few targets in this case.

The total collision rate *per unit volume*, Z, is simply the rate per molecule times the number density of those molecules. If there is only one component in the gas, this rate at equilibrium is

$$Z_{11} = \frac{1}{2} n z_1 = \frac{1}{2} \langle g\sigma \rangle n^2 = \frac{\sigma \langle v \rangle}{\sqrt{2}} n^2 \qquad (25.11)$$

where the factor of $\frac{1}{2}$ ensures that each collision is counted once, not twice. The final equality is strictly true only for constant cross sections, but it is a useful and accurate approximation in general.

If the molecules are different, the factor of $\frac{1}{2}$ is no longer needed, and we write

$$Z_{12} = n_1 z_{12} = n_2 z_{21} = \langle g\sigma \rangle n_1 n_2 \qquad (25.12)$$

where the approximation $\langle g\sigma \rangle \cong \langle g \rangle \sigma_{12}$ is also useful here. The mean relative speed depends explicitly on the nature of the colliding molecules through the reduced mass, μ_{12}. Table 25.1 summarizes these expressions, including versions of them that are useful for practical calculations.

TABLE 25.1 Collision Rate Expressions

Pure-gas mean collision rate per molecule:

$$z_1 = n \langle g\sigma \rangle = P\sigma \left(\frac{16}{\pi m k_B T}\right)^{1/2}$$
$$= 1.490 \times 10^{30} \text{ s}^{-1} (P/\text{bar}) (\sigma/\text{m}^2) (M/\text{g mol}^{-1})^{-1/2} (T/\text{K})^{-1/2}$$

Pure-gas collision rate per unit volume:

$$Z_{11} = \frac{n^2 \langle g\sigma \rangle}{2} = P^2 \sigma \left(\frac{4}{\pi m}\right)^{1/2} (k_B T)^{-3/2}$$
$$= 5.398 \times 10^{57} \text{ m}^{-3} \text{ s}^{-1} (P/\text{bar})^2 (\sigma/\text{m}^2)(M/\text{g mol}^{-1})^{-1/2}(T/\text{K})^{-3/2}$$

Gas-mixture collision rate per molecule:

$$z_{12} = n_2 \langle g\sigma \rangle = P_2 \sigma_{12} \left(\frac{8}{\pi \mu k_B T}\right)^{1/2}$$
$$= 1.054 \times 10^{30} \text{ s}^{-1} (P_2/\text{bar}) (\sigma_{12}/\text{m}^2) \{[M_1 M_2/(M_1 + M_2)]/\text{g mol}^{-1}\}^{-1/2} (T/\text{K})^{-1/2}$$

Gas-mixture collision rate per unit volume:

$$Z_{12} = n_1 n_2 \langle g\sigma \rangle = P_1 P_2 \sigma_{12} \left(\frac{8}{\pi \mu_{12}}\right)^{1/2} (k_B T)^{-3/2}$$
$$= 7.633 \times 10^{57} \text{ m}^{-3} \text{ s}^{-1} (P_1/\text{bar}) (P_2/\text{bar}) (\sigma_{12}/\text{m}^2) \{[M_1 M_2/(M_1 + M_2)]/\text{g mol}^{-1}\}^{-1/2} (T/\text{K})^{-3/2}$$

EXAMPLE 25.4

A sample of Ar is confined to a cube of 1.00 m edge length at 300 K and 1.00 bar pressure. How many collisions occur per second in this volume? How many collisions per second are there with the cube's walls?

SOLUTION The collision cross section is given in Example 25.3, and from Table 25.1, the collision rate per unit volume equals

$$Z_{11} = 5.398 \times 10^{57} \text{ m}^{-3} \text{ s}^{-1} (1.00)^2 (3.53 \times 10^{-19})(39.948)^{-1/2} (300)^{-3/2}$$
$$= 5.79 \times 10^{34} \text{ m}^{-3} \text{ s}^{-1}$$

while the total wall collision rate is the cube surface area, $A = 6.00 \text{ m}^2$, times the wall collision rate, J_x, given by Eq. (24.13):

$$AJ_x = \frac{AP}{(2\pi m k_B T)^{1/2}} = \frac{(6.00 \text{ m}^2)(10^5 \text{ Pa})}{[2\pi (6.634 \times 10^{-23} \text{ kg})(1.38 \times 10^{-23} \text{ J K}^{-1})(300 \text{ K})]^{1/2}}$$
$$= 4.57 \times 10^{26} \text{ s}^{-1}$$

There are $(5.79 \times 10^{34}/4.57 \times 10^{26}) \cong 1.3 \times 10^{8}$ gas-phase collisions for every wall collision.

➡ **RELATED PROBLEMS** 25.5, 25.6, 25.8

EXAMPLE 25.5

In a 50/50 mixture of He and Ar at a total pressure of 1.00 bar and 300 K, how many He–He collisions are there per Ar–Ar collision? per He–Ar collision?

SOLUTION We know the Ar–Ar hard-sphere diameter and collision cross section from Example 25.3, and we will need those for He–Ar and He–He collisions. We find a hard-sphere diameter for He as we did for Ar in Example 24.3: $d = R_e/2^{1/6}$. We find $d_{He} = (2.97 \text{ Å})/2^{1/6} = 2.67$ Å and $\sigma_{HeHe} = \pi d_{He}^2 = 2.20 \times 10^{-19} \text{ m}^2$. With $d_{Ar} = 3.35$ Å, $\sigma_{HeAr} = \pi(d_{He} + d_{Ar})^2/4 = 2.82 \times 10^{-19} \text{ m}^2$. The ratio of the He–He collision rate to the Ar–Ar collision rate is quite simple here, since the gases have the same temperature and partial pressures:

$$\frac{z_{HeHe}}{z_{ArAr}} = \frac{\sigma_{HeHe}}{\sigma_{ArAr}} \left(\frac{m_{Ar}}{m_{He}}\right)^{1/2} = \frac{2.20 \times 10^{-19} \text{ m}^2}{3.53 \times 10^{-19} \text{ m}^2} \left(\frac{39.948 \text{ g mol}^{-1}}{4.002\,6 \text{ g mol}^{-1}}\right)^{1/2} = 1.97 \;.$$

The mass ratio factor is roughly $\sqrt{10} = 3.16$ (helium moves much faster than argon), but the cross section ratio is only 0.62 (helium is much smaller than argon) so that there are about two He–He collisions per Ar–Ar collision. The ratio of He–He to He–Ar collision rates involves the He–Ar reduced mass:

$$\frac{z_{HeHe}}{z_{HeAr}} = \frac{\sigma_{HeHe}}{\sigma_{HeAr}} \left(\frac{2m_{Ar}}{m_{Ar} + m_{He}}\right)^{1/2} = \frac{2.20 \times 10^{-19} \text{ m}^2}{2.82 \times 10^{-19} \text{ m}^2} \left(\frac{2(39.948) \text{ g mol}^{-1}}{43.951 \text{ g mol}^{-1}}\right)^{1/2} = 1.05 \;,$$

and here the cross section ratio (0.780) offsets the mass ratio factor (1.348) so that, per unit time, a He atom is as likely to hit another He atom as to hit an Ar atom.

➡ **RELATED PROBLEMS** 25.5, 25.6, 25.9

We will return to the role played by collision rates in nonequilibrium phenomena in Section 25.3, and Chapter 27 will consider collision rates in the context of reactive collisions, but the next section considers macroscopic nonequilibrium phenomena. Just as thermodynamics can make useful equilibrium predictions without benefit of a microscopic theory of matter, *nonequilibrium thermodynamics* provides very general statements that overlay and guide microscopic theories of nonequilibrium dynamics.

25.2 PHENOMENOLOGICAL TRANSPORT—FORCES AND FLUXES

The Second Law gives a direction to the evolution of all nonequilibrium processes: any spontaneous process occurring in an isolated system continues, increasing entropy as it does, until the entropy is as large as the constraints imposed on the system allow. This suggests that *nonequilibrium* thermodynamics has a natural starting point: the *time dependence of entropy*. Nonequilibrium thermodynamics is a more advanced topic than we can consider here, but *entropy production* plays a central role in this branch of physical chemistry.

We used entropy maximization to define thermodynamic potentials in Chapter 5 where spontaneity was related to a minimization of the appropriate potential. Thus, *chemical potential gradients* drive systems toward equilibrium, and these are often simple gradients of macroscopic system parameters such as temperature, concentration, or density. For example, the ideal gas chemical potential expression at equilibrium, Eq. (5.36),

$$\mu(T,P) = \mu°(T) + RT \ln(P/1 \text{ bar}) \tag{5.36}$$

depends on pressure and temperature. If a closed system has a temperature gradient in the x direction, the pressure can be everywhere constant (or else "winds would blow") so that the chemical potential gradient is

$$\frac{\partial \mu}{\partial x} = \frac{\partial \mu}{\partial T}\frac{\partial T}{\partial x} = \frac{\partial(\mu° + RT \ln P)}{\partial T}\frac{\partial T}{\partial x}.$$

In mechanics, the spatial derivative of potential is the negative of *force*. This suggests that spatial derivatives of thermodynamic parameters, such as $\partial T/\partial x$, act as *the forces that drive transport phenomena*.

The phenomena we discuss here are familiar—thermal conductivity, diffusion, viscosity, and electrical conductivity. Everyone has a general feeling for them, recognizing what a viscous fluid is, how diffusion causes mixing, etc., and our first goal is to place these general feelings on a quantitative footing. Fortunately, it is easy to apply one mathematical formalism to all of them. We saw in Chapter 24 how the wall collision rate, J_x in Eq. (24.13), expressed a *flux* of molecules moving through a plane in the gas. Since a wall is impenetrable, the flux striking the wall is non-zero, but the *net* flux through any plane *in* the gas is zero at equilibrium. We will approach all transport phenomena in terms of the *vector flux of a transported quantity*:

$$\text{Flux} = \mathbf{J}(\text{something}) = \frac{\text{amount of something moving in a given direction}}{(\text{unit area})(\text{unit time})}.$$

The "something" that is transported can be any physical molecular property, but we will concentrate on the following four processes:

- *diffusion*, which transports *matter*,
- *thermal conduction*, which transports *energy*,
- *electrical conduction*, which transports *charge*, and
- *viscosity*, which transports *linear momentum*.

It is also common sense that any one flux increases if the system is moved farther from equilibrium. We will assume throughout this chapter that we are sufficiently close to equilibrium that "distance from equilibrium" is measured by the magnitude of *simple gradients of macroscopic parameters*. We write in general

$$\mathbf{J}(\text{something}) = -(\text{phenomenological coefficient}) \times (\text{gradient of a macroscopic parameter}).$$

For example, the z component of the flux of matter due to diffusion is

where D is the *diffusion coefficient* and $(\partial n/\partial z)$ is the particle density gradient in the z direction. Note the minus sign. The flux is in a direction *opposite* that of the gradient.

Systems that are very far from equilibrium have fluxes that do not follow such a simple proportionality. Only gentle gradients lead to *linear* flux–gradient relations. Each gradient, in turn, acts as a driving force, sustaining the flux but also forcing it to disappear as the system evolves towards equilibrium.[3] Table 25.2 summarizes the one-dimensional flux–force relations we use here, the phenomenological coefficients that appear in each, and their SI units.

Diffusion

Diffusion usually evokes a mental picture of, for example, a drop of ink diffusing through a glass of water until the entire solution is homogeneous. This is an example of *binary diffusion*—two distinguishable species diffusing through each other. In contrast, *self-diffusion* involves only one species. Experimentally, self-diffusion is less easily measured than binary diffusion, and often one uses innocuous isotopic labels to follow it. Less obvious, however, are measurements that use light scattering

TABLE 25.2 Phenomenological Transport Equations

Diffusion	density gradient transports matter
$J_{n,z} = -D\left(\dfrac{\partial n}{\partial z}\right)$	D = diffusion coefficient (m² s⁻¹)
Thermal conduction	temperature gradient transports energy
$J_{q,z} = -k\left(\dfrac{\partial T}{\partial z}\right)$	k = thermal conductivity (W m⁻¹ K⁻¹ = J m⁻¹ K⁻¹ s⁻¹)
Electrical conduction	electric potential gradient transports charge
$J_{e,z} = -\sigma_e\left(\dfrac{\partial \phi}{\partial z}\right)$	σ_e = electrical conductivity (S m⁻¹ = A² s³ kg⁻¹ m⁻³)†
Viscosity	velocity gradient transports momentum
$J_{mv,zx} = -\eta\left(\dfrac{\partial v_x}{\partial z}\right)$	η = viscosity (Pa s = kg m⁻¹ s⁻¹)‡

†1 S = 1 siemens = 1 Ω⁻¹ = 1 ohm⁻¹ = 1 A² s³ kg⁻¹ m⁻²
‡Viscosity is often expressed in poise units where 1 poise = 1 P = 10⁻¹ Pa s.

---

[3] In a *steady-state* nonequilibrium system, the surroundings maintain the gradient (and entropy production occurs only in the surroundings).

or NMR spectra to find diffusion coefficients, techniques that are discussed in this chapter's references.

The diffusion flux–gradient expression in Table 25.2 is known as *Fick's first law*, named after the 19th century German physiologist A. E. Fick (who also invented a method to measure cardiac output). The name suggests that he found more than one law, and Fick's second law (his only other one) is easy to derive. Imagine a density gradient in the z direction such that transport is in the direction of increasing z. The flux into a region from z to $z + dz$ is given by Fick's first law, evaluated at z:

$$J_{n,z} = -D\frac{\partial n(z)}{\partial z} \qquad (25.13)$$

and the flux out of this region, at $z + dz$, is

$$J_{n,z+dz} = -D\frac{\partial n(z+dz)}{\partial z} = -D\left[\left(\frac{\partial n(z)}{\partial z}\right) + \left(\frac{\partial^2 n(z)}{\partial z^2}\right)dz\right]$$

where we have used

$$n(z + dz) = n(z) + dn(z) = n(z) + \frac{\partial n(z)}{\partial z}dz \ .$$

The difference between the flux in and the flux out, per unit distance, is the time rate of change of concentration:

$$\frac{J_{n,z} - J_{n,z+dz}}{dz} = \frac{\partial n(z)}{\partial t} \ .$$

Substituting the expressions for $J_{n,z}$ and $J_{n,z+dz}$ from above leads to *Fick's second law*, which is also known simply as the *diffusion equation*:

$$\frac{\partial n(z,t)}{\partial t} = D\frac{\partial^2 n(z,t)}{\partial z^2} \ . \qquad (25.14)$$

Note that a perfectly linear concentration gradient would appear to be stable in time: $(\partial^2 n/\partial z^2) = 0$ so that $(\partial n/\partial t) = 0$, but spontaneous fluctuations make perfectly linear gradients impossible. A highly kinked or wrinkled concentration profile with large, local values of $\partial^2 n/\partial z^2$ changes rapidly in time as the system spontaneously smooths itself.

Specific solutions to Eq. (25.14) depend on the initial conditions and shape of the system. For example, consider concentrating species 1 along the yz plane at $x = 0$ and $t = 0$ and watching it diffuse toward $x = \pm\infty$ through a uniform concentration of species 2. Experimentally, we could coat a thin plate of area A with N_0 solute molecules and slide the plate into a solvent. Equation (25.14) in this case has the solution

$$n(x,t) = \frac{N_0}{A\sqrt{4\pi Dt}}e^{-x^2/4Dt} \qquad (25.15)$$

as you can verify by direct substitution and the normalization condition

$$\int_{-\infty}^{\infty} n(x,t)\, dx = \frac{N_0}{A} \ .$$

Equation (25.15) is a Gaussian function of the form $\exp[-(\text{distance/width})^2]$ where the width parameter measures the progress of diffusion: width $= (4Dt)^{1/2}$. Note how this depends on the square root of time. As diffusion proceeds, it slows as the concentration gradient decreases. Diffusion is ultimately a very slow mixing process.

Figure 25.3 plots Eq. (25.15) as a function of distance at various times, and Figure 25.4 simulates a movie of the diffusion process. Both of these are drawn for $D = 10^{-9}$ m² s⁻¹, a typical value for small molecule diffusion in a liquid. Note that after 25 s, the diffusing species are still well within 1 mm of where they started. Since diffusion's progress scales with $(Dt)^{1/2}$, we can use these figures to predict the rate of processes with other diffusion coefficients. Quantitatively, the rms spread at time t, $x_{rms} = \langle x^2 \rangle^{1/2}$, is

$$x_{rms} = \left[\int_{-\infty}^{\infty} \frac{x^2 e^{-x^2/4Dt}}{\sqrt{4\pi Dt}} dx\right]^{1/2} = \left[\frac{4Dt}{\sqrt{\pi}} \int_{-\infty}^{\infty} y^2 e^{-y^2} dy\right]^{1/2} = (2Dt)^{1/2} \ . \quad (25.16)$$

EXAMPLE 25.6

How long must one wait for the concentration profile shown at 25 s in Figure 25.3 if (a) the diffuser is a gas with $D = 10^{-5}$ m² s⁻¹ or (b) the diffuser is a metal with $D = 10^{-19}$ m² s⁻¹? At what later time will the profile of each have doubled in width?

SOLUTION The product Dt determines the shape of a profile. Thus, attaining the 25 s profile for $D = 10^{-9}$ m² s⁻¹ takes only $(10^{-9}$ m² s⁻¹$)(25$ s$)/(10^{-5}$ m² s⁻¹$) = 2.5$ ms for the gas but $(10^{-9}$ m² s⁻¹$)(25$ s$)/(10^{-19}$ m² s⁻¹$) = 2.5 \times 10^{11}$ s $\cong 8000$ yr for the metal. Since the profile width is measured by x_{rms}, which varies like $t^{1/2}$, doubling the width takes four times as long: 100 s, 10 ms, and 32 000 yr.

→ **RELATED PROBLEMS** 25.10, 25.11, 25.12

We will look further at the phenomenology of diffusion in Section 25.4, but we close this discussion with a general *microscopic* phenomenological discussion based on the *one-dimensional random walk*. Had Albert Einstein failed to shake the world of physics with his theories of relativity, his impact on physical chemistry would have assured his fame. We have already seen his mind at work in the photoelectric effect, solid heat capacities, and Bose–Einstein statistics. His explanation of *Brownian motion* in 1905 (Polish physicist Marian von Smoluchowski independently derived an explanation at about the same time) is a classic study of diffusion based on random collisions that produce the zig-zag motion characteristic of Brownian

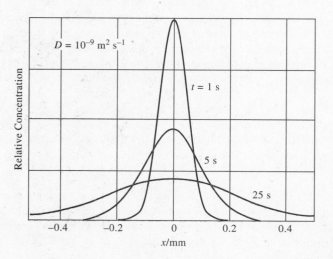

FIGURE 25.3 Concentration profiles at various times for molecules diffusing from the origin show how diffusion slows in time as the concentration gradient becomes smaller.

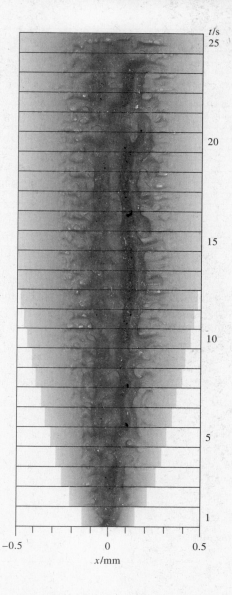

FIGURE 25.4 The diffusion process of Figure 25.3 is shown here as a series of movie frames taken at 1 s intervals.

motion.[4] This study was one of the convincing pieces of evidence supporting the existence of atoms, and Eq. (25.16) is sometimes called the *Einstein diffusion equation*.

The simplest random walk takes a particle and moves it randomly right or left a constant distance step after step. One such walk is shown in Figure 25.5. Elaborations include unequal step sizes and/or probabilities as well as multidimensions, but the essence of the problem comes from the simplest case. We ask for the probability that the particle has moved X integral steps from the origin after taking N random steps. Since any one step can be in either of two directions, N random steps can

---

[4]British botanist Robert Brown observed what became known as Brownian motion in 1827. He was observing pollen grains moving erratically in water, and the reason for that motion was a puzzle for the next 75 years.

FIGURE 25.5 A simulated 1-D random walk is represented as the unit step displacements, chosen here with equal probability in either direction, taken at each unit time interval. This is a 1-D version of Brownian motion.

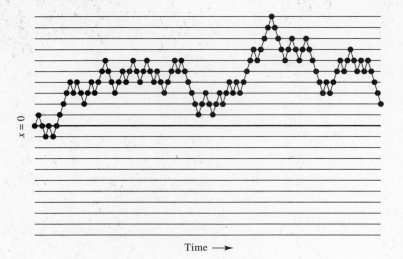

cover 2^N possible paths. The number of these with N_+ to the right and $N_- = N - N_+$ to the left is given by a familiar expression: the statistics of the two-level system, Eq. (4.5). The number, W, is

$$W = \frac{N!}{N_+!\,N_-!} = \frac{N!}{N_+!\,(N - N_+)!}\,. \tag{4.5}$$

Since each step covers unit length, the final position is

$$X = N_+ - N_- = N_+ - (N - N_+) = 2N_+ - N$$

or

$$N_+ = \frac{1}{2}(N + X)\,,$$

and since $N_+ + N_- = N$, we also have

$$N_- = N - N_+ = \frac{1}{2}(N - X)\,.$$

Equation (4.5) becomes

$$W = \frac{N!}{\left[\frac{1}{2}(N + X)\right]!\left[\frac{1}{2}(N - X)\right]!}$$

so that the probability, P, of ending a distance X integral steps to the right *or* left of the origin after N random steps is the number of ways of reaching X (which is W) divided by all the paths N steps could make (which is 2^N):

$$P = \frac{W}{2^N} = \frac{N!}{2^N\left[\frac{1}{2}(N + X)\right]!\left[\frac{1}{2}(N - X)\right]!}\,. \tag{25.17}$$

It is probably not obvious to you that this equation and Eq. (25.15), derived from Fick's second law, have anything in common, but they are in fact nearly identical. If we let the particle take very many steps so that N is large, we can use Stirling's approximation for the factorials (see Problem 25.13), and Eq. (25.17) becomes

$$P = \left(\frac{2}{\pi N}\right)^{1/2} e^{-X^2/2N}\,, \tag{25.18}$$

which has the same Gaussian form as Eq. (25.15). The difference between Eq. (25.15) and Eq. (25.18) is only one of normalization. The integral of Eq. (25.15) from $-\infty$ to ∞ equals 1, but the same integral of Eq. (25.18) equals 2, since it has a built-in bias: X is an even number only if N is an even number (see Figure 25.5 with $N = 100$ and $X = 2$) and X is odd if N is odd. Dividing Eq. (25.18) by 2 brings it into direct correspondence with Eq. (25.15):

$$\frac{1}{2}\left(\frac{2}{\pi N}\right)^{1/2} e^{-X^2/2N} = \left(\frac{1}{4\pi Dt}\right)^{1/2} e^{-x^2/4Dt},$$

and we can see that $X^2/2N$ corresponds to $x^2/4Dt$.

If each random step covers a distance x_0, so that $Xx_0 = x$, and if the time between steps is τ, so that $N = t/\tau$, we have

$$\frac{X^2}{2N} = \frac{x^2\tau}{2x_0^2 t} = \frac{x^2}{4Dt}$$

or

$$D = \frac{x_0^2}{2\tau} \qquad (25.19)$$

known as the *Einstein–Smoluchowski equation*. If we associate x_0 with λ, the mean free path, and τ with the mean time between collisions, $\langle \tau_{\text{coll}} \rangle = \lambda/\langle v \rangle$ where $\langle v \rangle$ is the mean speed, we find $D = \langle v \rangle \lambda/2$. We will return to this expression in the next section to see how it compares to a gas kinetic expression for D, and in the final section, we will find another expression for D based on the motion of ions in solution. Equation (25.19), however, shows not only that diffusion is faster the longer the step taken between collisions and the shorter the time between collisions (both fairly obvious), but also that it increases with the *square* of the step size (not so obvious).

The random walk is very easy to simulate on a computer with a good random number generator.[5] Figure 25.6 compares Eq. (25.18) to the results of a 2000-step random walk, repeated 2000 times. For each walk, the final displacement, X, was recorded, and the figure compares a histogram of those results to Eq. (25.18). The agreement is good, but the *disagreement* shows the importance of sampling over *very* many random walks in computer simulations. The difference between the analytic theory and the simulation varies roughly as $\sqrt{N}$ so that to decrease the error in the figure by a factor of ten requires 100 times as many simulations. Typically several million repetitions are used in research simulations.

A Closer Look at Concentration Gradients

While we have discussed diffusion as if only a concentration gradient can produce diffusion, we must remember that all transport properties are driven by *chemical potential gradients* which, in nonideal systems, can spontaneously *produce* concentration gradients. We have seen an example of this behavior in Chapter 6 in our

---

[5] No computer algorithm can generate truly random numbers, and writing an algorithm that does a very good job of simulating randomness, and does so quickly, is tricky. Computers do have the advantage that they can generate the *same* sequence of "random" numbers when asked to, a trait that makes testing computer code that depends on random numbers somewhat easier.

FIGURE 25.6 A histogram of the final displacement of a 2000-step 1-D random walk, repeated randomly 2000 times, is compared to Eq. (25.28), the limiting probability for an infinite number of steps (the solid line).

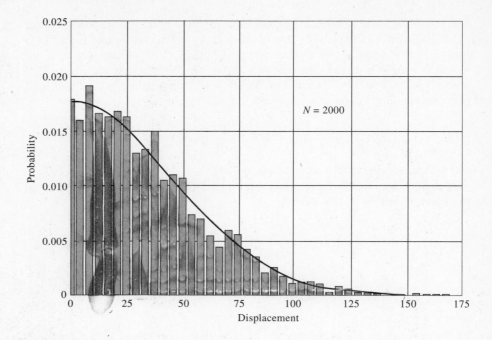

discussion of the Sn/Pb phase diagram. Freezing a homogeneous eutectic liquid solution spontaneously produces two mixed solid phases with very steep concentration gradients across phase boundaries, as shown in Figure 6.27.

Diffusion in solids is a major topic of materials science. Diffusion is often controlled by (or produces) defects of the type discussed in Section 16.4. Amorphous solids (such as polymers and glasses) have very high diffusion rates for small particles, such as light alkali atoms in glasses and H_2 and He in glasses and polymers. One can purify H_2 if the impure gas is allowed to diffuse through a thin sample of glass: the H_2 diffuses through rapidly, leaving the impurities (except for He) behind. Many metals (Pd is the best studied—see Problem 25.12) appear porous to H_2, and the metal often changes properties as a result of H_2 diffusion through it. Chemical reactions (metal hydride formation) can induce defects that make the metal brittle, for example.

When two pure metals or alloys are allowed to diffuse into each other and one diffuses much more rapidly than the other, unusual concentration gradients can result, called the *Hartley–Kirkendall effect*. Call the two samples A and B with B the faster moving. Initially, the A–B *diffusion couple* (as it is called by materials scientists) has a distinct interface:

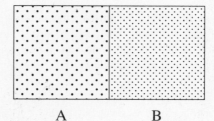

As the faster B atoms[6] diffuse, the interface appears to move toward them:

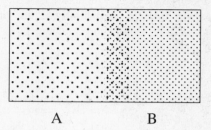

A B

and in some cases, the diffusion is so biased towards motion of B that significant voids appear in B:

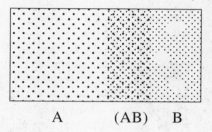

A (AB) B

In this extreme, it makes sense to consider atomic *vacancies* in the lattice as particles and to consider the diffusion coefficient of the vacancies.

The mutual diffusion coefficient in this case (and in general—the expression is not limited to solids) is proportional to $x_A D_B + x_B D_A$ where x_A and x_B are mole fractions and D_A and D_B are self-diffusion coefficients. As fast B ($D_B \gg D_A$) moves into pure A ($x_A \gg x_B$), diffusion is dominated by B, and $D_{AB} \cong D_B$. Slow A moving into pure B continues to move slowly ($x_A D_B + x_B D_A \cong D_A$). In effect, B gives A a boost, but A cannot keep up: the interface moves toward the B region, and ultimately B cannot keep up with its own desires and begins to deplete itself, forming voids.

Thermal Conduction

We encounter thermal conductivity in our everyday life far more often than we do diffusion. We don't drink hot drinks from silver chalices. We insulate our homes against both the winter cold and the summer heat. We understand why a down comforter is warmer than a silk sheet. Here, we consider the thermal conductivity, the number that makes these phenomena quantitative.

The flux–gradient expression in Table 25.2 is known as *Fourier's law*. Jean Fourier is perhaps better known for the mathematical techniques he discovered: the Fourier series and the Fourier transform, but these grew out of his study of heat. His 1822 book, *Théorie Analytique de la Chaleur*, (Analytic Theory of Heat), introduced not only his law but also much of the mathematics that has gained greater fame. At the time, heat was not well understood (Sadi Carnot's book that led to the Second Law was two years away), and Fourier's work was not widely believed.

---

[6]The diffusing species do not have to be atoms. Hartley first observed the effect between cellulose acetate and acetone, and Kirkendall used copper and brass.

The caloric theory of heat was still popular, and Fourier deftly side-stepped the controversy, refusing to speculate on the nature of heat.

Fourier's law makes the thermal conductivity, k, the proportionality constant between the energy flux and the temperature gradient:

$$J_{q,z} = -k \frac{\partial T}{\partial z}, \qquad (25.20)$$

where $J_{q,z}$ is the energy flux (J m^{-2} s^{-1}), k is the thermal conductivity, and $\partial T/\partial z$ is the temperature gradient.

and we consider the magnitude and general properties of k here. Thermal conductivities of the elements near 300 K span the range of k values found for most compounds, and, as you would expect, they reflect periodic trends. Figure 25.7 plots them on a logarithmic scale; they span nearly six orders of magnitude.

Metals generally have values in the range 10–300 W m^{-1} K^{-1}, peaking at Cu, Ag, and Au, but note that diamond has the highest value of all, around 1370 W m^{-1} K^{-1} near room termperature, compared to graphite's 129 W m^{-1} K^{-1} or silver's 430 W m^{-1} K^{-1}. Molecular solid thermal conductivities are significantly lower: 13.2 W m^{-1} K^{-1} for red phosphorus, 0.24 W m^{-1} K^{-1} for white phosphorus, 0.44 W m^{-1} K^{-1} for I$_2$, and so on. These are only representative values in many cases, since k depends on physical state of the sample—polycrystalline or single-crystal, large crystals or small, etc.—and on the direction through a single crystal for anisotropic solids—parallel to the carbon sheets in graphite or perpendicular to them, etc. Electrically insulating solids conduct heat through lattice-atom vibrations called *phonons*. Metals and semiconductors use this mechanism along with free-electron motion to conduct heat. The electron mechanism dominates when it is available.

Liquids have thermal conductivities in the range 0.03 W m^{-1} K^{-1} for ^{4}He(ℓ) at 4.2 K, to 0.10–0.15 W m^{-1} K^{-1} for most nonassociated molecular liquids, to

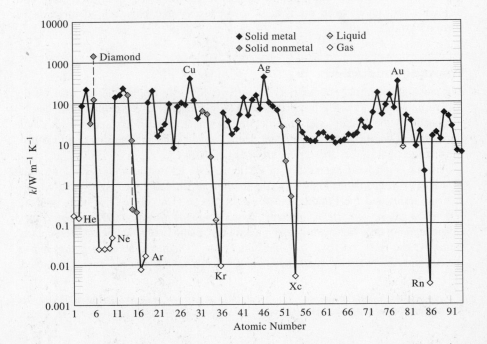

FIGURE 25.7 The thermal conductivities of the elements span nearly six orders of magnitude. Note the periodic variations, the general decrease with increasing mass, and the variations with phase.

0.61 W m^{-1} K^{-1} for water, to values >10 W m^{-1} K^{-1} for liquid metals. Thermal conduction in liquids (and in solids, to a degree) is related to the speed of sound, at least in mechanism. The speed of sound in a liquid is typically five to ten times larger than the corresponding gas speed given by Eq. (24.6). This increase can be explained if we assume that each collision in a liquid transfers energy instantaneously from one molecule to the next. This is also the mechanism of thermal conductivity. We will see in the next section that k for a gas is proportional to the heat capacity. This is also larger for liquids than for solids, providing another factor that leads to an increase in thermal conductivity over the gas value.

Gas values must depend on pressure, since "vacuum bottles" are good thermal insulators, but over wide ranges of pressure above the "vacuum" region, gas thermal conductivities are independent of pressure. We will say much more about gases in the next section and explain this observation.

Thermal conductivities also vary with temperature, generally increasing with increasing T above room temperature. Experimental values are often fit to a power series in T, and our model for gas thermal conductivity in the next section predicts a simple temperature dependence. Below room temperature, thermal conductivities of single crystal materials often rise dramatically to a maximum in the vicinity of 15–50 K, then fall rapidly at lower T.[7] For example, k(Cu) $\cong$ 400 W m^{-1} K^{-1} near room temperature, but around 20 K, k(Cu) $\cong$ 5000 W m^{-1} K^{-1}. This is a dramatic increase, but synthetic single-crystal sapphire (Al$_2$O$_3$) does even better. It has k = 37 W m^{-1} K^{-1} at room temperature (a modest, typical value for an ionic solid), but around 30 K, the sapphire thermal conductivity has risen to an enormous 6000 W m^{-1} K^{-1}. Alloys generally have much lower thermal conductivities than pure metals, and these dramatic differences are exploited in the design of low-temperature experimental apparatus.

EXAMPLE 25.7

You are designing an experiment at liquid He temperature (4.2 K), and you need to support your sample chamber with rods that run from it to walls that you can easily maintain at liquid-nitrogen temperature (77 K). You find that either pyrex glass, stainless steel, or copper (among the materials readily available in your shop) has the mechanical strength you need. You connect a N$_2(\ell)$ reservoir to a He(ℓ) reservoir (through vacuum!) with 10 cm × 1 mm^2 bars of each material and measure the rate of He evaporation due to thermal conduction from the N$_2(\ell)$ reservoir. You find the pyrex rate is 2.13 μmol s^{-1}, the stainless steel rate is 39.0 μmol s^{-1}, and the Cu rate is 8400 μmol s^{-1}. Given the molar enthalpy of evaporation of He, 84.5 J mol^{-1}, what is the thermal conductivity of each material?

SOLUTION The evaporation rates and $\Delta \overline{H}^\circ_{vap}$ tell us the heat-flow rate in each experiment: heat-flow rate/W = (evaporation rate/mol s^{-1}) × ($\Delta \overline{H}^\circ_{vap}$/J mol^{-1}). These rates are 1.80 × 10^{-4} W for pyrex, 3.30 × 10^{-3} W for stainless steel, and 0.71 W for Cu. Since the thermal conductivity is a function of temperature, we can calculate only the average k over the 77–4.2 K temperature range. The temperature gradient is $\Delta T/L$ where L is the length, 0.1 m, and ΔT is (77 K − 4.2 K) = 72.8 K, or $\Delta T/L$ = 728 K m^{-1}. Fourier's law multiplied by the sample cross-sectional area, A = 1 mm^2 = 10^{-6} m^2, is what we want:

$$\text{Heat flow} = J_{q,x} A = -kA \frac{\Delta T}{L}$$

---

[7]Debye, who advanced our understanding of low-temperature solid heat capacities, also contributed to an understanding of cryogenic thermal conductivity.

or, since we can ignore algebraic signs because we know the heat flow is in the opposite direction from the gradient,

$$k = \frac{(\text{Heat flow}) L}{A \Delta T}.$$

This expression yields $k(\text{pyrex}) = 0.25$ W m^{-1} K^{-1}, $k(\text{stainless}) = 4.5$ W m^{-1} K^{-1}, and $k(\text{Cu}) = 975$ W m^{-1} K^{-1}. One might guess that pyrex would be the best thermal insulator, but one might not guess that stainless steel is almost as good. Hold hot or cold water in both a stainless steel and a pyrex beaker to experience this.

⇒ *RELATED PROBLEMS* 25.16, 25.17

We will not go into solutions of Fourier's law in detail. The temperature dependence of k makes them difficult, and when we can assume k is constant (because we know it is or because the temperature span is relatively small), we usually have a simple equation to solve directly for either the temperature gradient or the energy flux. Methods that yield the entire temperature profile are discussed in this chapter's references.

Electrical Conduction

Metals conduct electricity well, insulators do not, and semiconductors are in between. This much you probably already know, and our discussion of metals in Section 16.2 described the microscopic mechanisms of conduction in terms of conduction electrons and band theory. The flux–gradient equation for charge transport is known as *Ohm's law*. You may have learned a variant of Ohm's law in your study of elementary electrical phenomena where it is often expressed in the practical form $E = IR$, "electromotive force (voltage), E, equals resistance, R, times current I." Resistance and conductance are reciprocals of each other. Electrical resistance is measured in ohms, Ω, and for many years, the unit for electrical conductance was the "mho"—"ohm" spelled backwards—symbolized ℧, an upside-down Ω. The modern SI unit for electrical conductance is the siemens,[8] symbolized S, where $1 \text{ S} = 1 \text{ }\Omega^{-1} = 1 \text{ A V}^{-1}$.

The bulk resistance of something depends on its size. Consider how one measures the resistance of a sample of cross-sectional area A as shown in Figure 25.8. A

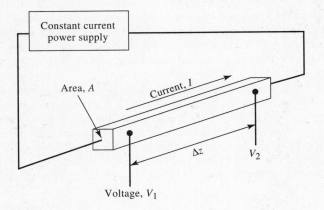

FIGURE 25.8 The resistance, R, of a sample is the measured voltage difference, $V_2 - V_1$, divided by the current, I. The resistivity is related to the resistance through $\rho_e = AR/\Delta z$.

---

[8] Note the final "s" in the spelling of siemens. The Siemens family produced several inventive brothers who founded a company in the mid-1800s that became the German industrial firm Siemens AG. One of the brothers, Ernst W. von Siemens, was the first to exploit, in 1861, the temperature dependence of resistivity with his platinum resistance thermometer.

constant current, I, is passed through the sample, and the voltage difference, $V_2 - V_1$, is measured between two points spaced a distance Δz along the current flow direction. The Ohm's law resistance is simply $R = E/I = |V_2 - V_1|/I$. Just as water flow through a pipe (analogous to electric current) increases at a constant pressure drop (analogous to a voltage difference) if the pipe is shortened or if its area is increased, electric current increases if the sample is made shorter or fatter. The current density in the sample, $J_{e,z}$, equals I/A, and the magnitude of the electric field is $|V_2 - V_1|/\Delta z$. Since electric field is the negative of the potential gradient, the flux-gradient version of Ohm's law is

$$J_{e,z} = \frac{I}{A} = \frac{\sigma_e |V_2 - V_1|}{\Delta z} \quad (25.21)$$

and the resistance, conductivity σ_e, and resistivity ρ_e, are related through

$$R = \frac{\Delta z}{\sigma_e A} = \frac{\rho_e \Delta z}{A}. \quad (25.22)$$

Electrical conductivities (units: S m^{-1}) of ordinary materials span over 24 orders of magnitude. If superconductors are added to the list, the span is infinitely greater, since superconductors have infinite conductivity (at least over limited current spans). Ordinary solids have σ_e values ranging from the high values of metals (Ag: 6.3×10^7 S m^{-1}; Cu: 5.85×10^7 S m^{-1}; stainless steel: 0.14×10^7 S m^{-1}) down through graphite ($\sim 10^5$ S m^{-1}), the semiconductors (diamond: $<10^{-8}$ S m^{-1}; Si: 2.52×10^{-4}; S m^{-1}; Ge: 1.45 S m^{-1}; gray Sn: 136 S m^{-1}), iodine (7.7×10^{-8} S m^{-1}), sulfur (5×10^{-16} S m^{-1}), and polymers such as polyethylene and polytetrafluoroethylene ($\sim 10^{-15}$–10^{-17} S m^{-1}). The conductivity of liquid electrolytes is a special topic we will consider in the final section. Gases have essentially zero conductivities at low electric fields, but they "break down" at fields sufficiently high to ionize the gas and then conduct quite well. The next section considers ionized gases, *plasmas*, in more detail.

Metal and semiconductor conductivities have different temperature dependencies. For metals above the Debye temperature, the resistivity ($\rho_e = 1/\sigma_e$) *increases* with increasing T in an approximately linear way. If we let ρ_0 be the resistivity at a reference temperature T_0 (such as room temperature), then $\rho_e(T) \cong \rho_0[1 + \alpha(T - T_0)]$ where α is the temperature coefficient of resistivity, usually in the range 0.002–0.008 K^{-1}. Above the Debye temperature, resistivity is controlled by free-electron scattering from vibrating lattice atoms. At very low T, the lattice imperfections and impurities control electron scattering, leading to a so-called residual resistivity. As in elementary electrical circuits in which the total resistance of two resistors connected in series is the sum of the individual resistances, the residual resistivity and the intrinsic resistivity add to produce the total resistivity. A simple measure of purity and structural perfection is the ratio of the room-temperature resistivity to the 4.2 K, He(ℓ), resistivity. At 4.2 K, (unless the material has become superconducting), only the residual resistivity remains. Since this quantity is smallest for highly pure and perfect materials, the $\rho_{300}/\rho_{4.2}$ ratio is largest for them. Values as large as 10^3 or more are possible for carefully purified metals, but values as low as 1 are observed for some alloys.

EXAMPLE 25.8

At room temperature, the resistivity of dilute Cu–Ni alloys increases approximately 1.2×10^{-8} Ω m for every 1% by weight increase in Ni concentration. The resistivity of pure Cu is 1.68×10^{-8} Ω m at 300 K and 0.35×10^{-8} Ω m at 100 K. What is the resistance at

250 K of a 10 cm length of a 1.0 mm² cross-sectional area wire of a Cu-Ni alloy that is 2.5% Ni?

SOLUTION We first find the temperature coefficient of resistivity of pure Cu. Since $\rho_{100} = \rho_{300}[1 + \alpha(100 \text{ K} - 300 \text{ K})]$, we find

$$\alpha = \frac{\rho_{100}/\rho_{300} - 1}{100 \text{ K} - 300 \text{ K}} = \frac{0.35/1.68 - 1}{100 \text{ K} - 300 \text{ K}} = 3.96 \times 10^{-3} \text{ K}^{-1}$$

so that $\rho_{250} = (1.68 \times 10^{-8} \, \Omega \text{ m})[1 + (3.96 \times 10^{-3} \text{ K}^{-1})(250 \text{ K} - 300 \text{ K})] = 1.35 \times 10^{-8} \, \Omega$ m. Alloying adds a residual resistivity directly to this intrinsic amount. We add $(1.2 \times 10^{-8} \, \Omega \text{ m \%}^{-1})(2.5\%) = 3.00 \times 10^{-8} \, \Omega$ m for a total resistivity $\rho_e = 4.35 \times 10^{-8} \, \Omega$ m. From Eq. (25.22), $R = \rho_e \Delta z/A$ where $A = 1.0$ mm² $= 1.0 \times 10^{-6}$ m² and $\Delta z = 10$ cm $= 0.1$ m. We find $R = (4.35 \times 10^{-8} \, \Omega \text{ m})(0.1 \text{ m})/(10^{-6} \text{ m}^2) = 4.35 \times 10^{-3} \, \Omega$. The resistance of the same size wire of pure Cu at 300 K is $1.68 \times 10^{-3} \, \Omega$, about 2.5 times less.

⇒ **RELATED PROBLEMS** 25.18, 25.19

Electrical conduction in semiconductors has two components: the motion of electrons with sufficient energy to occupy the *conduction band* and the motion (in the opposite direction) of the *holes* in the *valence band* left behind by those electrons. Pure semiconductors (also called *intrinsic* semiconductors) increase their conductivity with increasing T as thermal energy populates the conduction band, creating more electron–hole pairs. Doped semiconductors, in which an impurity is added to increase the number of holes or electrons through direct chemical stoichiometry—one Ga atom added to Ge adds one hole, and one As atom added to Ge adds one electron—also increase conductivity with increasing T. The increase roughly follows a simple Boltzmann expression, $\exp(-\Delta E_{\text{gap}}/k_B T)$, where ΔE_{gap} is the valence-band–conduction-band energy gap. Pure Ge (or Si) can be used as a low-temperature thermometer, since its resistivity can be accurately measured without dissipating much energy (the dissipated energy is $EI = I^2 R$), valuable if one does not want to heat the object in contact with the thermometer. Commercial carbon resistors are also used, even into the mK temperature range. Power dissipation as low as a few fW and an exponential dependence of ρ on T make these devices attractive.

Viscosity

Everyone understands qualitative differences in viscosity: Vermont maple syrup is more viscous than California white wine. Quantitatively, viscosity can be a difficult concept to grasp. There are reasons (which we will not consider) to define different types of viscosity (ours is sometimes called *dynamic* or *shear* viscosity), and the phenomenological flux–gradient equation for momentum transport, called *Newton's law of viscous flow*, is slightly more complicated than for other transport phenomena. As we explore this equation, we will learn not only about viscosity but also more about pressure.

The flux of momentum is intimately related to pressure. We begin to see this if we consider the dimensions of the quantities in Newton's viscous law:

$$J_{mv,zx} = -\eta \, \frac{\partial v_x}{\partial z} \quad (25.23)$$

↑ Flux of the x component of momentum in the z direction

↑ Coefficient of viscosity

↑ Gradient in the z direction of the x component of velocity

Consider the dimensions of the flux: momentum per unit time per unit area. From Newton's better known second law of dynamics, $F = d(mv)/dt$, momentum per unit time is a force, and a force per unit area is a pressure. Thus, pressure is another interpretation for momentum flux, but we seem to have a strange kind of pressure here. The area is perpendicular to the direction of the flux, but the pressure is *parallel* to the area.

We can write *nine versions* of Eq. (25.23) if we cycle x and z through the other Cartesian directions. For instance, $J_{mv,yz}$ represents the flux of the z component of momentum along the y direction, and among the nine versions, three have identical axes labels, such as $J_{mv,xx}$. A quantity that can be written in these nine different ways is said to form a *second-rank tensor*, a higher dimensional generalization of a simple three-component vector. In a fluid that is not flowing at all, the three fluxes $J_{mv,xx}$, $J_{mv,yy}$, and $J_{mv,zz}$ (called the *diagonal elements* of the tensor) are equal to each other and represent the equilibrium, static pressure.

From here, one can develop several powerful general expressions that are at the heart of hydrodynamics, but we will stop at the simple, one-dimensional Eq. (25.23). To see this equation in action, consider the experiment outlined in Figure 25.9. A large, thin plate is pulled through a fluid in a direction parallel to its face. As it moves, a thin layer of fluid coating it moves at its speed along the x direction. The similarly thin layer of fluid coating the walls of the container, however, is stationary. This means that between the plate and its container, along the z direction in the figure, each successive thin layer of fluid moving from the plate toward the wall must move at a slightly slower speed, creating the *velocity gradient* shown in the figure as a series of arrows of shorter length as one moves toward the walls.

Now imagine that the fluid is very viscous and the container is not held fixed. As we slowly move the plate, we drag the container along too, in the x direction. To do this, we have *transferred x momentum along the z direction to the container* to cause it to move. The velocity field in the fixed container shows us the mechanism of the momentum transport, from layer to layer. This is an example of a *shear force*, one perpendicular to the direction of motion, and consequently, the viscosity that characterizes this situation is called the *shear viscosity*.[9]

Further dimensional analysis shows that the coefficient of viscosity, η, has units of kg m^{-1} s^{-1}, or Pa s, the product of pressure and time. A viscosity of 1 Pa s

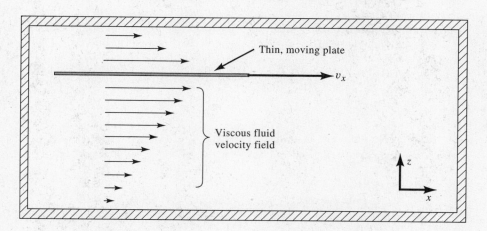

FIGURE 25.9 If a thin plate is pulled through a viscous fluid as shown, a fluid velocity gradient is established due to momentum transport through the fluid at right angles to the direction of the plate's motion. Momentum in the x direction is transported in the z direction here.

---

[9]The other kind of viscosity is called bulk viscosity and reflects the response of the system to uniform compression or expansion forces. See this chapter's references for further details.

represents, roughly, chilled Vermont maple syrup. (For a better comparison, the viscosity of glycerol is about 1.5 Pa s at 20 °C.) In contrast, California white wine (or water or ethanol) has a viscosity around 10^{-3} Pa s. The former unit for viscosity, the poise, (symbol: P), is 1 g cm^{-1} s^{-1} so that 1 P = 0.1 Pa s. Liquid viscosities typically fall in the range 0.1–100 cP (centipoise), a commonly tabulated unit, or 10^{-4}–0.1 Pa s. Gas viscosities are typically 100 or so times smaller (~100 μP or 10^{-5} Pa s). Several representative liquid values are given in Table 25.3. Many polymer solutions (including S(ℓ) above about 440 K) have very large viscosities, and amorphous or glassy materials can have huge, but measurable viscosities. Volcanic lava, for example, has a viscosity 10^4–10^6 Pa s.

There are several good experimental methods for measuring viscosity, but the arrangement in Figure 23.9 is not one of them. (In a close variation that is good, one measures the damping force on an oscillating disk.) A more useful method relies on viscosity to control the rate of flow of a fluid through a tube. A viscous force represents a friction, and when this force is balanced by the pressure difference

TABLE 25.3 Viscosities of Representative Liquids

Liquid	$\eta/10^{-3}$ Pa s	T/K
H_2	0.013	20.3
^{4}He	0.003†	4.2
^{3}He	0.001 9	3.2
Ne	0.125	27.1
Ar	0.257	87.3
Kr	0.431	119.8
Xe	0.507	165.0
N_2	0.155	77.3
O_2	0.193	90.2
F_2	0.238	85.0
Cl_2	0.492	239.1
$(C_2H_5)_2O$	0.233	293.0
NH_3	0.261	239.7
C_6H_{14} (isohexane)	0.306	293.0
C_6H_{14} (n-hexane)	0.326	293.0
CH_3OH	0.597	293.0
C_6H_6	0.652	293.0
Cs	0.676	302.0
$C_{10}H_{22}$ (n-decane)	0.92	293.0
CCl_4	0.969	293.0
Br_2	0.995	293.0
H_2O	1.002‡	293.0
C_2H_5OH	1.200	293.0
Hg	1.554	293.0
HCOOH	1.804	293.0
$C_{16}H_{34}$ (n-hexadecane)	3.19	293.0
H_2SO_4	25.4	293.0
$C_3H_5(OH)_3$ (glycerol)	1 490.	293.0

†The viscosity of superfluid ^{4}He is zero.
‡The following empirical expression reproduces the viscosity of water at 1 bar over its entire liquid range to better than 0.3%: $\eta/10^{-3}$ Pa s = 0.129 61 exp[343.28/(T/K − 168.55) − (2.424 4 × 10^{-3})(T/K)]

pushing the fluid through the tube, a steady-state flow results. The pressure difference and tube size must be chosen to avoid turbulent flow, but if they are, the flow properties are easy to analyze.

Flow through a tube has a *radial* velocity gradient: the flow is faster in the center than at the wall (where it is zero). Consider an incompressible fluid of viscosity η flowing through a tube of length L and radius R. The force driving the flow through a region in the tube of radius r is $\Delta P \pi r^2$ where ΔP is the pressure difference across the length of the tube. This flow feels viscous friction along the region's contact area, $2\pi r L$, and the ratio $\Delta P \pi r^2 / 2\pi r L = \Delta P r / 2L$, the driving force per unit friction-contact area, just balances the viscous radial flux of z momentum (flow momentum down the tube), $J_{mv,rz}$. Newton's viscous law becomes

$$J_{mv,rz} = -\frac{\Delta P r}{2L} = -\eta \frac{\partial v_z}{\partial r},$$

and if we separate variables and integrate in from the tube wall (at $r = R$, where $v_z = 0$) to a radius r, we find an expression for $v_z(r)$, the flow velocity at any radius:

$$\int_{v_z(R)}^{v_z(r)} dv_z = v_z(r) = \frac{\Delta P}{2L\eta} \int_R^r r \, dr = \frac{\Delta P}{4L\eta} (r^2 - R^2). \quad (25.24)$$

This expression shows that the flow velocity (a negative number, since it is in the opposite direction from the pressure gradient) decreases quadratically from a maximum at the center of the tube toward zero at the tube wall. Figure 25.10 plots

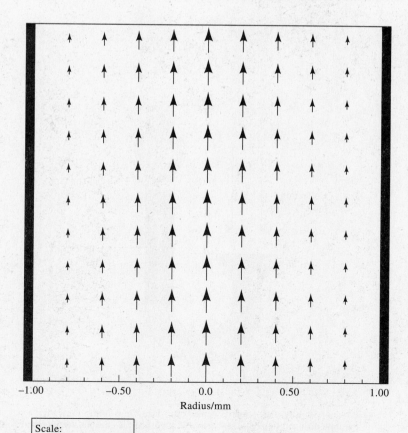

FIGURE 25.10 A fluid (water here) flowing slowly through a long capillary tube has a characteristic radial velocity field. The fluid flows faster in the center of the tube than it does at the tube wall (where it is stationary).

this equation as a flow velocity vector field for the case of water flowing (slowly!) through a 1 m long, 1 mm radius tube at 20 °C due to a 0.01 atm pressure difference.

Equation (25.24) can be used to derive an expression for the volume rate of flow through the tube (see Problems 25.20 and 25.21). For an incompressible fluid (a liquid), one finds

$$\frac{dV}{dt} = \frac{\Delta P \pi R^4}{8L\eta} \quad \text{(incompressible fluid)} \tag{25.25a}$$

while for a compressible fluid (a gas), one finds

$$\frac{dV}{dt} = \frac{\pi R^4}{16L\eta} \frac{P_i^2 - P_f^2}{P_0} \quad \text{(compressible fluid)} \tag{25.25b}$$

where P_i is the pressure at the tube's entrance, P_f is the pressure at the tube's outlet, and P_0 is the pressure at which the fluid's volume is measured. These are called *Poiseuille's equations*.[10]

The *Ostwald viscometer*, shown in Figure 25.11, exploits Poiseuille's equation to measure liquid viscosities. The collection bulb is filled with the liquid of interest, which is then drawn through the capillary until its upper meniscus is above the start mark. The liquid then drains under gravity through the capillary and back to the collection bulb. One measures the time, Δt, required for the liquid to drain from the start to the finish marks. As the liquid drains, the height difference between the upper and lower liquid levels, Δh, is changing with time. One can show that the viscosity is given by

$$\eta = \left(\frac{\pi R^4}{8L}\right) \frac{\rho g \langle \Delta h \rangle}{V} \Delta t \tag{25.26}$$

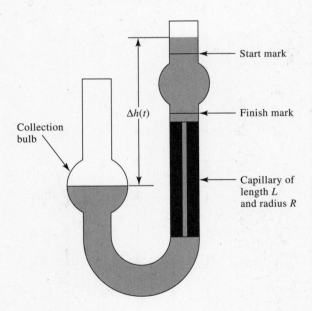

FIGURE 25.11 The Ostwald viscometer exploits the flow properties of a fluid through a capillary to measure the fluid's viscosity. The parameters of the viscometer are described in the text.

---

[10]The cgs system viscosity unit, poise, is named after Jean Louis Marie Poiseuille, a physiologist interested in the flow of blood through capillaries. He published his equation for incompressible flow in 1840. The unit name was suggested in 1913.

where V is the volume between the start and finish marks, ρ is the liquid density, g is the gravitational acceleration constant, and $\langle \Delta h \rangle$ is the time-average height difference given by

$$\langle \Delta h \rangle = \frac{\Delta h_{\text{start}} - \Delta h_{\text{finish}}}{\ln(\Delta h_{\text{start}}/\Delta h_{\text{finish}})} .$$

This device is usually used to compare the viscosity of one liquid to that of a standard. The draining times of the two liquids (when the viscometer is charged with equal volumes of each) and their densities are related to their viscosities through $\eta_1/\eta_2 = \rho_1 \Delta t_1/\rho_2 \Delta t_2$. This procedure eliminates the need to measure the capillary parameters where the R^4 dependence magnifies errors in the measurement of R.

EXAMPLE 25.9

A sample of water at 20 °C flows through an Ostwald viscometer in a time $\Delta t = 54.8$ s. The same amount of water at 50 °C takes 30.1 s. What is the viscosity of water at 50 °C, given the densities 998.204 1 kg m^{-3} at 20 °C and 988.036 3 kg m^{-3} at 50 °C?

SOLUTION From Table 25.3, the 20 °C viscosity is 1.002×10^{-3} Pa s. The 50 °C viscosity is then simply

$$\eta_{50} = \eta_{20} \frac{\rho_{50} \Delta t_{50}}{\rho_{20} \Delta t_{20}} = (1.002 \times 10^{-3}\text{ Pa s}) \frac{(988.036\,3)(30.1)}{(998.204\,1)(54.8)} = 0.545 \times 10^{-3}\text{ Pa s}$$

a value that agrees very well with the empirical expression for $\eta(\text{H}_2\text{O})$ given at the end of Table 25.3, which predicts $\eta_{50} = 0.5463 \times 10^{-3}$ Pa s. While this is a simple calculation, it makes an important point: viscosities of liquids generally *decrease with increasing temperature*, and they do so rather rapidly, roughly as an exponential function of $1/T$, as the empirical equation suggests.

⟹ RELATED PROBLEMS 25.22, 25.23, 25.24

We will look at another experimental method for measuring viscosities in Section 25.4, but now we move on in the next section to microscopic expressions for gas-phase transport coefficients. These expressions show us how the coefficients depend on macroscopic parameters such as temperature and pressure as well as microscopic parameters such as molecular mass and size.

25.3 MICROSCOPIC TRANSPORT COEFFICIENTS

In this section, we return to the approximate, linearized Boltzmann transport equation, Eq. (25.6), and learn how to write and solve it for simple transport phenomena in gases. The solutions, (the nonequilibrium number density distribution functions F^*), will not interest us as much as the expressions for the transport coefficients, but F^* is the key to each of the transport fluxes.

We start with expressions for these fluxes, working F^* into the flux–gradient expressions in Table 25.2. Since the units of F^* are molecules per unit volume per unit velocity element, each flux expression has the form

$$\text{flux} = \int \binom{\text{quantity transported}}{\text{per molecule}} \binom{\text{molecular speed in the}}{\text{transport direction}} F^*(\mathbf{v},\mathbf{r},t)\, dv_x dv_y dv_z .$$

Thus, for diffusion in the z direction, we have

$$J_{n,z} = \int v_z F^*(\mathbf{v},\mathbf{r},t)\, dv_x dv_y dv_z = -D\left(\frac{\partial n}{\partial z}\right) \qquad (25.27)$$

since matter is moving at the (local) speed v_z along the direction of the density gradient, and the net flux is just the density distribution-function weighted average of this speed.

For thermal conductivity, we write a similar expression, but we weight the integral by the transported quantity, the kinetic energy $mv^2/2$. (We assume for now that internal energies are absent—the molecules are monatomic—and that the gas is sufficiently dilute so that intermolecular potential energy transport is absent.) We have

$$J_{q,z} = \int \left(\frac{mv^2}{2}\right) v_z F^*(\mathbf{v},\mathbf{r},t)\, dv_x dv_y dv_z = -k\left(\frac{\partial T}{\partial z}\right) \quad (25.28)$$

for the flux of kinetic energy (heat) in the z direction.

Electrical conduction transports charge, and the current flux expression is very similar to the diffusive flux. We multiply by Q, the charge of the molecules represented by F^*, and write

$$J_{e,z} = \int Q\, v_z F^*(\mathbf{v},\mathbf{r},t)\, dv_x dv_y dv_z = -\sigma_e\left(\frac{\partial \phi}{\partial z}\right). \quad (25.29)$$

Finally, the viscous flux of x momentum in the z direction is

$$J_{mv,zx} = \int (mv_x)\, v_z F^*(\mathbf{v},\mathbf{r},t)\, dv_x dv_y dv_z = -\eta\left(\frac{\partial v_x}{\partial z}\right). \quad (25.30)$$

The experimental scenarios behind each of these flux expressions impose certain constraints on the Boltzmann equation. Together with other reasonable simplifying assumptions, they turn the four equations above into expressions that readily expose microscopic expressions for the transport coefficients.

Diffusion

We immediately attack the Boltzmann transport equation at three points. We assume

(1) a steady state has been reached so that $\partial F^*/\partial t = 0$,
(2) there are no external forces so that $\nabla_v F^* \cdot \mathbf{F}_{ext}/m = 0$, and
(3) the concentration gradient is in the z direction only so that $\nabla F^* \cdot \mathbf{v} = (\partial F^*/\partial z)\, v_z$.

The Boltzmann equation under these assumptions is

$$\frac{\partial F^*}{\partial z} v_z = -\frac{F^* - F}{\tau_{coll}}.$$

We also assume that F^* is not very different from F, the equilibrium function, but we will allow F to vary from point to point, since the density is not uniform. We write

$$F^*(\mathbf{r}) = F(\mathbf{r}) + F'(\mathbf{r})$$

where F' is a function that is everywhere small compared to F. If we also assume the spatial derivatives of F' are small compared to those for F so that $\partial F^*/\partial z \cong \partial F/\partial z$, we can write the Boltzmann equation as

$$\frac{\partial F}{\partial z} v_z = -\frac{F'}{\tau_{coll}},$$

which rearranges to

$$F' = -\tau_{coll} \frac{\partial F}{\partial z} v_z = -\tau_{coll}\, v_z \frac{\partial F}{\partial n}\frac{\partial n}{\partial z} = -\frac{\tau_{coll}\, v_z F}{n}\frac{\partial n}{\partial z} \quad (25.31)$$

since F depends on z only through the z dependence of the density n, and since $\partial F/\partial n = \partial(nf)/\partial n = f = F/n$. Notice how the density gradient, $\partial n/\partial z$, has appeared.

We now substitute $F + F'$ for F^* in the diffusive flux equation, Eq. (25.27). The integral over F vanishes (because it represents equilibrium), and we are left with just the integral over F', which, with Eq. (25.31), becomes

$$J_{n,z} = \int v_z F' \, d\mathbf{v} = -\left(\frac{\tau_{\text{coll}}}{n} \int v_z^2 F \, d\mathbf{v}\right)\left(\frac{\partial n}{\partial z}\right). \quad (25.32)$$

If we compare this equation to Eq. (25.27), we see that the quantity in the first set of parentheses on the right is the diffusion coefficient:

$$D = \frac{\tau_{\text{coll}}}{n} \int v_z^2 F \, d\mathbf{v} = \frac{\tau_{\text{coll}}}{n}\left(\frac{nk_B T}{m}\right) = \frac{\tau_{\text{coll}} k_B T}{m} \quad (25.33a)$$

where we have used $F = nf$ and $\langle v_z^2 \rangle = k_B T/m$ from the equipartition theorem. (See the discussion following Eq. (24.7).) If we relate τ_{coll} to the mean free path through $\tau_{\text{coll}} = \lambda/\langle v \rangle$ where the average speed, $\langle v \rangle$, is given in Table 24.1 and λ is given in Eq. (24.16), and if we assume a hard-sphere collision cross section and the ideal gas equation of state, we find

$$D = \frac{\lambda k_B T}{m \langle v \rangle} = \frac{1}{4n\sigma}\left(\frac{\pi k_B T}{m}\right)^{1/2} = \left(\frac{\pi}{m}\right)^{1/2}\frac{(k_B T)^{3/2}}{4P\sigma} \quad (25.33b)$$

where σ is the collision cross section and P is the pressure. In this expression, n represents the average concentration of the gas (N/V), since it appears through τ_{coll}, a quantity we have assumed to be the same throughout the gas.

These expressions support our intuition about diffusion. Larger molecules (larger σ) diffuse more slowly than smaller ones, everything diffuses more slowly as pressure is increased at constant T ($D \propto 1/P$), and everything diffuses faster as T is increased ($D \propto T^{3/2}$ at constant P, $D \propto T^{1/2}$ at constant V). They are also qualitatively in agreement with the Einstein–Smoluchowski expression, Eq. (25.19). If we neglect numeric factors on the order of unity, we have $\langle v \rangle^2 \cong k_B T/m$ so that $D \cong \lambda \langle v \rangle$ for the first equality in Eq. (25.33b). With $\langle v \rangle \cong \lambda/\tau_{\text{coll}}$, we have $D \cong \lambda^2/\tau_{\text{coll}}$, which is the same form as Eq. (25.19).

EXAMPLE 25.10

In 1950, E. B. Winn (no relation to the author of this book!) measured diffusion coefficients of Ar, Ne, N_2, O_2, CO_2, and CH_4 using isotopically enriched gases [*Phys. Rev.* **80**, 1024 (1950)] over a wide temperature range. The diffusion coefficient of one isotope into another was corrected for mass differences to yield self-diffusion coefficients at a constant 1.00 atm pressure. His data for Ar are tabulated below. How well do they follow the prediction of Eq. (25.33b)?

T/K	353.15	326.65	295.15	273.15	194.65	90.15	77.65
$D/10^{-5}$ m^2 s^{-1}	2.49	2.12	1.78	1.56	0.830	0.180	0.134

SOLUTION At constant pressure, Eq. (25.33b) predicts $D \propto T^{3/2}$ so that a graph of $D/T^{3/2}$ versus T should be a horizontal straight line. If we make this graph, we find the data do not fall on a horizontal straight line:

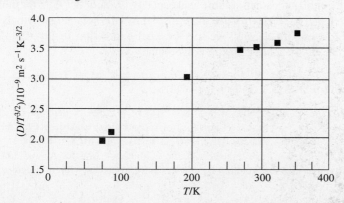

The deviation is most pronounced at the lowest temperatures, where we should expect a hard-sphere potential model to be least accurate. These data indicate that the attractive part of the intermolecular potential (which would enter our theory through a temperature-dependent cross section) plays an important role in transport, especially at low T. Problem 25.25 considers these data further.

⇒ **RELATED PROBLEMS** 25.25, 25.26, 25.27

Thermal Conductivity

We make the same assumptions as before, but we need to recognize that a temperature gradient at constant pressure implies a *density* gradient as well. If $PV = Nk_BT$ or $P = nk_BT$ (remember: $n = N/V$ here!), then if $\partial T/\partial z \neq 0$, $\partial n/\partial z$ must be non-zero as well to make P spatially uniform. The local equilibrium distribution function has an explicit z dependence to T and n:

$$F(z) = n(z)\left(\frac{m}{2\pi k_B T(z)}\right)^{3/2} e^{-m(v_x^2 + v_y^2 + v_z^2)/2k_B T(z)}.$$

The Boltzmann transport equation becomes

$$F' = -v_z \tau_{\text{coll}}\left(\frac{\partial F(z)}{\partial z}\right),$$

and from the expression for $F(z)$ and the equation of state $n = P/k_BT$ (for which $\partial n/\partial T = -n/T$), we can write

$$\frac{\partial F(z)}{\partial z} = \frac{F}{T}\left(\frac{mv^2}{2k_BT} - \frac{5}{2}\right)\left(\frac{\partial T}{\partial z}\right).$$

The energy flux expression, Eq. (25.28), becomes

$$J_{q,z} = \int \left(\frac{mv^2}{2}\right) v_z F' \, d\mathbf{v} = -\left[\frac{\tau_{\text{coll}}}{T}\int \left(\frac{mv^2}{2}\right) v_z^2 \left(\frac{mv^2}{2k_BT} - \frac{5}{2}\right) F \, d\mathbf{v}\right]\left(\frac{\partial T}{\partial z}\right) \quad (25.34)$$

and the quantity in square brackets is the thermal conductivity, k. If we carry out the integration, (see Problem 25.28), we find

$$k = \frac{5}{2}\frac{\tau_{\text{coll}} n k_B^2 T}{m}, \quad (25.35a)$$

which we can again write in other ways:

$$k = \frac{5}{2}\frac{\lambda n k_B^2 T}{m\langle v \rangle} = \frac{5}{2}\frac{k_B^{3/2}}{4\sigma}\left(\frac{\pi T}{m}\right)^{1/2} = \frac{5}{12}\frac{\overline{C}_V}{\sigma N_A}\left(\frac{\pi k_B T}{m}\right)^{1/2} \quad (25.35b)$$

where $\overline{C}_V$ is the molar constant-volume heat capacity ($3R/2$ or $3N_Ak_B/2$ for monatomic species). We can use $\overline{C}_V$ to introduce the transport of *internal energy* by polyatomics in at least an approximate way: we use the final expression in Eq. (25.35b) with the appropriate $\overline{C}_V$, σ, and m for the polyatomic.

EXAMPLE 25.11

At 300 K, $k(\text{Ar}) = 1.766 \times 10^{-2}$ W m^{-1} K^{-1}, $k(\text{CO}) = 2.524 \times 10^{-2}$ W m^{-1} K^{-1}, and $k(\text{N}_2) = 2.610 \times 10^{-2}$ W m^{-1} K^{-1}. Analysis of second virial coefficients for these gases yield the hard-sphere cross sections $\sigma(\text{Ar}) = 3.53 \times 10^{-19}$ m^2, $\sigma(\text{CO}) = 4.44 \times 10^{-19}$ m^2, and $\sigma(\text{N}_2) = 4.30 \times 10^{-19}$ m^2. Use Eq. (25.35b) with $\overline{C}_V = 3R/2$ for Ar and $5R/2$ for CO and N$_2$ (i.e., assume diatomic rotation contributes fully to $\overline{C}_V$, but vibration does not at all) to calculate the ratios $k(\text{Ar})/k(\text{CO})$ and $k(\text{N}_2)/k(\text{CO})$, compare your answers to the experimental ratio, and comment on the agreement or disagreement.

SOLUTION The theoretical ratios are

$$\frac{k(\text{Ar})}{k(\text{CO})} = \frac{\overline{C}_V(\text{Ar})}{\overline{C}_V(\text{CO})} \frac{\sigma(\text{CO})}{\sigma(\text{Ar})} \left(\frac{m(\text{CO})}{m(\text{Ar})}\right)^{1/2} = 0.633 \ ,$$

which is somewhat smaller than the experimental ratio, 0.700, and

$$\frac{k(\text{N}_2)}{k(\text{CO})} = \frac{\overline{C}_V(\text{N}_2)}{\overline{C}_V(\text{CO})} \frac{\sigma(\text{CO})}{\sigma(\text{N}_2)} \left(\frac{m(\text{CO})}{m(\text{N}_2)}\right)^{1/2} = 1.033 \ ,$$

which is almost exactly the experimental ratio, 1.034. There is uncertainty in the cross sections, since different kinds of measurements can yield cross sections that differ by 5% or so, and while the heat capacities are quite accurate, we can question whether or not a gas transports *kinetic* energy as well as it transports *internal* energy. Using $\overline{C}_V = 5R/2$ for CO assumes that it does. The agreement for $k(\text{N}_2)/k(\text{CO})$ suggests that internal energy is transported equally well by both (the heat capacity contributions cancel in this ratio), but the disagreement for $k(\text{Ar})/k(\text{CO})$ suggests that the full heat capacity does not model the transport of internal energy accurately. This point is discussed further near the end of this section.

⇒ **RELATED PROBLEMS** 25.29, 25.30, 25.31

The general intuitive behavior of k is again born out: k decreases for larger molecules, and increases with T (like $T^{1/2}$). Perhaps the most surprising prediction, however, and one that experiment confirms, is that the *thermal conductivity of a gas is independent of pressure*. If the pressure is so low that the mean free path is comparable to or larger than a characteristic size of the gas container, then molecular collisions no longer mediate thermal conduction, and our expression is invalid (k becomes proportional to P). We know that "vacuum bottles" have good thermally insulating walls, and this is the reason why. It is also the reason why a small air leak into the vacuum is bad news. A few mbar pressure is as effective as 1 bar in wrecking the thermal insulation. At very high P, some of our assumptions begin to fail (multiple collision events become important, etc.), and k begins to depend on pressure, but for wide variations around ordinary pressures, k is constant. The microscopic reason for this can be seen in the first expression for k in Eq. (25.35b), $k \propto \lambda n$. As P increases, n increases, increasing the number of molecules available to transport energy. But λ *decreases* at the same rate, decreasing the average distance any one molecule can carry energy before an interrupting collision. The two effects have exactly the opposite P dependence and cancel each other.

Electrical Conductivity

Here, we imagine a weakly ionized gas (a *plasma*) with n_+ positive ions per unit volume, each with charge Q_+, in thermal equilibrium with their parent neutral molecules. Free electrons or gaseous anions are also present, in order to assure that the plasma is electrically neutral, but we concentrate on the conductivity of one species at a time. The net current is due to the motion of all charged species of all types, of course, and we will say more about this later.

We impose on the plasma a small electric potential gradient, $\partial\phi/\partial z$, which is equal to $-E_z$, the z component of the external electric field. The flux–gradient equation becomes $J_{e,z} = \sigma_e E_z$. We also assume the ions and the neutrals have the same f^* (due to collisions), so that their F^* functions differ only through the density differences between the ions and neutrals: $n_+ \ll n$. We also assume these densities are constants, along with our usual assumption of a steady state and a small perturbation from equilibrium ($F^* = F + F'$). The Boltzmann equation now has no *spatial* gradient term, but due to the external force ($F_{\text{ext},z} = Q_+ E_z$), Eq. (25.6) becomes

$$\frac{\partial F}{\partial v_z} \frac{Q_+ E_z}{m} = -\frac{F'}{\tau_{\text{coll}}},$$

which rearranges to

$$F' = -\frac{\tau_{\text{coll}} Q_+ E_z}{m} \frac{\partial F}{\partial v_z}.$$

Since we know F, the equilibrium distribution, we can evaluate $\partial F/\partial v_z$:

$$\frac{\partial F}{\partial v_z} = -\frac{m v_z}{k_B T} F$$

and write F' as

$$F' = \frac{\tau_{\text{coll}} Q_+}{k_B T} v_z F E_z.$$

Since we want $F' \ll F$, this expression gives us an upper limit to the electric field, as discussed in Problem 25.33. The charge flux equation, Eq. (25.29), becomes

$$J_{e,z} = Q_+ \int v_z F' \, d\mathbf{v} = \left(\frac{\tau_{\text{coll}} Q_+^2}{k_B T} \int v_z^2 F \, d\mathbf{v}\right) E_z \quad (25.36)$$

and the expression in parentheses is the electrical conductivity:

$$\sigma_e = \frac{\tau_{\text{coll}} Q_+^2}{k_B T} \int v_z^2 F \, d\mathbf{v} = \frac{n_+ \tau_{\text{coll}} Q_+^2}{m}, \quad (25.37a)$$

which we can again write in other ways, if we assume the ions and neutrals have the same collision cross section:

$$\sigma_e = \frac{\lambda n_+ Q_+^2}{m \langle v \rangle} = \frac{n_+}{n} \frac{Q_+^2}{4\sigma} \left(\frac{\pi}{m k_B T}\right)^{1/2}. \quad (25.37b)$$

The Q_+^2 factor is reassuring. It tells us that the *sign* of the charge does not matter (and a *negative* conductivity would be a difficult beast to interpret!). To incorporate other ions of perhaps other charges, it is useful to introduce the *mobility, u*, which is the ions' drift velocity per unit electric field, or the conductivity per unit charge concentration:

$$u_+ = \frac{\text{conductivity}}{\text{unit charge concentration}} = \frac{\sigma_e}{n_+ Q_+} = \frac{\tau_{\text{coll}} Q_+}{m} = \frac{Q_+}{4n\sigma} \left(\frac{\pi}{m k_B T}\right)^{1/2}. \quad (25.38)$$

If there are two species of opposite charge with mobilities u_+ and u_-, the net conductivity is

$$\sigma_e = \sigma_+ + \sigma_- = n_+ Q_+ u_+ + n_- Q_- u_-. \quad (25.39)$$

Moreover, the ratio of mobility to the diffusion constant,

$$\frac{u_+}{D} = \frac{\tau_{\text{coll}} Q_+}{m} \frac{m}{\tau_{\text{coll}} k_B T} = \frac{Q_+}{k_B T} \quad (25.40)$$

is a very general result, known as the *Einstein mobility–diffusion equation*, that holds for ions in liquid solutions as well as in gases.

EXAMPLE 25.12

Consider a Xe atom at very low pressure in the center of a pair of parallel plates connected to a power supply so that they produce an electric field $E_z = 10^4$ V m^{-1}. The atom is ionized by a multiphoton laser process and begins to move toward the negative plate, which is 1.00 cm away. How long does it take to reach this plate if the pressure is so low that $\lambda \gg 1$ cm? How long does it take if the Xe in the cell is at 1.00 atm pressure?

SOLUTION In the low-pressure scenario, the Xe$^+$ moves under the constant force eE_z or the constant acceleration $a = eE_z/m$. The time to travel a distance $L = 0.01$ m is

$$t = \left(\frac{2L}{a}\right)^{1/2} = \left(\frac{2Lm}{eE_z}\right)^{1/2} = 1.65 \times 10^{-6} \text{ s} = 1.65 \text{ μs} .$$

At 1.00 atm, the ion has to make its way, via collisions, through the neutral background gas. It does so at a *constant* speed, v_z, equal to the product of the ion's mobility and the electric field:

$$v_z = u_+ E_z = \frac{e}{4n\sigma}\left(\frac{\pi}{mk_BT}\right)^{1/2} E_z$$

where $n = P/k_BT = 2.45 \times 10^{25}$ m^{-3} at 300 K and 1.00 atm. The hard-sphere cross section for Xe is 5.28×10^{-19} m^2, and u_+ is calculated to be 1.83×10^{-4} m^2 s^{-1} V^{-1}. The speed is $u_+ E_z = 1.83$ m s^{-1}, and the time to move 1.00 cm is $L/v_z = 5.47 \times 10^{-3}$ s, a considerably longer time. These calculations are important in designing experiments that use ion currents to monitor a photochemical event such as multiphoton ionization.

⇒ **RELATED PROBLEMS** 25.32, 25.33

Viscosity

We imagine the gas to have a slow *bulk flow* in the x direction, with a flow gradient in the z direction. The flow speed, u_x, and the random, fast, locally equilibrated molecular velocities add, component by component, but the local, equilibrated velocity distribution refers to just random motion in the reference frame of the flow. The distribution function depends on local values of these velocity components, U_x, U_y, and U_z, in the flowing frame of reference:

$$U_z = v_z, \quad U_y = v_y, \quad U_x = v_x - u_x .$$

We again assume a steady-state flow and write $F^* = F + F'$. The Boltzmann equation is

$$F' = -\tau_{\text{coll}} v_z \frac{\partial F}{\partial z} = -\tau_{\text{coll}} v_z \left(\frac{\partial F}{\partial U_x}\right)\left(\frac{\partial U_x}{\partial z}\right) = \tau_{\text{coll}} v_z \left(\frac{\partial F}{\partial U_x}\right)\left(\frac{\partial u_x}{\partial z}\right) ,$$

and Eq. (25.30) becomes

$$J_{mv,zx} = \int (mU_x) U_z F' d\mathbf{U} = m \int U_x v_z F' d\mathbf{U} = \left[\tau_{\text{coll}} m \int \left(\frac{\partial F}{\partial U_x}\right) U_x v_z^2 d\mathbf{U}\right]\left(\frac{\partial u_x}{\partial z}\right) . \quad (25.41)$$

The quantity in square brackets is the negative of the coefficient of viscosity:

$$\begin{aligned}\eta &= -\tau_{\text{coll}} m \int \left(\frac{\partial F}{\partial U_x}\right) U_x v_z^2 d\mathbf{U} = -\tau_{\text{coll}} m \int \left(-F\frac{mU_x}{k_BT}\right) U_x v_z^2 d\mathbf{U} \\ &= \frac{\tau_{\text{coll}} m^2}{k_BT} \int F U_x^2 v_z^2 d\mathbf{U} = \frac{\tau_{\text{coll}} m^2}{k_BT} n \left(\frac{k_BT}{m}\right)^2 = n\tau_{\text{coll}} k_BT ,\end{aligned} \quad (25.42\text{a})$$

which we again write in different ways:

$$\eta = \frac{n\lambda k_B T}{\langle v \rangle} = \frac{(\pi k_B T m)^{1/2}}{4\sigma}. \quad (25.42b)$$

We see that the *viscosity of a gas increases with temperature,* opposite the trend for simple molecular liquids, and, again perhaps surprisingly, η is *independent of pressure,* as was the thermal conductivity. The reason is the same: as the density goes up at constant temperature, the mean free path goes down at the same rate. More molecules are available to carry momentum, but they carry it a shorter distance, canceling any pressure dependence. The increase in η with T is approximately in accord with experiment, but since σ also varies with T for realistic intermolecular potentials, the $T^{1/2}$ dependence fails over wide temperature ranges. Figure 25.12 compares the experimental viscosity of Ar over a wide temperature range to the prediction of Eq. (25.42b) using the hard-sphere cross section from Chapter 24, 4.1×10^{-19} m². Also shown is the prediction of a refined hard-sphere theory that we discuss briefly below.

EXAMPLE 25.13

If Ar behaved as a hard sphere, and if our simplified treatment of transport phenomena was exact, then measured transport coefficients should yield the same hard-sphere cross section. The degree to which these cross-sections differ is a measure of the approximate nature of our expressions. Find σ from each of the following coefficients for Ar at 300 K: $\eta = 2.294 \times 10^{-5}$ Pa s and $k = 1.766 \times 10^{-2}$ W m⁻¹ K⁻¹.

SOLUTION From Eq. (25.35b), we have

$$\sigma = \frac{5k_B^{3/2}}{8k}\left(\frac{\pi T}{m}\right)^{1/2}$$

and from Eq. (25.42b), we have

$$\sigma = \frac{(\pi k_B T m)^{1/2}}{4\eta}.$$

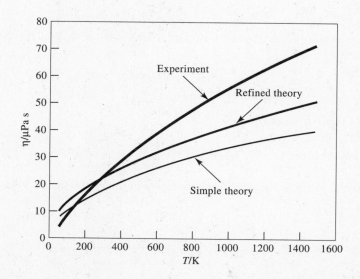

FIGURE 25.12 The experimental viscosity of Ar as a function of temperature (heavy line) is compared to the predictions of the simple hard-sphere theory and a refined hard-sphere theory, neither of which reproduces the experimental temperature dependence very well. If a more realistic intermolecular force is used in the refined theory, the experimental curve is reproduced almost exactly.

Substituting the data (with $m = 6.63 \times 10^{-36}$ kg, the mass of one Ar atom) into these equations yields $\sigma = 2.16 \times 10^{-19}$ m² from k and $\sigma = 3.20 \times 10^{-19}$ m² from η. Note too that our simple expressions predict $k/\eta = 5k_B/2m = 520$ J K⁻¹ kg⁻¹ for Ar, a number that is independent of T. The experimental ratio is 770.0 J K⁻¹ kg⁻¹ at 300 K, and it turns out to be nearly independent of T, as predicted. The disagreements point to both the simplicity of the derivation and the simplicity of the model intermolecular potential, but the derivation does support the observation that k/η is nearly independent of T, which suggests that better expressions for k and/or η differ from ours by simple constants.

⟹ RELATED PROBLEMS 25.34, 25.35, 25.36

We have now taken the simplified Boltzmann transport equation about as far as we can. The rigorous transport coefficients for hard spheres were first worked out in detail by David Enskog, a Swedish scientist, and Sydney Chapman, an English scientist, independently, starting around 1917. Maxwell, Boltzmann, and others had worked on this problem, and before the Enskog-Chapman results, there was general agreement about T, P, m, etc. dependencies, but confusion over numerical factors remained. Boltzmann compiled ten different expressions for thermal conductivity from the literature in 1909, indicating that there was both great interest in the subject and equally great confusion. The Enskog–Chapman results were both a relief and a warning: the calculations were very involved and complicated, and they indicated that extension to realistic intermolecular potentials would not be at all easy. When Chapman and Cowling published the first edition of their standard monograph in the field, *The Mathematical Theory of Non-uniform Gases*, in 1939, Chapman himself admitted that reading it was "like chewing glass."

Suffice it to say that the complicated theory *can* be extended to realistic potentials with excellent agreement with experiment. Even for hard spheres, the Enskog–Chapman corrections to our simple expressions are factors not far from unity (a factor of 1.25 in the case of viscosity, shown in Figure 25.12, and a factor of 1.5 for diffusion and thermal conductivity), and if one determines the hard-sphere cross section for a measurement of one property at one temperature (including our old friend, the second virial coefficient), that cross section gives reasonable results for other properties at the same temperature. Note also that we have *not* needed quantum mechanics to derive useful transport expressions. The only gases for which quantum corrections are important are the lightest ones, H_2, He, and Ne, at the lowest temperatures.

We close this section with a few miscellaneous comments on transport properties in gases. First, note the resemblance between the final expression for k in Eq. (25.35b) and for η in Eq. (25.42b). We can combine these equations and write

$$k = \frac{5}{3} \frac{\overline{C}_V}{N_A m} \eta ,$$

and the Enskog–Chapman theory predicts the same expression with 5/3 replaced by 5/2. For polyatomics, this factor is found to be somewhat less than 5/2. The reason is traced to the transport of internal energy: slower molecules can transport as much internal energy as faster ones, and thus internal and translational energy are not transported at the same rate. (See also Example 25.11.) Eucken found that a partition of the heat capacity into translational and internal components leads to a prediction that this factor (called f, the *Eucken factor*) should be $f = (9\gamma - 5)/4$ where γ is the ratio of heat capacities $\overline{C}_P/\overline{C}_V$. The Eucken factor reproduces $f = 5/2$ for monatomics ($\gamma = 5/3$), and for simple diatomics with $\gamma \cong 7/5$, $f \cong 1.9$.

Finally, we mention that Nature is remarkably adept at transporting physical quantities. In addition to the simple flux–gradient phenomena we have discussed here, numerous others are known involving so-called *cross-effects,* so that the general flux–gradient expression, $J_i = -c_{ii}\Phi_i$ where c_{ii} is a phenomenological coupling constant and Φ_i is a gradient, can be generalized to something like $J_i = -\Sigma c_{ij}\Phi_j$. For example, a density gradient (Fick's first law) *or* a temperature gradient, in the phenomenon known as *thermal diffusion* (the *Soret effect*), can produce a flux of matter. Similarly, a temperature gradient in a conductor produces an electrical potential difference between the hot and cold ends, a phenomenon exploited in the *thermocouple.* In addition, other external fields can induce transport. For example, if a magnetic field is applied to a gas in the x direction and a temperature gradient in the y direction, a heat flow is observed in the z direction as well as in the y direction. There are quite powerful, general arguments that can be made about which of the coefficients c_{ij} must be zero, and which must follow a so-called *reciprocal relation* of the form $c_{ij} = c_{ji}$. This latter relation will interest us when we consider the flux of chemical reactions and the general rates of reactions.

25.4 TRANSPORT IN CONDENSED MEDIA

Section 25.2 mentioned several general aspects of transport in condensed media, and in this section, we continue with a look at some of the inter-relations among diffusion, viscosity, and ion conductivity in liquid solutions. Conductivity is one of the hallmarks of electrolyte solutions, of course, and quantitative study of conductivity gives chemical information on not only ion concentrations but also on ion mobilities, ionization equilibrium constants, and hydrated ion structures. The hydrodynamics of motion in a viscous medium is exploited in centrifuge separation schemes (see also Chapter 9), and polymer and biochemical sciences take advantage of the size differences between macromolecules and simple solvent molecules in many ways that are controlled by viscous hydrodynamics.

Imagine a small sphere falling under its own weight through a very viscous liquid such as glycerol. Unlike free-fall through a vacuum, the sphere falls at a constant speed (after a brief spurt at the start of its trip). It is not constantly accelerated to greater and greater speed. If the flow is slow enough so that turbulence is avoided, the viscous force retarding its downward motion is directly proportional to its speed. If z is the "down" direction, we can write $F_z = -\zeta v_z$. The force, F_z, points in the opposite direction from the speed, v_z, and the two are related by a proportionality factor, ζ. For a sphere of radius R, this factor (called the *friction constant*) is given by *Stokes law:*

$$\zeta = 6\pi\eta R \tag{25.43}$$

where η is the solvent's viscosity. When viscous friction balances the gravitational force,

$$F_g = mg = \frac{4}{3}\pi R^3 (\rho_s - \rho_\ell)g$$

where ρ_s is the density of the sphere and ρ_ℓ is the density of the liquid (subtracted to account for the sphere's buoyancy), the sphere falls with a constant speed. If one times the fall through a distance L, the viscosity is found from

$$\eta = \frac{2R^2(\rho_s - \rho_\ell)gt}{9L} \tag{25.44}$$

where t is the fall time.

EXAMPLE 25.14

An Al sphere (ρ = 2698 kg m^{-3}) 2.5 mm in radius and a Pb sphere (ρ = 11 350 kg m^{-3}) 1.0 mm in radius are dropped at the same time into a 1.00 m deep vat of glycerol (ρ = 1261.3 kg m^{-3}). Which sphere reaches the bottom of the vat first?

SOLUTION Equation (25.44) can be manipulated into

$$\frac{t_{Al}}{t_{Pb}} = \frac{R_{Pb}^2 (\rho_{Pb} - \rho_\ell)}{R_{Al}^2 (\rho_{Al} - \rho_\ell)} = \frac{(1.0)^2 (11\,350 - 1261.3)}{(2.5)^2 (2698 - 1261.3)} = 1.12$$

so that the Pb sphere falls to the bottom first. The individual times, using Eq. (25.44) with η = 1.49 Pa s from Table 25.3, are t_{Al} = 76.1 s and t_{Pb} = 67.8 s. In contrast, the time to fall 1.00 m through a vacuum is $t = (2L/g)^{1/2} = 0.45$ s for both spheres.

If the radius of the sphere is not very small compared to the fall distance and/or if the radius is not very small compared to the radius of the liquid's container, corrections can be made to account for the effect the liquid flow field has on the sphere's motion. These corrections can be ignored if the same cell and sphere are used to measure the fall time in both a reference liquid of known viscosity and in a liquid of unknown viscosity.

➡ **RELATED PROBLEMS** 25.37, 25.38

If the sphere is charged and is moving under the influence of an electric field, the same force balance argument determines the sphere's constant, terminal drift speed. The electric force is QE_z for charge Q, and the force balance is $QE_z = 6\pi\eta R v_z$. If the sphere is a charged, large, roughly spherical molecule (large compared to a solvent molecule) such as a charged protein or other macromolecule, we have discovered the basis of *electrophoresis,* a method of separating mixtures of such molecules according to their drift speeds in an electric field. The details of electrophoresis are complicated by corrections for nonspherical shapes and uncertainties about charge and size (the net charge on a protein generally depends on pH, and "size" includes the solvent counter-ions that may be strongly bound to the protein), but as a separation method, electrophoresis is a powerful and routine tool.

The definition of an ion's mobility, Eq. (25.38), can also be written as the ratio of the drift velocity to the electric field: $u_+ = v_z/E_z = Q/6\pi\eta R$. If we combine this with Eq. (25.40), the Einstein relation between mobility and the diffusion constant, we find the *Stokes–Einstein relation:*

$$\frac{Q}{6\pi\eta R} = \frac{QD}{k_B T} \quad \text{or} \quad D = \frac{k_B T}{6\pi\eta R} \qquad (25.45)$$

relating the hydrodynamic friction constant $6\pi\eta R$ to the diffusion constant. While we have derived this relation from an ion mobility argument, it is independent of charge and holds for neutral species as well. Moreover, the friction constant can be generalized to ellipsoidal shapes, a better approximation for most molecules than spherical. This is discussed in references at the end of the chapter.

EXAMPLE 25.15

The diffusion constant of hemoglobin in water at 293 K is 6.9 × 10^{-11} m^2 s^{-1}. How large is hemoglobin?

SOLUTION Assuming hemoglobin is spherical (it almost is), we solve Eq. (25.45) for R, and use the viscosity of water from Table 25.3:

$$R = \frac{k_B T}{6\pi\eta D} = \frac{(1.38 \times 10^{-23} \text{ J K}^{-1})(293 \text{ K})}{6\pi(1.002 \times 10^{-3} \text{ Pa s})(6.9 \times 10^{-11} \text{ m}^2 \text{ s}^{-1})} = 31 \text{ Å}$$

The molar mass of hemoglobin is 68 000 g mol^{-1}, a fairly small value for a protein, and this radius is a reasonable one for such a molecule.

▶ RELATED PROBLEM 25.39

The conductivity (or mobility) of small ions is an important aspect of electrochemistry and electrolyte solution studies. It is tricky to measure aqueous ion conductivities, since electrolysis reactions and the phenomenon known as *polarization* that occurs near electrodes complicate direct current measurements. Conductivity equipment usually use an alternating current in the frequency range 100–3000 Hz and a standard electrolyte solution such as KCl(*aq*) to make relative conductivity measurements. Since the conductivity of a solution depends on ionic concentration, it is convenient to introduce the *molar conductivity,* Λ, of electrolyte A:

$$\Lambda = \sigma_e/[A] \tag{25.46}$$

where [A] is the molarity of A.[11] With σ_e in S m^{-1} units and [A] in M units (mol L^{-1} or mol dm^{-3}), Eq. (25.46) is

$$\Lambda/\text{S m}^2\text{ mol}^{-1} = \frac{\sigma_e/\text{S m}^{-1}}{(1000\text{ dm}^3\text{ m}^{-3})\,[A]/\text{mol dm}^{-3}}\ .$$

We know several reasons from Chapter 10 that explain the observation that Λ depends on concentration (or, said another way, that σ_e is not a simple linear function of [A]). First, even strong electrolytes such as KCl produce nonideal solutions in which ions interact strongly with each other. Weak electrolytes such as CH$_3$COOH have small equilibrium constants for dissociation, and this equilibrium controls the ion concentration as the weak electrolyte concentration is varied, as we saw in Chapters 7 and 10. In the limit of zero concentration, however, both effects—ion–ion interaction and weak electrolyte dissociation—are minimized. Ions are too far apart to interact, and entropy ensures complete dissociation at infinite dilution.

For dilute solutions of strong electrolytes, Kohlrausch proposed around 1916 that the equation

$$\Lambda = \Lambda° - \alpha[A]^{1/2} \tag{25.47}$$

should be used to extrapolate molar conductivities to their *limiting* value at infinite dilution, $\Lambda°$, and Onsager gave this expression theoretical justification about ten years later. In Eq. (25.47), α is a parameter that varies from substance to substance. This behavior is shown for several strong electrolyte chlorides in Figure 25.13. The straight lines are the predictions of Eq. (25.47), but the limiting intercepts, shown as crosses, were calculated independently. At infinite dilution, Kohlrausch argued, the solution conductivity should consist of independent conductivity contributions from each kind of ion. If we introduce *ion-specific limiting molar conductivities*, $\lambda_{\pm}$, we can write

$$\Lambda° = \nu_+\lambda_+ + \nu_-\lambda_-\ , \tag{25.48}$$

which is known as *Kohlrausch's law of independent conductivity*. The stoichiometric coefficients ν_+ and ν_- are defined in Eq. (10.2). The crosses in Figure 25.13 were calculated from this expression and the specific conductivities collected in Table 25.4.

---

[11]We will drop our usual notation of writing a bar over Λ to indicate that it is a molar quantity. This is the standard notation.

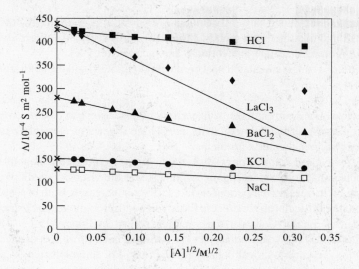

FIGURE 25.13 The molar conductivity for several chlorides is plotted versus the square root of the concentration, as suggested by Kohlrausch's law, Eq. (25.47), the straight lines. The law has less validity for more highly charged electrolytes, but it does have limiting, low-concentration validity for them all. The intercepts (the crosses) were calculated independently from limiting ion-specific molar conductivities.

These specific conductivities are interesting and tell us some of the secrets of ion motion in solution. Note first the huge value for H^+. The solvated proton seems to slip quite easily and quickly from water to water. In fact, protons are *passed* from one water to the next through hydrogen bonds (between H_2O and H_3O^+) so that a proton that lands on one side of a water molecule may become its chemically bonded H atom as the molecule frees and passes on a proton from its other side. One might expect that small size is a general indicator of large conductivity, but a glance through the alkali metal values shows that if this is true, it is not the crystal ionic radius that measures size. Strong *hydration* of small ions gives them

TABLE 25.4 Ion-Specific Limiting Molar Conductances at 298 K

Cation	$\lambda_+/10^{-4}$ S m² mol⁻¹	Anion	$\lambda_-/10^{-4}$ S m² mol⁻¹
H^+	349.65	OH^-	198
Li^+	38.66	F^-	55.4
Na^+	50.08	Cl^-	76.31
K^+	73.48	Br^-	78.1
Rb^+	77.8	I^-	76.8
Cs^+	77.2	HF_2^-	75
Be^{2+}	90.0	CN^-	78
Mg^{2+}	106.0	CH_3COO^-	40.9
Ca^{2+}	118.9	NO_3^-	71.42
Sr^{2+}	118.8	SO_4^{2-}	160
Ba^{2+}	127.2	HSO_4^-	100
Al^{3+}	183	CO_3^{2-}	138.6
La^{3+}	209.1	HCO_3^-	44.5
Ag^+	61.9	SCN^-	66
Cu^{2+}	107.2	$HCOO^-$	54.6
Ni^{2+}	100		
NH_4^+	73.5		

Note: These quantities are often tabulated *per unit charge* on the ion and listed, for example, as $\lambda(1/2 Ba^{2+}) = 63.6 \times 10^{-4}$ S m² mol⁻¹, which is one-half the value listed here.

hydrodynamic radii much larger than their crystal radii, accounting for the trends in the alkali and alkaline earth metals. Molecular complexity also affects hydrodynamic size (compare acetate and formate anions, for example).

Ion *mobilities* can be calculated in accord with Eq. (25.38) and the definition of $\lambda_\pm$:

$$\lambda_\pm = Q_\pm u_\pm N_A \,, \tag{25.49}$$

and our hydrodynamic arguments from earlier in this section can be used to estimate aqueous ion sizes, diffusion coefficients, etc. These aspects of electrolyte transport are explored in this chapter's problems.

Finally, we close with a brief discussion of the temperature dependence of transport phenomena in condensed media. In general, the viscosity of simple liquids decreases with increasing temperature while diffusion increases. Both of these trends spring from the difficult job a molecule faces when it tries to move past its neighbors in a condensed phase. Consider the situation shown in Figure 25.14. The light colored molecule finds itself adjacent to a transient *cavity* that has opened in the liquid due to spontaneous local density fluctuations. For this molecule to transport itself (or its momentum or energy) into the cavity, it has to get past the neighboring molecules numbered 1 and 2 in the figure. These, in turn, are held more or less in place by their neighbors, and so on throughout the liquid. The light colored molecule needs a higher than average energy to push its way into the cavity. If the average energy needed to push 1 and 2 aside is E_a, the probability that the light colored molecule can do the job is proportional to a Boltzmann factor, $\exp(-E_a/k_B T)$. As T increases, this probability increases, increasing the ability of the light colored molecule to diffuse (or transport its momentum or energy).

This is a simple view of an *activated process,* and E_a is called an *activation energy.* We will find activation energies for chemical reactions in later chapters, but the concept of an activated process is very general. We will also study chemical reactions in condensed phases where another aspect of Figure 25.14 will be important. Suppose the light colored molecule is about to dissociate (either because it just absorbed a photon to a dissociative state, or because the liquid is hot enough to induce thermal dissociation). As it tries to fall apart into fragments, the fragments find themselves trapped in a cage of neighbors. If they do not escape this cage quickly enough (and begin to diffuse from each other), they may simply give up trying and recombine into the original molecule (or an isomer of it, in some cases).

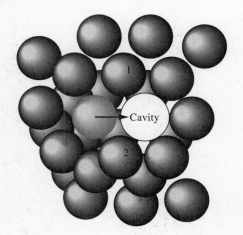

FIGURE 25.14 Diffusion and viscosity in condensed media are activated processes, which means that the crowded conditions of a condensed phase forces an atom (the lightly shaded one) to push its way past its neighbors to reach an adjoining cavity.

CHAPTER 25 SUMMARY

This chapter has introduced nonequilibrium dynamics, a topic we extend to chemical reactions in the following three chapters. We saw here how the nonequilibrium distribution function, F^*, representing a system not far from equilibrium, entered a simplified equation that described the return to equilibrium:

$$\text{Linearized Boltzmann transport equation:}$$
$$\nabla F^* \cdot \mathbf{v} + \nabla_v F^* \cdot \frac{\mathbf{F}_{\text{ext}}}{m} + \frac{\partial F^*}{\partial t} = -\frac{F^* - F}{\tau_{\text{coll}}}$$

We saw the meaning of the collision time, τ_{coll}, in more detail, and derived useful expressions for gas-phase collision rates, summarized in Table 25.1.

The phenomenological flux–gradient expressions summarized in Table 25.2 introduced the four transport coefficients of most interest. Diffusion was found to be a slow process at best, and one that continually slows as equilibrium is approached. The $t^{1/2}$ time dependence we found is characteristic of diffusion:

$$\text{Einstein diffusion equation: } x_{\text{rms}} = (2Dt)^{1/2}$$

and we also discussed diffusion in terms of a random walk process. We found several other fundamental relations among related transport coefficients and the microscopic parameters behind them:

$$\text{Einstein–Smoluchowski equation: } D = \frac{x_0^2}{2\tau}$$

$$\text{Einstein mobility-diffusion equation: } u_+ = \frac{DQ_+}{k_B T}$$

$$\text{Stokes–Einstein equation: } D = \frac{k_B T}{6\pi\eta R}$$

We discussed two experimental methods for measuring viscosity: the Ostwald viscometer, based on a viscous flow rate expression:

$$\text{Poiseuille equation: } \frac{dV}{dt} = \frac{\Delta P \pi R^4}{8L\eta} \text{ (incompressible)}$$
$$= \frac{\pi R^4}{16 L \eta} \frac{P_i^2 - P_f^2}{P_0} \text{ (compressible)}$$

and the falling-sphere method, based on the viscous force experienced by a sphere moving through a viscous medium:

$$\text{Stokes force equation: } F_z = -(6\pi\eta R)v_z$$

We applied the Boltzmann transport equation to gases in some detail and arrived at approximate expressions for gas transport coefficients:

$$\text{Diffusion coefficient: } D = \left(\frac{\pi}{m}\right)^{1/2} \frac{(k_B T)^{3/2}}{4P\sigma}$$

$$\text{Thermal conductivity: } k = \frac{5}{12} \frac{\overline{C}_V}{\sigma N_A} \left(\frac{\pi k_B T}{m}\right)^{1/2}$$

$$\text{Electrical conductivity: } \sigma_e = \frac{n_+}{n} \frac{Q_+^2}{4\sigma} \left(\frac{\pi}{m k_B T}\right)^{1/2}$$

$$\text{Coefficient of viscosity: } \eta = \frac{(\pi k_B T m)^{1/2}}{4\sigma}$$

For ions in solution, the ions move independently at low concentration, and we used limiting ion-specific molar conductances to express the limiting molar conductance of a solution:

$$\text{Kohlrausch's law: } \Lambda^\circ = \nu_+ \lambda_+ + \nu_- \lambda_-$$

FURTHER READING

The books covering the kinetic theory of gases referenced at the end of Chapter 24 also discuss transport phenomena, especially in gases, of course! One of the standard works on transport in general is *Transport Phenomena,* by R. Byron Bird, Warren E. Stewart, and Edwin N. Lightfoot (Wiley, New York, 1960), and the book *Chemical Kinetics and Transport,* by Peter C. Jordan (Plenum, New York, 1979) has several chapters that expand on the material here. A more detailed look at transport from a hydrodynamic point of view is found in *Physical Chemistry,* by R. Stephen Berry, Stuart A. Rice, and John Ross (Wiley, New York, 1980).

The linearized Boltzmann transport equation is discussed in *Fundamentals of Statistical and Thermal Physics,* by F. Reif (McGraw-Hill, New York, 1965) and in the article *Transport in Gases: An Alternative Treatment,* by Bruce H. Mahan [*J. Chem. Ed.* **55**, 23 (1978)]. Section 25.3 is closely based on this latter article.

Biophysical applications of transport are discussed in

Physical Chemistry, 2nd ed., by Ignacio Tinoco, Jr., Kenneth Sauer, and James C. Wang (Prentice-Hall, New York, 1985) and in *Biophysical Chemistry,* by Charles R. Cantor and Paul R. Schimmel (Freeman, San Francisco, 1980).

Access to Data

Most standard data handbooks, such as the *CRC Handbook of Chemistry and Physics,* referenced in Chapter 7, tabulate viscosities, thermal conductivities, and electrical conductivities of a wide range of materials. A particularly useful compilation of data on gaseous H_2O, O_2, N_2, H_2, CO, CO_2, Ar, and air is the *Tables of Thermal Properties of Gases,* by Joseph Hilsenrath, et al. (US National Bureau of Standards Circular 564, US Government Printing Office, Washington, 1955).

PRACTICE WITH EQUATIONS

25A What is the total time derivative of f^* for a system that is spatially uniform in the x and y directions and has no external force acting on it? (25.2b)

ANSWER: $df^*/dt = (\partial f^*/\partial z)(\partial z/\partial t) + \partial f^*/\partial t$

25B What is the linearized Boltzmann equation for a steady-state system in a gravitational field (along the z direction) with spatial gradients along z only? (25.6)

ANSWER: $(\partial F^*/\partial z)(\partial z/\partial t) + (\partial F^*/\partial v_z)g = -(F^* - F)/\tau_{coll}$

25C A beam of H^+ ions with a kinetic energy of 100 eV (and thus a speed $v_1 = 1.38 \times 10^6$ m s^{-1}) passes through a 1.00 cm long cell containing Xe at 300 K and 10 Pa (which is about 75 mtorr). The H^+–Xe collision cross section is 1.50×10^{-20} m^2. What is the average number of collisions each ion experiences? (25.8)

ANSWER: 0.48 collisions (assume $g = v_1$)

25D What is the mean H–H collision rate in the interstellar medium at $T = 50$ K and a density of one H atom per cm^3? Assume $\sigma = 1.0 \times 10^{-17}$ m^2. (25.9)

ANSWER: 1.4×10^{-8} s^{-1} (about one collision every two years!)

25E How many total collisions occur per second in 1.00 mol Ar at 300 K and 1.00 atm? (25.11), Example 25.3

ANSWER: 1.46×10^{33} s^{-1}

25F Natural Ar is 0.337% ^{36}Ar and 0.063% ^{38}Ar. In the Ar sample in 25E, how many ^{36}Ar–^{38}Ar collisions are there per second? (25.12)

ANSWER: 6.46×10^{27} s^{-1}

25G What is the diffusive flux at $z = 1.00$ cm in a system with the density gradient $n(z) = (3 \times 10^{23}$ m$^{-3}) e^{-z/(7.5 \text{ cm})}$ and a diffusion coefficient $D = 1.3 \times 10^{-5}$ m^2 s^{-1}? (25.13)

ANSWER: 4.55×10^{19} m^{-2} s^{-1}

25H How rapidly is the density changing at $z = 1.00$ cm in the system described in 25G? (25.14)

ANSWER: 6.07×10^{20} m^{-3} s^{-1}

25I The concentration profile of a solute diffusing from a plane is measured 980 s after diffusion starts. The rms spread of the profile is found to be 0.18 cm. What is the diffusion coefficient? (25.16)

ANSWER: 1.65×10^{-9} m^2 s^{-1}

25J What is the probability that a 1-D random walker is back at the start of the walk after 100 steps? (25.18)

ANSWER: 0.080

25K If a small liquid molecule diffuses as a random walk with a step size on the order of its diameter (8 Å), taking a step every 0.1 ns, what is its diffusion coefficient? (25.19)

ANSWER: 3.2×10^{-9} m^2 s^{-1}

25L A 1.2 m length of Al wire 0.50 mm in diameter has a measured resistance of 0.162 Ω. What is the conductivity of Al? (25.22)

ANSWER: 3.77×10^7 S m^{-1}

25M How fast would glycerol move at the center of a tube 0.1 cm in radius and 1.0 m long if driven by a 1.0 atm pressure difference? (25.24), Table 25.3

ANSWER: 1.7 cm s^{-1}

25N What is the flow rate through the tube in 25M? (25.25a)

ANSWER: 2.67×10^{-8} m^3 s^{-1}

25O At 25 °C and 1.00 atm, the diffusion coefficient of CH_4 is 2.40×10^{-5} m^2 s^{-1}. What is the hard-sphere cross section for CH_4? (25.33b)

ANSWER: 2.65×10^{-19} m^2

25P What collision time does the diffusion coefficient in 25O predict? (25.33a)

ANSWER: 0.155 ns

25Q At 300 K, the viscosity of H_2 is 8.96×10^{-6} Pa s. What is the viscosity of D_2 at this temperature? (25.42b)

ANSWER: 1.27×10^{-5} Pa s

25R What is the terminal speed of a 10^{-6} m radius drop of mist falling through air ($\eta = 1.85 \times 10^{-5}$ Pa s at 300 K)? (25.44)

ANSWER: 1.2×10^{-4} m s^{-1}

25S What is the diffusion constant of C_{60} ($R = 5$ Å) in n-hexane at 20 °C? (25.45), Table 25.3

ANSWER: 1.3×10^{-9} m^2 s^{-1}

25T What is the speed of $H^+(aq)$ in a 100 V m^{-1} electric field? (25.38), (25.49), Table 25.4

ANSWER: 3.62×10^{-5} m s^{-1}

PROBLEMS

SECTION 25.1

25.1 What is ∇f for the equilibrium distribution function, Eq. (24.7), including all three directions, x, y, and z? What is $\nabla_v f$ for this distribution?

25.2 What is the total derivative dF^*/dt for the following situations:

(a) steady state, concentration gradient in z direction, no external forces,
(b) steady state, concentration gradient in z direction, gravitational field along z direction,
(c) non-steady state, external force of the form $\mathbf{F}_{ext} = F_x \hat{\mathbf{x}} - F_y \hat{\mathbf{y}}$ where F_x and F_y are constants and $\hat{\mathbf{x}}$ and $\hat{\mathbf{y}}$ are unit vectors.

25.3 How long does it take the spike at $v_x = +250$ m s^{-1} in the distribution of Figure 25.1 to fall 95% of the way back to the equilibrium distribution? Derive first a general expression for this time in terms of τ_{coll}. The expression for τ_{coll}^{-1} for molecules with one specific velocity colliding with all other molecules (Eq. (25.8) weighted by f^* and integrated over $\mathbf{v}$ with $\mathbf{v}_1$ a constant) cannot be expressed in closed form. For Ar at 300 K and 0.10 bar, a numerical calculation leads to $\tau_{coll}^{-1} = 2.5 \times 10^8$ s^{-1}. Use this value to find the 95% relaxation time.

25.4 A beam of Ar^+ ions with a kinetic energy of 25 eV is directed into a cell of Ar(g) at 300 K and 10^{-3} torr pressure. What is the collision rate for these ions as they move through the gas? Note that the speed of the ions, v_+, is much greater than the typical speed of the target gas so that $g \cong v_+$. Assume the Ar^+–Ar collision cross section is 8×10^{-20} m^2.

25.5 Around an altitude of 40 km, in the stratosphere, the temperature is about 250 K. The O atom density is about 5×10^8 cm^{-3}, and the O_3 density is about 5×10^{11} cm^{-3}. If the radius of an O atom is 1.40 Å and that of O_3 is 2.0 Å, what are the collision rates per unit volume for (a) O–O collisions, (b) O–O_3 collisions, and (c) O_3–O_3 collisions?

25.6 On February 23, 1987, the first light from a blue supergiant star that had exploded as a supernova some 170 000 years ago reached earth. The light was studied intensely in the following weeks and months, and around 100 days later, the first evidence was found (from IR emission features) that CO molecules were being synthesized in abundance from the supernova's ejected atoms. At this point, the temperature of the material had cooled to around 3000 K and had a density around 1×10^{17} m^{-3}. The most abundant atoms in this gas are He and H, but C and O densities are significant, since before the star exploded, nuclear reactions had produced these atoms preferentially. If we assume 10% of the material was O and 2% was C, values suggested by supernova model calculations, what was the C–O collision rate per unit volume at the time CO was being formed? Assume a collision cross section $\sigma = 3 \times 10^{-19}$ m^2. The CO is thought to be formed directly in the reaction C + O $\rightarrow$ CO + IR radiation, and the data suggest that CO is formed at the rate 2×10^8 molecules m^{-3} s^{-1}. How many C–O collisions occur before one of them leads to CO formation?

25.7 The atmosphere on the surface of Saturn's largest moon, Titan, is largely N_2 at a temperature of 94 K and a pressure of 1.5 bar. What is the ratio of the N_2–N_2 collision rate on Titan's surface to that on Earth's surface?

25.8 The expression for the mean collision rate in a pure gas, $z_1 = P\sigma(16/\pi mk_B T)^{1/2}$, is very similar to the expression for the total rate of collisions with an area A of a container wall, $AJ_z = PA(1/2\pi mk_B T)^{1/2}$. The ratio z_1/AJ_z is $(\sigma/A) \cdot 4\sqrt{2}$, and if we neglect the physical differences between A and σ, this ratio is just the numerical factor $4\sqrt{2}$. Search through the derivations of z_1 and J_x, and locate the *physical* origins of this factor.

25.9 Example 25.5 considered a 50/50 mixture of He and Ar and showed that the He—He collision rate was about twice the Ar–Ar rate. What mixture would have equal rates? What would the He–Ar collision rate be for this mixture?

SECTION 25.2

25.10 Equation (25.15) described diffusion in two directions away from a plane source of molecules. If the diffusion is only along one direction from the plane (as when volatile components of fresh paint diffuse from a wall), the density profile, $n(x,t)$, is only slightly different from Eq. (25.15). Use the difference in normalization for one-sided versus two-sided diffusion to find this difference. In one-sided diffusion, the average distance traveled in time t is not zero. Show that $\langle x \rangle = 2(Dt/\pi)^{1/2}$.

25.11 Another simple solution to Fick's laws describes radial diffusion from a point source of N_0 molecules. One finds $n(r,t) = N_0 e^{-r^2/4Dt}/[(4\pi Dt)^{3/2}]$. Problem 14.33 introduced the silkworm moth attractant pheromone, bombykol ($C_{16}H_{29}OH$). As few as 200 bombykol molecules causes the male moth to flutter his wings in a characteristic frenzy. Suppose 1.0 mg of bombykol is placed in a small spot 1.0 m from a male moth. When will the moth start his frenzy? Assume the frenzy starts when the bombykol concentration at the moth reaches 200 cm^{-3}, and assume $D = 4.1 \times 10^{-7}$ m^2 s^{-1}. Do you conclude that diffusion alone is the major transport mechanism for pheromones?

25.12 Hydrogen gas diffuses far more rapidly through Pd than any other gas. The molecule dissociates on the Pd surface, diffuses as atoms, then recombines and pops off the other side of the metal. This is an important method of purifying H_2. The diffusion constant for H_2 in Pd at 1000 °C is about 10^{-7} m^2 s^{-1}. What is the flow rate (in mol s^{-1}) for H_2 diffusing through a 1.0 mm thick, 1.0 cm^2 area Pd foil at 1000 °C if the impure H_2 is at 1.0 atm and the pure H_2 is diffusing into a container so large that its pressure is negligible?

25.13 If one uses Stirling's approximation in the form

$$\ln n! \cong \left(n + \frac{1}{2}\right)\ln n - n + \ln(2\pi)^{1/2}$$

in Eq. (25.17) written as

$$\ln P = \ln N! - N\ln 2 - \ln\left\{\left[\frac{1}{2}(N+X)\right]!\right\} - \ln\left\{\left[\frac{1}{2}(N-X)\right]!\right\},$$

one can derive the approximation

$$\ln P \cong \ln\left(\frac{2}{\pi N}\right)^{1/2} - \frac{1}{2}[(N+X+1)\ln(1+r) + (N-X+1)\ln(1-r)]$$

where $r = X/N$. If $r \ll 1$, show that the approximation $\ln(1 \pm r) \cong \pm r$ leads to Eq. (25.18).

25.14 While Eq. (25.18) was derived for a large number, N, of random walk steps, plot this function along with Eq. (25.17) for $N = 10$, and show that they are quite similar even for small N.

25.15 The expression for 3-D radial diffusion quoted in Problem 25.11 is easy to derive from Eq. (25.15) for double-sided 1-D diffusion. We picture diffusion from a point as a 3-D random walk with each direction, x, y, and z, independent. The probability that the walk of one molecule from the origin ends at (x, y, z) is

$$\left(\frac{1}{\sqrt{4\pi Dt}}e^{-x^2/4Dt}\right)\left(\frac{1}{\sqrt{4\pi Dt}}e^{-y^2/4Dt}\right)\left(\frac{1}{\sqrt{4\pi Dt}}e^{-z^2/4Dt}\right) = (4\pi Dt)^{-3/2} e^{-r^2/4Dt},$$

which is the expression in Problem 25.11. Now consider the rms distance, $r_{rms} \equiv \langle r^2 \rangle^{1/2}$, traveled during time t. Since $dx\,dy\,dz = r^2 \sin\theta\,dr\,d\theta\,d\phi$, the probability of landing in an infinitesimal volume at a distance r at any angle is

$$P(r)\,dr = (4\pi Dt)^{-3/2}\left(\iint e^{-r^2/4Dt} r^2 \sin\theta\,d\theta\,d\phi\right)dr$$

$$= \frac{4\pi r^2\,dr}{(4\pi Dt)^{3/2}}e^{-r^2/4Dt}.$$

Show that this expression yields $r_{rms} = (6Dt)^{1/2}$, which is $3^{1/2}$ times larger than the 1-D expression in Eq. (25.16). Note that $r_{rms} = \int_0^r r^2 P(r)\,dr$. Now derive the same expression the easy way. Since the walk is random, x, y, and z are independent equivalent directions. Therefore, $\langle r^2 \rangle = \langle x^2 + y^2 + z^2 \rangle = \langle x^2 \rangle + \langle y^2 \rangle + \langle z^2 \rangle$. Show that this expression leads to $r_{rms} = 3^{1/2} x_{rms}$.

25.16 How efficient is a thermopane window? Calculate the heat flux, first, through a 1/8″ ($l = 3.175$ mm) thick pane of glass and then through two such panes separated by a 1/4″ ($L = 6.35$ mm) air space. The thermal conductivity of glass, k_1, is about 0.84 W m^{-1} K^{-1}. Assume the inside temperature is $T_1 = 20$ °C, the outside temperature is $T_2 = 0$ °C, and the thermal conductivity of air, k_2, is (at the average temperature of 10 °C) 24.1 mW m^{-1} K^{-1}. Show that the general expression for the heat flux can be written

$$J_{q,z} = -\frac{(T_1 - T_2)k_1 k_2}{Lk_1 + 2lk_2}$$

where the window parameters are defined in the sketch below. To do this, require the heat flux to be constant through the window (why must it be?), and assume the temperature gradients across each material have the form (temperature drop)/(material thickness). The temperatures T'_1 and T'_2 define the temperature drop across the air space.

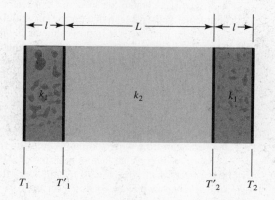

25.17 If a 1.00 cm² × 10.0 cm bar of Cu ($k = 400$ W m⁻¹ K⁻¹) has one end in contact with a 1.0 g sample of ice at 0 °C and the other in contact with a hot plate at 140 °C, how long does it take to melt the ice? The standard molar enthalpy of fusion for water is $\Delta \bar{H}°_{fus} = 6.008$ kJ mol⁻¹.

25.18 The electrical conductivity of pure Cd is $\sigma_e(\text{Cd}) = 1.38 \times 10^7$ S m⁻¹ and that of pure Bi is considerably less, $\sigma_e(\text{Bi}) = 8.67 \times 10^5$ S m⁻¹. The Cd/Bi phase diagram in Figure 6.25 shows that these metals are immiscible at low temperature so that when a homogeneous liquid solution is cooled, a randomized mixture of rather pure Cd and Bi microcrystals results. What would be the conductivity of such a mixture as a function of composition? Imagine a sample of some mixture in the shape of a rod of length L and area A, and suppose it has a measured end-to-end resistance $R = L/A\sigma_e$ as given by Eq. (25.22). Let f_{Cd} and f_{Bi} represent the volume fractions of each element in the sample. Show that the conductivity of the sample is given by

$$\sigma_e = \frac{1}{f_{Cd}/\sigma_{Cd} + f_{Bi}/\sigma_{Bi}} .$$

(*Hint:* It helps to think first about the resistivity and to remember that the total resistance of two resistors in series is the sum of the two individual resistances.) What is the conductivity of a sample containing 5.0 g of Cd and 5.0 g of Bi? The densities are $\rho_{Cd} = 8.65$ g cm⁻³ and $\rho_{Bi} = 9.80$ g cm⁻³.

25.19 The resistivity of Ag at 0 °C is 1.506 μΩ cm, and its temperature coefficient of resistivity is 6.12×10^{-3} μΩ cm K⁻¹. The resistivity of liquid Ag is 11.3 μΩ cm at 1000 °C and 15.3 μΩ cm at 1500 °C. What is the electrical conductivity of Ag(s) and Ag(ℓ) at the melting point, 1234 K? Do you conclude that electrical conductivity is a continuous function of T across a metal's solid–liquid phase transition?

25.20 The derivation of Poiseuille's equation for incompressible fluid flow, Eq. (25.25a), follows these steps. Consider the flow through a ring of area $2\pi r\, dr$ concentric with the tube. The volume flow rate at this radius is this area times the flow velocity at r, $v_z(r)$, as given by Eq. (25.24). The total volume flow rate is the integral of the flow at r over r from $r = 0$ to $r = R$. Show that this leads to Eq. (25.25a).

25.21 To derive the Poiseuille equation, Eq. (25.25b), for a compressible fluid, it helps to think of Eq. (25.24) in terms of the pressure gradient along the axis of the tube:

$$v_z(r) = \frac{1}{4\eta}\left(\frac{dP}{dz}\right)(r^2 - R^2) .$$

The volume flow rate is constant along the tube if the fluid is incompressible (the previous problem), but if the fluid is compressible, it is the *particle* flow rate that must be constant (in order to conserve matter). We still write the volume flow rate as

$$\frac{dV}{dt} = \int_0^R v_z(r)\, 2\pi r\, dr = \frac{\pi R^4}{8\eta}\left(\frac{dP}{dz}\right) ,$$

but if the gas is ideal, the particle flow rate is $dN/dt = (P/k_BT)(dV/dt) = (\pi R^4 P/8\eta k_BT)(dP/dz)$. Use the identity $2P(dP/dz) = [d(P^2)/dz]$ and the requirement that dN/dt is independent of z to complete the derivation of Eq. (25.25b).

25.22 In the course of your design of a high-performance liquid chromatograph, an HPLC, you begin with an estimate of the pressure required to force the liquid through the capillary chromatograph column. You plan to pump *n*-hexane through a column 1.0 m long with a bore 1.0 mm in diameter. What pressure must your pump generate to pass a sample injected at the high-pressure end of the column through the column in 120 s? You will find a very easily attained pressure, but a true HPLC column is packed with a chromatographic support, greatly restricting the flow through the column and requiring pumps capable of pressures on the order of 50–100 atm.

25.23 In vacuum engineering, one defines the *flow rate* of gas as $Q = P_0(dV/dt)$ and the *conductance* of a tube as $F = Q/(P_i - P_f)$. Find the conductance of a 0.60 cm inside diameter vacuum hose 25 cm long carrying air ($\eta = 1.85 \times 10^{-5}$ Pa s) at an average pressure of 100 torr.

25.24 Would you have the patience to measure the viscosity of glycerol in the Ostwald viscometer described in Example 25.9? The density of glycerol is 1 261.3 kg m⁻³ at 20 °C.

SECTION 25.3

25.25 The data discussed in Example 25.10 show that the collision cross section depends on temperature. Find the effective hard-sphere cross section at the highest and lowest temperatures quoted in that Example, convert them to effective diameters, and contrast them to the Ar–Ar nearest neighbor distances in solid Ar, 3.75 Å. Are the magnitudes and trend of your effective diameters consistent with this number?

25.26 If two isotopes of mass m_1 and m_2 interdiffuse, the mutual diffusion coefficient, D_{12}, is given by Eq. (25.33b) with m replaced by μ, the reduced mass of the isotopes. Show that this leads to the relation $D_{11} = [2m_2/(m_1 + m_2)]^{1/2} D_{12}$ between the mutual diffusion coefficient and the self-diffusion coefficient, D_{11}, for isotope 1. The paper cited in Example 25.10 also quotes data derived from the diffusion of ^{22}Ne into ^{20}Ne, expressed as D_{11} for ^{20}Ne at several temperatures. In particular, $D_{11} = 6.68 \times 10^{-5}$ m^2 s^{-1} at 80 °C and 1.00 atm. Find D_{12} and the Ne hard-sphere diameter at these conditions. Predict D_{11} at 0 °C, and compare it to the measured value, 4.35×10^{-5} m^2 s^{-1}.

25.27 Problem 25S considers the diffusion of C_{60} in a liquid, but the molecule itself is produced in a C arc in 300 torr He. Estimate the mutual diffusion coefficient (see also the previous problem) for C_{60} in He at this pressure and a temperature of 400 K. Take the radius of He to be 2.65 Å.

25.28 The integral that takes us from Eq. (25.34) to Eq. (25.35a) is most easily done in spherical polar velocity coordinates, $dv_x dv_y dv_z = v^2 \sin\theta\, dv\, d\theta\, d\phi$ and $v_z = v\cos\theta$. The distribution function in these coordinates is $F = nf = n(m/2\pi k_B T)^{3/2} \exp(-mv^2/2k_B T)$. Show that the integral equals $5nk_B^2 T^2/m$. You will need the following integrals:

$$\int_0^\pi \cos^2\theta \sin\theta\, d\theta = \frac{3}{2}, \quad \int_0^\infty x^8 e^{-x^2} dx = \frac{105\sqrt{\pi}}{32},$$

$$\int_0^\infty x^6 e^{-x^2} dx = \frac{15\sqrt{\pi}}{16}.$$

25.29 How well does Eq. (25.35b) predict the thermal conductivity of air quoted in Problem 25.16? Assume the cross section for "air" is 4.3×10^{-19} m^2, the N_2 cross section quoted in Example 25.11, and estimate the mass and heat capacity on the basis that air is 80% N_2 and 20% O_2.

25.30 Show that the diffusion constant and the thermal conductivity in the simplified hard-sphere theory are related through $D = 2kT/5P$. Test this expression using data for Ar at 300 K and 1.00 atm quoted in Examples 25.10 and 25.11, then test it for CH_4 under the same conditions using the experimental values $k = 3.42 \times 10^{-2}$ W m^{-1} K^{-1} and $D = 2.39 \times 10^{-5}$ m^2 s^{-1}. Explain why the expression works much better for Ar than for CH_4, and suggest a way to improve it, arguing as follows. The expression was derived for hard spheres, but Eq. (25.35b) shows $k \propto \overline{C}_V$. If the gas does not have the hard-sphere heat capacity, the expression can be corrected if we use an equivalent hard-sphere thermal conductivity defined by $k_{hs} = k_{exp}(\overline{C}_{V,hs}/\overline{C}_{V,exp})$ where the subscript exp refers to experimental values. Make this correction, estimating $\overline{C}_V$ for CH_4, and see if it improves the agreement.

25.31 The common vacuum gauge known as a "thermocouple gauge" works as follows. A constant current is passed through a wire suspended in the gas of interest. The current heats the wire, and a steady-state temperature is reached as the wire loses energy through radiation and thermal conduction through the gas to the gauge walls. A thermocouple records the temperature of the wire, and this is converted into a pressure reading. What factors limit the pressure range over which the gauge is useful? Does the pressure reading have to be corrected for the nature of the gas? In other words, would He at one pressure result in the same reading as Ar at the same pressure?

25.32 In the apparatus shown in cross-section below, $N_2(g)$ at 0.10 atm pressure and 300 K is contained in a tube the walls of which are coated with ^{241}Am, the same isotope used in home smoke detectors to generate small amounts of gaseous ions (^{241}Am, with a half-life of 432.2 y, emits 5.6 MeV α particles). The amount of Am in the apparatus below is sufficient to maintain 1 part in 10^8 of the gas ionized as singly charged ions. The metal plates at the ends of the tube are 1.0 cm^2 in area, 10 cm apart, and are connected to a 50 V power supply. What current of N_2^+ ions flows in the tube?

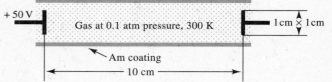

25.33 The derivation of the charge flux equation, Eq. (25.36), involved the equation $F'/F = \tau_{coll} Q v_z E_z/k_B T$. Show that if one approximates $\tau_{coll} v_z \cong \lambda$, the mean free path, this ratio can be written $F'/F = QE_z/\sqrt{2} P\sigma$. The theory requires $F'/F \ll 1$. What is the largest field that keeps $F'/F < 0.01$ for a singly charged ion at 0.1 atm? Use a value for the cross section that is representative of the simple gases we have discussed in this chapter.

25.34 In 1660, Robert Boyle measured the "damping time" of a pendulum in a vessel at two different air pressures. If a pendulum takes 15 min to come to rest swinging in air at 1.0 atm, will it come to rest faster or slower in air at 0.1 atm?

25.35 Show that our simplified hard-sphere transport theory predicts $\rho D = \eta$ where ρ is the gas density. Test this prediction using data for Ar at 300 K and 1.00 atm: $D = 1.833 \times 10^{-5}$ m^2 s^{-1} and $\eta = 2.294 \times 10^{-5}$ Pa s.

25.36 The viscosity of water vapor at 300 K is 9.81×10^{-6} Pa s, and at 1500 K, it is 5.17×10^{-5} Pa s. At the same two temperatures, the viscosity of CO is 1.79×10^{-5} Pa s and 5.44×10^{-5} Pa s. Calculate the hard-sphere cross sections these values imply, based on Eq. (25.42b). We know to expect σ to decrease with increasing temperature, since molecules are not really hard spheres, but can you explain why σ for H_2O decreases much more than for CO? Why does H_2O appear to be much larger than CO at 300 K in this model?

SECTION 25.4

25.37 One of the classic experiments in 20th century physics is the *Millikan oil-drop experiment* of 1909, which established the magnitude of the elementary charge. Millikan used real oil mist, but modern replicas of his experiment use plastic spheres of very uniform diameter. One version uses spheres with a radius of 0.90 μm and a mass of 2.9×10^{-15} kg. If one such sphere carried a single excess electron, what would be its terminal speed moving in air at 300 K (1.85×10^{-5} Pa s) under the influence of gravity alone? What electric field would be required to hold the sphere stationary? What is the terminal velocity if this field is reversed? What is the general expression for the terminal velocity in an arbitrary electric field? (It is an acceptable approximation to neglect the buoyant force of the air.)

25.38 The viscosity of seawater is a function of its composition. Oceanographers express seawater composition in terms of its *salinity*, an operationally defined quantity based on conductivity, but originally defined as the weight of inorganic salts contained in 1 kg seawater after bromides and iodides have been converted to chlorides and carbonates to oxides. It is usually expressed as a percentage, denoted S, and ocean water usually covers the range 33–38‰, but S in the Red Sea is as high as 45‰, and the relatively enclosed Baltic has S only 7–8‰ at the surface due to inflowing freshwater rivers. The viscosity as a function of salinity percentage at 20 °C is $\eta(S) = (1.002 \times 10^{-3}) \times (1 + 2.2 \times 10^{-3} S)$ Pa s. Is a falling-sphere viscometer useful as a salinity-measuring device? Imagine using the Pb sphere in Example 25.14, and calculate the sphere's terminal velocity at several salinities. How precisely must you measure this velocity to determine salinity to 10%? to 1%? to distinguish Red Sea water from Gulf of Finland water? (The density of seawater also depends on its salinity, but you can assume a constant density here for simplicity.)

25.39 The HIV virus is roughly spherical with a radius about 500 Å. How long does it take the virus to diffuse the length of a typical cell, 10^{-5} m, assuming it diffuses through water?

25.40 A weak electrolyte has a conductivity that depends strongly on concentration due to the ionization equilibrium. If the electrolyte is a weak acid, HA, the equilibrium is $HA + H_2O \rightleftarrows A^- + H_3O^+$ with an ionization equilibrium constant K_{eq}. If concentrations are so low that activity coefficients can be neglected and if $[HA]_0$ represents the total concentration of dissociated and undissociated acid at equilibrium (i.e., $[HA]_0 = [HA] + [A^-]$), we can define the degree of ionization, $\alpha = [A^-]/[HA]_0$. Show first that the equilibrium constant expression, written in terms of α and $[HA]_0$, can be manipulated into the form $1/\alpha = 1 + \alpha[HA]_0/K_{eq}$. The molar conductivity is related to the degree of dissociation and the limiting conductivity through the simple relation $\Lambda = \alpha \Lambda°$. Combine this and the previous equation to derive the *Ostwald dilution law,* first derived in 1888:

$$\frac{1}{\Lambda} = \frac{1}{\Lambda°} + \frac{\Lambda[HA]_0}{(\Lambda°)^2 K_{eq}}.$$

Manipulate the conductivity data below for formic acid so that they can be used in this law to produce a straight line. Plot them, deduce $\Lambda°$ and K_{eq} from the plot, and compare $\Lambda°$ to the value calculated from data in Table 25.4.

[HCOOH]/M	1.00×10^{-5}	5.00×10^{-5}	1.00×10^{-4}	5.00×10^{-4}	1.00×10^{-3}
Λ/S m^2 mol^{-1}	0.038	0.033	0.029	0.018	0.014

25.41 Conductance methods can be used to find end-points in so-called *conductometric* titrations. Consider first titration of HCl with NaOH. Sketch the qualitative curve of conductance versus volume of added NaOH. You might find it helpful to consider the conductances of each ion in solution during the course of the titration. The observed curve is the sum of curves for the individual ions. Assume the concentration of NaOH is sufficiently high that dilution effects can be neglected. Next, plot the titration curve for the titration of $AgNO_3$ with a solution of NaCl to an added volume equal to twice the equivalence point. Note that only three points determine the general shape of the plot: at the beginning, at the equivalence point, and at the end. Assume AgCl is completely insoluble.

25.42 Derive the *Nernst–Einstein relation* between the limiting molar conductance and the diffusion coefficient: $\lambda_\pm = Q_\pm^2 D N_A / k_B T$. If this is combined with the Stokes–Einstein relation, Eq. (25.45), one finds an expression for the so-called hydrodynamic radius of an ion. Calculate the aqueous hydrodynamic radii of the alkali metal ions, and compare them to the ionic solid crystal radii: Li^+–Cs^+: 0.59 Å, 1.02 Å, 1.38 Å, 1.49 Å, and 1.70 Å. You should find the trend is in the opposite direction. What physical reason or reasons can you give for this different trend and the surprising prediction that the hydrodynamic radius of the heavier ions is *less* than their crystal radius?

GENERAL PROBLEMS

25.43 The text discusses the Eucken factor, f, relating the thermal conductivity, viscosity coefficient, and constant volume heat capacity: $k = f\eta \overline{C}_V/N_A m$ where $f = (9\gamma - 5)/4$ with $\gamma = \overline{C}_P/\overline{C}_V$. Assuming $\overline{C}_P = \overline{C}_V + R$, calculate the Eucken factor for O_2, CO_2, and H_2O, and see how well it predicts k given, at 400 K, $\eta(O_2) = 2.56 \times 10^{-5}$ Pa s, $\overline{C}_V(O_2)/R = 2.621$, $\eta(CO_2) = 1.93 \times 10^{-5}$ Pa s, $\overline{C}_V(CO_2)/R = 3.984$, $\eta(H_2O) = 1.34 \times 10^{-5}$ Pa s, and $\overline{C}_V(H_2O)/R = 3.355$. The experimental values are $k(O_2) = 34.6$ mW m^{-1} K^{-1}, $k(CO_2) = 24.6$ mW m^{-1} K^{-1}, and $k(H_2O) = 26.0$ mW m^{-1} K^{-1}. Can you suggest a reason for the one glaring disagreement among these calculations?

25.44 The ancient water clock known as a *clepsydra*, which has been dated to at least 1600 BC in Egypt, operated on the basis of water dripping through a small hole. As the water level above the hole fell, it passed graduations that marked the passage of time. How should one calibrate the clepsydra illustrated below? It has a capillary 1/2 cm long and 1/4 mm in radius, and the reservoir is a cylinder 10 cm in diameter. Imagine charging the clock with 1 L of water, bringing the level to a height h_0 above the capillary entrance. Show that the pressure driving the water through the capillary is $\Delta P = \rho g h$ where ρ is the density of water, g is the acceleration of gravity, and h is the instantaneous position of the water level, a function of time. The volume of water delivered equals $A(h_0 - h)$ where A is the reservoir area so that $dV/dt = -A(dh/dt)$. Put all this together with the Poiseuille equation, and show that the level falls exponentially in time. You should find that this clock is good for about an hour. Contrast its operation on a hot day in Egypt ($T = 100$ °F $= 311$ K) to a winter's day farther north, at 33 °F $= 274$ K. The density of water at 100 °F is about 993 kg m^{-3}, and it is about 1000 kg m^{-3} at 33 °F. The viscosity of water at any T can be calculated from the formula in Table 25.3.

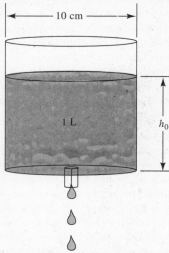

25.45 The temperature dependence of viscosity in a liquid follows an activated process relation of the form $\eta = \eta_0 \, e^{E_a/k_B T}$ where η_0 is a constant and E_a is the activation energy for viscosity. Find these quantities for liquid Hg and benzene from the data below.

Hg:

$T/°C$	−20	20	60	100	200	300
$\eta/10^{-3}$ Pa s	1.855	1.554	1.367	1.240	1.052	0.950

C_6H_6:

$T/°C$	0	10	20	40	60	80
$\eta/10^{-3}$ Pa s	0.912	0.758	0.652	0.503	0.392	0.329

A semiempirical expression due to Eyring states that $\eta = \eta_0 \, e^{3.8 T_b/T}$ where T_b is the normal boiling point. This expression is based on Trouton's rule (see Section 6.2), and you should find that it works much better for one of these liquids than for the other ($T_b = 353.3$ K for benzene, 680 K for Hg). Which one is it and why?

CHAPTER 26

The Phenomenology of Chemical Reaction Rates

REACTING compounds establish the most important nonequilibrium distributions in chemistry. When reactive species are mixed, they slide down a free-energy gradient toward an equilibrium mixture of reaction products and unspent reactants. Chapter 7 described that gradient and showed how to find its equilibrium point, and Chapter 25 used nonreactive concentration gradients to describe diffusion. Here, we put both ideas together. We ask how a nonequilibrium distribution of concentrations relaxes in time to the equilibrium distribution through chemical reactions. It will take this and the following two chapters to complete the story as we look at chemical reaction rates from macroscopic and microscopic points of view.

If thermodynamics is the soul of physical chemistry and quantum mechanics is its mind, reaction kinetics is its heart. This heart can beat at an astounding pace: reactions are studied today on a sub-picosecond time scale. It can also be languorous: Lavoisier stoked a furnace nonstop for twelve days to complete his celebrated experiment on the reaction of Hg with the oxygen in air. Physical chemistry not only tries to understand these time scales, it also tries to find ways to *control* them. Think how marvelous it would be if all reactions proceeded at the rate of our choosing, leading only to those products we want and minimizing undesirable products. As we progress through this and the next two chapters, we will learn the limits Nature imposes on us as we try to realize chemical reaction control.

26.1 Basic Ideas—Reactants Go, Products Appear

26.2 Mechanisms and Integrated Rate Expressions

26.3 The Effects of Thermodynamic Variables on Reaction Rates

26.4 The Attainment of Chemical Equilibrium

26.1 BASIC IDEAS—REACTANTS GO, PRODUCTS APPEAR

Chemical reaction kinetics has its own language, just as thermodynamics does, and our first task is to learn how to speak this language. If you studied chemical kinetics in a general chemistry course, some of this will be familiar to you, but some will be new.

The key macroscopic quantity of chemical kinetics is the time derivative of any reactant or product concentration, and for many simple reactions, stoichiometry governs the interrelationship among these derivatives. If we resurrect the general reaction we used to study equilibrium in Chapter 7:

$$\alpha A + \beta B \rightleftarrows \gamma C + \delta D$$

where α, β, γ, and δ are stoichiometric coefficients, whenever α moles of A disappear, sooner or later γ moles of C and δ moles of D must appear, and β moles of B must disappear as well. The reaction progress variable ξ relates amounts of reactants and products to their stoichiometric coefficients and initial amounts n^o through Eq. (7.1):

$$\frac{n_A^o - n_A}{\alpha} = \frac{n_B^o - n_B}{\beta} = \frac{n_C - n_C^o}{\gamma} = \frac{n_D - n_D^o}{\delta} = \xi \;, \quad (7.1)$$

and it is tempting to define the *reaction rate* as $d\xi/dt$, or this quantity converted to concentration units. Often, this is exactly the correct definition, and we write the rate $\mathcal{R}$ in concentration units as

$$\mathcal{R} = \frac{1}{V}\frac{d\xi}{dt} = -\frac{1}{\alpha V}\frac{dn_A}{dt} = -\frac{1}{\beta V}\frac{dn_B}{dt} = \frac{1}{\gamma V}\frac{dn_C}{dt} = \frac{1}{\delta V}\frac{dn_D}{dt} \quad (26.1a)$$

where V is the system's volume. Using the usual notation for molar concentration in mol dm^{-3} = mol L^{-1} = M units, we have

$$\mathcal{R} = -\frac{1}{\alpha}\frac{d[A]}{dt} = -\frac{1}{\beta}\frac{d[B]}{dt} = \frac{1}{\gamma}\frac{d[C]}{dt} = \frac{1}{\delta}\frac{d[D]}{dt} \;. \quad (26.1b)$$

When could this definition fail? Note that we were careful to say that when α moles of A disappear, *sooner or later* γ moles of C appear. In the *net* chemical reaction, the equilibrium reaction mixture's composition is governed by the thermodynamics of A, B, C, and D and the net reaction stoichiometry. Along the way to equilibrium, however, a series of *reaction intermediates* might exist with appreciable concentrations for appreciable times. The net reaction may occur through a *sequence* of reactions containing one or more of these intermediates. Depending on the way the net reactants convert to intermediates and the intermediates convert to net products, the full string of equalities in the reaction rate definition of Eq. (26.1) may or may not hold at all times.

A simple example makes this point clear. Consider the radioactive decay of an unstable isotope A. Suppose A decays to an isotope E that is also radioactive, and suppose E decays to a stable isotope C. If E decays slowly compared to A's decay rate, A quickly disappears, but C appears slowly. The net reaction is A $\rightarrow$ C, but $-d[A]/dt \neq d[C]/dt$ at all times.

However the reactant and product concentration derivatives are related, any one of them is an experimental handle to reaction rates. We can use any convenient method sensitive to a species' concentration to follow that concentration in time. Any physical property of the system—optical absorbance, electrical conductance, electrochemical potential, total pressure, fluorescence intensity, etc.—will do, as long as we can relate this property unambiguously to a species' concentration. Often, the reaction is probed continuously in time as it approaches equilibrium, but for slow reactions, we can sample the mixture periodically, *quench* the reaction in the sample (by changing the temperature, diluting the sample, etc.), and analyze it at our leisure. Any of these methods leads most directly to a plot of concentration versus time, and the *slope* of such a plot is related to the rate through Eq. (26.1b).

It is important to recognize that the rate is itself a function of time. We saw in Chapter 25 that diffusion slowed as the initial concentration gradient approached equilibrium. Similarly, a reaction rate slows as the system comes to chemical equilibrium. At equilibrium, concentrations no longer change in time, and the rate is zero. On the other hand, we might expect the rate to be greatest when the system is *farthest* from equilibrium, and often this is the case. The most common situation mixes pure reactants, placing the system as far from chemical equilibrium as it will ever be, and the *initial rate*, the concentration time derivatives at the moment of mixing, is the maximum rate.

EXAMPLE 26.1

The reactions among the various nitrogen oxides played a central role in the evolution of gas phase reaction kinetics, and one of the earliest studies was the thermal decomposition of N_2O_5 [F. Daniels and E. H. Johnston, *J. Am. Chem. Soc.* **43**, 53 (1921)].[1] The net reaction is

$$N_2O_5(g) \rightarrow 2NO_2(g) + \frac{1}{2}O_2(g) ,$$

which is complicated by the dimerization equilibrium $N_2O_4 \rightleftarrows 2NO_2$. Given the equilibrium constant for this dimerization, the amount of N_2O_5 that decomposes in time can be found from the *change in pressure* of the system. In one experiment at 35 °C, the data can be interpreted to give the following partial pressures of N_2O_5 as a function of time:

t/min	0.0	20.0	40.0	60.0	80.0
$P_{N_2O_5}$/torr	56.0	42.8	36.1	30.3	26.0

What is the average rate of this reaction over each 20 min period?

SOLUTION We convert partial pressures to molar concentrations through the ideal gas equation of state: $n_{N_2O_5}/V = P_{N_2O_5}/RT$, and express the time in seconds:

t/s	0.0	1200	2400	3600	4800
$[N_2O_5]$/mM	2.91	2.22	1.88	1.58	1.35

From Eq. (26.1b) and the net reaction, the instantaneous rate is $\mathcal{R} = -d[N_2O_5]/dt$, but for measurements made periodically rather than continuously, the *average* rate is defined in terms of incremental concentration changes divided by time intervals: $\mathcal{R} = -\Delta[N_2O_5]/\Delta t$. Here, $\Delta t = 1200$ s, and we have four intervals:

interval	0–1200 s	1200–2400 s	2400–3600 s	3600–4800 s
$\Delta[N_2O_5]$/mM	−0.69	−0.34	−0.30	−0.23
$\mathcal{R}/10^{-7}$ M s^{-1}	5.75	2.91	2.52	1.86

---

[1] It is interesting to note that this research was done while Daniels and Johnston were at the Fixed Nitrogen Research Laboratory of the (then) US War Department and the Bureau of Soils of the Department of Agriculture. Their paper starts, "In connection with the work of the Fixed Nitrogen Research Laboratory, it became desirable to measure the rate of decomposition of gaseous nitrogen pentoxide. ... such reactions are of great importance for the theory of chemical reactions." Fifty years later, this (and other) pure research studies of nitrogen oxide kinetics played a central role in the discovery and understanding of anthropogenic sources that attack atmospheric ozone.

Note how the rate is largest at the start of the reaction and is decreasing in time as equilibrium is approached. At 35 °C, the net reaction equilibrium constant is 130 so that there is very little N_2O_5 at equilibrium ($P_{N_2O_5}^{eq} = 0.025$ torr), and the data show that the reaction is still far from equilibrium after 80 min, as the graph below suggests:

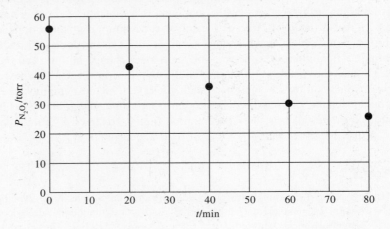

▶ **RELATED PROBLEMS** 26.2, 26.3, 26.4

The reason the rate varies with time is obvious: the *rate depends on concentrations* and the concentrations change in time as equilibrium is approached. One of the goals of chemical kinetics studies is establishing the relationship between the rate and the concentrations that affect it, an expression known as the *rate law*. This law holds many clues to the molecular nature of the reaction. Consider again our general reaction. The rate is potentially a function of the concentrations of *all* the species in the net reaction and perhaps of other species E, F, etc., that are not in the net reaction but are necessary for the reaction to proceed:

$$\mathcal{R} = f([A], [B], [C], [D], [E], [F], \ldots) \, .$$

We might find $\mathcal{R} = k[A][B]$, for example, where k is a proportionality constant, or perhaps $\mathcal{R} = k'[A]^2[B]$ or some other simple combination of powers of concentrations. This is often the form the rate law takes for the initial stage of a reaction begun with mixed reactants and no products. Neither of these simple laws can predict the equilibrium rate, $\mathcal{R} = 0$, and a more complicated expression is needed as the system approaches equilibrium, as discussed in Section 26.4.

But whatever the rate law may be, it is important to recognize that it cannot in general be predicted *a priori*. It must be gleaned from experimental rate data. We will see certain kinds of reactions for which the stoichiometry of the net reaction is enough to predict the rate law, and we will spend most of Chapter 27 studying these so-called *elementary reactions*. For more complicated reactions, the rate law and the net reaction often have no obvious connection.

There is an accepted terminology associated with rate laws. The *exponent* of each concentration in a law is called the *order* of the reaction with respect to that compound. For example, if $\mathcal{R} = k'[A]^2[B]$, we say the rate is *second order* with respect to A and *first order* with respect to B. Orders are generally integral, half-integral, or, rarely, third-integral in the range -1 to 2. Some rate laws do not include certain reactant (or product) concentrations, and the reaction is said to be *zeroth order* in those species, while compounds that are *not* in the net reaction but appear in the law are called *catalysts*. They may either increase or decrease the rate as their concentration increases. (If they decrease the rate, they are generally called *inhibitors* rather than catalysts.) Some catalysts (such as surfaces that catalyze certain reactions) formally have a zero order, but their presence or absence has a notable effect on the rate.

The sum of all the orders in the rate law is called the *reaction order,* and we say $\mathcal{R} = k'[A]^2[B]$ is a third-order reaction, for example. Sometimes, however, the law is sufficiently complicated that no overall reaction order can be identified. For example, a rate law of the form $\mathcal{R} = k[A][B]/(1 + \alpha[C])$ where α is a constant has no obvious order.

A term related to the reaction order is its *molecularity*. This term is usually reserved for only three cases: first order, called *unimolecular;* second order, called *bimolecular;* and third order, called *termolecular*. We will see how these terms relate to the molecular mechanism of reactions in the next chapter.

Finally, the proportionality constant between the concentration factors and the rate, (i.e., k in $\mathcal{R} = k[A][B]$ or k' in $\mathcal{R} = k'[A]^2[B]$), is called the *rate constant*. Since the rate has the dimensions of (concentration)/(time), the rate constant has variable dimensions depending on the order parameters in the rate law. For example, k in our second-order rate law has the dimensions (concentration)$^{-1}$ (time)$^{-1}$, and the third-order constant k' has the dimensions (concentration)$^{-2}$ (time)$^{-1}$. A first-order rate constant has the simple dimension (time)$^{-1}$.

EXAMPLE 26.2

Among the many reactions that govern the H_2–O_2 flame, the $H + O_2 \rightarrow O + OH$ reaction has the initial rate 15.5 M s^{-1} at 2000 K for [H] = 8.0×10^{-6} M and [O_2] = 6.0×10^{-4} M. The rate law is known to be bimolecular, first order in each reactant. What is the rate constant?

SOLUTION The rate law is $\mathcal{R} = k[H][O_2]$—first order in each reactant and bimolecular overall. The rate constant is

$$k = \frac{\mathcal{R}}{[H][O_2]} = \frac{15.5 \text{ M s}^{-1}}{(8.0 \times 10^{-6} \text{ M})(6.0 \times 10^{-4} \text{ M})} = 3.2 \times 10^9 \text{ M}^{-1} \text{ s}^{-1}$$

in molarity units for the concentration. Gas phase reaction rate constants are often expressed in molecule cm^{-3} concentration units instead of mol dm^{-3} units, and the conversion factor is $N_A/(1000 \text{ cm}^3 \text{ dm}^{-3}) \cong 6.02 \times 10^{20}$ molecule M^{-1} cm^{-3} so that we could also express this rate constant as $(3.2 \times 10^9 \text{ M}^{-1} \text{ s}^{-1})/(6.02 \times 10^{20} \text{ molecule M}^{-1} \text{ cm}^{-3}) = 5.3 \times 10^{-12}$ cm^3 molecule^{-1} s^{-1}.

⟹ **RELATED PROBLEMS** 26.4, 26.5, 26.7, 26.8

It is axiomatic that two molecules must collide before they can react. The rate constant holds, in a complicated way, the *dynamics* of these collisions, heavily averaged over all the collisions that can occur at a given total system energy. Reaction dynamics asks (and answers) the following questions: If reactant molecules are colliding at an enormous rate, why is the reaction rate often much slower? When two reactants collide, does their *internal* energy state affect the reaction rate? Does the relative *orientation* of the molecules as they collide have an effect on the rate? We will address some of these questions in this and the next chapter, and Chapter 28 considers chemical reaction dynamics exclusively.

26.2 MECHANISMS AND INTEGRATED RATE EXPRESSIONS

There are many tricks that kineticists exploit to uncover experimental rate laws. Most of them rely on clever choices of experimental conditions that simplify the kinetics as much as possible. These conditions are aimed at reducing the rate law to a simple form that leads to a characteristic time behavior during at least part of the system's progress toward equilibrium.

Consider, for example, a reaction that has a complicated general rate law involving many reactants and products. A technique known as *isolation* arranges concentrations at the outset of the reaction ("time zero," the time of initial, rapid reactant mixing or reaction initiation in general) so that only *one* compound changes concentration appreciably during the reaction. If no products are included in the initial mixture, then all terms involving product concentrations are zero initially and remain small during the early phase of the reaction. If the general rate law is reduced to, say, $\mathcal{R} = k[A]^2[B]$ by this trick, we can choose initial concentrations $[A]_0 \gg [B]_0$ and keep $[A]$ approximately constant as B disappears, reducing the rate law to $\mathcal{R} = k[A]_0^2[B] = k_A[B]$ where $k_A = k[A]_0^2$ is an effective rate constant. Similarly, we could make $[B]_0 \gg [A]_0$ so that $\mathcal{R} = k_B[A]^2$ where $k_B = k[B]_0$. Making one reactant the minor species isolates its effect on the observed rate and exposes its role in the rate law.

First-Order Reactions

If we have a reaction that follows a simple rate law, either in general or by experimental design, we can find an expression for the reactant concentration as a function of time. We integrate the differential rate law, using the initial concentrations as integral limits. The simplest case is a first-order reaction, A → Products, with the differential rate law

$$\mathcal{R} = -\frac{d[A]}{dt} = k_1[A] \tag{26.2}$$

where k_1 is the first-order rate constant. If $[A] = [A]_0$ at $t = 0$, we can rearrange this differential equation for $[A]$ and integrate it to find $[A]$ as a function of time:

$$\int_{[A]_0}^{[A]} \frac{d[A]}{[A]} = -k_1 \int_0^t dt = \ln\frac{[A]}{[A]_0} = -k_1 t \ ,$$

which can be written

$$[A] = [A]_0 \, e^{-k_1 t} \ . \tag{26.3}$$

We have seen this kind of expression before. In Chapter 18, spontaneous emission of an excited state led to Eq. (18.22) for the number of excited state species left after time t: $N(t) = N° e^{-t/\tau_{\text{rad}}}$, where τ_{rad} was the radiative lifetime. Here we can interpret τ_{rad} as the reciprocal of a first-order rate constant (the Einstein spontaneous emission rate constant) for the reaction (excited state) → (relaxed state). In Chapter 25, Example 25.2 showed how a nonequilibrium velocity distribution relaxed exponentially in time with a collision time scale τ_{coll} that also acts like the reciprocal of a first-order rate constant. Other important first order processes include radioactive decay of nuclei, many isomerization reactions such as $CH_3CN \rightarrow CH_3NC$, and some decomposition reactions such as that of N_2O_5 mentioned in Example 26.1.

A first-order process has a characteristic signature: exponential time dependence, and a plot of $\ln([A]/[A]_0)$ versus time is a straight line of slope $-k_1$. We can also measure the time for a fixed fraction of A to disappear, and for a first-order process, this time is *independent* of when in the course of the reaction we measure it.[2]

Consider, for example, the time for half of the amount of A present at any time to disappear, the so-called *half-life*, $t_{1/2}$. We write Eq. (26.3) as

$$\frac{[A]_0}{2} = [A]_0\, e^{-k_1 t_{1/2}} \quad \text{or} \quad t_{1/2} = \frac{-\ln(1/2)}{k_1} = \frac{0.6931}{k_1}. \tag{26.4}$$

After a time $t_{1/2}$ has passed, one half of A is gone, and in another $t_{1/2}$ period, half of what was left (three fourths in all) is gone, and so on. This constant internal clock is the basis for radioactivity dating methods.

EXAMPLE 26.3

Tritium, ^{3}H, was first synthesized in 1935 by bombarding deuterated phosphoric acid with deuterons. Tritium spontaneously decays to ^{3}He by emission of a β particle (an electron emitted from the nucleus) and an antineutrino. The decay has a 12.26 yr half-life. What fraction of the original ^{3}H synthesized in 1935 will still be around to celebrate the centennial of its discovery?

SOLUTION There are two equivalent ways to answer this question. The most direct turns the half-life into its corresponding first-order rate constant:

$$k_1 = \frac{0.6931}{12.26 \text{ yr}} = 0.056\ 53 \text{ yr}^{-1}$$

and uses Eq. (26.3) to find the fraction remaining, $[A]/[A]_0$:

$$\frac{[A]}{[A]_0} = \exp[-(0.056\ 53 \text{ yr}^{-1})(100 \text{ yr})] = 3.51 \times 10^{-3}\ .$$

A second approach says that after n half-lives, the fraction remaining equals 2^{-n}. (See Problem 26.11.) Here, $n = (100 \text{ yr})/(12.26 \text{ yr}) = 8.16$, and $2^{-8.16} = 3.51 \times 10^{-3}$, in agreement with the other calculation. Note that there is nothing magical about choosing 1/2 as the representative fraction. A first-order decay can be discussed in terms of the quarter-life, tenth-life, etc., just as easily. One does not have to wait for a half-life to pass to measure one, as is obvious if one consults a table of radioactive half-lives that are in some cases many millions of years or more!

→ RELATED PROBLEMS 26.7, 26.11

Second-Order Reactions

The next simplest rate law (aside from zero order, for which $[A] = [A]_0 - k_0 t$) is a second-order law. There are two major variants of this law. In the first, the typical reaction and its rate law are

$$A + A \rightarrow \text{Products} \qquad \mathcal{R} = -\frac{1}{2}\frac{d[A]}{dt} = k_2[A]^2 \tag{26.5}$$

while the second has

$$A + B \rightarrow \text{Products} \qquad \mathcal{R} = -\frac{d[A]}{dt} = k_2'[A][B]\ . \tag{26.6}$$

---

[2] If the products of a first-order reaction reach an appreciable concentration and begin to back-react rapidly to produce more of the initial reactant or reactants, this statement is no longer true.

There are many examples of each kind. The first represents, for example, the association of identical species, as in $2CH_3 \rightarrow C_2H_6$, $2NO_2 \rightarrow N_2O_4$, or $2Ar \rightarrow Ar_2$, while the second represents many types of bimolecular processes.

The first is the simpler to integrate. We again imagine $[A] = [A]_0$ at $t = 0$, and separate variables in the differential rate law, Eq. (26.5):

$$\int_{[A]_0}^{[A]} \frac{d[A]}{[A]^2} = -2k_2 \int_0^t dt$$

finding

$$\frac{1}{[A]} = \frac{1}{[A]_0} + 2k_2 t \quad . \tag{26.7}$$

Leaving the equation in this form (rather than writing it in the form "$[A] = \ldots$") emphasizes the characteristic signature of a second-order reaction: the *reciprocal* of the concentration is a linear function of time. We can again find a half-life if we set $[A] = [A]_0/2$ in Eq. (26.7):

$$t_{1/2} = \frac{1}{2[A]_0 k_2} \, , \tag{26.8}$$

which, unlike the first-order half-life, depends on the concentration $[A]_0$ at the start of the half-life measurement. This difference between first- and second-order behavior is illustrated in Figure 26.1. We imagine a second-order reaction begun under conditions such that it has an *initial* half-life equal to that for some reference first-order reaction. (Note in the figure that $[A]/[A]_0 = 1/2$ at the same time for both reactions.) If we wait for the first-order reaction to pass through two half-lives, we see that the second-order reaction has fallen behind and continues to do so. When the first-order reaction reaches $[A]/[A]_0 = 1/8$, the second-order reaction has reached only $1/4$.

The physical reason behind the relative slowing down of the second-order reaction is easy to understand. It takes *pairs* of A molecules to cause the second-order reaction, and the probability that one A finds another is falling as $[A]$ falls. The first-order reaction, in contrast, happens to A molecules in isolation.

For the other type of second-order reaction, $A + B \rightarrow$ Products, we integrate the rate law in Eq. (26.6). If we let x equal the concentration drop in A at time t,

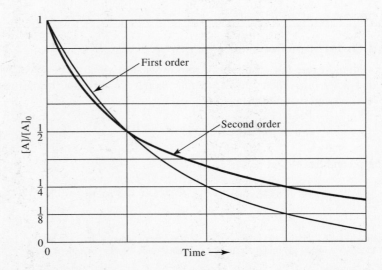

FIGURE 26.1 The half-life of a first-order reaction is constant in time, but the half-life for a second-order reaction *lengthens* in time. This is illustrated here for a second-order reaction chosen to begin with the half-life of a representative first-order reaction. (Both reach $[A]/[A]_0 = 1/2$ at the same time.)

so that $x = [A]_0 - [A]$, then by stoichiometry (every mole of A that reacts takes with it one mole of B)

$$x = [A]_0 - [A] = [B]_0 - [B] \quad \text{and} \quad \frac{dx}{dt} = -\frac{d[A]}{dt} \qquad (26.9)$$

so that the rate law becomes

$$\frac{dx}{dt} = k_2'([A]_0 - x)([B]_0 - x) ,$$

which we integrate (note that $x = 0$ at $t = 0$):

$$k_2' \int_0^t dt = \int_0^x \frac{dx}{([A]_0 - x)([B]_0 - x)} .$$

The first integral is simply $k_2' t$, and the second can be found in standard integral tables or evaluated directly if its integrand is simplified by a partial fraction technique:

$$\frac{1}{([A]_0 - x)([B]_0 - x)} = \frac{1}{([A]_0 - [B]_0)([B]_0 - x)} - \frac{1}{([A]_0 - [B]_0)([A]_0 - x)}$$

leading to the expression

$$k_2' t = \frac{1}{[A]_0 - [B]_0} \left(\ln \frac{[B]_0}{[B]_0 - x} - \ln \frac{[A]_0}{[A]_0 - x} \right) . \qquad (26.10a)$$

Introducing [A] and [B] through Eq. (26.9) and combining the logarithm terms gives

$$k_2' t = \frac{1}{[A]_0 - [B]_0} \ln \frac{[A]/[A]_0}{[B]/[B]_0} . \qquad (26.10b)$$

Both of these are fairly complicated expressions. Equation (26.10a) can be solved for x and thus for [A] or [B] as a function of time (see Problem 26.19), but if one has experimental data and wishes to see if they are consistent with second-order kinetics, one can plot the right-hand side of Eq. (26.10a) versus time and look for a straight line of slope k_2'.

Note that Eq. (26.10), in either form, seems to have a mathematical singularity—division by zero—if $[A]_0 = [B]_0$. Physically, however, this situation reduces to the A + A → Products case, and Eq. (26.7) applies without the factor of 2, which is in Eq. (26.7) only through our standard definition of the rate in Eq. (26.5).

EXAMPLE 26.4

A high-temperature study of the decomposition of SO_2Cl_2 into SO_2 and Cl_2, followed in time by measuring the system's total pressure at constant volume, gave the following results:

t/hr	0	3	6	9	12	15
P_{tot}/kPa	11.07	14.79	17.26	18.90	19.99	20.71

Is the rate law first order or second order in [SO_2Cl_2]?

SOLUTION Since each mole of SO_2Cl_2 that decomposes produces two moles of gas, the total pressure after the partial pressure of SO_2Cl_2 has fallen from an initial value P_0 to a value P_t at time t must be $P_{\text{tot}} = P_t + 2(P_0 - P_t) = 2P_0 - P_t$. We can turn the measurements into P_t values: $P_t = 2P_0 - P_{\text{tot}}$ and make the following table:

t/hr	0	3	6	9	12	15
P_t/kPa	11.07	7.35	4.88	3.24	2.15	1.43

At constant T and V, these partial pressures are directly proportional to molar concentrations. If the reaction is first order, a plot of $\ln P_t$ versus t will be a straight line; if second order, a plot of $1/P_t$ versus t will be straight:

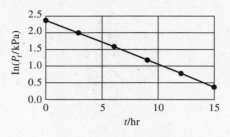
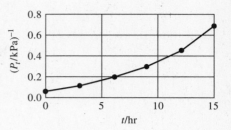

The plots show that the reaction is first order, but note that the second-order plot would look rather linear if data collection had stopped after only 4 or 5 hr. It is important to collect data over several reaction half-lives to make the distinction between these plots obvious.

➡ *RELATED PROBLEMS* 26.13, 26.14, 26.18

Suppose we arrange an "A + B → Products" reaction so that $[A]_0 \gg [B]_0$. This is the isolation method in action, and Eq. (26.10b) should reduce to an expression for [B] alone. If B is the minor species, then $[A]/[A]_0$ will be approximately 1 throughout the reaction, since very little A can disappear even if all of B does. Introducing this approximation along with $[A]_0 - [B]_0 \cong [A]_0$ into Eq. (26.10b) gives us

$$k_2' t = \frac{1}{[A]_0} \ln \frac{1}{[B]/[B]_0}$$

or

$$-[A]_0 k_2' t = \ln \frac{[B]}{[B]_0}$$

so that if we define an *apparent* or *pseudo first-order rate constant* $k_1' = k_2'[A]_0$, we have recovered the isolated, first-order behavior of B:

$$[B] = [B]_0\, e^{-k_1' t}\ .$$

EXAMPLE 26.5

The reaction between H atoms and O_2 mentioned in Example 26.2 was studied at 500 K as a function of O_2 pressure at very low [H], and the half-life for OH production was measured in each experiment, giving the data below:

$[O_2]_0$/mM	3.21	7.06	15.4	24.1
$t_{1/2}$/ms	21.8	9.72	4.48	2.83

Find the second-order rate constant for this reaction.

SOLUTION In this series of experiments, $[O_2]_0 \gg [H]_0$ so that the decay of [H] and the rise of [OH] follows pseudo first-order kinetics at any one $[O_2]_0$. We convert the measured half-lives into pseudo first-order rate constants k_1' with Eq. (26.4), and then use $[O_2]_0$ to turn these into true second-order rate constants $k_2 = k_1'/[O_2]_0$:

k_1'/s^{-1}	31.8	71.3	155.	245.
k_2/M^{-1} s^{-1}	9.9×10^3	1.01×10^4	1.00×10^4	1.02×10^4

The constancy of k_2 within experimental error shows that the reaction is second order overall and that the approximations behind the isolation method were valid.

➡ **RELATED PROBLEMS** 26.19, 26.22

There are several other rate laws for which one can find expressions for concentrations as a function of time, and Table 26.1 collects some of them along with the ones we have discussed. For many rate laws, however, an analytic expression is impossible, and a numerical solution by computer is used. There are many numerical techniques that integrate differential equations through a computer algorithm, and *kinetics modeling*, the technique used by, for instance, atmospheric chemists to predict the concentration profiles of many reacting species in the atmosphere, has become an established technique of the kinetics community. Analytic techniques are generally impossible to find or use if there are many consecutive or sequential reactions among many (often hundreds) of reacting species. We close this section with a look at some of the more common and important building blocks of these more complicated schemes.

The detailed steps taken as reactants evolve toward final stable products are together called the *reaction mechanism*. These steps may include formation of transient reaction intermediates that appear in one reaction early in the scheme but disappear in a later step. Sometimes, intermediates are potentially stable species that happen to find themselves produced in a hostile environment in which they are destroyed, but most often intermediates are chemically unique species that are inherently so reactive that we can never "put them in a bottle" by themselves. Such species include open-shell atoms, free radical molecular fragments, unstable oxidation states of ions in solution, molecular ions, and weakly bound molecular complexes.

Sequential Reactions

It is important to see how the concentration of intermediates changes in the course of a reaction. They are absent at the start, and they should be present in only very small amounts, if at all, at the end, but their time behavior *during* the reaction has

TABLE 26.1 Integrated Rate Expressions

Reaction	Rate law	Integrated form
A → Products	$\mathcal{R} = -\dfrac{d[A]}{dt} = k_0$	$[A] = [A]_0 - k_0 t$
A → Products	$\mathcal{R} = -\dfrac{d[A]}{dt} = k_1[A]$	$[A] = [A]_0 e^{-k_1 t}$
2A → Products	$\mathcal{R} = -\dfrac{1}{2}\dfrac{d[A]}{dt} = k_2[A]^2$	$[A] = [A]_0(1 + 2k_2 t[A]_0)^{-1}$
nA → Products	$\mathcal{R} = -\dfrac{1}{n}\dfrac{d[A]}{dt} = k_n[A]^n$	$[A] = [A]_0[1 + n(n-1)k_n t[A]_0^{n-1}]^{-1/(n-1)}$
A + B → Products	$\mathcal{R} = -\dfrac{d[A]}{dt} = k_2'[A][B]$	$[A] = [A]_0 \left[\dfrac{([B]_0 - [A]_0)e^{-([B]_0 - [A]_0)k_2' t}}{[B]_0 - [A]_0 e^{-([B]_0 - [A]_0)k_2' t}} \right]$

features that simplify analysis of the net reaction. With this in mind, consider the reaction sequence

$$A \xrightarrow{k_1} B \xrightarrow{k'_1} C$$

where each step is first order with rate constants k_1 and k'_1. This could represent a sequence of radioactive decays, or an isomerization through a structure B that was distinct from A and C but less stable than either. The rate laws for this sequence can be integrated exactly (see Problem 26.21) so that we can follow the concentrations of all three species under any hypothetical set of rate constants. Figure 26.2 shows the behavior under the conditions $k'_1 = k_1/3$. The initial reactant, A, disappears rapidly. If it decayed directly to C, [C] would follow the dashed line in the figure. Instead, the intermediate B appears rapidly at first, grows to an appreciable concentration, and then disappears as B converts to C. There is an appreciable time lag to C's appearance, compared to direct A → C conversion.

The characteristic rates of a sequence of reactions often differ by one or more orders of magnitude, and when this happens, we can invoke two important approximations that have simple physical interpretations. The first of these invokes the *rate-limiting step*, which is the slowest step in the sequence. The idea is very simple. Suppose you are in a classroom and hear the fire alarm sound. Your goal is to leave the building as quickly (and orderly!) as you can, and there is a sequence of steps you must follow: head for the exit from your seat, leave the room, then leave the building. Suppose, however, the door to the classroom is partly blocked so that only one person at a time can squeeze through. No matter how fast you run to the door, and no matter how fast you could run from the room once you passed through the door, your rate of exit is limited by the perhaps very long time it takes to get through the door. Exiting the room has become the rate-limiting step. So it is with many reaction sequences. Suppose B in our sequence is a free-radical fragment of A that is produced slowly but decays to C quickly once B is formed. Then, A → B is the rate-limiting step ($k_1 \ll k'_1$) and the overall rate of C production is governed by k_1, not k'_1.

The second approximation is known as the *steady-state approximation*. Returning to our fire alarm analogy, as you and your classmates discover the door is blocked, you try to figure out what to do next. Is there another way out? Can anyone break down the door? Is the door locked? As you work on this problem collectively, you

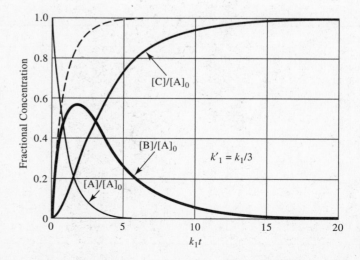

FIGURE 26.2 The exact time dependence for the reaction sequence A → B → C follows the solid lines in the figure. The intermediate, B, builds to a maximum concentration, then decays into the stable product, C. Without this intermediate, [C] would follow the dashed line.

will likely be running around the room, back to your seats, to a window, etc. At any one time, only one of you can be squeezing through the door: the rate of "people exiting the room" will be roughly constant from the time you first hear the alarm (the reaction starts) until you have all gotten through the door (the reaction is over, since once you are out the door, you can very quickly leave the building). This is the hallmark of the steady-state approximation: a concentration of an intermediate (people passing through the door) that is always small and roughly constant in time.

The mathematical version of the steady-state approximation was first suggested by David Chapman in a 1913 paper, and it was extensively employed and defended by Max Bodenstein, one of the founders of modern chemical reaction kinetics.[3] We can see it at work in the kinetic equations for our prototype reaction sequence. The exact expressions for [A], [B], and [C] under the assumption that $[A] = [A]_0$ and $[B]_0 = [C]_0 = 0$ are

$$[A] = [A]_0\, e^{-k_1 t} \qquad (26.11a)$$

$$[B] = \frac{k_1 [A]_0}{k_1' - k_1} (e^{-k_1 t} - e^{-k_1' t}) \qquad (26.11b)$$

$$[C] = [A]_0 \left(1 + \frac{k_1' e^{-k_1 t} - k_1 e^{-k_1' t}}{k_1 - k_1'}\right), \qquad (26.11c)$$

and these expressions were used to draw Figure 26.2. Now we impose the condition that makes the first step rate limiting: $k_1 \ll k_1'$. Equation (26.11a) is unchanged, since A still disappears at a rate governed by k_1. In Eq. (26.11b), we neglect k_1 in the denominator and set the second exponential equal to zero, since a large k_1' means this term will go to zero well before the first term does. For the same reasons, the second term in the parentheses in Eq. (26.11c) becomes $-e^{-k_1 t}$. The new expressions are

$$[A] = [A]_0\, e^{-k_1 t} \qquad \text{(unchanged)} \qquad (26.12a)$$

$$[B] = \frac{k_1}{k_1'} [A]_0\, e^{-k_1 t} \qquad (26.12b)$$

$$[C] = [A]_0 (1 - e^{-k_1 t}) \,. \qquad (26.12c)$$

Note how the only rate constant that multiplies time in all of these expressions is k_1. That is why we call it rate determining.

The steady-state approximation says that [B] is constant in time so that $d[B]/dt = 0$. The first step in the reaction sequence produces B at a rate $k_1[A]$, and the second step destroys B at a rate $k_1'[B]$ so that the net change in [B] is

$$\underset{\substack{\uparrow \\ \text{Net rate of} \\ \text{change of [B]}}}{\frac{d[B]}{dt}} = \underset{\substack{\uparrow \\ \text{Production rate} \\ \text{in first step}}}{k_1[A]} - \underset{\substack{\uparrow \\ \text{Destruction rate} \\ \text{in second step}}}{k_1'[B]} = \underset{\substack{\uparrow \\ \text{Steady-state} \\ \text{assumption}}}{0} . \qquad (26.13)$$

---

[3] Bodenstein was a colorful character. He insisted that his students learn to master the experimental methods of their work, and when one of his students had the laboratory technician construct a glass vacuum manifold for him, Bodenstein smashed it with his cane, admonishing the student to do it himself. The student was George B. Kistiakowsky, a physical chemist who went on to a distinguished career in kinetics and served as President Eisenhower's Special Assistant for Science and Technology from 1959–1961.

FIGURE 26.3 In the steady-state approximation to the A → B → C sequence, one takes advantage of the small and nearly constant value of [B] to simplify the more complicated exact integrated rate laws. The steady-state curve for [C] (dashed line) is an accurate representation of the exact behavior, and the agreement improves as the ratio of rate constants k_1'/k_1 increases.

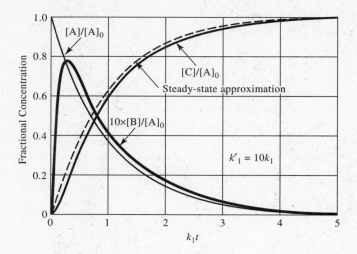

If we solve this equation for [B], we find

$$[B] = \frac{k_1}{k_1'}[A] \ . \tag{26.14}$$

The steady-state approximation uses this relation in the following way. The concentration of A is still governed by Eq. (26.11a), and substituting that expression for [A] into Eq. (26.14) gives Eq. (26.12b). The rate of formation of C is $d[C]/dt = k_1'[B]$, which Eq. (26.14) turns into $d[C]/dt = k_1[A]$. Integrating this from $t = 0$ (when [C] = 0 and [A] = [A]$_0$) to t (when [C] = [A]$_0$ − [A]) gives Eq. (26.12c).

The steady-state approximation leads to the same conclusions as did the rate-limiting step approximations to the exact equations, but the advantage of the steady-state assumption is that it *does not require us to have an exact integrated solution to the full rate law*. Instead, it simplifies the exact rate law, often into one of the forms we can integrate easily.

You may be troubled that Eq. (26.13), which says [B] is independent of time, leads to Eq. (26.12b), which clearly has [B] depending on time! If we differentiate Eq. (26.12b) with respect to time, we find, using Eq. (26.12a) also, that

$$\frac{d[B]}{dt} = \frac{k_1}{k_1'}(-k_1[A]_0 e^{-k_1 t}) = -\frac{k_1^2}{k_1'}[A]$$

and since $k_1 \ll k_1'$, this derivative is very small, if not exactly zero, and it is decreasing as [A] → 0. These conditions are at the heart of the steady-state assumption.

Figure 26.3 contrasts the exact expressions, Equations (26.11a, b, c), and the steady-state expressions, Equations (26.12a, b, and c) for the case $k_1' = 10k_1$. The solid lines are the exact expressions (with [B] multiplied by $10 = k_1'/k_1$). The dashed line is Eq. (26.12c) for [C], and Eq. (26.12b), when scaled by a factor of 10, would follow the exact [A] curve. Note that the steady-state curve is a very good approximation to the exact [C] curve, and if k_1' had been even larger than k_1, the agreement would have been closer.

Branching Reactions

Another common motif for reaction mechanisms involves *branching* from one reactant to more than one product, as in the bromination of toluene:

Toluene + Br_2 → o-bromotoluene (40%) + p-bromotoluene (60%)

or the spontaneous emission of an excited state to many lower energy states:

$$A^{***} \begin{cases} \xrightarrow{k_1} A^{**} + h\nu \\ \xrightarrow{k_1'} A^{*} + h\nu' \\ \xrightarrow{k_1''} A + h\nu'' \end{cases}$$

If the products are all produced through first-order processes, as in spontaneous emission, the rate law is easy to integrate. We have, for the example diagrammed above with three independent fates for the initial excited state A^{***},

$$\frac{d[A^{***}]}{dt} = -k_1[A^{***}] - k_1'[A^{***}] - k_1''[A^{***}]$$
$$= -(k_1 + k_1' + k_1'')[A^{***}] = -k_{net}[A^{***}] \quad (26.15)$$

where k_{net}, the sum of the first-order rate constants for the parallel paths, represents a net first-order rate constant for *all* parallel first-order processes. The initial state decays exponentially in time, and the apparent radiative lifetime of the state is $\tau_{rad} = 1/k_{net}$. The rate of production of any one product state, however, (assuming all such states are absent at $t = 0$ and that the products do not decay one into another, which is often *not* the case for spontaneous emission) follows, for example,

$$\frac{d[A^{**}]}{dt} = k_1[A^{***}] = k_1[A^{***}]_0 e^{-k_{net}t}$$

which integrates to

$$[A^{**}] = \frac{k_1[A^{***}]_0}{k_{net}}(1 - e^{-k_{net}t})$$

with similar expressions for $[A^*]$ and $[A]$ with k_1 replaced by the appropriate rate constant. Note that, somewhat surprisingly, the rate of appearance of *any* product is controlled by k_{net}, *not* by the rate constant of the reaction producing that product. The relative yield of any two products, $[A^{**}]$ and $[A^*]$ for example, *does* depend on individual rate constants:

$$\frac{[A^{**}]}{[A^*]} = \frac{k_1}{k_1'}$$

so that the *ratio* of the rate constants governs the *branching* of parallel first-order processes.

Complex Mechanisms—Chain Reactions

Often, reactions that one might expect to have simple, direct mechanisms have, for reasons we explore in later chapters, quite complex mechanisms. The gas phase reactions of hydrogen and the halogens are classic examples of this behavior that have been studied for roughly a century. It would seem at least plausible (until one looks more deeply at the problem) that the $H_2 + X_2$ reaction could proceed through a simple bimolecular encounter in which the two reactant bonds are broken as the two HX product bonds are formed:

$$\begin{array}{c} H \\ | \\ H \end{array} + \begin{array}{c} X \\ | \\ X \end{array} \longrightarrow \begin{array}{c} H\text{----}X \\ | \quad | \\ H\text{----}X \end{array} \longrightarrow \begin{array}{c} H\text{---}X \\ + \\ H\text{---}X \end{array}$$

The H_2X_2 intermediate in this scheme is called a *four-center intermediate* since there are four nuclear centers that interact and define it. For many years, the $H_2 + I_2$ reaction was thought to follow this microscopic mechanism, and the bulk kinetic rate law seemed to agree. The observed rate law is $d[HI]/dt = k[H_2][I_2]$, but it is now known that the mechanism is more complex and involves free I atoms from thermal I_2 dissociation. Note that the very different dissociation energies of H_2 and I_2, 4.48 eV and 1.54 eV, is a thermodynamic clue that I_2 thermal dissociation predominates over H_2 dissociation.

Evidence of a more complex mechanism came from experiments that followed Bodenstein's original 1894 study by 73 years. In an experiment that produced I atoms by *photodissociation* of I_2 at a temperature where thermal reaction was negligible, John Sullivan showed that the rate of production of HI at high temperatures could be extrapolated from his low-temperature rates. Further work has led to an overall mechanism that involves not only I atoms but also species such as IH_2, a weakly bound intermediate.

In contrast to the simple rate law but unexpectedly complex mechanism of the H_2/I_2 system, the $H_2 + Br_2 \rightarrow 2HBr$ reaction shows its complexity from its rate law. Again, Bodenstein published the first experimental study of the thermal reaction rate in 1907, and the rate law took the surprising form

$$\mathcal{R} = \frac{1}{2}\frac{d[HBr]}{dt} = \frac{k_{exp}[H_2][Br_2]^{1/2}}{1 + \alpha[HBr]/[Br_2]} \tag{26.16}$$

where k_{exp} is the experimentally observed rate constant and α is a dimensionless parameter that was found to be nearly independent of temperature over the ~470–570 K range of the experiments. Consider the early phase of a study begun with no HBr. The rate follows $k_{exp}[H_2][Br_2]^{1/2}$ and is first order in H_2, half order in Br_2, and $\frac{3}{2}$ order overall, since $[HBr] \cong 0$. As the reaction proceeds, however, [HBr] becomes appreciable (the experimental value for α is ~0.1) and *inhibits* the rate as the denominator of Eq. (26.16) increases above 1.

How can such simple reactants, as simple as the $H_2 + I_2$ system, lead to such a complicated rate law? To begin to answer this question, we start with the thermodynamics of the net reaction and all the plausible elementary reactions that could participate in the mechanism. Figure 26.4 is an enthalpy level diagram (drawn for the representative temperature 600 K) based on standard molar enthalpy of formation data from JANAF Tables.

We can tell at a glance that the net reaction is exothermic and that Br_2 dissociation is a much lower energy process than H_2 dissociation. Thus, we consider

$$Br_2 \rightarrow 2Br \, . \tag{26.17a}$$

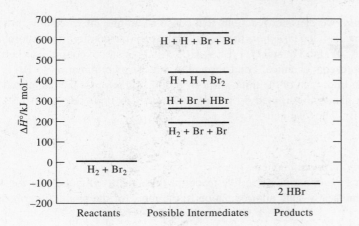

FIGURE 26.4 One good place to start when considering possible steps in a complex reaction mechanism is with an enthalpy level diagram relating the stable reactants and products to the various possible intermediate species. This is done here for the $H_2 + Br_2 \rightarrow 2HBr$ reaction.

We also see from the lower two levels in the center of the diagram that the reaction

$$Br + H_2 \rightarrow HBr + H \quad \quad \textbf{(26.17b)}$$

is endothermic while the middle two levels show us that

$$H + Br_2 \rightarrow HBr + Br \quad \quad \textbf{(26.17c)}$$

is exothermic. If we add these two reactions, we have the net reaction, and we can see how the atomic intermediates can churn through a mechanism, behind the scenes, in a process known as a *linear chain mechanism* that we can diagram schematically as

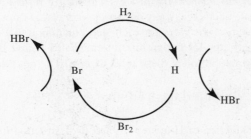

The first reaction uses one intermediate to produce a second intermediate that the second reaction uses to produce the first intermediate, and so on, spitting out two stable reaction products at each cycle.

What could lead to the rate inhibition by HBr as observed in the experimental rate law? Figure 26.4 shows us that the reverse of reaction (26.17b) is exothermic

$$H + HBr \rightarrow Br + H_2 \quad \quad \textbf{(26.17d)}$$

while the reverse of (26.17c) is substantially endothermic. The reaction enthalpy is a useful guide here because the energy to surmount a reaction endothermicity must come from somewhere, and in this reaction only those collisions in the improbable, high-energy tail of the Boltzmann distribution have sufficient energy to promote the reaction.[4] This leads us to neglect the reverse of (26.17c) and hypothesize that reaction (26.17d) serves as a *chain inhibition* step. It interrupts and reverses the direction of the chain.

---

[4]We leave unanswered for now the interesting question concerning the mechanical *type* of energy (translational, vibrational, electronic, etc.) that might be most efficient in surmounting the endothermic barrier. Chapter 28 considers this question in more detail.

Finally, we consider reactions that consume atomic intermediates, produce only stable species, and *terminate* the chain. There are three possibilities: recombination of Br to Br_2, of H to H_2, and H + Br → HBr. All three are, of course, exothermic, since each represents chemical bond formation, but Br recombination to Br_2 is arguably the most important of the three since Br production *initiates* the chain, and the chain itself is likely rate-limited by the endothermic step Br + H_2 → HBr + H that is fed by Br. Thus we call

$$2Br \rightarrow Br_2 \quad (26.17e)$$

the *chain termination* step.

We can summarize these five reactions, give them symbolic rate constants, and apply the steady-state approximation to the rates for the atomic intermediates:

Initiation:	$Br_2 \rightarrow 2Br$	k_a
Propagation:	$Br + H_2 \rightarrow HBr + H$	k_b
	$H + Br_2 \rightarrow HBr + Br$	k_c
Inhibition:	$H + HBr \rightarrow H_2 + Br$	k_d
Termination:	$2Br \rightarrow Br_2$	k_e

We will see in the next chapter that there is more to the initiation and termination steps than these simple reactions indicate, but for now, writing them as we have here is sufficient.

We can write rate laws for each of these steps because each is an elementary process. The hallmark of such processes is that *the reaction order of each species in the rate law for an elementary process equals its stoichiometric coefficient.* Moreover, the net rate of change of any one species appearing throughout a sequential mechanism is the sum of the rates for each step containing that species.

We start with net rates for the atomic intermediates H and Br, and invoke the steady-state approximation to set these rates to zero. The net rate for [H] is

$$\frac{d[H]}{dt} = k_b[Br][H_2] - k_c[H][Br_2] - k_d[H][HBr] = 0 \quad (26.18)$$

representing the production of H in step (b) and its destruction in steps (c) and (d). Similarly, the net rate for [Br] is

$$\frac{d[Br]}{dt} = 2k_a[Br_2] - k_b[Br][H_2] + k_c[H][Br_2] + k_d[H][HBr] - 2k_e[Br]^2 = 0 \quad (26.19)$$

where our definition of an individual reaction's rate in Eq. (26.1) introduces the factors of 2.

Note that the three middle terms in Eq. (26.19) are the negative of the three in Eq. (26.18) and must add to zero, reducing Eq. (26.19) to

$$2k_a[Br_2] = 2k_e[Br]^2 \quad \text{or} \quad [Br] = \left(\frac{k_a}{k_e}[Br_2]\right)^{1/2}, \quad (26.20)$$

which looks interesting, since it suggests a way to introduce $[Br_2]^{1/2}$ into the net reaction rate law. We can substitute this expression for [Br] into Eq. (26.18) and solve for [H]:

$$[H] = \frac{k_b(k_a/k_e)^{1/2}[H_2][Br_2]^{1/2}}{k_c[Br_2] + k_d[HBr]}. \quad (26.21)$$

Finally, we need an expression for the rate of the net reaction:

$$H_2 + Br_2 \rightarrow 2HBr, \quad \mathcal{R} = \frac{1}{2}\frac{d[HBr]}{dt}$$

and an expression for the net [HBr] rate from the reaction mechanism:

$$\frac{d[HBr]}{dt} = k_b[Br][H_2] + k_c[H][Br_2] - k_d[H][HBr] \quad (26.22)$$

reflecting the production steps (b) and (c) and the destruction step (d). These are the same three steps that contribute to $d[H]/dt$ in Eq. (26.18), and from that equation, we can write

$$k_b[Br][H_2] = k_c[H][Br_2] + k_d[H][HBr]$$

which turns Eq. (26.22) into

$$\frac{d[HBr]}{dt} = 2k_c[H][Br_2] = 2\mathcal{R} \ . \quad (26.23)$$

Substituting the steady-state expression for [H] into Eq. (26.23) gives us an expression for the net reaction rate in terms of the stable species:

$$\mathcal{R} = \frac{1}{2}\frac{d[HBr]}{dt} = k_c[H][Br_2] \quad (26.24)$$
$$= k_c\left(\frac{k_b(k_a/k_e)^{1/2}[H_2][Br_2]^{1/2}}{k_c[Br_2] + k_d[HBr]}\right)[Br_2] = \frac{k_b(k_a/k_e)^{1/2}[H_2][Br_2]^{1/2}}{1 + (k_d/k_c)[HBr]/[Br_2]} ,$$

which is exactly the form of the experimental rate law, Eq. (26.16). We identify

$$k_{\text{exp}} = k_b(k_a/k_e)^{1/2} \quad \text{and} \quad \alpha = k_d/k_c \ .$$

In the roughly 100 years since Bodenstein's first study of this reaction presented us with a puzzle for its rate law, continued study of it has brought us to a detailed picture of not only its mechanism, but of the microscopic details of each step in the mechanism. We will return to many of these details in later discussions, but each of the five mechanism step rate constants is known with good accuracy now, and this mechanism predicts the net rate law quantitatively. Figure 26.5 plots [HBr] in time from a numerical simulation based on this mechanism (at 570 K where $k_{\text{exp}} = 0.072$ cm$^{3/2}$ mol$^{-1/2}$ s^{-1} and $\alpha = 0.068$). The heavy line is the full rate expression, while the lighter line sets $\alpha = 0$, simulating no HBr inhibition. The inhibiting effect is fairly subtle in the early phases of the reaction, but at later times, it is quite noticeable.

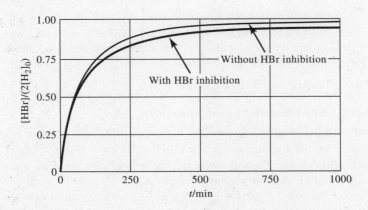

FIGURE 26.5 The rate of increase of [HBr] in the $H_2 + Br_2 \rightarrow 2HBr$ reaction shows the effect of *product inhibition*, expressed in the rate law through the appearance of the product concentration, [HBr], in the denominator. The curves here contrast the observed behavior with that expected with no inhibition.

EXAMPLE 26.6

The net reaction for the decomposition of ozone, $2O_3(g) \rightarrow 3O_2$, is notably exothermic: $\Delta_r \overline{H}° = -285$ kJ mol^{-1}. The reaction occurs in part on the surface of its container, but the homogeneous gas-phase reaction in the presence of a large excess of an inert gas such as argon follows the mechanism with associated rate constants shown below:

1. $O_3 + Ar \rightarrow O_2 + O + Ar \quad k_1$
2. $O_2 + O + Ar \rightarrow O_3 + Ar \quad k_1'$
3. $O + O_3 \rightarrow 2O_2 \quad k_2$

What is the net reaction rate law under the assumption of steady-state kinetics for the O atom concentration?

SOLUTION The net reaction rate is $\mathcal{R}_{net} = (1/3)d[O_2]/dt$, and thus we seek an expression for $d[O_2]/dt$ from the mechanism. Step 3 predicts

$$\mathcal{R}_3 = \frac{1}{2}\frac{d[O_2]}{dt} = k_2[O][O_3] \quad \text{or} \quad \frac{d[O_2]}{dt} = 2k_2[O][O_3]$$

while all three reactions give a net rate for [O] equal to

$$\frac{d[O]}{dt} = \underbrace{k_1[O_3][Ar]}_{\text{Step 1 production}} - \underbrace{k_1'[O_2][O][Ar]}_{\text{Step 2 destruction}} - \underbrace{k_2[O][O_3]}_{\text{Step 3 destruction}},$$

(Net rate)

which we set equal to zero in the steady-state approximation and solve for [O]:

$$[O] = \frac{k_1[O_3][Ar]}{k_1'[O_2][Ar] + k_2[O_3]}.$$

We substitute this into our expression for $d[O_2]/dt$ from step 3:

$$\frac{d[O_2]}{dt} = 2k_2\left(\frac{k_1[O_3][Ar]}{k_1'[O_2][Ar] + k_2[O_3]}\right)[O_3]$$

and finally divide by 3 to arrive at the net reaction rate:

$$\mathcal{R}_{net} = \frac{1}{3}\frac{d[O_2]}{dt} = \frac{2k_1k_2[O_3]^2[Ar]}{3(k_1'[O_2][Ar] + k_2[O_3])}.$$

Note that at any one Ar concentration, the mechanistic step rate constants and [Ar] can be combined into effective experimental rate constants, exposing the ozone and oxygen behavior more clearly:

$$\mathcal{R}_{net} = \frac{1}{3}\frac{d[O_2]}{dt} = \frac{k_{exp}[O_3]^2}{k_{exp}'[O_2] + k_{exp}''[O_3]}$$

where $k_{exp} = 2k_1k_2[Ar]$, $k_{exp}' = 3k_1'[Ar]$, and $k_{exp}'' = 3k_2$. At early stages of a reaction begun with pure ozone, $[O_2] \ll [O_3]$, and the rate appears to be first order in $[O_3]$ because the first term in the denominator is negligible. If the oxygen concentration dominates, however, then the rate is second order in ozone, and oxygen is an inhibitor.

➡ RELATED PROBLEMS 26.23, 26.26, 26.27, 26.28

Branching Chain Reactions—Explosions

On a windy Thursday in the first week of May in 1937, the *Hindenberg* airship exploded as it approached its mooring in Lakehurst, New Jersey. Newsreel film of the disaster, which killed 35 passengers and one ground crew member, showed that all 2×10^5 m^3 of H_2 in the ship burned in under 45 s. Nearly fifty years later, the space shuttle *Challenger* exploded shortly after launch when an O-ring seal in the solid fuel booster rocket failed, rupturing hydrogen and oxygen fuel tanks and igniting the mixed gases. The *Challenger* was completely destroyed, and its seven crew members were killed. Both of these tragedies were caused by the rapid and highly exothermic reaction between H_2 and O_2, but at room temperature, these gases can be mixed in any proportion and no reaction takes place. Add a spark, a flame, or suitable catalyst, however, and the reaction is complete and very fast. What triggers the reaction, and how does this trigger propagate a reaction mechanism that is so strikingly fast?

Figure 7.2 showed an enthalpy level diagram for the reaction $H_2(g) + \frac{1}{2}O_2(g) \rightarrow H_2O(g)$. The reaction is considerably exothermic, ($\Delta_r \overline{H}^\circ = -242$ kJ mol^{-1} at 298 K), but as we have seen for other reactions, exothermicity does not ensure a fast reaction rate. As with the H_2/X_2 systems, the "four-center" collision between H_2 and O_2 leads to no reaction, and the strengths of the H_2 and O_2 bonds are so great that there are essentially no H or O atoms in low-temperature thermal equilibrium. Sparks, flames, and other triggers initiate the reaction through local atomic radical production, and the speed of the reactions that follow is due to the *branching* nature of subsequent radical chain reactions. Unlike the hydrogen/halogen systems, in which mechanistic steps produced one chain-propagating radical for every one consumed, the hydrogen/oxygen mechanism includes steps that consume one radical but produce *two*. These are the branching steps that amplify the radical initiation step and lead to an explosively fast sequence of reactions.[5]

There are a surprising number of intermediate species involved in the complete mechanism: H, O, OH, HO_2, and H_2O_2, with perhaps more than one electronic state of importance, as Figures 14.26–14.29 suggest for the dissociation of H_2O. At low pressures, the important steps in the mechanism involve only H, O, and OH intermediates. Following the processes that initiate the reaction:

Initiation: $\quad H_2 \rightarrow 2H$
$\qquad\qquad\qquad O_2 \rightarrow 2O$

hydroxyl radical is produced in either of two chain-branching steps:

Branching: $\quad H + O_2 \rightarrow OH + O$
$\qquad\qquad\qquad O + H_2 \rightarrow OH + H$.

Note that these steps add to $H_2 + O_2 \rightarrow 2OH$, which creates two radicals at the expense of two stable species. The chain continues through

Propagation: $\quad OH + H_2 \rightarrow H_2O + H$,

and for each branching pair of reactions, the propagating step can occur twice, for a net change $3H_2 + O_2 \rightarrow 2H_2O + 2H$. Two stable product molecules are made, but two H atoms are produced as well to continue the chain branching. Chain termination can occur through several processes such as

Termination: $\quad 2H \rightarrow H_2 \quad$ (at the walls of the container)
$\qquad\qquad\qquad H + O_2 + M \rightarrow HO_2 + M$

where M represents any gas-phase species that can act as a "third body" to stabilize the HO_2 product. (The HO_2 radical is considerably less reactive.)

The nature of the net reaction—slow and controlled, rapid and explosive, or controlled rapid burning—depends on the total pressure of the system as well as the temperature. If you have ever operated a hydrogen/oxygen torch at atmospheric pressure, you have seen the controlled burning characteristic of that pressure region. If you stop the gas flows in a certain way to extinguish the flame, you will often hear a sharp "pop" as the flame dies. This is the chain-reaction explosion operating

---

[5]There are gentle ways to carry out this reaction if we leave the gas phase. See Problem 10.24.

at lower pressures for an instant. Similarly, the common lecture demonstration that captures H_2 and O_2 from water hydrolysis in soap bubbles or a balloon and then ignites the mixture with a loud pop is demonstrating the explosive chain branching regime.

We will encounter other generic types of mechanisms later in this chapter and again in the next, but now we turn from the effects of concentrations alone to the effects of other thermodynamic variables, especially temperature.

26.3 THE EFFECTS OF THERMODYNAMIC VARIABLES ON REACTION RATES

The rate of any reaction depends on the macroscopic conditions—temperature, pressure, volume, isothermal, adiabatic, etc.—to which it is constrained, and the dependence comes through two sources: the effects on concentrations and the effects on the rate constant. Usually, the concentration dependence is easy to recognize and account for. For example, a gas-phase reaction that has a different number of moles of gaseous reactants and gaseous products will have a rate that is different under constant volume versus constant pressure conditions simply because the reacting species' concentrations change in different ways under these two constraints. Similarly, reactions in condensed phases are generally insensitive to pressure because condensed phases usually have negligible compressibilities. (We will see in the next chapter that pressure can alter some condensed-phase reactions in an interesting way, but the effect is usually unimportant.)

The temperature dependence of reaction rates, however, is generally large and always due at least in part to the way the rate constant changes with T. These effects are commonplace: we "cook" food to speed the reactions of cooking and refrigerate food to inhibit many of the same reactions. The chirping rate of insects, the flash rate of fireflies, the crawling rate of ants, and our own metabolism rate are all affected by small changes in temperature. There is, of course, no single chemical reaction that accounts for a cricket chirp, and thus it is somewhat surprising that any elementary reaction between two simple molecules always has roughly the same functional temperature dependence to its rate as do cricket chirp rates. (See Problem 26.45 for the effect of temperature on a model chirping reaction.)

In spite of the common and strong dependence of rates on temperature in simple and complex reacting systems, our understanding of this dependence was slow in coming. Ostwald, one of the founders of physical chemistry, commented around 1900 that the temperature effect was "one of the darkest chapters in chemical mechanics," and two other founders of the subject, van't Hoff and Arrhenius, proposed theories around that time that we now know to be essentially correct. The simplest expression that reasonably relates the rate constant to temperature is still known as the Arrhenius equation.[6]

Arrhenius argued in an 1889 paper on reaction rates that temperature effects were generally too large to be explained by the role temperature plays in either molecular translational energies or solvent viscosities. Instead, he proposed that an

---

[6]Svante Arrhenius is best known as the founder of the correct view of electrolyte solutions as large numbers of dissociated ions, and it was this work that garnered him greatest fame—the 1903 Nobel Prize in Chemistry, the first awarded to a Swede, and the directorship of the Nobel Institute for Physical Chemistry from its founding in 1905 until his retirement shortly before his death in 1927. He also made important contributions to other areas of physical chemistry, including the effect of electrolyte dissociation on colligative properties, and in his later years, he wrote books and papers on subjects as diverse as the kinetics of enzymatic catalysis, immunochemistry, cosmology, and natural resources. His 1906 book, *Världarnas Utveckling* (*Worlds in the Making*), proposed a "Panspermia Theory" in which life was transported throughout the Universe by radiation pressure acting on living spores flying through the interstellar medium, a theory that attracted many followers and has proponents today.

equilibrium is established between nonreactive and reactive molecules (and he was coy in avoiding a detailed distinction between these two types of molecules) so that temperature effected a shift in this equilibrium along the lines proposed by van't Hoff for ordinary reaction equilibria, as we found in Chapter 7: $(\partial \ln K_{eq}/\partial(1/T))_P = -\Delta_r\overline{H}°/R$. In place of $\Delta_r\overline{H}°$, he introduced an energy parameter called the *activation energy*, E_a, which we now *define* as

$$E_a = -R\left(\frac{\partial \ln k}{\partial(1/T)}\right)_P \qquad (26.25)$$

where k is the rate constant.

This definition is easier to apply than it is to interpret, and thus we will postpone an interpretation until the next chapter.[7] Its application is simple. We measure rates, deduce rate laws, and calculate rate constants at a variety of temperatures. We plot $\ln k$ versus $1/T$ for all our data, and interpret the local slope of this plot as $-E_a/R$. Two experimental difficulties make this slightly more difficult than it sounds, however. The first is the problem of obtaining precise rate constants in the first place, since several manipulations of original data are required to extract them, data that themselves are subject to various uncertainties. The second is more severe, and led to the slow emergence of the Arrhenius expression out of many years of empirical data-fitting by early kineticists. This is the problem of a limited temperature span over which most reactions can be studied. Noisy data plotted as a function of $1/T$ over a brief span usually warrant no more than a *linear* interpretation: a plot of $\ln k$ versus $1/T$ either clearly looks like a straight line or else has sufficient scatter that higher order fits are undetermined.

A linear Arrhenius plot is equivalent to assuming that the activation energy is independent of temperature. When that is the case, we can integrate Eq. (26.25):

$$\int^k d \ln k = -\frac{E_a}{R}\int^{1/T} d\left(\frac{1}{T}\right)$$

and write

$$\ln k = -\frac{E_a}{RT} + \text{(constant)}$$

where (constant) represents combined integration constants. Exponentiating both sides gives

$$k = e^{-E_a/RT}\, e^{\text{(constant)}} = A\, e^{-E_a/RT}\,, \qquad (26.26)$$

which is known as the *Arrhenius equation*. The factor A is called the *pre-exponential factor*.

If the data warrant a nonlinear fit in an Arrhenius plot, the expression

$$k = A'T^m e^{-E/RT} \qquad (26.27)$$

where m and E are constants is often satisfactory. It has the temperature-dependent slope

$$\left(\frac{\partial \ln k}{\partial(1/T)}\right) = -mT - \frac{E}{R}\,,$$

which is equivalent to an activation energy that is a linear function of T. Most reactions, even those governed by complex mechanisms, have positive activation energies, but some reactions are known that have *negative* activation energies (they

---

[7]We have seen one simple interpretation of activated processes for nonreactive collisions in Figure 25.14 and its discussion of the temperature dependence of condensed phase transport coefficients. See Problem 25.45 as well.

slow with increasing T), others that have *no* activation energy, and others that have complicated temperature behavior that even Eq. (26.27) cannot describe. For example, the reaction

$$OH + CO \rightarrow H + CO_2 \;,$$

which is important in both combustion and atmospheric processes, has an activation energy that is nearly zero at 300 K, increasing to about 30 kJ mol^{-1} at 2000 K. The gas-phase reaction between hydroxyl radical and nitric acid,

$$OH + HNO_3 \rightarrow H_2O + NO_3 \;,$$

which is a major route to OH loss in the lower stratosphere, has a negative activation energy, about -6.3 kJ mol^{-1}, over the 220–440 K range.

EXAMPLE 26.7

The data below are measured rate constants at various temperatures for the reaction $C_2H_5I + OH^- \rightarrow C_2H_5OH + I^-$ in an ethanol solution. What is the rate constant at 320 K?

T/K	288.98	305.17	332.90	363.76
$k/10^{-3}$ M^{-1} s^{-1}	0.0503	0.368	6.71	119.

SOLUTION The first step is to extract the Arrhenius parameters from the data. Since there are only four data points over a brief temperature span, a linear fit will suffice (but see the Closer Look below). We plot ln k versus $1/T$ and use a least-squares analysis to calculate the slope and intercept of the line:

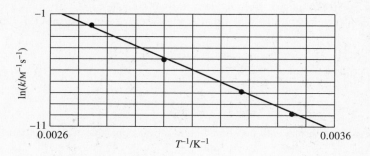

We find the slope is -1.09×10^4 K, and multiplying this by $-R$ gives us $E_a = 90.6$ kJ mol^{-1}. The intercept is 27.8, which equals ln A, and thus $A = e^{27.8} = 1.18 \times 10^{12}$ M^{-1} s^{-1}. With these parameters and $T = 320$ K, Eq. (26.26) predicts $k = 1.92 \times 10^{-3}$ M^{-1} s^{-1}. It is worth pointing out that a simple (but unfounded) *linear* interpolation between the data at 305.17 and 332.90 K predicts $k = 3.76 \times 10^{-3}$ M^{-1} s^{-1}, a value in considerable error even though the temperature span between the interpolated data is a modest 10% increase.

➡ RELATED PROBLEMS 26.29, 26.30, 26.31, 26.33, 26.36

A Closer Look at the Arrhenius Plot

A plot of "ln(something) versus $1/T$" pervades macroscopic physical chemistry, as we have seen in several places in this book. Such plots are often quite closely linear, and their most important parameter is their slope, which is interpreted as $-$(a characteristic molar energy)/R. In the Arrhenius plot, the characteristic molar energy is the activation energy, and we look here at the numerical methods used to extract this number from rate constant data.

Linear least-squares methods, which adjust the slope and intercept of the fit line so as to minimize the square of the deviations of the dependent variable from the line, are so well established that many personal computer programs and scientific calculators perform these calculations at the click of a mouse button or the push of a calculator key. This is certainly an advance over the days of graph paper and manual calculations, but it is *not* an advance if one fails to understand the validity of the fit parameters and the assumptions behind them. Consider, for example, the data in Example 26.7 above. The $\ln k$ versus $1/T$ plot certainly looks linear, and the line through the data was derived from a computer least-squares program asked to fit a linear function, $y = a_0 + a_1 x$, to them. In addition to the slope a_1 and intercept a_0, this particular program also calculated the statistically significant errors in them. It reported (stretching significant figures to make a point) a slope of $-1.089\,09 \times 10^4$ K with a statistical uncertainty at the 95% confidence level[8] of $\pm 4.235\,8 \times 10^2$ K. Now paying attention to significant figures, we can say that the slope is $-10\,900 \pm 400$ K.

We next ask a difficult question: do we expect the data to follow Eq. (26.25) or Eq. (26.26)? These two equations are apparently equivalent, since we can write Eq. (26.26) in the form of a straight line if we take the natural log of both sides: $\ln k = \ln A + (-E_a/R)(1/T)$ and identify this as a linear equation with $y = \ln k$, $x = 1/T$, $a_1 = -E_a/R$, and $a_0 = \ln A$. Equation (26.25) simply expresses the slope of this equation. The difficulty comes in the inner workings of the least-squares method. If we assume Eq. (26.26) is the physical model that the data should follow, least-squares mathematics (which we will not prove here) requires us to *weight each data point by the square of y*, that is, by the square of $\ln k$, emphasizing the high temperature (largest k) data over the low temperature data. Simple linear least-square routines do *not* include this weighting, since they assume the data are in fact based on a *linear* physical model, not an *exponential* one. Those programs that do include this weighting for fits to nonlinear model functions that can be written in a linear way, such as Eq. (26.26), correctly yield fit parameters that are *different* from those that do not. In our example, a correctly weighted fit to the exponential function $y = Ae^{B/x}$ yields B (which is still to be interpreted as $-E_a/R$) equal to $-1.127\,92 \times 10^4$ K $\pm 1.166\,1 \times 10^2$ K, or $-11\,300 \pm 100$ K, a value not only different from the simple linear fit, but one that only slightly overlaps it when the uncertainties of both are considered. One should never apply a numerical fitting technique without first understanding the method and the way experimental uncertainties affect the fitting parameters.

Control of temperature during a reaction is often a very important goal. While it is possible to measure rates as the temperature is changing in a known way, failure to maintain isothermal conditions can have catastrophic results. An exothermic reaction run under adiabatic or near-adiabatic conditions can lead to a *thermal explosion*. As the reaction proceeds, enthalpy is released, raising the temperature, increasing the reaction rate, releasing enthalpy at a greater rate, raising T at a greater rate, and so on in an exponentially sudden way, a phenomenon known as *thermal runaway*. In 1984, the thermal runaway that resulted from accidental water contami-

---

[8] A "confidence level" is a measure of the ability to predict the outcome of repeated measurements of a particular quantity, assuming systematic errors are constant. At the commonly used 95% level, there is only a 5% chance that a repetition of the entire experiment would have random errors that produce a slope outside the 95% confidence limit.

FIGURE 26.6 If the exothermic reaction $CH_3NC \to CH_3CN$ is begun at 500 K in an *adiabatic* container, the released enthalpy raises the system's temperature and increases the reaction rate. Initially, T rises slowly, but at ~7 s in this simulation, the reaction reaches *thermal runaway* and the temperature suddenly jumps to its final value around 3500 K.

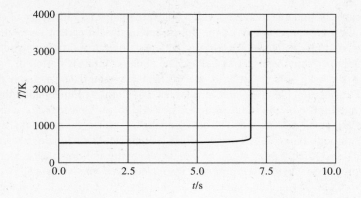

nation of a reaction tank in Bhopal, India, led to a massive leak of methyl isocyanate, the death of over 2000 people, and injury to hundreds of thousands.

Consider, for example, the simple exothermic isomerization

$$CH_3NC \to CH_3CN \qquad \Delta_r\overline{H}° = -100 \text{ kJ mol}^{-1}.$$

This is a very well-studied reaction, and it is first order with Arrhenius rate constant parameters $A = 3.98 \times 10^{13} \text{ s}^{-1}$ and $E_a = 161 \text{ kJ mol}^{-1}$. Since $\Delta_r\overline{S}° \cong 0$ so that $\Delta_r\overline{G}° \cong \Delta_r\overline{H}°$, we see that the equilibrium constant for this reaction will be large at almost any T, which tells us that a sample of pure isocyanide will be completely converted to the cyanide at equilibrium. We can purify the isocyanide only because the activation energy is so large that the reaction is very slow at low T. The discussion of adiabatic reactions in Chapter 7 along with a simple estimate of the heat capacity lets us calculate that complete conversion of the isocyanide to the cyanide under *adiabatic* conditions raises the temperature about 3000 K.[9] Figure 26.6 is a calculated temperature profile for this reaction, started at 500 K where the reaction is not fast, but allowed to run adiabatically at constant volume. Nothing much happens for the first 6 s or so until suddenly thermal runaway raises the temperature (and pressure) to the final value. The reaction covers the final 99% of its trip to equilibrium in its final second.

26.4 THE ATTAINMENT OF CHEMICAL EQUILIBRIUM

We have stressed that equilibrium is a dynamic, not a static, state of affairs. For a reaction at equilibrium, this means that the overall rate of the reaction in the *forward* direction must equal the overall rate in the *reverse* direction. This observation leads to a very important connection between forward and reverse rate constants and the reaction equilibrium constant. Once we establish this connection, we will be able to introduce the thermodynamic and statistical mechanic language we have developed for equilibrium constants in a reaction rate context. We will do this in detail in the next chapter, but in this section, we will make the connection, explore some of its consequences, and discuss experimental methods that probe kinetics by giving a system at equilibrium a small kick *away* from equilibrium and watching it relax back.

Detailed Balance

Consider again the first-order isomerization of methyl isocyanide. The reverse reaction is first order as well, and we can approach equilibrium (at a reasonable rate

---

[9]There is a component of the system's heat capacity due to the reaction enthalpy change (discussed in Problem 7.30), but it is quite small here.

only at high temperature) from either "end," starting with either the isocyanide or the cyanide. If we call the rate constants k_f and k_r for the forward and reverse directions of the equilibrium written as

$$CH_3NC \underset{k_r}{\overset{k_f}{\rightleftarrows}} CH_3CN$$

and consider that equilibrium means (compare Eq. (26.13))

$$-k_f[CH_3NC]_{eq} + k_r[CH_3CN]_{eq} = 0 \qquad (26.28)$$

$\uparrow$ Rate of CH_3NC destruction at equilibrium $\uparrow$ Rate of CH_3NC production at equilibrium $\uparrow$ Net reaction rate at equilibrium

where we are careful to specify equilibrium concentrations, then we see that

$$\frac{k_f}{k_r} = \frac{[CH_3CN]_{eq}}{[CH_3NC]_{eq}} = K_{eq} \qquad (26.29)$$

where K_{eq} is the equilibrium constant. This is a very important result. It expresses a general principle known as *detailed balance*.[10]

EXAMPLE 26.8

A 1.00 L adiabatic container is filled with CH_3NC so that when the temperature is raised to 500 K, the pressure is 800 torr. After 1.0 s, the concentration of CH_3CN has risen to 1.82×10^{-5} M. What are the forward and reverse rate constants for this isomerization at 500 K?

SOLUTION These are the conditions under which the simulation in Figure 26.6 was computed, and that figure shows that even under adiabatic conditions, the temperature stays close to 500 K during the first second. The initial rate is

$$-\frac{\Delta[CH_3NC]}{\Delta t} = \frac{\Delta[CH_3CN]}{\Delta t} = \frac{1.82 \times 10^{-5} \text{ M}}{1.00 \text{ s}} = 1.82 \times 10^{-5} \text{ M s}^{-1} = k_f[CH_3NC]$$

and the ideal gas equation of state tells us that $[CH_3NC]_0 = P/RT = 2.57 \times 10^{-2}$ M. Thus, the forward rate constant is

$$k_f = \frac{1.82 \times 10^{-5} \text{ M s}^{-1}}{[CH_3NC]_0} = \frac{1.82 \times 10^{-5} \text{ M s}^{-1}}{2.57 \times 10^{-2} \text{ M}} = 7.08 \times 10^{-4} \text{ s}^{-1}.$$

As discussed in the text, we can approximate the equilibrium constant as

$$K_{eq} = e^{-\Delta_r \overline{G}^\circ/RT} \cong e^{-\Delta_r \overline{H}^\circ/RT} = \exp\left[-\frac{-100 \text{ kJ mol}^{-1}}{(8.31 \text{ J mol}^{-1} \text{ K}^{-1})(500 \text{ K})}\right] = 2.8 \times 10^{10}$$

so that Eq. (26.29) tells us

$$k_r = \frac{k_f}{K_{eq}} = \frac{7.08 \times 10^{-4} \text{ s}^{-1}}{2.8 \times 10^{10}} = 2.5 \times 10^{-14} \text{ s}^{-1}.$$

▸ RELATED PROBLEMS 26.37, 26.38, 26.39, 26.40

Equation (26.29) seems to say that if we know one rate constant and the net reaction equilibrium constant, we know the other rate constant, and for elementary

---

[10]There are deep connections among three related principles: detailed balance (which involves rates at equilibrium), microscopic reversibility (which is based on the time-reversal invariance of dynamic equations of motion), and the reciprocal relations of nonequilibrium thermodynamics discovered by Lars Onsager (which relate coupled transport phenomena, including chemical reactions). We will explore some of these connections as we progress toward a more microscopic picture of reactive collisions.

processes, this is exactly right. If the net reaction is not elementary, however, detailed balance will hold for all the elementary steps in the net mechanism, and this can lead to an expression somewhat more complicated than Eq. (26.29) for the net reaction. Consider the sequence of first-order processes (written in a somewhat unusual way for reasons that will become clear)

$$C \underset{k_2}{\overset{k_{-2}}{\rightleftarrows}} A \underset{k_{-1}}{\overset{k_1}{\rightleftarrows}} B$$

leading to the net equilibrium $C \rightleftarrows B$. A study of the initial rate of disappearance of pure C leads directly to k_{-2}, and a study starting with pure B leads to k_{-1}, and we might conclude that $K_{eq} = [B]_{eq}/[C]_{eq} = k_{-2}/k_{-1}$, but a correct detailed balance analysis of this sequence shows that $K_{eq} = k_1 k_{-2}/k_{-1} k_2$.

Suppose the sequence above is initiated with pure A. A study of the relative rates of production of B and C (governed only by rate constants here, assuming first-order processes) gives us another way to view the approach to the $C \rightleftarrows B$ equilibrium. Suppose $k_1 = 1 \text{ s}^{-1} = 10 k_2 = 100 k_{-1} = 2000 k_{-2}$. Initially, A converts rapidly to B because $k_1 > k_2$, but the net equilibrium constant is $K_{eq} = k_1 k_{-2}/k_{-1} k_2 = 0.5$ so that the equilibrium system will contain twice as much C as B with less than 1% A remaining. Figure 26.7 plots the time evolution of this model system. Note

FIGURE 26.7 In some reaction mechanisms, the rate constants can be such that a product distribution very different from the equilibrium distribution is maintained for long times. Called *kinetic control* of a reaction, this phenomenon is illustrated in (a), in which $[B] \gg [C]$ for appreciable times, even though at equilibrium (shown at the longest times in figure (b)), $[C] \gg [B]$.

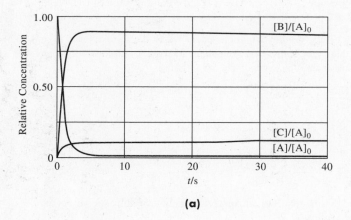

(a)

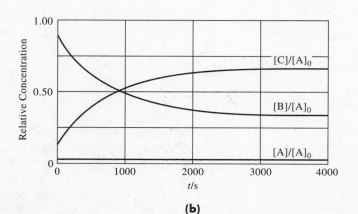

(b)

that in early phases of the reaction, shown in Figure 26.7(a), [A] decays very quickly, but for an appreciable time, [B]/[C] ≅ 10, the k_1/k_2 ratio. The system approaches thermodynamic equilibrium only after a much longer time, shown in Figure 26.7(b). When a reaction mechanism has this behavior—an appreciable period of time during which nearly constant but nonequilibrium concentrations are maintained—we say the reaction proceeds through *kinetic control*. The contrasting behavior—rapid decay toward the *true* equilibrium mixture—is known as *thermodynamic control*. Often, a mechanism that happens to operate under kinetic control can be exploited to isolate a reaction product that would be difficult to isolate otherwise. (Compound B is such a product in our example.)

Chemical Relaxation

We can also use detailed balance to our advantage in techniques known collectively as *relaxation kinetics* experiments. In these experiments, which were among the first to study very fast (sub-millisecond time scale) reactions, a system is allowed to reach equilibrium, then it is slightly *perturbed away from equilibrium* very quickly, and one studies the system's *return to a new equilibrium* (the system's "relaxation"). The perturbation must be very fast (faster than the relaxation time scale), and it is not difficult to do this. One can quickly release a compressed gas onto (or from) the system, creating a rapid *pressure* change (a *pressure jump* experiment), or a charged capacitor can be quickly discharged through a conducting system, raising the temperature (a *T-jump*), or a brief laser pulse can electronically excite an acidic species in solution, changing its proton dissociation equilibrium (a *pH jump*), or the laser or any brief light pulse can perturb a gas-phase equilibrium (a *flash photolysis* experiment). Any of these perturbations create almost instantaneous shifts away from equilibrium.

Recall from Chapter 7 that equilibrium corresponds to a reaction degree of advancement, ξ_{eq}, that minimizes the system's free energy. Figure 7.3 diagrams this minimum, and we can imagine perturbing a system in such a way that $G(\xi)$ shifts its equilibrium point slightly so that the system finds itself away from the new minimum:

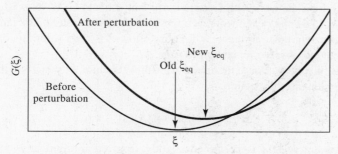

The full theory of nonequilibrium thermodynamics, which is beyond our scope to discuss in any detail, predicts that the reaction rate for perturbations such that $(\partial G/\partial \xi) \ll RT$ follows

$$\frac{\partial \xi}{\partial t} \propto \left(\frac{\partial^2 G}{\partial \xi^2}\right)_{eq} (\xi - \xi_{eq}) \qquad (26.30)$$

where ξ_{eq} is the equilibrium degree of advancement at which $(\partial G/\partial \xi) = 0$. Integrating this equation shows that the system relaxes back to equilibrium exponentially in time. (This is the reaction analog of the relaxation we found in Example 25.2 for a perturbed nonreactive system.) The characteristic parameter of an exponential time dependence is the relaxation time constant, τ, (or the relaxation rate, $1/\tau$), and an analysis of the kinetic mechanism relates τ to the rate constants of the rate law.

One of the most important reactions that has been studied this way is the autoionization of water. Anyone who has titrated an acid with a strong base using a phenolphthalein indicator (and that is probably everyone reading this book) has noticed that the transient pink color formed as the base first hits the solution disappears very quickly. In fact, the acid–base neutralization or *recombination* reaction

$$H^+(aq) + OH^-(aq) \rightarrow H_2O(l)$$

has one of the largest second-order rate constants of any condensed-phase reaction, $k_f = 1.12 \times 10^{11}$ M^{-1} s^{-1}. At 298 K for the reaction written in the direction above, the equilibrium constant is

$$K_{eq} = \frac{a_{H_2O}}{a_{H^+} a_{OH^-}} = 1.01 \times 10^{14} \tag{26.31}$$

where a_i is the activity of component i. Detailed balance tells us that the net rate is zero at equilibrium:

$$\mathcal{R} = 0 = k_f [H^+][OH^-] - k_r [H_2O] \; . \tag{26.32}$$

In pure water, $[H_2O] = 55.51$ M, $a_{H_2O} = 1$, and the equilibrium ion concentrations are so low that we can equate them to activities:

$$[H^+][OH^-] = \frac{1}{K_{eq}} = K_w = 9.9 \times 10^{-15} \tag{26.33}$$

where K_w is the traditional water ionic dissociation constant. We solve Eq. (26.32) for k_r:[11]

$$k_r = \frac{k_f [H^+][OH^-]}{[H_2O]} = \frac{(1.12 \times 10^{11} \; M^{-1} \; s^{-1})(9.9 \times 10^{-15})}{55.51 \; M} = 2.0 \times 10^{-5} \; s^{-1} \; . \tag{26.34}$$

The forward rate constant was first measured accurately in 1954 by Manfred Eigen, who pioneered and perfected fast relaxation methods and shared the 1967 Nobel Prize for chemistry with George Porter and Ronald G. W. Norrish, pioneers in the study of fast photochemical reactions. In his study of water dissociation, Eigen used a fast current pulse to create a T-jump followed by fast solution conductivity measurements to observe relaxation in time.

Analysis of a relaxation experiment with this stoichiometry starts with the rate law for the elementary reaction:

$$A + B \underset{k_r}{\overset{k_f}{\rightleftarrows}} C \qquad \mathcal{R} = k_f [A][B] - k_r [C] \; . \tag{26.35}$$

We imagine the system is perturbed slightly from equilibrium at $t = 0$. It has concentrations $[A]_0$, $[B]_0$, and $[C]_0$ at this time, and it relaxes to the new equilibrium characterized by concentrations $[A]_{eq}$, $[B]_{eq}$, and $[C]_{eq}$. The reaction stoichiometry (assuming also constant volume) tells us that at any time

$$[A]_{eq} = [A] - x, \quad [B]_{eq} = [B] - x, \quad \text{and} \quad [C]_{eq} = [C] + x \tag{26.36}$$

---

[11] Note that the thermodynamic equilibrium constant is *not* simply the ratio of the forward and reverse rate constants here. Simple dimensional analysis shows that it cannot be: the equilibrium constant is dimensionless, and the rate constants here have different units. The general connection between detailed balance and the equilibrium constant introduces the equilibrium constant and its associated concentrations into the net rate law set equal to zero, as in Eq. (26.34).

where x is the time-dependent reaction progress variable expressed as a concentration. Note that x could be either positive or negative, depending on the direction of the equilibrium shift. Introducing Eq. (26.36) into the rate law gives

$$\mathcal{R} = -\frac{dx}{dt} = k_f([A]_{eq} + x)([B]_{eq} + x) - k_r([C]_{eq} - x) \quad (26.37)$$

and expanding the right-hand side gives us three groups of terms:

$$-\frac{dx}{dt} = \underbrace{k_f[A]_{eq}[B]_{eq} - k_r[C]_{eq}}_{} + [k_f([A]_{eq} + [B]_{eq}) + k_r]x + k_f x^2$$

↑ Rate of relaxation   ↑ Zero by detailed balance   ↑ (Constant) × x ↑ Negligibly small

where the term quadratic in x can be neglected for small perturbations. Thus, we have

$$\frac{dx}{dt} = -[k_f([A]_{eq} + [B]_{eq}) + k_r]x = -\frac{x}{\tau} \quad (26.38)$$

where

$$\tau = \frac{1}{k_f([A]_{eq} + [B]_{eq}) + k_r} \quad (26.39)$$

is the relaxation time constant. Since $x = [A] - [A]_{eq}$, Eq. (26.38) has the form of Eq. (26.30). Rearranging Eq. (26.38) to $dx/x = -dt/\tau$ and integrating from $t = 0$ where $x = [A]_0 - [A]_{eq}$ to a later time t where $x = [A] - [A]_{eq}$ gives

$$[A] = [A]_{eq}(1 - e^{-t/\tau}) + [A]_0 e^{-t/\tau} . \quad (26.40)$$

Thus, an experimental measurement of [A] as a function of time gives us τ, which Eq. (26.39) relates to k_f and k_r. Since these are also related to K_{eq}, knowledge of K_{eq} and τ gives us both rate constants.

For the water dissociation reaction, the relaxation time constant is about 35 μs near room temperature, and K_w increases with increasing T, from 1.21×10^{-14} at 300 K to 2.04×10^{-14} at 305 K. Figure 26.8 plots the [H$^+$] behavior following a +5 K temperature jump from 300 K, based on Eq. (26.40). The change in the equilibrium H$^+$ concentration due to this jump is

$$\Delta[H^+]_{eq}/M = \sqrt{2.04 \times 10^{-14}} - \sqrt{1.21 \times 10^{-14}} = 3.3 \times 10^{-8} ,$$

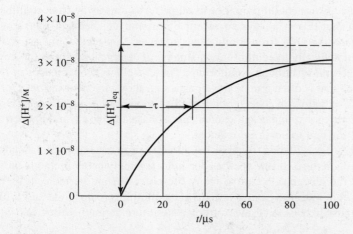

FIGURE 26.8 In a T-jump experiment in pure water (simulated for a 300 K to 305 K jump here), the equilibrium ion concentrations relax to their new value exponentially in time with a characteristic relaxation time constant, τ.

and the figure shows how the system relaxes exponentially toward its new equilibrium.

EXAMPLE 26.9

Hydrofluoric acid is an unusually weak acid when compared to HCl, HBr, or HI, but the recombination rate constant for $H^+ + F^- \rightarrow HF$ is 1.0×10^{11} M^{-1} s^{-1}, comparable to that for water. The equilibrium constant for this reaction is 1441 at 298 K. What is the rate constant for HF dissociation? What is the relaxation time constant in a 10^{-4} M HF solution?

SOLUTION Detailed balance tells us

$$k_f[H^+]_{eq}[F^-]_{eq} - k_r[HF]_{eq} = 0$$

and the equilibrium constant expression tells us

$$a_{HF} = K_{eq} a_{H^+} a_{F^-} \ .$$

At low concentrations, we can take activities to equal concentrations, so that the detailed balance expression becomes

$$k_f[H^+]_{eq}[F^-]_{eq} - k_r K_{eq}[H^+]_{eq}[F^-]_{eq} = 0$$

or $k_r = k_f/K_{eq} = 7.0 \times 10^7$ s^{-1}. (The equilibrium constant here adopts M^{-1} units.)

A 10^{-4} M HF solution has $[HF] + [F^-] = 10^{-4}$ M, and if we let $x = [H^+] \cong [F^-]$ (since the contribution to $[H^+]$ from the self-ionization of water is negligible), the equilibrium constant expression is

$$K_{eq} = \frac{[HF]}{[H^+][F^-]} = \frac{(10^{-4} - x)}{x^2} = 1441$$

and we find $x = 8.9 \times 10^{-5}$ M. The relaxation time constant follows from Eq. (26.39):

$$\tau = \frac{1}{k_f([H^+]_{eq} + [F^-]_{eq}) + k_r} = \frac{1}{2k_f x + k_r}$$
$$= \frac{1}{2(1.0 \times 10^{11} \, M^{-1} \, s^{-1})(8.9 \times 10^{-5} \, M) + 7.0 \times 10^7 \, s^{-1}}$$
$$= \frac{1}{1.8 \times 10^7 + 7.0 \times 10^7} s = 1.1 \times 10^{-8} \, s \ ,$$

a considerably smaller time than for pure water.

➡ *RELATED PROBLEMS* 26.41, 26.42

Chemical Oscillations

Usually, the path of any macroscopic nonequilibrium system is a monotonic and smooth one toward equilibrium. For small perturbations away from equilibrium, we have seen that this is always so in our discussion of chemical relaxation. For a system poised very far from equilibrium, however, this need not be so. Rather than decay monotonically to equilibrium, a number of chemical reactions are known that *oscillate* around or through equilibrium. These oscillations can occur in *time* with a frequency that is often remarkably constant, or they can occur in *space,* producing sustained oscillations in concentration profiles known as *chemical waves*. Oscillating reactions are known in gas and in condensed phases, in closed and in open systems, and in stirred and in unstirred reactors.

The study of oscillating reactions has attracted growing experimental and theoretical attention in the past few decades due to increasingly better methods for studying them and the recognition that so many life processes are periodic, from a heart's beat to a zebra's stripes. Living systems at equilibrium are no longer alive, of course, and life must be sustained away from equilibrium through eating and breathing, actions that sustain perhaps the most complicated set of reactions we will ever attempt to understand.

One of the first systematic theoretical studies of a reaction mechanism capable of showing oscillatory behavior appeared in the 1932 book with the provocative title *Leçons sur la Théorie Mathématique de la Lutte pour la Vie*, (*Lessons on the Mathematical Theory of the Battle for Life*), written by V. Volterra. He considered a scheme first introduced in the early 1920s by A. J. Lotka (whose papers had the less striking titles *Undamped Oscillations Derived from the Law of Mass Action* and *Analytical Note on Certain Rhythmic Relations in Organic Systems*). The scheme they discussed is known today as the *Lotka-Volterra* mechanism. It has three simple steps:

$$A + X \xrightarrow{k_a} 2X$$

$$X + Y \xrightarrow{k_b} 2Y$$

$$Y \xrightarrow{k_c} B \ .$$

While no chemical reaction is known that follows this mechanism, it has commonplace analogs. For example, let A be grain, X be ducks, Y be wolves, and B be dead wolves. Ducks eat grain, prosper, and reproduce (step 1). Wolves eat ducks, prosper, and reproduce (step 2), but wolves are mortal (step 3). Given a constant supply of grain, the net reaction is grain → dead wolves, but the population of ducks and live wolves oscillates in time. Abundant ducks provide a large food supply to the wolves, who multiply and rapidly deplete the duck population, leading to a duck scarcity, a collapse of the wolf population, and an opportunity for the duck population to build again, starting another cycle. Outside intervention, such as the hunters in the story of *Peter and the Wolf*, can intervene and radically alter these oscillations, and in more complicated mechanisms, seemingly innocuous steps are known that do just that.

Note that the first two steps could be written A → X and X → Y if we were interested in only the net reaction stoichiometry. The reaction rates, however, reflect the second-order nature of these steps. The rate of production of X depends on [X]: $k_a[A][X]$, and the rate of production of Y depends on [Y]: $k_b[X][Y]$. Processes for which the forward rate depends on the concentration of a product are called *autocatalytic*.[12] They represent a *positive feedback mechanism* that leads to a different signature to the time dependence of the product concentration. In contrast to a simple first-order growth, an autocatalytic step has a characteristic S-shaped curve:

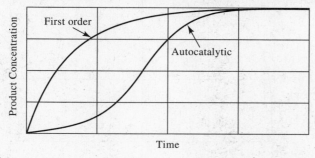

---

[12]Lotka, who was also a statistician, discovered that the number of scientific papers published per author followed an inverse square law: for every 100 authors of only 1 paper, there are roughly 25 who publish 2, 11 who publish 3, and so on. In contrast, the professor of political economy and sociological statistician Vilfredo Pareto found, in 1897, that society's distribution of wealth followed a $1/n^{1.5}$ law. Both phenomena have autocatalytic aspects to them, as expressed in the maxim, "it takes money to make money."

FIGURE 26.9 In the Lotka-Volterra mechanism, the concentrations of the intermediates [X] and [Y] oscillate in time as first [X] reaches a maximum, then falls rapidly as [Y] grows, which in turn falls, initiating a new cycle.

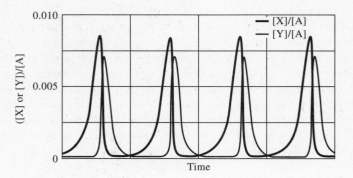

Autocatalytic steps are common in oscillating systems. In the Lotka-Volterra mechanism, a complete mathematical analysis (beyond our scope here) shows that the system can *never* attain stable equilibrium as long as the concentrations of the intermediates X and Y are finite and the reverse reactions (resurrection of wolves, wolves giving birth to ducks, and ducks turning into grain) have zero rates. (There is a unique steady-state solution for [X] and [Y], but it is unstable. Any perturbation from this state takes the system *away* from this point. See Problem 26.43.)

The mechanism can be studied by analytic and numerical methods, however. Figure 26.9 plots the relative X and Y concentrations versus time for a system constrained to have a constant amount of A. Note how X (the duck population) rapidly decays just as Y (the wolf population) begins to grow, followed almost immediately by collapse of the Y concentration. (The final product, B, is pulsing upward in steps at each cycle.) If we plot [Y] versus [X] parametrically in time, we get Figure 26.10. These concentrations loop around a closed curve as time goes on. In a system that decays normally to equilibrium, such a graph would be either a more or less smooth line toward the point that locates stable equilibrium or a spiral that closes in on the equilibrium point. The behavior shown in the figure, a

FIGURE 26.10 If the concentrations of X and Y in the Lotka-Volterra mechanism are plotted versus each other, parametrically in time, they are seen to follow a closed loop, characteristic of stable oscillatory behavior. If the system could approach stable equilibrium, the curve would be a spiral or other curved line ending at the equilibrium concentrations of X and Y.

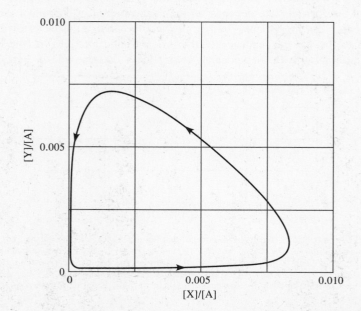

stable loop, is known as a *limit cycle*.[13] The size of this cycle depends on the details of the reaction (rate constants and initial amounts), but the general shape is characteristic of many oscillating reactions.

The so-called *Belousov-Zhabotinskii* reactions are among the best studied oscillating reactions. The original reaction, discovered by Belousov in 1951 but unpublished until 1958, and then in an obscure source, the Russian *Collection of Abstracts on Radiation Medicine,* was the Ce^{4+} catalyzed oxidation and bromination of citric acid by bromate ion. Other easily brominated acids such as malonic acid can be used in place of citric acid, and when the detailed reaction mechanism was deduced in 1972 by Field, Kőrös, and Noyes, it was found to involve 18 steps among 21 species. They were able to express the essence of the behavior of this complicated mechanism with a simplified 5 step scheme they called the *oregonator,* in honor of their home institution, the University of Oregon. The oregonator mechanism has reactants A and B, intermediates X, Y, and Z, and products P and Q:

$$A + Y \rightarrow X$$
$$X + Y \rightarrow P$$
$$B + X \rightarrow 2X + Z$$
$$2X \rightarrow Q$$
$$Z \rightarrow fY$$

where f is a variable stoichiometric factor. Note the autocatalytic third step.

Chemical waves, including perhaps the zebra's stripes and the leopard's spots, typically involve a complex interplay between diffusion and reaction. As we will discuss further in the next chapter, diffusion in condensed phases plays a critical role in reaction rate processes. Without mixing and diffusion (and thermal conduction), reaction can deplete reactant concentrations unevenly throughout a reactor. Usually, reactions are rapidly stirred to avoid these problems, but we are now learning that new phenomena appear and can be controlled in unstirred open reactors with variable reactant inflow rates. This will continue to be an area of interest, especially in models of biochemical systems that operate under these rules *in vivo*.

CHAPTER 26 SUMMARY

This chapter has discussed the macroscopic picture of chemical reaction kinetics: how we measure reaction rates, interpret the results, and predict rates under new conditions. We started with a definition of the reaction rate itself in terms of the reaction progress variable:

$$\text{Reaction rate: } \mathcal{R} = \frac{1}{V} \frac{d\xi}{dt}$$

and we discussed the reaction rate law, the equation relating $\mathcal{R}$ to interacting species' concentrations and rate constants. We examined several simple rate laws in detail:

$$\text{First order: } \mathcal{R} = k_1[A], \quad [A] = [A]_0 \, e^{-k_1 t}$$

Second order:
$$\mathcal{R} = k_2[A]^2, \quad \frac{1}{[A]} = \frac{1}{[A]_0} + 2k_2 t$$
$$\mathcal{R} = k_2'[A][B], \quad k_2' t = \frac{1}{[A]_0 - [B]_0} \ln \frac{[A]/[A]_0}{[B]/[B]_0}$$

and applied these simple laws to several basic reaction mechanisms: sequential reactions, branching reactions, and chain reactions. Two related approximations, the steady-state

---

[13] In general, a limit cycle is defined as a closed loop asymptotically reached by a system started anywhere in its vicinity. The Lotka-Volterra loops are unique and stable for any initial condition.

assumption and the concept of a rate-limiting step, simplified our analysis of reaction mechanisms.

The most important thermodynamic variable in reaction kinetics (aside from concentrations themselves) is temperature, which governs individual elementary reaction rate constants through the activation energy:

$$\text{Activation energy: } E_a = -R\left(\frac{\partial \ln k}{\partial (1/T)}\right)_P$$

leading to the Arrhenius expression:

$$E_a \text{ independent of } T: k = A\, e^{-E_a/RT}$$

or a simple variant of it:

$$E_a \text{ linear in } T: k = A'\, T^m\, e^{-E/RT}$$

The end-point of a reaction is its equilibrium state, and detailed balance tells us that at equilibrium, the forward rate equals the reverse rate for every elementary process. This allowed us to connect rate constants and equilibrium constants:

$$\text{Detailed balance: } A \underset{k_r}{\overset{k_f}{\rightleftarrows}} B, \quad K_{eq} = \frac{k_f}{k_r}$$

We saw how small perturbations away from equilibrium caused the system to move to a new equilibrium through an exponential relaxation on a characteristic time scale:

$$\text{Relaxation time constant:}$$
$$A + B \underset{k_r}{\overset{k_f}{\rightleftarrows}} C, \quad \tau = \frac{1}{k_f([A]_{eq} + [B]_{eq}) + k_r}$$

Finally, we pointed out mechanisms that did not move directly toward equilibrium, but instead oscillated in time. Autocatalytic steps were seen to be common features of these processes.

The next chapter continues the story as we begin to take apart a rate law on a microscopic scale. We have seen how statistical mechanics can turn molecular energies and symmetries into macroscopic equilibrium constants, and how it can turn simple dynamic arguments into macroscopic transport coefficients. Can it do the same for reaction rate constants?

FURTHER READING

There are several texts that supplement this and the next two chapters. In particular, *Chemical Kinetics and Dynamics,* by J. I. Steinfeld, J. S. Francisco, and W. L. Hase (Prentice Hall, Englewood Cliffs, NJ, 1989), *Chemical Kinetics,* by K. J. Laidler, 3rd ed. (Harper & Row, New York, 1987), and *Chemical Kinetics and Transport,* by Peter C. Jordan (Plenum, New York, 1979) treat this material in depth. Laidler's book includes interesting biographical sketches of many important contributors to the field of reaction kinetics. An interesting discussion of least-squares analysis and the use of weighting factors is given in a paper by R. de Levie, *J. Chem. Ed.* **63**, 10 (1986).

ACCESS TO DATA

Since there are so many more reactions than there are compounds, tables of evaluated rate data tend to focus on a small set of related reactions. The US National Institute of Standards and Technology has produced a computer-searchable database listing rate constants (in the form of Eq. (26.27)) for over 2000 gas-phase reactions [*NIST Chemical Kinetics Database,* NIST Standard Reference Database No. 17, (1989)]. This method of accessing large amounts of reaction rate data will certainly increase in time. The text by Steinfeld, *et al.,* referenced above contains an appendix listing many specialized compilations for areas such as combustion, atmospheric modeling, and high-temperature chemistry.

PRACTICE WITH EQUATIONS

26A At one point in a study of the reaction $2NO + O_2 \rightarrow 2NO_2$, it is found that $d[NO_2]/dt = 3.4 \times 10^{-6}$ M s^{-1}. What is the reaction rate at that time? (26.1)

ANSWER: $\mathcal{R} = 1.7 \times 10^{-6}$ M s^{-1}

26B The first-order rate constant for the reaction $NH_2NO_2(aq) \rightarrow N_2O(g) + H_2O(l)$ in alkaline solution is 9.3×10^{-5} s^{-1}. How long does it take for 5% of an initial amount of NH_2NO_2 to decompose? (26.3)

ANSWER: 550 s

26C What is the half-life in minutes for the reaction in 26B?

ANSWER: 124 min

26D At 400 K, the rate constant for the second-order reaction $2CF_3 \rightarrow C_2F_6$ is 2.51×10^{10} M^{-1} s^{-1}. Pulsed laser photolysis of perfluoroacetone produces an initial $[CF_3] = 2.7 \times 10^{-6}$ M. At what time after the laser pulse is $[C_2F_6] = 10^{-7}$ M? (26.7)

ANSWER: 0.6 μs

26E The rate constant at 40 °C for the ligand exchange reaction $Co(CN)_5(H_2O)^{2-} + N_3^- \rightarrow Co(CN)_5(N_3)^{3-} + H_2O$ is 5.0×10^{-5} M^{-1} s^{-1}, and the reaction is first order in each reactant. In an experiment with $[Co(CN)_5(H_2O)^{2-}]_0 = 0.002$ M and $[N_3^-]_0 = 0.1$ M, what is $[Co(CN)_5(N_3)^{3-}]$ after 24 hr? (26.10a)

ANSWER: 7.0×10^{-4} M

26F The initial discovery of element 98, Cf, produced about 5000 atoms of ^{245}Cf. This isotope decays to ^{245}Bk with a half-life of 43.6 min. In turn, ^{245}Bk decays to ^{245}Cm with a 4.94 d half-life, and ^{245}Cm, with a half-life of 8500 yr, is effectively stable on the time scale of its precursors. What fraction of the original Cf had turned into ^{245}Cm one week after its discovery? (26.11c)

ANSWER: 0.623

26G A cloud of 10^6 excited state atoms are produced at $t = 0$ by a laser pulse. These atoms emit to three lower energy states with Einstein spontaneous emission rates 5.3×10^8 s^{-1}, 2.3×10^8 s^{-1}, and 2.6×10^5 s^{-1}. What is the total photon emission rate from this excited state at $t = 0$? (26.15)

ANSWER: 7.6×10^{14} photons s^{-1}

26H By what factor does the rate constant change for a reaction with a 52 kJ mol^{-1} activation energy if the temperature is changed from 295 K to 305 K? (26.26)

ANSWER: 2

26I The activation energy for the reaction discussed in Examples 26.2 and 26.5 is 70.3 kJ mol^{-1}. What is the pre-exponential factor at 500 K? (26.26)

ANSWER: 2.2×10^{11} M^{-1} s^{-1}

26J The rate constant for the reaction $Cl + CH_4 \rightarrow HCl + CH_3$ over the range 200–500 K is accurately given by $(5180$ M^{-1} s$^{-1}) T^{2.11} e^{-(795 K)/T}$. What is the activation energy at 400 K? (26.25), (26.27)

ANSWER: 13.6 kJ mol^{-1}

26K The equilibrium constant for the interconversion of the cis and trans isomers of a particular MX_4Y_2 octahedral complex, cis $\rightleftarrows$ trans, is 0.23 at a temperature for which the half-life of the pure cis isomer is 2080 s. What is the half-life of the pure trans isomer at this temperature? (26.4), (26.29)

ANSWER: 478 s

26L The relaxation time constant for $H^+ + OH^- \rightarrow H_2O$ is 35 μs at room temperature in pure water. What is it at pH = 7.8? (26.34), (26.39)

ANSWER: 11 μs

PROBLEMS

SECTION 26.1

26.1 Suggest experimental methods that could follow the course in time of the following types of reactions. (There may be several ways to study any one of these, but try to suggest a way that is direct and unambiguous.)

(a) isotopic scrambling in the gas phase, as in $HF + DCl \rightarrow DF + HCl$

(b) hydrogenation of an unsaturated hydrocarbon, as in $C_2H_4 + H_2 \rightarrow C_2H_6$

(c) acid-catalyzed halogenation of acetone, as in $(CH_3)_2CO + I_2 \rightarrow CH_3COCH_2I + HI$

(d) identical ligand exchange, as in $Fe(CN)_6^{3-}(aq) + CN^-(aq) \rightarrow Fe(CN)_6^{3-}(aq) + CN^-(aq)$

(e) a disproportionation reaction, as in $3In^+(aq) \rightarrow 2In(s) + In^{3+}(aq)$

26.2 The graph below shows the relative amount of a reactant A as it disappears in time in an experiment begun with $[A]_0 = 2.5 \times 10^{-3}$ M. Estimate the reaction rate (assumed to be $-d[A]/dt$) at the following times: initially, at 5 s, and at 15 s.

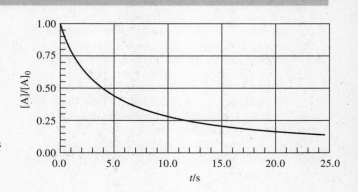

26.3 In an experiment in which one colored species is turned into another, a spectrophotometric probe of the reaction is often complicated if the absorption spectra of the two species overlap. One way around this problem uses a *diode array spectrophotometer* in which a linear array of many very small photodiodes monitors the absorption spectrum across a wide spectral range, one small wavelength increment per diode. Consider such a reaction of the form $A \rightleftarrows B$, started at $t = 0$ with $[A]_0 = 10^{-4}$ M and $[B]_0 = 0$, and monitored at two wavelengths. The table below shows absorbances at these wavelengths at three times early in the reaction and at a much later time at which the reaction has reached equilibrium. Given that the equilibrium constant for the reaction is $[B]_{eq}/[A]_{eq} = 4$, find the rate at $t = 5$ s.

t/s	0	5	10	∞
A_1	0.400	0.408	0.415	0.480
A_2	0.200	0.238	0.276	0.600

26.4 The rate of formation of phosgene in the reaction $Cl_2 + CO \rightarrow Cl_2CO$ has been extensively studied. Imagine an experiment in which Cl_2 and CO are rapidly mixed in a constant volume container. Derive an expression for the rate of this reaction in terms of the rate of change of the pressure in the container and any other experimental variables that are relevant.

26.5 Oxidation of bromide ion by hydrogen peroxide in acidic solution, $2Br^- + H_2O_2 + 2H^+ \rightarrow Br_2 + 2H_2O$, follows the rate law $\mathcal{R} = k[Br^-][H^+][H_2O_2]$. When bromide ion is destroyed at the rate 9.2×10^{-4} M s^{-1}, what is the rate of appearance of Br_2? What is the rate of disappearance of H_2O_2? Consider two experiments that start with the same initial amounts of reactants in stoichiometric proportion. In one, $[H^+]$ is buffered to remain essentially constant, but in the second, there is no buffer. Which experiment will reach equilibrium first?

26.6 The only isotope of radium that occurs naturally is ^{226}Ra, which decays through α-particle emission. A 0.1 mg sample emits 3.7×10^6 α particles per second, second after second, for many years. Why is this emission rate so constant in time? (*Hint:* What is the emission rate per single atom?)

26.7 The spontaneous radioactive decay of ^{40}K (to ^{40}Ar) follows the rate law $-d[^{40}K]/dt = k[^{40}K]$ with $k = 1.72 \times 10^{-17}$ s^{-1}. What is the rate of production of Ar (in atoms per second) in 1.00 L of a 0.100 M KCl solution? The natural fractional abundance of ^{40}K is 1.18×10^{-4} (and there are on the order of 10^{20} ^{40}K atoms in your body at any time, by the way).

26.8 The reaction between hypochlorite ion and iodide ion in alkaline solution:

$$ClO^-(aq) + I^-(aq) \rightarrow IO^-(aq) + Cl^-(aq)$$

has been studied in detail [Y. T. Chen and R. E. Connick, *J. Phys. Chem.* **63**, 1518 (1959)]. The table below gives initial rates for various initial reaction conditions based on this study. What is the rate law and the rate constant?

$[OH^-]_0$/M	$[ClO^-]_0$/M	$[I^-]_0$/M	$\mathcal{R}$/M s^{-1}
1.0×10^{-5}	4.7×10^{-3}	2.2×10^{-3}	62.0
1.0×10^{-5}	2.5×10^{-3}	4.7×10^{-3}	70.5
1.0×10^{-5}	1.9×10^{-3}	2.1×10^{-3}	23.9
4.5×10^{-6}	1.9×10^{-3}	2.1×10^{-3}	53.2

26.9 The initial rate of the reaction $2NO + 2H_2 \rightarrow N_2 + 2H_2O$ can be followed through the pressure change in the system at constant volume. The data below are from a study at 1099 K in which the initial partial pressures of the reactants were varied and the initial rate measured [C. N. Hinshelwood and T. E. Green, *J. Chem. Soc.* 730 (1926)]. What is the reaction rate law?

$P^0_{H_2}$/atm	P^0_{NO}/atm	$\mathcal{R}$/atm s^{-1}
0.380	0.526	2.11×10^{-4}
0.270	0.526	1.45×10^{-4}
0.193	0.526	1.04×10^{-4}
0.526	0.472	1.97×10^{-4}
0.526	0.395	1.36×10^{-4}
0.526	0.200	0.33×10^{-4}

SECTION 26.2

26.10 Prove that the integrated rate law for a reaction that is zero order in all species leads to a linear decrease in reactant with time if the reverse reaction is neglected. When is such a reaction over?

26.11 Consider a substance that undergoes a first-order reaction with a half-life $t_{1/2}$. Prove, starting from Eq. (26.3), that the fraction remaining after n half-lives equals 2^{-n}. The fraction remaining equals $[A]/[A]_0$, and Eq. (26.4) relates $t_{1/2}$ to k. (*Hint:* Recall that $x = e^{\ln x}$.) Show as well that the general "m^{th}-life," the time for the concentration to fall to $1/m$ times its original value, is given by $t_{1/m} = (\ln m)/k_1$.

26.12 One can find the rate constant for the first-order decomposition of N_2O_5 from the data in Example 26.1 in more than one way. One way applies Eq. (26.3) to each datum past $t = 0$, finds k_1 for each, and averages them. A second plots $\ln([N_2O_5]/[N_2O_5]_0)$ versus t and extracts k_1 from the slope of the resulting straight line. Try both ways, and see how well they agree. Daniels and Johnston reported $k_1 = 1.35 \times 10^{-4}$ s^{-1} in their paper, using many more measurements than were included in the example.

26.13 Find the rate constant and half-life for SO_2Cl_2 decomposition from the data in Example 26.4. What is the total pressure after 5 hr in that experiment? When did the total pressure reach 13.0 kPa?

26.14 The gas phase thermal decomposition of $(CH_3)_3COOC(CH_3)_3$ to ethane and acetone is first order. A sample of this peroxide is sealed in a cold container and rapidly taken to a high temperature where the rate constant $k_1 = 1.92 \times 10^{-2}$ s^{-1}. How long does it take for the pressure to double? (*Hint:* Be sure to write the net reaction!)

26.15 Isotopic ratios are used to establish ages of minerals on geologic time-scales through known radioactive decay patterns and rates. For example, ^{87}Rb decays through β + antineutrino emission to ^{87}Sr with a 4.90×10^{10} yr half-life, and ^{238}U decays through α emission to ^{234}Th with a 4.47×10^9 yr half-life. The ^{234}Th, in turn, rapidly decays through a series of many other short-lived isotopes, ending with ^{206}Pb, which is stable. The intermediate decays are so fast that the net process, ^{238}U $\rightarrow$ ^{206}Pb, has an effective 4.51×10^9 yr half-life, only slightly longer than that of the first step, ^{238}U $\rightarrow$ ^{234}Th. How old is a mineral that has the isotopic ratios $[^{206}$Pb$]/[^{238}$U$] = 0.71$ and $[^{87}$Sr$]/[^{87}$Rb$] = 5.1 \times 10^{-2}$, assuming the only sources of Pb and Sr are these two decay paths?

26.16 This problem explores the simplest numerical solution to a differential equation. Known as *Euler's method*, it makes the approximation $d[A] \cong \Delta[A] = (d[A]/dt)\Delta t$ for finite time steps Δt. At time $t + \Delta t$, we take $[A]_{t+\Delta t} = [A]_t + \Delta[A]$. Try this method on the first-order rate law $d[A]/dt = -k_1[A]$ (assuming $[A]_0 = 1$ M and $k_1 = 1$ s^{-1} for simplicity) and compare it to the exact expression. A computer spreadsheet program makes this exercise easy. We want Δt to be small; try $\Delta t = 0.2 t_{1/2}$ and find [A] at $2t_{1/2}$. Compare to the exact answer, which of course is 0.25 M here. Other algorithms that correct some of the approximations inherent in the Euler method are known, and since they are not difficult to program, they are usually used instead. (If you have a spreadsheet program available, you might try improving the Euler results by making Δt smaller. Checking the numerical stability and convergence of a differential equation integrating algorithm this way is always important to do.)

26.17 There are very few reactions that involve the simultaneous interaction of more than three molecules at a time, especially in gases or dilute solutions, simply because the probability that more than three molecules are simultaneously close enough to each other for a time long enough to do anything is very, very small. One classical case of three-body interaction, however, is well studied and is represented by the recombination reaction $3A \rightarrow A_2 + A$. (The next chapter explains in detail why $2A \rightarrow A_2$ alone does not, in general, work.) Integrate the rate law for this reaction, $d[A]/dt = -3k[A]^3$, and show that the reaction half-life is $t_{1/2} = (2k[A]_0^2)^{-1}$.

26.18 The data plotted in Problem 26.2 represent either a first-order or a second-order process. Which is it? What is the rate constant?

26.19 The rate constant for the reaction H + CH$_4$ $\rightarrow$ CH$_3$ + H$_2$ is 1.2×10^9 M^{-1} s^{-1} in an experiment in which CH$_4$, HI, and Ar are mixed with partial pressures 20 torr, 0.1 torr, and 500 torr, respectively, in a 1.50 L container at 320 K. A light pulse photodissociates some of the HI, creating H and I atoms with partial pressures of 1.3×10^{-3} torr at $t = 0$. What is the initial rate of appearance of H$_2$? How long does it take for [H] to fall to 5% of its initial value? (Assume that H reacts only with CH$_4$.)

26.20 Solve Eq. (26.10a) for x, and then use the definition of x in Eq. (26.9) to find expressions for [A] and [B] as functions of time. Explore the behavior of your expressions as follows: assume first $[B]_0 = 2[A]_0$ and verify that your expressions predict the expected equilibrium ($t = \infty$) values, $[A]_\infty = 0$ and $[B]_\infty = [A]_0$. Next, assume $[A]_0 \gg [B]_0$, and show that your expressions reduce to pseudo-first-order decay of [B]: [A] = constant and $[B] = [B]_0 \exp(-[A]_0 k_2' t)$.

26.21 Equations (26.11a–c) for the sequence $A \rightarrow B \rightarrow C$ are derived along these lines. Assume that at $t = 0$, only A is present with $[A] = [A]_0$. The rate law for [A] is $d[A]/dt = -k_1[A]$, which leads directly to Eq. (26.11a). Next, write the rate law for $d[B]/dt$, substitute Eq. (26.11a) for [A] into it, and show that Eq. (26.11b) is a solution to this differential equation. Finally, use the fact that $[A] + [B] + [C] = [A]_0$ at all times to deduce Eq. (26.11c) from Equations (26.11a–b).

26.22 In the Earth's stratosphere, around 30 km, there is always a trace amount of the excited ^{1}D state of atomic oxygen. During the day, there are only $\sim$10 atoms cm^{-3}, but these few atoms are very reactive, abstracting H from water and methane, forming OH which then enters a catalytic cycle destroying ozone. These O(^{1}D) atoms are formed naturally from photodissociation of ozone: $O_3 + h\nu \rightarrow O_2 + O(^1D)$, and while they have a long radiative lifetime, (see the discussion at the end of Section 18.3), they are relaxed to the ground ^{3}P state through collisions: $O(^1D) + M \rightarrow O(^3P) + M$ where M is either N$_2$ or O$_2$. This deactivation process has a rate constant that is about 1.8×10^{10} M^{-1} s^{-1}. The rate of production of O(^{1}D) can be written $d[O]/dt = j[O_3]$ where j is an effective first-order rate constant that depends on the intensity of the photodissociation radiation. The O$_3$ density is about 2×10^{12} molecules cm^{-3}, and the total [M] is about 5.8×10^{17} cm^{-3}. Apply a steady-state analysis to these two reactions, and use the data quoted above to calculate j.

26.23 The gas-phase conversion of para-hydrogen into ortho-hydrogen (see Chapters 19 and 23 for the differences between them) is usefully rapid at temperatures in the 900–1100 K range, and a landmark 1930 study followed the reaction through a change in the system's thermal conductivity [see A. Farkas, *Z. physik. Chem.* **10**, 419 (1930)]. The mechanism

$$H_2 \rightleftarrows 2H$$

$$H + p\text{-}H_2 \rightarrow o\text{-}H_2 + H$$

leads to the observed rate law for the net reaction $p\text{-}H_2 \rightarrow o\text{-}H_2$. What is this rate law?

26.24 The excited 3s $^2S_{1/2}$ state of C^+ has only two spontaneous radiative paths: to the $J = 1/2$ and $J = 3/2$ fine-structure levels of the $2s^2 2p\ ^2P^o$ ground state. The $J = 3/2$ level is only 64 cm^{-1} above the $J = 1/2$ level, and at any high-temperature thermal equilibrium, these two levels would have very nearly the same population. Suppose, however, we create the $^2S_{1/2}$ state selectively from laser excitation of C into an autoionizing state that forms $^2S_{1/2}$ preferentially. Given the Einstein A coefficients $A_1 = 4.2 \times 10^8$ s^{-1} for $^2S_{1/2} \rightarrow\ ^2P_{1/2}$ and $A_2 = 8.3 \times 10^8$ s^{-1} for $^2S_{1/2} \rightarrow\ ^2P_{3/2}$, find the population ratio for these lower states (i.e., [3/2]/[1/2]) that this process would produce. What is the rate constant for the disappearance of the excited state? Can one create a population inversion ([3/2] > [1/2]) this way?

26.25 In a high-energy electric discharge through a He/CO mixture, a number of processes involving excited and ground electronic states of CO and CO^+ are known to be important. In particular, the excited B $^2\Sigma^+$ state of CO^+ has two fates at low CO densities: spontaneous emission to the ground X $^2\Sigma^+$ state, or recombination with free electrons to form excited C atoms and ground-state O atoms. The radiative rate constant is $k_{rad} = 1.8 \times 10^7$ s^{-1}, and the second-order recombination rate constant is $k_2 = 1.8 \times 10^{-8}$ cm^3 molecule^{-1} s^{-1}. It is possible to have discharges in which $[e^-] \gg [CO^+(B)]$. Show that under these conditions, the CO^+ B state decays exponentially in time with an apparent first-order rate constant equal to $k_2[e^-] + k_{rad}$. You would like to measure k_{rad} alone in an experiment that excites a mixture with a brief pulse of electrons and then follows the CO^+ B fluorescence in time. Explain how you could do this, and suggest limits on the time duration of the electron pulse and the electron density.

26.26 The rates of formation and decomposition of phosgene, $CO + Cl_2 \rightleftarrows Cl_2CO$, were studied by Bodenstein in 1924. The observed rate law in the forward reaction was

$$\frac{d[Cl_2CO]}{dt} = k_f [Cl_2]^{3/2}[CO]$$

and for the reverse reaction, he found

$$\frac{d[Cl_2CO]}{dt} = -k_r [Cl_2]^{1/2}[Cl_2CO]\ .$$

One of the mechanisms he proposed (which is now believed to be correct) is a linear chain involving the three elementary steps

(1) $Cl_2 \rightleftarrows 2Cl$

(2) $Cl + CO \rightleftarrows ClCO$

(3) $ClCO + Cl_2 \rightleftarrows Cl_2CO + Cl\ .$

Identify the reaction intermediates in this scheme, and classify each step (in each direction individually) as to its role in the chain: initiation, propagation, inhibition, or termination. Assume the third step in its forward direction has the smallest rate constant, apply the steady-state approximation, and show that this mechanism predicts the observed rate law for the forward net reaction. Relate k_f to the rate constants of the mechanism. Similarly, assume that the reverse of step (3) controls the rate of phosgene decomposition, and show that this mechanism also predicts the observed reverse rate law.

26.27 The mechanism of N_2O_5 decomposition follows

(1) $N_2O_5 \rightleftarrows NO_2 + NO_3$

(2) $NO_2 + NO_3 \rightarrow NO_2 + O_2 + NO$

(3) $NO + N_2O_5 \rightarrow 3NO_2$

where the final two steps have negligible reverse rates. Apply a steady-state analysis to the NO and NO_3 intermediates, and show that first-order decomposition of N_2O_5 is predicted. Note that (2) and the reverse of (1) have the same reactants, but different products. Does your final expression for $-d[N_2O_5]/dt$ simplify if k_2, the forward rate constant for step (2), is much smaller than k_{-1}, the reverse rate constant of step (1)?

26.28 What would the $H_2 + Br_2 \rightarrow 2HBr$ rate law be if the important inhibition step was $Br + HBr \rightarrow H + Br_2$ instead of $H + HBr \rightarrow H_2 + Br$?

SECTION 26.3

26.29 The compound $ClCOOCCl_3$ decomposes irreversibly to phosgene, Cl_2CO, in a first-order reaction. When pure diphosgene is rapidly introduced into a container at 553 K, the pressure is found to be 0.0268 atm after 751 s, rising to a final value of 0.0396 atm. The same experiment at 578 K gave a pressure 0.0280 atm after 320 s, rising to 0.0351 atm. Find the rate constants at both temperatures as well as the activation energy and pre-exponential factor.

26.30 The decomposition of N_2O_5 discussed in Example 26.1 is found experimentally to be first order in N_2O_5. Use the data below to find the pre-exponential factor and activation energy for this reaction. Predict the rate constant at 35 °C, and compare to the value quoted in Problem 26.12.

T/K	273	298	318	338
$k/10^{-5}$ s^{-1}	0.0787	3.46	49.8	487

26.31 A second-order dimerization reaction, $2A \rightarrow A_2$, has a 48.5 kJ mol^{-1} activation energy and a 4.6×10^{-3} M^{-1} s^{-1} rate constant at 350 K. At what temperature should the reaction be run so that half of an initial sample of A, at $[A]_0 = 0.01$ M, dimerizes in one hour?

26.32 Consider a compound that decomposes irreversibly in a first-order reaction to either of two products, as in A → B and A → C. If formation of B has an activation energy E_1 and formation of C has E_2, how does the final product distribution, [B]/[C], depend on temperature?

26.33 Some of the data on which the expression in Problem 26J is based are given below, as reported by D. A. Whytock, J. H. Lee, J. V. Michael, W. A. Payne, and L. J. Stief, *J. Chem. Phys.* **66**, 2690 (1977). Plot them in the form ln k versus $1/T$. You should find that the data do not fall on a single straight line. Find a "low temperature" and a "high temperature" activation energy by using Eq. (26.26) twice: once for the low-T and once for the high-T data. How do these activation energies compare to the answer to Problem 26J? (The expression in Problem 26J is derived from measurements in the cited paper as well as other measurements. That paper's data alone lead to a slightly different expression.)

T/K	500	447	404	220	210	200
$k/10^6$ M^{-1} s^{-1}	546.	344.	227.	14.1	11.5	8.79

26.34 In a study of the temperature dependence of a reaction, why does it make sense to choose temperatures more closely spaced at the low end of one's experimental range than at the high end? How would you choose six experimental temperatures in the range 200–400 K that would be optimum?

26.35 Prove that Eq. (26.26) is equivalent to assuming that the activation energy is a linear function of temperature. How is dE_a/dT related to the parameters m and E of that equation?

26.36 Colorless NO(g) reacts rapidly with $O_2(g)$ to form the brown $NO_2(g)$, which contributes to the characteristic color of smog. The reaction is third order: 2NO + $O_2 \to 2NO_2$, $\mathcal{R} = k_3[NO]^2[O_2]$, and k_3 *decreases* with increasing T, as the experimental data below indicate. Derive Arrhenius parameters A and E_a from these data, then apply Eq. (26.27) to them, assuming $E = 0$ and plotting ln k_3 versus ln T to find m. Which fit do you prefer?

T/K	80	143	228	300	564
$k_3/10^3$ M^{-2} s^{-1}	41.8	20.2	10.1	7.1	2.8

SECTION 26.4

26.37 Problem 26.22 quotes the deactivation rate constant for $O(^1D) + M \to O(^3P) + M$ as $k_2 = 1.8 \times 10^{10}$ M^{-1} s^{-1}, which is valid for a temperature around 240 K, in the lower stratosphere. Calculate the equilibrium constant for this reaction using data in Table 13.4 and the statistical mechanical methods of Chapter 23, and then calculate the rate constant for the reverse reaction, collisional excitation of ground-state O atom. Note that the equilibrium constant is independent of the nature of M. Do you think that collisional excitation is an important source of $O(^1D)$ in this region of the atmosphere?

26.38 The two most abundant isotopes of Ne are ^{20}Ne (~91%) and ^{22}Ne (~9%), and both have spinless nuclei. For the elementary reaction $2\ ^{20}Ne^{22}Ne \rightleftarrows\ ^{20}Ne_2 + ^{22}Ne_2$, which is larger, the forward rate constant or the reverse rate constant? (*Hint:* Review Problem 7.34 and Example 23.10.)

26.39 Consider the isomerization equilibrium A ⇌ B with forward and reverse first-order rate constants k_f and k_r so that

$$-\frac{d[A]}{dt} = \frac{d[B]}{dt} = k_f[A] - k_r[B] .$$

Stoichiometry ensures $[A] + [B] = c$, a constant at all times in the reaction. Use this fact to show first that at $t = \infty$ (i.e., at equilibrium)

$$[A]_{eq} = \frac{k_r c}{k_f + k_r} \quad \text{and} \quad [B]_{eq} = \frac{k_f c}{k_f + k_r} .$$

Next, show that equilibrium is approached exponentially *from either direction* with an effective rate constant $k = k_f + k_r$. Do this by proving that

$$-\frac{d[A]}{dt} = -\frac{d([A] - [A]_{eq})}{dt} = k([A] - [A]_{eq})$$

and finding a similar expression for $[B] - [B]_{eq}$. Integrate from $t = 0$ where $[A] = [A]_0$ and $[B] = [B]_0$ to t and show that $[A] = [A]_0 e^{-kt} + [A]_{eq}(1 - e^{-kt})$ and similarly for [B].

26.40 Use the results of the previous problem to find the forward and reverse first-order rate constants for the isomerization of glucose (α-D-glucose ⇌ β-D-glucose, known as *mutarotation*) from the following data. At 291 K, pure α-D-glucose or pure β-D-glucose isomerizes in pure water with an apparent first-order rate constant 0.0116 min^{-1}. The reaction can be followed by *polarimetry*, the change in the plane of rotation of polarized light, since both isomers are chiral but rotate the polarization differently. One measures the so-called *specific rotation*, which is directly proportional to concentration. For pure α-D-glucose, the specific rotation (in standard units of degrees of rotation of the plane of polarized Na resonance radiation per decimeter path length per g mL^{-1} concentration) is +112.2°, and for pure β-D-glucose, it is +18.7°. The equilibrium mixture has a specific rotation of +52.6°. (By the way, an experiment of this type on saccharose "inversion" to glucose and fructose was performed in 1850 by Wilhelmy and

26.41 Note that the expressions derived in Problem 26.39 hold no matter how far one starts from equilibrium. They thus hold for relaxation experiments as well. Derive an expression for the relaxation time constant τ for this system. Write it first in terms of the forward rate constant k_f and the equilibrium constant K_{eq}, and then write it in terms of the reverse rate constant k_r and K_{eq}. Which quantity governs the *temperature dependence* of τ in each of the following cases: (a) $K_{eq} \ll 1$, (b) $K_{eq} \gg 1$, and (c) $K_{eq} \cong 1$, the forward activation energy, the reverse activation energy, the reaction enthalpy, or a combination of these?

26.42 Our derivation of Eq. (26.38) required us to neglect the term $k_f x^2$. How large a T-jump can we make and keep this term really negligible? Consider the following experiment: pure water at 60 °C is rapidly sprayed as small droplets onto a silver surface held at 10 °C, producing a -50 °C jump very quickly. At 60 °C, the water dissociation constant, K_w, is 9.4×10^{-14}, and at 10 °C, it is 2.9×10^{-15}. Compare $k_f x^2$ to $[k_f([H^+]_{eq} + [OH^-]_{eq}) + k_r]x$ for this experiment. Is our approximation valid? Repeat this calculation for the 5 °C jump from 300 K to 305 K discussed in the text. Is $k_f x^2$ negligible there?

26.43 Let the three steps in the Lotka-Volterra mechanism have rate constants k_a, k_b, and k_c. Write rate laws for the intermediates X and Y, and apply the steady-state approximation to them, assuming $[A] = [A]_0 = a$ constant at all times. You should find that the steady-state values for [X] and [Y] are independent of time. How are these values related to the rate constants? Where would you find the steady state on a graph such as Figure 26.10?

26.44 What is the net reaction for the oregonator mechanism, assuming $f = 1$? What are the rate laws for the intermediates?

GENERAL PROBLEMS

26.45 Suppose the periodic peaks in one of the intermediates of the Lotka-Volterra scheme as shown in Figure 26.9 produced a "chirp" in a simulated cricket. Would the chirp rate have an activation energy that had any simple relation to the activation energies of the individual steps of the mechanism? The data below were generated from computer simulations of the mechanism taking the first step to have an activation energy of 2 kJ mol^{-1}, the second to have 3 kJ mol^{-1}, and the third to have 1 kJ mol^{-1} (as if ducks eat rather well no matter how cold it gets, but wolves are sluggish in winter and die rather rapidly in hot weather). Find the chirping activation energy. Is it anywhere near the activation energies of the individual steps? Is the Arrhenius expression, Eq. (26.26), sufficient to represent these data?

T/K	250	290	300	320	380
chirp frequency (arb. units)	7.55	8.42	8.60	8.99	9.76

26.46 Consider the simplest autocatalytic reaction, A → B with a rate law $\mathcal{R} = k[A][B]$. Assume $[A] = [A]_0$ and $[B] = [B]_0$ at $t = 0$, let $x = [A]_0 - [A] = [B] - [B]_0$, and integrate dx/dt, ending with an expression for $x/[B]_0$ as a function of time. (You will need to consult an integral table or invoke integration by partial fractions.) Autocatalytic reactions are said to have an *induction time* during which their rate increases to a maximum. Find an expression for the time of maximum rate. (*Hint:* The rate is maximum at the inflection point of the expression for $x/[B]_0$ as a function of time, when $d^2x/dt^2 = 0$.)

26.47 Consider the first-order gas-phase reaction A → B + C run at constant temperature and *pressure* instead of constant temperature and *volume*. Show that $[A] = [A]_0 \, e^{-k_1 t}$, just as in the constant volume case. Imagine $[A] = [A]_0 = n_A^o/V = P/RT$ at $t = 0$ with $[B]_0 = [C]_0 = 0$, and neglect any reverse reaction. Introduce the degree of advancement variable $\xi = n_A^o - n_A = n_B = n_C$, and show that $[A]/[A]_0 = (n_A^o - \xi)/(n_A^o + \xi)$. Write the rate law in terms of ξ, integrate it, and show that the result is still $[A] = [A]_0 \, e^{-k_1 t}$.

CHAPTER 27

Elementary Processes and Rate Theories

THE previous chapter introduced elementary processes, the building blocks of net reaction mechanisms. These processes are the focus of this chapter. We first categorize them into classes based on the number of species that interact and the general physical process each represents. The second section introduces several physical ideas behind all elementary reaction rates through a simple collision model. There are advantages to the collision model, including the central role it gives a quantity known as the *reaction cross section*, but it is often not the most insightful way to view and understand the factors that govern chemical reaction rates.

The third section focuses on that transient species that fleetingly ties products to reactants, the *activated complex*. There, we see how the reaction *transition state* holds many of the secrets to reactivity and the control of rates. We derive and apply a statistical theory of rate constants called *activated complex theory* that shows us the role played by the transition state much more clearly than collision theory can, and the transition state is the unifying theme for the remainder of the chapter.

The final sections cover two specific topics: unimolecular reactions and catalysis. Unimolecular processes not only represent net reactions for single species' transformation, they also play a key role in the mechanisms of more complicated reactions such as those promoted

27.1 What Is an Elementary Process?

27.2 Simple Collision Theories of Reactions

27.3 The Transition State

27.4 Unimolecular Reactions—The Decay of Energy

27.5 Catalysts and the Design of Potential Energy Surfaces

by catalysts. Our discussion of unimolecular reactions thus leads naturally into a discussion of enzyme or surface-catalyzed reactions due to the physically similar processes that tie them together.

Catalysts control most life processes, and the biophysical chemistry of enzyme catalysis is a fascinating area of contemporary research. Industrial catalysts give economic life to many chemical processes, and attempts to design a catalyst for a particular reaction require detailed understanding of reaction mechanisms at the molecular level. We will see how catalysts interact with a reaction transition state and produce a new path for the reaction to follow. The reaction is much more likely to follow this new path due to its reduced activation energy, a hallmark of most catalytic processes.

27.1 WHAT IS AN ELEMENTARY PROCESS?

To answer the question posed in this section's title, it is helpful to consider first those processes that are *not* elementary. For example, the reaction that powers gasoline engines,

$$2C_8H_{18} + 25O_2 \rightarrow 16CO_2 + 18H_2O \; ,$$

is very fast, but it is unreasonable to imagine 25 oxygen molecules interacting simultaneously with two octane molecules to produce directly 16 carbon dioxide molecules and 18 water molecules. This net reaction cannot be elementary because it requires too many species interacting in the same vicinity at the same time. In contrast, the previous chapter discussed the reaction between H_2 and Br_2 to form HBr. Collisions between H_2 and Br_2 do not lead to HBr products, (nor do collisions between two HBr molecules produce H_2 and Br_2), and thus it is not correct to say that all imaginable bimolecular reactions are viable elementary processes.

An elementary process, therefore, must be reasonably probable for whatever reason. Such processes are conveniently categorized into classes based on the reaction *molecularity*, and traditional names are associated with different types of reactions within each class. Table 27.1 lists examples of many of these, and in this section, we survey characteristic features of each class. Throughout this chapter, we subscript elementary rate constants with 1, 2, or 3 to indicate that they refer to unimolecular, bimolecular, or termolecular elementary processes.

Unimolecular Processes

One of the simplest unimolecular processes is isomerization. For example, the internal rotation isomerization between staggered and eclipsed forms of ethane has a very low energy barrier and is probable at all temperatures, so much so that we cannot isolate the pure staggered isomer. In the methyl isocyanide–methyl cyanide isomerization, we imagine the CN group rotating from one orientation to the other, keeping the C≡N bond length and the methyl group geometry roughly constant:

As the CN rotates, the molecule's potential energy (in a Born–Oppenheimer sense) rises to a maximum, then falls again to the local potential energy minimum characteristic of the cyanide product. Section 27.4 will have much more to say about this reaction.

TABLE 27.1 Examples of elementary processes

	Unimolecular
Isomerization	$CH_3NC \rightarrow CH_3CN$
Elimination	$CH_3CF_3 \rightarrow CH_2CF_2 + HF$
	Bimolecular
Radiative association	$C + O \rightarrow CO + h\nu$†
Penning ionization (Chemi-ionization)	$He^* + H \rightarrow He + H^+ + e^-$
Associative ionization	$CH + O \rightarrow HCO^+ + e^-$
Atom abstraction	$F + H_2 \rightarrow HF + H$
Collision-induced dissociation	$Ar + I_2 \rightarrow Ar + I + I$
Charge transfer	$Ar^+ + Xe \rightarrow Ar + Xe^+$
Dielectric recombination	$Li^+ + F^- \rightarrow Li + F$
Collision-induced ionization	$Ar + Cs \rightarrow Ar + Cs^+ + e^-$
Ion-pair formation	$Ar + CsBr \rightarrow Ar + Cs^+ + Br^-$
Energy transfer	$Ar^* + N_2 \rightarrow Ar + N_2^*$
Chemiluminescence	$Ca + F_2 \rightarrow CaF + F + h\nu$
Associative detachment	$H + H^- \rightarrow H_2 + e^-$
Dissociative attachment	$HI + e^- \rightarrow H + I^-$
	Termolecular
Recombination	$Ar + I + I \rightarrow Ar + I_2$
	Photochemical
Electronic excitation	$I_2 + h\nu \rightarrow I_2^*$
Photoionization	$Cs + h\nu \rightarrow Cs^+ + e^-$
Photodissociation	$HI + h\nu \rightarrow H + I$
Multiphoton ionization	$Xe + nh\nu \rightarrow Xe^+ + e^-$
Multiphoton dissociation	$SF_6 + nh\nu \rightarrow SF_5 + F$

† The symbol $h\nu$ is shorthand for emission or absorption of a photon; similarly, $nh\nu$ is shorthand for emission or absorption of n photons in a multiphoton process.

In contrast, the interconversion of *optical* isomers is never observed to occur at a significant rate (in the absence of catalysts, the topic for Section 27.5), due to the generally enormous intramolecular potential energy barrier that exists between isolated optical isomers. The *height* of this barrier controls, in large part, the reaction rate.

Similarly, a unimolecular elimination (or unimolecular fragmentation in general) proceeds at a rate governed by the height of potential energy barriers between the isolated, ground-state reactant and the isolated products. Any internal excitation of the reactant that can help move atoms along the direction leading to products will raise the reaction rate. This excitation can be nonspecific, as happens with a temperature increase, or it can be designed to *direct* internal energy excitation in a favorable way, as many photochemical processes attempt to do.

Bimolecular Processes

Bimolecular elementary processes fall into many named types, as Table 27.1 indicates. Some bimolecular collisions cannot lead to a reaction due to energy and momentum conservation requirements. Thus, in the gas phase, a bimolecular recombination reaction such as $I + I \rightarrow I_2$ cannot happen in isolation. If the I atoms have enough energy to approach to a distance characteristic of the I_2 bond length, they have enough energy to separate back to atoms and will always do so. In a condensed phase, however, the I atoms remain close to each other long enough for solvent molecules to extract some of their relative energy, stabilizing the I_2 product.

FIGURE 27.1 The collision between two ground-state atoms, C and O here, leads to a bound molecule only if sufficient collisional energy is lost. The process called *radiative association* stabilizes the molecular product through spontaneous emission *during* the collision.

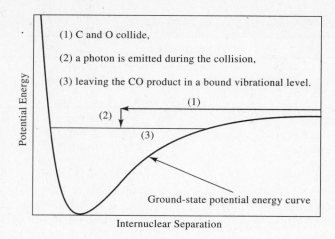

(1) C and O collide,

(2) a photon is emitted during the collision,

(3) leaving the CO product in a bound vibrational level.

The rare gas-phase process called *radiative association* uses spontaneous emission to remove collision energy and stabilize the product. In the example in Table 27.1, $C + O \rightarrow CO + h\nu$ where $h\nu$ represents a photon, we picture the reactant atoms approaching to some distance on the order of the CO bond length, as diagrammed in Figure 27.1.[1] They remain in this region for only a short time, typically less than a picosecond at ordinary temperatures. But while they are close, there is some probability for spontaneous photon emission (in the infrared, following something like the dipole emission rules discussed in Chapter 19 but modified because the initial state of the incipient CO is *unbound*). The photon removes energy from the colliding atoms, and if enough is removed, they no longer have sufficient energy to separate. The bond is made. As Problem 25.6 discussed, however, the short time available for emission compared to the small probability of emission per unit time makes any radiative association process very rare. Nevertheless, we conclude that gas-phase bimolecular elementary processes *must have at least two products*, even if one of them is a photon.

EXAMPLE 27.1

Several imaginable bimolecular elementary processes in a discharge through H_2 are listed below. Which are likely to be unimportant or impossible, and why?

(a) $H_2 + e^- \rightarrow H_2^+ + 2e^-$
(b) $H + e^- \rightarrow H^-$
(c) $H + H_2^+ \rightarrow H_3^+$
(d) $H_2 + H_2^+ \rightarrow H_3^+ + H$
(e) $H^- + H^+ \rightarrow H_2$
(f) $H_2^+ + e^- \rightarrow 2H$

SOLUTION Process (a), electron-impact ionization, is a common and important process in any gaseous discharge. Processes (b) and (c) are bimolecular associations that cannot lead to a bound, stable product as written. In (a), the electron cannot attach to an H atom without photon emission, a possible but improbable process. A more likely means of producing H^- is the *dissociative attachment* reaction, $H_2 + e^- \rightarrow H^- + H$. Similarly, process (c) cannot produce stable H_3^+ in a bimolecular association of H and H_2^+, but process (d), H atom

---

[1]This diagram is simplified somewhat in that it indicates only a single potential energy curve. In reality, ground-state C and O atoms can interact via several possible curves, some of which are repulsive and can never lead to stable products. Recall our discussion of the H_2 potential energy curves that dissociate to ground-state H atoms: the bound singlet and the repulsive triplet of Figure 14.10. Two colliding H atoms have only a one in four chance of interacting on the bound singlet curve.

abstraction, does lead to H_3^+. Process (e) cannot occur as written, since there is no mechanism to stabilize the H_2 product. The only likely reaction between H^- and H^+ is the charge-transfer reaction to form two free H atoms. Finally, process (f), *dissociative recombination,* is allowed and probable.

⟹ RELATED PROBLEM 27.1

We now have a puzzle. Why do some processes such as $H_2 + Br_2 \rightarrow 2HBr$ that satisfy rules for product stabilization fail to proceed at a measurable rate? There are many seemingly plausible bimolecular reactions that fail because of the *nature of the potential energy surface* that describes the collision between reactants. Understanding this surface is one of the goals of this and the next chapter. For now, we can say in a general way that the potential energy surface that governs the simultaneous interactions among two H atoms and two Br atoms must have features—large potential energy barriers—that preclude $H_2 + Br_2 \rightarrow 2HBr$. It is often the case that such barriers are not impossibly high but are nevertheless high enough that collisions with extraordinary energy, energy that could surmount a barrier, fail to do so simply because some lower energy option is available to them. In the $H_2 + Br_2$ case, collision-induced dissociation of Br_2 is such an option.

By far the most important chemical bimolecular process is simple atom abstraction or *metathesis.* The next two sections cover this type of reaction in more detail, but we already know many questions that must be posed in the course of understanding the rate of such a process: Is it exothermic or endothermic? Does it have a large or small activation energy? Is there more than one possible set of products? Can the reactants interact through more than one potential energy surface?

Termolecular Processes

Termolecular elementary processes in the gas phase often step in to provide reaction paths that are precluded from happening in a bimolecular way. Termolecular recombination is the most important of these. Since $I + I \rightarrow I_2$ cannot happen in isolation, but since the two I atoms remain in each other's vicinity for a finite time, there is some probability that a third atom (of any type) will collide with the transient $I \cdots I$ collision pair. As in the example in Table 27.1, if we imagine I atoms dilute in excess Ar, that third atom is most likely to be Ar. When Ar collides with $I \cdots I$, there is some probability that the rebounding Ar moves away from the collision faster than it approached. To do this, it must have extracted some energy from the $I \cdots I$ species, and if enough is extracted, the I_2 molecule finds itself with insufficient energy to separate into free atoms, and the bond is made. The Ar third body plays the role of the emitted photon in radiative association, but unlike the radiative process, this third body may also *increase* the internal energy of the $I \cdots I$ pair, *destabilizing* it further. This does *not* lead to a successful reaction, and thus the presence of a third collision partner is a necessary but not sufficient condition for recombination. In condensed-phase recombination, solvent molecules play the role of the third body.

From this discussion, it is clear that termolecular recombination is enhanced the longer the recombining fragments remain in close proximity. One way to do this is to lower the fragment's speed: lower the temperature (or have heavier fragments that move slowly), but at any one temperature, it is the nature of the intermolecular potential energy surface for the incipient molecule that governs the length of time for the encounter. Contrast, for example, I atom recombination with methyl radical recombination. The internal degrees of freedom of the methyl radicals provide places to temporarily store some of the collision energy during the encounter. As the radicals approach with some initial collision energy, the details of the collision (which is what we mean when we speak of the collision *dynamics*) can cause some of this energy to be transferred into the internal motion of the radicals.[2] This can temporarily divert energy from the dissociation coordinate ($H_3C \cdots CH_3$), keeping the radicals in close proximity for a longer time, increasing the chance that a third body collision can irreversibly transfer energy from them.

If combining radical pairs linger in each other's vicinity long enough, we can write the termolecular process as a sequence of bimolecular processes. If the radicals are A and the third body is M (a common symbol for a chemically inert third body), we can write

$$A + A \underset{k_{-1}}{\overset{k_2}{\rightleftarrows}} A \cdots A \qquad \text{(transient association)} \qquad (27.1a)$$

$$A \cdots A + M \overset{k_2'}{\rightarrow} A_2 + M \qquad \text{(collisional stabilization)} \qquad (27.1b)$$

The difference between this sequence and the imagined simultaneous interaction among 2A and M is simply one of time scales. If $A \cdots A$ cannot survive longer than the time of a typical A to A *vibrational* period, then M must interact with $A \cdots A$ during that brief time. This is the basic termolecular elementary process. If $A \cdots A$ lives for a longer time, characteristic of several $A \cdots A$ *rotational* periods, then we can designate it to be a separate species and imagine M colliding with it in a separate energy-transfer step, Eq. (27.1b).

In a variant of this scheme, the inert third body acts as a so-called *chaperone*. Suppose the $A \cdots A$ interaction does not last a long time, but the $A \cdots M$ interaction does. Then the analog of the sequence above becomes

$$A + M \underset{k_{-1}}{\overset{k_2}{\rightleftarrows}} A \cdots M \qquad \text{(chaperone complex formation)} \qquad (27.2a)$$

$$A \cdots M + A \overset{k_2'}{\rightarrow} A_2 + M \qquad \text{(bimolecular atom transfer)} \qquad (27.2b)$$

with the same net result (and net rate law—see Example 27.2 below) as in the previous sequence. Experimental evidence supports the choice of a chaperone mechanism in many cases. For example, the I atom recombination rate constant in the presence of excess I_2 is about 530 times faster at room temperature than it is in the presence of excess Ar. One expects the $Ar \cdots I$ complex to be less strongly bound than the I_3 (or $I \cdots I_2$) complex, and the increased lifetime of the I_3 complex increases the probability of the reaction.

EXAMPLE 27.2

Show that the chaperone mechanism and the long-lived radical–radical intermediate mechanism can lead to the same rate law as the pure termolecular elementary process.

SOLUTION We write the pure termolecular process as $2A + M \rightarrow A_2 + M$, with a rate law $\mathcal{R} = k_3[A]^2[M]$ where k_3 is the elementary termolecular rate constant. For the long-lived

---

[2] In the case of methyl recombination, one such motion suggests itself immediately. The geometry of a free methyl radical is planar, but in the ethane product, the geometry is roughly tetrahedral about the C atoms, suggesting that the out-of-plane bending motion of the radicals must play a role in the course of the reaction. Excitation of this motion can temporarily store some of the collision energy.

radical–radical intermediate mechanism, Equations (27.1a, b), we first write the rate law for [A · · · A]:

$$\frac{d[\text{A} \cdots \text{A}]}{dt} = k_2[\text{A}]^2 - k_{-1}[\text{A} \cdots \text{A}] - k_2'[\text{A} \cdots \text{A}][\text{M}] \ .$$

Using the steady-state approximation, we set this expression equal to zero and solve for [A · · · A], finding

$$[\text{A} \cdots \text{A}] = \frac{k_2[\text{A}]^2}{k_{-1} + k_2'[\text{M}]} \ .$$

The A_2 product is formed in the elementary bimolecular second step: $d[A_2]/dt = k_2'[\text{A} \cdots \text{A}][\text{M}]$, and substituting the steady-state expression for [A · · · A] gives

$$\frac{d[\text{A}_2]}{dt} = k_2'[\text{A} \cdots \text{A}][\text{M}] = k_2' \left(\frac{k_2[\text{A}]^2}{k_{-1} + k_2'[\text{M}]} \right) [\text{M}] \ .$$

There are two physically important limits that simplify this expression. If $k_2'[\text{M}] \gg k_{-1}$, we find $d[\text{A}_2]/dt = k_2[\text{A}]^2$. The recombination appears to be bimolecular because the net rate is controlled by A · · · A formation, and the A · · · A complex is immediately stabilized. If [M] is sufficiently low, $k_{-1} \gg k_2'[\text{M}]$, and $d[\text{A}_2]/dt = k_2'(k_2/k_{-1})[\text{A}]^2[\text{M}]$, which has the form of the elementary termolecular rate law. The experimental rate constant is the product of k_2' and the ratio k_2/k_{-1}, the equilibrium constant for the formation of A · · · A.

In the chaperone mechanism, we follow the same steps, starting with the rate law for [A · · · M] from Equations (27.2a, b):

$$\frac{d[\text{A} \cdots \text{M}]}{dt} = k_2[\text{A}][\text{M}] - k_{-1}[\text{A} \cdots \text{M}] - k_2'[\text{A} \cdots \text{M}][\text{A}] \ ,$$

applying steady-state:

$$[\text{A} \cdots \text{M}] = \frac{k_2[\text{A}][\text{M}]}{k_{-1} + k_2'[\text{A}]} \ ,$$

and substituting in the rate law for A_2 formation:

$$\frac{d[\text{A}_2]}{dt} = k_2'[\text{A} \cdots \text{M}][\text{A}] = k_2' \left(\frac{k_2[\text{A}][\text{M}]}{k_{-1} + k_2'[\text{A}]} \right) [\text{A}] \ .$$

Here, the usual experimental situation has [A] quite small so that $k_{-1} \gg k_2'[\text{A}]$, and we can make the approximation $d[\text{A}_2]/dt = k_2'(k_2/k_{-1})[\text{A}]^2[\text{M}]$, again leading to a third-order net reaction rate law.

➠ RELATED PROBLEMS 27.2, 27.3, 27.4

Photochemical Processes

Finally, we introduce the language of *photochemical* elementary processes. Grotthuss, in 1819, and Draper, in 1841, laid the foundation of photochemistry when they independently stated what today seems obvious: light must be absorbed to effect a chemical change.[3] Modern photochemistry was put on reasonably firm footing around the turn of the twentieth century when Stark and Einstein indepen-

---

[3]The modern way of stating this law is that a molecule must *exchange energy with the radiation field*. Absorption is the most common mechanism by far, but chemical changes are possible through a Raman process (see Chapter 19) in which a molecule *inelastically scatters* a photon, absorbing some, but not all, of the photon's energy.

dently stated what has become known as the Stark–Einstein *law of photochemical equivalence:* each absorbed photon leads to one photochemically active molecule. We treat a photon as another collision partner and say, for example, that the rate of photon absorption in a reaction such as $HI + h\nu \rightarrow H + I$ equals the rate of H atom production. A mole of photons is often called an *einstein,* a unit suggested by Bodenstein in 1929, and we convert among einstein units, light intensity, and irradiation time using expressions derived in Section 18.1. For example, the He/Ne laser discussed in Example 18.1 produced 3.2×10^{15} photons s^{-1}. Thus, this laser produces 1 einstein in $(6.02 \times 10^{23}$ photons$)/(3.2 \times 10^{15}$ photons s$^{-1}) = 1.8 \times 10^8$ s, or about 5.7 years, (which, coincidentally, is comparable to the typical operating lifetime of such lasers).

The amount of light absorbed is governed in most cases (but not for multiphoton processes) by the Beer–Lambert law, Eq. (18.15), which, for an incident intensity I_0 (with dimensions of energy per unit area per unit time), predicts energy absorption at a rate per unit illuminated area given by

$$I_a = I_0 (1 - e^{-abc}) \qquad (27.3)$$

where a is the absorbance coefficient at the illumination frequency, b is the "breadth" (optical path length) of the sample, and c is the absorber's concentration. The units of a depend on the units of b and c, of course, and if b is in cm and c in M units, then a is in M^{-1} cm^{-1} units. If c is in units of molecules per unit volume, then a has the dimensions of area, and the absorbance coefficient becomes an *absorption cross section.*

EXAMPLE 27.3

The 0.12 W output at 488 nm from an Ar$^+$ laser, a common continuous wave (CW) laboratory laser, is spread to illuminate uniformly the 1.0 cm^2 cross-sectional area of a 5.0 cm long cell. The cell is filled with a 3.5×10^{-4} M solution of a molecule A that produces exactly one product molecule B per photon absorbed by A. The absorption coefficient of A at 488 nm is 86 M^{-1} cm^{-1}. What is the rate of production of B?

SOLUTION From Eq. (27.3), with $I_0 = 0.12$ J cm^{-2} s^{-1}, $a = 86$ M^{-1} cm^{-1} (a modest value), $b = 5.0$ cm, and $c = 3.5 \times 10^{-4}$ M, we find that energy is absorbed at the rate

$$\begin{aligned} I_a &= I_0(1 - e^{-abc}) \\ &= (0.12 \text{ W cm}^{-2})\{1 - \exp[-(86 \text{ M}^{-1}\text{ cm}^{-1})(5.0 \text{ cm})(3.5 \times 10^{-4} \text{ M})]\} \\ &= 1.7 \times 10^{-2} \text{ J cm}^{-2} \text{ s}^{-1} \ . \end{aligned}$$

A photon at 488 nm wavelength has an energy $hc/\lambda = 4.1 \times 10^{-19}$ J. Thus, the absorbed photon flux is $(1.7 \times 10^{-2}$ J cm^{-2} s$^{-1})/(4.1 \times 10^{-19}$ J photon$^{-1}) = 4.1 \times 10^{16}$ photon cm^{-2} s^{-1}. This is the rate of B production per unit illuminated area, since each absorbed photon leads to one B, and the rate per unit volume is this number divided by the optical path length, or $(4.1 \times 10^{16}$ photon cm^{-2} s$^{-1})/(5.0$ cm$) = 8.2 \times 10^{15}$ cm^{-3} s^{-1}. Dividing this number by Avogadro's constant gives us the number of moles of B produced per cm^3 per second, and converting cm^3 to dm^3 gives us the rate of production of B in traditional molar concentration units:

$$\frac{d[B]}{dt} = \frac{8.2 \times 10^{15} \text{ cm}^{-3} \text{ s}^{-1}}{N_A (10^{-3} \text{ dm}^3 \text{ cm}^{-3})} = 1.4 \times 10^{-5} \text{ M s}^{-1} \ .$$

This is the *initial* rate. As A is converted to B, [A] falls, the amount of absorbed light falls as well, and the rate of production of B falls.

➠ RELATED PROBLEMS 27.5, 27.7, 27.8

Photochemical processes do not always have a one-to-one correspondence between photons absorbed and reaction products produced. The *photochemical product quantum yield* Φ is the experimental clue to many photochemical mechanisms:

$$\Phi = \frac{\text{number of molecules of a particular product produced}}{\text{number of photons absorbed}} . \quad (27.4)$$

(Compare to the fluorescence quantum yield, Φ_F, defined in Eq. (20.3).) The product quantum yield can be much less than 1 if fluorescence or collisional deactivation removes the absorbed energy before the reaction can occur. It can be much greater than 1 if the absorbed photon initiates a branching chain reaction, as in the photoinitiated $H_2 + Cl_2 \rightarrow 2HCl$ system where the quantum yield can be as high as 10^5. In multiphoton processes, $\Phi < 1$ simply because many photons are absorbed, as in the infrared multiphoton dissociation of SF_6 discussed in Section 18.4. Many dozen photons are absorbed by a single SF_6 molecule in an intense, brief laser pulse, leading to $SF_5 + F$ products. Similarly, multiphoton ionization requires typically two to five photons, absorbed simultaneously or in very rapid sequence, to produce one ionized molecule.

If there are physical processes that remove the absorbed energy faster than the excited molecule can produce products, continuous illumination produces a *photochemical stationary state*. Consider the simplest case: A absorbs light, producing an excited state A* which can only fluoresce with a first-order fluorescence rate constant $k_f = 1/\tau_{\text{rad}}$, the reciprocal of the A* radiative lifetime:

$$A + h\nu \underset{\text{fluorescence}}{\overset{\text{absorption}}{\rightleftarrows}} A^* .$$

Example 27.3 shows that the rate of A* production under constant illumination is

$$\frac{d[A^*]}{dt} = \frac{I_0(1 - e^{-abc})}{N_A bh\nu} = \frac{I_a}{N_A bh\nu} . \quad (27.5)$$

Spontaneous emission depletes [A*] at a rate equal to $-k_f[A^*]$. We add this term to Eq. (27.5), set the sum equal to zero to express a steady state, and solve for [A*]:

$$[A^*]_{ss} = \frac{\text{rate of A* production}}{\text{unimolecular rate constant for A* destruction}} = \frac{I_a}{N_A bh\nu k_f} . \quad (27.6)$$

This expression can be extended to include other physical relaxation processes, as discussed in Problems 27.11–13. Note that at thermal equilibrium, insignificant amounts of A* are usually present, but constant irradiation to produce the photochemical stationary state has the effect of *shifting* the equilibrium towards greater [A*]. If this excited state is significantly more reactive than the ground state, illumination has produced a way to enhance the reactivity of A. We distinguish the *primary* photochemical event, production of A*, from all subsequent *secondary* processes involving A* in this case. Photochemical initiation of the $H_2 + Cl_2 \rightarrow 2HCl$ reaction is an example of this idea: the light shifts the $Cl_2 \rightarrow 2Cl$ equilibrium from the thermally negligible amount of [Cl] necessary to initiate the net chain reaction to sufficient [Cl] to produce rapid reaction.

EXAMPLE 27.4

A gas at 300 K and 50 torr pressure is placed in a 100 cm³ volume. The cell length, 10 cm, is such that 5% of the light at a particular frequency is absorbed by the gas, producing the first excited state. The excited state undergoes only spontaneous emission back to the ground state with a unimolecular rate constant $k_f = 10^7$ s^{-1}. The excited state, however, can absorb the incident light as well as the ground state does. What is the probability of *sequential two photon* absorption if the gas is continuously illuminated with 10^{15} photons s^{-1}?

SOLUTION The light creates a steady-state molar concentration of first excited state molecules, A*, given by Eq. (27.6) with $I_a/N_A h\nu = (0.05 \times 10^{15} \text{ s}^{-1})/VN_A$:

$$[A^*]_{ss} = \frac{\text{rate of A* production}}{\text{rate constant for A* fluorescence}}$$

$$= \frac{(0.05 \times 10^{15} \text{ s}^{-1})/[(0.1 \text{ L})(6.02 \times 10^{23} \text{ mol}^{-1})]}{10^7 \text{ s}^{-1}}$$

$$= 8.3 \times 10^{-17} \text{ M},$$

which is a very small concentration. From the ideal gas equation of state, one can calculate that the number density of A molecules in the cell is about 1.6×10^{18} cm^{-3}. The steady-state number density of A* molecules is only 5×10^4 cm^{-3}. If these species can absorb a second photon with the same probability as the ground state, then the rate of sequential two photon absorption (i.e., $A + h\nu \rightarrow A^*$ followed by $A^* + h\nu \rightarrow A^{**}$) is less probable than the rate of ground-state absorption by a factor of $[A^*]_{ss}/[A] = (5 \times 10^4 \text{ cm}^{-3})/(1.6 \times 10^{18} \text{ cm}^{-3}) \cong 3 \times 10^{-14}$. This is a negligible probability. Problem 27.10 considers this system under intense pulsed laser illumination where the probability can be significant.

➡ **RELATED PROBLEMS** 27.8, 27.10

There are often several fates for the energy absorbed in a primary photochemical event. In fact, Stern and Volmer established in 1920 that the primary event often did *not* lead to dissociation or ionization directly, as had been suspected previously. If the primary event is electronic excitation of A to excited state A*, then A* can spontaneously fluoresce with a unimolecular rate constant k_f, as we have seen, or it can suffer *collisional quenching* with a bimolecular rate constant k_Q:

$$A^* + M \xrightarrow{k_Q} A + M$$

where M is a chemically inert collision partner.[4] Suppose we measure the fluorescence decay rate of A* as a function of [M], using a brief laser pulse to create an initial amount of A* at $t = 0$. In the absence of collisions, (i.e., at [M] = 0), this rate is governed by the fluorescence rate constant, k_f. In the presence of collisions, however, the measured rate is the *sum* of k_f and $k_Q[M]$:

$$k_{\exp} = k_f + k_Q[M] \tag{27.7}$$

so that a plot of $k_{\exp}$, the experimental fluorescence decay rate constant, should be a linear function of [M] (or of pressure, if T is constant). The slope of this line is k_Q, and its intercept is k_f. This is an example of one of several related quenching schemes that can be treated similarly, and the linear plot is known as a *Stern–Volmer plot*.

EXAMPLE 27.5

Example 27.3 mentioned the 488 nm emission of the Ar$^+$ laser. Ground state Na$_2$ absorbs this laser light in a transition to the B $^1\Pi_u$ excited electronic state. (See Table 19.3.) Experiments on the fluorescence decay rate of the $v = 9$, $J = 56$ vibration–rotation level of this state as

---

[4]Stern and Volmer pointed out in support of their arguments the known phenomena of molecular iodine and atomic mercury fluorescence quenching by added gases. They did not fully understand the collisional quenching mechanism (quantum mechanics was still six years away!), but they did deduce the correct phenomenon.

a function of Na atom pressure (the Na$_2$ partial pressure was always much less) gave the results below, expressed in terms of the experimental fluorescence rate constant, k_{exp}. What is the natural radiative lifetime of this state, and what is the rate constant for quenching by Na atoms?

P/torr	0.50	0.60	0.70	0.80	0.90
$k_{exp}/10^8$ s^{-1}	1.74	1.82	1.88	1.93	2.01

SOLUTION Since [Na] $\gg$ [Na$_2$] in these experiments, we can take P to be directly proportional to [Na] and write Eq. (27.7) as $k_{exp} = k_f + k_Q P$. A Stern–Volmer plot of these data, shown below, gives a least-squares fit intercept $k_f = 1.42 \times 10^8$ s^{-1}, or a radiative lifetime $\tau_{rad} = 1/k_f = 7.04$ ns (see Footnote k in Table 19.3) and a quenching rate constant $k_Q = 6.5 \times 10^7$ torr^{-1} s^{-1} from the slope.

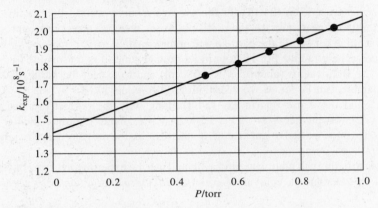

➠ RELATED PROBLEMS 27.13, 27.14, 27.15

27.2 SIMPLE COLLISION THEORIES OF REACTIONS

In this section, we take our first stab at predictive theories of bimolecular elementary rate constants. We consider gas- and solution-phase reactions, trying to expose and explain the dominant phenomena that govern bimolecular reaction rates. Our central theme is the *encounter rate,* the raw rate of bimolecular collisions whether they lead to reaction or not. On top of this rate, we add a factor for the probability that a collision is reactive. We have already mentioned several phenomena that go into this factor: *energy requirements* (that are caused by potential energy surface barriers), *reactant spin-state requirements* (that govern the electronic state in which the reactants interact), and *collision geometry requirements* (that express the need for reactants to have auspicious relative orientations in order to react). These are difficult to account for, but the encounter rate itself is straightforward to find. In the gas phase, we start with the hard-sphere binary collision rate from the kinetic theory of gases, and in solution, we invoke the diffusion rate of reactants. We start in the gas phase.

Gas-Phase Reactions

The elementary associative detachment reaction

$$H + H^- \rightarrow H_2 + e^-$$

is simple and well studied experimentally and theoretically. At 300 K, the rate constant is 1.17×10^{12} M^{-1} s^{-1}. How rapidly do H and H$^-$ collide at 300 K, no matter what the outcome of the collision? The answer is the total collision rate per unit volume, Eq. (25.12), a function of the relative collision velocity g, the collision cross section σ, and the number densities n_1 and n_2 (particles per unit volume) of the collision pair:

$$Z_{12} = <g\sigma> n_H n_{H^-} \ . \tag{25.12}$$

This is a traditional bimolecular rate law if we convert the number densities into molar concentrations: molar concentration = (particle number density)/$(N_A \times 1000 \text{ L m}^{-3})$, and it is our first expression for an elementary bimolecular rate constant:

$$k_2 = N_A <g\sigma>(1000 \text{ L m}^{-3}) = N_A \sigma_{12} \left(\frac{8k_B T}{\pi \mu}\right)^{1/2} (1000 \text{ L m}^{-3}) \tag{27.8}$$

where we have used Eq. (24.15) for the average relative speed, $<g>$, with μ the collision reduced mass and σ_{12} the collision cross section. To test this expression, we need a value for σ_{12}. In the hard-sphere approximation, $\sigma_{12} = \pi(r_1 + r_2)^2$ where r_i is a hard-sphere radius. For H atom in its ground state, it is reasonable to expect $r_H \sim a_0$, the Bohr radius (about 5.3×10^{-11} m) and to expect the radius of H$^-$ to be comparable, but larger. If we take both radii to equal a_0, $\sigma_{12} = \pi a_0^2 = 3.5 \times 10^{-20}$ m^2, and if we use this in Eq. (27.8) along with the correct reduced mass ($\mu = m_H/2$), we predict $k_2 = 7.5 \times 10^{10}$ M^{-1} s^{-1}, about 15 times smaller than experiment. Since we had to guess at hard-sphere radii, the rough order of magnitude agreement is encouraging.

We can turn this calculation around a bit, and use the experimental rate constant in Eq. (27.8) to calculate the equivalent *experimental thermal reaction cross section* σ_r:

$$\sigma_r = \frac{k_2}{N_A \left(\frac{8k_B T}{\pi \mu}\right)^{1/2} (1000 \text{ L m}^{-3})} = \frac{1.17 \times 10^{12} \text{ M}^{-1}\text{s}^{-1}}{2.14 \times 10^{30} \text{ m}^{-2}\text{M}^{-1}\text{s}^{-1}} = 5.5 \times 10^{-19} \text{ m}^2 \ .$$

This is in general a valid calculation—it simply makes our focus a little tighter, concentrating on the microscopic reaction cross section rather than the macroscopic rate constant. We can always go back and forth from thermally averaged reaction cross section to rate constant through this expression.

EXAMPLE 27.6

The rate constant at 300 K for the ion–molecule reaction Ar$^+$ + H$_2$ → ArH$^+$ + H is 6.6×10^{11} M^{-1} s^{-1}. What is the reaction cross section? What effective reaction distance does this cross section suggest?

SOLUTION We solve Eq. (27.8) for the cross section, using the collisional reduced mass

$$\mu = \frac{m_{Ar} m_{H_2}}{m_{Ar} + m_{H_2}} = 3.18 \times 10^{-27} \text{ kg} \ ,$$

which gives an average relative collision speed

$$<g> = \left(\frac{8k_B T}{\pi \mu}\right)^{1/2} = 1820 \text{ m s}^{-1}$$

and a reaction cross section

$$\sigma_r = \frac{k_2}{N_A <g>(1000 \text{ L m}^{-3})} = 6.0 \times 10^{-19} \text{ m}^2 \ .$$

If we express the cross section as πd^{*2} where d^* is an effective reaction distance, we find $d^* = 4.4 \times 10^{-10}$ m = 4.4 Å, a reasonable intermolecular distance, but one we do not yet know how to interpret in detail.

→ RELATED PROBLEMS 27.16, 27.18

Experimentally, the rate constant for H + H$^-$ → H$_2$ + e^- is nearly independent of temperature. There is no apparent activation energy. Equation (27.8) has a weak $T^{1/2}$ temperature dependence *at least*, since σ_r may depend on temperature in an unknown way, but there is no hint in our theory so far of the Arrhenius-like behavior predicted by Eq. (26.26) or (26.27). We need another reaction to explore the way to incorporate activation energy into the story.

Many ion–molecule reactions have very large rate constants because they have no activation energy. Consider, in contrast, the Cl + H$_2$ → HCl + H reaction. At 300 K, $k_2 = 9.4 \times 10^6$ M^{-1} s^{-1}, 70 000 times smaller than that for the isoelectronic Ar$^+$ + H$_2$ → ArH$^+$ + H reaction of Example 27.6. The Cl + H$_2$ reaction has an appreciable activation energy: $E_a = 18.3$ kJ mol^{-1}. If we compute a reaction cross section from Eq. (27.8), we find $\sigma_r = 8.6 \times 10^{-24}$ m^2. If we write $\sigma_r = \pi d^{*2}$ where d^* is an effective reaction distance, as used in Example 27.6, we find $d^* = 0.017$ Å, a physically absurd distance. It is unreasonable (and rather ridiculous) to imagine that Cl and H$_2$ must approach to within 0.017 Å before they react! What should we do? We can still retain the idea of a reaction cross section, but we need to reinterpret it. *A reaction cross section does not necessarily reveal the physical distance at which reaction occurs.* It is a more complicated thing.

The presence of an appreciable activation energy means that only those collisions with energy sufficient to surmount the activation barrier can lead to reaction, and perhaps not *all* of those with sufficient energy go on to reaction. Is it sufficient to tack the Arrhenius $\exp(-E_a/RT)$ factor onto Eq. (27.8)? This factor equals 6.5×10^{-4} for the Cl + H$_2$ reaction at 300 K, and the Arrhenius A factor is, from Eq. (26.26),

$$A = \frac{k_2}{e^{-E_a/RT}} = \frac{9.4 \times 10^6 \text{ M}^{-1}\text{ s}^{-1}}{6.5 \times 10^{-4}} = 1.45 \times 10^{10} \text{ M}^{-1}\text{ s}^{-1} ,$$

which is in the range of rate constants predicted by Eq. (27.8) alone.

This is encouraging, since it says the collision rate of Eq. (27.8) gives the A factor (with a suitable hard-sphere cross section) while the activation energy factor simply modulates the probability of reaction per collision. Now we ask what form of energy—relative translational energy, internal excitation energy, etc.—is effective in surmounting the activation barrier. One collision model assumes that only the *relative translational collision energy along a line connecting the centers of the collision pair* is used to surmount any potential energy barrier to reaction. It is beyond our scope here to discuss this model in detail, but it assumes hard-sphere collisions with a potential energy barrier of height E^* separating reactants and products. One still assumes a hard-sphere collision cross section and averages over all relative collision speeds. This theory yields the expression

$$k_2 = N_A \langle g \rangle \sigma_{12} (1000 \text{ L m}^{-3}) e^{-E^*/k_B T} = N_A \left(\frac{8 k_B T}{\pi \mu}\right)^{1/2} \sigma_{12} (1000 \text{ L m}^{-3}) e^{-E^*/k_B T} .$$

While this is an improvement, it is still not in quantitative agreement with experiment if one uses a temperature-independent cross section calculated from hard-sphere radii typically derived from transport phenomena. There is still a missing factor related to the collision dynamics, and for now, we express it in a time-honored *ad hoc* fashion. In his 1919 discussion of the H + HBr → H$_2$ + Br reaction's role in the mechanism of H$_2$ + Br$_2$ → 2HBr, Herzfeld called this missing factor the *steric factor,* symbolized p. He suggested that this reaction occurred only with certain *orientations* of the colliding species, a very reasonable suggestion.

We know today that effects other than orientation alone, such as the nature of the electronic state of the collision, enter the steric factor. For example, in the H + H$^-$ → H$_2$ + e^- reaction we first considered, ground-state H (1s ^{2}S) and ground-state H$^-$ (1s^2 ^{1}S) can interact along either of two molecular potential energy curves: $^2\Sigma_g^+$, which is repulsive and does not lead to stable H$_2^-$, and $^2\Sigma_u^+$, which does lead to stable H$_2^-$. At low collision energy, the reactants must interact along the $^2\Sigma_u^+$ curve to produce the final products, and each H + H$^-$ collision has a 50/50 chance of doing so. Since there is no "orientation" to the collision of two spherical species such as these, the steric factor here equals 0.5.

Thus, our final expression for a collision theory bimolecular rate constant has four pieces with a physical interpretation:

$$k_2 = N_A(1000 \text{ L m}^{-3}) \quad p \quad <g> \quad \sigma_{12} \quad e^{-E_a/RT} \ . \quad (27.9)$$

$$\uparrow \qquad\qquad\qquad \uparrow \quad\ \uparrow \quad\ \uparrow \quad\quad\ \uparrow$$

Rate constant · · · Steric factor · Relative speed · Collision cross section · Activation energy

We can calculate only one piece, the relative speed, with any certainty, but we can measure the activation energy experimentally.

EXAMPLE 27.7

Calculate the steric factor for the reaction Cl + H$_2$ → HCl + H, assuming hard-sphere radii of 1.82 Å for Cl (the Ar value) and 1.47 Å for H$_2$ (the value derived from transport phenomena).

SOLUTION We calculated the Arrhenius A factor for this reaction to be 1.45×10^{10} M^{-1} s^{-1}, and our collision theory says this should equal $N_A(1000 \text{ L m}^{-3})p<g>\sigma_{12}$. The collision cross section is

$$\sigma_{12} = \pi(r_{Cl} + r_{H_2})^2 = 34 \text{ Å}^2 = 3.4 \times 10^{-19} \text{ m}^2 \ ,$$

and the average relative collision speed at 300 K is

$$<g> = \left(\frac{8k_BT}{\pi\mu}\right)^{1/2} = 1825 \text{ m s}^{-1} \ .$$

Thus, the steric factor is

$$p = \frac{A}{N_A(1000 \text{ L m}^{-3})<g>\sigma_{12}} = 0.039 \ ,$$

a number that is difficult to interpret except to say that it indicates that, for whatever reason, only about 4% of those collisions with sufficient energy actually lead to products.

➡ **RELATED PROBLEMS** 27.19, 27.20

Solution Reactions

We consider next a collision theory for bimolecular elementary reactions in solution. Some of the gas-phase concepts still hold, such as activation energy, but the significant difference between solution and gas reactions is the rate of collisions and their nature. Gas molecules spend most of their time flying freely between occasional brief encounters while solution molecules are *always* interacting with *at least* solvent. Solution reactants must *diffuse* to reach each other. Once they do, the solvent cage around them can hold them in close proximity for a long time. As we discussed in Chapter 17 (see Figure 17.7 in particular), even hard spheres appear to have an effective attraction for each other due to this *cage effect,* and instead of the single collision characteristic of gas-phase encounters, solution encounters are a rapid sequence of many collisions of varying collision energy as the reactants jostle in the solvent, exchanging energy with it.

A little thought and a simple mechanistic picture lead to two distinct classes of solution reactions. If the activation energy is very low, the extra collisions between reactants caused by the solvent cage are not needed. Reaction occurs on the first collision, and the reaction rate is controlled by the rate at which reactants diffuse together. This is called, naturally enough, the *diffusion-limited* regime. On the other hand, a reaction with a significant activation energy can take advantage of the extra encounters provided by the cage. The cage molecules are constantly exchanging energy with the reactants causing their collision energy to fluctuate. An occasional fluctuation will be large enough to overcome the activation energy. In this, the *activation-limited* regime, diffusion does not control the reaction rate.

This mechanism has much in common with the sequence of Eq. (27.1) that we used to discuss termolecular processes. The species we symbolized A $\cdots$ A in that discussion is called the *encounter pair* or *encounter complex* in solution reactions, and the first step of the solution mechanism is bimolecular reactant diffusion to form this encounter pair and its possible unimolecular dissociation back to separated reactants A and B:

$$\text{A} + \text{B} \underset{k_{-1}}{\overset{k_2}{\rightleftarrows}} \text{A} \cdots \text{B} \quad \text{(reactant diffusion/encounter pair dissociation)} \quad \textbf{(27.10a)}$$

The encounter pair may also react, forming stable products (call them C and D), and we write this step as a unimolecular process, but recognize that it is in fact a pseudo-first-order process, since it is mediated by solvent present at a large and constant concentration:

$$\text{A} \cdots \text{B} \overset{k_1'}{\rightarrow} \text{C} + \text{D} \quad \text{(encounter pair reaction)} \quad \textbf{(27.10b)}$$

A steady-state analysis for [A $\cdots$ B] leads to the rate law for the net reaction A + B → C + D:

$$\mathcal{R} = \frac{d[\text{C}]}{dt} = k_1'[\text{A}\cdots\text{B}] = \frac{k_2 k_1'}{k_{-1} + k_1'}[\text{A}][\text{B}] = k_{\text{exp}}[\text{A}][\text{B}] \quad \textbf{(27.11)}$$

where k_{exp} is the experimental second-order rate constant.

In the diffusion-limited regime, k_1' (encounter pair *reaction*) $\gg k_{-1}$ (encounter pair *dissociation*), and we have the approximation $k_{\text{exp}} \cong k_2$, the diffusion rate constant. The other limit, slow reaction of the encounter pair or $k_1' \ll k_{-1}$, predicts $k_{\text{exp}} \cong k_1'(k_2/k_{-1}) = k_1' K_{\text{eq}}$ where K_{eq} is the equilibrium constant for Eq. (27.10a), diffusive formation of the encounter pair. This reaction usually has a negative reaction entropy change (since it is an association reaction in which two molecules form only one) and a positive or only slightly negative reaction enthalpy change (or else A $\cdots$ B could be the final product, stabilized as an AB product by solvent collisions) so that $\Delta_r \overline{G}^\circ$ is usually positive and $K_{\text{eq}} \ll 1$.

We can use ideas from Chapter 25 to find an expression for the diffusion rate constant k_2. Before we do, we should anticipate the factors that control it. Certainly the diffusion coefficients D_A and D_B for A and B will be important. These, in turn, are functions of temperature, and Eq. (25.45), the Stokes–Einstein relation $D = k_B T/6\pi\eta R$, will bring solvent viscosity η and molecular radii R into the story as well. If A and B are *ions*, electrostatic forces, governed by both the ions' charges and the relative permittivity of the solvent, play a role as well. In fact, once we recognize that electrostatic forces can affect reaction rates, we can begin to answer a question that may have entered your mind in the previous chapter. There, our discussion of detailed balance related rate constants to equilibrium constants in terms of reagent *concentrations* rather than *thermodynamic activities*. You may have asked yourself whatever happened to fugacity coefficients, activity coefficients, and the other nonideality corrections of thermodynamics. If detailed balance is general in the sense that equilibrium thermodynamics is general, then chemical kinetics *must* at some point invoke the nonideal corrections used in thermodynamics. Here is one place kinetics can do just that: the solution reactions of ions.

We will start with uncharged reactants first, however, since the diffusion problem alone takes some thought to solve. We picture an A molecule as a spherical entity of some characteristic radius R_A and ride along with it as spherical B molecules of radius R_B diffuse around us in a structureless, continuous solvent, as in Figure 27.2. We imagine the encounter complex A $\cdots$ B is formed when A and B are a distance R^* from each other. The conversion of B into A $\cdots$ B makes [B] = 0 at R^*, and this drop in [B] provides a *radial concentration gradient* that drives diffusion of B towards A.

The radial flux of B is related to its diffusion coefficient through Fick's first law, introduced in Section 25.2. For a radial concentration gradient, this law is (compare to Eq. (25.13) for a linear gradient)

$$J_{B,r} = -D_B \frac{\partial n_B}{\partial r} = -D_B \frac{d[B]}{dr} N_A (1000 \text{ L m}^{-3}) \tag{27.12}$$

where $J_{B,r}$ is the radial flux of B, D_B is the diffusion coefficient for B, and n_B is the B number density. Fick's second law says that the time rate of change of n_B, $\partial n_B/\partial t$, equals D_B times the *second* spatial derivative of n_B (see Eq. (25.14)). Differentiation of Eq. (27.12) with respect to r gives another expression for $\partial^2 n_B/\partial r^2$ which, when substituted into Fick's second law, shows that $\partial n_B/\partial t$ equals the negative of the *first*

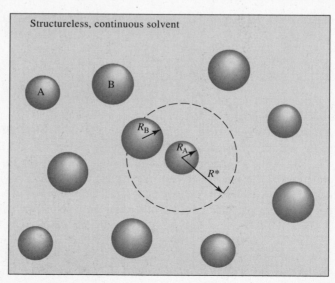

FIGURE 27.2 In solution, reactants diffuse together. We picture them as spheres diffusing through a continuous solvent, and we say the reactants form an *encounter pair* when they are within a distance R^* of each other.

derivative of the *flux*, $J_{B,r}$. If we assume a steady state, $(\partial n_B/\partial t = 0)$, Fick's second law shows that the *first derivative of the flux in the r direction equals zero*, which means the flux is the same through a spherical shell drawn at *any* radius r about A, including one at R^*.

The total radial flow of B through a sphere of arbitrary radius r must therefore equal the flux in Eq. (27.12) times the surface area of the sphere, $4\pi r^2$, and since the radial direction points *away* from the center of this sphere but we want the flow *toward* the center, we reverse the sign in Eq. (27.12) and write the total radial inward flow I_r as

$$I_r = -4\pi r^2 J_{B,r} = 4\pi r^2 D_B N_A (1000 \text{ L m}^{-3})\frac{d[B]}{dr} \ . \tag{27.13}$$

In order to conserve the number of B molecules diffusing toward A, this quantity must be a constant, independent of r. Thus we can write a simple differential equation for $d[B]/dr$:

$$r^2 \frac{d[B]}{dr} = \frac{I_r}{4\pi D_B N_A (1000 \text{ L m}^{-3})} = \text{constant} \ ,$$

which has the solution

$$[B] = [B]_\infty - \frac{I_r}{4\pi r D_B N_A (1000 \text{ L m}^{-3})}$$

where $[B]_\infty$ is the constant, bulk concentration of B at an appreciable distance from A. Since $[B] = 0$ at $r = R^*$,

$$[B]_\infty = \frac{I_r}{4\pi R^* D_B N_A (1000 \text{ L m}^{-3})}$$

or

$$I_r = [B]_\infty 4\pi R^* D_B N_A (1000 \text{ L m}^{-3})$$

so that the radial concentration profile for [B] is

$$[B] = [B]_\infty \left(1 - \frac{R^*}{r}\right) \ . \tag{27.14}$$

This profile is plotted in Figure 27.3 for a typical value of R^*, 5 Å, and we see that it extends many tens of ångströms away from the origin (where A is).

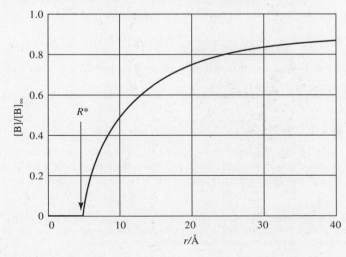

FIGURE 27.3 Reactive diffusion is driven by a radial concentration gradient centered on one reactant. This gradient produces a concentration profile of the other reactant as shown here. The encounter pair radius is R^*, assumed to be 5 Å, a typical value.

To account for the diffusive motion of A, we replace D_B with $(D_A + D_B)$, the sum of the A and B diffusion coefficients. Moreover, the total radial flow, by definition of R^*, is the rate of formation of $A \cdots B$ per A molecule, and thus the reaction rate is

$$\mathcal{R} = \frac{d[A \cdots B]}{dt} = -\frac{d[B]}{dt} = k_2[A][B] = [A]I_r \qquad (27.15)$$

or, using our expression above for I_r in which we write $[B]_\infty$ as simply $[B]$ and correct the diffusion constant, we find

$$k_2[A][B] = [4\pi R^*(D_A + D_B)N_A(1000 \text{ L m}^{-3})] [A][B] \ .$$

The quantity above in square brackets is k_2, the diffusion rate constant:

$$k_2 = 4\pi R^*(D_A + D_B)N_A(1000 \text{ L m}^{-3}) \ . \qquad (27.16)$$

If we take $R^* = 5$ Å and $D_A + D_B = 10^{-9}$ m^2 s^{-1} as representative values, we find $k_2 \cong 4 \times 10^9$ M^{-1} s^{-1}, which is comparable to the magnitude of the fastest bimolecular solution rate constants measured experimentally.[5]

It is interesting to contrast Eq. (27.16) to Eq. (27.8) for the gas-phase collision rate constant. Neglecting unit conversion factors, the gas rate constant is proportional to a relative speed times a cross section, while the solution constant is proportional to a diffusion constant times a length. The gas rate constant depends on the *square* of a characteristic reaction radius through the cross section, but the solution rate constant is a *linear* function of a similar radius.

EXAMPLE 27.8

The $I + I \rightarrow I_2$ recombination reaction in a CCl$_4$ solvent has a measured rate constant $k_{exp} = 8.2 \times 10^9$ M^{-1} s^{-1}, and the diffusion coefficient of I in CCl$_4$ is 4.2×10^{-9} m^2 s^{-1}. The magnitude of k_{exp} suggests that the reaction is diffusion limited. What reaction radius R^* do these numbers imply, and do they support the suspicion of diffusion-limited reaction?

SOLUTION For reaction between two *identical* reactants, as we have here, the rate law is $\mathcal{R} = -(\frac{1}{2})(d[I]/dt)$, but our derivation for a reaction between *different* reactants did not have the factor of $\frac{1}{2}$. Consequently, we divide Eq. (27.16) by 2 for reactions between identical reactants:

$$k_2 = \frac{4\pi R^*}{2}(D_A + D_A)N_A(1000 \text{ L m}^{-3}) = 4\pi R^* D_A N_A(1000 \text{ L m}^{-3}) \ .$$

If we equate k_2 to k_{exp} and solve for R^* using the data given, we find

$$R^* = \frac{k_{exp}}{4\pi D_A N_A(1000 \text{ L m}^{-3})} = 2.6 \times 10^{-10} \text{ m} = 2.6 \text{ Å} \ ,$$

which is a very reasonable length, equal to the I_2 bond length, 2.66 Å, from Table 19.2.

➡ *RELATED PROBLEMS* 27.21, 27.23

EXAMPLE 27.9

The aqueous solution reaction between O_2 and hemoglobin has a rate constant around 4×10^7 M^{-1} s^{-1}. Is this a diffusion-limited reaction? The diffusion coefficient of O_2 in water is 2.0×10^{-9} m^2 s^{-1}, and that for hemoglobin is 6.9×10^{-11} m^2 s^{-1} (from Example 25.15).

---

[5]It is worth pointing out that the *raw collision rate* in solution is roughly the same as that in the gas phase, but most of the collisions are with solvent molecules. Consequently, we call our diffusion rate constant an *encounter* rate constant rather than a *collision* rate constant.

SOLUTION The diffusion constant for hemoglobin is significantly less than that for O_2, as size considerations would indicate. Thus, we approximate the sum of diffusion constants in Eq. (27.16) by the O_2 value alone. The radius of hemoglobin, 31 Å in Example 25.15, is a good estimate for the critical radius R^* (or perhaps an ångström or two more to account for the size of O_2, but as we will see, we can answer the question with a crude R^* estimate). Substituting these values into Eq. (27.16) yields $k_2 = 4.6 \times 10^{10}$ M^{-1} s^{-1}, a value 1000 times greater than the experimental value. Here, the disagreement could be due to an activation energy *or* it could be due to a *steric factor* (or both). Hemoglobin is a much larger molecule than O_2, but reaction requires O_2 to find an active spot, the center of one of the four heme groups in the molecule. This argues in favor of at least a significant steric factor.

⇒ *RELATED PROBLEMS* 27.22, 27.26

We can write Eq. (27.16) in another approximate way if we take advantage of the Stokes–Einstein relation $D = k_B T/6\pi\eta R$ we mentioned earlier. We use it to write

$$D_A + D_B = \frac{k_B T}{6\pi\eta}\left(\frac{1}{R_A} + \frac{1}{R_B}\right)$$

and approximate R^* by $R_A + R_B$, the sum of the hydrodynamic radii of A and B. This turns Eq. (27.16) into

$$k_2 = \frac{2k_B T}{3\eta}\frac{(R_A + R_B)^2}{R_A R_B} N_A (1000 \text{ L m}^{-3}) ,$$

and if we further assume $R_A = R_B$ and write $k_B N_A = R$, the universal gas constant, we find

$$k_2 = \frac{8RT}{3\eta}(1000 \text{ L m}^{-3}) , \qquad (27.17)$$

an expression independent of molecular size. The size plays no role because while large species diffuse more slowly, they present a larger collisional target and encounter each other at the same rate as small species.

For water with a viscosity around 10^{-3} Pa s at 300 K (Table 25.3), Eq. (27.17) predicts $k_2 \cong 7 \times 10^9$ M^{-1} s^{-1}, similar to the value we found above with guesses for R^* and $D_A + D_B$. The viscosity dependence of Eq. (27.17) has been verified experimentally using, for example, water/ethanol solutions of variable viscosity.

The expressions for neutral diffusion-controlled rate constants were first deduced in 1917 by Smoluchowski in his study of the rate of colloidal particle growth. It was not until 1942 that Debye repeated the calculation for *charged* species and derived an expression for the diffusion-controlled rate constant of ion reactions. We expect ions with like charges to diffuse together more slowly than neutral species and those with opposite charges to diffuse together faster than neutrals of the same size. This means we must alter Fick's first law to account for the basic flux differences between charged and neutral species' diffusion. In addition to a concentration gradient, an *electrostatic potential energy gradient* also drives the diffusion of ions, since this gradient contributes to the chemical potential of the ions (their *electrochemical potential* as used in Chapter 10) and all transport processes are driven by chemical potential gradients.

Including this extra term driving diffusion (see Problem 27.50) leads to a simple modification of Eq. (27.16). One finds

$$k_2 = 4\pi R^*(D_A + D_B)N_A(1000 \text{ L m}^{-3})f \qquad (27.18)$$

where f is called the *electrostatic factor*:

$$f = \frac{R_0}{R^* (e^{R_0/R^*} - 1)} \qquad (27.19)$$

and R_0 is the distance at which the electrostatic energy equals $k_B T$:

$$\text{Coulomb energy at } R_0 = \frac{z_A z_B e^2}{4\pi\epsilon_0 \epsilon_r R_0} = k_B T = \text{Characteristic thermal energy}$$

or

$$R_0 = \frac{z_A z_B e^2}{4\pi\epsilon_0 \epsilon_r k_B T}. \qquad (27.20)$$

Here, $z_i e$ is the charge of ion i, ϵ_0 is the permittivity of vacuum, and ϵ_r is the solvent's relative permittivity (see Table 9.1). (Note that R_0 is negative if the charges have opposite signs; thus, only the magnitude of R_0 has physical meaning.)

At 25 °C in water, $\epsilon_r = 78.3$ and $R_0 = z_A z_B$ (7.16 × 10^{-10} m). If we take $R^* \cong 5$ Å, then $f = 1.88$ for $z_A z_B = -1$ (unlike singly charged ions), 0.45 for $z_A z_B = +1$ (like singly charged ions), 5.7 for $z_A z_B = -4$, 0.019 for $z_A z_B = +4$, and so on. Thus, singly charged ions have diffusion rate constants that are not much different from neutrals of the same size, but multiply charged ion rates are roughly an order of magnitude larger or smaller, depending on the signs of the charges.

If the total ion concentration is large ($\sim 10^{-4}$ M or larger), a nonideal correction to account for the effects of the electrostatic potential of the entire ionic atmosphere is needed. This correction is derived along the lines of the Debye–Hückel expression for an ion's activity coefficient, as discussed in Chapter 10, and leads to an expression for k_2 that depends on the solution's ionic strength. Increasing ionic strength increases the rate for reactions between ions of the *same* sign, and decreases the rate for reactions between ions of *opposite* sign. This behavior is called the *primary salt effect*.

This completes our first look at collisional effects in chemical reactions, but nearly all of the next chapter focuses on modern thoughts about the role of collisions in chemical reactions. In the next section, we take a very different point of view toward the rate constant. Instead of a focus on the probability of reaction per collision, we let statistical mechanics average over collisions in a way that has much in common with the statistical expression for the equilibrium constant.

27.3 THE TRANSITION STATE

Consider the first step someone interested in molecular electronic structure might take toward understanding a chemical reaction. That step might well be consideration of the forces among all the atoms at various geometries along the way from reactants to products. In the language of the Born–Oppenheimer approximation, one might start with the *potential energy surface* for the reaction. This surface is generally a complicated function of many variables. In principle *all* the relative spatial coordinates of all the atoms in the reactants govern it. Usually, however, "chemical intuition," that catch-all phrase that represents much study and experience with molecular structures and energies, points to a few and perhaps only two important geometric variables.

Reactions among only three atoms are geometrically simple, and we focus on them. Consider one of the simplest and best-studied reactions, H + H$_2$ → H$_2$ + H, which can be studied by isotopic substitution, as in D + H$_2$ → HD + H. This is a triatomic system with three geometric parameters, two bond lengths and one bond angle:

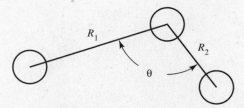

If we imagine the bond angle is fixed at 180° so that the atoms are collinear, the two bond lengths describe all possible configurations. For the H + H_2 system, it makes sense to consider collinear configurations first, since, as we found in Chapter 14 in our discussion of the electronic structure of H_3^+ and H_3, the linear configuration of H_3 has a lower energy than any bent configuration. We also found there that H_3 itself was not a stable species. Consequently, the association reaction H + H_2 → H_3 cannot occur, but somewhere along the way from reactants to products, we expect all three H atoms to be close together, and the forces among them tend to push the atoms toward a collinear configuration.

The *transition state* of a reaction is the somewhat ill-defined configuration that represents neither isolated reactants nor isolated products but something in between. The study of transition states is a very active and important area in contemporary research, and we look more closely at this research later, but first we need to specify some of the general properties of a transition state.

In the H + H_2 reaction, we know the H_2 interatomic potential energy function from quantum chemistry and molecular spectroscopy, and we can approximate it very well using, for example, a Morse potential function such as we used in Figure 19.15. Quantum chemistry provides equally accurate representations of the potential energy function for H_3 at any geometry, and Figure 27.4 shows a view of that function for a collinear geometry. In this diagram, R_2 is the *bound reactant* H—H internuclear distance, and R_1 is the distance between the *free atom reactant* and the *closest bound atom*. Note the heavy line that traces the surface along the R_2 direction at $R_1 = 3$ Å. This is essentially the H—H potential curve (a Morse curve) since $R_1 = 3$ Å places the free atom reactant far enough from H_2 to ensure very little perturbation of the isolated H—H interaction.

Several key points related to the reaction are noted on this figure. Point A corresponds to the start of the reaction, H + H_2, with R_2 at the equilibrium H_2 bond length and R_1 large. Point B is the *transition state*: both distances are comparable and close to chemical bond values. Point C corresponds to the reaction products (R_1 at the H_2 equilibrium value and R_2 large). Finally, point D, on a rather flat portion of the surface, corresponds to another fate of the H + H_2 collision, dissociation to three independent H atoms. This region is called the *dissociative plateau*, and both a flat potential (indicative of no interaction among the atoms) and large values for both R_1 and R_2 characterize it.

We can see the topography of the surface more clearly in the *contour map* of Figure 27.5. The zero of energy corresponds to all three atoms infinitely far apart (at the limits of the dissociative plateau, but as the figure shows, even at $R_1 = R_2 = 3$ Å, the potential energy is > -50 kJ mol^{-1}, which is close to zero for this system), and the minima at points A and B have energies corresponding to $-\mathcal{D}_e$, the negative of the H_2 equilibrium bond energy, which is about -458 kJ mol^{-1}. If

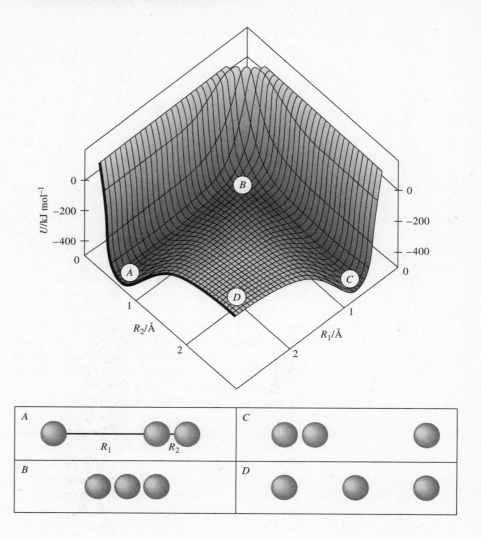

FIGURE 27.4 The interatomic potential energy function for the collinear H_3 system is shown here as a 3-D surface, $U(R_1, R_2)$. The points A–D locate important regions of the surface discussed in the text.

you think of this diagram as a geographical contour map and imagine walking on it from A to the transition state at B, you see that you *climb* in potential energy. The transition state is *higher* in energy than points A or B (but lower than the plateau at D).

Figures 27.4 and 27.5 were drawn using an approximate potential energy function known as the *LEPS model*, which stands for the last names of the four scientists who contributed to its development: London, Eyring, Polanyi, and Sato. In this model, the bound diatomic reactant and product potential curves are represented by Morse functions, and its interaction with the third atom is based on an early (1928!) quantum mechanical description of triatomic bonding put forth by London. Details of this method, largely superceded by modern quantum chemistry methods, are discussed in this chapter's references. Our H_3 LEPS surface has the transition state at $R_1 = R_2 = 0.93$ Å with an energy -424 kJ mol^{-1}, which is 34 kJ mol^{-1} higher than that at points A or C. Modern calculations place this state at -418 kJ mol^{-1}, 40 kJ mol^{-1} higher than points A or C. Thus, the LEPS method is reasonably accurate and considerably simpler to do than a full quantum mechanical calculation.

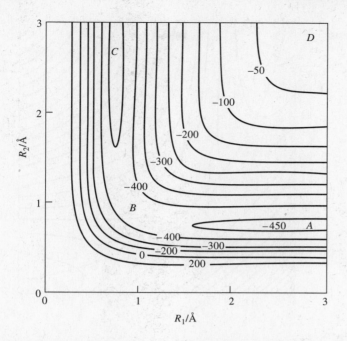

FIGURE 27.5 The collinear H_3 surface of Figure 27.4 is shown here as a contour map. The points A–D are the same as in that figure.

A potential energy barrier along *any* path from reactants to products suggests a microscopic origin for the activation energy, since any collision that has less than the barrier's energy cannot reach the transition state. This barrier height is *not*, however, equal to the activation energy. For the H + H_2 reaction, the experimental activation energy is $E_a = 32$ kJ mol^{-1}, notably less than the collinear barrier, and of course, experiments sample *all* orientations, not just collinear. (The barrier height increases for nonlinear geometries.) We do not yet have a clear and crisp definition of the activation energy on a microscopic level, but recognition of a potential energy barrier at the transition state brings us closer to a definition.

If we zoom in on point B in Figure 27.5, we can see more of the transition state's features. This is done in Figure 27.6. The dashed curve traces the *minimum energy path across the potential surface from point A, through the transition state, to point C*. Note that both R_1 and R_2 are changing along this path: the old bond is breaking as the new one is forming. This path is called the *classical reaction coordinate*: *classical*, since classical mechanics locates it (no zero-point energy or uncertainty in position) and *reaction coordinate*, since it connects the isolated reactants to the isolated products.

If we position collinear H_3 at the transition state, we can deduce something interesting about the *internal vibrations* of a transition state species. A triatomic constrained to be linear has only two normal modes of vibration, a symmetric and an antisymmetric stretch. (See Figure 19.6b for pictures of these motions in CO_2.) The two arrows labeled q_1 and q_2 in Figure 27.6 locate the *directions* of these vibrations. Along the q_2 direction, $R_1 = R_2$. This is the *symmetric stretch*. Along the q_1 direction, as R_1 increases, R_2 decreases the same amount. This is the *antisymmetric stretch*. It is q_1 that describes motion along the reaction coordinate at the transition state, but it is *not* a true vibration. The potential surface *curves the wrong way* along this direction.

The transition state locates a potential energy *minimum* along the q_2 direction. The surface has a positive second derivative in this direction, leading to a positive

FIGURE 27.6 The collinear H_3 transition state region, point B of Figures 27.4 and 27.5, is shown in greater detail here. The dashed line is the classical adiabatic reaction coordinate, the minimum energy path from reactants to products. The arrows q_1 and q_2, centered at the transition state, locate the two normal modes of motion for the system, the antisymmetric reaction coordinate motion q_1 and the symmetric stretch motion q_2.

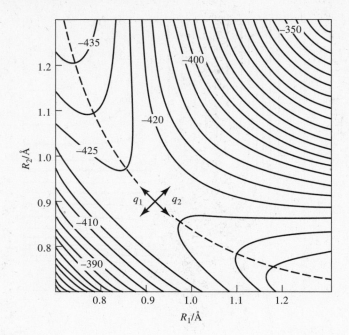

vibrational force constant in the usual way. Along the q_1 direction, the transition state locates a potential energy *maximum* with a *negative* second derivative. As Eq. (19.14) described, a negative second derivative of the intermolecular potential energy U means a *negative* harmonic force constant:

$$\frac{\partial^2 U}{\partial q_i^2} = \text{force constant for normal mode } q_i ,$$

and as Eq. (12.15) described, a negative force constant implies that the *harmonic vibration frequency is a purely imaginary number*:

$$\omega_i = \text{normal mode frequency} = \sqrt{\frac{\text{normal mode force constant}}{\text{normal mode reduced mass}}} .$$

Since the force constant is negative and the reduced mass is always positive, ω_1 is purely imaginary. This seemingly strange result simply means that *motion along the q_1 direction is not bound*, which is a characteristic of many transition states.

In another simple reaction, $H^+ + H_2 \rightarrow H_2 + H^+$, (which is also studied by isotopic substitution: $D^+ + H_2 \rightarrow HD + H^+$, for example), the transition state species, $H_3^{\ddagger}$, is quite different from H_3. As we found in Chapter 14 (see the discussion of Figure 14.17), $H_3^{\ddagger}$ is stable, and it is triangular. We can no longer limit ourselves to a collinear collision geometry, which makes the $H_3^{\ddagger}$ potential energy surface much more difficult to visualize, but along the reaction coordinate, the $H_3^{\ddagger}$ transition state represents a potential energy *minimum* rather than a *maximum*. This will always be the case for reactions in which the transition-state species represents a potentially stable molecule, as in $O + CO \rightarrow CO + O$ (stable CO_2) or $O + N_2 \rightarrow ON + N$ (stable N_2O).

Free atoms or radicals can often interact through more than one potential energy surface, as we have seen, and each potential energy surface generally has a unique transition state. Nevertheless, for each surface we can classify the *potential energy profile* along the reaction coordinate into several general categories. Figure 27.7 shows the most important of these. In Figure 27.7(a), the profile has a simple barrier.

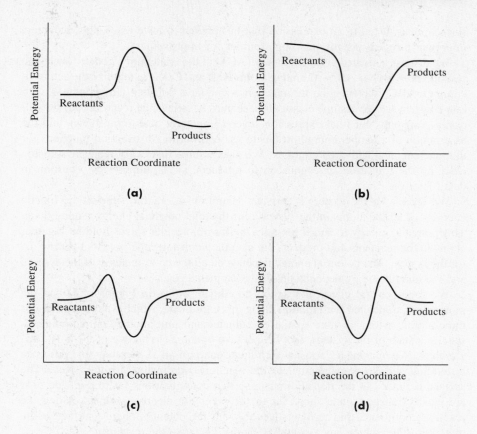

FIGURE 27.7 Typical energy profiles along the reaction coordinate of many atom exchange reactions look qualitatively like one of these sketches. In (a), there is a simple energy barrier between reactants and products, while (b) has a simple energy well. Schemes (c) and (d) are mixed barrier–well profiles with the barrier on the reactant side of the well in (c) and on the product side in (d).

The transition state cannot be stable. In Figure 27.7(b), the potential energy is *lowest* at the transition state, and the transition state species would form a stable compound if it lost energy through photon emission (radiative association) or collision with a third body (termolecular recombination). Figures 27.7(c) and (d) are common variants with both a potential barrier and a well; they differ only in the order in which colliding reactants encounter these features. In these, we locate the transition state at the top of the barrier, since that is the configuration along the reaction coordinate the reactants must reach to go on to products.

Activated Complex Theory

The transition-state species is usually called the *activated complex*. In 1935, Eyring and, independently but almost simultaneously, Evans and Polanyi published a theory of the bimolecular rate constant that today goes by the names *transition state theory, absolute rate theory,* or *activated complex theory*.[6] This theory recognized the role played by the reaction potential energy surface and its associated transition state, but it also mixed in the thermal distribution of collision energies characteristic of most kinetic experiments. Using a blend of statistical mechanics and potential energy

---

[6] Henry Eyring and Michael Polanyi had collaborated in 1931 on a paper describing the H + H_2 potential energy surface, which led to the LEPS model. Eyring's contributions to kinetics were enormous, and he would be on nearly anyone's short list of physical chemists whose contributions warranted a Nobel Prize but were never honored by one. Michael Polanyi worked in the area of chemical kinetics through the 1930s and early 1940s, but in 1948, he left physical chemistry and spent the remaining 28 years of his life in various areas of sociology. His son, John Polanyi, is an active researcher in the area of reaction dynamics, the topic of the next chapter. John Polanyi shared the 1986 Nobel Prize for this work.

surface ideas, it led to an expression that has proven to have reasonable accuracy, important physical insight, and opportunities for improvement.

We discuss activated complex theory (ACT) in the context of a generic exchange reaction A + BC → AB + C among atoms A, B and C. We symbolize the activated complex ABC‡ and imagine the reaction occurs at a constant temperature T with rate constant k_2. We assume classical mechanics describes the reaction (but we will easily graft onto our final expression several important quantum effects), and the assumption of a temperature allows us to use equilibrium distribution functions for the reactants and activated complex. We also assume that every reactant collision reaching the transition state goes on to products, an assumption we question in detail later on.

We start with the system's classical Hamiltonian, H, the sum of the kinetic energies of all the atoms in the system plus the total potential energy among them. This potential energy includes not only the *intra*molecular forces holding together atoms in the reactants and products but also the *inter*molecular potential energy, U, for the system. The potential energy depends on all the coordinates q_i of the system, and the kinetic energy depends on all the momenta p_i.

We focus on the classical reaction coordinate q_1 (as in Figure 27.6) and the momentum associated with motion along this coordinate, which we call p_1. For our three atoms, a total of nine spatial coordinates and nine momentum components specify a state of the system, and we can take q_1 and p_1 to be two of these 18 total variables. We imagine a Cartesian coordinate space of all 18 variables and postulate a dividing surface perpendicular to the reaction coordinate q_1 in this space. The surface is placed at the transition state so that configurations far to one side of it are clearly reactants, and those far to the other are clearly products. Along the reaction coordinate, this surface slices through the potential energy surface at the transition state, which has a potential energy $U^\ddagger$ above the reactants:

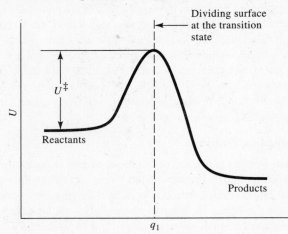

Classical statistical mechanics tells us that the fraction of a total of N systems found in any volume element $dq_1 dq_2 \ldots dq_9 dp_1 dp_2 \ldots dp_9$ of this space depends on the Hamiltonian and system's temperature in the following way:

$$\frac{d^{18}N}{N} = \frac{\text{(Boltzmann factor) (Volume element)}}{\text{Partition function}} = \frac{e^{-\beta H} dq_1 \ldots dq_9 dp_1 \ldots dp_9/h^9}{\int e^{-\beta H} dq_1 \ldots dq_9 dp_1 \ldots dp_9/h^9} \quad (27.21)$$

where $\beta = 1/k_B T$. We write $d^{18}N$ to remind us that we have an 18-dimensional differential on each side of the equation. This expression is exact at equilibrium, and we assume it holds during the evolution of a reacting system. We have used Eq. (23.17) for the classical partition function, and the factors h^9, Planck's constant raised to a power equal to three times the number of atoms in the system, are explained following our introduction of Eq. (23.17).

This expression is manipulated through several steps that are given a closer look below into an expression for the rate at which systems cross the dividing surface. In words, these steps are:

1. integrate both sides of Eq. (27.21) over positive values of p_1 to find the rate at which systems cross the dividing surface from reactants to products,
2. integrate the denominator of Eq. (27.21), recognizing that most of the time A and BC are far from each other so that we can neglect their interaction (the intermolecular potential energy component of H), which makes the denominator equal to the product of the A and BC partition functions, and
3. integrate the remainder of the numerator, recognizing that it has the form of a partition function for the activated complex $ABC^{\ddagger}$ with two corrections: the internal energy levels of $ABC^{\ddagger}$ are measured from the energy at the top of the barrier in U (which we denote $U^{\ddagger}$) and, most importantly, step 1 has *removed one degree of freedom* from $ABC^{\ddagger}$.

These steps turn Eq. (27.21) into

$$\frac{dN}{dt} = \frac{Nk_B T}{h} \frac{q^{\ddagger}}{q_A q_{BC}} e^{-U^{\ddagger}/k_B T}, \qquad (27.22)$$

where q_A and q_{BC} are partition functions for A and BC, $q^{\ddagger}$ is the partition function for the activated complex *with one degree of freedom, motion along the reaction coordinate, removed,* and $e^{-U^{\ddagger}/k_B T}$ acknowledges that energy levels of $ABC^{\ddagger}$ are measured from the top of the potential barrier at the transition state along the reaction coordinate. This equation represents the rate at which systems reach the dividing surface from the reactant side and, we assume, go on to products. A few clean-up steps turn this into an expression for the rate constant. First, we change it into a rate expression in molar concentration units:

$$\mathcal{R} = k_2[A][BC] = \frac{d[ABC^{\ddagger}]}{dt} = \frac{dN}{dt} \frac{1}{VN_A(1000 \text{ L m}^{-3})} \qquad (27.23)$$

where V is the system volume.

Next, we consider exactly what we mean by N in these equations. If U is zero everywhere (and it almost is, since A and BC have to be quite close to interact), we would write $d^{18}N/N$ in Eq. (27.21) as a product of two independent factors, one for A and one for BC, since the only part of H that couples A to BC is U (as used in step 2 above). Let the number of A and the number of BC molecules equal N_A (*not* Avogadro's constant!) and N_{BC}, respectively. Then

$$\frac{d^{18}N}{N} = \frac{d^6 N_A}{N_A} \frac{d^{12} N_{BC}}{N_{BC}}$$

so that N, the statistical number of ABC complexes possible, equals $N_A N_{BC}$. If we substitute Eq. (27.22) into Eq. (27.23), turn N into $N_A N_{BC}$, and turn this product of *numbers* of molecules into a product of *concentrations*, using $N_i = [i]VN_A(1000 \text{ L m}^{-3})$ *where here N_A is Avogadro's constant*, we find

$$\mathcal{R} = k_2[A][BC] = [A][BC] \frac{k_B T}{h} \left(\frac{q^{\ddagger}}{q_A q_{BC}} e^{-U^{\ddagger}/k_B T} \right) VN_A(1000 \text{ L m}^{-3})$$

so that the rate constant is

$$k_2 = \frac{k_B T}{h} \left(\frac{q^\ddagger}{q_A q_{BC}} e^{-U^\ddagger/k_B T} \right) V N_A (1000 \text{ L m}^{-3}) \;. \quad (27.24)$$

Finally, we note that the expression in parentheses in Eq. (27.24) looks very much like a statistical mechanical expression for an equilibrium constant, as in our basic statistical expression for K_{eq}, Eq. (23.29). Since we have assumed equilibrium conditions and identified the activated complex as a unique species, we should expect the equilibrium constant for

$$A + BC \rightleftarrows ABC^\ddagger$$

to appear somewhere in our theory. To put the expression in parentheses in the correct form of an equilibrium constant, we need to work the thermodynamic standard state into it. We can do this partition function by partition function, and then reassemble the partition function ratio. Consider the activated complex partition function $q^\ddagger$ first. There is a factor of the system volume V in $q^\ddagger$, as part of the ideal gas $ABC^\ddagger$ translational partition function, as in Eq. (23.5). If the $ABC^\ddagger$ standard state volume is $V^\circ_\ddagger$, we can write

$$q^\ddagger = q^\ddagger \frac{V^\circ_\ddagger}{V^\circ_\ddagger} = q^{\ddagger\circ} \frac{V}{V^\circ_\ddagger}$$

and introduce in this way the standard state partition function $q^{\ddagger\circ}$, which contains a factor of $V^\circ_\ddagger$. The standard state volume is related to the standard state pressure, $P^\circ = 1 \text{ bar} = 10^5 \text{ Pa}$, through

$$V^\circ_\ddagger = \frac{N_\ddagger RT}{N_A P^\circ} \;.$$

If we convert each partition function this way, we find

$$\frac{q^\ddagger}{q_A q_{BC}} e^{-U^\ddagger/k_B T} = K^\ddagger_{eq} \left(\frac{RT}{V N_A P^\circ} \right) \quad (27.25)$$

where $K^\ddagger_{eq}$ is the equilibrium constant for $ABC^\ddagger$ formation written in the form of Eq. (23.29):

$$K^\ddagger_{eq} = \frac{(q^{\ddagger\circ}/N_\ddagger)}{(q^\circ_A/N_A)(q^\circ_{BC}/N_{BC})} e^{-U^\ddagger/k_B T} \;.$$

If we substitute Eq. (27.25) into Eq. (27.24), we arrive at our final expression for the ACT bimolecular rate constant:

$$k_2 = \frac{k_B T}{h} \left(K^\ddagger_{eq} \frac{RT}{V N_A P^\circ} \right) V N_A (1000 \text{ L m}^{-3}) = \frac{k_B T}{h} K^\ddagger_{eq} \frac{RT}{P^\circ} (1000 \text{ L m}^{-3}) \;. \quad (27.26)$$

This expression has three distinct parts:

(1) $\dfrac{k_B T}{h}$, a factor with units of a *frequency*,

(2) $K^\ddagger_{eq}$, the equilibrium constant for formation of the activated complex, and

(3) $\dfrac{RT}{P^\circ} (1000 \text{ L m}^{-3})$, standard state and unit conversion factors.

Factor (1) appeared when we took one vibrational degree of freedom, the unbound reaction coordinate motion, away from the activated complex. Numerically,

$k_B/h = 2.08 \times 10^{10}$ s^{-1} K^{-1}, or $k_B T/h = 6.3 \times 10^{12}$ s^{-1} at 300 K. Factor (2) is an ordinary equilibrium constant, except for the loss of one degree of freedom in the activated complex. Factor (3) equals $0.0831 T$ L mol^{-1} K^{-1}, or 24.9 L mol^{-1} at 300 K, the standard state molar volume at this temperature. We can combine the two numerical factors and write k_2 as

$$k_2 = (1.73 \times 10^9 \text{ M}^{-1}\text{ s}^{-1}\text{ K}^{-2}) K_{eq}^\ddagger T^2 \qquad (27.27)$$

and, with the help of Table 23.2 and other expressions in Chapter 23, calculate $K_{eq}^\ddagger$ easily *if we know the structure and internal vibrational frequencies of the activated complex.* This is the tricky part, since spectroscopic or theoretical information on this complex is scarce, although modern experiments are making progress in this area.

A Closer Look at Activated Complex Theory

We derive here the mathematics behind the steps that take us from Eq. (27.21) to Eq. (27.22). Suppose we locate the 18-dimensional volume element in Eq. (27.21) somewhere slightly on the product side of our dividing surface. The speed at which systems move along the reaction coordinate into this volume element is dq_1/dt, and their associated momentum, p_1, is $m_1 dq_1/dt$ where m_1 is the effective mass associated with motion along q_1. (This mass does not appear in our final expression, and we do not need an explicit expression for it in terms of the A, B, and C masses.) Once a system reaches this element from the reactant side of the dividing surface, we declare it to have formed products.

Thus, we divide Eq. (27.21) by dt and rearrange it a bit:

$$\frac{d^{18}N}{dt} = \frac{N}{h}\frac{dq_1}{dt} \frac{e^{-\beta H} dq_2 \ldots dq_9 dp_1 \ldots dp_9/h^8}{\int e^{-\beta H} dq_1 \ldots dq_9 dp_1 \ldots dp_9/h^9} = \text{rate into } dq_2 \ldots dq_9 dp_1 \ldots dp_9 \ .$$

We first integrate over *positive* values of p_1, since only motion along q_1 from reactants to products leads to reaction. The piece of H associated with kinetic energy along q_1 is $p_1^2/2m_1$, and we use H' to represent the rest of H. Integrating over p_1 (writing $p_1 = m_1 dq_1/dt$ and dropping $d^{18}N$ down to $d^{17}N$ to account for one integration) gives us

$$\frac{d^{17}N}{dt} = \frac{N}{m_1 h} \frac{\left(\int_0^\infty p_1 e^{-p_1^2/2m_1 k_B T} dp_1\right) e^{-\beta H'} dq_2 \ldots dq_9 dp_2 \ldots dp_9/h^8}{\int e^{-\beta H} dq_1 \ldots dq_9 dp_1 \ldots dp_9/h^9} \ .$$

The integral in parentheses is known:

$$\int_0^\infty p_1 e^{-p_1^2/2m_1 k_B T} dp_1 = m_1 k_B T \ ,$$

and we have completed the first step in the derivation:

$$\frac{d^{17}N}{dt} = \frac{N k_B T}{h} \frac{e^{-\beta H'} dq_2 \ldots dq_9 dp_2 \ldots dp_9/h^8}{\int e^{-\beta H} dq_1 \ldots dq_9 dp_1 \ldots dp_9/h^9} \ .$$

The m_1 factor is gone, and we have 16 more integrations left in the numerator, plus the 18 in the denominator's integral.

For the second step, we note that the integral in the denominator should range over all that portion of coordinate and momentum space characteristic of the reactants, since we are considering only those systems moving from the reactant side of our dividing surface toward it. Since the interaction between A and BC is zero unless A is quite close to BC, most of the integration range of this integral covers *isolated* reactants. This means we can separate H into a sum of pieces for A and BC alone (i.e., we neglect U). This step lets us interpret the 18-dimensional integral as a product of two integrals, one for A alone and one for BC alone:

$$\int e^{-\beta H}\,dq_1\ldots dq_9 dp_1\ldots dp_9/h^9 =$$
$$\int e^{-\beta H_A}\,dq_{A1}\ldots dq_{A3} dp_{A1}\ldots dp_{A3}/h^3 \int e^{-\beta H_{BC}}\,dq_{BC1}\ldots dq_{BC6} dp_{BC1}\ldots dp_{BC6}/h^6$$

where H_A is the Hamiltonian for A and H_{BC} is that for BC. From the definition of the classical partition function, Eq. (23.17), we see that our 18-dimensional integral is the product of the A and BC partition functions:

$$\int e^{-\beta H}\,dq_1\ldots dq_9 dp_1\ldots dp_9/h^9 = q_A q_{BC}\;.$$

This completes the second step of the derivation and gives us

$$\frac{d^{17}N}{dt} = \frac{Nk_B T}{h}\frac{e^{-\beta H'}\,dq_2\ldots dq_9 dp_2\ldots dp_9/h^8}{q_A q_{BC}}\;.$$

For the third step, we remove one final piece of H'. If the intermolecular potential energy of the activated complex is higher than that of the isolated reactants by an amount $U^\ddagger$, we can separate this from H' and write $H' = U^\ddagger + H''$ so that

$$\frac{d^{17}N}{dt} = \frac{Nk_B T}{h}\frac{e^{-U^\ddagger/k_B T}}{q_A q_{BC}}e^{-\beta H''}\,dq_2\ldots dq_9 dp_2\ldots dp_9/h^8\;.$$

The remaining 16 differentials are located just to the product side of the dividing surface, and they represent an activated complex poised at the barrier in U with an internal Hamiltonian H''. Integration over them gives us the partition function for the activated complex, $q^\ddagger$, with motion along the reaction coordinate removed, giving Eq. (27.22).

We now explore some of the properties of the ACT expression for k_2. Suppose $K_{eq}^\ddagger \cong 1$ and $T = 300$ K. Then Eq. (27.27) predicts $k_2 \cong 1.6 \times 10^{14}$ M^{-1} s^{-1}, which is considerably larger than collision theory in Section 27.2 predicts. This is because $K_{eq}^\ddagger$ is *never* approximately 1. Two species are condensing to one in this equilibrium, and the corresponding entropy decrease plus the endothermicity of ABC‡ formation causes $K_{eq}^\ddagger \ll 1$.

We can see this if we assume A and BC have no internal structure, as if we had only atoms A and B interacting in their ground states. The AB‡ activated complex cannot become a stable product (unless we invoke radiative association, associative ionization, or a third body), but the exercise is instructive. We assume the transition state has no barrier ($U^\ddagger = 0$), but occurs at the AB equilibrium bond length, R_e. We also assume that A, B, and AB have electronically nondegenerate ground states.

We construct partition functions from expressions in Chapter 23 (see also Example 23.11, which considers K_{eq} for a similar reaction, H_2 dissociation). Those for A and B are easy to find, since these species have only translational components:

$$\frac{q_i^\circ}{N_i} = \frac{q_{tr,i}^\circ}{N_i} = \left(\frac{m_i k_B T}{2\pi\hbar^2}\right)^{3/2}\frac{RT}{N_A P^\circ}\;,\; i = \text{A or B}.$$

The AB‡ partition function has a translational part and a rotational part, but *no vibrational part*. This is the reaction coordinate motion in AB‡ that we removed. Thus we write

$$\frac{q^{\ddagger \circ}}{N_\ddagger} = \frac{q_{tr}^{\ddagger \circ \prime}}{N_\ddagger} q_{rot}^\ddagger = \left[\frac{(m_A + m_B)k_BT}{2\pi\hbar^2}\right]^{3/2} \frac{RT}{N_A P^\circ} \left(\frac{2\mu R_e^2 k_BT}{\hbar^2}\right)$$

where the final factor in parentheses is the diatomic rotational partition function, Eq. (23.11), written with the moment of inertia I expressed in terms of the $AB^\ddagger$ reduced mass $\mu = m_A m_B/(m_A + m_B)$ and bond length R_e.

If we substitute these expressions into Eq. (23.29), we find, after cancelling common factors,

$$K_{eq}^\ddagger = 2h \frac{N_A P^\circ}{RT} \left(\frac{2\pi}{\mu k_B T}\right)^{1/2} R_e^2 \qquad (27.28)$$

(which, for $\mu \cong 1$ amu, $T \cong 300$ K, and $R_e \cong 2$ Å, equals $\sim 10^{-3}$, supporting our claim that $K_{eq}^\ddagger \ll 1$). If we substitute Eq. (27.28) into Eq. (27.26) for k_2, a little algebra leads to a most remarkable result:

$$k_2 = \frac{k_B T}{h}\left[2h\frac{N_A P^\circ}{RT}\left(\frac{2\pi}{\mu k_B T}\right)^{1/2} R_e^2\right]\frac{RT}{P^\circ}(1000 \text{ L m}^{-3})$$

$$= N_A(\pi R_e^2)\left(\frac{8k_BT}{\pi\mu}\right)^{1/2}(1000 \text{ L m}^{-3}),$$

which is exactly our simplest collision theory expression for k_2, Eq. (27.8). We associate the collision cross section σ_{12} in that equation with πR_e^2 here.

Thus ACT has collision theory imbedded in it, and the mystery factors of collision theory—the phenomenological activation theory and the steric factor—are less mysterious in ACT due to the role played by the potential energy barrier and the reactant and activated complex *internal* partition functions. These still need some work in our ACT expression, since we have assumed classical mechanics in a quantum world. For internal partition functions, we simply abandon the classical expressions (which are the high-temperature limits of quantum expressions), and use the correct quantum expressions. For translations and rotations, the classical expressions are almost always valid.

For the potential barrier term, $e^{-U^\ddagger/k_BT}$, we have to be careful when we transcribe to quantum mechanics to include correctly *zero-point energy* effects. Figure 27.8 uses the energy profile from Figure 27.7(a) to illustrate this. We add the total zero-

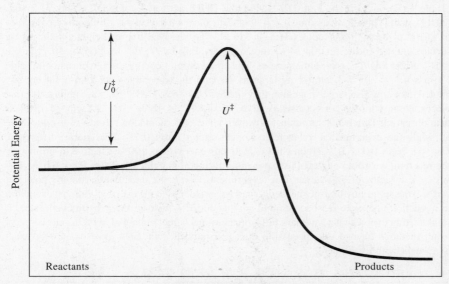

FIGURE 27.8 The classical potential energy of the activated complex, $U^\ddagger$, is corrected by the reactant and activated complex zero-point energies to yield $U_0^\ddagger$, the correct quantum-mechanical potential energy of the activated complex with respect to the reactants.

point energy of the reactants to U in the reactant region, add the zero-point energy of the activated complex (which is in general different from that for isolated reactants) to U at the transition state, and subtract these to find $U_0^\ddagger$, the barrier height corrected for zero-point energies. This quantity replaces $U^\ddagger$ in the exponential.

For atom A reacting with diatomic BC, we can express $K_{eq}^\ddagger$ in terms of partition functions from Chapter 23 and arrive at a working expression for k_2:

$$k_2 = \sqrt{2\pi}\, N_A (1000\text{ L m}^{-3}) \left(\frac{m_{ABC}}{m_A m_{BC} k_B T}\right)^{3/2} \frac{\sigma_{BC} q_{rot}^\ddagger \hbar^4}{2 I_{BC}} \frac{q_{vib}^\ddagger}{q_{vib,BC}} \frac{q_{el}^\ddagger}{q_{el,A} q_{el,BC}} e^{-U_0^\ddagger/k_B T}$$

$$= c \left(\frac{m_{ABC}}{m_A m_{BC} T}\right)^{3/2} \frac{\sigma_{BC} q_{rot}^\ddagger}{I_{BC}} \frac{q_{vib}^\ddagger}{q_{vib,BC}} \frac{q_{el}^\ddagger}{q_{el,A} q_{el,BC}} e^{-U_0^\ddagger/k_B T}$$

where c is a collection of constants:

$$c = 1.82 \times 10^{-75}\text{ kg}^{5/2}\text{ m}^2\text{ L s}^{-1}\text{ mol}^{-1}\text{ K}^{-3/2},$$

σ_{BC} and I_{BC} are the rotational symmetry number and moment of inertia of BC, respectively, and the other partition functions are discussed in Chapter 23 and Table 23.2 (where we choose the vibrational partition functions excluding zero-point energy).

EXAMPLE 27.10

The experimental rate constant at 310 K for the reaction between a deuterium atom and H_2 in its first vibrational state,

$$D + H_2(v = 1) \rightarrow HD + H,$$

is $k_2 = 1.1 \times 10^8$ M^{-1} s^{-1}. What does ACT and the LEPS potential energy surface predict? At the LEPS transition state, the activated complex moment of inertia is $I^\ddagger = 4.07 \times 10^{-47}$ kg m^2, and the DHH‡ zero-point energy is 3.47×10^{-20} J.

SOLUTION We first assemble the various pieces of our working formula for k_2. The ratio of electronic partition functions equals 1, since the H_2 ground state is nondegenerate, D is doubly degenerate, but so is DHH‡. For H_2, $\sigma_{BC} = 2$, $I_{BC} = \mu R_e^2 = 4.6 \times 10^{-48}$ kg m^2 (Table 19.2), and $q_{vib} = 1$ at 310 K. For the DHH‡ activated complex, $q_{rot}^\ddagger = 31.3$ (Eq. (23.11), using $I^\ddagger$), and we assume $q_{vib}^\ddagger = 1$ as well, since we expect the vibrational frequencies of DHH‡ to be high and not excited at 310 K. The classical barrier height for the LEPS surface, quoted in the text, is 34 kJ mol^{-1} or, dividing by N_A, $U^\ddagger = 5.65 \times 10^{-20}$ J. We must correct this for zero-point energies. The H_2 zero-point energy, using constants in Table 19.3, is 4.3×10^{-20} J, almost as large as $U^\ddagger$, and the transition state zero-point energy, quoted above, is also large. Together, they predict $U_0^\ddagger = 4.8 \times 10^{-20}$ J. Assembling all these pieces, along with the correct masses, gives $k_2 = 8.82 \times 10^5$ M^{-1} s^{-1}, a factor of 130 less than observed. This is not surprising, however, since our calculation refers to a *ground-state* H_2 while the experimental value refers to *vibrational excited* H_2. The energy difference, $\hbar\omega = 8.26 \times 10^{-20}$ J, is *greater* than $U^\ddagger$. If this energy was 100% effective in promoting the reaction, we could neglect the exponential factor in k_2, which leads to $k_2 = 6.6 \times 10^{10}$ M^{-1} s^{-1}, a factor of 580 larger than experiment. Given the uncertainties in the potential surface, the approximations in the theory, and uncertainty as to the use of vibrational excitation in a reaction, these are reasonable numbers that correctly bracket the experimental value. It is interesting to note that a landmark fully quantum-mechanical calculation of this rate constant using methods beyond our discussion here and published in 1994 gave essentially exact agreement with experiment.

➡ *RELATED PROBLEMS* 27.29, 27.30, 27.31, 27.32, 27.33

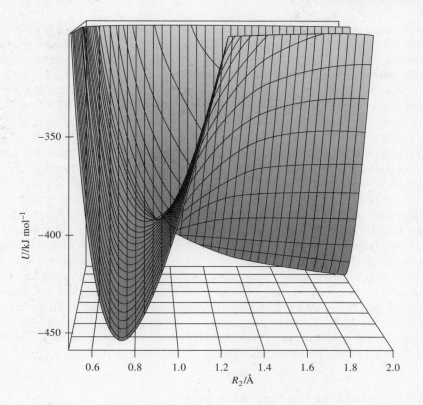

FIGURE 27.9 One of the assumptions of ACT is that every system that reaches the transition state goes on to products. The dynamics of the reactant's collision may be such that this is not always true. In this view of the H_3 surface, we look along the reactant entrance channel toward the transition state. To reach the product exit channel, the system's motion has to turn the corner in the potential energy surface. Some collisions fail to do this and reflect back along the entrance channel.

It is observed experimentally that elementary reaction rate constants have noticeable *isotope effects*, especially when the transferred atom is H or one of its isotopes. We can see several places in the ACT theory that predict such an effect: zero-point energy changes, moment of inertia changes, and mass changes, and the theory correctly predicts the general trends of these effects.

There are two remaining corrections to ACT that warrant discussion. The first is quantum-mechanical tunneling of the activated complex *through* the transition-state barrier. This is an important correction for H atom transfer reactions at low temperature where the majority of successful reactions are barely surmounting the barrier. In fact, several reactions are known that follow the Arrhenius expression as T is lowered, but below a certain temperature, their rates are nearly constant as they enter a regime dominated by tunneling.[7] There are several approximate ways to correct for tunneling (see Example 12.3 for one), but they require detailed knowledge of the shape of the barrier at the transition state.

The second correction comes from the assumption that every system reaching the transition state goes on to products. This is not correct, since some systems reach the transition state, but *reflect back* to reactants.

Figure 27.9 gives a view of the LEPS surface along the so-called *entrance channel,* the direction from point A to point B in Figure 27.5. The surface is shaded in proportion to the relative collision energy distribution at 2000 K. This is hot enough for there to be an appreciable number of collisions that pass over the transition state barrier, but the majority (darkly shaded regions) do not. Those with

---

[7]This may be sufficient reason to reject offers to have oneself frozen at death in hopes of revival at a later date far in the future.

sufficient energy to cross the barrier must also approach it in a way that allows the system to "turn the corner" and head out the *exit channel* toward isolated products. Imagine riding this surface on a skateboard.[8] If your aim is poor as you race toward the barrier, you might ride up one wall in such a way that you fall back toward the entrance channel from the transition state region rather than turning the corner. While this might earn you some style points in a skateboarding competition, it will not lead to a chemical reaction: the collision reaches the transition state, but fails to go on to products.

Reflection back into the entrance channel *always* has probability quantum mechanically (see the discussion of Figure 12.5 and the quantum phenomenology of 1-D motion across potential barriers and wells). If the transition state has a potential *well* rather than a barrier (as in Figure 27.7(b)), the collision complex can cross and recross the dividing surface many times, since the potential well provides forces that tend to keep the activated complex bound in all directions of motion. This *dynamic* failure to reach reactants is often expressed (along with tunneling corrections) by an extra factor in the ACT expression called the *transmission coefficient*, κ. Note that tunneling tends to *increase* κ, but reflection tends to *decrease* it. Neither effect is easy to calculate or even estimate without knowledge of the potential surface, and κ is somewhat like the collision theory steric factor, p: it is easy to see that it must be there, but it is not at all easy to quantify. Fortunately, it is not expected to be greatly different from 1, and the other assumptions and approximations of ACT are usually as severe as setting $\kappa = 1$ right from the start.

A Closer Look at the Transition State

Several new experiments allow close scrutiny of the transition state for simple reactions. Some of these rely on *molecular scattering* experiments, which are discussed in the next chapter, but others attempt spectroscopy on the activated complex itself. This is clearly a significant challenge, since gas-phase activated complexes are very transient species. Consider, for example, the $H + H_2$ system. The distance along the reaction coordinate from points A to C in Figure 27.5 is about 4 Å. A collinear collision that follows the reaction coordinate with 35 kJ mol^{-1} of relative translational energy will clear the transition-state barrier on this surface with 1 kJ mol^{-1} to spare. The system would traverse this 4 Å distance in only 4×10^{-14} s, or 40 fs, if there was no barrier along the reaction coordinate. Including the barrier slows the collision as relative kinetic energy is converted to potential energy at the transition state, but the collision is still over in about 65 fs. These very short times make it clear that the activated complex can never have an appreciable concentration, making direct spectroscopic observation very difficult.

Consequently, several very clever indirect methods have been devised that in one way or another enhance spectroscopic sensitivity. In one of the earliest of these, known as *laser assisted reactions*, light is absorbed *by the transition state but not by the reactants or products*. One of the first experiments of this type involved energy transfer from an excited Sr atom to a ground-state Ca atom, producing excited Ca and ground-state Sr:

$$Sr^* (5s5p\ ^1P_1) + Ca + h\nu \rightarrow Sr + Ca^{**} (4p^2\ ^1S_0).$$

---

[8]There is a coordinate transformation needed to make an exact analogy between sliding along U and the true dynamic response of the system, but the qualitative ideas of the next few sentences do not need this extra transformation.

The reaction was monitored through Ca** fluorescence to a lower excited state: Ca** → Ca* (4s4p $^1P_1$). Without the photon, the reaction is 227 kJ mol^{-1} (18 992 cm^{-1}) *endothermic* and has a negligible rate. The photon energy, however, was tuned near the *energy difference between the excited Sr* reactant and the excited Ca** product*. It was thus absorbed *only* by the transient Ca–Sr* collision pair. This idea has since been extended to a few molecular systems, and the variation in the reaction rate as a function of the photon energy carries information about the energy difference between the transition state of the reactants and that of the products (which, since the photon induces an electronic transition in the activated complex, are in different electronic states).

Other methods begin with a stable *precursor* to a transition state. For example, the stable linear anion IHI$^-$ is only one electron away from the activated complex for the I + HI reaction. Photoelectron spectroscopy of IHI$^-$, in which a fixed-energy photon detaches the extra electron and the energy of this electron is analyzed, reveals information about the structure and bound vibrational levels of the activated complex. In a variant of the anion precursor technique, weakly bound van der Waals molecules are prepared and photolyzed to initiate a reactive collision that starts in the vicinity of the transition state, but with a range of collision geometries limited by the geometry of the precursor van der Waals complex. (These experiments are sometimes called *half-collision* studies, since the interaction starts at the transition state and evolves from there.) For example, photolysis of the HI·CO$_2$ van der Waals complex in the region of the HI photodissociation absorption (see Figure 18.6) initiates the H + CO$_2$ → OH + CO reaction from the configuration of the HI·CO$_2$ complex. The OH product is detected through laser induced fluorescence, and the information on the transition state comes from a comparison of the OH internal energy distribution in this experiment to one performed in a traditional bulk kinetic experiment.

Over the past thirty years or so, the duration of laser pulses have fallen by a factor of 10^6 from the nanosecond through the picosecond and down to the femtosecond region. The femtosecond time scale, as we saw in the H + H$_2$ example, is fast enough to follow interactions through the transition state in real time. A typical experiment uses one laser pulse (the pulse duration is typically 10–20 fs) to initiate the interaction and a second probe pulse delayed from the first over a range variable from ~20 fs to several picoseconds. Typically, an evolving product absorbs the probe radiation, and fluorescence from a product excited state is detected. In one study, the HBr·I$_2$ complex was the precursor for the Br + I$_2$ → IBr + I reaction. The first pulse photodissociated HBr in the complex, and the light H atom moved quickly from the scene, leaving the Br ··· I ··· I activated state to evolve. The potential energy profile along the reaction coordinate of this reaction is similar to that in Figure 27.7(d). The reaction is exothermic by 28 kJ mol^{-1}, but the BrI$_2$ species represents a potentially stable species. There is a potential barrier about 12 kJ mol^{-1} above the isolated product energy, and this barrier together with the potential well causes the activated complex to have a fairly long lifetime. The experiment found the IBr product to appear over a 100 ps period.

Thermochemical Kinetics

We close our discussion of ACT with a thermodynamic analysis that shows us the meaning of the Arrhenius activation energy. Since $K_{eq}^{\ddagger}$ is an equilibrium constant, we can invoke thermodynamics if we write

$$K_{eq}^{\ddagger} = e^{-\Delta \overline{G}^{\ddagger\circ}/RT} \tag{27.29}$$

where $\Delta\overline{G}^{\ddagger\circ}$ is the standard molar *free energy of activation*. Like all free energies, we can express $\Delta\overline{G}^{\ddagger\circ}$ in terms of an *enthalpy and entropy of activation*:

$$\Delta\overline{G}^{\ddagger\circ} = \Delta\overline{H}^{\ddagger\circ} - T\Delta\overline{S}^{\ddagger\circ}. \tag{27.30}$$

The enthalpy is related to the temperature derivative of $K_{eq}^{\ddagger}$ in the usual way:

$$\left(\frac{\partial \ln K_{eq}^{\ddagger}}{\partial T}\right)_P = \frac{\Delta\overline{H}^{\ddagger\circ}}{RT^2}. \tag{27.31}$$

If we apply the Arrhenius definition of activation energy,

$$\left(\frac{\partial \ln k_2}{\partial T}\right)_P = \frac{E_a}{RT^2},$$

to Eq. (27.26) for k_2 and use Eq. (27.31), we find

$$\left(\frac{\partial \ln k_2}{\partial T}\right)_P = \left(\frac{\partial}{\partial T} \ln\left[\frac{k_B R T^2}{h P^\circ} K_{eq}^{\ddagger}(1000 \text{ L m}^{-3})\right]\right)_P$$
$$= \frac{2}{T} + \left(\frac{\partial \ln K_{eq}^{\ddagger}}{\partial T}\right)_P = \frac{2RT + \Delta\overline{H}^{\ddagger\circ}}{RT^2}$$

so that

$$E_a = \Delta\overline{H}^{\ddagger\circ} + 2RT. \tag{27.32}$$

Next, we work the potential energy barrier $U_0^{\ddagger}$ into the story. This quantity is the *internal energy of activation at absolute zero,* which thermodynamics writes as $\Delta\overline{U}_0^{\ddagger\circ}$. Since $\Delta H = \Delta U + \Delta(PV)$ in general, or $\Delta H = \Delta U + (\Delta n)RT$ for a reaction among ideal gases at constant T (see Eq. (2.34)), the enthalpy of activation at absolute zero, $\Delta\overline{H}_0^{\ddagger\circ}$, equals $\Delta\overline{U}_0^{\ddagger\circ}$. If we know the heat capacities of A, BC, and ABC‡ as functions of T, we can calculate a reaction heat capacity change $\Delta\overline{C}_P^{\ddagger\circ}$ and find $\Delta\overline{H}^{\ddagger\circ}$ at any T in terms of $\Delta\overline{H}_0^{\ddagger\circ}$ and thus $U_0^{\ddagger}$ through Eq. (7.9):

$$\Delta\overline{H}^{\ddagger\circ} = N_A U_0^{\ddagger} + \int_0^T \Delta\overline{C}_P^{\ddagger\circ} \, dT. \tag{27.33}$$

For example, suppose A is an atom, BC is a diatomic, ABC‡ is a linear triatomic, and each has only translational and rotational contributions to their heat capacities. Then, from our discussion of heat capacities in Chapter 23, these heat capacities are $5R/2$, $7R/2$, and $7R/2$, respectively, or $\Delta\overline{C}_P^{\ddagger\circ} = -5R/2$ so that (using N_A to convert the *molecular* energy $U_0^{\ddagger}$ into a *molar* energy)

$$\Delta\overline{H}^{\ddagger\circ} = N_A U_0^{\ddagger} - \frac{5RT}{2}$$

or, from Eq. (27.32)

$$E_a = N_A U_0^{\ddagger} - \frac{RT}{2}.$$

While this result is specific to our particular reaction and assumptions about its heat capacity, it is often the case that $N_A U_0^{\ddagger}$ is much larger than RT (i.e., $U_0^{\ddagger} \cong 100$ kJ mol^{-1} versus $RT \cong 3$–4 kJ mol^{-1}) so that the activation energy is usually close in magnitude to the potential energy barrier height.

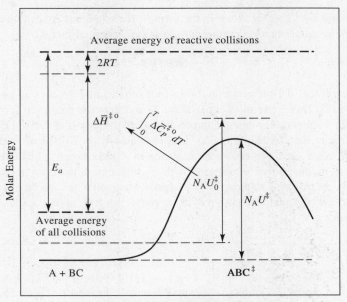

FIGURE 27.10 The Arrhenius activation energy E_a is the difference between the average energy of reactive collisions and the average energy of all collisions. The activation energy is related to other measures of the transition state energy as shown here and discussed in the text.

A more microscopic interpretation of activation energy was derived by Richard C. Tolman in the 1920s.[9] He invoked two average energies in his analysis: that of all the reactants, and that of those reactants with sufficient energy to react, a generally larger number. A simple statistical argument led him to conclude that the activation energy is the difference between these energies:

$$E_a = \begin{pmatrix} \text{Average energy of} \\ \text{reactive collisions} \end{pmatrix} - \begin{pmatrix} \text{Average energy of} \\ \text{all collisions} \end{pmatrix}.$$

Figure 27.10 summarizes the relationships between activation energy and the potential energy surface. While Eq. (27.32) is specific to a bimolecular elementary process, we nevertheless see that the activation energy of an elementary process is a fairly complicated quantity related through thermal averages to microscopic energies.

Next, we can express the rate constant in terms of thermodynamic activation quantities. We start with Eq. (27.26) for k_2, and successively invoke Equations (27.29), (27.30), and (27.32):

$$k_2 = \frac{k_B T}{h} e^{-\Delta \overline{G}^{\ddagger \circ}/RT} \frac{RT}{P^\circ} (1000 \text{ L m}^{-3}) \qquad \text{using Eq. (27.29)}$$

$$= \frac{k_B T}{h} e^{-\Delta \overline{H}^{\ddagger \circ}/RT} e^{\Delta \overline{S}^{\ddagger \circ}/R} \frac{RT}{P^\circ} (1000 \text{ L m}^{-3}) \qquad \text{using Eq. (27.30)}$$

$$= \frac{k_B T}{h} e^{-(E_a - 2RT)/RT} e^{\Delta \overline{S}^{\ddagger \circ}/R} \frac{RT}{P^\circ} (1000 \text{ L m}^{-3}) \qquad \text{using Eq. (27.32)}$$

$$= \left[e^2 \frac{k_B T}{h} \frac{RT}{P^\circ} (1000 \text{ L m}^{-3}) e^{\Delta \overline{S}^{\ddagger \circ}/R} \right] e^{-E_a/RT} \qquad \text{collecting factors}$$

---

[9] It is interesting to note that Tolman was the director of the US War Department's Fixed Nitrogen Research Laboratory from 1920–1922 at the time Daniels and Johnston did their seminal work on the unimolecular decomposition of N_2O_5 discussed in Example 26.1. Tolman moved to Caltech and eventually from physical chemistry to astrophysics, claiming he did so because "Chemistry is hard."

so that we end with an expression in the form of the empirical Arrhenius equation. The factors in square brackets represent the pre-exponential factor, A:

$$A = e^2 \frac{k_B T}{h} \frac{RT}{P^\circ} (1000 \text{ L m}^{-3}) \, e^{\Delta \overline{S}^{\ddagger \circ}/R} \qquad (27.34)$$

(an expression that depends on our choice of a bimolecular process as well as our standard state). This expression depends on T, and thus the empirical form of Eq. (26.27) with a temperature-dependent pre-exponential factor is closer to Eq. (27.34) than the simpler Eq. (26.26), but if $\Delta \overline{S}^{\ddagger \circ}$ depends slowly on T (as is often the case), we still predict a slow power-law dependence to A rather than a fast exponential dependence, in qualitative agreement with experiment. Since two reactants are forming one activated complex, $\Delta \overline{S}^{\ddagger \circ}$ is usually negative, as we have remarked earlier, and we see that the entropy factor plays the role of the steric factor in collision theory.

EXAMPLE 27.11

The pre-exponential factor for $\text{Cl} + \text{H}_2 \rightarrow \text{HCl} + \text{H}$ is $1.45 \times 10^{10} \text{ M}^{-1} \text{ s}^{-1}$ at 300 K, and the activation energy is 18.3 kJ mol^{-1}. (See also Example 27.7.) What are the thermochemical kinetic parameters $\Delta \overline{H}^{\ddagger \circ}$ and $\Delta \overline{S}^{\ddagger \circ}$ for this reaction?

SOLUTION From Eq. (27.32), we find

$$\Delta \overline{H}^{\ddagger \circ} = E_a - 2RT = 18.3 \text{ kJ mol}^{-1} - 5.0 \text{ kJ mol}^{-1} = 13.3 \text{ kJ mol}^{-1}$$

and from Eq. (27.34), we find $\Delta \overline{S}^{\ddagger \circ}$:

$$\Delta \overline{S}^{\ddagger \circ} = R \ln \left[\frac{AhP^\circ}{e^2 k_B RT^2 (1000 \text{ L m}^{-3})} \right] = -11.3 \, R = -93.8 \text{ J mol}^{-1} \text{ K}^{-1}.$$

The value for $\Delta \overline{S}^{\ddagger \circ}$ is negative, as expected, and we can check the calculation if we find $\Delta \overline{G}^{\ddagger \circ}$ from Eq. (27.30), calculate $K_{eq}^{\ddagger}$ from Eq. (27.29), and then find k_2 from Eq. (27.27):

$$\Delta \overline{G}^{\ddagger \circ} = \Delta \overline{H}^{\ddagger \circ} - T \Delta \overline{S}^{\ddagger \circ} = 41.5 \text{ kJ mol}^{-1}$$

$$K_{eq}^{\ddagger} = e^{-\Delta \overline{G}^{\ddagger \circ}/RT} = 6.06 \times 10^{-8}$$

$$k_2 = (1.73 \times 10^9 \text{ M}^{-1} \text{ s}^{-1} \text{ K}^{-2}) \, K_{eq}^{\ddagger} \, T^2 = 9.4 \times 10^6 \text{ M}^{-1} \text{ s}^{-1} ,$$

in agreement with the experimental value quoted in the text before Example 27.7.

→ RELATED PROBLEMS 27.35, 27.36

Finally, we can invoke another thermodynamic expression to find the role played by *pressure* in chemical kinetics. We can apply Eq. (5.36), $(\partial G/\partial P)_T = V$, to Eq. (27.26) with Eq. (27.29) expressing $K_{eq}^{\ddagger}$ and find

$$\left(\frac{\partial k_2}{\partial P} \right)_T = \left[\frac{k_B RT^2}{hP^\circ} (1000 \text{ L m}^{-3}) \right] \left(\frac{\partial}{\partial P} e^{-\Delta \overline{G}^{\ddagger \circ}/RT} \right)_T$$

$$= -\frac{k_2}{RT} \left(\frac{\partial \Delta \overline{G}^{\ddagger \circ}}{\partial P} \right)_T = -\frac{k_2 \Delta \overline{V}^{\ddagger \circ}}{RT}$$

or

$$\left(\frac{\partial \ln k_2}{\partial P} \right)_T = -\frac{\Delta \overline{V}^{\ddagger \circ}}{RT} \qquad (27.35)$$

where $\Delta \overline{V}^{\ddagger \circ}$ is called the *molar activation volume*. It represents the difference between the molar volume of the activated complex and the total molar volume of the isolated reactants. If $\Delta \overline{V}^{\ddagger \circ}$ is negative, implying an activated complex that is more compact than the molar volume sum of isolated reactants, increasing pressure increases the reaction rate.

If we assume ACT holds for solution reactions, as it does with only minor modifications, the wide range of pressures available to solution studies in contrast to gas studies allows Eq. (27.35) to be put to practical use. Both positive and negative values of $\Delta \overline{V}^{\ddagger \circ}$ are known, spanning the range from about -20 to $+20$ cm^3 mol^{-1}, and we can give rough physical interpretations to these numbers (see also Problem 27.39). For a unimolecular bond cleavage, the stable reactant is more compact than the dissociating transition state, and $\Delta \overline{V}^{\ddagger \circ}$ is generally positive. If the dissociation leads to ionized products from a neutral reactant, the transition state often has considerable charge separation, and the strong association of solvent to a charged transition state (the *electrostriction* effect mentioned in Section 9.3) usually makes $\Delta \overline{V}^{\ddagger \circ}$ negative.

In the next section, we consider some of the special properties of unimolecular reactions. While a topic unto itself, unimolecular decomposition is also a part of activated complex theory. It represents the last phase of the reaction in that products finally appear through unimolecular decomposition of the activated complex: ABC‡ → AB + C.

27.4 UNIMOLECULAR REACTIONS—THE DECAY OF ENERGY

Unimolecular processes were a considerable mystery for a long time. What caused an otherwise perfectly stable molecule to decompose or isomerize? Where did the energy needed for this reaction come from? For some time around the start of the 20th century, blackbody radiation was blamed in what was termed the *radiation hypothesis* of chemical kinetics. These were the early days of physical chemistry when chemists and physicists were struggling with many new ideas and phenomena, and we can forgive them if they embraced a mechanism that, as it turned out, invoked the wrong portion of the electromagnetic spectrum and also had the potential to violate the Second Law of thermodynamics![10] Once things were sorted out, the role of energy transfer via collisions again rose to the front, and thermal unimolecular decompositions became very well understood.

The correct foundation for thermal unimolecular reactions was proposed in 1921 at about the same time by Christiansen (in his PhD thesis, which also introduced the term "chain reaction") and Lindemann (who later served as science advisor to Winston Churchill during World War II). We assume the net reaction and experimental unimolecular rate law

$$A \xrightarrow{k_{uni}} \text{Products}, \quad \mathcal{R} = k_{uni}[A]$$

and imagine a sequential mechanism of two bimolecular and one unimolecular elementary processes. The first step is bimolecular collisional energy transfer to A

---

[10] One of the proponents of the radiation hypothesis throughout the 1920s was Richard C. Tolman, mentioned in the previous footnote. He set out many clear arguments in favor of and against this hypothesis, but he was unable to convince himself conclusively one way or the other for many years. Perhaps this helps us understand why he said, "Chemistry is hard."

from any collision partner M

$$A + M \xrightarrow{k_{-2}} A^* + M \qquad (27.36a)$$

where A* represents a reactant with somewhat greater than average internal energy. The second step is the reverse of the first, a bimolecular collisional deactivation of A*

$$A^* + M \xrightarrow{k_{-2}} A + M \; , \qquad (27.36b)$$

and the final step is the elementary unimolecular decomposition of A*

$$A^* \xrightarrow{k_1} \text{Products} \; . \qquad (27.36c)$$

If we apply a steady-state analysis to [A*] in this mechanism, we find

$$\mathcal{R} = k_{\text{uni}}[A] = k_1[A^*] = k_1 \left(\frac{k_2[A][M]}{k_1 + k_{-2}[M]} \right) = \left(\frac{k_1 k_2 [M]}{k_1 + k_{-2}[M]} \right)[A]$$

so that the experimental rate constant is

$$k_{\text{uni}} = \frac{k_1 k_2 [M]}{k_1 + k_{-2}[M]} \; . \qquad (27.37)$$

This expression has two clear pressure limits. At high pressure, collisions with M establish an equilibrium between A and A*, and we have

$$k_{\text{uni}} \xrightarrow{[M] \to \infty} k_\infty = \frac{k_1 k_2}{k_{-2}} = k_1 K^*_{\text{eq}} \qquad (27.38)$$

where K^*_{eq} is the A $\rightleftarrows$ A* equilibrium constant. At low pressure, such that $k_1 \gg k_{-2}[M]$, we find

$$k_{\text{uni}} \xrightarrow{[M] \to 0} k_0 = k_2[M] \qquad (27.39)$$

so that the reaction is governed by the rate of A energization.

Several physical ideas are behind this simple mechanism. Increased pressure means decreased time between collisions. If A* takes a finite time to decompose (which, of course, it must—nothing is truly instantaneous), then two critical quantities are the characteristic decomposition time of A*, $\tau_1 = 1/k_1$, and the time between *deactivating* collisions. Deactivation is the second step, Eq. (27.36b), and it is worth our while to stop a moment and consider collisions between A* and M. To make the discussion concrete, let A be a diatomic such as I_2 and let M be Ar. (We also image [Ar]$\gg$[I_2] so that we can neglect I_2–I_2 collisions, but this is a minor point.)

It is unlikely for dynamical reasons that a single Ar–I_2 collision leads to I_2 dissociation in *one step* unless the collision energy is extraordinarily high. Although the ArI_2 potential energy surface is very different from the H_3 surface in Figures 27.4 and 27.5 since ArI is not strongly bound, our skateboarding analogy helps explain why. Direct collisional dissociation requires a skateboarding trick that starts in the entrance channel (deep, since I_2 is reasonably strongly bound), rides up to the ArI$I^\ddagger$ transition state, and then *turns out toward the dissociative plateau*, avoiding the attractive forces that try to keep I_2 bound. A more likely consequence is reflection back out the entrance channel but with *vibrational excitation* of I_2, and such collisions of course serve to maintain the equilibrium distribution of I_2 vibrational level populations, as illustrated in Figure 23.2.

On occasion, an Ar encounters a vibrationally excited I_2. Such a collision can either de-excite the I_2 or excite it further. Both processes are required by detailed balance. Eventually, however, through collision-induced jumps up and down its vibrational manifold, the I_2 internal energy exceeds its bond energy. It may then dissociate in one vibrational swing, or it may hang around for several if its rotational energy is so large that a *centrifugal potential barrier* holds it together. Should this I_2^* species collide with Ar before it dissociates on its own, the collision could force dissociation, since it is hanging together by the thinnest of energy threads, but it could also deactivate I_2^* to a vibrationally excited I_2, but one now far from the dissociation limit. This is Eq. (27.36b). The so-called *strong-collision assumption* of unimolecular rate theory assumes that *every* I_2^*–Ar collision leads to substantial deactivation.

To return to the basic Lindemann expression, we can write Eq. (27.37) as

$$k_{\text{uni}} = \frac{k_\infty}{1 + k_\infty/k_0} = \frac{k_\infty}{1 + k_\infty/k_2[M]} \ . \quad (27.40)$$

If we measure k_{uni} as a function of [M], we can derive values for k_∞ and k_2 from a fit of Eq. (27.40) to the data. Such a fit is shown in Figure 27.11 for the classic $CH_3NC \rightarrow CH_3CN$ isomerization. The data are taken from the paper by Schneider and Rabinovitch (*J. Am. Chem. Soc.* **84,** 4215 (1962)) and represent experiments at 503.6 K begun with pure CH_3NC. (Only 10% of their data are plotted in the figure for simplicity.) While the qualitative trend predicted by Lindemann is correct, the low-pressure data fall less rapidly than Eq. (27.40) predicts as $P \rightarrow 0$.

Why should this be? The answer comes if we blend in some ideas from ACT. It is not sufficient to say that A has been energized to A* which immediately turns into products. One step is missing: A* must find its way to the activated complex transition state, $A^\ddagger$, of the potential energy surface. The rate at which A* turns into $A^\ddagger$ depends on the total internal energy in A* and the nature of A. If A is a large molecule, its many internal vibrational modes constitute a multidimensional maze through which energy travels, flowing from mode to mode through the coupling between them. The transition state is on one side of a special door in this maze (akin to our dividing surface in ACT), and finding it takes time, the more so if the energy in A* is not large or if A* has many vibrational modes.

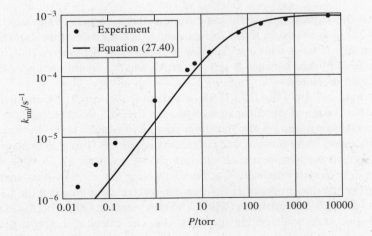

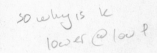

FIGURE 27.11 Experimental unimolecular rate constants for the isomerization of CH_3NC at 503.6 K as a function of pressure are contrasted to the simple Lindemann expression's prediction, Eq. (27.40). There is general qualitative agreement, but the theory falls more rapidly as $P \rightarrow 0$ than the experiment.

EXAMPLE 27.12

The data below represent some of the Schneider and Rabinovitch data for CH_3NC isomerization at 472.6 K. What high-pressure limiting rate constant k_∞ do they predict?

P/torr	2248	1500	570	208	72.3
$k_{uni}/10^{-5}$ s^{-1}	7.37	7.25	6.98	6.02	4.50

SOLUTION   If we write Eq. (27.40) in the form

$$\frac{1}{k_{uni}} = \frac{1}{k_\infty} + \frac{1}{k_2[M]} ,$$

recognize that M is CH_3NC in these experiments, and note that [M] is proportional to P since the volume is fixed, we see that a plot of $1/k_{uni}$ versus $1/P$ should be a straight line with intercept $1/k_\infty$. Such a plot is shown below along with a line representing a least-squares fit to the data.

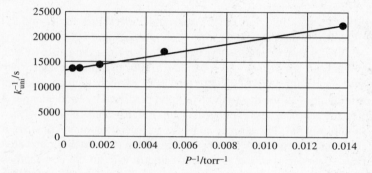

The calculated intercept, 1.33×10^4 s, corresponds to $k_\infty = 7.5 \times 10^{-5}$ s^{-1}. Note that the reaction is not very fast at these temperatures; the half-life is about 2.5 hr. Schneider and Rabinovitch ran each experiment to between 5% and 35% of completion.

➡ **RELATED PROBLEMS** 27.40, 27.41

If we apply ACT to the unimolecular process $A \to A^\ddagger \to$ products, we find an analog to the bimolecular ACT expression of Eq. (27.24):

$$k_{uni} = \frac{k_B T}{h} \frac{q^\ddagger}{q_A} e^{-U_0^\ddagger/k_B T} . \qquad (27.41)$$

Note that this expression has the same form as the Lindemann expression for k_∞, Eq. (27.38), if we consider $k_B T/h$ to represent k_1 and the rest of the expression to represent K_{eq}^*. It must represent the *high-pressure limit* because ACT assumes a statistical equilibrium between A and A* (or $A^\ddagger$) and this equilibrium is maintained only at high pressures.

Before we can apply Eq. (27.41) to a reaction such as the CH_3NC isomerization, we need structural and vibrational information on CH_3NC, which is readily available from spectroscopy, and on the transition state, which must come from quantum chemistry theory. An extensive study of this system in 1980 using *ab initio* quantum methods found the transition state structure shown in Figure 27.12. Note that the CN carbon is closer to the methyl carbon, indicating that the transition state looks a little more like the product CH_3CN than the reactant, and note that the CN bond is very slightly longer than that in CH_3NC (1.177 Å versus 1.170 Å), but qualitatively, the transition state holds no big surprises. One can calculate moments of inertia

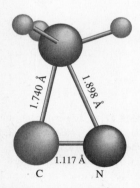

FIGURE 27.12 The structure of the $CH_3NC \to CH_3CN$ reaction's transition state, as predicted by a quantum chemical calculation, leads to activated complex vibrational frequencies, total energy, and moments of inertia that accurately predict the unimolecular rate constant at all pressures.

from this structure, quantities needed for the rotational part of $q^\ddagger$, but the difficult quantities to derive from theory are the normal mode vibrational frequencies. Theory more readily predicts a value for $U_0^\ddagger$, however, finding 2.84×10^{-19} J or 171 kJ mol^{-1}. At the temperature of Figure 27.11, this value gives

$$\frac{k_B T}{h} e^{-U_0^\ddagger/k_B T} = 1.93 \times 10^{-5} \text{ s}^{-1},$$

which is only a factor of 50 smaller than the experimental k_∞, 9.25×10^{-4} s^{-1}. This is the correct order of magnitude for the ratio $q^\ddagger/q_A$, and the same study that predicted the transition state structure calculated normal mode frequencies and moments of inertia that lead to a theoretical value for k_∞ only 50% greater than experiment, a remarkable result for a calculation based on no adjustable or guessed parameters.

Behind this accurate rate calculation lies a more complete theory of unimolecular processes that can be extended successfully to low pressures. Known as *RRKM theory* for its originators, Rice, Ramsperger, Kassel, and Marcus, it is currently the most complete theory of thermal unimolecular reaction rates, and it can be readily extended to nonthermal unimolecular processes such as those initiated through laser irradiation, a topic covered in the next chapter. The complete theory is beyond the scope of our survey here, and this chapter's references have discussions of it.[11] There are several basic assumptions in RRKM theory that should sound familiar: the reactant is assumed to be a collection of coupled harmonic oscillators, energy flows freely among the modes on a time scale faster than the reaction time scale, and all modes at the total energy of the molecule are equally accessible. In the high-pressure limit, the RRKM expression becomes the ACT expression, Eq. (27.41).

We close this chapter with a discussion of catalysis, starting with enzyme action. At first thought, enzyme catalysis might seem to have very little in common with unimolecular reaction theory, but we will find mechanistic interpretations for catalytic action that have a very close parallel with the Lindemann mechanistic ideas. Moreover, the role played by catalysts again focuses on the transition state, which has become the central species in all our understanding of reaction kinetics.

27.5 CATALYSTS AND THE DESIGN OF POTENTIAL ENERGY SURFACES

Chapter 26 introduced the idea of catalysis and discussed autocatalysis in some detail. Here, we focus on catalysts that are not in the net reaction. Catalysis plays a vital role in biochemical reactions (often mediated through *enzymes*), industrial processes (often mediated through *surface catalysis*), and synthetic methods (as in *acid–base catalysis*). We start our discussion with the catalytic activity of enzymes.

Enzyme Catalysis

The next time you have some fresh, uncooked chopped chicken liver that you would rather not cook, try sprinkling it with a hydrogen peroxide solution. You will see something most remarkable: the peroxide solution will bubble, froth, and foam vigorously. Sprinkle H_2O_2 on powdered MnO_2, and you will see the same thing. Both experiments demonstrate catalysis in action. Liver has an enzyme known as

---

[11] It is worth pointing out, however, that RRKM theory requires no new concepts beyond those discussed in this and previous chapters of this book! If you have covered all the major topics up to this point, you are well equipped to continue into more advanced areas of physical chemistry such as RRKM theory.

catalase that accelerates H_2O_2 decomposition by an astounding factor of 10^{15} over the rate in pure water. This is an example of *homogeneous catalysis*. Both the reactant (or reactants) and the catalyst are in the same phase. In contrast, catalysis by solid MnO_2 is an example of *heterogeneous catalysis*. The catalyst is in a different phase (almost always a solid) from the reactant (which can be in either liquid or gas phases).

In the last chapter, we encountered the contributions of Arrhenius to chemical kinetics, and here we have another great Swedish chemist to thank, Jöns Jakob Berzelius, who died in 1848, eleven years before Arrhenius was born. Berzelius was the first to use element *symbols* in chemical formulas, he discovered several elements, and he coined several now familiar terms such as *isomerism* and, to the point here, *catalysis*. The French chemist Louis Jacques Thénard had discovered hydrogen peroxide in 1818, and Berzelius, an indefatigable experimenter, discovered the effect MnO_2 (and Au, Ag, and Pt) had on its decomposition. Writing in the *Edinburg New Philosophical Journal* in 1836, he stated:

> The substances that cause the decomposition of H_2O_2 do not achieve this goal by being incorporated into the new compounds (H_2O and O_2); in each case they remain *unchanged* and hence act by means of an inherent force whose nature is still unknown ... So long as the nature of the new force remains hidden, it will help our researches and discussions about it if we have a special name for it. I hence will name it the catalytic force of the substances, and I will name decomposition by this force catalysis. The catalytic force is reflected in the capacity that some substances have, by their mere presence and not by their own reactivity, to awaken activities that are slumbering in molecules at a given temperature.

Alas, history does not record if Berzelius sprinkled H_2O_2 on chicken liver, but later in this paper, he argued that if we assume catalysts abound in living systems, "thousands of catalytic processes occur ... generating a multitude of substances of differing chemical compositions whose formation ... we could never understand before." This idea proved to be correct, of course, and over 1500 catalytic natural enzymes are now known.

From what we know now about the factors that influence a reaction rate, it is clear that a catalyst must participate in the reaction somewhere in the vicinity of the transition state, and it must do so in a way that lowers the potential energy barrier between reactants and products. A catalyst may also affect $\Delta \overline{S}^{\ddagger \circ}$, the entropy of activation, but this factor is difficult to change appreciably (since it is tied to the molecularity of the reaction and the basic structure of the transition state). We can see how a catalyst might lower an energy barrier if we first understand how a solvent can *inhibit* a reaction. We have discussed the *transport* differences between solution and gas-phase kinetics—diffusion versus free flight—but there are often important *energetic* differences as well. Rather few reactions have been studied in detail in both phases, but some general trends have emerged. Figure 27.13 considers hypothetical but reasonable potential energy profiles along the reaction coordinate for a reaction in which both the reactants and products are strongly solvated in solution but completely free as gases. In solution, the solvation energy *lowers* the energy of reactants and solvents below that of the gas phase. Since the solution activated complex necessarily has some of the coordinated solvent displaced so that the reactants can "touch," at least part of the solution activation energy is related to the energy needed to push solvent away. The gas-phase reaction does not have this component in its reaction energy profile.

An enzyme (and most catalysts in general) helps make up this energy component. It does so through *stabilization of the transition state*. An enzyme selectively binds the activated complex, and this binding energy helps take the place of reactant and

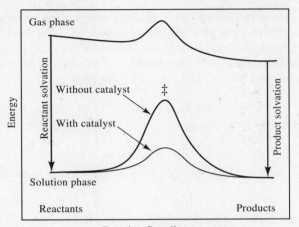

FIGURE 27.13 A reaction in the gas phase can be significantly faster than the same reaction in a solution due to the different amounts of solvation energy for the reactants and the transition state. A catalyst helps overcome these differences through increased stabilization of the transition state complex.

product solvation energy. Enzymes are amazingly engineered proteins that bind the reactant or reactants (called the *substrate*) in a lock-and-key fashion, an idea first suggested by Emil Fischer. The enzyme–substrate interactions are a scaffolding constructed from all sorts of intermolecular interactions: directional hydrogen bonds, charge–charge interactions, van der Waals interactions, and other electrostatic interactions. The scaffolding is designed to hold only one specific substrate with a selectivity that is remarkable. Some enzymes can pick out a particular peptide bond (—CO—NH—) for hydrolytic cleavage from among dozens of them in a particular protein substrate.

EXAMPLE 27.13

The jackbean enzyme *urease* catalyzes the hydrolysis of urea, $(NH_2)_2CO$, to ammonia and CO_2. In the absence of urease, hydrolysis at 21 °C has an activation energy around 125 kJ mol^{-1}. With urease, the activation energy falls to 46 kJ mol^{-1}. By what factor does urease increase the reaction rate? What temperature would give the uncatalyzed reaction the same rate as the catalyzed reaction at 21 °C?

SOLUTION The Arrhenius factor at 21 °C = 294 K without catalysis is

$$e^{-E_a/RT} = \exp\left[-\frac{125\,000 \text{ J mol}^{-1}}{(8.314 \text{ J mol}^{-1} \text{ K}^{-1})(294 \text{ K})}\right] = 5.0 \times 10^{-23}$$

and with catalysis, it is

$$e^{-E_a/RT} = \exp\left[-\frac{46\,000 \text{ J mol}^{-1}}{(8.314 \text{ J mol}^{-1} \text{ K}^{-1})(294 \text{ K})}\right] = 6.7 \times 10^{-9}$$

giving a rate increase by a factor of

$$\frac{\text{catalyzed rate}}{\text{uncatalyzed rate}} = \frac{6.7 \times 10^{-9}}{5.0 \times 10^{-23}} = 1.3 \times 10^{14}.$$

The uncatalyzed reaction rate would equal the 21 °C catalyzed rate at a temperature

$$(294 \text{ K}) \left(\frac{E_a(\text{uncatalyzed})}{E_a(\text{catalyzed})}\right) = (294 \text{ K}) \left(\frac{125 \text{ kJ mol}^{-1}}{46 \text{ kJ mol}^{-1}}\right) \cong 800 \text{ K or } 525 \text{ °C}.$$

This temperature is impossible in aqueous solution, of course, and thus the enzyme achieves a result that is otherwise impossible.

⟹ *RELATED PROBLEM* 27.43

The kinetic mechanism of enzyme catalyzed reactions focuses on the formation of the enzyme–substrate complex (which we denote ES). The basic scheme was formulated by Leonor Michaelis and Maud Menten in 1913 and further elaborated by Briggs and Haldane in 1925. We imagine a mechanism that starts with an elementary bimolecular process to form the ES complex:

$$E + S \xrightarrow{k_2} ES \qquad (27.42a)$$

and we allow this step to be reversible:

$$ES \xrightarrow{k_{-1}} E + S \qquad (27.42b)$$

The ES complex then undergoes a unimolecular step leading to products:[12]

$$ES \xrightarrow{k_1} E + P . \qquad (27.42c)$$

Note the similarity of this mechanism to the unimolecular mechanism of Eq. (27.36). Both start with a bimolecular step and end with a unimolecular step. A steady-state analysis for [ES] leads to

$$\mathscr{R} = \frac{d[P]}{dt} = k_1[ES]$$

$$\frac{d[ES]}{dt} = 0 = k_2[E][S] - k_{-1}[ES] - k_1[ES] .$$

In these expressions, [E] represents the amount of *free* enzyme, which is related to the total amount of enzyme $[E]_0$ through $[E]_0 = [E] + [ES]$. Since $[E]_0$ is the quantity we control experimentally, we write $[E] = [E]_0 - [ES]$ and substitute this into the steady-state expression above, finding

$$[ES] = \frac{k_2[E]_0[S]}{k_1 + k_{-1} + k_2[S]}$$

so that

$$\mathscr{R} = k_1 \frac{k_2[E]_0[S]}{k_1 + k_{-1} + k_2[S]} = \frac{k_1[E]_0[S]}{\left(\frac{k_1 + k_{-1}}{k_2}\right) + [S]} . \qquad (27.43)$$

In the enzyme kinetics literature, the ratio $(k_1 + k_{-1})/k_2$ is called the *Michaelis constant* K_M. At very large [S], the enzyme is saturated, $\mathscr{R} \cong k_1[E]_0$, and thus $k_1[E]_0$ is called the *maximum reaction velocity,* V_{max}. These definitions turn Eq. (27.43) into the *Michaelis–Menten equation:*

$$\mathscr{R} = \frac{V_{max}[S]}{K_M + [S]} . \qquad (27.44)$$

Note the similarity between this expression for the catalyzed rate and Eq. (27.37) for the unimolecular rate constant. Both exhibit limiting values—high pressure for k_{uni} and large [S] for catalysis—and both have similar hyperbolic shapes, as Figure 27.11 showed for k_{uni}. Note also that the reaction rate equals half its maximum when $[S] = K_M$. This is a convenient experimental measure of K_M, but a direct plot of rate versus [S] is often difficult to extrapolate to an accurate saturation value.

Two transformed versions of Eq. (27.44) are used to represent enzyme kinetic data in a more convenient way. One, known as the *Lineweaver–Burke plot*, writes Eq. (27.44) upside-down:

$$\frac{1}{\mathscr{R}} = \frac{K_M + [S]}{V_{max}[S]} = \frac{1}{V_{max}} + \frac{K_M}{V_{max}}\frac{1}{[S]} \qquad (27.45)$$

---

[12] An additional step, ES $\rightleftarrows$ EP where EP represents the enzyme–*product* complex can be added without changing the kinetic analysis.

so that a plot of $1/\mathcal{R}$ versus $1/[S]$ is a straight line of intercept $1/V_{max}$ and slope K_M/V_{max}. The other, called the *Eadie–Hofstee plot*, writes Eq. (27.44) as

$$\mathcal{R} = V_{max} - \frac{K_M \mathcal{R}}{[S]} \qquad (27.46)$$

and a plot of $\mathcal{R}$ versus $\mathcal{R}/[S]$ is linear.

Nature, on occasion, needs to regulate the otherwise voracious appetite of enzymes, and molecules known as *inhibitors* compete with the substrate for the enzyme's attention. The inhibitors undergo no reaction themselves, but they tie up enzyme active sites, effectively lowering the enzyme concentration. Problem 27.46 considers the kinetic consequences of inhibition.

Homogeneous Catalysis

Homogeneous gas-phase catalysis often takes the form of catalytic *cycles* in the net reaction mechanism, and the most important of these are the cycles implicated in the destruction of stratospheric ozone. For example, the chlorofluorocarbon CCl_2F_2, released on the Earth's surface, is sufficiently inert to survive the trip up to the stratosphere. Once there, photolysis in the UV region (λ about 180–220 nm) produces $CF_2Cl + Cl$ (the C—F bond is exceptionally strong—see Table 7.2) or reaction with atomic oxygen in the excited ^1D state produces $CF_2Cl + ClO$. The Cl and ClO radicals enter the following catalytic cycle (similar cycles operate with OH/HO_2 and with NO/NO_2) that has the net result of turning $O + O_3$ into $2O_2$.

$$\begin{array}{c} Cl + O_3 \rightarrow ClO + O_2 \\ ClO + O \rightarrow Cl + O_2 \\ \hline O + O_3 \rightarrow 2O_2 \text{ (net)} \end{array}$$

The activation energy for the uncatalyzed net reaction is 17.1 kJ mol^{-1}, but that for $Cl + O_3$ is only 2.1 kJ mol^{-1} and for $ClO + O$, it is 0.4 kJ mol^{-1}. Figure 27.14 contrasts the catalytic pathway to the uncatalyzed net reaction through an enthalpy level diagram. The transition states are located through activation energies, and we can see at a glance that the catalytic steps provide a much lower energy pathway for the net reaction.

Heterogeneous Catalysis

The advances in surface science methods over the past fifty years or so has led to an explosion of knowledge about and interest in heterogeneous catalysis. What was once a black art is now a grey art at worst as many solid catalysts have yielded their molecular catalytic mechanisms when exposed to the scrutiny of methods such

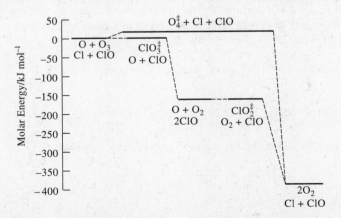

FIGURE 27.14 The homogeneous catalysis of the $O + O_3 \rightarrow 2O_2$ reaction by Cl and ClO carries the reaction through a pair of steps of much lower activation energy. This is a common feature to catalytic mechanisms.

as LEED, Auger spectroscopy, and scanning tunneling microscopy. Surface catalysts not only lower activation barriers, they can also *direct product formation* when more than one net reaction is possible. For example, H_2 and CO form methane and water with some catalysts, methanol and higher alcohols with other catalysts, and higher alkanes and alkenes with still others.

For a unimolecular reaction catalyzed on a surface, we imagine a mechanism that starts with an equilibrium between the reactant A, a surface site S, and the *adsorbed* reactant, AS:

$$A + S \rightleftarrows AS .$$

Since there are a finite number of binding sites on any surface (and on all but perfect surfaces, various sites with different binding strengths always exist), it is helpful to define a *surface coverage* parameter θ defined as

$$\theta = \frac{\text{number of occupied sites}}{\text{number of possible sites}} .$$

For a total of N sites, $N\theta$ is the number occupied, and $N(1 - \theta)$ is the number unoccupied, statements that depend on the assumptions that all sites have equal binding probabilities and that the binding probability of any site is independent of θ.

The surface behaves like a very large molecule at a finite temperature, and the "bimolecular" association represented above for the absorption step $A + S \rightarrow AS$ happens as A first strikes the surface, perhaps losing some energy to it, and then bounces around the surface a few times, losing more energy, until it has lost sufficient energy to remain bound.

Surface binding energies span a wide range. The term *physisorption* applies to species bound through noncovalent bonds with surface atoms, an interaction characterized by energies on the order of 10–80 kJ mol^{-1}, while much stronger *chemisorption* interactions involve true chemical bonds with the surface, on the order of 100–400 kJ mol^{-1}. Often, adsorption passes through a physisorption stage before reaching chemisorption, and *dissociative chemisorption* in which A dissociates on the surface and binds as radicals or atoms (as happens with H_2 on Pd, accounting for the semipermeability of Pd to H_2 discussed in Problem 25.12) is known for many species.

Fluctuations in the energy in the local physisorbed or chemisorbed bond drives unimolecular desorption, the $AS \rightarrow A + S$ step. Adsorption and desorption depend sensitively on temperature, and the technique known as *temperature programmed desorption* monitors the species evolved from a covered surface as a function of the surface temperature. Physisorbed species generally desorb at much lower temperatures than chemisorbed species.

If k_a and k_d are rate constants for adsorption and desorption, respectively, then the rate of change of occupied surface sites for A at a pressure P, which must equal zero at equilibrium, is

$$\underset{\underset{\text{rate of change}}{\text{Surface coverage}}}{\frac{d\theta}{dt}} = \underset{\underset{\text{rate}}{\text{Adsorption}}}{k_a PN(1 - \theta)} - \underset{\underset{\text{rate}}{\text{Desorption}}}{k_d N\theta} = \underset{\underset{\text{rate}}{\text{Equilibrium}}}{0} . \quad (27.47)$$

This gives an expression for the equilibrium surface coverage:

$$\theta_{eq} = \frac{k_a PN}{k_a PN + k_d N} = \frac{k_a P}{k_a P + k_d} . \quad (27.48)$$

If we define an *adsorption coefficient* as $K_{ad} = k_a/k_d$, the equilibrium constant for the adsorption/desorption equilibrium, we can write Eq. (27.48) in the form known as the *Langmuir isotherm*:

$$\theta_{eq} = \frac{K_{ad}P}{1 + K_{ad}P} . \qquad (27.49)$$

Note the similarity of this equation to the Michaelis–Menten rate, Eq. (27.44), and the k_{uni} expression in Eq. (27.37). All three have similar kinetic origins: mechanisms that start with an equilibrium producing an intermediate (AS here, ES for enzymes, and A* for unimolecular reactions) that is at the heart of subsequent chemistry. In fact, a catalytic surface is similar to a giant enzyme with a huge number of active sites.

Experimental adsorption data are often expressed in terms of the equivalent volume of gas at a reference temperature and pressure that was adsorbed to a particular surface (or this volume per unit area of surface). If we let V_m equal the gas volume that would adsorb as a monolayer on a particular surface, then $\theta = V/V_m$ where V is the measured volume at sub-monolayer coverage. The high-pressure limit of Eq. (27.49) is $\theta = 1$, corresponding to $V = V_m$, or surface saturation. This is a difficult limit to reach, since adsorption does not stop at a monolayer, but continues with a different binding energy (usually much smaller than to the naked surface) layer after layer. Introducing this ratio into Eq. (27.49) gives an expression that can be rearranged into a convenient form for data analysis:

$$\frac{P}{V} = \frac{1}{K_{ad} V_m} + \frac{P}{V_m} . \qquad (27.50)$$

One plots P/V versus P and extracts K_{ad} and V_m from the slope and intercept of the resulting straight line. Often, as in the following example, the fit is very satisfactory, but many systems do not follow the Langmuir isotherm very well. There are good physical reasons why the Langmuir isotherm might fail, and improved models of adsorption for more complicated surfaces exist.

EXAMPLE 27.14

Irving Langmuir, one of the few chemistry Nobel Prize winners to come from an industrial research setting, derived Eq. (27.49) in a paper published in 1918, *J. Am. Chem. Soc.* **40,** 1361 (1918). That paper also contains experimental data on the adsorption of several gases on mica, glass, and Pt surfaces. Some of his data for N_2 on a mica sample are listed below, expressed in modern units with volumes corrected to 298 K and 1 bar pressure. (Langmuir expressed his pressures in bar units, but to him and other chemists of the time, 1 bar equaled 1 dyne per cm^2. Our modern bar unit is 10^6 times larger.)

$P/10^{-2}$ torr	$V/10^{-8}$ m^3
2.55	3.65
1.79	3.41
1.30	3.12
0.98	2.82
0.71	2.64
0.56	2.39
0.46	2.10
0.38	1.88
0.30	1.67
0.26	1.48
0.21	1.33

SOLUTION A plot of P/V versus P, as suggested by Eq. (27.50), gives the straight line shown below:

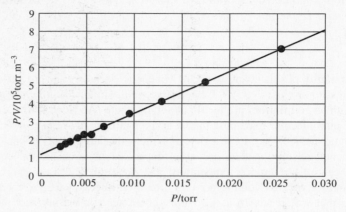

The slope is $1/V_m$, and the fit predicts $V_m = 4.3 \times 10^{-8}$ m^3. The intercept is $1/K_{ad}V_m$, and given V_m, the fit predicts $K_{ad} = 209$ torr^{-1} = 1.57 Pa^{-1}. If we substitute this value into Eq. (27.49), we can see how surface coverage saturates in a graph of θ_{eq} versus P:

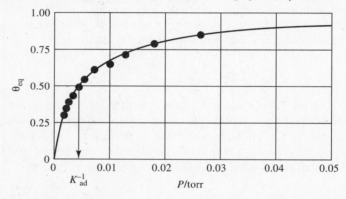

Note the slow approach to saturation, and note that $\theta_{eq} = 0.5$ when $P = 1/K_{ad}$.

➡ **RELATED PROBLEMS** 27.47, 27.48, 27.49

We now have species A adsorbed on a surface, and we return to the question of its catalyzed reaction rate. If A simply decomposes on the surface to products that rapidly desorb, the rate is the product of a unimolecular rate constant k_1 for reactive desorption

$$\text{AS} \xrightarrow{k_1} \text{Products} + \text{S}$$

times the number of occupied sites, $N\theta$:

$$\mathcal{R} = -\frac{d[A]}{dt} = k_1 N\theta = \frac{k_1 N K_{ad} P}{1 + K_{ad} P} = \frac{k_1 N K'_{ad}[A]}{1 + K'_{ad}[A]} \quad (27.51)$$

where K'_{ad} is the adsorption coefficient on a concentration rather than a pressure basis: $K'_{ad} = K_{ad}RT(1000$ L m$^{-3})$.

If the pressure is low, [A] is small, and the rate law is first order in [A]: $\mathcal{R} = k_1 N K'_{ad}[A] = k_{eff}[A]$ where k_{eff} is an effective first-order rate constant. The catalyzed net reaction's activation energy is now contained in k_{eff}. Imagine A binds to the surface much more strongly than the products. An energy profile for such a reaction, compared to the gas-phase reaction with transition state A‡, looks like this:

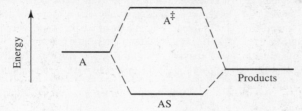

Binding has again produced a lower energy path for the reaction. This alone is not enough for catalysis, of course, since all binding is exothermic. The catalyst must also bind A in such a way that it *promotes rearrangement of A into a transition state structure that has a low desorption energy barrier to products*. This is the tricky part of catalyst design: finding those surface structures and binding energies that preferentially bind a transition state species strongly with respect to reactants but weakly with respect to products. Enzymes do just that, of course, and enzyme analogs with significant catalytic activity have been synthesized. Synthetic surface catalysts are a much greater challenge.

If the pressure is high, [A] is large, the surface active sites are saturated, and the rate approaches zeroth-order kinetics: $\mathcal{R} = k_1 N$. This is the usual case for industrial catalysts that operate at very high pressures. Note also that the rate is proportional to the number of active sites, N. Consequently, industrial catalysts try to achieve the greatest surface to volume ratio possible so as to expose the greatest number of active sites.

CHAPTER SUMMARY

This chapter has covered the way various elementary processes work. Photochemistry proceeds at a rate governed by the rate of photon absorption and the subsequent fate of the excited molecule. The product quantum yield Φ tells us how many product molecules are produced per photon absorbed by a reactant, and $\Phi < 1$ suggests nonreactive energy loss processes are important while $\Phi \gg 1$ suggests a photochemically initiated chain reaction. Under constant illumination, a photochemical reaction can reach a steady state that can be analyzed to yield information about various reactive and nonreactive processes such as fluorescence and collisional energy transfer (*quenching*) from the excited reactant.

A simple collisional model of gas reaction thermal rate constants points out the role of several factors in determining reaction rates: the activation energy, E_a, the average relative speed of collisions, $<g>$, the collision cross section, σ_{12}, and a steric factor, p, that expresses the role played by the relative orientation of the colliding reactants:

> Bimolecular gas-phase collision theory:
> $k_2 = N_A(1000 \text{ L m}^{-3}) p<g>\sigma_{12}e^{-E_a/RT}$

The collisional model in solutions focuses on the role diffusion plays in bringing reactants to a critical reaction distance R^*:

> Diffusion-controlled reactions:
> $k_2 = N_A(1000 \text{ L m}^{-3})4\pi R^*(D_A + D_B)f$

If the reactants are uncharged and of similar size, the electrostatic factor, f, equals 1, and this expression reduces to a useful approximation based on the solvent viscosity:

> Diffusion-controlled reactions:
> $k_2 = (1000 \text{ L m}^{-3})\dfrac{8RT}{3\eta}$

If the reactants are charged, the random, thermal diffusive force that brings reactants together is augmented by an electrostatic driving force expressed through the electrostatic factor:

> Electrostatic factor: $f = \dfrac{R_0}{R^*(e^{R_0/R^*} - 1)}$,
> $R_0 = \dfrac{z_1 z_2 e^2}{4\pi\epsilon_0\epsilon_r k_B T}$

Activated complex theory uses the reaction potential energy surface to locate a critical reactant configuration known as the transition state. Thermal equilibrium is assumed between the reactants and the activated complex, which is treated as a molecule with one unstable internal degree of freedom, the reaction-coordinate motion. For a bimolecular gas-phase reaction, ACT predicts a rate constant dependent on a thermal frequency factor, $k_B T/h$, times an equilibrium constant between the reactants and the activated complex:

$$\text{Activated complex theory:}$$
$$k_2 = \frac{k_B T}{h} K_{eq}^{\ddagger} \frac{RT}{P^\circ} \ (1000 \text{ L m}^{-3})$$

Thermochemical quantities are introduced into chemical kinetics through $K_{eq}^{\ddagger}$ and various thermodynamic activation quantities:

$$\text{Free energy of activation: } \Delta \overline{G}^{\ddagger\circ} = -RT \ln K_{eq}^{\ddagger}$$
$$\text{Enthalpy and entropy of activation:}$$
$$\Delta \overline{G}^{\ddagger\circ} = \Delta \overline{H}^{\ddagger\circ} - T\Delta \overline{S}^{\ddagger\circ}$$

These quantities give us a better picture of the Arrhenius activation energy (through the relation between features of the reaction potential energy surface and $\Delta \overline{H}^{\ddagger\circ}$) and preexponential factor (through the role played by $\Delta \overline{S}^{\ddagger\circ}$).

Unimolecular reactions require reactant excitation before products appear, and the role of collisional excitation, deexcitation, and product formation steps are expressed through a simple mechanism:

$$\text{Lindemann unimolecular mechanism:}$$
$$A + M \underset{k_{-2}}{\overset{k_2}{\rightleftarrows}} A^* + M$$
$$A^* \overset{k_1}{\to} \text{Products}$$

which leads to a simple expression for the observed unimolecular rate constant:

$$\text{Lindemann unimolecular rate constant:}$$
$$k_{uni} = \frac{k_1 k_2 [M]}{k_1 + k_2 [M]}$$

At high pressures, this expression approaches an experimentally observable limiting rate constant, k_∞, but at low pressures, as k_{uni} falls, constant values for k_1, k_2, and k_{-2} do not lead to good agreement with experiment, and more detailed theories are needed to interpret experiments.

Catalysis through an enzyme E and its reactive substrate S follows a mechanism similar to the Lindemann mechanism:

$$\text{Michaelis–Menten enzyme mechanism:}$$
$$E + S \underset{k_{-1}}{\overset{k_2}{\rightleftarrows}} ES$$
$$ES \overset{k_1}{\to} E + \text{Products}$$

The rate of the net reaction, S → products, depends on the substrate concentration, [S], and two parameters of the enzyme, the Michaelis constant, $K_M = (k_1 + k_{-1})/k_2$, and the maximum reaction velocity for a total enzyme concentration $[E]_0$, $V_{max} = k_1 [E]_0$:

$$\text{Michaelis–Menten rate law: } \mathscr{R} = \frac{V_{max}[S]}{K_M + [S]}$$

Adsorption on a surface, whether catalytic or not, depends on the number of surface binding sites available and the magnitude of the molecule's interaction energy with the surface. Adsorption and desorption exist in dynamic equilibrium, and the simplest description of the extent of surface coverage is based on molecule–surface binding up to a maximum of one monolayer coverage:

$$\text{Langmuir adsorption isotherm: } \theta_{eq} = \frac{K_{ad} P}{1 + K_{ad} P}$$

Coverage depends on the adsorbing gas pressure, P, and the adsorption coefficient, K_{ad}, which is the ratio of the adsorption rate constant, k_a, to the desorption rate constant, k_d.

FURTHER READING

In addition to the references at the end of Chapter 26, several specialized books cover topics from this chapter in some detail. Unimolecular reactions are treated in depth in *Theory of Unimolecular Reactions*, by W. Forst (Academic Press, New York, 1973) and in *Unimolecular Reactions*, by P. J. Robinson and K. A. Holbrook (Wiley, New York, 1972). Enzyme kinetics is discussed in *Physical Chemistry*, by I. Tinoco, Jr., K. Sauer, and J. C. Wang, 2nd ed. (Prentice-Hall, Englewood Cliffs, NJ, 1985), a textbook at the level of this one but with an emphasis on biophysical chemistry, in *Enzyme Structure and Mechanism*, by A. Fersht, 2nd ed. (W. H. Freeman, New York, 1985), and in *Enzymatic Reaction*

Mechanisms, by C. Walsh, (W. H. Freeman, New York, 1979). An excellent introduction to the history, chemistry, structure, and future of enzyme research is *Discovering Enzymes*, by D. Dressler and H. Potter, (Scientific American Library, New York, 1991). *An Introduction to Chemisorption and Catalysis by Metals*, by R. P. H. Gasser, (Oxford University Press, Oxford, 1985) and *Catalysis at Surfaces*, by Ian M. Campbell, (Chapman and Hall, London, 1988) both discuss heterogeneous catalysis at surfaces. Modern research into the nature of the transition state is discussed in a review article, *Transition State Spectroscopy*, by D. M. Neumark, (*Annu. Rev. Phys. Chem.* **43**, 153 (1992)).

PRACTICE WITH EQUATIONS

27A How many moles of H_2 are produced by photolysis of excess ethylene with 2×10^{19} photons at 185 nm if $\Phi = 0.42$? (27.4)

ANSWER: 1.4×10^{-5} mol

27B The rate constant for quenching the $^3P_1$ excited state of Hg by N_2 is 7×10^9 M^{-1} s^{-1} at 300 K. For $[N_2] = 1.6 \times 10^{-3}$ M (30 torr), the Hg* fluorescence decay rate is 2×10^7 s^{-1}. What is the Hg* radiative lifetime? (27.7)

ANSWER: $\tau_{rad} = 1/k_f = 114$ ns

27C What is the quenching cross section for the N_2–Hg* system of 27B? (27.8)

ANSWER: 6.3×10^{-21} m^2

27D The CH_4 hard-sphere radius is 1.9 Å from transport data, and the rate constant at 400 K for $Cl + CH_4 \rightarrow HCl + CH_3$ is 2.26×10^8 M^{-1} s^{-1} with an activation energy of 6.24 kJ mol^{-1}. What is the steric factor? (27.9), Example 27.7

ANSWER: 6.4×10^{-3}

27E What is the total radial flow I_r at a reaction radius $R^* = 5$ Å for a species with $[B]_\infty = 10^{-4}$ M and $D_B = 2 \times 10^{-9}$ m^2 s^{-1}? (27.13), (27.14)

ANSWER: 7.6×10^5 molecules s^{-1}

27F The $H + OH \rightarrow H_2O$ rate constant in aqueous solution is 7.0×10^9 M^{-1} s^{-1}. If the reaction is diffusion limited with R^* equal to the OH bond length, 0.97 Å, what is the diffusion constant sum, $D_H + D_{OH}$? (27.16)

ANSWER: 9.5×10^{-9} m^2 s^{-1}

27G A diffusion-limited reaction is run first in benzene and then in *n*-hexane. What is the ratio of the rate constants in the two experiments? (27.17), Table 25.3

ANSWER: $k_{benz}/k_{hex} = 0.5$

27H What is the rate constant at 298 K for $NH_4^+(aq) + OH^-(aq) \rightarrow NH_4OH(aq)$ if $R^* = 5$ Å, $D_{NH_4^+} = 1.9 \times 10^{-9}$ m^2 s^{-1}, and $D_{OH^-} = 5.08 \times 10^{-9}$ m^2 s^{-1}? (27.18), (27.19), (27.20), Table 9.1

ANSWER: 5×10^{10} M^{-1} s^{-1}

27I The rate constant for the stratospherically important reaction $NO + ClO \rightarrow Cl + NO_2$ at 298 K is 1.1×10^{10} M^{-1} s^{-1}. What is $K_{eq}^\ddagger$? (27.27)

ANSWER: 7.4×10^{-5}

27J What is the free energy of activation for the reaction in 27I? (27.29)

ANSWER: 23.6 kJ mol^{-1}

27K The pre-exponential factor for $H + F_2 \rightarrow HF + F$ is 8.9×10^{10} M^{-1} s^{-1} at 300 K. What is the reaction's entropy of activation? (27.34)

ANSWER: -79 J mol^{-1} K^{-1}

27L What pressure doubles the 1 atm reaction rate at 300 K if $\Delta \overline{V}^{\ddagger\circ} = -20$ cm^3 mol^{-1}? (27.35)

ANSWER: 854 atm

27M What is the high-pressure unimolecular rate constant at 500 K for a reaction with $U_0^\ddagger = 100$ kJ mol^{-1} and $q^\ddagger/q_A = 27$? (27.41)

ANSWER: 10^4 s^{-1}

27N The Michaelis constant for catalase is 1.1 M. At what $[H_2O_2]$ is the catalyzed reaction rate 25% of the maximum for a particular $[catalase]_0$? (27.44)

ANSWER: 0.37 M

27O What pressure gives 25% surface coverage for a catalyst that obeys the Langmuir adsorption isotherm with an adsorption coefficient 5.2 Pa^{-1}? (27.49)

ANSWER: 6.4×10^{-2} Pa

PROBLEMS

SECTION 27.1

27.1 Here are a few short questions about the elementary processes in Table 27.1.

(a) Why is the radiative association reaction $H + H \rightarrow H_2 + h\nu$ impossible for ground-state H atoms?

(b) Why does the Penning ionization reaction $H^* + He \rightarrow H + He^+ + e^-$ have a negligible rate at ordinary temperatures for *any* excited state of H?

(c) Why is the dissociative attachment reaction $HF + e^- \rightarrow H + F^-$ more probable than $HF + e^- \rightarrow H^- + F$?

(d) Which is more probable, the charge transfer reaction $O + H^- \rightarrow O^- + H$ or the associative detachment reaction $O + H^- \rightarrow OH + e^-$? (*Hint:* See Table 7.4.)

27.2 If the transient $A \cdots M$ species in the chaperone mechanism, Eq. (27.2), has a sufficiently large binding energy, collisions may stabilize it to a bound AM species, which then undergoes atom transfer with A to form the final A_2 product:

$A + M \rightleftarrows A \cdots M$ **(transient association/dissociation)**

$A \cdots M + M \rightleftarrows AM + M$ **(stabilization/destabilization)**

$AM + A \rightarrow A_2 + M$.   **(atom transfer)**

Assign rate constants to these processes, and use a steady-state analysis to show that at low [A], the rate law has the form $\mathcal{R} = k_{exp}[A]^2[M]$. Express the experimental rate constant k_{exp} in terms of your elementary rate constants.

27.3 The second step in the mechanism of the previous problem has a simple approximate form for its equilibrium constant. Assume AM and $A \cdots M$ have the same electronic state with $A \cdots M$ representing an AM molecule in the last bound vibrational level, and write the statistical-mechanical expression for this equilibrium constant in terms of the AM bond dissociation energy $\mathcal{D}_0$ and any other relevant molecular parameters. What role might entropy play in this equilibrium constant? (*Hint:* What is the average bond length of $A \cdots M$ compared to that for AM?)

27.4 Consider the formation of a weakly bound dimer such as Ar_2 in pure Ar. Show, using the chaperone mechanism adapted to a pure gas, that the rate of formation of Ar_2 follows a second-order rate law, $\mathcal{R} = k[Ar]^2$, as if the reaction $Ar + Ar \rightarrow Ar_2$ was an elementary process.

27.5 The N—O bond in N_2O is much weaker (161 kJ mol^{-1}) than the N—N bond (481 kJ mol^{-1}), but spin considerations show that photodissociation to ground-state $O + N_2$ is improbable: N_2O and N_2 are singlets, but ground-state O is a triplet. Consequently, photolysis of N_2O around 200 nm leads to the excited ^1D state of O. The N_2O absorption cross section at 200 nm is 4.1×10^{-24} m^2. What is the maximum number of O ^1D atoms that a single 0.1 mJ laser pulse at 200 nm could produce in a laser beam volume 10 cm long and 5 mm^2 in cross-sectional area through N_2O at 100 torr and 300 K?

27.6 Irradiation of HI in the UV leads to photodissociation with unit quantum yield of atomic H and I. Suggest a mechanism for photochemical conversion of HI into H_2 and I_2, taking into account the bond energies 294 kJ mol^{-1} for HI, 432 kJ mol^{-1} for H_2, and 149 kJ mol^{-1} for I_2. How many HI molecules are destroyed per absorbed photon in your mechanism?

27.7 You wish to carry out a photochemical reaction using a UV lamp filtered to generate 290 nm radiation with a 25 W power. Your reactants are sufficiently concentrated to absorb all the radiation from this lamp, and the reaction's photochemical product quantum yield is 0.28. How many moles of product will you generate per hour of illumination?

27.8 In 1825, Sir John Herschel invented an instrument for measuring the intensity of sunlight, and he named it the *actinometer*. Today, chemical actinometry refers to the measurement of light intensity through a photochemical reaction of known absorption and quantum yield. Derive a general expression for the intensity of a light source (in photon s^{-1} units) in terms of the rate of production of a stable product (in molecule s^{-1} units), the product quantum yield, the light path length, and the reactant absorbance coefficient and concentration. A common source for light in the vacuum ultraviolet (VUV, $\lambda < 200$ nm) is the fluorescence of rare gas atoms from their first excited states. In particular, Kr has a strong emission feature at 123.6 nm. These photons are sufficiently energetic to photoionize NO, and the photoionization current can serve as an actinometer for the Kr lamp. Due to subsequent chemistry of the NO^+ and e^- products, the observed quantum yield for ions in an NO actinometer is 0.77 at 123.6 nm. If a Kr source illuminates a cell 3.0 cm long containing NO at 1.0 torr pressure at 300 K and generates a current in the cell of 26 μA, what is the source intensity? The photoionization cross section for NO at this wavelength is 1.86×10^{-22} m^2. Don't forget that both the NO^+ ions *and* the photoelectrons contribute to the measured current!

27.9 The photochemical reaction between Br_2 and H_2 starts with the initiation step $Br_2 + h\nu \rightarrow 2Br$, the photochemical equivalent of Eq. (26.17a), and a chain reaction follows, as discussed in Chapter 26.

Find the net rate law for this reaction mechanism in terms of the rate constants k_b–k_e for Equations (26.17b–e) and the rate of light absorption in the photochemical initiation step.

27.10 Repeat Example 27.4, replacing the continuous 10^{15} photon s^{-1} lamp with a single 350 nm, 1.0 J, 10 ns duration laser pulse. By what factor is the probability of sequential two photon absorption greater for this intense, brief pulse compared to the weak, continuous source of the Example?

27.11 Chapter 20 defined the fluorescence quantum yield in Eq. (20.3) and discussed the radiationless transitions called intersystem crossing and internal conversion. This and the following three problems consider the kinetics of radiationless processes. Let $^1A^*$ represent the first excited singlet state of A, and let $\mathcal{R}_a = I_a/N_A bh\nu$, the rate of $^1A^*$ production from absorption, as in Eq. (27.5). Imagine two unimolecular loss processes for $^1A^*$, spontaneous singlet–singlet fluorescence to the singlet ground state:

$$^1A^* \xrightarrow{k_f} A + h\nu$$

and intersystem crossing to the lowest triplet state, $^3A^*$:

$$^1A^* \xrightarrow{k_{ISC}} {}^3A^* \ .$$

Derive an expression for the steady-state concentration of $^1A^*$ under constant illumination in terms of $\mathcal{R}_a$, k_f, and k_{ISC}, and derive an expression for the concentration of $^1A^*$ as a function of time after a brief excitation pulse. What is the half-life of $^1A^*$? The first excited singlet state of benzene has a 370 ns radiative lifetime in the gas phase, but in hexane, ^{1}benzene* fluorescence decays with a 26 ns lifetime. What is the fluorescence quantum yield for benzene in hexane?

27.12 Continuing from the previous problem, we consider the radiative and nonradiative fates of the lowest triplet state. This state can emit light through slow triplet–singlet fluorescence (*phosphorescence*):

$$^3A^* \xrightarrow{k_p} A + h\nu$$

or it can undergo intersystem crossing to vibrationally excited levels of the ground singlet state, represented as $A^\dagger$:

$$^3A^* \xrightarrow{k'_{ISC}} A^\dagger \ .$$

Derive an expression for the steady-state concentration of $^3A^*$ assuming constant illumination of A and using expressions from the previous problem when needed. Reactions of $^3A^*$ can be very different from those of $^1A^*$. What conditions among the rates and rate constants in this scheme make $[^3A^*] \gg [^1A^*]$?

27.13 Continuing from the previous two problems, we consider next the kinetic consequences of collisional quenching of the lowest triplet state, $^3A^*$. This quenching can be reactive or nonreactive, and thus we write a generic bimolecular step between $^3A^*$ and a quencher Q:

$$^3A^* + Q \xrightarrow{k_Q} \text{Products} \ .$$

Derive an expression for the steady-state concentration of $^3A^*$ including quenching as well as the processes of the previous two problems. Show that the ratio of the intensity of $^3A^*$ phosphorescence in the absence of Q to that when Q is at a finite concentration [Q] follows the Stern–Volmer relation

$$\frac{\text{Unquenched intensity}}{\text{Quenched intensity}} = 1 + k_Q \tau_3 [Q]$$

where τ_3 is the triplet phosphorescence lifetime. Find an expression for τ_3 as well.

27.14 A tunable laser is used to excite I_2 gas to a particular vibration–rotation level of the excited $^3\Pi_u$ electronic state. This state undergoes spontaneous emission and collisional quenching with ground-state I_2 molecules, which are always in great excess. In an experiment that populated the $v = 9$, $J = 33$ level in I_2 at 300 K, the following fluorescence decay rates were measured as a function of the I_2 pressure. Find the natural radiative lifetime of this level and the quenching rate constant. Express the rate constant in both torr^{-1} s^{-1} and M^{-1} s^{-1} units.

P/torr	0.023	0.051	0.081	0.093	0.121	0.125
$k_{exp}/10^6$ s^{-1}	1.79	1.93	2.09	2.14	2.29	2.30

27.15 In another experiment of the type described in the previous problem, the laser was tuned to the $v = 17$, $J = 27$ level of I_2^*. Analysis of the data found $\tau_{rad} = 1.146$ μs for this level with $k_Q = 4.75 \times 10^6$ torr^{-1} s^{-1}. What I_2 pressure would give a fluorescence decay rate of 2.0×10^6 s^{-1}?

SECTION 27.2

27.16 Problem 27.15 quotes a quenching rate constant for $I_2^* + I_2$ collisions as $k_Q = 4.75 \times 10^6$ torr^{-1} s^{-1} at 300 K. What quenching cross section does this correspond to? If we assume the cross section is independent of temperature, what is the rate constant at 400 K?

27.17 The reaction $H_2^+ + H_2 \rightarrow H_3^+ + H$ has a rate constant 1.3×10^{12} M^{-1} s^{-1} while that for the analogous deuterium reaction is 9.6×10^{11} M^{-1} s^{-1}. What does collision theory predict for the *ratio* of these two rate constants? How closely does experiment agree?

27.18 Find the reaction cross sections for the following reactions given their experimental rate constants at the indicated temperatures.

(a) $Br + H_2 \rightarrow HBr + H$
$k_2 = 1.8 \times 10^4$ M^{-1} s^{-1} at 600 K

(b) $O + HBr \rightarrow Br + OH$
$k_2 = 4.9 \times 10^7$ M^{-1} s^{-1} at 350 K

(c) $Cl + O_3 \rightarrow ClO + O_2$
$k_2 = 7.4 \times 10^9$ M^{-1} s^{-1} at 300 K

(d) $Cl + CH_4 \rightarrow HCl + CH_3$
$k_2 = 8.8 \times 10^6$ M^{-1} s^{-1} at 200 K

Which of these reaction cross sections is closest to a hard-sphere collision cross section?

27.19 The reaction $He^+ + CO \rightarrow C^+ + O + He$ is an important CO destruction reaction in stellar and interstellar gases. It has an effective reaction distance that is quite large, 6.3 Å. What rate constant does this distance predict for $T = 300$ K?

27.20 If the H_2 hard-sphere radius is 1.47 Å and that for Br is 2.05 Å, (the value for atomic Kr), what is the steric factor for the $Br + H_2$ reaction for which the pre-exponential factor is 1.7×10^{11} M^{-1} s^{-1} at 500 K? In contrast, the pre-exponential factor for the reaction $Kr^+ + H_2 \rightarrow KrH^+ + H$ is 3.0×10^{11} M^{-1} s^{-1} at 300 K. What is the steric factor for this reaction?

27.21 Example 27.8 quotes a rate constant $k_{exp} = 8.2 \times 10^9$ M^{-1} s^{-1} for I atom recombination in CCl_4. What value would you predict based on the viscosity of CCl_4 quoted in Table 25.3? Equation (27.17) is based on equal hydrodynamic radii for the reactants, a condition guaranteed to be true here, and the example quoted $D = 4.2 \times 10^{-9}$ m^2 s^{-1} for I in CCl_4. Calculate the I atom's hydrodynamic radius from this number and compare it to the reaction radius R^* found in the example. Is your value chemically reasonable? Does it suggest significant strong solvation of I?

27.22 Derive an expression more appropriate than Eq. (27.17) for the diffusion-limited rate constant for the reaction between a very large molecule such as a protein and a very small but very reactive species such as the OH radical. Imagine the reaction occurs when the small species touches the larger one so that $R^* = R_A + R_B$. Let the large molecule be hemoglobin (see Example 27.9) and give the small molecule a 1.0 Å radius. What rate constant do you predict for reaction in water at 300 K? How much larger is your prediction than one based on Eq. (27.17) alone?

27.23 Another way to think about diffusion-limited reaction rates invokes the Einstein–Smoluchowski equation, Eq. (25.19), for the diffusion constant. This equation is based on a random walk picture of diffusion as a series of jumps of length x_0 every τ seconds: $D = x_0^2/2\tau$. Work this expression into Eq. (27.16) for the case of I atom recombination assuming $x_0 = R^*$, and find a value for τ using k_2 and R^* from Example 27.8.

27.24 Problem 27F quotes the rate constant 7.0×10^9 M^{-1} s^{-1} for $H + OH \rightarrow H_2O$ in aqueous solution, and Chapter 26 quotes $k_2 = 1.12 \times 10^{11}$ M^{-1} s^{-1} for $H^+ + OH^- \rightarrow H_2O$, both at 298 K. Is the electrostatic attractive force in the second reaction sufficient to account for the rate constant difference? The diffusion constants for H^+ and OH^- are 9.3×10^{-9} m^2 s^{-1} and 5.3×10^{-9} m^2 s^{-1}. Assume first that $R^* = 0.97$ Å for the ion neutralization reaction, the same value assumed in Problem 27F, and calculate k_2. You should find that the electrostatic factor f, Eq. (27.19), is about 7.4, which is not large enough to account for the observed difference between the two reactions. What value for R^* in Eq. (27.18) does give the observed rate constant? You should find a rather large R^* and a significantly smaller electrostatic factor. Can you qualitatively justify this R^* in terms of the molecular structures of H_2O and $H^+(aq)$, which is, of course, really H_3O^+? (*Hint:* Diffraction studies of water show approximate tetrahedral coordination around each O with O–O distances around 2.6 Å.)

27.25 A modern study of the $H^+ + OH^- \rightarrow H_2O$ reaction used a laser pulse to perturb the system from equilibrium, following the system's relaxation through changes in conductivity (W. C. Natzle and C. B. Moore, *J. Phys. Chem.* **89**, 2605 (1985)). The 8 ns laser pulse was tuned into the first vibrational overtone of water, creating a quick temperature jump as the excited molecules relaxed. These authors measured recombination rate constants at various temperatures, and they analyzed their data using Eq. (27.18) to find $R^* = 5.8 \pm 0.5$ Å, independent of temperature. The rate constant itself is strongly dependent on T, of course. At the temperature limits of the study, k_2 varied from 6.34×10^{10} M^{-1} s^{-1} at 1.1 °C to 1.41×10^{11} M^{-1} s^{-1} at 42.5 °C. Calculate k_2 at these two temperatures using $R^* = 5.8$ Å and the 298 K values for ϵ_r and the diffusion constants quoted in Table 9.1 and the previous problem. Do you conclude that the temperature dependence of ϵ_r and the diffusion constants are important to take into account?

27.26 The diffusion constant for OH radical in water is 2.6×10^{-9} m^2 s^{-1} at 298 K, and the rate constant for OH recombination to H_2O_2 is 5.0×10^9 M^{-1} s^{-1}. Find the reaction radius R^* for this reaction, and decide whether or not the reaction is diffusion limited.

27.27 Is a diffusion-limited reaction between uncharged species faster in water or in ethanol? Use Eq. (27.17) to find out. What about a reaction between ions of charges $+e$ and $-e$ in water versus in ethanol? Tack the electrostatic factor of Eq. (27.19) onto Eq. (27.17), assume $R^* = 5.0$ Å, and take $T = 293$ K. You will need data from Tables 9.1 and 25.3.

SECTION 27.3

27.28 The contour plot below represents the collinear potential energy surface for an atom A interacting with a diatomic BC. The contours are spaced in 50 kJ mol^{-1} intervals, R_1 is the A—B distance, and R_2 is the B—C distance. What is the BC bond length and bond energy? What is the AB bond length and bond energy? What is the potential energy and ABC structure at the transition state?

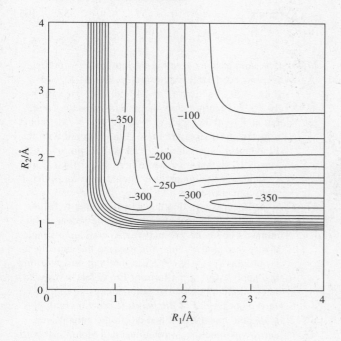

27.29 The collinear transition state for Br + H$_2$ → HBr + H has R_{Br-H} = 1.49 Å and R_{H-H} = 1.19 Å, and the transition state energy is about 83.9 kJ mol^{-1} above the energy of the isolated reactants. The H$_2$ bond energy is 458.4 kJ mol^{-1}, and the bond length is 0.741 Å. For HBr, the bond energy is 378.3 kJ mol^{-1}, and the bond length is 1.414 Å. Use these data to sketch a semi-quantitative potential energy contour map for the collinear Br—H—H system.

27.30 This and the next two problems build on the previous one toward a calculation of the ACT expression for the Br + H$_2$ → HBr + H rate constant. At 550 K, the experimental rate constant is 4.1 × 10^3 M^{-1} s^{-1} with about a 50% uncertainty. This is the number we will aim for, bit by bit. We start with the ACT expression neglecting rotational and vibrational partition function contributions. The energy at the transition state corrected for zero-point motion is $U_0^{\ddagger}$ = 1.29 × 10^{-19} J or 77.9 kJ mol^{-1}, and we will assume the electronic partition function factor equals 1, since the spin multiplicity of the reactants and the activated complex are the same and the temperature is low enough to neglect higher excited states. Calculate the contribution to k_2 from translational and $U_0^{\ddagger}$ components. You should find a number significantly lower than experiment, indicating the importance of internal degrees of freedom.

27.31 Continuing from the previous problem, we add rotational partition function contributions to k_2. The temperature is high enough that H$_2$ rotation is at the classical, high-temperature limit, and to save you a step, the expression in Table 23.2 gives q_{rot} = 3.14 at 550 K with the rotational constant from Table 19.2. Now we need $q_{rot}^{\ddagger}$, and for that we need the BrH$_2$ moment of inertia at the transition state, $I^{\ddagger}$, since $q_{rot}^{\ddagger}$ = $2I^{\ddagger}k_BT/\hbar^2$. In terms of atom masses and coordinates, $I^{\ddagger} = m_{Br}R_{Br}^2 + m_H R_{H1}^2 + m_H R_{H2}^2$ where the distances are measured from the center of mass:

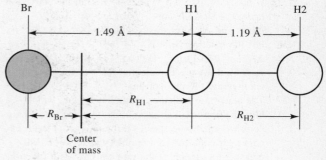

This is the transition state geometry mentioned in Problem 27.29, and the center of mass is 0.0513 Å from the Br atom. Calculate $I^{\ddagger}$, $q_{rot}^{\ddagger}$, and the rotational partition function factor to k_2. Multiply this factor and the previous problem's k_2 estimate, and you should find that nearly the entire rate constant is recovered.

27.32 Now we add vibration to the calculation running over the previous two problems. The H$_2$ vibrational partition function at 550 K is quite close to 1 due to the large H$_2$ vibrational constant. Calculations on the transition state indicate a symmetric stretch vibrational constant around 2340 cm^{-1} and a doubly degenerate bend constant around 500 cm^{-1}. Compute $q_{vib}^{\ddagger}$ from these constants, and complete the calculation of k_2. Does the calculation agree with experiment to within its uncertainty?

27.33 The experimental rate constant at 1000 K for H + H$_2$ → H$_2$ + H is 1.3 × 10^9 M^{-1} s^{-1} with an uncertainty around ±0.4 × 10^9 M^{-1} s^{-1}. Calculate the ACT rate constant at this temperature from the following data: H ··· H ··· H bond lengths of 0.93 Å at the linear transition state, H$_3^{\ddagger}$ frequencies of 2193 cm^{-1} (symmetric stretch) and 978 cm^{-1} (doubly degenerate bend), data on H$_2$ from Table 19.2, and $U_0^{\ddagger}$ = 36.1 kJ mol^{-1} or 5.99 × 10^{-20} J. Note that the

27.34 The reaction between H and HBr has two possible product channels, H atom abstraction to form H_2 + Br or H atom exchange. The exchange reaction is relatively unimportant since the potential energy barrier for abstraction is considerably lower. Nevertheless, exchange can be studied through isotopic substitution. What does ACT predict for the *ratio* of the rate constant for H + DBr $\rightarrow$ HBr + D to that for D + HBr $\rightarrow$ DBr + H? The transition state is linear, D $\cdots$ Br $\cdots$ H, but does this matter? Where have you seen this ratio before?

activated complex is a linear, symmetric triatomic so that the center of mass is at the central atom. This symmetry makes the moment of inertia particularly easy to calculate.

27.35 The pre-exponential factor for Br + H_2 $\rightarrow$ HBr + H is 1.7×10^{11} M^{-1} s^{-1} at 500 K, and the activation energy is 80.2 kJ mol^{-1}. What are $K_{eq}^{\ddagger}$, $\Delta \overline{G}^{\ddagger \circ}$, $\Delta \overline{H}^{\ddagger \circ}$, and $\Delta \overline{S}^{\ddagger \circ}$ for this reaction?

27.36 For the reaction Br + O_3 $\rightarrow$ BrO + O_2 at 300 K, $\Delta \overline{H}^{\ddagger \circ} = 330$ J mol^{-1} and $\Delta \overline{S}^{\ddagger \circ} = -102$ J mol^{-1} K^{-1}. Find the pre-exponential factor, the activation energy, and the rate constant.

27.37 The acid–base indicator bromphenol blue, which is a blue dianion in alkaline solution, slowly fades to a colorless trianionic species through reaction with excess hydroxide ion in a bimolecular reaction. The measured variation of the rate constant as a function of pressure at 298 K is tabulated below. (The "zero" pressure value is really for $P = 1$ atm, but the difference between 1 and zero atmospheres pressure is unimportant here.)

P/atm	0	272	544	816	1088
$k_2/10^{-3}$ M^{-1} s^{-1}	0.93	1.11	1.31	1.53	1.79

Find the molar activation volume from these data. You should find a negative number. Can you suggest a reason why it is negative?

27.38 The molar activation volume at 298 K for the dimerization of cyclopentadiene in *n*-butyl chloride solvent is -22.0 cm^3 mol^{-1}. What pressure doubles the 1 atm reaction rate?

27.39 Several unimolecular dissociations of the type $(CH_3)_3COOC(CH_3)_3 \rightarrow 2(CH_3)_3CO$ have been studied as a function of pressure in various solvents. These reactions typically have positive molar activation volumes due to the increased size of the transition state species as a bond is stretched before being broken. If you take the volume of a bond to be a cylinder of radius comparable to a typical atomic radius and imagine its length is stretched about 1 Å in the transition state, what $\Delta \overline{V}^{\ddagger \circ}$ do you predict for these types of reactions? For the example quoted here, $\Delta \overline{V}^{\ddagger \circ} = 6.7$ cm^3 mol^{-1}. How does your estimate compare?

SECTION 27.4

27.40 The data below are more of the Schneider and Rabinovitch data for CH_3NC isomerization mentioned in Example 27.12, but for 503.6 K here. What k_∞ do they predict? Use your value here and that found in Example 27.12 to predict k_∞ at 533.0 K assuming simple Arrhenius behavior, and compare your prediction to the measured value, 7.67×10^{-3} s^{-1}.

P/torr	3610	659	240	157	103
$k_{uni}/10^{-4}$ s^{-1}	9.06	7.74	6.82	6.84	5.70

27.41 The unimolecular reaction cyclopropane $\rightarrow$ propylene has been extensively studied. At 764 K, $k_\infty = 3.83 \times 10^{-4}$ s^{-1}. At $P = 84.1$ torr, $k_{uni} = 3.15 \times 10^{-4}$ s^{-1}, slightly below the infinite pressure value, but at $P = 0.067$ torr, $k_{uni} = 3.19 \times 10^{-5}$ s^{-1}, about a factor of 10 less. Are these data in accord with the Lindemann expression, Eq. (27.40)? A plot of $1/k_{uni}$ versus $1/P$ is not very useful here, but if Eq. (27.40) is valid, a single value of k_2 should hold for both measurements. Does it?

27.42 Experiment shows that the activation energy for the dissociation of a diatomic is always less than the diatomic's bond energy, and we can cook up a simple theory that predicts this. We imagine the unimolecular process $A_2 \rightarrow A_2^{\ddagger} \rightarrow 2A$ in the high-pressure limit and apply ACT in the form of Eq. (27.41). We take $U_0^{\ddagger} = \mathcal{D}_0$, the bond dissociation energy from the zero-point level. For the partition function ratio $q^{\ddagger}/q_{A_2}$, the translational and electronic parts cancel, and we assume the A_2 rotational and vibrational partition functions are at their classical limits: $q_{rot} = Ik_BT/\hbar^2$ and $q_{vib} = k_BT/\hbar\omega$. (The assumption for q_{vib} is justified because $\mathcal{D}_0 >> k_BT$ and the rate is appreciable only at very high temperatures.) The vibrational part of $q^{\ddagger}$ is the reaction coordinate that is removed in ACT, and the rotational part is tricky. The activated complex $A_2^{\ddagger}$ has very few closely spaced rotational states so that $q_{rot}^{\ddagger}$ is a constant at high temperature: all these states are equally populated. Put all this together to show that

$$k_2 = \frac{\text{constant}}{T} e^{-\mathcal{D}_0/k_BT},$$

identify the constant, and show that the activation energy is $E_a = \mathcal{D}_0 N_A - RT$.

SECTION 27.5

27.43 The activation energy for the uncatalyzed reaction $H_2O_2(aq) \rightleftarrows H_2O + \frac{1}{2}O_2(g)$ is 71 kJ mol^{-1}. Add HBr as a catalyst, and the activation energy falls to 50 kJ mol^{-1}. With the enzyme catalase, it falls to 8 kJ mol^{-1}. By what factors do HBr and catalase increase the forward reaction rate at 298 K? For the net reaction, $\Delta \overline{H}° = -94.66$ kJ mol^{-1}. Since a catalyst operates on the transition state, it affects the rate of the reverse reaction as well. What is the rate increase for the reverse net reaction with catalase? How does catalase affect the net reaction equilibrium constant?

27.44 If you start at the N-terminus end of the pancreatic enzyme α-chymotrypsin and walk along 195 amino acids, you reach a serine residue that plays a central role in the catalyzed hydrolysis of peptides. This is also the region of the enzyme attacked by several nerve gases. The details of the catalytic mechanism are described in a very readable article by R. M. Stroud (*Scientific American* **231**, 74 (1974)), and J. A. Hurlbut, *et al.*, describe an undergraduate laboratory experiment on its effect on the hydrolysis of the amide bond in the substrate *N*-glutaryl-L-phenylalanine-*p*-nitroanalide (*J. Chem. Ed.* **50**, 149, (1973)). These authors report the typical experimental data tabulated below for reactions run with a total enzyme concentration $[E]_0 = 4.0$ μM.

[S]/mM	0.25	0.50	1.0	1.5
$\mathcal{R}$/10^{-8} M s^{-1}	3.7	6.3	9.8	11.8

Use both the Lineweaver-Burke and Eadie-Hofstee plots to find V_{max}, K_M, and k_1 from these data.

27.45 Another way of analyzing enzyme kinetic data plots [S]/$\mathcal{R}$ versus [S] (called a *Dixon* or a *Woolf* plot). Show that this also leads to a straight line, and relate the line's slope and intercept to K_M and V_{max}. Plot the data from the previous problem this way, and find K_M and V_{max}.

27.46 Molecules that bind to the active site of an enzyme but do not react are called inhibitors, and several mechanisms for inhibition are known. The simplest is called *competitive inhibition:* inhibitor I and substrate S compete for the enzyme's active site:

$$E + S \rightleftharpoons ES \xrightarrow{k_1} E + P$$
$$E + I \rightleftharpoons EI$$

Let $K_I = [E][I]/[EI]$, the dissociation equilibrium constant for the enzyme–inhibitor complex and let $K_M = [E][S]/[ES]$, the dissociation constant for the enzyme–substrate complex. (How does this K_M differ from the Michaelis constant definition in the text?) Use $[E]_0 = [E] + [ES] + [EI]$ to show that the rate of product formation is

$$\mathcal{R} = \frac{k_1 [E]_0 [S]}{K_M (1 + [I]/K_I) + [S]}.$$

27.47 If an adsorbed molecule dissociates on the surface, the rate of adsorption is proportional to the pressure, the *square* of the number of unoccupied sites, and an adsorption rate constant k_a. Similarly, the desorption rate is proportional to a desorption rate constant k_d times the square of the number of occupied sites. Show that under these conditions, the equilibrium surface coverage can be written $\theta_{eq} = (K_{ad}P)^{1/2}/[1 + (K_{ad}P)^{1/2}]$. In Example 27.14, we saw that $\theta_{eq} = 0.5$ when $P = 1/K_{ad}$ if there is no dissociation. What pressure gives this coverage if there is dissociation?

27.48 Equilibrium adsorption studies of N$_2$ on 1 g of carbon powder at 0 °C and various N$_2$ pressures gave the data below (Zeise, *Zeit. physik. Chem.* **136**, 410 (1928)) with adsorbed gas volumes corrected to 298 K and 1 bar. Find the Langmuir adsorption isotherm parameters for this system, and find the range of equilibrium coverage these data span.

P/torr	V/10^{-7} m^3
0.43	1.21
1.21	3.25
3.98	10.8
12.98	33.2
22.94	55.5
34.01	76.9
56.23	113.
77.46	142.

27.49 Derive an expression for the rate of reaction between two species that only react when *both* are adsorbed to a surface so that the rate is proportional to the surface coverage of each. Assume the Langmuir isotherm describes the adsorption of each species, but take into account the competition between them for unoccupied sites.

GENERAL PROBLEMS

27.50 The derivation of the electrostatic factor for diffusion-limited rate constants, Eq. (27.19), starts with the introduction of diffusion driven by the directed electrostatic force between ions into Fick's first law of diffusion. This problem considers how that is done. We start with the electrostatic potential energy between ions A and B of charges $z_A e$ and $z_B e$ in a solvent of relative permittivity ϵ_r: $U(r) = z_A z_B e^2/4\pi\epsilon_0\epsilon_r r$. The speed v_i of an ion in an electric field $E(r)$ is related to its mobility u_i through $v_i = u_i E(r)$, the field is the force per unit charge: $E(r) = F(r)/e$, and the force is the negative derivative of the potential energy: $F(r) = -dU(r)/dr$. The flux is the product of the ion speed and density: $J^e_{i,r} = v_i n_i$. Incorporate the Einstein mobility-diffusion relation, Eq. (25.40), and show that

$$J^e_{i,r} = -D\left(\frac{n_i}{k_B T}\frac{dU}{dr}\right).$$

This term is added to the random thermal diffusion term of Eq. (27.12), and the derivation continues as discussed in the text. Mathematical details of the remaining steps are in this chapter's references.

27.51 In a mass spectrometer, a molecular beam is ionized by electron bombardment. The fraction of molecules that is ionized depends on the following four quantities:

(a) the speed of the molecules in the beam,
(b) the length of the molecular beam that is bombarded,
(c) the flux of electrons, and
(d) the ionization cross section.

A supersonic beam of H_2 is generated at a source temperature of 300 K under conditions that cause the most probable beam speed to be 1.2 times the most probable speed of the gas in the source. One in 10^3 of the molecules in this beam are ionized under a flux of 10^{18} electrons cm^{-2} s^{-1} when the electrons cross a 1.5 cm length of the beam. What is the ionization cross section? (*Hint:* Combine (a)–(d) above into an expression for the ionization fraction that is dimensionless and makes physical sense.)

27.52 Derive an approximate expression for the steric factor of collision theory starting from the ACT expression for the rate constant. Assume for simplicity an A + BC system with all atoms of equal mass and no activation energy. Assume as well that the BC equilibrium bond length, R_e, is the same as the A—B and B—C bond lengths in a linear A $\cdots$ B $\cdots$ C transition state, that there is no electronic partition function factor, and that the temperature is such that the vibration partition functions of BC and ABC‡ can be ignored but rotations are at their classical limit. Manipulate your expression into the form of Eq. (27.9) with πR_e^2 representing the collision cross section. You should find $p = 3\hbar^2/mR_e^2 k_B T$ where m is the atomic mass. How is this related to the rotational partition function for BC?

27.53 One often sees potential energy surfaces or contour maps plotted for collinear triatomic systems, as in Figures 27.4–27.6, but there is another configuration for which these plots can be drawn easily: the approach of an atom along the perpendicular bisector of a homonuclear diatomic. To get a feeling for the potential surface in this geometry, consider three hard spheres of radius r_{hs} in the coordinate system below in which atom A strikes diatomic B_2.

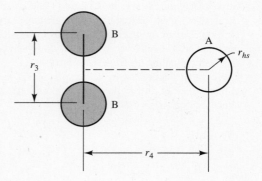

This geometry is not amenable to reactive collisions, but it is useful for energy transfer collision arguments. Sketch the hard-sphere repulsive wall in a coordinate system in which r_4 is the "x axis" and r_3 is the "y axis." Note that the potential is zero if $r_3 \geq 2r_{hs}$ and infinite if $r_3 < 2r_{hs}$. What is the minimum value for r_4 when $r_3 = 2r_{hs}$? What is a general expression for r_4 at any r_3 such that all three spheres are touching? (*Hint:* Note that r_4 can be zero if $r_3 \geq 4r_{hs}$. What does $r_4 = 0$ correspond to?)

CHAPTER 28

Chemical Reaction Dynamics

IMAGINE a three-dimensional game of pocket billiards. Some of the target balls are grouped together by springs, these groups are spinning in space, the cue ball itself might be part of a group, the cue stick wobbles with Heisenberg uncertainty, and the balls move at hundreds of meters per second. This is the game played by reaction dynamicists in the gas phase. The target group represents one reactant molecule rotating, vibrating, and translating through space, while the cue-ball group is the co-reactant. Unlike ordinary billiards, pieces of the cue-ball group might stick to a target group or exchange a ball or two with it. To win the game, one must figure out how to aim a cue ball group at a target group so that the desired exchange reaction is carried out and the product group ends in a particular target pocket, representing product formation in a particular quantum state.

We are now asking the most detailed questions possible about a chemical reaction, questions we first posed in Chapter 26 and began to answer in Chapter 27. We will see in this chapter how modern experiments work toward answers to these questions, and along the way, we will discuss methods that probe intermolecular interactions through analysis of molecular *scattering* patterns. If you throw one ball (the probe) at another ball (the target) and record only the *scattering direction* of the probe ball, after many throws you expect the pattern of these directions to be different from the pattern you find if the target

28.1	Where Does Chemical Energy Come From and Where Does It Go?
28.2	State-to-State Kinetics
28.3	Models for Elementary Reaction Dynamics
28.4	Is There Microscopic Control Over Reactions?

is a cube. Once you learn to recognize the two types of patterns as signatures of ball–ball or ball–cube scattering, you can throw the same probe repeatedly at a target of unknown shape and deduce its shape from the scattering pattern. This is the idea behind molecular beam scattering experiments designed to deduce *intermolecular forces*. As we have stressed in earlier chapters, the "size and shape" of a molecule is measured through its interactions with other molecules, and we have used this idea to discuss everything from the second virial coefficient of a nonideal gas in Chapter 1 to the reaction cross section and steric factor in Chapter 27.

A related concern beyond size and shape is the *energy requirements and consequences* of a reactive collision. A probe ball thrown very fast toward a target might scatter in a very different way from one thrown slowly but filled with a substance that explodes on contact. The fast throw concentrates the collision energy in *relative translational motion,* while the explosive probe concentrates energy in *internal degrees of freedom*. If our goal is a collision that causes the target to rupture (as in a chemical reaction), one type of energy concentration may be more effective than the other.

We start this chapter with an introduction to this question of energy requirements in reactive collisions. We consider the ways energy can be carried to a collision and the ways it appears in the collision products, whether the collision is reactive or not. Most interesting is the way we can distribute a given total energy among different degrees of freedom of the reactants and alter the collision outcome. Next, we consider the kinetic language needed to discuss molecular collisions in which we specify the initial state of the reactants and the final state of the products in as great a detail as quantum mechanics allows. Then, we consider simple collision models for reactive and nonreactive collisions. We will see that classical mechanics suffices to describe the motion of the colliding molecules, and that we need only classical Newtonian mechanics—the conservation of linear and angular momentum and energy—to understand the rudiments of these collisions. We will see how scattering patterns can be measured and interpreted to reveal intermolecular forces in some detail. In the final section, we consider ways to apply the ideas of the previous sections in attempts to direct the outcomes of reactive collisions. This is a tall order. The practical implications are potentially very big, comparable to the discovery of catalysts that also direct reactions, but the difficulties of controlling a reaction in any bulk near-equilibrium state are enormous.

28.1 WHERE DOES CHEMICAL ENERGY COME FROM AND WHERE DOES IT GO?

Thermodynamics measures chemical reaction energies through a few main techniques. For example, calorimetry measures enthalpy or internal energy differences between the equilibrium states of unmixed reactants and the final reaction mixture. Chemical kinetics, through the activation energy, measures a heavily averaged energy requirement for the collisions that lead to reaction. Quantum mechanics and spectroscopy measure the energy states allowed to individual molecules, and statistical mechanics links these energies to the energy of macroscopic samples at or very near to equilibrium.

The preceding paragraph is, roughly, a concise overview of all the chapters in this book except for this one. Here, we consider the way colliding molecules bring energy into a potentially reactive collision event and the fate of that energy. We consider the energy carried from the collision by the collision products, which may not be chemically different from the colliding reactants, and we begin to consider the *distribution* of the total collision energy among the various internal degrees of freedom of the products. The discussion here will be largely qualitative and pictorial. In later sections, we will derive some elementary models that have quantitative validity, but for now, we need to state the problem, develop some language, and discover some phenomena.

We begin with the collinear H + H$_2$ reaction system we introduced in Chapter 27. In our discussion there, we emphasized the *classical reaction coordinate,* the dashed line in Figure 27.6 that connects reactants to products along the path of lowest potential energy. No reaction follows this path exactly, and there are several reasons why. First, no atom–diatom collision is exactly collinear. (Quantum mechanics precludes this, as does classical mechanics, for reasons we will discover in Section 28.2.) Nevertheless, study of collinear collisions is simple to do and sufficiently realistic with classical mechanics to reveal the phenomena we need. Second, quantum mechanics demands *zero-point vibrational motion,* and any vibrational motion in either the diatomic reactant or the product produces a classical oscillatory motion of the system about the classical reaction coordinate path.

We can overlay a collinear potential-energy contour map with a plot of the two distances R_1 and R_2 (defined in Figures 27.4–27.6) that characterize a collinear collision, each calculated as a parametric function of time. Such a plot is called the *classical trajectory* of the collision. Classical trajectories reveal much of the dynamics of molecular collisions, and we will use them throughout this chapter to illustrate various phenomena. Figure 28.1 shows such a trajectory for the H + H$_2$ system, overlaid on the potential contours of Figure 27.5. The arrows along the trajectory point in the direction of increasing time as the collision evolves. As the H atom reactant approaches the closer end of the H$_2$ diatomic, coordinate R_1 is decreasing. The diatomic, however, is vibrating with its zero-point motion, and thus the R_2

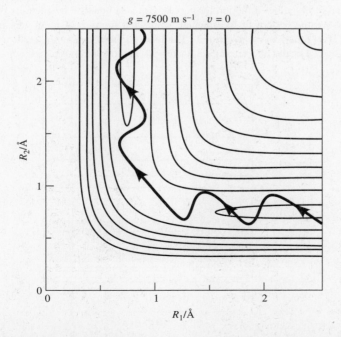

FIGURE 28.1 A classical trajectory for a collinear collision between H and H$_2$ plots the two interatomic distances R_1 and R_2 as parametric functions of time on the reaction's intermolecular potential-energy function, represented as a contour plot. Here we see a reactive collision for H hitting ground vibrational state H$_2$ with a relative speed $g = 7500$ m s^{-1}.

coordinate oscillates initially. Consequently, the trajectory snakes its way down the entrance channel of the potential energy surface toward the transition state. For the particular conditions of this collision, the total energy is great enough to surmount the transition-state potential-energy barrier, and the trajectory shows that these conditions allow the barrier to be crossed. As we will see, a total energy greater than the potential barrier energy is a necessary but *not* sufficient condition to ensure that the collision leads to reaction.

As this trajectory crosses the transition-state region, it happens to follow closely the classical reaction coordinate: R_1 is decreasing at about the same rate as R_2 is increasing, following the motion we denoted with a coordinate q_1 in Figure 27.6. Reaction results as the trajectory continues out the potential-surface exit channel, leading to an H_2 product vibrating with its zero-point motion. The trajectory tells us that the product is not vibrationally excited: note how the amplitude of the oscillations in the trajectory *along the R_2 direction in the entrance channel* are the same as the amplitude of the oscillations *along the R_1 direction in the exit channel*. *The greater the vibrational energy in a reactant or product, the greater the amplitude of the trajectory oscillations.*

EXAMPLE 28.1

What areas of the $H + H_2$ potential surface's entrance channel are allowed classically for an H_2 reactant in: (a) the $v = 0$ ground vibrational state, (b) the $v = 1$ first excited vibrational state?

SOLUTION Using the H_2 vibrational constants in Table 19.2 ($\omega_e = 4401.213$ cm^{-1} and $\omega_e x_e = 121.336$ cm^{-1}) and Eq. (19.26) for the vibrational energy of a diatomic, we can calculate that the H_2 zero-point energy ($v = 0$) is

$$G(0) = \frac{\omega_e}{2} - \frac{\omega_e x_e}{4} = 2170 \text{ cm}^{-1} = 26.0 \text{ kJ mol}^{-1} \ .$$

Similarly, the $v = 1$ energy is

$$G(1) = \frac{3\omega_e}{2} - \frac{9\omega_e x_e}{4} = 6329 \text{ cm}^{-1} = 75.7 \text{ kJ mol}^{-1} \ .$$

In the entrance channel, R_1 is large, and potential energy contours are parallel to the R_1 axis. The classically allowed motion for $v = 0$ is an area bounded by contours that are 26.0 kJ mol^{-1} above the bottom of the channel, while that for $v = 1$ is an area bounded by contours 75.7 kJ mol^{-1} above the bottom. The lightly shaded area in the contour plot below represents the classically allowed region for $v = 0$, and the darkly shaded region is that for

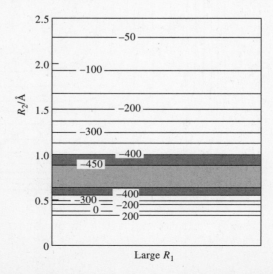

$v = 1$. (The bottom of the channel has a potential energy -477.4 kJ mol^{-1}, the negative of the H$_2$ Morse dissociation energy as discussed in Section 19.4.)

⟹ RELATED PROBLEMS 28.3, 28.7, 28.9

The initial conditions of the trajectory in Figure 28.1 were chosen to illustrate a collision that smoothly traverses the transition-state region and leads to an H$_2$ product with internal energy equal to that of the H$_2$ reactant. To define a collision's initial state classically, we specify the initial positions of the collision partners (at some initial separation so large that they do not interact) and their vector momenta. We express momenta in terms of masses and velocities, and while we can always express positions in terms of *laboratory coordinates,* the coordinates of a reference frame in which we sit motionless and watch the collision, two other coordinate systems are useful in discussions of collisions. These are the *relative* and *center-of-mass* coordinate systems.

If masses m_1 and m_2 are at laboratory position vectors r_1 and r_2, we define their *relative position vector* r through

$$r = r_1 - r_2 , \tag{28.1}$$

their *total mass* M through

$$M = m_1 + m_2 , \tag{28.2}$$

their *reduced mass* μ through

$$\mu = \frac{m_1 m_2}{M} , \tag{28.3}$$

and the position R of the *center of mass* of the collision through

$$MR = m_1 r_1 + m_2 r_2 . \tag{28.4}$$

A little algebra discussed in Problem 28.1 shows that we can write the total collision kinetic energy T as

$$T = \frac{1}{2} m_1 \left(\frac{dr_1}{dt} \right)^2 + \frac{1}{2} m_2 \left(\frac{dr_2}{dt} \right)^2 = \frac{1}{2} M \left(\frac{dR}{dt} \right)^2 + \frac{1}{2} \mu \left(\frac{dr}{dt} \right)^2 \tag{28.5}$$

where we have expressed velocities as explicit time derivatives of coordinates.[1]

One can also show that the center of mass of the system moves with a constant velocity throughout the collision (see Problem 28.2) so that the term $\frac{1}{2} M (dR/dt)^2$ is a constant. The more interesting term is $\frac{1}{2} \mu (dr/dt)^2 = \frac{1}{2} \mu g^2$ where $g = (dr/dt) = v_1 - v_2$ is the *relative velocity* and g is its magnitude, the relative collision speed introduced in Section 24.2. This term represents the *relative translational energy* E_{rel} of the collision:

$$E_{\text{rel}} = \frac{1}{2} \mu \left(\frac{dr}{dt} \right)^2 = \frac{1}{2} \mu g^2 , \tag{28.6}$$

and if we specify it at a point where r is so large that the intermolecular potential energy is zero, E_{rel} represents the *translational energy available to the collision.* The energy tied up in the uniform motion of the center of mass through space is *not* available to the collision, since it must remain a constant throughout the collision.

---

[1] Because velocities are vector quantities, the *square* of a vector, as in $(dr_1/dt)^2$, represents the *vector dot product* of the velocity with itself. See the discussion leading to Eq. (24.4). For collinear collisions, we will be concerned more with speeds than vector velocities.

For the trajectory plotted in Figure 28.1, the H_2 reactant was stationary while the H atom moved collinearly toward it with a speed $dr_1/dt = v_1 = 7500$ m s^{-1}. Since H_2 was stationary at the start of the collision, $v_{H_2} = 0$ and $g = v_H$. Thus, the relative collision energy was

$$E_{rel} = \frac{1}{2}\mu g^2 = \frac{1}{2}\left(\frac{m_H m_{H_2}}{m_H + m_{H_2}}\right)(7500 \text{ m s}^{-1})^2 = 3.14 \times 10^{-20} \text{ J} = 18.9 \text{ kJ mol}^{-1} \ .$$

From Chapter 27's discussion of the H_3 LEPS potential-energy surface, we learned that the barrier at the transition state is about 34 kJ mol^{-1} high. Thus, 18.9 kJ mol^{-1} of collision energy in relative translation is *not* sufficient to surmount the barrier all by itself. The remaining collision energy is in the vibrational motion of H_2. Since the 34 kJ mol^{-1} barrier height is measured from the bottom of the entrance channel to the top of the transition state barrier, the zero-point vibrational energy of the reactant is available (classically) to the collision. Example 28.1 found that this energy for H_2 is 26.0 kJ mol^{-1}. This is the *internal* energy of the collision, and when we add it to E_{rel}, we find that the *total* collision energy is greater than 34 kJ mol^{-1} and thus sufficient to exceed the transition-state barrier along the reaction coordinate:

$$E_{tot} = E_{rel} + E_{int} = 18.9 \text{ kJ mol}^{-1} + 26.0 \text{ kJ mol}^{-1} = 44.9 \text{ kJ mol}^{-1} \ .$$

We can always decompose the total laboratory energy of a collision into an internal energy component (that includes all internal degrees of freedom, not just vibration), a relative translational energy component, and a center-of-mass motion component.

EXAMPLE 28.2

What is the total laboratory energy of the collision in Figure 28.1? What fraction of that energy is tied up in the center-of-mass motion and thus unavailable to the collision? How would this fraction change for the collision $D + H_2(v = 0)$ with g still 7500 m s^{-1}?

SOLUTION The total energy in the laboratory frame of reference (with $H_2(v = 0)$ stationary and H moving at 7500 m s^{-1}) is the H translational kinetic energy:

$$E_{tr} = \frac{1}{2}m_H v_H^2 = \frac{(1.67 \times 10^{-27} \text{ kg})(7500 \text{ m s}^{-1})^2}{2}$$

$$= 4.71 \times 10^{-20} \text{ J} = 28.3 \text{ kJ mol}^{-1}$$

plus the 26.0 kJ mol^{-1} zero-point internal energy of the H_2, for a total laboratory energy $E_{lab} = 54.3$ kJ mol^{-1}.

We can find the speed of the center-of-mass of the collision, v_{cm}, if we differentiate Eq. (28.4) with respect to time:

$$\frac{dR}{dt} = v_{cm} = \frac{m_1}{M}\frac{dr_1}{dt} + \frac{m_2}{M}\frac{dr_2}{dt} = \frac{m_1}{M}v_1 + \frac{m_2}{M}v_2 \ ,$$

and with $m_1 = m_H$, $M = 3m_H$, $v_1 = v_H$, and $v_2 = v_{H_2} = 0$, we find

$$v_{cm} = \frac{m_H}{M} v_H = \frac{1}{3} v_H = 2500 \text{ m s}^{-1} ,$$

and thus the energy tied up in motion of the center of mass is

$$\frac{1}{2} M v_{cm}^2 = \frac{(5.02 \times 10^{-27} \text{ kg})(2500 \text{ m s}^{-1})^2}{2} = 1.57 \times 10^{-20} \text{ J} = 9.45 \text{ kJ mol}^{-1} .$$

Note that the total laboratory energy, 54.3 kJ mol^{-1}, equals the sum of the center-of-mass motion energy, 9.45 kJ mol^{-1}, and the available total collision energy, $E_{tot} = 44.9$ kJ mol^{-1}. The fraction of the total laboratory energy available to the collision is $44.9/54.3 = 0.83$.

If the reactant atom is changed from H to D, the laboratory translational energy is twice that of H moving at the same speed (or 56.6 kJ mol^{-1}), the H$_2$ internal energy is unchanged, and thus the total lab energy is $E_{lab} = 82.6$ kJ mol^{-1}. The center-of-mass velocity is now $v_{cm} = (m_D/M)v_D = v_D/2 = 3750$ m s^{-1}, and the energy of center-of-mass motion is $Mv_{cm}^2/2 = 28.3$ kJ mol^{-1}. The relative speed is still 7500 m s^{-1}, and the relative collision energy is

$$E_{rel} = \frac{1}{2} \mu g^2 = \frac{1}{2} \left(\frac{m_D m_{H_2}}{m_D + m_{H_2}} \right) (7500 \text{ m s}^{-1})^2 = 4.70 \times 10^{-20} \text{ J} = 28.3 \text{ kJ mol}^{-1} .$$

Note that $E_{rel} = Mv_{cm}^2/2$ here, but only because of the chance ratio of collision masses. The total collision energy is $E_{rel} + E_{int} = 28.3$ kJ mol^{-1} + 26.0 kJ mol^{-1} = 54.3 kJ mol^{-1} so that the fraction of the total laboratory energy available to the collision is only $54.3/82.6 = 0.66$. The heavy D atom ties up a greater portion of the laboratory energy in center-of-mass motion than does the light H atom.

RELATED PROBLEMS 28.1, 28.3, 28.9

We now ask how the outcome of the collision changes if we change either the speed of the attacking H atom or the internal energy of the target H$_2$. Suppose we increase only the internal energy, raising the H$_2$ vibrational quantum number from 0 to 1. The available collision energy is now the same 18.9 kJ mol^{-1} relative translational energy plus (see Example 28.1) 75.7 kJ mol^{-1} internal energy, or 94.6 kJ mol^{-1} total, far more than needed to exceed the transition-state barrier. Figure 28.2 plots the trajectory in this case, but we see that the collision is *nonreactive*: the transition state region is crossed, but the trajectory is reflected back into the entrance channel. This is an example of a collision that activated-complex theory assumes is so rare it can be neglected, as discussed in Chapter 27, and such collisions contribute to the ACT transmission coefficient κ discussed there.

Note in Figure 28.2 that the trajectory oscillates in the entrance channel along the R_2 direction with a greater amplitude on the way in than on the way out. This indicates that while the collision did not lead to reaction, it did lead to *energy transfer*. The H$_2$ internal energy *after* the collision is *less* than that *before* the collision. We have found a trajectory representing the *vibrational deactivation* process

$$H + H_2(v = 1) \rightarrow H + H_2(v = 0) .$$

Where did the internal energy of H$_2$ go? Since the center-of-mass motion cannot change, *any change in internal energy must appear as a change in relative translational energy in a nonreactive collision*. The lost vibrational quantum of energy

FIGURE 28.2 In contrast to the reactive collision of Figure 28.1, we see here a *nonreactive* collision at the same relative speed but with H_2 in its first excited vibrational state.

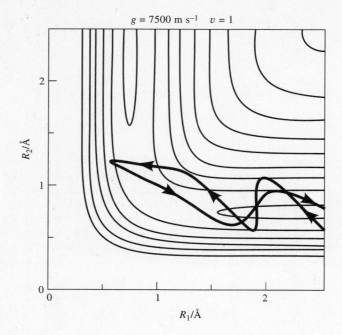

(49.7 kJ mol^{-1} from Example 28.1) increases E_{rel} after the collision. We will add a prime to quantities that refer to conditions after a collision so that here

$$E'_{rel} = E_{rel} + \Delta E_{int} = 18.9 \text{ kJ mol}^{-1} + 49.7 \text{ kJ mol}^{-1} = 68.6 \text{ kJ mol}^{-1}\ .$$

We can turn this number into a final relative speed g':

$$E'_{rel} = \frac{1}{2}\mu g'^2 \quad \text{or} \quad g' = \sqrt{\frac{2E'_{rel}}{\mu}} = 1.43 \times 10^4 \text{ m s}^{-1}\ .$$

This is nearly twice the initial relative speed, $g = 7500$ m s^{-1}.

Suppose we keep H_2 in $v = 0$ but increase the relative translational energy by increasing g from 7500 m s^{-1} to 11 200 m s^{-1}. This increases E_{rel} from 18.9 kJ mol^{-1} to 42.1 kJ mol^{-1}, and Figure 28.3 shows the trajectory. We do achieve reaction, but compare the exit channels of Figures 28.1 and 28.3. The H_2 product in Figure 28.3 is *vibrationally excited,* as we can tell from the amplitude of the trajectory oscillations along the R_1 direction.[2] The close spacing of the wiggles in the exit channel reflect a transfer of energy from relative translation to internal product excitation: the products move away from each other slower than they moved toward each other at the start of the collision.

We can, of course, increase both translational and internal energies. Figure 28.4 shows a trajectory in which H_2 has $v = 1$ as in Figure 28.2 and $g = 11\ 200$ m s^{-1} as in Figure 28.3. The result is reminiscent of Figure 28.1: smooth passage over the transition-state barrier leading to H_2 product with the same internal excitation as the H_2 reactant. Taken together, Figures 28.1–28.4 lead to an important conclusion:

---

[2] A close comparison of Figures 28.1–28.3 will show you that the exit-channel trajectory of Figure 28.3 has an oscillation amplitude along R_1 that is in between that of H_2 in $v = 0$ and H_2 in $v = 1$. This cannot happen, of course, because there are no fractional quantum numbers for vibration. But these are classical mechanics calculations, and a quantum interpretation of them would attribute a trajectory such as in Figure 28.3 to a collision that had some probability of leading to H_2 product in $v = 0$ and some probability for $v = 1$.

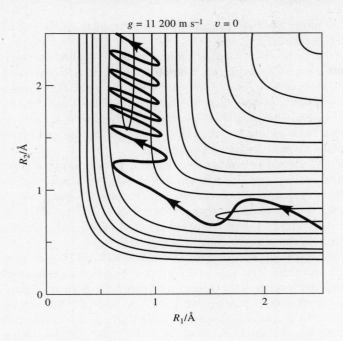

FIGURE 28.3 A reactive trajectory at a higher relative speed but with H_2 in $v = 0$ is shown here. From the extent of the trajectory along the R_1 direction in the exit channel, we can tell that this reaction leads to a vibrationally excited product.

the distribution of energy over collision product degrees of freedom is not simply related to the distribution over reactant degrees of freedom. This fact adds to the richness of chemical reaction dynamics and makes its study full of surprises. Nevertheless, we can establish collision distributions experimentally (using techniques discussed in previous chapters such as supersonic molecular beam tricks, selective laser excitation of internal motions, and so forth) and measure product distributions (using time-of-flight methods, laser-induced or spontaneous fluores-

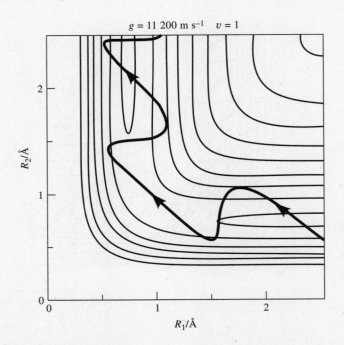

FIGURE 28.4 At very high total collision energy—enhanced translational and internal energies—collisions may still progress smoothly to products. Contrast this trajectory to the much lower total energy trajectory of Figure 28.1.

cence, and so forth). We will return to these experimental methods in more detail in Sections 28.2 and 28.3.

We close this section with a look at a trajectory that is qualitatively very different from any of the others we have seen. Computer techniques for generating classical trajectories have been used for over 25 years, and early in their development, it was recognized (primarily by the group of John Polanyi at the University of Toronto) that *model potential-energy surfaces*—surfaces that had representative features of real system surfaces but did not necessarily represent any one real system—could be used to hone our reaction dynamics intuition.

Our $H + H_2$ surface is particularly simple: it is thermoneutral with a simple barrier at the transition state. Suppose we alter the LEPS parameters of that surface so that we turn the barrier into a *transition-state potential-energy well*, along the lines of the reaction-coordinate energy profile in Figure 27.7(b). The general LEPS formalism allows one to do this easily enough, and the consequences this change makes on the collision are surprising.

Figure 28.5 shows a typical trajectory calculated with the $H + H_2$ masses on an LEPS surface with a transition-state well. Note how the trajectory enters the transition state smoothly, but once there, it spends a considerable amount of time deciding what to do. The transition state potential well has *held the collision partners in close proximity for a long time* (measured in terms of a characteristic vibrational period's heartbeat). This is characteristic of a *long-lived collision complex*, a true activated complex that lives long enough to be considered more than a fleetingly transient species. The other trajectories we have seen in this chapter crossed the transition state twice at most. They are said to be representative of *direct collisions*. The trajectory in Figure 28.5 is so snarled in the transition state region that it is difficult to tell which way time flows in a picture of the full trajectory. All three atoms vibrate in close proximity for many vibrational periods of all the vibrational modes of the activated complex.

Once we have a collision system capable of forming a long-lived collision complex, we can have greater confidence in *statistical theories* of reaction dynamics.

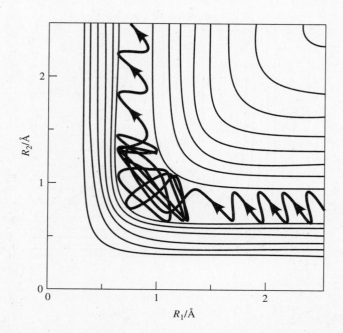

FIGURE 28.5 If the potential-energy surface has a well rather than a barrier at the transition state, it is possible that the trajectory spends a long time in the transition state region. This is the characteristic of *collisional complex* formation.

Long-lived complexes are formed, evolve over a long period of time, and effectively scramble the details of the energy distribution of the reactants. They explore all degrees of freedom of the activated complex, and the outcome of the reaction is best described as the *unimolecular decomposition* of the activated complex from some total energy state. Theories such as RRKM theory mentioned in Section 27.4 are particularly effective in these situations.

28.2 STATE-TO-STATE KINETICS

The previous section taught us how to decompose collision energy into internal and relative translational components and how to interpret reactive and nonreactive classical trajectory plots for collinear collisions. Here, we extend those ideas in two ways. We move into the real world of three dimensions, and we consider the *direction* of a scattering event as well as its energy. We also see how one probes single-collision dynamics experimentally.

Our discussion of chemical reaction kinetics in Chapters 26 and 27 focused on the reaction rate constant for a *bulk* system always so near to equilibrium that we could consider it a function of temperature alone. In contrast, reaction dynamics considers what is called the *state-to-state rate constant* (or state-to-state collision cross section—compare the discussion of the thermal rate constant and the thermal reaction cross section following Eq. (27.8) in the previous chapter). While a thermal reaction cross section expresses something about the reaction probability for a set of reactants in a particular thermally averaged energy distribution, a state-to-state cross section expresses reaction probability in terms of *unique quantum states of the reactants leading to unique quantum states of the products*. For example, the trajectory in Figure 28.1 refers to the system H atom in its ground state colliding collinearly with a relative speed of 7500 m s^{-1} with an H_2 that is also in its ground state, leading to a reaction in which the products are also in their ground states. It led to reaction, but to get a true state-to-state probability that such a collision is reactive, we would have to average over many trajectories starting from various initial H_2 vibrational *phases*, since not *every* H + $H_2(v = 0)$, $g = 7500$ m s^{-1} collinear collision leads to reaction.

Elastic Scattering

We will start with the simplest kind of scattering in three dimensions: *elastic, nonreactive scattering*. Collisions between rare gas atoms at low energies are elastic, for example. We will see how measurements of scattering directions allow us to find the *intermolecular potential-energy function* that governs the scattering. Also, since these are elastic collisions (they do not exchange energy between internal and translational degrees of freedom), they are particularly easy to analyze.

Experiments of this type are usually performed in a *crossed molecular beam* apparatus. Two molecular beams of known speed are directed through a vacuum so that they cross in a small collision volume. The beams usually move at right angles to each other, and they are so dilute and narrow that only about 1 in 100 atoms in one beam will strike an atom in the other. A particle detector such as a mass spectrometer is scanned in angle about the center of the collision volume, recording the intensity of the scattering as a function of the angle from one beam. While this geometry is experimentally convenient, it is slightly more difficult to analyze than our collinear systems in the previous section. The extra difficulties are covered in this chapter's problems, but for this section, we will assume that one atom (call it A) is moving in the lab towards a second atom (B) that is *stationary* in the lab.

FIGURE 28.6 There are three important dynamic frames of reference in which we can profitably discuss a collision event. Here, we see the same collision in: (a) the laboratory frame, in which A moves toward a stationary B target; (b) the center-of-mass frame, in which the collision evolves as if we were moving at the velocity of the collision's center of mass; and (c), the effective relative frame, in which we imagine a single particle with a mass equal to the real collision's reduced mass μ collides with a fixed point (located by a cross) from which the collision force originates. Note that the laboratory scattering angle θ is in general different from the center-of-mass or relative scattering angle χ.

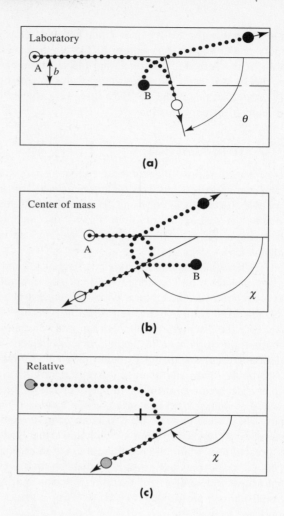

In this scenario, the lab picture of the collision starts with A far from B but moving toward it with some laboratory velocity v_A. This velocity equals the relative velocity g, since B is stationary. If A and B were both points that did not interact at all, A would move in a straight line at a constant speed. If we draw a line through the center of B parallel to the line A would follow in the absence of interaction, we can define a very important collision parameter, the *impact parameter b*. This quantity (also called the *aiming error*) is the perpendicular distance between the straight line path of A and the line through B parallel to this path. When the impact parameter is zero, the collision is "head on," as it was in our collinear collisions in the last section. When $b = \infty$, its upper limit, A has missed B as much as possible. Since real intermolecular forces (except those between charged particles) die rapidly with distance, an impact parameter greater than about 15 Å or so is usually equivalent (dynamically) to $b = \infty$. Thus, we can specify an impact parameter for a collision between interacting particles as long as we start the collision with A and B sufficiently far apart.[3]

In Figure 28.6(a), we show the evolution of a collision as we would see it in the lab with B sitting still and A approaching it at some non-zero, but not too large, impact

---

[3] While the impact parameter is a very useful concept, it is something we *cannot* control experimentally. We are very clever at manipulating the energies and momenta of molecules, but we cannot control their positions in space to anything near an ångström's precision.

parameter. Each dot represents the position of A and B at equal time intervals, and we assume A and B have the same mass. At first, B sits still as A approaches. They are too far apart to interact. Eventually, they start to interact notably through a typical attractive force. A starts to swing towards B, and B starts to move towards A; the collision has truly begun. They continue to approach, swing around each other, and head out to their final directions. At the end of the collision, A and B are (by definition of a collision's "end") so far apart that they no longer interact, and they continue to move through the lab at constant final velocities. As the figure indicates, we can describe this collision's outcome through the *laboratory scattering angle* θ of A, the angle through which A's velocity has turned as a result of the collision.

Suppose we were moving through the lab with the speed and direction of the center-of-mass velocity v_{cm} of this collision. With B initially still, v_{cm} points in the direction of A's velocity and is m_A/M times smaller in magnitude. In this frame of reference, we would see *both A and B moving towards us*. This frame is particularly simple, since it imposes a certain symmetry to the collision, and, as Problem 28.10 asks you to prove, the total linear momentum in this frame is zero. Figure 28.6(b) plots our representative collision in this frame of reference; A and B have mirror image trajectories. We describe the collision in this frame of reference in terms of the *center-of-mass scattering angle* χ, which is the same for A and B but, in general, different from θ.

Equation (28.5) showed us how to separate kinetic energy between relative and center-of-mass components, and we pointed out there that the center of mass moves with boring constancy; all the action is in relative motion. A further consequence of this separation of dynamics is that *two-body elastic scattering is dynamically equivalent to a single body of mass μ scattering from a fixed point with which it interacts through the real intermolecular potential*. Figure 28.6(c) shows our representative collision in these coordinates. The fixed point is located at the small cross, while the effective motion of the representative particle of mass μ is tracked from point to point. Note that in these coordinates, the scattering angle χ is *exactly the same as in the center-of-mass reference frame*.

The dynamic simplification of the relative frame makes it the easiest to consider, and the simplest intermolecular-potential function is our old friend the hard-sphere potential. Thus, we begin with a pair of hard spheres of radii r_1 and r_2 with $d = r_1 + r_2$, the hard-sphere diameter. The spheres will collide for any impact parameter $0 \leq b \leq d$; larger impact parameters guarantee a "miss," and the scattering angles are zero. A head-on collision ($b = 0$) means χ is 180°, its maximum value. Intermediate impact parameters lead to intermediate scattering angles, and we can use relative coordinates to relate χ to b.

Figure 28.7 shows a collision between two hard spheres for some $b < d$ in the relative frame of reference. The center of force is again at the cross, and the circle

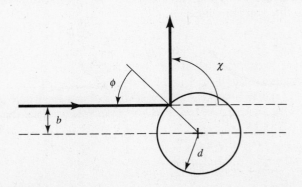

FIGURE 28.7 The scattering angle χ in the relative frame for hard spheres of effective diameter d and impact parameter b is found from the geometry shown here.

FIGURE 28.8 The deflection function $\chi(b)$ for hard spheres of effective diameter d is shown here. The three drawings show representative scattering trajectories at small, intermediate, and large impact parameters.

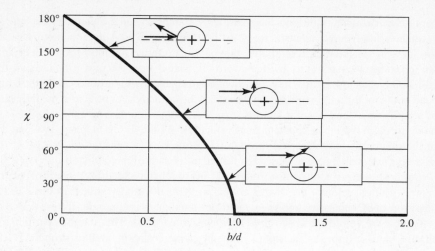

of radius d locates the hard-sphere diameter boundary. The trajectory is a simple pair of straight lines: the spheres approach with constant speeds until they instantaneously hit, scatter, and move away at constant speeds, just as two billiard balls do (that aren't spinning in some clever way so that they involve friction with the table surface to alter their otherwise straight-line trajectories). Simple geometry shows that χ is related to the angle ϕ defined in the figure through $\chi = \pi - 2\phi$. The angle ϕ is also related to b and the hard-sphere diameter through

$$\sin \phi = \frac{b}{d}, \quad b \leq d$$

so that the scattering angle as a function of impact parameter, a quantity called the *deflection function*, $\chi(b)$, is

$$\chi(b) = \begin{cases} 2 \cos^{-1} \frac{b}{d}, & b \leq d \\ 0, & b > d \end{cases} \quad (28.7)$$

where we have used $\pi - 2\sin^{-1}(b/d) = 2\cos^{-1}(b/d)$. Figure 28.8 plots this deflection function and also shows representative trajectories for several impact parameters.

Recall that the hard-sphere potential is purely repulsive. It reflects "size" only, not attraction. Note that χ from Eq. (28.7) is always *positive*. This is a general characteristic of a collision dominated by repulsion. In contrast, the collision shown in Figure 28.6 has a *negative* scattering angle; it happens to be dominated by the *attractive* forces of the real intermolecular potential. We will always draw scattering trajectories and define scattering angle algebraic signs following this convention, summarized in the diagram below:

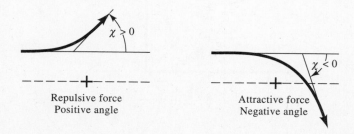

In any experiment, however, *only the absolute value of the scattering angle is measurable.* We can see this if we consider the diagram below in which a negative χ scattering event (the same one as shown above) has the same |χ| as a positive χ event in which the effective scatterer approaches the scattering center of force from the opposite side. The scattering has cylindrical symmetry about the line representing $b = 0$ that passes through the center of force:

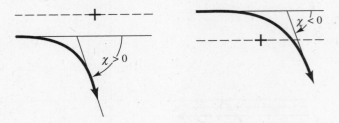

These two events are experimentally indistinguishable.

At small impact parameters, and any collision energy, the scattering is always dominated by the repulsive core of the intermolecular potential, and the scattering angle is positive. As b increases, the effects of the attractive part of the interaction become noticeable. If E_{rel} is very large compared to the intermolecular potential's well depth $\mathcal{D}_e$, attractive forces are too weak to have an effect, and the scattering approaches that of hard spheres. For the rare gases, with $\mathcal{D}_e$ values in the range 10–200 cm^{-1} (see Table 24.4), collisions with relative speeds based on most probable speeds at ordinary temperatures have E_{rel} comparable to $\mathcal{D}_e$.

EXAMPLE 28.3

What is the ratio of E_{rel} to $\mathcal{D}_e$ for a collision between a Ne atom moving at v_{mp}, the most probable speed, at 300 K and a stationary He atom? What is this ratio for Xe moving at v_{mp} at 300 K striking a stationary Xe? (From Table 24.4, $\mathcal{D}_e$ for Ne–He is 15.3 cm^{-1}, and for Xe–Xe, 196.2 cm^{-1}.)

SOLUTION Table 24.1 has the expression we need relating v_{mp} to mass and temperature: $v_{mp} = \sqrt{2k_B T/m}$. For Ne at 300 K, $v_{mp} = 497$ m s^{-1}, and for Xe, $v_{mp} = 195$ m s^{-1}. Since the other atom in each collision is stationary, $v_{mp} = g$. The corresponding relative collision energies are, with $\mu = m_{Ne} m_{He}/(m_{Ne} + m_{He})$ and $m_{Xe}/2$, respectively, 6.85×10^{-22} J for Ne–He and 2.07×10^{-21} for Xe–Xe. Converting from cm^{-1} to J units (1 cm$^{-1} \cong 2 \times 10^{-23}$ J), the intermolecular well depths are $\mathcal{D}_e = 3.04 \times 10^{-22}$ J (or $E_{rel}/\mathcal{D}_e = 2.26$) for He–Ne and $\mathcal{D}_e = 3.90 \times 10^{-21}$ J (or $E_{rel}/\mathcal{D}_e = 0.53$) for Xe–Xe.

→ RELATED PROBLEMS 28.12, 28.13

Collisions between Ar atoms are not only representative of all elastic collision events, they are also among the best studied. These studies have helped make the Ar–Ar intermolecular potential among the best known, as we discussed in Chapter 24. (See the multi-parameter Ar–Ar potential in Figure 24.11.) One can calculate trajectories at various impact parameters and relative collision energies based on this potential, and Figure 28.9 plots several such trajectories for $E_{rel} = \mathcal{D}_e, 2\mathcal{D}_e$, and $3\mathcal{D}_e$ along with hard-sphere trajectories for comparison. Note how the scattering angles for $b < 1$ Å are roughly the same for all four pictures: small b, large positive χ collisions are relatively insensitive to the attractive part of the interaction and the collision energy. Note as well how all the large b ($b > 6$ Å) trajectories for the

FIGURE 28.9 Shown here are representative elastic scattering trajectories for a typical realistic potential at three collision energies, expressed as multiples of the potential well depth, $\mathcal{D}_e$. The same trajectories for hard spheres are also shown for contrast. As E_{rel} increases, the real scattering approaches the hard-sphere limit.

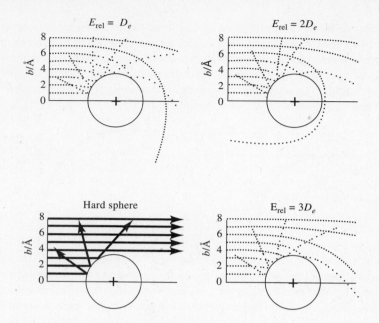

Ar–Ar collisions lead to negative χ collisions and are thus influenced by the attractive part of the interaction. Finally, compare the $b = 8$ Å trajectories for all three Ar–Ar collision energies. As E_{rel} increases, χ becomes less negative, and in general, as E_{rel} increases, *all* trajectories approach the hard-sphere trajectory limits.

If we measure χ for each impact parameter at any one E_{rel}, we can construct a deflection function plot for that E_{rel}. Figure 28.10 does this for the collision conditions shown in Figure 28.9. The hard-sphere curve is a repeat of Figure 28.8 using the Ar–Ar hard-sphere diameter suggested by the real potential's repulsive wall. Note again how the real deflection functions are approaching the hard-sphere curve as E_{rel} increases.

Most striking in Figure 28.10, however, are the sharp dips in each deflection function that extend toward large negative scattering angles. Locate the $b = 6$ Å

FIGURE 28.10 If very many trajectories of the type shown in Figure 28.9 are calculated, we can construct deflection-angle plots for any one collision energy.

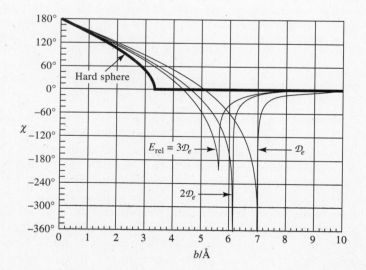

trajectory for $E_{rel} = 2\mathcal{D}_e$ in Figure 28.9. It turns around the scattering center and leads to scattering that reverses the initial direction; χ is about $-180°$. Figure 28.10 shows that slightly greater impact parameters at this E_{rel} lead to even more negative χ values, including $-360°$. This is the characteristic of an *orbiting collision*. The two atoms circle each other at least one full revolution during the collision. In fact, the deflection functions for the lowest two energies shown in Figure 28.10 actually extend *below* $-360°$; the graph has been chopped at $-360°$ in order to keep the majority of the curves on a reasonable scale. If the collision energy and the impact parameter are both just right, the scattering angle diverges to $-\infty$ as the two atoms enter a stable orbit around each other, an orbit that (classically) persists until the orbiting pair collides with something else.

Orbiting and near-orbiting collisions play two important roles. The first takes us all the way back to Chapter 1 and our discussion there of the origin of nonideal behavior in bulk gases. We argued how attractive forces caused the pressure of the gas to fall *below* the ideal value due to the way these forces *reduced* the wall collision rate below the ideal rate. Here we have found a detailed microscopic picture of this rate decrease. Orbiting and near-orbiting collisions slow the progress of gas atoms (or molecules) through the bulk gas as time is spent in long collisions during which the free atom speed is effectively slowed to the speed of the collision's center of mass.[4]

The second consequence of near-orbiting collisions is unexpected and, surprisingly, related to another natural phenomenon: the formation of a rainbow. What is the most amazing aspect of a rainbow? If you answered, "the beautiful colors," "the arc shape," or even "the pot of gold at its end," you have perhaps missed how amazing it is that you can *see* a rainbow at all! Think about the situation that leads to a visible rainbow: you look toward a fine mist or rain, but the *sun is behind you.* How do water drops *scatter* sunlight so that you see a *bright* rainbow *localized in space?* (The rainbow's colors are irrelevant here. They arise simply because sunlight is "white"—polychromatic—and the refractive index of water varies slightly with wavelength.)

The refraction of light of any one color was first described in detail by the French philosopher-scientist René Descartes (also of Cartesian coordinate fame) in his 1637 book *Dioptric*. He arrived at his explanation through a variety of analogies, including the action of a tennis ball when struck by a racket! The rainbow itself had long been a source of wonder and speculation, of course, as Descartes pointed out in his later book, *Météors,* in which he presented his explanation of the rainbow. He wrote the following:

> The rainbow is a marvel of nature that is so remarkable, and its cause has been so eagerly sought ... and is so little understood, that I could not choose a subject better suited to show how, by the method which I employ, we can arrive at knowledge which has not been attained by those whose writings we have.

---

[4]Consider a perfect orbiting collision: two Ar atoms collide in the lab reference frame in exactly opposite directions but with equal speeds (this ensures that the center-of-mass speed is zero) and at just the right collision energy and impact parameter to orbit. If we could watch this collision, we would see the two atoms approach, begin their collision, and then forever orbit *without moving further through space.* They would form an apparently bound Ar_2 molecule that would rotate but not translate. They would never hit the container wall and thus never contribute to the pressure. This type of collision seems to contradict our earlier statements in Chapter 27 that bimolecular association is impossible without, for instance, photon emission. Classically, there is only *one* precise impact parameter and collision energy that can lead to true orbiting, which makes this a very improbable event, and orbiting atoms are only metastably bound at best.

To Descartes, application of his "method" (by which he meant his entire philosophical approach to the acquisition of knowledge) to so striking and accessible a problem as the rainbow would be a *tour de force* that would help ensure his fame.

Realizing that raindrops were the intermediaries in rainbow formation, Descartes experimented with a large spherical glass ball filled with water.[5] He found that any one color of light when refracted once on entering the sphere, then reflected from the back side of the sphere, and then refracted a second time on leaving it would be strongly *focused* at a particular angle. For red light, the angle is about 42° from the incident light direction, and for blue, the angle is about 41°. The key is the focusing action, which he did not fully understand until, as he wrote, "I have taken my pen and calculated in detail all the rays falling on various points of a drop of water." He did dozens of calculations and found that a somewhat wide range of initial ray directions lead to final rays clustered at nearly the same angle—the focusing effect—whereas rays striking the drops in unfavorable positions or directions were scattered widely—not focused and thus not seen. We see a rainbow only because light is scattered strongly and preferentially back towards us over a small angular range, focused by the raindrops.

Now we make the connection between the optical rainbow and the elastic scattering of two atoms. The rainbow—its angular size and angular spread from red to blue—tells us something about water. If it ever rained small spherical diamonds, the rainbow would look very different due to the different refractive index of diamond and the different change of this index (called the *dispersion*) with wavelength. In other words, the angular position and extent of the rainbow tell us something about the nature of the scatterer. So it is for elastic scattering of atoms.

When we conduct a scattering experiment, we typically measure the flux of scattered particles that enters a small aperture on a particle detector pointing at some known laboratory angle from the initial direction of the scattering particles. A series of geometric and kinematic transformations that we need not discuss here converts that laboratory flux into the flux we would measure if we could do the idealized experiment represented by the fixed center-of-force reference frame. Since the scattering occurs in three dimensions, we express the scattering intensity in terms of the *number of molecules scattered into the solid angle*[6] *subtended by the detector per unit time per unit incident particle flux*. Note the dimensions of this intensity: (number of scattered particles)/[(solid angle) × (incident particle flux) × (time)]. The product of particle flux and time has the dimensions of particles per unit area, and dividing this into the number of scattered particles per unit solid angle reduces to something with dimensions of *area per unit solid angle*. This is called *the differential scattering cross section* $I(\chi)$ for scattering into the solid angle $d\omega$ located by some scattering angle χ. We recover the *total* scattering cross section σ if we integrate $I(\chi)$ over all solid angles:

$$\sigma = \int I(\chi)\, d\omega \ . \tag{28.8}$$

---

[5]This trick had been used, independently, by the Persian Kamāl al-Dīn al-Fārisī and the German Theodoric of Freiberg in the early 1300s. Both of these scientists hit on more or less the correct explanation, but there is no real evidence that Descartes knew of either of their experiments.

[6]A solid angle is to a sphere what a plane angle is to a circle. Solid angles are measured in *steradians* (in contrast to plane angles measured in *radians*) and range from 0 (a point) to 4π steradians (the full surface area of a sphere of unit radius). For example, if our detector has an aperture of small area A located a distance R from the point of collision, we say it subtends a solid angle $\omega \cong 4\pi$ (area of detector)/(area of sphere of radius R) = A/R^2. This relation is approximate because A is a *plane* area rather than the area of a *spherical* section, but if $A << R^2$, it is a very good approximation. See also the *Closer Look at an Effusive Molecular Beam* in Chapter 24.

We now need a connection between the measured differential cross section $I(\chi)$ and the deflection function $\chi(b)$. Because the scattering is cylindrically symmetric about the line $b = 0$ that passes through the scattering center of force, and since all trajectories start out parallel to this line at large distances, the fraction of the incident beam flux with impact parameters between b and $b + db$ is $2\pi b\, db$, the area of an annulus of radius b and width db. (Note that this fraction is zero if $b = 0$. This means that truly "head-on" collisions have zero probability, classically.) This fraction of the incident beam is scattered into an infinitesimal angular region between χ and $\chi + d\chi$ (if $d\chi/db > 0$) or between χ and $\chi - d\chi$ if $(d\chi/db < 0)$. Since $I(\chi)$ is the fraction scattered at χ per unit differential solid angle $d\omega$, we can equate the fractional flux before the collision to the fraction after the collision:

$$2\pi b\, db \;=\; I(\chi)\, d\omega \qquad (28.9)$$

$\uparrow$ Fraction of incident flux before collision $\qquad$ $\uparrow$ Fraction of incident flux after collision

and now all we need is a relationship between the solid-angle element $d\omega$ and the scattering angle χ. The solid angle $d\omega$ between cones defined by χ and $\chi + d\chi$ is $2\pi \sin\chi\, d\chi$ so that the elastic scattering differential cross section is

$$I(\chi) = \frac{2\pi b\, db}{d\omega} = \frac{2\pi b\, db}{2\pi \sin\chi\, d\chi} = \frac{b}{|\sin\chi\,(d\chi/db)|} \,. \qquad (28.10)$$

The absolute value sign has been added in the last step to reflect our inability to distinguish experimentally between scattering at $+\chi$ from scattering at $-\chi$.

EXAMPLE 28.4

Equation (28.7) gives the deflection function for hard-sphere scattering. What is the hard-sphere differential scattering cross section? Check that this $I(\chi)$ also yields the correct total scattering cross section for hard spheres, $\sigma = \pi d^2$.

SOLUTION The deflection function is defined only for $b \leq d$, which means that all trajectories with $b > d$ have no deflection angle and thus do not contribute to $I(\chi)$ at any $\chi \neq 0$. We first find $\sin\chi$ and $d\chi/db$ from Eq. (28.7) using some simple trigonometric identities:

$$\sin\chi = \sin\!\left[2\cos^{-1}\!\left(\frac{b}{d}\right)\right] \qquad \text{from Eq. (28.7)}$$

$$= \frac{2b}{d}\sin\!\left[\cos^{-1}\!\left(\frac{b}{d}\right)\right] \qquad \text{from } \sin 2x = 2\sin x\cos x$$

$$= \frac{2b}{d}\sqrt{1 - \frac{b^2}{d^2}} \qquad \text{from } \cos^{-1} x = \sin^{-1}\sqrt{1 - x^2}$$

$$\frac{d\chi}{db} = \frac{d}{db}\!\left[2\cos^{-1}\!\left(\frac{b}{d}\right)\right] \qquad \text{from Eq. (28.7)}$$

$$= -\frac{2}{d\sqrt{1 - b^2/d^2}} \qquad \text{from } \frac{d\cos^{-1} u}{dx} = -\frac{1}{\sqrt{1 - u^2}}\frac{du}{dx}$$

and then substitute into Eq. (28.10) and cancel common factors, yielding simply

$$I(\chi) = \frac{b}{|\sin\chi\,(d\chi/db)|} = \frac{d^2}{4}\,,$$

FIGURE 28.11 If the deflection functions as a minimum at a particular impact parameter, the scattering intensity at the angle corresponding to that minimum is enhanced. This is called *rainbow* scattering, and the angle is called the rainbow angle.

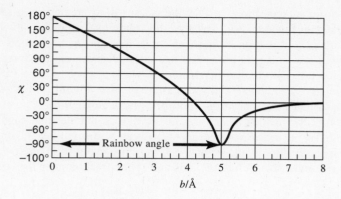

which is independent of scattering angle. One says that elastic hard-sphere scattering is *isotropic,* constant in intensity at all angles.

We find the total cross section from Eq. (28.8) using the solid angle element for cylindrically symmetric scattering $d\omega = 2\pi \sin\chi\, d\chi$:

$$\sigma = 2\pi \int_0^\pi I(\chi) \sin\chi\, d\chi = \frac{\pi d^2}{2} \int_0^\pi \sin\chi\, d\chi = \pi d^2$$

recovering the expected result.

⟹ RELATED PROBLEMS 28.14, 28.15, 28.16

Now we have what we need to discover the origin of rainbow scattering: a relationship between the measured scattered flux at any angle χ and the deflection function $\chi(b)$ through its derivative $d\chi/db$. Consider the deflection function shown in Figure 28.11. Since most collisions have large impact parameters (because $2\pi b\, db$ increases with b, or, put more simply, most of the time the colliding atoms "miss each other"), they have small scattering angles, and we expect $I(\chi)$ to be large at small χ.[7] Equation (28.10) and Figure 28.11 bear this out. At large b, $d\chi/db$ is small, χ is near zero (so that $\sin\chi$ is near zero), and $I(\chi)$ is large. For small impact parameters, b is small, and $I(\chi)$ is approaching the hard-sphere value derived in Example 28.4. But consider scattering with $b \cong 5$ Å in Figure 28.11, leading to $\chi \cong -90°$. That is the region of the *minimum* in the deflection function so that $d\chi/db = 0$, making $I(\chi) \to \infty$. This is rainbow scattering, or at least a large part of it, and we call the scattering angle at which the intensity is abnormally large the *rainbow angle,* χ_r. The minimum in the deflection function leads to a finite range of impact parameters that all produce roughly the same scattering angle. That is a focusing effect, just as in the optical rainbow.

Classically, $I(\chi)$ diverges to ∞ at the rainbow angle because the $d\chi/db$ factor in Eq. (28.10) has gone to zero at the rainbow angle. Note, however, that scattering with $b \cong 2.5$ Å leads to $\chi = +90°$, and since the sign of χ cannot be measured, scattering at this impact parameter also leads to intensity at χ_r. When elastic scattering is analyzed through quantum rather than classical mechanics, one finds that the intensity from the small impact parameter *interferes,* in a quantum wave-like way, with that from the larger impact parameter (and in quantum mechanics, impact

---

[7]More detailed classical theories of $I(\chi)$ than we can discuss here show that $I(\chi) \propto \chi^{-7/3}$ at small angles for an interaction potential with a long-range attraction that varies like R^{-6}. Classical mechanics always predicts $I(\chi) \to \infty$ as $\chi \to 0$, but quantum mechanics shows that $I(\chi)$ stays finite at $\chi = 0$.

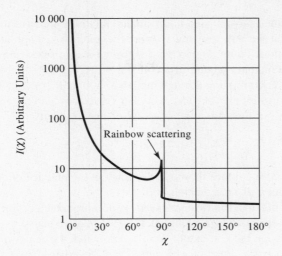

FIGURE 28.12 The differential elastic scattering cross section for a deflection function such as is shown in Figure 28.11 has this qualitative shape. Note that the $I(\chi)$ scale is logarithmic. Intensity is greatest at small scattering angles, since most collisions "miss" or are glancing at best, while at larger angles, the rainbow-scattered intensity appears prominently.

parameter is replaced by the *collisional angular momentum*, which equals μgb classically). This interference keeps the rainbow intensity large, but finite, and one also finds that interferences introduce other undulations in $I(\chi)$ as well.

Figure 28.12 plots the qualitative shape of $I(\chi)$ for a deflection function such as in Figure 28.11. Note that the differential cross section axis is logarithmic; rainbow scattering is rather bright! At scattering angles just past χ_r, the intensity drops suddenly to a nearly constant value from there to $\chi = 180°$. This is called the "dark side of the rainbow" (something optical rainbows have, too!), and the nearly constant $I(\chi)$ at large χ is caused by small impact parameter collisions following an $I(\chi)$ that is close to the isotropic hard-sphere value.

The rainbow angle (and, in fact, the detailed shape of the entire $I(\chi)$ curve) changes with collision energy. As E_{rel} is increased, χ_r decreases. If we measure $I(\chi)$ at several energies, we can attempt to *invert* these functions, first back to the deflection function (although this step is rarely done in practice) and then back to the intermolecular potential itself. If the differential cross section was featureless (as it is for hard spheres), we could not deduce much about the intermolecular potential, but by exploiting features such as the rainbow and other wiggles in $I(\chi)$ required by quantum mechanics, very precise intermolecular potential functions for not only rare gas atoms but also other spherical or nearly spherical systems (such as K–Hg, or H_2–He, or Ar–SF_6, to name a few) have been measured.

We have now spent some time establishing some of the language and phenomena of three-dimensional scattering, but since we have limited ourselves to elastic events, we have yet to consider reactive scattering and our goal of measuring state-to-state reaction cross sections. We turn next to *inelastic* scattering, which will take us almost all the way to our goal, and once we establish a few more phenomena, we can easily incorporate reactive scattering events.

Inelastic Scattering

An inelastic collision transfers relative translational energy into internal energy of one or both of the collision partners.[8] By conservation of energy, the amount transferred to internal excitation is lost to translation, and since the masses have

---

[8]If internal energy is transferred *to* relative translational energy, the collision is often called *superelastic*.

not changed, an inelastic collision has a smaller relative speed g' after the collision than before. What we will see here is how to exploit this simple consequence of energy conservation. To do so, we must return to elastic scattering to introduce a very clever diagrammatic picture of a collision.

Imagine a Ne atom moving through the lab at 500 m s^{-1} toward a stationary He atom, as pictured in Figure 28.13(a). This lab velocity is also the relative velocity g since the He is not moving, and the center-of-mass velocity is

$$\boldsymbol{v}_{cm} = \frac{m_{Ne}}{M}\boldsymbol{v}_{lab} = \frac{20}{24}(500 \text{ m s}^{-1}) = 417 \text{ m s}^{-1},$$

directed along g, as the figure shows. We will adopt the practice of drawing a cross along g at the center-of-mass velocity tip.

In an elastic collision, the relative *speed*, g, does not change, since the collision energy, $\mu g^2/2$, does not change. Consequently, the relative collision *velocity*, g, can only change *direction* but not length. We also know that the center-of-mass *velocity* cannot change, in *either* direction or length. Taken together, these constraints show us that g can only *rotate* (in any plane) *about the end of the center-of-mass velocity*

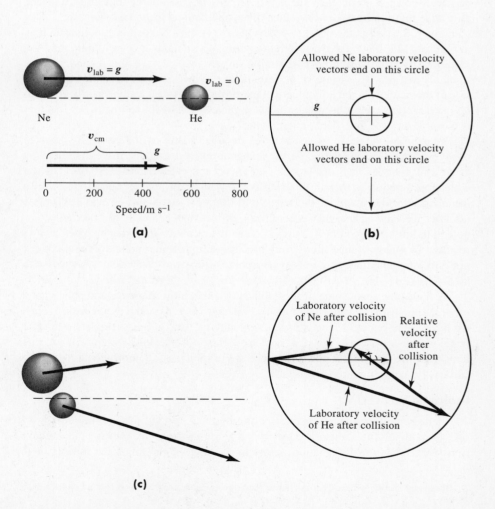

FIGURE 28.13 Imagine a collision between Ne moving at 500 m s^{-1} through the lab toward a stationary He atom, as in (a). The constraints of energy and momentum conservation allow us to draw the Newton diagram shown in (b), which locates the possible final laboratory velocity vectors of each atom after a collision. These vectors for a particular collision are shown on (c), along with the final relative velocity g', which we see is simply the initial g rotated about the tip of the center-of-mass velocity vector, located by a cross along g.

vector. Consequently, the final *laboratory* velocity of Ne *after* the collision must lie on a circle centered at the tip of $\boldsymbol{v}_{cm}$ with a radius equal to $g - v_{cm}$. Likewise, the He lab velocity after the collision must lie on a circle centered at the same place but of radius v_{cm}. Figure 28.13(b) shows these circles along with the initial relative velocity vector and the location of the center-of-mass velocity vector (the cross drawn on $\boldsymbol{g}$). This drawing is made in "laboratory velocity space," in which the origin (the end of *laboratory* vectors *without* an arrowhead) represents zero laboratory speed, and the circles are called *elastic circles* because they represent *all* possible outcomes of elastic scattering events.[9]

Now consider the collision of Figure 28.13(c) in which Ne strikes a glancing blow to He. Intuition tells us that the heavier Ne atom will continue to move more or less straight ahead—a heavy sphere thrown towards a light, stationary sphere will not scatter backwards towards us. Intuition also tells us that the light He atom will move forward at a *faster* speed than the Ne had originally if we hit the He nearly head-on, as is shown in Figure 28.13(c) by the arrows representing the lab velocities after the collision.

We transfer these arrows to our velocity-space construction in Figure 28.13(c). The lab-velocity arrows end on their respective elastic circles, and the arrow connecting them represents the final relative velocity, $\boldsymbol{g}'$. Note that $\boldsymbol{g}'$ passes through the end of $\boldsymbol{v}_{cm}$, as it must; the relative velocity is simply rotated about the end of the center-of-mass velocity.

The construction in Figure 28.13(c) is called a *Newton diagram,* since only simple Newtonian mechanics—the conservation of energy and of linear and angular momentum—is needed to draw it. Such diagrams summarize all *possible* scattering outcomes, but *not* their relative probability. That is still contained in the differential scattering cross section.

EXAMPLE 28.5

Suppose the Ne atom in Figure 28.13 hits He with zero impact parameter. What are the final laboratory velocities of the Ne and He atoms? What is the Ne scattering angle in the center-of-mass frame? in the laboratory frame?

SOLUTION For $b = 0$, we know that χ must equal 180°. That means the relative velocity vector is rotated 180° about the fixed end of $\boldsymbol{v}_{cm}$ so that $\boldsymbol{g}'$ points in the *opposite* direction from $\boldsymbol{g}$ (only portions of the Ne and He elastic circles are shown here):

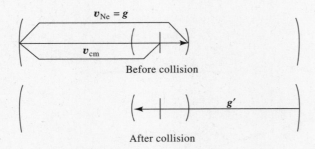

---

[9]In three dimensions, the circles are spheres with the same radius.

The final lab velocity of Ne, v'_{Ne}, extends from the origin to the $\chi = 180°$ point on the Ne elastic circle, and the final lab velocity of He extends to its elastic circle in the same direction:

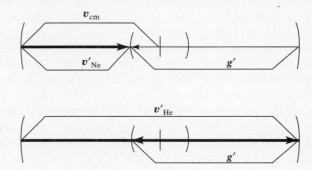

The corresponding speeds are easy to read from the diagrams:

$$v'_{Ne} = 2v_{cm} - g' = 2v_{cm} - v_{Ne}$$
$$v'_{He} = 2v_{cm}$$

where v_{cm} is the center-of-mass speed. Numerically, with $v_{Ne} = 500$ m s^{-1}, $v'_{Ne} = 333$ m s^{-1} and $v'_{He} = 833$ m s^{-1}.

Note that while the *center-of-mass scattering angle* is 180°, the *laboratory scattering angle is 0°*. The Ne atom is moving in the same direction through the lab after the collision as it was before the collision; only its speed has changed.

➠ RELATED PROBLEMS 28.20, 28.21, 28.22

EXAMPLE 28.6

An Ar atom moving 370 m s^{-1} through the lab strikes a stationary Ar atom so that the center-of-mass scattering angle is 90°. What are the final laboratory speeds and scattering angles of the two atoms?

SOLUTION If the colliding particles have equal masses and one of them is initially at rest, the Newton diagram is particularly simple: the center-of-mass velocity lies exactly half-way along the relative velocity and the two elastic circles lie on top of each other:

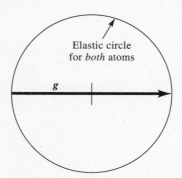

After the collision, g' is turned through $\chi = 90°$ about the center-of-mass velocity, and simple trigonometry gives us the laboratory speeds and scattering angles:

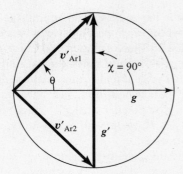

With $\chi = 90°$, the lab scattering angle θ of both atoms is 45°, and the speeds are both $g/\sqrt{2} = 262$ m s^{-1}. Since the atoms are identical, we cannot tell after the collision which atom was initially stationary. We have seen in other contexts (the electrons in He atom and the nuclei in homonuclear diatomics, for example) that quantum mechanics often has something new to say about the behavior of indistinguishable particles, and this is also true of scattering, as Problem 28.23 discusses.

➠ *RELATED PROBLEMS* 28.20, 28.21, 28.22, 28.23

There is one final experimental lesson we can learn from elastic scattering Newton diagrams that is invaluable to understanding inelastic and reactive scattering. *Particles scattered at one angle in the lab often represent more than one center-of-mass scattering angle.* For elastic scattering, "more than one" means simply "two at most" as we can see in Figure 28.14, which shows a Ne–He scattering event in which the Ne is scattered through the *same* lab angle as in Figure 28.13(c), but through a *smaller* χ. The He atom is moving in a very different direction and with a very different speed, but if our detector was set to detect *only* Ne and it was placed at this lab angle, we would measure scattering flux from *two* different types of collision events. Clearly, selecting the scatterer's mass and lab scattering angle alone is not enough to tell us all we need; we also must measure the scattered particle's *speed in the lab*.

This measurement is most often done through a time-of-flight measurement, as discussed in Chapter 24, or through a spectroscopic probe, such as laser induced fluorescence done with sufficient resolution to exploit the Doppler effect, as discussed in Chapter 19. In our Ne–He scattering experiment, we could place our mass

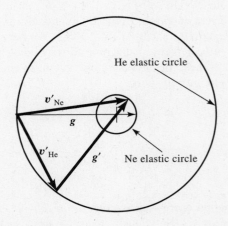

FIGURE 28.14 The laboratory scattering angle of one particle does not necessarily define the scattering event uniquely. Here, we show the Newton diagram for a collision in which the Ne lab scattering angle is the same as in Figure 28.13(c), but the center-of-mass scattering angle is smaller.

spectrometer detector at the lab angle θ of Figure 28.14, tune it to detect Ne, and place a rotating toothed wheel (a chopper) between the scattering volume and the detector entrance to initiate a time-of-flight measurement. The Ne time-of-flight signal would have a large peak at short times, representing the flux of the faster (in the lab) atoms and a smaller peak at longer times representing the flux of the slower atoms. The faster atoms' peak would be larger since, as the figure shows, these atoms are scattered through a *smaller* center-of-mass angle than are the slower atoms of Figure 28.13(c) and $I(\chi)$ falls with increasing χ (neglecting the possibility of rainbow-enhanced scattering).

Now we consider inelastic scattering. Suppose we substitute D_2 for He in our experiment. The elastic Newton diagram is unchanged, since He and D_2 have the same mass, but D_2 has internal vibration and rotation degrees of freedom that can absorb appropriately quantized amounts of energy from a collision's relative translational energy. Consider first the most extreme limit: a totally inelastic collision. All E_{rel} is absorbed by D_2. By conservation of energy, the final relative translational energy E'_{rel} must be zero, and thus so must be g'. The particles collide, energy is transferred, and since none is left to carry them away from each other, they travel as a pair through the lab *at the center-of-mass velocity*. They stay stuck together until some internal energy leaks back out from the D_2 and into relative translation, at which time they finally move apart. For the Ne–D_2 system, it is unlikely that much time would be needed to transfer this energy back, but for collisions between large polyatomics, the internal energy can take a significant time to wander first from translation, to internal motion, then back out again: a time long enough for the stuck-together pair to fly into our detector.

The dynamics of a totally inelastic event are particularly simple: the collision partners must both move through the lab with the speed and direction of the center-of-mass velocity. For inelasticities intermediate between total and none (i.e., elastic), this argument indicates that the scattered particle must have a final velocity vector that ends *somewhere inside its elastic circle*. For Ne moving at 500 m s^{-1} toward a stationary D_2, $E_{rel} = \mu g^2/2 = 6.86 \times 10^{-22}$ J. The vibrational constants for D_2 from Table 19.2 show that vibrational excitation of D_2 from $v = 0$ to 1 requires about 3000 cm^{-1} or about 5.9×10^{-20} J, more than is available in E_{rel}. The D_2 rotational constants show that rotational excitation from $J = 0$ to 1 requires about 60 cm^{-1} or 1.2×10^{-21} J, also more than E_{rel}. Consequently, in this experiment, we would *only* see elastic scattering, assuming all our D_2 molecules are in $v = 0$, $J = 0$.

Suppose we raise the Ne lab velocity to 1200 m s^{-1}, raising E_{rel} to 4.0×10^{-21} J. Now we have sufficient energy to excite D_2 rotationally not only to $J = 1$ but also to $J = 2$, which requires about 3.6×10^{-21} J. We next find the radii of the two *inelastic* circles in our Newton diagram corresponding to $J = 0 \rightarrow 1$ and $J = 0 \rightarrow 2$ inelastic events. For the first, conservation of energy tells us the final relative translational energy and speed:

$$E'_{rel} = E_{rel} - \Delta E_{int} = 4.0 \times 10^{-21} \text{ J} - 1.2 \times 10^{-21} \text{ J} = 2.8 \times 10^{-21} \text{ J}$$

and

$$g' = \sqrt{\frac{2E'_{rel}}{\mu}} = 1.0 \times 10^3 \text{ m s}^{-1}$$

where ΔE_{int} is the internal energy change due to the $J = 0 \rightarrow 1$ excitation of D_2. Since the center-of-mass velocity cannot change, $\boldsymbol{g}$ must *shrink about the end of*

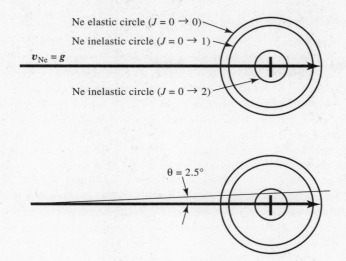

FIGURE 28.15 The Newton diagram for a nonreactive but possible inelastic collision system locates not only the elastic circles but also circles that locate scattering velocities for every possible excited quantum state of the collision partners that could be populated in a collision. In the lower part of the diagram, we see that Ne atoms moving at 2.5° in the lab after the collision could represent any of six different inelastic events: three different inelasticities with two unique center-of-mass scattering angles each.

v_{cm} to produce g'. The distance from the end of v_{cm} to the end of g is $(m_{D_2}/M)g = 200$ m s^{-1}, the elastic circle radius, and thus the radius of the $J = 0 \rightarrow 1$ inelastic circle is $(m_{D_2}/M)g' = 167$ m s^{-1}. Repeating these calculations for the $J = 0 \rightarrow 2$ excitation energy, $\Delta E_{int} = 3.6 \times 10^{-21}$ J, shows that this inelastic event has $g' = 379$ m s^{-1} and an inelastic circle radius of 63 m s^{-1}.

Figure 28.15 draws the Newton diagram for this experiment, showing only the Ne elastic and inelastic circles. The elastic and inelastic circles that locate allowed velocities for scattered D$_2$ are also centered at the tip of v_{cm} with radii given by m_{Ne}/M times the appropriate g'. Suppose, as suggested in the lower half of the figure, we place our Ne detector, equipped with time-of-flight capabilities, at a lab angle $\theta = 2.5°$ from the initial direction of Ne. What might we see? The line drawn at this angle intersects each of the three circles twice so that we could expect to find *six* different laboratory speeds at this angle. The fastest three would correspond to small center-of-mass scattering angles, one for elastic scattering and two slower speeds for the two inelastic events. The slowest three of the six speeds would correspond to large center-of-mass angles, one for elastic scattering and two somewhat faster for the inelastic events.

If we calculate from our six measured intensities the differential cross sections each corresponds to, we would have six state-to-state differential cross sections for our collision energy, such as that for Ne + D$_2(J = 0) \rightarrow$ Ne + D$_2(J = 2)$ at a particular scattering angle, that for Ne + D$_2(J = 0) \rightarrow$ Ne + D$_2(J = 1)$ at a different scattering angle, and so forth. If we repeat this experiment at various lab scattering angles, we can build up a complete picture of all the possible dynamical outcomes for Ne striking D$_2(J = 0)$ at that energy. (Our experimental *resolution*, both in angle and speed, would limit the precision of our measurements, of course.) All of our measurements—differential cross sections as functions of center-of-mass scattering angle—could be overlaid onto the Newton diagram as a *scattering-flux contour map* from which we can read at a glance the peaks and valleys of these quantities.[10]

---

[10] It is likely in this case that the $J = 0 \rightarrow 1$ cross section is very much less than that for $J = 0 \rightarrow 2$ at any scattering angle due to nuclear spin effects in homonuclear D$_2$. Changing J from 0 to 2 does not require a change in nuclear spin state, but changing J from 0 to 1 does, and a collision with Ne has a very small probability of flipping a nuclear spin.

Reactive Scattering

We now know how to treat energy changes in a collision, and to generalize to reactive collisions, we must learn how to treat *mass* changes as well. Suppose we change from Ne to F, keeping the D_2 target in our prototype experiment. The F mass is only slightly less than the Ne mass, which keeps the *nonreactive* scattering Newton diagram roughly the same, but the $F + D_2 \to DF + D$ reaction has an appreciable cross section. It is also among the best studied: one can generate a fluorine atom beam from thermal dissociation of F_2, and the system is sufficiently simple to permit highly accurate quantum chemistry calculations of the intermolecular potential energy surface. This reaction (and its H_2 variant) also has great practical interest in that the DF (or HF) product leads to *chemical lasing action* in the infrared. The reaction produces a distribution of internal energies that is so far from equilibrium that the DF or HF vibrational states are produced with a *population inversion* in which excited vibrational levels are produced at a greater rate than the ground level.

While this reaction is highly exothermic, imagine for a moment that the reaction occurs without any change in internal energy. How does the mass change affect the dynamics? If the internal energy is not changed, then the relative translational energy is not changed either, but the final relative speed changes because *the collisional reduced mass changes in a reactive collision*:

$$E_{\text{rel}} = \frac{1}{2}\mu g^2 = E_{\text{rel}} = \frac{1}{2}\mu' g'^2 \quad \text{thermoneutral, } \Delta E_{\text{int}} = 0 \;.$$

For $F + D_2 \to DF + D$, $\mu = m_F m_{D_2}/M = 76/23$ amu, but $\mu' = m_{DF} m_D/M = 42/23$ amu. The center-of-mass velocity vector is still unchanged in a reactive collision, and g' is still divided at the tip of $\boldsymbol{v}_{\text{cm}}$ into two segments, one $(m_{DF}/M)g'$ in length and the other $(m_D/M)g'$ in length.

Now we include the exothermicity (or endothermicity, to be general) of the reaction. If a reaction has an internal energy change ΔU_0, that number is negative for an exothermic reaction and represents energy *released* into the collision. Conversely, an endothermic reaction has a positive ΔU_0 representing energy *consumed* by the reactive collision. If we again assume that the reaction does not change internal energy other than through ΔU_0, we can write the energy balance expression

$$E_{\text{rel}} = E'_{\text{rel}} + \Delta U_0 \;,$$

which makes physical sense if we consider, for example, an endothermic reaction studied at E_{rel} equal to the endothermicity. In such a study, all of E_{rel} is consumed by the endothermicity (ΔU_0, a positive number) and no energy is left for product translation: $E'_{\text{rel}} = 0$, as the equation predicts.

Finally, we can add the internal energies of the reactants and products (which are usually both non-zero in practice) and write our most general energy balance expression for a reactive collision with arbitrary internal energies:

$$\underbrace{\frac{1}{2}mg^2}_{\substack{\text{Initial relative}\\ \text{translational energy}}} + \underbrace{E_{\text{int}}}_{\substack{\text{Reactants'}\\ \text{internal energy}}} = \underbrace{\frac{1}{2}\mu' g'^2}_{\substack{\text{Final relative}\\ \text{translational energy}}} + \underbrace{E'_{\text{int}}}_{\substack{\text{Products'}\\ \text{internal energy}}} + \underbrace{\Delta U_0}_{\substack{\text{Reaction internal}\\ \text{energy change}}} \quad (28.11)$$

This expression lets us calculate Newton diagrams for any reactive (or nonreactive, if $\mu = \mu'$) collision with known ΔU_0 and internal energies.

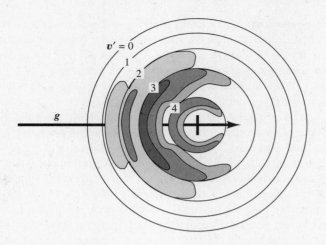

FIGURE 28.16 The reactive scattering of F and D_2 leading to detected DF product at one collision energy is shown here as a scattering flux contour map overlaid on the collisional Newton diagram. The intensity of shading within any contour indicates the qualitative intensity of the reactive flux, and we can see that DF product with $v' = 3$ vibrational excitation scattered at angles near 180° is the most probable outcome of the collision.

For the F + H_2 reaction, ΔU, the reaction internal-energy change neglecting zero-point energies, is about -2.21×10^{-19} J. Including zero-point energies $G(0)$ for D_2 and DF gives us ΔU_0 for F + D_2, but since these two zero-point energies are nearly identical and $\Delta U_0 = \Delta U + G_{D_2}(0) - G_{DF}(0)$, we find here that $\Delta U_0 \cong \Delta U$.

One study of this reaction had $E_{rel} = 1.26 \times 10^{-20}$ J and D_2 in predominantly its ground vibrational level and $J = 0, 1$, and 2 rotational levels.[11] To calculate internal energies for any vibrational quantum number, we also need vibrational constants for DF, and these are, of course, well known. Given these internal energies (and assuming the small D_2 rotational energy is negligible), we have enough information to apply Eq. (28.11) and construct the table below. The quantity $(m_D/M)g'$ is the radius of the DF product's limiting circle in the Newton diagram for a given product vibrational quantum number v', and we see that the total energy in this study was sufficient to produce DF with vibrational excitations in levels $v' = 0$ through 4. Using the data in the table, the initial relative velocity (2140 m s^{-1}) found from E_{rel}, and the center-of-mass velocity (1770 m s^{-1}), we can construct the reactive Newton diagram for this collision system, shown in Figure 28.16 along with contours of measured DF scattering flux.

v'	$E'_{int}/10^{-19}$ J	$E'_{rel}/10^{-19}$ J	$g'/10^3$ m s^{-1}	$\dfrac{m_D}{M} g'$/m s^{-1}
0	0.0	2.33	12.4	1081
1	0.58	1.75	10.7	937
2	1.14	1.19	8.84	773
3	1.68	0.65	6.53	572
4	2.20	0.13	2.90	254

---

[11] Atomic F has two low-lying electronic states of importance, $^2P_{3/2}$, the ground state, and $^2P_{1/2}$, lying 404 cm^{-1} higher. (See Example 13.5.) Since the F source was at a high temperature, both of these states had significant population, but, as discussed further in Section 28.4, only the lower state reacts significantly with D_2.

Figure 28.16 shows only a few of the experimentally derived contours for simplicity, but even with a few of them, we can reach several important conclusions about the *dynamics* of $F + D_2$ reactive scattering. The most obvious is that the DF product is scattered predominantly at large scattering angles, peaking at $\chi = 180°$. This indicates that nearly head-on collisions are most effective at promoting reaction. Second, we see the vibrational population inversion clearly in these data: there is *no* measurable DF produced in the ground vibrational state, and in fact, $v' = 3$ is the most probable vibrational state of the product. Finally, although the speed resolution in this experiment was not sufficient to resolve individual *rotational* states of the product, we can see from the way the contours hug the *inside* of each limiting circle that the products do not have large rotational excitations. A large rotational energy would place flux far from its corresponding vibrational circle, inwards toward the center of mass. Experiments at comparable energies with an H_2 or HD target show similar features.

If the experiment is repeated with a collision energy around twice that of Figure 28.16, something unexpected is found. While many general features of the dynamics remain the same, the flux for $v' = 4$ shows a peak at $\chi = 0°$ rather than 180°. This change and similar features in the HF from $F + H_2$ scattering flux have been interpreted in terms of a quantum mechanical dynamic *resonance,* a quantum feature of the scattering that focuses reaction products in a particular direction.

All of these experiments are in good over-all agreement with other types of dynamical studies of this reaction system. In particular, *infrared chemiluminescence* measurements have been used. In this experiment, the spontaneous infrared emission of the HF or DF products is dispersed and recorded under conditions in which collisions with background gas *after* the reactive collision were sufficiently infrequent to preserve the initial vibrational and most of the initial rotational distributions. Likewise, studies of the infrared chemical laser emission produced by this reaction's vibrational population inversion give good quantitative agreement for quantities such as the relative cross sections for various vibrational states.[12] Of equal importance to the experimental study has been the theoretical study. Quantum chemical calculations of the FH_2 intermolecular potential energy surface and various semiempirical calculations (such as elaborated LEPS methods) have been coupled to classical collinear and 3D trajectory calculations and to quantum reactive scattering theories (beyond our scope here) to produce a very complete picture of this system's dynamics.

It is, sadly, rather hopeless to expect that we can study every gas-phase reaction of interest with the same level of detail that the $F + H_2$ system has attained. Many other systems have been studied with comparable detail, of course, but one would like to find a set of *general* features exhibited by a small number of classes of reactions that would allow us to make reasonable expectations about chemically and dynamically similar reactions. This brings us to the realm of *reaction dynamics models,* the topic of the next two sections.

28.3 MODELS FOR ELEMENTARY REACTION DYNAMICS

The $F + H_2$ reaction falls into the general class of *direct* reactions. The colliding reactants do not linger in each other's vicinity for a long time. The other general class, mentioned at the end of Section 28.1, is the *long-lived collision complex*

---

[12]The crossed-molecular beam study of the $F + H_2$ and its isotopic variants is one of the classic studies in the reaction dynamics literature. The work was done over many years under the direction of Yuan T. Lee, who, along with another leader of the field, Dudley R. Herschbach, shared the 1986 Nobel Prize in Chemistry with John Polanyi, who was mentioned in Section 28.1 and was also instrumental in the $F + H_2$ study. Polanyi pioneered the use of infrared chemiluminescence in reaction dynamics studies.

model. These are usually easily distinguished dynamically. The direct reaction leads to products scattered predominantly in a rather narrow range of angles, such as the back-scattered DF product in Figure 28.16. In contrast, the collision complex has a scattering pattern that reflects its unimolecular dissociation: it has lost all memory of the direction of the initial relative velocity. This scattering is *isotropic* in the center-of-mass frame (or at least has forward-backward symmetry about the directions $\chi = \pm 90°$).

Here, we will look at some simple direct reaction models that have been found in real systems and that can be used to give simple expressions for the reaction rate constant. We will also explore one model that has features of a collision complex model—orbiting will play an important role in it—but also applies to reactions that are known to be direct. To get a feeling for how to proceed from a state-to-state cross section to a bulk rate constant, we will begin with a derivation of the expression quoted in Chapter 27 for the rate constant between hard spheres that must surmount a potential energy barrier of height E^* in order to react.

The expression we are aiming for is

$$k_2 = N_A \left(\frac{8 k_B T}{\pi \mu} \right)^{1/2} \sigma_{12} (1000 \text{ L m}^{-3}) \, e^{-E^*/k_B T} \qquad (28.12)$$

where k_2 is the bimolecular rate constant, E^* is the critical energy, and σ_{12} is the hard-sphere total cross section πd^2. To derive it from dynamics, we must begin at the beginning, with an expression for the conservation of energy *during* a collision. The kinetic and the potential energy sum to the total energy, of course, but for collisions in three dimensions, we must remember that *collisional angular momentum* plays a role and adds to the *real* intermolecular potential energy, $U(R)$, a term called the *centrifugal potential,* which we have encountered twice before: once in our discussion of the H atom in Chapter 12 (see Figure 12.13), and once in our discussion of diatomic molecule rotation and vibration (see Eq. (19.11)).

The centrifugal potential is $L^2/2\mu R^2$ where L is the *collisional* angular momentum. Imagine a collision with *no* force between collision partners. The collision has some impact parameter b and the trajectory is a straight line in the effective center-of-force frame. At some point, the representative particle of mass μ in this frame passes the center of force at a distance $R = b$. Note that this is the *minimum* distance between it and the center of force along the whole trajectory. The particle has linear momentum $p = \mu g$ directed at a right angle to a line of length b connecting it to the center of force:

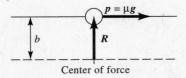

At this point, the angular momentum of the particle with respect to the center of force is, from the general definition of angular momentum (see Eq. (12.26) in Section 12.4),

$$\begin{aligned} L &= Rp \sin\alpha & &\text{from Eq. (12.26)} \\ &= bp & &\text{since } \alpha = 90° \text{ and } R = b \qquad (28.13)\\ &= \mu g b & &\text{since } p = \mu g. \end{aligned}$$

But since angular momentum is conserved in the absence of external forces, this must be *the* collisional angular momentum no matter *where* the particle is and no matter *what* the intermolecular potential is.

We can now write a simple expression for the total energy at any point in any collision between particles that have no internal energy or angular momentum and that interact through a potential that depends only on their separation, R:

$$E_{\text{tot}} = \frac{1}{2}\mu\left(\frac{dR}{dt}\right)^2 + U(R) + \frac{(\mu g b)^2}{2\mu R^2} \quad (28.14)$$

↑ Total collision energy, $\mu g^2/2$ ↑ Relative kinetic energy ↑ Intermolecular potential energy ↑ Centrifugal potential

This expression is useful since it contains dR/dt, and at the distance of closest approach in *any* collision, $dR/dt = 0$ as the trajectory instantaneously pauses as it switches between the "incoming" leg and the "outgoing" leg of the collision. If we solve Eq. (28.14) for dR/dt, we find

$$\frac{dR}{dt} = \pm\left\{\frac{2}{\mu}\left[E_{\text{tot}} - U(R) - \frac{(\mu g b)^2}{2\mu R^2}\right]\right\}^{1/2},$$

and at the distance of closest approach, R_c, we have $dR/dt = 0$ so that

$$E_{\text{tot}} - U(R_c) - \frac{(\mu g b)^2}{2\mu R_c^2} = 0. \quad (28.15)$$

Now we note that $E_{\text{tot}} = \mu g^2/2 = E_{\text{rel}}$, since the total collision energy is set at $R = \infty$ where $U(R)$ and the centrifugal potential are both zero and where $dR/dt = g$. This allows us to write Eq. (28.15) as

$$E_{\text{rel}} = U(R_c) + E_{\text{rel}}\left(\frac{b^2}{R_c^2}\right)$$

or

$$b^2 = R_c^2\left[1 - \frac{U(R_c)}{E_{\text{rel}}}\right]. \quad (28.16)$$

We now apply this expression to our "reactive hard spheres with a potential barrier" problem: in order to react, the colliding species must at least reach R_c at which their mutual potential energy is $U(R_c) = E^*$. Note that, in fact, this scenario is more general than for the hard-sphere potential. All we need is some $U(R_c) > 0$ to establish a barrier. We interpret b in Eq. (28.16) to be the *critical impact parameter for reaction*, b_c, since any collision with a smaller b will reach R_c. Thus, if we multiply Eq. (28.16) by π, we have the total reaction cross section for a collision with $E_{\text{tot}} = \mu g^2/2$:

$$\sigma = \pi b_c^2 = \pi R_c^2\left(1 - \frac{E^*}{E_{\text{rel}}}\right) = \pi R_c^2\left(1 - \frac{2E^*}{\mu g^2}\right), E_{\text{rel}} > E^*. \quad (28.17)$$

Note that this cross section is *zero* at and below threshold: $\sigma = 0$ for $E_{\text{rel}} \leq E^*$.

The reaction rate constant *for this collision energy* is just the product of the cross section and the relative speed (with conversion factors to yield traditional units, as we saw in Eq. (27.8)):

$$k_2(g) = N_A g \sigma (1000 \text{ L m}^{-3}).$$

The *thermal* reaction rate constant is $k_2(g)$ averaged over a Maxwell–Boltzmann distribution of relative speeds. This is given by Eq. (24.5) written to represent relative speeds and the collisional reduced mass:

$$f(g)\, dg = 4\pi\left(\frac{\mu}{2\pi k_B T}\right)^{3/2} g^2 e^{-\mu g^2/2k_B T}\, dg.$$

Thus, we must evaluate

$$k_2 = \int_{\sqrt{2E^*/\mu}}^{\infty} k_2(g) f(g) \, dg \qquad (28.18)$$

where the lower limit to the integral reflects the fact that the reaction cross section is zero for relative energies $\mu g^2/2 \leq E^*$ or for relative speeds $g \leq \sqrt{2E^*/\mu}$.

When the full expressions for k_2 and f are substituted into Eq. (28.18), one finds a rather formidable looking integral, but (see Problem 28.28) it can be evaluated in a straightforward way to yield

$$k_2(T) = N_A (1000 \text{ L m}^{-3}) \pi R_c^2 \left(\frac{8k_B T}{\pi \mu}\right)^{1/2} e^{-E^*/k_B T},$$

which is our target expression, Eq. (28.12), if we associate πR_c^2 with σ_{12}.

It is worth comparing this correct derivation to the slightly glib derivation implied in Chapter 27. There, we wrote $k_2 \propto \langle g \rangle \sigma_{12}$, which implies that the reaction cross section is independent of g (it is not) and that all values of g should enter the average (they should not, since any $g \leq \sqrt{2E^*/\mu}$ does not lead to reaction). Our derivation is more general than for the simple hard-sphere model.

We can apply this approach to other potential energy functions and derive rate constants appropriate for various classes of reaction. Problem 28.29 does this for the case of *dielectric recombination,* the mutual neutralization of positive and negative ions, and here we look at a related class, the reaction between a positive (or negative) ion and a neutral molecule. As we remarked in Chapter 27 (see Example 27.6 and the discussion following it), simple ion–molecule reactions often have no activation energy and very large rate constants. From Chapter 24 (see Table 24.3 and Example 24.10), we know that an ion of charge q_1 interacts with a molecule of polarizability volume α' through the purely attractive *ion–induced dipole* potential-energy function

$$U(R) = -\frac{q_1^2 \alpha_2'}{(4\pi\epsilon_0) 2R^4}. \qquad (28.19)$$

Since this potential is appreciable at fairly long range, we will neglect any short-range repulsion forces and assume the dynamics are entirely controlled by this attractive potential-energy term. We will also assume that the ion must approach the neutral to within a particular radial distance at least to react, and that any closer approach always leads to reaction. Unlike some other models we have encountered, we can give this distance a straightforward physical interpretation here. Suppose a collision has just the right impact parameter and E_{rel} to lead to *orbiting*. We will take the reaction distance to be the *orbiting radius* R_0 for such a collision. This makes sense for the following reason: if the impact parameter is smaller than the orbiting b, the attractive potential will ensure approach to $R < R_0$, and if b is larger, the collision will have a closest approach larger than R_0. We can see that this must be so if we plot the effective potential energy, $U_{\text{eff}}(R) = U(R) + (\mu g b)^2/2\mu R^2$, for various b at any one g.

To be specific, consider the $Ar^+ + H_2$ reaction mentioned in Example 27.6. We look up the average polarizability volume of H_2 in Table 15.3, give Ar^+ its charge $q_1 = e$, and assume for now g equals $\langle g \rangle$, the average relative speed, at $T = 300$ K. From Eq. (24.15), $g = 1820$ m s^{-1}.

Before we graph U_{eff}, we should find the orbiting impact parameter. At orbiting, $dR/dt = 0$, and all the collision energy has been converted to potential energy; there is no *radial* kinetic energy available to move the particles closer together or farther apart. Since the effective potential is a sum of an attractive part and a repulsive part (the centrifugal potential), $U_{\text{eff}}(R)$ has a maximum in R for any $b \neq 0$. When E_{rel} is equal to the value of U_{eff} at this maximum, orbiting occurs, and R_0 equals that R for which U_{eff} is maximum.

The condition $dU_{\text{eff}}(R)/dR = 0$ leads to an expression for R_0 in terms of b and g:

$$R_0^2 = \frac{2e^2\alpha'}{(4\pi\epsilon_0)\mu g^2 b^2} = \frac{e^2\alpha'}{(4\pi\epsilon_0)b^2 E_{\text{rel}}} . \quad (28.20)$$

Equating E_{rel} to U_{eff} at R_0 leads to a relation for the orbiting impact parameter b_0:

$$b_0^2 = \frac{2e}{g}\left[\frac{\alpha'}{(4\pi\epsilon_0)\mu}\right]^{1/2} . \quad (28.21)$$

If we multiply this expression by π, we have the so-called *Langevin cross section*, first derived in 1905:

$$\sigma_L = \pi b_0^2 = \frac{2\pi e}{g}\left[\frac{\alpha'}{(4\pi\epsilon_0)\mu}\right]^{1/2} = \left(\frac{\pi\alpha' e^2}{2\epsilon_0 E_{\text{rel}}}\right)^{1/2} . \quad (28.22)$$

EXAMPLE 28.7

What is the Langevin cross section for the $\text{Ar}^+ + \text{H}_2$ reaction for $g = \langle g \rangle$ at 300 K? How does this number compare to the cross section found in Example 27.6?

SOLUTION The parameters of Eq. (28.22) for this reaction are mentioned in the text and lead to a value $\sigma_L = 8.32 \times 10^{-19}$ m^2, which compares well with the value 6.0×10^{-19} m^2 found in Example 27.6 using just the 300 K bimolecular rate constant and the average relative speed. We should not expect outstanding agreement here for two reasons: first, we do not yet know if this reaction model is at all reliable, and second, σ_L is valid for only one relative energy. We must average it over a Maxwell–Boltzmann distribution to compare it to k_2.

We can also find the orbiting impact parameter from Eq. (28.21) and the orbiting radius from Eq. (28.20): $b_0 = 5.15$ Å and $R_0 = 3.63$ Å. These numbers tell us that the initial trajectory spirals in slightly to its stable orbit.

➡ *RELATED PROBLEMS* 28.31, 28.32, 28.33

Figure 28.17 plots $U(R)$ and $U_{\text{eff}}(R)$ for $b = b_0$, $0.9b_0$, and $1.1b_0$ for our chosen E_{rel}. The horizontal line locates the total collision energy, E_{rel}. For $b = b_0$ (the heavy

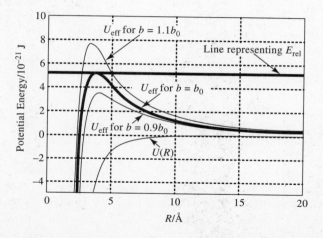

FIGURE 28.17 The effective potential $U_{\text{eff}}(R)$ for an ion–induced dipole potential is drawn for impact parameters at and near to the *orbiting* impact parameter b_0, all at one collision energy E_{rel}. The potential itself, $U(R)$, is also shown for comparison. The orbiting radius R_0 is located at the maximum of the U_{eff} curve for $b = b_0$.

curve), U_{eff} is maximum at R_0, and we see that $U_{eff}(R_0) = E_{rel}$. For smaller impact parameters, U_{eff} is always *below* E_{rel}, and the potential causes the trajectory to spiral in to $R < R_0$, while for larger b, trajectories have a *radial turning point* (distance of closest approach) at some $R > R_0$. For the $b = 1.1b_0$ curve shown in the figure, the radial turning point is about 5 Å, the point at which $E_{rel} = U_{eff}$ at largest R.

The thermal rate constant is again a Maxwell–Boltzmann average over g times the Langevin cross section. When this average is performed, one finds a very interesting expression known as the *Gioumousis–Stevenson rate constant*. We write

$$\begin{aligned}
k_2 &= N_A(1000 \text{ L m}^{-3}) \int g \, \sigma_L \, f(g) \, dg & &\text{by definition of } k_2 \\
&= N_A(1000 \text{ L m}^{-3}) \int \left(\frac{2E_{rel}}{\mu}\right)^{1/2} \left(\frac{\pi\alpha' e^2}{2\epsilon_0 E_{rel}}\right)^{1/2} f(g) \, dg & &\text{by definition of } g \text{ and } \sigma_L \quad (28.23)\\
&= N_A(1000 \text{ L m}^{-3}) \left(\frac{\pi\alpha' e^2}{\epsilon_0 \mu}\right)^{1/2} \int f(g) \, dg & &\text{collecting common factors} \\
&= N_A(1000 \text{ L m}^{-3}) \, e \left(\frac{\pi\alpha'}{\epsilon_0 \mu}\right)^{1/2} & &\text{since } \int f(g) \, dg = 1.
\end{aligned}$$

This expression is remarkable in that it represents a *thermal* rate constant, but it does *not* depend on temperature. This is approximately the observed behavior for many ion–molecule reactions. For the $Ar^+ + H_2$ reaction, it predicts $k_2 = 9.2 \times 10^{11} \text{ M}^{-1} \text{ s}^{-1}$, in reasonable agreement with the experimental value quoted in Example 27.6, $6.6 \times 10^{11} \text{ M}^{-1} \text{ s}^{-1}$.

The Langevin model takes the dynamics of the reaction up to the orbiting radius, but from then on, it fails to illuminate any further aspects of the reaction. Do the reactants need to form a long-lived collision complex in order to swap an atom, or is the swap a fast, direct process? Are the products scattered backwards from the initial direction of the reactants, as in $F + H_2$, or are they scattered in some other direction? The answers to these questions have been discovered in molecular beam experiments of ion–molecule collisions, performed much as neutral beam experiments are, but with typically one important difference: ion beams are difficult to generate with useful intensities at low laboratory speeds. A beam of ions naturally spreads due to the strong Coulombic repulsive forces that exist between ions of the same charge. Consequently (and for other reasons), ion beams usually have laboratory kinetic energies far above thermal. A typical thermal kinetic energy is on the order of $k_B T$ with T somewhere between 200 and 1200 K, which translates into 0.02–0.10 eV. In contrast, ion kinetic energies are more commonly in the range 1–100 eV.[13] While extrapolation of high relative energy dynamics down to ordinary thermal energy dynamics can be a tricky thing to do, the dynamics of some ion–molecule reactions have been studied in detail at energies more appropriate to thermal experiments. These lower energy experiments along with the higher energy studies show many common gross features for many simple systems, and simple dynamic models have been invented to describe them. We close this section with a look at one of the more common and important of these models.

We said that the DF product in $F + D_2$ was backscattered around $\chi = 180°$ because small impact parameter collisions must be required to surmount the potential-energy barrier to reaction. The $Ar^+ + H_2 \rightarrow ArH^+ + H$ reaction, however, shows

---

[13]There are other types of ion scattering experiments for which ion kinetic energies in the keV to MeV range are appropriate, but collision chemistry usually has too small a cross section at these higher energies.

dynamics that leads almost exclusively to product scattered at $\chi = 0°$. How can this be? We get an important clue when we look at the *speed* of the product as well as its scattering angle.

Imagine the following sequence of events during the collision: Ar^+ approaches a stationary H_2, the H_2 bond magically dissolves as the ArH^+ bond is formed, one H atom is transferred while the other is oblivious to the events around it, and the ArH^+ product moves in the same direction the Ar^+ had initially, requiring the H product to keep the initial direction of the H_2 reactant. This is the sequence embodied in the model known as *spectator stripping*. The H atom that becomes the product atom is a spectator to the reaction, while the other H atom is stripped from H_2 by Ar^+. The final velocity of the products is particularly easy to find in this model. Since the spectator H does nothing of great consequence, we can forget about it momentarily and think of the reaction as *the totally inelastic collision between* Ar^+ *and a single H atom*. This is proper to do because these species end the collision bound to each other, just as we pictured a totally inelastic event happening in the previous section.

If we again imagine the H_2 is stationary in the lab while Ar^+ moves toward it with velocity $\boldsymbol{v}_{Ar^+} = \boldsymbol{g}$, we can define the center-of-mass velocity of the collision in the usual way: $\boldsymbol{v}_{cm} = (m_{Ar}/M)\boldsymbol{g}$. In addition, we can define the center-of-mass velocity of Ar^+ with respect to *only one* of the H atoms in H_2. This quantity is $[m_{Ar}/(m_{Ar} + m_H)]\boldsymbol{g}$ and is equal to the *spectator stripping velocity*, $\boldsymbol{v}_{ss}$, the velocity at which we will find our ArH^+ product. For the general reaction $A + BC \rightarrow AB + C$, we have

$$v_{ss} = \frac{m_A}{m_A + m_B} g \ . \tag{28.24}$$

EXAMPLE 28.8

In an experimental study of the $Ar^+ + D_2 \rightarrow ArD^+ + D$ reaction, a beam of Ar^+ with a laboratory kinetic energy of 30.0 eV is created and directed toward a scattering cell containing D_2 gas at room temperature. What is the collision energy, E_{rel}, in this experiment? What are the laboratory speeds of the products, assuming the spectator stripping model?

SOLUTION A kinetic energy of 30.0 eV equals 4.81×10^{-18} J, and equating this to $m_{Ar}v_{Ar}^2/2$, the usual kinetic-energy expression, tells us that the Ar^+ lab speed v_{Ar} is 1.20×10^4 m s^{-1}. Room temperature D_2 has a most probable speed (see Table 24.1) $v_{mp} = 1100$ m s^{-1}, and this is sufficiently less than v_{Ar} that we can assume the D_2 is stationary in the lab. Consequently, $g = v_{Ar}$, and E_{rel} is given by

$$E_{rel} = \frac{1}{2}\mu g^2 = \frac{1}{2}\left(\frac{m_{Ar}m_{D_2}}{m_{Ar} + m_{D_2}}\right)g^2 = 4.40 \times 10^{-19} \text{ J} = 2.75 \text{ eV} \ .$$

This is a much higher collision energy than is common (or often even possible) in studies of neutral species reactions. For example, the $F + D_2$ experiment discussed in the previous section had $E_{rel} = 1.26 \times 10^{-20}$ J = 0.079 eV.

The ArD^+ laboratory speed is just the magnitude of the spectator stripping velocity:

$$v_{ss} = \frac{m_{Ar}}{m_{Ar} + m_D} g = 1.14 \times 10^4 \text{ m s}^{-1} \ .$$

Since we assumed D_2 was stationary in the lab, the spectator D atom that is the other reaction product will also be stationary.

Note that, due to the laboratory dynamics of this reaction (a heavy projectile striking a light target), the heavy reaction product is confined to a small range of laboratory speeds and scattering angles *no matter what the reaction dynamics might be.* The center-of-mass speed for the reaction is 1.09×10^4 m s^{-1}, only slightly less than the spectator stripping speed or the Ar$^+$ lab speed itself. Consequently, experiments of this type require good speed resolution to distinguish features in the scattering that might be closely spaced. For ions, a time-of-flight method is generally a less useful way to measure speeds than is any of several methods that use electric fields to measure an ion's laboratory kinetic energy.

⟹ RELATED PROBLEMS 28.34, 28.35, 28.36

Note that the spectator stripping velocity depends *only* on masses, not on collision energy or any other aspect of the reaction. This means that the *fraction* of collision energy that ends as internal energy in the diatomic product is constant, and thus, as the collision energy increases, the product internal energy increases. At some point, this internal energy will *exceed* the diatomic's bond energy. What happens then? Experiments have discovered two answers. First, at low energy, one finds that the product often appears at a speed slightly *greater* than v_{ss}, which means the product internal energy is slightly *less* than that predicted by spectator stripping. Some feature of the reaction's intermolecular potential energy helps to stabilize the product a bit. As the collision energy is increased to and beyond the point where spectator stripping puts more energy in the product than it can hold, one finds that product *still appears at $\chi = 0°$*, but at speed now significantly greater than v_{ss}, again indicating that the products somehow transfer product internal energy into relative translational energy as they separate, stabilizing the diatomic product.

The O$^+$ + H$_2$ → OH$^+$ + H reaction, however, switches from spectator stripping at low E_{rel} to a rather novel mechanism at high E_{rel}. The scattered product no longer appears at $\chi = 0°$, but at larger angles, and a dynamic analysis of the scattering has shown that the reaction follows a *sequence* of steps. First, O$^+$ strikes one of the H atoms, as if the other was not there. This is seen clearly in the *nonreactive* scattering of O$^+$, which looks very much like the *elastic* scattering of O$^+$ from *H atom*, not H$_2$. (In an O$^+$ + HD experiment, the nonreactive scattering looks like the *sum* of O$^+$–H and O$^+$–D elastic scattering, as if the H—D bond dissolves in the presence of O$^+$.) Once one H atom has been struck, it moves and, if the atoms are correctly aligned, strikes the other H atom. The O$^+$ then picks up and remains bound to whichever of the H atoms (and it is usually the first one struck) has a relative velocity with respect to O$^+$ that is less than the OH$^+$ bond energy.

This is called the *sequential impulse* mechanism. The word *impulse* implies the short time allowed for interactions due to the high collision energies of the colliding partners. In our discussion of elastic scattering, we saw that high collision energies take us to the hard-sphere scattering limit, and that is what we have here. The O$^+$ and each H atom behave very much like a set of three nonconnected hard spheres. Chemical bonding is almost an afterthought in this regime.

In this section, we have seen how simple reaction dynamics models allow us to connect the intimate details of bimolecular collisions to bulk thermal rate constants. We have found situations in which the rate constant depended on temperature (the potential energy barrier model of Eq. (28.12)), and in which it did not (the ion–molecule rate constant of Eq. (28.23)). We close this chapter, and this book, with a discussion of one of the most interesting possibilities raised by the study of reaction dynamics: the possibility that we can *control* dynamics in a way that is useful. Can we change reaction branching ratios between various possible products? Can we select reaction conditions that alter the state-to-state cross sections in a useful way, in order to, for example, find new reactions that lead to new chemical lasers?

In a sense, photochemistry has done this for years. Many reactions are known that are said to be "thermally forbidden but photochemically allowed," or the other way around. In your study of organic chemistry, you might have encountered what are known as the *Woodward–Hoffman rules,* a set of requirements for a reaction based on the evolution of molecular orbitals during the reaction. The next section explores the

28.4 IS THERE MICROSCOPIC CONTROL OVER REACTIONS?

There is nearly as much to be learned from the study of elementary reactions that *fail* to react as there is from the study of those that do react. If we can understand why a reaction has a very small cross section under ordinary conditions, then perhaps we can find a set of unusual conditions that allow it to proceed. Often the reasons behind a small cross section are easy to recognize: the reaction is very endothermic, or has a very large activation barrier, or is never reached because of competing reactions that are much more probable. In these cases, we know, in a general way, what to do. For instance, we raise the collision energy to defeat endothermicity, or search for a catalyst to lower the activation barrier.

Our concern here is with other *dynamical* reasons for small reaction cross sections. In the last section, we made use of the $Ar^+ + H_2 \rightarrow ArH^+ + H$ reaction as a paradigm of a fast, fairly easy to understand system. It is isoelectronic to the halogen atom plus H_2 reactions we have used elsewhere as models, such as $F + H_2$, and we might expect that any rare-gas ion would react readily with H_2. In fact, not all of them do. With Xe^+, the reaction is endothermic, but proceeds readily at sufficient relative energy. With Kr^+ and Ar^+, the reaction is exothermic and fast. But with Ne^+ and He^+, which also have exothermic reactions with H_2, *no* reaction is observed.

The He^+ and $Ne^+ + H_2$ systems are both sufficiently simple that we should be able to understand the changes in bonding induced by the reaction in some detail, and perhaps this will show us why they fail to react. After all, HeH_2^+ has only three electrons, just like $H + H_2$! We begin with some basic thermodynamics on this system. Figure 28.18 plots the energies of the various reactants and products for the HeH_2^+ system. The large exothermicity of $He^+ + H_2 \rightarrow HeH^+ + H$ is obvious

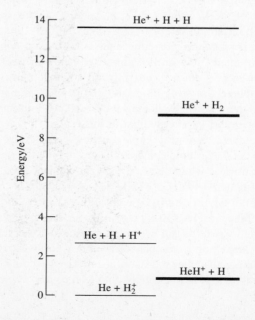

FIGURE 28.18 A diagram showing the relative energies of various possible products for the collision between He^+ and H_2 tells us important quantities such as exothermicities, but it does *not* tell us if or how the reactants can evolve to any of the product sets. That information comes from correlation diagrams.

here, and we also see that one possible reason the reaction fails is that the reaction dynamics forces this large exothermicity into the HeH$^+$ product, which cannot hold it all, leading to the dissociative charge-transfer products, He + H + H$^+$. We also note from the figure that the much larger ionization energy of He over H$_2$ makes the simple charge-transfer products He + H$_2^+$ also exothermic, and perhaps this is the outcome of collisions between He$^+$ and H$_2$.

While this analysis is always a help, it cannot be the entire story, since an energy-level diagram like this does not show us how various combinations of species are *connected by potential-energy surfaces*. In fact, to construct this diagram itself, we have to recognize that HeH$^+$ dissociates in its ground state to He + H$^+$ and *not* to He$^+$ + H, which are, since the ionization energy of He is greater than that of H, the dissociation products of an *excited* electronic state of HeH$^+$. We first encountered the notion of *correlation diagrams* in Section 14.5, where we used them to understand the various dissociation limits of H$_2$O in Figures 14.27–14.29. What we need here is a *reaction correlation diagram*.

Correlation diagrams can be drawn for *electronic states*, as in Figures 14.27–14.29, or for *molecular orbitals* (MOs). In the latter case, we seek the evolution of relevant occupied MOs (or atomic orbitals) in the reactants as they interact and evolve to products with their set of MOs. If these correlations connect occupied MOs of the reactants to MOs of the products characteristic of their ground states, we can expect reaction. If the correlation is to highly excited MOs of the products, then we have discovered either the source of a large potential-energy barrier to reaction or the source of an unexpected endothermicity, since the reactants most naturally lead to *excited* electronic states of the products in this case.

We will construct a collinear MO correlation diagram for the HeH$_2^+$ system. This is relatively easy to do, and it will expose the reason He$^+$ fails to react with H$_2$. Moreover, it will suggest which of the other possible outcomes of the energy-level diagram we are likely to observe. Figure 28.19 is the complete MO correlation

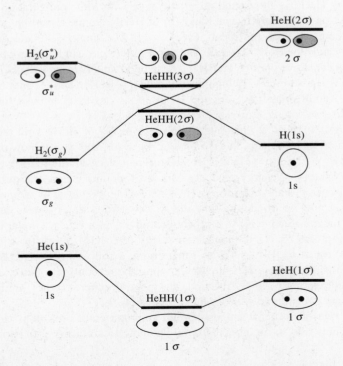

FIGURE 28.19 The MO correlation diagram for the HeHH system shows what is possible and what is difficult or impossible in He$^+$ + H$_2$ collisions. The MOs are qualitatively ordered in energy according to ionization and excitation energies, and sketches of the MO shapes are drawn at each level.

diagram, drawn in the following way. On the left, we order in energy the three important orbitals of the reactants (*without* specifying the number of electrons in each, so that the diagram is general for $He^+ + H_2$ *or* for $He + H_2^+$). The ionization and electronic excitation energies of these species is our guide: He 1s, with the largest ionization energy, is lowest, followed by the σ_g ground-state MO of H_2 and its σ_u^* first-excited MO at the top. On the right, we again use ionization and excitation energy to locate MOs of the products: HeH 1σ is the lowest MO, followed by ground-state H 1s and the first excited MO of HeH, 2σ. Schematic MO diagrams are drawn to show the rough symmetries and nodal patterns of these MOs, as we did in Chapter 14. In the center of the diagram, we locate the 1σ–3σ MOs of the HeHH collinear intermediate, along with the schematic diagrams of each.

The lines connecting these various MO energy levels constitute the correlations. The HeHH 1σ MO is predominantly made from the He 1s orbital when approached from $He + H_2$ *or* from the HeH 1σ MO when approached from $HeH + H$. It is a nodeless, bonding MO that is lower in energy than either of its predominant components: He 1s or HeH 1σ. The HeHH 2σ MO contains a single node near the central atom (but not directly on it, as is the case for the linear H_3 $1\sigma_u^*$ MO shown in Figure 14.17). As He and H_2 approach, the H_2 σ_g MO rises in energy as the node forms. Finally, the 3σ MO of HeHH has two nodes and is very antibonding. It arises from a combination of He 1s and H_2 σ_u^*. Initially, the σ_u^* orbital drops in energy, but at intimate He–H–H distances, the HeHH 2σ and 3σ orbitals interact (indicated by the correlation lines that cross between 2σ and 3σ). Similar arguments hold for the correlations starting from the product side.

Given this diagram, we can populate reactant MOs with electrons appropriate for either the $He^+ + H_2$ *or* the $He + H_2^+$ reaction. Consider $He + H_2^+$ first, which Figure 28.18 shows is slightly endothermic (about 0.8 eV) but which experiment shows to have an appreciable cross section if the collision energy exceeds the endothermicity. For this reaction, He 1s has two electrons, and the correlation diagram shows that they smoothly correlate through HeHH 1σ to the HeH^+ 1σ MO, giving HeH^+ in its ground electronic state with MO configuration $1\sigma^2$. The single electron in H_2^+ correlates through 2σ of $HeHH^+$ to the H-atom product. Now consider the electron arrangement of $He^+ + H_2$. The single He 1s electron moves to *singly occupy* the HeH^+ 1σ MO. The two H_2 electrons can have several possible fates. They can move through $HeHH^+$ 2σ and on to *doubly occupy* the H 1s product orbital. This leads to the very high-energy products HeH^{2+} and H^-, not at all what we want. Another possibility for the H_2 electrons has them interact through the $HeHH^+$ 2σ–3σ mixing region with one going on to H 1s, as we want, but the other going into HeH^+ 2σ, leading to an *excited state* of HeH^+ with the electron configuration $1\sigma 2\sigma$. Again, this is not what we want, and we can now understand why the $He^+ + H_2$ reaction fails to go: the reactant electrons are in the wrong place to lead to stable, low-energy products. Moreover, neither charge transfer to $He + H_2^+$ nor dissociative charge transfer to $He + H + H^+$ is predicted by the correlations.

You may now be wondering how the $Ar^+ + H_2$ reaction happens, and rightly so. While the nonreactive $Ne^+ + H_2$ correlation diagram is very similar to that for He^+ (see Problem 28.37), the ionization energy of Ar (15.759 eV from Table 7.3) is so close to that for H_2 (15.427 eV) that during a collision between Ar^+ and H_2, an electron can hop from H_2 onto Ar^+, fooling the reaction into thinking it is the smoothly correlating $Ar + H_2^+$ reaction. The highest occupied AO of Ar^+ has nearly the same energy as the lowest occupied MO of H_2.

This is one of the most interesting and common modes reactions can take to reach products that an initial analysis might show to be difficult or impossible to attain. If the reaction happened very quickly, the interactions that allow the electron to hop might not have time to do their job. This has been seen experimentally in another class of collisions. If an alkali halide is pulled slowly apart, it ends up as a neutral halogen and a neutral alkali metal atom, but if it is forced apart quickly, as it is in an energetic collisional dissociation event with an accelerated rare-gas atom, for example, it comes apart into the higher energy pair halide anion + alkali cation. (See Figure 14.25 for a view of the same phenomenon in LiH. There, the $Li^+ + H^-$ product pair lies well above the neutral Li + H pair.)

In electronic *state* correlation diagram language, one says the electron hopped (or otherwise rearranged itself into a new MO) through a *change from one potential-energy surface to another*. It is beyond our discussion here to consider in detail the types of interactions that cause a jump between potential-energy surfaces, but effects such as spin–orbit coupling, vibronic coupling, and other effects not treated explicitly by the Born–Oppenheimer approximation are the culprits. But whatever the reason for the jump, it will not happen unless the reaction trajectory takes the system into those atomic configurations for which the probability of the jump is large. *This* is one way we might be able to control reaction outcomes: we might find ways of *enhancing* the probability that a reaction reaches configurations that lead to favorable jumps (or enhancing access to regions that decrease the probability of *undesirable* jumps).

What sorts of things might we do? Theory must be our guide, since theory can locate auspicious configurations more easily than experimentation, in many systems. One might find that a vibrational excitation of a particular normal mode or two of a reactant moves atoms more closely into the coupling region, as the jump region is often called. Perhaps a totally different reactant electronic state is called for, as is commonplace in many photochemical reactions. Sometimes subtle changes in electronic state have profound consequences. For example, in the $F + H_2$ reaction, as mentioned in Footnote 11, the ground electronic state of F, $^2P_{2/3}$, does all the work, while the $^2P_{1/2}$ excited state, only 404 cm^{-1} = 0.050 eV higher, does *not* correlate to HF + H products and effectively fails to react at all. If the $^2P_{3/2}$–$^2P_{1/2}$ transition was strongly allowed, which, alas, it is not, we could shine 404 cm^{-1} infrared light on the $F + H_2$ chemical-laser reaction mixture and easily modulate its intense emission as we switched population between F $^2P_{3/2}$ and F $^2P_{1/2}$ states.

Even if a single potential-energy surface governs the reaction, there may be dynamic opportunities to modify the reaction rate. For example, consider a reaction with a modest potential-energy barrier at the transition state. We know that raising the system's temperature will increase the rate, but perhaps the dynamics indicate a more subtle approach. Systems are known for which selective vibrational excitation of one reactant increases the reaction rate many times more than would raising the temperature an amount equal to the vibrational energy input. Raising temperature spreads energy over many degrees of freedom, while selective excitation can place energy in modes that are most effective to promote a reaction.

We close with what is perhaps the most interesting opportunity. Theory suggests, but experiment has not fully shown, that it may be possible to design a light pulse that can selectively excite and perhaps break virtually any chosen bond in any molecule. This would be the ultimate manipulation of a molecule, and an analogy from more ordinary human experience may help you understand how difficult a task such a manipulation is.

Imagine you have just entered a concert hall. People around you are talking and moving to their seats, the orchestra is perhaps moving on stage and warming up, and what you hear is a rather incoherent signal: noise. This is the aural equivalent of "white" light, which has very nonspecific effects on molecules, if it is absorbed at all. Once people settle into their seats, often the oboe plays a single, pure note as a cue to the orchestra to check the tuning of their instruments. This is the equivalent of a laser, a monochromatic source that has a specific effect on those molecules sympathetic to its pure signal. The effect, however, is often no more exciting than that of the oboe: it "tunes-up" the molecule.

A graph of noise, plotted as sound intensity versus time, is an incoherent, random squiggle across the page. A graph of the oboe's tuning note is nearly a pure sine wave, tightly constrained to one frequency. Contrast either of those graphs to the one shown in Figure 28.20, which is a graph of the first second of a signal that has a profound and specific effect on a very complicated chemical system. It is music, and the chemical system is anyone listening to it. Theory suggests that a light signal that can produce the profound and specific effect of breaking a single selected bond in a polyatomic molecule may need to be as complicated as music. If we can learn how to generate such signals, then perhaps we can "play" molecules much as we play musical instruments and perhaps with as much pleasure.

The signal shown in Figure 28.20 is the first second of the overture to the musical adaptation of Voltaire's novel, *Candide,* composed by Leonard Bernstein. This

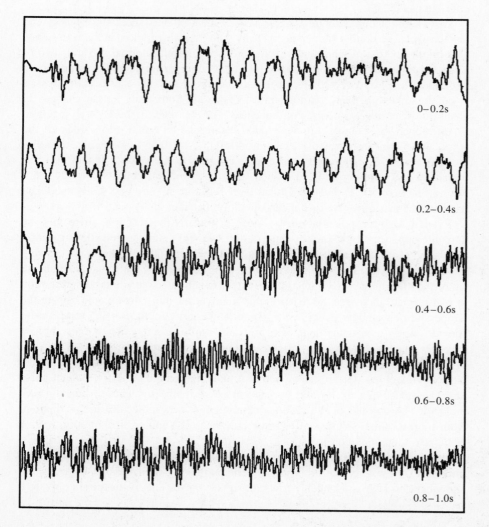

FIGURE 28.20 The acoustic signal as a function of time for one second of music is a complicated stimulus that evokes in us a similarly complicated and unique response. It is likely that evocation of a complicated response in a molecule will require a similarly complicated stimulus in the form of a light pulse. If such pulses can be designed, they may be able to direct chemical reactions in ways that are unknown now.

overture opens with a musical motif that is repeated throughout many other tunes in the show. It represents an oft-repeated phrase spoken (or sung) by the character Pangloss, the unrealistically optimistic teacher of the novel's hero, Candide, and it is the phrase spoken by him to end the musical. The phrase is also a good one with which to end this book. It is, "Any questions?"

CHAPTER 28 SUMMARY

This chapter has used classical mechanics, for the most part, to describe various aspects of isolated bimolecular collisions between molecules. The dynamics of these collisions depend on several quantities specified at the start of the collision. For a collision represented by the general reaction A + BC → AB + C, these quantities include:

$$\text{Collision reduced mass: } \mu = \frac{m_A m_{BC}}{m_A + m_{BC}} = \frac{m_A m_{BC}}{M}$$

$$\text{Collision relative velocity: } \boldsymbol{g} = \boldsymbol{v}_A - \boldsymbol{v}_{BC}$$

$$\text{Relative translational energy: } E_{rel} = \frac{1}{2}\mu g^2$$

$$\text{Center-of-mass velocity: } \boldsymbol{v}_{cm} = \frac{m_A}{M}\boldsymbol{v}_A + \frac{m_{BC}}{M}\boldsymbol{v}_{BC}$$

$$\text{Collisional energy balance:}$$
$$\frac{1}{2}\mu g^2 + E_{int} = \frac{1}{2}\mu' g'^2 + E'_{int} + \Delta U_0$$

These relations tell us the quantities we need to construct a Newton diagram, which shows us the dynamic limits imposed by conservation of energy and linear and angular momentum on any scattering product.

We looked at elastic scattering in detail, focusing on hard-sphere collisions as models of more realistic intermolecular interactions. For hard spheres of radii r_1 and r_2 such that $r_1 + r_2 = d$, the hard-sphere diameter, we derived the following quantities that describe their collisions in terms of the collisional impact parameter, b:

$$\text{Deflection function: } \chi(b) = \begin{cases} 2\cos^{-1}\frac{b}{d}, & b \leq d \\ 0, & b > d \end{cases}$$

$$\text{Differential scattering cross section:}$$
$$I(\chi) = \frac{b}{\left|\sin\chi \frac{d\chi}{db}\right|} \text{ (in general)}$$
$$= \frac{d^2}{4} \text{ (for hard spheres)}$$

We also considered two models that lead to expressions for the thermal rate constant in terms of quantities based on collision dynamics. In the first, the only assumption we made was that a collision had to possess energy in the relative translational coordinate in excess of a threshold amount E^* at some minimum collision separation R_c. We derived expressions for the total reaction cross section at any one relative translational energy and, by averaging this over a Maxwell–Boltzmann distribution of relative speeds, the thermal rate constant:

$$\text{Total reaction cross section for an energy threshold } E^*:$$
$$\sigma = \pi b_c^2 = \pi R_c^2 \left(1 - \frac{E^*}{E_{rel}}\right)$$

$$\text{Thermal rate constant for a threshold reaction:}$$
$$k_2(T) = N_A(1000 \text{ L m}^{-3}) \pi R_c^2 \left(\frac{8k_B T}{\pi\mu}\right)^{1/2} e^{-E^*/k_B T}$$

We then applied a related dynamic model that emphasized orbiting collisions to derive the Langevin cross section and the Gioumousis–Stevenson rate constant for ion–molecule reactions:

$$\text{Langevin cross section: } \sigma_L = \left(\frac{\pi\alpha' e^2}{2\epsilon_0 E_{rel}}\right)^{1/2}$$

$$\text{Gioumousis–Stevenson rate constant:}$$
$$k_2 = N_A(1000 \text{ L m}^{-3}) e \left(\frac{\pi\alpha'}{\epsilon_0 \mu}\right)^{1/2}$$

We closed with an introduction to reaction correlation diagrams, which guide our understanding of the possible products and their electronic quantum states that any one set of reactants can achieve.

FURTHER READING

In addition to the references at the end of Chapter 26, there are several books specific to the topic of chemical reaction dynamics. The most comprehensive is *Molecular Reaction Dynamics and Chemical Reactivity,* by Raphael D. Levine and Richard B. Bernstein (Oxford University Press, New York, 1987). It covers all the topics in this chapter in greater detail as well as several additional topics. A companion volume is *Chemical Dynamics via Molecular Beam and Laser Techniques,* by Richard B. Bernstein (Oxford University Press, New York, 1982). Both of these books contain extensive original literature references. A rather complete history of the optical rainbow, which makes for an interesting case study in the development of many branches of science, is *The Rainbow: from Myth to Mathematics,* by Carl B. Boyer (T. Yoseloff, New York, 1959).

PRACTICE WITH EQUATIONS

28A What is the collisional reduced mass for H colliding with H_2? (28.3)

ANSWER: $2m_H/3 = 1.12 \times 10^{-27}$ kg

28B What is the relative collision energy for H moving 3000 m s^{-1} toward a stationary H_2? (28.6)

ANSWER: 5.02×10^{-21} J

28C What speed would a D atom have to have to collide with H_2 at the same E_{rel} of 28B above? (28.3), (28.6)

ANSWER: 2450 m s^{-1}

28D What is the speed of the collision's center of mass in 28B above? (28.4), (28.5)

ANSWER: 1000 m s^{-1}

28E What fraction of the total translational energy in 28B is tied up in center-of-mass motion? (28.5), Example 28.2

ANSWER: 1/3

28F What impact parameter gives exactly $\chi = 90°$ for hard-sphere scattering with an effective diameter $d = 5$ Å? (28.7)

ANSWER: $5/\sqrt{2} = 3.54$ Å

28G What is the collisional angular momentum for the collision in 28B above at an impact parameter of 1.3 Å? (28.13)

ANSWER: 4.35×10^{-34} kg m^2 s^{-1}

28H What collision energy, expressed as a multiple of E^*, gives a reaction cross section 90% of its maximum for a reaction with a threshold energy E^*? (28.17)

ANSWER: 10

28I What is the Langevin cross section for the reaction between H$^+$ and H_2 at a relative collision energy of 0.10 eV? (28.22), Table 15.3

ANSWER: 4.8×10^{-19} m^2

28J What is the reaction rate constant for H$^+$ + H_2 in the Gioumousis–Stevenson theory? (28.23)

ANSWER: 1.54×10^{12} M^{-1} s^{-1}

28K What is the spectator stripping speed for a collision in which N$^+$ strips D from HD in a collision with $g = 3.4 \times 10^4$ m s^{-1}? (28.24)

ANSWER: 3.0×10^4 m s^{-1}

PROBLEMS

SECTION 28.1

28.1 Use the definitions for r, M, μ, and R in Equations (28.1)–(28.5) to prove that the decomposition of the total kinetic energy into a center-of-mass component and a relative component as expressed in Eq. (28.5) is valid. Derive general expressions for the center-of-mass energy and the relative collision energy in terms of the total translational energy for the special case we use frequently in this chapter: one mass is stationary in the laboratory while the other has non-zero velocity v_1. Can you find a real chemical system (rounding masses to integral amu amounts) for which $E_{rel} = T/2$ in this special case? (*Hint:* Example 28.2 has a situation in which $E_{rel} = Mv_{cm}^2/2$. Is this the same as $E_{rel} = T/2$?)

28.2 We stated in the text that the center-of-mass motion moves uniformly at a fixed velocity v_{cm} no matter what the collision dynamics do to the relative motion of the colliding partners. This problem establishes this important fact. Differentiate Eq. (28.4) twice with respect to time to arrive at an equation for the acceleration of the center-of-mass vector, d^2R/dt^2. Note as well that $m_1(d^2v_1/dt^2)$ is the *force* on mass 1 by Newton's second law, $F_1 = m_1a_1$, and that the force on mass 2 must be the *negative* of the force on 1, since there are no external forces acting on the collision. Show that this condition implies that the center-of-mass is subjected to no acceleration and thus moves uniformly.

28.3 What is the minimum relative speed that would give a collision between H and $H_2(v = 0)$ just enough energy to pass over the barrier in the LEPS potential energy surface? This is a *classical* threshold, and it neglects the *zero-point energy* at the transition state. The collinear H–H–H LEPS surface has a symmetric stretch vibrational frequency $\omega_e = 2108$ cm^{-1} at the transition state. Repeat your calculation including the zero-point energy of this motion.

28.4 *Collisional dissociation* is a possible result of atom–diatom collisions. Sketch a trajectory in (R_1, R_2) coordinates that represents collinear collisional dissociation between a diatomic in its $v = 0$ vibrational state and an atom.

28.5 Problem 27.28 shows a collinear triatomic potential-energy surface that has a barrier to reaction, but unlike the $H + H_2$ surface, the barrier is not located at $R_1 = R_2$. Instead, it is located in the entrance channel. Consider trajectories on this surface for two collisions with the same *total* energy, but with different partitionings of that energy. Let one collision have a minimum of reactant vibrational energy with the rest in relative translation, and let the second have considerable reactant vibrational energy and therefore much less relative translational energy. Sketch qualitative trajectories for both of these cases, taking into account the rise in potential energy at the transition state, and see if you can decide which partitioning would be more helpful in surmounting the barrier: excess relative translational energy or excess reactant vibrational energy.

28.6 One can make an interesting picture of a collinear collision if one plots separate curves for R_1, R_2, and $R_1 + R_2$, each as a function of time. Sketch qualitatively what these curves would be for the reactive trajectory shown in Figure 28.1. You should be able to tell at a glance which atom is bonded to which and the duration of the intimate period of the collision.

28.7 Use the Morse potential function for H_2 from Chapter 19, Eq. (19.47), to calculate the inner and outer classical turning points for the H_2 bond vibration with $v = 0$ and $v = 1$. Are your values in agreement with the figure shown in Example 28.1?

28.8 If all H atoms were hard spheres, and there was no bond between them, how would you describe a collinear collision in which one H collides with a pair of stationary H atoms that are separated by the H_2 bond length? Would this collision be "reactive"? If so, what is the internal energy of the product? All you need to answer this problem is a little intuition about collisions between equal masses.

28.9 Figure 28.3 shows a trajectory that leads to vibrationally excited H_2 product. Use the conditions for that trajectory along with spectroscopic constants for H_2 from Table 19.2 to calculate the *maximum* product vibrational quantum number that this collision could produce. Suppose H_2 was produced in this state. What would be the final relative speed of the collision, g'? Collinear collisions, of course, cannot involve reactant or product rotation, but in three dimensions, rotation can play an important dynamic role. What is the highest rotational quantum number that an H_2 product could have if it was formed also in its highest possible vibrational state?

SECTION 28.2

28.10 Consider an elastic collision between atoms A and B with B initially stationary in the lab (so that its velocity v_B is zero) and A moving toward B with velocity v_A so that the center-of-mass velocity is $(m_A/M)v_A$. What is the total *linear momentum* of the system in the laboratory frame? What is the linear momentum *of* the center of mass? What is the total linear momentum *in* the center-of-mass frame of reference? You might find it helpful to also find the velocities of A and B in the center-of-mass frame. These are simply the lab velocities minus the center-of-mass velocity.

28.11 He atoms are so weakly bound that their scattering is dominated by repulsive forces at almost any common collision energy. Use the high-temperature limit to the He second virial coefficient $B(T)$ in Table 1.1 along with the hard-sphere expression for $B(T)$ in Eq. (1.19) to calculate a hard-sphere radius for He. What range of impact parameters yields scattering angles $\chi \geq 90°$ for this size atom?

28.12 Problem 24.5 shows you how to calculate the fraction of molecules in an equilibrium gas that have kinetic energies between any two limits. This expression holds for relative collision energies as well. Use it along with data from Table 24.4 to find the fraction of collisions that have energies greater than $\mathcal{D}_e$ in Ar at 300 K, in He at 300 K, and in Xe at its boiling point, 166 K.

28.13 The equilibrium distribution of relative speeds follows a Maxwell–Boltzmann distribution, Eq. (24.5), with v replaced by g and m by μ. Use this expression to find the most probable relative collision speed in N_2 at 300 K and in H_2 at 4000 K. This latter scenario was used in Chapter 7 (see Eq. (7.28)) to explore the equilibrium composition of the $H_2 \rightleftarrows 2H$ dissociation reaction. How does your most probable relative speed, when converted into a relative collision energy, compare to the H_2 bond energy, 4.48 eV = 7.18×10^{-19} J? Do most collisions have sufficient energy to dissociate ground vibrational state H_2 at this temperature, or do only very few?

28.14 If you look closely at the highest energy, smallest impact parameter trajectories in Figure 28.9, you will see that they penetrate the circle drawn around the cross representing the center-of-force. The radius of this circle is that interatomic distance for which the interatomic potential energy has risen from its minimum

28.14 up an amount equal to $\mathcal{D}_e$ along its repulsive side, a distance we have previously used to approximate a hard-sphere diameter. Explain why the trajectories penetrate this circle, including why they do more and more as E_{rel} is increased and b is decreased.

28.15 Note that the deflection functions in Figures 28.10 and 28.11 predict $\chi = 0°$ for a finite impact parameter. What would a trajectory at this impact parameter look like in the relative frame of reference? Do any of the trajectories in Figure 28.9 come close to this one?

28.16 A particle detector with a 4.0 mm diameter opening is located 5.0 cm from the scattering volume defined by the intersection of two crossed molecular beams. What solid angle does this detector subtend? If one of the beams has very heavy and slow molecules and the other has very light and fast molecules, the lab scattering angle of the light molecules will always be very close to the center-of-mass scattering angle. Suppose the flux of these molecules is 6.5×10^{21} cm^{-2} s^{-1} and their interaction with the heavy molecules is approximated by hard-sphere collisions with $d = 3.5$ Å. How many light molecules enter the detector each second? The parameters of this experiment are fairly typical of real situations, and your answer should help you understand why molecular beam experiments are difficult to do!

28.17 Sketch trajectories in the relative frame for the two impact parameters in Figure 28.11 that lead to rainbow scattering: one at the minimum of the deflection function and the other at the smaller b for which $\chi = \chi_r$.

28.18 The table below lists calculated scattering angles as a function of impact parameter for elastic scattering by a realistic potential. Construct a deflection function plot from them, and estimate the rainbow angle.

b/Å	χ
0.5	159°
1.0	137°
1.5	114°
2.0	86°
2.5	57°
3.0	22°
3.5	−19°
3.6	−28°
3.7	−40°
3.8	−54°
3.9	−65°
4.0	−68°
4.1	−59°
4.2	−47°
4.3	−35°
4.4	−28°
4.5	−22°
5.0	−9°
5.5	−4°

28.19 Suppose the situation in Example 28.5 had the Ne atom stationary and the He atom hitting it at 500 m s^{-1}. Construct a Newton diagram for this case, and find the laboratory velocities for both atoms in the case of a head-on collision. Does your diagram make it clear why it is usually easier to detect the heavier collision partner in a scattering experiment?

28.20 The Newton diagram in Figure 28.13(c) can be used to find the maximum *laboratory* scattering angle at which the Ne atom could be found. Find this angle. What is the maximum lab angle at which scattered He could be found? What sort of collision would move He in that direction?

28.21 In a scattering experiment between Cs$^+$ ions moving through the lab at 2.0×10^4 m s^{-1} and striking a beam of Ar atoms that are moving so slowly that we can consider them stationary, Cs$^+$ ions are detected at a lab scattering angle of 5°. What lab speeds must these ions have, assuming only elastic collisions?

28.22 Identical particle scattering has some quantum mechanical features that arise when two different classical trajectories lead to the same scattering angle. Sketch two trajectories in the center-of-mass frame (along the lines of Figure 28.6(b)) that have the same χ if the particles are indistinguishable. One choice is a head-on trajectory and a large impact parameter, $\chi = 0°$ trajectory, but see if you can draw them for non-zero χ. Try $\chi = 90°$, as in Example 28.6.

28.23 The CsCl dissociation energy into neutral atoms is $\mathcal{D}_0 = 4.58$ eV. What relative speed must CsCl have with a Xe atom so that collision induced dissociation is just possible? What speed makes dissociation into Cs$^+$ and Cl$^-$ ions possible? You will need data from Tables 7.3 and 7.4.

28.24 In an experiment in which He$^+$ ions with a laboratory speed of 120.0 eV struck H$_2$ molecules moving so slowly that they could be considered stationary in the lab, it was found that He$^+$ scattered at $\chi = 180°$ had speeds indicating vibrational excitation and collisional dissociation of H$_2$. Find the He$^+$ laboratory speeds representing elastic scattering, inelastic scattering to H$_2$ $v = 1$, and H$_2$ dissociation, all at $\chi = 180°$.

28.25 The F + D$_2 \rightarrow$ DF + D reaction internal energy change, quoted in the text, is about -2.21×10^{-19} J. Although experiments show that this never happens, find the maximum laboratory speed DF in internal quantum state ($v' = 0$, $J' = 0$) could have if F moving at 2140 m s^{-1} struck a stationary D$_2$ with no internal excitation. The numbers you need are included in the table leading to Figure 28.16, but you should calculate them independently. Repeat this calculation for the observed reactive scattering flux maximum: DF in $v' = 3$, scattered at $\chi = 180°$.

28.26 The Newton diagram for photodissociation of a single molecule is particularly simple to find. The molecule moves through space at some velocity v, and since it has all the mass, this is also the center-of-mass velocity. After a photon of energy E_{ph} is absorbed, placing the molecule in an electronic state that spontaneously dissociates, the fragments fly apart, using whatever energy remains after a portion of the photon energy is used to break the bond and conserving linear momentum as they do. Suppose the molecule is HI. Explain how one can measure the speed of the I atom fragment to find its *electronic state*. (Recall that I has two low-lying electronic states, in analogy with F. See Example 13.5.) Can one find out the I atom's state by measuring the H atom's speed? This is the basis of the method known as *photofragment spectroscopy*.

SECTION 28.3

28.27 The integral in Eq. (28.18) looks more formidable than it is. Make the substitution $x = \mu g^2/2k_B T$ and use the integrals

$$\int x\,e^{-x}\,dx = -(1+x)e^{-x} \quad \text{and} \quad \int e^{-x}\,dx = -e^{-x}$$

to evaluate it and verify the result in the text.

28.28 Dielectric recombination is the mutual neutralization of ions, as in $A^+ + B^- \to A + B$. The ions attract through the very long-range Coulombic potential, $U(R) = -e^2/(4\pi\epsilon_0)R$, and these processes usually have very large rate constants. Assume that neutralization occurs if the ions approach within a distance R_n. Show that the cross section can be written

$$\sigma = \pi R_n^2 \left(1 - \frac{U(R_n)}{E_{\text{rel}}}\right)$$

and that the thermal rate constant is

$$k_2 = N_A(1000\,\text{L m}^{-3})\,\pi R_n^2 \left(\frac{8k_B T}{\pi\mu}\right)^{1/2}\left(1 + \frac{e^2}{4\pi\epsilon_0 R_n k_B T}\right).$$

Which term of the two in the last parentheses dominates for typical temperatures and neutralization radii (a few Å)? What is the predicted temperature dependence of this reaction model?

28.29 Suppose a reaction cross section follows $\sigma = 0$, $E_{\text{rel}} \leq E^*$, but $\sigma =$ constant, $E_{\text{rel}} > E^*$ where E^* is some threshold energy. Find the thermal rate constant in this case, (Problem 28.27 might help you), and compare your answer to the thermal rate constant based on Eq. (28.17). Does the one here follow an Arrhenius behavior with T?

28.30 The $H_2^+(v=0) + \text{He} \to \text{HeH}^+ + \text{H}$ reaction has an energy threshold $E^* = 0.8$ eV. Above that energy, the reaction cross section roughly follows Eq. (28.17) with $\pi R_c^2 \cong 0.15$ Å^2. Calculate the rate constant based on these data at 500 K. How much smaller is this than the rate constant based on a zero-threshold energy, Langevin orbiting reaction model? The polarizability volume of He is in Table 15.2.

28.31 The experimental rate constant for the exothermic reaction $\text{HeH}^+ + H_2 \to H_3^+ + \text{He}$ is $\geq 2.1 \times 10^{10}$ M^{-1} s^{-1} at 200 K. How well does this agree with the Gioumousis–Stevenson rate constant? What is the orbiting radius? Is it reasonable, given the H_2 and HeH$^+$ bond lengths, 0.74 Å and 0.77 Å, respectively? What does this theory predict for an isotope effect on the rate constant for $\text{HeH}^+ + D_2 \to HD_2^+ + \text{He}$?

28.32 Some steps leading up to the Langevin cross section expression, Eq. (28.22), were neglected in the text. Complete them by deriving Equations (28.20) and (28.21). Then consider collisions between positive and negative ions, as in Problem 28.29. Is an orbiting model appropriate here? How is U_{eff} changed if $U(R)$ is the Coulombic potential instead of the induced dipole potential?

28.33 Draw the Newton diagram for the $\text{Ar}^+ + D_2$ spectator stripping reaction discussed in Example 28.8. Make your drawing to scale, and indicate the circle corresponding to ArD$^+$ product moving at the spectator stripping velocity in the center-of-mass frame. (This is a circle of radius $v_{ss} - v_{cm}$.) What is the maximum laboratory angle at which ArD$^+$ product could be found with this center-of-mass speed? If an experiment finds ArD$^+$ product with center-of-mass speeds lying on this circle, suggest a model that is only a slight elaboration of spectator stripping that can explain how it got there. (*Hint*: The model is known as *elastic spectator stripping*.)

28.34 Consider a hypothetical reaction $A + BC \to AB + C$ that is thermoneutral but follows a spectator stripping mechanism. Find the maximum E_{rel} at which spectator-stripped AB product is stable if its bond energy is $\mathcal{D}_0$. Show as well that spectator stripping places a constant fraction of E_{rel} into product internal energy.

28.35 As mentioned in the text, the $O^+ + H_2 \to OH^+ + H$ reaction follows spectator stripping at low collision energies. Find the spectator stripping laboratory speeds for both OH$^+$ and OD$^+$ in an experiment using stationary HD as a target and O$^+$ moving with a laboratory kinetic energy of 40.0 eV.

SECTION 28.4

28.36 The MO correlation diagram for $\text{Ne}^+ + H_2$ is not much different from that for $\text{He}^+ + H_2$. The energies change a bit, and the Ne and NeH orbital designations change as well. Redraw Figure 28.19 to reflect these changes. Guess any energy changes you cannot look up in this book, but justify your guesses.

28.37 It is not difficult to draw an electronic *state* correlation diagram to go along with the MO diagram in Figure 28.19 for $\text{He}^+ + H_2 \to \text{HeH}^+ + H$. The relevant electronic ground states are He$^+$ $^2S_{1/2}$, H_2 $^1\Sigma_g^+$, H $^2S_{1/2}$, HeH$^+$ $^1\Sigma^+$, He $^1S_0$, H_2^+ $^2\Sigma_g^+$, and excited states with term symbols you should be able to figure out from information in Figure 28.19. Likewise, the energies of various ground state pairs are given in Figure 28.18, and excited state energies can be estimated from Figure 28.19's placement of MOs.

GENERAL PROBLEMS

28.38 There is a direct reaction dynamic model closely related to spectator stripping known as the *knockout model*. For the A + BC → AB + C reaction, this model assumes the following sequence of events. First, A strikes C and scatters from it elastically. A then picks up B to form the AB product while C continues on its way. Find the laboratory speed of AB assuming that BC is stationary and that A strikes C head on. If you recognize that after the A–C collision, product is formed as if A and B had a totally inelastic collision with B stationary and A moving with the speed it has after the A–C collision, you should be able to solve this problem quickly.

28.39 The remaining problems guide you through the construction and interpretation of Newton diagrams for the usual crossed molecular beam experiment in which one beam has velocity v_1 and the other has velocity v_2 directed at a right angle to the first. The relative velocity is still given by $g = v_1 - v_2$ and the center-of-mass velocity is $v_{cm} = (m_1/M)v_1 + (m_2/M)v_2$, but the directional properties of these vectors lead to the following diagram (assuming $m_1 > m_2$):

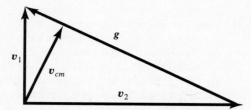

Prove that the tip of v_{cm} still lies on g and divides g into fractions $(m_1/M)g$ and $(m_2/M)g$. To show that v_{cm} just touches g, you might find it easiest to find an equation for the line along which g points and show that the tip of v_{cm} is a point on that line. Prove as well that the angle ϕ between v_2 and v_{cm} is given by $\tan\phi = (m_1v_1/m_2v_2)$.

28.40 Continuing from the previous problem, assume that beam 1 is Ne and beam 2 is He. Give the Ne beam a speed 500 m s^{-1} and give the He beam a speed 2000 m s^{-1}. Draw the Newton diagram for this experiment to scale. What is the relative speed, g? What is the collision energy, E_{rel}? How does the Ne elastic circle appear on your diagram? What is the *laboratory* scattering angle and speed of a Ne atom that strikes He head on and is scattered through a *center-of-mass* scattering angle $\chi = 180°$?

28.41 Finally, we consider a crossed-beam reactive scattering Newton diagram. The F + D$_2$ experiment that led to Figure 28.16 had the F beam's speed equal to 870 m s^{-1} and the D$_2$ speed equal to 1960 m s^{-1}. Show that this indeed implies $E_{rel} = 1.26 \times 10^{-20}$ J, as quoted in the text. Draw the Newton diagram (let F = beam 1, and your drawing should be identical to the sketch in Problem 28.39, which was drawn with this problem in mind), and use information quoted in the text to find the *laboratory* speed and direction of DF product scattered at $\chi = 180°$ into $v' = 3$ with no rotational excitation.

APPENDIX

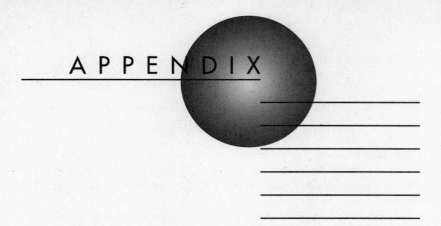

Units, Dimensions, and Constants

> Divers weights, and divers measures,
> both of them are alike abomination to the Lord.
> *Proverbs* XX:10
> The Bible, King James Version

The Syntax of Physical Chemistry

When you learn a new language, one of your first tasks is to learn the grammar and word order of the language so that you can understand and be understood in that language. In short, you learn its syntax. While the rest of this book teaches you the vocabulary of physical chemistry, this appendix is an introduction to its syntax. This syntax has relevance to all branches of science, of course, but the examples we use here are particularly common to the dialect of physical chemistry.

Our concern is with the syntax of *physical quantities.* Since most quantities in physical chemistry have associated physical dimensions, our first syntactical rule is that *no dimensioned physical quantity has meaning without a statement of the units used to express it.* There are a few important *dimensionless* quantities (such as the fine structure constant, the Madelung constant for a particular crystal structure, and the ratio of the proton mass to the electron mass) for which the numerical value alone suffices, but in general, units are *required,* and there are rules for their use and display.

Probably the single most common error made by students starting a course in physical chemistry is a failure to pay attention to units. It is vital to carry a quantity's units along in a calculation and to verify by simple unit algebra that any calculated quantity has the presumed units one desires. Common trouble spots include the universal gas constant R, often expressed in either J K^{-1} mol^{-1} or L atm K^{-1} mol^{-1} units, and molecular masses, often found or thought of in g mol^{-1} units, but needed in kg mol^{-1} units in many calculations.

Units are written in this text in a style that is common and generally accepted, but not universally so. Other styles are acceptable, but some are not. For example, we write the universal gas constant and the molar mass of H as

$$R = 8.314\ 51 \text{ J mol}^{-1} \text{ K}^{-1} \quad \text{(acceptable)}$$

$$M_\text{H} = 1.007\ 94 \text{ g mol}^{-1}\ . \quad \text{(acceptable)}$$

It is also acceptable to write

$$M_H = 1.007\ 94 \text{ g/mol} \quad \textbf{(acceptable)}$$

with a single solidus (as the "/" sign is known), but it is *not* acceptable to write

$$R = 8.314\ 51 \text{ J/mol/K} \quad \textbf{(unacceptable)}$$

since this style is ambiguous. Are the units (J/mol)/K = J mol^{-1} K^{-1} or J/(mol/K) = J K mol^{-1}? Consequently, units should have at most one solidus among them, but omitting the solidus completely and writing negative exponents where needed is never ambiguous.

It is useful to think of a physical quantity as the product of a pure number (8.314 51 in the case of R) and its unit symbols (i.e., J mol^{-1} K^{-1}). This allows us to tabulate or plot physical quantities in a compact and unambiguous way. Consider, for example, the second virial coefficients B in Table 1.1. At 20 K, B for He is -3.34 cm^3 mol^{-1}, which we can write as an equality

$$B = -3.34 \text{ cm}^3 \text{ mol}^{-1}$$

and then rearrange, following the usual rules of algebra, into a form that separates the pure number part of B from its unit symbols:

$$\underbrace{B/\text{cm}^3 \text{ mol}^{-1}}_{\substack{\uparrow \\ \text{Physical quantity} \\ \text{Unit symbols}}} = \underbrace{-3.34}_{\substack{\uparrow \\ \text{Pure number}}}$$

so that we tabulate the pure number and head the column of such numbers (or label the axis on a graph of them) with the physical quantity divided by its unit symbols.

If the most convenient or common units of a quantity lead to pure numbers that are many powers of ten away from 1, one can divide the quantity by unit symbols multiplied by some suitable power of ten to yield a pure number closer to 1. For example, molecular dipole moments tend to fall close to 10^{-30} C m. The HCl dipole moment is 3.697×10^{-30} C m, and thus Table 15.1 lists the pure number 3.697 under the heading "dipole moment/10^{-30} C m."

This convention is unambiguous and preferable to other display styles that one sometimes sees, such as "dipole moment (10^{-30} C m)," or "dipole moment in 10^{-30} C m," but no matter which style is chosen, the important rule is that both the units and the magnitude of a physical quantity must be clear to your reader in order for you to be understood.

Note as well that there is a space between the number and the following unit symbols. The only exception to this rule concerns angles, in which the degree sign (or the minute sign, ', or second sign,") follows the number immediately, as in 180° rather than 180 °. Celsius and Fahrenheit temperatures should be written with a space between the number and the degree sign, as in 100 °C and 98.6 °F rather than 100°C and 98.6°F.

A related convention followed in this book concerns the display of a number with many significant figures. Grouping makes long strings of digits easier to read. You are used to using a comma to separate groups of three digits to the left of the decimal point, but the convention followed in this book uses a space rather than a comma, and the convention applies to digits on both sides of the decimal point. If a number has *four or fewer digits on one side or the other (but not both) of the decimal,* no grouping is used. Thus, we write 1000, not 1 000, or 0.1234, not 0.123 4. If there are more than four digits, then they are grouped into threes on either side of the decimal, as in 0.123 45, 1.234 5, 12 345, etc. Note as well that numbers less than 1 should *always* have a leading zero: 0.123, not .123. A comma is not used to separate groups because it is the usual symbol for the *radix* (the integer–fraction separator—the English "decimal point") in European languages. What a native English speaker would naturally write as 0.12, a native French speaker, for example, writes 0,12.

> There is no need for the economist to prove ... that as a result of the adoption of a certain measure nobody is going to suffer.
> Welfare Propositions of Economics, *Economic Journal,* September, 1939
> Nicholas Kaldor, British economist

The SI System

The SI units (Système International d'Unités) scheme is a universally accepted and continually reviewed set of fundamental units for the base dimensions on which all physical measurements can be expressed in a compact and uniform way. While not the everyday units of the United States, they are in common use almost everywhere else in the world (including Britain, where it has been said, "Metrification is creeping in inch by inch."), and they pervade modern scientific literature. Nevertheless, older literature of great value is often expressed in non-SI units, and it is important to recognize these units and the conversion factors that take them into SI quantities. For example, the SI unit of pressure is the pascal (Pa) where $1 \text{ Pa} = 1 \text{ N m}^{-2}$ (1 newton per square meter), but the pressure units of atmosphere (atm), bar, torr, and mm Hg ("millimeter of mercury," which is not a measure of length, but a measure of the pressure exerted at the base of a column of Hg that many millimeters tall) are still widely used.

The SI scheme uses seven *base units,* and all other quantities can be expressed in terms of them.

SI Base Units

Physical quantity	Unit name	Unit symbol
Length	meter	m
Mass	kilogram	kg
Time	second	s
Electric current	ampere	A
Thermodynamic temperature	kelvin	K
Amount of substance	mole	mol
Luminous intensity	candela	cd

Each of these base units has a specific definition:

- meter: the length of path traveled by light in vacuum during a time interval of 1/299 792 458 of a second (so that the meter is based on the defined value for the speed of light),
- kilogram: the mass of the international prototype of the kilogram (an artifact maintained in Sèvres, France, constructed of a Pt–Ir alloy),
- second: the duration of 9 192 631 770 periods of the radiation corresponding to the transition between the two hyperfine levels of the ground state of the ^{133}Cs atom (as discussed in Chapter 21),
- ampere: the constant current which, if maintained in two straight parallel conductors of infinite length, of negligible circular cross-section, and placed

TABLE A1 Common Derived SI Units

Quantity	Unit	Symbol	SI base unit decomposition	Derived units equivalent
Force	newton	N	kg m s^{-2}	J m^{-1}
Energy	joule	J	kg m^2 s^{-2}	N m
Power	watt	W	kg m^2 s^{-3}	J s^{-1}
Pressure	pascal	Pa	kg m^{-1} s^{-2}	N m^{-2}
Electric charge	coulomb	C	A s	J V^{-1}
Electric potential	volt	V	kg m^2 s^{-3} A^{-1}	J C^{-1}
Electric resistance	ohm	Ω	kg m^2 s^{-3} A^{-2}	V A^{-1}
Electric conductance	siemens	S	kg^{-1} m^{-2} s^3 A^2	Ω^{-1}
Electric capacitance	farad	F	kg^{-1} m^{-2} s^4 A^2	C V^{-1}
Magnetic flux density	tesla	T	kg s^{-2} A^{-1}	N A^{-1} m^{-1}
Temperature	degree Celsius†	°C	K	
Frequency	hertz	Hz	s^{-1} (cycles per second)	

†The degree Celsius measures the same temperature *increment* as the kelvin, but to convert from one unit to the other, use $T/K = 273.15 + T/°C$ whenever T represents an actual temperature.

1 meter apart in vacuum, would produce between these conductors a force equal to 2×10^{-7} newton per meter of length (which cannot, of course, be realized exactly in practice),

- kelvin: the fraction 1/273.15 of the thermodynamic temperature of the triple point of water,
- mole: the amount of substance of a system which contains as many elementary entities as there are atoms in 0.012 kilogram of ^{12}C,
- candela: the luminous intensity in a given direction, of a source that emits monochromatic radiation of frequency 540×10^{12} hertz and that has a radiant intensity in that direction of (1/683) watt per steradian (and is a unit that is not needed in this book).

Note that temperature is symbolized K, not °K, and that one says, for example, "Room temperature is about three hundred kelvin," rather than "... three hundred *degrees* kelvin." Note also that when the mole is used, the "elementary entities" must be specified. They can be anything: atoms, molecules, ions, electrons, photons, or specified groups of such particles.

In addition to these base units, several *derived* units are common to physical chemistry. Each derived unit can be decomposed into the base units that define it. Some derived units are used in the definition of the base units, such as the watt and the newton, but the reasoning is not circular. The more important derived units are listed in Table A1. Note that the units are named after various scientists, but with the exception of the Celsius degree, they are *spelled* in English with lower case first letters: newton, not Newton, but *symbolized* with capital letters: N, not n.

Each SI unit, base or derived, can be prefixed with a standard notation to indicate various powers of ten multiplicative factors, as in 12 345 J = 12.345 kJ. A current list of these prefixes is shown on the inside back of this book, along with their origins. Only one prefix per unit is allowed: "kilomicrometer," for example, is just "millimeter."

> But the main thing is, does it hold good measure?
> Heaven soon sets right all other matters!
> 'Christmas-Eve' (1850) 1. 1311
> Robert Browning, English poet

Atomic Units

Chapter 14 introduced *atomic units,* and Table 14.1 lists the more common of them. The discussion in Chapter 14 points out the advantage of units based on fundamental atomic quantities. While the value of a quantity expressed in ordinary units might change slightly as fundamental measurements improve in time or as base units are redefined, a quantity expressed as a multiple of an atomic unit will *always* have the same value. *All* atomic units can carry the unit abbreviation au, which is confusing unless one is given a clear indication of the type of quantity under discussion.

Note that some of these units can be expressed in more than one way. For example, the atomic unit of time is not only the first Bohr orbit period, it is also $\hbar$ divided by the atomic unit of energy, the hartree.

Other atomic units have similar definitions, such as the atomic unit of electric dipole moment, which is simply ea_0, or of force, which is the atomic unit of energy divided by the atomic unit of length. Only the atomic units of mass, charge, angular momentum (or action), and length are independent. The others can always be expressed in terms of these.

> That man is the measure of all things.
> Plato's *Theaetetus*
> Protagoras, Greek sophist

Fundamental Constants

There is an international committee called CODATA, the Committee on Data for Science and Technology of the International Council of Scientific Unions, which is charged with the critical evaluation of new measurements of those quantities called the fundamental or universal constants. This committee periodically publishes recommended values for these constants (in 1965, 1969, 1973, and 1986, with a new listing due in 1995 or 1996) based on a critical evaluation of new measurements and a least-squares analysis of their uncertainties. Occasionally, the new values reflect fundamental definition changes, as was the case in 1986 when the speed of light became a defined quantity, $c = 299\,792\,458$ m s^{-1} exactly, as did the permeability of vacuum, $\mu_0 = 4\pi \times 10^{-7}$ N A^{-2} exactly.

These evaluations are not trivial to do, since measurements rarely lead directly to a new value of a *single* constant but rather to a combination of them, such as e/h. While some fundamental quantities have exact values, many others are based on somewhat unsatisfactory *artifacts* (such as the international kilogram standard) or physically impossible idealizations (such as the infinitely long conductors in the definition of the ampere). Over time, these artifacts or impossibilities have slowly been replaced, and there is great interest in finding ways to eliminate them all. Length, for example, was long based on an artifact. One of the first of international significance was an iron bar constructed in the mid-1700s called the "toise" and preserved today in the Paris Observatory. It was about 1.95 m long. It was followed by the meter bar, another artifact, and this bar was then replaced with a length standard based on atomic spectroscopy: 1 650 763.73 wavelengths in vacuum of the orange-red line of the spectrum of ^{86}Kr. This in turn was replaced with the current definition in terms of the defined speed of light.

Efforts are aiming toward a mass standard that would replace the artifact kilogram as well. Precision mass ratios (based on the ^{12}C standard) using *ion traps,* mass spectrometers that can hold as few as a single ion, have great promise.

The 1986 values of the more important fundamental constants for physical chemistry are gathered in the table on the inside front of this book. Uncertainties range from 128 ppm for the poorest known, the universal gravitational constant G, to 0.0012 ppm for the best known, the Rydberg constant. Just as you memorize certain key historical dates to a certain precision (such as July 4, 1776, for example, but do you know what time of day John Hancock signed the Declaration of Independence?), likewise you should memorize the important fundamental constants to two or three significant figures. The ability to estimate quantities to an order of magnitude or so without having to consult a reference is an important skill in any science.

> Bring out number weight and measure in a year of dearth.
> Proverbs of Hell, *The Marriage of Heaven and Hell*
> William Blake, English poet

Conversion Factors

Beyond concern for correct kinds of units is the concern for unit conversion from non-SI to SI units. As a simple example of the unit algebra called for in using a conversion factor (which itself is unitless, as we will see), consider the conversion factor needed to express the length 3.6 in. in meter units. By definition of the inch, 1 in. = 2.54×10^{-2} m exactly. (Note that, in order to avoid confusion with the English word "in," the inch is the only common unit with an abbreviation followed by a period.) We write the conversion factor $1 = 2.54 \times 10^{-2}$ m in.$^{-1}$. Note that this factor equals 1 and is dimensionless, since "m in.$^{-1}$" is a ratio of two units of length; a conversion factor is simply the pure number 1 written in disguise. To carry out the conversion, we write 3.6 in. = (3.6 in.) $\times$ (1) = (3.6 in.) $\times$ (2.54 $\times$ 10^{-2} m in.$^{-1}$) = 9.144×10^{-2} m (as displayed on a calculator), which we then round to the two significant-figure accuracy of the original quantity: 9.1×10^{-2} m.

A table on the inside back of this book lists a variety of conversion factors for quantities encountered in this book. To use them, note that the top row of the table is expressed as the denominator of the conversion-factor unit fraction while the table title expresses the numerator, always a fundamental SI quantity. The entries themselves are thus the factors that multiply a non-SI quantity and convert it to the fundamental SI quantity. Quantities in italics are exact.

REFERENCES

Several books cover the material of this chapter in greater detail. The Physical Chemistry Division of the Commission on Physicochemical Symbols, Terminology, and Units of the International Union of Pure and Applied Chemistry (IUPAC) has published *Quantities, Units and Symbols in Physical Chemistry* (Blackwell, Oxford, 1988), a very comprehensive and carefully considered guide to the syntax of physical chemistry. It includes several other standard tables, such as unit conversion factors and standard atomic weights of the elements. An earlier book, along the same lines as the IUPAC book but including numerous historical discussions, is *Physicochemical Quantities and Units,* by M. L. McGlashan (Royal Institute of Chemistry, London, 2nd ed., 1971). The most recent compilation of fundamental constants and a discussion of their sources and derivation is found in the article by E. R. Cohen and B. N. Taylor, *Rev. Mod. Phys.* **57,** 1121 (1987). An introduction to the whole field of metrology can be found in *The Fundamental Physical Constants and the Frontier of Measurement* by B. W. Petley (Adam Hilger, Bristol, England, 1985).

INDEX

Absolute zero, 89
   unattainability, 106
Absorption, 672
Absorption coefficient, 673
Absorption cross section, 1046
Action, 372, 868
Activated complex theory, 1063, 1105
Activation energy, 988, 1019, 1052, 1061, 1073
   Tolman expression, 1075
Activation volume, 1077
Activation-limited reaction, 1053
Activity, 165
Activity coefficient, 165
   mean ionic, 310
Adiabat, 50
Adiabatic reaction, 223
Adiabatic demagnetization, 297
Adiabatic depolarization, 285, 298
Adiabatic lapse rate, 273
Adiabatic process, 43
Adiabatic reversible process, 49
Adsorption, 628
Adsorption coefficient, 1087
Amagat's law, 163, 220
Amorphous solid, 625
Angular momentum, 415
   collisional, 1129
   coupling, 480
   electronic orbital, 506
   vibration-induced, 715
Anharmonic corrections, 409, 704
Annihilation operator, 405
Antibonding MO, 507
Antiferroelectric, 289
Arrhenius equation, 1019
Associated Laguerre polynomial, 427, 428
Asymmetric top, 717
Asymmetry parameter, 718
Atmosphere
   adiabatic, 273
   isothermal, 272
Atomic clock, 789
Atomic Energy Levels, 486
Atomic scattering factor, 623

Atomic units, 500, 1151
Auger process, 630, 1086
Autocatalysis, 1029, 1038
Autoionization, 488, 683
Average speed, 904
Azeotrope, 175

Band gap, 612
Band theory, 605, 611
Beer–Lambert law, 674, 1046
Belousov–Zhabotinskii mechanism, 1031
Bimolecular process, 1001, 1041
Binary diffusion, 958
Birefringence, 661
Birge–Sponer extrapolation, 736, 750
Blackbody, 352, 669, 690, 872, 875
Body-centered cubic, 589, 594
Bohr magneton, 577
Bohr radius, 426
Bohr theory, 356, 424, 430, 447, 449
Boiling point elevation, 176
Bolometer, 918
Boltzmann distribution, 93, 820, 831
Boltzmann H theorem, 846
Boltzmann transport equation, 951
   linearized, 952
Bond
   hydrogen, 104
Bond energy, 209, 735
Bond length
   Morse, 733
Bond order, 523
Bonding MO, 507
Born–Haber cycle, 246
Born–Mayer model, 600
Born–Oppenheimer approximation, 499
Bose–Einstein statistics, 821
Boson, 453, 821
Boyle temperature, 17, 936
   van der Waals, 20
Bragg reflection, 620
Branching reactions, 1010, 1017
Bravais lattice, 589
Brownian motion, 960
Bubble-point line, 173

Buckminsterfullerene, 613
Bulk compressibility, 21, 102, 601, 851

Cage effect, 1053
Calorimeter, 103
Canonical ensemble, 845
Canonical partition function, 839
Capillarity, 263
Carnot cycle, 67
Catalysis, 1001, 1081
Center of mass, 692, 1103
Centrifugal distortion, 704
Centrifugal field, 274
Centrifugal potential, 428, 1129
Centrifuge, 275, 277
Chain reactions, 1012
Chaperone mechanism, 1044
Charge-transfer reaction, 323
Charge-transfer transition, 762
Chemical equilibrium, 218
Chemical laser, 685, 866, 1126
Chemical oscillations, 1028
Chemical potential, 125, 841
   ideal gas, 129
   standard state, 130
Chemical relaxation, 1025
Chemical shift, 794
Chemiluminescence
   infrared, 1128
Chemisorption, 1086
Cholesteric phase, 657, 659
Chromophore, 755
Circular polarization, 668
Clapeyron equation, 151, 223
   integrated form, 155
Classical trajectory, 1101
   harmonic oscillator, 402
Classical turning point, 360
   harmonic oscillator, 402, 444
   radial, 428
Clausius–Mossotti equation, 286
Clepsydra, 996
Close packing, 589
   random, 664
Cloud chamber, 258

INDEX

Colligative properties, 170, 312
Collision complex, 1108, 1128
Collision rate, 953, 955
    mean, 954
    total, 955, 1050
    wall, 17, 912
Collision theory
    reactive, 1049, 1069
Color center, 626
Combination level, 716
Combination transition, 732
Combining rules, 932
Commutator, 363
    angular momentum, 417
Competitive inhibition, 1097
Complete set, 369
Complex conjugate, 358
Compressibility factor, 13
Concentration cell, 329
Conduction band, 970
Conduction electrons, 605
Conductivity, 969
Configurational energy, 645
Configuration integral, 920
Configuration interaction, 479, 515
Conjugation, 533
Consolute temperature, 169
Coordinate
    center of mass, 1103
    relative, 1103
Correlation diagram, 552
    orbital, 510
    reaction, 1137
Correspondence Principle, 357, 380, 408
Corresponding states, 25, 267, 932
Coulomb integral, 474, 477, 508
Covalent bond, 515, 525
Creation operator, 405
Critical opalescence, 254
Critical point, 23
Critical solution temperature, 169
Cross section
    absorption, 1046
    differential, 1116
    hard sphere, 913, 1050
    reaction, 1050, 1051
Cryoscopic constant, 178
Crystal defect, 625
Crystal field theory, 758
Crystallography, 616
Cubic closest packing, 591
Cubic site, 597
Curie temperature, 289, 296
Curie's law, 295
Curie–Weiss equation, 297
Curve crossing, 549

d-d transition, 759, 775
Dalton's law, 160
de Broglie wavelength, 355, 378
    thermal, 855
Debye equation
    polarizability, 287
Debye frequency, 873
Debye length, 317
Debye T-cubed law, 96
Debye temperature, 96, 873
Debye–Hückel limiting law, 315, 1058

Decoupling, 804
Deflection function, 1112
Degeneracy, 381
    angular momentum, 422
    hydrogen atom, 431, 439
    particle in a box, 412
Degree of advancement, 202, 218
Degree of freedom
    thermodynamic, 185
Delocalization, 292, 528, 534
Demagnetization, 297
    nuclear, 298
Density gradient, 276
Density of states, 448, 640, 754, 774, 856, 901
Depolarization, 285, 298
Desorption, 1086
Detailed balance, 1022, 1054
Determinant, 460, 534
    secular, 507, 529
    Slater, 471, 513
Dew-point line, 173
Diamagnetic, 291, 794
Diamond structure, 611
Dielectric constant, 280, 668
Dielectric recombination, 1131, 1145
Dielectric susceptibility, 280
Differential, 12
Differential scattering cross section, 1116
Diffraction
    X-ray, 617
Diffusion, 957, 976, 1053
    binary, 958
    coefficient, 958, 1054
    Einstein equation, 961
    self, 958
Diffusion equation, 959
Diffusion-limited reaction, 1053
Dipole approximation, 676
Dipole moment, 279, 561, 786
    function, 721
    induced, 569
    magnetic, 576
    operator, 565
    permanent, 574
    transition, 676
Direct collision, 1108, 1128
Dislocation, 626, 627
Dispersion, 1116
Dispersion energy, 927
Displacement field, 281
Dissociation
    multiphoton, 685
Dissociation energy, 210
    Morse, 733
Dissociative plateau, 1059
Distribution function, 820
    Boltzmann, 831
    configurational, 644
    Maxwell–Boltzmann, 903
    nonequilibrium, 948, 975
    Planck radiation, 669
    radial, 646
    speed, 903
    translational energy, 901
    velocity, 905
Doppler effect, 740, 741, 944, 1123
Doppler width, 740, 918

Drude model, 927
Dulong and Petit heat capacity, 95
Dunham expansion, 705
Dye laser, 771

Eadie–Hofstee plot, 1085
Ebullioscopic constant, 179
Effective charge, 469
Effective potential, 428, 701
Efficiency, 68
Effusion, 915
Ehrenfest's theorem, 368
Eigenfunction, 363
Eigenstate, 367
Eigenvalue, 363
Einstein diffusion equation, 961
Einstein mobility-diffusion equation, 980
Einstein–Smoluchowski equation, 963, 977, 1094
Elastic scattering, 1109
Electret, 289
Electric field
    thermodynamic, 278
Electric permittivity, 668
Electric potential, 278
Electric resonance
    molecular beam, 787
Electrical conductivity, 957, 958, 968, 979
Electrocaloric effect, 285
Electrochemical potential, 326, 1057
Electromagnetic spectrum, 666
Electron affinity, 212
Electron configuration, 451
    doubly excited, 489
    molecular, 513
Electron correlation, 479
Electron spin resonance (ESR), 809
Electron volt, 213, 353
Electronegativity, 527
Electrophoresis, 985
Electrostatic factor, 1058, 1097
Electrostriction, 285, 1077
Elementary process, 1000, 1014, 1040
Elliptical polarization, 668
Encounter pair, 1053
Encounter rate, 1049
Endothermic, 207, 223
Energy
    bond dissociation, 210
    configurational, 645
    internal, 43, 869, 1104
    most probable, 902
    relative translational, 1103
    rotational, 701
    vibrational, 701
Engel–Brewer model, 607
Ensemble, 844
    canonical, 845
    grand canonical, 845
    microcanonical, 844
Enskog–Chapman theory, 983
Enthalpy, 52
    activation, 1074
    atomization, 214
    bond, 209
    combustion, 214
    crystal, 214, 600
    formation, 206

fusion, 103
hydration, 215
lattice, 214
level, 215
solution, 215
specific, 207
sublimation, 104, 214
vaporization, 104
Entrance channel, 1071
Entropy, 66
 absolute, 105, 108
 activation, 1074
 Boltzmann, 66, 70, 90, 107
 phase transition, 105
 production, 957
 reaction, 204
 residual, 879
 statistical expression, 876
Enzyme catalysis, 1081
Equation of state, 5
 ideal gas, 2
 van der Waals, 17
 virial, 14, 646
Equilibrium, 3
 chemical reaction, 218
 gravitational, 270, 274
 sedimentation, 275
Equilibrium constant, 219, 337, 881
 pressure dependence, 225
 solubility, 226
 statistical expression, 887, 1066
 temperature dependence, 222
Equipartition theorem, 868
Ergodic hypothesis, 847
Eucken factor, 983, 996
Eutectic point, 189
Exact differential, 12
Excess functions, 165
Exchange integral, 474, 475, 477, 508, 528
Exchange operator, 453
Exchange symmetry, 452
Excluded volume, 18
Exclusion rule, 732
Exit channel, 1072
Exothermic, 207, 223
Expectation value, 368
Explosions, 1016
Exponential repulsion, 600, 922
Exponential-n potential, 929
Extended Debye–Hückel law, 320
Extensive variable, 4

F center, 626
Face-centered cubic, 589
Fermi energy, 607
Fermi resonance, 751
Fermion, 453, 821
Fermi–Dirac statistics, 821
Ferrimagnet, 296
Ferrites, 296
Ferroelectric, 288
Ferromagnetic, 291
Fick's laws, 959, 1054, 1097
Fine structure constant, 577
First Law of Thermodynamics, 44
First-order reaction, 1002
Fluctuations, 847
Fluorescence, 730, 762, 771

Force constant, 402, 701, 726
 transition state, 1062
Fourier transform, 625, 806, 807
Fourier transform infrared (FTIR), 740
Fourier's law, 965
Franck–Condon factor, 728, 750
Franck–Condon principle, 729
Free electron gas, 605
Free energy, 120
 activation, 1074
 formation, 233
 Gibbs, 119, 841
 Helmholtz, 118, 841
 interfacial, 254
 mixing, 161
Free energy function, 233, 890
Free expansion, 41, 46
Free induction decay, 808
Free jet expansion, 906
Free particle, 389
Freezing point depression, 176
Frenkel defect, 626, 639
Friction constant, 984
Fugacity, 130
 coefficient, 133
 partial, 163
 van der Waals, 132
Fullerene, 613
Fundamental constants, 1151
Fundamental transition, 724
Fusion, 103

g factor, 577
 nuclear, 578
Gaussian-type orbital, 472
Gerade, 506, 697
Giauque function, 233, 890
Gibbs free energy, 119, 841
Gibbs' paradox, 162
Gibbs–Duhem equation, 129
Gioumousis–Stevenson equation, 1133
Glass, 627
Glass transition temperature, 627
Gouy balance, 293
Graham's law, 915
Grain boundary, 626, 627
Grand canonical ensemble, 845
Graphite structure, 613
Gravity, 270
Ground state, 379

H theorem, 846
Half-life, 1003
Hamiltonian operator, 365
 matrix elements, 460
Hard-sphere fluid, 649
Hard-sphere model, 15, 647, 913, 922
 cross section, 1050
 rate constant, 1129
 scattering, 1111
 virial coefficient, 936
Harmonic oscillator, 402
Hartley–Kirkendall effect, 964
Hartree–Fock theory, 476
Heat, 43, 836, 880
 algebraic sign, 44
Heat capacity, 44
 configurational, 656

Debye, 96, 873
Dulong and Petit, 95
Einstein, 96, 872
electronic, 96
ideal gas, 45
liquid, 98
metallic, 96
polyatomic gas, 98
ratio, 50, 100
solid, 95, 872, 873
two-level system, 94
two-phase system, 103
van der Waals, 103
water, 98, 108
Heisenberg Uncertainty Principle, 808
Helium
 phase transitions, 153
 solid, 592
Helium atom, 455, 468
Hellmann–Feynman theorem, 574, 576, 586
Helmholtz free energy, 118, 841
Henry's law, 173
Hermite polynomial, 407
Hermitian, 362
Heterogeneous catalysis, 1082, 1085
Hexagonal closest packing, 592
Hildebrand's rule, 157
HOMO, 521
Homogeneous catalysis, 1082, 1085
Hooke's law, 402
Hückel theory, 528, 637
Hund's rules, 485, 524, 535
Hybrid, 434, 506, 542, 545, 609
Hydrogen
 ortho–para, 229, 744, 863, 1035
Hydrogen atom
 Bohr theory, 356, 424, 430, 447, 449
 degeneracy, 431, 439
 energy, 430
 ground state, 427
 ionization, 424, 426
 nodes, 431
 optical spectrum, 351
 radial wavefunction, 428
 size, 431
 spin-polarized, 517
Hydrogen bond, 104, 614, 655, 935
Hyperfine coupling, 578, 796, 810
Hyperpolarizability, 574

Ideal gas, 2
 adiabatic reversible path, 49
 chemical potential, 129
 heat capacity, 45
 temperature scale, 42
Ideal mixing, 159, 162
Ideal mixture
 reactive, 220
Ideal solubility, 226
Ideal solution, 165
Identical particles, 452
Immiscibility, 168, 190
Impact parameter, 1110
 reaction, 1130
Impulsive scattering, 1135
Independent electron approximation, 456
Index of refraction, 290

INDEX

Indistinguishability, 452
Inelastic scattering, 1119
Inexact differential, 12
Inhibition, 1001, 1014
   competitive, 1097
Inhibitor, 1085
Inhomogeneous broadening, 771
Initial rate, 999
Intensive variable, 4
Intercombination transition, 681
Intermolecular potential energy, 921
   dispersive, 927
   electrostatic, 924
   exponential-n, 929
   inductive, 925, 926
   Lennard-Jones, 445, 930
   repulsive, 922
Internal conversion, 764, 1093
Internal energy, 43, 869, 1104
   activation, 1074
Intersystem crossing, 763, 1093
Intramultiplet transition, 681
Inverse power repulsion, 636, 923
Inversion operator, 564
Inversion symmetry, 697
Inversion temperature, 55
Ion pair, 515
Ionic bond, 515, 526
Ionic character, 515, 568
Ionic solid, 597
Ionic strength, 309, 1058
Ionization potential, 212
Isenthalp, 55
Isentrope, 106
Isobar, 8
Isochore, 8
Isopleth, 173
Isopycnic point, 276
Isotherm, 7
Isotope effect, 1071
Isotopic substitution, 722, 726

Jahn–Teller distortion, 528, 760, 775
JANAF Thermochemical Tables, 228, 871, 888, 889
Joule coefficient, 61
Joule expansion, 41, 46, 66, 75
Joule-Thomson effect, 54
   inversion curve, 57
   coefficient, 55
   inversion temperature, 55
j–j coupling, 480, 489

Kelvin mist equation, 258
Kinetic control, 1024
Knockout model, 1146
Knudsen cell, 915
Knudsen number, 915
Kohlrausch's law, 986
Koopmans' theorem, 478
Kopp heat capacity, 95, 872

Lagrange undetermined multipliers, 830
Lamb dip, 741
Lambda transition, 187
Landé g factor, 782, 783
Langevin cross section, 1132
Langmuir isotherm, 1087

Laplace transform, 857
Laplacian operator, 410
Laporte's rule, 680
Laser
   carbon dioxide, 716
   chemical, 685, 866, 1126
   dye, 771
Laser assisted reaction, 1072
Laser induced fluorescence, 730
Laser magnetic resonance (LMR), 785
LCAO-MO method, 509
Le Châtelier–Braun principle, 227, 239, 251
LEED, 628, 1086
Legendre polynomial, 422
Legendrian operator, 417
Lennard-Jones potential, 445, 930
   virial coefficient, 937
LEPS model, 1060, 1108
Lever rule, 158, 173
Lewis–Randall rule, 163
Ligand field theory, 761
Limit cycle, 1031
Lindemann mechanism, 1077
Line defect, 626, 627
Linear combination, 369
Linear operator, 362
Linear polarization, 668
Lineshape, 678
Lineweaver–Burke plot, 1084
Liquid crystal, 657
Local mode, 713, 769
Localization, 392
London energy, 927
Lotka–Volterra mechanism, 1029
Low energy electron diffraction, 628, 1086
Lowering operator, 420, 800, 814
LUMO, 542
L–S coupling, 480, 489

Mach number, 908
Madelung constant, 601, 604
Magnetic dipole moment, 576
Magnetic field, 291
Magnetic flux density, 290
Magnetic moment, 781
   nuclear, 792
Magnetic permeability, 668
Magnetic resonance
   laser, 785
   molecular beam, 789
   nuclear, 791
Magnetic shielding, 794
Magnetic susceptibility, 291
Magnetization, 291, 293
   isothermal, 298
Magnetostriction, 294
Margules expression, 167
Maser, 722, 789
Matrix element, 460, 676
Matrix isolation spectroscopy, 770
Maxwell relations, 81, 110, 135, 266
Maxwell–Boltzmann distribution, 903
Maxwell–Boltzmann statistics, 827
Mean free path, 913, 952
Mean ionic activity coefficient, 310
Mean ionic molality, 310
Median speed, 904
Meissner effect, 296, 306

Mesophase, 657
Michaelis constant, 1084
Michaelis–Menten equation, 1084
Microcanonical ensemble, 844
Microscopic reversibility, 1023
Microstate, 64, 90, 107, 819, 823
Miller index, 620
Mobility, 980, 988
Molality, 173, 221
   mean ionic, 310
Molar conductivity, 986
Molar refraction, 290
Molar volume, 2
Molarity, 173
Mole fraction, 158
Molecular beam, 741
   effusive, 917
   supersonic, 906
Molecular beam electric resonance, 787
Molecular beam magnetic resonance, 789
Molecular crystal, 613
Molecular dynamics, 649
Molecular orbital, 507
Molecular partition function, 853
Moment of inertia, 376, 416, 693
Monte Carlo method, 650
Morse potential, 733, 734, 1059
Most probable speed, 904
Multiphoton dissociation, 685
Multipole expansion, 560

Nearest neighbors, 592
Néel temperature, 296
Nematic phase, 657
Nernst equation, 331
Nernst–Einstein equation, 995
Newton diagram, 1121
Newton's law of viscous flow, 970
Node, 361, 379
   planar, 431, 435
   spherical, 427
Non-crossing rule, 550
Nonideal mixing, 163
Nonradiative transition, 685
Normal mode, 713
Normalization, 359
Nozzle beam, 906
Nuclear g factor, 792
Nuclear magnetic moment, 792
Nuclear magnetic resonance (NMR), 791
Nuclear magneton, 578
Nuclear quadrupole coupling, 790
Nucleation, 259
   critical radius, 260

Oblate top, 717
Octahedral site, 597
Octupole moment, 562
Ohm's law, 968
Operator, 362
   annihilation/creation, 405
   dipole moment, 565
   exchange, 453
   Hamiltonian, 365
   inversion, 564
   Laplacian, 410
   Legendrian, 417
   permutation, 453

raising/lowering, 420, 800, 814
  symmetry, 504
Optoacoustic effect, 767
Orbiting collision, 1115, 1131
Order parameter, 658
Order-disorder transition, 188
Oregonator mechanism, 1031
Orthogonality, 369
Orthonormality, 369
Osmotic pressure, 179
Ostwald dilution law, 995
Ostwald viscometer, 974
Overlap integral, 508, 528
Overtone, 724, 766
Oxidation–reduction, 323

Packing fraction, 591
Pair potential, 600, 645
Paramagnetic, 291, 794
Parity, 564, 680
Partial molar quantity, 125
Partial molar volume, 128
Partial pressure, 160
Particle in a box, 394, 452, 822, 854
  multidimensional, 409, 606
Particle on a ring, 418
Particle on a sphere, 418
Partition function
  canonical, 839
  classical, 869, 919, 1068
  electronic, 859
  molecular, 830, 853
  nuclear, 859
  rotational, 859
  translational, 855, 856
  vibrational, 857
Paschen–Bach limit, 578
Pashen–Back effect, 785, 800
Path
  thermodynamic, 9
Pauli Exclusion Principle, 453, 821
Periodic boundary conditions, 418, 651
Permeability, 291
Permittivity, 280
Permutation operator, 453
Perturbation theory, 462
pH, 340
Phase diagram, 172, 186
Phase rule, 181
Phase space, 843
Phase transition
  order, 186
Phonon, 966
Phosphorescence, 730, 762, 1093
Photochemical process, 1045
Photodissociation, 682
Photoelectric effect, 354
Photoelectron spectroscopy, 520, 1073
Photofragment spectroscopy, 1145
Photoionization, 682
Photon gas, 875
Physisorption, 1086
Piezoelectricity, 285
Planck radiation distribution, 669
Plasma, 979
Point defect, 626
Poiseuille's equation, 974
Polarimetry, 1037

Polarizability, 282, 569, 574, 727, 925
Polarizability volume, 282
Polarization, 280
  tensor, 288
Polarized light, 668, 781
Population inversion, 93, 721, 789, 866, 1126
Positronium, 424, 448, 496, 558, 587
Potential
  electrochemical, 326
  reduction, 334
Potential energy surface, 933, 1058
  contour map, 1059, 1101
Potential of mean force, 654
Powder pattern, 618
Poynting correction, 164, 221, 258
Pre-exponential factor, 1019, 1076
Predissociation, 685
Pressure
  gas kinetic, 910
  internal, 18
  standard, 882
  statistical expression, 876
  tensor, 971
Primary salt effect, 1058
Principal quantum number
  effective, 472
Probability density, 360
Process
  adiabatic, 43
  adiabatic reversible, 49
  cyclic, 40, 207
  reversible, 41
  spontaneous, 77
  thermodynamic, 9
Prolate top, 717
Proton affinity, 214

Quadrupole coupling, 790
Quadrupole moment, 561
Quantum beats, 764
Quantum number, 379
  hydrogen atom, 427
  most probable rotational, 868
  particle in a box, 396
  principal, 427
Quantum yield
  fluorescence, 765, 1047, 1093
  photochemical, 1047
Quenching, 674
  collisional, 1048, 1093

Radial distribution function, 431, 447, 646
Radial turning point, 1133
Radiation intensity, 668
Radiationless transition, 762
Radiative association, 1042
Radiative lifetime, 677, 762, 1047
Radiative recombination, 778
Rainbow angle, 1118
Rainbow scattering, 1115
Raising operator, 420, 800, 814
Raman scattering, 696, 727, 1045
Ramsauer–Townsend effect, 399
Random walk, 960
Raoult's law, 171
Rate constant, 1001
  hard sphere, 1129
  state-to-state, 1109

Rate law, 1000
Rate-limiting step, 1008
Rayleigh scattering, 696, 727
Reaction
  adiabatic, 223
Reaction coordinate, 1061, 1101
Reaction cycle, 207
Reaction intermediate, 998
Reaction mechanism, 1007
Reaction progress variable, 202, 218, 998
Reaction rate, 998
Reaction rate order, 1000
Reactive scattering, 1126
Recombination, 778
Recurrence relation
  Hermite polynomial, 407
Recurrence time, 843
Redox reaction, 323
Reduced mass, 376, 402
  collisional, 1050
Reference state
  thermodynamic, 206
Reflection symmetry, 697, 698
Regular solution, 170
Relative permeability, 291
Relative permittivity, 280
Relative speed, 914
Relaxation time, 846, 1027
Reservoir, 78
Residual entropy, 879
Resistivity, 969
Resonance, 399, 1128
Resonance energy, 536
Resonance integral, 528
Rigid rotor–harmonic oscillator, 702
Rittner model, 944
Rms speed, 904
Rotational energy
  diatomic, 701
  polyatomic, 718
RRKM theory, 1081, 1109
Russell–Saunders coupling, 480
Rydberg state, 756

Sackur–Tetrode equation, 130, 347, 644, 877
Salt bridge, 333
Salt effect, 1058
Saturated calomel electrode, 336
Saturation
  spectroscopic, 742
Scanning tunneling microscope (STM), 632
Scattering
  elastic, 1109
  impulsive, 1135
  inelastic, 1119
  rainbow, 1115
  reactive, 1126
Scattering angle, 1111
Scattering flux, 1125
Schottky defect, 626, 639, 897
Schrödinger equation
  radial, 700
  time dependent, 365
  time independent, 366
Screening, 470
Second Law of Thermodynamics, 73

Second virial coefficient, 14, 921, 936, 937
　van der Waals, 20
Second-order reaction, 1003
Secular determinant, 529
Sedimentation, 275
Selection rules, 679
　electronic, 729
　NMR, 793
　rotational, 721, 730, 731
　vibrational, 724, 732
Self diffusion, 958
Self-consistent field theory, 476
Semiconductor, 611
　intrinsic, 613
Semipermeability, 161, 179
Separated atom limit, 507
Separation of variables, 410
Sequential reactions, 1007
Shear viscosity, 971
Shielding
　magnetic, 794
　nuclear charge, 470
SI units, 1149
Simon–Glatzel equation, 156
Simple cubic, 589
Singlet state, 475
Slater atomic orbital, 472, 546
Slater determinant, 471, 513
Slater diagram, 495
Smectic phase, 657, 659
Solubility, 240
　effect of pressure, 226
　equilibrium constant, 226
　ideal, 226
Solution
　ideal, 165
　ideal dilute, 173, 221
　regular, 170
　simple, 167
Solvus line, 190
Soret effect, 984
Spectator stripping, 1134, 1145
Spectrophone, 767
Speed of sound, 904, 967
Spherical harmonic, 422, 700
Spherical polar coordinates, 412
Spherical top, 717
Spin, 229, 294, 422, 435
　nuclear, 438, 743
　photon, 453
Spin–lattice relaxation, 792
Spin–orbit coupling, 438, 479, 578, 699, 748, 764, 796
　coupling constant, 480
　halogens, 486
Spin–spin coupling, 795
Spontaneous emission, 672, 1042
　coefficient, 673
Square-well potential, 647, 929
　second virial coefficient, 936
Standard hydrogen electrode, 334
Standard pressure, 103, 130, 882
Standard reduction potential, 334
Standard state, 130, 206
Standard volume, 882
Stark effect, 581, 786
Stationary state, 366
　photochemical, 1047

Statistical weight, 824
Steady state, 949, 1055
Steady-state approximation, 1008
Steric factor, 1052
Stern–Gerlach experiment, 581
Stern–Volmer plot, 1048, 1093
Stimulated emission, 672
　coefficient, 673
Stimulated emission pumping (SEP), 690, 768
Stirling's approximation, 92, 829
Stockmayer potential, 946
Stokes' law, 984
Stokes–Einstein equation, 985, 1054, 1057
Sublimation, 104
Substrate, 1083
Superconductivity, 296
Supersonic molecular beam, 906
　seeding, 909
Surface charge, 278
Surface coverage, 1086
Surface net, 628, 639
Surface reconstruction, 628
Surface tension, 255
Susceptibility, 280
Symmetric top, 717
Symmetry number, 860, 863
Symmetry operations, 504
Synchrotron radiation, 618

Temperature
　absolute, 2
　absolute zero, 89
　Debye, 96
　infinite, 92
　negative absolute, 93
　thermodynamic, 69
Temperature programmed desorption, 1086
Term symbols
　atomic, 484
　diatomic, 697
　polyatomic, 712
Termolecular process, 1001, 1043
Tetrahedral site, 599
Thermal conductivity, 957, 958, 965, 978
Thermal de Broglie wavelength, 855
Thermal diffusion, 984
Thermal expansivity, 83, 102
Thermal runaway, 1021
Thermal transpiration, 943
Thermocouple, 984
Thermodynamic control, 1025
Thermodynamic efficiency, 68
Thermodynamic temperature, 69
Thermodynamics
　master equation, 80
Thermoneutral, 207
Third Law of Thermodynamics, 106, 820
Tie line, 144
Time-of-flight method, 905, 916, 918, 1123
Total derivative, 7
Trajectory, 1101
Transition matrix element, 676
Transition moment, 728
Transition state, 1058
Transition state theory, 1063
Translational energy, 1103
Transmission coefficient, 1072, 1105

Transport phenomena, 957
Triple point, 3, 148
Triplet state, 475
Trouton's rule, 157
Tunneling, 378, 400, 1071
Two center integral, 508
Two-level system, 91, 229, 385, 819, 824
　heat capacity, 94
　perturbation theory, 464

Uncertainty Principle, 371
　energy–time, 376
Undetermined multipliers, 830
Ungerade, 506, 697
Unimolecular process, 1001, 1040, 1077, 1109
Unit cell, 589
United atom limit, 507
Unsöld's theorem, 446, 489

Valence bond theory, 515
van der Waals energy, 927
van der Waals equation, 17, 919
　chemical potential, 143
　critical point, 23
　fugacity, 132
　heat capacity, 103
　inversion curve, 57
　inversion temperature, 55
　phase diagram, 148
　phase transitions, 143
　vapor pressure, 144
　virial coefficient, 20
van der Waals molecules, 789, 1073
van't Hoff equation, 180
Vapor pressure, 144, 171
　equation, 155
Vaporization, 104
Variation theory, 456
Vector cross-product, 415
Velocity
　relative, 914
Velocity modulation spectroscopy, 742
Vibration-rotation coupling, 705
Vibrational deactivation, 1105
Vibrational energy
　diatomic, 701
　polyatomic, 716
Vibrational excitation, 1078
Vibronic coupling, 755, 764
Virial coefficient, 14, 646
　van der Waals, 20
Virial expansion, 14, 646, 919
Virial theorem, 368, 447, 448, 449, 494, 587, 646
　molecular, 518, 555, 932
Viscosity, 957, 958, 970, 981, 1054
　shear, 971
Volume
　activation, 1077

Walsh diagram, 538
Walsh's rules, 536
Wavefunction, 358
　complete set, 369
　curvature, 377
　free particle, 390
　hybrid, 434

hydrogen atom, 428
interpretation, 359
linear combination, 369
linear superposition, 370
node, 361, 379
normalization, 359
particle in a box, 396
orthogonal, 369
Wavenumber, 426, 666, 705
Wavepacket, 392
Wavevector, 390

Wetting, 263
Woodward–Hoffman rules, 1136
Work, 34, 880
adiabatic, 43
algebraic sign, 37
Wulff plot, 263

X-ray crystallography, 616
X-ray diffraction, 617

Zeeman effect, 579, 780
anomalous, 782

Zero-order reaction, 1003
Zero-point energy, 374, 381, 823, 858
harmonic oscillator, 404
multidimensional, 411
particle in a box, 396
transition state, 1069
Zero-point motion, 1101
Zeroth Law of Thermodynamics, 71
Zwitterion, 568

Δ notation, 6

Powers of Ten Prefixes

Factor	Prefix	Symbol	Origin
10^{-24}	yocto	y	Blend of *y* and Greek *octo*, eight ($24 = 3 \times 8$)
10^{-21}	zepto	z	Blend of *z* and Latin *septem*, seven ($21 = 3 \times 7$)
10^{-18}	atto	a	Danish or Norwegian *atten*, eighteen
10^{-15}	femto	f	Danish or Norwegian *femten*, fifteen
10^{-12}	pico	p	Spanish *pico*, small quantity
10^{-9}	nano	n	Latin *nanus*, dwarf
10^{-6}	micro	μ	Greek *mikros*, small
10^{-3}	milli	m	Latin *mille*, thousand
10^{-2}	centi	c	Latin *centum*, hundred
10^{-1}	deci	d	Latin *decimus*, tenth
10^{1}	deca	da	Greek *deka*, ten
10^{2}	hecto	h	Greek *hekaton*, hundred
10^{3}	kilo	k	Greek *khilioi*, thousand
10^{6}	mega	M	Greek *megas*, great
10^{9}	giga	G	Greek *gigas*, giant
10^{12}	tera	T	Greek *teras*, monster
10^{15}	peta	P	Greek *pente*, five
10^{18}	exa	E	Variant of Greek *hex*, six ($18 = 3 \times 6$)
10^{21}	zetta	Z	Variant of zepto
10^{24}	yotta	Y	Variant of yocto

"The only missing metric prefixes are groucho, chico, gummo, zeppo, and harpo. If that's not enough, they can use alto and soprano." J. W. Batchelder, *Metric Madness*, (Devon–Adair, Old Greenwich, CT, 1981).